13th International Conference on Radiation Shielding (ICRS-13) & 19th Topical Meeting of the Radiation Protection and Shielding Division of the American Nuclear Society (RPSD-2016)

EPJ Web of Conferences Volume 153 (2017)

Paris, France
3 - 6 October 2016

Part 1 of 2

Editors:

Fausto Malvagi
Fadhel Malouch
Cheikh M'Backe Diop

ISBN: 978-1-5108-4812-2

Printed from e-media with permission by:

Curran Associates, Inc.
57 Morehouse Lane
Red Hook, NY 12571

proceedings.com

Some format issues inherent in the e-media version may also appear in this print version.

Printed by Curran Associates, Inc. (2017)

For additional information, please contact EDP Sciences – Web of Conferences at the address below.

EDP Sciences – Web of Conferences
17, Avenue du Hoggar
Parc d'Activité de Courtabœuf
BP 112
F-91944 Les Ulis Cedex A
France

Phone: +33 (0) 1 69 18 75 75
Fax: +33 (0) 1 69 28 84 91

contact-edps@webofconferences.org

Additional copies of this publication are available from:

Curran Associates, Inc.
57 Morehouse Lane
Red Hook, NY 12571 USA
Phone: 845-758-0400
Fax: 845-758-2633
Email: curran@proceedings.com
Web: www.proceedings.com

TABLE OF CONTENTS

PART 1

PART 2

ICRS-13 & RPSD-2016

13th *International Conference on Radiation Shielding*
&
19th Topical Meeting of the *Radiation Protection & Shielding Division*
of the *American Nuclear Society* -2016

Welcome message of Dr. Cheikh M. DIOP

Research Director, CEA, France
General Chair

My first words are to wish you a very warm welcome in Paris after Nara in Japan in 2012 and Knoxville in the U.S. in 2014. It is symbolic that this conference is held almost 44 years ago to the day and at almost the same location near the Eiffel Tower, with the cooperation of the Nuclear Energy Agency, of the International Atomic Agency and of the American Nuclear Society.

What brings us together here today is our unity and our diversity. Unity because we draw on the same scientific heritage. The diversity is due to the specificity of our own fields of application.

The beneficial consequence is the mutual enrichment of our respective communities on scientific, technical, cultural and human levels. This enrichment supposes also the transmission of achievements and a continuity in this transmission.

This is the spirit in which we have requested the attendance of two eminent witnesses and actors of the radiation protection and shielding history: they are Dr. Enrico SARTORI and Jean-Claude NIMAL, who will speak not today but at the closing session of this conference.

Then this is the right time to recall the ambitious and long term objective of the Health Physics group composed of eight physicists – Bob COVEYOU, Carl GAMERTSFELDER, Jim HART, O. G. LANDSWERK, K.Z. MORGAN, L. A. PARDUE, Herb PARKER, Ernest WOLLAN – who met at the era of the first nuclear reactor at Chicago in USA :

"Health physics had its origin as a profession at the University of Chicago in 1942-43 as indicated in Figure I. the effects of ionizing radiation on matter with the immediate objective to maximize the ratio of benefits to hazards and the principal long-range goal was recognized as the development of a coherent theory of radiation damage. We today find health physics research programs employing physicists, chemists, engineers, biologists and mathematicians working together on theoretical and experimental problems. They are working at all levels--nuclear, atomic, molecular, gas, liquid, plasma, solid, crystal, living cell, animal and the ecosystem." (Karl Z., Morgan, "Education and Training", at the 121st Meeting of the Advisory Committee for the Division of Biology and Medicine. Oak Ridge. Tennessee, November *9.* 1967).

This leads to questioning the past and the present for the future:

- Theoretical, methodological, experimental breakthroughs?
- Innovative applications, new fields?
- What's new for solving the inverse problem?
- Finest dosimetry measurements & interpretation?

- The future of the computational methods and associated codes : deterministic versus Monte Carlo. The end of the semi-empirical methods?
- New optimization approaches?
- Impact of the High Performance Computing?
- New requirements for nuclear data?
- Towards a multiphysics approach for extreme situations?
- Validation and transposition?
- To refund the radiation shielding engineering occupation using CAD, virtual reality?
- ...

I am sure that the conference will provide elements of response to all these questions.

I would like to thanks here the Institution Representatives for their attendance:

- Dr. Andew C. KLEIN, President of the ANS and "the Named Editor of Nuclear Technology",
- Prof. Takashi NAKAMURA de Tohoku University, General Co-Chair of the conference, represented by Dr. Kimiaki SAITO from JAEA.
- Dr. Masashi HIRANO, Senior Coordinator for International Collaboration de la Nuclear Regulation Authority, Japan,
- Dr. Michele FERENCI from Penn State University, USA, General Co-Chair of the conference,
- Dr. Gianni BRUNA, Scientific Director of IRSN, France, General Co-Chair of the conference,
- Dr. Daniel IRACANE, Adjoint Director of OCDE/AEN, Chief Nuclear Officer,
- Dr. Sylvestre PIVET, Director of Innovation and Nuclear Support at the Nuclear Energy Division (DEN) of the CEA, France

Organized by

with the support of

co-organized by

and the partnership of

I thank all the committee members for their hard and valuable work that made this conference possible,

Thanks to the plenary speakers and all the attendees for your contributions and interest to this joint conference ICRS-13 & RPSD-2016.

I declare the ICRS-13 & RPSD-2016 conference open. Thank you for your attention.

Welcome to ICRS-13/RPSD2016 from the ANS

American Nuclear Society

ANS

Andrew C. Klein, PhD, PE(OR)

ANS President
Editor, *Nuclear Technology*
Professor of Nuclear Science and Engineering
Oregon State University

OSU Oregon State University

European Nuclear Conference 10 October 2016

ANS

Welcome to ICRS-13/RPSD2016 from ANS

- ANS highly values our strong relationship and partnership with the European Nuclear Society
- Very glad to be a collaborating Society in the European Nuclear Conference
- The importance of nuclear energy

ANS

Nuclear for CLIMATE

A Lead Partner at COP21 in Paris

ANS

Nuclear for
CLIMATE

**Public Information Materials Exchange (PIME)
Award for Communications Excellence 2016**

ANS NUCLEAR GRAND CHALLENGES

Dr. Andrew Klein
ANS President

October 2016
American Nuclear Society
www.ans.org

ICRS-13 & RPSD-2016

13th *International Conference on Radiation Shielding*
&
19th Topical Meeting of the *Radiation Protection & Shielding Division*
of the American Nuclear Society -2016

Speech of Prof. Takashi Nakamura

Tohoku University, Japan
General Co-Chair

To my great regret, I will not be able to attend the conference for reasons beyond my control.

Personally, I planned to have a special session, SS2 Neutron experimental studies under the quasi-mono-energetic neutron field in the energies from 100 to 400 MeV at Research Center of Nuclear Physics of Osaka University in Japan. Fortunately, seven oral presentations will be done three from Japan and four from Europe. I am very happy to have these special interesting oral presentations with some poster presentations in ICRS-13. I was also requested from French association of this conference to nominate two or three invited speakers from Japan on SS1, Fukushima special session after 5 years. I tried to find out appropriate persons and finally I could inform two invited speakers from Japan, Dr. Saito of Japan Atomic Energy Agency and Dr. Hirano of Nuclear Regulation Authority. Furthermore, nine oral wide-scope presentations on Fukushima accident will be given from Japan.

I really believe that the ICRS-13 conference will be a great success and I do hope for your fruitful and enjoyable meeting in Paris, although unfortunately I am very sorry that I will not be able to act as a general co-chair during the conference..

ICRS-13 & RPSD-2016

13th *International Conference on Radiation Shielding*
&
19th Topical Meeting of the *Radiation Protection & Shielding Division*
of the American Nuclear Society -2016

Speech of Dr. Giovanni Bruna

Scientific Director of IRSN, France
General Co-Chair

Ladies and Gentlemen,

It is a pleasure & a honor for me to welcome you at this ICRS 13-RPSD 2016 joint Conference on behalf of the IRSN, the French *Institute for Radioprotection and Nuclear Safety*, which is widely involved in its organization, cares of a stand, contributes to the technical sessions through Jean-Christophe Gariel's, along with Masashi Hirano's (from Nuclear Regulation Authority of Japan)]'s coordination of the special session on "Fukushima 5 years after", as well as through some chairmanships, and presents several papers in different fields of endeavor.

I feel that the choice of addressing Monte-Carlo computing and Radioprotection in this joint Conference is really a lucky one, because the links between the two disciplines are very tight and both can get big advantage from gathering, exchanging and working together.

Actually, Monte-Carlo computing techniques are widely adopted in the Radioprotection field for a large range of applications, including the evaluation of the particles propagation in NPPs - both in normal operation, incidental & accidental conditions - but also, and I would say even more, for Medical Facilities design, Radiotherapy & Medical Applications, with use spanning from the production of isotopes to the development of specific measurement devices and metrology.

In my youth, I've personally been involved in Monte-Carlo computing, in the field of reactor design and operation. From my personal past experience, I just remember:

- The computation of suitable cross-section data-set for design calculations (in this field, further improvements are expected to account for temperature feedback),
- The double heterogeneity cases, such as in the PB – Pebble Bed – graphite moderated, gas cooled reactors, where both the particle walking process and geometry are random. Micro-particles are distributed randomly and, in accidental states, the media have got fuzzy boundaries. The coupled Monte-Carlo depletion provides with an accurate description of the time behavior of the pebbles to optimize safety and operation,
- The evaluation of the cross-sections of the SPND - Self Powered Neutron Detectors - adopted for in-core on-line reactor monitoring, which require very accurate Monte-Carlo description to account for the space and time effect of shell depletion for compensation purposes.

In the field of Propagation of Particles and Radiation, I recall the accurate focusing techniques, which allow appreciating the contribution of very specific energy ranges to the activation & the damage of materials, to the measurement device counting rates, and so on... In the VR - Variance Reduction - the importance is identified with the adjoint flux. Among the way of providing an estimate of it, I have been familiar with a separate deterministic calculation. But it is likely using Monte Carlo itself. All these topics are extensively addressed in this Conference.

In the field of Radioprotection, I mention the wide contribution of the Monte-Carlo techniques to the simulation of irradiation effect on the tumors and the impact on their vicinity which allows predicting the collateral damage of irradiation, like radiation-induced secondary cancers, or the monitoring and the development of specific measurement devices and fakes.

All those very important topics and many others are discussed in this Conference, which moreover addresses the very crucial aspects of:

- The availability of experimental devices (which are often very expensive to build and operate) and sometimes shared, but which are definitely mandatory for the validation of the simulations in a wide range of topics addressed during this conference,
- The nuclear data, a crucial topic both for reactor and medical applications.

I am very impressed by the comprehensiveness and the exhaustiveness of the Conference program, and I wish to congratulate the General Chair, the Organizers, the Sponsors, the Contributors for their effort and commitment to make this Conference a success.

A last word to conclude: I am fully aware that innovation is a mandatory key for progress... and innovation can be achieved and speeded-up only through gathering forces and sharing knowledge among Organizations and Countries, worldwide.

That is the objective of the initiatives, platforms and R&D associations (such as, in Europe, NUGENIA for Reactors, OPERRA for Radioprotection), which are bourgeoning and growing-up everywhere in the world.
Among them, last but not least, I would mention the NI2050 initiative sponsored and supported by the NEA - Nuclear Energy Agency - which is presently underway. It intends to identify some high level challenges and main topics for R&D in the short & medium term, which can sustain the development of the peaceful applications of nuclear energy (not only for electricity production) and can increase its civil acceptance, which is a major issue today in many countries. Computation and radioprotection aspects are part of the initiative objectives.

I trust that attractive Conferences such as the present joint ICRS 13-RPSD 2016 one contribute efficiently to the emergency of new and innovative ideas and to the creation of collaboration frameworks which are to enhance and support effectively innovation in the reactor and radioprotection field.

I wish you all a very useful and profitable Conference and a pleasant stay in this charming end-of-summer Paris.

ICRS-13 & RPSD-2016

13th *International Conference on Radiation Shielding*
&
19th Topical Meeting of the *Radiation Protection & Shielding Division* of the American Nuclear Society -2016

Speech of Dr. Daniel Iracane

Deputy Director-General - Chief Nuclear Officer
OECD – NEA
General Co-Chair

Ladies and Gentlemen, Distinguished colleagues,

I am happy and honoured to share the introduction of this 13th International Conference on Radiation Shielding (ICRS-13) and the 19th Topical Meeting of the Radiation Protection & Shielding Division of the American Nuclear Society-2016 (RPSD-2016)

The OECD/NEA is a Forum for Co-operation for the Most Advanced Countries in the World. NEA has been founded in 1958 and gathers 31 member countries. The NEA is continuously supporting member countries in maintaining and further developing, through international co-operation, the scientific, technological and legal bases required for a safe, environmentally friendly and economical use of nuclear energy.

Through its 7 Technical Committees and 72 working parties and expert groups, the NEA is providing shared authoritative assessments on key issues, as input to government decisions on nuclear energy policy. I have seen that the International Radiation Shielding Conferences, the ICRS cycle, started also in 1958, like NEA. We are strong here of a powerful and bright legacy of 58 years built on an effective international cooperation. But looking at the future, I would say that the international cooperation is even more necessary today than yesterday.

Why?

Nuclear technology is a quite mature technology. We are far from the pioneering phase. Nevertheless **this technology is facing several important challenges**, all of them being related to the confidence building.

- Confidence of the local populations
- Confidence of the investors and policy makers
- Confidence of the international community

By putting the experts of the experienced countries around the table, the NEA produces authoritative technical output in a wide range of topics. And, by putting around the table, not only the experts but also the leaders of related agencies and public organisations, the NEA committees' structure allows to get an endorsement from the member countries. This process contributes strongly to the collective definition of the state of the art in numbers of sensitive topics. As such, the process continuously contributes to the confidence building. Let us point out few current examples where the confidence is at stake and where the international cooperation is bringing a valuable input for the policymakers.

- The Fukushima accident has created a significant loss of confidence in the Japanese population and a major concern in the international community.

  - The Committee gathering the regulatory body's leaders have deeply instigated the consequences of the Fukushima, including a self-assessment of the effectiveness of the regulatory bodies. The conclusion on the relevance of the defence in depth methodology is one of the several outputs that help the international community to keep confidence in the current regulatory process. Numbers of other topics have been addressed collectively to reach shared conclusion and actions have been implemented promptly in NEA member countries in a consistent way.
  - Regarding the radio-protection and public health, international discussion on the food safety and the management of the recovery process are very active topics of discussion within the international community.
  - Regarding the nuclear liabilities issues, the NEA and its member countries are quite committed in the monitoring of damage compensation in Japan as a practical implementation of principles set up by the international conventions in the real difficult situation created by an accident.

- But there are also other challenges for the decision makers, either the policy makers or the investors. The economy of the nuclear energy is challenged in different way.

  - Of course the fuel back-end and decommissioning are open issues that create uncertainties in the final cost of nuclear energy. Through its committees, NEA is strong investing these topics, with an increasing interest for decommissioning.
  - In addition, more recently the increased penetration rate of variable renewable energy raised a new concern in terms of the ability to invest in nuclear which is highly capitalistic. The renewable share increase has also an impact on the nuclear technology by requiring more flexibility.

These are few examples to illustrate the need of further exchanges at the international level to produce robust shared views that will help creating the necessary confidence to proceed. On a more scientific level, the NEA is also managing the Nuclear Data Bank and, together with the Science committee, produces a continuous input for the scientific community in terms of qualified nuclear data and computational codes. The trend in this field is dominated by the coupling between the different disciplines and the challenging experimental need for Multiphysics validation.

But the confidence is also built by maintaining and developing a high level of science and competences. And for that purpose, this conference is a cornerstone.

The agenda of the conference is very rich and illustrates the importance of the topic through numbers of stakes related to ALARA principles in several application fields such as nuclear energy systems, accelerator facilities, lasers, space, medical area, sensor optimisation or others.

We see also in the agenda that challenging scientific issues can still be overcome to provide more reliability and more performance to shielding calculations through faster computing, uncertainty control, improved capacities to handle singularities and extreme situations, etc.

So, such a conference appears to be an asset for the international community by promoting the best science in radiation transport, in order to support a robust radiological protection system that will sustain the right level of confidence among users and for the population.

Thank you for attention and I wish you a successful conference and fruitful outcomes

ICRS-13 & RPSD-2016

13th *International Conference on Radiation Shielding*
&
19th Topical Meeting of the *Radiation Protection & Shielding Division*
of the American Nuclear Society -2016

Speech of Dr. Sylvestre Pivet

Director of Innovation and Nuclear Support at the Nuclear Energy Division (DEN) of the CEA, France

Ladies and Gentlemen,

First of all, I want to welcome all of you again in Paris for the ICRS & RPSD joint conference. On behalf of the CEA I would like to thank the representative of the co-organizer and partner institutions:

- Dr. **Andy Klein**, President of the *American Nuclear Society* (ANS),
- Prof. **Takashi Nakamura** *Tohoku* University who could not come and is here represented by Dr. **Masashi Hirano** from the *Japan Nuclear Regulation Authority* and also Dr. **Kimiaki Saito** from *Japan Atomic Energy Research Institute*,
- Dr. **Gianni Bruna**, the Scientific Director of the *Institut de Radioprotection et de Sûreté Nucléaire* (IRSN) in France,
- Dr. **Daniel Iracane**, **NEA** Deputy Director-General and Chief Nuclear Officer of Nuclear Energy Agency of OECD.

I want to thank them both for their attendance and the messages they delivered.

Many thanks also to the two Honorary Chairs of this conference, that is Dr. **Enrico Sartori** formely *Head of Computer Program Service and Reactor Physics* at the NEA and Dr. **Jean-Claude Nimal** formely Research Director at the CEA.

The CEA applauds the format of joint ICRS & RPSD conferences that promotes the exchanges inside the international Radiation Protection and the Radiation Shielding communities.

During this presentation, I will now try to stress the importance of the radioprotection activity within CEA and particularly within its Nuclear Energy Division, and will try to underline that this topic has been a major preoccupation for CEA from its very creation.

1. General presentation of the CEA

As you probably know CEA is a key player in research, development and innovation in four main areas : defence and security, nuclear energy (fission and fusion), technological research for industry and fundamental research in the physical sciences and life sciences.

2. Presentation of the DEN activities

Within the CEA, the Nuclear Energy Division carries out research in three main fields:

- Developing nuclear systems of the future – dubbed "4th generation" reactors – and their fuel cycles,
- Optimizing the current nuclear industry;
- Developing and operating large experimentation and simulation tools needed for its research programs.

The detailed programs are defined with the French government and the industrial partners of the CEA, in order to provide them with technical expertise and innovation in nuclear power generation systems to develop sustainable nuclear energy that is both safe and economically competitive.

As a nuclear operator, the CEA/DEN also has to manage and upgrade its own fleet of nuclear facilities. It carries out numerous construction and refurbishment programs on these facilities, together with clean-up and dismantling programs for those that have reached the end of their service life.

3. A historical overview of the place of the radioprotection at the CEA

If we take a look in the past : on the 15th of December of the year 1948 in the Paris suburbs (Fort de Chatillon de Fontenay-aux-Roses) the first French nuclear reactor, called ZOE for *Zéro énergie (Zero power), Oxyde d'uranium (Uranium Oxyde), Eau lourde* (*Heavy Water*) achieves divergence.

A very recent book tells that from the very beginning the CEA primary concerns in the field of biology were centered on the protection against the ionizing radiations.

A health control service was created in February 1947 to follow the staff devoted to ZOE reactor operation. **Lew Kowarski**, a prominent ZOE team member, had previously made a sojourn at the Harwell Atomic Center in UK in order to learn about the best practices in radiation shielding.

In 1951, the CEA thus created the *Service of Protection against Radiations* (*Service de Protection contre les Radiations* SPR) under the direction of Dr. **Henri Jammet** who later became a member of the *International Commission on Radiological Protection* (ICRP).

Then the CEA defined a complete nuclear worker radiation protection policy and created various overseeing bodies.

At the same time, the CEA undertook extended research in the field of life sciences including *radiobiology, radiotoxicology, radiotherapy*. Today, this activity relies on digital and computational sciences but also on the experimental and computational facilities distributed over the different CEA centers.

Furthermore, the CEA has shared a long and fruitful common history with IRSN, since one of the original components of IRSN was the *Institut de Protection et de Sûreté Nucléaire* (IPSN) unit from CEA.

I will now focus on the genesis of CEA/DEN radioprotection activities more directly related to the ICRS & RPSD conference topics. It will be the opportunity to highlight the international opening of the "Radiation Shielding" domain, and the very important exchanges CEA has had, in particular with the Oak Ridge National Laboratory.

The first radiation shielding experiments and set of computational methods were developed inside the *Service d'Études de Protections des Piles* (SEPP) led by the mathematician and physicist **Pierre Lafore**, at the CEA center of *Fontenay-aux-Roses*. The first experiments were dedicated to the determination and validation of the water cross sections through the facility called *NAIADE* next to ZOE. Next, the reactor TRITON of 6.5 MWth, came into service in 1959. The SEPP unit has also developed a very important activity centered on dosimetry and instrumentation.

Starting in 1956, information exchanges were established between *Oak Ridge National Laboratory* in USA and the CEA/SEPP concerning both the experiments and the calculation methods in the field of Radiation Shielding.

At CEA, the first radiation shielding calculations using the Monte Carlo method were performed by three researchers: **Pierre Lafore**, **Robert Lattès** and **Jean-Paul Millot.** Their work was published in 1958, making references to that of **Henry C. Honeck** from ORNL.

These works were presented at the first edition of the ICRS (ICRS-1) or the *European Atomic Energy Society Symposium VI-58 Radiation Shielding* held at Cambridge *Caius College* in UK, from 26 to 29 August 1958.

In 1965, **Jean Rastoin** and **Jean-Claude Nimal** visited the *Radiation Shielding Information Center* (RSIC), at ORNL that had been created three years earlier. They met the RSIC director, the scientist **S. Keith Penny**, and **David Trubey** and **Fred C. Maienschein**, two other eminent specialists in the field of Radiation Shielding.

In 1966, the ongoing discussions were focused on theoretical aspects, in the frame of a seminar held at the CEA center of Fontenay-aux-Roses. The main topics discussed were geometry description and acceleration techniques dedicated to Monte Carlo particle transport simulations.

In the 60's CEA built a full body of knowledge on shielding computational methods and codes. At the same time radiation protection became a major subject in the nuclear world, both academic and industrial.
Indeed, in parallel to his experiment mockups, the CEA developed his first transport codes:

- NIOBE: a deterministic S_N code developed with the collaboration of **Everitt P. Blizard** from ORNL,
- The MERCURE and ORPHEE codes based on the straight line point kernel attenuation method, applied respectively for gamma rays and fast neutrons transport,
- The Monte Carlo simulation codes ZEUS and POKER written by **Jeanne-Marie Lanore**.

The CEA also went looking for radiation shielding calculation tools and nuclear data libraries in American, English and Italian nuclear centers, for instance:

- Two Monte Carlo codes SANE for the neutron propagation and SAGE for the photon propagation from ORNL.
- The 1D S_N transport code ANISN again from ORNL.
- The *Removal-Diffusion* neutron transport code SABINE code written by **Carlo Ponti** from ENEA Ispra center
- The cross section libraries from US (known as *Union National Carbide* libraries) or the *United Kingdom Nuclear Data Library* (UKNDL).

The result of the effort of the CEA to reach an international level was presented by **Christian Devillers**, **Pierre Lafore** and **Jean Rastoin** at the *Joint ENEA/IAEA Specialist meeting on the physics problems of reactor shielding*, held in Paris from 8 au 10 of December in 1970, under the title « *Reactor Shielding in the "Commissariat à l'Energie Atomique"* ».

The attention put to experimental code validation is shown by the measurements made on the HARMONIE facility: a 1 kWth source reactor built in CEA Cadarache in the frame of EURATOM-CEA cooperation. HARMONIE achieved it first divergence in 1965.

During this same period the TRIPOLI (*Monte Carlo Tridimensionnel Polycinétique*) code project was started with a great ambition.

This forward-thinking project has been decided jointly by **Pierre Lafore** and **Jean Rastoin**. The first TRIPOLI team was a small one composed of just two persons: **Jean-Claude Nimal** and **Shlomo Katz.** The latter left France shortly thereafter and **Thérèse Vergnaud** took his place.

The first version of TRIPOLI, TRIPOLI-1 was presented to the *ICRS-4* held in Paris in 1972. TRIPOLI-1 simulations were used to study the sodium activation of the PHENIX fast neutron reactor.

At CEA, the radiation shielding studies were the main driving force for the development of the Monte Carlo method and the associated computational codes.

The turn of the new decade 1970 could be considered as the entrance of the CEA in the international community of radiation shielding. The relationships with RSIC (that became RSICC: *Radiation Safety Information Computational Center* in 1996) became closer thanks to two persons: **Betty Maskewitz** from **RSICC** and **Enrico Sartori** from NEA. They were instrumental to the CEA allowing his own codes MERCURE, TRIPOLI and PEPIN (a depletion code) to be made available to NEA and RSIC: a rightful payback!

Another key moment was the seminar *"Monte-Carlo Methods and Applications in Neutronics, Photonics, and Statistical Physics"*, in the Château of Cadarache in the South of France, from 22th to 26th of April 1985. This meeting brought together the researchers from *Los Alamos National Laboratory* (LANL) and those from the *Commissariat à l'Énergie Atomique*. It has been organized respectively by **Robert Alcouffe** and **Robert Dautray**. **Daniel Verwaerde**, the current General Administrator of the CEA was present at that meeting.

The famed **Nicholas Metropolis**, the father of the modern Monte Carlo method with **John von Neumann** and **Stanislas Ulam,** was among the American delegates.

The CEA/DEN exposed there its theoretical and radiation shielding studies carried out with the TRIPOLI-2 code, which had been released in 1978.

In France, the Monte Carlo transport had finally demonstrated its ability to deal with complex, industrial scale radiation shielding problems. It was also the demonstration of its practical usefulness for nuclear engineering.

During the following decades, the CEA/DEN has pursued his R&D activity on the various types of methods traditionally used in the radiation shielding studies: deterministic methods, probabilistic methods and methods based on simplified models to compute dose rates, material structure activation, radiotoxicity and so on.

Our current calculation codes devoted to radiation shielding, TRIPOLI-4, DARWIN, MENDEL, NARMER, etc. are the heirs of this past briefly sketched.

Thanks to our nuclear data treatment system GALILEE still partly based on the NJOY American system, our calculations tools are connected to the most recent nuclear data evaluations, either American with ENDF/VII or European with JEFF-3 or TENDL, but also Japanese with JENDL-4, Chinese with CENDL-3, Russian with RUSFOND, and even international through IAIA with FENDL-2 or PDL.

Our simulation codes stand comparison with the best codes from other countries. In this respect, the SINBAD shielding benchmark data base, created by **Enrico Sartori** at NEA, plays a central role.

In order to deal with the present needs and to prepare the future, CEA continues to investigate the various radiation shielding problematics related to the different phases of the life of nuclear reactors: design, operation, dismantling/decommisionning, waste management. Its aim in to answer to EDF and AREVA industrial needs but also to the CEA internal needs as well.

The same tools are used to respond to the needs of research as those related to the MTR research reactors (OSIRIS, RJH...), related to the thermonuclear fusion devices (ITER project) or the power lasers.

In the field of code development, the advent of *high computing performance* opens new exciting challenges and perspectives both for the developers and the "end-users".

Furthermore, we observe a trend to extend the application scope of the tools developed, particularly towards nuclear instrumentation.

At last we can highlight the increasing involvement of the radiation shielding tools developed in the curricula of engineering school and of universities at the graduate level.

***

In conclusion, this brief focus on the CEA/DEN radioprotection activity, very much centered on the Monte Carlo simulation, must not overshadow the other branches of research and development related directly or indirectly to the radiation shielding.

Along those lines, we want to highlight the major contributions of CEA/DRF and of the CEA Bruyères-le-Châtel concerning the production of cross section data so essential for our numerical simulations as well as the radiation shielding studies around accelerators and power lasers

We are aware of the magnitude of the work that has been achieved and the progress made by the whole radiation protection and shielding international community aiming to respond to the crucial objective of human and environment protection. We can also add the objective to limit the material structure damages induced by particle irradiations.

If the radiation protection field has been led at the beginning by the nuclear reactor engineering for historical reasons, it is no longer the case today. This is shown by the diversity and the richness of your conference technical program and we think that that's a good thing.

You can be certain that the results of your works have been and will be carefully examined. They are an essential contribution to the orientations and planning of future R&D in our respective institutions.

Dear attendees, I wish you a wonderful and fruitful ICRS & RPSD conference. Having looked at the program, I have no doubt about it.

Thank you for your attention.

ICRS-13 & RPSD-2016

13th *International Conference on Radiation Shielding*
&
19th Topical Meeting of the *Radiation Protection & Shielding Division*
of the *American Nuclear Society* -2016

1st Plenary invited talk:

"Machine and radiation protection challenges of high energy/intensity accelerators: the role of Monte Carlo calculations"

Dr. Francesco Cerutti

CERN, Switzerland

The role of Monte Carlo calculations in addressing machine protection and radiation protection challenges regarding accelerator design and operation is discussed, through an overview of different applications and validation examples especially referring to recent LHC measurements.

Francesco Cerutti started his career on heavy ion reaction modeling at low energies, getting his degree in Physics in Milan and his PhD in Turin. After becoming a contributing author of the FLUKA code, he joined the CERN staff in 2006, as a member of the CERN FLUKA team, working both at the code development and at its application to various simulation studies concerning the CERN facilities, in particular the Large Hadron Collider. Since 2011, he leads the section of the Engineering Department in charge – among other activities - of beam machine interaction calculations. In the framework of the High Luminosity LHC project, he is the coordinator of the workpackage dedicated to the evaluation of particle showers and energy deposition. He is reviewer for several journals, including Physical Review, Nuclear Instruments and Methods in Physics Research, and Radiation Protection Dosimetry, and organizer of the series of Varenna Conferences on Nuclear Reaction Mechanisms.

MACHINE and RADIATION PROTECTION CHALLENGES of HIGH ENERGY/INTENSITY ACCELERATORS: The ROLE of MONTE CARLO CALCULATIONS

M.I. Besana, C. Bahamonde Castro, F. Cerutti, S. Damjanovic, L.S. Esposito, A. Ferrari, R. Froeschl, E. Iliopoulou, A. Lechner, A. Mereghetti, S. Roesler, E. Skordis, A. Tsinganis, V. Vlachoudis, J. Vollaire

ICRS 13
RPSD 2016

October 3|6, 2016
Paris, France

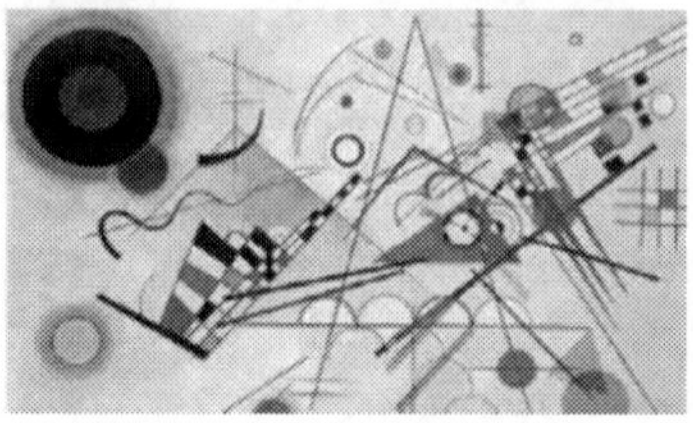

OUTLINE

- Beam losses and their consequences
- How to calculate them: Beam-machine interaction description
- Sightseeing through representative loss and machine cases

THE SOURCE OF THE PROBLEM: BEAM LOSSES

- *on beam intercepting devices (targets, collimators, dumps, stoppers, stripping foils, ...)*
  - *regular*
  - *accidental: smearing failures, injection mis-steering, asynchronous beam dump, top-off*
- *diffused*
  - *synchrotron radiation, beam-gas interaction, gas bremsstrahlung*
- *debris from regular collisions at the interaction points*
- *on unexpected obstacles (UFO, ULO, ...)*

CONSEQUENCES

	relevant macroscopic quantity
Heating	energy deposition (integral power)
Thermal shock	energy deposition (power density)
Quenching	energy deposition (power density)
Deterioration	energy deposition (dose), particle fluence, DPA
Oxidation, radiolysis, ozone production	energy deposition
Gas production	residual nuclei production
Single event effects in electronic devices	high energy hadron fluence [+ neutron fluence, energy deposition (dose)]
Shielding requirements	particle fluence (*prompt* dose equivalent)
Access limitations, radioactive waste, air activation	*residual* dose rate and activity
Beam Loss Monitors (BLM)	energy deposition
Radiation Monitors (RadMon)	thermal neutron and high energy hadron fluence
Tumor cell destruction	energy deposition (dose, biological dose)

PARTICLE INTERACTION WITH (ACCELERATOR) MATERIAL

A(n hadronic) shower implies a lot of different physics processes, touching a very broad energy [time-space] scale

Its description relies on the organic integration of diverse **theories and models**, and requires as essential pieces of **information**:

- reaction cross sections
- exclusive fragment production
- nuclide structure and decay data
- evaluated quantities of neutron induced reactions

Monte Carlo simulation is an effective way to calculate macroscopic quantities (like energy deposition, dpa, particle fluence, activation and residual dose rate) with an accuracy reflecting the quality of the critical processes implementation

Multipurpose widespread codes are available:
FLUKA, GEANT4, MARS15, MCNP6, PHITS, ...

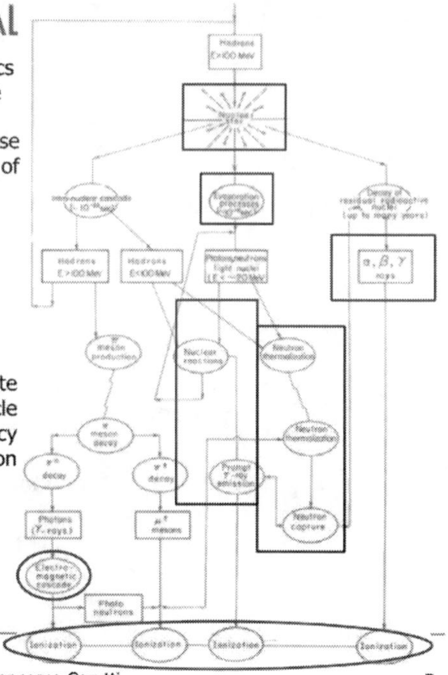

FLUKA

- hadron-hadron and hadron-nucleus interactions
- nucleus-nucleus interactions
- photon interactions (>100 eV)
- electron interactions (> 1 keV; including electronuclear)
- muon interactions (including photonuclear)
- neutrino interactions
- particle decay
- low energy (<20 MeV) neutron library
- ionization and multiple (single) scattering (including all ions down to 250 eV/u)
- combinatorial geometry and lattice capabilities
- voxel geometry and DICOM importing
- magnetic field
- analogue or biased treatment
- on-line buildup and evolution of induced radioactivity and dose

built-in scoring of several quantities (including DPA and biological dose)

✓ Accelerator design ✓ Shielding design ✓ Radiation protection
✓ Dosimetry ✓ Radiation damage ✓ Radiation to electronics effects
✓ Particle physics (calorimetry, tracking and detector simulation, ...)
✓ Cosmic ray physics ✓ Neutrino physics ✓ ADS systems, waste transmutation
✓ Medical applications, hadrontherapy ✓ Neutronics

LINE BUILDER

From DETAILED MODELS OF ACCELERATOR COMPONENTS WITH ASSOCIATED SCORING and the ELEMENT SEQUENCE AND RESPECTIVE MAGNETIC STRENGTHS, as given IN THE MACHINE OPTICS (TWISS) FILES, the **AUTOMATIC CONSTRUCTION OF COMPLEX BEAM LINES**, including collimator settings and element displacement (BLMs), is achievable, profiting from roto-translation directives and replication (lattice) capabilities

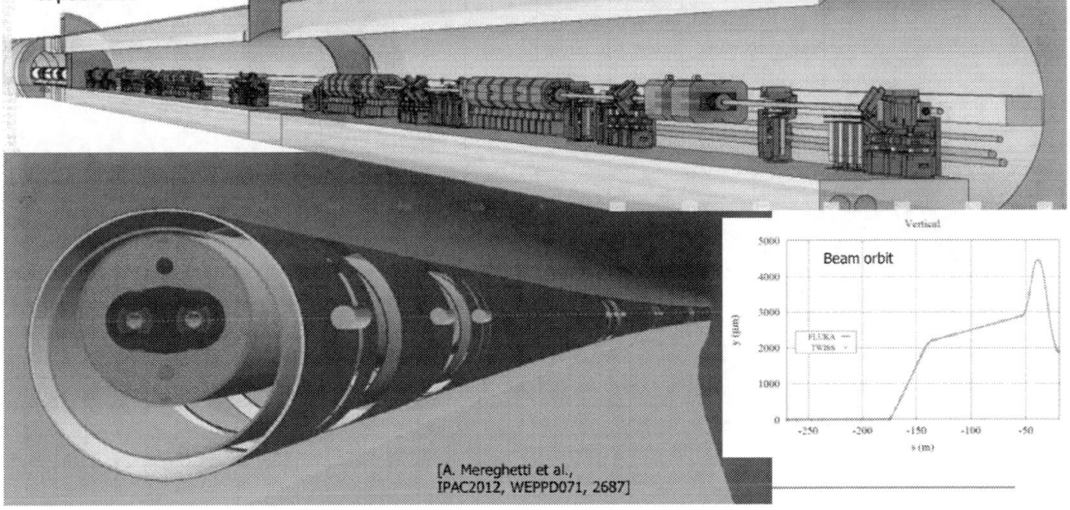

[A. Mereghetti et al., IPAC2012, WEPPD071, 2687]

CERN ACCELERATORS

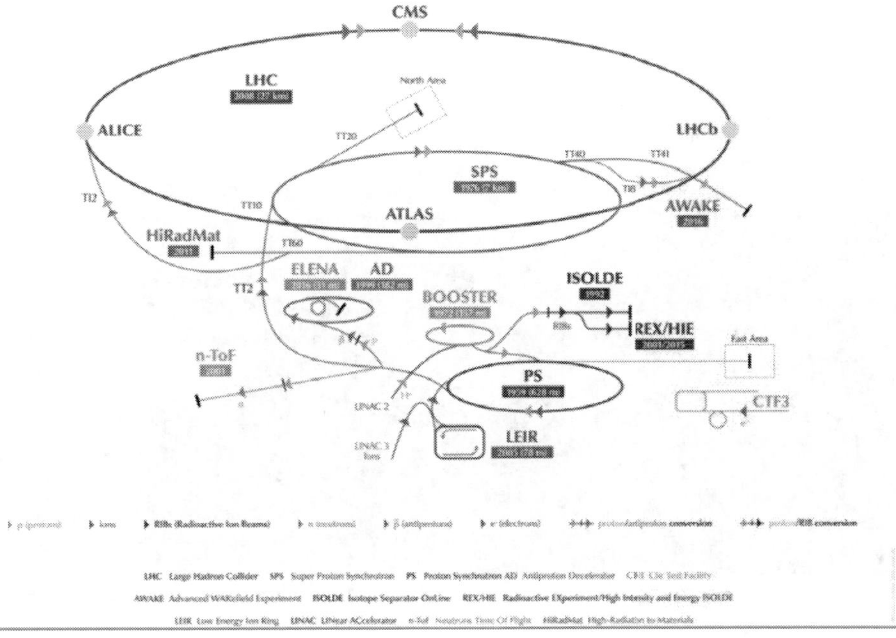

BEAM **L**OSS **M**ONITOR SYSTEM

Why to calculate the BLM response?

- Often BLM signals are the only measured quantity available when analysing loss events ... and they give absolute dose values!

The LHC BLM system:

- Ionisation chambers filled with N_2 gas at 1.1 atm, $\sim$1500 cm^3 sensitive volume
- More than 3000 BLMs are mounted along the ring
- The monitors trigger a beam extraction request in case of critical beam losses – machine protection, e.g. against magnet quench

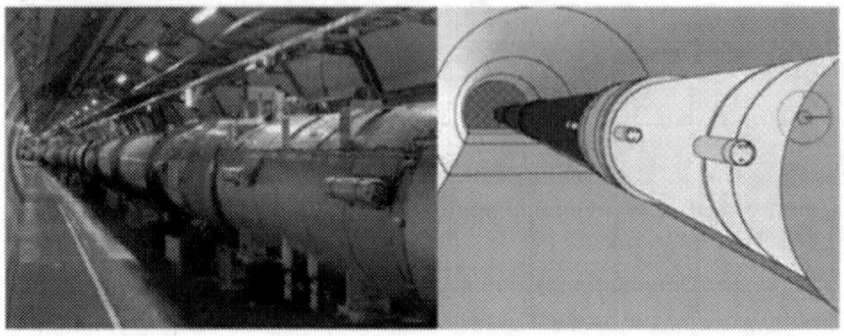

FLUKA BLM model

- Accurately reproduces circular electrodes, alumina spacers and stainless steel housing
- Response calculated by scoring energy deposition in active gas volume

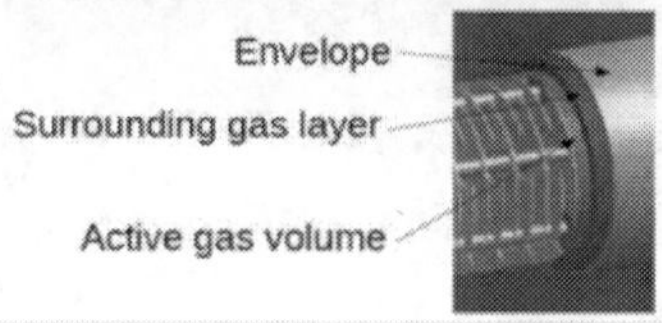

LUMINOSITY LOSSES

THE LHC 14 TeV CM p–p COLLISION DEBRIS

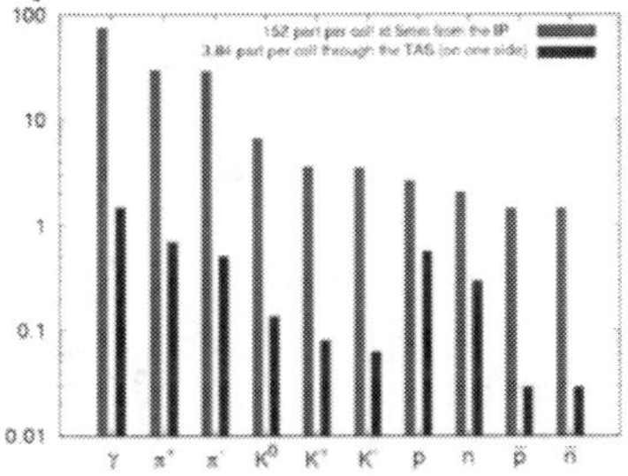

Instantaneous luminosity: 10^{34} cm^{-2} s^{-1}
(i.e. 8.5 10^{8} inelastic collisions)

950 W towards each (L&R) side

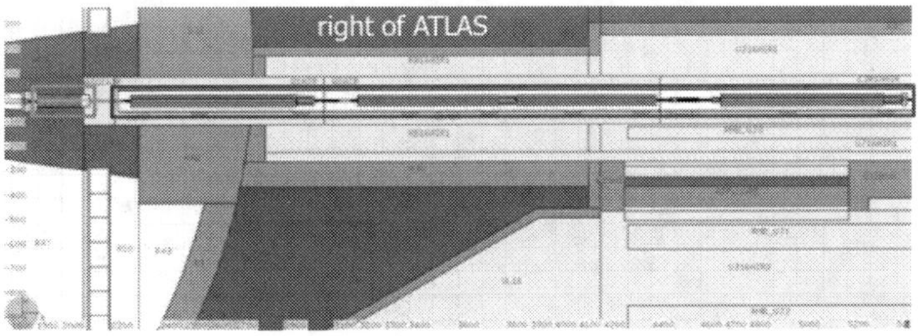

150 W absorbed in the TAS absorber

650 W going through the TAS

of which

150 W absorbed in the triplet

cold magnets

DEBRIS CAPTURE IN THE FINAL FOCUS "TRIPLET"

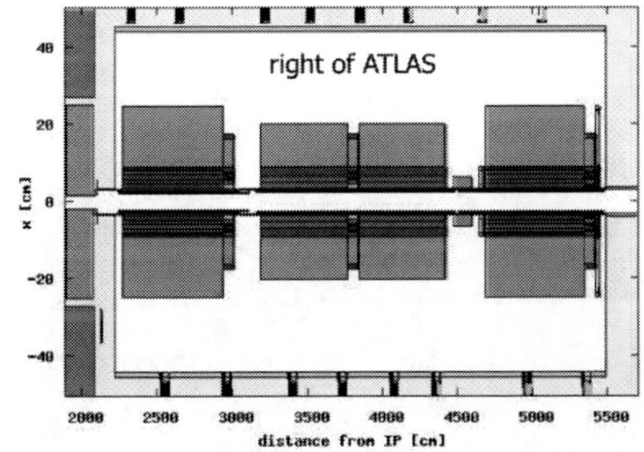

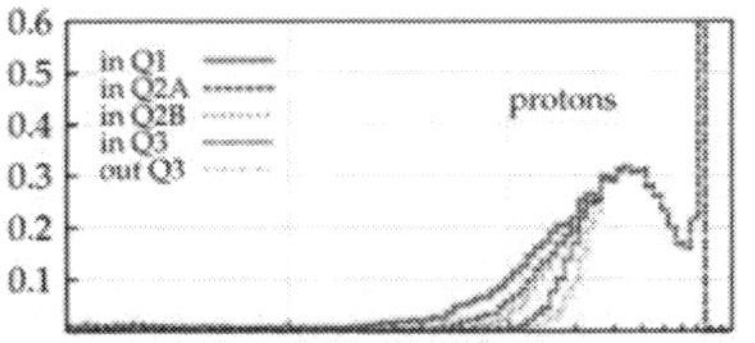

high energy spectra are depopulated by *charge particle* interception in the successive quadrupoles due to magnetic field bending

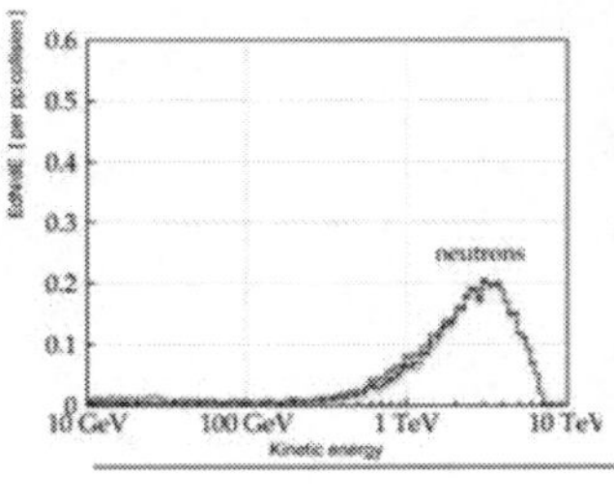

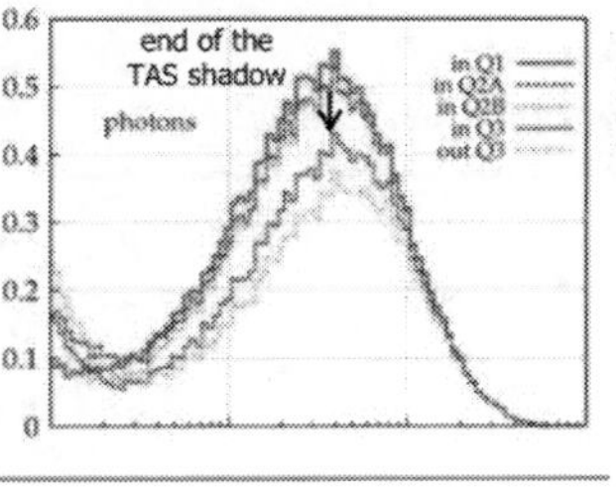

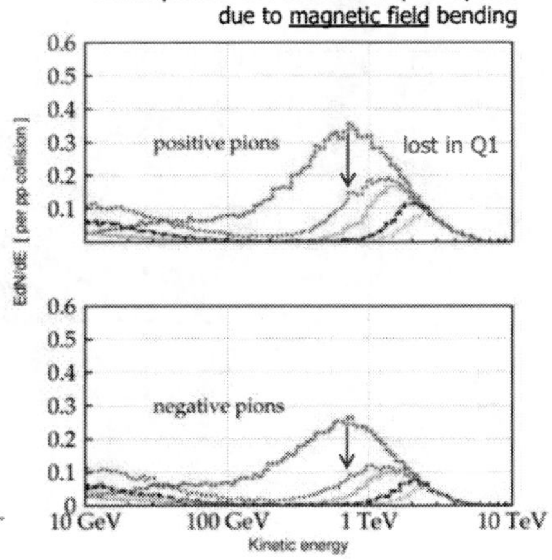

MARGIN TO QUENCH AND LIFETIME

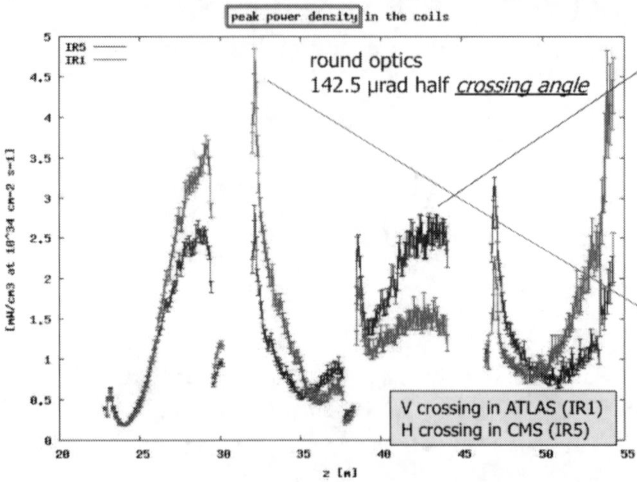

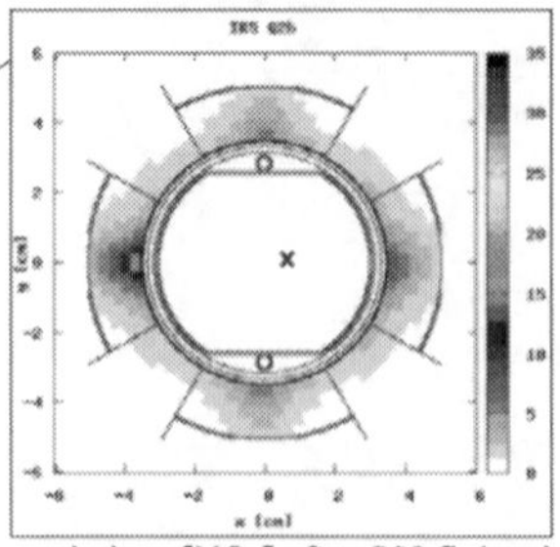

peak dose [MGy] after 300 fb^{-1}, relevant for triplet (i.e. coil insulator) lifetime

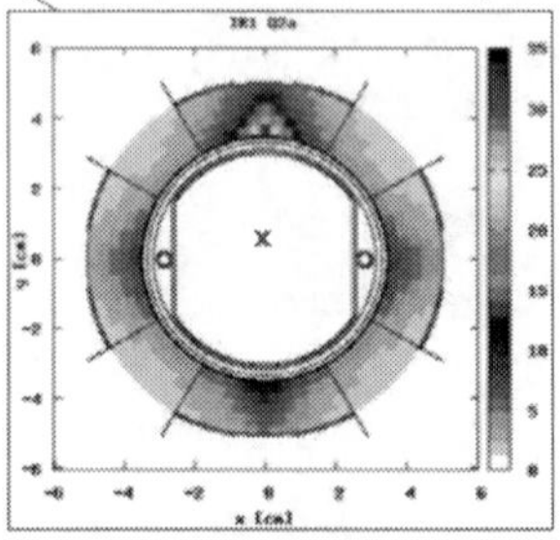

- different behavior reflecting the crossing plane variation for the same magnetic configuration of the triplet (FDF in the horizontal plane for positively charged particles coming from the IP)
- IR1 triplet (Q2a) reaches the design limit at nominal luminosity

RUN 1 (8 TeV CM): ATLAS

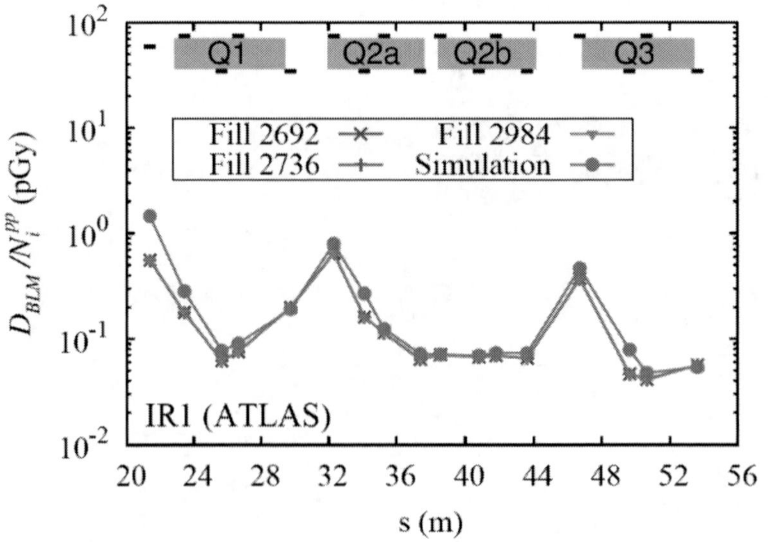

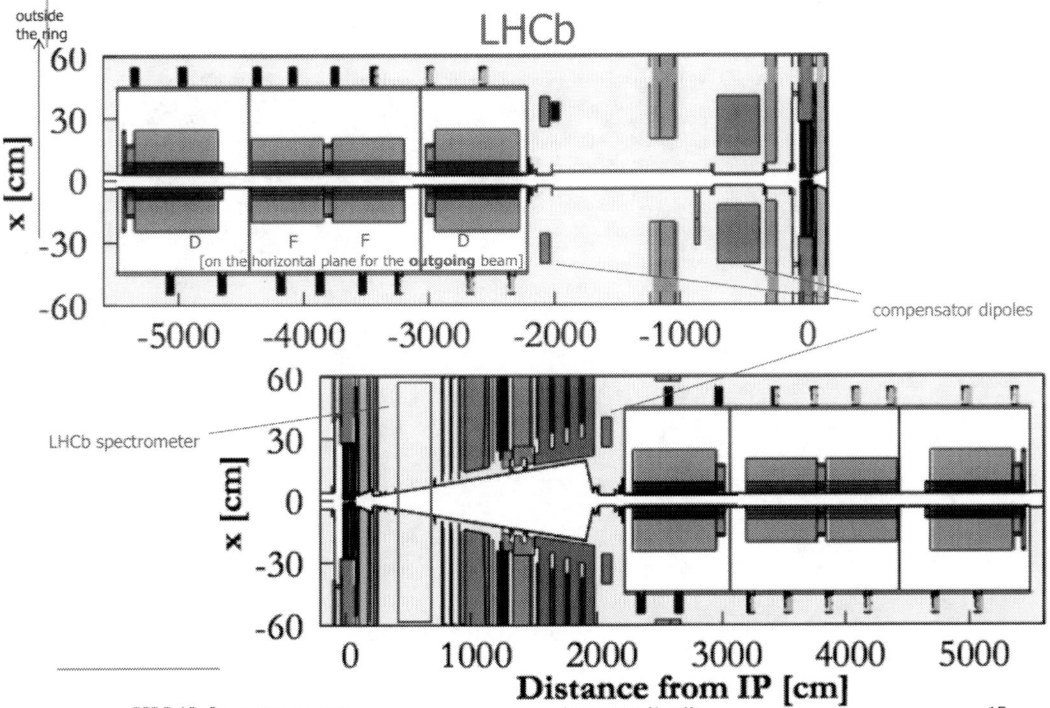

RUN 1 (4 TeV CM): LHCb

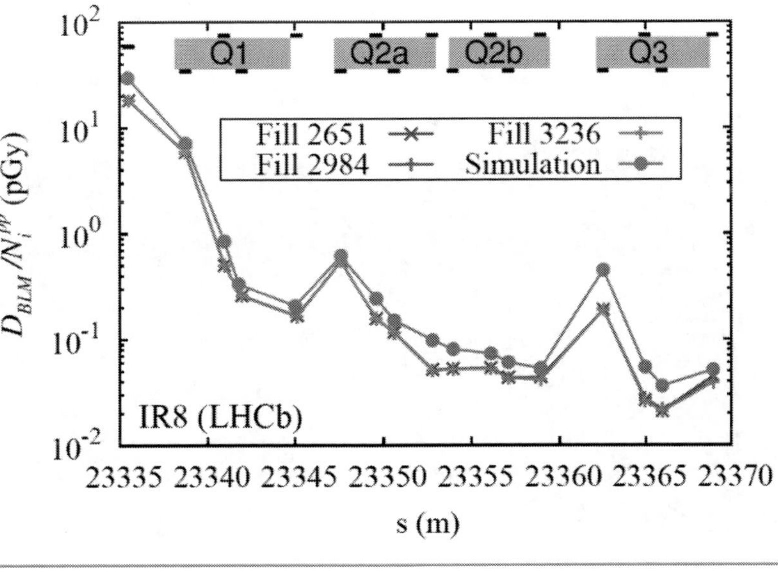

RUN 2 (13 TeV CM): ATLAS MATCHING SECTION

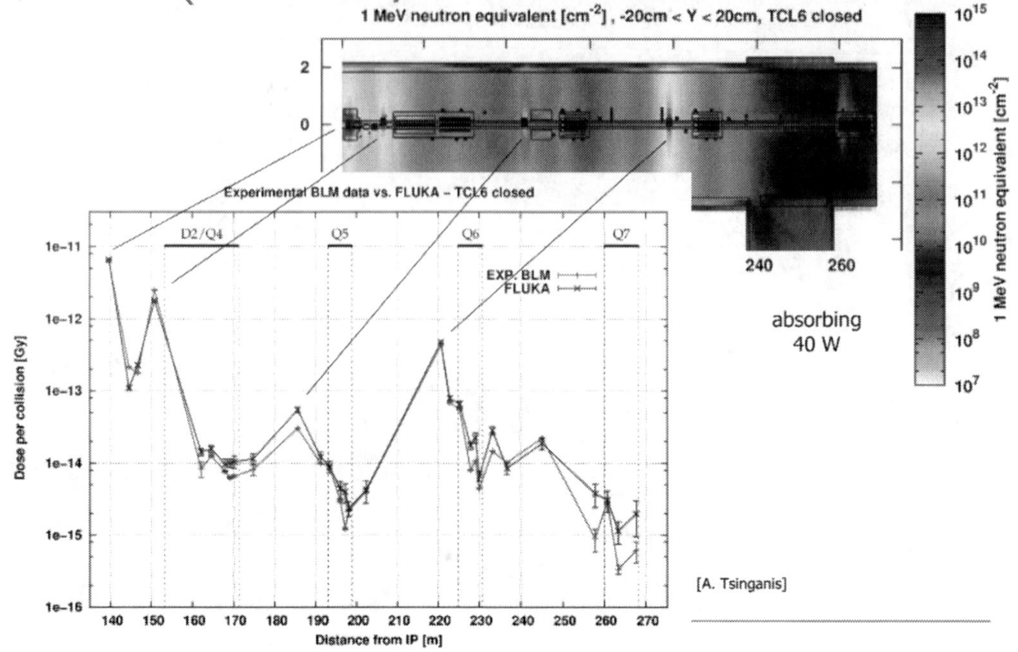

RADIATION TO ELECTRONICS

			relevant physical quantity the effect is scaling with
Single Event effects (Random in time)	***Single Event Upset (SEU)***	Memory bit flip (soft error) Temporary functional failure	High energy hadron fluence [cm^{-2}] (but also thermal neutrons!)
	Single Event Latchup (SEL)	Abnormal high current state Permanent/destructive if not protected	High energy hadron fluence [cm^{-2}]
Cumulative effects (Long term)	***Total Ionizing Dose (TID)***	Charge build-up in oxide Threshold shift & increased leakage current Ultimately destructive	Ionizing dose [Gy]
	Displacement damage	Atomic displacements Degradation over time Ultimately destructive	Silicon 1 MeV-equivalent neutron fluence [cm^{-2}] {NIEL -> DPA}

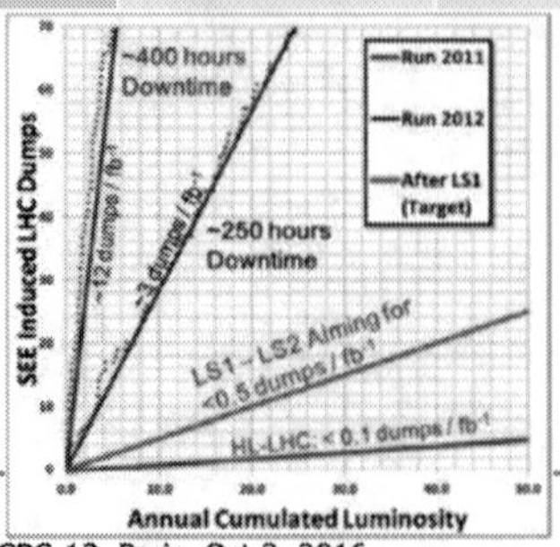

in 2016 3 R2E-related dumps up to 20 fb^{-1}

(i.e. 1.6 10^{15} inelastic collisions)

RUN 1: ELECTRONICS ALCOVE AROUND CMS (8 TeV CM)

RadMon

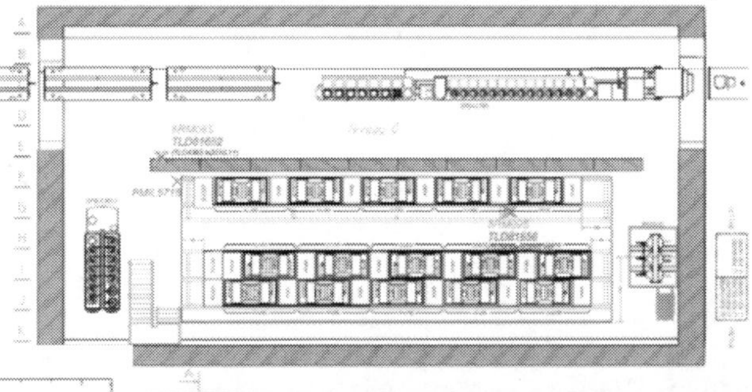

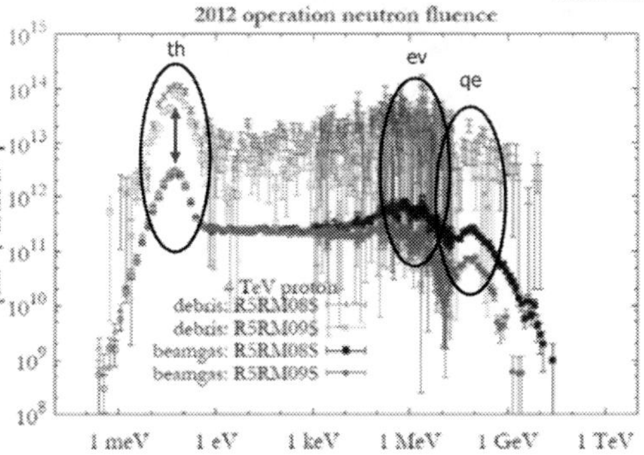

$F_{H>20MeV}$ [cm^{-2}] (L_{2012})	5RM08S	5RM09S
FLUKA	6.1 10^8	3.0 10^7
DATA	4.56 10^8 (256 upsets)	4.32 10^7 (25 upsets)

Agreement within 30%

(14 TeV CM) HL-LHC DESIGN: CODE INTERCOMPARISON

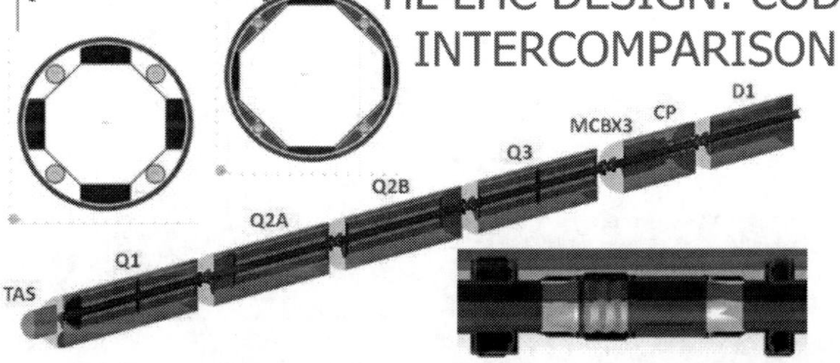

particle tracklength fraction in the Nb_3Sn coils [%]	
photons	88
electrons/positrons	7
neutrons	4
pions	0.45
protons	0.15

Instantaneous luminosity: 5 10^{34} cm^{-2} s^{-1}

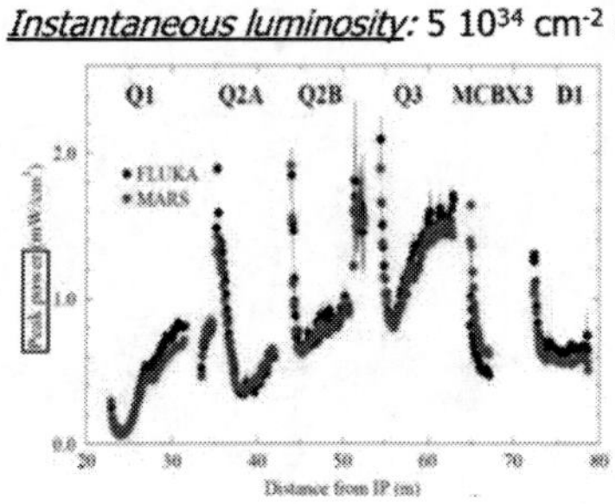

Integrated luminosity: 3000 fb^{-1}

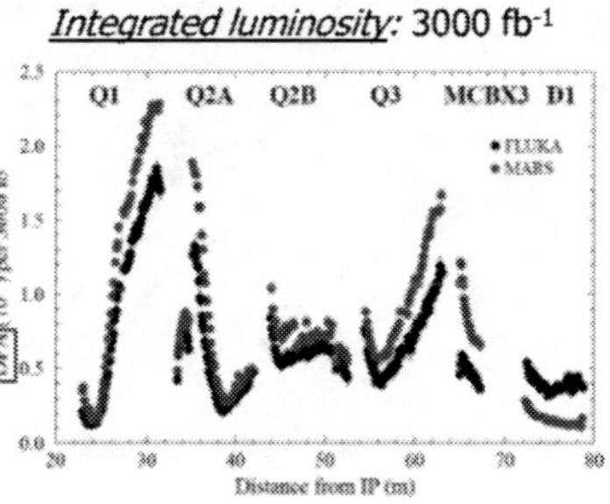

[MARS calculations by N. Mokhov and his team, FNAL]

5 A TeV CM Pb–Pb: BFPP LOSSES

Instantaneous luminosity: 10^{27} cm^{-2} s^{-1}

Bound Free Pair Production $Pb^{82+} + Pb^{82+} \rightarrow Pb^{82+} + Pb^{81+} + e^{+}$ 280 b (one side) (vs 8b nuclear inelastic)

23 W towards each (L&R) side

magnetic rigidity relative increase $\delta = 0.01235$

[C. Bahamonde]

Right DS of IP5

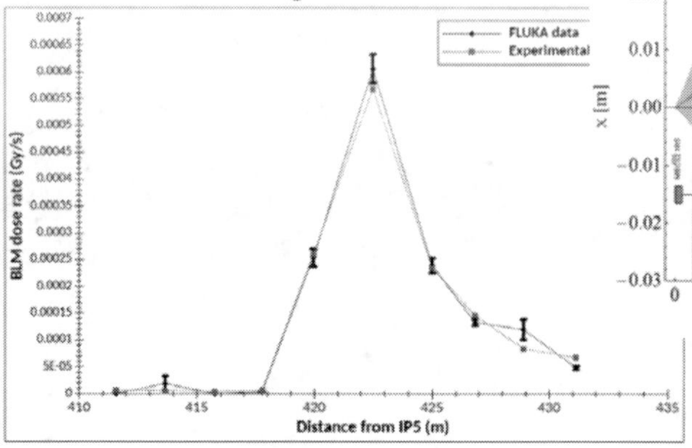

Secondary Beam Paths right of IP5

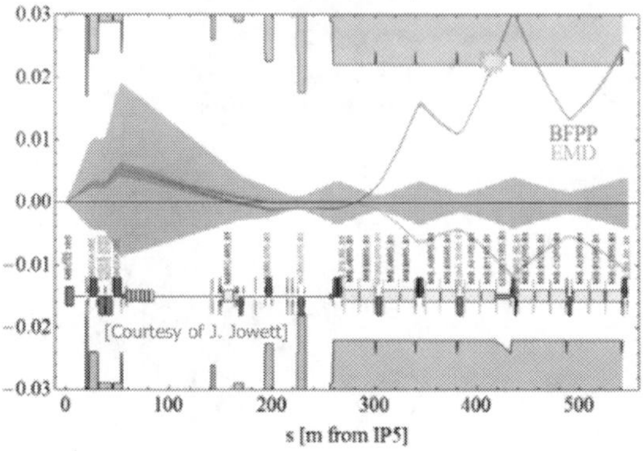

FUTURE CIRCULAR COLLIDERs

International FCC collaboration (CERN as host lab) to study:

- p-p collider *(FCC-hh)*, driving infrastructure requirements
  - 80-100 km tunnel infrastructure in Geneva area
  - 16 T dipoles → 100 TeV CM
- e^+-e^- collider *(FCC-ee)*, as intermediate step at different CM energies tailored to the production of
  - Z (45.5 GeV beam energy)
  - ZH (120 GeV beam energy)
  - $t\bar{t}$ (175 GeV beam energy)
- p-e^- *(FCC-he)*

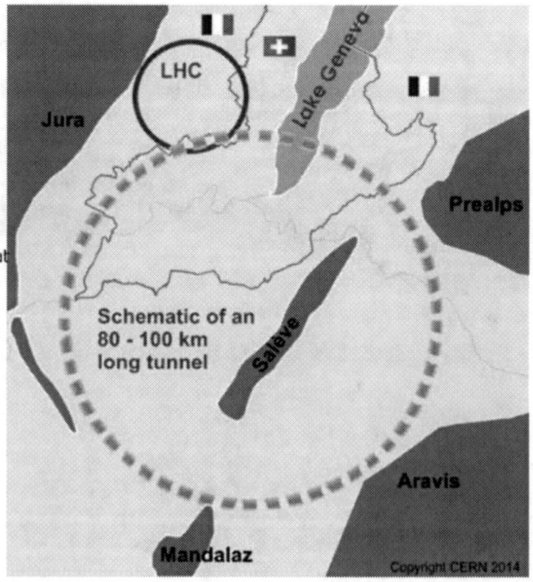

FCC-hh DETECTOR (100 TeV CM p–p)

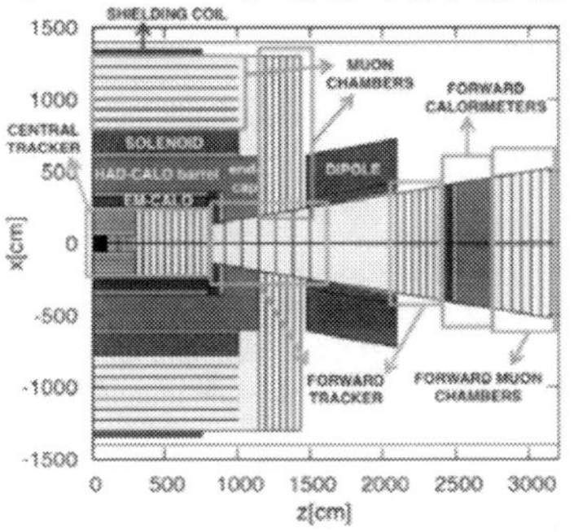

Instantaneous luminosity: 30 10^{34} cm^{-2} s^{-1}

[M.I. Besana]

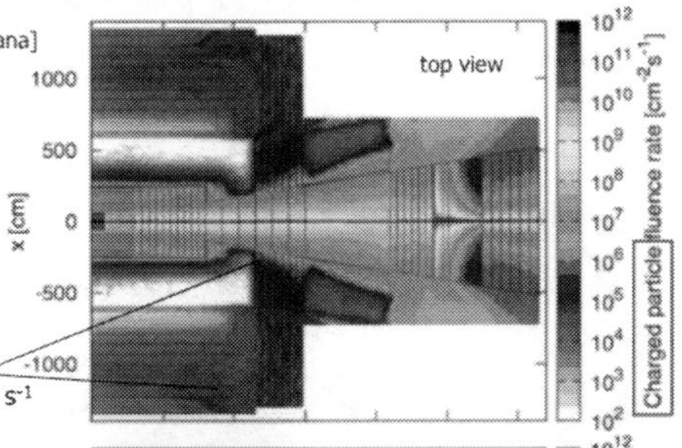

neutrons
@ 10^7–10^8 cm^{-2} s^{-1}

2m *iron* shielding with *polyethylene* and *lead* layers

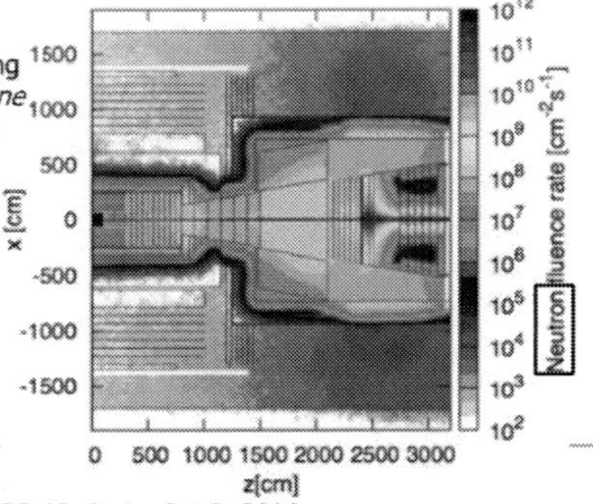

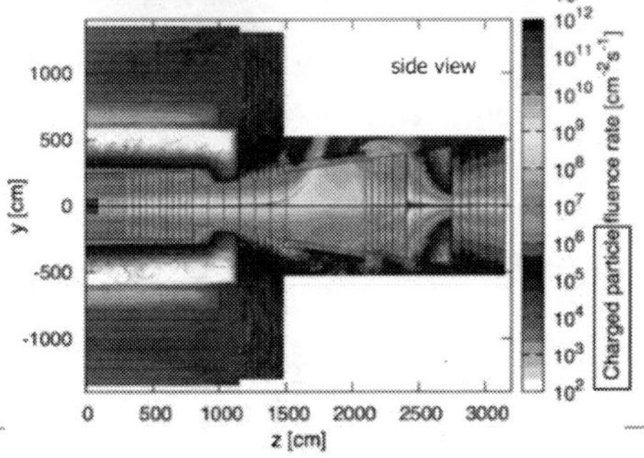

OBSTACLE-INDUCED LOSSES

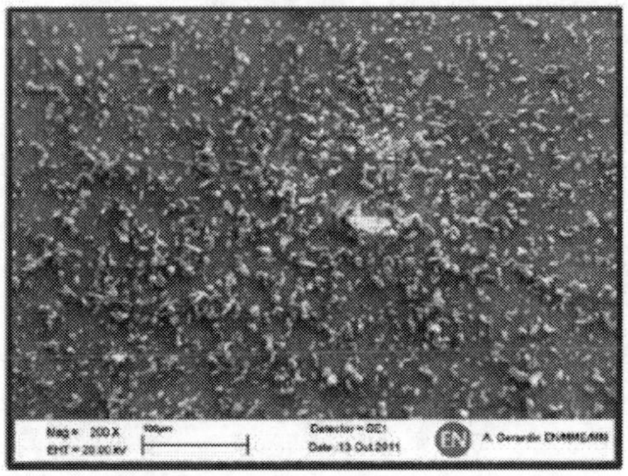

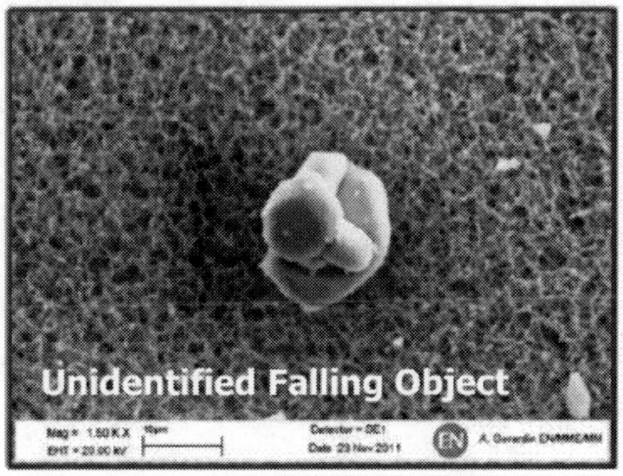

UFOs IN THE LHC (6.5 TeV/c p)

i. UFO locating

[A. Lechner]

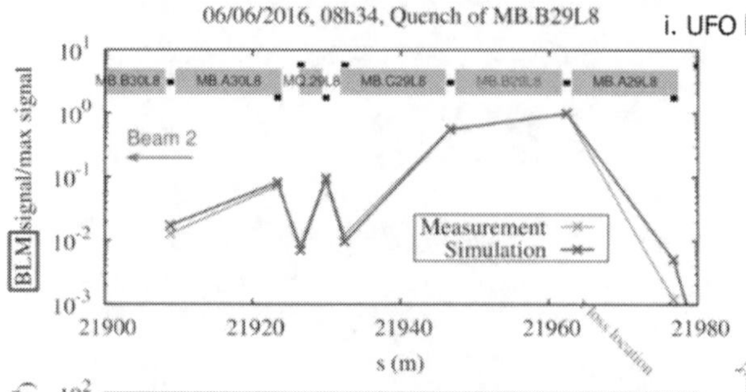

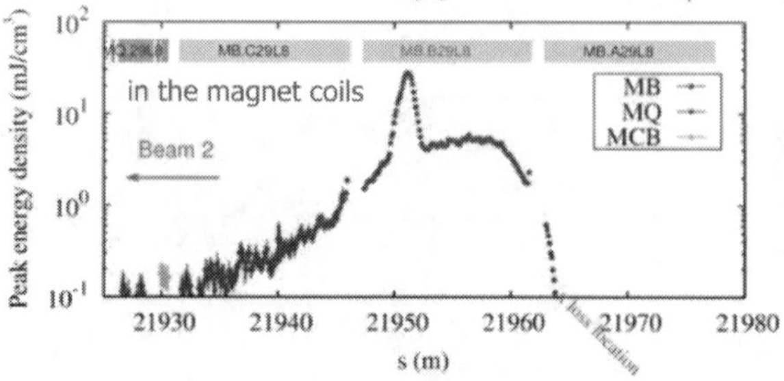

ii. quantification of the impact on the superconducting magnets

iii. quench limit understanding and BLM threshold setting

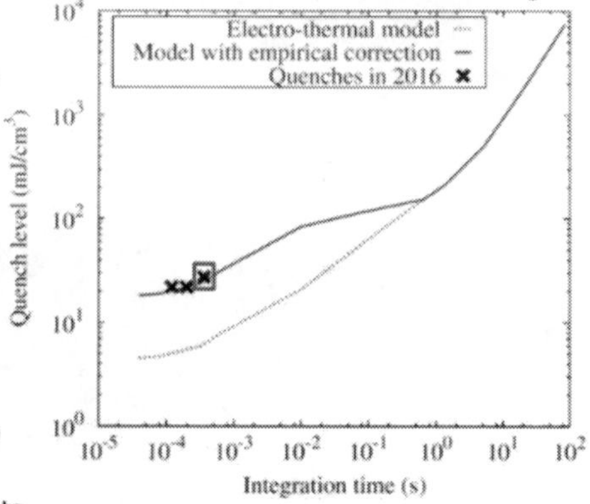

LOSSES ON BEAM INTERCEPTING DEVICES

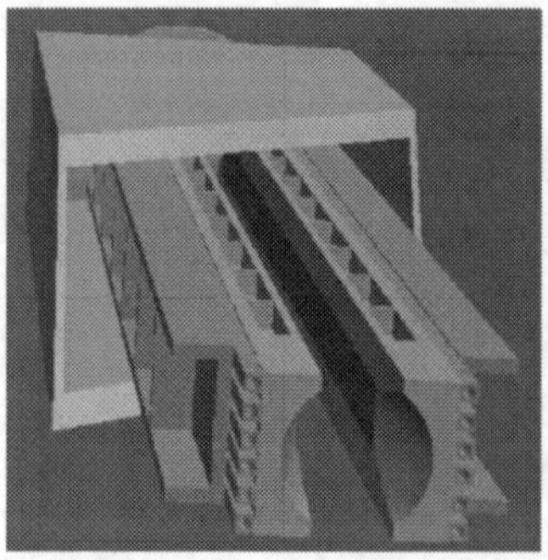

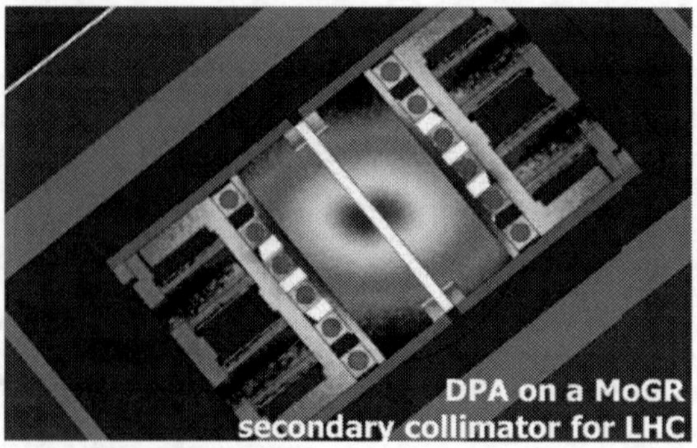

LHC COLLIMATION (4 TeV/c p)

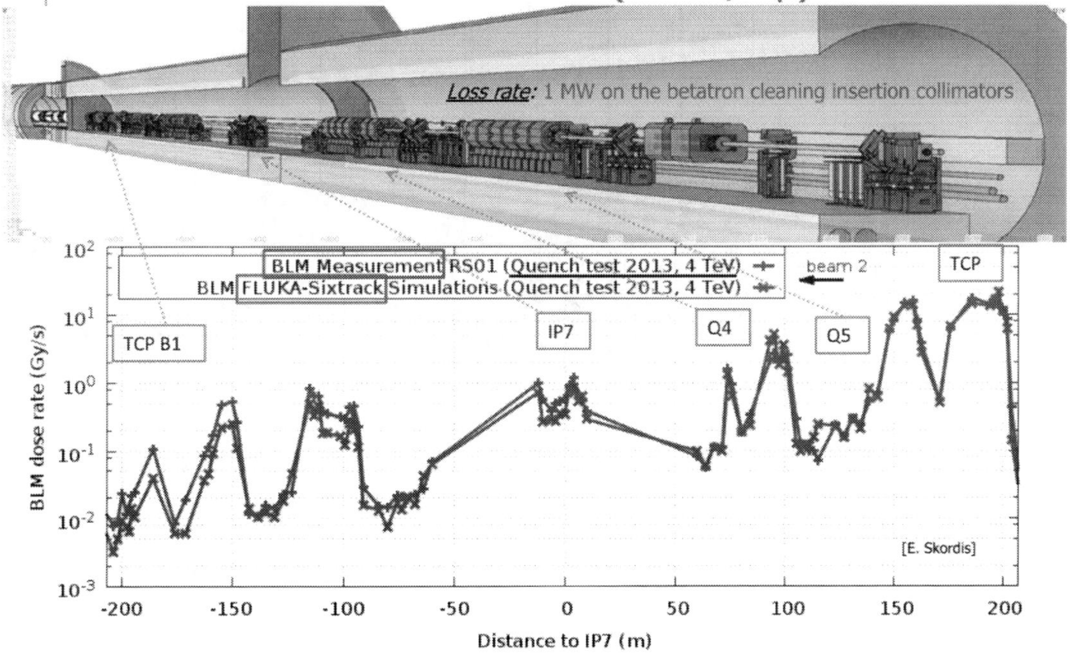

ANTIPROTON DECELERATOR TARGET AREA (26 GeV/c p)

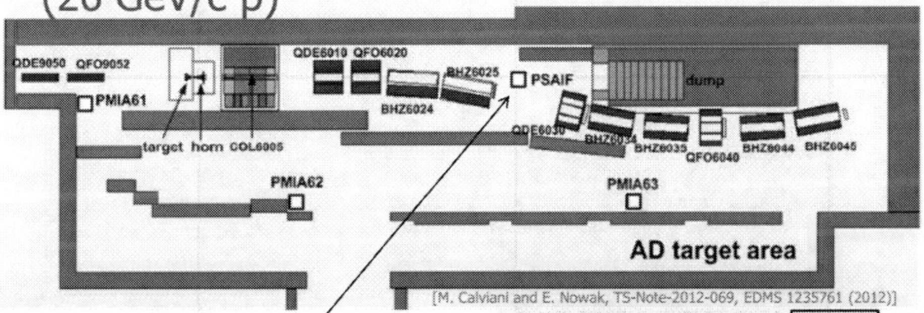

[M. Calviani and E. Nowak, TS-Note-2012-069, EDMS 1235761 (2012)]

66-80 mGy per pulse of 1.4 10^{13} protons

(1990-1991 measurements)

68-70 mGy/pulse (FLUKA simulation)

DOSE: 50 ± 15 Gy (meas.) vs. 60 Gy (FLUKA)

HEH fluence/thermal neutron fluence (R factor):

5 ± 40% (meas.) vs. 5 ± 10% (FLUKA)

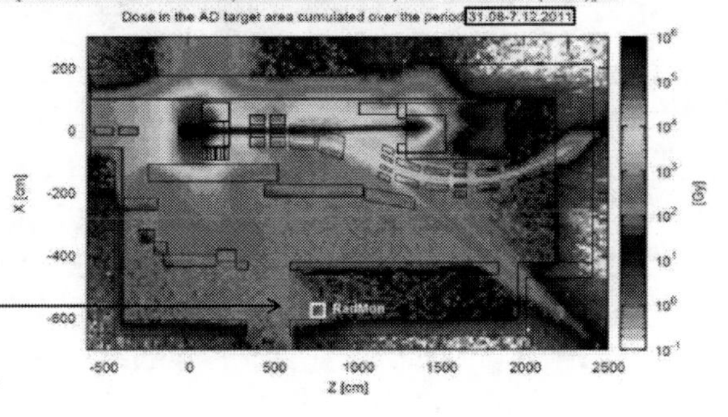

nTOF SHIELDING (20 GeV/c p)

top view of the ISR tunnel on the side of the *n_TOF EAR-2 vertical beam line*
(for neutrons from 1.6 10^{12} Hz 20 GeV/c protons on lead)

EAR2

Prompt dose equivalent
0.5 μSv/h classification limit

[I. Bergström, V. Vlachoudis, J. Vollaire]

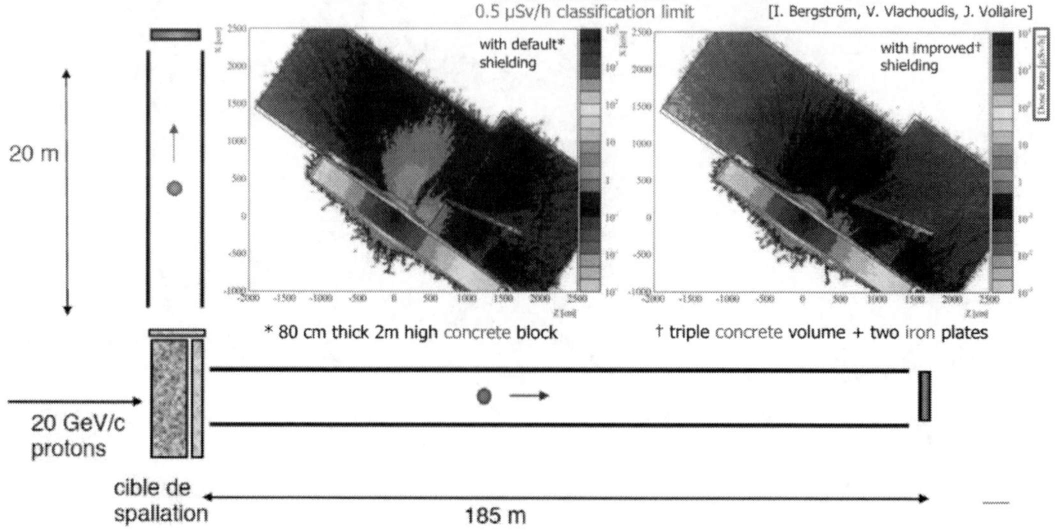

CERN SHIELDING BENCHMARK FACILITY (24 GeV/c p)

Situated laterally above the CHARM target
for deep shielding penetration studies (Detector calibration, Detector inter-comparison, Activation)

360cm of concrete and barite concrete
plus 80cm of cast iron

11 Bismuth and Aluminum samples
at different heights in CSBF and also inside CHARM (@ -80cm)

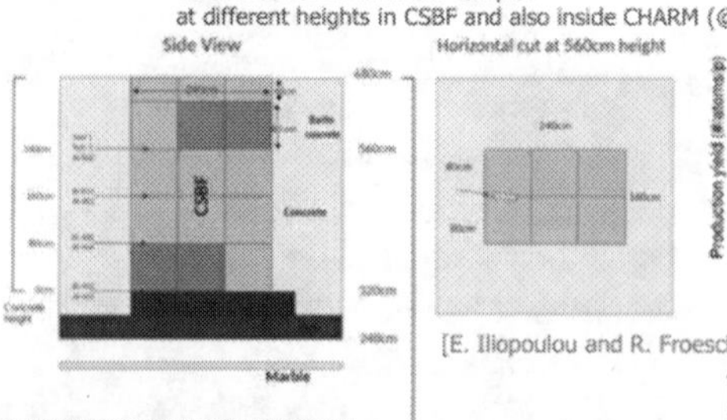

[E. Iliopoulou and R. Froeschl]

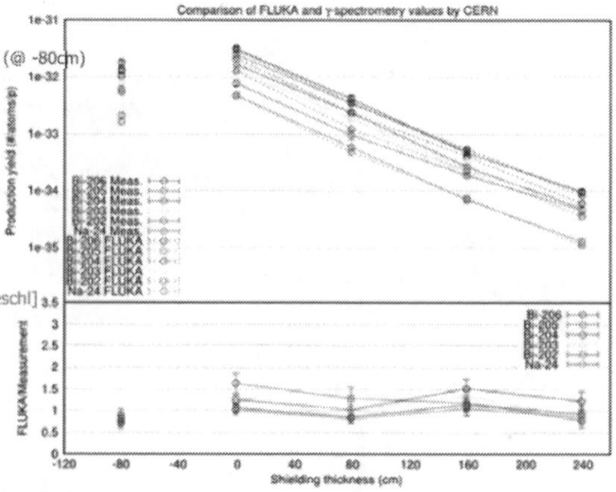

@ CHARM (CERN High energy AcceleRator Mixed field facility,
to study radiation effects on electronic components)
5 x 10^{11} protons/pulse, 350ms pulse length, max. average beam intensity 6.6 x 10^{10} p/s
three 50cm long 8cm diameter targets: Copper, Aluminum, Aluminum with holes

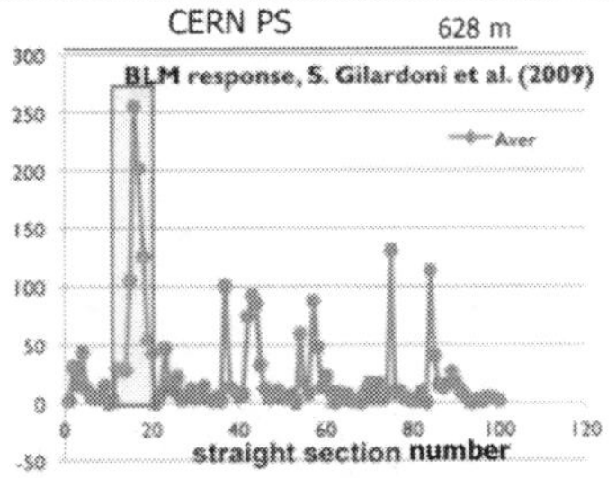

DIFFUSED LOSSES

2% beam loss, half of it in the extraction region at the SMH16 septum

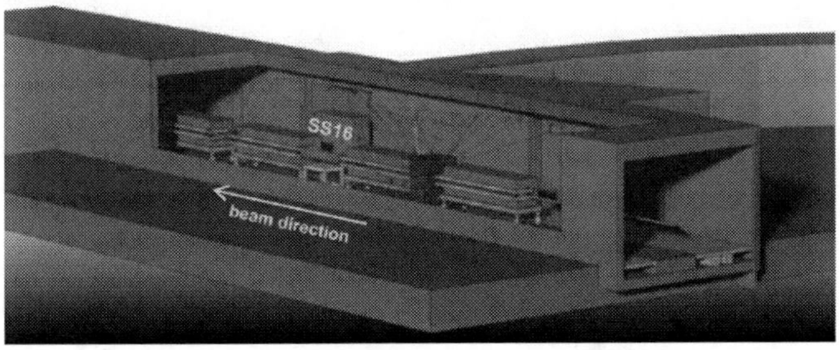

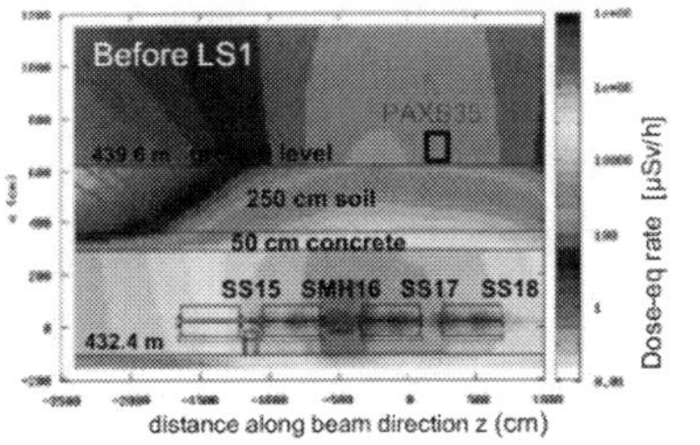

PS SHIELDING UPGRADE (14 GeV/c p)

FLUKA studies instrumental in defining the shielding improvements around the PS accelerator that were implemented in 2013-2014 (**L**ong **S**hutdown **1**). Simulation results systematically compared to data from radiation monitoring and a detailed measurement campaign (before and after the shielding re-enforcement).

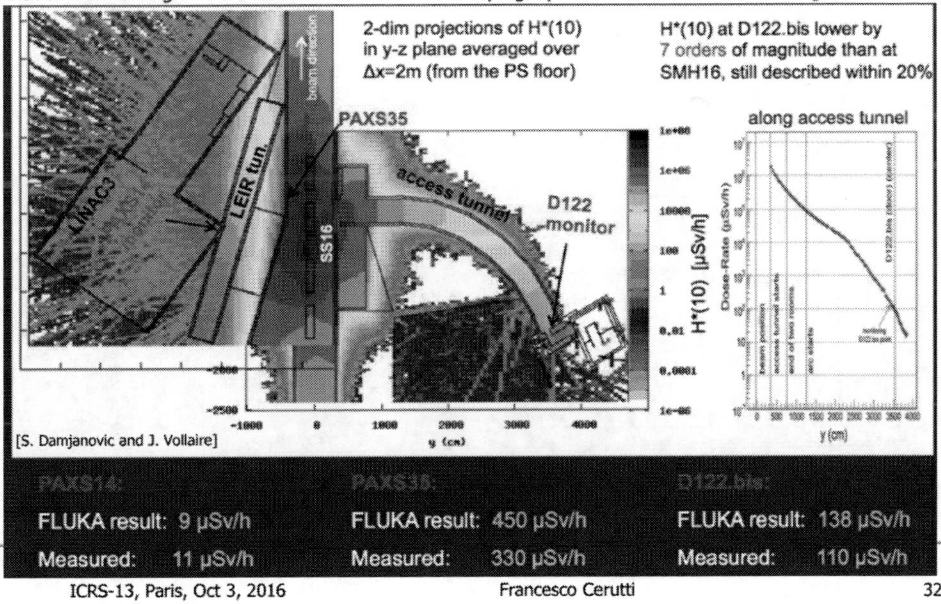

[S. Damjanovic and J. Vollaire]

PAXS14:	PAXS35:	D122.bis:
FLUKA result: 9 µSv/h	FLUKA result: 450 µSv/h	FLUKA result: 138 µSv/h
Measured: 11 µSv/h	Measured: 330 µSv/h	Measured: 110 µSv/h

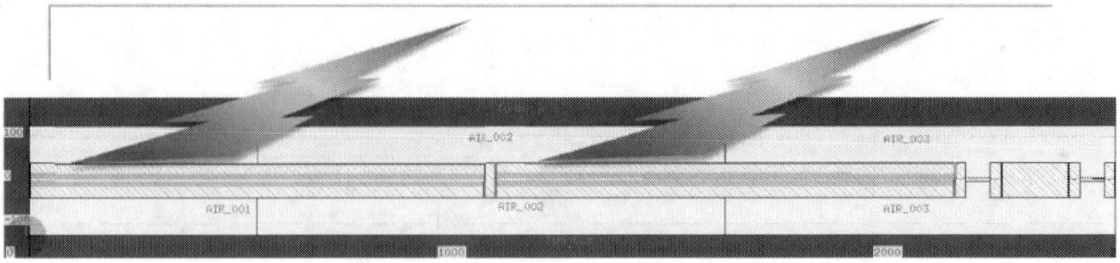

SYNCHROTRON RADIATION

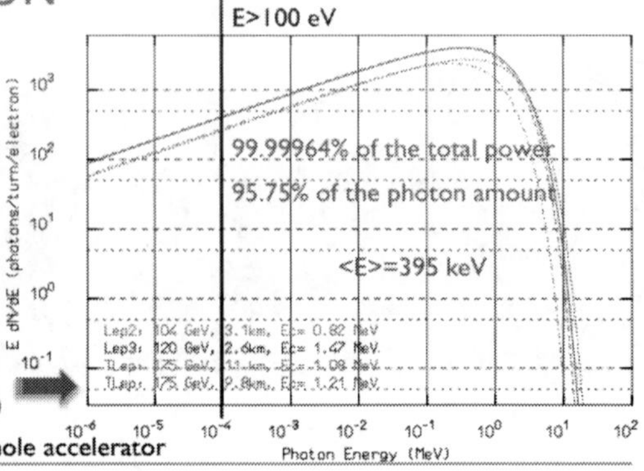

FCC-ee

- ΔE= 8.5 GeV/turn (dE/ds=1.375 keV/cm **in the dipoles**)
- P = 8.5 x I[mA] MW = 8.5 x <u>10mA</u> = 85 MW **in the whole accelerator**

(dP/ds= 1.375 x I[mA] W/cm **in the dipoles**)

FCC-ee

POWER INTERCEPTION

24 cm absorber concept (now obsolete)

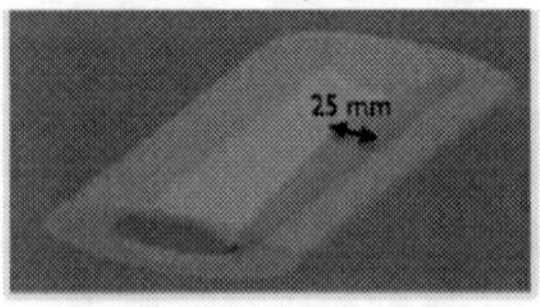

[L. Lari et al., IPAC 2014] power sharing

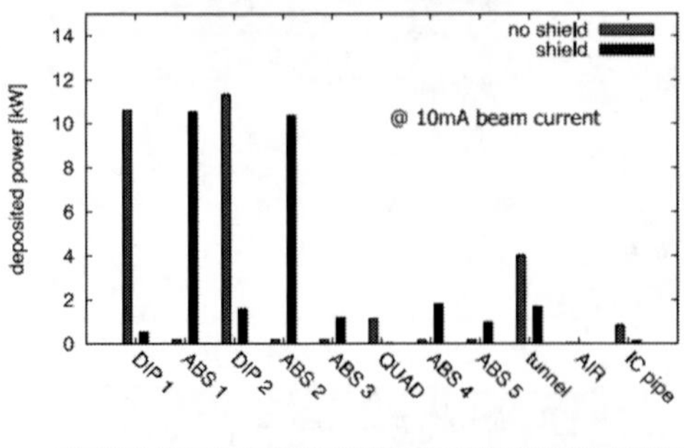

LONG TERM RADIATION LEVELS IN THE TUNNEL

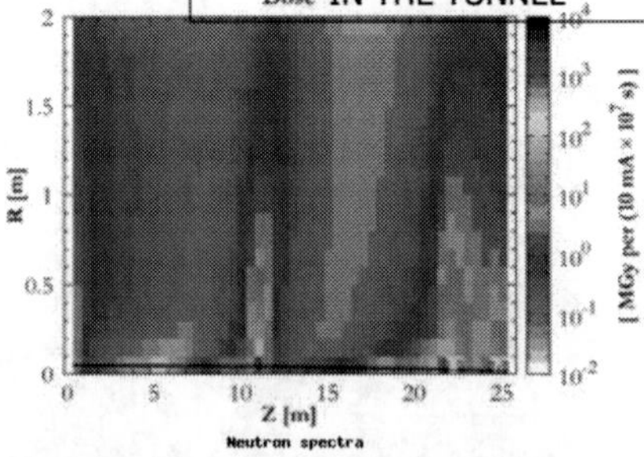

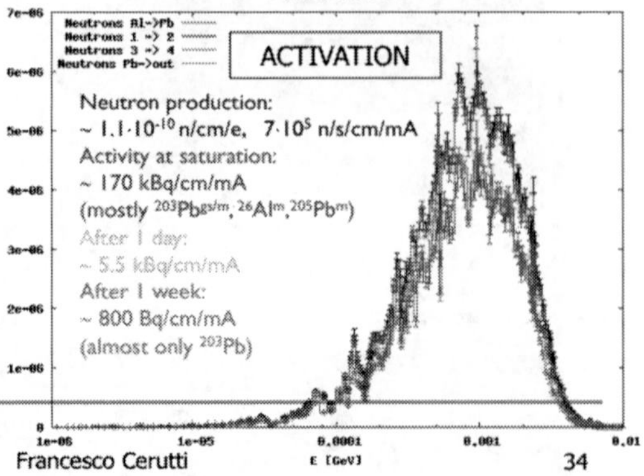

CONCLUSIONS

Monte Carlo has a role

ICRS-13 & RPSD-2016

13th ***International Conference on Radiation Shielding***
&
19th Topical Meeting of the ***Radiation Protection & Shielding Division***
of the ***American Nuclear Society*** **-2016**

2nd Plenary invited talk:

"Outline of the Fukushima Daiichi Accident. Lessons Learned and Safety Enhancements"

Dr. Masashi Hirano

Nuclear Regulation Authority, Japan

On March 11, 2011, an earthquake and subsequent tsunamis off the Pacific coastline of Japan's Tohoku region caused widespread devastation in Japan. As of June 10, 2016, it is reported that a total of 15,894 people lost their lives and 2,558 people are still unaccounted for. In Fukushima Prefecture, approximately 100,000 people are still obliged to live away from their homes due to the earthquake and tsunami as well as the Fukushima Daiichi accident. This paper briefly presents the outline of the Fukushima Daiichi accident and summarizes the major lessons learned having been drawn and safety enhancements having been done in Japan for the purpose of giving inputs to the discussions to be taken place in the Special Invited Session "Fukushima, 5 years after".

Dr. Masashi Hirano has worked in the area of nuclear safety at the Japan Atomic Energy Agency (JAEA) for more than 35 years. After the Fukushima Daiichi accident, he moved over to the Japan Nuclear Safety Organization (JNES) and then to the Nuclear Regulation Authority (NRA) of the Japanese Government, where he served as Director General for Regulatory Standard and Research. From 2015, he is Senior Coordinator for International Corroboration in the NRA.

NRA , Japan
Nuclear Regulation Authority

Fukushima Special Session
"Fukushima, 5 years after"

Outline of the Fukushima Daiichi Accident, Lessons Learned and Safety Enhancements

Masashi HIRANO and **Harutaka HOSHI**
Nuclear Regulation Authority (NRA)

The joint conference of
the 13th Int. Conf. on Radiation Shielding (**ICRS-13**), and
19th Topical Meeting of the Radiation Protection and Shielding Division of ANS - 2016 (**RPSD-2016**)
October 3 - 6, 2016, Paris, France

NRA , Japan
Nuclear Regulation Authority

Contents

- **Introduction**
  - Current Status of Nuclear Regulation in Japan
- **Outline of the Accident**
  - Earthquake and Tsunami
  - Accident Progression
  - Source Term Evaluation
- **Lessons Learned**
- **Safety Enhancements**
  - Creation of NRA
  - New Regulatory Requirements
- **Summary**

Current Status of Nuclear Regulation in Japan

- NRA was established in **Sep. 2012** and developed the **new regulatory requirements** for NPPs which came into force in **July 2013**.
  - **All the 48 units** had been shut down **since Sep. 2013**.
  - So far, a total of **26 units** (**17 PWRs** and **9 BWRs**) have applied for **conformance review**.
  - In **Sep. 2014**, NRA first approved the applications from **Sendai Units 1 and 2**.
  - As of Aug. 2016, NRA has granted permission to **7 PWRs** of which **four units have restarted** (**Sendai Units 1,2 and Takahama Units 3, 4**)
- The **IAEA IRRS mission's review** was taken place in **Jan. 2016**.

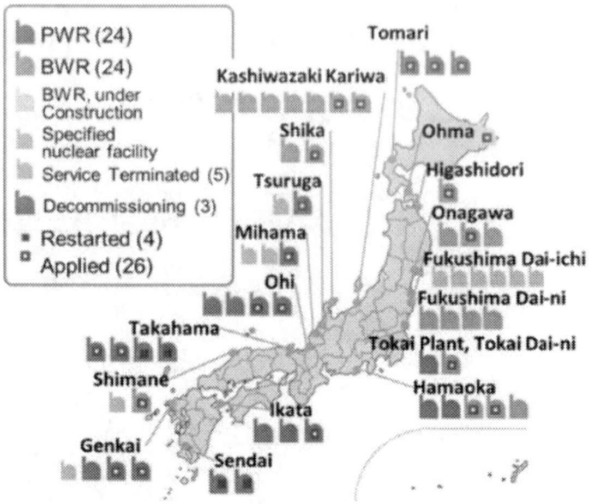

Y. Shimizu, IAEA Effective Nuclear Regulatory Systems (April 2016)

Outline of the Accident

NRA . Japan
Nuclear Regulation Authority

Tohoku-District off the Pacific Ocean Earthquake

- Occurred at **14:46** on **March 11, 2011**
- **9.0 Mw**: largest ever recorded in Japan
- Rupture duration: **170 sec**
- Epicenter: 38° 10''N and 142° 86''E, 23.7km in depth
- **Reverse fault type near the Japan Trench**

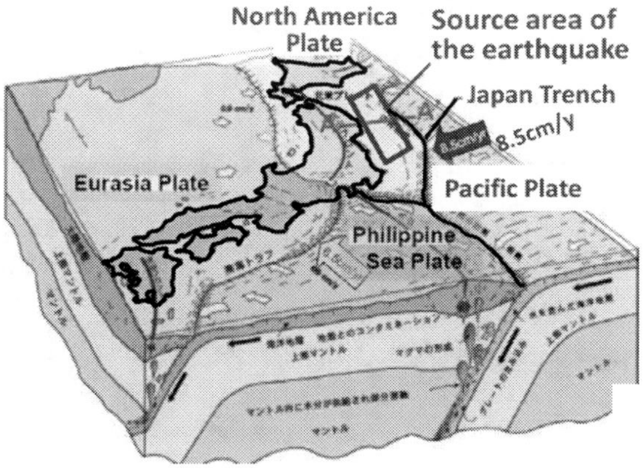

Edited to figure in "1st Government Report (June) 2011)"

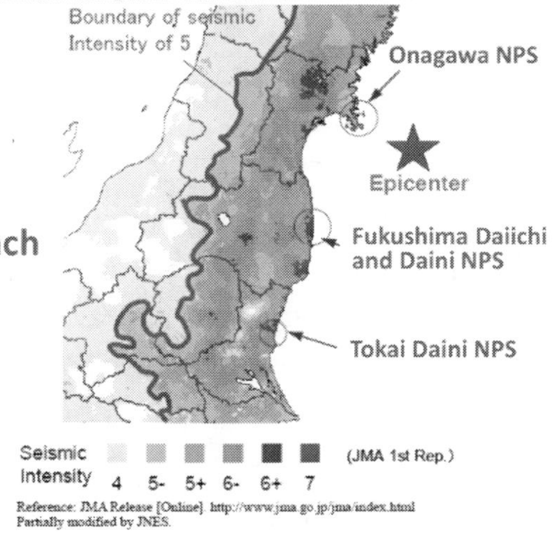

- Observed **max. acceleration** of seismic ground motion at **Fukushima-Daiichi** exceeded that of the **design basis by 26%**.

NRA . Japan
Nuclear Regulation Authority

Accompanied Tsunami

- A **huge tsunami** was generated due to a **large slip**, more than 56m, **near the Japan Trench**.

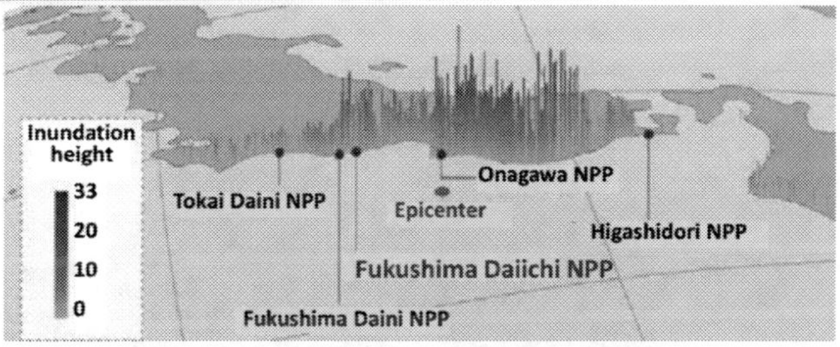

IAEA Fukushima report, Technical Volume 1/5 (2015)

Estimated slip distribution

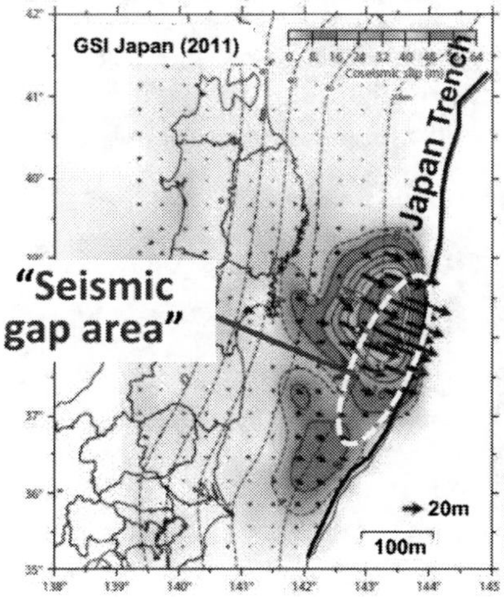

Sugino, et al., NRA, 24 Feb, 2016

- **Multi-segment rupture** including **"seismic gap area (no seismic record)"**

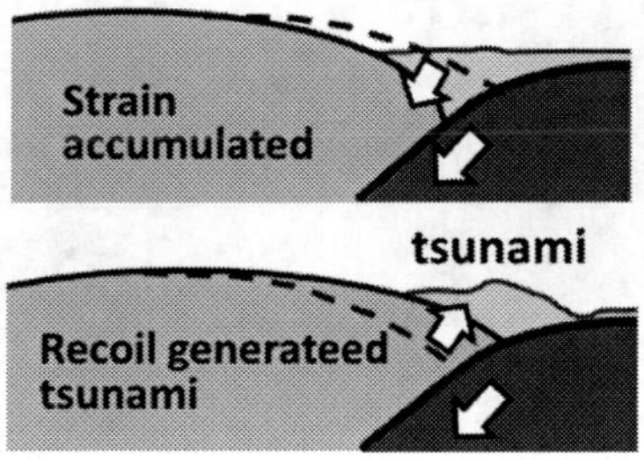

- Due to rupture of the **bonded interface**, the **North America Plate recoiled** and the strain energy was released, which generated the huge tsunami.

NRA. Japan Nuclear Regulation Authority

Consequences of Natural Hazards and the Accident

- **Casualties due to earthquake/tsunami** (as of Dec.1, 2011)
  - **Deceased**: 15,840
  - **Unaccounted for**: 3,547
  - **Injured**: 5,951
  - **Damaged buildings**: 1,009,074

Interim Report from Government's Investigation Committee (Dec. 2011)

- As of June 2016, the number of evacuees has been reduced from **470,000** to **155,000**. Still **51,000 people** live in **temporary housing**.
- In **Fukushima Prefecture**, approx. **100,000** people are still **obliged to live away from their homes** due to the earthquake/tsunami as well as the Fukushima Daiichi accident. (website of the Fukushima Prefecture)
- **Evacuation order**: (1) Return is difficult (red), (2) Habitation is restricted (yellow), (3) Preparing for lifting orders (Green)
  - **Yellow and Green areas** will be lifted no later than March 2017. Current Status of Reconstruction and Challenges, June 2016, Reconstruction Agency

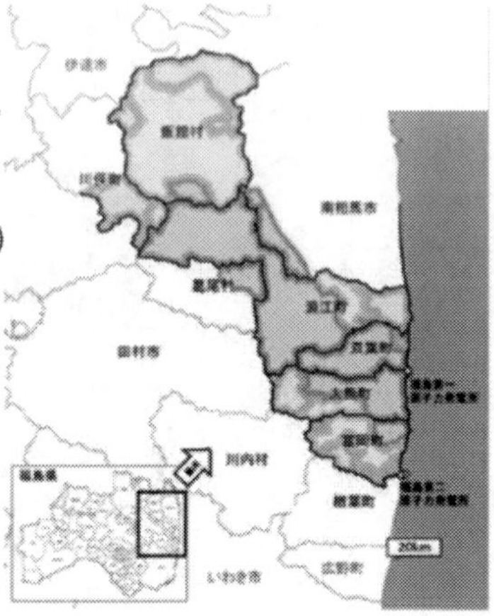

NRA. Japan Nuclear Regulation Authority

Tsunami Heights at four Different Sites

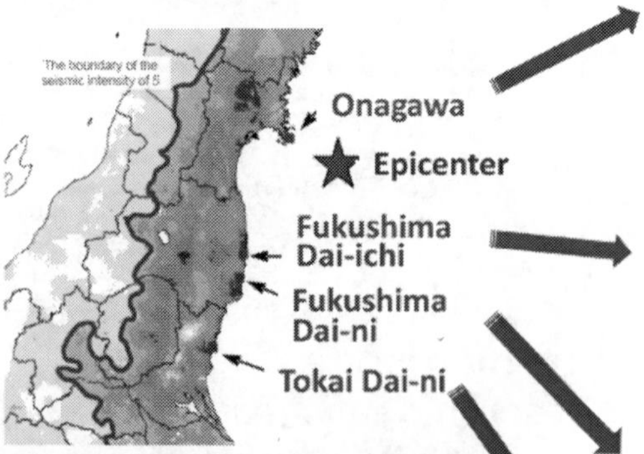

Tsunami at Fukushima Daiichi NPS

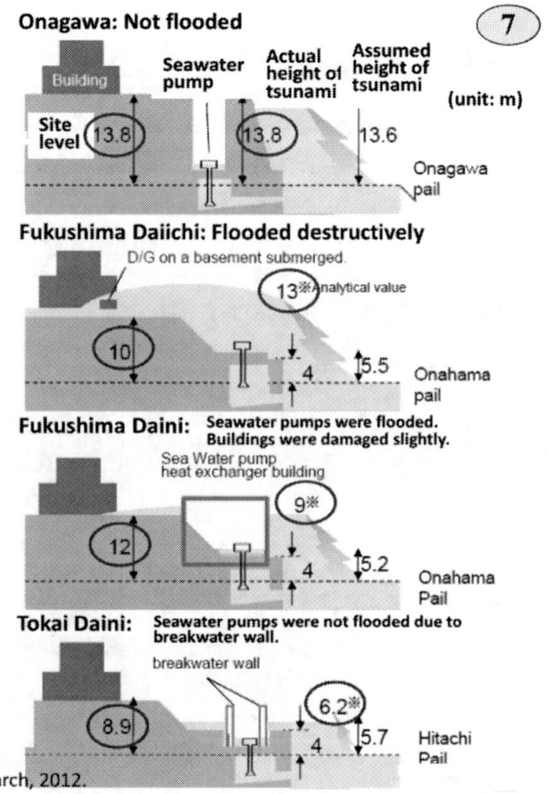

Yasui, METI, "Causes and Countermeasures", March, 2012.

NRA, Japan Nuclear Regulation Authority

Loss of Offsite Power by Earthquake and Common-Cause Failures by Tsunami

8

- Units 1 to 3 automatically scrammed on seismic reactor protection trips. The **earthquake** led to **loss of offsite power**. All the **emergency diesel generators (EDGs)** automatically started.
- Then, **about 40 min. later**, the **tsunami** arrived at the site.
  - In Units **1 to 4, all the EDGs became inoperable** (SBO) together with loss of DC power.* It also caused LUHS as well.
  - In Unites **5 and 6**, an **air-cooled EDG of Unit 6 survived** and it supplied **AC power** to both units.

*DC power was available in Unit 3.

Seawater pumps for residual heat removal became inoperable

⬇

Loss of Ultimate Heat Sink (LUHS)

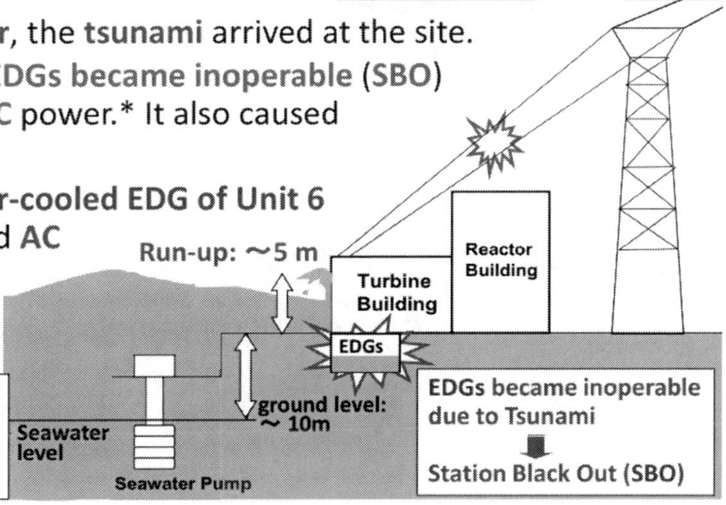

NRA, Japan Nuclear Regulation Authority

Major Chronology of Events

Source: IAEA DG Report and Technical Volumes 1 (2015):

9

Unit 1	Unit 2	Unit 3
14:46 on Mar. 11: Shutdown due to **earthquake**		
LOOP: Loss of off-site power ➡ **EDGs started automatically**		
ICs (Isolation Condensers) operated manually	RCIC (Reactor Core Isolation Cooling System) started manually	RCIC started manually
15:27 on Mar. 11: **Tsunami arrived**		
Loss of AC and DC power		**Loss of AC** power (DC power available)
ICs and HPCI were not operable **March 12** • **PCV venting at about 14:30** • **Hydrogen explosion at 15:36** • **Seawater injection at 19:04**	⬇ RCIC operated for about 70 hours without DC power ⬇ **March 14** • **RCIC stopped until ~12: 30** • **Operators opened SRV at 18:02** • **Seawater injection at 19:54** **March 15** • Possible PCV failure	**March 12** • RCIC stopped (11:36) and HPCI started (12:35) **March 13** • **HPCI was shut down remote-manually from MCR at 2:42** • **SRVs opened at 9:08** (ADS conditions met). • **Fresh water injection at 9:25** **March 14** • **Hydrogen explosion at 11:01**

Hydrogen Explosion at Units 1, 3

- Hydrogen leaked from RPV to PCV and then to the RB, and exploded there.
- Analyses were done by the former JNES with **FLUENT** and **AUTODYN**.
- It was assumed that hydrogen leaked at:
  - **Top flange gasket at Unit 1**
  - **Sealing of hatch or penetrations at 1st floor at Unit 3**

https://en.wikipedia.org/wiki/File:Browns_Ferry_Unit_1_under_construction.jpg

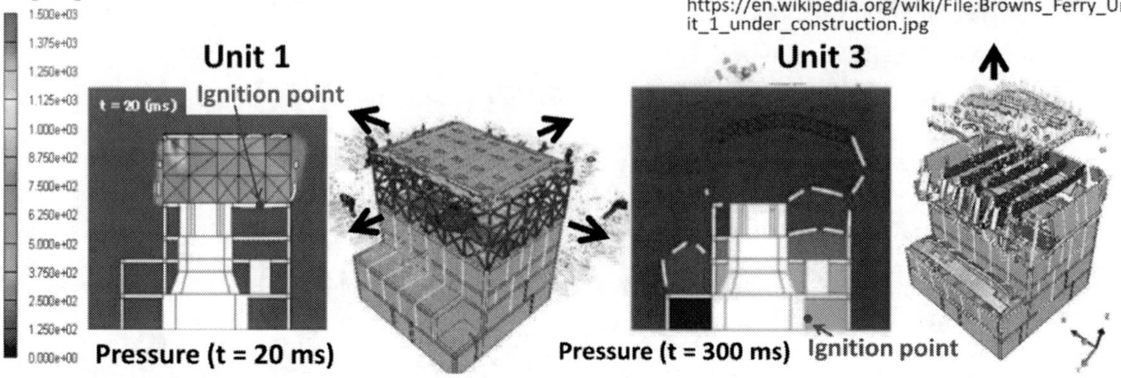

Pressure (t = 20 ms) **Pressure (t = 300 ms)**

Hydrogen Explosion at Unit 4

- **Hydrogen generated at Unit 3** might have been transported to Unit 4 **through SGTS piping and exhaust ducts.**
- The **exhaust stack was shared**. The valves of unit4 SGTS were open.

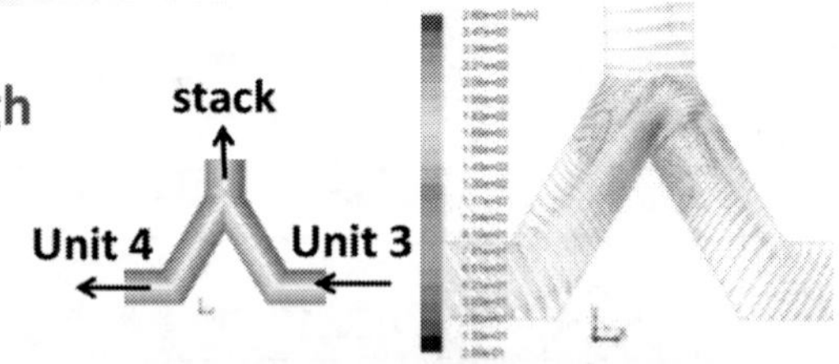

Analysis done by former JNES
WWW.nsr.go.jp/archive/jnes/content/000125907.pdf

Reactor Building of Unit 4

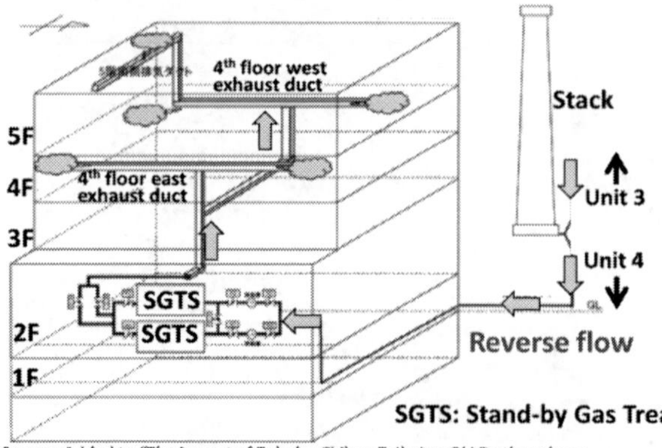

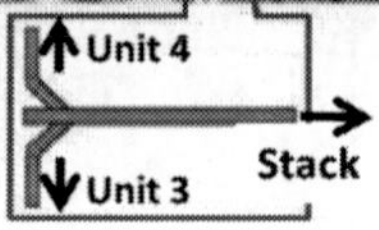

SGTS: Stand-by Gas Treatment System

Source: Added to "The impact of Tohoku-Chihou Taiheiyo-Oki Earthquake to Nuclear Reactor Facilities at Fukushima Daiichi Nuclear Power Station" (Sep. 9 2011, revised Sept. 28, 2011, Tokyo Electric Power Co., Inc.)

2nd Government Report

NRA, Japan
Nuclear Regulation Authority

End States of Units 1 to 3: OECD/NEA BSAF Project Phase I

OECD/NEA CSNI launched the **BSAF (Benchmark Study of the Accident at the Fukushima Daiichi NPP)** Project in **Nov. 2012**.

- **16 organizations** from **8 countries (France, Germany, Korea, Russia, Spain, Switzerland, the United States,** and **Japan)** participated.

Phase I Summary Report (March 2015)

- **Qualitative description of the plausible end states** after comparison of **the best estimate case analyses**

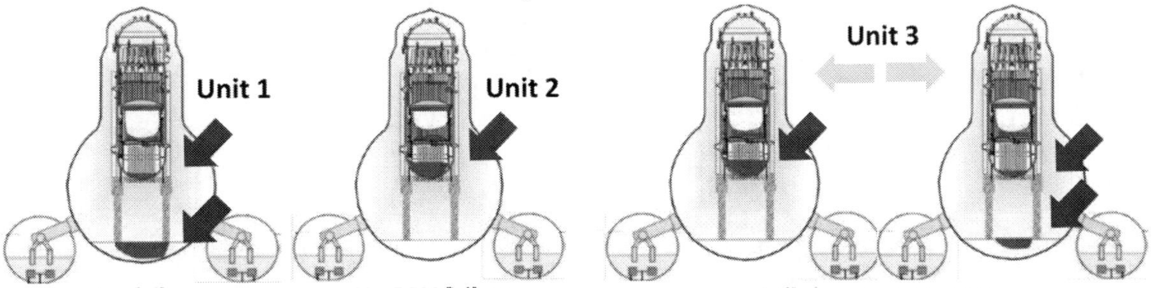

RPV failure
Every calculation predicted that extensive damage occurred.

No RPV failure
The majority of calculations do not predict RPV failure

Split into two groups:
a) No RPV failure
b) RPV failure and temporary MCCI scenario

NRA, Japan
Nuclear Regulation Authority

Source Term Evaluation

- The IAEA collected / compared the source terms evaluated by various organizations with different methods and assumptions.
- **Technical Volume 1 of the IAEA Fukushima report**: "**The total amount of aerosol based fission product releases** from all of the Fukushima Daiichi units was **about one order of magnitude smaller than that from the Chernobyl accident.**"

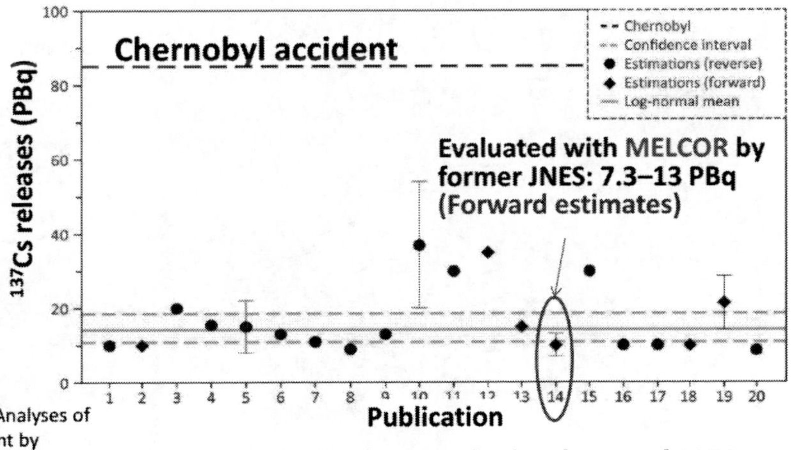

FIG. 1.4–4. Estimated atmospheric releases of 137Cs

IAEA Fukushima report, Technical Volume 1/5 (2015)

Publication [14]:
HOSHI, H., HIRANO, M., Severe Accident Analyses of Fukushima - Daiichi Units 1 to 3, Side event by Government of Japan at 56th IAEA General Conference, Vienna, 2012, JNES (2012),

SA Progression and Source Terms

Hirano, Hoshi and Homma, Presented at U.S.NRC RIC, March 2013.

- The former JNES has conducted a SA progression analysis with **MELCOR** developed by U.S.NRC. The NRA continues to do it.
  - **Release timings** are in good agreement with the monitoring data.
  - **The largest peak in the morning on March 15** was probably caused by **PCV failure at Unit 2**

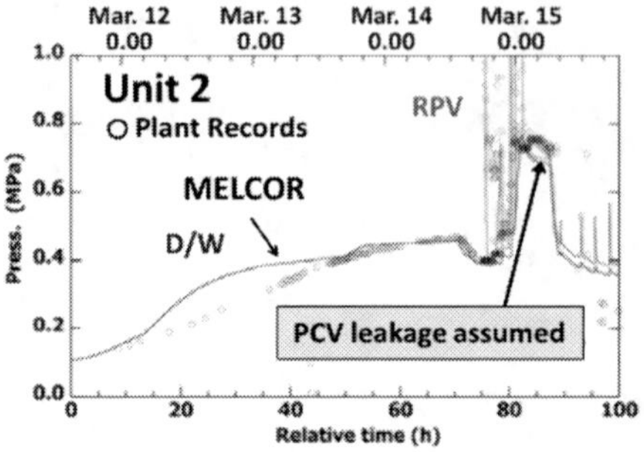

Comparison between the source terms by MELCOR and monitoring data

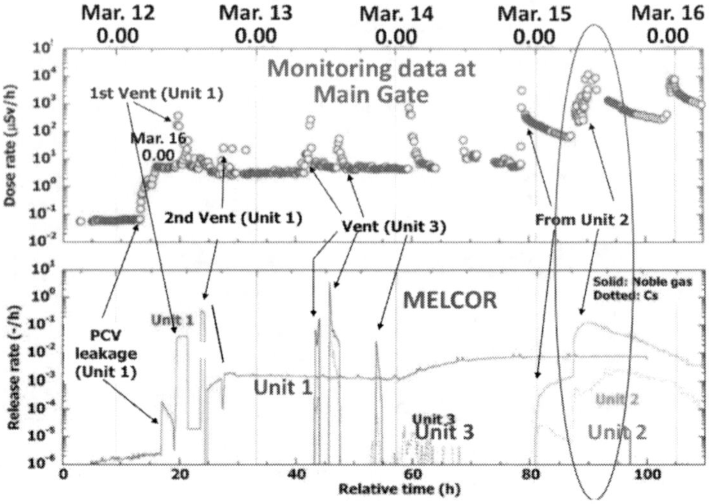

Environmental Consequence Analysis

Hirano, Hoshi and Homma, Presented at U.S.NRC RIC, March 2013.

- By using the **source terms with MELCOR**, an **environmental consequence analysis** was done in JAEA with **OSCAAR** developed there.

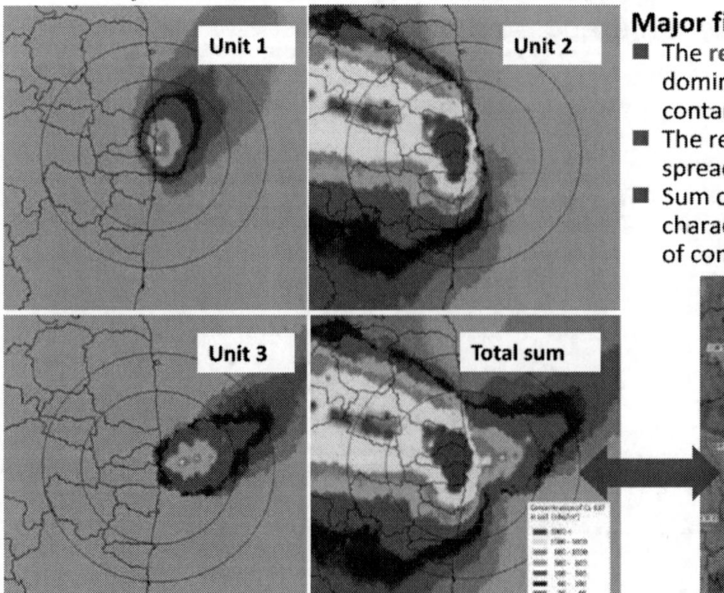

Major findings:

- The **release from Unit 2** on **March 15** had dominant contribution to the inland contamination.
- The release from **Units 1 and 3** mostly spread **to the ocean.**
- Sum of the release well reproduced the characteristics of the measured distribution of contamination.

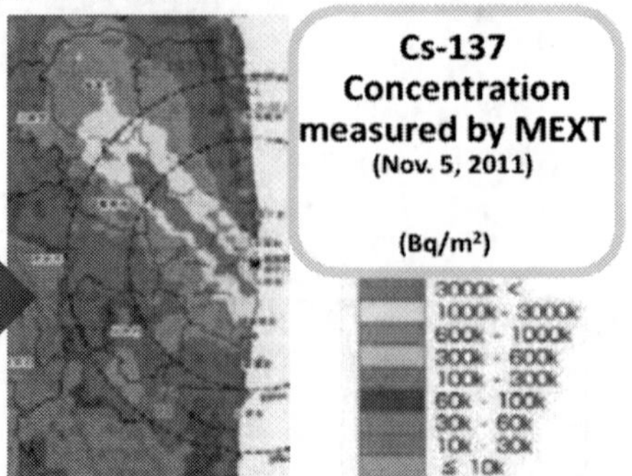

Wind was whirling clockwise on March 15, 2011.

Ref.: Hirano, Presented at IAEA IEM9, 20–24 April 2015

- During the accident, the **SPEEDI** simulation had been done every hour assuming a **unit release (1 Bq/hour)** since the source terms were not available. ➡ The results were equivalent to the **meteorological data**.

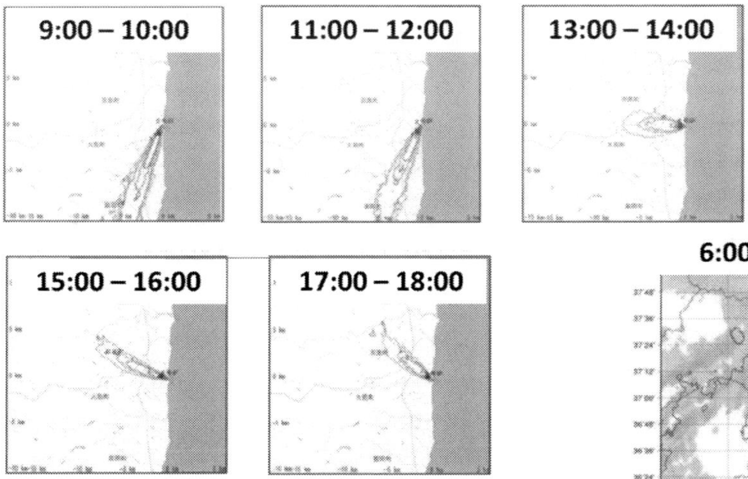

SPEEDI calculations Mar. 15, 2011

Source: M. Chino, JAEA, May 2013

Final Report from Government's Investigation Committee (July 2012)

Precipitation (Rainfall)

6:00 – 15:00 **15:00 – 21:00**

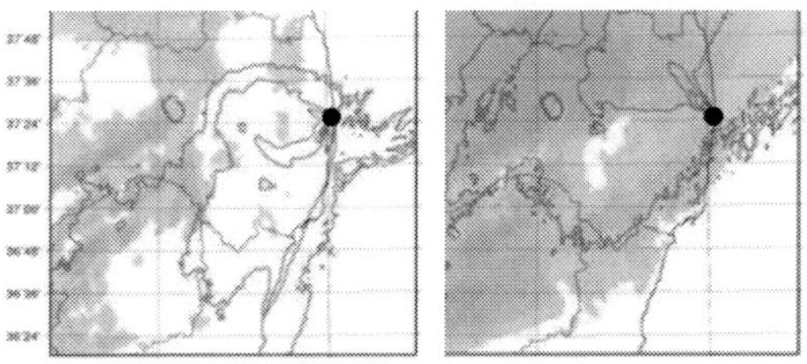

http://www.aec.go.jp/jicst/NC/iinkai/teirei/siryo2013/siryo18/siryo1.pdf

Some Lessons Learned

NRA, Japan Nuclear Regulation Authority

Lack of "Defense-in-Depth" against external hazards

- Large scale "common cause failures (CCFs)" by external hazards should have been prevented/mitigated.
  - We should have been prepared for "Low frequency, high consequence events" beyond design basis.
- Actually, the Defense-in-Depth (DiD) had not been taken into consideration for external events.
  - Watertight doors were not implemented, Seawater pumps were not protected. AM measures were not effective, ...
  - Training / drills assuming external events had not been conducted.
- **DiD has been and will continue to be an effective way to account for uncertainties.**
  - **Effective independence between different layers** of defense
  - More "diversity" and "independence" rather than "redundancy"

NRA, Japan Nuclear Regulation Authority

"Tsunami risk had been recognized": Lack of Agility

Diet`s Report: NAIIC : The National Diet's Fukushima Nuclear Accident Independent Investigation Commission, Reported to the Diet in July 2012

- The Seismic Design Review Guide was revised in **2006** and a requirement was newly introduced against tsunami. So-called "back-check" was in progress, but its progress was very slow.
  - **"Uncertainty allows for wishful thinking" + "Safety myth"**
  - **Lack of agility, lack of safety culture**

Diet Report

- As the regulatory agency was aware of TEPCO's delaying of countermeasures, but did not follow up with any specific instructions. Nor did they properly supervise the back-check progress.
- When new findings indicate the possibility of a tsunami exceeding previous assumptions, the operator is required to quickly implement countermeasures, rather than taking time to clarify the scientific basis for that possibility through studies of sediment ....

NRA, Japan Nuclear Regulation Authority

TEPCP's "Trial Calculations" of Tsunami

IAEA Fukushima report, Technical Volume 2/5 (2015)

Seismicity Gap Area (no record of large earthquake):

- **Major opinion**: Majority of scientists believed the shallow plate boundary **along the Japan trench** was **not able to store a large amount of strain** (**coupling is weak**).

New findings / minor opinion *Headquarters for Earthquake Research Promotion

- In **2002**, **HERP*** pointed out the **possibility of a tsunami earthquake occurring anywhere along the Japan Trench including the gap area could not be ruled out.**
- **In 2008, TEPCO** did hypothetical calculation assuming a **M 8.2 earthquake** occurring **off the coast of Fukushima Pref.** ➡ **Max. ~15.7 m**
- TEPCO also conducted a **Jogan** (869)-**type tsunami** simulation ➡ **Max. ~9.2 m**
- However, **TEPCO** decided only on a plan to ask JSCE* for investigation such as **tsunami deposit survey, rather than take any immediate measures.**

*JSCE : Japan Society of Civil Engineers

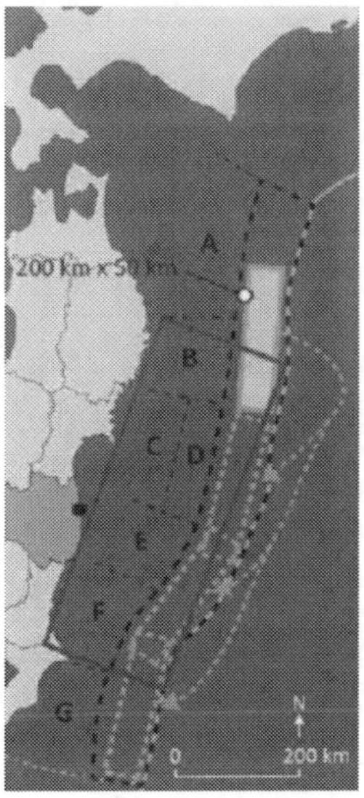

NRA, Japan Nuclear Regulation Authority

Regulatory Aspects: "Regulatory Capture" Pointed out in the Diet's Report

Diet's Report (July 2012)

Message from Chairman

- ... this was a disaster "**Made in Japan.**" Its fundamental causes are to be found in the ... **Japanese culture**: our **reflexive obedience**; our **reluctance to question authority**; ... and our **insularity**.

Organizational issues ...

- ... actual relationship **lacked independence** and **transparency**, ... In fact, it was a typical example of "**regulatory capture**," ...

Lack of expertise

- ... the two incorporated technical agencies advising NISA, namely, **JNES and JAEA**, have been **too rigidly tied to NISA** ....

Conclusions

- ... **The lack of expertise** resulted in "**regulatory capture**,"... They **avoided their direct responsibilities** by **letting operators apply regulations on a voluntary basis.**

NRA Japan
Nuclear Regulation Authority

Human / Organizational Aspects: "Myth of Safety" Pointed Out in the Government's Report

Final Report from Government's Investigation Committee (July 2012)

"Myth of Safety"

- ... a fundamental problem of the inability to capture such crises as a reality that could happen in our lives; this, in turn, is the result of a **myth of safety** that existed among **nuclear operators including TEPCO** as well as **the government**, that **serious severe accidents could never occur** in NPPs in Japan.

"Basic Assumption" in the **IAEA DG Report**

- Because of the **basic assumption** that **NPPs in Japan were safe**, there was a tendency for organizations and their staff **not to challenge the level of safety**.
- The reinforced **basic assumption** among the stakeholders about the robustness of the technical design of NPPs **resulted in a situation where safety improvements were not introduced promptly**.

NRA Japan
Nuclear Regulation Authority

"**Complacency**" together with "**Bureaucracy**" allowed "**Safety myth**" to prevail, having let "**Continuous improvement**" cease.

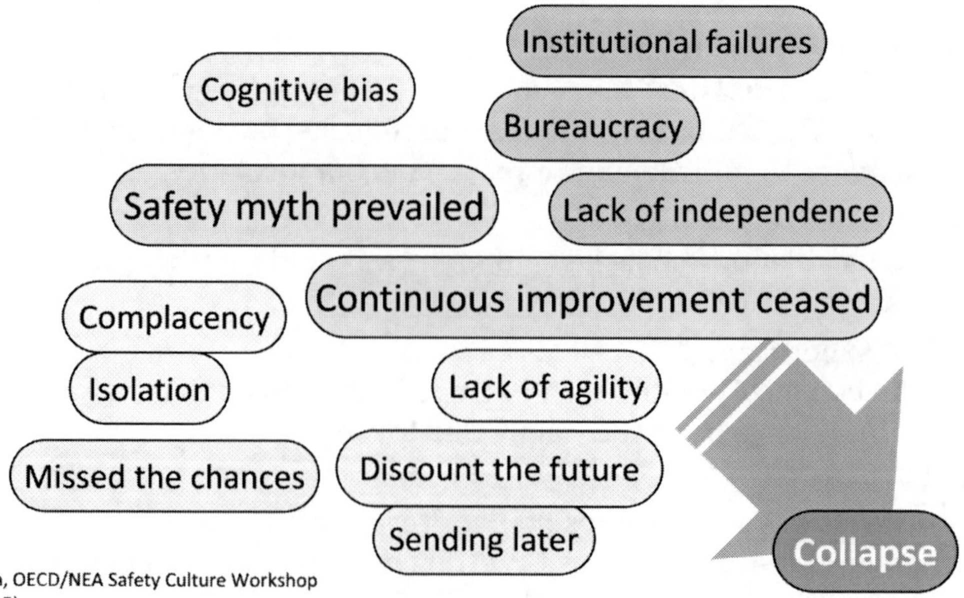

T. Fuketa, OECD/NEA Safety Culture Workshop (June 2015)

Safety Enhancements

NRA: Nuclear Regulation Authority

Established in Sept. 2012

- **Independence**

  Nuclear regulation and nuclear promotion were clearly separated, and the NRA was established as **an independent commission body** defined by law* affiliated with MOE (Minister of Environment).

  * a council-system organization based on Article 3 of the National Government Organization Act, ensuring its independence without any control or supervision by other organizations.

- **Integrated**

  Nuclear regulation functions regarding safety, security, safeguards, radiation monitoring and radioisotopes were integrated into the NRA.

Integrated and Independent

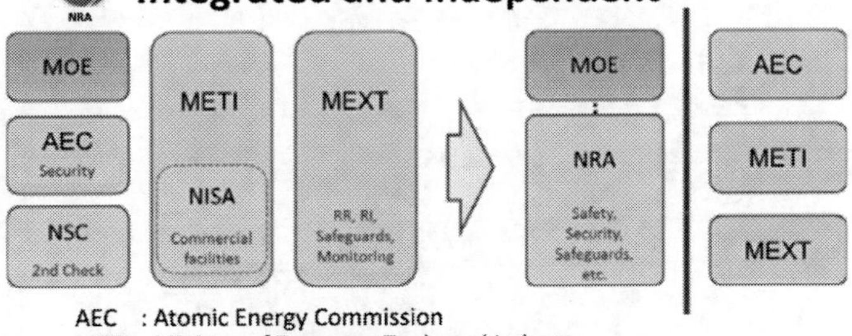

AEC : Atomic Energy Commission
METI : Ministry of Economy, Trade and Industry
MEXT: Ministry of Education, Culture, Sports, Science and Technology
MOE : Ministry of the Environment
NISA : Nuclear and Industrial Safety Agency (abolished)
NSC : Nuclear Safety Commission (abolished)

T. Fuketa, NRA, presented at U.S.NRC RIC2013, March 13, 2013: http://www.nsr.go.jp/data/000067048.pdf

NRA, Japan
Nuclear Regulation Authority

New Regulatory Requirements: Structure

http://www.nsr.go.jp/data/000067048.pdf

Requirements for B-DBA

- **DEC**: Design extension conditions defined in IAEA SSR-2/1

<New> (Severe Accident Measures)

4th Layer of DiD — NEW:
- Suppression of radioactive materials dispersal
- Specialized Safety Facility
- Prevention of CV failure
- Prevention of core damage

<Pre-existed>

<Pre-existed>		<New>	
Natural phenomena	3rd Layer of DiD	Natural phenomena	Reinforced
Fire		Fire	
Reliability		Reliability	
Reliability of power supply		Reliability of power supply	
Ultimate heat sink		Ultimate heat sink	
Function of other SCCs		Function of other SCCs	
Seismic/Tsunami resistance		Seismic/Tsunami resistance	Reinforced

NRA, Japan
Nuclear Regulation Authority

New Regulatory Requirements: Enhanced Measures against Tsunami

http://www.nsr.go.jp/data/000067048.pdf

More Stringent Standards on Tsunami ➡ It is required to define "design basis tsunami" that **exceeds the largest in the historical records** and to take **protective measures** such as **breakwater wall** based on it.

Enlarged Application of Higher Seismic Resistance ➡ SSCs for tsunami protective measures are **classified as Class S** equivalent to RPV etc. **of seismic design importance classification.**

Example of protective measures against tsunami (multiple measures)

- **Breakwater wall** for prevention of inundation to the site

- **Tsunami gate** for prevention of water penetration into the building

New Regulatory Requirements: Measures against Intentional Aircraft Crash, etc.

http://www.nsr.go.jp/data/000067048.pdf

"Specialized Safety Facility" is required to mitigate release of radioactive materials after core damage due to intentional aircraft crash.

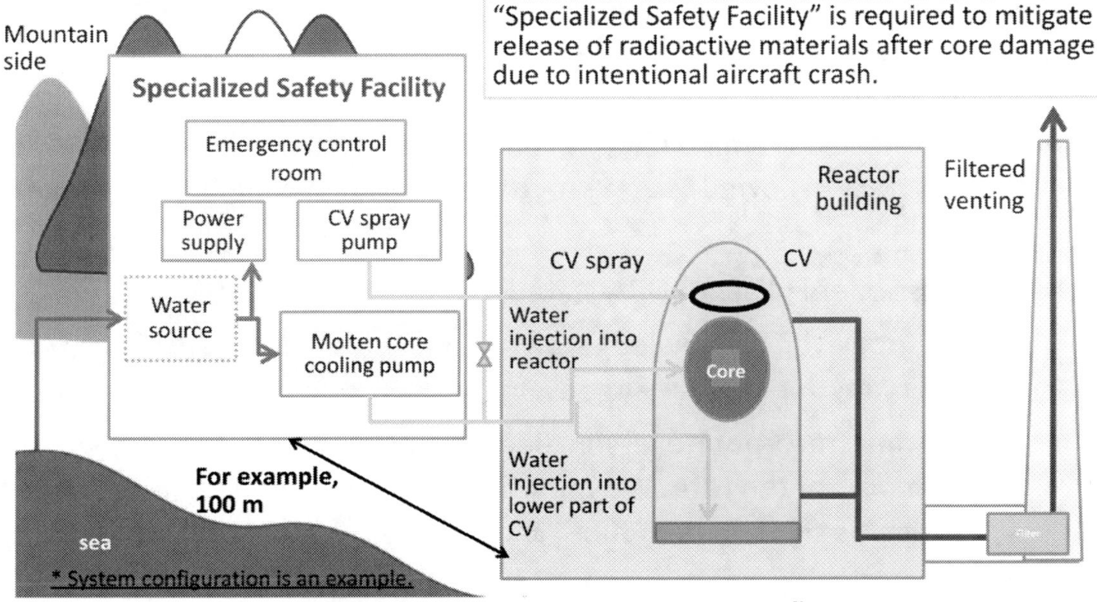

For BWR, one filtered venting for prevention of containment failure and another filtered venting of Specialized Safety Facility are acceptable solution.

NRA's Statement of Nuclear Safety Culture

Issued on May 27, 2015

T. Fuketa, OECD/NEA Safety Culture Workshop (June 2015)

- NRA's Mission Statements

The NRA's Core Values and Principles

The Statement of Nuclear **Safety Culture**	The Code of Conduct on **Nuclear Security**

- **Eight Traits** of Nuclear Safety Culture in NRA's Statement
  1. Priority to Safety ➡ Break with "**safety myth**"
  2. Decision-making prioritized by safety
  3. Fostering, sustaining and strengthening safety culture
  4. Learning organization ➡ Seek out "**opportunities for improvement**"
  5. Communication ➡ Get rid of "**isolation**"/"**self-righteousness**"
  6. Questioning attitude ➡ Avoid "**complacency**"
  7. Rigorous and prudent judgment and action with agility
  8. Harmonization with nuclear security

NRA, Japan

Summary

- One of the most important lessons learned is "**Lack of Defense-in-Depth against external hazards**". We need to continuously enhance DiD since **DiD has been and will continue to be an effective way to account for uncertainties.**
- We need to keep reminding that "**Complacency**" together with "**Bureaucracy**" allowed "**Safety myth**" to prevail, having let "**Continuous improvement**" **cease.**
- Commissioner Fuketa of NRA stated in the OECD/NEA SC Workshop*:
  - We, however, acknowledge we are oblivious. Numerous **sprouts of safety myth** reappear already.
  - We must incorporate lessons-learned into the "institutional memory" of the NRA. We must create an environment where **a gene letting us think "safety first" can survive**.

* T. Fuketa, OECD/NEA Safety Culture Workshop (June 2015)

NRA, Japan

References

- **1st Government Report (June 2011)**: Report of the Japanese Government to the IAEA Ministerial Conference on Nuclear Safety, Nuclear Emergency Response Headquarters, Government of Japan, June 2011
- **2nd Government Report (September 2011)**: Additional Report of the Japanese Government to the IAEA (Second Report), Nuclear Emergency Response Headquarters, Government of Japan, September 2011
- **Diet's Report (July 2012)**: The Official Report of the Fukushima Nuclear Accident Independent Investigation Commission (NAIIC), the National Diet of Japan, 2012
- **Interim Report from Government's Investigation Committee (Dec. 2011)**: Interim Report, Investigation Committee on the Accident at Fukushima NPS of TEPCO, Dec. 26, 2011
- **Final Report from Government's Investigation Committee (July 2012)**: Final report, Investigation Committee on the Accident at Fukushima NPS of TEPCO, July 23, 2012
- **IAEA DG Report and Technical Volumes 1 to 5 (2015)**: The Fukushima Daiichi Accident, Report by the Director General and Technical Volumes 1 to 5 (2015)
- **T. Fuketa, OECD/NEA Safety Culture Workshop (June 2015)**: OECD NEA/CNRA/ CSNI/CRPPH Joint Workshop on Challenges and Enhancements to Safety Culture of the Regulatory Body, Paris, June 3, 2015
- **Y. Shimizu, IAEA Effective Nuclear Regulatory Systems (April 2016)**: Y. Shimizu, "Lessons Learned from the Fukushima Daiichi Accident, Actions Taken and Challenges Ahead," presented at IAEA International Conference on Effective Nuclear Regulatory Systems, Vienna, Austria, April 11-15, 2016

Appendix — Key: It was difficult to open Air-Operated Valves (AOVs)

- For seawater injection by fire engines, it was necessary to depressurize RPV by opening SRVs. It was needed to open AOVs in the venting system to prevent PCV failure.
- AOVs need:
  - DC power to open the pilot valve (solenoid valve) ⟸ **Car batteries**
  - Compressed air ⟸ **Mobile compressor, nitrogen gas cylinders**

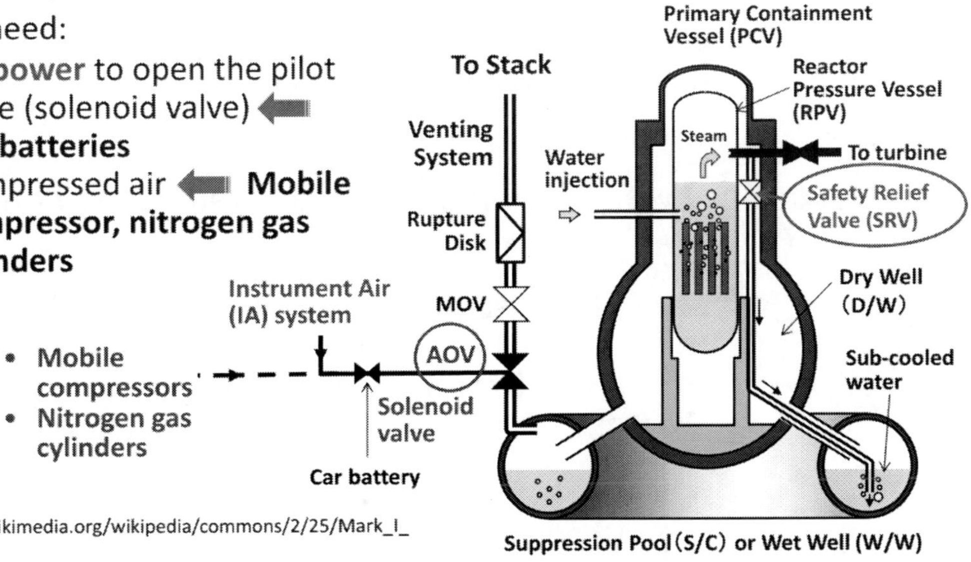

https://upload.wikimedia.org/wikipedia/commons/2/25/Mark_I_Containment.jpg

Appendix — Example of Use of Mobile Equipment: Measures against Station Blackout (SBO)

✓ Install both mobile alternate power sources (power vehicle, batteries, etc.) and permanent alternate AC power sources.

- Batteries for 8 hours without load shedding + 16 hours with load shedding
- Alternate onsite AC power supply for 7 days
- External support from offsite by the 6th day
- Alternative MC, PC, MCC

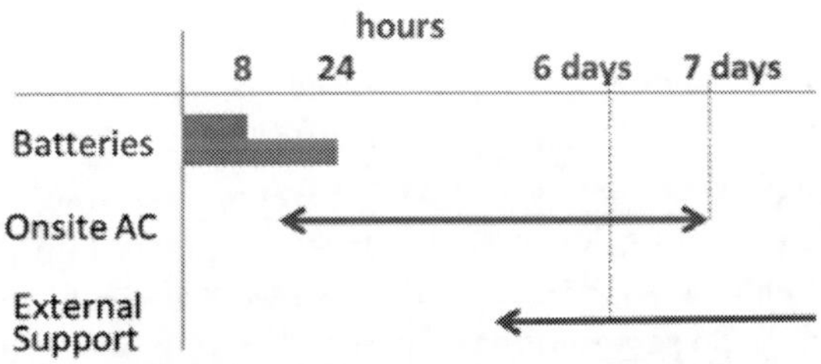

✓ It is also required to install the third station battery system.

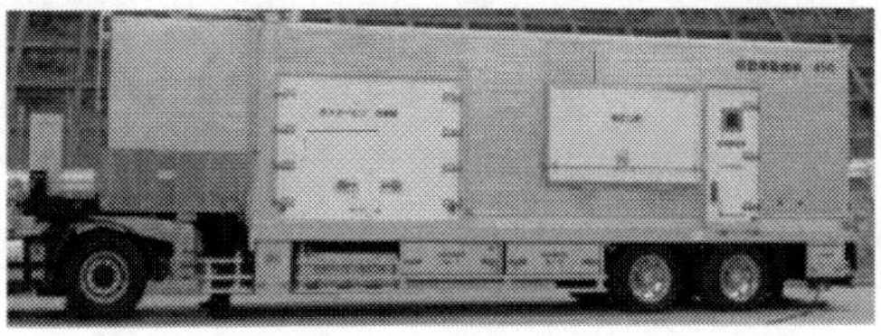

Alternative onsite AC power (Power Vehicle)

ICRS-13 & RPSD-2016

13th *International Conference on Radiation Shielding*
&
19th Topical Meeting of the *Radiation Protection & Shielding Division*
of the *American Nuclear Society* -2016

3rd Plenary invited talk:

Secondary Neutrons from Proton Radiation Therapy”

Dr. Rebecca M. Howell

University of Texas, USA

The author will present an overview of secondary neutrons in proton therapy. In particular, she will discuss the wide variability of data in the literature and focusing on the variability in measurement techniques used in different studies. Additionally the presentation will compare the secondary neutrons encountered in scanning beam and passive scatter proton therapy and high energy photon therapy.

Rebecca M. Howell, PhD, is a tenured Associate Professor at The University of Texas MD Anderson Cancer Center. Dr. Howell was elected Fellow of the American Association of Physicists in Medicine (AAPM) last year. She received her MS (2001) and PhD (2005) from the University of Texas Health Science Center San Antonio. Dr. Howell is certified by the American Board of Radiology and has 15 years of clinical experience. She has published 50 peer-reviewed papers and is Principal and Co-investigator on several funded research grants from the National Cancer Institute and other funding agencies. Her research focuses on out-of-field dose and late effects of radiation therapy.

Presented by:
Rebecca M. Howell
PhD, DABR, FAAPM

Secondary Neutrons in Proton Therapy

THE UNIVERSITY OF TEXAS
MD Anderson Cancer Center
Making Cancer History®

Background
Treatment Planning in Radiation Therapy

- Standard of care: commercial TPS used to calculate patients' treatment plans (all RT types)
- Accuracy
  - High-dose regions and areas within the primary beam path are typically well described (accuracy ± 2%)
  - Low-dose regions
    - Accuracy beyond a few centimeters outside the treatment field edge is usually poor
    - Secondary neutrons are not considered

Secondary Neutrons Photon Therapy

External Neutrons in Photon Therapy

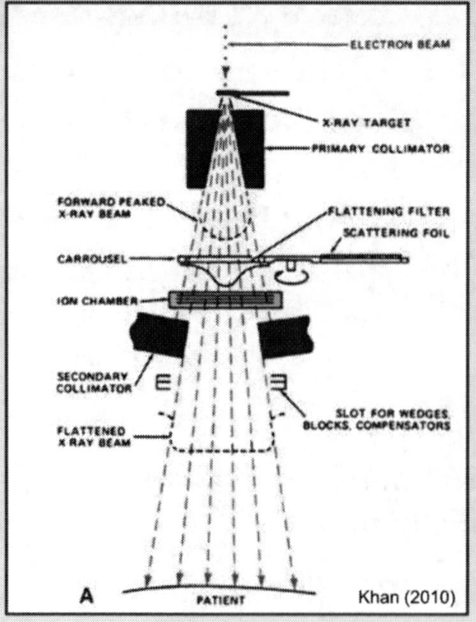

Khan (2010)

(γ,n) in High-Z Materials

- **Primary collimator**
- **Target**
- **Flattening filter***
- **Secondary jaws**
- **MLC****

Photon Beam Energy Effect on Neutron Production

- Bremsstrahlung photon beams
  - Wide distribution of energies
  - Max energy ≈ energy of electron beam on the target.

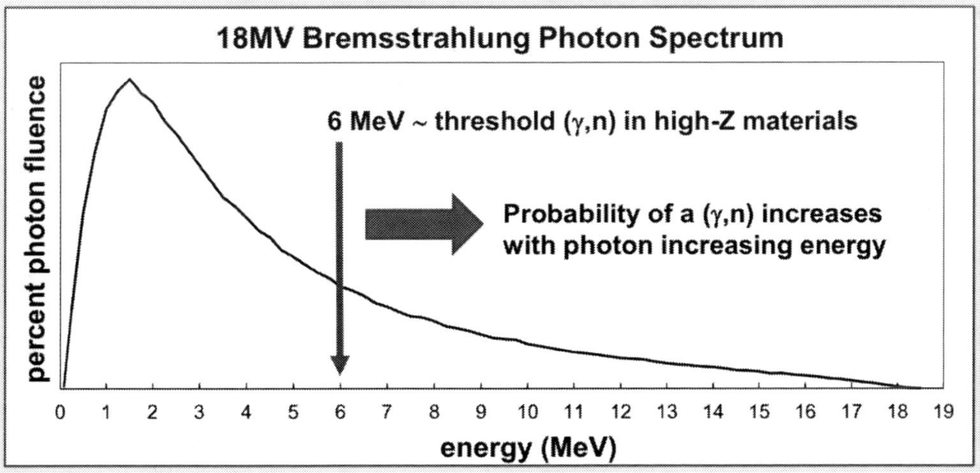

Secondary Neutron Spectra from Clinical Photon beams

- Energy distribution in the linac head ~ isotropic resembles a fission spectrum.
- Softens, as neutrons undergo interactions in the linac head, patient, and treatment vault.

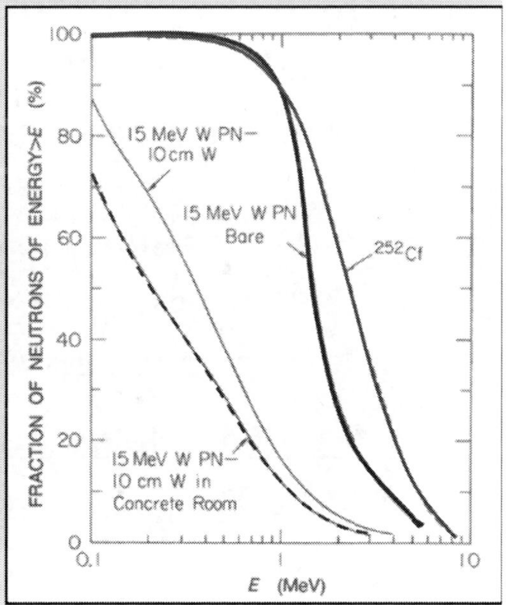

NCRP Report 79 (1984)

Secondary Neutron Spectra from Clinical Photon beams

- Similar energy distribution for all photon beams
- Fluence increases with increasing photon beam energy

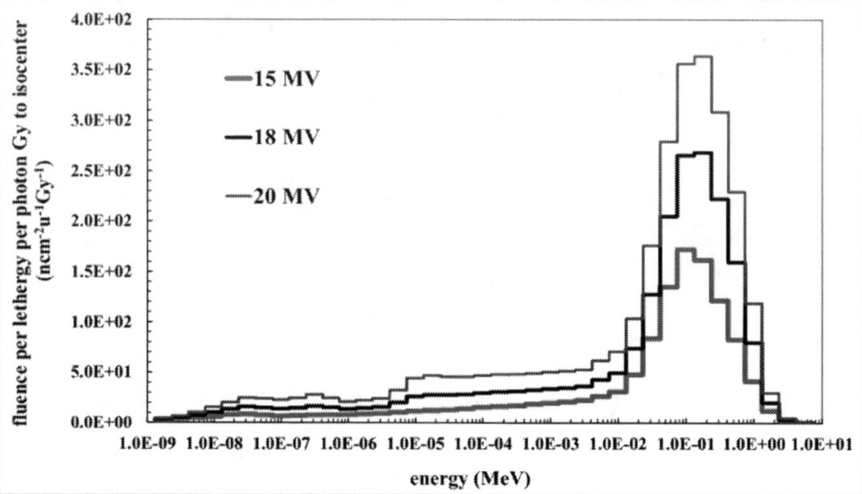

Howell *et al.*, Med Phys 36, (2009).

Internal Neutrons in Photon Therapy

- (γ,n) in low-Z Materials is negligible compared to high-Z materials
  - Higher threshold energies:
    - ^{16}O (17 MeV)
    - ^{12}C (19 MeV)
    - ^{14}N (8 MeV)
  - Lower x-sections (20X - 30X ↓)
- <u>Essentially no neutron production in patient</u>
  - *Exception high-z implants, e.g., hip prosthesis*

Secondary Neutrons
Proton Therapy

Background
External Neutrons in Proton Therapy

Passive Scatter Beamline

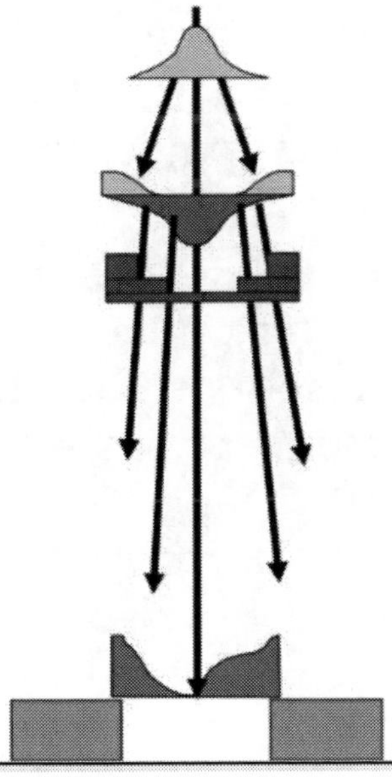

(p,n) in High-Z Materials

- **Double scatterer**
- **Modulation wheel (SOBP)**
- **Compensator**
- **Field aperture**

External Neutrons in Proton Therapy

Spot Scanned Beamline

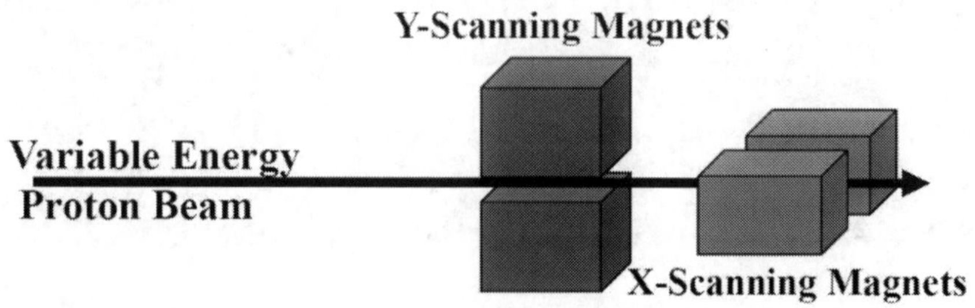

- **Less high-Z materials in beamline**
  - **Less neutron production**

Internal Neutrons in Proton Therapy

- **(p,n) in low-Z Materials is important and non-negligible downstream of the target and w/in 10 cm of field edge.**
  - **In passive scatter, can contribute ~ 50% of the dose equivalent.**
  - **In spot scanning, dominant source neutron dose.**
- **Internal neutron production ↑ with:**
  - **Beam energy due to higher (p,n) cross sections.**
  - **Field size because the proton beam is incident on a larger volume of tissue.**

Secondary Neutron Spectra from Clinical Proton beams

- Similar energy distribution for all proton beams

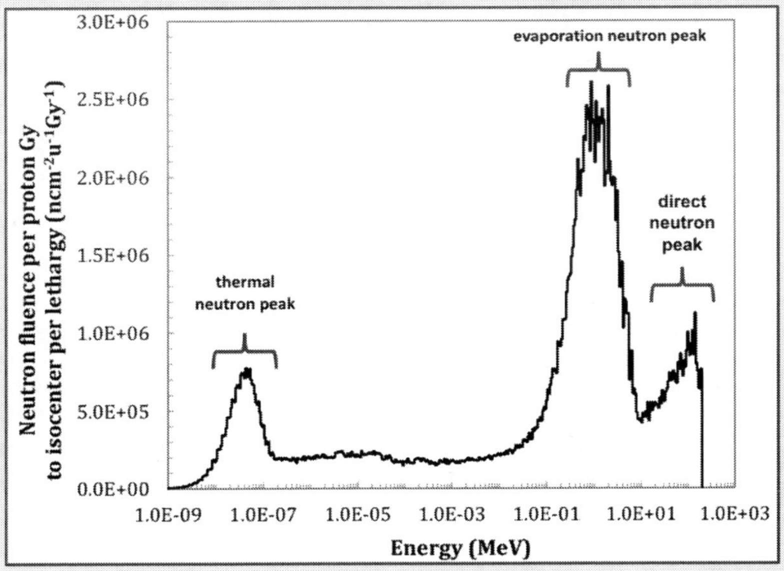

Howell and Burgett, Med Phys 41 (2014).

Location, location, location….

- Relative peak heights differ at different relative positions from isocenter, as beam softens.

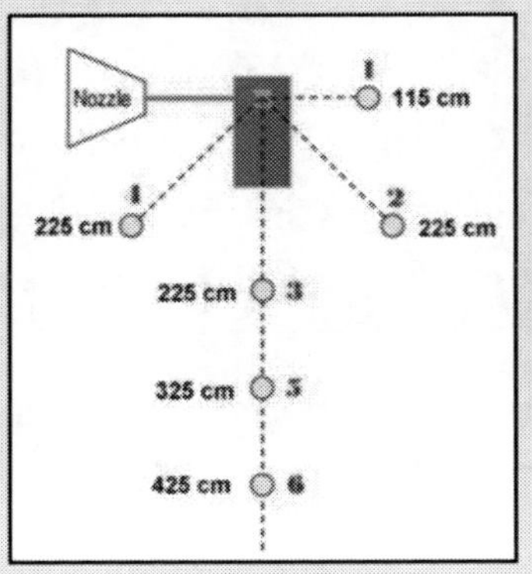

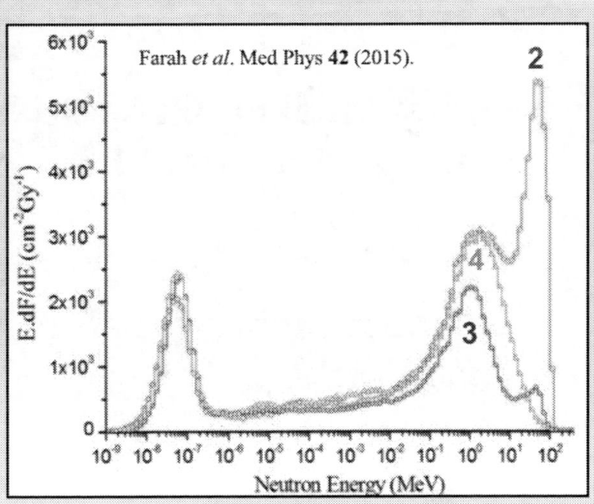

Scanning Beam Measurements - IBA

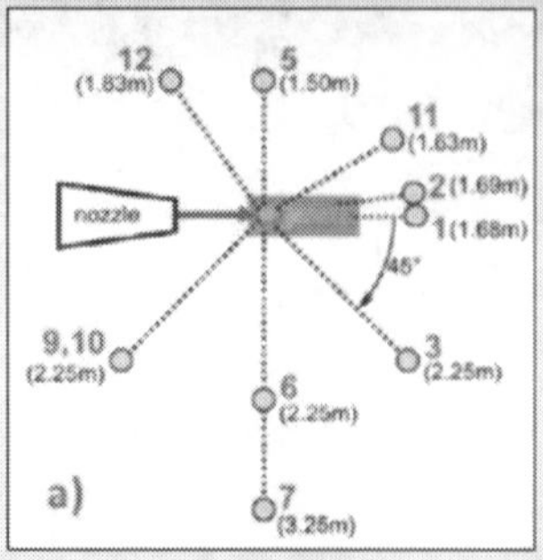

Mares *et al*,. 2016

- Greater high energy component
  - Along beam path (than off axis)
  - Closer to isocenter

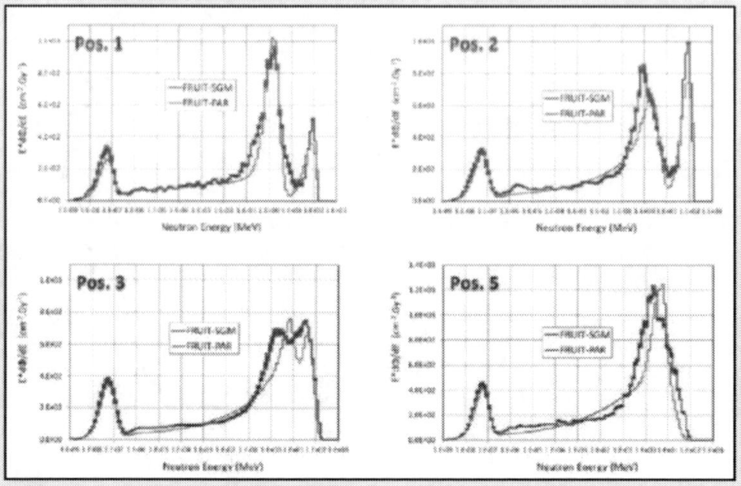

Published Data Secondary Neutrons in PRT

- How much secondary neutron dose results from PRT???
- Substantial published data on out-of-field neutron dose from PRT.
- but data highly variable spanning orders of magnitude….

Data Span Orders of Magnitude

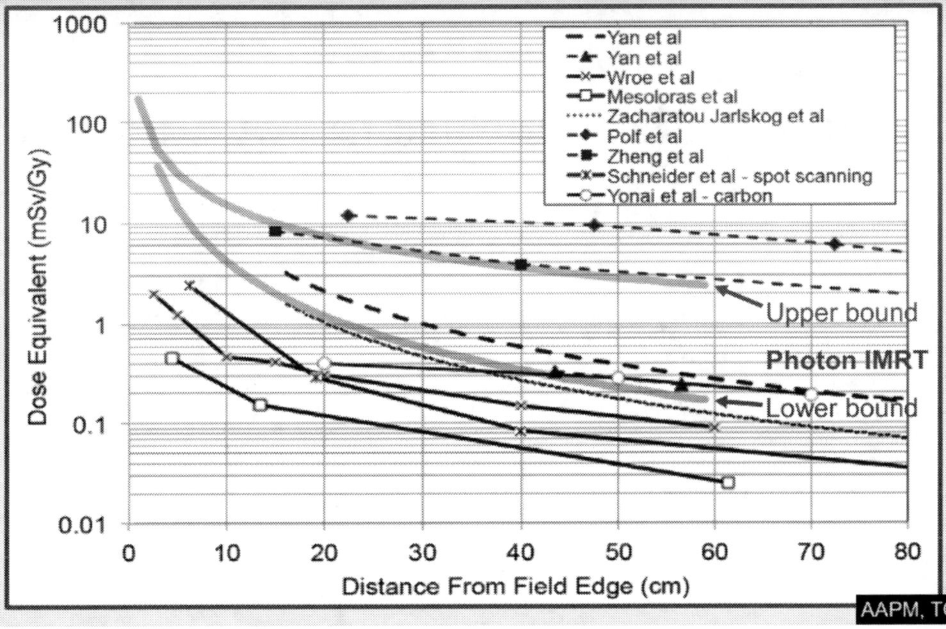

AAPM, TG-158

General Questions to Consider when Evaluating Neutron Data

Measurements

- **How were the data measured?**
- **Was the detector sensitive to neutrons over the entire energy range?**
- **If an active detector was used, was pulse pile-up a potential issue?**
  - **If yes, was it appropriately accounted for?**
- **If in phantom, was the change in spectrum with increasing depth taken into account?**

Simulations

- **Was the model appropriately validated against "good" measurements?**

Neutron Measurement Techniques

Detector Energy Response
In the Context of Spectra being Measured

Photon Therapy

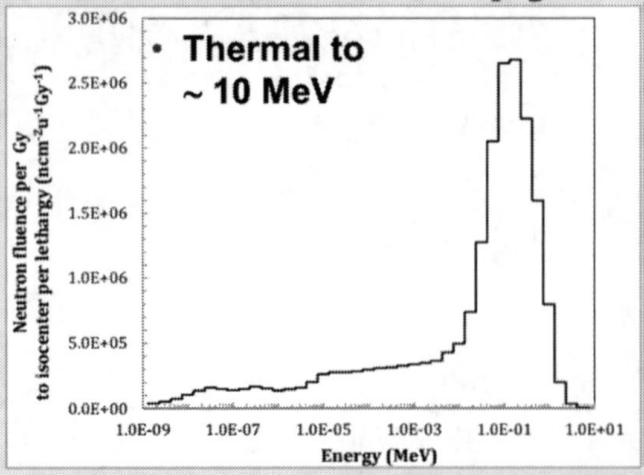

Proton Therapy

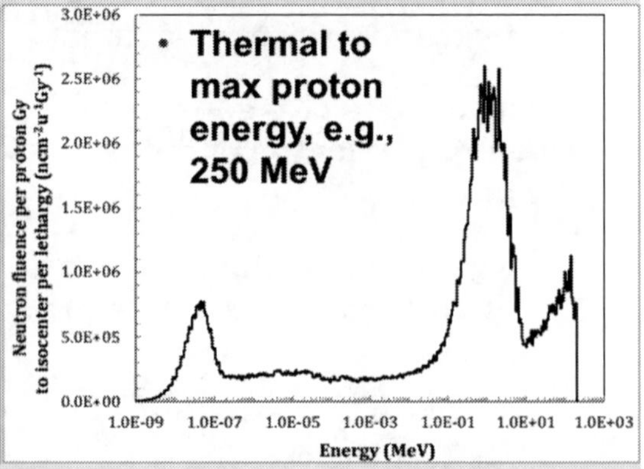

Neutron Detection Techniques

- **The vast majority of neutron detectors are only able to detect thermal neutrons.**

  - **Active Detection**
    - $^{10}B(n,\alpha)$
    - $^{6}Li(n,\alpha)$
    - $^{3}He(n,p)$
  - **Passive Detection**
    - Thermoluminescent Dosimeters (TLD)
    - Activation Foils (^{198}Au and ^{111}In)

- **Modified response based owhen embedded in various moderators (low-Z and high-Z).**

Systems of Moderators

- **By varying the thickness of this moderating material, the detector becomes optimized for specific neutron energy ranges.**

Tedious measurement system: repeat measurements with each sphere and then de-convolve data to determine spectrum.

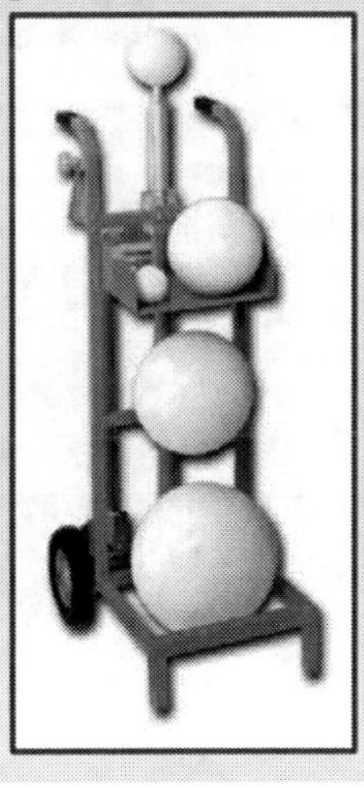

Response Functions for 2" to 12" Moderators

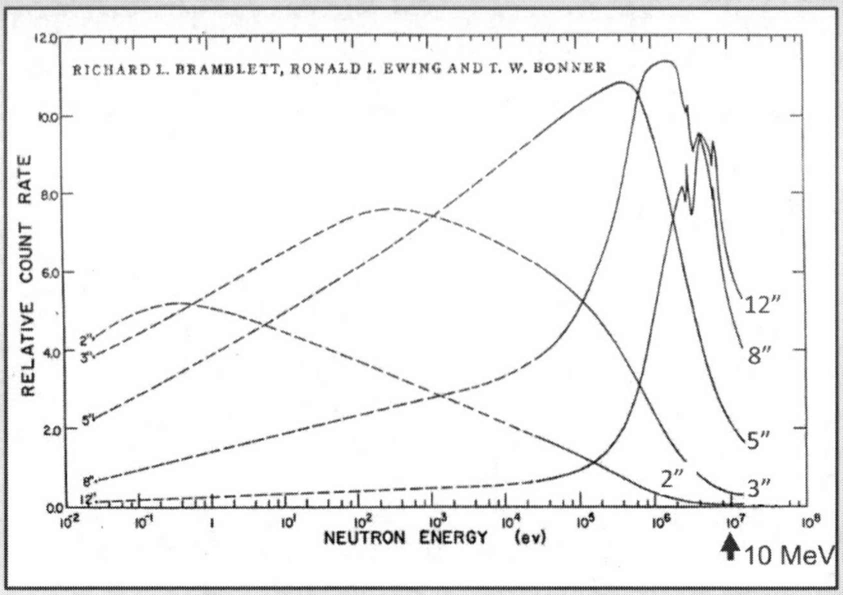

- For all spheres response begins to drop above 10 MeV.
- Suitable for measuring secondary neutrons from photon therapy.

What about PRT???

Extended Range Bonner Spheres

- Increase sensitivity to high energy neutrons by adding high-Z materials with large (n,xn) cross sections.

1. NEMUS

B. WIEGEL and A.V. ALEVERA. "NEMUS – the PTB Neutron Multisphere Spectrometer: Bonner spheres and more. *Nucl. Instr. and Meth. Phys. Res.* A476, 52 (2002)

2. CERN

C. BIRATTARI, P. CAPPELLARO, A MITAROFF, and M. SILARI. Development of an extended range Bonner sphere Spectrometer. Proceedings of Monte Carlo 2000 – Advanced Monte Carlo for Radiation Physics, Particle Transport Simulation, and Applications, Lisbon (Portugal), 1157-1162 (2000)

1. Bonner Sphere Extension

Howell RM, Burgett EA, Hertell NE, Kry SF, Wang Z and Salehpour M. Measurement of high-energy neutron spectra with a Bonner sphere extension. Nuclear Technology 168(2):333-339, 2009 and Howell RM, Burgett EA, Wiegel B, Hertel NE. Calibration of a Bonner sphere extension (BSE) for high-enegy neutron spectrometry. Rad Meas 45:1233-1237, 2010.

Bonner Sphere Extension

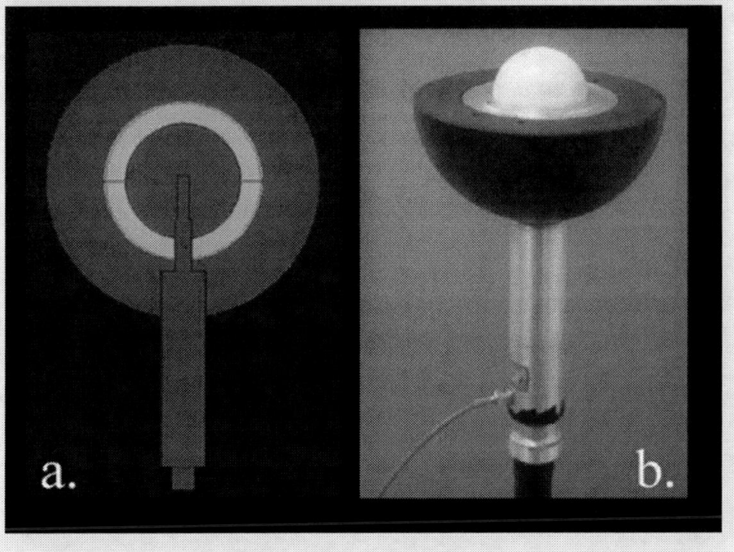

Response Functions for ERBSS

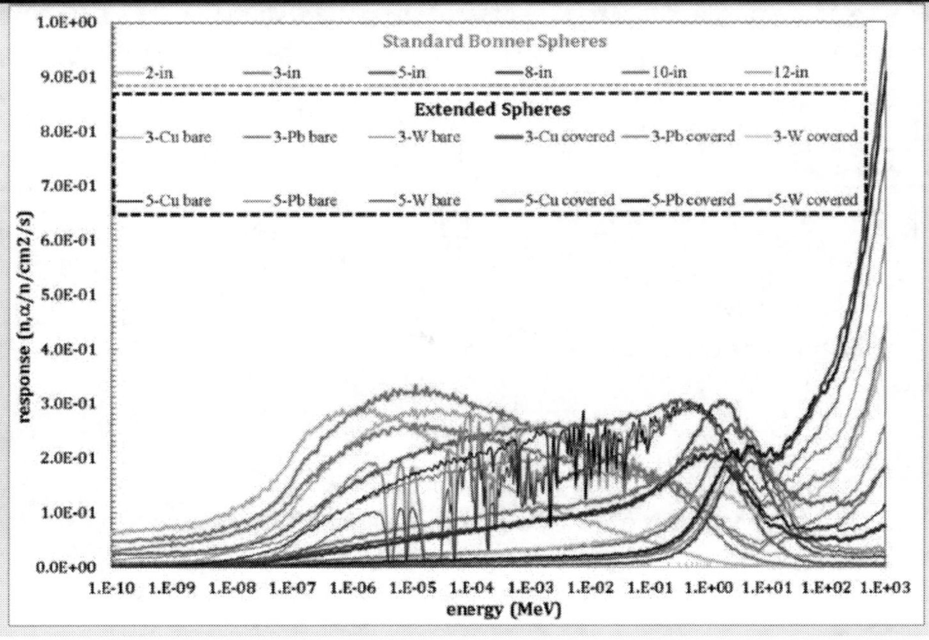

Neutron Survey Meters

- **Active thermal neutron detector (often BF_3) surrounded by a low-Z moderator.**

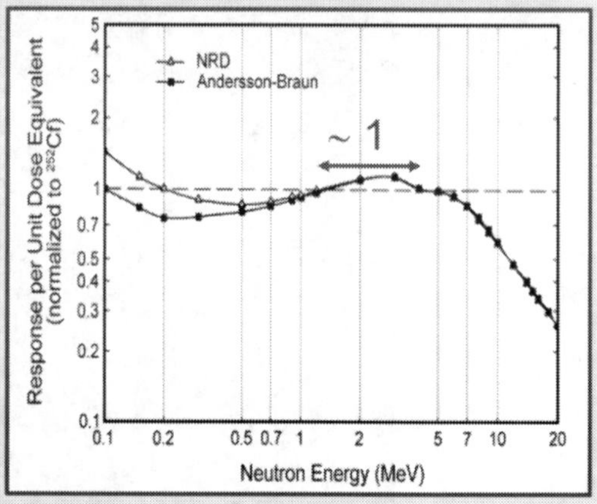

Energy Response

- **1 MeV, response ~1**
- **0.1 - 10 MeV, response 1.5 – 0.5**
  - **Under/over response within 50%**
- **> 10 MeV, response rapidly decreases with increasing energy.**
  - **underresponse ~6X at 100 MeV.**

For PRT, underestimate dose:

- **No response to neutrons from 10 MeV - max proton energy.**

Extended Range Neutron Survey Meters

- **Active thermal neutron detector surrounded by a low-Z moderator + Pb or W to extend high energy response.**

Energy Response

- **WENDI-II**
  - **2 – 10 Mev and 100 – 300 MeV, response ~ 1**
  - **10 – 100 MeV under-responds by as much as 40%**
- **PRESCILA generally over-responds**

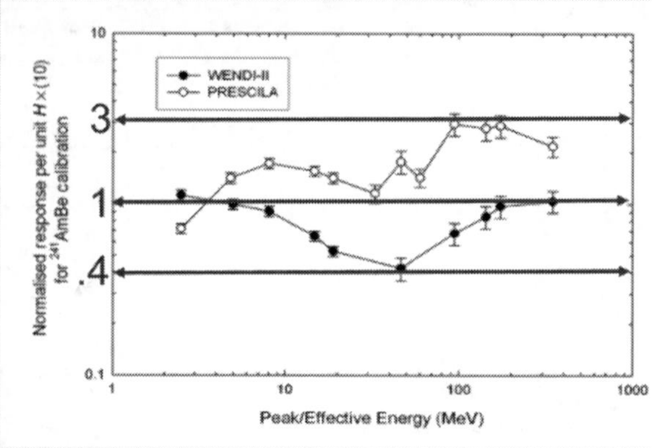

For PRT:

- **WENDI-II is reasonable detector, but still has uncertainty due to variable response, but generally within 50%.**

Fast Neutron Detectors
Bubble Detectors

- **Small sealed tube with superheated droplets of liquid dispersed throughout a polymer.**
- **When neutrons strike a droplet, 2°cp are emitted that vaporize the droplet, producing a bubble.**
- **Neutron dose determined by counting the bubbles and applying a calibration factor.**

Unirradiated detector

Bubbles form after neutron exposure.

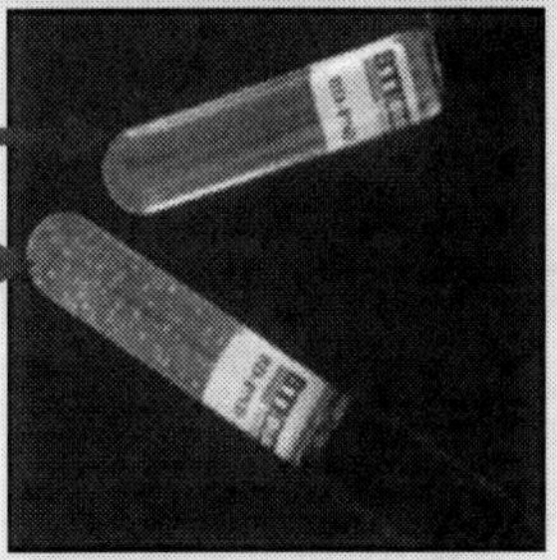

BD-PND (BTI Bubble Technology Industries, Ontario Canada)

Advantages
- Easy to use
- High sensitivities
- Reusable
- Instantaneous reading

Fast Neutron Detectors
Bubble Detectors

Energy Response:
- **Somewhat uniform response to neutrons over the energy range of 100 keV–10 MeV.**
- **At energies < and > this range, the response decreases rapidly.**

For PRT, underestimate dose:
- **No response to neutrons from 10 MeV - max proton energy.**
- **Further complicated In phantom neutrons lose energy rapidly limited response < 100 keV**

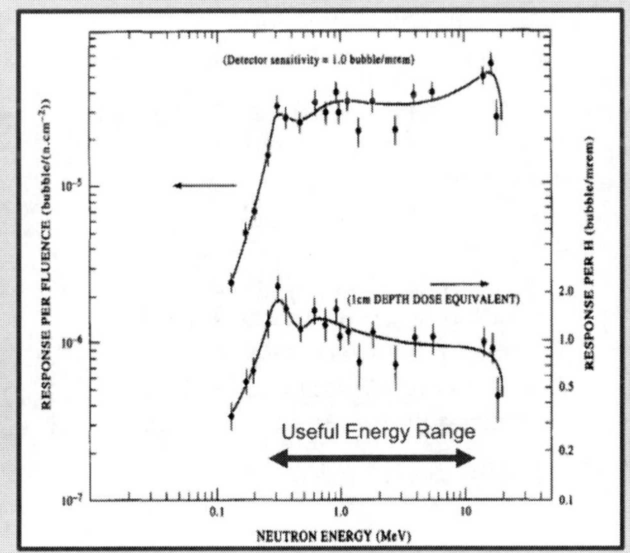

Fast Neutron Detectors Track Etch Detectors

- Solid-state nuclear track etch detectors are made of polymers that include mainly H, C, and O, e.g, CR-39 polymer.
- When a neutron interacts in these materials, the recoil nuclei from the constituent materials leave behind microscopic damage trails, or tracks.
- Chemical processing (etching) is used to enlarge the tracks.
- The track density can then be scored and neutron dose determined using a calibration factor.

Fast Neutron Detectors Track Etch Detectors

Energy Response:

- For most track etch detectors, response falls off above ~ 6 – 10 MeV.
- PN3 does respond above 10 MeV, but response is complexly energy dependent.

For PRT, underestimate dose:

- Most underestimate dose due to limited/no response to neutrons from 10 MeV - max proton energy.
- PN3 have successfully been used, but must account for energy dependence.

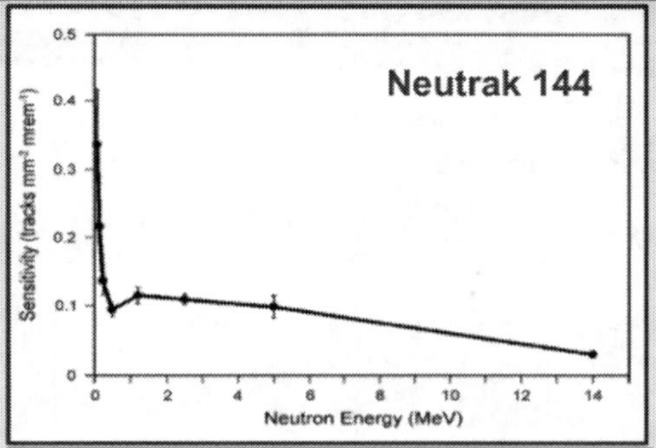

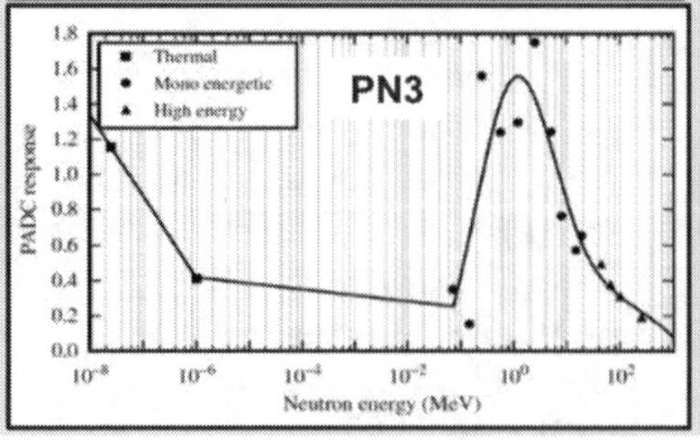

Special Considerations

In-phantom Measurement Challenges

- Neutron detectors with moderators are too large for phantom measurements.
- But for those that are small enough….
  - Neutrons rapidly lose energy in phantom due to scatter with low-Z media.
  - Detector responses to different neutron energies can vary by several orders of magnitude.
  - Thus, incomplete consideration of the detector response or neutron spectrum being measured can easily result in errors of several orders of magnitude.

Variations in the Neutron spectrum as a Function of Depth in Tissue

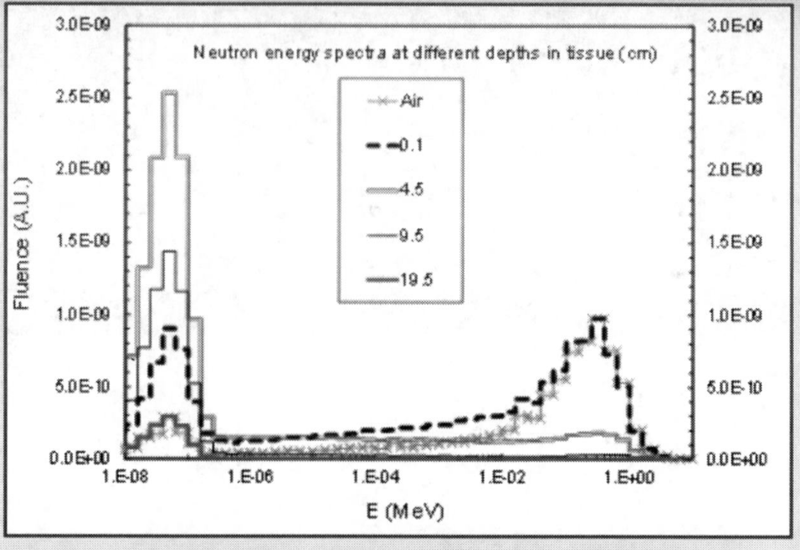

- **Let's consider the examples of TLD 600 & bubble detectors in the context of these data….**

Clinical Proton Beam Therapy Cranial Spinal Irradiation

What about Clinical Scenarios??

- **Dose from secondary neutrons is of particular concern in proton CSI because:**
  - almost universally utilized for children and adolescents,
  - who often survive many decades following diagnosis, and
  - are at significant risk of RT-related late effects.

Passive Scatter Beamline - Mevion

- **Measure Clinical Scenarios**

Field ty	ulation cm)
Whole br	16
Upper sp	8
Lower Sp	8

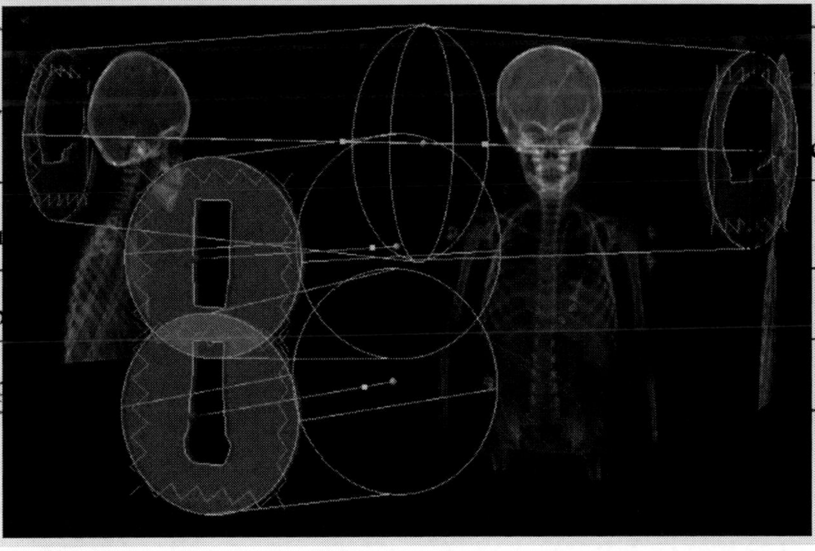

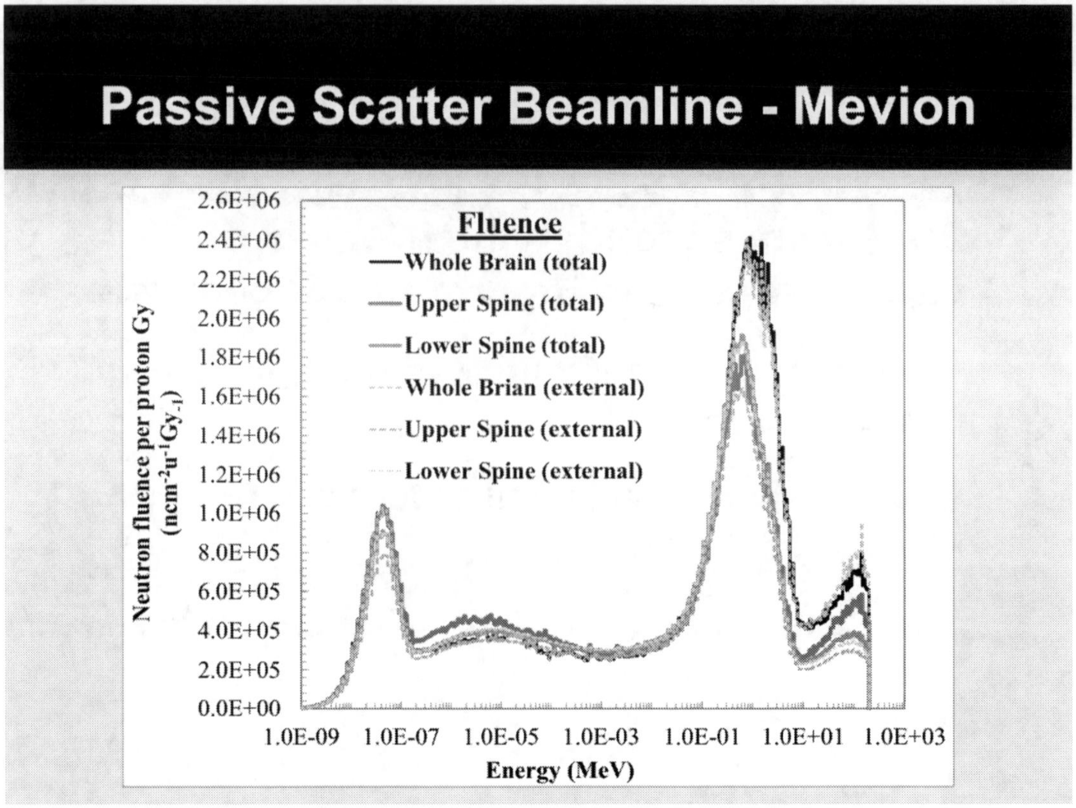
Passive Scatter Beamline - Mevion
Fluence
Whole Brain (total)
Upper Spine (total)
Lower Spine (total)
Whole Brian (external)
Upper Spine (external)
Lower Spine (external)
Neutron fluence per proton Gy
(ncm-2u-1Gy-1)
Energy (MeV)

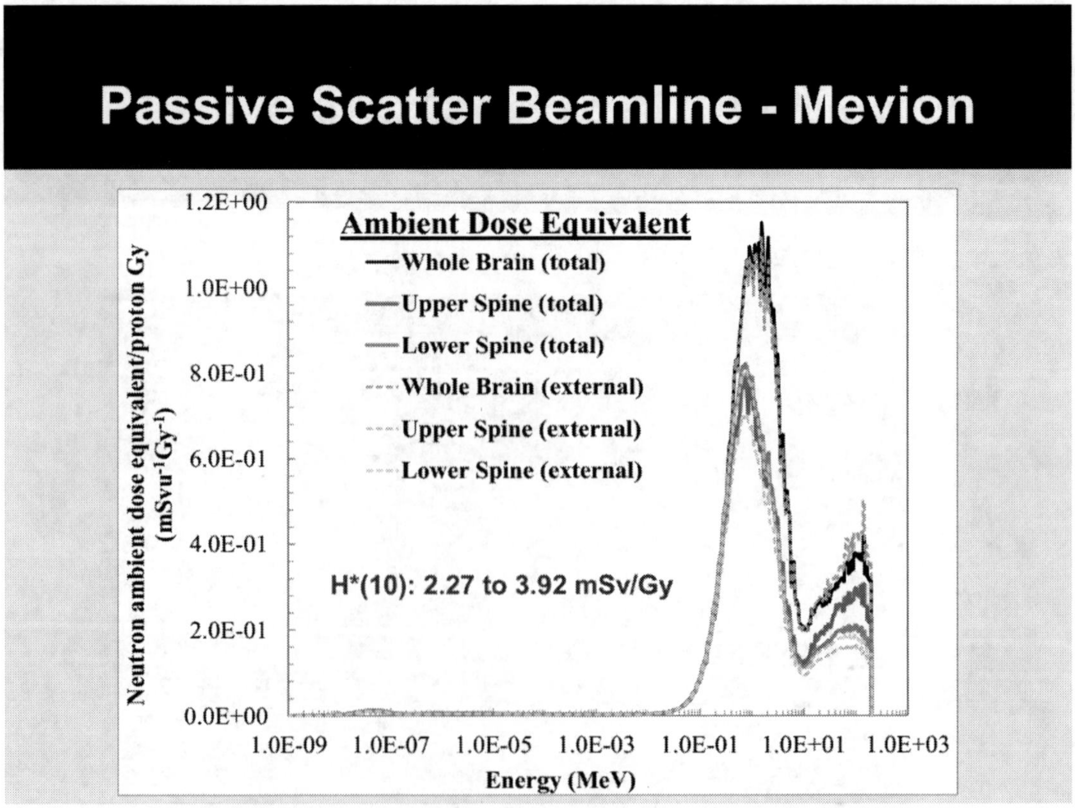
Passive Scatter Beamline - Mevion
Ambient Dose Equivalent
Whole Brain (total)
Upper Spine (total)
Lower Spine (total)
Whole Brain (external)
Upper Spine (external)
Lower Spine (external)
H*(10): 2.27 to 3.92 mSv/Gy
Neutron ambient dose equivalent/proton Gy
(mSvu-1Gy-1)
Energy (MeV)

What about Scanning Beam?

How much lower are the neutron doses compared to passive scatter?

- **EURADOS WG 9 Comprehensive Study**
- **IBA beamline (Krakow, Poland)**
  - **measured 10 spectra**
  - **Two proton energies 100 and 144 MeV**
  - **2 different ERBSS**
  - **High value: 2.7μSv/Gy at 1.6 m**
  - **Low value: 0.52 μSv/Gy at 3.35 m**

Passive Scatter versus Scanning Beam

External Neutrons

- Doses from neutrons generated external to the patient are highly dependent on the
  - proton beam incident on the patient, which depends on:
    - the design of the proton therapy machine itself and
    - treatment-specific devices within the beamline.

Internal Neutrons

- Doses from neutrons generated within the patient increase with:
  - beam range and
  - treatment volume irradiated (i.e., field size).

mSv/Gy versus μSv/Gy

Secondary Neutrons
Protons versus Photons

Photons versus Protons

Protons	Photons
• Narrow proton energy distribution	• Wide photon energy distribution
• Higher primary energies: 180 MeV, 200 MeV, 250 MeV	• Lower primary energies: 15 MV, 18 MV, 20 MV
• (p,n) in both high-Z and low-Z – External and Internal neutron production	• (γ,n) in high-Z – External neutrons
	• No (γ,n) in low-Z – No internal neutrons
• Patient dose: – μSv – mSv per Gy	• Patient does – mSv/Gy (varies over ~3X for CRT, IMRT, FFF)

The Bigger Picture

Proton CSI Photon CSI

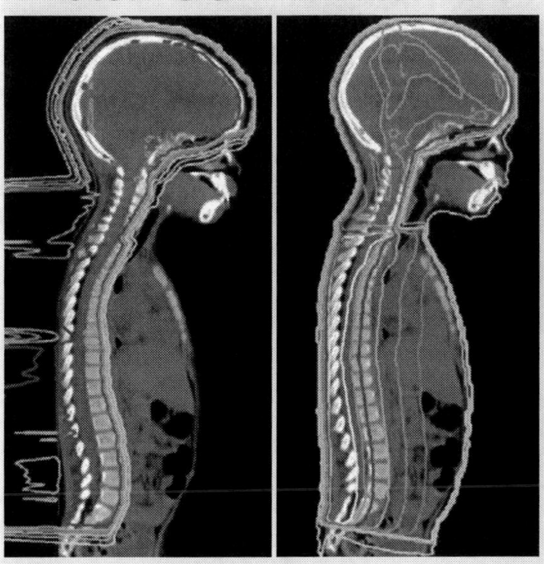

Dose (%)
110
105
100
90
80
70
50
25
5

Compare 21Gy RX

- For a PRT example of 5 mSv/Gy → **10.5cSv**
- For CRT FIF plan, entire abdomen is covered by 50% isodos line, i.e., **10 Gy**

Data Span Orders of Magnitude

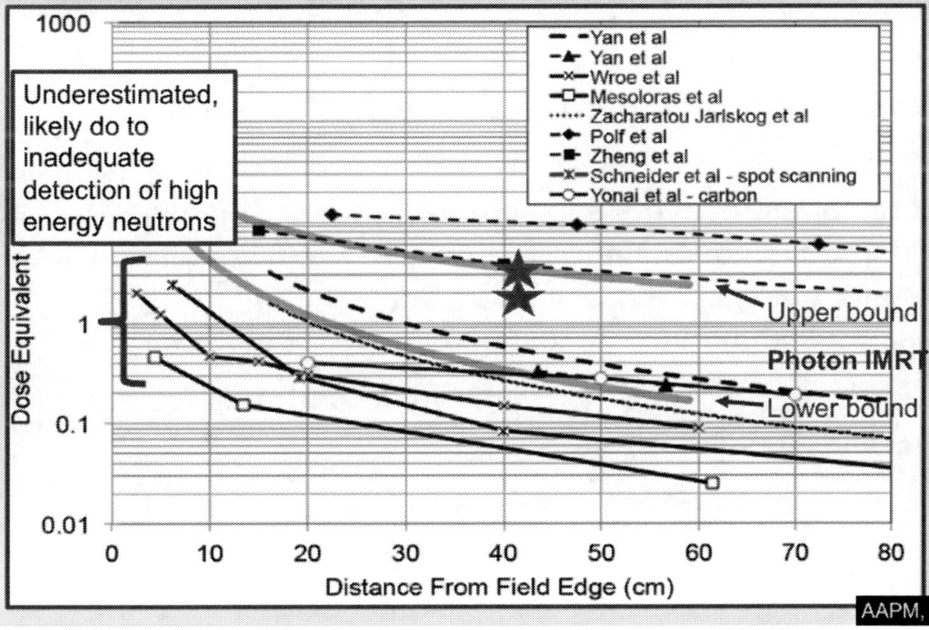

AAPM, TG-158

What's Recommended?
American Association of Physicists in Medicine

For Neutron Research

1. Optimal neutron dosimetry consists of in air spectra measurements
2. Monte Carlo is well-suited for in-patient/phantom neutron dose calculations, but the model must be first validated against in-air neutron measurements
3. Neutron measurements in-patient/phantom, particularly those based on thermal neutron detection, should account for the spatial variations in the neutron energy spectrum
4. **Low-energy neutron detectors (e.g., standard Bonner spheres, most track etch detectors, bubble detectors, etc.) should not be used in proton beams. Appropriate detectors include extended-range Bonner sphere systems, SWENDI detectors, and TEPCs**

What's Recommended?
American Association of Physicists in Medicine

For Clinical Concerns Regarding Neutrons

- Neutron dosimetry requires high levels of expertise and is prone to very large errors if techniques are performed incorrectly.
- Neutron measurements should only be conducted when there is a thorough understanding of neutron dosimetry and the dosimeter being used (particularly the spectrum being measured and the energy response of the detector).
- In lieu of this, **previously published data should be used to get a reasonable estimate.**

What's needed?

- **A comprehensive measurement campaign of the various PRT beamlines, several beamlines with similar (as possible) experimental design as possible…..**

Thank you

- **and stay tuned for……**
- AAPM TG 158: Measurement and calculation of doses outside the treated volume from external-beam radiation therapy

Memories from ICRS-10

References

- NCRP, Report 79 - Neutron Contamination from Medical Electron Accelerators. (National Council on Radiation Protection and Measurements, Bethesda, Md., 1984).
- R.M. Howell, S.F. Kry, E. Burgett, N.E. Hertel, D.S. Followill, "Secondary neutron spectra from modern Varian, Siemens, and Elekta linacs with multileaf collimators," Med Phys 36, 4027-4038 (2009).
- S.F. Kry, R.M. Howell, M. Salehpour, D.S. Followill, "Neutron spectra and dose equivalents calculated in tissue for high-energy radiation therapy," Med Phys 36, 1244-1250 (2009).
- R.M. Howell, E.A. Burgett, "Secondary neutron spectrum from 250-MeV passively scattered proton therapy: Measurement with an extended-range Bonner sphere system," Med Phys 41 (2014).
- F. Khan The Physics of Radiation Therapy 4th Ed. (2010)
- J. Fontenot, P. Taddei, Y. Zheng, D. Mirkovic, T. Jordan, W. Newhauser, "Equivalent dose and effective dose from stray radiation during passively scattered proton radiotherapy for prostate cancer," Phys Med Biol 53, 1677-1688 (2008).
- J. Farah, V. Mares, M. Romero-Exposito, S. Trinkl, C. Domingo, V. Dufek, M. Klodowska, J. Kubancak, Z. Knezevic, M. Liszka, M. Majer, S. Miljanic, O. Ploc, K. Schinner, L. Stolarczyk, F. Trompier, M. Wielunski, P. Olko, R.M. Harrison, "Measurement of stray radiation within a scanning proton therapy facility: EURADOS WG9 intercomparison exercise of active dosimetry systems," Med Phys 42, 2572-2584 (2015).
- PAPER
- V. Mares1, M. Romero-Expósito, J. Farah, S. Trinkl, C. Domingo, M. Dommert, L. Stolarczyk, L. Van Ryckeghem, M. Wielunski, A comprehensive spectrometry study of a stray neutron radiation field in scanning proton therapy
- Physics in Medicine and Biology, 61 (2016)

- B. Gottschalk, "Neutron dose in scattered and scanned proton beams: in regard to Eric J. Hall (Int J Radiat Oncol Biol Phys 2006;65:1-7)," International journal of radiation oncology, biology, physics 66, 1594; author reply 1595 (2006).
- H. Paganetti, T. Bortfeld, T.F. Delaney, "Neutron dose in proton radiation therapy: in regard to Eric J. Hall (Int J Radiat Oncol Biol Phys 2006;65:1-7)," International journal of radiation oncology, biology, physics 66, 1594-1595; author reply 1595 (2006).
- C.Z. Jarlskog, H. Paganetti, "Risk of developing second cancer from neutron dose in proton therapy as function of field characteristics, organ, and patient age," Int J Radiat Oncol 72, 228-235 (2008).
- D. Hecksel, V. Anferov, M. Fitzek, K. Shahnazi, "Influence of beam efficiency through the patient-specific collimator on secondary neutron dose equivalent in double scattering and uniform scanning modes of proton therapy," Med Phys 37, 2910-2917 (2010).
- C. Zacharatou Jarlskog, C. Lee, W.E. Bolch, X.G. Xu, H. Paganetti, "Assessment of organ-specific neutron equivalent doses in proton therapy using computational whole-body age-dependent voxel phantoms," Phys Med Biol 53, 693-717 (2008).
- U. Titt, Y.S. Zheng, O.N. Vassiliev, W.D. Newhauser, "Monte Carlo investigation of collimator scatter of proton-therapy beams produced using the passive scattering method," Phys Med Biol 53, 487-504 (2008).
- H. Jiang, B. Wang, X.G. Xu, H.D. Suit, H. Paganetti, "Simulation of organ-specific patient effective dose due to secondary neutrons in proton radiation treatment," Phys Med Biol 50, 4337-4353 (2005).

ICRS-13 & RPSD-2016

13th *International Conference on Radiation Shielding*
&
19th Topical Meeting of the *Radiation Protection & Shielding Division*
of the *American Nuclear Society* -2016

4th Plenary invited talk:

"From Earth to Moon, Mars and Beyond – Space Radiation and Implications for Human Exploration"

Prof. Dr. Robert F. Wimmer-Schweingruber

University of Kiel, Germany

High-energy radiation fills space and poses a short-term as well as long-term danger to humans in space, but is also of concern for electronics in spacecraft. The radiation environment on planetary bodies is quite different from that on Earth. On the airless Moon, space radiation is shielded by the massive Moon, but the interaction of radiation with it also generates new, secondary radiation. On Mars, the tenuous atmosphere provides some shielding, but also radiation interaction with it also creates secondary radiation, in addition to what is created in the soil. It is discussed measurements of the space radiation environment at Earth, Moon, en route to Mars and on Mars itself.

Robert Wimmer-Schweingruber („Bob Wimmer") received his PhD in experimental physics from the University of Bern, Switzerland, in 1994. He was asked to return to Bern after a PostDoc in the US, to lead two space projects. He obtained his venia docendi („Habilitation") in 2001 and is now a professor at the University of Kiel, Germany, since 2002, where he heads the Division of Extraterrestrial Physics. He is the Principal Investigator (PI) of three space instruments, and Co-PI and Co-I of many other space experiments.

From Earth to Moon, Mars, and Beyond – Space Radiation and Implications Human Exploration

Robert F. Wimmer-Schweingruber
Extraterrestrial Physics,
IEAP, University of Kiel, Germany
wimmer@physik.uni-kiel.de

Structure of my talk

- Radiation: Origin & transport in heliosphere
- Implications for human exploration

- Particle radiation at and on the Moon
- Particle radiation during the cruise to Mars
- Mars surface radiation environment
- Implications for human exploration

- ~~Why Mars, why Gale?~~
- ~~Was Mars habitable?~~

Radiation Sources Close to Earth

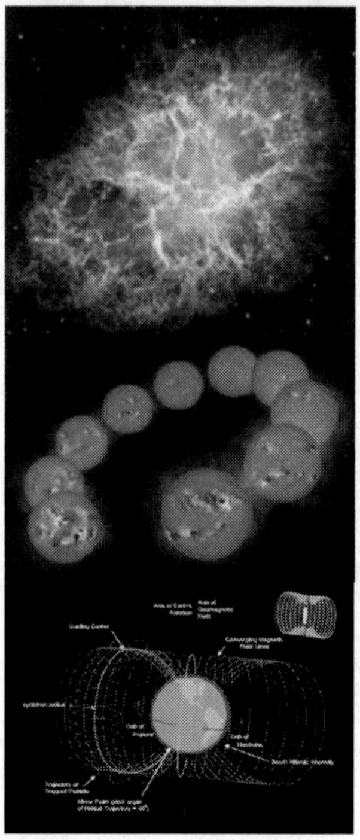

Galactic Cosmic Rays:
Very high energies, H through U
Solar modulation, predictable
Highly penetrating

Solar Energetic Particles:
High energies
Solar activity, unpredictble
High dose rate variability (up to ~1000)
Can be shielded

Trapped Radiation (Radiation Belts):
Very high energies, protons, electrons
Very high dose rate
Highly penetrating

Radiation Sources Close to Earth: GCR

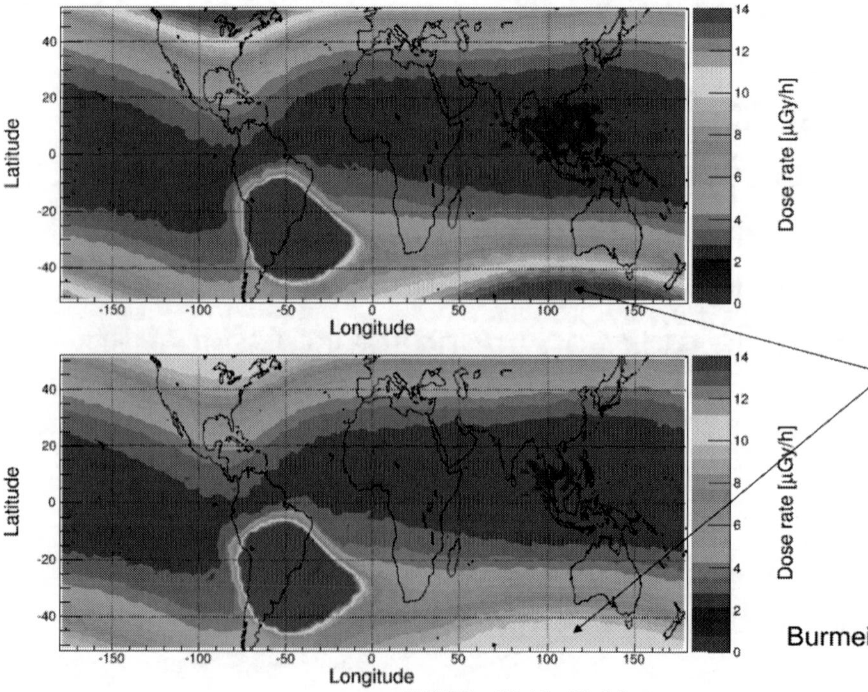

2009
closer to solar min

GCR

2013
closer to solar max

Burmeister et al., 2016

Radiation Sources Close to Earth: Trapped

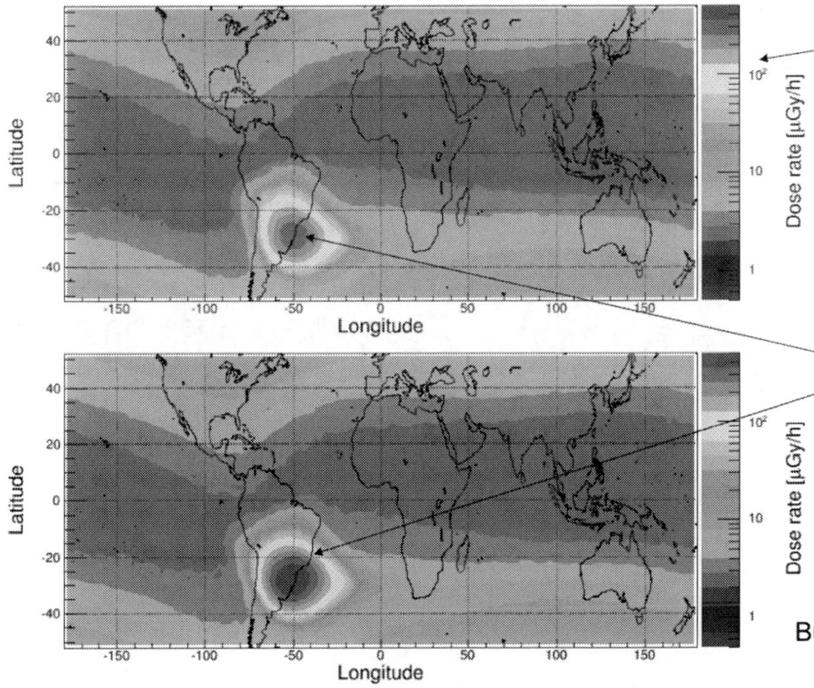

2009
@ 350 km

2013
@ 420 km

Burmeister et al., 2016

Radiation Sources Close to Earth: Solar

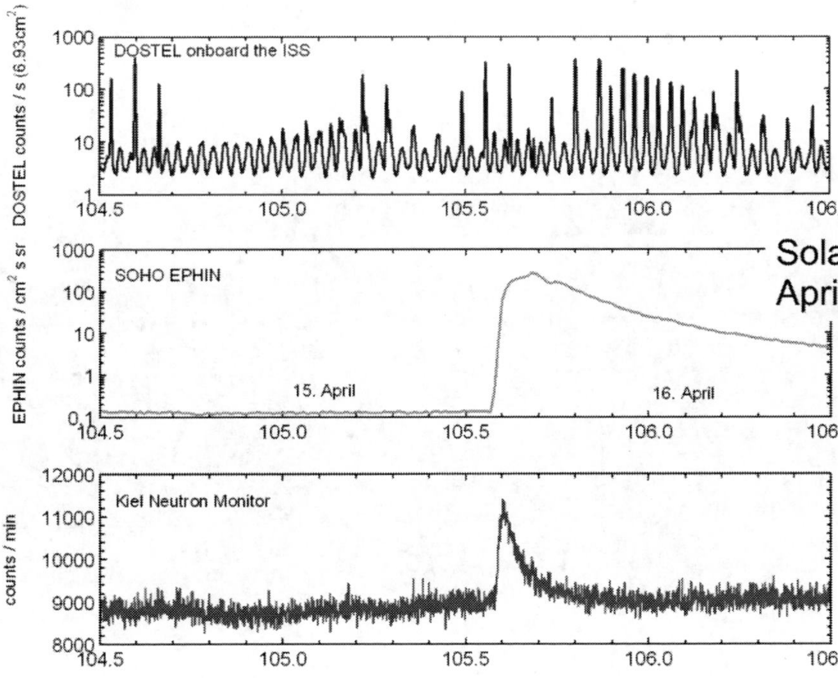

Solar Particle Event:
April 2001

Data:
Burmeister
Heber
NMDB

Radiation Sources Close to Earth: Comparison

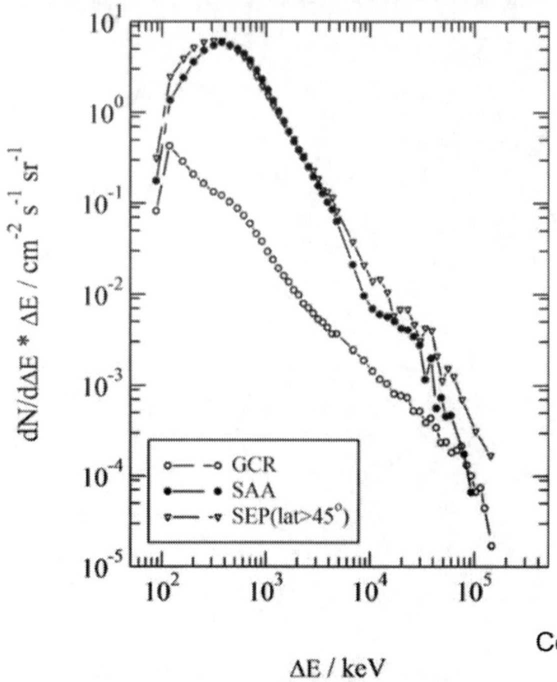

Nov 1997 SPE:

Comparison of the effect of the GCR, SPE, and trapped components

Measured spectrum of energy deposit in a 315 micron Si solid-state detector.

Courtesy S. Burmeister

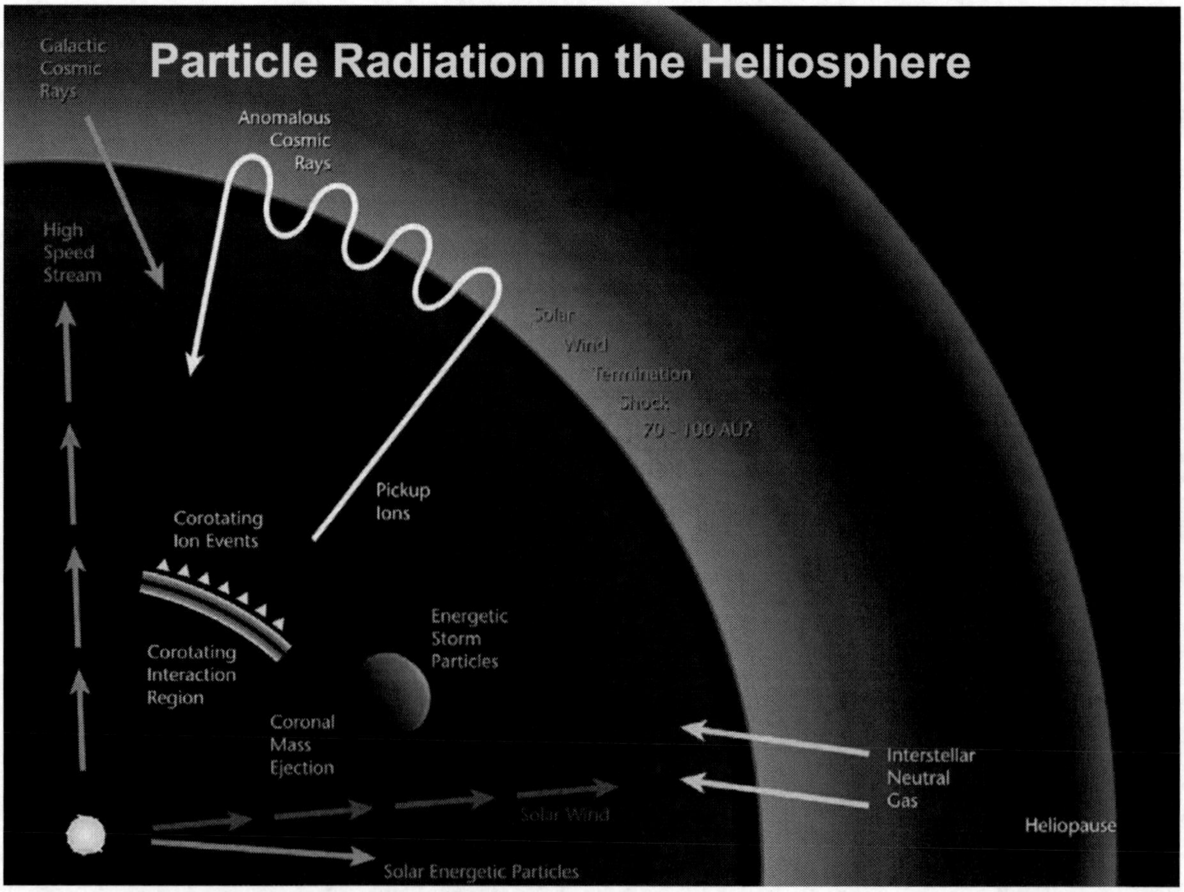

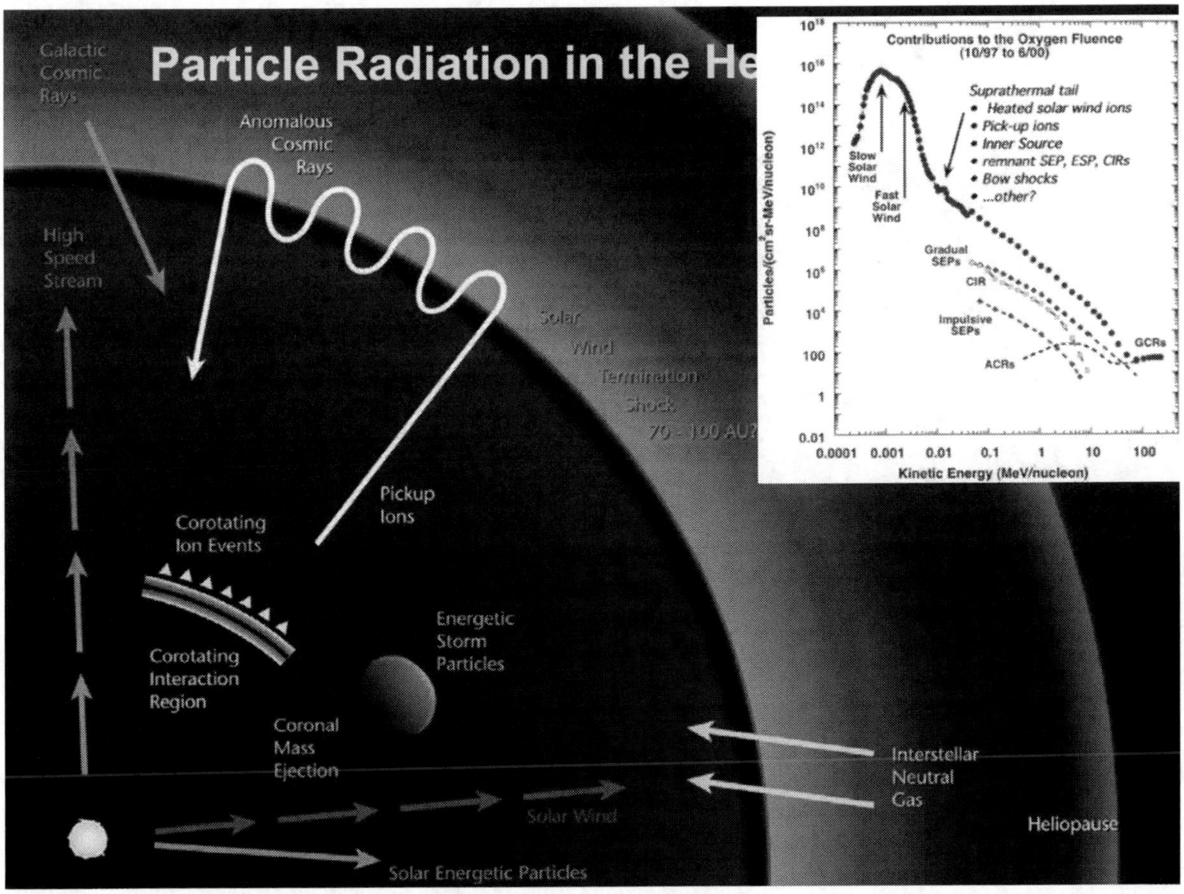
Particle Radiation in the He
Galactic Cosmic Rays
Anomalous Cosmic Rays
High Speed Stream
Solar Wind Termination Shock 70 - 100 AU?
Pickup Ions
Corotating Ion Events
Corotating Interaction Region
Energetic Storm Particles
Coronal Mass Ejection
Interstellar Neutral Gas
Solar Wind
Heliopause
Solar Energetic Particles
Contributions to the Oxygen Fluence (10/97 to 6/00)
Suprathermal tail
• Heated solar wind ions
• Pick-up ions
• Inner Source
• remnant SEP, ESP, CIRs
• Bow shocks
• ...other?
Slow Solar Wind
Fast Solar Wind
Gradual SEPs
CIR
Impulsive SEPs
ACRs
GCRs
Particles/(cm² sr-MeV/nucleon)
Kinetic Energy (MeV/nucleon)

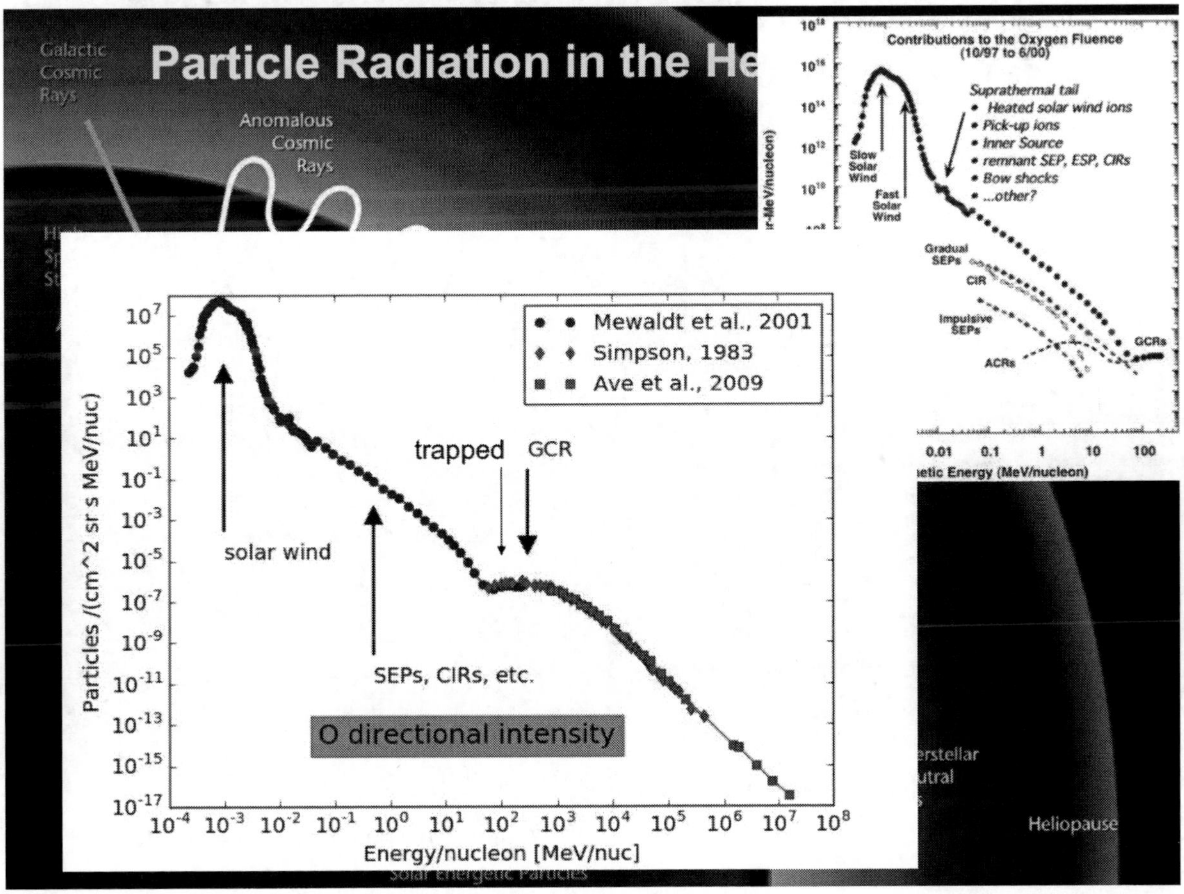
Particle Radiation in the He
Galactic Cosmic Rays
Anomalous Cosmic Rays
Contributions to the Oxygen Fluence (10/97 to 6/00)
Suprathermal tail
• Heated solar wind ions
• Pick-up ions
• Inner Source
• remnant SEP, ESP, CIRs
• Bow shocks
• ...other?
Slow Solar Wind
Fast Solar Wind
Gradual SEPs
CIR
Impulsive SEPs
ACRs
GCRs
Heliopause
Mewaldt et al., 2001
Simpson, 1983
Ave et al., 2009
solar wind
SEPs, CIRs, etc.
trapped
GCR
O directional intensity
Particles /(cm^2 sr s MeV/nuc)
Energy/nucleon [MeV/nuc]

Radiation Protection in Space

Personal Dosimeters

Radiation is *the* long-lasting risk for astronauts. Radiation damage can persist after the end of a long-duration space flight.

Radiation must be measured!

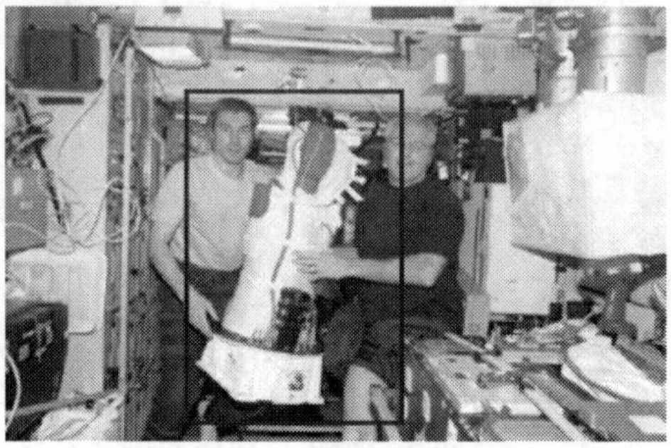

Phantom experiments
MATROSHKA

(Photos courtesy G. Reitz)

Radiation Protection in Space

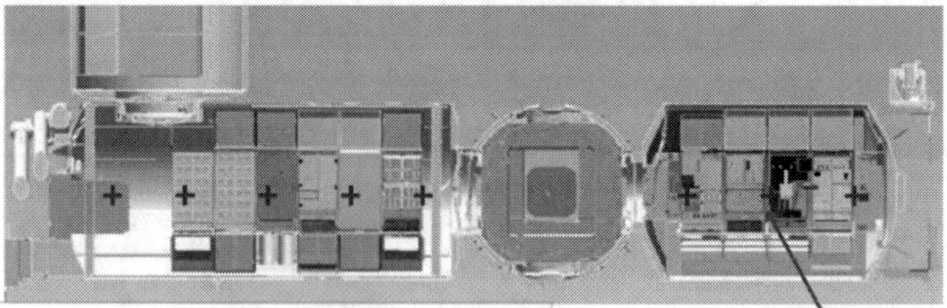

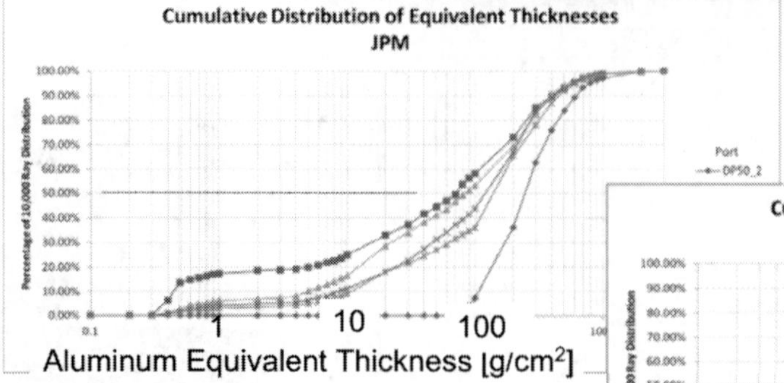

COLUMBUS
median shielding ≈100 g/cm2

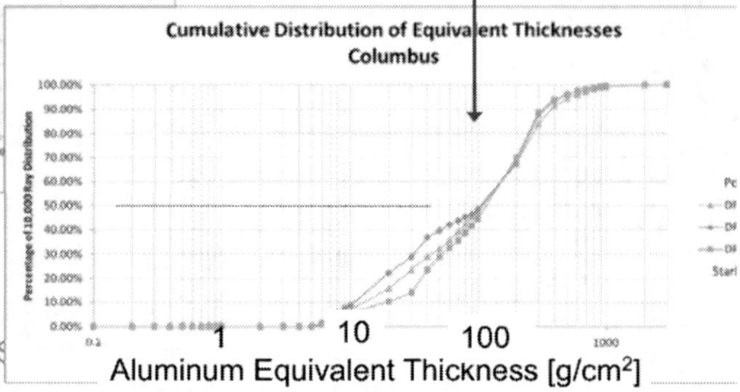

(courtesy Daniel Matthiä)

CAU Institute of Experimental and Applied Physics
Extraterrestrial Physics

Radiation Protection in Space

GCR radiation eposure on ISS
Dec. 2009 (max. GCR intensity)

- Dose rates in a water sphere with radius 20 cm
  - surface (depth < 1cm)
  - whole sphere
- Spherical aluminium shielding
- Columbus Module:
  - ≈ 100 g/cm², Stoffle et al. (2012)
  - 158 µGy/d, 448 µSv/d, Q = 2.84, Semones et al. (2009)
  - 150 - 157 µGy/d, 496 – 517 µSv/d, Q = 3.30, Burmeister et al. (2012)
- Zvezda Service Module:
  - 32 – 47 g/cm², Jardrnickova et al. (2009)
  - 100 – 110 µGy/d, Lishnevskii et al. (2012)
  - 125 µGy/d, Semones et al. (2009)

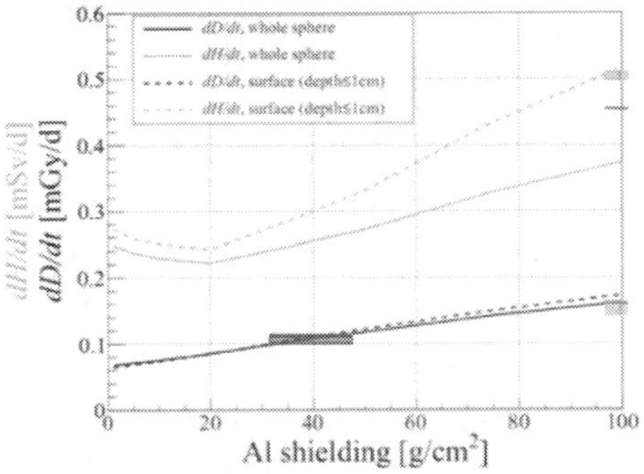

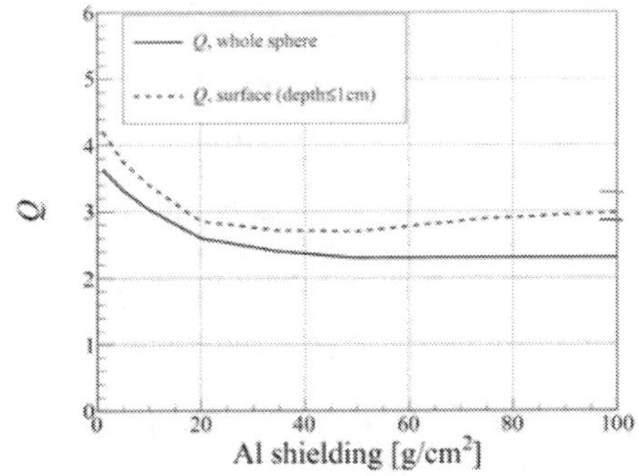

(courtesy Daniel Matthiä)

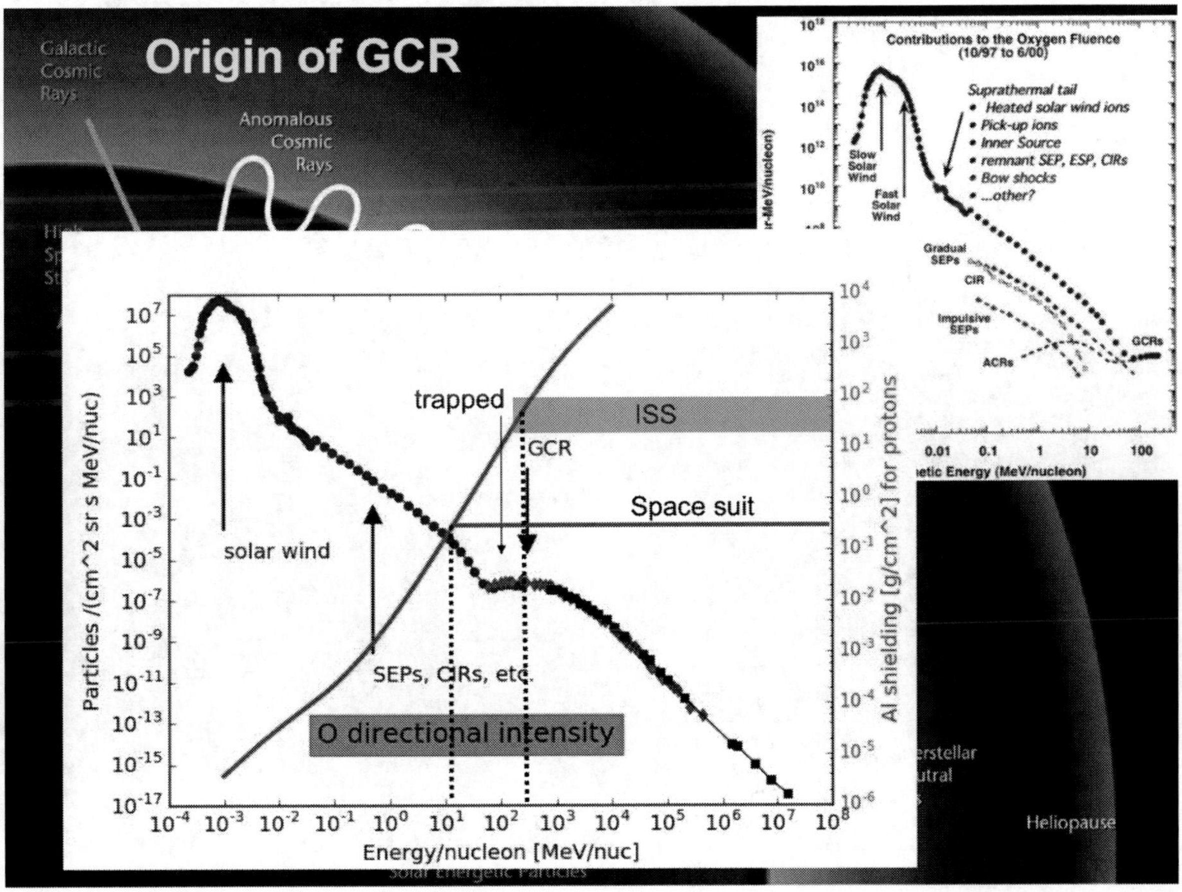

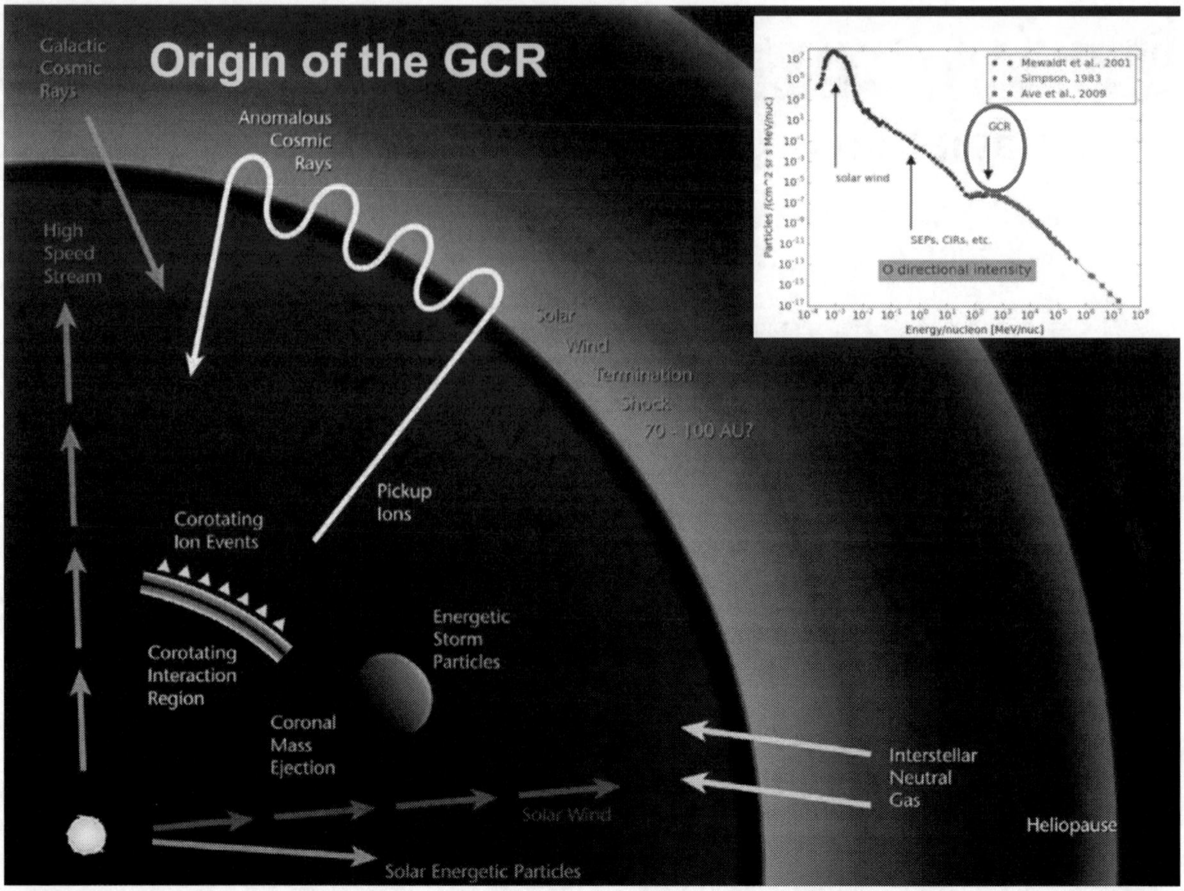
Origin of the GCR
Galactic Cosmic Rays
Anomalous Cosmic Rays
High Speed Stream
Solar Wind Termination Shock 70 - 100 AU?
Pickup Ions
Corotating Ion Events
Corotating Interaction Region
Energetic Storm Particles
Coronal Mass Ejection
Interstellar Neutral Gas
Solar Wind
Heliopause
Solar Energetic Particles
Mewaldt et al., 2001
Simpson, 1983
Ave et al., 2009
GCR
solar wind
SEPs, CIRs, etc.
O directional intensity
Energy/nucleon [MeV/nuc]

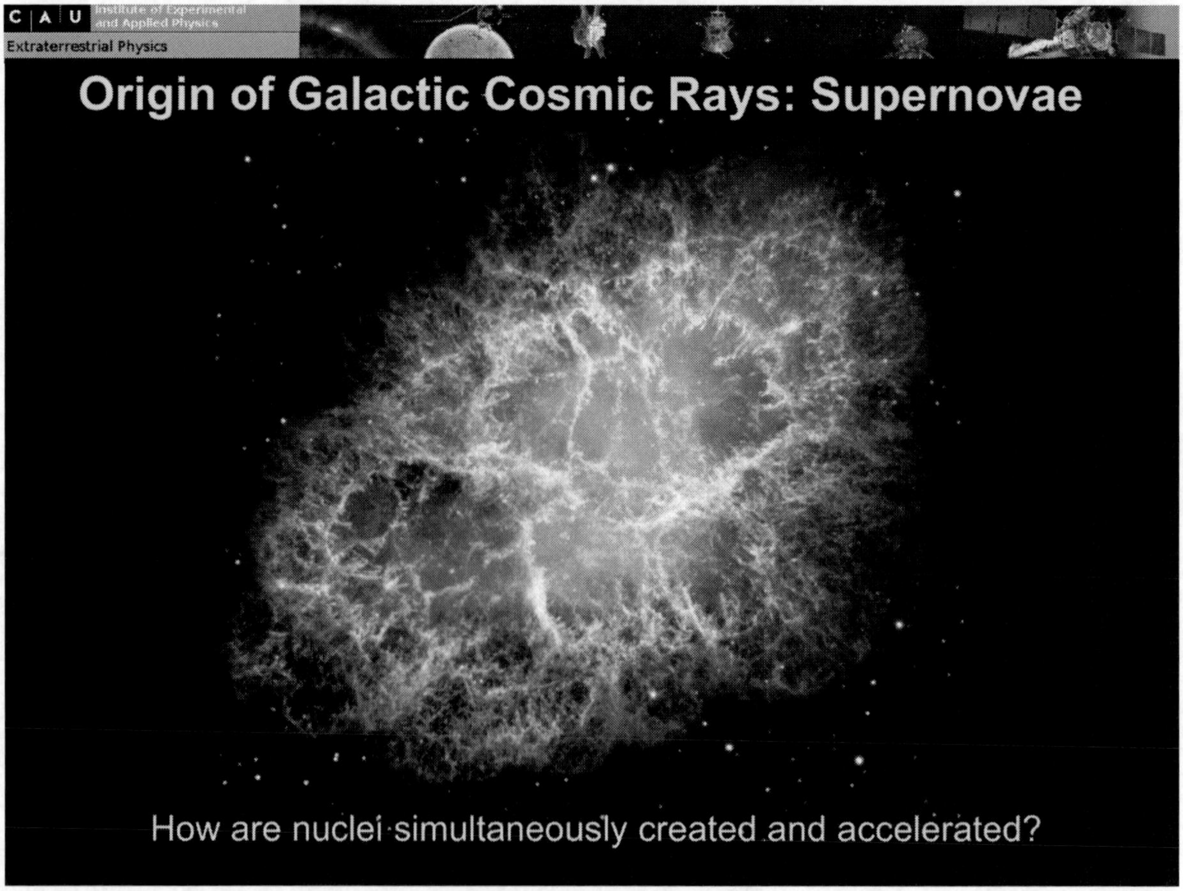
C A U Institute of Experimental and Applied Physics
Extraterrestrial Physics
Origin of Galactic Cosmic Rays: Supernovae
How are nuclei simultaneously created and accelerated?

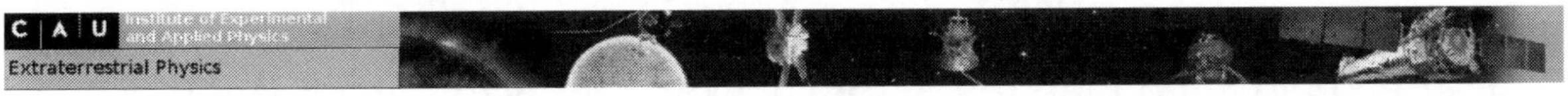

Origin of Galactic Cosmic Rays: OB-Associations

^{59}Ni decays into ^{59}Co via electron capture $T_{1/2}$ = 76,000 years

Measured ^{59}Ni/^{59}Co < 0.055 implies acceleration > 100,000 years after creation of ^{59}Ni

^{60}Fe β-decays into ^{60}Co with $T_{1/2}$ = 2.62 Myrs. Its measurement implies acceleration within few Myrs of nuclei creation

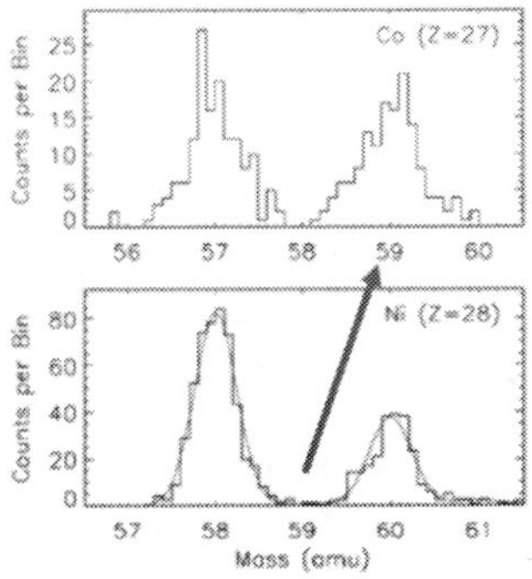

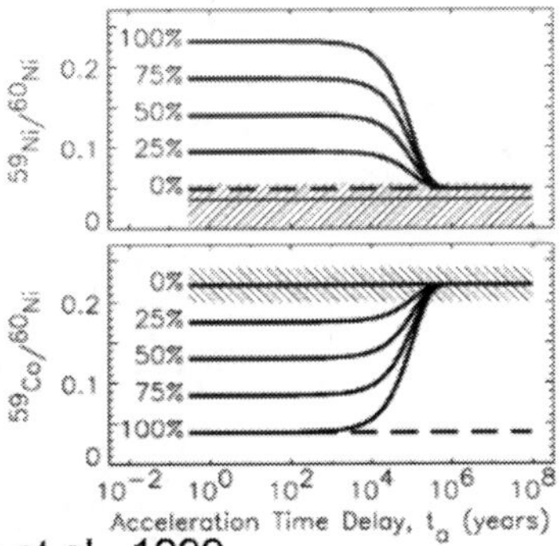

Wiedenbeck et al., 1999

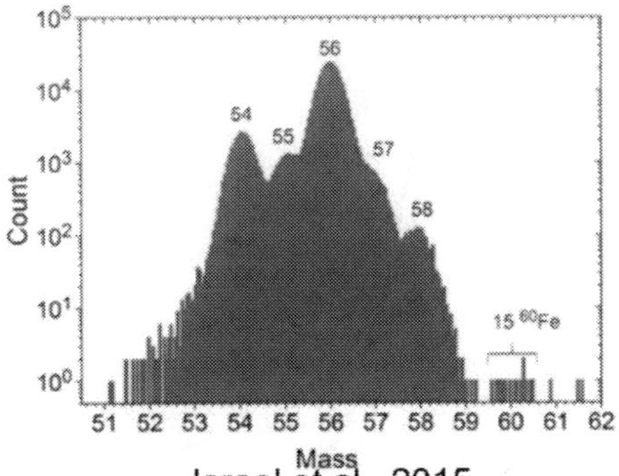

Israel et al., 2015

One supernova creates GCR nuclei, a second one accelerates them

==> OB Associations

Radiation Transport in the Heliosphere

- Electrically charged ions and electrons
- Interplanetary magnetic field controls propagation

$$r_c = \frac{|p_\perp|}{q\,B}$$

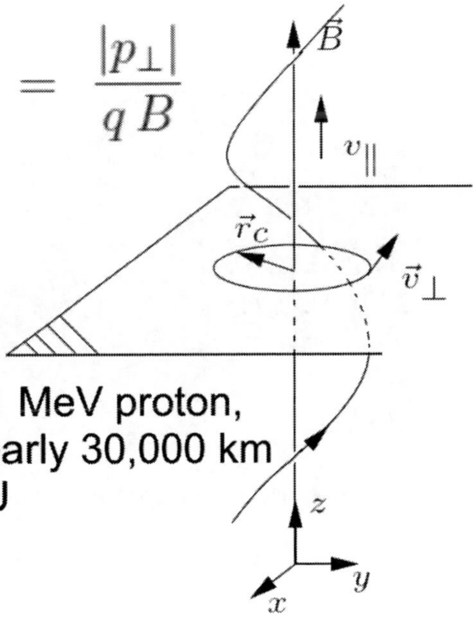

For a 1 MeV proton, r_c is nearly 30,000 km at 1 AU

Large-scale interplanetary magnetic field organized along Parker spiral

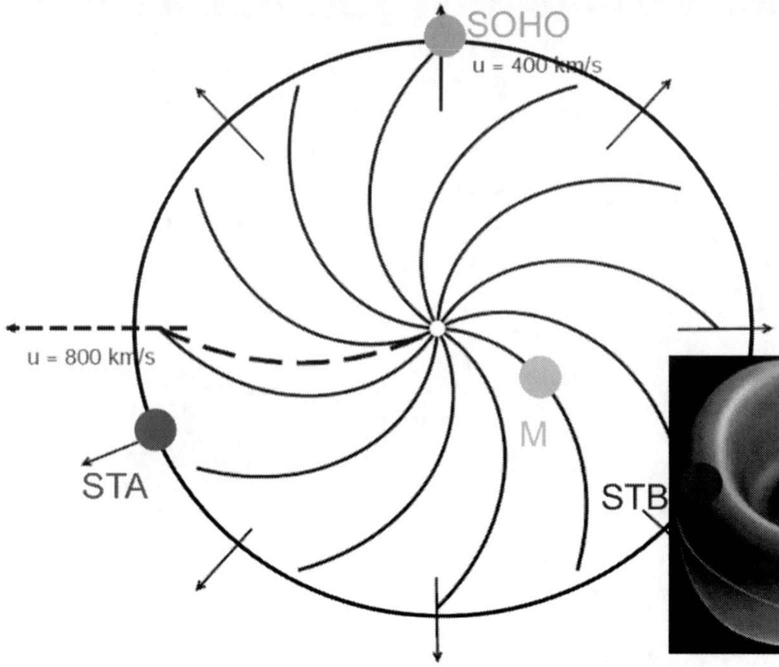

Charged energetic particles are tied to magnetic field...

SOHO, STEREO, Messenger widely separated in Nov‘ 11

Connection Sun-Heliosphere for November 2011

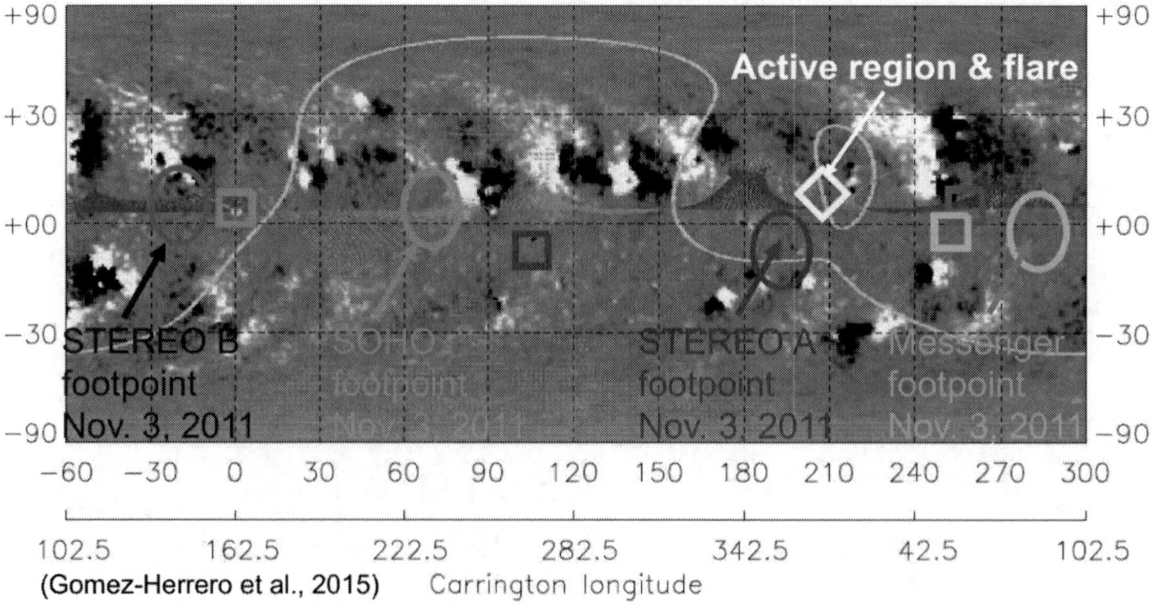

(Gomez-Herrero et al., 2015)

You don't really expect all spacecraft to see the event, do you?

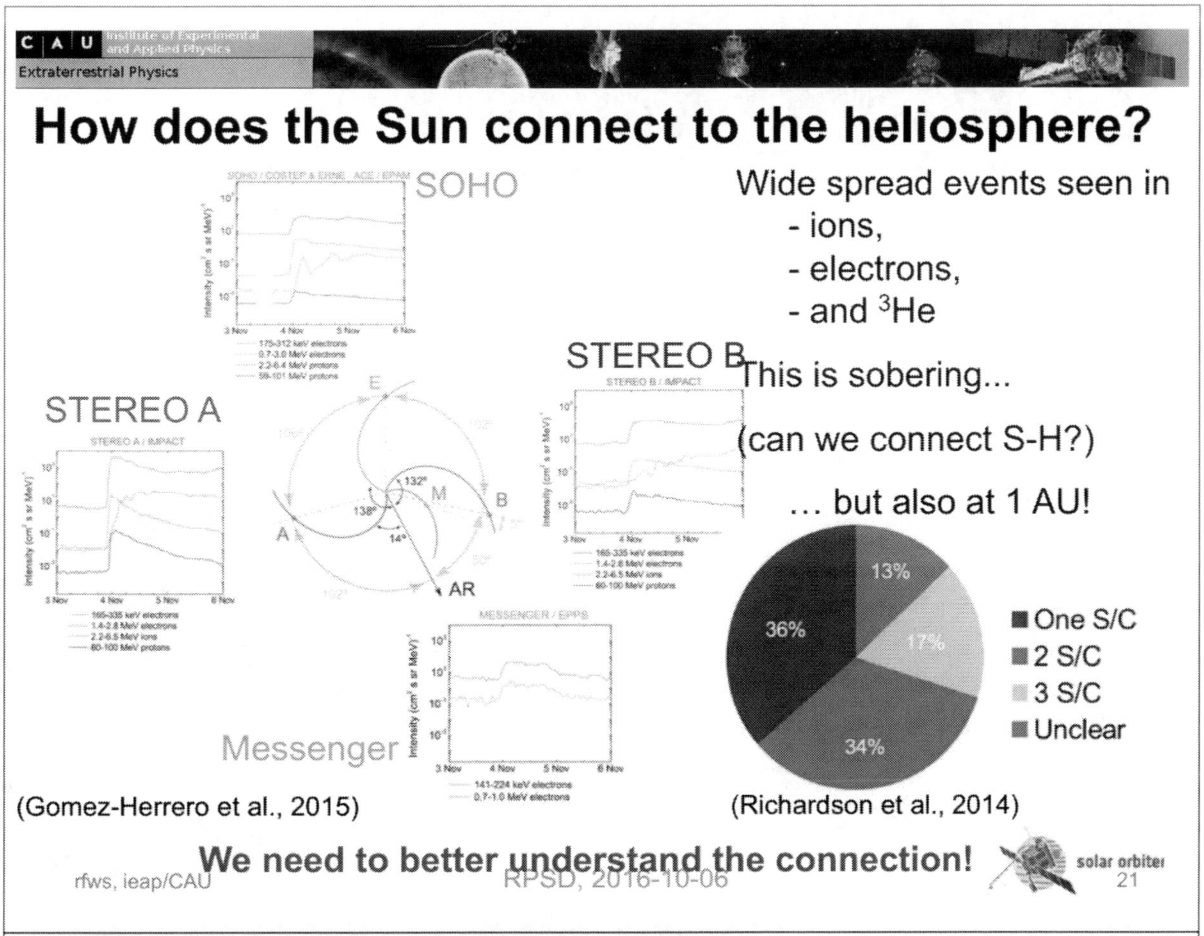
C A U Institute of Experimental and Applied Physics
Extraterrestrial Physics
How does the Sun connect to the heliosphere?
SOHO
STEREO A
STEREO B
Messenger
AR
Wide spread events seen in
- ions,
- electrons,
- and 3He
This is sobering...
(can we connect S-H?)
… but also at 1 AU!
13%
17%
34%
36%
One S/C
2 S/C
3 S/C
Unclear
(Gomez-Herrero et al., 2015)
(Richardson et al., 2014)
We need to better understand the connection!
rfws, ieap/CAU
RPSD, 2016-10-06
solar orbiter
21

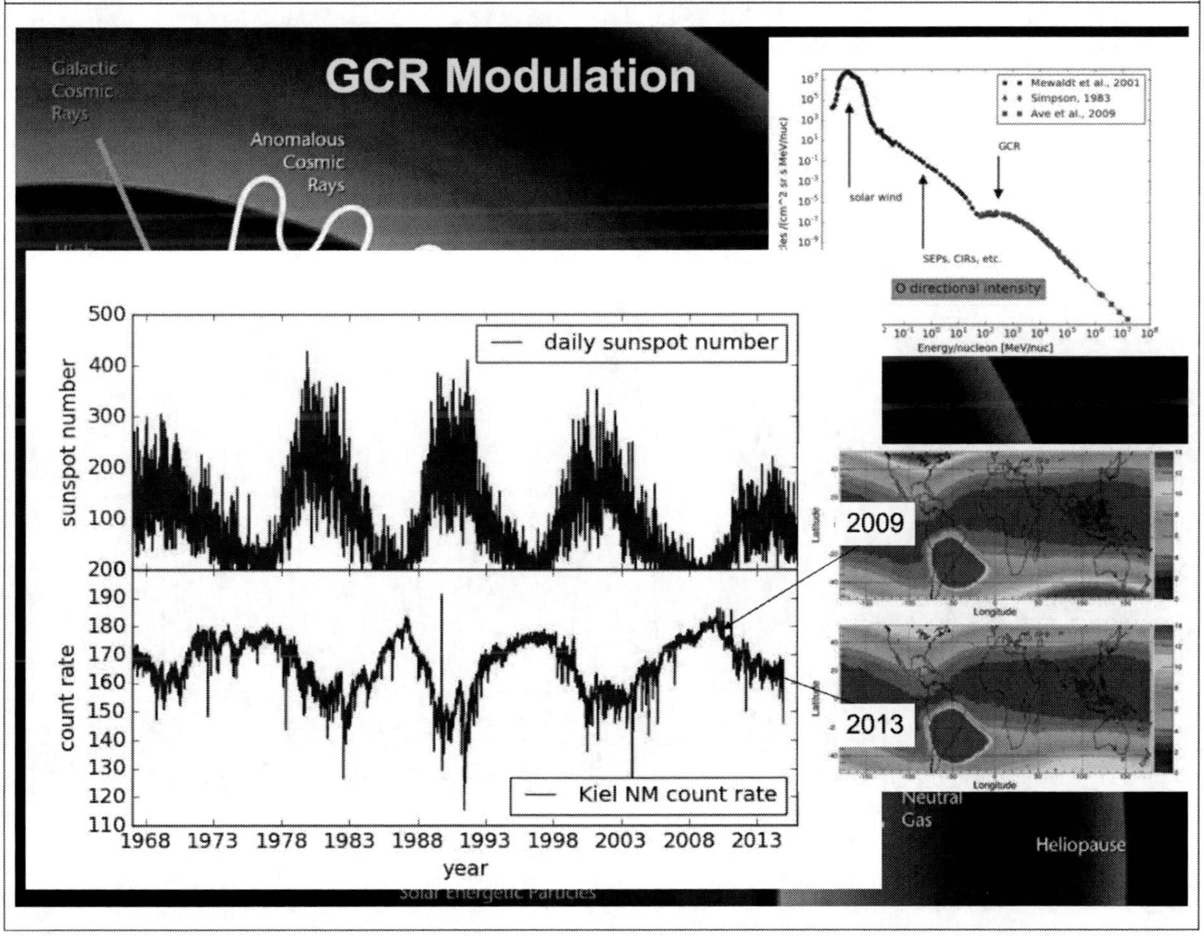
GCR Modulation
Galactic Cosmic Rays
Anomalous Cosmic Rays
solar wind
SEPs, CIRs, etc.
GCR
Mewaldt et al., 2001
Simpson, 1983
Ave et al., 2009
Energy/nucleon [MeV/nuc]
daily sunspot number
sunspot number
Kiel NM count rate
count rate
year
2009
2013
Neutral Gas
Heliopause

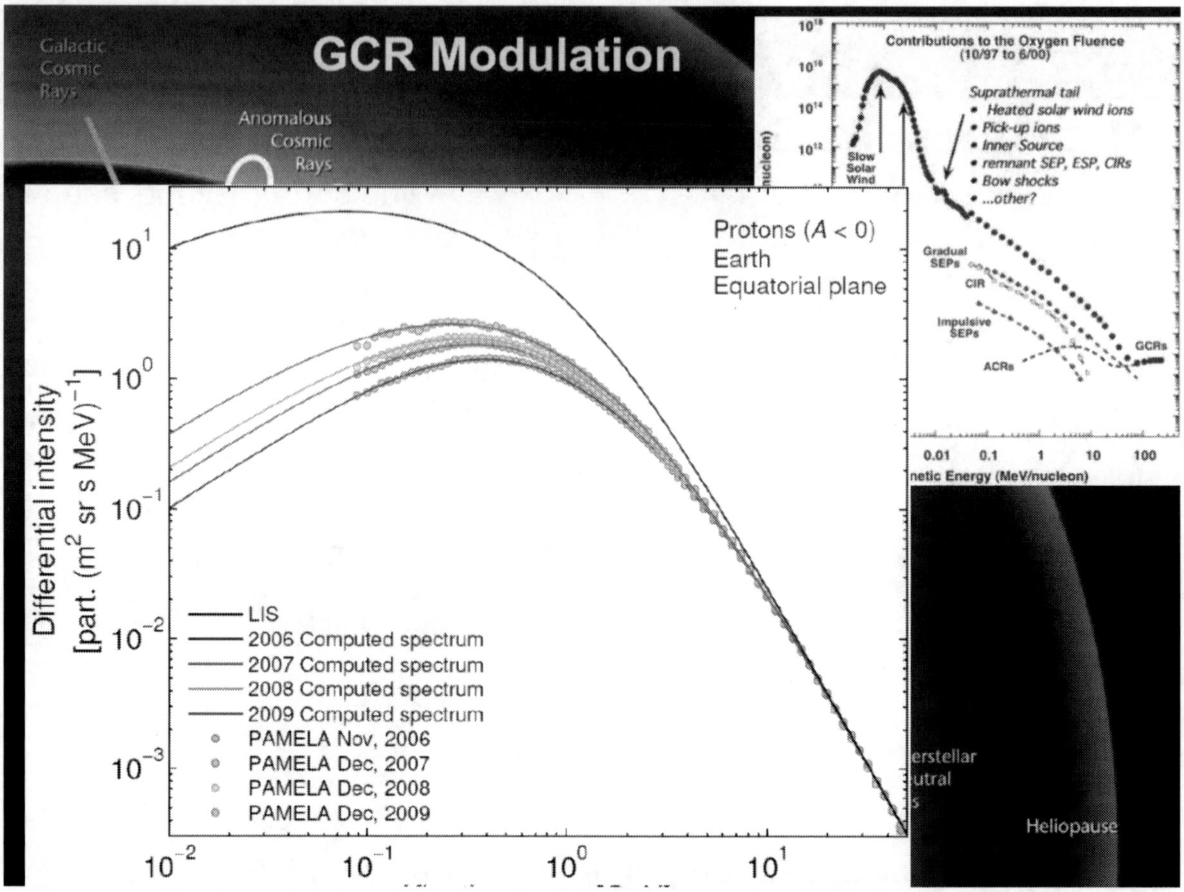

GCR Modulation
Galactic Cosmic Rays
Anomalous Cosmic Rays
Protons (A < 0)
Earth
Equatorial plane
Differential intensity [part. (m2 sr s MeV)-1]
LIS
2006 Computed spectrum
2007 Computed spectrum
2008 Computed spectrum
2009 Computed spectrum
PAMELA Nov, 2006
PAMELA Dec, 2007
PAMELA Dec, 2008
PAMELA Dec, 2009
Contributions to the Oxygen Fluence (10/97 to 6/00)
Slow Solar Wind
Suprathermal tail
Heated solar wind ions
Pick-up ions
Inner Source
remnant SEP, ESP, CIRs
Bow shocks
...other?
Gradual SEPs
CIR
Impulsive SEPs
ACRs
GCRs
netic Energy (MeV/nucleon)
Heliopause

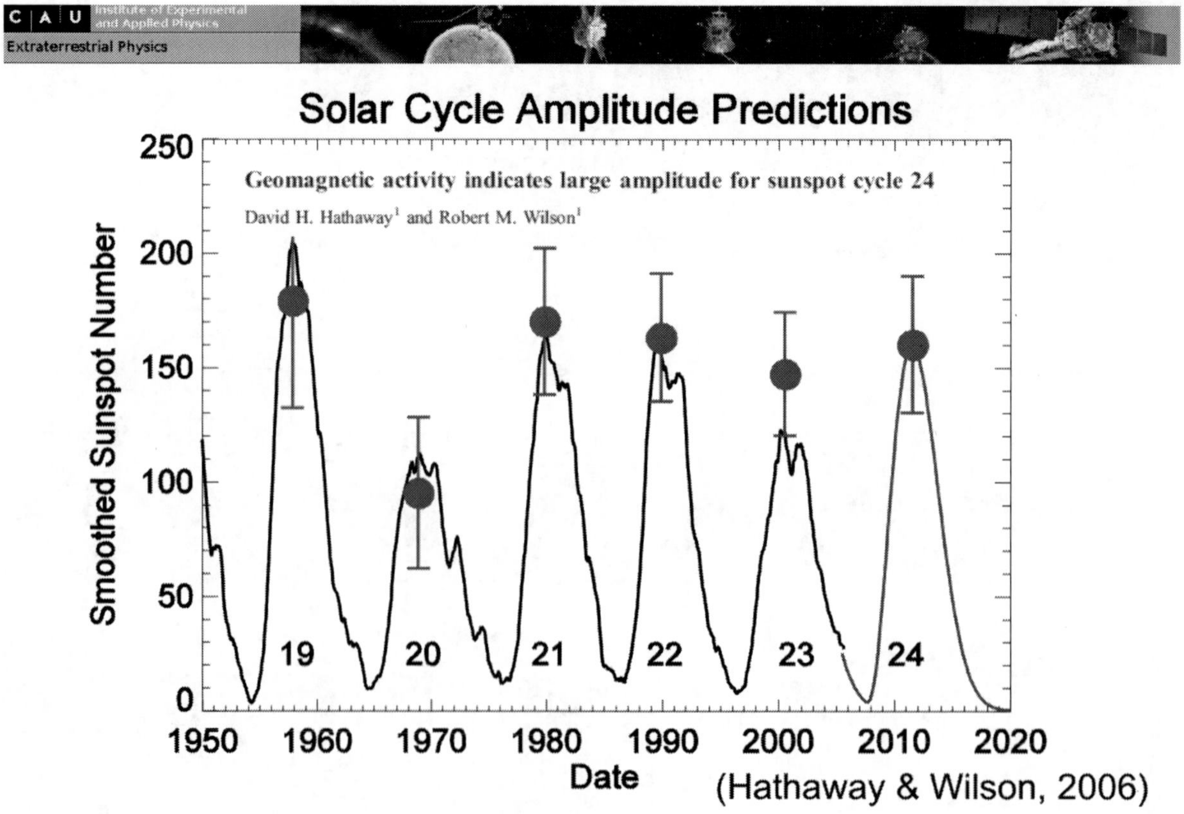

C A U Institute of Experimental and Applied Physics
Extraterrestrial Physics
Solar Cycle Amplitude Predictions
Geomagnetic activity indicates large amplitude for sunspot cycle 24
David H. Hathaway and Robert M. Wilson
Smoothed Sunspot Number
19
20
21
22
23
24
Date
(Hathaway & Wilson, 2006)

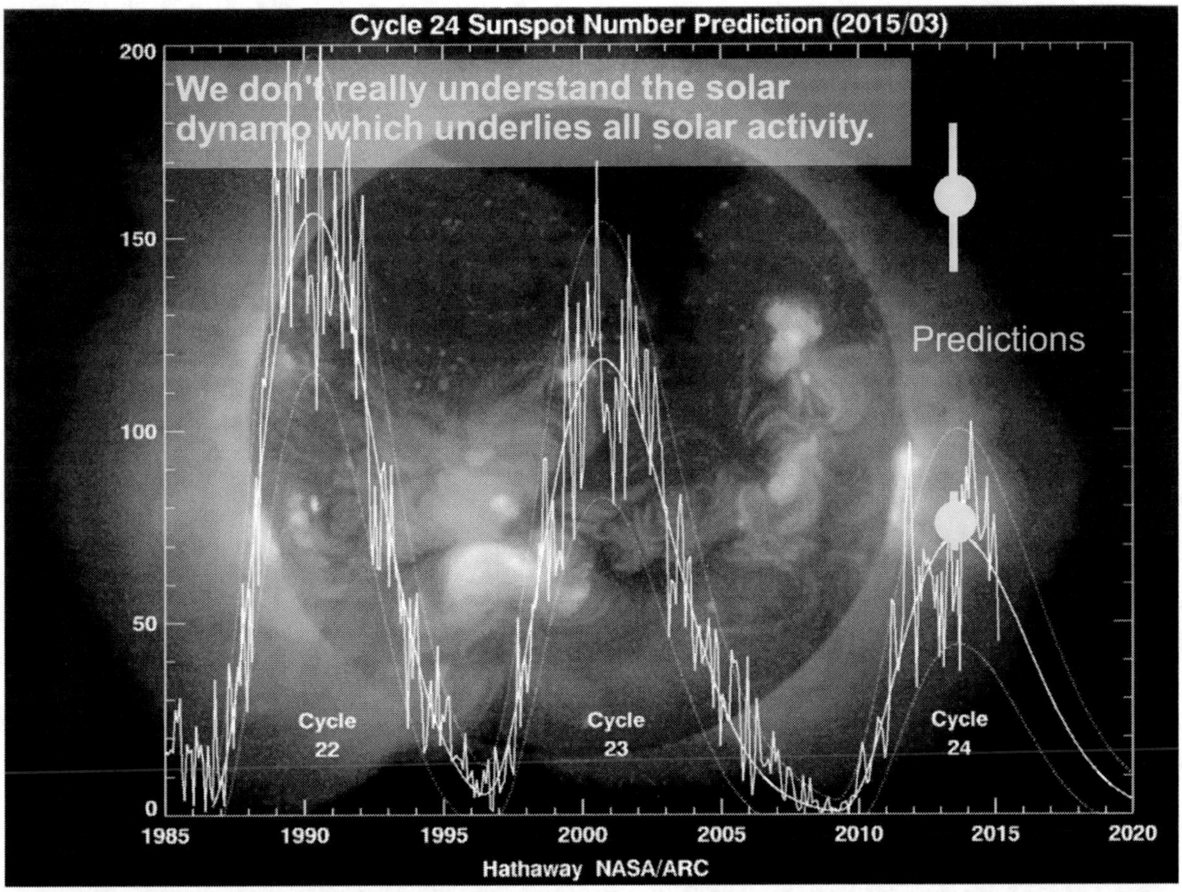
Cycle 24 Sunspot Number Prediction (2015/03)
We don't really understand the solar dynamo which underlies all solar activity.
Predictions
Cycle 22
Cycle 23
Cycle 24
200
150
100
50
0
1985
1990
1995
2000
2005
2010
2015
2020
Hathaway NASA/ARC

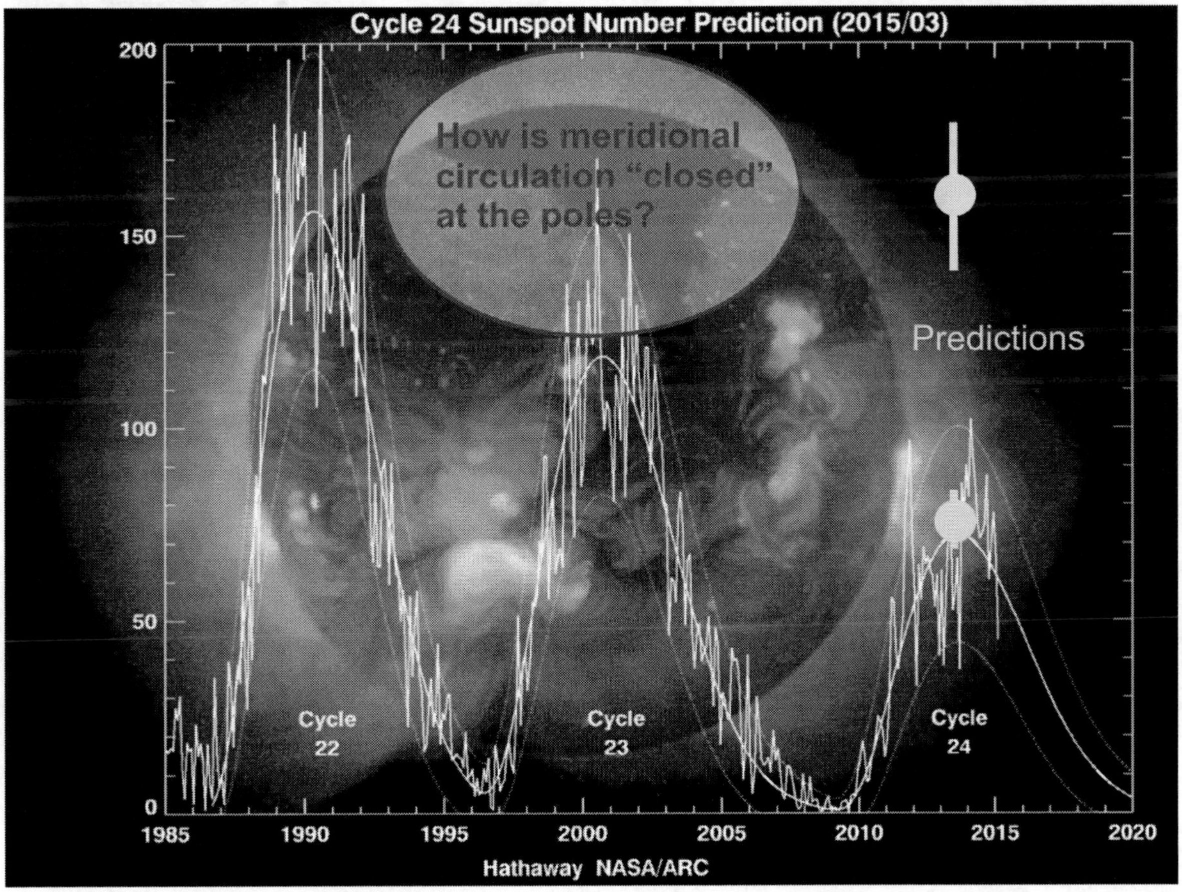
Cycle 24 Sunspot Number Prediction (2015/03)
How is meridional circulation "closed" at the poles?
Predictions
Cycle 22
Cycle 23
Cycle 24
200
150
100
50
0
1985
1990
1995
2000
2005
2010
2015
2020
Hathaway NASA/ARC

We don't even know how many polar convective cells there are!

solar orbiter

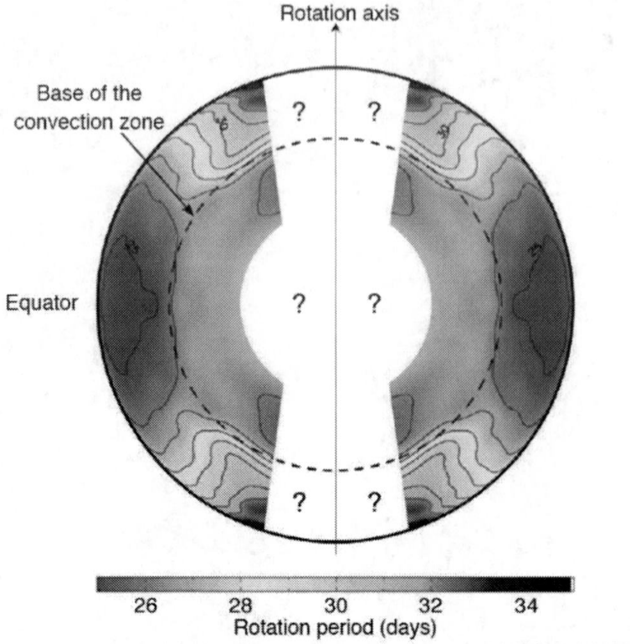

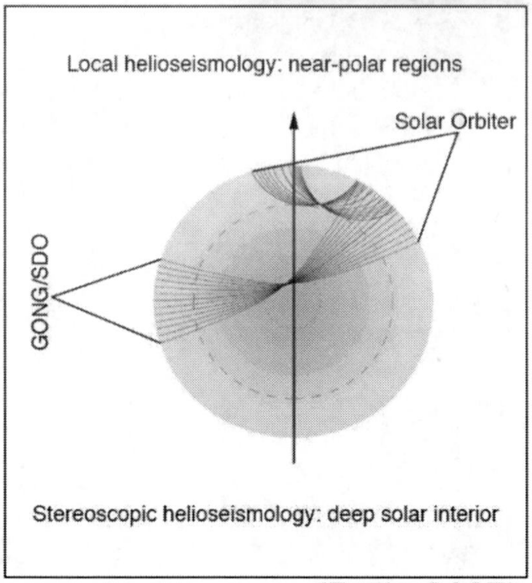

(Roth, 2007)

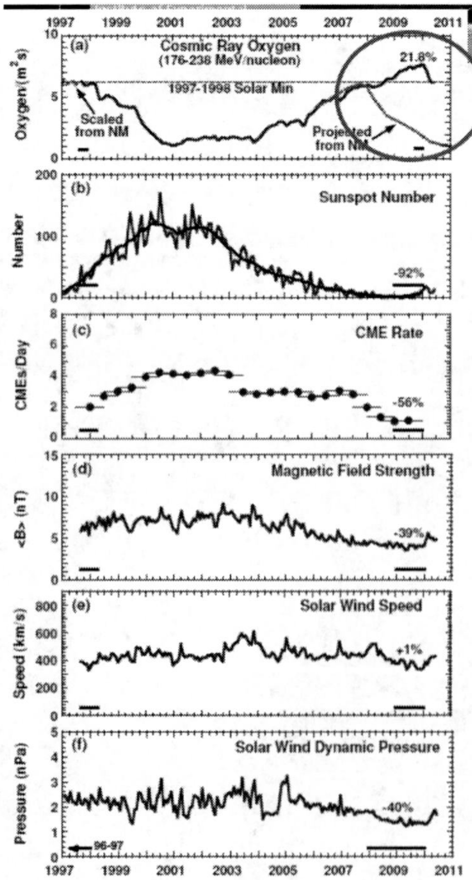

The Consequence:

RECORD-SETTING COSMIC-RAY INTENSITIES IN 2009 AND 2010

R. A. Mewaldt[1], A. J. Davis[1], K. A. Lave[2], R. A. Leske[1], E. C. Stone[1], M. E. Wiedenbeck[3], W. R. Binns[2], E. R. Christian[4], A. C. Cummings[1], G. A. de Nolfo[4], M. H. Israel[2], A. W. Labrador[1], and T. T. von Rosenvinge[4]

- Lowest solar wind dynamic pressure
- Smallest total heliospheric flux
- Smallest CME rate

… were all not predicted.

solar orbiter

We need to better understand the Sun!

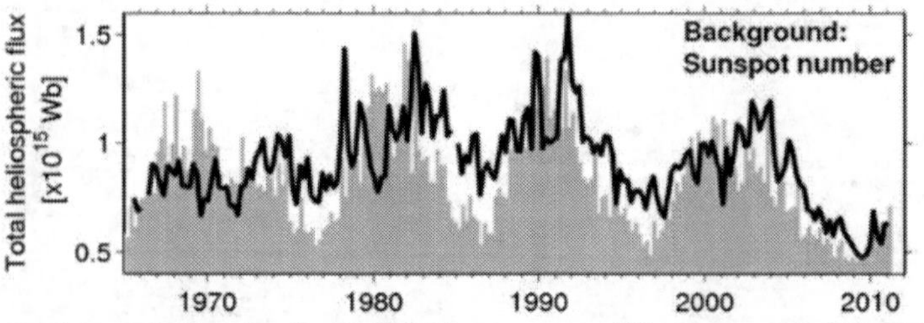

(Mewaldt et al., 2010) (Owens, 2008)

CAU Institute of Experimental and Applied Physics
Extraterrestrial Physics

Risks Assessments for all DRMs (Jan 2015)

Human Spaceflight Risks Human Spaceflight Risks 01/12/15	In Mission Risk - Operations						Post Mission Risk – Long Term Health					
	Low Earth Orbit 6 Months	Low Earth Orbit 12 Months	Deep Space Sortie 30 Days	Lunar Visit/Habitation 1 year	Deep Space Journey/Habitation 1 Year	Planetary 3 years	Low Earth Orbit 6 Months	Low Earth Orbit 12 Months	Deep Space Sortie 30 Days	Lunar Visit/Habitation 1 year	Deep Space Journey/Habitation 1 Year	Planetary 3 years
VIIP	A	A	A	A	RM	RM	A	A	A	A	RM	RM
Renal Stone Formation	A	A	A	A	RM	RM	RM	RM	RM	RM	RM	RM
Inadequate food and nutrition	A	A	A	A	RM	RM	A	A	A	A	A	RM
Risk of Space Radiation Exposure	A	A	A	A	A	A	A	A	A	RM	RM	RM
Medications Long Term Storage	A	A	A	A	A	RM	A	A	A	A	A	RM
Acute and Chronic Carbon Dioxide	A	A	A	A	RM	RM	A	A	A	A	A	A
Inflight Medical Conditions	A	A	A	RM	RM	RM	A	A	A	A	RM	RM
Cognitive or Behavioral Conditions	A	A	A	A	RM	RM	A	A	[illegible]	RM	RM	RM
Risk of Bone Fracture	A	A	A	A	A	A	A	[illegible]	[illegible]	A	A	RM
Team Performance Decrements#	A	A	A	A	RM	RM	A	[illegible]	[illegible]	A	A	A
Reduced Muscle Mass, Strength	A	A	A	A	A	RM	[illegible]	[illegible]	[illegible]	[illegible]	A	A
Reduced Aerobic Capacity	A	A	A	A	A	RM	A	[illegible]	[illegible]	A	A	A
Sensorimotor Alterations	A	A	A	A	A	RM	A	[illegible]	[illegible]	A	A	RM
Human-System Interaction Design#	A	A	A	RM	RM	RM	A	A	A	A	A	A
Injury from Dynamic Loads	A	A	RM	RM	RM	RM	A	A	RM	RM	RM	RM
Sleep Loss	A	A	A	A	RM	RM	A	A	A	A	RM	RM
Altered Immune Response	A	A	A	A	RM	RM	A	A	A	A	A	RM
Celestial Dust Exposure	N/A	N/A	A	TBD	TBD	TBD	N/A	N/A	A	TBD	TBD	TBD
Host-Microorganism Interactions	A	A	A	A	RM	RM	A	A	A	A	A	RM
Injury due to EVA Operations	A	A	A	RM	A	RM	A	A	A	RM	RM	RM
Decompression Sickness	A	A	A	A	RM	A	A	A	A	RM	A	RM
Toxic Exposure	A	A	A	A	A	A	A	A	A	A	A	A
Hypobaric Hypoxia	A	A	A	A	A	A	A	A	A	A	A	A
Space Adaptation Back Pain	A	A	A	A	A	A	N/A	N/A	N/A	N/A	N/A	N/A
Urinary Retention	A	A	A	A	A	A	A	A	A	A	A	A
Hearing Loss Related to Spaceflight	A	A	A	A	A	A	A	A	A	A	A	A
Orthostatic Intolerance	A	A	A	A	A	A	A	A	A	A	A	A
Injury from Sunlight Exposure - retired	A	A	A	A	A	A	A	A	A	A	A	A
Risk of electrical shock - Retired	A	A	A	A	A	A	A	A	A	A	A	A

Radiation!

A – Accepted RM- Requires Green – Mitigated Yellow – partially controlled Red – uncontrolled

29

CAU Institute of Experimental and Applied Physics
Extraterrestrial Physics

GCR models show substantial differences in their predictions

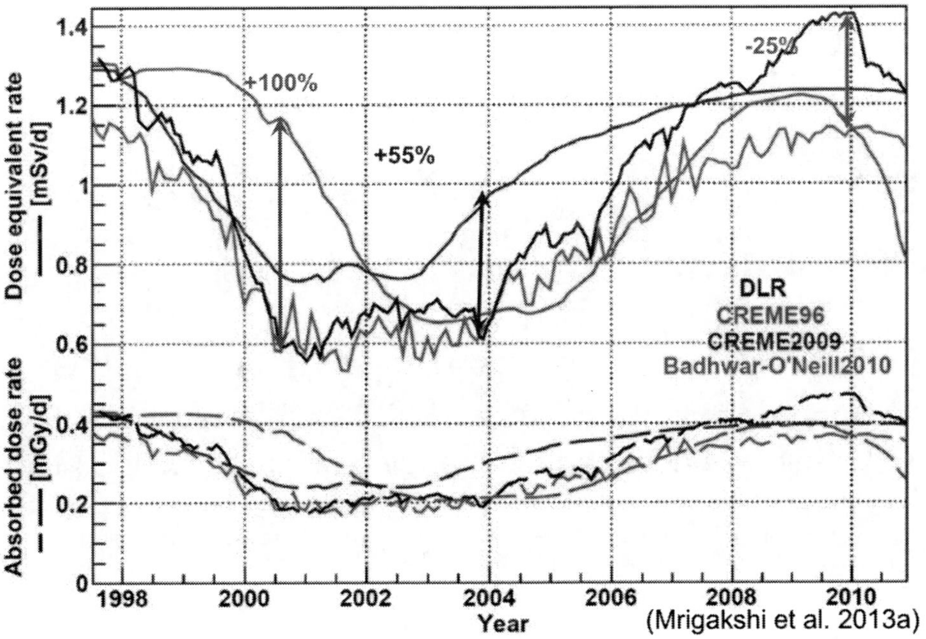

(Mrigakshi et al. 2013a)

Dose Rate Measurements at the Moon?

-Only results from two missions in Lunar orbit avaible (CRATER and RADOM) with dose values of a

T.P. Dachev et al. *An overview of RADOM results for earth and moon radiation environment on Chandrayaan-1 satellite, Advances in Spa*
Spence H. E. (CRaTER Science Team)*, An Overview of Results from the Lunar Reconnaissance Orbiter (LRO) Cosmic Ray Telescope f*
J.E. Mazur et al. *New Measurements ot total ionizing dose in the lunar environment, Space Weather, 9, S07002, doi:10.1029/2010SW00*

-Calculations show an Effective Dose of 600 µSv/day taking into account Solar Minimum condition

NASA-TP-2003-212158, "Analysis of a Radiation Model of the Shuttle Space Suit"
D. Matthiae*, T. Berger, G. Reitz: DLR – Updated GCR Model / Moon calculations, 2012*

-„Safe haven“ to prevent exposures due to SPE needs approx. 80 cm of lunar regolith

D. Matthiae*, T. Berger, G. Reitz: DLR – Updated GCR Model / Moon calculations, 2012*

NO MEASUREMENTS on the LUNAR surface available --> needed as input
a) benchmarking of radiation transport codes
b) prerequisite for future human presence on the Moon

Dose Rate Measurements at the Moon?

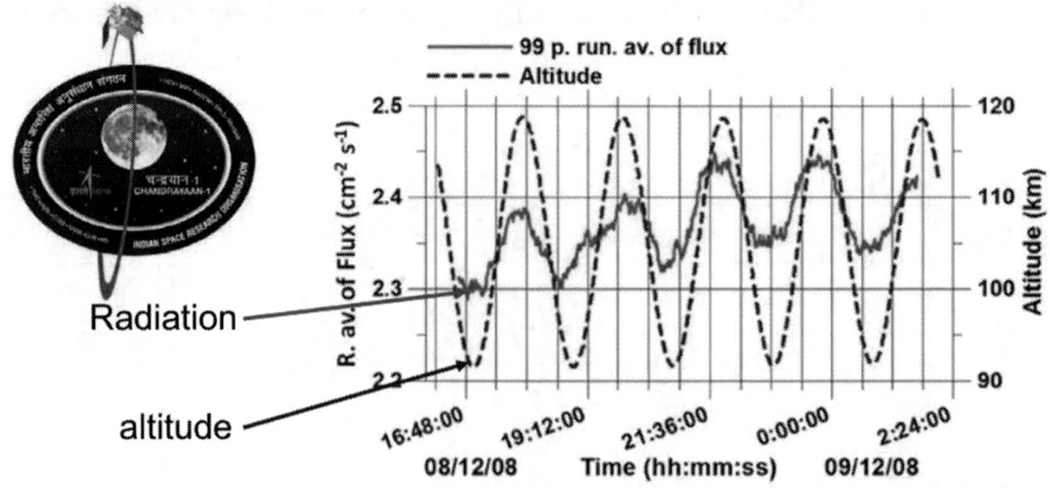

Radiation is modulated by changing shielding by the Moon

T.P. Dachev et al. An overview of RADOM results for earth and moon radiation environment on Chandrayaan-1 satellite, Advances in Space Research 48 (2011) 779-791

Dose Rate Measurements at the Moon?

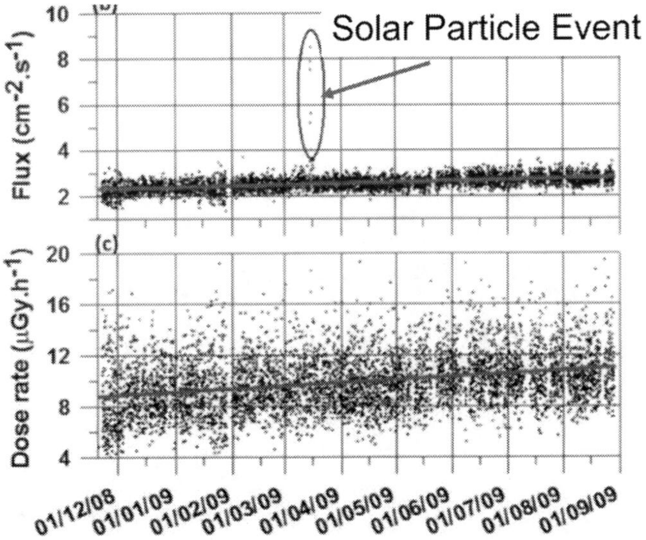

Dose rates :

~ 100 km: 227 µGy/d

~ 200 km : 257 µGy/d

T.P. Dachev et al. An overview of RADOM results for earth and moon radiation environment on Chandrayaan-1 satellite, Advances in Space Research 48 (2011) 779-791

C A U Institute of Experimental and Applied Physics
Extraterrestrial Physics

Particle Radiation Measurements at the Moon

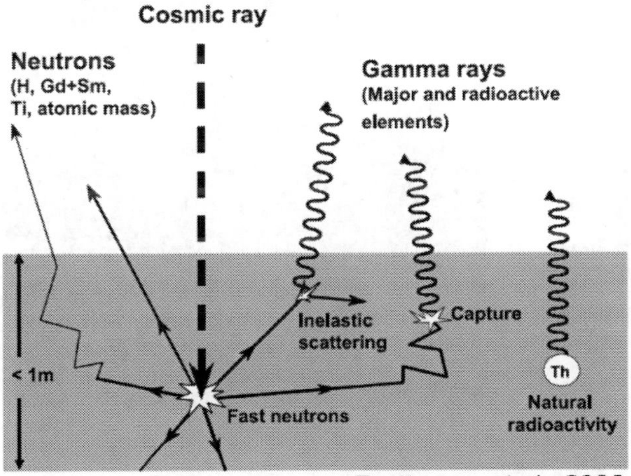

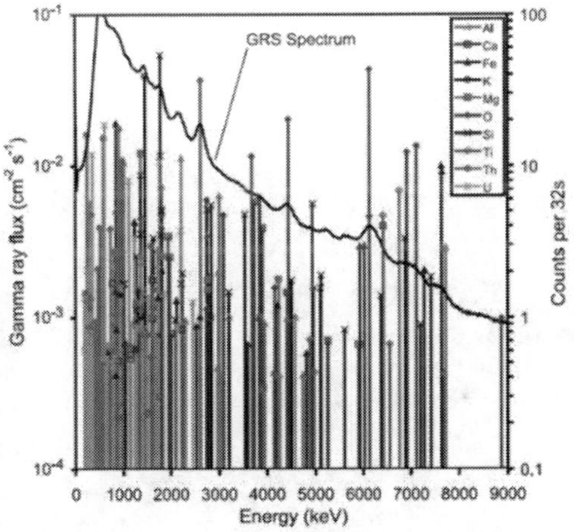

Prettyman et al., 2006

Particle Radiation Measurements at the Moon

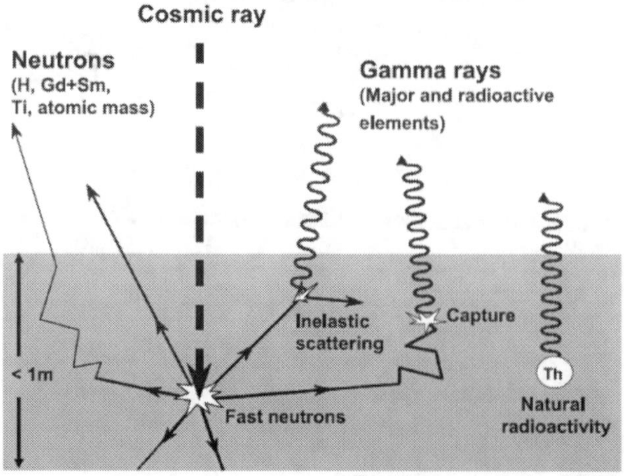

Prettyman et al., 2006

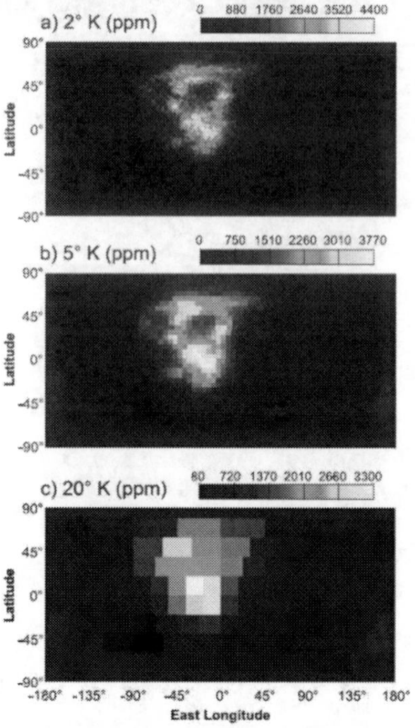

Particle Radiation Measurements at the Moon

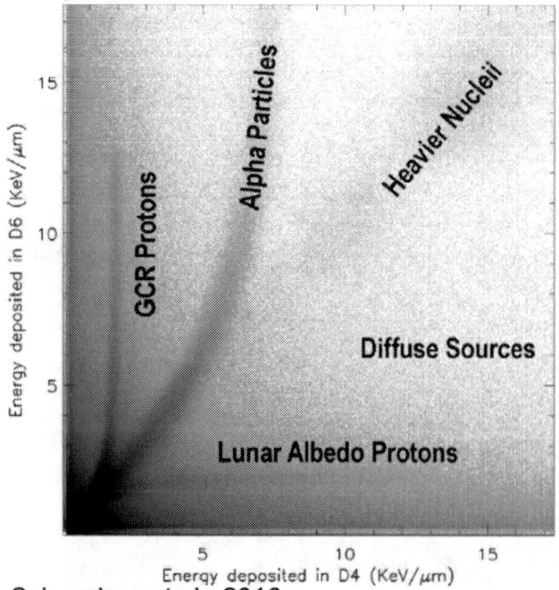

Schwadron et al., 2016

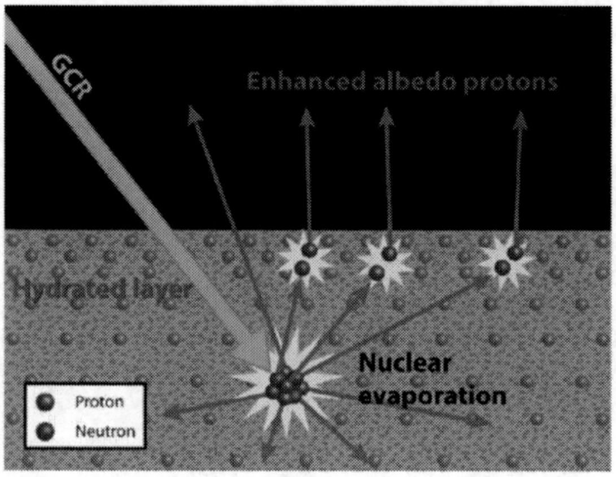

Particle Radiation Measurements at the Moon

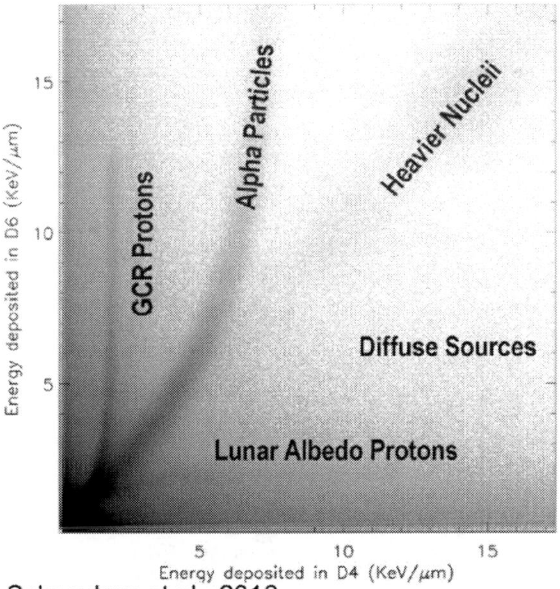

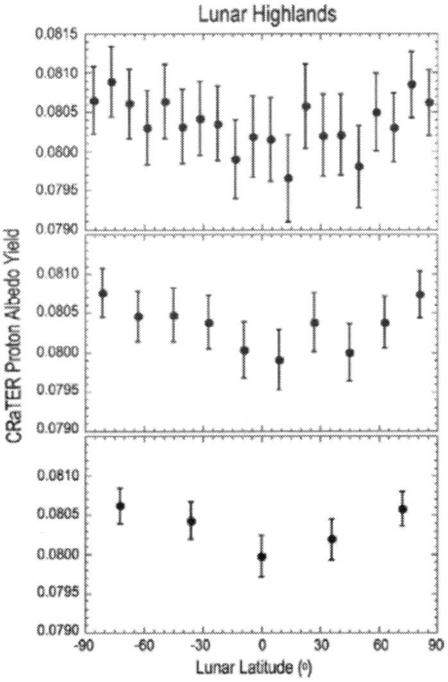

Schwadron et al., 2016

Dose Rate Measurements on the Moon?

Large solar particle event took place between Apollo 16 and Apollo 17

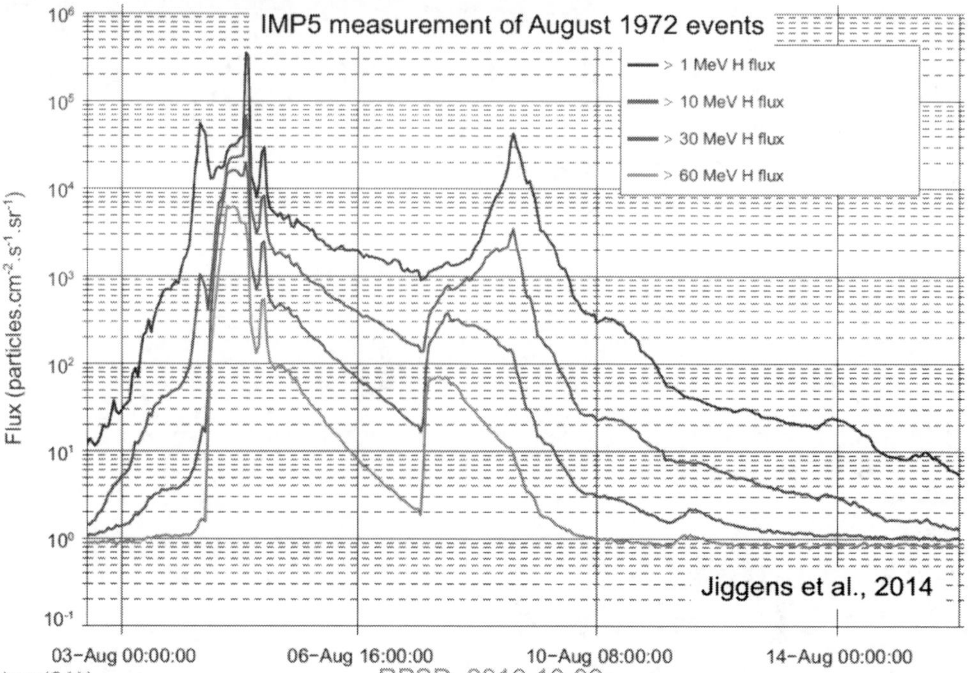

Efffects of solar particle events

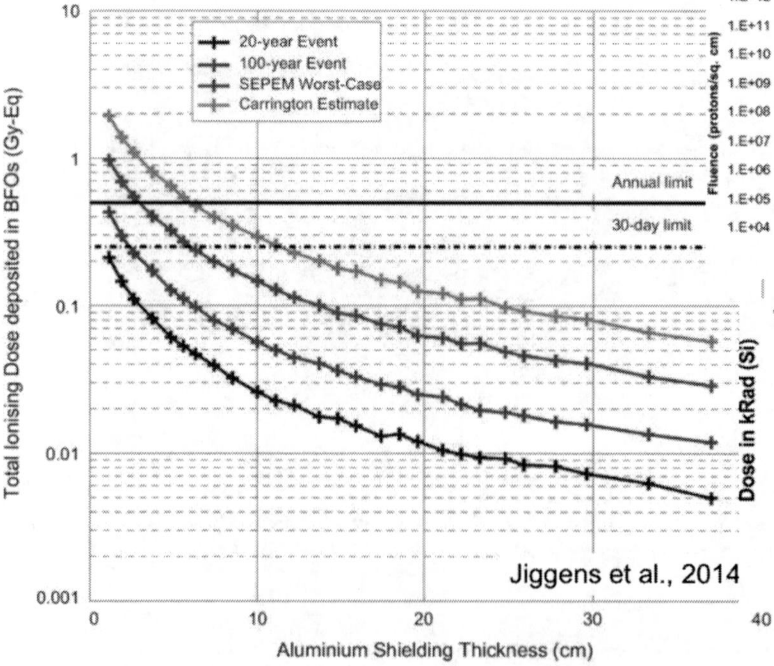

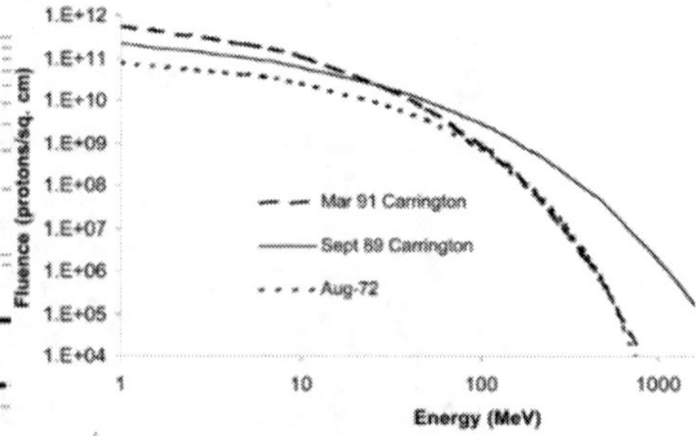

Townsend et al., 2003

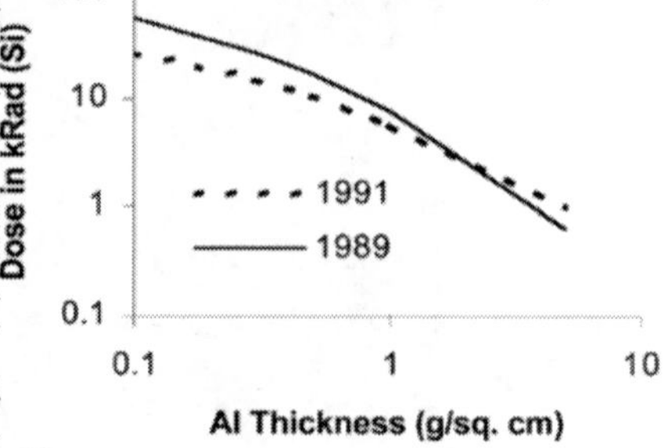

Dose Rate Measurements on the Moon?

C A U Institute of Experimental and Applied Physics
Extraterrestrial Physics

Dose Rate Measurements on the Moon?

None!

Dose Rate Measurements on the Moon?

Lunar Neutron and Dosimetry (LND) Experiment on Chang'E-4 will provide such measurements (launch 2018)

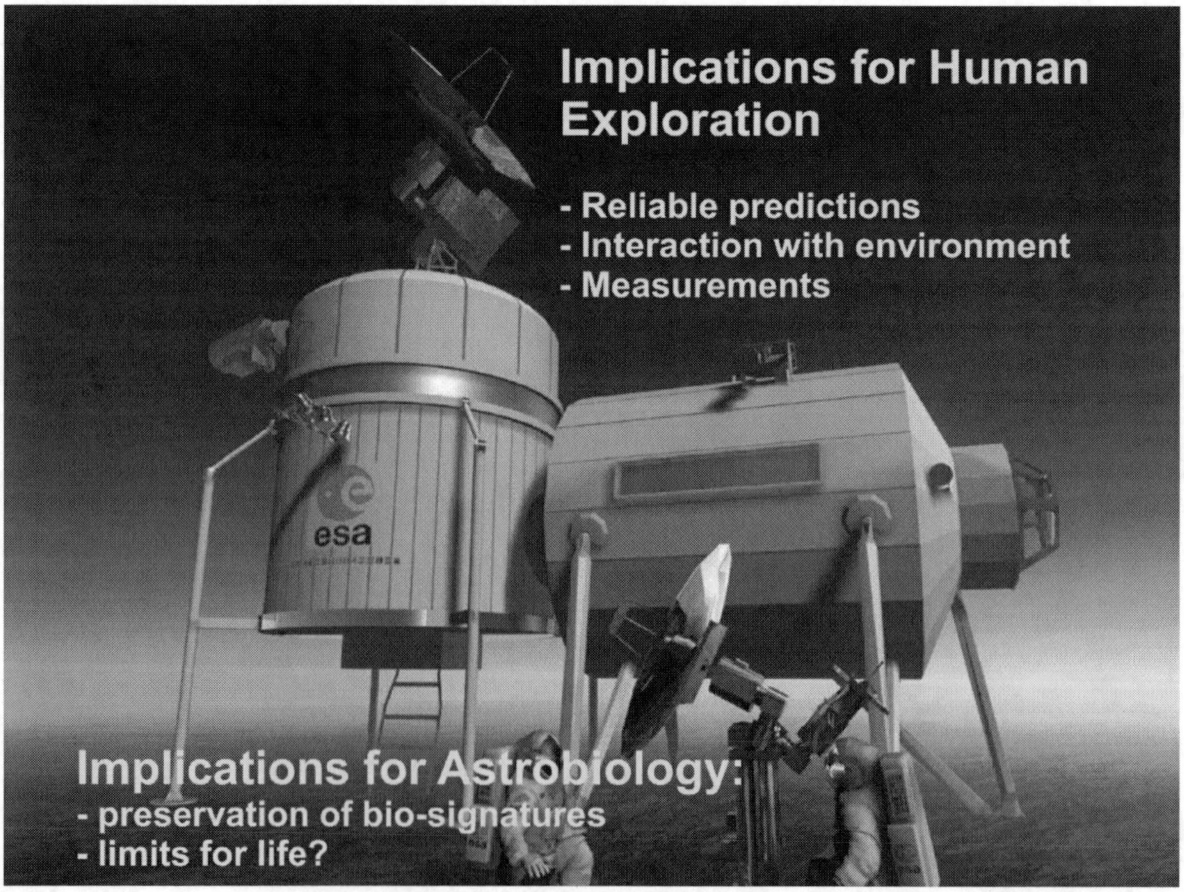

Radiation Measurements on Mars

Requirements
:

- Charged particles (1 < Z < 27) up to 100 MeV/nuc
- Neutral particles (n, γ) up to 100 MeV
- LET
- Composition
- Time series
- Autonomous operations

?

The Radiation Assessment Detector (RAD)

Requirements:

- Charged particles (1 < Z < 27) up to 100 MeV/nuc
- Neutral particles (n, γ) up to 100 MeV
- LET
- Composition
- Time series
- Autonomous operations

Solution:

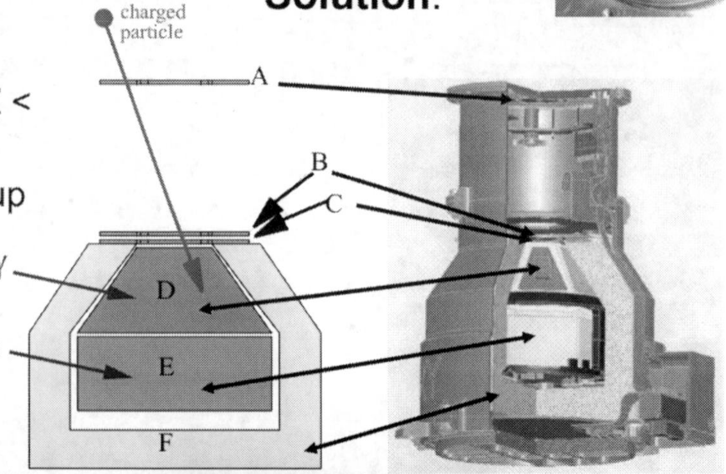

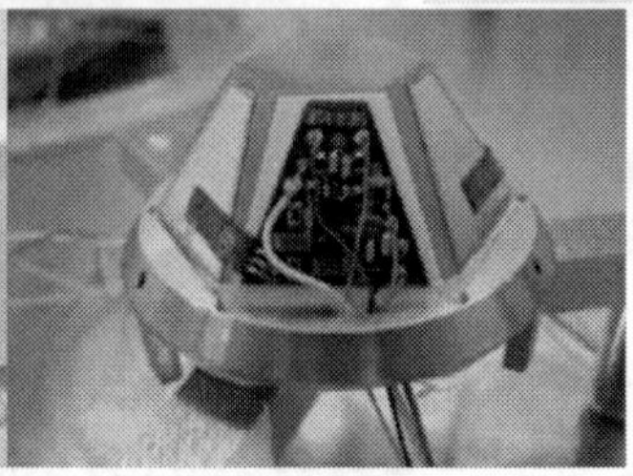

RAD:

- Mass: 1,56 kg
- Power: 4,2 W
- Data rate: ~130 bps
- RSH: FM & FS

Summary of solar particle events seen by MSL

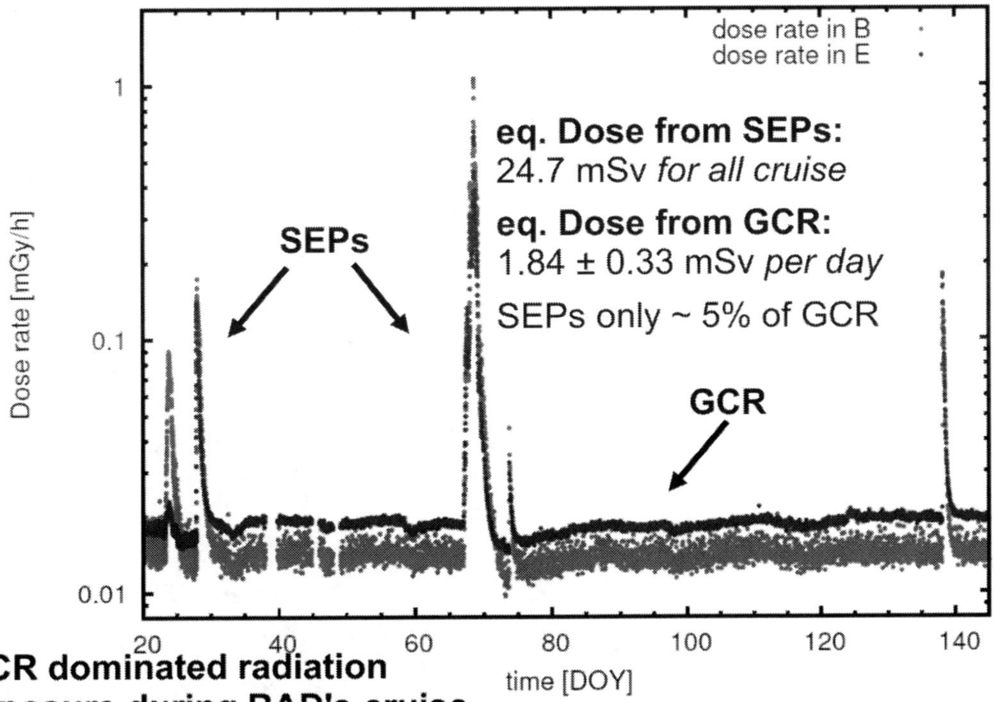

GCR dominated radiation exposure during RAD's cruise

CAU Institute of Experimental and Applied Physics
Extraterrestrial Physics

During cruise, RAD was heavily shielded by inhomogeneous and anisotropic structures.

1 g/cm^2 < shielding < 80 g/cm^2

This complicates interpretation of data.

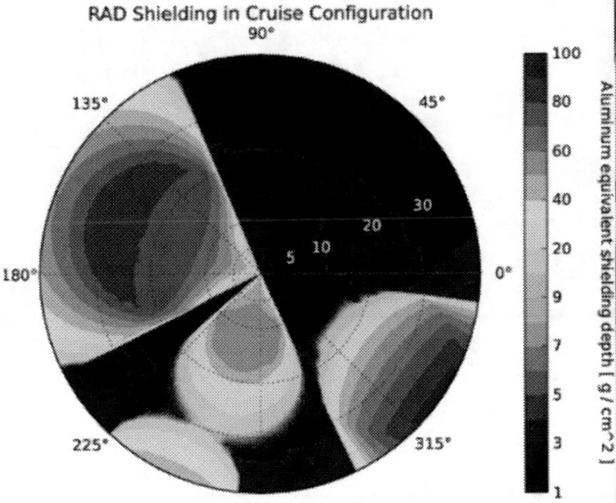

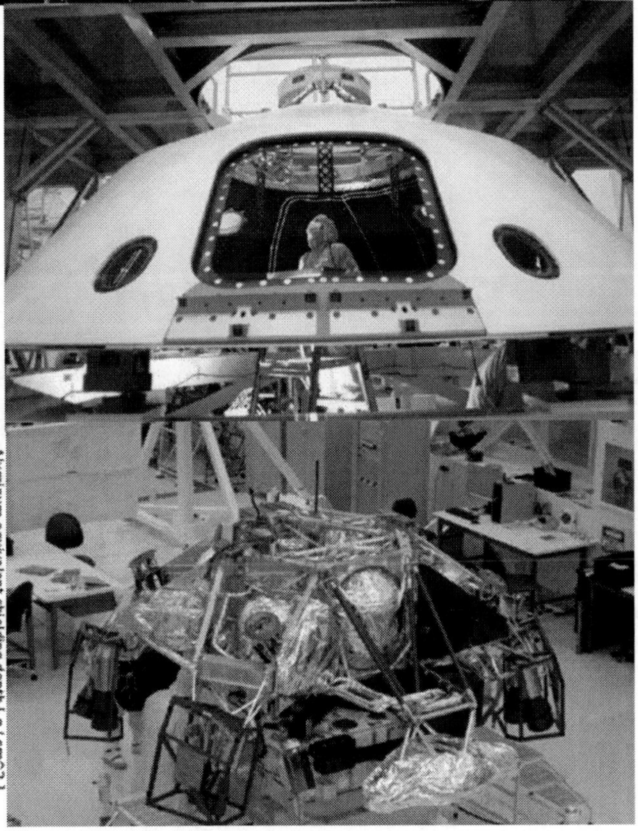

During cruise, RAD was heavily shielded by inhomogeneous and anisotropic structures.

1 g/cm^2 < shielding < 80 g/cm^2

This complicates interpretation of data.

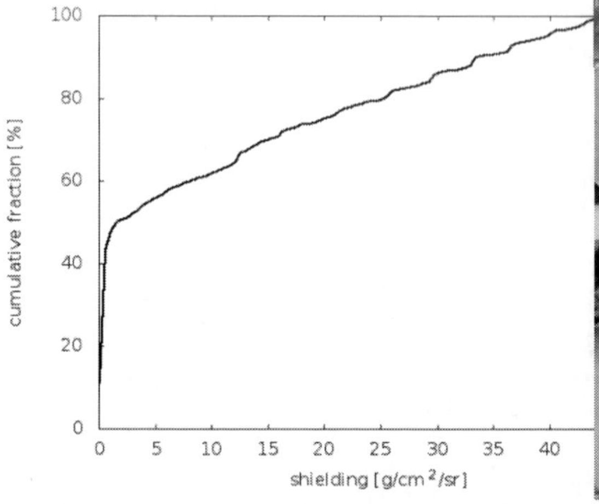

Inversion towards MSL long-term average spectr

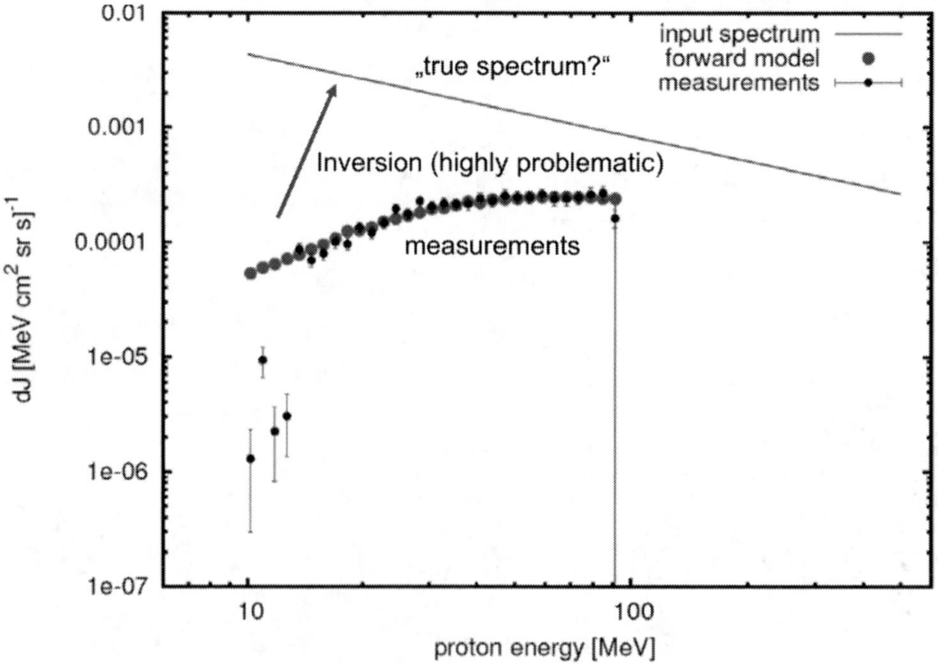

Summary of cruise radiation environment of MSL (This can be measured accurately!)

		Estimated variability	Two 180-day legs return trip
RAD cruise measurement SEPs	**24.7 mSv (5% of all)**	**Orders of magnitude**	**?**
RAD cruise measurement GCR	**1.84 mSv/d**	**0.33 mSv/d ± 20%**	**662 ± 108 mSv ± 20%**
6-month stay of astronaut on ISS	75-90 mSv/(a/2)	20,00%	150-180 mSv
Radiation worker limit (ICRP)	20 mSv/a	n/a	
Average exposure of normal population	4 mSv/a	Wide range, radon!	
Allowable additional exposure norm. pop.	1 mSv/a	n/a	

How variable are solar particle events?

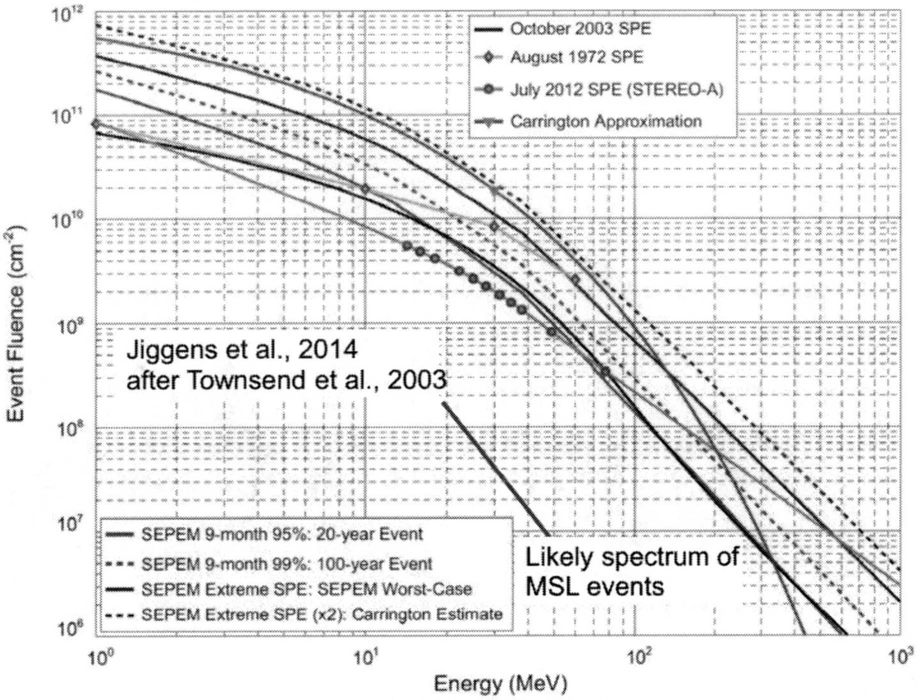

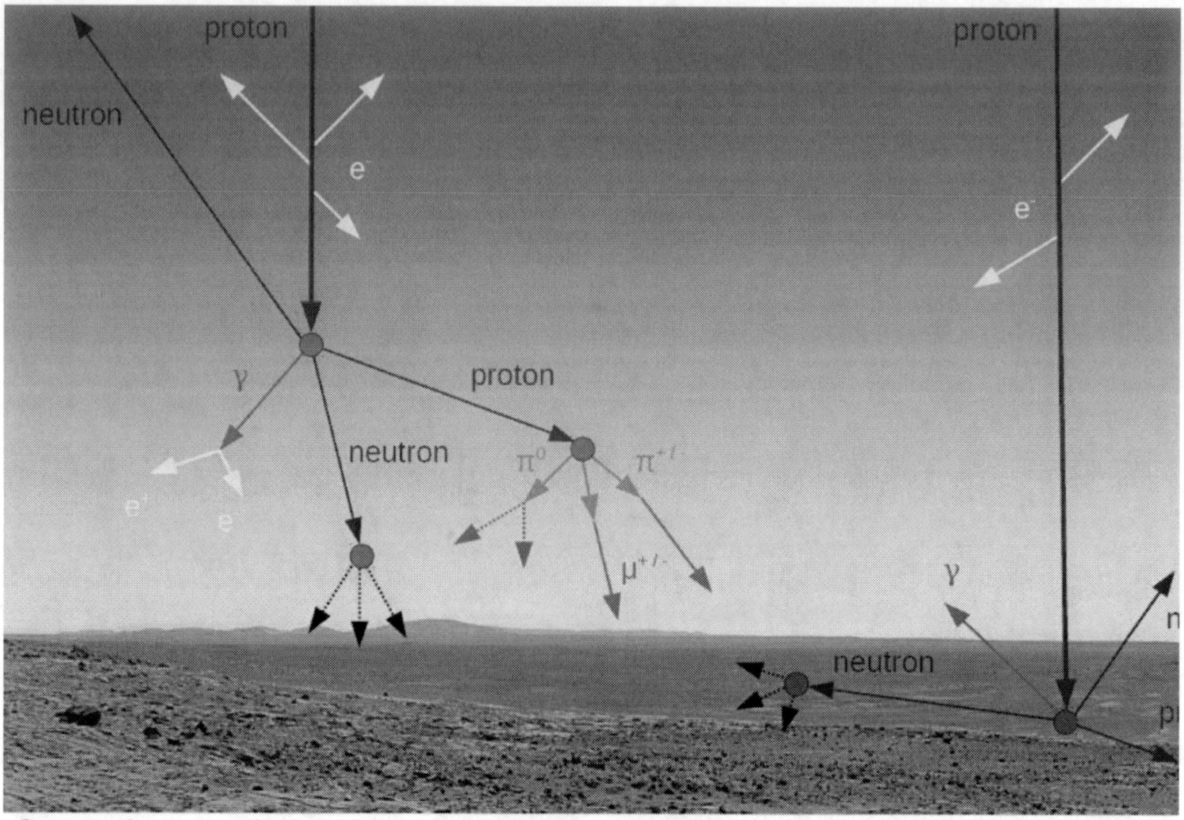

Secondary radiation (neutrals!) plays an important role! (Ehresmann, 2011)

CAU Institute of Experimental and Applied Physics
Extraterrestrial Physics

RAD Surface Measurements

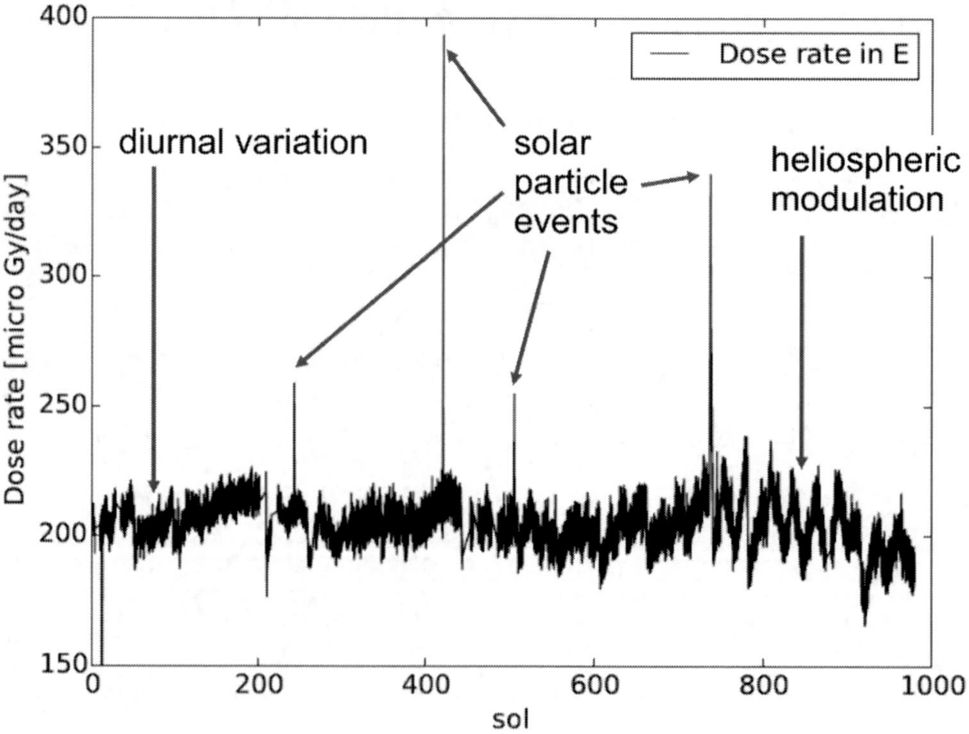

RAD Surface Measurements

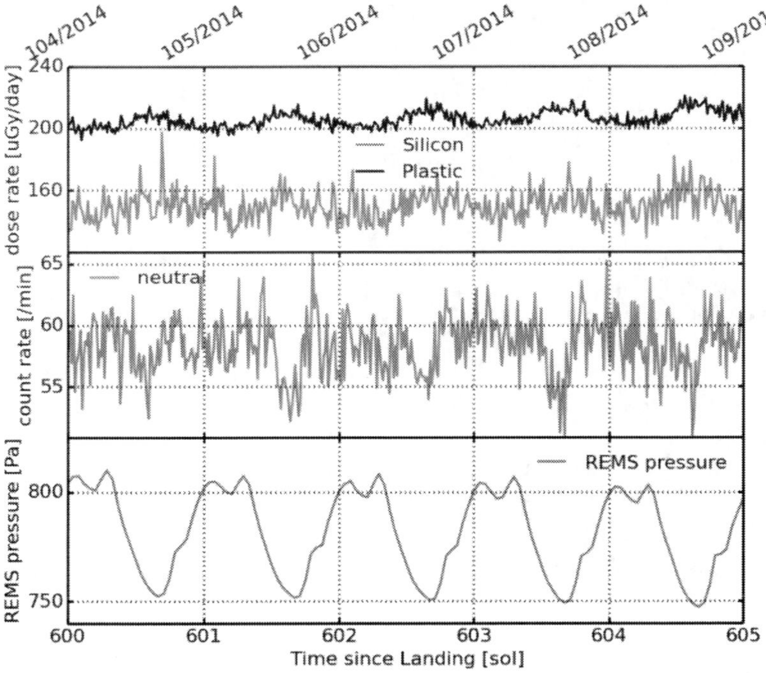

RAD Surface Measurements

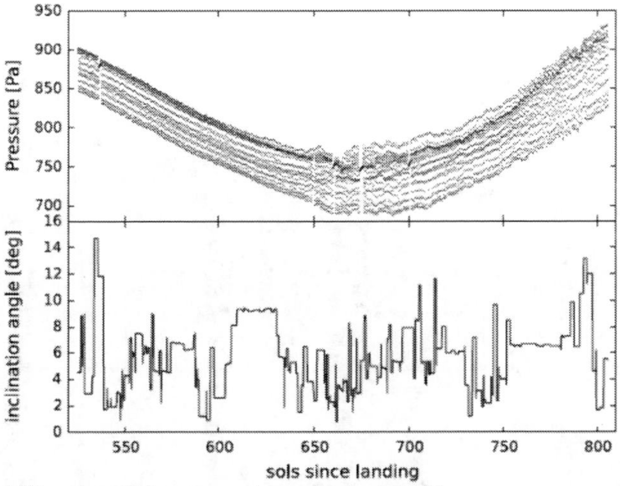

(Wimmer-Schweingruber et al., 2015)

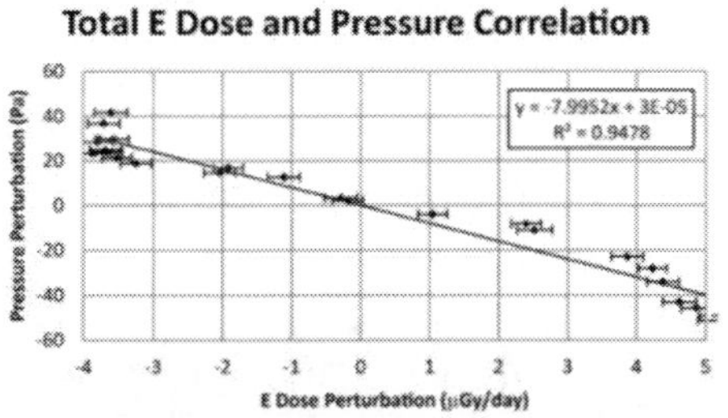

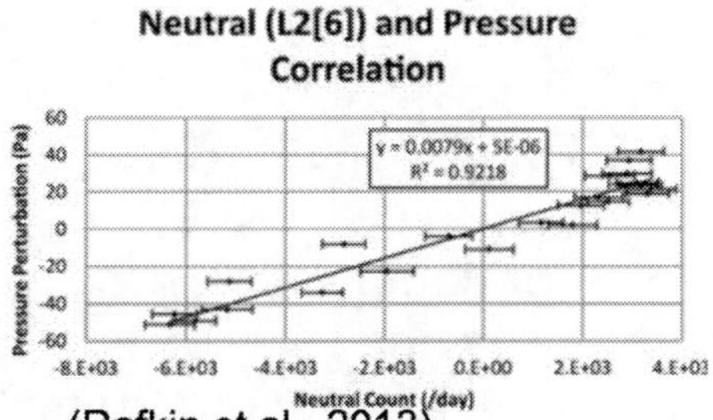

(Rafkin et al., 2013)

- RAD measurements consistent with isotropic Martian surface radiation
- Very thin Martian atmosphere provides only a small amount of shielding
- Shielding effect partially compensated by generation of secondary particles

RAD Surface Measurements

D/E

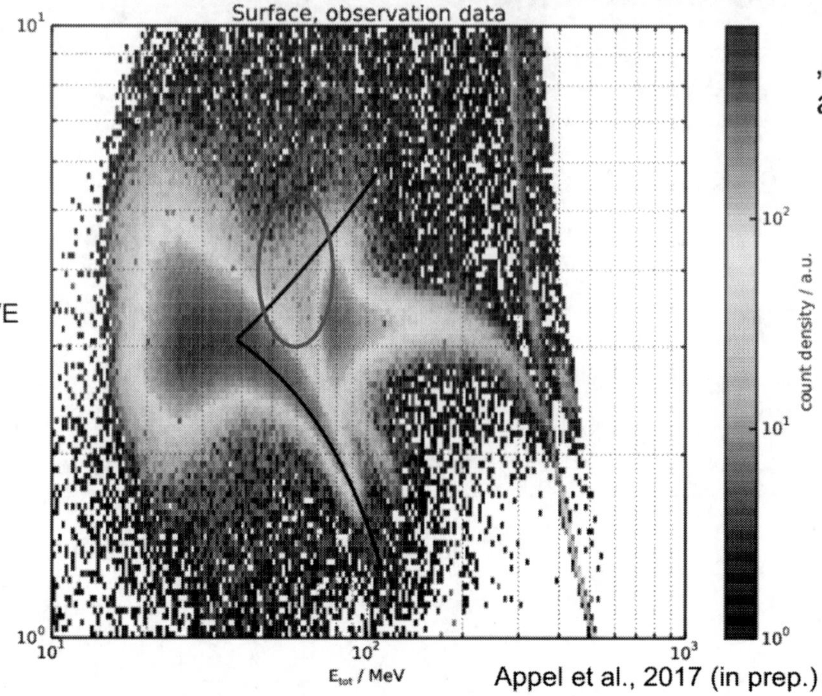

„Albedo"-protons also on Mars

Appel et al., 2017 (in prep.)

Summary of RAD measurements

Radiation exposure on a mission to Mars:

Cruise: 662 +/- 108 mSv
Mars: 320 +/- 50 mSv

Total ~ 1000 mSv

<u>For comparison:</u>
6 months ISS: 75-90 mSv
radiation worker: 20 mSv/y
CT-scan: 8 mSv

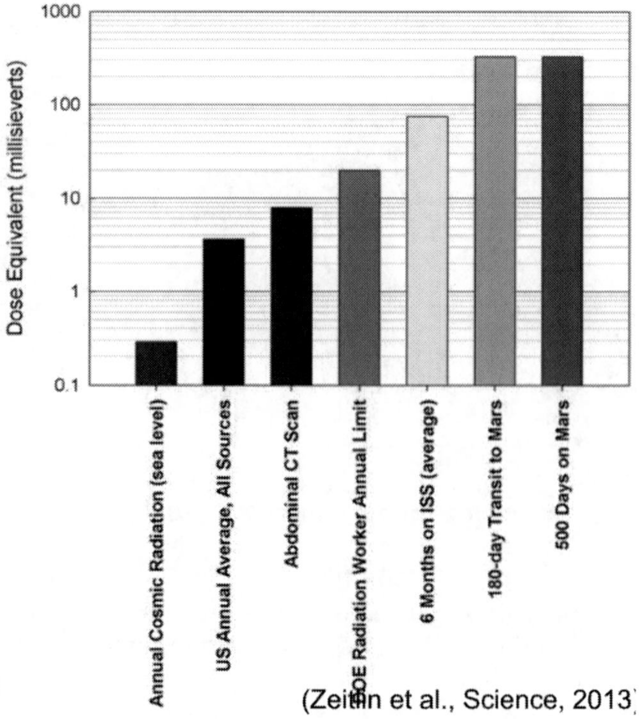

(Zeitlin et al., Science, 2013)

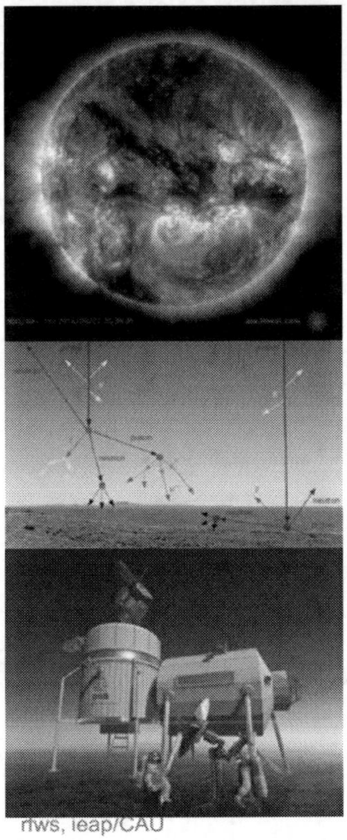

Implications for human exploration

- Space particle radiation is complex!
- “Shielding“ primarily “magnetic“
- Space weather predictions are still very difficult
- Large variability (solar, heliospheric, seasonal, diurnal)
- Secondary radiation important (n/γ)
- Where should we live on Mars?

Implications for non-terrestrial life?
Exo-, astrobiology?

Preservation of bio-signatures?

Summary, Conclusions, and Thanks

- Radiation in heliosphere very variable
- Models are even more variable...
- First radiation measurements on another planet
- Radiation environment on Mars quite different than on Earth and important for future human exploration
- Modulation is an important „shielding“ factor

CAU Christian-Albrechts-Universität zu Kiel

Supported by: Federal Ministry of Economics and Technology on the basis of a decision by the German Bundestag

DLR esa NASA

Destruction of Organic Compounds by Radiation

Ionizing radiation breaks chemical bonds and produces radicals and oxidants.

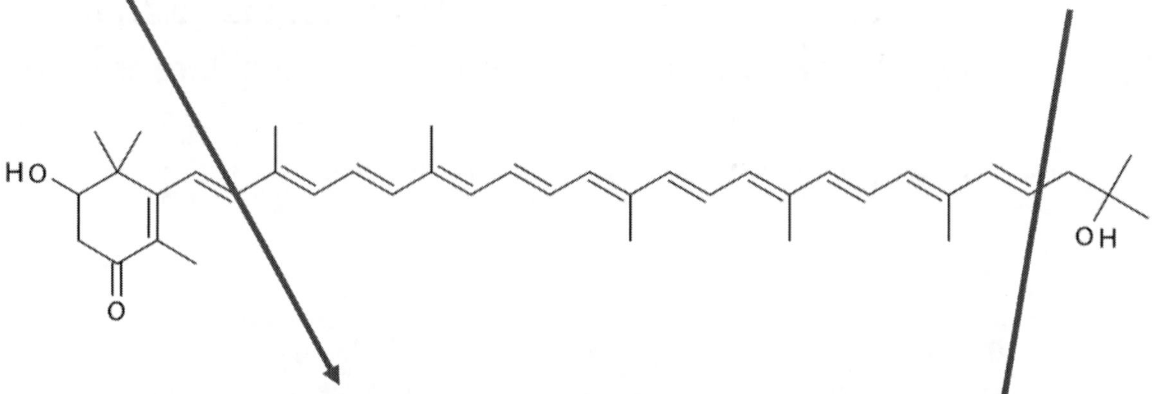

Result: Destruction of large organic molecules (if there is no repair mechanism)

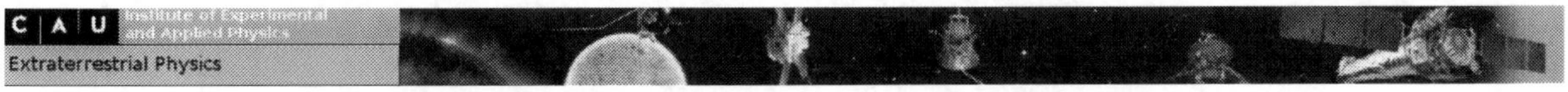

How long could organic molecules survive ionizing radiation environment?

Previous models: 50 – 150 mGy/y
RAD measurements: 76 mGy/y

Organic molecules are efficiently destroyed at a depth of 4-5 cm. In 650 million years only 1/1000 survives.

How many after 3.8 Gy?

==> Half of the organics should still be around if the soil were only exposed for 65 million years.

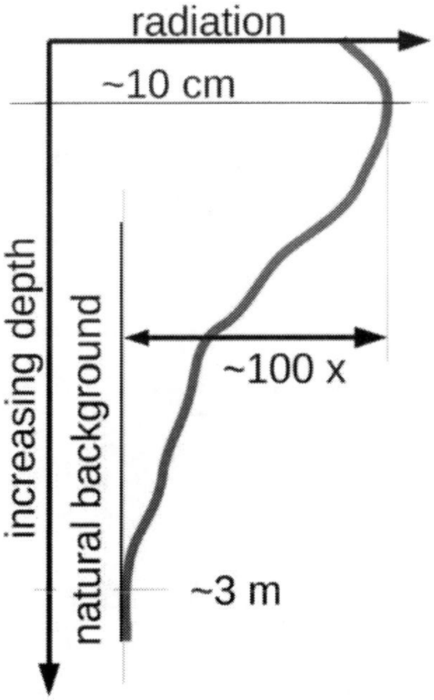

Age determination with cosmogenic isotopes

Sheepbed was exposed for only 80 ± 30 million years!

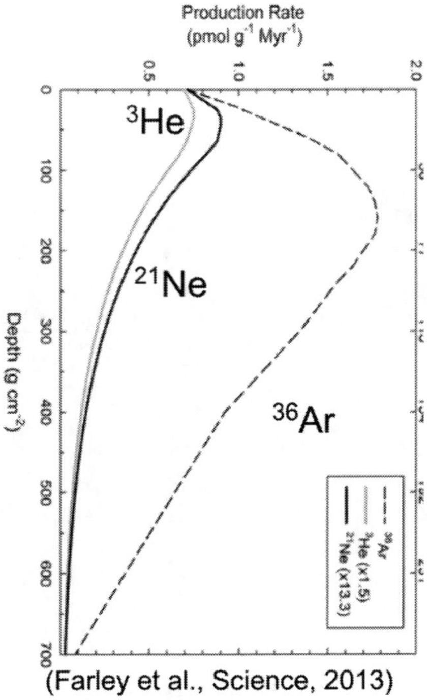

(Farley et al., Science, 2013)

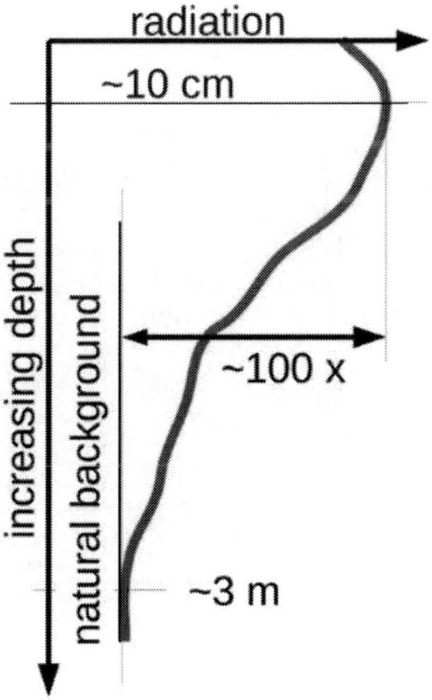

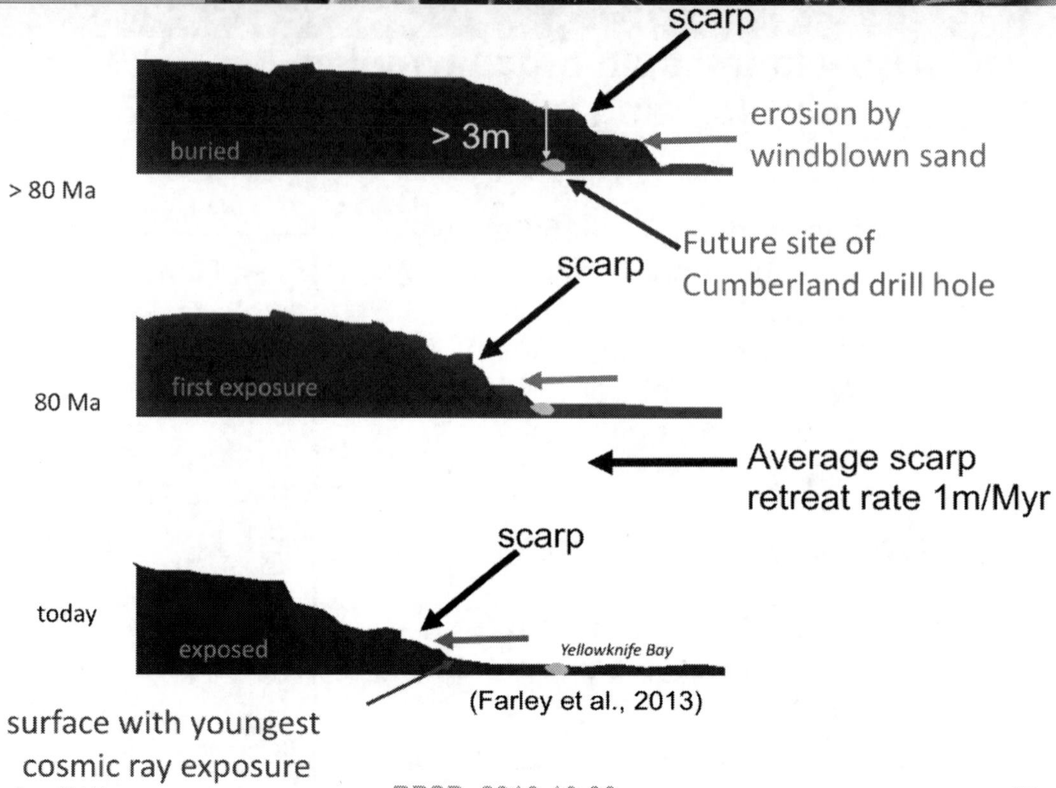

Some of the SAM Data

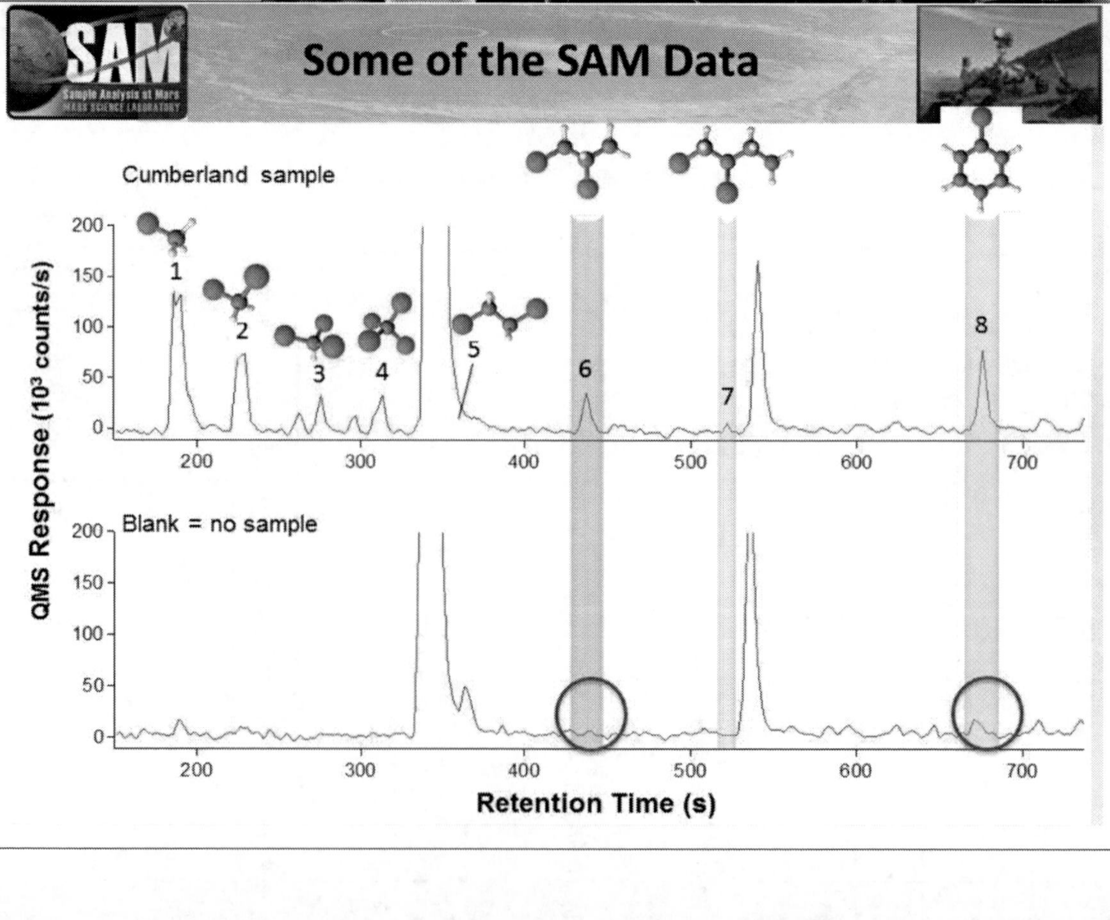

Summary, Conclusions, and Thanks

- Radiation in heliosphere very variable
- Models are even more variable...
- First radiation measurements on another planet
- Radiation environment on Mars quite different than on Earth and important for future human exploration
- Organics can be and have been detected on Mars
- Mars was habitable – but is it still so?

C A U Christian-Albrechts-Universität zu Kiel

Supported by: Federal Ministry of Economics and Technology on the basis of a decision by the German Bundestag

DLR esa NASA

Additional Slides

Yes, there are planets everywhere!

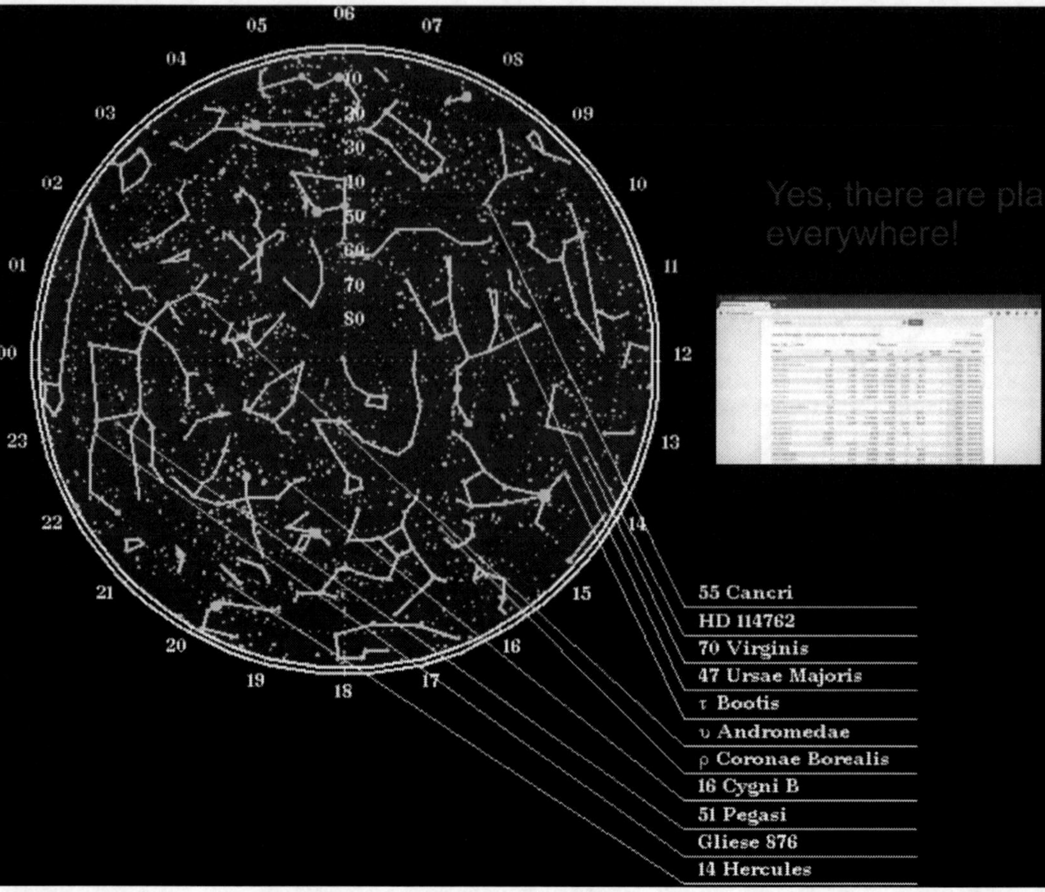

C A U Institute of Experimental and Applied Physics

Extraterrestrial Physics

exoplanet.eu/catalog

Search

Catalog

Download VOTable | CSV | DAT

All planets | Filter

Showing 2107 planets / 1349 planetary systems / 511 multiple planet systems

All fields

Show 100 entries   Planet Search   Show / hide columns

Planet	Mass (M_{Jup})	Radius (R_{Jup})	Period (day)	a (AU)	e	i (deg)	Ang. dist. (arcsec)	Discovery
WASP-121 b	1.184	1.865	1.2749255	0.02544	0	87.6	—	2015
K2-19 b	0.187	0.691	7.9194	0.077	0.119	88.83	—	2015
K2-19 c	0.031	0.434	11.90715	0.1032	0.095	89.91	—	2015
EPIC 201505350 d	—	0.102	2.50856	—	—	—	—	2015
HD 9174 b	1.11	—	1179	2.2	0.12	—	—	2016
HD 68402 b	3.07	—	1103	2.18	0.03	—	—	2016
HD 72892 b	5.45	—	39.475	0.228	0.423	—	—	2016
HD 128356 b	0.89	—	298.2	0.87	0.57	—	—	2016
HD 224538 b	5.97	—	1189.1	2.28	0.464	—	—	2016
HD 165155 b	2.89	—	434.5	1.13	0.2	—	—	2016
HD 147873 b	5.14	—	116.596	0.522	0.207	—	—	2016
HD 147873 c	2.3	—	491.54	1.36	0.23	—	—	2016
Pr 0211 b	1.88	—	2.14609	0.03184	0.017	—	—	2012
Pr 0201 b	0.54	—	4.4264	—	—	—	—	2012
Pr 0211 c	7.95	—	5300	5.8	0.7	—	—	2016
TYC-3667-1280-1 b	5.4	—	26.468	0.21	0.036	—	—	2016

What is life?

Life multiplies | Life is self-organizing | Life adapts to its environment

What is life?

~~Life multiplies~~ | ~~Life is self-organizing~~ | ~~Life adapts to its environment~~

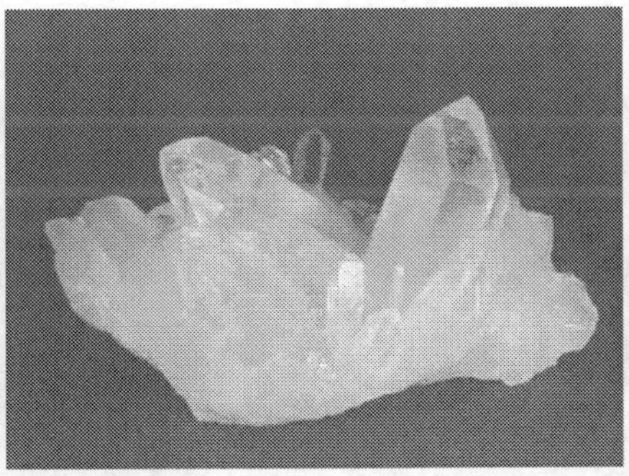

Life must be described differently:
What does life need?

What is life?

~~Life multiplies~~ ~~Life is self-organizing~~ ~~Life adapts to its environment~~

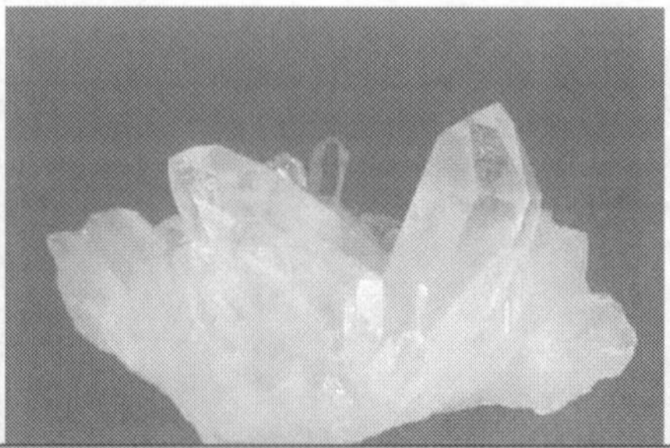

Life – as we know it – requires:

Liquid Water	stable environment	trace elements	energy

Habitability in the solar system

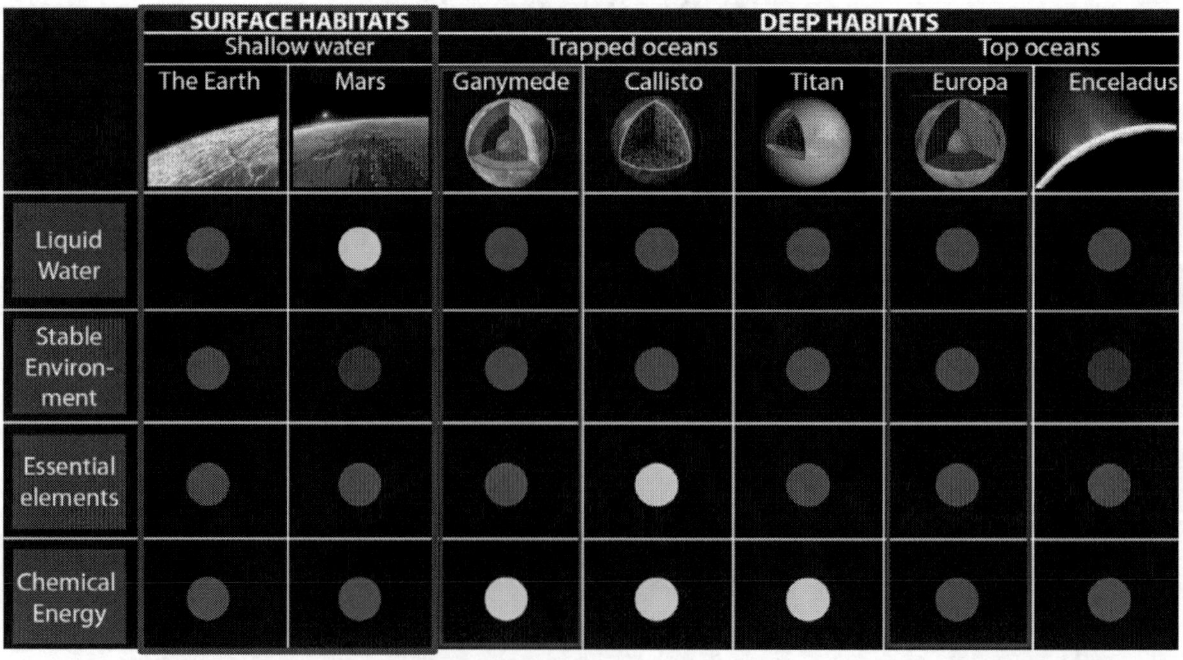

Earth **Mars**

CAU Institute of Experimental and Applied Physics
Extraterrestrial Physics

Mars – our neighbor planet

Mars – our neighbor planet

Mars – our neighbor planet

Mars – our neighbor planet

Mars – our neighbor planet

Mars atmosphere:

Surface pressure:	approx. 0,6% Earth value
Temperature at surface:	approx. -130 to + 30°C
Composition:	95% carbon dioxide; 2,7% nitrogen
Earth (for comparison):	78% nitrogen; 21% oxygen

Mars – our neighbor planet

CAU Institute of Experimental and Applied Physics
Extraterrestrial Physics

Despite Mars' very thin atmosphere, dust storms can cover Mars nearly completely.

'Everything is dusty'

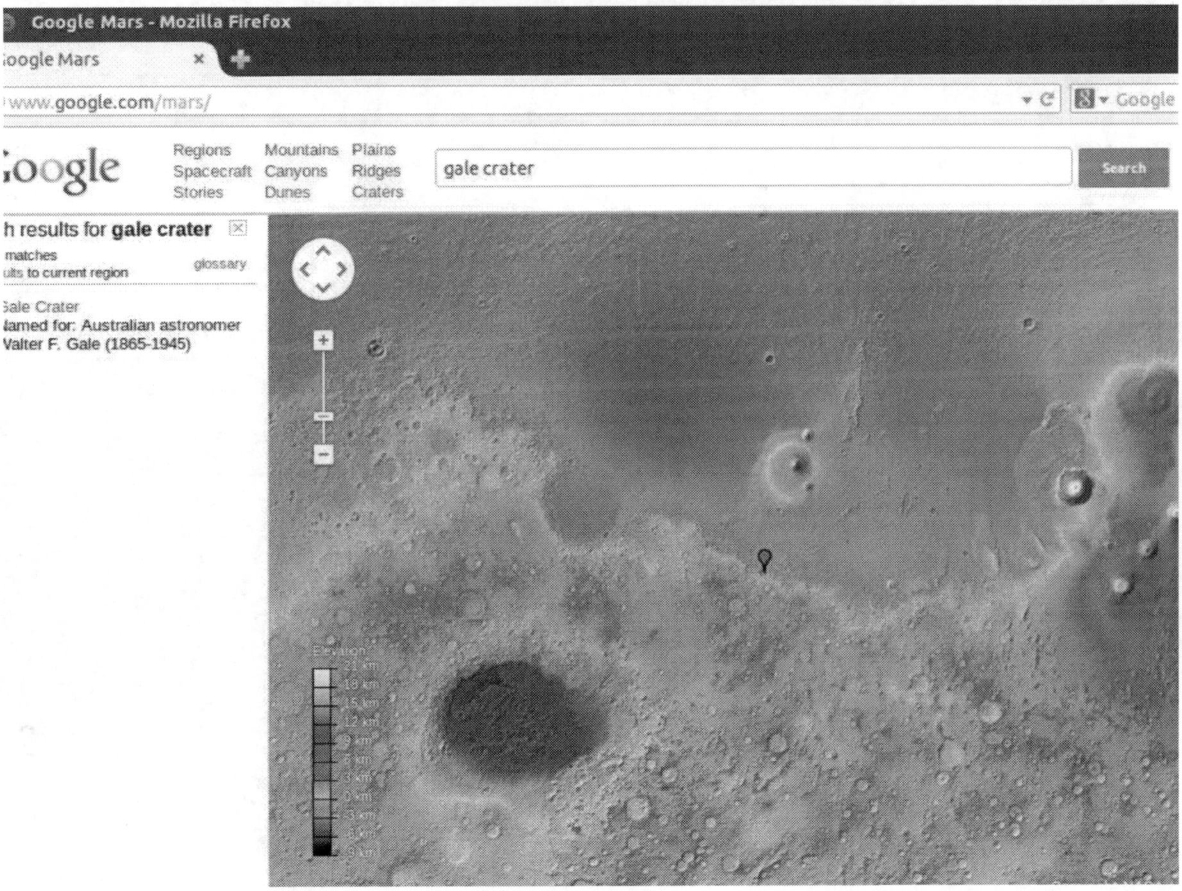
Google Mars - Mozilla Firefox
www.google.com/mars/
Regions
Spacecraft
Stories
Mountains
Canyons
Dunes
Plains
Ridges
Craters
gale crater
Search
results for gale crater
matches
glossary
Gale Crater
Australian astronomer
F. Gale (1865-1945)
Elevation

Why Gale crater?

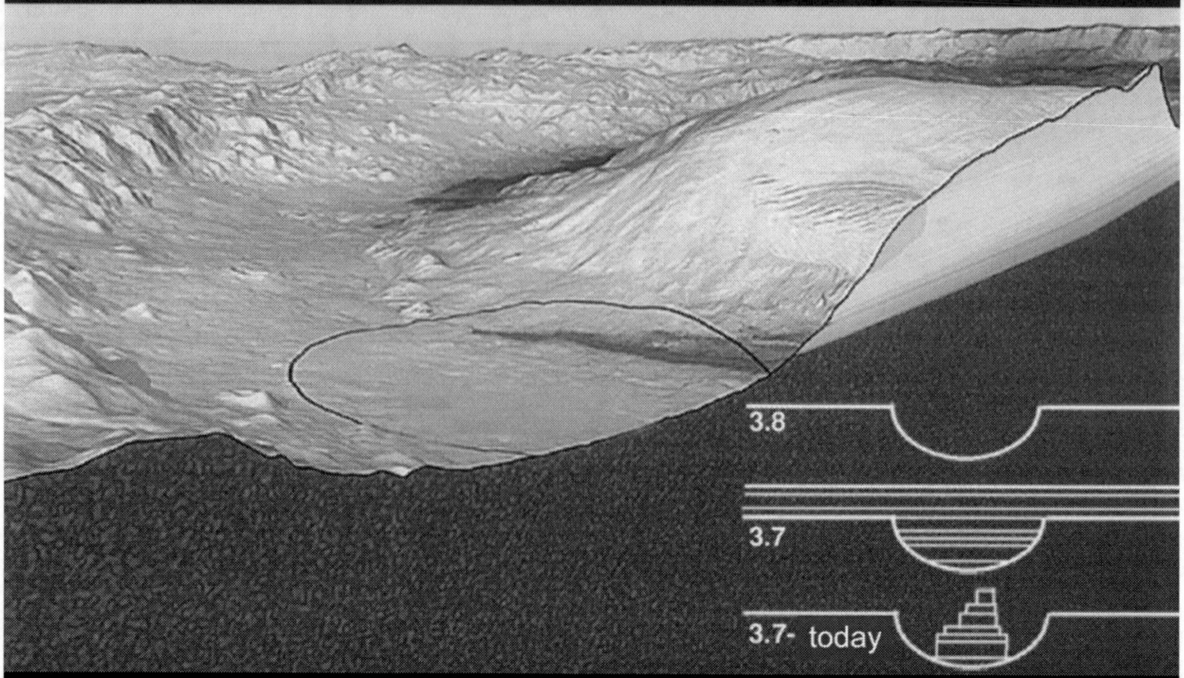

Sediments were deposited some 3.7 billion years ago. Most of them have been eroded away by wind and sand.

CAU Institute of Experimental and Applied Physics
Extraterrestrial Physics

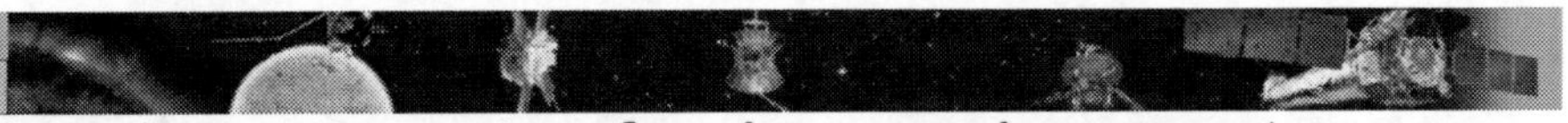

Curiosity landed in the remnants of a river environment...

… and drove on into the remnants of an early lake!

Drill results from SAM

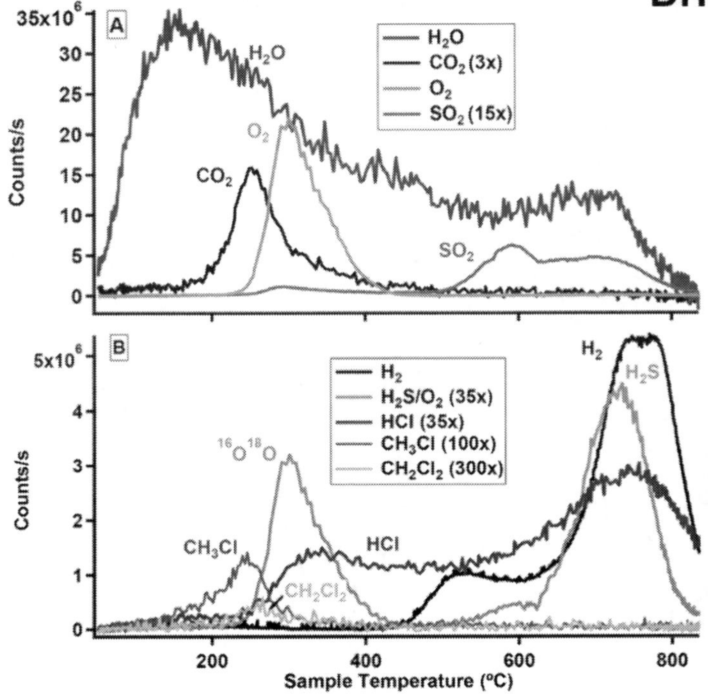

(Ming et al., Science, 2013)

Measurements show that the drilled mudstone contained carbon, nitrogen, oxygen, and sulfur. These elements are needed for life to form.

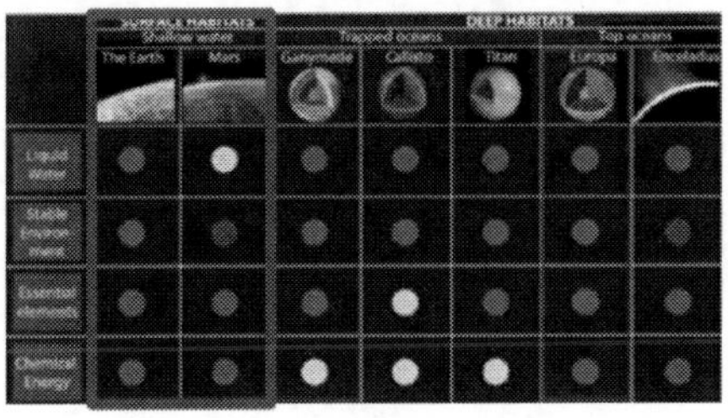

Destruction of Organic Compounds by Radiation

Ionizing radiation breaks chemical bonds and produces radicals and oxidants.

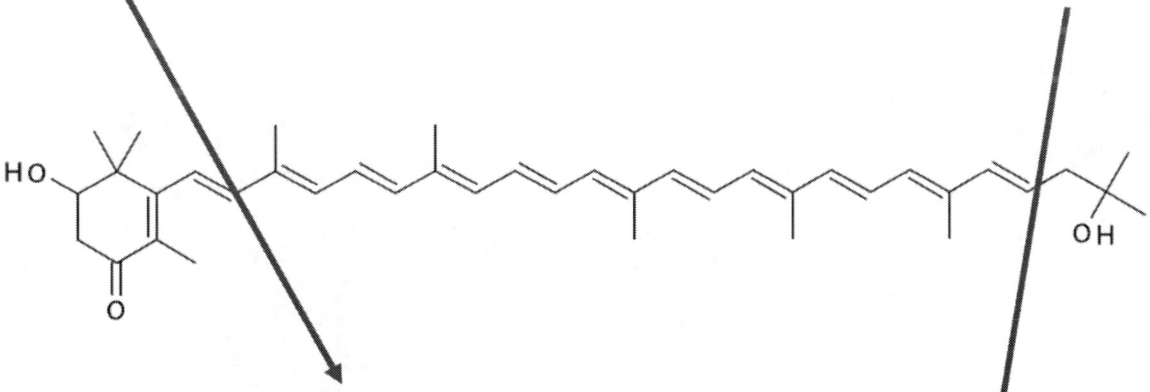

Result: Destruction of large organic molecules (if there is no repair mechanism)

How long could organic molecules survive ionizing radiation environment?

Previous models: 50 – 150 mGy/y
RAD measurements: 76 mGy/y

Organic molecules are efficiently destroyed at a depth of 4-5 cm. In 650 million years only 1/1000 survives.

How many after 3.8 Gy?

==> Half of the organics should still be around if the soil were only exposed for 65 million years.

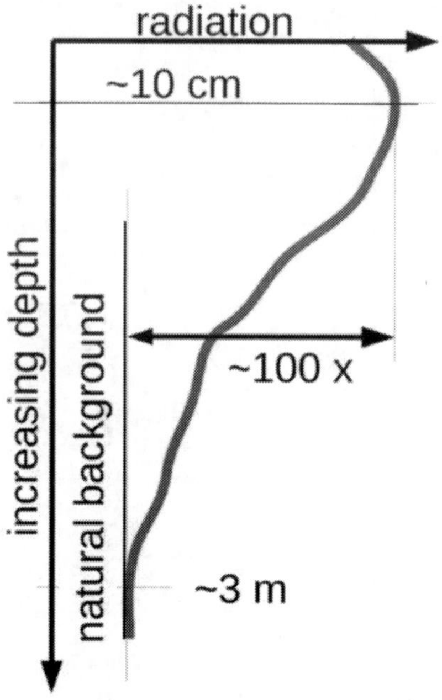

Age determination with cosmogenic isotopes

Sheepbed was exposed for only 80 ± 30 million years!

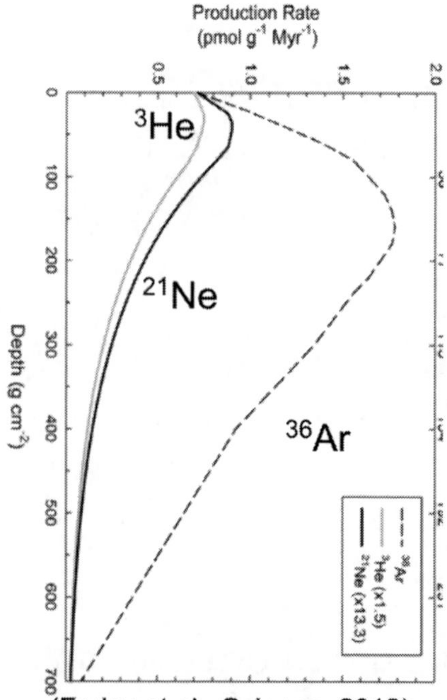

(Farley et al., Science, 2013)

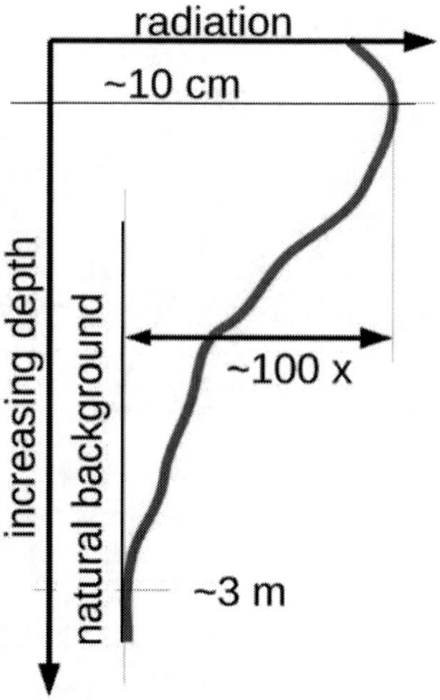

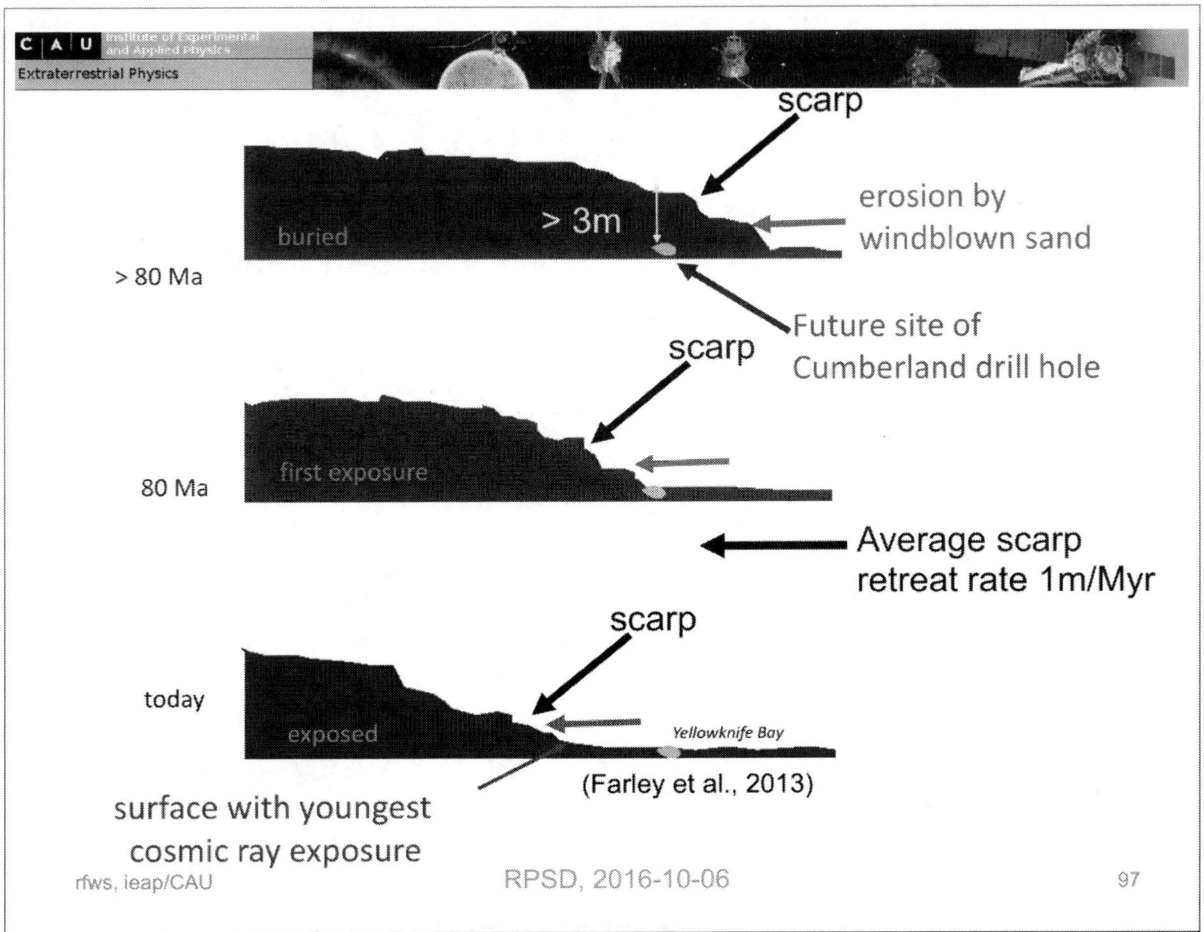
C A U Institute of Experimental and Applied Physics
Extraterrestrial Physics
scarp
> 3m
buried
erosion by windblown sand
> 80 Ma
Future site of Cumberland drill hole
scarp
first exposure
80 Ma
Average scarp retreat rate 1m/Myr
scarp
today
exposed
Yellowknife Bay
(Farley et al., 2013)
surface with youngest cosmic ray exposure
rfws, ieap/CAU
RPSD, 2016-10-06
97

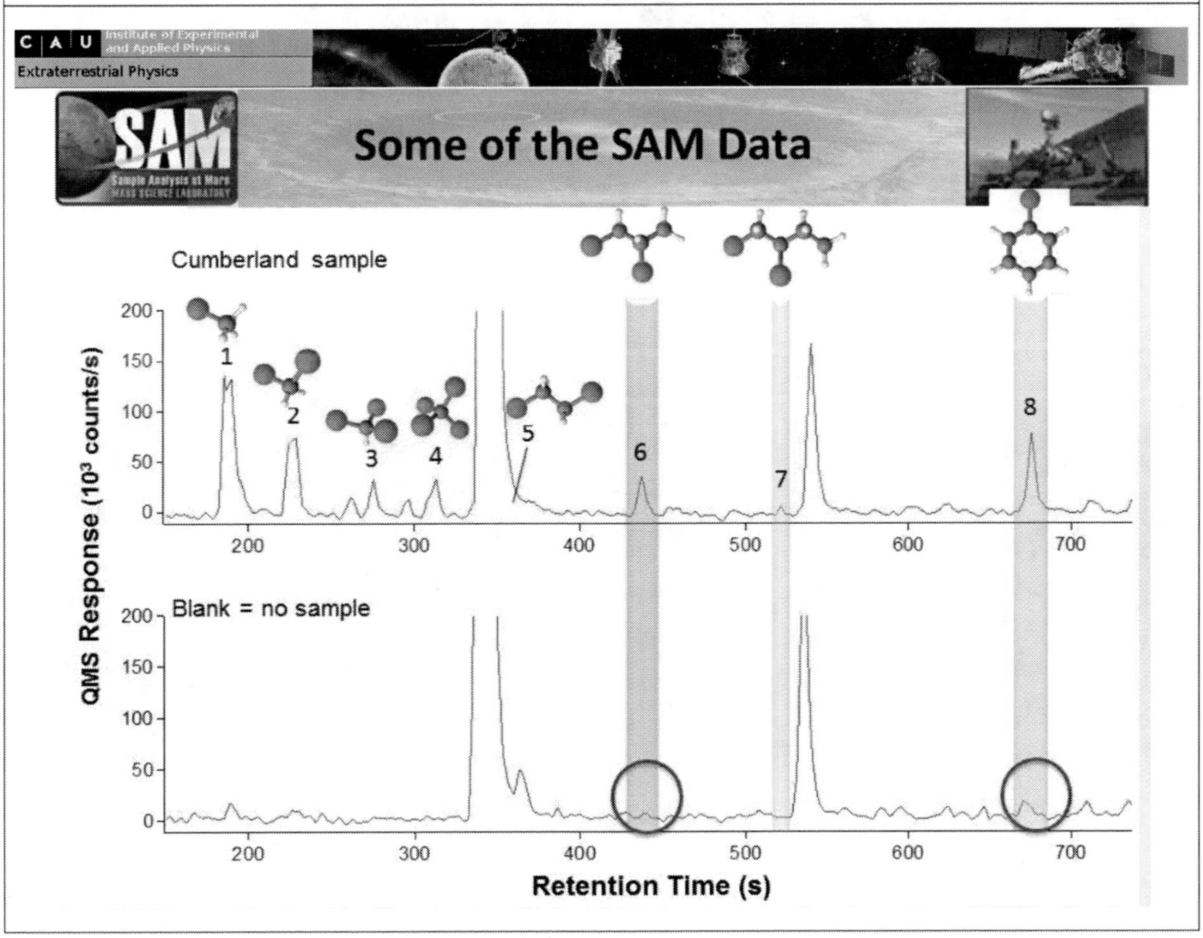
C A U Institute of Experimental and Applied Physics
Extraterrestrial Physics
SAM
Sample Analysis at Mars
Some of the SAM Data
Cumberland sample
1
2
3
4
5
6
7
8
Blank = no sample
QMS Response (10³ counts/s)
Retention Time (s)

Summary, Conclusions, and Thanks

- Radiation in heliosphere very variable
- Models are even more variable...
- First radiation measurements on another planet
- Radiation environment on Mars quite different than on Earth and important for future human exploration
- Organics can be and have been detected on Mars
- Mars was habitable – but is it still so?

C A U
Christian-Albrechts-Universität zu Kiel

SwRI

JPL

Supported by:
Federal Ministry of Economics and Technology
on the basis of a decision by the German Bundestag

DLR

esa

NASA

Backup Slides

Why do we go to the trouble of predicting radiation?

Radiation health primer:

Average exposure: **few mSv/year**
Allowed additional exposure: **1 mSv/year**
Variability of exposures: **from ~1mSv/year to 440 (!) mSv/year**

Guarapari, Brasil

Variability mainly due to radon. CH: factor 240!

NASA: Cancer risk shall not increase by more than 3%.

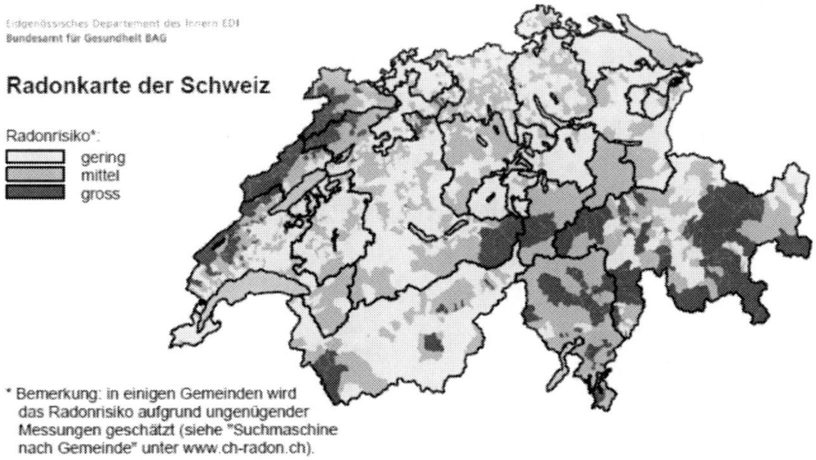

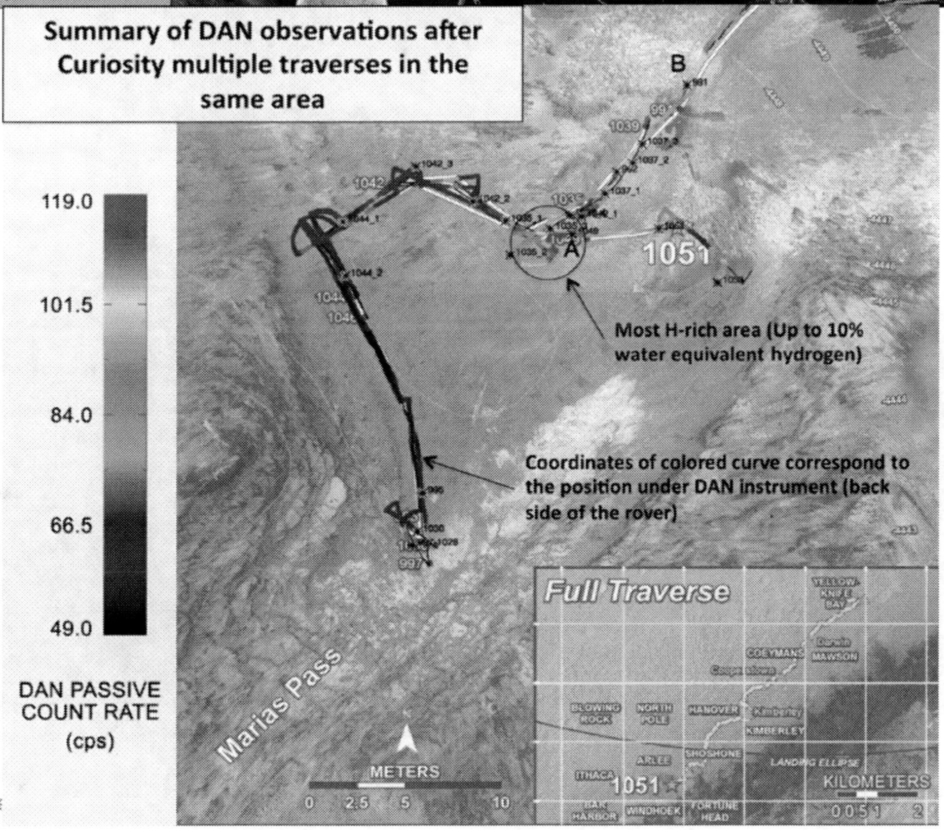

Why does ionization increase with height?

Measurements show count rate increases with height.

Air pressure decreases with height.

If Earth's radiation were ionizing, count rate should decrease with height because there is less air to ionize.

Can only be explained by ionizing agent coming in from outside the Earth ---> extraterrestrial origin! (Millikan)

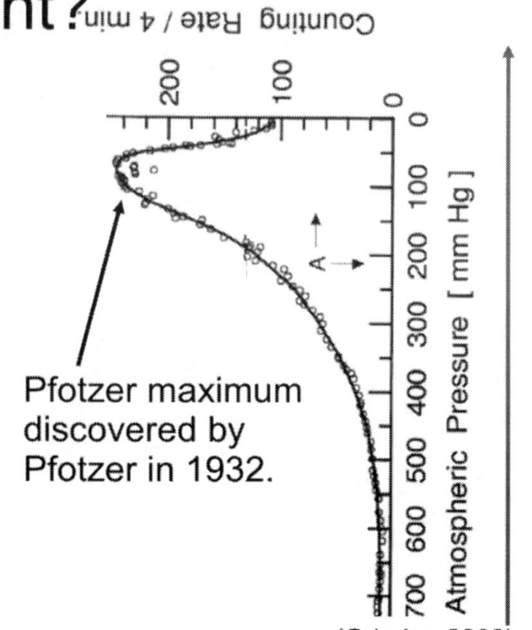

(Grieder, 2002)

Where is Curiosity?

Where is Curiosity?

Where is Curiosity?
Marias Pass
... and what did we find?

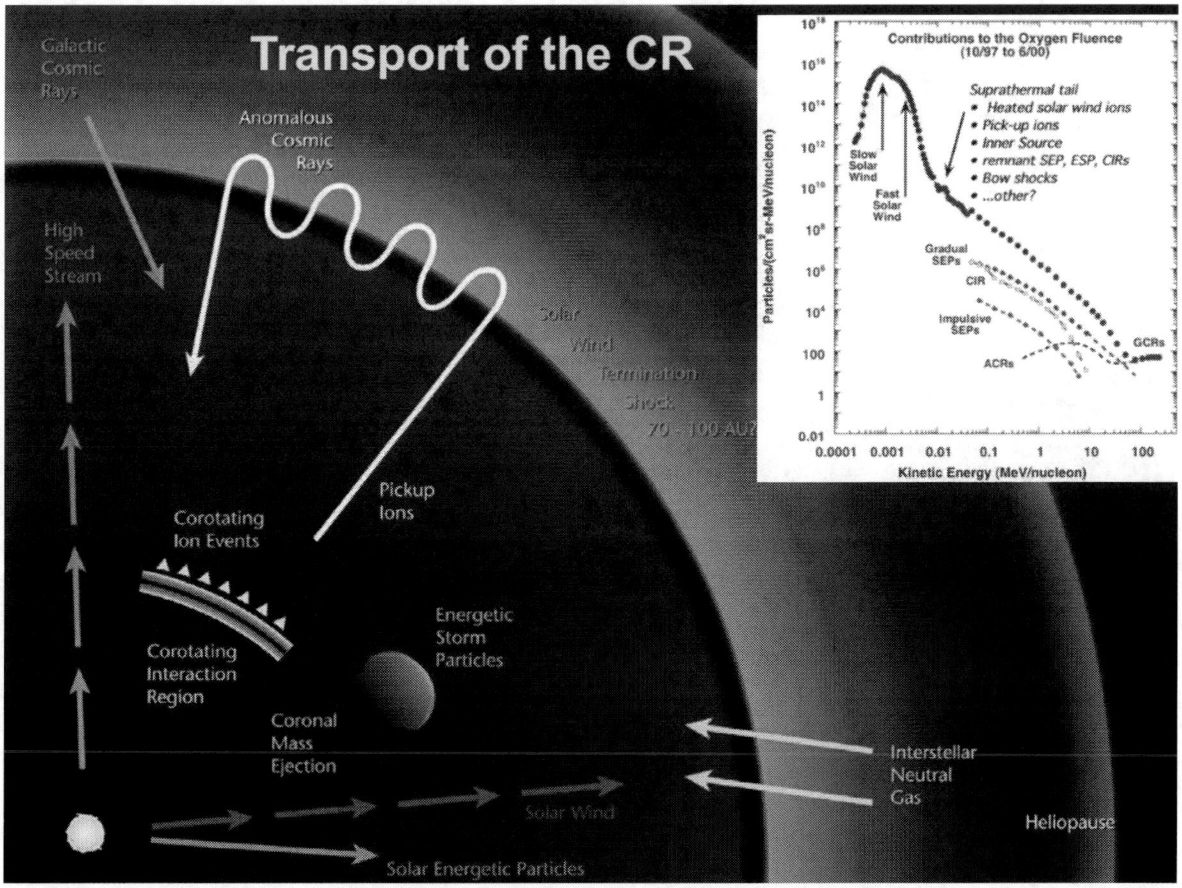
Transport of the CR
Galactic Cosmic Rays
Anomalous Cosmic Rays
High Speed Stream
Solar Wind Termination Shock 70 - 100 AU?
Pickup Ions
Corotating Ion Events
Corotating Interaction Region
Energetic Storm Particles
Coronal Mass Ejection
Interstellar Neutral Gas
Solar Wind
Heliopause
Solar Energetic Particles
Contributions to the Oxygen Fluence (10/97 to 6/00)
Suprathermal tail
• Heated solar wind ions
• Pick-up ions
• Inner Source
• remnant SEP, ESP, CIRs
• Bow shocks
• ...other?
Slow Solar Wind
Fast Solar Wind
Gradual SEPs
CIR
Impulsive SEPs
ACRs
GCRs
Particles/(cm² sr-MeV/nucleon)
Kinetic Energy (MeV/nucleon)

CAU Institute of Experimental and Applied Physics
Extraterrestrial Physics
From Kiel to Space
Space missions with contributions from CAU Kiel
Ulysses (1990-2009)
Advanced Composition Explorer (ACE, 1997-today)
Galileo (1989-2003)
SOlar and Heliospheric Observatory (SOHO, 1995-today)
Space Shuttle (1981-2011)
STEREO A and B (2006-today)
Long Duration Exposure Facility (1984-1990)
International Sun-Earth Explorer (ISEE-2, 1977-1987)
Helios 1 and 2 (1974-1986)
Azur (1969-1970)
Mir (1986-2001)
International Space Station (1998-today)
Chandra X-Ray Observatory (1999-today)
Solar Orbiter (2018-?)
Balloon Flights (ongoing)

Solar Orbiter

Exploring the Sun-Heliosphere Connection

Remote-sensing windows (10 days each)

High-latitude Observations

Mission Summary

Launch: October 2018
Cruise Phase: 2.5 years
Nominal/Extended Mission: 3.5/2.5 years
Orbit: 0.28–0.91 AU (P=150-180 days)
Out-of-Ecliptic View:
Multiple gravity assists with Venus to increase out-of-ecliptic inclination to >33°

Perihelion Observations

Science Focus

How does the Sun create and control the Heliosphere – and why does solar activity change with time ?

1) What drives the solar wind and where does the coronal magnetic field originate?
2) How do solar transients drive heliospheric variability?
3) **How do solar eruptions produce energetic particle radiation that fills the heliosphere?**
4) How does the solar dynamo work and drive connections between the Sun and the

High-latitude Observations

C A U Institute of Experimental and Applied Physics
Extraterrestrial Physics

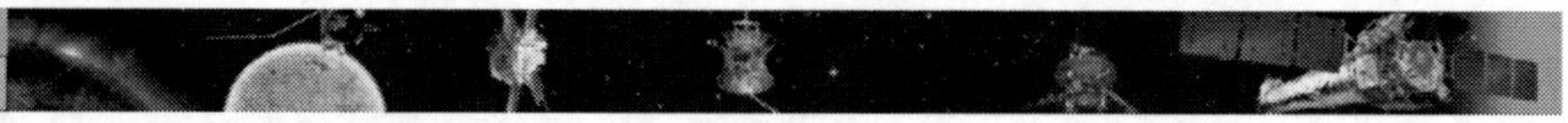

EPD - The Team

Led by University of Alcala, Spain,
PI: Javier Rodriguez-Pacheco (Alcala)
Co-PI: Robert Wimmer-Schweingruber (Kiel)
Hardware contributions from:

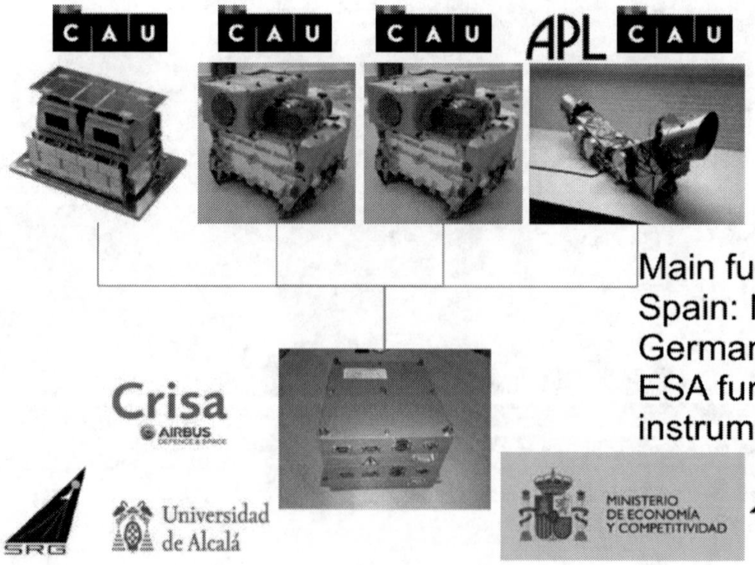

System Engineering:
SENER

Main funding agencies:
Spain: Mineco
Germany: DLR
ESA funds SIS as a facility instrument

MINISTERIO DE ECONOMÍA Y COMPETITIVIDAD
DLR
esa

Background: Radiation on Earth

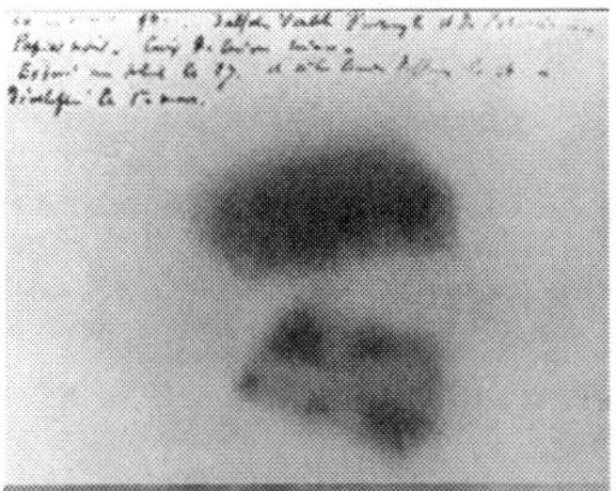

In 1896 Henri Bequerel and others discovers non-chemical process capable of penetrating black paper and darkening photo plates. It also excites fluorescence.

Marie Curie discovers that uranium salt also ionizes air, strictly proportional to the amount of material.

Pitchblend is four times more effective, therefore, it must contain a new element. Discovery of polonium and radium.

Invention of the word 'radioactivity'.

Something can ionize air and penetrate paper.

Ionization is a good measure of radioactivity.

Other early discoveries

In 1912 Victor Hess discovers 'Höhenstrahlung' at large heights. Degree of ionization increases with height.

Image credit: NY Times

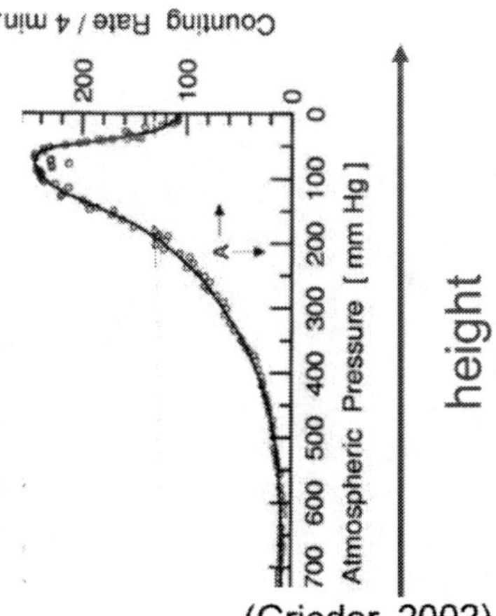

(Grieder, 2002)

Why does ionization increase with height?

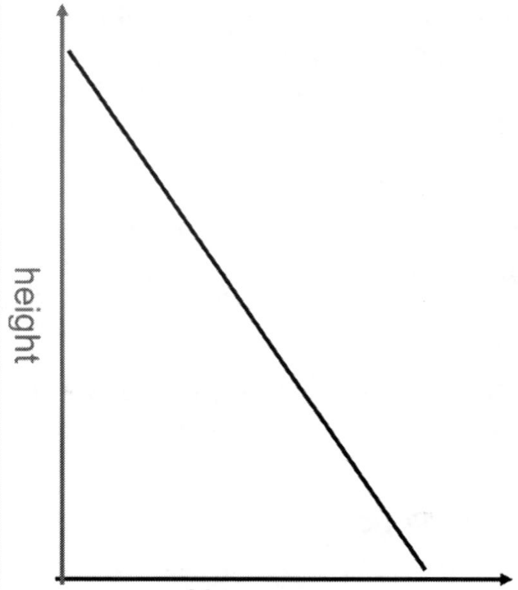

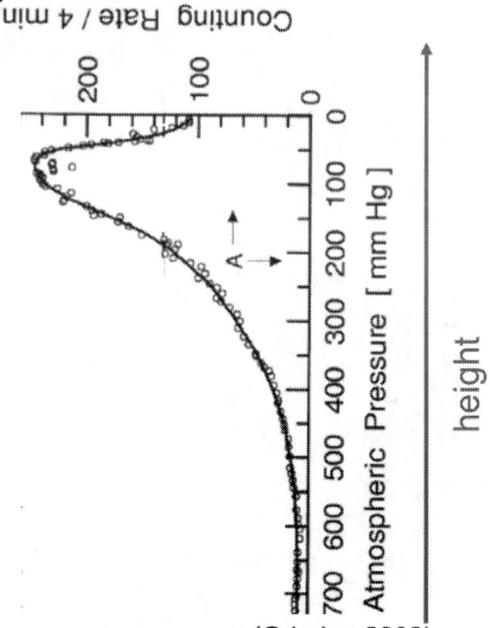

(Grieder, 2002)

Why does ionization increase with height?

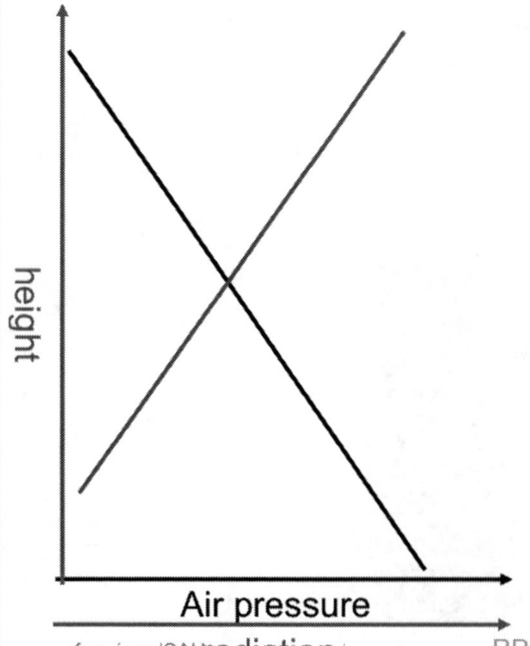

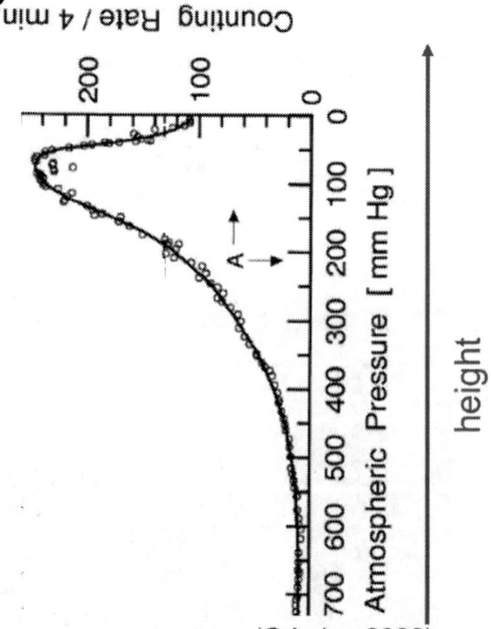

(Grieder, 2002)

Why does ionization increase with height?

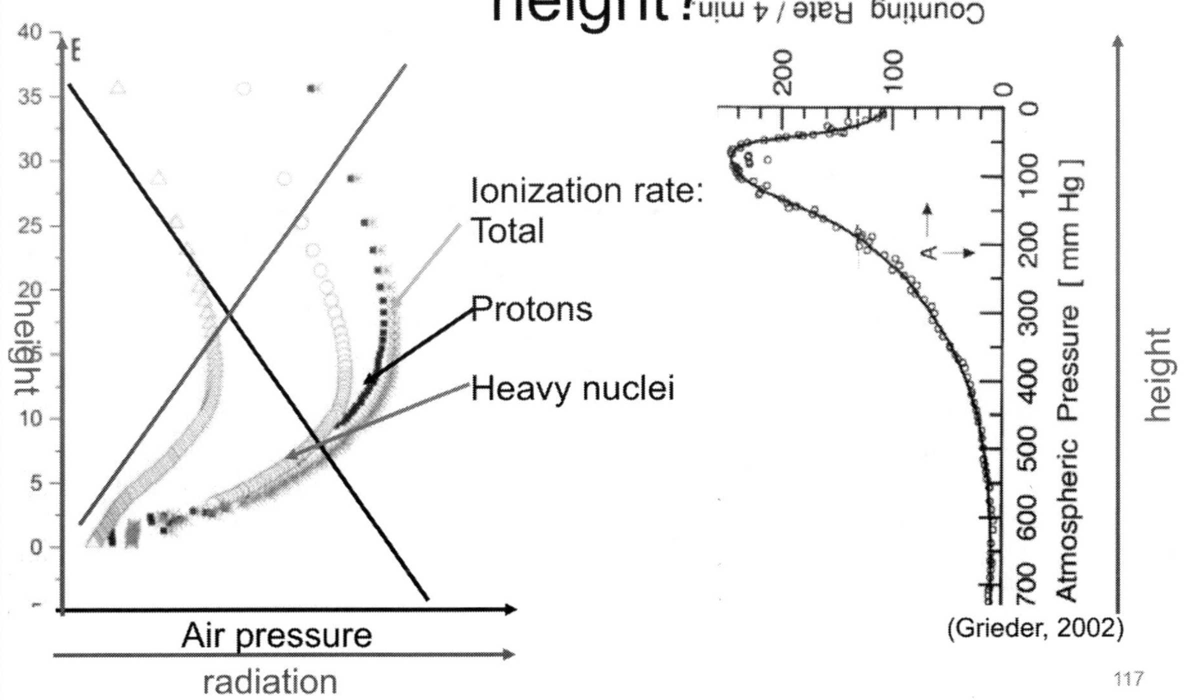

C A U Institute of Experimental and Applied Physics
Extraterrestrial Physics

Why does ionization increase with height?

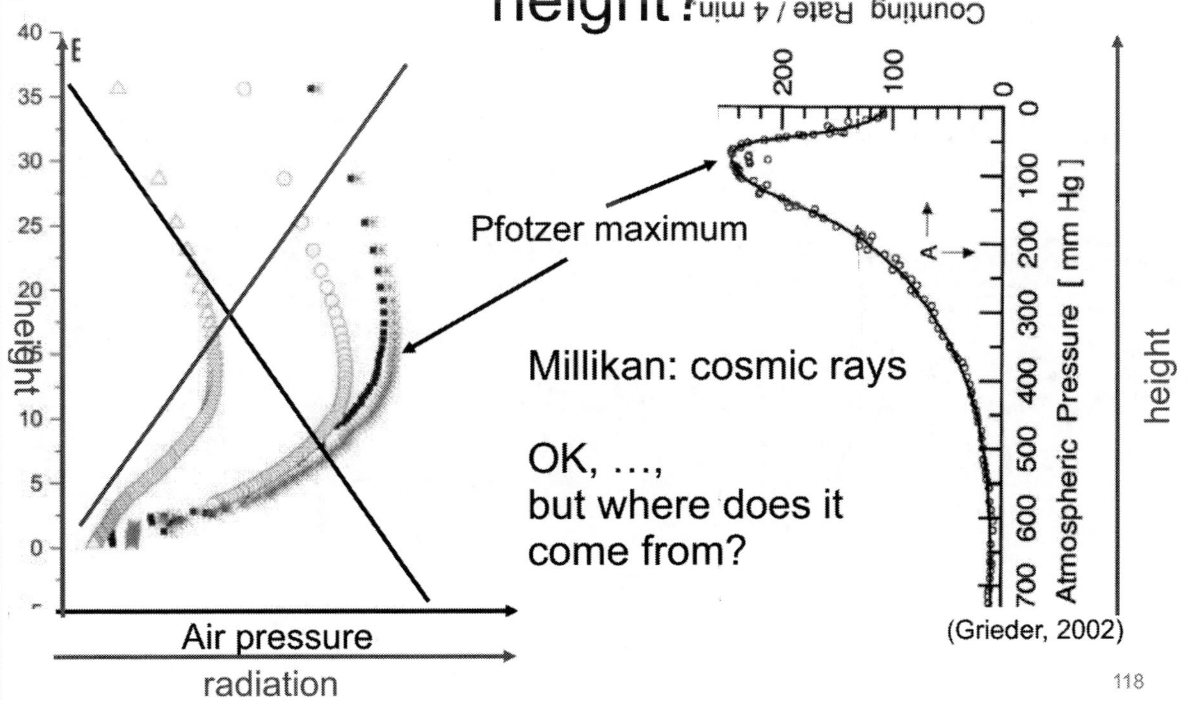

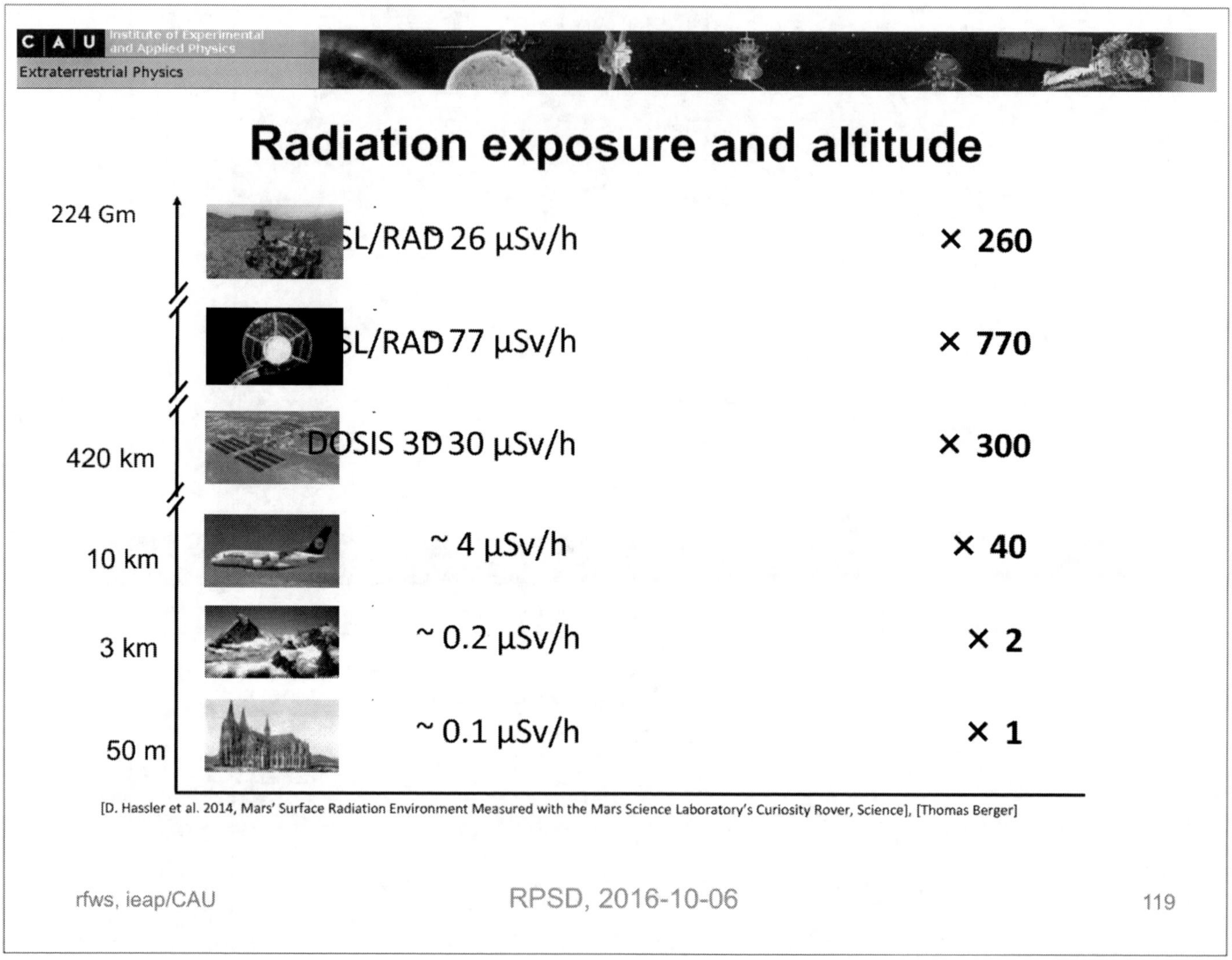

CAU Institute of Experimental and Applied Physics
Extraterrestrial Physics
Radiation exposure and altitude
224 Gm
SL/RAD 26 µSv/h
× 260
SL/RAD 77 µSv/h
× 770
420 km
DOSIS 3D 30 µSv/h
× 300
10 km
~ 4 µSv/h
× 40
3 km
~ 0.2 µSv/h
× 2
50 m
~ 0.1 µSv/h
× 1
[D. Hassler et al. 2014, Mars' Surface Radiation Environment Measured with the Mars Science Laboratory's Curiosity Rover, Science], [Thomas Berger]
rfws, ieap/CAU
RPSD, 2016-10-06
119

ICRS-13 & RPSD-2016

13th *International Conference on Radiation Shielding*
&
19th Topical Meeting of the *Radiation Protection & Shielding Division*
of the *American Nuclear Society* -2016

Plenary keynote:

"AAEA contribution to strengthening radiation protection in the Arab countries"

Dr. Abdelmajid Mahjoub

Arab Atomic Energy Agency, Tunis

Dr. Abdelmajid Mahjoub received a master's degree from Tulane University in New Orleans and a PhD from the University of Nebraska-Lincoln in the USA.

Dr. Mahjoub was a Professor at the University of Tunis (1985-1997). He held the position of Director General of the National Center for Nuclear Sciences and Technologies of Tunisia (1997-2001). He worked at the International Atomic Energy Agency (IAEA) in the Department of technical cooperation for seven years.

Dr. Mahjoub is currently Director General of the Arab Atomic Energy Agency (AAEA) based in Tunis since February 2009.

AAEA CONTRIBUTION TO STRENGTHENING RADIATION PROTECTION IN THE ARAB COUNTRIES

Abdelmajid Mahjoub

Arab Atomic Energy Agency

Tunis

aaea@aaea.org.tn

ICRS 13
RPSD 2016
3-6 Octobre2016, paris, France

ANNuR MEMBRE STATES

22 member states in LAS

African Countries

Egypt
Libya
Sudan
Tunisia
Mauritania
Morocco
Algeria
Comoros
Djibouti
Somalia

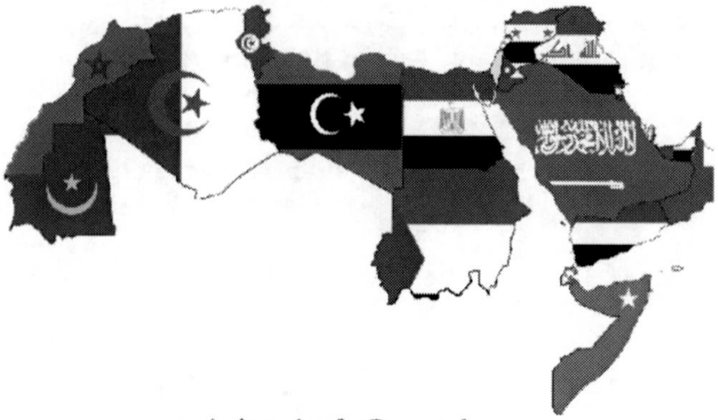

Gulf Countries

Bahrain
KSA
Kuwait
Oman
Qatar
UAE

Asian Arab Countries
Iraq, Jordan, Lebanon ,
Palestine, Syria, Yemen,

AAEA Activities

- AAEA translate many IAEA guides and technical documents related to Radiation Protection into Arabic like:
  - Safety series, International Basic Safety standards for protection from Ionizing radiation and safety of radioactive sources BSS
  - Assessment of occupational exposure (Safety guide)
  - Occupational radiation protection
  - Technical document of radioactive waste disposal
  - Decommissioning of research reactors and NPP
  - ICRP- Publications

AAEA spreads the safety and security culture among Arab speaking nuclear community

AAEA Missions

- Coordinates among member states in the field of peaceful applications of atomic energy,
- Assists in human resources development and transfer of knowledge and information concerning nuclear sciences & technologies,
- Sets up unified Arab regulations for radiation protection, nuclear safety and security and safe handling of radioactive materials. Building the SS culture
- Raise the public awareness of benefits and hazards of ionized radiation.,

Need for Arab Safety Networks

- To ensure that safety, security and knowledge, experience, and lessons learned are made visible and available, through links between Arab RB,
- To enable and support interaction and collaboration between competent people and organizations relevant to nuclear regulation.
- Therefore, The Arab Network of Nuclear Regulators (ANNuR) has been established

ANNuR's THEMATIC WORKING GROUPS

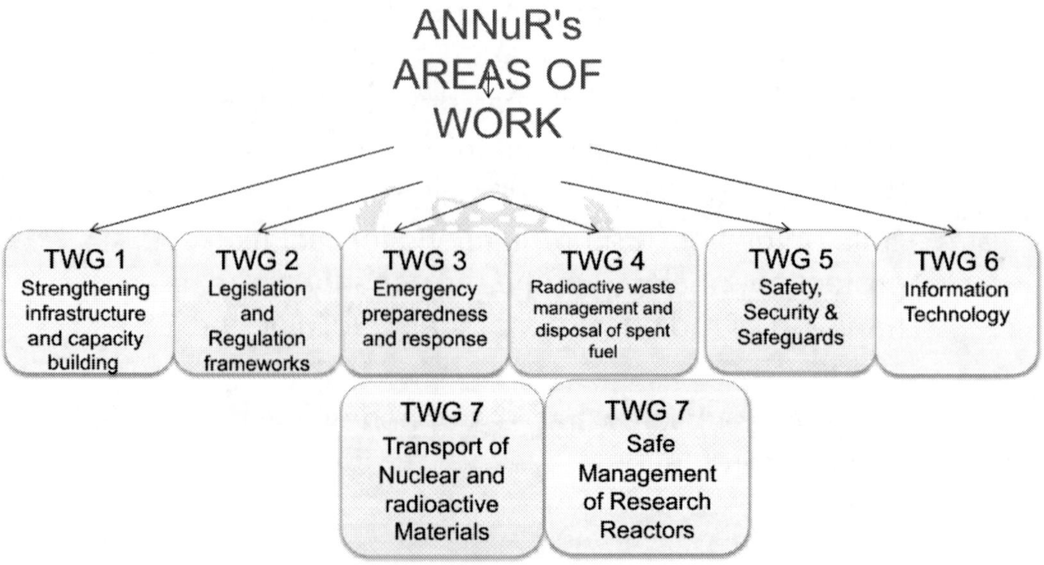

The Thematic Working Groups

Thematic Working Group	coordinator
strengthening infrastructure and capacity building for regulatory bodies	Jordan
Legislation and Regulation frameworks	Tunisia
Emergency preparedness and response	Iraq
Radioactive waste management and disposal	Libya
Safety, Security and Safeguards	Egypt
Information Technology	IAEA/AAEA
Transport of Nuclear and Radioactive Materials	Tunisia
Safety Management of Research Reactors	Morocco

ANNuR Meetings

January 2010 +++++→ First meeting
May 2011 +++++→ Second meeting
April 2012 +++++→ Third meeting
March 2013 +++++→ Fourth meeting
March 2014 +++++→ Fifth meeting

March 2015 +++++→ Sixth meeting

March 2016-----------------→ Seventh Meeting

ICRS13 - RPSD 2016, Paris, France, 3-6 October 2016.

Cooperation with IAEA

IAEA assisted in establishing ANNuR

IAEA provides expert support to different ANNuR activities

IAEA funds many ANNuR activities, (TC, Wkps, EM,...)

ANNuR and IAEA signed a practical agreement on cooperation in the area of strengthening the nuclear safety and security frameworks

Cooperation with KINS

- ***MOU has been signed between KINS and AAEA***
- ***Co- sponsoring of ANNuR activities***
- ***Attended all six ANNuR Annual Plenary meetings***
- ***MsC grants in ISNS for ANNuR member states.***

- ***Joint IAEA-KINS Basic Professional Training Course on Nuclear Safety for Regulators, (Tunis, Tunisia, 4-15 May 2015)***
- ***Workshop on Safety and Protection of radiation sources, (Daejong, Korea, 1-11 June 2015)***
- ***Emergency preparedness and Radiation protection measures, (Daejong, Korea, 5-9 Sept. 2016)***

ICRS13 - RPSD 2016, Paris, France, 3-6 October 2016.

US-NRC support to ANNuR

- Nuclear Executive Workshop for Decision Makers in Nuclear Power Programs, (March 2012, New Orleans, USA).
- Workshop on the Physical Protection and Security of Radioactive Sources, (June 2012, Tunisia).
- Site Application Review, (February 2013, Tunisia).
- Workshop on Physical Protection of Nuclear Materials and Installations, (November 2013, Kuwait).
- Regulatory Aspect of Power and Research reactors siting, (September 2013, Jordan).
- Radiological and Nuclear Emergencies, (August 2013, Tunisia).
- Workshop on Construction Permit for Power and Research Reactors, (October 2014, Rockville, USA).
- Workshop on Probabilistic Risk Assessment, (November 2014, Kuwait).
- Workshop on Advanced Steps on Safety of Radioactive Sources and Nuclear Materials (8-12 June 2015, Marrakech, Morocco).
- Workshop on SMRs Technologies and Regulations (January 2016, Vienna, Austria).

European Commission

- **EU representatives participate at all ANNuR meetings**
- **EU supports ANNuR activities since 2013**

Actions planned/implemented:

- EU-Acquis, legal and regulatory framework, regulatory role and responsibilities for licensing and enforcement (Hammamet, Tunisia, 20-24 April 2013)
- **Radiation protection principles and system at Nuclear Facilities:** (Hammamet,Tunisia 21-24/4/2015)
- Radiation Protection and Regulatory Emergency Preparedness (Bizerte, Tunisia, 19-23 Sept. 2016)
- Regulatory functions, licensing management, basic safety concepts and decision making. (To be implemented).

Lawrence Livermore National Laboratory

➢ AAEA had an agreement with DoE-LLNL to support three activities annually in the field of Nuclear SSS

➢**SUBJECTS:**

➢NPP Project Management

➢Basic Requirements of Nuclear Regulatory Bodies

➢Radiation Protection, Nuclear Safety and Nuclear Security

➢Nuclear Security and Safeguards

➢Regulation, Legislation and Security of Radioactive Sources

➢Radiation Monitoring and Export Control

VENUES

- Kuwait, Jordan, Morocco, Saudi Arabia Bahrain.

ANNuR Number of training events per Year

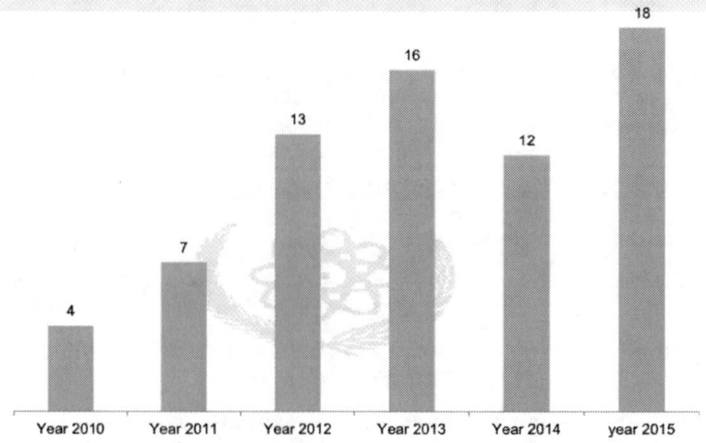

Number of trainees per year

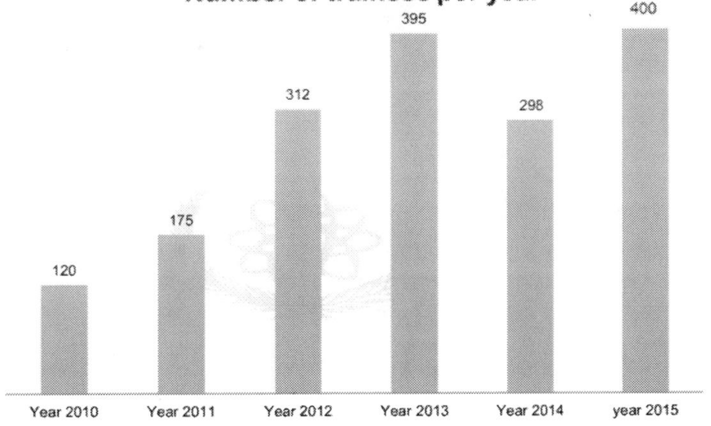

ICRS13 - RPSD 2016, Paris, France, 3-6 October 2016.

Number of Participants per Country 2010-2015

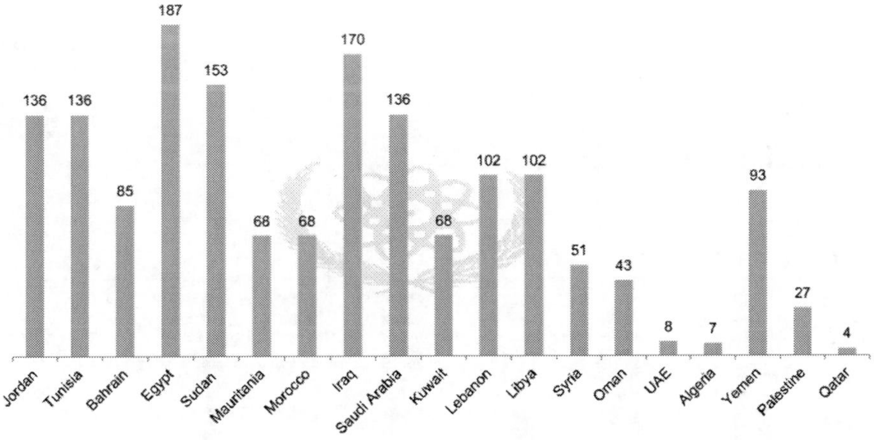

ICRS13 - RPSD 2016, Paris, France, 3-6 October 2016.

For best browsing experience, please use Internet Explorer 7 or a later version.
IAEA NUCLEUS Site Actions
Daw Mosbah EN Help Sign Out
IAEA | ANNuR Arab Network of Nuclear Regulators
Home About ANNuR Thematical Working Groups Steering Committee IT Support Group Terms of Reference GNSSN
ANNuR Documents Overview
ANNuR Contact List
Thematic Working Groups
ANNuR Blog
ANNuR Discussion Board
National Nuclear Regulatory Portal List of Countries
Arab Network of Nuclear Regulators
ANNUR Regis
Related Links

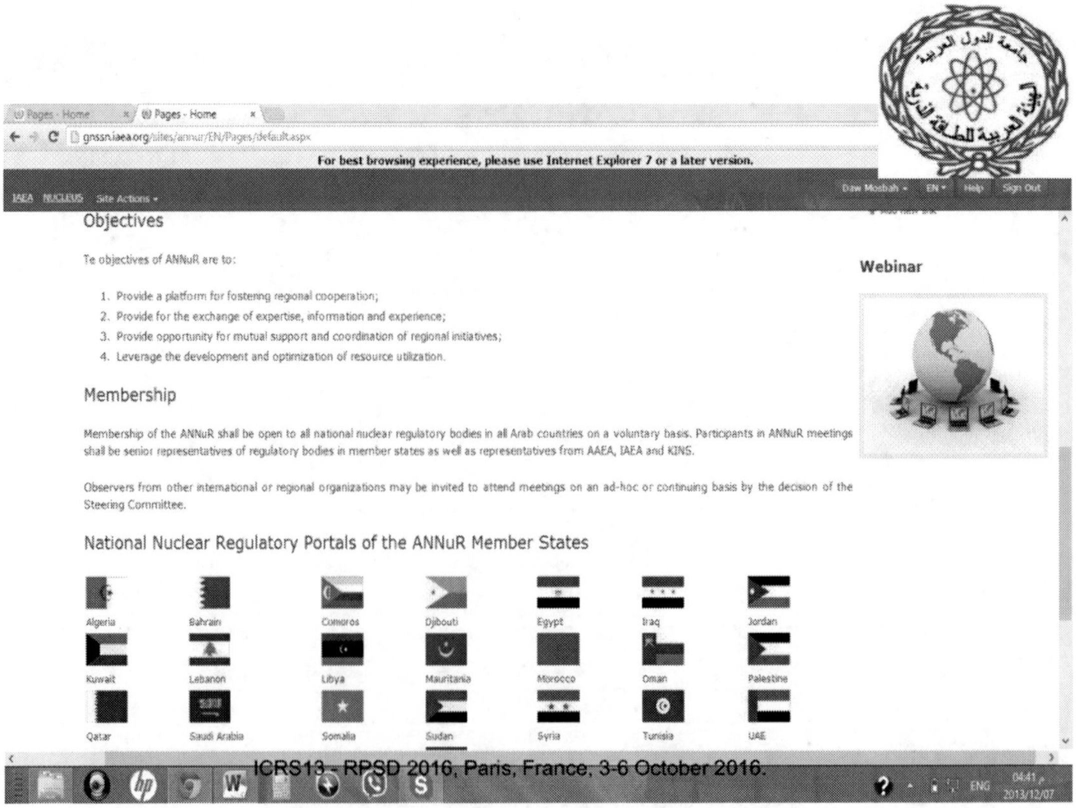

For best browsing experience, please use Internet Explorer 7 or a later version.
IAEA NUCLEUS Site Actions
Daw Mosbah EN Help Sign Out
Objectives
Te objectives of ANNuR are to:
1. Provide a platform for fosterng regional cooperation;
2. Provide for the exchange of expertise, information and experience;
3. Provide opportunity for mutual support and coordination of regional initiatives;
4. Leverage the development and optimization of resource utilization.
Membership
Membership of the ANNuR shall be open to all national nuclear regulatory bodies in all Arab countries on a voluntary basis. Participants in ANNuR meetings shall be senior representatives of regulatory bodies in member states as well as representatives from AAEA, IAEA and KINS.
Observers from other international or regional organizations may be invited to attend meetings on an ad-hoc or continuing basis by the decision of the Steering Committee.
National Nuclear Regulatory Portals of the ANNuR Member States
Algeria Bahrain Comoros Djibouti Egypt Iraq Jordan
Kuwait Lebanon Libya Mauritania Morocco Oman Palestine
Qatar Saudi Arabia Somalia Sudan Syria Tunisia UAE
Webinar

ANNuR
الشبكة العربية للهيئات الرقابية

ANNuR is an effective mechanism to address regional challenges and shifts in the member states priorities

ANNuR has proven its capabilities by providing the framework for conducting the first inter-regional inspection programme for a Research Reactor Facility (Cairo, September 2014)

ANNuR
الشبكة العربية للهيئات الرقابية

ANNuR is developing an action plan for all member states operation a research reactor or willing to implement a RR, using the milestone approach. In May 2015, three IAEA departments jointly are implementing the final step of this project.

ANNuR is working with the IEC/IAEA to further develop the Emergency preparedness self-assessment for the region and a regional emergency action plan is expected.

ANNuR
الشبكة العربية للهيئات الرقابية

In order to respond to the ANNuR member States needs in a systematic and proactive manner, ANNuR is working with the ANSN to use the ANSN regional self assessment and to adopt it for the ANNuR region

As requested by partners during the GNSSN coordination meeting, ANNuR is developped a capacity building framework, in order to promote the challenges and the priority areas in the region and to build strategic partnerships with donors.

ANNuR
الشبكة العربية للهيئات الرقابية

ANNuR is coordinating closely with FNRBA and ANSN to strengthen the Global nuclear safety framework.

ANNuR with the IAEA secretariat is promoting the National Nuclear Safety Knowledge management Platforms hosted under GNSSN. To this date 9 Countries have developed their NNRP.

13th Arab Conference on the Peaceful Uses of Atomic Energy

Hammamet, Republic of Tunisia
18-22 December 2016

Organized by

Arab Atomic Energy Agency

in Collaboration with

National Center for Nuclear Sciences and Technologies – Tunis

ICRS13 - RPSD 2016, Paris, France, 3-6 October 2016.

Closing session

- General Chair comments and some figures
- Honorary Chairs messages
- Honorary Chairs distinctions
- Student Poster Awards
- Final word by General Co-Chair Michele FERENCI

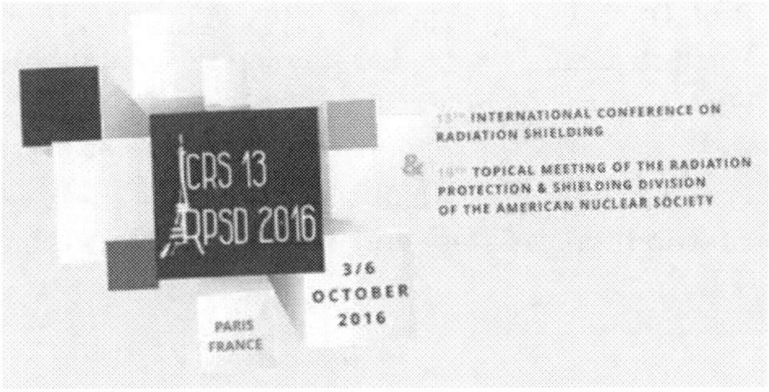

Participation

- 303 participants
- 37 students
- 26 countries:

  Austria (1), Australia (1), Belgium (5), Canada (1), China (17), Czech Republic (3), Germany (14), France (100), Grèce (1), Israël (1), Italia (11), Japan (45), Korea (23), Netherland (1), Poland (1), Portugal (1), Russia (2), Slovenia (3), Spain (5), Sweden (2), Switzerland (9), Tunisia (1), Turkey (3), Taïwan (3), UK (17), USA (32)

4 plenary talks & 1 key note

Dr. Francesco Cerutti, *CERN, Switzerland*, "Machine and radiation protection challenges of high energy/intensity accelerators: the role of Monte Carlo calculations"

Dr. Masashi Hirano, *Nuclear Regulation Authority, Japan,* "Outline of the Fukushima Daiichi Accident. Lessons Learned and Safety Enhancements"

Dr. Rebecca M. Howell, *University of Texas, USA*, "Neutron Measurements in Proton Radiation Therapy"

Prof. Dr. Robert F. Wimmer-Schweingruber, *University of Kiel, Germany*, "From Earth to Moon, Mars and - Beyond Space Radiation and Implications for Human Exploration"

Prof. Abdelmajid Mahjoub, *Arab Atomic Energy Agency, Tunisia*, "AAEA contribution to Safety enhancement in the Arab countries"

2 special sessions

"Fukushima five years after": 13

"Neutron experimental studies under the quasi-monoenergetic neutron field in the energies from 100 to 400 MeV": 7

4 workshops

SuperMC Tutorial
Jing Song - "Key Laboratory of Neutronics and Radiation Safety,
Chinese Academy of Sciences, Institute of Nuclear Energy Safety Technology,
Chinese Academy of Science . FDS Team"

Attila4MC-Software for Simplifying Monte Carlo
Greg Failla, Sr. Manager - Attila Product Line Varian Medical Systems -
Imaging Components

MRT Methodologies for Real-Time Particle Transport Simulation of Nuclear Systems
Prof. Alireza Haghighat Virginia Tech Transport Theory
Group (VT3G) Director of Nuclear Engineering and
Science Lab (NSEL) at Arlington Nuclear Engineering Program,
Mechanical Engineering Department Arlington, VA, USA

RayXpert - 3D radiation modeling software with Monte Carlo solver
Hugo Canovas -TRADE - Sales Manager

129 oral presentations in // sessions

T1_Nuclear Data, Radiation Detection, Measurements & Dosimetry: 27

T2_Shielding Experiments & Benchmarks: 10

T3_Accelerators & Fusion Facilities: 12

T4_Medical Facilities, Radiotherapy & Medical Applications, Space Dosimetry & Shielding: 20

T5_Fission Facilities, Fuel Cycle & Waste Management Facilities, Decommissioning: 25

T6_Calculation methods Monte Carlo & Deterministic: 35

Parallel sessions

60 Chair persons, 44 sessions

Monday October 3, 2016			
T1 : Nuclear Data, Radiation Detection, Measurements & Dosimetry	T3 : Accelerators & Fusion Facilities	T5 : Fission Facilities, Fuel Cycle & Waste Management Facilities, Decommissioning	T6 : Calculation methods Monte Carlo & Deterministic
Zhi Chen; Mariya Brovchenko	Alfredo Ferrari; Toshiya Sanami	Yurdunaz Celik; David Blanchet	David Dixon; Andrea Zola
T1 : Nuclear Data, Radiation Detection, Measurements & Dosimetry	T3 : Accelerators & Fusion Facilities	T5 : Fission Facilities, Fuel Cycle & Waste Management Facilities, Decommissioning	T6 : Calculation methods Monte Carlo & Deterministic
Zhi Chen; Mariya Brovchenko	Alfredo Ferrari; Toshiya Sanami	Yurdunaz Celik; David Blanchet	David Dixon; Andrea Zola
T1 : Nuclear Data, Radiation Detection, Measurements & Dosimetry	T2_Shielding Experiments & Benchmarks	T4_Medical Facilities, Radiotherapy & Medical Applications, Space Dosimetry & Shielding	T6 : Calculation methods Monte Carlo & Deterministic
Thomas Miller; Cedric Jouanne	Ivan Kodeli; Joachim Miss	Lawrence Heilbronn; Vladimir Mares	H. Grady Hughes; Eric Dumonteil

Tuesday October 4, 2016			
T1 : Nuclear Data, Radiation Detection, Measurements & Dosimetry	SS1 : Fukushima special session "Fukushima five years after"	T5 : Fission Facilities, Fuel Cycle & Waste Management Facilities, Decommissioning	
Stefano Agosteo; Jean-Christophe Trama	Jean Christophe Gariel; Masashi Hirano	Arkady Serikov; Vincent Léger	
T1 : Nuclear Data, Radiation Detection, Measurements &	SS1 : Fukushima special session (continued) "Fukushima five years after"	T5 : Fission Facilities, Fuel Cycle & Waste Management Facilities, Decommissioning	T6 : Calculation methods Monte Carlo & Deterministic
Stefano Agosteo; Jean-Christophe Trama	Jean Christophe Gariel; Masashi Hirano	Arkady Serikov; Vincent Léger	Anil Prinja; Odile Petit
T1 : Nuclear Data, Radiation Detection, Measurements & Dosimetry	SS1 : Fukushima special session (continued) "Fukushima five years after"	T2_Shielding Experiments & Benchmarks	T6 : Calculation methods Monte Carlo & Deterministic
Yoshihiro Asano; Mansour Fadil	Jean Christophe Gariel; Masashi Hirano	Irina Popova; Yannick Peneliau	Yasushi Nauchi; Jing Song
T3 : Accelerators & Fusion Facilities	SS1 : Fukushima special session (continued) "Fukushima five years after"	T4_Medical Facilities, Radiotherapy & Medical Applications, Space Dosimetry & Shielding	T6 : Calculation methods Monte Carlo & Deterministic
Chang-Ho Shin; Jean-Charles Jaboulay	Jean Christophe Gariel; Masashi Hirano	Isao Murata; Cheick Thiam	Yasushi Nauchi; Jing Song

Wednesday October 5, 2016			
T1 : Nuclear Data, Radiation Detection, Measurements & Dosimetry	T3 : Accelerators & Fusion Facilities	T5 : Fission Facilities, Fuel Cycle & Waste Management Facilities, Decommissioning	T4_Medical Facilities, Radiotherapy & Medical Applications, Space Dosimetry & Shielding
Hee-Seock Lee; Pierre Leconte	Rong-Jiun Sheu; Celine Guenaut	Alireza Haghighat; Yi-Kang Lee	Michele Ferenci; François Xavier Hugot
T1 : Nuclear Data, Radiation Detection, Measurements & Dosimetry	T3 : Accelerators & Fusion Facilities	T5 : Fission Facilities, Fuel Cycle & Waste Management Facilities, Decommissioning	T2_Shielding Experiments & Benchmarks
Hee-Seock Lee; Pierre Leconte	Rong-Jiun Sheu; Celine Guenaut	Alireza Haghighat; Yi-Kang Lee	Enrico Sartori; Frederic Damian

SS2 : "Neutron experimental studies under the quasi- monoenergetic
Yosuke Iwamoto

Thursday October 6, 2016			
T1 : Nuclear Data, Radiation Detection, Measurements & Dosimetry			T6 : Calculation methods Monte Carlo & Deterministic
Stephane Bourganel; Thomas Miller			Xie George Xu; Jeremy Sweezy
T3 : Accelerators & Fusion Facilities	T4_Medical Facilities, Radiotherapy & Medical Applications, Space Dosimetry & Shielding	T5 : Fission Facilities, Fuel Cycle & Waste Management Facilities, Decommissioning	T6 : Calculation methods Monte Carlo & Deterministic
Toshiya Sanami; Davide Mancusi	Michele Ferenci; Carmen Villagrasa	Vladimir Smutny; Matthieu Longeot	Paul Warner; Daniele Tomatis
T3 : Accelerators & Fusion Facilities	T4_Medical Facilities, Radiotherapy & Medical Applications, Space Dosimetry & Shielding	T5 : Fission Facilities, Fuel Cycle & Waste Management Facilities, Decommissioning	T6 : Calculation methods Monte Carlo & Deterministic
Toshiya Sanami; Davide Mancusi	Rebecca Howell; Laurence Lebaron-Jacobs	José Rodenas Diago; Laurence Pangault	Kenneth Burn; Shengpeng Yu

66 Posters
17 student posters

T1_Nuclear Data, Radiation Detection, Measurements & Dosimetry: 23

T2_Shielding Experiments & Benchmarks: 8

T3_Accelerators & Fusion Facilities: 10

T4_Medical Facilities, Radiotherapy & Medical Applications, Space Dosimetry & Shielding: 3

T5_Fission Facilities, Fuel Cycle & Waste Management Facilities, Decommissioning: 10

T6_Calculation methods Monte Carlo & Deterministic: 11

SS2 _Neutron spallation source session: 1

Some trends

✓ *Methods* and *codes*

Monte Carlo transport evidently more and more used both for applications and for fundamental research:

- analogue simulation for nuclear instrumentation
- completeness of representation of radiation phenomena
- coupling of physical phenomena: transport/depletion, neutron/photon/charged particles...

✓ *Deterministic* still very useful w.r.t.:

- performance
- 3D adjoint information for Monte Carlo (variance reduction)
- well adapted for coupling particle transport and depletion.

✓ *Simplified methods*:

- heuristic approach,
- meta models *e.g.* for optimization,
- Coupling (*e.g.* simplified charged particle slowing down)

✓ Powerfull *pre- and post- treatment tools*

- input data *e.g.* geometry/materials 2D & 3D
- results of calculations, intermediate information (*e.g.* particle tracks)

Some trends

✓ Extension of the needs of *nuclear & atomic data*

- to answer to safety requirements
- for high energy ranges (above 20 MeV) for both application & research fields
- Investigation of very low energy range

✓ Development of *multiscale* & *multiphysics approaches* (physics/biology/chemistry...)

✓ Emergence of *holistic approach* (particularly for safety in extreme situations)

✓ Uncertainties?

ICRS 13
RPSD 2016
ICRS-13 & RPSD-2016 Award
To Dr. Jean-Claude NIMAL
"In recognition of his outstanding scientific contribution
in the field of Computational Radiation Shielding"
Paris, October 3-6, 2016

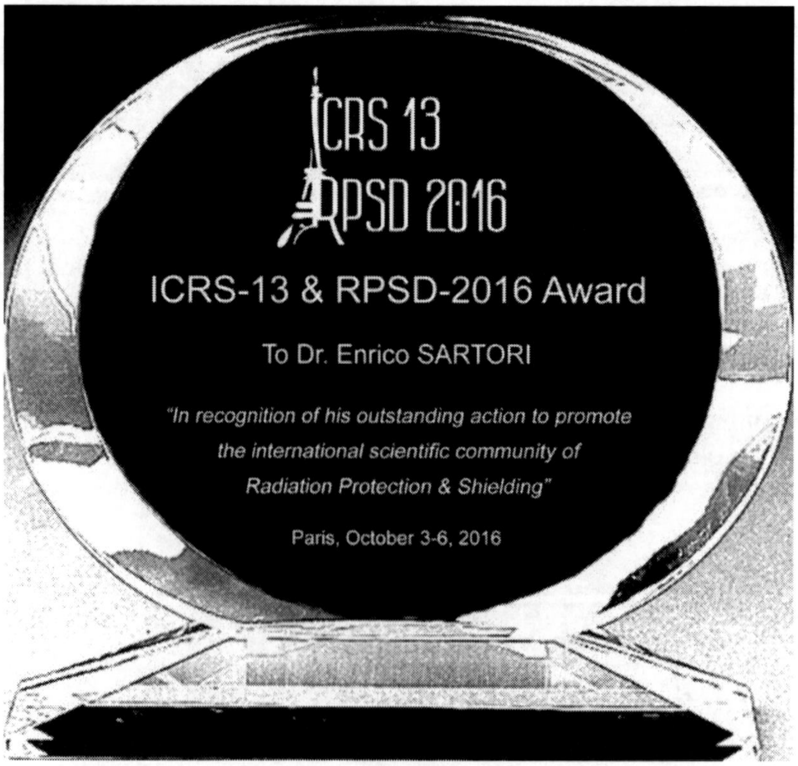
ICRS 13
RPSD 2016
ICRS-13 & RPSD-2016 Award
To Dr. Enrico SARTORI
"In recognition of his outstanding action to promote
the international scientific community of
Radiation Protection & Shielding"
Paris, October 3-6, 2016

Student Awards

ICRS-13 & RPSD-2016

POSTER AWARD

To

Karen Arlet GUZMAN-GARCIA

for the contribution entitled

10B+ZnS(Ag) as an alternative to 3He-based detectors for Radiation Portal Monitors

Dr. Michele FERENCI
General co-Chair

Dr. Thomas MILLER
Technical co-Chair

Dr. Toshiya SANAMI
Technical co-Chair

ICRS-13 & RPSD-2016 AWARDS CEREMONY - OCTOBER 6 - PARIS

ICRS 13 RPSD 2016 3/6 OCTOBER 2016 PARIS FRANCE

ICRS-13 & RPSD-2016

POSTER AWARD

To

Sylvain MEYLAN

for the contribution entitled

Modeling of the human genome with DnaFabric for the calculation of radio-induced DNA damages

Dr. Michele FERENCI
General co-Chair

Dr. Thomas MILLER
Technical co-Chair

Dr. Toshiya SANAMI
Technical co-Chair

ICRS-13 & RPSD-2016 AWARDS CEREMONY - OCTOBER 6 - PARIS

ICRS 13 RPSD 2016 3/6 OCTOBER 2016 PARIS FRANCE

Thanks to all the attendees

Thanks to Patricia, Michèle & Lise from SFEN

ICRS-13 & RPSD-2016

Organized by

with the support of

co-organized by

and the partnership of

ICRS-13 & RPSD-2016

13th *International Conference on Radiation Shielding*

&

19th Topical Meeting of the *Radiation Protection & Shielding Division* of the American Nuclear Society -2016

Closing session
Speech of Dr. Enrico Sartori

retired formerly OECD – NEA
Consultant for the CEA, IRSN, Government of Norway
Conference Honorary Chair

ICRS 13 RPSD 2016

What has a person of my age to add to what was shown here already? All the important advancements and issues in radiation protection and shielding have already been shown in the hundreds of presentations and posters.

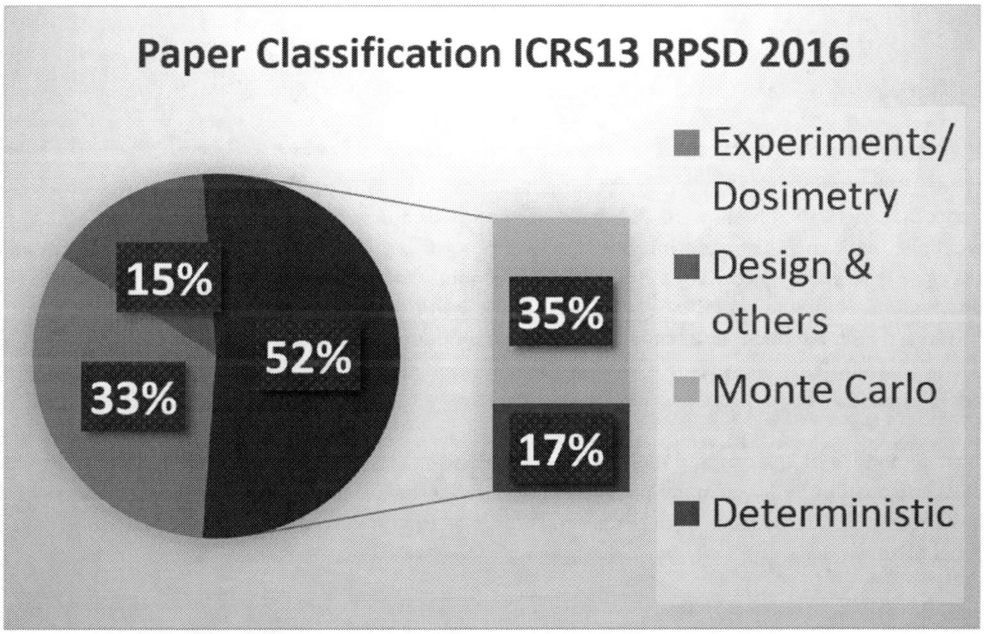

This pie chart shows an approximate distribution of presentations by general topics: numerical simulation, experiments/dosimetry, measurements and benchmarks and other issues such as design. There are of course correlations among the topics which are not shown here. But this gives you a rough idea.

You see that among the simulation methods the stochastic ones represent 2/3 of the total while the deterministic methods the remaining 1/3. Also here there is some overlap because some use hybrid methods, others carry out multi-physics / multiscale simulations where both methods may be present.

In the following picture you see most of the equations used in nuclear applications – all of them are deterministic.

So, we use for mostly stochastic phenomena, deterministic equations which in majority are solved with stochastic methods. What's wrong about that? There is nothing wrong, it works. Also the deterministic ones work.

Equations used in the computer codes for nuclear power and non-power applications

$$\left(\beta mc^2 + \sum_{k=1}^{3} \alpha_k p_k c\right) \psi(\mathbf{x},t) = i\hbar \frac{\partial \psi(\mathbf{x},t)}{\partial t}$$

$$i\hbar \frac{\partial}{\partial t}\Psi(\mathbf{r},t) = -\frac{\hbar^2}{2m}\nabla^2\Psi(\mathbf{r},t) + V(\mathbf{r})\Psi(\mathbf{r},t)$$

$$\left[\frac{\partial^2}{\partial \kappa^2} + \frac{2}{\kappa}\frac{\partial}{\partial \kappa} - \frac{\partial^2}{\partial \chi^2} + \frac{2}{\chi}\frac{\partial}{\partial \chi}\right] V(\kappa, \chi) = 0$$

$$\frac{\partial u}{\partial t} = k\nabla^2 u$$

$$\oint_{\partial S} \mathbf{E} \cdot d\mathbf{l} = -\frac{\partial \Phi_{B,S}}{\partial t} \qquad \nabla \times \mathbf{E} = -\frac{\partial \mathbf{B}}{\partial t}$$

$$\frac{\partial N(\vec{r},\vec{\Omega},E,t)}{\partial t} = -v\vec{\Omega}.\vec{\nabla}N(\vec{r},\vec{\Omega},E,t) - \Sigma_t(\vec{r},E)vN(\vec{r},\vec{\Omega},E,t) + Q(\vec{r},\vec{\Omega},E,t) + \int_{4\pi} d\Omega' \int_0^\infty dE' \Sigma_s(\vec{r},\vec{\Omega}' \to \vec{\Omega}, E' \to E) v' N(\vec{r},\vec{\Omega}',E',t) \ldots$$

$$\frac{dQ_i}{dt} = S_i + k_{Q_{i-1},Q_i} \cdot Q_{i-1} - k_{Q_i} \cdot Q_i$$

$$\rho\left(\frac{\partial \mathbf{v}}{\partial t} + \mathbf{v} \cdot \nabla \mathbf{v}\right) = -\nabla p + \nabla \cdot \mathbb{T} + \mathbf{f},$$

Dirac, Schrödinger, nuclear matter, heat conduction, Maxwell, Boltzmann, Bateman and Navier-Stokes

Poor Ludwig Boltzmann, about 130 years ago, he has shown that the macroscopic behaviour of a large amount of microscopic stochastic phenomena can be described with a deterministic equation which is dissipative and not reversible. With this, he made many enemies among the physicist of those times. They argued that the particles were interacting according to the laws of mechanics which was the base on which the whole universe was functioning. Knowing an initial status would allow you to predict in detail all future events. The only reason it could not be done is because of lacking complete knowledge of initial conditions. Philosophers moved in with arguments whether our will was then free or not as everything had in this case been fixed from the start.

There was a mixture of science and ideology, --- it's a bad marriage that leads mostly to divorce. Boltzmann was attacked so violently by so many physicists and intellectuals that some say he committed suicide for that reason others say it was because of depression. Later, Boltzmann was proven to be right. Nowadays he would be a Nobel prize winner.

Somehow the argument was about:

what is reality and what is a model?

When did the little monkey of about 50 000 generations ago become human? And how?

She started to perceive the signals coming from her environment, then to observe them and their pattern, she found analogies among different events, gathered experience which lead to synthesizing the whole into a concept, into an idea and as she discovered that she could modulate sounds the result is what we call today the word, the synthesis of the whole process.

A Word? Or was it a first model of reality constructed through this process, the first acquired knowledge? These conceived words were linked together and a natural language emerged that permitted communication with the others and she could start to simulate what happens if, and transmit it to others. She started then to represent the different ideas with signs, a new abstraction was born.

This was the moment when the little monkey became human: it all started with the process of observation and simulation. She transmitted her genes and capacities to the descendants who became increasingly stronger, more intelligent, and because of this they did survive.

But words and language were not efficient enough to make predictions, although it had helped to produce some useful survival tools. To improve them, quantities needed to be defined, and so she invented the numbers, and in order to manipulate them she invented the universal language of mathematics and she put aside into a corner the magic formulae of hokuspokus and abracadabra.

That permitted the development of technology and 50 000 generations later here we are after this first experience of modelling we call today the word; an incredible development followed, algorithms, computers and ever more sophisticated modelling and an incredible explosion of technological developments have emerged. It is modelling and the following development of technology that made us human and special compared to the other living on our planet.

Modelling started with perception and observation and then experimentation, by using the filtered signals coming from the real world. --- In a dream this connection does not exist and there the images are not the reflections of the signals from the real world.

Our human detectors, our senses that are perceiving the signals from the real world are limited in scope and capability just so much as required to survive in our environment here on earth. With technology we have constructed new detectors of additional signals, which we cannot perceive directly, but which are unfolded through instruments and computers into new signals that our senses can then perceive and interpret. This allows us to construct new, more complex models of our real world.

If instead we make simulations with models using computing only, it is like dreaming: the image produced may not correspond to the real world.

It is for that reason that I am so pleased with this conference:

1/3 of the papers are about capturing, unfolding, interpreting signals coming from the real world, from experimentation, dosimetry, benchmark studies, look at the enormous amount of measurements from Fukushima. Only when your simulation is validated against the signals from the real world can you trust it.

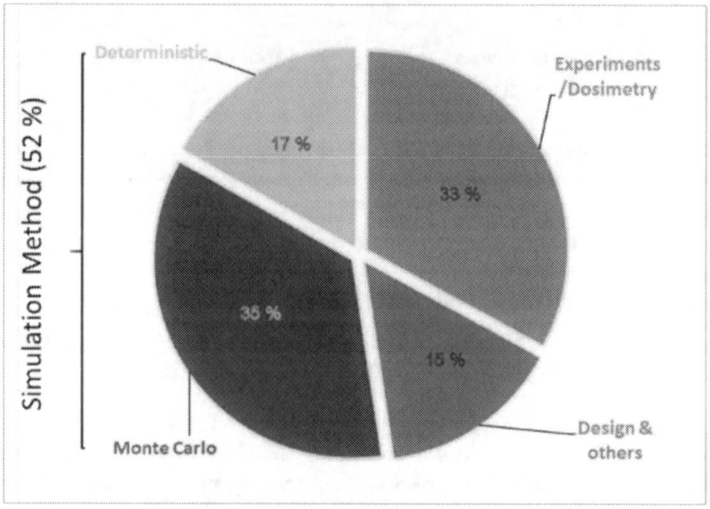

There is a tendency though today, to believe too much in computer modelling results, with the beautiful, colorful graphs and videos that are produced which seem more real than reality. They mostly represent a model, an objective oriented, rough approximation of the world.

A person of my age has only this to add: do not trust your results unless you can prove them right against signals from the real world and do not follow one person I have met saying:

"You know, Monte Carlo results represent the real world better than measurements as they result from the use of the exact and clean laws of physics".

Thank you all for listening to me and I wish you all good luck for the future.

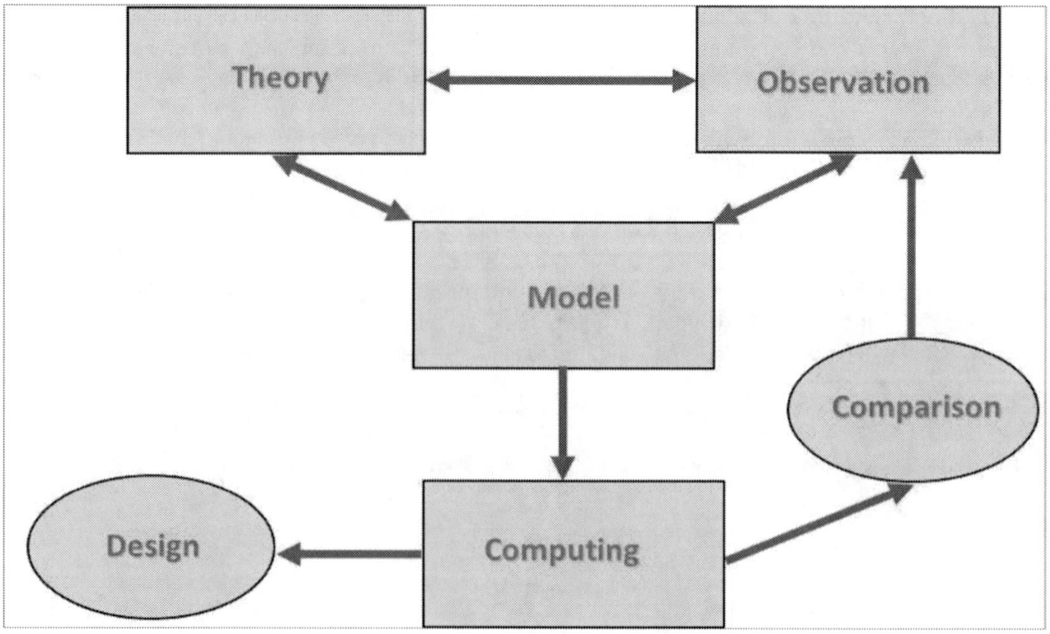

ICRS-13 & RPSD-2016

13th *International Conference on Radiation Shielding*
&
19th Topical Meeting of the *Radiation Protection & Shielding Division*
of the American Nuclear Society -2016

Closing session
Speech of Dr. Jean-Claude Nimal

retired formerly CEA
Consultant for the CEA
Conference Honorary Chair

I will talk about the evolution of radiation shielding studies over the last sixty years mainly from my French point of view. I will also offer some indications for future developments.

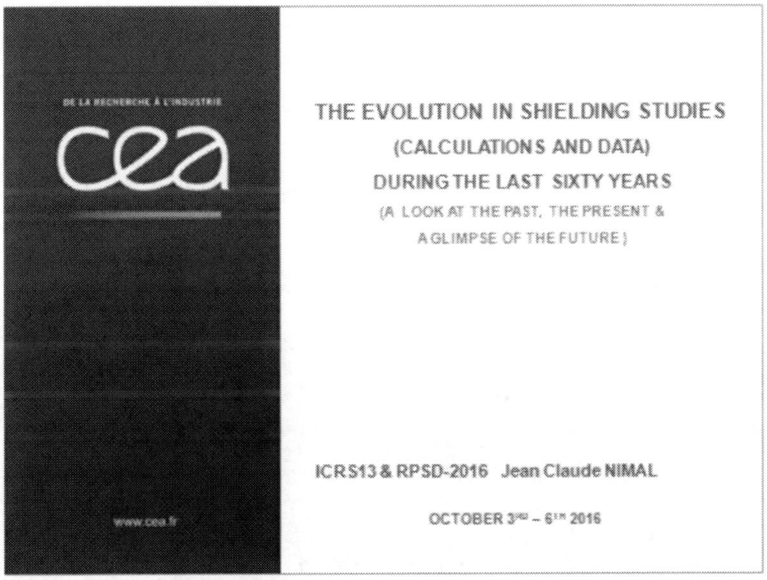

As a preliminary remark, please note that the guiding thread of my talk stems from two observations:

The first observation concerns the three following improvements made during this period:

- The first one is better nuclear data knowledge: accurate values and their uncertainties.

- The second one is the exponential explosion of computer power with the capability of treating huge amounts of data and with increased calculation speed thus allowing precise and complex calculations using realistic physical models.

- The third one is to note the evolution in the field of radiation shielding studies which now involve the high energy range and many types of particles. There are medical and spatial applications, accelerator shielding studies, etc. A number of papers presented during this conference have been devoted to these new fields of application.

The second observation concerns the interdisciplinary approach bringing together different disciplines: computer sciences and data processing, fundamental physics and theoretical mathematics. We can observe that codes are put together in “calculation systems”. Concerning the oral presentations let me cite two examples presented at this conference: TRIPOLI 4 with GEANT4 and FIFRELIN-TRIPOLI 4 coupling. Many presentations in the poster sessions have also shown the associating of several disciplines like the one which received an award today.

It must also be noted that great “international collaboration” exists within the “shielding community” and this is true since the beginning of radiation shielding studies. From the sixties many exchanges occurred with the USA essentially with RSIC (radiation shielding information center) and with several centers (ORNL, Los Alamos national laboratory) or Universities. The OEDC’s Nuclear Energy Agency played and still plays a very important part in the shielding community. Also, many European shielding engineers meet annually in Zurich, as they have for many years, to exchange their points of view concerning software improvements, design studies, etc.

I will now distinguish four sections in my talk going from the bacterial era to the future.

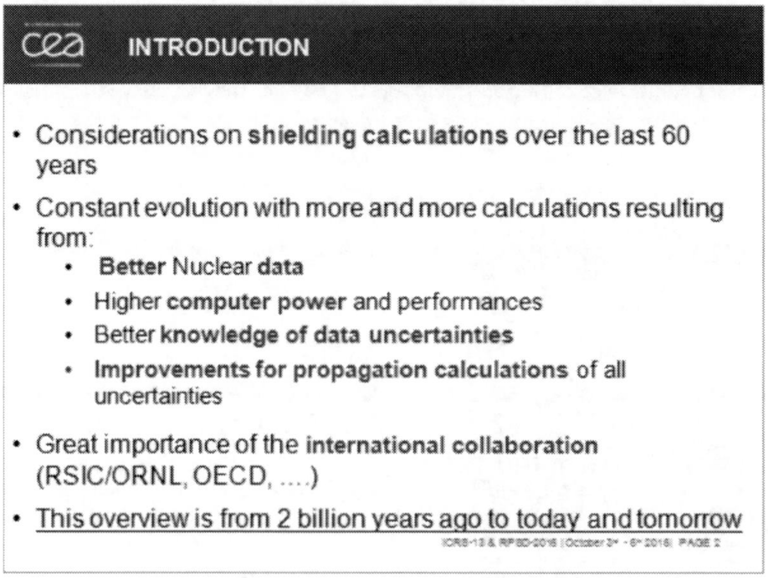

I. The bacterial age.

You were not born or even me when the first reactor we know of diverged about 2 billion years ago in Gabon, Africa. It was located in the uranium mine, called “Oklo mine”. This reactor operated for only several centuries. It is now a fossil reactor. Isotopic analyses were performed in 1972 by the French Atomic Energy Commission. These isotopic measurements involved stable fission products such as Nd, Sm, Eu and essentially U235. It was possible to prove the existence first of the chain fission reactions and secondly the conversion from U238 to Pu239. OKLO was thus the first thermal nuclear reactor using fuel recycling. The features of this reactor were fully presented at the fourth ICRS in Paris in October 1972. This reactor did not need sophisticated calculations or nuclear data like cross sections which we need today. OKLO is now the subject of many discussions and many hazardous assumptions since its discovery by Professor El Albany (Poitiers University, CNRS). In 2010, he and his collaborators extracted close to 250 fossils from the Oklo site. The dating of sediments allows asserting that these fossils are 2,1 billion years old. This means that they date back to 1,5 billion years before the oldest previously known trace of evolved life on Earth. The most interesting fact of the Oklo site is the simultaneous spatial and temporal presence of bacteria, oxygen, radioactivity and high grade uranium metal ore.

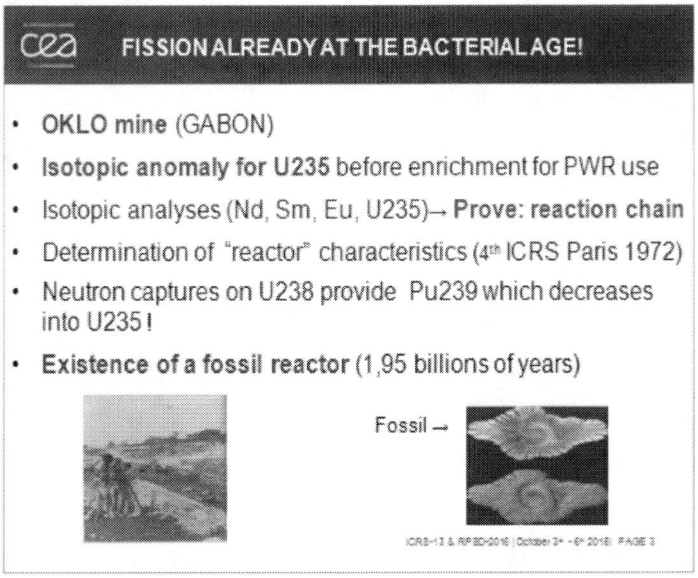

II. "The first try"

We will now take a giant step in time, leaving the prehistoric era to reach the Enrico Fermi reactor located near Chicago, USA at its first occurrence of divergence initiated by physicists in December 1942. It was also at this time that John von Neumann was a precursor in the development and the use of the simulation method for solving complex physical problems. His first applications were devoted to calculations in neutronics with the criticality approach and the explosion height determination. A name was attributed to the corresponding classified document: "Monte Carlo". As a reminder, the essential notion of this method is the game theory. But before going forward, let us take a small step backwards to the naturalist Buffon (1707-1788). This scientist presented a game (1833) for a literary salon which permitted to simply calculate the value of "π". The experiment consisted in dropping needles on a floor made of parallel and equidistant strips to estimate the π value. John von Neumann, along with others, developed theoretical studies in physics and on mathematics. This Monte Carlo method is now applied in other fields like economics, theoretical physics for multi-dimensional integral calculations and other numerical applications. For physics applications, two conditions must exist simultaneously:

- First an exact knowledge of all physical laws for the studied phenomena.
- And second good computing capacity coupled eventually with variance reduction techniques to obtain sufficient statistical accuracy.

The two preceding requirements dictated the orientation of work related to the nuclear field. The execution of radiation transport calculations (shielding and core studies) precise and representative is at the same time a consequence and a motivation of the joint evolution of nuclear data improvements and of the exponential growth of computing powers. If the numerical methods greatly rose during this half-century, the use, a priori not very realistic, of the Monte Carlo methods became common. The Monte Carlo method, based on simulation, has indeed two particular inherent characteristics: the slowness of statistical convergence according to calculation time but also the great consistency in the description of physical phenomena. The increasing interest of the Monte Carlo method is due to these last two features and to the nuclear data accuracy combined with the computing power improvement.

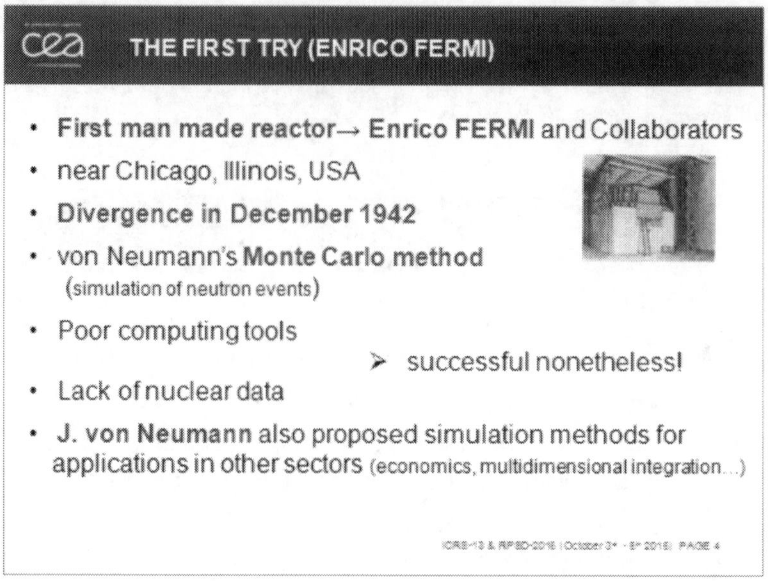

The following slides show the evolution of radiation shielding studies over time.

III. The age of modeling using mock ups for radiation shielding calculations

For this third section, we make a second jump in time to reach the 1950s and 60s. At this time, French shielding engineers employed formulas and abacuses from handbooks like the "Handbook Shielding Design Manual" by Theodor Rockwell III published in the USA (1956) and the "Handbook for the Calculations of Reactor Protections published in France in 1963 (CEA report prepared by C. Devillers).

Some years later, shielding calculations were performed using specific codes written by the CEA. These simplified codes were based on diffusion or removal approximations and on the point attenuation kernel method. The NARMER software and MERCURE 6 code belong to these series of codes using synthetic models. The still used MERCURE 6 code is still also distributed in several countries. For calculations involving particle spectrum use, numerical codes like NIOBE, ANISN and DOT were imported from the United States and used in France since the sixties. This is an important example of the RSIC's international dissemination. At the same time, the diffusion/removal software SABINE from Italy's ISPRA center was used as well.

The simplified codes using removal cross sections or the diffusion theory with few energy groups needed synthetic data. These data were absorption, transfer cross sections, diffusion coefficients. In France, these data were determined by attenuation measurements in bulk shields placed in the NAÏADE facility. It was during this same period that a number of nuclear shielding facilities were developed in other countries (USA, UK, Italy ...). I will only mention here the British facility located in WINFRITH coupled to the NESTOR reactor core.

The first version of the French facility, NAÏADE 1, was coupled to the ZOE reactor core (first occurrence of divergence in December 1948). The NAÏADE 1 and the WINFRITH facilities both have a large experimental area containing various mock ups. This area was located behind a converter plate irradiated by the thermalization of fission neutrons issued from a reactor core. For the British facility, the converter plate was very thin and consisted of enriched Uranium. In the French case, the converter plate was composed of natural Uranium and thus had to be 2 centimeters thick meaning more complicated power plate calculations. Political reasons were at the origin of these constraints.

In the ZOE environment, calorimeters were used to measure the residual heat for spent fissile isotopes like U235 and Pu239, both thermal neutrons, and fast neutrons like U238. These measurements result in heat decay curves after "the elementary fission" on several fissile isotopes. These synthetic data were used by convolution software to calculate the residual core power after shutdown.

A second version of the French facility, NAÏADE 2, was constructed with a subcritical fuel assembly located behind the swimming pool reactor core (TRITON). The neutron emission power of this second facility was greatly increased when compared to NAÏADE 1's fission neutron emission power.
The French facility was used in the sixties and seventies to carry out three types of applications:

- Multilayer bulk shields for synthetic data determination.
- Lacunar mock ups with annular spaces, gaps or pipes, were implanted essentially for shielding studies involving graphite gas cooled reactors used in France at that time.
- Fission product decay analyses using calorimetric methods.

The measurements from WINFRITH and FONTENAY are still used today to create shielding benchmarks. These benchmarks and those provided by other countries have been put together by the OECD/AEN in the SINBAD data base for cross section analysis and code qualification. Papers on these benchmarks have been presented during this conference.

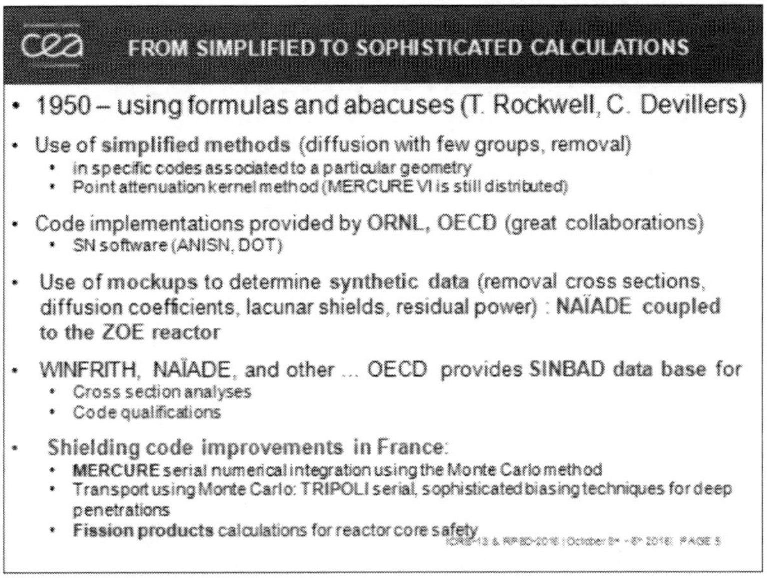

We leave this period with simplified codes to move to the period of sophisticated software improvements and uses.

IV. From simplified to high computer calculations with uncertainty analysis.

To show the evolution of high computer use, I will present two typical aspects of the TRIPOLI software creation. These aspects were the outcome of many deep reflections since 1965!

The first aspect concerns biasing techniques. It was a successful idea developed during the sixties. Particular attention was provided by the decision-makers (P. Lafore and J. Rastoin) concerning the choice of the forward-looking software features: "a 3D polycinetic Monte Carlo neutron transport code for deep penetration calculations". Thus the non-analog game was studied and implemented using the exponential transform on the neutron path simulation according to a given importance distribution. Associated collision biasing completed the successful spatial biasing. Two types of collision biasing were implemented: the first one devoted to the "normal" bulk shields and the second one for the pipes and experimental reactor channels. The primary importance definition was performed using hand-made drawings. To make TRIPOLI easier to use, the first improvement we made was to automatically determine the spatial importance variations. Based on the initial manual graphic method, we developed the INIPOND module using the Dijkstra algorithm. This algorithm allows calculating the spatial importance by minimizing for each spatial point the shortest optical pathway to reach the result area. The Monte Carlo biasing techniques offered the opportunity for many theoretical discussions on "the zero variance game" and its associated "figure of merit". The INIPOND formulation is in accordance with a quote from Seneca:

"The wind is only favorable for he who knows where he is going".

(SENECA, Sénèque in French, was born ~4 years before Christ). This sentence appears on the first page of the TRIPOLI 3 report. It seems to me to be excellent advice not only valuable for neutron propagation!

The aim of the second aspect was to "simulate the neutron thermalization in some moderators, like light or heavy water, directly from the frequency spectra of the molecules without using the well-known S (α, β) functions." It was overambitious at the time and was a complete failure. But could it be possible in the future with the improvement of computer capabilities? It is nonetheless possible to directly simulate thermalization with the free gas model.

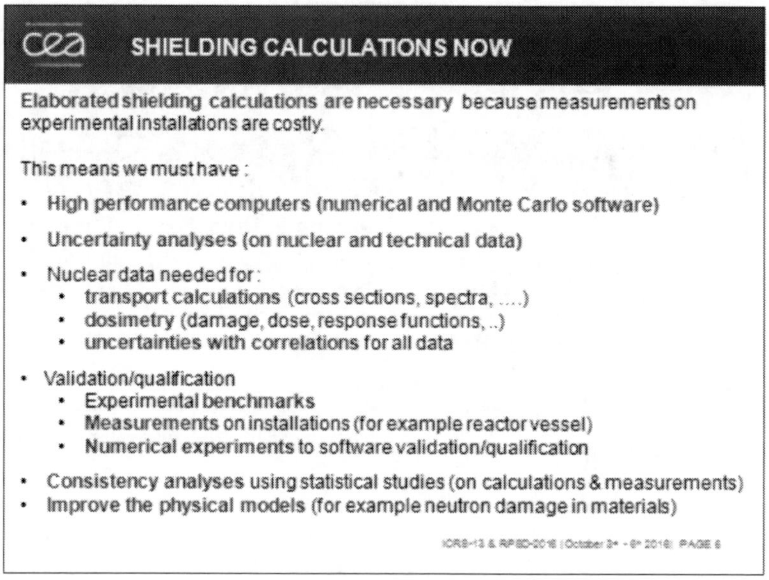

The following slide 7, presents the four stages we have seen and I will briefly make two remarks:

<u>Remark 1:</u> in 1968, the number of simulated particles in some Monte Carlo calculations was a few thousand for a benchmark calculation in the sodium tank placed in the Harmonie reactor. In the 1980s, 2 calculations using 30,000 neutrons were necessary to predict the control rod efficiency. The difference between these two calculations allows to appreciate the control rod efficiency to avoid criticality experiments. Such sophisticated calculations using a simulation technique can reduced the utilization of the experimental way if and only if first the nuclear data are accurate and reliable and secondly if there are sufficiently accurate uncertainty calculations. Now with TRIPOLI 4 we simulate 1 to 3 billion particles to calculate the damage rate for the pressure vessel in a PWR!

<u>Remark 2</u>: in 1968 the memory size of 300 kilo bytes available for the sodium activation calculation (PHENIX reactor) obliged us to represent the double differential cross sections with "half words" (2 octets). The memory size can now reach eight giga bytes or more and some hundreds of processors can easily be used with Monte Carlo software. Monte Carlo methods are naturally well adapted to a parallel treatment. The memory capacities now allow the fine treatment of a large amount of physical data.

The uncertainty calculations on calculated values are now reliable. Initially for the uncertainty calculations they were limited to linear evaluations using sensitivities and partial derivatives. Several software were then improved like SWANLAKE, DUCKPOND and SUSD3D. The nonlinear approaches have been improved with the "correlated sampling method" in the Monte Carlo software. For these approaches, geometrical perturbations can't be treated. More general methods now exist: codes using probability tables or determination from simulations using a lot of random values of data. The aim of these methods is to calculate the probability density of all results from which we can deduce the interval of confidence. These general methods allow the treatment any perturbations without approximation but require important calculation power.

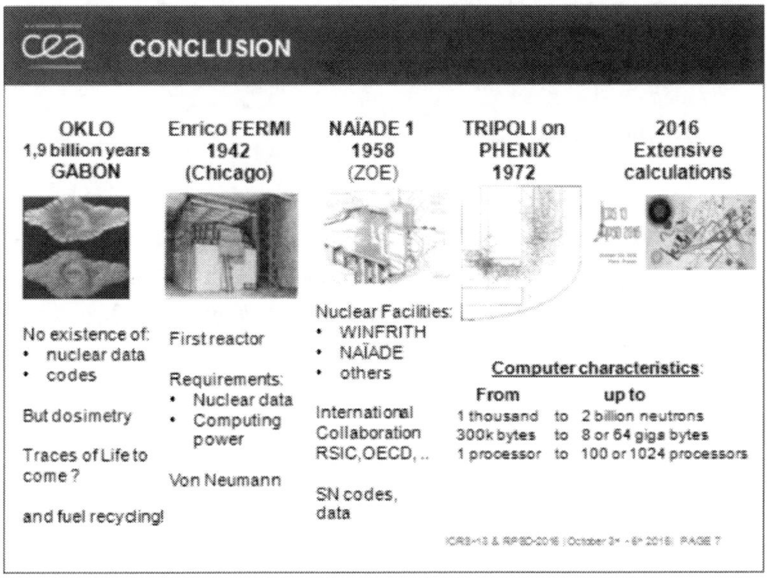

What could be for the future?

For me, the following two remarks are important:

- Improvements in new application fields. At the beginning, the Monte Carlo method was only applied to neutron transport calculations like criticality conditions or sky shine effects. Later it was extended to reliability studies in many fields such as nuclear reactor safety. It could be useful in cell biology, medicine, fundamental research in physics, physical-chemistry and so on. These applications will require more and more calculation power.

- Some applications involve very rare events such as particle transmissions through deep shields or accidents in a nuclear plant (reliability). To reduce computing time used for simulations we will need more powerful biasing techniques. The biasing techniques were already used for deep penetration of neutrons. Always for rare event estimations, despite the high performances of the computer, it will be essential to develop new techniques such as Adaptive Multilevel Splitting, or to improve biasing parameter values using simulated annealing or artificial intelligence. It would be useful to take into account new knowledge as it is acquired along the way during the simulation.

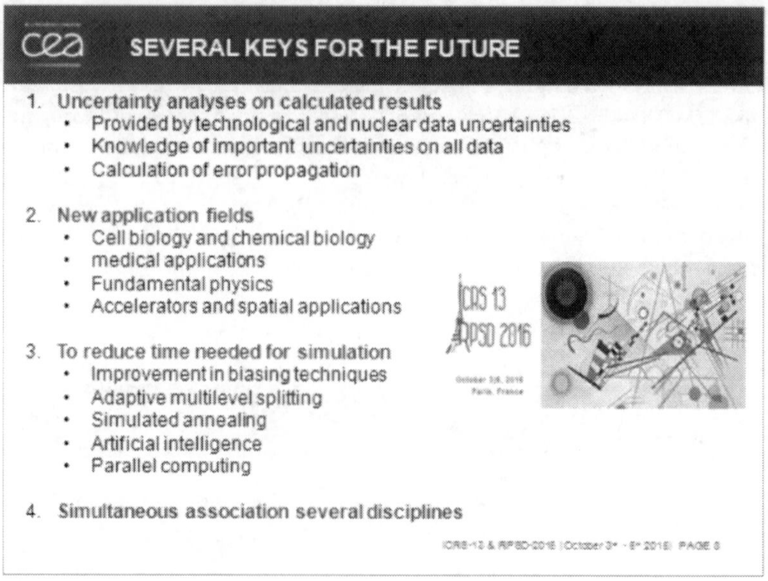

I wish to thank the organizers for the excellent quality of this 5-day conference and also to extend my thanks to Elizabeth COUVERT (*Cedc Consultancy*, devlin.couvert@gmail.com) for our various discussions on these subjects and for helping me to prepare this presentation in English.

ICRS-13/RPSD 2016

Closing

ICRS Meetings

1	1958	Cambridge,	UK
2	1961	Studsvik,	Sweden
3	1967	Oxford,	UK
4	1972	Paris,	France
5	1977	Knoxville,TN,	USA
6	1983	Tokyo,	Japan
7	1988	Bournemouth,	UK
8	1994	Arlington, TX,	USA
9	1999	Tsukuba,	Japan
10	2004	Funchal, Madiera,	Portugal
11	2008	Pine Mountain, GA	USA
12	2012	Nara,	Japan
13	2016	Paris,	France

ICRS-14

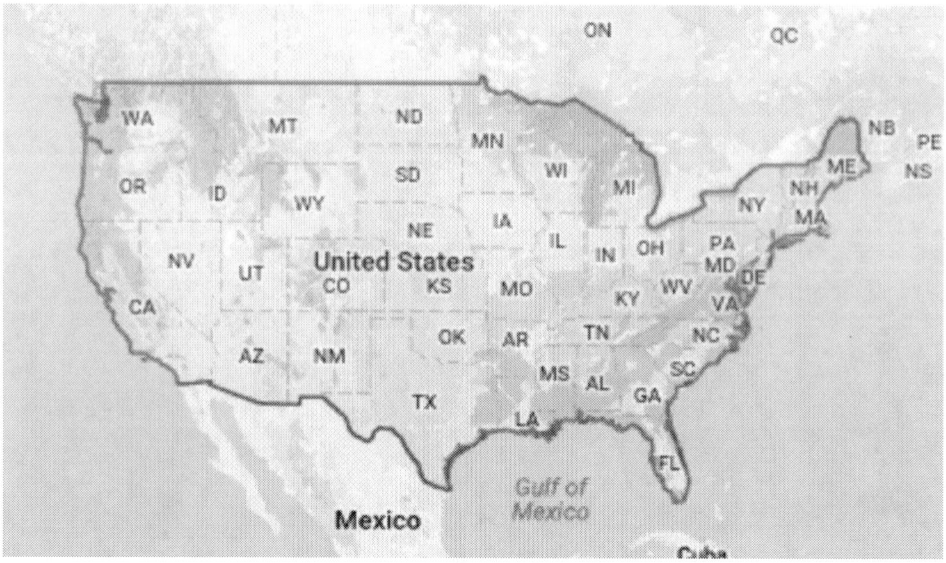

THANK YOU!

See you in 2020!

edp sciences

EPJ Web of Conferences

Statement of Peer review

In submitting conference proceedings to EPJ Web of Conferences, the editors of the proceedings certify to the Publisher that they adhere to the **Policy on Publishing Integrity** of the journal in order to safeguard good scientific practice in publishing.

1. All articles have been subjected to peer review administered by the proceedings editors.
2. Reviews have been conducted by expert referees, who have been requested to provide unbiased and constructive comments aimed, whenever possible, at improving the work.
3. Proceedings editors have taken all reasonable steps to ensure the quality of the materials they publish and their decision to accept or reject a paper for publication has been based only on the merits of the work and the relevance to the journal.

Title, date and place of the conference

ICRS-13 & RPSD-2016, 13th *International Conference on Radiation Shielding* & 19th Topical Meeting of the *Radiation Protection* and *Shielding Division* of the *American Nuclear Society* -2016
Paris, 3-6 October, 2016
Novotel Paris Centre Eiffel Tower Hotel

Proceedings editor(s):

Fausto MALVAGI, Fadhel MALOUCH, Cheikh M'Backé DIOP

Date and editor's signature

11/08/2017 1/08/2017

17, avenue du Hoggar -PA de Courtabœuf – BP 112 -91944 Les Ulis Cedex A (France)
Tél. : 33(0)169 1875 75 -Fax : 33(0)1 69 07 45 17 – www.edpsciences.org

Uranium, radium and thorium in soils with high-resolution gamma spectroscopy, MCNP-generated efficiencies, and VRF non-linear full-spectrum nuclide shape fitting

Robert Metzger[1,*], *Kenneth* Van Riper[2], and *George* Lasche[3]

[1]Radiation Safety Engineering, Inc, 3245 North Washington Street, Chandler, AZ, USA
[2]White Rock Science, PO Box 4729, Los Alamos, NM, USA
[3]Snakedance Scientific, LLC, 13509 Quaking Aspen Place NE, Albuquerque, NM, USA

Abstract. A new method for analysis of uranium and radium in soils by gamma spectroscopy has been developed using VRF ("Visual RobFit") which, unlike traditional peak-search techniques, fits full-spectrum nuclide shapes with non-linear least-squares minimization of the chi-squared statistic. Gamma efficiency curves were developed for a 500 mL Marinelli beaker geometry as a function of soil density using MCNP. Collected spectra were then analyzed using the MCNP-generated efficiency curves and VRF to deconvolute the 90 keV peak complex of uranium and obtain ^{238}U and ^{235}U activities. ^{226}Ra activity was determined either from the radon daughters if the equilibrium status is known, or directly from the deconvoluted 186 keV line. ^{228}Ra values were determined from the ^{228}Ac daughter activity. The method was validated by analysis of radium, thorium and uranium soil standards and by inter-comparison with other methods for radium in soils. The method allows for a rapid determination of whether a sample has been impacted by a man-made activity by comparison of the uranium and radium concentrations to those that would be expected from a natural equilibrium state.

1 Introduction

Rapid determination of uranium and radium concentrations in bulk soil samples is important for waste characterization for naturally-occurring radioactive materials (NORM) rules, determination if soils on a site are impacted by man-made activities, safeguards activities, and accident response. Gamma spectroscopy with high-purity germanium detectors is an attractive approach, however many of the lines needed to determine the uranium and radium concentrations are convoluted and commercially available software packages cannot deconvolute them. Further, there is significant self absorption of photons in the soil sample matrix that affect the efficiency curves needed to quantify results. The changes in photon efficiency vary with the soil density and the sample counting geometry chosen.

To address these deficiencies, detector efficiencies were calculated using MCNP [1] modeling and deconvolution of interfering peaks was accomplished with non-linear full-spectrum nuclide shape fitting using VRF, which is short for "Visual RobFit", and is described in more detail in [2]. These methods were then applied to radium, thorium and uranium soil standards and results were compared to those of commonly-used other methods.

2 Method

2.1 Efficiency curves

Efficiency curves as a function of soil density for the detector and sample geometry chosen were determined using an MCNP model of the detector and the 500 mL Marinelli beaker counting geometry used in this work. The model [3] consisted of the crystal, dead layer, detector cover, and the Marinelli beaker. The thickness of the dead layer was estimated by varying the thickness of the layer until the MCNP-generated efficiency values matched the empirically-determined values using a NIST-traceable multiline standard. Once a good agreement was obtained, the matrix in the beaker in the model was changed to a standard soil and MCNP was run to obtain efficiency curves

*Corresponding author: rlmetzger@radsafe.com

 DOI: 10.1051/epjconf/201715301001

for soil densities from 1.1 to 2.0 g/cm^3. The efficiency curves are shown in Figure 1. The values for full-energy peak efficiency that were calculated with MCNP at 60, 88, 122.1, 169.5, 279.2, 391.7, 661.7, 898, 1173.2, 1332.5, and 1836.1 keV were then fitted with a 5-parameter efficiency function using VRF. An example of the efficiency fit to the MCNP values for soil density of 1.3 g/cm^3 that was used for the thoriated soil analysis in this work is shown in Figure 2.

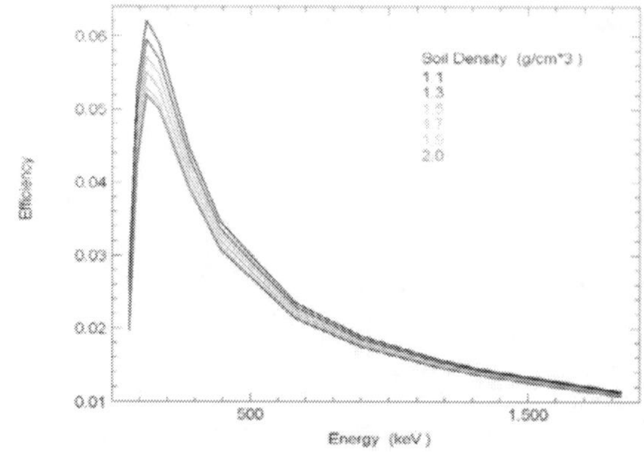

Figure 1. Efficiency curves generated with MCNP for soil in a 500 mL Marinelli beaker.

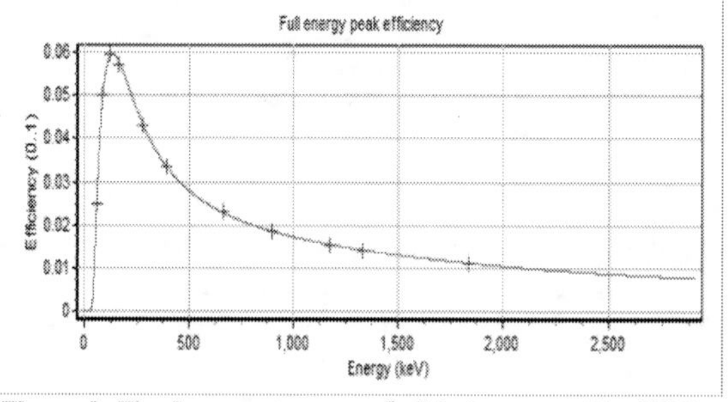

Figure 2. The 5-parameter curve fit (blue) generated by VRF to the efficiency points (red) that were generated with MCNP for a soil density of 1.3 g/cm^3. This efficiency curve was used in the analysis of standard thoriated soil.

2.2 Spectral analysis

To deconvolute the overlapped spectra generated by natural sources the VRF gamma spectroscopy software was chosen. This analysis program differs from other commercial software in that it forms, at each of many automated iterations, a spectrum-wide shape for each nuclide and, also at each iteration, it adjusts the activities of each nuclide, as well as user-enabled parameters of energy calibration, attenuation, peak width as a function of energy, coincidence summing, and efficiency until no better fit to the data can be obtained. Other commercial softwares in wide use simply attempt to find each peak in the spectrum and then use tables to identify nuclide activities, with the result that minor peaks that are masked by larger, overlapping peaks cannot be identified. For this reason, VRF is capable of direct measurement of ^{226}Ra (whose peak at 186.2 keV appears to be merged with the peak of ^{235}U at 185.7 keV), while simple peak-search methods fail. Once the spectrum has been fit, then the spectrum is reassembled and the relative contributions of different lines in overlapped peaks can be seen. Figure 3 shows a segment of the deconvoluted spectrum of a sample of water that was contaminated with depleted uranium. Finally, quantification of the activities in the sample is done using the MCNP-generated efficiency curve that best matches the density of the sample under test.

The resulting method can measure ^{238}U, ^{235}U, ^{226}Ra and ^{228}Ra in soil samples and can further distinguish man-made contributions of uranium and radium in the soil or water sample to natural concentrations by comparing the measured values to expected equilibrium values based on other isotopes in the chain. The method can evaluate ^{226}Ra by directly measuring the deconvoluted 186.2 keV line, or by the activities of the short-lived radon daughters, if the equilibrium status of the sample is known. Uranium enrichment may also be estimated if the counting statistics for the ^{235}U lines are sufficient. Similarly, ^{232}Th and ^{228}Th concentrations may also be determined when they are present.

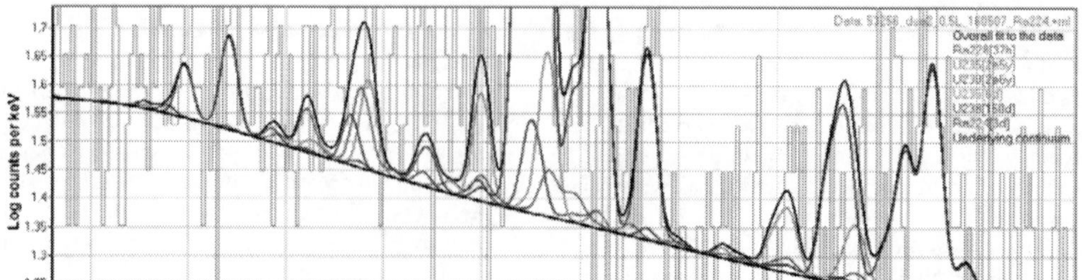

Figure 3. The 90 keV region of a spectrum of water contaminated with depleted uranium and analyzed with VRF. The spectrum has been decomposed into a best fit of full-spectrum shapes of six equilibrium decay chains: geologically aged ^{238}U (green) and ^{235}U (pink), recently processed ^{238}U (purple) and ^{235}U (orange), and fresh ^{224}Ra (red) and ^{228}Ra (blue). The sum of all fits is shown in black. The decay-chain shapes include decay x-rays of the constituent nuclides in addition to their primary, escape, and coincidence sum peaks.

3 Validation

The method was validated by counting soil standards with known activities of uranium, thorium and radium in them. For the radium isotopes, further validation was made by inter-comparison with other methods for a set of routine samples with low concentrations of radium in them.

3.1 Uranium

A uranium standard solution was used to create water and synthetic soil sample standards. The standards were counted on the characterized spectrometer and evaluated as described above. Results are shown in Table 1 below. Results for the ^{238}U were within counting error. The ^{235}U results needed to determine enrichment were within 25% of the actual values. It should be noted that extended counting times are necessary to obtain an accurate measure of the uranium enrichment at natural and depleted abundances due to the low concentrations of the ^{235}U in environmental samples. At natural abundance, the ^{235}U is 0.71% of the ^{238}U by mass.

3.2 Radium

A radium standard in synthetic soil at 1.3 g/cm^3 was made using a NIST-traceable radium standard, and the standard was counted on the characterized spectrometer. The concentration of ^{226}Ra and ^{228}Ra was determined with the new method and the results are shown in Table 2 below. Results show that the new method was within counting error for the two isotopes. The ^{226}Ra concentration was determined directly from the deconvoluted 186 keV line, while the ^{228}Ra was estimated from the ^{228}Ac 911 keV line after the ^{228}Ac had been allowed to grow in for 48 hours.

Finally the new method was compared to two existing methods for determining radium in soils for a series of normal soil samples with relatively low natural radium concentrations. The acquired spectra were evaluated using the new method, the Rosson [4] method which uses a direct calibration and two lines each to estimate the ^{226}Ra and ^{228}Ra values, and the Canberra Genie [5] software. Since the equilibrium status of the samples were known (full in-growth), the lines from the ^{214}Pb and ^{214}Bi were used to determine the ^{226}Ra concentration directly from the Genie analysis software. The MCNP-generated efficiency curves accounted for self-absorption in the different soil densities and were used for all of the methods. Results of this inter-comparison are shown in Figures 4 and 5. Results show relatively good agreement across the different methods.

3.3 Thorium

A thorium standard in synthetic soil at 1.3 g/cm^3 was made using calibrated thorium ore, and the standard was counted on the characterized spectrometer. The concentration of ^{232}Th and ^{228}Th was determined with the new method and the results are shown in Table 3 and Figure 6 below. Results show that the new method was within counting error for the two primary thorium chain isotopes.

4 An application

The new method was used in conjunction with traditional alpha spectroscopy to evaluate a water sample suspected of being impacted by depleted uranium (DU). The water sample was evaluated using SM 7500U-C [6] and the alpha spectrometer spectrum is shown in Figure 7. The spectrum shows a strong disequilibrium between the ^{238}U peak and

Table 1. Uranium in water and uranium in synthetic soil standards by gamma spectroscopy. Only results from VRF are shown because peak-search analyses by Genie and Rosson were unable to deconvolve or resolve critical uranium peaks at this low level of activity.

	Actual (Bq/g)	VRF (Bq/g)	% Error
^{238}U in H_2O	0.0069	0.0061	9.7
^{235}U in H_2O	0.00031	0.00044	26
^{238}U in Soil	1.07	1.08	0.59
^{235}U in Soil	0.054	0.050	8.25

Table 2. Radium by gamma spectroscopy. Radium standard in synthetic soil at 1.3 g/cm^3.

	Actual (Bq/g)	Genie (Bq/g)	% Error	Rosson (Bq/g)	% Error	VRF (Bq/g)	% Error
^{226}Ra	0.019	0.020	4.6	0.021	10.5	0.019	0.03
^{228}Ra	0.014	0.013	2.7	0.013	8.1	0.0145	3.45

the ^{234}U peak and a much lower enrichment (0.2%) than would be expected in a natural source (0.71%). While 234/238 disequilibrium is not uncommon in natural sources, it is normally the slightly more soluble ^{234}U that has the higher activity. The strong disequilibrium in favor of the ^{238}U is characteristic of depleted uranium, a waste product from the enrichment process that has excellent shielding properties, and is also used as an anti-tank projectile.

The sample was also evaluated using the new method and the deconvoluted spectrum of the crucial 90 keV peak complex of uranium was shown earlier in Figure 3. The enrichment estimate by the gamma spectrometer method is shown in Table 4, and also shows that the uranium in the sample is primarily depleted uranium. The other gamma spectroscopy methods were unable to determine uranium activity or enrichment as they cannot deconvolute the heavily overlapped spectrum.

This method was also able to estimate the extent to which the sample was impacted by the DU. Natural uranium, of course, was also present in the sample. The other natural chain isotopes in the sample were used to estimate expected values of natural uranium that should have been present, and the DU to natural ratio was found to be 49:1, clearly indicating the sample was primarily DU.

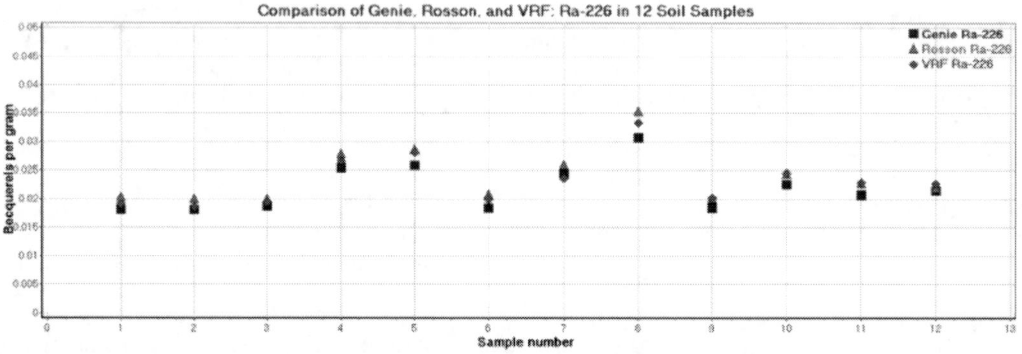

Figure 4. Comparison of analysis results from Genie, the method of Rosson, and VRF of specific activities of ^{226}Ra in 12 soil samples with densities of 1.3 g/cm^3 in 500 mL Marinelli beakers. The efficiency curve was modeled with MCNP.

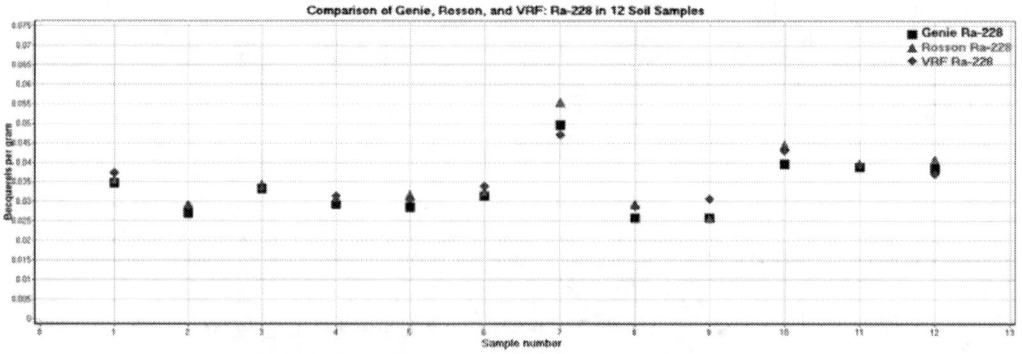

Figure 5. Comparison of analysis results from Genie, the method of Rosson, and VRF of specific activities of ^{228}Ra in 12 soil samples with densities of 1.3 g/cm^3 in 500 mL Marinelli beakers. The efficiency curve was modeled with MCNP.

Table 3. Thorium in synthetic soil standards.

	Actual (Bq/g)	VRF (Bq/g)	% Error
^{232}Th	1.58	1.64	3.42
^{228}Th	1.58	1.64	3.42

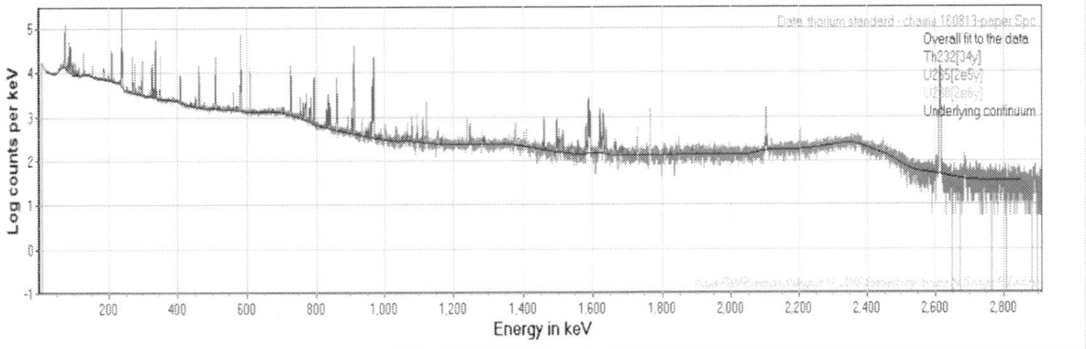

Figure 6. Graphical results of analysis with VRF of thoriated standard soil. In this case VRF was used to fit entire equilibrium decay chains as one emitting source. The equilibrium decay chain of ^{232}Th is shown in red and is evidenced primarily by peaks from ^{228}Ac, ^{212}Pb, and ^{208}Tl. The equilibrium decay chain of ^{238}U is shown in green and is evidenced primarily by peaks from ^{214}Bi, ^{214}Pb, ^{226}Ra, and ^{234m}Pa.

Table 4. Analysis of impacted water sample in good agreement assessing the uranium as DU. Uranium enrichment by alpha spec and VRF (natural abundance of uranium is 0.71%).

Alpha spec (%)	VRF (%)
0.2 +/- 0.001	0.3 +/- 0.084

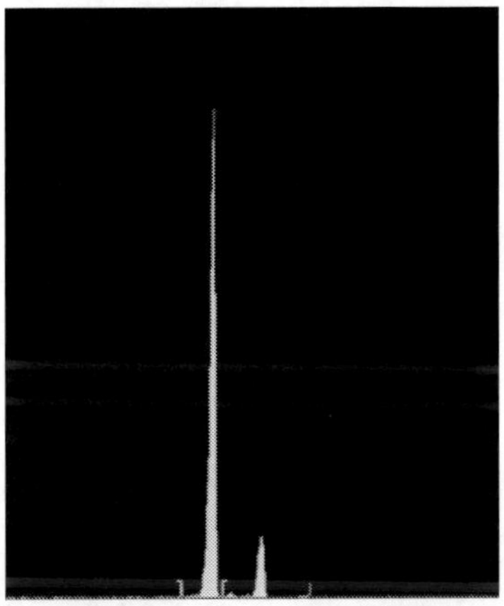

Figure 7. Alpha spectrum of impacted uranium in water sample, showing a strong disequilibrium between the ^{238}U peak and the ^{234}U peak and a much lower enrichment (0.2%) than would be expected in a natural source (0.71%).

5 Conclusion

A new method has been developed that is especially useful for determining uranium and radium concentrations in soil or water samples by gamma spectroscopy using conventional P- or N-type HPGe spectrometers. The method is less expensive and requires less time than traditional methods that solubilize the soil sample through acid leaching or various fusion methods, and which then perform radiochemical separations and alpha spectroscopy. The new spectroscopic analysis techniques of VRF can directly measure ^{226}Ra concentrations without the usual in-growth time for its daughter nuclides. The method can determine uranium enrichment and whether soils are impacted by some man-made activity or are natural.

References

1. T. Goorley, M. James, T. Booth, F. Brown, J. Bull, L. J. Cox, J. Durkee, J. Elson, M. Fensin, R. A. Forster, J. Hendricks, H. G. Hughes, R. Johns, B. Kiedrowski, R. Martz, S. Mashnik, G. McKinney, D. Pelowitz, R. Prael, J. Sweezy, L. Waters, T. Wilcox, T. Zukaitis, "Initial MCNP6 Release Overview", *Nuclear Technology*, **180**, pp.298–315,(2012)
2. G.P. Lasche, R.L. Coldwell, R.L. Metzger, "VRF – nuclear spectral analysis with non-linear whole-spectrum nuclide shape fitting", to be published in the proceedings of the ICRS-13 & RPSD-2016, Paris, France, Oct 3-6, 2016.
3. R. Metzger, K. Van Riper, "A Monte Carlo Approach to Food Density Corrections in Gamma Spectroscopy", Journal of Radioanalytical and Nuclear Chemistry, DOI: 10.1007 (2015)
4. B. Kahn, R. Rosson, J. Cantrell, "Analysis of Ra-228 and Ra-226 In Public Water Supplies by a Gamma Ray Spectrometer," Health Physics, **59**, 125-131, (July 1990)
5. *Genie 2000 Operations Manual* (3.2 ed.), Canberra Industries. (2009)
6. L. Clesceri, A. Greenberg,, A. Eaton, *Standard Methods for the Examination of Water and Wastewater* (20th ed., pp. 7-47 - 7-51). Washington: APHA American Public Health Association (1998)

VRF ("Visual RobFit") — nuclear spectral analysis with non-linear full-spectrum nuclide shape fitting

George Lasche[1,*], *Robert* Coldwell[1], and *Robert* Metzger[2]

[1]Snakedance Scientific, LLC, 13509 Quaking Aspen Place NE, Albuquerque, NM 87111, USA
[2]Radiation Safety Engineering, Inc, 3245 North Washington Street, Chandler, AZ 85225, USA

Abstract. A new application (known as "VRF", or "Visual RobFit") for analysis of high-resolution gamma-ray spectra has been developed using non-linear fitting techniques to fit full-spectrum nuclide shapes. In contrast to conventional methods based on the results of an initial peak-search, the VRF analysis method forms, at each of many automated iterations, a spectrum-wide shape for each nuclide and, also at each iteration, it adjusts the activities of each nuclide, as well as user-enabled parameters of energy calibration, attenuation by up to three intervening or self-absorbing materials, peak width as a function of energy, full-energy peak efficiency, and coincidence summing until no better fit to the data can be obtained. This approach, which employs a new and significantly advanced underlying fitting engine especially adapted to nuclear spectra, allows identification of minor peaks that are masked by larger, overlapping peaks that would not otherwise be possible. The application and method are briefly described and two examples are presented.

1 Introduction

Applications for nuclear spectral analysis that are now in wide use for commercial applications typically begin with a peak search (such as the Mariscotti [1] method of second derivatives), and then match the energies and areas of the peaks that were found to nuclear tables to determine the activities of nuclides evident in the spectrum, taking into account calibrations for energy, resolution (or "peak-width"), and detector efficiency. This approach has difficulty resolving peaks that lie closely together unless they are sufficiently well separated in energy, have comparable areas, and have sufficient counting statistics.

These limitations can be overcome with a different approach using minimization of the chi-squared statistic, as with the method of Marquardt [2] and detailed in Bevington [3], of full-spectrum shapes for all the nuclides under consideration. This requires, however, that the second derivative matrix can be successfully inverted, which is often not possible with standard techniques such as described by Press et al. [4] due to the limitations of the numerical precision, computer memory, and processor speed. An extension of the method of Marquardt, known as RobFit [5], has been developed and is capable of inverting large, sparse matrices successfully and rapidly enough to meet the requirements of sophisticated nuclear spectral analysis. An early attempt at applying less-developed versions of RobFit was described by Lasche [6], but the graphical implementation lacked many essential capabilities for practical analysis, the library was severely limited, and the interface that applied nuclear spectral data to RobFit had only limited capabilities for calibration, no gamma emission line search capability, crude efficiency fitting methods, inefficient continuum fitting methods, and no means of identifying or resolving coincidence summing peaks. All of these limitations have been overcome with an entirely new implementation of a greatly improved interface to RobFit called "Visual RobFit", or "VRF" for short.

2 Description

In the recent four years the authors have created an entirely new procedural interface to analyze nuclear spectral data with an improved RobFit engine and a completely new and comprehensive graphical user interface. During analysis, full-spectrum shapes of user-selected emitters are generated by referencing a library consisting of every peak of every nuclide listed in the ENSDF [7], all of the elemental thermal neutron capture peaks and elemental x- rays listed in the TORI [8], and the mass absorption data given by Berger et al. [9] for all of the elements up to atomic number 98, as well as some common charged-particle reactions. Users are able to add to or edit any of the library data. Equilibrium decay

* Corresponding author: George.Lasche@alum.mit.edu

chains can be selected to unify all of the emissions of the nuclides in equilibrium -- this feature is useful not only to apply more known physics for better results, but also to expose impacted nuclides not in equilibrium. Any combination of these emitters can be grouped as subject to up to three differing attenuations by intervening matter.

Each analysis step consists of a series of iterations seeking a best fit while varying the activities of each emitter and the user-enabled parameters of energy calibration, resolution calibration, detector full-energy peak efficiency, and attenuation by intervening compounds or mixtures of material. Also, a single parameter can be allowed to vary for the best fit to all random (or "accidental") coincidence peaks in the spectrum, and another single parameter for all of the cascade (or "true") coincidence peaks in the spectrum. When no further improvement can be made, the analysis step terminates. The values of all activities and other fitted parameters are then reported together with their uncertainties (or "errors") which include the effects of covariance with all other variables in the fit.

3 Validation

In a companion paper [10], automated analysis results for ^{226}Ra and ^{228}Ra from 12 soil samples, and from a sample of synthetic soil spiked with radium from a certified standard, are compared to results from well-known commercial software with excellent agreement. In cases of low-level uranium concentrations, the commercial software was unable to give useful results due to peak interferences and low counting statistics, so results of analysis from VRF were compared to concentrations from certified standards or results from alpha spectrometry, again with excellent agreement. These include impacted uranium in a water sample and certified uranium standard in water at only 0.014 Bq/g. Further validation was conducted with samples of standard thoriated soil, also with excellent agreement with the known values of the NIST-traceable source.

4 Example analysis #1 – uranium metal

A good example well-suited to illustrate the capabilities of VRF is the analysis of a spectrum from a small flat piece of uranium. The spectrum was collected without the benefit of a low-background chamber and the analysis was carried out without the benefit of a local background spectrum. The task was to determine if the sample was natural, depleted, or enriched uranium. With VRF the usual subtraction of a background spectrum for *in situ* measurements is neither needed nor desired — instead, the natural background radiation environment that is mixed with the radiation from the sample gives valuable information about the energy and resolution calibrations and the detector efficiency, and is easily deconvoluted from the peaks caused by radiation from the sample of interest. Furthermore, the use of a single spectrum (rather than subtracting a background spectrum as is common practice) avoids the errors associated with background subtraction and preserves Poisson statistics.

Two groups of emitters were simultaneously fitted during the analysis. The first group, for the natural radiation environment, consisted of ^{40}K, the annihilation reaction peak at 511 keV, and fully-aged equilibrium decay chains of ^{232}Th, ^{238}U, and ^{235}U; the second, for the sample, consisted of uranium x-rays and equilibrium decay chains of ^{238}U and ^{235}U aged more than 5 months (to allow significant growth of ^{234}Th) but less than 5500 years (before significant growth of ^{231}Pa). The first group was designated to be self-absorbed by a typical optical depth of 10 cm of typical soil; the second, self-absorbed by uranium metal an estimated 0.5 cm thick. Four parameters of a cubic polynomial for energy calibration were allowed to vary, and four for resolution also. The thickness of the uranium metal was also allowed to vary during the best-fit minimization process. The fits to all of the peaks of each of the emitters in the region of 778-812 keV is shown in Figure 1, and in the region of 186 keV in Figure 2. In these figures the advantage of whole-spectrum nuclide fitting for resolving overlapping peaks becomes clear.

Because the detector efficiency was not well known, a default efficiency typical of HPGe detectors was used and the activities were reported only as relative to each

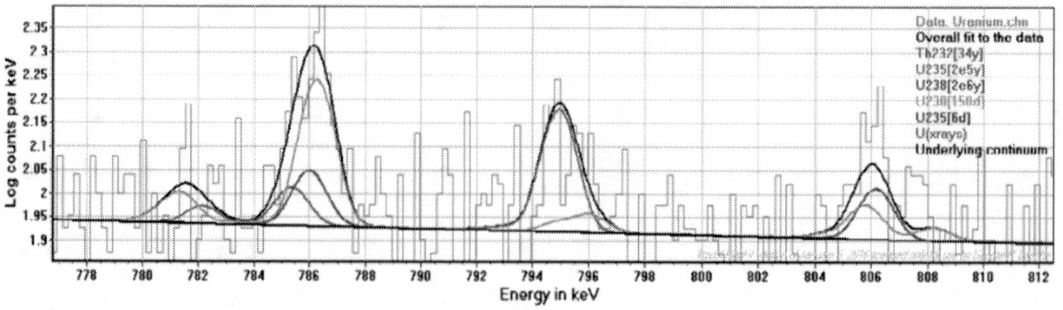

Figure 1. The analyzed spectrum of uranium in the region of 776-811 keV showing the overall fit (black), the fit to recently-processed ^{238}U (orange), the fit to fully-aged ^{232}Th (blue), and the fit to fully-aged ^{238}U (green). Shapes that appear to be skewed are sums of peaks from the same emitter. The 786.27 keV of recently-processed ^{238}U appears shifted to higher energy than the 785.96 keV of fully-aged ^{238}U because the peak at 785.96 keV is due to ^{214}Pb, which is only evident in fully-aged ^{238}U due to the long growth times of ^{234}U and ^{230}Th. On the other hand, the peak at 786.27 keV is due to ^{234m}Pa, which is the dominant emitter in ^{238}U that is only recently processed.

other. However, relative activities are sufficient to answer the question of enrichment or depletion. The results of relative activities and their uncertainties for the second group (the uranium sample) are shown in Figure 3, as well as the fitted thickness of 0.33 cm. Also, in Figure 3, the calculation of the weight-percent ratio of ^{235}U to (^{235}U + ^{238}U) is shown as 0.2567 +/- 0.0416 percent, with the conclusion that the uranium sample was clearly depleted uranium, which is typically about 0.2% ^{235}U by weight [11], compared to 0.72% for natural uranium, or more for enriched uranium..

5 Example analysis #2 – pipe scale

In this example pipe scale from a new process in a copper mine was found to be radioactive and was sent to us for analysis. Unlike the earlier example of the piece of uranium metal, this spectrum was acquired in a low-background chamber and quantitative results were requested.

This example also gives an excellent illustration of the use by VRF of "equilibrium decay chains" in which advantage is taken of the additional physical requirement that all radionuclides in such a chain have fixed activities in relation to each other. VRF considers such chains as if they were single "emitter" whose activity is to be evaluated as if it were a single individual radionuclide.

VRF takes advantage of covariance terms from the matrix inversions that are needed for its non-linear fits. The inclusion of covariance terms, if present, always reduces the statistical uncertainty of the results. VRF uses the ratio of results for activity to the uncertainty of the activity, or "significance", as a measure of confidence that an emitter is present or not. A significance of 5.0 is normally taken as a minimum indication that an emitter is present.

As an *ansatz*, or trial assumption, an analysis was made assuming only geologically-aged nuclides in secular equilibrium. From the result it was clear that all the radiation emitters in this spectrum are found in the natural environment. However, the excessive residuals from the initial fit with the equilibrium decay chains of fully-aged ^{232}Th, ^{238}U, and ^{235}U showed that some of the nuclides were clearly not in equilibrium, so as a second *ansatz* these large decay chains were broken into a continuous sequence of either smaller chains with shorter equilibration times or individual radionuclides. In this second trial assumption the longest equilibration time of the smaller chains was 150 days to allow for the in-growth of ^{234}Th. The test for validity of trial assumptions comes after the analysis by observing if there are no excessive residuals and if the chi-squared statistic of the final fit is satisfactorily low.

Equilibrium decay chains in VRF are labeled with a time to equilibrium in square brackets following the parent nuclide symbol, typically about six half-lives. For example, the fully-aged ^{238}U equilibrium decay chain is represented as U238[2e6y]. For samples such as the pipe

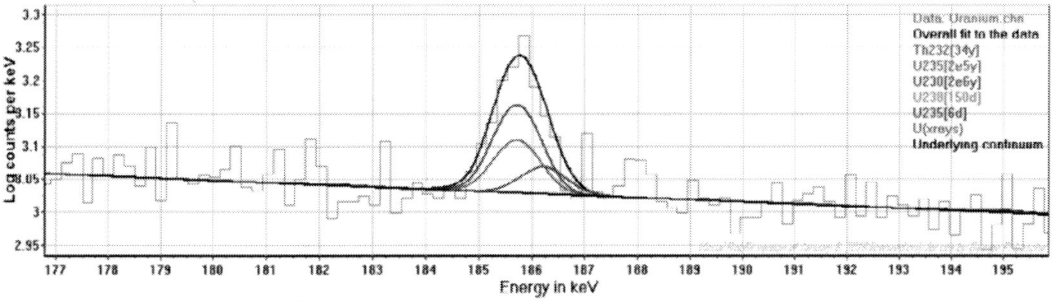

Figure 2. The analyzed spectrum of uranium in the region of 177-195 keV showing the overall fit (black), the fit to recently-processed ^{235}U (purple), the fit to fully-aged ^{235}U (fuchsia), and the fit to fully-aged ^{238}U (green). The peak at 186.2 keV is actually due to ^{226}Ra, a decay daughter of fully-aged ^{238}U. The power of full-spectrum nuclide shape fitting is well-illustrated here. Simple peak-search methods cannot resolve the 186.2 of ^{226}Ra from the 185.7 of ^{235}U, nor can they resolve the 185.7 of ^{235}U due to fully-aged ^{235}U from the 185.7 of recently-processed ^{235}U. Both are critical for determining isotopic ratios of a uranium sample with gamma spectroscopy.

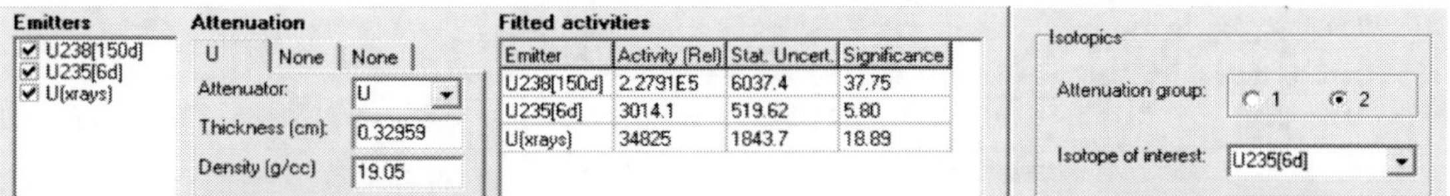

Figure 3. The results of the analysis for the attenuation group of the uranium sample.

scale example, where fully-aged equilibrium clearly cannot be justified for nuclides with very long half lives compared to the age of the sample, VRF provides decay chain segments in which the terminal nuclide of the decay chain segment has a half-life much greater than the half-lives of the other decay daughters of the parent.

The use of decay chain segments is valid if the age of the sample is much greater than the longest half-life of the decay daughters of the parent nuclide of the chain segment, not counting the terminal nuclide.

For example, for processed uranium the ^{238}U decay chain effectively stops at ^{234}U, which has a half-life of 245,490 years. The decay daughter with the longest half-life after the ^{238}U parent and before the ^{234}U terminal nuclide is ^{234}Th, which has a half-life of 24.1 days. If the processed uranium has aged for at least six of the ^{234}Th half-lives, the ^{234}Th and ^{234m}Pa daughters will have essentially the same activity as the ^{238}U parent. However, there is no relation between the activity of the ^{234}U terminal nuclide and the activities of the other members of the chain segment. For the case of processed uranium, VRF represents the decay chain segment for ^{238}U as U238[150d].

For analysis of the pipe scale spectrum, the fully-aged ^{238}U decay chain was broken into four parts:

1. $^{238}U \rightarrow ^{234}Th \rightarrow ^{234m}Pa \rightarrow ^{234}U$: This chain segment reaches equilibrium in about 150 days due to the in-growth of ^{234}Th, which has a half-life of 24.1 days. This chain was evident with high significance.

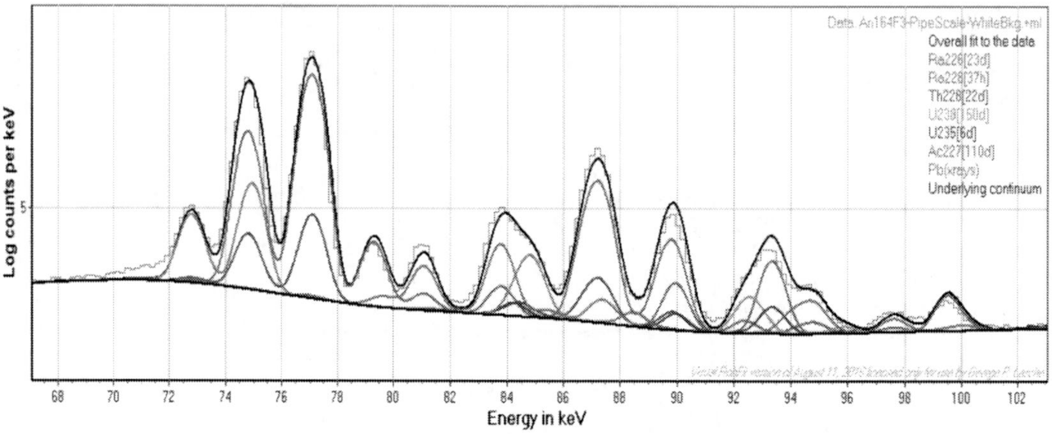

Figure 4. The low-energy region of the spectrum of radioactive pipe scale, as fitted with VRF. Note the deconvolution of the many collided peaks from different emitters in the 92-96 keV region that would not be possible with conventional methods that are based on simple peak-searches.

Table 1. The results from VRF of fitted activities and minimum detectable activities for emitters with low significance. Emitters with text in square brackets in the Emitter column are parent nuclides of an equilibrium decay chain. The equilibration time of the chain is shown in square brackets. The equilibration times of the chains are much less than the half-life of the terminal member of the chain.

Fitted activities

Emitter	Bq/g	Uncertainty	Significance
e+e-	0.016624	0.00091857	18.10
K40	0.08065	0.0072095	11.19
Ra226[23d]	9.2849	0.0063479	1462.68
Ra228[37h]	2.1831	0.0040017	545.55
Th228[22d]	0.81893	0.0019725	415.17
U238[150d]	0.55968	0.01071	52.26
U235[6d]	0.18917	0.0016698	113.29
Th230	0.33051	0.27287	1.21
U234	0.00034723	1.3301	0.00
Ac227[110d]	0.349	0.0015604	223.65
Pb(xrays)	0.49855	0.0025445	195.93
Pa231	0.053006	0.010273	5.16
Th232	1.1852	0.45939	2.58

Fitted minimum detectable activities Group 1

Emitter	Bq/g	Uncertainty	Significance
Th230	1.3201	0.26503	5.0
U234	5.5999	1.1233	5.0
Th232	2.2081	0.44411	5.0

2. ^{234}U: No equilibrium could be assumed due to the long half-life of 245,490 years. It was not detected with minimum significance due to having only low-yield, low-energy gammas, so only its minimum detectable activity (MDA) could be usefully reported.
3. ^{230}Th: No equilibrium could be assumed due to the long half-life of 75,386 years. It was not detected with minimum significance due to having only low-yield, low-energy gammas, so only its minimum detectable activity (MDA) could be usefully reported.
4. ^{226}Ra→^{222}Rn→^{218}Po→^{214}Pb→^{214}Bi→^{210}Po→^{210}Pb→^{210}Bi→^{206}Po→^{206}Pb: This chain segment reaches equilibrium in about 23 days due to the in-growth of ^{222}Rn, which has a half-life of 3.82 days. This chain was evident with very high significance.

The fully-aged ^{235}U decay chain was broken into three parts:

1. ^{235}U→^{231}Th→^{231}Pa: This chain segment reaches equilibrium in about 6 days due to the in-growth of ^{231}Th, which has a half-life of 1.06 days. This chain was evident with high significance.
2. ^{231}Pa: No equilibrium could be assumed due to the long half-life of 32,765 years. This nuclide was evident, however, with minimal significance.
3. ^{227}Ac→^{227}Th→^{223}Ra→^{219}Rn→^{215}Po→^{211}Pb→^{211}Bi→^{207}Tl→^{207}Pb: This chain segment reaches equilibrium in about 110 days due to the in-growth of ^{227}Th, which has a half-life of 18.68 days. This chain was evident with very high significance.

The fully-aged ^{232}Th decay chain was broken into three parts:

1. ^{232}Th: No equilibrium with its ^{228}Ra decay daughter could be assumed because of the evident enhancement of radium in the sample due to an unknown chemical process. It was not detected with minimum significance due to having only low-yield, low-energy gammas, so only its minimum detectable activity (MDA) could be usefully reported.
2. ^{228}Ra→^{228}Ac→^{228}Th: This chain segment reaches equilibrium in about 37 hours due to the in-growth of ^{228}Ac, which has a half-life of 6.15 hours. This chain was evident with very high significance.
3. ^{228}Th→^{224}Ra→^{220}Rn→^{216}Po→^{212}Pb→^{212}Bi→(^{212}Po or ^{208}Tl)→^{208}Pb: This chain segment reaches equilibrium in about 22 days due to the in-growth of ^{224}Ra, which has a half-life of 3.66 days. This chain was evident with very high significance.

With these ten emitters and the addition of emitters for positron annihilation, ^{40}K, and Pb x-rays a satisfactorily low reduced chi-squared of 3.2 was achieved and no excessive residuals were observed, indicating completeness of the analysis. Attempts to achieve a lower chi-squared resulted in what visually appeared to be possible over-fitting, so the analysis was considered complete at that point.

Figure 4 shows the fits to the data in a low-energy region of the spectrum crowded with x-rays and some gammas. Note the deconvolution of the many collided peaks from different emitters in the 92-96 keV region that would not be possible with conventional methods that are based on simple peak-searches.

Table 1 shows the activities of the fitted emitters and the MDA of those with low significance. For each decay chain segment, each member nuclide of the decay chain segment is reported with the same activity, uncertainty, and significance as those of the decay chain segment.

With VRF, MDA is calculated in a different but simple and straightforward manner. The user selects a significance threshold (in this example, a threshold of 5.0 was used). After analysis, MDA is calculated for each emitter that had a significance below the threshold (except for elemental x-ray emitters, thermal neutron capture emitters, and the positron annihilation emitter). For each emitter below threshold VRF adds counts from that emitter's full-spectrum shape to the original spectrum until that emitter can be detected at the desired threshold significance. The resulting activity is the MDA.

6 Conclusion

A new application that uses non-linear fitting of full-spectrum emitter shapes for nuclear spectral analysis has been demonstrated and validated. This application has significant advantages compared to standard peak-search methods, in particular when the resolution of overlapping peaks from different emitters is important.

Emitter shapes can be formed from a single nuclide's gammas and decay x-rays or from an equilibrium decay chain of many nuclides. The non-linear fitting process consists of many automated iterations in which adjustments are made to the activities of each nuclide or decay chain, as well as to user-enabled parameters of energy calibration, attenuation by up to three intervening or self-absorbing materials, peak width as a function of energy, full-energy peak efficiency, and coincidence summing, until no better fit to the data can be obtained.

References

1. M.A. Mariscotti , Nuclear Instrumentation Methods 50 (1967), 309.
2. D.W. Marquardt, Journal of the Society for Industrial and Applied Mathematics, 11 (1963) 431-441.
3. P.R. Bevington, D.K. Robinson, Data Reduction and Error Analysis for the Physical Sciences, McGraw-Hill, New York, 1969, Chapter 11.
4. W.H. Press, S.A. Teukolsky, W.T. Vetterling, B.P. Flannery, Numerical Recipes, Cambridge University Press, New York, 1992, (Sec 15.5).
5. R.L. Coldwell, G.J. Bamford, 1991, The Theory and Operation of Spectral Analysis Using ROBFIT, American Institute of Physics, New York, 1991.

6. G.P. Lasche, R.L. Coldwell, Analysis of Nuclear Spectra with Non-Linear Techniques and its Implementation in the Cambio Application, Journal of Radioanalytical Nuclear Chemistry (2009) 282:211-215.
7. Evaluated Nuclear Structure File (ENSDF), National Nuclear Data Center (NNDC), Brookhaven National Laboratory, www.nndc.bnl.gov/ensdf
8. S.Y.F. Chu, L.P. Ekstrom and R.B. Firestone, WWW Table of Radioactive Isotopes, database version 1999-02-28, http://nucleardata.nuclear.lu.se/nucleardata/toi/
9. M.J. Berger, J.H. Hubbell, S.M. Seltzer, J. Chang, J.S. Coursey, R. Sukumar, and D.S. Zucker, XCOM: Photon Cross Sections Database, National Institute of Standards and Technology, NBSIR 87-3597, 1998.
10. R.L. Metzger, K.A. Van Riper, G.P. Lasche, "Uranium, radium and thorium in soils with high-resolution gamma spectroscopy, MCNP-generated efficiencies, and VRF non-linear full-spectrum nuclide shape fitting", to be published in the proceedings of the ICRS-13 & RPSD-2016, Paris, France, Oct 3-6, 2016.
11. B. Rostker, Properties and Characteristics of DU, U.S. Department of Defense, Dec 13, 2000, http://www.gulflink.osd.mil/du_ii/index.htm#TABC_Properties and Characteristics of DU

A Rapid Method for Measuring Strontium-90 Activity in Crops in China

Lingjing Pan Pan[1], *Guobing* Yu[2], *Deyun* Wen[2], *Zhi* Chen[1], *Liusi* Sheng[1], *Chung-King* Liu[1], and *X. George* Xu[1,3]*

[1] *University of Science and Technology of China, Hefei, China*
[2] *Environmental Radiation Surveillance Center, Environmental Protection Department of Anhui Province, Hefei, China*
[3] *Rensselaer Polytechnic Institute, Troy, New York, USA*

Abstract. A rapid method for measuring Sr-90 activity in crop ashes is presented. Liquid scintillation counting, combined with ion exchange columns 4', 4''(5'')-di-t-butylcyclohexane-18-crown-6, is used to determine the activity of Sr-90 in crops. The yields of chemical procedure are quantified using gravimetric analysis. The conventional method that uses ion-exchange resin with HDEHP could not completely remove all the bismuth when comparatively large lead and bismuth exist in the samples. This is overcome by the rapid method. The chemical yield of this method is about 60% and the MDA for Sr-90 is found to be 2.32 Bq/kg. The whole procedure together with using spectrum analysis to determine the activity only takes about one day, which is really a large improvement compared with the conventional method. A modified conventional method is also described here to verify the value of the rapid one. These two methods can meet different needs of daily monitoring and emergency situation.

1 Introduction

Sr-90 is an artificial radionuclide produced by spontaneous fission of U and Pu [1], which is associated with nuclear accidents, nuclear facility operations, and nuclear weapon explosion [2, 3]. With historical nuclear explosion and accidents, a large number of radionuclides such as fission product strontium have been released to the atmosphere [4, 5]. Due to its long half-life of about 28.5 years, it still exists in the environment. Once the fallouts were absorbed by crops, Sr-90 can eventually get into the human body and accumulate in the skeleton as the result of its chemical similarity with the calcium, and its daughter radionuclide Y-90 emits high-energy β particles delivering a relatively high radiation dose to the bone and bone narrow tissues [6-7]. Thus it's extremely critical to determine the activity of Sr-90 in crops. As China is rapidly developing the nuclear power industry, establishing associated daily monitoring and emergency determining systems of Sr-90 are really urgent.

Sr-90 is a pure beta-emitter with a maximum energy of $0.5 MeV$. When using the traditional determination method, chemical pretreatments are needed to separate and purify Sr-90 from other elements by selective precipitations, liquid-liquid extraction (LLE), extraction chromatography (EC) and ion exchange (IE) [8]. One of the widely used precipitation methods is based on the solubility difference of strontium and other cations in fuming nitric acid, which is more accurate than other methods. However, numerous steps are required before separating because the fuming nitric acid is harmful to the operator and environment. Other methods such as LLE, EC and IE usually have to wait for about 14 days till the Sr-90 sample reaches equilibrium with the Y-90 [9], and then isolate Y-90 for the indirect determination of Sr-90. The long waiting time of these methods makes them inapplicable to emergency response. In addition, for the IE, the ion exchange resin does not have a higher selectivity for Sr-90 either, so that the larger the sample, the more complex the isolation, because large amount of other elements may interfere separating small amount of strontium. LLE is suitable for the matrix of simple composition. A lot of work has been done to find simpler, faster and more accurate methods. Nowadays, extraction chromatographic resin has been developed for isolation of radionuclides from various kinds of matrix [10-13]. In China, the conventional EC method for analyzing strontium-90 in crops is based on the $GB11222.1-89$ [14] which uses ion-exchange resin with di (2-ethylhexyl) phosphoric acid (HEDHP) to isolate Y-90, counting Y-90 with low-background proportional counter, and then the radioactivity of Sr-90 can be evaluated. Horwitz et. al reported that strontium could be extracted in nitric acid using a crown ether [15], dicyclohexano-18-crown-6 (DCH18C6). Right now, this substance is commercially available called Sr•Spec. The resin enables a rapid and simple separation for strontium. It's used more and more widely to separate Sr-90 from different matric, such as milk, soil, water and biological sample [16-19]. However, before passing through the resin a suitable chemical pretreatment is needed to eliminate some interfering nuclides since the resin has saturated adsorption capacity. For milk samples, a large amount of Ca^{2+} interfere the resin from absorbing

*e-mail: xgxu@ustc.edu.cn

Sr because of their chemical similarities. Thus special procedures for eliminating Ca^{2+} should be taken before the sample is passed through the resin. For sediment samples, Ba^{2+} may be the interference. For biological samples, the organic matter need to be removed before analyzing. In this paper, a rapid method and a modified conventional method for analyzing the Sr-90 with background level in crops are described. The modified conventional method is based on the $GB11222.1-89$ in China. The conventional method for analyzing Sr-90 could not completely remove Bi-210 which interferes the measurement results. Some crops samples have a large amount of lead, bismuth and potassium. In those cases, the conventional method would not work. The rapid method uses liquid scintillation counter (LSC) together with Eichrom Sr•spec extraction resin as an agent to separate strontium and applies spectrum analysis [20-22] for measuring the Sr-90 activity. Compared to the similar methods in former literatures, a special co-precipitation method is applied before samples passed through the resin in order to prevent the competition of other ions in the rapid method. The modified conventional method and the rapid method for analyzing Sr-90 in crops cover all the situations of emergency response, daily monitoring.

2 Experiments

2.1 Reagents

Sr•Spec Resin (particle size: $100-150\ \mu m$) was obtained from *EiChrom Industries, Inc.* *Nitric* and hydrochloric acids of various concentrations were used to dissolute precipitation and other functions. All chemical reagents are of analytical grade quality. The Sr-90 (Sr-90/Y-90 in equilibrium) and Pb-210 (Pb-210/Bi-210/Po-210 in equilibrium) as certified standard aqueous solution are obtained from textitEckert & Ziegler.

2.2 Instrumentations

- *Quantulus* 1220^{TM} liquid scintillation counter (Perkin-Elmer, Inc. USA).
- Low-background proportional counter.
- Centrifugal machine.
- Water bath kettle.
- Analytical balance, 0.1 mg accuracy.

2.3 Sample pretreatment

The purpose of sample pretreatment is to obtain a homogenous sample. The crops were collected in China. All the samples were homogenized and dried on a hot plate until constant weight. Afterwards, the samples were ashed to 600 in muffle furnace.

2.4 Rapid method for Sr-90 separation from crops samples

Based on the previous researches, it's possible to directly isolate strontium from the crops. This method was investigated and developed here. The mainly procedures of rapid method for analyzing Sr-90 include three parts: radiochemical procedure, source preparation & measurement, and activity determination. The flow diagram of this method is shown in figure 1.

2.4.1 Detailed radiochemical procedures

After pretreatment, white ashes were obtained. 5 grams crops ashes were taken for the analysis. 30 mg Sr^{2+} carrier and 10 mg Y^{3+} were added for chemical yield determination. The ashes were mixed with hydrochloric acid 5M for leaching about one hour. The insoluble substances were filtered to get clear filtrate. After leaching and filtering, oxalic acid was added to the filtration and Sr was separated from the other alkaline earth metals by oxalate precipitation at pH = 4. Compared to the strontium, a large amount of potassium was in the crops, so oxalic acid was added to the solution to separate strontium from potassium avoiding ion competition. Ion competition may lead to lower chemical yield, which interferes the measuried results. Afterwards, the precipitation was dissolved with 20 mL $8M$ nitric acid and the solution was passed through the column filled with *Eichrom* Sr•Spec resin.

The resin should be soaked in the water for a long time (several hours) to remove impurities before being transferred into the column. The Sr•Spec column was equilibrated with 15mL 3M nitric acid, and then the sample solution was passed through. The column was washed successively with 20 mL $8M$ nitric acid and 15 mL $3M$ nitric acid, Bi, Y, K, Ca and other elements passed the column, strontium was retained by it. Then 40 mL deionized water was used to elute Sr-90. It is important to realize the washing solution volume should be controlled, otherwise it would elute strontium as well.

2.4.2 Source preparation and measurement

The Sr-90 eluting solution was adjusted to pH = 9.0 with ammonium hydroxide. After adding 5mL saturated ammonium carbonate, filtering and heating, the yields of strontium were determined by gravimetric method. The precipitation was dissolved with 8 mL 0.1M nitric acid, and then mixed with 12 mL liquid scintillation cocktail *HiSafe TMIII* for the measurement using LSC.

The measurement was performed with a *Quantulus* 1220^{TM} counter. This apparatus uses low coincidence bias to reduce the background. The multichannel analyzer (MCA) configuration was set in the Alpha/ Beta option and the pulse shape analysis (PSA) was set to 100. For the determination of activity, the efficiency calibration for different radionuclides should be performed. A certain amount of Sr-90/Y-90 standard solution with increasing amounts of CCl_4 as quenching agent was used. Then the quenching correction curve was

DOI: 10.1051/epjconf/201715301003

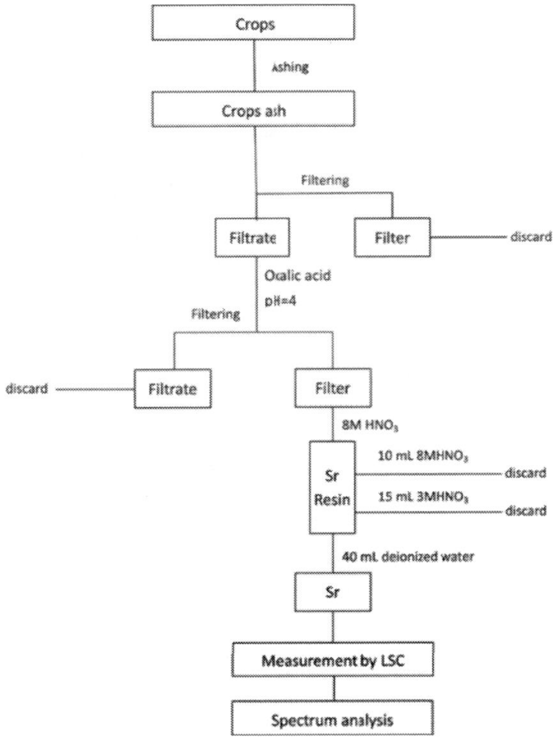

Figure 1: Flow diagram for rapid method determination of Sr-90 in crops.

obtained by using an Eu-152 source. Sr-90 as a β radionuclide with high energy normally has little influence on chemical quenching, which keeps the quench correction curve away from obvious changes. Finally, background correction and time correction were performed.

2.4.3 Activity determination

LSC is a detector with spectrum properties. Therefore, the spectrum analysis method can be applied to determine the Sr-90 activity. Impurities can be detected when using spectrum analysis method performing different spectra, especially for radionuclides mixed with their isotopes which cannot be separated with chemical treatments. The activities of all the radionuclides can be calculated using spectrum analysis, as long as the standard spectra of all the radionuclides are known.

There are totally 1024 channels in LSC and the Sr-90 spectrum covers from channel range [$5th$ – $600th$]. The spectrum obtained from the LSC shows that high energy windows (channel range [$150th$ – $600th$]) should be focused on to avoid the interference from low energy radionuclides, such as Pb-210. Then spectrum unfolding based on the least squares algorithm (as shown in Eq. 1) is applied to determine the activities. Once all the radionuclides in samples are known, the activity of each nuclide can be easily calculated using the spectrum analysis method, without any chemical treatments to the samples.

$$Y_i = \sum_n C_n \cdot a_{n,i} \quad (1)$$

Where y_i is the counting number of each channel. C_n is the activity of the unknown nucleus n in the mixed sample, and $a_{n,i}$ is nuclide n's counting responding in the ith channel.

In this rapid method, the Sr resin can separate Sr^{2+} from other interference substance directly. In an emergency situation, this rapid method can be used to determine the activity of Sr-90 together with other strontium isotopes and oxalic acid cannot be added to the filtrate due to the small sampling. Even though the rapid method works for all the situations, the main problem of this method is that the Sr resin is really expensive. Regarding to the daily monitoring, a much cheaper resin separating Y-90 can be used for samples in which Sr-90 already reaches equilibrium with the Y-90. Some of these conventional methods using other resins, like the Chinese national standard method for determining Sr-90 activity, may have shortcomings in the presence of large amounts of lead and bismuth. To meet the needs for daily monitoring, a modified method was described as below.

2.5 Procedures for modified conventional method

Ion-exchange resin with HDEHP is used in the modified conventional method, together with low-background proportional counter (LLP) to determine the activity of Y-90.

And then Sr-90 activity is derived from some calculation formulas.

A certain amount of sampling ash was taken for the analysis. 50mg Sr^{2+} carrier and 20 mg Y^{3+} were added for chemical yield determination. The ash samples were dissolved by leaching with hydrochloric acid 5M. Oxalic acid was added and the pH was adjusted to 3 after leached. Yttrium and other alkaline earth elements were precipitated as oxalates, a process in which the potassium and some other interferences were removed. Oxalate precipitation was destructed by ashing and then was dissolved by nitric acid. The filtrate was passed through the HDEHP column to remove Sr, Cs, Ce and other elements. Y-90 was eluted with 6M nitric acid. 10mg Bi carrier was added to the eluent solution to purify Y-90 from bismuth sulfide precipitation. Thereafter, the purification solution was taken to determine the chemical yield by gravimetric method and the activity of Y-90 was measured by a LSC or LLP. The activity of Sr-90 was then determined. The flow diagram for the separation of Y-90 from crops ash is presented in figure 2.

3 Results and discussion

The spiked samples for conventional method were added with Sr-90 standard solution, Pb-210 standard solution and Sr-90 mixed with Pb-210 standard solution respectively and then measured by LSC. The results of the LSC were shown in figure 3. Comparing the spectra in figure 3, the spectrum spiked with Pb-210 obtained from the conventional method has an obvious peak observed between channel range $[200th - 400th]$. The spectra of Sr-90 mixed with Pb-210 standard solution and Sr-90 standard solution are obviously different. Comparison of the three figures implies that Bi-210 was not removed thoroughly in the conventional method which leads to the interference.

The same samples were analysed by modified conventional method and conventional method respectively, the results were shown in figure 4 and figure 5. The spectrum from the conventional method has obvious peaks near the center of the energy channels, but the spectrum using modified measuring method does not. The results imply that the modified method can separate Bi from the sample to some extent.

Figure 6 and figure 7 show the spectra of the crops sample and the spiked crops sample operated by rapid method. A good consistency with the strontium standard spectrum can be observed. From figure 4 to figure 6, there are obvious peaks in low energy channels when the sample has low-level radioactive contamination. It may be caused by the interference from environmental radon and its daughters. That is why we shifted to the high energy windows while using the spectrum analysis method to calculate the activity of Sr-90.

3.1 Spectrum method for determining Sr-90 activity

Figure 8 shows a spectrum corresponding to Sr-90 and Y-90 in secular equilibrium. Both Sr-90 and Y-90 present a continuous spectrum from 0 to Emax since they are pure β nuclides. Y-90 has a shorter half-life so that Sr-90 decays to Y-90 rapidly leading to an overlap between their individual spectra even though Y-90 was eliminated completely. Spectrum analysis method can be used to eliminate the interference from Y-90. Appropriate windows should be chosen based on the rules that window A contains pulses both from Sr-90 and Y-90 whereas window B only contains Y-90 pulses.

From Y-90's standard spectrum in figure 3, which spiked Sr-90 standard solution obtained from conventional method, window B could be chosen from channel range [600th-850th], in which only Y-90 pulses exist. With respect to window A, the window was selected with the purpose of getting a high efficiency for Sr-90 and low interference from other nuclides. Because of the influence from environmental radon and cosmic rays, the window A could be chosen from channel range [150th-600th], which gets rid of low channel parts. The expression for calculating the Sr-90 activity is shown as below:

$$C_A = A_1\epsilon + A_2\phi_A \tag{2}$$

$$C_B = A_2\phi_B \tag{3}$$

$$A_1 = \frac{C_A - \frac{\phi_A}{\phi_B}C_B}{\epsilon} \tag{4}$$

Where A_1 denotes the Sr-90 activity in the sample, ϵ is the counting efficiency of Sr-90 in window A, A_2 is the activity of Y-90 in the sample, C_A and C_B are the count rate in the window A and window B, and ϕ_A and ϕ_B are the counting efficiencies of Y-90 in window A and window B. The value of these parameters could be obtained in such a way described as follow. Sr-90 was separated from a certain standard solution and then this sample was measured several times after the separation date. The count rate in the window A and window B at different time would be recorded. And then least square method was used to determine each parameter. We chose the optimal window A to be in the channel range [150th-600th] and window B to be in the channel range $[600th - 850th]$, and the value of the Sr-90 counting efficiency ϵ was 88.4% in window A and the ratio of efficiencies for Y-90 in the two windows $\frac{\phi_A}{\phi_B}$ was 0.458.

3.2 Minimum detectable activity (MDA)

MDA is one of the most important value to estimate the radioactivity method. The formula of MDA was evaluated as follows:

$$MDA(Bq/kg) = \frac{2.71 + 4.65\sqrt{B}}{ETYmf} \tag{5}$$

Where B is the total number of background measured separately with the same counting time, E is the counting efficiency, T is the sample counting time, Y is the chemical yield, m is the mass of sample and f is the correction coefficient. According to the equation, the value of MDA for rapid method is $2.32Bq/kg$. When the calculation method

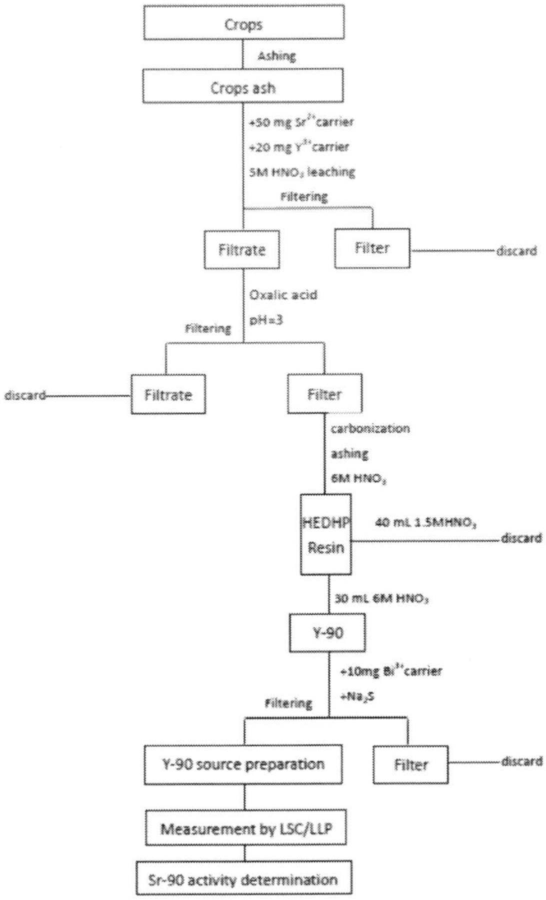

Figure 2: Flow diagram for modified conventional method of Y-90 in crops

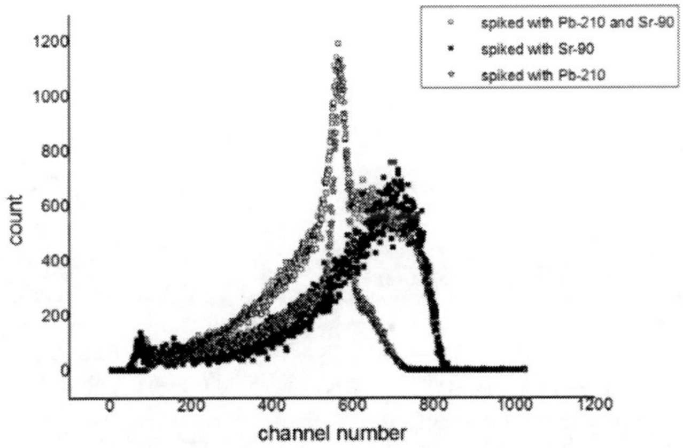

Figure 3: Spectra of spiked samples obtained from the conventional method.

DOI: 10.1051/epjconf/201715301003

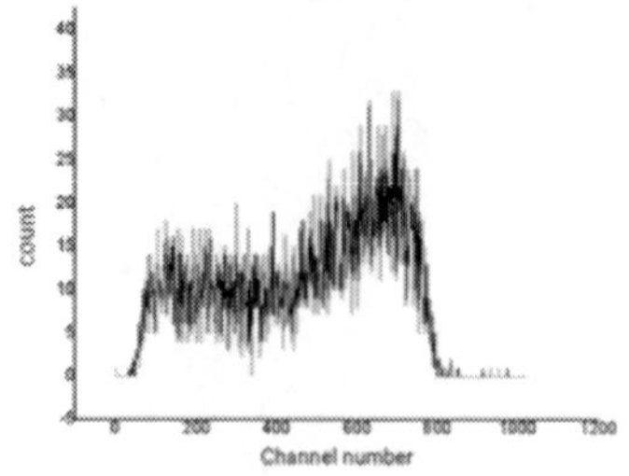

Figure 4: Spectrum of crops ashes obtained from the modified conventional method.

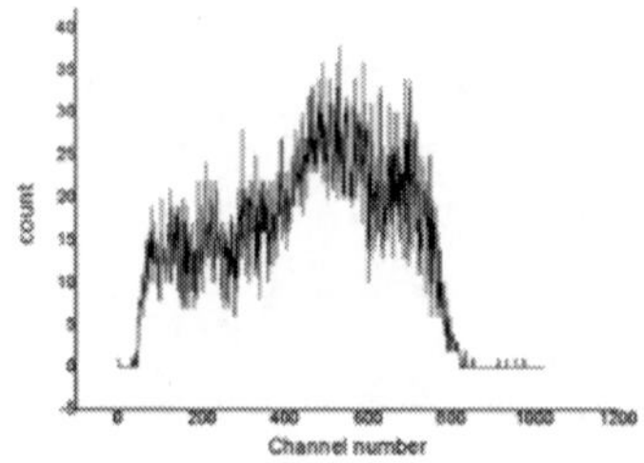

Figure 5: Spectra of spiked samples obtained from the conventional method.

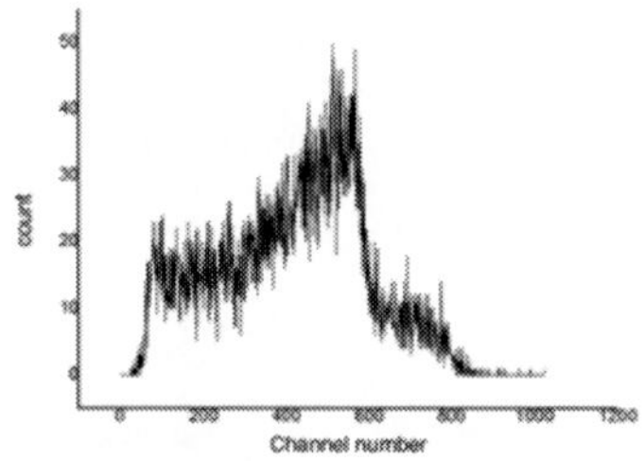

Figure 6: Spectrum of crops ashes obtained from the modified conventional method.

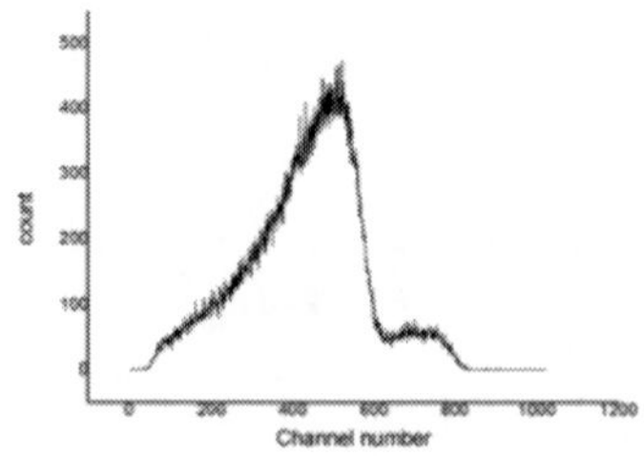

Figure 7: Spectra of spiked samples obtained from the conventional method.

is applied to the environmental samples, the MDA should be estimated for each sample due to the different factor.

As shown in Table 1, the results of the rapid method are slightly and systematically smaller than those of the modified conventional method. The reasons for this may be the modified conventional method for the sample with a large amount of bismuth not eliminating bismuth completely in the error-allowed range and the system error between different detectors. However, the difference of the

Table 1: Comparison of analytical results of Sr-90 activity concentration. (CM: conventional method; MCM: modified conventional method; RM: rapid method)

Sample	CM(Bq/kg)	MCM(Bq/kg)	RM(Bq/kg)
A	23.0±2.5	18.0±1.8	16.0±1.7
B	8.0±0.9	6.8±0.7	5.4±0.5
C	11.0±1.2	9.1±1.0	7.2±0.7
D	12.0±1.3	8.9±0.9	7.4±0.8
E	8.3±1.1	4.7±0.8	4.1±0.4
F	7.9±1.2	3.4±0.6	3.4±0.3
G	14.2±1.7	11.1±1.0	10.5±1.1
H	4.0±0.8	4.0±0.9	3.6±0.3

values between the modified method and rapid method are actually considered to be acceptable.

4 Conclusions

The rapid method proposed in this paper for chemical separation and purification of Sr-90 was found to be simpler, faster and more accurate than the conventional method. The activity of Sr-90 can be calculated without considering Y-90, so it no longer requires waiting for Sr-90 and Y-90 reaching equilibrium. As a result, the whole analysis procedure can be accomplished within one day. Furthermore, Sr-89 and Sr-90 can be detected simultaneously using this spectrum analysis method,so it can be used in nuclear emergency monitoring. For daily monitoring with large amount of samples and low radioactivity, the much cheaper modified conventional method would be a better choice. The two methods do cover different situations.

Acknowledgements

The work was funded in part by a grant from National Natural Science Foundation of China (*No.*11575178).

References

[1] H. Isnard, M, Aubert, P, Blanchet, et al. Spectrosc Spect Anal. **61(2)**, 150-156 (2006)
[2] C. Landstetter, G. Wallner. J Environ Radioactiv. **87(3)**, 315-324(2006)
[3] I. Lopes, M. J. Madruga, A. Mourato, et al. J. Radioanal. Nucl. Chem. **87(3)**, 315-324 (2006)
[4] E. Kabai, L Hornung, B. T. Savkin, et al, Sci. Total Environ. **410**, 235-240 (2011)
[5] P. Neis, R. Hille, M, Paschke, et al. Forensic Sci Int. **99(1)**, 47-51(1999)
[6] E. P. Hardy, M. W. Meyer, J. S. Allen, L. T. Alexander, Nature 219. 584-587 (1968)
[7] F. Goutelard, R. Nazard, C. Bocquet, et al. Radiat. Isotopes. **53(1)**, 145-151 (2000)
[8] N. Vajda, C. K. Kim. Radiat. Isotopes. **68(12)**, 2306-2326 (2010)
[9] M. Heilgeist. Radioanal Nucl Chem. **245(2)**, 249-254 (2000)

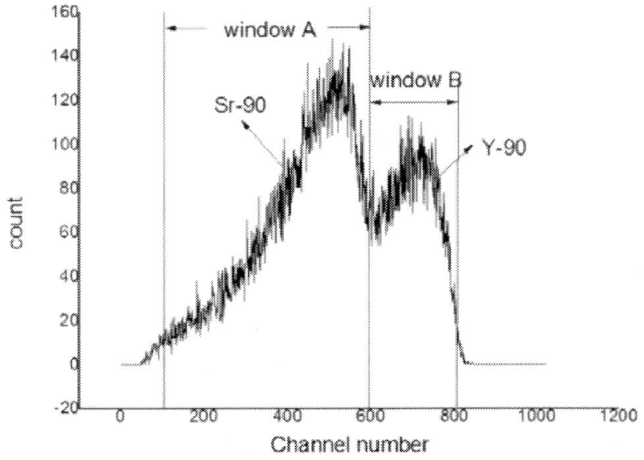

Figure 8: Spectrum of Sr-90/Y-90 secular equilibrium standard solution

[10] Z. Grahek, N. Zecevic, et al, Anal Chim Acta. **399**, 237 (1999)

[11] IAEA Tehnical report series 295. 70-89 (1989)

[12] Z. Grahek, I. Eskinja, K. Kosutic, et al. Anal Chim Acta. **379**, 107 (1999)

[13] Z. Grahek, K. Kosutic, S. Lulic, et al. J Radioanal Nucl Chem. **242**, 33 (1999)

[14] Ministry of Environmental Protection of China, National Standard GB 11222.1–89 (1989) (in Chinese)

[15] R. Bojanowski, D. Knapinska-Skiba. J Radioanal Nucl Chem. **138(2)**, 207-218 (1990)

[16] R. Bojanowski, Z. Radecki, M. J. Campbell, K. I. Burns, A. Trinkl. IAEA-326 and 327, IAEA Vienna (2001)

[17] S. Brun, S. Bessac, D. Uridat. B, Boursier. J Radioanal Nucl Chem. **252(2)**, 191-197 (2002)

[18] ASTM Standard Test Method for Strontium-90 in water. D5811-08 (2008)

[19] E. Blasius, W. Klein. J Radioanal Nucl Chem. **89(2)**, 307-310 (1985)

[20] L. J. Pan, Z. Chen, G. B. Yu, et al. Nuclear Electronics & Detection Technology. **34(7)**, 812-815 (2014) (in Chinese)

[21] M. Heilgeist. J Radioanal Nucl Chem. **245(2)**. 249-254 (2000)

[22] K.H. Hong, Y. H. Cho, M. H. Lee, et al. Radiat. Isotopes. **54**. 299-305 (2001)

Fast Computation of Pulse Height Spectra Using SGRD Code

Philippe Humbert, *Boukhmès* Méchitoua*

CEA DAM Ile-de-France, Bruyères-le-Châtel, 91297 Arpajon cedex, France

Abstract. SGRD (Spectroscopy, Gamma rays, Rapid, Deterministic) code is used for fast calculation of the gamma ray spectrum produced by a spherical shielded source and measured by a detector. The photon source lines originate from the radioactive decay of the unstable isotopes. The emission rate and spectrum of these primary sources are calculated using the DARWIN code. The leakage spectrum is separated in two parts, the uncollided component is transported by ray-tracing and the scattered component is calculated using a multigroup discrete ordinates method. The pulsed height spectrum is then simulated by folding the leakage spectrum with the detector response functions which are pre-calculated using MCNP5 code for each considered detector type. An application to the simulation of the gamma spectrum produced by a natural uranium ball coated with plexiglass and measured using a NaI detector is presented.

1 Introduction

Real time applications require fast and accurate calculation of the detected gamma-ray spectra produced by shielded sources. For this purpose the SGRD (Spectroscopy, Gamma rays, Rapid, Deterministic) code [1] that was used to calculate the leakage spectra of one-dimensional spherical assemblies has been updated in order to take into account the response function of various types of detectors.

In the following we recall the methods used in SGRD to calculate the leakage spectra, then we describe the pre-calculation of detector response functions (DRF) using Monte Carlo code MCNP5 [3] simulations and finally we show some numerical results concerning the simulation of a natural uranium ball coated with plexiglass and measured using a NaI detector.

2 Gamma-Ray Spectra

The gamma-ray spectrum simulation is the result of successive steps. The first step is the computation of the primary gamma source emission rate and spectrum. The second step is the photon transport through the shielding materials up to the external surface of the detector in order to derive the leakage spectrum. The third step is the photon transport into the detector in order to evaluate the distribution of the photon energy deposited inside the detector's sensitive volume.

When the source can be modeled as one-dimensional spherical shells, the second step can be handled very effectively using deterministic solvers [1,4]. The first and third steps can be solved using pre-calculated sources and DRF.

The SGRD input/output flow diagram for gamma spectra simulation is presented in figure 1.

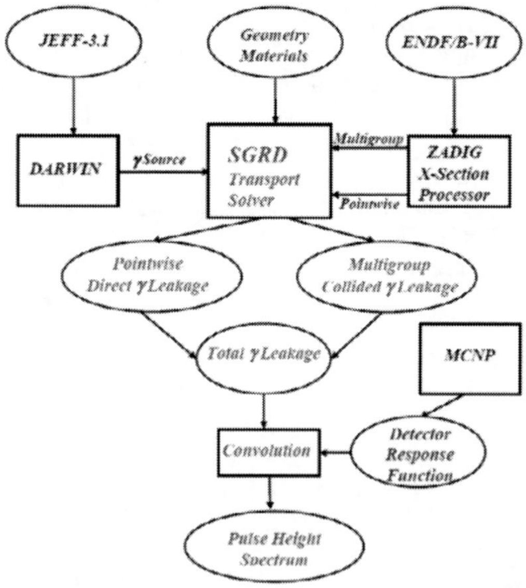

Fig. 1. SGRD input/outplut flow diagram for gamma spectra calculation.

2.1 Primary source

The primary photon source has several components, the gamma lines due to the radioactive decay of unstable isotopes, the gamma resulting from spontaneous fission, the bremsstrahlung radiation produced by charged

* Corresponding author: boukhmes.mechitoua@cea.fr

particles slowing down and the neutron induced gamma production.

The radioactive decay source line spectrum and emission rate are pre-calculated using DARWIN code [2]. The treatment with SGRD of the other components of the primary source terms are under study.

2.2 Leakage spectra

The leakage L is the number of particles leaving the external surface of the source per unit time and solid angle. Considering a spherical source with an external radius R and an angular flux $\psi(r,\mu)$ we have:

$$L = 2\pi R^2 \int_0^1 \mu\psi(R,\mu)d\mu \quad (1)$$

The angular flux is solution of the transport equation. It has a discrete part due to the uncollided transport of gamma lines and a continuous part due to the scattering. The discrete component is calculated using a ray-tracing transport solver. The continuous component is transported using a multigroup discrete ordinates (S_N) solver.

The flux computation is performed using three transport calculations.

The first step is the uncollided transport of each source line using a ray-tracing algorithm (cf.figure 2).

$$\vec{\Omega}\cdot\vec{\nabla}\psi^{UNC}(E_\gamma) + \sigma_T(E_\gamma)\psi^{UNC}(E_\gamma) = q(E_\gamma) \quad (2)$$

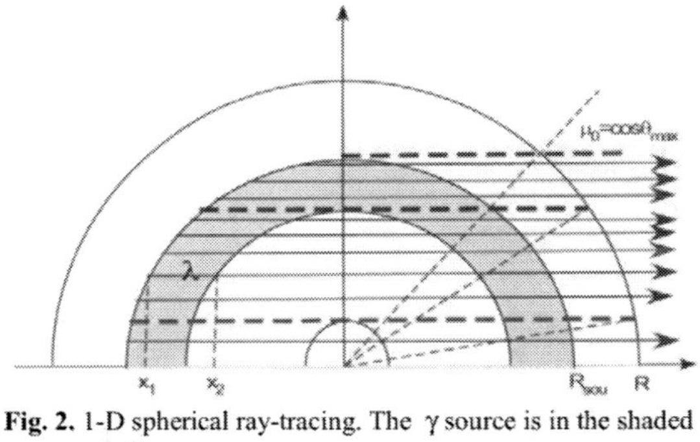

Fig. 2. 1-D spherical ray-tracing. The γ source is in the shaded shell.

The second step is the total multigroup S_N transport taking into account the scattering term. The multigroup source is obtained by condensing the gamma lines into groups (cf. figure 3).

$$\vec{\Omega}\cdot\vec{\nabla}\psi_g^{TOT} + \sigma_g^T\psi_g^{TOT} = \sum_{g'=1}^{G}\int_{4\pi}\sigma_{g'g}(\vec{\Omega}'\cdot\vec{\Omega})\psi_{g'}^{TOT}(\vec{\Omega}')d\Omega' + q_g \quad (3)$$

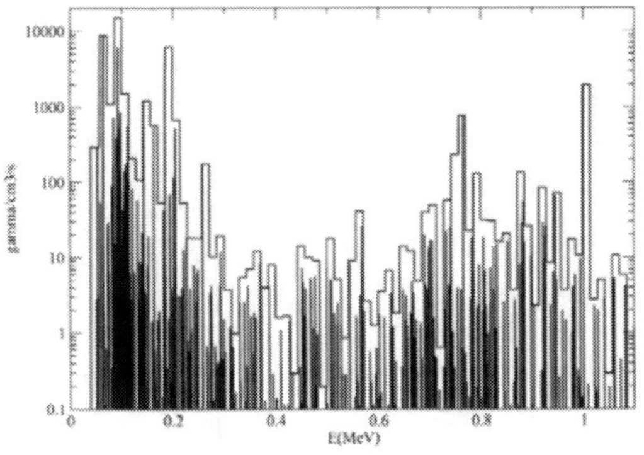

Fig. 3. Gamma source lines (blue) are condensed into groups (red) for multigroup S_N calculations.

The third step is the uncollided multigroup S_N transport of the multigroup source.

$$\vec{\Omega}\cdot\vec{\nabla}\psi_g^{UNC} + \sigma_g^T\psi_g^{UNC} = q_g \quad (4)$$

The scattered component is given by subtracting the uncollided multigroup flux to the total multigroup flux as shown in figure 4.

$$\psi_g^{COL} = \psi_g^{TOT} - \psi_g^{UNC} \quad (5)$$

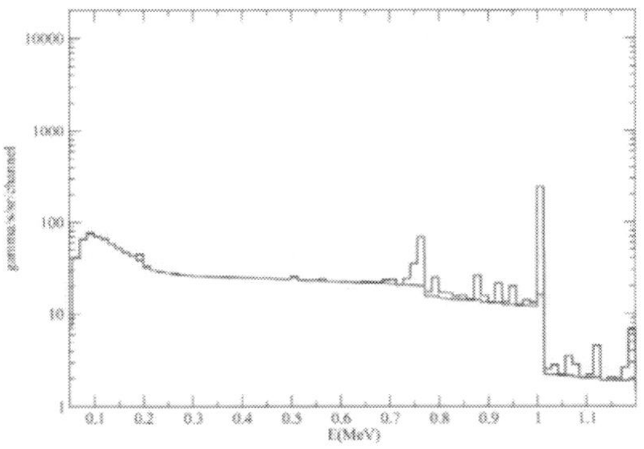

Fig. 4. Multigroup SN leakage spectra for both uncollided component in blue and scattered background in red.

Finally, the total leakage spectrum is the superposition of the discrete uncollided and multigroup scattered components on a very fine multigroup grid (cf. figure 5).

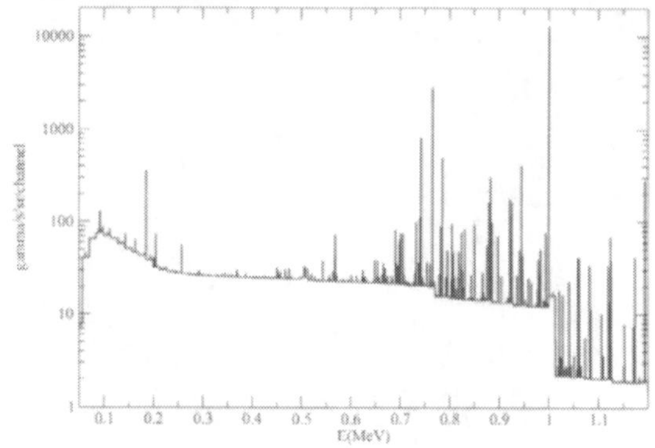

Fig. 5. Discrete uncollided lines with multigroup scattered background.

2.3 Detector response Function

The DRF is used to convert the leakage spectrum into a pulse height spectrum using equation (6) assuming that the pulse height and leakage spectra have the same energy discretization into *n* channels.

Notations

Δt : Duration of the measurement.

$\Delta\Omega$: Solid angle of the detector viewed from the center of the source.

L_j : Leakage spectrum = Number of gamma particles leaving the source with energy in channel *j* per unit time and solid angle.

N_i : Pulse height spectrum = Number of counts in channel *i*.

R_{ij} : Detector response matrix = Number of counts in channel *i* due to one gamma entering the detector with energy in channel *j*.

$$N_i = \Delta\Omega\Delta t \sum_{i=1}^{n} R_{ij} L_j \tag{6}$$

The response matrix is obtained by using Monte Carlo simulations of the energy deposition within the detector using the same methodology as in [5].
The experimental spectra have a Gaussian distribution shape for the photons energy lines. We take this effect into account by modifying the MCNP5 simulations results, we use a fitting technique to approximate the resolution of the detector which is an experimentally measured data.
With MCNP5, we use a "FT8 Gaussian Energy Broadening" card so we can simulate the full width at half maximum (FWHM) around the peak.
An example of NaI detector response functions calculated with MCNP5 are presented in figure 7.

3 Application to Scintillation NaI Detector

As an example we present the simulation of a pulse height spectra obtained with a scintillation sodium iodide detector and a spherical source made of a natural uranium ball surrounded by a plexiglass shell as shown in figure 6.

Fig. 6. Detector and source geometry.

The example geometry is detailed in table 1. The distance between the center of the source and the detector is 10 cm.

Table 1. Example specification.

	SOURCE		DETECTOR	
Materials	Natural uranium	Plexiglass	NaI	Aluminum
Density (g/cm^3)	18.9	1.2	3.7	2.7
Dim. (cm)	R=0.5017	R=1.0	R=1.0 H=3.0	Thickness 0.2cm

The results shown in this paper are given for guidance.

The leakage spectra for the gamma particles leaving the source is calculated using SGRD. As shown in figure 7 there is a good agreement with a corresponding MCNP5 calculation.

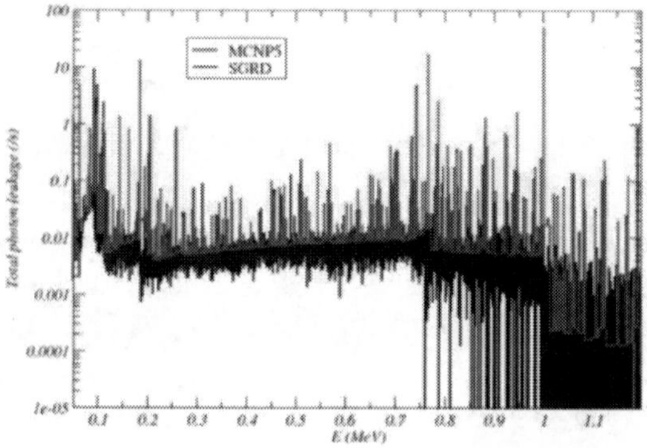

Fig. 7. Source leakage spectra calculated using MCNP5 and SGRD.

The pulse height spectrum is obtained by folding the SGRD leakage spectrum with the pre-calculated MCNP5 DRF. They are obtained using MCNP5 pulse height tally and Gaussian broadening to take into account the detector resolution.

A non-linear function was applied to estimate the values of the triplet (a, b, c) used to fit the FWHM in function of the photon energy.

Let $FWHM = a + b\sqrt{E + cE^2}$ (E = incident gamma rays energy in MeV, (a, b, c) : constants from a fitting or from a system of a few equations).

The triplet used was : (a, b, c) = (-0.002, 0.05, 2.86).

A set of response functions is presented in figure 8 for different incident energy of the gamma particles entering the NaI detector. The DRF which are between two calculated ones are interpolated.

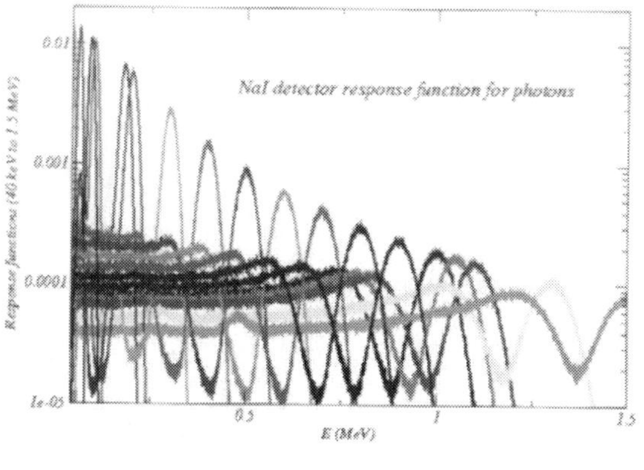

Fig. 8. Scintillation NaI detector response functions

For verification purpose the same calculation was performed using MCNP5 only. Both spectra are presented in figure 9. They show a very good agreement although the SGRD calculation is much faster and less prone to statistical artifacts. The typical computer time for SGRD is of several seconds compared to one hour or more for a classical Monte Carlo calculation.

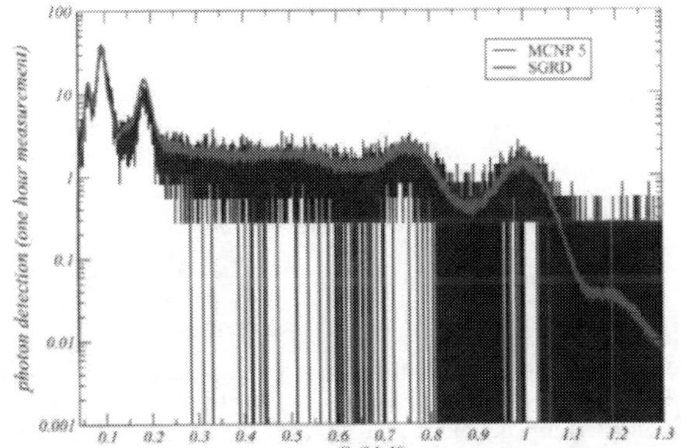

Fig. 9. Pulse height spectra calculated using MCNP5 and SGRD.

4 Conclusion

SGRD code which is used for fast calculation of uncollided and total leakage spectrum out of spherical shielded sources has been updated in order to simulate the detector pulse height spectra. For this purpose the detector response function calculated using MCNP5 is folded with the leakage spectrum. The methodology is illustrated on a test problem with a gamma source and a scintillation NaI detector. The results are in good agreement with a full Monte Carlo calculation.

Future developments will include the extension of the source term with neutron induced gamma rays and electron bremsstrahlung radiation.

References

1. Ph. Humbert, B. Méchitoua, "Fast Gamma Ray Leakage Spectra Simulation,"*Proc. M&C 2009*, Saratoga Springs, New York, May 3–7, 2009, American Nuclear Society (2009) (CD-ROM).
2. A. Tsilanizara et al., "DARWIN: an evolution code system for a large range of applications," Proc. Of the 9th Int. Conf. on Radiation Shielding, Tsukuba, Japan, Oct. 17-22, p. 845 (1999).
3. X-5 Monte Carlo Team 2003, "MCNP A General Monte Carlo N-Particle Transport Code Version 5," LA-UR-03-1987, (2003).
4. D.J. Mitchell, J. Mattingly, "Rapid Computation of Gamma-ray Spectra for One-Dimensional Source Models", *Trans. Am. Nuc. Soc.*, **98**, pp565-568 (2008).
5. C.M. Salgado et al., "Validation of a NaI(Tl) detector's model developed with MCNP-X code," *Progress in Nuclear Energy*, **59**, 19-25 (2012).

Visual Simultaneous Localization And Mapping (VSLAM) methods applied to indoor 3D topographical and radiological mapping in real-time

Felix Hautot[1,3], Philippe Dubart[2], Charles-Olivier Bacri[3], Benjamin Chagneau[2], Roger Abou-Khalil[4]

[1]*AREVA D&S, Technical Department, 1 route de la Noue 91196 Gif-sur-Yvette, France*
[2]*AREVA D&S Technical Department, Marcoule, France*
[3]*CSNSM (IN2P3/CNRS), Bat 104 et 108, 91405 Orsay, France*
[4]*AREVA Corporate, Innovation Department, 1 place Jean Millier, 92084 Paris La Défense, France*

Abstract. New developments in the field of robotics and computer vision enables to merge sensors to allow fast real-time localization of radiological measurements in the space/volume with near-real time radioactive sources identification and characterization. These capabilities lead nuclear investigations to a more efficient way for operators' dosimetry evaluation, intervention *scenarii* and risks mitigation and simulations, such as accidents in unknown potentially contaminated areas or during dismantling operations

1 Introduction

Nuclear back-end activities such as decontamination and dismantling lead stakeholders to develop new methods in order to decrease operators dose rate integration and increase the efficiency of waste management. One of the current fields of investigations concerns exploration of potentially contaminated premises. These explorations are preliminary to any kind of operation; they must be precise, exhaustive and reliable, especially concerning radioactivity localization in volume.

Furthermore, after Fukushima nuclear accident, and due to lack of efficient indoor investigations solutions, operators were conducted to find new methods of investigations in order to evaluate the dispersion of radionuclides in destroyed zones, especially for outdoor areas, using Global Positioning Systems (GPS) and Geographical Information Systems (GIS) as described in [1]. In both cases, nuclear dismantling and accidents situations, the first aim is to explore unknown potentially contaminated areas and premises so as to locate radioactive sources. Previous methods needed Geographical Information systems and Global positioning systems or placement of markers inside the building before localization of measurements, but plans and maps are often out-dated or unavailable.

Since the end of 2000's years, new emergent technologies in the field of video games and robotics enable to consider fast computations due to new embedded GPU and CPU architectures. Since the Microsoft Kinect™ has been released in 2010, a lot of developers "hacked" the 3D camera system in order to use 3D video streams in many fields of use such as robotics, motion capture or 3D imaging processing algorithms development. During the few following years, Light and low power consuming 3D cameras enabled to consider new 3D reconstruction of environment methods such as Simultaneous Localization And Mapping (SLAM) based on visual odometry [2].

This paper will present new progresses in merging SLAM and nuclear measurement in motion methods in order to detect, locate, and evaluate the activity of radioactive sources in 3D.

These new methods enable to reconstruct indoor areas and eventually out-door areas in real-time and 3D and also reconstruct 3D radioactive sources in volume.

In 2013, AREVA D&S started an R&D program for developing new investigations technics based on autonomous sensing robotics and localization apparatus in order to provide new efficient exploration and characterization methods of contaminated premises and areas. This work corresponds to MANUELA™ system developments by Areva D&S. Part of this work is protected by a patent (number WO2015024694) [3].

2 Material and Methods

2.1 General method

The presented method is based on two completely different technics. The first one, which is called SLAM, is well known in the field of robotics, it constitutes a specific branch of computer perception R&D. The second one, as described in [4], concerns radioactive sources localization and activity quantification from *in-situ* measurements and data acquisitions. The usual method for these acquisition are time consuming for operators and, in consequence, integrated dose of workers during these investigation could be decreased.

* Corresponding author: hautot@csnsm.in2p3.fr

Chen *et al.* [5] described such a mapping system based on merging RGBD-camera with radioactive sensors. The presented system automatically detects radioactive sources by estimating their intensities during acquisition with a deported computer. Our aim was to build a complete autonomous system for being totally independent of any external structures, including GPS.

A great problem for an autonomous apparatus (such as robot) is to locate itself in unknown environments, in order to compute appropriate motions and trajectories in volume. A simple formulation of this problem is that the apparatus must know its position in a map in order to estimate its trajectory. Using sensors, the system will build a map and compute its next position (translation and orientation in volume) at each acquisition step.

In order to compute SLAM inherent calculation in autonomous and light device development context, hardware specifications investigations are particularly important, due to required software performances.

2.2 Hardware

The presented radiological mapping system is embedded and designed for real-time investigations inside contaminated areas or premises. The whole system is enclosed and autonomous, and needs no external marker or network for being active. However, the operator's real-time intervention requires real-time reconstruction and visualization, which is very performance-consuming.

2.2.1 Sensors

The system software input uses different sensors:

- 3D camera based on active stereoscopy. As shown in figure 1, this camera's output consists in two different kinds of frame, a normal colour pixels image, and a depth map. The depth map is based on active stereoscopy technique and provides each colour pixel distance to sensor.

Figure 1: Outcoming data from 3D sensor (left: depth-map, right: RGB image)

- Nuclear measurements sensors including a dose rate meter and a micro CZT spectrometer (Figure 2)

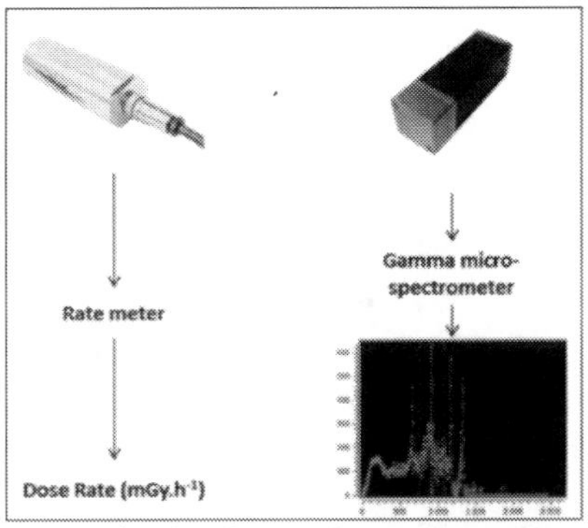

Figure 2: Outcoming data from nuclear measurements sensors

2.2.2 Computing unit

The 3D reconstruction and nuclear measurements are performed fully embedded, in real-time, due to operator interactions and acquisition time optimization in contaminated environment. Furthermore, the computing hardware must be fan-less in order to avoid nuclear contamination. So as to satisfy these constraints, the embedded CPU must be enough powerful for supporting parallel processing.

2.3 Simultaneous localization And Mapping

SLAM concept (simultaneous localization and mapping) can be performed by merging different kind of sensors, such as Inertial Measurement Units (IMU), accelerometers, sonars, lidars, cameras for example. In our method, only 3D cameras are used. We propose to merge two different kinds of algorithms so as to reconstruct the environment in 3D, compute the trajectory with 6 degrees of freedom (translation, pitch, roll, and yaw) in volume, and merge measurements with the device's poses.

Two problems appear during that kind of acquisition. First of all, slight error during the odometry computation causes a non-regular drift of the trajectory. The second problem concerns the memory management of acquisitions in real time. Indeed, 3D video gross data can quickly cost a considerable amount of active memory during the acquisition. Then, implementing a circular buffer is necessary for increasing the scanning volume up to hundreds of cube meters.

In order to develop our measurement method, we modified the RtabMap software [6-7-8] provided by IntroLab (Sherbrooke). By this way, we are able to use visual odometry with 3D cameras in order to reconstruct the environment and compute the device trajectory at 25Hz.

Pose-graph visual Slam is based on the principle that each acquisition step is a combination of constraints links between observations. These constraints are established using features detection and extractions of each processed image. This kind of SLAM problem is represented with graphs of constraints. Each observation of the robot creates node, new links and constraints. This method allows fast node recognition and including loop & closure based-on optimization methods.

2.3.1 Visual Odometry

The goal of visual odometry is to detect the relative motion between two poses of the camera, and then to back-project the 3D and RGB streams in a computing reconstruction volume. This problem can be exposed as equation 1. This equation describes the transformation of each pixel of the camera to a 3D point, depending on intrinsic and extrinsic camera parameters (Equation 1).

$$z\begin{pmatrix}u\\v\\1\end{pmatrix}=\begin{pmatrix}f_x & 0 & c_x & 0\\0 & f_y & c_y & 0\\0 & 0 & 1 & 0\end{pmatrix}\begin{pmatrix}R_{1,1} & R_{1,2} & R_{1,3} & T_1\\R_{2,1} & R_{2,2} & R_{2,3} & T_2\\R_{3,1} & R_{3,2} & R_{3,3} & T_3\\0 & 0 & 0 & 1\end{pmatrix}\begin{pmatrix}x\\y\\z\\1\end{pmatrix} \quad [1]$$

pixel — Camera intrinsic parameters — Camera extrinsic parameters — 3D point

z : depth at pixel (u,v)
u, v : coordinates of the considered pixel
f_x : focal length along x
f_y : focal length along y
c_x, c_y : lens centering on the optical axis
$R_{a,b}$: element of the rotation matrix
T_a: element of the translation vector
x, y, z : projected point coordinates in volume.

Visual odometry is processed on a real-time RGB data stream to detect the motions of the device in volume and get the colour for reconstruction. Simultaneously, the corresponding depth stream is used for calculating the rotation/translation matrix between two successive frames.

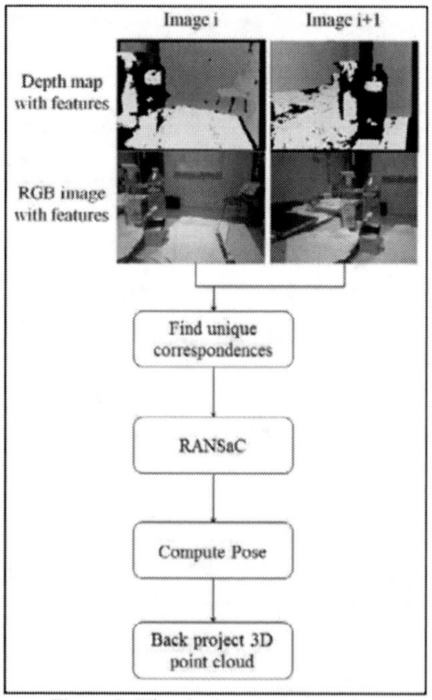

Figure 3: Visual odometry processing

Visual odometry is features extraction based-on. Each RGB image is processed in order to extract interest points. These interest points are formed by image corners. The corresponding pixel in the depth map is also extracted.

Depending on the pine-hole model, features are then back-projected in volume.

As described in figure 3, unique correspondences are researched between two sets of consecutive images. If enough unique correspondences are detected, then, odometry is processed. A RANdom SAmple Consensus (RANSAC) algorithm calculates the best spatial transformation between input images (rotation and translation).

2.3.2 Loop and closure

Errors during the pose matrix computation cause a non-regular drift of the trajectory. Graph-Slam and constrained optimization methods based on loop-and-closure correct this drift when a previously scanned zone is reached by adding constraints to previous acquired constraint graph as described in figure 4 [6].

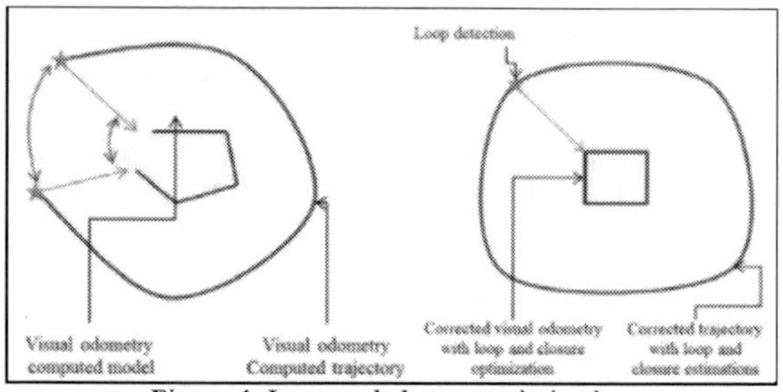

Figure 4: Loop and closure optimization

2.4 Nuclear measurement Management and location

Measurements in motion related work [9] by F. Panza, describes a system used within motions in two dimensions with collimated measurements probes. In this case, using leads collimator could be possible, but our case concerns a handheld system measuring in a near 4pi sphere, and moving with six degrees or freedom.

All the data (nuclear measurement and positioning, 3D geographical and trajectory reconstruction) are performed in real-time while the device can have different kinds of status: moving or motionless. All the nuclear measurements are considered isotropic.

Each set of measurement (integrated or not) is attached to the graph-slam geographical constraints structure. This allows performing trajectory optimizations and measurement positioning optimization simultaneously.

In order to satisfy the "real-time" constraint, a user interface displays every current measurement and process step in real-time in order to provide all pertinent and essential information to the operator (figure 5).

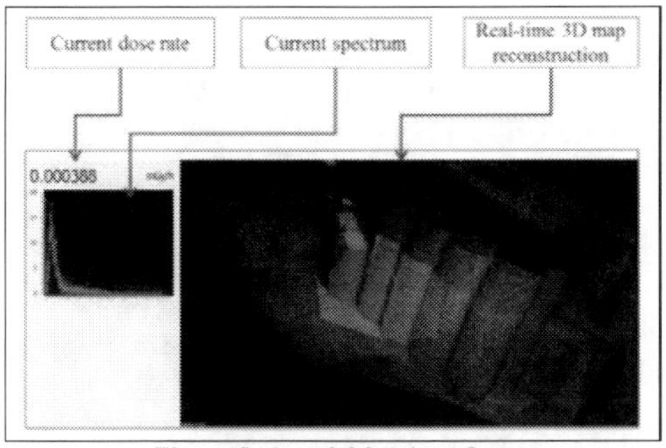

Figure 5: Acquisition interface

2.4.1 Continuous measurement

Dose rate measurements are processed during the 3D reconstruction with a lower frequency (around 2Hz) than video processing (around 20-25Hz). In order to manage the nuclear measurement positioning, we had to find a compromise between positioning uncertainty, which depends on the counting time and counting uncertainty that depends on the inverse counting time. So, first and foremost, dose rate measurement is positioned at half path distance during integration (Figure 6).

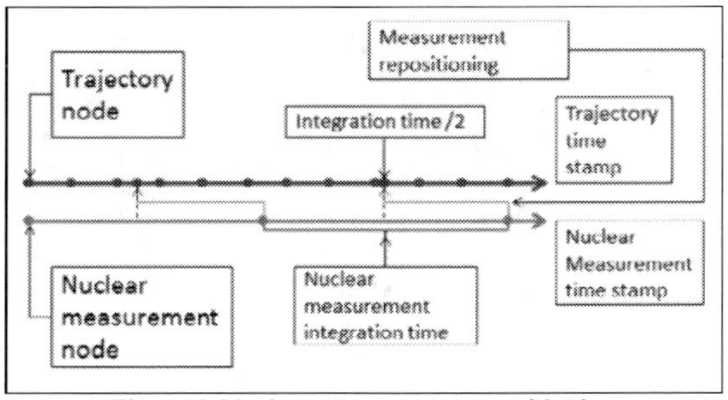

Figure 6: Nuclear measurements positioning

Gamma spectrometry measurements are processed with an even lower frequency (around 0.3-0.5Hz) than the dose rate measurements. Consequently, the uncertainty on spectrum positioning is more important, compared to dose rate positioning. So as to compensate this error, dose rate values will help to distribute weighted spectrums for the acquired one (Figure 7).

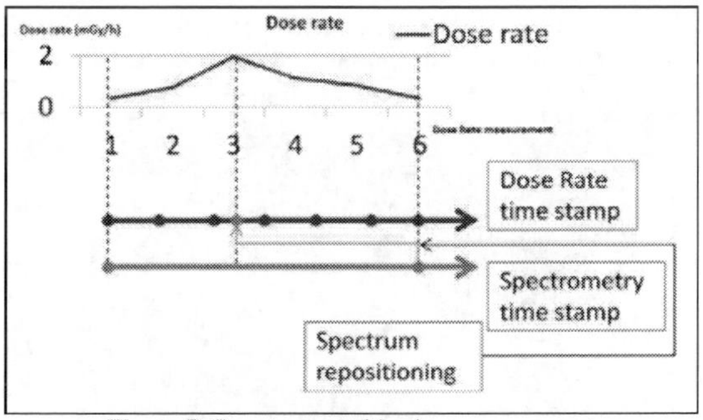

Figure 7: Spectrum positioning management

2.4.2 Integrating measurement

In some case, very precise measurements are required to build a representative map of the environment. The fast pose calculation method we use allows considering the device as a 3D accelerometer with a higher frequency than nuclear measurements. While the 3D video stream is being acquired, the acceleration of the device is estimated and if the device is motionless, measurements can be integrated at the current pose (Figure 8)

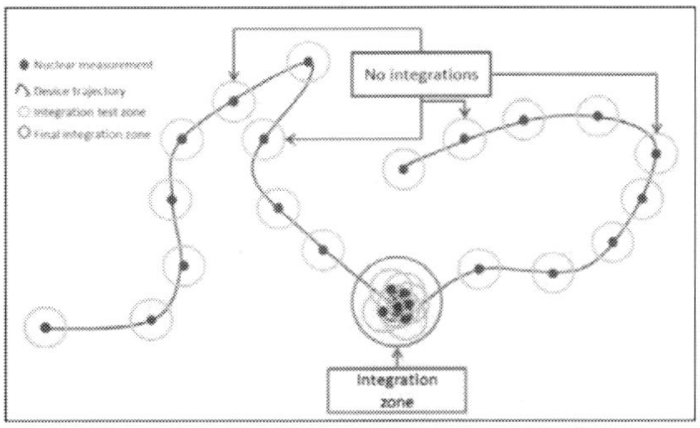

Figure 8: radioactive measurements positioning

2.5 Near real-time post-processing, sources localization

At the end of acquisition, radioactive source localization computation methods are available with a set of algorithms that provide interpolations and back-projections of measured radioactive data in volume. The algorithms are optimized for providing results in a few seconds, even if uncertainties could be reduced by more accurate methods.

2.5.1 Measurements 3D interpolation

For interpolating measurements in 3D, we use a simple deterministic Inverse Distance Weighting (IDW) method, which is accurate enough considering the usual radioprotection operating accuracy. Furthermore, this fast computed method allows operators to consider the operating room state of contamination very quickly with this embedded method. The used IDW method is described within equation 2 and 3.

$$v(x) = \frac{\sum_{i=0}^{n} w_i(x) \cdot v_i}{\sum_{i=0}^{n} w_i(x)} \quad [2]$$

with:

$$w_i(x) = \frac{1}{D_{x,x_i}{}^{P}} \quad [3]$$

v(x): interpolated value at x
$\mathbf{w_i}$**:** weight of the measurement point i
$\mathbf{D_{x,x_i}{}^{P}}$**:** distance between current interpolated point and measurement point i.
n: number of measured points

2.5.2 Dose rate back projection

Back projection method is also deterministic and uses the 3D reconstruction to compute radiation emission zones in volume. This method is described within equations 4 and 5.

$$B(x) = \frac{\sum_{i=0}^{n} \frac{e^{-\mu_a D_{x,x}}}{D_{x,x}{}^{2}}}{n} \cdot w(x) \quad [4]$$

$$w(x) = f(var(B(x))) \quad [5]$$

n : Number of nuclear measurement point.
$B(x)$: Back projection value ($mGy.h^{-1}$)
x : Location (x_1,y_1,z_1) of back-projected value
X: Location (x_2,y_2,z_2) of nuclear measurement point
μ_a : Linear attenuation coefficient of air (cm^{-2})
$w(x)$: Weight associated to x location.
$D_{X,x}$: Distance between X and x location

The back-projection algorithm inputs are:
- 3D reconstruction decimated point cloud
- Nuclear measurements and position data.

Each point of the 3D point cloud (x in eq.4) is considered as a possible radioactive source, then, emerging mean fluency or dose rate at the 3D reconstruction point (B(x) in eq.4) is computed for every measured point (X in eq.4). Then, variance distribution of the back projected value enables to evaluate the possibility of radioactive source presence in volume at the back projected point.

2.5.3 Topographical study

Topographical measurements can be performed as soon as the acquisition is terminated. This function gives instant information on the situation of premises.

2.6 Offline post-processing

The device output data can be processed in back office with a set of tools for estimating accurate gamma-emitting sources localization and quantification in volume. It also provides tools for estimating effects of a dismantling or decommissioning operation on dose rate distribution and allows the user to estimate the exposure of operators during interventions.

Next subparagraphs will present these different tools such as radioactive sources quantification, operators' avatars, and topographic studies.

2.6.1 Topographical measurements

The dedicated post-processing software provides two kinds of topographical study tools, according to figure 9: a global grid containing the scanned volume for global intervention prevision and a drag and drop tool for specific structure measurements (volumes, length, and thickness).

Figure 9: topographical study interface presenting dimensioning tools.

2.6.2 3D Nuclear Measurements interpolation

Nuclear measurement interpolation characterization tool (Figure 10) is based on IDW, and uses the same principle than the near real time post-processing interpolation method, however, slights modifications of scales allows user to refine the computation and then locate low emitting sources. Moreover, spectrometry can be exploited by interpolating user's defined region of interest of the spectrum. This enables specific studies concerning radionuclides diffusion in the investigated area, such as ^{137}Cs or ^{60}Co containers localization for example.

Figure 10: Radioactive measurements 3D interpolation

2.6.3 3D Back-projection and avatar dose integration simulation

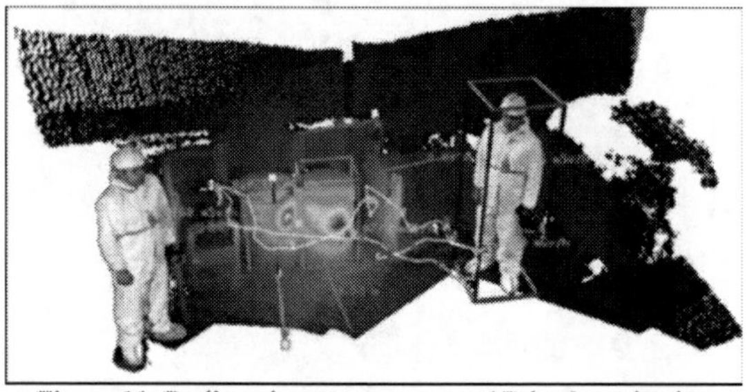

Figure 11: Radioactive measurements 3D back-projection

Back-projection algorithm also benefits of an improved interface in order to locate accurately radionuclides in volume using spectrometry (Figure 11).

Avatars of operators can be used for estimating previsions of their exposure before operations. This dose rate integration estimation is performed by extrapolating dose rates from measurements points.

2.6.4 Sources activities estimations

Radioactive sources activities are estimated with a set of algorithms combining 3D transfer functions calculations and minimization methods (Figure 12).

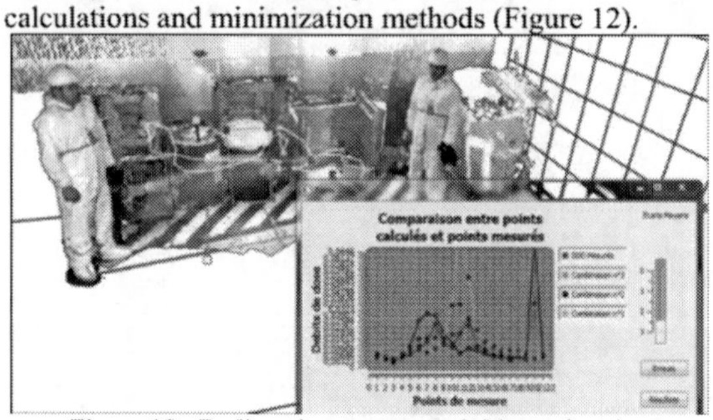

Figure 12 : Radioactive sources activities estimation

- Transfer function calculation

The transfer function will quantify the relation between volume radioactive source activities and resulting dose rate for a specific radionuclide.

The first step of radioactive sources estimation consists in modelling them according to available data provided by the results of acquisition on a hand, and by operating documents on the other hand (volume, enclosure type, shielding, materials, radionuclides) (Figure 13).

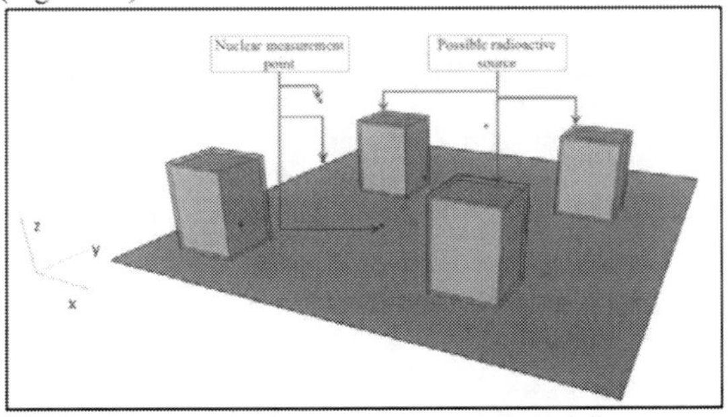

Figure 13: Radioactive sources scene modeling

In order to satisfy the nearest real-time calculation constraint, the transfer function calculation method is deterministic and based on ray-tracing and radioactive kernels point distribution in volume. Potential radioactive sources are designed by the user with the help of localization algorithm.

Equation [6] describes the transfer function calculation method.

$$\dot{D}_{Bu} = \frac{A_v}{4\pi} \cdot C \iiint_V \prod_{i=0}^{n} \frac{Bu_i e^{-\mu_i d_i}}{d_i^2} \cdot dV \qquad [6]$$

n : Number of attenuating volumes on the ray path

$\dot{D}_{Bu}$: Dose Rate at Measurement point (considering the Build-up factor)

A_v : Volume activity of the radioactive source.

C : Dose Rate – gamma fluency conversion coefficient

Bu : Build-up factor for the "i" attenuating volume

μ_i : Linear attenuation coefficient of the « i » attenuating volume

d_i : Path length in the « i » attenuating volume.

The numerical integration method for source kernels distribution in the source is a Gauss-Legendre integration based-on method.

- Radioactive sources activities minimization method.

The minimization method is based on iterative technique for which each step consists in considering a different combination of radioactive sources.

The equation system resolution method is based on the most important transfer function selection at each step of calculation.

The following figure (figure 14) presents the whole algorithm process for computing sources activities.

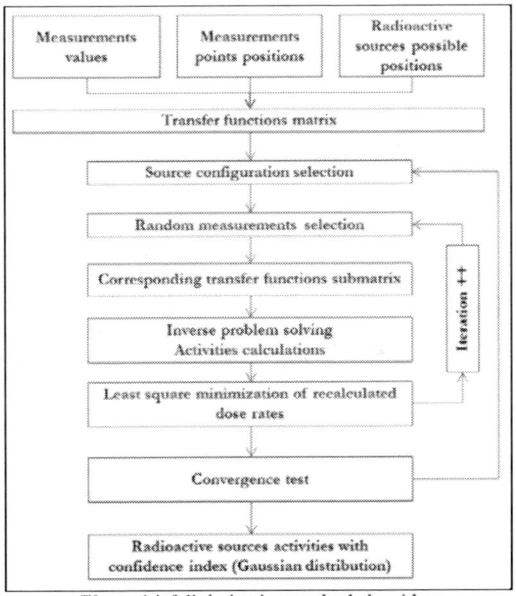

Figure 14: Minimization method algorithm

3. Benchmarks

Since June 2016, a set of benchmarks is performed in order to compare the performances of this acquisition system with different systems that can be considered as references. Different parts of the system are compared such as:

- 3D volumetric reconstruction
- 3D trajectory reconstruction
- Dose Rate measurements
- Spectrometry measurements
- 3D nuclear measurements interpolation
- 3D radioactive source localization without collimator.

3.1 3D volumetric reconstructions

This part of the benchmark has been realized with a 3D scanner and is based on point cloud registration techniques such as Iterative Closest Points (ICP) method. We used "Cloud compare", developed by Telecom ParisTech and EDF R&D department.

The reference point cloud was acquired with a 3D laser scanner from Faro Corporation (Table 1).

Table 1: Comparison data

Device	Faro laser scanner	MANUELA Device
Error	6mm/10meters	?
Acquisition time	~11mn	~3mn
Dynamic/ static	Static	Dynamic
Number of vertex	2.157E+3	1.470E+3

3.1.1 Benchmark example

As an experimental test, we used a storage room "as it is" (Figure 15)

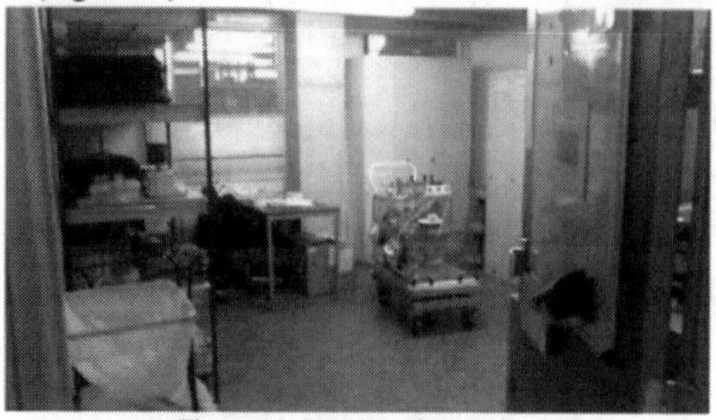

Figure 15: Benchmark situation

Room dimensions: 25 square meters, 2.5 meters high.

3.1.2 Results:

Following clouds superposition prepositioning, the ICP registration algorithm is performed in order to compute the best cloud-to-cloud matching. Then, point-to-point distance is computed for the whole reference point cloud.

Figures 16 to 18 show a part of the benchmark procedure.

The first step in figure 16 concerns Faro and MANUELA point clouds superposition in order to prepare the matching process.

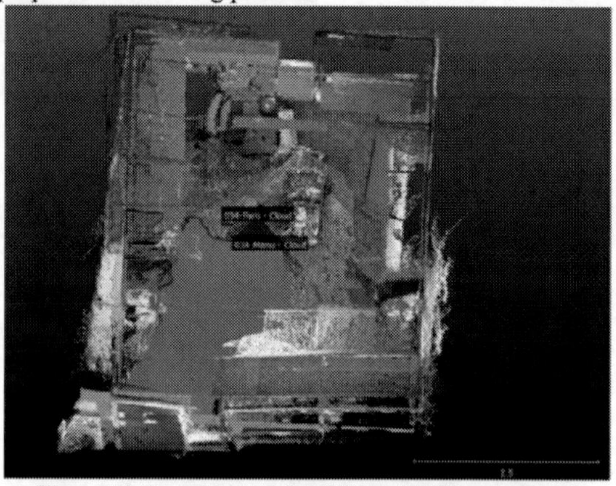

Figure 16: Faro (blue) / MANUELA (RGB) point cloud superposition and registration

The second step consists in using ICP matching, and then computing point-to-point distances in the best matching case. In figure 17, the colour scale displays the distribution of point-to-point distances between point clouds, and the chart shows the distribution of these distances in the whole scene.

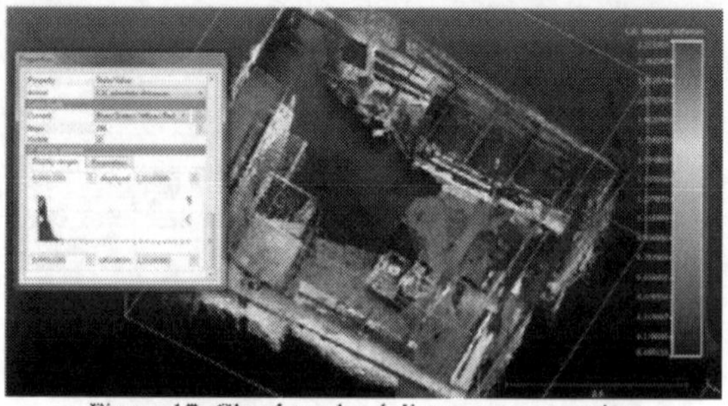

Figure 17: Cloud-to-cloud distance computation

The last graph displays the point-to-point distances distributions as a Gaussian (Figure 18 shows the chart of figure 17 with a different binning).

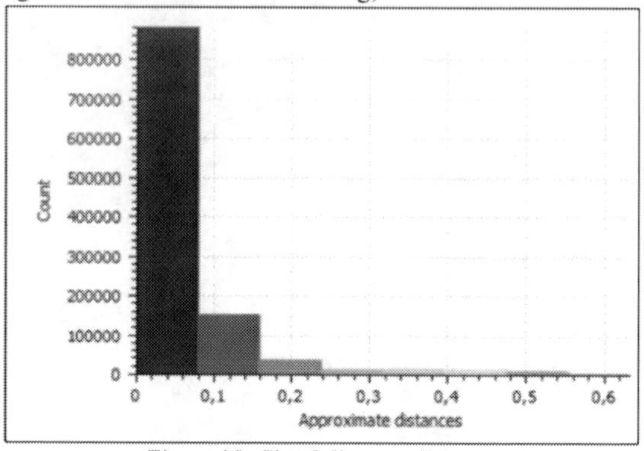

Figure 18: Cloud distance diagram

General conclusion of 3D point cloud reconstruction Benchmark shows a Gaussian distribution of cloud-to-cloud distance. Global cloud-to-cloud mean distance is around $7.3e^{-2}$m (σ $=10.2e^{-2}$m).

4 Uncertainties estimation

4.1 General considerations

Current work is going on estimating uncertainties on each step of the acquisition.

Uncertainties estimation and propagation in such kind of acquisition and processing system is necessary for interpreting the results, estimating the relative probability of presence of radioactive sources, and building sensibility analysis in order to increase the system performance, by detecting the most uncertainty generator step in the process.

The whole process chain depends on four simple entries, the RGB image, the depth map, the dose rate measurements and the spectra measurements. Each one of these entry elements is tainted by systematic and stochastic uncertainties. Moreover, each step of process generates uncertainties and amplifies the input uncertainties.

In this paragraph, we will describe all parts of the acquisition process and present the current results on a new method we use for propagating uncertainties in real-time. We compared this new method to a Monte-Carlo calculation that will be considered as the reference.

In the case of input detectors, we will only consider the model generated or counting (nuclear) measurement uncertainties.

4.1.1 Uncertainty estimation, general description

The acquisition and processing devices is decomposable in a few building blocks (sensor or processing technic). They will be considered separately in order to quantify each block uncertainty (Figure 19).

Each block will be considered as a linear system such as described in equation [7-8].

$$\mathrm{O} = \mathrm{T} \cdot \mathrm{I} \qquad [7]$$

$$\begin{pmatrix} \mathrm{u} \\ \mathrm{v} \\ \mathrm{w} \end{pmatrix} = \begin{pmatrix} \mathrm{a}_{1,1} & \mathrm{a}_{1,2} & \mathrm{a}_{1,3} \\ \mathrm{a}_{2,1} & \mathrm{a}_{2,2} & \mathrm{a}_{2,3} \\ \mathrm{a}_{3,1} & \mathrm{a}_{3,2} & \mathrm{a}_{3,3} \end{pmatrix} \begin{pmatrix} \mathrm{x} \\ \mathrm{y} \\ \mathrm{z} \end{pmatrix} \qquad [8]$$

$\mathrm{O}(\mathrm{u}, \mathrm{v}, \mathrm{w})$: Output vector
$\mathrm{T}(a_{x,x}, \ldots, a_{x,x})$: Transformation matrix
$\mathrm{I}(\mathrm{x}, \mathrm{y}, \mathrm{z})$: Input vector

Each of these blocks generates an intrinsic uncertainty and amplifies inputs errors. Then, the final $\|dO\|$ will represent the global uncertainty on the acquisition and processing chain.

Each system uncertainty is considered as:

- $\|dO\| = f(\|dT\|)$ error intrinsic generation estimation (Equation 9)

$$\begin{pmatrix} \mathbf{u + du} \\ \mathbf{v + dv} \\ \mathbf{w + dw} \end{pmatrix} = \begin{pmatrix} \mathbf{a}_{1,1} + \mathbf{da}_{1,1} & \mathbf{a}_{1,2} + \mathbf{da}_{1,2} & \mathbf{a}_{1,3} + \mathbf{a}_{1,3} \\ \mathbf{a}_{2,1} + \mathbf{da}_{2,1} & \mathbf{a}_{2,2} + \mathbf{da}_{2,2} & \mathbf{a}_{2,3} + \mathbf{da}_{2,3} \\ \mathbf{a}_{3,1} + \mathbf{da}_{3,1} & \mathbf{a}_{3,2} + \mathbf{da}_{3,2} & \mathbf{a}_{3,3} + \mathbf{da}_{3,3} \end{pmatrix} \begin{pmatrix} \mathbf{y} \\ \mathbf{y} \\ \mathbf{z} \end{pmatrix} \qquad [9]$$

- $\|dO\| = f(\|dI\|)$ error amplification estimation (Equation 10)

$$\begin{pmatrix} \mathbf{u + du} \\ \mathbf{v + dv} \\ \mathbf{w + dw} \end{pmatrix} = \begin{pmatrix} \mathbf{a}_{1,1} & \mathbf{a}_{1,2} & \mathbf{a}_{1,3} \\ \mathbf{a}_{2,1} & \mathbf{a}_{2,2} & \mathbf{a}_{2,3} \\ \mathbf{a}_{3,1} & \mathbf{a}_{3,2} & \mathbf{a}_{3,3} \end{pmatrix} \begin{pmatrix} \mathbf{x + dx} \\ \mathbf{y + dy} \\ \mathbf{z + dz} \end{pmatrix} \qquad [10]$$

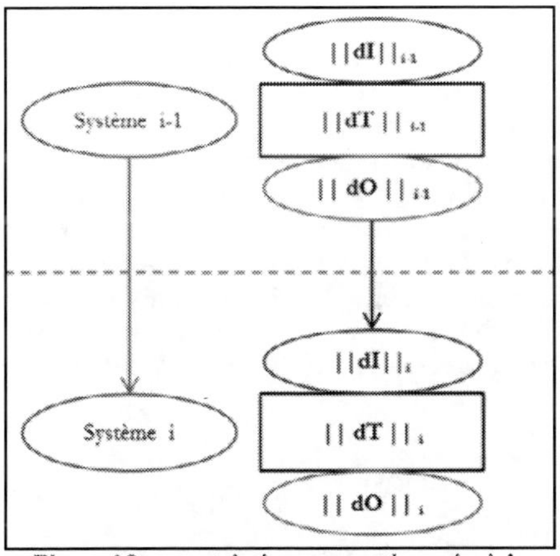

Figure 19: uncertainties propagation principle

We use Frobenius matrix norm (Equation 11-12) in order to quantify the global variation of matrix terms.

$$A = \begin{pmatrix} a_{0,0} & \cdots & a_{0,m} \\ \vdots & \ddots & \vdots \\ a_{n,0} & \cdots & a_{n,m} \end{pmatrix} \qquad [11]$$

$$\|A\| = \sqrt{\sum_{i=0,j=0}^{i=n,j=m} a_{i,j}} \qquad [12]$$

4.1.2 Monte Carlo methods

Monte Carlo methods are efficient for propagating and estimating models uncertainties accurately. The main problem in such method is computing time. This method will be considered as reference for comparison with the new method we developed. Furthermore, Monte-Carlo method will provide accurate results on each acquisition and process steps and will help to estimate how performances will be increased.

4.1.3 Interval arithmetic method

Interval arithmetic methods have been initially developed due to rounding errors of floating values in computing calculation. The principle is to describe a scalar value with an interval. This help to consider not finite floating values as a duo of floating scalars with specific rounding policies. For example, $\pi \in [3.14; 3.15]$.

This description of values will change arithmetic rules. For example, mathematical product becomes:

$$[-5; -4.5] * [-0.5; 2] = [-10; 2.5]$$

With this arithmetic, we developed a method for propagating uncertainties as boundaries of a noisy system. We will compare this method with Monte-Carlo uncertainties estimations method. We choose, as an example case, the 3D camera projection Matrix (pine-hole model).

4.2 3D Camera calibrations

The 3D camera output data is a duet of frames, a RGB one and a depth map. Each of them can be described as pinholes. Then, pixels 3D back-projections in volume are considered in figure (13)

$$\mathrm{z}\begin{pmatrix} \mathrm{u} \\ \mathrm{v} \\ 1 \end{pmatrix} = \begin{pmatrix} \mathrm{f_x} & 0 & \mathrm{c_x} \\ 0 & \mathrm{f_y} & \mathrm{c_y} \\ 0 & 0 & 1 \end{pmatrix} \begin{pmatrix} \mathrm{x} \\ \mathrm{y} \\ \mathrm{z} \end{pmatrix} \qquad [13]$$

z : Depth at pixel (u,v)
u, v : Coordinates of the considered pixel
$\mathrm{f_x}$: Focal length along x
$\mathrm{f_y}$: Focal length along y
$\mathrm{c_x}, \mathrm{c_y}$: Lens centering on the optical axis
x, y, z : Projected point coordinates in volume.

Values of interest in this models are (x,y,z) coordinates of the projected pixel. These coordinates, depending on features detection, will be processed in the VSLAM algorithm.

So as to provide interpretable results, the pinhole must be calibrated, and the calibration process will give interpretable back-projection uncertainties on (x,y) coordinates. Uncertainty on z coordinate will be estimated in function of the depth, material and radioactivity level. Indeed, depth map is computed with active stereoscopy, which involves infrared coded grids projection and interpretations, which can be sensitive to external interferences.

4.2.1 Uncertainties estimations on camera calibration coefficients

The calibration of the camera consists in minimizing the pine hole projection matrix elements, using a well-known projection pattern.

The calculation of back-projections errors is performed with a Monte-Carlo random selection of calibration values c_x, c_y, f_x, f_y on a hand, and then, these back-projection errors are compared with intervals estimations. Deviations estimations methods are described with equation 14 and 15.

$$\begin{pmatrix} du+u \\ dv+v \\ 1 \end{pmatrix} = \begin{pmatrix} df_x+f_x & 0 & dc_x+c_x \\ 0 & df_y+f_y & dc_y+c_y \\ 0 & 0 & 1 \end{pmatrix} \begin{pmatrix} x \\ y \\ 1 \end{pmatrix} \quad [14]$$

$$\begin{pmatrix} du+u \\ dv+v \\ 1 \end{pmatrix} = \begin{pmatrix} f_x & 0 & c_x \\ 0 & f_y & c_y \\ 0 & 0 & 1 \end{pmatrix} \begin{pmatrix} dx+x \\ dy+y \\ 1 \end{pmatrix} \quad [15]$$

4.2.2 Monte-Carlo estimation:

This method gives exhaustive results on the camera response, but it takes around 2 minutes to be computed for each frame, which is not suitable with real-time computations.

The Figure 20 shows results of Monte-Carlo simulation varying the pinhole projection matrix elements; each matrix element could vary from -10 to +10%. 1e+6 random selections have been performed. The resulting error due to the lenses system will not cause a larger deviation than 80 pixels on the photographic sensor.

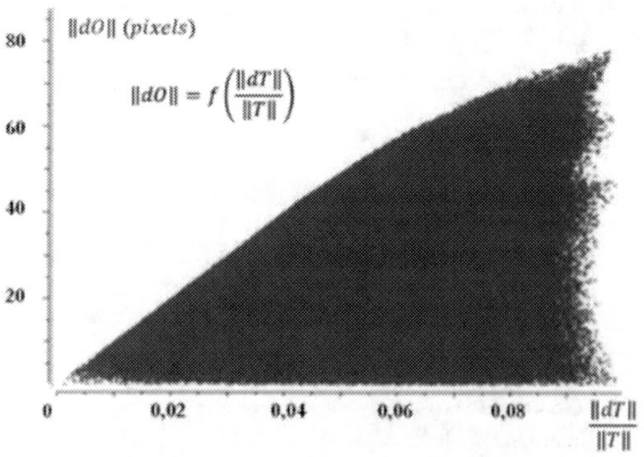

Figure 20: Monte-Carlo estimation of pinhole model uncertainties generation

The second diagram (Figure 21) shows the error amplification by the projection matrix. In this specific case, it doesn't have physical sense because it would represent the variation of a vertex coordinates as a function of pixel coordinate variation, which has no physical sense.

The pinhole model is linear, consequently, variation of output variation norms are distributed along a straight line.

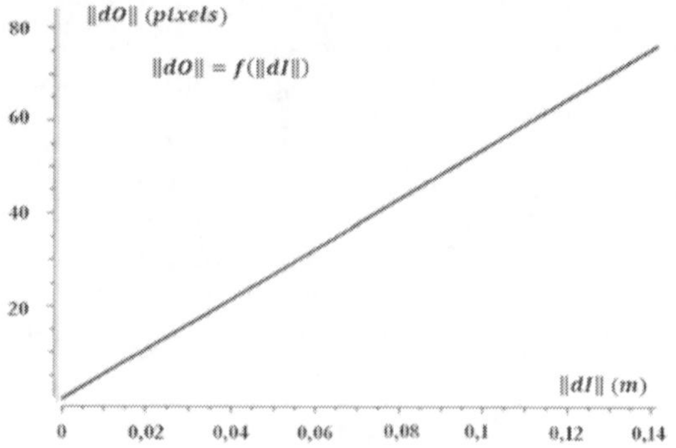

Figure 21: Linear system error amplification by Monte-Carlo methods

The last diagram (Figure 22) combines errors generated and amplified by the pinhole projection matrix. And this error will be considered as input error by the next processing block of the system. (for each value of $\|dI\|$, $\frac{\|dT\|}{\|T\|} < 0.1$)

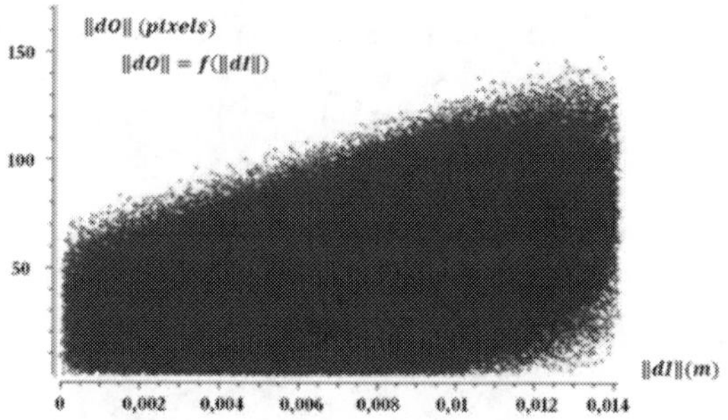

Figure 22: Global uncertainties generated by the pinhole model

4.2.3 Intervals estimation.

Provided results are less exhaustive than with Monte-Carlo methods, nevertheless, these results are bounding phase space provided by Monte-Carlo method. Furthermore, executing time is around 5ms per frame, which is compliant with real-time analysis and processing.

The figures 23-24-25 show comparisons between uncertainties computed by Monte-Carlo and by intervals. This graphical representation represents the bounding of Monte-Carlo computed values by interval arithmetic method. Interval arithmetic results have been added to Monte-Carlo charts in order to figure both methods results (interval arithmetic results have been encircled to remain visible).

The simulation has been performed considering exactly the same matrix element variations. Indeed, matrix element have been defined as intervals (equation 16)

$$\begin{pmatrix} a_{0,0} & \cdots & a_{0,j} \\ \vdots & \ddots & \vdots \\ a_{i,0} & \cdots & a_{i,j} \end{pmatrix} \rightarrow X = 0.1 \cdot a_{i,j} \quad [16]$$

$$\begin{pmatrix} [a_{0,0}-X, a_{0,0}+X] & \cdots & [a_{0,m}-X, a_{0,m}+X] \\ \vdots & \ddots & \vdots \\ [a_{n,0}-X, a_{n,0}+X] & \cdots & [a_{i,j}-X, a_{i,j}+X] \end{pmatrix}$$

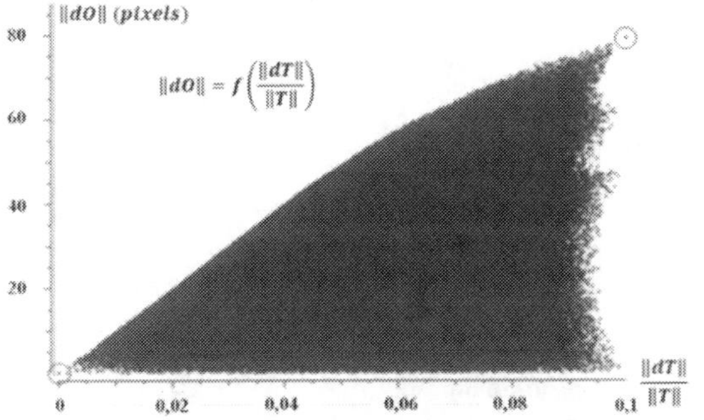

Figure 23: Comparison between Monte-Carlo and Intervals for estimating uncertainties of pinhole model

Results in figure 23, show a good concordance of both methods, intervals arithmetic provides limits of Monte-Carlo method for intrinsic uncertainties generation.

Concerning the amplification of input system uncertainties, the input vector will be defined as a vector of intervals (equation 17)

$$\begin{pmatrix} x \\ y \\ z \end{pmatrix} \rightarrow \begin{pmatrix} [x-dx, x+dx] \\ [y-dy, y+dy] \\ [z-dz, z+dz] \end{pmatrix} \quad [17]$$

Depending on this definition of the input vector, intervals of the output vector will be estimated, as figuring on figure 24.

In this case, intervals also provide bounds of Monte-Carlo simulation.

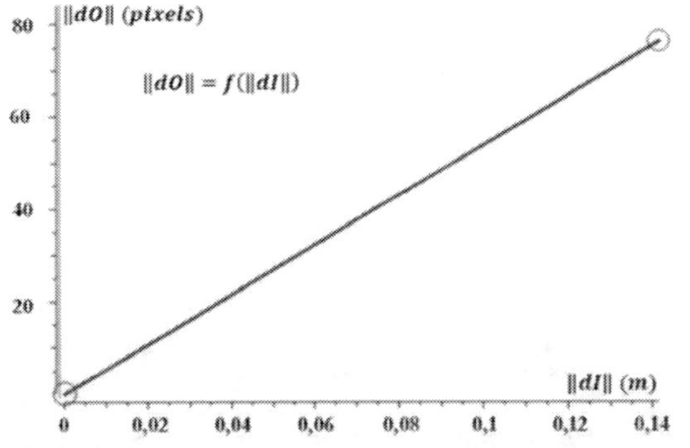

Figure 24: Comparison between Monte-Carlo and Intervals for estimating uncertainties Amplification by the pinhole model

Finally, by merging the system extrinsic and intrinsic parameters, a global estimation of generated uncertainties can also be provided with intervals technics, by bounding Monte-Carlo values (Figure 25).

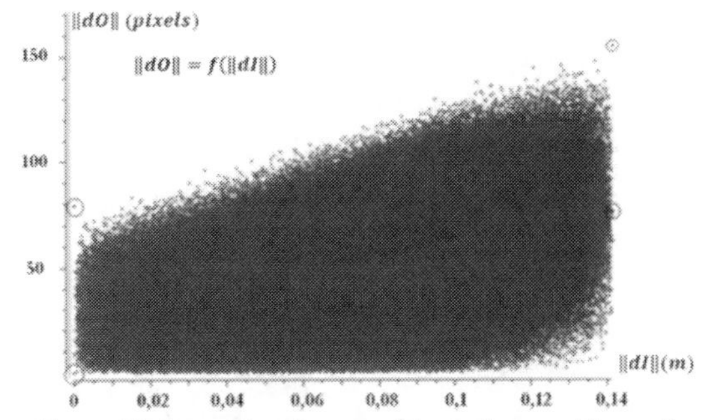

Figure 25: Comparison between Monte-Carlo and Intervals method for propagating the global uncertainties of the model

4.2.4 Performances benchmark

In this case, both methods can provide clear uncertainties results about input data uncertainties amplification and intrinsic uncertainties generation.

These methods can have different use cases depending on each method pros and cons.

The Monte-Carlo method will describe accurately the different systems, but interval arithmetic method can provide pertinent accuracy indices on each acquisition, in real-time or near real time. Indeed, Monte-Carlo algorithm runs in around two minutes, and interval arithmetic one runs in almost 5ms in the same study case.

4.3 Radiological measurement

The second kind of algorithms input data are nuclear probes outcoming ones.

4.3.1 Dose rate measurements

Dose rate measurements uncertainties are linked to counting errors. Equations [18-19-20] describe the global dose rate counting error.

Count to dose rate pre-processing:

$$\dot{D} = \frac{C_{pm} \cdot H}{(1 - C_{ps} \cdot \tau) \cdot S_d} \quad [18]$$

With:

$$C_{pm} = \frac{60 * C}{t} \quad [19]$$

And Relative uncertainty is given by:

$$u = \frac{k \cdot \sqrt{\frac{N}{t}}}{N} \quad [20]$$

$\dot{D}$: Dose rate (mGy.h^{-1})
C_{pm} : Count per minute
H: Conversion factor mR.h^{-1} to mGy.h^{-1}
C_{ps} : Count per second
τ: Detector's dead time constant (s)
S_d: Detector sensibility coefficient (Cpm.mR.h^{-1})
t: Acquisition time (s)
C: Count during the acquisition time t in seconds
N: Result of counting
u: Relative uncertainty (%)

4.3.2 Spectrum measurements

Gamma spectrometry also depends on counting uncertainties. In this case, counting uncertainties are considered for each spectrometry channel. Spectrometry uncertainties also depend on energy/efficiency calibration quality and acquisition parameters.

4.4 SLAM method uncertainties

During SLAM acquisitions, each pose computed by odometry and re-estimated following to loop and closure algorithm is associated to a variance, due to statistical way of calculating spatial transforms which gives a confidence index on each pose computation. Concerning the 3D reconstruction, uncertainty will combine poses computations and 3D projection uncertainties.

4.5 Ongoing work

Current work concerns the estimation of each building-block uncertainties generation and amplification, such as depth map variations as a function of the environment conditions, spatial transforms variance due to descriptors variations, numerical applications of back-projection or geostatistical interpolation. Radioactive sources localization uncertainties generation and propagation will also be estimated in order to provide a complete method for estimating uncertainties and errors in such kind of system that combines acquisition of physical measurements and automatic processing in near real-time.

Also, a full benchmark of the system is performed in order to correlate its results with real-time uncertainties estimations. Each building-block will be compared to well-known references systems. For example, in order to compare 3D reconstructions, we will use a Faro scanner as reference, and concerning trajectory computation, Vicon™ camera system will provide reference of spatial motions of the device.

5 Conclusions

The new presented method for 3D Indoor / Real-time Topographical and Radiological Mapping (ITRM), with Visual Simultaneous Localization And Mapping (VSLAM) has been developed to perform near real time acquisitions and post processing for radiological characterization of premises. Such method allows optimizations of operational time, personnel dose and waste minimization. One of the major challenges consists in optimizing the mobile experimental apparatus in order to satisfy the desired performances, by generating optimized programming code, selecting high performance computer units and measuring instrument. Such an optimization needs also a careful study of uncertainties introduced by the used instrumentation and developed algorithms. Indeed, physical measurement must be provided with associated uncertainty, in order to estimate its pertinence. Also, this uncertainties study method will finally enable performing a sensibility analysis of the whole system. Consequently, it will help to define the building blocks to improve and increase the system's performances.

Finally, this new approach that we developed and applied for radiological measurements within the nuclear field could be of interest in other domains by adapting the sensors for the required measurements (Chemical, environmental...).

This ongoing work is funded by ANRT (Agence Nationale de la Recherche et des Technologies) under a CIFRE contract (number 2013/1542) between AREVA/STMI and CNRS.

References

1. M. Morichi, H. Toubon, R. Venkataraman, J. Beaujoin, Ph. Dubart, "Nuclear measurement technologies & solutions implemented during nuclear accident at Fukushima" in Proceedings of the IEEE/ Advancements in Nuclear Instrumentation Measurement and their Applications (ANIMMA) international conference, 2013
2. T. Whelan, H Johannsson, M. Kaess, J.-J.Leonard, J.McDonald, "Robust real-time visual odometry for dense RGB-D mapping, in proceedings of the IEEE International conference on robotics and Automation (ICRA), 2013, pp. 5724-5731.
3. Ph. Dubart, M. Morichi, "3D topographic and radiological modeling of an environment", STMI International Patent Application WO2015024694 published Feb. 26, 2015
4. F. Hautot, Ph. Dubart, R. Abou-Khalil, M. Morichi, "Novel real-time 3D radiological mapping solution for ALARA maximization" IEEE/ Advancements in Nuclear Instrumentation Measurement and their Applications (ANIMMA) international conference, 2015 Lisbon
5. C. Liang-Chia, T. Nguyen Van, S. Hung-Fa, L. and Hsien-I, "In-situ clouds-powered 3-D radiation detection and localization using novel color-depth-radiation (CDR) mapping". In "advanced robotics", vol. 28, no. 12, pp. 841-857,2014
6. M. Labbé and F. Michaud, "Online Global Loop Closure Detection for Large-Scale Multi-Session Graph-Based SLAM", in Proceedings of the IEEE/RSJ International Conference on Intelligent Robots and Systems, 2014
7. M. Labbé and F. Michaud, "Appearance-Based Loop Closure Detection for Online Large-Scale and Long-Term Operation," in IEEE Transactions on Robotics, vol. 29, no. 3, pp. 734-745, 2013.
8. M. Labbé and F. Michaud, "Memory management for real-time appearance-based loop closure detection," in Proceedings of the IEEE/RSJ International Conference on Intelligent Robots and Systems, 2011, pp. 1271–1276.
9. F. Panza, « Développement de la spectrométrie gamma in situ pour la cartographie de site ». thèse de doctorat. Université de Strasbourg, 2012.

DOI: 10.1051/epjconf/201715301006

RadFET dose response in the CHARM mixed-field: FLUKA MC simulations

M. Marzo[1,*], *S.* Bonaldo[1†], *M.* Brugger[1†], *S.* Danzeca[1†], *R.* Garcia Alia[1†], *A.* Infantino[1†], and *A.* Thornton[1†]

[1] *CERN, European Organization for Nuclear Research, Geneva, Switzerland.*

Abstract. This paper focuses on Monte Carlo simulations aiming at calculating the dose response of the RadFET dosimeter, when exposed to the complex CHARM mixed-fields, at CERN. We study how the dose deposited in the Gate Oxide (SiO_2) of the RadFET is affected by the energy threshold variation in the Monte Carlo simulations as well as the materials and sizes of scoring volumes. Also the characteristics of the input spectra will be taken into account and their impact on the final simulated dose will be studied. Dose variation as a function of the position of the RadFET in the test facility will be then examined and comparisons with experimental results will be shown. The contribution to the total dose due to every single particles of the mixed-field, under different target-shielding configurations, will be finally presented, aiming at a complete characterization of the RadFETs dose response in the CHARM mixed-fields.

1 Introduction

We present Monte Carlo calculations of the dose response of the RADiation sensing Field Effect Transistors (RadFET) [3] [4] [6] [9], when they are exposed to the radiation environment at the CERN High Energy Accelerator Mixed-field (CHARM) test facility [8].

The main purpose of the facility is to replicate different radiation environments (space, atmospheric, accelerator complexes, for instance) for radiation effects testing on electronic components and systems. RadFETs at CHARM are then used as online dosimeters during the radiation tests, to characterize the complex mixed-fields reproduced in the test area.

In this regard, we are interested in investigating the dose deposited in the Gate Oxide (GO) of the RadFET, a very small sensitive volume where the increase of the accumulated charge δQ_{tot}, as a consequence of the exposure to the radiation fields of interest, implies a change of the source-drain channel thickness of the MOSFET. This causes a drift in the threshold voltage δV_{th} of the device and a modification of its electric response, which can therefore be correlated to the deposited dose.

Measurements and FLUKA [1] [2] simulations of the dose response of RadFETs in the mixed-field at CHARM had been already performed to understand the radiation field in the test area, but discrepancies in the 25-45% range had been identified between simulated and experimental values. In this paper we want to set up more accurate simulations to investigate this mismatch.

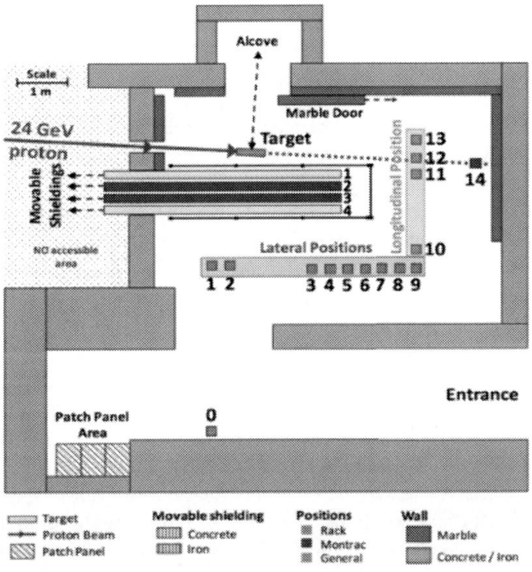

Figure 1: CHARM test facility model, top view.

2 CHARM and its mixed-field: measurements *VS* FLUKA simulations

The CHARM test facility [8] is located in the Meyrin site of CERN, specifically in the Proton Synchrotron (PS) East Area hall. The main source of the radiation field reproduced at CHARM is in fact the shower developing from the interaction between the 24 GeV/c protons beam extracted from the PS and a metallic target.

*Author's e-mail: matteo.marzo@cern.ch

†Please contact the author, if you need the coauthors' mail addresses

2.1 Facility variables

Due to the high energetic particles treated at CHARM, particular attention has been given to the protection of the personnel, by using concrete, marble and iron as materials surrounding the central part of the facility itself and separating the irradiation chamber, from the technical area [8] [10]. The irradiation chamber is a 7x7x3 m^3 room, available for the irradiation tests.
The 4 most relevant variables at CHARM, needed to obtain and properly tune the mixed-field of the facility, are:

- The 24 GeV/c beam from the PS;
- A target- Copper (cp), Aluminum (al) or Aluminum with holes (alh), depending on the wanted mixed-field intensity and composition, as it will be clear in the following- placed along the beam direction (Fig. 1);
- 13 rack test locations in lateral and downstream positions with respect to the beam direction, averagely 3÷5 m distant from the target (Fig. 1);
- A shielding between the target and the lateral positions created by using 4 movable blocks of concrete (C) and iron (I) (Fig. 1).

Entering the CHARM facility, in fact, the the 24 GeV/c protons beam impinges against the above mentioned target. From the interaction between the two, a shower of particles takes place. This shower can further interact with the shielding (if any), possibly generating other secondaries and consequently producing (or stopping) other particles: this is the way we create and tune the mixed-field at CHARM.
Changing the target material, the shielding configuration and having the possibility of testing the electronics at many test locations, different radiation fields in terms of composition and intensity can be reproduced. In particular, in this paper, two target-shielding layouts will be examined: copper target and no shielding (movable plates in OFF position), namely cp_0000, and copper target and concrete-iron-iron-concrete shielding, i.e. cp_CIIC. Other mixed-fields can be originated using different shielding configurations and different target materials.
FLUKA Monte Carlo simulations have already been performed to characterize the different radiation fields: the knowledge of the field at any rack test locations in terms of particles spectra (coming from the previous simulations) is the first, preliminary step to run the dedicated simulations on the RadFET.

2.2 FLUKA simulations settings to calculate the mixed-field at CHARM

A FLUKA model of CHARM was built to simulate the radiation field inside the test facility and retrieve all the most relevant quantities for radiation to electronics testing.
The NEWDEFA default in FLUKA (characterized by particles transport thresholds at 10MeV, except for neutrons at 10^{-5}eV and delta ray production with threshold at 1MeV) had been used as a first approximation to get reasonable results, saving CPU time, given the size of the geometry. The use of NEWDEFA was crucial to reduce the computational time, given the large geometry of the CHARM test facility (the higher the thresholds for production and transport of particles, the lower the computational time).
For the same reasons, an air cube of 20cm side had been chosen as sensitive volume: scoring, for instance, dose inside a volume having the same size as the actual size of the Gate Oxide (GO) of the RadFET -whose characteristic dimension is several order of magnitude lower than the CHARM's size-, would have led to a disproportional increase of the CPU time (a larger amount of primaries to get the same statistics).
An example of the characterization of the mixed-field at CHARM in terms of simulated particle energy spectra (lethargy), for the cp_0000 and cp_CIIC configurations is presented- for the position 1- in Fig. 2. It is clearly visible how the field is populated by secondary electrons, positrons, muons ($\mu\pm$), photons, neutrons and charged hadrons (protons, kaons ($\kappa\pm$) and pions ($\pi\pm$)). Those spectra have been obtained from FLUKA, making the 24 GeV/c proton beam interact with a copper target and simulating the entire CHARM geometry, reproducing the two different shielding layouts.

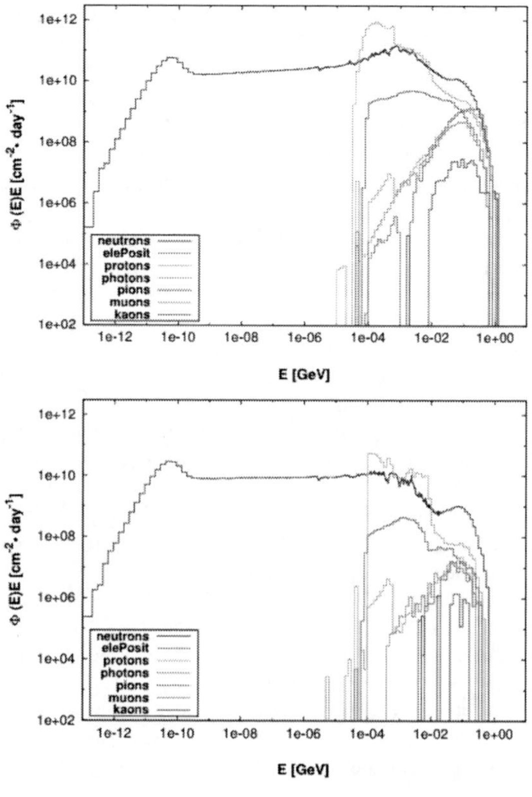

Figure 2: Mixed-field: simulated spectra (lethargy), position 1, cp_0000 and cp_CIIC configurations.

It is clear for instance how, from Fig. 2, in the shielded configuration the lethargy (cm^{-2}·day^{-1}, considering that nominal daily value of protons on target- pot- is $1.15 \cdot 10^{15}$, during the runs) is overall lower for all particles of the mixed-field, if compared to the cp_0000 configuration: the CIIC walls are stopping a certain amount of particles of the shower, preventing them from reaching the rack position 1, even if they are generating other secondaries.

2.3 Experimental measurements and simulated dose: benchmark

Dose at CHARM is experimentally detected by an integrated dosimetry system, called RadMon (Radiation Monitoring system) [5] [7]. The RadMon executes an electronic readout of the threshold voltage of the RadFETs and sends the voltage value to the CERN database. The conversion from voltage (V_{gs}) to dose (Gy) is post-processed by means of a calibration curve, obtained for each RadFET batch in a Co60 source. The experimental results presented in this paragraph refer to the data measured during 2015 run period of CHARM, for the copper target and no shielding configuration, measured by 100 nm p-channel RadFET on RadMon in different test positions. The experimental doses in Tab. 1 are averages of the calibration factors retrieved during different runs of 2015 plus minus total associated errors, which take into account primary proton beam intensity error, RadFET error and standard deviation with respect to the average value [10].
It's worth noticing that the previous FLUKA simulations results (obtained by scoring the dose in a cubic volume of air, 20 cm side, placed in the different test positions, considering the beam-target interaction and the whole CHARM geometry, and setting energy thresholds for transport and production of secondaries at 1MeV) in Tab. 1 are systematically overestimating the measured doses of 25% and 43%.

Position	Exp. dose	1st step dose	% diff.
1	0.776 ± 23.0%	1.101 ± 0.8%	30%
3	1.510 ± 23.0%	2.281 ± 0.6%	34%
5	1.190 ± 22.0%	2.110 ± 0.6%	43%
7	1.460 ± 23.0%	2.260 ± 0.5%	35%
10	1.550 ± 29.0%	2.401 ± 0.5%	35%
13	3.420 ± 28.0%	4.510 ± 0.3%	24%

Table 1: Experimental and 1st step simulated doses in 10^{-14}Gy/pot, in different positions at CHARM, cp_0000 configuration.

Fig. 3 shows a plot of the trend of dose simulated in different positions in the test area. The errors associated to the simulated values (as also clear from Tab. 1) are low if compared to the experimental ones, since they take into account only statistical uncertainties and no systematic errors, like in the case of measured doses.

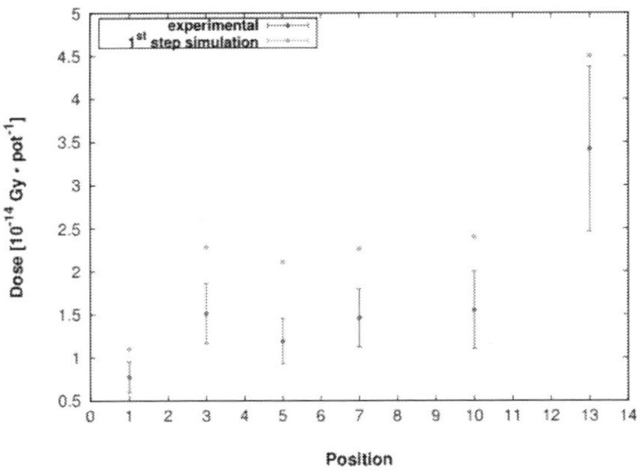

Figure 3: Experimental and simulated doses at different test positions, cp_0000 target-shielding layout.

3 The RadFET model and the second step simulations: benchmarks

Starting from the mismatch between the performed FLUKA simulations and the measurements detected by the RadMon, focusing on position 1 and cp_0000 layout, we want now to investigate these discrepancies and evaluate if it's possible to better simulate the radiation dose levels at CHARM. If we can improve simulations settings, aiming at obtaining doses that are more realistic and in line with experimental measurements, we could use our calculations to predict the radiation field at each position and for each target-shielding layout at CHARM, even out of the 14 ordinary predefined positions.
In order to improve the reliability of our MC simulations and the agreement with respect to the experimental data, we decide to implement a more realistic second step simulation acting on:

- Geometry and materials of the sensitive volume;
- Energy thresholds for production and transport of particles;
- Energy cutoffs for input spectra.

We want to conduct different parametric analyses to find out which are the most accurate FLUKA simulation setting, aiming at simulating the impact of the CHARM mixed-field on the RadFET oxide, through a two step simulation.
In the following paragraph, the first step dose calculated in correspondence of the position 1, for the copper target and no shielding configuration (Sect. 2.2), is taken as reference and its value is $1.101 \cdot 10^{-14} \pm 0.8\%$ Gy/pot (protons on target, meaning each proton from the PS beam and interacting with the target- $1.15 \cdot 10^{15}$ pot/day).
Before going through the details of the parametric simulations that aim at finding the best settings for the second step, we need to put in relation the two steps: if the geometry and the size of the sensitive volume is going to change,

also the overall geometry and scale of the problem have to change accordingly.

3.1 Link between the first and second simulations step

As already specified (Sect. 2.1), particles spectra characterizing the CHARM mixed-field had already been retrieved in the first step simulation (Sect. 2.2). They are now available and will be used as input to perform the second step simulations: the RadFET will be irradiated by a rectangular beam carrying information about the mixed-field in a given position and for a specific target-shielding layout. The beam has the same surface as the RadFET crossing section ($1mm^2$). In order to give an example, if we want to simulate the dose response of the RadFET, we will irradiate the actual geometry of the MOSFET using a beam that is carrying with itself the spectra of protons, electrons, positrons, muons, ... calculated in the first step simulation, in wanted position target-shielding configuration, available as text files.
The approximation of the radiation field at CHARM as a mono-directional beam, is realistic as the real mixed-field is in good approximation radial. The particles shower originating from the target (and possibly interacting with the shielding, if present) reaches all the rack test positions moving almost along radial directions.

3.2 The RadFET geometry and materials: their role in the second step

As it is shown in the Fig. 4, where the yz (z is assumed to be the beam direction) and the xy views are shown, the RadFET is mainly composed of a 250μm kovar lid, a vacuum region of 250μm and a 500 μm Silicon substrate; the crossing plane (xy) surface is $1x1mm^2$. A 1.4μm Silicon die is then located on the Silicon substrate: the SiO_2 gate oxides we are interested to study are deposited on this thin layer (Fig. 4). The dioxides, 4 in total in our RadFET (named ox1, ox2, ox3, ox4), can be 100 or 400 nm thick and both the configurations are studied in this paper. For our simulations we took into account just the ox1 Gate Oxide, having dimensions 300μm x 50μm x 400nm (or 100nm), centered in x = 0.000cm and y = -0.0135cm, with respect to the RadFET center.
The second step approach has been then validated by reproducing the first step configuration (air cube of 20 cm side) for position 1 in copper target and shielding OFF configuration. We got a deposited dose in the air rack of $1.101 \cdot 10^{-14}$ ±0.7% Gy/pot. This is in perfect agreement with the dose calculated in the 1st step, obtained from a more realistic and complex geometry model of the entire test facility, with a more realistic mixed-field. We are therefore confident that the approximation we introduced to reduce the CPU time, given the microscopic dimensions of our sensitive volume (mixed-field given through a directional beam), is good enough to faithfully reproduce the 1st step.

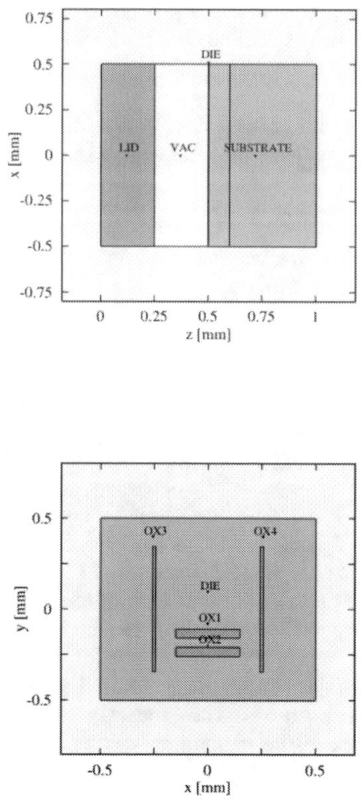

Figure 4: RadFET geometry and materials: FLUKA model.

After this validation of our model, using the 20cm-side volume of air as scoring volume, we decide to calculate dose in the $6 \cdot 10^{-6} mm^3$ gate oxide (SiO_2) of the RadFET, in kovar lid configuration, using the same energy thresholds as before. The simulated dose in this case is $0.895 \cdot 10^{-14}$ ±2.5% Gy/pot, showing a reduction of 19%, if compared to the dose in the air volume.
As clearly visible in Tab. 2, the deposited dose shows

Scoring vol.	1MeV th.	1keV th.	% diff.
air cube	1.101± 0.7%	0.840± 0.5%	24%
radfet	0.895± 2.5%	0.757± 2.0%	15%

Table 2: Dose in 10^{-14}Gy/pot deposited within the air rack and the RadFET gate oxide in SiO_2, using 1MeV (with NEDEFA defaults) and 1keV (using PRECISIO defaults) energy thresholds.

a further decrease if we use the PRECISIO physics settings plus 1keV energy thresholds: in this case the dose is $0.757 \cdot 10^{-14}$ ± 2.0% Gy/pot, 15% lower than the one obtained with the NEWDEFA settings and thresholds at 1MeV and 31% smaller than what calculated on the air

volume, using the same settings as in the 1st step simulation.
This evidently indicates that if we set a more accurate simulation in terms of geometry, materials and energy thresholds, the deposited dose shows a significant decrease with respect to what obtained in the 1st step simulation.

3.3 Update of the energy thresholds

As already discussed, the use of the NEWDEFA default in the first step simulation was crucial to reduce the CPU time, given the large geometry of the CHARM test facility. On the other hand, considering that the range of secondary electrons in air is 376cm [12] and that the characteristic dimension of the air scoring volume we are using is 20cm, lowering thresholds represents an improvement of the accuracy of our simulations (first row in Tab. 2). It is important to point out, in fact, that in the case of 1MeV thresholds, we are clearly overestimating the energy deposited, introducing an artifact: all the particles having energies lower than 1MeV will be treated by the FLUKA Monte Carlo algorithm as depositing their energy on spot, without producing other particles and being transported.
Following the previous considerations, the second step simulations will be then mostly conducted using the PRECISIO default, mainly characterized by particles transport thresholds set at 100keV (except for neutrons, 10^{-5}eV) and delta ray production threshold at 100keV. In addition, by using the EMF-CUT and DELTARAY cards, thresholds for transport and production of secondary electrons, positrons and photons and delta ray generated by muons and charged hadrons will be set at 1 MeV, 100keV, 10keV and 1keV. The PART-THR card will be instead employed to lower energy transport cut-offs for protons, pions, muons, kaons and α particles to 1keV.
Given the 15% difference in the dose deposited in the RadFET gate oxide, going from 1MeV to 1keV (last row in Tab. 2), it's worth investigating now in which thresholds energy range this main change occurs. The results of this parametric analysis can be found in Tab. 3.

Energy threshold	Dose
1MeV	0.895 ± 2.5%
100keV	0.768 ± 1.5%
10keV	0.787 ± 1.7%
1keV	0.757 ± 2.0%

Table 3: Parametric analysis changing the energy thresholds of the simulations (PRECISIO defaults): dose in the SiO_2 GO of the RadFET, in 10^{-14}Gy/pot.

It is clearly visible that the main change in dose occurs between 1MeV and 100keV, showing a decrease of 14%. Going down to 10keV and 1keV does not considerably change the scenario, the differences are in the range of the statistical uncertainties. This behavior can be explained considering the range of electrons in SiO_2: 1MeV electrons have a range of 1.96mm, 100keV electrons have a range of 66.15μm, while 10keV electrons have a range of 1.23μm. By comparing these ranges of secondary electrons with the characteristic lateral dimensions of the gate oxide of the RadFET (300μm x 50μm), it is evident that thresholds lower than 100keV are the most appropriate to simulate and faithfully reproduce the energy deposition in the gate oxide of the RadFET.
In this specific case, since the beam surface has the same size as the crossing surface of the RadFET and the ox1 sensitive volume is completely irradiated by the beam itself, we have a good statistics using a reasonable number of primaries: we decide to lower the thresholds down to 1keV to have most precise simulated doses, without loosing big amounts of CPU time.

3.4 Input spectra cutoffs and oxides thickenss

After evaluating the dose dependency on materials, scoring volumes sizes and energy thresholds, we decide now to study what is the impact of the oxide thickness and how input spectra cutoffs could affect the final dose.
We simulate RadFET having 100nm and 400nm thick gate oxides. We reproduce, on the other hand, input spectra characterizing the mixed-field, through a dedicated and independent first step simulation, by cutting them at 1MeV and 100keV, to see how lower energy particles of the spectra could affect the final deposited dose. Going down to energies lower than 100keV is counterproductive: the lower the thresholds, the lower the ranges and that particles can stopped by the kovar lid of the RadFET. As clearly visible from Tab. 4, the oxide thickness does not have a large impact on the simulated deposited dose. On the other hand, using spectra cut at 1MeV or 100keV makes a difference that is of the order of 10%.

Thickness	1MeV input	100keV input	% diff.
400nm	0.757 ± 2.0%	0.820 ± 4.3%	8%
100nm	0.719 ± 3.8%	0.816 ± 2.4%	12%

Table 4: Effect of Gate Oxide thickness and input spectra thresholds on the RadFET dose response: simulated dose in 10^{-14}Gy/pot (PRECISIO defaults set).

4 Second step simulations: the cp_0000 target-shielding configuration

The analyses conducted in Sect. 3 led us to identify the following simulation settings as the most suitable for our second step simulations:

- 400nm thick oxide;
- PRECISIO default in FLUKA;
- Energy thresholds at 1keV;
- Input spectra cutoffs at 100keV.

This is the starting point to test the second step algorithm on other relevant positions inside the test area, concerning the cp_0000 target-shielding configuration.

4.1 Doses simulated and measured in different positions at CHARM

We set the RadFET to positions 3, 5, 7, 10 and 13, by using as input the spectra of the 1st step simulations, calculated for the new locations of interest, as described in Sect. 2.2.

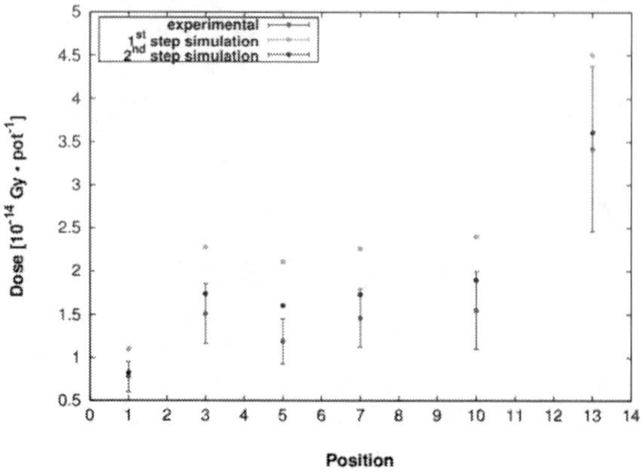

Figure 5: Experimental and simulated doses- 1st and 2nd step simulations- at different test positions, cp_0000 configuration.

Fig. 1 shows how the new considered locations are differently exposed to the mixed-field, mostly because of geometric reasons: we expect the position 13 to be actually the most solicited one, since directly exposed to the flux and almost in the beam direction. In addition, we want to compare simulated doses to experimental data (Sect. 2.3), to show the improvements we got by performing a more accurate second step simulation.
Tab. 5 clearly shows how the difference between experimental data and simulated values is now lower if compared to what obtained in Sect. 2.3, Tab. 2. The mismatch measured-simulated is 5÷25%- compatible with experimental uncertainties instead of 25÷43%, showing a substantial reduction (20÷25%) of the simulated dose (Tab. 6), in the direction of experimental data.
It is then evident that the new simulation allowed us to obtain most accurate simulated values, in better agreement with the measured doses. This is also clear from Fig. 7, where experimental, 1st and 2nd step doses are plotted: the second step simulation has represented an important improvement in simulating the radiation field at CHARM in the copper target and shielding OFF configuration.

Position	Exp. dose	2nd step dose	% diff.
1	0.776 ± 23.0%	0.820 ± 0.6%	5%
3	1.510 ± 23.0%	1.740 ± 0.4%	13%
5	1.190 ± 22.0%	1.603 ± 0.8%	25%
7	1.460 ± 23.0%	1.733 ± 0.6%	15%
10	1.550 ± 29.0%	1.900 ± 0.2%	18%
13	3.420 ± 28.0%	3.610 ± 0.3%	5%

Table 5: Experimental and 2nd step simulated doses in 10^{-14}Gy/pot, in different positions at CHARM.

Position	1st step dose	2nd step dose	% diff.
1	1.101 ± 0.8%	0.820 ± 0.6%	25%
3	2.281 ± 0.6%	1.740 ± 0.4%	24%
5	2.110 ± 0.6%	1.603 ± 0.8%	24%
7	2.260 ± 0.5%	1.733 ± 0.6%	23%
10	2.401 ± 0.5%	1.900 ± 0.2%	21%
13	4.510 ± 0.3%	3.610 ± 0.3%	20%

Table 6: 1st and 2nd simulated doses in 10^{-14}Gy/pot, in different positions at CHARM.

4.2 Contribution of each mixed-field's particles to the total dose

After validating the 2nd step simulation settings for the cp_0000 target-shielding configuration for different positions inside the test facility, the simulations results of the single particles (of the mixed-field) contribution to the total dose will be presented. The dose coming from all the particles of the mixed-field is going to be tabulated for the position 1, 3, 10 and 13; particular attention will be given to the weight of each particle over the total dose.
As showed in Fig. 6, downstream positions (10÷13) are the most exposed to the particle shower developing after the beam-target interaction. Speaking in terms of doses deposited by the different particles of the mixed-field, the energies deposited by charged hadrons, electrons and positrons, muons increase going from location 1 to 13 (Fig. 6). On the other hand, doses coming from neutrons and photons are almost constant. The explanation has to be found in the topography of our radiation field: fluences of charged hardons, electrons and positrons an muons are in fact characterized by a forward directionality, while, on the contrary, photons and neutrons fluences are more isotropic.
Considering the relative contribution of each particle of the field over the total dose (Fig. 7), it's interesting to note how, going from lateral to downstream positions, from 1 to 13:

- Charged Hadrons (protons, pions, kaons) contribution rises from 34% to 48%;
- Electrons and positrons contribution rises from 30% to 40%;
- Muons doses are stable around the 3 – 4%, representing an almost negligible portion of the total;
- Photons doses decrease from 24% to 7% in terms of contribution to the total;
- Neutrons contribution goes down from 8.5% to 2.5% in terms of contribution to the total.

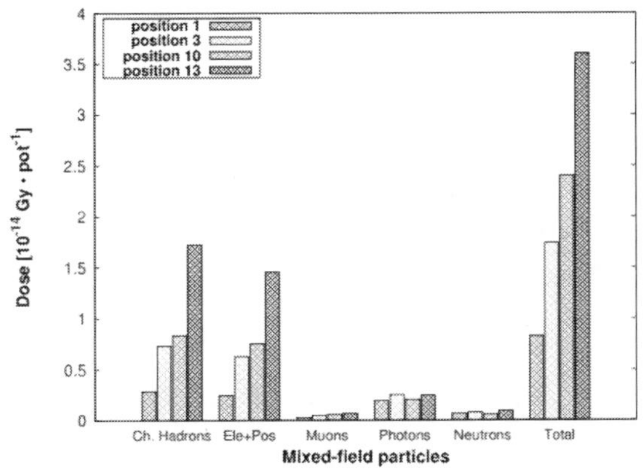

Figure 6: Particles contribution to the total dose: dose values.

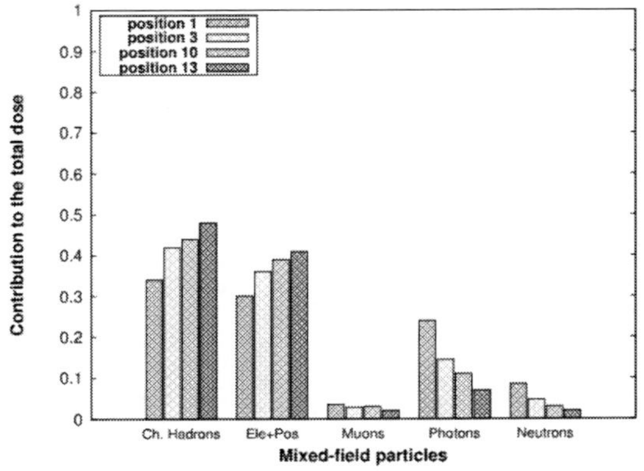

Figure 7: Particles contribution to the total dose: relative contribution.

As already said, this is a specular way to see how going from lateral to upstream positions, the contributions of the primary particles of the mixed-field to the total dose change and in particular charged hadrons, electrons and positrons become dominant, while photons and neutrons weights loose relevance.

5 cp_CIIC target-shielding configuration: the thermal neutrons issue

In this paragraph we will use the same simulation settings employed in the frame of the cp_0000 target-shielding configuration, for the cp_CIIC layout.
As for the previous case, we will irradiate the RadFET with a beam carrying information about the different spectra of the mixed-field, starting from position 1.

5.1 Doses simulated and measured in position 1

We want to compare the two step simulations results. As visible in Tab. 7, where doses for position 1 and cp_CIIC are reported, the mismatch between first and second step doses is now 52%. This could mean that probably something is missing in our model.
Given the unexpected discrepancy, we decide to evaluate

2nd step	1st step	%diff.
0.0327 ± 0.6%	0.0678± 0.8%	52%

Table 7: Experimental, 2nd and 1st step simulated doses in 10^{-14}Gy/pot, in position 1, cp_CIIC configuration.

the single contributions to the total dose, due to all the particles populating the CHARM mixed-field, in cp_CIIC configuration, in order to investigate this issue. In this regard, Fig. 8 shows a comparison between the two step in the air cube of 20cm side (taken as a reference) and the RadFET gate oxide, using the same settings as in Sect. 3.1. It's worth noticing that the doses deposited by charged hadrons, muons and electrons+positrons+photons are perfectly comparable in both cases (air volume and RadFET). What really changes is the dose contribution due to neutrons: the dose simulated in the large air volume is a factor 5.3 higher than what scored in the RadFET.

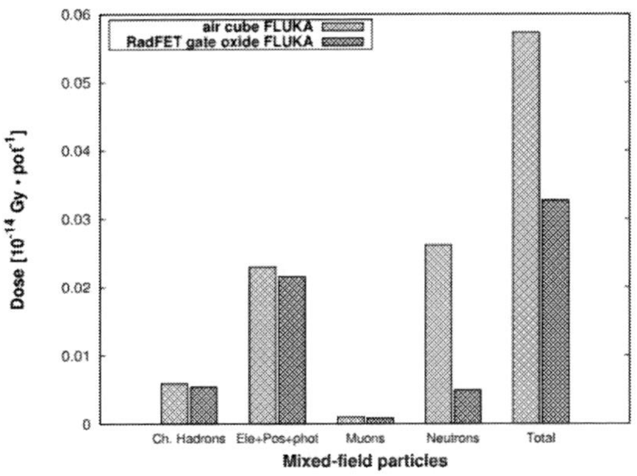

Figure 8: Particles contribution to the total dose: dose values.

5.2 Splitting neutrons spectrum: contribution at different energies

For the reasons exposed in the previous paragraph, we decide to split the spectrum of neutrons in different energy ranges, to further investigate where this mismatch could be originated. From Tab. 8 it's clear that the discrepancy we got is mainly due to thermal neutrons, which are depositing in the RadFET gate oxide a dose that is a factor 50 lower than expected (dose simulated in the air volume and measurements). This is, anyway, something in line with expectations: thermal neutrons undergo np nuclear interactions with the nuclei of Nitrogen atoms present in air and these interactions give birth to protons at 300÷ 400 keV that are actually depositing energy in the air cube volume.

Energy range	Air Vol. dose	RadFET dose	ratio
$E_n \leqslant 0.1$eV	0.0151 ± 0.5%	0.0003 ± 8.5%	50.0
0.1eV < E_n < 20MeV	0.0087 ± 2.5%	0.0029 ± 4.9%	3.0
$E_n \geqslant 20$MeV	0.0024 ± 5.5%	0.0017 ± 5.5%	1.4
Full spectrum	0.0262 ± 1.5%	0.0049 ± 4.0%	5.4

Table 8: Contributions to the total dose deposited in the air cube and the RadFET GO, in 10^{-14}Gy/pot, in position 1, due to neutrons at different energies.

On the other hand, this is evidently not the case if we consider the dose simulated in the RadFET gate oxide: before reaching the oxide, thermal neutrons have interacted with the Kovar (Ni-Co) lid (possibly developing a further particles shower) and finally deposited energy in SiO_2, without interacting with any air molecule.
This is clearly only partially explaining the discrepancy we found and further investigations are already ongoing, laying the basis for the next step of research. In the next future, in fact, we want to build a most accurate model of the RadFET in terms of materials, thinking, for instance, that the doping could be a potential source of this discrepancy. In parallel, new test runs at CHARM will give us more information and most reliable measurements about the dose absorbed by RadFETs in the cp_CIIC configuration. Also a test campaign to measure the RadFET sensitivity to low energy neutrons will take place in the early 2017.

6 Summary and conclusions

By implementing the real geometry of the RadFET in FLUKA, using most accurate physics settings and input spectra from the first simulation step, we are able to simulate the RadFET dose response in the CHARM mixed-field- copper target and shielding OFF configuration- with a high accuracy. Through the second simulations step, we got a reduction of the mismatch between measured and simulated doses of 20÷25%. Equivalently, simulated doses are definitely closer to the experimental data (5÷25% difference between the two), than what had already been done with the previous calculations, which showed a difference of 25÷45% between simulated and measured values.
On the other hand, the same simulation settings do not seem to be entirely applicable when simulating the dose response of the RadFET under the mixed-field generated in copper target and concrete-iron-iron-concrete shielding configuration. From the simulations we conducted, this is apparently due to the thermal neutrons component of the radiation field at CHARM (dominant in the shielded case and evidently negligible in th cp_0000 configuration) and requires further investigation both from the experimental and simulation point of view. In this regards, this is the starting point for the next future research, aiming at finding unambiguous simulation settings to be used under different target-shielding configurations at CHARM and confirmed by experimental measurements.

References

[1] A. Ferrari *et al.*, "FLUKA: a multi-particle transport code", CERN-2005-10 (2005), INFN/TC05/11, SLAC-R-773

[2] T.T. Böhlen *et al.* "The FLUKA Code: Developments and Challenges for High Energy and Medical Applications", Nuclear Data Sheets 120, 211-214 (2014)

[3] P. Beck *et al.*, "Investigation on Photon Energy Response of RadFET Using Monte Carlo Simulations", IEEE Transactions on nuclear science, Vol. 54, no. 4, pp. 1151-1157, August 2007.

[4] M. Wind *et al.*, "Investigation of the Energy Response of RadFET for High Energy Photons, Electrons, Protons, and Neutrons", IEEE Transactions on nuclear science, Vol. 56, no. 6, pp. 3387-3392, December 2009.

[5] G. Spiezia *et al.*, "The LHC Accelerator Radiation Monitoring System-RadMon", 2011.

[6] J. Mekki *et al.*, "Mixed Particle Influence on RadFET Responses Using Co-60 Calibration", IEEE Transactions on nuclear science, Vol. 60, no. 4, pp. 2435-2443, August 2013.

[7] G. Spiezia *et al.*, "A New RadMon Version for the LHC and its injection lines", IEEE Trans. Nucl. Sci, vol. 61, no. 6, pp. 3424-3431, 2014.

[8] J. Mekki *et al.*, "A Mixed Field Facility At CERN for Radiation Test: CHARM", RADiation and its Effects on Components and Systems (RADECS), 14-18 September 2015.

[9] A. Holmes-Sieldle and L. Adams, "Handbook of Radiation Effects", Second Edition, Oxford University Press, 2002.

[10] S. Bonaldo, "CHARM Una nuova facility del CERN per test di elettronica con spettri misti di radiazione ionizzante", 2016.

[11] F. Ravotti *et al.*, "Response of RadFET Dosimeters to High Fluences of Fast Neutrons", IEEE Transactions on nuclear science, Vol. 52, no. 4, pp. 959-965, August 2005.

[12] Martin J. Berger *et al.*, "ESTAR, PSTAR, and ASTAR: Computer Programs for Calculating Stopping-Power and Range Tables for Electrons, Protons, and Helium Ions (version 1.21)", Citation: http://physics.nist.gov/Star

Pulsed neutron fields measurements around a synchrotron storage ring

Marco Caresana[1*], *Marcello* Ballerini[2], *David* Garf Ulfbeck[4], *Niels* Hertel[3], *Giacomo Paolo* Manessi[2], *Carsten* Søgaard[5]

[1] Politecnico di Milano, Department of Energy, Via Lambruschini 4, 20156 Milan, Italy
[2] Else Nuclear srl, Via Tito Vignoli 44, 20146 Milan, Italy
[3] Department of Physics and Astronomy, Institute for Storage Ring Facilities, 8000 Aarhus, Denmark
[4] National Institute for Radiation Protection, 2300 Copenhagen, Denmark
[5] eScience group, Niels Bohr Institute, University of Copenhagen, Denmark

Abstract. A measurement campaign was performed for characterizing the neutron ambient dose equivalent, H*(10), in selected positions at ISA, Aarhus, Denmark, around the ASTRID and ASTRID2 storage rings. The neutron stray radiation field is characterized here by very intense radiation bursts with a low repetition rate, which result in a comparatively low average H*(10) rate. As a consequence, devices specifically conceived for operating in pulsed neutron fields must be employed for efficiently measuring in this radiation environment, in order to avoid severe underestimations of the H*(10) rate. The measurements were performed with the ELSE NUCLEAR LUPIN 5401 BF3-NP rem counter, a detector characterized by an innovative working principle that is not affected by dead time losses. This allowed characterizing both the H*(10) and the time structure of the radiation field in the pre-selected positions.

1 Introduction

The need of radiation detectors capable of measuring efficiently in pulsed neutron fields is attracting widespread interest for an increasing number of applications, such as radiation protection monitoring and beam diagnostics around medical and research particle accelerators. Reliable measurements cannot be carried out with conventional rem counters operating in pulse mode, as they underestimate the true neutron interaction rate when exposed to high intensity and short duration radiation bursts. The underestimation is due to dead time losses. Recently [1] it was demonstrated that rem counters operating in pulse mode can properly measure radiation bursts of intensity up to about 10 nSv (H*(10)). Above this threshold, the dose underestimation starts to be important and devices specifically conceived for pulsed neutron fields must be used.

An innovative rem counter, called LUPIN (Long interval, Ultra-wide dynamic, PIle-up free, Neutron rem counter) [2,3], has been specifically designed to work in pulsed neutron fields as its working principle is not characterized by any dead time. LUPIN has been employed to characterize the neutron ambient dose equivalent, H*(10), around the Aarhus Storage Ring in Denmark (ASTRID) and ASTRID2 storage rings at the Centre for Storage Ring Facilities (ISA) at Aarhus University, Denmark. The stray radiation field at ISA is characterized, on average, by a low neutron H*(10) rate.

However, the neutron stray field is generated only during the injection in ASTRID and the extraction phase from ASTRID to ASTRID2, which lasts for a few seconds and happens once every 200 s. This causes a relatively high neutron H*(10) rate to be measured during the short amount of time.

2. MATERIALS AND METHODS

The time structure of the stray neutron field around the ASTRID and ASTRID2 storage rings represents the typical measuring situation for the LUPIN, which has been conceived for coping with very intense radiation bursts. In this paragraph both the experimental facility and the working principle of the LUPIN are discussed.

2.1 ASTRID and ASTRID2 storage rings

ASTRID is a 40 m circumference electron storage ring operating since 1990. It operates as a 580 MeV electron storage ring, injected from a 100 MeV racetrack microtron. ASTRID2 is a 45.7 m circumference storage ring located next to ASTRID (see Figure 1) that operates at 580 MeV and is optimised for photon energies from the visible to around 1 keV [4,5].

ASTRID is used as an injector for ASTRID2, which maintains an almost constant circulating current fluctuating around 200 mA. As the current falls below about 150 mA the beam is injected into ASTRID2 to

* Corresponding author: marco.caresana@polimi.it

 DOI: 10.1051/epjconf/201715301007

restore the nominal current. The injection occurs every about 150 – 200 s.

In ASTRID, a fast injection mode can also be operated, with an injection frequency from the microtron of 2 Hz. The neutron stray field around the two storage rings could not be characterised with passive methods (TLDs and CR39) due to the very low expected average H*(10) rate, which induces very high fluctuations on the results of the readings.

A measurement campaign with the ELSE NUCLEAR LUPIN 5401 BF3-NP rem counter was then planned in agreement with the Danish National Institute for Radiation Protection in order to characterize the neutron H*(10) rate in 7 selected positions around the storage rings, see Figure 1.

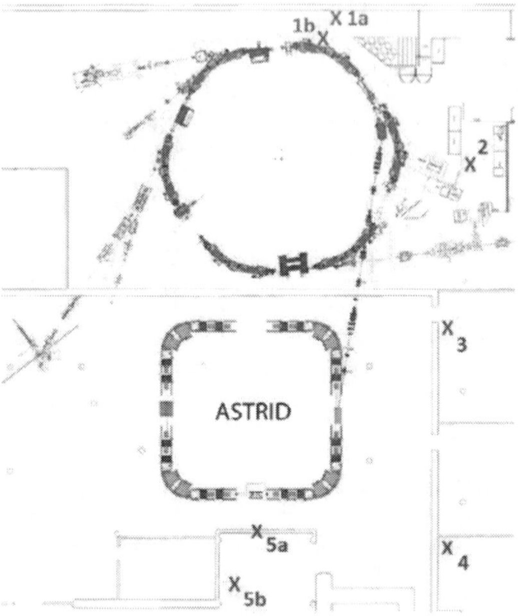

Fig. 1. Map of the measuring positions around the accelerators.

In particular, some of the positions were located in offices and laboratories close to the accelerators and on a platform recently built to allow external visitors to have a look at parts of the ASTRID2 ring. (Position 1b, see Figure 2)

2.2 LUPIN

The LUPIN is based on a standard rem counter moderating assembly and is available in two configurations: either a spherical one [2], which embeds a ^{3}He proportional counter, or a cylindrical one [3], which embeds a BF_3 proportional counter.

Regardless to the particular configuration, the main difference distinguishing the LUPIN from a standard rem counter is the signal acquisition technique.

It is well known that commercial rem counters can suffer of dead time losses and this effect is particularly severe during an intense and short radiation burst. This effect can lead to important dose underestimation, up to three orders of magnitude [6].

The dead time can be reduced by reducing the shaping time of the output signal, but such an approach can just attenuate the dead time losses, thus reducing the underestimation, but does not fix the problem. Besides there is a physical limit to the reduction of the shaping time defined by the charge mobility inside the proportional counter.

Fig 2. Measuring position 1b on top of the visitor platform.

2.2.1 Working principle of the LUPIN

The approach adopted in the LUPIN is completely different. The classical acquisition chain composed by preamplifier – amplifier – discriminator – counter is replaced by the electronics shown in figure 3

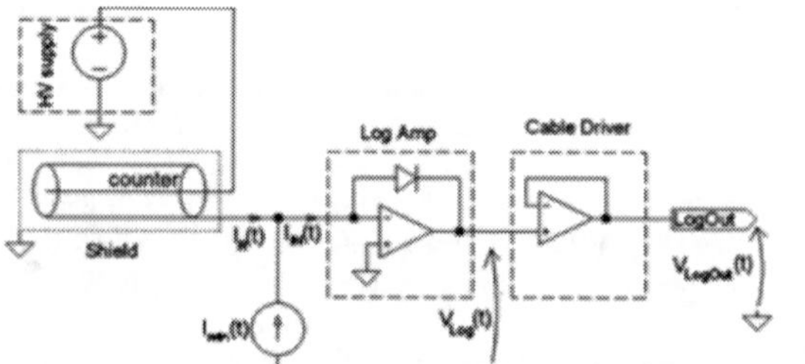

Fig 3. General scheme of the detector electronics.

A user-adjustable high voltage module on board the instrument supplies the voltage bias at the electrodes of the proportional tube. The acquisition electronics is based on a current-to-voltage logarithmic amplification of the current signal produced in the proportional counter. The choice of a logarithmic amplifier is driven by the need to measure the output current over few order of magnitude, in order to achieve the same dynamic range also in terms of burst intensity.

From a theoretical point of view the logarithmic amplifier can accept only a positive input current. From a practical point of view, this causes that an offset current I_{min} of minimum 100 pA must be supplied in input to avoid negative saturation.

In order to provide a positive current at the input of the logarithmic amplifier the polarization of the proportional counter is positive at the anode while the cathode is grounded.

Because the signal is collected at the cathode, the proportional counter must be electrostatically shielded in order to avoid noise pick up. The electrostatic shield is made of an aluminium cylinder that completely encases the counter.

The output of the logarithmic amplifier passes through a cable driver that permits to connect long cables, up to about 200 m.

More details about the theoretical operating principles of the circuit can be found in [7].

The output voltage of the cable driver is processed in two different ways:

1) via an oscilloscope (PicoScope442 by PicTechnology) and monitored by a LabVIEW program running on a PC that converts the voltage back to current.
2) Via a custom board based on a FPGA model Cyclone III produced by ALTERA.

In the oscilloscope acquisition the signal is sampled at a rate of 10MHz. This method is mainly used to visualize the current profile in time, but is unfit for a continuous measurement because of the rather slow processing time of the software running on the PC.

In the FPGA acquisition the signal is sampled via a 12 bit ADC converter again at 10 MHz, but the analysis is performed on-line by the FPGA that send to a remote PC the dose value every second.

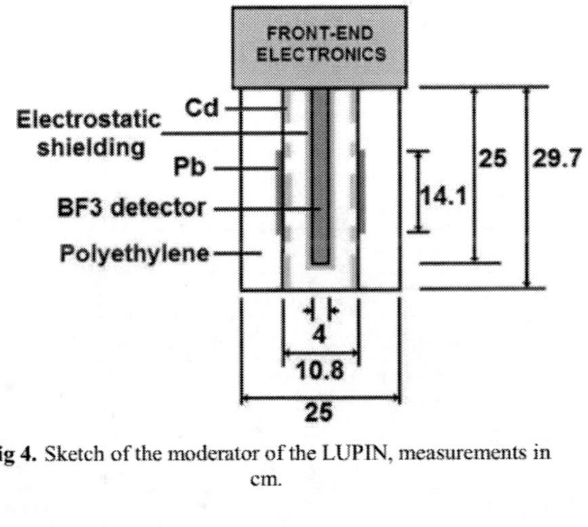

Fig 4. Sketch of the moderator of the LUPIN, measurements in cm.

2.2.2 Design of the moderator

Figure 4 shows a sketch of the LUPIN moderator. It consists of a polyethylene cylinder of 25 cm diameter with lead and cadmium inserts. The lead layer permits to extend the energy response range up about 10 GeV [8], while the cadmium inserts permit to tune the response to slow and thermal neutron. An electrostatic shielding, constituted by an aluminium cylinder of 1.5 mm thickness, contains the detector and two polyethylene inserts, which are used to fill the void around the detector.

The moderator design is optimized so that the response function of the instrument reproduces sufficiently well the curve of the conversion coefficients from neutron fluence to H*(10) over a wide energy range, from thermal neutrons up to 10 GeV.

2.2.3 LUPIN calibration

In order to measure the number of neutrons interacting in the proportional counter during a neutron burst the LUPIN software integrates the current over a user settable time window and calculates the total charge generated in the proportional counter.

Dividing the integral charge by the mean collected charge (MCC) due to the interaction of a single neutron, the LUPIN can calculate the number N of interacting neutrons. See equation (1)

$$N = round\left(\frac{\sum_{i=1}^{n} I_n \cdot \Delta t}{MCC}\right) \quad (1)$$

where N is the number of acquired neutrons, ''round'' is a function which rounds up the number in brackets to the nearest integer number, n is the number of samples in a single acquisition, I_i is the current value of each sample, Δt is the time width of the sampling (e.g. for 10MHz sampling, Δt =100 ns) and MCC is a fundamental parameter characterizing the LUPIN.

The MCC was measured with an Am-Be source tuning the source to detector distance to obtain an interaction rate of about 10 s^{-1} and setting four different integration times: 0.5 ms, 1 ms, 2 ms and 4 ms. The proportional counter bias voltage was 1180 V.

Figure 5 shows the frequency distribution, the main peak corresponding to the full energy deposition of the reaction products is clearly visible. The continuum below the main peak is due to the wall effect.

The MCC corresponding to an integration time t=0.5 ms is 547 fC. For higher integration times the MCC reaches a steady value of 615±3 fC, indicating that an integration time of 1 ms is enough to avoid any ballistic deficit.

The calibration in terms of H*(10) has been done with a Pu-Be source at the calibration facility at CERN and cross checked at the Politecnico di Milano with an Am-Be source. The sensitivity is 2.13 ± 0.2 nSv^{-1}.

2.2.4 Photon rejection capability

It is possible, especially around electron accelerators, that the radiation burst is composed by a photon and a neutron component. Usually both radiation components have a pulsed time structure. Of course also the photon component can generate charge inside the proportional counter.

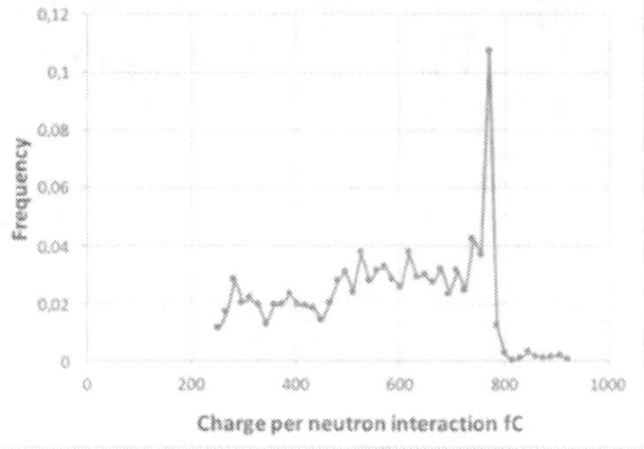

Fig 5. Frequency distribution of the charge per neutron interaction for a 4 ms integration time.

The separation of the two contributions can be done on a time basis considering that the photon component immediately interacts with the detector, while the interaction of the neutron component is delayed because fast neutrons take some time for the moderation and, once thermalized, a further delay due to the diffusion time is introduced.

The average thermalization and diffusion time (TDT) is typically in the 350 – 700 µs range [9].

Figure 6 shows an example of the exploitation of the TDT to separate the photon and neutron component. The measurement was performed at PSI around the injector test facility used as principal test and demonstration plant for the SwissFEL project. [10]

Data shown in Figure 6 was taken close to the beam dump where the photon contribution is particularly intense, compared to the neutron one.

The graph shows the typical fingerprint of a pulsed and mixed radiation field, composed by a very intense initial peak due to the photon interaction followed by neutron interactions spread over hundreds of μs.

The integral charge of the photon peak is about 1100 fC, equivalent to the interaction of two neutrons, and its FWHM is around 200 ns. The gamma rejection is obtained starting the charge integration about 5 μs after the photon peak that can be used as a trigger.

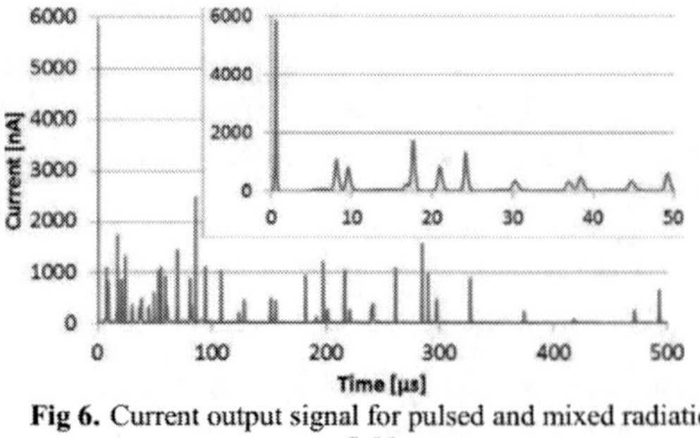

Fig 6. Current output signal for pulsed and mixed radiation fields.

3. RESULTS AND DISCUSSION

The results of the measurements are given in Table 1, which shows both the average H*(10) rate and the H*(10) per single burst.

The presence of a photon contribution to the neutron signal was checked for every position by sampling the signal via oscilloscope. Figure 7 shows an example of signal taken in position 5a. A comparison with figure 6 permits to conclude that the photon contribution is negligible.

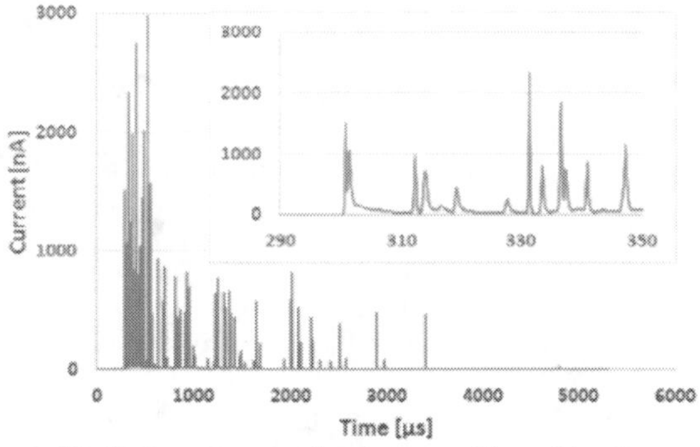

Fig 7. Current output signal measured in position 5a.

For some positions two sets of measurements were performed: the first one with ASTRID operating in slow extraction mode, the second one in fast injection mode. The high sensitivity of the device and its wide operating range allowed measuring very low average H*(10) rate as well as comparatively high H*(10) per burst, like the ones measured in position 5a. Here the time structure of extraction phase is composed by four different sub-phases, three injection sub-phases from the microtron to ASTRID and one extraction sub-phase from ASTRID to ASTRID2.

Each of the injection phase has a time duration of 800 ns while the extraction is over one revolution in ASTRID which is 133 ns.

Table 1. Results of the measurements. (n.a. = not applicable, * = fast injection mode). Position 5a reports four values of H*(10) according to the injection-extraction time structure (see figure 8)

Position	H*(10) per burst [nSv]				Average H*(10) rate [µSv/h]
1a	5.6				0.39 ± 0.11
1b	31.7				1.7 ± 0.2
2	1.9				0.27 ± 0.08
3	2.8				0.09 ± 0.01
3*	n.a.				2.6 ± 0.8
4	n.a.				0.005 ± 0.001
4*	n.a.				0.16 ± 0.05
5a	27.7	23.3	23.5	43.7	2.4 ± 0.1
5a*	n.a.				422 ± 63
5b	n.a.				43.2 ± 10.1

The innovative working principle of the LUPIN allowed the distinguishing of the time structure of the stray radiation field and the contribution of each sub-phase to the total integrated H*(10) can be separated very well.

The average H*(10) per burst is given separately for each injection sub- phase and for the extraction sub-phase.

Figure 8 shows the H*(10) data acquired in position 5a.

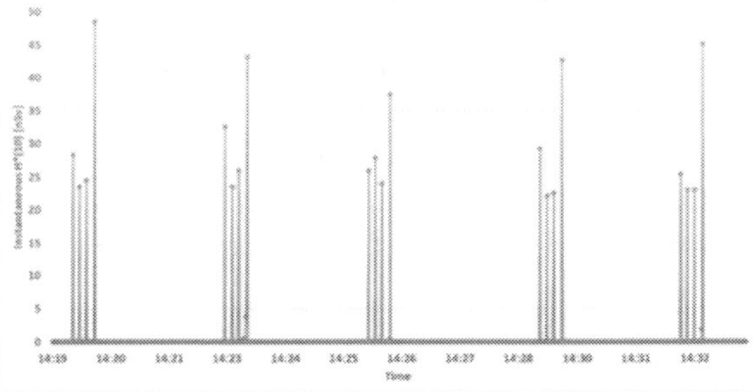

Fig 8. H*(10) data acquired with the LUPIN rem counter in position 5a.

In order to cross check the consistency of the results it was performed a statistical analysis on the measurements in position 1b aiming at evaluating the distribution of the bursts intensity. As the accelerator is characterized by a very high working stability, it is assumed that the fluctuation of the signal detected by the LUPIN in terms of interacting neutrons per burst is dominated by the Poisson statistic. Figure 9 shows the time profile on the number of neutron interactions recorded by the LUPIN.

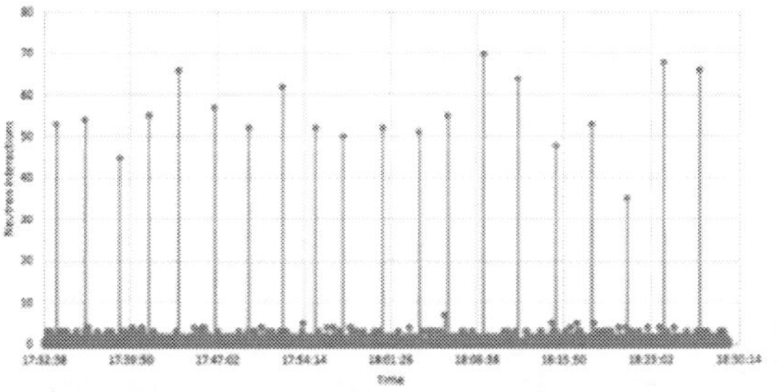

Fig 9. Data acquired with the LUPIN rem counter in position 1b in terms of neutron interactions.

The injection phase from ASTRID to ASTRID2, occurring every 150 - 200s is clearly visible. Similar time distributions have been measured for the positions listed in table 2. The same table reports also the reduced χ^2 of the distribution of the number of neutrons detected by the LUPIN during each injection, calculated assuming the Poisson distribution as target distribution.

The result is clearly shows that data are Poisson distributed with a high level of confidence. The result permits to conclude that the injection process is characterized by a very good repeatability and that the LUPIN is not affected by any signal loss.

Table 2. Statistical analysis to check the compliance to the Poisson distribution for repeated measurements in different positions. (n.a. = not applicable, * = fast injection mode).

Position	Reduced χ^2	Degree of freedom	90% confidence limits
1a	0.991	22	0.53 –1.58
1b	1,05	19	0.53 –1.59
2	1,09	22	0.53 –1.58
3	0.985	25	0.58 – 1.51
3*	1.09	450	0.89 – 1.11
4	0.997	4750	0.96 – 1.03
4*	1.15	150	0.82 – 1.2
5a injection 1	0.62	4	0.18 – 2.37
5a injection 2	0.49	4	0.18 – 2.37
5a injection 3	0.15	4	0.18 – 2.37
5a extraction	0.88	4	0.18 – 2.37
5a*	n.a.	n.a.	n.a.
5b	1.08	79	0.75 – 1.27

The value of the reduced χ^2 for position 5a* (fast injection mode) has not been reported as the signal has a time distribution as reported in figure 10. The fast injection mode produces an almost continuous neutron stray field with superimposed peaks due to the extraction form ASTRID to ASTRID2, occurring every 12 s. Of course this measurement cannot be Poisson distributed. However considering just the number of neutron interactions in the peak, a value of χ^2(reduced)=1.29 is obtained, with a number of degree of freedom υ=19. This value is within the statistical limit. (see table 2)

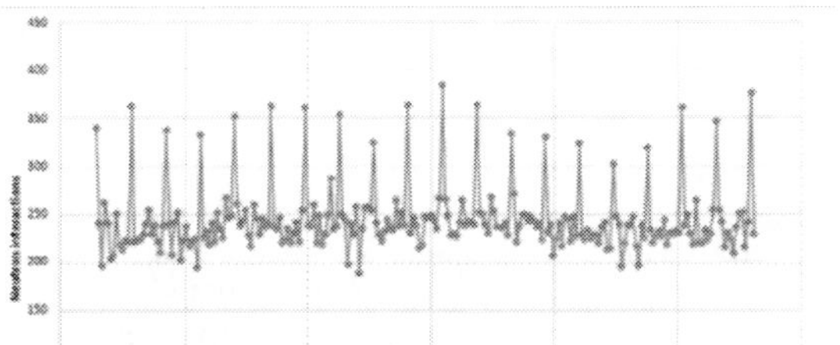

Fig 10. Data acquired with the LUPIN rem counter in position 5a in terms of neutron interactions. Accelerator operating in fast injection mode.

4. Conclusion

The measurement campaign performed with the ELSE NUCLEAR LUPIN 5401 BF3-NP rem counter at ISA allowed characterising the neutron H*(10) rate in selected positions around the ASTRID and ASTRID2 storage rings.

The stray field was characterised by intense pulsed bursts of radiation, in the range of 20 – 50 nSv. In a previous work of benchmark of rem counters in pulsed neutron field [1], it was demonstrated that in this burst intensity range almost all the rem counters are affected by important signal loss (up to a factor 5). The LUPIN is an exception because, thanks to the front end electronics, the linearity extends up to hundreds of nSv per burst.

The radiation fields around the ASTRID accelerator complex, in spite of the very high instantaneous dose rate, results in a pretty low average dose rate, in the range of few μSv/h. This means that these areas can be attended by workers of even general public (In position 1b there is the visitor platform). As a consequence of that that, environmental dosimetry plays a key role in the radiation protection and the need of a specific device capable to cope with pulsed neutron fields is even more pressing.

The pulsed photon component associated to the stray neutron field was negligible, in the sense that its intensity that cannot affect the neutron measurements.

The particular time structure of the stray radiation field defined by the working parameters of the accelerators is coherent with the measurement and the statistical distribution of repeated measurement is in agreement with the Poisson distribution.

References

1. M. Caresana et al., NIM A **737**, 203–213 (2014)
2. M. Caresana, M. Ferrarini, G.P. Manessi, M. Silari and V. Varoli, NIM A, **712**, 15-26 (2013).
3. M. Caresana, C. Cassell, M. Ferrarini, E. Hohmann, G.P. Manessi, S. Mayer, M. Silari and V. Varoli, Rev Sci Instrum, **85**, 065102 (2014).
4. J.S. Nielsen, N. Hertel and S.P. Moller, Proceedings of IPAC2011, San Sebastian, Spain, THPC003, 2909-2911 (2011).
5. N. Hertel , S. Vrønning Hoffmann, **24-1**, (2011)
6. A. Klett and A. Leuschner, in Proceedings of the IEEE 2006 Nuclear Science Symposium and Medical Imaging Conference, San Diego, CA, 29
7. M. Ferrarini, V. Varoli, A. Favalli, M. Caresana, B. Pedersen, NIM A, **613**, 272 (2010).
8. C. Birattari, A. Esposito, A. Ferrari, M. Pelliccioni, T. Rancati, M. Silari, Radiat Prot Dosim, **76**, 135 (1998)
9. A. L. Justus, Health Phys, **102**, 9 (2012)
10. M. Pedrozzi, SwissFEL Injector Conceptual Design Report, PSI Report, 10-05, (2010)

Application of the Monte Carlo method to the analysis of doses and shielding around an X-ray fluorescence equipment

José Ródenas[1,*], *Belén* Juste[1], *Sergio* Gallardo[1], and *Andrea* Querol[1]

[1]Departamento de Ingeniería Química y Nuclear, Universidad Politécnica de Valencia, Spain

Abstract. An X-ray fluorescence equipment is used for practical exercises in the laboratory of Nuclear Engineering of the Polytechnic University of Valencia (Spain). This equipment includes a compact X-ray tube, ECLIPSE-III, and a Si–PIN XR-100T detector. The voltage (30 kV), and the current (100 µA) of the tube are low enough so that expected doses around the tube do not represent a risk for students working in the laboratory. Nevertheless, doses and shielding should be evaluated to accomplish the ALARA criterion. The Monte Carlo method has been applied to evaluate the dose rate around the installation provided with a shielding composed by a box of methacrylate. Dose rates calculated are compared with experimental measurements to validate the model. Obtained results show that doses are below allowable limits. Hence, no extra shielding is required for the X-ray beam. A previous Monte Carlo model was also developed to obtain the tube spectrum and validated by comparison with data from manufacturer.

1 Introduction

X-ray fluorescence analysis (XRFA) is a standard technique [1] to obtain the composition of a sample. Any photon beam can produce fluorescence. The X-ray fluorescence analysed in this work is produced by an X-ray tube. The X-ray fluorescence equipment includes an X-ray tube and a detector with the appropriate software to analyse fluorescence photons produced in the sample.

Students of different courses such as Radioactive Contamination, Radioprotection, Nuclear Materials, etc. use this equipment for practical exercises in the laboratory of Nuclear Engineering of the Polytechnic University of Valencia (Spain).

The MCNP code [2], based on the Monte Carlo (MC) method is applied to obtain the spectrum of the tube, but also to estimate doses produced by the tube at certain distances and consequently to determine the shielding necessary for a safe work around the equipment. Therefore, two MC models have been developed.

The model developed to obtain the X-ray spectrum was validated by comparison with data provided by the manufacturer.

The MC method has been also applied to develop a second model to evaluate the dose rate around the X-ray equipment provided with a shielding composed by a box of methacrylate. Results of doses estimated with MCNP simulations are compared with experimental measurements to validate the MC dose model.

The voltage and the current of the tube are low enough so that expected doses around the tube do not represent a risk for students working in the laboratory. Nevertheless, doses and shielding should be evaluated to accomplish the ALARA criterion.

2 Materials and methods

2.1 Eclipse III

An ECLIPSE-III X-ray tube [3] was acquired by the Department of Nuclear Engineering of the Polytechnic University of Valencia (Spain) for the development of X-ray fluorescence analysis. The ECLIPSE-III is a self-contained miniature X-ray tube system, which includes the X-ray tube and the High Voltage power supply in a single compact cylindrical enclosure.

The X-ray generator consists of an axial beam emitted through a Be window. The nominal size of the focal spot is 1.5 mm diameter. An Ag target (anode) is used. The voltage is limited to 30 kV, which is the limit of exemption for X-ray tubes according to the Spanish legislation [4]. The maximum value of the current for this tube is 100 µA. The equipment is completed with a Si–PIN XR-100T detector offering a good resolution at low energy, specially designed to obtain spectra from XRFA.

A portable Controller is provided, powered from 80 - 240 V AC. It generates all the voltages needed to operate the X-ray tube and provides both voltage (kV) and current

[*] Corresponding author: jrodenas@iqn.upv.es

(μA) display and control. Both the X-ray tube and the Controller can be seen in figure 1.

The ECLIPSE-III is designed to replace radioisotope sources in XRFA applications. Using ECLIPSE-III, a safer operation is ensured, which is the optimal situation in any environment, but especially in a teaching laboratory.

Fig. 1. X-Ray tube and Controller.

The main features of ECLIPSE-III are its small size, a low power, with adjustable current and voltage, and a lifetime over 2000 hours. It is a portable system, appropriated in XRFA applications for both research and teaching. Especially for a teaching environment due to the mentioned features of the equipment.

The ECLIPSE-III is intended to generate X-ray radiation during normal operation. The ECLIPSE-III has been designed to focus radiation in the designated output direction. However, radiation in other directions is possible and it should be considered with shielding and/or monitoring in the final application.

2.2 X-ray fluorescence (XRF)

When a primary excitation source from an X-ray tube or a radioactive source strikes a sample, the X-ray (excitation particles in general) can be either absorbed by the atom or scattered through the material. The process in which an X-ray or a photon in general, is absorbed by the atom transferring all of its energy to an innermost electron is called photoelectric absorption. During this process, if the primary excitation source (the X-ray photon in our case) has sufficient energy, electrons are ejected from the inner shells, creating vacancies. These vacancies present an unstable condition for the atom. As the atom returns to its stable condition, electrons from the outer shells are transferred to the inner shells and in the process give off a characteristic X-ray whose energy is the difference between the two binding energies of the corresponding shells. The emitted X-rays produced from this process are called X-ray fluorescence (XRF). The process of detecting and analysing the emitted X-rays is called X-ray fluorescence analysis (XRFA).

In most cases, the innermost K and L shells are involved in XRF detection. A typical X-ray spectrum from an irradiated sample will display multiple peaks of different intensities.

The characteristic X-rays are labelled as K, L, M or N to denote the shells they originated from. Another designation alpha (α), beta (β) or gamma (γ) is made to mark the X-rays that originated from the transitions of electrons from higher shells. Hence, a Kα X-ray is produced from a transition of an electron from the L to the K shell, and a Kβ X-ray is produced from a transition of an electron from the M to a K shell, etc. Since within the shells there are multiple orbits of higher and lower binding energy electrons, a further designation is made as α_1, α_2 or β_1, β_2, etc. to denote transitions of electrons from these orbits into the same lower shell.

The XRF method is widely used to measure the elemental composition of materials. Since this method is fast and non-destructive to the sample, it is the method of choice for field applications and industrial production for control of materials. Depending on the application, XRF can be produced by using not only X-rays but also other primary excitation sources such as alpha particles, protons, high-energy electron beams or gamma radiation.

2.3 Using ECLIPSE-III in a teaching laboratory

Therefore, XRFA is a standard technique to obtain the composition of a sample. A qualitative analysis is quite simple based on the fluorescence spectrum recorded in a detector and analysed by the appropriated software.

The ECLIPSE-III equipment is used by students of different courses of the Polytechnic University of Valencia to determine the composition of different materials, as well as to identify impurities contained in a sample.

It is not expected that doses around the tube do represent a risk for people working in the laboratory, due to low voltage and current operating values of the ECLIPSE-III tube. On the other hand, a flashing red LED and a beeper warns the user while the X-ray tube is working. However, as the tube will be mainly used for practical exercises performed by students of Radioprotection courses, or having basic Radioprotection concepts among their matters, doses and subsequent shielding should be evaluated. A layout of the equipment at the laboratory is shown in figure 2.

Fig. 2. XRFA layout at the laboratory.

As mentioned above, it is not necessary a shielding of the equipment due to the low values of voltage and current and taking into account that the maximum permanence time of students near the equipment is about 2 hours. Furthermore, they do not stay in the direct path of the beam. For more precautions a radiation monitor is placed close to the laptop where students manage the XRFA software.

However, the application of the ALARA criterion, well studied by our students, advises the use of a shielding around the equipment as well as other cautions such as focusing the tube towards the wall where it is impossible to find anybody.

Therefore, four methacrylate plates of 1 cm thickness have been disposed around the equipment, at the top, front, right and left sides. It is not necessary a shielding at the back because of the 25-cm concrete wall of the laboratory. Moreover, the equipment is placed on a laboratory table, so there is no need to shield the bottom side.

Another dosimeter is placed next to the shielding. The front side of the shielding box can be opened to manage samples. It can be seen in figure 2 the disposition of the X-ray tube and the detector, as well as methacrylate plates.

To prepare the exercise a small piece of gafchromic film is placed in the sample holder. The X-ray tube is switched on for warming during some minutes and it produces a dark spot on the film. After switching off the tube, two laser lights focus on the film so that both lights match the mark. Afterwards, laser lamps are fixed with their screws. Then to correctly situate the sample both laser lights must coincide on the same point in the sample. This procedure can be seen in figure 3.

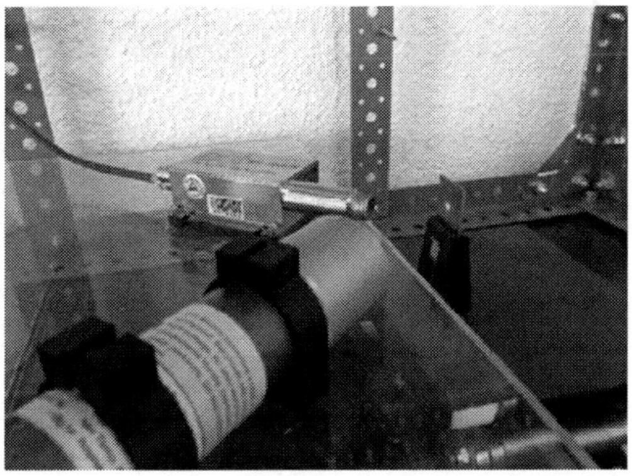

Fig. 3. Placement of the sample with laser lights.

2.4 Monte Carlo models

2.4.1 Simulation of the X-ray spectrum

The MCNP5 code, based on the MC method, is suitable to simulate the production of X-ray spectra. A model of the ECLIPSE-III X-ray tube has been developed for the MCNP5 code. For simulation of X-ray spectra, the MCNP5 code has been run in photon and electron mode (mode: P, E).

To improve the efficiency of electron and photon transport, two cards (PHYS: P and PHYS: E) are considered in MCNP5 for biasing some physical parameters such as production of secondary electrons by photons (IDES), coherent scattering (NOCOH), bremsstrahlung angular distribution (IBAD) and production of characteristic X-rays (XNUM).

A continuous slowing down model is used for electron transport. For photon transport, the code takes into account incoherent and coherent scattering, the possibility of fluorescent emission after photoelectric absorption and bremsstrahlung. To follow an electron through a significant energy loss, the MCNP5 code breaks the electron path into many steps. One way to achieve acceptable computing times is to reduce the time spent on tracking electrons. This can be done by increasing the cut-off energy for electrons, so the value of this cut-off in MCNP5 has been modified for these calculations. It has been set to 5 keV.

MCNP5 is suitable for modelling the flux (tally F5) at a point emitted by the X-ray tube using a semi-deterministic method to improve the variance of results. According to the MCNP5 user manual, the F5 tally requires a statistical uncertainty lower than 5% to produce a generally reliable confidence level [1].

X-ray tubes generate X-rays by accelerating electrons across a vacuum and bombarding them into a metal target. A portion of the electron energy as it strikes the target is converted into X-ray radiation. All X-ray exposures produce photons of different energy. The distribution of these photons makes the X-ray spectrum.

The procedure to simulate X-ray production consists in tracking a large number of incident electrons on the anode target material until they are absorbed or emerge from it. When the electrons strike the target, the code transports electrons inside the target material until they are stopped after losing their kinetic energy. During the transport of electrons, all bremsstrahlung and characteristic X-ray production is considered.

Bremsstrahlung X-rays have a broad energy distribution, with the greatest number of X-rays being produced at lower energies.

A limited number of X-rays are produced as characteristic X-rays. Because characteristic X-ray production involves transitions of electrons between various orbital shells of the target atoms, characteristic X-rays appears as discrete lines at certain fixed energy in the X-ray spectrum. The fixed energy levels represent the differences between the binding energies of the electron orbital shells of the target atoms.

In these cases the electron source, which has been defined as a point source, emits electrons with a given energy within a solid angle ψ. The electric field has not been simulated, so the maximum energy reached by electrons is the energy specified in the source definition. The simulated X-ray spectrum has been recorded at the exit of the X-ray tube using a point detector tally F5 ($1/cm^2$ per emitted particle).

The X-rays propagate, within a solid angle ψ, from the point of production in the direction towards the beryllium window. After passing throw this window, the X-ray beam passes through the air until the measurement point.

To study the effect of the electron energy distribution on the X-ray spectrum, it has been considered in the MCNP5 simulations the ripple factor with different energy distributions (uniform and normal). The ripple factor corresponds to the amount of voltage fluctuation, represented as a percentage of the maximum voltage, present in the waveform.

DOI: 10.1051/epjconf/201715301008

Therefore, the MC model includes a point electron source with a directional beam along the axis of the tube, hitting the Ag target where the X-rays are generated.

The X-ray beam is transported in air passing through the Be window. A point detector (F5 tally) is used to estimate the photon flux produced by electrons impinging the target. With adequate bins, the spectrum can be reconstructed.

The simulated spectra with uniform and normal electron energy distributions for 15% of ripple are compared with specifications given by AMPTEK, the tube manufacturer. Comparison is shown in figure 4.

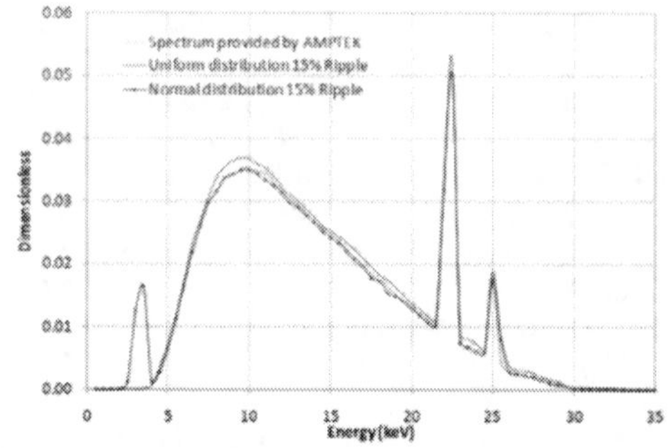

Fig. 4. Comparison of spectra provided by AMPTEK and spectra simulated with two electron energy distributions for 15% of ripple.

As it can be seen in figure 4, there are not important differences between the two types of distributions simulated, although normal distribution shows more fluctuations than uniform distribution. Another fact that can be observed is that the ripple factor produces a reduction in the fluence of the X-ray spectra and in the characteristic lines.

It can be also seen in figure 4 a peak at 3.5 keV, which appears only for simulated spectra. This peak does not exist in the actual tube probably due to an aluminium filter. However, it has not been considered in the MNCP5 model due to the lack of technical specification regarding the filter.

The area under the curve represents the number of photons with an energy between 4 and 30 keV produced by each electron emitted per cm^2. The value obtained by MATLAB calculating the integral of the spectrum is equal to $2*10^{-4}$ photons/e$^-$ emitted cm^2 [5].

2.4.2 MC model for dose calculations

Another MC model has been developed to evaluate the dose rate around the XRFA installation for an optimal calculation of the radioprotection.

Dose rates calculated with the MCNP code are compared with experimental measurements.

Isodoses around the XRFA installation have been obtained with an F4 tally, which is a fluency tally that can be modified by DE, DF cards to obtain dose rate values.

The DE and DF cards allow modelling an energy-dependent dose function that is a continuous function of energy from a table whose data points are given by DE (energy values) and DF (values of the function). In this case, DF gives conversion factors from fluency to dose rate.

The DF card was evaluated using the mass energy transfer coefficient μ_{en}/ρ (cm^2/g) extracted from NIST [6]. This coefficient multiplied by the energy (MeV), gives Mev-cm^2/g and using the appropriate constant for conversion of units, permit to express DF in μSv/cm^2. Therefore, the tally F4 (cm^{-2}) will be expressed in μSv per photon emitted at the source.

The detailed gamma physics has been used, taking into account coherent and incoherent scattering, photoelectric effect (with fluorescence photon production).

A number of 500 million of particles (NPS) have been launched to improve statistics.

For the dose distribution calculation, a mesh grid has been super-imposed to the installation and the surrounding media geometry (FMESH). F4 tally with FMESH card allows the user to define a mesh tally superimposed over problem geometry.

Fluency in nodes of this mesh (in cm^{-2}) is obtained. If the source is expressed in photons/s, the tally will be obtained in particle flux (cm^{-2} s^{-1}). Using DE, DF cards the tally is obtained in μSv/s per photon emitted at the source. Therefore, it is necessary to estimate the number of photons/s emitted at the source.

With this goal in mind, we come back to the previous model. The area under the ECLIPSE spectrum was calculated as $2*10^{-4}$ photons/cm^2 per emitted electron. To estimate the number of emitted electrons, we should consider the current given to the filament, 100 μA, that is, 10^{-4} C/s, and the equivalence of one Coulomb in electrons ($6.2422*10^{18}$ e$^-$/C) obtaining $6.2422*10^{14}$ e$^-$/s.

Multiplying the area under the spectrum times the area of the tube focus times the number of electrons per second, one can obtain the number of photons/s emitted at the source, the X-ray tube focus.

This value multiplied per the tally result gives the dose rate in μSv/s and applying a constant (3600 s/h) for conversion of units, in μSv/h.

2.5 Measurements

Dose rate measurement around the XRFA installation have been done with an ionization chamber ROTEM RAM ION [7]. It is a battery operated, auto ranging portable ion chamber survey meter designed for highly stable and accurate measurement of dose rates and integrated dose of gamma, X-ray and beta radiation.

A picture of the chamber showing the screen and the measuring probe, can be seen in figure 5.

Main specifications of this ion chamber are the following:

- *Measuring Range:* 1 μSv/hr to 500 mSv/hr
- *Display Range:* 0.1 μSv/hr to 500 mSv/hr
- *Accuracy:* ±10% of reading within measuring range
- *Ion Chamber Volume:* 500 cc
- *Chamber Wall and Cover Thickness:* 300 mg/cm^2 (tissue equivalent)
- Window Thickness: 7 mg/cm^2

- *Response Time:* 2 sec. for readings above 1 mR/h 5 sec. for auto-ranging change, from Low to High Range
- *Power Source:* Two 1.5V C-type Alkaline cells - 100 hours of continuous operation
- *Display :* DigiLog (3 digits and 2 decades of analog bar graph)
- *Data Logging :* 347 data records (1415 with ex-tended memory)
- *Temperature Range:* Operation: -10ºC to +50ºC (15ºF – 122ºF) Storage: -20ºC to +60ºC
- *Humidity Range:* Up to 95% RH (non condensing)
- *Dimensions:* Width: 10cm, length 25cm, height 19cm
- *Weight:* 1,1 kg

Fig. 5. ROTEM ionization chamber.

Measurements were done at different distances from the X-ray tube, inside and outside the methacrylate box. Dose rate was also measured at the other side of the concrete wall of the laboratory, but only background dose (order of nSv/h) was found.

3 Results

First, a conservative approximation was adopted to simplify the model calculating dose rates. In this approximation, the tube is placed parallel to the methacrylate folding lid and perpendicular to the right methacrylate plate. Dose rates are calculated in the axis of the beam.

However, measurements in those points along the axis of the beam. are not reliable due to the pile-up produced by photons. The monitor should be placed far away of the device being very difficult the measurement due to dimensions of the room. Another solution is to use a scattering bar, but this implies a complicated calculation involving unfolding of values measured.

The second version of this model is the actual layout of the equipment for XRFA in the laboratory of Nuclear Engineering of the Polytechnic University of Valencia. The tube forms 45º with the plate walls, so the beam is not perpendicular to any of them. Now measurements are reliable out of the shielding box.

Nevertheless, inside the box, again, pile-up is produced in the detector and measurements are not reliable.

For this second version of the model that, on the other hand, is a more real solution, MCNP simulations and measurements have been performed at 3 points in a plane containing the X-ray focus:

- (P1), at 15 cm on the perpendicular to the front plate (lid)
- (P2), inside the box, at 15 cm from the X-ray focus
- (P3), at 7.5 cm on the perpendicular to the right plate.

For these three points, results from MCNP simulations and measurements are listed in Table 1.

Table 1. Comparison of results.

Position	MCNP µSv/h	Relative error	measured (µSv/hr)	Ratio M/m	Disagr.
P1	7.4	0.0195	10.2	0.725	27.5%
P2	$1.3*10^3$	0.0013	**Not reliable**	-------	
P3	52.5	0.0064	53.2	0.987	1.3 %

The disagreement for P1 is not significant because the value measured (340 nSv in 30 sec) is practically the background (150-200 nSv).

4 Discussion

MCNP models cannot be completely validated because of the problems with pile-up in the detector.

However, results from the second version of the MCNP simulation for doses are acceptable for those points around the box containing the XRFA installation according to values listed in the Table 1.

The nominal allowable limit for normal working conditions is 50 mSv/yr. Considering 2000 working hours/year, the limit would be 25 µSv/hr. However, the maximum time for students around the device can be estimated as 2 hours per year that means a limit equal to 25 mSv/hr, while for teachers this time is not higher than 20 hours per year, that is, the limit would be equal to 2.5 mSv/hr.

Therefore, all values calculated and measured are under limits.

On the other hand, the working point, that is, the control point, is not placed into the path of the X-ray beam while the tube is working and where dose values are measured and calculated by MCNP. It is located next to the laptop where the XRFA software is managed, at a distance about 2.3 m from the box containing the XRFA installation. At this point, the dose measured is about 200 nSv/h a bit higher than background (150 nSv/h).

Doses are higher than limits inside the methacrylate box, but it is forbidden to introduce hands into the box while the tube is working.

5 Conclusions

The Monte Carlo method is a powerful tool to obtain the spectrum of the tube in an XRFA device.

An MC model was developed for the MCNP code to obtain the ECLIPSE-III X-ray tube spectrum.

The spectrum obtained with MCNP shows no important discrepancies when compared with specifications given by AMPTEK, the tube manufacturer.

Therefore, the first MC model is validated.

Another MC model was developed to evaluate doses produced by the tube and the thickness of shielding necessary to attenuate these doses, applying the ALARA criterion.

This model can be validated only for points out of the X-ray beam axis because dose measurements are not reliable into the beam.

The validation of the models developed by means of comparison of results will permit their application to other similar situations.

Calculated and measured doses outside the shielding (methacrylate box) are lower enough than limits.

Shielding provided to the XRFA installation at the laboratory of Nuclear Engineering of the Universidad Politécnica de Valencia (UPV) is sufficient to protect students and teachers working in the laboratory.

These calculations and measurements can be proposed to students as a complementary work to their practical exercises in courses related to Radioprotection.

It is highly interesting for students at the laboratory of Nuclear Engineering and Radiation Protection to assess doses and shielding checking that ALARA criterion is accomplished and the working situation is safe enough.

References

1. B. Beckhoff, B. Kanngießer, N. Langhoff, R. Wedell, H. Wolff (Eds.), *Handbook of Practical X-Ray Fluorescence Analysis*, Springer, Berlin, (2006)
2. X-5 Monte Carlo team, *MCNP – A General Monte Carlo N Particle Transport Code, Version 5*, LA-UR-03-1987, Los Alamos National Laboratory, (April 2003)
3. ECLIPSE-III X-ray tube. Amptek. www.amptek.com (2011)
4. DIRECTIVA 2013/59/EURATOM DEL CONSEJO de 5 de diciembre de 2013 por la que se establecen normas de seguridad básicas para la protección contra los peligros derivados de la exposición a radiaciones ionizantes.
5. A. Querol, S. Gallardo, J. Ródenas and G. Verdú, *Analysis of the effect of the Electron Energy Distribution on the X-Ray Spectra Produced* International Nuclear Atlantic Conference - INAC 2011 ISBN:978-85-99141-04-5 (2011)
6. NIST (Physical Measurement Laboratory), *NIST: X-ray Mass Attenuation Coefficients*, physics.nist.gov/PhysRefData
7. http://www.gammadata.se/products/radiation-detection/portable-instruments/dose-rate-instruments/ram-ion/

 DOI: 10.1051/epjconf/201715301009

Organ and Effective Dose Coefficients for Cranial and Caudal Irradiation Geometries: Neutrons

K. G. Veinot[1,2,a], K.F. Eckerman[2], N.E. Hertel[2,3], M.M. Hiller[2]

[1] *Y-12 National Security Complex P.O. Box 2009, M.S. 8206 Oak Ridge, TN, USA 37831*
[2] *Center for Radiation Protection Knowledge Oak Ridge National Laboratory Oak Ridge, TN, USA 37831*
[3] *Georgia Institute of Technology Atlanta, GA, USA 30332-0745*

Abstract. With the introduction of new recommendations by ICRP Publication 103, the methodology for determining the protection quantity, effective dose, has been modified. The modifications include changes to the defined organs and tissues, the associated tissue weighting factors, radiation weighting factors, and the introduction of reference sex-specific computational phantoms (ICRP Publication 110). Computations of equivalent doses in organs and tissues are now performed in both the male and female phantoms and the sex-averaged values used to determine the effective dose. Dose coefficients based on the ICRP 103 recommendations were reported in ICRP Publication 116, the revision of ICRP Publication 74 and ICRU Publication 57. The coefficients were determined for the following irradiation geometries: anterior-posterior (AP), posterior-anterior (PA), right and left lateral (RLAT and LLAT), rotational (ROT), and isotropic (ISO). In this work, the methodology of ICRP Publication 116 was used to compute dose coefficients for neutron irradiation of the body with parallel beams directed upward from below the feet (caudal) and directed downward from above the head (cranial). These geometries may be encountered in the workplace from personnel standing on contaminated surfaces or volumes and from overhead sources. Calculations of organ and tissue absorbed doses for caudal and cranial exposures to neutrons ranging in energy from 10^{-9} MeV to 10 GeV have been performed using the MCNP6 radiation transport code and the adult reference voxel phantoms of ICRP Publication 110. At lower energies the effective dose per particle fluence for cranial and caudal exposures is less than AP orientations while above about 30 MeV the cranial and caudal values are greater.

1 INTRODUCTION

The International Commission on Radiological Protection (ICRP) Publication 103 [1] established a revised method for determining the protection quantities for organs and tissues (equivalent doses) and the effective dose. Changes to the defined organs and tissues, their associated tissue weighting factors, radiation weighting factors, and the use of sex-specific phantoms from ICRP 110 [2] required a recomputation of dose coefficients. The ICRP DOCAL task group reported these dose coefficients in ICRP Publication 116 [3] which replaces ICRP 74 [4] and the International Commission on Radiation Units and Measurements Report 57 [5]. In ICRP 116 coefficients for the six standard irradiation geometries (anterior-posterior (AP), posterior-anterior (PA), right and left lateral (RLAT and LLAT), rotational (ROT), and isotropic (ISO)) are provided.

Perhaps the most commonly used orientation in the application of the coefficients in practice is the AP geometry since this corresponds to the standard geometry used for personal dosimeter and instrument calibrations. In addition, the ROT irradiation geometry has obvious uses in the computation of effective dose to workers moving about in a radiation field and the ISO geometry also would have application in practical situations. However, in many operational scenarios these standard geometries may not be appropriate and angular effects, both on the dosimetric quantities as well as the dosimeter response, must be considered. The irradiation geometries of ICRP 116 also do not directly account for the possibility of a person being exposed from sources either beneath the feet (caudal) or from overhead (cranial). In operational settings personnel may perform work on contaminated surfaces or with sources positioned overhead. Since the effective dose is dependent upon the doses delivered to organs and tissues, and because these organs and tissues are shielded differently by the body for different angles of irradion, the caudal (CAU) and cranial (CRA) geometries may result in significantly different effective doses compared to the six standard geometries addressed in ICRP 116.

[a] Corresponding author: veinotkg@y12.doe.gov

2 MATERIALS AND METHODS

2.1 Protection Quantity Calculations

As in previous ICRP recommendations, such as ICRP 26 [6] and ICRP 60 [7], the method for determining protection quantities begins with calculating doses to organs and tissues using mathematical phantoms. In ICRP 103 methodology, the organ and tissue absorbed dose coefficients are determined for the male and female reference voxel phantoms separately, the appropriate radiation weighting factor applied, then the sex-averaged equivalent dose coefficients obtained by averaging the values from the male and female phantoms. The radiation weighting factors, w_R, are given in ICRP 103 for various radiations and energies and for neutrons as a continuous function of neutron energy. Once the sex-averaged equivalent doses are determined, they are weighted by the tissue weighting factors and summed to determine the effective dose coefficient. In the ICRP 103 methodology, the specific organs used to define the remainder are prescribed.

ICRP Publication 103 defines effective dose, E, as

$$E = \sum_T w_T \left[\frac{H_T^M + H_T^M}{2} \right] \quad (1)$$

where H_T^M is the equivalent dose to organ T of the male phantom, H_T^F is the equivalent dose to organ T of the female phantom, and w_T is the organ or tissue weighting factor. The equivalent dose to the organ or tissue is

$$H_T = \sum_R w_R D_{T,R} \quad (2)$$

where w_R is the radiation weighting factor and $D_{T,R}$ is the absorbed dose to tissue or organ T from incident radiation R.

2.2 Reference Phantoms

The reference phantoms of ICRP 110 [2] were used for all calculations presented herein. These three-dimensional phantoms are volumized-pixel (voxel) representations of male and female subjects based on high-resolution scans. The male phantom contains almost two million voxels with each rectangular solid being 8 mm X 2.137 mm X 2.137 mm. The female phantom contains almost four million voxels with each being 4.84 mm X 1.775 mm X 1.775 mm. Tissue compositions and densities are based on ICRP 89 [8] information. The male phantom is 1.76 m in height and has a mass of 73 kg while the female is 1.63 m tall and has a mass of 60 kg. Both subjects were scanned while laying prone so the phantoms do display some flattening. Additionally, the female head position slightly favors a head down orientation. Because of the voxel resolution, exact calculations to sensitive regions of the skin (e.g. a depth of 0.07 mm) and to the lens of the eye are not possible, however, for the calculations reported here the dose to those voxels defining the skin were used. Lens of the eye dose coefficients are not reported. Differences in phantom size and organ mass impact the dose coefficients for certain organs because of shielding effects.

2.3 Reference Radiation Field Orientations

For caudal and cranial calculations, planar monoenergetic neutron sources were oriented either below the feet or over the head with particles directed parallel to the long-axis of the phantom. Source areas were sufficiently large to ensure the entire cross-sectional area of the phantom was within the radiation field.

2.4 Calculational Method

Calculations were performed using the MCNP6 [9] radiation transport code along with ENDF-B VII cross sections. Both the male and female voxel phantoms from ICRP Publication 110 were modeled and irradiated with planar sources oriented upward from below (CAU) or downward from above (CRA) the phantoms in a vacuum. Organ equivalent doses were computed by tracking the secondary particles via the modified kerma tally (+F6 in MCNP nomenclature). This provides a more accurate computation of the absorbed dose since it does not assume all energy from interactions is deposited at the interaction site and tracks the secondaries created in the interactions. The dose coefficient energy grid duplicated that used in ICRP Publication 116 and ranged in energy from 10 keV to 10 GeV. Organ and effective dose coefficients were normalized to particle fluence and are presented in units of pSv-cm^2.

Because of the segmentation of the cortical (bone surface) and spongiosa (red bone marrow) regions these doses were tallied individually and mass weighted to determine the total bone surface and red bone marrow doses. This method is in accordance with that used for ICRP Publication 116.

Tally uncertainties were minimized, although some organs that received small doses had large tally uncertainties. In cases where these organs and tissues did not significantly contribute to the effective dose this uncertainty was accepted as a practical limitation on computation time. When the organs or tissues were significant contributors to effective dose (the weighted organ equivalent dose, $w_T H_T$, was greater than 1% of the effective dose), additional particle histories were executed to reduce the uncertainty to less than 3%. In the cases of the bone surface and the red bone marrow the contributor tally errors were propagated using mass weighting. The effective dose error was determined by propagating the sex-averaged, organ weighted equivalent doses in quadrature.

3 RESULTS AND DISCUSSION

Effective dose coefficients for cranial (CRA) and caudal (CAU) exposures are shown in Fig. 1 along with AP irradiation geometry values from ICRP Publication 116. Ratios of the CAU and CRA effective dose coefficients to AP are shown in Fig. 2. Tabulated values of the effective dose and all organs and tissue equivalent dose coefficients are available for download from the Oak

Ridge National Laboratory Center for Radiation Protection Knowledge website (http://crpk.ornl.gov/resources/) or can be obtained by contacting the authors.

The effects of caudal and cranial exposures on effective dose are evident, and not unexpected. Differences in organ doses between the male and female were noted and are a consequence of organ location within the body (effects of different body shielding of internal organs for different angles of incidence), phantom mass and size, and phantom posture itself. For example, the gonads and brain dose are two organs that exhibit the highest dependence on irradiation orientation. Ratios of the organ equivalent dose for the two geometries for the brain and gonads are shown in Fig. 2.

4 CONCLUSION

Dose coefficients for caudal and cranial irradiation geometries have been calculated in accordance with ICRP Publication 103. The coefficients differ significantly from AP values for both organ and tissues of ICRP Publication 116, as would be expected. The differences can lead to significant overestimation of personnel doses if the AP irradiation geometry is assumed.

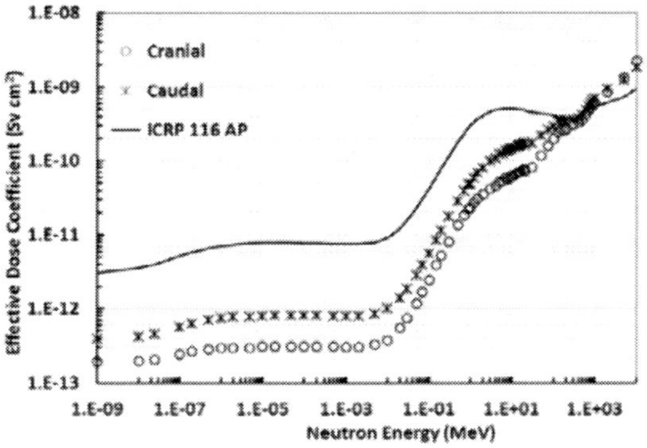

Figure 1. Effective dose coefficients for cranial and caudal irradiations and for AP as listed in ICRP 116.

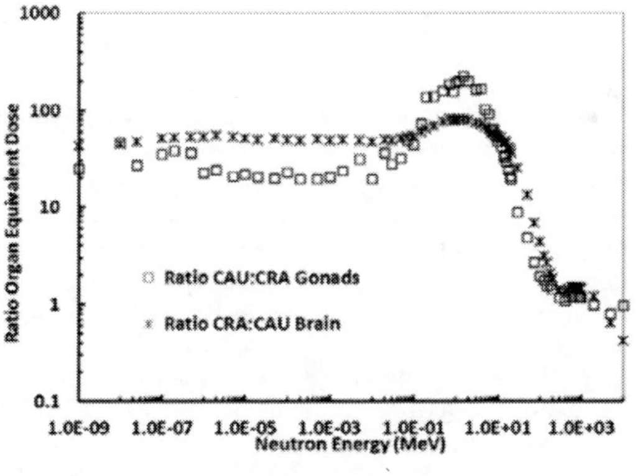

Figure 2. Ratio of organ equivalent doses for cranial and caudal geometries for the brain and gonads..

References

1. INTERNATIONAL COMMISSION ON RADIOLOGICAL PROTECTION (ICRP), "The 2007 Recommendations of the International Commission on Radiological Protection," ICRP Publication 103. Ann. ICRP 37 (2-4). Elsevier (2007).
2. INTERNATIONAL COMMISSION ON RADIOLOGICAL PROTECTION (ICRP), "Adult Reference Computational Phantoms," ICRP Publication 110. Ann. ICRP 39 (2). Elsevier (2009).
3. INTERNATIONAL COMMISSION ON RADIOLOGICAL PROTECTION (ICRP), "Conversion Coefficients for Radiological Protection Quantities for External Radiation Exposures," ICRP Publication 116. Ann. ICRP 40 (2-5). Elsevier (2010).
4. INTERNATIONAL COMMISSION ON RADIOLOGICAL PROTECTION (ICRP), "Conversion Coefficients for Use in Radiological Protection Against External Radiation," ICRP Publication 74. Ann. ICRP 26 (3-4). Pergamon Press (1996).
5. INTERNATIONAL COMMISSION ON RADIATION UNITS AND MEASUREMENTS (ICRU), "Conversion Coefficients For Use In Radiological Protection Against External Radiation," ICRU Report 57, Bethesda, MD; ICRU (1998).
6. INTERNATIONAL COMMISSION ON RADIOLOGICAL PROTECTION (ICRP), "Recommendations of the International Commission on Radiological Protection," ICRP Publication 26. Ann. ICRP 1 (3). Pergamon Press (1977).
7. INTERNATIONAL COMMISSION ON RADIOLOGICAL PROTECTION (ICRP), "1990 Recommendations of the International Commission on Radiological Protection," ICRP Publication 60. Ann. ICRP 21 (1-3). Pergamon Press (1991).
8. INTERNATIONAL COMMISSION ON RADIOLOGICAL PROTECTION (ICRP), "Basic Anatomical and Physiological Data for Use in Radiological Protection Reference Values," ICRP Publication 89. Ann. ICRP 32 (3-4). Pergamon Press (2002).
9. MCNP6 USER'S MANUAL Version 1.0, Denise B. Pelowitz, Ed., LA-CP-13-00634 Rev. 0, Los Alamos National Laboratory Los Alamos, New Mexico (2013).

Activation experiment for concrete blocks using thermal neutrons

Koichi Okuno[1*], *Seiichiro Tanaka*[1]

[1]Hazama-Ando Corporation Technical Research Institute, Karima 515-1, Tsukuba, Ibaraki, 305-0822, Japan

Abstract. Activation experiments for ordinary concrete, colemanite-peridotite concrete, B_4C-loaded concrete, and limestone concrete are carried out using thermal neutrons. The results reveal that the effective dose for gamma rays from activated nuclides of colemanite-peridotite concrete is lower than that for the other types of concrete. Therefore, colemanite-peridotite concrete is useful for reducing radiation exposure for workers.

1 Introduction

In recent years, applications of neutrons have been studied in various fields. Concrete is widely used for the neutron shields. Although concrete is inexpensive, it requires massive thickness to make enough shields to neutrons and to occupy the room spaces.

Boron carbide is a good neutron absorber and often used as the containing material for the concrete. But the cost of its shield has risen steeply because boron carbide has a very high cost.

Colemanite is an example of the natural rock that contains boron. It is produced in several places in the world, but most is produced in Turkey. The active research on concrete using colemanite as an aggregate was done in the 1950s and 1960s.(1)-(5) Research on concrete blocks using colemanite was carried out.(6)-(9) But, they cannot use facility.

Hydrogen-containing material is used for neutron shielding. Serpentinite and peridotite rocks contain much hydrogen. Serpentinite was used for neutron shielding concrete on the pedestal concrete shield on the reactor vessel in the fast breeder demonstration reactor Monju,(10) but it is not desirable for outer shields now because serpentinite is also a raw material for asbestos.

To realize a compact shield satisfying the radiation dose criteria even in a limited space such as in a hospital, neutron shield concrete using colemanite and peridotite has been developed (11). Such concrete has been used in four beam lines at the Materials and Life Science Facility (MLF) of the Japan Proton Accelerator Research Complex (J-PARC) and in the shielding wall of the RIKEN accelerator-driven compact neutron source (RANS)(12) (13).

In many cases, it is necessary to perform estimations of neutron activation for construction materials and research instruments. The influence of the activation is mainly divided into effects caused by the short half-life nuclides and those caused by the long half-life nuclides. An increase in the ambient dose in a room after irradiation is caused by short half-life nuclides, and increased radioactive waste is caused by long half-life nuclides. In past research, many activation experiments for concrete have been carried out by concrete power that weighs around 100 mg because the restriction of the irradiation port in research reactors. However, this method is not very realistic.

In this research, an activation experiment using various types of concrete blocks is carried out using thermal neutrons with an accelerator neutron source.

2 Concrete blocks

Ordinary concrete, colemanite-peridotite concrete (11), B_4C-loaded concrete, and limestone concrete are used in this research. Each concrete sample was made by us. The element analysis of each concrete was carried out by chemical analysis specifically ICP-AES (Inductively Coupled Plasma Atomic Emission Spectroscopy).

2.1 Colemanite-peridotite concrete

Colemanite-peridotite concrete uses colemanite and peridotite rocks as the aggregates. Colemanite is a natural mineral that is rich in boron in the form of B_2O_3. Peridotite is a natural mineral that is rich in hydrogen atoms in the form of H_2O. These minerals are used as aggregates, and together with ordinary Portland cement, form the colemanite-peridotite concrete. Colemanite forms 5–15 wt% of the concrete composition for improved shielding performance. The density of colemanite-peridotite concrete is the same as that of ordinary concrete.

The elemental analysis data for ordinary concrete and colemanite-peridotite concrete used in this research are presented in Tables 1 and 2.

* Corresponding author: okuno.koichi@ad-hzm.co.jp

Table 1. Ordinary concrete (wt%)

SiO_2	H_2O	MgO	MnO	Al_2O_3	CaO
62.4	4.73	1.36	0.06	10.4	10.2
Na_2O	K_2O	TiO_2	Fe_2O_3	SO_3	
2.06	2.48	0.42	3.19	0.29	

Table 2. Colemanite-peridotite concrete (wt%)

SiO_2	H_2O	MgO	MnO	Al_2O_3	B_2O_3
34.8	14.9	31.6	0.12	1.81	2.4
CaO	Na_2O	K_2O	TiO_2	Fe_2O_3	SO_3
9.31	0.08	0.06	0.08	7.08	0.28

2.2 B_4C-loaded concrete

B_4C (boron carbide)-loaded concrete contains 3 wt% of boron carbide powder loaded into ordinary concrete. The elemental analysis data for B_4C loaded concrete used in this research are presented in Table 3.

Table 3. B_4C loaded concrete (wt%)

SiO_2	H_2O	MgO	MnO	Al_2O_3	CaO
49.6	3.07	1.34	0.07	9.46	15.4
Na_2O	K_2O	TiO_2	Fe_2O_3	C	B
2.49	1.38	0.39	2.87	2.47	3.40

2.3 Limestone concrete

Limestone concrete contains limestone as the coarse aggregate of the concrete. The fine aggregate used is ordinary sand. The elemental analysis data for limestone concrete used in this research are presented in Table 4.

Table 4. Limestone concrete (wt%)

SiO_2	H_2O	MgO	MnO	Al_2O_3	CaO
33.3	4.84	0.70	0.03	4.68	7.99
Na_2O	K_2O	TiO_2	Fe_2O_3	SO_3	$CaCO_3$
0.92	0.99	0.17	1.47	0.29	46.0

3 Experimental Method

Neutrons from the cyclotron accelerator with a beryllium target were used. Proton energy was 18MeV; beam current was 17μA.Thermal neutrons were obtained from the 90° direction of the beryllium target with a polyethylene moderator and collimator. The size of each concrete sample was 45 mmφ × 50 mm thickness, and the samples were piled up to a height of 150 mm. To remove thermal neutrons other than those in the beam direction, a cadmium sheet was used around and at the bottom of the concrete sample. Thermal neutron flux irradiating the concrete sample was measured using gold foil. The experimental setup is shown in Figure 1. All the various concrete samples were irradiated at once, for the same time, i.e., 16 h. In the BNCT treatment room, Neutron flux at the concrete wall that calculated by PHITS code is the order of 10^6 (n/cm^2 s), and treatment time is 1 h, Therefore, integrated neutron flux is 3.6×10^9 (n/1h cm^2). On the other hand, Neutron flux in experiment facility used in research is the order of 10^5 (n/cm^2 s). So, The irradiation time equivalent to BNCT treatment is 16 h. After irradiation, the gamma spectra of the concrete samples were measured by a Ge detector. Specification of Ge detector is 15% of efficiency, 1.9keV of resolution at ^{60}Co. Then, quantitative analysis was carried out for all samples by using calibration data measured for a radioactive standard volume source. The measured activated nuclides are ^{56}Mn and ^{24}Na, which are the most detected nuclides.

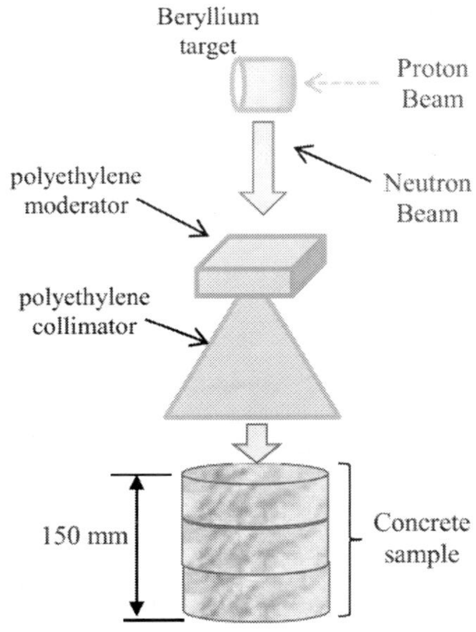

Experimental setup (sketch)

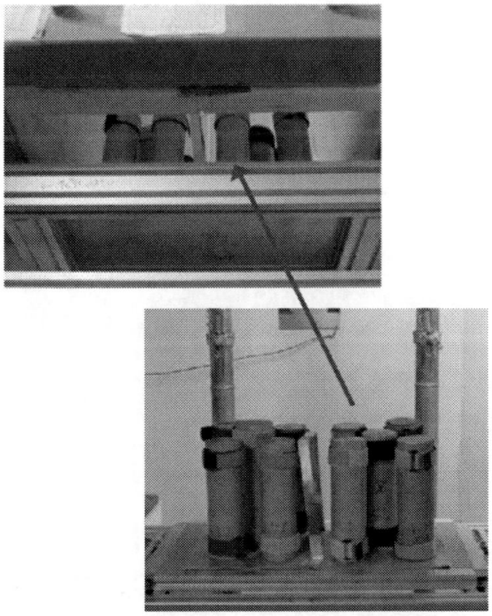

Experimental setup (picture)

Fig. 1. Experimental setup

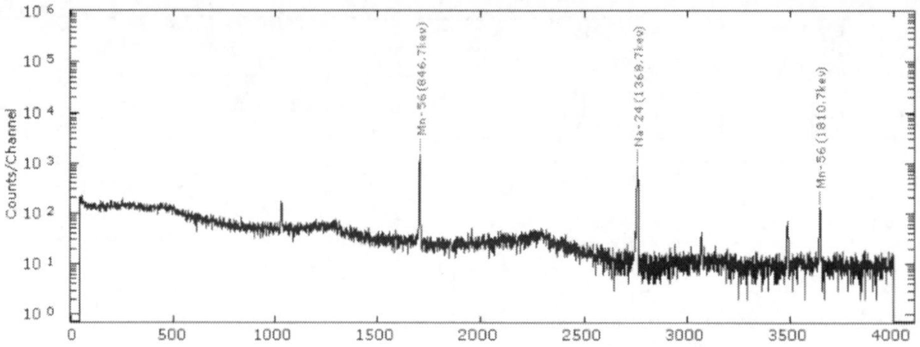

Fig. 2. Gamma spectrum for ordinary concrete (most closer to the neutron source)

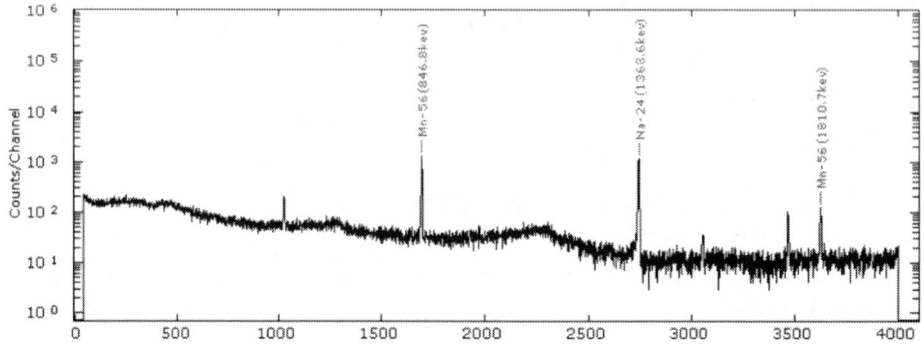

Fig. 3. Gamma spectrum for limestone concrete (most closer to the neutron source)

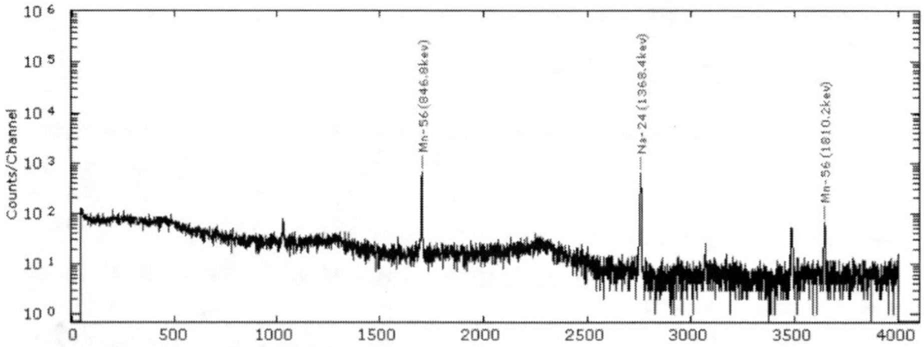

Fig. 4. Gamma spectrum for colemanite-peridotite concrete (most closer to the neutron source)

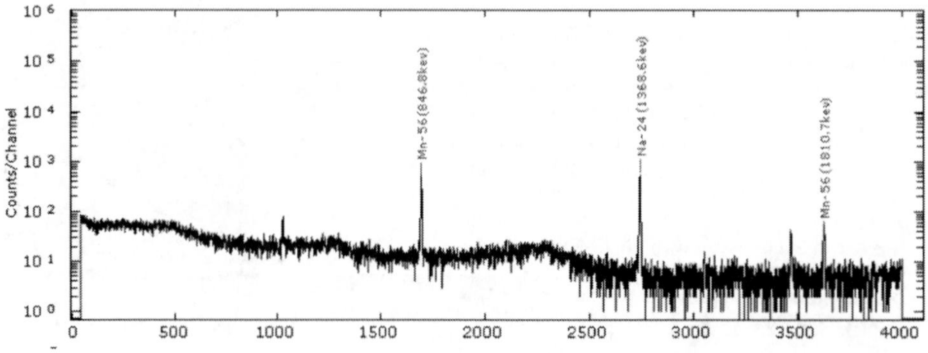

Fig. 5. Gamma spectrum for B_4C-loaded concrete (most closer to the neutron source)

Table 5. Results of quantitative analysis

Ordinary concrete		
Depth	^{56}Mn (Bq/g)	^{24}Na (Bq/g)
5cm	10.56	8.82
10cm	1.31	1.02

Table 6. Results of quantitative analysis

Limestone concrete		
Depth	^{56}Mn (Bq/g)	^{24}Na (Bq/g)
5cm	4.15	5.93
10cm	0.55	0.70

Table 7. Results of quantitative analysis

Colemanite-peridotite concrete		
Depth	^{56}Mn (Bq/g)	^{24}Na (Bq/g)
5cm	5.92	0.93
10cm	0.39	0.47

Table 8. Results of quantitative analysis

B_4C loaded conctete		
Depth	^{56}Mn (Bq/g)	^{24}Na (Bq/g)
5cm	2.94	4.26
10cm	0.30	0.30

4 Results

The irradiated thermal neutron flux measured by the gold foil is 1.6×10^5 (n/s cm^2) and the cadmium ratio is 5.4. Neutron spectra were not measured in this research because quantities of total thermal neutron are important for neutron activation in application facilities of neutrons. The measured gamma spectra at the top block of the concrete samples are shown in Figures 2–5. The live time in Figures 2–5 is 600 s because measurement error is low enough. In Figures 2–5, a sharp peak of ^{56}Mn and ^{24}Na is seen for all concrete samples. The counts of gamma spectrum for concrete with boron are lower than that of concrete without boron. The results of quantitative analysis for each sample are shown in Tables 5–8. The amount of activation for ordinary concrete is the largest. The production amount of ^{24}Na and ^{56}Mn for limestone concrete, which has no boron content, was lower than that for ordinary concrete because elemental analysis data in table 4 were lower than ordinary concrete. For the concrete containing boron, i.e., colemanite-peridotite concrete, the Mn content was twice that of ordinary concrete, and four times that of limestone concrete. However, the amount of ^{56}Mn in colemanite-peridotite concrete was 1.4 times that of limestone concrete, and 44% lower than that of ordinary concrete. This is the effect of the neutron capture of boron. For the B_4C-loaded concrete, the amount of Mn and Na are almost the same as that of ordinary concrete. However, the amount of ^{56}Mn and ^{24}Na was respectively 3.6 and 2 times lower than that of ordinary concrete. This is also the effect of the neutron capture of boron.

5 Conclusion

Activation experiments for ordinary concrete, colemanite-peridotite concrete, B_4C-loaded concrete and limestone concrete were carried out. From the view of radiation protection, dose calculations for activated concrete shown in Tables 5-8 were carried out by PHITS code Ver.2.88[(14)]. Nuclear data was used JENDL-4[(15)]. Calculation model was assumed as shown in Figure 6.

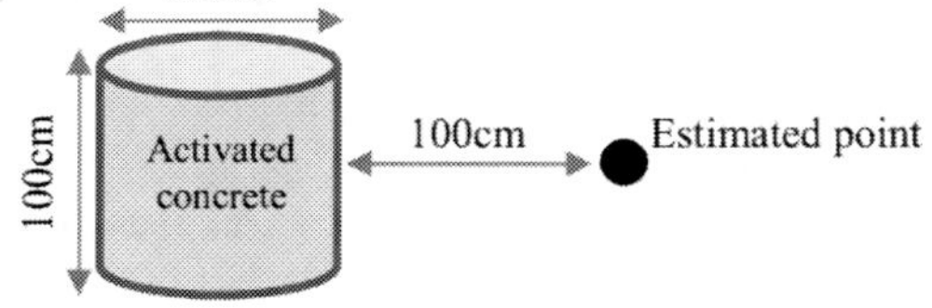

Fig. 6. Calculation model

Estimated point is 100cm far from the activated concrete. Gamma ray energy data produce from activated nuclei were used the RI-source function in PHITS code. Flux to dose conversion factor was used built-in data of PHITS code. The effective doses at a distance of 100cm from a activated concrete are listed in Table 9.

Table 9. Effective dose (μSv/h)

Ordinary Concrete	Limestone Concrete
0.121	0.068
Colemanite-peridotite concrete	B_4C loaded concrete
0.031	0.050

Results reveal that the dose for gamma ray from activated nuclides of colemanite-peridotite concrete is lower than that for the other concrete types. The reason for this is considered to be derived from amount of ^{24}Na. Gamma ray energy of ^{24}Na are 1.37MeV and 2.75MeV, therefore, influence to dose is high. From the reslut, colemanite-peridotite concrete will be useful for reducing radiation exposure for workers.

References

1. R. B. GALLAHER and A. S. KITZES, "Summary Report on Portland Cement Concretes for Shielding," ORNL-1414, Oak Ridge National Laboratory (1953)
2. A. N. KOMAROVSKII, Shielding Materials for Nuclear Reactors, Pergamon Press, New York (1961)
3. H. S.DAVIS, "High Density Concrete for Shielding Atomic Energy Plants," J. Am. Concrete. Inst., 54, 5, 1965 (1958)
4. K. OKADA et al., "A Few Experiments for Shielding Concrete," in "Technical Annual Report for Cement," ,452 (1957) in Japanese
5. J. O. HENRIE, "Properties of Nuclear Shielding Concrete,", J. Am. Concrete. Inst., 56, 7, 37 (1959)
6. Y. YARAR and A. BAYULKEN, "Investigation of Neutron Shielding Efficiency and Radioactivity of Concrete Shielding Containing Colemanite," J. Nuclear. Materials., 212–215, 1720 (1994)
7. D. E. VOLKMAN and P. L. BUSSOLINI, "Comparison of Fine Particle Colemanite and Boron Frit in Concrete for Time-Strength Relationship," J. Test. Eval., 20, 92 (1992)
8. Berna Oto et al., Photon attenuation properties of some concretes containing barite and colemanite in different rates. Annals of Nuclear Energy 51 (2013) 120–124
9. Turgay Korkut et al., Neutron dose transmission measurements for several new concrete samples including colemanite. Annals of Nuclear Energy 37 (2010) 996–998
10. S. USAMI, K. SASAKI, T. DESHUMARU, and F. NAKASHIMA," Measurement of Shielding Characteristics in the Prototype FBR Monju," J. Nucl. Sci. Technol., Suppl. 1, 83(2000)
11. Koichi Okuno, Masayoshi Kawai, Hitoshi Yamada, "Development of Novel Neutron Shielding Concrete", Nuclear Technology, 168, 545–552 (2009).
12. Masayoshi Kawai,Masao Yonemura,Shuki Torii,Takashi Mutoya,Kazuhiro Mori,Toshiharu Fkunaga,Koichi Okuno,Kazuo Takeda,Koji Niita and Takashi Kamiyama, "Novel monolayer shields of a neutron powder diffractometer SPICA at BL09 of J-PARC", Progress in Nuclear Science and Technology, 4, 156–159 (2014).
13. Koichi Okuno,Masayoshi Kawai,Hitoshi Yamada,Takenao Shinohara,Shin-ichi Takata,Jun-ichi Suzuki,Kentaro Suzuya and Kazuya Aizawa, "Application of neutron shield concrete to neutron scattering instrument TAIKAN in J-PARC", Progress in Nuclear Science and Technology, 4, 619–622 (2014).
14. T. Sato, K. Niita, N. Matsuda, S. Hashimoto, Y. Iwamoto, S. Noda, T. Ogawa, H. Iwase, H. Nakashima, T. Fukahori, K. Okumura, T. Kai, S. Chiba, T. Furuta and L. Sihver, Particle and Heavy Ion Transport Code System PHITS, Version 2.52, J. Nucl. Sci. Technol. 50:9, 913-923 (2013)
15. K. Shibata, O. Iwamoto, T. Nakagawa, N. Iwamoto, A. Ichihara, S. Kunieda, S. Chiba, K. Furutaka, N. Otuka, T. Ohsawa, T. Murata, H. Matsunobu, A. Zukeran, S. Kamada, and J. Katakura: "JENDL-4.0: A New Library for Nuclear Science and Engineering," J. Nucl. Sci. Technol. 48(1), 1-30 (2011).

A novel avalanche-confinement TEPC for microdosimetry at nanometric level

D. Bortot [1,2*], S. Agosteo [1,2], P. Colautti [3], V. Conte [3], M.V. Introini [1], M. Lorenzoli [1], S. Pasquato [1], A. Pola [1,2]

[1]*Politecnico di Milano, Dipartimento di Energia, via La Masa 34, Milano, Italy*
[2]*Istituto Nazionale di Fisica Nucleare - INFN, Sezione di Milano, via Celoria 16, Milano, Italy*
[3]*Laboratori Nazionali di Legnaro LNL-INFN, viale dell'Università 2, Legnaro (Padova), Italy*

*Corresponding Author: davide.bortot@polimi.it

The tissue equivalent proportional counter (TEPC) is the most accurate device for measuring the microdosimetric properties of a particle beam, showing to properly assess the relative biological effectiveness by linking the physical parameters of the radiation with the corresponding biological response. Nevertheless no detailed information on the track structure of the impinging particles can be obtained, since the lower operation limit of the common TEPCs is about 0.3 μm. On the other hand, the pattern of particle interactions at the nanometer level, which demonstrated to have a strong correlation with radiation-induced damages to the DNA, is directly measured by only three different nanodosimeters worldwide: practical instruments are not yet available.

The gap between microdosimetry and track-nanodosimetry can be filled partially by extending the TEPC response down to the nanometric region. A feasibility study of a novel TEPC designed to simulate biological sites in the nanometric domain was performed. The present paper aims at describing the design, the development and the characterization of this avalanche-confinement TEPC. Irradiations with photons, fast neutrons and low-energy carbon ions demonstrated the capability of this TEPC of measuring in the range 0.3 μm - 25 nm.

KEYWORDS: microdosimetry, nanodosimetry, tissue equivalent proportional counter (TEPC), carbon ions

I. Introduction

In the last three decades a great interest has grown for hadron therapy, which consists in a radiation therapy modality based on charged particles (protons and other ions such as carbon ions) for treating radio-resistant cancers and malignant tumors close to critical organs[1]. At present, 43 proton and 8 carbon ion therapy facilities are in operation worldwide, and many other are being planned or constructed[2]. The assessment of the biological effective dose of clinical hadron beams is based on the measurements of the absorbed dose, which is a macroscopic quantity that demonstrated to be not adequate to describe the energy deposition process at micrometric level, because it does not take into account neither the stochastic of particle interaction in the target volume nor the track structure of ionizing charged particles, which is fundamental for initiating of the radiation damage[3]. A more comprehensive physical knowledge of the local energy deposition can be accomplished with high accuracy by innovative methodologies and instruments provided by microdosimetry, which aims at characterizing the statistical fluctuations of the local energy imparted at the micrometric level, and track-nanodosimetry, devoted to the description of the pattern of particle interactions at the nanometric level.

The tissue equivalent proportional counter (TEPC) is the most accurate device to measure the microdosimetric properties of a particle beam, and it is capable of simulating site sizes in the micrometric domain. At present, transportable devices are available, allowing to characterize radiation fields. Nevertheless, since the lower operation limit of standard TEPCs is about 0.3 μm, no detailed information on the track structure can be obtained.

On the other hand, the pattern of particle interactions at the nanometer level is measured by track-nanodosimetry, which derives the single-event distribution of ionization cluster size for site dimensions from a few nanometers up to tens of nanometers. Nanodosimetric probability distributions have demonstrated to show a trend similar to the cellular inactivation cross-sections[4]. Anyway, only three detectors are available worldwide for measuring these distributions. In order to fill the gap between standard TEPCs and nanodosimeters, an avalanche-confinement TEPC capable of simulating biological sites down to the nanometric region was designed and constructed. This detector was calibrated with an internal alpha source. Irradiations with photons emitted by a Cs-137 source demonstrated the good performances of the device and the reproducibility of the calibration procedure. Finally its response to fast neutrons and to low-energy carbon ions was assessed experimentally.

 DOI: 10.1051/epjconf/201715301011

II. Methods

1. Design and development of the counter

A TEPC can simulate site sizes in the nanometric domain only if the electronic avalanche is confined in a defined region inside the sensitive volume in order to guarantee a proper energy resolution of the device.

For this reason, an innovative TEPC based on a three-electrodes structure and capable of performing microdosimetric measurements from 0.3 μm down to 25 nm in simulated site size, was designed basing on a prototype[5]. The peculiarities of the new design are mainly a thinner wall chamber, which allows the microdosimeter to measure low-energy (and high-LET) hadron beams. Moreover, a removable internal alpha source and a very compact solid state detector were inserted inside the sensitive zone for energy calibration. The detector sensitive volume is cylindrical, 13 mm in diameter and length and houses three electrodes independently biased: a central anode wire (graphite, 1 mm in diameter), a cylindrical cathode shell (conductive plastic A-150 type, 13 mm in internal diameter and 1 mm in thickness) and a helix (gold-plated tungsten, 100 μm in diameter) made of 19 coils, 6 mm in diameter, which surrounds the anode wire and subdivides the sensitive volume into a drift and a multiplication region. Two field tubes (stainless steel, 6 mm in diameter) are employed for both sustaining the helix and defining the sensitive volume, thus avoiding any distortion of the electric field, while two insulating Rexolite caps enclose the chamber. An orifice on the basis cap permits the gas inlet and outlet (Figure 1).

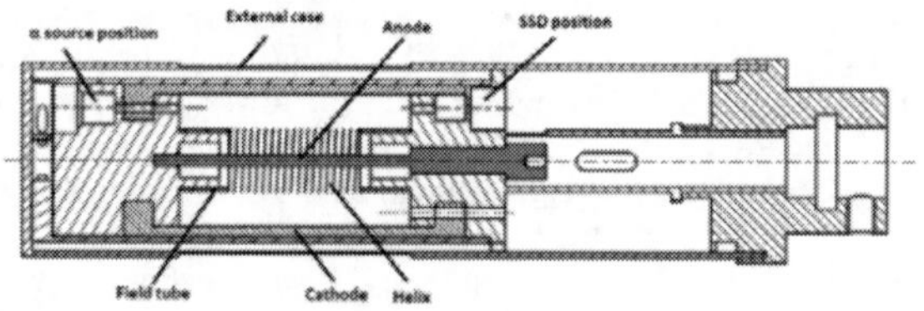

Figure 1: **Cross sectional view of the new avalanche-confinement TEPC.**

Two aligned holes were also designed in order to contain a thick removable Cm-244 alpha source, sealed by a mylar layer, and a miniaturized solid state detector (SSD). This configuration allows to determine the response of the TEPC as a function of i) the simulated site size and ii) the polarization of the three electrodes. It guarantees that only signals due to alpha particles with a straight path inside the sensitive volume, i.e. the drift region, are collected.

Figure 2 shows the main components of the microdosimeter before its assembly and the assembled TEPC without the external case: the SSD cables and the small white hose for the gas inlet can be recognized.

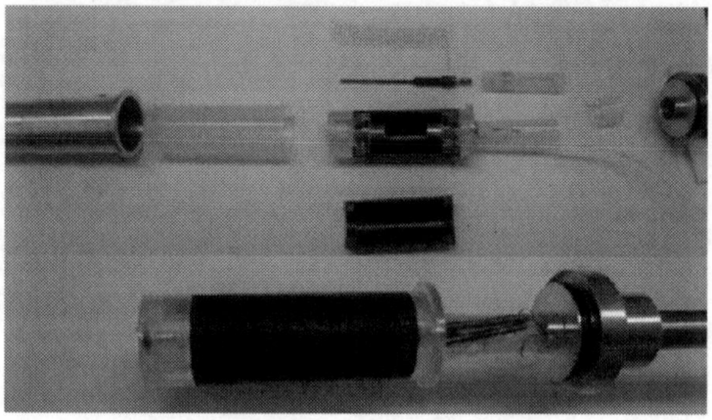

Figure 2: Pictures of the components which compose the TEPC and of the assembled device (without the external case).

2. Vacuum and gas flow system

A customized and transportable vacuum and gas flow system was developed in order to guarantee vacuum conditions and ensure a continuous replacement of tissue equivalent gas inside the chamber[6]. In such a way, the high purity required for stable gas gain during the irradiation is preserved and the ageing and outgassing effects are minimized.

Dimethyl ether (DME: $(CH_3)_2O$) is the selected filling gas for this avalanche-confinement TEPC: it can be considered as a tissue-equivalent gas, apart from the lack of nitrogen[7]. With respect to other tissue-equivalent gases, DME was chosen for its high gain (that leads to a better signal to noise ratio), its low diffusion coefficient (which increases the transparency of the helix to the drifted electrons), the presence of the chemical bond between carbon and oxygen (which lowers the polymerization rate, which is the aging effect of the anode wire) and its compatibility with all the materials of the TEPC.

3. Operating conditions and energy calibration

The characterization was devoted to optimize the operating conditions of the detection system by determining the proper bias voltage and pressure combinations, the stability and the performances of the whole system. This was accomplished through the determination of the so-called "electron transmission windows", empirical charts reporting the optimum operating conditions of the device in terms of polarization of the three electrodes at a selected simulated site size. A procedure for the calibration of the microdosimetric spectra based on the internal alpha source was also set up and exploited.

Irradiations with photons emitted by a Cs-137 isotopic source confirmed that the tested operating parameters of the device, i.e. the bias voltages of the electrodes which regulate the multiplication process, do not affect the microdosimetric information measured by the detector, which depends on the cavity size and on the radiation field only.

DOI: 10.1051/epjconf/201715301011

III. Applications

The response of the novel TEPC to fast neutrons and to low-energy carbon ions was experimentally evaluated.

1. Description

Irradiations with a high-energy neutron field produced at the CN Van de Graaff accelerator of the INFN-Laboratori Nazionali di Legnaro (LNL-INFN) were performed. Fast neutrons up to 20 MeV were generated through the ^{7}Li(d,n)^{8}Be reaction with 5.5 MeV deuterons striking a thick Lithium target. This experimental characterization was carried out for simulated site sizes in the range 300 nm - 25 nm at different bias voltages.

Moreover, the response of this novel avalanche-confinement TEPC to a 62 AMeV carbon ion beam was assessed at the carbon ion facility of the INFN-Laboratori Nazionali del Sud (LNS-INFN) in Catania. Several configurations were tested by varying both the simulated site size and the depth along the Bragg peak of the dose distribution of the delivered beam.

2. Results with high energy neutrons

Figure 3 (on the left) shows a neutron microdosimetric spectrum measured at a simulated site of 300 nm. Three contributions can be observed: at low lineal energy values, ionization events are primarily due to electrons produced by gamma rays, while the central peak is mainly given by proton recoils produced through neutron elastic scattering. At higher y-values, the events are due to low-energy recoils (mainly carbon, nitrogen and oxygen ions) from the A-150 cathode wall. A contribution is also given by the ^{4}He ions produced by the neutron capture reaction on carbon.

Figure 3 (on the right) shows neutron spectra measured at 300, 100, 50, 35, 30 and 25 nm. Because of different values in the minimum detectable lineal energy, the spectra were normalized in order to obtain equal integrals for lineal energies above about 23 keV μm^{-1} (threshold at 25 nm). The main proton peak at 300, 100 and 50 nm is slightly shifted at higher lineal energies with respect to the other simulated sites. The distributions at 35, 30 and 25 nm show a very similar shape, but different acquisition thresholds due to different values of the respective gain.

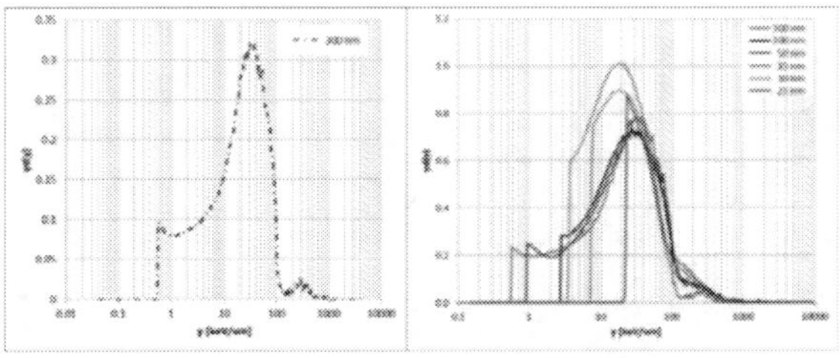

Figure 3: Neutron microdosimetric spectrum at 300 nm (left) and comparison between all measured neutron distributions at different simulated site sizes (right).

At 30 nm the main proton peak is still visible, even if located just above the threshold, while at 25 nm the limit condition is reached and only a small portion of the proton peak and of the contribution of light ions are present.

The reason of the discrepancy in the proton edge of the distributions measured at 300, 100 and 50 nm with respect to those derived at lower simulated sites is still under investigation. A future work will be devoted to irradiations with monoenergetic neutron fields in order to compare directly the experimental results with Monte Carlo simulations.

3. Results with 62 AMeV carbon ions

A set of PMMA foils was placed in front of the detector in order to reproduce different depths across the Bragg peak.

Figure 4 shows the microdosimetric distributions measured at different simulated site sizes at the reference position, which corresponds to a total depth of 6.76 mm PMMA equivalent (in the distal part of the Bragg peak).

This comparison allows to appreciate a small change in the shape of carbon peak when the site size reduces from 300 nm to 100 nm. Owing to different minimum detectable lineal energy, spectra at 50, 35 and 25 nm were normalized in order to obtain equal integrals for lineal energies above about 23 keV μm^{-1} (threshold at 25 nm).

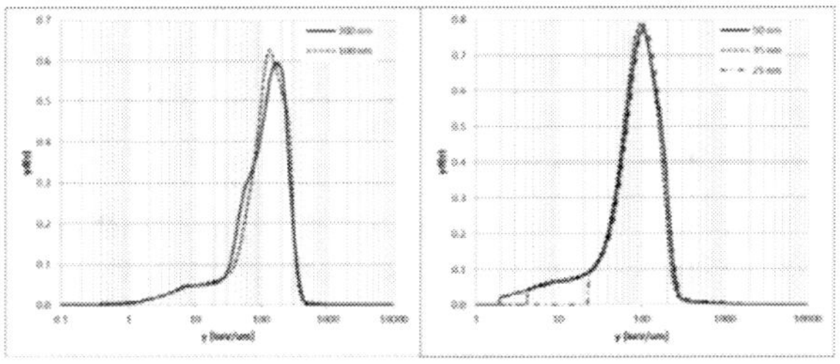

Figure 4: Spectra obtained by irradiating the TEPC at the reference depth (6.76 mm) with a 62 AMeV carbon ion beam at 300 nm and 100 nm (left) and 50 nm, 35 nm and 25 nm (right).

Figure 5 (left) shows the comparison between the lineal energy spectra obtained for a simulated site of 300 nm at a depth 6.50 mm, 6.76 mm (reference point) and 7.09 mm in PMMA. At 6.50 mm the carbon peak is slightly shifted at lower lineal energies with respect to the one at 6.76 mm, but the edge is the same. At 7.03 mm, the spectrum changes completely: the primary carbon contribution is no more present and that of high energy fragments, which is barely visible at 6.50 mm and 6.76 mm, becomes predominant. This depth is beyond the Bragg peak.

The microdosimetric spectra obtained at eight different points along the depth dose profile for a simulated site size of 25 nm are shown in the right part of Figure 5. The distributions shift towards higher y values as depth increases (from 1.76 mm to 6.76 mm), according to the beam slowing-down. In contrast, at 7.29 mm in depth, which is beyond the Bragg peak, the distribution changes dramatically due to the contribution of high energy fragments.

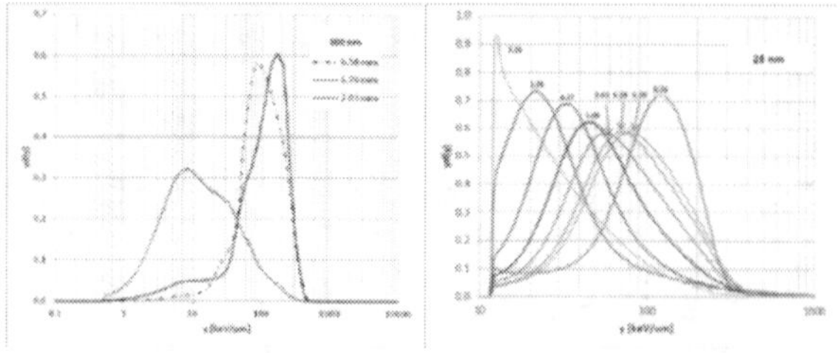

Figure 5: microdosimetric spectra obtained for a simulated site of (left) 300 nm at a depth 6.50, 6.76 and 7.03 mm in PMMA and (right) 25 nm at eight depths from 1.76 mm to 7.29 mm.

IV. Conclusion

The novel avalanche-confinement TEPC showed to be capable of measuring in the range 0.3 µm - 25 nm when irradiated with low-energy carbon ions. It should be stressed that microdosimetric measurements of hadron beams at the nanometric scale with portable devices have never been performed before.
A direct and systematic comparison with reference devices both in the micrometric and nanometric domains has to be performed and will be matter of future work. Moreover, a new facility constituted by a track-nanodosimeter coupled with a wall-less avalanche-confinement TEPC will be designed and constructed, for correlating directly the microdosimetric spectra with the track-nanodosimetric distributions at the same experimental conditions. The microdosimetric spectra measured by such a device consist in the convolution of core and penumbra events, the two components of a particle track measured through a nanodosimeter. A systematic comparison between the microdosimetric and track-nanodosimetric distributions for the same particle beam will allow to investigate the possibility of unfolding the microdosimetric spectrum into its different track components. In such a way, the beam characteristics measured by a portable TEPC could be transferred to a track-nanodosimetric description, much more relevant in the framework of the biological damages to the nanometric targets, such as the DNA and the chromosome fiber.

Acknowledgment

This work has been supported by the MITRA project (Scientific Commission V) of the Italian National Institute for Nuclear Physics-INFN.

References

1) M. Durante, J.S. Loeffler, *Charged particles in radiation oncology*, Nature Reviews Clinical Oncology 7: 37-43 (2010).
2) Particle Therapy Co-operative Group PTCOG: http://www.ptcog.ch/index.php/facilities-in-operation.
3) K. Weyrather, G. Kraft, *RBE of carbon ions: Experimental data and the strategy of RBE calculation for treatment planning*, Radiotherapy and Oncology 73 (2): S161-S169 (2004).
4) V. Conte, D. Moro, P. Colautti, B. Grosswendt, *Nanodosimetric descriptors of the radiation quality of carbon ions*, Radiation Protection Dosimetry 166 (1-4): 214-218 (2015).
5) V. Cesari et al., *Nanodosimetric measurements with an avalanche confinement TEPC*, Radiation Protection Dosimetry 99(1-4): 337-342 (2002).
6) D. Bortot et al., *Development of a Vacuum and Gas Flow System for a New Avalanche Confinement TEPC*, INFN Website, LNL Annual Report 2013.
7) International Commission on Radiation Units and Measurement (ICRU), *Microdosimetry*, Report 36 (1983).

Mini-TEPC Microdosimetric Study of Carbon Ion Therapeutic Beams at CNAO

V.Conte[1], P.Colautti[1], S.Chiriotti[1,2], D.Moro[1], M.Ciocca[3], A.Mairani[3]

1. INFN Laboratori Nazionali di Legnaro, viale dell'Università 2, 35020 Legnaro, Italy
2. Belgian Nuclear Research Centre, SCK•CEN, Boeretang 200, 2400 Mol, Belgium
3. Fondazione CNAO, strada Campeggi 53, 27100 Pavia, Italy

*Corresponding Author: paolo.coautti@lnl.infn.it

Mono-energetic carbon ion scanning beams of 195.2 MeV/u at the Italian National Centre for Oncological Hadrontherapy (CNAO) have been used to study the microdosimetric features of an "active" carbon ion beam used in hadrontherapy. A 30x30 mm^2 area has been scanned by a 6 mm beam with scanning steps of 2 mm. A mini TEPC of 0.57 mm^3 has been used to perform measurements in a water phantom at different depths on the beam axis. The detector small size allowed for measuring, with good spatial resolution, also inside the relatively small Bragg peak region and inside the distal edge, where the radiation quality varies quickly. In spite of the high event rate (up to ~ 10^5 s^{-1}), no pile-up effects were observed. Results showed that the frequency-mean lineal energy scaled well with the absorbed dose. Moreover, the dose-mean lineal energy itself seemed to be a good descriptor of the radiation quality.

KEYWORDS: microdosimetry, mini tissue-equivalent proportional-counter (TEPC), carbon ions

I. Introduction

The ion beam radiotherapy is a technique that uses fast light ions (protons and carbon ions) to treat solid tumours. Its success is mainly due to the good ballistic properties of ions with respect to electrons, photons and neutrons. The ion finite range allows maximizing the tumour dose, while sparing the dose to the surrounding healthy tissues. Therefore, such a radiation therapy has therapeutic indications for solid tumours that are located close to critical healthy tissues. Nowadays, forty-nine therapeutic centres use ion beams to treat their patients [1]. Most of them use fast protons. At present, eight centres (2 in Europe, 4 in Japan and 2 in China) use carbon ions. Carbon ions have high relative biological effectiveness (RBE) and low oxygen enhancement ratio (OER) [2, 3]. These features make carbon ions suitable to treat tumours, which are low-LET radiation resistant. However, both the RBE the OER values change with the radiation quality, namely with the ion charge and velocity. Therefore, the radiation-field biological action varies with the spatial coordinates inside the treated volume. As far as the RBE and OER can be assessed from the radiation field quality, it should be known with high spatial resolution in order to optimise the therapeutic plan and its success. That is of specific utility when complex radiation fields, namely radiation fields with several radiation quality components, are used (IMPT for instance). The only available technique for such an aim is the experimental microdosimetry. Microdosimetric detectors measure the energy imparted ε by a single particle in a site of finite size, for instance 1 µm site (~ the chromosome thickness). In microdosimetry, the ratio $\varepsilon/\bar{l}$, where $\bar{l}$ is the microdosimeter sensitive-volume mean-chord length, is called lineal energy y. In a mixed radiation field, y events originate the frequency distribution $f(y)$, the mean value of which is called frequency-mean lineal energy $\bar{y}_F$. $\bar{y}_F$ is proportional to the absorbed dose through the equation:

$$D = \frac{k}{d^2} N \bar{y}_F \quad (1)$$

where N is the event number and d the sensitive volume size and k a constant factor depending on the sensitive volume shape. The average value of the weighted distribution $d(y)$ is called dose-mean lineal energy, $\bar{y}_D$. The two microdosimetric averages can be used as a radiation quality signature.

Microdosimetric detectors can be made of silicon [4], diamond [5] or organic gas. Gas counters have the advantage to amplify in proportional mode the ionization events produced by the imparted energy. Therefore, they are potentially able to detect a single ionization event, namely all the imparted energy ε.

Various microdosimetric measurements of carbon ion beams were performed in the past with TEPCs (tissue-equivalent proportional counters). The most of them used the commercial gas counter LET ½ (Far West Technology, Inc., Goleta, CA, 93117), which is a relatively large detector of 1.07 cm^3 (a sphere of 12.7 mm of diameter). Therefore, in order to prevent pile-up effects, these counters were used in low-intensity ion beams or with rather cumbersome trigger and veto detection systems. They are not suitable for scanning beams used in the hadrontherapy clinical practice, where the fluence rates span a 10^4 –10^7 mm^{-2} s^{-1} range. The mini-TEPC developed by INFN Legnaro laboratories [6], which has a cylindrical sensitive volume of only 0.57 mm^3 (0.9 mm of diameter and height), was already successfully tested in intense therapeutic proton beams [7]. Therefore, in principle it could be used also in therapeutic carbon ion beams. Moreover, the small counter cavity allows to perform measurements in radiation fields with high dose-gradients with a spatial accuracy of about 0.1 mm.

The aim of this research was to test the ability of mini-TEPCs to properly measure microdosimetric spectra in an active carbon ion beam at different depths in a water phantom. For this study, the scanning therapeutic beam of the Italian national centre for oncological hadrontherapy (CNAO) [8] has been used.

II. Instruments and Methods

1. The CNAO carbon ion beam

At CNAO a fully active scanning delivery system is used. A treatment planning system (TPS) is used to calculate how to adequately cover the entire tumour volume, while sparring normal tissues and organs, using a very large number of beam spots. Every spot (6 mm of diameter at FWHM as an average over the whole energy spectrum) is characterized by the particles energy, the particles number and the position where it has to be placed. All spots with the same energy are grouped in slices and a synchrotron cycle is used to generate a beam with the right energy and dimension. With a couple of scanning magnets the beam is sent in the right position and, when the requested particles number is delivered for the first spot the beam is moved to the following spot position. When the last spot belonging to a slice has been delivered, the beam extraction is stopped and a new synchrotron cycle is generated. The sequence continues slice by slice until all spots are delivered. The control of the treatment is done in real time by the "dose delivery system", which is composed mainly by the monitoring system, that measure the beam intensity and position, and the scanning system. More detailed have been published elsewhere [9].

In order to simulate a TPS single-slice irradiation, a mono-energetic carbon ion beam of 195.2 MeV/u, which was enlarged by using two thin PMMA ripple filters, was used. The emerging beam has been calculated to have an average energy of 189.5 MeV/u and a FWHM (energy spread) of 0.28 MeV/u at the isocentre. The multi-spot scanning system was used to uniformly irradiate a 30 x 30 mm^2 surface. The surface was irradiated with equally-separated 225 spots, approximately lasting 0.04 s each. This procedure was repeated about 90 times, for a total of ~ 2,250 spots, to obtain the total absorbed dose to water of 10 Gy at the reference depth of 2 cm. The maximum fluence rate in a single spot was calculated to be about $2 \cdot 10^5$ $s^{-1}mm^{-2}$.

2. The microdosimetric detection system

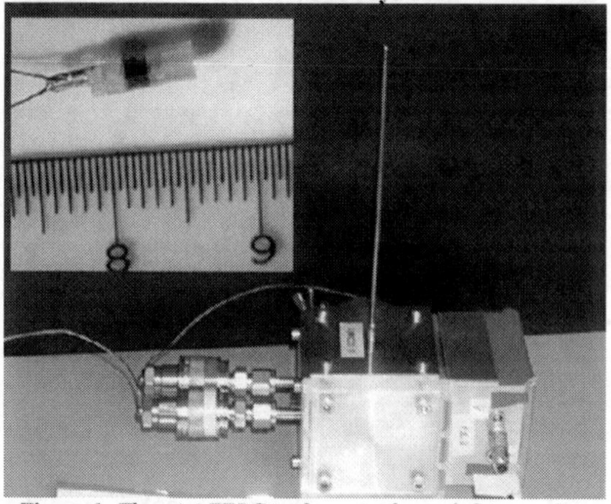

Figure 1: *The mini TEPC is shown in the up-left insert. It is placed inside the thin aluminum sleeve (the long stick in the figure).*

Measurements were performed with a cylindrical detector made of A-150 plastic, the cathode, and Rexolite® plastic, the insulators. The detector stands inside 2.7 mm diameter 16 cm long aluminium sleeve, which has a thickness of 0.2 mm (see figure1).

Measurements were performed with pure propane gas to prevent the response inaccuracy due to the uncertainty about the propane-based tissue-equivalent gas-mixture composition. The gas pressure value was set to simulate 1 µm-equivalent propane-TE site size. All the measurements were performed in gas-flow modality (1 cm^3/min) with the mini-TEPC filled with 454 mbar of propane at 21.8 °C of temperature.

The mini-TEPC is equipped with a homemade charge preamplifier of ~ 300 equivalent electrons of rms noise and a dynamic range of four and a half order of magnitude. The preamplifier is placed inside the aluminium box, on which the aluminium sleeve is inserted. Three spectroscopy linear amplifiers were used for conditioning and simultaneously amplifying the pulses coming from the preamplifier. The shaping times were set at 250 ns. The analogic pulses were digitized with three CAMAC-standard ADCs of 16k, 8k and 8k channels respectively. The CAMAC system was driven by a portable computer with Kmax® software.

3. The irradiation set up

Measurements were performed in a liquid water phantom, at the isocentre. The detector was placed on a rigid support, which could be moved along the beam axis with a precision of 0.1 mm. The thin rod containing the mini-TEPC was inserted in a PMMA cylinder of 24 mm outer diameter (see figure 2). The overall position uncertainty, mainly due to the PMMA window and cylinder thickness uncertainties, has been assessed to be 0.6 mm. The relative depth-dose profile has been measured with the Advanced Markus parallel-plate ionization chamber (model 34045, PTW, Freiburg, Germany).[8].

Figure 2: *The water phantom used during measurements at CNAO. The detector was placed upside down inside a PMMA cylinder that is immerged in the water.*

4. Data processing

Every microdosimetric spectrum is the result of two measurements: one measurement at low gas gain (600 V) and the second one at higher gas gain (750 V). The two spectra were calibrated with a ^{137}Cs source. More details about the calibration procedure have been published elsewhere (10). The spectra at different gas gains were joined together superimposing the common spectral parts. This "photon calibrated" spectrum has been eventually re-calibrated by using the carbon-edge value, which has been determined with the same procedure explained in reference 10. The carbon edge corresponds to the maximum energy transferred in 1 µm site of liquid water, namely 931 keV/µm, which corresponds to a lineal energy value of 1,397 keV/µm.

III. Results and discussion

Because of the electronic noise, the experimental spectra had a low detection threshold of 0.2-0.3 keV/µm. Therefore, all the spectra were extrapolated down to 0.01 keV/µm by using the linear best-fit of 0.2-0.5 keV/µm data .

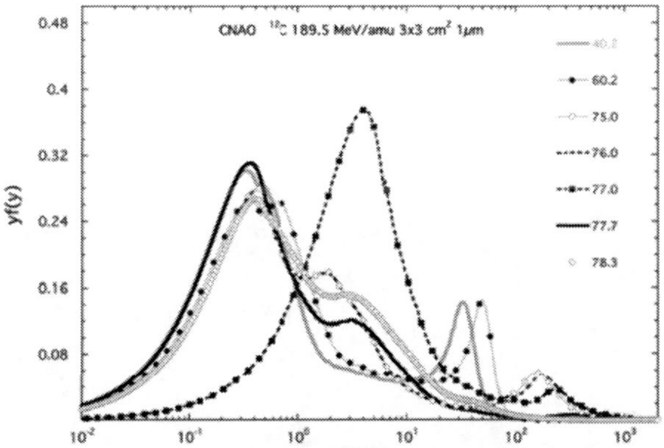

Figure 3: *Microdosimetric frequency distributions at different depths in the water phantom.*

In figure 3, the frequency distribution multiplied by the y is plotted against log(y). In such a representation, equal visual areas under the curve mean equal relative contribution to the total event number. In the proximal edge, the carbon ion events give rise to roughly Gaussian peaks in the 10 – 100 keV/µm region. In the Bragg peak region, carbon ion events give rise to broader and lower peaks in the region 10^2 – 10^3 keV/µm. Figure 3 shows also that at any depth the event majority has sizes less than 10 keV/µm. Such small events are likely due to lighter ions emerging from nuclear reactions.

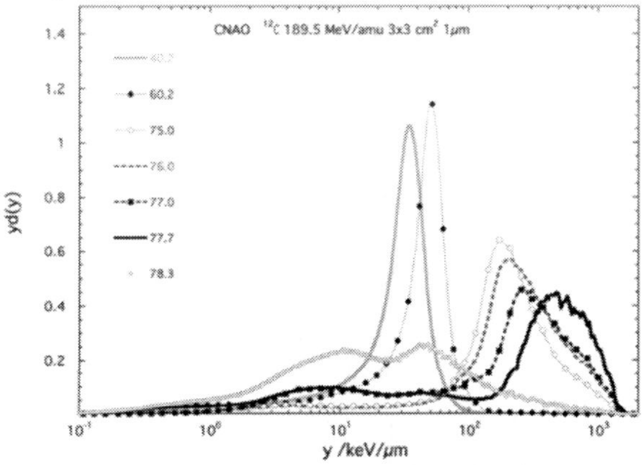

Figure 4: *Microdosimetric dose-weighted distributions at different depths in the water phantom.*

In figure 4, the dose-averaged lineal energy multiplied by y-value is plotted against log(y). In such a representation, equal visual areas under the curve mean equal contribution to the total absorbed dose. Differently from the frequency distributions, the dose distributions are almost everywhere dominated by carbon ion events, which give rise to roughly-Gaussian peaks of 10 – 100 keV/µm in the proximal edge and to broader peaks of 100-1000 keV/µm in the Bragg peak region. However, in the Bragg peak distal edge (78.3 mm of depth), where the primary carbon ion beam is quickly vanishing, low y-values events due to light ion from nuclear reactions become dominant.

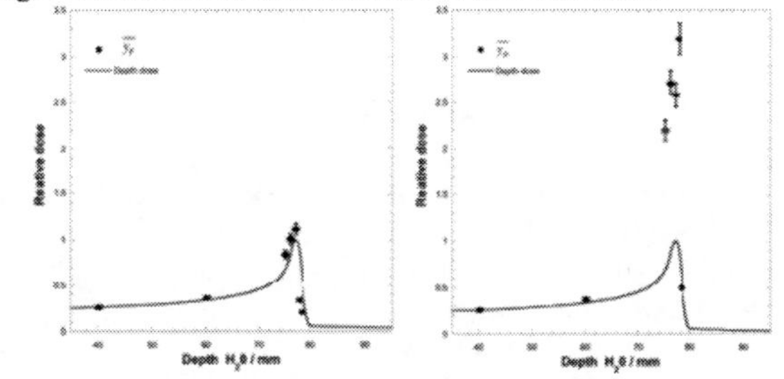

Figure 5: *Comparison of relative absorbed dose (line) with the microdosimetric relative mean-values (full dots). Left side: frequency-mean lineal energy. Right side: dose-mean lineal energy.*

In figure 5, the frequency-mean and the dose-mean lineal energy values are superimposed to the relative absorbed dose curve. For sake of comparison, the microdosimetric values have been scaled to match the absorbed-dose relative value at 40.2 mm of depth. Figure 5 shows that the frequency-mean values scale rather well with the absorbed dose, according to equation 1. On the contrary, the does-mean values are much higher than the relative absorbed-dose values in the Bragg peak region. At the Bragg peak, the $\overline{y}_D$ value is 3.2 times higher.

IV. Conclusion

For the first time, microdosimetric measurements have been performed in a carbon ion scanning beam, at the CNAO hadrontherapy facility. In spite of the high counting rate, no spectral distortion, due to pile-up events, were observed. Moreover, the frequency-mean values at different depths scales rather well with the relative absorbed dose values, this being a further indication of the accuracy of measured spectra. The $\overline{y}_D$ values do not scale with the absorbed dose. They are much higher in the Bragg peak region. The increase factor is not far from the expected RBE values in the Bragg peak [12]. Discussing about the use of microdosimetric data in radiobiological models is not the aim of this paper. However, CNAO microdosimetric data suggest that mini-TEPCs are able to monitor the quality of carbon-ion active-beams used in the modern hadron therapeutic centres.

Acknowledgment

This work has been supported by the MITRA project (Scientific Commission V) of the Italian National Institute for Nuclear Physics INFN.

References

1) Hadrontherapy facilities worldwide in operation. www.ptcog.ch,updated 09-Nov-2014)
2) Raju MR, Amols HI, Bain E, Carpenter SG, Cox RA, Robertson JB, "A heavy particle comparative study. Part III. OER and RBE", Br. J. Radiol., **51**, 712-719 (1978)
3) Curtis SB, "The OER of mixed high- and low-LET radiation", Radiat Res **65**, 566-572 (1976).
4) S. Agosteo, G.A.P. Cirrone, P. Colautti, G. Cuttone, G. D'Angelo, A. Fazzi, M.V. Introini, D. Moro, A. Pola , V. Varoli. "Study of a silicon telescope for solid state microdosimetry: Preliminary measurements at the therapeutic proton beam line of CATANA" Radiat. Meas. **45**, 1284-1289 (2010).V. Cesari et al., *Nanodosimetric measurements with an avalanche confinement TEPC*, Radiation Protection Dosimetry 99(1-4): 337-342 (2002).
5) S.Rollet, M.Angelone, G.Magrin, M.Marinelli, E.Milani, M.Pillon, G.Prestopino, C.Verona, G.Vrrona-Rinati. "A novel microdosimeter based upon arificial single crystal diamond", Nucl.Sci. IEEE Trans. 59, 2409-2415 (2012).Meas. **45**, 1284-1289
6) L. De Nardo, V.Cesari, G.Donà, G.Magrin, P.Colautti, V.Conte and G.Tornielli, "Mini-TEPCs for radiation therapy", Rad. Prot. Dos. **108**, 345-352 (2004).
7) L. De Nardo, V. Cesari, N. Iborra, V. Conte, P. Colautti, J. Héreault, G. Tornielli, P. Chauvel. "Microdosimetric assessment of Nice therapeutic proton beam biological quality, Physica Medica **XX**, 71-77 (2004).
8) S. Rossi, "The National Centre for Oncological Hadrontherapy (CNAO): Status and perspectives", Physica Medica **31**, 333-351 (2015). Doi: 10.1016/j.ejmp.2015.03.001.
9) S. Giordanengo, M.A. Garella, F. Marchetto, F. Bourhaleb, M. Ciocca, A. Mirandola, V. Monaco, M.A. Hosseini, C. Peroni, R. Sacchi, R. Cirio, M. Donetti, "The CNAO dose delivery system for modulated scanning ion beam radiotherapy" Med Phys. **42**, 263-275 (2015). doi: 10.1118/1.4903276
10) Mirandola A, Molinelli S, Vilches Freixas G, Mairani A, Gallio E, Panizza D, Russo S, Ciocca M, Donetti M, Magro G, Giordanengo S, Orecchia R. *Dosimetric commissioning and quality assurance of scanned ion beams at the Italian National Center for Oncological Hadrontherapy*. Med Phys. 2015 42(9):5287-300.
11) Conte, D. Moro, B. Grosswendt, P. Colautti. *Lineal energy calibration of mini tissue-equivalent gas-proportional counters (TEPC)*, in Multidisciplinary Applications of Nuclear Physics With Ion Beams. AIP Conference collection, 2013, pp. 171 -178.
12) IAEA 2008. *Relative Biological Effectiveness in Ion Beam Therapy*. Technical report 461

Optimization of a neutron dosimeter for the high energy accelerators

Alexey Sokolov[1,*], *Georg* Fehrenbacher[1], *Ekaterina* Kozlova[1], *Torsten* Radon[1], and *Izabella* Zahradnik[1]

[1] *GSI, Darmstadt (Germany)*

Abstract. In high energy accelerator facilities the neutron radiation should be continuously measured during operation to control the ambient dose. This requires a reliable neutron dosimeter in a wide energy range. In this work we present an optimization of a compact cylindrical passive neutron dosimeter for the usage in wide energy neutron fields.

1 Introduction

For various high energy physics experiments, ions are accelerated to high energies and in collision with a target or losses in a beam line produce neutrons. These neutron fields are possibly harmful not only to the personnel behind the shielding, but also to electronics located in the vicinity of a radiation source. The energy range of neutrons at accelerator complexes, such as the future Facility for Anti-proton and Ion Research (FAIR) [1] can vary from thermal up to GeVs. The position of the ion losses in the beam line are often unpredictable and time dependent thus an accumulative dose measurement of the ambient dose is preferred for monitoring over a wide area. This and necessity in a reliable measurements during a long period, for instance one year, can be provided with thermoluminescence detector (TLD) based dosimeters [2, 3]. In the TLD cards the neutrons are detected via a $^6Li(n,\alpha)^3H$ reaction whose cross section increases for low energy neutrons. To perform neutron fluence to equivalent ambient dose conversion an appropriate combination of a neutron moderator and absorber layouts should be used. In this work we present the results of simulations and test measurements of a recently developed compact dosimeter layout [4] and continue to search for new materials and layouts for further improvements. The choice of materials and dimensions of different layers were tested with simulations using the FLUKA code [5], results were processed using FLAIR [6].

2 Dosimeter reading calculations

The TLD cards signal is proportional to the absorbed dose in the TL chips. In the simulations the energy absorbed in TL chips under irradiation in isotropic neutron fields was scored. The fields used are simple parallel neutron beams of certain energy and additionally a ^{241}AmBe source spectrum which was also used for calibration purposes. The TLD card of Harshaw type consists of two pairs of chips, 'TLD-600' based on ^{6}Li, sensitive to gamma and neutron radiation and 'TLD-700' based on ^{7}Li, sensitive to the gamma radiation. So the simulated relative reading or response of the dosimeter to a certain neutron energy was obtained as a sum of the absorbed dose in the TLD-600 crystals minus the dose in the TLD-700 crystals and divided by the same value for the neutrons of a ^{241}AmBe source spectrum. The calculated relative reading is further compared to the fluence-to-dose values according to Pellicioni's publication [7] normalized to 391 pSv·cm^2, which is a reference for ^{241}AmBe source provided by ISO Standard 8529-1 [8]. The closer are the relative readings to the mentioned values, the closer are the ambient dose readings of the dosimeter to the real equivalent dose H*(10).

FLUKA is also capable to predict an amount of produced isotopes in reactions of neutrons with matter. Several simulations were performed to count the number of ^{3}H and α-particles produced in TLD-600 chips during the detection time and the relative dosimeter reading was obtained as:

$$R_R = N_E(^3H,\alpha)/N_{AmBe}(^3H,\alpha) \quad (1)$$

where $N_E(^3H,\alpha)$ is either the number of produced tritium or helium nuclei for a certain neutron energy E and $N_{AmBe}(^3H,\alpha)$ is the number of those produced under irradiation with neutrons of a ^{241}AmBe spectrum. Such an approach provided the same response characteristics as in the first method, which takes into account the absorbed energy. This shows that the neutron capture process releases the energy which contributes mostly to the absorbed dose in the TL chips even for fast neutrons.

The method calculating the amount of e.g. ^{3}H – particles is very suitable to estimate the dose reading of active detectors, where the dose is calculated as a number of pulses. Each pulse corresponds to a neutron capture event. As a couple of tests we have also performed some simulations for active ^{3}He dosimeters where neutrons are detected in the $^3He(n,p)^3H$ reaction. Again we used also the

* e-mail: Al.Sokolov@gsi.de

DOI: 10.1051/epjconf/201715301013

absorption energy for detection and as usually both methods of isotope counting and energy deposition calculations gave the same response characteristics of the investigated dosimeter layouts.

The energy dependence of the capture cross sections for ^{3}He and ^{6}Li are pretty similar and our simulations showed that exchange of a TLD card with a ^{3}He proportional counter is possible. Exchanging the sensitive element of the dosimeter in studied layouts should provide close relative read-out. Of course the response characteristics are influenced by the size of the ^{3}He tube as well as by the gas pressures of the counter. The active dosimeter prototypes are to be tested in the near future.

3 Compact cylindrical dosimeter

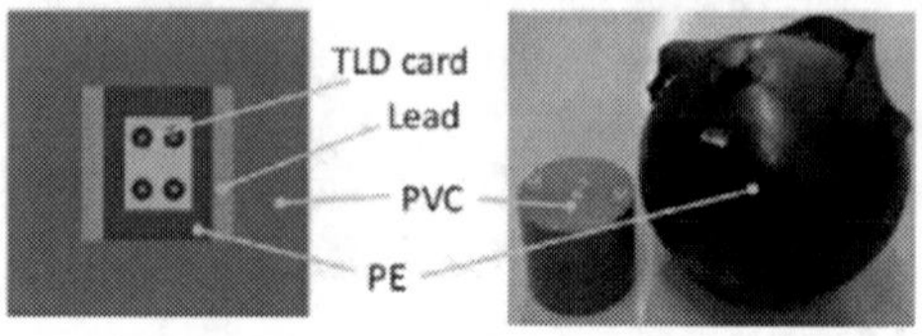

Figure 1. (left) Neutron dosimeter layout with a sensitive element – TLD card, PVC, lead and PE layers. Upper green squares in the TLD card depict TLD-700 chips and lower blue ones TLD-600 chips. (right) Prototype 2 of the dosimeter and the 12" sphere [3] currently used at GSI for ambient dose measurements.

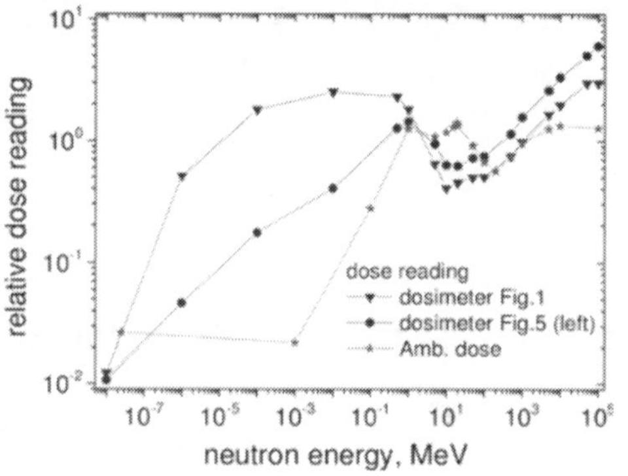

Figure 2. The relative dose readings for different dosimeter layouts from Fig. 1 and Fig. 5 and their comparison to the expected ambient dose. The values are normalized to the reading for the ^{241}AmBe source neutrons.

Table 1. The ratios of the measured at CERF ambient dose equivalent to the reference value and to the expected from FLUKA simulations with and without correction of the calibration to the scattered neutrons.

irradiation direction	CERF	FLUKA	CERF corrected	FLUKA corrected
front	120%	100%	145%	121%
top	129%	109%	155%	131%

As it was shown by us previously [4, 9] compact sizes of broad energy range neutron dosimeters are possible, introducing a polyvinylchloride (PVC) layer, which serves for both absorption and moderation. This material also reduces the production costs and makes the dosimeter design simple. In Fig. 1 (left) one can see a schematic layout of a cylinder dosimeter and a photo of a built prototype Fig. 1 (right). The dosimeter has an outer PVC layer, afterwards a lead layer used to increase the response for high energy neutrons [3]. In the middle are an additional moderating high density polyethylene (PE) layer and the TLD card, as a sensitive element. The developed design with height and diameter of only 14 cm and weight of 4.5 kilogram provided a very good response of the dosimeter in high energy region (MeVs - GeVs) critical at high energy accelerators. Still the over-response in medium (eVs-keVs) energy region leads to a dose overestimation. The relative dosimeter reading calculated using FLUKA simulations can be found in Fig. 2. In further sections we describe a possibility to improve the response in the eVs-keVs energy interval. This is necessary even if the measured neutrons are of higher energies, because the over- or underestimation in the low energy region can influence the calibration and lead to a incorrect dose reading.

The readings of the described compact dosimeter layout with the PVC layer were tested at the CERN-EU high-energy Reference Field (CERF) facility [10] and can be found in Table 1. CERF is the facility providing neutron fields in a wide energy range typically produced by cosmic rays interacting with the atmosphere and by ions at high energy accelerators. The first column 'CERF' are the ratios of the value measured with a dosimeter to the provided reference ambient dose equivalent value [11]. The measured value applied a calibration done with a ^{241}AmBe source under an assumption of only direct neutrons from the source with average fluence-to-dose conversion equal to 391 pSv·cm^2 mentioned in calculation section. Two values of a column are for the irradiation perpendicular to the TLD card surface and for the irradiation from the top of the dosimeter, both indicating angular dependence of the dosimeter response. Also to be mentioned is that the values in the table are the results averaged over several experimental runs with only slight deviations of few percent between the runs. In the second column 'FLUKA' are the ratios between the measured value and the one obtained folding the provided neutron spectrum during irradiation at the CERF facility and the calculated dosimeter response from simulations. The reference equivalent ambient dose value is a result of folding between the given spectrum and Pellicioni's fluence-to-dose coefficients. So the 'CERF' value shows the precision of the fluence-to-dose conversion of the dosimeter and the 'FLUKA' value shows the precision of the simulations. In the first approximation the dosimeter overestimated the H*(10) value only about 20 %. FLUKA simulations of the dosimeter response are defined to be 100 % precise for the frontal irradiation.

The calibration of the neutron dosimeters, typically done using ^{241}AmBe or ^{252}Cf sources, requires the knowledge of the neutron spectrum, including the scattered neutrons. For this purpose a shadow cone technique [12] or simulations should be used. If the scattered neutrons contribute significantly to the neutron spectrum and are not taken into account the calibration can be underestimated and can provide lower measured values.

Another pair of columns in Table 1 marked as 'CERF corrected' and 'FLUKA corrected' are the ratios mentioned above, taking into account the experimentally verified neutron scattering.

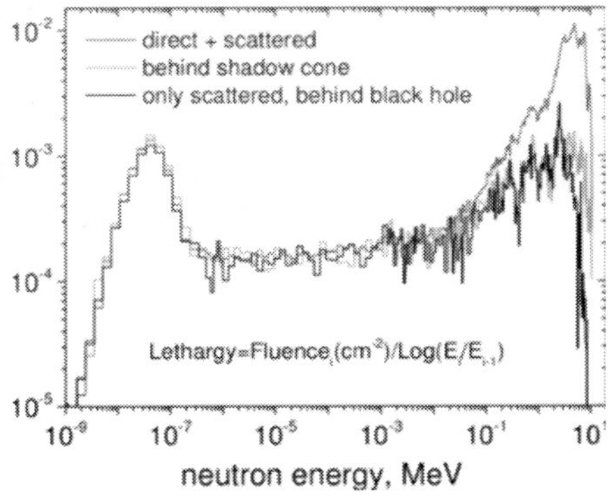

Figure 3. Simulated spectra of neutrons from the ^{241}AmBe source at the calibration point, behind the 'shadow cone' and behind the 'blackhole' providing only scattered neutrons.

The neutron scattering for the calibration procedure was simulated using the FLUKA code. The ^{241}AmBe source was positioned in FLUKA geometry inside the room, 1.6 m above the floor, 2.5 m below the ceiling, 3 m away from concrete walls in one direction and 6 m in another direction. To separate the scattered neutrons from those coming directly from the source so called 'blackhole' material was inserted into the FLUKA geometry between the source and the dosimeter. This FLUKA 'material' is introduced to stop the transport of particles and thus very suitable for our purposes. Instead of 'blackhole' in another simulation a paraffin wall was inserted with varying thickness. It was found that a 25 cm wall thickness is sufficient enough to stop the direct flux of neutrons from the source so having almost the same effect as the 'blackhole'. In Fig. 3 one can see the spectra for the described simulations. After folding the neutron spectra with and without the paraffin wall, or so called shadow cone, with Pellicioni's values it was found that the contribution from the scattered part to the dose is almost 20% of the initially expected. To test the simulation results the GSI ball dosimeter [13], with approved close to ideal response characteristics, was used. The measured readings of the dosimeter with and without the wall also showed 17 % difference for the doses absorbed by the TLD cards and thus around 17 % difference for H*(10) value, confirming the simulation results.

Coming back to the Table 1, the 20 % contribution to the equivalent ambient dose from the scattered neutrons, taken into account for calibration, changes the dosimeter reading. The ambient dose measured at CERF for the frontal irradiation is increased from 120 % to 145 %. Still the 45 % overestimation of the dose for the compact cylindrical dosimeter is a rather good agreement, which happens due to the good fluence-to-dose response in the region of interest for the accelerator complexes, i.e. thermal and from MeVs to few GeVs intervals [4]. Still the overestimation can be reduced if the response of the dosimeter will be improved in eVs-keVs energy region. The overestimation also occurs for the comparison between FLUKA simulations and measurements, if the scattered neutrons are taken into account. Thus our simulation instead of almost perfect fit shows a 20 % overestimation for the calculated dose values, which is also quite acceptable and approves our simulation approach. As it is also clear from

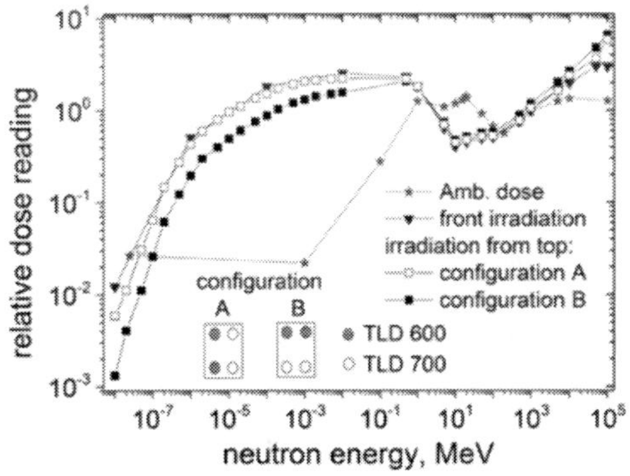

Figure 4. The relative dose readings of the compact 14 cm large dosimeter layout for irradiations from aside and from top for various configurations of TL chips in TLD card.

the table the response of the dosimeter is almost the same for irradiation from aside and from the top, despite the lack of the top and bottom caps of the lead cylinder. Thus the dosimeter read-out is quite angular independent though having only a cylindrical shape and rather small dimensions. The only requirement is set to the location of the chips in the card, they should be vertically the same in configuration A and not B as depicted in Fig. 4 with the responses for different irradiation directions and different TL chips configurations. For any irradiation perpendicular to the symmetry axis the readings are almost identical to the front irradiation and are independent on chips configuration. For the irradiation from top for the configuration B the material thickness pathed by neutron towards the TLD-600 chips is bigger and response of the dosimeter in low energy region decreases as it will be discussed below. In our measurements the configuration A was used and though the response for the irradiation from top is close to frontal irradiation (see Fig. 4) the simulations show bigger deviation from the measurements as can be seen in Table 1.

4 Modification of the Andersson-Braun dosimeter

A recently investigated layout showed that the compact size of the broad energy range neutron dosimeters is possible, but requires an improvement of the response in the low energy region. A solution could be a separation of the outer layer into 2 as in the modified Andersson-Braun type [14]. In Fig. 5 (left) one can see a schematic layout of a cylinder dosimeter, which has an outer 16 cm in diameter and 16 cm in height moderating high density polyethylene layer, afterwards a 10% borated PE or rubber to increase the absorption, then a lead layer and again a PE layer used to increase the response in the high energy region. In the middle of the dosimeter a TLD card is used

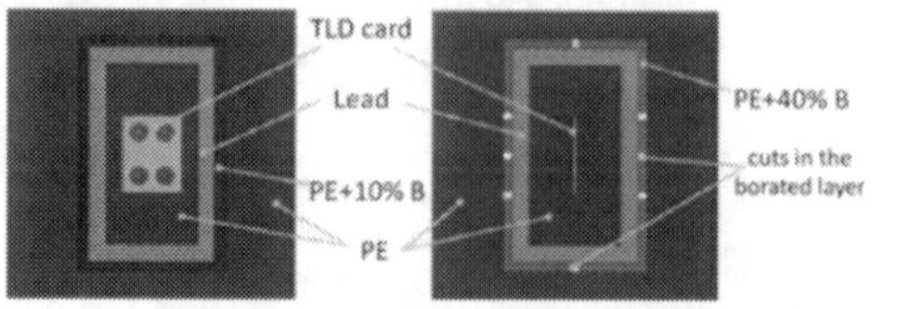

Figure 5. The modified Andersson-Braun dosimeter (left) and its further modification with concentric openings in the borated PE layer, side view relative to the TLD card (right).

as a sensitive element. Detailed description of the dimensions for the mentioned geometry and materials composition and densities can be found in [4]. The relative reading of the dosimeter can be found in Fig. 2 in comparison to the dosimeter from Fig. 1. The improved dose reading is clearly seen.

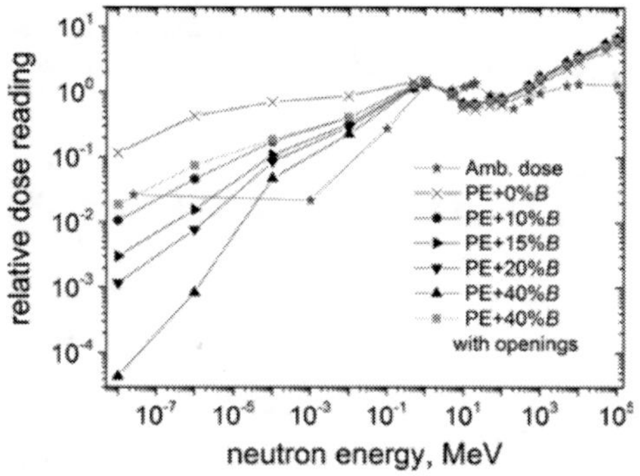

Figure 6. Dose readings of the dosimeter with varying borated polyethylene layer composition.

In Fig. 6 one can see the response of the dosimeter if the borated PE layer is modified. Using just a PE without any absorptive admixtures increases the response of the dosimeter at low and medium neutron energies, while an addition of 10 % boron to the layer drastically improves the response. Further increase of the boron doping increases the underestimation of the dose at lower energies and for 40% of the mass the response at thermal energy is more than two orders of magnitude less than required. Following an idea in [15] to increase the response for low energy neutrons holes can be drilled in the dosimeter to provide more transparent flow of neutrons towards the TLD cards. Modifying the dosimeter with 40% doped PE layer as depicted in Fig. 5 (right), introducing concentric openings has the same influence as a decrease of the boron admixture.

5 Modification of the compact dosimeter

Another solution to improve the response of the compact dosimeter with a PVC layer could be changing the outer diameter of the dosimeter. For different neutron energies the relative ambient dose readings of the dosimeters with varying PVC layers can be seen in Fig. 7. For consistency in the figure the neutron energies are limited to 10 MeV, the red curve represents the expected values according to [7]. Decrease of the dimension leads to the reduced absorption and moderation, which makes it unfavourable to decrease further the size of the dosimeter. Still in high energy region the readings of more compact layouts are close to the one from Fig. 2 and thus even smaller dosimeters can be used for detection in the energy region above 1 MeV. The neutron spectra typically have a peak in the vicin-

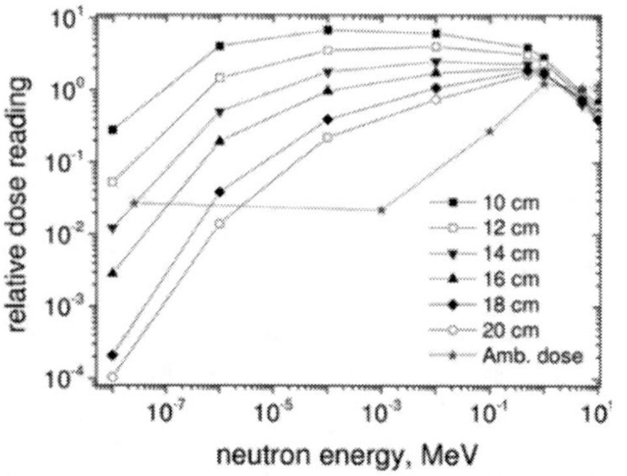

Figure 7. Readings of the dosimeter in low and medium energy region for different diameters of the PVC layer from 10 cm to 20 cm.

ity of thermal energies requiring an appropriate reading around 25 meV. A diameter of 14 cm of the current compact PVC cylinder is optimal for such energies. Increase of the dosimeter dimension bends the response curve in such a way that the dose reading is underestimated around 25 meV and for eVs-keVs region it is better due to more efficient moderation. For the dosimeter with the diameter and height of 20 cm the reading in the thermal region is almost two orders of magnitude smaller due to strong absorption and thus not acceptable, though in eVs-keVs region the reading is drastically improved in comparison to the 14 cm layout. Following our investigations with the dosimeter layout from Fig. 5 the openings were cut through the PVC layer. In Fig. 8 the simulated relative readings are shown when additional openings 1 mm and 0.5 mm in width are introduced in PVC layer, as depicted in the insertion of the figure. Such dosimeters would have very angular dependent response and are not usable in practice.

Extensive simulations were performed in order to investigate the influence of different openings configurations, their widths and depths. Finally it was found that an

DOI: 10.1051/epjconf/201715301013

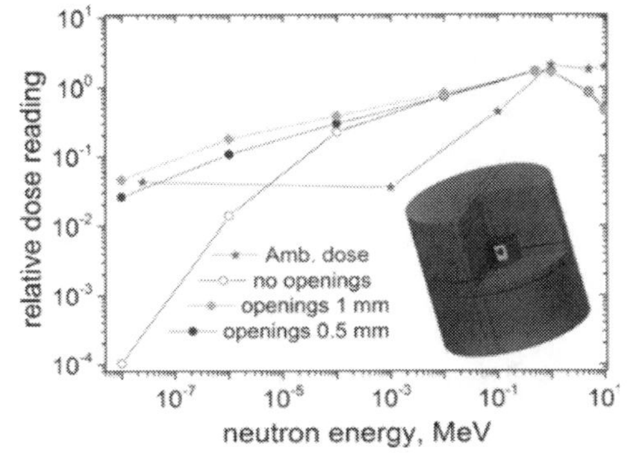

Figure 8. Responses of dosimeters having 20 cm diameter with and without openings and comparison to the required ambient dose values. Insertion shows the layout with 1 mm wide openings.

optimal configuration providing nearly angular independent reading is achieved for a dosimeter layout shown on Fig. 9. The initial 6.7 cm in diameter (d) and 7.1 cm in

Figure 9. Neutron dosimeter in 3D view (obtained using Flair [6]) with openings in PVC layer. Material colours correspond to those in Fig. 1.

height (h) lead cylinder containing high density polyethylene cylinder (d=4.7 cm and h=7.1 cm) is located in futuristic structure of PVC layer cut to a depth of 5.5 cm from the side and to 5 cm from the top and bottom with openings, each 2 mm in width. The openings release some of material making the dosimeter up to 0.5 kg lighter. The inner PVC part and openings can be varied in such a way that increase of the openings width or their amount can be compensated with the increase of the non-cut PVC region.

6 Summary and conclusion

The final neutron dosimeter layout shown in Fig. 9 improves the fluence-to-ambient dose conversion and provides almost angular independent equivalent dose reading. In Fig. 10 one can find the simulated reading of the dosimeter in wide energy range, green curve with square symbols. Folding the response of the dosimeter from Fig. 9 with the CERF spectrum should provide almost 100 % fit with the expected equivalent ambient dose value. Taking into account our comparison between FLUKA simulations and measurements would most probably lead to a dose overestimation of around 20 %. The

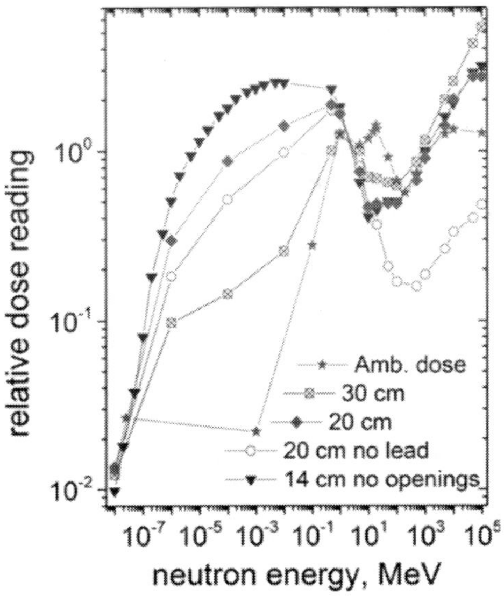

Figure 10. Readings for the dosimeters: (green curve) layout from Fig. 9; (brown curve) the same layout with PE instead of lead; (blue curve) layout from Fig. 1; (black curve) example of a dosimeter with d=h=30 cm; (red curve) expected ambient dose equivalent.

dosimeter can be further increased in dimensions and provided with the openings will have even more improved reading, example of such dosimeter response can be also found in Fig. 10, black curve for a 30 cm diameter dosimeter, though such a dosimeter is already quite bulky, heavy and complicated for production. Additionally a reading for the dosimeter with the same layout as in Fig. 9, but with a polyethylene instead of lead, is also depicted in Fig. 10 and provides even better response for low and medium energies.

The final design from Fig. 9 consumes an appropriate response from the initial design in high energy neutron fields and improves it for medium energies. Thus the dosimeter response is covering almost 12 orders of magnitude of neutron energies and is rather sufficient for the usage at high energy accelerators.

References

[1] https://www.gsi.de ; https://www.fair-center.de.
[2] M. Awschalom and R. S. Sanna, Applications of Bonner sphere detectors in neutron field dosimetry, Radiation Protection Dosimetry, **Vol.10** No. 1-4 pp. 89-101 (1985).
[3] G. Fehrenbacher, F. Gutermuth, E. Kozlova, T. Radon and R. Schuetz, Neutron dose measurements with the

GSI ball at high-energy accelerators, Radiation Protection Dosimetry (2007), Advance Access publication 16 September 2007, **Vol. 125**, No. 1-4, pp. 209-212.

[4] A. Sokolov, G. Fehrenbacher, T. Radon , Design Development of a passive neutron dosemeter for the use at high-energy accelerators, Radiation Protection Dosimetry 2015; doi: 10.1093/rpd/ncv489

[5] A. Fasso', A. Ferrari, J. Ranft and P. R Sala, FLUKA: a multi-particle transport code, CERN-2005-10 (2005), INFN/TC 05/11, SLAC-R-773.

[6] V. Vlachoudis, FLAIR: A Powerful But User Friendly Graphical Interface For FLUKA Journal Name, Saratoga Springs, New York, 2009.

[7] M. Pelliccioni, Radiation Protection Dosimetry, Nuclear Technology Publishing, 88, 279-297 (2000).

[8] ISO, International Standardization Organization, Neutron Reference Radiations for Calibrating Neutron Measurement Devices Used for Radiation Protection-Purposes. ISO 8529, 1st Edition, 1989.

[9] I.A. Zahradnik, Development of the passive neutron dosimeter, ANDREA, for high energy accelerators, verification measurements and simulation of the dose response using the Monte Carlo code , bachelor thesis, THM Giessen, 2015.

[10] A. Mitaroff and M. Silari, The CERN-EU high-energy Reference Field (CERF) facility for dosimetry at commercial flight altitudes and in space, Radiation Protection Dosimetry 102, 7-22, 2002.

[11] http://cerf-dev.web.cern.ch/content/fluka-reference-values.

[12] H. Kluge, A.V. Alevra, S. Jetzke, K. Knauf, M.Matzke, K. Weise and J. Wittstock, Scattered neutron reference fields produced by radionuclide sources, Radiation Protection Dosimetery, Nuclear Technology Publishing, 70, Nos. 1-4, pp. 327-330 (1997).

[13] F. Gutermuth, T. Radon, G. Fehrenbacher and J.G. Festag, The response of various neutron dosemeters considering the application at a high energy particle accelerator. Kerntechnik 68, 172–179 (2003).

[14] V. Mares, A.V. Sannikov and H. Schraube, Response functions of the Andersson–Braun and extended range rem counters for neutron energies from thermal to 10 GeV, Nuclear Instruments and Methods in Physics , **Vol. 476**, Issues 1–2, 1 Jan 2002, p. 341–346.

[15] F. H. Attix (Editor), *Topics in Radiation Dosimetery: Radiation Dosimetery, Band 1*, Academic Press 1972

Ionization Quenching Factor measurement of 1 keV to 25 keV protons in Isobutane gas mixture

Benjamin Tampon[1], Daniel Santos[2], Olivier Guillaudin[2], Jean-François Muraz[2], Lena Lebreton[1], Thibaut Vinchon[1] and Philippe Querre[3]

[1]*Institut de Radioprotection et de Sureté Nucléaire (IRSN), PRP-HOM/SDE/LMDN, 13115 Saint Paul Lez Durance, France*
[2]*Laboratoire de Physique Subatomique et de Cosmologie, 38000 Grenoble, France*
[3]*Institut de Radioprotection et de Sureté Nucléaire (IRSN), PSN-RES/SA2I/LIE, 13115 Saint Paul lez Durance, France*

Abstract. The French Institute for Radiation protection and Nuclear Safety (IRSN) is providing reference neutron fluence energy distribution at its standard monoenergetic neutron fields, produced at the AMANDE facility. The neutron energy is assessed by measuring the recoil nuclei energy in a μTPC detector, the LNE-IRSN/MIMAC detector. The knowledge of the ionization quenching factor (IQF) is fundamental to determine the kinetic energy of the recoil nuclei. For some various gases and pressures, discrepancies of about 15% were observed between IQF calculations using the SRIM software and experimental measurements. No data are available for the iC_4H_{10} + 50% CHF_3 gas mixture which are used for measurements from a few keV up to 565 keV neutron energies in the μTPC detector. The experimental determination of the IQF is of primary importance to provide reference neutron fluence energy distribution. After a short description of the experimental set-up, this paper presents the first results of the IQF measurements in a iC_4H_{10} + 50% CHF_3 gas mixture in the energy range 1 keV – 25 keV.

1 Introduction

The Laboratory for Neutron Metrology and Neutron Dosimetry (LMDN) of the French Institute for Radiation protection and Nuclear Safety (IRSN), is equipped with several neutron facilities. The AMANDE facility, in particular, aims at producing monoenergetic standard neutron fields as recommended by the ISO 8529 [1]. As an associated laboratory to the French National Metrological Institute (LNE), the LMDN has to develop, improve, and maintain the reference quantities associated to its standard neutron fields. To that end, the laboratory has undergone the development of a recoil nuclei-based gas spectrometer, called LNE-IRSN/MIMAC [2], (MIcro-tpc MAtrix of Chambers) in order to provide reference values of the neutron fluence, and neutron fluence energy distribution in the energy range of a few keV up to 5 MeV. For the lowest energies, between 8 keV and 565 keV, a iC_4H_{10} (isobutane) + 50% CHF_3 gas mixture, at 30 and 50 hPa gas pressure, could provide promising results [3][4]. The development of the LNE-IRSN/MIMAC detector is a collaborative effort with the Laboratory of Subatomic Physics and Cosmology (LPSC), which has developed the first MIMAC prototype for directional dark matter search [5].

The neutron fluence energy distribution is determined from the nuclear recoils produced by the neutron elastic scattering in the gas mixture. The neutron energy is determined, event by event, by measuring the scattering angle θ and the kinetic energy of the gas nucleus recoil. Preliminary measurements have shown that the recoil nuclei release only a part of their energy through the ionization process. Therefore, the measured energy in the detector does not correspond to the kinetic energy of the recoil nuclei. To assess this quantity, the Ionization Quenching Factor (IQF) has to be determined. The IQF, which is the ratio of the ionization energy and the kinetic energy (Eq. 1), depends on the gas mixture, the pressure, and the energy of the recoil nuclei,

$$IQF = \frac{E_{ionization,ion}}{E_{kinetic,ion}} \quad (1)$$

In the energy range of our interest, and for the chosen gaseous composition, no experimental data are available. The only way to estimate the IQF in our working conditions is to do simulations using SRIM [6] software. To perform experimental measurements of the IQF the LPSC has developed a dedicated facility, called COMIMAC [10].

2 The COMIMAC facility

COMIMAC is a table-top ion and electron beams facility, which produces beams in the energy range from few tens of eV up to 50 keV. A picture and a schematic view of the device are shown respectively in Figures 1 and 2. This facility is mainly composed of a COMIC ion source, (Compact Microwave Coaxial) [7], and a measurement chamber. This measurement chamber is a gaseous detector for measuring the ionization energy (Figure 2). A plasma mainly composed of mono-charged ions and electrons is created and maintained in the resonant cavity of the COMIC ion source. Then two electrodes extract and focus the beam. Depending of the polarity and the voltage applied on these electrodes, extracted particles are electrons or ions. A voltage of 1 kV up to 50 kV can be applied to the electrodes, which allows thus the generation of mono-charged beams from 1 keV up to 50 keV. The beam passes then through a micrometric hole [8] which separates the measurement chamber, filled with gas (10 hPa up to 950 hPa), from the ion source, pumped at a secondary vacuum (~10^{-5} hPa). Once entering in the measurement chamber, the beam's particles lose their energy by ionising the gas. The collection of ionization electron induces an electric current on the anode of a micromegas-type detector [9]. The charges collected on the micromegas are proportional to the ionization energy released in the gas by the particle.

Figure 1. Picture of the COMIMAC ion and electron beam generator.

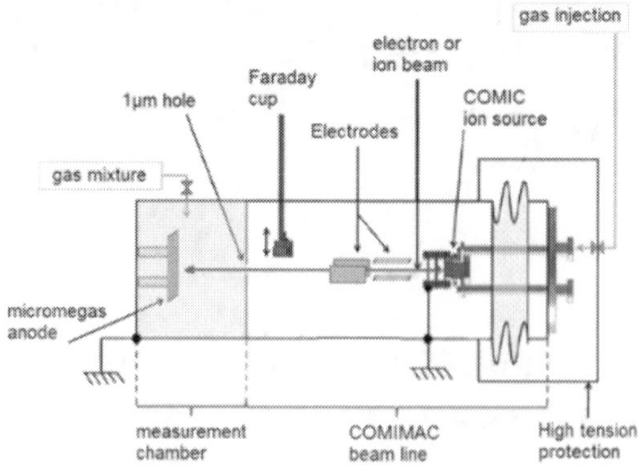

Figure 2. Schematic view of the COMIMAC facility.

3 IQF measurements in C_4H_{10} + 50% CHF_3

3.1 Gas purity

The gas used to create the plasma in the COMIC ion source is dihydrogen (H_2[a]). Despite the high quality of the gas, some impurities were still present. They were mainly coming from air and water outgassing from the one meter long "plastic" pipe, transporting the gas from bottle to the COMIC ion source. The plasma was thus composed of several atoms, including protons which were well identified, as explained in section 4.2.2.

Concerning the gas mixture, the available C_4H_{10}[b] and CHF_3[c] gases were provided and then mixed in the laboratory, in a dedicated buffer volume, as explained later. In case of pollution, electronegative atoms, like oxygen atoms, will capture electrons, and the number of collected charges will be reduced. Therefore, before using the gas mixture, one has to ensure as much as possible, a clean enough chamber and gas pipes. For that, all gas pipes and volumes are preliminary pumped (pressure around 10^{-5} hPa) and gas pipes from bottle to volumes are heated for out-gassing. These precautions allow limiting the gas mixture pollution.

To further limit the pollution, the measurement chamber is completely emptied and then refilled every 4 hours. In this case, the challenge is to keep constant the proportion of both gases in the mixture during the whole measurement campaign. For that, a buffer volume was set up, where gases could be mixed and stored (see Figure 3). A sufficient high quantity of gas mixture is prepared in this buffer at the beginning of the measurement campaign, and then is used to refill the measurement chamber.

---

[a] Air Liquid - H2 type N55 - ref P1416S01B0D001
[b] Messer - quality 3.5 (iC_4H_{10} >99.95%)
[c] Air Liquid - CHF3 type N50 – ref P0737S05R0A001

DOI: 10.1051/epjconf/201715301014

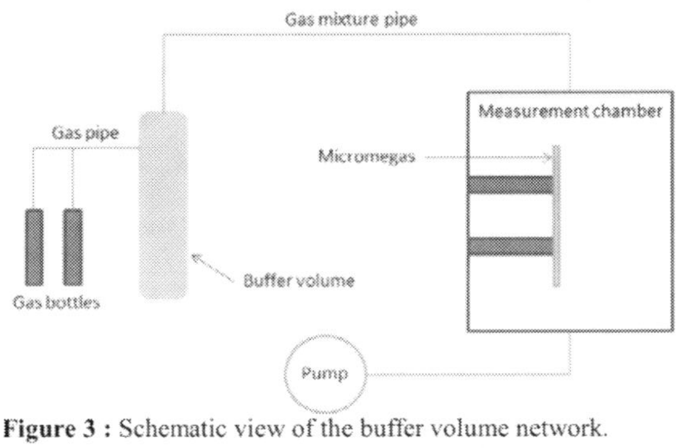

Figure 3 : Schematic view of the buffer volume network.

3.2 Data acquisition

3.2.1 Electronics and acquisition system

The fast electronics as well as the data acquisition software use to perform this experiment has been developed at the LPSC [11]. In the detector, each particle releases its energy by ionizing the gas; this induces a charge current on the grid of the Micromegas [12] which is proportional to the ionization energy deposited. This charge current is then treated by a preamplifier and the output signal is further sent to the FPGA, located on the printed circuit board. The value of the collected charge is recorded and finally sent to the acquisition software. This allows reconstructing for each event a charge collection profile whose amplitude is proportional to the energy deposited by the particle in the detector.

3.2.2 Energy calibration

As shown in the Equation 1, the Ionization Quenching Factor depends on the measured ionization energy and the particle kinetic energy. For electrons, in the energy range of 1 keV up to 50 keV, the ionization energy measured corresponds to their kinetic energy, leading to IQF values equal to 1. For that reason, electron beams are ideal to perform the detector energy calibration, as it will be detailed in section 4.2.1.

For a reliable energy calibration, electrons have to lose all their kinetic energy in the active volume of the measurement chamber. This implies that, in our low gas pressure conditions (30 or 50 hPa), only electrons with energy lower than 10 keV can be used.

The other way used to perform the detector energy calibration is to use X-ray sources which are permanently located inside the measurement chamber (see Figure 4). The X-rays interact through photoelectric effect with atoms in the gas. Photoelectrons produced by these interactions lose their kinetic energy by ionizing the gas and electrons are collected by the Micromegas anode. The X-ray sources used are ^{55}Fe and ^{109}Cd which produced electrons at 5.92 keV and 3.04 keV respectively [2]. Third electron energy of 1.486 keV, emitted by excited ^{27}Al atoms, is also used for the calibration. These electrons come from the interaction between the 22 keV electrons of ^{109}Cd and the ^{27}Al present in the detector components.

Once the calibration is performed using these three energy points, the X-ray sources, which are sticked onto a ferrite plate, are moved and hidden behind the Micromegas, by handling a magnet, as shown on Figure 4.

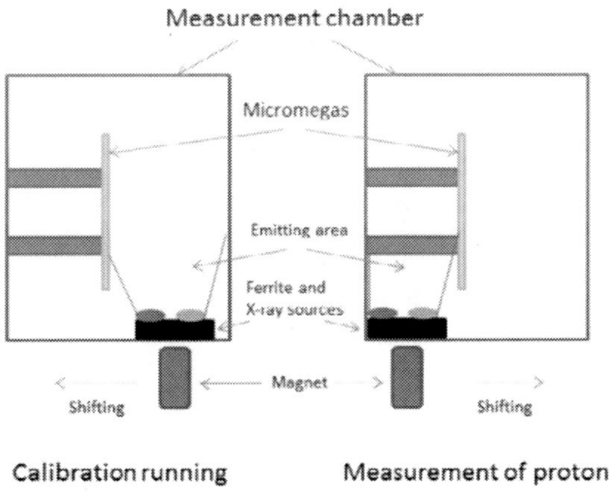

Figure 4. Schematic view of measurement chamber with position of X-ray sources.

3.2.3 Proton measurements

The kinetic energy of proton beams is determined by the extraction voltage applied on the electrodes see Figure 2. This method has been validated by comparison of X-ray sources with electron beams. In this experimental campaign, measurements were performed for eight different proton energy beams, between 1.5 keV and 25 keV. For each proton beam a minimum of 5000 counts under the proton peak was achieved with a data acquisition time of about 5 minutes each.

3.2.4 Monitoring of gas mixture with electrons

Even though all efforts made to keep the pollution as low as possible, as detailed in 3.1, pollution atoms can still be present in the gas mixture. In order to prevent pollution the gas is regularly renewed in the ionization chamber. In addition, during the experiment, the monitoring of the gas mixture degradation is possible by performing frequent energy calibrations. In such a way, one has a quantitative degradation of the gas mixture allowing deciding when the mixture should be renewed.

From this monitoring data, a correction of the Ionization Quenching Factor is computed as a function of this degradation, as it will be explained in section 4.2.3.

4 Data analysis

4.1 Ballistic effect correction

Due to the charge collection time of the charge preamplifier, which depends on the RC time constant, with R = 10 MΩ and C = 1 pF, the total charge collection may be reduced. This reduction, called also ballistic effect, is an exponential function expressed as exp(-t/τ) where τ = R*C. As a consequence, the measured collected charge does not completely correspond to the ionization energy released by the particles in the detector, which is thus underestimated, as illustrated in Figure 5 for a proton event. In this figure, the experimental collected charge of a proton event (represented by star), which reaches its maximum after about 450 ns, is decreasing with time due the ballistic effect. The second curve represents the collected charge of the same proton if there was no ballistic effect. A correction of this effect is necessary for the determination of the total integration of the collected charges.

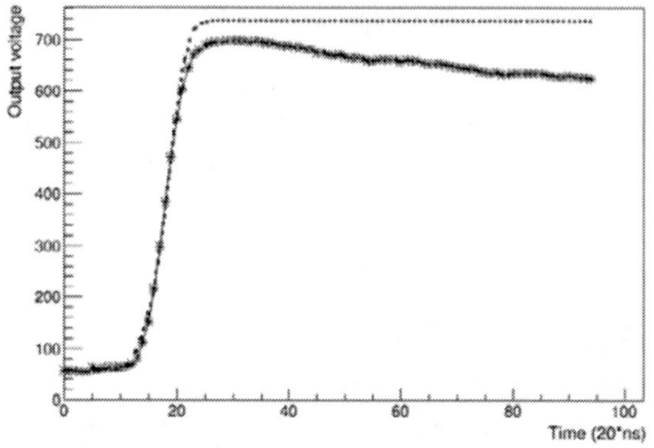

Figure 5. Charge collection of a proton event. Black stars is the data measurements and the blue curve represents the ideal charge collection (without ballistic effect).

As explained before, the electronic decreasing is estimated as exp(-t/τ) where τ = R*C (with R = 10 MΩ and C = 1 pF). The total charge integration of a given event is the difference between the maximum and the minimum values of its charge collection profile. These minimum and maximum values are measured at times T_{min} and T_{max} (respectively around 0 ns and 450 ns in Figure 5). The time difference T, given by T_{max}-T_{min} corresponds to the duration of the charge collection of this event. The ballistic effect during this event is exp(-T/τ), so the associated ballistic correction is exp(+T/τ).

This ballistic correction has been applied to all events presented hereafter.

4.2 Determination of energy

4.2.1 Energy calibration

By analysing each electron event; it is then possible to plot a histogram about the number of events as a function of the integrated charge value measured on the charge preamplifier profile, which corresponds to the ionization energy released in ADC channels. The spectrum of an electron calibration is shown in Figure 6.

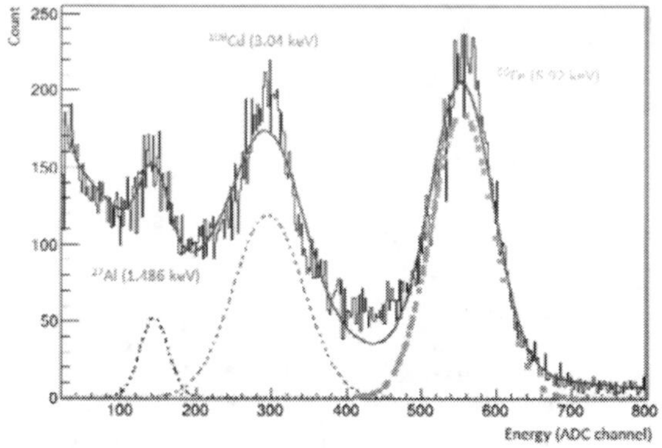

Figure 6. Spectrum of calibration running with ^{55}Fe and ^{109}Cd in the detector. The red curve is the fit of the spectrum by three Gaussians.

The spectrum shows 3 peaks and a continuum background. The peaks correspond respectively to the three energies of electrons produced by X-rays used to calibrate. The continuum observed is induced by the incomplete collection of electrons with energy greater than 10 keV (like X-ray at 22 keV of ^{109}Cd).

The analysis of this spectrum enables to build a calibration curve between ionization energies and ADC channel numbers. In order to determine the parameters associated to this calibration curve, the software MCARE [13] is used. It has been developed by the "Collège Français de Métrologie" and enables computing the equation of a calibration curve with different methods. The most popular is GGMR (General Gauss Markov Regression) which computes the curve parameters with uncertainties on X (number of ADC channel) and Y (energy) as shown on Figure 7.

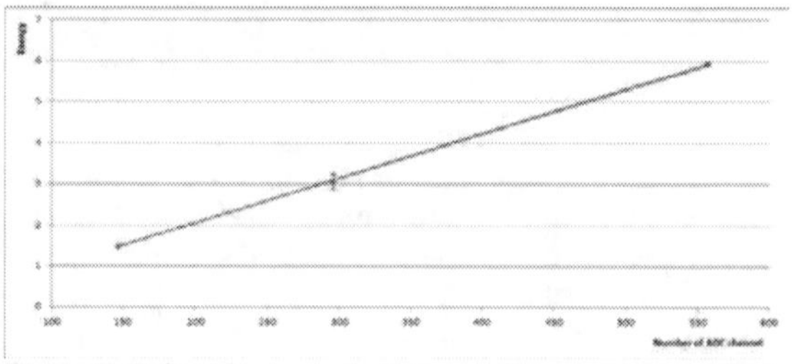

Figure 7. Calibration curve built from measurements presented on Figure 6.

Uncertainties of X-ray energies of ^{27}Al, ^{55}Fe and ^{109}Cd are taken from ENDF VII.1 data base. Indeed, the measurement of the energy mean (1,486 keV and 3,04 keV and 5,92 keV) and the uncertainty associated (0,01 keV and 0,2 keV and 0,04 keV respectively) are realized with the probability distribution of X-ray for these three energies.

As explained before, the software MCARE allows calculating parameters for the energy calibration curve. Moreover, it gives the uncertainty associated to each point on the curve too. This is needed to perform the calculation of uncertainty on IQF measurements, as it will be explained in section 5.

4.2.2 IQF measurements

The same kind of spectrum shown for electrons is created for proton events. In Figure 8 is presented the spectrum obtained for the protons with a kinetic energy of 25 keV.

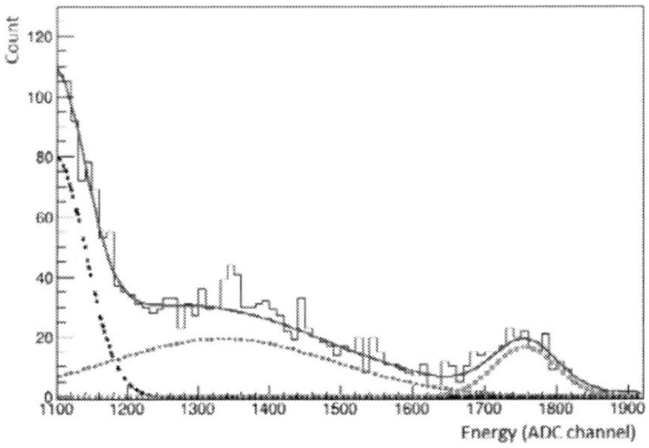

Figure 8. Spectrum obtained for a proton beam with 25 keV kinetic energy. The red line corresponds to the fit of the spectrum. The green Gaussian fit is the proton peak and the two other Gaussians are peaks of pollutions (other nuclei at 25 keV kinetic energies).

On a given spectrum for protons, there are, in general, several well-defined peaks, which correspond to atoms present in the gas constituting the plasma in the ion source, see section 2. These peaks are even more important when a Wien filter is not used, as it was the case for these measurements. Whatever the ionization energy of the particle, the proton peak is always at the highest ionization energy. This is because the quenching in ionization increases with the mass of the atoms. This means that heavier the atoms smaller are their IQF. According to equation 1, the heavier than the proton atoms have smaller released ionisation energy than the proton. That property explains that protons, which are the lightest elements, will be always on the right of spectra.

In Figure 8, the spectrum obtained with a 25 keV proton beam is shown. The position of the proton peak, which is the mean of the Gaussian, was combined with the energy calibration on MCARE to determine the ionisation energy released by protons at this energy. Then, according to the equation 1, the Ionisation Quenching Factor could be calculated.

Uncertainties associated to these IQF measurements for each energy were calculated from the uncertainty of the proton peak and the ionisation energy released (related to the energy calibration uncertainties like explained in part 4.2.1). The proton peak is assumed to be a Gaussian so the associated uncertainty was sigma/$(N)^{1/2}$, where sigma is normalized by the mean and N is the number of events issued from the Gaussian curve integral.

As explained in 3.2, energy calibrations are repeated before and after each measurement of protons at different energies to monitor the degradation of the gas mixture. In order to correct this degradation, a linear interpolation is realized for each proton measurement. After the calculations of IQF for the two energy calibrations (performed just before and after this measurement), the corrected IQF is defined as the linear combination function of time of the two calibrations and of the time of the proton measurement.

Table 1. Notation definitions for IQF interpolation.

Measurements	Time	IQF (t1)	IQF(t2)	IQF(t)
Calibration 1	t1			
proton	t	IQF1	IQF2	IQF
Calibration 2	t2			

In the situation described in **Table 1**, the interpolated value of IQF will be:

$$IQF = \frac{\left(\frac{t-t1}{t2-t1}*IQF2+\frac{t2-t}{t2-t1}*IQF1\right)}{2} \tag{2}$$

With this method, the gas degradation is no longer an issue for the measurement of IQF, as long as a linear degradation of the gas can be assumed.

5 Results

Before this experiment, no experimental data on Ionization Quenching Factors were available in literature in our energy range of interest. As seen in the section 1, the only way to estimate the IQF was to use the SRIM simulation software. In Figure 9 are presented the IQF values as a function of proton energy in a 50% C_4H_{10} + 50% CHF_3 gas mixture, computed by SRIM simulation in red and those deduced from the data obtained during this experimental campaign.

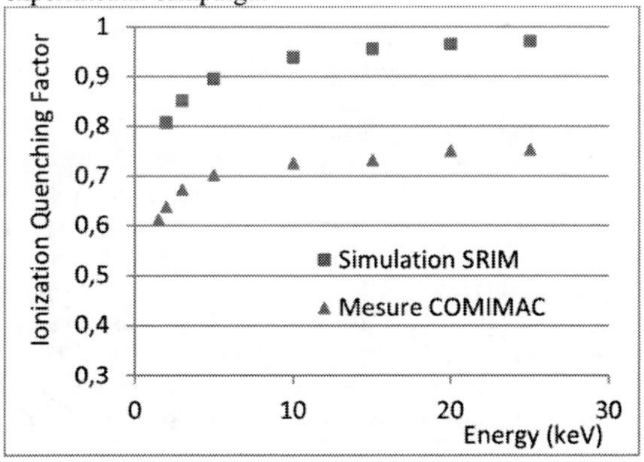

Figure 9. IQF measurements.

Uncertainties of experimental data are not visible on the Figure 9 because they do not exceed 0,01, which correspond to 1.16% of uncertainties at 1.5 keV.

The COMIMAC facility allows measuring the Ionization Quenching Factor of a gas mixture in-situ taking into account the pollution in the chamber with low uncertainties. These results will become the reference for the IQF of proton in 50% C_4H_{10} + 50% CHF_3 gas mixture at 50 mbar. The IQF depends on the amount of pollution and on the pressure of the gas. Indeed, the SRIM simulations on the Ionization Quenching Factor do not show this dependency on the gas pressure. The SRIM simulation, shown in figure 8, overestimates the IQF of protons in this case. This difference represents the limitations of SRIM calculations on the IQF estimation in this low energy range. Indeed, this software is only optimized and validated for calculations around a few MeV per nucleon when this work concerns energies around a few keV.

It is theoretically expected that at higher energies the IQF will be closer than 1. But it appears that the limitations in the energy range of COMIMAC (E < 50 keV) prevent to explore the IQF in the MeV range. However, for fast neutron fields characterization in the MeV range an exploration of the IQF in such a range of energy will be performed in a future work.

Acknowledgement

This work is partially funded by the French National Metrology Institute (LNE) under contract LNE/DRST 15 7 004.

Bibliography

[1] International Organization for standardization 2001 *Reference Neutron Radiations - Part 1: Characteristics and methods of production* ISO8529-1

[2] D. Maire, *Développement d'une chambre à projection temporelle utilisant une anode pixélisée pour les références en énergie et en fluence de champs neutroniques de basses énergies (quelques keV à 5 MeV)*, PhD Thesis (in french) (2015).

[3] C. Golabek et al., *A μTPC detector for the characterization of low energy neutron fields*, NIM A, vol. 678, pp. 33-38. 2012.

[4] D. Maire et *al.*, *Development of a μTPC detector as a standard instrument for low energy neutron field characterization*, Rad. Prot. Dos., vol. 161, N°s 1-4, pp. 245-248, 2014.

[5] D. Santos et *al.*, *MIMAC, Micro-tpc Matrix of Chambers for dark matter directional detection*, Proceedings of the 4th International Conference on Directional Dark Matter Detection CYGNUS2013, held in Toyoma (Japan), 2013.

[6] J.F. Ziegler et al., *SRIM – The stopping and range of ions in matter (2010)*, NIM B 268 (2010) 1818-1823.

[7] P. Sortais et al., *Ultra compact / Ultra low power ECRIS for multi-purpose applications*, Rev. Sci. Instrum. 81, 02B314, 2010.

[8] D. Santos et al., *Ionization Quenching Factor Measurement of 4He*, arXiv :0810.1137v1.

[9] I. Giomataris et *al.*, *Micromegas in a bulk*, NIM A, vol. 560, pp. 405-408, 2006.

[10] J.F.Muraz et al., *A table-top ion and electron beam facility for ionization quenching measurement and gas detector calibration*. NIM A 832 (2016) 214-218

[11] O. Guillaudin et al., *Quenching factor measurement in low pressure gas detector for directional dark matter search*, arXiv:1110.2042v1

[12] G.F. Knoll. *Radiation and Measurement Third Edition*, Wiley and sons, 2000

[13] M-CARE v1.0: C4GH10, retrieved 20/12/2014.

Calibration of a FDS-LC long counter for monitoring the neutron fluence at the HINEG facility

Bing Hong, Yanan Li, Mohamed Mazuga, Taosheng Li*, Yongfeng Wang, Xiang Ji, Chao Liu, Jieqiong Jiang, FDS Team [1*]

[1]*Institute of Nuclear Energy Safety Technology, Chinese Academy of sciences, Key Laboratory of Neutronics and Radiation Safety, Hefei, Anhui, 230031, China*
*Corresponding Author, E-mail: taosheng.li@fds.org.cn

Abstract: A FDS-LC long counter was developed to monitor the neutron fluence at the HINEG facility. The response function and the effective centre of long counter were calculated by using SuperMC code. The calibration of this long counter was performed in Research Facility, China institute of atomic energy. The results of experimental and simulative characterization of the FDS-LC long counter are persented in this paper. The comparison of the measured results with the calculated results shown a good agreement within 5%.

KEYWORDS: Long counter, Response, The effective centre, Calibration

I. Introduction

The High Intensity D-T Fusion Generator (HINEG) with yield of 10^{12} n/s is constructed by institute of nuclear energy safety technology, chinese academy of science(INEST,CAS). It can be used for the neutronics, radiation technology, and other characteristic technology for advanced nuclear energy system-such as fusion/fission reactor and the calibration of measurement devices. In order to monitor the neutron fluence or the neutron emission at the HINEG facility, the neutron fluence monitoring system needs to be developed.

The long counter is insensitive to γ-rays, high efficiency, and has a flat response function over a wide neutron energy range, which make it an ideal neutron detector for the measurements of parallel neutron fields [(1)]. Such a neutron detector is chosen to monitor the neutron fluence or neutron emission at the HINEG facility. In previous work, our team has designed a long counter having a relatively flat response from neutron energies 1 keV to 20 MeV by adding tungsten material and the maximal relative variation is 7.8% [(2)]. The design is then further optimized by our team to extend its energy range from 1 keV to 150 MeV with response fluctuation less than 9%. A long counter called FDS-LC based on this design is developed by our team considering actual technology at manufacturing process.

II. Description of FDS-LC long counter

The geometry of the FDS-LC long counter is based on the De Pangher long counter[(3)]. Figure 1 shows the structure of this long counter. This long counter is about 50.4 cm in diameter and 52.2 cm in length. It mainly consists of a BF_3 thermal neutron counter, the inner and outer polyethylene, borated polyethylene absorption layer, chromium and lead metal (n,xn) converter. The outer polyethylene annulus and the borated-polyethylene are used to reduce the effect of scattered neutron. The BF_3 proportional counter, manufactured by LND.INC, is installed in the centre of the cylinder of inner polyethylene. The external diameter of this proportional counter is 50.8 mm, the nominal active is 215.7 mm and the pressure of the BF_3 gas is about 150 torr. The cadmium sheet in the front face of long counter is used to absorb thermal neutron of incident direction.The polyethylene cylinder of 25 mm long is placed in front the BF_3 counter to flatten the response of the low energy neutron.To improve the response in the intermediate neutron energies, the air annulus with 12 cm depth and 0.4 cm width is added on the inner cylinder moderator. To improve the response function at higher neutron energies, a batch of 12 of the chromium bar of dimension 1.7 cm diameter and 31 cm length are inserted in inner polyethylene, and a thick 0.5 cm lead layer is placed in outside of the BF_3 tube.

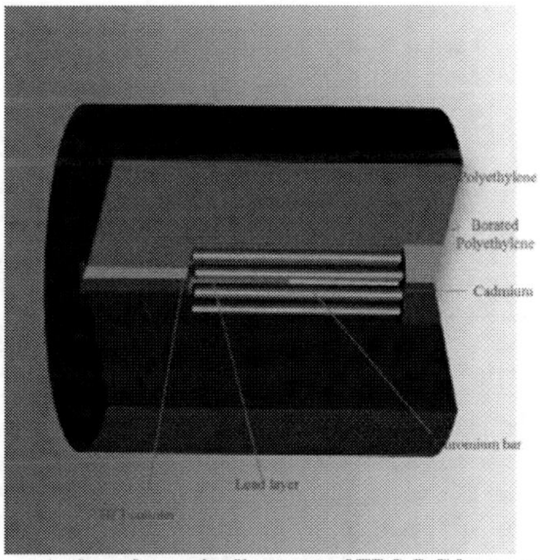

Figure 1. the schematic diagram of FDS-LC long counter

III. The characteristics of FDS-LC

According to the actual geometric structure and material parameters of FDS-LC long counter, the long counter was modeled using the Super Monte Carlo (SuperMC) code[(4)]. The response function and the effective centre of FDS-LC long counter were calculated by using SuperMC code. The response function of FDS-LC long counter was calculated for the radioactive source (^{241}Am-Be ,^{252}Cf) and for about hundreds of monoenergetic neutron in the range 1 keV to 150 MeV. The calculated results revealed that the responses were almost flat from 1 keV to 150 MeV with the mean value 1.1 cm^2. The effective centre of FDS-LC long counter was also calculated. The relationship between the effective centre of FDS-LC long counter and the neutron energy in the range 0.1 MeV to 20 MeV was also estimated.

IV. The Calibration of FDS-LC

The FDS-LC long counter developed in this study was shown in Figure 2. It was calibrated using ^{241}Am-Be source and T(d,n)^{3}He source in Research Facility, China institute of atomic energy. The FDS-LC long counter signal was acquired by the experimental acquisition system. The acquisition system consisted of an amplifier, a discriminator and a counter. The discriminator was used to discriminate between the neutron-induced signal and the γ-induced signal. The experimental results shown that the response of FDS-LC long counter was 1.11 cm^2 for the T(d,n)^{3}He source and 1.06 cm^2 for the ^{241}Am-Be source, respectively. The calculated and measured results were also shown in Table 1.

Figure 2. the calibration of FDS-LC long counter in Research Facility, China institute of atomic energy

Source	Energy (MeV)	Calculated efficiency (cm$^{2)}$	Measured Efficiency (cm^2)
^{241}Am-Be	4.16	1.01	1.06
T(d ,n)^{3}He	14.8	1.11	1.11

Table 1 .Comparison of the calculated and measured effeciency for 241Am-Be source and T(d,n)3He source

V. Conclusion

The good agreement within 5% was confirmed between the measured and calculated response for the ^{241}Am-Be source and T(d,n)^{3}He source, when uncertainties with k=2 are 5% for ^{241}Am-Be source and 7% for T(d,n)^{3}He source, respectively. The results demonstrated that the design of the FDS-LC long counter was valid and consistent with the design requirements. It is now available for monitoring the neutron fluence at the HINEG facility

Acknowledgment

This work was supported by the insititute of nuclear safety and technology , Chinese Academy of Scineces. And Thank Members of FDS team for their helpful advice.

References

1) H.Tagziria and D.J.Thomas, "Calibration and Monte Carto modelling of neutron long counters,"Nucl.Instruments Methods Phys.Res.Set. A Accel.Spectrometers.Detect.Assoc.Equip., vol.452,no. 3,pp.470-483,2000
2) Yanan Li,Taosheng Li,Gang Song ,et al. Response improved for neutron long counter.Radiation Protection Dosimetry.2015,164:93-96.
3) De Pangher, J., Nichols L L. "precision long counter for measuring fast neutron flux density.," *BNWL-260,1966.*
4) L.Song,J.,Sun,G.,Chen,Z.,Zheng,H. and Hu," Benchmarking of CAD-based SuperMC with ITER benchmark model," Fusion Eng.Des,vol.89.pp.2499-2503.2014.

Validation of PHITS Spallation Models from the Perspective of the Shielding Design of Transmutation Experimental Facility

Hiroki Iwamoto[1,a] and Shin-ichiro Meigo[1,b]

[1] *J-PARC Center, Japan Atomic Energy Agency, 2-4, Shirakata, Tokai-mura, Naka-gun, Ibaraki 319-1195, Japan*

Abstract. The impact of different spallation models implemented in the particle transport code PHITS on the shielding design of Transmutation Experimental Facility is investigated. For 400-MeV proton incident on a lead-bismuth eutectic target, an effective dose rate at the end of a thick radiation shield (3-m-thick iron and 3-m-thick concrete) calculated by the Liège intranuclear cascade (INC) model version 4.6 (INCL4.6) coupled with the GEM code (INCL4.6/GEM) yields about twice as high as the Bertini INC model (Bertini/GEM). A comparison with experimental data for 500-MeV proton incident on a thick lead target suggest that the prediction accuracy of INCL4.6/GEM would be better than that of Bertini/GEM. In contrast, it is found that the dose rates in beam ducts in front of targets calculated by the INCL4.6/GEM are lower than those by the Bertini/GEM. Since both models underestimate the experimental results for neutron-production double-differential cross sections at 180° for 140-MeV proton incident on carbon, iron, and gold targets, it is concluded that it is necessary to allow a margin for uncertainty caused by the spallation models, which is a factor of two, in estimating the dose rate induced by neutron streaming through a beam duct.

1 Introduction

To promote research, development, and demonstration of elemental technologies for accelerator-driven systems (ADS) [1–5], the Japan Atomic Energy Agency (JAEA) is planning to construct the Transmutation Experimental Facility (TEF) at J-PARC [6, 7]. In the construction of TEF, the Particle and Heavy Iron Transport code System, PHITS [8], plays an important role in the design of TEF shielding. Since the development of PHITS version 2.52, the Liège intranuclear cascade (INC) model version 4.6 (INCL4.6 [9]) has been employed as a default spallation model for the INC calculation in PHITS instead of the Bertini INC model [10]. Although, within the framework of the International Atomic Energy Agency (IAEA) benchmark program [11], a benchmark survey of spallation models demonstrated that the INCL model was the most accurate, further validation is still required to employ the INCL4.6 implemented in PHITS for specific applications such as ADS (see Ref. [5]).

This study investigates the impact of different spallation models, i.e., INCL4.6 and Bertini INC, on the TEF shielding characteristics. Here, the Generalized Evaporation Model (GEM) code [12] was

[a] e-mail: iwamoto.hiroki@jaea.go.jp
[b] e-mail: meigo.shinichiro@jaea.go.jp

employed for the de-excitation calculation following the intranuclear cascade calculation. Among various parameters on the shielding design such as effective dose rate, absorption dose rate, and radioactivity, we focus on the effective dose rate for representative apparatuses arranged in TEF; (1) a lead-bismuth eutectic (LBE) target and (2) a 10-W beam dump for beam commissioning. Another purpose of this study is to validate these models by comparing with experimental results, which were not included in the IAEA benchmark, from the perspective of the TEF shielding design.

2 Outline of TEF and calculation conditions

TEF utilizes a 250-kW proton beam accelerated with a 400-MeV LINAC accelerator at J-PARC. The proton beam is injected to the LBE target, in which intense spallation neutrons are produced. Basic concept of the TEF shielding design is the same as existing spallation neutron source facilities such as JSNS [13] at J-PARC, namely, the LBE target is surrounded by iron blocks, and an ordinary concrete shield is arranged outside the iron shield. To utilize the proton beam for various reactor-physics experiments with a subcritical assembly in TEF, a proton beam with powers up to 10 W is extracted from the 250-kW proton beam, and the 10-W beam dump for beam commissioning is installed in the TEF facility. Since a proton-beam duct is attached to the upstream side of each apparatus, estimation of the dose rate by neutron streaming through the beam duct is of importance.

Main characteristics of the LBE target and the 10-W beam dump are listed in Tables 1 and 2, respectively.

Table 1. Main characteristics of the LBE target.

	unit	value
Proton beam		
Energy	MeV	400
Maximum power	kW	250
LBE target		
Dimension	cm	D15 × L60
Density	g/cm^3	10.21

Table 2. Main characteristics the 10-W beam dump.

	unit	value
Proton beam		
Energy	MeV	400
Maximum power	W	10
Beam dump		
Material	—	graphite, iron, and ordinary concrete
Dimension	cm	20 × 20 × 40 (graphite)
Beam duct		
Material	—	titanium
Inner diameter	cm	10

Figures 1 and 2 illustrate the PHITS calculation models of the LBE target and 10-W beam dump, respectively. The calculation was performed by PHITS version 2.76, and the nuclear data library JENDL-4.0 [14] was employed for the calculation of neutron-induced reactions below 20 MeV. The effective dose rates were evaluated by the use of dose conversion coefficients [15] implemented in PHITS.

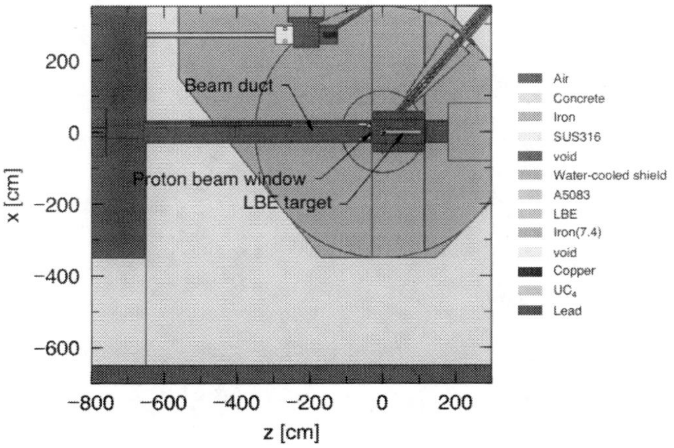

Figure 1. Calculation model of the LBE target.

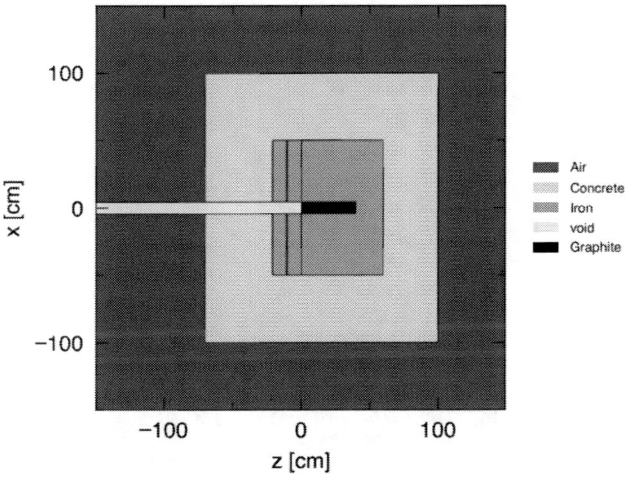

Figure 2. Calculation model of the 10-W beam dump.

3 Results and discussion

3.1 LBE target

Figure 3 shows the effective dose rate in 3-m-thick iron and 3-m-thick concrete shields at 90° from the beam direction. Here, the neutron contribution is divided into three components with resepct to its energy; thermal (below 0.4 eV), intermediate (0.4–1 MeV), and high-energy (above 1 MeV) neutrons.

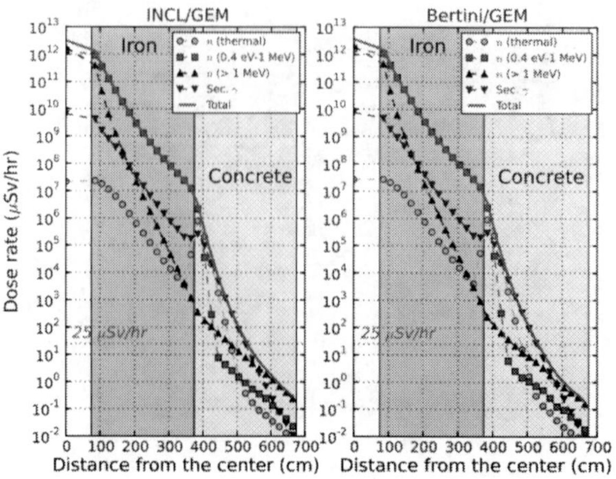

Figure 3. Effective dose rate in iron and concrete shields at 90° from the beam direction.

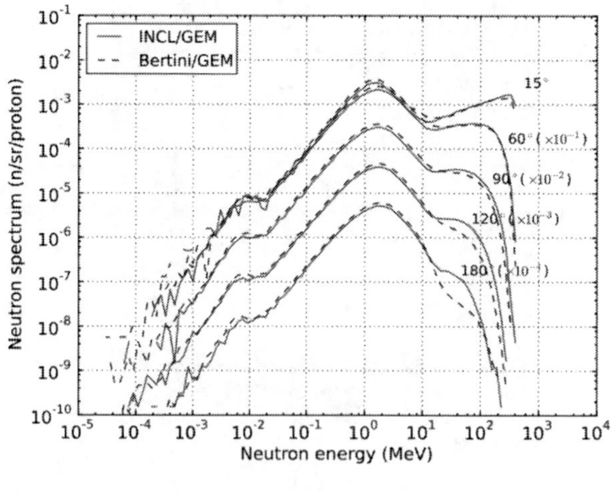

Figure 4. Neutron spectra produced from 400-MeV proton incident on the LBE target.

The dose rate calculated by the INCL4.6/GEM is higher than that calculated by the Bertini/GEM. At the end of the iron and concrete shield, the INCL4.6/GEM calculation yields about twice as high as the Bertini/GEM, and the total values are dominated by the contribution of neutrons with energies above 1 MeV. Since both models uses the same total cross sections in simulating the particle transport, it appears that this difference mainly comes from the difference in neutrons produced from the LBE target.

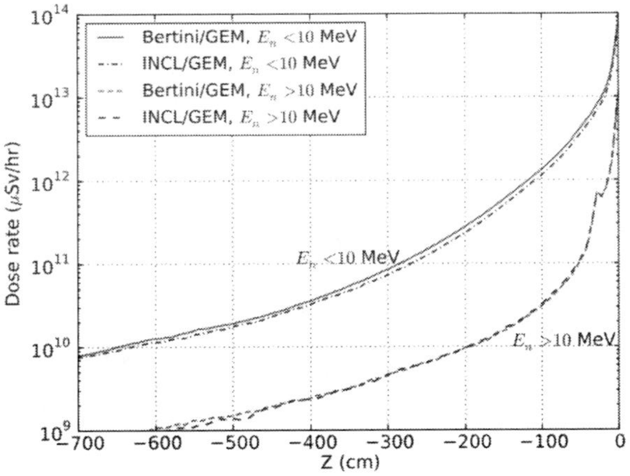

Figure 5. Effective dose rate at 180° from the beam direction. The values of a coordinate on the horizontal axis are the same as those in Figure 1.

Figure 4 shows calculated neutron spectra produced from 400-MeV proton incident on the LBE target for angles of 15°, 60°, 90°, 120°, and 180°. The INCL4.6/GEM gives higher neutron spectra than the Bertini/GEM for energies above 10 MeV. This is the reason for the higher dose estimation by INCL4.6/GEM in Fig. 3.

Figure 5 shows the effective dose rate at 180° from the beam direction. In contrast to the above results, the dose rate at 7 m away from the target calculated by the INCL4.6/GEM is slightly lower than that by the Bertini/GEM. This difference is attributable to the difference in low-energy neutrons produced from the LBE target (see Fig. 4).

3.2 Beam Dump

Figure 6 shows the effective dose rate around the beam dump for the beam direction. Figure 7 shows the double-differential cross sections at 180° for 400-MeV proton incident on a carbon target. As is the case with the LBE target, the INCL4.6/GEM gives lower values than the Bertini/GEM at the backward direction, which is due to the difference in neutron spectra for energies below 10 MeV as shown in Fig. 7.

3.3 Validation

Figure 8 compares the thick target neutron yield for 500-MeV protons on a 20-cm-thick lead target. Here, the experimental data are taken from Ref. [16]. INCL4.6/GEM is better suited for the spectral shape, whose prediction is within 1σ experimental uncertainty over a wide energy range, whereas the Bertini/GEM underestimates the experiments for angles from 90° to 150°.

Figure 9 compares the double-differential cross sections at 180° for 140-MeV protons on carbon, iron, and gold targets. The experimental data are taken from Ref. [17]. Although the incident energy is

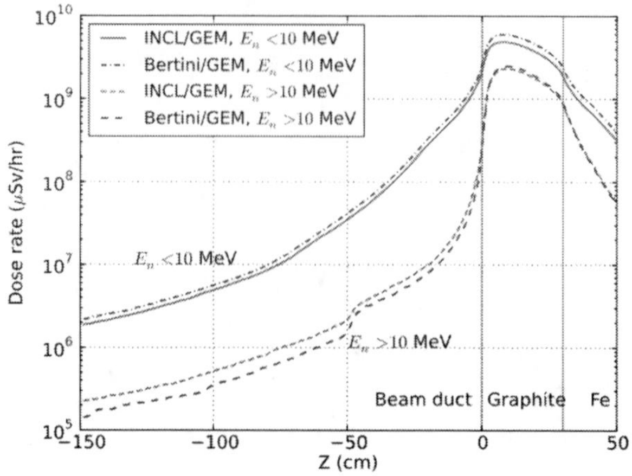

Figure 6. Effective dose rate around the beam dump for the beam directions. The values of a coordinate on the horizontal axis are the same as those in Figure 2.

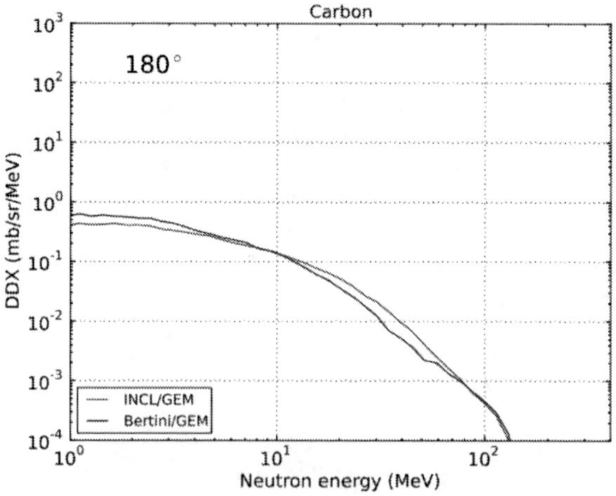

Figure 7. Double-differential cross sections at 180° for 400-MeV proton incident on a carbon target.

different, the trends of spectral shape are the same as those for the 400-MeV proton incident reaction (see Fig. 7). It was found from Fig. 9 that both models underestimate the experiments overall and fail to reproduce the spectral shape. This underestimation, especially at energies below 10 MeV, would cause an underestimation of the effective dose rate for backscattered neutrons. Since neutrons with energies below 10 MeV at the backward directions are mainly produced by the de-excitation

DOI: 10.1051/epjconf/201715301016

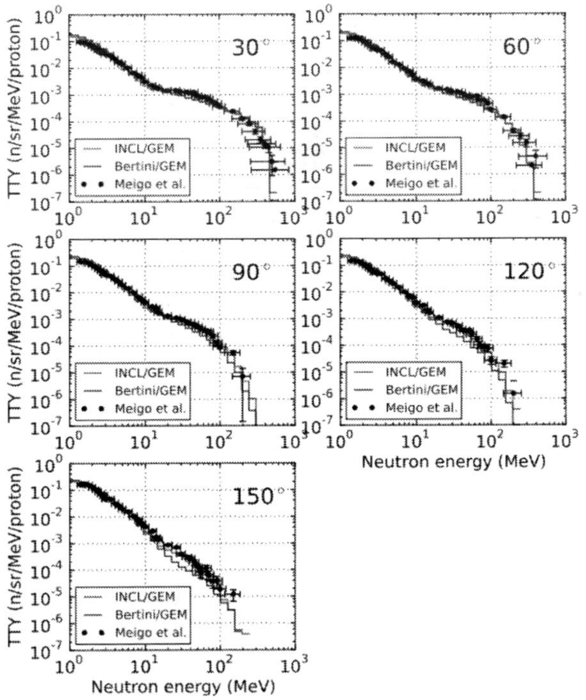

Figure 8. Comparison of thick target neutron yield for 500-MeV protons on 20-cm-thick lead.

process, the results indicate that modification or improvement of the de-excitation model, GEM, will be necessary.

4 Conclusion

In this study, the impact of different spallation models (i.e., INCL4.6/GEM and Bertini/GEM) on the TEF shielding design was investigated. The results showed that the INCL4.6/GEM estimates the dose rate at the end of the radiation shield ~1.5 times higher than the Bertini/GEM, whereas the INCL4.6/GEM calculation gives lower dose rate than the Bertini/GEM at 180° for the LBE target and the 10-W beam dump. The comparison with the experiment of Ref. [16] showed that the INCL4.6/GEM calculation is more reliable for the TEF shielding design. In contrast, for energies below 10 MeV, both spallation models underestimate the neutron spectra at 180° in Ref. [17]. This indicates that the PHITS spallation models could underestimate the dose rate for the backward directions. While it is difficult to derive a conclusion only from the limited experimental results, it appears

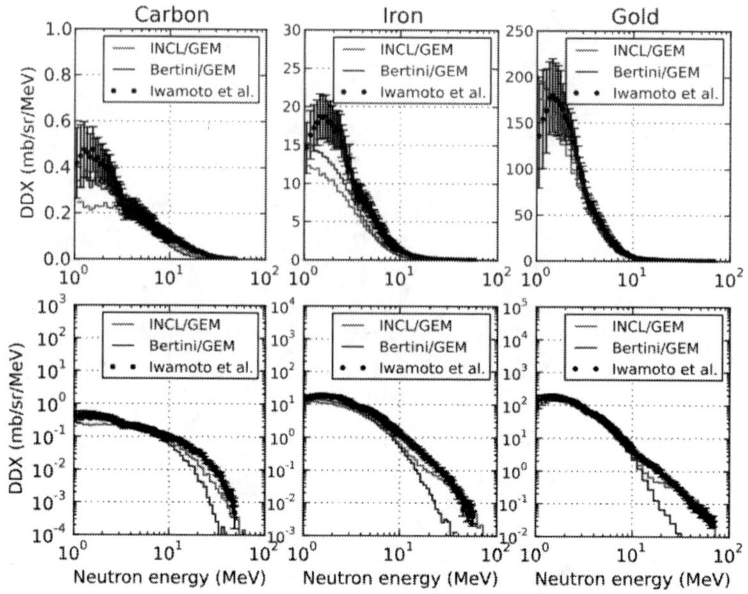

Figure 9. Comparison of double-differential cross sections at 180° for 140-MeV proton incident on carbon, iron, and gold targets. Upper and lower panels indicate linear and logarithmic scales on the vertical axis, respectively.

that uncertainty in the effective dose rate at the 180° direction induced by spallation models would be within a factor of two. Therefore, in calculating the dose rate by backscattered neutrons, a margin of this uncertainty should be considered. Thus, the PHITS spallation models still have room for improvement especially for neutron emissions at the backward angles. To achieve more reliable dose estimation, systematic experiments on the backward neutron data is needed.

References

[1] K. Tsujimoto *et al.*, Nuclear Technology **161**, 315–328 (2008)
[2] K. Nishihara *et al.*, Journal of Nuclear Science and Technology **45**, 812–822 (2008)
[3] T. Sugawara *et al.*, Journal of Nuclear Science and Technology **47**, 953–962 (2008)
[4] H. Iwamoto *et al.*, Journal of Nuclear Science and Technology **50**, 856–862 (2013)
[5] H. Iwamoto *et al.*, Journal of Nuclear Science and Technology **53**, 1585–1594 (2016)
[6] K. Tsujimoto *et al.*, Journal of Nuclear Science and Technology **44**, 483–490 (2007)
[7] T. Sasa and H. Oigawa, Plasma and Fusion Research **9**, 4401113 (2014)
[8] T. Sato *et al.*, Journal of Nuclear Science and Technology **50**, 913–923 (2013)
[9] A. Boudard *et al.*, Physical Review C **87**, 014606 (2013)
[10] H. W. Bertini, Physical Review **131**, 1801–1821 (1963)
[11] J. C. David *et al.*, Progress in Nuclear Science and Technology **2**, 942–947 (2011)

[12] S. Furihata, Nuclear Instruments and Methods in Physics Research, Section B **171**, 251–258 (2000)

[13] F. Maekawa *et al.*, Nuclear Instruments and Materials in Physics Research, Section A **620**, 259–265 (2010)

[14] K. Shibata *et al.*, Journal of Nuclear Science and Technology **48**, 1–30 (2011)

[15] Y. Sakamoto *et al.*, *Dose conversion coefficients for high-energy photons, electrons and protons* (Japan Atomic Energy Research Institute, Tokai, 2003) JAERI-1345

[16] S. Meigo *et al.*, Nuclear Instruments and Methods in Physics Research, Section A **431**, 521–530 (1999)

[17] Y. Iwamoto *et al.*, Journal of the Korean Physical Society **59**, 1753–1756 (2011)

Depth profiles of production yields of $^{nat}Pb(p, xn)^{206,205,204,203,202}Bi$ reactions using 100-MeV proton beam

Leila Mokhtari Oranj[1], Nam-Suk Jung[2], Joo-Hee Oh[2], Arim Lee[2], Dong-Hyun Kim[2], Oryun Bae[2] and Hee-Seock Lee[2,a]

[1]*Division of Advanced Nuclear Engineering, POSTECH, Pohang 37673, Republic of Korea*
[2]*Pohang Accelerator Laboratory, POSTECH, Pohang 37673, Republic of Korea*

Abstract. In this study, results of the experimental study on the depth profiles of production yields of $^{206,205,204,203,202}Bi$ radio-nuclei in the natural Pb target irradiated by a 100-MeV proton beam are presented. Irradiation was performed at proton linac facility (KOMAC) in Korea. The target, irradiated by 100-MeV protons, was arranged in a stack consisting of natural Pb, Al, Au foils and Pb plates. The proton beam intensity was determined by activation analysis method using $^{27}Al(p, 3p1n)^{24}Na$, $^{197}Au(p, p1n)^{196}Au$, and $^{197}Au(p, p3n)^{194}Au$ monitor reactions and also using dosimetry method by a Gafchromic film. The production yields of produced Bi radio-nuclei in the natural Pb foils and monitor reactions were measured by gamma-ray spectroscopy. Monte Carlo simulations were performed by FLUKA, PHITS, and MCNPX codes and compared with the measurements in order to verify validity of physical models and nuclear data libraries in the Monte Carlo codes. A fairly good agreement was observed between the present experimental data and the simulations by FLUKA, PHITS, and MCNPX. However, physical models and the nuclear data relevant to the end of range of protons in the codes need to be improved.

1 Introduction

The development of high-intensity and high-energy accelerator is gaining interest in Korea, in recent years. Rare Isotope Science Project (RISP), Pohang Accelerator Laboratory X-ray Free Electron Laser (PAL-XFEL), Korea Heavy Ion Medical Accelerator (KHIMA) facilities are being designed as well as Korea Multi-purpose Accelerator Complex (KOMAC) and Korea National Cancer Center (KNCC) which they have started their operations for various applications. In the frame of above projects, for the safety issue and shielding analysis of the accelerator facility, extensive studies including experimental and simulation on the production yields of residual nuclei induced in the accelerator materials such as Pb, Cu and Bi by protons and heavy ions are in progress.
Many experiment studies have been performed on the production yields of radio-nuclei induced in natural Pb target by a wide range of proton energies in the recent years. However, in case of 100-MeV proton energy few experiments have been performed so far [1,2].
In this work, we extend experimentally and theoretically earlier works. We measured production yields of $^{nat}Pb(p, xn)^{206,205,204,203,202}Bi$ radio-nuclei at three different depth points of the ^{nat}Pb target irradiated by the 100-MeV protons. The measured data were compared with the results simulated using the FLUKA, PHITS, and MCNPX codes to verify validity of nuclear models and nuclear data libraries.

[a] Corresponding author: lee@postech.ac.kr

2 Material and method

2.1 Target and irradiation conditions

A layout of the experimental set-up, designed by simpleGeo 4.3 software, is illustrated in Figure 1. Target was arranged in a stack consisting of an Au (30 μm), Al (100 and 200 μm) and ^{nat}Pb (125 μm) foils together with natural Pb (5 mm) plates. Natural Pb activation foils were placed between two ^{nat}Pb foils to avoid cross contamination and recoil effects. Similar to the ^{nat}Pb activation foils discussed above, Al monitor foils also were placed between two Al foils. The size of the target was about 5 cm × 5 cm. Natural Pb plates were inserted between Pb activation foils to degrade proton beam energy. The total thickness of target was considered to be higher than the range of proton beam.
The irradiation was performed at the high-intensity 100-MeV proton linac facility (KOMAC). The target was irradiated during 3.4 minutes with 103-MeV protons. The beam shape at the target can be described by Gaussian distribution in both x and y coordinates. The initial proton energy was calibrated to be 103 ± 0.13 MeV using multi-energy degrading techniques [3].
After irradiation, the gamma-ray spectra of Au, Al, and Pb foils were measured using an HPGe detector with a relative efficiency of 15% at 5 cm from the surface of detector.

Figure 1. A layout of the experimental set-up, designed by simpleGeo 4.3 software.

2.2 Simulation of experiment

Monte Carlo codes FLUKA, version 2c.0, PHITS, version 2.64, and MCNPX, version 2.7 were used to simulate the experiment. The production yields of Bi radio-nuclei were estimated using MCNPX and activation code SP-FISPACT, version 2010 and PHITS and activation code DCHAIN, version 2001. The measured and simulated production yields of Bi radio-nuclei were compared after end of irradiation. The model of simulation was quite similar to experiment, consisting of the proton beam, Al_4Be_6 window and target.

3 Results and discussion

3.1. Measurement of beam intensity

Beam intensity were measured using the ^{27}Al(p, 3p1n)^{24}Na, ^{197}Au(p, p1n)^{196}Au, ^{197}Au(p, p3n)^{194}Au monitor reactions. And also a Gafchromic, film with a dimension of 7 cm × 7 cm which was placed in front of the target to measure beam intensity. Table 1 shows results of measured beam intensity. It can be seen that a satisfactory agreement was also observed between the beam intensity measurements using activation analysis method and the dosimetry method using Gafchromic film

3.2 Depth profiles of production yields of ^{nat}Pb(p, xn)^{206,205,204,203,202}Bi

In case of ^{206}Bi, production yields of ^{206}Bi ($T_{1/2}$ = 6.243 d) were detected using photon peaks at 803.1 and 881.01 keV that were well-defined in the spectrum of each ^{nat}Pb irradiated foils. The measured production yields include the contribution from the decay of the simultaneously produced short half-life isomeric state ($T_{1/2}$ = 0.89 ms, 100% IT). Figure 2 shows the measured depth profile of production yield of ^{nat}Pb(p, xn)^{206}Bi reaction together with the results of simulations. It can be seen that from Figure 2 a good agreement was observed between the measured production yield and simulated production yield up to the depth of 0.44 range of protons in the natural Pb target. However, MCNPX and PHITS slightly overestimate the measurement within a factor of 1.5 at the depth of 0.44 range of protons which is still good agreement. At the depth of 0.83 range of protons, considerable difference was observed between the results of simulations and the measurement.

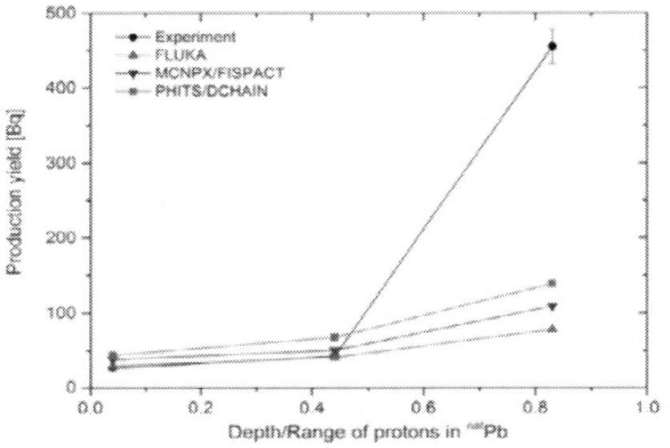

Figure 2. Depth profile of production yield of ^{nat}Pb(p, xn)^{206}Bi

Table 1. Measurement of beam intensity

Method		Ep [MeV]	Beam intensity [protons/s]
Activation analysis	^{197}Au(p, pn)	100.9±0.1	7.72E+09
	^{197}Au(p, p3n)	100.9±0.1	8.05E+09
	^{27}Al(p, 3p1n)	100.9±0.1	8.93E+09
Dosimetry	Gafchromic	---	7.30E09

4 Conclusions

Natural lead target was irradiated with 100-MeV proton beam and depth profiles of production yields of ^{206,205,204,203,202}Bi radio-nuclei were measured using gamma-ray spectroscopy. Experimental data were compared with results of simulation using FLUKA, PHITS and MCNPX. Generally, a good agreement was observed between the measurements and simulations at the beginning of range of protons in the natural Pb target and a satisfactory agreement was observed between the measurements and simulations at the middle of range of protons in the natural Pb target. Lastly, a considerable difference between the results of simulations and measurement was observed. This work demonstrated that physical models and nuclear data relevant to the end of range of protons in the all codes need to be improved.

References

1. Y.E. Titarenko, V.F. Bataev, R.D. Mulambetov, V. M. Zhivun, V.S. Barashenkov, S.G. Mashnik, Y.N. Shubin, A.V. Ignatyuk, Nucl. Instr. Meth. Phys. Res. A **562** (2006)
2. M. Gloris, R. Michel, F. Sudbrock, U. Herpers, P. Malmborg, B. Holmqvist, Nucl. Instr. Meth. Phys. Res. A **463** (2001)
3. K.R. Kim, N.S. Jung, B.G. Park, J. Nucl. Sci. Technol, **5** (2008)

Experimental approach to measure thick target neutron yields induced by heavy ions for shielding

N.D. Trinh[1,*], M. Fadil[1,*], M. Lewitowicz[1,*], C. Brouillard[1], T. Clerc[1], S. Damoy[1], V. Desmezières[1], E. Dessay[1], M. Dupuis[1], G.F. Grinyer[1], J. Grinyer[1], B.Jacquot[1], X. Ledoux[1], A. Madeline[1], N. Menard[1], M. Michel[1], V. Morel[1], F. Porée[1], B. Rannou[1] and A. Savalle[1]

[1] *Grand Accélérateur National d'Ions Lourds, Bd Henri Becquerel, 14076 Caen, France*

Abstract. Double differential (angular and energy) neutron distributions were measured using an activation foil technique. Reactions were induced by impinging two low-energy heavy-ion beams accelerated with the GANIL CSS1 cyclotron: (^{36}S (12 MeV/u) and ^{208}Pb (6.25 MeV/u)) onto thick ^{nat}Cu targets. Results have been compared to Monte-Carlo calculations from two codes (PHITS and FLUKA) for the purpose of benchmarking radiation protection and shielding requirements. This comparison suggests a disagreement between calculations and experiment, particularly for high-energy neutrons.

1 Introduction

Nowadays, particle accelerators are used in many different fields from fundamental research to applications in medicine and in industry. In order to maintain a good level of knowledge concerning the radiation protection and shielding requirements and to answer to emerging needs in these fields with next-generation particle accelerators, characterization of potential radiation hazards like neutron yields induced by particle beams must be well known [1,2]. At GANIL, operation of its existing cyclotrons and the recent installation of a new high-intensity LINAC for SPIRAL 2 imply that any potential safety risk must be accurately evaluated.

Neutron yields are usually estimated from Monte-Carlo transport codes such as FLUKA or PHITS. Due to the lack of the data concerning neutron yields induced by heavy-ion reactions around 10 MeV/u [1,2], we performed an experimental campaign at GANIL with the purpose of benchmarking the physics models in transport codes around this low energy limit.

2 Experimental method

2.1 Experimental setup

Two experiments were carried out in 2015 using the GANIL CSS1 cyclotron with two low-energy beams: ^{36}S at 12 MeV/u and ^{208}Pb at 6.25 MeV/u.

Since copper is one of the prominent elements in accelerators and beam-line components, a thick ^{nat}Cu target was chosen to study neutron yields under nominal operating conditions. The target was 5*5 cm^2 and had a thickness of 0.1cm to completely stop the incident beams.

To measure neutrons, the activation-foil technique was employed. During the irradiation of the target, the resulting neutrons induced various threshold reactions in a series of activation foils that were placed around the target position (see Fig. 1). In this work, indium, nickel, cobalt, iron, aluminum, niobium and bismuth were chosen as activation foils. This choice permits neutron energies from 0.5 MeV to 70 MeV to be measured. The foils were 2.5*2.5 cm^2 with a thickness of 0.1 cm in order to minimize the attenuation and the diffusion of neutrons. After irradiation of the ^{nat}Cu target, the measurement of the resulting activities in the activation foils was used to reconstruct the incident neutron spectra [3,4,5] using an unfolding method.

The experimental setup is shown in Fig. 1. The ^{nat}Cu target is mounted on a 1.6 cm thick steel backing. On the other side of this backing, activation foils were placed from 0° to 75° on an aluminum support. At each position, different foils were stacked in a sandwich form.

* To whom all correspondence should be addressed: trinh@ganil.fr, fadil@ganil.fr and lewitowicz@ganil.fr

 DOI: 10.1051/epjconf/201715301018

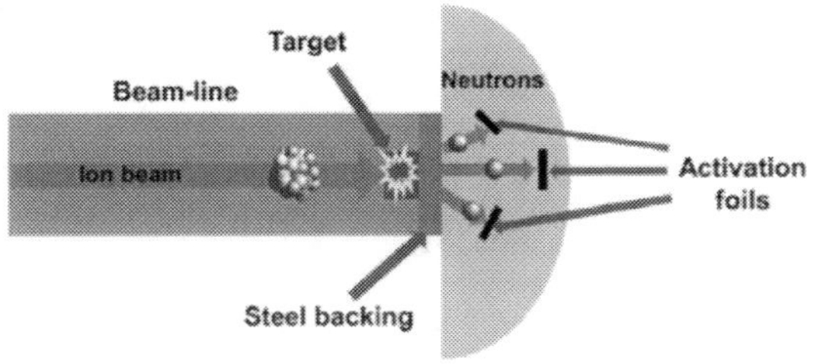

Figure 1: Experimental setup for the ion-beam irradiation and foil-activation measurements

In Table 1, a summary of the beam characteristics for both experiments is provided. During the irradiation, the beam intensity was monitored and measured in real time.

Table 1: Beam characteristics

Beam	Energy (MeV/u)	Irradiation time (h)	Average intensity (pps)
^{36}S	12	12	$1*10^{12}$
^{208}Pb	6.25	15	$1*10^{11}$

2.2 Analysis method for the activation foils

2.2.1 Measurement and unfolding method

Two Ortec HPGe detectors were used to measure the γ-ray spectra from the activation foils. The measurements were performed successively and the ordering was chosen according to the half-life of each nuclide. Gamma-ray spectra were recorded using the Interwinner software and were analyzed manually with the Fityk program [6]. After the activity for each isotope of interest in the foils was determined, the corresponding reaction rate was calculated.

In this work, we developed an unfolding program to reconstruct the incident neutron spectrum. This program adopts the SAND-II algorithm that is coupled to an applied Monte-Carlo technique [7,8,9]. The input of the unfolding program requires reaction rates, cross sections for each isotope, as well as their relative uncertainties. The cross sections used were those from the evaluated neutron cross-section libraries EAF-2010 [10].

2.2.2 Uncertainties

There were three main sources of uncertainties:

1. The uncertainty on the measured activities is about 5% - 10% due mainly to the efficiency calibration of the HPGe detector.
2. The uncertainty of the beam intensity and total irradiation time was evaluated to be about 5% for these two experiments.
3. Uncertainties associated with the cross sections range from a few to up to 10% for high energy threshold reactions [10].

3 Monte-Carlo simulation codes

Two nuclear Monte-Carlo transport codes, PHITS and FLUKA, were used to simulate the neutron spectrum. Comparison to our experimental data then provides a means to benchmark the simulated results.

3.1 FLUKA

FLUKA (FLUktuierende KAskade) is a Monte-Carlo transport code developed by INFN and CERN. In FLUKA, the Boltzmann Master Equation (BME) model is used to simulate the dynamic processes of reactions from ions lower than 100 MeV/u [11, 12]. Fluka also uses the Weisskopf-Ewing model to compute the resulting compound nuclei and their evaporation products [11, 12].

3.2 PHITS

PHITS (Particle and Heavy Ion Transport code System) is a Monte-Carlo code developed by a Japanese collaboration between JAEA, RIST, KEK and other institutes. This code uses the JAERI Quantum Molecular Dynamics (JQMD) model to simulate the dynamic processes of reactions. JQMD can be used to simulate induced heavy-ion reactions from 10 MeV/u to 3.5 GeV/u. A Generalized Evaporation Model (GEM) is adopted to simulate the evaporation of the compound nuclei [13].

4 Results

In this section, we compare the unfolded spectra to the simulated ones obtained from PHITS and FLUKA for neutrons emitted at 0° with respect to the beam axis.

The unfolded spectra of the ^{36}S (12 MeV/u) + Cu reaction and comparison to FLUKA and PHITS estimates are shown in Fig.2. In this experiment, the highest threshold reaction observed was at 40 MeV. Above this energy, threshold reactions were not statistically observable at this beam intensity (10^{12} pps). The uncertainty of unfolded spectra is about 15 % below this threshold (40 MeV).

For low neutron energies (<20 MeV), FLUKA agrees rather well with the experimental spectrum while PHITS underestimates the experimental data by a factor of 2. The JQMD model does not simulate heavy-ion reactions below 10 MeV/u. This could explain why PHITS underestimates the low-energy neutrons that are mainly produced from low-energy heavy ions that slow down in the target.

At higher neutron energies, there is good agreement between FLUKA and PHITS. However, these calculations both overestimate the experimental data by up to a factor of 3 at 40 MeV.

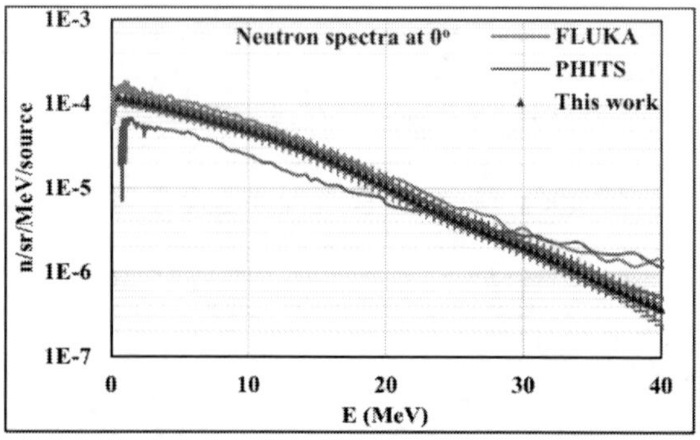

Figure 2: Unfolded spectrum of ^{36}S (12MeV/u) + ^{nat}Cu compared to FLUKA and PHITS simulations.

The unfolded spectrum of the second experiment (^{208}Pb (6.25 MeV/u) + ^{nat}Cu) and its comparison to FLUKA and PHITS calculations is shown in Fig.3. In this experiment, the highest threshold reaction measured was about 15 MeV due also to statistics. Uncertainties in this experiment were also about 15%.

For this reaction FLUKA calculates a spectrum that is about 4 times larger than the experimental unfolded one. In this case, PHITS was not used to simulate this reaction because of its inherent limitations at low energies as described above.

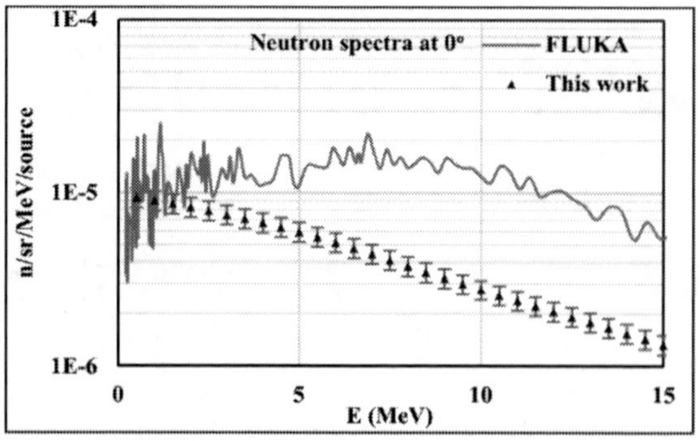

Figure 3: Unfolded spectrum of ^{208}Pb (6.25 MeV/u) + ^{nat}Cu compared to the FLUKA simulation.

These results show clear limitations of transport codes to estimate neutron yields from the interactions of low-energy and heavy ions, especially for high-energy neutrons.

5 Conclusion

In this work, we measured the double differential neutron spectrum from a thick ^{nat}Cu target bombarded by ^{36}S (12 MeV/u) and ^{208}Pb (6.25 MeV/u) beams using the foil-activation method. An unfolding program using the SAND-II algorithm coupled to a Monte-Carlo method has been developed to reconstruct the incident neutron spectra. Two Monte-Carlo transport codes (FLUKA and PHITS) were used to calculate the neutron spectrum and to benchmark them with experiment.

For the ^{36}S (12 MeV/u) + ^{nat}Cu reaction, PHITS underestimates the results for low-energy neutrons (<20 MeV) at 0° while FLUKA seems to be in good agreement with the experiment. At higher energies, both codes agree with each other but systematically overestimate the experimental data by up to a factor of 3.

For the ^{208}Pb (6.25 MeV/u) + ^{nat}Cu reaction, the spectrum at 0° calculated by FLUKA is higher than unfolded experimental one by up to a factor of 3.

The results presented here are still preliminary and are being checked and improved by our team. After the final verification, we will apply the method described here to other angles. We are also studying the physics models at this low-energy limit for heavy ions in order to better understand the disagreement between calculation and experiment.

References

[1] C. Sunil et al, Thick target neutron yield from 145 MeV ^{19}F+^{27}Al system, Nuclear Instruments and Methods in Physics Research A 721 (2013) 21-25.
[2] V. Suman et al, Thick target double differential neutron energy distribution from ^{12}C + ^{27}Al at 115 MeV, Nuclear Instruments and Methods in Physics Research A 800 (2015) 29-33.
[3] G. Lhersonneau et al, Neutron yield from carbon, light- and heavy-water thick targets irradiated by 40 MeV deuterons, Nuclear Instruments and Methods in Physics Research A 603 (2009) 228-235.
[4] G. Lhersonneau et al, A facility for fast-neutron irradiations at Jyväskylä and its use for nuclide cross-section measurements in fission, Nuclear Instruments and Methods in Physics Research A 698 (2013) 224-233.
[5] G. F. Knoll, Radiation Detection and Measurement – Third edition, John Wiley & Sons, Inc, 2000.
[6] M. Wojdyr, Fityk: a general-purpose peak fitting program, J. Appl. Cryst. (2010). 43, 1126-1128.
[7] W. N. McElroy et al, SAND-II Neutron Flux Spectra Determinations by Multiple Foil Activation Iterative Method. RSIC Computer Code Collection CCC-112. Oak Ridge National Laboratory, Information Center, May 1969.

[8] É. M. Zsolnay and E. J. Szondi, Neutron spectrum determination by multiple foil activation method, Nuclear Training Reactor of the Technical University Budapest, 1982.
[9] S. Maeda et al, Fundamental Study on Neutron Spectrum Unfolding using Maximum Entropy and Maximum Likelihood Method, Progress in Nuclear science and technology, Vol. 1, p.233-236 (2011).
[10] J.-Ch. Sublet et al, The European Activation File: EAF-2010 neutron-induced cross section library, EASY Documentation Series CCFE-R (10) 05
[11] A. Ferrari et al, FLUKA: a multi-particle transport code, CERN-2005-10 (2005), INFN/TC_05/11, SLAC-R-773.
[12] T.T. Böhlen et al, The FLUKA Code: Developments and Challenges for High Energy and Medical Applications, Nuclear Data Sheets 120, 211-214 (2014).
[13] T. Sato et al, Particle and Heavy Ion Transport Code System PHITS, Version 2.52, J. Nucl. Sci. Technol. 50:9, 913-923 (2013).

Measurement of neutron spectra for photonuclear reaction with linearly polarized photons

Yoichi Kirihara[1,*], *Toshiro* Itoga[2], *Toshiya* Sanami[1,4,5], *Hiroshi* Nakashima[3,4], *Yoshihito* Namito[1,5], *Shuji* Miyamoto[6], *Akinori* Takemoto[6], *Masashi* Yamaguchi[6], and *Yoshihiro* Asano[4,6]

[1]High Energy Accelerator Research Organization, 1-1 Oho, Tsukuba, Ibaraki 305-0801, Japan
[2]JASRI, 1-1-1 Kouto, Sayo-cho, Sayo-gun, Hyogo 679-5198, Japan
[3]Japan Atomic Energy Agency, 2-4 Shirakata, Tokai-mura, Naka-gun, Ibaraki 319-1195, Japan
[4]RIKEN SPring-8 Center, 1-1-1 Kouto, Sayo-cho, Sayo-gun, Hyogo 679-5198, Japan
[5]The Graduate University for Advanced Studies (SOKENDAI), 1-1 Oho, Tsukuba, Ibaraki 305-0801, Japan
[6]University of Hyogo, 1-1-2, Koto, Kamigori-cho, Ako-gun, Hyogo 678-1205, Japan

Abstract. Spectra of neutrons produced by a photonuclear reaction from a ^{197}Au target were measured using 16.95 MeV linearly and circularly polarized photon beams at NewSUBARU-BL01 using a time-of-flight method. The difference in the neutron spectra between the cases of a linearly and circularly polarized photon was measured. The difference in the neutron yield increased with the neutron energy and was approximately threefold at the maximum neutron energy. In a direction perpendicular to that of the linear polarization, the neutron yields decreased as the neutron energy increased.

1 Introduction

In high-energy electron-accelerator facilities, electrons hitting components of the accelerators produce high-energy photons through bremsstrahlung. The photons generate fast neutrons through photo-nuclear reactions. The neutrons induce issues in radiation protection, for example, increase of ambient dose behind a shielding wall and production of activity in air and surrounding materials. To evaluate these issues, the data on not only neutron production but also neutron energy and angular distribution are important for photo-nuclear reactions.

For instance at the International Linear Collider (ILC) facility, polarized photons with hundreds of kW impinge on a target to produce positrons [1]. Thus, from the target, significant amount of neutrons are produced as a result of photo-nuclear reactions. The neutron production form the reaction should be estimated reasonably well for radiation shielding design of the ILC.

Experimental data for neutron production of photo-nuclear reaction, until now, have been taken intensively by Livermore and Saclay groups for giant dipole resonance (GDR) region [2]. Recently, the anisotropy of neutron azimuthal angular distributions with linearly polarized photon beams was measured by Horikawa et al. in GDR region [3].

Nevertheless, the experimental data on neutron energy and angular distribution are still not sufficient since most of the data were obtained with low intensity photon sources. Furthermore, effects on neutron energy distribution due to polarization are not obvious since Horikawa et al. provided data based on relative neutron yield [3]. They didn't take data for un-polarized photon to reveal enhancement on neutron production due to polarization.

On the other hand, reaction model of photo nuclear reaction was proposed by Agodi [4]. Thee consequence of the model was partially confirmed by the data of Horikawa et al. [3]. This model, however, has not been included in the particle-transport codes that are used for shielding design of high-energy accelerators. Thus, to develop and evaluate a reaction model to treat the effect of polarization in particle transport codes, experimental data are desired on neutron energy and angular distribution for linear and circular polarization photons.

Recently, intense mono-energy photon beam up to 70 MeV become available at NewSUBARU-BL01. University of Hyogo, Japan, using laser Compton scattering (LCS) technique [5, 6]. The polarization of LCS photon can be easily controlled by changing polarization of laser system. The usage of the LCS photon enables us to take systematic data of neutron energy and angular distribution for mono-energetic photons with considering effects from polarization.

In this study, as a part of the systematic data taking, we obtain the data of neutron energy and angular dependency to reveal effect of polarization. We use not only linearly polarized photon beam but also circularly polarized to obtain difference between them. To obtain absolute neutron spectra for the photons, incident number of photons and neutron detector efficiency are carefully determined experimentally, which are not considered in ref. 2. As a material of the neutron production target, ^{197}Au was chosen since it showed large polarization dependence [3].

* Corresponding author: kyoichi@post.kek.jp

2. Experimental

2.1. Experimental set up

An overview of BL01 and how to obtained LCS photons are described elsewhere [3, 5], only outline is summarized in this section.

Figure 1 shows schematic drawing of BL01. An LCS photon beam was generated by scattering laser photons and electron bunches in a storage ring. The storage ring was operated in the single bunch with top-up mode. The typical stored beam current was 30 mA during the experiment. The electron energy was 982.4 MeV. An $NdYVO_4$ laser system provided laser photons with a wavelength for the laser photon of 1.064 μm. The laser power was 20 W. Under these conditions, the maximum energy of the LCS photon beam was 16.95 MeV at a scattering angle of 180°. The LCS photon beam was collimated using a 3-mm-inner-diameter collimator in an enclosure of the storage ring, 13.82 m away from the collision point, and a 2-mm-inner-diameter collimator in optics hutch 1, 16.67 m from the collision point. After passing through these collimators, the beam is led to an optics hutch 2.

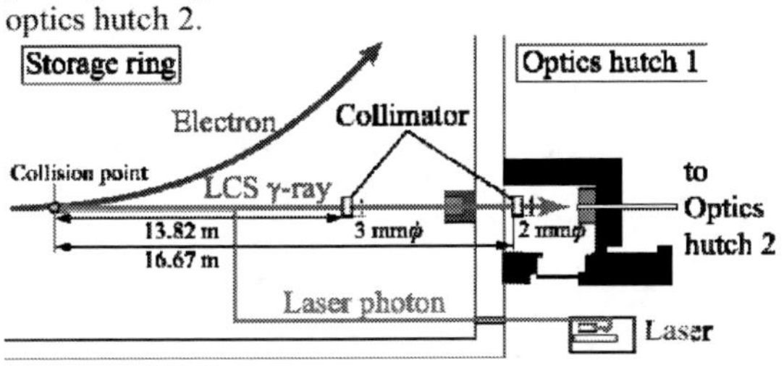

Fig. 1. Schematic of NewSUBARU-BL01.

Figure 2 shows the experimental setup in the optics hutch 2. At the entrance of the hutch, a plastic scintillator plate with thickness of 5 mm coupled with photo-multiplier tube was placed on the path of LCS photon beam to monitor relative intensity of the beam (PLS in Figure 2). 1.4 m downstream from the PLS, a cylindrical shaped ^{197}Au target was placed. The axis of the target was parallel to the beam line. The diameter and thickness of the target were 10 and 40 mm, respectively. Although the neutron attenuates within the target, this attenuation was not considered in this paper.

We used three NE213 organic liquid scintillators, 12.7 cm in diameter and 12.7 cm in thickness, coupled with photo-multiplier tubes as neutron detectors. The positions of the NE213 detectors are shown in Figure 2 and listed in Table 1. The pulse height NE213 detectors were calibrated using γ-ray sources prior to the beam experiment. The detection efficiencies for each detector were determined as a function of the pulse height bias (see section 3.3 in detail).

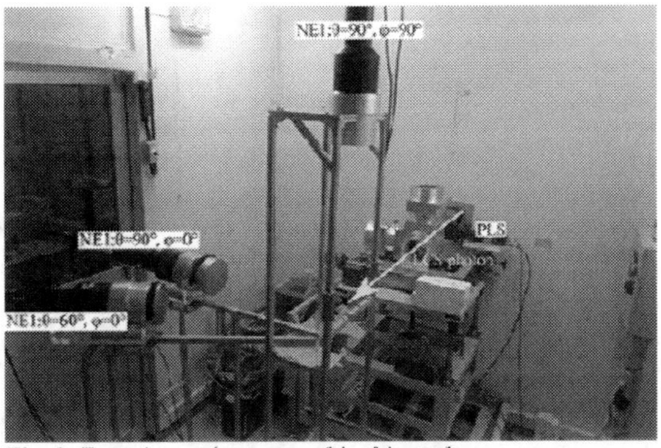

Fig. 2. Experimental setup used in this study.

Table 1. Positions of the NE213 detectors.

Detector label	Flight path from target [cm]	Polar angle θ [degrees]	Azimuthal angle ϕ [degrees]
NE1	70.9	60	0
NE2	64.7	90	0
NE3	65.9	90	90

2.2. Readout electronics

Figure 3 shows a block diagram of the readout electronics. Analogue signals from NE213 detectors were divided to three to obtain pulse integral data (PH) with total and slow gates for pulse shape discrimination (PSD) and time difference from RF signal for time-of-flight (TOF) measurement. Only beam coincident events were selected using 4 fold 1 veto coincidence module to generate the gate and TDC start signal.

The RF signal was also used to trigger laser to generate LCS photon [7] (not shown in this figure) and TDC stop signal.

The analogue signal from PLS was divided to two signals to obtain total pulse integral.

The data acquisition system is A3000 system that consists of QDC of A3200, TDC of A3300 modules and VME controller (Nikiglass Co.Ltd.). The modules were operated free-run list mode that accumulates data from these modules with adding time stamp with XX ns resolution event by event. The time stamp was used to find out a pair of PH and TOF data in offline analysys. The data were stored in HDD of Linux PC that was connected with VME controller via gigabit Ethernet interface.

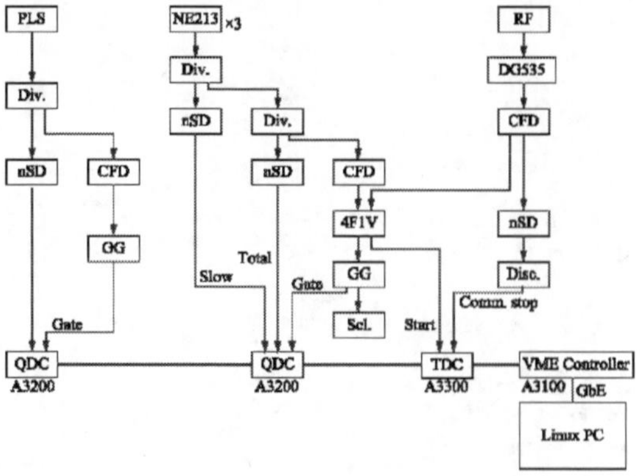

Fig. 3. Block diagram of the electronics circuit. PLS: plastic scintillator; NE213: organic liquid scintillator; Div.: divider; DG535: digital delay; nSD: nano second delay; CFD: constant fraction discriminator; 4F1V: 4-fold 1-veto coincidence; GG: gate and delay generator; Disc.: discriminator; Scl.: scaler; QDC: charge-to-digital converter; TDC: time-to-digital converter; GbE: giga bit ethernet.

2.3. Data taking procedure

In prior to neutron data taking, at first, alignment of the collimators and the target was carefully checked using the LCS photon beam. For collimator alignment, gadolinium orthosilicate sintillator (GSO), rectangular parallel-piped shape with 76 mm × 76 mm × 180 mm was placed instead of the target. The photon intensity was monitored using the GSO with changing positions and angles of the collimators. The positions and angles were chosen to provide maximum count rate of the LCS photon.

To confirm the target position, a piece of imaging plate was attached on the front face of the target. The beam image of the plate was read by a scanner. The position of image was exactly at the center of the target cylinder. The beam size at the target was 2.3 mm in diameter.

The polarization of the laser was confirmed using transmitted laser. The results were that the polarization degree was 99.3%.

After these confirmations, neutron measurements were performed to linear and circular polarized beams. The laser system provides vertical polarized beam. The circular polarized beam was prepared using 1/4 λ plate in front of the laser. The running time of the measurement was 47 hours.

For each beam condition, count rate of PLS was calibrated using the GSO. Figure 4 shows setup for the calibration. The GSO detector was placed on the beam line to count.

Stability of neutron detectors was also checked using photon check sources.

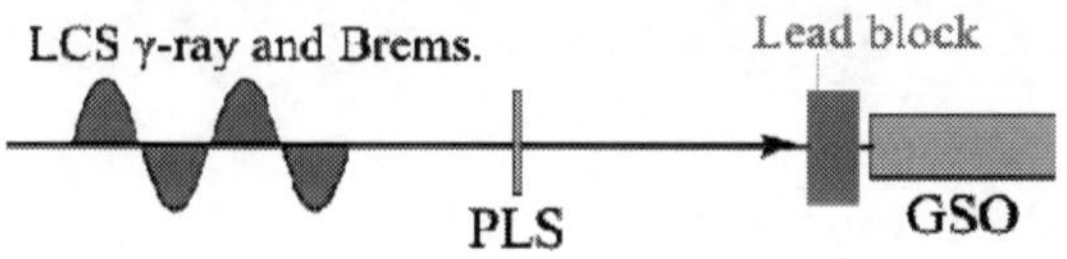

Fig. 4. Experimental geometry used in that the PLS detector is calibrated with the GSO detector.

3. Data analysis

3.1. Number of LCS photon

To obtain the number of photons, the count rates of the PLS was calibrated with the GSO. Figure 4 shows experimental setup for this calibration. As shown in this figure, a lead block with various thicknesses was placed in front of the GSO to multi-gate pileup of pulses.

Figure 5 (a) and (b) show pulse height spectra of the GSO and the PLS for laser on and off condition. The GSO pulse height spectrum is for 5 cm in thickness lead. From the data for laser off, contribution from bremsstrahlung can be seen since the electron bunches interact in the storage ring. Therefore, the laser off events was subtracted from laser on to obtain LCS photon contribution. After this subtraction, the counts of the PLS and the GSO were summed for the pulse height channels from 100 ch to 200 ch and from 100 ch to 8192 ch, respectively.

Figure 6 shows thickness of the lead block versus a ratio of count for the GSO to that for the PLS. The data were fitted by an exponential function. As a result, an attenuation coefficient is 0.62 in this measurement. Figure 7 shows the counts for the PLS versus that for the GSO. The counts of the GSO were evaluated with the attenuation coefficient. The data were fitted by a linear function, and the conversion factor from the counts for the PLS detector to the counts for the GSO detector was 111.02±1.15.

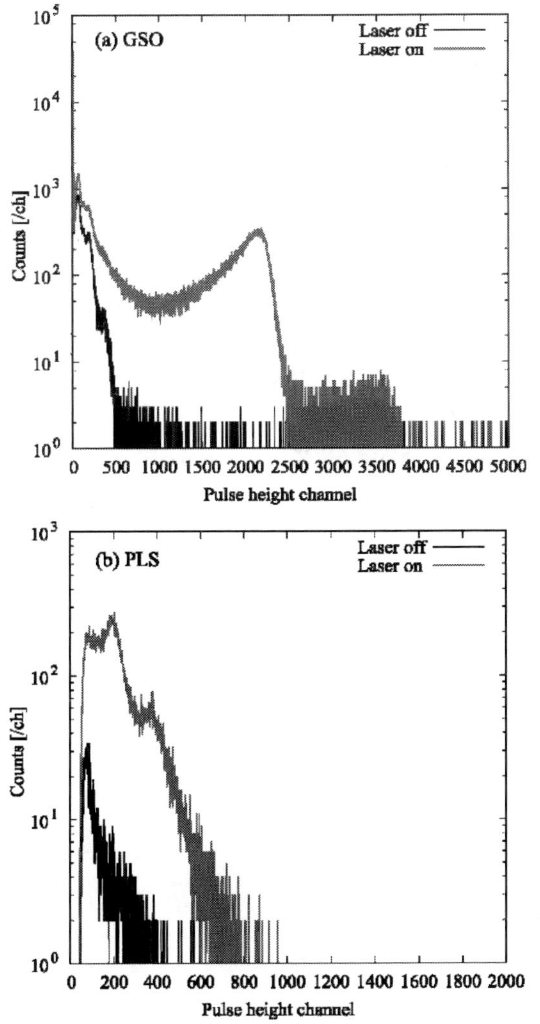

Fig. 5. Pulse height spectra of (a) the PLS detector and (b) the GSO detector for a lead block of 5 cm thickness. Red line: laser on; black line: laser off.

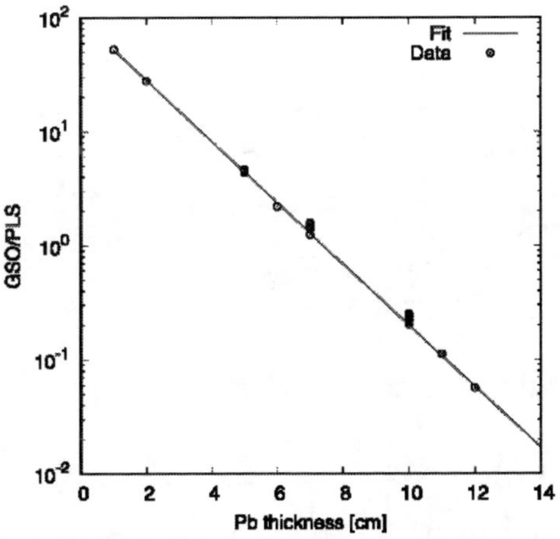

Fig. 6. Thickness of the lead block versus ratio of the count for the GSO to those for the PLS. Red line is the fit by the exponential function.

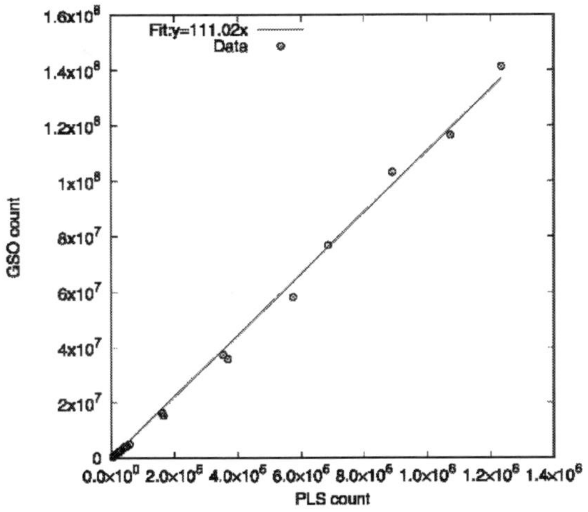

Fig. 7. Counts for the PLS detector versus counts for the GSO detector. The red line is the fit by the linear function.

3.2. Neutron and gamma separation

Figure 8 shows a 2D plot for the total component of the pulse-height versus the slow component of the same. The events having data greater than 0.25 MeVee (equivalent to a half of the Cs bias) were chosen. Clear neutron-gamma separation was obtained for three detectors.

Figure 9 shows the TOF spectrum of raw data and data with neutron-gamma separation. The neutron component and the peak of the γ-ray events are clearly separated. From the FWHM of the γ-ray peak, we determined that the time resolution was 0.81 ns.

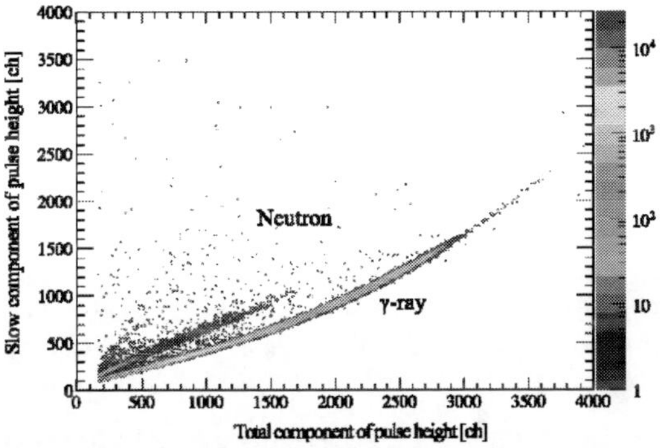

Fig. 8. 2D plot for a total component of the pulse height versus a slow component of the same for the NE1 with the linearly polarized photon.

DOI: 10.1051/epjconf/201715301019

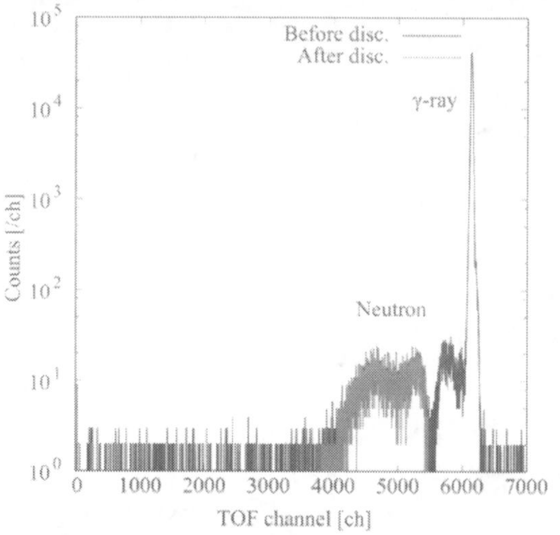

Fig. 9. Time-of-flight spectra for the NE3 detector. Black line: TOF spectrum of raw data; red line: TOF spectrum of neutron event after the pulse-shape discrimination.

3.3. Neutron detection efficiency

The efficiency of the NE213 detectors required to transforming the TOF spectrum into the energy spectrum was derived using a SCINFUL-QMD code [8]. Generally, the efficiency of NE213 detectors in low energy region is difficult to estimate purely by calculation because of detector-dependent pulse-height resolution. Thus, the efficiencies of the NE213 detectors for the energies ≤3 MeV were derived experimentally using a ^{252}Cf neutron source, the spectrum of which is well known. To estimate the efficiencies of the NE213 detectors, the neutron spectra of the spontaneous fission of ^{252}Cf were measured using a TOF method, and compared with the evaluated spectrum by the National Bureau of Standards (NBS) [9]. The NBS spectrum $X_{Cf}(E)$ is obtained as

$$X_{Cf}(E) = 0.6672(E)^{\frac{1}{2}} \exp\left(-\frac{E}{1.42}\right)\mu(E) \qquad (1)$$

where the values of $\mu(E)$ are listed in Table 2 [10]. The relative efficiency was deduced by dividing the observed energy distribution of ^{252}Cf fission neutrons $Y(E)$ by $X_{Cf}(E)$:

$$\varepsilon(E) = C\frac{Y(E)}{X_{Cf}(E)}, \qquad (2)$$

where C is the normalization factor. Figure 10 shows the detector efficiencies determined by ^{252}Cf measurement and the SCINFUL-QMD code for a 0.25 MeVee bias. To estimate the efficiencies by ^{252}Cf measurement for energies ≤3 MeV, the efficiencies taken by ^{252}Cf measurement were normalized with those determined by the SCINFUL-QMD code between 3 and 5 MeV.

Table 2. The values of $\mu(E)$ in Eq (1) [10].

Energy interval [MeV]	$\mu(E)$
0-0.25	$1+1.20E-0.237$
0.25-0.8	$1-0.14E+0.098$
0.8-1.5	$1+0.24E-0.0332$
1.5-6.0	$1-0.00062E+0.0037$
6.0-20	$1.0\exp[-0.03(E-6.0)]$

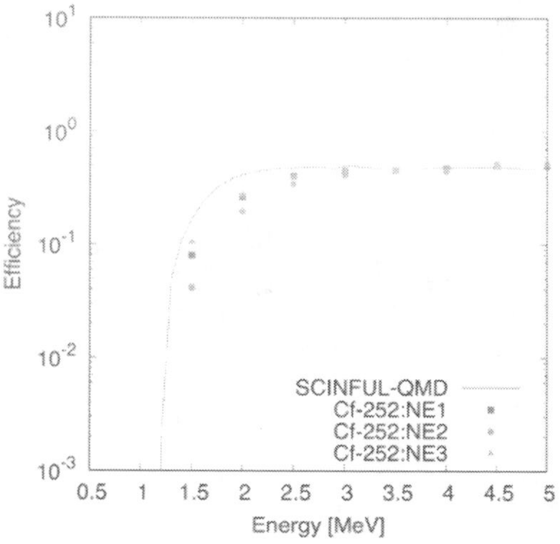

Fig. 10. Detector efficiencies determined by ^{252}Cf measurement and SCINFUL-QMD code for a 0.25 MeVee bias.

3.4. Uncertainty

The uncertainties were estimated as shown in Table 3. The uncertainty of the measured number of neutron was evaluated with the statistical uncertainty with each energy. The uncertainty of number of photon was evaluated with determination accuracy of the counts for the GSO to that for the PLS. The uncertainty of solid angle was evaluated with the uncertainty of the detector position. The uncertainty of detector efficiency was the uncertainty of SCINFUL-QMD code. The total uncertainty was estimated with a root mean square of these uncertainties.

Table 3. The values of uncertainties.

Uncertainty elements	Uncertainty [%] (1σ)
Uncertainty of the measured number of neutron	2-15
Uncertainty of number of photon	5
Uncertainty of solid angle	0.2
Uncertainty of detector efficiency	5
Total uncertainty	7-17

4 Results and discussion

4.1. Energy spectra

Figure 11 shows the energy spectra of neutron from the ^{197}Au target produced by the photonuclear reaction with the circularly and the linearly polarized photons. The spectra were divided with the photon flux and solid angles. At energies ≤4 MeV, the neutron yield decreased exponentially as the neutron energy increased. Between 4 MeV and 9 MeV, the neutron yield was flat.

For the linearly polarized photons, the spectrum, which is took at the same direction of the linear polarization (ϕ = 90°), is higher than the spectra of the other directions between neutron energies of 4 MeV and 9 MeV. This difference was due to the influence of the linear polarization, which was absent for the circularly polarized photons.

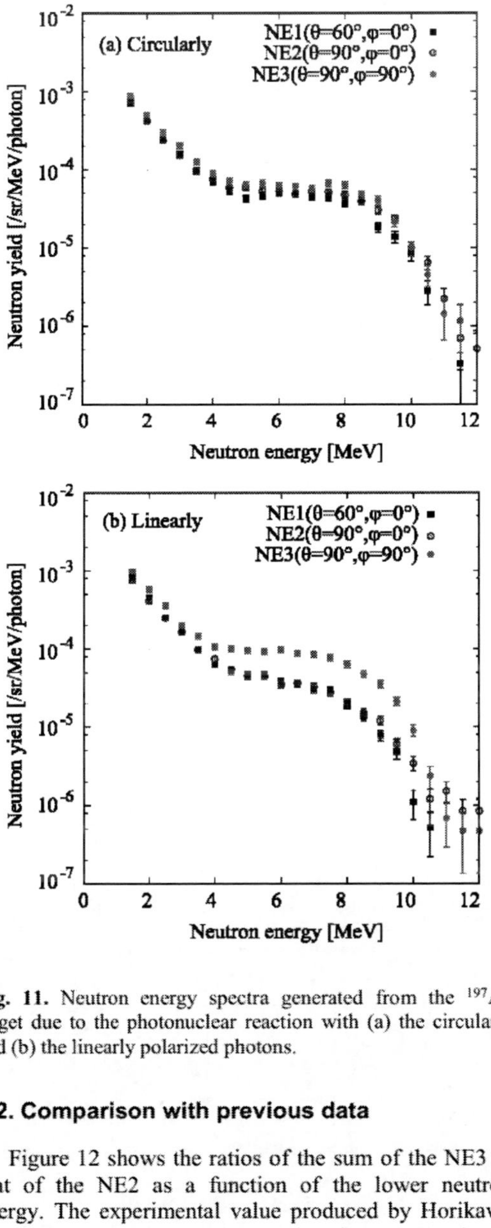

Fig. 11. Neutron energy spectra generated from the ^{197}Au target due to the photonuclear reaction with (a) the circularly and (b) the linearly polarized photons.

4.2. Comparison with previous data

Figure 12 shows the ratios of the sum of the NE3 to that of the NE2 as a function of the lower neutron energy. The experimental value produced by Horikawa et al. shows a black point in this figure. The lower energy of the experimental value produced by Horikawa et al. was evaluated from a TOF spectrum in their paper [3]. The ratios in this work reproduce the experimental value produced by Horikawa et al. roughly.

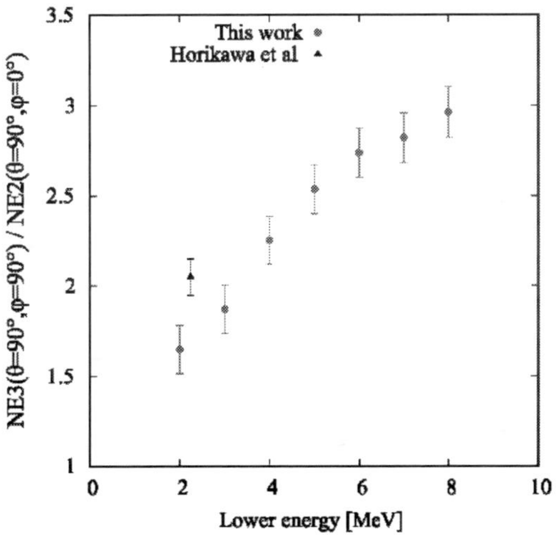

Fig. 12. Ratio of summation for the NE3 to that for the NE2 as a function of lower neutron energy. A black point is the experimental value by Horikawa et al.

4.3. Energy dependence

To estimate the energy dependency of ratios of the neutron yield, the ratios of the neutron yield of NE3 (ϕ = 90°) to that of NE2 (ϕ = 0°) were calculated, as shown in Figure 13. For the circularly polarized photons, the ratio was approximately constant. For the linearly polarized photons, the ratio increased with the neutron energy, up to approximately three times its initial value.

To estimate the energy dependency between the linearly polarized photon and the circularly polarized photon, ratios of the neutron yields of the linearly polarized photon to those of the circularly polarized photon were calculated, as shown in Figure 14. For the NE3, the ratio was greater than one for the entire neutron energy. For NE1 and NE2, both of which were set perpendicular to the direction of the linear polarization, the ratios decreased as the neutron energy increased.

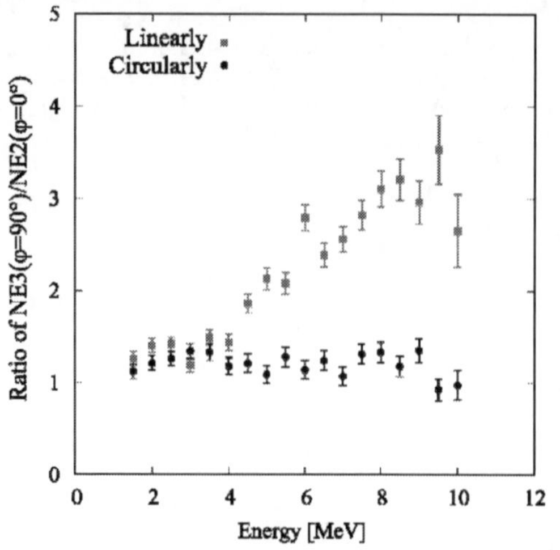

Fig. 13. Ratio of the neutron yield of NE3 to those of NE2.

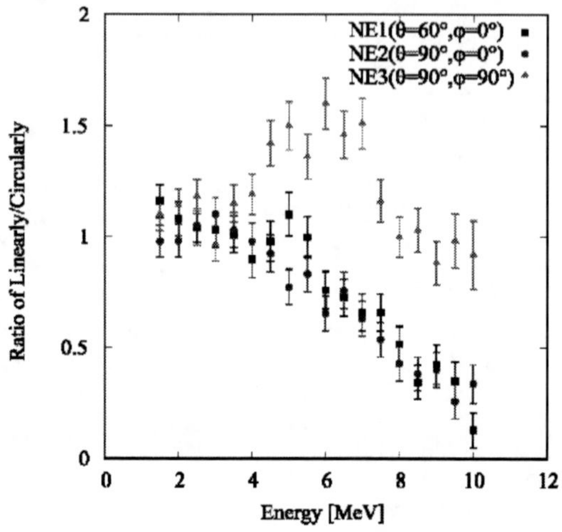

Fig. 14. Ratio of the neutron yield of the linearly polarized photons to that of the circularly polarized photons.

5 Conclusions

To estimate the influence of a photonuclear reaction with polarized photons, the neutron spectra generated from a ^{197}Au target by the photonuclear reaction caused by 16.95 MeV polarized LCS γ-ray beams were measured using a TOF method at NewSUBARU-BL01. The neutron yield that was measured at the same direction as the linear polarization was higher than the yields in the other directions. This difference in the neutron yield increased with the neutron energy, up to a factor of approximately three. The ratios of the sum of the NE3 to that of the NE2 as a function of the lower neutron energy were compared with the experimental value produced by Horikawa et al. The ratios in this work reproduce the experimental value produced by Horikawa et al. roughly. To estimate the energy dependency between the linearly polarized photon and the circularly polarized photon, ratios of the neutron yields of the linearly polarized photon to those of the circularly polarized photon were calculated. In a direction perpendicular to that of the linear polarization, the ratio of the neutron yield decreased as the neutron energy increased.

References

1. International Linear Collider Technical Design Report, https://www.linearcollider.org/ILC/Publications/Technical-Design-Report
2. V.V. Varlamov, B.S. Ishkhanov, "Study of Consistency Between (γ, xn), [γ, n) + (γ, np)] and (γ, 2n) Reaction Cross Sections Using Data Systematics", IAEA, Vienna, Austria, INDC (CCP)-433 (2002)
3. K. Horikawa et al., Phys. Lett. B **737**, 109 (2014)
4. A. Agodi, Nuovo Cimento, **5**, 21, (1) (1957)
5. S. Miyamoto et al., Radiat. Meas. **41**, S179 (2007)
6. Y. Asano et al., Prog. Nucl. Sci. Tech. **4**, 254 (2014)
7. S. Miyamoto, "NewSUBARU Laser Compton Gamma-ray Source and Study on Photo-nuclear Reaction", Japan Forum of Nuclear Astrophysics, (2009.7.27-7.29)
8. D. Satoh, et al., Nucl. Sci. Tech. **2**, 657 (2002)
9. T. Win, "Experimental Studies on Fast Neutron-Induced Fission Spectra for ^{237}Np and ^{233}U", PhD thesis, Tohoku University (1999).
10. W. A. G. Sauerwein et al., "Neutron Capture Therapy", Springer Verlag Berlin Heidelberg, (2012).

DOI: 10.1051/epjconf/201715301020

Neutron Productions from thin Be target irradiated by 50 MeV/u ^{238}U beam

Hee-Seock Lee[1*], Joo-Hee Oh[1], Nam-Suk Jung[1], Leila Mokhtari Oranj[2], Noriaki Nakao[3], and Yoshitomo Uwamino[4]

[1]*Pohang Accelerator Laboratory, POSTECH, Pohang 37673, Republic of Korea*
[2]*Division of Advanced Nuclear Engineering, POSTECH, Pohang 37673, Republic of Korea*
[3]*Institue of Technology, Shimizu Corporation, Tokyo, Japan*
[4]*RIKEN, Wako, Japan*
*Corresponding Author, E-mail: lee@postech.ac.kr

Neutrons generated from thin beryllium target by 50 MeV/u ^{238}U beam were measured using activation analysis at 15, 30, 45, and 90 degrees from the beam direction. A 0.085 mm-thick Be stripper of RIBF was used as the neutron generating target. Activation detectors of bismuth, cobalt, and aluminum were placed out of the stripper chamber. The threshold reactions of ^{209}Bi(n, xn)$^{210-x}$Bi(x=4~8), ^{59}Co(n, xn)$^{60-x}$CO(x=2~5), ^{59}Co(n, 2nα)^{54}Mn, ^{27}Al(n, α)^{24}Na, and ^{27}Al(n,2nα)^{22}Na were applied to measure the production rates of radionuclides. The neutron spectra were obtained using an unfolding method with the SAND-II code. All of production rates and neutron spectra were compared with the calculated results using Monte Carlo codes, the PHITS and the FLUKA. The FLUKA results showed better agreement with the measurements than the PHITS. The discrepancy between the measurements and the calculations were discussed.

***KEYWORDS:** ^{238}U beam, Be, Activation analysis, Neutron spectrum*

I. Introduction

Neutron production is one of important subjects for shielding calculation of particle accelerator facility, such as RIBF/RIKEN. However the number of high energy heavy ion accelerators is small in a world. The accuracy of shielding calculation of such facilities has been studied until now. Especially the information of neutron production is very rare if the accelerated particles are heavy ions like uranium. The study to measure the neutron production yields was done using 50 MeV/u ^{238}U beam of RIBF for the shielding design of the Rare Isotope Science Project (RISP) ion accelerator in Korea.
The interesting energy of 238U beam is lower than 100 MeV/u because another group had already performed similar study using 345 MeV/u at RIBF[1]. The produced neutrons were measured when 50 MeV/u ^{238}U beam hit thin Be using activation analysis. The production rate of radionuclides were compared with the Monte Carlo calculation and neutron spectra were also obtained using the unfolding method.

II. Material and Method

1. Experiments

The charge stripper chamber was chosen as the experimental station as shown in figure 1. The produced neutrons were measured when 50 MeV/u ^{238}U beam hit thin Be stripper (0.085 mm) in the chamber. For this measurement, the activation analysis was applied using Bi, Co, and Al samples which were attached on the outer surface of the stripper chamber. The samples have a diameter of 2 cm (or 2.5x2.5 cm2) and a thickness of 5 mm (or 2 mm). We chose four different emission angles (15, 30, 45, and 90 degrees) of neutron productions as shown in Fig. 1. The irradiation of 238U beam was done during normal operation for the other experiments at the RIBF target station. The beam intensity of ^{238}U ions was monitored by a phase probe (PPM04) and calibrated by the current of Faraday cup (FCM04). The gamma-rays from activated samples were measured using an HPGe detector and analyzed using HYPEGAM[2] software. Radionuclide yields in the samples were obtained. The gamma spectroscopy system was calibrated using multi-lines standard source and verified by the MCNPX simulation[3].

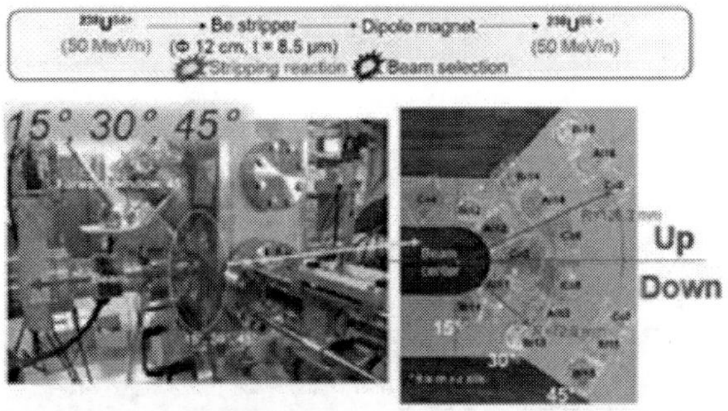

Figure 1: Experimental setup at a charge stripper chamber and the arrangement of activation samples at the forward direction. (Charge stripping process is described up)

2. Monte Carlo calculation and unfolding

Monte Carlo codes FLUKA[4], version 2c.0, PHITS[5], version 2.64, and MCNPX, version 2.7 were used to simulate this experiment. The radionuclide yields in Bi, CO, and Al samples were estimated using the FLUKA or the PHITS and activation code, DCHAIN[5], version 2001. The flux and energy distribution of all kinds of particles produced from Be stripper target were estimated. In figure 2, the part of calculated PHITS results, fluence distribution of neutrons and nucleus including 238U ions, were presented. All devices and real geometry were considered in this simulation in order to figure out the contribution of those to sample activation, especially a big dipole magnet downstream of the stripper chamber.

The energy distributions of produced neutrons were calculated by the unfolding method. The SAND-II code[6)] was applied for several threshold reactions of Bi, Co, and Al elements. The measured data were compared with the results simulated using the FLUKA, and the PHITS.

III. Results and Discussion

The radionuclide production rates of Bi, Co and Al samples were measured for different angles, 15, 30, 45, and 90 degrees as shown in figure 2. The rate of Bi samples are presented in figure 3. The angular distribution of production rate regenerated well the theoretical trend. The comparison between the measured and the calculated data by two codes showed reasonable agreements but the PHITS results underestimated more than the FLUKA results relatively.

All neutron-induced reactions of Bi, Co, and Al samples have each threshold energy. The neutron spectra above 10 MeV were obtained from the production yields of each reactions through the unfolding process7). The contamination of neutron energy spectrum by surrounding devices except of the Be stripper target was found as very small at the energy range higher than 20 MeV. The calculated spectra using the FLUKA had good agreements with the unfolded one for every emission angle. Figure 4 shows one spectra. However the PHITS results showed big discrepancy. The reason of the discrepancy has been discussed at present. The different physics model of two codes at the energy range has been reviewed.

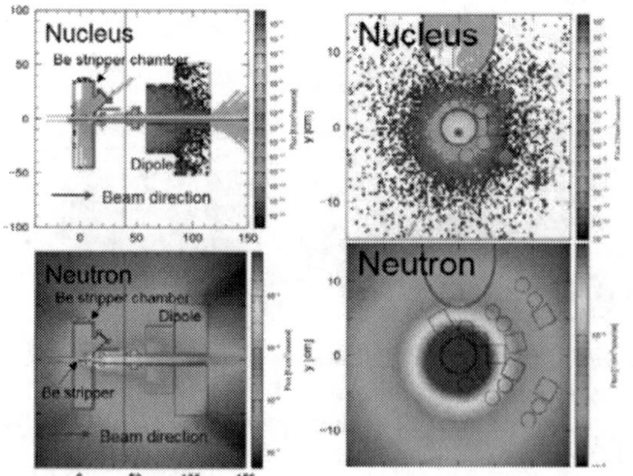

Figure 2: **Fluence Distribution of nucleus (up) and neutrons (down) at stripping reaction area.**

IV. Conclusion

These experimental results are very important to compensate the fact that there is no proved data below about 300 MeV/u for benchmarking 238U induced neutron production. The enhanced analysis is ongoing to confirm the discrepancy between the PHITS and the FLUKA results or with the experimental results.

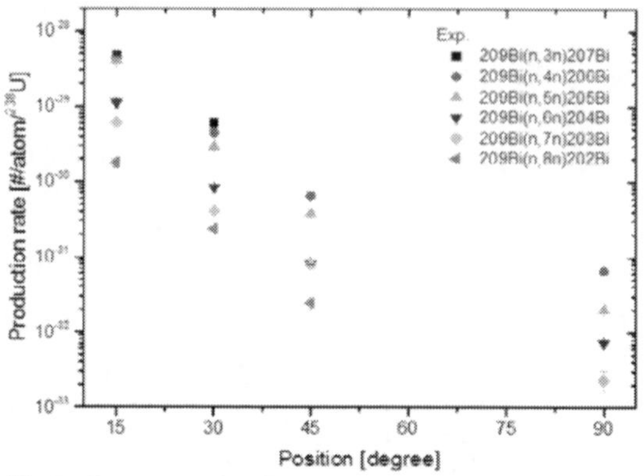

Figure 3: **Angular distribution of radionuclide yields in Bi samples.**

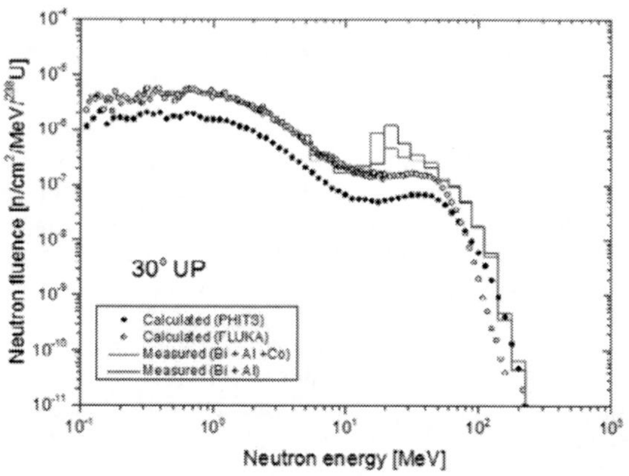

Figure 4: **Differential neutron yields from thin Be target by 50 Mev/u ^{238}U beam at 30 degree: comparison of the measured using two or three elements with resulted calculated by PHITS and FLUKA.**

References

1) N. Nakao, et al.: Proc. IPAC2014, (2015) p.1811.
2) H.D. Choi, et al.: Nucl. Eng. Tech. 44, (2012) p.781.
3) G.W. McKinney, et al.: LA-UR-12-25775, LANL (2012).
4) A. Ferrari, et al.: CERN-2005-10, CERN (2005).
5) K. Nitta, et al.: Phys. Rev. C 52, (1995) p.2620.
6) P.J. Griffin, et al.: SAND93-3957, Sandia Nat. Lab. (1994).
7) J.H. Oh, et al.: presented at ISORD-8, Jeju (2015). N.S. Jung, B.G. Park, J. Nucl. Sci. Technol, 5 (2008)

Measurement and Analysis of Neutron Leakage Spectra from Pb and LBE Cylinders with D-T Neutrons

Size Chen[1], Leting Gan[1, 2], Taosheng Li[1], Yuncheng Han[1], Chao Liu[1,a], Jieqiong Jiang[1] and Yican Wu[1,2]

[1]*Key Laboratory of Neutronics and Radiation Safety, Institute of Nuclear Energy Safety Technology, Chinese Academy of Sciences, Hefei, Anhui, 230031, China*
[2]*University of Science and Technology of China, Hefei, Anhui, 230027, China*

Abstract. For validating the current evaluated neutron data libraries, neutron leakage spectra from lead and lead bismuth eutectic (LBE) cylinders have been measured using an intense D-T pulsed neutron source with time-of-flight (TOF) method by Institute of Nuclear Energy Safety Technology (INEST), Chinese Academy of Sciences (CAS). The measured leakage spectra have been compared with the calculated ones using Super Monte Carlo Simulation Program for Nuclear and Radiation Process (SuperMC) with the evaluated pointwise data of lead and bismuth processed from ENDF/B-VII.1, JEFF-3.1 and JENDL-4.0 libraries. This work shows that calculations of the three libraries are all generally consistent with the lead experimental result. For LBE experiment, the JEFF-3.1 and JENDL-4.0 calculations both agree well with the measurement. However, the result of ENDF/B-VII.1 fails to fit with the measured data, especially in the energy range of 5.5 and 7 MeV with difference more than 80%. Through sensitivity analysis with partial cross sections of ^{209}Bi in ENDF/B-VII.1 and JEFF, the difference between the measurement and the ENDF/B-VII.1 calculation in LBE experiment is found due to the neutron data of ^{209}Bi.

1 INTRODUCTION

Lead bismuth eutectic (LBE) alloy plays an important role in many advanced nuclear applications such as accelerator driven subcritical system (ADS)[1-3]. It is necessary to perform a new integral experiment on Pb and Bi to validate the current evaluated neutron data libraries for the design of ADS.

2 EXPERIMENT PROCEDURE

This experiment was performed at a Cockcroft-Walton-type neutron generator installed at China Institute of Atomic Energy (CIAE) [4]. It can be used for both continuous and pulsed beam operations to generate neutrons for various kinds of experiments. In this work, the pulsed deuteron beam was accelerated up to 300 keV and bombarded an air-cooling tritium target to yield 14.8 MeV neutrons with 3 ns FWHM (full width at half maximum). The repetition rate was 1.5 MHz and the average deuteron current was 30 μA, which produced a neutron intensity of about 10^9 n/s.

A stilbene crystal detector was positioned at 7.05±0.01 m from the neutron source at forward direction to check the accelerator pulsed mode operation. The leakage neutron spectra were measured using a BC501A (Ø5.08 × 2.54 cm) liquid scintillation detector by time of flight (TOF) technique.

The process of the leakage neutron spectra measuring included the sample in and sample out measurements. The sample in measurement and sample out survey obtained the foreground and background data, respectively. The net TOF spectra were achieved by subtracting the background data from the foreground data. The uncertainties of the experimental data contain the statistical uncertainty (<5%), the neutron detection efficiency (~3%) and the neutron source yield (~2%).

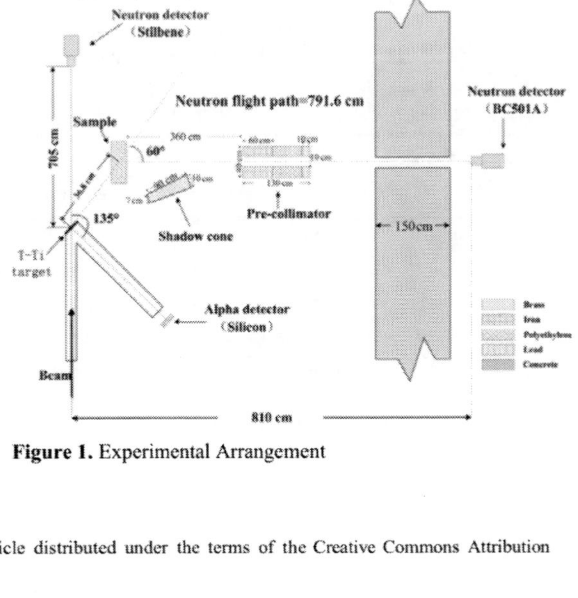

Figure 1. Experimental Arrangement

[a] Corresponding author: chao.liu@fds.org.cn

3 METHOD OF CALCULATION

The calculated leakage spectra were gained by using SuperMC[5] with lead and bismuth pointwise data from the ENDF/B-VII.1[6], JEFF-3.1[7] and JENDL-4.0 [8] libraries.

In this simulation, source neutron was assumed to be a point source. And a point detector was used at the position of 791.6 cm from the sample (Pb/LBE). The angle dependent neutron energy spectra of the neutron source, the energy efficiency of the BC501A detector and the width of the beam pulse were considered in order to accord with the actual situation of the experiment. The total neutron histories in the model were 109, and the average error is about 0.5%.

4 RESULTS AND DISCUSSION

Figures 2-3 show the comparisons between the measured and simulated spectra on Pb and LBE samples, respectively.

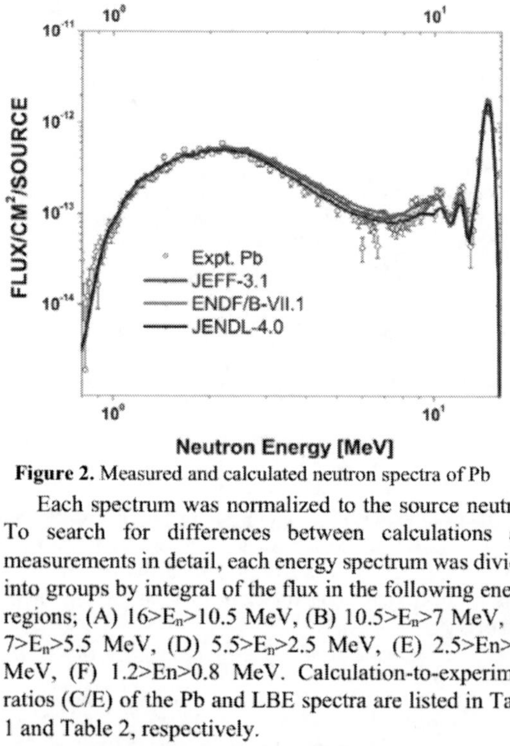

Figure 2. Measured and calculated neutron spectra of Pb

Each spectrum was normalized to the source neutron. To search for differences between calculations and measurements in detail, each energy spectrum was divided into groups by integral of the flux in the following energy regions; (A) $16>E_n>10.5$ MeV, (B) $10.5>E_n>7$ MeV, (C) $7>E_n>5.5$ MeV, (D) $5.5>E_n>2.5$ MeV, (E) 2.5>En>1.2 MeV, (F) 1.2>En>0.8 MeV. Calculation-to-experiment ratios (C/E) of the Pb and LBE spectra are listed in Table 1 and Table 2, respectively.

Table 1: C/E values for Pb in different energy intervals

Energy region (MeV)	C/E		
	JEFF-3.1	ENDF/B-VII.1	JENDL-4.0
16>En>10.5	1.039	1.080	0.963
10.5>En>7	1.092	1.044	0.871
7>En>5.5	1.124	1.158	0.944
5.5>En>2.5	1.113	1.188	1.013
2.5>En>1.2	0.995	1.027	1.003
1.2>En>0.8	0.963	0.943	1.020

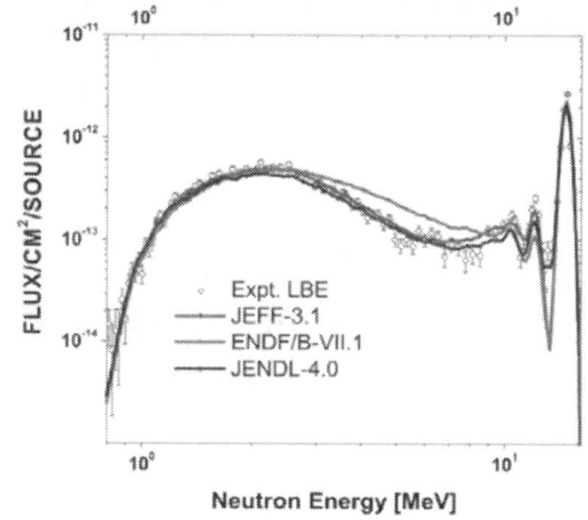

Figure 3: Measured and calculated neutron spectra of LBE

As shown in Figure 2 and Table 1 for Pb sample, in the energy range of 10.5-16 MeV, discrepancies between calculations of three libraries and experiment exceed no more than 8%. In 7-10.5 MeV region, the result of JENDL-4.0 underestimates nearly 13%, while the C/E values of JEFF-3.1 and ENDF/B-VII.1 are 1.092 and 1.044 respectively. The difference between the partial spectrum calculated with JENDL-4.0 and obtained from experiment doesn't exceed 6% in the energy range between 2.5 and 7 MeV, and calculated results from JEFF-3.1 and ENDF/B-VII.1 in the same energy range overestimate that from measurement 10% and 15%, respectively. When neutron energy is less than 2.5 MeV, all of three calculations exceed the measurement less than 6%. In general, calculated results obtained from ENDF/B-VII.1, JEFF-3.1 and JENDL-4.0 agree well with experimental result and the conclusion consists with Chikara Konno's evaluation on JENDL-4.0, while small discrepancies existed in partial energy interval.

Table 2: C/E values for LBE in different energy intervals

Energy region (MeV)	C/E		
	JEFF-3.1	ENDF/B-VII.1	JENDL-4.0
16>En>10.5	1.012	1.050	1.011
10.5>En>7	1.120	1.195	0.913
7>En>5.5	1.144	1.803	1.051
5.5>En>2.5	1.020	1.265	0.899
2.5>En>1.2	0.968	0.893	0.880
1.2>En>0.8	1.026	0.892	1.021

Figure 3 and Table 2 give the result of LBE measurement. It is found that the simulation of JEFF-3.1 overestimates the experimental result 12-15% in the energy region of 5.5-10.5 MeV. In other energy range, the differences are within 4%. The result with JENDL-4.0 library underestimates the measurement by 10-12% in the 1.2-5.5 MeV range. For other energy regions, the difference between the calculation and the experiment is less than 10%. However, the discrepancy between ENDF/B-VII.1 calculation and the experimental spectrum is more than 10% in the energy group, 0.8-10.5 MeV, and nearly 80% in the interval 5.5-7 MeV. In 10.5-16 MeV energy range, the C/E value of ENDF/B-VII.1 is 1.050. From the above observation, it can be seen that the simulations of JEFF-3.1 and JENDL-4.0 agree with the experimental data much better than calculation of ENDF/B-VII.1 in general. Drastic overestimation is observed in the ENDF/B-VII.1 calculation by a factor of 1.8 as compared to the experimental result in the 5.5-7 MeV range.

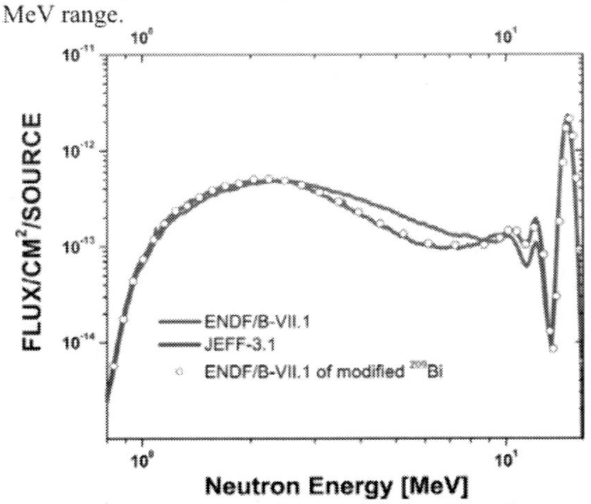

Figure 4: Calculation spectra dependence of modified ^{209}Bi cross section from ENDF/B-VII.1 to JEFF-3.1

To clarify the big difference between ENDF/B-VII.1 and measurement in the 2.5-7 MeV energy region on LBE experiment, more calculations using JEFF-3.1 and ENDF/B-VII.1 were made. Calculation was taken with modified ENDF/B-VII.1 library which replaced the data of ^{209}Bi, the only nature nuclide of bismuth, from ENDF/B-VII.1 to JEFF-3.1. The result is similar to the JEFF-3.1 calculation as shown in Figure 4. From the above comparison, we find that the difference between the measurement and the ENDF/B-VII.1 calculation in LBE experiment is really due to the neutron data of ^{209}Bi as the calculation with JEFF-3.1 agrees well with the experimental result.

ACKNOWLEDGMENT

This work was supported by the Natural Science Foundation of China, Grant No. 91026004. We further thank the great help from members of FDS Team and from researchers and engineers of CIAE in this work.

References

1. C. Bowman, E. Arthur, P. Lisowski, et al., Nucl. Instrum. and Meth. A, **320.1**, 336-367 (1992)
2. S. Kapoor, Pramana, **59.6,** 941-950 (2002)
3. P. Gokhale, S. Deokattey, V. Kumar, Prog. in Nucl. Energy, **48.2**, 91-102 (2006)
4. J.Q. Lu, D.L. Xie, S.W. Quan, et al., Atom. Energy Sci. and Tech., **37.4**, 376−380 (2003)
5. Y.C. Wu, J. Song, H. Zheng, et al., Ann Nucl. Energy, **82**, 161-168 (2015)
6. M. Chadwick, M. Herman, P. Obložinský, et al., Nucl. Data Sheets, **112.12**, 2887-2996 (2011)
7. A. Santamarina, D. Bernard, P. Blaise, et al., JEFF Report, **22.10.2**, 2 (2009)
8. K. Shibata, O. Iwamoto, T. Nakagawa, et al., J. Nucl. Sci. and Tech., **48.1**, 1-30 (2011)

EPJ Web of Conferences **153**, 01022 (2017)
ICRS-13 & RPSD-2016
DOI: 10.1051/epjconf/201715301022

Measurement of activation of helium gas by ^{238}U beam irradiation at about 11 A MeV

A. Akashio[a], K. Tanaka, H. Imao, and Y. Uwamino

RIKEN Nishina Center 2-1 Hirosawa, Wako, Saitama, Japan

Abstract. A new helium-gas stripper system has been applied at the ~11 A MeV uranium beam of the Radioactive Isotope Beam Factory of the RIKEN accelerator facility. Although the gas stripper is important for the heavy-ion accelerator facility, the residual radiation that is generated is a serious problem for maintenance work. The residual dose was evaluated by using three-layered activation samples of aluminium and bismuth. The γ-rays from produced radionuclides with in-flight fission of the ^{238}U beam and from the material of the chamber activated by neutrons were observed by using a Ge detector and compared with the values calculated by using the Monte-Carlo simulation code PHITS.

1 Introduction

The Radioactive Isotope Beam Factory (RIBF) of the RIKEN accelerator facility is used to generate intense beams of radioactive isotopes for use in studies on nuclear physics and nuclear astrophysics. The superconducting-ring cyclotron produces beams of heavy ions of nuclei ranging from hydrogen to uranium. The energy of the primary beam is about 350 MeV per nucleon, and the targeted intensity is as high as one particle microampere, corresponding to 6.2×10^{12} particles/s [1].

To accelerate a beam of uranium ions for use in generating a variety of radioactive nuclei, it is necessary to strip electric charges from the uranium ions by passing them through materials at an intermediate energy of 10–20 A MeV. Serious radiation damage caused by the high-intensity uranium beam causes problems with the lifetimes of solid materials; consequently, it is difficult to use solids in the electron stripper. This is common problem at high-intensity heavy-ion accelerator facilities.

At the RIBF, a helium-gas stripper system that changes the charge on the uranium ions from U^{35+} to U^{71+} has been successfully developed [2]. Because the stripper setup has to be changed for each ion nuclide, such as uranium, xenon, etc., the residual dose is a serious problem with respect to maintenance work. The intensity of the uranium beam at the target in operation in 2015 was 50 particle nanoamperes, and this will become more intense in the future. It is important to understand the origin of the high residual dose and to make plans to reduce the radiation dose during maintenance work.

Because gas stripper systems might become popular solutions for use in high-intensity accelerator facilities, the activation around the gas stripper chamber was measured by using activation samples. The activation of the stripper system, intended to provide a low activation of the setup, good shielding, and an effective maintenance process for future design, was evaluated by using the Monte-Carlo simulation code PHITS [3] and the results were compared with those obtained experimentally.

2 Experiment

Figure 1 shows the place of the helium gas stripper at RIBF, and Figure 2 is a schematic view of the helium-gas stripper setup used in the PHITS calculation. The energy of the uranium beam in the helium gas region was 10.75 MeV per nucleon, and the beam intensity was 0.86 particle microamperes. The helium-gas pressure in the region of highest pressure was 7 kPa, and the thickness of the helium layer was about 500 mm, which corresponds to 0.7 mg/cm^2. The beam intensity was measured nondestructively by a using phase probe [4]. The helium gas was irradiated with 5.4×10^{18} uranium ions (±20% accuracy) during about 18 days of operation of the accelerator.

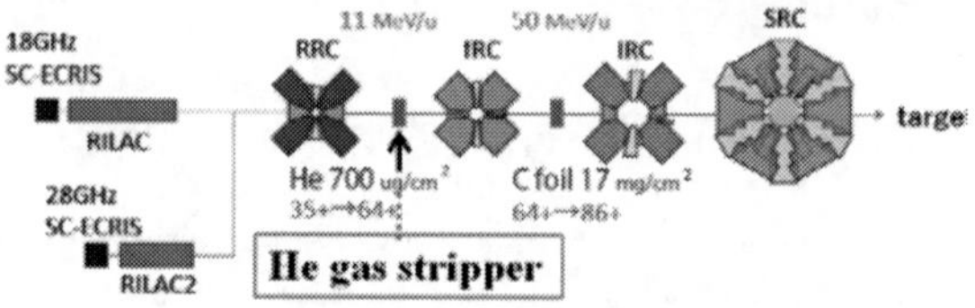

Figure 1. Place of the helium gas stripper setup at RIBF.

[a] Corresponding author: horigome@riken.jp

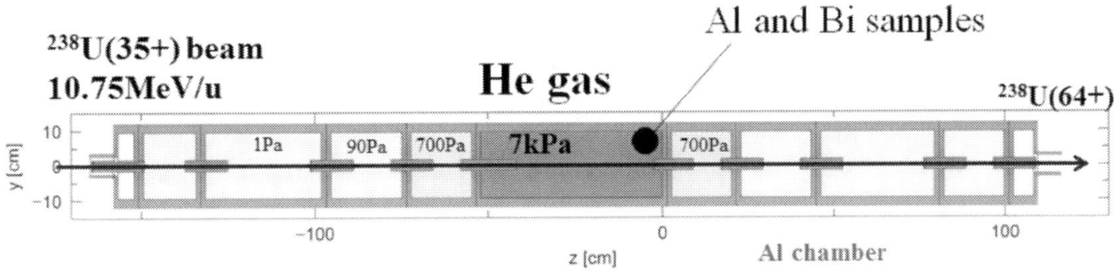

Figure 2. Schematic side-cross sectional view of the helium-gas stripper setup. The helium gas is contained in the gas cells. The uranium beam penetrates the centre of the chamber. Each cells are connected each other at the beam pass without any window. The highest pressure of the helium was 7k Pa.

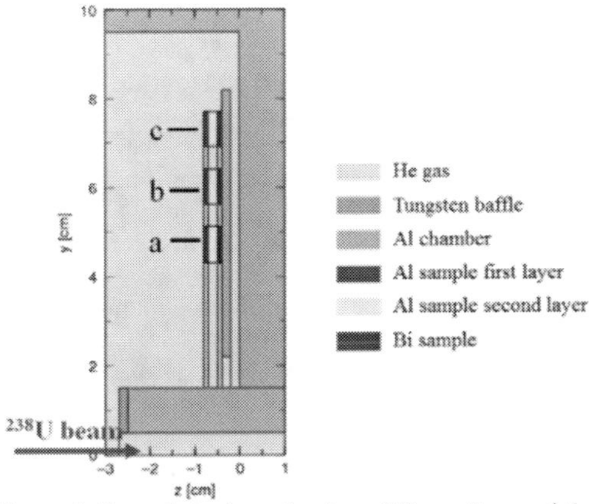

Figure 3. Expanded schematic view of Figure 2 around the sample position. Each samples consisted of three layers of the same three materials. The symbols a–c are assigned with respect to the beam line. The positions are indicated as Al-1a, Al-2a, and Bi-a for positions in the first, second, and third layers, respectively.

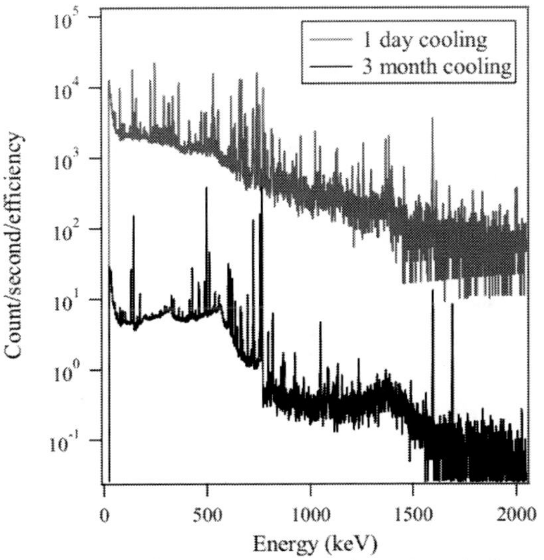

Figure 4. Typical γ-ray spectra of an Al sample set in the gas stripper chamber. The spectra were normalized with respect to the measurement time and the efficiency of the Ge detector. The red spectrum was recorded following 1 day of cooling after beam irradiation ceased, and the black spectrum was recorded after 3 months of cooling. Many of the energy peaks for short-lived radioactive isotopes generated by fission of ^{238}U showed a decrease after 3 months.

The activation samples were located inside the gas stripper chamber. Each sample consisted of three layers, the structure of which is shown in Figure 3. The first layer, consisting of a 1-mm-thick, 10 × 10 mm specimen of aluminium, was intended to catch the products of in-flight fission of the uranium beam. The second layer consisting of a 2-mm-thick, 10 × 10 mm specimen of aluminium was activated by neutrons only, because most of the fission products from the uranium beam were captured by the first layer. These aluminium specimens were used to evaluate the activation of the aluminium chamber of the gas stripper. The third layer of the sample consisted of 1-mm-thick, 10 × 10 mm specimen of bismuth. Bismuth isotopes generated from natural ^{209}Bi are sensitive to neutron energy, because of the low neutron-energy thresholds of their reactions. The uranium beam passed though the helium gas. The sample closest to the source of the beam inside the chamber was at a distance of 47.25 mm.

Figure 4 shows typical γ-ray spectra from the Al-1a sample, measured following about 1 day and 3 months of cooling time.

Because evaluation of radiation from long-lived isotopes is more important than that for short-lived isotopes, we performed our measurements after three months, so that the background level was reduced. The radioactive nuclei generated in the samples were identified and counted from the spectra, taking into account the efficiency and half-life. Table 1 lists the radioactive nuclei identified by γ-ray spectroscopy after the three-month cooling period, and mass distribution of the identified nuclei are shown in Figure 5. Typical fission-product nuclei from the uranium beam were observed. The long-lived isotope ^{22}Na was also observed; this was generated from the aluminium sample by neutron-induced reaction. Table 2 lists the observed bismuth isotopes, with their neutron-energy thresholds. 22.55 MeV neutrons were detected, whereas 28.64 MeV neutrons, which can generate ^{205}Bi from ^{209}Bi, were not detected.

Table 1. Typical radioactive nuclei observed in the aluminium sample Al-1a from the stripper chamber after cooling for about 3 months.

Nuclide	Half-life	Nuclide	Half-life
^{91}Y	58.5 day	^{134}Cs	2.06 year
^{95}Zr	64.0 day	^{137}Cs	30.1 year
^{103}Ru	39.3 day	^{141}Ce	32.5 day
^{106}Ru	371 day	^{144}Ce	285 day
^{124}Sb	60.2 day	^{155}Eu	4.75 year
^{125}Sb	2.76 year		

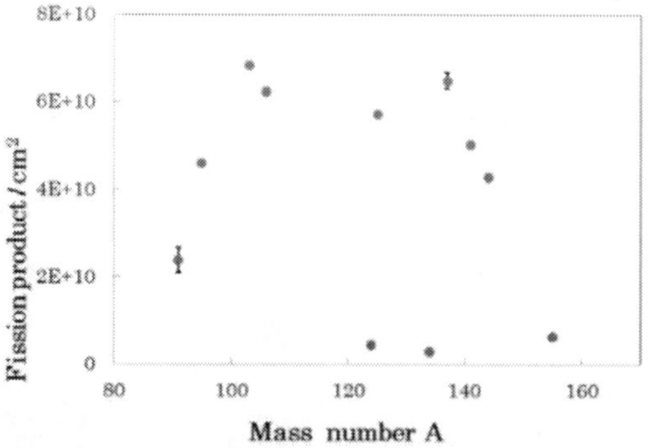

Figure 5. Mass distribution of the fission products observed in the aluminium sample Al-1a after the cooling for about 3 months. The low production of A = 124 and 134 were produced directly. The others are the summation of produced nuclei of the same mass because of the beta-decay, respectively as explained in Figure 8.

Table 2. Threshold of the neutron energy corresponding to the production of radioactive bismuth isotopes. ^{206}Bi, ^{207}Bi, and ^{208}Bi were observed in this study.

Nuclide	Half-life	Reaction	Threshold (MeV)
^{208}Bi	1.4 × 10^5 year	^{209}Bi(n,2n)^{208}Bi	
^{207}Bi	31.6 year	^{209}Bi(n,3n)^{207}Bi	14.12
^{206}Bi	6.4 days	^{209}Bi(n,4n)^{206}Bi	22.55
^{205}Bi	15.31 day	^{209}Bi(n,5n)^{205}Bi	29.62

3 PHITS calculation

The calculation was performed by using PHITS ver. 2.76 for the geometry shown in Figure 2. PHITS uses the JAERI Quantum Molecular Dynamics model for the nucleus–nucleus reaction of ^{238}U with helium. The Generalized Evaporation Model was used to simulate the evaporation and fission processes of the excited states, the INCL model was used for nucleon–nucleus collisions at energies above 20 MeV [3], and the JENDL-4 library of evaluated nuclear data was used to calculate collisions below 20 MeV [5].

Figure 6 shows the results of calculations of the fluxes of generated nuclei and neutrons. The fission products reached the position of the aluminium sample.

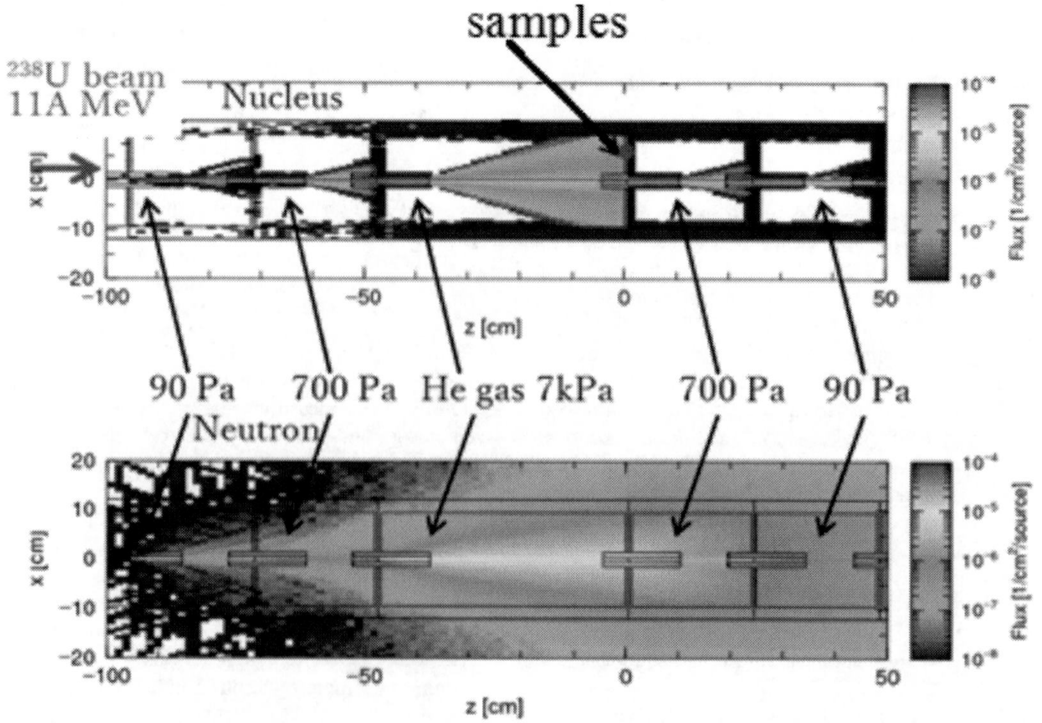

Figure 6. Cross-sectional view from above of the flux of generated nuclei (upper) and neutrons (lower), as calculated by PHITS

The sample sizes were small and it was difficult to obtain enough statistics of fission products and neutrons by PHITS calculation. To obtain enough statistics for the samples, ring-type samples were set in the PHITS geometry by the area as explained by D. Satoh et al [6]. Figure 7 shows the neutron spectrum irradiated to the aluminium sample Al-1a. The peak energy is the same as the uranium beam energy. About 90 % of the neutron were less than 20 MeV thus aluminium and bismuth activation by neutron was calculated by using the nuclear data, mainly.

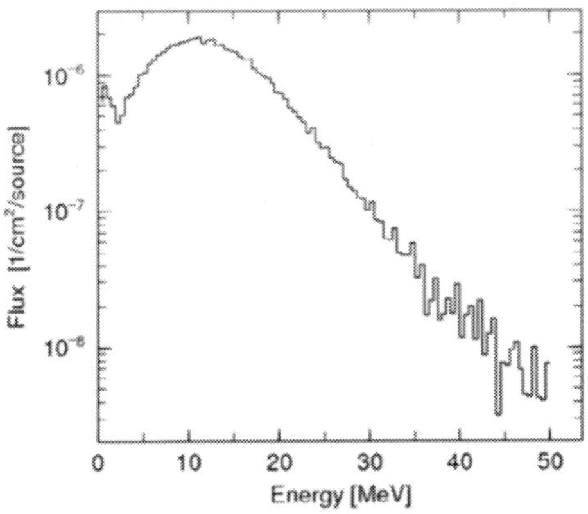

Figure 7. Neutron spectrum impinged to the sample Al-1a.

The numbers of nuclei that were stopped by the samples were counted, and the results were compared with the measured value. Although short-lived nuclei were also observed, we chose to examine long-lived nuclei in the evaluation of the activation because of their contribution to the residual dose in the experimental setup.

Figure 8 shows a nuclear chart of the PHTIS calculation results for radioactive nuclei produced in sample Al-1a for irradiation by 8×10^9 uranium ions. Short-lived nuclei were eliminated from the spectra measured after three months of cooling. Therefore, the numbers of short-lived nuclei deduced by PHITS were counted and summed to give the long-lived daughter nuclei shown in Figure 8.

Because the helium gas was about 50-cm thick, it was difficult to determine the angle of fission products against the samples, experimentally. From the calculation result, the average angles of the fission products to the first-layer specimens were 6°, 8°, and 10° for Al-1a, Al-1b, and Al-1c, respectively. Therefore, the angular property was evaluated with thin target case by PHITS calculation. Figure 9 shows angular distributions of neutron, cesium isotopes, and ruthenium isotopes which are typical fission products. The helium target was assumed as 1-mm thick and 0.7 mg/cm^2 as same as the stripper in this study. Because the angular difference of the samples estimated by PHITS calculation were small, angular dependence of the fission products yield were also small.

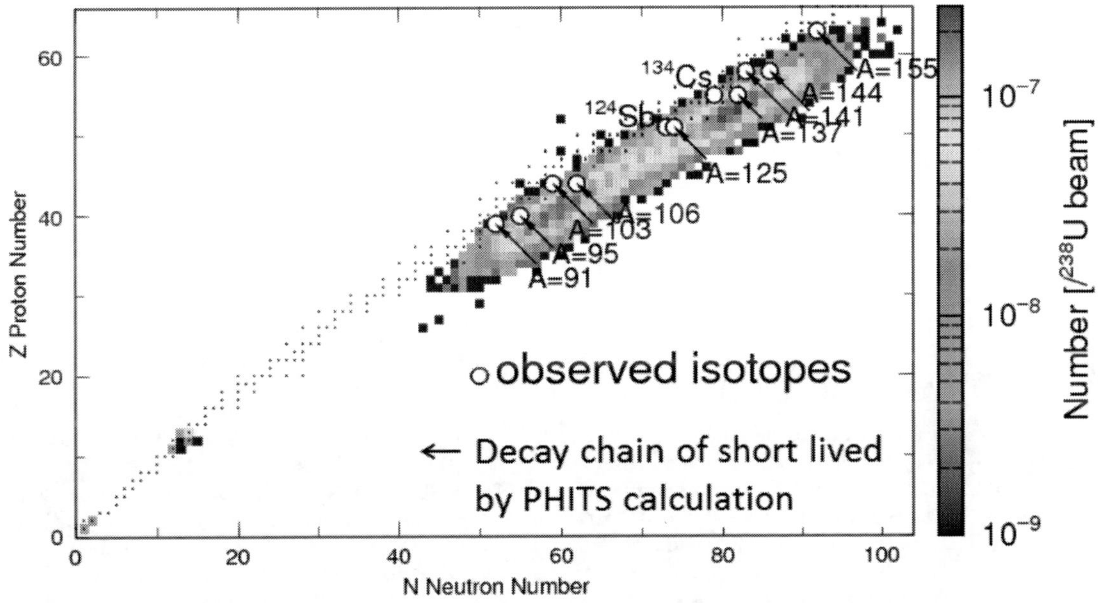

Figure 8. Nuclear chart of produced nuclei stopped at the Al-1a sample. The black dots indicate stable nuclei. The open circles show observed long-lived daughter nuclei, as listed in Table 1. The arrows indicate decay chains to the daughter nuclei evaluated by PHITS. Long-lived ^{124}Sb and ^{134}Cs do not have parent nuclei, because their $Z = +1$, $N = -1$ nuclei are stable. The ^{124}Sb and ^{134}Cs nuclei were therefore used to make a direct comparison between the results of the PHITS calculation to the measured values.

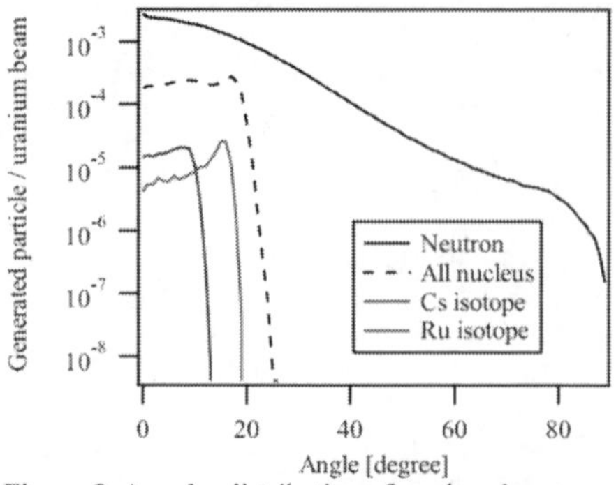

Figure 9. Angular distribution of produced neutron and fission product nuclei calculated by PHITS. 0.7 g/cm^2 helium target of 1-mm thickness was set for the calculation.

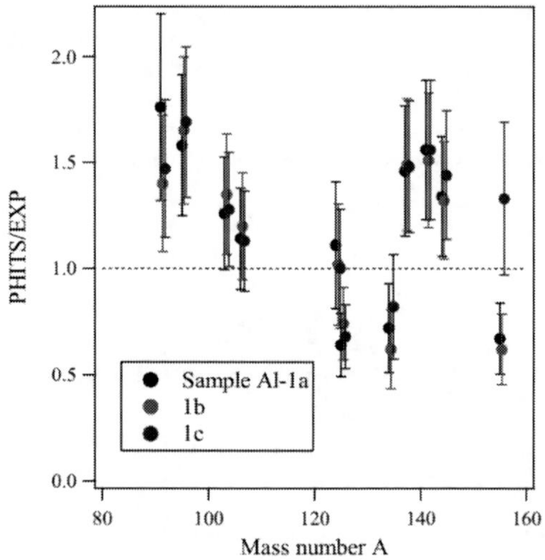

Figure 10. Ratios of calculated fission products to observed them.

4 Conclusion

Figure 9 shows the ratios of the numbers of nuclei calculated by PHITS calculations to the observed numbers of nuclei at the samples. The errors in the plots arose mainly from statistics of the γ-ray and PHITS calculations, and the 20% variation in accuracy of the beam intensity. The fission products stopped inside the aluminium sample first layers are shown. Evaluations of the neutron activation of the aluminium and bismuth samples are ongoing.

In this study, the results of the PHITS evaluation for fission products from a uranium beam at around 11 A MeV agreed within a factor of two with the measured values for all mass regions. Previously, K. Tanaka et al. measured sample activation by neutrons around a copper beam dump irradiated by a uranium beam at 345 A MeV at RIBF [7]. N. Nakao also measured neutron production through irradiation of a beryllium target at RIBF by a uranium beam at 345 A MeV by using bismuth samples placed from 60 to 90 degree of the target [8]. The results of their PHITS calculation for the activation ratio also agreed with the measured values. K. Kusaka et al. measured heat load by neutron and charged particles from the target on the first superconducting quadrupoles magnet at RIBF. The measured result of radiation heat load by a uranium beam at 345 A MeV also agreed with that of PHITS calculation [9]. D. Sato et al. and K. Niita et al. studied other beam nuclides for a beam of more than 100 A MeV [6, 10], and PHITS reproduced these measurements within a factor of two. For the case of low-energy reactions, Y. Iwamoto et al. measured the angular distribution of neutrons from a ^{9}Be target irradiated with a 10 MeV beam of protons and deuterons [11], and the neutron production in the forward angles calculated by PHTIS agreed with the experimental value. In the current uranium study, the calculated result for uranium fission agreed with the results of these previous studies for other nuclides and beam energies. Further evaluations of neutron activation for aluminium and bismuth samples are necessary.

Acknowledgement

The PHITS calculation was performed by using the RIKEN HOKUAI–GreatWave system, which permits rapid large-scale computations. The authors thank RIKEN for supporting the computing systems essential for the completion of this work. We also grateful to the accelerator operation staff and the experimental group of RIBF for helpful support in the parasitic experiment.

References

1. Y. Yano, Nucl. Instrum. Methods Phys. Res., Sect. B, **261** 1009 (2007); DOI: 10.1016/j.nimb.2007.04.174.
2. H. Imao, H. Okuno, H. Kuboki, et al., *Proceedings of IPAC2013,* 12–17th May 2013, Shanghai China, 3851–3853 (2013); http://accelconf.web.cern.ch/accelconf/IPAC2013/papers/thpwo038.pdf (Accessed 24th Oct. 2016).
3. T. Sato, K. Niita, N. Matsuda, et al., J. Nucl. Sci. Technol. (Abingdon, U. K.) **50**, 913 (2013); DOI: 10.1080/00223131.2013.814553
4. R. Koyama, N. Sakamoto, M. Fujimaki, et al., Nucl. Instrum. Methods Phys. Res., Sect. A, **729**, 788 (2013); DOI: 10.1016/j.nima.2013.08.056.
5. K. Shibata, O. Iwamoto, T. Nakagawa, N. Iwamoto, A. Ichihara, S.Kunieda, S. Chiba, K. Furutaka, N.Otsuka, T. Ohsawa, T. Murata, H. Matsunobu, A. Zukeran, S. Kamada, and J. Katakakura, J. Nucl. Sci. Technol. (Abingdon, U. K.) **48**, 1 (2011); DOI: 10.1080/18811248.2011.9711675
6. D. Sato , T. Kurosawa, T. Sato, A. Endo, M. Takada, H. Iwase, T. Nakamura, K. Niita, Nucl. Instrum.

Methods Phys. Res., Sect. A, **583** 507 (2007); DOI: 10.1016/j.nima.2007.09.23.

7. K. Tanaka, N. Inabe, K. Yoshida, T. Kubo, Prog. Nucl. Sci. Technol. **4**, 201 (2014); DOI: 10.15669/pnst.4.201.
8. N. Nakao, Y. Uwamino, K. Tanaka, *Proceedings of IPAC2014*, 15–20th June 2014, Dresden, Germany, 1811–1813 (2014); http://accelconf.web.cern.ch/AccelConf/IPAC2014/papers/tupri101.pdf?n=IPAC2014/papers/tupri101.pdf (Accessed 24th Oct 2016).
9. K. Kusaka M. Ohtake, K. Tanaka, K. Yoshida, and T. Kubo, IEEE Transactions and Applied Superconductivity, **25**, Issue3, 4100404 (2015); DOI: 10.1109/TASC.2014.2365957.
10. K. Niita, T. Sato, H. Iwase, H. Nose, H. Nakashima, L. Sihver, Radiat. Meas., **41** 1080 (2006); DOI: 10.1016/j.radmeas.2006.07.013.
11. Y. Iwamoto, Y. Sakamoto, N. Matsuda, et al. Nucl. Instrum. Methods Phys. Res., Sect. A, **598** 687 (2009); DOI: 10.1016/j.radmeas.2006.07.013.

 DOI: 10.1051/epjconf/201715301023

On the feasibility to perform integral transmission experiments in the GELINA target hall at IRMM

Pierre Leconte[1,a], Cyrille De Saint Jean[1], Benoit Geslot[2], Arjan Plompen[3], Francesca Belloni[3], Markus Nyman[3]

[1]*CEA Cadarache, DEN/CAD/DER/SPRC/LEPh, 13108 Saint Paul Lez Durance Cedex, France*
[2]*CEA Cadarache, DEN/CAD/DER/SPEx/LPE, 13108 Saint Paul Lez Durance Cedex, France*
[3]*JRC-IRMM, Geel, Belgium*

Abstract. Shielding experiments are relevant to validate elastic and inelastic scattering cross sections in the fast energy range. In this paper, we are focusing on the possibility to use the pulsed white neutron time-of-flight facility GELINA to perform this kind of measurement. Several issues need to be addressed: neutron source intensity, room return effect, distance of the materials to be irradiated from the source, and the sensitivity of various reaction rate distributions through the material to different input cross sections. MCNP6 and TRIPOLI4 calculations of the outgoing neutron spectrum are compared, based on electron/positron/gamma/neutron simulations. A first guess of an integral transmission experiment through a ^{238}U slab is considered. It shows that a 10 cm thickness of uranium is sufficient to reach a high sensitivity to the ^{238}U inelastic scattering cross section in the [2-5 MeV] energy range, with small contributions from elastic and fission cross sections. This experiment would contribute to reduce the uncertainty on this nuclear data, which has a significant impact on the power distribution in large commercial reactors. Other materials that would be relevant for the ASTRID 4th generation prototype reactor are also tested, showing that a sufficient sensitivity to nuclear data would be obtained by using a 50 to 100cm thick slab of side 60x60cm. This study concludes on the feasibility and interest of such experiments in the target hall of the GELINA facility.

1 Introduction

Shielding experiments concern the measurement of neutron and gamma flux attenuation through large thicknesses of material, using either point-like sources (^{252}Cf for instance) or neutron beams from test reactors. Most of these experiments were initially driven by the need to validate calculation codes related to radiation protection. They also have an interest for nuclear data evaluators as they provide a convenient way to validate neutron inelastic scattering cross sections which are often difficult to measure by differential experiments.

SINBAD [1] is an international effort to store, evaluate and share shielding experiments that were made by fourteen different countries, among them USA, Russia, UK and France. Materials covered include air, graphite, concrete, N, O, H_2O, Al, Be, Cu, C, Fe, Pb, Li, Ni, Nb, SiC, Na, Stainless Steel, W and V. Most of these experiments are more than thirty years old and not all of them have the sufficient quality assurance to be used for nuclear data validation. Moreover, the list of materials does not include heavier materials than tungsten. Uranium 238 for instance belongs to a high priority request list at OECD/NEA to improve the inelastic scattering cross section, due to its major impact on both thermal and fast reactor estimates. Indeed, a paper published at the PHYSOR conference [2] showed that the *a priori* uncertainty on the radial power map for LWR-UOx cores (and its swing between the center and the periphery) is 5-6% (1σ).

This is why CEA and IRMM have considered the possibility to perform new shielding experiments to meet the target accuracies which are required for reactor physics studies (typically less than 5% at 1σ). Such an experiment was recently performed in the CALIBAN critical facility at Valduc [3], in the framework of collaboration between CEA/DEN (Nuclear Energy Division) and CEA/DAM (Military Application Division).

In this paper, we are considering the feasibility of using the pulsed white neutron source GELINA, located at Geel in Belgium, to perform shielding experiments within the target hall. Firstly, we present comparisons of MCNP6 and TRIPOLI4 calculations of the outgoing neutron source, generated by the slowing down of 70-150 MeV electrons within the depleted Uranium target. Secondly, we present the main characteristics of the neutron flux at 70 to 120 cm from the source. Thirdly, we present preliminary calculations of a transmission experiment based on several disks of depleted Uranium available at IRMM and the use of several activation dosimeters based on threshold reactions. Finally, some quantitative results are presented for the measurement of other materials of interest for the ASTRID 4th generation prototype reactor: iron, sodium and magnesium oxide.

[a] Corresponding author: pierre.leconte@cea.fr

2 Monte-Carlo calculation of the GELINA neutron source

The MCNP input file of GELINA was provided by IRMM and is described in Ref. [4]. An equivalent model was built for TRIPOLI4. It consists of a mono-directional electron beam, with homogeneously distributed energies from 70 to 140 MeV, reaching a depleted Uranium torus, cooled by a mercury flow. The electron slowing down produces high energy bremsstrahlung photons which generate neutrons by (γ,n), (γ,xn) and (γ,f) reactions. A scheme of the accelerator and target hall is presented in Figure 1.

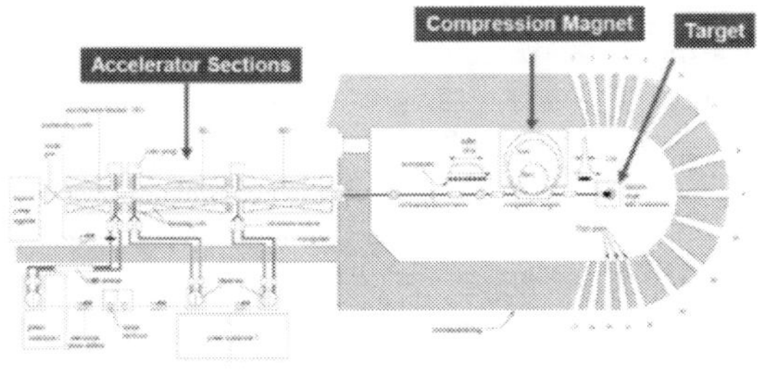

Figure 1. The GELINA electron accelerator

For this simulation, four types of particle are followed: electrons, positrons, gammas, and neutrons. In this first step, the environment of the electron gun was not included in the model, as the purpose of this calculation is only to get the source term. The calculated neutron spectrum is illustrated in Figure 2. It is compared to the neutron spectrum that was calculated for the CALIBAN experiment [3].

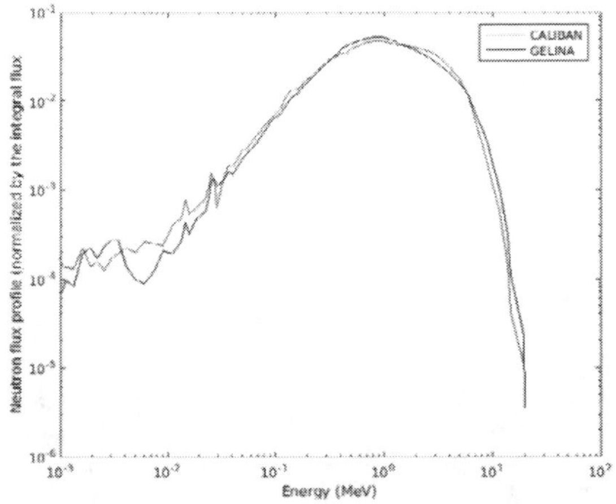

Figure 2. MCNP6 calculation of the neutron flux spectrum in GELINA, compared with the neutron spectrum from the CALIBAN critical facility.

The GELINA spectrum matches the one from CALIBAN very well in the fast energy range. The maximum neutron emission rate in GELINA is 3.2 10^{13} n/s. It is equivalent to a 300W irradiation in CALIBAN (maximum allowed is 1 kW).

An equivalent TRIPOLI4 input file of GELINA was prepared as well, in order to cross check the two codes, based on the same nuclear data for neutrons, photons and electrons. The comparison of neutron spectra is showed on the Figure 3. No significant difference can be observed in the fast energy range between the two codes, except a slightly harder shape in the energy range above 10 MeV for MCNP6. The change from ENDF/B-VII.1 nuclear data file to JEFF-3.2 does not have any impact, except in the lower part of the spectrum, below 10 keV where the slowing down contribution has almost a one order magnitude difference between two libraries. It should be noticed also that the MCNP spectrum has a smoother shape in this region than the TRIPOLI one, a difference that is still under investigation.

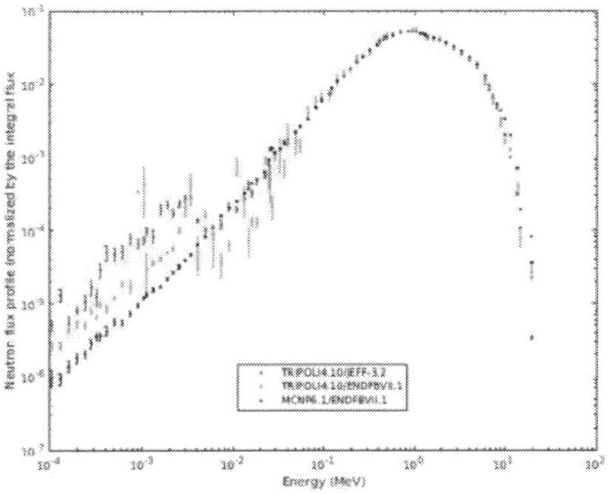

Figure 3. MCNP6.1 vs TRIPOLI4.10 calculation of the neutron flux spectrum in GELINA at the U disk position.

3 Characterization of the neutron flux within the target hall

The simulated neutron flux is used in a second step as an input source in a TRIPOLI calculation where the environment of the target hall is taken into account. The geometry is described in a rather simple way: only the concrete walls are modelled, as they should be the highest contributor to the thermal neutron background due to the scattering on light elements (H_2O, Si, Ca…). The neutron flux is calculated at several distances from the target and is characterized in terms of amplitude and energy spectrum. Figure 4 illustrates calculated neutron flux for three distances from the target: 40, 70 and 100 cm.

One can notice that the thermal neutron background is independent of the distance from the source, while the fast neutron component is approximatively decreasing by $1/d^2$, with d the distance from the target. As a consequence, to prevent the influence of thermal background, the material to be irradiated should be positioned as close to the target as reasonably achievable, to maximize the fast over thermal ratio. The main issue when reducing the distance is linked to the dose rate of operators who will prepare the device and retrieve the activated dosimeters after the irradiation. A compromise will have to be found to satisfy both aspects. Some tests are currently ongoing to evaluate the dose rate within the target hall. A short time irradiation of gold, indium, nickel and rhodium foils will be prepared either to evaluate the expected activity, in order to prepare more accurate design calculations.

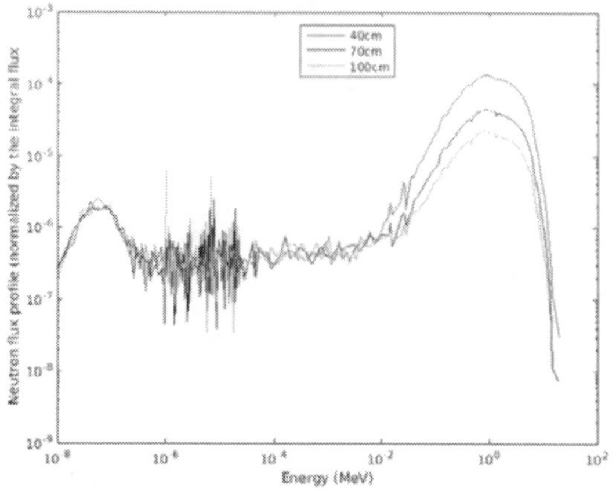

Figure 4. TRIPOLI4 neutron flux for several distances from the Uranium target.

4 Feasibility of an integral transmission experiment through a Uranium slab

We have tested the possibility and the interest of an integral transmission experiment through a set of different depleted Uranium disks (see Figure 5).

Figure 5. Depleted Uranium disks available at IRMM

The disks have a diameter of 21 cm and various thicknesses of 0.5 to 2 cm. They can be piled to build a cylinder of total thickness 15 cm, interleaved with Al sheets for containment and dosimeter holders. γ-spectrometry measurements were performed on one of them in order to record the 186 keV and 1001 keV peaks corresponding respectively to ^{235}U and ^{238}U decay. The ^{235}U mass enrichment of the disk was found to be (0.264±0.012)% (1σ).

Calculations were performed to evaluate the integral transmission of the neutron flux, using various threshold dosimeters. For this study, we have considered a mean distance of 100cm from the source but the conclusions should not be changed if we revise this value. We present the results hereafter.

4.1 Integral transmission coefficients T

Firstly, we evaluate the integral transmission coefficient $T(x)$ which is defined as follows:

$$T(x) = R(x) / R(0) \quad (1)$$

where R(x) defines the microscopic activation rate on the dosimeter placed behind a thickness of Uranium Δx and R(0) is the microscopic activation rate on the dosimeter placed on the front size of the Uranium slab.

Table 1 presents the calculated coefficients for three kinds of threshold reactions (E50% represents the energy corresponding to half of the total reaction rate in a pure ^{252}Cf spectrum). They have been obtained with TRIPOLI4.9 and the JEFF-3.1.1 nuclear data library. For a qualitative appreciation of the influence of the room return effect, we have computed the same transmission coefficients without the concrete walls of the target hall.

Table 1. Integral transmission coefficients for different threshold dosimeters and different penetration length (uncertainties are given at 1σ)

Threshold reaction	E50%	x (cm)	*Integral transmission coefficient T(x)*	
			Geometry with concrete walls	Geometry without concrete walls
^{115}In(n,n')	2.7	3	0.620 ± 0.017	0.645 ± 0.005
		6	0.350 ± 0.011	0.351 ± 0.003
		9	0.192 ± 0.017	0.184 ± 0.002
		12	0.105 ± 0.015	0.091 ± 0.001
		15	0.073 ± 0.004	0.043 ± 0.001
^{54}Fe(n,p)	4.3	3	0.600 ± 0.025	0.631 ± 0.008
		6	0.339 ± 0.016	0.355 ± 0.005
		9	0.196 ± 0.012	0.192 ± 0.003
		12	0.105 ± 0.008	0.102 ± 0.002
		15	0.063 ± 0.006	0.052 ± 0.002
^{24}Mg(n,p)	8.3	3	0.617 ± 0.051	0.629 ± 0.051
		6	0.338 ± 0.031	0.383 ± 0.058
		9	0.213 ± 0.029	0.211 ± 0.033
		12	0.115 ± 0.018	0.123 ± 0.063
		15	0.051 ± 0.011	0.070 ± 0.006

The dependence of transmission coefficients with Δx is very close to an exponential function *exp*(-ax), which is consistent with the H. Bethe empirical formulation [5]. The influence of concrete walls is only significant for transmission with x = 15 cm, as the relative abundance of slowing down neutrons become significant, compared to the low transmission rate of fast neutrons. Moreover, as it is very unlikely that scattered neutrons could have high energies, the wall return effect will decrease with the activation threshold. To mitigate this issue, it is planned to add a 0.5 cm disk at the back side of the 15 cm cylinder. Acting as a shield to the thermal neutron background, it would significantly reduce the call return contribution, especially for the low energy dosimeters.

4.2 Sensitivity of integral transmission coefficients to ^{238}U partial cross sections

Secondly we have computed integral sensitivity coefficients to evaluate how the transmission coefficients will be related to reaction cross sections. Three reactions have been tested by changing three different input cross sections by 1% over the full energy range: elastic, total inelastic (continuum + discrete) and fission. Preliminary calculations have showed that other reactions (i.e. radiative capture and (n,xn) reactions) can be neglected.

To do so, we have used the correlated sampling method [6][7] in TRIPOLI4 which evaluates the change in reaction rates thanks to a modified weight, according to the change in the cross section.

Table 2 gives the sensitivity coefficients $S(\sigma_i, \Delta x)$ defined as follows:

$$S(\sigma_i, \Delta x) = [\delta T(\Delta x) / T(\Delta x)] / [\delta\sigma_i / \sigma_i] \quad (2)$$

The absolute uncertainty is typically ±0.05 on $S(\sigma_{el}, \Delta x)$, ±0.02 on $S(\sigma_{inel}, \Delta x)$ and $S(\sigma_f, \Delta x)$.

Table 2. Sensitivity coefficients of transmission coefficients T to different ^{238}U cross section

Threshold reaction	E50%	Δx (cm)	$S(\sigma_i, \Delta x)$		
			Elastic scattering	Total inelastic	Fission
$^{115}In(n,n')$	2.7	3	0.03	-0.36	-0.09
		6	-0.04	-0.77	-0.16
		9	-0.16	-1.13	-0.25
		12	-0.26	-1.39	-0.30
		15	-0.42	-0.99	-0.21
$^{54}Fe(n,p)$	4.3	3	-0.03	-0.29	-0.11
		6	-0.08	-0.63	-0.21
		9	-0.14	-0.97	-0.34
		12	-0.25	-1.24	-0.41
		15	-0.45	-1.11	-0.34
$^{24}Mg(n,p)$	8.3	3	-0.11	-0.07	-0.16
		6	-0.22	-0.26	-0.29
		9	-0.09	-0.38	-0.49
		12	-0.20	-0.56	-0.57
		15	-0.38	-0.66	-0.65

For threshold reactions occurring in the [2-5 MeV] energy range, the sensitivity coefficient mostly depends on the plateau value of the inelastic scattering cross section. The optimal compromise between room return effect and the relative contributions of elastic and fission cross sections is 9 or 12 cm attenuation where the inelastic is 3-10 times higher than other reaction. For reactions occurring at higher energies than 5 MeV where the fission cross section increases while the inelastic decreases, the sensitivity becomes equivalent between the two reactions. The uncertainty of the fission cross section is <1%, so that C/E differences are expected to reflect differences in inelastic cross section [8].

Based on these preliminary results, a project proposal have been submitted in June 2016 and accepted by the EUFRAT PAC (European facility for nuclear reaction and decay data measurements) for a one week experiment on ^{238}U scheduled in September 2016.

5 Application for other materials: Fe, Na and MgO

Iron, Sodium and Magnesium Oxide are other relevant materials for the ASTRID 4th generation prototype reactor: they are respectively used in assemblies and claddings, for the fuel cooling and to prevent radial leakage to preserve the core vessel. In the fast energy regions, elastic and inelastic scattering cross sections are important nuclear data to validate for design calculations. Moreover, covariance matrices, which are required for uncertainty analysis, are missing for ^{16}O and ^{24}Mg in the current JEFF-3.2 library.

In this framework, we have investigated the possibility of using large blocks of these materials within the target hall of GELINA, in order to make integral transmission measurements as well. For the block dimension definition, one of our constraints was to adapt the block to the support table that has been manufactured for the experiment on ^{238}U: the maximum material size was imposed to be a rectangular block of maximum 60x60x200cm (largest dimension in the direction towards the source). The positions for computing the activation rates are the following ones: 0, 5, 10, 15, 20, 25, 30, 40, 50, 60, 70, 80, 100, 120, 140, 160, 180 and 200cm (0cm being the front side position inside the block). The list of dosimeters is : $^{103}Rh(n,n')$, $^{115}In(n,n')$, $^{58}Ni(n,p)$, $^{24}Mg(n,p)$.

A general scheme is represented in Fig.6 (the actual scale is not respected).

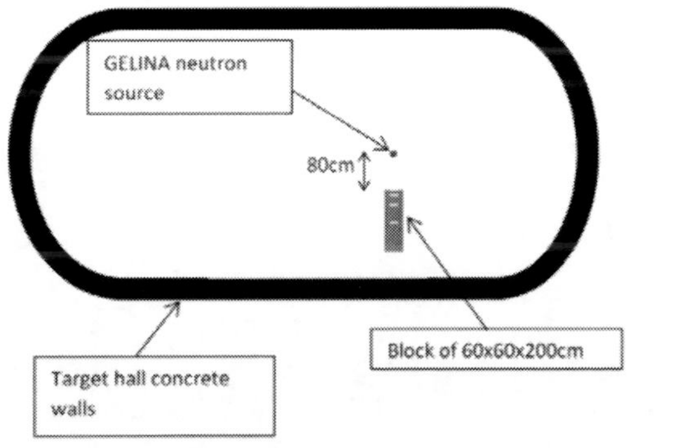

Figure 6. Scheme of the GELINA target hall with the block of material to be investigated

5.1 Results for Na

We are representing hereafter the relative activation rate distribution within a pure sodium block of 60x60x200 cm (Figure 7) and the sensitivity coefficients for the $^{103}Rh(n,inel)$, $^{115}In(n,inel)$, $^{58}Ni(n,p)$ and $^{24}Mg(n,p)$ threshold reactions (Figures 8 to 11).

 DOI: 10.1051/epjconf/201715301023

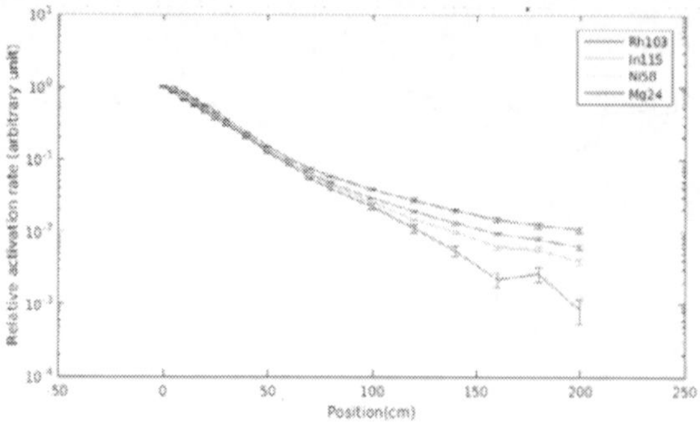

Figure 7. Activation rate relative distribution as a function of dosimeter position within the sodium block

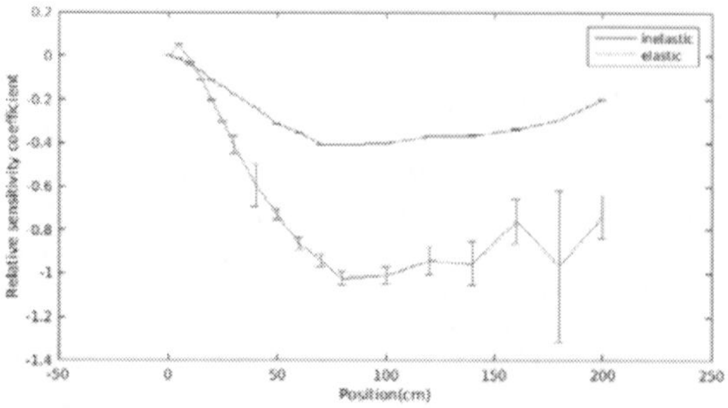

Figure 8. Sensitivity coefficients of the integral transmission rate of ^{103}Rh(n,n') for the ^{23}Na scattering cross sections

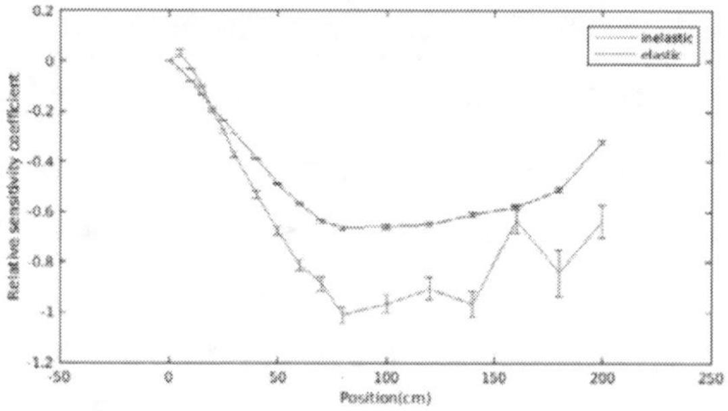

Figure 9. Sensitivity coefficients of the integral transmission rate of ^{115}In(n,n') for the ^{23}Na scattering cross sections

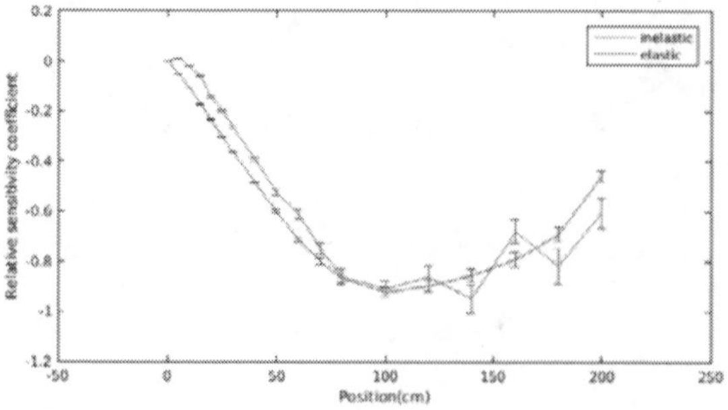

Figure 10. Sensitivity coefficients of the integral transmission rate of ^{58}Ni(n,p) for the ^{23}Na scattering cross sections

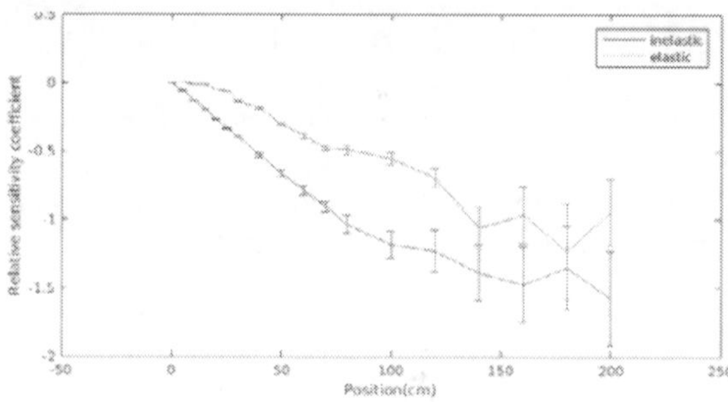

Figure 11. Sensitivity coefficients of the integral transmission rate of ^{24}Mg(n,p) for the ^{23}Na scattering cross sections

The lowest transmission rates vary from 10^{-3} to 10^{-2} depending on the E50% threshold energy of the dosimeter. The shape is roughly linear in log-scale for positions below 60cm.

Sensitivity coefficients are increasing almost linearly from 0 to 80cm then they all tend to a saturated value above 100cm. The breakdown between the elastic and inelastic contributions of ^{23}Na shows the first one is dominant for low energy reactions (Rh), while the second one becomes dominant for high energy reactions (Mg).

Based on these different elements, a maximum thickness of 100cm seems to be the best compromise for an integral transmission experiment with this material.

5.2 Results for Fe

We are representing hereafter the relative activation rate distribution within a pure iron block of 60x60x200cm (Figure 12) and the sensitivity coefficients for the ^{103}Rh(n,inel), ^{115}In(n,inel), and ^{24}Mg(n,p) threshold reactions (Figures 13 to 16).

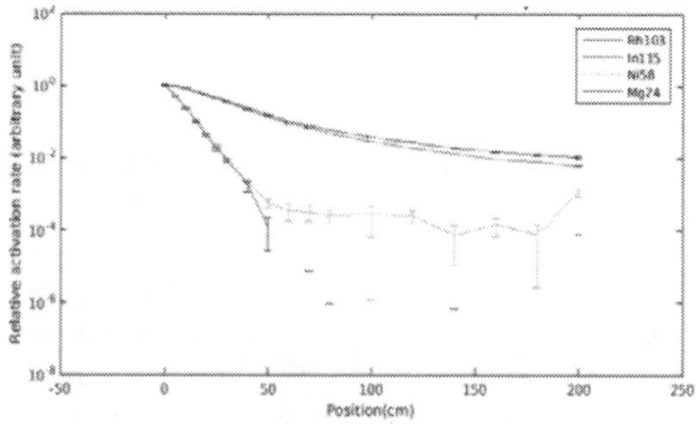

Figure 12. Activation rate relative distribution as a function of dosimeter position within the iron block

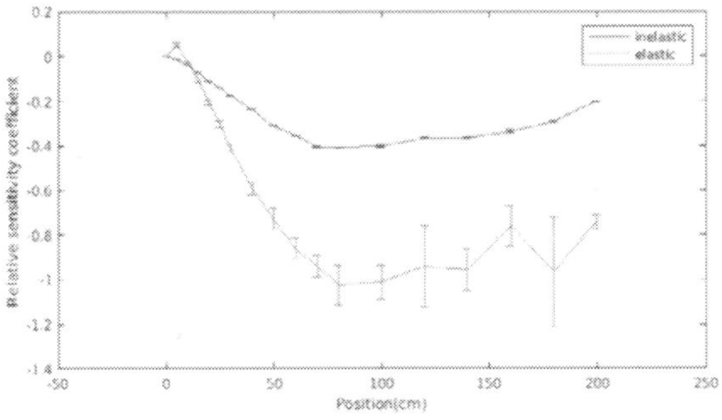

Figure 13. Sensitivity coefficients of the integral transmission rate of $^{103}Rh(n,n')$ for the ^{56}Fe scattering cross sections

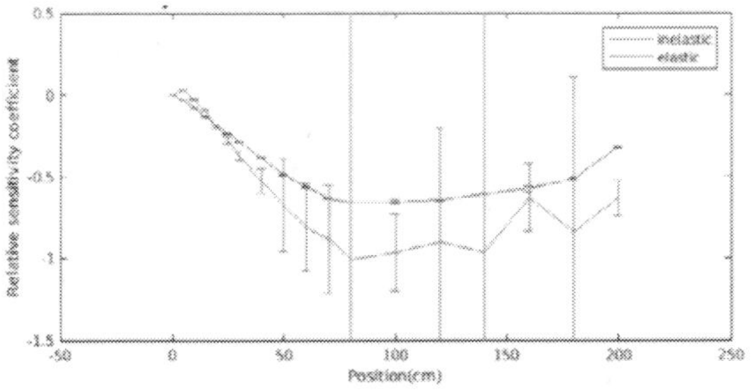

Figure 14. Sensitivity coefficients of the integral transmission rate of $^{115}In(n,n')$ for the ^{56}Fe scattering cross section

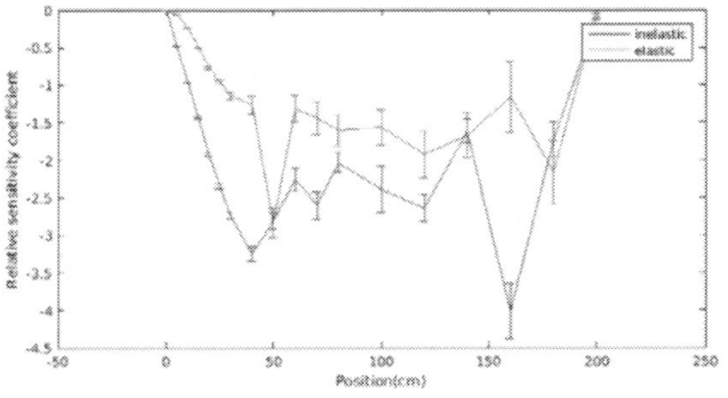

Figure 15. Sensitivity coefficients of the integral transmission rate of $^{58}Ni(n,p)$ for the ^{56}Fe scattering cross sections

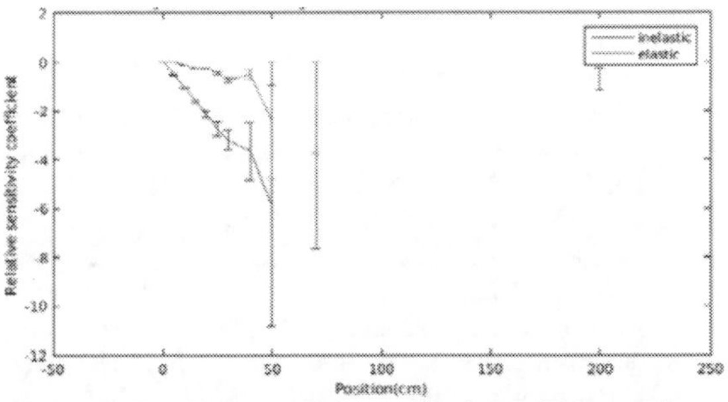

Figure 16. Sensitivity coefficients of the integral transmission rate of $^{24}Mg(n,p)$ for the ^{56}Fe scattering cross sections

The lowest transmission rates vary from 10^{-4} to 10^{-2} depending on the E50% threshold energy of the dosimeter. The shape is roughly linear in log-scale for positions below 60 cm.

Sensitivity coefficients are increasing almost linearly from 0 to 80 cm for Rh and In, but only from 0 to 40 cm for Ni and Mg, then they all tend to a saturated value. The breakdown between the elastic and inelastic contributions of ^{56}Fe shows the first one is dominant for low energy reactions (Rh), while the second one becomes dominant for high energy reactions (Mg).

Based on these different elements, a maximum thickness of 50 cm seems to be the best compromise for an integral transmission experiment with this material.

5.3 Results for MgO

We are representing hereafter the relative activation rate distribution within a pure magnesium oxide block of 60x60x200 cm (Figure 17) and the sensitivity coefficients for $^{103}Rh(n,inel)$, $^{115}In(n,inel)$, $^{58}Ni(n,p)$ and $^{24}Mg(n,p)$ threshold reactions (Figures 18 to 21).

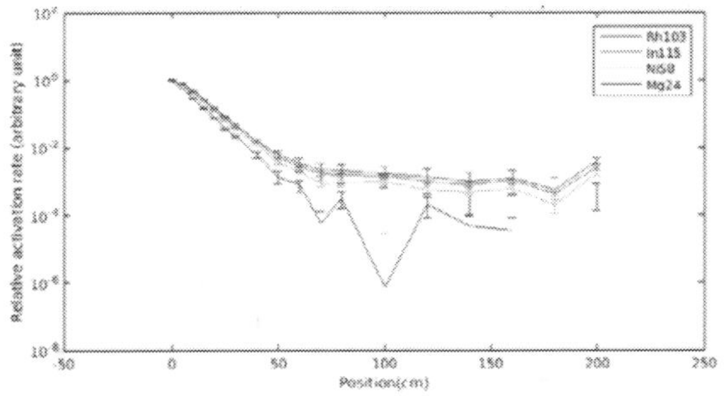

Figure 17. Activation rate relative distribution as a function of dosimeter position within the magnesium oxide block

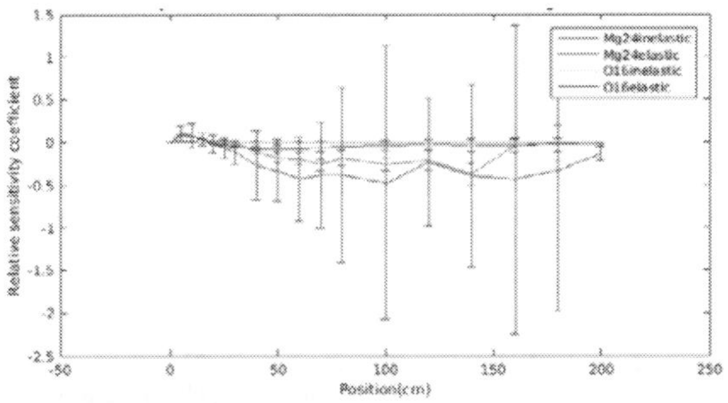

Figure 18. Sensitivity coefficients of the integral transmission rate of $^{103}Rh(n,n')$ for the Mg and O scattering cross sections

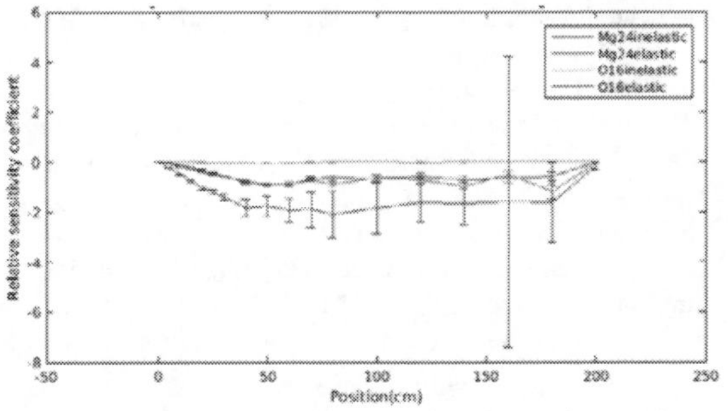

Figure 19. Sensitivity coefficients of the integral transmission rate of $^{115}In(n,n')$ for the Mg and O scattering cross sections

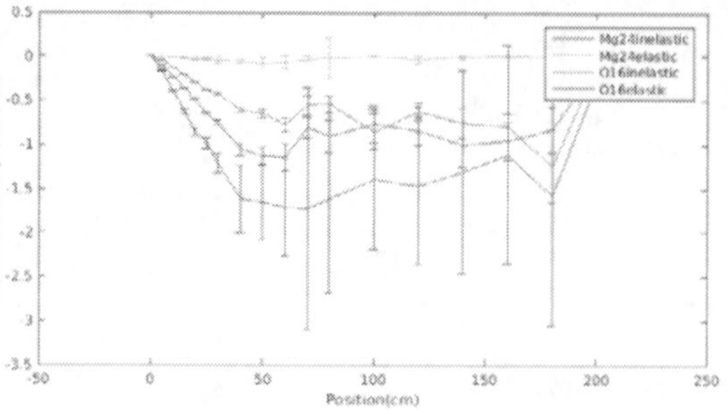

Figure 20. Sensitivity coefficients of the integral transmission rate of ^{58}Ni(n,p) for the Mg and O scattering cross sections

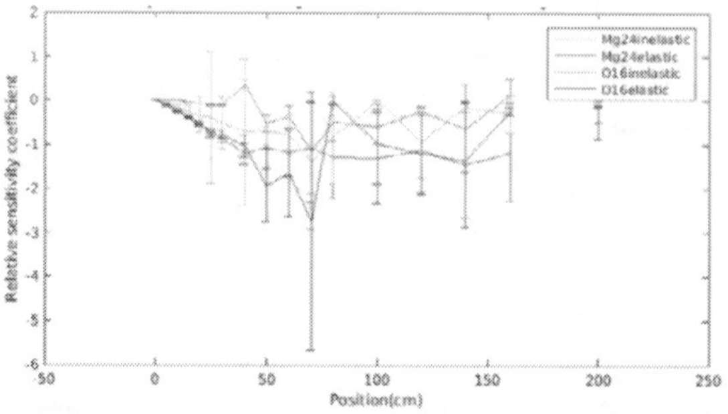

Figure 21. Sensitivity coefficients of the integral transmission rate of ^{24}Mg(n,p) for the Mg and O scattering cross sections

The lowest transmission rate varies from 10^{-4} to 10^{-2} depending on the E50% threshold energy of the dosimeter. The shape is roughly linear in log-scale for positions below 50cm.

Sensitivity coefficients are increasing almost linearly from 0 to 50cm then they all tend to a saturated value. The sensitivity breakdown reveals a major contribution of ^{16}O elastic scattering, followed by ^{24}Mg elastic and inelastic scattering. Sensitivities reach a maximum for thickness of about 50cm.

Based on these different elements, a maximum thickness of 50cm seems to be the best compromise for an integral transmission experiment with this material.

6 Conclusions

Preliminary calculations were performed to assess the feasibility of shielding experiments in the target hall of the GELINA facility. The main concerns were the neutron flux shape and intensity, the influence of the room return effect, and how the integral transmission coefficient would be sensitive to elastic and inelastic scattering cross sections.

From this preliminary study, it can be concluded that GELINA has the potential to perform such experiments. The total neutron emission rate is equivalent to the one that was used in the CALIBAN experiments [3]. The depleted Uranium disks available at IRMM have the suitable characteristics to obtain a targeted feedback on the plateau value of the ^{238}U inelastic cross section, with low contributions of elastic and fission cross sections for a selection of relevant dosimeters.

Other applications have been considered based on materials of interest for 4th generation-type reactors, like ASTRID, including iron, magnesium oxide and sodium. It can be concluded that sensitivity coefficients close to unity can be reached with material thickness of 50cm for Fe and MgO and 100cm for Na. An appropriate selection of different activation dosimeters with threshold energies covering the range of interest for reactor applications would provide feedback on both elastic and inelastic scattering cross sections. This would be useful to identify possible improvements on the cross sections and/or to add constraints in the evaluation process.

Additional design calculations would be required to optimize the experiment: for instance, the addition of a thin boron shelter covering the device to be irradiated would help reducing the influence of the room return effect and to get feedback on the neutron cross sections in the lower energy range (typically less than 1MeV). Moreover, the analysis will have to assess the influence of the calculated neutron flux on the transmission coefficient and also the impact of angular distributions, sensitivity coefficients that will require new developments in the TRIPOLI4 Monte-Carlo code.

References

1. I. Kodelia, A. Milocco, Pedro Ortego and Enrico Sartori, 20 Years of SINBAD (Shielding Integral Benchmark Archive and Database), Progress in Nuclear Science and Technology, Volume 4 (2014) pp. 308-311.
2. A. Santamarina, P. Blaise, N. Dos Santos, C. Vaglio-Gaudard, "Nuclear data uncertainty propagation on Power Maps", PHYSOR 2014, Kyoto, Japan, September 28 – October 3, 2014.
3. D. Bernard, P. Leconte, et al., EXCALIBUR at CALIBAN: a neutron transmission experiment for ^{238}U(n,n'$_{continuum}$γ) nuclear data validation, Proceedings of the International Conference ANIMMA 2015, Lisboa (Portugal), April 2015.
4. M. Flaska, A. Borella, D. Lathouwers, L.C. Mihailescu, W. Mondelaers, A.J.M. Plompen, H. van Dam, T.H.J.J. van de Hagen, Modeling of the GELINA neutron target using coupled electron-photon-neutron transport with the MCNP4C3 code, NIM A, **531** (2004) 392-406.
5. H.A. Bethe, J.R. Beyster, and R.E. Carter, Inelastic Cross Sections for Fission Spectrum Neutrons I, II, III, IV, Journal of Nuclear Energy I, 1956 (Vol. 3), 1957 (Vol. 4).
6. G. Dejonghe, "Etude d'Effets Différentiels par la Méthode de Monte Carlo dans la Cadre de l'Equation du Transport – Applications aux calculs de protection et de neutronique," PhD. Thesis, University of Paris XI, April, 1982.
7. Perturbation Calculations by the Correlated Samples Method, Proc. of the Joint LANL – CEA Meeting on Monte Carlo Method and Applications in Neutronics, p. 311, April, 1985.

8. A. Carlson et al., "International evaluation of neutron cross section standards", Nuclear Data Sheets 110 (2009) 3215.

EPJ Web of Conferences **153**, 01024 (2017) DOI: 10.1051/epjconf/201715301024
ICRS-13 & RPSD-2016

JENDL-4.0/HE Benchmark Test with Concrete and Iron Shielding Experiments at JAEA/TIARA

Chikara Konno[1*], *Norihiro* Matsuda[1], *Saerom* Kwon[2], *Masayuki* Ohta[2], and *Satoshi* Sato[2]

[1]Japan Atomic Energy Agency, Nuclear Science and Engineering Center, 319-1195 Tokai-mura, Japan
[2] National Institutes for Quantum and Radiological Science and Technology, Fusion Energy Research and Development Diretorate, 039-3212 Rokkasho-mura, Japan

Abstract. As a benchmark test of JENDL-4.0/HE released in 2015, we have analyzed the concrete and iron shielding experiments with the quasi mono-energetic 40 and 65 MeV neutron sources at TIARA in JAEA by using MCNP5 and ACE files processed from JENDL-4.0/HE with NJOY2012. As a result, it was found out that the calculation results with JENDL-4.0/HE agreed with the measured ones in the concrete experiment well, while they underestimated the measured ones in the iron experiment with 65 MeV neutrons more for the thicker assemblies. We examined the ^{56}Fe data of JENDL-4.0/HE in detail and it was considered that the larger non-elastic scattering cross sections of ^{56}Fe caused the underestimation in the calculation with JENDL-4.0/HE for the iron experiment with 65 MeV neutrons.

1 Introduction

A Japanese Evaluated Nuclear Data Library (JENDL) special purpose file, JENDL-4.0 High Energy File (JENDL-4.0/HE) [1], is an extended version of JENDL-4.0 for incident neutrons up to 200 MeV for 130 nuclei, and a nuclear data library for incident protons up to 200 MeV for 133 nuclei, which are relevant to high-energy accelerator applications. It was released in November, 2015.

More than 20 years ago shielding experiments at the Takasaki Ion Accelerators for Advanced Radiation Application (TIARA) in Japan Atomic Energy Agency (JAEA) were carried out for concrete and iron with quasi-mono energetic 40 or 65 MeV neutrons produced by bombarding 43 or 68 MeV protons to a ^{7}Li target, which were collimated and injected to a test shield of 1.2 m x 1.2 m as shown in Fig. 1 [2]. The neutron spectra above 10 MeV just behind the test concrete or iron shield on beam axis were measured with a BC501A scintillation detector. These experiments are very useful for benchmarking higher energy neuron incident nuclear data libraries [3]. So far we benchmarked FENDL-3.0 [4], JENDL/HE-2007 [5] and ENDF/B-VII.1 [6] with these experiments and deduced the following conclusions [7].

1) Concrete experiments : The calculation result with JENDL/HE-2007 agrees with the measured one best. Those with FENDL-3.0 and ENDF/B-VII.1 drastically overestimate the measured data. It is confirmed that this overestimation is due to both the larger elastic and smaller non-elastic scattering data of ^{16}O in FENDL-3.0 and ENDF/B-VII.1.
2) Iron experiments : The calculation result with FENDL-3.0 agrees with the measured one best. That with

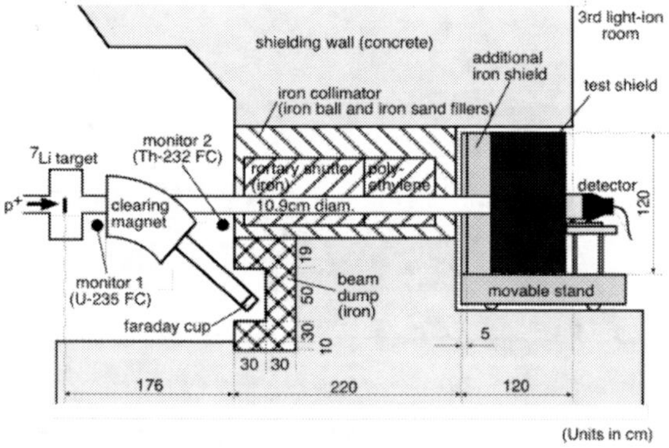

Fig. 1. Experimental configuration of TIARA shielding experiments.

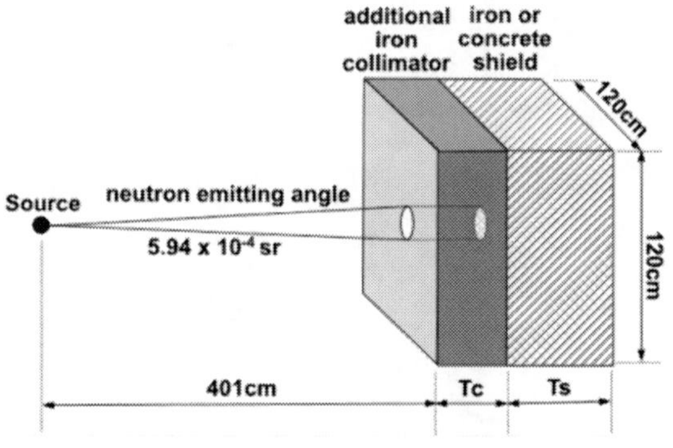

Fig. 2. Calculation model.

* Corresponding author: konno.chikara@jaea.go.jp

JENDL/HE-2007 fairly agrees with the experimental data. That with ENDF/B-VII.1 drastically overestimates the measured one, which is confirmed to be due to the smaller non-elastic scattering data of ^{56}Fe in ENDF/B-VII.1.

In this study we carry out a benchmark test of JENDL-4.0/HE with the concrete and iron shielding experiments at TIARA in JAEA.

2 Analysis

The measured spectrum data of quasi-mono energetic 40 or 65 MeV neutrons were adopted as the source neutron in the analysis. Only the collimated neutron beam and test shield (iron or concrete) with additional shields were modeled in the analysis as shown in Fig. 2. The measured neutron spectra above 10 MeV just behind the test shield on beam axis were compared with the calculated ones in this study.

The Monte Carlo code MCNP-5.14 [8] and the ACE file processed from JENDL-4.0/HE with the NJOY2012.50 [9] code by ourselves were used in the analysis. We also used the official ACE files of JENDL/HE-2007 [10] and FENDL-3.1b (the revised FENDL-3.0) [11]. FENDL-3.1b was used only for the iron experiment, because it was pointed out that ^{16}O in FENDL-3.0 had some problems [7] and the ^{16}O was not revised in FENDL-3.1b. ENDF/B-VII.1 was not used because of defects of the ^{56}Fe and ^{16}O data [7].

3 Results and discussion

3.1 Concrete experiment with 40 MeV neutrons

Figure 3 shows the measured and calculated neutron spectra in the concrete experiment with 40 MeV neutrons. The agreement between the measured and calculated spectra is roughly good. For more quantitative comparison of the overall tendency, Figs. 4 and 5 show the ratios (C/Es) of the calculated peak (35 - 45 MeV) and continuum (10 - 35 MeV) neutron fluxes to the measured ones, respectively, where "Expt. error" means an experimental error estimated from the errors of the measured neutron spectra by assuming no correlation among the measured flux errors in Fig. 3 because no correlation data among the measured flux errors were not specified in Ref. 2. The calculation results with JENDL/HE-2007 and JENDL-4.0/HE agree with the measured one well, though the calculation result with JENDL-4.0/HE is slightly better than that with JENDL/HE-2007.

3.2 Concrete experiment with 65 MeV neutrons

The measured and calculated neutron spectra in the concrete experiment with 65 MeV neutrons are shown in Fig. 6. The calculated spectra generally represent the measured ones. Figures 7 and 8 show more quantitative

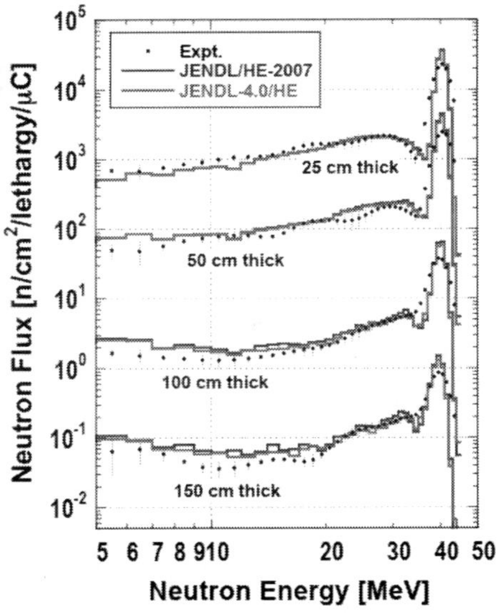

Fig. 3. Measured and calculated neutron spectra in concrete experiment with 40 MeV neutrons.

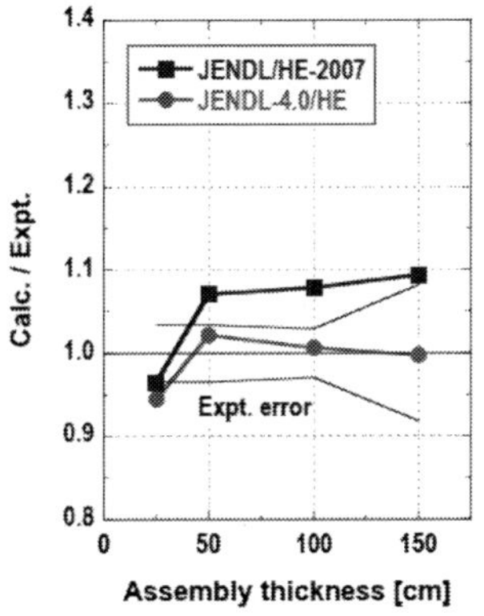

Fig. 4. C/Es of peak (35-45 MeV) neutron fluxes in concrete experiment with 40 MeV neutrons.

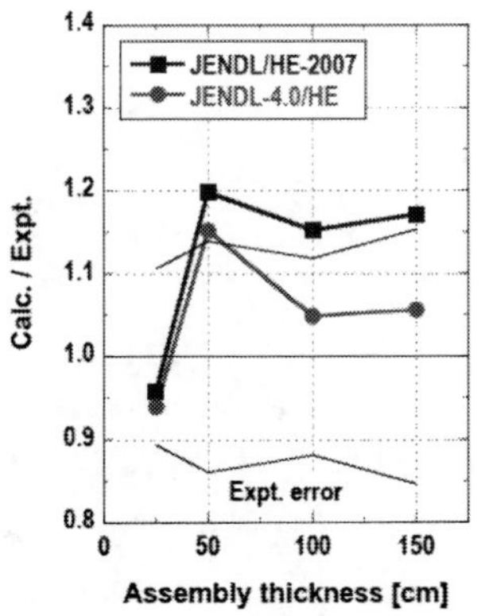

Fig. 5. C/Es of continuum (10-35 MeV) neutron fluxes in concrete experiment with 40 MeV neutrons.

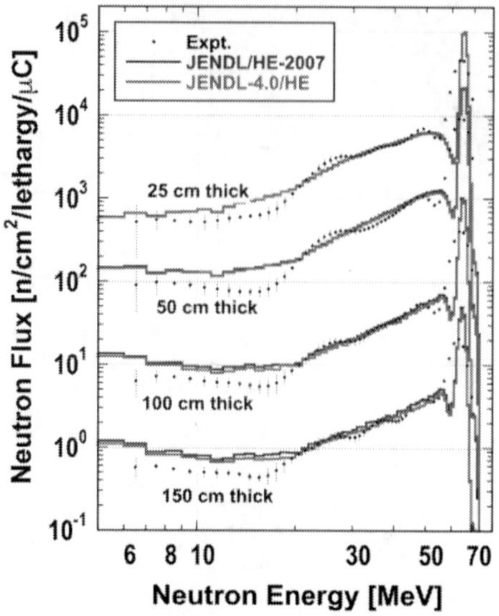

Fig. 6. Measured and calculated neutron spectra in concrete experiment with 65 MeV neutrons.

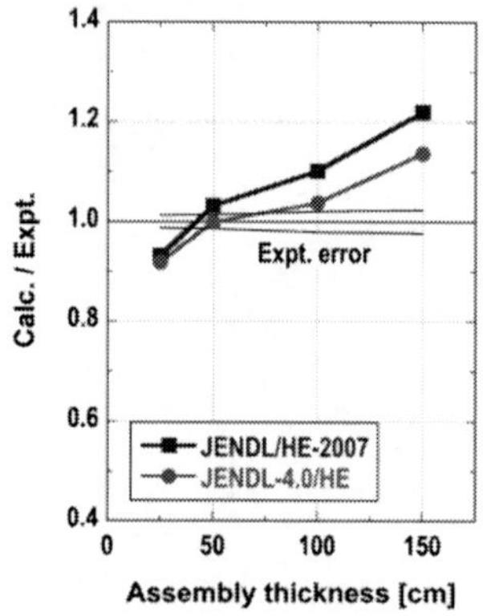

Fig. 7. C/Es of peak (60-70 MeV) neutron fluxes in concrete experiment with 65 MeV neutrons.

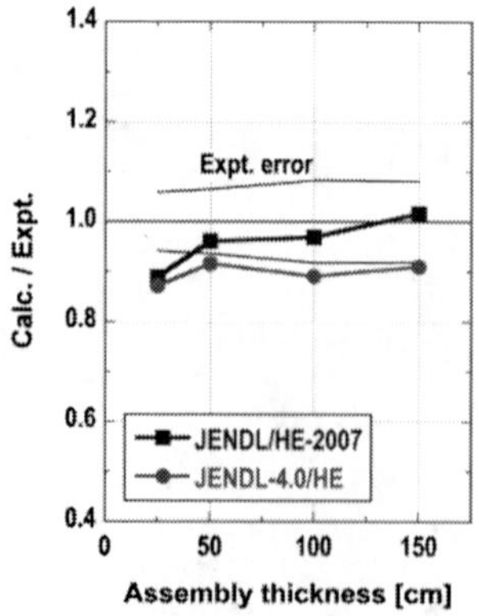

Fig. 8. C/Es of continuum (10-60 MeV) neutron fluxes in concrete experiment with 65 MeV neutrons.

comparison between the measured and calculated neutron fluxes, i.e., the C/Es of the peak (60 - 70 MeV) and continuum (10 - 60 MeV) neutron fluxes, respectively. The calculation results with JENDL/HE-2007 and JENDL-4.0/HE agree with the measured one well.

It is concluded that the JENDL-4.0/HE data are good for concrete.

3.3 Iron experiment with 40 MeV neutrons

Figure 9 shows the measured and calculated neutron spectra in the iron experiment with 40 MeV neutrons. The calculated neutron spectra with FENDL-3.1b, JENDL/HE-2007 and JENDL-4.0/HE agree with the measured one well, though a small strange step, which is probably caused by discontinuity between data below and above 20 MeV, appears around 20 MeV in that in FENDL-3.1b. The quantitative comparison between the measured and calculated neutron fluxes also indicates that the three nuclear data libraries have no problem for iron as shown in Figs. 10 and 11, though JENDL-HE/2007 is slightly better than FENDL-3.1b and JENDL-4.0 for the peak neutron fluxes, while JENDL-4.0 is slightly better than FENDL-3.1b and JENDL-HE/2007 for the continuum neutron fluxes.

3.4 Iron experiment with 65 MeV neutrons

The measured and calculated neutron spectra in the iron experiment with 65 MeV neutrons are shown in Fig. 12. In this logarithmic scale the calculated neutron spectra seem to agree with the measured ones. However from the quantitative comparison between the measured and calculated neutron fluxes in Figs. 13 and 14, it is noted that the C/E tendencies are clearly different among the three nuclear data libraries : the calculated neutron fluxes with FENDL-3.1b agree with the measured ones the best, while those with JENDL-4.0/HE are smaller by ~ 40 % and ~ 20 % at maximum than those with FENDL-3.1b and JENDL-HE/2007, respectively.

3.5 Problems of ^{56}Fe data in JENDL-4.0/HE

In order to specify reasons for the larger underestimation of the calculation result with JENDL-4.0 in the iron experiment with 65 MeV neutrons, the cross section data of ^{56}Fe, which is a main iron isotope, above 20 MeV in FENDL-3.1b, JENDL/HE-2007 and JENDL-4.0/HE are compared. Note that the cross section data of ^{56}Fe above 20 MeV are only non-elastic scattering cross sections and elastic scattering cross sections as shown in Figs. 15 and 16. The non-elastic scattering cross section data in JENDL-4.0/HE are larger than those in FENDL-3.1b and JENDL/HE-2007, while the elastic scattering cross section data in JENDL-4.0/HE are smaller than those in FENDL-3.1b and JENDL/HE-2007.

Thus we temporarily produce two nuclear data, JENDL-4.0/HE non-elas and JENDL-4.0/HE elas, where non-elastic scattering and elastic scattering data above 20 MeV of the ^{56}Fe data in JENDL-4.0/HE are replaced with those in FENDL-3.1b, respectively, and process them

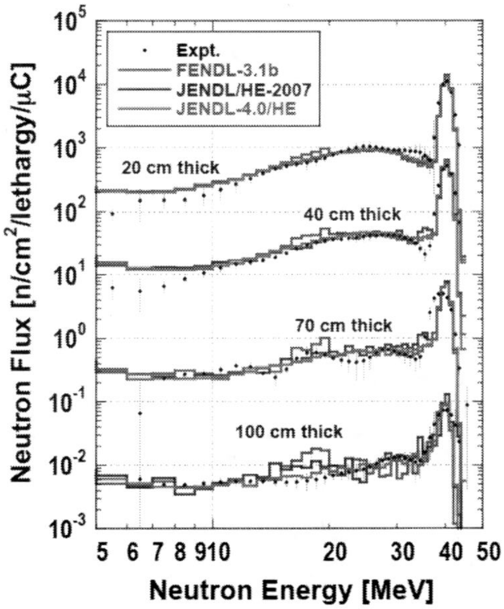

Fig. 9. Measured and calculated neutron spectra in iron experiment with 40 MeV neutrons.

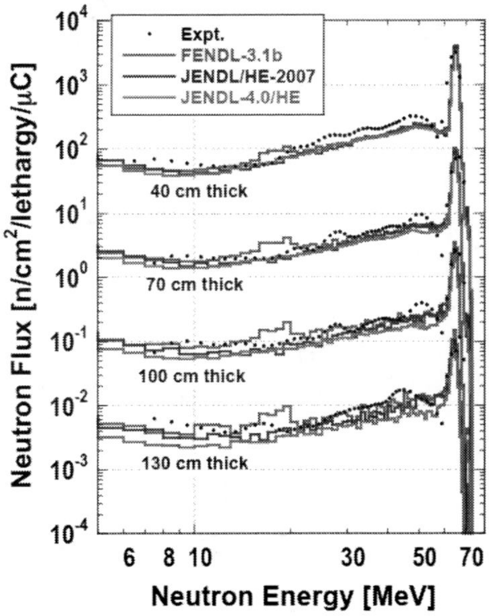

Fig. 12. Measured and calculated neutron spectra in iron experiment with 65 MeV neutrons.

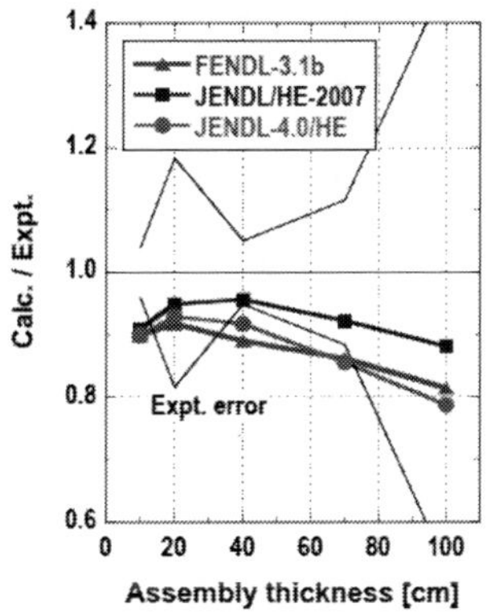

Fig. 10. C/Es of peak (35-45 MeV) neutron fluxes in iron experiment with 40 MeV neutrons.

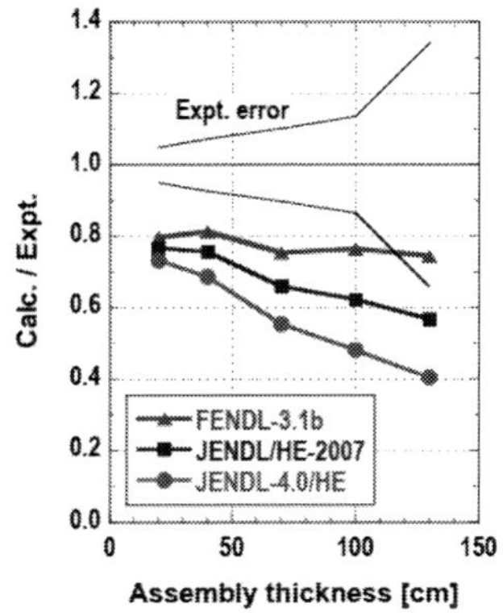

Fig. 13. C/Es of peak (60-70 MeV) neutron fluxes in iron experiment with 65 MeV neutrons.

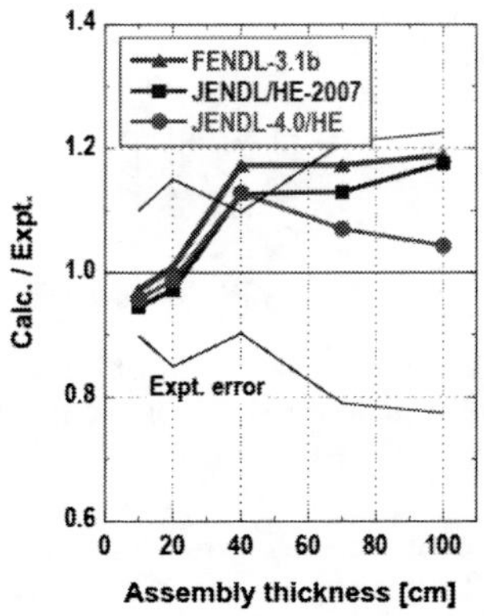

Fig. 11. C/Es of continuum (10-35 MeV) neutron fluxes in iron experiment with 40 MeV neutrons.

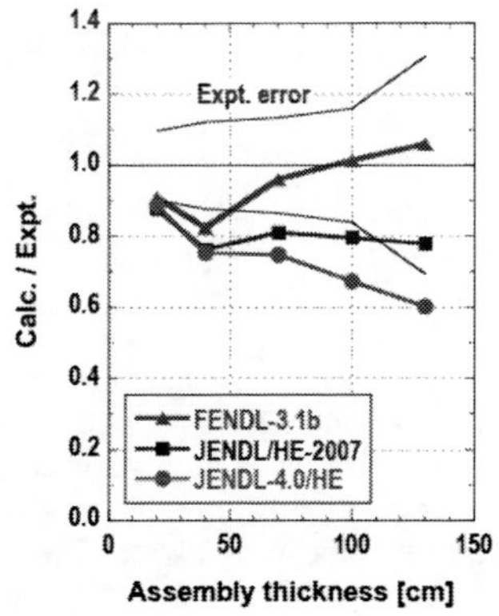

Fig. 14. C/Es of continuum (10-60 MeV) neutron fluxes in iron experiment with 65 MeV neutrons.

DOI: 10.1051/epjconf/201715301024

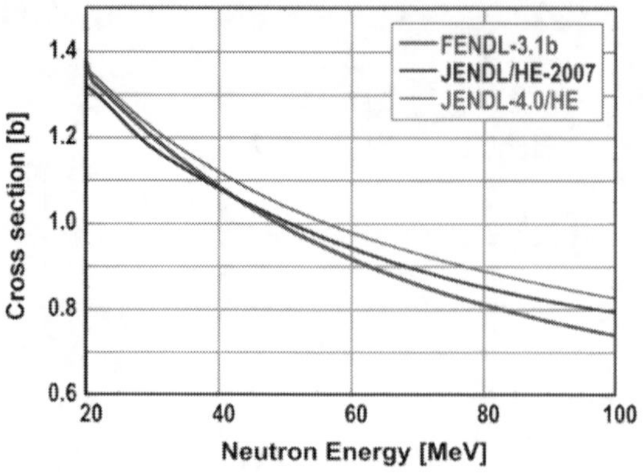

Fig. 15. Non-elastic scattering cross sections of ^{56}Fe.

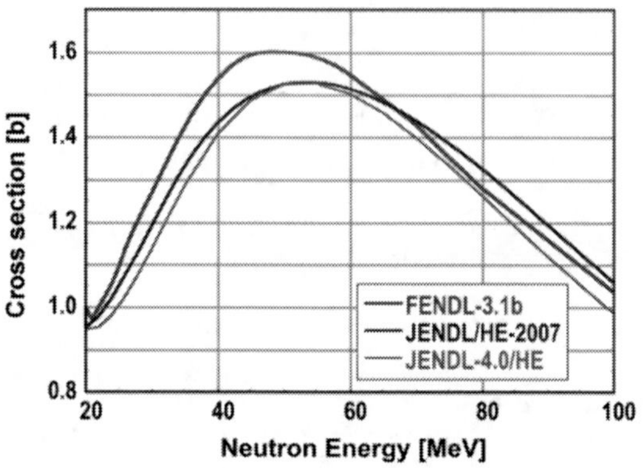

Fig. 16. Elastic scattering cross sections of ^{56}Fe.

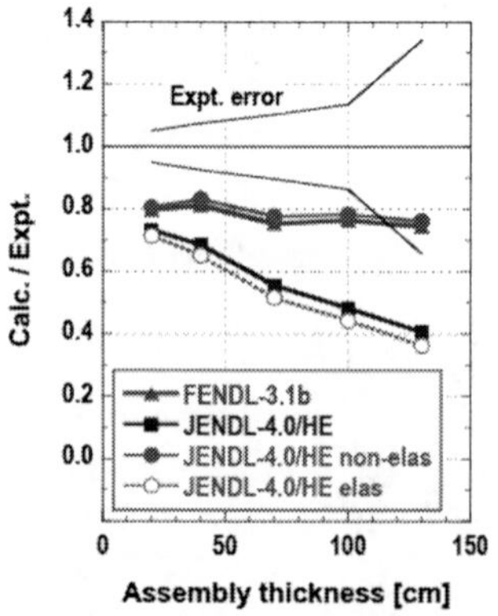

Fig. 17. C/Es of peak (60-70 MeV) neutron fluxes with JENDL-4.0/HE non-elas and JENDL-4.0/HE elas in iron experiment with 65 MeV neutrons.

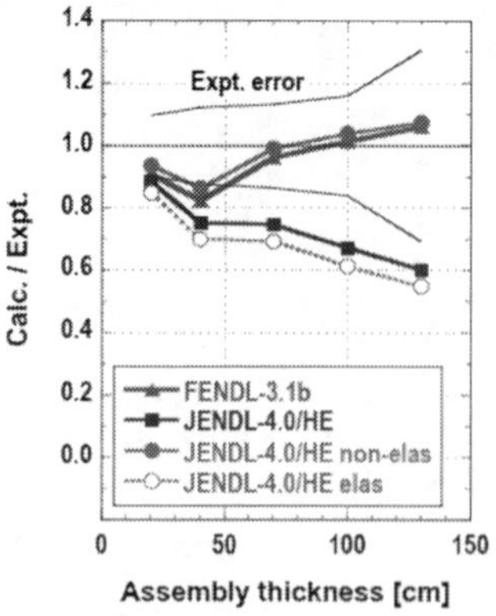

Fig. 18. C/Es of continuum (10-60 MeV) neutron fluxes with JENDL-4.0/HE non-elas and JENDL-4.0/HE elas in iron experiment with 65 MeV neutrons.

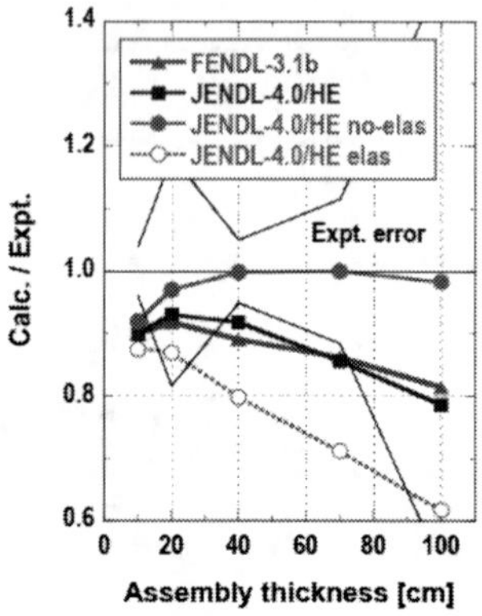

Fig. 19. C/Es of peak (35-45 MeV) neutron fluxes with JENDL-4.0/HE non-elas and JENDL-4.0/HE elas in iron experiment with 40 MeV neutrons.

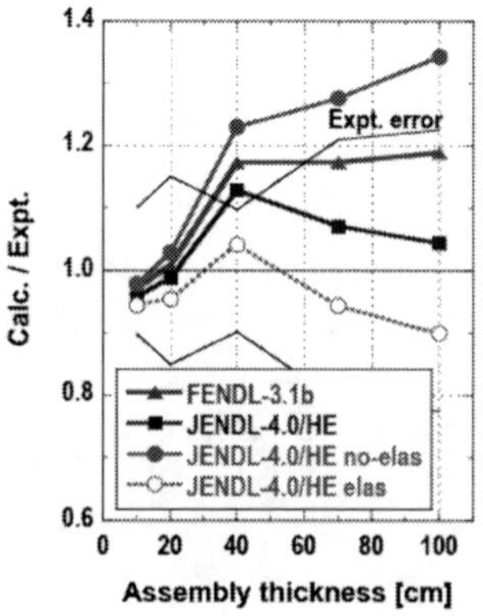

Fig. 20. C/Es of continuum (10-35 MeV) neutron fluxes with JENDL-4.0/HE non-elas and JENDL-4.0/HE elas in iron experiment with 40 MeV neutrons.

with NJOY2012.50. The calculation results with JENDL-4.0/HE non-elas and JENDL-4.0/HE elas for the iron experiment with 65 MeV neutrons are shown in Figs. 17 and 18. The calculation result with JENDL-4.0/HE non-elas is almost the same as that with FENDL-3.1b. On the other hand, the calculation result with JENDL-4.0/HE elas is almost the same as that with JENDL-4.0/HE. It is considered that the smaller neutron fluxes with JENDL-4.0/HE in the iron experiment with 65 MeV neutrons are due to the larger non-elastic scattering cross sections of

^{56}Fe around 65 MeV.

However the results for the iron experiment with 40 MeV neutrons are different as shown in Figs. 19 and 20. The calculation result with JENDL-4.0/HE non-elas is larger than those with FENDL-3.1b and JENDL-4.0/HE and that with JENDL-4.0/HE elas is smaller than those with FENDL-3.1b and JENDL-4.0/HE. It is considered that the calculation result with JENDL-4.0 agrees with that with FENDL-3.1b as a result of accidental cancellation of the overestimation and underestimation.

4 Conclusions

We analyzed the concrete and iron shielding experiments at JAEA/TIARA with JENDL-4.0/HE, JENDL/HE-2007 and FENDL-3.1b for benchmarking JENDL-4.0/HE. The calculation result with JENDL-4.0/HE agreed with the measured one for the concrete experiments. It also agreed with the measured one for the iron experiment with 40 MeV neutrons, but it underestimated the measured neutron fluxes in the iron experiment with 65 MeV neutrons. The comparison of the ^{56}Fe data among JENDL-4.0/HE, JENDL/HE-2007 and FENDL-3.1b indicated that the non-elastic scattering cross section data in JENDL-4.0/HE were larger than those in JENDL/HE-2007 and FENDL-3.1b around 65 MeV, which caused the underestimation of neutron fluxes in the iron experiment with 65 MeV neutrons. The ^{56}Fe data in JENDL-4.0/HE should be revised based on this study.

References

1. S. Kunieda et al., Japan Atomic Energy Agency report, JAEA-Conf 2008-004, 41 (2016)
2. Y. Nakane et al., Japan Atomic Energy Research Institute report, JAERI-Data/Code 96-029 (1996)
3. K. Kondo et al., Prog. Nucl. Sci. Technol. **4**, 322 (2014)
4. R.A. Forrest et al., International Nuclear Data Committee report, INDC(NDS)-0628 (2012)
5. Y. Watanabe et al., J. Korean Phys. Soc. **59**, 1040 (2011)
6. M.B. Chadwick et al., Nuclear Data Sheets **112**, 2887 (2011)
7. C. Konno, et al., Fusion Eng. Design **98-99**, 2178 (2015)
8. X-5 Monte Carlo Team, Los Alamos National Laboratory report, LA-UR-03-1987 (2005)
9. A.C. Kahler et al., Los Alamos National Laboratory report, LA-UR-12-27079 (2012)
10. T. Sasa et al., Japan Atomic Energy Agency report, JAEA-Data/Code 2008-022 (2008)
11. https://www-nds.iaea.org/fendl/

Determination of a non-measurable quantity using information from calculations and experimental measurements: application to the damage rate determination

Stéphane Bourganel[1,a], and Jean-Claude Nimal[1]

[1]*CEA/Saclay – DEN/DANS/DM2S/SERMA – Bâtiment 470, PC 57 – 91191 Gif sur Yvette Cedex - France*

Abstract. This article presents a method dedicated to the determination of the best value of a required quantity which is estimated by calculation, using information closely related, obtained by measurements and calculations. This best value, called thereafter the "target", is not measurable in most cases. DPA and high energy neutron fluence (typically higher than 1 MeV) involved in vessel surveillance programs, using measurements of dosimeters, are some examples of application of this methodology. This methodology is applied without spectrum adjustment, but the spectrum shape is implicitly taken into account.

In this article, an example is presented based on the FLUOLE-2 experimental program, which is developed and conducted by CEA. Neutron information is derived from a set of different kinds of neutron dosimeters. The objective is to estimate the best value of reaction rate values for each kind of dosimeters. All calculations are carried out using TRIPOLI-4 3D pointwise Monte Carlo code, and DARWIN/PEPIN2 depletion code.

This article presents a method dedicated to the determination of the best value of a required quantity which is estimated by calculation, using information closely related, obtained by measurements and calculations. This best value, called thereafter the "target", is not measurable in most cases. DPA and high energy neutron fluence (typically higher than 1 MeV) involved in vessel surveillance programs, using measurements of dosimeters, are some examples for which this methodology can be applied. In this methodology, spectrum adjustment is not used, but the spectrum shape is implicitly taken into account.

1 General case

Let C_k and M_k be a calculated and a measured value of the physical quantity k respectively (a reaction rate for instance; k spans 1 to NK), and C_0 an estimate of the target value also obtained by calculation. C_0 is a physical quantity close to C_k. First, C_0 may be enhanced, using each of the NK measured and calculated information $<C_k, M_k>$, providing NK improved values called observances, what will be noted R_k further. These values represent imperfect results of the target value deduced from each dosimeter. These R_k values can be linearly combined to determine the best estimate target value R_0. R_0 represents an improvement of the first estimated target value C_0. Indeed, C_0 is degraded by imperfections in the modelling (lack of knowledge or uncertainties on modelling data, cross sections ...). By contrast, measurements are perfect with regard to the reality, even if they are biased by experimental uncertainties.

[a] Corresponding author: sbourganel@cea.fr

The determination of the R_0 value requires a weighting of the contribution of each observance R_k. To achieve this goal, two kinds of elements are used:

- The concept of proximity between each couple $<C_k, M_k>$ and the imperfect target C_0, which incorporates the representativeness of the couples $<C_k, M_k>$ with regard to the target.
- The uncertainty associated to each calculated and measured values ($<C_k, M_k>$).

2 Application of the methodology: neutron flux > 1 MeV calculation

Some details to carry out a numerical application of the methodology described above, is presented in this section. The aim is to determine a damage rate, based on neutron flux with energy greater than 1 MeV: ϕ_1. Here, this high energy neutron flux is the required target value. Required information comes from measurements and calculations which are linked to a set of NK neutron dosimeters:

- M_k: The reaction rate values which are deduced from the measured dosimeter activities. M_k values are strongly linked to the reality of the irradiation. They are also independent of the system modelling.
- C_k: The calculated reaction rate values using

TRIPOLI-4 Monte Carlo code [1].

- $\phi_{1,T4}$: The neutron flux with energy greater than 1 MeV, also calculated with TRIPOLI-4® code. $\phi_{1,T4}$ (called C_0 in the previous section) represents an estimate of the target value.

NK observances $\phi_{1,k}$ (called R_k in the previous section) are defined, representing NK values of the neutron flux with energy greater than 1 MeV which are "seen" by dosimeters. Each of these observances is deduced by measurement and calculation. Impact of measurements is taken into account in the $\phi_{1,k}$ determination in the following manner: the reaction rate values M_k deduced from measurements are divided by the calculated cross-section value > 1 MeV ($\sigma_{k,1\ MeV}$) as shown in the formula (1). This cross-section value is the ratio between the calculated reaction rate C_k, and the calculated flux > 1 MeV value $\phi_{1,T4}$. $\sigma_{k,1\ MeV}$ is therefore independent of the power level of the system modelling.

$$(1) \qquad \phi_{1,k} = \frac{M_k}{\sigma_{k,1\ MeV}} = \frac{M_k}{C_k} \times \phi_{1,T4}$$

Formula (1) is valid only if dosimeters were irradiated with a single uniform cycle. This formula must be changed when several cycles are considered [2].

Besides this methodology is applied to the determination the neutron fluence, it can be generalized to others physical quantities as DPA.

The best estimate target value $\phi_{1,estimate}$ is obtained by linear combination of the NK observances $\phi_{1,k}$:

$$(2) \qquad \phi_{1,estimate} = \sum_{k=1}^{NK} W_k \times \phi_{1,k}$$

The importance of the observance $\phi_{1,k}$ in the determination of the target value $\phi_{1,estimate}$ is quantified by the weight value W_k. This weight is obtained by minimizing the variance associated to the best target value $\phi_{1,estimate}$. The best estimate of the target value $\phi_{1,estimate}$ and its associated variance value $\sigma^2(\phi_{1,estimate})$ are given in formulae (3) and (4).

$$(3) \qquad \phi_{1,estimate} = \frac{\overrightarrow{1^t} \times \|CV\phi_1\|^{-1} \times \overrightarrow{\phi_1}}{\overrightarrow{1^t} \times \|CV\phi_1\|^{-1} \times \vec{1}} = \sum_{k=1}^{NK} W_k \times \phi_{1,k}$$

$$(4) \qquad \sigma^2(\phi_{1,estimate}) = \frac{1}{\overrightarrow{1^t} \times \|CV\phi_1\|^{-1} \times \vec{1}}$$

In formulae (3) and (4), $\|CV\phi_1\|$ is the variance-covariance matrix associated to the NK observances $\phi_{1,k}$. It is a square matrix of order NK. By definition, this matrix is positive-definite and thus, it is invertible. $\overrightarrow{\phi_1}$ is the $\phi_{1,k}$ vector of NK observances.

This matrix is obtained by the simple matrix expression for error propagation:

$$(5) \qquad \|CV\phi_1\| = \|S\|^t \times \|CV_{niv3}\| \times \|S\|$$

Formula (5) is based on a linear approximation. ||S|| is a sensitivity matrix which is easily deduced from formula (1).

$\|CV_{niv3}\|$ is a square matrix of order (2×NK+1). It represents the variance-covariance matrix of the NK measured values M_k and the (NK+1) calculated values C_k. It represents uncertainties associated to measured values, modelling data, nuclear data, etc. Paragraph 3 is dedicated to the determination of the $\|CV_{niv3}\|$ matrix.

3 determination of the $\|CV_{niv3}\|$ matrix

The $\|CV_{niv3}\|$ matrix comprises 3sub-matrices:

- One square sub-matrix of order NK, representing uncertainties on measured values. These sub-matrix elements are given by experimenters.
- One square sub-matrix of order (NK+1), representing variance and covariance values associated to calculated values.
- Two rectangular sub-matrices NK × (NK+1) representing correlations between measured and calculated values. We should note that these two kinds of values are usually uncorrelated.

This paragraph is dedicated to the determination of sub-matrix representing calculated values. This is the most difficult point to deal with the problem of uncertainties. Each element of this sub-matrix is given by formula (6).

$$(6) \qquad M(CV_{niv3})_{k,k'} = r_{k,k'}\sigma_k\sigma_{k'}$$

$r_{kk'}$ is the correlation coefficient between dosimeters k and k'. k is between 0 an NK. k=0 corresponds to the calculated target value, and 1≤k≤NK to calculated reaction rate values associated to dosimeters. Data for completing the $\|CV_{niv3}\|$ matrix may have different origins:

- A methodologically uncertainty propagation method. In this way, variance and covariance values are rigorously linked to uncertainties associated with the problem (technological data, nuclear data…). $r_{k,k'}$ values are then known. The URANIE platform [3] developed by CEA is one example to address this need.

- Results from ESTIMA computer program when only variance values are available. ESTIMA allows calculating covariance values according to a methodology presented in chapter 4. Variance values are known through feedback. A practical application of this way is presented in chapter 5 where $r_{k,k'}$ values are deduced from neutron spectra.
- Results from CORRELA computer program (see chapter 4) when reliable uncertainty data are not available. $r_{k,k'}$ values are also deduced from neutron spectra.

4 ESTIMA computer program

A computer program has been developed to carry out numerical applications. It consists of a main program called ESTIMA, dedicated to matrix calculation. It can be used for any kinds of problems. It comprises a function named CORRELA, dedicated to the calculation of the $r_{k,k'}$ correlation factors. ESTIMA allows obtaining the best target value $\phi_{1,estimate}$, and the associated variance $\sigma^2(\phi_1)$. This package allows to process a set of covariance matrices that might be generated randomly.

5 Numerical applications

Numerical applications are presented in this paper. They are based on FLUOLE-2 program [4], and concern reaction rate calculations.

5.1 FLUOLE-2 program

FLUOLE-2 [4] is a benchmark-type experiment dedicated to neutron attenuation analysis with the aim of improving the TRIPOLI-4® [1] Monte Carlo code validation. This two-year program has been developed by CEA (Commissariat à l'Energie Atomique et aux Energies Alternatives), with the financial support of EDF, to be representative of 900 and 1450 MWe Pressurized Water Reactors. In this perspective, different stainless steel structures have been designed and appropriately positioned inside the EOLE facility [5]. In FLUOLE-2 program, a set of dosimeters were irradiated and the activity was measured. Calculated activity of each dosimeter was obtained with a specific calculation scheme based on TRIPOLI-4® Monte Carlo code, and DARWIN/PEPIN2 depletion code.

5.2 TRIPOLI-4®

TRIPOLI-4® [1] is a 3D transport code using full pointwise cross section data. It is dedicated to radiation protection and shielding, nuclear criticality safety, fission and fusion reactor design, and nuclear instrumentation. It is used as a reference tool by CEA, EDF, and several other industrial or institutional partners. In this study, TRIPOLI-4® is used to calculate the neutron source distribution in the core, the neutron propagation through the different structures, and the fission rate values inside fission chambers.

5.3 DARWIN/PEPIN2

DARWIN/PEPIN2 [6] solves Bateman's generalized differential equations governing the time dependence of isotope concentrations. This code may be coupled with the TRIPOLI-4® code. In this analysis, the DARWIN/PEPIN2 code is dedicated to dosimeter activation calculations.

5.4 Applications: general

For the numerical application, 8 dosimeters have been selected from the FLUOLE-2 analysis result database. These dosimeters were irradiated in the same position from the core [4]. The analysis of 8 dosimeters shows relatively good agreement between calculated and measured activity values, as presented in Table 1. Average C/M values for all dosimeters grouped by type are similar to those results.

Table 1. C/M results from FLUOLE-2 analysis

Dosimeters	C/M
^{27}Al(n,α)	1.00 ±0.03
^{54}Fe(n,p)	1.00 ±0.03
^{115}In(n,n')	0.98 ±0.03
^{58}Ni(n,p)	1.00 ±0.03
^{103}Rh(n,n')	0.86 ±0.05
^{117}Sn(n,n')	1.05 ±0.03
^{46}Ti(n,p)	1.00 ±0.03
^{51}V(n,α)	0.98 ±0.04

Numerical applications have the objective to show that it is possible to enhance the target value using sometimes unsatisfying results, and running ESTIMA program with calculated and measured results. This improvement consists of a significant reduction of the variance associated to the target value. The $\|CV_{niv3}\|$ matrix is filled as described in chapter 3: experimental uncertainties are given by measurement laboratory, and calculation uncertainties are derived through feedback on PWR uncertainty analysis.

Dosimeters can be divided into two categories:

- Dosimeters with poor C/M ratio. These results may be due to inadequate knowledges of modelling data (nuclear data, technological data …) or incorrect measurement. Rhodium and tin dosimeters are part of this category.
- Dosimeters with very good C/M ratio. All others dosimeters are classified in this group.

Further on, the main reaction rate for nickel, rhodium, and tin dosimeters will be considered as target values.

5.5 Numerical applications

5.1.1 General

Numerical applications have the objective to demonstrate the robustness of the methodology proposed in the paper. The possibility to improve target values is shown, including when data involved in the analysis process (C/M calculations) are imperfect.

In our test, we will assume that uncertainty on the neutron spectrum and reaction rates for each dosimeter, both calculated with TRIPOLI-4®, is altered by a lack of knowledge of nuclear data. The impact of this on results given by ESTIMA will be checked. To achieve this goal, the calculated spectrum is deliberately distorted by the use of a set of multiplying factors. For each case, the target value is then calculated as well as the associated uncertainty, called residual uncertainty.

5.1.2 Altering spectrum

The calculated spectrum is altered using a multiplying factor $F_p(E)$ defined as follows:

- 0.04 MeV $\leq$ E $\leq$ 8 MeV: $F_p(E)$ increases linearly, with $F_p(0.04\ \text{MeV}) = a_p$ and $F_p(8\ \text{MeV}) = 1$.
- 8 MeV $\leq$ E $\leq$ 20 MeV: $F_p(E)$ increases linearly, with $F_p(8\ \text{MeV}) = 1$ and $F_p(20\ \text{MeV}) = 1/a_p$.

The a_p factor values will be specified further. Beside this example of a voluntary degradation of the spectrum, other combinations are possible.

5.1.3 Application #1: nickel dosimeter

In this test, ^{58}Ni(n,p) reaction rate is the target value. A high level of confidence on calculated and measured values associated to this reaction is provided by a very good agreement between these two values for this dosimeter (C/M = 1.00 ± 0.03). Besides, this reaction in deemed to be a reference reaction in the field of dosimetry nuclear data.

Impact of altering spectra on the target value (reaction rate) given by ESTIMA is gathered in Table 2: first, C/M ratios considering altering spectra are calculated, then, C/M ratios from ESTIMA are shown.

Table 2. Application #1: C/M ratios for nickel dosimeter

a_p factors	Results from altering spectra	Results from ESTIMA
0.75	0.865	1.004
0.80	0.892	1.005
0.90	0.947	1.005
0.95	0.974	1.005
1.00	1.001	1.005
1.05	1.028	1.006
1.10	1.056	1.006
1.20	1.110	1.006
1.25	1.137	1.007

ESTIMA provides stable C/M values, close to the FLUOLE-2 reference result (Table 1), whereas the altering spectra produce erroneous reaction rates heavily dependent on the altering factors a_p.

5.1.4 Application #2: rhodium dosimeter

In this application, ^{103}Rh(n,n') reaction rate is the target value. Measured results for rhodium dosimeters are subject to some uncertainties: X-rays (around 20 keV) emitted by ^{103m}Rh are largely absorbed by the dosimeter itself though it is very thin (50 µm). This involves significant corrective factors. Furthermore, the X-rays emission probabilities are known with an uncertainty of 7%.

As in application #1, Table 3 shows impact of the perturbation of the altering spectra, and the results of ESTIMA

Table 3. Application #2: C/M ratios for rhodium dosimeter.

a_p factors	Results from altering spectra	Results from ESTIMA
0.75	0.795	0.908
0.80	0.808	0.900
0.90	0.833	0.886
0.95	0.846	0.887
1.00	0.859	0.874
1.05	0.872	0.869
1.10	0.885	0.863
1.20	0.910	0.853
1.25	0.923	0.850

As in the previous case, ESTIMA provides stable C/M results. This trend confirms that the measured reaction rate value needs to be revised.

5.1.5 Application #3: tin dosimeter

In this third application, ^{117}Sn(n,n') reaction rate is the target value. It should be noted that the use of tin dosimeters is fairly new, and the inelastic scattering cross section for ^{117}Sn is probably not well known. Indeed, depending on the origin of nuclear data (JEFF3.1.1, Endf/B-VII ...), the calculated reaction rate can vary by twice as much. In addition, this reaction is available neither in the International Reactor Dosimetry File (IRDF2002) nor in the new Dosimetry library IRDFF. For the FLUOLE-2 analysis, inelastic scattering cross section data for ^{117}Sn(n,n') reaction come from the JENDL/A-96 nuclear library which performs a little better than other libraries. By contrast, the activity measurement doesn't have a particular technical difficulty.

As in applications #1 and #2, Table 4 shows impact of altering spectra on C/M results obtained from altering spectrum, and using ESTIMA.

Table 4. Application #2: C/M ratios for tin dosimeter.

a_p factors	Results from altering spectra	Results from ESTIMA
0.75	0.939	1.029
0.80	0.962	1.038
0.90	1.003	1.052
0.95	1.032	1.058
1.00	1.055	1.063

1.05	1.078	1.067
1.10	1.101	1.071
1.20	1.147	1.077
1.25	1.170	1.080

ESTIMA provides, once again, stable values. The nuclear data problem is confirmed by this result.

6 Conclusion

In light of the above considerations, an estimate of uncertainties associated to calculated reaction rate results can be given. It is the same with uncertainties associated to measured reaction rate values even if they are usually given by the measurement laboratories. Rhodium dosimeter is an exception: an uncertainty of 13% for measurement seems to be more adequate. For calculated results, uncertainty is obtained by CORRELA, except for tin dosimeter. For that kind of dosimeter, cross section data seem to be incorrect because C/M ratio is not perfect, and results from ESTIMA remain unchanged, regardless of the altering spectrum.

The methodology presented in this article allows to make up for a lack of information to determinate and to enhance a requested result, using an appropriate weighting of available information (typically calculation and/or measure data). This methodology also alerts when any incorrect data are used in a calculation. Although a good neutron flux is provided by ESTIMA, agreement between measurement and calculation cannot be achieved when external data are incorrect (measurement results or cross section data for instance).

The authors would like to thank Mr. Jean-Marc Martinez (CEA/Saclay) for many fruitful discussions [6].

References

1. "TRIPOLI-4® version 8.1, 3D general purpose continuous energy Monte Carlo Transport code", NEA-1716 TRIPOLI-4 VERS. 8.1, https://www.oecd-nea.org/tools/abstract/detail/nea-1716
2. Bevilacqua, R. Lloret, J.C. Nimal, S.H. Zengh, C. Rieg, "Special dosimetry at Saint Laurent B1A", ASTM-EURATOM 1993.
3. F. Gaudier, "URANIE : The CEA/DEN Uncertainty and Sensitivity platform", Procedia Social and Behavioral Sciences 2 (2010) 7660–7661
4. N. Thiollay et al, "FLUOLE-2: an Experiment for PWR Pressure Vessel Surveillance", Fifteenth International Symposium on Reactor Dosimetry (ISRD 15), Aix-en-Provence, France, May 18-23, 2014.
5. JC. Bosq, M. Antony, J. Di-Salvo, J.C. Klein, N. Thiollay, P. Blaise, P. Leconte, "The use of EOLE and MINERVE critical facilities for the Generation 3 light water reactors studies", Proc. Of ICAPP 2011, Nice, France, May 2-5, 2011
6. A. Tsilanizara et al, "Experimental validation of the DARWIN2.3 package for fuel cycle applications", PHYSOR 2012, Knoxville, Tennessee, USA, April 15-20, 2012.
7. J.M. Martinez (CEA/Saclay), Private communication, 2014 September 15th.

Generation of an activation map for decommissioning planning of the Berlin Experimental Reactor-II

Janis Lapins[1,a], Nicole Guilliard[1] and Wolfgang Bernnat[1]

[1]*Institute of Nuclear Technology and Energy Systems (IKE), University of Stuttgart, 70569 Stuttgart, Germany*

Abstract. The BER-II is an experimental facility with 10 MW that was operated since 1974. Its planned operation will end in 2019. To support the decommissioning planning, a map with the overall distribution of relevant radionuclides has to be created according to the state of the art. In this paper, a procedure to create these 3-d maps using a combination of MCNP and deterministic methods is presented. With this approach, an activation analysis is performed for the whole reactor geometry including the most remote parts of the concrete shielding.

1 Introduction

The BER-II is a light water cooled pool type 10 MW reactor with UAl-plate fuel elements of MTR-type [1, 2] The core consists of 30 fuel elements, in a 6 x 7 grid plate. Some positons are filled with beryllium reflectors. There are 6 control rods and a total height of 60 cm. Also there is a permanent beryllium reflector around the core, and one in-core irradiation channel, see Figure 1 . The core is submerged in a pool of 3.5 metres diameter and a depth of 11 metres. Next to the reactor pool, there is a second pool for discharged fuel elements. The pools are surrounded by massive concrete structures; barite concrete around the core pool, normal concrete for the other pool (Figures 2 and 3).

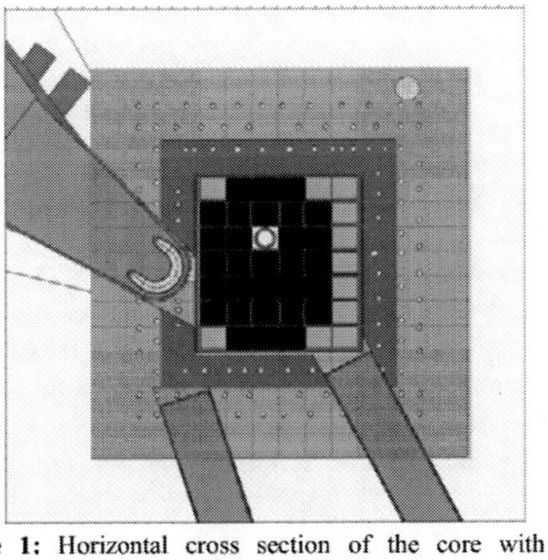

Figure 1: Horizontal cross section of the core with fuel elements (black), beryllium (green and magenta), ends of irradiation channels (light pink)

The barite concrete structures are penetrated by multiple irradiation tubes to allow for neutron experiments. With its 10 tangential, diagonal and radial irradiation channels and one conical beam tube housing the cold neutron source, see Figure 4, a cylindrically symmetric approach cannot be applied and the reactor is fully modelled in 3 dimensions.

The reactor will operate until the end of 2019. The activation analysis comprises all load periods until the planned shutdown.

2 Methods and models

In order to have an accurate description of all the components with their real geometric complexity, MCNP6 [3] is used to create the geometry model. With MCNP6, all components can be input explicitly without discretisation. However, for MCNP6 it is rather time consuming, if not impossible, to calculate the neutron flux distribution for the most remote parts of the concrete. Given that, a different approach was applied. MCNP6 has an automatic geometry export option which can be used for the discrete ordinates transport code PARTISN [4]. Since PARTISN needs a discretisation, a mesh grid subdivision in x,y,z-direction has to be provided within the MCNP6 input before applying the export option. When using the export function, the reactor geometry is subdivided into volumes according to the subdivision done by the user with up to 6 different materials from MCNP6 being mixed together with their respective volume fractions. For all different materials appropriate

[a]Janis.lapins@ike.uni-stuttgart.de

macroscopic cross sections (XS) have to be generated for PARTISN and subsequently used for determination of the neutron flux distribution. To create these XS, use is made of the SCALE6 package [5]. In this package the ENDF/B-VII.1 library [6] is available in 238 energy group structure and different σ_0-background cross sections to account for self-shielding effects in both fuel and structure materials. The self-shielding was treated with the BONAMI module of SCALE6 applying the Bondarenko method. The 238 group library was then used for radial 1-d spectrum calculations with ANISN [7] to get 238 group spectra for the different material regions of interest. The 1D- calculations were performed with the S_N order 16 and the Legendre expansion P_5 to account for anisotropic scattering. The cross sections for all different materials of the material specifications of the MCNP6 model and the PARTISN model were then condensed from 238 to 65 groups to limit the computational effort for the 3D-PARTISN calculations. With the spectrum weighted macroscopic 65 group XS, the neutron transport equation was solved for the whole reactor geometry with PARTISN. The solution is a detailed flux distribution in 65 groups for a fine mesh grid. The calculation was performed as eigenvalue problem. For the range of the core a flat burnup distribution was assumed and an average operational control-rod bank position to reach an effective multiplication factor near to one. A control with the MCNP6 model showed a good agreement with the multiplication factor calculated by PARTISN.

For the activation analysis, more data have to be available, e.g. the power history over the whole lifetime has to be accounted for; also the detailed compositions of all materials for which activation has to be determined. This comprises not only the main nuclides necessary to compute the neutron flux, but also the traces of manufacturing impurities such as uranium, cobalt, caesium, europium or gadolinium etc. The activation calculations itself were performed with the program FISPACT [8]. The FISPACT program package contains activation cross section for a large number of nuclides for the most important (n,X) reactions and corresponding decay data. The cross section libraries are available in different group structures. The group structures used here were 69 groups (WIMS structure) for PARTISN and 315 groups for MCNP6. All spectra with the 65 group structure of PARTISN were converted to the 69 group structure of the FISPACT library, the 292 group spectra of the MCNP results to the 315 group structure. The weighted activation XS for the calculation of the nuclide build-up and decay finally were generated by collapsing the 69 or 315 group XS via the converted spectra.

This whole procedure is depicted in Figure 5. On the left side of this figure, there is also the MCNP6 transport calculation step included that is especially performed if certain details have to be calculated more accurately without discretization, e.g. the activity of certain screw nuts. To get a flux distribution around the beam tubes based on continues energy cross sections, special MCNP6 transport calculations were performed for fine mesh grids (mesh tallies in environment of beam tubes). For these calculations, weight windows were generated derived from adjoint fluxes calculated with PARTISN.

2.1 The BER-II geometry model in MCNP6

The geometry of the BER-II was modelled with MCNP6. From a previous project that concerned the optimisation of the geometry of the cold neutron source container, the model of the core up to the aluminium liner was already available. In this work, the components surrounding the reactor pool were complemented. This includes the spent fuel pool, the concrete shielding, and the irradiation tubes with their respective penetrations within the concrete to coincide with the irradiation tubes within the reactor pool. The model comprises 17 different materials: the fuel elements, $AlMg_3$, water, beryllium, helium, CO_2, air, stainless steel, steel, grey cast iron, boron carbide, bismuth pellets, lead shot, normal concrete, magnetite concrete, barite concrete, and concrete bricks with a steel coating. However, the barite concrete also contains steel reinforcements which are denser in the region close to the reactor pool and the region next to the reactor hall (see Figure 6, three subdivision of the barite concrete). All, materials are assumed at 300 K, except the hydrogen chamber in the conical irradiation tube which is at 25 K.

Using the export option from MCNP6 to PARTISN the geometry is discretised according to the user-provided subdivisions. The material specification in every mesh volume of the grid is a homogeneous mixture of materials sharing the mesh according to its fraction. While the mesh grid is finer in the core region (approx. 3.5 cm), the mesh size increases towards the periphery (5+ cm). The coarsest mesh size is 11 cm. The overall number of computational cells for the whole geometry in PARTISN is 2,790,000 (155 x 120 x 158)

Material specifications for the activation calculations are provided by the Helmholtz-Zentrum Berlin (HZB). However, for some components there are only unprecise material specifications, e.g. Co < 10 ppm which are not measured afterwards. Since cobalt is not an alloy element, the respective content is estimated according to either values from literature or from other reactor activation analysis. Here, measured values for impurities of uranium, caesium, gadolinium, cobalt, and europium content could be used for this analysis.

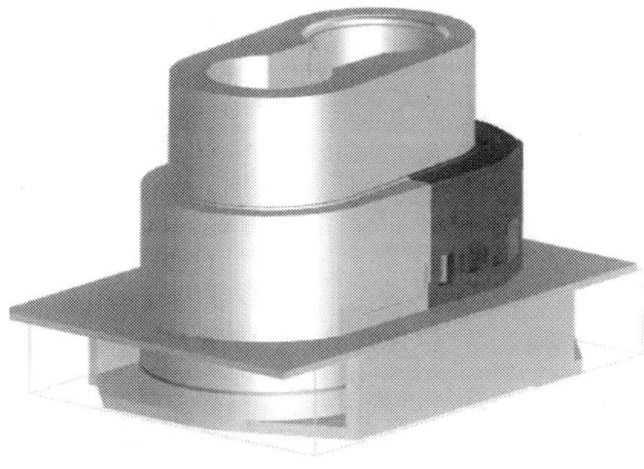

Figure 2: 3D-Model of the BER-II reactor /1 /

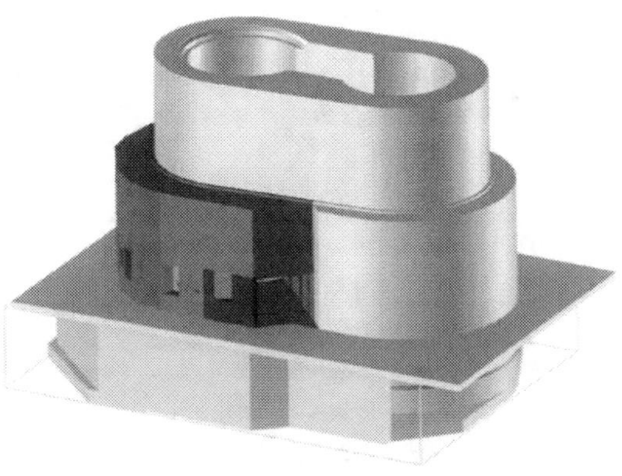

Figure 3: 3D-Model of the BER-II reactor /1 /

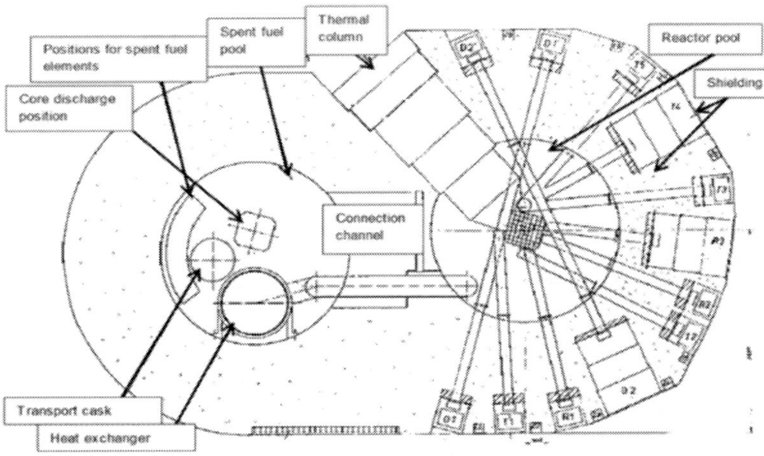

Figure 4: Horizontal cross section of the BER-II with irradiation channels before retrofitting in 1985 – 91, there is a graphite thermal column instead of a conical irradiation tube housing a cold neutron source /1/.

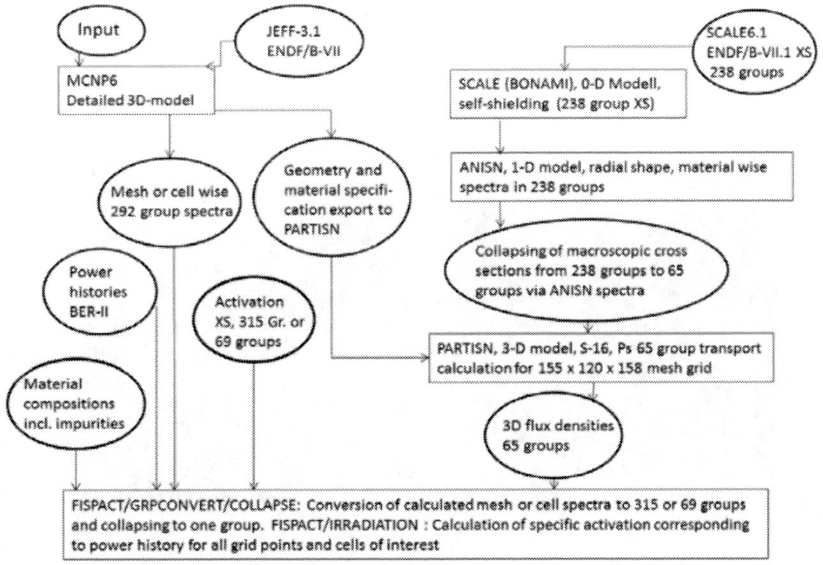

Figure 5: Flowchart for transport calculations with MCNP and PARTISN and activity calculations with FISPACT.

DOI: 10.1051/epjconf/201715301026

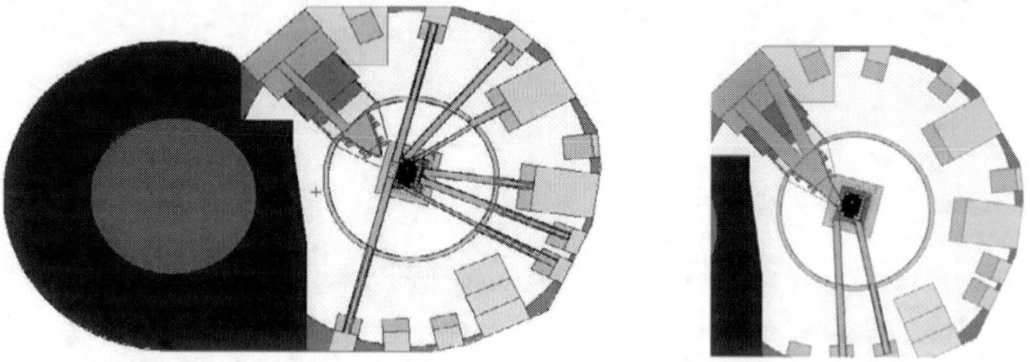

Figure 6: Horizontal sections at the plane with 7 out of 9 beam tubes (left) and the cold neutron source within the conical beam tube and two beam tubes (right). The barite concrete structure is subdivided into three parts according to their steel reinforcement content.

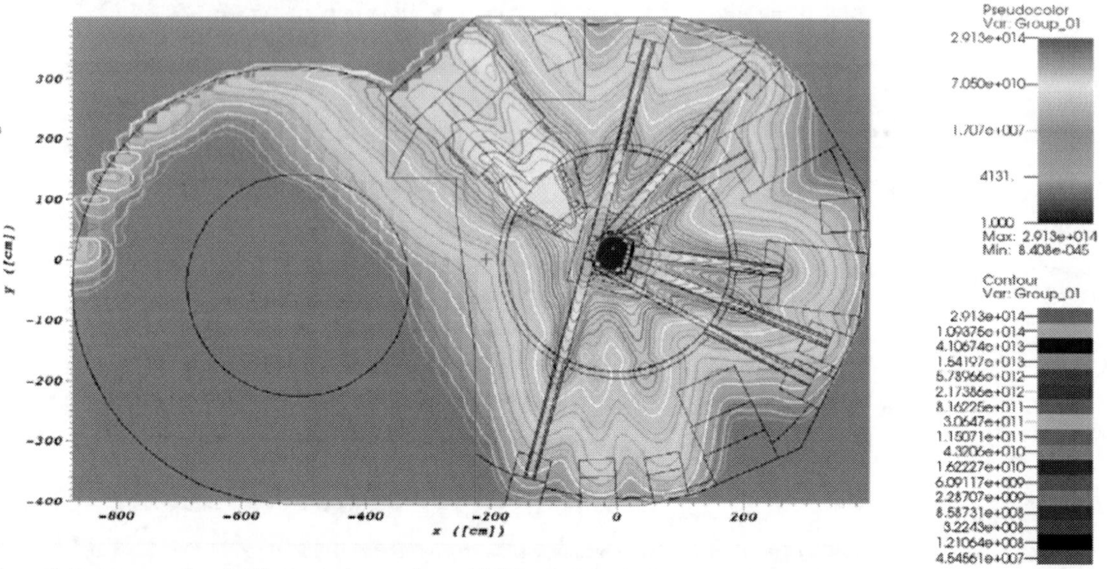

Figure 7: Fast neutron flux distribution in the x,y-plane with 7 irradiation tubes.

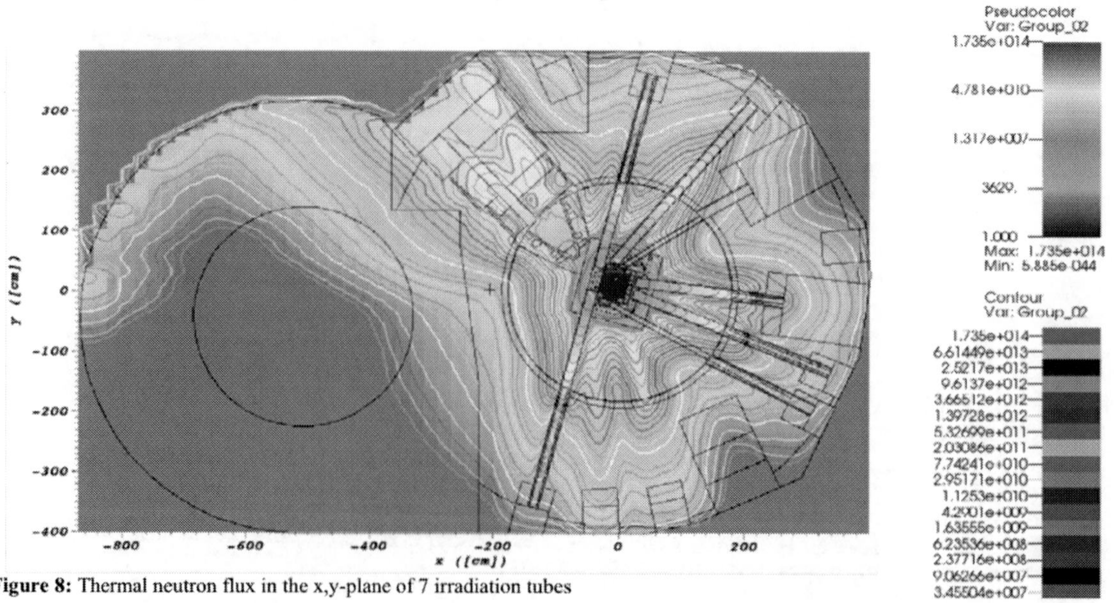

Figure 8: Thermal neutron flux in the x,y-plane of 7 irradiation tubes

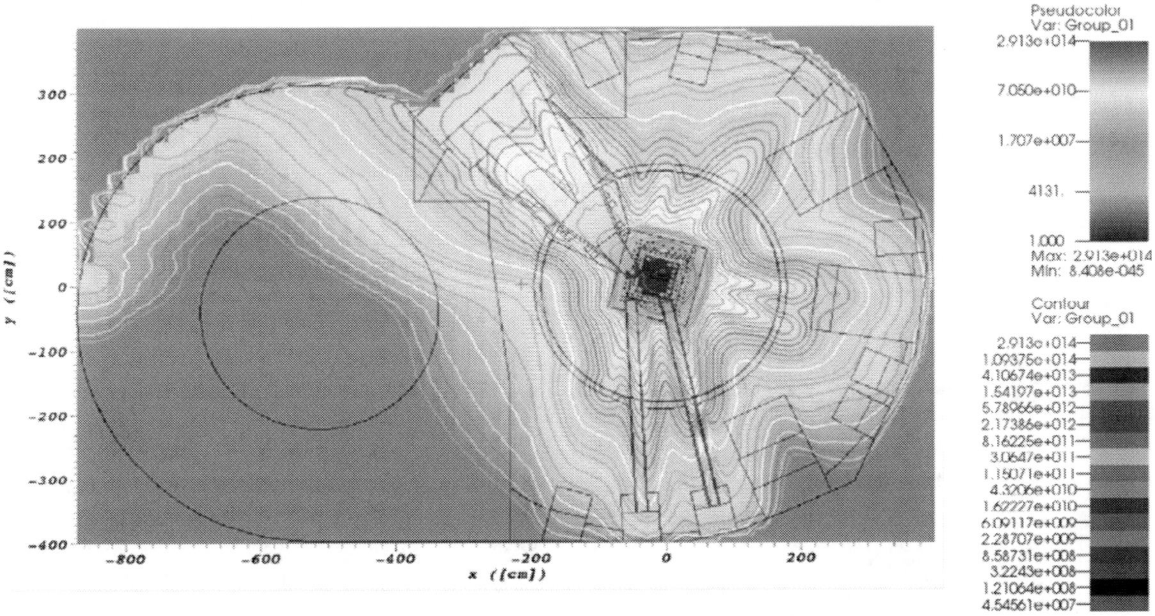

Figure 9: Fast neutron flux in the x,y-plane of 2 irradiation tubes and the conical tube of cold neutron source

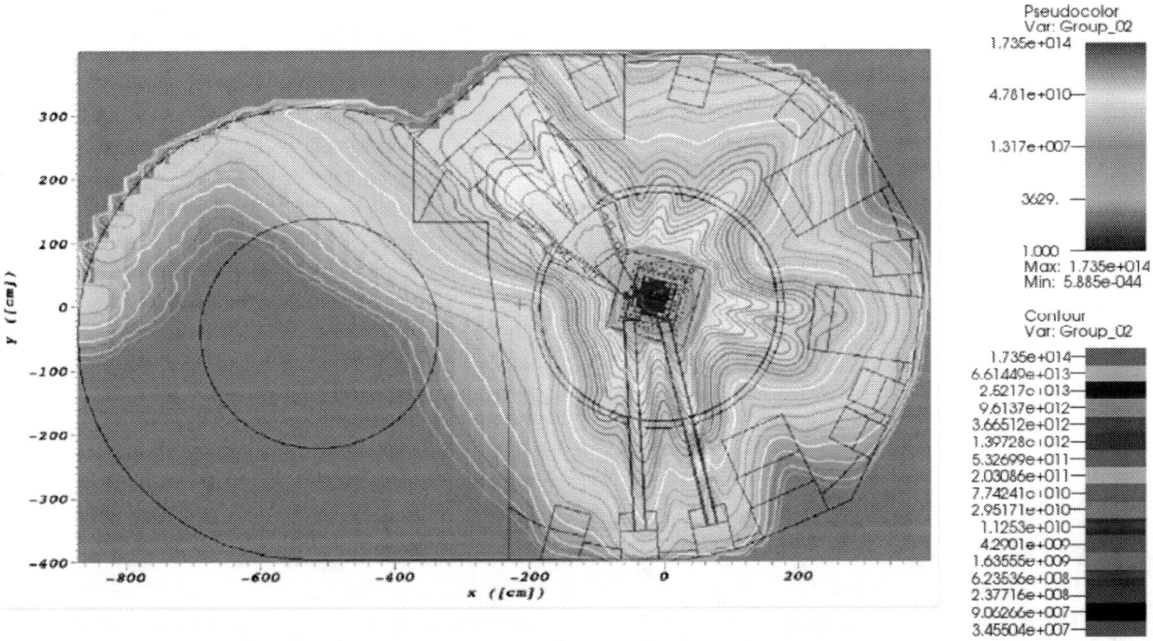

Figure 10: Thermal neutron flux in the x,y-plane of 2 irradiation tubes and the conical tube of cold neutron source

3 Results of the Neutron Flux Calculation

The neutron flux distribution of the BER-II calculated with PARTISN for the full geometry model is shown for two horizontal sections of major interest, i.e. the heights of the two different levels of the irradiation tubes. In one section there are the seven irradiation tubes, in the other section there are two irradiation tubes and the conical beam tube housing the cold neutron source. The corresponding fast and thermal fluxes are shown in Figures 7-10, respectively. The figures show that neutrons escape through the large conical beam tube containing the neutron guides and through the surrounding concrete structure and increase the fast and thermal flux density next to it and even at the leftmost (upper) side by travelling through the concrete structure of the pool for discharged fuel elements,

Since the geometric structure of the conical beam tube housing the cold source and the neutron guides was very complicated, a separate flux density calculation for this region was made with a fine mesh-grid by MCNP6. The weight windows for this MCNP6 run were prepared by an adjoint PARTISN calculation. The thermal flux distribution for this region is shown in

Figure 11:

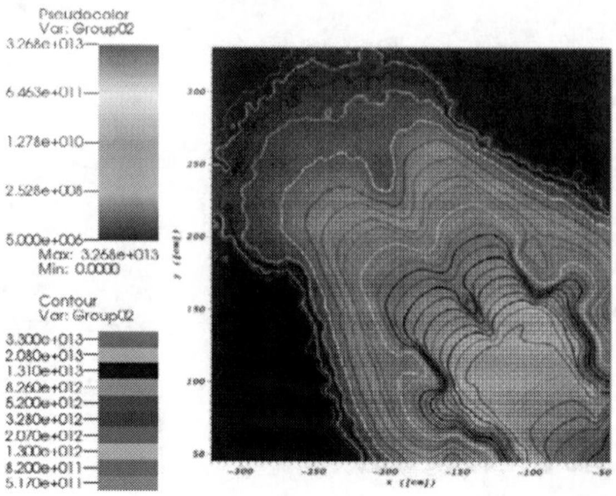

Figure 11: Thermal flux distribution in the conical beam tube and surrounding concrete

4. Power Histories used for the activation calculations

For the activation, it is important to know the neutron flux distribution throughout the whole operational life-time. In this analysis, it is assumed that the neutron flux distribution throughout the whole operation time keeps the shape, but the amplitude is adjusted with respect to the increase in burn-up in a certain time interval. For the years 1974 – 1985, the increase of burn-up is only given for a whole year. Starting from 1991, the increase in burn-up is given in weeks. From there on, the power level is adjusted such that the power produced matches the burn-up increase. This is done for the 1991 – 2015. For the years 2016 – 2019 there are only prognoses of the planned power production. As only the planning for the year 2016 is available, the years 2017 – 2019 are assumed to have the same load periods. For this analysis there are 274 time intervals considered over the whole life-time, see Figure 12. Each time interval consists in a full power period and a decay period. For components with different irradiation histories for which measurements were performed the appropriate operational history was regarded and the decay periods from the time of removal until the time of measurement are accounted for, e.g. some screw nuts of the flange of the conical tube containing the cold neutron source container.

These screw nuts were removed and the activities of corresponding chip samples were measured.

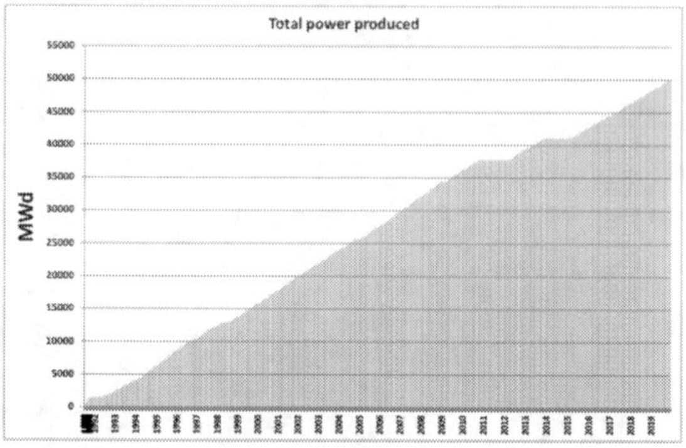

Figure 12: Total power production of the BER-II during operational lifetime including planned power production until 2019 (the year of 2016 is used for extrapolation).

5 Results of the activation calculation and comparison with measurements at calibration points

With the mesh wise neutron flux density distributions presented in the previous chapter, the activation calculations were performed with the burn-up and activation code FISPACT. For the determination of the activity distribution of important nuclides in the activated structures outside the core and reflector for all PARTISN or MCNP6 meshes containing irradiated material a FISPACT run was performed for the material of the mesh and the power history connected to the calculated flux spectrum. The power history of the BER-II is taken as shown in chapter 4. The spatial and energetic flux shape remains the same during the whole irradiation period, only the flux amplitude varies with respect to the power produced in each irradiation interval. Effects of burn-up or moving of control rods are not accounted for.

As an example, the resulting Co-60 activity distribution in Bq/g within the reactor geometry is presented in Figure 13 for the x,y-section at the height of the seven irradiation tubes. The values are given for the planned final shutdown time point. The complete set of 3D specific activity distributions are stored in a special data file in so called silo structure [9] which allow to show activity distributions for distinct 2D slices in x-, y- or z-direction. With the graphical software VISIT [10] visualizations of activity distributions or single values are possible.

The typical permitted limits for the most important nuclides in decommissioning are presented in Table 1 below [11]. The figure shows that the activation is strongly influenced by the neutron streaming through the beam tubes. The lower limit of the colour bar defines the unrestricted release value for the Co-60 activity. For the verification of the activation calculation, some measurements of components were provided by HZB. These measurements were performed at the irradiation channel inside the core and at the screw nuts on the flange of the conical irradiation tube, see Figure 14. The nuts were in the reactor from 1991 until week 40 of 2011. To account for spectral changes due to the adjacent steel bolts, the back flange is modelled with two bolts next to the observation point. The screw nuts were measured Nov 23rd 2015. Another calibration point was an AlMg3 sample from the in-core irradiation channel. The results are presented in Table 2, and are in sufficient agreement with the measurement as a first estimate, even though the first calibration point (front flange) is somewhat underestimated. However, due to the fact that rod movement was not considered, and the height of the screw nuts on the front flange practically coincided with the lower tip of the control rod facing the cold neutron source, the deviations could be explained. More accurate values could be achieved if the realistic burnup distribution would be taken into account.

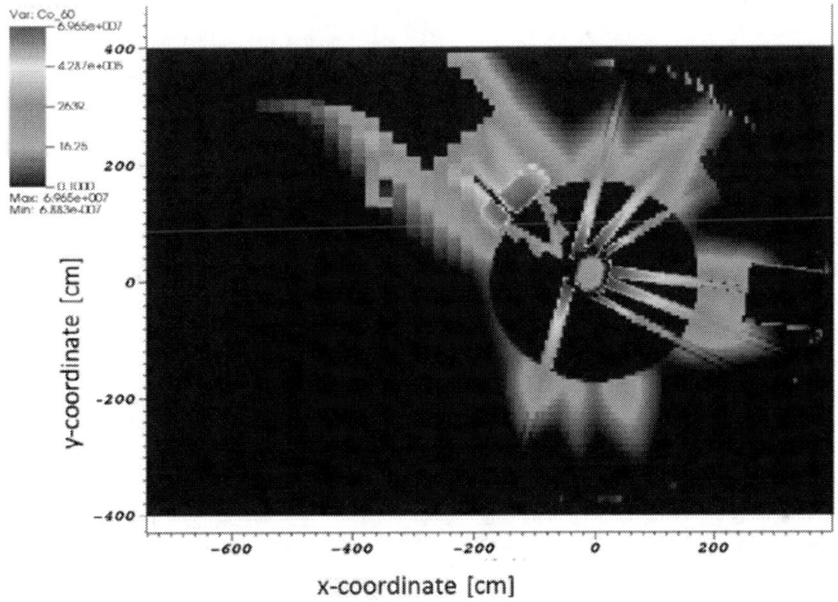

Figure 13: Distribution of Co-60 in Bq/g per gram of material mixture within a cell, x,y-section of the plane with 7 irradiation tubes at the end of operation

Table 1: Unrestricted release limits for different isotopes from the "Strahlenschutzverordnung"(German Radiation Protection Ordinance) [11]

Isotope	Unrestricted	Restricted	Unit
Co-60	1E-01	6E-01	Bq/g
Fe-55	2E+02	1E+04	Bq/g
Eu-152	2E-01	1E+01	Bq/g
H-3	1E+03	6E+04	Bq/g
C-14	8E+01	4E+03	Bq/g

Table 2: Comparison of the specific Co-60 activity, measured and calculated [2]

Component	Position	Measured [Bq/g]	Calculated [Bq/g]
Screw nut (stainless steel)	front flange	5.84E+07	3.81E+07
Screw nut (stainless steel)	back flange	9.9E+06	1.12E+07
Irradiation channel (AlMg3)	inside the core	2.4E+06	4.12E+06

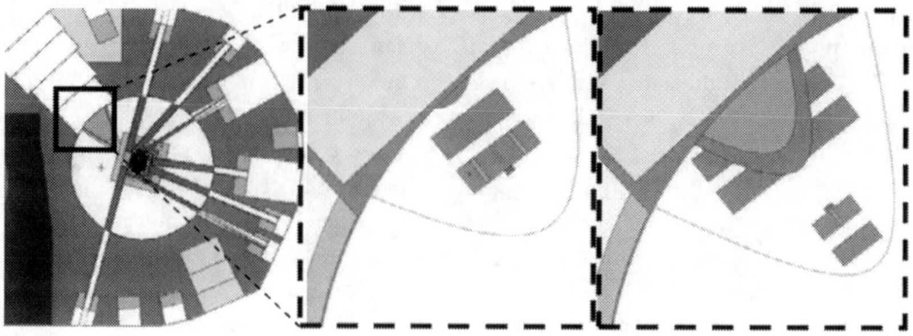

Figure 14: Location of calibration points for the activation analysis, measured values provided BER-2 operational staff

6 Conclusion

In this paper, a procedure for an activation calculation of the pool structure of the BER-II research reactor including the concrete shielding is presented. With the combination of the options offered by MCNP6 and PARTISN, an effective method to compute the neutron flux for a reactor geometry model with 2,790,000 cells could be achieved. The fact that for calculations with PARTISN some extra effort to generate the XS is needed is more than compensated by the fact that for all volumes of the model, neutron fluxes are produced that can be subsequently used for the activation calculation. With the power history provided together with the material compositions, an activation map was created for the whole structure, even for the most remote parts. The activity can be calculated for all nuclides of the FISPACT library. Dependent from material and irradiation history, however, only a limited number of nuclides must be regarded to reduce the size of activation data base.

The results for the activation at calibration points give an example that the use of detailed calculation methods allows reliable estimation of the activity of relevant nuclides, here Co-60. Especially the mean and the maximum activity can be determined. For a more detailed analysis, it is desirable to provide some more calibration points and realistic material specifications including impurities where the results of the activation can be validated, especially for more remote regions of the concrete structures.

References

1. H. Buchholz, C. O. Fischer, K. Waßerroth & INTERATOM, „Sicherheitsbericht für den Reaktor BER-II“ (engl.: Safety report for the BER-II reactor), (June 1972).
2. D.B. Pelowitz (Ed.), "MCNP6™ User's Manual Version 1.0," LA-CP-13-00634, Rev. 0 (2013).
3. R. E. Alcouffe, R. S. Baker, J. A. Dahl, S. A. Turner, R. C. Ward: "PARTISN: A time dependent, Parallel Neutral Particle Transport Code System", LA-UR-08-07258, (November 2008).
4. R. E. Alcouffe, R. S. Baker, J. A. Dahl, S. A. Turner, R. C. Ward: "PARTISN: A time dependent, Parallel Neutral Particle Transport Code System", LA-UR-08-07258, (November 2008).
5. SCALE: A Comprehensive Modelling and Simulation Suite for Nuclear Safety Analysis and Design," ORNL/TM-2005/39, Version 6.1 (2011).
6. M.B. Chadwick, et. Al. : "ENDF/B-VII.1: Nuclear Data for Science and Technology: Cross Sections, Covariances, Fission Product Yields and Decay Data", Nucl. Data Sheets **112**(2011)2887.
7. W. W. Engle, Jr., "A USER MANUAL FOR ANISN, A One Dimensional Discrete Ordinates Transport Code with Anisotropic Scattering", K-1693, ORNL, (March 1967).
8. R. A. Forest: "FISPACT-2007: User manual", UKAEA FUS 534, (March 2007).
9. Silo Users Guide, Version 4.10, LLNL-SM-654357
10. VISIT Users Manual, Version 1.5: UCRL-SM-220449, October 2005.
11. Verordnung über den Schutz vor Schäden durch ionisierende Strahlen §29 Anhang 3 (engl.: German Radiation Protection Ordinance §28 appendix 3), (Stand 11.12.2014).

ACTIVATION OF DOSIMETERS USED IN QA OF MEDICAL LINEAR ACCELERATORS

Kinga Polaczek-Grelik
Department of Medical Physics
University of Silesia, Katowice, Poland
NU-MED Cancer Diagnosis and Treatment Centre Katowice, Poland
kinga.polaczek-grelik@us.edu.pl

Magdalena Nowacka
Department of Medical Physics
University of Silesia, Katowice, Poland

Maciej Raczkowski
Department of Medical Physics
Lower Silesian Centre of Oncology, Wrocław, Poland

ABSTRACT

This paper presents the first results of a project intended to investigate γ-radiation activity induced in dosimeters used in clinical practice during routine quality assurance of high-energy photon beams emitted by electron linear accelerators. Two aspects of the activation via photonuclear reactions (X,n) of therapeutic beam and subsequent capture of secondary neutrons (n,γ) are under considerations: the influence of activation on intrinsic background of the dosimeters and exposure of dosimetrists who operate this equipment. The activation of several types of ionization chambers as well as the silicon diodes was studied after long-time exposure (10 000 MUs) of the 15 MV photon beam (Elekta Synergy). Photon fluxes obtained from spectra of γ-rays registered by HPGe spectrometer were subsequently converted to equivalent doses using appropriate coefficients. The main contribution to the induced activity comes from the neutron capture process on Al, Mn and Cu, therefore it decays quite fast with the half-lives of the order of 15 minutes. Nevertheless, the activation of chlorine was also observed. The estimated equivalent doses to skin and eye lens were in the range 0.19 – 0.62 μSv/min. However, no influence on intrinsic background signal of all studied dosimeters was observed. The preliminary results indicate that induced radioactivity of dosimeters is strongly influenced by therapeutic beam quality and neutron source strength of particular linac. This dependence will be studied deeper in order to quantify it more precisely.

Key Words: **Induced radioactivity, gamma radiation, medical accelerators, ionization chambers, silicon diodes, equivalent doses.**

1. INTRODUCTION

The phenomenon of inducing radioactivity in the construction materials of medical linear accelerators working in high-energy (above ~9MV) photon mode has been widely studied over the last decades experimentally as well as using the calculations, either a Monte Carlo simulations or the equations of activation and depletion processes, e.g. [1-4]. The induced radioactivity is the consequence of nuclear photo effect (X,n) as well as neutron capture reactions (n,γ), being dependent on linac beam emission time. Therefore, it seems to be especially important during linac commissioning and performance of periodical quality assurance dosimetric tests.

Nuclear photo effect, e.g. (γ/X,n) is observed when energy of photons exceeds approximately 6 MeV in giant dipole resonance (GDR) region. Up to 30 MeV of photon beam generated by linacs, neutrons and protons are the main reaction products [2,5]. However, only neutrons are measured all over the accelerator room [6,7] due to much more limited range of charged particles [4,8].
Computer simulations reported e.g. by Sheikh-Bagheri and Rogers [9] have shown that significant part of high-energy therapeutic photon beam spectrum (up to even 10%) could have energy above the photonuclear reaction threshold for many construction materials, what is exemplary shown in Figure 1.

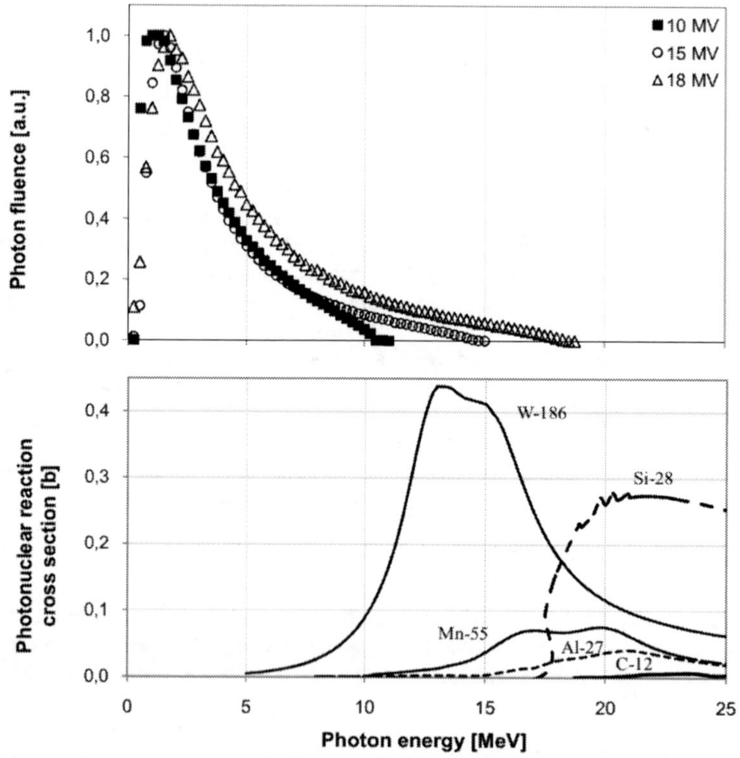

Figure 1. Comparison of the spectra of high-energy therapeutic photon beams generated by linacs obtained in MC simulations by Sheikh-Bagheri and Rogers [9] with the photonuclear reaction cross sections [10] of the most abundant isotopes of the elements found in construction materials of linacs and dosimeters used in clinical practice.

The process of originating radioactivity via photonuclear interaction (γ/X,n) is accompanying by production of secondary particles, mostly – neutrons. The main source of photoneutrons in teleradiotherapy is the accelerator head [1], which is made of heavy materials having lower energy thresholds for this phenomenon. As linac head constitute no shielding against neutrons, these particles are present all over the therapeutic room as it previously has been shown by several authors, e.g. [2,6,7,11] . Moreover, neutron flux can induce radioactivity itself. Cross sections of slow neutron nuclear reactions for many nuclides have high and sharp resonance structure [10]. Generally, the lower neutron energy the higher probability of capture reaction (n,γ). From the in-phantom measurements in QA tests of linac beams point of view, the thermalization process taking place in hydrogen-rich water phantom provides slow neutrons what could gain the activation of dosimeters, e.g. ionization chambers and silicon diodes.

All mechanisms of inducing radioactivity mentioned above can produce unstable isotopes. As a consequence of their decay, secondary gamma radiation is able to appear. The time scale of γ radiation as a consequence of nuclear reactions is directly connected with half-lives of activated radioisotopes. Therefore, it is a process which is extended in time after the end of high-energy beam emission. Consequently, this activity is able to accumulate and finally result in not negligible dose, what is reported for example in [5,12] with respect to patients and workers. However, up to now there is no study concerning the effects of radiation emitted as a consequence of activation of dosimetric equipment, however, some investigations concerning the activation of metal-based ionization chambers were already published, e.g. [13].

The development of modern radiotherapeutic techniques, aimed at improving the conformance of dose distribution in target volume, is usually connected with increased energy emission, e.g. intensity-modulated IMRT techniques or flattening filter-free (FFF) beams. These highly specialized modalities require extended and more sophisticated dosimetric tests to be performed for a given linac before approval for clinical use. During these tests usually beams are emitted for a long time, allowing the induced radioactivity to rise even to the saturation level in the case of short-lived radionuclides.
This paper contains our preliminary results of the study aimed at:

- the assessment of the influence of the activation of commonly used dosimeters on their intrinsic background signal;
- the evaluation of exposure of the staff who use these equipment (hand held carriage, vicinity of eye lens).

2. MATERIALS AND METHODS

Dosimeters commonly used in radiotherapy with external photon beams were investigated in presented study. Among the detectors, characterized in Table 1, were 3 types of ionization chambers as well as silicon diodes dedicated to photon radiation measurements.
The elemental composition of these dosimeters includes mostly non-metals, i.e. C, O, H, F, N, which in general have lower activation cross sections. Nevertheless some of metallic elements are also present, like Al and steel (electrodes), Si (active volume of diodes), W (for homogeneous energy response) and copper wires.

Table 1. Basic characteristics of dosimeters (IBA Dosimetry GmbH) used in presented study.

Type	Model	Active volume [cm^3]	Construction materials	
			Inner electrode	Outer wall
Farmer	FC65-G	0.65	Al	Graphite
Compact	CC13	0.13	C552*	C552*
Compact	CC01	0.01	Steel	C552*
Plane-parallel	PPC05	0.05	PEEK#	C552*
Si diode	PFD^{3G}	0.001	pSi	W, epoxy
Si diode	SFD	0.0004	pSi	epoxy

* C552 – air-equivalent plastic, ρ = 1.76 g/cm^3, (2.5% H, 50.2% C, 0.5% O, 46.2% F, 0.4% Si)
\# PEEK – Polyether Ether Ketone, ρ = 1.32 g/cm^3, type of thermoplastic resins, (C, H, O)

2.1. Experimental Procedure

Activation of dosimetric equipment has been performed in conditions of beam quality assurance tests, i.e. exposition of ionizing chambers and silicon diodes to the treatment X-ray beam and accompanying neutron radiation was performed in the set-up presented in Figure 2, which includes:

- position at central beam axis (CAX),
- irradiation field of 30×30cm^2,
- source-to-surface distance (SSD) of 90cm,
- depth of 10 cm in water phantom (48L × 48W × 41H cm^3 of effective scanning volume).

This procedure allowed for taking into considerations real contributions of primary and scattered photons as well as the fast and water-slowed down neutrons in activation process. Since each time whole dosimeter was inside the irradiation field, applying such arrangement one could achieve the maximal activation signal in comparison with partial dosimeter irradiation, i.e. applying lower field size. Each dosimeter was irradiated with the high-energy photon beam with a nominal potential of 15 MV, generated by the Elekta Synergy linear medical accelerator. 10 000 monitor units (MU) with the rate of 600 MU/min was emitted in each case, which means that the activation time was about 17 min.

a)  b) 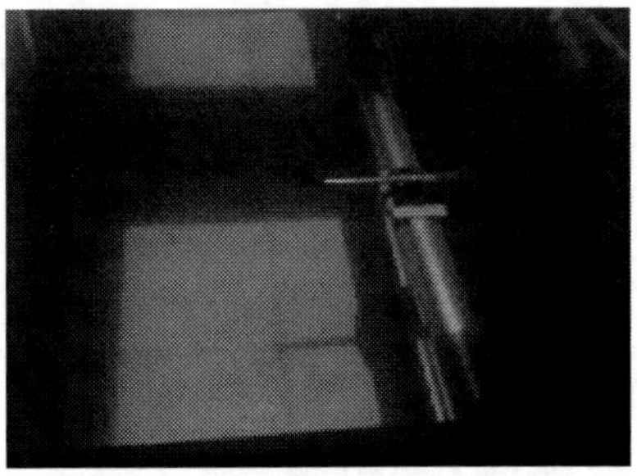

Figure 2. Exemplary view (Farmer-type ionization chamber) of the positioning of dosimeters before activation: a) general view with water phantom and linac head visibility, b) in-field positioning with respect to CAX.

2.2. Intrinsic Signal

The instrumentation issue of presented study has been focused on monitoring of dosimeters' intrinsic charge signal, to check whether should it be periodically compensated during long-time measurements.
The background signal of each dosimeter (in nC) has been investigated before the irradiation as well as immediately after the end of beam emission.
Charge signal of each ionization chamber was determined with the use of a standard electrometer (Dose1, IBA Dosimetry), whereas in case of silicon diodes CCU unit coupled with OmniPro Accept7 software (IBA Dosimetry) was used.
Before the irradiation 10-second measurement was repeated 10 times, then averaged and referred as intrinsic background signal. After the end of activation procedure, 10 second-lasting measurements of the charge were carried out for one hour, to investigate potentially occurring decay of intrinsic signal connected with the activated radionuclides.

2.3. Induced Radioactivity

The induced radioactivity of each studied dosimeter has been investigated using gamma spectrometry system consists of:

- portable coaxial high-purity Ge detector with reverse electrodes (REGe) (Canberra Inc.);
- multichannel analyser InSpector 2000 (8194 channels);
- Genie2000 software, v.3.2.1.

The gamma-energy spectrum in the energy range of 6-3150 keV of each studied dosimeter has been registered twice:

1) before the activation process, for 30 minutes and
2) immediately after the end of irradiation, for one hour.

The first spectrum is considered as an internal background of the investigated equipment (containing also the natural radiation of the place in which the measurements were carried out). The second spectrum was analyzed in terms of induced radionuclides.
The set-up for induced radioactivity registration is presented in Figure 3.

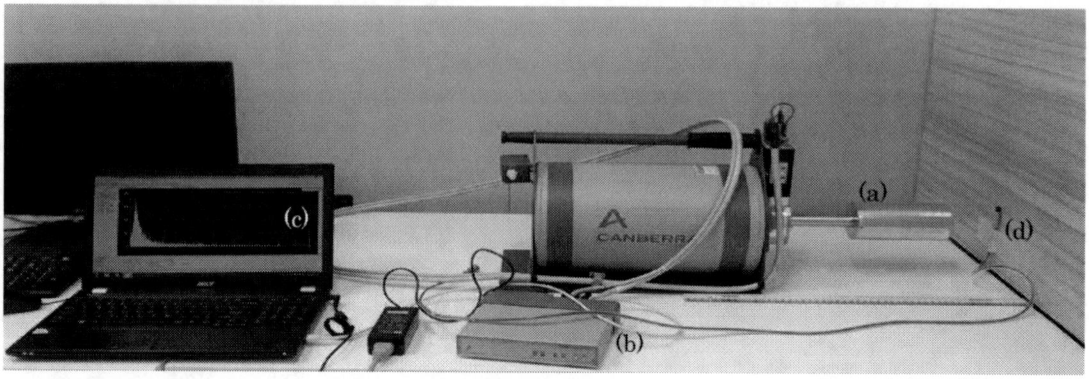

Figure 3. Spectrometric system, which contain REGe detector (a), multichannel analyzer (b), registration software (c) in arrangement with activated ionization chamber (d) during induced radioactivity measurement.

2.3.1. Spectrometric analysis

The spectrometric analysis of radionuclides has been shown by us previously [14] to be good method of low dose assessment in comparison with the standard dose rate meters, giving furthermore the additional information of the elements, which are the most intense sources induced radiatoactivity and secondary doses.
Identification of radionuclides was done on the base of the energy at the centre of photopeak, therefore, this method required exact energy calibration of spectrometric system. This calibration was performed with the use of point-like sources (^{22}Na, ^{54}Mn, ^{57}Co, ^{60}Co, ^{65}Zn, ^{109}Cd, ^{133}Ba and ^{137}Cs) with the activity of the order of 1 μCi (comparable with experimental once), which energies covered most of the useful energy range.

The quantitative analysis of gamma radiation flux (required for dose estimation) at a distance of 10 cm from the dosimeters was based on net count rate under the photopeak around the energy of photons emitted by particular radioactive nuclide (E_γ). The efficiency (ε) calibration of spectrometric system is described by the Equation (1).

$$\varepsilon = 4.74 \cdot E_\gamma^{-1.05} \cdot \exp\left[-\left(\frac{E_\gamma}{-0.63}\right)^{132.27}\right] \tag{1}$$

2.3.2. Dose calculation

Equivalent doses to skin and eye lens were estimated on the base of the gamma-ray flux of particular energies obtained from registered spectra, applying fluence-to-dose conversion coefficients according to ICRP report no 74 [15]. Since this report supplies conversion coefficients for discrete values of energies, according to the suggestions given therein, values required in our analysis were calculated using Lagrange interpolation formula of 3rd degree (4-point). The dependency of conversion coefficients on photon energy is exemplary shown in Figure 4.

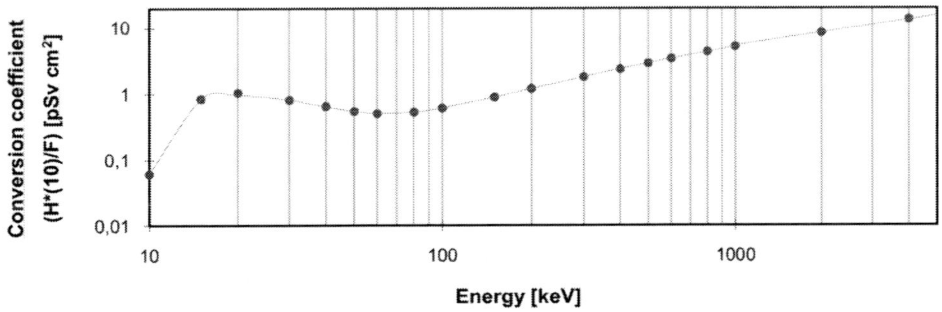

Figure 4. Photon fluence-to-quivalent dose conversion coefficients obtained on the base of [15]. The points indicate discrete values given in [15], which serves for interpolation with the use of Lagrange interpolation formula.

2.3.3. Uncertainty of the method

The major source of the uncertainty in applied technique is the statistical error of gamma-ray counting performed by spectrometry system. This contribution to the total uncertainty may be reduced by applying longer irradiation time, allowing to reach the saturation activity for majority of nuclides. In applied activation and spectrum registration time conditions this results in average counting uncertainty of 10 %.

The discrepancies in conversion coefficient values specified in report 74 of ICRP [15] of 10% are another source of total dose uncertainty.

Additional uncertainty in determination of the activity, and afterwards in estimation of photon flux and subsequently – equivalent doses, could appear due to the use of point sources in calibration process, whereas extended volume sources were under study. However, as our previous measurements for higher activity level have shown [14], the results obtained using presented method in comparison with the use of a standard dose rate meter agree within the accuracy of 4 %.

3. RESULTS

3.1. Dosimeters Reading

Typical values of intrinsic background current signal for studied ionization chambers with applied operation voltage of 300V is 0.0010 ± 0.0004 nC. The leakage current of silicon diodes under investigation is 0.0040 ± 0.0008 nC. These values were obtained as an average from 10 consecutive measurements before irradiation.

Activation process of ionization chambers does not increase the intrinsic charge signal (average value is around 0.0008 ± 0.0003 nC), i.e. it remains statistically dispersed over time, what is presented exemplary in Figure 5. In case of silicon diodes the signal was increased about 3 times to the value of 0.011 ± 0.001 nC. Even after 10 minutes from the end of irradiation the background current signal for diodes was still rising (see Figure 6), what suggests that it was not affected by decay of activated nuclides.

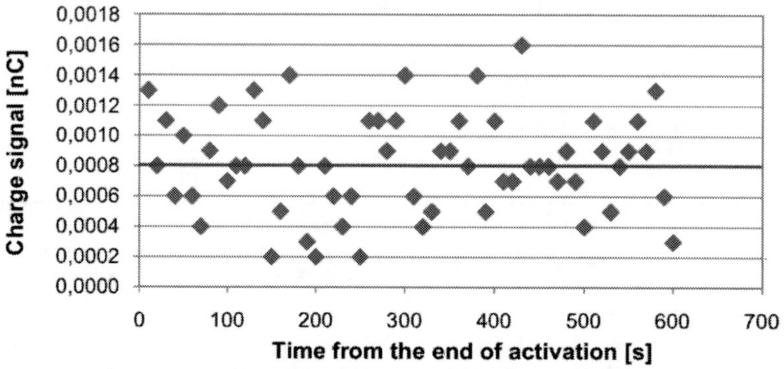

Figure 5. Charge signal of CC01 ionization chamber after irradiation with 10 000 MUs 15MV photon beam. The blue line represents the average signal.

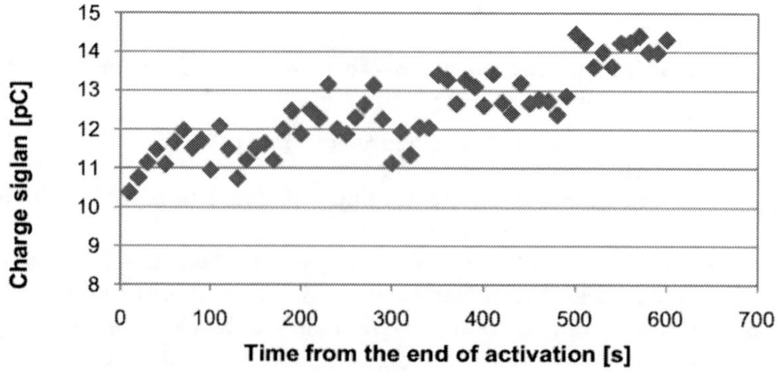

Figure 6. Evolution in time of charge signal for silicon diode irradiated with 10 000 MUs 15MV photon beam.

3.2. Activated Radionuclides

The identification of activated radionuclides (presented in Table 2) has been done using the spectra of activated dosimeters registered for one hour after irradiation, to achieve good counting statistics.

Table 2. Gamma energy-ordered analysis of activation process occurring in studied dosimeters during the irradiation with 15 MV therapeutic photon beam.

Radionuclide origin	Energy of γ-ray [keV]	Half-life $T_{1/2}$	Dosimeter
$^{186}W(n,\gamma)^{187}W$	479.5	23.7 h	PFD
$^{186}W(n,\gamma)^{187}W$	551.5	23.7 h	PFD
$^{186}W(n,\gamma)^{187}W$	618.4	23.7 h	PFD
$^{186}W(n,\gamma)^{187}W$	685.8	23.7 h	PFD
$^{186}W(n,\gamma)^{187}W$	772.9	23.7 h	PFD
$^{55}Mn(\gamma/X,n)^{54}Mn$	834.9	312.1 d	CC01
$^{55}Mn(n,\gamma)^{56}Mn$	846.8	2.6 h	CC01, CC13, FC65-G, SFD, PFD
$^{65}Cu(n,\gamma)^{66}Cu$	1039.3	5.1 min	CC01, CC13, FC65-G, PPC05, SFD, PFD
$^{59}Co(n,\gamma)^{60}Co$	1173.2	5.3 y	CC01
$^{58}Fe(n,\gamma)^{59}Fe$	1291.6	44.5 d	CC01
$^{59}Co(n,\gamma)^{60}Co$	1332.5	5.3 y	CC01
$^{65}Cu(\gamma/X,n)^{64}Cu$	1345.8	12.7 h	CC01, FC65-G
$^{37}Cl(n,\gamma)^{38}Cl$	1642.7	37.2 min	CC01, CC13, FC65-G
$^{27}Al(n,\gamma)^{28}Al$	1779.0	2.2 min	FC65-G, PPC05
$^{55}Mn(n,\gamma)^{56}Mn$	1810.8	2.6 h	CC01, FC65-G, SFD, PFD
$^{55}Mn(n,\gamma)^{56}Mn$	2113.1	2.6 h	CC01, PFD
$^{37}Cl(n,\gamma)^{38}Cl$	2167.4	37.2 min	CC01, CC13, FC65-G, PPC05, SFD, PFD

The analysis presented in Table 2 has shown, that in case of compact chamber CC01 and Farmer-type chamber FC-65G, the most intense activation occurs in the materials of inner electrodes, i.e. stainless steel (Fe, Mn, Co) and aluminium, respectively. In case of shielded photon diode PFD, material (W) aimed at making the response uniform over all photon energies, is the most susceptible on activation process. In all plastic materials activation concerns only chlorine in applied experimental conditions (15MV beam). This could be different for higher therapeutic beam energies and/or more intense neutron fluxes, i.e. linacs from other manufacturers. Copper wires connected with all studied dosimeters are also activated.

The exemplary comparison of the spectra registered before and after 15 MV photon beam irradiation, during which the activation occurs is shown in Figure 7. It shows that the studied effects are clearly visible only for radionuclides which were able to reach the saturation activity during irradiation, i.e. short-lived once (e.g. ^{28}Al and ^{66}Cu).

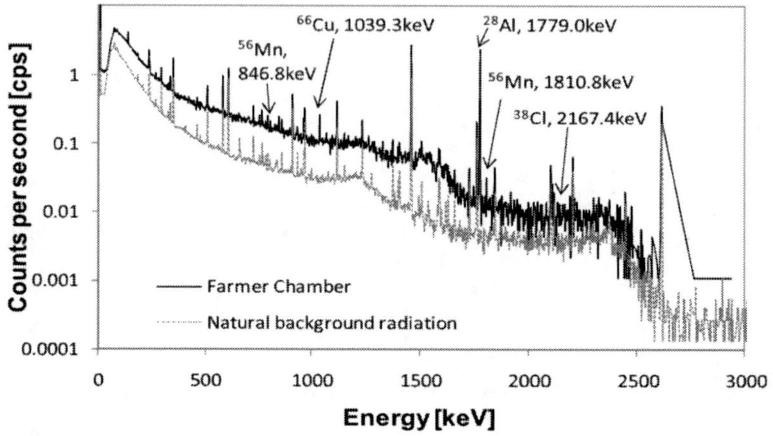

Figure 7. Spectra of gamma radiation of activated Farmer-type ionization chamber FC-65G (solid line) and natural background radiation (dotted line) for comparison.

3.3. Equivalent Doses

Equivalent doses to skin and eye lens were calculated due to higher risk for these organs during operation of dosimetric equipment in teleradiotherapy. The results are presented in Table 3. Since the limited time of direct contact with this equipment as well as the fact, that the most activated radionuclides are short-lived, i.e. ^{28}Al ($T_{1/2}$ = 2.2 min) and ^{37}Cl ($T_{1/2}$ = 37.2 min), the doses are expressed in μSv/min. The estimated values are relevant for the distance from radiation source to the skin/eye lens of 10 cm, since in such conditions the detection efficiency curve was determined.

Table 3. Equivalent doses to skin and eye lens, dominant components of these doses and effective half-lives estimated on the base of registered gamma-energy spectra of activated dosimeters.

Dosimeter	Equivalent dose [μSv/min]		Dominant component	Second contributor	Effective half-life [min]
	Skin	Eye lens			
FC65-G	0.38 ± 0.06	0.41 ± 0.08	^{37}Cl(n,γ)^{38}Cl	^{27}Al(γ/X,n)^{26}Al	16 ± 3
CC13	0.22 ± 0.05	0.24 ± 0.05	^{37}Cl(n,γ)^{38}Cl	^{55}Mn(n,γ)^{56}Mn	12 ± 2
CC01	0.56 ± 0.13	0.62 ± 0.13	^{55}Mn(n,γ)^{56}Mn	^{37}Cl(n,γ)^{38}Cl	16 ± 3
PPC05	0.19 ± 0.04	0.20 ± 0.04	^{37}Cl(n,γ)^{38}Cl	^{27}Al(n,γ)^{28}Al	19 ± 4
PFD^{3G}	0.37 ± 0.10	0.41 ± 0.10	^{55}Mn(n,γ)^{56}Mn	^{37}Cl(n,γ)^{38}Cl	16 ± 3
SFD	0.22 ± 0.05	0.24 ± 0.05	^{37}Cl(n,γ)^{38}Cl	^{27}Al(γ/X,n)^{26}Al	12 ± 2

4. SUMMARY AND CONCLUSIONS

Presented study have shown that the process of radionuclides activation takes place in materials of commonly used dosimeters, e.g. ionization chambers and silicon diodes, at a measurable level. However, it has been shown that this induced activity does not affect the intrinsic signal of these dosimeters. Therefore, it seems to be unnecessary to periodically compensate the background signal during the performance of long-lasting medical linac commissioning or QA tests for that reason. Nevertheless, in case of silicon diodes, typical behavior of this type of dosimeters was demonstrated, i.e. the intrinsic signal increases rapidly with collected dose, which should be keeping in mind by dosimetrists.

The analysis of gamma-energy spectra of activated dosimeters have shown, that in this case neutron capture (n,γ) process is the main mechanism of inducing radioactivity. Radionuclides originated this way are rather short-lived, what has been pointed out by the assessment of effective half-lives of apparent radioactivity induced in each dosimeter under study, being of the order of 15 minutes. Furthermore, since intensity of neutron production depends on therapeutic beam end-point energy and linac construction (manufacturer), the studied phenomenon is beam-quality dependent, and is currently the subject of ongoing research project.
As suggested by the overview of photonuclear reaction cross sections (see Figure 1), silicon in diodes undergoes no activation in the 15 MV therapeutic beam. Nevertheless, applying higher voltage, the results could be different.

Skin of hands and eye lens – organs which are the most likely to be exposed to radiation during the operation and in-phantom positioning of dosimeters, are able to receive equivalent doses of the order of a tenth of μSv/min, in which the activation of chlorine and inner electrode material contribute the most. It is worth noticing that such an exposition is usually not detected by the personal dosimeters worn by the staff, which are calibrated in personal dose equivalent $H_p(10)$.

ACKNOWLEDGMENTS

The study was partially support by the program for young scientists (KPG) realized in the Institute of Physics, University of Silesia also aimed at supporting the cooperation with business sector.

REFERENCES

[1] J. Becker, E. Brunckhorst, R. Schmidt, "Photoneutron production of a Siemens Primus linear accelerator studied by Monte Carlo methods and a paired magnesium and boron coated magnesium ionization chamber system", *Physics in Medicine and Biology*, **52(21),** pp.6375 – 6387. (2007)

[2] H.W. Fisher, B.E. Tabot, B. Poppe, "Activation process in medical linear accelerators and spatial distribution of activation products", *Physics in Medicine and Biology*, **51**, pp. N461 – N466. (2006)

[3] K. Polaczek-Grelik, B. Karaczyn, A. Konefał, "Nuclear reactions in linear medical accelerators and their exposure consequences", *Applied Radiation and Isotopes*, **70**, pp. 2332 – 2339. (2012)

[4] W. Abdel-Rahman, E.B. Podgorsak, "Neutron-activation revisited: the depletion and deplete-on-activation models", *Medical Physics*, **32(2)**, pp. 326 – 336. (2005)

[5] I. Gudowska, A. Brahme, P. Andreo, W. Gudowski, J. Kierkegaard, "Calculation of absorbed dose and biological effectiveness from photonuclear reactions in a bremsstrahlung beam of end point 50 MeV", *Physics in Medicine and Biology*, **44**, pp. 2099 – 2125. (1999)

[6] J.-P. Lin, T.-Ch. Chu, S-Y. Lin, M.-T. Liu, "The measurement of photoneutron in the vicinity of Siemens Primus linear accelerator", *Applied Radiation and Isotopes*, **55**, pp. 315 – 321. (2001)

[7] A. Konefał, A. Orlef, M. Dybek, Z. Maniakowski, K. Polaczek-Grelik, W. Zipper, „Correlation between radioactivity induced insi de the treatment room and the undesirable thermal/resonance neutron radiation produced by linac", *Physica Medica*, **24**, pp. 212 – 218. (2008)

[8] P.D. Allen, A. M. Chaudhri, " Charged photoparticle production in tissue during radiotherapy", *Medical Physics*, **24**, pp. 837 – 839. (1997)

[9] D. Sheikh-Bagheri, D.W.O. Rogers, "Monte Carlo calculations of nine megavoltage photon beam spectra using the BEAM code", *Medical Physics*, **29**, pp. 391 – 402. (2002)

[10] ENDF/B-VII.1 (USA, 2011) data base, available online. URL http://www.nndc.bnl.gov/sigma/search.jsp (2016)

[11] K. Polaczek-Grelik, B. Karaczyn, M. Grządziel, M. Pieńkos, A. Konefał, W. Zipper, „The map of thermal and resonance neutron distribution insi de the treatment room for akcelerator therapy", *Polish Journal of Environmental Studies, Series of Monographs*, **1**, pp. 139 – 145. (2010)

[12] L. L. Donadille et al., "Radiation protection of workers associated with secondary neutrons produced by medical linear accelerators", *Radiation Measurements*, **43**, pp. 939 – 943. (2008)

[13] Y.-H. Liu et al., "The activation contamination in the metal-based ionization chambers as gamma dosimeters in the mixed field dosimetry", *Radiation Measurements*, **45**, pp. 1427 – 1431. (2010)

[14] M. Janiszewska, K. Polaczek-Grelik, M. Raczkowski, B. Szfron, A. Konefał, W. Zipper, „Secondary radiation dose during high-energy total body irradiation", *Strahlentherapie und Onkologie*, **190(5)**, pp. 459 – 466. (2014)

[15] H. Smith (ed.), *Conversion coefficients for use in radiological protection against external radiation*, ICRP Publication 74, Pergamon Press, Oxford, United Kingdom, (1996)

Determination of power density in VVER-1000 Mock-Up in LR-0 reactor

Michal Košťál *, Vojtěch Rypar, Davit Harutyunyan, Martin Schulc, Evžen Losa

Research Center Rez Ltd., 250 68 Husinec-Rez 130, Czech Republic

Abstract. The pin power density is an important quantity which has to be monitored during the reactor operation, for two main reasons. Firstly, it is part of the limits and conditions of safe operation and, secondly, it is source term in neutron transport calculations used for the adequate assessing of the state of core structures and pressure vessel material. It is often calculated using deterministic codes which may have problems with an adequate definition of boundary conditions in subcritical regions. This may lead to overestimation of real situation, and therefore the validation of the utility codes contributes not only to better fuel utilization, but also to more precise description of radiation situation in structural components of core. Current paper presents methods developed at LR-0 reactor, as well as selected results for pin power density measurement in peripheral regions of VVER-1000 mock-up. The presented data show that the results of a utility diffusion code at core boundary overestimate the measurement. This situation, however satisfactory safe, may lead to unduly conservative approach in the determination of radiation damage of core structures.

1 Introduction

The pin power density distribution cannot be measured directly, but it can be determined indirectly via fission density. Such approach is possible due to very low disproportionality between fission and power density. The fission density can be determined by means of gamma spectroscopy of irradiated fuel, where the amount of radioactive fission products and fission density are proportional. This criterion has to be met also in the cases with different spectra, especially near baffle or control rods [1, 2, 3]. Gamma spectroscopy is suitable also for burnup determination [4].

Semiconductor gamma spectrometry with an HPGe coaxial detector in horizontal orientation was used to measure net peak areas (NPA, the area under selected gamma energy peak) of chosen fission products induced in the fuel during its irradiation. Detector is placed in a thick Pb cylindrical shield with various types of collimator. Especially in axial measurement the thin collimator (2×1 cm) has to be used. Measured gamma spectra were analyzed with the Genie 2000 software (Canberra).

2 LR-0 reactor

The LR-0 reactor, operated by the Research Centre Rez (the Czech Republic), is a pool type zero power light water reactor. For the layout of the reactor see Figure 1. All described experiments were performed with the VVER-1000 Mock-Up. This mock-up core consists of 32 shortened VVER-1000 type fuel assemblies with different enrichments of ^{235}U [3]. The fuel length is optimized for the LR-0 reactor and the fissile column is 125 cm long. The core layout with marked measured fuel pins is shown in Figure 2.

3 Experimental and calculation methods

The project, carried out on LR-0 reactor, was focused on fission rates determination in different fuel pins in VVER-1000 Mock-Up core. Positions of selected pins, together with pin numbering, are highlighted on Figure 2. There were carried out 3 irradiation sets to determine the fission density based on both short-living (^{92}Sr, ^{91}Sr, ^{97}Zr, ^{135}I, ^{88}Kr; they can be observed immediately after a short irradiation on low power level) and week-living (^{140}Ba, ^{103}Ru, ^{131}I, ^{95}Zr, ^{140}La; they can be observed after a long-lasting irradiation on higher power level and after appropriate decay time) fission products. Short living products were measured after 5 irradiation batches on power about 10 W lasting 2.5 hours each. To induce measurable activities of long living products, the fuel was irradiated for 100 hours on power about 700 W. Absolute reactor power was monitored by activation foils. The fission reaction rates measurements were then compared with computational models.

* Corresponding author: Michal.Kostal@cvrez.cz

 DOI: 10.1051/epjconf/201715302001

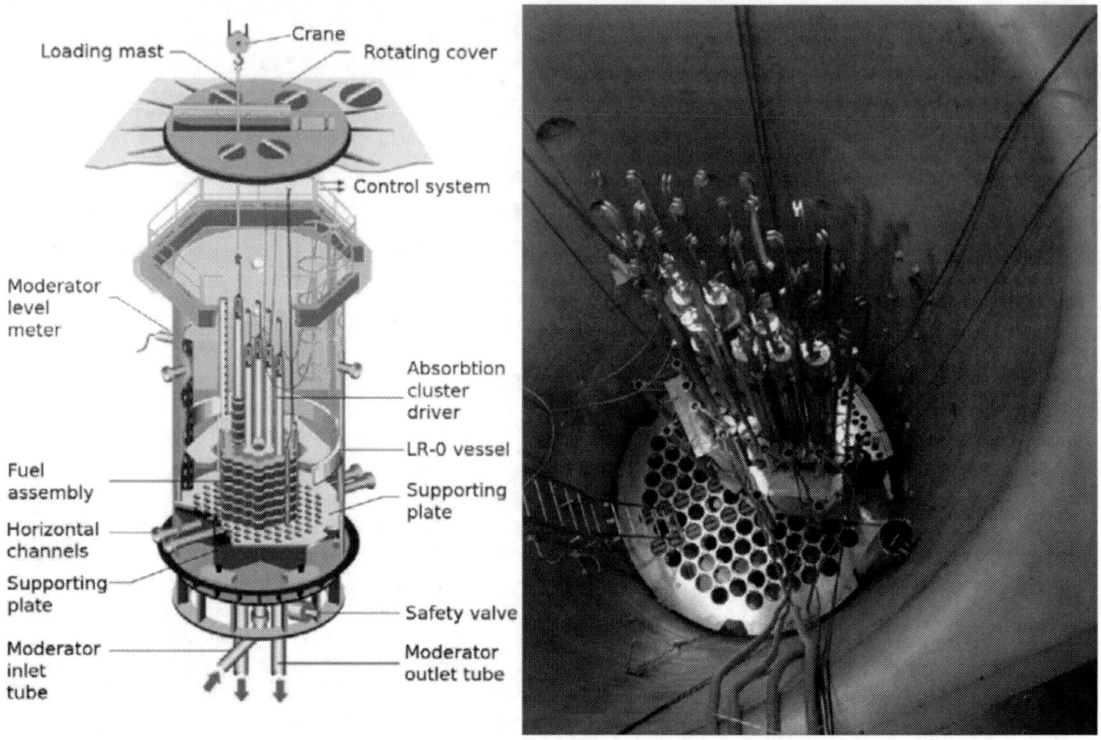

Figure 1. General view on LR-0 and Mock-Up core with barrel and baffle simulators

3.1. Fission rates measurement

The measurement was performed in the centre of the fissile column. The measured NPA's were corrected to selected reference date. In case of short living fission products, it is the end of irradiation, in other cases selected reference date. That is chosen to the time when all mother nuclides of studied radioisotopes are decayed. In case of ^{140}La measurement, the reference time is set to transient equilibrium between ^{140}Ba and ^{140}La. This correction is applicable only when the photons of the analysed peak, emitted during the decay, occur in the sample [4, 5, 6]. Observed photons can also be extracted in case of mixed peak [7]. This approach of extracting peak of interest from mixed peak was used only in case of ^{97}Zr.

Net peak areas in different energy peaks of selected fission products induced in the fuel during its irradiation were measured by means of semiconductor gamma spectrometry with an HPGe coaxial detector in a streamline horizontal configuration (Ortec GEM70, resolution approximately 2.1 keV at Eγ = 1333 keV). Gamma spectra were measured using multichannel analyser DSA 2000 controlled by computer using Genie 2000 spectroscopy software via the in-built Ethernet interface. Genie 2000 software with gamma analysis package also analysed acquired gamma-spectra.

To suppress the effect of non-homogenous fission products distribution in the fuel, the pins axially rotate during the measurement. This effect could play its role in the pins adjacent to the reactor baffle model and reflects the strong gradient in neutron flux caused by a strong absorption in the steel which forms the reactor baffle model. The theoretical prediction states significant heterogeneities in the distribution of fission products. Namely there are 20% more fission products in the pellet region opposite to the baffle, than in the region near the baffle (see Figure 3) The layout of the experimental arrangement is shown in Figure 4.

3.1.1 Short living fission products

In this experiment, the fission products induced within 2.5 h irradiation on power level ~10 W were measured. Under these conditions, the observed fission products have relatively short half-life, thus the pins were measured after 5 irradiation steps. The fuel in this experiment was measured immediately after each irradiation, therefore it was highly radioactive. Due to this fact the dead time for HPGe measurement was very high, thus the signal had to be suppressed by a Pb-Cu plate (3.3 mm of Pb followed by 1 mm of Cu) placed between the measured pin and the HPGe front end cap.

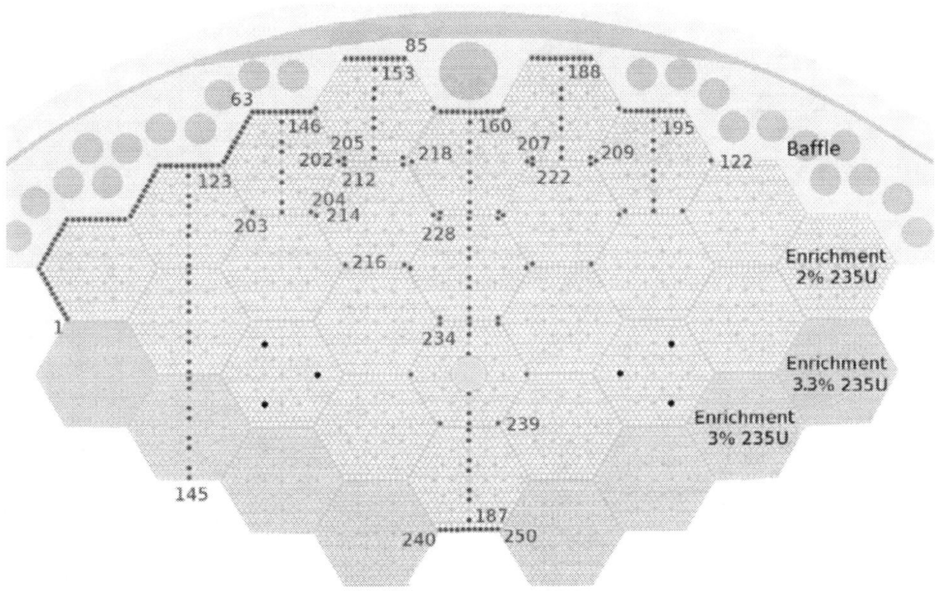

Figure 2. Positions of measured pins

Many peaks can be found in the fuel shortly after the end of irradiation. It reflects high activity of short living fission products after short irradiation. Most notable peaks are presented in Table 1, but it is worth noting that only few of reported fission products are suitable for fission density determination. The comparison between calculated and measured NPA for selected nuclides is plotted in Figure 5. The satisfactorily agreement shows, that not only ^{92}Sr, but also ^{91}Sr is suitable for fission density determination. More about this method is presented in [5] and [9].

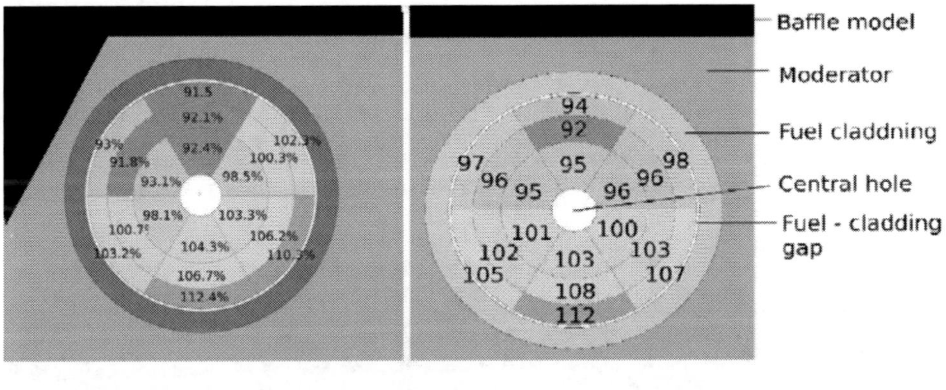

Figure 3. The theoretical prediction of fission distribution in pin near baffle model (% of average fission density)

3.1.2 Week living fission products

Net peak areas of selected longer-living fission products induced in the fuel during 100 hours irradiation and at least 16 days of decay time were measured. The peaks in the energy interval below 1 MeV were analysed. Specifically these were 537 keV peak for ^{140}Ba, 498 keV peak for ^{103}Ru, 365 keV peak for ^{131}I and 724 keV peak for ^{95}Zr. The count times were chosen to obtain over 10^4 counts in net peak area of ^{140}Ba peak, thereby achieving statistical uncertainty less than 1%. The analyzed NPAs varies around 10^4-10^5 except measured nuclides ^{131}I with 5.10^3 pulses in minimal NPA values. The effect of true summation is neglected, because of the measurement geometry, the low gamma branching ratio of the coincidence peaks, or low energy of the coincidence peaks, as well as low HPGe sensitivity. It was observed good agreement between calculated and measured NPA of studied nuclides. Thus it seems, they are suitable for fission density determination (see Figure 6). The more detailed results of week-living fission products measurement are presented in [6].

Table 1. Radionuclides selected from fuel pin gamma spectra measured 1.2 h after irradiation (irradiation period = 2.5 h, reactor power ≈ 9.5 W). Activity values were corrected for radioactive decay to the time of reactor shutdown [1].

Radioisotope	Activity (%)	Uncertainty (%)	Key line (keV)	Half-life (h)	Decay scheme
^{134}I	100.0	1.0	847.0	0.88	^{134}Te → ^{134}I
^{138}Cs	98.5	0.8	1435.8	0.56	^{138}Xe → ^{138}Cs
^{134}Te	68.2	1.1	767.2	0.70	^{134}Te → ^{134}I
^{142}La	44.1	0.9	641.3	1.52	^{142}Ba → ^{142}La
^{133m}Te	29.1	1.3	912.7	0.92	^{133m}Te → ^{133}Te
^{92}Sr	25.1	0.8	1383.9	2.71	^{92}Sr → ^{92}Y
^{87}Kr	19.2	3.9	402.6	1.27	^{87}Kr → ^{87}Rb
^{88}Kr	13.9	2.1	2392.1	2.84	^{88}Kr → ^{88}Rb
^{135}I	13.6	1.2	1260.4	6.57	^{135}I → ^{135}Xe
^{91}Sr	8.5	2.0	1024.3	9.63	^{91}Sr → ^{91}Y
^{97}Nb	8.4	2.2	657.9	1.20	^{97}Zr → ^{97}Nb
^{132}Te	3.5	5.9	228.2	3.2	^{132}Te → ^{132}I
^{97}Zr	5.7	2.1	743.4	16.9	^{97}Zr → ^{97}Nb
^{133}I	4.9	2.2	529.9	20.8	^{133}Te → ^{133}I
^{135}Xe	3.2	5.7	249.8	9.14	^{135}I → ^{135}Xe

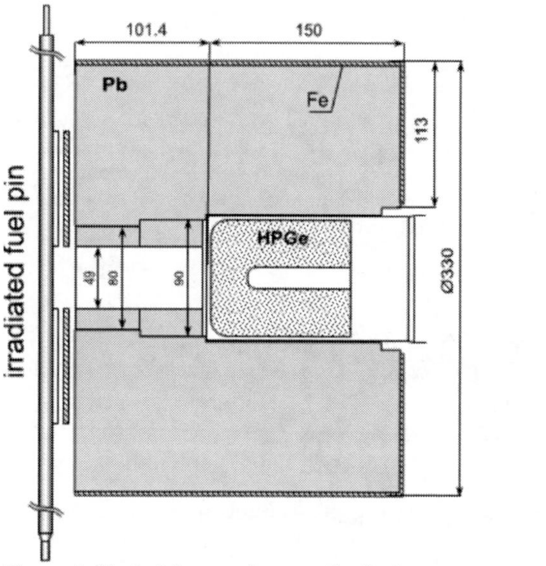

Figure 4. Vertical layout of power distribution measurement (Dimension are in mm]

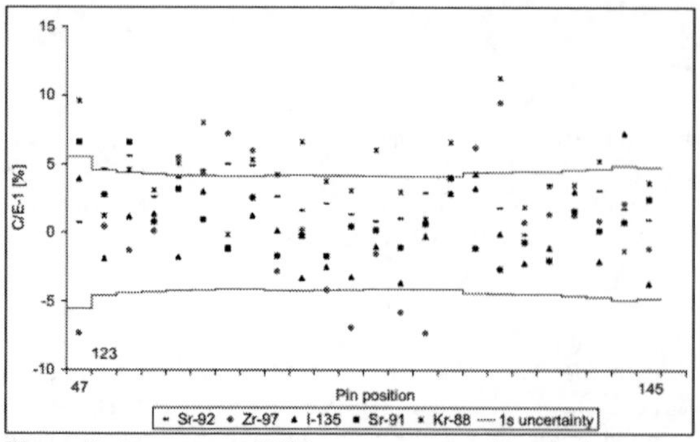

Figure 5. Agreement of NPA measurement with calculation for selected short living fission products.

3.1.3. ^{140}La Measurement

The HPGe measurement of pins started 15 days after the reactor shut-down and 1597 keV (^{140}La) energy peak was assessed. Such time is needed for establishing of transient equilibrium between mother ^{140}Ba and daughter ^{140}La nuclides. In this transient equilibrium, the decay correction of both radionuclides can be simply calculated by means of the ^{140}Ba decay half-life (12.75 days). The uncertainties in realized experiment are below 10 %. The power level during irradiation was not defined, that is the reason why the measured data have relative character. Such approach simplifies the evaluation process of experimental data, because otherwise the measured gamma activity of ^{140}La must be corrected for true summation effect, due to the many coincidence peaks present in its decay scheme.

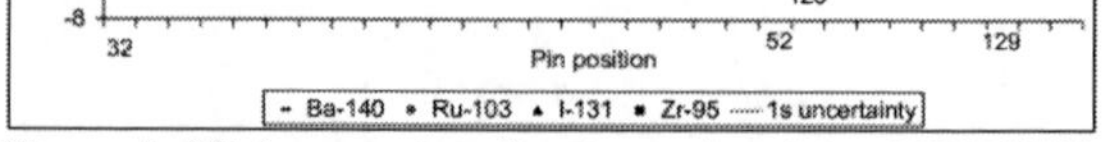

Figure 6. C/E-1 comparison for observed nuclei and ENDF VII calculation

In relative comparison, on the other hand, such correction is not necessary, because all measurements are done in the same geometry and the correction factor is already the same. More details can be found in [6].

3.2. Detector arrangement for measurement of activation foils

The value of the mock-up power, which is used for normalization of absolutely measured fission products (^{92}Sr, ^{91}Sr, ^{97}Zr, ^{135}I, ^{88}Kr, ^{140}Ba, ^{103}Ru, ^{131}I, and ^{95}Zr), was determined by means of reaction rate determination in the activation foil using reaction ^{197}Au(n,g)^{198}Au. The gold, in mass about 30 µg per foil, was in an aluminium alloy with 1% of Au. The foil diameter was 3.6 mm and its thickness was 0.1 mm. The ^{198}Au peak at 412 keV was measured immediately after the irradiation by the coaxial HPGe detector in vertical configuration (Ortec GEM35). For background radiation suppression, the detector and measured foil are located in the sandwich shielding box. The experimental uncertainties were estimated to be below 3%. More details can be found in [5].

3.3. Calculations

The pin-by-pin fission density distribution in the reactor core was calculated by the MCNP6 [10] transport code in critical mode [11] and ENDF/B-VII.0 library [12].

In case of steel reactor structures and thermal neutron transport, the free gas approach for ^{56}Fe was used.

The HPGe responses used in evaluation of fission densities of selected fission products were also determined by calculation using the MCNP6 code. The parameters of the detector producer were used for computational model compilation. The yields of the individual nuclides, which contribute to the production of selected isotopes, were used directly from the ENDF/B-VII.0 nuclear data library.

4 Results

The fission densities from nuclides induced during defined irradiation were derived using equation (1).

The ^{140}La measurement from 2008 has character of relative distribution of fission rates. The selection of pins for relative measurement contains same 12 pins as selection for absolute measurement, namely pins in position 53, 63 - 73 (see Figure 2). Due to this overlap, the normalization of relative ^{140}La distribution to fission densities can be realized.

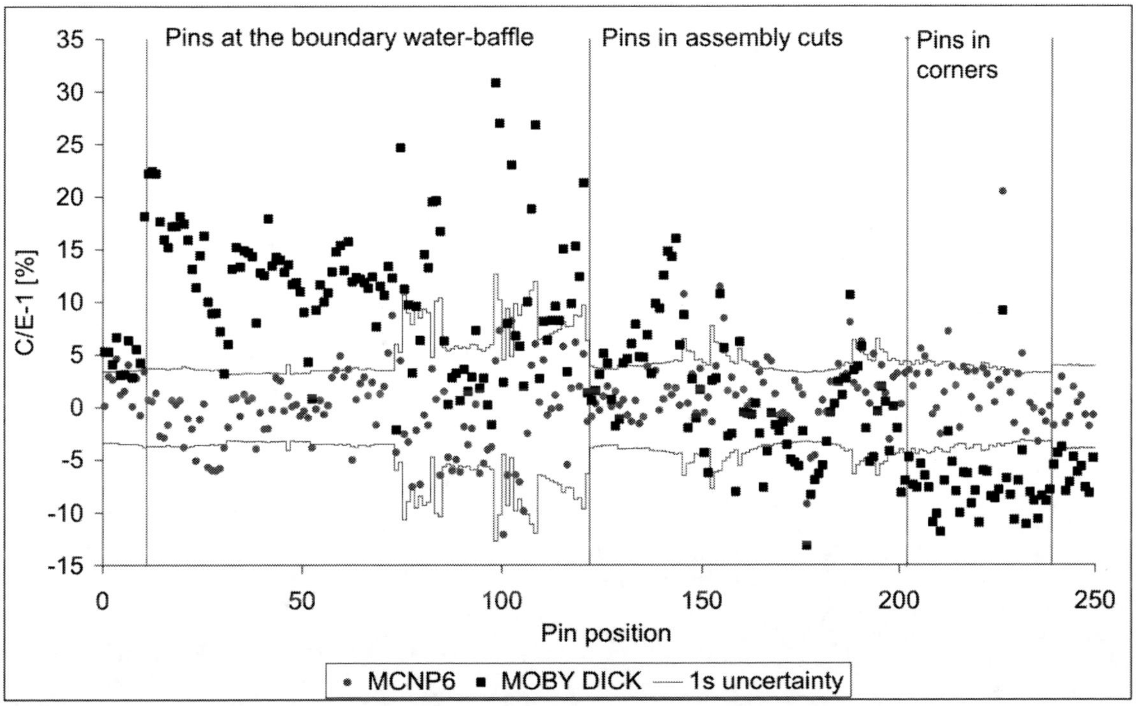

Figure 7. Experiment to calculation comparison of fission densities in VVER-1000 Mock-Up by means of C/E-1

$$F_j^i = \frac{NPA_j^i(t)}{\eta^i.\varepsilon^i.\lambda^i}.\frac{1}{N^i(t)} = NPA_j^i(0).\frac{1}{\eta^i.\varepsilon^i.\lambda^i.N^i(0)} \quad (1)$$

$$NPA_j^i(0) = NPA_j^i(t).\frac{1}{e^{-\lambda^i.t}} \times \frac{\lambda^i.\Delta T}{1-e^{-\lambda^i.\Delta T}} \quad (2)$$

$$N^i(0) = \frac{q^i}{\lambda^i}.(1-e^{-\lambda^i.T_{irr}}) \quad (3)$$

F^i_j fission rate determined via the i-th nuclei and j-th pin; $N^i(t)$ calculated number of observed nuclei in fuel pin when 1 fission/s occurs, in time t after irradiation end; $NPA^i_j(t)$ measured Net Peak Area of the observed nuclei and selected peak in i-th pin (equation (2); λ^i decay constant of selected nuclide; η^i efficiency of HPGe for the selected gamma line of the i-th nuclide; ε^i gamma branching ratio of the selected peak from observed nuclei i; t start i-th pin measurement; ΔT length of i-th pin HPGe measurement; q^i is fission yield of i th isotope, T_{irr} is time of irradiation. The resulting fission densities are compared with Monte Carlo MCNP6 and diffusion code MOBY DICK [13] calculations, in form of C/E-1 comparison in, see Figure 7. The comparison covers especially the pins in positions with expected C/E discrepancies, i.e. near the core and baffle boundary, between the assemblies (where water gaps are present) and in the assemblies corners. Especially the diffusion codes may have problems with adequate definition of boundary conditions near the steel internal structures of nuclear reactor. It can be said, that the expectations were fulfilled, and notable discrepancies can be observed in case of diffusion code MOBY DICK results.

Generally, the diffusion code results overestimate experiment in positions near baffle, while in positions in assembly corners they underestimate experiment. It can be explained by the fact, that as steel does not have so good moderation properties as fuel cell (i.e. pin surrounded by water) and the real backscatter from steel is lower than prediction based on parameters derived for fuel cell. In the corners, the situation is opposite; the real moderation properties are better than theoretical prediction for fuel cell, which reflects the overestimation of diffusion calculational approach. Fuel pins of positions on core-stainless steel baffle border have numbering 11–122 and those ones on central axis of fuel assemblies 123 – 201.

5 Conclusions

The methodology, developed at LR-0 reactor, is suitable for experimental determination of fission density. The experimental measurement is in satisfactory agreement with Monte Carlo simulation. In case of diffusion calculation, notable variations in regions in corners and baffle can be observed. Diffusion code MOBY DICK underestimates experiment in corner pins, while near baffle it overestimates. The rate of overestimation can be as high as 25%. It's worth noting, such kind of overestimation is on safe side, because the calculational dose in the RPV overestimates true value. On the other hand the residual lifetime prediction of RPV, based on diffusion code might be unduly conservative.

It can be concluded that fission source described by Monte Carlo code is more reliable than description by diffusion code, therefore it should be used for the serious issues concerning the topics such as RPV lifetime assessment or estimation of the flux behind the reactor vessel.

Acknowledgements

The presented work was financially supported by the Ministry of Education, Youth and Sport Czech Republic Project LQ1603 (Research for SUSEN).

References

1. M. Švadlenková et al., Gamma spectrometry of short living fission products in fuel pins, Nucl. Inst. and Meth. in Physics Research Section A, 739, (2014), pp 55-62,
2. H. Kröhnert et al., Gamma-ray spectrometric measurements of fission rate ratios between fresh and burnt fuel following irradiation in a zero-power reactor, Nucl. Inst. and Meth. in Physics Research Section A, 698, (2013), pp 72-80
3. M. Košťál et al., Effect of inhomogeneities inserted into fuel assembly in the VVER-1000 mock-up on the LR-0 research reactor, Ann. of Nucl. En., 69, (2014), pp 168-174
4. M. Koleška et al., Determination of IRT-2M fuel burnup by gamma spectrometry, Appl. Rad. and Isot., 107, (2016), pp 92-97
5. M. Košťál et al., Comparison of various hours living fission products for absolute power density determination in VVER-1000 mock up in LR-0 reactor, Appl. Rad. and Isot., 105, (2015), pp 264-272
6. M. Košťál et al., Determination of critical assembly absolute power using post-irradiation activation measurement of week-lived fission products, Appl. Rad. and Isot., 89, (2014), pp 18-24
7. M. Košťál, V. Rypar, M. Švadlenková, The pin power distribution in theVVER-1000 mock-up on the LR-0 research reactor, Nucl. Eng. and Des., 242, (2012), pp 201-214
8. M. Košťál, M. Švadlenková, J. Milčák, The application and comparison of 97Zr and 92Sr in the absolute determination of the contribution of power density and cladding activation in a VVER-1000 Mock-Up on the LR-0 Research Reactor, Nucl. Inst. and Meth. in Physics Research Section A, 738, (2014), pp 87-92
9. U.C. Bergmann et al, Optimised non-invasive method to determine 238U-captures-to-total-fissions in reactor fuel, Nucl. Inst. and Meth. in Physics Research Section A, 556, 1, (2006), pp 331-338
10. D.B. Pelowitz (Ed.), MCNP6™ User's Manual, Los Alamos National Laboratory Report, LA-CP-13–00634 (2013)
11. T. Goorley, et al., "Initial MCNP6 Release Overview", Nuclear Technology, 180, pp 298-315 (Dec 2012).
12. M.B. Chadwick et al, ENDF/B-VII.0: Next Generation Evaluated Nuclear Data Library for Nuclear Science and Technology, Nuclear Data Sheets, 107, 12, (2006), pp 2931-3060
13 C. Guler, S. Levine, K. Ivanov et al, Development of the VVER core loading optimization system, Ann. of Nucl. En., 31, 7, (2004), pp 747-772

Characteristic Analysis of Concrete for Radiation Shielding Evaluation of Gamma Irradiation Facilities

Ki-tek Han[1] , Sang Mook Kang[1], Jeho Min[1] and Han Ki Jang[1,a]

[1]KARA, Quality Assurance Center for Medical Radiation, 04790 77 Seongsuil-ro Seongdong-gu Seoul, Korea

Abstract. Physical and chemical properties of concrete sample was obtained using various methods such as density analysis, elemental analysis and WD-XRF analysis to get accurate characteristics of concrete collected in our construction site in Korea. Using these results, in this study, we perform the simulation to compare the dose evaluation results between NBS concrete of ANSI/ANS 6.4.3 and characteristics of concrete samples using MCNPX. As a result, simulation results of collected concrete samples in our facility were underestimated up to 54.01%. Through these results, we verify the necessity of accurate properties of materials.

1 Introduction

Monte Carlo radiation transport computer codes such as MCNPX, Geant4 are widely used in radiation shielding evaluation(RSE)[1].

Most case of concrete in RSE with Monte Carlo radiation transport computer code, National Bureau of Standards(NBS) concrete values in ANSI/ANS 6.4.3 is widely used[2-3]. However, there are many differences between NBS concrete properties in ANSI/ANS 6.4.3 and real concrete. So that reason, RSE results may overestimate or underestimate. To improve this problem, exact characteristics of material such as density and weight fraction is necessary.

There are many advantages to using real properties of materials. For example, it helps prevent overestimate or underestimate, thereby calculating accurate construction cost.

In this study, we collected concrete samples from our construction site of gamma irradiation rooms in Quality Assurance Center for Medical Radiation of the Korean Association for Radiation Application(KARA). Using these concrete samples, we carried out physical and chemical analysis such as density analysis, elemental analysis(EA) and X-ray fluorescence(XRF) analysis to get accurate properties for characteristics of concrete.

Finally, we perform the simulation using MCNPX Monte Carlo radiation transport computer code to compare the gamma-ray shielding performance between NBS concrete and concrete samples collected in our construction site.

2 Materials and Methods

2.1 Preparation of Concrete Samples

For analyzing characteristics of real concretes, we prepare concrete samples as shown in figure 1. These concrete samples are collected from various part of gamma irradiation rooms in Quality Assurance Center for Medical Radiation. And collected concrete samples were made by two types, cylinder and rectangular type.

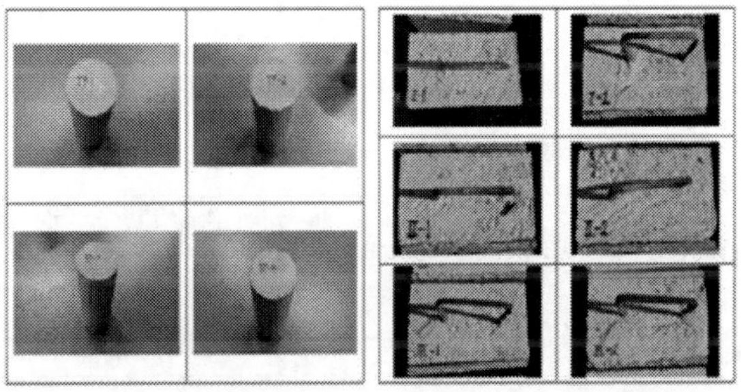

(a) (b)

Figure 1. Concrete samples(a: cylinder type, b: rectangular type).

2.2 Characteristics Analysis

As mentioned, we carried out concrete sample analysis to obtain physical and chemical properties. Concrete samples were analyzed in three ways, density analysis, elemental analysis(EA) and wavelength dispersive X-ray fluorescence (WD-XRF) analysis using five cylinder type samples.

[a] Corresponding author: hkjang@ri.or.kr

DOI: 10.1051/epjconf/201715302002

2.2.1 Density Analysis

Density of collected concrete samples were measured by Korea Conformity Laboratory (KCL) following KS F 2459:2002[4].

2.2.2 Elemental Analysis

Generally, EA method is used in quantitative analysis for organic properties such as hydrogen, carbon, oxygen, nitrogen and sulfur. This test was measured by Korea Basic Science Institute (KBSI). using elemental analyzer.

2.2.3 WD-XRF Analysis

XRF analysis using WD-XRF is widely used in quantitative analysis of inorganic properties of material. For analyzing samples using WD-XRF, pretreatment is necessary as shown in figure 2.

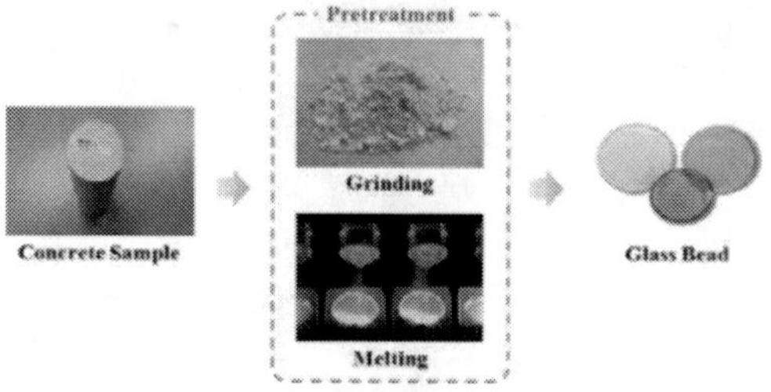

Figure 2. Pretreatment process of sample.

3 Experimental Results

3.1 Results of Characteristics Analysis

In this study, we analyze characteristics of concrete samples to obtain physical and chemical properties.

Table 1 shows the results of density analysis. Average density of 2.208 g/cm^3 and a relative standard deviation(RSD %) of 1.044 was obtained.

Table 1. Measurement of .concrete density.

Samples	Concrete Samples (Test Piece, TP)				
	TP-1	TP-3	TP-2	TP-4	TP-5
Density (g/cm^3)	2.199	2.217	2.179	2.202	2.241
Average		2.208 g/cm^3			

In case of EA, properties of carbon, hydrogen and oxygen were obtained as shown in table 2. And the WD-XRF results obtained from five collected concrete samples in oxide form were shown in table 3.

Table 2. Elemental analysis results.

Element	C	H	O	N	S
Weight Fraction (wt %)	1.07 ±0.01	0.05 ±0.01	6.09 ±0.05	None	None

Table 3. WD-XRF analysis Results.

Oxygen Compounds	Weight Fraction (wt %)
Al_2O_3	12.49 ± 0.32
CaO	11.53 ± 0.40
Fe_2O_3	2.13 ± 0.42
K_2O	3.00 ± 0.19
MgO	1.39 ± 0.39
MnO	0.08 ± 0.05
Na_2O	3.05 ± 0.23
P_2O_5	0.13 ± 0.10
SiO_2	58.77 ± 0.70
TiO_2	0.34 ± 0.17
L.O.I	6.29 ± 0.36
Total	99.20 ±0.28

Table 4 shows comparison of ANSI/ANS 6.4.3 and experimental results. Experimental results were obtained from calculation of EA and WD-XRF results. As a result, element properties are similar. However some elements are little different such as carbon, phosphorus, sulfur and titanium.

Table 4. Comparison of ANSI/ANS 6.4.3 and experimental results.

Elements	Weight Fraction	
	ANSI 6.4.3	This Study
Hydrogen	0.006	0.005
Carbon	-	0.011
Oxygen	0.498	0.493
Sodium	0.017	0.020
Magnesium	0.002	0.016
Aluminum	0.046	0.060
Silicon	0.316	0.258
Phosphorus	-	0.001
Sulfur	0.001	-
Potassium	0.019	0.025
Calcium	0.083	0.084
Iron	0.012	0.024
Titanium	-	0.003
Total	1	1

3.2 Comparison of Shielding Performance

To compare gamma-ray shielding performance between ANSI/ANS 6.4.3 NBS concrete and concrete sample collected in our construction site, we perform the simulation through the MCNPX Monte Carlo radiation transport computer code work.

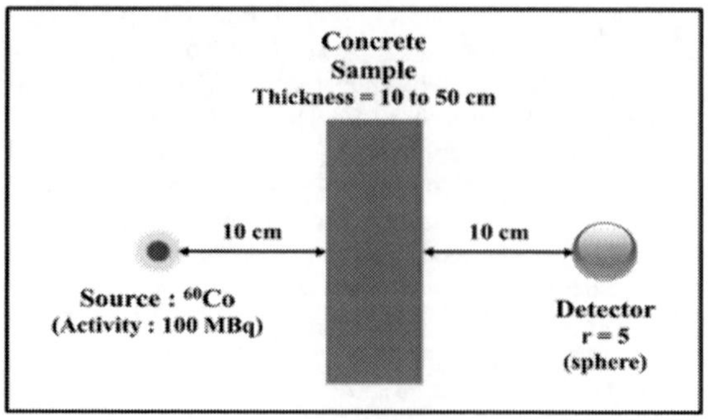

Figure 3. Geometry of MCNPX simulation

Figure 3 shows geometry of MCNPX simulation. RSE performed according to thickness of concrete samples using 100 MBq ^{60}Co radiation source. The thickness of concrete increases by 5 cm increments from 10 to 50 cm. In case of radiation detector, sphere type with a radius of 5 cm was used.

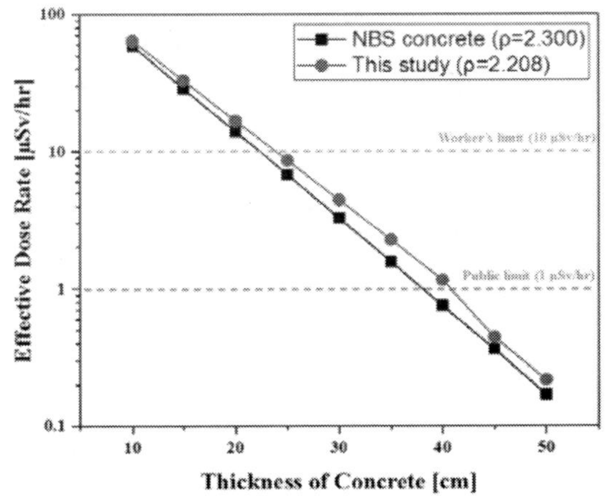

Figure 4. MCNPX simulation results according to thickness of concrete samples.

Figure 4 shows MCNPX simulation results, according to thickness of concrete samples. These results are calculated by tally 4 in MCNPX and ICRP 74 DCF(dose conversion factor) was used to convert effective dose per photon fluence. Relative errors of all calculated results are less than 1%. As a result, simulation results of collected concrete samples in our facility were underestimated up to 54.01% against NBS concrete. Therefore, accurate properties of material is necessary to prevent overestimate or underestimate.

4 Conclusion

In this study, we obtained physical and chemical properties of concrete samples collected from our construction site. Using this properties, shielding performance compared between NBS concrete and collected concrete sample using MCNPX. In conclusion, we verify that accurate properties of material are necessary to prevent overestimate or underestimate in RSE.

Further studies will be carried out experiments using collected concrete samples and various gamma irradiation systems such as ^{60}Co, ^{137}Cs, and ^{192}Ir irradiation systems.

References

1. A.M. El-khayatt, Ann Nucl Energy **38**, 128 (2011)
2. Oak Ridge National Laboratory, *ORNL/RSIC-49* (1988)
3. K.S. Mann, T. Korkut, Ann Nucl Energy **41**, 81 (2013)
4. Korea Agency for Technology and Standards, *KS F 2459* (2002).

SCALE/MAVRIC calculation of dose rates measured for a gamma radiation source in a thick-walled transport and storage cask of ductile cast iron with lead inserts

Werner Baumgarten[1,a], Holger Thiele[1], Benjamin Ruprecht[1], Peter-W. Phlippen[2], Luc Schlömer[2]

[1]*Bundesamt für kerntechnische Entsorgungssicherheit (BfE), Postfach 100149, 38201 Salzgitter, Germany*
[2]*WTI Wissenschaftlich-Technische Ingenieurberatung GmbH, Karl-Heinz-Beckurts-Straße 8, 52428 Jülich, Germany*

Abstract. Dose rate calculations are important for judging the shielding performance of transport casks for radioactive material. Therefore it is important to have reliable calculation tools. We report on measured and calculated dose rates near a thick-walled transport and storage cask of ductile cast iron with lead inserts and a Co-60 source inside. In a series of experiments the thickness of the inserts was varied, and measured dose rates near the cask were compared with SCALE/MAVRIC 6.1.3 and SCALE/MAVRIC 6.2 calculation results. Deviations from the measurements were found to be higher for increased lead thicknesses. Furthermore, it is shown how the shielding material density, air scattering and accounting for the floor influence the quality of the calculation.

1 Introduction

The calculation of dose rates behind multiple shielding layers of material is highly relevant for radiation shielding assessment tasks within package design approval procedures. Thus, it is important that the underlying calculation methods work as accurate as possible and that they are validated against experimental data. A series of experiments was performed and evaluated by WTI (Wissenschaftlich-Technische Ingenieurberatung GmbH) allowing to check the accuracy of such calculation tools (cp. [3]). The paper reports on modelling and dose rate calculations with SCALE/MAVRIC 6.1.3 [1] and SCALE/MAVRIC 6.2 [4] by BfE and how the results compare to the measured data.

2 Experimental Setup

The experimental setup consisted of a freely moveable gamma radiation detector and a cylindrical transport and storage cask of ductile cast iron standing on a concrete floor (Fig. 1). Through small holes in the lid of the closed cask a Co-60 gamma source was placed inside. The shielding properties of the cask were changed by inserting hollow lead cylinders with different thicknesses between 20 mm and 80 mm.
Gamma energy dose rates (kerma-free-in-air) were measured using different ionization chambers connected to a universal dosimeter of type PTW UNIDOS. By varying the position of the source as well as the thickness of the inserts different configurations were considered.

The gamma energy dose rates were measured at radial distances of 16.5 cm, 50 cm, 100 cm, 150 cm, and 200 cm from the surface of the cask body in planes at the top, the middle and the bottom. Due to the high source activity of about 1 TBq, dose rates could be measured accurately even for the highest lead thickness.
In order to ensure homogeneous shielding properties of the cast iron wall and the lead between source and detector these structures were checked for material errors, *e.g.* cavities, by gamma radiography before the experiments were carried out. Additionally, the real cast iron and lead density were measured after the measurement series had been conducted.

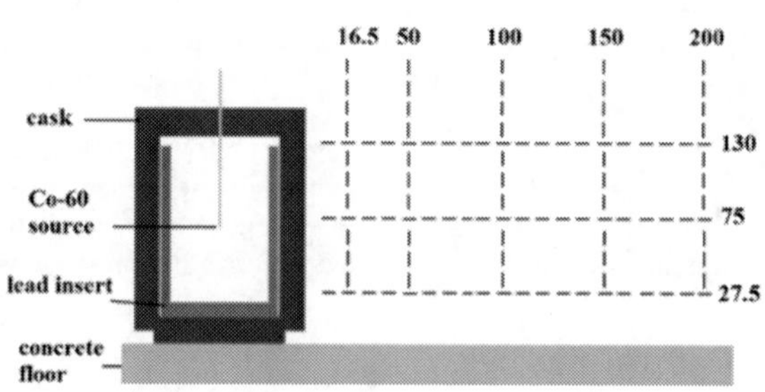

Fig. 1: Cross-section view of the storage cask and the detector positions (dimensions in cm).

[a] Corresponding author: Werner.Baumgarten@BfE.bund.de

3 Calculation Methods and Model

Co-60 predominantly emits gamma photons with energies of 1.17 MeV and 1.33 MeV, which can only be modelled directly in programs working with continuous energies. Thus, in order to model the Co-60 source in multi-group calculations an energy conserving gamma spectrum was calculated for the 47 gamma group library available in the SCALE 6 package with ORIGEN-ARP 6.1. Dose rates were then calculated with the MAVRIC [2] sequence of SCALE 6.1.3 using point tallies and the v7-200n47g shielding library based on ENDF/B-VII data.
However, SCALE/MAVRIC 6.2 offers the possibility to use continuous energy cross sections in the transport calculation which allowed us to model the Co-60 source without any pre-processing of the spectrum. In our model the source definition relies on the implemented single-nuclide spectrum.
A cross section of the cask model and the different detector positions are shown in Fig. 1. The grid geometry for the calculation in MAVRIC that covers the cask consisted of 60 bins for each of the x- and y-direction, and 100 bins for the z-direction (to the top), while the concrete floor was modelled with 10 bins in x, y and z direction. The source itself was modelled by a small cylinder of 0.4 cm in diameter and height. Monte-Carlo calculations with SCALE/MONACO code were conducted with 1600 batches of 1000 particles resulting in typical relative standard deviations of up to 0.7 % of the calculated dose rates.
The dose rates gained from the model are influenced by several parameters. In this paper we focus on the lead thickness, the floor, the density of the shielding cask material, air in the free volume and the type of the energy library used to model the Co-60 source and the photon transport. For these model simulations, if not stated otherwise, the following setup has been used. The source was placed in the centre of the cask at 75 cm height and the tallies (virtual detector positions) were placed in the middle plane also at 75 cm height (Fig. 1). Energy dose rates were calculated using the appropriate ICRU-57 flux to dose rate conversion factors integrated in SCALE. The uncertainties of the Co-60 source activity (2.4 %) and the calibration of the dosimeter (2.0 %) result in a systematic error of 4.4 %. The relative error of 1.3 % is composed of the standard deviation of the repeated measurements (0.5 %), the exact positioning of the source (1.0 %), and the Monte-Carlo simulation results (0.7 %).

4 Results

The relative deviation of the calculated dose rates from the measured dose rates changes only negligibly when the detector distance from the cask is varied. This is shown in Fig. 2 for the results obtained using SCALE/MAVRIC 6.1.3 with the multi-group shielding library. However, the deviation from the measured dose rate significantly increases when thicker lead insets are used, which can also be seen in Fig. 3 where the distance-averaged deviations are plotted vs. the lead thickness. The multi-group model underestimates the dose rate by up to 25% at 80 mm lead thickness (Fig. 2) whereas the underestimation is as low as about 7 % when no lead is present. For the SCALE/MAVRIC 6.2 model with continuous energy library a similar behaviour is observed, see Fig. 3. Yet, an underestimation of the dose rate is only observed for lead thicknesses above 40 mm. Without lead and for the 20 mm inset the dose rate is even overestimated.

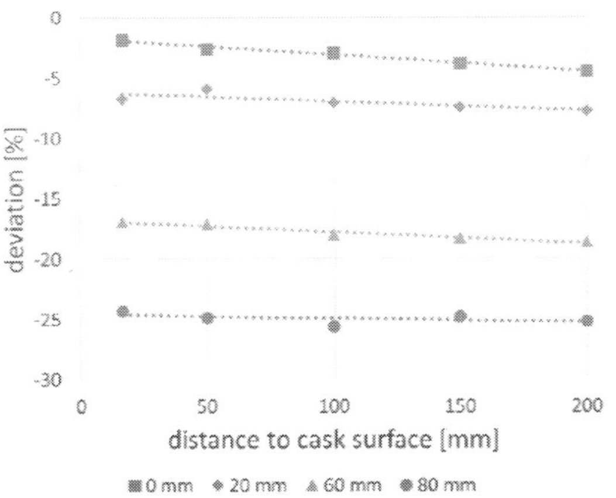

Fig. 2: Deviations from the measured dose rates in SCALE/MAVRIC 6.1.3 multi-energy group calculations with different lead thicknesses (0 mm, 20 mm, 60 mm and 80 mm) and detector positions.

Nevertheless, the slopes of the trending lines in Fig. 3 are comparable. Additional tests have clearly shown that SCALE 6.2 with multi-group energy model settings leads to exactly the same results as SCALE 6.1.3 (not shown here). All in all, the clear trend for underestimation of the gamma dose rates when thick-walled shields are employed has to be considered in future studies with both SCALE 6.1.3 and SCALE 6.2 calculations.

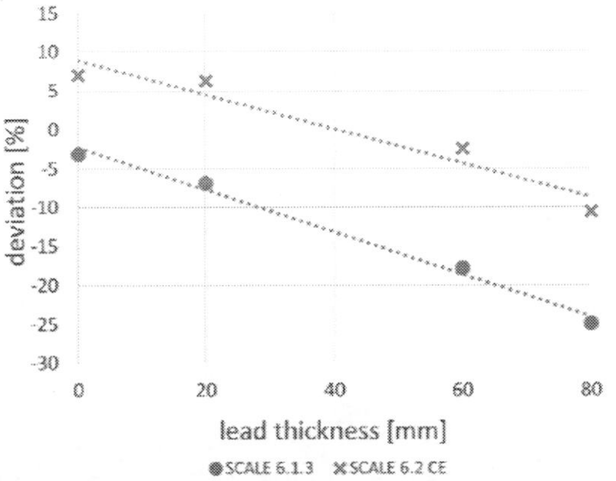

Fig. 3: Average deviation from measured dose rates by a multi-group energy model (SCALE 6.1.3, blue line) and a continuous energy model (SCALE 6.2, orange line).

The uncertainty in the exact knowledge of the density of the shielding material influences the results of the

DOI: 10.1051/epjconf/201715302003

simulation. To quantify this influence, the density of the shielding material of the cask was changed from 7.07 g/cm³ to 7.00 g/cm³, *i.e.* was reduced by 1%. Fig. 4 shows that on average this 1% decrease of the density results in an increase of the dose rate of about 5.0 %, regardless of the thickness of the lead insert. Thus, the exact knowledge of the density of the shielding material is crucial to gain correct dose rate results in the model.

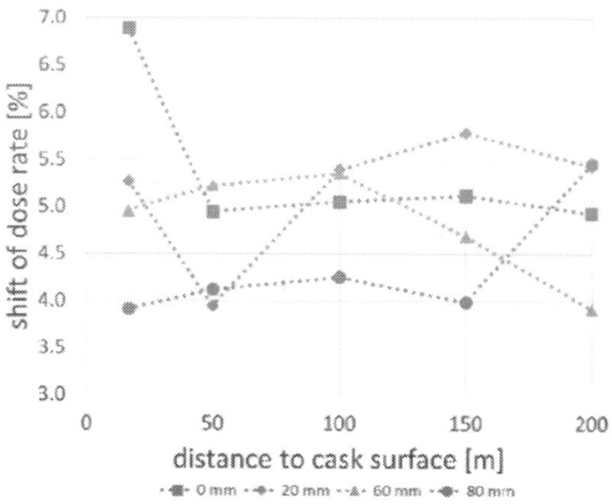

Fig. 4: Influence of changing the shielding material density from 7.07 g/cm³ to 7.00 g/cm³ on the calculation results for different lead thicknesses (0 mm, 20 mm, 60 mm and 80 mm). Calculated with SCALE 6.1.3.

Air, as a weak absorber and scatterer of gamma radiation, may also affect the resulting dose rate. However, our calculations show that at short distances of not more than 200 cm, air has only little influence of about 1% on the results, see Fig. 5, which is not significant within two standard deviations of the Monte-Carlo calculations.

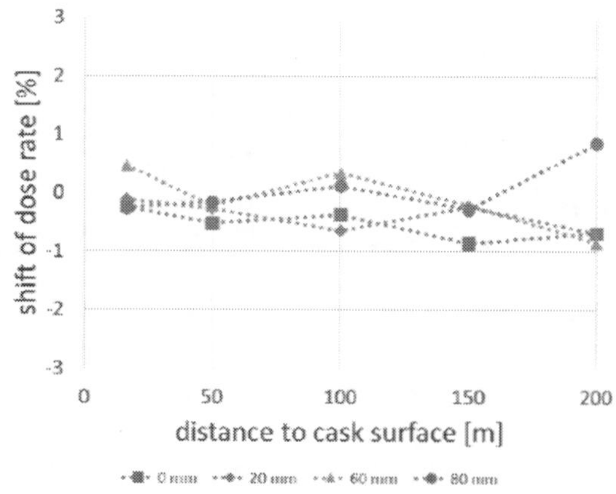

Fig. 5: The effect of including air into the models at various lead insert thicknesses (0 mm, 20 mm, 60 mm and 80 mm) is between -1 % and 1 %. Calculations performed with SCALE 6.1.3.

We also studied to which extend the concrete floor affects the dose rate results in the model. Whereas the floor was included in all previous models, we ran the same simulations as before but without the floor, and compared the dose rates. For the detector positions in the middle plane the results show that accounting for the floor increases the dose rates by up to 4% (see Fig. 6). In an additional set of models, we also included the detector positions right above the floor on the lower plane in Fig. 1.

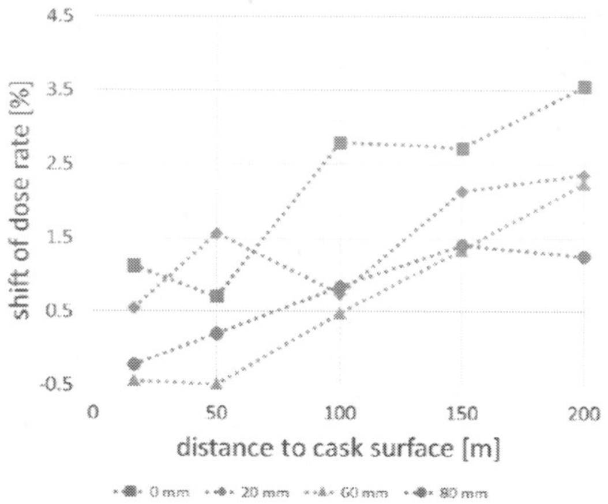

Fig. 6: Increase of dose rates in the middle plane (Fig. 1) when accounting for the concrete floor, under various lead insert thicknesses (0 mm, 20 mm, 60 mm and 80 mm). Calculations performed with SCALE 6.1.3.

At these positions the dose rate turns out to be much more sensitive to the absence of the floor. From Fig. 7 it can be seen that the presence of the floor contributes up to about 6 % of the value of the dose rates, as measured by the virtual detectors.

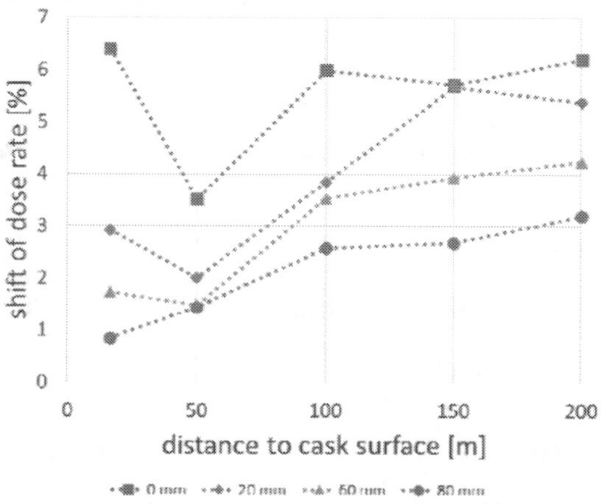

Fig. 7: Increase of dose rates in the lower plane (Fig. 1) when accounting for the concrete floor, studied for various lead insert thicknesses (0 mm, 20 mm, 60 mm and 80 mm). Calculations performed with SCALE 6.1.3.

The large influence of the floor on the calculated dose rates can be explained by considering that there is a scattering of the gamma radiation that interacts with the concrete.

5 Discussion and Conclusion

The goal of the study was to determine the accuracy of dose rate calculations for thick-walled transport and storage casks using two versions of SCALE/MAVRIC. Precise modelling of the different configurations and dose rate comparisons were possible due to an elaborate experimental setup and well-documented measurement series. Overall, the results of our SCALE 6.1.3 model fit well the results described in an earlier study of the same cask, where analyses identify the variability of the group wise modelling of the Co-60 source as one significant parameter causing the observed deviations to the measurements and the MCNP-results presented in [3]. The results of the present work do not fully support this finding. Our analysis has shown that continuous energy calculations with the release version of SCALE 6.2 does not completely resolve the problem of the observed underestimations. For a Co-60 source switching from a SCALE/MAVRIC full multi-group model to a continuous energy model shifts the dose rate deviations by about 10% to the positive direction, but does not significantly mitigate the trend for underestimation. It is worth noting that this trend is less pronounced in calculations conducted with the continuous energy code MCNP [3]. Furthermore, additional SCALE 6.2 models with continuous energy cross sections used in the transport calculation but a multi-group source definition yield dose rates between those resulting in the deviations shown in Fig. 3. Hence, both, the change from the multi-group (ENDF/B-VII.0) to the continuous energy cross section library (ENDF/B-VII.1) and from the multi-group to the continuous energy source definition contribute to these differences with similar magnitude.

These effects must be accounted for in shielding assessments with SCALE/MAVRIC in multi-group and continuous mode for package design approval procedures. Careful code validation for the application case is still essential when using SCALE/MAVRIC for dose rate calculations in geometries with thick-walled gamma radiation shields and an appropriate quantitative correction of the results may be necessary. Apart from that, the present paper has shown that parameters such as accounting for the floor, which scatters the gamma radiation and the density of the cask shielding material crucially affect the deviation of the measured dose rates, up to 6 % for the floor and 5 % for a hypothetical change of 1 % in the density of the cast iron. At the short distances used in this measurement setup, air does not play an important role for the dose rates observed.

References

[1] SCALE: A Comprehensive Modeling and Simulation Suite for Nuclear Safety Analysis and Design, ORNL/TM-2005/39, Version 6.1, Oak Ridge National Laboratory, June 2011

[2] D. E. Peplow, Monte Carlo Shielding Analysis Capabilities with MAVRIC, Nucl. Technol. 174(2), 289-313, May 2011.

[3] L. Schlömer, S. Tittelbach, P.-W. Phlippen, J. Pillath, W. Filbert, Validation and Benchmarking of Calculation Methods for Photon and Neutron Transport at Cask Configurations, PHYSOR 2014, 28. September - 3. October 2014, Kyoto, Japan

[4] SCALE Code System ORNL/TM-2005/39, Version 6.2, Oak Ridge National Laboratory, April 2016

Discussion on the Standardization of Shielding Materials – Sensitivity Analysis of Material Compositions

Tomohiro Ogata[1,*], *Ken-ichi* Kimura[2], *Mikihiro* Nakata[3], *Koichi* Okuno[4], *Takayuki* Hirouchi[5], *Naofumi* Kawano[6], *Koji* Oishi[7,8], *Ken-ichi* Tanaka[9], *Toshio* Amano[10] and *Yoshihiro* Hirao[11]

[1]Mitsubishi Heavy Industries, Ltd., 1-1-1 Wadasakicho, Hyogo-ku, Kobe 652-8585, Japan
[2]Fujita Corporation, 2025-1 Ono, Atsugi 243-0125, Japan
[3]MHI Nuclear Solutions and Engineering Co., Ltd., 3-3-1 Minatomirai, Nishi-ku, Yokohama 220-8401, Japan
[4]HAZAMA ANDO Corp., 515-1 Karima, Tsukuba 305-0822, Japan
[5]Toshiba Corporation, 8 Shinsugitacho, Isogo-ku, Yokohama 235-8523, Japan
[6]Hitachi-GE Nuclear Energy, Ltd., 3-1-1 Saiwaicho, Hitachi 317-0073, Japan
[7]Japan Environment Research Co., Ltd., 6-24-1 Nishi Shinjuku, Shinjuku-ku, Tokyo-160-0023, Japan
[8]Giken Kogyo Co., Ltd., 3-7-2 Asagayaminami, Suginami-ku, Tokyo 166-0004, Japan
[9]The Institute of Applied Energy, 1-14-2 Nishi Shinbashi, Minato-ku, Tokyo 105-0003, Japan
[10]ITOCHU Techono-Solutions Corporation, 3-2-5 Kasumigaseki, Chiyoda-ku, Tokyo 100-6080, Japan
[11]National Maritime Research Institute, 6-38-1 Shinkawa, Mitaka-shi, Tokyo 181-0004, Japan

Abstract. The overview of standardization activities for shielding materials is described. We propose a basic approach for standardizing material composition used in radiation shielding design for nuclear and accelerator facilities. We have collected concrete composition data from actual concrete samples to organize a representative composition and its variance data. Then the sensitivity analysis of the composition variance has been performed through a simple 1-D dose calculation. Recent findings from the analysis are summarized.

1 Introduction

In radiation shielding design and safety analysis, it is necessary to employ appropriate calculation conditions of shielding material, e.g., material density and composition. Concrete, lead and iron are widely used as shielding materials for nuclear power plants and accelerator facilities. Among them, the concrete composition cannot be universally specified in design phase, because it varies greatly depending on the locality of aggregate and a mixture ratio of aggregate and cement.

In Japan, the composition data listed in American National Standards Institute[1] or other reports[2,3] are often quoted for concrete composition. However, it is not clear whether they can represent the actual samples at Japanese construction sites. In fact, the importance and difficulties of this issue has long been widely shared in Japan.

In 2014, a working group was established in order to standardize shielding materials under the Standards Committee of Atomic Energy Society of Japan. Then we have discussed representativeness of the material composition and implication of the composition variance in shielding performance, especially for concrete. In this paper, our approach to standardize material composition is described, and calculation models and results of sensitivity analysis of concrete compositions are shown.

2 Basic approach for standardization of material composition

Our approach to standardize concrete composition is as follows. First, we collect concrete composition data used in the present shielding calculation and actual sample data to organize the representative composition and variance of composition. Then neutron and gamma ray dose calculations are performed to evaluate influences of the composition variation on shielding performance. Lastly, influences of other variations, such as a moisture within concrete, mixture non-uniformity and trace elements contributing to induced activity, are considered.

3 Sensitivity analyses

It is needed to prepare sensitivity analyses data so that shielding designers can reffer the influence of the fluctuation of concrete composition and water content on the shielding design. Fig. 1 shows sensitivity analysis method in this paper.

3.1 Method

Influences of variation of concrete composition on a shielding calculation were evaluated. Table 1 shows

* Corresponding author: tomohiro_ogata@mhi.co.jp

basic sensitivity analysis conditions. Neutron and gamma dose calculations were performed by using one dimensional S_N code ANISN[4] with MATXSLIB-J33 library[5]. As shown in Fig.2, 1-D simple sphere geometry was used for the sensitivity analysis so that influences of concrete composition can be shown clearly. A hollow concrete sphere having inner radius 500cm was used to avoid appearing the distance attenuation nearby source. A radiation source of 1cm diameter is placed at the center of the sphere. Representative concrete composition and its elemental variance data derived from actual samples was used in this calculation. Composition of F02HT is shown in table 2. As radiation sources ^{60}Co for gamma and ^{235}U for neutron were used in this paper. In neutron dose calculation, it is also considered dose by secondary gamma rays.

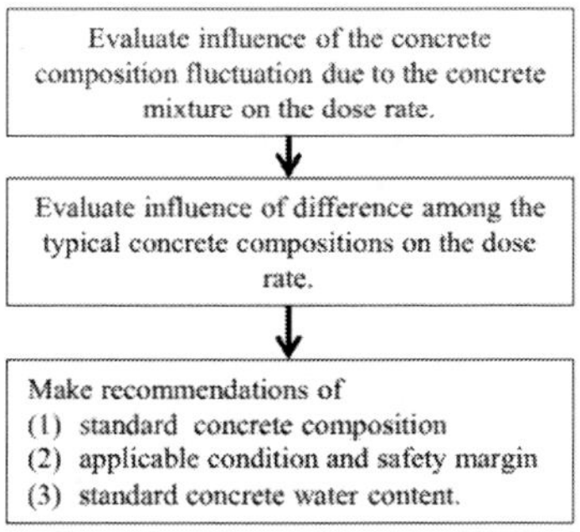

Fig.1. Flow chart of the sensitivity analysis.

Table 1. Basic conditions of sensitivity analysis

Item	Condition
Calculation code	ANISN (Doors3.2)
Evaluated nuclear data file	JENDL-3.3
Library	MATXSLIB-J33
Model	1 dimensional shell model (See Fig.2)
Maximum order of scatter	P3
Order of angular quadrature	S8
Boundary condition	Left: reflection Right: vacuum
Accuracy	0.001

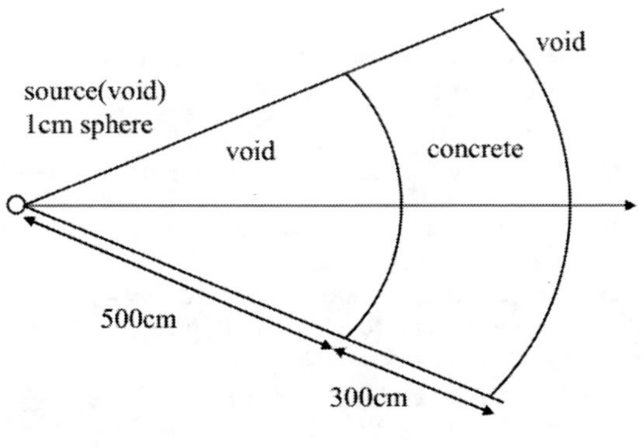

Fig.2. ANISN calculation model for the sensitivity analysis.

Table 2. Concrete compositions (atomic density) of F02HT

x Avogadro Number

Element	-3sigma	average	+3sigma
H	6.31E-03	6.31E-03	6.31E-03
C	5.89E-03	5.89E-03	5.89E-03
O	3.92E-02	4.08E-02	4.23E-02
Na	3.81E-04	3.81E-04	3.81E-04
Mg	2.93E-04	2.93E-04	2.93E-04
Al	1.23E-03	1.34E-03	1.45E-03
Si	6.37E-03	7.50E-03	8.63E-03
P	1.17E-05	1.17E-05	1.17E-05
K	3.80E-04	3.80E-04	3.80E-04
Ca	6.89E-03	7.85E-03	8.81E-03
Ti	2.26E-05	2.26E-05	2.26E-05
Mn	5.86E-06	5.86E-06	5.86E-06
Fe	1.20E-04	1.44E-04	1.68E-04
density	2.05g/cm^3	2.21g/cm^3	2.38g/cm^3

3.2 Influence of the concrete composition

Fig.3 shows the calculation results of neutron dose attenuation rates of F02HT, 'F02HT+3σ'and 'F02HT-3σ' with U-235 neutron source. 'F02HT±3σ' indicates a representative average density plus/minus 3 times of standard deviation for each of the nuclides. Similarly, Figures 4 and 5 show the secondary gamma and the total dose attenuation rates, respectively. Each curve were normalized by total dose rate of F02HT average concrete at inner concrete surface The results suggest that concrete density variation is dominant attenuation factor compared with the variation of elemental composition.

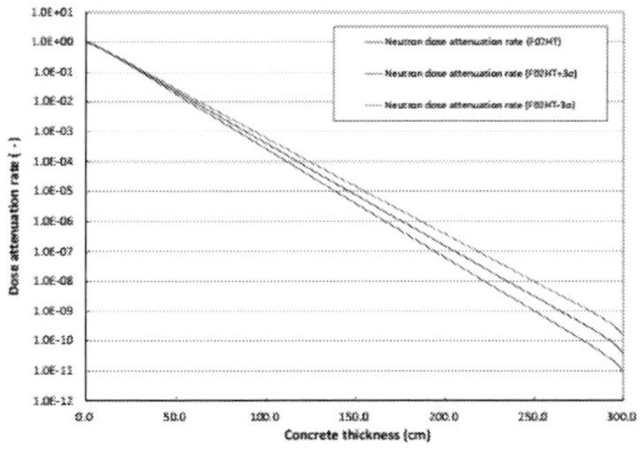

Fig.3. Neutron dose attenuation rates of F02HT with ^{235}U source.

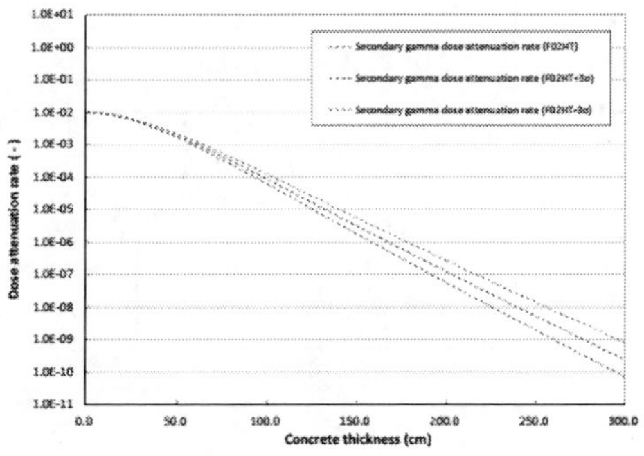

Fig.4. Secondary gamma dose attenuation rates of F02HT with ^{235}U source.

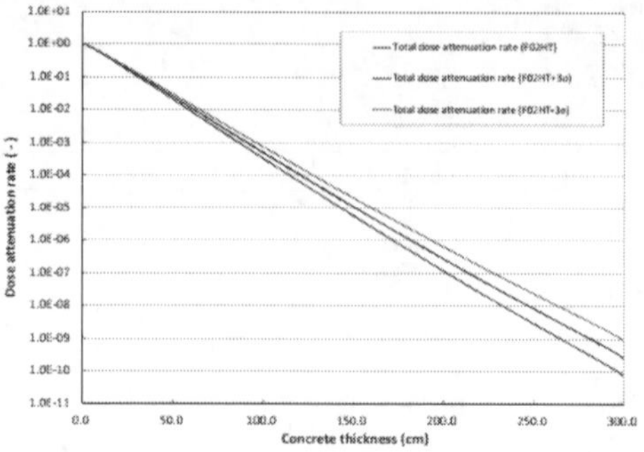

Fig.5. Total dose attenuation rates of F02HT with ^{235}U source.

3.3 Influence of the concrete composition with realistic density

Although calculation with measured elemental composition with ±3sigma carried out, further calculation was carried out with restricted density between ±3sigma of the analyzed an actual concrete density since concrete densities shown in table 1 were too varied (about 8%) to evaluate effect of concrete composition. Concrete density used for radiation shielding was usually controlled to be desinged value. In this calculation, density restricted concrete composition derived with actual concrete data was emplyed. Table 3 shows density restricted concrete composition of F02HT (about 2.2%). Dose attenuation calculation model is shown in Fig.2 (same as first calculation).

Fig.6, Fig.7 and Fig.8 show neutron, secondary gamma and total dose atteuation rate with U-235 source. Neutron dose attenuation rate at 250cm depth of -3sigma concrete is about 2 times as large as that of +3sigma. As a result of this calculation, it can be concluded that neutron dose rate variation due to the density fluctuation by concrete mixing is at most within 2 times.

Table 3. Concrete compositions (atomic density) of F02HT by density reconfigured with analysed concrete density data.

x Avogadro Number

Element	-3sigma	average	+3sigma
H	6.31E-03	6.31E-03	6.31E-03
C	5.89E-03	5.89E-03	5.89E-03
O	4.03E-02	4.08E-02	4.13E-02
Na	3.81E-04	3.81E-04	3.81E-04
Mg	2.93E-04	2.93E-04	2.93E-04
Al	1.31E-03	1.34E-03	1.37E-03
Si	7.16E-03	7.50E-03	7.84E-03
P	1.17E-05	1.17E-05	1.17E-05
K	3.80E-04	3.80E-04	3.80E-04
Ca	7.56E-03	7.85E-03	8.14E-03
Ti	2.26E-05	2.26E-05	2.26E-05
Mn	5.87E-06	5.86E-06	5.87E-06
Fe	1.37E-04	1.44E-04	1.51E-04
density	2.16g/cm^3	2.21g/cm^3	2.26g/cm^3

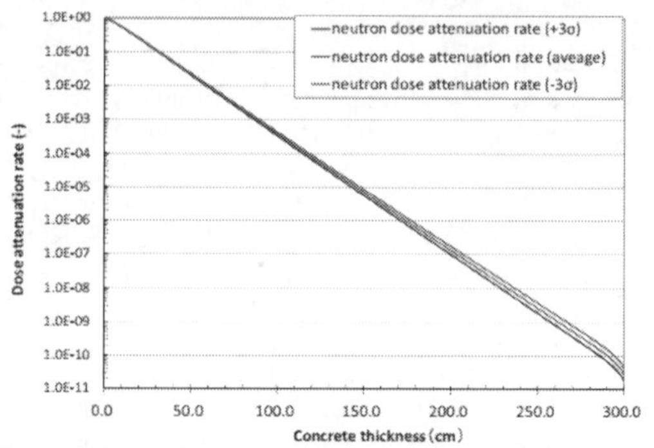

Fig.6. Neutron dose attenuation rate of F02HT with dinsity reconfigured.

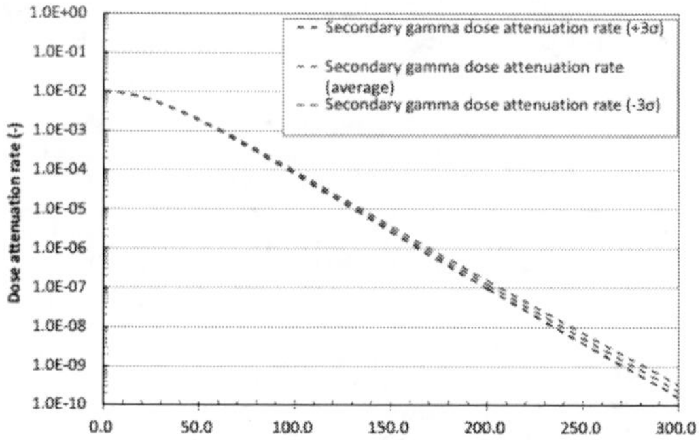

Fig.7. Secondary gamma dose attenuation rate of F02HT with dinsity reconfigured.

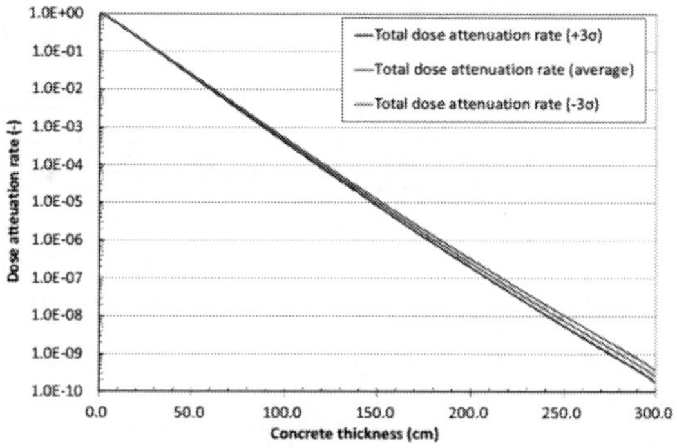

Fig.8. Total dose attenuation rate of F02HT with dinsity reconfigured.

3.4 Influence of the concrete composition with water content

3.4.1 Sensitivity analysis with U-235 neutron source

It is important to set water content in concrete appropriately when neutron transport calculation is carried out because neutron is easily slowing down by low atomic numbered elements (typically hydrogen in water). In this paper, three cases of water content in concrete were evaluated. The cases were absolute dry (0g/cm^3 free water), 0.04g/cm^3 free water and 0.1g/cm^3 free water in concrete. Table 4 shows concrete composition considering water content. Dose attenuation calculation model is shown in Fig.2 (same as first calculation)

Fig.9, Fig.10 and Fig.11 show the result of this calculation. Neutron dose attenuation rate is worse with decreasing the water content in concrete because neutron is easily slowing down in rich water content. Secondary gammas generation rate increases with increasing water content in concrete but dose rate of poor water content is higher than dose rate of rich water content because of neutron slowing down by water.

Table 4. Concrete compositions (atomic density) of F02HT with water content sensitivity analysis data.

x Avogadro Number

Element	average	-3sigma	+3sigma
H	6.31E-03	8.98E-03	1.30E-02
C	5.89E-03	5.89E-03	5.89E-03
O	4.08E-02	4.21E-02	4.41E-02
Na	3.81E-04	3.81E-04	3.81E-04
Mg	2.93E-04	2.93E-04	2.93E-04
Al	1.34E-03	1.34E-03	1.34E-03
Si	7.50E-03	7.50E-03	7.50E-03
P	1.17E-05	1.17E-05	1.17E-05
K	3.80E-04	3.80E-04	3.80E-04
Ca	7.85E-03	7.85E-03	7.85E-03
Ti	2.26E-05	2.26E-05	2.26E-05
Mn	5.86E-06	5.87E-06	5.87E-06
Fe	1.44E-04	1.44E-04	1.44E-04
density	2.21g/cm^3	2.25g/cm^3	2.31g/cm^3

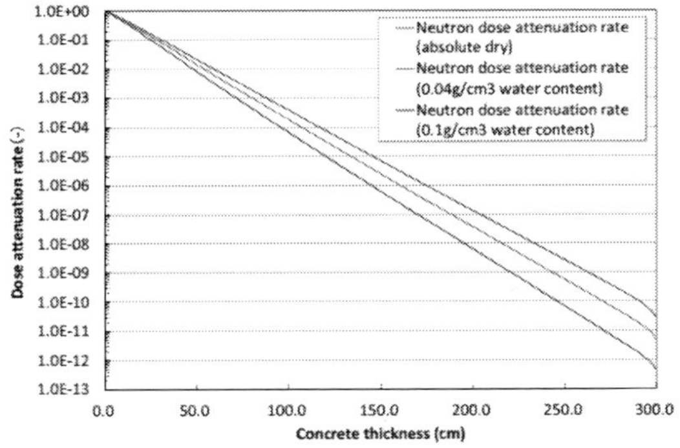

Fig.9. Neutron dose attenuation rate of F02HT with dinsity reconfigured.

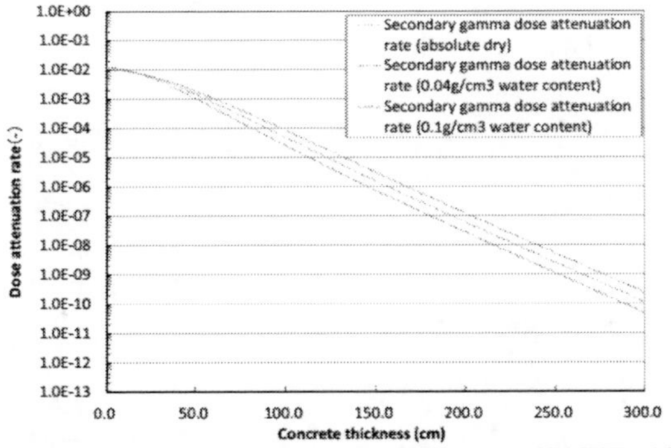

Fig.10. Secondary gamma dose attenuation rate of F02HT with dinsity reconfigured.

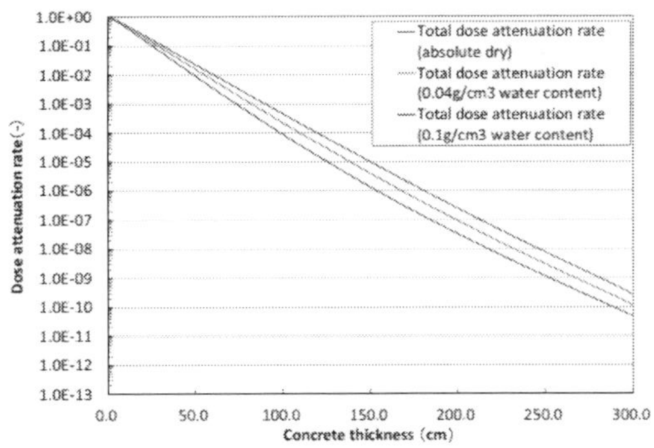

Fig.11. Total dose attenuation rate of F02HT with dinsity reconfigured.

3.4.2 Sensitivity analysis with gamma source

Dose rate sensitivity analysis for water content with gamma source was carried out. Co-60 was set as a gamma source. Fig. 12 shows the dose attenuation rate for these gamma rays. Since increase of water content brings increasing concrete density, gamma dose rate is well attenuated with rich water content concrete. Fig.13 shows the gamma dose attenuation for concrete mass thickness of F02HT. Fig.13 indicates that dose attenuation rate depends on its density since water content in concrete make a little contribution to the gamma dose attenuation rate by mass thickness.

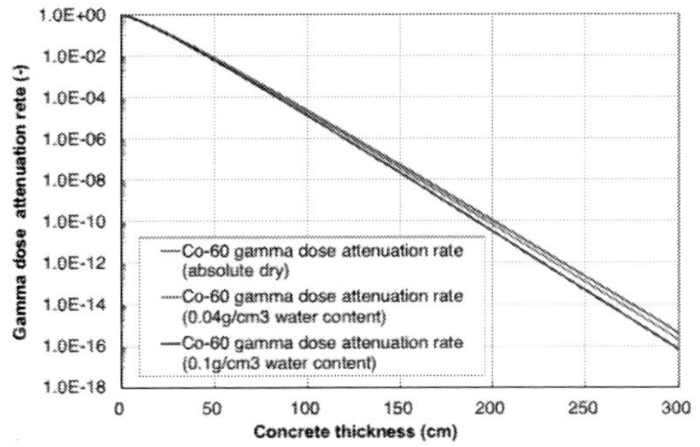

Fig.12 Co-60 gamma dose attenuation rate for the F02HT.

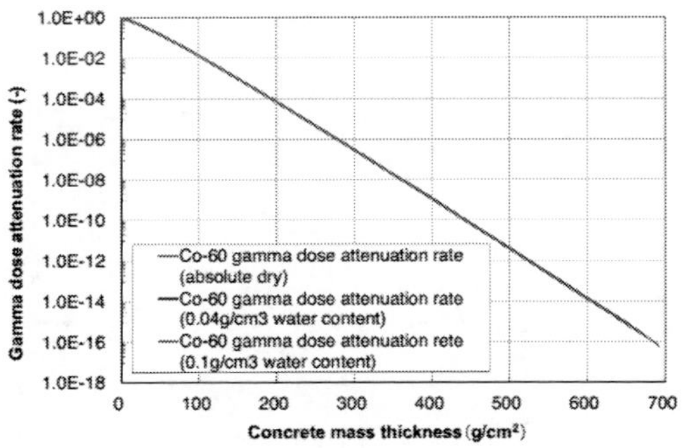

Fig.13 Co-60 gamma dose attenuation rate for the F02HT concrete mass thickness.

4 Conclusions

The overview of standardization activities for shielding materials was described. As a first step, important data to determine radiation shielding concrete composition was obtained. Variation and behaviour of dose attenuation rate for gamma ray and neutron caused by variances of the concrete and the water content of the concrete indicates that neutron dose rate is influenced by light elements and gamma dose rate is influenced by concrete density. In the determination of the standard material composition, it has to present the consideration for variation of composition, the effect and dose caused by concrete density for neutron and gamma ray, consideration for water content and conservativeness of the shielding calculation with the standard concrete composition.

The authors wish to acknowledge the contribution of Y. Sakamoto, M. Taniguchi, T. Tsukiyama, S. Ishikawa, H. Sakamoto, H. Kawano, M. Yoshida and K. Kosako.

References

Online references will be linked to their original source, only if possible. To enable this linking extra care should be taken when preparing reference lists.

1. American Nuclear Society, *Nuclear analysis and design of concrete radiation shielding for nuclear power plants*, ANSI/ANS-6.4.1-1997 (1997)
2. Argonne National Laboratory, *Reactor physics constants*, ANL-5800 (1963)
3. R. L. Walker and M. Grotenhuis, *A Summary of shielding constants for concrete*, ANL-6443 (1961)
4. Oak Ridge National Laboratory, *Doors3.2: one, two- and three-dimensional discrete ordinates neutron/photon transport code system,* Radiation Safety Information Computational Center computer code collection, Oak Ridge, TN, CCC-650 (1998)
5. K. Kosako, N. Yamano, T. Fukahori, K. Shibata and A. Hasegawa, *The libraries FSXLIB and MATXSLIB based on JENDL3.3*, JAERI-Data/Code 2003-011 (2003)

A Monte-Carlo Benchmark of TRIPOLI-4® and MCNP on ITER neutronics

David Blanchet[1,a], Yannick Péméliau[1], Romain Eschbach[1], Bruno Fontaine[1], Bruno Cantone[2], Marc Ferlet[2], Eric Gauthier[2], Christophe Guillon[2], Laurent Letellier[2], Maxime Proust[2], Fernando Mota[3], Iole Palermo[3], Luis Rios[3], Frédéric Le Guern[4], Martin Kocan[5], Roger Reichle[5]

[1]*CEA,DEN/DER, Cadarache, 13108 Saint-Paul-lez-Durance, France*
[2]*CEA, IRFM, Cadarache, 13108 Saint-Paul-lez-Durance, France*
[3]*CIEMAT, Avda.Complutense, 40, Madrid 28040, Spain*
[4]*F4E, Josep Pla 2, Torres Diagonal Litoral B3, 08019 Barcelona, Spain*
[5]*ITER Organization, Route de Vinon sur Verdon, CS 90 046, 13067 Saint-Paul-lez-Durance, France*

Abstract. Radiation protection and shielding studies are often based on the extensive use of 3D Monte-Carlo neutron and photon transport simulations. ITER organization hence recommends the use of MCNP-5 code (version 1.60), in association with the FENDL-2.1 neutron cross section data library, specifically dedicated to fusion applications. The MCNP reference model of the ITER tokamak, the 'C-lite', is being continuously developed and improved. This article proposes to develop an alternative model, equivalent to the 'C-lite', but for the Monte-Carlo code TRIPOLI-4®. A benchmark study is defined to test this new model. Since one of the most critical areas for ITER neutronics analysis concerns the assessment of radiation levels and Shutdown Dose Rates (SDDR) behind the Equatorial Port Plugs (EPP), the benchmark is conducted to compare the neutron flux through the EPP. This problem is quite challenging with regard to the complex geometry and considering the important neutron flux attenuation ranging from 10^{14} down to 10^{8} n•cm^{-2}•s^{-1}. Such code-to-code comparison provides independent validation of the Monte-Carlo simulations, improving the confidence in neutronic results.

1 Introduction

One of the most critical issues for ITER neutronics analysis concerns the assessment of radiation levels behind the Equatorial Port Plugs. The Port Plugs are water-cooled stainless steel structures designed to host a wide variety of diagnostic systems. Among them, the Equatorial Visible/Infra-Red Wide Angle Viewing System (WAVS) [1] plays a key role in machine protection by providing on-line monitoring of the surface temperature of plasma facing components, in order to avoid excessive heating. In addition to support diagnostics, the Port Plugs shall provide sufficient radiation shielding to attenuate the propagation of the intense, high-energy plasma-neutron flux. However, the Port Plug itself in the Port and the diagnostics create feedthroughs and streaming gaps, allowing the neutron flux to penetrate deeply into the Port Plug structures. This neutron irradiation could lead to significant activation of structural materials, with subsequent decay gamma emissions, contributing to the shutdown dose rate.

In order to limit the occupational radiation exposure during maintenance, the shutdown dose rate behind the Port Plug, in the Port Interspace area (Figure 1), should not exceed 100 μSv.h^{-1} after 12 days shutdown. Actually, recent studies [2] have shown that this project criterion is particularly challenging. Improvements are sought in the design of the shielding of the radiation streaming through the gaps and penetrations (adopting optical labyrinths) and in the choice of materials (use of boron carbide as neutron absorber). High density materials should be avoided because of severe constraints on weight limits. Moreover, the problem cannot be solved locally without considering all contributions, because of the cross-talk interferences (contributions from neighbouring components), mainly from the Lower Ports. Hence, there is a need to consider complete models with a lot of details included and to develop them in a modular and flexible approach allowing regular updates during the design phase. The convergence of Monte-Carlo simulations on these large-scale models requires the use of biaising techniques for variance reduction. This last point highlights the importance of cross-checking the simulations. There is an interest in proposing alternative models for independent validations that will improve the confidence in the neutronic results. For this reason, the present study describes the development of an equivalent 'C-lite' model but for the Monte-Carlo code TRIPOLI-4® [3] as explained in Section 2. A benchmark study is also defined to test the TRIPOLI-4® ITER model as detailed in Section 3. The results of such code-to-code comparison are provided in Section 4.

[a] Corresponding author: david.blanchet@cea.fr

 DOI: 10.1051/epjconf/201715302005

2 Development of the ITER C-lite TRIPOLI-4® model

ITER organization recommends the use of MCNP-5 [4] code (version 1.60), in association with the FENDL-2.1 [5] neutron cross section data library, specifically dedicated to fusion applications. The MCNP reference model of the ITER tokamak, known as the 'C-lite', is a representation of the machine in a 40° torus sector, with specular reflections at the ±20° boundaries. The C-lite model integrates generic descriptions of the components, but depending on the level of interest, it can also incorporate very detailed and sophisticated descriptions of the components under study. In the present study, we consider a MCNP C-lite model composed of about 10'000 volumes.

2.1 Methodology for Geometry Creation

The TRIPOLI-4® model is created, component-by-component in order to preserve the original structure of the C-lite based on Universes. The MCNP Universes can be translated in TRIPOLI-4® by using fictitious volumes. These fictitious volumes can be seen as containers, to be filled with detailed component descriptions using volume intersection operator. This way of creating the geometry ensures the required flexibility for regular update of a particular component if it is needed or if a refined model is available. This is the case of the Diagnostic Generic Equatorial Port Plug (DGEPP) models which are of particular interest in the present study. The length of the EPP represented in figure 1 is about ~3 meters.

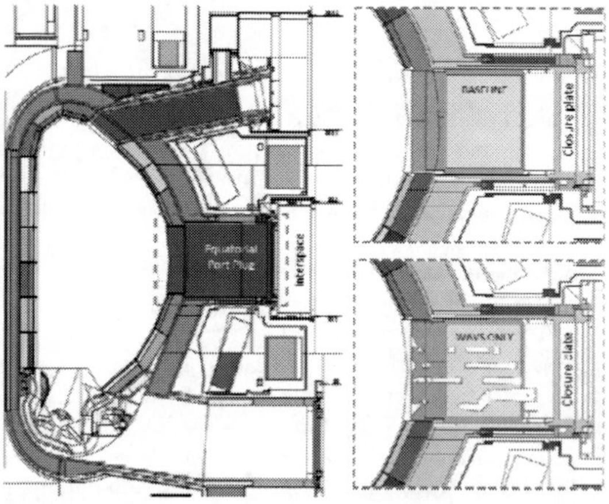

Figure 1. RZ (poloidal) view of the generic EPP in MCNP geometry (left); detailed views of the "baseline" (top right) and "WAVS only" (bottom right) DGEPP models.

Among the various diagnostics to be housed by the EPP, this study only focuses on the WAVS diagnostics. The optical system of the WAVS diagnostics is represented by 4 lines-of-sight (respectively for the Upper, Left, Right and Divertor views) creating apertures and feedthroughs in the Diagnostic Shield Module (DSM) structure. In order to avoid direct streaming corridors for neutrons, dogleg configurations are highly recommended; the number and the position of doglegs are still under optimization. The optical paths of the WAVS system are illustrated in Figure 2.

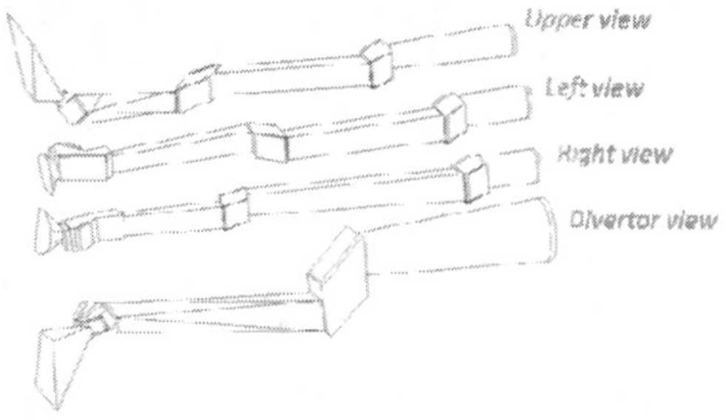

Figure 2. CAD model of the WAVS optical paths (with dogleg configurations) through the EPP (length ~3m, diameter < 16 cm).

The CAD models of the detailed components are first simplified using SpaceClaim Engineer 2015 CAD tool [6]. This simplification basically consists in: removing all unnecessary elements (which do not influence the transport analyses of the high energy neutrons) and preserving only the most important ones; substituting the sp-lines with surfaces understandable by the code (i.e. second-degree surfaces and toroidal fourth-degree surfaces); splitting the large and complex components into simple ones, and finally filling the little voids because all the space must be defined to ensure particle transport doesn't stop in a void medium. Besides, Monte-Carlo codes require void cell descriptions, which are generally lacking in the CAD models, hence complementary cells have to be defined. This can be done by modifying the original CAD geometry, either directly with SpaceClaim or later on with MCAM.

During this complex simplification process, care should be taken to preserve mass balance when the volumes are modified: this is done by adjusting material densities. An illustration of this simplification process is proposed in Figure 3.

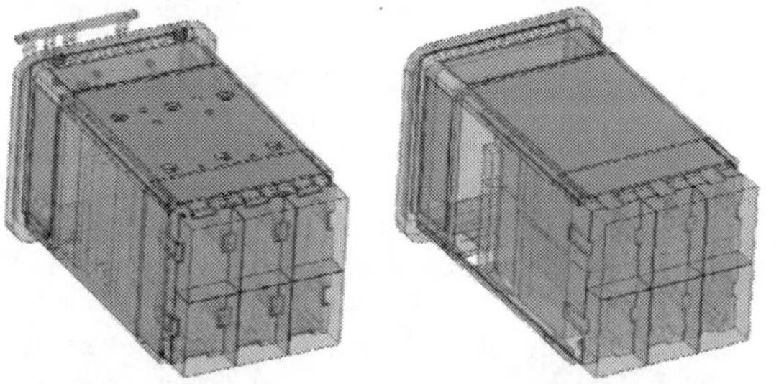

Figure 3. Simplification of the DGEPP CAD model with SpaceClaim (on left: the original design, on right: the simplified design).

Once simplified, the CAD geometry can be converted to MCNP-5 or TRIPOLI-4® geometry using MCAM (Professional Version 5.2) software [7]. The created Monte-Carlo geometries are finally included into the C-lite model.

In Figure 4, part (b) (c) and (d) represent the successive steps in the TRIPOLI-4® geometry creation. These illustrations were generated using the embedded

 DOI: 10.1051/epjconf/201715302005

plot utility named T4G [8]. This tool allows considerably fast display and navigation into the geometry.

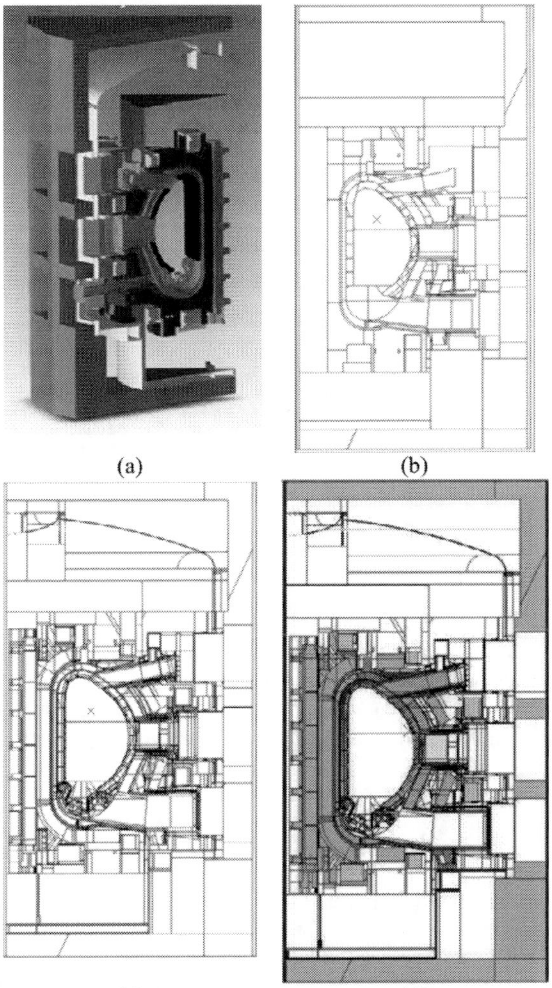

Figure 4. Development of the ITER 'C-lite V1' model in TRIPOLI-4®. (a) Original CAD model in step format. (b) Definition of fictitious volumes representing Universes in TRIPOLI-4®. (c) Definition of detailed Components. (d) Geometry to material association.

The TRIPOLI-4® geometry is tested by counting lost particles when all cells are filled with void material. It is mandatory to get zero lost particles in simulations of at least 10^7 source particle histories run. Another interesting geometrical test consists in the stochastic volume estimation by ray tracing performed by setting a uniform neutron source distribution inward-directed, on a plane surface, voiding out all materials in the problem and setting all nonzero importances to one. In this way the calculation of the neutron flux tallies (using the track-length estimator, *F4* in MCNP and *TRACK* in TRIPOLI-4®) provide Monte-Carlo estimates of all cell volumes. These checks were successfully done in both MCNP-5 and TRIPOLI-4® models showing a perfect agreement of the volumes calculated by the two codes when they are compared with the direct CAD volumes. Following this analysis, the volumes and the mass balance of DGEPP cells are deduced and pointed out in Table 1.

Table 1. Volume and mass Balance in dgepp-0-wavs model (estimations by TRIPOLI-4®).

	Density	Volumes	Masse
Material	**t.m^{-3}**	**m^3**	**t**
SS316L(N)	7,93	2,12	16,782
SS316L(N)+10%H2O	7,19	0,81	5,845
SS316L(N)+5%H2O	7,58	2,48	18,833
B4C	1,89	3,03	5,731
TOTAL DGEPP	5,59	8,45	47,191

Considering fully shielded DSM is a rather conservative assumption for the total EPP weight estimate which is already exceeding the 45 tons limit imposed by the project. According to the above table, the shielding material in the EPP could not exceed ~3.5 tons. It is clear that high density materials should be avoided.

2.2 Verification of the D-T neutron source

The ITER standard neutron source is defined by the nominal inductive operation scenario, which assumes a deuterium-tritium plasma pulse with a total fusion power of 500 MW. The energy distribution of the 14 MeV neutrons emitted by D-T reactions is described by the built-in Gaussian fusion energy distribution, centred at E = 14.0791 MeV, with a spread of σ = 0.53415 MeV. The total neutron emission rate is normalized to $1.9718 \cdot 10^{19}$ neutrons per second. Figure 5 is an illustration of the toroidally-averaged poloidal distribution of the neutron source, as sampled in analog (unbiased) simulations by the Monte-Carlo TRIPOLI-4® code.

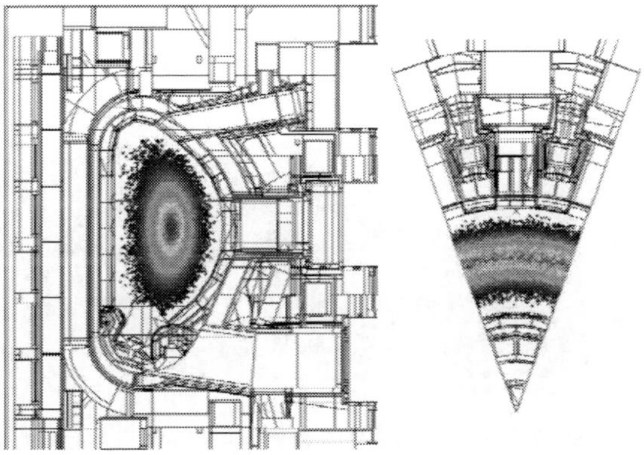

Figure 5. Source Density Sampling by TRIPOLI-4®.

2.3 Assessment of Variance Reduction Techniques

Variance-reduction techniques are used in both MCNP-5 and TRIPOLI-4® to improve statistics (Weight Windows in MCNP, Exponential Transform in TRIPOLI-4® [9] [10]).

In MCNP, the automatic weight windows generator is used on fictitious geometry with cells having artificially reduced material densities (by a factor 0.1). This technique is widely recognized to easier the production of importance maps on large geometries, especially when there are strong neutron flux attenuations. Neutron importance maps are generated on a rather broad energy mesh in 3 groups (defined with boundaries at 10^{-11}, 10^{-3}, 0.1 and 20 MeV), which is efficient for hard neutron spectra. An irregular spatial mesh is defined to match the various components (mesh sizes might be varying from 10 cm up to few meters). The resulting maps are plotted in Figure 6, where blue zones are standing for high importance regions. These maps were obtained by specifying a target detector at the rear of the Equatorial Port Plug. They were improved by several iterations in a so-called learning approach.

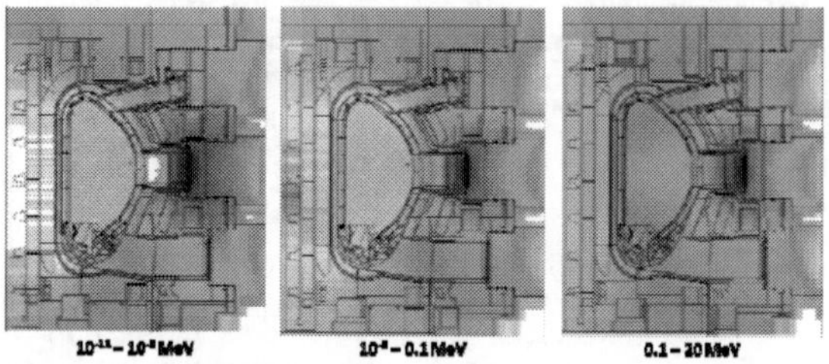

Figure 6. Weight Windows map for MCNP.

In a first attempt, the same spatial and energy meshes are used for the automatic biasing scheme implemented in TRIPOLI-4®. As opposed to the user-friendly generation of weight windows in MCNP which relies on brute force to be efficient (thanks to the automatic learning approach), the "automatic" ponderation in TRIPOLI-4®, requires much efforts and expertise in preliminary tuning to be efficient.

The preliminary importance maps generated by TRIPOLI-4® using a cylindrical detector are illustrated in Figure 7. These importance maps are not yet very satisfactory and would require further optimization. This will be investigated in a future work.

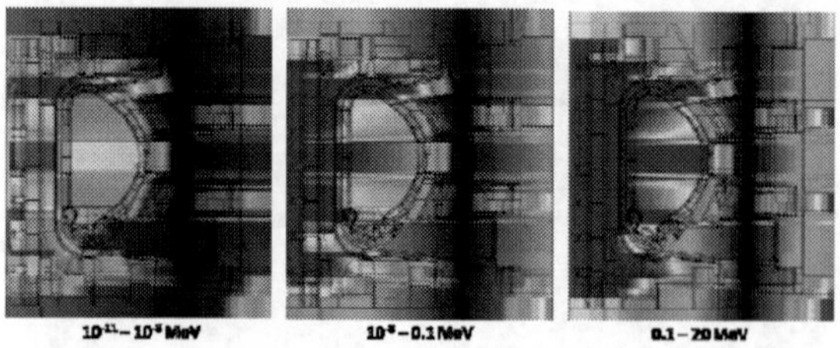

Figure 7. Importance maps obtained with automatic variance reduction technique in TRIPOLI-4®.

The efficiency of the variance reduction is assessed by comparison with analog simulations (i.e. simulations not using any variance reduction, except for the implicit capture). General characteristics of these simulations are reported in Table 2. The improvement induced by the variance reduction technique is measured by a statistical indicator denoted FoM and defined as the Figure of Merit:

$$FoM = \frac{1}{R^2 \cdot T} \tag{1}$$

Where R is the relative standard deviation and T is the computer running time. Results having the highest FoM are preferred. According to Table 2, weight windows are especially improving the statistical convergence of neutron flux results at the target location.

Table 2. Example of statistical indicators (figure of merit) associated to neutron flux obtained either in analog simulations or in simulations using variance reduction techniques. These computations were run on parallel machines at the CEA supercomputing center, using 256 cores from the Curie nodes (Intel® Nehalem-Ex X7560 at 2.26 GHz).

	Analog Simulation		**Variance Reduct°**
	T-4	**MCNP**	**MCNP**
Time (h)	501	468	302
Particles (#)	10^9	10^9	$5 \cdot 10^9$
Std-Dev(%)	2.0%	1.6%	0.7%
FoM	5.0	8.9	67.6

Such a comparison with analog simulations is not only meant to measure the performance of the variance reduction, but it provides also an additional check on the absence of deviation introduced by the biasing techniques. Values obtained with variance reduction were found in good agreement with those coming from analog simulations according to their statistical error uncertainty.

3 Description of the benchmark study

This study is conducted to assess the radiation level at the rear of the EPP, the neutron fluxes being calculated at the Closure Plate (CP) and at the surrounding Closure Flange (CF) of the EPP. The activation of the closure plate is one of the most important contributions to the shutdown dose rate in the Interspace. For this reason, the calculation of the neutron flux at the closure plate can be considered as a preliminary step before the determination of shutdown dose rates in the Interspace.

The following two DGEPP models are being considered for the benchmark study:

- a model denoted as "dgepp-0" representing the *initial design*;
- a model denoted as "dgepp-1" representing the *optimized design,* with improved neutron shielding features.

The optimized design ("dgepp-1") incorporates some neutron labyrinths in the surrounding gaps between the Port Plug and the Vessel; and some neutron stoppers made of natural boron carbide in the inter-drawer gaps. These neutron shielding features are well-illustrated in Figure 7. In this figure, it can also be observed that the DGEPP is divided into 3 drawers filled with Diagnostic

DOI: 10.1051/epjconf/201715302005

Shield Modules (DSM) made of powdered natural B_4C at 1.89 g/cm^3 density. Two configurations are under study:

- The "baseline" configuration, which does not house any diagnostic inside, represents a fully dense EPP, having 3 drawers completely filled by the DSM.
- The "wavs" configuration represents the EPP with the 4 optical paths of the WAVS diagnostics included in one of the 3 drawers (the other ones being full of shielding material).

The study measures the impact that the WAVS diagnostics have on the flux level at the back-end of the EPP. In this approach, all possible contributions from other diagnostics are voluntary ignored. Some illustrations of these four models are proposed in the Figure 8.

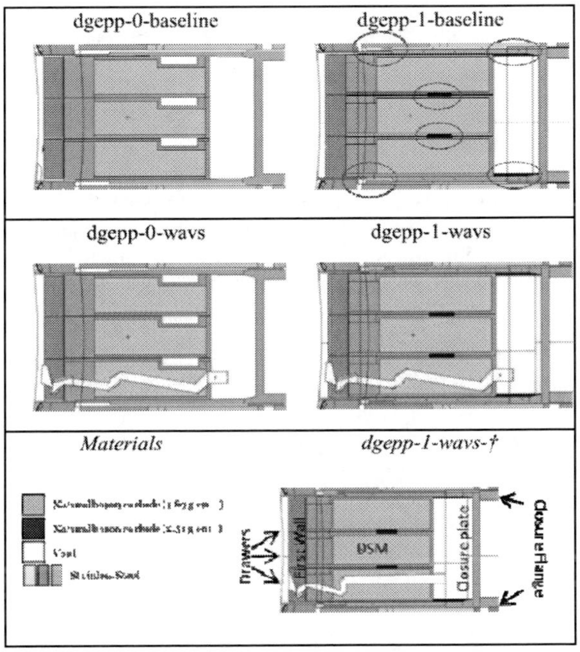

Figure 8. Benchmark cases. (†Alternative optical paths defined by only 1 dogleg instead of 2).

There are on-going studies to investigate alternative designs for the optical paths, by looking at the impact of adopting one dogleg instead of two, as illustrated in the additional configuration denoted as "dgepp-1-wavs-†" in Figure 8. However this case is presented only for illustration purpose and will not be considered in the rest of the study.

4 Results

4.1 Neutron Flux at the Equatorial Port Plug

Neutron spectra are calculated at the closure plate by the use of MCNP-5 and TRIPOLI-4®. Some of these results are depicted in Figure 9. The energy-integrated neutron flux values are reported in Table 3. All comparisons between MCNP-5 and TRIPOLI-4® have revealed an excellent agreement; according to Table 3 for instance, there is no difference at all between the results obtained by the two codes in consideration of the statistical errors.

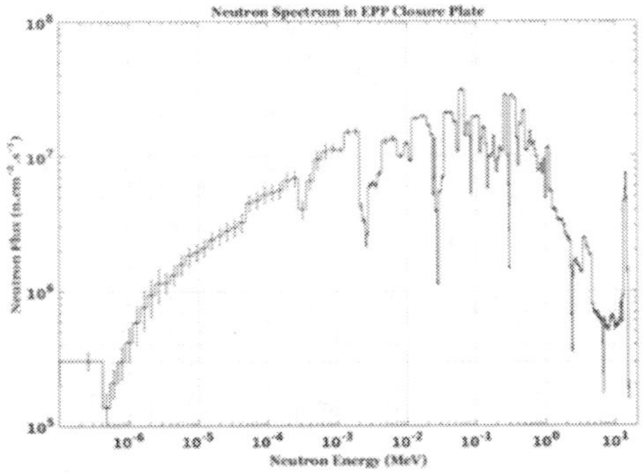

Figure 9. Neutron Spectrum in VITAMIN-J Energy Structure (175 groups) at the EPP Closure Plate (± 2σ) by MCNP-5.

Table 3. Comparison of neutron flux obtained at Closure Plate (CP) and Closure Flange (CF) by analog simulations with TRIPOLI-4® and MCNP-5, based on the "dgepp-0-wavs" model.

FENDL-2.1	Flux at CP		Flux at CF	
	$n.cm^{-2}.s^{-1}$	σ %	$n.cm^{-2}.s^{-1}$	σ %
TRIPOLI-4.10®	$8,14 \cdot 10^9$	2,0%	$6,66 \cdot 10^9$	3,0%
MCNP-5.1.60	$8,24 \cdot 10^9$	1,6%	$6,70 \cdot 10^9$	2,4%
(T4-M5)/T4 %	-1,3%	2,5%	-0,6%	3,8%

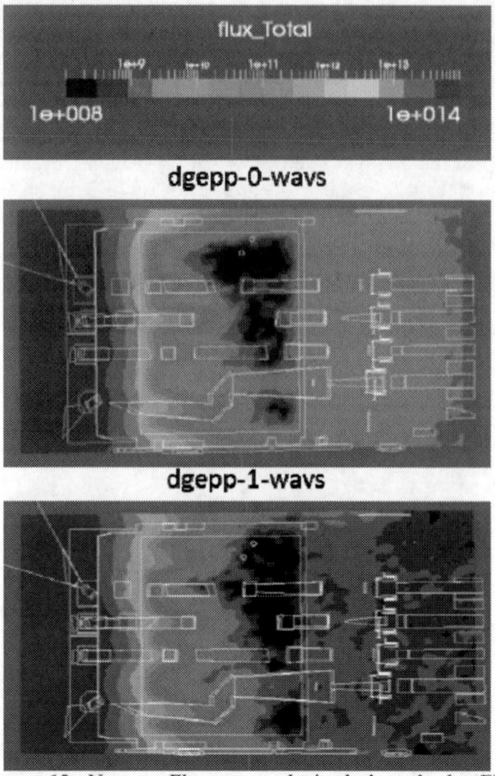

Figure 10. Neutron Flux maps obtained through the EPP (dgepp-0-wavs and dgepp-1-wavs models) by MCNP-5.

Table 4 summarizes the relative neutron flux results obtained for the different benchmark cases previously defined. Neutron flux maps obtained in the EPP area are depicted in Figure 10. These results were obtained with MCNP-5, using variance reduction techniques (based on the Weight Windows maps defined in section §2.3).

Table 4. Neutron flux results (in relative units) obtained at the closure plate for the different cases under study (σ ~2%).

	dgepp-0	dgepp-1
baseline	4	1
wavs	10	1,1

In figure 10, the dark blue regions inside the EPP can be explained by the strong neutron absorption in the boron carbide shielding modules. The shielding options adopted in dgepp-1 model (i.e. neutron labyrinths and neutron stoppers) are efficient in decreasing the neutron flux levels at the CP: a reduction factor of ~10 can be obtained in the "wavs" case (see Figure 10). However this factor is only ~4 for the "baseline" cases. This non-linearity could be explained by the existence of cross-talk phenomenon between the various streaming paths: from the inter-vessel gap into the optical path of the Divertor view for instance. This assumption should be verified: it could be further analysed using tally-tagging features.

4.2 Sensitivity to Nuclear Data Library

A sensitivity study is conducted to assess the impact of different nuclear data libraries on the results. Three libraries are used: FENDL-2.1, JEFF-3.1.1 [11], ENDFB-VII.0. Results of neutron flux are reported in Table 5.

Table 5. Neutron Flux obtained at the EPP Closure Plate, results of TRIPOLI-4® simulations based on different nuclear data libraries.

	Φ (n.cm^{-2}.s^{-1}) $\pm$ 1σ	Rel. Diff. (%)
FENDL-2.1	$8.14 \cdot 10^9 \pm 2\%$	-
JEFF-3.1.1	$7.92 \cdot 10^9 \pm 2\%$	-2.7%
ENDFB-VII.0	$7.86 \cdot 10^9 \pm 2\%$	-3.3%

In terms of neutron shielding and streaming through the Equatorial Port Plug, no significant differences have been found between ENDFB-VII.0, JEFF-3.1.1 and FENDL-2.1.

5 Conclusion and Perspectives

A C-lite integrated neutronic model of ITER is developed for the Monte-Carlo code TRIPOLI-4®. This model is equivalent to the C-lite version 1 model officially released by ITER Organization for MCNP-5. This equivalence is verified by cross-checking the geometries and by doing the present benchmarking study.

It is noteworthy to mention that the newly developed model is originally produced for local shielding analysis, in the frame of design studies that are conducted for the WAVS diagnostics in the Equatorial Port Plugs. The scope of the benchmarking between TRIPOLI-4® and MCNP-5 is thus limited to the assessment of neutron flux levels at the closure plate of the EPP. The neutron-induced activation of the closure plate (with subsequent decay-gamma emissions) is one of the most important contributions to the shutdown dose rate in the Interspace, where personnel access for maintenance is foreseen. The assessment of radiation levels behind the Equatorial Port Plugs is therefore a major concern for the Project.

The TRIPOLI-4® model has kept the same flexibility as the original MCNP-5 model to allow easy integration of detailed components when it is needed. Various DGEPP configurations were integrated in order to investigate shielding options. These integrations are based on CAD models and rely on MCAM, a CAD-to-Monte-Carlo geometry conversion tool.

Based on the analysis of the various cases investigated, a few conclusions can be drawn: firstly, there is a significant neutron leakage in the gaps between the Vacuum Vessel and the Port Plug, and in the inter-DSM gaps. These streaming paths can be effectively shielded by the use of neutron labyrinths and neutron stoppers (absorbers). Secondly, the existence of cross-talk phenomenon between the various streaming paths makes it difficult to do separate-effect analysis, for instance by considering the impact of only one diagnostics among the others. For this reason, it is preferable to consider the problem as a whole, with all diagnostics included so far.

These preliminary results constitute a first step before the calculation of the shutdown dose rates in the

Interspace. Shutdown dose rate calculations will be performed in the near future on the newly developed model. To be able to perform such calculations, the authors will rely on the on-going implementation of the so-called "rigorous-two-step" methodology [12] into TRIPOLI-4® [13]. Future work should also consider the optimization of the variance reduction scheme in TRIPOLI-4®, necessary to obtain reasonable statistical accuracy in the neutron flux meshes.

Finally, one of the main interests in developing alternative models for ITER is to provide independent validation of the results by code-to-code comparisons, thereby strengthening the confidence in Monte-Carlo simulations. This is especially important when variance reduction techniques are employed.

Acknowledgments

Part of this study was done with the financial support from Fusion for Energy in the frame of the contract F4E-FPA-407. The views and opinions expressed herein do not necessarily reflect those of Fusion for Energy nor those of the ITER Organization.

References

1. S. Salasca et al., "The ITER Equatorial Visible/Infra-Red Wide Angle Viewing System: Status of design and R&D", Fusion Engineering and Design, March 2015.

2. R. Juárez, R. Pampin, B. Levesy, F. Moro, A. Suarez, J. Sanz, "Shutdown dose rates at ITER equatorial ports considering radiation cross-talk from torus cryopump lower port", Fusion Eng. Des., 100 (2015), pp. 501–506.

3. J.-C. Trama, E. Brun, F. Damian, C. Diop, E. Dumonteil, F.-X. Hugot, "TRIPOLI-4® version 9 Shielding for the fusion community: overview, relevant bench-marking for fusion and licencing policy", ISFNT-11, Barcelona, Spain, September 16-20, 2013.

4. X-5 Monte Carlo Team, "MCNP – A General Monte Carlo N-Particle Transport Code Version 5", LA-CP-03-0245, Los Alamos National Laboratory, 2005.

5. D. Lopez Al-dama and A. Trkov, " FENDL-2.1: update of an evaluated nuclear data library for fusion applications Report INDC(NDS)-46 (Vienna: IAEA)", 2004.

6. SpaceClaim CAD Software: www.spaceclaim.com.

7. Y. Wu, "CAD-based interface programs for fusion neutron transport simulation FDS Team", Fusion Engineering and Design. vol. 84 (2009) pp. 1987 – 1992.

8. F.-X. Hugot and Y.-K. Lee, "A New Prototype Display Tool for the Monte-Carlo Particle Transport Code TRIPOLI-4®", Progress in Nuclear Science and Technology, Vol 2, pp.851-854, 2011.

9. J.-P. Both, J.-C. Nimal, T. Vergnaud, "Automated Importance Generation and Biasing Techniques for Monte-Carlo Shielding Techniques by the TRIPOLI-3 Code", Progress in Nuclear Energy, Vol 24, pp.273-281, 1990.

10. O. Petit, Y.-K. Lee, C.M. Diop, "Variance Reduction Adjustment in Monte-Carlo TRIPOLI-4® Neutron Gamma Coupled Calculations", Progress in Nuclear Science and Technology, Vol 4, pp.408-412, 2014.

11. A. Santamarina et al., "The JEFF-3.1.1 nuclear data library, JEFF report 22, validation results from JEF-2.2 to JEFF-3.1.1", vol. 6807. OECD, NEA, 2009.

12. Y. Chen, U. Fischer, "Rigorous MCNP Based Shutdown Dose Rate Calculations: Computational Scheme, Verification Calculations and Application to ITER", Fusion Engineering and Design, Vol. 63-64, (2002).

13. F. Malouch et al., "Recent developments in the TRIPOLI-4® Monte-Carlo code for fusion applications", in proceedings of SOFT 2016, Prague, Czech Republic.

Residual activity evaluation: a benchmark between ANITA, FISPACT, FLUKA and PHITS codes

Gabriele Firpo[1,*], *Carlo Maria* Viberti[1], *Anna* Ferrari[2], *Manuela* Frisoni[3]

[1]Ansaldo Nucleare, Corso Perrone 25, Genova, Italy, 16152
[2]Helmholtz-Zentrum Dresden-Rossendorf e. V., Bautzner Landstraße 400, Dresden, Germany, 01328
[3]Agenzia Nazionale per le Nuove tecnologie, l'Energia e lo Sviluppo economico sostenibile, Via Martiri Monte Sole 4, Bologna, Italy, 40129

Abstract. The activity of residual nuclides dictates the radiation fields in periodic inspections/repairs (maintenance periods) and dismantling operations (decommissioning phase) of accelerator facilities (*i.e.*, medical, industrial, research) and nuclear reactors. Therefore, the correct prediction of the material activation allows for a more accurate planning of the activities, in line with the ALARA (As Low As Reasonably Achievable) principles. The scope of the present work is to show the results of a comparison between residual total specific activity versus a set of cooling time instants (from zero up to 10 years after irradiation) as obtained by two analytical (FISPACT and ANITA) and two Monte Carlo (FLUKA and PHITS) codes, making use of their default nuclear data libraries. A set of ~40 irradiating scenarios is considered, *i.e.* neutron and proton particles of different energies, ranging from zero to many hundreds MeV, impinging on pure elements or materials of standard composition typically used in industrial applications (namely, AISI SS316 and Portland concrete). In some cases, experimental results were also available for a more thorough benchmark.

1 Background and Scope

The evaluation of the residual nuclide activity in irradiated materials is an important issue in radiological analysis. The activity of residual nuclides dictates the radiation fields in periodic inspections/repairs (maintenance periods) and dismantling operations (decommissioning phase) of accelerator facilities (i.e., medical, industrial, research) and nuclear reactors. Therefore, the correct prediction of the material activation allows for a more accurate planning of the activities, in line with the ALARA (As Low As Reasonably Achievable) principles.

Typically, the irradiation source, impinging on the material during operation, is known in terms of particle type (potentially a combination of neutrons and other hadrons), flux, energy spectrum and time evolution. While the composition of irradiation samples is usually well known, radiation protection practically refers to standard compositions for structural materials; from an engineering point of view, steel and concrete materials are of particular interest in many nuclear applications.

Starting from the characterized source and material composition, the time evolution of the nuclide inventory, and the corresponding residual activity, can be predicted by solving the set of Bateman equations [1].

Nowadays, many analytical and Monte Carlo computer codes include a Bateman solver routine, offering suitable simulation capabilities. The external irradiation source specification, the material composition, the irradiation history and cooling time instants represent the input data set, which is combined with half-life parameters and activation cross sections of nuclides to calculate the isotope inventories.

Different geometrical approaches are followed by the computer codes, depending on the selected methodologies, whether purely analytical (i.e., Bateman equations solver in a so called "0-dimensional" approach) or Monte Carlo (i.e., Bateman equations solver integrated in a multi-dimensional radiation transport code). These approaches are synthetically described both here below and in Table 1.

The "0-dimensional" approach of the analytical codes assumes an homogeneous, isotropic and energy-dependent flux of the irradiating particle defined in an infinite medium of specified density and isotopic composition.

A fully "3-dimensional" approach is instead implemented in the Monte Carlo codes: the modelled geometry and materials are irradiated by a user defined source; the simulation particles are transported, taking into account any possible interaction with the material nuclides (i.e., absorption, down and up scattering, build-up,…).. The irradiating flux can be averaged in tallied regions (e.g. through track length estimators), and then passed to the Bateman solver routine.

* Corresponding author: Gabriele.Firpo@ann.ansaldoenergia.com

Table 1. Comparison between the "0-dimensional" and "3-dimensional" approaches for the evaluation of the residual activity or nuclide inventory by computational codes.

		0-dimensional	**3-dimensional**
Method of solution		Solution of the Bateman equations	Full Monte Carlo calculation or mixed Monte Carlo and solution of the Bateman equations
Parame-ters of the Bateman equations	*Initial material composi-tion*	Input by user (infinite medium assumed, mass and volume treated as normalization factors)	Input by user (real geometry modelled)
	Transmu-tation cross section	Evaluated data libraries	Evaluated data libraries and/or models
	Irradiating flux	Input by user (homogeneous and isotropic flux assumed)	Track-length averaged by Monte Carlo calculation starting from user-defined real impinging flux (more realistic representation)
Graphical representation		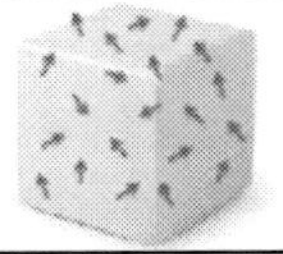	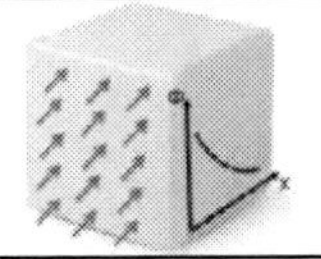
Limitations		No build-up*; No self-shielding and flux attenuation*	No build-up*

* Estimation by models and/or iterative approaches available in some codes.

The scope of the present work is to show the results of a comparison between residual total specific activity versus a set of cooling time instants (from zero up to 10 years after irradiation) as obtained by two analytical (FISPACT [2] and ANITA [3]) and two Monte Carlo (FLUKA [4][5] and PHITS [6]) codes, making use of their default nuclear data libraries.

2 Methodology Description

A set of several tens irradiating scenarios was considered, *i.e.* neutron and proton particles of known energies, ranging from zero to many hundreds MeV, impinging on a restricted set of defined materials. In many of the considered scenarios, called "calculational case studies", no corresponding experimental results were available, and the benchmark was performed only among the results of the calculations. In some other cases, called "experimental case studies", experimental results were also available for a more thorough benchmark.

In general, the following parameters must be fixed in order to completely set an irradiation scenario:

- The irradiation particle type and intensity;
- The spatial, directional and energy distribution of the irradiation particle;
- The initial material composition of the irradiated sample;
- The sample geometry;
- The irradiation time(s);
- The cooling time instant(s) after irradiation at which the total specific activity has to be evaluated.

Details on how these parameters were fixed, slightly depending both on the case study type and on the analytical or Monte Carlo method of solution, are reported hereafter.

2.1 The Calculational Case Studies

In this set of case studies, comparison was made among total specific activity values obtained by analytical (FISPACT and ANITA) and Monte Carlo (FLUKA and PHITS) codes.

2.1.1 The irradiation geometry and parameters

First of all, the irradiation geometry and the geometrical model was fixed as simple as possible as shown in Figure 1: the irradiating flux, monochromatic in energy, impinges perpendicularly and uniformly on the basis of a cylindrical sample. It had fixed radius equal to 1 cm and variable height, being 0.1 or 3 times the irradiating particle mean free path and defining the "thin" and "thick" cases, respectively[a]. A set of two typical structural materials used in industrial application, namely the AISI SS316 and the Portland concrete, was chosen; their isotopic compositions and densities were taken from [7].

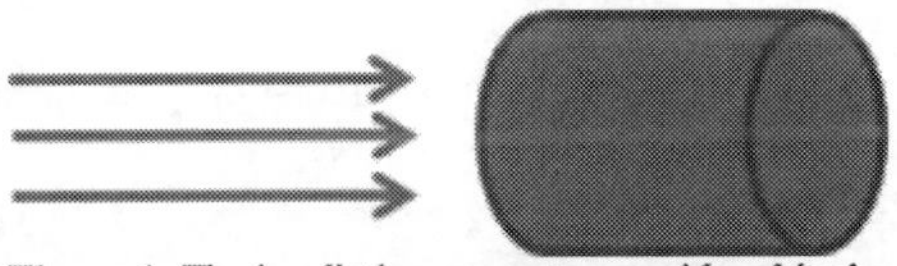

Figure 1. The irradiation geometry considered in the "calculational case studies".

The irradiation parameter values considered in the calculations are summarized in Table 2. Any combination of the irradiation particle energy and type, sample material and sample thickness values defined a different irradiation scenario.

[a] The thin and thick cases were defined in order to consider both the scenario where the irradiating energy spectrum and intensity are negligibly or slightly modified in the sample (thin case) and the scenario where they are considerably modified and attenuated in the sample (thick case).

Table 2. List of the irradiation parameter values considered in the present work for the "calculational case studies". Any combination of the values of the highlighted parameters defines a different irradiation scenario; the other parameters remained unchanged for each scenario.

Parameter	Values
Irradiation particle and energy [MeV]	Neutron: 2.5e-8, 50, 200
	Proton: 10, 50, 600
Sample material	Steel (AISI SS316)
	Concrete (Portland)
Sample thickness (height)	~λ/10 (thin case)
	~3λ (thick case)
Flux intensity [particle/cm^2/s]	1e10
Irradiation time	1 y
Cooling time instants	1 s, 1 h, 1 d, 1 week, 1 month, 6 months, 1 y, 5 y, 10 y

* λ is equal to the mean free path of the irradiating particle as taken from literature or dedicated Monte Carlo evaluation for each material and energy value defined above.

The total number of irradiation scenarios considered for the Calculational case studies was then equal to:

$$(\#_{neutron\ energies})\ (\#_{materials})\ (\#_{thicknesses})\ +\ (\#_{proton\ energies})\ (\#_{materials})\ (\#_{thicknesses})\ =\ 3{\cdot}2{\cdot}2 + 3{\cdot}2{\cdot}2 = 24$$

2.1.2 The calculational methodology

In the Monte Carlo approach with FLUKA and PHITS codes, the calculation of the residual specific activity at the requested cooling time instants starting from the irradiation scenarios beforehand described were performed automatically during the Monte Carlo run via the RESNUCLEI card or via the dedicated post-Monte Carlo processing code "dchain-sp", respectively.

A two-step method was instead necessary for the simulations with the analytical codes FISPACT and ANITA: firstly, the averaged flux (track-length estimation, discerned in few hundreds of logarithmic-interpolated energy bins) in the sample material was evaluated by a radiation transport simulation with a Monte Carlo code, making use of the same irradiation source and geometry of the Monte Carlo approach; secondly, the deterministic calculation was performed with FISPACT and ANITA, defining the track-length averaged particle flux evaluated in the first step as the irradiating flux.

All the Monte Carlo calculations were performed setting an appropriate number of starting particles so that the output results (total activity and track-length estimator values) had a relative error well below 1% in all the cases.

The information on the code versions and nuclear dataset used in the calculations, as well as the information on the computable scenarios for each code—due to nuclear data availability—, is reported in Table 3.

Table 3. Code name, version and nuclear dataset (transmutation cross sections) used in the calculations of the "Calculational case studies".

Code name	Version	Nuclear dataset	Irradiation scenarios
FLUKA	2011.2c	Proprietary	All the scenarios calculated
PHITS	2.82	JENDL 4.0	All the scenarios calculated
FISPACT	2010	EAF2010	All except 600 MeV protons and 200 MeV neutrons
	II	TENDL2014	Only 200 MeV neutrons
ANITA*	4M	EAF2010	Only 2.5e-8 MeV neutrons

* Proprietary version of the ANITA code developed by ENEA Italy.

2.2 The Experimental Case Studies

In this set of case studies, comparison was made among total specific activity values following neutron irradiation obtained by analytical codes (FISPACT and ANITA), Monte Carlo codes (FLUKA and PHITS) and experimental data taken from activity measurements performed at the Karlsruhe Isochronous Cyclotron ("IFMIF cases" in the following [8]) and at the Frascati Neutron Generator ("FNG cases" in the following [9]).

2.2.1 The irradiation geometry and parameters

The irradiation parameters, materials and geometry were defined referring to the experimental specification reported in [8] and [9]. The irradiating neutron spectra on thin cylindrical samples for the IFMIF and FNG cases are shown in Figure 2 and Figure 3, respectively.

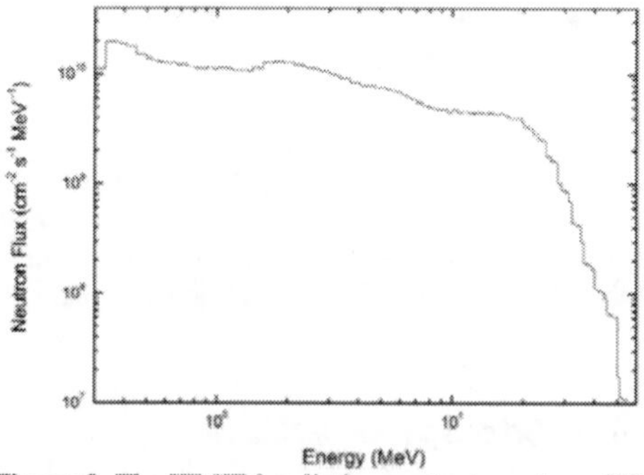

Figure 2. The IFMIF irradiating neutron spectrum [8].

In the IFMIF case, an energetic neutron spectrum (upper energy ~55 MeV) is present, whereas the typical fusion neutron spectrum, peaked at ~14 MeV, is present in the FNG case. The irradiation and cooling times, as well as sample material compositions of the samples, are here summarized in Table 4. The total number of irradiating scenarios considered for the Experimental case studies was equal to:

$$\#_{FNG\_Cases} + \#_{IFMIF\_Cases} = 13 + 4 = 17$$

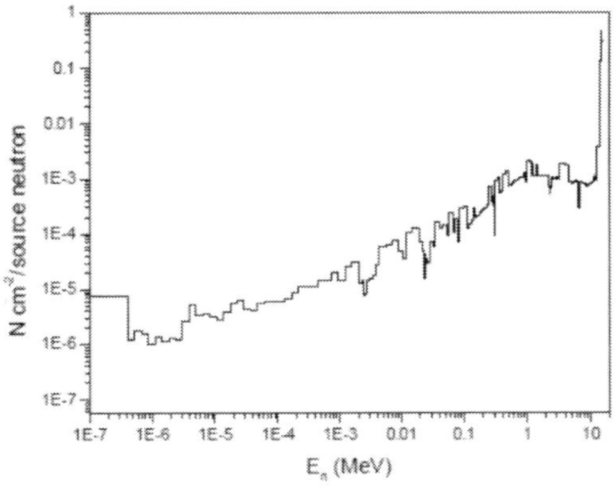

Figure 3. The FNG irradiating neutron spectrum [9].

Table 4. List of the irradiation parameter values [8] [9] considered in the present work for the "experimental case studies". Each line in the table represents an irradiation scenario.

Case	Sample Material	Flux intensity [n/cm^2/s]	Irradiation time [s]	Cooling time instants [s]
IFMIF	SS316	4.10e11	7525	5965, 18848, 93333, 178860, 432539, 2520240, 12962640
	F82H	4.14e11	7525	4173, 18011, 92043, 158665, 424416, 2922060, 13036200
	V-alloy	4.22e11	7525	5477, 17189, 90710, 417962, 2254080, 13341780
	V-pure	4.27e11	7525	4983, 16396, 90001, 412474, 3100500, 13146900
FNG	Mo	3.13e8	330	148, 320, 625, 926
	Cu	2.46e8	161	172, 348, 880, 1443
	Hf	4.71e8	300	136, 318, 638, 963
	Mg	4.64e8	300	136, 316, 616, 958
	Ni	5.72e8	600	139, 318, 617, 918
	Cd	4.23e8	300	136, 317, 618, 917
	Tin	3.70e8	300	137, 316, 616, 917
	Re	5.56e8	1500	136, 318, 619, 919, 1141
	Ti	4.97e8	600	137, 317, 617, 918
	W	5.63e8	1500	139, 320, 619, 923, 1085
	Ag	2.42e8	300	138, 319, 621, 923
	Al	4.73e8	600	138, 323, 621, 922
	Al	3.24e8	901	4547, 6324

2.2.2 The calculational methodology

Since all the irradiation scenarios in the Experimental case studies had very thin samples (sample thickness < 25 μm in FNG cases [9], similar conditions are assumed for the IFMIF cases), in the present work it was assumed that the irradiating neutron fluxes shown in Figure 2 and Figure 3 were not significantly modified or attenuated in the samples. Hence, the samples were assumed to have an homogeneous and isotropic distribution of these neutron fluxes inside their volumes.

By consequence, in the Monte Carlo approach with FLUKA and PHITS codes the irradiating source was defined differently with respect to the "Calculational case studies", since here the neutron energy spectra were defined as an homogeneous and isotropic volume source term. Again, the calculation of the residual specific activity at the requested cooling time instants were performed automatically during the Monte Carlo run via the RESNUCLEI card or via the dedicated post-Monte Carlo processing code "dchain-sp", respectively.

In the analytical approach, the two-step method followed in the "Calculational case studies" was no more necessary since the irradiating neutron fluxes could be just defined as the input irradiating flux without any modification, thus performing directly the single step method of the deterministic calculation.

It is here underlined that, apart from the experimental results, in [8] and [9] the results of corresponding calculations with ANITA (both for FNG and IFMIF cases) and FISPACT (for FNG cases) were already available; hence, these calculations were not repeated in this work, and results are here reported as taken from these references.

The information on the code versions and nuclear dataset used in the calculations is reported in Table 5.

Table 5. Code name, version and nuclear dataset (transmutation cross section) used in the calculations of the "Experimental case studies".

Code name	Version	Nuclear dataset	Irradiation scenarios
FLUKA	2011.2c	Proprietary	All the scenarios
PHITS	2.82	JENDL 4.0	All the scenarios
FISPACT ǂ	2010	EAF2010	All the scenarios
ANITA*	IEAF	IEAF2001	IFMIF
	4M	EAF2010	FNG

* Proprietary versions of the ANITA code developed by ENEA Italy. The corresponding calculations were not performed for the present work and the results were taken from [8] (IFMIF case) and [9] (FNG case).

ǂ The calculations with this code were performed for the present work only in the IFMIF case. The results of the FNG case were taken from [9].

2.3 Uncertainty Estimation

A detailed analysis of the uncertainties was considered out of the scope of the present work. When available from the codes, the errors on the total specific activities were evaluated and reported also in the graphs as error bars. Depending on the code, the error associated to the total specific activity value is either due to statistics (for Monte Carlo codes) or estimated by evaluated data libraries (for analytical codes). As stated before, all the Monte Carlo calculations were performed setting an appropriate number of starting particles so that the output results (total activity and track-length estimator values) had a relative error well below 1% in all the cases. The eventual contribution of secondary particles (*i.e.*, primarily, neutron particles in case of proton irradiating particles and vice versa) was considered during the definition of the methodology and the execution of the calculations; indeed, in some of the considered codes the evaluation of this contribution is embedded or estimated in the calculation by dedicated models. Anyway, some sensitivities were preliminary performed in several case studies considered in the present work in order to discern the percent contribution to the total specific activities of the secondary particles: in all the considered cases, it was found that this contribution is no more than few percent of the one of the primary particles. Hence, for simplicity, the contribution of the secondary particles was neglected—by consequence, a systematic underestimation error ranging from about zero up to few percent of the total specific activity can be roughly considered in the final results reported in the present paper.

3 Results and Conclusions

In the present Section the results of the calculations and the comparison among them are reported both for the calculation and for the experimental case studies.

3.1 The Calculational Case Studies

The behaviour of the total specific activity of the SS316 and Portland Concrete samples for each irradiation scenarios are reported hereafter.

Looking at the results from Figure 4 through Figure 27, in general it can be concluded that a good agreement among the results obtained with the different codes is observed. On average, differences of the order of factor of a few are present. Very few cases, in particular for high cooling time instants, do present big differences (factor 10 and more) among the codes.

In particular, the results of SS316 are in better agreement among the codes than the Portland concrete ones. It is suggested that this is due to some difference on the dominating activated isotopes in Portland among the nuclear dataset used in the calculations. Moreover, the results in thin samples are in better agreement among the codes than the thick ones. It is suggested that it is due to the fact that in the thick cases the averaged irradiating energy spectra are no more monochromatic in energy as the source is, and a wider range of energy values is spanned: the results are then again potentially more sensitive to differences on the cross section shape versus energy among the different nuclear dataset.

3.1.1 Thermal (2.5e-8 MeV) neutrons

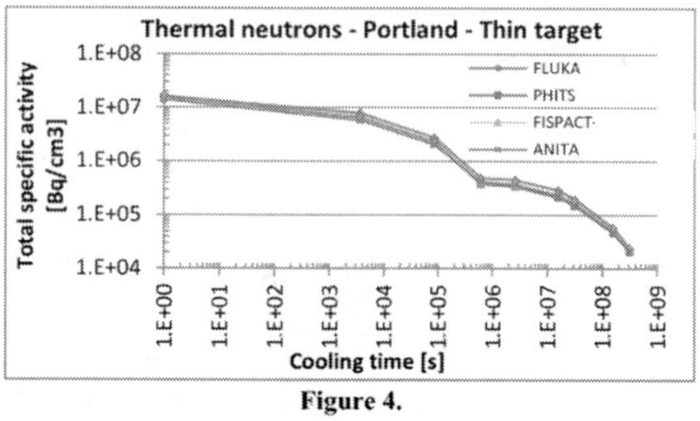

Figure 4.

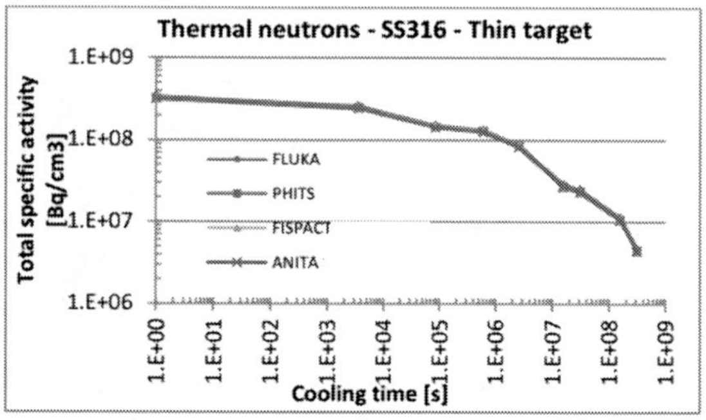

Figure 5.

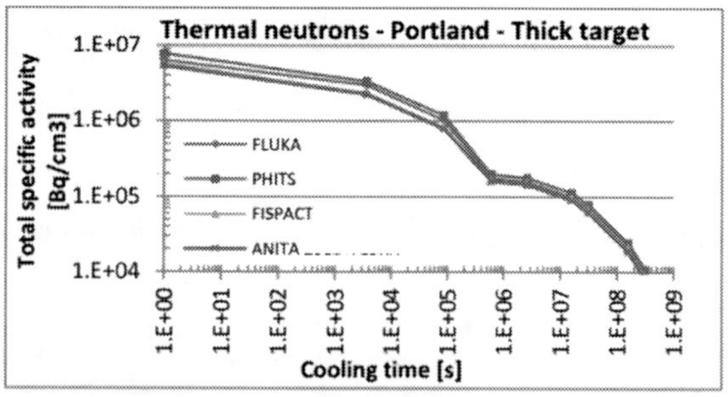

Figure 6.

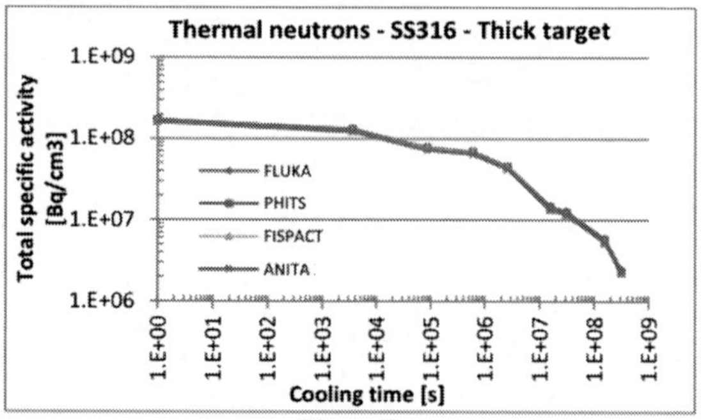

Figure 7.

3.1.2 10 MeV protons

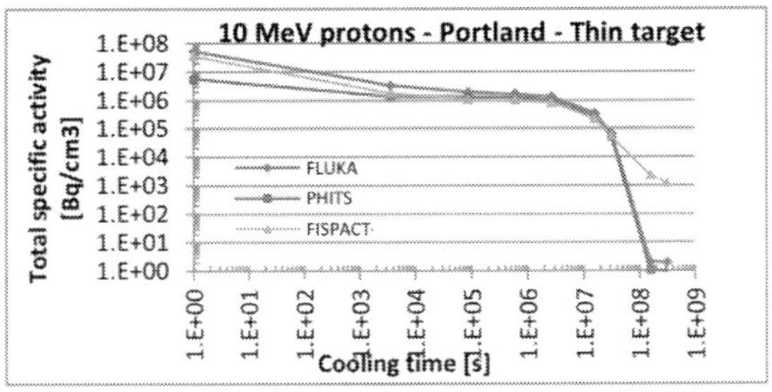

Figure 8.

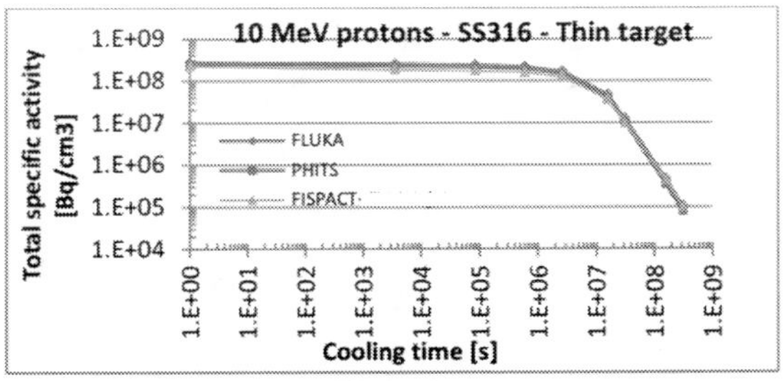

Figure 9.

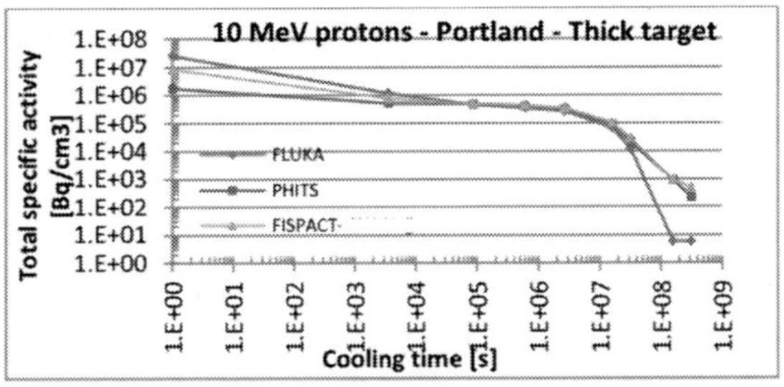

Figure 10.

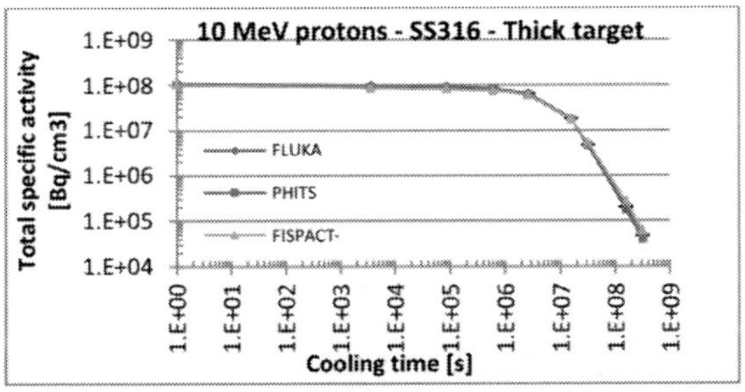

Figure 11.

3.1.3 50 MeV neutrons

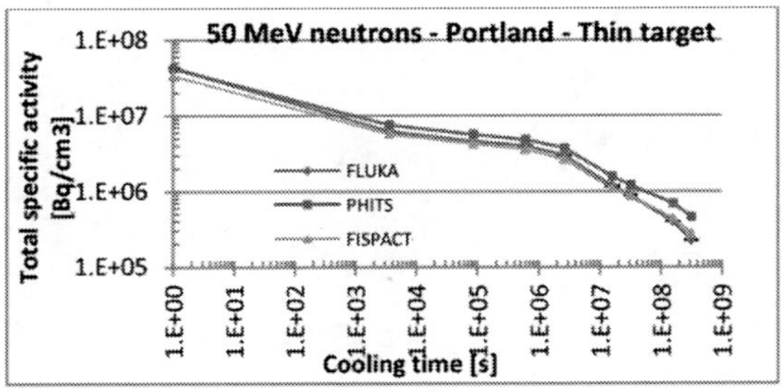

Figure 12.

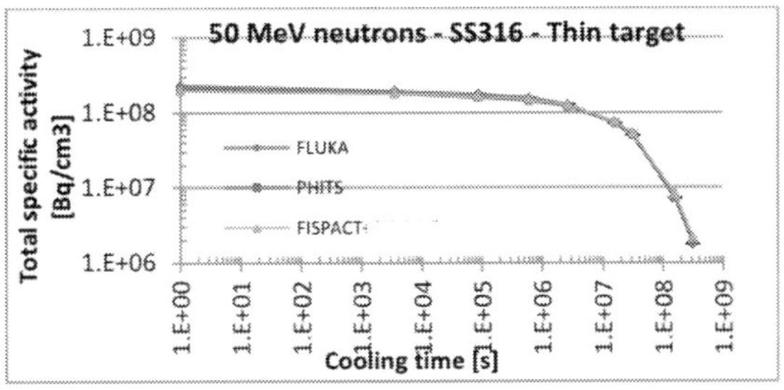

Figure 13.

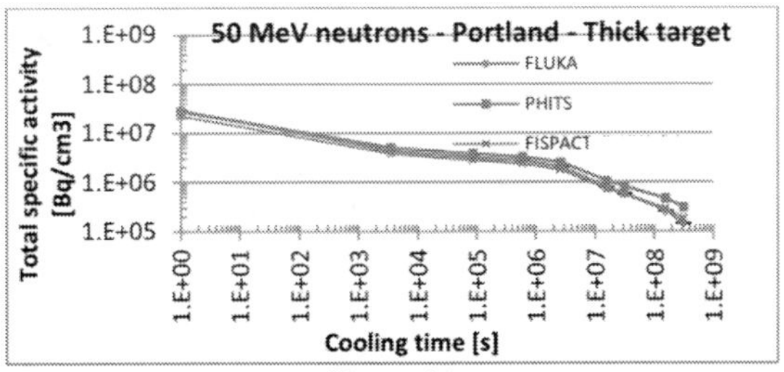

Figure 14.

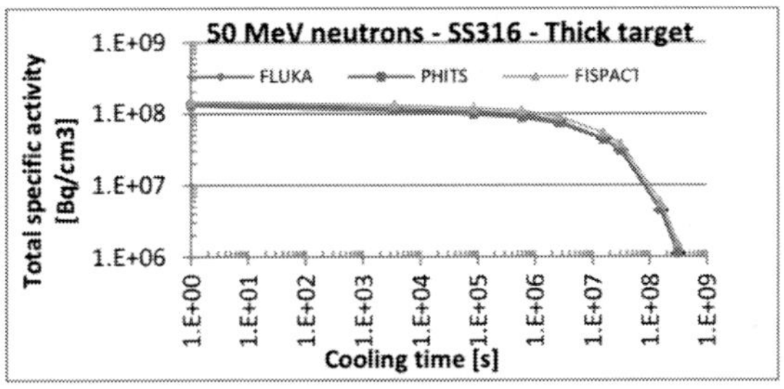

Figure 15.

3.1.4 50 MeV protons

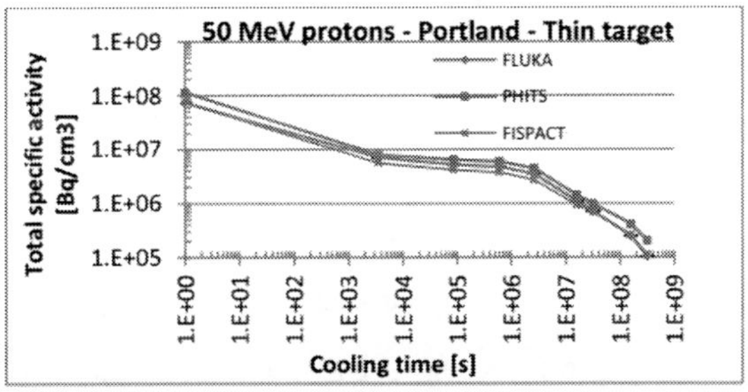

Figure 16.

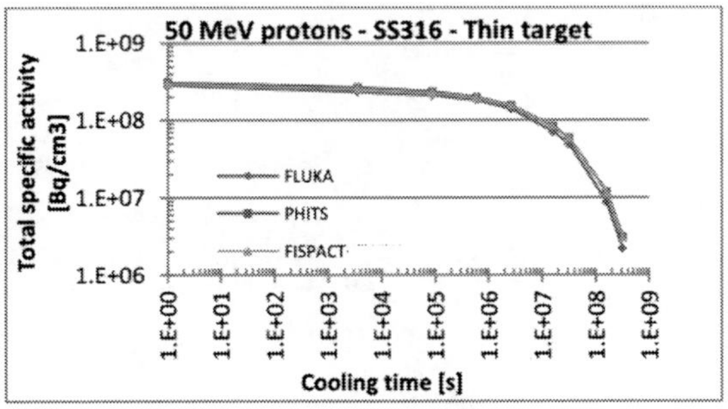

Figure 17.

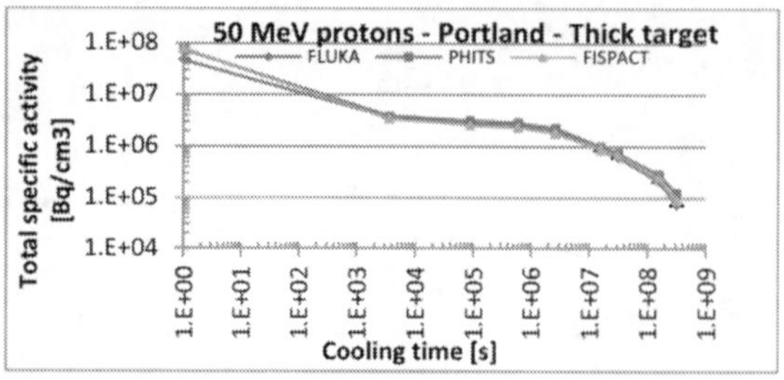

Figure 18.

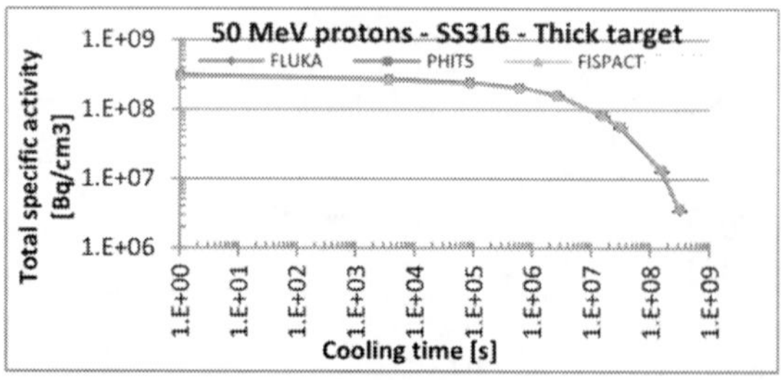

Figure 19.

3.1.5 200 MeV neutrons

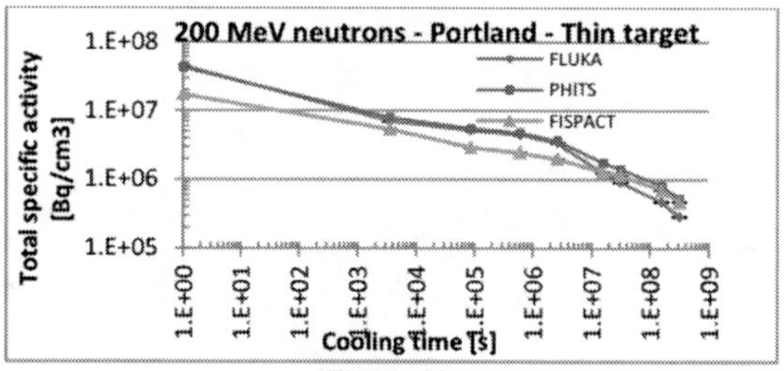

Figure 20.

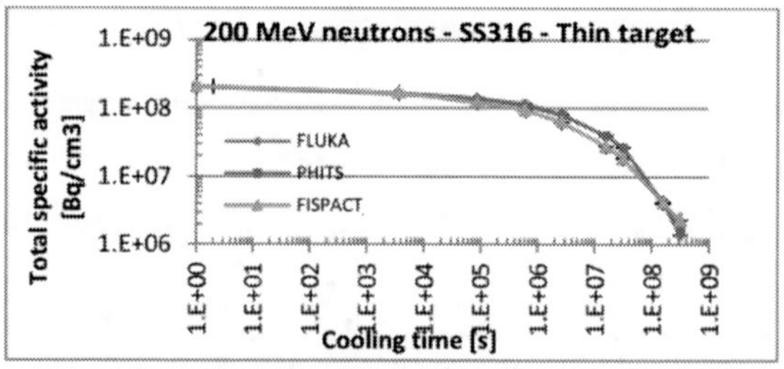

Figure 21.

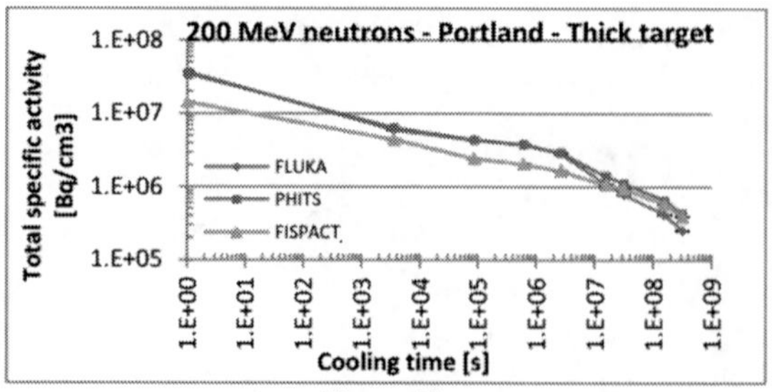

Figure 22.

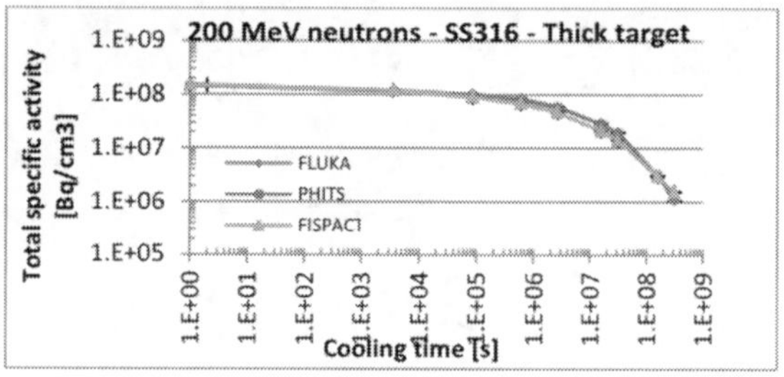

Figure 23.

3.1.6 600 MeV protons

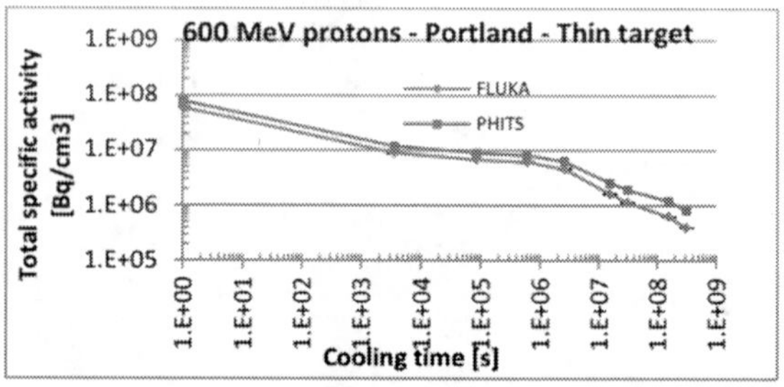

Figure 24.

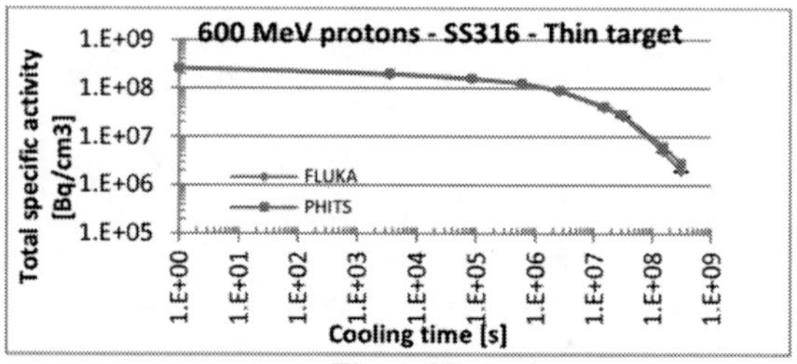

Figure 25.

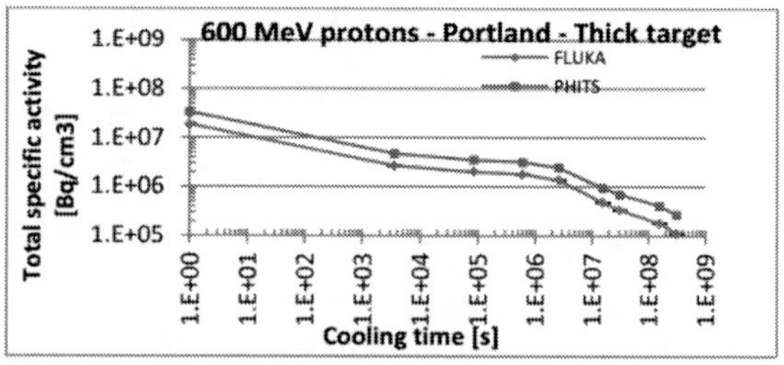

Figure 26.

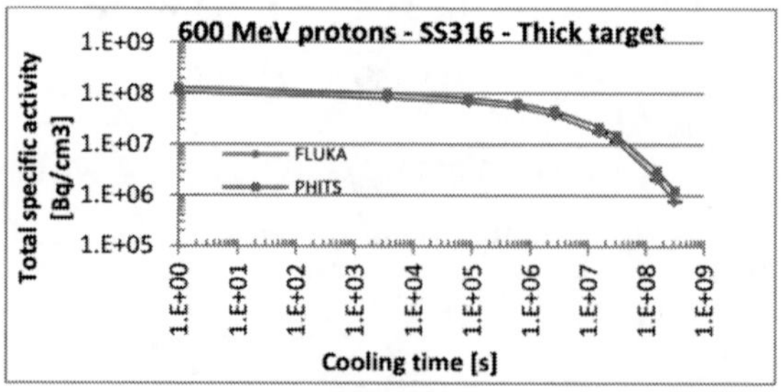

Figure 27.

 DOI: 10.1051/epjconf/201715302006

3.2 The Experimental Case Studies

The behaviour of the total specific activity of some pure elements and of SS316 thin samples for each irradiation scenarios are reported hereafter as obtained from the calculations. The comparison with experimental data and the corresponding C/E (calculation over experiment) values are also reported. Their error values (1σ) are not reported in the graphs for convenience, and reported in Table 6 and Table 7.

Looking at the results from Figure 28 through Figure 53, it can be concluded that in almost all the FNG cases the results of the different codes are in good agreement (few tens percent difference among them) and in agreement within a factor of 2 at maximum with respect to the experiment values. Only in the Hafnium case, probably because of issues on the nuclear dataset, very big differences are present between the calculations and the experiment.

Finally, looking at the results from Figure 54 through Figure 61, also for the IFMIF cases a good agreement among the calculation results was observed, and an acceptable agreement within a factor 4 at maximum was found against the experimental values in all the cases except the Vanadium ones, where the disagreement went up to more than a factor 10 for long cooling time instants ("Fluka_3"), again probably because of nuclear data issue. Also the calculation results obtained with FLUKA in the Vanadium cases at long cooling time instants were found not to be in good agreement with the other codes: it was found that it was related to an overestimation of the ^{65}Zn isotope contribution: by eliminating it, as shown in Figure 58 and Figure 60 ("Fluka_3 – Zn65" label), the FLUKA results are closer both the experimental and the calculated results of the other codes[b].

3.2.1 FNG – Molybdenum sample

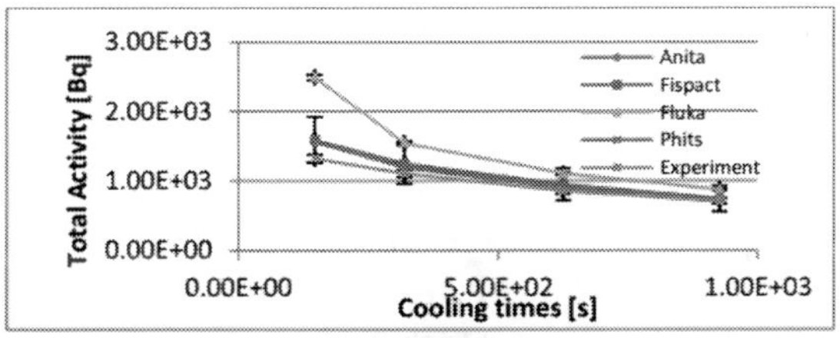

Figure 28.

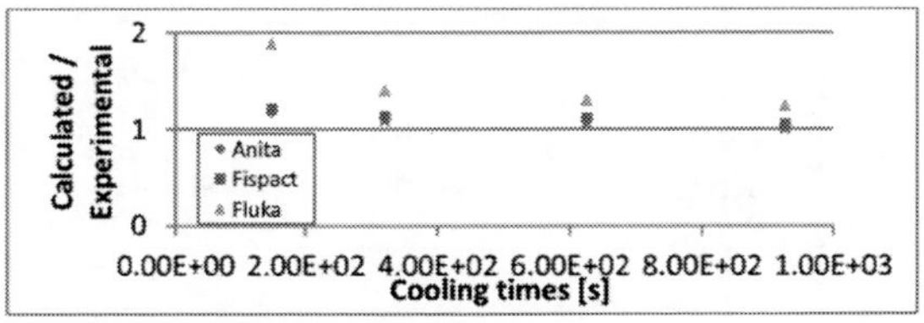

Figure 29.

[b] This issue was discussed with the FLUKA Developers. It was found that there was a minor bug since the ^{65}Zn libraries were wrongly linked by the software. A new FLUKA version was then released in October 2016 (*i.e.* FLUKA 2011.2c.5). The Vanadium cases were then computed again, showing that this issue was solved (see Figure 58 and Figure 60, label « Fluka_5 »).

3.2.2 FNG – Copper sample

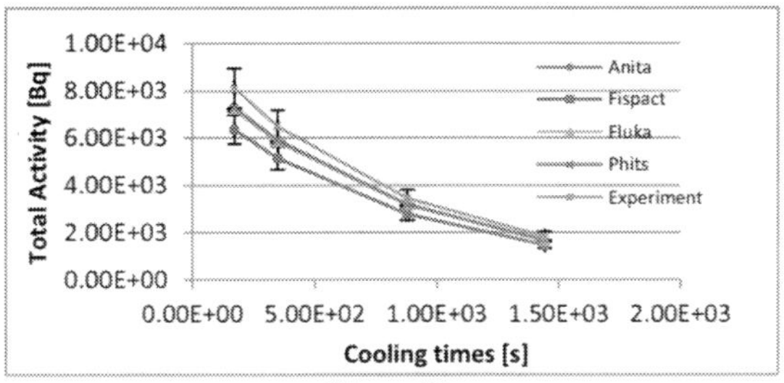

Figure 30.

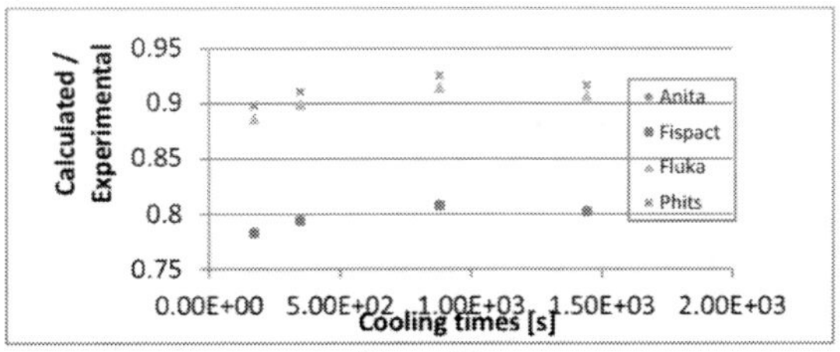

Figure 31.

3.2.3 FNG – Hafnium sample

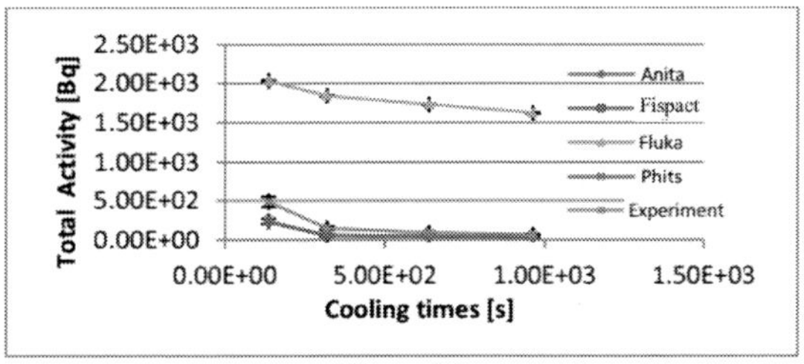

Figure 32.

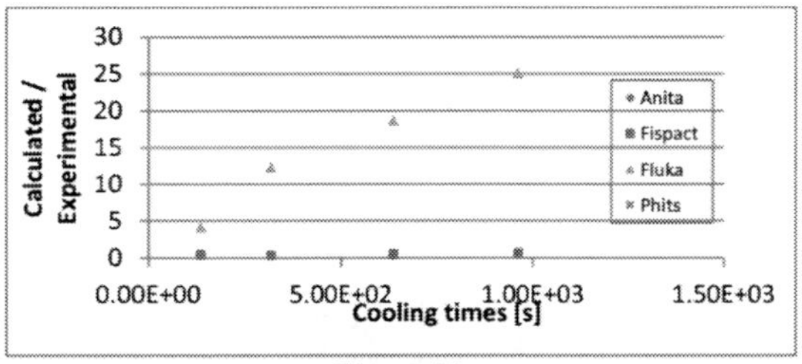

Figure 33.

3.2.4 FNG – Magnesium sample

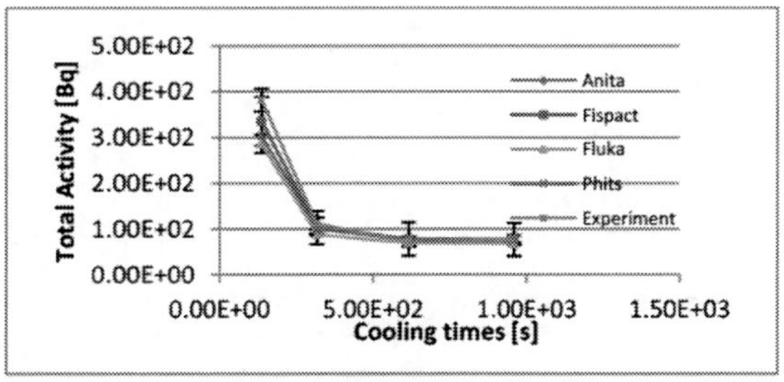

Figure 34.

 DOI: 10.1051/epjconf/201715302006

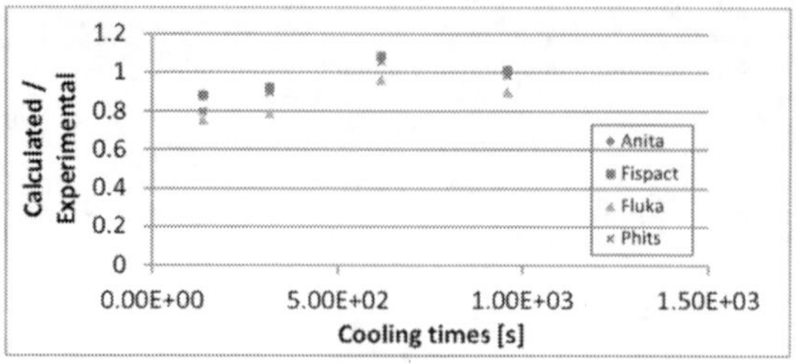

Figure 35.

3.2.5 FNG – Nichel sample

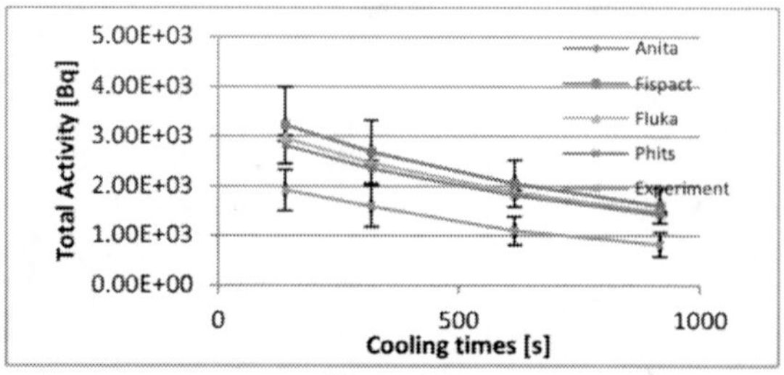

Figure 36.

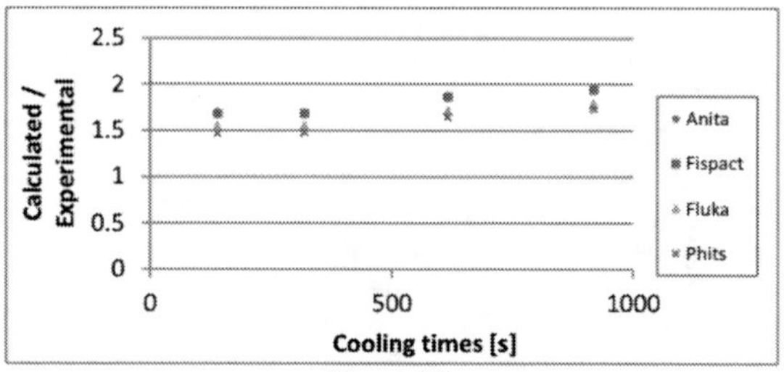

Figure 37.

3.2.6 FNG – Cadmium sample

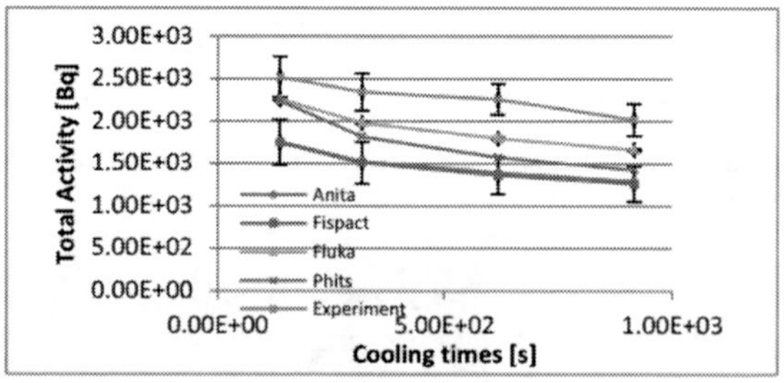

Figure 38.

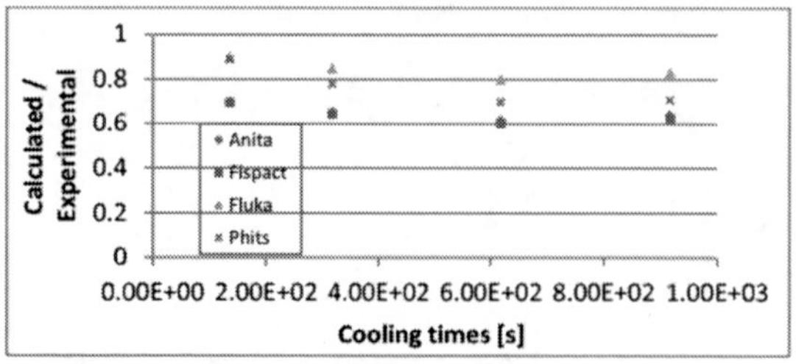

Figure 39.

3.2.7 FNG – Tin sample

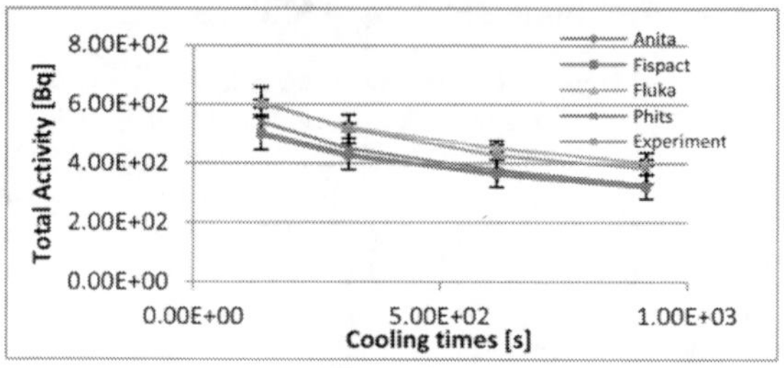

Figure 40.

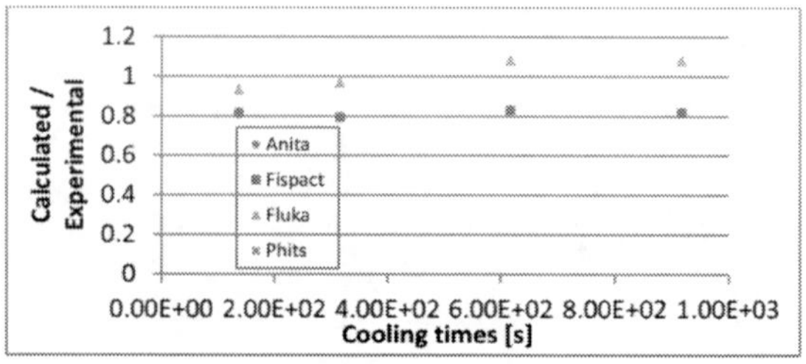

Figure 41.

3.2.8 FNG – Rhenium sample

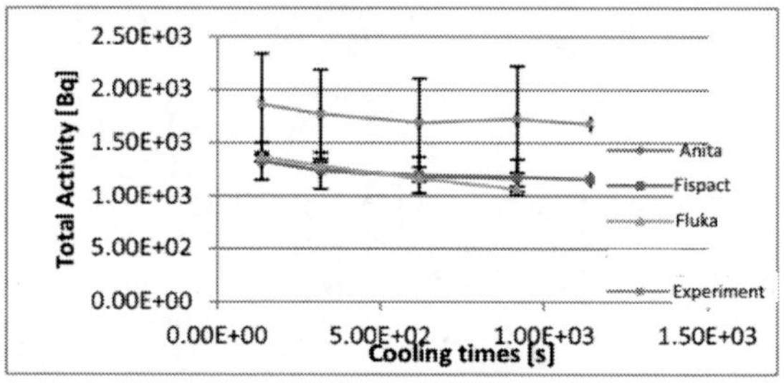

Figure 42.

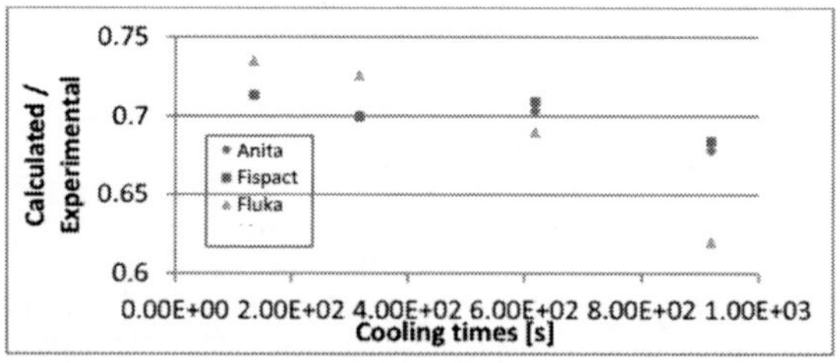

Figure 43.

3.2.9 FNG – Titanium sample

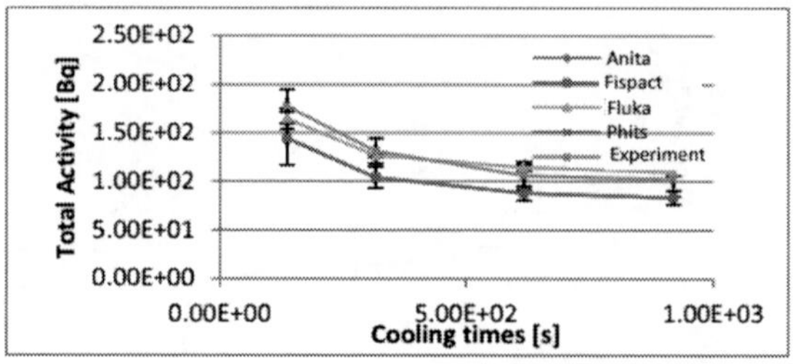

Figure 44.

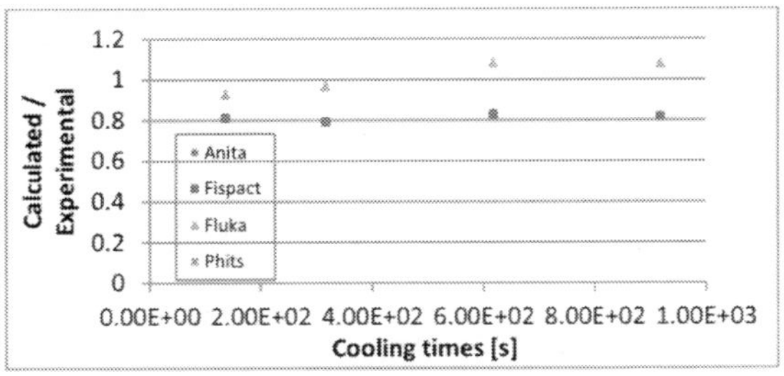

Figure 45.

3.2.10 FNG – Tungsten sample

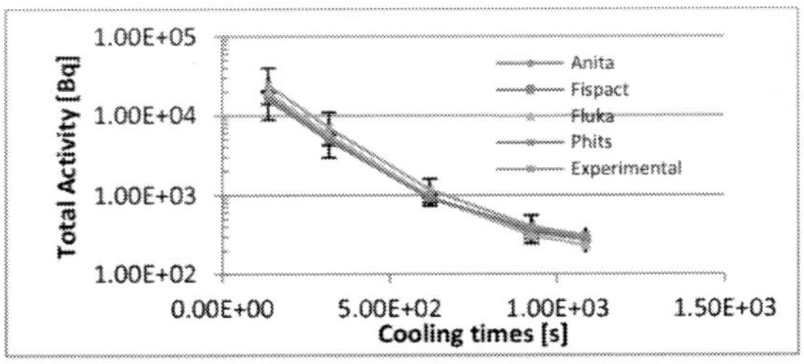

Figure 46.

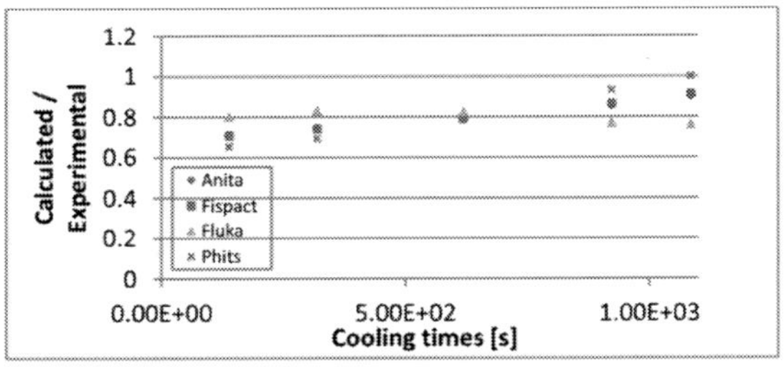

Figure 47.

3.2.11 FNG – Silver sample

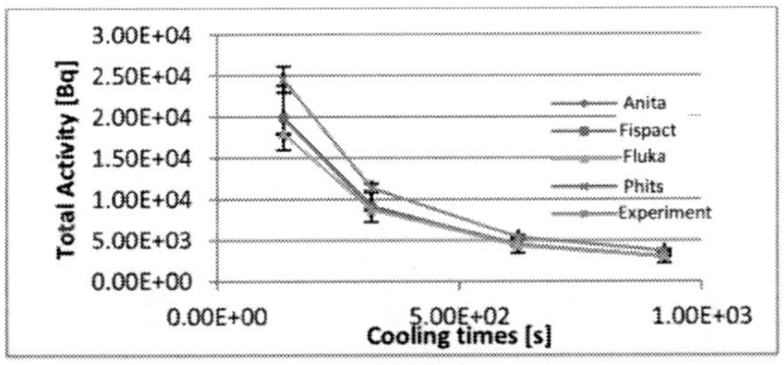

Figure 48.

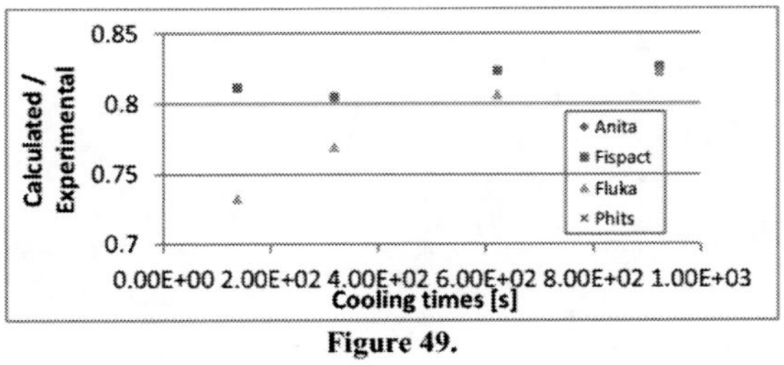

Figure 49.

3.2.12 FNG – Aluminium sample number 1

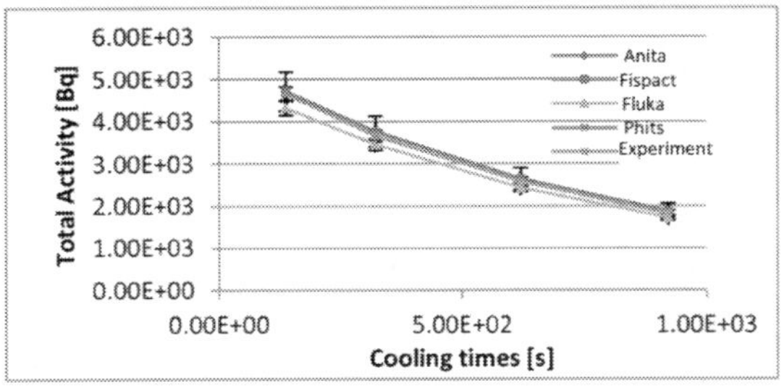

Figure 50.

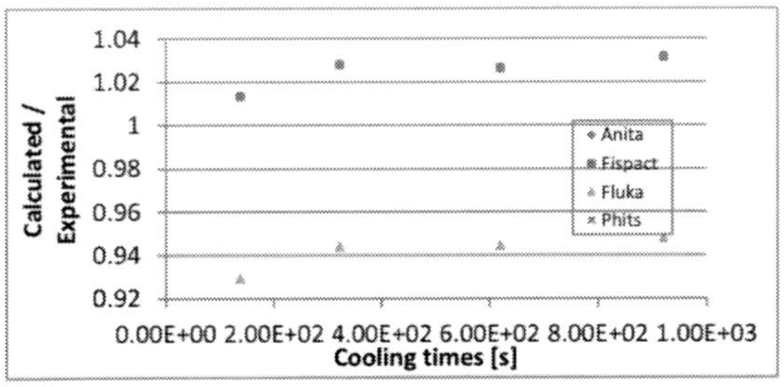

Figure 51.

3.2.13 FNG – Aluminium sample number 2

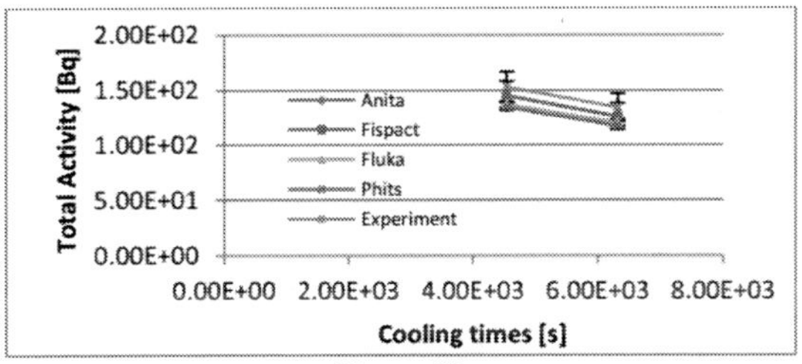

Figure 52.

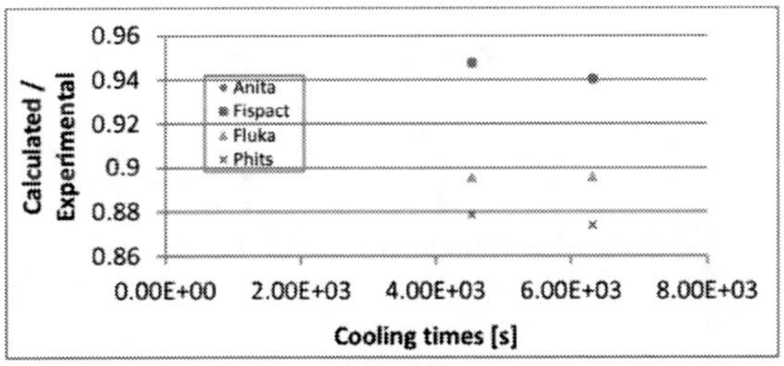

Figure 53.

DOI: 10.1051/epjconf/201715302006

3.2.14 IFMIF – SS316 sample

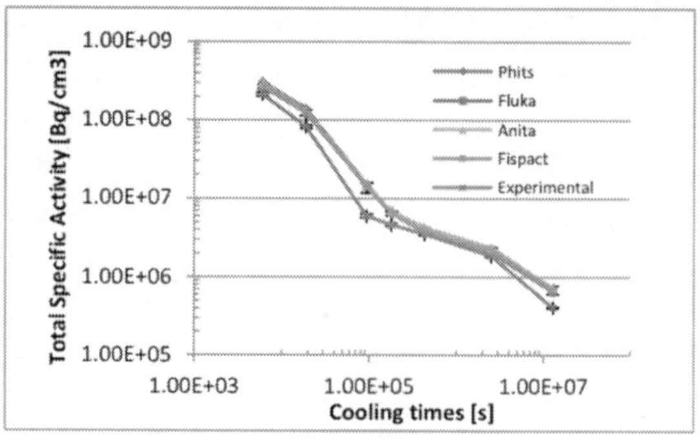

Figure 54.

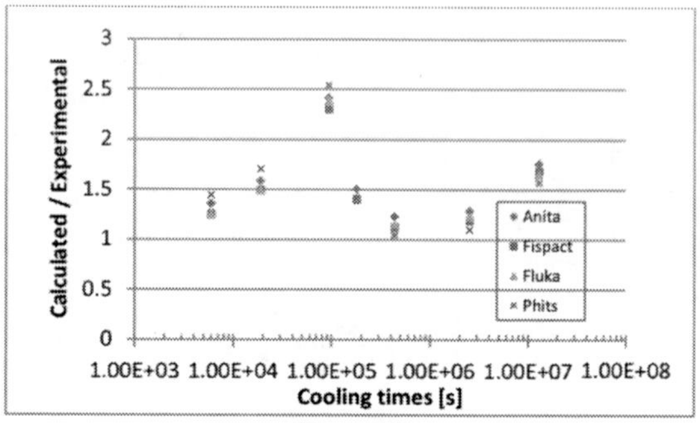

Figure 55.

3.2.15 IFMIF – F82H sample

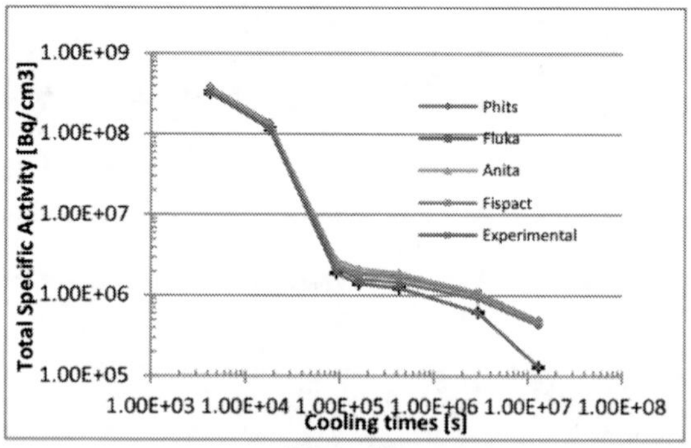

Figure 56.

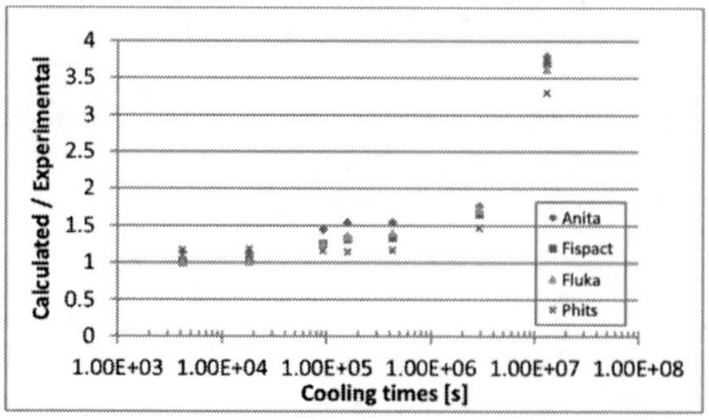

Figure 57.

3.2.16 IFMIF – V-alloy sample

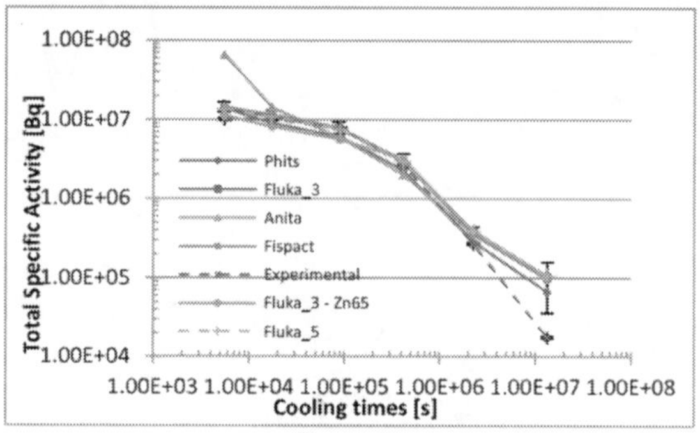

Figure 58.

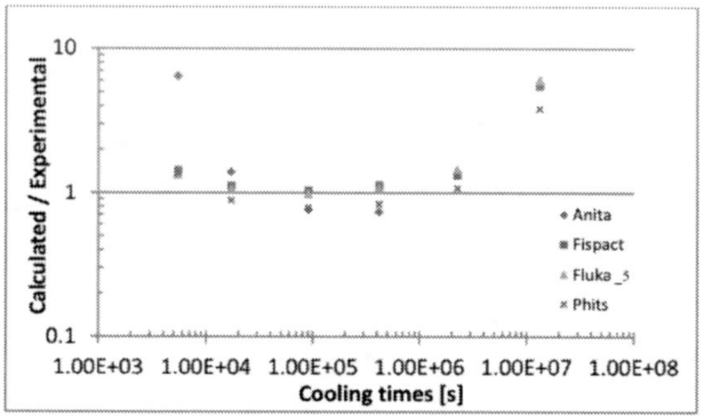

Figure 59.

3.2.17 IFMIF – V-pure sample

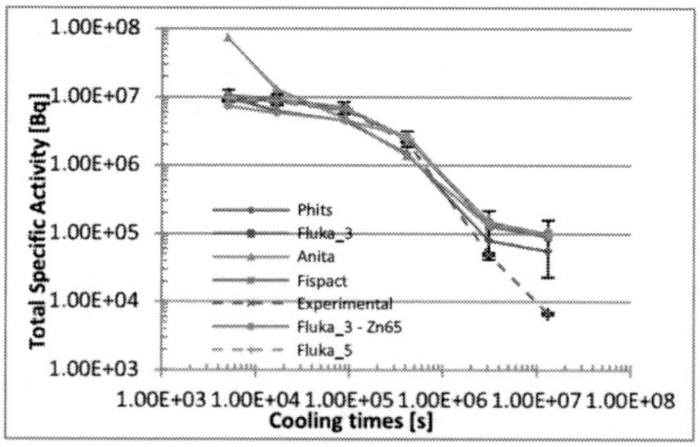

Figure 60.

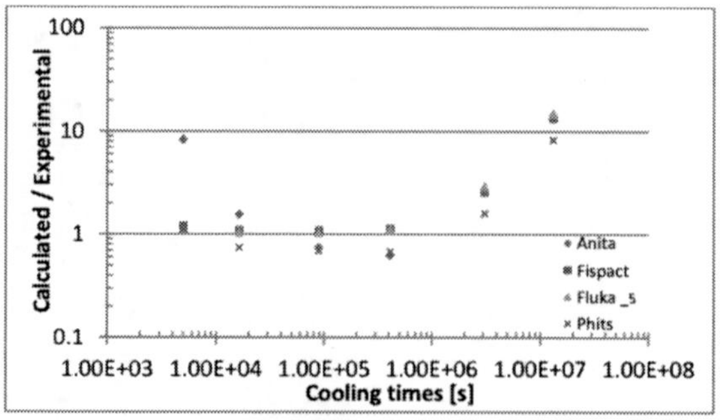

Figure 61.

3.3 Additional Notes

Apart from the total specific activity values, in principle it could have been also possible to compare the detailed list of nuclide inventories obtained by the different codes; anyway, for simplicity this analysis was not systematically performed in the present work, and the corresponding results are not reported. It is just underlined here that this analysis was performed in some cases, looking at the list of the nuclides in decreasing order of their contribution to the total specific activity. In general it was noted that, from the qualitative point of view, the dominating 4 or 5 nuclides in the isotopic inventories are typically the same for all the codes—just some position exchange in this list was observed among the codes—whereas, going on through the list, where the contribution of the nuclides goes well below 1% of the total specific activity, the scenarios are completely different among the codes. It is suggested that this is due primarily to the different nuclear dataset used by the codes.

Table 6. Summary of the C/E values for the IFMIF cases.

Sample	Cooling time [s]	Anita		Fispact		Fluka		Phits	
		C/E	$\sigma_{C/E}$	C/E	$\sigma_{C/E}$	C/E	$\sigma_{C/E}$	C/E	$\sigma_{C/E}$
SS316	5965	1.36	0.07	1.26	0.09	1.24	0.06	1.44	0.07
	18848	1.59	0.08	1.50	0.14	1.49	0.07	1.71	0.08
	93333	2.42	0.13	2.30	0.36	2.39	0.13	2.54	0.13
	178860	1.50	0.06	1.40	0.11	1.42	0.06	1.41	0.06
	432539	1.23	0.04	1.12	0.09	1.15	0.04	1.04	0.03
	2520240	1.28	0.04	1.19	0.11	1.23	0.04	1.10	0.04
	12962640	1.76	0.05	1.69	0.21	1.65	0.04	1.57	0.04
F82H	4173	1.14	0.06	1.02	0.05	1.00	0.05	1.17	0.06
	18011	1.12	0.06	1.04	0.06	1.01	0.05	1.18	0.06
	92043	1.45	0.05	1.26	0.15	1.26	0.04	1.16	0.04
	158665	1.54	0.06	1.32	0.19	1.36	0.05	1.15	0.05
	424416	1.54	0.06	1.34	0.18	1.40	0.06	1.17	0.05
	2922060	1.77	0.06	1.66	0.22	1.72	0.06	1.47	0.05
	13036200	3.79	0.12	3.70	0.75	3.62	0.12	3.30	0.11
V_Alloy	5477	6.45	0.25	1.43	0.21	1.34	0.05	1.38	0.05
	17189	1.40	0.05	1.12	0.17	1.06	0.04	0.88	0.03
	90710	0.76	0.03	1.04	0.18	0.98	0.03	0.78	0.03
	417962	0.73	0.02	1.13	0.23	1.08	0.03	0.83	0.03
	2254080	1.36	0.04	1.33	0.31	1.45	0.05	1.08	0.04
	13341780	5.58	0.18	5.53	3.48	6.01	0.19	3.80	0.12
V_pure	4983	8.28	0.31	1.21	0.25	1.10	0.04	1.11	0.04
	16396	1.55	0.06	1.10	0.21	1.04	0.04	0.74	0.03
	90001	0.74	0.03	1.09	0.22	1.04	0.04	0.69	0.03
	412474	0.62	0.02	1.14	0.29	1.10	0.03	0.68	0.02
	3100500	2.60	0.12	2.56	1.74	2.91	0.13	1.59	0.07
	13146900	13.50	0.53	13.51	10.06	14.84	0.58	8.31	0.32

Table 7. Summary of the C/E values for the FNG cases.

Sample	Cooling time [s]	Anita		Fispact		Fluka		Phits	
		C/E	$\sigma_{C/E}$	C/E	$\sigma_{C/E}$	C/E	$\sigma_{C/E}$	C/E	$\sigma_{C/E}$
Mo	148	1.18	0.05	1.20	0.26	1.88	0.08	1.18	0.05
	320	1.11	0.05	1.13	0.27	1.40	0.07	1.08	0.05
	625	1.09	0.05	1.11	0.28	1.30	0.06	1.05	0.05
	926	1.03	0.06	1.05	0.27	1.23	0.08	0.99	0.06
Cu	172	0.78	0.08	0.78	0.11	0.89	0.09	0.90	0.09
	348	0.79	0.08	0.79	0.11	0.90	0.09	0.91	0.09
	880	0.81	0.08	0.81	0.11	0.91	0.10	0.92	0.10
	1443	0.80	0.09	0.80	0.11	0.91	0.10	0.92	0.10
Hf	136	0.47	0.06	0.49	0.10	4.15	0.56	0.38	0.05
	318	0.35	0.05	0.41	0.10	12.30	1.77	0.58	0.08
	638	0.45	0.07	0.54	0.14	18.61	3.00	0.70	0.11
	963	0.55	0.10	0.66	0.18	25.00	4.50	0.81	0.15

Table 7.(Contd)Summary of the C/E values for the FNG cases.

Sample	Cooling time [s]	Anita		Fispact		Fluka		Phits	
		C/E	$\sigma_{C/E}$	C/E	$\sigma_{C/E}$	C/E	$\sigma_{C/E}$	C/E	$\sigma_{C/E}$
Mg	136	0.88	0.06	0.88	0.15	0.75	0.07	0.80	0.05
	316	0.92	0.11	0.92	0.34	0.79	0.09	0.89	0.10
	616	1.07	0.15	1.08	0.52	0.96	0.14	1.06	0.15
	958	1.01	0.13	1.01	0.49	0.90	0.12	0.98	0.13
Ni	139	1.68	0.36	1.68	0.54	1.54	0.33	1.47	0.32
	318	1.68	0.44	1.68	0.59	1.55	0.40	1.47	0.38
	617	1.86	0.49	1.86	0.65	1.71	0.45	1.64	0.43
	918	1.94	0.58	1.94	0.72	1.78	0.54	1.73	0.52
Cd	136	0.70	0.07	0.69	0.13	0.90	0.08	0.89	0.08
	317	0.65	0.06	0.64	0.12	0.85	0.08	0.78	0.07
	618	0.61	0.05	0.60	0.11	0.80	0.06	0.70	0.06
	917	0.64	0.06	0.62	0.12	0.83	0.08	0.71	0.07
Tin	137	0.81	0.07	0.83	0.12	0.99	0.09	0.89	0.08
	316	0.82	0.08	0.83	0.13	1.01	0.10	0.87	0.08
	616	0.84	0.10	0.86	0.15	1.06	0.12	0.88	0.10
	917	0.83	0.12	0.84	0.16	1.05	0.15	0.85	0.12
Re*	136	0.71	0.18	0.71	0.21	0.73	0.19	n.a	n.a
	318	0.70	0.17	0.70	0.19	0.73	0.17	n.a	n.a
	619	0.70	0.17	0.71	0.20	0.69	0.17	n.a	n.a
	919	0.68	0.20	0.68	0.22	0.62	0.18	n.a	n.a
	1141	0.69	0.16	0.69	0.19	0.60	0.14	n.a	n.a
Ti	137	0.81	0.08	0.81	0.18	0.93	0.11	0.94	0.09
	317	0.79	0.08	0.79	0.11	0.97	0.10	0.94	0.09
	617	0.83	0.11	0.83	0.13	1.08	0.15	1.00	0.14
	918	0.82	0.14	0.82	0.16	1.08	0.19	0.99	0.17
W	139	0.71	0.45	0.71	0.47	0.80	0.51	0.65	0.42
	320	0.74	0.43	0.74	0.45	0.83	0.48	0.69	0.40
	619	0.79	0.30	0.79	0.32	0.82	0.31	0.79	0.30
	923	0.86	0.33	0.87	0.34	0.77	0.30	0.93	0.36
	1085	0.90	0.22	0.91	0.23	0.76	0.20	1.00	0.24
Ag	138	0.81	0.05	0.81	0.17	0.73	0.05	0.79	0.05
	319	0.81	0.04	0.81	0.17	0.77	0.04	0.78	0.04
	621	0.82	0.03	0.82	0.19	0.81	0.03	0.81	0.03
	923	0.83	0.03	0.83	0.22	0.82	0.03	0.82	0.03
Al_1	138	1.01	0.04	1.01	0.10	0.93	0.05	0.92	0.03
	323	1.03	0.04	1.03	0.11	0.94	0.05	0.93	0.03
	621	1.03	0.04	1.03	0.11	0.94	0.05	0.93	0.04
	922	1.03	0.04	1.03	0.11	0.95	0.05	0.94	0.04
Al_2	4547	0.95	0.09	0.95	0.12	0.90	0.08	0.88	0.08
	6324	0.94	0.09	0.94	0.13	0.90	0.09	0.87	0.08

*Calculation with PHITS not available because of lack in nuclear data for Re.

References

[1] J. Cetnar, Ann. Nucl. En. **33**, issue 7, 640 (2006)
[2] J. C. Sublet, J. Eastwood, G. Morgan, A. Koning, D. Rochman, Progr. in Nucl. Sc. and Tech. **4**, 349 (2014)
[3] M. Frisoni, EPJ Web of Conferences **111**, 11004 (2016)
[4] T.T. Bohlen, F. Cerutti, M.P.W. Chin, A. Fasso`, A. Ferrari, P.G. Ortega, A. Mairani, P.R. Sala, G. Smirnov, and V. Vlachoudis, Nucl. Data Sheets **120**, 211-214 (2014)
[5] A. Ferrari, P.R. Sala, A. Fasso`, and J. Ranft, **CERN-2005-10** (2005), **INFN/TC_05/11**, **SLAC-R-773**
[6] T. Sato, K. Niita, N. Matsuda, S. Hashimoto, Y. Iwamoto, T. Furuta, S. Noda, T. Ogawa, H. Iwase, H. Nakashima, T. Fukahori, K. Okumura, T. Kai, S. Chiba, L. Sihver, Ann. Nucl. Energy **82** (2014)
[7] RJ McConn Jr, CJ Gesh, RT Pagh, RA Rucker, RG Williams III, **PNNL-15870 Rev. 1** (2011)
[8] D.G. Cepraga, M. Frisoni, G. Cambi, **FUS-TN-SA-SE-R-70 Rev. 1** (2003)
[9] Courtesy of M. Frisoni, ENEA, Italy

Shielding and Activation Analyses for BTF Facility at SNS

Irina Popova[1,*] and *Franz* X. Gallmeier[1]

[1]ORNL, PO box 2008, MS 6466, Oak Ridge, TN 37831-6474, USA

Abstract. The beam test facility (BTF), which simulates front end of the Spallation Neutron Source (SNS), has been built at the SNS, and is preparing for commissioning. The BTF has been assembled and will operate in one of service buildings at the site. The 2.5 MeV proton beam, produced in the facility, will be stopped in the beam dump. In order to support BTF project from radiation protection site, neutronics simulations and activation analyses were performed to evaluate the necessary shielding around the facility and radionuclide inventory of the beam stop.

1 Introduction

The beam test facility (BTF) will be a copy of the accelerator front end of the Spallation Neutron Source (SNS) facility and will be used for test and development of new front-end equipment, for conducting research in accelerator physics (in a further build-out) for SNS second target station moderator development (tests). The BTF is assembled and will be operating in the radio frequency (RF) on-site annex building. Negatively charged hydrogen ions (H^-), generated at the ion source, will be accelerated to 2.5 MeV. The facility is capable of producing one-ms-long pulses at 60Hz repetition rate with up to 50mA peak current. This beam will be safely terminated in the beam stop.

In order to support the BTF project from radiation protection point of view, neutronics simulations and activation analyses were performed to develop shielding around the beam stop, evaluate dose rates in case of an accident and to evaluate the radionuclide inventory at the facility end-of-life for decommissioning.

The challenge of the analyses was to consider the low energy charges particles reaction physics to fit the calculations into our routine transport and activation methods as practiced for the much higher energy spallation environment.

2 BTF layout

BTF is a "room sized accelerator", which is located in a technical support building on SNS site. BTF consists of a 65keV H^- ion source, a Low Energy Beam Transport (LEBT), a 2.5 MeV Radio Frequency Quadrupole Accelerator (RFQ), a Medium Energy Beam Transport (MEBT) diagnostic systems, a high power beam stop and support systems. A V shaped beam stop is located at the end of MEBT, which is being re-used from SNS commissioning. The water-cooled beam stop is made from TZM, which is an alloy of 0.50% titanium, 0.08% zirconium and 0.02% carbon with the balance molybdenum, with density of 10.2g/cc, and it is water-cooled. The beam stop from SNS Frontend commissioning was re-used in this setup. A radiation level of 300 mrem/hour at 30 cm distance from the surface is expected at full beam power, according to measurements performed during the commissioning, which requires radiation shielding to be installed.

For the neutronics analyses of dose rates in the vicinity of the BTF beam stop and to design appropriate shielding, a very simple model of the beam stop in MCNPX language was developed. This model was used later for activation calculations as well.

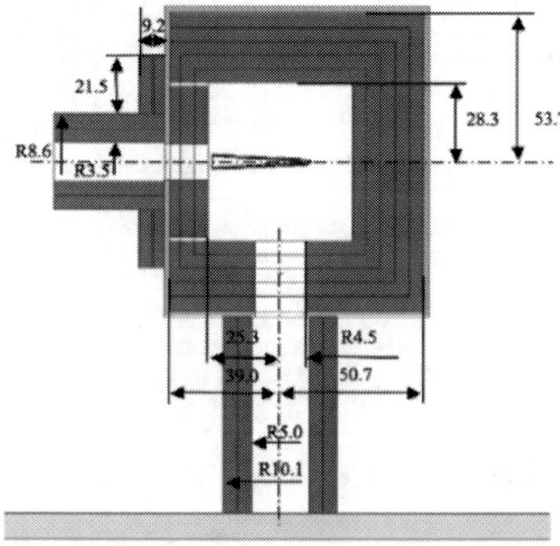

Figure 1. Vertical cross section of the beam stop shielding configuration, dimensions are in cm.

The model for the analyses is presented in Figure 1 with the water-cooled beam stop, is V shaped assembly in the middle of Figure 1 and is shown in greater detail in Figure 2. The calculational model of the beam stop consists from 4 cells - inner cell, outer cell, end cell and water between inner and outer cells. The layers of TZM,

* Corresponding author: popovai@ornl.gov

where the protons impact, is too small to see, and located on the inner surface of the water layer.

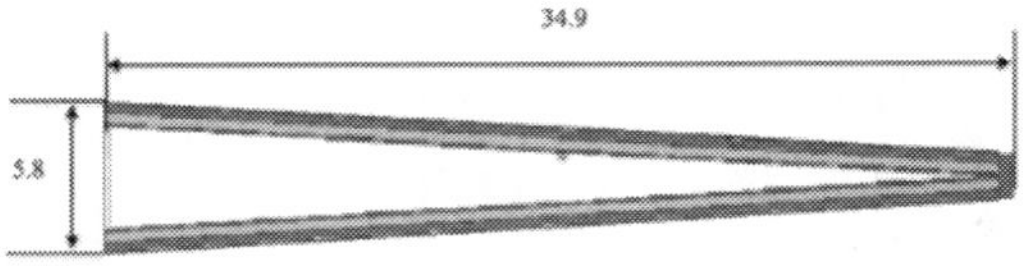

Figure 2. Vertical cross section of the beam stop, dimensions are in cm.

The H$^-$ beam is delivered by the beam pipe from the left side of the figure and intercepts the beam stop. The beam stop is surrounded by a shielding configuration consisting of 5% borated polyethylene, red color, enclosed in 1.9 cm steel, green color, resting on a concrete floor, yellow color, under the beam stop shielding assembly.

There are two penetrations at the beam stop shielding – one is on the upstream edge of the shielding for the beam pipe and the second is on the bottom portion of the shielding, under the beam stop for wiring and piping. It was shown in the analyses that the dose rates were high in penetration vicinity and, in order to reduce dose rates, additional borated polyethylene shielding was suggested around both penetrations.

3 Methods for the analyses and source terms

3.1 Dose rates evaluations

Shielding analyses at SNS are routinely performed using the Monte Carlo transport code MCNPX [1]. Evaluated proton reaction cross sections are sparse, and physics models not applicable for energy range in interest, which is up to 2.5MeV. A hand estimate of secondary neutrons and gammas yield resulting from protons reactions with TZM beam stop was performed.

The negatively charged hydrogen ions (H$^-$), generated at the ion source, are accelerated in the BTF assembly up to 2.5 MeV before they interact with the beam stop material. The source is a H$^-$ beam corresponding to the SNS accelerator full beam at 50mA, 1 ms, 60Hz, The averaged beam current is 3 mA, which is converted into number of H$^-$ per second - $1.87 \cdot 10^{16}$ H$^-$/s.

The simplified model for the hand estimation considers a massive material structure (TZM) hit by the 2.5 MeV proton beam. For simplicity, all secondary particles are considered to be distributed isotropically.

The reaction rate for producing neutrons or gammas from proton interactions is:

$$R = \Phi^{p} \sigma N V \qquad (1)$$

where σ is average cross section for reaction (p,n) or (p,g), N is the nuclear concentration of the target isotope (nucleus/barn*cm), V is the target volume and Φ^{p} is the proton flux. Proton cross sections for (p,n) and (p,g) reactions are taken from Experimental Nuclear Reaction Data (EXFOR) [2]. The proton flux may be determined as the number of protons per second (n) per cm^3 (V) of material multiplied by the track length through the material or stopping length (d).

$$\Phi^{p} = n\, d / V \qquad (2)$$

Then the reaction rate will be:

$$R = n\, d\, \sigma N \qquad (3)$$

Because the proton energy is low (2.5 MeV) the neutrons and gammas will be born very close to the point where the proton beam hits the material. Hence, the produced neutrons or gammas may be considered as a point source. Using the number of produced neutron or gamma reaction rates, the fluxes have been calculated by dividing the reaction rate by the area of the sphere with radii (r) corresponding to the distance from the material.

Finally dose rates are calculated by multiplying the calculated fluxes by flux to dose conversion factors. In order to be conservative, for neutron interactions the flux to dose conversion factor is taken for 2.5 MeV both neutrons and gammas. The non-attenuated neutron dose rate at 30 cm for full beam is expected to be approximately 324 mrem/h according to the results of the analyses, which are shown at Table 1 and compare well to the measurements during the commissioning (300mrem/h).

Table 1. Dose rates due to various isotopes from the unshielded beam stop.

Isotope and reaction	Cross section, mb	Reaction rate at full beam, particle/s	Dose Rate at 30cm, mrem/hr
Mo92(p,g)	2.50E-02	1.16E+07	2.24E+00
Mo94(p,g)	5.00E-02	1.44E+07	2.80E+00
Mo95(p,g)	6.00E-02	2.98E+07	5.77E+00
Zr94(p,n)	0.11	5.04E+04	6.24E-01
Ti49(p,n)	15	2.55E+07	3.15E+02

As shown in Table 1 dose rate is dominated by the neutron production from the (p,n) reaction on titanium isotope - Ti49. Thus we assume the entire source is coming from isotropically emitted 1.5MeV neutrons. This choice was made on the basis of TENDL-2012 [3] for Ti49.

3.2 Activation analyses

The standard way of SNS activation analyses, using the activation script [4], assumes that analyses are driven by high-energy proton beams. However, this procedure is not suitable for assessing problems driven by low-energy protons, which depend largely on correct treatment of threshold reactions. For this reason a custom procedure for analyzes was developed, covering the proton-induced and neutron-induced activities.

The larger part of the beam stop material activation is proton-induced, the calculation of which, is described first. Due to the relatively low energy of the source protons of 2.5 MeV and the slowing down of protons within micrometers of track length, the probability of the (p,x) reaction, where x is any out-coming particle, is low, and there are no activation cross section libraries coming with MCNPX transport code that would allow a direct isotope production analysis. In order to calculate isotope production rates, isotope production cross section for the stables isotopes in the beam stop were extracted from TENDL-2014 nuclear data library [5]. Macroscopic cross sections are obtained by multiplying cross sections with atomic density of the isotope in the material. The obtained macroscopic cross sections have been transformed to group macroscopic cross sections in the group structure shown in Table 2 assuming a flat proton flux energy distribution.

Proton fluxes in this energy group structure were calculated using MCNPX. The full proton beam, described above, was used as a source. Due to the low incoming proton energy, proton fluxes only in the inner cell of the beam stop were found to be nonzero. Fluxes for this cell were extracted.

Then macroscopic cross sections were folded with proton fluxes, in order to obtain isotopes production rates that were collected in the file "splprods" for consequent transmutation analyses by CINDER'90 code [6].

Table 2. Group structure.

Lower energy boundary, MeV	Upper energy boundary, MeV
1	1.2
1.2	1.5
1.5	1.75
1.75	2
2	2.2
2.2	2.5

Neutrons originating from the proton reactions in the beam stop contribute a small fraction to the activity. Their source strength was assessed for the dose rate analyzes analyses above and applied in a MCNPX transport calculation, which provided neutron fluxes in the beam stop geometry (cells).

The cell based activation analysis using the standardized activation script considered the neutron part by feeding the neutron fluxes into the CINDER90 code (the folding of neutron fluxes and neutron activation cross section is performed by CINDER90 internally) and by feeding the proton part as isotope production rates through the "splprods" file.

4 Results

Dose rate and activation analyses were performed for the facility operation at full beam power in order to develop proper shielding around the beam stop and to predict isotope inventory of the TZM beam stop in the end of the life time for decommissioning purposes The accident case to estimate dose rate around the facility fence in case of beam mis-steering was considered as well.

4.1 Normal beam operation

According to the provided drawing of the BTF beam stop, shielding analyses of the dose rates were performed and suggestions for additional shielding in order to reduce dose rates to below 0.25 mrem/h were made.

The first set of analyses was performed with no shielding near the penetrations, just general shielding around the beam stop. A second set of analyses was performed to reduce the dose rate around the beam stop below 0.25 mrem/h at 30 cm distance from the surfaces. Shielding was added under the beam stop, around the beam pipe and on the upstream surface of beam stop shielding. The dose rate maps in the beam stop vicinity with shielding only around the beam stop, and with added shielding as on Figure 1 (extended shielding) are shown in vertical cross section on Figure 2 and 3 respectively.

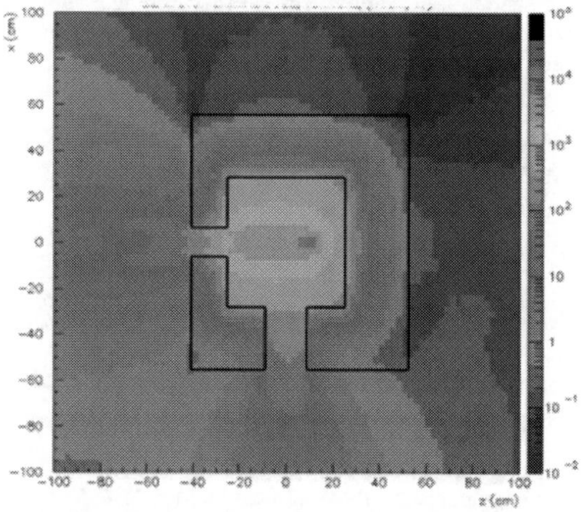

Figure 3. Total dose rate map in the vicinity of the beam stop with the general shielding, vertical cross section, mrem/h.

Table 3 shows dose rates at key locations at 30 cm distance from the surfaces, locations for the dose rate on the floor are directly on the floor surface.

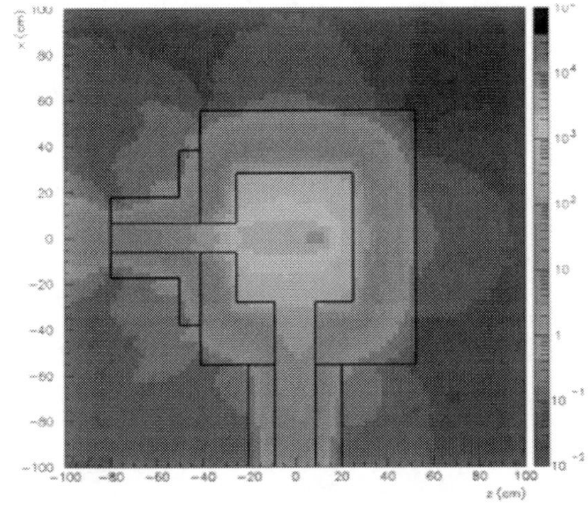

Figure 4. Total dose rate map in the vicinity of the beam stop with the added shielding, vertical cross section, mrem/h.

Table 3. Neutron dose rates during at key locations at 30 cm distance from the surfaces

Location	Shielding only around the beam stop	Extended shielding
Above the shielding	0.1	0.1
Behind the shielding	0.14	0.14
Aside the shielding	0.1	0.1
Upfront shielding	3.3	0.11
Beam pipe	4.3	0.12
On the floor under the beam stop	30	30
On the floor at the downstream edge of the beam stop shielding	0.8	0.08

4.2 Accident case

The accident scenario, when beam is mis-steered, was analyzed as well. The possible accidents are when the proton beam will intercept the stainless steel beam pipe or the copper structures. Hand estimates of secondary neutrons and gammas yield resulting from protons reactions with steel and copper were performed and then converted to the dose rates as described in the methods section.

According to the analyses the expected dose rate is 7600 mrem/h and 13700 mrem/h neutron dose at 30 cm from the accident point on stainless steel and on copper, respectively. The expected gamma dose rate is about two orders of magnitude less than neutron dose rate.

The BTF facility is located inside a fence enclosing the radiation area. In order to predict the dose rates at the different places near the fence for unshielded section of the facility, hand estimates, based $1/R^2$ attenuation were used. Table 4 shows dose rate at the different distances from beam interception point during the accident on stainless steel, when 2.5 MeV proton beam is mis-steered and hits the unshielded beam pipe. For accident on the copper the dose rates will be 1.8 times higher.

Table 4. Neutron dose rates during the accident near the radiation fence.

Distance, cm	Dose rate, mrem/h
30	7600
100	684
232	1
300	72
400	43
600	19
700	14
1000	6.9

4.3 Activation analyzes

Isotope inventory for TZM beam stop after 100 hours of operation spread over 10 years was analyzed for decay times for up to 10 years. In order to make simulation closer to the real scenario it was assumed that there is a proton beam on beam stop for 25 hours at an experiment campaign and 730 days of break between campaigns for four campaigns. The nuclide of highest long-term activation (up to 10 years) is V49, and those contributing most at short decay times (up to 1 day) are Tc95 and Tc96. Activity in Ci in both TZM beam stop and cooling water during 10 years decay time followed 10 years of operation is shown in Figure 5. Isotope inventory was extracted from transmutation code output and results were compared to DOE CAT-3 threshold [7]. Hazard values, ratio of isotope activity over the threshold values, were estimated for each isotope; the summed up hazard values over all isotopes are under 3e-08 even for 1 minute after the beam termination.

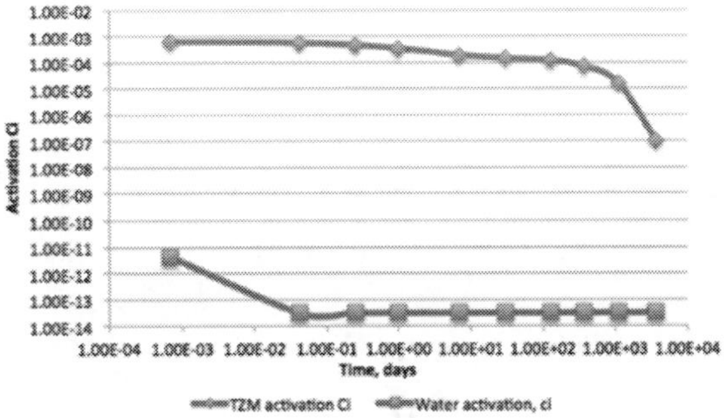

Figure 5. Activity in Ci in both TZM beam stop and cooling water during 10 years decay time followed 10 years of operation.

5 Conclusions

Shielding and activation analyses are performed to address radiation protection questions for BTF Dose rates near the beam stop were analysed and proper shielding was suggested. Accident cases, when beam was mis-steered and hits steel or copper, were analysed.

Activation analyses for beam stop material and cooling water were performed

References

1. C D. Pellowitz, ed., "MCNPX User's Manual, Version 2.6.0," LA-CP-07-1473, Los Alamos National Laboratory, Los Alamos, New Mexico (April 2008). (also see http://mcnpx.lanl.gov).
2. Experimental Nuclear Reaction Data (EXFOR) Database Version of May 17, 2007. See http://www.nndc.bnl.gov/exfor/exfor.htm.
3. TENDL-2012 TALYS-based evaluated nuclear data library, 2012, see http://www.talys.eu/tendl-2014/
4. F. X. Gallmeier and M. Wohlmuther, Activation Script Version 1.0 User Guide, ORNL-TM-2008/031, Oak Ridge National Laboratory, August 2008.
5. By A.J. Koning[1], D. Rochman[1], S.C van der Marck[1], J. Kopecky[2], J. Ch. Sublet[3], S. Pomp[4], H. Sjostrand[4], R. Forrest[5], E. Bauge[6], H. Henriksson[7], O. Cabellos[8], S. Goriely[9], J. Leppanen[10], H. Leeb[11], A. Plompen[12], R. Mills[13] and S. Hilaire[6] , TENDL-2014: TALYS-based evaluated nuclear data library, [1] NRG, [2] JUKO Research, [3]CCFE, [4]Uppsala Univ., [5]IAEA, [6]CEA,[7]Vattenfall, [8]NEA, [9]ULB, [10]VTT, [11]ATI, [12]IRMM, [13]NNL. See (ftp://ftp.nrg.eu/pub/www/talys/tendl2014/tendl2014.html), December 2014
6. W. B. Wilson, S.T. Cowell, T. R. England, A.C.Hayes, P. Möller, A Manual for Cinder'90 Version 07.4, LA-UR-07-8412, Los Alamos National Laboratory, Los Alamos, (2007).
7. DOE STANDARD, HAZARD CATEGORIZATION AND ACCIDENT ANALYSIS TECHNIQUES FOR COMPLIANCE WITH DOE ORDER 5480.23, NUCLEAR SAFETY ANALYSIS REPORTS, DOE-STD-1027-92, December 1992.

Rigorous-two-Steps scheme of TRIPOLI-4® Monte Carlo code validation for shutdown dose rate calculation

Jean-Charles Jaboulay[1,a], Emeric Brun[1], François-Xavier Hugot[1], Tan-Dat Huynh[1], Fadhel Malouch[1], Davide Mancusi[1], Aime Tsilanizara[1]

[1]*CEA-Saclay, DEN, DM2S, SERMA, F-91191 Gif-sur-Yvette, France*

Abstract. After fission or fusion reactor shutdown the activated structure emits decay photons. For maintenance operations the radiation dose map must be established in the reactor building. Several calculation schemes have been developed to calculate the shutdown dose rate. These schemes are widely developed in fusion application and more precisely for the ITER tokamak. This paper presents the rigorous-two-steps scheme implemented at CEA. It is based on the TRIPOLI-4® Monte Carlo code and the inventory code MENDEL. The ITER shutdown dose rate benchmark has been carried out, results are in a good agreement with the other participant.

1 Introduction

To safely conduct maintenance operation or repair in a nuclear reactor (fission or fusion) the radiation dose map in the reactor building must be well known. These doses are due to the decay gamma emitted by the activated structures and the fuel in fission reactor. For fission reactor a strong feedback (measures) is available to address this problem. For, the futures fusion devices like ITER the shutdown dose map must be calculated. Two methods have been developed to perform shutdown dose rate (SDR) mapping in the complex geometry of fusion reactors: the Rigorous Two Steps (R2S) and the Direct One Step (D1S) method. They are well described in [1]. A summary of the SDR mapping tools currently existing in Europe is given in [2]. There is one based on direct-one-step (D1S) treatment [3] and three following the rigorous-two-step (R2S) coupling methodology [4], [5] and [6].

In this context, CEA develops his own SDR scheme with its codes TRIPOLI-4® [7] and MENDEL [8]. This development is supported by a EUROfusion [9] task: *Monte Carlo code development.* This new tool will support the CEA activities in ITER and DEMO nuclear analysis and decommissioning preparation.

2 R2S Scheme developed with TRIPOLI-4® and MENDEL

Two activation schemes are commonly proposed for the shutdown dose rate calculations: D1S use only one coupled neutron-photon Monte Carlo transport calculation, prompt photons are replaced by decay photons. R2S use two Monte Carlo transport calculation one for the neutrons to calculate the activation and one for photons emitted by activated materials. The methodology developed in TRIPOLI-4® follows the R2S approach. The scheme of Figure 1 details these two steps.

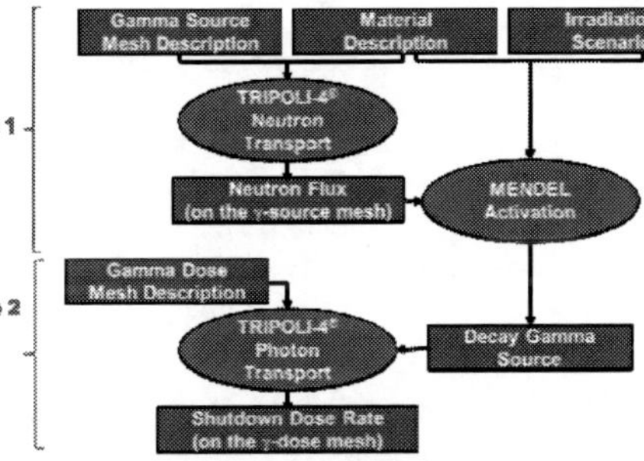

Figure 1. Flow chart of TRIPOLI-4® R2S scheme

2.1 Presentation of TRIPOLI-4® and MENDEL codes

2.1.1 *TRIPOLI-4®*

TRIPOLI-4® is a three-dimensional and continuous-energy Monte Carlo particle transport code dedicated to shielding, reactor physics with depletion, criticality safety and nuclear instrumentation. TRIPOLI-4® is currently able to simulate four kinds of particles: neutrons from 20 MeV down to 10-5 eV, photons from 50 MeV down to 1 keV, electrons and positrons from 100 MeV down to 1 keV. TRIPOLI-4® verification and validation operations are mainly based on ICSBEP [10] criticality cases

[a] Corresponding author: jean-charles.jaboulay@cea.fr

(around 1000), SINBAD [11] shielding cases (ASPIS, NESDIP, REPLICA and FNG). More particularly, numerous benchmarks on fusion applications were recently conducted [12-16].

2.1.2 MENDEL

MENDEL is a depletion code developed at CEA dedicated to radioactivity calculations related to neutron irradiation and based on deterministic depletion solvers. MENDEL is the successor of the DARWIN/PEPIN2 code [17]. MENDEL is developed in C++ with an object-oriented architecture allowing other codes to be easily coupled. For example, the depletion solvers of MENDEL are currently used in the deterministic transport code APOLLO3® [18] and in the Monte Carlo transport code TRIPOLI-4®, in order to perform depletion calculations. Recently, a new capability has been added to handle the propagation of the nuclear data uncertainties by a probabilistic approach [8] based on URANIE [19], the CEA uncertainty and sensitivity platform.

2.2 General principle of TRIPOLI-4® R2S scheme

The implemented R2S scheme is divided into three steps. In **step1a**, TRIPOLI-4® performs a neutron transport calculation in order to compute the flux and/or the energy integrated reaction rates in each region susceptible to produce decay gammas; in **step1b** MENDEL computes the decay photon sources for each region based on the neutron fluxes and/or the reaction rates calculated in step1a, as well as the specified irradiation and cooling times. Finally, in **step2** TRIPOLI-4® transports the decay photons and computes the dose rates in each region of interest.

The geometrical regions where decay gamma can be produced, corresponding to spatial regions where the neutron flux and/or the reaction rates are calculated during the first step can be defined cell by cell or by a mesh superposed to the geometry. In the current version of the code, an important hypothesis is done: each mesh cell should contains only one homogeneous material.

3 ITER Shutdown Dose Rate Benchmark

The ITER Shutdown Dose Rate (SDR) Benchmark was defined to compare the different SDR calculation schemes developed at UNED, CCFE, KIT and ENEA. The case studied is relatively simple but relevant for fusion applications. It aims to represent the typical configuration of a port plug in ITER. Focusing on a streaming path that contribute to activate a steel chamber.

The geometry considered is cylindrical (Figure 2). The radius of the outermost cylinder is 100cm. All radiation is lost beyond this cylinder. The source cell is 10cm thick. Neutron source is isotropic and emitted at 14 MeV, the source intensity is given in Table 2. There is then a gap of 100cm. The material (Table 1) section consists of an outer steel (density is equal to 7.93 g/cm^3) cylinder that is 550cm long with a 50cm radius hole through it. The first 210cm of this hole is nearly filled with a mixed steel-water (density is equal to 6.536 g/cm^3) cylinder, which itself has a 7.5cm radius hole through its centre. There is a 2cm gap between the steel-water cylinder and the outer steel cylinder. There is a 15cm thick steel plate at the end of outer steel cylinder. There is a 2cm gap between this plate and the outer cylinder. The tally cells are four concentric circular cells in a void at the rear of the material. These cells begin 30cm from the back of the steel and are 10cm thick (i.e. the centre of the tally cells is 35cm from the plate's rear face). The outer radii of these tally cells are 15, 30, 45, and 60cm respectively.

The steel and water region represent an experimental device of ITER typically steel cooled by water. The rear plate model the closure plate between the port plug and the inter-port region. In ITER SDR are calculated in the inter-port and the closure plate activation play a major role in the dose rate. The irradiation scenario (table 2) is typical to ITER experiment: low power at the beginning and high power cycles at the end of reactor life.

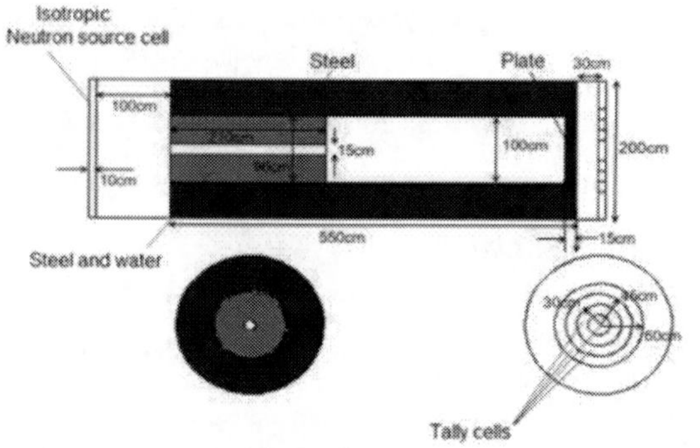

Figure 2. ITER SDR Benchmark.

Table 1. Materials definition in atom fraction.

	Steel and Water	Steel		Steel and Water	Steel
H	1,46E-1	-	Mn	1,42E-2	1,82E-2
B	4,02E-5	5,14E-5	Fe	5,03E-1	6,44E-1
C	8,14E-4	1,04E-3	Co	3,68E-4	4,71E-4
N	2,17E-3	2,78E-3	Ni	9,06E-2	1,16E-1
O	7,29E-2	6,95E-5	Cu	2,05E-3	2,62E-3
Al	8,04E-4	1,03E-3	Zr	9,52E-6	1,22E-5
Si	7,73E-3	9,89E-3	Nb	4,67E-5	5,98E-5
P	3,50E-4	4,48E-4	Mo	1,13E-2	1,45E-2
S	1,02E-4	1,30E-4	Sn	7,31E-6	9,36E-6
K	5,55E-6	7,10E-6	Ta	2,40E-5	3,07E-5
Ti	1,36E-3	1,74E-3	W	2,36E-6	3,02E-6
V	3,41E-5	4,36E-5	Pb	1,68E-6	2,14E-6
Cr	1,46E-1	1,87E-1	Bi	1,66E-6	2,13E-6

Table 2. Irradiation scenario.

Source Intensity in n/s	Duration	Number of cycles
1.0714×10^{17}	2 years	1
8.25×10^{17}	10 years	1
0	0.667 years	1
1.6607×10^{18}	1.33 years	1
0	3920 sec	17
2.0×10^{19}	400 sec	
0	3920 sec	4
2.8×10^{19}	400 sec	

4 Results

The ITER SDR benchmark was modelled with the TRIPOLI-4® R2S scheme. The FENDL-2.1 [**Erreur ! Source du renvoi introuvable.**] nuclear data library was used for the transport (neutron and photon) and the EAF2003 [21] library for the depletion calculation. The figure 3 shows the total neutron flux distribution over the ITER SDR benchmark geometry. Variance reduction techniques have been used (the neutron flux is attenuated by five decades between the front face and the rear face). The INIPOND module of TRIPOLI-4® automatically produces an importance map (figure 4). Classical techniques are used with this map: Russian roulette and splitting plus the exponential transform method.

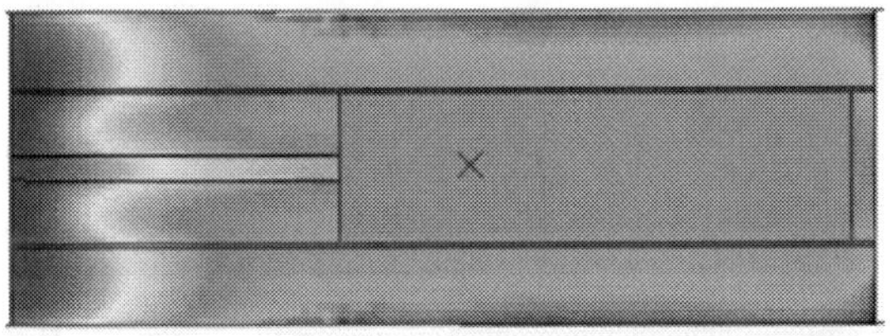

Figure 3. Total neutron flux distribution over the geometry

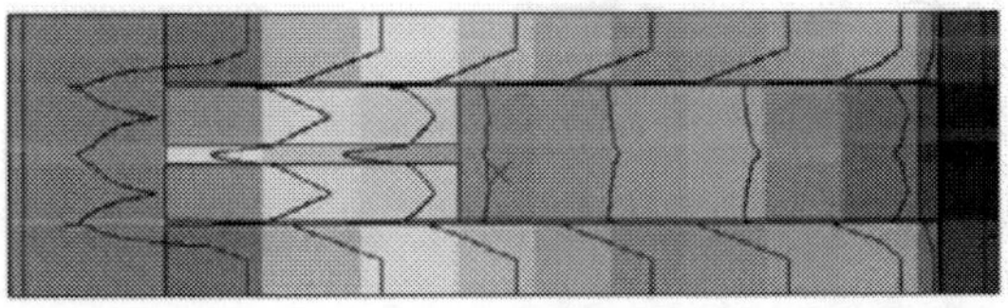

Figure 4. Importance map calculated by the INIPOND module (back curves are the iso-importance profile)

The neutron flux at the front face (facing the source) and at the rear face (near the tally region) were compared between the TRIPOLI-4® code and the MCNP5 code. The figure 5 shows a very good agreement around 0.4% for the front face and [-2%;+10%] for the rear face. These discrepancies are included in the statistical error.

Dose rate were calculated with the R2S scheme of TRIPOLI-4® and compared with R2SUNED [22] ones. Results are gathered in table 3, there is a good agreement discrepancies ranges between -2% up to + 3%. The statistical uncertainty could explain these discrepancies. The gamma dose rate is well converged in both calculation but neutron flux statistical error are around [2%; 5%] in both case. A better estimation in statistical error propagation would be necessary to conclude (§5). Another fact could explain the discrepancies: the activation library EAF2003 for TRIPOLI-4®, EAF2007 for R2SUNED. EAF2010 processing for its use in MENDEL is underway to estimate the effect of the activation library.

The dose rate contributors were investigated. The rear plate activation contributes to 99.7% except for the outermost tally (93.7%) placed in front of a streaming gap. In terms of isotopes ^{58}Co, ^{54}Mn, ^{60}Co contribute to 96% of the dose rate, respectively 58%, 28%, 10%.

Table 3. Shutdown dose rates.

Inner Radius [cm]	Outer Radius [cm]	R2SUNED dose rate [mSv/h]	R2S TRIPLI-4® dose rate [mSv/h]
0	15	16.9 ± 0.7%	17.4 ± 0.7%
15	30	13.7 ± 0.4%	13.8 ± 0.4%
30	45	9.3 ± 0.4%	9.1 ± 0.4%
45	60	5.8 ± 0.4%	5.8 ± 0.4%

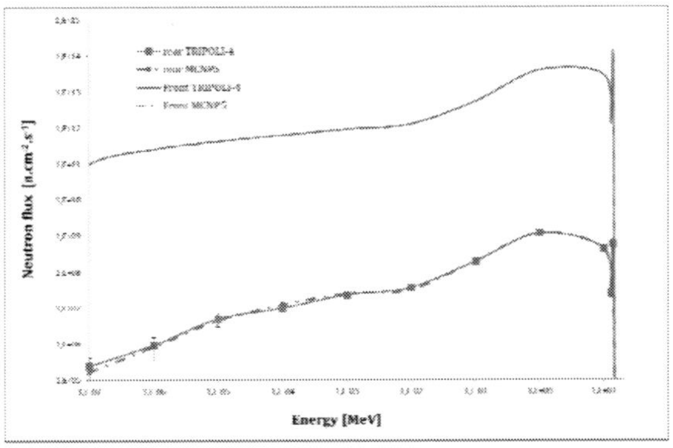

Figure 5. Neutron flux spectrum calculated with TRIPOLI-4® and MCNP5 at the front face and the rear face of the geometry

5 Futures developments

One important issue of the classical R2S approach is the materials handling inside a mesh. Currently, in each mesh only one material can be treated if not some bias may occur [22]. The mesh tallies of TRIPOLI-4® will be improved to calculate the flux and other quantities in each material of in a mesh-cell, this development will enable the materials handling inside mesh.

To estimate the doses uncertainties due to statistics (statistical error of the decay gamma simulation plus propagation of the statistical error on neutron flux) independent simulations will be used (making N complete R2S calculations with a different seed initialisation, Figure 6).

 DOI: 10.1051/epjconf/201715302008

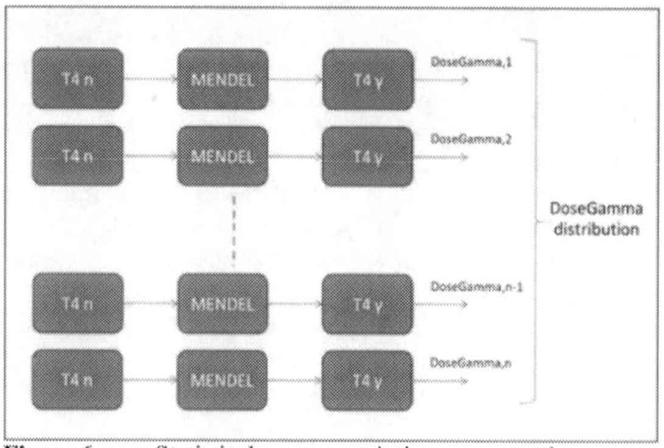

Figure 6. Statistical uncertainties propagation scheme (independent replicas)

Finally the MENDEL calls will be parallelized to accelerate the depletion calculation.

6 Conclusions

A shutdown dose rate calculation scheme was implemented in the TRIPOLI-4® code. It is based on the rigorous-two-step methodology (first step: neutron transport + depletion calculation, second step: decay gamma transport). The MENDEL inventory code was used for depletion calculation. This paper presents a numerical benchmark (ITER SDR benchmark) to validate the TRIPOLI-4® R2S. Results are in good agreement with the other participant (UNED). Improvement of the R2S scheme are planned: to accelerate the calculation (MENDEL parallelization), to propagate the statistical error of the first step and to handle different materials in a mesh-cell. Extended validation exercises will be carried out with the Frascati Neutron source Generator SDR experiment [11] and the JET SDR measurement [23]. TRIPOLI-4® R2S should be used in near-term for ITER neutronic analysis at CEA [24]. The R2S feature should be available in the next version of TRIPOLI-4® (v11, 2018). Beyond fusion applications, this new feature will be by very helpful for fission applications like reactor decommissioning analysis, dosimetry and activation calculation.

Acknowledgments

This work has been carried out within the framework of the EUROfusion Consortium and has received funding from the Euratom research and training programme 2014-2018 under grant agreement No 633053. The views and opinions expressed herein do not necessarily reflect those of the European Commission. TRIPOLI® is a registered trademark of CEA. We gratefully acknowledge EDF for their long term partnership and AREVA for their support. The authors warmly thanks the researcher of UNED (J.P. Catalan, J. Sanz, P.Sauvan) for their support on the ITER SDR benchmark and E. Dumonteil who started the implementation of the R2S scheme in TRIPOLI-4®.

References

1. L. Petrizzi, Fus. Eng. & Des. **81**, 1417 (2006)
2. R. Pampin, Fus. Eng. & Des. **88**, 454 (2013)
3. L. Petrizzi, Proc. ANS MC-2000, Lisbon, Oct. 2000
4. A. Davis, Fus. Eng. & Des. **85**, 87 (2010)
5. M. Majerle, Fus. Eng. & Des. **87**, 443 (2012)
6. J.P. Catalan, Fus. Eng. & Des. **88**, 2088 (2013)
7. http://www.oecd-nea.org/tools/abstract/detail/nea-1716/
8. A. Tsilanizara, Ann. of Nucl. Ener. **90**, 62 (2016)
9. https://www.euro-fusion.org/
10. http://www.oecd-nea.org/science/wpncs/icsbep/handbook.html
11. https://www.oecd-nea.org/science/wprs/shielding/
12. Y. K. Lee, Fus. Eng. & Des. **85**, 1125 (2010)
13. Y. K. Lee, Fus. Eng. & Des. **86**, 2246 (2011)
14. C. Fausser, Fus. Eng. & Des. **86**, 2135 (2011)
15. J.C. Jaboulay, Fus. Eng. & Des. **89**, 1195 (2014)
16. J.C. Jaboulay, Fus. Eng. & Des. **98**, 1800 (2015)
17. A. Tsilanizara, J. of Nucl. Sci. & Tech. **1**, 845 (2000)
18. H. Golfier. Proc. M&C 2009, Saratoga Springs
19. F. Gaudier, Proc.-Soc. & Behav. Sci. **2**, 7660 2010
20. D. Lopez, IAEA, Rep. INDC (NDS)-467, Dec. 2004
21. EAF2003
22. P. Sauvan, TNS, 2015
23. R. Villari, Fus. Eng. & Des. **109-111**, 895 (2016)
24. D. Blanchet, this conference

Validation of Shielding Analysis Capability of SuperMC with SINBAD

Chaobin Chen, Qi Yang, Bin Wu, Yuncheng Han, Jing Song*, FDS Team

Key Laboratory of Neutronics and Radiation Safety, Institute of Nuclear Energy Safety Technology, Chinese Academy of Sciences, Hefei, Anhui, 230031, China

* Corresponding Author, E-mail: jing.song@fds.org.cn

Abstract: The shielding analysis capability of SuperMC was validated with the Shielding Integral Benchmark Archive Database (SINBAD). The SINBAD was compiled by RSICC and NEA, it includes numerous benchmark experiments performed with the D-T fusion neutron source facilities of OKTAVIAN, FNS, IPPE, etc. The results from SuperMC simulation were compared with experimental data and MCNP results. Very good agreement with deviation lower than 1% was achieved and it suggests that SuperMC is reliable in shielding calculation.

KEYWORDS: Monte Carlo, SuperMC, SINBAD, Validation

I. Introduction

SuperMC is a general purpose, easy-to-use Monte Carlo software for modeling and simulation of nuclear facilities. The latest version of SuperMC can perform the simulation of neutron, photon and coupled neutron and photon transport, and also equipped with the function of automatic modeling and visualization[1-2].

For neutron transport, elastic scattering, inelastic scattering, and various absorption reaction processes includes (n, xn), (n, f), (n, γ), etc., with neutron energies from 0 MeV to 150 MeV, are considered. For photons, the code accounts for Compton scattering, coherent scattering, possibility of fluorescent emission after photoelectric absorption, electron-positron pair production and photonuclear processes with photon energies from 1 keV to 100 GeV. SuperMC uses high-resolution pointwise cross-section databases created from ENDF/B-6 format evaluated nuclear data libraries.

SINBAD is a compilation of 77 fission and fusion shielding experiments, most of which have been modeled with Monte Carlo (McBEND, TRIPOLI, MCNP) or deterministic codes (ANISN and DOT3.5). Shielding analysis is a typical application of SuperMC, and the primary objective for the radiation shielding validation is to assess the impact of SuperMC methodology and its neutron and photon data libraries on attenuation factors and dose rates. D-T fusion neutron source benchmark cases were selected to validate the shielding analysis capability of SuperMC.

II. Validation Method

SuperMC has been extensively tested and validated in-house over the years with benchmark cases from criticality safety experiments[3] and some typical radiation shielding cases, like ITER benchmark problems[4]. In this work, SuperMC transport functions were validated against benchmark experiments from three D-T neutron source facilities compiled in SINBAD. The problems with time-of-flight spectra were chosen to test several features of the code, such as neutron and photon transport, secondary photon production, coupling of the neutron-photon transport, point detector, boundary crossing detector or surface detector, and angular flux. In the next work, dose rates from neutrons and photons would be concerned.

A brief description of selected experiments was given below, and the calculation results were presented in the next section.

1. Osaka experiments

Neutron leakage spectra from different diameters of iron, aluminum, silicon, nickel and tungsten sphere were measured between 10 keV and 15 MeV by time-of-flight techniques. Some measurements were carried out for two energy regions, the high energy region for high energy spectrum and the low energy region for low energy spectrum, such as iron and nickel sphere experiments. In some cases, gamma-ray leakage spectra were measured, such as aluminum, silicon and tungsten sphere experiments.

2. FNS experiments

Materials including graphite, oxygen, vanadium, iron, tungsten, lead, etc., had been performed on the FNS facility. The neutron spectra emerging from slabs of material of varying thickness were measured at five different angles. The slabs were placed at 20 cm distance from a 14 MeV D-T neutron source.

3. IPPE experiments

Time-of-flight measurements of neutron leakage spectra from iron, nickel, vanadium, bismuth, uranium and thorium spheres with central 14MeV neutron sources and from uranium and thorium sphere with ^{252}Cf sources were performed.

III. Results

Benchmark experiments were modeled by modeling function of SuperMC, and all the problems were run on single CPU computers with ENDF/B-VII.1 neutron cross section library and ENDF/B-VI.8 photon cross section library. Integrated results over energy were dicussed and typical graphs were presented in this section.

1. Osaka benchmarks

SuperMC was used to simulate Osaka benmarks to get the neutron and photon leakage spectra from Al, Si, Fe, Ni and W spheres. The experiments were modeled with a simple spherical geometry, and two isotropic point sources, a neutron source and a photon source, located in a spherical air cavity at the center of the spheres. The results were compared with measured data and calculated values from MCNP. The integrated leakage fluxes over energy were in good greement with experimental data and within 0.1% with MCNP results. Leakage spectrums from 50cm iron sphere and 16cm nickel sphere are shown in Fig. 1.

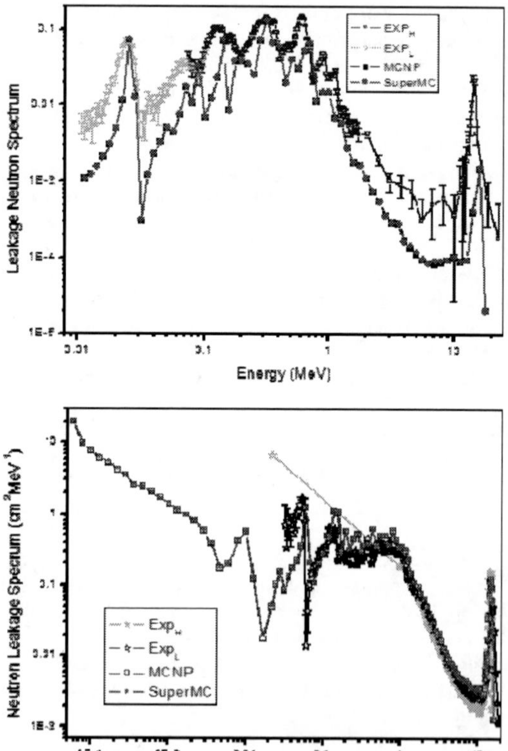

Fig.1 Comparison of neutron leakage currents of Osaka benchmark experiments (Up: Iron, Down: Nickel)

2. FNS benchmarks

The FNS time-of-flight benchmark specifications are given on the IAEA web site[5] and in SINBAD. Experiments with materials of C, O, V, W were simulated using SuperMC, with neutron source spectrum taken from the reference[6] and an isotropic neutron distribution assumed. The differences of point integrated flux over energy are within 1% compared with calculated results by MCNP and the point flux spectra of 20cm oxygen slab in two angles are shown in Fig. 2.

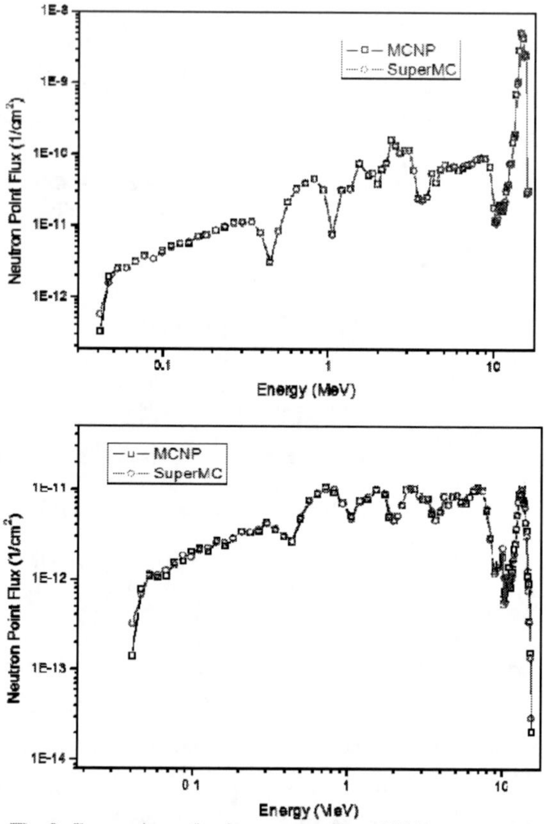

Fig. 2. Comparison of point neutron flux of FNS oxygen slab benchmark experiments (Up: 0 degree, Down: 66.8 degree)

3. IPPE benchmarks

V and Th spheres were simulated using SuperMC with 14MeV neutron source. The differences of point integrated flux over energy were about 0.1% compared with calculated results by MCNP and the point flux spectra of thorium sphere are shown in Fig. 3.

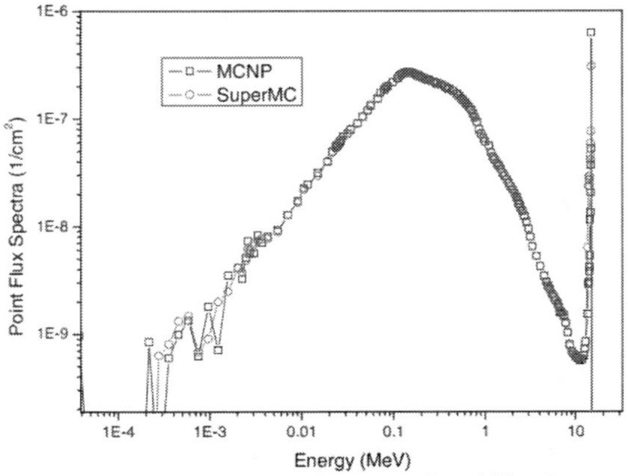

Fig. 3. Comparison of point neutron flux of Th sphere experiments with 14MeV neutron source

IV. Conclusion

SuperMC had been verified with D-T fusion neutron source benchmark cases from SINBAD, and excellent agreement has been obtained between SuperMC and MCNP. The correctness and reliability of SuperMC transport function and high accessibility of modeling function were fully validated.

Acknowledgment

The work was supported by the Strategic Priority Research Program of Chinese Academy of Sciences (No. XDA03040000), the National Special Program for ITER (No. 2014GB112001), the National Natural Science Foundation of China (No.11305205).

References

1) WU YC, SONG J, ZHENG HQ, et al, CAD-based Monte Carlo Program for Integrated Simulation of Nuclear System SuperMC, *Annuals of Nuclear Energy*, DOI: 10.1016/j.anucene.2014.08.058.
2) WU YC, FDS TEAM, CAD-based Interface Programs for Fusion Neutron Transport Simulation, *Fusion Engineering and Design*, 84, 1987 (2009).
3) Binhang Zhang, Criticality Validation of SuperMC with ICSBEP, *Annals of Nuclear Energy*, 87, 494–499 (2016).
4) Jing Song, Benchmarking of CAD-based SuperMC with ITER benchmark model, *Fusion Engineering and Design*, 89, 2499–2503 (2014).
5) www-nds.iaea.or.at/fendl2/validation/benchmarks/jaerim94014/fns-tof
6) M.Z. Youssef, Verification of ITER shielding capability and FENDL data benchmarking through analysis of bulk shielding experiment on large SS316/water assembly bombarded with 14 MeV neutrons, *Fusion Engineering and Design*, 42, 235–245 (1998).

Examples of Use of SINBAD Database for Nuclear Data and Code Validation

Ivan Kodeli[1,a], Gašper Žerovnik[1,b] and Alberto Milocco[2]

[1]*Jožef Stefan Institute, 1000 Ljubljana, Slovenia*
[2]*University of Milano Bicocca, Piazza della Scienza 3, 20126,Milan, Italy*

Abstract. The SINBAD database currently contains compilations and evaluations of over 100 shielding benchmark experiments. The SINBAD database is widely used for code and data validation. Materials covered include: Air, N. O, H_2O, Al, Be, Cu, graphite, concrete, Fe, stainless steel, Pb, Li, Ni, Nb, SiC, Na, W, V and mixtures thereof. Over 40 organisations from 14 countries and 2 international organisations have contributed data and work in support of SINBAD. Examples of the use of the database in the scope of different international projects, such as the Working Party on Evaluation Cooperation of the OECD and the European Fusion Programme demonstrate the merit and possible usage of the database for the validation of modern nuclear data evaluations and new computer codes.

1 Introduction

Benchmark experiments can serve several purposes and the availability of suitable experimental data is fundamental to ensure that modelling tools meet the requirements of nuclear industry and licensing needs. Essentially, they are used for:
- validation of nuclear cross-section data,
- validation of nuclear codes and models used,
- qualification and training of users through benchmarking and code comparison exercises.

The Shielding INtegral Benchmark Archive and Database (SINBAD) [1-3] project was started in the early 1990's at the OECD Nuclear Energy Agency Data Bank (NEA DB) and the Radiation Safety Information Computational Center (RSICC) as a collection of quality experiments for validating and benchmarking computer codes and data for radiation transport and shielding applications. Currently, the SINBAD database comprises over 100 benchmark compilations and evaluations of relevance to:
- reactor shielding, pressure vessel dosimetry (48)
- fusion blanket neutronics (31)
- accelerator shielding (23)

In order to facilitate their use, the data are organized in the standardized format structure. The index and the master document (called abstract) use the Hyper Text Markup Language (HTML) format with links to text files, figures, Portable Document Format (pdf) documents and computer code inputs. The format allows easy access and forward/backward navigation among the data and has proved to be suitable, general enough and above all easy (and low-cost) to maintain.

In addition to the characterization of the radiation source, description of the shielding set-up, the instrumentation and the relevant detector measurements, most sets in SINBAD contain also the deterministic or Monte Carlo radiation transport computer model used for the interpretation of the experiment and, where available, results from uncertainty analysis. The set of primary documents used for the benchmark compilation and evaluation are provided in computer readable form.

The database is intended for different types of users, including nuclear data evaluators, computer code developers, experiment designers and university students. SINBAD is available from the NEA Data Bank and RSICC (see https://www.oecdnea.org/science/wprs/shielding/sinbad/). Up to now several hundreds of SINBAD packages were distributed from the NEA DB and RSICC.

Since 2011 the maintenance and further development of the database is coordinated in the scope of the OECD NEA Nuclear Science Committee (NSC) Working Party on Scientific Issues of Reactor Systems (WPRS) Expert Group on Radiation Transport and Shielding (EGRTS). A key objective of the group is to identify, evaluate and preserve experimental data on shielding benchmarks. The Expert Group maintains close links with the International Reactor Physics Experiment Evaluation (IRPhE) and International Criticality Safety Benchmark Evaluation (ICSBEP) projects, as well as in co-ordination with the JEFF and EFF NEA Data Bank projects and with the SATIF community of accelerator shielding specialists. In the last years, much effort was devoted to the quality review of the available benchmark information. Setting up of the official SINBAD Quality Evaluation Group

[a] Corresponding author: ivo.kodeli@ijs.si
[b] Currently at EC-JRC-IRMM, Retieseweg 111, B-2440 Geel, Belgium

such as the groups established for the criticality benchmarks (ICSBEP) and reactor physics experiments (IRPhE) is planned. It would be based on the pilot evaluation group already actively performing the quality evaluations of SINBAD. The exact details on the peer review process for the future quality evaluations is still to be agreed upon.

2 Examples of nuclear data and code validation using SINBAD

The paper covers a few examples of recent analysis of the SINBAD benchmarks, demonstrating how the SINBAD data can be used for modern neutron/gamma nuclear data and computer code validation. The activities were done in the scope of several international projects such as the Fusion for Energy, the Working Party on Evaluation Cooperation of the OECD (WPEC) [4] subgroup SG40 (CIELO), WPEC SG39 and the International Atomic Energy Agency (IAEA). WPEC-SG40 was set up to focus on the evaluated nuclear data of the major nuclides in reactor technology, namely ^{1}H, ^{16}O, ^{56}Fe, ^{235}U, ^{238}U and ^{239}Pu. WPEC-SG39 entitled "Methods and approaches to provide feedback from nuclear and covariance data adjustment for improvement of nuclear data files", also expressed interest in using SINBAD iron benchmarks to study nuclear data adjustment methods. Fusion for Energy (F4E) is the European Union's Joint Undertaking responsible for providing Europe's contribution to ITER and the Development of Fusion Energy.

Several examples of recent nuclear data validation are presented, including materials such as iron and oxygen, for both neutrons and photons. The data were used also to for computer code validation. In some cases, the analysis was complemented with the sensitivity and uncertainty analyses to provide an insight into the role of the basic data used and help in the interpretation and understanding of the physical phenomena involved.

2.1 Iron cross-section validation

Due to their importance and complexity, the reaction cross sections of the iron isotopes traditionally attracted considerable attention and have been evaluated and measured several times. Altogether 25 SINBAD benchmarks include iron or steel as shielding material. The agreement between measurements and calculations of integral benchmarks using the present evaluations is reasonable, except at certain energy ranges. The comparison is however hindered by the fact that many of the experiments are old and the authors were not able (or willing) to provide additional information to improve the quality of the integral experiments, which would allow detailed explicit modelling of the experimental setup and reduce the uncertainties. Furthermore, much of the data testing refers to energies below 20 MeV.

A recently evaluation is ongoing in the scope of the Working Party on Evaluation Cooperation of the OECD set up a subgroup WPEC-SG40 (alias CIELO). Iron cross-section evaluation and validation is also part of the recent F4E activities. Furthermore, the ASPIS-Iron88 iron benchmark (see Fig. 1) study was also used in the scope of the WPEC WG39 [5, 6] including complete sensitivity and uncertainty analysis combined with data adjustment.

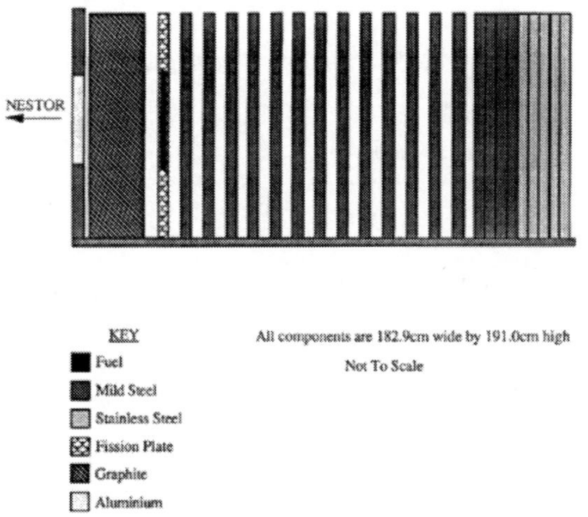

Figure 1. Schematic view of the Aspis Irion88 benchmark experimental shield.

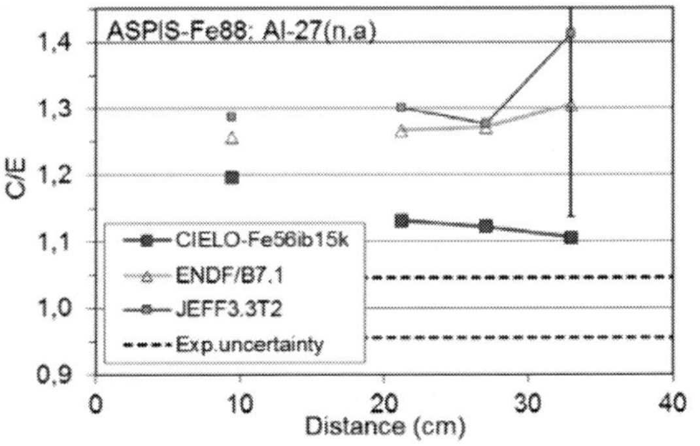

Figure 2. Relative axial ^{27}Al(n,α) reaction rate distribution for Aspis Iron88 benchmark. Comparison between experimental data and Monte Carlo simulation using different cross sections for ^{56}Fe. CIELO results refer to the "ib88" evaluation. Dashed lines delimit the ± 1 σ standard deviations of the measurements. ± 1 σ computational uncertainty due to the nuclear data is also shown with the ENDF/B-VII.1 results (at 35 cm).

Among the recently revised SINBAD integral benchmarks sensitive to iron cross sections the ASPIS-Iron88, JANUS Phase 1, NESDIP-3 and EURACOS Fe were selected for the validation of the recent iron evaluations. Comparison between the measured and calculated ^{32}S(n,p) reaction rate distributions for the Aspis Iron88 and JANUS-1 iron shielding benchmarks are presented in Figs. 2 – 4. Different ^{56}Fe cross sections (ENDF/B-VII.1 [7], JEFF-3.3T2, and two recent CIELO evaluations "ib04s" and "ib88" [8]) were used in the MCNP-6 simulation of the benchmark. For all other materials, ENDF/B-VII.1 was used. In particular, compared to the ENDF/B-VII.1, the version "ib04s" includes more detailed angular distributions of the first-level inelastic scattering, while "ib88" includes a new optical model evaluation of the fast energy.

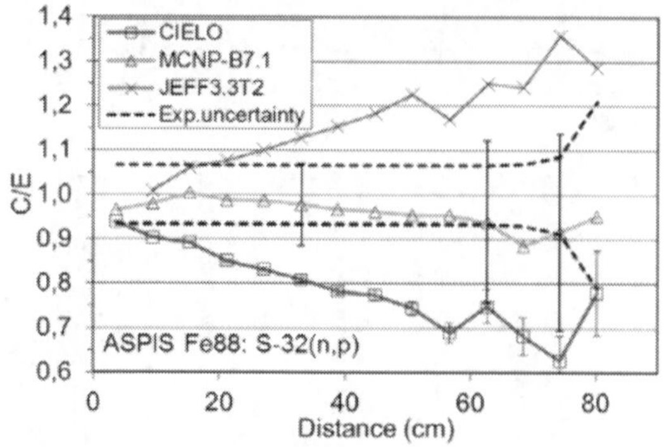

Figure 3. Relative axial $^{32}S(n,p)$ reaction rate distribution for Aspis Iron88 benchmark. Comparison between experimental data and Monte Carlo simulation using different cross sections for ^{56}Fe. CIELO results refer to the "ib88" evaluation. Dashed lines delimit the ± 1 σ standard deviations of the measurements. ± 1 σ computational uncertainty due to the nuclear data is also shown with the ENDF/B-VII.1 results.

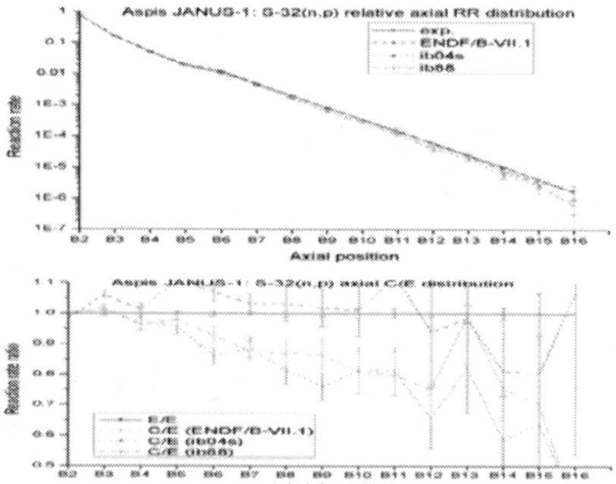

Figure 4. Relative axial $^{32}S(n,p)$ reaction rate distribution for Aspis JANUS-1. Comparison between experimental data and Monte Carlo simulation using different cross sections for ^{56}Fe.

2.2 Oxygen cross-section validation

The FNS Liquid Oxygen benchmark [9] performed at the 14 MeV D-T neutron facility at FNS/JAERI was considered as suitable for the validation of the new oxygen evaluations prepared in the scope of the CIELO project. In the benchmark the angular neutron leaking spectra from a 20 cm slab of liquid oxygen in the 0.05 - 15 MeV energy range at different angular directions (0, 12.2, 24.9, 41.8 and 66.8 degrees) were measured using the Time of Arrival (TOA) technique.

In 2010 the benchmark was re-analysed and SINBAD compilation updated including complete new revision of time vs. energy domain computational models. The benchmark was found to be of benchmark quality, however more information would be needed on the uncertainty in the neutron effective flight path parameter, on detector efficiency function, details on conversion of experimental TOA to energy spectra.

The analysis was performed using the MCNP-6 Monte Carlo code in the energy and time domain for the oxygen cross sections taken from ENDF/B-VII.1 and new CIELO evaluations by Luiz Leal and Gerry Hale [8]. Results of calculations performed in time domain and converted to energy are presented in Fig. 5. In the past the benchmark was also analysed using DORT with first collision source (GRTUNCL). This approach is suitable for cross-section sensitivity-uncertainty analysis.

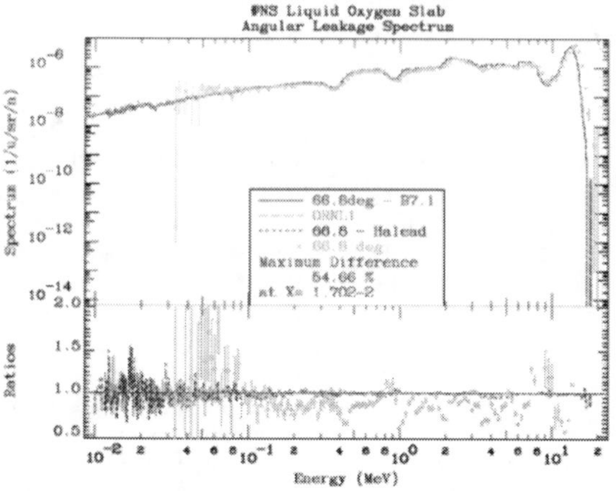

Figure 5. Neutron spectra calculated using the ENDF/B-VII.1 and new evaluations by Luiz Leal (file ORNL1) and Gerry Hale (file "O16_halead") [6] compared to the measured at the FNS-O benchmark at the 66^0 angle.

The main conclusions are the following:

- Relatively good agreement was found between the measured and calculated spectra for Leal, Hale and ENDF/B-VII.1 evaluations, with no significant trends with increasing angle.
- Little difference observed between spectra calculated using MCNP6 with ENDF/B-VII.1, Luiz and Hale files suggests that the benchmark may not be sufficiently selective in this particular case.

2.3 Photon data validation

In the scope of the F4E project the photon libraries available in the MCNP-6 code package were studied, in particular the following sets [10]:

- 04p: ENDF/B-VI.8 Photoatomic Data (MCNP standard library),
- 84p: Update of MCPLIB04 Photon Compton Broadening Data,
- 12p: new library (extended to lower energies).

SINBAD includes two sets of Photon Leakage Spectra benchmarks measured at RFNC-VNIITF (Zababakhin Russian Federal Nuclear Center – All-Russian Scientific Researching Institute of Technical Physics), Snezhinsk [11]:

- NEA-1517/74: Photon Leakage Spectra from Al, Ti, Fe, Cu, Zr, Pb, ^{238}U Spheres with a Central 14-MeV Neutron Source

• NEA-1517/80: RFNC Photon Spectra from H_2O, SiO_2 and NaCl with a Central 14-MeV Neutron Source.

Both benchmarks used 14-MeV neutron source placed in the center of spherical samples of inside diameter 100 mm and outside diameter 200 mm (see Fig. 6). Absolute measurements were conducted using the activation method and the reaction $^{27}Al(n,\alpha)^{24}Na$. The photon spectrum measurements were conducted using a scintillation detector having a stilbene crystal with dimensions of 60 × 60 mm. Measured energy range is 0.3 - 8.0 MeV. Energy resolution for ^{60}Co lines (1.17 and 1.33 MeV) was about 10%. At energies higher than 3 MeV it was 6-7% and at energies less than 0.5 MeV it was 15-20%.

The results are shown on Fig. 7 indicating good consistency between the measured and calculated spectra. Almost no differences were observed between the 04p, 84p and 12p data for the studied cases (iron and water).

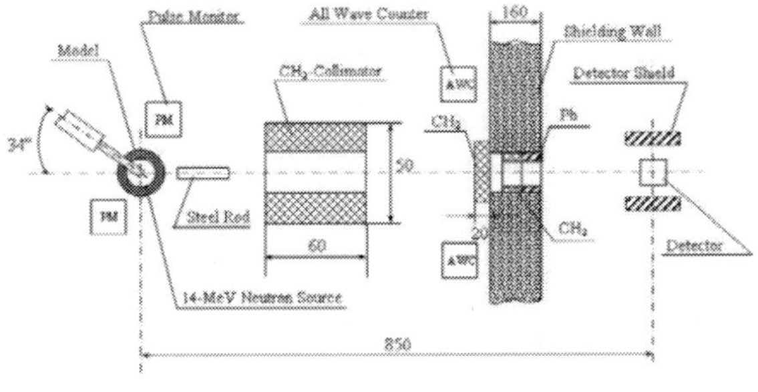

Figure 6. Experimental arrangement of the RFNC photon spectra experiment.

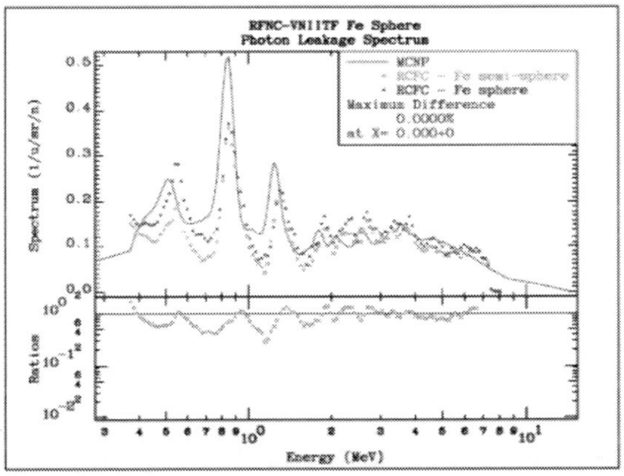

Figure 7. Measured and calculated gamma spectra for the RFNC-VNIITF Iron sphere measurements.

2.4 IRDF-2002 and IRDFF dosimetry data validation

The IAEA Research Coordination Project on "Testing and Improving the International Reactor Dosimetry and Fusion File (IRDFF)" [12] is aimed at updating the library of dosimetry cross sections. Several benchmark experiments performed at the ENEA Frascati Neutron Generator (FNG), and in ASPIS, AEA Technology, Winfrith included in SINBAD were analysed to compare the performance of the new IRDFF and previous IRDF-2002 libraries, to check for improvements between measured and calculated reaction rates and removal of some inconsistent trends in the results for different monitors. An example of results is shown in Fig. 8 below.

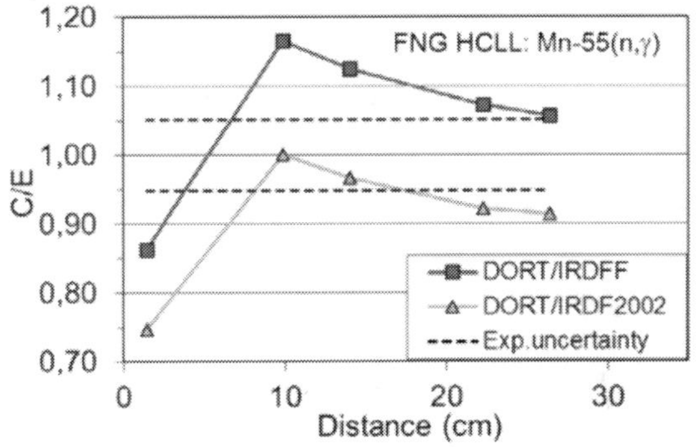

Figure 8. Calculated/Experimental (C/E) $^{55}Mn(n,\gamma)$ detector responses for the FNG Bulk-shield benchmark based on calculations with IRDFF and IRDF-2002 libraries. Dashed lines delimit the ± 1 σ standard deviations of the measurements.

2.5 Computer code testing

The MCNP model of FNG benchmarks presently included in SINBAD makes use of the ENEA-JSI source subroutine. This approach requires a recompilation of MCNP, therefore access to MCNP Fortran source code. As an alternative the replacement of the MCNP subroutine with the MCUNED [13] model was studied, allowing an explicit modelling of DT reactions (see Fig. 9).

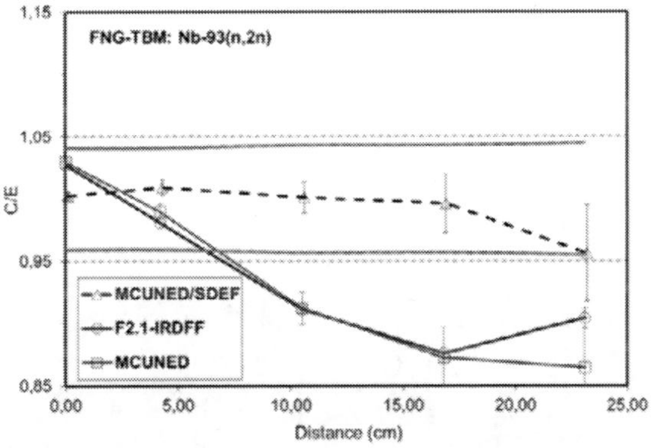

Figure 9. Validation of the MCUNED using the FNG HCPB benchmark experiment.

3 Conclusions

The SINBAD database currently contains compilations and evaluations of experiments for 48 reactor shielding problems, 31 for fusion neutronics shielding and 23 for accelerator shielding cases. 17 experiment re-evaluations to be updated. Revision and classification of benchmark experiments according to the completeness and reliability of information is ongoing in order to provide the users more detailed information on experimental setup, measurements and corresponding uncertainties and in this way facilitate the use of these data.

Results of the analysis of several SINBAD benchmark experiments demonstrate that these data can be useful for modern nuclear data and code validation, provided additional effort is invested in obtaining additional information on the measurements and in developing more detailed computational models for transport calculations. Sensitivity and uncertainty analysis provide a valuable insight into the importance of different nuclear data and reactions involved in the measurements.

Acknowledgments

The work leading to this publication has been funded partially by Fusion for Energy under the Specific Grant Agreement F4E-FPA-168-02. This publication reflects the views only of the authors, and Fusion for Energy cannot be held responsible for any use, which may be made of the information contained therein. Authors would also like to express their gratitude to the European Commission for the partial financial support of the CHANDA project (FP7-Fission-2013 – 605203),

References

1. I. Kodeli, A. Milocco, P. Ortego, E. Sartori, 20 Years of SINBAD (Shielding Integral Benchmark Archive and Database), *Progress in Nuclear Science and Technology*, Volume **4** (2014) pp. 308-311.
2. I. Kodeli, A. Milocco, A. Trkov, Lessons Learned from the TOF-Benchmark Intercomparison Exercise within EU Conrad Project (How Not to Misinterpret a TOF-Benchmark), *Nuclear Technology*, **168** (2009) 965-969.
3. A. Milocco, I. Kodeli, A. Trkov, The 2010 Compilation of SINBAD: Quality Assessment of the Fusion Shielding Benchmarks, Proc. NEMEA-6 Scientific workshop on Nuclear Measurements, Evaluations and Application, Krakow, Poland, 25-28 October 2010.
4. Working Party on International Nuclear Data Evaluation Co-operation (WPEC), https://www.oecd-nea.org/science/wpec/
5. I. Kodeli, SINBAD ASPIS-IRON88 benchmark results using different iron nuclear data including JEFF3.3T2, Processing, Verification & Benchmarking, Validation, OECD/NEA, Nov. 30, 2016, Paris.
6. G. Palmiotti1, M. Salvatores, M. Hursin, I. Kodeli, F. Gabrielli, A. Hummel, New approaches to provide feedback from nuclear and covariance data adjustment for effective improvement of evaluated nuclear data files, ND2016 Conference, Bruges, Belgium, 11-16/09/2016.
7. M.B. Chadwick et al, ENDF/B-VII.1 Nuclear Data for Science and Technology: Cross Sections, Covariances, Fission Product Yields and Decay Data, *Nucl. Data Sheets* **112** (2012) 2887.
8. IAEA CIELO web page: https://www-nds.iaea.org/CIELO/
9. Y. Oyama, S. Yamaguchi, H. Maekawa: "Experimental Results of Angular Neutron Flux Spectra Leaking from Slabs of Fusion Reactor Candidate Materials (I)," JAERI-M 90-092 (1990).
10. B. Colling, I. Kodeli, L.W. Packer, Benchmarking comparison and validation of MCNP photon interaction data, ND2016 Conference, Bruges, Belgium, 11-16/09/2016.
11. A.I. Saukov, E.N. Lipilina, V.D. Lyutov, “Measurements of Neutron and Photon Leakage from Spherical and Hemispherical Samples with a Central 14-MeV Neutron Source as a Possible Type of Benchmarks", Proc. Int. Conf. Radiation Safety, ICRS10–RPS-2004, May 9-14, 2004, Madeira, Portugal.
12. Capote, R.,Zolotarev, K.I., Pronyaev, V.G., Trkov, A., Journal of ASTM International (JAI)- Vol. 9, Issue 4, April 2012, JAI104119.
13. MCUNED: A MCNPX Extension for Using light Ion Evaluated Nuclear Data library- NEA-1859.

Neutron Deep Penetration Calculations in Light Water with Monte Carlo TRIPOLI-4® Variance Reduction Techniques

Yi-Kang Lee [1*]

[1]*DEN, Service d'étude des réacteurs et de mathématiques appliquées, CEA, Université Paris-Saclay, F-91191 Gif-sur-Yvette, France*
*Corresponding Author, E-mail: yi-kang.lee@cea.fr

Abstract

Nuclear decommissioning takes place in several stages due to the radioactivity in the reactor structure materials. A good estimation of the neutron activation products distributed in the reactor structure materials impacts obviously on the decommissioning planning and the low-level radioactive waste management. Continuous energy Monte-Carlo radiation transport code TRIPOLI-4 has been applied on radiation protection and shielding analyses. To enhance the TRIPOLI-4 application in nuclear decommissioning activities, both experimental and computational benchmarks are being performed. To calculate the neutron activation of the shielding and structure materials of nuclear facilities, the knowledge of 3D neutron flux map and energy spectra must be first investigated. To perform this type of neutron deep penetration calculations with the Monte Carlo transport code, variance reduction techniques are necessary in order to reduce the uncertainty of the neutron activation estimation. In this study, variance reduction options of the TRIPOLI-4 code were used on the NAIADE 1 light water shielding benchmark. This benchmark document is available from the OECD/NEA SINBAD shielding benchmark database. From this benchmark database, a simplified NAIADE 1 water shielding model was first proposed in this work in order to make the code validation easier. Determination of the fission neutron transport was performed in light water for penetration up to 50 cm for fast neutrons and up to about 180 cm for thermal neutrons. Measurement and calculation results were benchmarked. Variance reduction options and their performance were discussed and compared.

KEYWORDS: TRIPOLI-4 Monte Carlo Code, Light water shielding, Neutron transport, Variance reduction

I. Introduction

Nuclear decommissioning is one of the most important activities in current nuclear industry. Generally it takes place in several stages due to the radioactivity in the nuclear facilities. The decontamination work can be actively performed to reduce the removable radioactive materials from the contaminated zones. The radioactivity caused by the neutron activation of shielding and structure materials can be actively removed or passively treated with the cooling time. A good estimation of the neutron activation products distributed in the shielding and structure materials impacts obviously 1) the decommissioning techniques, planning, and costs, 2) the radiation protection implementation, and 3) the low-level radioactive waste management [1, 2].

The continuous-energy Monte-Carlo radiation transport code TRIPOLI-4 has been used for radiation protection and shielding analyses [3]. To develop its application on nuclear decommissioning activities, both experimental and computational benchmarks are being developed and performed. For the neutron activation calculation of shielding and structure materials of nuclear facilities, the knowledge of 3D neutron flux map and the energy spectra must be first investigated. In this kind of neutron deep penetration calculation in shielding, variance reduction techniques are necessary in order to reduce the uncertainty of the Monte-Carlo calculations [3, 4].

Different water shielding benchmarks are available from the OECD/NEA SINBAD database. The NAIADE 1 water benchmark was essential because of its important water thickness [5, 6]. In this study, the fission neutron deep penetration in water shielding was investigated with TRIPOLI-4 code for NAIADE 1 water benchmark. In order to make the code validation easier, a simplified shielding model of the NAIADE 1 water experiment was first developed in this work from the 2007 published fixed-source sub-criticality model [6]. Calculations were performed in water for penetration up to 50 cm for fast neutrons and up to 180 cm for thermal neutrons. Measurement and calculation results for equivalent fast and thermal neutron flux of ^{31}P(n,p), ^{32}S(n,p), and ^{10}B(n,α) were benchmarked. Different variance reduction techniques based on the INIPOND module of TRIPOLI-4 code were investigated. Their performances were compared and discussed.

II. NAÏADE 1 light water benchmark

NAÏADE 1 shielding experiments were realized at Fontenay -aux-Roses (France) during the 1960s and 1970s. These experiments were designed to study the fission and thermal neutron flux attenuation in various shielding materials including water, graphite, iron and concrete. The related calculation benchmarks were reassessed with TRIPOLI-4 Monte Carlo code by J.-C. Nimal [5-6]. Fig. 1 shows the configuration of the NAIADE 1 experiments. Neutron source in NAIADE 1 was produced from a natural uranium fission plate converter irradiated by a collimated thermal neutrons beam coming from the graphite reflector of the ZOE heavy water reactor. The thermal column of ZOE reactor is shown in the right side of Fig. 1. The fission plate and its Boral screen were installed in the central part. The removable shielding tank filled with light ordinary water is presented in the left side of Fig. 1.

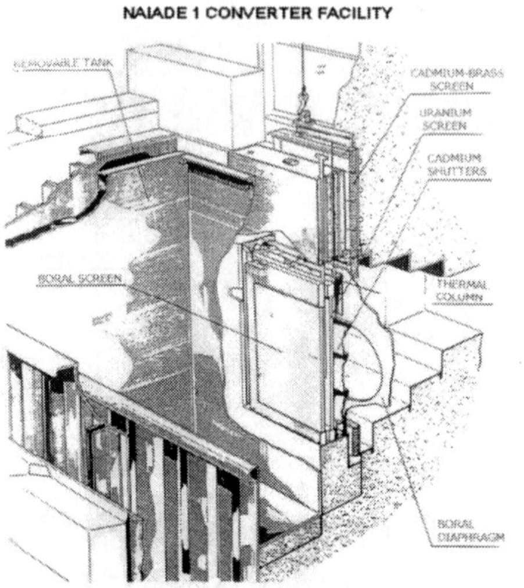

Figure 1: NAIADE 1 shielding experiment facility (Right - Thermal neutron column from ZOE reactor core, Middle - Natural Uranium fission plate and Boral screen, Left – Removable shielding tank for light water) [5, 6]

The water along the fission plate axis was 250 cm deep. The square section of the tank was 300 cm x 300 cm. The tank was surrounded laterally by an ordinary concrete shield of thickness 50 - 80 cm. This sub-critical system was first evaluated with the TRIPOLI-4 sub-critical option [6]. The following detectors were used in experiments: ^{31}P(n,p), ^{103}Rh(n,n'), and ^{32}S(n,p) dosimeters for fast neutron flux measurements, ^{115}In(n,γ), ^{55}Mn(n,γ), and ^{197}Au(n,γ) foil covered by cadmium for epithermal neutron, and bare ^{55}Mn(n,γ) and BF_3 chamber for thermal neutron measurements.

III. TRIPOLI-4 modeling of NAÏADE 1 water shielding benchmark

Fig. 2 shows the TRIPOLI-4 modeling of the NAÏADE 1 light water shielding experiment. The fission plate converter was modelled in the left side in red color. The 250 cm-thick light water in blue color was modelled in the center of the gray concrete wall. Using this shielding experiment to validate different options of TRIPOLI-4, two calculation models were developed. As shown in Figure 3, these two models presented different geometry representations of the experiment structure and tally. Table 1 presents different details in modeling and calculation of these two models.

Figure 2: TRIPOLI-4 modeling of NAIADE 1 light water shielding experiment (Red – natural uranium fission plate, Blue – light water shielding 250 cm x 300 cm x 300 cm, and Gray – ordinary concrete shielding wall) [5, 6]

The earlier 2006 model simulated the collimated thermal neutron source and the fission neutron attenuation in water via a fixed-source sub-criticality mode. This mode calculates both the neutron multiplication effect of the central fission plate and the neutron transport in water shielding. However in this study, a simplified shielding model of the NAIADE 1 water experiment was developed in order to make the validation work easier on variance reduction techniques and recent options of the code. That means, the present calculations concentrate on the neutron transport from the fission plate to the water by taking into account different TRIPOLI-4 variance reduction options.

IV. TRIPOLI-4 variance reduction techniques

Basically, the TRIPOLI-4 built-in variance reduction techniques such as implicit capture (survival biasing) and Russian roulette, are involved by default in the shielding mode calculation but they are not sufficient for neutron deep penetration transport calculations.

The importance map generator INIPOND module of TRIPOLI-4 can also be used to improve the variance reduction performance and to reduce the user involvement, especially for the neutron deep penetration cases [3-8]. The generated important map is then used by TRIPOLI-4 in the same run to perform importance sampling through the exponential transform method, quota sampling, splitting and Russian roulette, and source biasing.

Different options under INIPOND module were investigated in this study. Generally the importance function from INIPOND is assumed to be factorized in space, energy, and angle variables, with a coupling between space and energy variables. A space mesh and an energy grid were separately defined by the author for the initialization of the importance map by the code.

Benchmark models	SINBAD [6] (2006)	This work (2016)
Mode	Fixed-source sub-criticality	Fixed-source shielding
Simulation	Neutron transport only	Coupled Neutron-photon transport
Neutron source	Thermal neutron from graphite (Mono –directional)	Fission neutron from U plate (Isotropic)
Source distribution	R - θ - Z based on Mn-55 (n,γ) measurements	X-Y-Z based on P-31 (n,p) measurement
Geometry	Analytical surfaces	Combinatorial volume cells
Tally	Surface	Mesh & thin volume
Variance reduction (INIPOND)	Point attractor	Spherical surface attractor
Data library	ENDF/B-VI.r4 & IRDF 1990	JEFF3.2 (CEAV6) & IRDF 2002

Table 1: **NAIADE 1 Water shielding benchmark and TRIPOLI-4 calculation models**

In this initialization step of the TRIPOLI-4 shielding calculation with the INIPOND module, the simulation of the first batches of neutron histories is specific. Case-dependent information is first collected and subsequently used to improve the multi-group and 3D mesh-based important map. As each case may have different geometry volume cells, materials distributions, neutron source energy and space zones, tally energy and space zones, so the particle transport information of the sampled neutrons need to be collected in order to know the visited volume cells and the neutron energy slowing-down in space. The purposes of this batch-based iteration procedure are to regulate and improve the neutron population distribution in geometry and energy space. At the end of this initialization step, the importance map from this learning-by-simulation procedure helps guiding the neutron transport in all next sets of neutron histories.

The "monitoring" option was used by default in the TRIPOLI-4 variance reduction run. The adjustments of the importance map were performed from energy group to energy group by taking advantage of statistics on discrepancies between particle weights and reference weights given by the importance map, when particles slowing down in energy from group to group. Of course the performance of the associated variance reduction depends importantly on the input data dedicated to the INIPOND module.

The options for displaying of collision points and iso-importance curves were helpful to understand the particle propagation in water and useful to improve the importance map calculation. Considering the measurement positions in water, fictitious detector(s) can be placed in deep water of the calculation model so as to act as ''attractor(s)'' for simulated particles.

The β and K parameters of the INIPOND module were usually set in order to adjust the global strength of the "biasing" by the exponential transform techniques [4, 7]. When using the INIPOND AUTO option, the values K were calculated in the code according to the β input by solving the Placzeck equation, after collapsing the pointwise cross-sections in each composition and each energy group. As each mesh can contain different compositions, the initialization of importance map may be built using graph theory with the user-estimated or code-generated K values. Generally the shorter paths in the 3D graph between the neutron source and the "attractor" were also the guiding ways in neutron transport simulation. The optional energy factor α (with a set of parameter values given for each energy group) was also useful to adjust neutron population for high energy neutrons in this work as the uranium thermal fission spectrum has a peak zone around 2 MeV.

Due to the mesh-based variance reduction method used in INIPOND point(s) attractor, a file per thread calculation line is generated to store the importance map data. The size of this file depends on the meshes dimensions in space and energy. To reduce the memory requirement for the importance map calculations of INIPOND, the spherical analytical surface attractor was also chosen for the spatial importance calculation. This option is useful in massively parallel calculations.

V. Results

TRIPOLI-4 calculation results are divided in three parts. The first part shows the intermediate graphic results. These results are helpful to understand the variance reduction effects on fission neutron transport in water, to adjust the variance reduction parameters, and to improve the calculation performance. The second part presents the numerical calculation results of the NAIADE 1 water benchmark including the fast neutron and thermal neutron equivalent flux as a function of water shielding thickness. The third part shows the variance reduction performance results.

The graphic intermediate results including the collision sites maps and the importance maps generated in this study are presented in Fig. 4 and 5. Two different input data sets for the INIPOND variance reduction module, one with discrete attractors and the other with the spherical attractor surface [7], were applied to qualitatively demonstrate the variance reduction effects.

The direction of interest of simulated particles is calculated with the gradient of the importance lines/surfaces and it is perpendicular to the iso-importance surfaces (see Fig. 5). According to the position of the measurement, the neutrons were guided from the fission source plate and the collision sites to the pre-defined attractor(s). For the INIPOND variance reduction, the space mesh dimensions were 7.5 x 7.5 x 7.5 cm and three energy groups were set for fast, intermediate, and slow neutrons. Using the point attractor, initial computational resources are necessary. They depend on the mesh size and the number of energy groups. Several point attractors can be assigned to guide the neutron transport. Using the spherical attractor surface in INIPOND module, the importance map calculation time was shorter and very limited computer memory was necessary. But only one spherical attractor surface can be set.

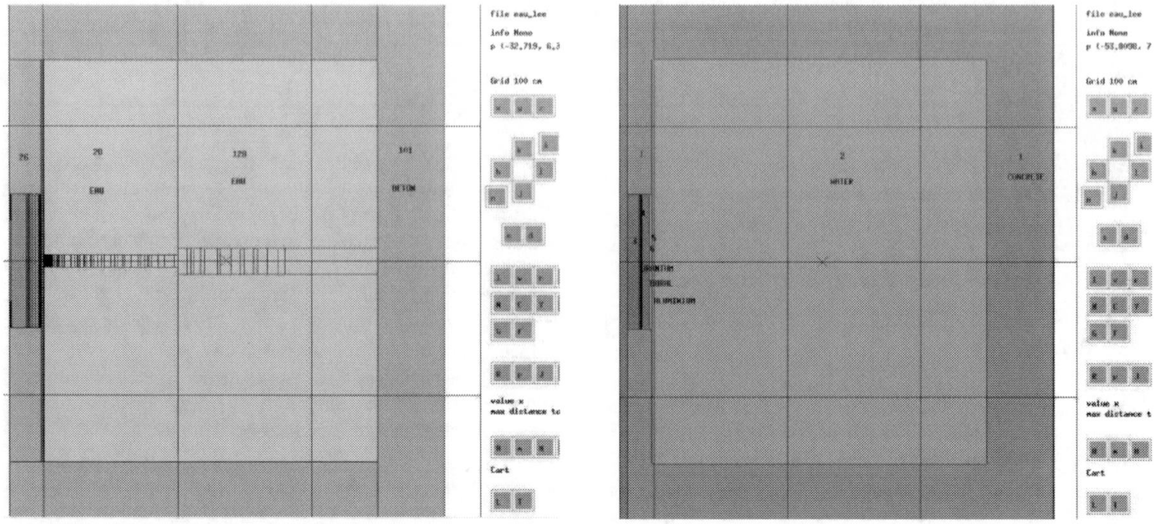

Figure 3: TRIPOLI-4 modeling of NAIADE 1 light water shielding experiment
Left: fixed-source sub-criticality model from SINBAD ([6], 2006), Right: related simplified shielding model (2016)

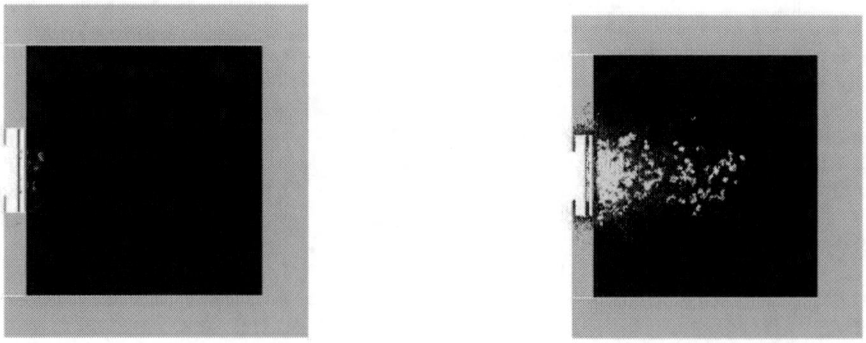

Figure 4: TRIPOLI-4 calculated projection maps of neutron collision sites for NAIADE 1, 250 cm x 300 cm x 300 cm, light water shielding surrounded by an ordinary concrete and irradiated by fission neutrons from a fission plate
Left: INIPOND option not used, Right: Variance reduction with INIPOND - Yellow points show the collision sites of slow neutrons.

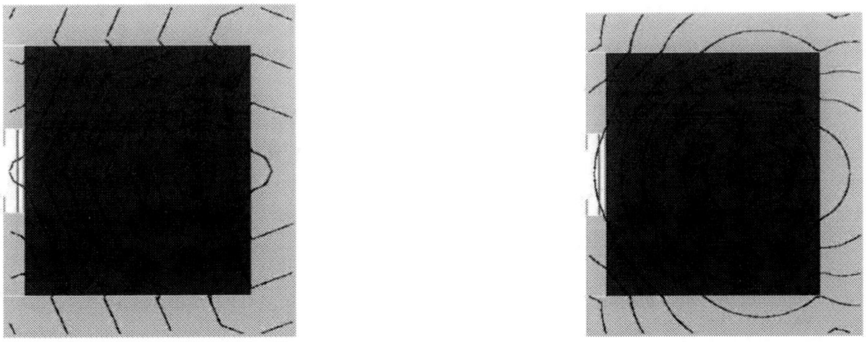

Figures 5: TRIPOLI-4 calculated iso-importance maps for NAIADE 1, 250 cm x 300 cm x 300 cm, light water shielding surrounded by an ordinary concrete and irradiated by fission neutrons from a fission plate
Left: Variance reduction with discrete attractor, Right: Variance reduction with a spherical attractor surface

The numerical calculation results are presented in Figures 6 to 9. Figure 6 shows the TRIPOLI-4 calculated and measured equivalent fast neutron flux based on $^{31}P(n,p)$ dosimeters. The calculation results of this work are generally in good agreement with the previous calculations of 2006 and the measurements of 1965 [6].

It should be noted that the average cross section of $^{31}P(n,p)$ based on the ^{235}U thermal fission spectrum was used to get the equivalent neutron flux from the calculated reaction rates. A 25% difference of this $^{31}P(n,p)$ average cross section has been found between 2.858e-2 barns from IRDF-1990 and 3.586e-2 barns from the recent evaluation of IAEA-NDS-0668.

In Figure 6 the calculated fast neutron flux attenuation in water was also shown as a function of penetration thickness. Both the above 1 MeV and the above 0.1 MeV fast neutron flux were included. Generally, they were lower than the equivalent fast neutron flux based on $^{31}P(n,p)$ dosimeters.

Using the $^{32}S(n,p)$ dosimeter measurement results, a similar benchmark trend was found in Figure 7. The fission neutron transport in light water for penetration up to 50 cm for fast neutrons was benchmarked. For water thicknesses larger than 30 cm, the calculated fast neutron flux and the equivalent fast neutron flux diverged because of the degradation of the fission neutron spectrum in water and the constant average cross-section of the dosimeter.

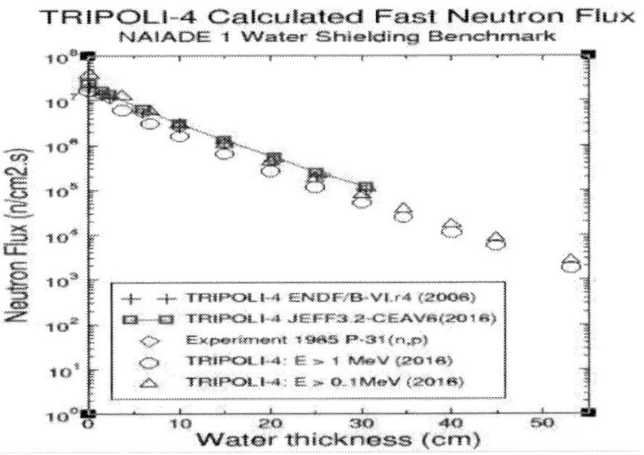

Figure 6: TRIPOLI-4 calculated fast neutron flux and equivalent fast neutron flux based on $^{31}P(n,p)$ measurements

In Figure 8, the calculated equivalent thermal neutron flux attenuation in water was shown as a function of the penetration thickness up to 180 cm. The equivalent thermal neutron flux was obtained from the calculated $^{10}B(n,\alpha)$ reaction rate. The $^{10}B(n,\alpha)$ average cross-section based on the 2200 m/s thermal neutron was 3840 barns [6].

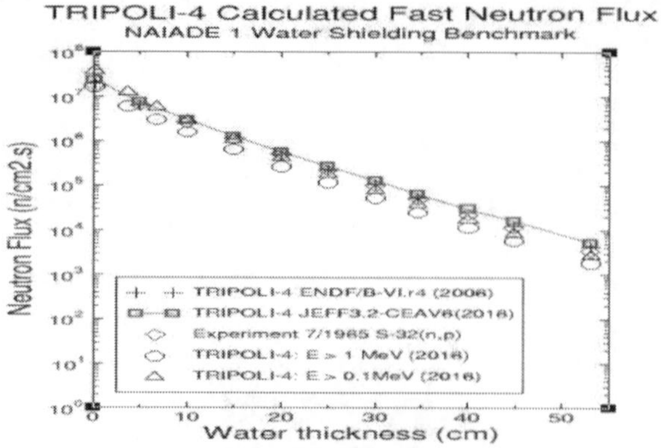

Figure 7: TRIPOLI-4 calculated fast neutron flux and equivalent fast neutron flux based on ^{32}S(n,p) measurements

The typical build-up of thermal neutron flux in the first water zone was observed for water thicknesses from 0 to 5 cm in Figure 8. The thermal neutron flux attenuation factor of 180 cm light water was about ten decades. The TRIPOLI-4 INIPOND variance reduction module was really helpful. The calculation results of this work are in good agreement with the previous calculations of 2006 and the measurements of 1957-1960 [6].

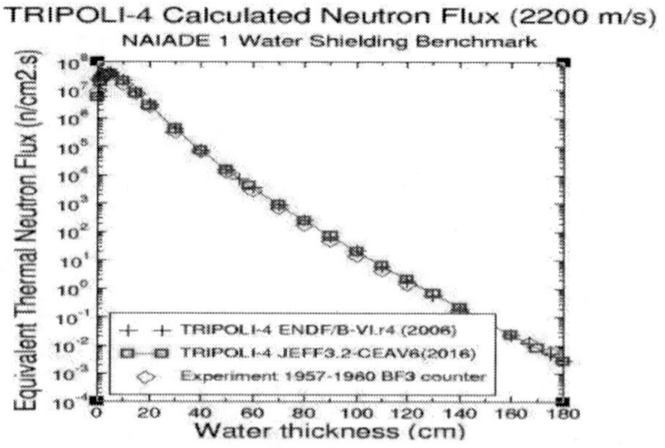

Figure 8: TRIPOLI-4 calculated equivalent thermal neutron flux based on the ^{10}B(n,α) reaction rate of BF3 counter measurements

The TRIPOLI-4 calculated equivalent dose rates in water as a function of the penetration thickness up to 90 cm was shown in Figure 9. For neutron dose rate, the calculation results of this work were in good agreement with the previous calculations of 2006 [6]. The dose rates from secondary gamma-ray contributions were also presented in this Figure. The main neutron absorption in water is based on the ^{1}H(n,γ) capture reaction. The 2.2 MeV secondary gamma dose was build-up in the first water zone and its domination in dose contribution appeared from 40 cm depth in water.

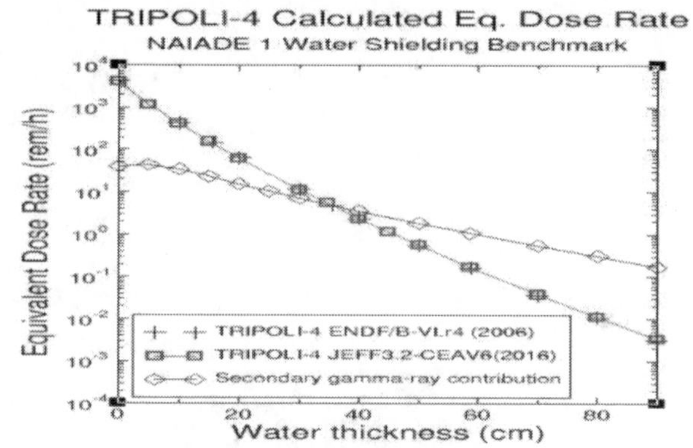

Figure 9: TRIPOLI-4 calculated equivalent dose rates of neutron and secondary gamma-ray generated in water

The performance study of the INIPOND variance reduction is presented in Table 2. Different options of the TRIPOLI-4 INIPOND variance reduction were used in this study to demonstrate and compare their performance on the fission neutron deep penetration calculation in water. Both the ^{31}P(n,p) equivalent fast neutron flux and the ^{10}B(n,α) equivalent 2200 m/s thermal neutron flux were taken into account to benchmark against the measurements. As the mean free path of slow neutrons is small and the neutron absorption in water based on the ^{1}H(n,γ) capture reaction is owing to the 1/v nature, the initial effort in adjusting the INIPOND variance reduction parameters was aimed at pushing fission neutron forward. In the same time, tally positions as a function of water thicknesses were assigned. An over push forward of fission neutron was not worthy. From fast to slow energy groups in runs (2) to (5) of Table 2, different K values of INIPOND module were adjusted according to the TRIPOLI-4 output of the initial runs.

FOM	**^{31}P(n,p) Eq. fast flux**	**^{10}B(n,α) - 2200 m/s Eq. thermal flux**
Eq. neutron flux in water depth	30 cm	50 cm
(1) TRIPOLI-4 Analog run	12	1
(2) INIPOND Point attractor	75	300
(3) (2) + Monitoring 0	3470	46
(4) INIPOND Surface attractor	100	137
(5) (4) + Monitoring 0	350	113
(6) (5) + INIPOND AUTO	3900	800

Table 2: TRIPOLI-4 variance reduction performance for equivalent neutron flux calculation of NAIADE 1 Water shielding benchmark

The performance of the calculation with variance reduction was evaluated with the factor FOM (figure of merit). The FOM is defined as FOM = $1 / (\sigma^2 t)$, where the standard deviation σ and the run time t are available from the code output. The FOM values in Table 2 were calculated by the author and their uncertainties were estimated to be less than 5 %.

The no monitoring feature "Monitoring 0" [7] in runs (3), (5), and (6) of Table 2 deactivated the monitor of importance and the regulation of neutron population between neutron energy groups as presented in section IV. That is why the FOM values of runs (3) and (5) show an improvement in equivalent ^{31}P(n,p) fast neutron flux, compared to runs (2) and (4) with the "monitoring" option. In the same time, the regulation of neutron population between energy groups was deactivated so the FOM values of equivalent thermal neutron flux in runs (3) and (5) decreased, compared to runs (2) and (4). When the option INIPOND AUTO in run (6) was activated with the spherical surface attractor option, the best FOM of this test was obtained for both fast and thermal neutron flux. In fact the attenuation of neutron flux in NAIADE 1 water benchmark was in important pure water. Neutron streaming was negligible. Different materials and void zones in attenuation way were absent. Under these conditions, the INIPOND AUTO option worked perfectly and produced converged results on both fast and thermal neutron flux.

```
VARIANCE_REDUCTION     NEUTRON       // Particle type

  GRID 4   20.  2.  0.005.  1.E-11     // Three groups in energy mesh

  ALFA      0.2   0.      0.           // Adjust group population size
  DETECT_RHO 1    1.      0.5          // Sphere surface attractor & β

  MESH     45      53     53           // VR geometry mesh no.
        7.247   7.551   7.551          //  Mesh dimension (cm)
          FRAME  CARTESIAN
       -25.9  -200.1  -200.1           // Mesh geometry origin (x, y, z)
     1.  0.  0.   0. 1. 0.   0. 0. 1.  //  vectors X Y Z
  END_MESH

  INIPOND  AUTO
    OMEGA SET_OMEGA SPHERE 50.  0.  0.  // Sphere center
  END_INIPOND

END_VARIANCE_REDUCTION
```

Table 3: TRIPOLI-4 variance reduction (VR) input of run (6) for equivalent neutron flux calculation of NAIADE 1 Water shielding benchmark

Table 3 shows the run (6) variance reduction input data. Three energy groups were set from 20 MeV to 1E-11 MeV to cover the fission neutron and the thermal neutron zones. A cartesian mesh geometry with 45 x 53 x 53 meshes in X, Y, and Z was set to cover the transport simulation geometry zone 326 x 400 x 400 cm^3 with a mesh size around 7.5 cm. The center of fission plate was placed at (-10, 0, 0) (see Fig. 2, red zone) and that of the spherical surface attractor was placed at (50, 0, 0) (Fig. 2, blue water zone) for this variance reduction performance study. The radius of the attractor sphere was set to 1 cm and the β value to 0.5. An α value was set only in the fast neutron group to improve the neutron population distribution above 2 MeV because only fast neutrons have a significant probability to penetrate into deep water.

VI. Conclusions

To support the nuclear decommissioning activities on reactor structure materials activation calculations, the TRIPOLI-4 Monte Carlo transport code capabilities on fission neutron shielding calculations were demonstrated by using the NAIADE 1 water benchmark experiments. Both the shielding mode and the fixed-source sub-criticality mode of the code were used. Collision sites projection maps were generated to show the neutron migration and the neutron energy slowing down along with the depth of the neutron transport routes. Iso-importance maps were generated to verify the direction dependent variance reduction effects and the shielding material dependent importance gradient. Neutron penetration in water up to 50 cm for fast neutrons and up to 180 cm for thermal neutrons was taken into account. Calculation results of equivalent fast neutron flux and thermal neutron flux were validated against the measurements. The variance reduction options of TRIPOLI-4 code were helpful to perform these calculations with the INIPOND point attractor and the INIPOND spherical surface attractor. A simplified NAIADE 1 water shielding model was proposed in this work. It allowed performing the code validation easier. New developed options for variance reduction in TRIPOLI-4 will be investigated with this simplified model in a future work.

Acknowledgment

TRIPOLI-4® is a registered trademark of CEA. The author gratefully acknowledges EDF for their long term partnership and support.

References

1) Y. K. Lee, "Démantèlement de Niveau 3 de L'installation TRITON: Etude d'Activation de Béton et de Ferraillage de la Piscine Liée á l'Exploitation des Réacteurs TRITON et NEREIDE (1959-1981)," Report CEA/DEN/DM2S/SERMA/ LEPP/ RT/01-3028, Mars 2002.

2) Y. K. Lee et C. M. Diop, "Expertise sur la note d'ATR Ingénierie intitulée - Calculs d'activation des structures de la piscine MELUSINE," Report 1607/VIL/NOT/001, 27/03/2006.

3) E. Brun, F. Damian, C.M. Diop et al., "TRIPOLI-4®, CEA, EDF and AREVA reference Monte Carlo code," *Annals of Nuclear Energy*, **82**, 151-160 (2015).

4) J.-P Both, J.-C. Nimal, T. Vergnaud, "Automated importance generation and biasing techniques for Monte Carlo shielding techniques by the TRIPOLI-3 code," Prog. Nucl. Energy **24**, 273–281 (1990).

5) J.-C. Nimal, "Status of NAIADE 1 Shielding Benchmarks," *Nuclear Technology*, **168**, 2, 405-410 (2009).
6) J.-C. Nimal, "NAIADE 1 light water benchmark (60cm)," SINBAD NEA-1517/79 (2007).
7) O. Petit, F. X. Hugot, Y. K. Lee, C. Jouanne et al., "TRIPOLI-4 user guide," CEA Report CEA-R-6169, (2008).
8) O. Petit, Y. K. Lee, C. M. Diop, "Variance reduction adjustment in Monte Carlo TRIPOLI-4 neutron gamma coupled calculations," *Progress in Nuclear Science and Technology*, Volume **4**, pp. 408-412 (2014).

DOI: 10.1051/epjconf/201715303001

The CERN-EU high-energy Reference Field (CERF) facility: applications and latest developments

Marco Silari[1,*], and *Fabio* Pozzi[1]

[1]CERN, 1211 Geneva 23, Switzerland

Abstract. The CERF facility at CERN provides an almost unique high-energy workplace reference radiation field for the calibration and test of radiation protection instrumentation employed at high-energy accelerator facilities and for aircraft and space dosimetry. This paper describes the main features of the facility and supplies a non-exhaustive list of recent (as of 2005) applications for which CERF is used. Upgrade work started in 2015 to provide the scientific and industrial communities with a state-of-the-art reference facility is also discussed.

1 Introduction

Neutron calibrations often need to be performed with neutron energies or spectral distributions very much different from those generated by radioactive sources such as Am-Be and Cf-252, employed in standard calibration laboratories. This is also highlighted by refs. [1] and [2], which call for the simulation of the neutron energy spectra encountered in the vicinity of high-energy accelerator facilities and in high-flying aircrafts, since high-energy neutrons can deliver a significant fraction of the ambient dose equivalent.

The CERF facility [3], set-up at CERN in the early nineties, provides mixed radiation fields representative of those found outside the shielding of high-energy hadron accelerators and sufficiently similar to the cosmic ray field encountered at 10-20 km altitude. This allows instrumentation to be tested, inter-compared and calibrated at CERN and subsequently used for either in-flight measurements on aircrafts or radiation protection surveys around particle accelerators. CERF has also been employed for a variety of other applications such as radiobiology studies, cross section measurements, investigation of activation of accelerator materials, intercomparison of individual dosimeters, and benchmarking of Monte Carlo codes against experimental data (see Section 3). This work describes the facility, lists its main applications, provides some representative results and illustrates the upgrade program started in 2015 to provide the scientific and industrial communities with an improved, state-of-the-art reference workplace facility.

2 Description of the facility

2.1 Set-up

Figure 1 shows the detailed geometry of CERF as simulated in the FLUKA Monte Carlo code [4, 5]. CERF is located in the North Experimental Area on the CERN Prévessin site. The CERF hadron beam is produced by the primary 450 GeV c^{-1} proton beam from the Super Proton Synchrotron (SPS) on the T4 production target. The secondary beam originating from the T4 target is composed of 61% positive pions, 35% protons and 4% positive kaons [6] with a momentum of 120 GeV c^{-1}. To calculate the beam composition at the CERF irradiation area, one must take into account that 4.2% and 27% of the pions and kaons, respectively, decay over the approximately 410 m long transfer line between T4 and CERF. Hence, the actual beam composition is 60.7% positive pions, 36.3% protons and 3.0% kaons. The hadron beam impinges on a copper target, 7 cm in diameter and 50 cm in length, which can be installed under either a concrete roof 80 cm thick or an iron roof 40 cm thick. The secondary particles produced in the target (mostly neutrons and photons, with a few percent contribution from electrons, protons, pions and muons) traverse these roof shields producing almost uniform radiation fields over two areas of 2x2 m^2 located at approximately 90° with respect to the incoming beam direction. Each 2x2 m^2 area is divided into 16 squares of 50x50 cm^2, and each element of this grid represents a reference exposure location. Additional reference positions are available behind the lateral shielding of the irradiation area, at the same angles with respect to the target as for the two roof positions. The nominal measurement locations are at the centre of each square at 25 cm height above the roof [3] or 25 cm distance from the lateral shielding, respectively. The copper target can also be removed to allow instrumentation or other targets to be directly placed in the beam, as for cross section measurements (see Section 3.5).

* Corresponding author: Marco.Silari@cern.ch

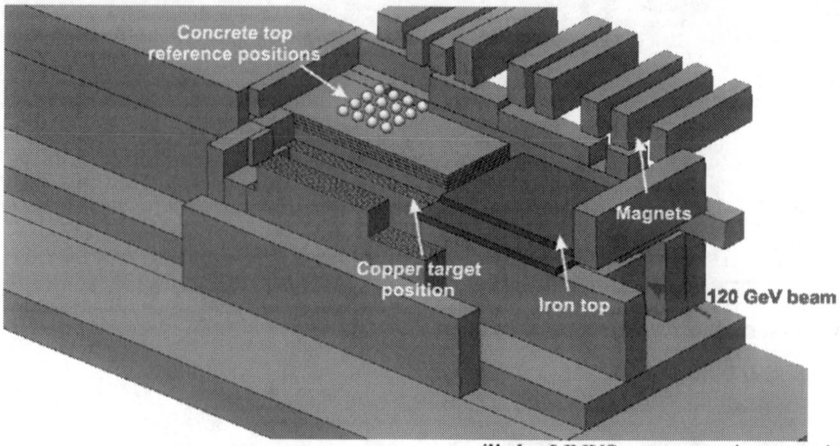

Figure 1. Geometry of CERF as simulated with FLUKA. Here, the copper target is placed under the concrete roof.

2.2 Beam monitoring and imaging

The beam is delivered to CERF with a typical intensity in the range 10^6 to 10^8 particles per SPS beam extraction (spill), with two to three beam extractions of about 5 s duration over an SPS cycle that can vary in the range 30-45 s. During the extraction time the beam intensity is constant resulting in a constant radiation field at the exposure positions. The beam monitoring, on which the normalization of all measurements relies, is provided by an air-filled, parallel-plate, transmission-type ionization chamber (IC); the IC calibration factor was recently verified by means of the multi-foil activation technique using two reactions: the well-known $^{27}Al(p,3pn)^{24}Na$ monitor reaction and the alternative $^{nat}Cu(p,x)^{24}Na$ reaction [7, 8]. These experimental values were compared with the calibration factor obtained simulating with FLUKA the expected charge collected on the plates of the IC; the agreement is very satisfying [7]. One IC-count corresponds to 22000±2200 particles impinging on the target.

Typical values of ambient dose equivalent rates are 0.2-0.3 nSv per IC-count over the 16 reference exposure locations on the concrete roof and 1-1.5 nSv per IC-count on the iron roof. By assuming an average of three 5-second spills per minute and adjusting the beam intensity on the target, one can obtain ambient dose equivalent rates approximately from 5 μSv h^{-1} (30 nSv per spill) to 250 μSv h^{-1} (1.5 μSv per spill) on the concrete roof and side, and from 18 μSv h^{-1} (100 nSv per spill) to 360 μSv h^{-1} (2 μSv per spill) on the iron roof. To provide an independent reference and to monitor the beam stability, the CERN rem counter LINUS [9, 10, 11, 12] is placed in a fixed position on either the concrete or iron roof (according to where measurements are being performed). The steady value of H*(10) per IC-count measured by the LINUS indicates stable beam conditions. The example in Figure 2 shows how the statistical fluctuations of the LINUS reading correctly decrease with increasing beam intensity while the average remains constant. At about 100 IC-counts per spill the LINUS count rate is approximately 20 counts per spill (for a LINUS sensitivity of 0.92 counts per nSv and an H*(10) value of about 0.2 nSv per IC-count on concrete top, CT, position 14) and it is 50 times higher at 5000 IC-counts per spill. A change in the LINUS reading would point to a mis-steering of the beam on the target.

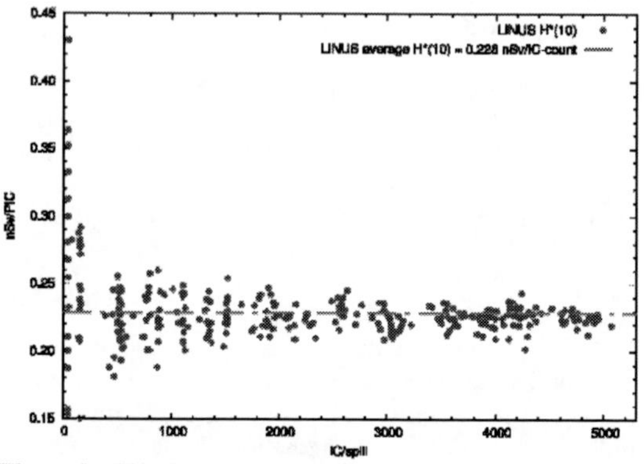

Figure 2. H*(10) rate (normalized to the IC-count rate) measured by the LINUS placed in CT14 as a function of the IC count rate and the average value.

The target alignment is made by exposing Polaroid films for a few minutes to the beam in order to integrate a fluence of about 10^4 IC-counts (Figure 3). The beam is Gaussian and its rectangular shape, mostly contained in an area of about 3.5 cm wide and 3 cm high (at the maximum beam intensity) or smaller, is defined by a set of collimators in the T4-CERF beam line. After the target has been aligned, during measurements the beam profile can be routinely checked by a multi wire proportional chamber (MWPC) installed between the two target positions (Figure 4). Recently, a GEM detector was installed to monitor the alignment of the beam with respect to the target (see Section 4.1).

DOI: 10.1051/epjconf/201715303001

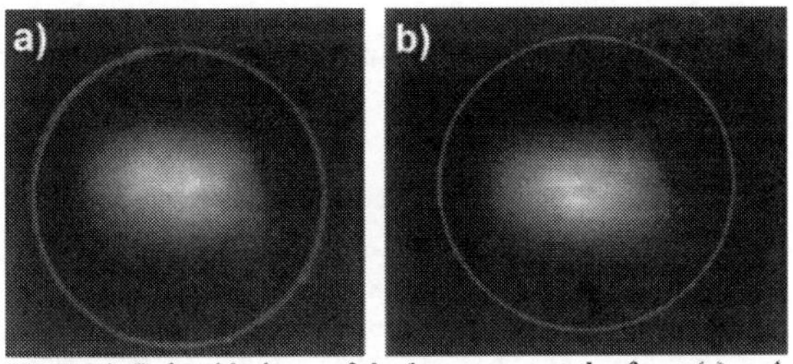

Figure 3. Polaroid views of the beam spot at the front (a) and exit (b) face of the target. The target diameter is indicated by the red circle.

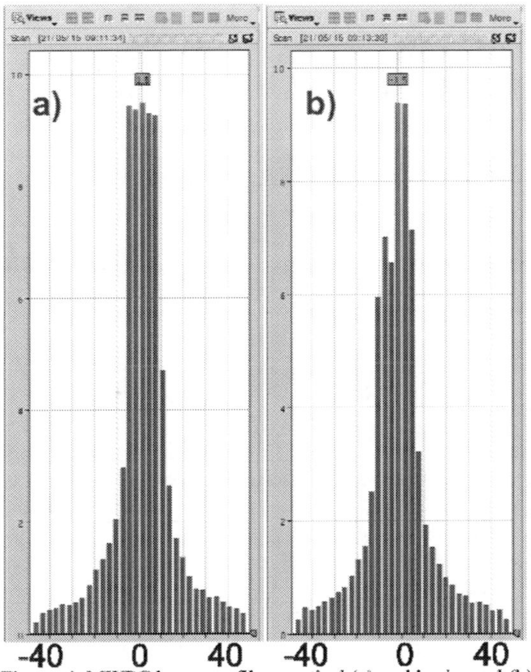

Figure 4. MWPC beam profiles, vertical (a) and horizontal (b). The x-axis scale is in mm.

2.3 Muon background

A significant low-LET background radiation field is present on the concrete and iron roof-shields. This background is mainly due to high-energy muons originating from the pion decay in the beam line. It can result in an apparent non-linearity of a detector response when measurements are carried out at different intensities of the incident hadron beam. This effect must be taken into account when measurements are performed by means of ionisation chambers, TEPC and other active devices that have considerable sensitivity to low-LET radiation. No significant muon background is present at the side concrete shielded positions. For more information on the muon background component and how to subtract it, the reader can refer to [3] and the references there-in quoted.

3 Applications

To show the range of uses of CERF, this section gives a brief, non-exhaustive overview of the most important and recent applications (as of 2005), together with a few representative results. An overview of earlier experiments can be found in ref. [3].

3.1 Test, workplace field calibration and intercomparison of radiation protection instrumentation

In 2007 the performances of conventional and extended-range rem-counters and of a tissue-equivalent proportional counter (TEPC) were tested for their use around high-energy accelerators [13]. The results confirmed the reliability of the extended-range rem-counters and TEPC, which are specifically designed to operate in these conditions. The conventional rem-counters, designed for neutron energies up to 15 MeV, substantially underestimated the neutron H*(10). However, a field calibration factor for each detector can be deduced at CERF; this should be used when these devices are employed in a similar high-energy field.

A similar experiment was conducted in 2012 by Caresana *et al.* [14] to investigate the performances of conventional and extended-range rem-counters including a novel instrument called LUPIN, two TEPCs and bubble detectors. The results were also compared with the FLUKA reference values [3]. The conventional rem-counters were in good agreement within their uncertainties and underestimated the H*(10) as measured by the extended-range instruments and as predicted by FLUKA. The TEPCs slightly overestimated the FLUKA value and indicated that the non-neutron part of the stray field accounts for about 30% of the total H*(10).

Unlike the two above-mentioned studies, which were carried out on the concrete roof, the following experiments were performed inside the irradiation cave to evaluate the capabilities of potential beam monitors. Vincke *et al.* [15] experimentally and theoretically studied the response to several types of incident particles of a PMI ionisation chamber, designed for use in the Large Hadron Collider (LHC) to obtain an indication of the machine performance via a measurement of radiation under beam operation conditions. This study also demonstrated that the FLUKA code is well suited to predict the response of ionisation chambers in a mixed radiation field.

Aza *et al.* [16] investigated the use of a triple GEM detector designed for low-energy neutrons as secondary on-line monitor, measuring the low-energy neutron flux scattered from the target inside the CERF irradiation area.

Ytre-Hauge *et al.* [17] investigated the capability of detecting neutrons by a monitor based on the registration of radiation effects in Static Random Access Memories (SRAMs). The SRAMs were irradiated in beams lines at PTB, Braunschweig, Germany, at the Oslo Cyclotron Laboratory, Norway, at the Svedberg Laboratory in Uppsala, Sweden, at the Institute for Energy Technology, Kjeller, Norway, and at CERF.

Haninger *et al.* [18] tested and calibrated a new albedo neutron dosemeter based on TLD, which was introduced by the individual monitoring service of the Helmholtz Zentrum Munchen. The field calibration revealed that this kind of dosemeter can also be used for personal monitoring at workplaces of high-energy particle accelerators.

3.2 In-flight and in-space measurements based on CERF calibration

Since the CERF mixed field is similar to the one encountered at aircraft altitudes, several studies and instrument calibrations were performed over the past 20 years, of which we provide a few examples. Vuković *et al.* [19] investigated the use of a combination of track etch detectors (CR-39, LR-115) with a plastic converter or boron foil. These detectors were calibrated at CERF as well as in a nuclear reactor before their use on aircraft.

Jadrníčková and Spurný [20] tested a LET spectrometer based on track detectors on the International Space Station (ISS), comparing the results with the measurements performed at CERF. The LET distributions of dose equivalent obtained at CERF and on-board ISS agree very well within their uncertainties, making CERF a suitable calibration facility for in-space dosimetry. The Timepix detector, which is currently employed on the ISS for the measurements of the dose to astronauts [21], was also tested at CERF for its use as a single dosimeter and LET spectrometer in mixed radiation fields created by heavy ions [22].

3.3 Dosimeter intercomparison

In 2014 a dosimeter intercomparison was organized at CERF. Seven U.S. laboratories participated sending their dosimeters to CERN (Table 1). Four prior comparisons were performed in the past to evaluate the state of personnel dosimetry implemented at accelerator facilities at the U.S. Department of Energy (USDOE) laboratories. Since the last test in 1999, the dosimetry technology has changed to some degree and USDOE regulations in the U.S. have adopted dose equivalent quantities specified in the ICRP 60. In addition, many laboratories have moved to the use of vendor-provided dosimetry. The aim of this study was to evaluate the current state of dosimetry related to high-energy neutron personnel monitoring [23].

Table 1. List of the US laboratories participating in the 2014 individual dosimetry intercomparison.

Laboratory
Brookhaven National Laboratory
Thomas Jefferson National Accelerator Facility
Pacific Northwest National Laboratory
Idaho National Laboratory
Los Alamos National Laboratory
Nevada National Security Site
Landauer, Inc.

3.4 Spallation cross section measurements

Recently, the CERF facility has been employed for the measurements of spallation cross sections in thin Fe, Ti and Cu targets [24, 25]. Thin foils were exposed to the CERF hadron beam and irradiated for about 8 hours to obtain measurable activities. The spallation cross sections for protons and pions at 120 GeV c^{-1} were derived from γ-spectrometry measurements and FLUKA simulations; they were compared with the existing values in the literature and the fragmentation and factorization hypothesis were evaluated.

Additional measurements of spallation cross sections on other target materials are planned for the near future. In order to obtain absolute cross sections, the possibility of using a 100% negative pion beam is currently under investigation; this will avoid having to resort to the Monte Carlo method to obtain the proton and pion cross sections, which are otherwise derived from mixed pion/proton irradiation.

3.5 Other applications

Over the years CERF has also been used for a number of other applications:

- Shielding experiments [26], which allowed measuring the inelastic attenuation length of high energy neutrons in a thick iron shield.
- Performance test of beam loss monitors, which are essential tools to protect accelerator components from damage due to unforeseen, critical beam losses [27].
- Measurements of Single Event Upsets (SEUs) for the ESA SEU Monitor [28] and Single Event Effects (SEEs) for a new platform dedicated to the long-term characterization of the atmospheric natural radiative environment at mountain altitude [29].
- Benchmark of Monte Carlo codes against experimental data [30].

4 Latest upgrades

Since 2015 an upgrade work has been undertaken, starting with the installation of improved target supports to make the target alignment to the beam easier and more accurate. A new up-to-date website [31] is currently maintained; this provides important information concerning the facility, access and safety rules, beam time and schedule, and reference articles.

4.1 New beam imaging and monitoring systems

A new beam imaging system employing a triple-GEM detector was tested in view of substituting the Polaroid films for target alignment [32] (section 2.2). The GEM system is mounted on both the downstream and upstream end of a hollow tube of the same dimensions of the copper target. This device allows a real time imaging of the beam spot on the target (Figure 5) providing a 2D image of the beam (Figure 6). This system thus allows a more accurate and rapid alignment of the beam on the copper target as compared to the films.

DOI: 10.1051/epjconf/201715303001

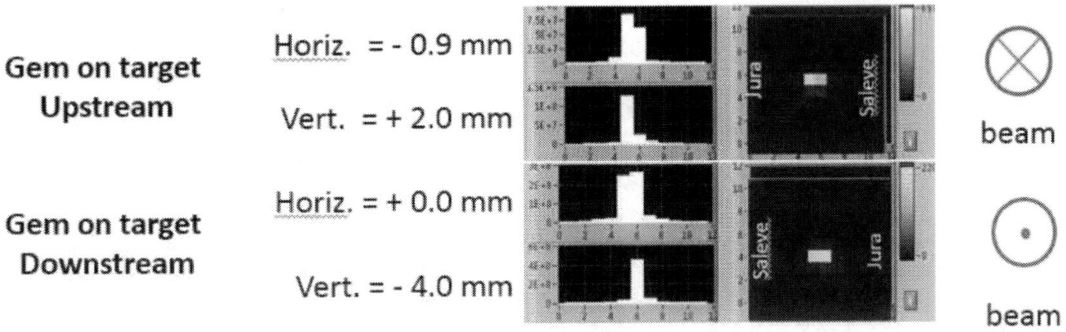

Figure 5. Position of the beam with respect to the target axis as measured by the triple-GEM detector.

In addition, the use of the GEM detector as beam intensity monitor was investigated, showing a satisfactory agreement with the measurements provided by the IC [32]. However, the triple-GEM has not yet shown to provide the same accuracy and stability of the IC as reference monitor to count beam particles. Since the IC is a single point of failure, options for a back-up solution are currently under investigation.

4.2 New FLUKA reference simulations

Measurements at CERF rely on the accurate characterization of the radiation field. FLUKA simulations of the entire facility providing reference neutron H*(10) values and spectral fluence at each exposure location were performed in 1992 (quoted in ref. [3]) and have been used since that time. Improvements of the physics models (especially concerning the neutron transport) were included in FLUKA since then, so that some variations in the simulation results can be expected with the latest version of the code.

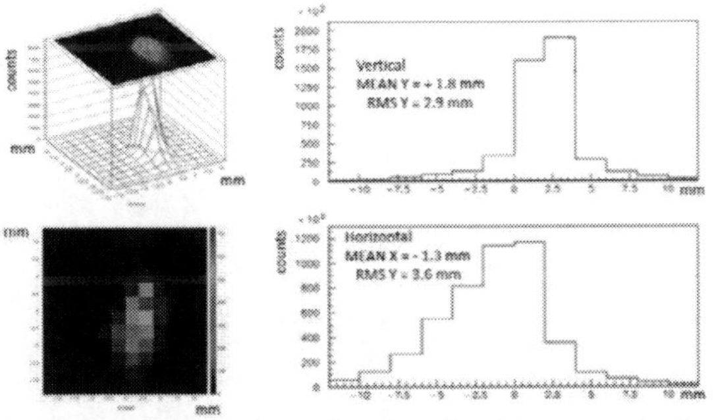

Figure 6. GEM 2D beam imaging showing a non-circular structure.

In addition, some details of the CERF geometry not included in the original simulations might result in small variations in the reference field. For instance, structures like the thin iron roof of the North Experimental Area and the shielding of the beam line adjacent to CERF can cause neutron scattering, which has some impact on the thermal and epithermal parts of the neutron spectrum. The elemental composition of the concrete and iron shielding were also investigated. All of these elements were taken into account in the new simulations currently ongoing with the latest FLUKA version (the development version available at CERN) and will be published in the coming months. In addition to neutrons, all other components of the radiation field (photons, protons, electrons) are being scored.

5 Conclusions

Since many years CERF at CERN provides a simulated workplace reference field similar to the ones encountered in the proximity of high-energy accelerators and at cruising altitudes in civil aviation. Its importance is underlined by the ISO standard dealing with simulated workplace neutron fields [1] as well as by an extensive scientific literature reporting on results obtained at CERF. Its wide range of applications extend from the calibration/test of radiation protection instrumentation, to nuclear physics, studies of radiation damage to electronics and more. The accurate characterization of the mixed radiation field by means of Monte Carlo simulations and experimental measurements ensures the reliability of the calibrations and experiments there performed. The undertaken upgrade work has the scope to provide the scientific and industrial community with a further improved and rather unique high-energy reference radiation field.

References

1. International Standard 12789-1
2. H. Schuhmacher, Radiat. Prot. Dosim. **110**, 33 (2004)
3. A. Mitaroff, M. Silari, Radiat. Prot. Dosim. **102**, 7 (2002)
4. T. T. Bohlen, F. Cerutti, M. P. W. Chin, A. Fasso, A. Ferrari, P. G. Ortega, A. Mairani, P. R. Sala, G. Smirnov, V. Vlachoudis, Nucl. Data Sheets **120**, 211 (2014)
5. A. Ferrari, P.R. Sala, A. Fassò, J. Ranft, CERN-2005-10/INFN/TC_05/11, SLAC-R-773, (2005)
6. H.W. Atherton, C. Bovet, N. Doble, G. Von Holtey, L. Piemontese, A. Placci, M. Placidi, D.E. Plane, M. Reinharz, E. Rossa, CERN Yellow Report 80-07 (1980)

7. A. Ferrari, F.P. La Torre, G.P. Manessi, F. Pozzi, M. Silari, Nucl. Instrum. Meth. A **763**, 177 (2014)
8. A. Curioni, R. Froeschl, M. Glaser, E. Iliopoulou, F. P. La Torre, F. Pozzi, F. Ravotti, M. Silari, submitted for publication in Nucl. Instrum. Meth. A
9. C. Birattari, A. Ferrari, C. Nuccetelli, M. Pelliccioni, M. Silari, Nucl. Instrum. Meth A **297**, 250 (1990)
10. C. Birattari, A. Esposito, A. Ferrari, M. Pelliccioni, M. Silari, Radiat. Prot. Dosim. **44**, 193 (1992)
11. C. Birattari, A. Esposito, A. Ferrari, M. Pelliccioni, M. Silari, Nucl. Instrum. Meth A **324**, 232 (1993)
12. C. Birattari, A. Esposito, A. Ferrari, M. Pelliccioni, T. Rancati, M. Silari, Rad. Prot. Dosim. **76**, 135 (1998)
13. S. Mayer, D. Forkel-Wirth, M. Fuerstner, H. G. Menzel, M. J. Mueller, D. Perrin, C. Theis, H. Vincke, Radiat. Prot. Dosim. **125**, 289 (2007)
14. M. Caresana, M. Helmecke, J. Kubancak, G. P. Manessi, K. Ott, R. Scherpelz, M. Silari, Radiat. Prot. Dosim. **161**, 67 (2014)
15. H. Vincke, D. Forkel-Wirth, D. Perrin, C. Theis, Radiat. Prot. Dosim. **116**, 380 (2005)
16. E. Aza, M. Magistris, F. Murtas, S. Puddu, M. Silari, J. Instrum. **9**, T11005 (2014)
17. K. S. Ytre-Hauge, A. Velure, E. F. Larsen, C. H. Stokkevåg, D. Röhrich, N. Brekke, O. H. Odland, Nucl. Instrum. Meth. A **804**, 64 (2015)
18. T. Haninger, P. Kleinau, S. Haninger, Radiat. Prot. Dosim. Doi: 10.1093/rpd/ncw166 (2016)
19. B. Vuković, M. Poje, M. Varga, V. Radolić, I. Miklavčić, D. Faj, D. Stanić, J. Planinić, Appl. Radiat. Isotopes **68**, 2398 (2010)
20. I. Jadrníčková, F. Spurný, Radiat. Meas. **43**, 683 (2008)
21. M. Kroupa et al., Life Sci. Space Res. **6**, 69 (2015)
22. O. Ploc, J. Kubancak, L. Sihver, Y. Uchihori, J. Jakubek, I. Ambrozova, A. Molokanov, L. Pinsky, J. Radiat. Res. **55**, 141 (2014)
23. R. K. Piper, J. C. McDonald, M. Silari, F. Pozzi, G. P. Manessi, F. P. La Torre, N. Dinar, to be submitted for publication in Radiat. Prot. Dosim.
24. A. Ferrari, F. P. La Torre, G. P. Manessi, F. Pozzi, M. Silari, Phys. Rev. C **89**, 034612 (2014)
25. F. Pozzi, A. Curioni, N. Dinar, F. P. La Torre, G. P. Manessi, M. Silari, submitted for publication in Phys. Rev. C
26. S. Agosteo, F. Pozzi, M. Silari, L. Ulrici, Nucl. Instrum. Meth. B **312**, 36 (2013)
27. L. Sarchiapone, M. Brugger, B. Dehning, D. Kramer, M. Stockner, V. Vlachoudis, Nucl. Instrum. Meth. A **581**, 511 (2007)
28. R. Garcia Alia, M. Brugger, S. Danzeca, V. Ferlet-Cavrois, C. Poivey, K. Roed, F. Saigne, G. Spiezia, S. Uznanski, F. Wrobel, IEEE T. Nucl. Sci. **60**, 4142 (2013)
29. A. Cheminet, G. Hubert, V. Lacoste, R. Velazco, D. Boscher, IEEE T. Nucl. Sci. **60**, 2411 (2013)
30. www.oecd-nea.org/science/wprs/shielding/ sinbad
31. www.cern.ch/cerf
32. E. Aza, D. Celeste, N. Dinar, E. Iliopoulou, F. P. La Torre, F. Murtas, F. Pozzi, S. Romano, M. Silari, CERN Technical Note, EDMS 1558978, (2015)

Clearance of materials from accelerator facilities

Sayed H. Rokni[1], Scott L. Davis[2], Ryan Ford[1], James C. Liu[1], Elaine Marshall[3], Scott O Schwahn[4] and Keith Welch[5]

[1]*SLAC National Accelerator Laboratory, Menlo Park, CA 94025, U.S.A.*
[2]*US Department of Energy, Germantown, MD 20874, U.S.A.*
[3]*Sandia National Laboratories, Albuquerque, NM 87185, U.S.A.*
[4]*Oak Ridge National Laboratory, Oak Ridge, TN 37831, U.S.A.*
[5]*Thomas Jefferson National Accelerator Facility, Newport News, VA 23606, U.S.A.*

Abstract. A new Technical Standard that supports the clearance of materials and equipment (personal property) from U.S. Department of Energy (DOE) accelerator facilities has been developed. The Standard focuses on personal property that has the potential to be radiologically impacted by accelerator operations. It addresses material clearance programs and protocols for off-site releases without restriction on use. Common metals with potential volumetric activation are of main interest with technical bases provided in Appendices of the Standard. The clearance protocols in the Standard include three elements: 1) clearance criteria, 2) process knowledge, and 3) measurement methods. This paper presents the technical aspects of the new Standard, discusses operational experience gained in clearance of materials and equipment from several accelerator facilities at SLAC and examples as to how this Standard can be applied to benefit the entirety of the DOE Accelerator Complex.

1 Introduction

Clearance of materials from accelerators is subject to many regulatory requirements demonstrated by the existence of international, national and regional clearance regulations or standards [1-3]. Thus, unrestricted release of materials from radiological controls can be a prohibitively restrictive process. Restrictions are generally due to: lack of accepted clearance criteria, lack of established processes to demonstrate compliance with the clearance criteria, and policies that further restrict release of materials due to public concerns, or inadequate practices. Accelerator materials and components are often subject to these restrictions despite the fact that the majority of components in accelerator facilities are not impacted by beam operations.

A new U.S. DOE Technical Standard [4] has been developed for use in DOE accelerator facilities that address these issues. The purpose of the Standard is to provide an acceptable approach for managing and dispositioning personal property that may be radiologically impacted by accelerator operations. The Standard provides criteria and guidance for accelerator facilities to develop site-specific material clearance programs (including management, technical, and operational aspects) that are in compliance with the myriade of regulatory requirements for protection of public and environment [5]. Both surface and volumetric radioactivity issues are addressed with focus on the volumetric radioactivity issues by establishing three-tier clearance criteria related to ANSI N13.12-2013 volume screening levels [3]. The appendices of the Standard provide general process knowledge, technical bases, and measurement methods for clearance of common metals with potential volumetric activation in electron and proton accelerators. The Standard leverages the large body of technical expertise and experience that exist in design and operation of accelerators [6-9], which include results of simulations by Monte Carlo radiation generation and transport codes and dedicated benchmark experiments. The technical bases and approaches used in this Standard may be applied to other types of accelerator, or even non-accelerator facilities.

2 Activation of materials in accelerators

Beam operation has the potential to generate radioactivity in various components in accelerators. However, in most accelerators, normal beam losses which cause material activation occur only on a limited number of components designed to intercept the full beam power (beam dumps or targets) or a fraction of the beam (e.g., collimators and septa). Other abnormal beam losses may occur at a few locations infrequently, due to mis-steering. Therefore, a large majority of materials and components in accelerator (magnets, support structures, sections of vacuum chamber) are not impacted by direct or stray radiation fields in accelerators and are good candidates for clearance

Processes leading to induced radioactivity in accelerators are well established and have been studied in detail with the use of well-benchmarked Monte Carlo codes such as FLUKA [10-11]. General process knowledge that is applicable to many accelerators can be derived from these studies. For example, the main characteristics of activation at high-energy (GeV) electron accelerators are:

- The most abundant radionuclides are those with a half-life that is on the order of the irradiation time,
- No activation can be produced by photons with energies < 1 MeV, such as those present around klystrons and synchrotron radiation beam lines,
- No alpha emitters are produced,

- No surface contamination is produced in metals and other solid materials, unless they are submitted to machining, drilling or grinding,
- Induced activity in any object from activation is volumetric and presents its maximum value at the surface that faces the beam loss point (surface maximum concept)
- Radioisotopes that are difficult to detect are always accompanied by "proxy" radioisotopes that can be clearly measured (proxy radionuclides concept).

The concepts of "proxy radionuclides" and "maximum surface activity" are essential for the development of measurement methods for clearance of materials with potential volumetric radioactivity (as explained in section 3.3).

The Standard provides the types of process knowledge that may allow for evaluation of volumetric induced radioactivity and its characteristics in accelerator facilities. Appendices C and D of the Standard provide the technical basis for the material clearance at high energy electron and proton accelerator facilities, respectively.

Clearance criteria and measurement methods for materials with surface radioactivity are available and standardized. Therefore, clearance of activated materials that may contain residual volumetric radioactivity in the main scope of the Standard and the main subject discussed in the following sections.

3 Clearance Protocols

The clearance protocols in the Standard include three elements: 1) a 3-tiered clearance criteria, 2) process knowledge, and 3) measurement methods. Process knowledge and/or measurements are used to demonstrate that the materials released meet the selected clearance criteria. Materials that have been characterized based on process knowledge evaluation and/or measurements to meet one of the clearance criteria may be released.

3.1 Clearance criteria

DOE Order 458.1 [5] specifies a Total Effective Dose constraint of 10 μSv above background in any calendar year for each specific clearance of personal property with potential residual radioactivity. Based on a dose constraint of 10 μSv/y, the ANSI N13.12 Standard [3] derives and establishes the screening levels (SLs) for various radionuclides (divided into 5 groups) for the clearance of materials that contain or may contain residual levels of surface or volume radioactivity of radioactive materials from radiological control.

The ANSI N13.12 volume SL for proxy radionuclides represents an activity level that is detectable with common, sensitive survey meters. Therefore, for clearance of property with potential volumetric radioactivity, this Standard adopts a 3-tiered clearance criteria that can be related to ANSI N13.12 volume SLs:

1) Indistinguishable from Background (IFB) at a level lower than the ANSI N13.12 volume SLs,
2) Non-IFB, but no more than the ANSI N13.12 volume SLs,
3) Higher than the ANSI N13.12 volume SLs. Release based on this clearance criterion would follow the site or operation specific standards provided for in existing DOE regulations. It may include use restrictions or designated end use for the property.

3.1.1 Indistinguishable from Background (IFB) Clearance Criterion

The IFB clearance criterion (i.e., material has no detectable radioactivity other than background) is the lowest and most restrictive criterion. Materials meeting the IFB clearance criterion need not be subject to radiological control and may have unrestricted release. Release based on IFB criterion requires the least justification and regulatory review and should be the preferred clearance criterion. Measurements and clearance based on IFB criterion are more straight-forward to be conducted than those with non-IFB clearance criteria.

The IFB level of a measurement method depends on the detection capability of the chosen method. When IFB is used as the clearance criterion, the measurement method needs to have sufficient detection capability (which is less than the ANSI N13.12 volume SLs) to provide reasonable assurance that materials and equipment released on the basis of this criterion satisfy the 10 μSv/y dose constraint. The measurement methods described in the Standard are designed to meet these objectives.

3.2 Process knowledge and its application

General process knowledge such as beam particle type, beam energy, average power, etc., is applicable to accelerator facilities with similar characteristics and operational modes (in terms of producing induced radioactivity). Specific process knowledge includes operational information such as beam energies and beam losses that are relevant to the specific facility, location and use of equipment within the facility as well as any temporary or long-term storage decisions. Specific process knowledge should be used to support the clearance measurements and decisions.

3.3 Measurement methods

Radionuclides that can be produced in common metals or concrete from activation such as ^{22}Na, ^{54}Mn, ^{60}Co, ^{65}Zn and ^{152}Eu are in Group 1 radionuclides in ANSI N13.12-

2013 Standard, as they emit high-energy and high-yield gamma rays. These proxy radionuclides have a volume SL of 0.1 Bq/g, that is 1/10 of the previous ANSI N13.12-1999 value. For the measurements of potential volumetric radioactivity in this Standard, many of these radionuclides may serve as "proxy" radionuclides for hard to detect radionuclides, because they are easy to be measured with common survey instruments.

The concepts of "measurements of proxy radionuclides" and "maximum surface activity" serve as the technical bases for the surface survey method for volumetric radioactivity recommended by this Standard. These concepts allow for efficient material release surveys in the fields via the following measurement protocols:

- Measurements for proxy radionuclides are necessary and sufficient.
- Surface surveys for proxy radionuclides using portable instruments and techniques with high sensitivity in an environment with low background are adequate and sufficient.
- Surface surveys of all surfaces of an object, regardless of its shape and size, are sufficient. Measurement of only the surface of an object that faces the beam loss point is also acceptable, if the beam loss geometry is known.

The detection capability needs to be evaluated and documented based on the instrument's sensitivity to proxy radionuclides in the materials and the acceptable background in the measurement environment. The sensitivity depends on factors such as the material type, types of radionuclides, volumetric radioactivity distribution, the detector-sample geometry, and the measurement mode such as scanning or fixed-position, etc. The minimum sensitivity and appropriate background level should be used such that the detection capability can be estimated conservatively. The Standard provides examples of appropriate instruments and techniques and estimations of associated detection capabilities. Many sensitive, portable instruments that are commercially available (e.g., scintillator-based survey meters) have capability to detect the proxy radionuclides at levels that are less than the corresponding ANSI N13.12 SL values.

4 Operational experience

The clearance program requirements and protocols prescribed by the DOE Technical Standard have been used for recycling and reuse of large amounts of metals at several DOE accelerator facilities. At SLAC, approximately 2,000 tons of metals have been recycled from Positron Electron Project accelerators and the BaBar detector, SPEAR-II and FFTB since 2011. The types of materials released include cables, pipes, magnets, vacuum chambers and support structures for the accelerators and detectors. These recycling projects create values to DOE and SLAC and avoid costs of disposing the metals as radioactive waste.

Only materials that meet the IFB clearance criterion have been released for recycling at SLAC. Materials from accelerator housings are surveyed for both surface contamination and volumetric activation using practical and sensitive field survey instruments with detection capabilities that are a few times lower than the ANSI N13.12 volume SLs. The measurements using sensitive field survey instruments are supplemented by more sensitive confirmatory measurements such as high resolution field gamma-ray spectrometry. SLAC measurement instruments and methods have sufficiently low detection capabilities that satisfy the requirements in this Standard and are lower than the ANSI N13.12 Screening Levels which correspond to a dose based limit of less than 10 μSv/y.

References

1. IAEA, "Clearance Levels for Radionuclides in Solid Materials", International Atomic Energy Agency (IAEA), Vienna, Austria, IAEA-TECDOC-855, (1996)
2. NCRP, "Managing Potentially Radioactive Scrap Metal," National Council on Radiation Protection and Measurements (NCRP), Bethesda MD, NCRP Report No. 144, (2002)
3. ANSI, "Surface and Volume Radioactivity Standards for Clearance", American National Standard Institute (ANSI), ANSI N13.12, (2013)
4. U.S. DOE, "Clearance and Release of Personal Property from Accelerator Facilities", U.S. Department of Energy Standard, in review and approval process (2016)
5. U.S. DOE, "Radiation Protection of the Public and the Environment", Department of Energy Order 458.1, Chg. 3, (2013)
6. SLAC, "SLAC Material Release Program Manual", SLAC National Accelerator Laboratory (SLAC), Radiation Protection Department procedure RPD-010, (2011)
7. JLAB, "Technical Basis for Release of Solid Materials from Radiological Control when Residual Radioactivity Levels Are Indistinguishable from Background," Thomas Jefferson National Accelerator Facility (JLAB), Radiation Control Department, (2009)
8. A. H. Sullivan, "A Guide to Radiation and Radioactivity Levels near High Energy Particle Accelerators," Nuclear Technology Publishing, ISBN 1-870965-18-3, Ashford, Kent, England, (1992)
9. K. Masumoto, H. Matsumura, K. Bessho, and A. Toyoda, "Role of Activation Analysis for Radiation Control in Accelerator Facilities." Journal of Radioanalytical and Nuclear Chemistry, **278**, No.2, 449–453, (2008)

10. T. Böhlen, F. Cerutti, M.P.W. Chin, A. Fassò, A. Ferrari, P.G. Ortega, A. Mairani, P.R. Sala, G. Smirnov and V. Vlachoudis,, The FLUKA Code: Developments and Challenges for High Energy and Medical Applications", Nuclear Data Sheets **120**, 211-214 (2014)

11. A. Ferrari, P.R. Sala, A. Fasso`, and J. Ranft, "FLUKA: a multi-particle transport code", CERN-2005-10 (2005), INFN/TC_05/11, SLAC-R-773

*This work is supported by the U.S. DOE, Office of Science, Office of Basic Energy Sciences, under Contract No. DE-AC02-76SF00515.

Radiation protection design for the Super-FRS and SIS100 at the international FAIR facility

Ekaterina Kozlova[1,*], *Alexey* Sokolov[1], *Torsten* Radon[1], *Rupert* Lang[1], *Inna* Conrad[1], *Georg* Fehrenbacher[1], *Helmut* Weick[1] and *Martin* Winkler[1]

[1]GSI Helmholtzzentrum für Schwerionenforschung GmbH, Planckstrasse 1, 64291 Darmstadt, Germany

Abstract. The new accelerator SIS100 and the Super-FRS will be built at the international Facility for Antiprotons and Ion Research FAIR. The synchrotron SIS100 is a core part of the FAIR facility which serves for acceleration of ions like Uranium up to 2.7 GeV/u with intensities of $3x10^{11}$ particles per second or protons up to 30 GeV with intensities of $5x10^{12}$ particles per second. The Super-FRS is a superconducting fragment separator, it will be able to separate all kinds of nuclear projectile fragments of primary heavy ion beams including Uranium with energies up to 1.5 GeV/u and intensities up to $3x10^{11}$ particles per second. During operation activation of several components, especially the production target and the beam catchers will take place. For handling of highly activated components it is foreseen to have a hot cell with connected storage place. All calculations for the optimisation of the shielding design of the SIS100, the Super-FRS and the hot cell were performed using the Monte Carlo code FLUKA, results are presented.

1 Introduction

FAIR [1, 2], Facility for Antiproton and Ion Research, will be a new international accelerator facility for the research with antiprotons and ions. It will be based upon an expansion of the GSI Helmholtz Centre for Heavy Ion Research which is situated in Darmstadt, Germany.

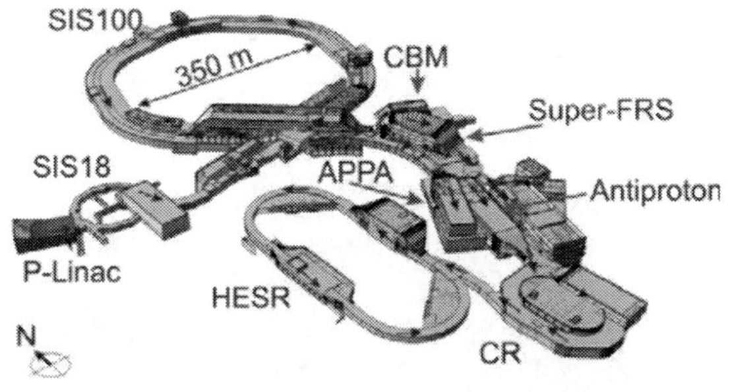

Fig. 1. The FAIR accelerator facility with the synchrotron SIS100, the experiments CBM, Super-FRS, APPA and PANDA, as well as the storage rings HESR and CR. The injected beam for FAIR comes from the GSI accelerator SIS18.

The existing accelerator facility at GSI after some appropriate modifications will be used as a pre-accelerator for the FAIR facility. Extracted from the pre-accelerator, high-intensity ion beams will be injected into the SIS100 and accelerated to the energy demanded by the experiments. Fig. 1 shows an overview of the FAIR facility with the SIS100 accelerator, the Super-FRS as well as the storage rings CR (Collector Ring), HESR (High Energy Storage Ring) and other experimental areas of different collaborations.

In this paper the radiation protection aspects of the SIS100 and the Super-FRS are discussed. They are the crucial parts of the facility in terms of radiation production.

2 Prompt dose rate calculations

The main role of radiation protection consists in avoiding unwanted exposure of man and environment to ionising radiation. The German radiation protection ordinance has defined dose limits which affect the design of the planned accelerator facilities. The effective dose rate must not exceed 1 mSv per year for all areas and accelerator installations, which are within the premises and are freely accessible for workers and visitors. Since a working year is counted in this respect as 2000 working hours, the dose rate limit is 0.5 μSv/h. Thus for the shielding design the dose rate outside the shield must not be larger than this limit.

All calculations of the prompt dose rates for the SIS100 and Super-FRS were done with the Monte Carlo code FLUKA [3, 4]. It is able to perform simulations of interactions with matter and propagations in matter for about 60 different particles. The transport of heavy ions was added in 2004 employing the RQMD and DPMJET models [5, 6]. FLUKA has the capability of scoring fluences, energy depositions, activities, physical doses,

* Corresponding author: e.kozlova@gsi.de

etc. Dose rates were calculated with the help of fluence-to-dose conversion factors of [7, 8] linked to the code.

In the calculations the electro-magnetic cascade was switched off because photon and electron contributions to the integral dose value play a minor role in the concept of the shielding design here.

The use of biasing techniques was essential to obtain results with reasonable statistical significance. To apply Russian roulette and splitting at boundaries of certain volumes the shielding was divided into many layers of 50-cm thickness in case of concrete.

2.1 SIS100

The SIS100 with circumference of 1083.60 m is five times longer than the injector SIS18 and can reach a magnetic rigidity of Bρ = 100 Tm. It is located in the underground Tunnel 110 (shown on Fig.2). The SIS100 accelerates high energy and high intensity proton and ion beams by the synchrotron principle, in which charged particles are accelerated in high-frequency cavities and bend on a circular path with a constant radius by magnetic fields. The extracted primary particles from the SIS100 are protons and heavy ions up to uranium from which secondary particle beams of antiprotons and rare isotopes can be generated. These ions are applied for experiments in the fields of atomic, nuclear and plasma physics, and also for biophysics and materials research [1, 2].

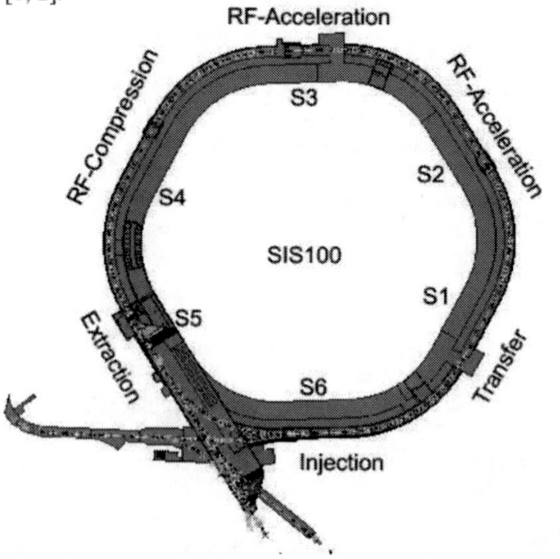

Fig. 2. Schematic layout of the SIS100 with injection, extraction, acceleration and beam compression on the straight sections of the sectors S1 - S6 along the circumference of 1083.60 m.

The key parameters for the primary beams are summarised in Table 1:

Table 1. Maximum energies and intensities for accelerated particles in SIS100.

Type of particle	Energy	Intensity [s^{-1}]
Protons	29.0 GeV	$5x10^{12}$
$Uranium^{28+}$	2.7 GeV/u	$3x10^{11}$

Such parameters can be achieved be applying new challenging features for the SIS100 synchrotron:

- operation at a very low base pressure of 10^{-12} mbar,
- careful control of beam losses,
- superconducting synchrotron magnet operation at a high ramp rate of 4 T/s,
- bunch compression of high intensity proton and uranium ion beams.

Tunnel T110, in which the SIS 100 is located, has many levels, three of them are underground (U10, U20 and U30), the other two (E10 and E20) are above. The level U30 basically consists of a ring-shaped beamline tunnel with the SIS 100 and a supply tunnel, separated by a total of 9.7 m wide shielding package as it is shown in Fig. 3 and Fig. 4. The space in the support tunnel is mostly designated for the electrical cabinets, which will provide the electrical support of the accelerator. The upper levels of the support tunnel house the extensive ventilation systems as well as supply rooms for cooling agents (water and liquid helium). Dedicated tracks connect the supply area with the accelerator. It is also planned to have three decay rooms in the support tunnel for the internal storage of not anymore needed or defective accelerator components. After sufficient decay time the activated components can be transported to the building for radioactive waste (Building 21). Building 21 is situated on the FAIR site and is used for the storage of the activated components and radioactive waste which will be produced during the facility operation.

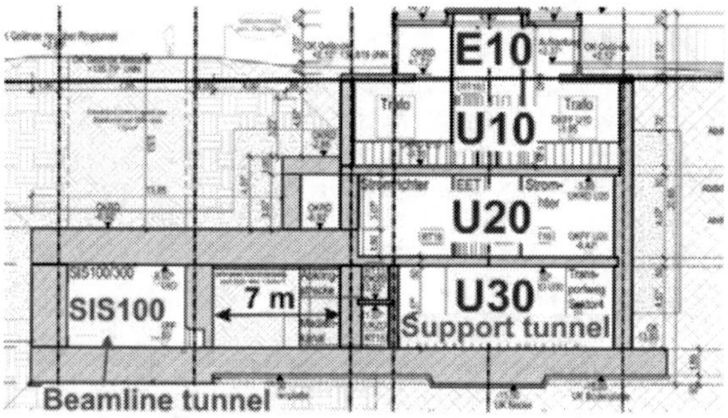

Fig. 3. Building plan for the SIS100 building showing cross section of the SIS100 beamline tunnel and the support tunnel. Both tunnels are divided by an earth layer.

For the calculation of the prompt dose rates the worst scenario with a 2.7 GeV/u uranium beam and local beam losses of $1x10^{10}$ ions per second (3% of the total beam intensity) are assumed. The result of the calculation is shown on Fig. 4. The dose rates above the soil layer on the public ground are below 50 nSv/h. During synchrotron operation the dose rates in the supply tunnel and rooms above are less than 0.5 μSv/h. The access to the areas is allowed during beam operation. The calculations for $1x10^{11}$ protons per second with the energy of 30 GeV were also done. The resulting dose rates are close to the case of uranium losses. The assumed proton losses are approximately 2 % of the total beam intensity, this should be very conservative for this part of the synchrotron ring as such losses are not expected according to beam loss estimations [9].

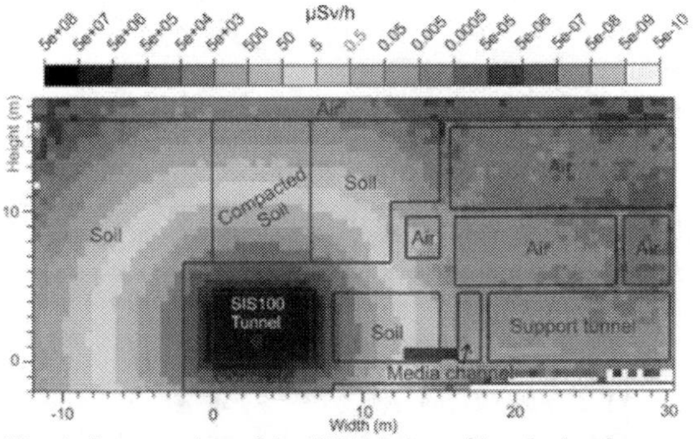

Fig. 4. Cross section of the SIS100 ring with calculated prompt dose rate in the SIS100 beam tunnel together with support tunnel, media channel and adjacent rooms.

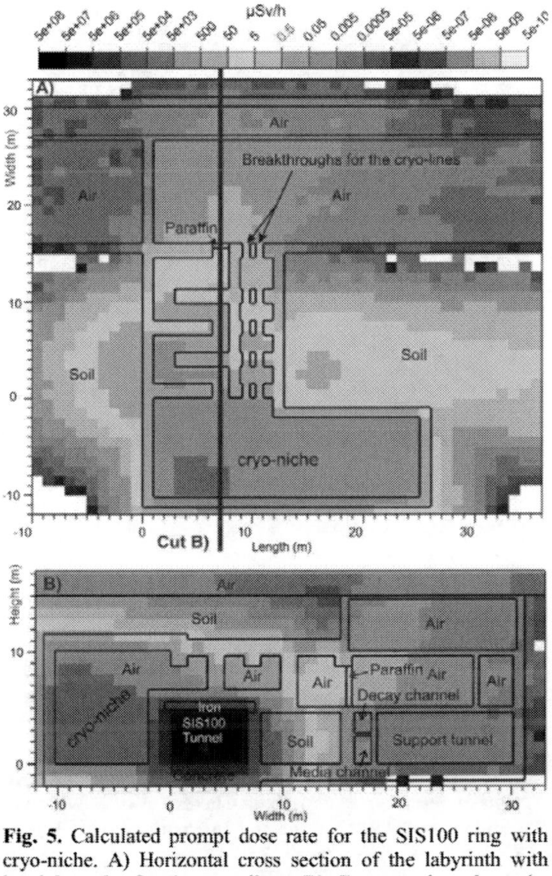

Fig. 5. Calculated prompt dose rate for the SIS100 ring with cryo-niche. A) Horizontal cross section of the labyrinth with breakthroughs for the cryo-lines. B) Cross section along the labyrinth, SIS100 beamline tunnel, support tunnel, cryo-niche and other rooms at different height levels.

On the outer side of the beamline tunnel three cryo-niches are planned. The niches will provide the connection of the synchrotron via cryo-lines with the cryo-stations, which are situated on level U20. On Fig. 5 (A-top view) the cross section of the U20 level with the maze and additional access labyrinth for the cryo-lines are shown. Due to technical requirements of the cryo-system the cryo-lines should go straight to the cryo-niches via breakthroughs in the additional access labyrinth. A location of 1 meter thick iron plate above the beamline tunnel and a 0.5 meter paraffin wall at the end of the labyrinth (as presented on Fig. 5 B) helps to reduce the dose rates above the building down to below 50 nSv/h.

2.2 Super-FRS

The Superconducting magnetic FRagment Separator Super-FRS represents a major component of the planned FAIR [10]. The Super-FRS is an in-flight facility similar to the existing FRS at GSI [11] and serves for the production of rare isotopes at relativistic energies via fragmentation or fission in flight. The Super-FRS uses the Bρ-ΔE-Bρ method, where a two-fold magnetic rigidity analysis is applied in front of and behind a shaped energy degrader. Compared to the existing separator the Super-FRS has more separation stages and twice larger magnet apertures made possible by use of superconducting coils.

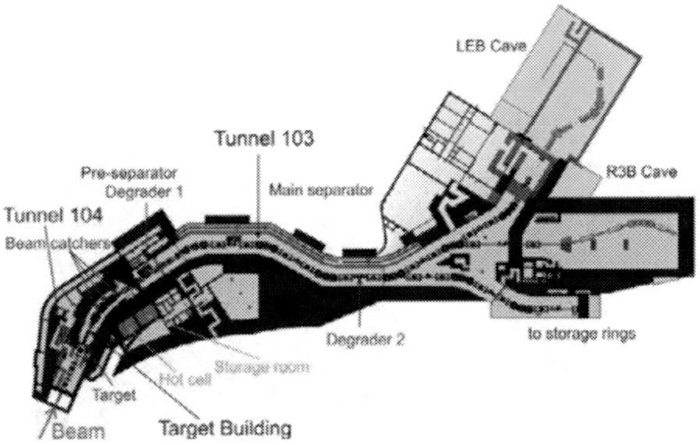

Fig. 6. Super-FRS layout, including the buildings and tunnels.

The SIS100 synchrotron will provide all kinds of heavy ion beams up to ^{238}U with maximum intensities of $3x10^{11}$/s and beam energy up to 1.5 GeV/u for the Super-FRS. As it is shown in Fig. 6 the accelerated beam hits the production target after the focusing system. The created fragments leave the production target in forward direction, and then the isotope of interest can be selected depending on the setting of the separator.

For cleaner separation the Super-FRS applies the Bρ-ΔE-Bρ method twice in the pre- and in the main-separator, both equipped with a degrader in the beam. After separation the rare-isotopes can be directed to one of the three experimental areas:

- The High-Energy Branch (HEB) where reactions of high-energy heavy nuclei will be investigated,
- The Low-Energy Branch (LEB) where properties of nuclei such as decay modes and energy levels can be explored using gamma and decay spectroscopy as well as trapped ions,

- The Ring Branch (RB) where exotic nuclei are collected, cooled and stored in the Collector Ring CR. In the CR masses and life-times of unknown nuclei will be measured [2].

The production of rare isotopes in the target by fragmentation of heavy nuclei is accompanied by creation of neutrons of similar high velocities as that of projectile [12, 13]. In Fig. 7 the pre-target focusing system of the ion-beam coming from the SIS 100 is shown, followed by the Super-FRS production target. The total beam losses can amount to $1x10^{8}$ ^{238}U/s in the quadrupoles before the target in addition to the unavoidable beam losses in the target ($3x10^{11}$ ^{238}U/s). The beam energy is 1.5 GeV/u. The dose rates around the quadrupole triplet just in front of the production target are still high due to the strong neutron backscattering from the production target. For the access to the tunnel an entrance maze is foreseen. It is designed in such a way, that dose rates outside of the shielding and close to the maze in the stairwell, during Super-FRS operation are below 0.5 μSv/h.

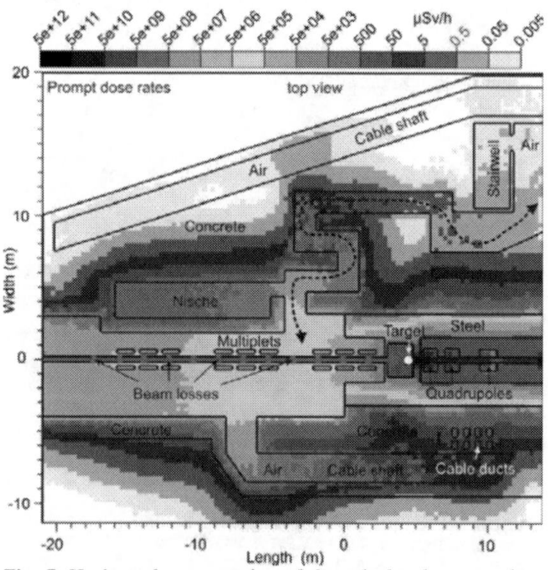

Fig. 7. Horizontal cross section of the calculated prompt dose rate distribution in the area upstream of the Super-FRS production target. The focusing system is situated in Tunnel 104, the production target is located inside Building 18.

In the target and beam catchers area the shielding concept foresees a close compact shielding of the beamline to reduce the overall volume required due to the vicinity of other beamlines of the FAIR complex. To achieve a compact shielding, a multi-layer shield of steel followed by concrete is used. In the inner steel layer the high-energy neutrons are attenuated by elastic scattering and inelastic interaction (spallation reactions). In the outer regions of this steel layer the dose rate is dominated by contributions of low-energy neutrons and in the concrete region low-energy neutrons are absorbed due to the elastic scattering effect of hydrogen contained in the concrete. In Fig. 8 the distribution of dose rates towards the vertical direction in the plane intersecting the beam catchers is shown. The beam catchers are foreseen to absorb the primary beam at positions where its diameter is larger than at the target. In total six of these beam catchers are planned [12]. They are located in the first half of the Super-FRS pre-separator always following one of the 11° dipole magnets as it is shown in Fig. 6. Depending on the experiment each beam catcher can absorb up to 85% of the non-interacted primary beam. In Fig. 8 the beam line is surrounded by the steel core and followed by the concrete shielding. The dose rates outside the shielding are less than 0.5 μSv/h. The drives, connections and supplies for the target, beam catchers and many other components are done from above through the iron shielding. Therefore, on top of the iron shielding a service tunnel is provided for faster access. Otherwise the whole area is covered with a mobile ceiling consisting of 5 layers of concrete bars that can be taken off by the hall crane of the building.

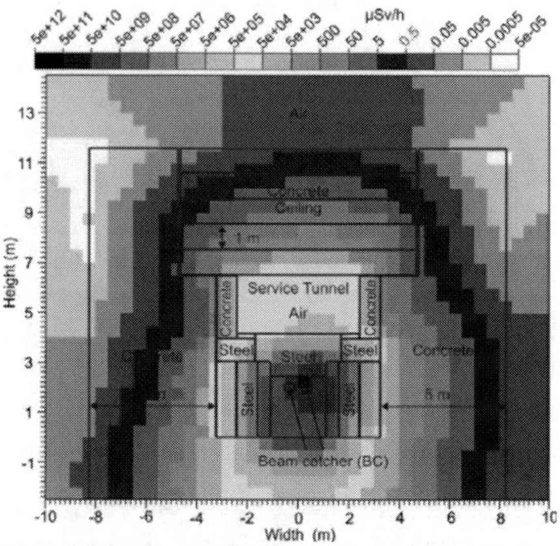

Fig. 8. Prompt dose rate calculations for 1.5 GeV/u Uranium beam of intensity of 3E11 U/s in a cross section perpendicular to the beam at the beam catcher position.

As all primary beam and also the majority of the created fragments will be stopped in the target and beam catchers area the following part of the separator can be in an open tunnel with concrete shielding or a combination of concrete with soil as shown in Fig. 9 for the main separator. For the calculations of the dose rates in the main separator a ^{114}Pd beam as a typical fragment of the primary uranium beam was taken. ^{114}Pd will have energies up to 1.3 GeV/u and intensities up to $5.5x10^{10}$ ions/s. The ions interact with two energy degraders and two slits. After a degrader ions have an energy lowered by a different amount depending on their atomic number and after the following separator stage the unwanted elements can be stopped in the thick tungsten slits. The second stage of the pre-separator and main separator is located in Tunnel 103.

In Fig. 9 all beam losses are marked. The number of ions hitting a degrader, due to their function, is higher

DOI: 10.1051/epjconf/201715303003

than on the slits. If Degrader 1 is hit by $3x10^{10}$ ions/s and Degrader 2 by 10^{10} ions/s, in the front slits 10^{10} ions/s and in the back slits $5x10^{9}$ ions/s are deposited. As the ions are not lost in the aluminium degraders plates but only converted to some fraction or produce secondaries in the material in the FLUKA simulation they were modelled as plates of maximal useful thickness in a beam. The shielding in the Tunnel 103 is a composition of concrete walls with a thickness of 0.5 m and 2.5 m of compacted soil in between, followed by at least 7 m of soil outside the building. Along the main tunnel with the separator there is the supply tunnel with control racks for the cryo- and vacuum support, data acquisition, quench detection and protection electronics, etc. The supply tunnel is subdivided from the main tunnel by a concrete wall with 0,5 m thickness.

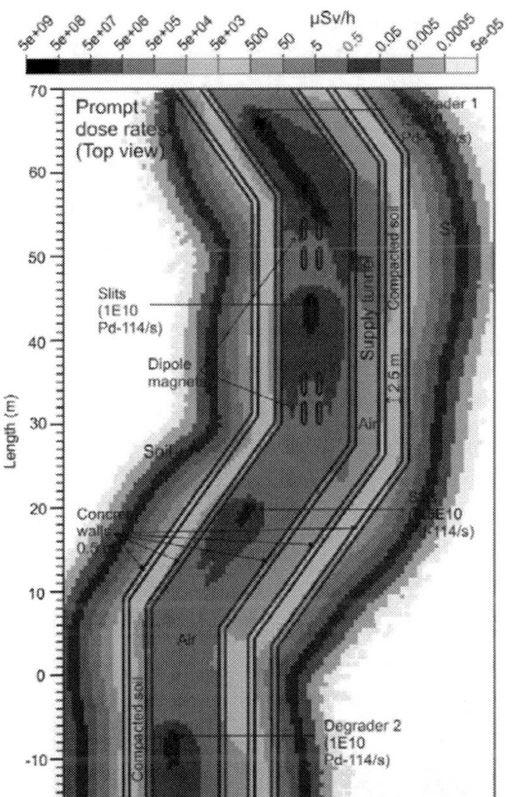

Fig. 9. Horizontal cut showing the prompt dose rate distributions in Tunnel 103 with four interaction points, two degraders and two slits.

3 Hot cell

The operation of the Super-FRS is inevitably associated with significant radioactive activation of components and targets where the primary beam interacts directly with materials. Activated components which have to be exchanged or maintained after an irradiation period cannot be handled by personnel directly. Therefore, it is obligatory to have a hot cell for handling these components using remote handling methods.

The use of the hot cell is planned as follows: The component (e.g. the production target) will be moved from the beam line into a shielding flask, which - on its part - will be moved to the roof position of the hot cell. The component within the flask will be lowered to the handling position in the hot cell, where it can be repaired or cut, separated and packed in barrels and moved over to a shielded storage position for radioactive decay (see Fig. 10). Maintenance work of the components will be carried out by means of remote manipulators. The hot cell with the storage place will be situated close to the Super-FRS target area as indicated in Fig. 6. The shielding layout of the hot cell was developed on the basis of activation calculations for components of the Super-FRS. The calculated radionuclide distributions within the components were used as radiation sources for estimating the shielding of the transport flask, the hot cell and the storage places in the neighbouring cell. The shielding design was done by using the FLUKA code in two steps. In the first step a simulation of production of the radioactive nuclides with build-up and decay for a certain irradiation pattern and different cooling times were performed. The simulations of the radioactive decay and the transport of the produced radiation were done in the second step with a new shielding geometry.

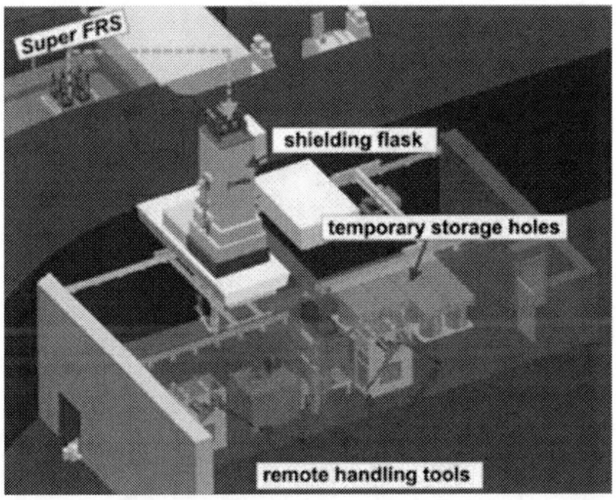

Fig. 10. Layout of the Super-FRS target and beam catchers area with the hot cell and the storage cell.

The largest and the most activated parts that have to be handled with the shielding flask are the beam catchers [12, 13]. Fig.11 shows as an example of the plug with the beam catcher. Beryllium and graphite beam absorbers are used to stop the heavy-ion beam or to convert it to lighter fragments and to absorb most of the beam energy. The beryllium absorber is for the slow beam extraction and graphite absorber is for the fast beam extraction. The beam absorbers are coupled to a copper heat sink and mounted to an iron shielding plug with a size of 60 cm x 80 cm x 174 cm with drives, vacuum seals and feedthroughs on top. The total weight that has to be handled by the overhead crane is 7.5 t.

DOI: 10.1051/epjconf/201715303003

After a long production period at maximum uranium beam intensity the activation reaches up to $3x10^{13}$ Bq [12, 13]. To obtain a realistic scenario for the estimation of the required wall thickness of the shielding flask, in a first step the beam catcher and plug were irradiated including the shielding environment of the Super-FRS. All information on the produced radionuclides is written to external files via a FLUKA user-routine (usrrnc.f). In the second step the beam catcher and plug are moved in the shielding flask and the information on radionuclides is read in from the file created in the first step via another user-routine (source.f). This method provides an accurate distribution of the nuclides in the device. The result is presented in Fig. 12. As realistic scenario, 3 days after four periods of 90 days of irradiation with an intensity of $3x10^{11}$/s with a cooling time of 120 days between the blocks are assumed. In the example case only the graphite part was irradiated. The dose rates outside the shielding flask are below 10 μSv/h. During the transport of the activated components in the flask to the hot cell the target hall is specified as a controlled area.

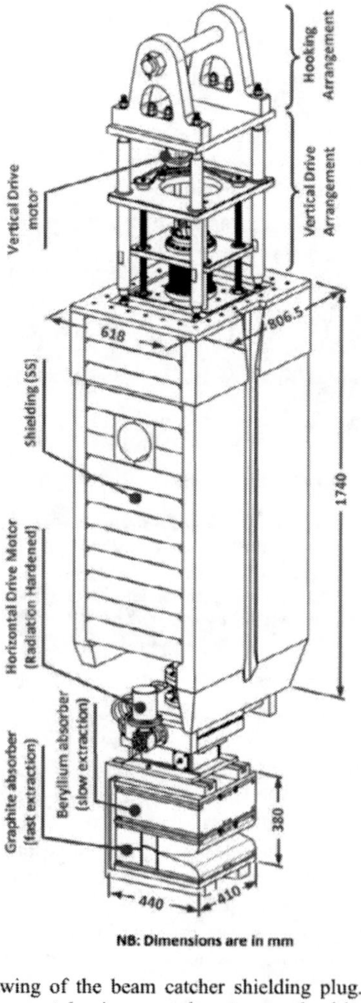

Fig. 11. Drawing of the beam catcher shielding plug. At the bottom the beam catcher is mounted, on top are the drives and a hook for removing. The catcher consists of the graphite and beryllium absorber.

The two step method was also applied for the optimization of the wall thickness in the hot cell. The activated beam catcher with the plug was situated in the center of the hot cell. In this case to achieve the dose rate outside the shielding below 0.5 μSv/h at the concrete wall with a thickness of 1 meter is required.

For the design of the storage place also the activated beam catchers were taken in the calculations. It is foreseen to have 20 places for the 200-l barrels with two levels (10 barrels on each level). In Fig. 13 four activated beam catchers after cooling times of 1 week, 1 month, 0.5 year and 1 year are situated in the storage holes with a concrete wall and iron cover. It is seen from Fig. 13 that the shielding wall thicknesses are sufficient to lower the dose rates outside the storage place to an appropriate level. The holes are situated in the storage room with the concrete wall thickness of 1 m. Thus the dose rates outside the storage room will be below 0.5 μSv/h.

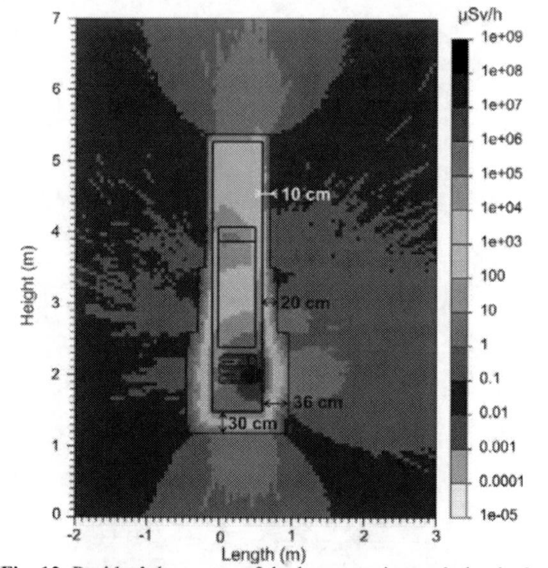

Fig. 12. Residual dose rates of the beam catcher und plug in the shielding flask.

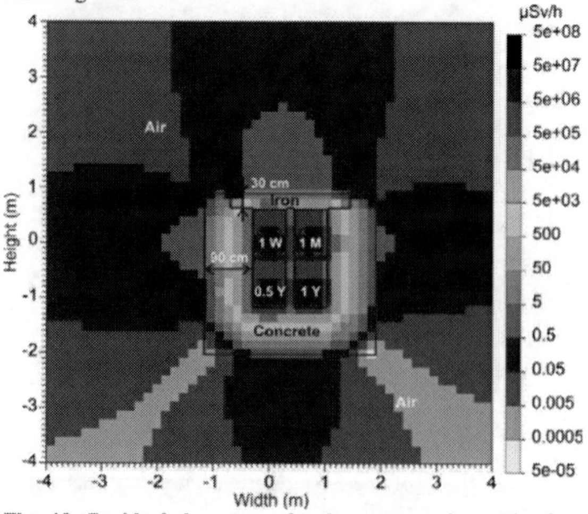

Fig. 13. Residual dose rates for the storage place. The four activated beam catchers are situated in the holes

4 Conclusions

This paper presents the radiation protection studies for the SIS 100 and Super-FRS including the hot cell. Monte Carlo calculations were used for approving the shielding thickness and estimation of the propagation of neutrons through the mazes, shafts, cable ducts in the accessible areas. The calculations show that the shielding design for the SIS100 and Super-FRS is adequate from a radiation protection standpoint. The design value, to keep dose rates below 0.5 μSv/h outside of the shielding and in all supervised areas nearby, is achieved.

A set of FLUKA calculations for the hot cell, the storage place and the flask was also performed to assess the residual dose rates inside and outside the shielding. In this case the calculations were done with a two steps method. In the first step the activation of some components and targets after several periods of Super-FRS operation and for different decay times was estimated. In the second step the residual dose rates for the hot cell, the storage place and the shielding flask from the most activated components, beam catchers, were calculated. As the result a suitable shielding design was developed on the basis of the FLUKA calculations.

The engineering design of Beam Stoppers for Super-FRS in FAIR project is currently ongoing at CSIR-CMERI, Durgapur, India under DST-DAE grant (GAP-098212) through BI-IFCC, Govt. of India. The authors acknowledge the contribution of Dr. Avik Chatterjee and Abhijit Mahapatra, Advance Design and Analysis Group, CSIR-CMERI for their contribution in engineering design.

References

1. An International Accelerator Facility for Beams of Ions and Antiprotons, Conceptual Design Report, GSI November 2001.
2. FAIR Baseline Technical Report, Darmstadt (2006). (https://www-alt.gsi.de/documents/DOC-2006-Jul-40-1.pdf)
3. A. Fassò, A. Ferrari, J. Ranft, and P.R. Sala, FLUKA: a multi-particle transport code, CERN 2005-10 (2005), INFN/TC_05/11, SLAC-R-773
4. A. Fassò, A. Ferrari, S. Roesler, P.R. Sala, G. Battistoni, F. Cerutti, E. Gadioli, M.V. Garzelli, F. Ballarini, A. Ottolenghi, A. Empl and J. Ranft, The physics models of FLUKA: status and recent developments, Computing in High Energy and Nuclear Physics 2003 Conference (CHEP2003), La Jolla, CA, USA, March 24-28, 2003, (paper MOMT005), eConf C0303241 (2003), arXiv:hep-ph/0306267
5. H. Sorge, Flavor production in Pb(160A GeV) on Pb collisions: Effect of color ropes and hadronic rescattering, Physical Review C **52** (1995) 3291-3314.
6. J. Ranft, Phys. Rev. D51 (1995) p.64; New features in DPMJET version II.5, hep- ph/9911213 and DPMJET version II.5, code manual, hep-ph/9911232.
7. International Commission on Radiological Protection, "Conversion Coefficients for use in Radiological Protection against External Radiation", ICRP Publication 74, Pergamon Press (1996).
8. A. Ferrari and M. Pelliccioni, Fluence to Dose Equivalent Conversion Data and Effective Quality Factors for High Energy Neutrons, Rad. Prot. Dosimetry **76** (1998) 215.
9. Beam Loss in SIS18/100/300/HEBT. Status 24.1.2009 – Version 7.0. FAIR Synchrotrons.
10. H. Geissel, M. Winkler, H. Weick, et al., Technical Design Report on the Super-FRS, 2008
11. H. Geissel, et al., The GSI projectile fragment separator (FRS): a versatile magnetic system for relativistic heavy ions, Nucl Instr. And Meth. in Phys. Res. B **70**, 286-297 (1992)
12. E. Kozlova, H. Weick et al., Layout of the Super-FRS target hall, Nucl Instr. And Meth. in Phys. Res. B **266**, 4275-4279 (2008)
13. T. Radon, E. Kozlova et al., Shielding of the target area of the fragment separator Super-FRS. Int. Conf. on Radiation Shielding (ICRS-11) **168**, 492-496 (2009)

Preliminary design of CERN Future Circular Collider tunnel: first evaluation of the radiation environment in critical areas for electronics

Angelo Infantino*, *Rubén* García Alía, *Maria Ilaria* Besana, *Markus* Brugger and *Francesco* Cerutti

European Organization for Nuclear Research (CERN), Geneva 23, Geneva CH-1211, Switzerland

Abstract. As part of its post-LHC high energy physics program, CERN is conducting a study for a new proton-proton collider, called Future Circular Collider (FCC-hh), running at center-of-mass energies of up to 100 TeV in a new 100 km tunnel. The study includes a 90-350 GeV lepton collider (FCC-ee) as well as a lepton-hadron option (FCC-he). In this work, FLUKA Monte Carlo simulation was extensively used to perform a first evaluation of the radiation environment in critical areas for electronics in the FCC-hh tunnel. The model of the tunnel was created based on the original civil engineering studies already performed and further integrated in the existing FLUKA models of the beam line. The radiation levels in critical areas, such as the racks for electronics and cables, power converters, service areas, local tunnel extensions was evaluated.

1 Introduction

1.1 From the LHC era to FCC

The main goal of CERN's mission is pushing back the frontiers of human knowledge and, in doing so, drive innovation, develop forefront technologies, stimulate international collaboration and inspire tomorrow's leaders in science and technology. This mission is reached particularly through fundamental research in physics. In the last years, the Large Hadron Collider (LHC) allowed to push back the frontiers of human knowledge as anyone before bringing to the discovery of the Higgs boson in 2012, from the ATLAS and CMS experiments. The LHC and its high-luminosity upgrade, the HL-LHC, have an exciting physics program which extends through the mid 2030's [1].

The Future Circular Collider (FCC) is an integral conceptual design study for post-LHC particle accelerator options in a global context. The study is conducted within a collaboration that is open to scientific institutes and companies of any size from all nations and it currently counts 70 institutions spread in 26 countries. The FCC study has an emphasis on proton-proton collider (FCC-hh) but includes also a 90-350 GeV lepton collider (FCC-ee), seen as a potential intermediate step, as well as a lepton-hadron option (FCC-he) [2]. The FCC study is exploring the potential of hadron and lepton circular colliders, performing an in-depth analysis of infrastructure and operation concepts and considering the technology research and development programs that would be required to build a future circular collider. The need for building such large circular hadron collider is to reach energy levels far beyond the range of the LHC to provide access to the direct production of new particles with masses up to tens of TeV, as well as to obtain much increased rates for phenomena in the sub-TeV mass range, with the corresponding greatly improved precision [1].

The goal of FCC-hh is to provide proton-proton collisions at a centre-of-mass energy of 100 TeV. Assuming a nominal dipole field of 16 T, such a machine would have a circumference of the order of 100 km (Figure 1).

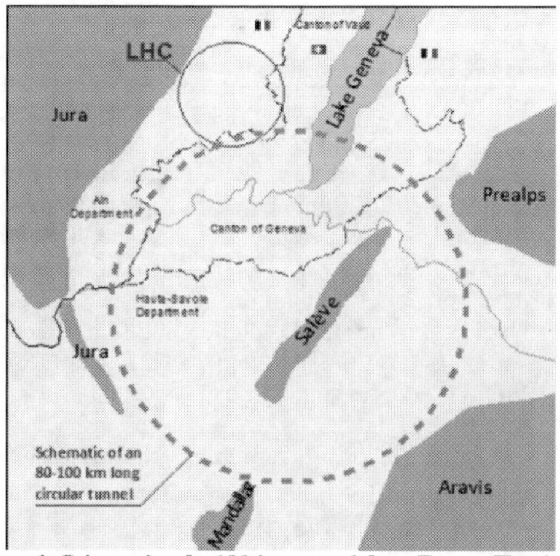

Figure 1. Schematic of a 100 km tunnel for a Future Circular Collider in the Lake Geneva basin [1].

The machine will accommodate two main proton experiments that are operated simultaneously but the

* Corresponding author: angelo.infantino@cern.ch

layout allow for two additional special-purpose experiments. The machine will deliver a peak luminosity of 5 - 25·10^{34} cm^{-2} s^{-1} [3]. The machine is compatible with ion beam operation. The main baseline features of FCC-hh are reported in Table 1. The realisation of a machine with 100 km circumference poses an enormous challenge in the availability and reliability of all kind of systems, since interventions and repairs would imply long downtime.

Table 1. FCC-hh baseline parameters compared to LHC and HL-LHC [3].

	LHC	*HL-LHC*	*FCC-hh*
c.m. Energy [TeV]	14		100
Circumference [km]	26.7		100
Dipole Field [T]	8.33		16
Number of interaction points	2+2		2+2
Injection Energy [TeV]	0.45		3.3
Peak Luminosity [10^{34} $cm^{-2}s^{-1}$]	1	5	5
Optimum run time [h]	15.2	10.2	12.1
Bunch Population [10^{11}] (25ns)	1.15	2.2	1.0
Number of bunches (25 ns)	2808		10600
Beam Current [A]	0.584	1.12	0.5
Stored Energy per beam [GJ]	0.392	0.694	8.4

1.2 Radiation to electronics studies for FCC

Radiation to Electronics (R2E) represents a crucial issue to be taken into account as design criteria of any high energy and intensity machine. Radiation effects in electronic devices can be divided into two main categories: cumulative effects and Single Event Effects (SEE). Cumulative effects, proportional to Total Ionizing Dose (TID), are due to damage induced by ionizing radiation and displacement atoms: the accumulation of defects causes measurable effects that can ultimately lead to device failure. On the other hand, SEE, proportional to High Energy Hadrons fluence (HEH, i.e. hadrons > 20 MeV), are due to the direct or indirect ionization by a single particle, able to deposit sufficient energy in order to disturb the operation of the device. They can only be characterized by their probability to occur, given their stochastic nature, which strongly depend on the device, the intensity and the kind of radiation filed [4]. The risk of failure of the before mentioned SEE becomes high as electronic components (e.g. logic, power devices) are present in a growing number of equipment. It is important to note that the radiation environment encountered in a high energy and intensity accelerator, the high number of electronic systems and components exposed to radiation, as well as the actual impact of radiation-induced failures on the machine operation, pose challenges not common in any other field.

In this context, in 2007 CERN decided to create a dedicated R2E task force (R2E-project) which consists of experts in various fields related to electronics damage: indeed, the study of the electronics sensitivity to radiation requires a multi-disciplinary approach, spanning from the knowledge of the electronic components, to the radiation environment, and to the physics models that describe the interaction of the radiation with matter.

In the past, a considerable amount of work has been done in R2E-oriented studies for LHC [4-7]. In this context, thanks also on the extensive experience maturated for LHC, the CERN's R2E project is currently involved in the analysis of the critical areas for electronics for FCC. As required for the LHC, FCC will need a significant amount of electronic components in the accelerator tunnel and in the side galleries to control and monitor the various infrastructures and systems: given that, an a priori evaluation of the radiation environment, the technology that would be required, the failure rate and the possible mitigation actions become strategic for a long-term planning.

For a high intensity and energy machine like FCC, typical sources of radiation are luminosity debris, direct losses on collimators and dumps, and beam interactions with the residual gas inside the vacuum chamber all along the accelerator; in addition, as regards FCC-ee, the main source of radiation in the tunnel is represented by the synchrotron radiation [8]. Moreover, the shape of the tunnel and the shielding thickness and configuration affects the radiation environment and consequently the effects on the electronic components. For these reason, the design of the FCC tunnel and the interaction regions becomes crucial for the choice of the technology to be used in the most important electronic components, e.g. control units or safety systems.

1.3 The FLUKA code and the FCC model: scoring capabilities important for shielding and radiation damage to electronics studies

In order to evaluate the impact of the radiation on the machine equipment, MC simulation represents an indispensable tool but it needs to rely both on a refined implementation of physics models of the particle interaction with matter and an accurate 3D-description of the region of interest [8].

In this context, FLUKA [9,10] is a well benchmarked, multi-purpose and fully integrated particle physics Monte Carlo code for calculations of particle transport and interactions with matter, covering an extended range of applications like for example proton and electron accelerator shielding, target design, calorimetry, activation and dosimetry, cosmic ray studies, and radiotherapy.

FLUKA is constantly employed in the majority of CERN technical and engineering applications such as machine protection issues, energy deposition calculations and material damage to accelerator elements, or shielding design. In R2E context, FLUKA has been extensively employed in the past for studies of the radiation levels in critical areas for electronics in the LHC environment [4-8] as well as in modelling SEE [11,12] demonstrating its capabilities in this field; recently it has been used in the FCC study [13,14] for characterization of the radiation field in the FCC-hh detector.

2 Aim of the work

A conceptual design report for FCC will be delivered before the end of 2018, in time for the next update of the European Strategy for Particle Physics. In this context, this work will provide a first evaluation of the radiation environment in critical areas for electronics, based on FLUKA Monte Carlo (MC) simulation, for the FCC tunnel. In particular, a model of the tunnel in the arc section was created based on the original civil engineering (CE) studies already performed and further integrated in the existing FLUKA models of the beam line. Furthermore, the radiation environment were evaluated in critical areas for electronics such as electronic racks, power converter location and auxiliary tunnel.

Finally, first considerations on the impact of the radiation environment on electronics have been conducted for an a priori evaluation of possible relocation or mitigation actions.

3 Materials and Methods

FCC-hh aims to provide, compared to LHC, proton collisions a factor ~7 and ~3 higher in energy and bunches number respectively. The radiation levels are expected to scale with the beam intensity for direct losses, with luminosity for collision debris, and with both beam current and residual gas density for beam-gas interactions. At the same time the higher energy will lead to a further increase. While direct losses or collisions debris will be typically the dominating sources of radiation for specific FCC areas, a significant contribution from beam-gas interactions is expected in the large part of the accelerator due to the extension of the arc segments.

An analytical scaling of the radiation levels in a given section of the machine may result generally unreliable given the competition of multiple sources of radiations. In the arc section, a tentative analytical scaling can be performed given that beam-gas interaction represents the dominant source of radiation: a projection of the radiation levels in the FCC arc cell can be performed, in first approximation, by a scaling from LHC data taking into account the different energy, the beam current and the residual gas-density profile within the vacuum chamber. The energy can be scaled using the approach proposed by Stevenson [15] where the energy scales as a power of the ratio of the energy to a reference energy (i.e. LHC beam energy). The current, as reported in Table 1, is almost the same for the two machines. The residual gas-density profile of LHC, in the arc section, is unknown: generally, the gas-density in the LHC arc is considered to be $\leq 10^9$ H_2 molecules per cm^3. Due to the lack of knowledge of the residual gas-density profile in the LHC arc, we decided to go through a detailed MC simulation of the FCC machine. Indeed, in opposition to analytical methods, MC simulation allows for an accurate modelling of the particle transport but it needs to rely both on a refined implementation of physics models of the particle interaction with matter and an accurate 3D-modelling of the region of interest [8].

In the present work, FLUKA simulation was used for a detailed modelling of the radiation environment in the FCC arc taking into account all the main parameters that can affect the radiation field such as the expected gas-density profile along the beam orbit inside the vacuum chamber; the beam optics; the accurate geometric and material reproduction of all the main elements of the generic arc cell including dipoles, quadrupoles, tunnel elements.

At CERN, a very detailed library of materials and geometry descriptions is available for accelerator elements such as dipoles, quadrupoles, collimators, etc. Based on this library, the baseline parameters and the most up-to-date studies on the beam optics, the generic arc cell was built, as shown in Figure 2. The generic arc cell is made of 12 bending dipoles and 2 quadrupoles (six dipoles followed by one quadrupole). Each dipole is 14.3 m long while the quadrupoles are 6.9 m long: the full arc cell has a total length of 213 m. A detailed model of the tunnel was created based on the current layout designed by the CE. The cross section of the tunnel, 6 m in inner diameter, is shown in Figure 3.

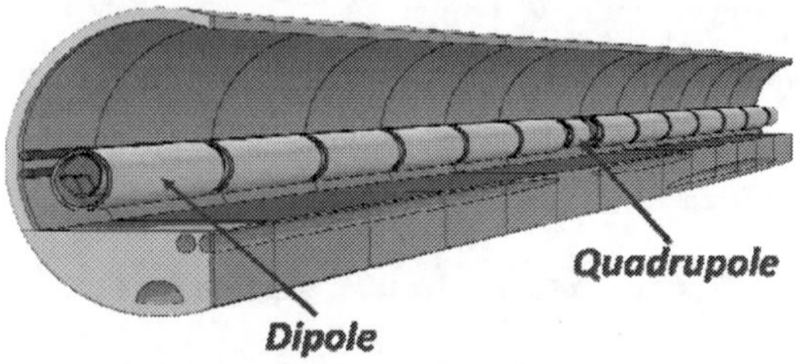

Figure 2. 3D-modelling of the FCC FLUKA arc cell: the cell is made of 12 bending dipoles and 2 quadrupoles, labelled in the following B1-B12 and Q1-Q2 respectively.

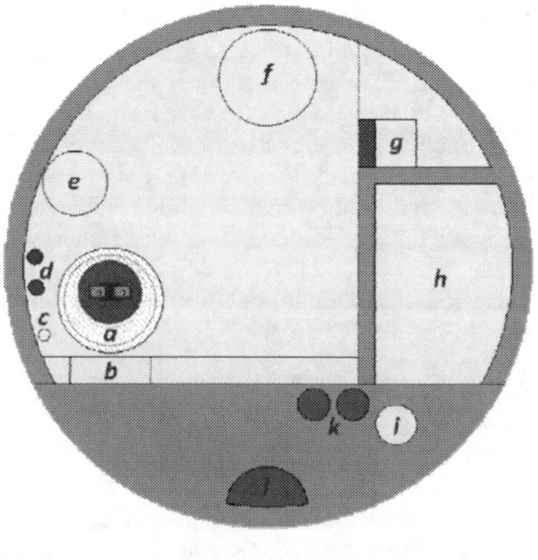

Figure 3. Cross section of the FLUKA model of current version of the FCC tunnel. The main elements included in the tunnel are a) dipole, b) power converter location, c) compressed air pipe, d) Helium recovery lines, e) cryo line, f) emergency extraction duct, g) electronics rack location, h) safe passage tunnel, i) safe passage ventilation, k) cooling water pipes, l) drainage.

Beam-gas interactions are considered the main source of secondary radiations in the arc section. A special source routine, already used in the past from our group [13], was used to allow for beam-gas interactions along the beam pipe. A tentative gas-density profile along the beam orbit inside the vacuum chamber, based on the most up-to-date studies from the CERN's vacuum group (R. Kersevan, private communication), was implemented as shown in Figure 4: the gas-density increases at beam vacuum interconnects while it was considered substantially constant within the magnets.

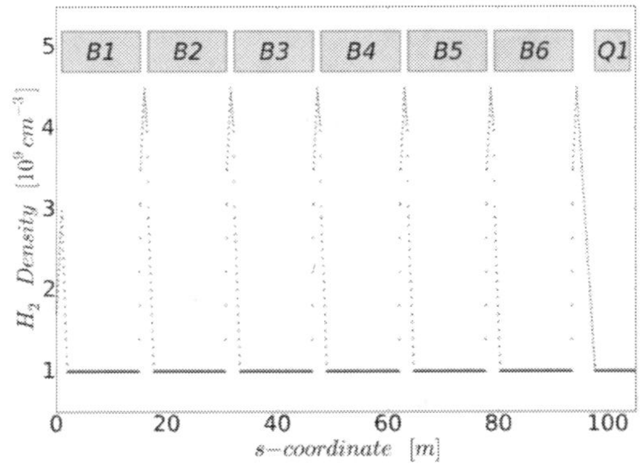

Figure 4. Tentative gas-density profile sampled in the FLUKA model. The gas-density profile is given in H_2 molecules per unit volume as a function of the curvilinear s-coordinate. The picture shows only half-cell, including the first six bending magnets and the first quadrupole.

Both the clockwise and anticlockwise 50 TeV/c proton beam were simulated to consider real operation conditions. High energy hadron-nucleus interaction (> 20 TeV/c) are treated in FLUKA through the DMPJET [16] event generator; the full electromagnetic transport allow for an accurate scoring of the energy deposition in the areas critical for electronics. Due to the beam energy the calculation time needed was a bottleneck. Indeed, a hadronic nuclear interaction at FCC energies, like for LHC, can end in a very complex particle shower with hundreds of secondary: thus to simulate both the hadronic cascade and the electromagnetic shower in bulk matter may require a huge CPU effort. As the focus of this paper is on radiation effects on electronics, the simulations was optimized, in terms of physical and transport parameters, based on what is reported in [13]. To allow for a reasonable compromise between the CPU time per primary particle and the final statistical uncertainty, leading particle biasing on the electromagnetic shower ($e^-/e^+/\gamma$ < 1 GeV) and multiplicity biasing on the hadronic cascade were enabled. In addition to the optimization options mentioned above, the simulation was split into multiple jobs and distributed, on different CPUs, to the CERN's cluster: a total of 120.000 primary protons were simulated.

As mentioned in Section 1.2, both cumulative and stochastic effects lead to possible radiation damage to electronics. TID was scored in our model through a dedicated score in the location *b*, *g*, *h* of Figure 3. The SEE rate is estimated by calculating the HEH fluence distribution in energy and convoluting it with a measured (or estimated) probability of observing SEE in the device of interest. FLUKA allows for a direct scoring of HEH fluence above 20 MeV [4]. Different track-length density estimators were used to assess the differential particle fluence distributions in energy in the location *b*, *g*, *h* of Figure 3: estimators were set in order to score HEH, protons, neutrons, pions, kaons and muons.

Finally, TID and HEH fluence were scored along the whole tunnel using a coarse Cartesian mesh 5cm x 5cm x 100 cm.

4 RESULTS

Figure 5 and Figure 6 show the assessment of the HEH fluence and the dose respectively from the coarse mesh set on the entire tunnel. The Cartesian mesh allows for a qualitative analysis of the radiation levels which, depending from the location of interest, can be evaluated in the range 10^{10}-10^{11} HEH cm^{-2} y^{-1} and 10^1-10^2 Gy y^{-1} respectively.

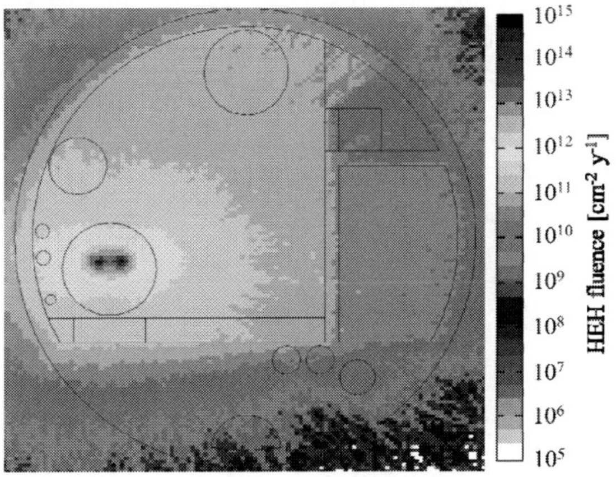

Figure 5. Assessment of the HEH fluence in the FCC tunnel. The slice is taken at the middle point of the bending magnet B1. A nominal FCC operational year of 10^7 s was considered.

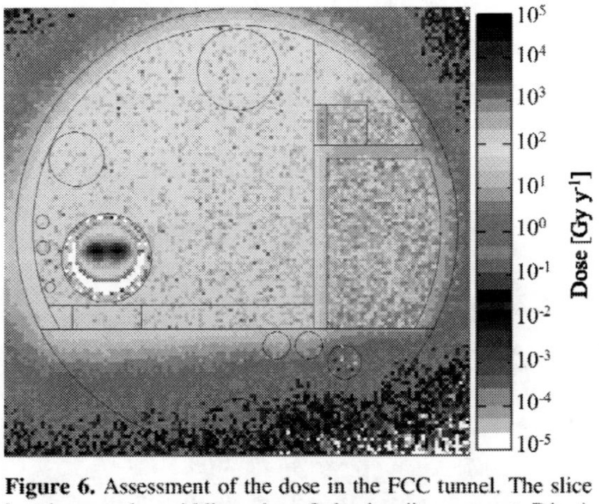

Figure 6. Assessment of the dose in the FCC tunnel. The slice is taken at the middle point of the bending magnet B1. A nominal FCC operational year of 10^7 s was considered.

The dose and HEH fluence profile along the arc are shown in Figure 7 and Figure 8. Figure 7 shows how the dose increases at beam vacuum interconnects between two magnets where the gas-density increases and the macroscopic cross section of a beam-gas interaction is high. A similar trend can be found for the LHC [17], despite the fact that in [17] the gas-density profile was considered constant.

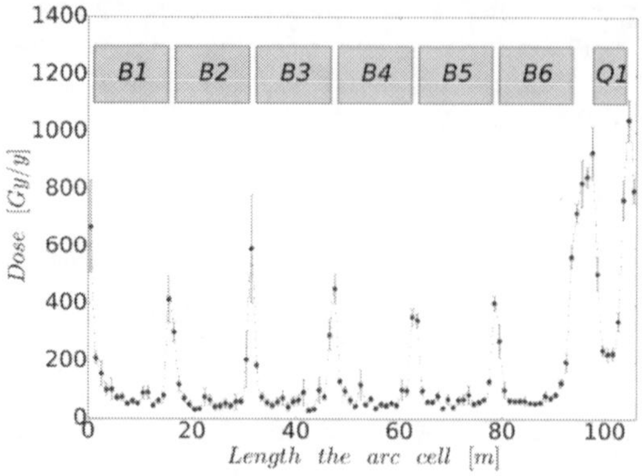

Figure 7. Dose profile (both beams) averaged along the arc in the power converter location (b). The picture shows only half-cell, including the first six bending magnets and the first quadrupole.

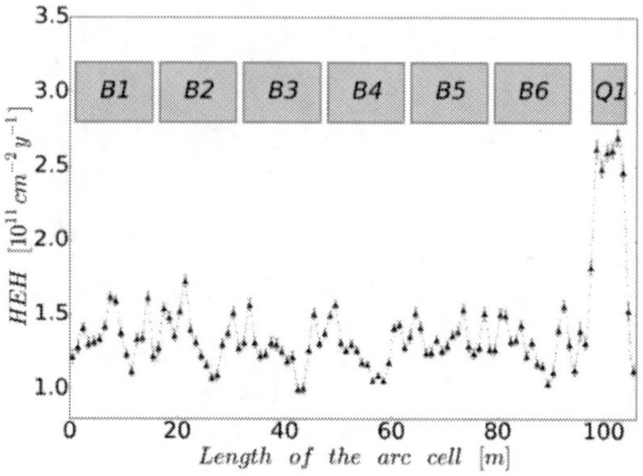

Figure 8. HEH fluence profile (both beams) averaged along the arc in the power converter location (b). The picture shows only half-cell, including the first six bending magnets and the first quadrupole.

Table 2. Radiation levels in the power converter location (*b*), electronic rack location (*g*), Safe Tunnel (*h*).

Power Converter (b)	
Dose [Gy y^{-1}]	(1.02±0.03)E+02
Protons [cm^{-2} y^{-1}]	(5.5±0.1)E+09
Kaons [cm^{-2} y^{-1}]	(6.5±0.3)E+08
Pions [cm^{-2} y^{-1}]	(6.2±0.1)E+09
Muons [cm^{-2} y^{-1}]	(1.17±0.05)E+09
Neutrons [cm^{-2} y^{-1}]	(1.25±0.01)E+13
HEH [cm^{-2} y^{-1}]	(1.25±0.01)E+11
Electronic Rack (g)	
Dose [Gy y^{-1}]	(2.04±0.09)E+01
Protons [cm^{-2} y^{-1}]	(7.4±0.5)E+08
Kaons [cm^{-2} y^{-1}]	(7.1±0.1)E+07
Pions [cm^{-2} y^{-1}]	(5.1±0.3)E+08
Muons [cm^{-2} y^{-1}]	(3.5±0.4)E+08
Neutrons [cm^{-2} y^{-1}]	(3.74±0.04)E+12
HEH [cm^{-2} y^{-1}]	(2.00±0.06)E+10
Safe Tunnel (h)	
Dose [Gy y^{-1}]	(1.63±0.04)E+01
Protons [cm^{-2} y^{-1}]	(1.60±0.06)E+09
Kaons [cm^{-2} y^{-1}]	(1.8±0.1)E+08
Pions [cm^{-2} y^{-1}]	(1.65±0.06)E+09
Muons [cm^{-2} y^{-1}]	(9.8±0.04)E+08
Neutrons [cm^{-2} y^{-1}]	(1.60±0.02)E+12
HEH [cm^{-2} y^{-1}]	(3.21±0.09)E+10

The bi-dimensional dose and fluence map of Figure 5 and Figure 6 do not allow for a detailed analysis of the radiation levels: indeed, a quantitative analysis of the radiation levels can be performed through the dose and the particle fluence averaged over the regions of interest nearby a bending magnet (Table 2).

Finally, Figure 9, Figure 10 and Figure 11 show the detailed particle spectra in the region *b*, *g*, and *h* respectively. The knowledge of the differential fluence distribution in energy allow, as mentioned above, to calculate the SEE rate: indeed, the SEE rate is estimated by calculating the HEH fluence distribution in energy and convoluting it with a measured (or estimated) probability of observing SEE in the device of interest [18]. Currently, we cannot calculate the SEE rate due to the lack of knowledge about the electronics that we are going to use in the future machine and its SEE cross section. Nevertheless, the analysis of the spectra can be useful to understand the particle environment at the different energies as well as the leading particles for SEE. With regard to protons, kaons, pions (and muons) the scoring was truncated at 20 MeV, as it possible to see from the unphysical cut in the spectra, given that for R2E purpose we are in interested in hadrons > 20 MeV: indeed, below this threshold the Single Event Upset (SEU, i.e. the change of the state of a memory bit from 0 to 1 and vice versa) cross section falls down. The particle population is dominated by neutrons which can induce SEE in semiconductor devices through inelastic interactions in the sensitive volume. Furthermore, thermal neutrons, mainly generated through the slowing down of neutrons due to the interaction with the shielding materials, can contribute to SEE in electronics [19].

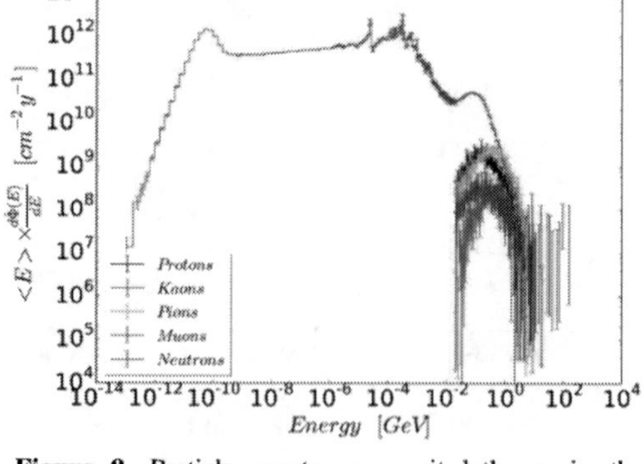

Figure 9. Particle spectra per unit lethargy in the power converter location (*b*).

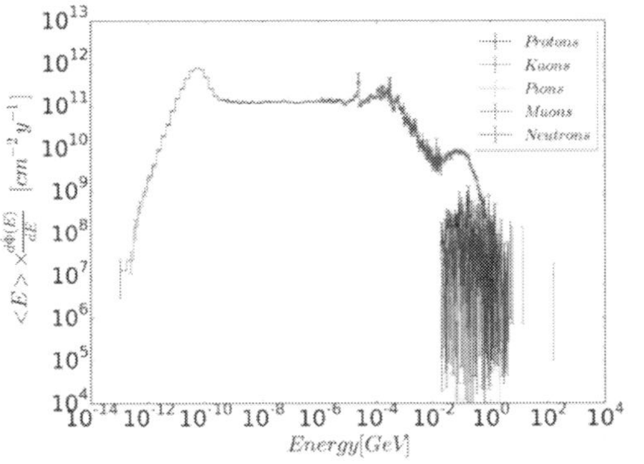

Figure 10. Particle spectra per unit lethargy in the electronic rack location (*g*).

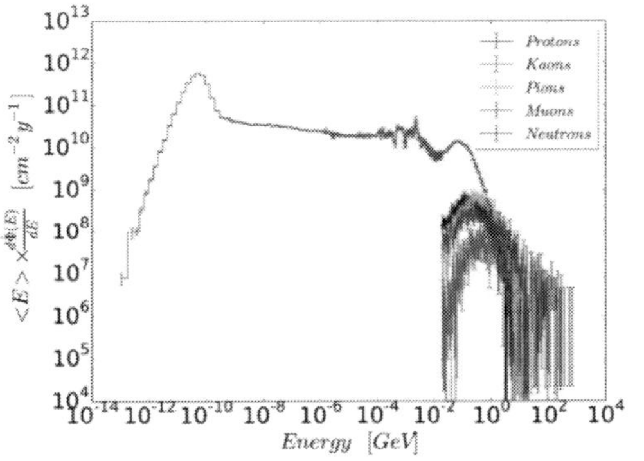

Figure 11. Particle spectra per unit lethargy in the safe tunnel passage (*h*).

5 CONCLUSION

The FCC study is exploring the potential of hadron-hadron and lepton-lepton circular colliders, for the post-LHC era. The goal of FCC-hh is to provide proton-proton collisions at a centre-of-mass energy of 100 TeV. The need for building such large circular hadron collider is to reach energy levels far beyond the range of the LHC to provide access to the direct production of new particles with masses up to tens of TeV, as well as to obtain much increased rates for phenomena in the sub-TeV mass range, with the corresponding greatly improved precision.

In this work, a first evaluation of the radiation levels in the FCC arc was discussed. In particular, the evaluation of the radiation levels was performed for critical areas for electronics.

An analytical scaling from LHC data could be performed in first approximation from the knowledge of the beam energy, the beam current and the residual gas-density profile along the beam orbit in the vacuum chamber. The former, which is unknown for the LHC arc, cannot allow to perform a proper analytical scaling of the radiation levels. Moreover, such kind of scaling from the current machine may not be fully representative of the operational conditions of the future machine: indeed, unavoidable differences in the machine lattice, gas-density profile, and physical mechanisms that might be accessible at higher energies are not taken into account and may lead to significant difference.

In opposition to analytical methods, MC simulation allows for an accurate particle transport, taking into account the actual geometry, the exact source term and all the main physical mechanisms. In this work, a FLUKA MC model of the FCC arc was developed considering the most up-to-date design of the tunnel layout, the beam optics, a tentative residual gas-density profile. MC simulations allowed to evaluate the TID and the HEH fluence in areas critical for electronics, namely the power converter location below the beam pipe, a dedicated electronic rack location and the safety tunnel. Considering the power converter area, the most critical of the three locations considered in this work, the expected TID is (1.02 ± 0.03)E+02 Gy y^{-1} while the HEH fluence is (1.25 ± 0.01)E+11 cm^{-2} y^{-1}.

As a general conclusion, we do not know yet the technology that we are going to use in more than 20 years but considering that we expect an increment of the radiation levels of factors compared to LHC (i.e. increasing probability of SEE), the electronic components are going to be much more integrated (i.e. increasing sensibility to radiations), the number of electronics devices in FCC will be larger than in LHC (i.e. increasing probability of SEE) the need of dedicated shielded areas for the arc become strategic to increase the FCC reliability of factors.

Moreover, the results presented in this paper refer to the tentative residual gas-density profile used in the simulation: we can expect an improvement of the vacuum quality in time, as from the experience of LHC, which will reduce the TID and HEH.

Further considerations and a comparison with the LHC data for the arc imply more details about the actual residual gas-density profile along the beam orbit in the vacuum chamber. New simulations and a detailed data collection is ongoing.

References

1. M. Benedikt, F. Zimmermann, *CERN-ACC-2016-005*, (2016)
2. R. Tomás, M. Benedikt, A.V. Bogomyagkov, L. Bottura, F. Cerutti, A. Ferrari, B. Haerer, B. Holzer, E. Jensen, M. Koratzinos, R. Martin, L. Medina, L.S. Esposito, D. Schulte, E. Todesco, J. Wenninger, S. White, F. Zimmermann, Nucl. Phys. B **00**, 1–8 (2014)
3. A. Ball, M. Benedikt, L. Bottura, O. Dominguez, F. Gianotti, B. Goddard, P. Lebrun, M. Mangano, D. Schulte, E. Shaposhnikova, R. Tomas, F. Zimmermann, *CERN FCC-1401101315-DSC*, (2014)
4. G. Battistoni, V. Boccone, F. Broggi, M. Brugger, M. Campanella, M. Carboni, F. Cerutti, A. Empl, A. Fassò, A. Ferrari, A. Ferrari, E. Gadioli, M. V. Garzelli, D. Kramer, M. Lantz, E. Lebbos, A. Mairani, A. Margiotta, A. Mereghetti, C. Morone, S.

Muraro, K. Parodi, V. Patera, M. Pelliccioni, L. Pinsky, J. Ranft, K. Roeed, S. Roesler, S. Rollet, P. R. Sala, M. Santana, L. Sarchiapone, M. Sioli, G. Smirnov, F. Sommerer, C. Theis, S. Trovati, R. Versaci, R. Villari, H. Vincke, H. Vincke, V. Vlachoudis, J. Vollaire, N. Zapp, Prog. Nucl. Sci. Tech. **2**, 948-954 (2011)

5. B. Bellesia, M. Brugger, A. Ferrari, D. Kramer, R. Losito, S. Myers, M. Pojer, S. Roesler, M. Solfaroli, A. Vergara, S. Weisz, T. Wijnands, *Proceedings of Chamonix 2009 workshop on LHC Performance*, 160-170 (2009)
6. K. Røed, V. Boccone, M. Brugger, A. Ferrari, D. Kramer, E. Lebbos, R. Losito, A. Mereghetti, G. Spiezia, R. Versaci, IEEE Trans. Nucl. Sci. **58**, 932-938 (2011)
7. K. Røed, M. Brugger, G. Spiezia, *CERN-ATS-Note-2011-077* (2012)
8. L. S. Esposito, M. Brugger, F. Cerutti, A. Ferrari, R. Losito, *Proceedings of HF2014*, 157-159 (2014)
9. T.T. Böhlen, F. Cerutti, M.P.W. Chin, A. Fassò, A. Ferrari, P.G. Ortega, A. Mairani, P.R. Sala, G. Smirnov, V. Vlachoudis, Nucl. Dat. She. **120**, 211-214 (2014)
10. A. Ferrari, P.R. Sala, A. Fassò, J. Ranft, *CERN-2005-10, INFN/TC_05/11, SLAC-R-773* (2005)
11. A. Infantino, R. García Alía, M. Brugger. IEEE Trans. Nuc. Sci. DOI: 10.1109/TNS.2016.2621238.
12. R. García Alía, B. Biskup, M. Brugger, M. Calviani, C.Poivey, K. Røed, F. Saigné, G. Spiezia, F. Wrobel. IEEE Trans. Nuc. Sci. **60** (4), 2469-2476 (2013).
13. M. I. Besana, F. Cerutti, L.S. Esposito, A. Ferrari, A. Lechner, A. Mereghetti, E. Skordis, *Proceedings of the first Annual Meeting of the Future Circular Collider study* (2015)
14. M. I. Besana, F. Cerutti, A. Ferrari, W. Riegler, V. Vlachoudis. *Proceedings of IPAC2016*, 1414-1417, (2016)
15. G. R. Stevenson, Rad. Prot. Dos. **96**, 359-371 (2001)
16. S. Roesler, R. Engel, J. Ranft, *Proc. Monte Carlo 2000 Conference*, 1033-1038 (2001)
17. F. Cerutti. Radiation levels today and in the future. Available: https://indico.cern.ch/event/308246/contributions/710485/attachments/588449/809908/R2EAvailWork.pdf [Accessed 25 October 2016]
18. JEDEC. JEDEC Standard No. 89A (JESD89A), (2006).
19. R. García Alía, S. Bonaldo, M. Brugger, D. Danzeca, F. Ferrari, A. Infantino, Y. Iwamoto, J. Mekki, C. Theis, A. Thorton. *Proceedings of ICRS13 & RPSD2016*, Unpublished data, (2016)

Shielding Design for the South Pole nToF Diagnostic at the NIF

Hesham Khater*, *Shiva* Sitaraman, *James* Hall, *Robert* Hatarik, *Joseph* Caggiano, and *Cory* Waltz

Lawrence Livermore National Laboratory, Livermore, CA 94550, USA

Abstract. Neutron time of flight (nToF) detectors are fielded at the National Ignition Facility (NIF) to measure neutron yield, ion temperature, and downscattering in the cold fuel for D-T implosions. Anisotropically assembled cold fuel may generate different nToF data when measured by detectors located at the Target Chamber equator and poles. A collimated nToF line of sight has been fielded near the Target Chamber South Pole (SP) to examine any possible anisotropy in the cold fuel. The SP nToF detector is located in the lowest floor level of the NIF's Target Bay and at a distance of ~ 18 m from the Target Chamber Center. The detector utilizes a solid bibenzyl scintillator and four photomultiplier tubes. The line of sight includes a port collimator that is attached to the Target Chamber and a bore hole collimator in the concrete floor above the detector. In addition, a beam line get lost hole is constructed in the Target Bay floor to minimize the backscattered radiation at the detector location. Initial measurements indicated the need for installation of additional shielding to eliminate gamma background during the period before arrival of the 14.1 MeV neutrons to the detector. A set of MCNP Monte Carlo simulations with the full Target Bay model were conducted to provide an estimate of the expected neutron and gamma backgrounds during D-T shots. A new shielding scheme is designed to reduce the gamma background by an order of magnitude.

1 Introduction

The National Ignition Facility (NIF) at Lawrence Livermore National Laboratory is the world's largest and most powerful laser system for inertial confinement fusion. The NIF is a 192 laser beam facility capable of producing 1.8 MJ, 500 TW of ultraviolet light. The NIF is configured for "indirect drive" approach where high-energy laser beams are focused to enter a small metal can (hohlraum) held at the center of the Target Chamber (TC). The laser beams heat the interior surface of the hohlraum to create intense flux of thermal x-rays that ablates the surface of a fusion capsule located at the center of the hohlraum. The capsule interior contains a solid cryogenic Deuterium-Tritium (D-T) fuel layer surrounding a central low-density D-T gas. Ablation of the exterior layer of the capsule compresses the remaining capsule and the fuel. In addition to the 14.1 MeV neutrons produced by the D-T fusion, inelastic collisions of these neutrons with the compressed fuel and/or ablator results in "down-scattered" (DS) neutrons in the 10-12 MeV spectral range.

A smaller fraction of "reaction-in-flight" (RIF) neutrons (with energy up to ~ 30 MeV) are produced by up-scattered deuterons or tritons undergoing D-T reaction with thermal ions. The full spectrum of neutrons can be measured using neutron time of flight (nToF) detectors. Anisotropically assembled cold fuel may generate different neutron time of flight data when measured by detectors located at the TC equator and poles. Several nToF detectors at different lines of sight are fielded on the NIF to measure the neutron spectrum with high accuracy [1-3]. The nToF diagnostics measure the neutron yield as a function of neutron energy by recording the detector signal strength as a function of time. For a specific distance from the source, a neutron arrival to the detector correlates to the neutron energy. The nToF diagnostics utilize scintillation detectors coupled to photomultiplier tube (PMT). The scintillators used at the NIF are characterized with low afterglow and are surrounded by light materials to reduce background from neutron scattering. Reducing scattering from near-by masses is important for accurate measurements of the DS neutrons. The detector signal is recorded by using a high bandwidth multi-channel digital oscilloscope. The nToF detectors are used to obtain spectral information such as ion temperature and the ratio of DS neutrons relative to the primary neutron yield.

There are seven main floor levels inside the NIF Target Bay (TB). The main floors are located at elevations of -10.29 m (-33'9"), -6.63 m (-21'9"), -1.07 m (-3'6"), 5.33 m (17'6"), 8.99 m (29'6"), 12.19 m (40'0"), and 15.39 m (50'6") with respect to the ground level. The TC is located in the middle of the TB with its center at an elevation of a 7.01 m (23'0") from the ground level. A collimated nToF line of sight (LOS) has been fielded near the TC South Pole (SP) using the TC port (θ=161°, ϕ=56.25°). The goal of the LOS is to examine any possible anisotropy in the cold fuel. The SP nToF detector is located in the lowest floor level of the TB and at a distance of ~ 18 m from the Target Chamber Center (TCC). The detector utilizes a solid bibenzyl

* Corresponding author: khater1@llnl.gov

DOI: 10.1051/epjconf/201715303005

scintillator and four photomultipliers (PMT). Figure 1 shows the SP nToF system layout.

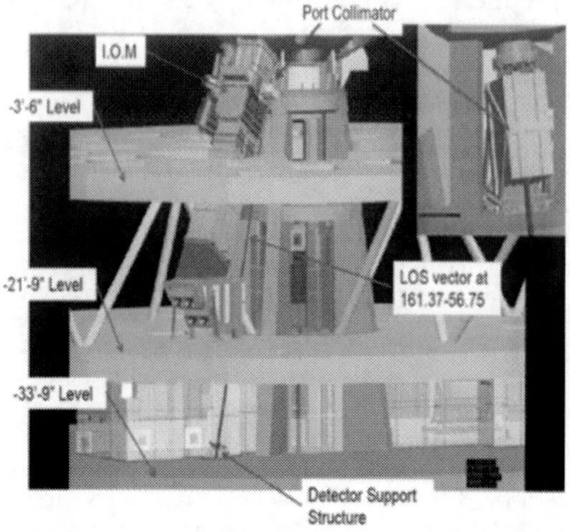

Figure 1. South Pole nToF system layout.

The LOS is collimated through the use of apertures to minimize contributions from neutrons and gammas scattered off the target positioner and diagnostic instrument manipulator (DIM) instruments from reaching the detector. The target positioner and the DIM instruments are inserted inside the TC at a few centimeters from the TCC. Contribution from radiation scattered off the TC wall is collimated through the use of a port collimator that is attached to the TC port. A two-dimensional view of the port collimator's MCNP model is shown in Fig. 2.

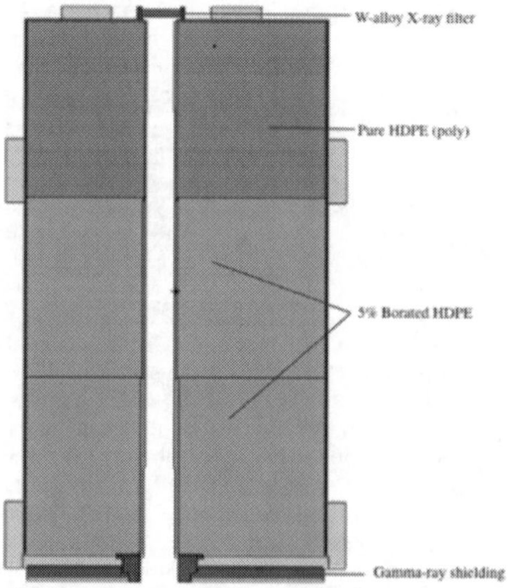

Figure 2. MCNP model of the TC port collimator.

Additional collimation is achieved through the installation of another collimator inside the 11.43 cm diameter bore hole in the -21'9" concrete floor above the detector. Finally, as shown in Fig. 3, a beam line get lost is constructed in the TB -33'9" floor to minimize/delay backscattered radiation at the detector location.

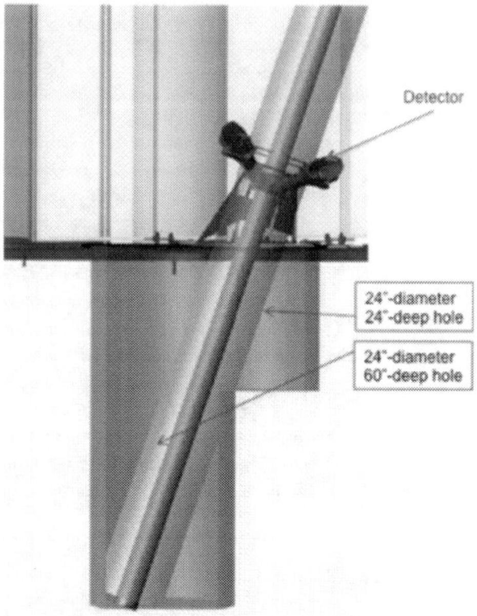

Figure 3. Get lost hole at the -33'9" floor.

2 Simulation of radiation backgrounds inside the Target Bay

Fusion neutrons generated by fuel burn and secondary neutrons and gammas resulting from fusion neutron interactions with structures present inside and outside the TC contribute to the radiation background. Inside the TC, neutrons interact with the Cryogenic Target Positioner, and three DIM instruments. Neutrons also interact with the stainless steel first wall panels, aluminum TC, gunite shield, port covers, and the Final Optics Assemblies. Gammas induced by neutron interactions with different structures inside and outside the TC contribute to the gamma background. A detailed MCNP [4] model has been developed to simulate the expected neutron and gamma backgrounds at different locations inside the NIF facility. The model includes most of the major equipment inserted in the TC and the major diagnostics and structures inside the TB. A 3-D visualization [5] of the detailed TB model is shown in Fig. 4.

The as-built facility drawings were used to create the MCNP model. The TC is made of a 10 cm-thick aluminum shell surrounded by 40 cm of borated concrete "gunite". The TC includes 48 indirect-drive and 24 direct-drive laser beam ports as well as 120 diagnostics ports. The NIF is currently configured for "indirect-drive", which as mentioned before focuses laser beams onto the interior wall of the hohlraum, thereby generating x-rays that heat the capsule contained within.

DOI: 10.1051/epjconf/201715303005

The NIF beampath would need to be reconfigured to operate in "direct-drive" mode, where the laser beams focus directly onto the capsule and a hohlraum is not used. All unused diagnostics and direct-drive ports are unshielded and only covered with ~ 5 cm-thick port covers made of aluminum alloy. Indirect-drive ports are connected to full models of the Final Optics Assemblies (FOA). The TB shielding walls are made of 1.83 m-thick concrete.

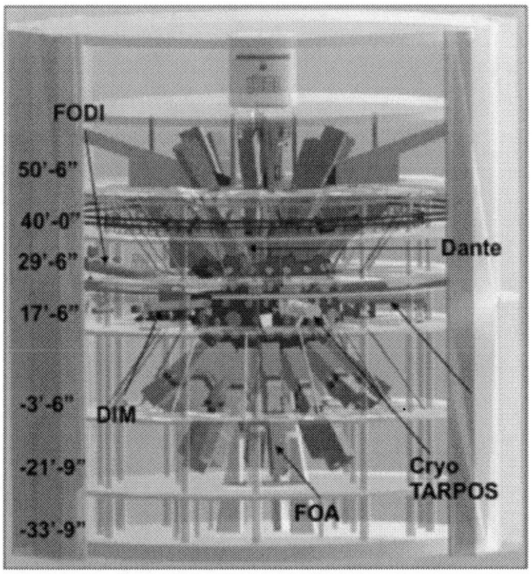

Figure 4. MCNP model of the Target Bay.

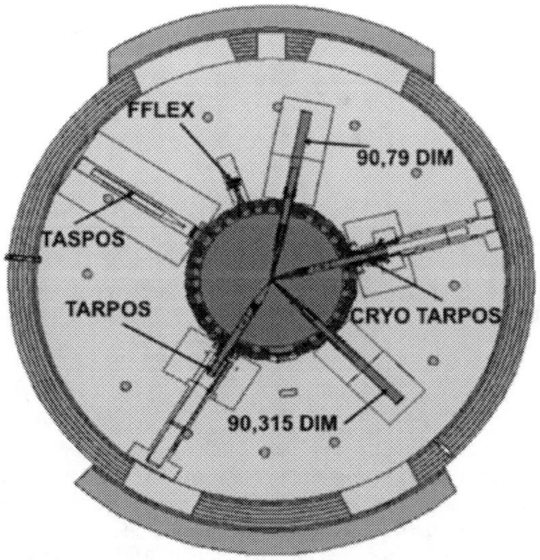

Figure 5. Equatorial view of the Target Bay model.

A two-dimensional view of MCNP model of the TB is shown in Fig. 5. In this view of the TC center, all positioners and DIM instruments are inserted inside the chamber during the shot. Finally, a detailed MCNP model of the nToF detector was separately developed and added to the facility model to be utilized in the background calculations. A 3-D visualization of the detector's MCNP model is shown in Fig. 6.

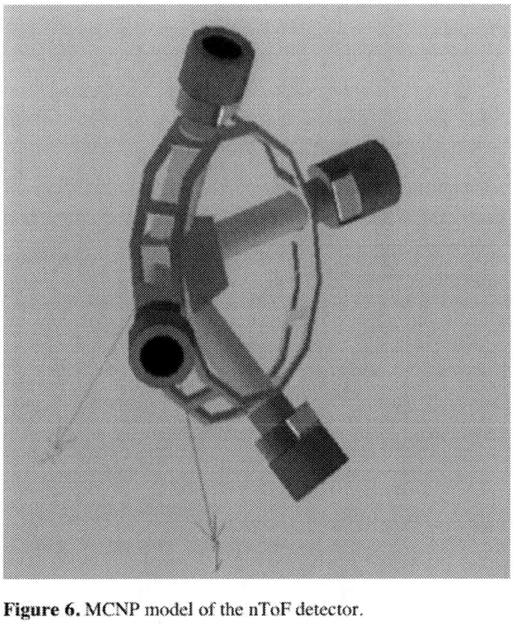

Figure 6. MCNP model of the nToF detector.

The TB model has been utilized in the simulation of expected neutron and gamma fluences throughout the TB [6]. The D-T neutron source is modeled as a volumetric source filling the hohlraum. The neutron cross section data were taken from the FENDL-2.1 cross section database [7]. Photon cross sections are derived from the ENDF/B-VI.8 data library. The neutron and gamma fluence maps are created for each of the TB floors. A Cartesian mesh size of 30 cm x 30 cm x 30 cm is used to create the maps. To reduce the statistical error associated with the simulation, particle splitting and Russian roulette are used throughout the geometry. The typical statistical error associated with the calculated neutron and gamma fluences inside the TB was less than 10%.

The calculated fluences were also used to create a set of dose maps for all floor levels. Silicon kerma factors were used to calculate the dose [8]. Figures 7 and 8 show the simulated neutron and gamma doses (in mSv to Si) for the -21'9" floor level during a 20 MJ shot (7.1×10^{18} neutrons), respectively. This floor level contains the final mirrors that are used to focus the laser beams onto the target in the center of the TC. The 192 laser beams enter the TC through 48 indirect-drive beam ports. Following fusion reactions, neutrons and gamma radiation streaming through the lower 24 beam ports passes through the final mirrors with minimal scattering and end up colliding with the -21'9" concrete floor. The concrete floor is 30.48 cm-thick, and for all practical purposes, it acts as a beam dump for neutrons and gamma streaming out of the TC. The lower 24 beam ports form two rings. The 8 hot spots shown on Fig. 7 in red near the TCC represent dose due to neutrons streaming through the inner lower ring of beam ports. Dose due to radiation streaming through the outer lower ring can be seen in the 16 spots near the TB wall.

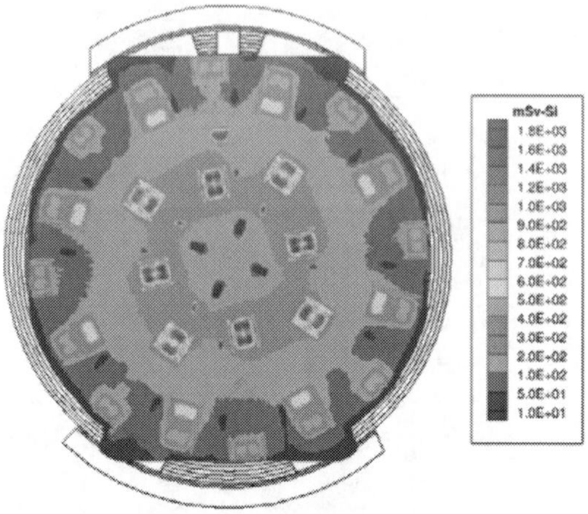

Figure 7. Neutron dose in the -21'9" TB floor.

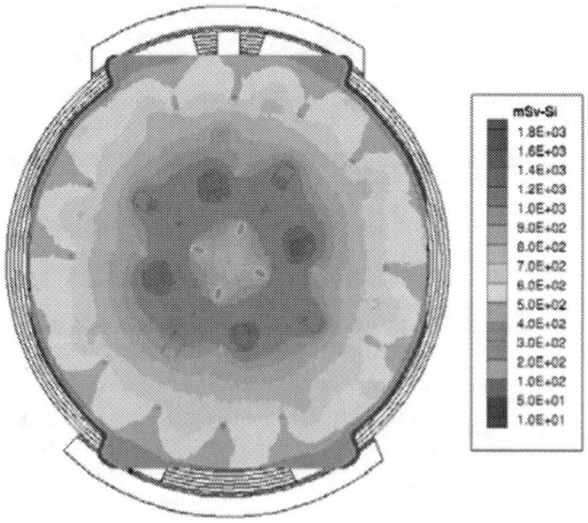

Figure 8. Gamma dose in the -21'9" TB floor.

Neutrons and gamma passing through the -21'9" concrete floor create hot spots in the -33'9" floor level, in the vicinity of the SP nToF detector location. Figures 9 and 10 show the expected neutron and gamma doses in the -33'9" floor level, respectively. As can be seen, the location of the detector experiences a moderate neutron dose but a relatively high gamma dose. The gamma radiation is mostly produced by the neutron interactions with the -21'9" concrete floor. The gamma dose is problematic because it arrives at the detector location well before the primary neutron signal. This early arrival would have an impact on the measured signal from the high energy RIF neutrons that are expected to arrive early at the detector location during the same time as the gamma radiation.

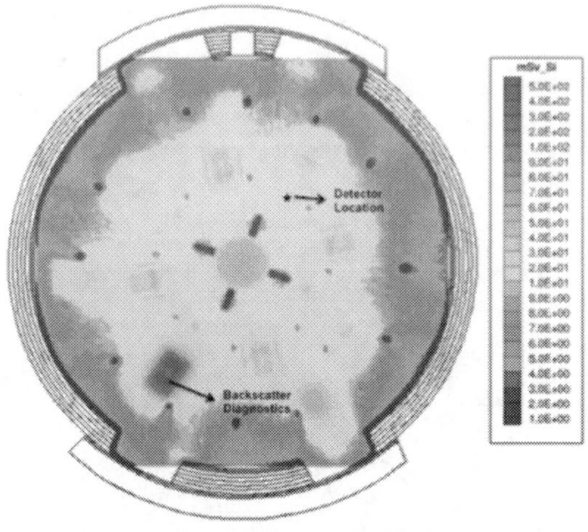

Figure 9. Neutron dose in the -33'9" TB floor.

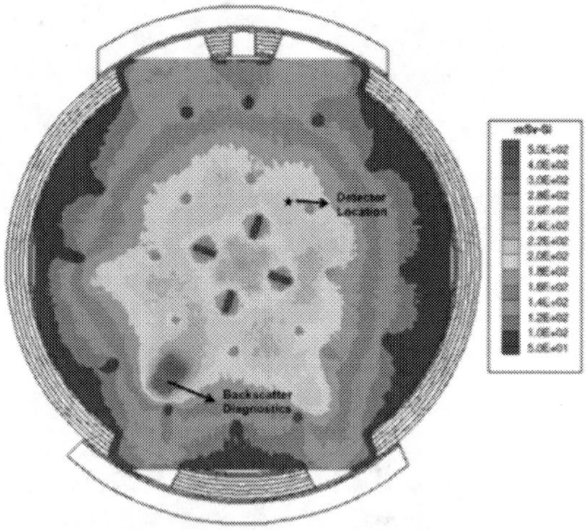

Figure 10. Gamma dose in the -33'9" TB floor.

3 Reducing detector background

The SP nToF detector is located at a distance of about 1.4 m above the -33'9" floor, and the front face of the bibenzyl scintillator is at a distance of 17.88 m from the TC center [9]. The detector has been installed inside the NIF facility and started taking data during shots with lower neutron yield (10^{15} to 10^{16}). During cryogenic shots, failure to identify the primary neutrons scattering back from the get lost hole could interfere with the measured signal due to the 10-12 MeV DS neutrons (due to the primary 14.1 MeV neutron interactions with the target fuel). To solve this problem, data were taken during shots with non-cryogenic, D-T gas-filled "exploding pusher" targets. During these shots, no DS neutrons are expected and hence it is possible to identify and separate the measurement due to primary neutrons passing through the scintillator and scattering back from the get lost hole in the TB floor. On the other hand, early measurements indicated the need to develop a shielding solution to significantly reduce the dose due to gammas arriving at earlier times, before the main 14.1 MeV neutron pulse.

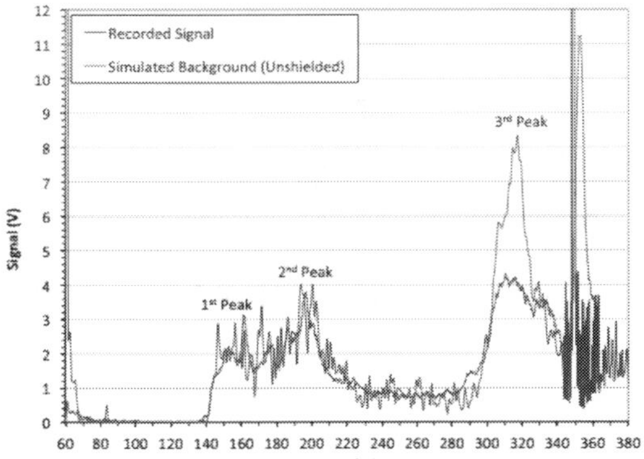

Figure 11. Comparison between simulated background and recorded signal during a low yield shot.

The energy deposited into the SP nToF bibenzyl scintillator has been simulated for a 14.1 MeV neutron source. A comparison between the detector recorded signal and simulated gamma background is shown in Fig. 11. The data were taken during a shot with a yield of ~ 10^{16} neutrons. As shown in the figure, there is a very good agreement between the recorded data and simulated background. The measured data shows a large gamma signal recorded by the scope at ~ 60 ns following the shot. The gamma is generated by neutron interactions with masses near the TCC. The hohlraum, target positioner and DIM snouts represent the masses near the TCC. The gamma signal associated with the primary neutron interactions with the scintillator is recorded at ~ 350 ns following the shot. The measured data shown on the figure were adjusted to match the arrival time simulated by the MCNP simulations. The actual 14.1 MeV neutron signal is usually recorded by the scope at ~ 370 ns, which is about 20 ns more than the time it takes the neutrons to reach the front face of the scintillator. The additional time represents the time periods between firing the laser and when it hits the hohlraum in addition to the time to nuclear burn of the D-T fuel.

Both of the recorded signal and simulation results pointed out to a problem with the unexpected arrival of early gammas before the 14.1 MeV neutron signal. The early gammas arrive at the detector between ~140-340 ns. There are three different gamma peaks, the first peak between ~140-170 ns, the second peak is between 170-220 ns with the final and largest peak is between 300-340 ns. A careful examination of potential sources for the early gamma radiation identified them as a product of fusion neutron collisions with several structures inside the TB. The first peak is due to gammas generated by neutron interactions with the TC first wall steel panels, the TC aluminium wall, and the gunite. The second peak is due to gammas generated by neutron interactions with the TB concrete floor at the -3'6" floor level. The third and largest peak is due to gammas generated by neutrons colliding with the -21'9" concrete floor. As mentioned in the previous section and shown in Fig. 7, the concrete floor at the -21'9" floor level acts as a beam dump for neutrons and gammas streaming out of the TC and traveling through the open laser beam ports. Even though the gammas are generated at different times by neutron collisions with different structures, they all stream through the lower rings of the laser beam ports before colliding with the -21'9" floor. The early gammas arriving before the 14.1 MeV neutron signal travel through the inner lower ring (8 beams). On the other hand, neutrons and gammas passing through the outer lower ring (16 beams) travel for longer distances before colliding with the -21'9" floor. They are mostly responsible for much smaller gamma peaks that arrive at the detector after the arrival of the 14.1 MeV neutrons.

The early arrival of the gamma radiation interferes with measurement of the RIF neutrons. The RIF neutrons can have energies up to ~ 30 MeV and are produced by up-scattered deuteron or triton undergoing D-T reaction with thermal ions. The RIF neutrons are very difficult to measure as they represent ~ five orders of magnitude lower than the primary 14.1 MeV neutrons. They are a very important measure of the cold fuel density [10]. It quickly became clear that there is a need for shielding the detector to eliminate the gamma background during the period before the arrival of the 14.1 MeV neutrons at the detector.

A set of MCNP simulations with the full TB model were conducted to facilitate the design of a new shielding scheme with the goal of reducing the early gamma background by about an order of magnitude. The options of using shielding made of tungsten or carbon steel were examined. A picture of the proposed shielding design is shown in Fig. 12. The shielding box fully surrounded the detector except for the 11.43 cm diameter hole on the top to allow for clear passage of the primary neutron beam.

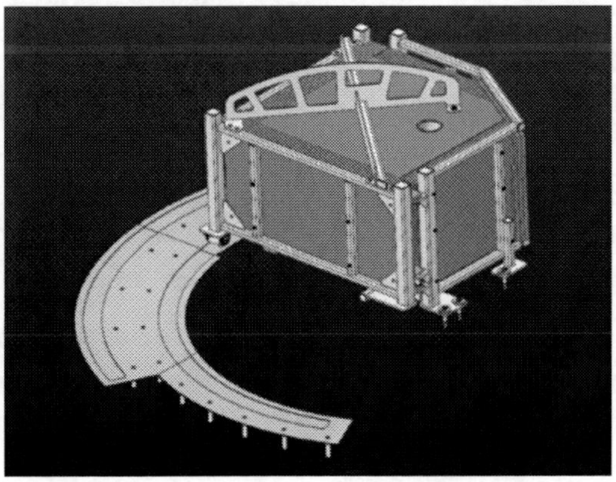

Figure 12. Conceptual design of the nToF detector shielding.

As shown in Fig. 13, a 5 cm-thick tungsten shield option successfully reduced the energy deposition in the scintillator during the period of concern by about an order of magnitude. Tungsten is an excellent gamma shield but has the known disadvantage of a higher weight and cost than steel. As an alternative to tungsten, a similar reduction in the gamma dose could be achieved through the use of a carbon steel shield.

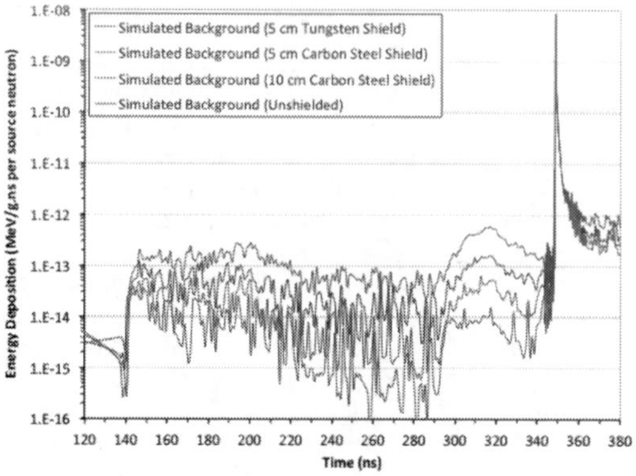

Figure 13. Energy deposition in the nToF scintillator for different shielding options.

As shown in Fig. 14, simulations performed with the proposed shielding options significantly reduced the expected detector signals due to early gamma background. Using a 5 cm-thick tungsten shield would eliminate the signal arriving at the detector as a result of gamma production by neutron collisions with the different structures inside the TB. A somewhat similar reduction in the recorded signal could also be achieved by using a 10 cm-thick carbon steel shield. The proposed shielding solution will allow the detector to successfully record signals associated with the RIF neutrons.

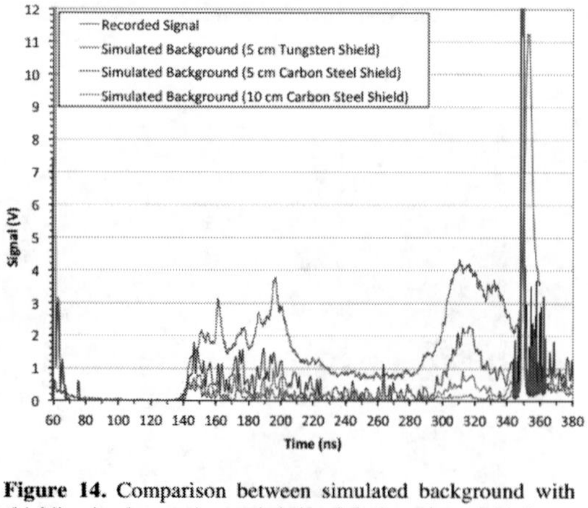

Figure 14. Comparison between simulated background with shielding in place and recorded signal during a low yield shot.

4 Summary

A South Pole (SP) neutron time of flight (nToF) detector has been fielded in the National Ignition facility (NIF) to measure neutron yield, ion temperature, and neutron downscattering in the cold fuel for D-T implosions. The SP nToF is located in the lowest floor level of the NIF's Target Bay and at a distance of ~ 18 m from the Target Chamber Center. The detector utilizes a solid bibenzyl scintillator and four photomultiplier tubes. Initial measurements indicated the need for installation of additional shielding to reduce the gamma background during a period before the arrival of the primary 14.1 MeV neutrons to the detector. Gamma arrival during this time period interferes with measurement of the higher energy "reaction-in-flight" (RIF) neutrons. The RIF neutrons can have energies up to ~ 30 MeV and are produced by up-scattered deuteron or triton undergoing D-T reaction with thermal ions. The RIF neutrons are very difficult to measure as they are at about five orders of magnitude lower level than the primary neutrons.

The early gammas arrive at the detector between ~140-340 ns. They are the product of fusion neutron collisions with the TC walls as well as TB concrete floors at the -3'6" and the -21'9" levels. Even though the gammas are generated at different times by neutron collisions with different structures, they all stream through the lower rings of the laser beam ports before colliding with the -21'9" floor. A new shielding that will fully surround the detector has been designed. The gamma background has been reduced by about an order of magnitude during the period of concern through the use of a new 5 cm-thick tungsten or 10-cm thick carbon steel shield.

This work performed under the auspices of the U.S. Department of Energy by Lawrence Livermore National Laboratory under Contract DE-AC52-07NA27344.

References

1. V. Glebov et al., Rev. Sci. Instrum. **81**, 10D325 (2010)
2. R. Lerche et al., Rev. Sci. Instrum. **81**, 10D319 (2010)
3. T. Clancy et al., Proc. SPIE 9211, 92110A (2014)
4. X-5 MONTE CARLO TEAM, Los Alamos National Laboratory, LA-UR-03-1987 (2003)
5. MORITZ, http://www.whiterockscience.com
6. H. Khater and S. Brereton, Prog. in Nuclear Sci. and Tech., **2** (2011)
7. D. Aldama and A. Trkov, International Atomic Energy Agency, INDC(NDS)-467 (2004)
8. M. Singh, Lawrence Livermore National Laboratory, UCRL-52850 (1979)
9. R. Hatarik et al., Rev. Sci. Instrum. **83**, 10D911 (2012)
10. A. Hayes et al., Journal of Physics: Conference Series **717**, 012022 (2016)

Development of High Intensity D-T fusion NEutron Generator (HINEG)

Yican Wu*, Chao Liu, Gang Song, Yongfeng Wang, Taosheng Li, Jieqiong Jiang, Yong Song, Xiang Ji, FDS Team

Key Laboratory of Neutronics and Radiation Safety, Institute of Nuclear Energy Safety Technology, Chinese Academy of Sciences, Hefei, Anhui, 230031, China
*Corresponding Author, Email: yican.wu@fds.org.cn

Abstract: A high intensity D-T fusion neutron generator (HINEG) is keenly needed for the research and development (R&D) of nuclear technology and safety of the advanced nuclear energy system, especially for the radiation protection and shielding. The R&D of HINEG includes two phases: HINEG-I and HINEG-II. HINEG-I is designed to have both the steady beam and pulsed beam. The neutron yield of the steady beam is up to 10^{12} n/s. The width of pulse neutron beam is less than 1.5 ns. HINEG-I is used for the basic neutronics study, such as measurement of nuclear data, validation of neutronics methods and software, validation of radiation protection and so on. HINEG-II aims to generate a high neutron yield of 10^{13} n/s neutrons by adopting high speed rotating tritium target system integrated with jet/spray array enhanced cooling techniques, and can further upgrade to obtain neutron yield of 10^{14}~10^{15}n/s by using of accelerators-array in a later stage. HINEG-II can be used for fundamentals research of nuclear technology including mechanism of materials radiation damage and neutronics performance of components, radiation shielding as well as other nuclear technology applications.

KEYWORDS: Fusion, Accelerator Based Neutrons Source, High Intesity, Radiation Shielding

I. INTRODUCTION

Fusion energy becomes essential to solve the energy problem with the increase of energy demands. Although the recent studies of fusion energy have demonstrated the feasibility of fusion power, it commonly realizes that more hard work is needed on neutronics and safety before real application of fusion energy, especially on radiation protection and radiation biological effect of fusion neutron [1]. However the intensity of D-T neutron generators currently on operation around the world is lower than 10^{13}n/s, which is severely restricting the research capability.

The Institute of Nuclear Energy Safety Technology (INEST), Chinese Academy of Sciences (CAS)/ has launched the High Intensity fusion Neutron Generator (HINEG) project to develop an accelerator-based D-T fusion neutron generator with the neutron yield higher than $10^{14\sim15}$ n/s. HINEG project is divided into two stages. At the first stage, a D-T fusion neutron generator with both steady and pulse neutrons (HINEG-I) will be developed. The intensity will reach 10^{12}n/s. At the second stage, a high intensity D-T fusion neutron generator will be developed. The intensity will reach $10^{14\text{-}15}$n/s. In this paper, the design and progress of HINEG will be described.

II. DEVELOPMENT OF HIGH INTENSITY D-T FUSION NEUTRON SOURCES

HINEG-I has two beam lines. The intensity of the steady beam could reach 10^{12}n/s. For the pulse beam, the full width at half maximum (FWHM) is smaller than 1.5 ns. HINEG-I can be used for the basic research for fusion neutron science. HINEG-I can also be utilized in nuclear technology applications, such as neutron radiography, neutron activation analysis, neutron radiation detection and so on. At present, all components of HINEG-I have been developed. The installation has been finished (see Figure1 and Figure 2) and is commissioning in December, 2015.

HINEG-I will focus on the basic research of neutronics, including nuclear data measurement, verification and validation (V&V) of the methods and code, etc. For example, it can be used for the measurement of the (n, p), (n, α) cross sections of transmutation reactions that produce gases and affect the service performance of structure materials in fusion reactors, and the (n, γ) reaction which contributes to the nuclear heating deposition in superconducting coils [2-4]. It is also a significant platform to perform the V&V for the neutronics simulation software, such as SuperMC, and its applicability to fusion neutron environment [5]. Note that the SuperMC code was developed by INEST/FDS Team and has been validated by ~2000 benchmarking cases for both fission and fusion reactors.

Figure 1: Steady beam line

DOI: 10.1051/epjconf/201715303006

Figure 2. Ion injector of HINEG-I.

During the second phase (HINEG-II), in order to reach a neutron yield higher than 10^{14} ~10^{15}n/s, several key technologies have been developed, such as the spraying impingement cooling for the target as well as a high intensity ion source. The feasibility of technological route is demonstrated. Furthermore, the neutron yield of 10^{15}~10^{16}n/s is expected to be achieved through adopting accelerator-array and upgrading the rotating target. The design and development of key components of HINEG-II is ongoing.

HINEG-II will focus on fundamental phenomena of nuclear technology, including mechanism of material irradiation damage and neutronics performances of key components. For example, a series of material irradiation experiments will be performed to study irradiation effects on the mechanical properties of the China Low Activity Martensitic steel (CLAM), the primary candidate structural material for CN ITER TBM(Test Blanket Module), in high energy and high intensity neutron environment [6]. A series of experiments for fusion TBM, such as the Dual Function Lithium Lead liquid TBM and Helium Cooled Ceramic breeder TBM, will be performed to confirm whether the neutronics performance, such as tritium breeding ratio, nuclear heating deposition and material activation behavior, is consistent with the design or not.

III. RESEARCH AND DEVELOPMENT OF KEY TECHNOLOGIES

In order to produce high current ion beams, an optional concept is proposed, the accelerator is equipped with a ion source extracted a high intensity beam, which is installed (with its associated LEBT) inside a HV cabinet and connected to a HV accelerator column. At present, we have started to study the development difficulties and feasibility analysis of the whole accelerator.

During the transportation of HINEG-II, various of beam instabilities occurred due to space charge effect, which is partially compensated by the interaction of the beam with the residual gas. Such space charge compensation can be calculated to allow simulating beam trajectories.

During the operation of HINEG-II, the power of D^+ ion beams hitting on the tritium target can be higher than 50~100kW/cm^2. The temperature must be controlled bellow 200 °C in order to prevent the diffusion of tritium. To enhance the heat transfer capacity of tritium target system, jet array, spraying impingement, and nano-fluid cooling enhancement technologies are explored. A high heating target experiment platform have been developed (see Figure 3), and the technology verified experimental results show that the temperature of tritium target can be control under 180 °C, while the beam current on the tritium target is bigger than 500 mA. The comparison of the experimental and the theoretical results (see Figure 4) indicated that the heat transfer capability can reached to 57 kW/(m^2·K), which has verified the feasibility of the cooling technologies.

Figure 3: Physical map of experiment platform for cooling technology of high heating target

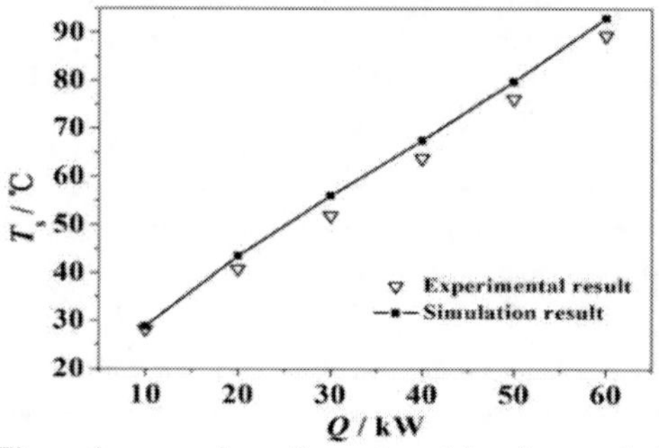

Figure 4: comparison of experimental and numerical simulation results

V. RESULTS

In this paper, the design and progress of HINEG project is presented. HINEG-I has both steady and pulse beams. The intensity of the steady beam is up to 10^{12} n/s. The FWHM of pulse beam will less than 1.5 ns. The installation of HINEG-I has been finished and is commissioning in December, 2015. HINEG-II is a steady D-T fusion generator. The intensity is $10^{14\text{-}15}$ n/s. In order to achieve so high intensity, multi-accelerators array and high heating load rotating tritium target will be used.

HINEG-I and HINEG-II could be used in the areas of fusion nuclear data and fusion neutron transport in complicate medium, irradiation damage and activation test of fusion material, and so on. HINEG can also be utilized for radiation protection. For example, in biological research field, HINEG can be applied to neutron irradiation breeding, neutron cancer treatment, biological protection, etc... Recently, the DNA lesions and cell transformation which were induced by neutron irradiation were test, and further the chromosome aberration, tumorigenesis were examined in animal experiments to explore how the neutron irradiation act on living body and its mechanisms. The results can contribute on developing new antitumor treatment, and further research will be sustained carried on HINEG.

Acknowledgment

This work is supported by the Strategic Priority Science & Technology Program of the Chinese Academy of Sciences (NO. XDA03040000) and ITER 973 Program (No. 2014GB112001). We would further like to thank the great help from the other members of FDS Team in this research.

References

1) A.Serikov, U.Fischer, C.S.Pitcher, A.Suarez, and B.Weinhorst, Computational Challenges for Fusion Neutronics for ITER Ports. SNA+MC 2013, Paris, France, (2013).
2) http://www-nds.iaea.org.
3) R.A. Forrest, "Nuclear Science and Data Need for Advanced Nuclear Systems," Energy Procedia, 7, 540-552, (2011).
4) TANG Jing-yu, JING Han-tao, XIA Hai-hong et al., "Key Nuclear Data Measurements for advanced Fission Energy and White Neutron Source at CSNS," Atomic Energy Science and Technology, 47, 1089-1095, (2013).
5) Y. Wu, J. Song, H. Zheng, et al., "CAD-Based Monte Carlo Program for Integrated Simulation of Nuclear System SuperMC," Ann. Nucl. Energy, 82: 161-168, (2015).
6) Q. Huang, FDS Team, "Development status of CLAM steel for fusion application," J. Nucl. Matter. 455, 649-654, (2014).

Neutronic design of MYRRHA reactor hall shielding

Yurdunaz Celik[1,*], *Alexey Stankovskiy*[1], and *Gert Van den Eynde*[1]

[1]SCK•CEN, Institute for Advanced Nuclear Systems, Boeretang 200, B-2400 Mol, Belgium

Abstract. The lateral shielding of a 600 MeV proton linear accelerator beam line in the MYRRHA reactor hall has been assessed using neutronic calculations by the MCNPX code complemented with analytical predictions. Continuous beam losses were considered to define the required shielding thickness that meets the requirements for the dose rate limits. Required shielding thicknesses were investigated from the viewpoint of accidental full beam loss as well as beam loss on collimator. The results confirm that the required shielding thicknesses are highly sensitive to the spatial shape of the beam and strongly divergent beam losses. Therefore shielding barrier should be designed according to the more conservative assumptions.

1 Introduction

The Monte Carlo simulation technique is commonly used in radiation shielding analysis. It is based on tracking each individual particle and is considered a very reliable method. However, calculations of huge shields are computationally expensive and require special methods, such as a variance reduction, to obtain statistically reliable results. The limitations of available computing resources often restrict the use of Monte Carlo methods for thick shielding barriers. In this case the results obtained for moderate shield thicknesses can be propagated to thicker shields using analytical models.

This paper deals with the design of lateral shielding of MYRRHA reactor hall. Multi-purpose hYbrid Research Reactor for High-tech Applications (MYRRHA) is an accelerator-driven system (ADS) being designed at SCK•CEN in Belgium [1,2].The 600 MeV, 4 mA proton beam from a high-power continuous wave superconducting linear accelerator penetrates to the reactor from the top to drive the subcritical core. The bending magnet diverting the beam into vertical direction is located above the ceiling of 35 m–high reactor hall. The beam travelling from bending magnet through the reactor hall could experience certain proton losses on the beam focusing equipment such as collimators.

To evaluate the shielding barrier, both continuous and accidental beam losses have been considered.

Various spatial profiles of the proton beam entering the reactor hall have been studied to define the specific requirements for the reactor hall shielding. The required minimum shielding thickness was determined according to the dose rate limits for the public and controlled area behind the shielding. The sensitivity of the required shielding thickness to the proton source distributions and analytical model is discussed as well.

The induced radioactivity and residual doses in reactor hall have also been calculated in order to better evaluate the required shield thicknesses.

2 Geometry and Method

The structure of a 35 m high vertical part of the beam line below the last bending magnet and penetrating the reactor hall is shown in Fig. 1. The inner diameter of 13.75-mm thick ferritic-martensitic T91 stainless steel (SS) beam tube is 84.3 mm. The distance between the beam tube and inner wall of concrete biological shieldingis 8 m. This volume is filled with air in the model. A 3 m thick standard concrete cylinder models the biological shielding. The components of the reactor cover (stainless steel plates enclosing borated water with penetrations for primary heat exchanger, primary pumps and in vessel fuel handling machine) were also included in the geometry in order to simulate their influence on the dose distribution.

After the bending magnet the divergent beam must be collimated by the collimator for an optimal beam profile. This will protect the beam line and reactor cover components in the reactor hall from protons deviated from their prescribed trajectory. Therefore, the computational model contains the collimator 1 m below the reactor hall ceiling.

Since different beam loss scenarios may lead to completely different dose distributions, the effects of proton beam spatial shapes on a required shielding thickness during a routine accelerator operation have been investigated.

A reference profile models the Gaussian beam ($\sigma = 9$ mm) swept around the central axis with the sweeping radius being 21.5 mm [3]. To simulate the beam loss, the source was diverted by 0.07° at the top of 35m-high vertical beam line. The beam profile is shown in Fig. 2.

* Corresponding author: ycelik@sckcen.be

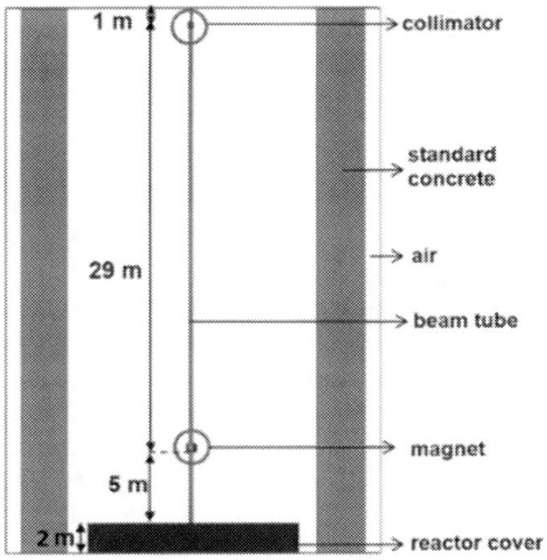

Figure 1. Vertical beam line geometry.

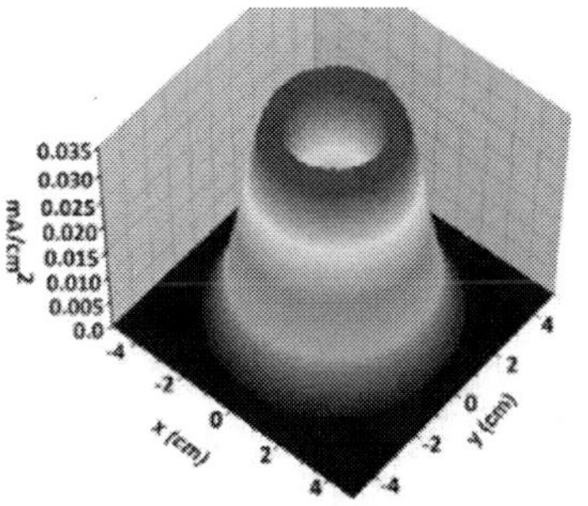

Figure 2. Profile of the sweeping proton beam at the top of the vertical beam line (reactor hall ceiling).

Another beam profile, a virtual one, is used to assess the dose rates in the situation when all the protons hit the beam tube close to the reactor cover. The source protons were linearly distributed in a 1 m long segment of the central axis and directed downwards to hit the beam tube at the last meter above the reactor cover.

To account beam losses on the collimator in a most conservative way, another virtual beam profile has been developed, representing a concial shape where protons launched from a point source above the hall ceiling with a divergence of 7.41° from the original beam direction hit the collimator.

The attenuation doses in the shielding were fitted by double-exponential functions [4-6], which assume a uniform beam loss along a linear proton source (henceforth denoted as "function 1") and a localized point source (henceforth denoted as "function 2"), see Eq.1 and Eq.2, respectively.

$$H(d) = \frac{H_1}{r}\exp\left(-\frac{d}{0.94\lambda_1}\right) + \frac{H_2}{r}\exp\left(-\frac{d}{0.94\lambda_2}\right) \quad (1)$$

$$H(d) = \frac{H_1}{r^2}\exp\left(-\frac{d}{\lambda_1}\right) + \frac{H_2}{r^2}\exp\left(-\frac{d}{\lambda_2}\right) \quad (2)$$

where $H(d)$ is the dose equivalent at depth d in shield, (H_1, λ_1) and (H_2, λ_2) are source terms and the attenuation lengths of the small thickness and large thickness of shield, respectively, d is the thickness of the shield and r = a + d, where a is the distance from the radiation source to the inner wall of the shield.

Shielding calculations have been performed with the general purpose radiation transport code MCNPX-2.7.0 [7]. The ambient equivalent doses have been calculated in MCNPX by collapsing the particle fluxes with energy-dependent dose conversion factors. Flux-to-ambient dose conversion factors were taken from the ICRP report 74 [8] for neutrons up to 200 MeV. For neutrons of higher energies as well as for protons and photons of all energies the functions from Ref. [9] have been used. Collimator activation and residual dose calculations have been carried out with the ALEPH2 depletion code developed at SCK-CEN [10] to determine the final disposal options and to assess the residual dose rate, which the workers or the equipment will be subject to during occasional maintenance and/or repair of collimator.

According to the IAEA recommendations [11] the dose equivalent rate limit for normal operation is 0.1 μSv/h for public areas and 10 μSv/h for controlled areas. When the operating time is assumed to be 270 day per year, i.e., 6480 h y^{-1}, these values will ensure that the annual dose limits of 1 mSv for the public and 20 mSv for workers will not be exceeded.

3 Main results

3.1 Normal operation losses on beam tube

For normal operation conditions we assume a loss of 10 W along every meter of the vertical beam line [12], with a total length of 37 m (35 m of beam tube in reactor hall and 2 m in the reactor cover). This value corresponds to beam losses of 3.84×10^{12} protons per meter and per second (37 m × 1.04×10^{11} p/s per meter).

The spatial distributions of the dose rates in the whole geometry for two proton source distributions are shown in Fig. 3.

The dose rates in the air around the beam tube rise up to 10^7 μSv/h in case of the reference Gaussian beam profile (a). Because of the spatial distribution of the proton flux in case of linear source distribution, the ambient equivalent dose is concentrated in the region around the last meter of the beam line (b). Dose rates decrease by almost two orders of magnitude in the reactor hall air if the linear proton source profile is considered. On the other hand, using linear source one gets the dose values behind the biological shielding lower than the control area dose limit (10 μSv/h), while in case of Gaussian source the dose values are around 200 μSv/h. Provided that access to this area is allowed only when the accelerator is not operational, the dose rate in these areas does not represent any danger.

DOI: 10.1051/epjconf/201715303007

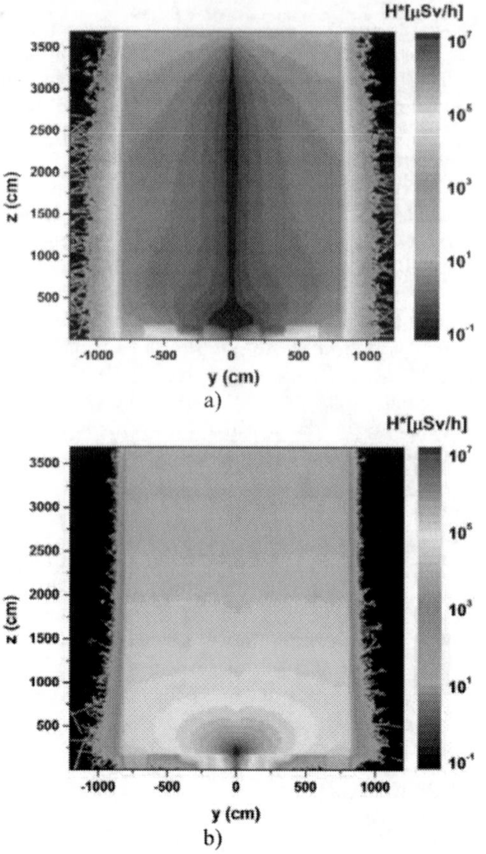

Figure 3. Spatial distribution of the dose rate for a) reference Gaussian source and b) line source.

Fig. 4 (a) displays the ambient equivalent doses calculated as a function of shield thickness. This plot allows to determine the required thickness according to the classification of the area behind the shielding. The calculated doses were fitted and extrapolated for larger depths by dose function 1. The first term of the function 1 describes the attenuation of dose below 0.45 cm. The rest of the data was extrapolated for second term of function 1. This value sreflects the equilibrium state when the neutron spectrum shape does not change with the shield depth and the neutron attenuation is purely exponential.

Clearly, there is a significant difference in the required shield thicknesses to reach the safety limits for the public and controlled areas. The dose curves are highly sensitive to the primary proton source distribution. In case of a Gaussian source, biological shieldings with thicknesses of about 4.07 m and 5.95 m are needed to remain within the controlled area and public area dose limits, respectively, while for a linear source distribution thicknesses of only 2.35 m and 4.10 m are sufficient. This huge difference reflects the differences in model assumptions for the beam losses: the Gaussian beam profile provides uniformly distributed losses along the height of the reactor hall, while the linear beam profile results in losses concentrated in the last meter of the beam tube above the reactor cover.

The neutron attenuation profiles obtained using the line source distribution were also fitted by function 2 assuming a localized radiation point source. A comparison between the raw data and fit of the neutron dose rate attenuation for point and linear source equations is shown in Fig. 4 (b). The usual approach that can be found in the literature [13-15] is either to use a line source in calculations and then fit the calculated values with equation (1) or to use a point source in calculations and fit the calculated values with equation (2). We demonstrated here that the data points obtained using linear source distribution are in very good agreement with both functions 1 and 2 up to 4 m of shield thickness. Above this thickness the two fits begin to deviate. The fit by function 1 guarantees a conservative estimation of the required thickness. This comparison was also performed for the Gaussian source and these two functions provide a similar required shielding thicknesses through the whole fit line.

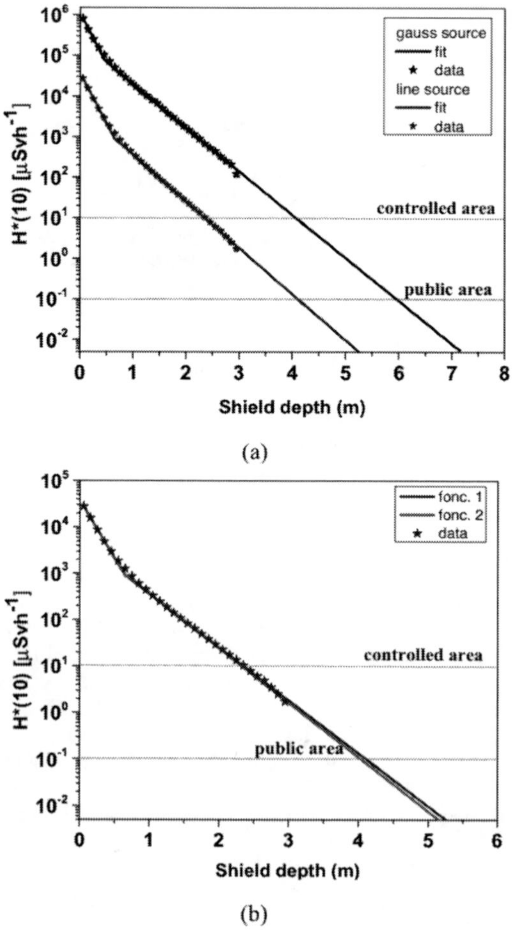

Figure 4. a) Neutron dose rate attenuation curves in standard concrete used as biological shielding for Gaussian and linear proton source distributions. b) Comparison between the fits by functions 1 and 2 to the data obtained with the line source distribution. The dose rate limits for controlled and public areas are indicated by the dashed lines.

DOI: 10.1051/epjconf/201715303007

3.2 Normal operation losses on collimator

The collimator is of particular concern due to exposure by the strongly divergent beam from the bending magnet. It was assumed that the collimator absorbs 1 % of the total proton current (4 mA). However, it is an important source of secondary particles, which can cause induced radioactivity and prompt doses. Therefore, to evaluate the shield thickness in the presence of collimator in the reactor hall the doses must not be underestimated.

A simple cylinder was used to simulate the collimator geometry. The collimator height was determined by a range of 600 MeV protons in the material. The projected range estimated with the SRIM code [16] was found to be 30 cm for copper. The diameter of the collimator made of copper (without the beam tube) equals to the height. The elemental composition of copper with account of impurities necessary for accurate activation and secondary doses calculations is shown in Table 1.

Table 1: Elemental composition used in [17] for MCNPX calculations. The density is 8.96 g/cm³ at room temperature.

Element	wt. %	Element	wt. %	Element	wt. %
Cu	99.9769	Ca	0.00025	Mo	0.0003
Li	0.00005	Ti	0.0001	Ag	0.0013
B	0.00005	V	0.00015	Sn	0.0005
N	0.001	Cr	0.0001	Cs	0.0001
Na	0.0001	Mn	0.00005	Ba	0.00004
Mg	0.00005	Fe	0.006	Eu	0.00002
Al	0.001	Co	0.00005	Ho	0.0001
Si	0.0001	Ni	0.0004	Hg	0.0002
P	0.004	Zn	0.0006	Pb	0.0008
S	0.0015	Ga	0.0002	Bi	0.0001
Cl	0.003	As	0.0001	Th	0.00005
K	0.00005	Rb	0.0001	U	0.00005

Fig. 5 shows the axial distributions of the energy deposited by neutrons, protons and photons. One can see that protons determine the largest part of the total heating. The proton heating density decreases with increasing depth of collimator thickness. Between depth of 23 and 25 cm there is a sharp decrease in proton heating since the primary protons are fully stopped by electromagnetic interactions. Residual proton heating at higher depth is due to secondary protons produced by neutrons in copper. The second largest contribution to the total heating density comes from photons. Contribution of neutrons is negligible.

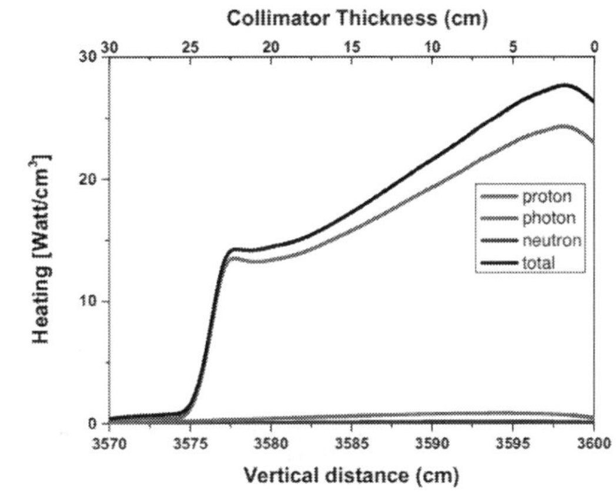

Figure 5. Energy deposition in the copper collimator.

The spatial distribution of the neutron ambient equivalent dose in the entire geometry is shown in Fig. 6. Large quantities of neutron doses, which are represented by the red colour, are produced by interactions with copper and the plot shows the doses exceed 10^7µSv/h. It is clearly seen that the collimator itself behaves as a radiation source in the reactor hall and causes higher doses complementary to ones produced during routine accelerator operation.

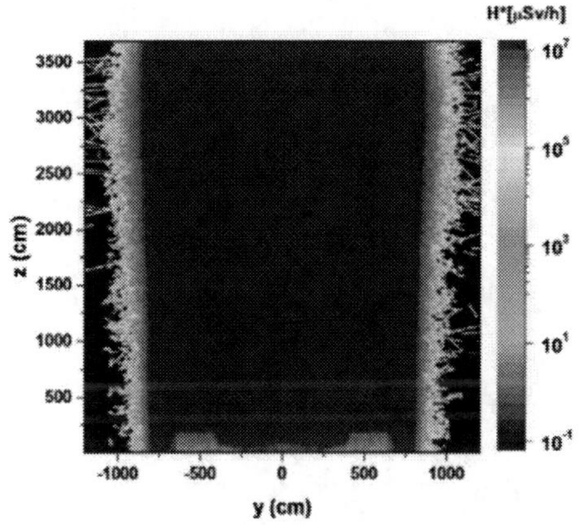

Figure 6. Dose rate spatial distributions in case of beam loss on collimator.

Detailed dose rates in the biological shielding are shown in Fig. 7. To identify the required shield thickness, dose values were fitted to function 2. According to the calculation results, the shield thicknesses determined by normal operation beam losses (see Fig. 4 (a)) do not meet dose limits due to additional doses produced by collimator. In this case, to reach the safety limits, 6.33 m and 8.25 m of shielding for the controlled and public area are needed, respectively.

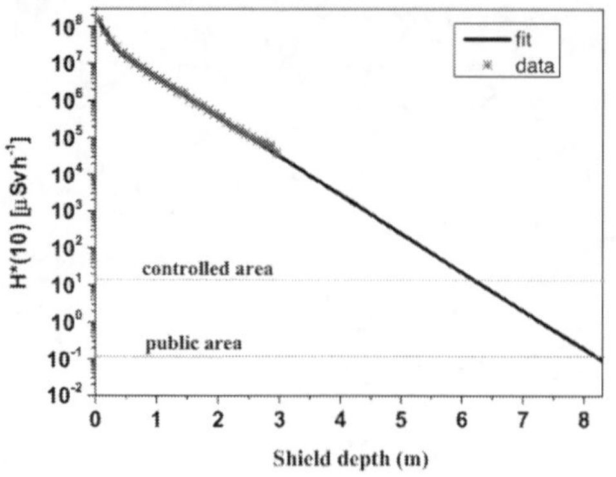

Figure 7. Neutron dose rate attenuation curve in biological shielding for beam loss on collimator.

3.3 Accidental full beam losses

An accidental scenario of full loss of a 4 mA proton beam (2.5×10^{16} protons per second) was considered to define the maximum acceptable integrated dose that can be received by personnel who may be present in the area. To simulate full beam loss in this case, the primary protons have been launched from both linearly distributed source and Gaussian source. The loss from the linear source produces an instantaneous equivalent dose rate of 2.4×10^{6} $\mu Sv{\cdot}h^{-1}$ after a minimum of 2.24 m biological shielding in order to remain below the controlled area limit. When the accelerator is interrupted within 100 μs, the integrated dose equivalent is about 0.07 μSv for the controlled area, as is demonstrated in Figure 3. The shield thickness obtained for normal operation beam loss is thus also acceptable for the full loss if the accelerator power is cut within 100 μs. The integrated dose from full beam loss for the Gaussian source distribution is 0.002 μSv behind 4.07 m of biological shielding that is below the controlled area limit. If full beam loss is assumed on the collimator, integrated doses will be 3×10^{-5} μSv for the control area behind 6.35 m of shield barrier.

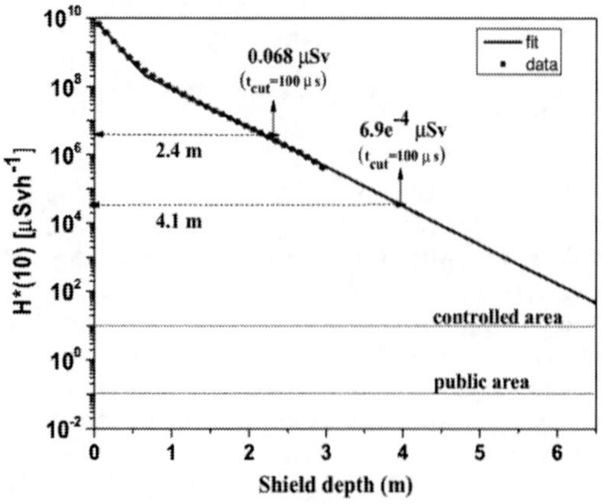

Figure 8. Neutron dose rate attenuation in the biological shielding at full beam loss.

3.4 Activation of the Collimator

Even if the collimator is designed to withstand the radiation, it becomes highly radioactive due to the proton beam exposure. The dominant isotopic activities in the collimator are presented as a function of time in Fig. 9 for 40 years of operation and 10^5 years of cooling.

During irradiation the dominant radioisotopes produced in the copper are ^{57}Co, ^{55}Fe and ^{49}V. They both decay quickly after the end of operation due to their relatively short half-lives. ^{53}Mn, ^{36}Cl, ^{41}Ca, ^{59}Ni and ^{14}C are major contributors in the total activity until 10^5 years after shut down. It is clear that copper collimator will remain highly activated long after irradiation. Therefore alternative low-activated materials such as graphite are suggested.

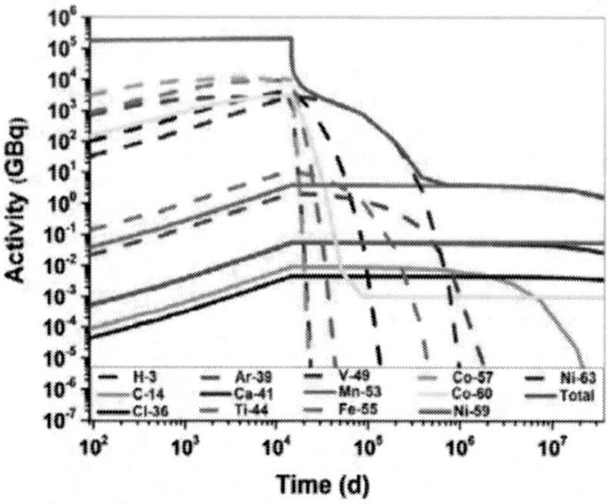

Figure 9: Time behaviour of copper collimator activity.

3.5 Residual gamma doses

After its operational lifetime, the collimator will be removed from the reactor hall and placed in the storage. It is important to know the dose rates in the close proximity of the collimator. Therefore, additional calculation was performed to determine residual gamma doses in the air at 1 m from the collimator. A separate MCNPX photon transport problem with a simple geometry including the collimator surrounded by air was has been developed. The energy spectrum of residual gamma rays emitted from the collimator at different cooling times was used as photon source distribution.

Fig. 10 shows residual gamma doses during 10^5 years of cooling after 40 years of operation. There is observed a slow decrease in dose rates since copper produce significant amounts of long living isotopes such as ^{59}Ni, ^{53}Mn, ^{41}Ca . After 200 years of cooling dose rates reach controlled area dose limit 10 μSv/h. In order to reduce the radiation exposure during maintenance of the collimator, the collimator must be covered with sufficient shielding.

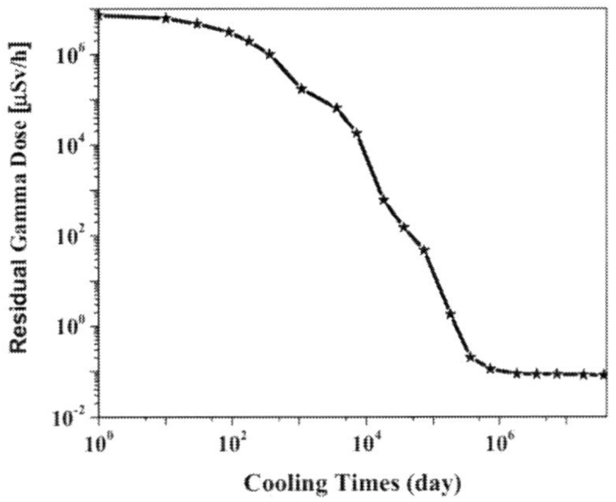

Figure 10: The residual dose rate in the air at 1 m distance from collimator.

4 Conclusion

The minimum required shielding thicknesses depend on the primary proton source spatial distributions for the fixed beam line geometry, spatial location of beam losses and their absolute values. The results demonstrate that the minimum shielding thicknesses defined for normal operation conditions meet the radiation dose limits in case of full beam loss. However, the integrated doses at full beam loss could be high, especially when the shield thicknesses are designed according to optimistic assumptions. To ensure the integrated doses at accidental conditions to be below the dose limits, the most conservative source distribution and strong divergent beam losses must be used for simulations, providing highest dose rates.

The obtained results also show that the copper collimator produces high dose levels even after shut down. From the radiation protection point of view it should be shielded also in the storage area to protect workers and equipment.

References

1. R. Fernandez, et al. MYRRHA Primary System Design Technical Description Rev.1.6 , SCK•CEN/1088304 (2014)
2. E. Malambu and A. Stankovskiy, Revised core design for MYRRHA – Rev. 1.6, SCK•CEN/3958903 (2014).
3. S. Keijers et al., "Design of the MYRRHA spallation target assembly", Proc. 13th Information Exchange Meeting on Actinide and Fission product Partitioning and Transmutation, Seoul, Korea, 23-26 September 2014.
4. S. Agosteo, T. Nakamura, M. Silari, Z. Zajacova, Attenuation curves in concrete of neutrons from 100 to 400 MeV per nucleon He C, Ne, Ar, Fe and Xe ions on various targets, Nucl. Instr Meth .B 217, 221–336 (2004)
5. S. Agosteo, G. Fehrenbacher, M. Silari, Attenuation curves in concrete of neutrons from 1 GeV/u C and U ions on a Fe target for the shielding design of RIB in-flight facilities, Nucl. Instr. Meth. B 226, 231–472 (2004).
6. A. H. Sullivan, A Guide to Radiation and Readioactivity Levels Near High-Energy Particle Accelerators, Nuclear technology Publishing (1992).
7. D .B. Pelowitz, et al., "MCNPX 2.7.0 Extensions", LA-UR-11-02295 (2011).
8. International Commission on Radiation Protection, "Conversion coefficients for use in radiological pro-tection against external radiation," ICRP Publication 74, Pergamon Press, Oxford (1997).
9. M. Pelliccioni, Overview of fluence-to-effective dose and fluence-to-ambient dose equivalent conversion coefficients for high energy radiation calculated using the fluka code, Radiation Protection Dosimetry 88 (2000) 279–297.
10. A. Stankovskiy, G. Van den Eynde, Sci. Thechnol. Nucl. Install, 2012;2012. 545103.
11. IAEA, Radiation Protection Programmes for the Transport of Radioactive Material, IAEA Safety Standards Series No. TS-G-1.3, Vienna (2007).
12. D. Vandeplassche, private communication, SCK•CEN (2014).
13. S. Agosteo, G. Fehrenbacher, M. Silari, Attenuation Curves in Concrete of Neutrons from 1 GeV/u C and U Ions on a Fe target for the Shielding Design of RIB in-Flight Facilities, Nucl. Instrum. Methods B 226 (2004a) 231–472.
14. S. Agosteo, T. Nakamura, M. Silari, Z. Zajacova, Attenuation Curves in Concrete of Neutrons from 100 to 400 MeV per Nucleon He C, Ne, Ar, Fe and Xe Ions on Various Ttargets, Nucl. Instrum. Methods B 217 (2004b) 221–336.
15. D. Ene et al., Radiation Protection Studies for ESS Superconducting Linear Accelerator, Progress in Nuclear Science and Technology 2 (2011) 382-388.
16. www.SRIM.org.
17. M.Wohlmuther, et al., Validation of Activation Calculations with MCNPX with Samples from a Copper Beam Dump, Proc. AccApp'07, Pocatello, ID AIP Press (2007), p.259.

Shielding design and radioprotection for Andromede and ThomX accelerators

Jean-Michel Horodynski[1,*], and *Sebastien* Wurth[2]

[1]iRSD – ingénierie Radioprotection sûreté démantèlement, UPS3364, CNRS/INP, 91898 Orsay, France
[2]IPNO – Institut de physique nucléaire d'Orsay, UMR8608, CNRS/IN2P3, Université Paris-Sud, 91898 Orsay, France

Abstract. The *Institut de physique nucléaire* – Institute of Nuclear Physics (IPN) and the *Laboratoire de l'accélérateur linéaire* – Laboratory of the linear accelerator (LAL) are about to build two accelerators at the University Paris-Sud, Orsay. Andromede is a new imaging mass spectrometry instrument for surface analysis: the electron positron emission microscope (EPEM) will analyse impacts on a surface made by nano-particles (Au404) in the keV to MeV range to study ion-surface interactions. ThomX is a compact X-ray source (energy up to 90 keV – flux up to 10^{13} photons.s^{-1}) that will produce X-rays using Compton backscattering effect between an electron beam and a laser. In order to use these machines in the same area, hutches are designed to comply with French regulations for radiation protection. This paper presents the methods used to design the hutches, complying with safety objectives, technical issues and budget.

1 Introduction

Andromede [1] and ThomX [2] are new accelerators conceived by two French research laboratories, respectively IPN (*Institut de physique nucléaire*) and LAL (*Laboratoire de l'accélérateur linéaure*). Based at the *Université Paris-Sud* at Orsay, these two facilities are going to be settled in the same building called "Igloo", which previously sheltered the synchrotron storage ring DCI of LURE, dismantled three years ago [3].

1.1 Purpose of IGLEX

IGLEX is a multidisciplinary center for technologies and applied sciences. The main goal is the research and development on accelerators and their use in various application fields: bio-chemical, astrophysics, materials science, medical imagery & therapy and cultural heritage.

Andromede is a Secondary Ion Mass Spectrometry (SIMS) for nano-domains and nano-objects. A Van der Graaf generator will accelerate nano-particles of Au (Au4004+ up to Au10000n+) and C_{60}^{n+} fullerenes between 1 to 4 MeV. Ions will be produced by two sources, a Liquid Metal Ion Source (LIMS) as a NanoParticle Ion Source (NIPS) provided by Orsay Physics and an Electron Cyclotron Resonance source (ECR) for multi-charged molecular ions, provided by Pantecknik.

ThomX is a source of X-rays of high flux and energy, produced by backscattering Compton interaction. The linear accelerator will produce a 70 MeV electron beam which will be injected in the compact storage ring (greater diagonal of about 7 m length). The X-rays will be produced with the interaction between electrons and photons coming from an amplified laser within a Fabry-Perot cavity [4]. The purpose of ThomX is to demonstrate that historical heritage study and Xray imaging can be affordable for a museum, or a large hospital, due to the compactness of the installation.

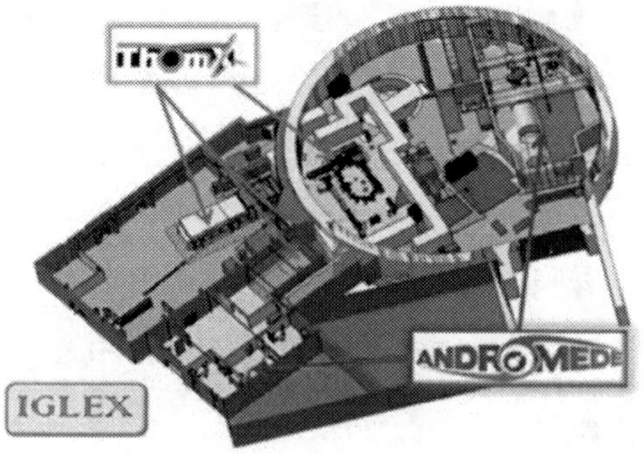

Fig. 1 IGLEX with the two accelerators, Andromede (IPN) and ThomX (LAL).

1.2 Limitations – specificities for the radiation protection

iRSD (*ingénierie Radioprotection Sûreté Démantèlement* – engineering in Radiation protection, Safety and Dismantling) and IPN radiation safety teams, due to their experience and expertise, are in charge of the studies leading to a high level of performance in operational and occupational safety, including radiation protection. On this purpose, national regulations and the French standard NF M 62-105 [5] for industrial accelerators have to be implemented.

As shown in fig. 2, the two accelerators will be placed in the same area but each one will be operated independently. Therefore, one bunker has to be built for each accelerator.

In the same time, workers could be present in one hutch or in the area between the two bunkers, called "circulation area" while one or both accelerators will work. In order to reduce radiation exposure while complying with technical and economic limitations, both teams proceed with a process of radiation protection optimization.

In particular maximum load of the ground limits the weight of the radiological protection used to design the bunkers. So, following the French regulations, circulation area will be defined as a "*Zone Surveillée*" or "Monitored Area", considering that the highest radiation level allowed is 7,5 μSv per hour for a worker. To avoid unexpected interferences, radiation levels around each shielding hutch are limited to 3 $\mu Sv.h^{-1}$ at the most.

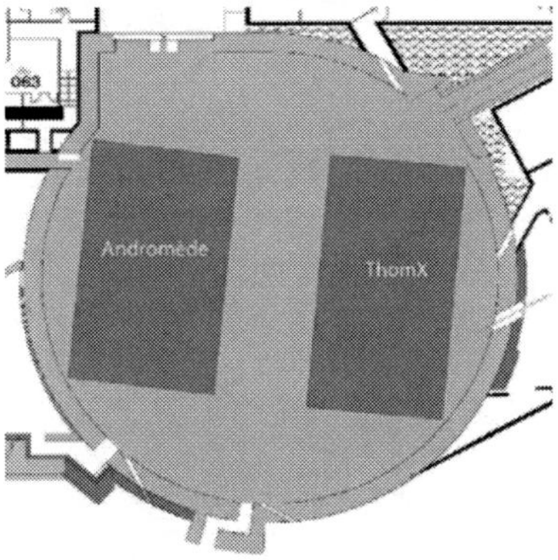

Fig. 2. Radiation protection zoning in the "Igloo" during the exploitation of Andromede and ThomX. In gray, the circulation area classified as "Zone Surveillée" (Monitored Area). In dark grey, the two hutches classified as "Zone Interdite" (Prohibited Area).

Access control, radiation monitoring system, command control and safety need to be installed considering the bunkers and specifications from AFNOR standard [5]. Interaction between components from the two machines had been designed to protect workers and to minimize operational constraints (Fig. 3). For example, radiation monitoring system interlocked with both accelerators stopped them if the level of radiation above the limit is detected. In the same time, access in hutches will be independent, except for the door leading to the circulation area (if the door stays opened during a time superior to a defined limit, both accelerators will stop).

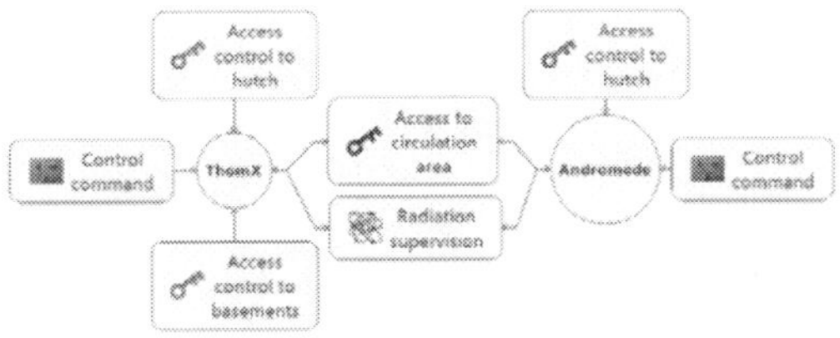

Fig. 3 : Simplified diagram about interlocks between safety components and accelerators in IGLEX plant.

Both project-teams need a strong coordination for risks assessment and operational safety of two plants in one building. LAL, supported by iRSD, and IPN work together to conduct radiation protection and operational safety studies. At the end, for the first time, both laboratories will provide the French nuclear authority, ASN (*Autorité de sûreté nucléaire*), two applications files with some common parts: operational and occupational safety, health and safety, radiation protection (Fig. 4).

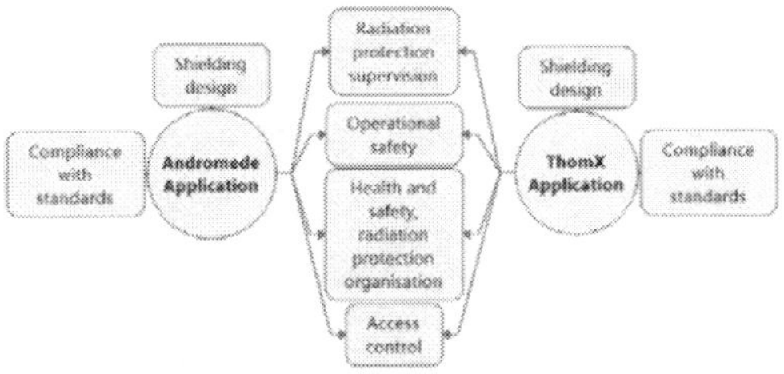

Fig. 4. Common and specific parts of the two application files in order to obtain the license to use the two accelerators.

The health and safety organization needs to be reliable and efficient. In this way, an agreement between three different partners, LAL, IPN and iRSD, is currently drafted to define responsibilities and operational scope of each laboratory.

2 Method used for shielding designs

Monte-Carlo codes of transport and interaction of particles are the most powerful tools to assess ionising radiation risks for accelerators. From shielding design to radiation protection and activation assessment, their capabilities increased over the past years. Choosing which code to use lies on multiple criteria, specifically: the kind of study to do, the results of benchmarking and comparisons tests, the regular updates of the code, the interactions between users and the accessibility of the code.

2.1 FLUKA as a multi-purpose code for radiation protection and radioactive waste management

FLUKA [6], [7] is a Monte Carlo code simulating the transport and the interactions of particles. Created in 1962, the third main version has been developed and maintained by INFN and CERN. This code has been chosen by the IPN and the iRSD to conduct radiation protection studies [8]. The powerful tools developed, like the 3D modelling and visualizing interface FLAIR [9], as the new features like the on-line activation calculation give FLUKA great capabilities to conceive new installations in an optimized way regarding radioprotection, environment and waste management.

2.2 Benchmarks on ions accelerators

As developed within the CERN, FLUKA has always been suitable for hadrons accelerators studies, for high energies before and then for lower energies with the successive implementations of the code.

2.3 Benchmarks on electrons accelerators

Improvements of low energy neutrons and electrons/photons modules for transport and interactions [10] during the last years are a strong justification to use FLUKA to make radiation protection or activation studies for electron accelerators.
Several plants used FLUKA for radiation protection studies: ARIEL – TRIUMF (Vancouver, Canada) or Spring-8 (Saitama, Japan) [11] for example.

2.4 Statistical

In electron accelerators, *Bremsstrahlung* and photonuclear interactions are predominant. Use of methods of variance reduction for shielding designs is mandatory to optimize the factor of merit (FOM). In the case of ThomX, importance biasing was applied to photonuclear interactions and inelastic interactions with electrons, and the mean free path in materials has been reduced to increase the probability of interactions with photons and electrons. The same techniques were used for Andromede as well as multiplicity biasing (based on importance biasing) for beam particles (proton or alpha).

2.5 Defining the sources

Both accelerators have different final scientific purposes, so their operational modes are not related. Sources must be defined considering these specificities.

Andromede will use ion beams to study interactions with targets. Therefore, a two-step method based on the assessment of the production rate for neutrons/photons and X-rays is used.

ThomX will produce X-rays in a storage ring on top-up mode. Several leak points for the primary beam are identified all along the accelerator.

2.5.1 Andromede

Two kinds of particles productions need to be taking account: X-rays losses within the accelerator tank and the neutrons and gammas produced in Faraday cups.

While conditioning, some X-rays can be emitted through the last titanium electrode tube within the accelerator tank. The model is considered conservative enough: 200 keV electron produced along the electrode within the accelerator's tank at 1% of maximum intensity (200 µA).

For ion beams, neutron and gammas produced in Faraday cups have been assessed using a two-step method with FLUKA:

- First, use of a reduced geometry in order to define neutron (n) and photon (g) yields from 4 MeV p^+ and 8 MeV He^{2+} reactions on natural C. Results are shown in table 1.

Table 1 Neutron and photon yields from p+ and 4He2+ beams on natural carbon

Projectile / E (MeV) / I (µA)	Target	Secondary radiation / E (MeV)[1]	Production Yield	References [12] [13]
p^+ / 4 / 10	Natural carbon	n / 0,8	$6*10^{-6}$ n/p	~ 10^{-6} n/p
		g / 3,1	$5*10^{-7}$ g/p	-
$^4He^{2+}$ (a) / 8 / 2,5	Natural carbon	n / 1,5 and 6,5	$2*10^{-5}$ n/a	~ $8*10^{-6}$ n/a
		g / 4,5	$2*10^{-3}$ g/a	-

(1): the given energy value is the most probable one in a radiation safety approach.

- The second one in another FLUKA calculation with the whole geometry and those yields as entry data to define the different local shielding thickness.

2.5.2 ThomX

ThomX is a standard storage ring using electron as the primary beam. Several feedbacks from exploitation of synchrotron electron rings can be used to localize and define leak points along the machine.

For example, the main leak point in storage rings, like SPEAR3 (Stanford - USA) [14], DELTA (Dortmund – Germany) [15] [16], SOLEIL (Saclay – France) [17] or ESRF (Grenoble – France) [18] identified the "*septum*", component used to inject electron beam into the ring, as the main leak point.

A scraper will be used in the transfer line to improve the quality of the primary beam. It removes electrons located outside the maximum diameter allowed. As a "target with a hole", continuous losses will occur at this point.

Two beam-dumps will be used. The one situated at the end of the LINAC will totally stop the beam in setting mode. Another one will be located at the end of the ring to stop the extracted beam, when X-rays will be produced (Top-up mode). The design of these components are the same as those used at SOLEIL Synchrotron.

With the cooperation of the team in charge of the beam dynamics, five losses points (Fig. 5) have been defined, as presented in table 2.

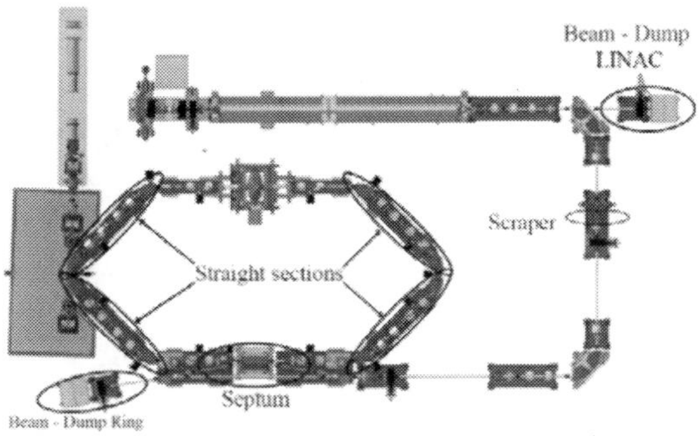

Fig. 5. Losses points localized all along the ThomX machine.

Losses in the scraper have been modelled in this way: the primary beam is directed in the jaws of the scraper. But width of the jaws (copper) are not sufficient to stop the primary beam. In order to stop the propagation of the electrons from the primary beam, a little area situated downstream stops all the electrons leaving the jaws (Fig. 6).

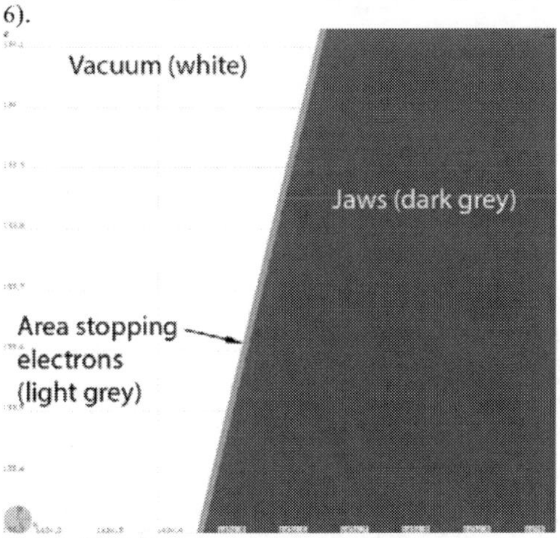

Fig. 6. Modelling scraper losses area. Downstream the jaws (dark grey), an area (light grey) stopped all leaving electrons.

Losses in the straight sections came from electrons situated at the periphery of the bunch [19] [20]. They are modelled as a cylinder starting at the center and propagating along the beam pipe (Fig. 7.)

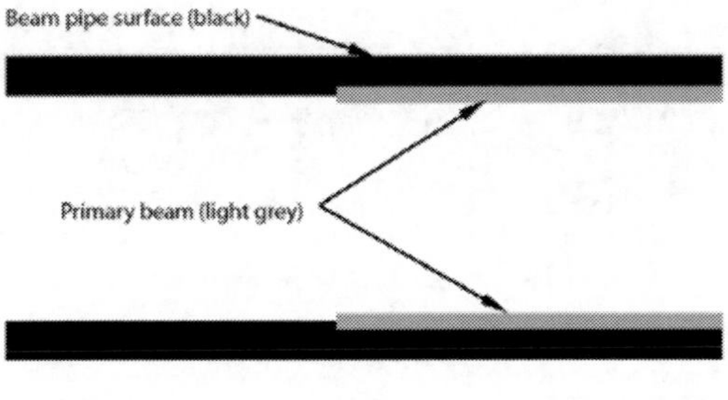

Fig. 7. Modelling straight section losses (side view). The primary beam (light grey) is a cylinder which diameter is slightly larger than the beam-pipe (black).

Table 2. Losses points defined all along ThomX

Localization	Percentage of loss (%)	Current load of the primary beam (nC)
Beam - dump LINAC	100	1,5
Beam – dump ring	100	0,9
Septum – loading	10	1
Septum – emptying	10	1
Straight sections of the ring	4	1
Scraper	15	1,5

2.6 Geometrical modelling

2.6.1 Andromede

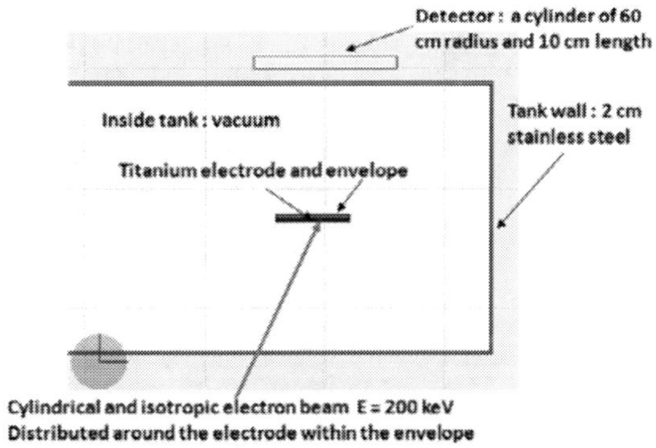

Fig. 8. Modelling X-rays losses along the last titanium electrode tube within the stainless steel accelerator tank.

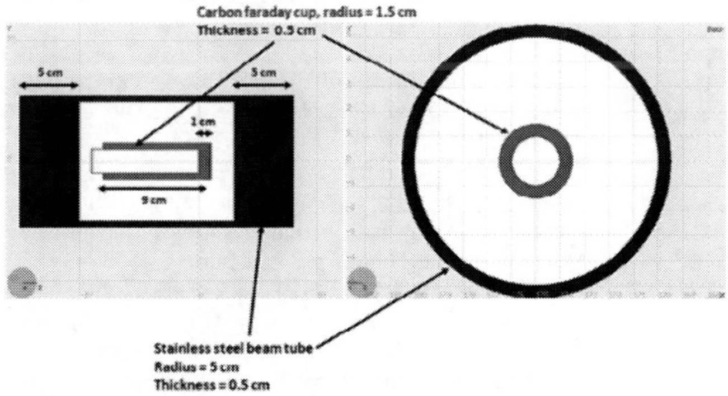

Fig. 9. Modelling carbon Faraday cup within stainless steel beam tube

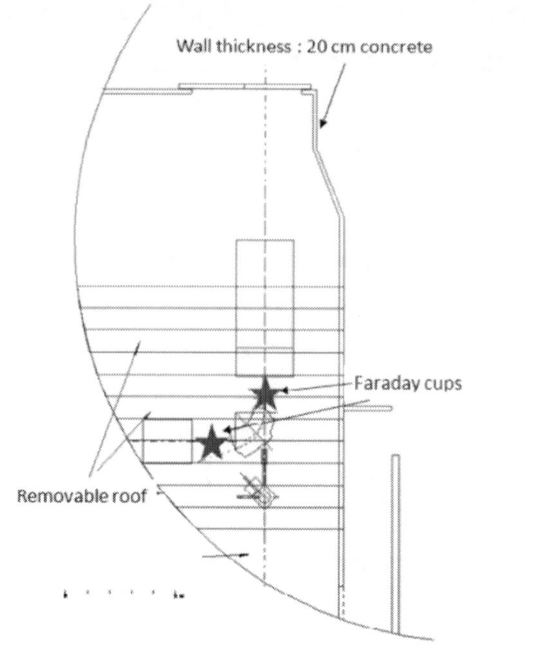

Fig. 10. Modelling the whole geometrical situation. The two stars show the location of the two Faraday cups as beam dump.

2.6.2 ThomX

At first, ThomX was intended to be the sole device in the "Igloo" building. First batch of numerical simulations was made to confirm the efficiency of the walls of the building (Fig. 11).

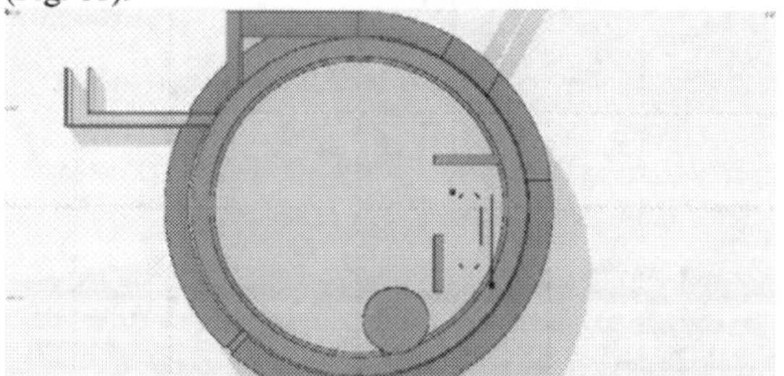

Fig. 11. First version of the geometrical modelling containing only the walls of the Igloo Building. Accesses and basements are missing.

When Andromede and ThomX merged in the IGLEX project, numerical simulations were used to design the hutches. The first step consisted in defining the minimum thickness of the hutch walls considering the radiation safety objectives and specific constraints (technical and financial). After that, a first design of the hutch was provided by the ThomX project team.

Therefore, new batches of numerical simulations have been done in an iterative way (Fig. 12). Each non-conformity with the safety and technical objectives was corrected by creating local shielding to reduce radiation leaks at the source of production or by modifying the design of the hutch or the materials used for the walls for example.

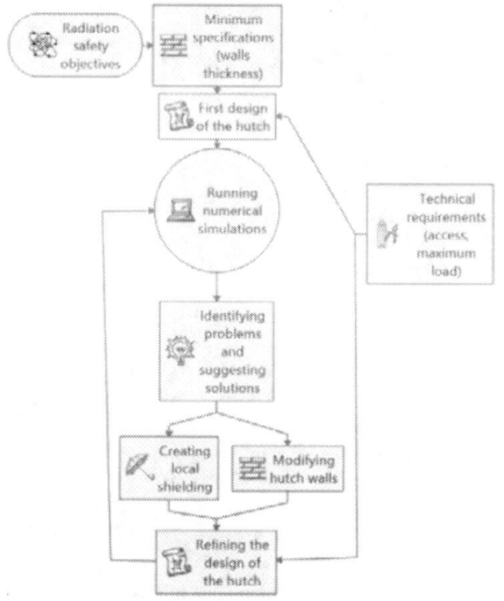

Fig. 12. Geometrical modelling process during the conception phase for ThomX.

The great flexibility of this process complied with the continuous conception of the ThomX machine: modifications of the machine, adapted integration in the hutch. The counterpart was the large amount of time needed to run each batch of numerical simulations (changing geometries, debugging, setting up the physics, running time, data analysis).

Concerning the machine, only parts involved in beam losses have been modelled (Fig. 12). Dipoles yokes of the ring have been included because they could alter the secondary radiations. Septum was modelled following indications provided by the "pulsed devices" team.

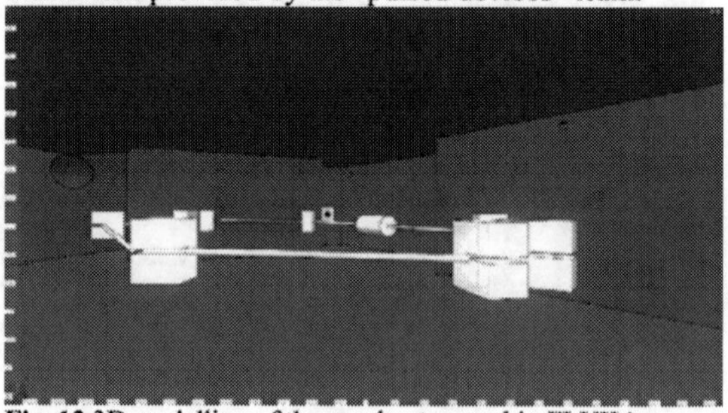

Fig. 13 3D modelling of the accelerator used in FLUKA.

3 Andromede shielding design

3.1 Description

Most of the time this accelerator facility will not produce radiation at all as the studied physics will only lead to sub-Coulomb reactions. But in case of light ions beams (proton or alphas), there will be some rare nuclear reactions while interacting with light material like carbon (typical Faraday cups). Identified radiation points are Bremsstrahlung X-rays from tank while conditioning beam, gamma rays and neutrons produced in nuclear reaction in the vacuum chamber or the Faraday cups while the light ion beam is stopped.

To keep produced radiation in that particular perimeter the approach is two-folded approach:

- A global shielding made of concrete (20 cm thick) and a half roof (45 cm thick) for the Andromede hutch.
- Two local shielding around the point of interaction (the Faraday cups or beam dumps) with classic shielding materials (high density polyethylene (PE) for neutrons and lead or stainless steel for photons).

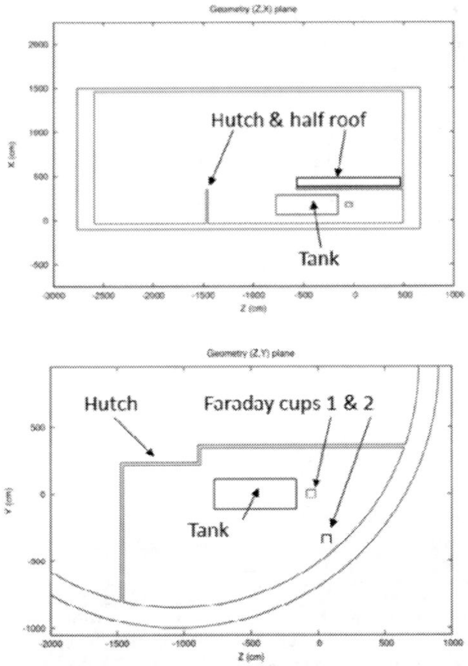

Fig. 14 Display of Andromede situation, vertical (left) and horizontal (right) cuts.

3.2 X-ray radiation from the tank

The plots show that in this case the shielding is sufficient to insure an exposure under the limits outside the hutch.

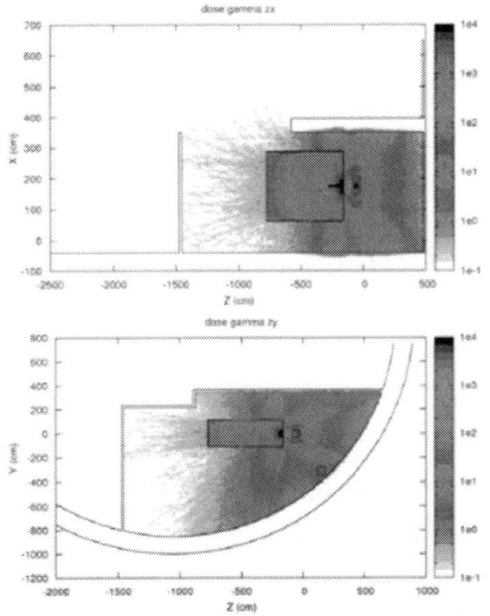

Fig. 15 Display of ambient dose rate in $\mu Sv.h^{-1}$ for X-ray production within the tank, vertical (left) and horizontal (right) cuts.

3.3 Faraday cups

For alphas: 5 cm PE & 10 cm lead, gamma dose showed in the case of Faraday cup n°1.

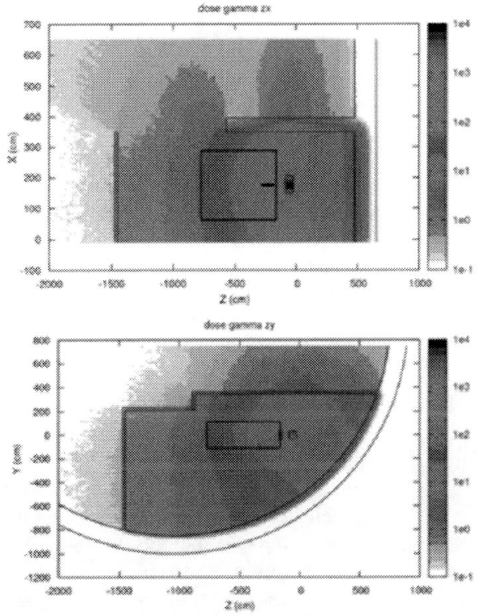

Fig. 16 Display of ambient dose rate in $\mu Sv.h^{-1}$ for an 8 MeV / 2,5 μA $^4He^{2+}$ beam interacting in Faraday cup n°1, vertical (left) and horizontal (right) cuts (gamma dose).

For protons: 15 cm PE, neutron dose showed in the case of Faraday cup n°2.

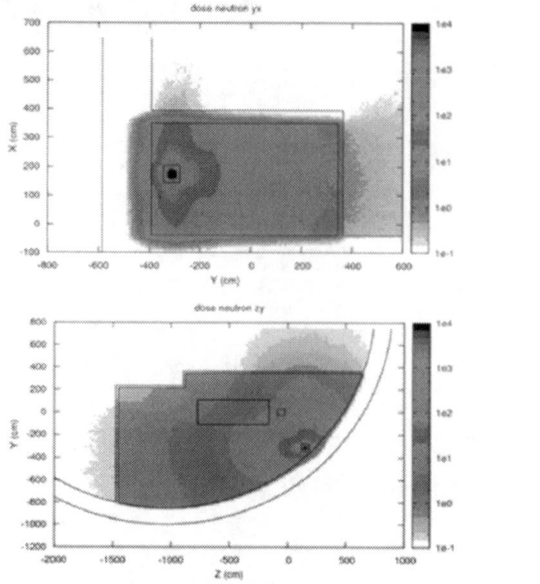

Fig. 17 Display of ambient dose rate in μSv.h^{-1} for a 4 MeV / 10 μA p$^+$ beam interacting in Faraday cup n°2, vertical (left) and horizontal (right) cuts (neutron dose).

4 ThomX design shielding

4.1 Accelerator Hutch

The main objective is to classify radiologically the circulation area as a monitored area. At first approximation, using analytical solutions, the minimum thickness for the walls of the hutch needs to be equal to 150 cm of ordinary concrete for high-dispersion straight sections losses. Some beam-loss locations produce higher level of radiation and need thicker walls. The floor load and budget for the infrastructure being limited, thickness of some walls needs to be optimized.

In a first time, the minimum thickness of the roof of the accelerator hutch was proportioned to ensure an ambiant dose rate $\dot{H}^*(10)$ inferior to 7.5 μSv.h^{-1}, limit value to define a monitored area. One meter of ordinary concrete was required. It had to be reduced because of technical issues, as the total weight or the supporting. Therefore, the project team, in accordance with the radiation safety team, decided to forbid the access to the upper part of the hutch during the operation of ThomX accelerator. After, a step-by-step method was used with numerical simulations to optimize the thickness of the roof, removing 10 cm of ordinary concrete by iteration. Considering the height between the roof of the hutch and the one of the Igloo building (more than 2 meters), the skyshine effect is negligible. Finally, the thickness of the roof will be 55 cm without altering the radiation zoning on the common area below the top of the roof.

Mazes were designed for the North and the South accesses of the hutch. Thickness of the South walls has been reduced and some parts removed to decrease the floor load. A prohibited area must be created outside the hutch to forbid presence of people during the work of the accelerator (Fig. 18).

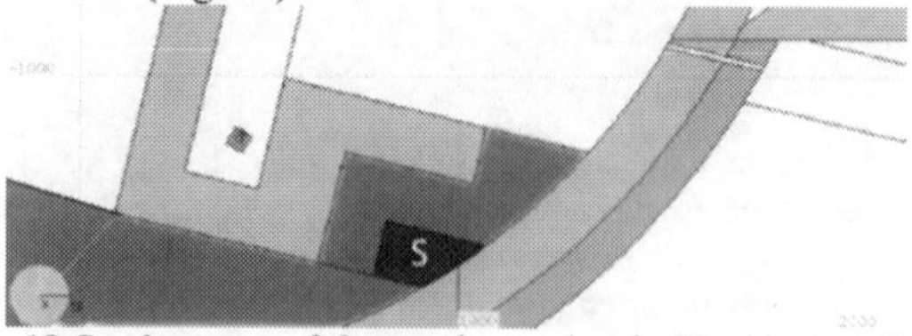

Fig. 18 South access of the accelerator hutch. The black wall (with an "S") has been removed in the final design. A prohibited area (middle gray) is created when the accelerator runs.

In the North maze, the thickness of one of the wall has to be proportioned according to the width needed to pass through the corridor leading to the control room toward the circulation area. Considering the first specifications for shielding, 1.5 meter of ordinary concrete was needed so the width available is around 30 cm (Fig. 19). Then, the thickness of the wall has been reduced using barite concrete (d = 3.2). In addition, a local shielding has been designed for the beam-dump of the LINAC (see 4.2).

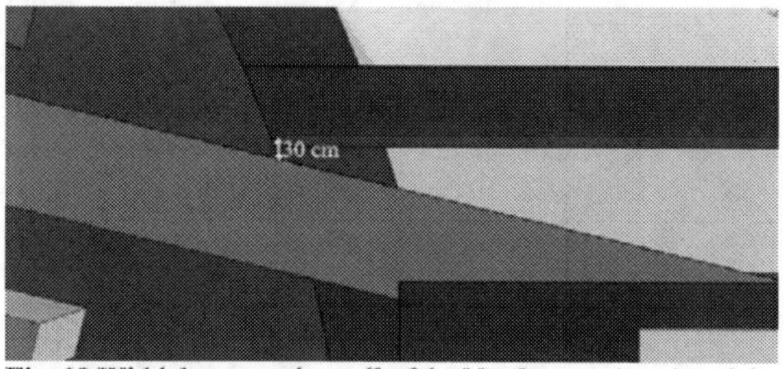

Fig. 19 Width between the wall of the North maze (gray) and the corridor between Igloo and control room (black) with ordinary concrete and 1.5 meter of thickness.

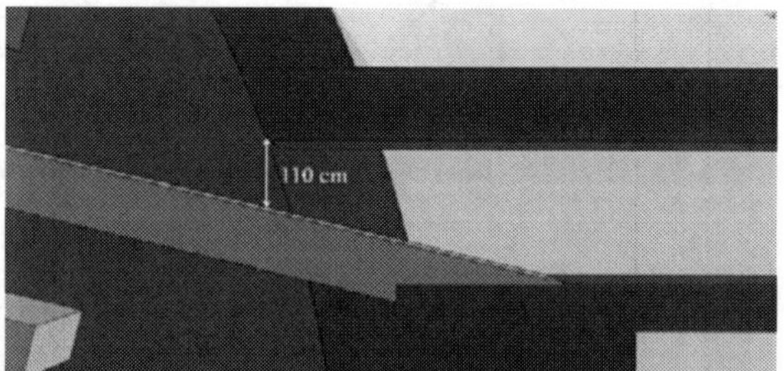

Fig. 20 Width between the wall of the North maze (gray) and the corridor between Igloo and control room (black) with barite concrete and 80 cm of thickness.

Other improvements have been made to increase the efficiency of the shielding: the different holes needed have been positioned to reduce risks of radiation leaks (heights, angles, slope) and local shielding have been designed. The final version of the accelerator hutch (Fig. 21) is currently under construction until the end of September 2016 (Fig. 22).

DOI: 10.1051/epjconf/201715303008

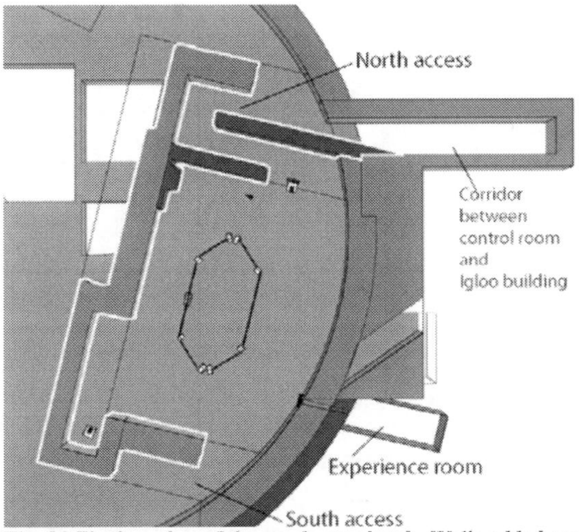

Fig. 21 Final version of the accelerator hutch. Walls added are outlined in white.

Fig. 22 View from the North maze of the accelerator hutch under construction (July 2016).

4.2 Local shielding

As seen in 4.1, the width of the accelerator hutch walls has been optimized using local shielding on critical elements as the scraper for example. Local shielding is needed too for the use of X-rays, produced. So, a beam-shutter for the X-ray line has been designed.

The scraper is the main beam-loss point, excepted beam-dumps. To comply with the radiation protection objective, a local concrete shielding has been designed on a wall of the accelerator hutch. Using of barite concrete and a two-part shield optimized the volume of concrete used (Fig. 23).

Another shielding has been designed too, to reduce exposure to radiations when the accelerator is off. Jaws in copper will be highly activated after the shutdown. So, the system will be placed in a lead and tungsten box (Fig. 24). The tungsten increases the efficiency of the shielding and protects the lead from activation.

As explain in 2.6.2, many changes were made during the conception phase and in particular, the diameter of the beam-line was increased. So a lead and tungsten pipe has been added to surround the vacuum pipe located at the front of the scraper. It will collimate the secondary particles produced when accelerator is on.

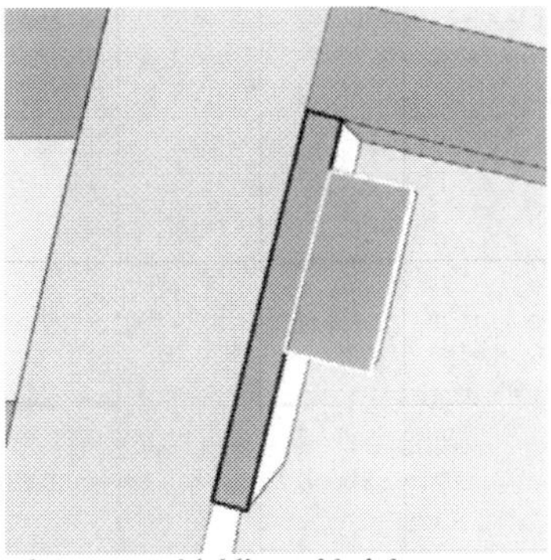

Fig. 23 Local concrete shielding added downstream the scraper (perpendicular). In black line, the first part (300*25*277 cm). In white line, the second one (130*60*130 cm).

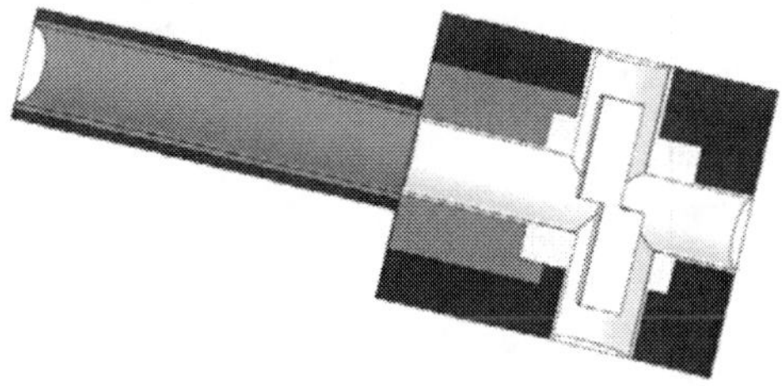

Fig. 24 Top view of the shielding of the scraper. The tungsten parts are in dark grey, the lead ones in black. The pipe shield is located downstream the scraper to collimate secondary particles produced when accelerator is on.

A hole is needed to transport X-rays through the Igloo wall toward the experience room, situated in the contiguous building. The opening is situated in the beam plan, at about 5.5 m from the interaction point. Two issues needed to be resolved: development of applications depends on the energy spectrum of the X-rays, itself depending on the opening angle of the beam created (Fig. 25). Furthermore, the diameter of the hole needs to be enough to exploit the largest spectra of X-rays.

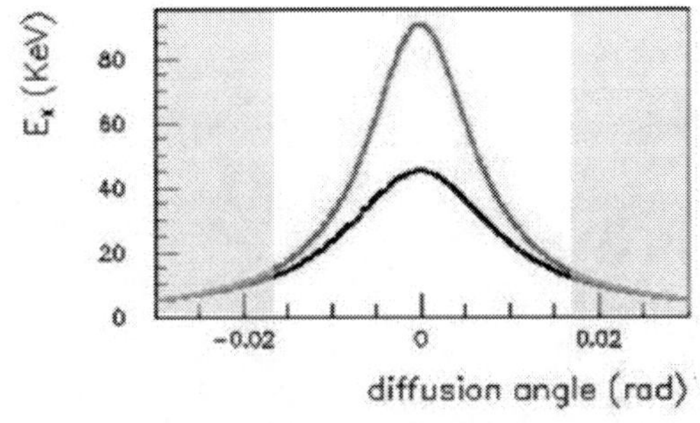

Fig. 25 Simulation of the energy of the X-rays produced by Compton backscattering effect versus diffusion angle. The weakest energy is produced in the largest diffusion angle **[2]**.

Nevertheless, users need to work in the experience room without stopping the accelerator (setting up the beam-line, changing samples…) and without being exposed to radiation. A balance between radiation protection and scientific performances has been made designing a beam-shutter coupled with a shielding inside the accelerator hutch (Fig. 26): the beam-shutter have been designed to stop X-rays and to reduce the scattered radiation coming from the accelerator. The lead shielding, situated inside the accelerator hutch, surrounds the beam-shutter and the pipe transporting the X-rays through the walls. We can see in fig. 26 that the hole in the wall has a larger diameter (25 cm) than the hole in the shielding (20 cm). The smallest one is sufficient for scientific exploitation, but if radiation leaks are weaker than anticipated, a new shielding will be designed to use a wider X-rays energy spectrum.

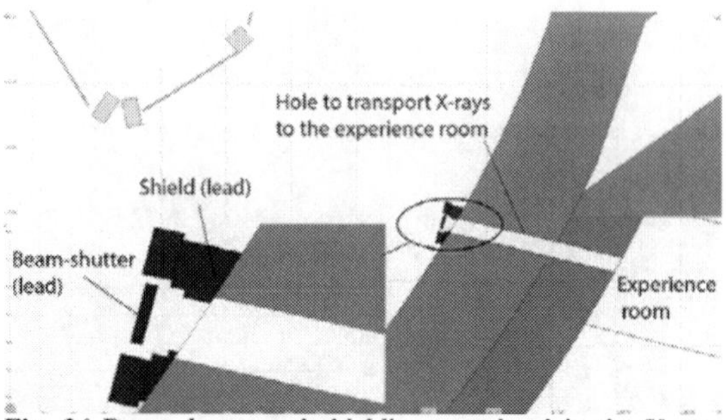

Fig. 26 Beam-shutter and shielding associated in the X-rays transport line.

4.3 Results

Balance between radiation protection, technical and financial optimization is obtained using full capacities from FLUKA. Use of Monte-Carlo code allows many improvements in shielding design by identifying and refining critical areas.

For ThomX accelerator hutch, additional local shielding has been designed to reduce thickness of concrete walls, in particular for the roof and the mazes (Table 3). Weight gains were important as were cost-savings: more than 250 tons of concrete corresponding to more than 100 k€ were saved.

The shielding complies with the radiation protection objectives, where the ambient dose rate needs to be lower than 3 μSv.h^{-1} in the common area opened when the machines will work (Fig. 27). However, some places will be closed to forbid the presence of workers during the operation of ThomX: the roof of the hutch and the south of the Igloo building.

The South access of the hutch will be delimited by a fence. A safety door interlocked with the safety system will control the access: possible when ThomX is off, prohibited when it is on.

A fixed ladder with safety cage, located in the South access, will be used to climb on the roof of the accelerator hutch. Safety barriers will be placed all along the roof to prevent falls or climbing up. The use of the travelling crane will be prohibited when one accelerator is working.

Table 3 Reductions of the thickness of concrete walls applied on the accelerator hutch. Weight of concrete removed and money saved have been assessed. Costs for local shielding designed and concrete added for the scraper are estimated too.

Zones	Reduction of the thickness of concrete walls	Weight removed (t)	Money saved (k€)
Roof	Thickness reduced by 50 cm	248	108
North maze	Thickness of some walls reduced by 50 cm (normal concrete)/80 cm (barite concrete)	33	14
South maze	One wall removed and thickness reduced by 50 cm	69	30
Total		350	152
Estimated cost for local shielding (LINAC beam-dump, scraper, beam-shutter for the X-rays line)			20 - 30
Cost for barite concrete added for scraper			66

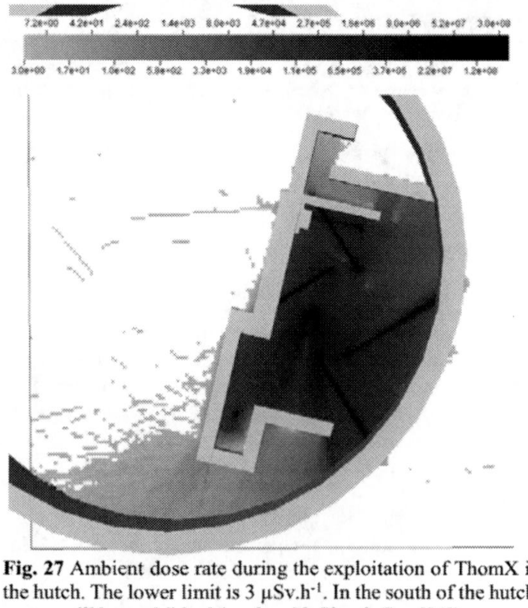

Fig. 27 Ambient dose rate during the exploitation of ThomX in the hutch. The lower limit is 3 μSv.h^{-1}. In the south of the hutch, access will be prohibited (made with SimpleGeo **[21]**).

5 Conclusion

Those simulations were determining to define the nature and dimensions of the shielding design (local and global ones) as well as to demonstrate the compliance with respect of public and worker regulations in order to plan the building of the infrastructures. In particular, those studies are valuable assets to obtain the authorization to exploit those facilities from the French nuclear authority.

The two project teams are collaborating to define a common safety and radiation protection supervision, taking into account various interactions (accesses, radiation supervision system, etc.) to comply with the NF M 62-105 [5] French standard for industrial accelerators.

As explain above (see 1), each accelerator will be operated by different laboratory (Andromede-IPN; ThomX-LAL). Thus, occupational health & safety and radiation protection needs a specific organization, taking account of the interaction between the exploitation of the two plants. Each safety radiation team from the LAL and the IPN will intervene for it accelerator. iRSD team will probably collaborate with the two laboratories for the coordination and the management of the radiation protection.

Acknowledgment

This work is supported by the French "*Agence Nationale de la Recherche*" as part of the program "Investing in the Future" under reference ANR-10-EQPX-51. This work was also supported by grants from *Région Ile-de-France.*

References

[1] M. J. Eller, E. Cottereau, V. Verkhoturov, A. Delobbe, J. Renaud, E. A. Schweikert and S. Della-Negra, *Andromede Project : Surface analysis and nano-particle surface interactions in the keV to Mev Energy Range,* Leuven, 2014.

[2] A. Variola, J. Haissinski, A. Loulergue and F. Zomer, "ThomX : Technical design report," LAL, Orsay, 2013.

[3] N. Pauwels, J.-M. Horodynski, P. Robert and A. Tadjeddine, "Démantèlement de l'installation nucléaire de base 106 (LURE)," *Radioprotection,* vol. 48, no. 4, pp. 545-561, October 2013.

[4] J. Bonis, R. Chiche, R. Cizeron, M. Cohen, E. Cormier, P. Cornebise, N. Delerue, R. Flaminio, D. Jehanno, F. Labaye, M. Lacroix, R. Marie, B. Mercier, C. Michel, Y. Peinaud, L. Pinard, C. Prevost, V. Soskov, A. Variola and F. Zommer, "Non-planar four-mirror optical cavity for high intensity gamma ray flux production by pulsed laser beam Compton scattering off GeV-electrons," *Journal of instrumentation,* January 2012.

[5] AFNOR, "Energie nucléaire - Accélérateurs Industriels : Installations," AFNOR, La Plaine Saint-Denis, 1998.

[6] T. T. Bölhen, F. Cerutti, M. P. Chin, A. Fasso, A. Ferrari, P. G. Ortega, A. Mairani, P. R. Sala, G. Smirnov and V. Vlachoudis, "The FLUKA Code: Developments and Challenges for High Energy and Medical Applications," *Nuclear Data Sheets,* vol. 120, pp. 211-214, June 2014.

[7] A. Ferrari, P. R. Sala, A. Fasso and J. Ranft, "FLUKA : a multi-particle transport code," CERN, Geneva, 2011.

[8] CNRS-IN2P3-IPNO, "IPN Orsay - Radioprotection Dosimétrie: SPR," IPN Orsay, [Online]. Available: http://ipnwww.in2p3.fr/Radioprotection-dosimetrie-SPR. [Accessed January 2016].

[9] V. Vlachoudis, "FLAIR : a powerful but user friendly graphical interface for FLUKA," in *Proceedings of international conference on mathematics, computationnal methods and reactor physics*, Saratoga Springs, 2009.

[10] A. Fasso, P. R. Sala and A. Ferrari, "Designing electron accelerator shielding with FLUKA," in *8th Int. Conf. on Radiation Shielding*, Arlington (Texas), 1994.

[11] Y. Asano, "Shielding design for the electron transport line of XFEL facility, SACLA to SPring-8 storage ring," in *7th International workshop on radiation safety at Synchrotron Radiation Sources*, Brookhaven, 2013.

[12] National Council on radiation Protection and Measurements, "Radiation Protection for Particle Accelerator Facilities : Recommendations of the National Council on Radiation Protection and Measurements," NCRPM, Bethesda, Md, 2003.

[13] T. D. Cradduck, "NCRP REPORT NO. 51: Radiation Protection Design Guidelines for 0,1-100 MeV Particle Accelerators Facilities," *Journal of Nuclear Medecine Technology,* vol. 6, no. 1, p. 40, March 1978.

[14] S. H. Rokni and J. C. Liu, *Radiation safety considerations for design of the SPEAR3 storage ring,* Stanford: SLAC, 2003.

[15] G. Schmidt, E. Kasel, K. Wille, M. Körfer, H. Henschel and J. Kuhnhenn, "Optical fibre beam loss monitors for storage rings at DELTA," in *Proceedings of EPAC 2002*, Paris, 2002.

[16] G. Schmidt, *Status of the DELTA machine operation ESLS 200,* Dortmund: Universität Dortmund, 2003.

[17] M.-A. Tordeux, P. Lebasque, P. Gros, J. Da Silva and P. Feret, "General performances of the injection scheme into the SOLEIL storage ring," in *Proceedings of EPAC 2004*, Lucerne, 2004.

[18] P. Berkvens and P. Colomp, *Measurements of radiation levels inside the ESRF storage ring,* Grenoble: ESRF.

[19] H. Lengeler, *Radiation protection of measuring set-ups for LEP 4-cell cavity,* Geneva: CERN, 1989.

[20] H. Lengeler, *X-ray radiation from a module of 4 superconducting cavities,* Geneva: CERN, 1989.

[21] C. Theis, K. H. Buchegger, M. Brugger, D. Forkel-Wirth, S. Roesler and H. Vincke, "Interactive three dimensional visualization and creation of geometries for Monte Carlo simulations," *Nuclear instruments and methods in physics research,* vol. 562, pp. 827-829, 2006.

A Study on the Establishment of Radiation Dose Estimation Procedure for Accumulated Radioactive Ions for RAON ISOL System

Do Hyun KIM[1], Song Hyun KIM[1] Jong Woo KIM[1], Chang Ho SHIN[1, a] and Shin Woo NAM[2]

[1]*Department of Nuclear Engineering, 04763 Hanyang University, Korea*
[2]*Facility Construction & Infrastructure Division, Rare Isotope Science Project, 34047 Institute for Basic Science, Korea*

Abstract. For purposes of various experiments, RAON heavy ion accelerator facility has been designed in Korea. ISOL is one system of RAON accelerators to generate and separate rare isotopes. Radioactive ions generated from target-proton reactions are separated and accumulated at separation devices. The accumulated isotopes release the gamma radiations; therefore, the radiation safety must be clearly estimated. In this study, a process to evaluate radiations from the accumulated ions was proposed by modifying FISPACT code. The proposed process was validated by comparing a solution of single element decay problem. Using the process, a preliminary study for radiation doses were performed in a virtual separation devise.

1 Introduction

In Korea, RAON accelerator facility is under construction for various experiments using heavy ions. It includes the In-flight Fragment (IF) and Isotope Separation On-Line (ISOL) systems. In the ISOL system, 70 MeV proton beam, which was designed to have 70 kW maximum beam power, is induced into UCx target to generate rare isotopes. Lots of unstable isotopes are generated from the UCx-proton reactions. Specific isotopes of the generated isotopes are accumulated at various separation devices in ISOL system during operation. Radiations are generated from the accumulated radioactive ions. In previous study [1], a study using ^{132}Sn isotope was evaluated for ISOL system of SPIRAL 2 [2]. However, the radiation from various radioactive ions should be individually performed for shielding design and operation plan. As a preliminary study of the radiation safety, process to evaluate radiation doses generated from accumulated ions is proposed. In addition, a preliminary study of radiation doses according to individual mass number of accumulated ions was performed.

2 Method and results

2.1 Overview of ISOL system of RAON

Figure 1 is a schematic design of ISOL system for RAON accelerator. ISOL system has lots of devices for generating and separating radioactive ions as follows: (A) Cyclotron to accelerate protons (1 mA and 70 MeV); (B) ISOL target system to generate various secondary particles using UCx-proton reaction; (C) Pre-separator to separate ions extracted from UCx target (In pre-separator, specific isotopes having a mass number can be separated); (D) diagnostic device used to check beam information. (E) RF-cooler which is used to slow down the beam; (F) HRMS is a devise to further separate ions. (G) ECR-CB is used to breed charges of ions from the +1 to +n charge state; (H) RF-Cooler/Buncher is a devise to cool and bunch the ions; (I) EBIS-CB having similar function of ECR-CB; and, (J) A/q separator is a remover of contaminants from charge breeders.

In these beam line, ions have low energy about 50 keV. Therefore, it is impossible to generate secondary particles from ions and beam line reactions; however, they are accumulated at the specific locations. Hence, it is important to estimate the radiation safety from the accumulated ions in the beam line. Table 1 shows the maximum beam transport rate for each device. Also, the locations of the devices are given as shown in Figure 1.

Table 1. Transport rate of major devices to separate isotopes.

Location	Device Name	Maximum Transport Rate
A	Cyclotron	-
B	Target System	20 % (extraction rate)
C	Pre-separator	5 %
D	Diagnostic Device	0% (if use)
E	RF-Cooler	100 %
F	HRMS	15 % (only Z-1, Z,Z+1)
G	ECR-CB	60 %
H	RF-Cooler /Buncher	45 %
I	EBIS-CB	100 %
J	A/q separator	25 %

[a] Corresponding author: chshin@nural.hanyang.ac.kr

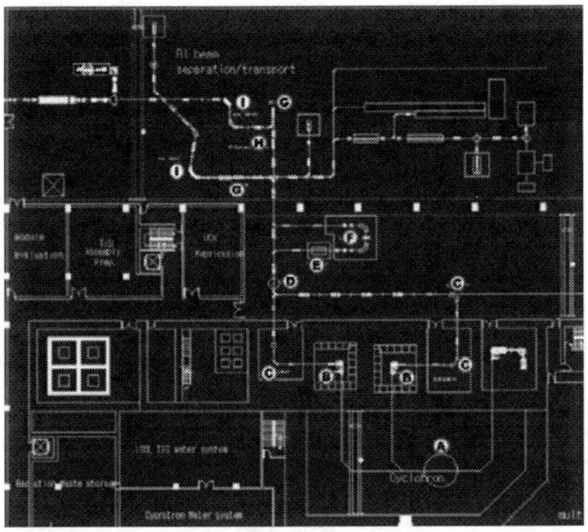

Figure 1. ISOL system and the device locations.

2.2 Calculation Method

To calculate residual radiation, BATEMAN equation is used as follows:

$$\frac{dN_m(t)}{dt} = -\beta_m N_m(t) + \sum_{k \neq m} N_k(t)\gamma_{k\rightarrow m} \quad (1.1)$$

$$\beta_m = \lambda_m + \Phi\sigma_a^m \; and \; \gamma_{k\rightarrow m} = \lambda_{k\rightarrow m} + \Phi\sigma_{k\rightarrow m} \quad (1.2)$$

where N_m = number of '*m*' nuclide, λ_m is decay constant of '*m*' nuclide, Φ is neutron flux, σ_a^m is absorption cross section of '*m*' nuclide, $\lambda_{k\rightarrow m}$ is decay constant from '*k*' to '*m*' nuclide and $\sigma_{k\rightarrow m}$ is reaction cross section from '*k*' to '*m*'. In our case, the neutron flux is not considered. Also, addtional term, which is ion generation rate accumulated at each device, is required. The general procedure of FISPACT code is briefly reviewed in section 2.2.1. To estimate our case, a process was proposed in section 2.2.2 using modified FISPACT 2010 [3-4] code.

2.2.1 Review of procedure for FISPACT code

To solve the Eq. (1) in FISPACT code, 3 steps are performed as shown in Figure 2. The first step is to produce a collapsed library called 'COLLAPX'. In the process, EAF-2010 reaction cross section library is weighted by neutron fluxes and the 'COLLAPX' is generated. The secondary step is to produce a library called as 'ARRAYX' condensed from collapsed library and decay library. Last step is a standard running module to solve the Eq. (1) with interest time.

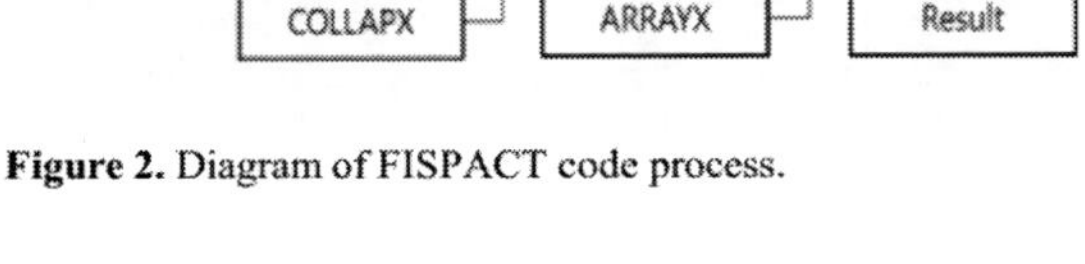

Figure 2. Diagram of FISPACT code process.

2.2.2 Proposed process for calculating accumulated ions

In FISPACT procedure, it is impossible to directly apply the production term caused by accumulations of the ions. To apply the production rate of accumulated ions, virtual condition, which is that all accumulated ions are produced to stable nuclides, was assumed as follows: ^{1}H is fully filled in an activation system with 1 cm^3 volume. The density is set to 1×10^{50} atoms/cm^3 to assume an enough amount to prevent the change of the number of ^{1}H atoms. Then, virtual reaction was used by modification of 'COLLAPX' file.

To verify the result using this modification, a simple test was performed by using ^{67}Ni with following conditions:

- Production Rate (*K*): 1.754 $\times10^9$ atoms/s
- Decay Constant of ^{67}Ni: 0.033 s^{-1}
- Period: production time = 400 sec and decay time = 400 sec.

This case is solved by following equation.

$$\frac{\partial N}{\partial t} = -\lambda N + K \quad (2)$$

where *N* is the number of ^{67}Ni, λ is decay constant of ^{67}Ni and *K* is production rate of ^{67}Ni. The solution of Eq. (2) is given by using Eq. (3).

$$\begin{cases} N = \frac{K}{\lambda}(1 - e^{-\lambda t}), & (if < 400\, s) \\ N = Ce^{-\lambda(t-400)}, & (if > 400\, s) \end{cases} \quad (3)$$

Where $C = \frac{K}{\lambda}(1 - e^{-\lambda 400})$. Table 2 and Figure 3 are results of the simple test using exact solution and proposed process. The result shows that the proposed process can accurately estimate the number of accumulated ions.

Table 2. Results from exact solution and proposed process.

Time [sec]	1	200	400	600	800
Proposed Process [atoms]	1.06 $\times10^9$	3.25 $\times10^{10}$	3.26 $\times10^{10}$	4.43 $\times10^7$	6.01 $\times10^4$
Solution [atoms]	1.06 $\times10^9$	3.25 $\times10^{10}$	3.26 $\times10^{10}$	4.43 $\times10^7$	6.03 $\times10^4$

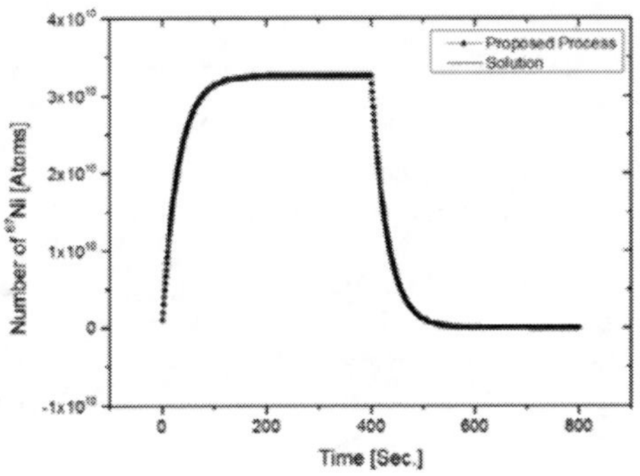

Figure 3. Results from exact solution and proposed process during all time periods.

2.3 Calculation results

2.3.1 Results of ion production rate

The ion production rates in the ISOL system were calculated by rigorous-two-step (R2S) method [5] with following step:

- MCNPX 2.7 [6] code simulation is pursued to get neutron spectra using JENDL/HE 2007 cross sections and HISTP option [7].
- Amounts of radioactive nuclides are estimated by modified FISPACT 2010 code [4].

The UCx target was assumed as Table 3, which generate a largest number of radioactive nuclides, to get secure result for radiation safety.

Table 3. Target information to calculate production rate of ions.

Element	Value
Composition	UC_2 (Natural abundance)
Length	3.77 cm (stopping length of proton)
Density	2.5 g/cm^3
Diameter	6 cm
Shape	Cylinder

Production rates using results of activation were calculated by Eq. (4) from solution of Eq. (2).

$$K = \frac{N\lambda}{1 - e^{(-\lambda t)}} \tag{4}$$

Figure 4 shows results of the ions production rates depend on mass number. The ^{235}U and ^{238}U nuclides were excepted because those are target material. Total 153 kind of isobar were estimated.

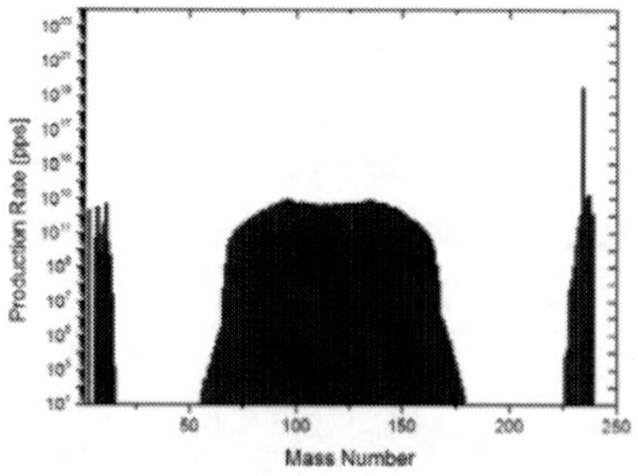

Figure 4. Production rates of ISOL system.

2.3.2 Results of radiation doses from accumulated ions

ISOL system is planned to use a single target for 14 days. After that, the target is cooled for 14 days. In this study, the operation time was assumed as 14 day, 1 year, 10 year, and 30 year with 0 second, 1 minute, 1 hour, 30 day, 180 day, and 1 year decay time for each operation time. To perform a preliminary study, the accumulation rate of the ions was assumed to 1 % (20 % × 5 %) for a virtual device after passing through pre-separator.

After passing the pre-separator, lots of isobar nuclides excepting target ions can be accumulated in the separation devise. It cannot be predicted to use which ions will be used as a target ions in experiments, and thus, some conservative assumption is necessary. Therefore, in this study, radiation doses of the isobars were estimated to confirm ions having maximum radiation dose. And, it was assumed that the isobar showing maximum radiation dose is accumulated in the devise. To evaluate the radiation doses, a simple model, as shown figure 4 was used with ICRP-116 [8] dose conversion factor for Antero-posterior.

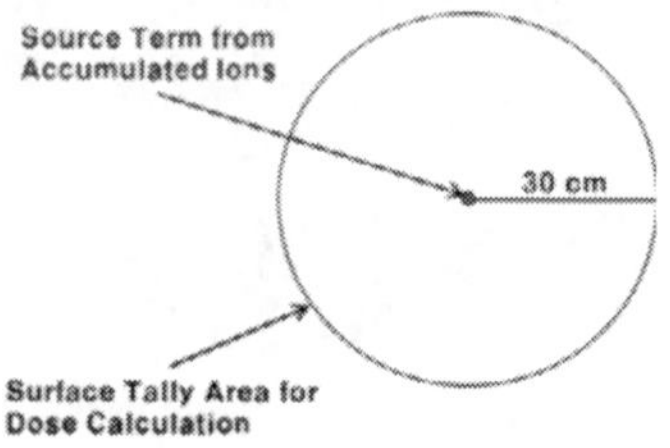

Figure 5. Model for dose estimation of each accumulated isobar.

Figure 6-7 are the results of decay gamma doses. The results show that doses for 0 decay time have similar dose distributions in cases of 2 week and 30 year operation periods. However, doses after 1 year decay time gives a considerable differences. It was analysed that the differences is caused by the differences of the decay constants. Thus, for the radiation safety, an additional study to select the isobar having maximum radiation dose should be performed as the decay times.

Figure 8 shows maximum gamma doses among various selected isobars for each decay time. The analysis of the results shows that following phenomenon should be considered for the radiation safety analysis: i) at each decay time, the isobar which shows maximum radiation dose is different; ii) for the conservative analysis of the radiation safety, it is recommended to select maximum accumulation time of the isobars.

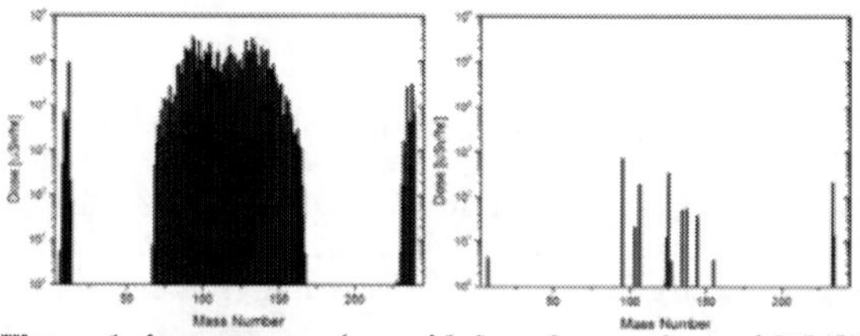

Figure 6. decay gamma dose with 2 week operation and 0 (left) 1 year (right) decay time.

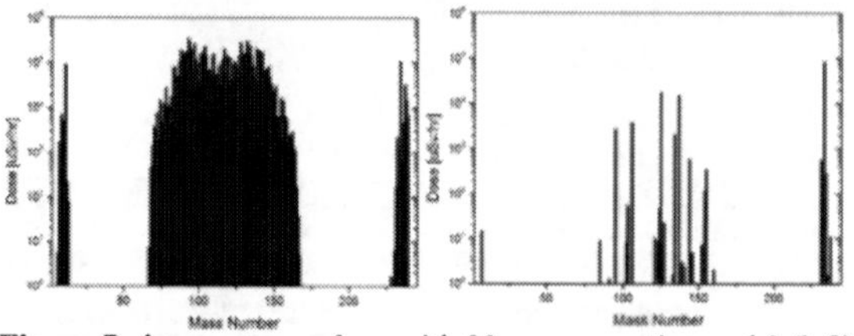

Figure 7. decay gamma dose with 30 year operation and 0 (left) 1 year (right) decay time.

 DOI: 10.1051/epjconf/201715303009

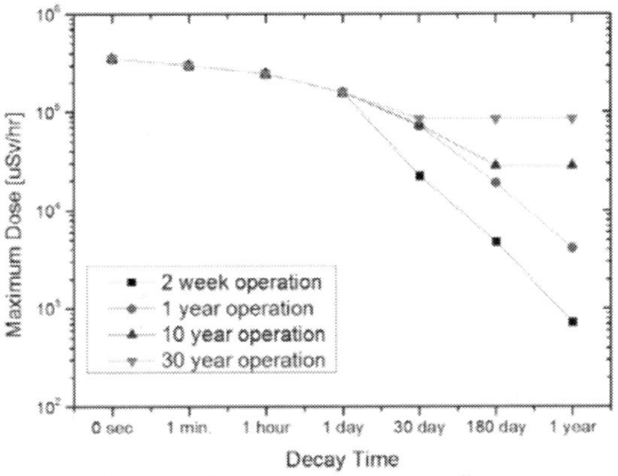

Figure 8. Maximum decay gamma dose according to operation and decay times for all accumulated isobars.

3 Conclusion

In the ISOL system, lots of separation systems are installed to select specific ion beam. Lots of ions generated by proton-target reactions are unstable nuclides. In this study, an estimation procedure for the radiation safety near the separation devises was proposed. First, modification of 'COLLAPX' file of FISPACT code was performed. This process was validated by solving a test problem, and the results estimated with the proposed method were compared with an exact solution. Second, the properties on the radiation doses of the isobars separated by ion separation devices were analysed. It is planned that the proposed method and procedure will be used for the radiation safety analysis in ISOL system.

References

1. F. Osswald, E. Bouquerel, D. Boutin, A. Dinkov, N. Kazarinov, L. Perrot, and A. Sellam, Rev. Sci. Instrum. **85**, 123301, (2014)
2. S. Gales, Nucl. Phys. A, **834**, 717c, (2010)
3. R. Forrest, *FISPACT-2007: User Manual*, (AKAEA Fusion Association, Abingdon UK, 2007)
4. D.H. Kim et al., *Transactions of the Korean Nuclear Society Spring Meeting*, **15s**, 150 (2015)
5. Y. Chen, U. Fischer, Fusion Eng. Des., 63, 107, (2002)
6. D. B. Pelowitz, et al., *MCNPXTM User's Manual, Version 2.7*, (LANL, USA, 2011)
7. B. J. Micklich, et al., Nucl. Technol., **168.3**, 700 (2009)
8. C.H. CLEMENT, *ICRP PUBLICATION 116*, (ICRP, 2010)

CAD-Based Shielding Analysis for ITER Port Diagnostics

Arkady Serikov[1,*], *Ulrich* Fischer[1], *David* Anthoine[2], *Luciano* Bertalot[2], *Maartin* De Bock[2], *Richard* O'Connor[2], *Rafael* Juarez[3], and *Vitaly* Krasilnikov[2]

[1]Karlsruhe Institute of Technology (KIT), Institute for Neutron Physics and Reactor Technology, Hermann-von-Helmholtz-Platz 1, 76344 Eggenstein-Leopoldshafen, Germany
[2]ITER Organization, Route de Vinon-sur-Verdon, CS 90 046, 13067 St. Paul Lez Durance Cedex, France
[3]Departamento de Ingeniería Energética, ETSII-UNED, Calle Juan del Rosal 12, Madrid 28040, Spain

Abstract. Radiation shielding analysis conducted in support of design development of the contemporary diagnostic systems integrated inside the ITER ports is relied on the use of CAD models. This paper presents the CAD-based MCNP Monte Carlo radiation transport and activation analyses for the Diagnostic Upper and Equatorial Port Plugs (UPP #3 and EPP #8, #17). The creation process of the complicated 3D MCNP models of the diagnostics systems was substantially accelerated by application of the CAD-to-MCNP converter programs MCAM and McCad. High performance computing resources of the Helios supercomputer allowed to speed-up the MCNP parallel transport calculations with the MPI/OpenMP interface. The found shielding solutions could be universal, reducing ports R&D costs. The shield block behind the Tritium and Deposit Monitor (TDM) optical box was added to study its influence on Shut-Down Dose Rate (SDDR) in Port Interspace (PI) of EPP#17. Influence of neutron streaming along the Lost Alpha Monitor (LAM) on the neutron energy spectra calculated in the Tangential Neutron Spectrometer (TNS) of EPP#8. For the UPP#3 with Charge eXchange Recombination Spectroscopy (CXRS-core), an excessive neutron streaming along the CXRS shutter, which should be prevented in further design iteration.

1 Introduction

The most (~80%) of ITER diagnostic systems are hosted by the Port Plugs which are arranged at 3 levels (Upper, Equatorial, and Lower) depicted in Fig. 1 as UPP, EPP, and LP. The plugs have dual purpose: provide the access for diagnostic instruments to measure the plasma characteristics (density, energy, etc.), and at the same way plug the ports against the neutron and gamma radiation and vacuum leakage. Therefore, the designing of the port diagnostics must be supported by the radiation shielding analysis for each of port system. At ITER the design developing process is performing in the Computer Aided Design (CAD) models with ENOVIA software platform. In order to follow all the important from neutronic point of view changes in the CAD models, the nuclear analyst should be capable to use the CAD model in creation of the neutronics models in the CAD-based shielding analysis. The main objective of the nuclear analyses of the ports is to guarantee radiation shielding of the personnel access to the Port Interspace (PI) areas behind the ports shown in Fig. 1 with compliances to the ITER project restrictions of nuclear safety, weight and consistency with port interfaces to the blanket and vacuum vessel. For the shielding optimization of the ports, the objective function was the Shut-Down Dose Rate (SDDR) at PI with the target of the SDDR minimization following the ALARA (As Low As Reasonable Achievable) principle. Because according to the ITER project requirements, the SDDR in PI have to be ALARA and below the limit of 100 μSv/hr after ~12 days of cooling for hands-on maintenance activities. Currently, many port systems have predicted SDDR in PI above this limit, therefor shielding design improvements should be investigated. Several shielding approaches to bring these values down and to minimize the dose to a worker have been proposed in the paper.

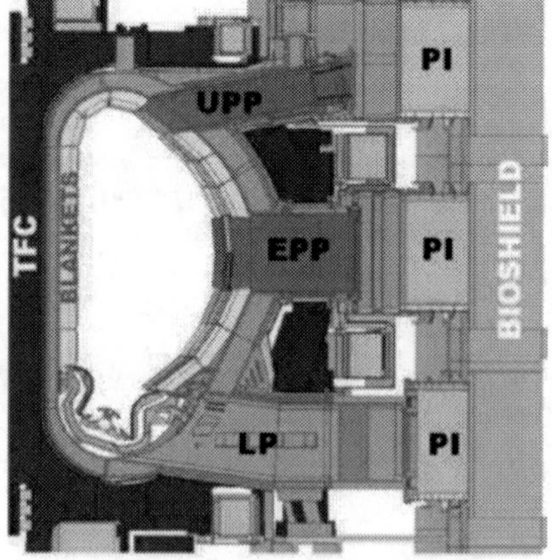

Fig. 1. CAD model of ITER with 3 levels of ports with Port Plugs at Upper (UPP), Equatorial (EPP), and diagnostic rack at Lower Port (LP). Port Interspace (PI) is between Port Plugs and Bioshield. Blankets and Port Plugs are used to protect the superconductive Toroidal Field Coils (TFC).

* Corresponding author: arkady.serikov@kit.edu

This paper presents new results of CAD-based neutronics analyses performed with the Monte Carlo MCNP5 code [1] and the Rigorous 2-Step mesh (R2Smesh) method [2] for SDDR-calculations. The ITER tokamak geometry complexity and multiple diagnostic instruments inside the Upper and Equatorial Port Plugs (UPP and EPP) shown in Fig. 2, make the MCNP simulation challenging. Some of challenges are discussed in work [3].

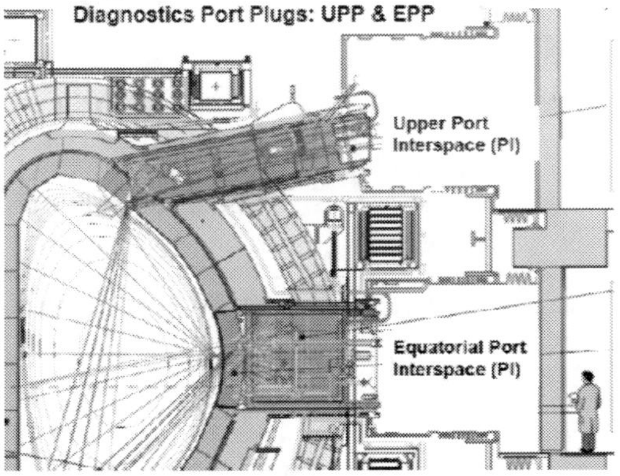

Fig. 2. Drawing of ITER with multiple diagnostic instruments integrated inside the Diagnostics Port Plugs and Port Interspace (PI) behind the plugs where the personnel access is expected for the diagnostics maintenance.

The results for three particular ports are discussed in Section 3 with the following diagnostic systems: (1) Tritium and Deposit Monitor (TDM) integrated in EPP#17; (2) Tangential Neutron Spectrometer (TNS) with 6 optical diagnostic systems inside EPP#8; and (3) Charge eXchange Recombination Spectroscopy (CXRS-core) in UPP#3, and (4) Diagnostic Generic Upper Port Plug (DGUPP) of ITER.

2 Methodology

Fusion neutronics calculations in general are conducted by Monte Carlo (MC) codes such as MCNP using the geometry representations which are well adapted from the engineering CAD model. In approach of this paper, the geometry of CAD model is converted to MCNP model. Reflecting boundary conditions are implemented on the poloidal vertical planes which are located at ±20° from the center of the mid-plane port in order to create a 40 degree toroidal sector of the MCNP model of ITER machine. This sector is symmetrically duplicated 9 times to simulate full 360 degree machine shown in Fig. 3, the CAD model of 40 degree sector is presented in Fig. 4. The reflective planes cut through the center of the ports at the ±20 lateral sides, therefore only symmetrical ports can be modelled on those sides and modelled as port halves. The geometry of central port is modelled completely, and any design features could be reproduced in the model. This 40 degree model includes all the major components of ITER such as blanket, vacuum vessel, toroidal and poloidal magnets, 3 levels of the ports, bioshield, and so on. Therefore, it is calls "ITER global MCNP model". It exists in several modifications and the latest MCNP ones are C-lite v.2 released in 2015 and C-Model 2016.

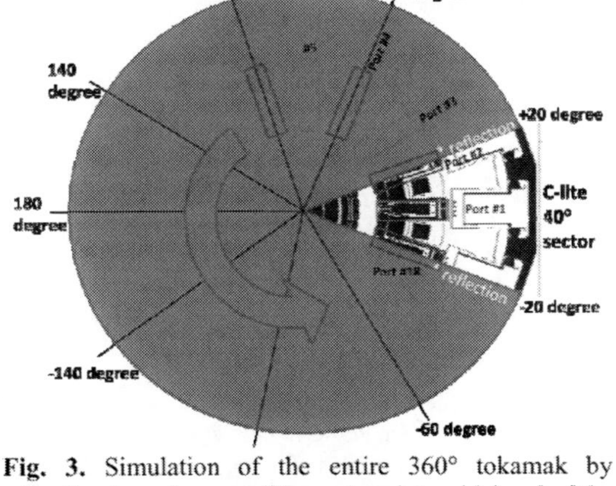

Fig. 3. Simulation of the entire 360° tokamak by reflective boundary conditions at each toroidal end of the 40° torus sector with reflective planes at ±20°.

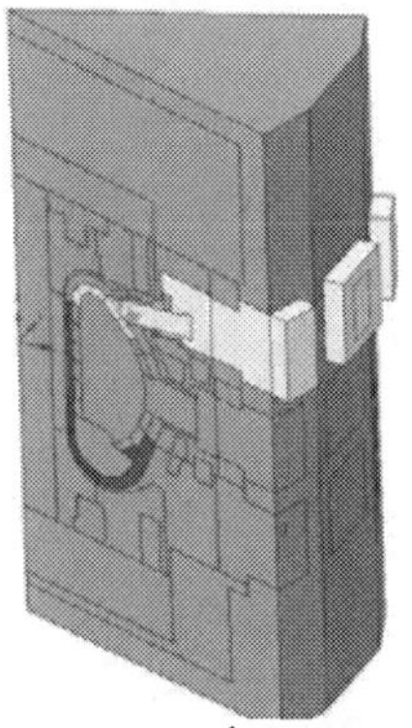

Fig. 4. CAD model of ITER 40 degree sector corresponded to the MCNP global C-Model 2016 – Diagnostics Generic Upper Ports (DGUPP) are highlighted.

Methodological and scientific aspects of shielding and activation analyses for the ITER diagnostic port systems are discussed in [4]. The original engineering CAD models usually are too complicated in geometry to be used directly in neutronics models, and CAD pre-processing should be performed. Geometry simplification should be introduced because some CAD details such as small bolt holes, chamfers, grooves and surface rounding are excessive for the neutron transport tasks. Problem also arises due to free-form and splines surfaces of complex CAD models. The MCNP5 code still supports only limited number of the geometry types, as defined in the MCNP User's Guide [1]. Therefore unsupported surfaces (mostly splines) must be approximated to series of planes and cylinders. The SpaceClaim 2015 CAD-modeling software (http://www.spaceclaim.com) has been used for these pre-processing operations. Then, after the CAD models simplification and surface approximations, in order to use CAD geometry for MC particle transport in the Constructive Solid Geometry (CSG) format one has to convert a solid given by a boundary representation into a cell given by a semi-algebraic description. The conversion process could be performed automatically by means of the CAD-to-MCNP converter programs MCAM [5] and McCad [6]. In this paper, the most conversions have been carried out with MCAM.

The methodology of SDDR calculations consisted in use of recently updated Rigorous 2-Step mesh-based (R2Smesh) method [2] which allows studying decay gamma sources in particular materials with estimations of isotopic contributions to SDDR. The maximal decay sources were produced in steel of Diagnostic First Wall (DFW) and steel structure of Diagnostic Shield Module (DSM). The following nuclides were found to dominate in SDDR at 12 days inside the port interspace: ^{60}Co, ^{182}Ta, ^{58}Co, ^{54}Mn, and ^{59}Fe. These radioactive isotopes are generated in the steel structure of the port rear side, mostly its closure plate, rear flange and port walls.

The visualization capabilities are based on the ParaView code http://www.paraview.org, which allows overlying the MCNP mesh tally distributions with the CAD geometry written in STEP-file. The MCNP parallel runs with message passing interface (MPI) were done on the Helios supercomputer (http://www.iferc.org).

3 Results

The described CAD-based methodology is routinely applying for neutronics analyses of the ITER diagnostics. The results of particular diagnostic systems integrated inside the two Equatorial Port Plugs (EPP #17 and #8) and one Upper Plug (UPP #3) are addresses in this Section. Neutronics results for Diagnostic Generic Upper Port Plug (DGUPP) are also presented at the end of this Section.

3.1 Tritium and Deposit Monitor (TDM) integrated in EPP#17

Shielding analysis of Tritium and Deposit Monitor (TDM) is performed with the local MCNP models of EPP#17 which comprises two adjacent diagnostics: TDM and Core-Imaging X-ray Spectrometer (CIXS) elaborated in works [4, 7]. With the applied approach of local modelling all contributions from other ITER systems and ports are neglected. The MCNP 3D model was converted from the corresponding CAD model. The result of such CAD-to-MCNP conversion is visualized in Figs. 5 and 6.

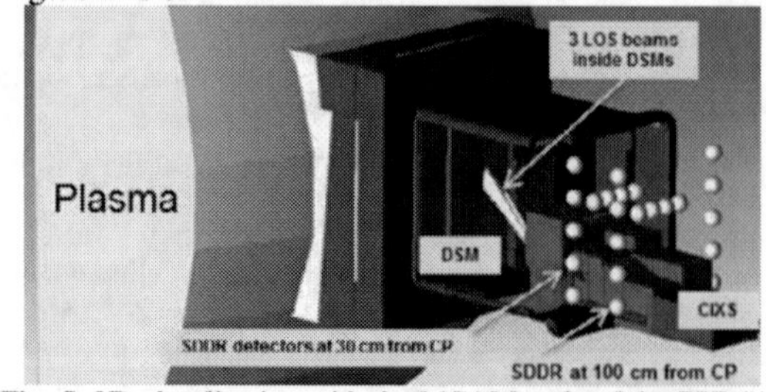

Fig. 5. 3D visualization with the MCAM code of the EPP #17 MCNP local model with inclusion of only the CIXS system with 3 Line-Of-Site (LOS) x-ray beams inside the DSMs #2 and #3.

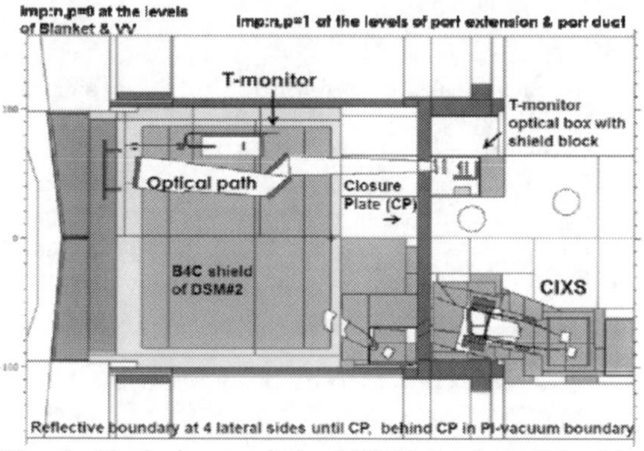

Fig. 6. Vertical cut of the MCNP local model with Diagnostics apertures of two systems: **Tritium & Deposit monitor & CIXS**. Neutron and photon importances and boundary conditions are specified.

The introduced modeling assumptions (e.g. homogenization in material specification) and the level of geometry details correspond to the ITER Conceptual Design Review (CDR) level of the design development. Critical neutronics issues of TDM related to radiation streaming have been resolved. For instance, it was proposed to increase the length of one segment between two particular mirrors (M4 and M5 depicted in Fig. 7) in the labyrinth optical mirror system to prevent direct neutron streaming from the plasma to the optical box in Port Interspace (PI), where maintenance is planned ~12 days after shutdown. To provide the possibility of personnel access to the PI area, Shut-Down Dose Rate (SDDR) has been calculated with the results presented as map distributions and estimates in spherical detectors. Shield block made of boron carbide behind the optical box was added to reduce the SDDR in PI.

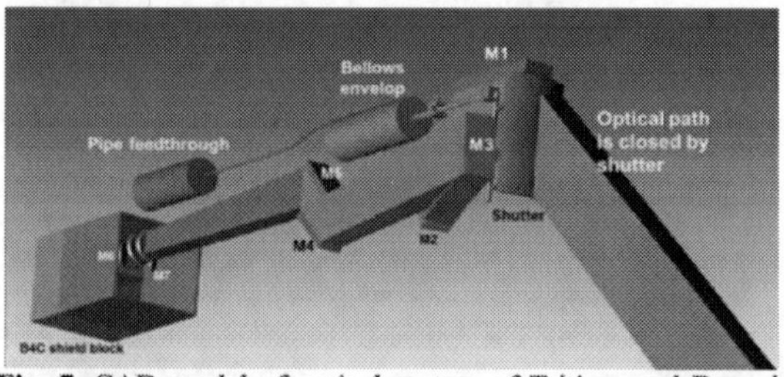

Fig. 7. CAD model of optical system of Tritium and Deposit Monitor (TDM)

All the space of MCNP local model has been covered with a fine mesh grid having 2x2x2 cm^3 of element size for tallying the neutron flux distribution. The mesh-tally maps are visualized with ParaView code (http://www.paraview.org) as shown in Fig. 8 with a distribution of total neutron flux from Diagnostics First Wall (DFW) till PI. The highest flux integrated by all the neutron energy range (total n-flux) is ~1e14 n/cm^2/s at the DFW. On the T-monitor (TDM) first mirror the flux is reduced to ~ 3e12 n/cm^2/s, and at the Port Interspace it is attenuated to ~ 1e7 n/cm^2/s. A detailed analysis [7] with several additional cuts of neutron flux maps reveals that mirror M5 pointed in Figs. 7, 8 does not present enough shift in the vertical segment M4-M5 of the bent optical pathway. Part of neutron flux can penetrate directly or through the thin shield at M5, resulting in excessive neutron streaming pathway to the PI. Therefore, we proposed to move mirror M5 up to increase the shift.

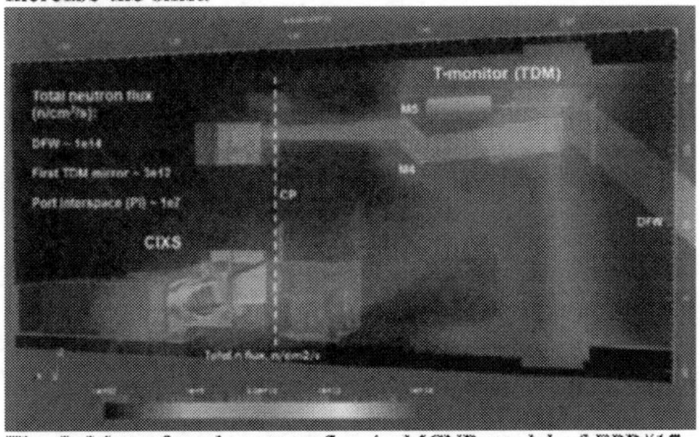

Fig. 8. Map of total neutron flux in MCNP model of EPP#17 included TD-Monitor & CIXS.

The R2Smesh method of SDDR calculations allowed us to distinguish different decay gamma sources in forming the SDDR. The SDDR results are presented for two variants of the EPP #17-CIXS models, with and without the TD-Monitor. Therefore the contribution to SDDR from the monitor was deduced. This work provides only the relative values of SDDR. The absolute values will be obtained after the finalizing of the designing work for all the EPP #17 diagnostics. This is a forthcoming task of EPP #17 diagnostics integration into the ITER-global C-lite MCNP model.

3.2 Tangential Neutron Spectrometer (TNS) with 7 optical diagnostic systems inside EPP#8

The Tangential Neutron Spectrometer (TNS) is devised to measure spectra of neutrons flying in tangential direction as a collective D-T plasma rotation. The TNS detectors should allow estimating the Doppler energy shift in the D-T source neutron spectrum emission. Such neutron spectrometer with tangential view to the plasma (TNS) has been modeled inside the Equatorial Port Plug #8 (EPP #8) of the ITER global MCNP model C-lite v.2 with 7 systems inside the port shown in Fig. 9.

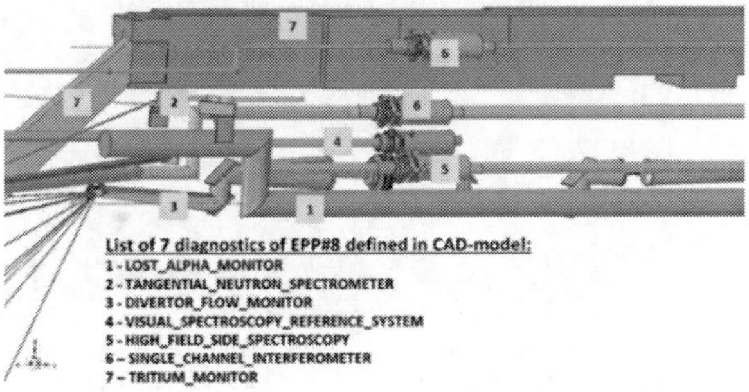

Fig. 9. CAD model of 7 diagnostics have been integrated into the EPP#8, among them two TNS detectors are listed by #2.

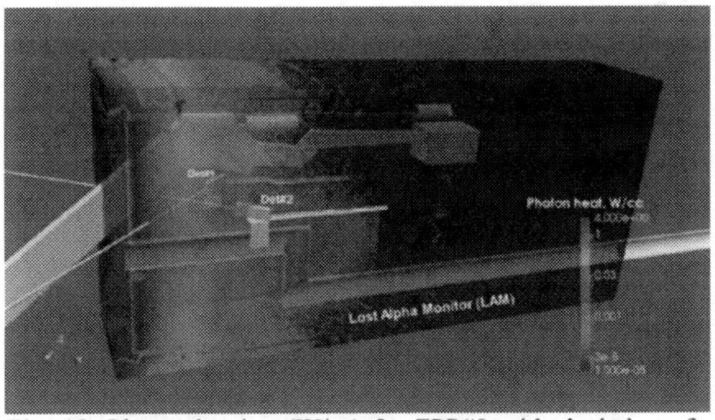

Fig. 10. Photon heating (W/cc) for EPP#8 with depicting of two TNS detectors and Lost Alpha Monitor (LAM)

The nuclear heating in the EPP#8 materials is dominated by photons. Map of photon heating distribution with the cut through the central Detector#2 of TNS and the channel of the Lost Alpha Monitor (LAM) is plotted in Fig. 10. In approximation of this neutronic analysis, the geometry of the LAM diagnostics was modelled as Z-shaped cylindrical channel and its model could be generalised as simple representation of Fast Ion Loss Detector (FILD). In this paper LAM and FILD are synonyms. It is noticeable from Fig. 10 that the LAM channel introduces radiation streaming deep inside the port plug shielding. This streaming disturbs the neutron energy spectrum detecting in the central TNS Detector #2. Such neutron spectral effect in TNS has been analyzed in the paper by modification of the MCNP C-lite model consisted in filling the LAM in EPP#8. As an alternative design option, instead of totally filling the LAM (that means LAM removing), we have proposed to turn upside-down the Z-shape channels from the original pathway in Fig. 11a to the turned pathway in Fig. 11b. The turning was implemented by symmetrical exchange of the shifted legs positions. In the MCNP model with the original shape of the LAM (FILD) channels the distance between TNS and LAM was 10 cm, by the turning the was increased to 60 cm.

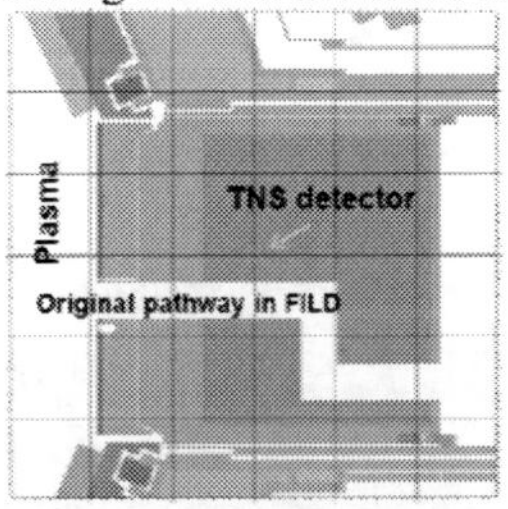

Fig. 11a. MCNP cut through the original orientation of the LAM (FILD) channel.

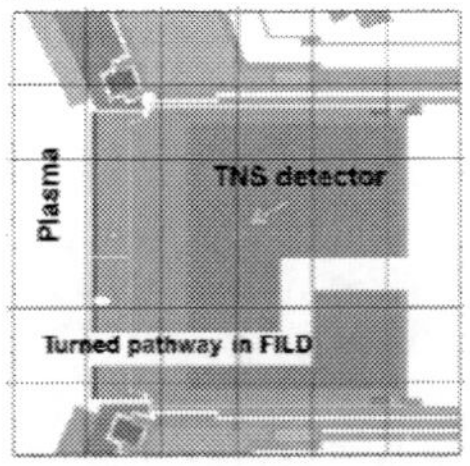

Fig. 11b. MCNP cut through the turned upside-down LAM (FILD) channel.

The comparisons of neutron spectra calculated with several TNS detectors and with the Radial Neutron Camera (RNC) in Equatorial Port are presented in Fig. 12. In its legend "TNS-no-LAM" corresponds to the modification of the MCNP C-lite model consisted in completely filling the LAM (FILD) channel. The LAM channel is filled with steel at the FW and boron carbide (B_4C) – inside the DSM part. By filling the LAM the

 DOI: 10.1051/epjconf/201715303010

neutron flux is reduced in the TNS central Detector#2, with higher 14 MeV peaking factor, which is explained by eliminating the noise of neutrons of low and intermediate energies. This is due to stronger moderation and neutron absorptions by isotope ^{10}B in the shield material inside the LAM channel.

The neutron spectra results indicated that the turned LAM option is an equivalent to one of its absence – option of totally filled LAM ("TNS-no-LAM"). As a conclusion of this analysis, it is recommended to introduce stronger moderators and neutron absorbers around the TNS detectors in order to detect only the neutron passing through the TNS collimators.

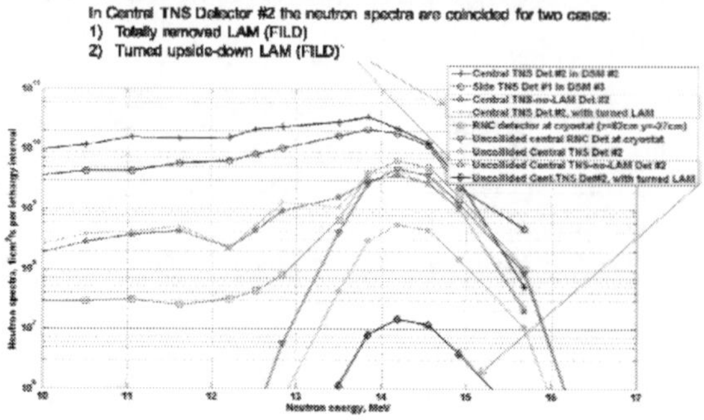

Fig. 12. Equivalence of neutron spectra in central TNS detector for two cases: (1) totally removed LAM, and (2) turned upside-down LAM.

3.3 Charge eXchange Recombination Spectroscopy (CXRS-core) in UPP#3

The Charge Exchange Recombination Spectroscopy (CXRS) diagnostic system measures photon line emissions of several impurity isotopes in the ITER plasma. This system is going to be installed inside the ITER Upper Port Plug (UPP) #3 together with Glow Discharge Cleaning (GDC) system. To satisfy the ITER radiation requirements, nuclear analysis appropriate for the CDR stage of CXRS development has been performed for the Upper Port (UP) environment including its Port Interspace (PI). Three cases of the UPP configurations have been analyzed: Case #1 is characterized by inclusion of both systems: CXRS and GDC, in Case #2 the GDC was removed, keeping shield in its place and the CXRS, and Case #3 has no any apertures at all in the Generic Upper Port Plug (GUPP). For the GUPP design, the 3D B-lite MCNP model with more detailed description of the blanket modules around the UP has been used in the state of GUPP Final Design Review (FDR) 2013 [8] with only one modification – introduction of single labyrinth at the bottom gap of GUPP. The UPP apertures of Cases #1 and #2 were inserted in the bulk shield of the GUPP Case #3.

The CAD model files of CXRS-core and Glow Discharge Cleaning (GDC) system were prepared for neutronics analysis by means of simplification and substitution of all spline surfaces with cylinders or planes keeping the original volumes of the CXRS components unchanged. The shape of the GDC model was updated to the newest design and its definition was simplified to the two-part geometry (front-end electrode and transmission line) with homogeneous material compositions. The final CAD model of the CXRS-core and GDC developed for the neutronics analysis is presented in Fig. 13. The properly prepared CAD files of CXRS and GDC were converted to the MCNP model by means of MCAM [5].

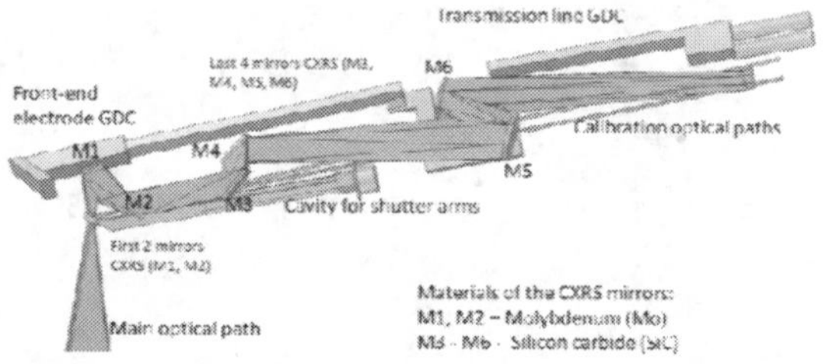

Fig. 13. CAD neutronics model of the CXRS-core and GDC

For calculation of neutron flux distributions over the whole space of the Upper Port Plug (UPP) and Port Interspace (PI) with the MCNP5 code, a fine mesh grid has been set up with the mesh element size of 1x1x1 cm^3. The comparison of the maps of total neutron fluxes for Cases #1 and #3 is illustrated in Fig. 14 in contrast for the cases with most difference in number of apertures: Case #1 with 4 types of radiation openings (upper picture of Fig. 14), and Case #3 with opening in for of gaps around the UPP (bottom picture of Fig. 14). Radiation streaming was expected along the following pathways: 1–Gaps around the UPP; 2–CXRS shutter; 3–CXRS main optical path; 4 – GDC electrode. The map of total neutron flux plotted in Fig. 15 emphasized the neutron streaming along the CXRS shutter to the mirror M5 located deep inside the port and farther to the Port Interspace (PI) area.

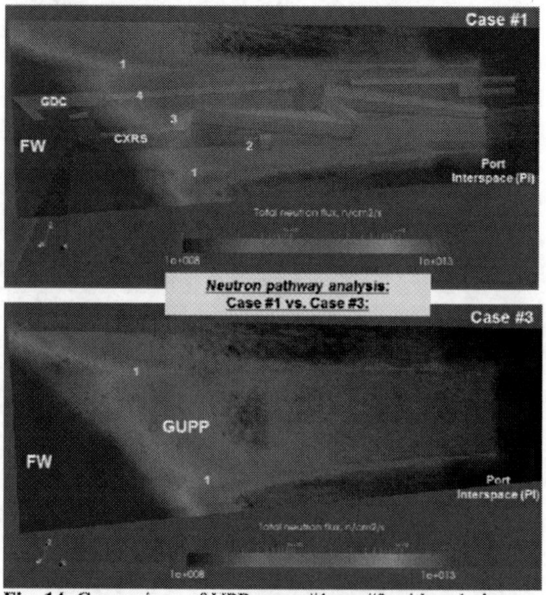

Fig. 14. Comparison of UPP cases #1 vs. #3 with pointing out the number of the major streaming pathways on the maps of total neutron flux (n/cm^2/s).

Fig. 15. Total neutron flux (n/cm^2/s) distribution with thresholds between (1e9 – 1e10) n/cm^2/s to demonstrate strong impact of neutron streaming along the CXRS shutter to mirror M5 and then farther to Port Interspace (PI) behind the UPP.

For the MCNP calculation of nuclear heating distribution inside the UPP we have used MCNP mesh-tallies with separate contributions from neutron (n) and photons (p). The dominant contribution to the heating is coming from photon heating deposition. The maximum (n+p) heating was ~5.5 W/cm^3 on the DFW, with strong attenuation to 0.06 W/cm^3 at 60 cm from the DFW surface, at which distance the DSM is beginning as shown in Fig. 16.

Based on the performed neutronics analysis, it is recommended to add shielding material behind the shutter to reduce the streaming to the mirror M5 and PI. The obtained SDDR in PI is between 124 μSv/hr (behind the port closure plate) and 91 μSv/hr (PI at the bioshield) with a contribution from CXRS-core of 25 μSv/hr (at closure plate) and 10 μSv/hr (at bioshield) correspondingly, and no signification contribution to SDDR from GDC was found.

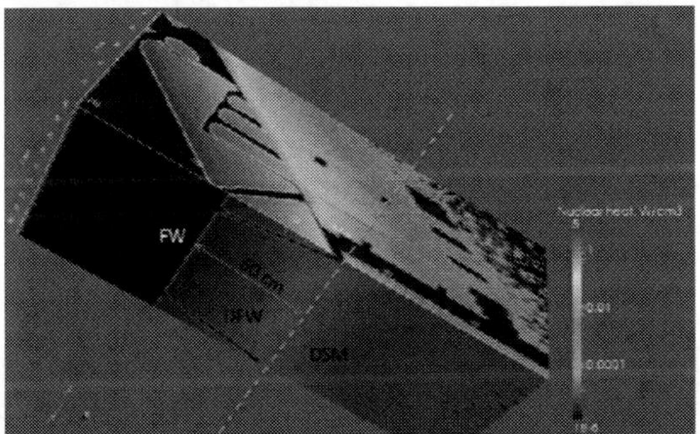

Fig. 16. Total nuclear heating (neutron and photon) in 100% steel for the Case #1 with attenuation from 5 W/cm^3 at the FW down to 0.06 W/cm^3 at 60 cm where DSM begins.

3.4 Diagnostic Generic Upper Port Plug (DGUPP) of ITER

The result of MCNP 3D model development of Diagnostic Generic Upper Port Plug (DGUPP) is shown in Fig. 17. That model will be used as standard neutronics model for ITER-related neutronics calculations and distributed among the Domestic Agencies and ITER contactors. Its geometry was converted from the actual CAD models of ITER Upper Port, inter-space structures (ISS), and bioshield port plug. The model was created in format of the MCNP code which is suitable for fusion neutronics to perform radiation transport calculations with the Monte Carlo stochastic methodology. The model will be used for analysing any other ITER systems and ports such as Equatorial or Lower ports and it is ready for integration of any particular Diagnostic system inside its Diagnostic Shielding Module (DSM). Using this new DGUPP model certain series of neutronics analyses has been performed in order to test its shielding performance and recognize potential problems. By conducting the Shut-Down Dose Rate (SDDR) analysis one of the problems was recognized in excessive neutron streaming along the gaps around the DGUPP depicted in Fig. 18 and it draws attention of the ITER engineers to mitigate such streaming by arranging additional shielding. We have also evaluated the contribution of neutron radiation (neutron cross-talk) from Equatorial Port (EP) and Lower Port (LP) to Upper Port interspace of DGUPP as depicted in Fig. 19. The analysis performed has been recognized as an important step for investigation of shielding improvement of the Upper Port design and other ITER systems interfacing with DGUPP. The analysis outcomes result in establishing new project tasks directed to the reduction of SDDR at port inter-space area following the ALARA principle because this area is supposed to be accessible by personnel for port maintenance activities.

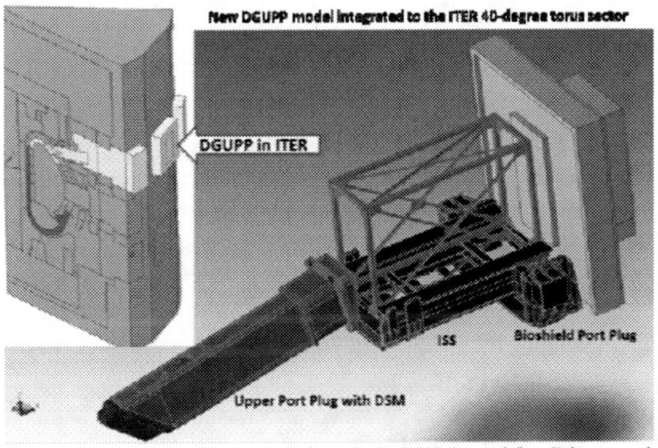

Fig. 17. Neutronics model of DGUPP with Diagnostic Shielding Modules (DSM), Inter-Space Structure (ISS), and Bioshield Port Plug created and integrated in ITER torus sector

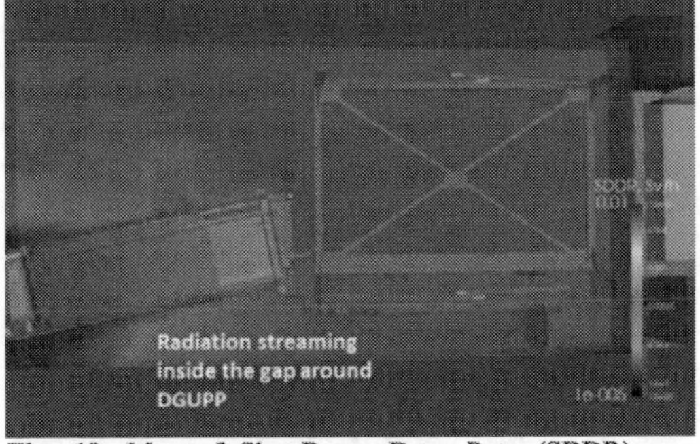

Fig. 18. Map of Shut-Down Dose Rate (SDDR) reveals excessive radiation streaming inside the gap around DGUPP.

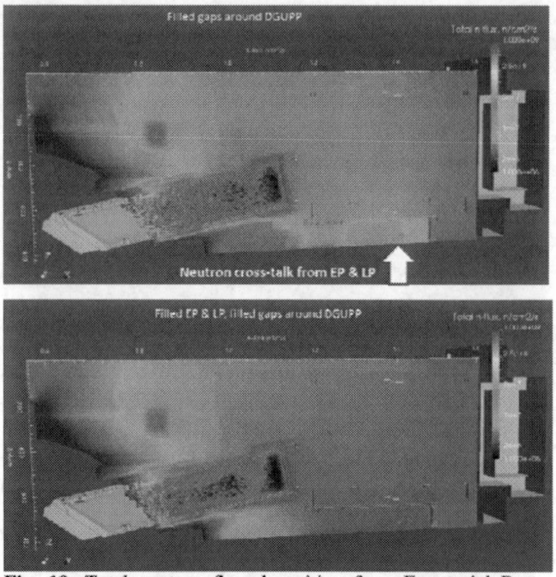

Fig. 19. Total neutron flux deposition from Equatorial Port (EP) and Lower Port (LP) to Upper Port interspace of DGUPP: model with radiation streaming from EP & LP (top picture), and variant with closed EP & LP ports (bottom picture).

Acknowledgment

This work has been funded by the ITER Organization under the ITER service contract Nr. IO/4300000896 using an adaptation of the C-lite MCNP models which were developed as a collaborative effort between the FDS team of ASIPP China, University of Wisconsin-Madison, ENEA Frascati, CCFE UK, JAEA Naka, and ITER Organization. The views and opinions expressed herein do not necessarily reflect those of the ITER Organization.

This work was carried out using an adaption of the Diagnostics Generic Equatorial Port Plug MCNP model which was developed as a collaborative effort between Universidad Nacional de Educacion a Distancia (UNED) and ITER Organization.

References

1. X-5 Monte Carlo Team, MCNP—A general Monte Carlo N-particle Transport Code, Version 5, LA-UR-03-1987, Los Alamos National Laboratory (2003)
2. P. Pereslavtsev, et al., Fusion Eng. Des., **88**, pp. 2719-2722 (2013), http://dx.doi.org/10.1016/j.fusengdes.2013.02.091
3. A. Serikov, et al., *SNA+MC 2013*, 01502 (2014) http://dx.doi.org/10.1051/snamc/201401502
4. A. Serikov, et al., *M&C+SNA+MC 2015 conference*, Nashville, TN, April 19-23, 2015, Volume 2, 2015, pp. 951-962 (2015)
5. Y. Wu, FDS Team, Fusion Eng. Des., **84**, pp. 1987-1992 (2009), http://dx.doi.org/10.1016/j.fusengdes.2008.12.041
6. D. Grosse, et al., Fusion Eng. Des., **88,** pp. 2210–2214 (2013), http://dx.doi.org/10.1016/j.fusengdes.2013.02.146
7. A. Serikov, et al., Fusion Eng. Des., **109–111**, pp. 848–854 (2016), http://dx.doi.org/10.1016/j.fusengdes.2016.01.063
8. S. Pak, et al., Fusion Eng. Des., **102**, pp. 21-27 (2016), http://dx.doi.org/10.1016/j.fusengdes.2015.11.016

Shielding calculations for industrial 5/7.5MeV electron accelerators using the MCNP Monte Carlo Code

Eyal Peri[1,*] and *Itzhak* Orion[2]

[1]*Nuclear Research Centre Negev, P.O.B 9001, Beer-Sheva, 84190, Israel*
[2]*Ben-Gurion University of the Negev, Beer Sheva, Israel*

Abstract. High energy X-rays from accelerators are used to irradiate food ingredients to prevent growth and development of unwanted biological organisms in food, and by that extend the shelf life of the products. The production of X-rays is done by accelerating 5 MeV electrons and bombarding them into a heavy target (high Z). Since 2004, the FDA has approved using 7.5 MeV energy, providing higher production rates with lower treatments costs. In this study we calculated all the essential data needed for a straightforward concrete shielding design of typical food accelerator rooms. The following evaluation is done using the MCNP Monte Carlo code system: (1) Angular dependence (0-180°) of photon dose rate for 5 MeV and 7.5 MeV electron beams bombarding iron, aluminum, gold, tantalum, and tungsten targets. (2) Angular dependence (0-180°) spectral distribution simulations of bremsstrahlung for gold, tantalum, and tungsten bombarded by 5 MeV and 7.5 MeV electron beams. (3) Concrete attenuation calculations in several photon emission angles for the 5 MeV and 7.5 MeV electron beams bombarding a tantalum target. Based on the simulation, we calculated the expected increase in dose rate for facilities intending to increase the energy from 5 MeV to 7.5 MeV, and the concrete width needed to be added in order to keep the existing dose rate unchanged.

1 Introduction

Industrial electron accelerators are used for many applications such as: irradiation of medical devices, food and wires, etc. [1]. In this study we calculated essential data needed for a straightforward concrete shielding design of an industrial electron accelerator, using 5 MeV and 7.5 MeV X-Rays. These energies are mainly used to irradiate food ingredients. Food irradiation using ionizing radiation is a common process to prevent the growth and development of several types of biological organisms in food. In the past, the main process to provide the desired dose required a ^{60}Co irradiation facility. Nowadays, due to their inherent safety, electron accelerators are more in use. When deeper penetration is needed, the electron beam produces bremsstrahlung (X-rays) by using targets with high Z, acting as convertors.

Due to the low range of electrons in matter, irradiation by the electron beam is limited to small packages. To irradiate larger packages, higher penetration is needed, therefore the electron beam has to be converted to bremsstrahlung X-rays. The targets used to convert the electron beam to bremsstrahlung are tantalum, tungsten, gold and aluminum. Aluminum is used as an electron beam absorber.

High energy photons can cause food activation due to (γ,n) reactions. Until 2004, to eliminate the possibility of food activation, the electron energy was limited to 5 MeV X-rays for food irradiation. In 2004, the FDA approved the usage of up to 7.5 MeV, but only with tantalum and gold targets [2].

Higher X-ray energy results an increased flux of X-rays in the forward direction, increased penetration, and higher photon dose rate due to better electron-to-photon conversion. These improvements could decrease the irradiation time and allow irradiation of larger packages, thereby providing higher production rates with lower treatment cost.

Electron accelerators emit high currents of high energy electrons and X-Ray. The x-ray energies are determined by the beam energy; the higher the energy, the better the X-ray penetration. The magnitude of the accelerator dose rates is mainly a function of the beam current. The typical dose rates in medical accelerator are quite low, up to 100 Gy h^{-1}, compared to the dose rates in industrial accelerators, which are much higher: 100 kGy h^{-1} in X-Ray mode, and up to 360,000 kGy h^{-1} in electron irradiation mode [1]. Due to the difference in dose rates, the shielding thickness of industrial accelerator is much higher. Therefore, accurate shielding values such as angular dependence of dose rates and concrete attenuation ratios, is very important for designing the optimal shielding.

Medical accelerators usually work with 6-18 MV electron energy with a tungsten target to convert the electron beam to X-rays. In order to protect the patients, the accelerator head is protected with a heavy lead shielding; therefore, the bremsstrahlung radiation is emitted only in the forward direction. Shielding thickness calculations are based on available bremsstrahlung spectrum data in the forward direction (primary shielding). Secondary shielding is designed to protect against photon scattering from the primary shielding or leakage from the accelerator head, which are two orders of magnitude smaller than the primary beam [3]. There are many publications and standards that guide how to design optimal shielding for medical accelerator rooms [3-5]. The shielding data for medical

* Corresponding author: eyalperi22@gmail.com

accelerators is not applicable for industrial accelerators, since the data is for different conversion targets, different X-Ray energies, and only for the forward direction.

In general, collimators are not in use in industrial accelerators, since human protection is not needed inside the room (e.g. patients), and therefore bremsstrahlung photons can be emitted in all directions. The bremsstrahlung spectrum and dose rate change as a function of the emission angle. The dose rate decreases from maximum in the forward direction (0°) to minimum at 180° by 1-2 orders of magnitude. In order to design and calculate optimal shielding for food accelerator rooms, there is a need to have the bremsstrahlung spectrum data and dose rates in all emission directions, which are different for each conversion target and energy (5 MeV or 7.5 MeV). There are publications and standards available such as: NCRP 144 [6], IAEA Report 188 [7], ANSI N43.3 [8], to guide how to design shielding for industrial accelerator rooms. The shielding data in these guides is suitable for most industrial accelerators, but does not contain data for 5 MeV and 7.5 MeV X-rays, such as used in food irradiation facilities.

One publication written specifically to give data for the shielding design of food accelerator rooms was published by Barkova [9]. The angular dependence of the KERMA rates from aluminum, iron, and tungsten targets irradiated by 5 MeV electrons were calculated using Monte Carlo simulations with the use of the program package SPIN-PC, presented by Barkova [9]. The concrete attenuation ratio was evaluated analytically using Monte Carlo simulations to obtain the bremsstrahlung spectrum for each angle.

Petwal [10] performed MCNP calculations to optimize the design of the conversion target for maximizing the X-ray output. They also presented calculated 5 MeV and 7.5 MeV relative bremsstrahlung spectra, but only for the forward direction, and relative angular dose rate up to 60°. The authors point out that from the three possible targets (gold, tantalum, tungsten), tantalum is the most suitable for reasons of photon yield, residual radioactivity, mechanical properties, and fabrication simplicity.

2 Objective

The following evaluations were done using the MCNP Monte Carlo code system. Simulations for angular dependence calculations of photon dose rate in units of Gy h^{-1} were carried out at the distance of one meter from several common converting targets as a function of the beam current for 5 MeV and 7.5 MeV electron beams. Photon attenuation ratios in concrete were calculated for a range of 0-180 cm for a tantalum target at selected photon emission angles (0°, 90°, 135°, 180°) for the 5 MeV and the 7.5 MeV electron beams. Based on the simulation, we calculated the expected increase in dose rate for facilities intending to increase the energy from 5 MeV to 7.5 MeV.

3 Materials and methods

The simulations were carried out using the MCNP version 4C2 code package [11]. Initiation of MCNP requires an input file that contains complete information about the radiation source, the geometry of the system to be simulated, and both the density and the elemental composition of all interacting materials. The Concrete composition was taken as an ordinary concrete 2.3 g cm^{-3} in density [12].

In addition, the output tallies to be calculated need to be specified in the MCNP input file. The tally F6 in a cell volume was used for absorbed dose calculations. The dose was calculated by employing the MCNP 'F6:p' track length tally and making the KERMA approximation. Since most of the dose is by photons of less than 3 MeV, electron equilibrium is satisfied, than the terms 'dose' and 'KERMA' are thus used synonymously in this work. Tally *F4 of photon fluence was used for the photon spectra graphs.

Simulation of the angular dependence of photon dose rate (Gy h^{-1}) at one meter from the targets as a function of beam current (mA) and beam energy were done by irradiating gold, tantalum, and tungsten targets with a 5/7.5 MeV pencil electron beam. In real accelerators the beam dimension is not narrow, but the distance from the source to the exterior side of the building is more than a few meters. Therefore, for shielding purposes it is possible to simulate the electron beam as a narrow beam to simplify the simulation. The targets depth and radial dimensions were chosen to be about 20% less the CSDA range using ESTAR database [13], which is approximately 20% more than the penetration range of the electrons in the target [14]. Target thickness of Al, Fe, Ta, W and Au were 9.4, 3.2, 1.5, 1.3 and 1.3 mm for 5 MeV electrons, and target thickness of Ta and Au were 2.3 and 2.0 mm for 7.5 MeV electrons.

The three-dimensional space was subdivided into conical shells to 18 sectors, from 0° (forward direction) to 180° (backward), 10° for each sector. At a radius of 1 meter around the target, 18 volumes of "detectors" were made by generating two concentric spheres at 99 cm and 101 cm, and dividing them by the 18 sectors. All the space around the target was covered by 18 cells, defined as "detectors" filled with standard air. The bremsstrahlung photons were collected by the 18 "detectors". The results were converted to units of (Gy mA^{-1} h^{-1}) at a distance of 1 m from the target, for each 10° sector. Since every photon was collected by one of the 10° "detectors", the number of photons reaching each detector was large. Therefore the statistics was adequate, with a small standard deviation (less than 2%) obtained with very short run time (e.g., 1 hour). For shielding purposes, averaging the dose rate at a 10° angle does not consider significant error. In order to validate the correctness of the results, simulations of aluminum, iron, and tungsten were done and compared to the data of Barkova [9]. As can be seen in Figure 1, the results are in good agreement, except in aluminum and iron targets, in the forward directions the difference that our results are higher up to 20% than the previous study results.

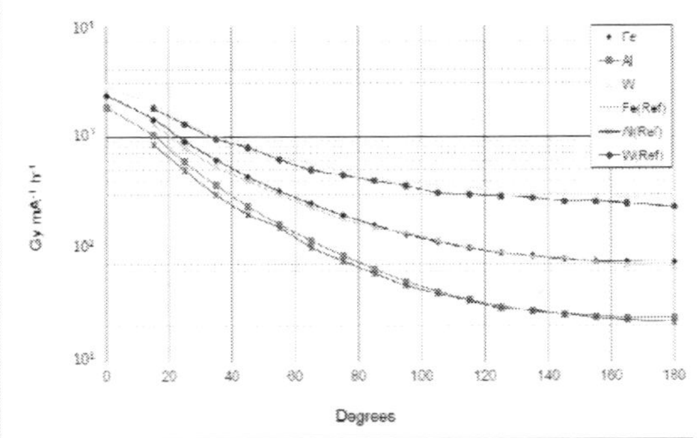

Fig. 1. Angular dependence of photon dose rate (Gy h^{-1}) at one meter from Fe, Al, W targets as a function of beam current (mA) for 5 MeV electron beam. MCNP (our result) vs. Barkova [9].

Simulation of the energy distribution of bremsstrahlung emission for gold, tantalum, and tungsten targets by 5 MeV and 7.5 MeV electron beams were performed. Energy distribution of bremsstrahlung intensity was calculated using Tally *F4 as kdn dk^{-1} $d\Omega^{-1}$ (MeV MeV^{-1} Sr^{-1} $electron^{-1}$), by dividing each bin by its energy interval and multipling it by the square of the distance. The plots are presented as the number of bremsstrahlung photons of energy k within an increment of energy dk, emitted at angle θ within a solid angle dΩ, by one electron whose initial energy is E_0. The same geometry configuration like the previous section was provided. The bremsstrahlung photons were collected in the "detectors" cells. The energy bins were 0.2 MeV at the following angles: 0°, 10°, 20°, 45°, 60°, 90°, 120°, 150°, and 180°. Each angle is the average of ±5° around it. In order to validate the results, the first simulation was compared to the spectrum of 2.8 MeV electrons hitting an iron target as measured by Dance [15]. Three emission angles were selected for comparison (10°, 45°, 120°); as can be seen in Figure 2, the results showed a good agreement.

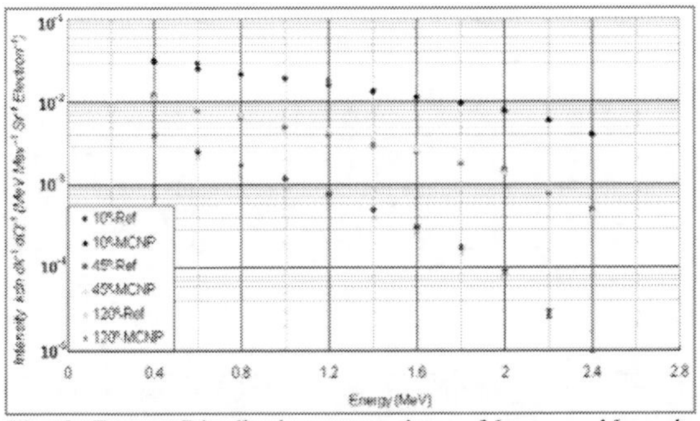

Fig. 2. Energy Distribution comparison of bremsstrahlung in Iron target 2.8 MeV. MCNP (our result) vs. Fig. 11 in Dance [15].

Simulation of photon attenuation in concrete of selected photon emission angles (0°, 90°, 135°, and 180°) were performed for 5 MeV and 7.5 MeV electron beams. The same configurations as the previous section were done. A spherical concentric concrete shield was added between the target and detectors from a thickness of 10 cm up to 180 cm, hence the detectors were moved to 200 cm. Every 10 cm addition required a separated run; when simulating concrete shielding thicker than 80 cm, only a few photons were tallied and there was a need to use variance reduction techniques in order to get improved statistics. The attenuation data were calculated for four emission angles. In order to improve the statistics, broad angles were chosen (up to ±20°). Since there is an increased concentration of bremsstrahlung photons in the forward direction, the collection angle was set to be larger in the backward direction than in the forward direction. The emission angles used were: 0° (0-10°), 90° (80-100°), 135° (120-150°), and 180° (160°-180°) (Figure 3). For shielding purposes, averaging the backward radiation dose rate in a ±20° range does not lead to significant uncertainties. The variance reduction technique that was used is geometry splitting: As illustrated in Figure 3, the concrete was divided to 10 cm sections. The importance of each section was gradually increased from the inner to the outer sections, and the importance of the backwards direction was greater than in the forward direction, so the photon population will be approximately equal. Figure 3 is a cross-section illustration of the geometry, and illustrates a single run with 80 cm of concrete.

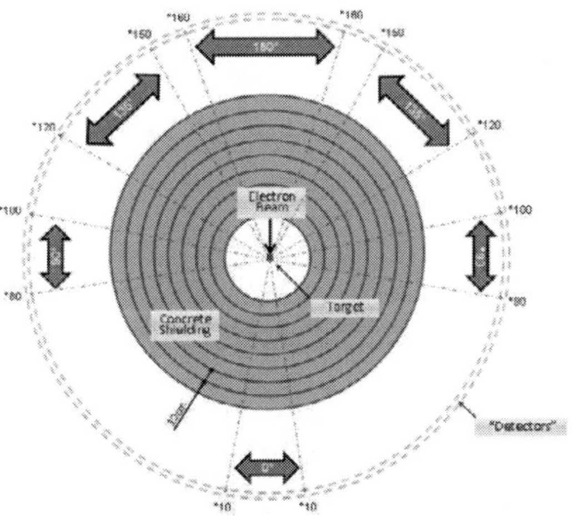

Fig. 3. Illustration of concrete shielding between the target and the detectors at 0°, 90°, 135°, and 180°.

In order to validate the results, the attenuation simulations results of an aluminium target with emission angles of 90° and 180° were compared to the data published in Barkova [9]. As can be seen in Figure 4, the results showed a good agreement.

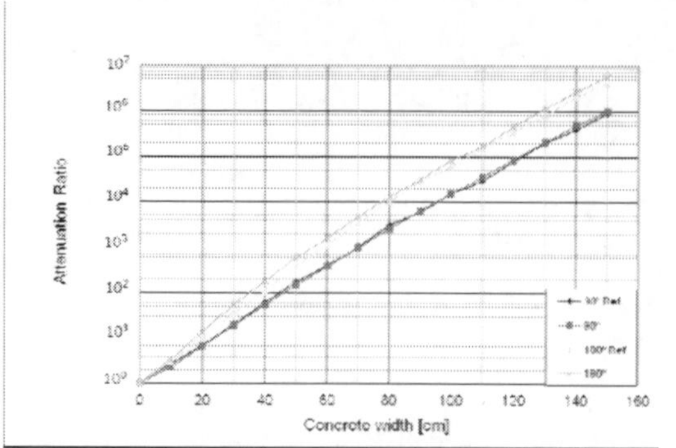

Fig. 4. Dose attenuation by concrete of the bremsstrahlung emission from Al target for beam energy of 5 MeV. MCNP (this study) vs. Barkova [9].

4 Results

Figure 5 presents the angular dependence (0-180°) of photon dose rate (Gy h^{-1}) at 1 m from gold, tantalum, and tungsten targets as a function of beam current (mA) for 5 MeV and 7.5 MeV electron beams. 10^6-10^7 histories were used to obtain relative statistical uncertainties lower than 2%. Error bars for less than 5% are too small to be shown in the graphs. The results of tantalum and tungsten for 5 MeV are almost identical. The reason is the close atomic number of the two targets. Due to the larger atomic number of gold its dose rate is slightly higher, but only for the backward angles (>90°). This is one of the reasons why a gold target is not cost effective. The 7.5 MeV dose rates curves show almost the same shape of the ones relating to 5 MeV and are higher by a factor from 1.5 up to 2.6.

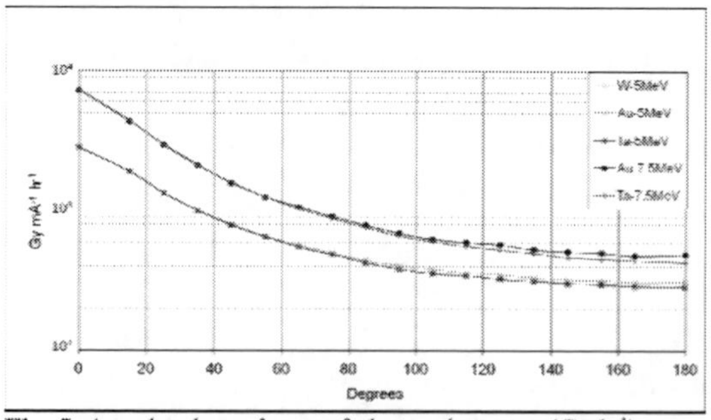

Fig. 5. Angular dependence of photon dose rate (Gy h^{-1}) at one meter from Au, Ta, W targets as a function of beam current (mA) for 5/7.5 MeV electron beam.

Figures 6 and 7 present the energy distribution of bremsstrahlung emission for gold and tantalum targets by 5 MeV and 7.5 MeV electron beams. 10^7-10^8 histories were used to usually obtain relative uncertainties lower than 5%. Due to the small difference in the atomic numbers of the targets, the spectra are almost identical for each energy. The spectrum of the forward direction photons is higher and "harder" than the spectrum in the backward emission. In the lower energy bins, the fluence in the forward directions is higher by an order of magnitude than the backward emission. As the energy bins increased, the gap increases up to three orders of magnitudes in the higher energy bins. From these figures we can understand why the dose rate is decreasing as the photon emission angle increases. The fact the spectrum of the backward is softer explains why the concrete shielding is more effective to attenuate backward photon emission.

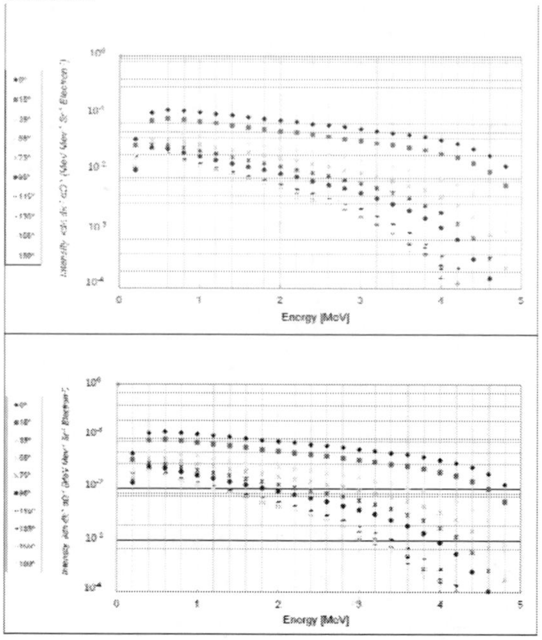

Fig. 6. Energy distribution of 5 MeV bremsstrahlung in Au (Upper) and Ta (Lower) Targets.

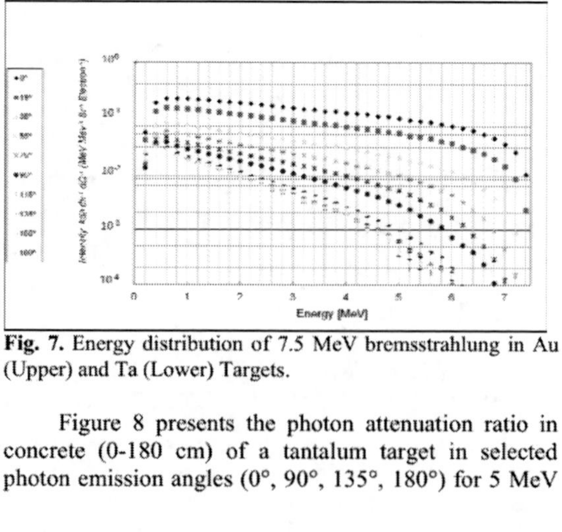

Fig. 7. Energy distribution of 7.5 MeV bremsstrahlung in Au (Upper) and Ta (Lower) Targets.

Figure 8 presents the photon attenuation ratio in concrete (0-180 cm) of a tantalum target in selected photon emission angles (0°, 90°, 135°, 180°) for 5 MeV and 7.5 MeV electron beams. The attenuation ratio was obtained as the ratio between the dose without shielding to the dose behind the concrete shielding. 10^6-10^8 histories were used to obtain relative uncertainty less than 5%. As with the dose rate and bremsstrahlung results, the data for gold and tungsten should be similar, and therefore were not calculated here. The photon attenuation in concrete for 5 MeV is higher than for 7.5 MeV by up to an order of magnitude. The attenuation is larger for the backward emitted photons compared to the forward ones by up to an order of magnitude.

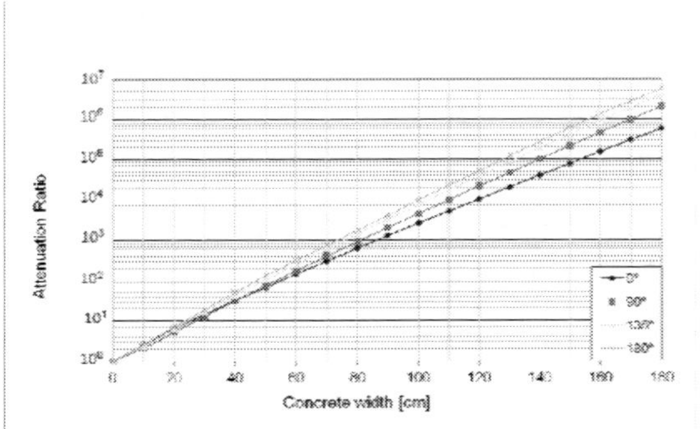

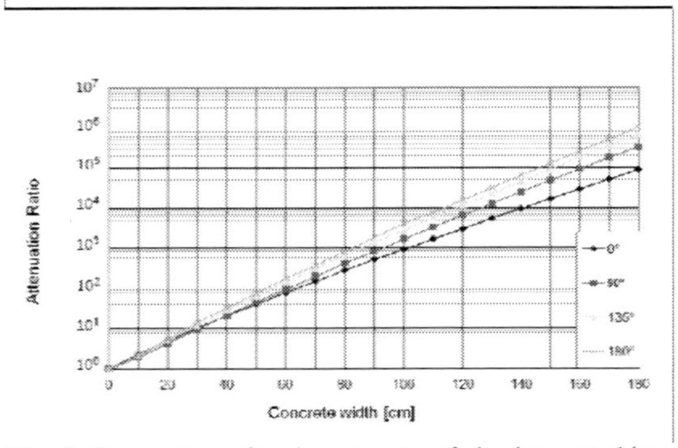

Fig. 8. Dose attenuation in concrete of the bremsstrahlung emission from Ta target for electron beam energy of 5 MeV (Upper) and 7.5 MeV (Lower).

Table 1 is an example of how to calculate the expected dose rate behind concrete shielding using the graphs given in this study. The example was set for an accelerator of 1 mA current and a tantalum target. The listed energy, emission angle, and concrete width and distance from the target are reasonably chosen settings. The dose rate at 1 m from the target is taken from Figure 5, and the attenuation ratio is taken from Figure 8; both are based on the settings data given in Table 1. The final dose rate behind the shielding at distance L from the target is obtained by the following equation:

$$\dot{D}_{out}[\frac{\mu Gy}{mA \cdot h}] = \frac{\dot{D}_{in}[\frac{Gy}{mA \cdot h}] \cdot 10^6}{(Atten.Ratio_{concrate})} \left(\frac{1[m]}{L[m]}\right)^2 \quad (1)$$

$\dot{D}_{in}$ - The dose rate one meter from the conversion target; 10^6 is a factor from Gy to μGy.

Table 1. Example of shielding calculation for 1mA current with Ta target.

Point	Energy [MeV]	Emission Angle [deg.]	Concrete width [cm]	Distance [m]	Dose Rate[2] [Gy*mA^{-1}*h^{-1}]	Concrete Attenuation ratio[3]	Dose Rate Behind Shielding [μGy*h^{-1}]
1	5	90	170	7	400	1.0 10^6	8
2	5	180	120	8	285	5.3 10^4	84
3	7.5	135	150	7	500	8.8 10^4	116
4	7.5	90	160	11	700	9.2 10^4	53

[2]From Figure 5 (before shielding).
[3]From Figure 8.

The quantities of concrete required to be added in a case of upgrading a 5 MeV beam to 7.5 MeV in a facility, in order to keep the 5 MeV dose rate are listed in Table 2. The expected difference in dose rate between 5 MeV and 7.5 MeV emitted from a tantalum target at 0°, 90°, and 180° are presented in Table 2. The presented results are taken from Figure 5 (2.6, 1.7, 1.5, respectively), multiplied by the difference in the concrete attenuation, taken from Figure 8. The data are presented for three emission angles (0°, 90°, 180°) as a function of the existing concrete thickness in the facility. The data are not dependent on the distance between the target and the interest point outside the facility.

Table 2. Dose rate ratios of bremsstrahlung emission between 5 MeV and 7.5 MeV beams Ta target as a function of existing concrete for a given angle (left), and the quantity of required concrete to be added for 7.5 MeV facility in order to keep the 5 MeV dose rate (right).

Existed Concrete (cm)	5 MeV and 7.5 MeV Dose Rate Ratio			Required Concrete addition (cm)		
	0°	90°	180°	0°	90°	180°
0	2.6	1.7	1.5	8	5	0.5
10	2.8	1.9	1.6	9	6	2
20	3.3	2.0	1.7	12	7	2
30	3.5	2.3	1.8	13	9	3
40	4.0	2.5	2.0	15	10	4
50	4.2	2.8	2.1	16	12	5
60	4.8	3.3	2.2	18	14	5
70	5.4	3.6	2.5	19	15	7
80	5.9	3.8	2.7	21	16	8
90	6.5	4.3	2.8	23	18	9
100	7.6	4.9	2.9	25	20	9
110	8.2	5.1	3.5	26	20	12
120	9.0	6.0	3.8	28	22	13
130	9.7	6.3	4.1	29	23	14
140	11.1	6.9	4.4	31	24	15
150	12.2	7.6	4.9	33	26	16
160	13.7	8.6	5.2	35	28	17
170	15.3	9.4	5.8	36	29	18
180	16.5	10.5	6.1	38	31	19

By using Table 2, facilities that plan upgrading their accelerator to 7.5 MeV can easily estimate the expected increase in dose rate around the shielded facility. The additional concrete thickness ensuring the same dose rate around the facilty can also be evaluated, thus maintaining the same safety status. For example, a

DOI: 10.1051/epjconf/201715303011

facility interested in the dose rate over a 120 cm shield located 90° from the electron beam direction, can use Table 2 to find that the expected dose rate rises by a factor of 6, or an additional 22 cm of concrete thickness to that area should lower the dose rate back to the original rate.

5 Conclusions

Angular dependence MCNP simulations of dose rate, bremsstrahlung, and concrete attenuation ratio were carried out for 5 MeV and 7.5 MeV accelerators for gold, tantalum, and tungsten targets. Since the targets atomic numbers are similar, there was only a slight difference in dose rates and bremsstrahlung spectra of the targets. The dose rate of 7.5 MeV accelerators is higher than the 5 MeV dose rate by factor of about 2.6 in the forward direction, and a factor of 1.5 in the backward direction. The difference in dose rates from the forward to backward directions is almost two orders of magnitude. Photon attenuation ratios in concrete of 7.5 MeV beam is lower than that of 5 MeV beam by almost an order of magnitude for thick shielding for all emission angles. There is a significant difference in the bremsstrahlung spectrum between the forward emission angles to the backward angles. Since the bremsstrahlung spectrum of the backward angles is softer than in the forward direction, the concrete shielding is more effective in attenuating the backward direction. This difference is small for thin shielding, but for thick shielding of more than 150 cm, the attenuation of the backward emissions is larger than for forward direction by an order of magnitude, for the same concrete thickness.

References

1. IAEA, Industrial Radiation Processing with Electron Beam and X-Rays - Revision 6, 1 May 2011.
2. Federal Register. **69**(246):76844; 2004.
3. IAEA, Safety Report Series No.47, Radiation Protection in the design of Radiotherapy Facilities, 2006.
4. NCRP Report No.49, Structural Shielding Design and Evaluation for Medical Use of X-Rays and Gamma-Rays up to 10 MeV, 1976.
5. NCRP Report No.151, Structural Shielding Design and Evaluation for Megavoltage X and Gamma Ray Radiotherapy Facilities, 2005.
6. NCRP Report No.144. Radiation Protection for Particle Accelerator Facilities; 2005.
7. IAEA, Technical Reports Series No.188, Radiological Safety Aspects of the Operation of Electron Linear Accelerators 1979.
8. ANSI/HPS N43.3-2008, For General Radiation Safety – Installation Using Non-Medical X-ray and Sealed Gamma-Ray Sources, Energy Up to 10 MeV, 2008.
9. Barkova VG, Kiselev AV, Chudaev VYa. Evaluation of Efficiency of Concrete Shielding Against Bremmsstrahlung of 5 Mev Electrons at Pre-Commissioning of the Accelerator ILU-10, Proceeding of RuPAC, Novosibirsk, Russia; 2006.
10. Petwal VC, Bapna SC, Sanda RS, Kotaiah S, Subbaih KV. Bremsstrahlung Converter for High Power EB Radiation Processing Facility, APAC, RRCAT. Indore, India; 2007.
11. Briesmeister JF, ed. MCNP A General Monte Carlo N-Particle Transport Code, version 4C, LA-13709-M. Los Alamos National Laboratory, Los Alamos, NM; April 2000.
12. Williams R.G., Compendium of Material Composition Data for Radiation Transport Modeling, DOE, April 2006.
13. http://physics.nist.gov/PhysRefData/Star/Text/EST AR.html
14. Attix F H, Introduction to Radiological Physics and Radiation Dosimetry, pp. 184-185, Wiley 2004.
15. Dance WE, Rester DH, Farmer BJ, Johonson JH. Bremsstrahlung Produced in Thick Aluminum and Iron Targets by 0.5 to 2.8 MeV Electrons. J Appl Phys **39**(6):2881-2889; 1967. DOI: http://dx.doi.org/+10.1063/1.1656689

EPJ Web of Conferences **153**, 03012 (2017)
ICRS-13 & RPSD-2016
DOI: 10.1051/epjconf/201715303012

Determination of neutron spectrum outside the lateral shielding of Taiwan Photon Source using high-sensitivity Bonner cylinders

Yu-Chi Lin[1,2], *Ang-Yu* Chen[1], *Joseph* Liu[1], *Kuo-Wei* Lee[2], *Rong-Jiun* Sheu[2,3,a]

[1]National Synchrotron Radiation Research Center, 101 Hsin-Ann Road, Hsinchu Science Park, Hsinchu, Taiwan
[2]Institute of Nuclear Engineering and Science, National Tsing Hua University, 101, Sec. 2, Kuang-Fu Road, Hsinchu, Taiwan
[3]Department of Engineering and System Science, National Tsing Hua University, 101, Sec. 2, Kuang-Fu Road, Hsinchu, Taiwan

Abstract. A homemade Bonner cylinder spectrometer was used to determine the neutron spectrum from thermal energy up to several GeV at a heavily-shielded light source facility. The spectrometer, similar to the design of Bonner spheres, features high sensitivity of neutron detection because of using a long cylindrical ^{3}He proportional counter. The spectrum measurement was performed during the facility commissioning by intentionally parking the injected electrons at the septum of the storage ring. Based on a high-fidelity FLUKA simulation, the predicted neutron spectrum at the location of measurement was adopted as the initial guess of spectrum unfolding. The unfolded result indicated an underestimation of the calculated neutron spectrum in the high-energy portion, leading to a substantial revision of the neutron dose rate at the location.

1 Introduction

Taiwan Photo Source (TPS), situated at the National Synchrotron Radiation Research Center in Hsinchu, is a 3 GeV light source facility with a circumference of 518 m, aiming to provide synchrotron light with extremely high brilliance and low emittance [1]. Electrons are generated and pre-accelerated to 150 MeV through a linear accelerator (LINAC), with a typical beam output of 2.25 W corresponding to 5 nC per pulse and 3 Hz repetition rate. These electrons are further accelerated to 3 GeV in a booster synchrotron and then injected into a storage ring. Fig. 1 shows a schematic layout of the main accelerator components of TPS and the bulk concrete structure. The storage ring and concentric booster were installed in a shared tunnel made of 1 m thick concrete walls, and the shielding in the injection area and ratchet end walls were enhanced to 1.2 m thick concrete against injection beam loss and forward-peaked bremsstrahlung, respectively.

According to the facility safety analysis report [2], the dose contributions from the neutron and gamma-ray components outside the lateral wall of the TPS tunnel are roughly comparable. In addition, neutrons with energies higher than 10 MeV contribute more than 50% of the total neutron dose. These estimates were obtained based solely on Monte Carlo simulations. Because of the detector limitation, conventional neutron dosemeters show almost no response to those high-energy neutrons and the extended-range neutron detectors should be used in this environment for proper dose responses. How accurate are the simulation results? Various factors may affect their accuracies, including beam loss assumptions, geometry and material modelling, cross section data and underlying physics models. Therefore, experimentally determining the neutron energy distribution outside the accelerator tunnel is of great interest, in particular the high-energy portion of the spectrum.

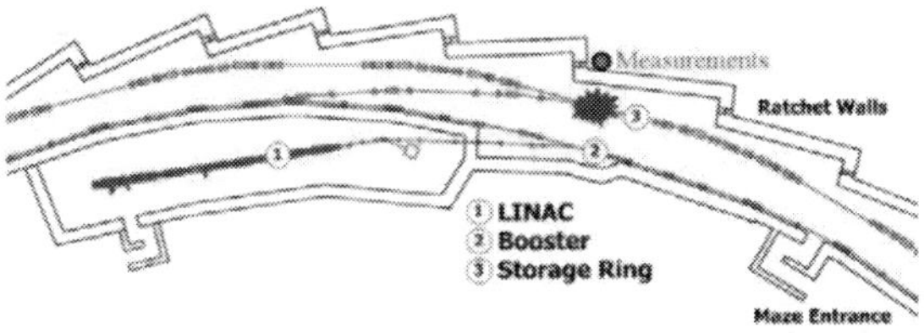

Fig. 1. A schematic layout of the TPS facility, showing the main accelerator components and the bulk concrete structure as well as the location of measurement (green spot).

2 Materials and methods

2.1 Spectrometer and experimental setup

Bonner spheres are widely used in neutron spectrometry because of the advantages of isotropic and wide-energy responses to incoming neutrons. However, the detection efficiency of a Bonner sphere is limited by the small active volume of its central probe and could become impractical in determining neutron spectra of low intensity. Thus, as shown in Fig. 2, a set of Bonner cylinders was assembled by warping polyethylene (PE) of various thicknesses around a long cylindrical ^{3}He proportional counter. The adopted neutron probe had an

[a] Corresponding author: rjsheu@mx.nthu.edu.tw

active volume of 2.54 cm in diameter and 45.72 cm in length that filled with 4 atm helium-3 gas, exhibiting a sensitivity of ~18 times higher than the typical probe used in the famous Bonner sphere spectrometer (BSS) NEMUS [3]. The Bonner cylinder spectrometer (BCS) consisted of 16 configurations in total, including the bare tube and 11 combinations with PE cylinders of thicknesses ranging from 1.5 to 10 cm. In addition to these standard configurations, four sets of special cylinders labelled 2C2.5_6.5, 2P2.5_6.5, 2P3_6.5, and 3P4_6.5 were the extended-range versions. The naming, taking 2P2.5_6.5 for example, indicated the ^{3}He tube being surrounded by consecutive shells of 2-cm-thick PE, 0.5-cm-thick lead and another 4-cm-thick PE. The character P in the naming distinguished the material of the embedded metal in moderator, P was lead and C stood for copper.

Fig. 2. A photo of the homemade high-sensitivity Bonner cylinders used in this study.

Both Monte Carlo simulations and measurements were performed to investigate the characteristics of the neutron field at the TPS. In order to well define the beam loss location, we intentionally parked all the injected electrons from the booster at the injection septum by turning off the power supply of the magnet. The detectors were arranged at the location outside the lateral shielding wall of the beam loss point, as illustrated in Fig. 1. Because it took time to switch and complete the measurements involving 16 configurations of the spectrometer, a counting rate normalization was necessary to account for the variation of the injected beam. For this reason, an independent neutron detector was installed in a close proximity of the Bonner cylinder to provide a continuous monitoring of the neutron field, which can ensure all the measurements corresponded to the same strength of irradiation. The reference neutron detector was extended-range type and its setup is shown in Fig. 3.

During the experiment, the injected beam current was kept stable and continuously monitored by checking the reading of the booster current. The reading is shown in Fig. 4, indicating an average current of approximately 0.027 mA, corresponding to a beam output of about 40 pC per pulse. Fig. 4 also shows the dose rate history of two nearby neutron and gamma-ray detectors during that period of time. The neutron and gamma-ray dose rates were indeed comparable as expected in the safety analysis report. Note that the level of dose rates was low, only approximately 0.2–0.4 μSv/h, which hinders the use of conventional BSS in neutron spectrum measurement.

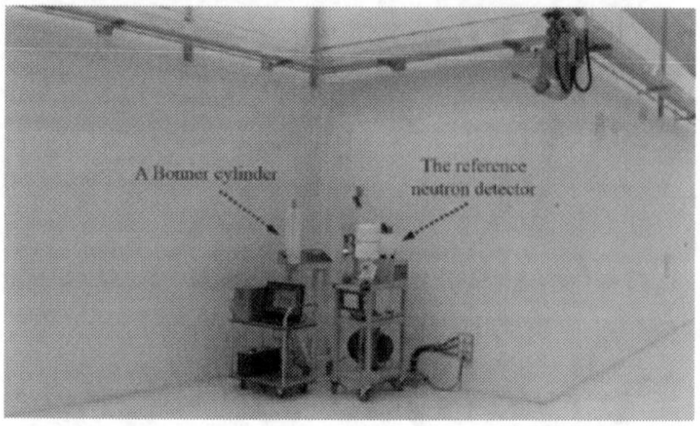

Fig. 3. A photo of the experimental setup at the TPS (See Fig. 1), showing the locations of a Bonner cylinder and the reference neutron detector.

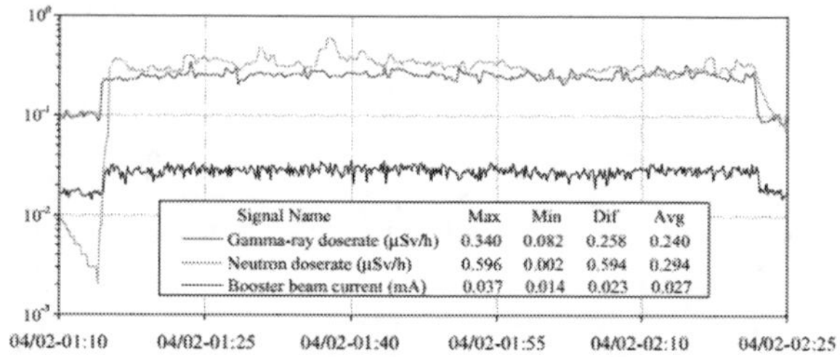

Fig. 4. Monitoring of the booster beam current during the experiment and the dose rate history of two nearby detectors.

2.2 Monte Carlo simulations

Two types of Monte Carlo simulations were conducted to provide information regarding the response functions of detectors and the characteristics of radiation field at the TPS. The response functions of 16 configurations of the BCS were calculated using MCNPX [4] with evaluated cross-section data for neutron energies below 150 MeV and the Bertini nuclear model for neutron interactions above 150 MeV. Full details of the neutron probe and moderator configurations were included in the calculation model. The resulting response functions were checked and verified by a series of integral tests performed in a standard calibration room using Cf-252 neutron source. For more details on the calculation model and verification of response functions, interested readers are referred to the paper [5].

Radiation environment outside thick shielding of a high-energy electron accelerator mainly consists of bremsstrahlung and secondary neutrons, resulting from the dominant electromagnetic cascade and subsequent photonuclear reactions, respectively. To acquire a quantitative description of the radiation field at the location of measurement, the multi-particle Monte Carlo transport code FLUKA [6] was employed to simulate the shower development and radiation transport around the beam loss spot. A rather detailed geometry model of the TPS injection section, including main components of the accelerator system and the surrounding shielding structure, was built aiming to make the simulation as realistic as possible. In addition to dose maps in the region, gamma-ray and neutron spectra were scored at the location of interest. By folding the neutron spectrum with detector response functions, we could predict the neutron counting rates of 16 Bonner cylinders at the location, which can be directly compared with the measurements.

3 Results and discussion

3.1. Detector response functions

Fig. 5 shows the calculated response functions of 16 Bonner cylinders under a parallel-beam irradiation of neutrons moving perpendicular to the tube axis. The response matrix indicates that the bare tube is essentially sensitive to thermal neutrons only. The 11 standard configurations comprised of PE in various thicknesses can detect neutrons in a wide range of intermediate energies up to several MeV. In contrast, the four extended-range detectors exhibit improved detection efficiencies to high-energy neutrons above 10 MeV due to a neutron multiplication effect caused by the embedded heavy metal. Compared with a conventional BSS, the main feature of this BCS was its high sensitivity, as indicated by the overall magnitude of these response functions. It was essential for our purpose of determining the neutron spectrum at a location with low intensity.

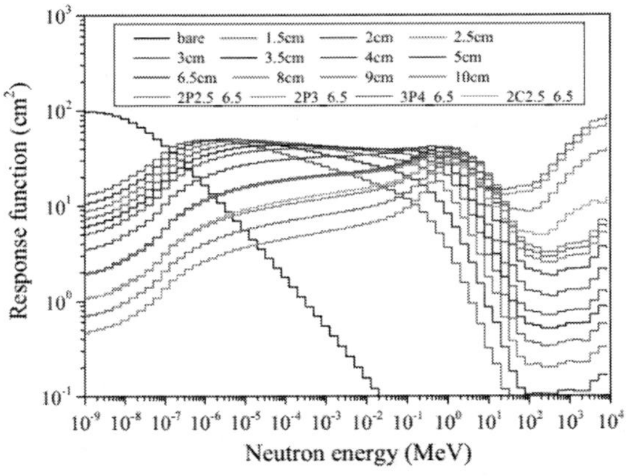

Fig. 5. The MCNPX-calculated response functions of 16 Bonner cylinders for a parallel-beam irradiation.

A problem of the BCS was the loss of an isotropic response to incident neutrons because the cylindrical-shaped detectors are not symmetric in the polar angle with respect to the tube axis. The authors prepared another set of response matrix to explore the effect of this problem on spectrum unfolding. In contrast to the parallel-beam irradiation, the second response matrix of the BCS was calculated by considering an isotropic irradiation of neutrons. Fig. 6 shows the calculated results. The two response matrices in Figs. 5 and 6, respectively, have very similar characteristics but with small differences when observing in details. The isotropic response functions are somewhat lower in magnitude than those calculated under parallel broad beam geometry because of the effect of slant incidence.

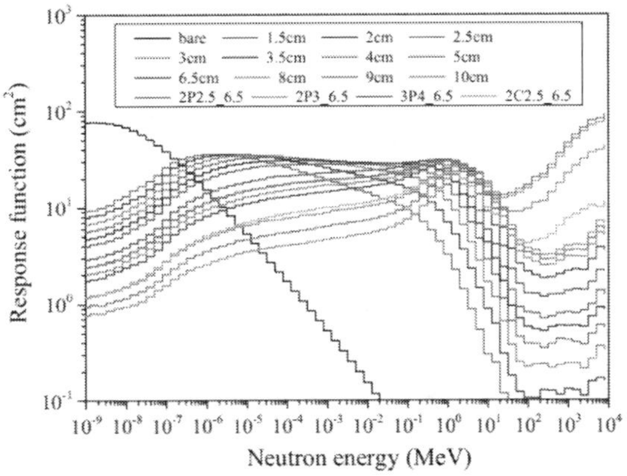

Fig. 6. The MCNPX-calculated response functions of 16 Bonner cylinders for an isotropic irradiation.

3.2 Characteristics of radiation field

According to the beam loss scenario and the ICRP-74 ambient dose equivalent conversion factors [7], Fig. 7 shows the FLUKA-predicted neutron and gamma-ray dose distributions around the injection area. In the TPS tunnel, the gamma-ray doses were apparently higher than neutron doses. However, neutrons are generally more penetrating than gamma rays. Outside the lateral shielding wall, neutron doses were observed comparable to or even higher than gamma-ray doses. Note that gamma rays induced by high-energy electrons are significantly forward-peaked, but neutrons lack of a strong directional dependence in this case.

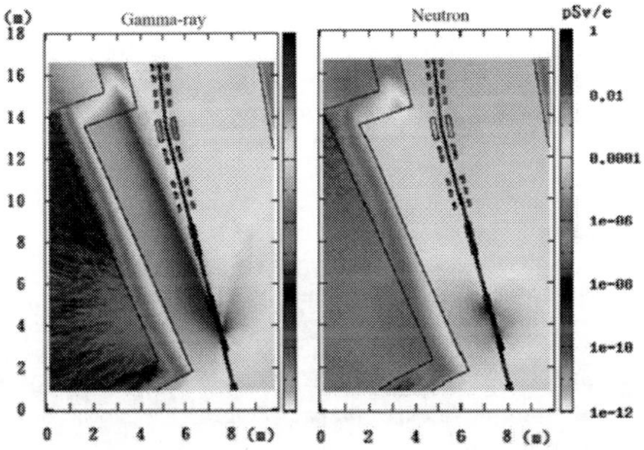

Fig. 7. The FLUKA-predicted neutron and gamma-ray dose distributions around the beam loss point, in unit of pSv per electron loss.

The neutron and gamma-ray spectra at the location of measurement were of particular interest because the information was important to the detector response and dose assessment. Fig. 8 shows the FLUKA-calculated results, where a multiplication factor of 10 was applied for the neutron spectrum to fit in the same scale with the

DOI: 10.1051/epjconf/201715303012

gamma-ray spectrum. The energies of most gamma rays are between 0.1 and 10 MeV, while the energies of neutrons have a wide range from thermal up to several hundreds of MeV. In addition, high energy neutrons account for an appreciate portion of the total neutrons. Considering their high values of fluence-to-dose conversion coefficients, those abundant high-energy neutrons may cause significant and even higher neutron doses than gamma rays.

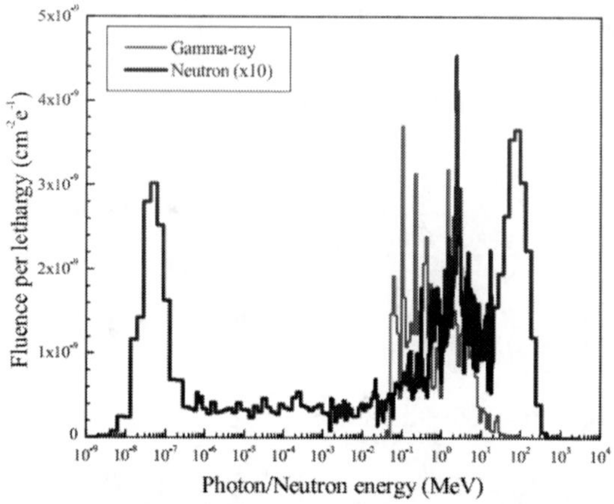

Fig. 8. Neutron and gamma-ray energy spectra scored at the location of measurement.

3.3 Measurements and spectrum unfolding

Fig. 9 summarizes the measured neutron counting rates of 16 Bonner cylinders, along with two sets of the FLUKA prediction derived by folding the calculated neutron spectrum with the response functions of detectors. The measured counting rates ranged from approximately 6 to 13 cps depending on the configuration of neutron moderator and material of neutron multiplier. These counting rates were not high and therefore the associated counting fluctuations were obvious because of limited time available for this experiment. On the other hand, the FLUKA-predicted neutron counting rates to some extent agree with the measurements, which in turn strongly indicates that both the Monte Carlo simulations and neutron measurements conducted in this study were reliable because the two approaches share similar trend in an absolute scale. Among these measurements, the counting rates of three lead-embedded detectors (2P2.5_6.5, 2P3_6.5, and 3P4_6.5) were the highest, suggesting a significant percentage of high-energy neutrons presented in the radiation field at the location of measurement.

For an underdetermined problem such as spectrum unfolding, an appropriate initial guess is requisite for most unfolding algorithms. Our best guess was the neutron spectrum in Fig. 8 obtained from the FLUKA simulation. The MAXED deconvolution code [8] based on the maximum entropy principle was used for spectrum unfolding. Because we are not sure of the angular distribution of neutrons incident on the detector, two sets of response matrices in Figs. 5 and 6, respectively, were considered in the unfolding process. As shown in Fig. 10, two unfolded spectra are similar to each other for neutron energies below 10 MeV, and without surprise consistent with the FLUKA-predicted result. However, the high-energy neutron peak around 100 MeV was substantially revised by the unfolding code for better agreement with measurements.

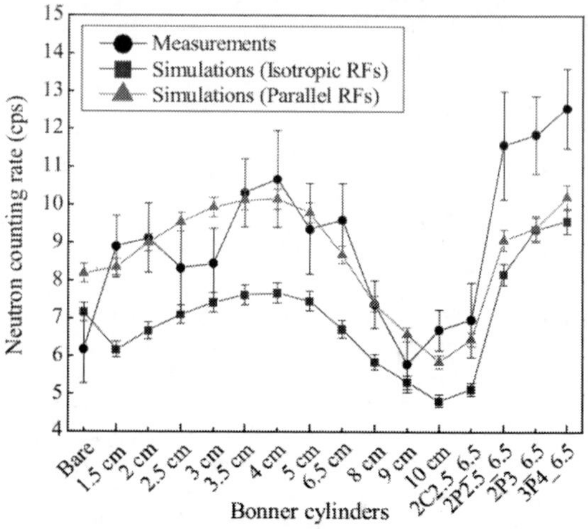

Fig. 9. The FLUKA-predicted neutron counting rates of all Bonner cylinders (with parallel and isotropic response functions) and the measured data.

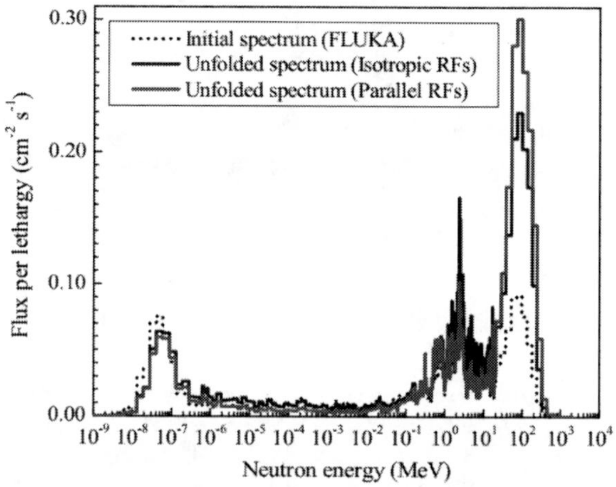

Fig. 10. The FLUKA-calculated initial spectrum and two unfolded spectra based on measurements.

The revision of the high-energy peak in the neutron spectrum has a significant impact on the dose assessment. For the parallel-beam case, the percentage of high-energy neutrons was increased to approximately 61% of the total neutron flux compared with the initial estimation of 30% in the calculated spectrum at the location. Taking the fluence-to-dose conversion coefficients into account, the neutron dose rate at the location was estimated to be 0.91 μSv/h, which was about 2 times higher than the FLUKA-predicted dose rate of 0.46 μSv/h because of a significant revision of the neutron spectrum in the high-energy portion. Assuming an isotropic irradiation, as indicated by the comparison of two unfolded spectra in Fig. 10, the high-energy peak was also revised by increasing its importance to better meet the measured data as the parallel-beam case. But the degree of the revision was less than the previous case. High-energy neutrons then accounted for approximately 48% of the total neutron flux and the neutron dose rate at the location was estimated to be 0.85 μSv/h, still significantly higher than the FLUKA prediction.

4 Conclusions

A high-sensitivity BCS was assembled, studied, and tested at the TPS, aiming to determine the neutron spectrum outside the lateral shielding wall of a 3-GeV electron accelerator. For a controlled beam loss scenario, FLUKA was used to simulate the radiation field around the experimental area and calculate the neutron spectrum at the location of measurement. Neutron counting rates of all Bonner cylinders were evaluated by folding the resultant spectrum with corresponding detector response functions. Overall, the calculated results were in satisfactory agreement with the measurements, indicating the response functions of detectors, the characteristics of radiation field, and the measured data were reasonable and reliable.

Regarding the radiation field outside the lateral shielding wall of the TPS, high-energy neutrons are important and dominate the dose contribution. The FLUKA simulation predicted those high-energy neutrons contribute approximately 55% of the total neutron dose. The unfolded neutron spectrum based on the BCS measurements revised the estimation by raising the percentage of high energy neutrons to 78% if the parallel-beam response functions were used and 67% if the isotropic response functions were considered. Both cases suggested an overall correction of the total neutron dose rate by almost doubling the FLUKA prediction. The authors currently are not sure which case can better suit for the actual measuring environment; however, it is reasonable if the answer lies in between these two desperate cases.

Acknowledgments

The authors would like to thank Yi-Chih Liu for his great help in accelerator operation during measurements. This work was supported by the Ministry of Science and Technology in Taiwan (MOST 104-2623-E-007-004-NU).

References

1. National Synchrotron Radiation Research Center, *Taiwan Photon Source design handbook* (2009)
2. R.J. Sheu, J. Liu, J.P. Wang, K.K. Lin, G.H. Luo, Nucl. Technol. **168**, 417 (2009)
3. B. Wiegel, A.V. Alevera, Nucl. Instrum. Meth. A **476**, 52 (2002)
4. D.B. Pelowitz, Ed., *MCNPX user's manual, version 2.7.0*, LA-CP-11-00438 (2011)
5. K.W. Lee, M.C. Yuan, S.H. Jiang, R.J. Sheu, Radiat. Prot. Dosim. **161**, 233 (2014)
6. A. Fasso`, A. Ferrari, J. Ranft, P.R. Sala, *FLUKA: a multi-particle transport code*, CERN-2005-10, INFN/TC_05/11, SLAC-R-773 (2005)
7. ICRP Publication 74, *Conversion coefficients for use in radiological protection against external radiation* (Pergamon Press, 1996)
8. M. Reginatto, P. Goldhagen, Health Phys. **77**, 579 (1999)

Radiation Monitoring System of 30 MeV Cyclotron

Jin-Woo Lee[1,2], Min-Goo Hur[1] , Gyosung Jeong[2,3] and Jongil Kim[2]

[1]*Advanced Radiation Technology Institute, KAERI, Jeongeup-si, Korea*
[2]*Chonbuk National University, Jeonju-si, Korea*
[3]*Corporation of Sean Technology, Seoul-si, Korea*

Abstract. A state-of-the-art radiation monitoring system was implemented at KAERI for a 30-MeV cyclotron. This system consists of several types of radiation measuring systems for ambient dose equivalent rate measurements of outside photon and neutron areas as well as inside the cyclotron, and monitors the alpha and beta particulates released from a stack, as well as the results of worker contamination at the portal of the cyclotron. In addition, an automatic alarm system is also mounted if there are alarms in the measuring systems.

1 Introduction

ARTI (Advanced Radiation Technology Institute) is one of the research centers of KAERI (Korea Atomic Energy Research Institute). The main role of ARTI is to research the radiation usage for a variety of technologies. ARTI has large radioisotope irradiation facilities, and thus many laboratories using isotopes and a 30MeV cyclotron for research and development. In Korea, many cyclotrons are being operated for hospitals at low energy levels compared to the research cyclotron of ARTI. A 30MeV cyclotron has been utilized for the research of new diagnostic isotope production, and utilization of the beam and target. The main facilities of a 30MeV cyclotron are the main body, target room, hot cell room, and affiliated laboratory. In addition, over 20 researchers are working in the controlled radiation area of the facilities of the cyclotron. Thus, it is necessary to measure periodically the radiation level and inspect the radioactive remains of the facilities of the cyclotron.

2 The need to operate RMS

It is necessary to check the radiation exposure unexpected during the radioisotope production in a cyclotron, and manipulate the isotopes in a hot cell and the contamination of radiation workers after radioisotope handling. In particular, it is very important for the radiation safety staff to monitor the radiation data during an unsealed radioisotope production and usage of the radiation in real time because particulates and gases from the radioisotopes might be vented into the atmosphere.
The facility of the 30MeV cyclotron has several radiation measuring instruments for radiation safety and protection. The on-line radiation monitor instruments that are currently operating in the ARTI cyclotron are gamma and neutron area monitors, stack monitors, and portal monitors. In addition, all of the measuring instruments have their own specific measurement methods, shown in Tables 1 and 2. Each instrument has a separate monitoring system [1].
A variety of communication methods such as Ethernet and serial connections are applied depending on the time of the installation and the instrument manufacturer.

Table 1. Area monitor

Instrument	Detector	Quantity	Communication
Gamma Detector	Proportional counter	6	Serial (RS-232)
Neutron Detector	He-3 counter	3	Ethernet (TCP/IP)
Portal Monitor	Plastic Scintillator	2	Ethernet (TCP/IP)

If radiation safety staff members operate all monitoring systems of the instruments, they cannot do anything but monitor all systems. Thus, the combined radiation monitoring system is essential for the radiation staff because of the limited radiation personnel and budget.
Thus, it is necessary to develop a combined RMS (radiation monitoring system), which includes all measuring data from each instrument for real-time remote monitoring and the measurement information database management of these instruments [2].

Table 2. Stack air monitor

Instrument	Detector	Quantity	Communication
Particulate Monitor	Proportional counter	2	Ethernet (TCP/IP)
Iodine Monitor	NaI(Tl) scintillator	1	Ethernet (TCP/IP)
PET Monitor	Plastic Scintillator	1	Serial (RS-485)

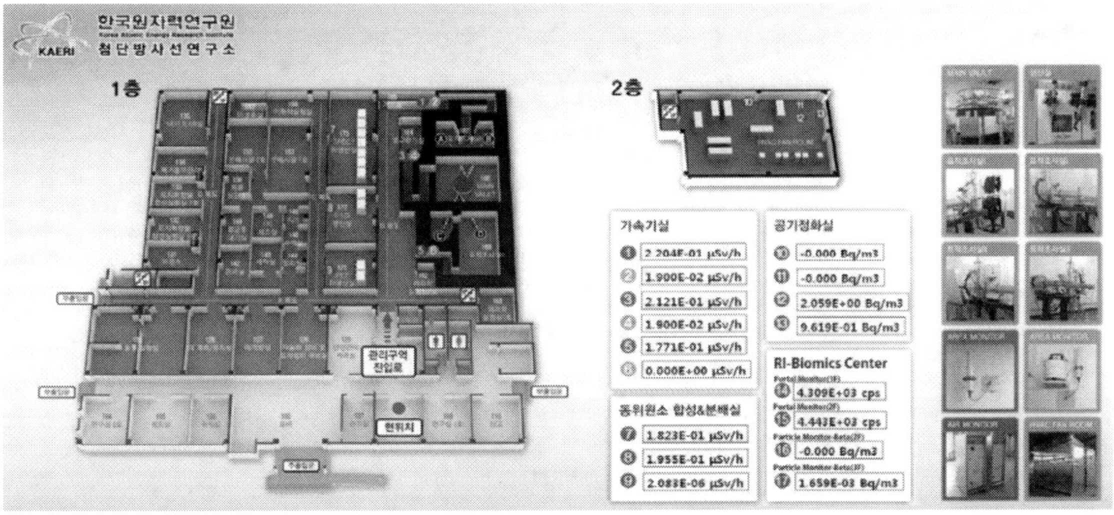

Figure 1. Main Portal Screen of RMS

3 Function of RMS

The configuration of the RMS of a 30 MeV cyclotron is composed of radiation measuring systems, Agent, DBMS (Database Management System) and an administrator program. Agent receives real-time data provided by the radiation measuring systems and communication, and performs the storage the reception of data to the DBMS. In addition, it has a function of setting the raw data from each radiation measuring system in units such as the dose rate (μSv/h), air concentration of radioactive materials being released (Bq/m^3), and the count rate of the portal monitor (cps). In addition, radiation staff members can give each instrument the calibration constants if needed, and be notified through an alarm from each radiation measuring system any transmission error between the RMS and radiation measuring systems in real time.

Figure 2. The basic configuration of RMS

The DBMS uses Microsoft SQL Server 2008 R2 Version. Each instrument is managed by classifying through a data table, and the table of the alarm history has been configured independently to facilitate the expansion of the system. An administrator program collects data in real time on the screen to query the database by the Agent. The main menu of the administrator program utilized by radiation safety manager is the query of real-time data and historical data and the alarm data The report of the administrator program are the measuring data in real-time and for a daily, weekly, and monthly basis. In addition, it is possible for radiation safety managers to make reports for specific periodic doses.

4 Conclusions

The recent developments of radiation measurement technology have occurred faster than before owing to the rapid development of information technology based on the MMI (Man-Machine Interface) technique. In this paper, we introduced the combined RMS as a part of the optimal radiation safety management of ARTI for radiation safety officers and management staff. In addition, it is also the basic infrastructure of the radiation safety program. In particular, the real-time display of the dose rate, air concentration of radioactive materials being released, and the count rate of the portal monitor of the radiation controlled area enables radiation workers and radiation safety staff members to achieve good side effects to check the availability of the radiation measuring instrument remotely on one screen. In the future, the contents of RMS will be applicable to other sites and other organizations within KAERI.

References

1. KAERI Manual of operating radiation measuring instrument (2015)
2. C. Kim, J. Shin, J. Cha J..The journal of Institute of Internet, Broadcasting and Communication (JIWIT 2012-5-32), A Study for Protocol for Heterogeneous Interface in Sensor Networks within Water Restore Facilities (2012)

Radiological Environmental Protection for LCLS-II High Power Operation

James Liu*, Jan Blaha, Maranda Cimeno, Stan Mao, Ludovic Nicolas, Sayed Rokni, Mario Santana, Henry Tran

SLAC National Accelerator Laboratory, Menlo Park, CA, 95125, USA

Abstract. The LCLS-II superconducting electron accelerator at SLAC plans to operate at up to 4 GeV and 240 kW average power, which would create higher radiological impacts particularly near the beam loss points such as beam dumps and halo collimators. The main hazards to the public and environment include direct or skyshine radiation, effluent of radioactive air such as ^{13}N, ^{15}O and ^{41}Ar, and activation of groundwater creating tritium. These hazards were evaluated using analytic methods and FLUKA Monte Carlo code. The controls (mainly extensive bulk shielding and local shielding around high loss points) and monitoring (neutron/photon detectors with detection capabilities below natural background at site boundary, site-wide radioactive air monitors, and groundwater wells) were designed to meet the U.S. DOE and EPA, as well as SLAC requirements. The radiological design and controls for the LCW systems [including concrete housing shielding for ^{15}O and ^{11}C circulating in LCW, ^{7}Be and erosion/corrosion products (^{22}Na, ^{54}Mn, ^{60}Co, ^{65}Zn, etc.) captured in resin and filters, leak detection and containment of LCW with ^{3}H and its waste water discharge; explosion from H_2 build-up in surge tank and release of radionuclides] associated with the high power beam dumps are also presented.

1 Introduction

SLAC National Accelerator Laboratory (SLAC) is designing the LCLS-II hard X-ray free-electron-laser facility, which utilizes a new superconducting (SC) electron linac for operation with a maximum beam of 4 GeV, 240 kW average power and MHz capability. The LCLS-II high power operation will create higher radiological impacts to the workers, public and environment, particularly near the beam loss points such as beam dumps and halo collimators.

The main impacts to the public and environment include direct or skyshine neutron and photon radiation, effluent of radioactive air (e.g., ^{13}N, ^{15}O and ^{41}Ar), and activation of groundwater (mainly tritium). Radiological design and controls for Low Conductivity Water (LCW) systems is one of the main subjects for protection of the workers, public and environment. The assessment, controls and monitoring of these hazards to satisfy the U.S. Department of Energy (DOE) and SLAC requirements are described.

2 Protection of Public and Environment

2.1 LCLS-II facility

SLAC has an existing 3-km-long underground linac tunnel. Figure 1 shows the LCLS-II layout from SC linac along Sector 0 to 10 to beam switching in BSY (via a spreader line), either dumping all or partial of the electrons at the BSY dump or directing beams through two variable-gap undulators (SXU and HXU) to generate hard and soft X-rays. The electrons beams are then bended to two main dumps (EBD). The BSY dump and EBD dumps, as well as halo collimators along linac and BSY, are the main beam loss points.

Table 1 summarizes the normal electron beam losses estimated by LCLS-II [1]. Maximum power and average power (considering use factors) are used to evaluate the radiological hazards to workers and public/environment, respectively.

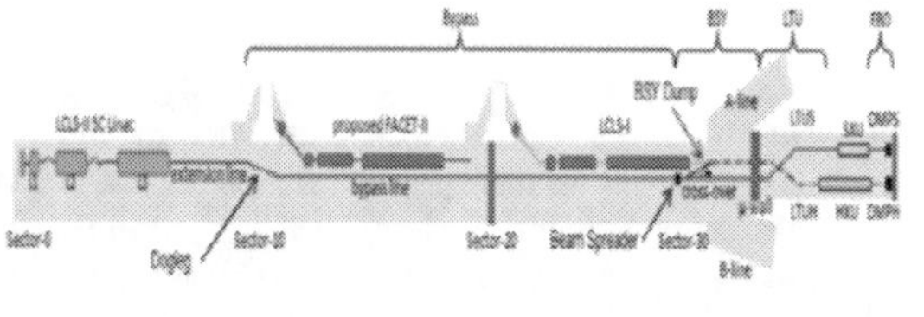

Figure 1. LCLS-II accelerator and beamline layout.

Table 1. LCLS-II main electron beam loss points.

	Linac-BSY	**BSY**	**Dump Hall**
Component	Each Halo Collimator	Dump	2 EBD dumps
Maximum Power	32 W	250 kW	2x120 kW
Average Power	13 W	90 kW	27+48 kW

2.2 Skyshine radiation

Because most of LCLS-II accelerator housings are deep underground, skyshine radiation to the public residing near SLAC site boundary is dominated by high-energy neutrons penetrating the 1.2-1.8 meter thick concrete roofs of the only aboveground section BTH.

With a total beam loss limited to 75 W by SLAC Long Ion Chambers inside BTH, the maximum skyshine dose to the public was calculated to be 6 μSv/y at 500 m from BTH with the SKYSHINE code [2]. This meets the SLAC shielding design goal of 50 μSv/y per facility within SLAC.

The skyshine radiation will be monitored continuously with eight stations around SLAC site boundary with radiation detectors sensitive to detect 1 nSv/h and 10 nSv/h for neutrons and photons, respectively [3].

* Corresponding author: james@slac.stanford.edu

2.3 Radioactive air effluent

The dose to the public from the exposure pathway of radioactive air effluent is dictated by the radioactive air with ^{11}C, ^{13}N, ^{15}O and ^{41}Ar released from the accelerator housing. For LCLS-II, high beam loss points are heavily shielded locally for the protection of groundwater and workers from induced activity. In this case, FLUKA calculations [4] show that the production of radioactive air is dominated by the small air volume between dump (or collimator) and local shielding, that should be minimized. Table 2 gives the saturation activity of radioactive air at LCSL-II, which is dominated by 240 kW beam loss at BSY dump with local shielding [4].

SLAC accelerator housing will have no forced ventilation during beam on, but the housing is not tightly sealed and it will have natural air leakage. With the calculated saturation activity (which can be reached with a few hours of beam on) and the natural air leakage rates, annual activity releases (Bq/y) can be estimated. The doses to the Maximum Exposed Individual (MEI) of public can then be calculated using the DOE-approved CAP-88 analytic code, considering the population data and annual average meteorological data in Bay area [5]. Table 3 shows that the maximum annual dose to the MEI from LCLS-II is 390 nSv/y from BSY dump operation. This is lower than the design goal of 1 μSv/y to MEI from each air release point (such that there are no requirements for continuous air monitoring) and is much lower than the design limit of 100 μSv/y.

Periodic confirmatory measurements of the radioactive air production and concentration inside housing will be conducted in-situ using Eberline AMS-4 noble gas monitors, which can detect down to 0.1 Derived Air Concentration [3] of ^{13}N and ^{41}Ar.

Table 2. Production of radioactive air at LCLS-II.

Saturation Activity (MBq)	Linac Collimator	BSY Dump	EBD Dumps
Maximum Power (kW)	1 (no shielding)	240	2x120
^{11}C	33	225	13.5
^{13}N	57	496	40.7
^{15}O	61	326	18.0
^{41}Ar	26	537	2.41

Table 3. Annual doses to the MEI of public from radioactive air effluent at LCLS-II.

Dose to MEI (nSv/y)	Linac Collimator	BSY Dump	EBD Dumps
Average Power (kW)	1 (no shielding)	90	75
^{11}C	9.3	64	0.57
^{13}N	12.7	111	1.38
^{15}O	3.9	22	0.27
^{41}Ar	13.5	193	0.17

2.4 Groundwater activation

The main concern of radiological impacts to environmental media is the potential activation of soil and groundwater near the beam loss points from high-energy neutrons (over 100 MeV) penetrating through heavy local and housing shielding. The regulatory requirement for groundwater protection is that there should be no impact to groundwater. Therefore, the shielding design goal for LCLS-II is that there should be no detectable activity in groundwater. Because of its high production cross section and high leachability from soil to groundwater, tritium is the main radionuclide of concern for groundwater protection. The design goal [3] is therefore that potential tritium concentration in groundwater should be less than the detection limit of 37 Bq/L for ^{3}H required by the U.S. Environmental Protection Agency.

Extensive local shielding has been added to BSY and EBD dumps to reduce the groundwater activation [6,7]. Due to limited space, the shielding was optimized based on calculations using FLUKA Monte Carlo code.

Figure 2 shows the groundwater activation at the LCLS-II EBD dumps. The groundwater table is located at about 10 m below the dump such that direct activation of groundwater is smaller than the indirect path in which the most irradiated rain water 1-m^3 volume flows at the worst drift speed from earth surface to groundwater table. With a 30% leachability and conservatively assuming no diffusion and no dilution in soil and groundwater, the ^{3}H concentration in groundwater table is estimated to be 26 Bq/L for 75 kW average power. For ALARA purpose, geo-membrane is considered as an option to further reduce the tritium concentration.

Several groundwater monitoring wells will be added near the BSY and EBD dumps, as well as Linac collimators with beam losses more than 10 W, to the existing ~100 wells at SLAC [3].

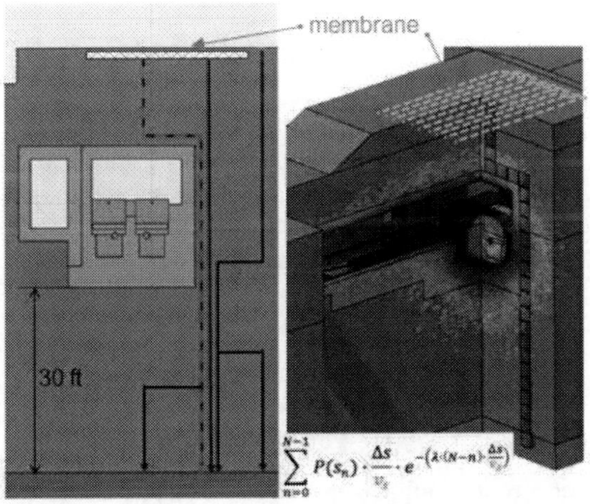

Figure 2. Activation of Groundwater located at 10 m below the LCLS-II EBD dumps.

3 Radiological design and controls for LCW system

A large amount of LCW is needed to cool high-power dumps and collimators, which will activate the LCW and any erosion/corrosion products flowing with LCW. Figure 3 shows a typical LCW system for a high power dump at SLAC, which include the LCW and its piping, surge tank, resin tanks, filter bottles, heat exchanger inside a concrete housing [8]. Main radiological hazards for a LCW system that need to be addressed include: ^{3}H in water, ^{15}O and ^{11}C in water, ^{7}Be and erosion/corrosion products (^{22}Na, ^{54}Mn, ^{60}Co, ^{65}Zn, etc.) captured in resin and filters, and potential explosion from H_2 build-up in surge tank. The mitigation measures include [8]: shielding of concrete housing and underground piping from ^{15}O and ^{11}C in LCW, shielding of concrete housing from ^{7}Be and erosion/corrosion products in resin/filters, access controls to housing (locked door and interlocked to radiation monitoring), detection and containment of LCW leakage, management of discharge of waste water with ^{3}H into sewer, using hydrogen recombiner to reduce H_2 build-up, and utilization of remote or automatic systems for the exchanges and disposal of resin and filters to minimize worker exposure and accelerator operational impacts.

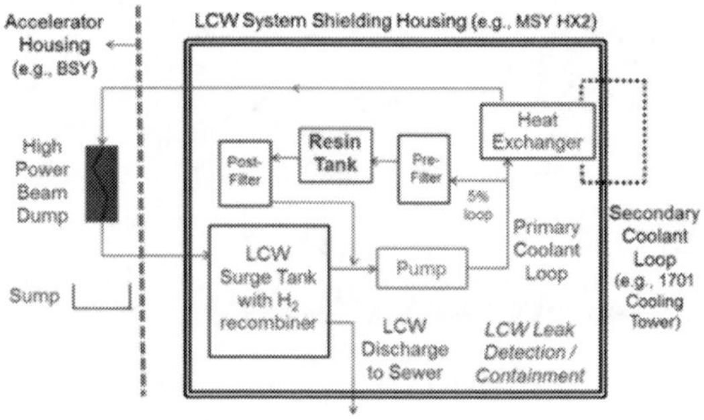

Figure 3. LCW system for a LCLS-II high power dump.

Table 4 summarizes the induced activities in the LCW system components for BSY and EBD dumps, estimated using FLUKA code or SLAC empirical data [9]. The BSY dump has small aluminum balls mixed in 39-liter of water followed by a solid Cu end-cap. Each EBD dump is an aluminum core plus tungsten end-cap that is surface cooled with 6.6-liter water. It is clear that the BSY dump (a volume-cooled dump) will have much higher induced activity than that of the EBD dump (a surface-cooled dump); ~150 times in this design.

To meet the shielding design goal of 1 mSv/y, the concrete housing needs to have 60 cm concrete walls (45 cm ceiling) and the underground pipes need to shielded with 100 cm soil; which is dictated by ^{15}O and ^{11}C flowing with LCW, that would decay within 1-2 hours after beam is off. For erosion/corrosion products (dictated by ^{22}Na or ^{60}Co) and ^{7}Be captured in resin, 45 cm concrete walls are needed for shielding when workers access the housing after beam is off. Other key LCW components include a locked door interlocked with a photon radiation detector, three 100-gallon resin tanks with remote resin exchange features, vented surge tank with hydrogen recombiner, LCW pressure sensors to detect leak and sump to contain leak, and a controlled, direct LCW discharge piping to sewer.

Table 4. Induced activities in LCW system components.

	BSY Dump	**EBD Dumps**
^{3}H saturation activity	510 GBq at 90 kW	3.3 GBq at 76 kW
^{7}Be saturation activity at 250 kW	200 GBq	1.4 GBq
^{15}O and ^{11}C saturation activity at 250 kW	2850 GBq ^{15}O, 700 GBq ^{11}C & 0.56 Sv/h at 1 m	19 GBq ^{15}O, 3.7 GBq ^{11}C & 3.5 mSv/h at 1 m
^{22}Na, ^{54}Mn, ^{60}Co, ^{65}Zn maximum activities estimated from historical data of similar LCW systems at SLAC	5 mSv/h at 1 m from two 400-L resin tanks (equivalent to 15 GBq ^{60}Co)	50 μSv/h at 1 m from two 400-L resin tanks (equivalent to 0.15 GBq ^{60}Co)

4 Conclusions

LCLS-II beam power is at least 2 orders of magnitude higher than LCLS-I. Radiological protection for the public, environment and workers is more challenging, considering that there are less safety margin and the constraints of building LCLS-II on existing LCLS-I accelerator housing and infrastructure configuration. The main issues of hazard evaluation and controls for BSY dump, EBD dumps, and halo collimators are being addressed to meet the regulatory and SLAC requirements.

References

1. J. Welch, SLAC LCLS-II Physics Requirement Document, LCLSII-2.7-PR-0079-R1, (2015)
2. S. Mao, SLAC Memo, RP-RPG-131105-MEM-01, (2013)
3. J. Liu, SLAC LCLS-II Physics Requirement Document, LCLSII-1.2-PR-0265-R0, (2014)
4. L. Nicolas, SLAC Memo, RP-RPG-140806-MEM-01, (2014)
5. H. Tran, SLAC Technical Note, RP Note 15-05, (2015)
6. L. Nicolas, SLAC Technical Note, RP Note 15-06, (2015)
7. M. Santana, SLAC LCLS-II Physics Requirement Document, LCLSII-1.2-PR-0100-R0, (2014)
8. J. Liu, SLAC LCLS-II Physics Requirement Document, LCLSII-1.2-PR-0211-R0, (2014)
9. J. Blaha, SLAC Technical Note, RP Note 15-19, (2016)

*This work is supported by the U.S. DOE, Office of Science, Office of Basic Energy Sciences, under Contract No. DE-AC02-76SF00515.

 DOI: 10.1051/epjconf/201715303015

Radiation containment at a 1 MW high energy electron accelerator:

Status of LCLS-II radiation physics design

M. Santana Leitner[1,a], J. Blaha, M. W. Guetg, Z. Li, J. C. Liu, S. X. Mao, L. Nicolas, S. H. Rokni, S. Xiao, and L. Ge

[1] SLAC National Accelerator Laboratory, 2575 Sand Hill Road, 94085-CA, USA

Abstract. LCLS-II will add a 4 GeV, 1 MHz, SCRF electron accelerator in the first 700 meters of the SLAC 2-mile Linac, as well as adjustable gap polarized undulators in the down-beam electron lines, to produce tunable, fully coherent X-rays in programmable bunch patterns. This facility will work in unison with the existing Linac Coherent Light Source, which uses the legacy copper cavities in the last third of the linac to deliver electrons between 2 and 17 GeV to an undulator line. The upgrade plan includes new beam lines, five stages of state of the art collimation that shall clean the high-power beam well up-beam of the radio-sensitive undulators, and new electron and photon beam dumps. This paper describes the challenges encountered to define efficient measures to protect machine, personnel, public and the environment from the potentially destructive power of the beam, while maximizing the reuse of existing components and infrastructure, and allowing for complex operational modes.

1 Introduction

LCLS-II will bring high-power electron operations back to SLAC. Such endeavor entails a broad span of challenges, some of which are presented here. The addition of superconducting cryomodules means potential generation of field-emission dark current, the mitigation of which needs to be designed by using developing and applying new models. Also LCLS-II is expected to have higher beam losses than its predecessor LCLS, and those will need to be cleaned to ensure its undulators are not damaged by radiation, which means that dark current will be strategically shredded in other areas, where shielding will be required. Not only will normal losses be higher, but also accidents can be more severe, and thus, shut-off mechanisms ought to be fast, and containment has to be designed to capture any high-power errant beams. Finally, the machine will use existing buildings, many of which were designed for low power beams, and which typically cannot be retro-fitted, or have no room for additional local shielding.

2 Beam Losses

2.1 Field emission from superconducting cavities

The high-power of LCLS-II electron beam will be supplied at 35 cryomodules distributed in four cold sections, and separated by bunching chicanes with halo cleaning collimators. Each cryomodule has eight nine-celled niobium cavities followed by a quadrupole, all cooled with liquid helium at ~2 K. The first cryomodule will operate at a high frequency of 3.9 GHz, while all others will run at 1.3 GHz. The nominal acceleration at cryomodules is set at 16 MV/m, although higher gradients may be used if cavities are not found to emit substantial dark current.

Because the aperture through the cold sections is large (70 mm), normal beam losses are not expected there. However electrons field-emitted in high gradient RF areas of the cavities will hit the iris of the cavities and other components, sometimes after having captured sufficient energy to the generate subsequent showers, thereby potentially damaging superconducting components, insulators, etc. Such effects were extensively studied [1] by recursively transferring field-emission and EM transport data and the generated radiation transport cascades between Track3P [2] and FLUKA [3, 4] on a 3D model with up to nine fully detailed back-to-back cryomodules, fig. 1.

It was found out that for nominal 10 nA/cryomodule 'captured' field emission, i.e. dark current generated in a cryomodule and escaping at each of its ends, prompt dose rates at the upper Klystron Gallery[1] were insignificant, while residual dose rates during access to the linac aisle after one hour cool-down would be just under the 50 μSv/h goal, meaning that the Linac aisle is not expected to fall into the *Radiation Area* classification.

As for *machine protection*, results of energy deposition at the niobium cavities indicate that heat generation from field-emitted electrons hitting the cavities will be negligible towards RF heating, and also that fundamental alterations to the crystalline structure of Nb should not be expected, as peak dose levels are far below any of the established damage thresholds for metals. Regarding sur-

[a] e-mail: msantana@slac.stanford.edu

[1] For LCLS-II the Klystron Gallery will host the Solid State Amplifiers, which will supply the RF energy to the cryomodules through waveguides that will use existing vertical penetrations into the Linac.

Figure 1. Flair [5] rendering of the FLUKA model for nine full cryomodules. Outer total fluence (un-normalized) from field-emission is overlaid in the vertical plane. The dose pattern on a cavity is shown in the lower right corner.

vival of cables, of insulators and of electronic components which could be installed in several positions within the cryomodules or outside in the Linac, a new function for FLUKA (SIDO) was developed [6] to compute *absorbed dose fields*, i.e. maps that display what the absorbed dose would be in each voxel should a light material be placed any of those positions. That function, which is folded real-time with the fluence histogram in a similar fashion as how biological dose is computed via run-time effective dose conversions, can be set for millimetric polyethylene type objects such as cables and insulators or for sub-millimetric silicon, representing electronics.

The resulting absorbed dose maps indicated that certain light materials (less robust to radiation) shall be avoided near the cavities. Indeed dose rates may exceed 100 kGy/y in some locations inside the cryomodules, while the level drops below 30 kGy/y under the cryomodules and to 0.2 kGy/y above those due to the attenuation provided by the large central Helium pipe above the cavities.

Two findings were made for the superconducting quadrupoles. First, its coils are expected to last for more than 20 years of operation. Second, their magnetic rigidity was found to play a major role in the distribution and intensity of field emission losses. At medium energies (~700 MeV), dose rates in the cryomodules from field emission induced showers are highest, while maximal transmission of field-emitted electrons to warm sections occurs for switched off quads, reaching integrated intensities of 25 W/nA for a string of 20 cryomodules. Such dark current beam resulting from accidentally powering off all cryomodule quadrupoles, would be cleaned up in the collimation sections, which also remove other undesired terms, as described next.

2.2 Additional beam loss terms: collimation shielding

LCLS-II halo collimation base-line design includes a first energy collimator followed by four transverse collimators (in bunch compressors), a second energy collimator in the dogleg, more transverse collimators in the by-pass line and both energy and transverse collimators in the transfer lines to the undulators. LCLS-II collimators do not have a line of sight with down-beam undulators, thereby reducing irradiation of those by secondaries.

Energy collimators are especially effective in the upstream end of the machine due to the large energy spread of the gun dark current, while down-beam collimators mainly serve to suppress gas-scattered electrons in the by-pass line. ASTRA [7] was used to simulate the gun dark current up to the first cryomodule, and ELEGANT [8] from there on [9]. No dark current from either the cathode or cavity field emissions has been found to lead to appreciable losses at the undulator, according to Top-up tracking. Intra-beam scattering losses is below half a milliwatt in the undulator for the worst conditions (300 pC/120 kW). Currently, higher order field errors are being studied, suggesting only small corrections.

With respect to LCLS, the effective collimator thickness has been increased to 15 radiation lengths by selecting Tungsten, rather than Titanium as material, based on Monte Carlo simulations of the collimator inefficiency.

Once the locations, sizes, materials and beam losses at the collimators had been established it was time to evaluate what would be the required shielding around every collimator. The concerns that needed to be addressed were the residual dose rate near those at Linac access times (one hour cool-down), and the radio-activation of soil and groundwater surrounding the Linac. The first criteria dictates the shielding between the collimator and the most accessed areas (i.e. aisle side, and above or below the collimator if the beam-line is low or high, respectively), while the second rules the shielding between the collimator and the closest/thinnest walls. A summary of these aspects, is presented below.

Simulations for the various energy/power combinations with local shielding of several thicknesses were performed to find the shielding requirements in terms of *local activation*. Results showed that at 98 MeV, beam losses of 32 W would need to be shielded by a 8-cm thick *cast iron* plate, while this thickness could be reduced to 6 cm by using *lead* plates. Also it was observed that, regardless of beam energy, below 20 W losses, shielding is not generally required. However, most collimators should be equipped with thin wire meshes or plastic barriers at 40 cm from the jaws to keep personnel away from the relatively high residual dose rates on the collimator surfaces, even for low loss rates. Also, it should be pointed out that *cast iron* residual dose shielding plates should *not* be replaced by some steel alloy, since some of the components of the latter could even render residual dose rates in the tunnel higher that they would be without any local shielding.

Tritium concentration at the soil and groundwater adjacent to the Linac is mostly due to spallation reactions induced by high-energy neutrons in silica and oxygen. Here we summarize the method that was used to compute the amount of environmental local shielding to contain the leaking neutrons from each of the halo collimators of LCLS-II.

First, the *production rate* of various radio-isotopes was simulated in FLUKA for a broad array of configurations, including several target materials, beam energies, and local shielding media / thicknesses. As expected, the *attenuation function for the production of radio-isotopes* seemed to be a decaying exponential function of the shielding thickness. However, a single exponential was not enough to properly fit the functions. Instead, two exponentials should be used to better describe the different attenuations of high and mid-energy neutrons. Equation 1 links the *effective attenuation for radiation to generate a given radioisotope* (α) with the thickness (r) and density (ρ) of the shielding through three parameters c, λ_H and λ_M:

$$\alpha(\rho, t) = (1 - c) \cdot exp\left(-\frac{\rho}{\lambda_H} \cdot r\right) + c \cdot exp\left(-\frac{\rho}{\lambda_M} \cdot r\right) \quad (1)$$

For all simulated combinations of incident beam energy, target material, shielding media and scoring media (soil or groundwater), the coefficients of equation 1 were found by fitting the simulated data to the functions. Fitting was carried out using an algorithm based on genetics, where mutations occur towards solutions that improve the outcome, i.e. reduce the fitting error. The formulae were then inverted numerically in order to obtain the required shielding thickness. In that process, the attenuation factor was modified to account for isotope leaching from soil to groundwater, as well as radioactive decay along the percolation path to the water-table. Since the distance between the aquifer and the tunnel fluctuates along the linac, two extreme scenarios of 3 m and 30 m depth, with associated high (30%) and low (10%) were considered. Moreover, direct activation of groundwater was also verified, as well as radio-activity at the soil itself 10 years after the facility final shut-off (for decommissioning).

Figure 2 summarizes the parameters relevant for calculations and the four applied criteria. It should pointed out that for each collimator, (1) direct activation of *soil* was verified in all four transverse directions, (2) direct activation of the *aquifer* and (3, 4) indirect activation of the aquifer through percolating humidity were taken into account for all directions except for the *ceiling*.

The logic and equations described above were implemented in a program that performed all calculations for a table listing the conditions of figure 2 for every halo collimator. Results can be summarized as follows. For 45 cm thick walls at 1 m from the collimator jaws, an environmental protection shielding between the two is necessary above 8 W beam losses on low-Z jaws such as aluminum (this was one of the materials which had been initially considered), while for a heavy-Z material like tungsten, beam losses need to exceed 50 W at 98 MeV and 15 W at 4 GeV in order for the environmental shielding to be necessary.

Other related aspects such as activation of surrounding air and of the collimator cooling water are described in [10].

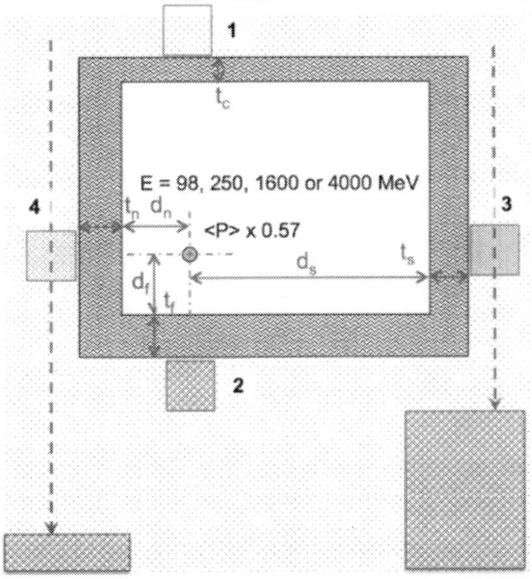

Figure 2. Scheme of LCLS-II linac showing the *d*istance of the collimator to the walls, the *t*hickness of each of the *c*eling, *f*loor, *n*orth and *s*outh walls, the respective scoring volumes for radio-isotopes and the transport decay paths. The orientation of the Linac is *west* → *east*

3 Mis-Steering. Fast beam shutoff and burn-through monitors

Besides normal beam-losses, beam can be accidentally steered out of the beam-stay clear, thereby hitting collimator jaws, magnets and other components. Such failures may damage equipment and, if unmitigated, they could also lead to high dose rates in some areas. Thus, several protective layers of engineering controls (Machine Protection System, Beam Containment System and Personal Protection System) are being designed or re-designed, including changes or additions to the current available instrumentation detectors, software and hardware architecture, policies, shut-off schemes, etc. Two outstanding examples of ongoing studies are presented in this section.

3.1 Hazard shut-off, transient thermal analysis

Because of the high beam power, hazardous conditions should be shut-off very rapidly, as otherwise BCS hardware such as *protection collimators*, *stoppers*, etc. may fracture or burn-through in the course of the event or at a later similar failure, thus not offering their presumed protection.

In order to examine what is to be demanded from detectors and shut-off mechanisms, energy deposition and heat transfer calculations have been performed for several beam-power levels, target materials and geometries, as well as beam irradiation conditions. For that purpose, an FEA code (FHeat3D) was developed to work in unison with FLUKA-generated dose maps, allowing for quick turnout of results even for relatively non-standard heat simulations that included pulsed source terms, complex power distributions, variable conditions, etc. From those, it was concluded that at maximum beam powers of 2 MW (this is the 'so-called' *maximum credible beam for LCLS-II*), shut-off should occur within 200 μs to avoid excessive stress in the irradiated safety component, while for lower beam powers, shut-off was driven by burn-through considerations, e.g. 1 s response to avoid melting with 1 kW beams.

3.2 Establishing and trimming the phase-space of mis-steered events

The phase-space of errant electron orbits compatible with magnet failure scenarios in different parts of LCLS-II is being projected over the accelerator layout to define where supplemental shielding panels and protection collimators are needed to contain the powerful beam. In the past such task would have been performed 'manually' by computing analytically or graphically the maximum kicks from dipoles and from other strong magnets.

Here we have applied the concept outlined in [12], based on Monte Carlo tracking of events over controlled magnet perturbations. Once configured, such method is flexible to changes and it can cope with the complex layout of beam-lines in LCLS-II, accounting also for secondary effects (e.g. over-powered dipole over-kicks the beam and well-behaved down-beam defocusing quadrupole sends it out of the beam-line due to the excessive transverse offset), incorporating pre-existing alignment tolerances and even imposing linked failure restrictions due to common power supplies.

This technique has been paired with MadFLUKA code [11], which converts MAD survey output files into a beam-line geometry file that can be incorporated in FLUKA, and also generating the *nominal* optics. The optics has then been modified to reflect the *failure* logics so that ray-traces can be drawn over the existing accelerator geometry by simulating a high number of events.

Figure 3 shows the phase-space of mis-steered rays in the beam-switchyard region for the three beam-lines in that area. In that simulation, custom latching and scoring were coded to identify which magnets caused beams to reach a specific section of the geometry. For example, the non-allowed events, marked with a red box, come from over-powering the dipole circled in green. In that case a collimator of optimized dimensions was immediately prescribed.

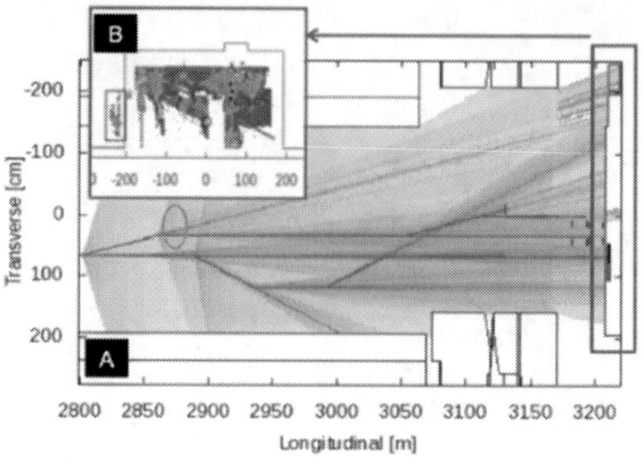

Figure 3. Ray-Trace for LCLS-II beam switchyard. **A)** Top projection of all rays, and **B)** cross-section projection at the end of BSY.

4 Re-use of current facilities

LCLS-II will mostly fit as-is in the existing SLAC infrastructure, i.e. the historic 2-mile linac followed by the beam transfer hall (BTH), undulator hall (UH), main dump hall (MDH) and by the experimental halls that were built for LCLS. The latter sections were designed for 5 kW beams, while LCLS-II will send up to 240 kW through the same enclosures. Moreover, although the Linac is underground and has occasionally hosted high-power beams, LCLS-II accelerator will be installed near the vertical plane of most penetrations to the Klystron Gallery, which means that beam-losses in that segment of the machine could lead to substantial exposure in the upper occupied area.

4.1 Linac penetrations

There are about 450 penetrations and shafts connecting the Klystron Gallery where personnel will be present, to the underground linac, 8 m below. The axis of some penetrations are almost aligned with the future vertical beam plane, and therefore they offer a direct line of sight for radiation to stream upwards. Moreover, most of these penetrations cannot be plugged as waveguides, helium supply/return lines and cables will go through them, and because some will be used to vent the Linac in case of accidental helium spills during access.

Thus, shielding configurations blocking just a fraction of the open penetration were studied. Simulation tests were numerous and included solutions with partial filling of the penetration cross-sections, or with local shielding over the beam-line, or with shielding caps at the klystron gallery, as well as combinations of the three concepts.

Finally, the last approach, i.e. cylindrical caps at the Klystron Gallery with exit routes for cables and waveguides, were adopted as the shielding solution for most used penetrations. The caps, shown in figure 4 are 1 m in diameter and have ~22 cm of iron plates and ~30 cm of concrete. With that shielding, accidental point-type full beam losses of up to 2 MW by any penetration would lead to dose rates in the gallery below 250 mSv/h, which is the design Limit at SLAC. For normal beam losses at a rate of 1 kW (this limit is controlled by the BCS), dose rates would not exceed the 5 μSv/h design limit for the Klystron Gallery.

It should be pointed out that dose rates fall very rapidly with distance from the penetration axis, which means that personnel performing any kind of installation or service work at the Klystron Gallery (there are no offices in that area), should not get any appreciable dose.

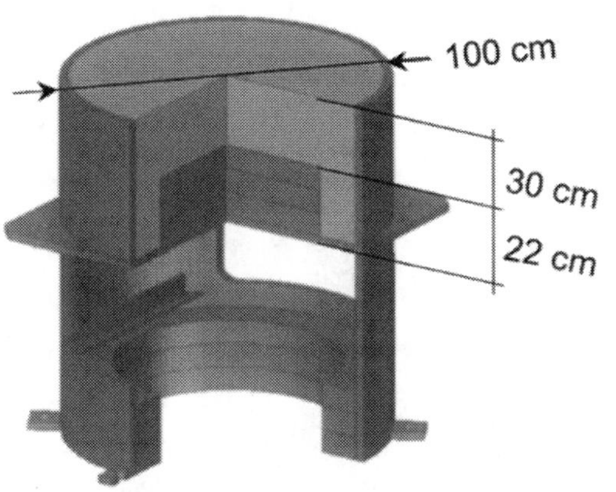

Figure 4. Shielding caps for used penetrations

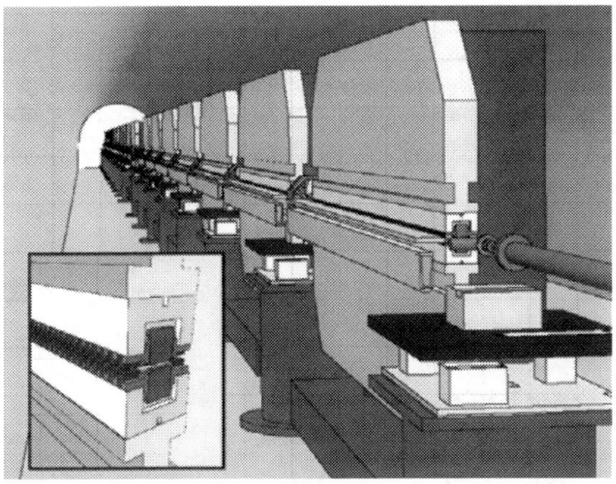

Figure 5. Detailed implementation of LCLS-II SXR for undulator damage studies

4.2 Overground beam transfer hall

The Beam Transfer Hall (BTH) crosses SLAC research yard at ground level. This building, made of reinforced concrete in 2006 for LCLS transfer line, has 1.8-m thick walls and a 1.2-m thick roof (except under the three service buildings, where it is also 1.8-m thick). Long ion chambers inside the building would shut the beam off for losses in excess of 5 W in any 35 m section. This protection guarantees dose rates by the external walls under 5 μSv/h.

In LCLS-II normal beam losses are expected to top at about 1 W/m, and in accident cases, up to 2 MW could be lost at a point. Instead of retrofitting the long building accordingly, fences at 5.5 m from the outer walls will be installed, equipped with gates with access controls. Access to the service buildings will be restricted to LCLS operation only, and tight monitoring will be deployed to ensure that sky-shine dose to other buildings is not significant.

4.3 High-power beam dumps

Three high-power beam dumps rated at 250 kW (1) and 120 kW (2) will be installed at the beam switchyard and at the end of the undulator lines.

As for the 250 kW dump, it will be installed inside of the existing 17-m long iron wall (the so-called muon shielding) that separates the switch-yard from the BTH. Samples of its (old) shielding blocks were sent out for chemical analysis, as being such a thick shielding, realistic composition becomes essential to predict the residual dose rate for access of personnel during beam-off, as well as to compute the neutron fluence outside of the building enclosure. Moreover the muon wall will require limited reconfiguration to host the dump, including features that ease eventual dump retrieval and substitution, with minimal exposure to personnel and without generating contamination.

The two 120 kW dumps will be installed in the existing LCLS pits at the end of the narrow main dump-hall (MDH). Due to the substantial average power increase that represents LCLS-II, the walls would need to be augmented to prevent groundwater activation by high-energy neutrons generated at the two dumps, but neither is there free space inside MDH, nor is it feasible to patch the walls from the outside, as this building is underground. Instead, the location of the dumps within the pits was shifted, i.e. they will sit 30 cm higher and closer to the aisle. This will be achieved by lifting and rolling the dump-lines. More details about this can be found in [10].

The average power on the dumps could be allowed to raise (e.g. in case of future Linac upgrades) if some amount of the steel in the dump pits is replaced by a tungsten alloy. Such upgrade should be performed before any beam is sent to the high-power dumps, as in the future the area will be too activated to perform those works.

Substantial shielding will be added on top of the dumps, including not only steel but also an intermediate plate of concrete that will limit activation of MDH. With these modifications, the residual dose rate inside MDH is expected to remain below 50 μSv/h after one hour cool-down time while the prompt dose to nearby accessible areas will be a 100 times lower.

5 Further studies, machine protection vs. radiation protection

In a high-power machine like LCLS-II there are many cross implications between machine protection and radiation protection. Indeed, high-power beams can damage and destroy machine components, and such incidents are not only costly, but also they may lead to temporary high prompt dose rates and also they may entail collective doses during the repair or replacement of those components. Examples of this are safety devices like stoppers and dumps, where radiation protection considerations are part of the design, and also components like kickers/septa, in which excessive beam losses may not only generate radiation, but could also alter their magnetic performance, thereby leading to higher losses. Finally, in FEL machines, collimators and undulators need special attention, as beam losses in the first are desired to avoid damage of the latter. Extensive simulations are required to verify that high-power can be handled and shielded in the collimators and that doses to the fragile undulators are minimal (fig. 5).

Acknowledgement

This work was supported by U.S. Department of Energy contract DE-AC02-76SF00515.

References

[1] M. Santana, L. Ge, Z. Li, C. Xu, C. Adolphsen, M. Ross, M. Carrasco, *Studies of Radiation Fields of LCLS-II Super Conducting Radio Frequency Cavities*, International Journal of Modern Physics: Conference Series, Vol. **44** (2016) 1660209; DOI: 10.1142/S201019451660209X

[2] K. Ko et. al., *SciDAC and the International Linear Collider, Petascale Computing for Terascale Acelerator*, Proceedings of SciDAC 2006, Denver, Colorado

[3] T.T. Böhlen, F. Cerutti, M.P.W. Chin, A. Fassò, A. Ferrari, P.G. Ortega, A. Mairani, P.R. Sala, G. Smirnov and V. Vlachoudis, *The FLUKA Code: Developments and Challenges for High Energy and Medical Applications*, Nuclear Data Sheets **120**, 211-214 (2014)

[4] A. Ferrari, P.R. Sala, A. Fassò, and J. Ranft, *FLUKA: a multi-particle transport code*, CERN-2005-10 (2005), INFN/TC_05/11, SLAC-R-773

[5] V. Vlachoudis, *FLAIR: A Powerful But User Friendly Graphical Interface For FLUKA*, Proc. Int. Conf. on Mathematics, Computational Methods & Reactor Physics (M&C 2009), Saratoga Springs, New York, 2009

[6] M. Santana Leitner, *A method to efficiently simulate absorbed dose in radio-sensitive instrumentation components*, Journal of Instrumentation YYY

[7] K. Flöttmann, *ASTRA Manual*, http://www.desy.de/~mpyflo/Astra_manual/

[8] M. Borland, *Elegant: A flexible SDDS-compliant code for accelerator simulation*, Technical Report, Argonne National Lab., 2000, http://www.osti.gov/bridge/product.biblio.jsp?osti_id=761286

[9] M. Borland, *Elegant: A flexible SDDS-compliant code for accelerator simulation*, Technical Report, Argonne National Lab., 2000, http://www.osti.gov/bridge/product.biblio.jsp?osti_id=761286

[10] J. C. Liu et al, *Radiological Environmental Protection for LCLS-II High Power Operation*, these proceedings

[11] M. Santana-Leitner, Y. Nosochkov and T. Raubenheimer, *MadFLUKA Beam Line 3D Builder. Simulation of Beam Loss Propagation in Accelerators*, Proceedings of the 5th International Particle Accelerator Conference, **MOPME040** / ISBN 978-3-95450-132-8, Dresden (2014)

[12] M. Santana Leitner, T. Liang, *Monte Carlo simulation of beam mis-steering at electron accelerators*, Proceeding of SATIF-11: Shielding Aspects of Accelerators Targets and Irradiation Facilities, Tsukuba (2012)

Radioprotection shielding for neutrons induced by the reaction (^{2}H (40 MeV), ^{12}C)

M. Fadil[1,a] and SPIRAL2 team collaboration

[1]*Grand Accélérateur National d'Ions Lourds, Bd Henri Becquerel, 14076 Caen, France*

Abstract. In the framework of design studies for SPIRAL2, the simulation of the neutron flux generated by 40 MeV deuterons on a thick ^{12}C target was performed and compared to experimental data. The calculation of the dose rate of these neutrons allowed to compare four materials being considered for radioprotection shielding: barites, gypsum, ordinary concrete and heavy concrete. The simulated map of the neutron dose rate in the production building shows a very high dose rate around the neutron source and in the environment of some of the accelerator equipment.

1 Introduction

SPIRAL2 is a major extension of GANIL. It was designed to produce high quality beams of exotic nuclei for several applications: fundamental nuclear physics (especially far from stability), astrophysics, isotope production for medical imaging and radiotherapy, material properties, and radioactive waste studies [1]. With its high-intensity neutron beams, this new facility will also be ideally suited for applications that require high neutron flux in the range from 1 to 40 MeV. Such studies include the measurement of induced neutron cross sections with different techniques namely time-of-flight, materials irradiation and fission-yield experiments.

SPIRAL2 adds a linear accelerator to the already five existing cyclotrons in use at GANIL. This new high-current accelerator is dedicated to the acceleration of light ions (up to 20 MeV/u) and heavy ions (up to 14.5 MeV/u) [2,3,4]. Irradiated by 40 MeV deuterons with an intensity of 5 mA, a graphite converter will be used as a neutron source of high flux. Located in the production building, the converter is coupled to a uranium carbide target (UC_2) which will produce high rates of fission fragments [5].

Safety studies and the definition of the radiological environment of the new facility require a detailed study in order to quantify the impact of these neutrons and their behaviour in the attenuation of the shielding. We investigated these issues using a simulation of related nuclear reactions and neutrons transport as well as by the modelling of the production building and some of its components.

2 Neutron production

The 40 MeV deuteron beam from the linear accelerator will nominally be stopped in a 4 mm thick graphite converter. After deuteron stripping, neutrons emerge as a continuous spectrum.

A neutron distribution measurement was performed by Lhersonneau et al. at the Physics Department in Jyväskylä, Finland, on behalf of the SPIRAL2 project. In this experiment, several foils (Al, Ni, Co, In and Bi) were used to generate neutron reactions with different thresholds, which allowed the unfolding of the neutron spectra. Figure.1 , taken from Ref. [6], shows the spectra.

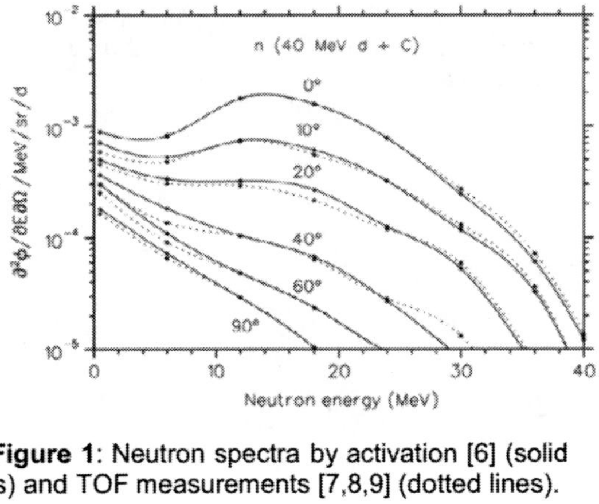

Figure 1: Neutron spectra by activation [6] (solid lines) and TOF measurements [7,8,9] (dotted lines).

[a] Corresponding author: fadil@ganil.fr

DOI: 10.1051/epjconf/201715303016

A comparison between these measurements and an MCNP calculation performed by the author of this abstract showed a ratio MCNP/experiment of 1.45 for the neutron yield generated by the interaction of deuterons with the graphite converter within a cone of half-angle of 50° and above 4 MeV (figure 2). The experimental flux is namely 0.0096 per deuteron and the MNCP value 0.0139. We took into account this conclusion in the reconstruction of the neutron source in all MCNP modelling of the shielding studies.

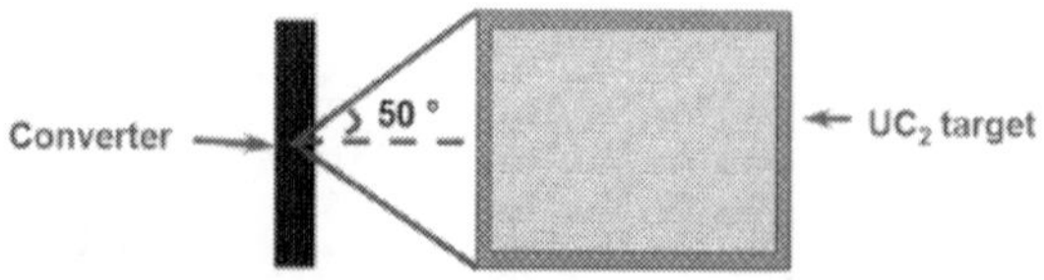

Figure 2: Schematic layout of the converter and the uranium carbide target.

3 Neutron attenuation

3.1. Configuration at 0° angle

Several materials, potentially interesting for our case, were chosen to study the attenuation of these neutrons in the protection shielding: barites, gypsum, ordinary concrete and heavy concrete. The characteristics of these materials are provided in table 1. The calculations were performed with the MCNP code [10].

We calculated the neutron dose rate throughout a shielding protection situated at 300 cm along the beam line axis in the front of the graphite converter and within 300 cm of thickness (figure 3).

We refer to the attenuation factor, μ, as the absolute value of the negative exponential slope ($\Phi(x) = \Phi_0 e^{-\mu x}$) of the curves in figure 3. The μ values in 10^{-2} /cm are 4.5 (barites), 7.6 (gypsum), 6.2 and 7.1 for normal and heavy concretes, respectively. Both gypsum and heavy concrete seem to be excellent materials for the attenuation of neutrons.

Table 1: Composition and density of the shielding materials.

material (density)	Composition (Ratio by mass (%))
Barites (2.6)	Ag(3.2e-4);Ba(53.9);Cd(4.0e-4);Ce(2.2e-3);Cu(1.5e-2);Fe(1.4);Mn(4.7e-2);P(2.3e-2);Pb(4.7e-2);S(12.7);Sb(2.6e-3);Si(2.1);Sr(1.2);Ti(1.2e-2);Zn(5.6e-2);O(28.5)
Gypsum (2.5)	H(2.3);O(55.8);S(18.6);Ca(23.3)
Ordinary concrete (2.3)	H(0.5);C(2.2);O(50.9);Na(1.0);Mg(0.5);Al (3.8);Si(26.7);Cl(0.3);K(1.1) ;Ca(10.9) ;Mn(0.03) :Fe(1.8) ;Pb(0.2)
Heavy concrete (3.3)	H(0.2);Na(36.7);Na(0.01);Mg(0.25);Al(2.2);Si(9.5);P(0.64);S(0.15);Cl(0.007);K(0.7);Ca(3.7);Ti(0.2);Mn(0.1);Fe(45.6)

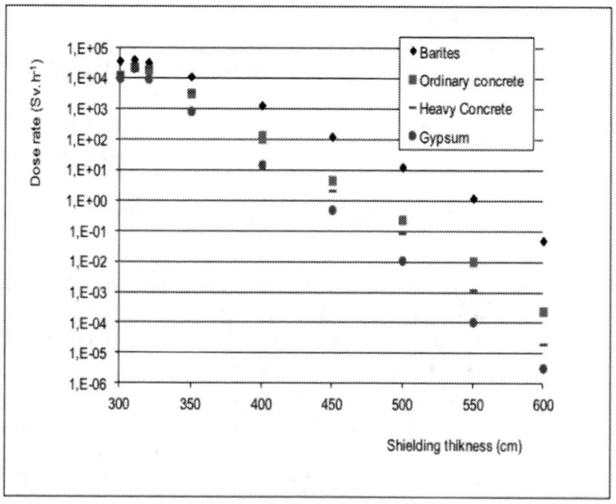

Figure 3: Calculated neutron dose rate and its attenuation within various shielding materials.

3.2 Canal configuration of the production building

A canal configuration was one of the options that were studied for the production building of SPIRAL2 [figure 4]. This building would use primarily ordinary concrete. The MCNP modelling of the building as well as some of its components (quadrupoles, diaphragms, dipole, graphite converter, uranium target) was used to obtain a neutron dose map of the whole building [figure 5].

The results show that canal 1 is the place where the dose reaches its highest rate: ~10^4 $Sv.h^{-1}$ around the neutron source and 6 10^3 $Sv.h^{-1}$ in the middle of the canal. The dose rate in canal 3 is about 10^3 $Sv.h^{-1}$ while it is about 5 10^3 $Sv.h^{-1}$ in canal 2 where some equipment is planned to be installed.

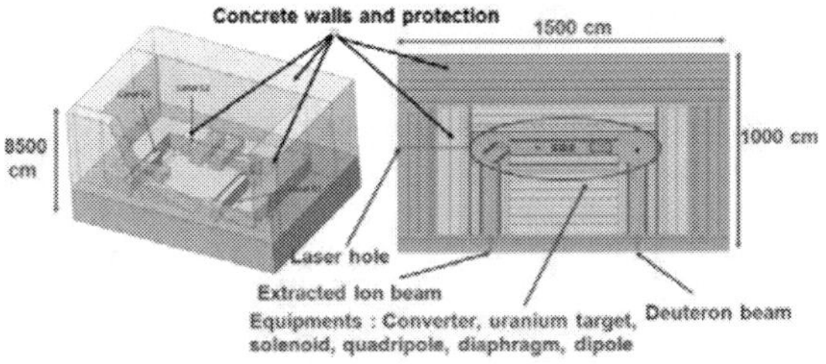

Figure 4: Canal configuration for the production building.

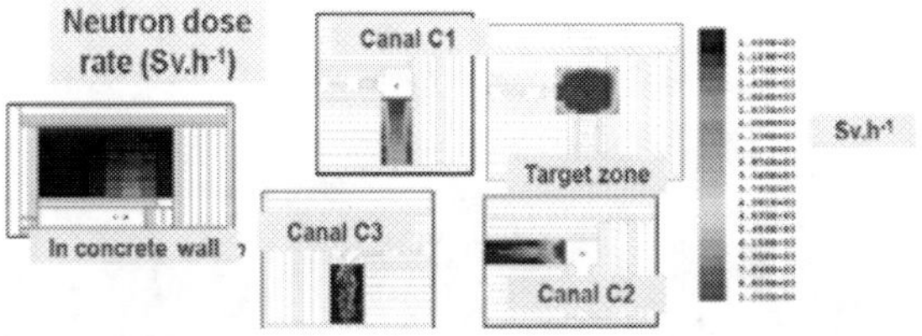

Figure 5: Neutron dose map in the production building

4. Conclusion

The MCNP simulation, together with the experimental neutron spectrum allowed to reconstruct a realistic neutron source based on 40 MeV deuteron stripping on a thick graphite converter target.

The calculated dose rates showed that gypsum and heavy concrete are good materials for attenuation of very fast neutrons.

The simulation of one of the configurations of the production building, the canal configuration, allowed the neutron dose rate to be calculated throughout the building. The dose rate for nominal conditions (5 mA of 40 MeV deuteron) is between 10^3 and 10^4 $Sv.h^{-1}$.

References

[1] S. Galès, "SPIRAL2 at GANIL: Next Generation of ISOL Facility for Intense Secondary Radioactive Ion Beams", Nuclear Physics A 834 (2010) 717c–723c.
[2] T. Junquera et al, "The high superconducting LINAC for the SPIRAL2 project at GANIL" Proceedings of SRF2013, Paris, France.
[3] R. Ferdinand et al, "Status and challenges of SPIRAL2 SRF LINAC" Proceedings of SRF2013, Paris, France.
[4] R. Ferdinand et al, "The SPIRAL2 superconducting LINAC" Proceedings of LINAC2008, Victoria, France
[5] M. Fadil, "About the production rates and the activation of the uranium carbide target for SPIRAL 2", Nuclear Instruments and Methods in Physics Research B 266 (2008) 4318–4321.
[6] G. Lhersonneau et al, "Neutron yield from carbon, light- and heavy-water thick targets irradiated by 40 MeV deuterons", Nuclear Instruments and Methods in Physics Research Section A: Accelerators, Spectrometers, Detectors and Associated Equipment, Volume 603, Issue 3, 21 May 2009, Pages 228-235.
[7] Z. Radivojevic et al, "Neutron yields from thick ^{12}C and ^{9}Be targets irradiated by 50 and 65 MeV deuterons", Nucl. Instr. and Meth. B 183 (2001) 212.
[8] K. Shin et al, "Neutron and photon production from thick targets bombarded by 30-MeV p, 33-MeV d, 65-MeV ^{3}He, and 65-MeV alpha ions: experiment and comparison with cascade Monte Carlo calculations", Phys. Rev. C 29 (1984) 1307.
[9] M. Hagiwara et al, "Experimental studies on the neutron emission spectrum and activation cross-section for 40 MeV deuterons in IFMIF accelerator structural elements", J. Nucl. Mater. 329–333 (2004) 218.
[10] Monte Carlo N-Particle Transport Code System for Multiparticle and High Energy Applications, Oak Ridge National Laboratory (ORNL/RSICC), USA.

SNS Proton Beam Window Disposal

Irina Popova[1,*], *Franz* X Gallmeier[1], *Steven* Trotter[1]

[1]*ORNL, PO box 2008, MS 6466, Oak Ridge, TN 37831-6474, USA*

Abstract. In order to support the disposal of the proton beam window assembly of the Spallation Neutron Source beamline to the target station, waste classification analyses are performed. The window has a limited life-time due to radiation-induced material damage. Analyses include calculation of the radionuclide inventory and shielding analyses for the transport package/container to ensure that the container is compliant with the transportation and waste management regulations. In order to automate this procedure and minimize manual work a script in Perl language was written.

1 Overview

The Spallation Neutron Source (SNS) in Oak Ridge, Tennessee, is an accelerator driven neutron scattering facility for materials research. Presently SNS operates at 1.4 MW proton beam power incident on a mercury target with a proton beam energy of 1 GeV and 60 Hz repetition rate. The SNS consists of accelerator facility, target facility, and a world-class suite of neutron scattering instruments.

Some of the components of the facility are scheduled for replacement, because they reach their end-of-life due to radiation induced material damage or burn-up, or because of mechanical failure, or design improvements. All these components must be safely removed, placed in a container for temporary on-site storage, and ultimately transported off-site to a nuclear waste disposal site.

In order to characterize and classify spent components, accurate estimates of radionuclide inventory are performed for components that are exchanged on regular basis such as target vessel and proton beam window module (PBW). A bounding case considering the maximum possible radionuclide inventory, based on a scenario with maximum possible irradiation exposed during the life-time, is established for the respective component. Using these data, the spent component is characterized and classified, and an appropriate container for temporary storage on-site and subsequent transport off-site is suggested. Once the container has been selected, transport calculations for the Department of Transportation (DOT) package are performed to ensure that the transport package is compliant with transportation and waste management regulations. Analyses for the radionuclide inventory and dose rates are performed for each off-site shipment with PBW realistic irradiation history.

The neutronics support analyzes and methods for spent PBW disposal are described in this paper.

2 PBW layout

The PBW module is mounted in the target monolith, approximately 2.3 meters upstream from the target, and establishes the boundary between the accelerator and target environments. Prior to intercepting the target, full beam power passes through the window, and therefore, the PBW receives a significant radiation exposure, leading to high activation in the materials.

The PBW module, which is shown in Figure 1 is comprised from the Inconel-718 window and is encased in a 316 stainless steel holder, and water-cooled collimator assembly. The holder is attached to a steel shield plug. The module weighs 3140 kg. Water lines for cooling the assembly exit the shield plug on top, and are shear-cut during PBW module removal activities. These lines along with the interior of the water jacket are dried to a dew point of –35 F with compressed air prior to removing the PBW module for disposal.

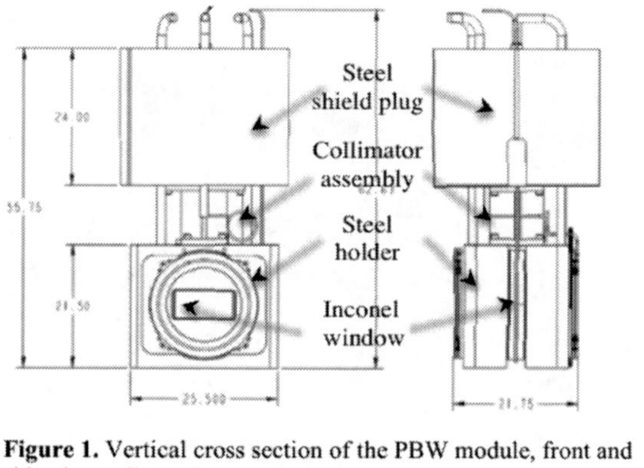

Figure 1. Vertical cross section of the PBW module, front and side views, dimensions are in inches.

The maximum allowable peak damage of steel and Inconel structures is 12dpa. In order to avoid excessive degradation of the materials, periodic replacement of the

* Corresponding author: popovai@ornl.gov

PBW during maintenance outages is scheduled when it approaches its estimated lifetime. Depending on the facility power, these exchanges occur every two or three years. The PBW is safely removed, placed into a handling liner and temporarily stored on site for cool-down. Later, the loaded liner is placed into a transport cask for transport off-site to a nuclear waste disposal facility.

3 Methods and codes

Neutronics analyses are performed to support spent PBW deposition. Full three-dimensional radiation transport calculations, using the state-of-the-art code MCNPX Version 2.6.0 [1] and the latest as-built target station model, are performed to simulate the radiation environment of the PBW module, and to calculate isotope production rates due to spallation reactions and the below-20-MeV neutron fluxes in areas of interest. The calculational model is shown in Figure 2. Isotope reaction rates and neutron fluxes are fed into the CINDER90 code [2] using the standardized ACTIVATION_SCRIPT [3] to calculate the PBW radionuclide inventory. Realistic irradiation history and decay time are used in the analyses. In order to obtain radionuclide inventory distributions in the proton beam window, the PBW module is subdivided into 82 cells, and activation calculations are performed based on cell - specific isotope production rates and neutron fluxes. Then, to obtain total isotope inventory and decay gammas, the post-processing code ALLCODE is used. This code is part of CINDER90 package and accumulates quantities from all material regions.

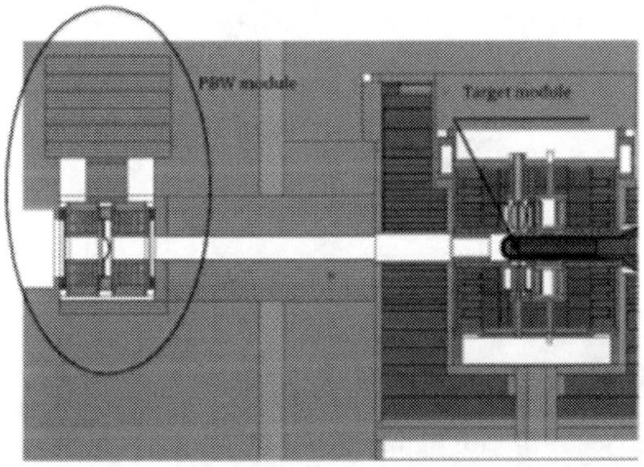

Figure 2. Vertical cross section of the target station including PBW module.

Decay gamma sources for a defined history of build-up and decay are extracted from each cell and compiled to a source term in MCNPX language by running the GAMMA_SOURCE_SCRIPT [4]. The decay gamma source is prepared based on forecasted for transport to the waste disposal facility. The source is utilized in photon transport calculations of the PBW module in the liner and also of the liner containing PBW module once it is placed inside the transport cask for prospective off-site transport time. Furthermore, gamma dose rates at requested locations around both the liner and the cask are provided to assist radiation protection staff planning the respective transport and disposal. For analyses of the residual dose rates, next-event point and ring detectors were applied as well as dose rate mesh tallies to allow for dose rate contours in and around the liner and cask geometries.

For transportation and waste characterization and classification, a report containing isotope inventory, dose rate and transport package description is generated on the basis of the component operation history.

Based on the estimated radionuclide inventory for full beam power (2MW) for 2 years beam exposure at 5000h of operations per year, it was found that the PBW waste profile fits on a per radionuclide comparison into the target vessel waste profile. As such the PBW is handled just as a spent target vessel and transported to off-site disposal using the TN-RAM. A model of the spent PBW module in the liner inserted into a TN-RAM cask is shown in Figure 3.

Figure 3. Model of spent PBW module in the liner inserted in TN-RAM cask.

TN-RAM cask model incorporates an inner steel shell of radial thickness 0.75 inches and of 0.5 inches thickness at both the bottom and the top were included as well as the outer steel shell of 1.5 inches radial walls, 2.5 inches walls at the bottom and 2.5 inches walls on the top. The PBW liner is a cylindrical liner with 0.25 inches thick steel sidewalls, a bottom plate of 1 inch thick steel, and a top lid of 2 inches thick steel.

From a transportation and waste management perspectives, supporting analyses with an accurate estimate of the radionuclide inventory and dose rates, assuming realistic irradiation history and decay time, are required for each shipment going to the disposal site. In order to simplify analyses, expedite work, generate a report and avoid errors arising from manual preparation of the calculations, a script in Perl language was written.

4 Script for the analyses

Script named PBWDISPOSAL.PL is written on Perl language, runs on LINUX machine and, using fluxes and reaction rates from MCNPX output file for transport calculational simulation of the radiation environment of the PBW module, performs neutronics and activation analyses for spent PBW, sorts out data and prepares the final report containing:

- History of irradiation;
- Radionuclide inventories in the PBW module;
- Activity in the PBW module structures after the beam termination versus decay time;
- Gamma contour plots of residual radiation coming from the spent PBW module;
- Predicted dose rates from the surface to the distance 3000 ft from the PBW module;
- Distances from the side and the bottom of the spent PBW module in the liner for which the dose rates fall below 100 mrem/hr and 5 mrem/hr, respectively.

Script uses 3 input files and supporting files for the analyses. Main input file provides general information such as report number, spent PBW module number, decay history and so on. Second input contains irradiation history. Third input contains cell numbers of PBW to be used in the analyses, which gives flexibility to use different model of the window.

This script:

- Reads input files;
- Creates input files for CINDER analyses, feeds isotope reaction rates and neutron fluxes to the script, and runs activation script for transmutation analyses;
- Creates input file and links for ALLCODE, that combines information by material region;
- Creates input file for GAMMA_SOURCE script and runs it to obtain SDEF card file with residual sources for liner and TN-RAM cask transport analyses for the time of off site transport;
- Runs MCNPX decay gamma transport analyses;
- Prepares tables and figures for the final report using transmutation and transport output files (additional post-processing script, which extracts data from MCNPX generated mctal file, then formats these data into gnuplot plotting format);
- Creates PDF report using latex, generated report contains information required by waste management.

5 Example of analyses

The last PBW that was replaced was PBW #4. It was in service from August 2012 to July 2014. During that time it was exposed to 8674 MWh proton beam. Off-site transport was tentatively scheduled to occur after 504 days of cooling down. The shut-down activity in the PBW's steel and Inconel structures is shown in Figure 4. Steel structures have higher overall activation because Inconel structures are less than 0.4% in the PBW module volume.

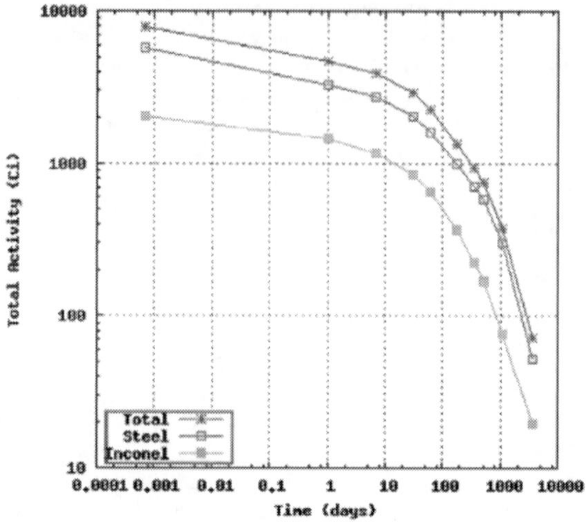

Figure 4. Activity (Cu) in spent PBW module, steel and Inconel structures.

Dose rate contours in and around the PBW module #4 in the liner after 504 days decay are shown in Figure 5.

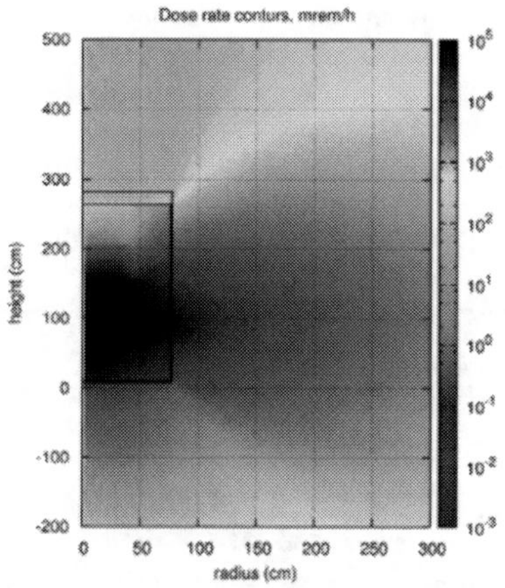

Figure 5. Dose rate contours in and around the PBW module in the liner after 504 days decay.

Dose rate contours in and around the TN-RAM cask loaded with PBW module #4 in the liner after 504 days decay are shown in Figure 6.

Predicted residual dose rates at specified locations around the PBW module in the liner and the TN-RAM cask loaded with PBW module in the liner after 504 days decay are presented in Table 1. Calculated values have less than 5% standard deviation.

DOI: 10.1051/epjconf/201715303017

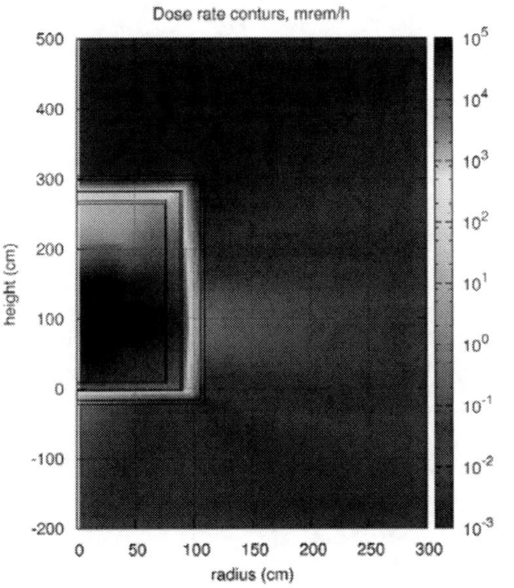

Figure 6. Dose rate contours in and around the TN-RAM cask loaded with PBW module in the liner after 504 days decay.

Table 1. Residual dose rates at specified locations around the PBW module in the liner and the TN-RAM cask loaded with PBW module in the liner after 504 days decay, mrem/h.

Location	Dose Rate TN-RAM	Dose Rate Liner
0 ft from bottom	2.01e-01	1.41e+04
1 ft from bottom	1.13e-01	7.00e+03
3 ft from bottom	4.68e-02	2.74e+03
10 ft from bottom	9.02e-03	5.06e+02
25 ft from bottom	1.87e-03	1.03e+02
50 ft from bottom	5.13e-04	2.89e+01
125 ft from bottom	8.81e-05	5.37e+00
0 ft from side	5.60e-01	2.33e+04
1 ft from side	3.03e-01	1.33e+04
3 ft from side	1.32e-01	5.70e+03
10 ft from side	2.77e-02	1.12e+03
25 ft from side	5.87e-03	2.32e+02
50 ft from side	1.60e-03	6.73e+01
125 ft from side	2.59e-04	1.06e+01

6 Sensitivity analyses

Residual dose rates analyses were performed to evaluate the sensitivity to impurities, geometry model precision and high-energy physics models. Transport analyses to calculate reactions rates and fluxes in the PBW were done using CEM model and Bertini model for high-energy interactions. Two geometry models were used – one is an old, physics geometry model and another is a new one,that was developed according to the engineering drawings. Cobalt (Co) impurity is introduced into stainless steel as 0.1% in the initial MCNPX run. One run with CEM model was done without Co added to stainless steel. Dose rates, produced with corresponding model on the liner in key locations are shown in Table 2.

Table 2. Residual dose rates for different models, mrem/h.

Location	30-cm from bottom	30-cm from side
Bertini model, physics geometry, with Co	1.06E+01	1.58E+01
CEM model, no Co	4.21E+00	7.00E+00
CEM model, with Co	5.79E+00	1.08E+01
Bertini model, with Co	7.03E+00	1.33E+01

Adding 0.1% cobalt to stainless steel increases dose rate about 30%. Bertini versus CEM high energy physics model increases dose rates about 23. Bertini model is used in the analyses.

7 Conclusions

Transport and activation analyses are performed to support SNS spent components waste characterization and handling. In order to simplify generating the waste handling report for PBW module, a script in Perl language running on LINUX platform, was written to automate the whole process.

References

1. C D. Pellowitz, ed., "MCNPX User's Manual, Version 2.6.0," LA-CP-07-1473, Los Alamos National Laboratory, Los Alamos, New Mexico (April 2008). (also see http://mcnpx.lanl.gov).
2. W. B. Wilson, S.T. Cowell, T. R. England, A.C.Hayes, P. Möller, A Manual for Cinder'90 Version 07.4, LA-UR-07-8412, Los Alamos National Laboratory, Los Alamos, (2007).
3. F. X. Gallmeier and M. Wohlmuther, Activation Script Version 1.0 User Guide, ORNL-TM-2008/031, Oak Ridge National Laboratory, August 2008.

Machine and radiation protection challenges of high energy/intensity accelerators: the role of Monte Carlo calculations

F. Cerutti[1,a] on behalf of the FLUKA collaboration and several colleagues of the CERN Accelerator and Technology sector

[1] *CERN, CH-1211 Geneva 23, Switzerland*

Abstract. The role of Monte Carlo calculations in addressing machine protection and radiation protection challenges regarding accelerator design and operation is discussed, through an overview of different applications and validation examples especially referring to recent LHC measurements.

1 Introduction

The interaction between particle beams and accelerator components is at the origin of a wealth of effects requiring a careful evaluation at the stages of machine design, operation, and dismantling. These include for instance heating, thermomechanical stress of beam intercepting devices (BID), gas production and material damage, superconducting magnet quenching, single event effects in electronic devices, and activation.

Such an interaction is triggered by different scenarios, i.e. either regular or accidental beam losses on BID (targets, collimators, dumps, stoppers, stripping foils, etc.), dust fall into circulating beam, particle debris generation inside the detectors of a collider, nuclear reactions on the residual gas in the vacuum chamber, synchrotron radiation, gas bremsstrahlung.

In order to quantify the above effects starting from the listed loss terms, multipurpose Monte Carlo codes like FLUKA [1, 2] represent an essential tool, allowing to evaluate macroscopic quantities through the microscopic description of particle transport and interaction in matter. This implies tracking in magnetic field as well as the account of all relevant electromagnetic and nuclear processes over an extremely wide energy range. The code reliability is verified through individual benchmarking of physics models against exclusive data. Moreover, a profitable calculation demands to implement to a challenging degree of accuracy the machine geometry, including material information.

This paper presents an overview of FLUKA studies instrumental in CERN accelerator design and operation, with special emphasis on the validation opportunities already offered by the first Large Hadron Collider (LHC) runs.

2 Energy deposition in sensitive components and monitor signals

At CERN, FLUKA is the reference tool to address machine protection aspects as well as the complementary radiation protection scope. It is regularly and extensively used for the whole accelerator

[a] e-mail: Francesco.Cerutti@cern.ch

chain, from the beam dump design of low energy injectors as Linac4 [3] up to LHC collimation [4] and High Luminosity upgrade of the LHC experimental insertions [5]. Also in the context of the Future Circular Collider (FCC) perspective, it plays a crucial role since the early stage of the project, both for accelerator and detector design and both for the hadron and lepton machines. This means that simulations have to deal with protons of some million TeV (the equivalent energy with target at rest of 100 TeV center-of-mass collisions) down to the lowest transport limit of 100 eV photons (for the study of the lepton ring synchrotron radiation). Such a task calls for a continuous improvement of different interaction models, having in mind that, despite the very high energy of beams of interest, several quantities, like for instance those related to radionuclide inventory, are extremely sensitive to low energy nuclear physics ingredients ruling the reaction fragment de-excitation. Together with this physics oriented effort, significant technical developments made possible to automatize the construction of consistent geometry models of several hundred meter accelerator portions [6, 7].

The LHC operation during Run-I (2010-2013) and Run-II (started in 2015) gave as intriguing byproduct a significant amount of data allowing to test the degree of reliability of simulations performed in the long course of the LHC design phase [8]. In particular, the Beam Loss Monitor (BLM) system, consisting of a few thousand ionization chambers all along the 27 km beam line, provides on-line measurement of the energy released by the particle shower originated by beam particle interactions, and triggers beam aborting if the detected values exceed pre-defined thresholds. Particle shower calculations permit to predict BLM signals for the relevant loss scenarios, correlating them at the same time with the energy deposition levels in the most exposed or sensitive elements. A quantity typically investigated is the power density in the superconducting coils of LHC magnets [9]. Its distribution is used to calculate the quench limit, and its absolute value, compared to the latter, indicates the likelihood of a quench event, implying accelerator downtime. While occasional quenches are to some extent natural in a cold machine, despite the BLM protection system in place to prevent them, steady heating from regular collisions in the experiment locations, scaling with the target instantaneous luminosity, has to be maintained at acceptable levels.

Among the various examples that will be included in the full contribution, the following subsection focuses on a very recent case referring to ion operation, enlarging instructively the picture with respect to more usual proton beams.

2.1 BFPP losses in LHC ion operation

Bound Free Pair Production (BFPP) was indicated since long ago [10] as a threatening electromagnetic process occurring in lead-lead collisions at the interaction points and causing specific losses hundreds of meters downstream, in the Dispersion Suppressor (DS) magnets. In fact, the creation of an electron-positron pair, followed by electron binding to one of the two fully stripped lead ions, has a considerable cross section and alters critically the ion magnetic rigidity due to the charge state change to 81+. This secondary beam travels on a peculiar orbit, making it impact the beam screen delimiting the mechanical aperture in the DS.

Using as input the impact distribution determined by independent tracking studies [11], a FLUKA calculation yielded the expected BLM pattern, which turned out to be in excellent agreement with late 2015 measurements, as shown in Fig. 1. As discussed above, the same simulation estimates the peak power density in the magnets coils.

A subsequent test pushing the luminosity to more than twice the design value led the impacted dipole to quench [12], demonstrating that the planned luminosity upgrade requires the already envisaged mitigation measures, that are the displacement of BFPP losses outside the magnet or, where this is not possible, the installation of a dedicated collimator upstream [13].

DOI: 10.1051/epjconf/201715303018

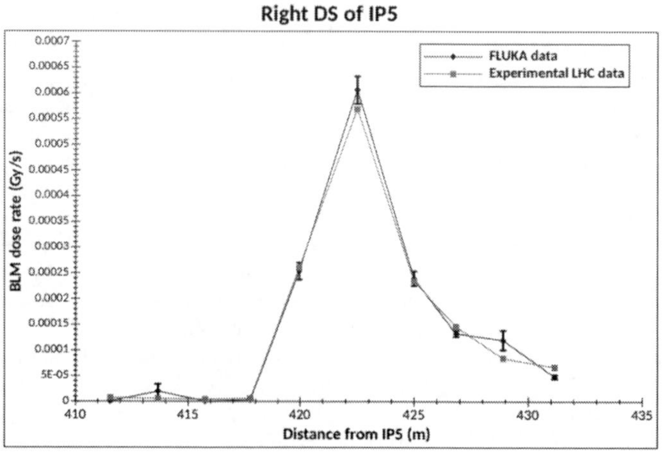

Figure 1. BLM pattern in the DS right of CMS for lead-lead collisions at 10^{27} cm^{-2} s^{-1}. For each monitor in the considered 20 m region, reported values are the absolute dose rate as measured (pink) and calculated by FLUKA (blue). In the latter case, statistical errors are also indicated.

References

[1] A. Ferrari, P.R. Sala, A. Fassò, and J. Ranft, *FLUKA: a multi-particle transport code* (CERN 2005-10, INFN/TC_05/11, SLAC-R-773, 2005)
[2] G. Battistoni et al., Annals of Nuclear Energy **82**, 10-18 (2015)
[3] C. Maglioni, *The Linac4 commissioning dump* (https://edms.cern.ch/document/1182648/1.0, 2013)
[4] R. Bruce et al., Phys. Rev. Special Topics - Accelerators and Beams **17**, 081004 (2014)
[5] N.V. Mokhov et al., Phys. Rev. Special Topics - Accelerators and Beams **18**, 051001 (2015)
[6] A. Mereghetti et al., *Proc. IPAC 2012* (New Orleans, 2012) 2687
[7] V. Vlachoudis, *Proc. M&C 2009* (Saratoga Springs, 2009)
[8] V. Boccone et al., Nuclear Data Sheets **120**, 215-218 (2014)
[9] B. Auchmann et al., Phys. Rev. Special Topics - Accelerators and Beams **18**, 061002 (2015)
[10] J.M. Jowett, *Proc. LHC Performance Workshop* (Chamonix, 2003) 84-90
[11] T. Mertens, private communication
[12] M. Schaumann, *LHC BFPP Quench Test with Ions (2015)* (CERN-ACC-NOTE-2016-0024, 2016)
[13] J.M. Jowett, *Proc. LHC Performance Workshop* (Chamonix, 2016) in preparation

Measuring space radiation shielding effectiveness

Amir Bahadori[1,2,⋆], Edward Semones[2], Michael Ewert[3], James Broyan[3], and Steven Walker[4]

[1] *Kansas State University, Department of Mechanical and Nuclear Engineering, Manhattan, KS, USA 66506 (current address)*
[2] *NASA Johnson Space Center, Space Radiation Analysis Group, Houston, TX, USA 77058*
[3] *NASA Johnson Space Center, Crew & Thermal Systems Division, Houston, TX, USA 77058*
[4] *Old Dominion University, Department of Mathematics & Statistics, Norfolk, VA, USA 23529*

Abstract. Passive radiation shielding is one strategy to mitigate the problem of space radiation exposure. While space vehicles are constructed largely of aluminum, polyethylene has been demonstrated to have superior shielding characteristics for both galactic cosmic rays and solar particle events due to the high hydrogen content. A method to calculate the shielding effectiveness of a material relative to reference material from Bragg peak measurements performed using energetic heavy charged particles is described. Using accelerated alpha particles at the National Aeronautics and Space Administration Space Radiation Laboratory at Brookhaven National Laboratory, the method is applied to sample tiles from the Heat Melt Compactor, which were created by melting material from a simulated astronaut waste stream, consisting of materials such as trash and unconsumed food. The shielding effectiveness calculated from measurements of the Heat Melt Compactor sample tiles is about 10% less than the shielding effectiveness of polyethylene. Shielding material produced from the astronaut waste stream in the form of Heat Melt Compactor tiles is therefore found to be an attractive solution for protection against space radiation.

1 Introduction

Human protection from space radiation is a challenging problem that must be addressed using various strategies, including passive shielding, active shielding, improved environment prediction, operational planning and actions, biological uncertainty reduction, and biological countermeasures. In the past, passive shielding studies have focused on using additional mass for the primary purpose of space radiation protection [1, 2]. However, adding mass for no purpose other than radiation shielding is unpalatable due to the limited launch capability of rockets and the mass required to sustain astronauts for extended periods of time for missions with destinations outside of low-Earth orbit. Therefore, alternative passive shielding strategies, such as using advanced structural materials [3], food [4], or astronaut waste [5] have been suggested. While computational and experimental methods to determine water equivalent thickness for use in proton therapy calculations have been published [6], no standardized procedure has been developed to determine the effectiveness of composite materials for space radiation shielding using measurements without detailed knowledge of the constituents. The purpose of the present study is to develop such a procedure, apply the procedure to sample tiles from the NASA Heat Melt Compactor, and determine whether the Heat Melt Compactor tiles are a viable alternative to dedicated passive shielding constructed from polyethylene.

⋆e-mail: bahadori@ksu.edu

2 Problem Description

The space radiation environment is comprised of radiation from three sources: geomagnetically-trapped particles, galactic cosmic rays, and solar particle events. Each of these sources presents unique challenges from the perspective of space radiation protection, including the efficacy of passive shielding as a means by which to mitigate space radiation exposure. Since mass launch capability is limited for missions beyond Earth orbit, compacted astronaut waste is attractive as an option for passive shielding against space radiation. No additional launch mass would be required to implement such a shielding strategy. However, the shielding effectiveness of the astronaut waste must be characterized to understand the level of protection afforded by the material. The hypothesis to be tested in the present study is that the Heat Melt Compactor tiles have a shielding effectiveness similar to that of polyethylene due to the high content of plastic and other hydrocarbons in the astronaut waste stream.

2.1 Space Radiation Environment

For missions beyond Earth orbit, the contribution of geomagnetically-trapped particles to overall mission radiation exposure can be minimized through judicious selection of trajectories that avoid high intensity regions of Earth's radiation belts. NASA's experience to date with crewed beyond Earth orbit missions consists of the Apollo

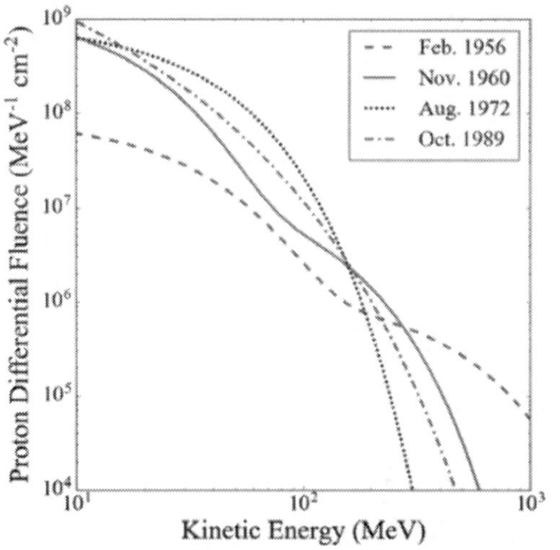

Figure 1. Selected historical solar particle event fluence spectra (reproduced from [9])

program, in which Earth orbits at low altitudes were completed prior to trans-lunar injection [7]. The radiation exposure from geomagnetically-trapped sources was largely limited to fast transits through the particle belts.

Galactic cosmic rays consist of heavy charged particles, ranging in atomic number from 1 (protons) to 58 (nickel ions) and above, with protons comprising about 90% and alphas comprising about 9% of the galactic cosmic ray spectrum [8]. The kinetic energies range from fractions of one megaelectron-volt per nucleon (MeV n^{-1}) to thousands of gigaelectron-volts per nucleon (GeV n^{-1}) [8], although the vast majority of ions have kinetic energies less than 50 GeV n^{-1}. The flux density spectra for galactic cosmic ray ions peak at kinetic energies between 100 MeV n^{-1} and 1 GeV n^{-1}.

Solar particle events are a concern for human space flight in the low cutoff rigidity regions of low-Earth orbit (e.g., near the geomagnetic poles) and outside of low-Earth orbit, such as missions to the Moon or Mars. Several major solar particle events have been observed in the Space Age via balloon measurements or satellites. Several historic solar particle event fluence spectra [9] are shown in Figure 1. These major solar particle events have spectral characteristics that can vary substantially. For example, the February 1956 solar particle event had a substantially larger high energy component than the August 1972 solar particle event [10]. The fluence spectrum impacts the radiation transport and ultimately the exposure and resulting radiation risk to humans within a vehicle or habitat [11, 12].

2.2 Space Radiation Shielding

Reducing exposure time, increasing distance from the source, and increasing shielding are frequently cited as the three major ways by which radiation exposure levels may be reduced. In space, reducing exposure time is only possible by completing missions with shorter durations or developing faster transit capabilities. Increasing distance from the source is not meaningful since the sources are nearly isotropic. Therefore, shielding remains the only option for reducing space radiation exposure for a given mission.

Galactic cosmic rays are difficult to shield due to the high kinetic energies of the particles and the tendency for ions of higher atomic number to fragment, creating lighter, secondary charged particles. The stopping power of passive shielding is smaller for the lighter ion fragments, which are created with velocities similar to the incident ions, allowing the secondary charged particles to more readily penetrate the shielding. Solar particle events tend to be easier to shield than galactic cosmic rays, as they are comprised mostly of protons with fluence distributions weighted towards lower kinetic energies. However, solar particle events are not easily predicted and can be very intense, resulting in high absolute exposure levels. For both galactic cosmic rays and solar particle events, polyethylene has been shown to have superior shielding characteristics to aluminum [13, 14], which is the dominant constituent in most space vehicle structures. Unfortunately, due to mass constraints for missions beyond Earth orbit, there are no plans to include material for the sole purpose of radiation shielding. Therefore, a desirable passive shielding strategy is to use material required for other purposes as radiation shielding.

2.3 NASA Heat Melt Compactor

The NASA Heat Melt Compactor technology has been under development for several years with the goal of developing a device that compacts trash by a factor of seven or more, recovers water from the trash, and produces a stable tile suitable for radiation shielding. Ground testing has demonstrated these capabilities. Plastic in the astronaut waste stream, primarily from food packaging, is melted in the Heat Melt Compactor process and helps encapsulate other items in the trash. The Heat Melt Compactor heats as it compresses in order to sterilize and limit odor as well as to vaporize water for subsequent condensation and recovery.

The first generation Heat Melt Compactor unit at NASA Ames Research Center was used to produce five round tiles for this evaluation with approximate diameter of 20.3 cm and thickness of 1.9 cm. Figure 2 shows one of the round tiles tested in the present study. The composition of the tiles was estimated based on the constituents of simulated astronaut trash that was processed, which included items such as food packaging materials, wasted food, clothing, and disposable wipes. The second generation Heat Melt Compactor ground unit, which is now being tested, produces square tiles with 22.9 cm side length.

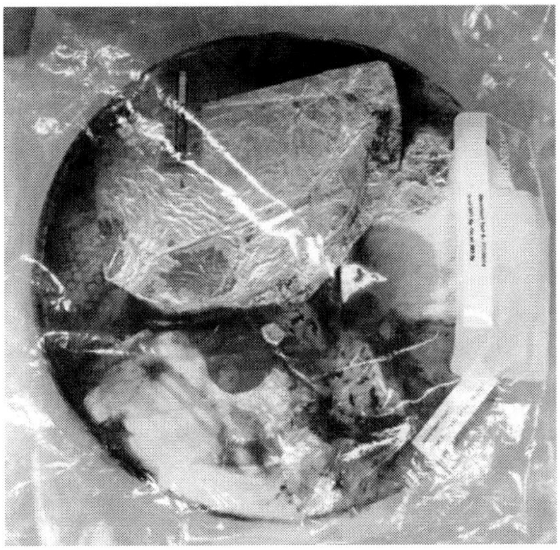

Figure 2. Photograph of Heat Melt Compactor tile tested at NSRL

3 Methods and Materials

First, the relative shielding effectiveness of the material-under-test with respect to the reference material must be defined in terms of measurable or known quantities of both materials using an accelerated charged particle beam. A measurement procedure is then developed to acquire the data necessary to calculate the relative shielding effectiveness.

3.1 Relative Shielding Effectiveness Definition

Previously, Guetersloh et al. [15] defined a shielding effectiveness by the ratio of measured absorbed doses with and without shielding material present in the beam. While this is a meaningful definition for a space radiation environment with a broad energy spectrum, it is difficult to interpret when used with a monoenergetic ion beam delivered at an accelerator facility. For certain combinations of ion, energy, and shielding, the measured absorbed dose increases with increasing shielding thickness because of particles slowing within the shield. Therefore, a new definition of shielding effectiveness, defined in terms of a reference material and applicable to accelerator beam measurements, is proposed.

The relative shielding effectiveness f is defined as the ratio of the mass stopping powers for the material-under-test (MUT) and the reference material (RM), as shown in Equation 1. Note that implicit dependence on the particle type and energy is suppressed in this formulation.

$$f = \frac{\left[-\frac{1}{\rho}\left(\frac{dE}{dx}\right)\right]_{MUT}}{\left[-\frac{1}{\rho}\left(\frac{dE}{dx}\right)\right]_{RM}} \tag{1}$$

Now, it is assumed that the charged particles have negligible energy loss while traversing the material-under-test sample and the equivalent thickness of reference material, such that the linear stopping power may be calculated as the change in energy resulting from travel through the thickness of each material. Furthermore, the changes in energy for the charged particles traversing each material must be equal. Therefore, Equation 1 reduces to

$$f = \frac{t_{RM}\rho_{RM}}{t_{MUT}\rho_{MUT}} \tag{2}$$

where t_{RM} is the equivalent reference material thickness for the material-under-test sample, ρ_{RM} is the density of the reference material, t_{MUT} is the thickness of the material-under-test, and ρ_{MUT} is the density of the material-under-test.

Finally, to account for density and thickness variations in the material-under-test, the areal density of the material-under-test, which is equal to the denominator of Equation 2, is simplified to the quotient of the material-under-test sample mass and cross-sectional area. The the relative shielding effectiveness calculation further simplifies to

$$f = \frac{t_{RM}\rho_{RM}}{\frac{m_{MUT}}{A_{MUT}}} \tag{3}$$

where m_{MUT} is the mass of the material-under-test sample and A_{MUT} is the cross-sectional area of the material-under-test sample. All values in Equation 3 required for calculation of the relative shielding effectiveness can be measured or are known prior to experiment.

3.2 Measurement Procedure

Four quantities must be measured or known to calculate the relative shielding effectiveness for a given material-under-test sample: the equivalent reference material thickness of the material-under-test sample, the density of the reference material, the mass of the material-under-test sample, and the cross-sectional area of the material-under-test sample.

The equivalent reference material thickness of the material-under-test sample must be determined using experimental measurements at a charged particle accelerator beam line. First, a Bragg peak measurement is performed using reference material to stop the charged particle beam in the detector. The thickness of reference material required is recorded. Next, a Bragg peak measurement is performed with the material-under-test sample proximal to the charged particle beam outlet, and the difference in reference material thickness between the Bragg peak measurement with no material-under-test sample and the Bragg peak measurement with the material-under-test sample is calculated.

The density of the reference material is either known from experience or can be measured by calculating the quotient of the mass and volume. The mass of the material-under-test sample is measured, and the cross-sectional area of the material-under-test sample is calculated using measurements of the sample dimensions. It is important that the beam area and detector area envelop the material-under-test sample cross-sectional area to accurately characterize the entire material-under-test sample.

4 Data Collection

For the present study, equivalent reference material thickness data were collected at the NASA Space Radiation Laboratory (NSRL) at Brookhaven National Laboratory in Upton, NY, USA. The ions available for measurement were limited to alphas and ^{56}Fe ions due to scheduling constraints. Measurements using alphas were determined to be most appropriate, as alphas are less likely to fragment than ^{56}Fe ions. Two considerations in addition to applicability to the space radiation environment dictated the selection of the ion kinetic energy. First, the ion kinetic energy must be high enough such that the change in kinetic energy over the Heat Melt Compactor tile is small; and second, the ion kinetic energy must be low enough to permit Bragg peak measurements with the available amount of degrader material. Since measurements for the present study were made in concert with measurements for other projects, a limited selection of kinetic energies was available. Of these, the nominal kinetic energy of 200 MeV n^{-1} was chosen. Polyethylene is used for performing Bragg peak measurements at the NSRL and therefore served as the reference material for this experiment.

Bragg peak measurements made using only the polyethylene degrader indicated that the alphas had a nominal range of 26.875 cm, corresponding to a kinetic energy of 207.8 MeV n^{-1}. Bragg peak measurements were next performed with each of the five Heat Melt Compactor samples (set-up shown in Figure 3), and the equivalent polyethylene thickness, t_{PE}, was found. Here, it was estimated that the Bragg peak measurements have an uncertainty of 0.1 cm. The equivalent polyethylene thicknesses and Heat Melt Compactor sample masses, m, are shown in Table 1. The area of each Heat Melt Compactor sample was calculated using a measured diameter of 20.0 cm, yielding a cross-sectional area of 314.2 cm^2. The density of the polyethylene used at the NSRL is 0.97 g cm^{-3} [16].

Figure 3. Heat Melt Compactor sample (circular disk in plastic bag) on beam line

5 Analysis

The relative shielding effectiveness for each Heat Melt Compactor tile as formulated in Equation 3 was calculated

Table 1. Heat Melt Compactor sample equivalent polyethylene thicknesses and masses

Tile No.	t_{PE} (cm)	m (kg)
1	1.2 ± 0.14	0.398
2	1.3 ± 0.14	0.438
3	1.4 ± 0.14	0.450
4	1.3 ± 0.14	0.435
5	1.3 ± 0.14	0.447

and the results are shown in Table 2. Also shown is the estimated uncertainty for the relative shielding effectiveness. The mean relative shielding effectiveness for the group of samples was found to be 0.91 ± 0.044, indicating that the Heat Melt Compactor samples are 91% ± 4.4% as effective as polyethylene at shielding against alpha particles with a kinetic energy of 207.8 MeV n^{-1}.

Table 2. Heat Melt Compactor sample relative shielding effectiveness experimental values

Tile No.	f (unitless)
1	0.92 ± 0.11
2	0.91 ± 0.095
3	0.95 ± 0.095
4	0.91 ± 0.098
5	0.89 ± 0.096

To check the accuracy of the measurements, the Heat Melt Compactor sample mass stopping power for alpha particles with kinetic energy of 831.2 MeV, corresponding to the beam kinetic energy of 207.8 MeV n^{-1}, was calculated using SRIM-2013 [17] for the assumed composition shown in Table 3. The Heat Melt Compactor sample mass stopping power was found to be 0.01700 MeV cm^2 mg^{-1}, while the polyethylene mass stopping power was found to be 0.01864 MeV cm^2 mg^{-1}. The results from SRIM-2013 indicate a material with the assumed composition is 91.2% as effective as polyethylene at shielding against alpha particles with kinetic energy of 207.8 MeV n^{-1}, exhibiting excellent agreement with the experimentally-determined relative shielding effectiveness.

Table 3. Heat Melt Compactor sample assumed composition

Element	Mass fraction
H	0.08
C	0.54
N	0.01
O	0.24
F	0.04
Na	0.01
Al	0.05
Cl	0.02
Fe	0.01

Both galactic cosmic ray and solar particle event environments are dominated by protons as previously described. While the measurements performed in the present study used alpha particles, the relative shielding effectiveness results are applicable protons of the same velocity (i.e., protons with kinetic energy of 207.8 MeV).

The ratios of the mass stopping powers for protons and alphas were calculated for the Heat Melt Compactor material and polyethylene using SRIM-2013 to demonstrate applicability to protons. These ratios are shown in Figure 4 as a function of kinetic energy per nucleon. Excellent agreement is observed for the ratios between 2 MeV n^{-1} and 10 GeV n^{-1}. Additionally, the ratios of the mass stopping powers increase with kinetic energy per nucleon, indicating that the shielding effectiveness of the Heat Melt Compactor material with respect to polyethylene would be even more favorable for ions with greater kinetic energy than that tested in the present study.

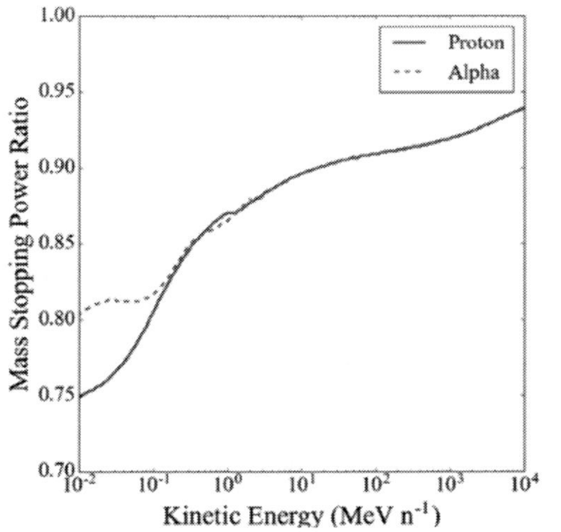

Figure 4. Ratios of Heat Melt Compactor material to polyethylene mass stopping powers for protons and alphas

Simulations of the August 1972 solar particle event using the King model [10] and the 1977 solar minimum galactic cosmic ray environment [18] have been performed using the assumed composition for the Heat Melt Compactor samples. The NASA On-Line Tool for the Assessment of Radiation In Space [19] was used to simulate the effective dose using the FAX female astronaut computational phantom [20] shielded by Heat Melt Compactor material in a spherical shell configuration for shield thicknesses between 0 and 100 g cm^{-2}. It is noted that the effective dose calculation uses both the linear energy transfer-dependent quality factor and tissue weighting factors defined in ICRP Publication 60 [21] and NCRP Report No. 132 [22] without gender averaging. Simulations were also performed for aluminum and polyethylene shielding for comparison.

The results for the August 1972 solar particle event and the 1977 solar minimum galactic cosmic ray spectrum are shown in Figures 5 and 6, respectively. The simulation results for both environments indicate that the performance of the Heat Melt Compactor material is superior to aluminum and similar in trend, although inferior, to polyethylene.

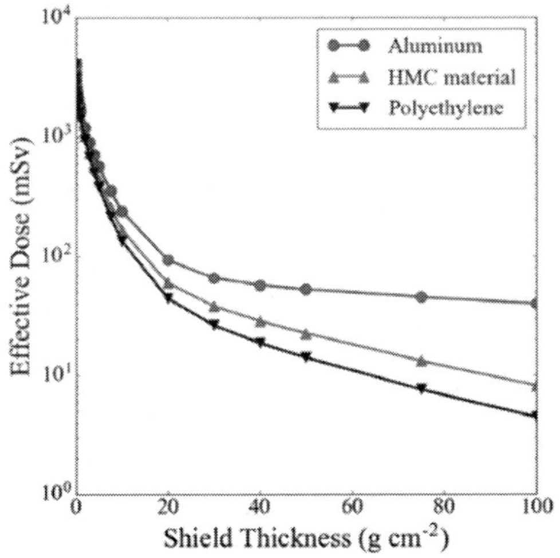

Figure 5. Female effective dose for August 1972 solar particle event environment

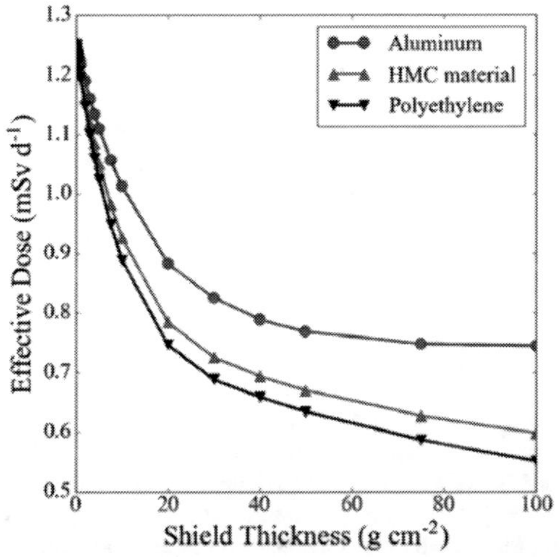

Figure 6. Female effective dose for 1977 solar minimum galactic cosmic ray environment

6 Conclusions

The results of the present study indicate that the Heat Melt Compactor material could be used as an acceptable replacement for polyethylene shielding. Simulations of space radiation spectra show that a material with the assumed composition exhibits favorable shielding characteristics. The Heat Melt Compactor material is particularly attractive since it requires no dedicated mass for radiation shielding. The hypothesis was therefore proven correct, since the space radiation shielding characteristics were shown to be similar to polyethylene.

There are several opportunities for future work investigating the use of the Heat Melt Compactor tiles for space radiation shielding. First, there is the problem of how to incorporate the tiles into the vehicle shielding. They are rigid and not easily conformed to the hull of a spacecraft. Also, there must be some method of tethering them in a particular location. Optimization of placement, including whether they placed externally or internally with respect to the vehicular structural shielding, should be performed. Next, there is a time dependence in the availability of astronaut waste for use in the Heat Melt Compactor. At the beginning of a mission, other consumable materials, such as water or food, that have the reverse time-dependent pattern of availability, should be included in the concept of operations for radiation shielding to make up for the lack of astronaut waste if augmented shielding is required early in a mission of extended duration. Finally, to avoid selection bias and to more completely quantify the shielding effectiveness against solar particle events, simulations should be performed using a proper sample of the historical database of solar particle events.

One limitation of the present study is that neutron contributions are not explicitly considered. Supplementing the measurements with neutron spectrum measurements behind equivalent thicknesses of Heat Melt Compactor material and polyethylene would indicate the degree to which neutron production is enhanced or suppressed in the Heat Melt Compactor material. Another important limitation of the proposed method is that the maximum kinetic energy that can be tested is limited by the amount of reference material available for Bragg peak measurements. Additionally, the kinetic energy must be high enough such that the change in energy experienced by the charged particles traversing the material-under-test sample is small. It is also difficult to test many kinetic energies due to practical limitations such as accelerator beam time availability and the scarcity of funds to pay for accelerator testing. As a result of these limitations, computational methods will continue to play a prominent role in characterizing space radiation shielding effectiveness, especially for very low and very high kinetic energies. Measurements of selected kinetic energies should be performed as validation of the results generated using computational methods, ensuring that astronaut waste stream variability does not result in unacceptably large variation in shielding characteristics among the tiles.

This work was supported by the NASA Advanced Exploration Systems (AES) program RadWorks project. We thank Bobbie Gail Swan and Catherine Mcleod, AES RadWorks Project Management, for supporting this study. We also thank Michael Sivertz at the NSRL for his work and advice during beam line data acquisition. Finally, we thank J. Kenneth Shultis for reviewing the draft paper and providing helpful comments. Funding from the Kansas State University Office of Engineering Research and Graduate Programs and Department of Mechanical and Nuclear Engineering supported data analysis, paper preparation, and conference travel.

References

[1] M. Shavers, N. Zapp, R. Barber, J. Wilson, G. Qualls, L. Toupes, S. Ramsey, V. Vinci, G. Smith, F. Cucinotta, Adv Space Res **34**(6) 1333-1337 (2004)

[2] R. Singleterry, K. Bollweg, T. Martin, S. Westover, R. Battiston, W. Burger, R. Meinke, *NASA/TP-2015-218689* (2015)

[3] A. Emmanuel, J. Raghavan, Adv Space Res **56**, 1288-1296 (2015)

[4] S. Walker, M. Clowdsley, H. Abston, A. Gallegos, M. Simon, *43rd ICES* (2013)

[5] J. Broyan, M. Ewert, P. Fink, *AIAA SPACE 2014* (2014)

[6] R. Zhang, P. Taddei, M. Fitzek, W. Newhauser, Phys Med Biol **55**(9) 2481-2493 (2010)

[7] R. English, R. Benson, J. Bailey, C. Barnes, *NASA TN D-7080* (1973)

[8] J. Simpson, Annu Rev Nucl Part S **33** 323-382 (1983)

[9] F. Badavi, J. Wilson, A. Hunter, *NASA/TP-2005-213941* (2005)

[10] J. King, J Spacecr Rockets **11**, 401-408 (1974)

[11] A. Bahadori, M. Van Baalen, M. Shavers, C. Dodge, E. Semones, W. Bolch, Phys Med Biol **56**(6) (2011)

[12] A. Bahadori, T. Sato, T. Slaba, M. Shavers, E. Semones, M. Van Baalen, W. Bolch, Phys Med Biol **58**(20) 7183-7207 (2013)

[13] J. Wilson, F. Cucinotta, J. Shinn, L. Simonsen, R. Dubey, W. Jordan, T. Jones, C. Chang, M. Kim, Radiat Meas **30**, 361-382 (1999)

[14] G. Badhwar, F. Cucinotta, Radiat Res **153**, 1-8 (2000)

[15] S. Guetersloh, C. Zeitlin, L. Heilbronn, J. Miller, T. Komiyama, A. Fukumura, Y. Iwata, T. Murakami, M. Bhattacharya, Nucl Instrum Methods Phys Res B **252** 319-332 (2006)

[16] U.S. Department of Energy, NSRL User Guide Part III. Technical Data, https://www.bnl.gov/nsrl/userguide/bragg-curves-and-peaks.php

[17] J. Ziegler, SRIM-2013 Software, http://www.srim.org/

[18] P. O'Neill, IEEE Trans Nucl Sci **57**(6) 3148-3153 (2010)

[19] R. Singleterry, S. Blattnig, M. Clowdsley, G. Qualls, C. Sandridge, L. Simonsen, T. Slaba, S. Walker, F. Badavi, J. Spangler, A. Aumann, E. Zapp, R. Rutledge, K. Lee, R. Norman, J. Norbury, Acta Astronaut **68**(7-8) 1086-1097 (2011)

[20] R. Kramer, H. Khoury, J. Vieira, E. Loureiro, V. Lima, F. Lima, G. Hoff, Phys Med Biol **49**, 5203-5216 (2004)

[21] ICRP, Ann ICRP **21**(1-3) 1-201 (1990)

[22] NCRP, *NCRP Report No. 132* (2000)

Double differential neutron yields from thick targets used in space applications

Natalie A. McGirl[1,*], *Luis* A. Castellanos[1], *Ashwin* P. Srikrishna[1], *Lawrence* H. Heilbronn[1], *Chiara* La Tessa[2], *Adam* Rusek[2], *Michael* Sivertz[2], *Steve* Blattnig[3], *Martha* Clowdsley[3], *Tony* Slaba[3], and *Cary* Zeitlin[4]

[1]Department of Nuclear Engineering, University of Tennessee, 1004 Estabrook Rd, Knoxville, TN, USA, 37996
[2]NASA Space Radiation Laboratory, Brookhaven National Laboratory, Upton, NY, USA, 11973
[3]NASA, Langley Research Center, Hampton, VA, USA, 23681
[4]Lockheed Martin, Information Systems and Global Solutions, Houston, TX, USA, 77285

Abstract. In March 2016, secondary neutron production from thick-target shielding experiments were conducted at the National Aeronautics and Space Administration's (NASA) Space Radiation Laboratory at Brookhaven National Laboratory. Ion beams of proton, helium, and iron projectiles were aimed at aluminum targets with areal densities of 20, 40, and 60 g/cm^2. The ion beams were extracted at energies of 400 and 800 AMeV and neutron yields were measured with liquid scintillators at 10°, 30°, 45°, 60°, 80°, and 135° off the beam axis. A second 60 g/cm^2 aluminum target was placed 3.5 m downstream from the middle of front target to study backscattered neutrons. Double differential thick-target neutron yields for various combinations of projectile, projectile energy, target material, target thickness, and detector location were produced using the time-of-flight technique. These measurements will help NASA perform uncertainty analyses on their transport codes and contribute to shielding design studies for future space applications.

1 Introduction

The future of manned, deep-space missions includes the need to sufficiently protect astronauts from the ionizing radiation effects caused by Solar Energetic Particles (SEP) and Galactic Cosmic Rays (GCR) for an extended length of time. Habitats and other transit vehicles often contain areas of thick (30-40 g/cm^2 or greater) shielding materials such as aluminum or high density polyethylene (HDPE) [1]. While an enclosed, thickly-shielded environment may decrease an astronaut's exposure to the primary radiation field, the creation of a secondary radiation field, which includes neutrons and light charged ions, still poses a risk [2]. In particular, secondary neutrons are a concern due to their highly-penetrative nature and large dose equivalent conversion factors [3]. The purpose of this study is to determine the double differential thick-target yields for neutrons produced directly from GCR-like heavy ion interactions with aluminum or HDPE shielding. These results will be compared with transport model calculations and incorporated into the uncertainty analysis for transport codes developed by the National Aeronautics and Space Administration (NASA). This paper presents a selection of results from an experiment conducted in March 2016.

2 Experiment Overview

Secondary neutron production measurements for the 100-hour March 2016 experiment occurred at Brookhaven National Laboratory's (BNL) NASA Space Radiation Laboratory (NSRL). BNL's Booster accelerator was used to deliver the 400- and 800-MeV protons, 400-AMeV helium, and the 400 and 800-AMeV iron beams to the 20, 40, and 60 g/cm^2 thick aluminum targets (100 x 30 cm^2). A back aluminum target with an areal density of 60 g/cm^2 (100 x 100 x 22.2 cm^3) was placed 3.5 m downstream from the middle of the front target. The addition of this target allowed for the study of neutrons scattered from or produced in the back target, which will occur at a later date. Additionally, two beam-defining EJ-228 plastic scintillator detectors were placed directly in front of the forward target and identified valid beam particles from coincidence events.

Neutrons were detected with three EJ-301 liquid scintillators placed at 10°, 30°, 45°, and three EJ-309 liquid scintillators placed at 60°, 80°, and 135° off the beam axis, with flight paths ranging from 2 to 3 m from the center of the front target to the front face of the liquid scintillator. Each liquid scintillator was 12.7 cm in diameter and 12.7 cm tall. Two thin EJ-304 solid plastic scintillators (12.7 x 12.7 x 0.635 cm^3) covered the front face of each liquid scintillator to distinguish between incident neutral and charged particle events. Finally, a shadow bar system was used to allow for full background characterization. This system consisted of two iron bars with lengths of 1 and 2 m. These shadow bars fully blocked the front face of a liquid scintillator and prevented neutrons originating directly from the front target from entering the scintillator. Liquid scintillator

* Corresponding author: nmcgirl@vols.utk.edu

locations, target locations, and flight path values are shown in Figure 1.

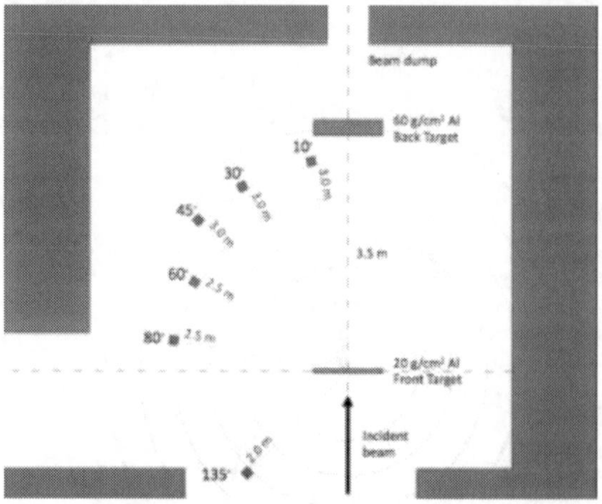

Fig. 1. March 2016 NSRL target room experimental setup for 20 g/cm² front and 60 g/cm² back aluminum targets. Grey blocks represent concrete walls.

3 Neutron Time of Flight Analysis

Several steps were taken to determine the double differential thick-target neutron yields for the March 2016 100-hour experiment. First, the charge deposited in the beam-defining "start" scintillators were plotted against each other. A noticeable beam spot was identified and a graphical ROOT cut was taken around it [4]. This ensured that only beam particles that did not deviate too far from the center of the beam were taken as source particles incident upon the front target. On average, approximately 30% of the beam particles were eliminated from the analysis with the good beam cut.

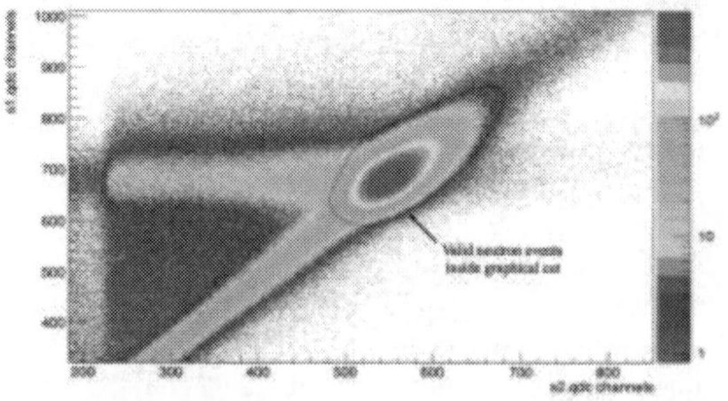

Fig. 2. Charge deposited in start scintillators for a 400-AMeV iron beam with good beam cut boundaries displayed.

Next, the neutral and charged particle events that set the liquid scintillator trigger were gated on in ROOT by selecting the self-time peak in the time-to-digital conversion (TDC) spectrum. The contributions to the liquid scintillator signals were then separated using the thin EJ-304 "veto" detectors located in front of the scintillators. Events that registered in both the veto and liquid scintillator detectors were taken to be the result of a charged particle interaction, due to the low probability of a neutron or gamma interaction in the veto detector [5]. By examining the veto detector's charge-to-digital converter (QDC) output, the neutral and charged particle information was separated and the neutral particle data were refined further to determine the neutron and gamma signal contributions.

Pulse shape discrimination (PSD) was used to separate neutrons and gammas based on the differences between the charge contained in the first 20 to 35 ns of a signal (h_qdc) and in the total signal (t_qdc) [6]. An example of a typical PSD plot is shown in Figure 3. The lowest energy of neutron-gamma separation was examined by taking one-channel-wide slices of the PSD plots and projecting them onto the Y-axis. The resulting neutron and gamma peaks were considered well-separated if their 2σ values did not overlap. The channel number at the separation threshold was converted to incident neutron energy (MeV) using the experimentally determined calibration curves of the liquid scintillators and a light output calibration [7]. Because a neutron may deposit all of its energy in a single interaction, the energy at the separation threshold represented the lowest energy neutron detectable by the scintillator. A 4.0 MeV neutron was detected at the separation threshold for the scintillators at 10°, 30°, and 45°, while a 2.0 MeV neutron was detected at the separation threshold for the scintillators at 60°, 80°, and 135°. Afterwards, neutron and gamma event data were extracted from the PSD plots using ROOT graphical cuts.

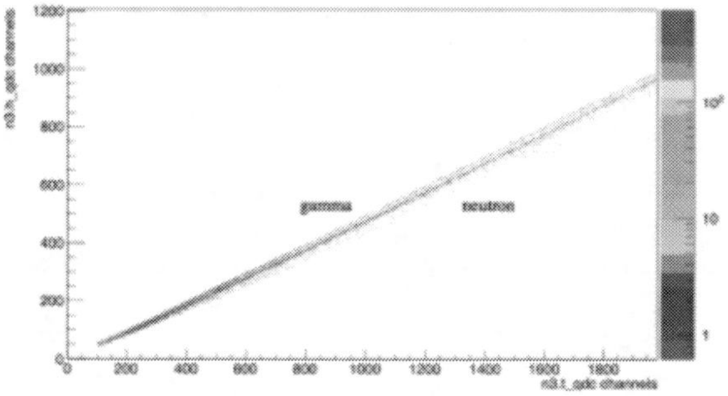

Fig. 3. Neutron and gamma PSD at 45° for 400-MeV protons on 20, 40, and 60 g/cm² aluminum front targets; 60 g/cm² back aluminum target.

To convert the start scintillator's time-to-digital converter (TDC) outputs to neutron energy spectra, the time-of-flight (TOF) method was utilized. TDC spectra collection started with a coincidence between the beam-defining scintillators and a neutron detector, and stopped once a delayed signal from the beam-defining scintillators was registered. To determine the neutron times of flight, a reference channel number, which corresponded to the gammas that were created in the projectile-target interactions and subsequently detected in the liquid scintillators, was identified on the TDC spectra. This channel number was found by plotting the TDC spectra gated on the gammas identified in PSD. The centroid of the resulting prompt gamma peak was used as the reference channel number, which was different for each beam-target combination.

Before the neutron TOF could be calculated, the background neutron contributions were taken into consideration. During the experiment, multiple runs were taken for each beam-target configuration with and without a shadow bar covering the front face of the liquid scintillators. Because runs varied in duration, system live time, and source particle exposure, counts ($N_{measured}$) in each covered run were combined and normalized to the recorded number of source particles ($s_1 \cdot s_2$), good beam fraction (GB), and live time (LT) per equation 1. This process was then repeated for the uncovered runs. After normalization, the sum of the covered runs for a given beam-target combination were subtracted from the sum of the uncovered runs. This meant the final spectra only contained information for neutrons coming directly from the front target, and not background neutrons that entered from the sides of the liquid scintillator or background neutrons that bypassed the shadow bars by skipping off the floor and entered the front of the scintillator.

$$Yield = \frac{\sum_{i=1}^{n} \frac{N_{i(measured)}}{LT_i}}{\sum_{i=1}^{n} GB_i \times (s_1 \cdot s_2)_i} \quad (1)$$

Neutron times of flight (t_n), velocities ($\beta_n = v_n / c$), and kinetic energies (T_n) were calculated using equations 2 through 5, respectively. The TDC calibration (k) varied from 0.228 ns/channel to 0.238 ns/channel, depending upon the liquid scintillator. Neutron flight paths (d) also varied depending upon the liquid scintillator. The prompt gamma peak channel number (tdc_g) was dependent upon the beam-target combination, while gamma velocity (c) and the neutron rest mass (m_n) were constant values. Finally, the TDC channel number of neutron events (tdc_n) were taken from the TDC output spectra.

$$t_n = (d / c) + k\,(tdc_g - tdc_n) \quad (2)$$

$$v_n = d / t_n \quad (3)$$

$$T_n = (\gamma - 1)\, m_n \quad (4)$$

$$\gamma = (1 - \beta_n^2)^{-1} \quad (5)$$

After converting channel number to neutron energy, an energy-dependent neutron detection efficiency correction factor was applied to each bin. Efficiencies were calculated for a 12.7 cm in length, 12.7 cm in diameter cylindrical liquid scintillator placed 2.0, 2.5, and 3.0 m away from a neutron source using the SCINFUL-QMD Monte Carlo code [8]. Double differential thick-target neutron yields were then calculated by dividing the efficiency-adjusted neutron yields by the solid angle of the liquid scintillator and the energy bin width for units of neutrons per source particle per steradian per energy (neutrons / S.P. / Ω / MeV).

Finally, the uncertainties in the counts were taken into consideration. Statistical uncertainties were calculated in ROOT and varied for each bin. Systematic uncertainties included uncertainties in solid angle, neutron detection efficiency, and graphical cuts. Solid angle uncertainties varied from 6.5% to 13.8% depending upon the flight path of the liquid scintillator and the front target thickness. The uncertainty in neutron detection efficiency was estimated at 15% using published experimental and SCINFUL-QMD calculated results [8]. Finally, the uncertainty in the ROOT graphical cuts was estimated at 5%. The summary of these systematic uncertainties organized by front target thickness is included in Table 1.

Table 1. Fractional systematic uncertainties for front aluminum target thicknesses of 20, 40, and 60 g/cm^2.

Type	Fractional Systematic Uncertainties at 10°, 135°		
	20 g/cm^2	**40 g/cm^2**	**60 g/cm^2**
Solid Angle	6.5, 8.7%	8.9, 11.3%	11.4, 13.8%
Detection Efficiency	15%		
ROOT Graphical Cuts	5%		

4 Selection of Experimental Results

A selection of background-subtracted results from the March 2016 experiment are provided below in Figures 4 through 8. Each figure contains double differential thick-target neutron yields at 10° and 135° for a single beam species on the 20, 40, and 60 g/cm^2 front aluminum targets. The 10° yields were multiplied by a factor of 100 to prevent the overlap of the 10° and 135° low-energy yields. Only statistical uncertainties were shown and they varied widely depending upon the ion beam. In general, statistical uncertainties from this experiment decreased with an increase in front target thickness. As the front target thickness increased, the number of nuclear interactions between the incident beam and target material also increased, resulting in a larger production of secondary neutrons. Additionally, an increase in beam energy typically led to lower statistical uncertainties because of the increased number of beam-target interactions. However, this was not seen with the 800-AMeV iron beam, as discussed later. Statistical uncertainties also improved as the neutron energy increased for the 10° liquid scintillator, but the uncertainties worsened at the highest neutron energies (about 60 to 90 MeV) for the 135° liquid scintillator.

Double differential thick-target neutron yields at 10° mostly consisted of neutrons produced during projectile breakup. The characteristic forward angle peak was located at approximately 70% of the incident beam energy and the yields rapidly decreased after this point, as expected. The thick-target neutron measurements and uncertainties at 10° suffered from background subtraction at low energies due to the liquid scintillator's proximity to the back target. Neutrons produced by other secondary ions or reflected off the back target typically had energies less than 100 MeV. These neutrons entered through the sides of the 10° liquid scintillator instead of the front face, which meant they were tagged as background events during the shadowed runs. Additionally, many of the beams were high enough in energy to punch through the front target, which resulted in secondary neutron production by primary beam interactions in the back target. The higher than expected background yields below 100 MeV resulted in lower-than-expected background-subtracted yields as well as poor statistics from error propagation. While the average fractional uncertainties below 100 MeV varied widely depending upon the ion beam, all average uncertainties above 100 MeV were less than 20%.

Yields at 135° were dominated by low-energy, isotropically emitted neutrons from target evaporation. Detected neutrons at this angle typically had an energy less than 100 MeV and yields exponentially decreased with energy. Statistical uncertainties were best for neutron measurements at 135° when compared to the other liquid scintillator locations due to the low neutron background at back angles. The average fractional uncertainties for all beams at all neutron energies ranged from 3.5% to 10%.

Double differential thick-target neutron yields for the 400-AMeV iron beam are displayed in Figure 4. This beam was stopped in all aluminium target thicknesses, indicating that only secondary particles were emitted from the front target. At 10°, the average fractional statistical uncertainties below 100 MeV for the 20, 40, and 60 g/cm^2 thick beam-target configurations were 12%, 9.5%, and 9.0%, respectively. Above 100 MeV, fractional uncertainties were less than 3.5%.

Figure 5 contains double differential yields for 800-AMeV iron on 20, 40, and 60 g/cm^2 aluminum front targets. Yields generated with the 800-AMeV iron beam were generally higher than with the 400-AMeV iron beam, except with the 20 g/cm^2 beam-target configuration. The 800-AMeV iron beam stopped in the 40 and 60 g/cm^2 aluminum targets, but punched through the 20 g/cm^2 target. This meant that the incident ion beam also interacted in the back 60 g/cm^2 aluminum target, increasing the number of neutrons that entered through the sides of the liquid scintillator at 10°. Double differential yields at 20 g/cm^2 were lower than expected for all secondary neutron energies because of this increase in the "neutron background" close to the back target.

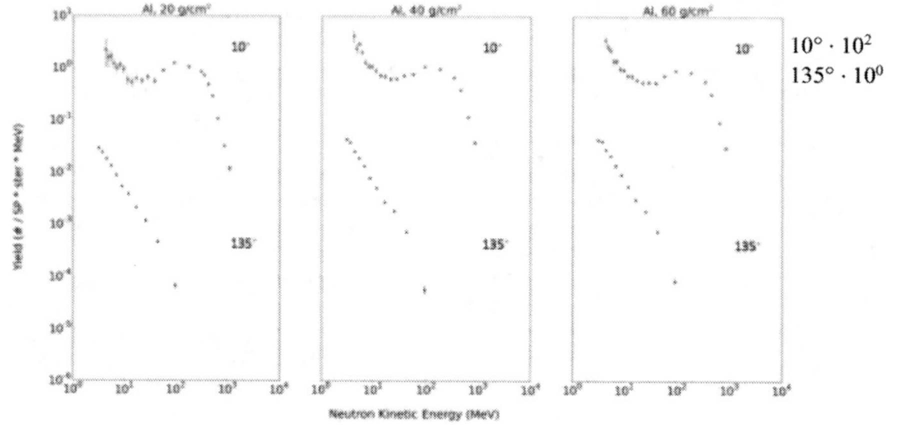

Fig. 4. Background subtracted double differential thick-target neutron yields at 10° and 135° for 400-AMeV iron projectiles on 20, 40, and 60 g/cm^2 front and 60 g/cm^2 back aluminum targets. The 10° yields were multiplied by a factor of 100 to prevent overlap.

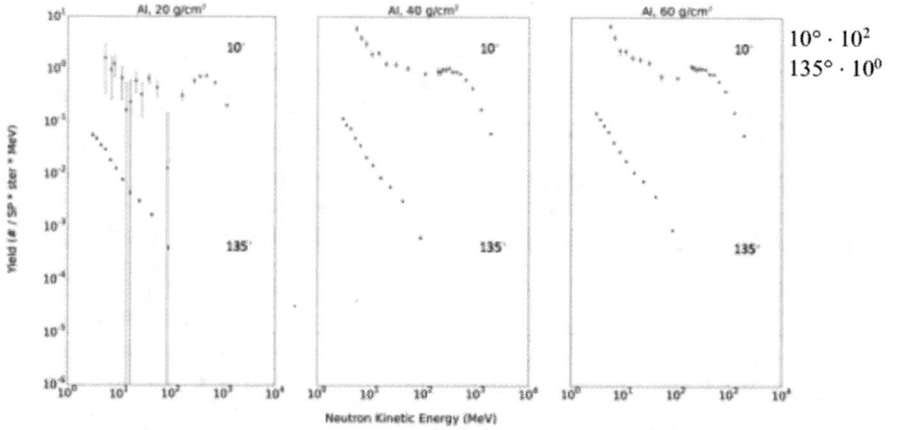

Fig. 5. Background subtracted double differential thick-target neutron yields at 10° and 135° for 800-AMeV iron projectiles on 20, 40, and 60 g/cm^2 front and 60 g/cm^2 back aluminum targets. The 10° yields were multiplied by a factor of 100 to prevent overlap.

Adding the shadowed and unshadowed uncertainties in quadrature resulted in average uncertainties over 100% below 100 MeV. However, the average uncertainty was less than 8.5% above neutron energies of 100 MeV. For the 40 and 60 g/cm^2 target, the average uncertainties below 100 MeV were less than 13%, and less than 5% at higher energies.

Double differential neutron yields for the 400- and 800-MeV proton beams are displayed in Figures 6 and 7. Both beams punched through all front target thicknesses and interacted in the back target, resulting in poor statistics below 100 MeV, similar to what was seen with the 800-AMeV iron beam on 20 g/cm^2 aluminum. Due to the increase in beam-target interactions and subsequent increase in secondary neutrons with a higher beam energy, average uncertainties were lower for the 800-MeV proton system at all target thicknesses when compared with the 400-MeV proton system. Depending upon the target thickness, average uncertainties below 100 MeV for the 400- and 800-MeV proton beams ranged from 32% to 54% and 15% to 31%, respectively. Above 100 MeV, average uncertainties were less than 22% (400-MeV protons) and less than 10% (800-MeV protons).

Finally, the 400-AMeV helium beam-target double differential yields are displayed in Figure 8. While the helium beam also punched through all target thicknesses, the uncertainty issues below 100 MeV were less pronounced than what was seen with the proton beams and the 800-AMeV iron beam on 20 g/cm^2 aluminum. Average uncertainties below 100 MeV ranged from 12% to 27%, depending upon the target thickness, while uncertainties above 100 MeV were less than 5%.

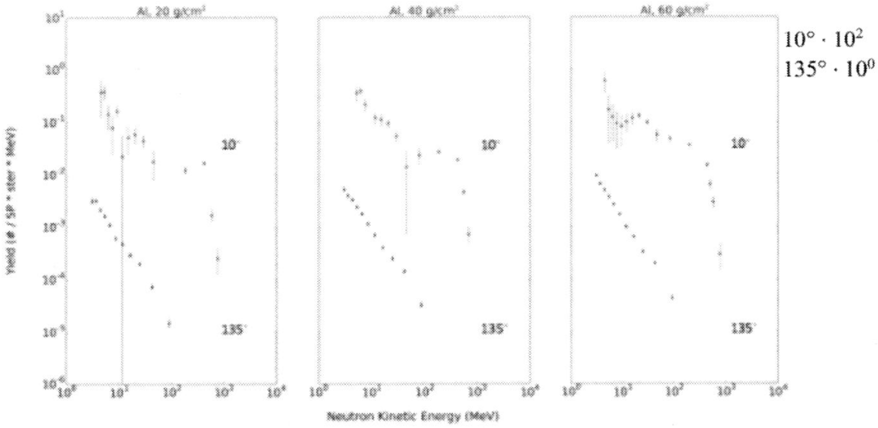

Fig. 6. Background subtracted double differential thick-target neutron yields at 10° and 135° for 400-MeV proton projectiles on 20, 40, and 60 g/cm^2 front and 60 g/cm^2 back aluminum targets. The 10° yields were multiplied by a factor of 100 to prevent overlap.

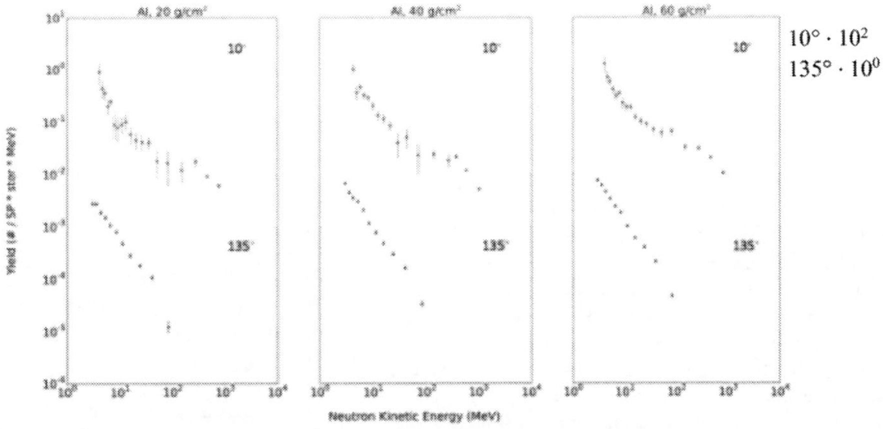

Fig. 7. Background subtracted double differential thick-target neutron yields at 10° and 135° for 800-MeV proton projectiles on 20, 40, and 60 g/cm^2 front and 60 g/cm^2 back aluminum targets. The 10° yields were multiplied by a factor of 100 to prevent overlap.

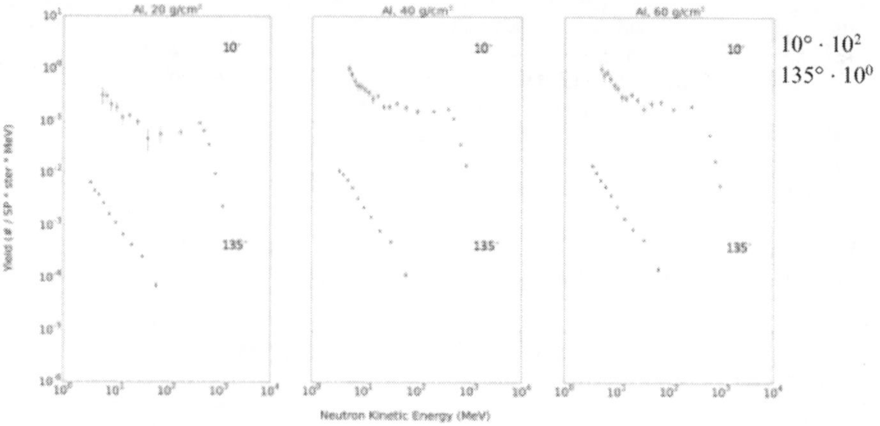

Fig. 8. Background subtracted double differential thick-target neutron yields at 10° and 135° for 400-AMeV helium projectiles on 20, 40, and 60 g/cm^2 front and 60 g/cm^2 back aluminum targets. The 10° yields were multiplied by a factor of 100 to prevent overlap.

5 Future Work

Further analysis will include quantifying the energy resolution of each neutron event using the method described in [9]. ROOT graphical cut uncertainties will be improved by reanalyzing the data with new graphical cuts. A fit to the background data at the 10° liquid scintillator will be considered to improve statistical uncertainties below 100 MeV. Additionally, these experimental results will be compared to MCNP-, PHITS-, and GEANT-simulated double differential thick target yields and their differences will be explored. In addition to this experimental dataset, two 200-hour experiments are scheduled for late 2016 and 2017 at BNL's NSRL. These experiments will follow the same setup and execution of the March 2016 experiment, but will test different beam-target configurations. Targets considered will include 20, 40, and 60 g/cm^2 aluminum and HDPE targets, and ion beams will include proton, helium, carbon, silicon, and iron projectiles at 0.4, 0.8, 1.0, and 2.5 AGeV. The analysis procedure laid out in this paper will be repeated for the new datasets.

6 Summary

100 hours of thick-target experiments were conducted in March 2016 at BNL's NSRL facility. Five ion beam species (400- and 800-AMeV iron, 400 and 800-MeV protons, and 400-AMeV helium) were bombarded on three aluminum upstream targets of varying areal thicknesses (20, 40, or 60 g/cm^2). A second 60 g/cm^2 aluminum target was placed 3.5 m downstream from the front target. Neutrons and light charged ions were measured at six angular locations off beam axis using EJ-301 and EJ-309 liquid scintillators. Neutron events was separated from the gamma and light charged ion events using a series of cuts in ROOT and then normalized to the number of source particles. Data was then background subtracted for each beam-target configuration, and the TDC channels were converted to energy using the time-of-flight method. Finally, neutron yields were corrected for neutron detection efficiency and then normalized to solid angle and energy bin width to produce double differential thick-target yields in units of neutrons / source particle / steradian / MeV. Neutron yields at 10° suffered from a high neutron background due to the liquid scintillator's proximity to the back target, while yields at 135° had the best uncertainties of all the angles. Additional analysis is needed to explore other methods of quantifying background neutrons at 10°, as well as to compare the experimental results presented here to Monte Carlo simulations. Ultimately, the experimental double differential thick target yields will be incorporated into NASA's uncertainty analysis for their radiation transport codes.

This work was supported by Grants NNX12AL51A and NNX15AD89A from the United States National Aeronautics and Space Administration. Special thanks to Matthew Beach, Ph.D. for his assistance with ROOT. Additional thanks to the team at Brookhaven National Laboratory for their extensive assistance with the setup and execution of the experiment.

References

1. P.-E. Tsai, *et. al.*, *2015 IEEE Aerospace Conference*, 1-7 (2015).
2. L. Heilbronn, *et. al.*, Nucl. Sci. Eng., **132**, 1-15 (1999).
3. P.-E. Tsai, "Study of secondary particles produced from heavy-ion interactions," Ph.D. dissertation, Dept. of Nucl. Eng., Univ. of TN, Knoxville, TN (2015).
4. R. Brun, F. Rademakers, Nucl. Instr. Meth. Phys. Res. A, **389**, 81–86 (1997).
5. T. Nakamura, L. Heilbronn, *Handbook on Secondary Particle Production and Transport by High-Energy Heavy Ions* (Singapore: World Scientific Publishing Co. Pte. Ltd., 2006).

6. J. H. Heltsley, *et. al.*, Nucl. Instr. Meth. Phys. Res. A, **263**, 441-445 (1988).
7. V. V. Verbinski, *et. al.*, Nucl. Inst. Meth., **65**, 8–25 (1968).
8. D. Satoh, *et. al.*, J. Nucl. Sci. Tech., **39**, 657–660 (2002).
9. Y. Iwata, *et. al.*, Phys. Rev. C, **64** (2001).

A Radiation Analysis of 500W RTPV System having Cylindrical Sources

Seong Jae Cheon [1], *Ser Gi Hong* [1,*] *Young Suk Nam* [2] and *Jung Hun Lee*[2]

[1]Dept. of Nuclear Engineering, Kyung Hee University, 1732 Deokyoungdaero, Giheung-gu, Yongin, Gyeonggi-do 446-701, Republic of korea

[2]Dept. of Mechanical Engineering, Kyung Hee University, 1732 Deokyoungdaero, Giheung-gu, Yongin, Gyeonggi-do 446-701, Republic of korea

Abstract. In this work, the radiation shielding analysis for a RTPV (Radioisotope Thermo-Photo-Voltaic) device using $^{238}PuO_2$ was performed to estimate the radiation dose distribution over source region and radiation doses in a specified tally zone. The radiation source intensities and spectra were evaluated with ORIGEN-S and the detailed shielding analyses were performed with MCNP6 and MONACO/MAVRIC (in SCALE 6.1). In particular, comparative shielding analysis using two different types of RTPV (cubic homogeneous source type and cylindrical heterogeneous source type) were performed to show their relative performances. The results show that the RTPV device using the heterogeneous cylindrical source type has lower dose values by 5~11% depending on the packing factor at the measurement cell than the one using the homogeneous cubic source type. The additional thermal analysis using COMSOL shows that the temperature distribution of the heterogeneous cylindrical source type is more flat than that of a homogeneous cubic source type.

1 INTRODUCTION

Radioisotope Thermoelectric Generators (RTG) has been considered as a power supply system for long-lived operation in space. RTG is an essentially nuclear battery that converts the heat resulted from the radioactive decay into electricity. RTG has the advantage of high energy density and long life operation due to very long half-lives of the radioisotopes. However, the RTG systems have relatively low thermoelectric energy conversion efficiency of 3-7% because RTGs rely on the thermos-couples to convert heat to electricity. On the other hand, in a RTPV, radiations such as beta or alpha particles heat an emitter material which emits photon to a PV cell converting the photons into electricity. Therefore, a RTPV (Radioisotope Thermo-Photo-Voltaic) system has typically much higher energy conversion efficiency of 15~30% than RTG system [1].

Recently, we have worked on a project for designing a small RTPV device (500W) and simulating its thermal performances by using a small scale experiment. In this project, the kinetic energy of alpha particles (~5MeV) from $^{238}PuO_2$ is considered as a long-term power supplier for the RTPV device. The alpha decay of ^{238}Pu accompanies additional neutrons by (α,n) reactions of ^{17}O and ^{18}O, and also the spontaneous fission of ^{238}Pu generates neutrons. In addition to these neutrons, these radioactive decay and reactions are accompanied by gammas. Therefore the shielding of these radiations is very important for radiation safety.

The objective of this work is to perform radiation dose analysis of heat source region for the RTPV system. In particular, a comparative shielding analysis using two different types of RTPV (cubic homogeneous source type and cylindrical heterogeneous source type) was performed to show their relative performances. In addition, we estimated the temperature distribution for thermal safety analysis of RTPV devices.

2 RADIATION SAFETY ANALYSIS

2.1 Radiation shielding analysis procedure

Fig. 1 shows the calculation procedure for the safety analysis of RTPV system. The radiation safety analysis consists of two steps. First, the ORIGEN-S code was used to evaluate the intensities and spectra of neutrons and gamma rays emitted from the $^{238}PuO_2$ [2]. Then, the doses are estimated by performing neutron and gamma transport simulations using two Monte Carlo codes and the source intensities and spectra estimated with ORIGEN-S. In this work, we used MCNP6 and MONACO/MAVRIC to perform the neutron and gamma transport calculations [3,4].

* Corresponding author: sergihong@khu.ac.kr

DOI: 10.1051/epjconf/201715304003

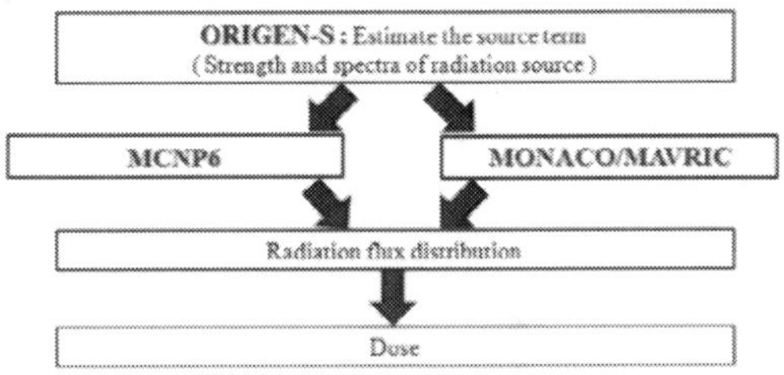

Fig. 1. The calculation procedure for a safety analysis.

2.2 Source term evaluation with ORIGEN-S

ORIGEN-S which is a part of SCALE6.1 is a computer code for evaluating the radioactive decay of radionuclide and the amount of nuclides after the neutron irradiation and cooling [2]. In this work, we used ORIGEN-S code for evaluating the intensities and spectra emitted from the source of RTPV system. The radiation source terms should be described prior to the detailed radiation safety analysis and their accuracies are quite important for accurate estimation of radiation dose.

The alpha decay of ^{238}Pu releases ~5.5MeV energy and also additional neutrons are emitted as result of the reactions (i.e., (α,n) reactions) of ^{17}O and ^{18}O with the alpha particles. In addition, the spontaneous fission of ^{238}Pu generates neutrons. Table 1 summarizes the neutron source intensities estimated using ORIGEN-S for a 500W heat generation.

Table 1. Neutron source intensities (neutrons/sec) for 500W $^{238}PuO_2$ source.

Components	Values
Spontaneous fission	$2.39\text{x}10^6$
Delayed neutrons	Negligible
(α,n) reactions - (mainly from ^{18}O)	$1.253\text{x}10^7$
(α,n)reactions - (from ^{17}O)	$9.996\text{x}10^5$
Total (α,n) reactions	$1.253\text{x}10^7$
Total	$1.492\text{x}10^7$

For gamma source, the most dominant contribution is from the radioactive decay of ^{238}Pu while the spontaneous fission of ^{238}Pu and (α,n) reactions generate additional gammas [5]. The gammas from the decay of ^{238}Pu have peaks at 43 keV, 99 keV, and 152 keV energies These spectra are obtained by using ORIGEN-S with the 200 and 47 energy group structures for neutron and gamma, respectively.

2.3 Radiation safety analysis with MCNP6 and MONACO/MAVRIC (SCALE6.1)

2.3.1 Computer codes for radiation dose analysis

We performed radiation safety analysis by using MCNP6 and MONACO/MAVRIC codes with the specified intensities and spectra of the radiation source determined with ORIGEN-S. MCNP6 is an advanced general purpose (continuous-energy and general geometry) simulation code that transports 37 different particles for criticality, shielding, dosimetry and so on. It is the latest version of the famous MCNP code [4]. MONACO is a new 3-D Monte Carlo code being developed within SCALE for shielding calculations. It is a multi-group Monte Carlo transport code with fixed sources for shielding applications. The tallies available in MONACO include point detectors, region-based flux tallies and mesh tallies. Specially, mesh tally values can be viewed with the mesh file viewer in SCALE6.1. The MAVRIC sequence is completely automated for shielding analysis [3].

2.3.2 Description of RTPV using homogeneous cubic source

In our previous work, a homogeneous cubic source sequentially surrounded by a tungsten material region and a thin emitter region of tantalum was considered for RTPV device [6]. In this work, the effects of the packing factor which is the ratio of heat source region (i.e., $^{238}PuO_2$ region) volume to the total region enclosed by the emitter on the radiation dose are analysed. In particular, a cladding is considered to enclose the source region in order to prevent releasing of $^{238}PuO_2$ under accident. The size of the source is fixed because the release of heat source is fixed to 500W. The total region enclosed by emitter consists of a 4.4910cm x 4.4910cm x 4.4910cm central cubic source region and its three surrounding regions. The central source region is first surrounded by an iridium (Ir) cladding region. The second surrounding region is a tungsten (W) radiation shielding region that is followed by a 0.5cm thick tantalum (Ta) outer emitter region. The thickness of the shielding material is determined by the packing factor. Fig. 2 shows the configuration of the cubic heat source region for RTPV device.

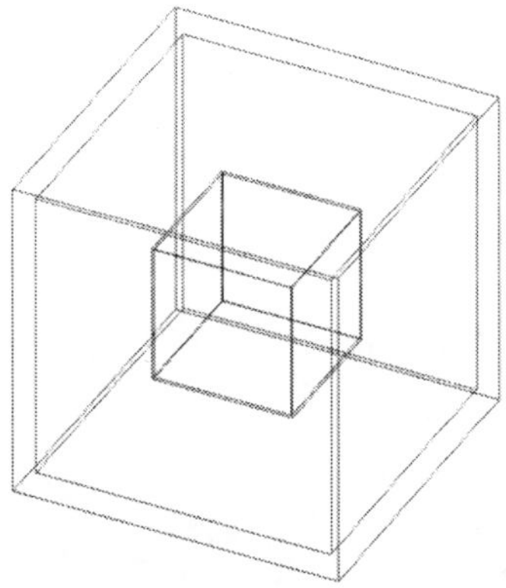

Fig. 2. Geometric model for the source of a cubic design

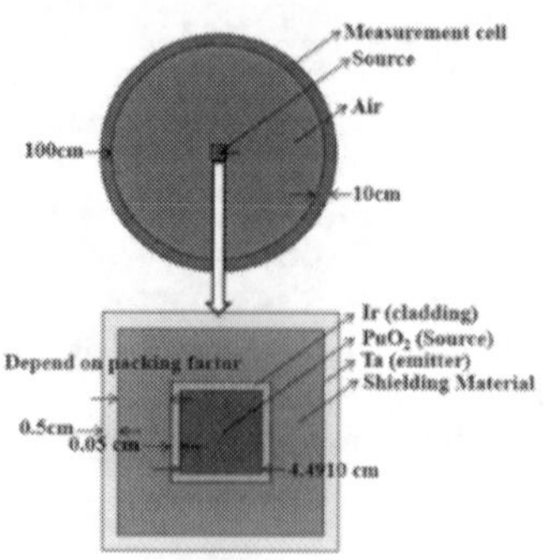

Fig. 3. Geometric model for the RTPV device of the cubic source design

The radiation dose is estimated in a 10cm thick spherical measurement shell which is located at 100cm distance from the center of the source. The region between RTPV device and the measurement spherical shell is assumed to be filled with air. Fig. 3 shows MCNP6 and MONACO/MAVRIC geometric modelling for radiation safety analysis. The radiation dose was converted from the radiation fluxes using the ICRU-57 and the ICRP-21 dose conversion factors for SCALE6.1 [7] and MCNP6 [4], respectively.

2.3.3 Description of RTPV using cylindrical sources

In this work, we also considered RTPV device where four separate cylindrical source zones having iridium cladding are symmetrically located and the dose values in a 10cm thick spherical measurement shell which is located at 100cm distance from the center of the source are estimated with MONACO/MAVRIC. In particular, the dose values for this RTPV device having cylindrical sources are compared with those for the one having cubic source. The reason for the consideration of separated cylindrical sources is to flatten the neutron flux and temperature distributions. As in the previous design using a single cubic source, the total volume of the source is fixed and it is determined to generate 500W heat. Fig. 4 shows the geometric model of the RTPV device having separated cylindrical sources for radiation safety analysis. A cylinder source whose radius and height are 1.217cm and 4.867cm, respectively releases 125W heat.

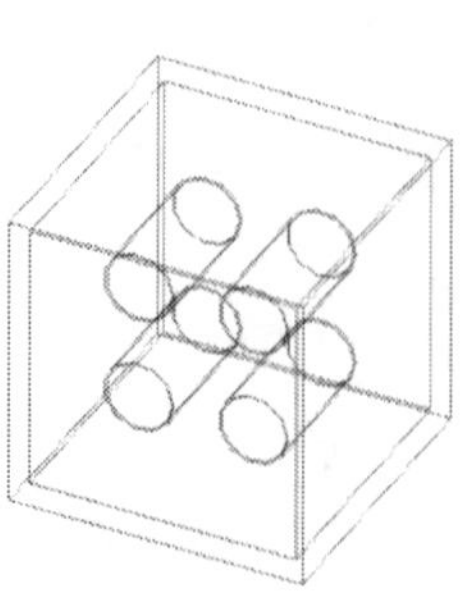

Fig. 4. Geometric model for the source of the cylindrical design

Fig. 5 shows MCNP6 and MONACO/MAVRIC geometric model for the RTPV device of the cylindrical source design. The distances between the cylindrical sources were adjusted to keep the same distance between the outer surface of the source cladding and the inner surface of the emitter for different packing factors.

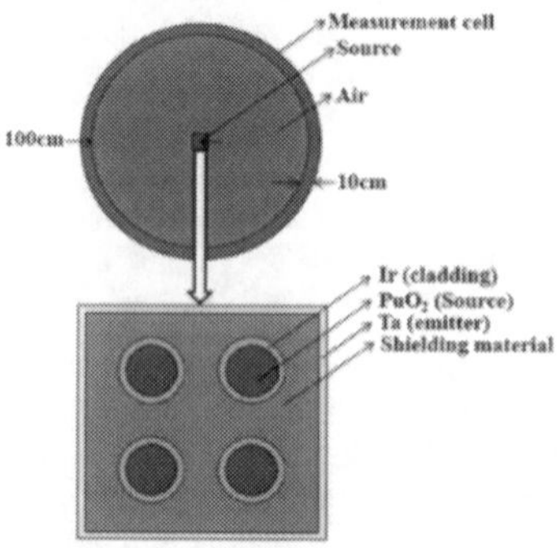

Fig. 5. Geometric model for the RTPV device of the cylindrical source design

2.3.4 Evaluation of the radiation doses

The thermal-to-electric energy conversion efficiency of RTPV strongly depends on the packing factor and the increase of packing factor usually leads to the higher conversion efficiency. On the other hand, the effect of the packing factor on the radiation dose is also needed to be estimated. In this work, the effects of the packing factor on the radiation dose are comparatively estimated for the RTPV devices having a single cubic source and separated cylindrical sources described above. Table 2 compares the total radiation doses in a 10cm thick measurement shell that is located at 100cm distance from the center of the source for the RTPV devices having a single cubic source and the separated cylindrical sources. As shown in Table 2, the dose values for the separated cylindrical sources are smaller than that for the single cubic sources for all the packing factors. In particular, it is noted that the difference in the dose values between these two different source configurations has the largest value of ~11.1% for the smallest packing factor (i.e., 10%). As expected, these differences in the dose values decrease as the packing factor increase. Fig. 6 compares these dose values graphically.

Table 2. Doses in the measurement sell (mSv/hr)

Packing factor (%)	Total dose value	
	Cubic source	Cylindrical sources
10	1.79E-01	1.59E-01 (11.1[a])
30	1.94E-01	1.78E-01 (8.2)
50	1.99E-01	1.87E-01 (6.2)
55	1.93E-01	1.89E-01 (5.3)

[a] Difference (%) between the dose values for the different source configurations

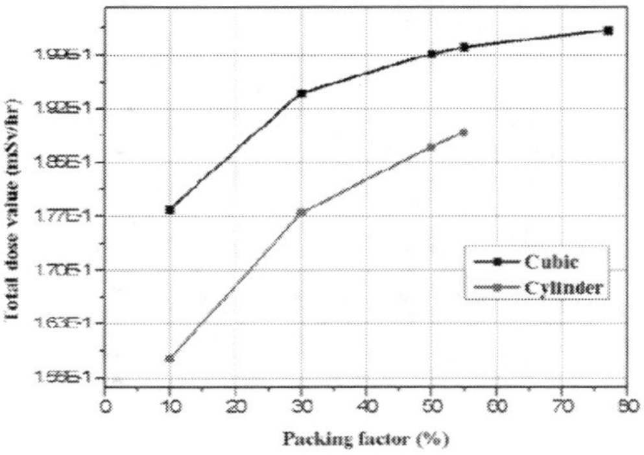

Fig. 6. Comparison of the total doses for a 500W heat generation between two source designs

Actually, the neutron dose is much more dominant than the gamma dose. Figures 7(a) and 7(b) compare the neutron dose distributions at the center planes for the RTPV devices having a cubic source and separated cylindrical sources, respectively. As expected, it is shown that the neutron dose distribution for the separated cylindrical sources is much more flat and the dose at outer surface for this case is lower than those for the single cubic source.

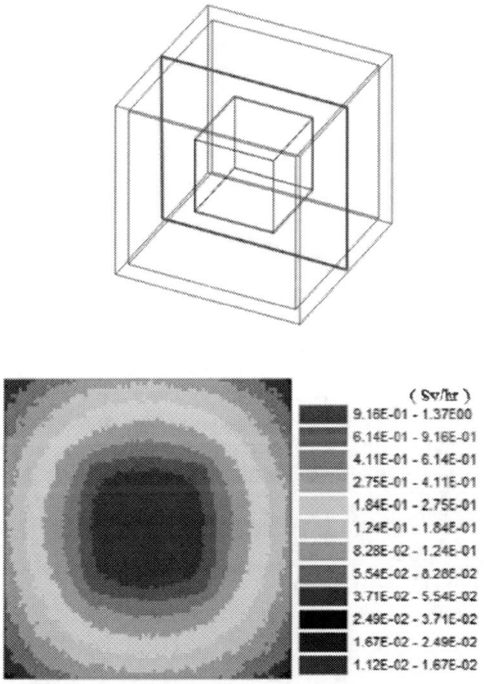

(a) A cubic source design

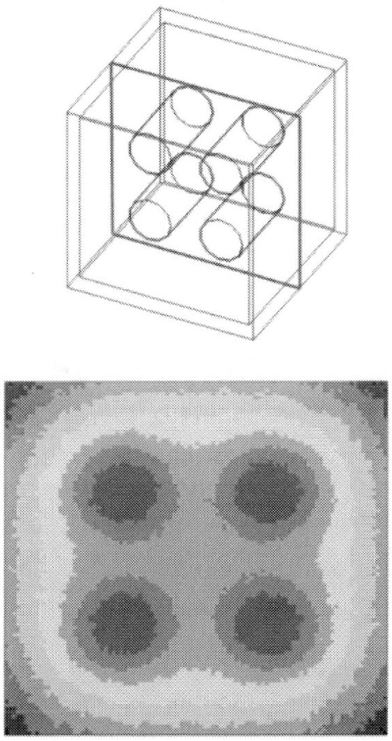

(b) Cylindrical sources design

Fig. 7. Comparison of the neutron dose distributions at the central planes (with same scale in dose, Sv/hr)

3 Thermal analysis

3.1 Modelling with COMSOL for thermal analysis

Next, we performed thermal safety analysis by using the COMSOL program. The COMSOL Multiphysics program is a powerful interactive environment for modeling and solving all kinds of scientific and engineering problems based on partial differential equations [8]. We analyzed temperature distributions using COMSOL and tetrahedral meshes. The geometrical modelings for the safety analysis are same as those for the radiation safety analysis. Table 3 shows the input physical parameters for thermal analysis of the RTPV devices.

Table 3. Input parameters for thermal analysis for the RTPV devices

Parameter	PuO_2	Ir	W	Ta
Thermal conductivity (W/m·K)	6.3	147	173	57.5
Heat capacity (J/kg·K)	240	233	130	140
Density (g/cm^3)	11.5	22.42	19.25	16.69

Eq. (1) shows the radiation boundary condition for estimating temperature distribution for the RTPV devices.

$$Q = \varepsilon\sigma T^4, \qquad (1)$$

where Q is the outward heat flux (W/m^2), ε is the emissivity, σ is the Stefan-Boltzmann constant (5.67 x 10^{-8}) and T is temperature (K). Figure 8 shows the emissivity as a function of temperature which is from the experiment data for the RTPV device.

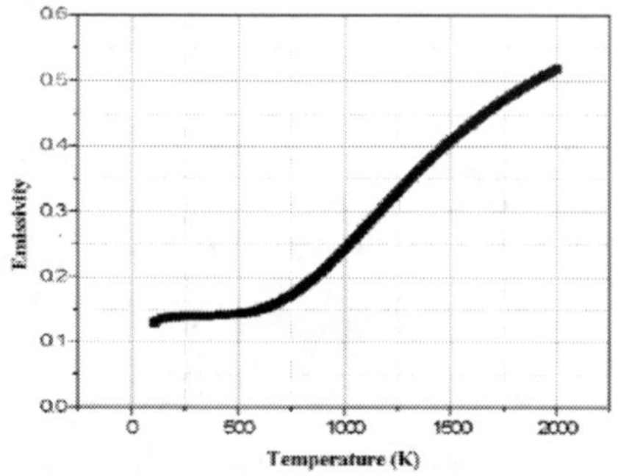

Fig. 8. Emissivity (ε) by temperature from the experiment data

3.2 Result of the thermal analysis

Figures 9(a) and 9(b) compare the temperature distributions at the axially central planes for the different source designs. Figure 9 shows the temperature distribution for the cylindrical sources design is much more flat than that for the cubic source design. Table 4 shows the maximum temperatures and the average temperatures on the outer faces for the two RTPVs having different source shapes. For the RTPV having cubic source, the maximum temperature is about 993.4 K which is higher by ~69 K than that of the one having the four cylindrical sources while the average temperatures on their outer faces are almost the same. Therefore, the temperature gradients for the cylindrical sources are smaller than that those for the cubic source design, which means that the cylindrical sources design will enhance the thermal integrity of the RTPV system.

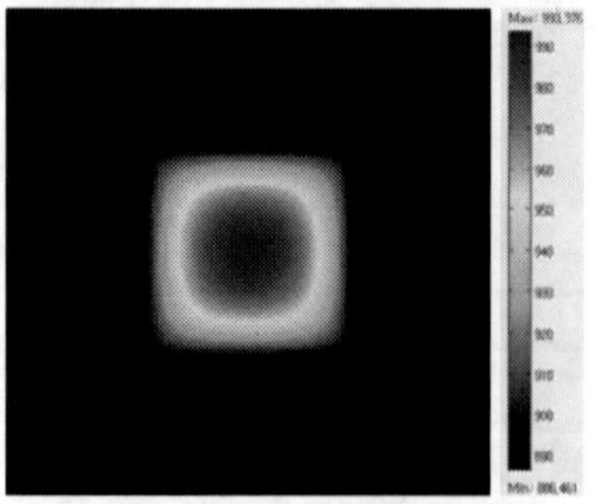

(a) RTPV with cubic source design

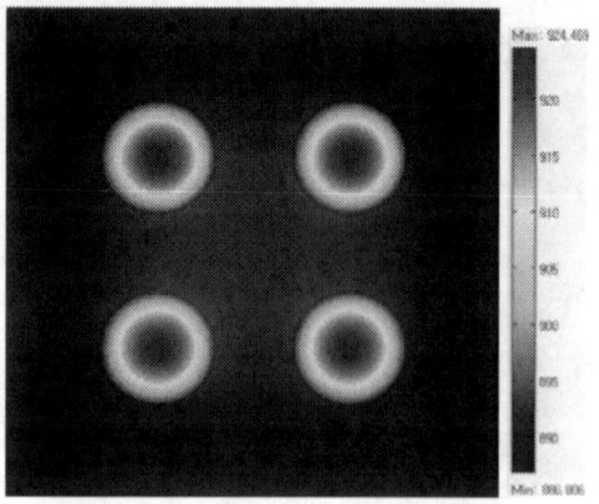

(b) RTPV with four cylindrical sources

Fig. 9. Comparison of temperature distributions at the central planes (with scale in temperature, K)

Table 4. Maximum and outer face average temperatures for RTPVs having a cubic source and four cylindrical sources (K)

Case	Maximum	Average on the outer face
Cubic source	993.4	887.5
Four cylindrical sources	924.5	887.5

4 Conclusion

In this work, the radiation shielding analysis was performed for 500W RTPVs having different source designs. In particular, a comparative shielding analysis using two different types of source design (i.e., homogeneous cubic source type and heterogeneous cylindrical sources type) was performed to show their relative performances. The results show that the RTPV device using the heterogeneous cylindrical sources has lower dose values by 5~11% depending on the packing factor at the measurement cell than the RTPV device using the cubic source type. In addition, we evaluated and analysed the temperature distributions for these two RTPVs having different source designs. The results of the thermal analysis showed that the RTPV having the separated cylindrical sources has much smaller temperature gradients than the one having a cubic homogeneous source. Therefore, it can be concluded that the RTPV design having separated cylindrical sources can be considered as a candidate for the future RTPV design from view point of the radiation dose and the thermal integrity.

Acknowledgement

This work was supported by Space Core Technology Program through the National Research Foundation of Korea (NRF) funded by the Ministry of Science, ICT & Future Planning (No. 2014M1A3A3A02034818).

References

1. J. Crepeau, *Design of a radioisotope thermo-photovoltaic power source by laser driven isotope surrogate assembly*, 8 (December 2013)
2. I. C. Gauld, *ORIGEN-S: Depletion Module to Calculate Neutron Activation, Actinide Transmutation, Fission Product Generation, And Radiation Source Terms* (June 2011)
3. D. E. Pelplow, S. M. Bowman, James E. Horwedel, and J. C. Wagner. *Monaco/MAVRIC: Computational Resources for Radiation Protection and Shielding in SCALE, Oak Ridge National Laboratory*, MS 6172.
4. D. B. Pelowitz, *MCNP6 User's Manual Version 1.0*, Manual Rev.0. (May 2013)
5. NASA Ames Research Center, *RTG/Science Instrument Radiation Interactions Study for Deep Space Promes*, Contract No. 2-5222. (1969)
6. S. J. Cheon, S. G. Hong, Korean Nuclear Society, *A Radiation Shielding Analysis for 500W Radioisotope Thermo-Photo-Voltaic(RTPV) System with MCNP6 and MONACO/MAVRIC,* Transaction of the Korean Nuclear Society Autumn meeting (Octobor 2015)
7. Oak Ridge National Laboratory, *Scale: A Comprehensive Modeling and Simulation Suite for Nuclear Safety Analysis and Design*, ORNL/TM-2005/39, Version 6.1. (June 2011)
8. COMSOL 3.5a, User's Guide

Secondary light-ion transport from intermediate-energy hadron experiments

Ashwin P. Srikrishna[1,a], *Luis* A. Castellanos[1], *Natalie* A. McGirl[1], *Lawrence* H. Heilbronn[1], *Chiara* La Tessa[2], *Adam* Rusek[2], *Michael* Sivertz[2], *Steve* Blattnig[3], *Martha* Clowdsley[3], *Tony* Slaba[3], and *Cary* Zeitlin[4]

[1]*Department of Nuclear Engineering, University of Tennessee, 1004 Estabrook Rd, Knoxville, TN, USA, 37996*
[2]*NASA Space Radiation Laboratory, Brookhaven National Laboratory, Upton, NY, USA, 11973*
[3]*National Aeronautics and Space Administration, Langley Research Center, Hampton, VA, USA, 23681*
[4]*Lockheed Martin, Information Systems and Global Solutions, Houston, TX, USA, 77285*

Abstract. The aim of this research is to produce double differential thick target yields, angular distributions and integrated yields for the inclusive production of neutrons, protons, deuterons, tritons, ^{3}He, and ^{4}He from intermediate heavy-ion interactions on thick targets of aluminium, polyethylene and other targets of interest to the radiation shielding program as specified by the National Aeronautics and Space Administration (NASA). In tandem with the experimental research, transport model calculations of these thick target yields were also performed. The first such experimental run was conducted in May 2015, with the expectation of improved experimental results at a following March 2016 run at the NASA Space Radiation Laboratory (NSRL) on the campus of Brookhaven National Laboratory (BNL). The May 2015 commissioning run served to test the electronics of the experimental setup, as well as the various detectors and other equipment under the conditions in which the following measurements will be run. The series of future accelerator-based experiments will rely on the inclusion of two separate upstream and downstream targets. Analysis of the data from both sets of detectors – liquid scintillator and sodium iodide – using both pulse height and time-of-flight methods will allow NASA to perform uncertainty quantification and sensitivity analysis on their transport codes and future shielding studies.

1 Introduction

There are two primary sources of ionizing radiation in the solar system – solar energetic particles (SEP) and galactic cosmic rays (GCR). Any future manned mission into space requires a more complete understanding of the radiation effects of exposure to these two types of ionizing radiation. SEP derive from energy released via solar flares in the form of highly energetic nuclei, so an increase in the frequency or intensity of solar flares (i.e. mimicking the solar activity cycle) roughly corresponds with an increase in SEP flux [1]. And similarly, GCR derive from the ever-present high energy nuclei present in the cosmos. Naturally, dose reduction is a major target for the future success of deep-space missions, and the current strategies rely on the use of thick (areal density of 30 g/cm^2 or greater) shielding materials, including aluminium and high density polyethylene [1]. NASA is attempting to model these types of shielding setups through the use of both stochastic Monte Carlo codes (FLUKA, PHITS, etc.) and deterministic codes (HZETRN), but there are issues involving both code accuracy and computational runtime [2]. Running experiments designed to calculate double differential (spatial, energy) thick target yields at the NASA Space Radiation Laboratory (NSRL) at Brookhaven National Laboratory (BNL) will allow for a collaboration with NASA to improve the quality of the shielding codes used by NASA. These results will allow for the opportunity to practically study the output of these simulations, as well as provide input in the form of uncertainty quantification and sensitivity analysis in the development of NASA codes.

Figure 1 illustrates one such example of the various models (HZETRN, Geant4, FLUKA and PHITS) providing very interesting results from a simulation of a GCR field spectrum directed at a water target with both front and back shields of infinite lateral dimensions [2]. There are differences between the models, but all seem to imply that at a particular point, increasing the shielding actually increases the calculated dose equivalent. Figure 1 represents the dose equivalent from all charged particles and charged ions – but if specific segments are looked at, the specific deviations can be found. For instance, looking only at the differential flux for deuterons, tritons, helium-3 and helium-4 in this simulated setup, there are statistically-significant disagreements amongst all of the codes. For the heavier ions ($Z > 8$) and heavy ion fragments ($2 < Z < 8$), there are reasonable levels of agreement on dose contributions. As such, this experiment is designed to study the $Z = 1$ & 2 particles - their production, their various energy- and spatial-based fluxes, and their dose contributions inside the target area.

[a] Corresponding author: asrikris@vols.utk.edu

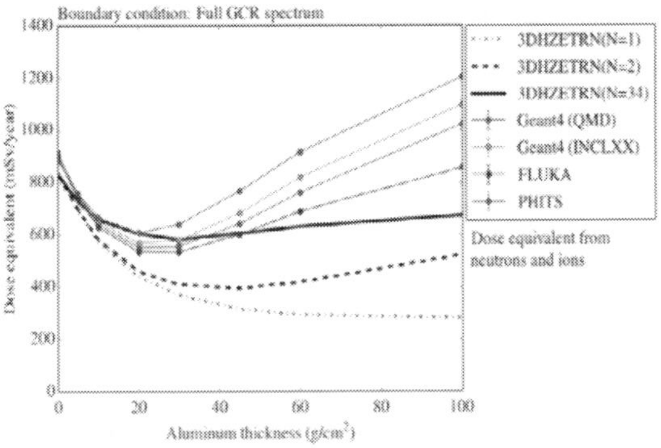

Figure 1. NASA code output for a simulated water target with front and back shielding struck by a full GCR spectrum [2].

This seems to suggest that at a point, while the shielding decreases the exposure concerns with respect to the primary GCR and SEP radiation fields, a secondary radiation field is created posing an additional risk [3]. This secondary radiation field – composed primarily of neutrons and charged light ions – is derived from the primary ionizing radiation interacting with the shielding material. This is further exacerbated by the secondary field interacting with the back (downstream) target and producing a tertiary field. The highly penetrative nature and large dose equivalent factors of the secondary radiation field necessitate that a better understanding of this field is needed for both the safety of the astronauts and the improvement of these transport codes [2,4]. This experiment is designed to study and investigate these light ions by simulating the effects of GCR and SEP on varying aluminium and HDPE shielding thicknesses.

2 Experiment Design

Secondary light ion production (including neutrons and charged light ions ranging from protons up through ^{4}He) were measured during a commissioning run at the NSRL facility at BNL in May of 2015. The experiment focused on four separate beams – 400 and 2500 MeV protons, and 400 MeV/nucleon and 1000 MeV/nucleon iron – striking a single upstream target of 30 g/cm^2 aluminum

The sodium iodide array was composed of four 4 x 2 x 16 in^3 (10.16 x 5.08 x 60.64 cm^3) sodium iodide detectors. These detectors were the best available at the time of the commissioning run, and have since been replaced by larger 4 x 4 x 16 in^3 (10.16 x 10.16 x 60.64 cm^3) sodium iodide detectors provided by the Department of Homeland Security on long-term loan to the University of Tennessee (UT-Knoxville). These detectors originate from portal monitors no longer in use, and allow for the implementation of additional sodium iodide arrays. They are twice as thick as the detectors used in the commissioning run, meaning only two such detectors are necessary to form a similarly sized array. The new detector setup for the March 2016 experimental run can be seen in Figure 2.

The locations for the organic liquid scintillators (ELJEN Model EJ-301) has slightly changed (each arc represents a radial increase of 0.5 m), due to the inclusion of more sodium iodide arrays, as well as the introduction of shadow bars. The large sections of iron shadow bars (lengths of 1 m and 2 m) are placed between the upstream target and the liquid scintillators to help separate neutrons coming from the upstream target and those coming from other sources (primarily the downstream target and room-scattered neutrons).

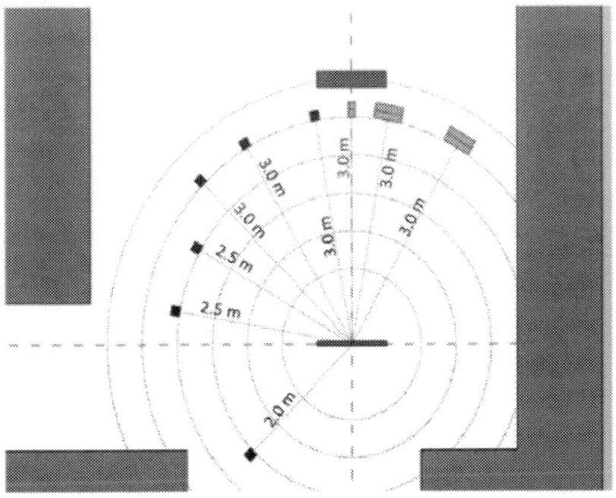

Figure 2. Detector types and locations, as well as relative target thicknesses and locations, for 2016 experiment at BNL.

Among other changes for the experimental run occurring in March 2016 include a second downstream target, as well as an increase in beam species and energies: the front target varies between 20, 40 and 60 g/cm^2 aluminum and the beam species include helium, in addition to iron and protons, at energies of 400 and 800 MeV/nucleon. The back target is placed 3.5 m downstream from the front target, and is composed of either aluminum or HDPE with a surface density of 60 g/cm^2. A detailed schematic showing features of the irradiation room is presented in Figure 3, illustrating the locations of the aforementioned targets and detectors with regards to the physical beamline at NSRL. Table 1 summarizes the beam species/energy/shielding configurations used in both the commissioning run and the March 2016 follow-up run.

Table 1 – Summary of experimental setups

Beam Species	Beam Energy (MeV)	Shielding (g/cm^2)
Helium (He)	400	Al: 20, 40, 60 + 60
Hydrogen (H)	400	Al: 20, 40, 60 + 60
Hydrogen (H)	800	Al: 20, 40, 60 + 60
Hydrogen (H)	400	Al: 30
Hydrogen (H)	2500	Al: 30
Iron (Fe)	400	Al: 20, 40, 60 + 60
Iron (Fe)	400	Al: 30
Iron (Fe)	800	Al: 20, 40, 60 + 60
Iron (Fe)	1000	Al: 30

Not present in any of these schematics is the illustration of the thin solid plastic scintillators, all of which cover the faces of the particle detectors. These scintillator paddles are important, for they allow for neutron/charged-particle discrimination, and identifying event occurrence (i.e. detection/triggering). These veto paddles are placed facing the target (the liquid scintillators have an additional paddle covering the detector face) – two were used for each neutron EJ-301 scintillators, and six (three in the front and three in the back) for the sodium iodide array. The paddles used in conjunction with the sodium iodide arrays also allow for increased angular dependencies (by focusing on particles triggered by specific configurations of the six paddles) for energy deposition data and measuring particle flight times. Additionally, two scintillating paddles were used in front of the upstream target to help detect incoming beam particles, which allowed for the discrimination of detected events by focusing only on the valid beam particles. Furthermore, these paddles were used within the logic units to set triggers for particle detection within the detectors – by splitting the signal into a real-time signal and a delayed signal, a time gap between the real-time (start) and delayed (stop) is created. This time gap is used to correlate coincidences between these "Start" paddles and the initial signals registered by the paddles in front of the sodium iodide detectors. If a correlation is confirmed, a true signal is registered and the data point is deemed 'valid'.

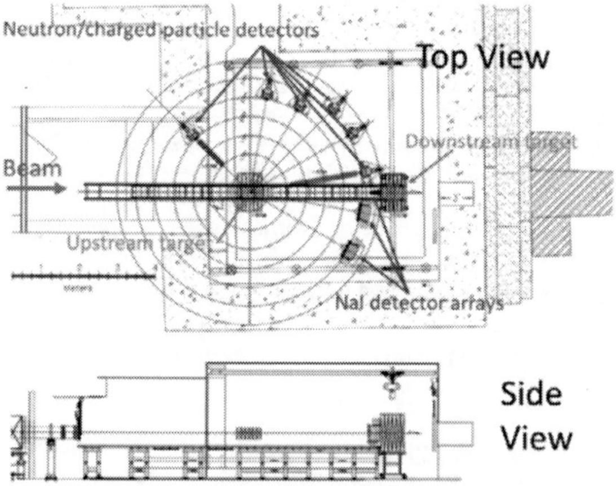

Figure 3. Schematic of experimental setup to be used for thick-target measurements at NSRL/BNL, illustrating location of beam direction, target location and detector location.

3 Experimental Analysis

The analysis process required several steps to produce the relevant particle spectra. The first such step involved the two scintillator paddles in front of the upstream target. The secondary paddle has a slightly smaller active area than the former, and as such by plotting the charge deposition against each other, the "good beam" can be identified. This region can be seen in Figure 4 – the central oval is the good beam. Figure 4 is an example from the iron 800 MeV/nucleon run on 60 g/cm^2 – these types of plots were used for each set of runs. By focusing on the good beam, extraneous events that set incorrect timing information can be removed, as well as ignoring any particles that were off-axis relative to the beamline.

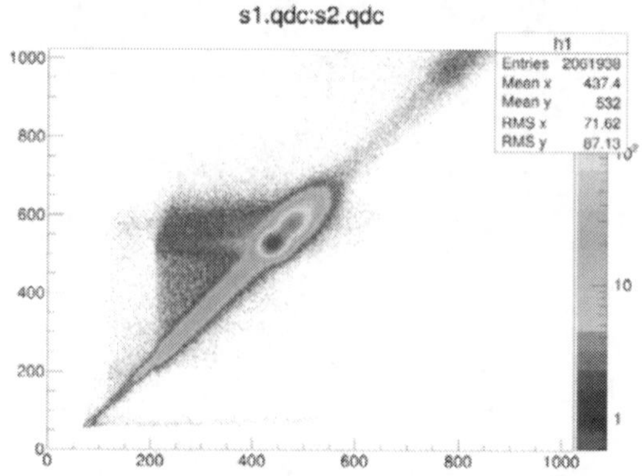

Figure 4. Iron 800 MeV/nucleon charge deposition plot used to identify the good beam. The front paddle (Y-axis) is plotted against the back paddle (X-axis) in terms of channel number.

Once this initial beam cutoff is made, the timing information from the veto paddles is necessary. There are essentially two signals registered by the timing modules for the paddles – a self-time peak that represents the timing coincidence between the veto paddle and the Start paddles, and an overflow peak that represents all of the data when a different detector set the trigger. By further gating the data on the self-time peaks for the front three paddles in front of each sodium iodide array in inclusive disjunction, the data from the sodium iodide arrays is only due to particles who have been born as a result of setting off timing-based triggers in both the Start scintillator paddles (in front of the upstream target) and the veto paddles (immediately in front of the sodium iodide arrays).

The next step in the analysis is charged particle identification. There are two methods used within the sodium iodide array, with varying levels of success. The time-of-flight (ToF) method involves the timing information from the Start paddles compared with the energy deposited within the front sodium iodide detector. An example of a ToF plot from the front 10° sodium iodide detector is shown in Figure 5. The channel numbers for flight time are inversely related – a higher channel number represents a shorter flight time. Each individual curve represents a specific charged particle ion, and the peak point of each curve represents the "punch-through" point. The punch-through point represents the maximum amount of energy that can be deposited before the particle ranges out in the forward detector. The dynamic energy range of the detector can be summarized by how much of each right leg can be accurately ascribed to a particular isotope. The punch-through point can be used as a point to calibrate the timing data – converting the channel number into a time scale (nanoseconds) – via the distances between the start paddles, the target (i.e. where the secondary particles are born) and the detector. As can be seen in Figure 5, the left leg (relative to the punch-through point) for each particle species can be cleanly identified, but there is a significant overlap of the right legs (as well as extraneous particle background– that is, particles that cannot be segmented out of the relevant data). As such, the use of a second method – comparing energy deposition in each detector – is necessary, because of poor particle identification in that second (right) leg, as well as inconsistencies with the timing data from the veto paddle triggers.

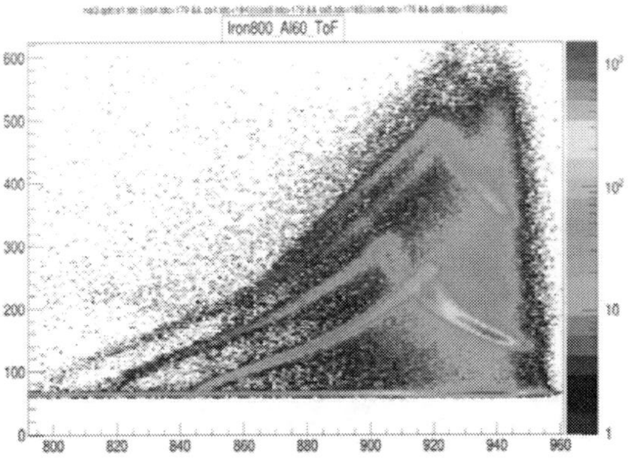

Figure 5. Iron 800 MeV/nucleon on a 60 g/cm^2 Aluminum target in the 10° sodium iodide detector array. The plot represents timing data (x-axis) plotted against energy deposition (y-axis).

Figure 6 – from the same iron 800 MeV system with the detectors from the 10° sodium iodide detectors – shows the charge deposited in the forward detector (y-axis) against the charge deposited in the back detector (x-axis). A similar pair of "legs" and a "punch-through point" can be seen in Figure 6. The legs represent the energy deposited in each detector – the top leg represents particles that deposited their energy in the front detector whilst the bottom represents particles with enough energy to make it through the front and deposit the remainder in the back. A second calibration point (the first being the punch-through point) is found in the bottom left of Figure 6 – this bright red spot represents zero energy deposition, and by assuming a linear relationship, the zero energy and the punch-through point can be used to convert the channel numbers into energy deposition in MeV.

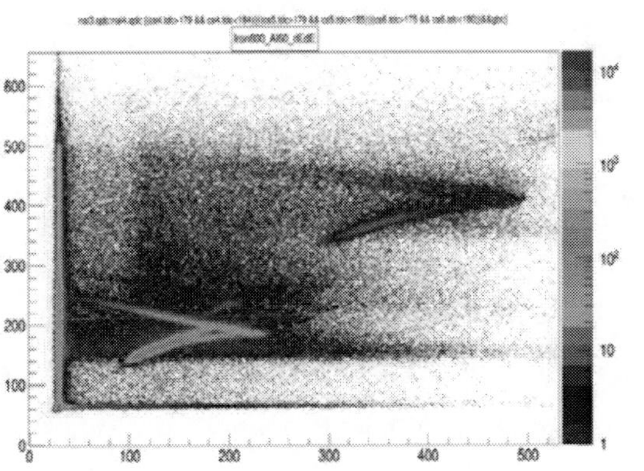

Figure 6. Iron 800 MeV/nucleon on a 60 g/cm^2 Aluminum target in the 10° sodium iodide detector array. The plot represents energy deposition in the front (y-axis) versus back (x-axis) detector.

Figure 6 has clearer particle identification than the ToF data in Figure 5. Each set of curves represents a unique Z value. In the example provided in Figure 6, three unique Z = 1 lines can be seen (from bottom to top: proton, deuteron and triton) and a Z = 2 (helium-4) line. By graphically gating on each set of legs, the specific data corresponding to each particle can be identified. The data is then projected onto the y-axis (which is the front detector of each array), creating a particle histogram.

It is important to note that for the Z = 1 data, there are visible regions of overlap between protons-deuterons, deuterons-tritons and so on. In an effort to avoid "double counting" (i.e. including deuteron results in a proton spectrum and vice versa), data interpolation in the overlap regions was done. By comparing the ratio of deuterons/protons in other data, an estimation for the expected ratio can be calculated and then used to ensure that the interpolated data is calculated correctly.

As mentioned earlier, the channel numbers were converted to energy deposition using the punch-through point. And using NIST tables as well as the Bethe-Bloch equation (Equation 1), for each particle species the energy deposition can be reverse-calculated into the incoming energy of each particle. Energy deposition is directly correlated with the energy of the incoming particle, and the NIST table with the Bethe-Bloch equation are used to make this calculation [6].

$$-\frac{dE}{dx} = \frac{5.08\times 10^{-31}z^2 n}{\beta^2}\left[\left(\ln\frac{1.02\times10^6\beta^2}{1-\beta^2} - \beta^2\right) - \ln I_{eV}\right] \quad (1)$$

In terms of uncertainty analysis, there were a few sources of error – systematic and statistical. The systematic uncertainties came from uncertainties in the solid angle, detector efficiency and graphic cuts performed on the data. The statistical errors came from the number of measured events. The errors in energy were also determined to be a property of the data as well – the width of the particle species identifiers (the legs from Figures 5 and 6) were determined to best represent the uncertainty in the deposited energy.

4 Experimental Results

Depending on the experimental setup (incoming source particle, source energy and shielding), different particles could be identified at the various detectors. Table 2 summarizes what particles could be identified for each setup (and with which sodium iodide array block - 10° or 30°), and the resulting figures are an illustrative example of the calculated results. The italics represent those particles that were detected, but for which a statistically significant amount of data is unavailable. For ease of comparison, only proton results from the 10° sodium iodide array are presented for each beam species-energy configuration at all three upstream target thicknesses. Additional plots from the proton runs will also be presented to show the angular dependence on the data. Figures 7-9 illustrate the proton yield [#/(source particle-MeV-steradian)] on a logarithmic scale plotted against the incoming particle energy.

Table 2 – Summary of particle identification

Beam Species/Energy [MeV]	Particles (Angle)
Helium (He) 400	Proton (10) Deuteron (10) Helium (10) Proton (30) *Helium-4 (30)*
Hydrogen (H) 400	Proton (10) *Deuteron (10)* Proton (30)
Hydrogen (H) 800	Proton (10) *Deuteron (10)* Proton (30)
Iron (Fe) 400	Proton (10) Deuteron (10) *Triton (10)* Helium-*4* (10) Proton (30) *Deuteron (30)*
Iron (Fe) 800	Proton (10) Deuteron (10) Triton (10) Helium (10) Proton (30) Deuteron (30) *Helium-4 (30)*

Figures 7-9 display the detected proton flux at 10° as a function of their incoming energy for the following incoming species: helium at 400 MeV/nucleon (Fig. 7), iron at 400 MeV/nucleon (Fig. 8), and iron at 800 MeV/nucleon (Fig. 9).

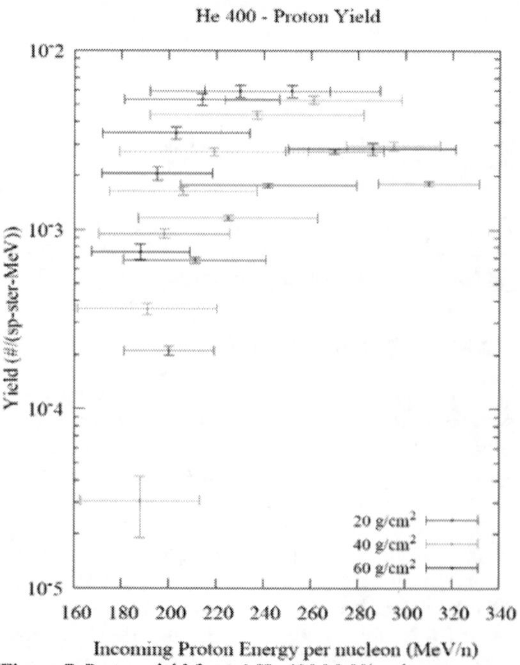

Figure 7. Proton yield from 4-He 400 MeV/nucleon on varying aluminum shield areal densities.

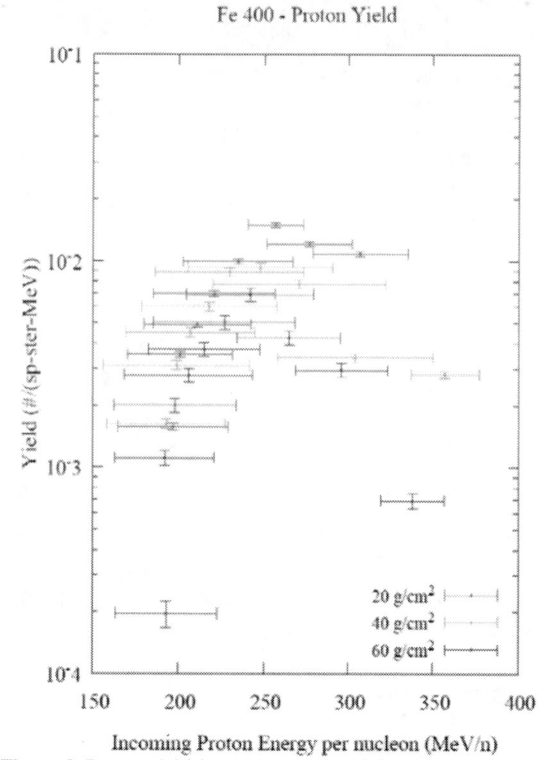

Figure 8. Proton yield from 56-Fe 400 MeV/nucleon on varying aluminum shield areal densities.

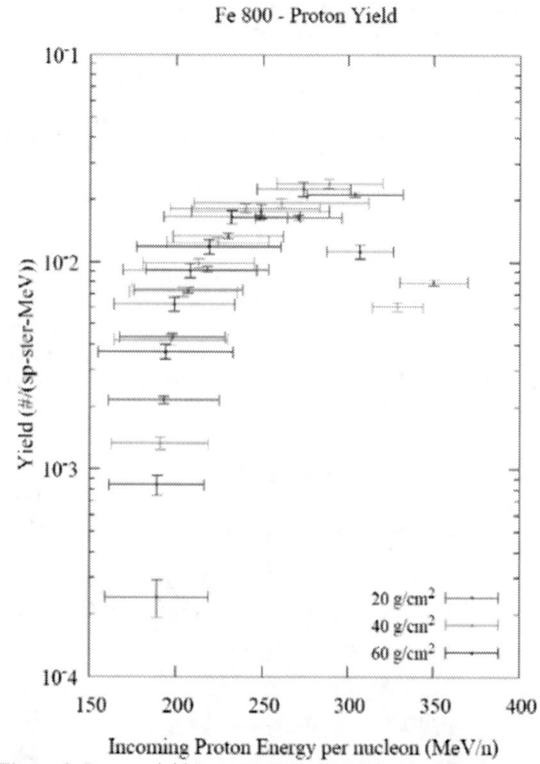

Figure 9. Proton yield from 56-Fe 800 MeV/nucleon on varying aluminum shield areal densities

The most interesting conclusion from the data presented in Figures 7-9 is that as the areal density of the upstream aluminium target increases, the proton yield (normalized to incoming source particle) actually increases, which seems to support the simulations (Figure 1). Another interesting takeaway is that as the incoming particle energy decreases, so does the associated errors with the calculated yield. These secondary particles are created when the incoming particle undergoes nuclear interactions within the upstream shield, so the statistics at higher energies would be worse, as the physics suggests that less secondary light ions would be created at higher energies. Thus, the statistical errors (both statistical and systematic errors are summed in the illustrated error bars) would be inherently higher as a result of fewer measured events at those energy ranges. The thicker target also inherently would also provide more material for the source particles to traverse through, so it supports the fact that there are higher yields of these light ions at higher shielding thicknesses.

Figures 10-11 show the proton yield from hydrogen nuclei at 400 MeV/nucleon (Fig. 10) and 800 MeV/nucleon (Fig. 11) for each shielding thickness, comparing the yield at the respective sodium iodide detector blocks (10° and 30°).

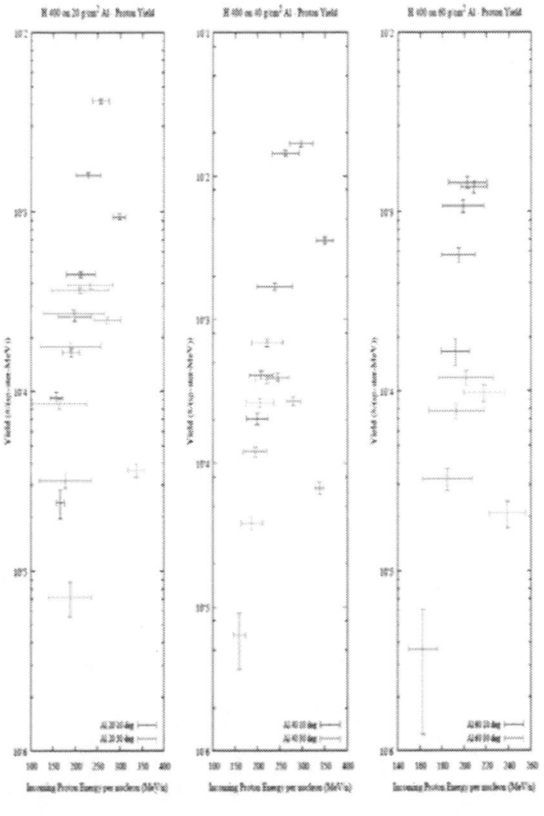

Figure 10. Spatially-dependent proton yields from a 400 MeV/nucleon upon 20 g/cm^2 (left) 40 g/cm^2 (middle) 60 g/cm^2 (right) aluminum target.

Generally, there is increased yield at every thickness for the 10° detector block as compared to the 30° block. Considering that the light ions are created from beam interacting within the shielding, it is presumed that most are initially produced along the beamline, so the closer the detector is to the center of the upstream and downstream targets, the higher the expected yield. The breakup of the projectile is also a source of these light ions, and as such the projectile fragments maintain the projectile's momentum in the forward direction. In conjunction with trends from Figures 7-9, errors tend to decrease with incoming particle energy, and that the thicker shielding results in higher yields. Furthermore, the yields from increased thicknesses result in more of the data being in lower energy ranges than the thinner shields.

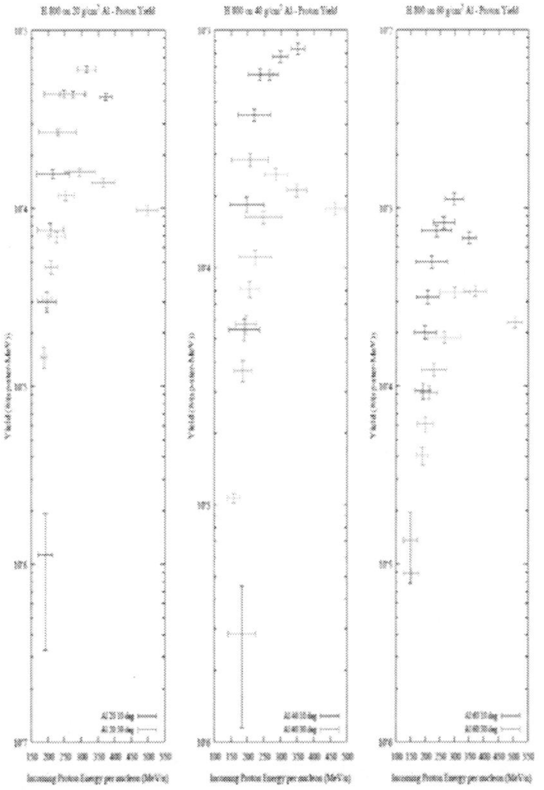

Figure 11. Spatially-dependent proton yields from a 800 MeV/nucleon upon 20 g/cm^2 (left) 40 g/cm^2 (middle) 60 g/cm^2 (right) aluminum target.

There are two sources of these secondary particles – breakup of the source particle, and breakup of the target. Figures 10 and 11 suggest that the wider the angle, the less yield measured and that even when the incoming particle energy [significantly] increases, a majority of the secondary particles come at energies lower than the source particle energy. This is primarily due to energy loss through the target after production.

5 Future Work

The next experiment is scheduled to take place in December 2016 at the NASA Space Radiation Laboratory at Brookhaven National Laboratory with the intent of adding new beam species and energies to the already-studied systems (Table 1). Table 3 summarizes the new beam species/energies and targets that are expected to be studied in the December 2016 run.

Further analysis is expected within this current data set as well – there are certain light ions for which there is data (the italicized species in Table 2) but it needs further graphical analysis to remove background data, and the errors associated with the incoming particle energy calculations will be reduced.

Table 3 – Upcoming (Dec. 2017) experimental setups

Beam Species	Beam Energy (MeV)	Shielding (g/cm²)
Helium (He)	800	Al: 20, 40, 60 + 60
Helium (He)	1500	Al: 20, 40, 60 + 60
Silicon (Si)	400	Al: 20, 40, 60 + 60
Silicon (Si)	800	Al: 20, 40, 60 + 60
Silicon (Si)	1500	Al: 20, 40, 60 + 60
Carbon (C)	400	Al: 20, 40, 60 + 60
Carbon (C)	800	Al: 20, 40, 60 + 60
Carbon (C)	1500	Al: 20, 40, 60 + 60
Iron (Fe)	1500	Al: 20, 40, 60 + 60

6 Summary

A series of thick-target experiments has been designed to aid NASA interests in deep-space manned missions by providing experimental data to assist with improving the transport models via uncertainty quantification. A commissioning experiment at the NSRL facility at BNL was performed in May 2015, with the preliminary results presented here. The methodology was studied and improved upon from the commissioning run, and the first set of data was taken in March 2016 runs. Double differential thick-target yields were calculated (as a function of incoming particle energy) for the light ions of interest. Between the March 2016 run and future runs at NSRL, the chosen beam species and energies will – as a whole – simulate the important components of the primary GCR field.

Acknowledgments

This work was supported by Grants NNX12AL51A and NNX15AD89A from the United States National Aeronautics and Space Administration.

References

1. D.F. Smart, M.A. Shea, *Handbook on Geophysics* (1985)
2. T. Slaba, *2015 Advanced Radiation Protection – EOY Review*, 1-15 (2015).
3. L. Heilbronn, R.S. Cary, M. Cronqvist, F. Deak, K. Frankel, A. Galonsky, K. Holabird, A. Horvath, A. Kiss, J. Kruse, and others, Nucl. Sci. Eng., **132**, 1-15 (1999)
4. P.-E. Tsai, "Study of secondary particles produced from heavy-ion interactions," Ph.D. dissertation, Dept. of Nucl. Eng., Univ. of TN, Knoxville, TN (2015)
5. T. Nakamura, L. Heilbronn, *Handbook on Secondary Particle Production and Transport by High-Energy Heavy Ions* (2006)
6. International Commission on Radiation Units and Measurements, *ICRU Report 49, Stopping Powers and Ranges for Protons and Alpha Particles* (1993)

DOI: 10.1051/epjconf/201715304005

Shielding design of the Mayo Clinic Scottsdale cyclotron vault

Kenneth A. Van Riper[1,*], *Robert L.* Metzger[2], and *Kevin* Nelson[3]

[1]White Rock Science, P. O. Box 4729, Los Alamos, NM 87547, USA
[2]Radiation Safety Engineering, 3245 N. Washington, Chandler, AZ 85225, USA
[3]Scottsdale Mayo Clinic, 13400 E. Shea Blvd., Scottsdale,AZ 85259, USA

Abstract. Mayo Clinic Scottsdale (Scottsdale, Arizona) is building a cyclotron vault containing a cyclotron with adjacent targets and a beam line leading to an external target. The targets are irradiated by high energy (15 to 16.5 MeV) protons for the production of radioisotopes. We performed Monte Carlo radiation transport simulations to calculate the radiation dose outside of the vault during irradiation of the cyclotron and external targets. We present the Monte Carlo model including the geometry, sources, and variance reduction methods. Mesh tallies surrounding the vault show the external dose rate is within acceptable limits.

1 Introduction

A cyclotron facility is under construction at the Scottsdale campus of the Mayo Clinic. The General Electric (GE) cyclotron's primary use is for production of ^{18}F for PET imaging. The PET isotopes are produced in a target adjacent to the accelerator. The proton beam can be directed to an external target in an adjoining room for production of other clinical and research isotopes including ^{11}C, ^{13}N, ^{15}O, ^{63}Zn, and ^{68}Ga. We refer to the former as the *cyclotron target* and the latter as the *external target*.

2 Monte Carlo Model

2.1 Geometry

The geometry model consists of the entrance maze, the cyclotron room containing the cyclotron target and collimator material, and the external target room with the external target. Details of the cyclotron are not included; we assume the shielding supplied by the vendor will reduce the leakage of any internal radiation from the accelerator to values low in comparison to the sources we consider: a fractional loss of the proton beam passing through a collimator, the interaction of the beam with the target, and the photons emitted by ^{18}F.

Figure 1 shows a floor plan of the model overlaid on an architect's drawing of the vault. Thick pink lines outline the Monte Carlo regions except for the interior and exterior air spaces. The concrete walls and ceiling are 6 feet thick. The floor is 3 feet thick. An air space approximately 3 1/2 feet (1.1 meter) thick surrounds the vault above the floor level. Below the floor line, the exterior consists of concrete.

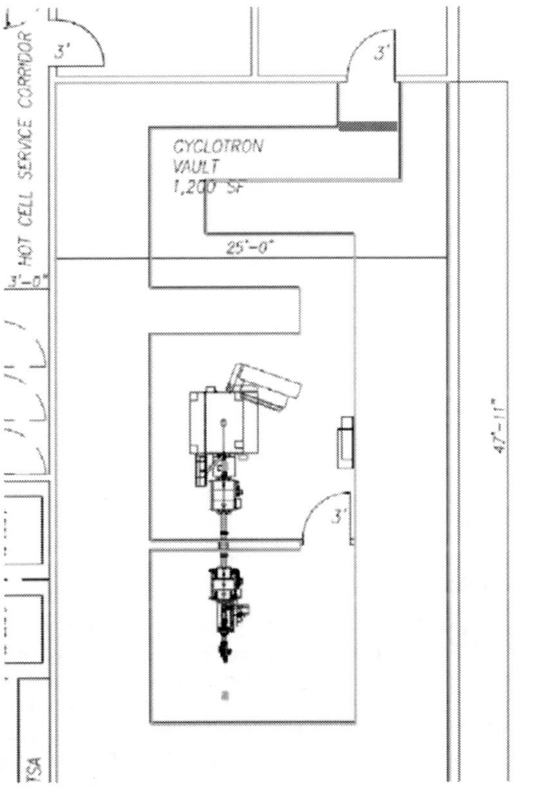

Figure 1. Monte Carlo model overlaid on the architectural drawing.

Also shown in pink in Figure 1 are the outlines of the targets and the collimator as modeled. The external target is located further from the end of the beam line shown in the drawing. The exact location will not affect

* Corresponding author: kvr@rt66.com

the radiation produced and its penetration through the walls.

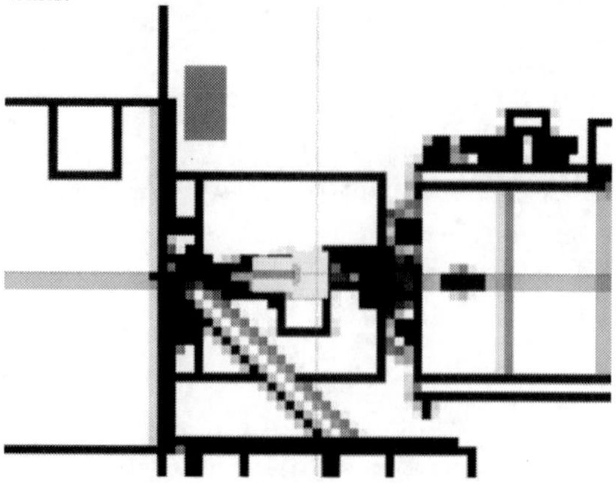

Figure 2. Detail of Figure 1 showing the cyclotron target and collimator block.

Figure 2 shows an enlarged section of Figure 1 around the cyclotron target. The blue block offset from the beam line is a graphite cube used to model the beam fraction (assumed to be 10 %) that is lost when passing though the collimator. It is offset from the beam line to prevent any particles that penetrate the block from reaching the target. We refer to this block as the *collimator* even though it is not a detailed model of that component. The cyclotron and beam line components were not included in the model.

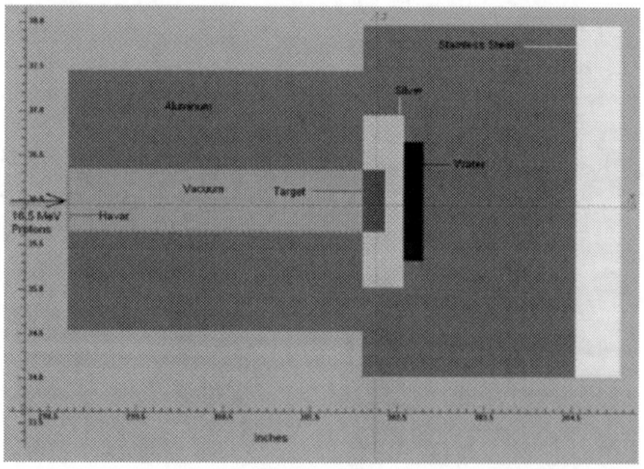

Figure 3. Target Model

Figure 3 shows the target model. It was adapted from the GE PETTRACE document[1]. The same model is used in both target locations. All components are concentric cylinders centered 36 inches above the floor. The dimensions are in inches. A beam of 16.5 MeV protons is incident on the 75 μM thick Havar foil covering a vacuum (brown in Fig. 3) downstream from the target. The target (magenta) is ^{18}O enriched water. It is surrounded upstream by a silver backing plate (grey) with cooling water further upstream. These components are encased in aluminum (green) and stainless steel (yellow).

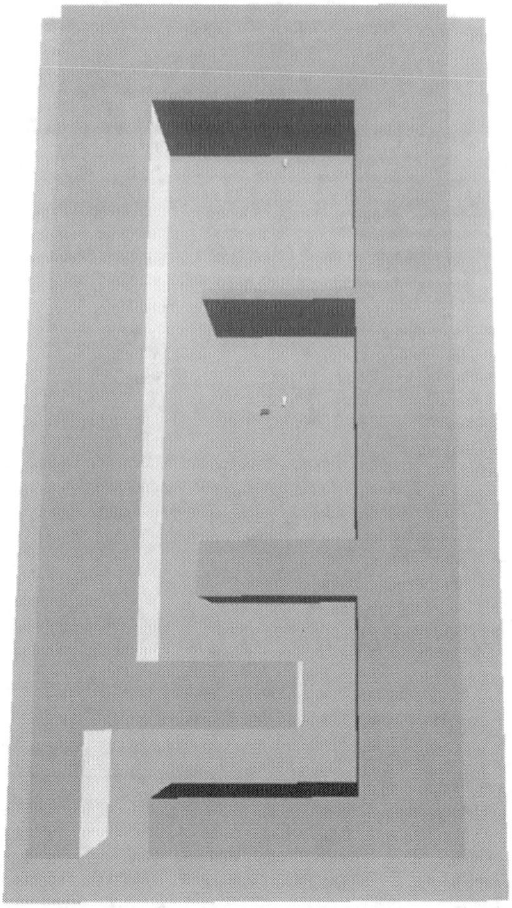

Figure 4. 3D Cutaway view of the cyclotron vault.

Figure 4 is a 3 dimensional (3D) view of the cyclotron vault with the ceiling cut away. The air is invisible, the concrete walls are gray, and the concrete floor is orange. The targets are yellow and the collimator block is blue.

2.2 Source terms

The cyclotron emits a beam of 16.5 MeV protons. The maximum current used in the cyclotron room is 130 μA (micro Amperes) after a 10% loss upon passing through the collimator. The 130 μA is split into 65 μA on two targets. For the Monte Carlo model, we use 130 μA on a single target. We also assume a fully irradiated target emitting 2 x 3500 mCi of 0.511 MeV photons. Table 1 summarizes the cyclotron room target source components. The total rate of emission is 8.927×10^{14} particles / second.

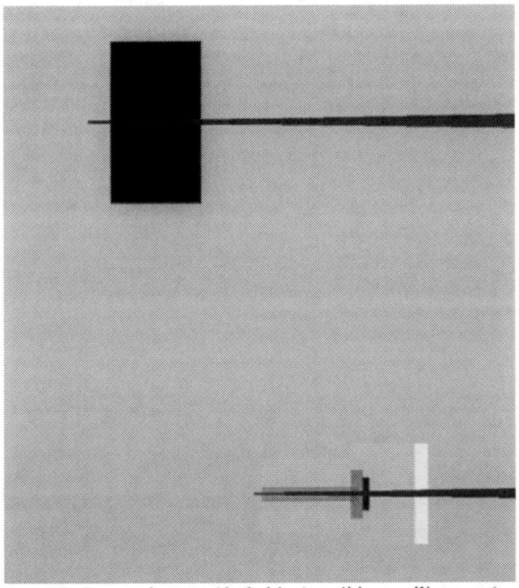

Figure 5. Proton beams (dark blue) striking collimator (top) and cyclotron target (bottom).

Two proton beams are directed towards the target and collimator with a small divergence angle. The beam origins are just downstream of the target and collimator. Figure 5 shows the two beams for the cyclotron room model. The figure is from a test run with no interactions; with interactions, the beams are stopped in the target and collimator.

Table 1. Source components for cyclotron target model.

Component	Particle	Energy (MeV)	Particles / Second
130 μA on Target	Proton	16.5	8.1133 x 10^{14}
13 μA on Collimator (10% loss)	Proton	16.5	8.1133 x 10^{13}
7000 mCi Photons	Photon	0.511	2.59 x 10^{11}

The photons are emitted isotropically. The photon source is distributed uniformly throughout the target water ($H_2{}^{18}O$) volume.

Table 2. Source components for the external target model

Component	Particl e	Energy (MeV)	Particles / Second
80 μA on Target	Proton	16.5	4.9928 x 10^{14}
8 μA on Collimator (10% loss)	Proton	16.5	4.993 x 10^{13}
3500 mCi Photons	Photon	0.511	1.255 x 10^{11}

The target room model assumes a 80 μA proton beam impinging on the target in the external target room, a 8 μA beam loss on the collimator in the cyclotron room, and single 3500 mCi load of irradiated target material. Table 2 summarizes the source components. The total rate of emission is 5.493 x 10^{14} particles / second.

2.3 Mesh tallies

The results are shown in the form of mesh tallies where the dose equivalent rate and its statistical error (relative error in Monte Carlo parlance) are calculated in each cell of a mesh overlaid on the geometry. Each mesh cell in the plane of the mesh is approximately 5 cm square and is 50 cm thick. Each mesh tally extends 1 meter outside of the vault. There are separate mesh tallies for the neutron, photon, and proton dose equivalent rates. These three were summed in post processing to give the total dose equivalent rate. Figures 6 and 7 show the mesh tally locations.

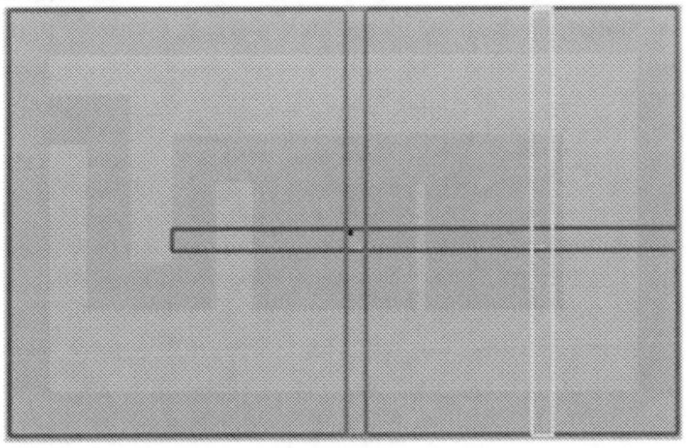

Figure 6. Mesh tally locations in the XY floor plan view.

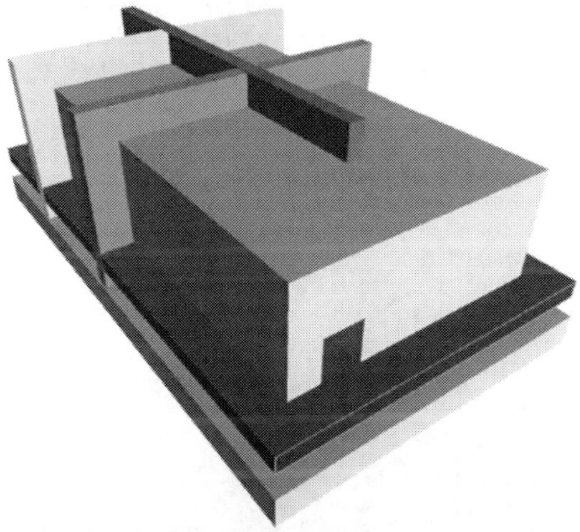

Figure 7. Mesh tally locations in the 3D view.

The red mesh tally covers the entire model, centered on the height of the targets and collimator. The vertical dark blue tally also passes through the targets and collimator. The red and blue tallies are used by both models (cyclotron targer and external target). The dark green tally is used in the cyclotron room target model and passes through the cyclotron target and collimator. The dark green tally is not used in the external target model. That model uses the yellow mesh tally that passes through the external target.

2.4 Monte Carlo items

The transport of protons, neutrons, and photons was calculated with the Monte Carlo code MCNPX version 2.7.0 [2].

Dose response functions are used to convert the calculated flux, in particles/cm^2/second, to a dose equivalent rate in Sv/hour. The dose equivalent is the absorbed dose at a point in tissue weighted by a distribution of quality factors related to the LET distribution of radiation at that point. For neutrons, the response function is taken from NCRP–38 1971, ANSI/ANS 6.1.1–1977. The quality factors in the NCRP-38 dose response function match those listed in 10 CFR Part 20. For photons, the values in ICRP–21 1971 are used. For protons, the values given in a NASA report were used.

Variance reduction, or biasing, is required to calculate penetration of the thick walls and ceiling and to direct the radiation through the entrance maze. We used the technique of importance splitting. When a particle traverses from a region of lower to high importance, it is split into two equivalent particles so that there is a greater chance of one of those particles avoiding absorption. The model uses 26 importance layers through the walls and ceiling and 31 layers between the source and entrance.

A calculation was run for each of the two source configurations. Each followed at least 5 x 10^7 source particles. The cyclotron target model had some large (> 0.1) relative errors in the external target room and far wall. An additional cyclotron target model with biasing adjusted towards the external target room was run to correct those deficiencies. Combined mesh tallies show the mesh cells with the lowest relative error from the two cyclotron target runs.

3 Vault shielding results

The mesh tally results are presented as color wash figures overlaid with an outline of the geometry. Contours are shown at 1 x 10^{-2}, 1 x 10^{-3}, 1 x 10^{-4}, 1 x 10^{-5}, 1 x 10^{-6}, 1 x 10^{-7}, and 1 x 10^{-8} Sv/hour. Values greater than 100 Sv/hr are colored red. Values less than 10^{-10} Sv/hr are dark blue. Different color schemes are used for the dose rate and for the relative error.

3.1 Cyclotron target model

Figures 8, 9, 10, and 11 show the neutron, photon, proton, and total dose equivalent rates, respectively, in the horizontal mesh tally for the cyclotron target model. External to the vault, the neutrons and photons give nearly equal contributions to the total. The only significant proton contribution is near the target and collimator from protons in the beam and protons scattered off components along the beam line. Figure 10 shows the beams are completely stopped in the target and collimator. Additional protons are produced by interactions in the concrete. These secondary protons contribute a very small fraction to the total dose rate.

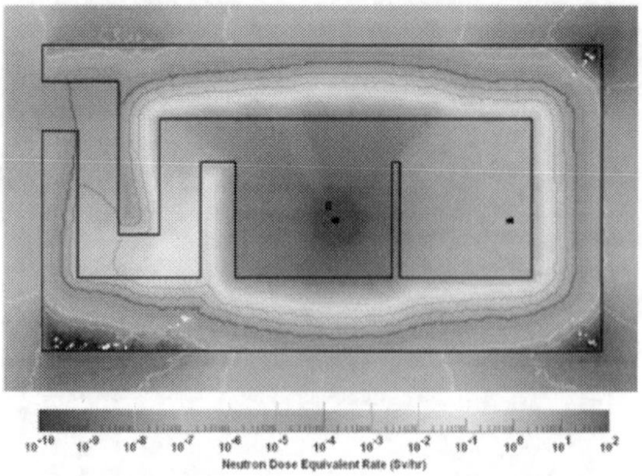

Figure 8. Neutron dose equivalent rate in the horizontal mesh tally of the cyclotron target model.

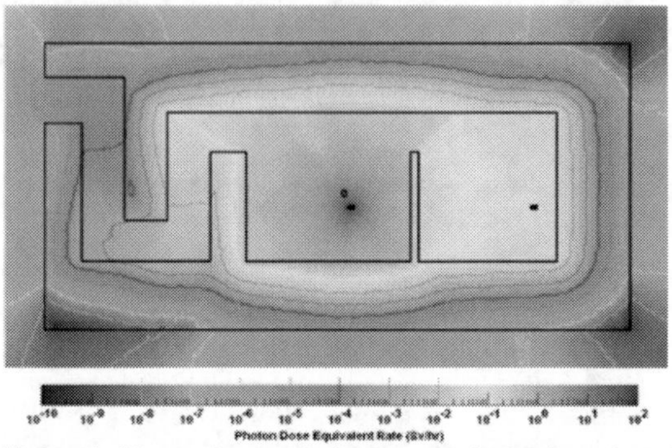

Figure 9. Photon dose equivalent rate in the horizontal mesh tally of the cyclotron target model.

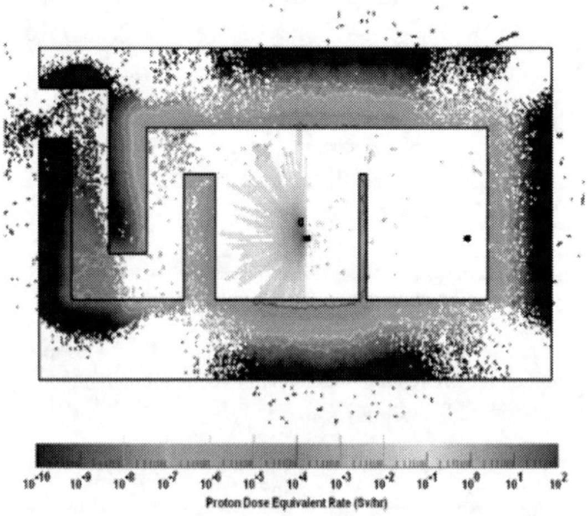

Figure 10. Proton dose equivalent rate in the horizontal mesh tally of the cyclotron target model.

Figure 12 shows the relative error for the total dose rate. Except for some regions within the walls, the relative error is less than 0.1 in areas of interest, an indication that the results are valid.

Figure 13 plots the dose equivalent rates extracted from the mesh tallies of Figures 8 through 10 along a line from top to bottom passing through the target and collimator. The proton contribution peaks at the target and collimator.

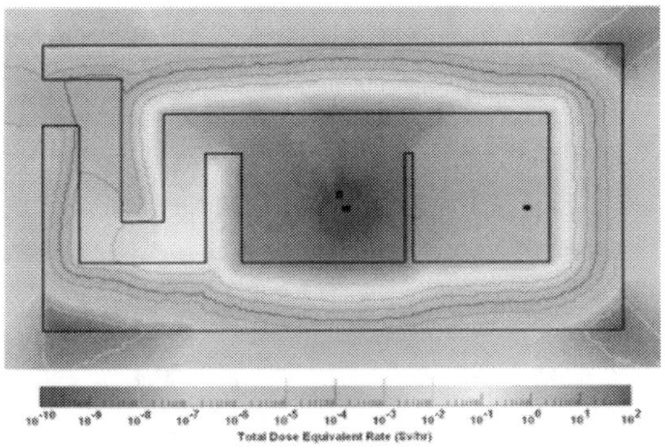

Figure 11. Total dose equivalent rate in the horizontal mesh tally of the cyclotron target model.

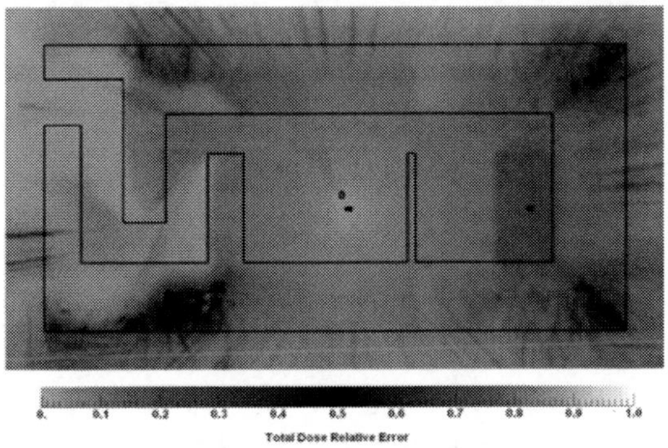

Figure 12. Total dose equivalent rate relative error in the horizontal mesh tally of the cyclotron target model.

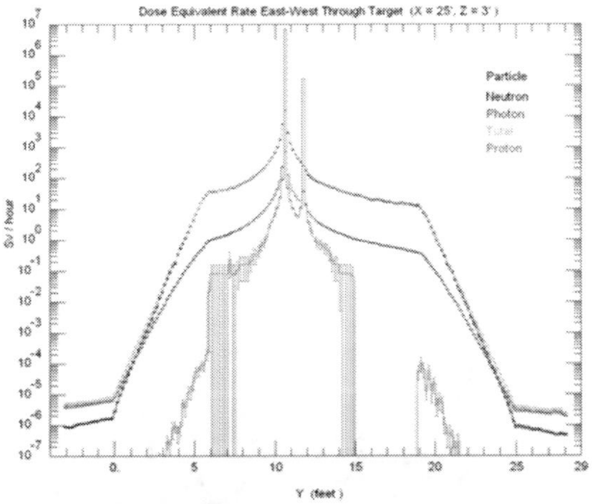

Figure 13. Dose rate profiles from top to bottom through the target in Figures 8-10.

Figure 14 shows the horizontal total dose rate mesh tally of figure 11 in a 3D view where the third dimension is proportional to the tally value. A different color scheme is used than in Figures 8 through 12 in order to better show the decline through the entrance maze.

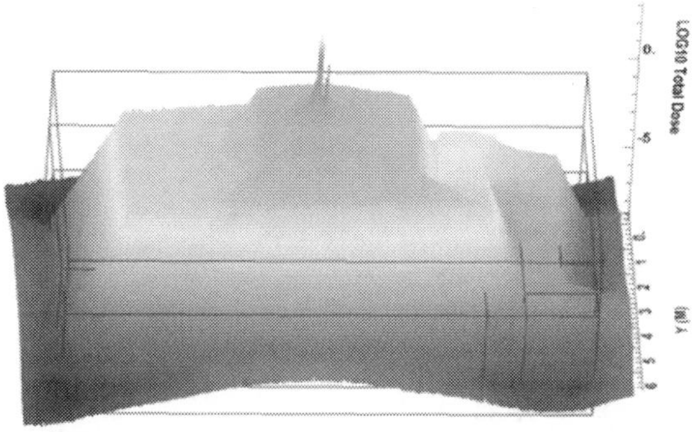

Figure 14. A 3D view of the total dose equivalent rate mesh tally of Figure 11.

3.2 External target model

Figures 15 and 16 show the total dose equivalent rate from the external target model. The relative neutron, photon, and proton contributions and the relative errors are similar to those in the cyclotron target model. The peak in the cyclotron room is due to the fraction of protons that strike the collimator.

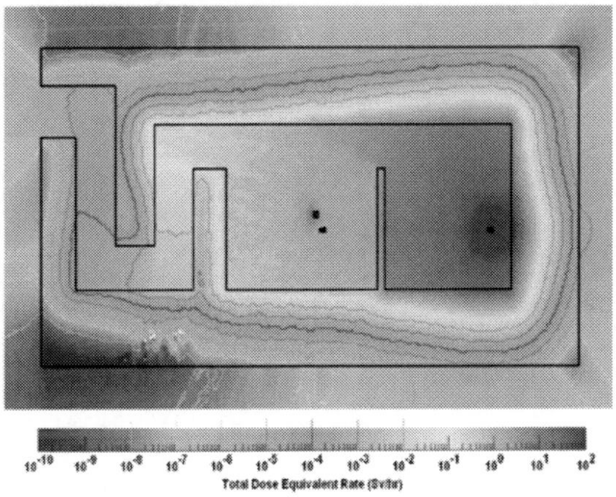

Figure 15. Total dose equivalent rate in the horizontal mesh tally of the external target model.

Figure 16. A 3D view of the total dose equivalent rate mesh tally of Figure 15.

4 Maze entrance door

The vault shielding models described above did not include a door at the maze entrance. We made several studies of the attenuation through entrance doors of different thicknesses and compositions. Except for the door and modifications near the door, the geometry, source, and materials are the same as in the cyclotron target model described above. Because the dose rate at the maze entrance is larger using the main cyclotron target, we assumed irradiation of that target for the door calculations.

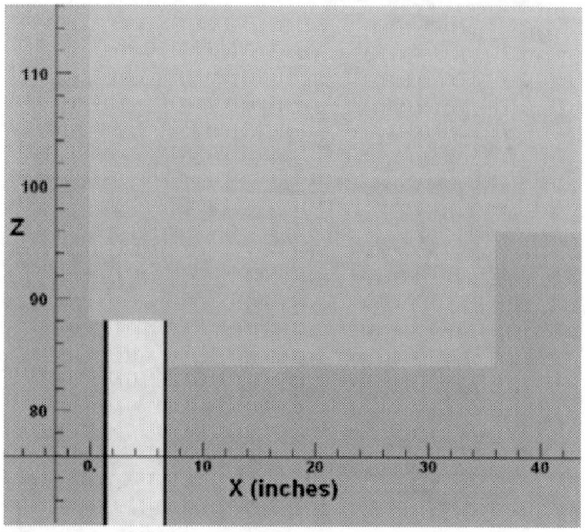

Figure 17. Vertical slice through the top of the maze entrance door. Both scales are in inches.

4.1 Surface source calculations

The calculation was made in two stages using the MCNP surface source write (SSW) and surface source read (SSR) features. Without the need to transport particles from the target through the maze, the SSR calculations require much less computation time. The shorter SSR runs permitrd exploration of a number of door designs.

The initial SSW calculation had irradiation of the main target and collimator by the proton beam as the source. All neutrons and protons crossing a surface spanning the air space in the maze coincident with the first inner wall (the green line in Figure 1) were saved to a file. Only outward (upwards in Figure 1) crossings were saved. To reduce calculation time, biasing to enable penetration of the side and rear walls and the ceiling was not included. Biasing through the maze was retained.

The SSW calculation followed 5,761,845 histories, approximately 1/10 of the histories followed in the vault shielding calculations. The small relative errors around the maze entrance found in the previous results justified using the smaller number of histories. The run resulted in a 1.9 GByte surface source file. Comparison of mesh tallies showed that the SSW calculation gave the same results in the cyclotron room and through the maze as the previous shielding calculation.

An SSR run in the same geometry was made to verify that the same results were obtained. Plots of the ratios of the horizontal neutron and photon mesh tallies near the maze entrance showed that the results agree within 20% in the air volumes downstream of the surface source. A vertical mesh tally outside of the door opening was used to examine any vertical variation of the neutron and photon dose rates. No significant variation was seen.

4.2 Door model

The door consists of a polyethylene slab sandwiched between steel plates. It fits into recesses in the outer maze entrance. Figure 17 is a horizontal cut through the model. Figure18 is a vertical cut through the door. In the figures, concrete is grey, air is light blue, polyethylene is yellow, and steel is dark blue. A horizontal mesh tally extends 25 feet in front of the door.

The side and top recesses are 6 ⅞ inches deep (in the +X direction in Figures 17 and 18). The east recess (towards the top in Figure 17) extends 9 inches beyond the maze opening. The west recess extends 5 inches beyond the opening. The top overlap is 4 inches high. The door bottom is flush with the concrete floor. The height of the opening is 84 inches, less than the 96 inch height of the maze. Except for the opening through the outer north wall, the height in the maze was kept at 96 inches.

The outer steel layer is ⅜ inch thick; the inner ¼ inch. A 5 inch thick polyethylene layer was used for the initial calculations. For the 3 inch polyethylene model, the inner thinner steel layer remained as shown and the outer steel layer was moved 2 inches inward (+X direction).

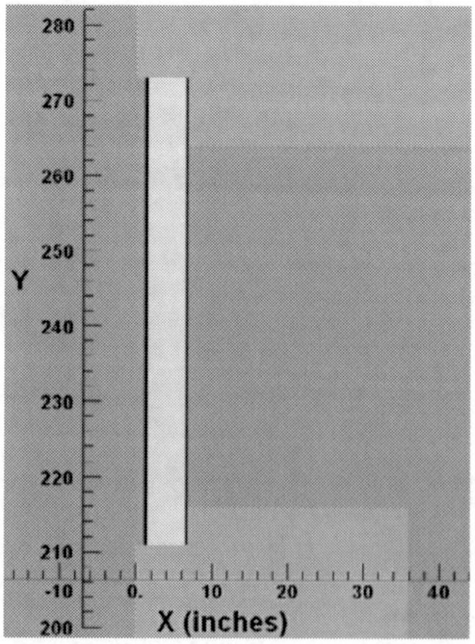

Figure 18. Horizontal slice through the maze entrance and door. Both scales are in inches.

The SSR door calculations followed neutrons and protons from the surface source. These calculations did not include penetration of the outer wall other than at the maze opening. The full shielding calculations showed some penetration, especially opposite the second turn in the maze.

4.3 Polyethylene composition

A model without a door was run to establish a baseline against which to measure the shielding efficacy of the door options. In the *No Door* model, the polyethylene and steel layers were replaced by air.

Several compositions were used for polyethylene. The *Natural Polyethylene* model did not contain any dopants, only ⅔ H and ⅓ C by number. The *Borated Polyethylene* model contained 5% boron by weight added as boric oxide (B_2O_3). Trace amounts of iron (0.00058% by weight), sulfate (SO_4, 0.0051%), and water (1.1%) found in an analysis of the boric oxide were ignored.

In MCNP, a material constituent can be specified as an element or as the fractions of isotopes making up the element. In the element specification, MCNP chooses the isotopes naturally occurring the element. For boron, we compared runs with natural boron (element specification, Natural B in Figure 11) and with explicit fractions for boron 10 and boron 11 (B10_B11). The results were the same within the statistical uncertainty

4.4 Maze entrance door results

The curves in the Figures 19 through 22 show the dose equivalent rates along a north-south line passing through the center of the door (240 inches in Figure 18). The peaks at 91.44 cm mark the location of the surface source. The curves to the right (larger X) of the peak should be ignored. The error bars in Figures 19 and 20 are omitted for clarity.

Figures 19 and 20 show the neutron and photon contributions to the dose rate with no door (19) and with a 5 inch borated polyethylene door. Figure 21 shows that boron doping does not change the neutron contribution. Figure 22 shows that the difference between 5 and 3 inch doors is small.

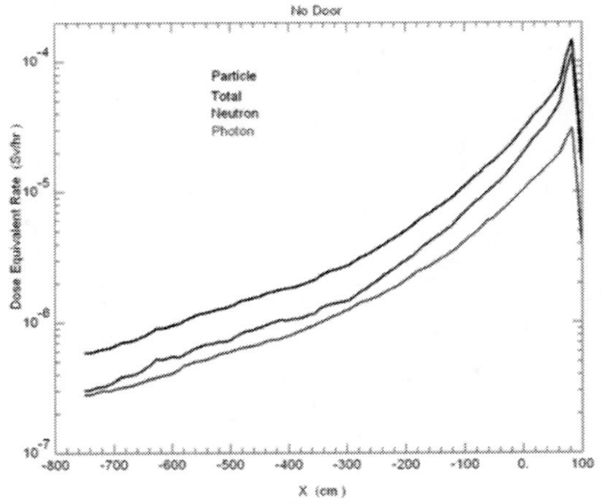

Figure 19. Dose equivalent rate profiles with no door.

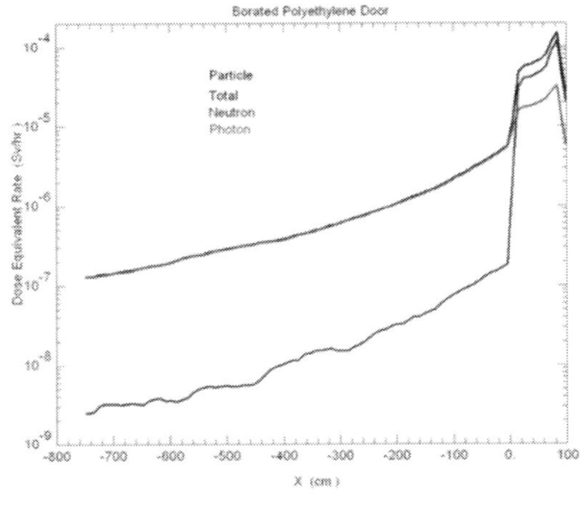

Figure 20. Dose equivalent rate profiles with a borated polyethyline door.

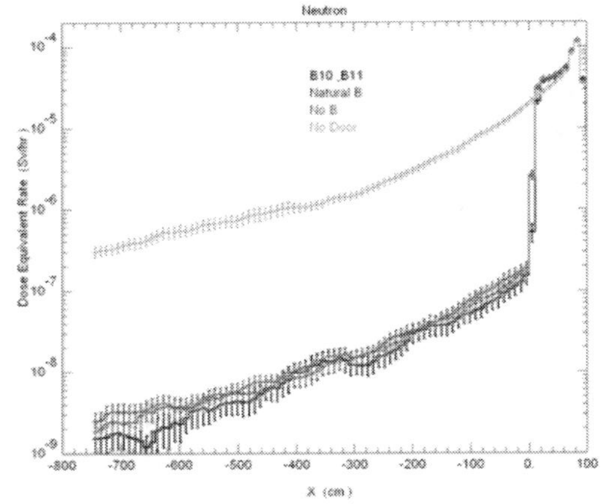

Figure 21. Neutron dose equivalent rate profiles.

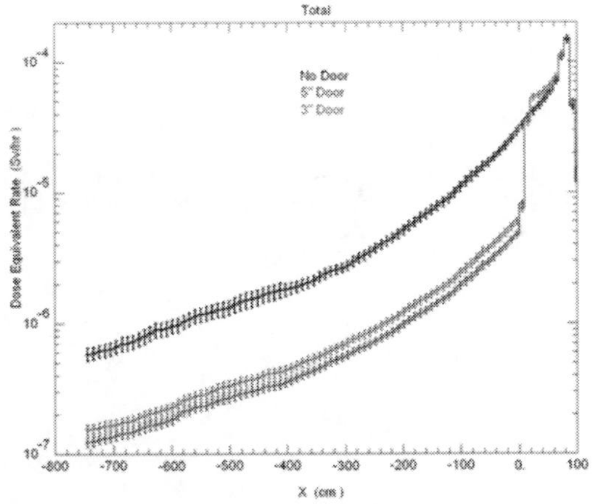

Figure 22. Total dose equivalent rate profiles.

5 Steel Rod Penetrations

A few dozen steel rods were used to hold the concrete forming framework together. These 7/8 inch diameter rods were left in place after the pour. We wanted to ensure that the presence of the rods did not affect the efficacy of the concrete sheilding wall. We prepared a modified model of the cyclotron target room that omitted the entrance maze and the external target room. Eight inches of the wall between the cyclotron room and the external target room and 8 inches of the the opposite wall were retained. The outer walls included 12 rods on each side. Horizontal mesh tallies at the height of the target also covered a row of three rods in each outer wall. Runs with and without the rods showed no difference in the external photon and neutron dose rates.

6 Conclusion

We have calculated the dose equivalent rates throughout and just outside of the cyclotron vault. The dose rate external to the vault is less than regulatory limits. In particular, we used the pregnant worker limit of 5 mSv over the course of the pregnancy for occupationally controlled areas and the 1 mSv per year limit for non-occupational areas. These limits factored in a workload (or use factor) of 0.97.

Photon measurements were taken above the cyclotron room during irradiation of the main target. The highest dose rates at the roof surface were 0.9 μSv/hour and 0.4 μSv/hour at 1 meter above the roof. Our calculations predicted 5 and 3 μSv/hour at these locations.

Acknowledgement

We thank Richard Wentz and Uno Zetterberg of GE Healthcare for their patience in answering numerous questions about the cyclotron and target design and source terms.

References

1. J. O. Bergstrom, *PETTRACE - unshielded machine : Summary of source terms, radiation fields and radwaste production*, GE Healthcare Document Ref: RP001181(0.4).doc, (2004)
2. D. Pelowitz (ed.), *MCNPX User's Manual*, Version 2.7.0, Los Alamos National Laboratory, LA-CP-11-00438 (2011).

A Shielding Concept for the MedAustron Facility

L. Jägerhofer*[1], E. Feldbaumer[2], S. Roesler[2], C. Theis[2] and H. Vincke[2]

[1] *EBG MedAustron GmbH, Austria*
* Lukas.Jaegerhofer@medaustron.at
[2] *CERN, HSE-RP, Geneva 23, CH-1211 Geneva, Switzerland*

Abstract. MedAustron is a synchrotron based accelerator facility for cancer therapy and research in Wiener Neustadt, 50 km south of Vienna. The facility will provide protons up to kinetic energies of 250 MeV and carbon ions up to 400 MeV/n for ion beam therapy. Additionally, protons up to 800 MeV kinetic energy will be used in a dedicated room for non-clinical research. In order to obtain a shielding concept for this facility a detailed geometry of the accelerator facility was implemented into the Monte-Carlo code FLUKA and shielding simulations were performed. In the course of these simulations the contributions of different particle types to the mixed fields around the accelerator and behind shielding were analysed. In an iterative process with the architect the final design of the shielding concept was developed until it was capable of reducing the effect of secondary radiation on humans and the environment below Austrian legal limits.

KEYWORDS: EBG MedAustron GmbH, Shielding Calculations, Hadron Therapy, FLUKA

1 Introduction

MedAustron is a synchrotron based accelerator facility which was built in Wiener Neustadt, Austria. It will provide proton beams up to kinetic energies of 250 MeV and ion beams up to 400 MeV/n for cancer treatment. It is planned to treat 1200 patients per year as an ambulatory. Apart from clinical treatment, the facility will provide one target room with up to 800 MeV protons for non-clinical research in the fields of medical radiation physics, radiation biology and experimental physics. The medical application has defined the accelerator design for MedAustron but from a radiation protection point of view the protons with 800 MeV play an important role. Figure 1 shows the patients' entrance of the facility.

Figure 1. Patients'entrance of the MedAustron facility (1)

Beam operation in the MedAustron facility produces secondary radiation fields due to beam losses along the accelerator chain and the beam transfer line as well as due to intended irradiation of objects or persons in the treatment rooms. Therefore it is required to shield the facility sufficiently. The Monte Carlo Code FLUKA (2) (3) was used to determine the shielding concept which reduces the effect of this secondary radiation on humans and environment below acceptable values, according to the legal limits in Austria.

In order to account for uncertainties of the Monte-Carlo simulations and the fact that incorporation of radionuclides is contributing to the accumulated effective dose, all simulation results are evaluated with a safety factor of 10. This results in dose limits of 0.1 mSv in publicly accessible areas, 0.6 mSv for supervised areas and 2 mSv for controlled areas which are accessible during operation. It has to be mentioned that the whole process of shielding design was iterative. The wall thicknesses were adapted to our demands, according to the simulation results. Then the geometry was adapted and simulations were repeated until the dose map met the requirements of legal dose limits outside as well as inside radiation areas.

2 The MedAustron Accelerator Facility

2.1 Ion beam therapy

MedAustron offers an innovative form of radiation therapy (ion beam therapy) using protons and carbon ions. Compared to conventional radiation therapy, this treatment is able to reduce radiation exposure to adjacent healthy tissue and to spare the tissue behind the tumour almost entirely. Therefore, ion beam therapy is an optimal treatment for tumours close to radiosensitive organs, like the brain and the spinal cord, eyes, liver and lungs. Since tissue in growth is more sensitive to radiation, proton radiation is particularly suitable for cancer in children and adolescents. The focus lies on reducing the risk of long-term consequences by minimising dose to healthy tissue. (4)

2.2 Irradiation rooms

MedAustron has four irradiation rooms. Three will be used

for patient treatment and one for non-clinical research. Though the rooms differ in terms of flexibility of the particle beam, they all have one thing in common: each of the four rooms is equipped with an innovative robotic positioning system. This will allow direct comparisons and applications of the results drawn from non-clinical research for the patient treatment. (4)

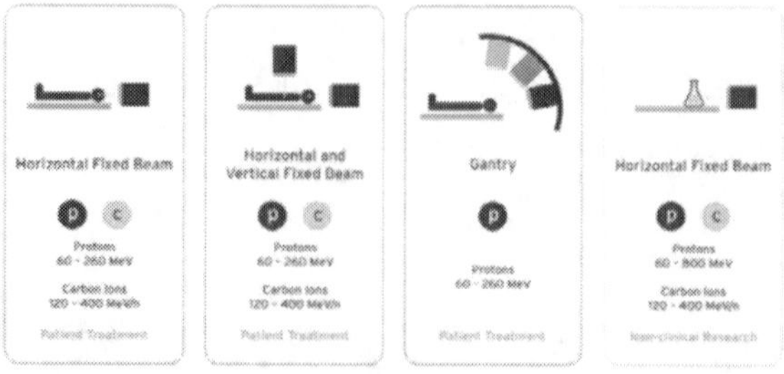

Figure 2. Four different irradiation rooms (4)

2.3 Base parameters of the MedAustron beams

Although the medical applications defined the accelerator design, the beam parameters for the experimental operation and the medical operation are different. Table 1 shows possible beam parameters for protons and carbon ions.

Table 1. Beam parameters at extraction from the synchrotron.

Protons	
Kinetic energy (min. – max.)	60 - 800 MeV
Beam intensity	$\leq 2x10^{10}$ particles per pulse
Extraction time (min. – max.)	0.1 – 10 s
Repetition rate (max.)	1 Hz
Carbon ions	
Kinetic energy (min. – max.)	120 - 400 MeV/n
Beam intensity	$\leq 1x10^{9}$ particles per pulse
Extraction time (min. – max.)	0.1 – 10 s
Repetition rate (max.)	1 Hz

3. Monte Carlo simulations

3.1 Setup of the simulation geometry

To assess the necessary shielding, simulations with the Monte-Carlo code FLUKA were carried out. For the verification of sufficient shielding a three-dimensional geometry was developed and implemented in FLUKA. Figure 3 shows a three dimensional view of the geometry visualized in SimpleGeo. (5) Ceilings are not shown in this picture for better understanding. The interior and exterior of this model are filled with air. Simplified models of dipoles and quadrupoles are placed in crucial positions to take the shielding effect of their massive iron yokes into account. The emergency exits, the air ducts and some representative cable ducts are included in the model.

Along the beam line numerous places can be defined where particles will be lost due to beam optic devices like dumps or magnets. Also at the end of the beam line in each of the 4 treatment rooms particles are delivered according to irradiation plans. All these particles contribute to the annual ambient dose equivalent in that area and build the basis for the simulation of a detailed dose map. Due to the fact that the contribution of the particles to the ambient dose equivalent is strongly related to the energy of the particles, only losses with maximum energies of 250 MeV and 800 MeV for protons and 400 MeV/n for carbon ions were taken into account for the simulations.

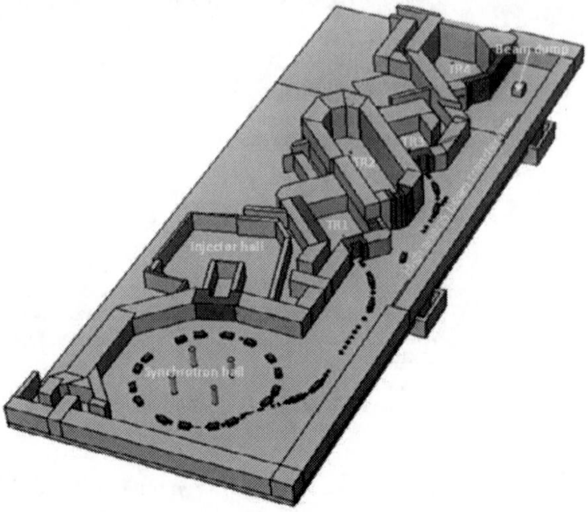

Figure 3. Schematic view of the 3-D FLUKA model of the MedAustron accelerator complex.

3.2 Method for the reduction of variance

To reduce simulation time significantly biasing methods were introduced. These mathematical methods are used in Monte Carlo simulations to reduce the variance of the results in relevant areas behind thick shielding. For the following simulations those areas, which are adjacent to the room during treatment, are considered to be relevant and may be accessed. A so called "importance biasing" method was applied. With this method importance factors are assigned to different regions to increase the statistical weight of each particle entering the region. For this purpose all walls and the ceilings as well were divided in 40 cm slabs, which is approximately the inelastic scattering length of the high energy cascade caused by 800 MeV protons in the MedAustron concrete. These sectors (40 cm slabs) were then assigned an importance factor from the inside to the outside according to the formula

$$importancefactor = startvalue * e^{number\ of\ sector} \quad (1)$$

Figure 4 shows irradiation room 1 as an example for importance biasing. The air in the maze has to be biased to assure increasing importance factors along the paths of each particle. By these means statistical uncertainties below 20% could be obtained at relevant places outside of the shielding.

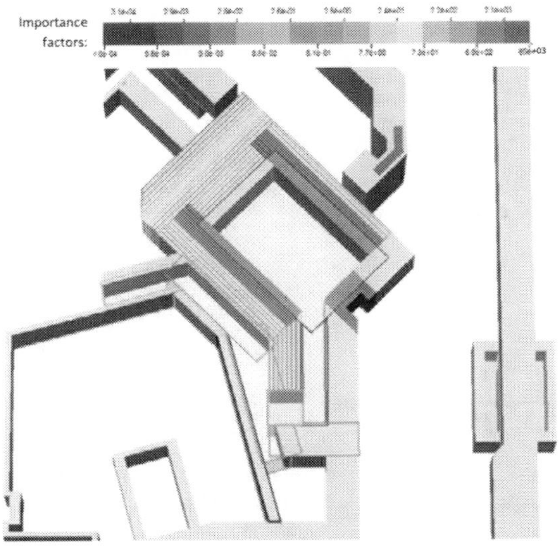

Figure 4. Importance biasing for irradiation room 1. The colours represent the importance factors.

3.3 Results of the Monte Carlo Simulations

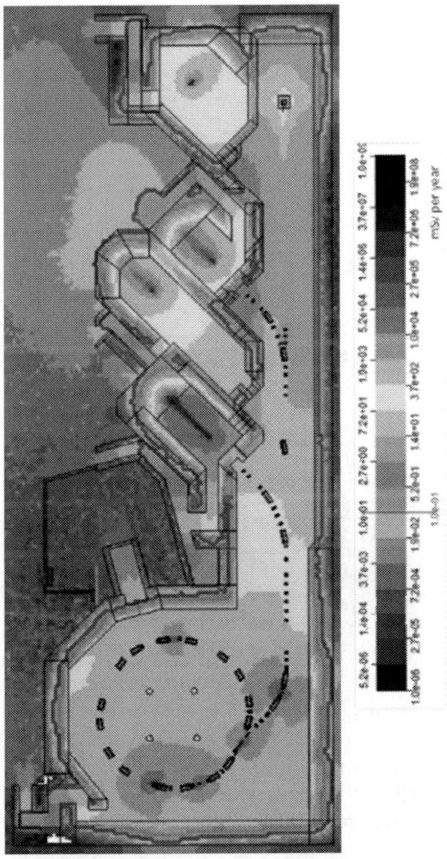

Figure 5. Annual ambient dose equivalent map of the full MedAustron facility at beam height.

A full annual ambient dose equivalent map of the MedAustron accelerator facility as well as close surroundings can be obtained by summation of the ambient dose equivalent originating from each simulated loss points and the treatment rooms. The resulting dose map is shown in Figure 5 and provides a clear indication that the shielding concept is capable of reducing the effect of secondary radiation on humans and the environment below Austrian legal limits. The purple line indicates the self-introduced 0.1 mSv limit.

3.4 Analysis of the radiation environment

Considering that different types of particles in mixed radiation fields contribute differently to the total ambient dose equivalent rates inside and around the MedAustron facility at neuralgic points, it is of great interest to quantify these contributions accurately. Furthermore, for the conception of a detector system it is crucial to know the composition of the radiation environment at the position of the detector to be able to install an appropriate detector and furthermore make correct measurements of ambient dose equivalent rates.

3.4.1 Spectra originating from the main loss points in the Synchrotron hall

To obtain the spectra at the positions shown in Figure 6 particle losses of the 800 MeV protons at the main loss points in the Synchrotron hall, the horizontal beam dump, the vertical beam dump, the electrostatic extraction septum and the magnetic extraction septum were summed up and weighted accordingly to the respective annual particle losses at each loss point. The following spectra represent only particle losses of 800 MeV protons, since these beams as well as their secondary radiation represent the main contribution to the annual ambient dose equivalent rate inside and outside of shielding at MedAustron. The spectra are scored inside air filled boxes of the dimensions 100x100x200 cm^3 standing at the floor level. Scoring volumes are distributed over the whole facility, here the synchrotron is chosen as an example.

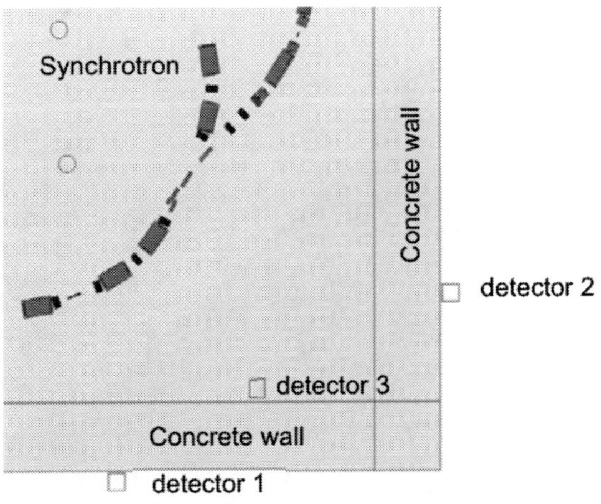

Figure 6. Scoring volumes inside and around the synchrotron hall.

The total fluence Φ in the detector volumes consists of the sum of the single fluences Φ_i created by each of the four loss points:

$$\frac{d\Phi}{dE} = \sum_i w_i \frac{d\Phi_i}{dE} \quad (2)$$

with

$$w_i = \frac{\alpha_i}{\sum_i \alpha_i} \quad (3)$$

where w_i represents the fracture of the particles loss sum at each loss point where i = 1 to 4 is the number of the loss point. α_i is the number of 800 MeV protons lost annually.
The absolute statistical error, originating from the simulations, was weighted and then propagated as independent error.
Figure 7 to Figure 9 show the spectra originating from particle losses at specific loss points around the synchrotron.

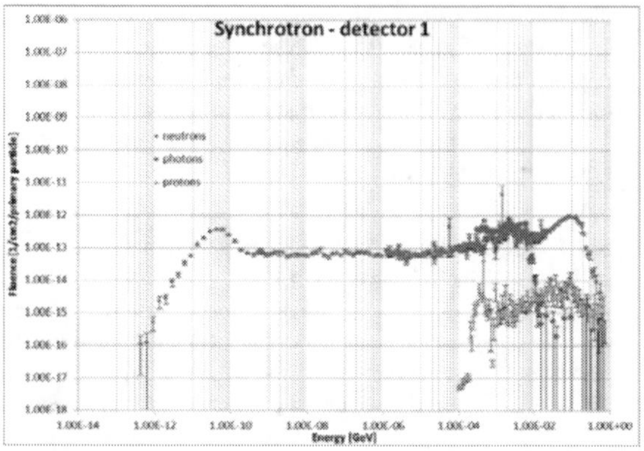

Figure 7. Particle fluence spectra for detector 1

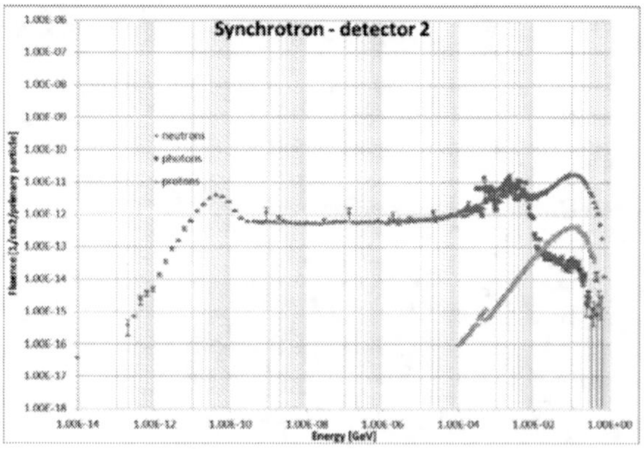

Figure 8. Particle fluence spectra for detector 2

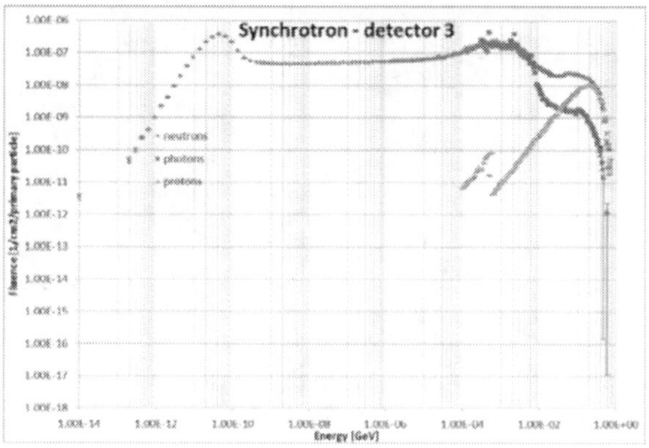

Figure 9. Particle fluence spectra for detector 3

3.4.2 Contributions to annual ambient dose equivalent values by different particle types

Table 2 shows the relative ambient dose equivalent contributions by neutrons, photons and protons, which are the main contributors to ambient dose equivalent rates in and around the synchrotron. The values are derived by convolution of the fluence spectra with the respective ambient dose equivalent conversion function by Pelliccioni.[6)] The conversion functions were linearly interpolated in order to be able to perform the convolution.

Table 2. Annual ambient dose equivalent contributions in the detectors of the synchrotron hall.

Synchrotron hall						
Ambient dose equivalent rate from	detector 1 [%]	+/- [%]	detector 2 [%]	+/- [%]	detector 3 [%]	+/- [%]
Neutrons:	96	0,15	91	0,08	88	0,01
Photons:	1	0,11	1	0,05	1	0,01
Protons:	3	0,55	8	0,12	11	0,09
Total dose: [mSv/a]	**0.001**	**0.10**	**0.017**	**0.05**	**211**	**0.01**

4. Summary and Conclusion

Extensive Monte-Carlo simulations were performed and for this purpose a detailed geometry of the MedAustron facility, based on the architectural plans, was developed and implemented in FLUKA simulations. The premise of these simulations is a map of foreseen particle loss points along the accelerator chain. Simplified models of magnets were implemented close to these loss points to consider the shielding effect of the massive iron yokes. The information about the annually lost particles was then used

as source terms for the FLUKA simulations. The results of the single FLUKA simulations were then converted to annual ambient dose equivalent rates (H*(10)) and summed up to build a full annual dose equivalent rate map which shows that the shielding concept of the MedAustron facility is capable of reducing the effect of secondary radiation on humans and the environment below acceptable values, according to the legal limits in Austria.
The simulation was an iterative process and input was given to the architect if shielding dimensions were too small or if a dump was needed in specific walls. The architect was then adapting the plans according to the results of the FLUKA simulations.
A more detailed analysis of the radiation field compositions clearly shows that neutrons are the main contributors to ambient dose equivalent rates outside but also inside the shielding walls of MedAustron. Furthermore the neutrons can reach up to primary beam energies with a clear quasielastic peak at around 100 MeV. These facts have to be taken into consideration for the design of a radiation monitoring system.

References

1) © Thomas Kästenbauer
2) A. Fassò, A. Ferrari, J. Ranft, and P.R. Sala, *FLUKA: a multi-particle transport code*, CERN Yellow Report, INFN/TC_05/11, SLAC-R-**773**, (2005).
3) G. Battistoni, S. Muraro, P.R. Sala, F. Cerutti, A. Ferrari, S. Roesler, A. Fassò, J. Ranft, The FLUKA code: Description and benchmarking, *Proceedings of the Hadronic Shower Simulation Workshop 2006*, Fermilab 6--8 September 2006, M. Albrow, R. Raja eds., AIP Conf. Proc., **896**, pp. 31-49. (2007)
4) http://www.medaustron.at (25.01.2016)
5) C. Theis, K.H. Buchegger, M. Brugger, D. Forkel-Wirth, S. Roesler, H. Vincke: *Interactive three dimensional visualization and creation of geometries for Monte Carlo calculations,* Nuclear Instruments and Methods in Physics Research A **562**, pp. 827-829 (2006).
6) M. Pelliccioni, *Overview of Fluence-To-Effective Dose and Fluence-To-Ambient Dose Equivalent Conversion Coefficients for High Energy Radiation calculated using the FLUKA Code*, Radiation Protection Dosimetry, **Vol. 88**, No. 4, pp. 279-297, (2000).

Neutron transport calculation for Activation Evaluation for Decommissioning of PET cyclotron Facility

Fumiyoshi Nobuhara[1,*], *Makoto* Kuroyanagi[1], *Kazuyoshi* Masumoto[2], *Hajime* Nakamura[2], *Akihiro* Toyoda[2], and *Katsuhiko* Takahashi[3]

[1]*Tokyo Nuclear Services Co. LTD, 1-3-5, Taito, Taito-ku Tokyo, 110-0016, Japan;*
[2]*KEK, the High Energy Accelerator Research Organization, 1-1 Oho, Tsukuba, Ibaraki , 305-0801 Japan;*
[3]*Dokkyo Medical. University, 880 Kita-Kobayashi, Mibu-machi, Shimotsuga-gun Tochigi , 321-0293, JAPAN;*

Abstract. In order to evaluate the state of activation in a cyclotron facility used for the radioisotope production of PET diagnostics, we measured the neutron flux by using gold foils and TLDs. Then, the spatial distribution of neutrons and induced activity inside the cyclotron vault were simulated with the Monte Calro calculation code for neutron transport and DCHAIN-SP for activation calculation. The calculated results are in good agreement with measured values within factor 3. Therefore, the adaption of the advanced evaluation procedure for activation level is proved to be important for the planning of decommissioning of these facilities.

1 Introduction

About 150 cyclotrons have been installed in the hospitals and clinics in Japan to supply radioisotopes for PET diagnosis of cancer and aging [1]. Several cyclotron facilities are in decommissioning stage recently. In these facilities, it is important to estimate the volume of radioactive wastes in advance.

There are two causes for activations. In case of the activation of charged particles, activated area is limited in a small area. On the other hand, neutrons are also emitted from the target by nuclear reactions and are widly distributed in a cyclotron vault. The evaluation of neutron activation of concrete around the cyclotron affects the cost of decommissioning. The purpose of this study is to estimate degrees and ranges of activation in a cyclotron vault.

In the previous report [2], we used the Monte Calro [MC] calculation code such as PHITS [3] or MCNP [4] for calculation of neutron spectrum emitted by the $^{18}O(p,n)^{18}F$ reaction [5] and the induced radioactivity calculation code DCHAIN-SP [6]. The results were compared with the measurement results in case of the PET Center of Dokkyo Medical University. In this work, we attempt to use PHITS with the revised nuclear reaction mode, in order to re-evaluate the activation results.

2 Materials and Methods

2.1 Facility

The cyclotron of Dokkyo Medical University is a CYPRIS-HM18 made by Sumitomo Heavy Industries, Ltd (Figure 1). This cyclotron has a target box to shield the radiation from several taget chambers. Target box is composed of polyethylene, lead and iron. In this facility, ^{18}F is mainly produced by the typical operation condition of the 18-MeV protons of 50μA for one hour in every morning on weekday. Two target (A and B) ports are used alternatively. Cyclotron has been operated about 200 days per year.

Fig. 1. Outview of the cyclotron CYPRIS-HM18 (Sumitomo Heavy Industries Co. Ltd.) .

* Corresponding author: nobuhara@tokyo-nucl.co.jp

2.2 Measurements

The gold (Au) foils (6mm in diameter and 20μm in thickness) with and without the cadmium (Cd) cover (1mm in thickness) were used for measuring the thermal and epi-thermal neutron flux(Figure 2) and the aluminium (Al) plates were used to detect a fast neutron flux by the $^{27}Al(n,\alpha)^{24}Na$ reaction(Figure 3) .

TLD UD-813PQ4 (Panasonic Co. Ltd.) (Figure 4) were used for measuring the neutron dose. The TLD has four phosphor elements. Two of the four elements (elements 1 and 2) are $^{6}Li_2{}^{10}B_4O_7(Cu)$, which is sensitive for thermal neutrons and gamma rays, and the others (elements 3 and 4) are $^{7}Li_2{}^{11}B_4O_7$ (Cu), which is sensitive for γ-rays only. A neutron dose for a TLD can be obtained by subtracting the average reading of elements 3 and 4 from that of elements 1 and 2.

Three TLDs were set inside the neutron monitor case UDS893-P1 for measuring the Ambient Dose Equivalent H*(10), respectively. The neutron monitor case is a polyethylene cube (outer size:225×225×220mm). This case is designed to moderate a fast neutron to thermal neutron and to increase a TLD reaction rate for measuring neutron ambient equivalent dose. The outside view of the neutron monitor case is shown in Figure 5.

These detectors were placed on the floor, wall, target box shielding and cyclotron surface. Irradiation of 18 MeV proton was performed for 55min in the target port A. The location of cyclotron components and detectors for neutron measurements in the cyclotron vault are shown in Figure 6.

In addition, for the verification of the activity calculation, several metal samples, bolts and screws, were obtained in the room (Figure 7). These samples are located at the same place all the time during operation period. The location of activity samples were shown in Figure 6.

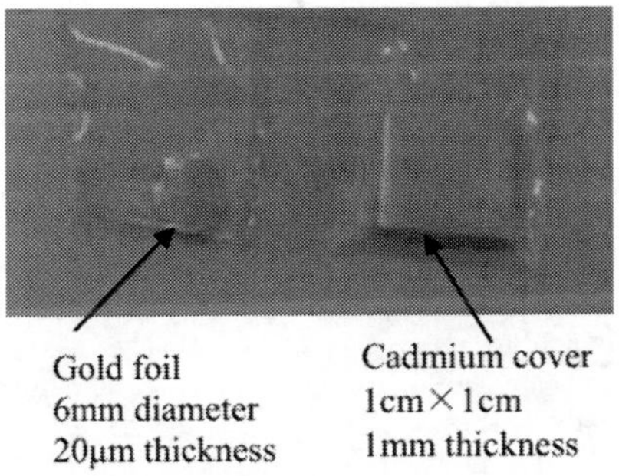

Fig. 2. A set of gold foils with and without Cd cover.

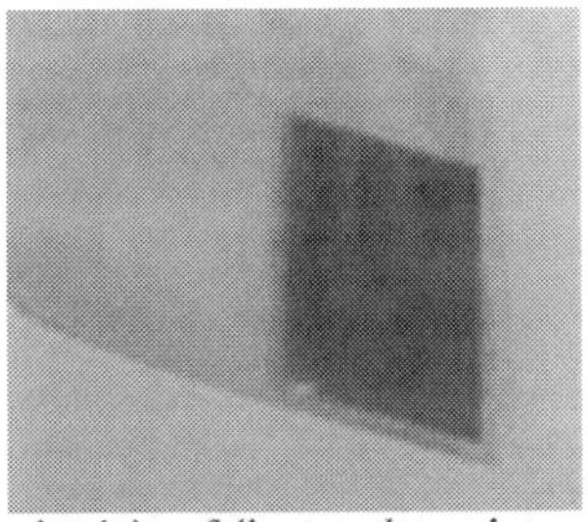

Fig. 3. An aluminium foil set on the cyclotron surface.

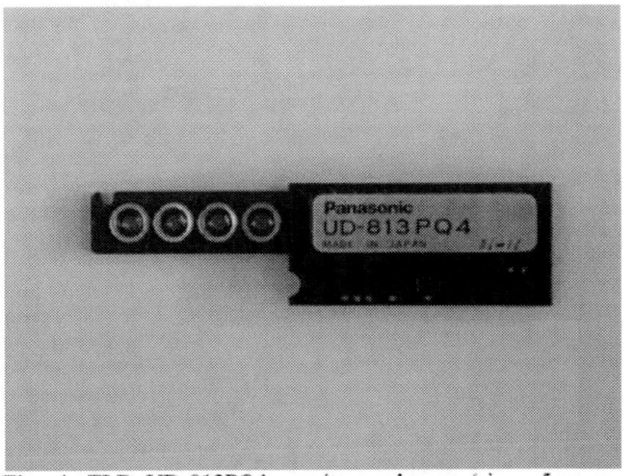

Fig. 4. TLD UD-813PQ4 monitor and case (size : 5cm × 2:5cm×0.5cm).

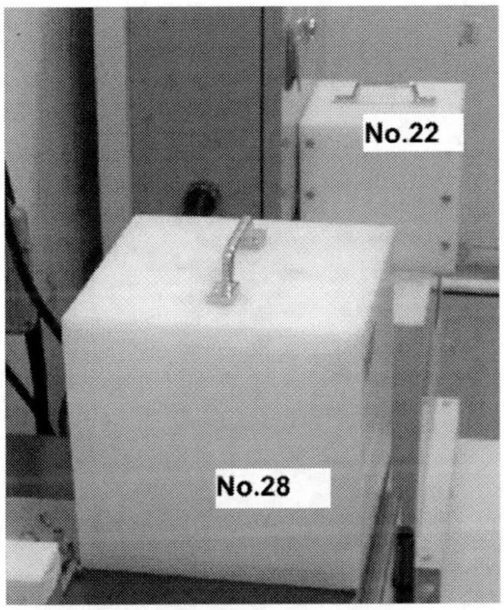

Fig. 5. Photo of two neutron monitor cases.

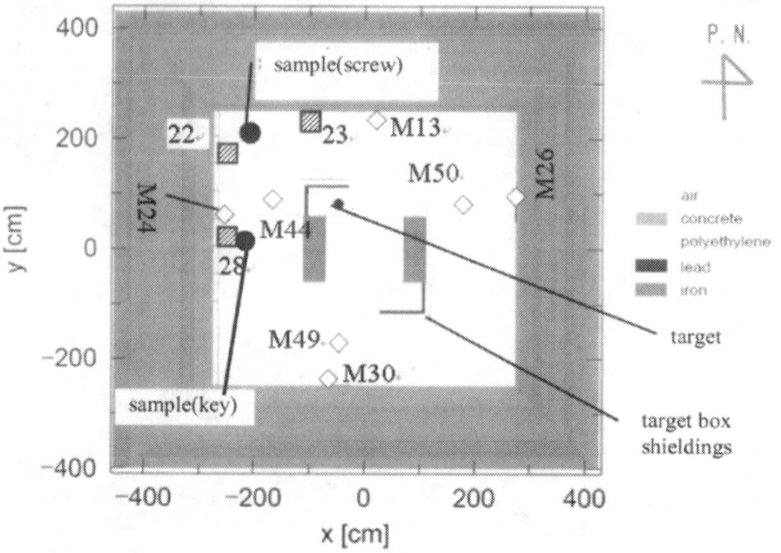

Fig. 6. Setting positions of monitors and neutron distribution calculation geometry in the cyclotron vault.
No. 22,23,28 means the position number of TLD dose monitor (for H 1cm)
Mxx stands for the position number of gold and aluminium foils

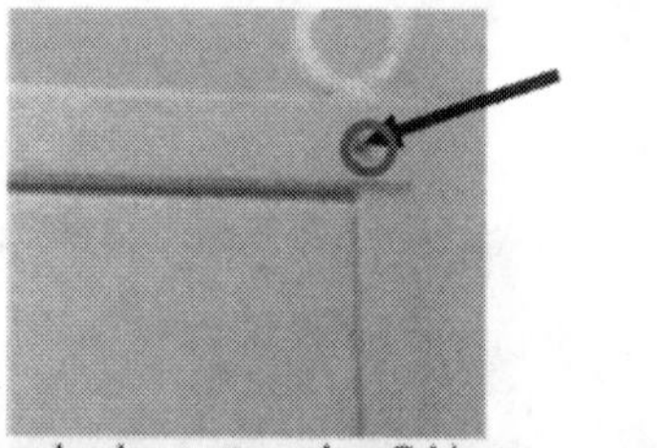

(1) A pan head screw tapped on Cabinet

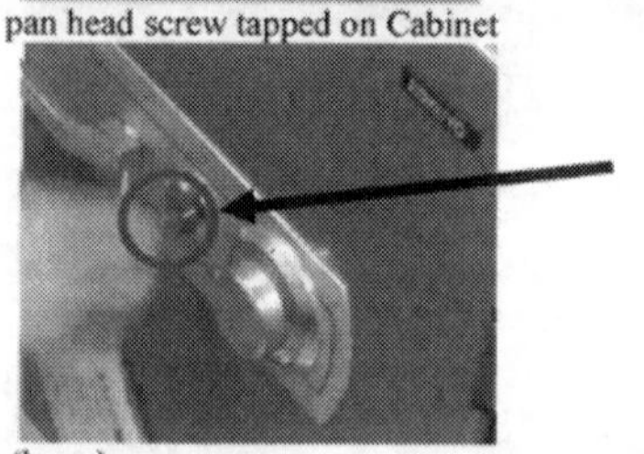

(2)key(brass)
Fig. 7. Metal samples for neutron flux monitoring

2.3 Neutron calculation

The PHITS ver. 2.82 code [3] is used for particle transport calculation with TENDL-2014 library [8] for the protons and JENDL-4.0 library [9] for the neutrons.

The calculation geometry for the neutron distribution in the cyclotron vault is shown in Figure 6. The target material is ^{18}O-enriched water in a niobium vessel with Harvar® foil window.

In our calculation, the cyclotron components were simply modelled as the structure of a target, a target box and a yoke. Data of the material compositions shown in the reference [10] are used.

We calculated the induced activity of activation foils and neutron dose in the case of the target port A.

To calculate the neutron dose, neutron fluxes are calculated by Track Length method [T-Track] where tallies are multiplied by dose conversion factors [11].

To calculate the thermal neutron fluxes (by gold foils), it is also calculated by Track Length method [T-Track] with low energy range (E<0.414eV). The thermal neutron fluxes are also corrected by the self-shielding and effective cross-section factors. In this study, the self-shielding and effective cross-section factors for thermal neutrons are calculated by PHITS neutron induced calculation and the value was estimated about 0.76 for a gold with 20 μm in thickness for on normal conditions.

To confirm the acuracy of flux of high energy part of neutrons in our calculation, a specific activity of ^{24}Na at the aluminium plates was calculated by Eq.(1) and compared with the that of measurements.

$$R = \int N\phi\,(E)\,\sigma(E)\,\mathrm{d}E\,(1-\exp(-\lambda t)) \qquad (1)$$

where, N (atoms/g) ; aluminium atomic density,
$\phi(E)$ (n/cm^2/sec) ; neutron flux at energy E,
$\sigma(E)$; the ^{27}Al (n, α) ^{24}Na reaction cross section (Fig.8) of JENDL/D-99 library[12],
λ ; decay constant and
t; experiment irradiation time.
In the PHITS caclculation, $\int \phi(E)\ \sigma(E)\ dE$ is calculated by Track Length method [T-Track] .

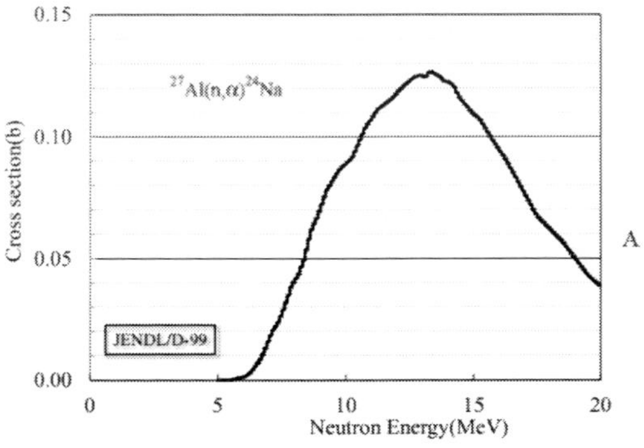

Fig. 8. The cross section of the ^{27}Al (n,α)^{24}Na reaction.

2.4 Activation calculation

Radioactivitiy induced in samples is calculated by using the DCHAIN-SP code. DCHAIN-SP code needs the neutron spectra which were calculated by the PHITS [T-Dchain] method for the activation calculation. A flow chart of linkage calculation between PHITS and DCHAIN-SP is shown in Figure 9. The [T-Dchain] tally method in PHIS code generates an input file for DCHAIN-SP. This is a merit to use the PHITS code and the DCHAIN code.

For the calculation of activity, the elemental composition (especially parent elements of radioactive nuclide) of samples were measured by an X-ray fluorescence analysis(Table 1). These samples have been set at the sampling positions since the start of this facility. In this facility, two target ports have been used alternatively. But in order to calculate induced activity simply, we supposed that the target port A has been only used, because the neutron fluxes during irradiation between target port A and B were alomost same level at the wall of cyclotron vault.

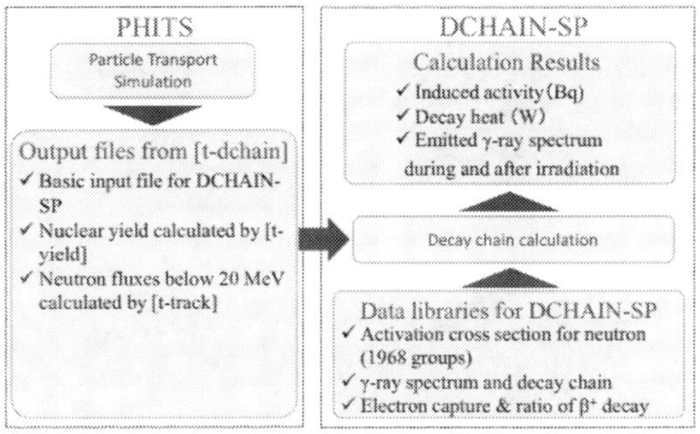

Fig. 9. Flow chart of connection calculation between PHITS and DCHAIN-SP

Table 1. material composition for activilty samples calculation

Element	Composition(%)	
	A pan head screw tapped on Cabinet (stainless steel)	key (brass)
Fe	69.8	-
Cr	18.8	-
Co	0.17*	-
Mn	1.19*	-
Ni	9.03*	-
Zn	-	34*

*based on JSS Standard stainless value

3 Results and Discussions

3.1 Spatial distribution of neutrons

The spatial distribution of neutron dose is shown in Figure 10. At the target port A, neutrons are produced by protons and neutrons are partially shielded by the target box shielding and the magnet yoke. For the estimating a radioactivitive volume, understanding the location of the maximum neutron flux at the wall is important.

In this facility, the location of maximum neutron flux is not occurs near the target because of the target box shielding.

Results of neutron dose calculation and TLD measurement are shown in Table 2. The results of calculated neutron dose are in good agreement with the measured results to within the factor (C/M; calculation / measurements) 0.79 to 0.99.

The thermal neutron flux distribution in the cyclotron vault is shown in Figure 11. The results of calculation and measurement obtained by activation of gold foils are shown in Table 3. The calculated results of thermal neutron flux are in good agreement with the in-room measured results to within the factor (C/M) 1.1 to 3.0.

Thermal neutron flux on the concrete surface may vary according to the concrete composition like as moisture content.

The calculated and measured values of ^{24}Na with activity of R(Bq/g) were shown in Figure 12. The calculation results are smaller than measurement values. The cause of this discrepancy might be came from the simple modelling of the ^{18}O target. We only considered the proton beam attack ^{18}O target and a Havar foil only. Any other proton beam loss point such as a vacuum foil may slightly contribute the emission of high energy part of neutrons.

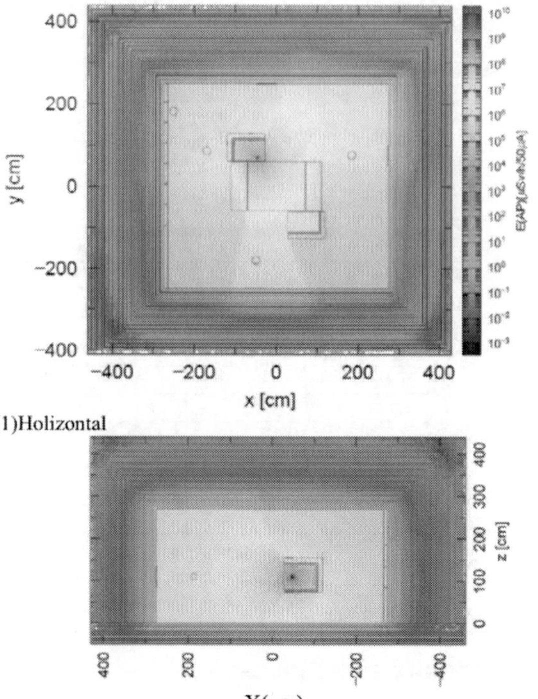

(1)Holizontal

(2)Vertical

Fig. 10. Neutron dose distribution in the room.

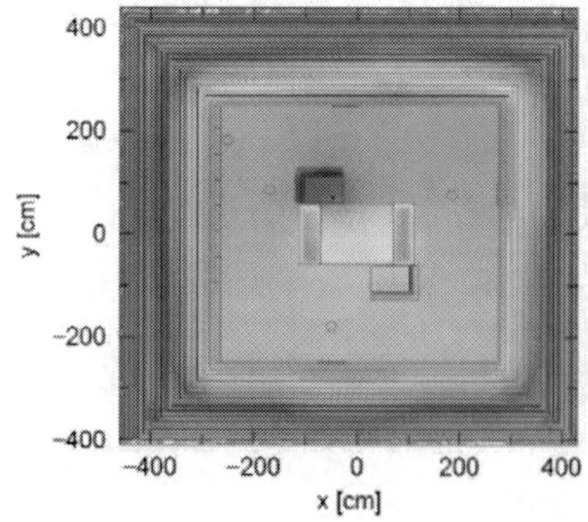

Fig. 11. Thermal neutron flux disttibution in the room (Holizontal).

Table 2. Neutron dose obtained by calculation and measurement.

Measurement Point No. (TLD)	Measurement (M) (μ Sv/h)	PHITS(C) (μ Sv/h)	C/M
22	1.6E+6	1.5E+06	0.94
23	2.6E+06	2.5E+06	0.99
28	1.0E+06	8.2E+05	0.79

Table 3. Thermal neutron flux obtained by calculation and measurement.

Measure-ment point No.	Measurement (M) ($n/cm^2/sec$)	PHITS(C) ($n/cm^2/sec$)	C/M
(1)In target box			
M01	1.0E+07	2.2E+07	2.1
M11	3.6E+06	5.8E+06	1.6
(2)In Room			
M44	1.1E+06	2.3E+06	2.2
M49	4.0E+05	1.1E+06	2.8
M50	8.6E+05	2.1E+06	2.4
(3)Wall Surface			
M13	2.5E+06	2.7E+06	1.1
M24	8.0E+05	1.9E+06	2.4
M26	7.9E+05	2.2E+06	2.7
M30	3.5E+05	1.2E+06	3.5

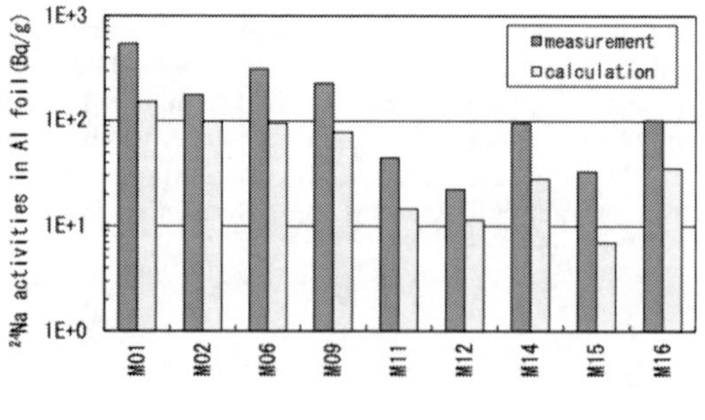

Fig. 12. The comparison of the results of calculation and measuremet value of ^{24}Na activity in aluminium plates which were set in the cyclotron vault

3.2 Activity calculation of samples

For the activity metal samples in the cyclotron vault, the activity calculation results are shown in Table 4 and Table 5. These measured and calculated data are in good agreement with each other. Detected nuclides in a screw were Cr-51, Fe-59 and Co-60, which are produced by thermal neutron capture reactions. The C/M ratios of the activities are comparable to the C/M ratios of thermal neutron fluxes. In this activity calculation, samples are very small, so activity self-shielding factor is not so large.

Table 4. Radioactivity induced in a screw tapped on a cabinet obtained by calculation and measurement.

Nuclide	**Measure-ment (M)** (Bq/g)	**PHITS+ DCHAIN-SP (C)** (Bq/g)	C/M
Fe-59	5.1E-01	6.5E-01	1.3
Cr-51	2.4E+01	4.5E+01	1.9
Co-60	3.8E+00	7.5E+00	2.0

Table 5. Radioactivity induced in a key set at the safety box obtained by calculation and measurement.

Nuclide	**Measure-ment (M)** (Bq/g)	**PHITS+ DCHAIN-SP (C)** (Bq/g)	C/M
Zn-65	1.2E+01	2.9E+01	2.4

4 Conclusion

It was found that a newer version of PHITS/DCHAIN-SP with a proton induced cross section library can be applicable to estimate radioactivity for PET cyclotron facility. The combination of MC calculation code and DCHAIN-SP code is suitable to estimate the induced activity and to depict the 3 dimensional map inside the cyclotron vault and concrete wall.

References

1. Japan Radioisotopes Association ,"*Statics on The Use of Radiation in Japan 2014*"
2. F Nobuhara et al. ,*Activation Evaluation for Decomissioning of PET cyclotron*", AESJ 2008 Fall Meeting, Sept. 4-6,2008 Kochi University of Technology H44
3. T. Sato, K. Niita, N. Matsuda, S. Hashimoto, Y. Iwamoto, S. Noda, T. Ogawa, H. Iwase, H. Nakashima, T. Fukahori, K. Okumura, T. Kai, S. Chiba, T. Furuta and L. Sihver, *Particle and Heavy Ion Transport Code System PHITS, Version 2.52*, J. Nucl. Sci. Technol. **50:9**, 913-923 (2013)
4. X-5 Monte Carlo Team, *CCC-710/MCNP - A General Monte Carlo N-Particle Transport Code, Version 5 Volume I: Overview and Theory*, LA-UR-03-1987 (April 2003 Revised 2/1/2008)
5. M. Hagiwara, T. Sanami, Y. Iwamoto, N. Matsuda, Y. Sakamoto, Y. Nakane, H. Nakashima, K. Masumoto, Y. Uwamino, *Spectrum Measurement of Neutrons and Gamma-rays from Thick $H_2{}^{18}O$ Target Bombarded with 18 MeV Protons*, Prog. in Nucl. Sci. Techn. 2011 (accepted)
6. T. Kai, F. Maekawa, K. Kosako, Y. Kasugai, H. Takada, and Y. Ikeda, *DCHAIN-SP 2001: High Energy Particle Induced Radioactivity Calculation Code*, JAERI-Data/Code 2001-016, 82 (2001)
7. Sumitomo Heavy Industries Co. Ltd. WWW Page, http://www.shi.co.jp/
8. A.J. Koning, et al.,*TENDL-2014: TALYS-based evaluated nuclear data library*, www.talys.eu/tendl-2014.html
9. K. Shibata,*JENDL-4.0: A New Library for Nuclear Science and Engineering*, J. Nucl. Sci. Technol. **48:1**, 1-30 (2011).
10. Nuclear Safety Technology Center , *Practical Manual for Shielding Calculation in Radiation facilities 2015* (Housyasen-sisetsu no Shahei Keisan Jitsumu Manual 2015,2015(Japanese only)
11. ICRP, 1996. *Conversion Coefficients for use in Radiological Protection against External Radiation*. ICRP Publication 74. Ann. ICRP 26 (3-4)
12. K.Kobayashi, et al, "*JENDL-Dosimetry File 99(JENDL/D-99)* ", JAERI-1344 (2002)

Development of An Epi-thermal Neutron Field for Fundamental Researches for BNCT with A DT Neutron Source

Yuta Osawa [1,], Shoichi Imoto[2], Sachie Kusaka[1], Fuminobu Sato[1], Masahiro Tanoshita[3], Isao Murata[1]*

[1]*Graduate School of Engineering Osaka University, Yamadaoka 2-1, Suita, Osaka, Japan*
[2]*City government of Kobe, Kanou 6-5-1, Tyuoku, Kobe, Hyogo, Japan*
[3]*Department of engineering Fukui University of Technology, Gakuen 3-6-1, Fukui, Fukui, Japan*

Abstract. Boron Neutron Capture Therapy (BNCT) is known to be a new promising cancer therapy suppressing influence against normal cells. In Japan, Accelerator Based Neutron Sources (ABNS) are being developed for BNCT. For the spread of ABNS based BNCT, we should characterize the neutron field beforehand. For this purpose, we have been developing a low-energy neutron spectrometer based on ^{3}He position sensitive proportional counter. In this study, a new intense epi-thermal neutron field was developed with a DT neutron source for verification of validity of the spectrometer. After the development, the neutron field characteristics were experimentally evaluated by using activation foils. As a result, we confirmed that an epi-thermal neutron field was successfully developed suppressing fast neutrons substantially. Thereafter, the neutron spectrometer was verified experimentally. In the verification, although a measured detection depth distribution agreed well with the calculated distribution by MCNP, the unfolded spectrum was significantly different from the calculated neutron spectrum due to contribution of the side neutron incidence. Therefore, we designed a new neutron collimator consisting of a polyethylene pre-collimator and boron carbide neutron absorber and confirmed numerically that it could suppress the side incident neutrons and shape the neutron flux to be like a pencil beam.

1 Introduction

Recently, Boron Neutron Capture Therapy (BNCT) is known to be a new promising cancer therapy. In BNCT, after ^{10}B is accumulated in cancer cells thermal or epi-thermal neutrons are irradiated from outside of a human body, and then α and ^{7}Li are produced. These particles have high LET and the ranges are as long as the size of one cell. As a result, tumor cells can selectively be destroyed, simultaneously suppressing influence against normal cells. Figure 1 shows the principle of BNCT.

Up to now, only nuclear reactor has been used as a neutron source for BNCT because BNCT requires a strong low-energy neutron source. In Japan, Accelerator Based Neutron Sources (ABNS) which can be constructed in medical facilities such as hospitals are being developed instead of nuclear reactor. However, characteristics of neutron field such as neutron spectrum and neutron intensity depend on kinds of accelerators or their parameters. Therefore, evaluation of neutron fields obtained in accelerators is important to know and control the effect of therapy, treatment time, and dose of patients. In BNCT, thermal or epi-thermal neutrons are used for the treatment. In particular, estimation of epi-thermal neutron spectrum is known to be difficult because the shape of thermal neutron spectrum is fixed for the atmospheric temperature and fast neutrons can be identified by nuclear reactions such as charged particle emission reactions, e.g., elastic scattering, and activation reactions having threshold energies. For these reasons, a neutron spectrometer is required which can measure thermal and epi-thermal neutron energy and intensity.

In the author's group, a low-energy neutron spectrometer based on a ^{3}He position sensitive proportional counter has been developed [1].

Figure 2 shows the detection principle of the new spectrometer. We measure the detection depth distribution Y of a nuclear reaction by an incident neutron entering the spectrometer parallel to axis of the neutron detector. We can estimate the neutron energy spectrum X by using an unfolding process with the detector response function R that was estimated as the reaction depth probability density function for neutron energy.

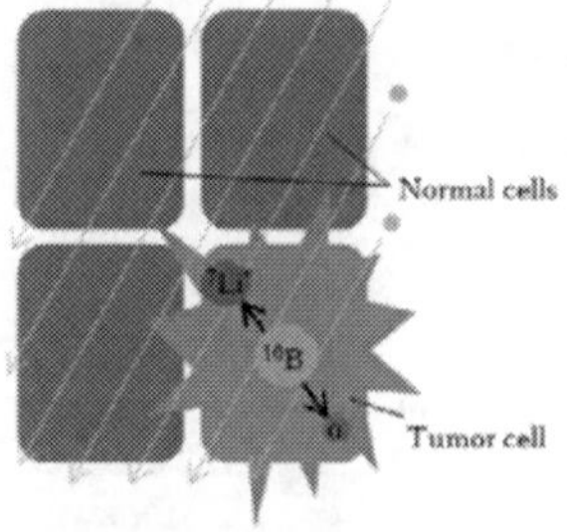

Figure 1 Principle of BNCT.

* Corresponding author: y-osawa@qr.see.eng.osaka-u.ac.jp

DOI: 10.1051/epjconf/201715304008

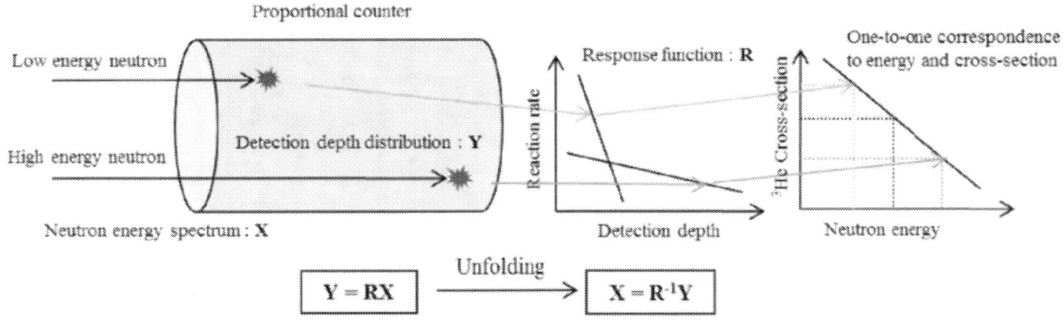

Figure 2 Principle of the neutron spectrometer.

In the previous study, verification of the spectrometer validity has been carried out by using an epi-thermal neutron field made with an AmBe neutron source [2]. However, there was a serious problem we faced, that is, weakness of the utilized AmBe neutron source intensity, leading that the statistical accuracy was seriously deteriorated. As a result, it took a very long time to complete one measurement. Therefore, we developed a new intense epi-thermal neutron field by using the Intense 14 MeV Neutron Source Facility (OKTAVIAN) of Osaka University specifically for fundamental researches of BNCT [3]. After examining the characteristics of the produced neutron field, we have designed and constructed a beam shaping assembly to realize a beam epi-thermal neutron so that we can verify the developed spectrometer experimentally.

2 Epi-thermal neutron column

As mentioned in Chapter 1, an intense beam epi-thermal neutron field is required for fundamental researches of BNCT, especially validity verification of the low-energy neutron spectrometer. In this study, we designed and constructed an intense epi-thermal neutron field having a uniform neutron field with as wide range as possible with a DT neutron source instead of an AmBe neutron source. Two design targets of the present epi-thermal neutron field were given as epi-thermal neutron flux $\Phi_{epi\text{-}thermal}$ and ratio of the epi-thermal neutron flux to the fast neutron flux η. In case of the AmBe neutron source, $\Phi_{epi\text{-}thermal}$ = several tens n/sec/cm^2 had been achieved by the author's group in the previous study. In this study, hundred times larger epi-thermal neutron flux was aimed, i.e., $\Phi_{epi\text{-}thermal}$ > 1000 n/sec/cm^2, considering the DT neutron source intensity which is several ten-times stronger than the AmBe neutron source. In addition, we defined η as $\Phi_{epi\text{-}thermal} / \Phi_{fast}$, which is an index to indicate the mixing degree of fast neutrons in the epi-thermal neutron field. In the present study, the target value of η is more than 10.

2.1 Design of epi-thermal neutron column

Design of the epi-thermal neutron field was carried out using Monte Carlo N-Particle Transport Code (MCNP-5) and evaluated nuclear data library of JENDL-4.0 [4]. Differently from the previous design using an AmBe neutron source (4.0×10^6 n/sec), the Intense 14 MeV Neutron Source Facility OKTAVIAN of Osaka University was used as a strong neutron source. In the OKTAVIAN Facility, deuteron is accelerated to about 350 keV with a Cockcroft-Walton type electrostatic accelerator and then 14MeV neutrons are generated by deuteron-triton (DT) reaction. The designed maximum neutron source intensity is around 1.0×10^9 n/sec in a pulse mode. In this study, we designed the epi-thermal neutron field as in the following steps under the assumption that the DT neutron intensity was 4.0×10^8 n/sec at 5 μA, which was a common value at present in the facility. Figure 3 shows neutron spectra obtained after each design step.

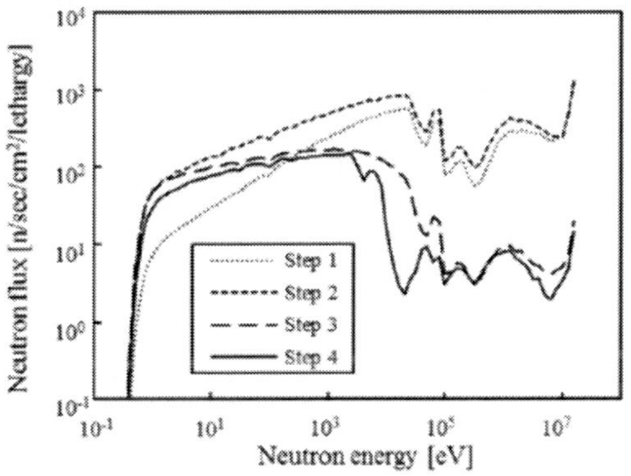

Figure 3 Neutron spectrum for each step.

Table 1 Simuration result of Step 1.

Material	$\Phi_{epi\text{-}thermal}$ [n/sec/cm^2]	η
Al	2.1	0.00
AlF_3	5714.7	0.58
C	3025.3	0.20
CF_2	5487.5	0.44
D_2O	5515	0.55
Fe	220.3	0.01
Pb	22.9	0.00
PE	345.7	0.07

2.1.1 1st moderator of 14 MeV neutrons

First of all, we selected the first moderator material and determined the moderator thickness for moderating 14 MeV neutrons effectively (Step 1). The moderator area was 50 cm × 50 cm seeing from the source and several thick moderators were investigated. As a result, we decided the optimum thickness of 40 cm which was able to moderate high energy neutrons keeping the epi-thermal neutron flux intensity sufficiently high. In this column, thermal neutrons are not removed by neutron absorbers for other applications. However, thermal neutrons below 0.5 eV could be cut by using a cadmium sheet. Table 1 shows $\Phi_{epi\text{-}thermal}$ of each material we investigated under the same condition. From the result of Table 1, aluminum fluoride shows the best performance as the first moderator. However, because production of an aluminum fluoride block of 50 cm × 50 cm × 40 cm was technically difficulty, we substituted an aluminum fluoride cylinder of 42 cm length and 20 cm in diameter which was successfully produced so far, and the rest space surrounding the cylinder was filled with Teflon pellets to supplement the function of the aluminum fluoride in an actual column.

2.1.2 Reflector to increase neutron flux intensity

Next, we determined the thicknesses of side reflectors to increase the neutron flux intensity (Step 2). Lead and graphite, which are known as a good neutron reflector, were selected as candidates. Because normally lead is known to be better than graphite, we examined the performance with lead reflector. As shown in Figure 4, the thicker the lead reflector is, the larger $\Phi_{epi\text{-}thermal}$ becomes. However, η saturates for the thickness more than 15 cm. Therefore, we decided the lead thickness was 15 cm. Since lead is much heavier than graphite, we decided to partly combine graphite with lead under the condition that the total thickness of the reflector was 15 cm. From the results of additional analysis shown in Figure 5, combinations of 5 cm lead and 10 cm graphite (Pb+C (B)) or 10 cm lead and 5 cm graphite (Pb+C (C)) have similarly high performance to the 15 cm lead reflector case. Finally, we decided that the combination of Pb+C (B) was employed as the present reflector configuration, because Pb+C (B) combination has a larger η value and the weight was lighter as compared to Pb+C (C) or Pb 15 cm reflector.

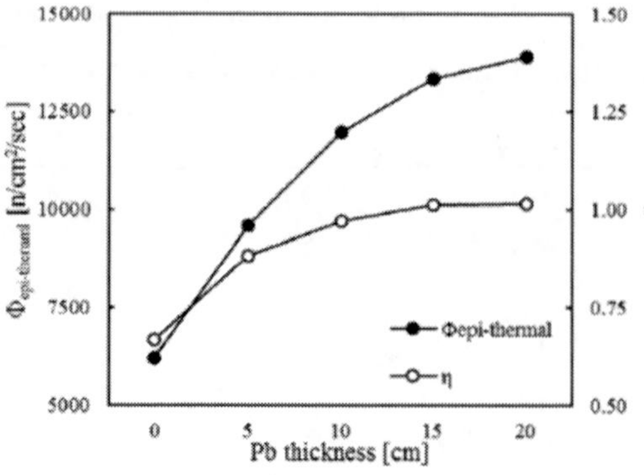

Figure 4 Simulation result of Step 2.

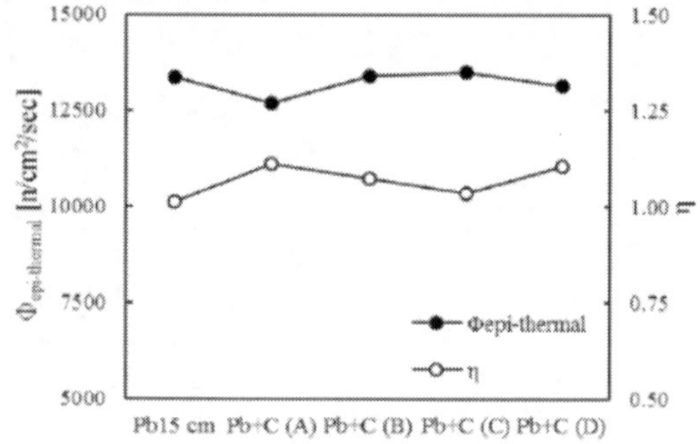

Figure 5 Simulation result of Step 2.
(Pb+C (A): Inside C 10 cm + Outside Pb 5 cm, Pb+C (B): Inside Pb 5 cm + Outside C 10 cm, Pb+C (C): Inside Pb 10 cm + Outside C 5 cm, Pb+C (D): Inside C 5 cm + Outside Pb 10 cm)

2.1.3 Reduction of fast neutron contribution

Up to Step 2, as shown in Figure 3, direct contribution of high energy neutrons around 14 MeV neutrons clearly remains. We thus arranged an iron shield as an attenuator in front of the moderator and between the moderator and the neutron source to suppress high energy neutron contribution as small as possible compared to the previous steps (Step 3). Table 2 shows the $\Phi_{epi\text{-}thermal}$, Φ_{14MeV} and η values as a function of the iron shield thickness. The thicker the iron shield becomes, the smaller Φ_{14MeV} is and the larger η is, while the $\Phi_{epi\text{-}thermal}$ decreases. In this study, we assumed that both of $\Phi_{epi\text{-}thermal}$ and η have equal importance. And then, we made a product of $\Phi_{epi\text{-}thermal}$ and η to examine the optimum thickness of the iron shield. The result is shown in Figure 6. From the Figure 6, we determined the thickness of the iron shield attenuator to be 25 cm because the product showed the highest value.

Table 2 Simulation result of Step 3.

Thickness of iron [cm]	$\Phi_{epi\text{-}thermal}$ [n/sec/cm^2]	$\Phi_{14\,MeV}$ [n/sec/cm^2]	η
0	12609.4	1276.8	0.95
5	11844.8	534.7	1.83
10	9976.8	238.7	2.95
15	8060.3	106.6	4.28
20	6363.1	42.9	5.82
25	5036.3	19.4	7.53
30	3984	9	9.16
35	3152.1	4.3	10.87
40	2485.5	1.9	12.22
45	1965.1	0.5	13.64
50	1564.3	0.3	14.91

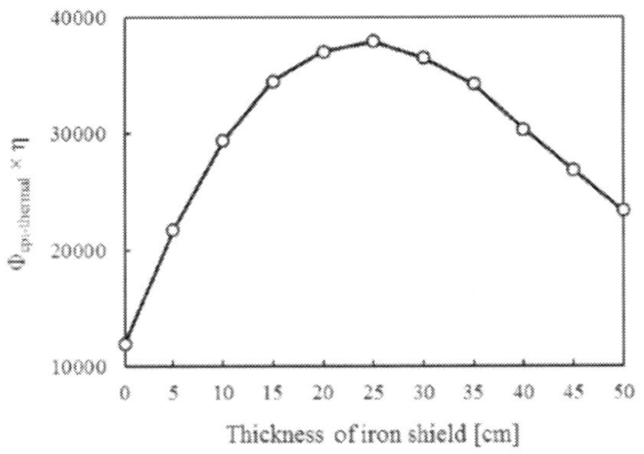

Figure 6 Product of epi-thermal neutro flux and η.

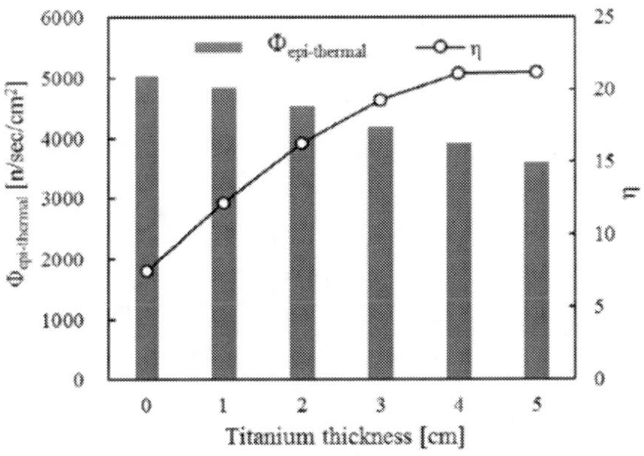

Figure 7 Simulation result of Step 4.

2.1.4 2nd moderator to cut higher energy neutrons over 10 keV

Finally, we utilized titanium, which has large resonance above 10 keV region, as the second moderator to cut higher energy neutrons over 10 keV and increase η value (Step 4). The size of a titanium plate was 50 cm × 50 cm and set up in the rear side of the first moderator. We then changed the titanium thickness to decide the optimum thickness. The η values saturated over 4 cm thickness as shown in Figure 7, we thus determined the titanium thickness of 4 cm. However, in the actual epi-thermal neutron column, we used the size of 30cm × 30cm × 4cm titanium due to its cost.

Figure 8 shows the finally designed neutron column after the 4 steps. From the spectrum after Step 4 in Figure 3, the designed neutron spectrum has an excellent spectrum shape having a clear peak in the epi-thermal energy region and at the same time supresses fast neutrons very low. Moreover, the designed epi-themal neutron field could accomplish the design target, i.e., $\Phi_{epi\text{-}thermal}$ = 3915 n/sec/cm^2 (> 1000) and η = 21.2 (> 10). We set this column in a steel container, which is the size of 80 cm × 80 cm × 100 cm and 700 kg in weight. Total weight of the column and the container was about 3000 kg.

2.2 Construction of the epi-thermal neutron column

The epi-thermal neutron column was constructed by filling designed materials listed in Table 3 in the container of 80 cm × 80 cm × 100 cm. Figure 9 is an inside photo of the column in the middle step of construction. A cylinder of aluminum fluoride moderator was set in the center of an aluminum box. Then Teflon pellets were poured in this box to fill up around the cylinder. Five Iron shield plates and titanium plates were set in the front and rear sides of the box, respectively. Reflector blocks of graphite and lead were arranged so as to cover the moderators and shields entirely.

Table 3 Materials of the epi-thermal neutron column.

Material	Size [cm]	Weight [kg]
AlF_3 cylinder	φ20×42	40.5
CF_2 pallets	50×50×43	115.6
C blocks	thickness 10	470.4
Pb blocks	thickness 5	900.6
Fe layer	50×50×25	492.0
Ti plate	30×30×4	16.2

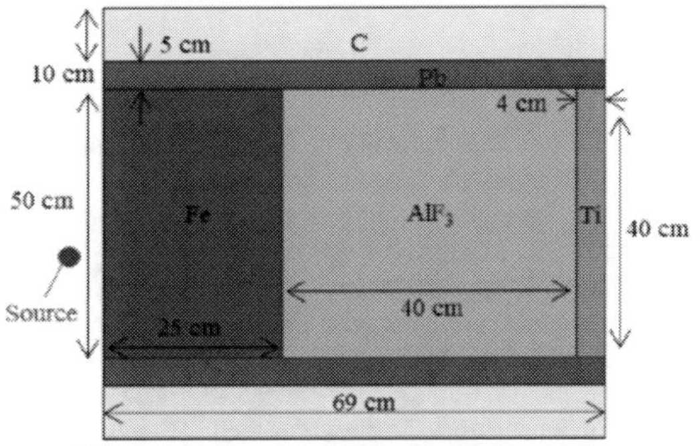

Figure 8 Designed epi-thermal neutron column.

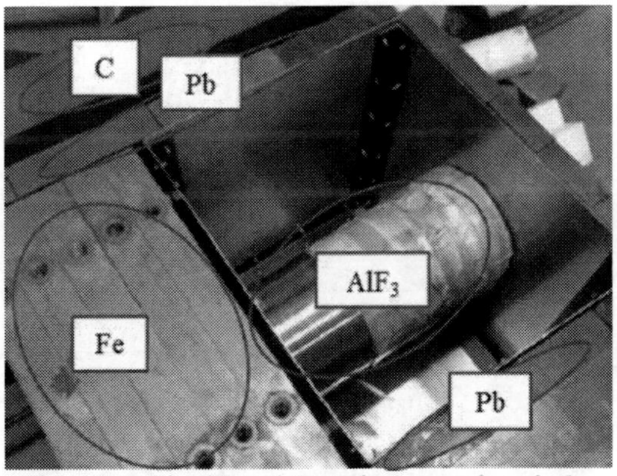

Figure 9 Inside of the epi-thermal neutron column in the middle stage.

3 Evaluation of the epi-thermal neutron column

3.1 Experiment with activation foils

For characterization of the constructed epi-thermal neutron column, we have carried out measurements of neutron flux intensity and spectral index with activation foils as shown in Table 4. $^{197}Au(n,\gamma)^{198}Au$ reaction is sensitive especially to low energy neutrons. In this experiment, we used gold foils with and without a cadmium absorber and measured the reaction rates to confirm whether the epi-thermal neutron field was formed properly. Indium and aluminum foils were used for an evaluation of the amount of fast neutrons. $^{115}In(n,n')^{115m}In$ reaction and $^{27}Al(n,\alpha)^{24}Na$ reaction are high threshold energy reactions for 1.2 MeV and 7.2 MeV neutron, respectively. Thus we can examine existence of fast neutrons. Indium foil was covered with a cadmium absorber because $^{115}In\ (n,\gamma)\ ^{116m}In$ reaction has a large cross section with a resonance in the thermal energy region. Figure 10 shows positions of activation foils in this experiment. We set the foils at the same height as the DT neutron source that is 5 cm lower than the center of the column. In this experiment, the irradiation time was 5 hours and the absolute neutron source intensity was determined using a niobium foil set in front of the neutron source.

3.2 Experimental result

The absolute DT neutron intensity was determined to be 1.08×10^{8} n/sec from a niobium foil in this experiment. After the irradiation, we measured decay gamma-rays of activation foils with two HpGe detectors. Reaction rates of the foils are shown in Table 5 compared with the calculations. Gold foils with and without cadmium were activated sufficiency, and we confirmed that a good agreement was obtained between the calculation and experimental values as shown in the C/E values. Next, as for the evaluation of fast neutrons, in the measurement we could not detect gamma-rays from the indium and aluminum foils. It means that higher energy neutron contribution could substantially be suppressed. From these results, we confirmed that an excellent epi-thermal neutron field was constructed having a sufficiently intense epi-thermal neutron flux simultaneously removing fast neutrons.

Table 4 Activation foils used.

	Nuclear reaction	γ-ray energy [keV]	Half life
^{197}Au	$^{197}Au(n,\gamma)^{198}Au$	411.8 (95.7 %)	2.69 d
^{115}In	$^{115}In(n,n')^{115m}In$	336.3 (45.8 %)	4.49 h
^{27}Al	$^{27}Al(n,\alpha)^{24}Na$	1369 (100 %)	14.7 h

Table 5 Comparison of experiment and calculation.

Activation foil	Reaction rate (calculation) [/sec/source]	Reaction rate (experiment) [/sec/source]	C/E
Au	2.67×10^{-6} ($\pm$3.94%)	2.48×10^{-6} ($\pm$2.14 %)	1.08
Au + Cd	2.28×10^{-6} ($\pm$4.08 %)	2.17×10^{-6} ($\pm$3.59 %)	1.05
In	2.97×10^{-9} ($\pm$13.2 %)	Not detected	
Al	5.50×10^{-10} ($\pm$23.4 %)	Not detected	

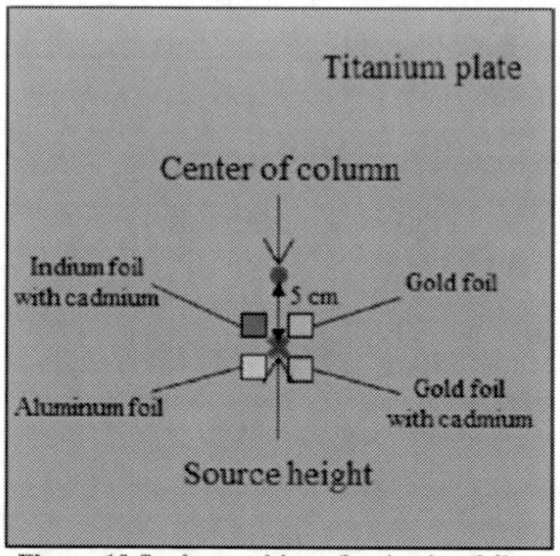

Figure 10 Setting position pf activation foils.

4 Experimental verification of the low energy neutron spectrometer

4.1 Principle of detection

There is no direct method to measure the low energy neutron spectrum. In the author's group, a new low-energy neuron spectrometer was developed using a position sensitive ^{3}He proportional counter [2]. The principle is as follows: ^{3}He has a large reaction-cross section in lower energy regions, in addition, the cross-section monotonically increases as the velocity of neutron deceases. Thus it shows a clear one-to-one correspondence between neutron energy and ^{3}He cross-section value. Therefore, when neutrons are incident to the spectrometer in parallel to the axis of the neutron detector, we will measure many nuclear reactions near the entrance of the detector because low-energy neutrons have a large reaction cross-section. By contrast, higher-energy neutrons will be measured at deeper positions. Namely, if we measure a neutron detection depth distribution in the detector, the distribution is expected to change depending on the neutron energy. The reaction depth probability density function for neutron energy can thus be used as the detector response function to estimate the neutron energy. The response matrix R(E,r)dr can ea-

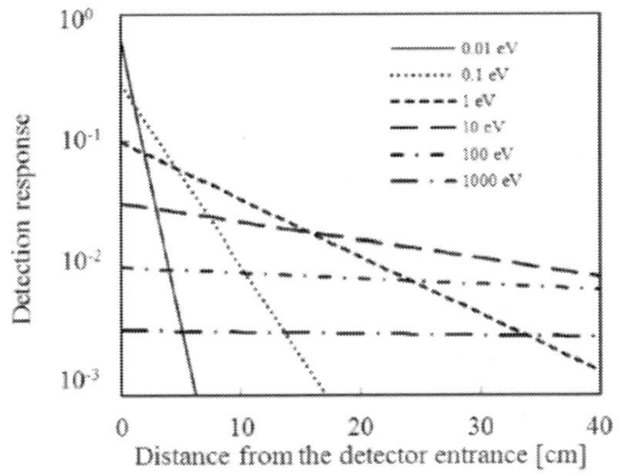

Figure 11 Response function of detector: R(E,r). Each line corresponds to incident neutron energy from 0.01 eV to 1000 eV.

sily be deduced as $\sum(E)\exp(-\sum(E)r)\cdot dr$ if neglecting the effect of a casing, where $\sum(E)$ is macroscopic cross section of ^{3}He(n,p) reaction, E is neutron energy and r is depth. Figure 11 illustrates some examples of the detector response function for the presently used ^{3}He detector. If measured detection depth distribution is Y(r) and neutron energy spectrum is X(E), they are related by $Y(r)=\int R(E,r)X(E)dE$. The neutron energy spectrum is estimated by using unfolding process from the measured reaction position distribution, Y(r). For the unfolding process, we adopted the Bayes theorem [5].

4.2 Detector

Figure 12 shows the ^{3}He position sensitive proportional counter we used in this study. The length is 40 cm, the diameter of the detector is 2.5 cm and the ^{3}He gas pressure is 0.5 MPa. The core wire resistance we used as anode is 4.6 kΩ. The position sensitive proportional counter has two output BNC connectors at both ends to identify where the incident neutron reacts with inner gaseous material. When the emitted charged particles induce secondary electrons in the counter and they reach the anode, the induced charge is separated and conveyed to two directions and different voltage pulses are generated according to the reaction position because of a high resistance level of the core wire. Two output signals are amplified adequately by two pairs of pre-amplifiers and main-amplifiers, and then fed to a multi-parameter system (MPS-1600). Figure 13 shows the block diagram of the measurement circuit. The measurable energy range of this proportional counter was estimated to be from 0.01 eV to 1 keV in the previous study [6].

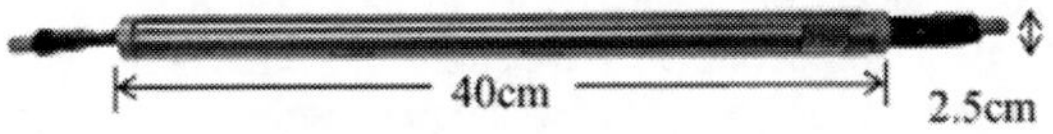

Figure 12 ^{3}He position sensitive proportional counter.

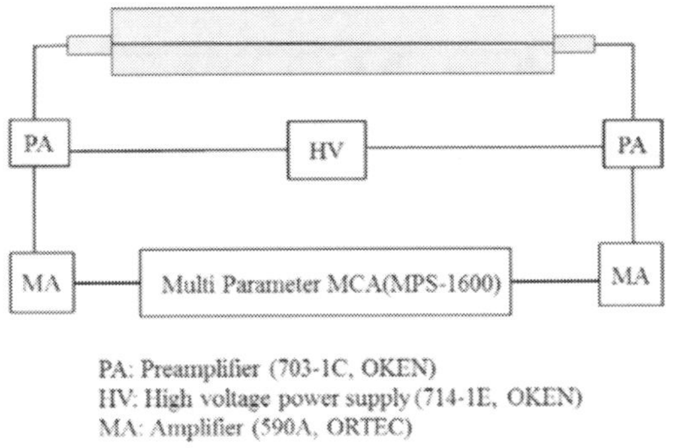

Figure 13 Schematic block diagram of measuring system.

4.3 Verification

4.3.1 Experiment

We carried out experimental verification of the neutron spectrometer with the epi-thermal neutron column designed and constructed in Chapter 2. Figure 14 shows the experimental setup. The DT neutron source was positioned at 8.5 cm distant from the column and shifted 5 cm vertically and 2.3 cm horizontally from the center of the column. The detector was set on the deuteron beam axis and at 20 cm away from the column. From the standpoint of the detection principle of the present neutron spectrometer, incident neutrons should be incident in parallel to the axis of the detector, that is, the neutron beam shaping is required. We covered the detector using cadmium sheet as a neutron absorber and polyethylene as a neutron moderator to collimate neutrons and to suppress influence of neutrons entering from the side of the detector. Practically, the neutrons entering from the side was moderated to thermal neutrons by polyethylene and cadmium can cut thermal neutrons. Accordingly, we shaped the neutron flux of the epi-thermal neutron column to be a neutron beam. The irradiation time was 5 hours and we used a niobium foil to determine the absolute neutron source intensity by setting it in front of the neutron source.

4.3.2 Result

In the experiment, the absolute DT neutron source intensity was determined to be 8.9×10^8 n/sec from a niobium foil. Figure 15 shows the measured detection depth distribution. In the distribution, there was no count at both ends of the detector, because proportional counter has dead regions at the both ends. From the result of the detection depth distribution, the count number was large near the entrance of the detector and was smaller at deeper positions. It means that low energy neutrons entered the detector from one side of it and then much more neutrons reacted with internal ^{3}He gas near the detector entrance. In Figure 15, the calculated detection position distribution by MCNP-5 was shown. From the figure, it was found that the calculation result agreed with the measurement result very well. However, the unfolded neutron spectrum was significantly different from the calculated spectrum by

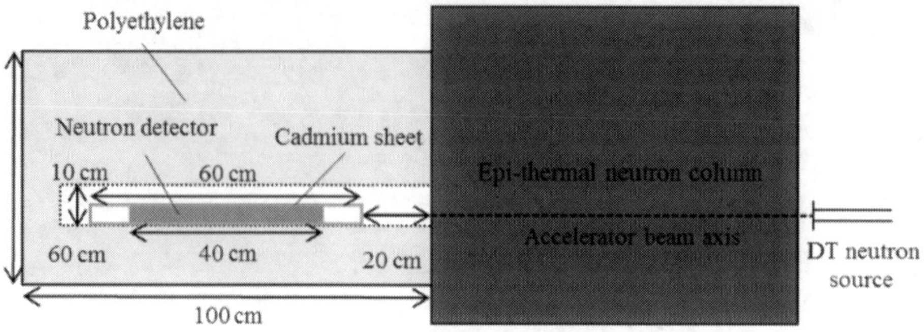

Figure 14 Experimental setup.

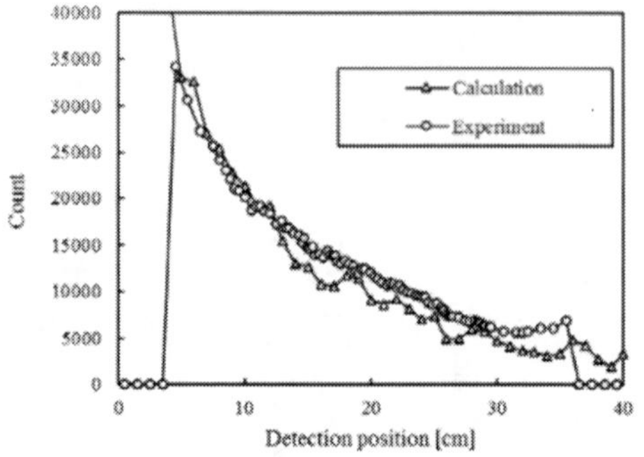

Figure 15 Detction depth distribution.

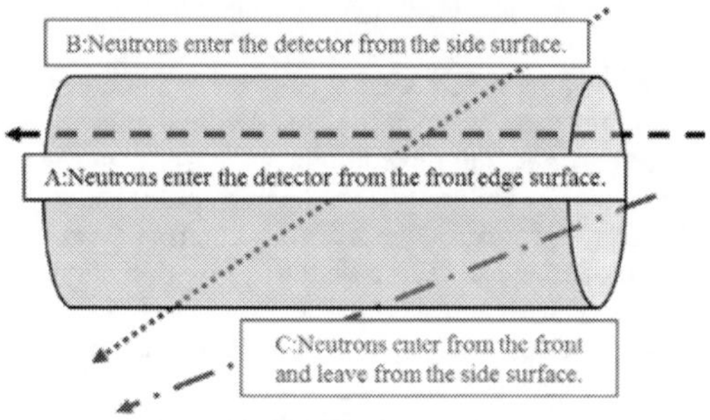

Figure 16 Incident neutrons.

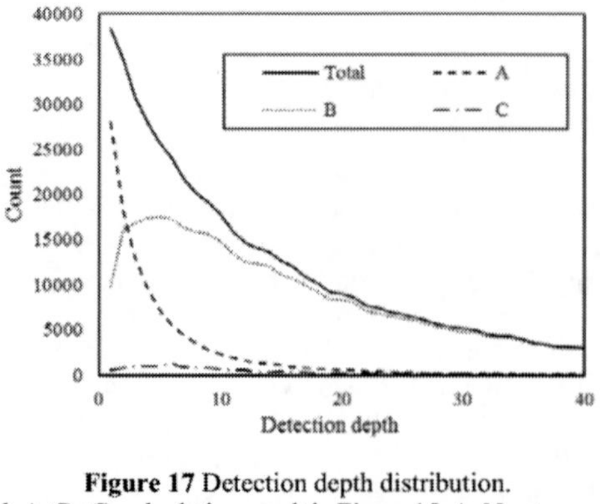

Figure 17 Detection depth distribution.
Total: A+B+C, calculation result in Figure 15, A: Neutrons enter the detector from the front edge surface, B: Neutrons enter from the side surface, and C: Neutrons enter from the front and leave from the side surface.

MCNP. To examine the discrepancy, we analyzed the contribution of incident neutrons to the detector separately, resolving the incident neutrons into three components; A, B and C as shown in Figure 16. As a result of Figure 17, it is made clear that not incident neutrons from the front edge (A) but those from the side surface (B) dominantly cause the detection depth distribution of. Namely, the cadmium sheet and polyethylene moderator could not suppress neutron's entering from the side of the detector sufficiently.

5. Design of a new neutron collimator

The experimental result of Chapter 4 suggested that a spectrometer shield having a collimator role made of cadmium and polyethylene should be improved. It is expected that it can shape the neutron flux to be a clear angular flux and at the same time to shield neutron incidence from the side of the detector, in order to carry out the experimental verification of the present neutron spectrometer. In this chapter, we design a new neutron collimator consisting of a pre-collimator and neutron shields.

5.1 Design of pre-collimator

To collimate the neutron flow, we set a pre-collimator additionally in front of the detector as show in Figure 18. The performance of pre-collimator saturated over 20cm. Therefore, the length of the pre-collimator was determined with 20 cm because the intensity was the highest in the thickness more than 20 cm and the diameter was 4 cm. We investigated suitable materials by examining angular distributions of neutron incident angles to the detector entrance surface as shown in Figure 19. Borated polyethylene or polyethylene pre-collimator could shape the neutron flux to be like a pencil beam. Although both of them have almost the same performance, we decided to use polyethylene in consideration of costs.

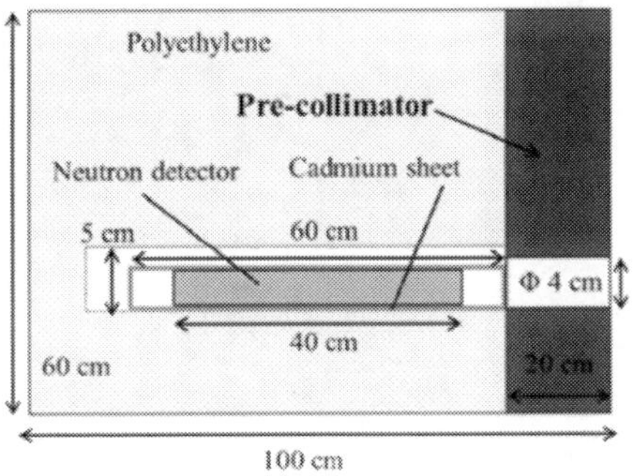

Figure 18 Pre-collimator design.

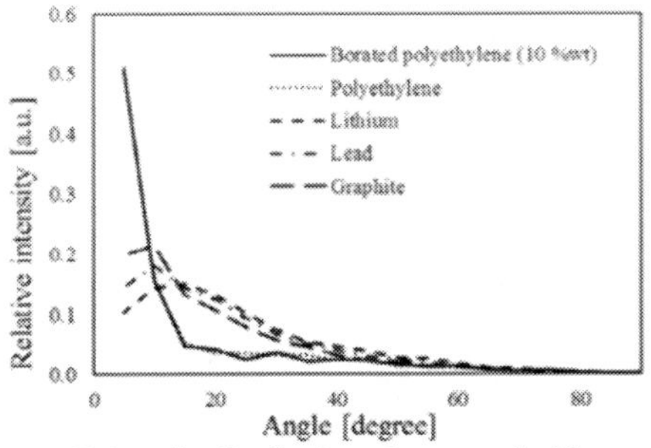

Figure 19 Angular distribution of neutrons incident to the detector entrance surface for several materials.

5.2 Design of neutron shield

For suppressing the side incident neutrons, we installed a neutron absorber surrounding the neutron detector in addition to a polyethylene layer covering the neutron shield as a neutron moderator as shown in Figure 20. We selected lithium and solid boron carbide as the neutron absorber material because these materials are known as a good neutron absorber, so that we could cut neutrons for energies smaller than the cadmium cutoff energy of 0.5eV. In order to determine the material and thickness, we changed their thickness and analyzed the contribution of incident neutrons; A, B and C in Figure 16. Figure 21 shows each contribution at 20 cm depth from the detector entrance. As shown in Figure 21, the performance of lithium becomes greatest at 20 cm thick, while the performance of solid boron carbide saturates over 5 mm. Comparing 20 cm lithium and 5 mm boron carbide, boron carbide has better performance. Therefore, we decided to use boron carbide of 5 mm thick for the neutron shield.

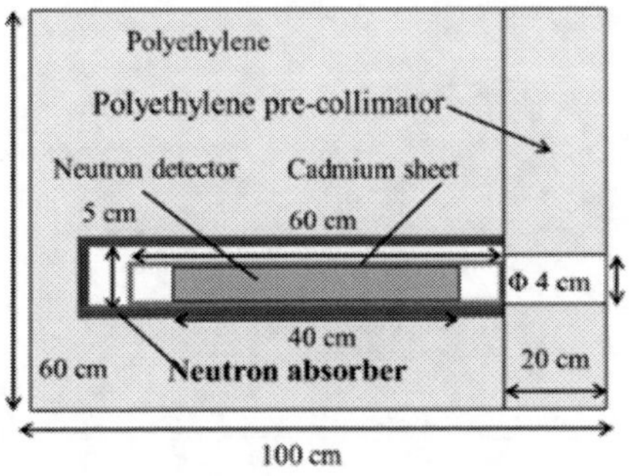

Figure 20 Neutron shield design around the detector.

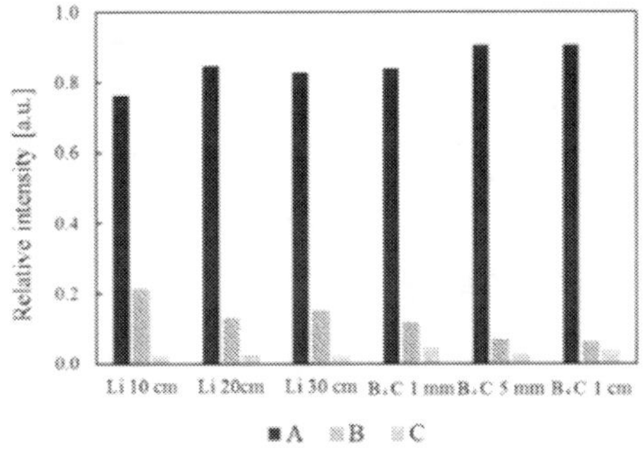

Figure 21 Performance of neutron absorbers of Li or B_4C.

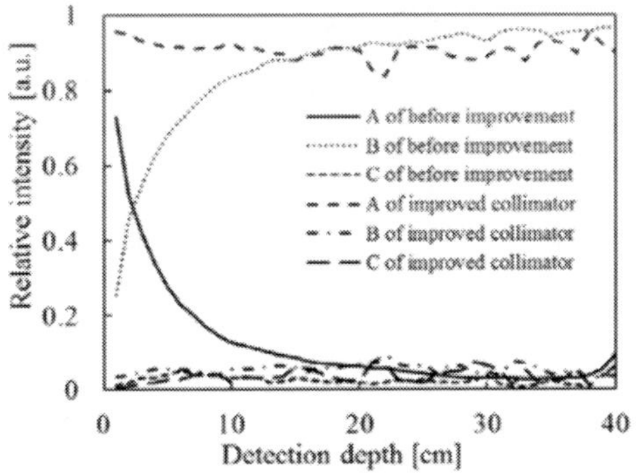

Figure 22 Presently obtained collimator performance.

Finally, the performance of the designed neutron collimator is shown in Figure 22. The designed neutron collimator could suppress influence of the side incident neuron to be sufficiency low and shape the neutron flux to be an excellent pencil beam.

6. Conclusion

We designed and constructed an epi-thermal neutron field with a DT neutron source and then evaluated the characteristics of the produced neutron field by using activation foils. Additionally, the experimental verification of the developed neutron spectrometer was carried out. As a result of the evaluation of the neutron field, we confirmed the epi-thermal neutron field was successfully formed so as to supply a high epi-thermal neutron flux field simultaneously suppressing fast neutrons substantially. The present neutron field is expected to give an epi-thermal flux intensity of over 1×10^4 n/sec/cm^2 and would be available for fundamental researches of BNCT, if the OKTAVIAN facility provides the spec value of 1×10^9 n/sec. In the experimental verification of the neutron spectrometer, the measured detection depth distribution agreed well with the calculation. However, the unfolded neutron spectrum was different from the neutron spectrum predicted by MCNP due to contribution of neutron incidence from the side surface of the detector. It means that the present beam shaping assembly with cadmium and polyethylene should be improved in order appropriately to carry out the

DOI: 10.1051/epjconf/201715304008

experimental verification of the neutron spectrometer. Therefore, we designed a new neutron collimator consisting of polyethylene pre-collimator and boron carbide neutron absorber with a polyethylene shield, so that it can shape the neutron flux to be an excellent pencil beam suppressing neutron incidence from the side surface of the detector substantially.

References

[1] I. Murata, H.Miyamaru, *Low-energy neutron spectrometer using position sensitive proportional counter-Feasibility study based on numerical analysis*, Nuclear Instruments and methods in Physics Research A **589**,445-454(2008).

[2] K. Tsubouchi, *Design calculation of epi-thermal neutron field with DT neutron source for low energy neutron spectrometer developed for BNCT*, Master's thesis, Osaka University (2014) [in Japanese].

[3] K. Sumita, A.Takahashi, T. Iida, J. Yamamto, S. Imoto, K. Matuda, *Osaka University 14Mev Intense Neutron Source and Its Utilizations for Fusion Studies (OKTAVIAN Program)*, Fusion Technology **1**, 675-680 (1982)

[4] X-5 Monte Carlo Team, *MCNP-A General Monte Carlo N-Particle Transport Code Vesion 5*, (2003).

[5] S. Iwasaki, *A new approach for unfolding PHA problems based only the Bayes' Theorem,* Proc. 9 th International Symposium on Rector Dosimetry, Prague, Chezk Republic, 245-252 (1996).

[6] T. Obata, *Development of keV region neutron spectrometer using 3He proportional counter*, Master's thesis, Osaka University (2013) [in Japanese]

Gamma-Ray Dose Measurement with Radio-Photoluminescence Glass Dosimeter in Mixed Radiation Field for BNCT

K. Hiramatsu [1], *S. Yoshihashi* [2], *S. Kusaka* [1], *F. Sato* [1], *E. Hoashi* [1], *I. Murata* [1]

[1]Graduate School of Engineering, Osaka University, Yamada-oka 2-1, Suita, Osaka 565-0871, Japan
[2]Department of Applied Nuclear Technology, Fukui University of Technology, Gakuen 3-6-1, Fukui-city, Fukui 910-8505, Japan

Abstract. Accelerator based neutron sources (ABNS) are being developed as the next generation neutron irradiation system for BNCT. From the ABNS, unnecessary gamma-rays will be generated by neutron capture reactions, as well as fast neutrons. To control the whole-body radiation dose to the patient, measurement of gamma-ray dose in the irradiation room is necessary. In this study, the objective is to establish a method to measure gamma-ray dose separately in a neutron/gamma mixed field by using RPL glass dosimeter. For this purpose, we proposed a lead filter method which uses a pair of RPL glasses with and without a lead filter outside. In order to realize this method, the basic characteristics of glass dosimeter was verified in the gamma-ray field, before adapting it in the mixture field. From the result of the experiment using the lead filter, the simulation result especially for the case with a lead filter overestimated the absorbed does obtained from measurement. We concluded that the reason of the discrepancy is caused by existence of gradient of the dose distribution in the glass, and the difference of sensitivity to low-energy photon between measurement and theory.

1. Introduction

The number of people who die of cancer is increasing all over the world. Especially in Japan, according to the cause-specific ranking of death cancer has become first place since 1981 [1]. Furthermore, as aging of population progresses, the number of cancer morbidity and/or death is expected to increase in the future. Therefore, in modern medical science, development of cancer therapies is an urgent task. Mainly, cancer treatment can be classified into surgical therapy, chemotherapy and radiotherapy. There are some advantages and disadvantages in each treatment, and the optimal treatment method of individual has thus been selected by considering each case of cancer. Further, it is possible to enhance the therapeutic effect by combining these treatment methods.

Recently, radiotherapy has attracted attention with development of accelerator technology. In radiotherapy, cancer cells will be destroyed by ionization effect of radiation, which injure the DNA in living cells. Because it is possible to irradiate around the cancer cells, radiation therapy can be regarded as a local treatment. Since forms or functions of organs will be preserved in radiotherapy, invasion to the normal tissues can be suppressed and quality of life can consequently be maintained fairly well after treatment compared with surgical therapy. In addition, general anesthesia of surgery or chemotherapy with an anti-cancer agent will become a heavy burden to a human body, especially for elderly patients, while it is possible to perform radiotherapy almost regardless of age of the patients because radiotherapy is a minimally invasive treatment. Therefore, radiotherapy is expected to be a treatment that can maintain QOL at high levels.

Ideally, as for radiotherapy, only target cancer cells will be irradiated, while normal cells surrounding the target will not be irradiated at all. In fact, however, since the target cells are irradiated from outside of the body, not only cancer cells but also normal cells will possibly be affected by radiation. Generally, radiation damage of cancer cells is almost the same as that of normal cells, while it takes more time to recover from the damage for cancer cells compared to normal cells. Therefore, actual treatments are performed with a special devise so as to destroy only cancer cells while suppressing the damage to normal cells, e.g., irradiation of the target is repeatedly carried out with a certain time interval for which normal cells will recover from the damage.

Radiation therapy is divided into two categories for the type of incident radiation; indirectly ionizing radiation such as X-rays and gamma rays, and directly ionizing radiation such as particle beam which normally uses heavier particles than electrons. In the former, the dose is at maximum on the surface of the body and is attenuating with increase of the depth in the body, so that the therapeutic effect can hardly be obtained if cancer cells are in a deep position inside the body. Furthermore, if we try to give a sufficient dose to cancer cells located deep inside the body, it would significantly affect the normal cells shallower than the cancer cells. Therefore, in a cyber knife and gamma knife using X-rays and gamma rays, respectively, cancer cells are irradiated with relatively weak dose radiation from multiple directions. As the result, only cancer cells will be irradiated with a sufficient dose, while the radiation effects on the normal cells can be suppressed.

On the other hand, as for the particle beam therapy, the pass length of incident charged particles is determined by the incidence energy, and the particles transfer a very large energy to surrounding cells, when stopping after losing the whole energy (Bragg peak). Therefore, the Bragg peak can be adjusted to match the cancer cell depth by changing extracted energy from the accelerator. For that reason, irradiation effects on normal cells existing between the body surface and cancer cells can be minimized. At the actual treatment, the Bragg peak is expanded with various energetic particles to cover the whole cancer cells (Spread Out Bragg Peak ; SOBP).

Recently, a new radiotherapy, BNCT (Boron Neutron Capture Therapy), is known as a quite new cancer treatment by which damage to normal cells can be substantially smaller than other radiation therapies. Since boron has a specific property of easily reacting with neutrons, cell selective radiotherapy is realized.

1.1 Boron Neutron Capture Therapy

BNCT is one of the particle beam therapies which uses a nuclear reaction of boron and neutron. Equation.a indicates the reaction of boron (B) and neutron [2].

$$^{10}B + n \rightarrow \begin{cases} ^{4}He + ^{7}Li(2.79MeV) \cdots 6\% \\ ^{4}He + ^{7}Li(2.31MeV) + \gamma(0.48MeV) \cdots 94\% \end{cases} \quad (a)$$

Table 1-1 shows the reaction cross-sections of thermal neutrons for boron-10 and other elements constituting the human body. From the table, it is seen that boron-10 has a large reaction cross-section for thermal neutrons compared with the other elements. Figure 1-1 shows a schematic diagram of BNCT.

In this radiotherapy, high concentration of boron drug; 20 μg/g, will be accumulated in cancer cells before irradiation of neutrons from outside of the body. After the irradiation, alpha particles and lithium nuclei generated in the reaction of Eq. (1), will destroy cancer cells. Since alpha particles and Li-7 nuclei have very short range of about 4-9 μm, this radiotherapy can destroy only cancer cells selectively without affecting normal cells around the cancer cells.

Table 1-1. Reaction cross-sections of thermal neutrons for some materials.

Nuclide	Reaction	Cross section [barn]
^{1}H	$^{1}H(n,\gamma)^{2}H$	0.332
^{12}C	$^{12}C(n,\gamma)^{13}C$	0.0035
^{14}N	$^{14}N(n,p)^{14}C$	1.83
^{16}O	$^{16}O(n,\gamma)^{17}O$	0.00019
^{10}B	$^{10}B(n,\alpha)^{7}Li$	3838

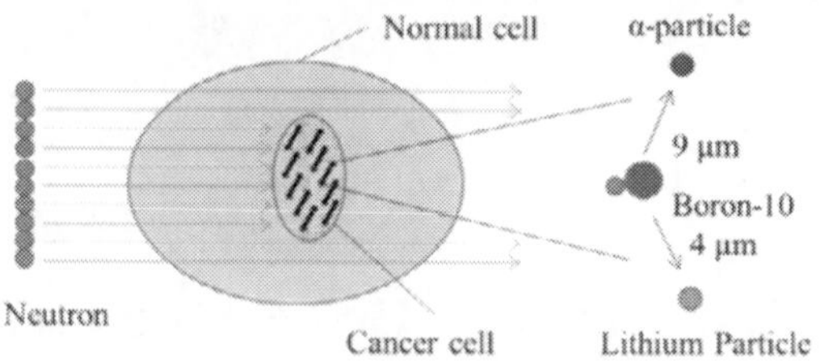

Fig 1-1. Schematic diagram of BNCT.

For advances in the treatment technology of BNCT, further investigation is required in several fields, i.e., drug delivery techniques that can administer boron to cancer cells, development of the neutron source instead of nuclear reactors and radiation dosimetry at the time of treatment. In particular, the development of the neutron source is an essential issue for the development of BNCT. Because BNCT requires a high flux neutron field, treatment has been performed only with research reactor until now. However, there are several problems for reactor based BNCT. At first, the number of research reactors that can perform neutron irradiation is not so many. As a result, patients are required to move to the nuclear reactor consuming a long time. It is a severe burden on the patient's body. Further, since the regulation of the nuclear reactor is very strict and hard, it is prohibited to build inside or adjacent to hospitals from the standpoint of safety. Therefore, accelerator based neutron sources (ABNS) are being developed as the next generation irradiation system for BNCT instead of nuclear reactors.

ABNS is a facility that can irradiate neutrons generated by nuclear reaction of accelerated particles and target materials. The neutron beam characteristics are controlled by a so-called beam shaping assembly containing neutron moderator and radiation shields. ABNS can be designed compactly compared with the nuclear reactor, and its inherent safety enables to install an ABNS in a hospital. However, for commercial use of the ABNS, further investigation is necessary such as accelerator, target and beam shaping assembly developments.

At present, liquid lithium, solid lithium and solid beryllium have been considered as the target materials of ABNS. Each material has advantage and disadvantage. In Osaka University, development of liquid lithium target ABNS is underway.

1. 2 Objectives

In the ABNS, neutrons generated by the reaction of protons and target are moderated in the beam shaping assembly. In the process, unnecessary gamma-rays are produced by neutron capture reaction of neutrons with materials arranged on the transport route to the beam exit. In radiology, it is very important to reduce impact on normal cells by secondary radiation as much as possible. It can affect the whole body exposure dose of patient. Because development of ABNS has been just started, the characteristics of the secondary radiations in the irradiation field of the ABNS such as neutrons and gamma-rays can change depending on the type of accelerators and thus have not been understood. Therefore, measurement of gamma-ray dose given at the same time is necessary for controlling the whole body radiation dose of patients.

There are several detectors which are used for measurement of gamma-rays. For example, Geiger-Müller counter, ionization chamber, scintillation detector, semiconductor detector, thermoluminescent dosimeter (TLD), radio-photoluminescence glass dosimeter (RPLDG) etc. are known so far. Among these detectors, most of detectors can measure time dependent dose rate of gamma-rays. On the other hand, TLD and RPLGD are used for measuring the cumulative dose at the measurement position. Until now, TLD is widely used for dosimetric applications to determine absorbed dose in radiotherapy. TLD has good reproducibility and small energy dependence [3]. However, it is quite difficult to handle because it has fading characteristics and its dose information will be deleted by dose measurement. Thus, RPLGD has recently been used for dosimetry instead of TLD. There is almost no fading characteristics in RPLGD, and it can be used easily.

Generally, for the dosimetry in the field where neutrons and gamma-rays exist, it is known that it is difficult to measure each radiation dose separately, because the detectors have sensitivity to both of radiations. This is also the case of using a RPLGD. Normally, RPLGD is used for environmental monitoring at nuclear reactor facility and its surrounding. Therefore, it is not designed to apply in the mixed field of neutron and gamma-ray. In other words, for the measurement in a strong neutron field, since gamma-rays are also existing, it is very difficult to measure gamma-ray dose separately. Until now, several detectors are proposed for measurement in the mixed field of neutrons and gamma-rays such as TLD and ionization chamber. However, these detectors still have problems. TLD has fading characteristics and it is impossible to be readout repeatedly. Ionization chamber also has some problems such as the size and difficulty of handling. In BNCT project of Osaka University, RPLDG is adopted for measurement of gamma-rays in the irradiation room from the view point of its size and ease of use. Therefore, the aim of this study is to establish a new measurement method with RPLDG which can evaluate only gamma-ray dose from dose values measured in a mixed field of neutrons and gamma-rays in BNCT.

To evaluate the gamma-ray dose from the measured dose, we have devised a new measurement method which uses a lead shield. The key is that neutrons can more-or-less penetrate the lead shield, on the other hand, gamma-rays will be attenuated partly by the lead. We call it "A lead filter method."

In this method, bare RPLGD and RPLGD covered with a lead filter were irradiated at the same irradiation field. By making difference of both measured values, the effect of neutrons will be cancelled, and only the gamma-ray dose can be shielded by the lead shield and a part of the dose will remain. The real gamma-ray dose can be estimated from the measured dose of RPLGD and the correction coefficient evaluated beforehand.

In the present study, since the reaction of neutrons and gamma-rays differ with the elements, correction coefficient may vary depending on the energy variation. In this study, to determine the correction coefficient used for this method, we evaluate the energy dependence of the measurement by RPLGD covered by a lead filter in the gamma-ray field as the first step of this method before adapting in mixed field.

2. Materials and Methods

2.1 RPLGD

RPLGD is one of the integrated-type individual dosimeters which is currently put into practical use. RPLGD is composed of silver activated phosphate glass. The glass irradiated with radiation emits orange light by stimulation of ultraviolet beam. This phenomenon is referred to as radio-photoluminescence (RPL). By utilizing the characteristics, that is, the amount of light emission is proportional to exposed dose, RPLGDs are mainly used for measurement of gamma-ray dose. Figure 2-1 shows a schematic diagram of the generation mechanism of an RPL center in a silver activated phosphate glass. As an initial condition, a silver (Ag) atom contained in the silver activated phosphate glass is stable in the state of Ag^{+}. After irradiation, a pair of electron and hole is generated inside the glass. The generated electron is trapped in Ag^{+} to create Ag^{0}. On the other hand, the generated hole is once captured by a PO_4 tetrahedron, and after that, the hole charge is shifted to Ag^{+} to finally make a stable Ag^{2+}. This phenomenon is called RPL center build-up [4]. Figure 2-2 indicates the schematic diagram of RPL emission process of silver active phosphate glass due to ultraviolet excitation.

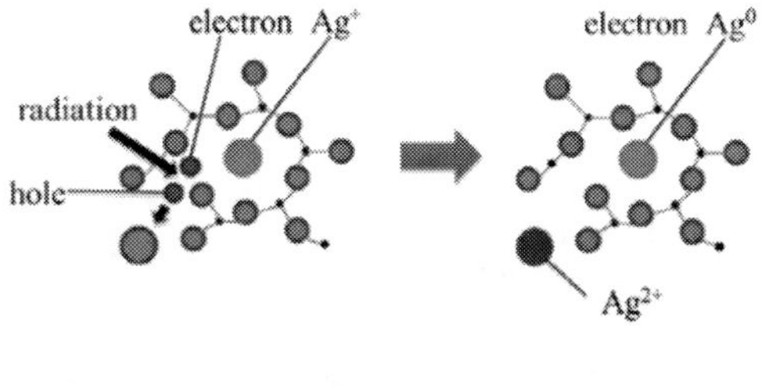

Fig 2-1. Schematic diagram of the generation mechanism of an RPL center.

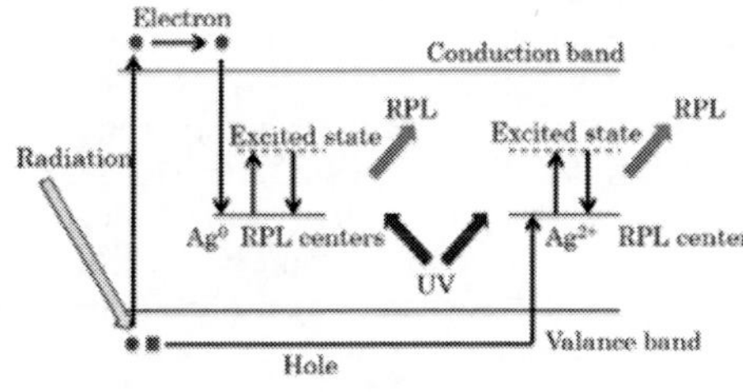

Fig 2-2. Schematic diagram of RPL emission process.

The RPLGD measurement system (Dose Ace, AGC Asahi Glass Company, Japan) is composed of RPLGD (GD-301, AGC Asahi Glass Company) and a readout system (Dose Ace FGD-1000, AGC Asahi Glass Company). Figure 2-3 is a photo of RPLGD. As shown in this figure GD-301 is 1.5 mm in diameter and 8.5 mm in length. The weight composition is P (31.55 %), O (51.16 %), Al (6.12 %), Na (11.0 %) and Ag (0.17 %), and the density is 2.61 [5]. For the readout, ultraviolet excitation light source is required to induce RPL. As for the ultraviolet source, LD-pumped ultraviolet solid-state laser (Explorer 349, Spectra-Physics) is implemented in FGD-1000. The laser is collimated in 1mm in diamiter through an aperture. The amount of RPL measured by the readout device is calibrated by that of a refference RPLGD which was irradiated beforehand and already known the accurate absorbed dose.

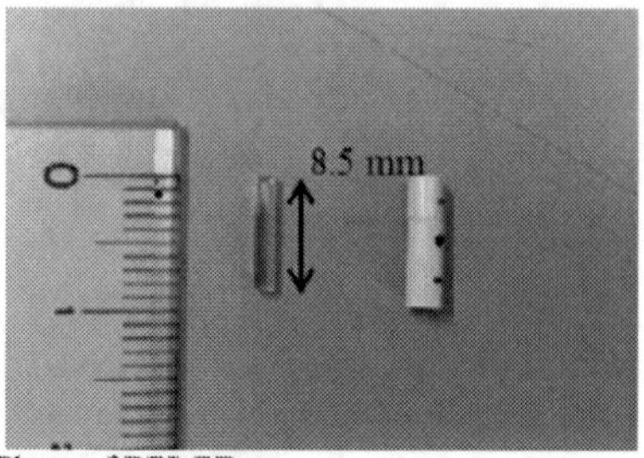

Fig 2-3. Photo of RPLGD.

2.2 Lead Filter Method

In this study, we aimed to establish a new measurement method; "lead filter method", to realize a separate measurement of gamma-ray dose in a mixed field of neutrons and gamma-rays. This method utilizes the characteristics that lead transmits neutrons, on the other hand, shields gamma-rays if the thickness of the lead is appropriately thin. Under this assumption, the measured value of absorbed dose D of the bare RPLGD in the mixed field can be derived by the following equation.

$$D = d\gamma + dn + dn\gamma \tag{1}$$

On the other hand, the measured value of absorbed dose D_{Pb} of the RPLGD covered with a lead filter can be derived by the following equation.

$$D_{Pb} = d^*\gamma + dn + dn\gamma \tag{2}$$

Here, dn and dγ are the dose of neutrons and gamma-rays, respectively. dnγ is the dose of gamma-rays induced by neutron capture reaction. As the reaction of gamma-ray, photonuclear reaction; (γ, n) is also considered. However, since this nuclear reaction will not occur unless the incident photon energy is higher than the threshold energy of each element, this nuclear reaction can be ignored in this method. Further, d*γ is dose of gamma-ray attenuated by the lead filter, and can be written as equation (3) with transmittance factor: ε.

$$d^*\gamma \fallingdotseq \varepsilon d\gamma \tag{3}$$

By making difference of the two measured values, the neutron effect can be removed, as shown in equation (4).

$$\begin{aligned} D - D_{Pb} &= d\gamma - d^*\gamma \\ &= (1-\varepsilon) d\gamma \end{aligned} \tag{4}$$

Finally, the absorbed dosed which we want to know can be deduced with a correction factor by equation (5).

$$d\gamma = (D - D_{Pb})/(1-\varepsilon) \tag{5}$$

The gamma-ray dose can be estimated by correcting the measured values of bare and lead filtered RPLGDs with the correction factor estimated in advance.

Figure 2-4 shows the lead filter used in this study. It is a cylinder the size of which is 2.2 cm in diameter and 8.5 cm in length. The thickness of the lead is designed as 1 cm, which can shield gamma-rays partly, on the other hand, will not affect the neutron dose.

Fig2-4. Lead filter used in this study.

2.3 Simulation Method

In order to analyse the presently carried out experimental results PHITS (Particle Heavy Ion Transport code System) was used. PHITS is a general purpose Monte Carlo particle transport simulation code developed in Japan, in which EGS5 (Electron Gamma Shower 5) is implemented for precise gamma-ray and electron transport. As for neutron reaction cross section library, JENDL-4.0 was used. In the simulation, the experimental system was modelled precisely [6].

2.4 Experimental Method

It is conceivable that dosimetry using RPLGDs with a lead filter depends on various factors, such as kind of radiation and its energy. In fact, when this method is applied to treatment field of BNCT, it is a mixed field of mainly gamma-rays and neutrons. Therefore, it is essential to evaluate the energy dependence of RPLGD in a gamma-ray field, neutron field and mixed field of gamma-ray and neutron. In this study, we evaluate the energy dependence of RPLGD in the gamma-ray field as the first step of development of this method. To realize this, irradiation experiments with standard gamma-ray sources have been carried out.

As for standard gamma-ray sources, ^{60}Co, ^{137}Cs and ^{22}Na were used. Table 2-1 summarizes the details of the sources. Figure 2-5 shows the size and photo of the standard gamma-ray source. This type of a radiation source is a point source the size of which is normally smaller than 6mm in diameter. Aluminium foil of about 0.1 mm thick and acrylic resin are used for the source window and casing, respectively. In the analysis of experiments, the structure of the sources was accurately modelled.

Table 2-1. Details of the sources.

Nuclide	Intensity [kBq]	Half Life time [year]	Energy [MeV]	Emission Ratio[%]
^{60}Co	1060	5.27	1.173	100
			1.333	100
^{137}Cs	1000	30.17	0.662	85.1
^{22}Na	867	2.60	1.275	100
			0.511 an.	180

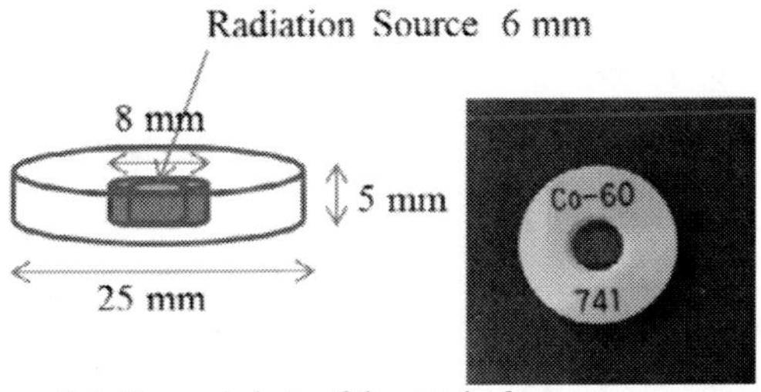

Figure 2-5. Size and photo of the standard gamma-ray source.

Figure 2-6 shows the experimental setup. The experimental setup was made on an aluminium plate of which the thickness is 5 mm. Both RPLGDs with and without a lead filter were put on 5 cm from the radiation source and an irradiation experiment was performed. Irradiating three pairs of both RPLGDs, the averaged values were measured. RPLGDs without a lead filter were put on a styrofoam foundation in order to fix the measurement position. Further, measurement results were compared with calculation results which simulate the experimental system exactly. The background dose in the irradiation room was measured by Geiger Muller survey meter showing that the air dose rate was less than 1 μSv/hr.

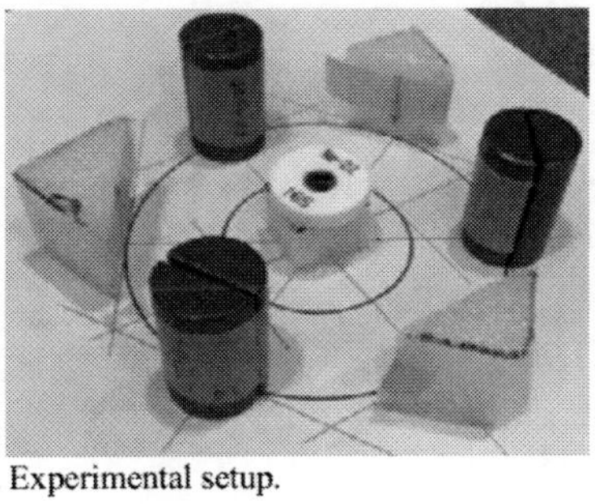

Figure 2-6. Experimental setup.

3. Result

We performed irradiation experiments with standard gamma-ray sources listed in Table 2-1. Figure 3-1 shows the comparison results of absorbed dose of RPLGD between measurement and simulation for each radiation source. Figures 3-1.a and 3-1.b are the results with and without lead filters, respectively. From the results, in the case of the RPLGD without a lead filter, measurement and simulation result agree well except ^{60}Co source. For the ^{60}Co source, the simulation value becomes smaller than the measurement value. In contrast, the simulation values show a little larger than the measurement for the RPLGDs with lead filters for all radiation sources.

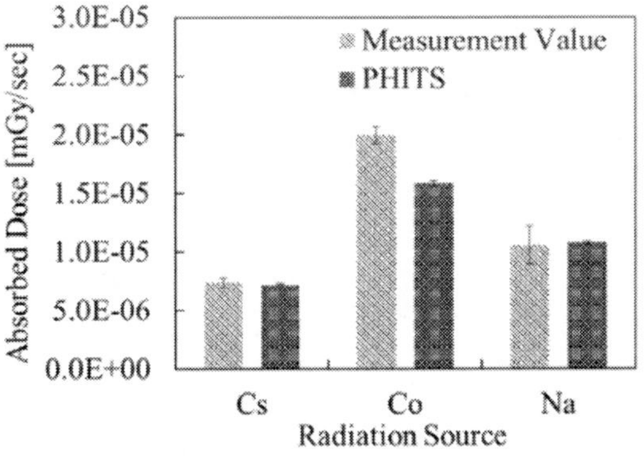

Fig 3-1.a. Comparison result of absorbed dose between measurement and simulation for bare RPLGDs.

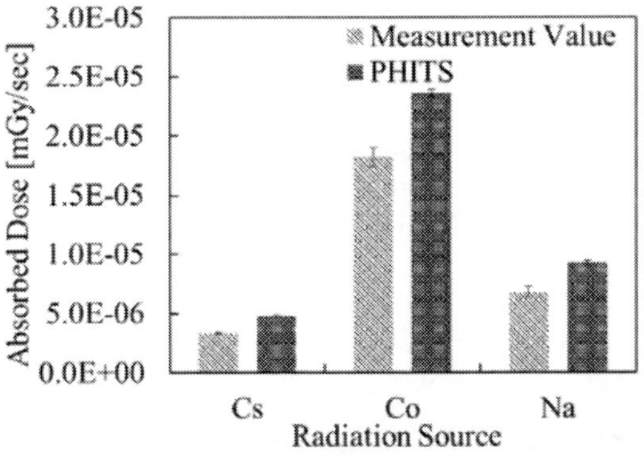

Fig 3-1.b. Comparison result of absorbed dose between measurement and simulation. RPLGDs were covered with a lead filter.

To clarify the discrepancy, we focused on two problems, dose distribution inside a glass and sensitivity to low-energy photon. As for the first problem, it should be noted that the readout volume of the glass is fixed to be 6 mm in length and 1 mm in diameter inside the glass due to the technical restriction of the reader and the diameter of ultraviolet beam. Figure 3-2 shows the size of readout volume inside the RPLGD. In the measurement, the estimated absorbed dose is the averaged value only inside the readout volume. It means, if the dose distribution inside a glass is not uniform, absorbed doses of the entire glass and only of the readout volume will not agree. For accurate measurement, it is necessary to evaluate distribution of absorbed dose inside a glass. The details of the dose distribution inside a glass are discussed a little later in Sec. 3.1.

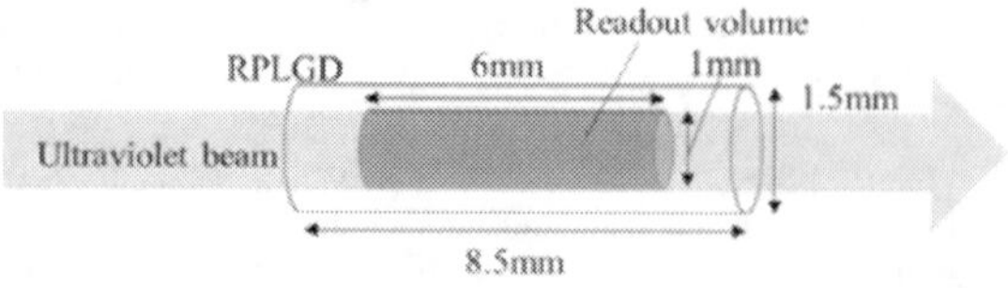

Fig 3-2. Size of readout volume inside RPLGD.

For another reason for the discrepancy of measurement and simulation, the sensitivity of RPLGD for radiation, especially for low-energy photon was examined. When high-energy gamma-rays are incident to lead, low-energy characteristic X-rays will be generated. Figure 3-3 shows the photon flux calculated by PHITS, which was incident on the glass from the lead filter. The source was ^{137}Cs (662 keV). From this figure, it was found that many low-energy photons can be generated by interaction of gamma-rays and lead atoms.

It is known that RPLGD has high sensitivity to low-energy photons [7]. Fig 3-4 shows the theoritical value of the mass energy absorption coefficient ratio of the air to the glass calculated from the composition ratio of the glass elements [8]. From the figure, it is found that RPLGD has high sensitivity to photon of which the energy is less than 100 keV. Since in this energy region, a lot of characteristic X-rays are produced as shown in Fig. 3-3, we presumed that there would be discrepancy of sensitivity to photons between measurement and simulation. In order to comfirm this, we performed experiments to examine whether the theoritical value of the mass energy absorption coefficient ratio calculated by PHITS is consistent with the dose response of actual RPLGD. The detailed results will be discussed in Sec. 3.2.

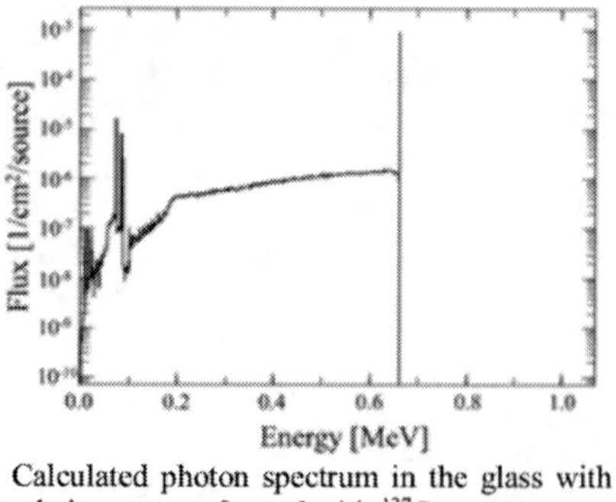

Fig 3-3. Calculated photon spectrum in the glass with a lead filter. Simulation was performed with ^{137}Cs source.

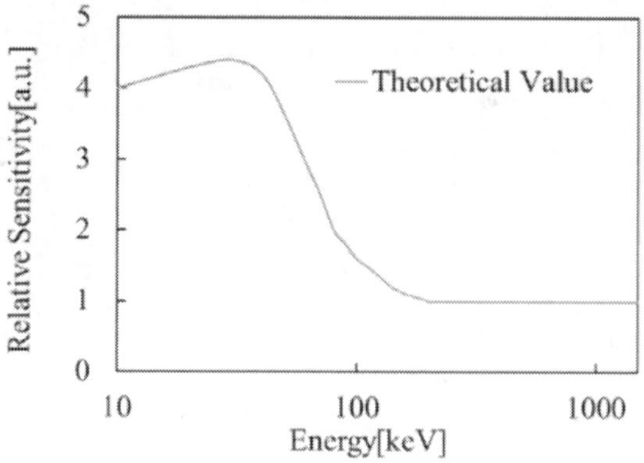

Fig.3-4. Theoritical value of the mass energy absorption coefficient ratio of the air to glass calculated from the composition ratio of the glass element.

3.1 Dose Distribution

In this section, the absorbed dose distribution in the glass is discussed. In the next section, the sensitivity examination to low energy photons is detailed based on the experimenatl results. To evaluate distribution of absorbed dose inside a glass, cross section of the glass was subdivied, and absorbed dose of each voxel was calculated by PHITS for the experimental setup of Figure 2-6. Figure 3-5 shows the calculation model of the cross section of a glass. Shaded area indicates the readout volume of RPLGD. The size of each voxel is $0.075 \times 0.075 \times 8.5mm^3$. The distribution of absorbed dose was calculated in the parallel and vertical directions with respect to the gamma-ray inicidence direction as indicated in red frame. By simulating ^{60}Co and ^{137}Cs gamma-rays as a radiation source, energy dependent dose distribution can be discussed.

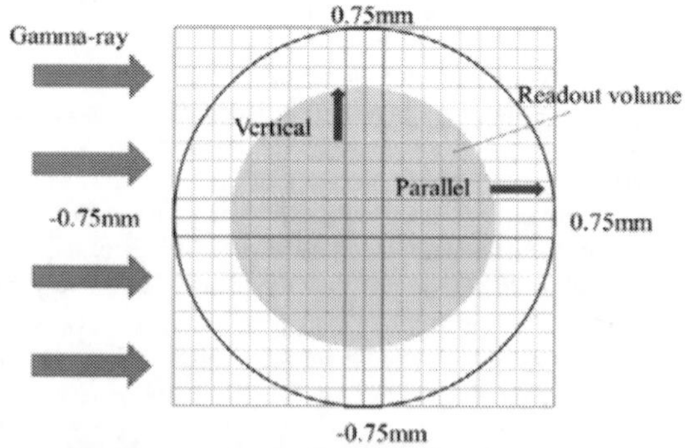

Fig 3-5. Calculation model of cross section of a glass.

Figures 3-6 and 3-7 show the simulation results of dose distribution for parallel and vertical directions inside a glass for ^{137}Cs and ^{60}Co gamma-ray sources, respectively. Shaded areas on the graph indicate the readout volume of RPLGDs. As shown in the figures, in the case for RPLGDs without a lead filter, the absorbed doses near the center of the glass are higer than those of around the peripheral region. On the other hand, in the case for

RPLGDs with a lead filter, absorbed doses around the periphery become higher compared with the doses near the center of the glass. Furthermore, as focusing on the parallel direction, the shape of the curve of dose distribution differs depending on the energy (^{137}Cs and ^{60}Co), in the case for RPLGDs with a lead filter. As for ^{137}Cs, the absorbed dose of the nearest position to the radiation source was the highest. On the other hand, in the case of ^{60}Co, the deeper the position is from the radiation source, the more the absorbed dose increases. In the measurement, the real absorbed dose is the averaged value only inside the readout volume indicated as the shaded part in the graph. It means absorbed dose in the peripheral region of RPLGDs were removed in the measured absorbed doses. This is the reason why descrepancy is observed between measurement and simulation.

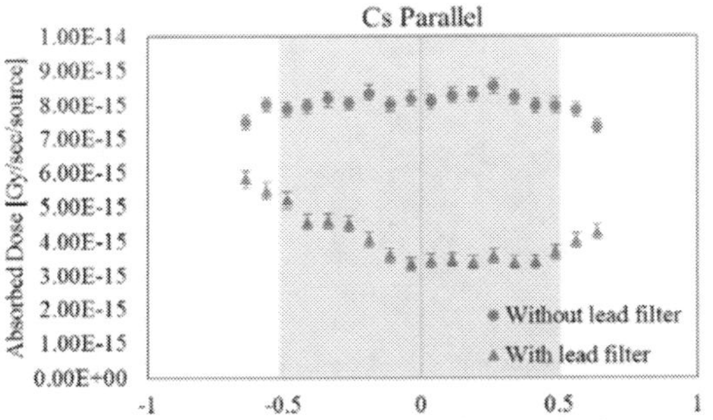

Fig 3-6.a. Simulation results of dose distribution of parallel direction. ^{137}Cs was simulated as a radiation source.

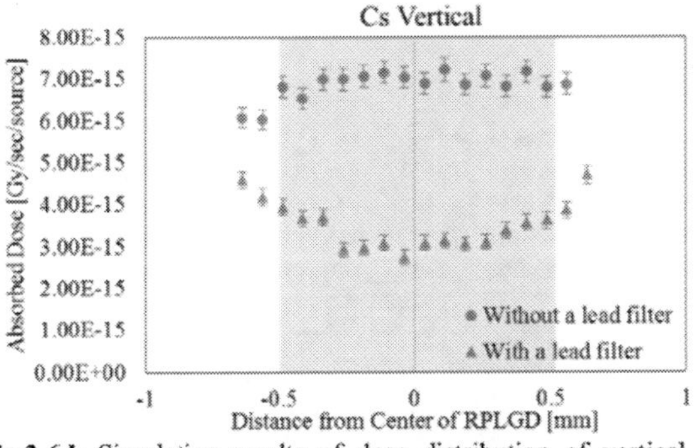

Fig.3-6.b Simulation results of dose distribution of vertical direction. ^{137}Cs was simulated as a radiation source.

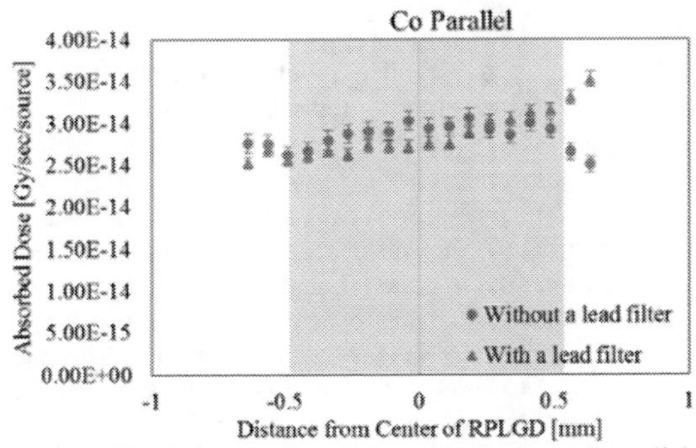

Fig 3-7.a. Simulation results of dose distribution of parallel direction. ^{60}Co was simulated as a radiation source.

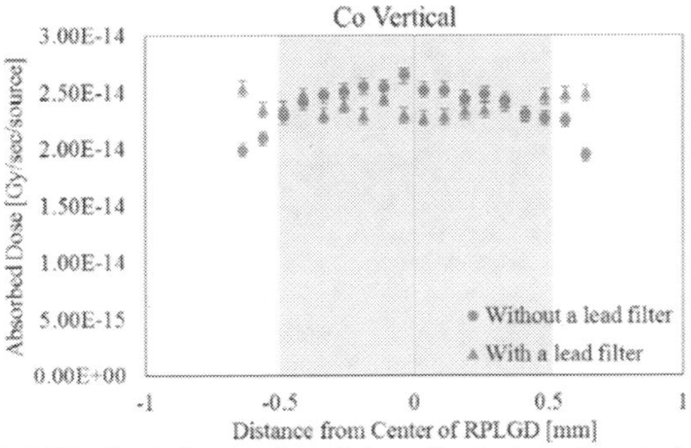

Fig.3-7.b Simulation results of dose distribution of vertical direction. ^{60}Co was simulated as a radiation source.

3.2 Dose Sensitivity

We carried out irradiation experiments using low-energy X-rays to compare the theoretical sensitivity of RPLGDs to photons with the actual dose sensitivity of the glass. From the results in the previous experiments using gamma-ray source and RPLGDs without a lead filter, measurement and simulation agreed fairly well. We thus focused on the sensitivity especially for low-energy photons of which energy is less than 100 keV, comparable to energy of characteristic X-rays generated by interaction of gamma-rays and lead atoms. Figures 3-8.a and 3-8.b are the photos of an X-ray irradiation system (MBR-1520R-4, Hitachi Power Solutions) and the experimental arrangement of irradiation, respectively. By measuring the spatial dose rate at the irradiation position beforehand, it is possible to adjust the actual irradiation dose accurately. Table 3-1 shows the irradiation condition. Irradiating three pairs of RPLGDs at the same time, the obtained values were averaged to be the measured value. Dose sensitivity of RPLGD was calculated by making the ratio of irradiation dose and measured absorbed dose.

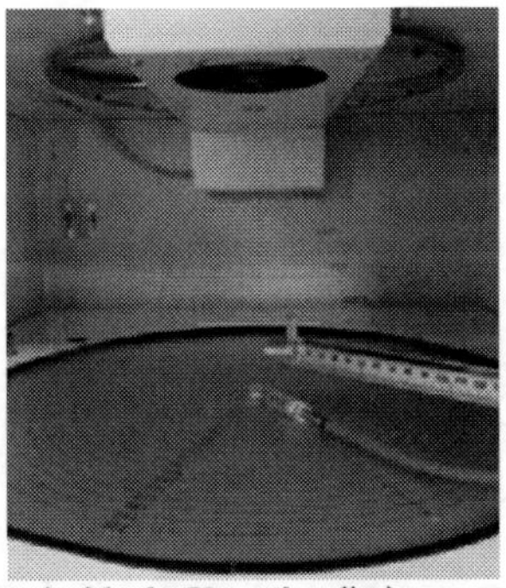

Fig 3-8.a. Photo inside the X-ray irradiation system.

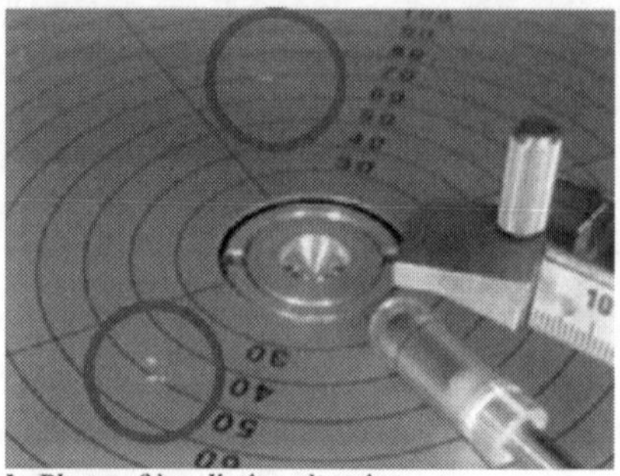

Fig 3-8.b. Photo of irradiation situation.

Table 3-1. Irradiation condiation.

Tube Current	10 mA
Tube Voltage	150 kV
Energy	10, 25, 39, 45, 55, 65 keV
Irradiation Dose	100 mGy

Figure 3-9 shows a comparison result of sensitivity to photon between measurement and theoretical value calculated by mass energy absorption coefficient. As the result, the measured sensitivity tends to become lower than the theoretical value. Since many characteristics X-rays will be produced if using the lead filter, we concluded that the reason of the disagreement between experimental values and simulation in Figure 3-1.b has been caused by the decline of the sensitivity to the low-energy photons in measurement.

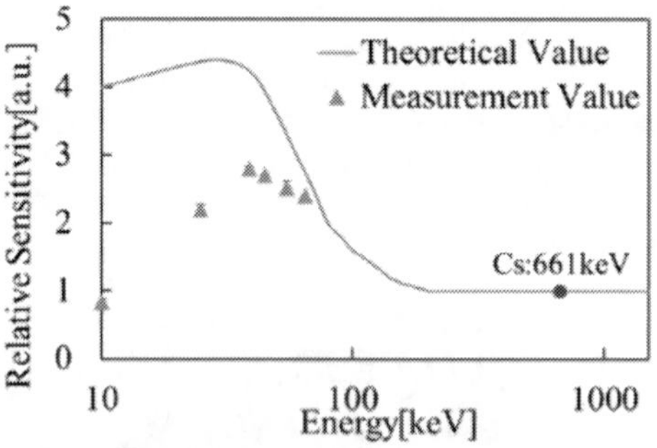

Fig 3-9. Comparison result of sensitivity to photon between measurement and theoretical value.

3.3 Recalculation Result

Based on the results above, absorbed dose of glass dosimeter was recalculated taking into account the dose distribution in a glass and the sensitivity for low-energy photons. In the recalculation, tally region in the simulation model was set to fit to the actual readout volume in the glass. Figure 3-10 shows the results of recalculation. From the figure, in both cases with/without a lead filter, the difference of values between measurement and simulation is reduced by the correction. In detail, for RPLGD without a lead filter the simulation value increases. On the other hand, for RPLGDs with a lead filter, the simulation value decreases by the correction. However, gamma-ray dose could not be reduced sufficiently even if using a lead filter in case of ^{60}Co. It means, if the energy is too large, the strength of gamma-rays may not be lowered enough by the lead filter. In other words, the lead filter method may not be utilized if the gamma-ray dose could not be attenuated sufficiently with a lead filter.

By the current structure of the filter, this method can be applied to gamma-rays the energy of which is about 1MeV or less. At the treatment field of BNCT higher energy gamma-rays than ^{60}Co, up to about 10 MeV, will be produced by neutron capture process. We thus need to re-examine gamma-ray attenuation process, i.e., the structures of the lead filter and glass in the next step.

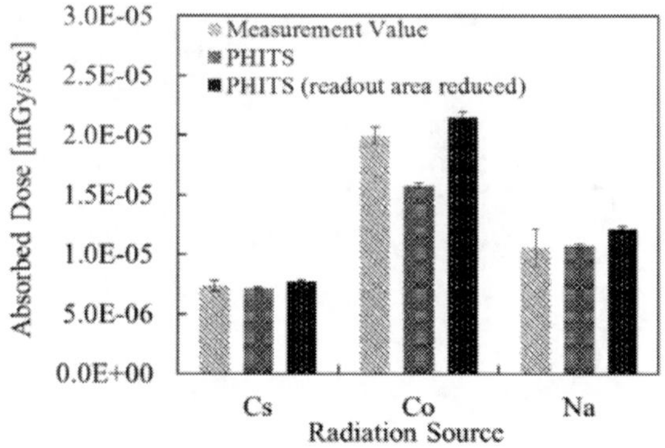

Fig 3-10.a. Comparison result of absorbed dose with recalculated value. RPLGDs were covered without a lead filter.

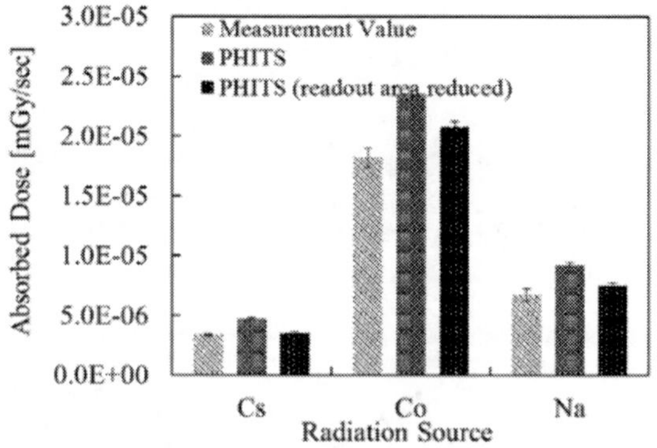

Fig 3-10.b. Comparison result of absorbed dose with recalculated value. RPLGDs were covered with a lead filter.

4. Conclusion

In this study, to establish a method of measuring gamma-ray dose separately at a neutron/gamma-ray mixed field, we proposed "Lead filter Method." As the result of the series experiments in a gamma-ray field, the simulation result especially for the case with a lead filter overestimated the absorbed dose obtained from the measurement. We concluded that the reason of the discrepancy is caused by existence of gradient of the dose distribution in the glass, and the difference of sensitivity to low-energy photons between measurement and theory.

However, gamma-ray dose could not be reduced sufficiently even if using a lead filter for gamma-rays of which the energy is higher than 1 MeV. In the future, to adapt to the actual treatment field of BNCT, we need to re-examine the structure of the lead filter to shield high-energy gamma-rays sufficiently and the size of the glass.

References

1. Ministry of Health, Labour and Welfare, Japan, Abridged Life table for Japan 2015, http://www.mhlw.go.jp/english/database/db-hw/lifetb15/index.html, (2015)
2. Wolfgang. A.G. Sauerwein, A. Witting, R. Moss, Y. Nakagawa, Neutron Capture, Radiology, Springer, p.1-3 (2012)
3. J.H. Lee, M.S. Lin, S.M. Hsu, I.J. Chen, W.L. Chen, C.F. Wang, Radiation Measurements, **44**, 1 (2009)
4. Y. Miyamoto, Y. Takei, H. Nanto, T. Kurobori, A. Konnai, T. Yanagida, A. Yoshikawa, Y. Shimotsuma, M. Sakakura, K. Miura, K. Hirao, Y. Nagashima, T. Yamamoto Radiation Measurements, **46**, 12 (2011),
5. Asahi Techno Glass Corporation, "RPL Glass Dosimeter / Small Element System, Dose Ace" (2014)
6. T. Sato, K. Niita, N. Matsuda, S. Hashimoto, Y. Iwamoto, S. Noda, T. Ogawa, H. Iwase, H. Nakashima, T. Fukahori, K. Okumura, T. Kai, S. Chiba, T. Furuta and L. Sihver, J. Nucl. Science Technol. **50**, 9 (2013)
7. N. Kadoya,K. Shimomura, S. Kitou, Y. Shiota, Y. Fujita, S. Dobashi, K. Takeda, K. Jingu, H. Matsushita, Y. Namito, S. Ban, S. Koyama, K. Tabushi, Med Phys.,**39**, 10(2012)
8. M.J. Berger, J.H. Hubbell, S.M. Seltzer, J. Chang, J.S. Coursey, R. Sukumar, D.S. Zucker, and K. Olsen, NIST Standard Reference Database, Xcom; Photon Cross Sections Database, National Institute of Standard Technology (2003)

Modeling of Radiotherapy Linac Source Terms Using ARCHER Monte Carlo Code: Performance Comparison for GPU and MIC Parallel Computing Devices

Hui Lin[1], Tianyu Liu[1], Lin Su[2], Bryan Bednarz[3], Peter Caracappa[1], X. George Xu[1*]

[1]*Nuclear Engineering Program, Rensselaer Polytechnic Institute, Troy, NY, USA, 12180*
[2]*Radiation Oncology, John Hopkins University, Baltimore, MD , USA, 21218*
[3]*Department of Medical Physics, University of Wisconsin, Madision, WI, 53705*

Abstract. Monte Carlo (MC) simulation is well recognized as the most accurate method for radiation dose calculations. For radiotherapy applications, accurate modelling of the source term, i.e. the clinical linear accelerator is critical to the simulation. The purpose of this paper is to perform source modelling and examine the accuracy and performance of the models on Intel Many Integrated Core coprocessors (aka Xeon Phi) and Nvidia GPU using ARCHER and explore the potential optimization methods. Phase Space-based source modelling for has been implemented. Good agreements were found in a tomotherapy prostate patient case and a TrueBeam breast case. From the aspect of performance, the whole simulation for prostate plan and breast plan cost about 173s and 73s with 1% statistical error.

1 Introduction

As a modern radiotherapy technique, Intensity modulated radiation therapy (IMRT) is designed to deliver superior dose conformity and uniformity compared to traditional three-dimensional conformal therapy [1]. However, few treatment planning systems (TPS) employ sophisticated dose calculation algorithms such as Monte Carlo (MC) method. Approximate algorithms are widely used in many TPSs, such as convolution/superposition method, are known to be fast but only able to generate approximate dose results in cases when the treatment site involves complex and heterogeneous tissue structures [2]. In comparison, Monte Carlo method for dose calculations in radiation therapy is well recognized for its accuracy. The issue of lengthy calculation times used to be the main bottleneck of MC method being widely applied in the medical physics community as a clinically feasible approach.

Recently, with the development of MC codes optimized for radiotherapy calculations as well as the availability of much faster and affordable computer accelerator technologies, such as the general-purpose Graphical Processing Units (GPU) by NVIDIA and AMD and the Xeon Phi coprocessors by Intel, decent acceleration factors have been observed compared to conventional CPU-based calculations. These significant advances have led to the clinical use of MC algorithms at some treatment centers and the promised availability of MC photon/electron planning modules among several commercial treatment planning vendors. With this in mind, we developed the ARCHER Monte Carlo code that takes advantage of a new class of parallel computing devices — GPU and Xeon Phi coprocessors [3-5]. Our motivation for developing the ARCHER code is to perform a thorough performance evaluation of modern computing devices in the context of Monte Carlo simulations, while providing the community a practical, efficient, and reliable dosimetry tool that can utilize all those devices. Previously we have showed that, by utilizing the GPU, ARCHER was able to finish a treatment planning dose calculations for Tomotherapy in about 70 seconds using single-precision floating point while the same plan took the GEANT4 code 10 CPU hours [6]. This paper demonstrates the latest effort to develop and apply the source modelling of Radiotherapy Linac photon source in the ARCHER GPU and Xeon Phi-based MC dose engine.

2 Methods and Materials

2.1 Accelerator Treatment Head Modelling

Generally, source modelling in the context of MC treatment planning contains multiple beam components. Energy, location and direction distributions of each source particle are explicitly expressed for each component of the accelerator treatment head. Over the years, some routes have been employed including direct use of phase space data from the MC treatment head simulations, measurement-based models and virtual, multi-source models from the treatment head simulation. Among them, phase space sources have the potential to provide the most accurate physical descriptions of the source, and they

* Corresponding author: xug2@rpi.edu

have been accepted as inputs by many well-known MC dose calculation tools such as EGSnrc and GEANT4 [7,8].

The simulation of a radiotherapy Linac treatment head could be divided into several methods. The first and most straightforward method is to perform the simulation of the patient-dependent phase-space sources. In ARCHER, two commonly used phase-space file formats defined in EGS code and the International Atomic Energy Agency (IAEA) [9] are supported. In this method, the phase space files were generated with the effects of secondary collimators. When the patient-dependent phase space files are available, this method is efficient. However, in most cases, the generation of patient-dependent phase spaces depends on other Monte Carlo codes like BEAMnrc, which is time-consuming and clinically impractical. In the following section, we described a method utilizing patient-independent phase space files.

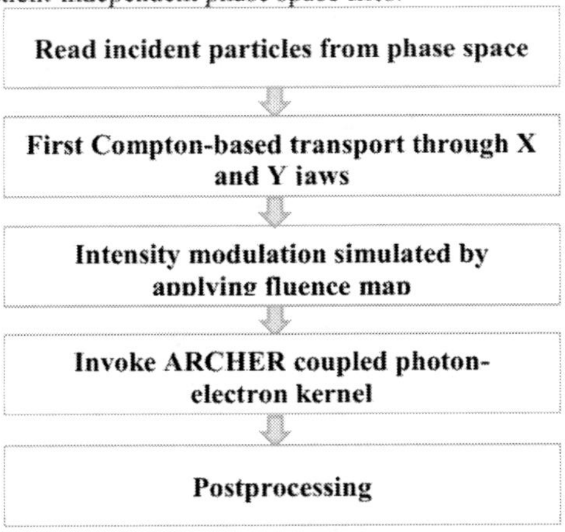

Fig. 1. Workflow of ARCHER Linac source modelling.

Fig. 1 demonstrated the workflow of the photon Linac source modelling in ARCHER. The namely patient-independent phase space file is generated by performing the simulation of the patient-independent structures of treatment head and storing a phase-space file at a plane just below the fixed elements of the accelerator head. The advantage of this approach is that once this part of phase space files is scored, it could then be reused as often as necessary. Then ARCHER takes care of the modelling of beam collimation and intensity modulation. For the simulation of secondary jaws, First Compton-based approximate transport was implemented. The assumption was built on the fact that the principle photon interaction in tungsten from 0.5 to 5MeV is the Compton interaction. The location of the interaction as the photon traverses thickness t is randomly sampled over [0,t) assuming exponential attenuation. The interaction probability is evaluated using the ratio of the Compton and total attenuation coefficients as shown in Equation (1), where μ indicates the attenuation coefficient of photons and t indicates the jaws thickness as particle transverses through.

$$f(t) = Ce^{-\mu t}\mu dt \qquad (1)$$

As Equation (2) demonstrated, the particle weight was then modified based on scattered photons' energy and direction, while the emerged Compton electrons are not continued.

$$w_f = w_i(1 - e^{-\mu(E)t})\frac{\mu_c}{\mu}e^{-\mu(E')t'} \qquad (2)$$

Intensity modulation is achieved by supplying fluence map for each treatment field. Phase-space particles that pass through the jaw openings are projected to the MLC plane. A weighting factor determined by the particles position in the fluence map is then assigned to the particle. If the particle already carries a weighting factor in the patient-independent phase-space, the two factors are multiplied. The final factor is carried throughout the subsequent transport process and is used to adjust the dose deposition accordingly.

2.2. Helical Tomotherapy accelerator

Helical tomotherapy is a sophisticated intensity-modulated radiation therapy (IMRT) delivery modality developed at the University of Wisconsin-Madison and commercialized by Tomotherapy, (now owned and distributed by Accuray Inc). The Hi-Art Tomotherapy system utilizes a unique mechanical structure that resembles a helical CT scanner. Different from the traditional IMRT, the treatment head in a Tomotherapy system is mounted on a slip ring gantry and can rotate continuously around the isocenter. During the treatment, the head rotates continuously while the couch is translated concurrently, delivering the dose in a helical manner. In our model, all particles defined in the phase space files lie on a cylindrical surface formed by the gantry rotation and couch translation—such that these particles can be used in the next-stage MC simulations involving patient-specific parameters of the treatment plan. Contaminant electrons are not included in PSFs because a previous study had showed that the electron contamination from a helical tomotherapy system could be ignored without introducing measurable error. After benchmarking the code, one clinical prostate case is calculated.

2.3. Electron-Photon Coupled Transport Kernel

The main physics models implemented in ARCHER were similar to production MC code DPM. The original sequential DPM code was developed for fast dose calculations in radiotherapy treatment planning [11]. It aims for simulating coupled photon-electron transport with a set of approximations valid for the energy range considered in radiotherapy. Specifically, for photon transport, we considered photoelectric effect, Compton scattering, and pair production. Rayleigh scattering is disregarded in this study because it has negligible impact on dose distribution for photon energy range used for radiotherapy. The photon transport is handled by using the Woodcock tracking method which eliminates frequent boundary checking for photon transport in heterogeneous geometries [12]. As for the electron transport, DPM implements a condensed history technique. Step-by-step simulation is used for inelastic collisions and bremsstrahlung emission involving energy losses above certain cutoffs. It also employs new transport mechanics and electron multiple scattering distribution functions to allow long transport steps. These mechanisms enable the electron cross a few heterogeneity boundaries in one step and hence increase the simulation efficiency. The continuous slowing down approximation is employed for energy losses below some preset energy thresholds. Positron transport is treated as electron and two photons are created at the end of the positron path to account for the annihilation process. The accuracy of DPM has been demonstrated to be within for both clinical photon and electron beams [13]. In ARCHER, this DPM-based physics kernel is coupled with other modules, such as the Constructive Solid Geometry (CSG), DICOM decoder and phase space file manager, to enable accurate, fast, and patient-specific dose calculations.

2.4. Code Design and micro-optimization

ARCHER uses "symmetric execution" model where the CPU and MIC work concurrently. The code is written in hybrid MPI-OpenMP. Three types of processes were created: root process, I/O process and compute process. The root process is launched on the host CPU and acts as an overall coordinator. Its job is to distribute the task and perform tally reduction. The I/O process, also on the host, is used to read the phase space files from the hard drive to the system memory and send them to the compute process. The compute process resides in the MIC coprocessor and performs transport simulation in parallel. For a heterogeneous system with 1 CPU and n MICs, we used 1 root process, and n I/O processes paired with n compute processes. On each MIC, the compute process spawned a total of 240 threads to fully saturate the hardware resource. To hide I/O and data transfer with computation, the double buffering method was implemented, which lets the I/O process to read the next phase space file and copy it to the compute process before the simulation using the current phase space file is completed.

To ensure the reliability of our MIC and GPU codes, we decided to use double-precision for all calculations. However, before the Pascal architecture, NVIDIA GPUs don't support FP64 natively. NVIDIA has offered some workarounds, but is very slow. To solve this problem, we implemented this fast warp-based floating point 64 atomic add algorithm. According to our test, our implementation is ~600x faster than Nvidia's original workarounds.

For CPU and MIC, localized vectorization and software-based data prefetch were implemented. Although MC is known to have many branch statements which is hard to vectorize on global scale, local vectorization for tight for-loops still exist. Thus for CPU, SSE4 is used to vectorize the instruction set, and IMCI is the instruction set for vectorization on the MIC.

The current MICs use legacy processors that are subject to stall when the data are being loaded from the DRAM, resulting in long latency. The data prefetch method is able to reduce this latency by loading data several loops ahead so that they are readily available in the cache when actually needed.

2.5. Phase Space Source Implementations

For the prostate plan, the size of the phase space file is 6.4 GB, which is too large to load into Xeon Phi's memory. Another concern is of efficiency. It takes at least tens of seconds to transfer a large amount of particle data from a hard drive to Xeon Phi memory. Although tens of seconds is relatively short for CPU-based MC simulations, it consumes a large portion of the total computational time for Xeon Phi-based MC dose calculations, posing a bottleneck to further improve efficiency. To circumvent the above problems, the entire PSF file is partitioned into smaller pieces based on their type and energy and then simulated in batches. Typically, the batch number is set to be 12, reducing the sub-PSF to a more manageable size of 530MB. A sorting of particles from the phase-space source has also been employed before launching them for dose calculations. Specifically, the code works as following: the CPU launches the particles sequentially and the CPU reads the first sub-PSF and copies the data to the Xeon Phi memory. The Xeon Phi kernel is then launched asynchronously to perform the MC calculations. The CPU reads in the next sub-PSF immediately after the first kernel is launched. In this way, the simulation and PSF reading are performed concurrently, thus enhancing the code execution efficiency.

2.6. Patient CT Phantom

The CT dataset is converted into mass density and material composition using the Hounsfield unit (HU)-to-density conversion curve. Using a commercial treatment planning system at UW–Madison, the conversion is calibrated in accordance with recommendations by Verhaegen and Devic [14]. In these simulations, four materials, each having densities specified by the HU, are used for the patient phantom: water, dry air, compact bone (defined by ICRU), and lung (defined by ICRP) (15). Although the code is capable of more materials, these four materials provide enough accuracy for the MC investigation. The prostate phantom consists of 260×256×108 voxels, with a voxel size of $0.1875\times0.1875\times0.3 cm^3$.

3 Results

3.1. Commissioning Fields Validation

Fig. 2 illustrates the percentage depth dose (PDD) between simulated phase space source modelling and measurement in $60\times60\times40cm^3$ water phantom for Varian TrueBeam 6MV beam at SSD 100 cm. The dose was normalized to different scale for better visualization. The voxel size of the phantom is $0.4\times0.4\times0.4cm^3$ for field size of 4×4, 10×10, and $40\times40cm^2$. A good agreement was observed for all fields. The dose difference was better than 2% except in build-up region where the difference is up to 5.87%.

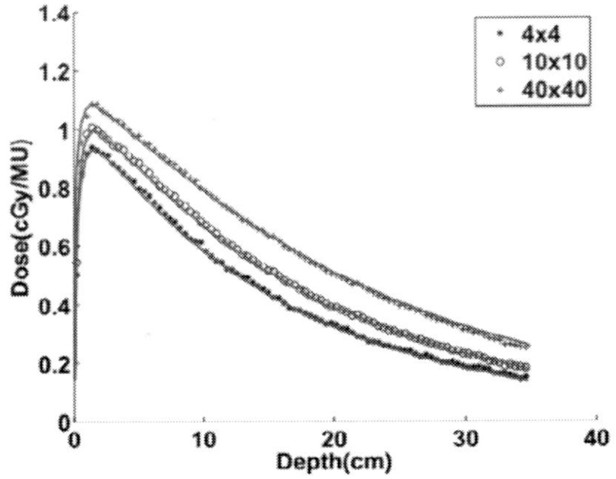

Fig. 2. Experimental (red solid lines) and ARCHER calculated depth dose curves. The Monte Carlo dose was averaged over 3×3 voxels.

Fig. 3 illustrates the comparison of cross profile at depth 2.5 cm for field sizes 4×4, 10×10, $40\times40cm^2$ between our simulation and measurement at SSD 100 cm. It is shown that the profiles at depth 2.5 cm matched well in all fields.

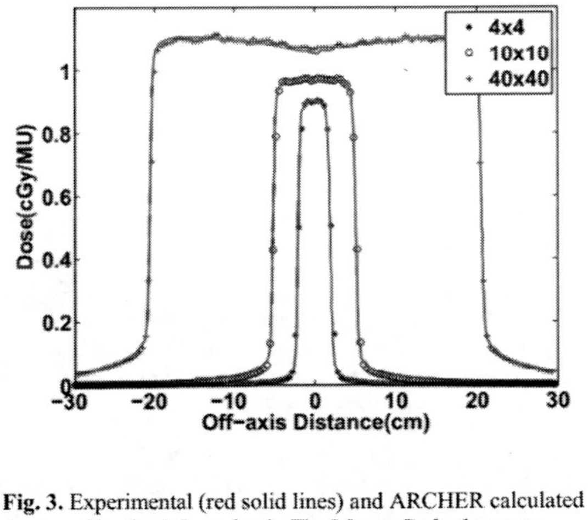

Fig. 3. Experimental (red solid lines) and ARCHER calculated dose profiles for 2.5 cm depth. The Monte Carlo dose was averaged over three voxels with a total length of 1.2 cm.

3.2. Dosimetric Results for Clinical Cases

One realistic Tomotherapy and TrueBeam plan were selected for this study. They are optimized in commercial TPS and then simulated with phase space based source modelling. The calculation results of the prostate tomotherapy case and the TrueBeam breast treatment using ARCHER were shown in Fig. 4 and 5. For the voxel doses in the PTV, relative statistical error (1σ) is kept to be about 1% considered in this study. The simulated dose distribution was shown in the first column. For comparison, the dose results from DPM are provided in the second column. It could be observed that these two sets of isodose lines overlapped with each other at most regions for both cases.

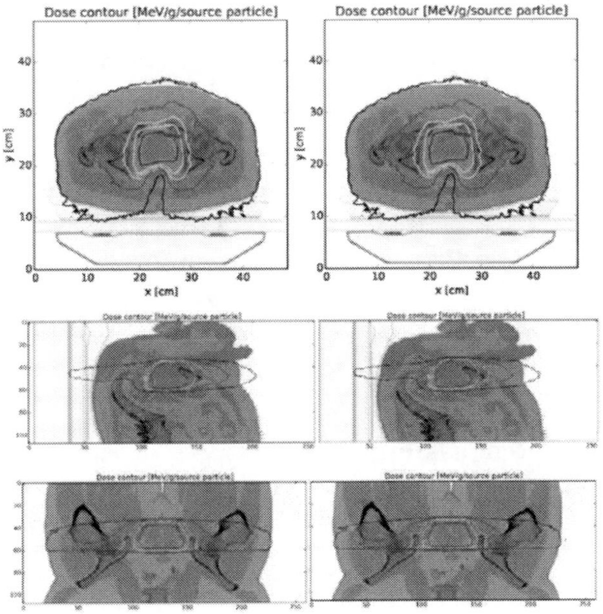

Fig. 4. Isodose distribution of a Tomotherapy plan between ARCHER simulated results (left) and DPM results (right). The isodose curves were shown in the first column, with transverse, coronal and sagittal views in three columns, respectively.

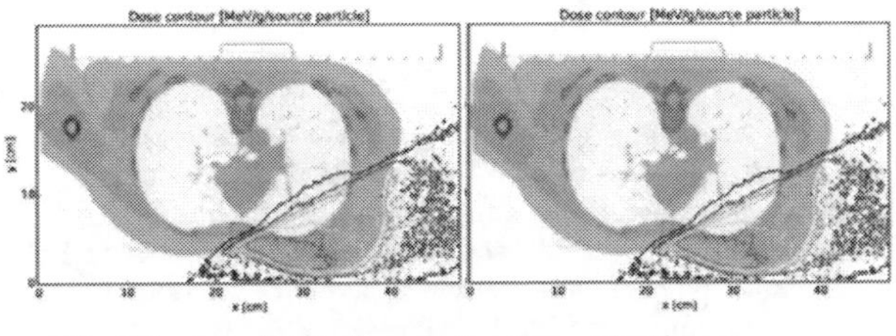

Fig. 5. Isodose distribution of a TrueBeam 6MV breast treatment plan between ARCHER simulated results (left) and DPM results (right).

3.3. Performance Studies

For fast MC simulations using Phase Space as source input, it is found that reading PSF from the hard drive costs a large fraction of code execution time. On the other hand, using the batch simulation explained before, file reading works concurrently with kernel execution, thus avoiding explicit file read time. With this improvement, the preprocess time includes loading 600 million particles and CT phantom is found to be only about 3s for the prostate case. Table 1 summarizes the performance results for comparison among CPU, Xeon Phi and GPU. To be noted, all the results were calculated under double-precision, and no aggressive compiler options were used since that may trade the accuracy for speed. It could be seen that by employing Intel Xeon Phi coprocessors and GPUs, comparing with Intel x5650 CPU, about 9-12 times speedup has been achieved.

Table 1. Timing comparisons among CPU, MIC and GPU for Tomotherapy and TrueBeam cases.

Processor	No. of particles	Total wall time [sec]	
		Tomotherapy	TrueBeam
1 X5650 CPU(one thread)	600 million	2057	758
1 5110P MIC	600 million	187	81
3 5110P MICs	600 million	84	40
1 Kepler K40 GPU	600 million	173	73
2 Kepler K40 GPU	600 million	89	39

4 Discussions

The treatment head model is crucial to the Monte Carlo dose simulation in external radiotherapy. Some previous studies have shown good agreements between simulation and measurement using MC simulated phase space as the head model. As the most accurate way of head model, direct and full MC simulation of accelerator head components can provide us a phase space file which contained nearly realistic particles' fluence distribution for the rest phantom simulation. The GPU implementation using Phase space file [16] were also carried out in several different methods.

In this work, we took the concept of approximate transport through patient-dependent collimators, where the jaws used First-Compton based model and then intensity modulated by using the fluence distribution. This model is efficient to implement on MIC/GPU while its accuracy is validated through the comparison between the simulation result and measurement of several open fields range from 4×4 cm^2 to 40×40 cm^2.

As for speed, compared to a highly optimized CPU implementation, the MIC and GPU have shown comparable performance gain of 9 to 12 times when simulating the real clinical cases. The reasons we optimized CPU code are mainly for code portability and fair comparison. To compare the accelerator's performance with CPU, the sequential code should be optimized, multi-thread parallelized. Additionally, although single precision floating point operations are faster than double precision, it's also more error-prone, since it could result in dose underestimate, especially for the voxels surrounding a radiation source. This importance of using double-precision for radiotherapy dose calculation was proven by Magnoux et al. in their paper [18].

5 Conclusions

In this study, we have successfully developed and extended ARCHER to Xeon Phi-based fast and accurate dose calculation engine for radiotherapy. The simulated results were validated with experimental measurement and one clinical TomoTherapy treatment plan of the prostate and one TrueBeam treatment plan of the breast were studied to further benchmark against DPM to show the clinical utility. Optimization techniques have been applied, includes sorting the source particles on-the-fly before transport to Xeon Phi and GPU based on particle type and energy, and batch method transferring the Phase Space source to reduce the preprocess time. The future work includes the development of explicit approximate transport of linac MLCs and extension to other energy levels and other popular vendor machines such as Varian TrueBeam Flattening Filter Free machine.

Acknowledgement

This research is funded by the US National Institute of Biomedical Imaging and Bioengineering (R01EB015478). We would also like to thank Intel and Nvidia for the generous hardware donation.

References

1. T. Bortfeld. Phys. Med. Biol.51, R363–R379 (2006)
2. A. Fogliata, E. Vanetti, D. Albers, C. Brink, A. Clivio, T. Knöös, G. Nicolini, and L. Cozzi. Phys. Med. Biol.52, 1363–1385 (2007)
3. XG Xu, T. Liu, L. Su, et al. Annals of Nuclear Energy (2014)
4. T Liu, XG Xu, CD Carothers. Annals of Nuclear Energy (2014)
5. XG Xu, T Liu, L Su, FB Brown et al. Trans. Am. Nucl. Soc (2013)
6. L Su, Y Yang, B Bednarz, XG Xu et al. Medical physics (2014)
7. I. Kawrakow and D. W. O. Rogers. NRCC Report No. PIRS-701 (2011)
8. S. Agostinelli, J. Allison, K. E. Amako, J. Apostolakis, H. Araujo, P. Arce, M. Asai, D. Axen, S. Banerjee, and G. Barrand. Nucl. Instrum. Methods, Phys. Res A506(3), 250–303 (2003)
9. Capote R. Radiother. Oncol.84S217 (2007)
10. Schmidhalter D, Manser P, Frei D, Volken W and Fix M K. Med. Phys.37492–504 (2010)
11. Sempau J, Wilderman S J and Bielajew A F. Physics in Medicine and Biology 45 2263-91 (2000)
12. Woodcock E, Murphy T, Hemmings P, et al. Applications of Computing Methods to Reactor Problems: Argonne National Laboratorie s Report) pp ANL -7050 (1965)

13. N. Tyagi, A. Bose, and I. J. Chetty. Med. Phys.31, 2721–2725 (2004)
14. F. Verhaegen and S. Devic. Phys. Med. Biol.50, 937–946 (2005)
15. I. Kawrakow and B. R. B. Walters. Med. Phys.33, 3046–3056 (2006)
16. Townson, Reid W., et al. Physics in medicine and biology 58.12 (2013)
17. Siebers, Jeffrey V., et al. Physics in medicine and biology 47.17 :3225 (2002)
18. Magnoux, Vincent, et al. Physics in medicine and biology 60.13 : 5007 (2015)

EPJ Web of Conferences 153, 04011 (2017) DOI: 10.1051/epjconf/201715304011
ICRS-13 & RPSD-2016

Inter-comparison of Dose Distributions Calculated by FLUKA, GEANT4, MCNP, and PHITS for Proton Therapy

Zi-Yi Yang[1,†], *Pi-En* Tsai[1,2,*,†], *Shao-Chun* Lee[1], *Yen-Chiang* Liu[1], *Chin-Cheng* Chen[3], *Tatsuhiko* Sato[2], and *Rong-Jiun* Sheu[1]

[1]Institute of Nuclear Engineering and Science, National Tsing-Hua University, Hsinchu, Taiwan
[2]Japan Atomic Energy Agency, Tokai, Japan
[3]ProCure Proton Therapy Center, Somerset, NJ, USA

† Contributed equally to this work with: *Zi-Yi* Yang and *Pi-En* Tsai

Abstract. The dose distributions from proton pencil beam scanning were calculated by FLUKA, GEANT4, MCNP, and PHITS, in order to investigate their applicability in proton radiotherapy. The first studied case was the integrated depth dose curves (IDDCs), respectively from a 100 and a 226-MeV proton pencil beam impinging a water phantom. The calculated IDDCs agree with each other as long as each code employs 75 eV for the ionization potential of water. The second case considered a similar condition of the first case but with proton energies in a Gaussian distribution. The comparison to the measurement indicates the inter-code differences might not only due to different stopping power but also the nuclear physics models. How the physics parameter setting affect the computation time was also discussed. In the third case, the applicability of each code for pencil beam scanning was confirmed by delivering a uniform volumetric dose distribution based on the treatment plan, and the results showed general agreement between each codes, the treatment plan, and the measurement, except that some deviations were found in the penumbra region. This study has demonstrated that the selected codes are all capable of performing dose calculations for therapeutic scanning proton beams with proper physics settings.

1 Introduction

Proton radiotherapy, compared to the conventional radiation therapy using photons, has the advantages of a more pronounced dose profile, as the so called Bragg peak, and a sharper distal dose falloff (DDF) followed by the Bragg peak. As such, doses can be delivered more restrictively to tumour regions, and less damages are caused to surrounding normal tissues. However, also due to proton's sharp Bragg peaks, the accuracy of the proton beam delivery is of critical importance in clinical treatments.

In the regulation for clinical treatment in radiotherapy, it is often required an additional independent methodology other than the semi-empirical method used in treatment planning systems (TPS) for verification and validation of the TPS. Monte Carlo (MC) simulations by using the general-purpose radiation transport codes, such as FLUKA [1], GEANT4 [2, 3], MCNPX [4, 5], MCNP6 [6], and PHITS [7], are able to calculate the dose profiles delivered by energetic protons. Hence, this study aims to investigate the applicability of these four MC codes in the proton therapy using the pencil beam scanning technique. The simulation results with respect to the accuracy, computing efficiency, and capability of performing complicated source setup of each code were inter-compared and discussed in this study. The simulation results were also validated by the experimental data.

2 Materials and Methods

2.1 Test Cases

The following describes the details of three test cases in this study: two for the integrated depth dose curves (IDDCs) in water from a mono-directional proton pencil beam but with different energy distributions, and one for the three-dimensional (3D) volumetric dose distribution from the pencil beam scanning.

Case (a) is a mono-energetic proton pencil beam incident on a water phantom with a depth of 40 cm and an infinite area. Two proton energies, respectively at 100.0 MeV and 226.0 MeV, were tested.

Case (b) is a proton pencil beam with a Gaussian energy distribution centred at the energy slightly shifted away from 100 MeV or 226 MeV. A pure mono-energetic proton beam like case (a) is an ideal case. In reality, the proton beam is not able to remain mono-energetic due to the complex beam transport and

* Corresponding author: tsai.pien@jaea.go.jp

delivery process. Our previous study [8] showed that the adjustments in the beam energy and its energy distribution in MC calculations can better match the experimental data, and the degree of energy spread-out decreases with the increasing proton energy. As such, the proton beam originally at 100 MeV was shifted to 100.5 MeV with 1.62-MeV full width at half maximum (FWHM) (standard deviation σ = 0.690 MeV) of a Gaussian distribution. For the original 226.0-MeV proton beam, it became a Gaussian distribution centred at 227.5 MeV with 0.54-MeV FWHM (σ = 0.023 MeV). The amount of the energy shift and the degree of the spread-out were determined by the FLUKA calculations with the same methodology used in [8].

Case (c) is to simulate the volumetric dose distribution with various proton energies between 116.1 MeV and 181.5 MeV by using the pencil beam scanning. The planning target volume (PTV) is a 10 × 10 × 10 cm^3 cube centred at depth of 15 cm inside a 40 × 40 × 40 cm^3 water phantom. How each pencil beam is delivered to each spot was determined by the RayStation treatment planning system (TPS) (version 4.0.3.4, RaySearch Laboratories AB, Stockholm Sweden).

2.2 Monte Carlo Simulations

The FLUKA, GEANT4, MCNP, and PHITS codes were employed for this MC simulation study owing to their established and widespread use for particle transport calculations. A brief description on the physics setting for each code and the simulation setup will be given in this section. The parameters set in each code followed either the recommended setting or the default value.

2.2.1 FLUKA

The FLUKA code version 2011.2c.4 was utilized for dose calculations. We selected the particle transport physics card HADROTHErapy, which sets the neutron transport threshold at 10^{-5} eV for neutrons. Transport and production cuts were set at 100 keV for electrons/positrons, photons, and protons by EMFCUT and THRESHOLD. Also, delta ray production by charged hadrons with a 100-keV threshold is activated in this physics option. The hadron-nucleus nuclear interactions were treated by the PEANUT (PreEquilibrium-Approach-to-Nuclear-Thermalization) package. The neutron transport below 20 MeV is handled by FLUKA's own group-wise cross section library, which was oriented from E. Cuccoli et al. [9] and updated based on the evaluations like ENDF/B, JEF, and JENDL.

2.2.2 GEANT4

The GEANT4 code version 10.1.1 was employed for this study. The physics lists adopted in this study were based on the suggestions from the user guide and recommended setting in other studies [10, 11]. The reference physics list Standard EM option 3 in the extended example TestEm7 of electromagnetic, G4Binary and G4LElastic were selected. The range cutoffs for electrons/positrons, photons, and protons were respectively set at 143.1 µm, 34 cm, and 1 mm, which all correspond to cutoff energy at about 100 keV (the default range cut is 1 mm for the above-mentioned particles), and were not specified for the remaining particle species. The neutron transport below 20 MeV is based on the default cross section data library, G4NDL4.5, whose data comes largely from the ENDF/B-VI library [12]. Note that the delta ray production is simulated in the electromagnetic processes for both electrons (and positrons) and hadrons above the range cuts.

2.2.3 MCNP

The MCNPX code version 2.6.0 as well as the MCNP6 code version 1.0 were both used in this study, since they both have a large number of users. Both codes employ the tabulated data library for nucleon-induced reactions for projectile energies less than 150 MeV. In this study, we used the same data libraries for both versions of the code, which are LA150N [13] sourcing from ENDF/B-VI.6 [12] for neutrons, and LA150H [13] for protons. For energies above, MCNPX uses the Bertini model as default [14], whereas MCNP6 changes the default model to the cascade-exciton model (CEM) as incorporated in the event generator CEM03.03 [15]. For both of MCNPX and MCNP6, the Coulomb diffusion and the energy straggling of protons as well as elastic and inelastic scatterings were taken into account, while the photonuclear interaction was not considered. The cutoff energies were both set to 100 keV for electrons, photons, protons, and no thresholds for neutrons. Also, both codes consider the delta-ray production only from electron-electron collisions by PHY:E card, but not from heavy charged hadrons.

2.2.4 PHITS

The PHITS code version 2.86 was employed for the inter-code comparison. ATIMA [16] was used to calculate stopping powers for protons and nucleus as the default setting since this version. The options for the Landau Vavilov energy straggling and the Coulomb diffusion by Lynch's formula [17] were turned on. The nucleon-induced nuclear reactions were modelled by the INCL4.6 model [18]. For neutron energies below 20 MeV, the transport is based on the cross section data library, JENDL-4.0 [19]. Photon and electron transports were calculated by the PHITS built-in EGS5 [20] code. The cutoff energies were 100 keV for electrons/positrons, photons, and protons, and 10^{-5} eV for neutrons. Knock-on electrons are treated in the simulation, and the option for delta rays produced from heavy charged particles with a threshold of 100 keV was turned on in order to compare the CPU time with the other codes.

2.2.5 Simulation

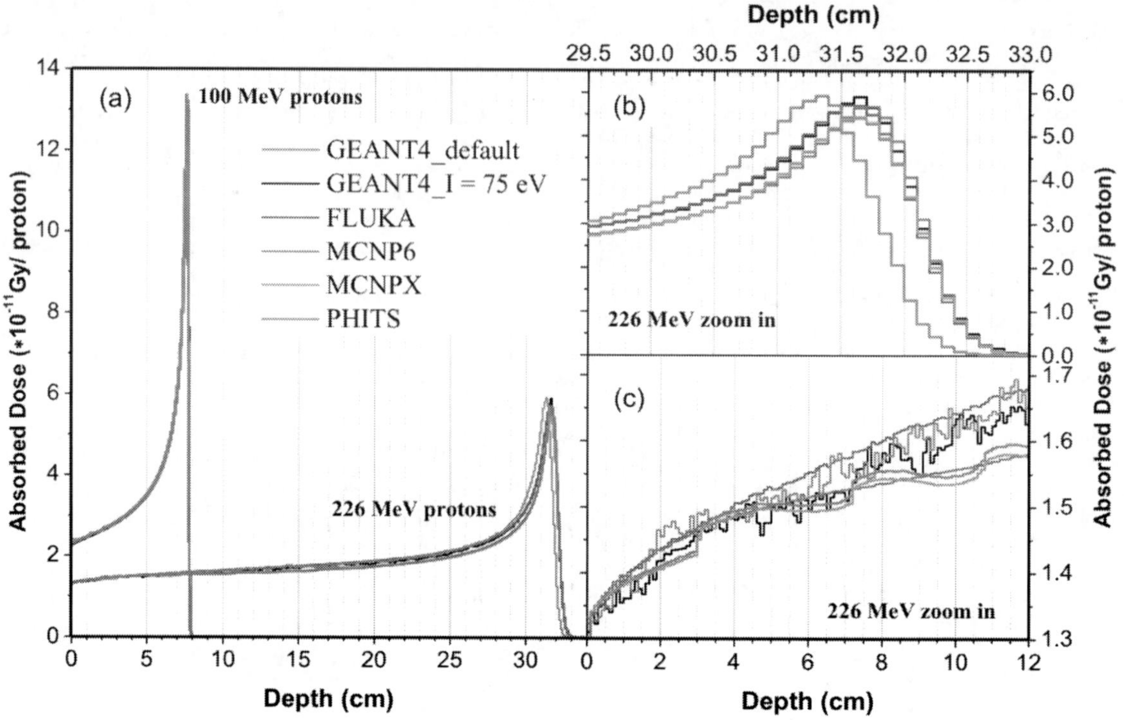

Fig. 1. (a) The IDDCs in water calculated by FLUKA, GEANT4, MCNP, and PHITS for mono-energetic protons, respectively at 100 MeV and 226 MeV; zoom in of **(b)** the Bragg peaks, and **(c)** the entrance and plateau region from 226-MeV proton.

A proton pencil beam impinged a water phantom from 200 cm upstream in vacuum. The water phantoms for case (a) and (b) are identical: both are 40-cm deep with an infinite cross sectional area. The absorbed doses were scored by disks with a 4.08-cm radius and a 1-mm depth along the beam direction, whose geometry is the same as the Bragg peak chamber (model 34070, PTW-Freiburg, Germany) used in the measurement.

The phantom for case (c) is a 40 × 40 × 40 cm^3 cubic water phantom. Two different tallies were scored for this case: (i) the central-axis depth dose distribution. The tallied voxel size is 0.3 cm in thickness and 1.77 × 1.77 cm^2 in cross sectional area (to mimic the collecting electrode dimension of the multiple layer ionization chambers, (MLIC, IBA Dosimetry GmbH, Germany). (ii) The two-dimensional (2D) dose distributions perpendicular to the central axis respectively at the depths of 5, 11, 13, 15, 17, and 19 cm. The size of each scoring volume is 0.1 × 0.1 × 0.1 cm^3.

The treatment plan for the dose delivery in case (c) was designed by RayStation TPS. In this treatment plan, there are 28 different incident energies ranging from 116.1 MeV to 181.5 MeV, and more than 10,000 scanning spots in total. Beam characteristics including proton energy, measured spot sizes, virtual and effective source-to-axis distance, and absolute dose correction [8] were applied in the simulation.

2.3 Measurements

The simulated results of case (b) and (c) were validated by the measurements performed at the ProCure Proton Therapy Center in NJ, USA.

For case (b), the IDDCs were measured with a Bragg peak ionization chamber (model 34070, PTW-Freiburg, Germany) [21]. The central-axis depth dose distribution in case (c) was measured by a multi-layer ionization chamber (MLIC, IBA Dosimetry GmbH, Germany) [22]. The 2D dose distributions at specific depths were measured with a 2D ion chamber array (MatriXX PT, IBA Dosimetry GmbH, Germany) [23]. More details of the measurements and the experimental uncertainties are described in the references [21-23].

3 Results and Discussions

3.1. Case (a): IDDCs from mono-energetic protons

Prior to comparing the proton ranges calculated by different MC codes, it is necessary to check the mean ionization potential (*I*) of the medium used in each MC code, since the stopping power described by the Bethe-Bloch formula [24-26] is dependent on I^{-1}, and since range is determined by integrating the reciprocal stopping power over energy from the projectile energy down to zero. FLUKA, MCNP [27], and PHITS all use 75 eV for their default *I* value, which is the same as in the ICRU Report 37 [28]. Only the GEANT4 code version 10.1.1 employs 68.998 eV for its default ionization potential. Fig. 1 (a) shows the comparisons of the IDDCs calculated by GEANT4 (with the default *I* value, and *I* = 75 eV), FLUKA, MCNPX, MCNP6, and PHITS, respectively for 100-MeV and 226-MeV protons. The calculated peak locations for 100-MeV protons show no prominent differences, except that the GEANT4 with the default setting has the Bragg peak location 0.1 cm shallower than the others. The difference in peak location becomes significant as the proton energy increases to 226 MeV as shown in Fig. 1(b), the zooming-in of the 226-MeV protons' IDDCs around the Bragg peak. The result of GEANT4 with the *I* value of 75 eV showed better agreement with the others. Based on this fact, the ionization potential was set at 75 eV for later calculations in GEANT4 to be consistent with the other codes.

In MCNP simulations, it is observed that, in the entrance region, their IDDCs look stepwise and are not as smooth as the other codes'. An example is shown in Fig. 1 (c) with 226-MeV protons incident on a water phantom. Such smoothness can be adjusted by the *efac* parameter in the PHYS:H card in MCNP6 (MCNPX does not offer this option to users). More details of the effects of this parameter will be discussed in section 3.2.2 along with its computation time.

3.2 Case (b): IDDCs from proton beams with energy shift and energy spread-out

The MC simulations and the measured relative IDDC data were compared here for case (b): proton beams with a Gaussian energy distribution centred at the energy slightly higher than case (a).

3.2.1 Depth dose distribution

Table 1 lists the total energy depositions in the cylindrical scoring region calculated by each codes for

Table 1. The total energy deposition in the cylindrical scoring region (radius = 4.08 cm) for two proton beam energies calculated by each MC codes.

Code	100.5 MeV (FWHM: 1.62 MeV)	227.5 MeV (FWHM: 0.54 MeV)
FLUKA	98.60 MeV	204.52 MeV
GEANT4	98.18 MeV	203.91 MeV
MCNP6	97.21 MeV	201.96 MeV
MCNPX	96.84 MeV	200.90 MeV
PHITS	98.22 MeV	204.20 MeV
Average (Measurement)	97.81 MeV	203.10 MeV
Maximum difference*	1.82%	1.80%

* (maximum − minumum)/minumum × 100%

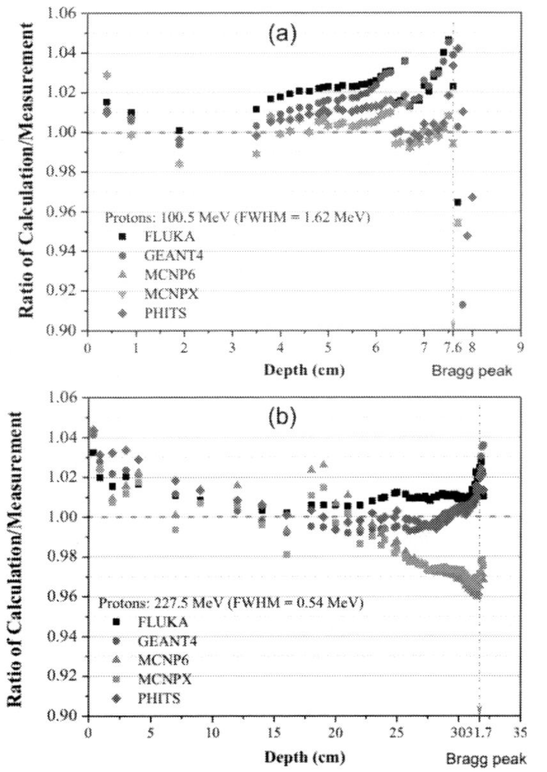

Fig. 2. The C/M ratios of each code for proton beams with a Gaussian energy distribution: **(a)** 100.5 MeV (1.62-MeV FWHM), and **(b)** 227.5 MeV (0.54-MeV FWHM).

two proton beam energy distributions. The values were obtained by integrating the doses deposited in each 4.08-cm-radius disk along the central axis over the entire phantom depth, i.e. the area under calculated IDDCs. Because the experimental data are available only in relative intensity, in order to compare the measurement and the simulations, the area under the measured IDDC was scaled to the average area under the MC calculated IDDCs, which is 97.81 MeV for the 100.5-MeV proton beam, and 203.10 MeV for the 227.5-MeV proton beam.

Table 1 also shows that, for both proton energies, MCNPX deposits the least amount of the total energy in the water phantom among all MC codes, while FLUKA deposits the largest amount of the total energy. The maximum differences among all the simulations are around 1.8%.

The ratios of calculated doses to measured doses (C/M ratios) over the entire range along the central axis are shown in Fig. 2 (a) and (b), respectively for two proton beam energies. In the entrance regions for both incident proton energies (depths < ~1 cm for 100.5-MeV protons, and < ~12 cm for 227.5-MeV protons), Fig. 2 shows that the C/M ratios for each codes are greater than 1.0, indicating that the physics models might underestimate the neutron production cross sections such that more energies are deposited locally by protons. As the depth becomes deeper, in general, the FLUKA's simulated results deposit more energies in the water phantom than other codes, whereas MCNPX and MCNP6 deposit the least amount of energies among all the codes. Such trends in the deeper region are in consistent with the total energy depositions listed in Table 1.

Fig. 2 also shows that the discrepancies also occur in the ascending region before the Bragg peaks, as well as the DDF region (here the distal dose falloff is defined as the distance from the 80% to the 20% dose level). Before the pristine Bragg peak, these inter-code differences in the C/M ratios are mainly resulted from (i) different stopping powers adopted in each code, and (ii) different physics models implemented in each code. However, as the depth in the water phantom increases, proton energy is continuously slowed down, and the inter-code differences are contributed by both factors, and thus need further study to investigate which factor has more impact on the differences.

3.2.2 Computation time

The computation time of each code was also compared for these two proton energies by using a PC equipped with a 3.4-GHz i7-3770 CPU, and the results were listed in Table 2. Note that PHITS does not recommend turning on delta ray production from heavy charged particles in calculations for particle therapy, as the transport of delta electrons is CPU expensive but only contributes negligible dose (less than 1%) to the final IDDCs due to the extremely short ranges of those delta rays. Thus the computation time for PHITS without considering delta rays from hadronic interactions is also tested and listed in Table 2. In FLUKA, though the default setting in the physics card HADROTHErapy considers the delta ray production from charged hadrons, the option can be turned off by users. Thus the run time was also tested, respectively with activating and deactivating this option, in order to investigate its impact on the computation time.

Table 2. The computation time (minutes) for 10^7 particles using different MC codes with different parameter settings.

Code	Option/Setting	100.5 MeV (FWHM = 1.62 MeV)	227.5 MeV (FWHM = 0.54 MeV)
FLUKA	HADROTHErapy default	133	416
	HADROTHErapy with delta ray off	130	200
GEANT4	default	800	4102
MCNP6	efac = 0.80 (minimum)	58	102
	efac = 0.917 (default)	160	209
	efac = 0.99 (maximum)	1649	2036
MCNPX	default	85	115
PHITS	With delta ray	136	806
	With delta ray off (recommended)	46	172

From Table 2, it is shown that, if the calculation of delta rays from heavy charged particles is taken into account (in FLUKA, GEANT4, and PHITS), then FLUKA and PHITS require about the same least amount of time for 100.5-MeV protons. If the proton energy increases to 227.5 MeV, FLUKA's computation time is about half of PHITS', indicating that the transport of delta electrons in FLUKA is more efficient than in PHITS. In contrast, for both proton energies, the computation time for GEANT4 is much longer than FLUKA and PHITS, and should be further improved.

When the transport of delta ray produced from charged hadrons is not considered, the difference in computation time required among all codes (excluding GENAT4 and MCNP6 with non-default *efac* settings) is not as significant as the time difference whilst the detla ray produced from charged hadrons is taken into consideration. The run time is least for PHITS with the 100.5-MeV proton beam, and for MCNPX with the 227.5-MeV proton beam, whereas the MCNP6 code with the default *efac* requires the longest time. Note that for both of FLUKA and PHITS, the computation time is much shorter when delta rays produced heavy charged particles are not considered, especially for higher incident proton energies.

It is worth mentioning that in MCNPX or MCNP6, the short calculation time is partially owing to its relatively rough energy spacing in the stopping power calculation by the default setting for "*efac*" in the PHYS:H physics card. Such default setting also results in its IDDCs in the entrance region coarse-stepwise and not as smooth as the other codes', as shown in Fig. 1 (c). In MCNP6, users can determine the *efac* value within the range of 0.80 and 0.99 (default is 0.917); a larger *efac* value produces more points in the stopping power tables, and thus a smoother IDDC. In this study, we also investigated how this setting affects the computation time and IDDCs; the result is shown in Fig. 3. With a larger *efac* value, (i) the IDDC is smoother. (ii) The range is slightly shorter. For instance, the range calculated with *efac* = 0.99 is about 2 mm shorter than the one with *efac* = 0.80. (iii) The required computation time, as shown in Table 2, is much longer. (iv) The area under the IDDCs (the total energy deposition) remains about the same (≪ 1.0%) regardless the *efac* setting. As

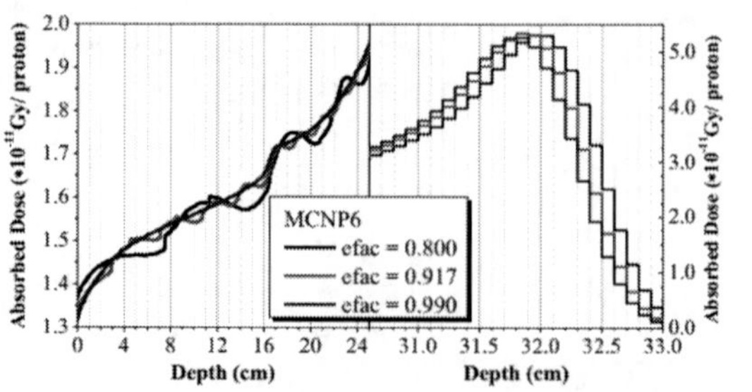

Fig. 3. The IDDCs calculated by MCNP6 with *efac* at 0.80 (minimum), 0.917 (default), and 0.99 (maximum) with the 227.5-MeV proton beam.

DOI: 10.1051/epjconf/201715304011

such, this parameter setting depends on users' requirement and the compensation between computation time, precision in range calculation, and smoothness of IDDCs.

3.3 Case (c): Volumetric dose distributions

In this test case, the source setup based on the TPS was to deliver a uniform volumetric dose distribution by using scanning pencil beams with energies from 116.1 MeV to 181.5 MeV. The calculation results of FLUKA, GEANT4, MCNP6, and PHITS along with the treatment plan and the measurement are compared in this section.

For the central-axis depth dose curves, all the measured and calculated results were normalized to 110 at the depth of 15 cm (the isocentric plane), because the dose distribution is quite flat in the plateau region. The MC simulations, the experimental data, and the TPS calculation are shown in Fig. 4(a); the calculation to measurement ratios for each MC codes are shown in Fig. 4(b). As seen in Fig. 4, though the deviations of the TPS to the measurement or the MC simulations to the measurement can be as large as ~5% at the phantom entrance window and the beginning region of the optimized target volume, they mostly agree with each other with C/M within 1.00 ± 0.02. Besides, the MC simulations can well overlap with TPS result. Considering the distal dose fall-off edge and lateral penumbra, the differences among the calculations, TPS, and the measurements were within 1 mm.

The calculation results of total dose and the total energy deposition per incident proton were also inter-compared for each MC code and listed in Table 3. The

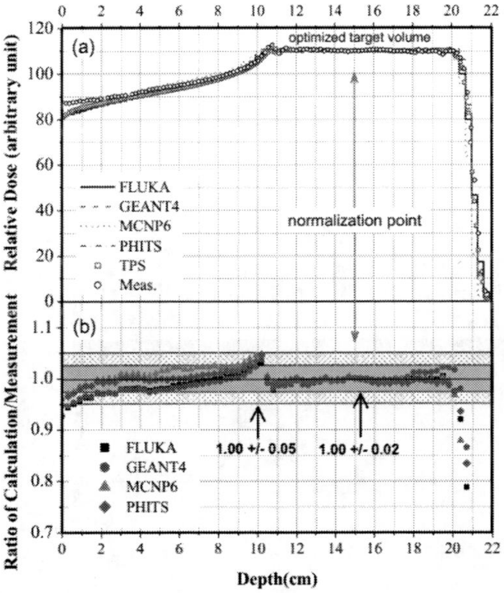

Fig. 4. **(a)** The relative depth dose distributions along the central axis for case (c) from measurement, TPS, as well as FLUKA, GEANT4, MCNP6, and PHITS simulations. **(b)** The C/M ratios for each MC code.

Table 3. The total dose and energy deposition per proton, respectively in the PTV (10 × 10 × 10 cm^3) and in the entire water phantom (40 × 40 × 40 cm^3).

Code	PTV		Entire water phantom	
	Total dose[1]	Total energy deposition[2]	Total dose[1]	Total energy deposition[2]
FLUKA	90.06	56.22	3.941	157.45
GEANT4	92.69	57.86	3.891	155.44
MCNP6	91.58	57.17	3.865	154.41
PHITS	93.46	58.34	3.901	155.86
Maximum difference[3]	3.77%		1.96%	

1. Unit in 10^{-13} Gy/proton
2. Unit in MeV/proton
3. (maximum − minumum)/minumum × 100%

trend in total energy deposition in the entire water phantom is consistent with Table 1, which shows the energy deposition is the greatest with the FLUKA calculation, followed by PHITS, GEANT4, and then MCNP6. The maximum inter-code difference is about 2.0%. Nevertheless, the trend in the total dose per proton delivered to the PTV is slightly different. PHITS estimates the largest amount of dose in the PTV region, and is then followed by GEANT4, MCNP6, and FLUKA in order. The maximum inter-code difference, 3.8%, is also slightly larger. Since the proton beam scanning spots are not far from the boundaries of the PTV, the relatively large inter-code differences and inconsistent trend with the total dose in the entire phantom are likely resulted from the different models employed in each MC code for angle straggling.

The 2D dose distributions of the measurements, TPS, and MC simulations were performed or tallied at the depths of 5 cm, 11 cm, 13 cm, 15 cm, 17 cm, and 19cm inside the water phantom, which include both upstream and downstream of the isocentric plane (at the depth of 15 cm in this case). To obtain the relative 2D dose distributions, the doses inside the water phantom at different depths were all normalized to the average value of the central 0.6 cm × 0.6 cm area at the isocentric plane.

The ratio of the TPS or calculation over the measurement within 1.00 ± 0.05 are considered in good agreement with the data, as the statistical errors of the experimental data are about 5% within the PTV. As shown in Fig. 5, for the 10 cm × 10 cm target areas, the TPS and the MC calculations, overall, have agreement within 5% with the measurement at all of the depths, although there are some spots that slightly underestimating the delivered does for (i) FLUKA and GEANT4 in the 5-cm-deep plane, and (ii) MCNP6 and PHITS in the 19-cm-deep plane.

However, in contrast to the good agreement in PTV, in the penumbra area (between ±5 cm and ±7 cm of the x or y axis), the TPS and MC simulations start to disagree with the experimental data. It can also be observed in Fig. 5 that the trend of underestimation by the TPS was similar to the MC simulations, and such underestimation becomes more significant as the distance is further from

 DOI: 10.1051/epjconf/201715304011

the PTV. It is likely that such underestimation is resulted from the background contribution of room scattering during the measurement, which was disregarded in neither the TPS nor the MC simulations.

To quantify the agreement of results, the gamma evaluation method [29, 30] was utilized here. This evaluation method provides a quantitative index γ from a composite analysis of dose-difference (DD) and distance-to-agreement (DTA, defined as the distance between a measured data point and the nearest point in the simulated dose distribution that has the same dose). The smaller the value of γ, the better the agreement between the reference and the calculation. When $\gamma \leq 1$, it means calculations pass the evaluation criteria; when $\gamma > 1$, it means calculations fail in the evaluation. Table 4 lists the γ passing rates for each code at six different depths, respectively with the measurement and the TPS as the reference. The criteria to distinguish the differences were DD = 2% and DTA = 2 mm, as suggested by the ICRU report [31]. Higher passing rates over all points in a certain distribution indicates better agreement for this comparison. As seen in Table 4, all of the results had the γ passing rates greater than 90%. However, the results versus the measurement were lower than the results versus the TPS, which is due to (i) the disregard of background contribution in neither the TPS nor the MC simulations, and (ii) the statistics in the measurement becomes poor as the distance from the

Table 4. The gamma passing rates with 2% DD and 2mm DTA criteria for each code compared with the measurement and the TPS at the depths of 5 cm, 11 cm, 13 cm, 15 cm, 17 cm, and 19 cm for case (c).

	Depth	*FLUKA*	*GEANT4*	*MCNP6*	*PHITS*
vs. Meas.	5 cm	92.39 %	92.03 %	91.68 %	90.45 %
	11 cm	93.93 %	93.83 %	93.72 %	90.06 %
	13 cm	90.06 %	92.31 %	93.30 %	92.77 %
	15 cm	99.85 %	98.77 %	90.80 %	99.08 %
	17 cm	93.72 %	93.72 %	99.01 %	91.55 %
	19 cm	94.10 %	91.72 %	91.81 %	91.63 %
vs. TPS	5 cm	96.18 %	94.63 %	93.74 %	95.55 %
	11 cm	99.44 %	99.36 %	98.10 %	98.16 %
	13 cm	99.91 %	99.22 %	99.72 %	98.67 %
	15 cm	99.85 %	98.77 %	96.71 %	99.15 %
	17 cm	99.73 %	97.00 %	95.24 %	98.47 %
	19 cm	98.79 %	98.26 %	94.08 %	96.36 %

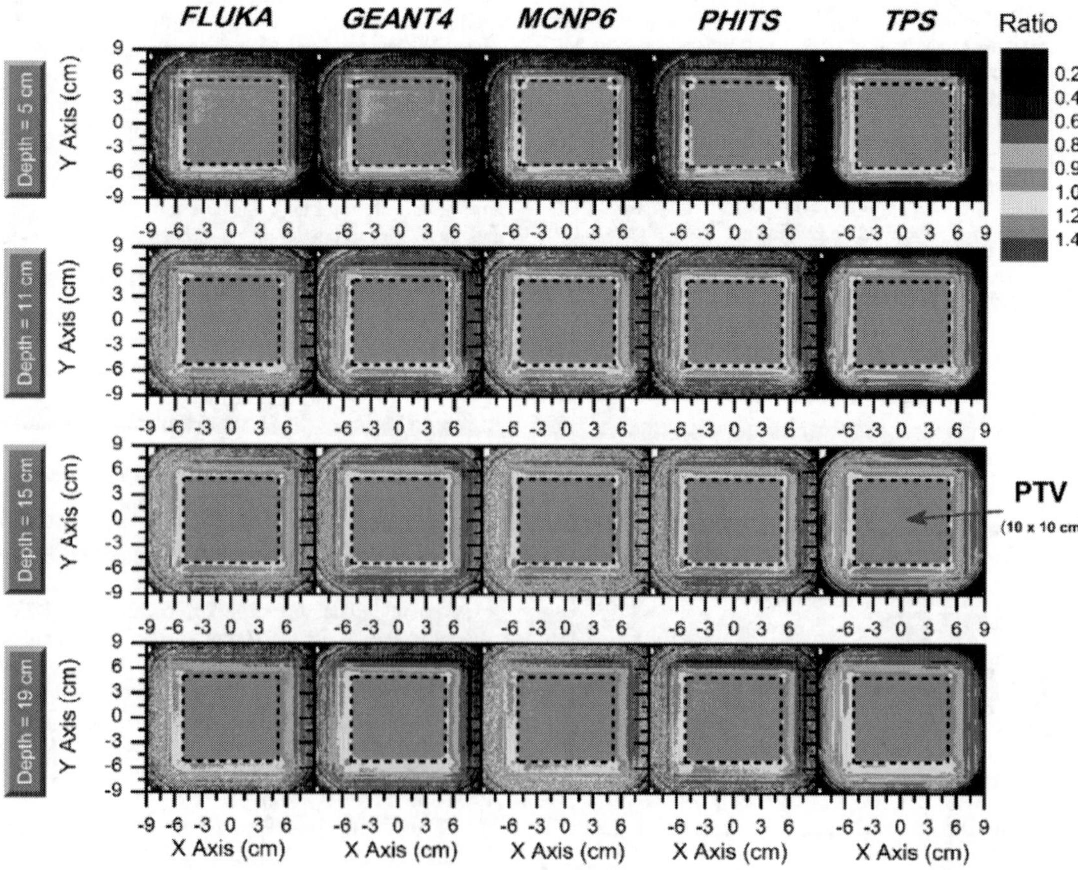

Fig. 5. The representative C/M ratios of 2D dose distributions at depths of 5 cm (1st row), 11 cm (2nd row), 15 cm (3rd row), and 19 cm (4th row) for case (c). From left to right in order are the ratios of FLUKA/measurement, GEANT4/measurement, MCNP6/measurement, PHITS/measurement, and TPS/measurement.

PTV increases, and such statistical fluctuation causes the drop of the γ passing rates versus the measurement. For this case (c), in summary, the TPS as well as the MC simulations by FLUKA, GEANT4, MCNP6, and PHITS are all able to reproduce the measured results in the PTV. Although the disagreement becomes more significant in the high dose gradient regions, the dose contribution to the phantom also becomes smaller due to the sharp distal falloff of the scanning proton beams as the distance was further away from the PTV. As such, the differences in the total dose to the entire water phantom are also small.

4 Conclusions

This study has demonstrated that the five MC codes – FLUKA, GEANT4, MCNPX, MCNP6, and PHITS, are capable of performing dose calculation for therapeutic proton beams, even for complex source setup with more than 10,000 scanning spots. Several physics settings were probed, including the ionization potential of water in GEANT4, the energy spacing for stopping power calculations in MCNP6, and the delta ray production from heavy charged particles in FLUKA and PHITS. It is suggested these setting or parameters should be carefully determined or selected, based on the users' requirements for computational time, smoothness of IDDCs, as well as preciseness and accuracy of range calculations. If users require a rough and quick calculation, MCNPX or MCNP6 with large energy spacing in stopping power calculations is suitable for this purpose. If users consider both computation time and smoothness of IDDCs, it is recommended to use the PHITS code without considering delta rays produced from heavy charged particles.

The results of this study has also shown that, with proper physics parameter settings in the MC codes, the simulations, measurements and TPS are generally in a few percent agreement with each other. As such, these MC codes can be further developed as an independent tool for verification of TPS in clinical use.

References

1. K. Parodi, A. Ferrari, F. Sommerer, H. Paganetti, Phys. Med. Biol., **52**, 3369-87 (2007)
2. H. Paganetti, H. Jiang, K. Parodi, R. Slopsem, M. Engelsman, Phys. Med. Biol., **53**, 4825–53 (2008)
3. T. Akagi, T. Aso, G. Iwai, A. Kimura, S. Kameoka, S. B. Lee, Y. Maeda, N. Matsufuji, T. Nishio, C. Omachi, S. Park, T. Sasaki, T. Toshito, W. Takase, T. Yamashita, Y. Watase, Prog. Nucl. Sci. Technol., **4**, 896-900 (2014)
4. A. E. Torshabi, A. Terakawa, K. Ishii, H. Yamazaki, S. Matsuyama, Y. Kikuchi, M. Nakhostin, H. Sabet, A. Ishizaki, W. Yamashita, T. Togashi, J. Arikawa, H. Akiyama, K. Koyata, World Acad. Sci. Eng. Technol., **3**, 1124-29 (2009)
5. G. O. Sawakuchi, D. Mirkovic, L. A. Perles, N. Sahoo, X. R. Zhu, G. Ciangaru, K. Suzuki, M. T. Gillin, R. Mohan, U. Titt, Med. Phys., **37**, 4960-70 (2010)
6. J. M. Verburg, H. A. Shih, J. Seco, Phys. Med. Biol., **57**, 5459-72 (2012)
7. T. Sato, K. Nitta, N. Matsuda, S. Hashimoto, Y. Iwamoto, S. Noda, T. Ogawa, H. Iwase, H. Nakashima, T. Fukahori, K. Okumura, T. Kai, S. Chiba, T. Furuta, L. Sihver, JNST, **50**, 913-23 (2013)
8. Z. Yang, C. Chen, D. Mah, R. Sheu, Int. J. Particle. Ther., **2**, 298, (2015)
9. E. Cuccoli, A. Ferrari, G.C. Panini, JEF-DOC-340, **91** (1991)
10. C. Z. Jarlskog, H. Paganetti, IEEE Trans. Nucl. Sci., **55**, 1018-25 (2008)
11. L. Grevillot, T. Frisson, N. Zahra, D. Bertrand, F. Stichelbaut, N. Freud, D. Sarrutet, Nucl. Instr. Meth. Phys. Res. B, **268**, 3295-3305 (2010)
12. P.F. Rose, *Report BNL-NCS-17541* (ENDF-201), (USA, 1991)
13. M. B. Chadwick, P. G. Young, P. Moller, G. M. Hale, A. J. Koningt, Satoshi Chiba, *Los Alamos National Laboratory report LA-UR-99-1222* (1999).
14. D. B. Pelowitz et al., MCNPX™ User's manual Version 2.6.0 (2008)
15. D. B. Pelowitz et al., MCNP6™ User's manual Version 1.0 (2013)
16. https://web-docs.gsi.de/ weick/atima/
17. G. R. Lynch, O. I. Dahl, Nucl. Instr. Meth. Phys. Res. B, **58**, 6-10 (1991).
18. A. Boudard, J. Cugnon, J. -C. David, S. Leray, D. Mancusi, Phys. Rev. C, **87**, 014606 (2013)
19. Shibata, Keiichi, et al. JNST, **48**, 1-30 (2011).
20. H. Hirayama, Y. Namito, A. F. Bielajew, S. J. Wilderman, W. R. Nelsonet, *SLAC-R-730* (2005) and *KEK Report 2005-8* (2005)
21. X. Zhang, W. Liu, Y. Li, X. Li, M. Quan, R. Mohan, A. Anand, N. Sahoo, M. Gillin, X. R. Zhu, Phys. Med. Biol., **56**, 7725-35 (2011)
22. S. Dhanesar, N. Sahoo, M. Kerr, MB. Taylor, P. Summers, X. R. Zhu, F. Poenisch, M. Gillin, Med. Phys., **40**, 092102 (2013)
23. B. Arjomandy, N. Sahoo, X. Ding, M. Gillin, Gillin, Med. Phys., **35**, 3889-94 (2008)
24. H. Bethe, AdP, **397**, 325-400 (1930)
25. F. Bloch, Z. Phys., **81**, 363-76 (1933)
26. F. Bloch, AdP, **408**, 285-320, (1933)
27. R. M. Sternheimer, S. M. Seltzer, M. J. Berger, Phys. Rev. B, **26**, 6067-76 (1982)
28. ICRU, *ICRU Rep. No. 37*, Bethesda, MD (1984)
29. D. A. Low, W. B. Harms, S. Mutic, and J. A. Purdy, Med. Phys. **25**, 656-61, (1998).
30. D. A. Low and J. F. Dempsey, Med. Phys. **30**, 2455-64, (2003).
31. ICRU, *ICRU Rep. No. 42*, Bethesda, MD (1987)

VARIAN CLINAC 6 MeV Photon Spectra Unfolding using a Monte Carlo Meshed Model

S. Morató[1,a], B.Juste[1] , R.Miró[1] and G. Verdú[1]

[1]*ISIRYM, Institute for Industrial, Radiophysical and Environmental Safety. Universitat Politècnica de València,. Camí de Vera s/n 46022 Valencia.*

Abstract. Energy spectrum is the best descriptive function to determine photon beam quality of a Medical Linear Accelerator (*LinAc*). The use of realistic photon spectra in Monte Carlo simulations has a great importance to obtain precise dose calculations in Radiotherapy Treatment Planning (RTP). Reconstruction of photon spectra emitted by medical accelerators from measured depth dose distributions in a water cube is an important tool for commissioning a Monte Carlo treatment planning system. Regarding this, the reconstruction problem is an inverse radiation transport function which is ill conditioned and its solution may become unstable due to small perturbations in the input data. This paper presents a more stable spectral reconstruction method which can be used to provide an independent confirmation of source models for a given machine without any prior knowledge of the spectral distribution. Monte Carlo models used in this work are built with unstructured meshes to simulate with realism the linear accelerator head geometry.

1 Introduction

Detailed knowledge of the spectrum emitted by the linear accelerator is essential, due to the fact that dosimetric factors depend directly on the beam energy. The principal objective of accurate dose calculation is delivering a prescribed dose to a target volume while minimizing radiation damage to the surrounding healthy tissues.

However, the energy spectrum of a *Bremsstrahlung* photon beam used in cancer treatments is difficult to be characterized by direct measurements because the emitted photon flux is very high and it produces the detector`s saturation [1].

For this reason, a methodology which guarantees a realistic shape of the spectrum emitted by *LinAc* is necessary. This work presents a robust and practical procedure to obtain reconstructed spectra by using depth dose measurements in a water phantom and Monte Carlo simulations [2].

A first order *Fredholm* integral equation allows to relate the depth dose curve measurements in the water phantom`s central axis with the *Bremsstrahlung* spectrum that forms it, but the solution to this equation is not simple due to the ill-conditioned nature of the mathematical problem.

The work is focused on solving the inverse problem based on reconstructing a spectrum from its experimental depth dose curve. The reconstructed spectrum will be used as source spectrum in a Monte Carlo simulation to compare depth dose curves between experimental measures and simulated results, validating the performed procedure.

It is here where it becomes important to have a detailed, accurate and realistic Monte Carlo model of the linear accelerator head to run a simulation which takes into account each material and geometric measures according to manufacturer data.

To that, this study has used the last version of *MCNP6.1.1* Monte Carlo code to simulate the LinAc head and water phantom irradiation. MCNP v.6 is the first Monte Carlo code which allows introducing meshed geometries into simulations [3].

The use of meshed geometries to generate the three-dimensional models presents the advantage of a higher accuracy in the geometry modelling, which is reflected in a precision improvement of the absorbed dose estimation.

According to this, the developed procedure presents an alternative method to calculate a linear accelerator photon beam spectrum based on indirect measurements. Specifically, the study has been focused on the reconstruction of a 6 MeV primary photon beam issued by the medical linear accelerator *Varian Clinac* available at the *Hospital Universitari i Politècnic La Fe, Valencia, Spain.*

[a] Corresponding author: smorato@iqn.upv.es

2 Methods and Materials

2.1 Spectrum reconstruction basic theory

The unfolding methodology used in this work for spectra reconstruction was developed and applied in previous works obtaining accurate results in other LinAc machines and for different energy beams [4].

2.1.1 Linear Problem and Discretization

The photon attenuation from the treatment beam is governed by an exponential law whose attenuation coefficients vary with the energy of these photons, therefore the first order equation of *Fredholm* can relate the energy deposition to different depths of photons in the water phantom. This equation can be expressed as:

$$d(x) = \int_0^{E_{max}} r_E(x)\ \chi(E)dE \qquad (1)$$

where *d(x)*, is the depth dose, *x* is the water phantom depth, $\chi(E)$ the incident energy spectrum and $r_E(x)$ a response matrix that contains simulated depth dose curves for different consecutive mono-energetic beams.

If the preceding equation is inverted, it makes it possible to calculate the spectrum which produces the depth dose curve. Nevertheless, the inverse matrix problem is ill-conditioned due to the nature of the problem. For this reason, to solve the equation it is necessary to discretize the problem into different intervals of depth positions.

$$d_i = \int_{x_{i-1}}^{x_i} d(x)dx, \quad i = 1, \dots, M \qquad (2)$$

The energy is discretized as well in different energy bins of the spectrum $\chi(E)$, where χ_j is the photon spectrum with energy between E_{j-1} and E_j.

$$\chi_j = \int_{E_{j-1}}^{E_j} \chi(E)dE, \quad j = 1, \dots, N \qquad (3)$$

On the other hand, $r_E(x)$ can be approximated by a M x N matrix, *A*:

$$A_{ij} = \int_{x_{i-1}}^{x_i} \int_{E_{j-1}}^{E_j} r_E(x)\ dE\ dx \qquad (4)$$

where A_{ij} is the deposited dose in a position *i*, for a given energy spectrum *j*.

Thus the equation becomes as follows:

$$d_i = \sum_{j=1}^{N} A_{ij}\,\chi_j\ , \quad i = 1, \dots, M \qquad (5)$$

where $\vec{\chi} = (\chi_1, \dots, \chi_N)$ is the original unknown beam and $\vec{d} = (d_1, \dots, d_M)^T$ is the depth dose curve which has been obtained experimentally at the Hospital Universitari i Politècnic La Fe .

To obtain the A_{ij} matrix (M × N) sequential Monte Carlo (MCNP6.1.1) simulations were executed acquiring depth dose curves for different mono-energetic incident beams with different energies. These doses have been calculated as a sum of the dose due to electrons and the dose due to photons.

Figure 1 shows the simulated results which configure the response matrix. This response matrix (dim. 300x350) with energy range between 0-7 MeV (energy bin 20 KeV) and the depth stepped by millimeter were calculated by simulation taking the contribution of photons and electrons.

A small line of noise can be seen in this figure, it is probably due to the fact that these values have been obtained from a simulation using low number of simulated particles and this produces statistic errors higher than rest. This fact had no relevance or significant influence in the calculation of the reconstructed spectrum.

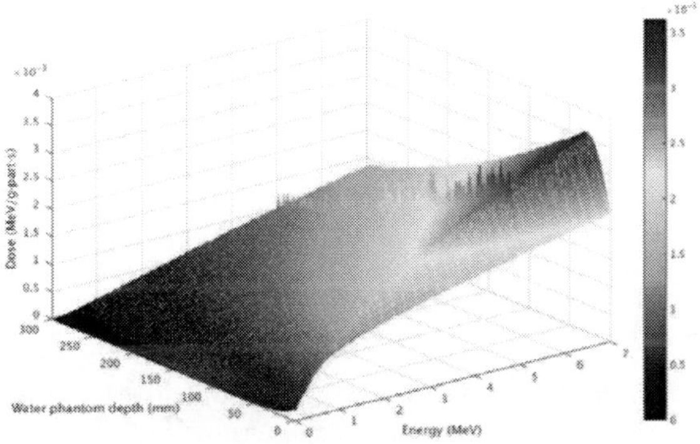

Fig. 1. Response matrix with energy range between 0 - 7 MeV, A ϵ $R_{M\times N}$

For the 6 MeV energy range we are considering, depth dose curves from mono-energetic spectra were calculated using the MCNP6 Monte Carlo code. The Source Surface Distance (SSD) from a point source on the central axis to a homogeneous water phantom has been 100 cm. The used field size has been 10 cm × 10 cm. Each Monte Carlo simulation was run until the uncertainty in all evaluated points of computed depth dose curves was less than 0.3%. To register the contribution of photons and electrons, the FMESH tally has been used with its respective flux-to dose conversion factors, implemented as the DE/DF target in MCNP.

Once the response matrix is obtained, the mathematical procedure is applied to reconstruct spectrum. The discretization of the problem offers a linear system equation matrix:

$$\begin{aligned} d_1 &= a_1 \cdot x_1 + a_2 \cdot x_2 + \ldots + a_M \cdot x_M \\ d_2 &= b_1 \cdot x_1 + b_2 \cdot x_2 + \ldots + b_M \cdot x_M \\ &\;\;\vdots \\ d_N &= e_1 \cdot x_1 + e_2 \cdot x_2 + \ldots + e_M \cdot x_M \end{aligned} \quad (6)$$

where $A \in R_{M\times N}$, $\chi \in R_M$ and $d \in R_N$.

The original photon spectrum has been defined here (Eq. 6) as the solution vector χ.

2.1.2 Singular Value Decomposition Method (SVD)

In this case, the matrix equation $A \cdot \chi = d$ allows to obtain the original spectrum if the response matrix is known but several error affect the spectrum´s determination as explained above and it can´t be obtained directly, hence we use the singular value decomposition method (*SVD*):

$$A = \sum_{i=1}^{n} u_i \sigma_i {v_i}^T \quad (7)$$

where u_i and v_i are the orthonormal singular vectors and σ_i the singular values. The same equation can be expressed by matrix form:

$$A = U\Sigma V^T \quad (8)$$

where Σ is a diagonal matrix with its singular values σ_i in the diagonal and the rest zeros, while U and V are the matrix of the singular vectors.

The SVD algorithm works very well for noise-free data. For data affected by noise, direct SVD reconstruction is extremely inaccurate but there are several regularization schemes that improve its performance.

Tikhonov methodology has been used to solve this matrix system [5], [6]. With this numeric algorithm, an optimum solution χ can be obtained by generating a new response matrix, removing the parts of the solution corresponding to the smallest singular values.

Tikhonov regularization is the most common and well-known form of numerical stabilization. This methodology corresponds to making the normal system of the least-squares problem $A^T Ax = A^T d$ better conditioned.

The idea is to define the general regularization solution x_λ as the minimum of the following weighted combination of residual norms and constraint:

$$x_\lambda = \min\left(\|A \cdot x - d\|_2^2 + \lambda^2 \|L(x - x^*)\|_2^2\right) \quad (9)$$

where L is the discrete derivative operator and x^* is the residual solution. L is a general matrix with column rank complete, and which is usually related to some derivation operator. The classic standard form of Tikhonov regularization considers that L is the identity matrix, so the solution is:

$$x_\lambda = \min\left(\|A \cdot x - d\|_2^2 + \lambda^2 \|x\|_2^2\right) \quad (10)$$

Hence, the equation is equivalent to solving the linear system:

$$\left(A^T A + \lambda^2 I\right) x_\lambda = A^T d \quad (11)$$

where the regularization parameter λ controls the weight granted to the minimization of the restriction of the part relative to the minimization of the residual norm. Clearly, a high λ (equivalent to a large amount of regularization) favors a small semi-norm solution and a large residual norm, while a small λ (i.e. a small amount of regularization) has the opposite effect.

Thus, the regularization parameter is an important parameter that controls the properties of the regularized solution and therefore should be carefully selected.

One way to determine the optimal regularization parameter, λ, is to plot the L curve, which in an ideal environment has a relatively vertical segment and a relatively horizontal line. For this curve, the norm-2 of the solution vector is drawn, $\|x\|_2$ (ordinate), against the norm-2 of the residual vector, $\|A \cdot x - d\|_2$ (abscissa) for different values λ. As it has been explained, it is recommended to place the truncation parameter λ, at the value closest to the point of maximum curvature of this curve, that is, by selecting the value corresponding to the corner that conforms to *L*.

2.2 Monte Carlo Simulation

The *Varian CLINAC* head was modeled in detail according to the manufacturer's specification and each element is assumed to have the typical natural abundance of different nuclides.

The *LinAc* of the present study is the *Varian CLINAC* 6 MeV photon beam configuration installed at the *Hospital Universitari i Politècnic La Fe* which is equipped with an adjustable jaw photon collimator as well as with an multi leaf collimator (*MLC*) with 120 leafs. The accelerator geometry has been accurately modeled as shown in *figure* 2.

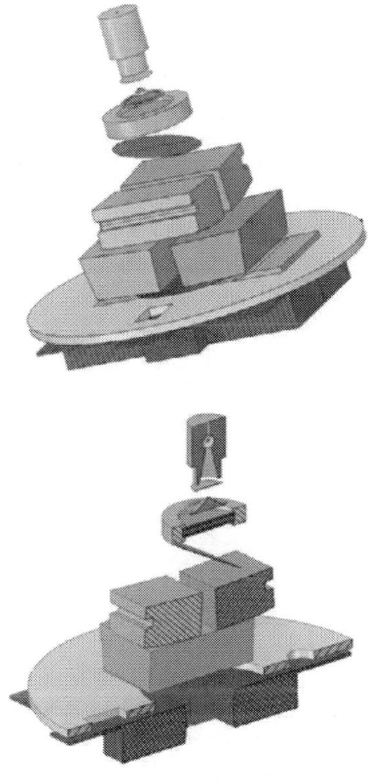

Fig. 2. Three-dimensional model of the *Varian CLINAC*

The model has been generated using 3D Modeling Software for Engineering *ANSYS SpaceClaim* creating quickly and easily the solid model *CAD*.

Later, the 3D solid model was meshed with *Abaqus*/CAE [7], which is a software used for finite element modeling and analysis of mechanical components and assemblies. It has been used to define the meshed geometry that will be introduced in the *MCNP6.1.1* simulation.

Meshed geometry modeling tool is a novelty of MCNP version 6.1.1, (older MCNP versions and other Monte Carlo codes do not present this capability) and it is an important advantage when modelling complex geometries with precision.

Fig. 3. *Varian CLINAC meshed with Abaqus/CAE*

The use of meshed geometries, presents the advantage of a higher accuracy in the geometry modelling which is reflected in an increased absorbed dose estimation precision. Unstructured meshes enable the use of cells of different types and/or different size, allowing complex geometries and optimizing the number of cells used according to the needs. That is why, for the present work, we have used unstructured grids imported from *Abaqus/CAE*.

Finally, the *Varian CLINAC* simulation has been performed using the reconstructed spectrum as a source.

The simulation was divided in two parts in order to speed up the calculations. The first simulation corresponds to the *Varian CLINAC* head part and the particles are stored in a surface source (*phase space*) located behind the *MLC*. This phase space was generated by MCNP6.1.1 simulation. The second simulation corresponds to the water phantom and starts from the previous surface source generated with more than 10^7 collected particles. [8]

To speed up the calculations without compromising the accuracy of the results, the MCNP6 (version 6.1.1) code has been parallelized in SENUBIO ISIRYM research group´s cluster Quasar using the MPI protocol with 26 processors. Finally, the computational real time real time was about 48 hours.

Concerning radiation transport, a detailed physics treatment for a coupled photon and electron mode simulation has been considered, taking into account the following physical processes: photoelectric effect with fluorescence production, Compton and Thomson scattering, and pair production in the energy range between 0.001 and 7 MeV.

Figure 4 shows the photon flux obtained by simulation passing through the whole Varian CLinac head.

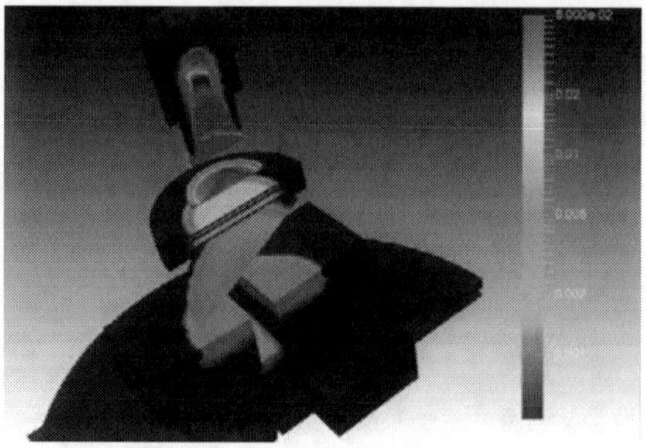

Fig. 4. MCNP6 flux results visualization with software *ParaView*.

An important mesh advantage is the easy results visualization. The results supplied in ASCII by means of huge matrix can be converted easily using some code tools or *FORTRAN* routines to obtain a compatible three-dimensional visualization format that can be opened with most of 3D visualization software such as *Paraview* or *Visit* [9]. By this way, an interactive and complete three-dimensional view is achieved and dose or flux results can be shown quickly and easily in any point of the simulated geometries.

Figure 5 shows 3D dose distribution in a water phantom and isodose curves obtained by MCNP6 simulation of a 6 MeV treatment beam transport.

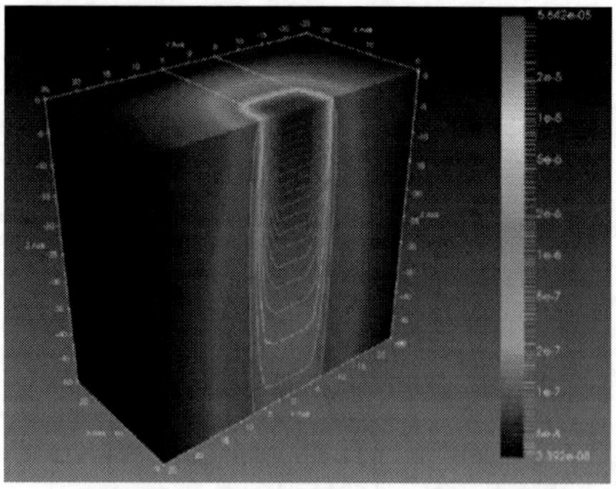

Fig. 5. MCNP6 water phantom dose and its isodose curves visualized with Paraview.

2.3 Experimental procedure

In order to validate the Monte Carlo simulation and the unfolded spectrum, experimental measures were obtained using a water phantom to determine the real depth dose curve and compare it with the simulated depth dose curve.

These experimental measures were taken at the *Hospital Universitari i Politècnic La Fe* in Valencia and were performed with a high resolution detector (registering the dose contribution of both electrons and photons) placed in a motorized guide inside the 50 cm side phantom ("*PTW 31010 Semiflex*").

The semiflex chambers are designed for therapy dosimetry, mainly for dose distribution measurements in motorized water phantoms. They have a short stem for mounting and a flexible connection cable. The nominal useful energy range is from 140 kV to 50 MV photons and 10 ... 45 MeV electrons.

The wall material is graphite with a protective acrylic cover. The guard rings are designed up to the measuring volume. Air density correction is required for each measurement, and a radioactive check device is available as an option.

Chambers are shaped cylindrically with an inner diameter of 5.5 mm. The 0.125 cm^3 chamber is ideal for 3D dosimetry in a water phantom, since the measuring volume is approximately spherical resulting in a flat angular response over an angle of ± 160° and a uniform spatial resolution along all three axes of a water phantom.

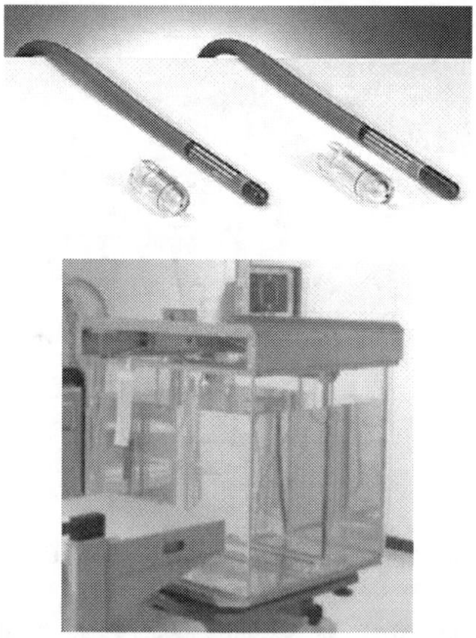

Fig. 6. Semiflex Chambers 31010 and 31013 and water phantom.

The water cube was irradiated with a 10 cm x 10 cm field size using a 6 MeV photon beam maintaining the source to surface distance (SSD) equal to 100 cm, source to collimator distance (SCD) equal to 37.9 cm, collimator angle was zero degrees, temperature was 20 degrees Celsius and 1013.25 mbar of pressure.

Depth dose curves and profile curves can be measured accurately in this phantom because precise positioning of high resolution detectors can be easily accomplished using a guide driven by reinforced toothed belts.

Only have been used depth dose curves to validate the spectrum and profile curves have not been used because the curves obtained by mono-energetic beams used in the unfolding process are depth dose curves too.

3 Results

The reconstructed photon spectra obtained using unfolding Tikhonov method for 6 MeV photon beams is shown in *Figure 7*.

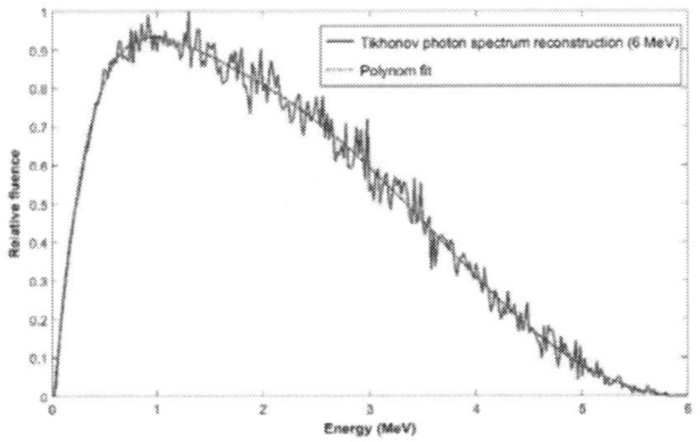

Fig. 7. Reconstructed 6 MeV spectrum obtained with Tikhonov method

The figure above represents the reconstructed 6 MeV spectrum and its respective polynomial fit. It has been calculated according deconvolution procedure explained. The method is based on a previously simulated response matrix, as well as an experimental depth dose curve.

The reconstructed spectrum has been validated comparing depth dose curves obtained by Monte Carlo simulations including the Tikhonov reconstructed spectrum as a source with experimental depth dose curves measured at the hospital.

This comparison is shown in *Figure 8*; it shows that simulated dose curve fits accurately the measured depth dose curve, being the average relative error is 1.89 %.

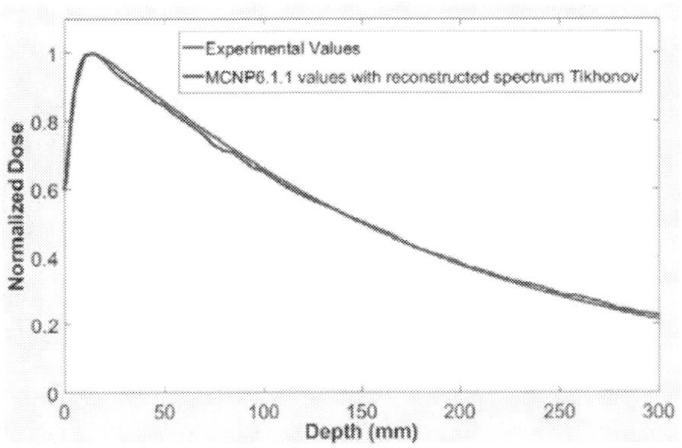

Fig. 8. Depth dose curve obtained with simulation of 6 MeV photon spectrum by Tikhonov method

Experimental values have been compared with the results of the MCNP simulation using as a source the Skeikh-Bagheri spectrum, spectrum commonly publicized and used in several works [10]. It can be appreciated in figure 9, that results are worst, the peak is not coincident, and the dose curve using Sheikh-Bagheri Spectrum is below the experimental spectrum.

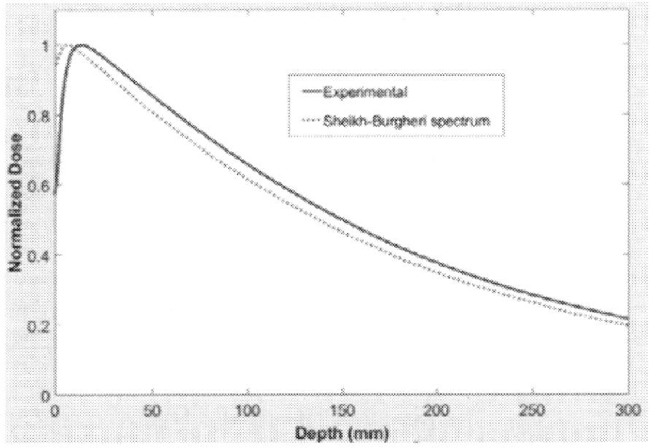

Fig. 9. Depth dose curve obtained with simulation of 6 MeV photon spectrum from Skeikh-Bagheri by Tikhonov method

The average relative error in this case is 7.1, higher than the one obtained using the spectrum reconstructed with the presented unfolding technique.

The importance of obtaining a reconstructed spectrum and not using a theoretical spectrum extracted from the literature is because each radiotherapy unit will have slight variations in its spectrum due to minor modifications in the installation of its components and the life of the machine. Thus, if we want a real calculation of the dose given to the patient, we will need a precise calculation of the real spectrum used in the treatment.

4 Conclusions

The procedure of this study is an important approach that allows characterizing the primary photon spectrum emitted by a LinAc. A computer model of a *Varian CLINAC* head has been generated to be used by *MCNP6.1.1*. The model was used to calculate the depth dose curves generated by reconstructed spectra and validate them by comparison with experimental depth curves.

This paper also presents the innovation of using unstructured mesh geometries in Monte Carlo calculations which means a direct impact on calculations because it allows to define the geometry with more accuracy.

Obtained results prove that the methodology described estimates accurately the beam energy spectrum emitted by LinAc. The validation was performed comparing depth dose curve in a water tank between simulated and measurements.

Acknowledgment

The authors wish to thank the *Hospital Universitari i Politècnic, La Fe de Valencia (Spain)* for their kind support in doing this work and *Varian Medical Systems* for providing us with all the necessary information.

References

[1] P.M. Charland, L. D. Paniak, "*Enhanced bremsstrahlung spectrum reconstruction from depth-dose gradients*". Physics in Medicine & Biology,14, pp. 3245-3262, (2005)

[2] Gui LI, Huaqing ZHENG, Guangyao SUN and Yican WU, "Photon Energy Spectrum Reconstruction Based on Monte Carlo and Measured Percentage Depth Dose in Accurate Radiotherapy", Progress in NUCLEAR SCIENCE and TECHNOLOGY, Vol. 2, pp.160-164 (2011)

[3] Roger L. Martz, "*The MCNP6 Book On Unstructured Mesh Geometry: User's Guide*", LA-UR- 11-05668 Rev. 8, MCNP6 code release to RSICC, Oak Ridge, TN and general distribution.

[4] B. Juste, R.Miró, G. Verdú, S.Díez, J.M. Campayo "*6 and 15 MeV photon spectra reconstruction using an unfolding depth dose gradient methodology*", Conf Proc IEEE Eng Med Biol Soc. 2011;2011:129-32.

[5] C. Hansen. "REGULARIZATION TOOLS*: A Matlab package for analysis and solution of discrete ill-posed problems*" Numerical Algorithms. Volume 6, Issue 1, pp 1-35. March 1994.

[6] D. Calvetti, S. Morigib, L. Reichel, F. Sgallarid, "*Tikhonov regularization and the L-curve for large discrete ill-posed problems*", Journal of Computational and Applied Mathematics, Volume 123, Issues 1–2, 1 November 2000, Pages 423–446.

[7] Dassault Systèmes, 3DExperience Company, "*ABAQUS 6.14, ABAQUS/CAE User's Guide", 2014.*

[8] B.Juste, R.Miró, V.Abella, A.Santos, G.Verdú, "*Use of MOSFET dosimeters to validate Monte Carlo radiation treatment calculation in an anthropomorphic phantom*", Radiation Physics and Chemistry Volume 116, November 2015, Pages 208–213

[9] Kenneth Moreland, "The ParaView Tutorial Version 4.2" Sandia National Laboratories

[10] Daryoush Sheikh-Bagheri, D. W. O. Rogers, "*Monte Carlo calculation of nine megavoltage photon beam spectra using the BEAM code*" Medical Physics 29, 391 (2002); doi: 10.1118/1.1445413

Re-Shielding of Cobalt-60 Teletherapy Rooms for Tomotherapy and Conventional Linear Accelerators using Monte Carlo Simulations

Yiğit Çeçen[1,2,a] , Çağrı Yazgan[1,2]
[1] *Akdeniz University, Nuclear Sciences Application and Research Center, 07058, Antalya, Turkey*
[2] *Akdeniz University, School of Medicine, Department of Radiation Oncology, 07058, Antalya, Turkey*

Abstract

Purpose. Nearly all Cobalt-60 teletherapy machines were removed around the world during the last two decades. The remaining ones are being used for experimental purposes. However, the rooms of these teletherapy machines are valuable because of lack of space in radiotherapy clinics. In order to place a new technology treatment machine in one of these rooms, one should re-shield the room since it was designed only for 1.25 MeV gamma beams on average. Mostly, the vendor of the new machine constructs the new shielding of the room using their experience. However, every radiotherapy room has different surrounding work areas and it would be wise to shield the room considering these special conditions. Also, the shield design goal of the clinic may be much lower than the International Atomic Energy Agency (IAEA) or the local association accepts. The study shows re-shielding of a Cobalt-60 room, specific to the clinic, using Monte Carlo simulations.

Materials & Methods. First, a 6 MV Tomotherapy machine, then a 10 MV conventional linear accelerator (LINAC) was placed inside the Cobalt-60 teletherapy room. The photon flux outside the room was simulated using Monte Carlo N-Particle (MCNP6.1) code before and after re-shielding.

For the Tomotherapy simulation, flux distributions around the machine were obtained from the vendor and implemented as the source of the model. The LINAC model was more generic with the 10 MeV electron source, the tungsten target, first and secondary collimators. The aim of the model was to obtain the maximum (40x40 cm^2) open field at the isocenter. Two different simulations were carried out for gantry angles 90° and 270°. The LINAC was placed in the room such that the primary walls were A' (Gantry 270°) and C' (Gantry 90°) (figure 1).

The second part of the study was to model the re-shielding of the room for Tomotherapy and for the conventional LINAC, separately. The aim was to investigate the recommended shielding by the vendors. Left side of the room was adjacent to a LINAC room with 2 meters thick concrete wall (figure 1). No shielding was necessary for that wall. Behind wall A-A' there was an outdoors forbidden area; behind wall B-B' was the contouring room for the doctors; and the control room was behind wall C-C' (figure 1). After some modifications, the final shielding was designed.

Results. The photon flux distributions outside the room before and after the re-shielding were compared. The re-shielding of Tomotherapy reduced the flux down to 1.89 % on average with respect to pre-shielding (table 1). For the conventional LINAC case; after re-shielding, the photon flux in the control room -which corresponds to gantry 90°- decreased down to 0.57% with respect to pre-shielding (table 2). The photon flux behind wall A' -which corresponds to gantry 270°- decreased down to 2.46%. Everybody was all safe behind wall B' even before re-shielding.

[a] Corresponding author: ycecen@yahoo.com

Table 1. Highest photon fluxes behind the walls for the Tomotherapy model before and after re-shielding.

Wall	Relative Photon Flux Before re-shielding	Relative Error (%)	Relative Photon Flux After re-shielding	Relative Error (%)	Decrease in photon flux (%)
A	2.71E-06	0.24	5.74E-08	0.42	97.88
B	6.94E-06	0.31	1.11E-07	0.29	98.41
C	2.86E-06	0.67	5.11E-08	0.37	98.21

Table 2. Highest photon fluxes behind the walls for the conventional LINAC model before and after re-shielding.

Wall	Relative Photon Flux Before re-shielding	Relative Error (%)	Relative Photon Flux After re-shielding	Relative Error (%)	Decrease in photon flux (%)
A' (G270^0)	1,81E-05	0.54	4,45E-07	0.34	97,54
B'	0	0	0	0	0
C' (G90^0)	2,31E-05	0.59	1,31E-07	0.29	99,43

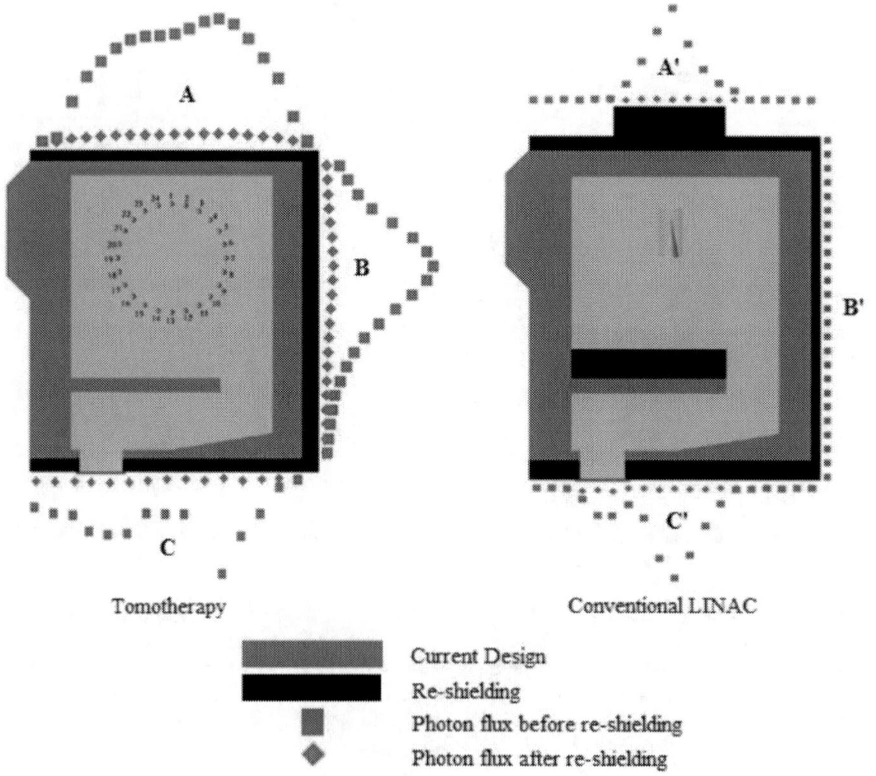

Figure 1. The photon flux distributions outside the Cobalt-60 room before and after re-shielding.

DOI: 10.1051/epjconf/201715304013

Conclusion. This study assumes that flux and dose relation is linear and 1 Gy dose corresponds to the flux at the isocenter for the LINAC case. For the Tomotherapy case, the dose distributions were given along with the flux distributions by the vendor. The fluxes calculated by the Monte Carlo simulations should give a hint for the instantaneous dose rates (IDR). More work is needed to calibrate the findings with the design goals and achieve correct comparison for the dose behind the walls.

In order to place new technology treatment machines inside old Cobalt-60 teletherapy rooms which were designed to shield only 1.25 MeV gamma radiation, it is necessary to design radiation shielding for high energy photons. However, the re-shielding should not solely be left to the vendor's experience. The re-shielding calculations should be made according to International-National Committee on Radiation Protection (ICRP-NCRP) protocols and, if possible, Monte Carlo simulations should be used specifically to meet the clinics needs.

Development and application of a set of mesh-based and age-dependent Chinese family phantoms for radiation protection dosimetry: Preliminary Data for external photon beams

Yifei Pi[1], *Lian* Zhang[1], *Wanli* Huo[1], *Mang* Feng[1], *Zhi* Chen[1] and *X. George* Xu[1,2,*]

[1]University of Science and Technology of China, Hefei, China
[2]Rensselaer Polytechnic Institute, Troy, New York, USA

Abstract. A group of mesh-based and age-dependent family phantoms for Chinese populations were developed in this study. We implemented a method for deforming original RPI-AM and RPI-AF models into phantoms of different ages: 5, 10 ,15 and adult. More than 120 organs for each model were processed to match with the values of the Chinese reference parameters within 0.5%. All of these phantoms were then converted to voxel format for Monte Carlo simulations. Dose coefficients for adult models were counted to compare with those of RPI-AM and RPI-AF. The results show that there are significant differences between absorbed doses of RPI phantoms and these of our adult phantoms at low energies. Comparisons for the dose coefficients among different ages and genders were also made. it was found that teenagers receive more radiation doses than adults under the same irradiation condition. This set of phantoms can be utilized to estimate dosimetry for Chinese population for radiation protection, medical imaging, and radiotherapy.

1 Introduction

A main and indispensable task in radiation protection, medical imaging, and radiotherapy is to estimate organ doses. Computational phantoms, especially whole body computational phantom for estimations of dosimetry have become widely used [1,2]. Such phantoms are typically based on reference anatomical values, for the population of interests. The ICRP firstly used so called "Reference man" to define them in the ICRP Publication 23 in 1975. Since the 1960s, three categories of computational phantoms have appeared, including stylized phantoms, voxel phantoms, and boundary representation (BREP) phantoms [1]. Compared with stylized phantoms, voxel phantoms utilize huge building blocks and mix them together. Each block has their own density data and material information based on medical images such as CT or MRI. Further, variability in shape and mess define introduced by motion of human body can be a significant factor affecting organ dose. However, changing phantoms in the voxel format is extremely inefficient. BREP phantoms, in the form of either Non-Uniform Rational B-Spline (NURBS) or polygonal meshes, possess attractive features in flexibility and anatomical realism [3,4,5]. China is rapidly developing its nuclear power industry and its radiation protection system follows those recommended by the ICRP. However, the ICRP Computational Phantoms are based on western populations and are not representative of the Chinese. Most reported Chinese phantoms are designed for Chinese adults (CNMAN、VCH and CVP) [6,7,8], there are currently no phantoms for Chinese minors. This paper reports our preliminary efforts in developing a set of mesh-based and age-dependent Chinese family phantoms for the purposes of deriving radiation protection dosimetry data.

2 Materials and methods

Figure 1 illustrates steps of the developing process. The process starts with the collection of initial input data including human dimensions of Chinese adults and minors [9,10], reference organ volume information and the RPI-AM and RPI-AF mesh phantoms. The deforming process was divided into two parts: human dimensions scaling and organ volume deformation. The first part was finished by some scripts in Matlab form automatically and the organ volume deformation step was done manually through Rhinoceros, a CAD software. We utilized an in-house C# based voxelization tool [11] to convert the meshed-based age-dependent phantoms into voxel formatted phantoms since polygonal mesh cannot be able to handle straight forward by MCNPX [12]. The organ doses and effective doses of our USTC adult phantoms are compared with those from RPI-AM and RPI-AF with six classic source directions (AP, PA, LLAT, RLAT, ROT, ISO) at energies from 20keV to 10 MeV [13]. This paper also reports our attempts on comparisons of the dose coefficient for phantoms with different ages and genders.

* Corresponding author: xgxu@ustc.edu.cn

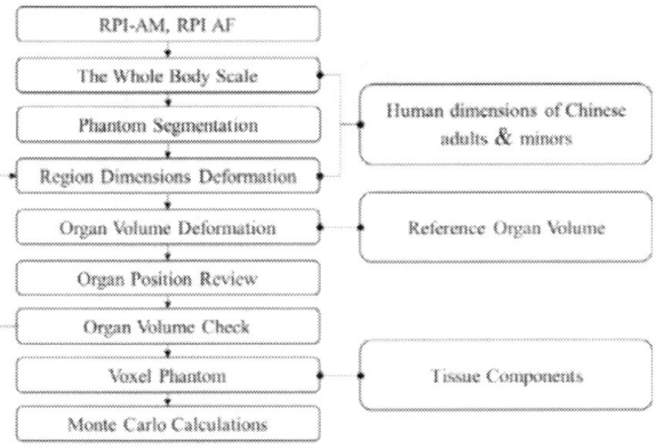

Figure 1. Flowchart of construction of mesh-based and age-dependent family phantoms

2.1 Original models and reference parameters

RPI-AM and RPI-AF, which were developed at RPI, are a pair of adult male and adult female computational phantoms and are compatible with anatomical parameters for the 50th percentile population as specified by the ICRP publication 89 [14]. Up to 140 inner organs or tissues from original 500 available structures were filtrated and designed to represent the whole human body. The organ volume were automatically adjusted to match the reference value. And they were all designed entirely with polygonal mesh surface that affords the ability to sufficiently deform the dimensions of phantoms and the shape of organs. The pair of phantoms are fully validated for their accuracy [4]. Furthermore, researchers in RPI have resoundingly extended the original reference phantoms to represent different populations and obese patients [11,15]. For all these reasons, we selected RPI–AM and RPI-AF as our initial phantoms.

Corresponding to ICRP reports, Chinese government issued reference individuals for radiation protection, a national standard to standardize Chinese "Reference Man" [10], whose organ size and mass information are contained. Some important data are listed in Table 1. However, we could not find all data for every organ or tissue in published Chinsed standards. Organ contents and some radial insensitive tissue, such as muscles, vessel, are missed. We took notice of that the IAEA (International atomic energy agency) group published some anatomical, physiological and metabolic characteristics for "Reference Asian Man" with the age group as new-born, 1 year, 5 years, 10 years and adult from 20 to 50 as the ICRP done for white people [16,17,18]. Apparently, we could adopt these data for Chinese people as a member of them. Particularly, we made use of the ICRP proportion data of the bones to calculate the mass of the red bone marrow (RBM), yellow bone marrow (YBM), trabecular and cortical part for every part of phantom bones [19,20]. It was because that we cannot find enough useful information from Chinese standards and Asian reference data.

Table 1. Reference values (g) for masses of important organs/tissue for Chinese

organs/ tissue	5 years	10 years	15 years		20-50 years	
			male	female	male	female
Testicles(2)	3.1	4.7	33	-	40	-
Ovaries(2)	0.5	1.4	-	9.8	-	11
RBM	305	710	900	750	1100	800
Colon	102	170	291	225	310	240
Lungs	360	580	940	720	1250	960
Stomach	47	75	120	95	145	110
Urinary bladder	13	21	38	30	40	30
Mammary gland	-	38	-	200	-	300
Liver	575	850	1170	1050	1410	1290
Esophagus	13	25	30	28	40	30
Thyroid	3.4	7.9	12	12	20	17
Skin	775	1200	2200	1700	2400	1800
Bone surface	-	-	-	-	100	90

Table 2. Reference values (g) for masses of other organs/tissue for Chinese

organs/ tissue	5 years	10 years	15 years		20-50 years	
			male	female	male	female
Body fat	1900	5000	6500	9000	9000	12500
LBW	17100	27000	48500	41000	54000	41500
Skeleton	2200	4500	7300	5700	8000	6000
Brain	1200	1350	1480	1360	1460	1330
Heart	95	150	240	200	325	290
Kidney(2)	115	175	230	220	290	260
Spleen	70	100	140	120	165	150
Sialaden	26	45	77	59	82	62
Gall bladder	3	4	8	6	9	7
Small intestine	190	325	540	420	620	450
Pancreas	40	60	90	75	120	100
Eyes	13	14	15	12	15	12
Eye lens	0.35	0.35	0.4	0.35	0.4	0.35
Adrenal gland	5	6	10	10	14	13
Thymus	33	37	37	32	30	27
Pituitary gland	0.3	0.4	0.53	0.61	0.7	0.8
Body weight	19000	32000	55000	50000	63000	54000

Table 3. Reference values (g) for masses of organs/tissue for Asian Reference Man

organs/ tissue	5	5	10	10	15	15	Adult	Adult
	M	F	M	F	M	F	M	F
Colon contents	120	120	190	190	340	260	360	280
Stomach contents	80	80	130	130	230	170	240	180
Bladder contents	33	32	54	54	96	78	100	85
Heart contents	130	130	210	220	380	290	400	300
Gall bladder contents	16	16	27	29	47	36	50	38
Small intestine contents	190	190	190	190	330	260	350	270
Rectum	6	6	10	10	19	19	20	20
Prostate	1	-	1.5	-	11	-	12	-
Prostate contents	0.34	-	0.51	-	3.6	-	4	-
Tongue	64	50	67	51	67	51	67	51
Tonsil	1	0.98	1	1	4	3	4	3
Bronchus	8.5	8.3	14	15	24	18	26	20
Trachea	3.2	2.9	5	5	8	6	9	6.8
Ureter	5.1	4.9	8	8.6	13	14	14	15
Uterus	-	5.9	-	8.9	-	63	-	70
Spinal cord	20	20	30	30	30	30	30	30
Teeth	45	34	45	34	45	34	43	34

2.2 Phantom adjustment

To obtain a phantom with a given height, we developed a uniform scaling factor adjustment method for adult phantoms and different factors for different parts in teenagers' deformation. The growth rate is fairly slow after 18 years old for boys and 16 years old for girls. It is reasonable to deform phantoms among adults using uniform scaling [15,21]. But for adolescents, growth and development occurs more rapidly for arms, legs and the trunk. For that reason, we split the original models into 4 parts, the head part, the arms part, the trunk part and the leg part. Figure 2 illustrates our distributing standard. As is shown in the picture, the head part are those structures from top of the model to gnathion and organs or tissue in the section from gnathion to perineal constitute the trunk part. According to the length, width and heightof every part, we applied the uniform scaling factor adjustment method. Scale factors were calculated by using:

$$WFactor = {AIMWidth}/{Width}$$
$$HFactor = {AIMHeight}/{Height} \quad (1)$$
$$LFactor = {AIMLength}/{Length}$$

$$PointNew = (Point.x * WFactor, Point.y * HFactor, Point.z * LFactor) \quad (2)$$

where $AIMWidth$ is the reference width of our model part, $Width$ represents the actual width value. $AIMHeight$, $Height$, $AIMLength$ and $Length$ have similar significances. $Point.x$, $Point.y$ and $Point.z$ are coordinates of vectors on our original models. $PointNew$ is the newly calculated point.

After this step, we assembled all the parts together to build one whole body model. Some reparative attempts were finally took in order to smooth discrete commissures. All of these steps for scaling were completed automatically with some in-house MATLAB R2014B[22] scripts except for the reparative works. We finished human dimension adjustment automatically and obtained rough phantoms.

However, those elementary models could not be utilized because of their rough anatomic details. In order to match internal organ volume with the Chinese reference parameters, a 3D model software named Rhinoceros 5 was used. By carrying out mesh offset, mesh scale and box edit operations, we can change organs' volume and shape. Figure 3 displays the deformation methods. We manually took efforts to adjust organs parameters to agree within 0.5% with the reference people data of China. Finally, all of these regions were assembled together.

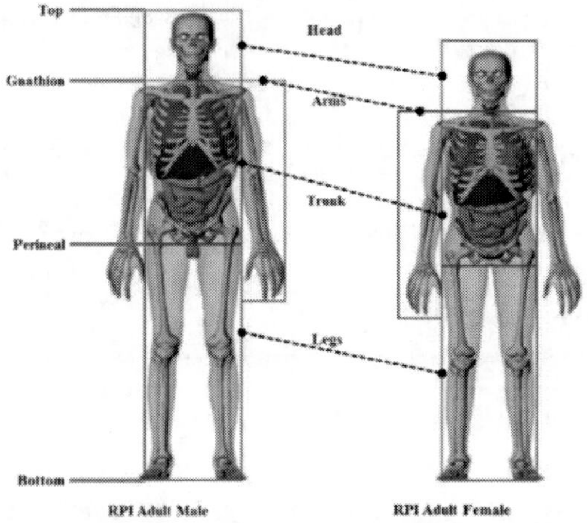

Figure 2. Different scale factors for different parts

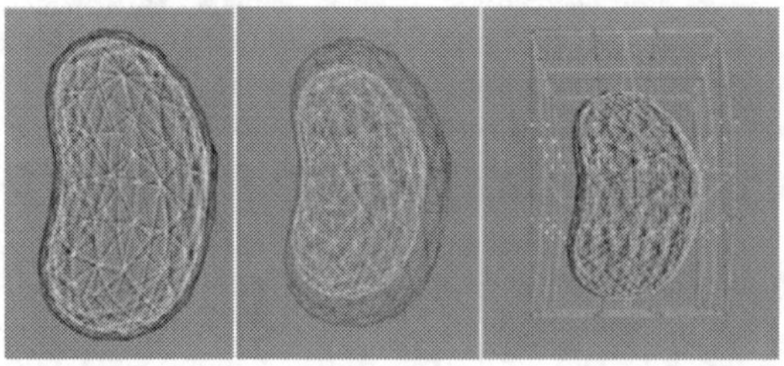

Figure 3. Mesh offset (left), mesh scale (middle) and box edit (right) operations in Rhinoceros

2.3 Tissue composition and Voxelization

The tissue element compositions were defined for the purposes of radiation transport simulations involving

Monte Carlo methods. Organ-specific element compositions were based on reference value of the Chinese standard and the ICRU report 46 [23,24]. A C++/C# based voxelization tool was used to convert the meshed-based age-dependent phantoms into voxel formatted phantoms. The organ volumes after the voxelization, were found to agree within 1% with the reference parameters.

2.4 Dose calculation

In this project, MCNPX code was used to calculate dose coefficients. We defined "photon and electron"mode in input files for photon simulations. The electron cross-section library el03 and photon cross-section library mcnplib04 were used. Monoenergetic and parallel photon beams from 6 normal source geometries: AP, PA, LLAT, RLAT, ROT and ISO, were took into consideration. We simulated incident energies covering 20 values from 20keV to 10MeV. The deposited energy for each organ was counted using the *F8 card. Three correction factor methods, noted by Lee et al [25] and applied by many other works [4,26], were adopted to compute the red bone marrow doses from the bone spongiosa doses. To keep statistical error less than 1%, up to 1e8 particles were used for every instance. All simulations were finished on a 24-core Linux server with 32GB RAM memory.

The stochastic efforts on the whole body could be evaluated by the effective dose, E, defined by the ICRP [27]. E is computed as weighted organ equivalent doses involving sex-averaged value according to:

$$E = \sum_T \omega_T \left[\frac{H_T^M + H_T^F}{2}\right] \quad (3)$$

where ω_T is the weighting factor for specified tissue or organ T defined in ICRP-103, $H_T^M and\, H_T^F$ are the equivalent dose of T in the male and female phantoms, respectively.

3 Results and discussion

3.1. Mesh-based and age-dependent family phantoms

Eight new mesh-based and age-dependent Chinese phantoms were developed: USTC-AM, USTC-AF, USTC-15M, USTC-15F, USTC-10M, USTC-10F, USTC-5M, USTC-5F. A total of 70 internal organs, 45 bone components and 4 muscle structures are contained in each single newly developed phantom. Figure 4 displays 3D rendering pictures of these adult phantoms. Figure 5 represents the geometries for teenager phantoms.

Table 4 and 5 illustrate the dimensions of the mesh-based and age-dependent Chinese phantoms. They agree well with the references parameters.

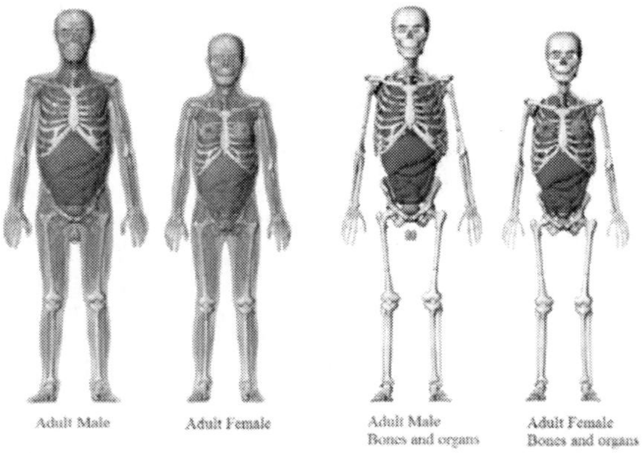

Figure 4. USTC - adult male and female mesh phantoms

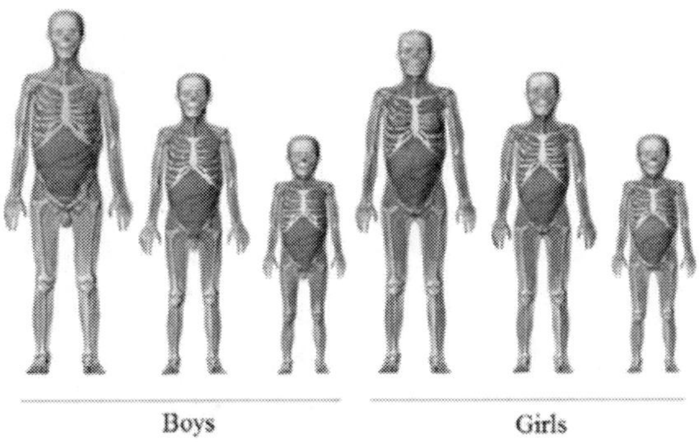

Figure 5. USTC – teenage mesh phantoms

Table 4. The value of stature for USTC phantoms.

	Real(cm)	Reference(cm)	Relative Error(%)
USTC-AM	170	170	0
USTC-AF	158	158	0
USTC-15M	168	168	0
USTC-15F	158	158	0
USTC-10M	139	139	0
USTC-10F	139	139	0
USTC-5M	110	110	0
USTC-5F	110	110	0

Table 5. The head height for USTC phantoms.

Name	Real(cm)	Reference(cm)	Relative Error(%)
USTC-AM	23.78	23.69	0.38
USTC-AF	22.95	22.82	0.57
USTC-15M	23.43	23.30	0.57
USTC-15F	22.44	22.70	1.16
USTC-10M	22.00	21.85	0.67
USTC-10F	21.40	21.65	1.18
USTC-5M	20.18	20.60	2.02
USTC-5F	20.18	20.20	0.08

3.2 Absorbed dose and effective dose of USTC phantoms

Using standard radiation protection dosimetry methods, monoenergetic and parallel photon beams from AP, PA,

LLAT, RLAT, ROT and ISO geometries were simulated and calculated doses were normalized by kerma free-in-air. Absorbed dose conversion coefficients for every organ and tissue in USTC phantoms were compared with those from RPI-AM and RPI-AF under the same conditions.

Figure 6 summarizes the relative absorbed doses to the male testicles and figure 7 shows the relative doses to the female ovaries. For low-energy photons, considerable differences are observed between USTC phantoms and RPI phantoms. Because of the anatomical dependency for the relative absorb dose to a particular organ or tissue, these differences are expected. Generally, the dimensions of USTC adult phantoms are less than those of RPI adult phantoms, and the mass value for a single organ is smaller than that in the Caucasian. Therefore, relative absorb doses in USTC adult phantoms are higher. As for high-energy photons, the differences of dose coefficients are relatively small. In the case of the ratio of ovaries absorbed dose, large differences are illustrated in figure 7. USTC-AF and RPI-AF phantoms have the same ovaries' mass, but the ovaries in USTC-AF are further to the posterior side than RPI-AF. And the abdominal width of USTC-AF is smaller than the value of RPI-AF phantom.

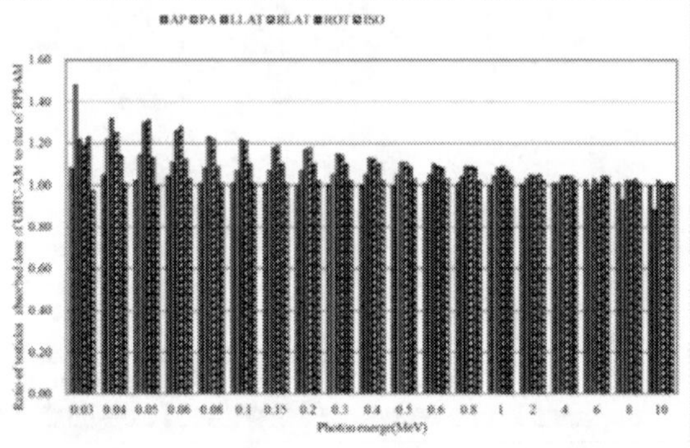

Figure 6. Ratio of testicles absorbed dose for USTC-AM to that of RPI-AM from photon beams.

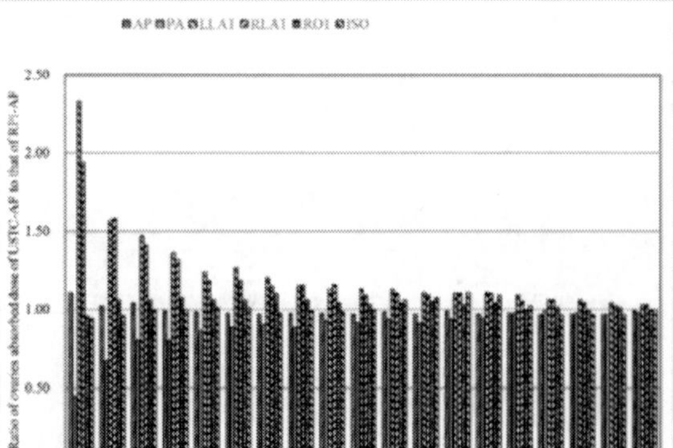

Figure 7. Ratio of ovaries absorbed dose of USTC-AF to that of RPI-AF showing large differences caused by the location of organs and the abdominal width

Figure 8 presents the results of the normalized effective doses for adult, 15-year old, 10-year old and 5-year old phantoms. For each phantom, the AP irradiation geometry shows the largest effective dose. When comparison difference ages, it is found that the 5-year old phantoms have the largest effective dose, due to the fact that the dimensions of smaller phantoms and organs are located closer to the body surfaces.

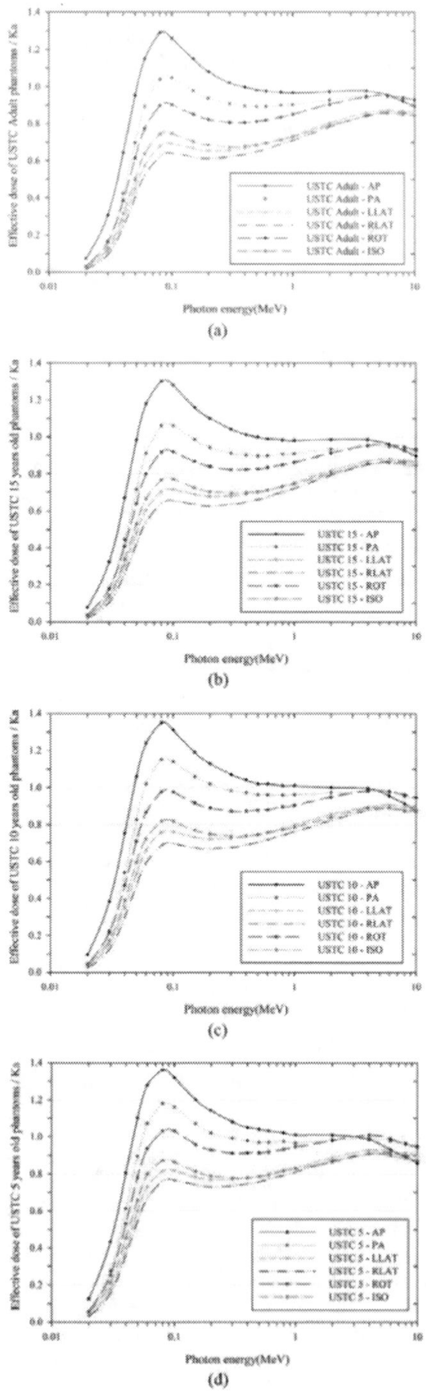

Figure 8. The normalized effective doses for (a) USTC adult, (b) 15 years old, (c) 10 years old and (d) 5 years old phantoms

4 Conclusion

A set of mesh-based and age-dependent adult and adolescent Chinese phantoms have been developed. Automated and manual methods were implemented to match the phantom parameters with Chinese reference values. All phantoms were voxelized for the purposes of Monte Carlo dose calculations. For low-energy photons, considerable differences are observed between USTC phantoms and those of RPI phantoms, suggesting that it is important to consider population related anatomical information in radiation protection dose calculations.

Acknowledgment

This work was supported in part by grants from the National Natural Science Foundation of China (Grant No. 11375181).

References

1. X. G. Xu, Physics in medicine and biology, **59**, R233 (2014)
2. X. G. Xu, K. F. Eckerman, *Handbook of anatomical models for radiation dosimetry* (CRC Press, 2009)
3. L. Su, B. Han, X. G. Xu, Radiation protection dosimetry, ncr194 (2011)
4. J. Zhang, Y. H. Na, P. F. Caracappa, X. G. Xu, Physics in medicine and biology, **54**, 5885 (2009)
5. J. Zhang, X. G. Xu, C. Shi, M. Fuss, Journal of applied clinical medical physics/American College of Medical Physics, **9**, 2700 (2008)
6. J. Li, R. Qiu, Z. Zhang, L. Liu, Z. Zeng, L. Bi, W. Li, Radiation protection dosimetry, ncp087 (2009)
7. B. Zhang, J. Ma, L. Liu, J. Cheng, Radiation protection dosimetry, **124**, 130-136 (2007)
8. G. Zhang, Q. Liu, S. Zeng, Q. Luo, Physics in medicine and biology, **53**, 3697 (2008)
9. GB/T 26158-2010 (Chinese Standard Press, Beijing, 2010)
10. GBZ/T200.1-2007 (People's Medical Publishing House, Beijing, 2007)
11. A. Ding, M. M. Mille, T. Liu, P. F. Caracappa, X. G. Xu, Physics in medicine and biology, **57**, 2441 (2012)
12. D. B. Pelowitz, Los Alamos National Laboratory, **76**, (2005)
13. C. O. R. P. International, *ICRP Publication 74: Conversion Coefficients for Use in Radiological Protection Against External Radiation* (Elsevier Health Sciences, 1997)
14. J. Valentin, Annals of the ICRP, **32**, 1-277 (2002)
15. Y. H. Na, B. Zhang, J. Zhang, P. F. Caracappa, X. G. Xu, Physics in medicine and biology, **55**, 3789 (2010)
16. H. KAWAMURA, G. TANAKA, K. SHIRAISHI, Y. HONDA, M. NISHIMUTA, T. KOYANAGI, F. KARIM, M. MOLLA, A. RAB, J. WANG, Journal of radiation research, **39**, 350 (1998)
17. G. Tanaka, H. Kawamura, R. V. Griffith, M. Cristy, K. F. Eckerman, Radiation protection dosimetry, **79**, 383-386 (1998)
18. J. X. Wang, R. Chen, H. Zhu, Y. Z. Zhou, R. W. Ma, *Data of anatomical physiological and metabolic characteristics for Chinese reference man* (Atomic Energy Press, Beijing, 1998)
19. M. Zankl, K. F. Eckerman, W. E. Bolch, Radiation protection dosimetry, **127**, 174-186 (2007)
20. C. J. Watchman, D. Hasenauer, W. E. Bolch, Physics in medicine and biology, **52**, 3133 (2007)
21. V. F. Cassola, V. J. de Melo Lima, R. Kramer, H. J. Khoury, Physics in medicine and biology, **55**, 133 (2009)
22. MathWorks,
23. GBZ/T200.5-2009 (Beijing, People's Medical Publishing, 2009)
24. J. Deasy, Medical Physics, **21**, 709-710 (1994)
25. C. Lee, C. Lee, A. P. Shah, W. E. Bolch, Physics in medicine and biology, **51**, 5391 (2006)
26. L. Liu, Z. Zeng, J. Li, B. Zhang, R. Qiu, J. Ma, Physics in medicine and biology, **54**, 6675 (2009)
27. ICRP, *The 2007 Recommendations of the International Commission on Radiological Protection* (Elsevier, Oxford, 2007)

DOI: 10.1051/epjconf/201715304015

Simulation of irradiation exposure of electronic devices due to heavy ion therapy with Monte Carlo Code MCNP6

Janis Lapins[1], Nicole Guilliard[1,a], Wolfgang Bernnat[2] and Arnulf Buck[3]

[1]University Stuttgart, Institute for Nuclear Energy and Energy Systems, 70569 Stuttgart, Germany
[2]University Stuttgart, KE Technology GmbH, 70569 Stuttgart, Germany
[3]Graz University of Technology, Master Student, Austria

Abstract. During heavy ion irradiation therapy the patient has to be located exactly at the right position to make sure that the Bragg peak occurs in the tumour. The patient has to be moved in the range of millimetres to scan the ill tissue. For that reason a special table was developed which allows exact positioning. The electronic control can be located outside the surgery. But that has some disadvantage for the construction. To keep the system compact it would be much more comfortable to put the electronic control inside the surgery. As a lot of high energetic secondary particles are produced during the therapy causing a high dose in the room it is important to find positions with low dose rates. Therefore, investigations are needed where the electronic devices should be located to obtain a minimum of radiation, help to prevent the failure of sensitive devices. The dose rate was calculated for carbon ions with different initial energy and protons over the entire therapy room with Monte Carlo particle tracking using MCNP6. The types of secondary particles were identified and the dose rate for a thin silicon layer and an electronic mixture material was determined. In addition, the shielding effect of several selected material layers was calculated using MCNP6.

1 Introduction

During the last years the treatment of patients with tumours with heavy ions became a big issue. The main advantage compared to the standard treatment with photons is the precise energy deposition in the tissue of the patient [1]. Healthy tissue is barely affected as the heavy ions deposit the majority of their energy at the end of their range in a very sharp Bragg-Peak. The mutated tissue can be scanned step by step to irradiate all tumour regions. But that means that the patient himself has to be immobilized throughout the therapy and a movement of the patient table in the range of millimetres is necessary. BEC Engineering [2] was involved in the development of such a table in the MedAustron Project [3] which allows precise location of the patient with exact minimal movement in any direction. Originally, it was planned to put the electronics for the table outside the room to avoid failure of the electronics due to high irradiation during the therapies. The high dose in the room is a result of secondary particle production. But this is not desirable from a construction point of view. The easiest way of construction would be to keep the electronic control at the table or at least next to it. The question arose how much dose the electronic devices will receive and if this dose rate affects the functions of the systems and which kind of shielding can prevent a high dose deposition in the electronics? The present work focuses on therapies with either carbon ions or protons.

2 Validation of MCNP6 for use with high energy heavy ion simulation

2.1. Validation calculations for carbon ions

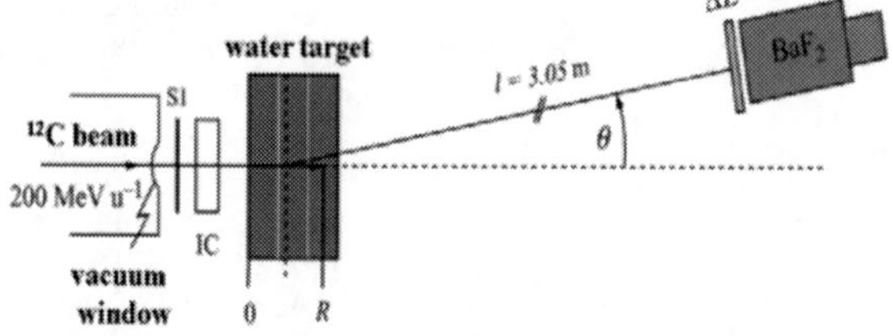

Figure 1: Experimental setup for validation calculations with MCNP6. A carbon ion beam hits a water target producing secondary particles.

[a] Nicole.guilliard@ike.uni-stuttgart.de

To make sure that the production of secondary particles during irradiation is correctly simulated with the selected models implemented in MCNP6 [4] for high energy particles an experiment [5] was recalculated. The experimental setup can be seen in Fig.1. A carbon ion beam with 200 MeV/amu was focused on a water target. The target has a width of about 12 cm. The Bragg peak occurs in the target. The secondary particles were detected in 3.05m behind the Bragg-Peak with a BaF2 detector. The angle between the beam direction and the detector was varied in the range from 0° up to 30°.

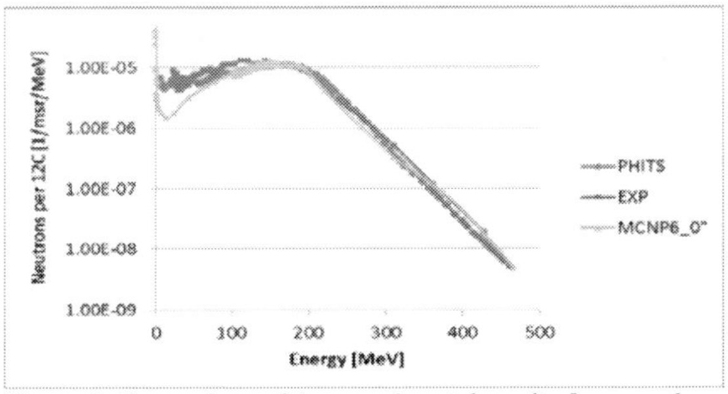

Figure 2: Comparison of the experimental results for secondary neutron production in beam direction (0°) with the simulations in PHITS and MCNP6.

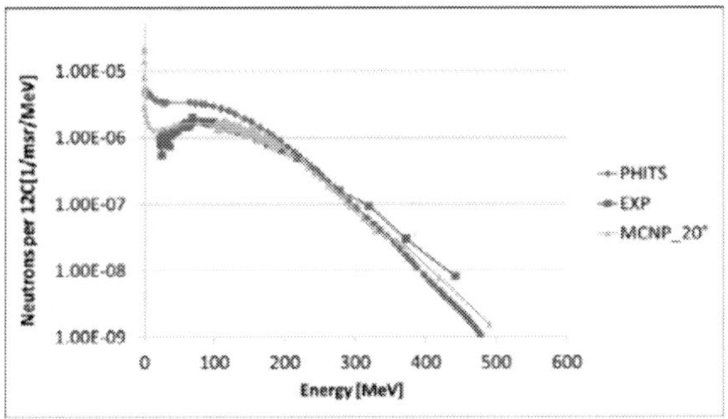

Figure 3: Secondary energy dependent neutron spectrum produced by a C12 ion beam hitting a water target. The experimental results measured at an angle of 20° to the original beam direction are compared with the simulations in PHITS and MCNP6.

In Figure 2 and Figure 3 the experimental results for the production of secondary neutrons are compared to the simulations with the Monte Carlo Code PHITS [5,6] (particle and heavy ion transport code system) and the results obtained with MCNP6 for different angles referred to the incident beam direction. In the high energy region the simulation reproduces the experimental values very accurately. The difference of the detector and the MCNP6 results below 20 MeV are remarkable. The strong rise calculated was not detected. But the detector efficiency in that range is very poor and the experiment does not cover that range properly [7]. The diagrams for other angles show approximately the same behaviour as well as the investigations for other secondary particles like Deuterons, Tritons or Alpha particles (see Figure 4).

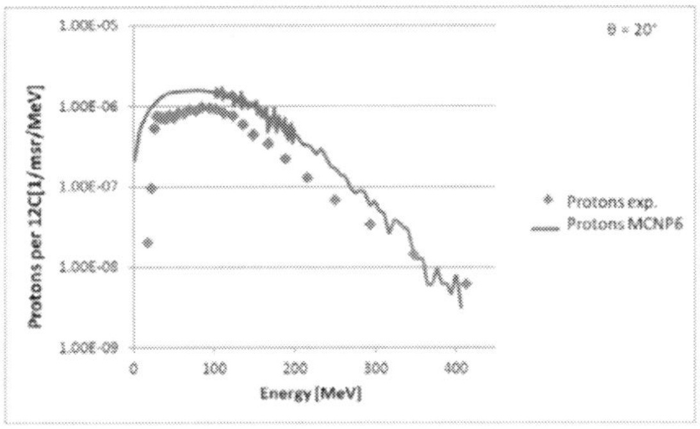

Figure 4: Comparison of experimental and MCNP6 results for secondary proton production in 20° direction.

2.1. Validation calculations for protons

Beside therapies with carbon ions it is planned in the MedAustron project to treat patients with protons. To make sure that the used models in MCNP6 provides reliable results for proton source particles and the production of secondary particles additional validation calculations were performed. Due to the leak of open experimental data a code to code benchmark was conducted. The comparative simulation was taken from Agosteo et al. [7]. In this work the secondary particle production for several types of ions on different target materials was investigated. The simulations were performed with the Monte Carlo Code FLUKA [8]. We will compare the results for 215 MeV/u protons impinging a cylindrical ICRU tissue like target [9]. The target length was chosen in such a way that the Bragg peak occurred in the material. The secondary neutrons were detected on a shell surrounding the target and which was divided in 12 angular bins with a width of 15°. Figure 5 shows the results for secondary neutron production as a function of the neutron energy for both the calculation from Agosteo et al [7]. with FLUKA and the ones made at IKE with MCNP6. The results for four different angle bins are plotted (0°-15°, 30°-45°, 60°-75°, 165°-180°, relative to the incident beam). The spectra in the different directions do not differ significantly and show more or less the same behaviour. A slight decrease of neutron production can be observed in the high energy region with increasing angles to the beam direction with both codes, especially for backscattered particles with an angle larger than 165°. In this energy region the results of the simulations do not agree as well as in the very low energy region. As both tools are Monte Carlo Codes and as the total amount of back scattered particles is much lower than the amount of particles in forward direction the statistical error increases with larger angles. The large fluctuations in the thermal energy region indicate that the results have large statistical errors as the amount is very low. But all together the code to code benchmark validates that the used MCNP6 models are an appropriate tool for the analysis of secondary particle production in proton therapy.

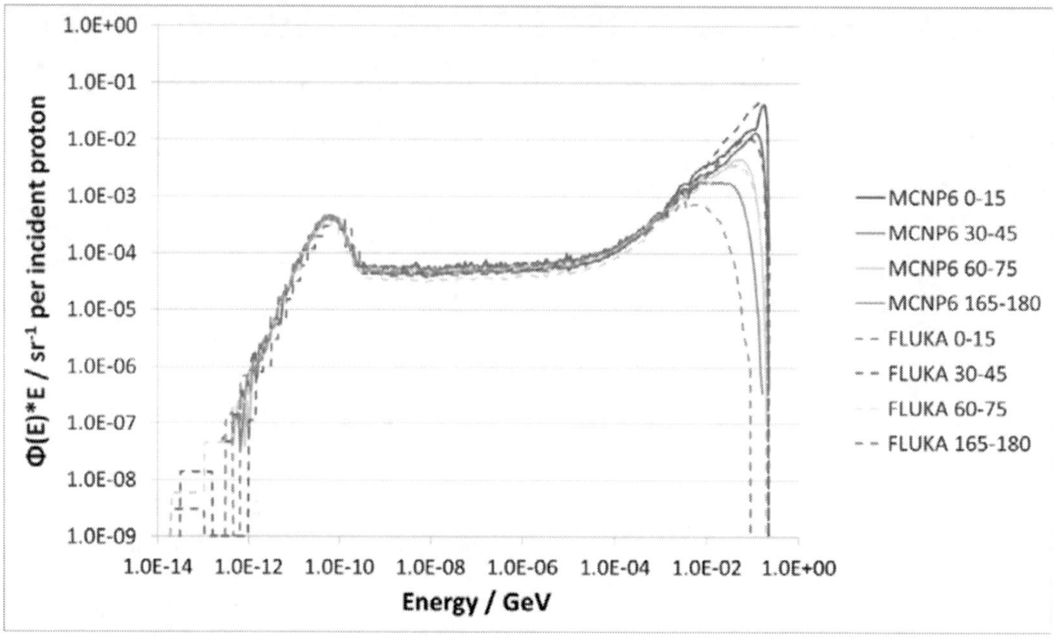

Figure 5: Spectrum of secondary neutrons for certain angles after irradiation of a tissue target with 215 MeV/u protons.

3 Response functions

The MCNP6 simulation for heavy ion irradiations with production of secondary particles tracks the particle way through the geometry and gives an energy dependent flux value Φ for every point of the geometry and for each particle type. To estimate a dose D which is deposited in a certain material the user needs to define response functions *f(E)* for all interesting particle types. With the simulated flux for different particles *i* the total deposition in this material can be calculated for every location in the room.

$$D_x = \sum_{i=1}^{n} \phi_i(E) f_{x,i}(E) dE \tag{1}$$

The deposited energy depends also on the material thickness *x* as particles like protons and alphas have sharp Bragg-Peaks and release the majority of their energy at a certain depth or high energy particles can cross thin layers without any interaction. To determine the response function for a certain particle type, simulations with an isotropic point source and a distinct energy of the source particles are performed for the energy region of interest. By varying the source energy an energy dependent response function can be generated. To obtain a complete set of response functions the calculations were made for several particle types such as neutrons (n), protons (h), deuterons (d), tritons (t), electrons (e) photons (p), He-3 (s) and He-4 (a).

Fig. 6 shows the response function for neutrons (n) for Silicon layers with different thicknesses s. The function is normalized to one gram of silicon. The response function depends on the target thickness. As the way through a thick target is longer, the reaction possibility is higher. Same applies to the secondary particles that travel a longer distance inside the sample. Even though very high energetic neutrons can pass the target without any interaction the response function increases in the high energy region. This is due to the fact that the deposition per interaction rises. So although the number of events decreases the energy transfer of a single event deposits an enormous amount of energy. The peak in the thermal energy region arises from non-elastic reactions. In the high energy region the energy transfer is dominated by elastic scattering and spallation reactions.

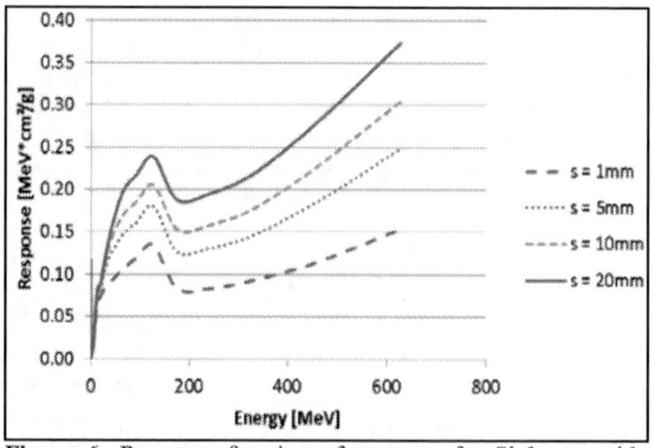

Figure 6: Response function of neutrons for Si layers with different thickness.

DOI: 10.1051/epjconf/201715304015

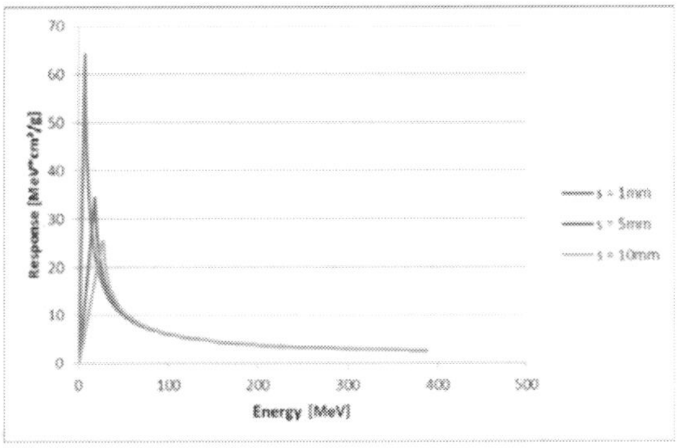

Figure 7: Response function of protons for Si layers with different thickness.

Figure 7 shows the response functions for protons (p) hitting Silicon layers of different thicknesses. Compared to the results for neutrons, the response of protons differs by a power of ten although the particles have comparable masses. Generally, the interaction possibility is strongly affected by the charge of the particles. For heavy particles (compared to the electron mass) the energy transfer can be described by the Bethe-Bloch formula. The formula says that the energy transfer *–dE/dx* is proportional to the square of the charge *z* and reciprocal proportional to the square of the particle velocity *v*:

$$-\frac{\mathrm{dE}}{\mathrm{dx}} \propto \frac{z^2}{v^2} \qquad (2)$$

Figure 7 also shows that the maximum of the response tends to higher energies for thicker targets. Above a certain energy the Bragg peak occurs beyond the length of the thin layer and the particles pass the target without strong interactions. The decreasing maximal value can be explained by the fact that the values are normalized to one gram of Silicon. In fact, the total energy release in a thick layer is higher than in a thin layer. In the energy region where all neutrons pass even the thick layer without interaction the curves coincide as the energy transfer is constant along the way and therefore proportional to the mass.

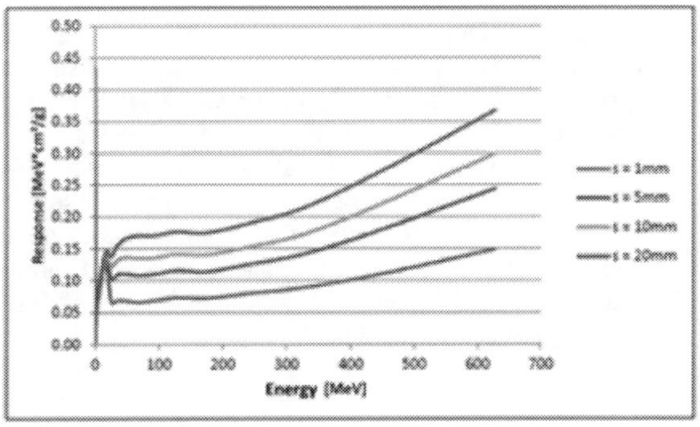

Figure 8: Response function of neutrons for electrical device with different thickness.

The functions were also determined for a material mix consisting of typical materials used for electronic devices (see figure 8). But only small differences were identified. The higher response in the thermal energy region can be explained by the larger capture cross sections of iron and copper. For high energies the response function is very similar to the one calculated for Silicon Hence, the following simulations and results apply to depositions in Silicon.

4 Simulations of radiation exposure to the electrical devices in a therapy room with MCNP6

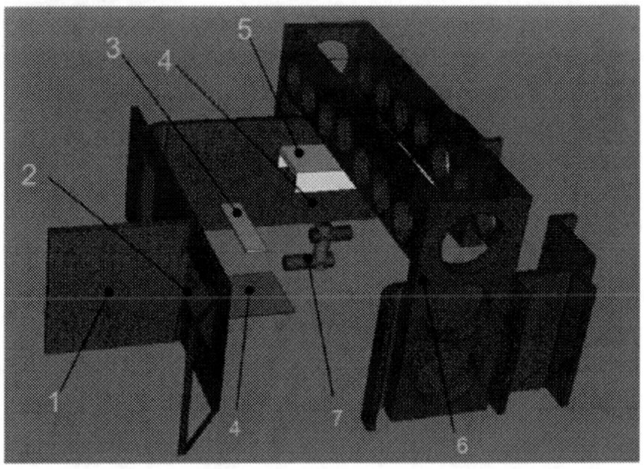

Figure 9: 3-D model of the room for MCNP6 without floor, ceiling and concrete walls. 1) wooden wall, 2) Lead window, 3) table, 4) Si plate for deposition, 5) linear axle, 6) supporting structure, 7) robot arm.

A typical therapy room was modelled with MCNP6 Figure 9 shows the interior of the room without the surrounding concrete wall, the floor and the ceiling. The most important components are modelled as simple geometric bodies. Next to the patient table (no.3) is the robot arm (no.7) for positioning which is connected to the linear axle (no.5) and the supporting structures (no.6). For deposition simulations a patient model is taken from former investigations of the Argonne National Laboratory (ANL) [11] consisting of a simplified body with typical human tissue and skeleton material.

Figure 10 is a cross section of the room at the level of the patient during therapy sessions. The red arrow shows the beam direction of the heavy ions in case of Carbon therapy. The orange dot next to the patient represents the robot arm. The thin blue lines are the wooden partition panels. The cross section also shows the thick concrete walls surrounding the room in light blue. During the carbon ion therapy the beam comes from the left side (figure 9) whereas in case of proton therapy the beam can be adjusted for irradiation from top to bottom. Below the table is a cavity in the floor where a camera for positioning of the patient should be located.

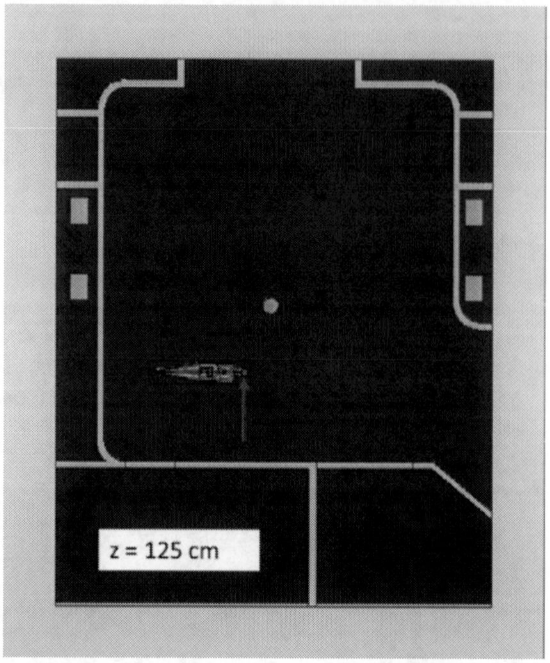

Figure 10. Cross section of the simulated therapy room in MCNP6 at a height of 125cm.

To estimate the deposition during the irradiations with a cell tally a plate of Silicon is simulated below the table. Additional Si-cells for cell tally measurements were located next to the robot arm (between position 4 and 7) and above the linear axle (position 5) as those places are preferred positions for electrical devices. The results of the cell tally measurements can be compared to those obtained by a flux mesh tally covering the whole room using the response functions. The simulations were performed using the response functions for a 0.1 mm thick Si layer. Considered particles are those mentioned above: neutrons, electrons, photons, deuterons, tritons, helions, alphas and heavy ions. To normalize the deposited dose during one year of operation some assumption were made. These are summarized in table 1.

Table 1. Assumptions for normalization of the cumulated dose after one year of irradiation.

	Carbon ions	Protons
Number of patients per year	800	800
Intensity	$4*10^8$ Ions/pulse	$4*10^{10}$ Ions/pulse
Pulse frequency	0.5 Hz	0.5 Hz
Duration of one irradiation	45 min	45 min
Starting particles per year	$4.32*10^{14}$	$4.32*10^{16}$

Carbon ions with 200 MeV/u

Figure 11 gives an overview of the deposited dose for irradiation with 200MeV/amu carbon ions after one year in the layer containing the initial ion beam. The highest dose occurs in beam direction, as expected. The robot arm shields part of the secondary particles and a lower dose is expected behind the apparatus. Another decrease of dose can be seen behind the human body on the left side of Figure 10 due to the absorption of secondary particles in the patient.

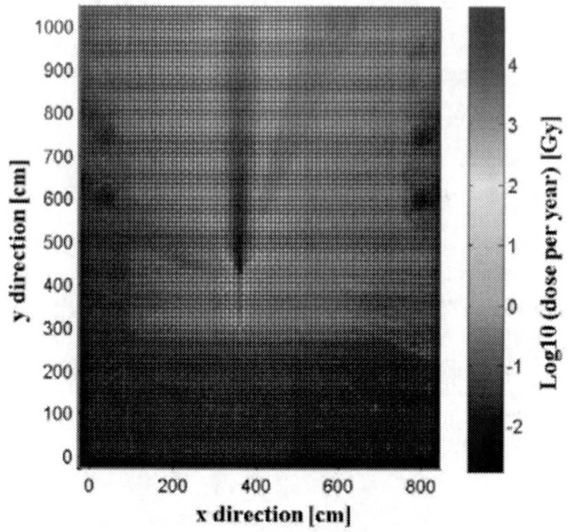

Figure 11. MCNP6 calculation of the dose deposited in a 0.1 cm thick Si layer after one year Carbon ion therapy with 200 MeV/u.

The cross section of the mesh tally in yz-direction is shown in figure 12. The dose per year upon the linear axle and on the left side of the room (beam entry) is more than a factor of 10^4 smaller than the dose expected in beam direction. The heavy particles are mainly scattered in forward direction. Therefor the installation of sensitve electrical devices on top of the axle is more advisable than put it directly next to the robot arm. Assuming a limit of 10^2 Gy/year it is recommended to avoid the red and orange locations for very sensitve devices.The wooden panel at y=300cm shields the backscattered particles and leads to a very small dose rate in the anterroom. To know which particle type is mostly responsible for the deposition the percental fraction of the different types was identified. It was found that the deposition in forward direction near the patient originates predominantely from heavy ions, alphas and protons (figure 13). The deposition caused by photons, electrons and neutrons can be neglected.

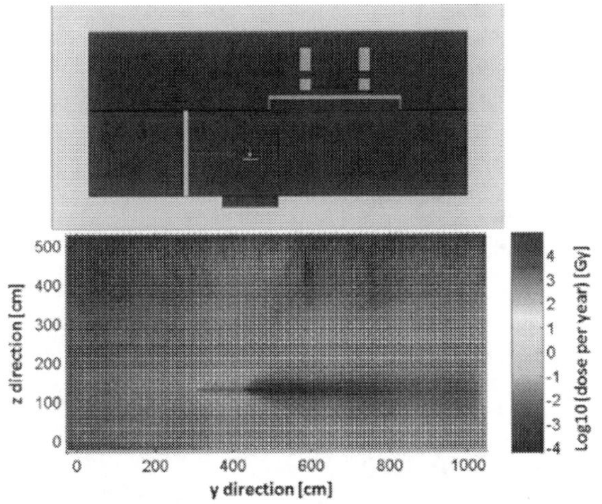

Figure 12. *top*: yz-cross section of the therapy room. *bottom*: MCNP6 calculation of the dose deposited in a 0.1 cm thick Si layer after one year Carbon ion therapy with 200 MeV/u.

In the upper and lower part of the room with an increasing angle to the original beam the influence of the heavy particles gets lower and most of the deposition comes from protons, electrons and photons (figure 14). The results of the MCNP6 cell detectors were used for comparison and validation of the mesh tally method and showed good agreement with the mesh tally results. The deposited energy during one year above the linear axle calculated with the cell tally was $2.17*10^{-1}$ Gy and the dose for one year at the position next to the robot arm was about $1.88*10^{3}$Gy. This confirms the results obtained by the mesh tally in figure 11 and figure 12.

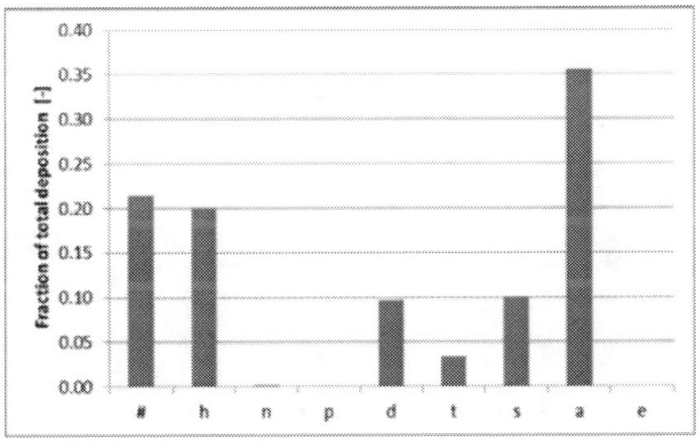

Figure 13. Fraction to the total deposition in a Si-layer of different particle types during a heavy ion therapy with C12 ions at the robot arm. (#: heavy ions, h: protons, n: neutrons, p: photons, d: deuterons, t: tritons, s: He-3, a: He-4, e: electrons).

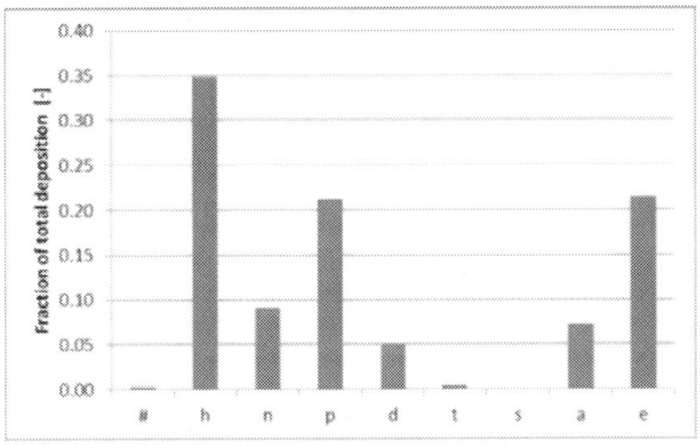

Figure 14. Fraction to the total deposition in a Si-layer of different particle types during a heavy ion therapy with C12 ions above the linear axle.

Protons with 250 MeV/u

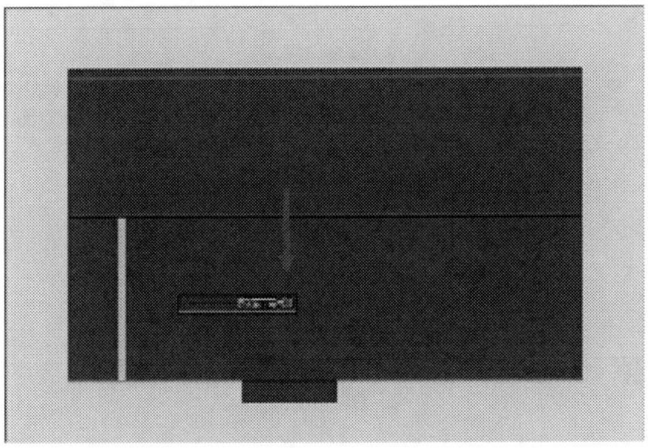

Figure 15. Axial cut of the therapy room in the beam plane simulated with MCNP6.

Analogue calculations were made for irradiation therapy with 250MeV protons. The assumptions for normalization of the accumulated dose after one year of therapy can be found in table 1. The geometry modeling of the room and the patient are the same as the one used for the former calculation with carbon ions. The proton beam direction is changed and goes from top to bottom (see figure 15) as forseen in the planning of the irradiation centre. The energy of the protons is 250Mev/amu. figure 15 shows the cumulated dose after one year of proton therapy deposited in a 0.1mm thick Silicon layer. The absolute deposited energy is about 100 times higher than for carbon ions due to the higher intensity of the protons during irradiation (see table 1). Again the region above the linear axle seems to be most favourable as location for electrical devices. The asymmetric distribution at z=125cm on the left side of the beam shows the position of the phantom. Below the table a cavity can be seen. The camera necessary for the exact positioning of the patient should be placed at this

 DOI: 10.1051/epjconf/201715304015

location. To check the results, standard track tallies were used that are implented in MCNP6. The flux in a cell volume is estimated by track length. Results with cell detector in the cavity was $2.72*10^3$Gy. This value seems to be very low compared to the meshtally results in figure 16. But the cell tally covers the whole cavity. The average dose of the mesh tally calculation over the whole cavity is in acceptable agreement (as can be seen in figure 17). Although the dose rate in the cavity is very high the camera has to be placed there for technical reasons. But it might help to avoid the position in direct beam direction. A slight shift may helb to decrease the deposition.

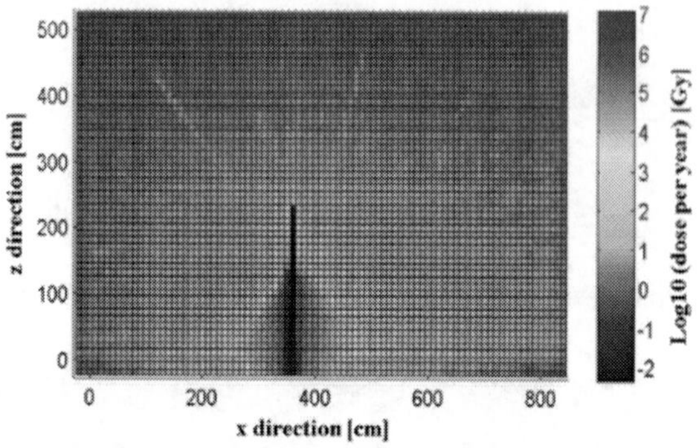

Figure 16. MCNP6 calculated deposition in a 0.1cm thick Si layer after one year therapy with 250MeV protons.

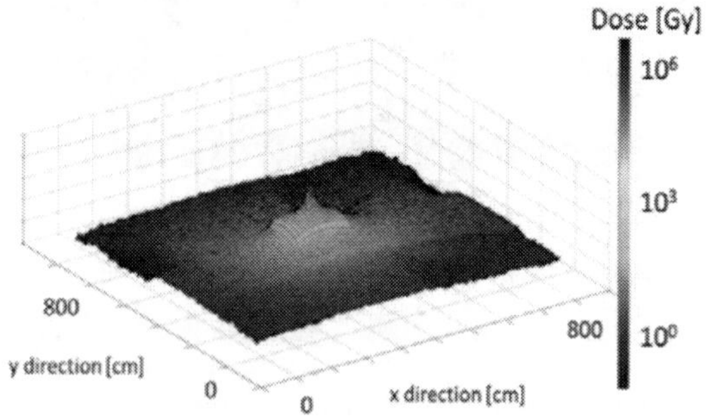

Figure 17. MCNP6 calculated deposition in a 0.1cm thick Si layer after one year therapy with 250MeV protons. The region around the cavity under the table is shown at z level of the cavity.

5 Shielding

To investigate which shielding materials can protect the electronic devices simulations with a simplified model were performed. A tissue like target is pounded with ^{12}C ions with an energy of 200 MeV/u and protons with an energy of 250 MeV/u. The dose deposited in a one cm thick Silicon layer behind a shielding layer was compared to the dose deposited without any shielding. The tissue target has a thickness of about 18cm (see figure 18). A set of different materials were determined. Beside the typical neutron absorbers like boron (B) and Gadolinium (Gd), polyethylene (PE) as well as heavy materials like concrete, lead (Pb) and iron (Fe) were used. Figure 19 gives a summary of the results in the case of carbon source particles. As figure 13 already indicated, most of the deposition is caused by heavy charged particles. Therefore, the best shielding effect is obtained with heavy elements, whreas the light shielding materials like boron granulate (boron mixed with carbon) or Polyethlene reduced the original dose without shielding only by a factor of 2/3.

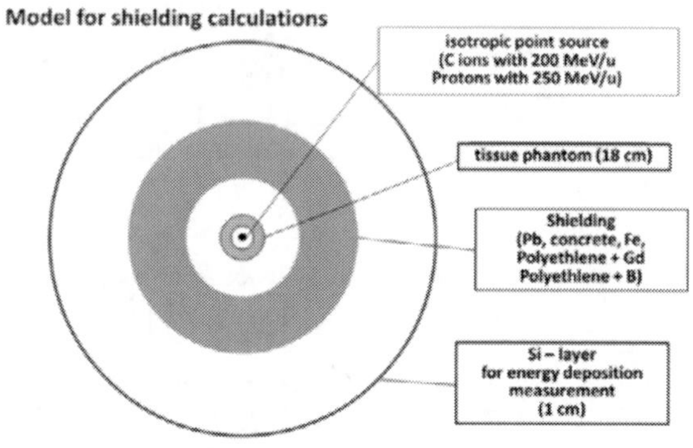

Figure 18. Setup for the simulation of a C12 ion beam impinging a tissue target. Different particle types are measured in a Si layer with and without shielding.

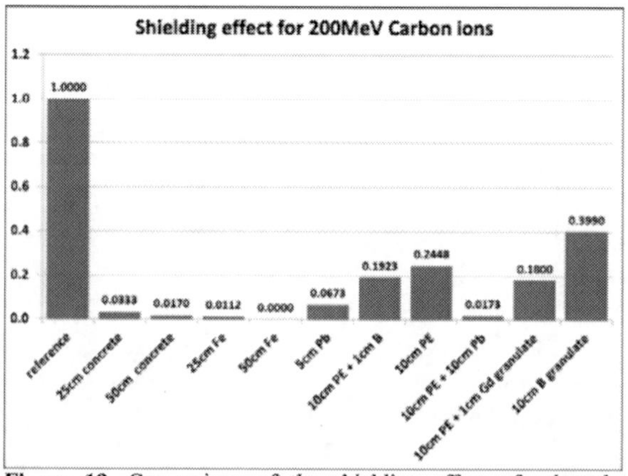

Figure 19. Comparison of the shielding effect of selected materials for the secondary particles produced by carbon ions with 200 MeV/u impinging a tissue target.

The same model and shielding materials were used for 250 MeV/u protons as source particles. Again the thick heavy shielding materials provides the best shielding effect (see figure 20). An interesting behaviour can be

DOI: 10.1051/epjconf/201715304015

observed using exclusively light materials. The relative dose after the shielding is increased. It seems as there are a lot of new secondary particles produced when the protons pass the Silicon layer.

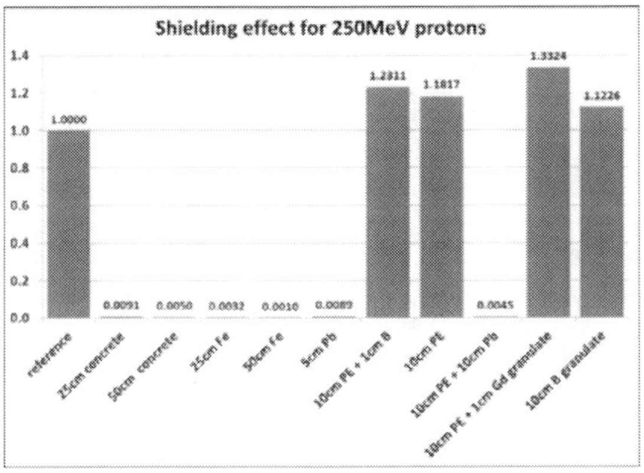

Figure 20. Comparison of the shielding effect of selected materials for the secondary particles produced by protons with 250 MeV/u impinging a tissue target.

6 Conclusion

It was shown that MCNP6 is an adequate tool for estimating the production of secondary particles and to estimate the energy deposition in strucutres during heavy ion therapy. The dose rate in beam direction is very high and it is questionable to put sensitve electrical devices there. Increasing the angle between the beam direction and the point of intrest can help to reduce the dose significantly. Additional shielding installation can protect sensitive electrical devices. The best shielding effect is achieved by heavy thick material which comes along with unhandy constructions. Therefore it would be much easier to use light material. But for light material shieldings it is important to know the exact position and particle type to get an optimized shileding. The impact on dose reduction may be sufficient for Carbon ions but can be a bad choice in case of proton therapy.

Acknowledgment

We thank the colleauges from Buck Engineering & Consulting GmbH who initiates this interesting collaboration and provided technical data for the calculations.

References

1. G. Kraft, "Tumor therapy with heavy charged particles", Prog. Part. Nucl. Phys. **45**, p473-544,(2000).
2. Buck Engineering & Consulting GmbH, 72764 Reutlingen (Germany)
3. E. Griesmayer, T. Schreiner und M. Pavlovic, „The MedAustron project," Nucl. Instr. and Meth. in Phys. Res. **B 258**, pp. 134-183, (2007).
4. D.B. Pelowitz (Ed.), "MCNP6™ User's Manual Version 1.0," LA-CP-13-00634, Rev. 0 (2013).
5. K. Gunzert-Marx, H. Iwase, D. Schardt, R. S. Simon, „Secondary beam fragments produced by 200 MeV/u ^{12}C ions in water and their dose contributions in carbon ion radiotherapy," New Jour. of Phys. (2008) 075003 (21pp).
6. H. Iwase, K. Niita, T. Nakamura: "Development of general-purpose particle and heavy ion transport Monte Carlo code", J. Nucl. Sci. and Technol. 39, 1142 (2002).
7. S. Agosteo, A. Mereghetti, E. Sagia, M. Silari, "Shielding data for hadron-therapy ion accelerators: Attenuation of secondary radiation in concrete", Nucl. Instr. and Meth. in Phys. Res. **B 319**, pp. 154-167 (2014).
8. A. Ferrari, P.R. Sala, A. Fassò, J. Ranft, FLUKA: a multi-particle transport code, CERN-2005-10, 2005, INFN/TC_05/11, SLAC-R-773.
9. International Commission on Radiation Units and Measurements, Report No. **33**, Bethesda, Maryland, USA, 1980 p. 20.
10. K. Gunzert-Marx, "Nachweis leichter Fragmente aus Schwerionenreaktionen mit einem BaF2-Teleskop-Detektor", Diss. D17, TU Darmstadt (2004)
11. D. Krstic und D. Nikezic, „Input files with ORNL-mathematical phantoms of the human body for MCNP-4B",Computer Physics communications **176**, pp. 33-37 (2007).

Monte Carlo calculations for reporting patient organ doses from interventional radiology

Wanli Huo[1], *Mang* Feng[1], *Yifei* Pi[1], *Zhi* Chen[1], *Yiming* Gao[2], *X. George* Xu [1, 2*]

[1]*University of Science and Technology of China, Hefei, China*

[2]*Rensselaer Polytechnic Institute, Troy, New York, USA*

Abstract. This paper describes a project to generate organ dose data for the purposes of extending VirtualDose software from CT imaging to interventional radiology (IR) applications. A library of 23 mesh-based anthropometric patient phantoms were involved in Monte Carlo simulations for database calculations. Organ doses and effective doses of IR procedures with specific beam projection, filed of view (FOV) and beam quality for all parts of body were obtained. Comparing organ doses for different beam qualities, beam projections, patients' ages and patient's body mass indexes (BMIs) which generated by VirtualDose-IR, significant discrepancies were observed. For relatively long time exposure, IR doses depend on beam quality, beam direction and patient size. Therefore, VirtualDose-IR, which is based on the latest anatomically realistic patient phantoms, can generate accurate doses for IR treatment. It is suitable to apply this software in clinical IR dose management as an effective tool to estimate patient doses and optimize IR treatment plans.

1 INTRODUCTION

VirtualDose is a cloud-based software tool for assessing and reporting patient organ doses from radiological imaging procedures [1-7]. VirtualDose was designed to improve the patient dose estimation with considering the latest ICRP (International Commission on Radiological Protection) tissue weighting factors, several validated CT scanner models, scanner-specific correction factors, and a family of anatomically realistic phantoms which include female and male adults, children with different ages, pregnant females at different gestational stages and obese people of different body weights [8-13]. The first application-specific module for patients undergoing CT examinations, VirtualDose-CT, was released in 2013 and is currently used by hundreds of medical physicists and radiologists. This paper describes a project to generate organ dose data for a new application – interventional radiology (IR). The database generated from this study will be used in VirtualDose-IR which is designed for reporting patient organ doses and effective dose during an interventional radiology (IR).

IR has been used for treatment and diagnose purposes since 1960s and the number of procedures has grown rapidly recently. According to NCRP (national council on radiation protection) Report No. 160 the exposure to ionizing radiation of the U.S. population has nearly doubled over the past two decades [14]. More than 17 million IR procedures were performed in the U.S. in 2006 and these figures will continue to grow [15]. Although often clinically justified, the radiation doses from X-ray examinations contribute significantly to the total radiation exposure. The NCRP stated that IR contributed 14% of the total yearly radiation exposure from medical imaging [15]. In complicated IR procedures, patients are exposed to high radiation doses up to 0.5 Gy to some organs, causing a real concern for radiation injury [16]. In order to avoid the deterministic effect and mitigate the risk of unnecessary radiation exposure, it is necessary to assess and report organ doses from these medical procedures.

Estimating organ doses for patients undergoing IR can be done using either experimental measurements or computational simulations. Experimental measurements can be done using dosimeters embedded in a physical phantom. This method is time-consuming and costly, often suffering from the lack of precision in dosimeter reading [17, 18]. For IR procedures, several studies have been performed to calculate organ doses and effective doses based on P_{ak} (air kerma-area product) and precalculated conversion factors. Some researchers employed the

[a] Corresponding author: xug2@rpi.edu

relationship between dose and source-to-surface distance (SSD) in IR to calculate peak skin doses (PSD) and skin dose maps [19-21]. Chugh et al and Khodadadegan et al used stylized phantoms and SSD relationship to calculate the skin dose of patients [19, 20]. Johnson et al employed some more anthropomorphic patient-dependent phantoms and SSD relationship to calculate PSD for patients [21]. These methods can only be used to calculate skin doses, but not internal organ doses.

The computational method, on the other hand, takes advantage of advanced Monte Carlo simulation methods and anatomically accurate models of the patients, thus showing a number of advantages in medical physics applications. Several groups used Monte Carlo methods to simulate the patient doses under IR procedures [22-24]. FAX06 and MAX06 phantoms were applied by Kramer et al to calculate organ doses for the most common examinations in X-ray diagnosis [22]. Bozkurt and Bor used VIP-Man to simulate the organ doses in an interventional cardiological procedure [23]. Garzón et al employed Monte Carlo code with FASH and MASH phantoms to stimulate the organ doses of patients undergoing hepatic chemoembolization procedures [24]. Those studies used only a few phantoms and cannot represent patient population accurately.

Radiation-induced organ injuries are correlated with the absorbed doses to organs and tissues. The absorbed dose for a specific organ is related to the radiation type, radiation energy, shielding between organ and radiation source. To quantify the whole-body radiation-induced health effects, the effective dose, a radiation protection quantity which is the sum of each absorbed organ dose multiplied with its specific tissue weighting factor, is recommended by ICRP [25].

The positions of irradiation area, patient's body mass indexes and ages for adolescent patients can affect the effective doses and specific organ doses significantly since the tissue thicknesses between source and organs changed dramatically with these parameters. Even the radiation type and the mean absorbed dose for whole body are the same, effective dose varies with irradiation position. While IR procedures can be operated at any parts of the body, previous researches only studied a few body part IR operations. There is no previous study about organ doses for IR operations of all part of body.

To address the needs for the development of a user friendly software tool, a comprehensive organ database deriving from extensive Monte Carlo simulations was established using a set of 23 anatomically realistic patient phantoms including adults, children at different ages, pregnant females at different gestational stages, and obese patients with different body mass indexes (BMIs). We adopted various X-ray energies, absorbers, beam projections, field of views and all parts of the body irradiated in this study, considering most potential IR procedures at all potential parts of the body. This paper describes the calculations and analyses of preliminary data.

2 METHODS

2.1 A Set of 23 phantoms

Computational human phantoms play a vital role in the field of medical physics patient dose simulations. It can be divided into three generations during its development: the stylized phantoms are combined with simple geometric elements such as cuboids, cylinders spheres and cones; the voxelized phantoms are constructed by massive of tiny voxels representing anatomical structures; the latest generation computational phantoms are the mesh-based phantoms, which are consisted of Non-Uniform Rational B-Splines (NURBs) or polygonal mesh [26-31]. Mesh-based phantoms have powerful abilities of great geometrical deformation and shape adjustment. These advantages make it suitable to change outside geometry and inside organ shape and to develop a series of phantoms with realistic anatomical structures from single mesh-based phantom.

A total of 23 whole-body phantoms were used in this project, including children, male and female adults, obese people and pregnant females in a joint effort by Rensselaer Polytechnic Institute (RPI) and the University of Science and Technology of China (USTC). The group of age-dependent Chinese adults and adolescences phantoms include males and females with age of new born, 5-year, 10-year, 15-year and adult (named USTC-Adult-Male and USTC-Adult-Female) [32-34], all the anatomical parameters adjusted to agree within 0.5% with the reference people data of China [35]. Other phantoms include pregnant females at 3-month, 6-month and 9-month gestational stages developed from a 30-week pregnant woman CT imaging [27], reference male and female adults according to ICRP 89 report (named RPI-Adult-Male (RPI-AM) and RPI-Adult-Female (RPI-AF)) [26], a set of obese males and females with body sizes from normal weight to morbidly obese [30].

The geometries for different type of phantoms are displayed in Figure 1. Those phantoms were originally developed using triangular surface meshes, and were converted to voxel-based phantoms by an in-house voxelization algorithm for the purpose of Monte Carlo dose calculations. In order to calculate PSDs accurately, we defined a 2.5 mm-thick layer of air in front of the phantoms and divided it into cubes, each having a 2.5 mm side length to calculate $Hp_{(0.07)}$. The arms of phantoms were removed when the simulated patients' arms are raised to the overhead position in lateral beam projections [11].

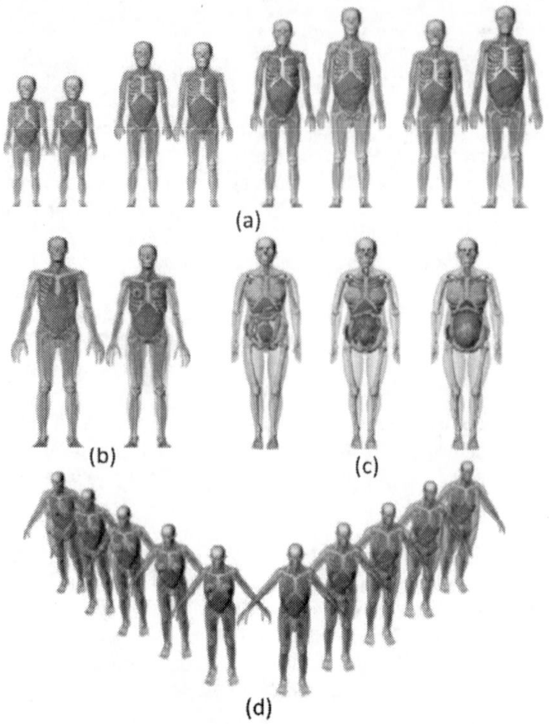

Figure 1. Phantoms in VirtualDose-IR: (a) adolescent and adult phantoms (from left to right: 5, 10, 15, and adult females and males) which are based on Chinese reference standards; (b) Adult male (RPI-AM) and female (RPI-AF) based on ICRP-89 report; (c) Pregnant females at three different gestational stages (3, 6, 9 month); (d) Obese patients in different obesity categories

2.2. VirtualDose Architecture

VirtualDose-IR was designed as a SaaS (Software as a Service) application which hosted all its associated data and up-to-date resources centrally on a remote computer server and allow multiple users to access the software functions at the same time via Internet. To implement this architecture, a "Service-Orientated Architecture (SOA)" design was adopted, as shown in figure 2. Programming languages, including Hypertext Markup Language (HTML), Cascading Style Sheets (CSS) and JavaScript, were used to provide an interactive graphical user interface (GUI). For the server-side scripting, C# was used as the primary programming language.

The completed VirtualDose-IR contains the client-side and server-side interfaces. The client-side interface consists of parameter selection panel, a patient model display, and a dose result display. These allow users to specify the operating conditions of a particular IR treatment for a particular patient.

The server-side interface stores data of 23 different phantoms and their organ doses and effective doses with various IR procedures produced by Monte Carlo simulations. Based on the user-specified IR parameters, VirtualDose-IR fetches and calculates the patient-specific organ dose and effective dose data from the remote server-side database. The results are displayed with a table and a figure on user's browser.

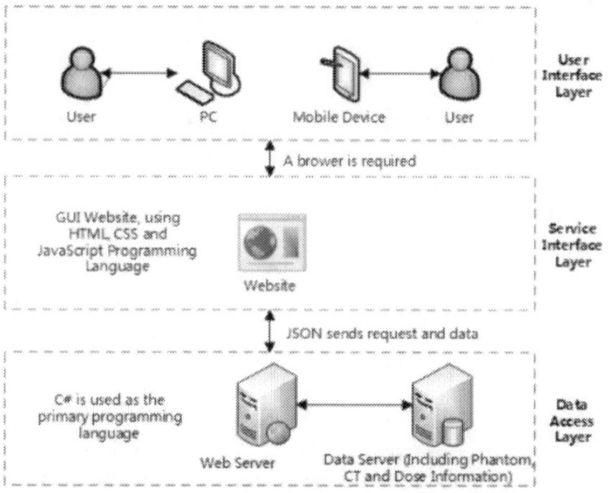

Figure 2. Schematic diagram of the SOA architecture for VirtualDose-IR

2.3 The Monte Carlo simulations

The Monte Carlo method is a well-established technique as well as the most accurate way to calculate radiation dose for medical physics applications. In this research, the software MCNPX (Monte Carlo N-Particle eXtended) version 2.7.0 was used for Monte Carlo simulations [36]. This production code designed at the Los Alamos National Laboratory (LANL) can handle the transportation and interaction of photons, electrons, neutrons, protons and heavy ions for complex three-dimensional geometries in a wide range of energies. Coupled with anatomical realistic computational patient phantoms, a Monte Carlo radiation simulation can produce a detailed distribution of radiation dose across various organs and tissues.

A 2.15 mm thick Al bed represents the operating bed in these simulations with the consideration of attenuation of the bed. Patient lays on his back at the bed and X-ray irradiated the patient from specific position with certain direction in each simulation. In order to report organ doses, each phantom was deconstructed into several parts. For each part, a series of IR exposure procedures were simulated using the specific tube voltage, tube current, absorber, field of view (FOV) and beam projection in the MCNPX code [37-38]. An air ball with constant distance to source was placed above the source as the detector to calculate the air kerma in the center of field in each simulation. Organ doses and effective doses of certain parts of body corresponding to a specific IR protocol were derived from integrating data.

In this research, several different IR exposure situations were studied, the following aspects were considered.

- Beam projections:

posterior anterior (PA), left anterior oblique, 45° (LAO), right anterior oblique, 45° (RAO), left lateral (LLAT), right lateral (RLAT) and cranial, 45° (CRAN).

- Field of views (FOV):

40 cm × 40 cm, 30 cm × 30 cm, 20 cm × 20 cm, 10 cm × 10 cm.

- Tube voltages:
  55 kVp, 60 kVp, 70 kVp, 80 kVp, 90 kVp, 100 kVp, 110 kVp, 120 kVp.
- Filtrations:
  3.5 mm Al + 0 mm Cu, 3.5 mm Al + 0.1 mm Cu, 3.5 mm Al + 0.2 mm Cu, 3.5 mm Al + 0.3 mm Cu.

The geometries of those projections are shown in figure 3.

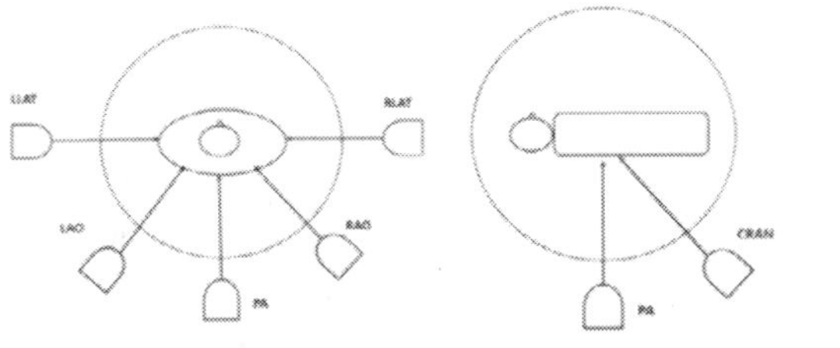

Figure 3. The diagrammatic sketch of beam projections

Considering different patient phantoms with different body sizes, we used different tube voltages and filtrations for every single phantom. While different parts of the body with different thicknesses, the body have been divided into three major sections: head, body and legs, as shown in the figure 4. Each section has different tissue structures and different body thicknesses, but for all the parts of body in every section, the body thicknesses are similar. For each body section of each phantom, the tube voltages and filtrations employed in IR simulations were the same. The voltages and filtrations employed in this study are summarized in table 1. The ranges of tube voltage and filtration change with body thicknesses according to the relation between body thickness and X-ray quality from AAPM (The American Association of Physicists in Medicine) 125 report [39].

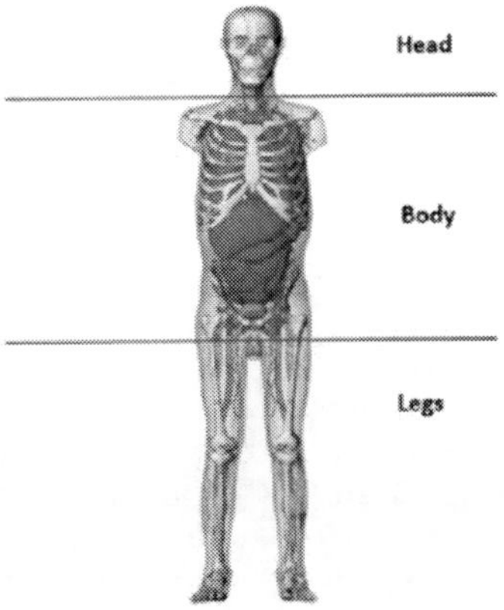

Figure **4**. Scheme of dividing patient phantom into three sections: head, body and legs

A conversion factor (CF) was used to convert the Monte Carlo simulation result to absorbed dose per unit integrated tube current in the units of mGy/100 mAs. These CFs were unique to each combination of tube voltage and absorber thickness. The CF can be obtained according to equation (1).

$$CF = \frac{\text{Kerma}_{\text{in-air}}^{\text{Measured}}}{\text{Kerma}_{\text{in-air}}^{\text{Simulated}}} \quad (1)$$

Where $\text{Kerma}_{\text{in-air}}^{\text{Measured}}$ is the measured kerma value in air in the units of mGy/100 mAs by applying the ionization chamber 44 cm above the source in air at center of the field; $\text{Kerma}_{\text{in-air}}^{\text{Simulated}}$ is the corresponding air kerma value in the units of MeV/gram·source particle obtained by simulating the ionization chamber in the MCNPX code under the same condition. The units of CF is (mGy·gram·source paticle)/ (MeV·mAs).

The simulated results from the MCNPX code can be easily converted to absorbed doses by employing these CFs according to equation (2).

$$D_{\text{absorbed}} = D_{\text{simulated}} \times CF \times (\text{Total mAs}/100) \quad (2)$$

Where D_{absorbed} is the absorbed dose in unit of mGy, $D_{\text{simulated}}$ is the MCNPX simulation result in the units of MeV/ (gram·source particle).

The effective dose was used to evaluate the whole body radiation damage during an IR procedure in this paper. The effective dose can be calculated by using organ doses and organ's tissue weighting factor accounting for equation (3)

$$E = \sum_T w_T \sum_R w_R D_{T,R} \quad (3)$$

where $D_{T,R}$ is the absorbed dose in tissue T for radiation type. w_R is the radiation weighting factor accounting for the relative biological damage of radiation type. w_T is the tissue weighting factor for specific tissue according to tissue's relative radio sensitivity. The tissue weighting factors used in this paper are derived from ICRP 103 report.

The radiation doses of 23 different patient phantoms undergoing IR procedures with various X-ray quantities, beam projections and all the potential parts of body were calculated in this study. These data will be converted to organ absorbed doses and effective doses and integrated into a database. In VirtualDose-IR, all the detailed distributions of radiation doses of different organs/tissues for a specific type of patient during an IR operation were derived from this large organ dose database.

3 Results

3.1 Comparison of organ dose data for different beam qualities

To illustrate how the software tool can be used, here we simulate a series of liver IR treatments for adult male patients with the same imaging parameters except for the tube voltages and absorbers. The beam projections and FOVs involving in those simulations are PA and 40 cm×40 cm. The results are plotted in figure 5. Six pairs of beam qualities (70 kVp + 0.1 mm Cu, 70 kVp +0.2 mm Cu, 80 kVp + 0.1 mm Cu, 80 kVp + 0.2 mm Cu, 90 kVp + 0.1 mm Cu and 90 kVp + 0.2 mm Cu) with a constant 3.5mm Al absorber were considered to generate those data for the comparison. As shown in figure 5 organ dose values increase as the tube voltages increase and decrease with the thickness of absorbers increases. The doses for 90 kVp

Table 1. Summary of tube voltages and filtrations used in IR Monte Carlo simulations for different body sections of type of patient phantom

	Body Sections					
	Head		**Body**		**Legs**	
	tube voltages (kVp)	**Filtrations (mm)**	**tube voltages (kVp)**	**Filtrations (mm)**	**tube voltages (kVp)**	**Filtrations (mm)**
Adolescences						
5 year male	60, 70, 80	3.5 Al + 0.2 Cu/0.3 Cu	60, 70, 80	3.5 Al + 0.2 Cu/0.3 Cu	55, 60, 70	3.5 Al + 0.2 Cu/0.3 Cu
5 year female	60, 70, 80	3.5 Al + 0.2 Cu/0.3 Cu	60, 70, 80	3.5 Al + 0.2 Cu/0.3 Cu	55, 60, 70	3.5 Al + 0.2 Cu/0.3 Cu
10 year male	60, 70, 80	3.5 Al + 0.2 Cu/0.3 Cu	60, 70, 80	3.5 Al + 0.2 Cu/0.3 Cu		3.5 Al + 0.2 Cu/0.3 Cu
10 year female	60, 70, 80	3.5 Al + 0.2 Cu/0.3 Cu	60, 70, 80	3.5 Al + 0.2 Cu/0.3 Cu		3.5 Al + 0.2 Cu/0.3 Cu
15 year male	70, 80, 90	3.5 Al + 0.1 Cu/0.2 Cu	70, 80, 90	3.5 Al + 0.1 Cu/0.2 Cu	60, 70, 80	3.5 Al + 0.2 Cu/0.3 Cu
15 year female	70, 80, 90	3.5 Al + 0.1 Cu/0.2 Cu	70, 80, 90	3.5 Al + 0.1 Cu/0.2 Cu	60, 70, 80	3.5 Al + 0.2 Cu/0.3 Cu
Pregnant females						
3 month pregnant	70, 80, 90	3.5 Al + 0.1 Cu/0.2 Cu	70, 80, 90	3.5 Al + 0.1 Cu/0.2 Cu	60, 70, 80	3.5 Al + 0.2 Cu/0.3 Cu
6 month pregnant	70, 80, 90	3.5 Al + 0.1 Cu/0.2 Cu	70, 80, 90	3.5 Al + 0.1 Cu/0.2 Cu	60, 70, 80	3.5 Al + 0.2 Cu/0.3 Cu
9 month pregnant	70, 80, 90	3.5 Al + 0.1 Cu/0.2 Cu	70, 80, 90	3.5 Al + 0.1 Cu/0.2 Cu	60, 70, 80	3.5 Al + 0.2 Cu/0.3 Cu
Average adults						
USTC-Adult-Male	70, 80, 90	3.5 Al + 0.1 Cu/0.2 Cu	70, 80, 90	3.5 Al + 0.1 Cu/0.2 Cu	60, 70, 80	3.5 Al + 0.2 Cu/0.3 Cu
USTC-Adult-Female	70, 80, 90	3.5 Al + 0.1 Cu/0.2 Cu	70, 80, 90	3.5 Al + 0.1 Cu/0.2 Cu	60, 70, 80	3.5 Al + 0.2 Cu/0.3 Cu
RPI-AM	70, 80, 90	3.5 Al + 0.1 Cu/0.2 Cu	70, 80, 90	3.5 Al + 0.1 Cu/0.2 Cu	60, 70, 80	3.5 Al + 0.2 Cu/0.3 Cu
RPI-AF	70, 80, 90	3.5 Al + 0.1 Cu/0.2 Cu	70, 80, 90	3.5 Al + 0.1 Cu/0.2 Cu	60, 70, 80	3.5 Al + 0.2 Cu/0.3 Cu
Obese patients						
Normal body-weight male	70, 80, 90	3.5 Al + 0.1 Cu/0.2 Cu	80, 90, 100	3.5 Al + 0.1 Cu/0.2 Cu	60, 70, 80	3.5 Al + 0.2 Cu/0.3 Cu
Normal body-weight female	70, 80, 90	3.5 Al + 0.1 Cu/0.2 Cu	70, 80, 90	3.5 Al + 0.1 Cu/0.2 Cu	60, 70, 80	3.5 Al + 0.2 Cu/0.3 Cu
Over-weight male	70, 80, 90	3.5 Al + 0.1 Cu/0.2 Cu	90, 100, 110	3.5 Al + 0.1 Cu/0.2 Cu	60, 70, 80	3.5 Al + 0.2 Cu/0.3 Cu
Over-weight female	70, 80, 90	3.5 Al + 0.1 Cu/0.2 Cu	80, 90, 100	3.5 Al + 0.1 Cu/0.2 Cu	60, 70, 80	3.5 Al + 0.2 Cu/0.3 Cu
Obese level-I male	70, 80, 90	3.5 Al + 0.1 Cu/0.2 Cu	100, 110, 120	3.5 Al + 0 Cu/0.1 Cu	60, 70, 80	3.5 Al + 0.2 Cu/0.3 Cu
Obese level-I female	70, 80, 90	3.5 Al + 0.1 Cu/0.2 Cu	90, 100, 110	3.5 Al + 0.1 Cu/0.2 Cu	60, 70, 80	3.5 Al + 0.2 Cu/0.3 Cu
Obese level-II male	70, 80, 90	3.5 Al + 0.1 Cu/0.2 Cu	100, 110, 120	3.5 Al + 0 Cu/0.1 Cu	60, 70, 80	3.5 Al + 0.2 Cu/0.3 Cu
Obese level-II female	70, 80, 90	3.5 Al + 0.1 Cu/0.2 Cu	100, 110, 120	3.5 Al + 0 Cu/0.1 Cu	60, 70, 80	3.5 Al + 0.2 Cu/0.3 Cu
Morbidly-Obese male	70, 80, 90	3.5 Al + 0.1 Cu/0.2 Cu	100, 110, 120	3.5 Al + 0 Cu/0.1 Cu	60, 70, 80	3.5 Al + 0.2 Cu/0.3 Cu
Morbidly-Obese female	70, 80, 90	3.5 Al + 0.1 Cu/0.2 Cu	100, 110, 120	3.5 Al + 0 Cu/0.1 Cu	60, 70, 80	3.5 Al + 0.2 Cu/0.3 Cu

DOI: 10.1051/epjconf/201715304016

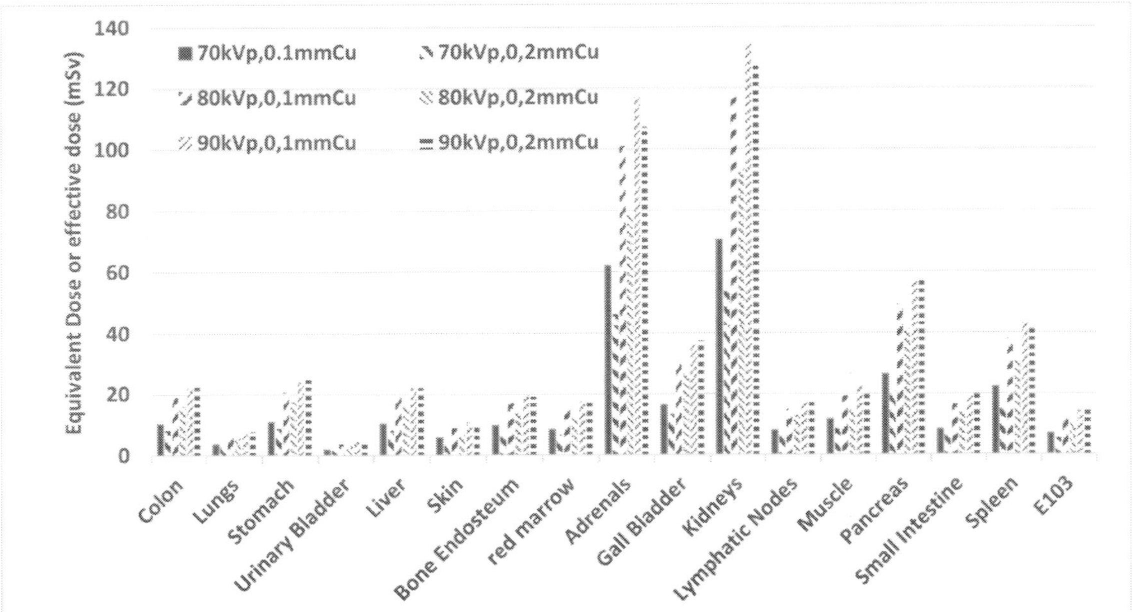

Figure 5. Comparison of organ dose data for different beam qualities.

+ 0.1 mm Cu are as almost two times high as those of 70 kVp +0.2 mm Cu.

This is mainly due to the fact that an increasing tube voltage can significantly enhance the beam energy while the increasing thickness of absorber would shield the low energy X-ray and thus reduce the X-ray intensity. The tube voltage determines the energy spectrum of the X-ray emitted from the source, the mean energy of X-ray emitted from source enhancing with increasing the tube voltage. The filtration impacts energy spectrum of X-ray reaching patient body, the mean energy of X-ray reaching patient also enhancing with increasing the thickness of filtration while the number of X-ray reaching patient body decreasing. Changing X-ray quantity can directly influence radiation –induced healthy effect of patient. Lower tube voltage and thicker filtrations should be employed at clinic to reduce patient's radiation risk as soon as possible.

3.2 Comparison of organ dose data for different beam projections

Figure 6 shows the results for different beam projections. The doses of adult male patients undergoing liver IR treatments with 90 kVp + 0.1 mm Cu, 40 cm×40 cm FOV and six beam projections (PA, LAO, RAO, CRAN, LLAT and RLAT) were generated from VirtualDose-IR. Differences among the effective doses of all six projections are not far from each other, while some obvious divergences of organ doses among different projections have been found, especially for organs locating at the field.

The first reason is that the irradiation regions inside the body are not completely equivalent for each projection even the target positions are the same. Secondly, the attenuation distances for X-ray reaching internal organs change with beam projections. Couple with projections change, some parts of organs slightly move out the radiation field while other parts of organs move in, some organs are shielded by thicker tissues whereas others are shielded by thinner ones. The organs that distributed throughout the body are hardly impacted by these little field changes and receiving similar absorbed organ dose. Those reasons leading effective doses do not fluctuate tremendous following the projection changes.

Furthermore, the most significant divergences were observed for organ doses between LLAT and RLAT, especially for the asymmetric organs and tissues. Because shielding tissues thicknesses of these asymmetric organs and tissues can be changed dramatically with X-ray shifted to opposite side. In clinical operations, doctors can employ suitable beam projections to reduce doses for certain organ at risk.

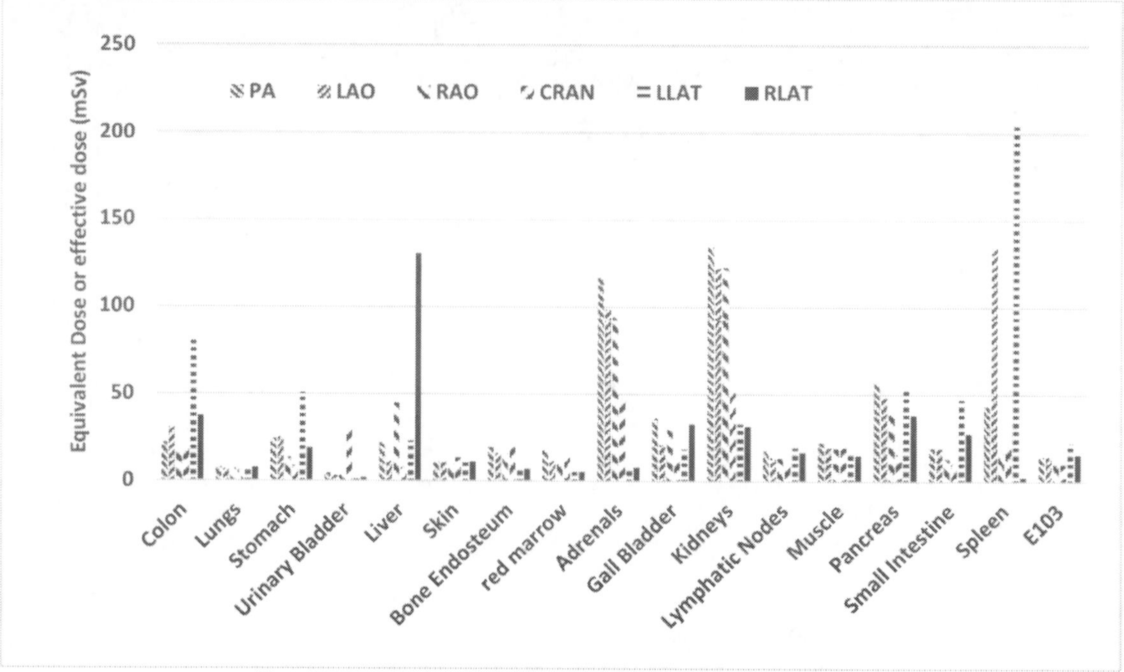

Figure 6. Comparison of organ dose data for different beam projections

3.3 Comparison of organ dose data for patients with different ages

A total of 4 liver IR treatments were independently performed on 5-year, 10-year, 15-year and adult male patients with the same conditions (70 kVp + 0.2mm Cu beam quality, 40 cm×40 cm FOV, PA beam projection). The results were summarized in figure 7. The effective doses and organ doses increase tremendously with age decreasing under this IR condition. Moreover, all the organ doses of 5-year male are much larger than others.

This is because the small size of paediatric patients will lead more organs and tissues to receive direct irradiations. Organs irradiated directly will receive far more radiation doses than organs which outside the irradiation region, because only a little parts of X-ray are scattered out the irradiation region. Moreover, the body sizes of young children also get smaller with age decreasing. Lacking other tissue shielding, the inner organs in a small body will receive more radiation dose.

As the adolescent are more sensitive to radiation rather than adult, especially for young children, the radiation doses for adolescence IR operations should be particularly concerned. The tube voltages performed in children IR treatments need to be set lower and filtrations employed need to be set thicker than adult treatments according to patient's age and body size.

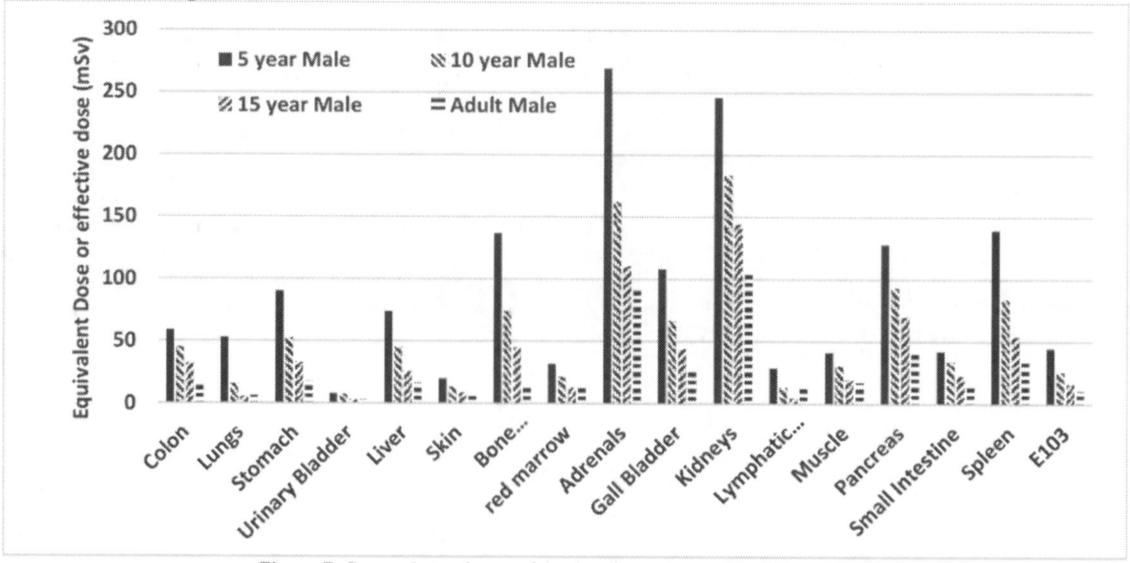

Figure 7. Comparison of organ dose data for patients with different ages

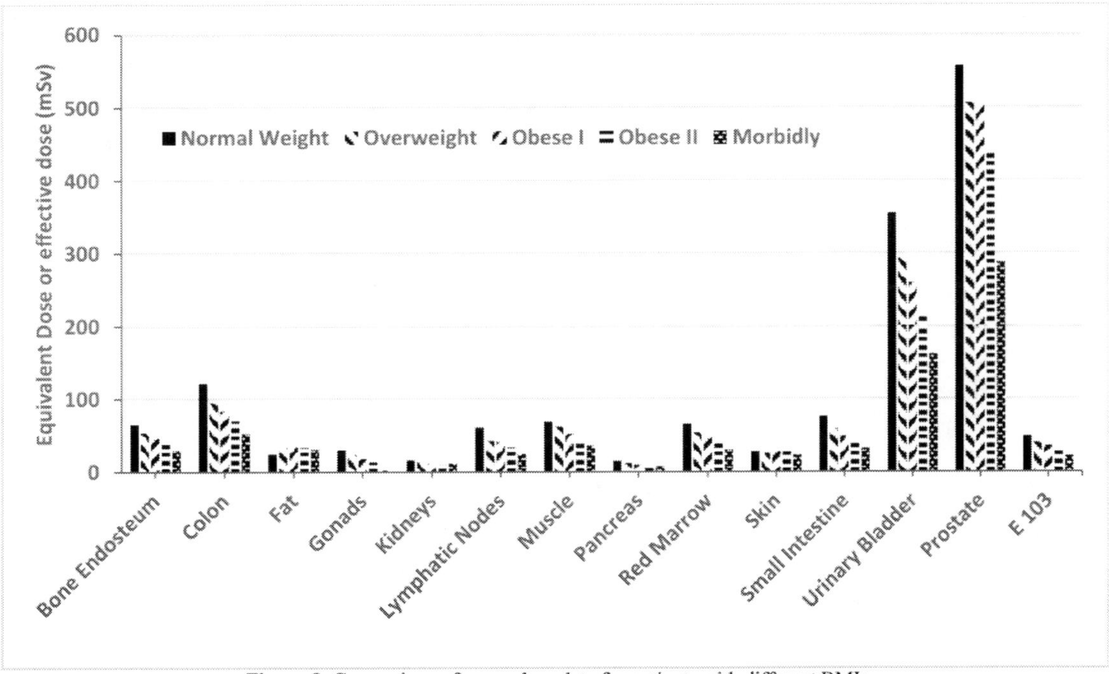

Figure 8. Comparison of organ dose data for patients with different BMIs

3.4 Comparison of organ dose data for patients with different BMIs

A small intestine IR treatment with 100 kVp + 0.1 mm Cu tube voltage, 40 cm×40 cm FOV and PA beam projection was independently performed on 5 male patients with different obesity (Normal weight, Over weight, Obese I, Obese II and Morbidly). Figure 8 shows that all the organ doses and effective doses are reducing with the enhancing of BMI. The morbidly obese patients receive minimum radiation doses compared to other type of patients in the same IR condition. Although on account of the effect that visceral fat occupies the original locations of some internal organs and changes their positions, the locations of those organs are not identical in those phantom. It still demonstrate the fact that BMI have great impact on doses of patient for an abdominal IR procedure.

These discrepancies are attributed to the fact that the abdominal fat has a great shielding effect against X-ray for internal organs. The abdominal fat is the major factor contributing to BMI increasing and has a great influence on body size, which will result in less radiation damage appear at obese patient compare with average adult and children.

But this shielding effect usually reduces image quality, so doctors need to enhance beam energy and intensity to get satisfied images during clinical operations. This result reveals that the BMI can impact patient doses and image quality markedly. Doctors are supposed to choose different IR plans for patient with different BMIs.

4 Conclusion

Based on a family of 23 anatomically realistic patient phantoms of both genders and various ages, organ dose database for the development of the VirtualDose-IR software has been reported successfully. The X-ray quality, beam projection, patient's age, patient's BMI and irradiation position have great influence on radiation dose of patients. When compared with previous researches, this database is found to be more suitable and flexible to evaluate radiation doses of patients in IR procedures because more IR parameters and irradiation positions were considered. Complying to the ALARA (As Low As Reasonably Achievable) principle in IR treatments, it is necessary to reduce the risk from the IR radiation exposure and optimize the medical benefit of IR procedures to patients. Clinical staffs can access to this database though VirtualDose-IR software to estimate radiation damage of patient before IR operation.

Acknowledgment

This project was supported in part by the National Institute of Biomedical Imaging and Bioengineering (NIBIB) (1R42EB010404). And by National Natural Science Foundation of China (Grant No. 11375182).

References

1. A. Ding, Y. Gao, H. Liu, P. F. Caracappa, D. J. Long, W. E. Bolch, B. Liu, X. G. Xu, PHYS. MED. BIOL **60**, 5601-5625 (2015)
2. A. Ding, J Gu, P. F. Caracappa, X. Xu, MED. PHYS **37**, 3463-3463 (2010)
3. A. Ding, Y. Gao, P. F. Caracappa, B. Liu, X. Xu,

MED. PHYS **39**, 3876-3877 (2012)
4. A. Ding, J Gu, M. Mille, X. Xu, B. Liu, J NUCL MED **51**, 520-520 (2010)
5. B. Liang, Y. Gao, Z. Chen, X. G. Xu, RADIAT. PROT. DOSIM , 1-10 (2016)
6. A. Ding, Y. Gao, H. Liu, P. F. Caracappa, D. J. Long, W. E. Bolch, B. Liu, M. Kalra, X. G. Xu, MED. PHYS **40**, 458-458 (2013)
7. A. P. Ding, *Development of a radiation dose reporting software for X-ray computed tomography (CT)* (PhD Thesis Rensselaer Polytechnic Institute, 2012)
8. A. P. Ding, J. W. Gu, A. V. Trofimov, X. G. Xu, Med. Phys. **37** 6199–6204 (2010)
9. J. Gu, B. Bednarz, P. F. Caracappa, X. G. Xu, Phys. Med. Biol. **54** 2699–2717 (2009)
10. J. W. Gu, X. G. Xu, P. F. Caracappa, B. Liu, Radiat. Prot. Dosim. **155** 64–72 (2013)
11. H. Liu, Y. Gao, A. Ding, P. F. Caracappa, X. G. Xu, Radiat. Prot. Dosim. **164** 368–375 (2015)
12. H. Liu, J. Gu, P. F. Caracappa, X. G. Xu, Phys. Med. Biol. **55** 1441–1451 (2010)
13. P. F. Caracappa, J. Gu, X. G. Xu, Med. Phys. 37 3100 (2010)
14. D. A. Schauer, O. W. Linton, HEALTH. PHYS **97**, 1-5 (2009)
15. US Food and Drug Administration. *Initiative to reduce unnecessary radiation exposure from medical imaging* (FDA, US, 2010)
16. F. A. Stewart, A. V. Akleyev, M. Hauer-Jensen, J. H. Hendry, N. J. Kleiman, T. J. MacVittie, B. M. Aleman, A. B. Edgar, K. Mabuchi, C. R. Muirhead, R. E. Shore, W. H. Wallace, Annals of the ICRP **41**, 1-322 (2012)
17. J. C. Waite, M. Fitzgerald, RADIAT. PROT. DOSIM, **94**, 89-92 (2001)
18. C. Theodorakou, J. A. Horrocks, BRIT. J. RADIOL, **76**, 546-552 (2014)
19. K. Chugh, P, D. R. Bednarek, D. Wobschall, S. Rudin, K. Hoffmann, R. Peterson, M. Zeng, Medical Imaging 2004 Conference **5367**, 464-473 (2004)
20. Y. Khodadadegan, M. Zhang , W. Pavlicek, R. G. Paden, B. Chong, B. A. Schueler, K. A. Fetterly, S. G. Langer, T. Wu, J. DIGIT. IMAGING **24**, 626-639 (2011)
21. P. B. Johnson, D. Borrego, S. Balter, K. Johnson, D. Siragusa, W. E. Bolch, MED. PHYS **38**, 5490-5499 (2011)
22. R. Kramer, H. J. Khoury, J. W. Vieira, PHYS. MED. BIOL **53**, 6437-6459 (2008)
23. A. Bozkurt, D. Bor, PHYS. MED. BIOL **52**, 317-330 (2007)
24. W. J. Garzón, R. Kramer, H. J. Khoury, V. S. M. D. Barros, G. Andrade, J. RADIOL. PROT **35**, 629-647 (2015)
25. ICRP, *The 2007 Recommendations of the International Commission on Radiological Protection* (Oxford: Pergamon, 2007)
26. J. Zhang, Y. H. Na, P. F. Caracappa, X. G. Xu, Phys. Med. Biol. **54** 5885–5908 (2009)
27. X. G. Xu, V. Taranenko, J, Zhang, C. Shi, Phys. Med. Biol. **52** 7023–7044 (2007)
28. X. G. Xu, K. F. Eckerman, *Handbook of Anatomical Models for Radiation Dosimetry* (Boca Raton, FL: CRC/Taylor and Francis Group, 2009)
29. X. G. Xu, Phys. Med. Biol. **59** R233–302 (2014)
30. A. P. Ding, M. M. Mille, T. Y. Liu, P. F. Caracappa, X. G. Xu, Phys. Med. Biol. **57** 2441–2459 (2012)
31. Y. H. Na, B. Zhang, J. Zhang, P. F. Caracappa, X. G. Xu, Phys. Med. Biol. **55** 3789–3811
32. Y. Pi, M. Feng, W. Huo, L. Zhang, T. Liu, H. Lin, L. Yang, F. Zheng, H. Tan, F. Pan , Z. Chen, X. G. Xu, 5th International workshop om computational human phantoms, (2015)
33. M. Feng, A. Ding, Y. Gao, Z. Chen, X. G. Xu, 5th International workshop om computational human phantoms, (2015)
34. W. Huo, Y. Wang, X. Zhuang, Y. Pi, Z. Chen, X. G. Xu, 5th International workshop om computational human phantoms, (2015)
35. Ministry of Health, The People's Republic of China, *Dose conversion coefficients for use in radiological protection against photon external radiation* (The People's Republic of China National Standard of Occupation Health, 2002)
36. D. B. Pelowitz, *MCNPX User's Manual Version 2.7.0* (LLNL, Los Alamos, 2011)
37. C. Koukorava, J. Farah, L. Struelens, I Clariand, L. Donadille, F. Vanhavere, P. Dimitriou, J. Radiat. Prot. **34** 509-528 (2014)
38. E. Carinou, P. Ferrari, C.Koukorava, S. Krim, L. Struelens, Radiat. Prot. Dosim **144**, 492-496 (2011)
39. AAPM, *Functionality and Operation of Fluoroscopic Automatic Brightness Control/Automatic Dose Rate Control Logic in Modern Cardiovascular and Interventional Angiography Systems* (The American Association of Physicists in Medicine, 2012)

Validation of Monte Carlo simulation of mammography with TLD measurement and depth dose calculation with a detailed breast model

Wenjing Wang[1, 2], Rui Qiu[1, 2, a], Li Ren[1, 2], Huan Liu[1, 2], Zhen Wu[3], Chunyan Li[3] and Junli Li[1, 2]

[1]Department of Engineering Physics, Tsinghua University, Beijing, China
[2]Key Laboratory of Particle & Radiation Imaging (Tsinghua University), Ministry of Education, Beijing, China
[3]Nuctech Company Limited, Beijing, China

Abstract. Mean glandular dose (MGD) is not only determined by the compressed breast thickness (CBT) and the glandular content, but also by the distribution of glandular tissues in breast. Depth dose inside the breast in mammography has been widely concerned as glandular dose decreases rapidly with increasing depth. In this study, an experiment using thermo luminescent dosimeters (TLDs) was carried out to validate Monte Carlo simulations of mammography. Percent depth doses (PDDs) at different depth values were measured inside simple breast phantoms of different thicknesses. The experimental values were well consistent with the values calculated by Geant4. Then a detailed breast model with a CBT of 4 cm and a glandular content of 50%, which has been constructed in previous work, was used to study the effects of the distribution of glandular tissues in breast with Geant4. The breast model was reversed in direction of compression to get a reverse model with a different distribution of glandular tissues. Depth dose distributions and glandular tissue dose conversion coefficients were calculated. It revealed that the conversion coefficients were about 10% larger when the breast model was reversed, for glandular tissues in the reverse model are concentrated in the upper part of the model.

1 Introduction

The average dose to the glandular tissue (mean glandular dose, MGD, or average glandular dose, AGD) is used to estimate the risk of breast cancer induced by ionizing radiation by ACR [1], EUREF [2] and IAEA [3]. MGD is estimated by the product of the measured incident air kerma at the upper surface of a breast and the conversion coefficient calculated using Monte Carlo codes. Conversion coefficients for breasts of different glandular contents and different compressed thicknesses (CBTs) are presented by Dance et al [4-7], which are adopted by the IAEA and EUREF protocols. Conversion coefficients presented by Wu et al [8, 9], which are adopted by the ACR protocol, are also listed by glandular content and CBT. The calculations are all based on a simple model of breast developed by Hammerstein et al [10].

However, MGD depends not only on the glandular content and the CBT, but also on the distribution of glandular tissues within the breast [11]. Depth dose distribution in breast in mammography, which is expressed as percent depth dose (PDD) or relative depth dose, has been widely concerned as the glandular dose decreases rapidly with increasing depth. Relative exposure was plotted as a function of depth in a phantom by Hammerstein et al [10], which was measured with Thermo Luminescent Dosimeters (TLDs) TLD-100. Comparison between depth exposure measured by Hammerstein et al [10] and that calculated with Monte Carlo codes was carried out by Wu et al [8].

A set of PDD curves were derived by a wide range of different combinations of target materials, filters and tube voltages with water phantom based on Monte Carlo simulation by Delis et al [12]. The PDD curves in 4.5 cm homogeneous breast phantom of 30% glandular and 70% adipose composition, 50% glandular and 50% adipose composition, and 70% glandular and 50% adipose composition were measured with TLD GR-200F by Tsai et al [13]. The PDD curves in 7 cm PMMA phantom for Mo/Mo, Mo/Rh and Rh/Rh target-filter combinations were measured with TLD-100 by Camargo-Mendoza et al [14]. Comparison of Monte carlo simulations and TLD-100H measurements of PDD in a 4 cm homogeneous breast phantom of 50% glandular and 50% adipose tissue by weight was carried out by Nigapruke et al [15] for verification of the EGSnrc MC code. The PPD curve was also measured with TLD-100 by Maria et al [16] in a phantom with composition of an average breast with 50% adipose and 50% glandular tissue, which was enveloped inside an adipose equivalent tissue case of 0.5 cm thickness. The TLD measurements were used to validate the dose values calculated using Monte Carlo programs of MCNPX and PENELOPE.

In this study, TLD measurements with simple breast phantoms are used to validate the simulations with Geant4. And a detailed breast model constructed in

[a] Corresponding author: qiurui@mail.tsinghua.edu.cn

DOI: 10.1051/epjconf/201715304017

previous work is used to study the effects of the distribution of glandular tissues in breast on MGD in mammography.

2 Materials and methods

2.1 Dose measurements with TLD

The GE Senographe DS Mammomat was used to perform the measurements. A Mo target with a 0.03 mm Mo filter was chosen. Half value levels (HVLs) at different X-ray tube voltages (25 kV, 28 kV, 30 kV, 32 kV) were obtained from the measurements with TLD GR200A.

PDD values in breast phantoms of 3 cm, 4.5 cm and 6 cm thicknesses in mammography were obtained from the measurements. The CIRS breast phantoms were used to perform the measurements, which include tissue equivalent slabs enabling a range of thickness from 0.5 cm to 7 cm. PDD values at 28 kV were obtained from the measurements with placing TLDs at different depth values in the breast phantoms. Figure 1 shows the sketch of depth dose measurements with a 4.5 cm breast phantom.

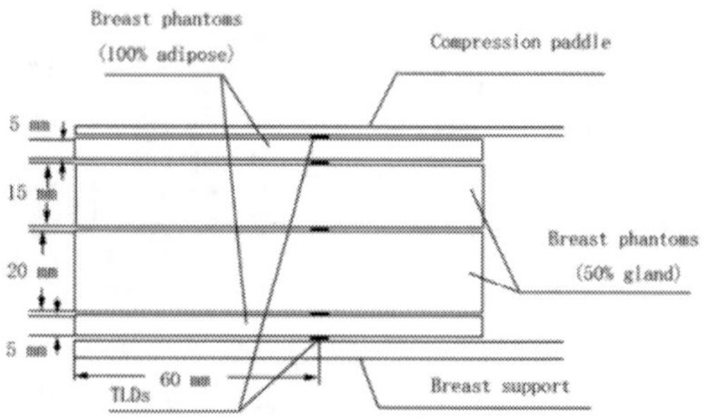

Figure 1. Sketch of depth dose measurements

2.2 Monte Carlo simulations with Geant4

Geant4 was used to perform the simulations. Figure 2 shows the sketch of the geometry in the simulations. The X-ray spectra data were taken from Boone et al [17].

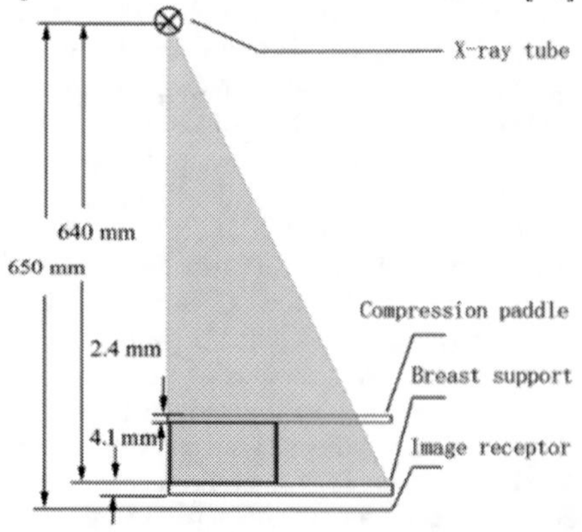

Figure 2. Sketch of the geometry in the simulation

The comparison between HVLs measured and simulated is shown in figure 3. The comparison of PDD values is shown in figure 4. The values obtained from simulations are well consistent with these obtained from measurements.

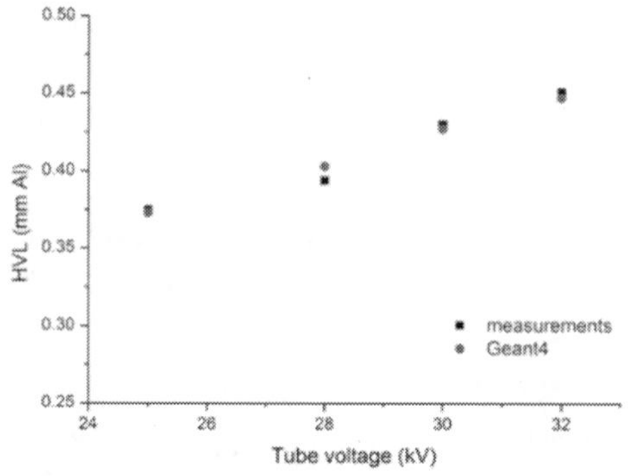

Figure 3. Comparison between HVLs measured and simulated

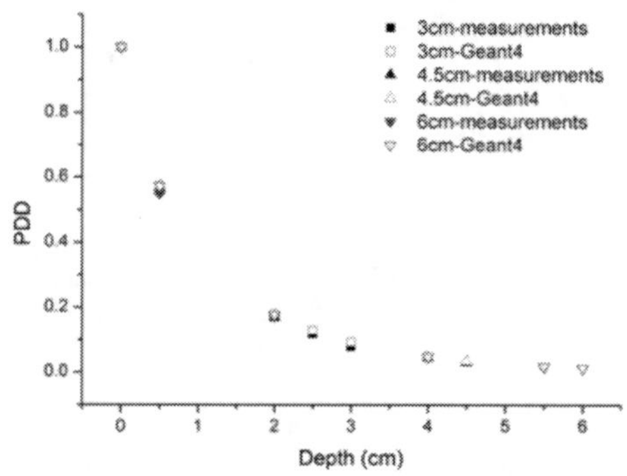

Figure 4. Comparison between PDD values measured and simulated

3 Results and discussions

The average value of compressed breast thickness (CBT) obtained from 4226 CC views from 2450 Taiwanese women is 4.1 cm, and the average glandularity obtained from 3910 CC views is 54% [18]. Therefore a 3D detailed breast model with a CBT of 4 cm and a glandularity of 50%, which has been constructed in previous work, is used to study the effects of the distribution of glandular tissues in breast on MGD.

The detailed compressed breast model was constructed for a more precise breast dose estimation in mammography in previous work, which included skin, subcutaneous fat, Cooper's ligaments, intraglandular fats, ductal trees and lobules, retromammary fat and so on. It is a voxel model with a voxel size of 0.2 mm × 0.2 mm × 0.081 mm, which is shown in figure 5.

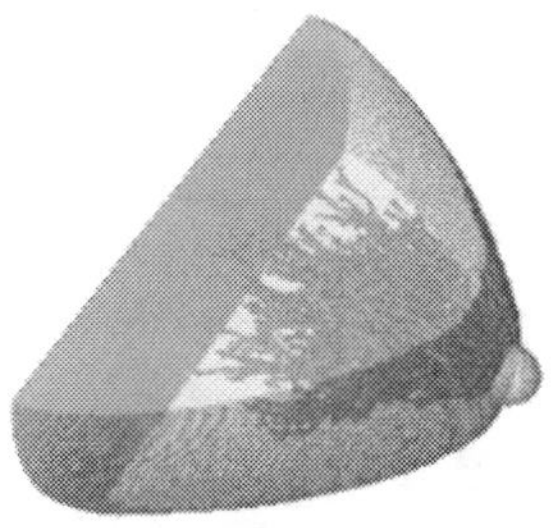

Figure 5. A detailed breast model

To study the effects of the distribution of glandular tissues, the breast model is reversed in direction of compression so that it has the same CBT and glandularity but a different distribution of glandular tissues. The distributions of glandular tissues in the detailed breast model (Breast) and the reverse model (Breast-rev) are shown in figure 6. It shows that glandular tissues in the reverse model are concentrated in the upper part of the model. The difference between the distributions of glandular tissues in the central part is not obvious.

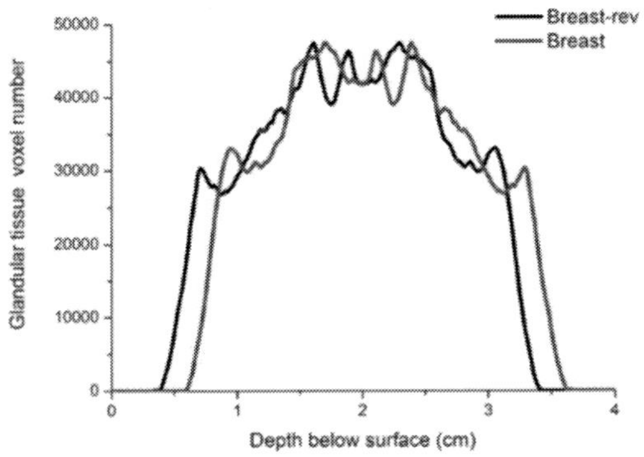

Figure 6. Distributions of glandular tissues in the detailed breast model and in the reverse model

The sketch of the geometry in the simulations is shown in figure 2. Glandular tissue dose conversion coefficients calculated with the detailed breast model (cc) and the reverse model (cc-rev) are shown in figure 7. The conversion coefficients obtained from Dance et al [5] are also shown (Dance). Relative depth doses at different HVLs are shown in figure 8.

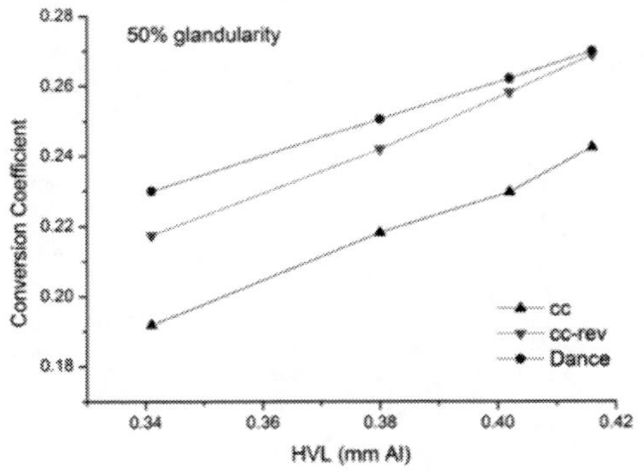

Figure 7. Glandular tissue dose conversion coefficients

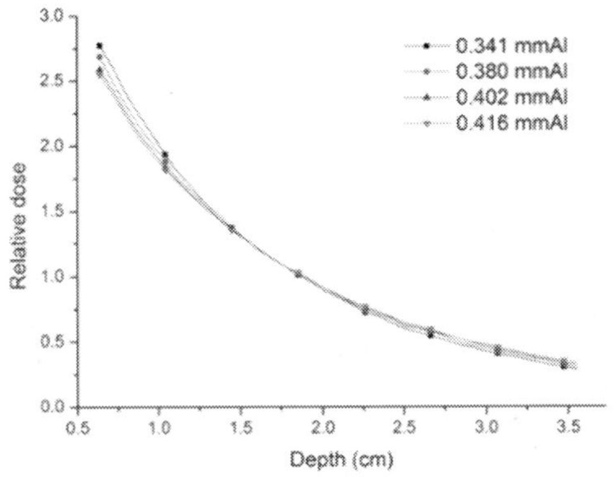

Figure 8. Relative depth doses at different HVLs

It shows that glandular tissue dose conversion coefficients are about 10% larger when the breast model is reversed. It is because that glandular tissues in the reverse model are concentrated in the upper part of the model from figure 6. Glandular tissue dose conversion coefficients calculated with the reverse model are better consistent with these obtained from Dance et al [5] and it consists better when the HVL is larger. It may be attributed to two factors. One factor that the adipose fat layer in the reverse model is about 5 mm (figure 6), which is better consistent with the simple breast model used by Dance et al [5]. Another factor is that glandular tissues in the reverse model are concentrated in depth from 1.5 cm to 2.5 cm while glandular tissues in the simple breast model distribute evenly in depth from 0.5 cm to 3.5 cm. Relative depth doses change less rapidly for a larger HVL from figure 8 therefore it consists better for a larger HVL.

4 Conclusions

The simulations with Geant4 are validated through an experiment with TLDs in this study. The values obtained from simulations are well consistent with these obtained from measurements. A detailed breast model constructed in previous work is used to study the effects of the distribution of glandular tissues in breast on MGD. The breast model is reversed in direction of compression to get a reverse model with the same CBT and glandular content but a different distribution of glandular tissues. Glandular tissue dose conversion coefficients calculated with the reverse model are better consistent with these obtained from Dance et al and it consists better when the tube voltage is larger. It is because that glandular tissues in the reverse model are concentrated in the upper part than the detailed model and in the more central part than the simple model.

Acknowledgment

This work was supported by the National Natural Science Foundation of China [grant number 11375103, 11275110].

References

1. ACR, *Mammography Quality Control Manual*, America: American College of Radiology (1999)
2. EUREF, *European guidelines for quality assurance in breast cancer screening and diagnosis*, The Netherlands: European Communities (2003)
3. IAEA, *Dosimetry in diagnostic radiology: an international code of practice*, Technical Reports Series no. 457 (2006)
4. Dance D R, Phys. Med. Biol., **35**, 1211-1219 (1990)
5. Dance D R, Skinner C L, Young K C, Beckett J R, Kotre C J, Phys. Med. Biol., **45**, 3225-3240 (2000)
6. Dance D R, Skinner C L, Young K C, Van Engen R E, Phys. Med. Biol., **54**, 4361-4372 (2009)
7. Dance D R, Young K C, Phys. Med. Biol., **59**, 2127-2137 (2014)
8. Wu X, Barnes G T, Tucker D M, Radiology, **179**, 143-148 (1991)
9. Wu X, Gingold E L, Barnes G T, Tucker D M, Radiology, **193**, 83-89 (1994)
10. Hammerstein G R, Miller D W, White D R, Masterson M E, Woodard H Q, Laughlin J S, Radiology, **130**, 485-491 (1979)
11. Zankl M, Fill U, Hoeschen C, Panzer W, Regulla D, Radiat. Prot. Dosim., **114**, 406-409 (2005)
12. H. Delis, G. Spyrou, G. Tzanakos, G. Panayiotakis, Radiat. Meas, **39**, 149-155 (2005)
13. H.Y. Tsai, N.S. Chong, Y.J. Ho, Y.S. Tyan, Radiat. Meas., **45**, 726-728 (2010)
14. Raúl E. Camargo-Mendoza, Martin E. Poletti, Alessandro M. Costa, Linda V.E. Caldas, Radiat. Meas, **46**, 2086-2089 (2011)
15. Khatayut N., Patana P., Nakorn P, Wiwat Y, J. Radiat. Res., **51**, 441-448 (2010)
16. S. Di Maria, S. Barros, J. Bento, P. Teles, C. Figueira, M. Pereira, P. Vaz, G. Paulo, *Radiat. Meas*, **46**, 1103-1108 (2011)
17. Boone J M, Fewell T R, Jennings R J, Med. Phys., **24**, 1863–1874 (1997)
18. S. L. Dong, *The development and application of a model breast and equivalent phantoms in mammography*, Ph.D. Dissertation: Department of Biomedical Engineering and Environmental Sciences National Tsing Hua University (2010)

A preliminary Monte Carlo study for the treatment head of a carbon-ion radiotherapy facility using TOPAS

Hongdong Liu[1], Lian Zhang[1], Zhi Chen[1], Xinguo Liu[2], Zhongying Dai[2], Qiang Li[2], and Xie George Xu[1,3]*

[1]School of Nuclear Science and Technology, University of Science and Technology of China Hefei 230027, China
[2] Institute of Modern Physics, Chinese Academy of Sciences, Lanzhou 730000, China
[3] Rensselaer Polytechnic Institute, Troy, New York, USA

Abstract. In medical physics it is desirable to have a Monte Carlo code that is less complex, reliable yet flexible for dose verification, optimization, and component design. TOPAS is a newly developed Monte Carlo simulation tool which combines extensive radiation physics libraries available in Geant4 code, easy-to-use geometry and support for visualization. Although TOPAS has been widely tested and verified in simulations of proton therapy, there has been no reported application for carbon ion therapy. To evaluate the feasibility and accuracy of TOPAS simulations for carbon ion therapy, a licensed TOPAS code (version 3_0_p1) was used to carry out a dosimetric study of therapeutic carbon ions. Results of depth dose profile based on different physics models have been obtained and compared with the measurements. It is found that the G4QMD model is at least as accurate as the TOPAS default BIC physics model for carbon ions, but when the energy is increased to relatively high levels such as 400 MeV/u, the G4QMD model shows preferable performance. Also, simulations of special components used in the treatment head at the Institute of Modern Physics facility was conducted to investigate the Spread-Out dose distribution in water. The physical dose in water of SOBP was found to be consistent with the aim of the 6 cm ridge filter.

1 INTRODUCTION

Over the past 30 years, ion beam radiotherapy using particles like proton, helium, carbon ions has been gradually and slowly increasingly in popularity in radiation oncology [1-3].The inverted depth dose profile —the increase of dose along the penetration depth (the Bragg Peak) and a sharp drop after the Bragg Peak — make ion beams an ideal tool for treatment of the deep-seated tumours [1-7]. For carbon ions, the excellent dose profile is further boosted by an additional increase in the relative biological effectiveness (RBE) along the depth till the end of the particle range [2-4, 6, 8].Because of the distinguished physical and biological properties, carbon ion radiotherapy (CIRT) has become a unique research focus around the world. In China, after almost 10 years of research about heavy ion therapy at the Institute of Modern Physics (IMP), Chinese Academy of Sciences (CAS), sophisticated and advanced techniques of beam delivery and treatment planning system have been established [5, 6]. Since 2006, a total of 213 patients have been treated using carbon ions and most of them are observed with impressive therapeutic effect [5, 6]. Encouraged by the superior clinical results, two dedicated heavy-ion therapy facilities are now under construction in Lanzhou and Wuwei, China, respectively. Both of them are expected to start treatment in the next few years [6].

Monte Carlo technique has become ubiquitous in medical physics [8].Monte Carlo particle transport codes are practical and useful in calculating the radiation fields and dose distributions. They have also been widely used in hadron therapy as an alternative to fast but less accurate analytical algorithms or to provide data when it is impossible to conduct a physical experiment [9-11]. A number of studies have shown that Monte Carlo simulation is the most accurate method to calculate doses in radiotherapy [9-12]. Indeed, general purposed Monte Carlo codes such as Geant4, FLUKA, PHITS, and MCNPX are reportedly used for radiotherapy to validate the clinical treatment plans that are mostly calculated by using analytical algorithms [12, 13]. Clearly, the accuracy of the Monte Carlo codes is of the foremost importance as it will ultimately impact the quality of the treatment plan. In reality, however, the special components of the treatment head used in ion beam radiotherapy are too complicated to be faithfully modelled using those general purpose Monte Carlo codes [12]. TOPAS ("TOol for PArticle Simulation") was developed as a new Monte Carlo research platform that is based on the toolkit Geant4 ("GEometry ANd Tracking 4"), and was specially designed for particle radiotherapy simulations, calculations and visualizations [12, 14, 15]. The aim of TOPAS code is to provide an easier-to-use, efficient but still accurate applications for medical physicists [12]. Although this tool has been widely tested and verified in simulation of proton

* Corresponding author: xgxu@ustc.edu.cn

therapy, TOPAS utilization in carbon ion therapy has not been reported [12, 14, 15].

To evaluate the feasibility and accuracy of TOPAS simulations in carbon ion therapy, a licensed TOPAS code (version 3_0_p1) was used to perform dosimetric study for carbon ion therapy.

2 Method

2.1 Structure of TOPAS code

TOPAS is a particle transport simulation code developed with the specific aim of making Monte Carlo simulations user-friendly for users in the particle therapy community. Currently in the 3_0_p1 version, TOPAS is designed as a "user code" layered on top of Geant4, which includes the standard Geant4 toolkit, plus additional codes to access Geant4 functionality [12]. TOPAS can model quantities such as passive scattering or scanning beam treatment heads, model a patient geometry based on CT (computed tomography) images, score dose and fluence, save and restart phase space [12]. It provides advanced graphics and is fully four-dimensional (4D) views to handle variations in the beam delivery and patient geometry during treatment [12]. Initial TOPAS validation results show agreement with measured data at the University of California San Francisco Comprehensive Cancer Center involving the eye treatment head, MGH stereotactic alignment in radiosurgery treatment head and MGH treatment heads, which are simulated in scattering and scanning modalities [12].

The heart of TOPAS is the parameter control system which allows the user to control nearly everything by constructing txt-files without the need of detailed knowledge of Geant4 [12, 16]. In TOPAS, it is the user input "parameter files" that specify everything: geometry, particle source, fields, motion, scoring, graphical output and physics settings [12, 16]. Every command line in a txt-file has the same structure: "Parameter-Type: Parameter-Name = Parameter-Value" and the order of command lines does not matter [16]. With the correct format of "parameter files", one can run TOPAS as a command-line program with the name of the top level parameter file. That file includes whatever other parameter files are needed. Each parameter file is a simple text file consisting of one or more lines, specifying either an include file or a parameter definition. Each parameter definition line has the same easily mastered format that specifies a parameter type, parameter name, parameter value [16].

2.2 Physics model for ion simulations in TOPAS

TOPAS provides a list of default physics models for proton simulation and it has been carefully validated by clinical proton beam measurements at Massachusetts General Hospital [12, 14, 17].Now for the new version of TOPAS (3_0_p1), the physics models in the default list includs: "g4em-standard_opt4" "g4h-phy_QGSP_BIC_HP" "g4decay" "g4ion-binarycascade" "g4h-elastic_HP" and "g4stopping" [16]. This list includes models that handle not only protons but also all secondary particles (neutrons, helium ions, deuterons, tritons, photons, electrons) produced by proton interaction with targets. The default physics list provided by TOPAS is the kind 'modular physics list' in Geant4, which allows that multiple physics 'processes' and 'models' can be assigned as alternatives to the same physical process in the user's application, the actual physics setting being decided upon at run time [17]. Unlike Geant4, the users do not have to specify their choice of physics models and settings in a piece of C++ code. There is no single default physics in Geant4 and the variety of physics models and many adjustable settings within each model may lead to the dilemma that various groups in different fields are using their own physics lists which may or may not be properly validated. Of course, users can also further adjust settings from their parameter file, turning off various processes, adjusting the step size or the range cut, etc. Adjustments may be set independently in different geometry components, allowing, for example, more detailed simulation in an ionization chamber [12]. Parameters also allow the user to replace the TOPAS physics list with any of the reference physics lists included in Geant4. As a result, TOPAS provides user an accurate but still flexible default physics models for proton therapy simulation in case some users are not expert enough. But can the default physics models be directly used for carbon ion simulation? Maybe corresponding changes and adjustments are needed when it comes to carbon ions with higher energy, since the default physics models have only been verified for therapeutic proton simulation.

The default modular physics list can be mainly characterized by addressing the following types of interactions: (1) electromagnetic process, (2) elastic scattering, (3) inelastic scattering of protons and neutrons and (4) inelastic scattering of heavier ions [17]. For example, the standard electromagnetic package is an analytical model that derives directly from QED calculations and describes the interactions of photons and all charged particles down to 1 keV [17, 18]. The energy loss of hadrons is calculated by the Bethe-Bloch formula down to 2 MeV, below which a parameterization based on ICRU 49 is implemented [17]. Scattering is considered by a condensed history algorithm with functions to calculate angular and spatial distributions that match the Lewis theory, validated with an electron scatter benchmark [12, 19]. After development by years, some of these models have recently undergone major improvements and corrections. For example, according to a benchmark work about hadron therapy both in measurements and Monte Carlo simulation, it is revealed that G4QMD model is preferable over the BIC model for the correct prediction of fragment yields [10]. G4QMD and simultaneous inclusion of de-excitation models will result in a better agreement with the experimental data compared with the BIC model [10]. While in TOPAS, the intranuclear cascade propagating interactions is addressed by the Geant4 BIC model. As a result, for carbon ion radiotherapy simulation in TOPAS, it still needs to investigate if the BIC model in default physics models should be replaced by the G4QMD model.

2.3 Description of the treatment head at IMP

After verification and adjustment in physics models for carbon ion simulation, a study was conducted to model the treatment head which are practically used at IMP treatment terminal. At the IMP carbon ion oncology center, the ion beam is firstly deflected by scanning magnet to form a uniform irradiation field. Then carbon ions will go through the passive scattering modality treatment head in order to produce the desired dose distribution in PTV (Planning Target Volume). Figure 1 shows the simplified layout of the passive scattering treatment head at IMP. The treatment head is a complicated system which usually consists of various kinds of special components such as collimator, range shifter, ridge filter, ion chambers, aperture, etc. [11, 20]. For the dose calculation and verification, four important components (As seen in figure 1) are practically constructed in this simulation. The primary collimator is a component made of 4 identical copper bars with each thickness of 7.5cm, whose function is blocking the unnecessary particle and controlling the original size of the treatment field. Range shifter, which is constituted by various dimensioned Lucite slices (mass density ρ= 1.19kg/m3), is used to modulate the depth of the pristine Bragg Peaks. The ridge filter is a stationary plate shaped device made of aluminum typically (mass density ρ=2.7 kg/m3), and consists of about a total of 40 ridge bars abreast placed every 5 mm in this modelling. The beam energy will be degraded by bring in the ridge filter [21, 22]. After the ridge filter, the Spread-Out Bragg peak will be produced to cover the target in depth with a relatively homogeneous RBE-weighted dose [2-4]. The SOBP dose distribution span depends on the period of the ridge filter structure and the angular straggling of the ridge filter placement. Multi-Leaf Collimator (MLC), which is made of the tungsten copper alloy (W 90%, Cu 10% and mass density ρ= 16.75kg/m3), is a component specially chosen and customized for the individual case to block the unwanted particles. In this study, the MLC arrangement of each slice is consistent with the previous treatment case at IMP. All these special components have been carefully optimized and refined over years at IMP to achieve a high degree of dose conformation to the target volume while protect the OAR (Organs At Risk).

Fig. 1. Layout diagram for the simulation of the treatment head (the four components are primary collimator (No.1), range shifter (No.2), ridge filter (No.3) and multi-leaf collimator (No.4), respectively). The photographs are the multi-leaf collimator and range shifter routinely used at IMP.

3 Results

3.1 Verification of TOPAS default physics model for proton and carbon ions simulation

Figure 2 shows the pristine Bragg Peaks for mono-energetic proton and carbon ions calculated using TOPAS. Figure 2a is the calculation results of proton beam with energy from 50 to 250 MeV. Figure 2b is the analogous results for carbon ions. Both of the simulation were performed based on the TOPAS default physics models. As can be seen from the Figure 2a, when the primary energy of proton beam is lower, for example less than 150 MeV, the proton Bragg Peak in water is sharp and the dose level is higher at the peak region. But for proton with higher energy of 200 MeV or greater, the

Bragg Peak region will be broadened due to severe energy straggling for protons with such high energies. But for carbon ions of 400 MeV/u, the pristine Bragg Peak is still sharp confirming that carbon ions are much heavier than protons and therefore the energy straggling effects are less important.

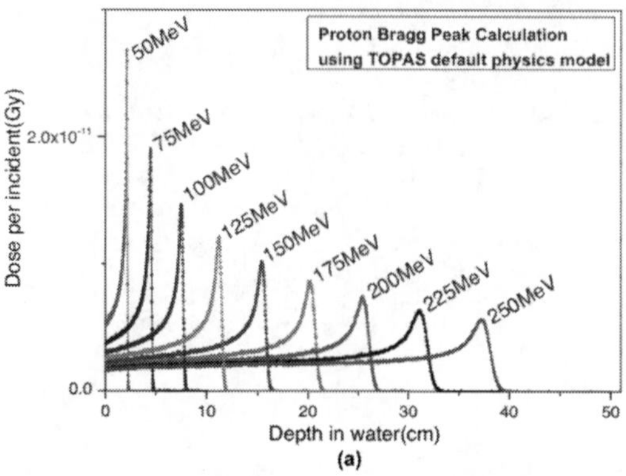

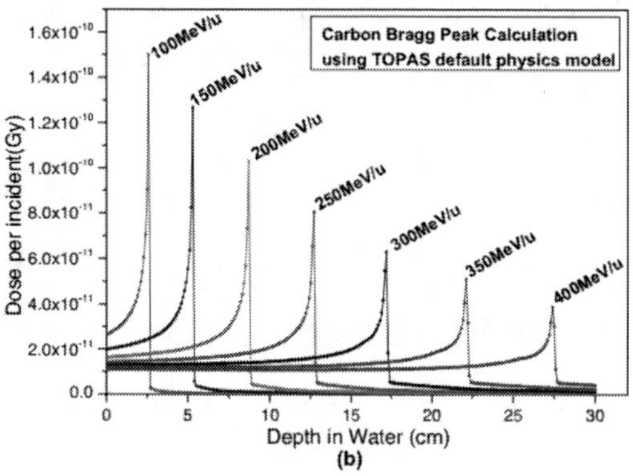

Fig. 2. Depth dose distribution of proton (a) and carbon ions (b) based on TOPAS default physics models.

Although the TOPAS default physics models for proton therapy simulation have been testified and validated by many studies [14, 15]. As discussed earlier in section 2.2, TOPAS will evolve following the Geant4. In TOPAS v3_0_p1, the default physics models have been slightly changed. For example, the electromagnetic model has been modified from opt3 to opt4. For comparison, Figure 3 provides the verification for proton ranges using the PSTAR range data from National Institute of Standard and Technology [23]. The mean projected range (or range) of a proton beam, R_0, is defined as the penetrating depth in the material where half of the protons have been stopped by undergoing electromagnetic interactions only. It also corresponds to 80% of the maximum dose after the Bragg peak [24]. It can be seen from the Figure 3 that the default physics models are fairly accurate for proton calculations with negligible differences (less than 2%).

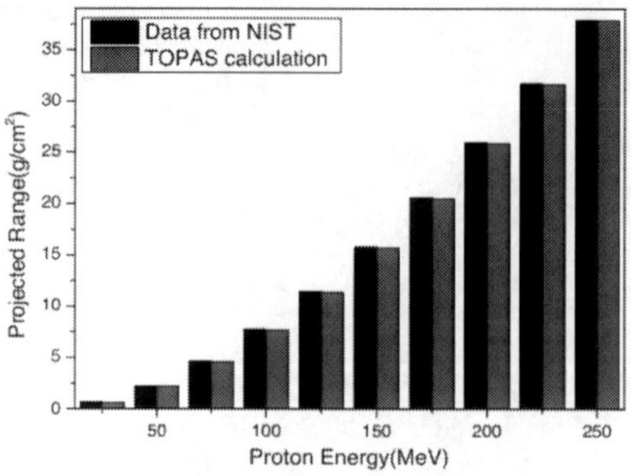

Fig. 3. Comparison of proton ranges calculated using TOPAS and NIST data

For carbon ions simulations, TOPAS does not provide corresponding physics models, so the precision of result using the TOPAS default physics models still needs further investigation. Furthermore, several studies reported that the QMD model are preferable than the BIC model, while the TOPAS default physics models utilized the BIC model and the QMD model is abandoned [10, 16]. As a result, it is necessary to compare and determine the more accurate physics model. Figure 4 compares the results obtained by QMD and BIC models. Both calculations are verified by the measurements data from IMP experiment. It was found that both the QMD and BIC models can give highly accurate results, especially for carbon ions with energy less than 300 MeV/u. But when carbon ion energy reaches 400 MeV/u, slight differences are observed before and after the Bragg Peak. As seen from Figure 4 (d), the QMD result curve agrees better with the measurement, suggesting that the QMD model is more accurate than BIC model when carbon ions are at a high energy. Noted that the vertical axis of Figure 4 is the relative dose rather than the absolute dose.

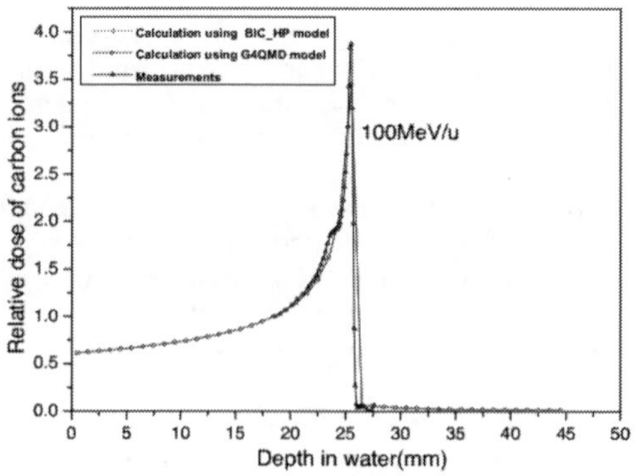

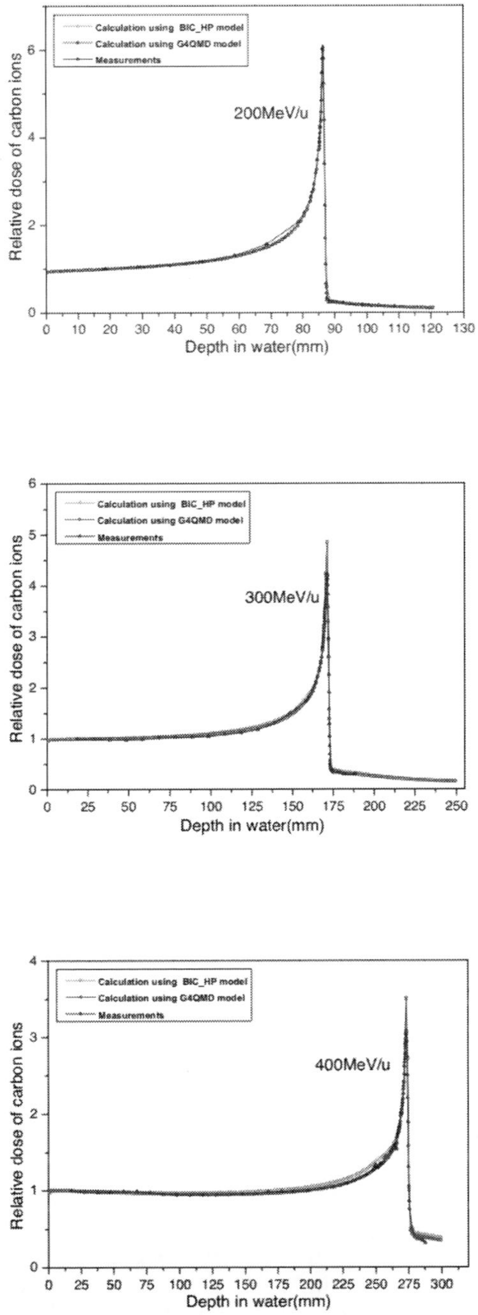

Fig. 4. Comparison of QMD and BIC model with the measurement for carbon ion from 100 MeV/u to 400 MeV/u.

3.2 Treatment head modelling and SOBP calculation

As a preliminary study, a simplified model of the IMP treatment head was constructed to verify the spread-out dose distribution in water. But even in a real treatment case, only the ionization chamber, collimator and the compensator are included, which would have a slight effect on the SOBP study. The full modelling of a more realistic treatment head is now under progress. Figure 5 is the visualization for the special components of the treatment head, and the detailed sketches of the ridge filter and multi-leaf collimator are supplied.

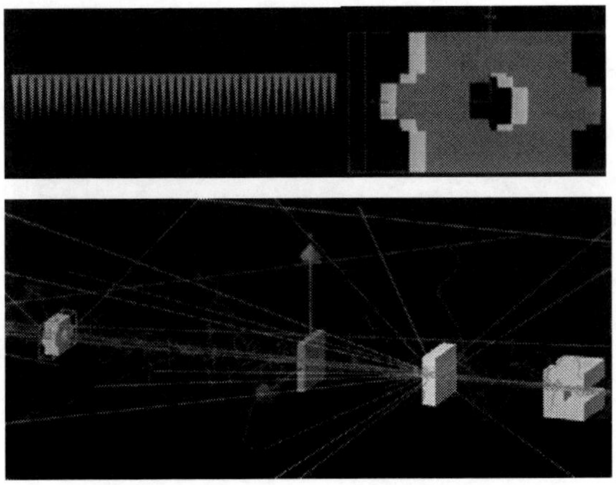

Fig. 5. Visualization of the whole simulation.

Figure 6 is the calculation of SOBP by 300 MeV/u carbon ions using the default physics models. Noted that this result is obtained by 300 MeV/u carbon ions getting through a 6 cm ridge filter and 5 cm range shifter, which is also the reason that the SOBP end is not consistent with the pristine Bragg Peak of 300 MeV/u carbon ions. The function of the ridge filter has also been verified in figure 6, as can been seen, the SOBP region spans from nearly 11 cm to 17 cm, which exactly accords with the design aim of spreading the peak for 6 cm. SOBP calculated by TOPAS is the physical dose, the relative biological effectiveness has not been considered. According to the design of the ridge filter, the biological effective dose, which is the product of physical dose and RBE, should be uniform and with the plateau at the SOBP region. Studies related to biological effective dose of carbon ions are still the on-going effort.

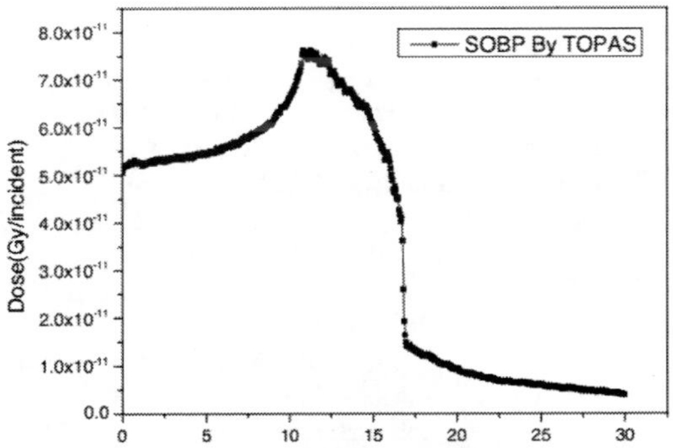

Fig. 6. TOPAS calculated SOBP of 300MeV/u carbon ions with 6cm ridge filter and 5cm range shifter.

4 Conclusion

To conclude, TOPAS is well-designed to make the simulation for Ion Beam Radiotherapy, both proton and carbon ions, easier and more efficient. The default physics models in TOPAS are elaborately chosen and have been validated with a high accuracy. However, when it comes to the high energetic carbon ions like 400 MeV/u or higher, the G4QMD model is recommended instead of the BIC model. Especially when using the TOPAS in other research field and the object is the heavier ions like neon, iron or uranium, etc., the G4QMD model is the preferable choice. Other outstanding features supported by TOPAS are including the import of XIO format phantom, DICOM files and CAD models. Time related feature and the support of electromagnetic field simulation are two other highlight features in TOPAS. But, still, the calculation capability based completely on CPU is limited even with the parallel run of multi-thread. One of the possible solution is transplanting the computing to a GPU heterogeneous environment where the Monte Carlo method can have a chance to be more efficient and practical for clinical usage [25].

This project was supported by the NSFC (Natural Science Foundation of China) with grant (Grant No. 11575180).

References

1. M. Scholz. Nuclear Instruments and Methods in Physics Research Section B: Beam Interactions with Materials and Atoms. **161**: 76-82. (2000)
2. D. Schardt, T. Elsässer, D. Schulz-Ertner. Reviews of modern physics, **82**(1): 383 (2010).
3. Ugo Amaldi, Kraft Gerhard. Reports on progress in physics **68**.8: 1861. (2005).
4. H. Tsujii, J. Mizoe, T. Kamada, et al. Radiotherapy and Oncology, **73**: S41-S49 (2004).
5. X. Liu, Q. Li, Q. Wu, et al. Chinese Physics C **33**.2: 151. (2009).
6. Q. Li, X. Liu, Z. Dai, etc. *AIP Conference Proceedings*. 1533, 174 (2013).
7. H Suit, T DeLaney, S Goldberg, et al. Radiotherapy and Oncology. **95**(1): 3-22. (2010)
8. Pedro Andreo. Physics in medicine and biology **36**.7: 861. (1991).
9. H Paganetti H Jiang, Parodi K, et al. Physics in medicine and biology **53**.17: 4825. (2008).
10. F Cerutti, M Dosanjh, A Ferrari et al. Physics in medicine and biology **55**.19: 5833. (2010).
11. H Paganetti, H Jiang, S Y Lee, et al. Medical physics, **31**(7): 2107-2118. (2004).
12. J Perl, J Shin, J Schümann, et al. Medical physics, **39**(11): 6818-6837. (2012).
13. C Robert, G Dedes, G Battistoni et al. Physics in medicine and biology **58**.9: 2879. (2013).
14. M Testa, J Schümann, H M Lu, et al. Medical physics, **40**(12): 121719. (2013)
15. J R Méndez, J Perl, J Schümann, et al. Physics in medicine and biology. **60**(13): 5019. (2015).
16. J Perl. User Guide for TOPAS Version 3.0 (rev. 20160522b) 72–79, 106–110. (2016).
17. Jarlskog, Christina Zacharatou, H Paganetti. IEEE Transactions on Nuclear Science **55**.3: 1018-1025. (2008).
18. Geant4 Collaboration. "Physics reference manual." *Version geant4.10.p01* 101-123, 203-210 (2015).
19. H. W. Lewis, Phys. Rev. **78**, 526–529 (1950).
20. K Chung, J Kim, D H Kim, et al. Journal of the Korean Physical Society **67**.1: 170-174. (2015).
21. M Sakama, T Kanai, Y Kase, et al. Physics in medicine and biology, **57**(20): 6615 (2012).
22. R Fujimoto, T Takayanagi, S Fujitaka. Physics in medicine and biology, **54**(13): N273 (2009).
23. NIST National Institute of Standards and Technology. PSTAR stopping power and range tables for protons.
24. H Paganetti. Physics in medicine and biology **57**.11: R99. (2012).
25. L Su, Y Yang, B Bednarz, et al. Medical physics **41**.7: 071709. (2014).

Geant4-DNA simulation of DNA damage caused by direct and indirect radiation effects and comparison with biological data.

Carmen Villagrasa[1,a], Sylvain Meylan[1] , Geraldine Gonon[1], Gaëtan Gruel[1], Ulrich Giesen[2], Marta Bueno[1] and Hans Rabus[2]

[1] *Institut de Radioprotection et Sûreté nucléaire (IRSN), BP-17, 92262-Fontenay-aux-Roses, France;*
[2] *Physikalisch-Technische Bundesanstalt (PTB), D-38116 Braunschweig, Germany*

Abstract. In this work we present results obtained in the frame of the BioQuaRT project. The objective of the study was the correlation between the number of radiation-induced double strand breaks (DSB) of the DNA molecule and the probability of detecting nuclear foci after targeted microbeam irradiation of cells with protons and alpha particles of different LET. The former were obtained by simulation with new methods integrated into Geant4-DNA that permit calculating the number of DSB in a DNA target model induced by direct and indirect radiation effects. A particular focus was laid in this work on evaluating the influence of different criteria applied to the simulated results for predicting the formation of a direct SSB. Indeed, these criteria have an important impact on the predicted number of DSB per particle track and its dependence with LET. Among the criteria tested in this work, the case that a direct radiation interaction leads to a strand break if the cumulative energy deposited in the backbone part of one nucleotide exceeds a threshold of 17.5 eV leads to the best agreement with the relative LET dependence of number of radiation induced foci. Further calculations and experimental data are nevertheless needed in order to fix the simulation parameters and to help interpreting the biological experimental data observed by immunofluorescence in terms of the DSB complexity.

1 Introduction

In cancer treatment using ionising radiation (radiotherapy), dosage is quantified by the absorbed dose to water expressed in the derived SI unit gray (Gy). Several radiotherapy modalities such as proton and ion beams require an additional weighting factor to account for increased tumor cell mortality for a given absorbed dose (as compared to a high-energy photon beam). The growing use of radiotherapy modalities whose biological effectiveness differ from that of conventional high-energy photon beams raises the need for establishing a new dosimetric concept expressed in terms of new quantities, which allow a transparent separation of the physical processes (dependent on modality) from the biological ones (independent of modality).

The joint research project "Biologically weighted quantities in radiotherapy" (BioQuaRT [1], 2012-2015) aimed at laying the foundation for these biologically-weighted radiation quantities based on a multi-scale approach. This approach used measurement and simulation techniques to determine the physical properties of ionising particle tracks on different length scales (from about 2 nm to 10 μm), which were then correlated with biological effects of radiation.

Presented here are the development of simulation tools and the quantitative comparison of the simulation results with experimental radiobiological data. The aim of this study was to evaluate the probability of structural DNA damage originating from particle irradiation as a function of the particle energy. From the simulations, the pattern at nanometric scale of the energy deposited by protons and alpha particles of different energies was obtained and different criteria for these energy depositions to break the DNA backbone were tested. The corresponding experimental biological data on early DNA damages were produced by targeted irradiation of cells at an ion microbeam and subsequent analysis of the formation of radiation-induced nuclear foci, such as γ-H2AX and 53BP1, observed by immunofluorescence.

2 Materials and methods

The evaluation of radiation-induced DNA damage in this work considers single strand breaks (SSB) and double strand breaks (DSB) of the DNA molecule. Monte Carlo simulations were used to compute both directly and indirectly induced breaks. An important aspect in this evaluation is the geometrical description of the DNA target. Indeed, the separation between neighbouring ionisations in the track of a densely ionising particle (such as low-energetic secondary electrons or ions in the Bragg peak region) is in the order of some nanometers and corresponds to the same order of magnitude as the size of the different molecules constituting the DNA. Hence, it is necessary to use a realistic description of the

[a] Corresponding author: Carmen.villagrasa@irsn.fr

target at nanometric scale in order to properly determinethe energy depositions that are located in the backbone region of the DNA which can give rise to direct single strand breaks (direct SSB).

On the other hand, a molecular description of the different components of the DNA was also needed in this work to extend the simulation beyond the physical interactions, i.e. in order to include the chemical reactions between the radicals produced by radiolysis of water in the vicinity of the target and the DNA. These chemical reactions could also break the structure of DNA and are accounted as indirectly induced breaks (indirect SSB).
Both types of damages, direct and indirect SSB were analysed using the DBSCAN clustering algorithm [2] in order to determine the relative distance between them and to calculate the number of predicted DSBs to be compared to the biological experimental data.
More precisely, the simulations modeled the ionising particle interactions on a virtual phantom of a chromatin fiber. The results obtained in the simulation were then scaled to the whole DNA (~ 6 Gbp) contained in the nucleus of a HUVEC (Human umbilical vein endothelial) cell of similar shape and volume as those HUVEC cells used in the radiobiological microbeam experiments. Then, the expected number of radiation-induced foci (RIF) was calculated from the number of DSB per track by taking into account the characteristics of the microbeam experiment. Finally the results predicted for the different radiation qualities were compared to the respective experimental radiobiological data.

2.1 Cell irradiation using a microbeam and DNA damage detection

Primary HUVEC cells cultures, predominantly in G0/G1 phase of cell cycle, were exposed to ion microbeam irradiation using the facility at the Physikalisch-Technische Bundesanstalt (PTB) [3]. The microbeam cell irradiation system enables hitting individual cell nuclei with a pre-determined number of particles in a given pattern of irradiation. In this experiment, each cell nucleus was hit by 5 particle tracks placed at the corners and the center of a square of 4 µm side length. The middle of the square was positioned at the barycenter of the targeted cell nucleus. Ten or thirteen minutes after the irradiations, the cells were fixed and the immunofluorescence protocol was followed.

Four types of ion/energy combinations were used in the experiment: α particles of 8 MeV, 10 MeV and 20 MeV (nominal beam energy) with an LET in the center of the cell nucleus of about 160 keV/µm, 90 keV/µm and 37 keV/µm, respectively, as well as 3 MeV protons with an LET of about 23 keV/µm. The karyotype and the genomic stability of cells were evaluated to ensure an equivalent DNA content in all cells of a population.

The foci analysis was based on a platform of high-throughput microscopy, with a powerful and robust infrastructure necessary for a massive image analysis (Scan-R software, Olympus). It enables the efficient measurement of numerous topological parameters on foci such as area, shape (i.e. circularity factor, elongation factor, etc.), relative positions in the cell nuclei, density, etc. With automated high-speed microscopy, the statistical evaluation of these measurements can be done in a large population of cell nuclei and provides an objective way to estimate a probability of foci formation related to a particle traversal.

2.2 Monte Carlo simulation of the early DNA damage for proton and alpha particle irradiation

2.2.1 Geometrical model for DNA target

The simulations were performed within a DNA fibre [4] of 34 nm diameter composed of 90 linked nucleosomes as depicted in figure 1. For each nucleosome, the histone proteins were represented by a cylinder of 2.4 nm radius and were surrounded by a DNA double helix composed of 200 pairs of nucleotides. Each of these nucleotides was composed of 3 adjacent spherical volumes of 0.091 nm^3, 0.060 nm^3 and 0.093 nm^3 representing the sugar, phosphate and base molecules, respectively. A liquid water shell of 12 molecules per nucleotide around the DNA structure was also taken into account as a part of the target for the direct effects calculations. Throughout the whole geometrical structure, liquid water cross-sections were used for the physical radiation interactions.

Fig. 1. Geometrical description of the DNA target used in this work. Upper part: (Two adjacent nucleosomes linked with the DNA double helix. Each nucleosome consists of a part of the DNA double helix wound around histone proteins (white cylinders). The DNA double helix is described at the molecular level. Lower part: 90 nucleosomes forming a 34 nm diameter chromatin fibre. Red arrows represent the random directions of the projectiles.

In the simulation, the fibre was surrounded by liquid water and was irradiated with helium or hydrogen ions impinging with random directions in order to simulate random positions of the chromatin fibre within the cell nucleus. The results obtained in this target of 18,000 base pairs (bp) were then extrapolated to the total DNA content of a eukaryotic cell ($6{\cdot}10^9$ bp) by following the method described in Ref. [5] and taking into account the shape and mean-size characteristics of the irradiated HUVEC cell nuclei: cylinder with an ellipsoidal basis with semi-axes a=9.5 µm and b=5.5 µm, and a height h=2 µm.

2.2.2 Physical interactions and evaluation of direct SSB

The Monte Carlo track structure simulation was computed with the Geant4 code [6]. In particular, the Geant4-DNA [7,8] processes included in the Low Energy Package were used. These processes allow the transport of protons, alpha particles and all other charged states of hydrogen and helium as well as of electrons down to a few eV. All interactions with the target medium leading to either an energy loss or a non negligible change in the particle direction were taken into account.

The simulation of the interactions was purely discrete, calculating all the elementary interactions on an event-by-event basis down to the eV scale and not using a condensed history approximation. This method allows obtaining a nanometric description of the energy deposited by radiation in a target of liquid water material within a complex geometry as detailed below.

Each physical interaction leading to an energy deposition in one of the DNA elements during the Monte Carlo simulation was registered for further analysis.

In order to calculate from these data the number of strand breaks (SSB) that these interactions could cause, the following scenarios were considered:

1. Each ionisation located in a sugar or a phosphate volume is directly converted to a direct SSB ("all ionisations")
2. The energy deposited by different interactions in the sugar and the phosphate volumes of the same nucleotide is added up. Then a threshold value of 17.5 eV is applied. If the cumulated energy is higher that the threshold, a direct SSB is registered at that location of the DNA molecule [9] ("threshold")
3. The energy deposited in the nucleotide backbone is obtained as before but linear probability between 5 eV and 37.5 eV is applied to determine the SSB formation. No direct SSB is formed for an energy lower than 5 eV and an SSB is always formed for energies higher than 37.5 eV [10] ("linear probability")

2.2.3 Chemical interactions in the DNA target

The simulation of the radicals created after the ionisation or excitation of the water molecules around the DNA target was performed using the processes implemented in Geant4-DNA since December 2014 [8, 11]. The simulation takes into account the creation of 7 different radical species, their diffusion in liquid water and their chemical reactions. The implementation of the DNA target within this simulation was made by taking into account the position of each molecule composing the target: deoxyribose, phosphate, base and histone volumes. These positions were maintained fixed during the simulation of the chemical stage and new chemical reactions were included in the list of possible reactions given by default in the code (see Table 1). Therefore, in this simulation, an $OH^{\bullet}$ radical was able to chemically interact with a deoxyribose molecule of the target giving raise to an indirect SSB in ~40% of the cases. The reason of this 40% probability comes from the fact that the deoxyribose molecular structure only leads to a break of the DNA backbone when a chemical attack occurs in certain parts of the molecule. Since our geometrical DNA model represents the whole deoxyribose structure as a sphere, we took that fact into account by applying a 40 % uniform random selection of the interactions leading effectively to a strand break. Chemical reactions between the $OH^{\bullet}$ radical and the DNA bases were also simulated but they were never considered to lead to a strand break. Instead, they were considered as base damages and were not included in the comparison with biological data.

Table 1. Chemical reactions included in the simulation of the chemical stage performed with Geant4-DNA [11]. Reaction rate values were taken from [12]

Chemical reaction	**Reaction rate ($m^3\,mol^{-1}{\cdot}s^{-1}$)**
Deoxyribose + $OH^{\bullet}$	$2.5{\cdot}10^6$
Adenine + $OH^{\bullet}$	$6.1{\cdot}10^6$
Guanine + $OH^{\bullet}$	$9.2{\cdot}10^6$
Thymine + $OH^{\bullet}$	$6.4{\cdot}10^6$
Cytosine+ $OH^{\bullet}$	$6.1{\cdot}10^6$

2.2.4 Double strand breaks (DSB) evaluation and RIF simulation

Double strand breaks were calculated from computed single strand breaks obtained either by direct or indirect effect in the chemical stage. The simplest definition of a DSB is a DNA damage formed by two SSB separated by less than 10 bp. Nevertheless, in this work, the DBSCAN algorithm [2] was used to identify clustered DNA damages within the simulated SSB results. Therefore, the DSB presented here can be composed by two or more SSBs (also called complex DSB) only if all of them were separated by less than 10 bp from another SSB within the cluster.

In order to be able to compare the simulated results on DSB per track of a given projectile to the experimental biological data of probability of RIF formation per particle track, a simple assumption was made based on the characteristics of the microbeam irradiation. The particles irradiating the cell nucleus crossed perpendicularly to the ellipsoidal surface (x-y plane) of the cell nucleus as shown in figure 2. While traversing the cells, the hydrogen and helium ions are only slightly deviating from this initial linear trajectory.

The image of experimentally detected RIF is obtained in the x-y plane and is not able to resolve two possible RIFs formed in very close (x,y) values along the z axis. For this reason, whatever the number of DSB created by the projectile (and their complexity) along the z axis and even though each of them could potentially lead to a visible RIF, the experimental image in the x-y plane will only show one single RIF. Therefore, the number of DSB per track obtained by Monte Carlo calculation was translated in a probability going from 0 to a maximum of 1 by considering the probability that at least one DSB is formed.

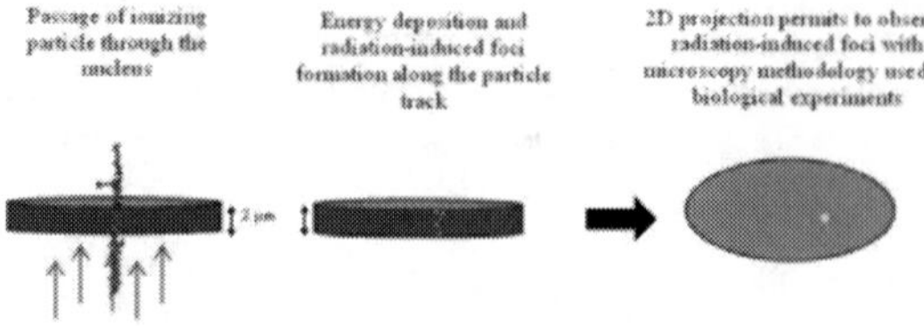

Fig. 2. Representation of the microbeam irradiation with ions traversing an endothelium cell nucleus. The experimental foci are obtained in a 2D projection losing information on the number of foci or DSB formation along the track.

3 Results and discussion

The mean number of DSBs per particle track traversing the endothelium cell nucleus was calculated using the three different criteria for the direct SSB formation. The method described in Ref. [5] to extrapolate the simulated data on a chromatin fibre to the whole trajectory within the cell nucleus results in a probability distribution on the number of traversed fibres with a mean number of 7.77. From this assumption, the results for the 4 radiation qualities in terms of mean number of DSB per track are given in table 2 with the uncertainty corresponding to the standard deviation of the distribution.

As it can be seen from these values, the criteria used for determining the formation of a direct SSB in the DNA backbone has a strong influence on both the absolute value of the number of DSB per track and also the dependence of DSB formation on LET. Moreover, these different criteria also have a high impact on the proportion between direct and indirect damages. Indeed, in the case that all ionisations in the backbone region are taken as direct breaks, the number of direct damages is higher than the number of indirect damages for all the LET values considered here: $SSB_{indirect}$ over SSB_{direct} decreases with increasing LET from 0.9 at 23 keV μm^{-1} to 0.5 at 160 keV μm^{-1}. On the contrary, for the linear probability or the energy threshold, the indirect damages are more frequent than the direct ones and the ratio $SSB_{indirect}$ over SSB_{direct} decreases with increasing LET: from 3.4 to 1.7 for the linear probability and from 16.5 to 3.2 for the threshold case.

Table 2. Simulation results on the mean number of DSB per track produced by the ion projectile when traversing the 2 μm thickness of the endothelium cell nucleus.

	All ionisations	Threshold 17.5 eV	Linear probability (5 eV-37.5 eV)
Protons (23 keV μm^{-1})	0.73±0.11	0.14±0.05	0.20±0.08
Alpha particles (37 keV μm^{-1})	1.01±0.11	0.30±0.09	0.45±0.01
Alpha particles (90 keV μm^{-1})	4.6±0.3	1.17±0.13	1.88±0.17
Alpha particles (160 keV μm^{-1})	14.8±0.5	5.3±0.2	6.9±0.2

Nevertheless, in order to calculate the probability of RIF formation in the cell nucleus per track, the important simulated result is the cumulative probability of obtaining at least 1 DSB in the track. Those results are presented in table 3 for the different criteria used in this work and compared to the experimentally observed 53BP1 foci in figure 3.

Table 3. Simulation results on the probability of RIF formation per particle track for the three different direct SSB formation criteria.

	All ionisations	Threshold 17.5 eV	Linear probability (5 eV - 37.5 eV)
Protons (23 keV μm^{-1})	0.50	0.13	0.19
Alpha particles (37 keV μm^{-1})	0.68	0.26	0.35
Alpha particles (90 keV μm^{-1})	0.94	0.63	0.79
Alpha particles (160 keV μm^{-1})	0.99	0.48	0.69

The different slopes in the increasing number of DSB with LET in the results presented in table 2 translate into different probabilities of RIF formation with LET. If all the ionisations are considered as responsible of a direct SSB, the probability of RIF formation increases rapidly with LET and for projectiles of LET of 90 keV/μm or higher, this probability is very close to unity. On the contrary, in the case of the threshold criterium or the linear probability criterium, the resulting probability of RIF formation shows a maximum around 100 keV/μm that is lower than 1 and that corresponds to what is also seen in the biological experimental data.

In particular, the threshold criterium of 17.5 eV [9] seems to be the one that better reproduces the relative dependence of the biological data on LET even though the absolute values deviate significantly.

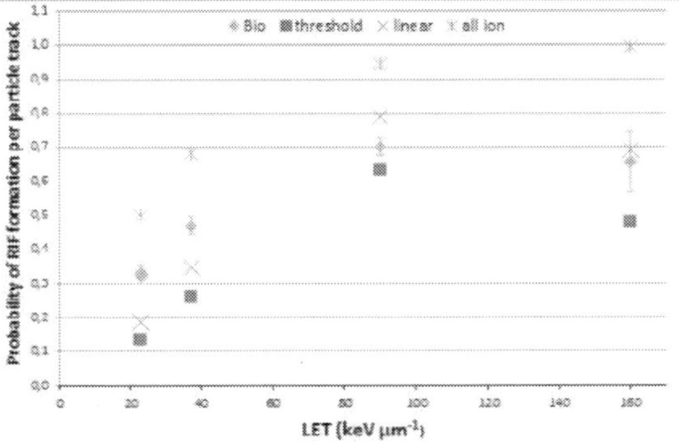

Fig. 3. Comparison of the probability of RIF formation simulated in this work for the three presented criteria on the direct SSB formation (blue stars, purple crosses and red squares) compared to the mean number of 53BP1 foci detected per particle track (green circle for protons and green diamonds for alpha particles).

The simulated results in this work are not given with associated uncertainty estimates. The reason for this is the difficulty of estimating these uncertainties owing to the simulation procedure being composed of several stages, each of them based on hypotheses and parameters that are still under investigation and do not offer a simple approach for estimating their associated uncertainty. Nevertheless, the main contributors to the uncertainty budget of the results presented here can be highlighted.

Concerning the absolute value of the number of DSB produced in the endothelium nucleus for a given track, the most important uncertainty comes from the scaling procedure adopted in this work to extend the simulated results in the chromatin fibre to the whole nucleus [5]. The chromatin domains size or the compaction of the chromatin fibre within the domains used in the procedure, can change the mean value (and the distribution) of the number of traversed fibres per track by more than 50%. For this reason, the relevant feature of the results presented in figure 3 is the dependency on LET and not their absolute value.

Besides the scaling procedure, the total uncertainty associated with the simulation performed in this work is the result of combining the uncertainties related to several factors: First, the calculation of physical interactions using Monte Carlo transport and liquid water cross-sections (especially for very low energy electrons). Second, the geometrical description of the DNA target at G0/G1 phase and the potential geometrical variability of the whole nucleus that is not taken into account here. Third, the limited number of chemical reactions taken into account and their reaction rates and probabilities. Fourth, the maximum time simulated for the chemical stage. Here the simulation time was limited to 2.5 ns in order to take into account the scavenging of the $OH^{\bullet}$ radicals by other entities in the nucleus which are not explicitly simulated in the present work. Finally, the statistical uncertainty related to the number of simulated projectiles (here 900 per energy), to cite the most important uncertainty sources.

The preceding is not an exhaustive list of parameters and hypotheses in the simulation that will influence the outcome of the calculated energy deposited in the DNA molecule during the physical stage and the mean number of chemical reactions between the radiation-induced radicals and the DNA.

In particular, in order to obtain from the simulated results a number of strand breaks for comparison with the experimental biological data, it is still necessary to determine which of these physical or chemical interactions will lead to a break. In this frame, and more specifically concerning the physical interactions, there is still debate within the scientific community, which is the reason why different criteria have been tested in this work. From the obtained results, using an energy threshold around 17.5 eV for the energy deposited in the backbone region of a nucleosome of our geometrical DNA model seems to give the best agreement with the experimental biological data obtained in this project.

4 Conclusions

Monte Carlo simulations using Geant4-DNA have been performed in order to calculate the number of DSB produced in a chromatin fibre target described with molecular detail. In these calculations, both, the physical stage and the chemical stage have been taken into account resulting in direct and indirect DNA damage. Concerning the direct effects, different criteria have been tested for the induction of a single strand break, based on the energy deposited in the backbone or the type of physical interaction. From these results, an evaluation of DSB in a HUVEC cell and the probability of detecting nuclear foci after irradiation with protons or alpha particles of different energy has been calculated. Besides all the hypotheses and parameters used in our simulation, the aforementioned criteria used to determine the formation of direct breaks of the DNA molecule have a great impact on the number of DSB and its dependence on the LET of the projectile. From the results presented in this work, the use of a threshold of 17.5 eV for the cumulated energy deposited in a single nucleosome in the DNA backbone seems to reproduce quite well the LET dependence of the biological data obtained in this work. Nevertheless, absolute values of the RIF formation probability simulated in this work do not coincide with the experimental data.

There are potentially two main reasons for explanation of this discrepancy: On the one hand, in this work a scaling procedure was used to extrapolate from the simulated results in a chromatin fibre to the whole cell nucleus. This could give rise to a significant bias in the results. In order to fix this potential problem, calculations are ongoing that simulate proton and alpha particle tracks in a geometrical model describing the complete DNA genome in the endothelium nucleus. The results of the very computer time consuming simulations

 DOI: 10.1051/epjconf/201715304019

will eventually also be analysed using the criteria studied in this work.

On the other hand, one should also keep in mind that the measured 53BP1 foci, may not represent 1:1 all the DSB produced by the radiation. Therefore, the simulation method developed in this work also offers the possibility of investigating the complexity and position in the chromatin of the DNA damages that can correspond to those detectable by an immunofluorescence technique.

Acknowledgements

This work is carried out within EMRP (European Metrology Research Programme) Joint Research Project SIB06 'BioQuaRT', which has received funding from the European Union on the basis of decision number 912/2009/EC. The EMRP is jointly funded by the EMRP participating countries within EURAMET and the European Union. The EMRP is jointly funded by the EMRP participating countries within EURAMET and the European Union.

References

1. URL: http://www.ptb.de/emrp/bioquart.html
2. M. Ester, H-P. Kriegel, J. Sander, X. Xu. In: E. Simoudis, J. Han, U. M. Fayyad (Eds.). *Proceedings of the Second International Conference on Knowledge Discovery and Data Mining*, AAAI Press, pp. 226–231 (1996)
3. K. D. Greif, H.J. Brede, D. Frankenberg, U. Giesen, Nucl Instrum Methods Phys Res **B217**, 505-512(2004)
4. S. Meylan, U. Vimont, S. Incerti, I. Clairand and C. Villagrasa. Comp. Phys and Comm. In press.
5. B. Lee, PhD thesis, Georgia Institute of Technology, 2014. URL: http://hdl.handle.net/1853/53106
6. Agostinelli, S. et al., Nucl. Instrum. Meth. Phys. Res. A **506** pp. 250-303(2003)
7. S. Incerti et al, Med. Phys. **37** pp. 4692-4708 (2010)
8. M. A. Bernal et al., Phys. Medica **31,** pp. 861-874 (2015)
9. D.T. Goodhead and H. Nikjoo, Int. J. of Rad. Biol. **55**, pp. 513-529 (1989)
10. W. Friedland, P. Jacob, P. Bernhardt, H. G. Paretzke and M. Dingfelder, Rad. Res. **159**, pp 401-410 (2003)
11. M. Karamitros et al., J. Comput. Phys. **274**, pp. 841-882 (2014)
12. B. Aydogan, W.E. Bolch, S.G. Swarts, J.E. Turner and D.T. Marshall, Rad. Res. **169**, pp. 223-231 (2008)

Radiation-Induced Second Cancer Risk Estimates From Radionuclide Therapy

Bryan Bednarz[1,*] and Abigail Besemer[1]

[1]Department of Medical Physics, University of Wisconsin, Madison, WI, 53705

The use of radionuclide therapy in the clinical setting is expected to increase significantly over the next decade. There is an important need to understand the radiation-induced second cancer risk associated with these procedures. In this study the radiation-induced cancer risk in five radionuclide therapy patients was investigated. These patients underwent serial SPECT imaging scans following injection as part of a clinical trial testing the efficacy of a ^{131}Iodine-labeled radiopharmaceutical. Using these datasets the committed absorbed doses to multiple sensitive structures were calculated using RAPID, which is a novel Monte Carlo-based 3D dosimetry platform developed for personalized dosimetry. The excess relative risk (ERR) for radiation-induced cancer in these structures was then derived from these dose estimates following the recommendations set forth in the BEIR VII report. The radiation-induced leukemia ERR was highest among all sites considered reaching a maximum value of approximately 4.5. The radiation-induced cancer risk in the kidneys, liver and spleen ranged between 0.3 and 1.3. The lifetime attributable risks (LARs) were also calculated, which ranged from 30 to 1700 cancers per 100,000 persons and were highest for leukemia and the liver for both males and females followed by radiation-induced spleen and kidney cancer. The risks associated with radionuclide therapy are similar to the risk associated with external beam radiation therapy.

I. Introduction

Radionuclide therapy is a form of therapy involving the systemic administration of a radioactive compound in a patient. The primary basis for successful radionuclide therapy is the preferential uptake and retention of the radiopharmaceutical in the tumor without inducing acute or latent side effects in the patient. Radionuclide therapy has a proven track record in the clinical setting for treating a broad spectrum of solid and hematological cancers. The use of radionuclide therapy is expected to significantly increase in the next decade owing to the rapid development of a diverse set of novel radiopharmaceuticals.

While the risk of second cancer development from external beam radiation therapy (EBRT) has been investigated extensively [1-3], less is known about the radiation-induced cancer risk from radionuclide therapy. Because this risk is largely dependent on the absorbed dose and dose rate in the body one would expect sizeable differences in the risk predictions between these two treatments. The absorbed dose distribution resulting from radionuclide therapy is often heterogeneously distributed throughout the body of the patient unlike the absorbed dose from EBRT that falls off rapidly as a function of distance from the tumor. The absorbed dose from radionuclide therapy is delivered continuously over an extended period of time as governed by the radionuclides half-life, which is in stark contrast to EBRT where sizable doses are delivered to a tumor volume in multiple fractions. It is impossible to relate the risks between these treatment options without accounting for these aforementioned differences.

In order to provide adequate risk comparisons between EBRT and radionuclide therapy, new computational tools are needed. Currently, the only internal dosimetry software approved by the US Food and Drug Administration (FDA) is OLINDA/EXM (Organ Level Internal Dose Assessment for EXponential Modeling)[4]. OLINDA calculates absorbed doses using the formalism developed by the Medical Internal Radiation Dose (MIRD) committee of the Society of Nuclear Medicine [5]. OLINDA uses tabulated activity-to-dose scaling factors, or organ S-values (OSVs), generated based on standard phantoms [6]. Tumors are modeled separately as a unit-density sphere isolated in an infinite unit density medium. The major limitations of the OSV method implemented in OLINDA are that (1) phantoms that are not an accurate representation of each patient's unique geometry, (2) the radioactivity uptake and dose deposition are assumed to be uniformly distributed within each organ, and (3) the tumor not being modeled as part of the phantom leads to an underestimate of the normal tissue and tumor dose. Despite these limitations, OLINDA remains the only commercial internal dosimetry software utilized clinically because of its simplicity and ease of use.

The purpose of this work is to investigate radiation-induced second cancer risks from radionuclide therapy using an advanced Monte Carlo (MC) 3D dosimetry platform called RAPID (Radiopharmaceutical Assessment Platform for Internal Dosimetry).

II. Methods and Materials

2.1 Dose Calculation Procedure

The RAPID framework and workflow is demonstrated in Fig. 1. Serial 3D images (SPECT/CT) are acquired to map the agent's pharmacokinetics. Following acquisition, the images

*Corresponding author: bbednarz2@wisc.edu

are coregistered, resampled, rescaled, corrected for partial volume effects, and masked. First, the PET/SPECT images are coregistered to the CT at each time point, a using normalized mutual information affine transformation with 12 degrees of freedom (3 rotational, 3 translational, 3 scaling, and 3 sheering). The SPECT data was then up-sampled to match the higher CT resolution. Given that Monte Carlo dosimetry relies both on accurate quantification of the activity distribution by the SPECT and the material composition/density by the CT, each resampling option has both advantages and disadvantages. Resampling is performed using the Mitchell filter kernel that offers a good compromise between ringing and blurring artifacts.

The MC simulations in this work were done using the MC code Geant4 version 9.6 [7]. Geant4 is a versatile object-oriented simulation toolkit that allows for the modeling of complex geometries, radiation sources, and detectors. Geant4 has been benchmarked for a variety of different medical physics applications [8-10]. The CT and the SPECT images are used in the Monte Carlo simulation to define the geometry and source distribution, respectively. The CT images define the material composition and mass density of the simulation geometry. For this work, the entire HU range was segmented into 27 materials based on the human tissue compositions defined by Schneider et al [11]. Each HU was assigned a unique mass density according to the conversion defined in Schneider et al [11]. The SPECT images were used to define the radionuclide activity in each voxel. The decay was s simulated using the G4RadioactiveDecay module which allows the simulation of the decay of more than 2000 radioactive nuclei using the decay information from the ENSDF database [12]. The point of the radionuclide decay is sampled uniformly throughout each voxel.

The statistical relative error calculated on the voxel and ROI-level. Sufficient particles was simulated such that the average organ-level relative error was less than 1% and the average voxel-level relative error within the patient volume was less than 10%. Note that the relative error only refers to the statistical precision of the Monte Carlo dose calculation and does not take in to account other sources of uncertainty such as the activity quantification, material/density quantification, coregistration, resampling, kinetic fitting, or ROI contouring.

To speed up the simulation time, the simulation jobs were parallelized into individual slices. The whole patient CT was used to define the patient geometry and calculate the dose deposition but each job only contains one slice of the SPECT defined source distribution. For example, a whole patient simulation would run 1 job containing 80 slices while a simulation parallelized into slices would run 80 jobs each containing 1 slice of the PET/SPECT defined activity distribution. The simulation efficiency of the slice parallelization method was compared with the whole patient method by calculating the absorbed dose rate distribution at a single time point based on a SPECT/CT image. It was shown that the slice parallelization simulations were ~250-400 times faster and 3-5 times more efficient than the whole patient simulations. The image contained 80 slices with a matrix size of 512 × 512 voxels. All simulations were run on the UW Center for High Throughput Computing (CHTC) cluster.

Following the MC simulation at each time point the absorbed dose rate is co-registered to a single CT and resampled. The absorbed dose is calculated by time-integrating the absorbed dose rate in kinetic fitting module. Radiobiological dose metrics can also be calculated. The dosimetry output includes 3D dose distributions, biological equivalent dose (BED) distributions and equivalent uniform dose values in regions of interest.

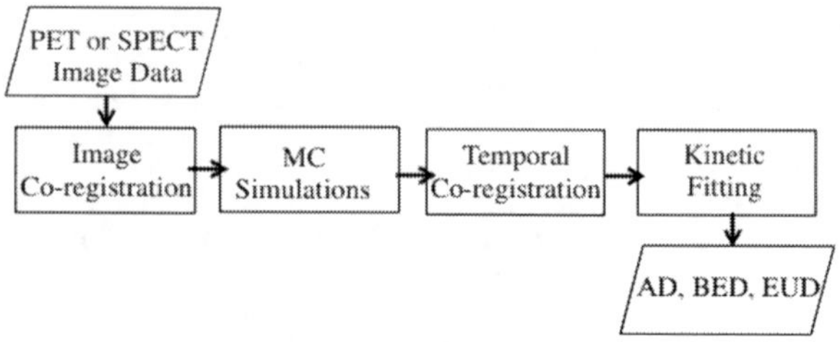

Figure 1. The RAPID framework and workflow.

For this investigation 5 patient datasets were analyzed from a Phase 1 dose escalation clinical trial investigating a ‘theranostic/diapeutic’ molecular agent, which can be radiolabeled with either ^{124}I for PET imaging or ^{131}I for radiotherapy or SPECT/planar imaging [13]. Each patient underwent serial SPECT/CT ^{131}I scans following treatment. Information about dosing and scan times are provided in Table 1. Following the workflow illustrated in Figure 1, the total committed absorbed dose distribution in each patient was determined. In addition committed absorbed doses were also calculated using the OLINDA for comparison.

Table 1. Injected ^{131}I dosage and scanning times for each patient considered in this investigation.

Patient	Dose Group (mCi/m^2) / Inj. Act. (mCi)	Scan Time Post Injection			
		Day 03	Day 06	Day 14	Day 21
1	25 / 48.6	□	□	□	□
2	37.5 / 84.5	-	□	□	□
3	37.5 / 64	-	□	□	□
4	31.25 / 53.8	□	□	□	□
5	31.25 / 77.74	□	□	□	□

2.2 Radiation Risk Estimation

Second cancer risks to dose limiting organs from these treatments were calculated using the BEIR VII methodology [14]. Details of this methodology can be found in Chapter 12 of the BEIR VII report, but will be described below. The relationship between the risk of second cancer induction in

DOI: 10.1051/epjconf/201715304020

exposed populations and the unexposed populations can be quantified by their difference known as the excess absolute risk (EAR) or their ratio known as the relative risk (RR). Typically, the excess relative risk (ERR) is used instead of RR, where the ERR is equivalent to RR−1. That is the EAR and ERR are defined as the additive and multiplicative increase in incidence related to an exposure, respectively, where the incident rate is number of new cases of disease occurring in a specified amount of time among previously unaffected individuals. In this study, the age-dependent organ-specific ERRs were calculated for a variety of sites.

The BEIR VII report defines the ERR for solid tumors as a function of attained age (a) for patients over 30 as (Ref. 28):

$$ERR = \rho(D) \cdot \beta_s \cdot (a/60)^{\eta} \quad (1)$$

where e is the age at treatment, D is the total dose equivalent received and s is the gender of the patient. For solid tumor sites, the parameter $\rho(D)$ is a linear function of dose. The model parameters β_s and η are provided for several organs in the BEIR VII report and are typically an order of magnitude different for the ERR.

The ERR for leukemia is defined by BEIR VII for patients over 30 as:

$$ERR = \beta_s \cdot D(1 + \theta D)\exp[\delta \log(t/25] \quad (2)$$

where t is the time since the treatment and the parameter θ indicates the degree of curvature of the linear quadratic model to represent the dose-response relationship for leukemia. The model parameters β_s and δ are provided for both ERR and EAR in BEIR VII. The absorbed dose to the bone marrow, which was determined within a small region of interest within the pelvic bone, was used to determine the ERR of inducing leukemia in these patients.

The lifetime attributable risk (LAR) is the probability that an exposed patient will develop a cancer during their lifetime. The BEIR VII report recommends taking a weighted average (log-scale) between the ERR and EAR to derive the LAR given as:

$$LAR\,(D,e) = \left(\int_{a+L}^{100} M(D,e,a)_{ERR} \cdot S(a)/S(e)\, da\right)^{0.7} \times \left(\int_{a+L}^{100} M(D,e,a)_{EAR} \cdot S(a)/S(e)\, da\right)^{0.3} \quad (3)$$

where,

$$M(D,e,a)_{ERR} = ERR(D,e,a) \cdot \lambda_I^c \text{ and } M(D,e,a)_{EAR} = EAR(D,e,a) \quad (4)$$

where $S(a)/S(e)$ is the probability of a patient surviving from their exposure age (e) to their attained age (a) and were derived from life-span tables provided by Anderson and DeTurk [15]. The recommended latent periods by BEIR VII of 5 years for solid cancers and 2 years for leukemia were used. The parameters λ_I^c can be considered the baseline cancer risk for the exposed individual and are based on data provided in the SEER Cancer Statistics Review. The LAR coefficients derived from Eq. 1-4 and provided in BEIR VII were used for this work. The EAR values for the kidney and spleen were determined using the generic "Other solid cancer" coefficients set forth by BEIR VII. These coefficients were rescaled by a factor of 13 given that the "Other solid cancer" category combines risk for 13 remainder organs. Organ-specific coefficients were used for all remaining organs. The risks derived for leukemia assumed that the age of the exposure was 30 years. All other solid cancer risks assumed the ERR was independent of the age at exposure following recommendations from the BEIR VII report. The risk coefficients were rescaled by a factor of 1.5 to eliminate the dose and dose rate effectiveness factor (DDREF) used by BEIR committee to adjust for low dose rate exposures. This is based on the assumption that ^{131}I therapy can be considered a high dose and dose rate therapy due to biological clearance of CLR1404. A more appropriate approach, albeit out of the scope of this work, would be to determine DDREF with a value between 1 and 1.5 specific to this type of therapy.

III. Results

Mean absorbed doses to the kidneys, spleen, liver, and bone marrow were calculated using RAPID from these treatments are provided in Figure 2. Also included are mean doses calculated by OLINDA. As seen in Figure 2 the mean absorbed dose calculated by RAPID ranged between 1.66 - 3.44 Gy. There are clear differences between absorbed doses by RAPID using patient-specific MC simulations and OLINDA where these differences can be attributed to the aforementioned weaknesses of the OLINDA software. RAPID calculates the committed absorbed dose in each voxel using patient-specific images as illustrated in the dose color washes provided in Figure 3 for Patient 2. One would expect similar variations in the ERR values calculated for these patients using absorbed dose values from these two sources.

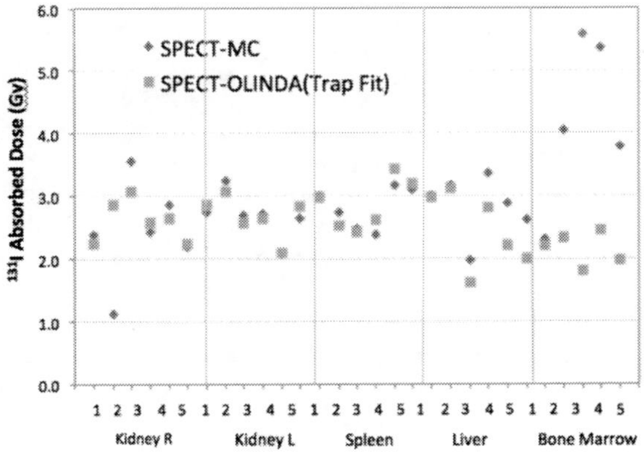

Figure 2. Mean committed absorbed doses in the kidneys, spleen, liver, and bone marrow for each patient calculated with RAPID and OLINDA.

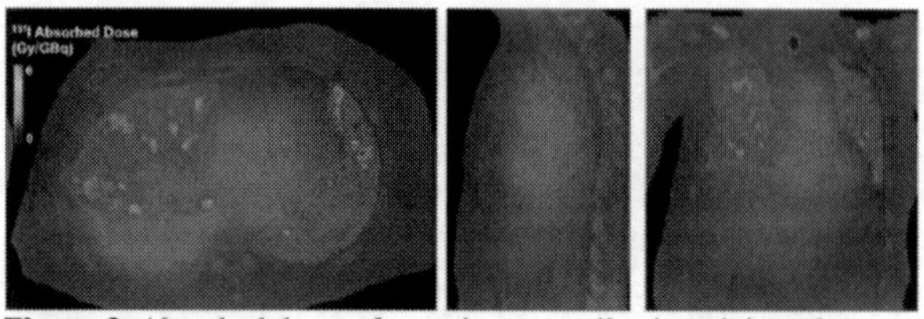

Figure 3. Absorbed dose colorwashes normalized per injected activity shown in the transverse, sagittal and corona l plane for Patient 2.

Figure 4 (a–e) provides sex-averaged ERR values for radiation-induced cancers in the kidney, spleen, liver, and blood. The highest ERR values are associated with leukemia and the lung whereas the liver, kidneys and spleen all had lower ERR values. Relatively large patient risk variations are seen within different cancer groupings.

Table 2 provides the LAR for inducing cancer in the kidney, spleen, liver and blood as a function of age derived using the ERR values provided above and the EAR values (data not shown). LARs for both males and females are included.

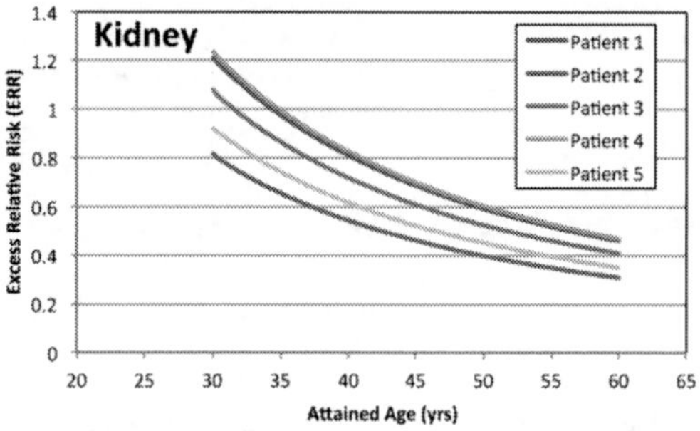

(a.)

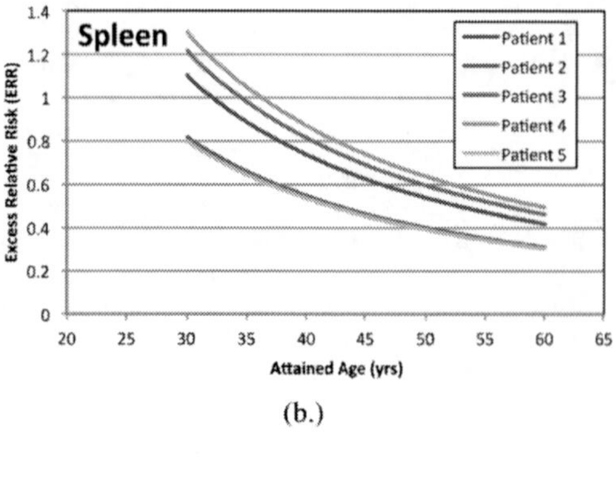

(b.)

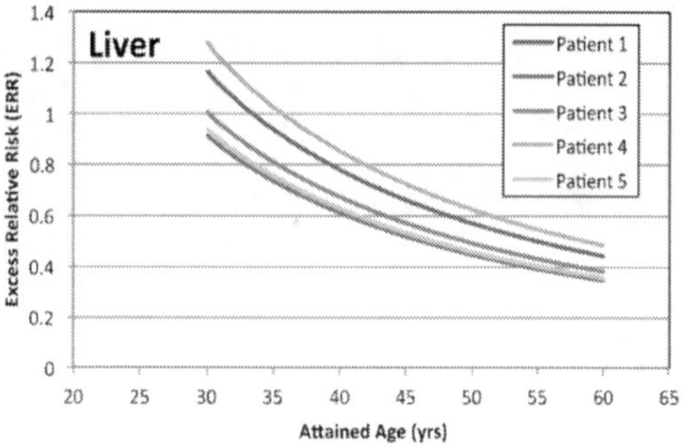

(c.)

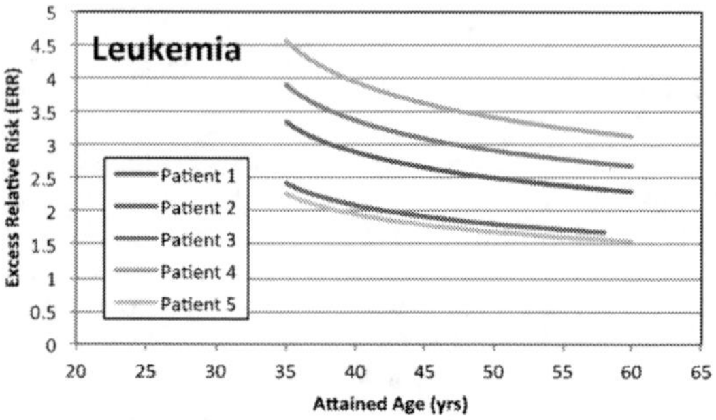

(d.)

Figure 4. Sex-averaged ERR values for (1.) kidney, (b.) spleen, (c.) liver, and (d.) leukemia for the five patients considered in this study.

IV. Discussion and Conclusion

This work investigated the risk of radiation-induced second cancers in radionuclide therapy patients. The committed absorbed dose to multiple sensitive structures was calculated using RAPID. The sex-averaged ERR and LAR of developing a radiation-induced cancer in these structures was then derived following the recommendations set forth in the BEIR VII report [7].

Our results indicate that the sex-averaged ERR in these patients range from 0.5 to 4.5 depending on the organ site, absorbed dose to the site, and obtained age of the patient following the exposure. The LARs associated with these exposures ranged roughly from 30 to 1700 cancers per 100,000 persons. The LAR risks were highest for radiation-induced leukemia risks for both males and females followed by radiation-induced liver cancer. This likely due to the relatively slow clearance of CLR1404 from the bloodstream.

There has been heightened interest and concern about radiation-induced cancer risk in external beam radiation therapy in recent years. The initial results from this work indicate that the risk of radiation-induced second cancers from radionuclide therapy are similar to those predicted for external beam radiation therapy [16]. However, the risk of radiation-induced leukemia is potentially much higher given the differences in dose distributions involved. This is supported by several clinical studies on radiation-induced leukemia in patients treated with ^{131}I for thyroid cancer [17]. Given these findings, it is imperative that more attention be afforded to investigating the cancer risks associated with radionuclide therapy.

Acknowledgments

We would like to thank Dr. Joseph Grudzinski and Mr. Alex Newell for help with data collection and analysis. We would like to thank the UW Center for High Throughput Computing (CHTC) for the use of their cluster and their computational support.

References

[1] X. G. Xu, B. Bednarz, and H. Paganetti, Phys. Med. Biol. 53, R193–R241 (2008)
[2] E. J. Hall and C. S. Wuu, Int. J. Radiat. Oncol., Biol., Phys. 56, 83–88 (2003)
[3] M. Tubiana Radiother. Oncol. 91, 4–15 (2009)
[4] M.G. Stabin, R.B. Sparks, E. Crowe, J Nucl Med, 46(6), 1023-1027 (2005)
[5] W.E. Bolch, K.F. Eckerman, G. Sgouros, S.R. Thomas Journal of Nuclear Medicine 50, 477-484 (2009).
[6] M.G. Stabin, E. Watson, M. Cristy, J. Ryman, K. Eckerman, J. Davis, D. Marshall, M. Gehlen ORNL Report (1995)
M. Cristy, K. Eckerman, ORNL/TM-8381 6 (1987)
[7] Agnostinelli, Nuclear Instruments and Methods in Physics Research Section A: Accelerators, Spectrometers, Detectors and Associated Equipment 506, 250 - 303 (2003).
[8] B.A. Faddegon, I. Kawrakow, Y. Kubyshin, J. Perl, J. Sempau, L. Urban, Phys. Med. Biol 542009).
[9] T. Aso, A. Kimura, S. Tanaka, H. Yoshida, N. Kanematsu, T. Sasaki, T. Akagi, Nuclear Science, IEEE Transactions on 52, 896-901 (2005).
[10] C. Thiam, V. Breton, D. Donnarieix, B. Habib, L. Maigne, Phys. Med. Biol. 53, 3039 (2008).
[11] W. Schneider, T. Bortfeld, W. Schlegel Phys. Med. Biol. 53, 459 (2000)
[12] S. Hauf, M. Kuster, M. Batic, Z.W. Bell, D.H. Hoffman, P.M. Lang, et al Nuclear Science IEEE Trans. 60, 2966-2983 (2013)
[13] J. Grudzinski, B. titz, K. Kozak, W. Clarke, et al et al PLOS one 9 (11) 1-11 (2014)
[14] BEIR, Health Risks from Exposure to Low Levels of Ionizing Radiation: BEIR VII, Phase 2. National Academy of Science, Washington, DC, (2006)
[15] R.N. Anderson and P.B. DeTurk, National Vital Statistics Report, 50(6), 1-12 (2002)
[16] Shuryak I et al. J Natl Cancer Inst. 98(24) 1794-1806 (2006)
[17] Fard-Esfahani A, Emamu-Ardekani A, Fallahi B, et al Nucl Med Commun 35(8), 808-817 (2014)

Table 2: LAR as a function of cancer site and age at exposure (yr) for the five patients considered in this study.

	LAR (cases per 100,000 persons)											
	Age at exposure (yr)											
	30		40		50		60		70		80	
Organ												
Patient 1	Male	Female	Male	Female	Male	Female	Male	Female	Male	Female	Male	Female
Kidney	291	304	253	266	206	218	144	160	84	100	34	44
Liver	455	207	434	207	393	186	290	145	165	103	62	41
Spleen	265	277	230	242	187	198	131	146	76	91	31	40
Leukemia	1353	1015	1353	999	1353	999	1321	918	1176	822	773	596
Patient 2												
Kidney	195	204	170	179	138	146	97	108	56	67	23	30
Liver	356	162	340	162	307	146	226	113	129	81	49	32
Spleen	197	206	171	180	139	147	97	108	57	68	23	30
Leukemia	1090	817	1090	804	1090	804	1064	739	947	662	623	480
Patient 3												
Kidney	258	269	224	236	182	193	128	142	74	89	30	39
Liver	356	179	340	179	307	161	226	125	129	89	49	36
Spleen	293	307	255	268	207	219	145	161	84	101	34	44
Leukemia	1501	1125	1501	1108	1501	1108	1465	1018	1304	911	858	661
Patient 4												
Kidney	297	310	258	271	210	222	147	163	85	102	34	45
Liver	498	227	476	227	431	204	317	159	181	113	68	45
Spleen	312	326	271	285	221	233	154	172	90	107	36	47
Leukemia	1668	1251	1668	1231	1668	1231	1628	1132	1450	1013	953	735
Patient 5												
Kidney	221	232	192	202	157	166	110	122	64	76	26	34
Liver	365	166	348	166	315	149	232	116	133	83	50	33
Spleen	193	202	167	176	136	144	95	106	56	66	22	29
Leukemia	1031	773	1031	761	1031	761	1006	700	896	626	589	454

DOI: 10.1051/epjconf/201715305001

Shielding and activation calculations around the reactor core for the MYRRHA ADS design

Anna Ferrari [a,1], Stefan Mueller[1], J. Konheiser[1], D. Castelliti[2], M. Sarotto[3] and A. Stankovskiy[2]

[1]Helmholtz-Zentrum Dresden-Rossendorf, PF 910119, 01314 Dresden, Germany
[2]SCK·CEN, Boretang 200, 2400 Mol, Belgium
[3]ENEA, Via Martiri di Monte Sole 4, 40129 Bologna, Italy

Abstract. In the frame of the FP7 European project MAXSIMA, an extensive simulation study has been done to assess the main shielding problems in view of the construction of the MYRRHA accelerator-driven system at SCK·CEN in Mol (Belgium). An innovative method based on the combined use of the two state-of-the-art Monte Carlo codes MCNPX and FLUKA has been used, with the goal to characterize complex, realistic neutron fields around the core barrel, to be used as source terms in detailed analyses of the radiation fields due to the system in operation, and of the coupled residual radiation. The main results of the shielding analysis are presented, as well as the construction of an activation database of all the key structural materials. The results evidenced a powerful way to analyse the shielding and activation problems, with direct and clear implications on the design solutions.

1 Introduction

Accelerator-driven systems (ADS) are one of the options studied for the transmutation of nuclear waste in the international community. The MYRRHA [1] facility at SCK·CEN in Mol (Belgium) aims to demonstrate efficient transmutation of high level waste and associated ADS technology. The heart of the system is a lead-bismuth eutectic (LBE) cooled reactor, working both in critical and in sub-critical operation modes. The neutrons needed to sustain fission in the sub-critical mode are produced via spallation processes by a 600 MeV, ≤ 4 mA proton beam, which is provided by a linear accelerator and hits, along a vertical beamline, a LBE spallation target located inside the reactor core. The combination with a nuclear reactor core makes the shielding problem an issue, being the protection from the prompt radiation and the spent beam handling the main points. Due to the backscattered neutron radiation on the spallation target, the operation in sub-critical mode drives the shielding in all the vertical structures, and requires more innovative solutions. This work is therefore focused on this operation mode.

2 Method

A systematic analysis has been done using two state-of-the-art Monte Carlo codes: MCNPX [2], version 2.6.0, and FLUKA [3, 4], version 2011.2. As first step, an accurate MCNPX model of the reactor sub-critical core has been developed, in agreement with the most recent MYRRHA design. To simulate high energy processes, the Cascade-Exciton model, CEM03 [5], has been used. Neutron radiation fields have been then fully characterized on suitable surfaces around the core barrel and at the top of the fuel elements and used to build complex source terms, to be used in a second row of FLUKA simulations. A precious advantage of FLUKA is the possibility to evaluate, in the same simulation, the radiation fields due to the system in operation and the ones due to the coupled residual radiation. This is performed in the code by a half-analytic/half- Monte Carlo method, which is shortly summarized in the following. For a given irradiated material, the time evolution of its isotopic densities, N_i, is given by the set of the Bateman equations:

$$\frac{dN_i}{dt} = - \sum_{j \neq i} \left[\lambda^d_{ji} + \bar{\sigma}_{ji} \bar{\varphi} \right] N_i + \sum_{j \neq i} \left[\lambda^d_{ij} + \bar{\sigma}_{ij} \bar{\varphi} \right] N_j \quad (1)$$

where λ^d_{ji} is the decay probability of the radionuclide i in the radionuclide j and σ_{ji} is the particle induced cross section for transmutation from the isotope i to the isotope j, and where $\bar{\varphi}$ and $\bar{\sigma}_{ji}$ are the average spectrum and the spectrum averaged, particle induced cross-section, respectively:

$$\bar{\varphi} = \int \varphi(E)dE \quad , \qquad \bar{\sigma}_{ji} = \frac{1}{\bar{\varphi}} \int \varphi(E)\sigma_{ji}(E)dE \quad (2)$$

In the Eq. (1) second terms in brackets represent the radionuclide production due to the irradiation, and in FLUKA are evaluated runtime via Monte Carlo. First terms represent on the other hand the decay contribution

[a] Corresponding author: a.ferrari@hzdr.de

to the radionuclide equilibrium, and are analytically computed. Once the Bateman equations are solved, in the same run FLUKA performs the generation and transport of the emitted residual radiation with a full Monte Carlo method. This means that in the same run we can obtain the production of the residuals, their time evolution and the residual dose due to their decay.

To characterize the structures around the core, the main primary system components have been represented with a simplified geometry and a more homogeneous material composition. Minor internal elements have been neglected and all components have been described through a radial and axial homogenization. The resulting structure maintains a high level of complexity and includes the reactor vessel, the diaphragm and the cover model, the LBE and the gas plenum above, a homogenization of the structures above the core, the fuel handling machinery, the primary heat exchangers and the primary pumps. Beyond the reactor cover, a simplified geometrical model extends to include all the vertical part of the proton beamline, until the final roof.

3 Results

3.1 Shielding

Neutron and photon radiation fields and ambient dose equivalent distributions have been computed in all the structures around the core. Fig. 1, Fig. 2 and Fig. 3 show an example of neutron and photon fluence rate distributions, as well as of the H*(10) rate. The photon contribution of Fig. 2 is not due to photons coming from the reactor, but to the interactions of neutrons in the in-vessel structures, and – if we neglet the secondary radiation possibly coming from the beam losses inside the proton beam-pipe - is the only photon contribution that affects the dose rate far from the core. The present study had an impact on the design of the cover, which is in borated water contained in a steel structure. It can be observed that neutrons are very well absorbed in the borated water, while on the other hand the neutron capture processes on hydrogen and on boron generate

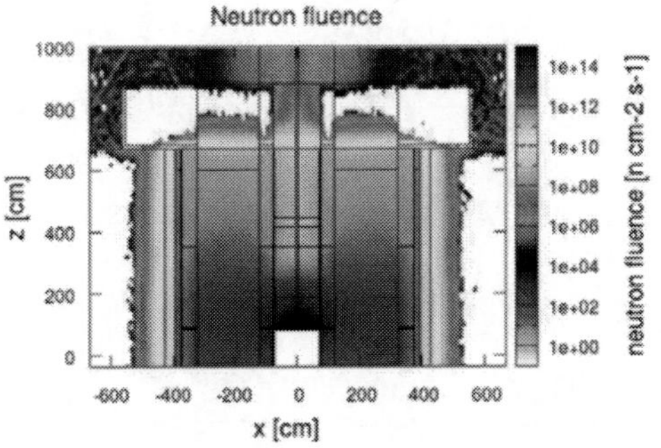

Figure 1. Neutron fluence rate [neutrons cm^{-2}s^{-1}] in the in-vessel structures, until the cover plate.

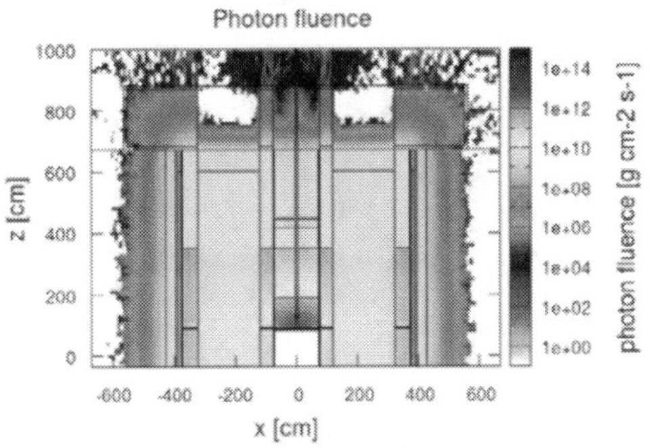

Figure 2. Photon fluence rate [photons cm^{-2}s^{-1}] in the in-vessel structures.

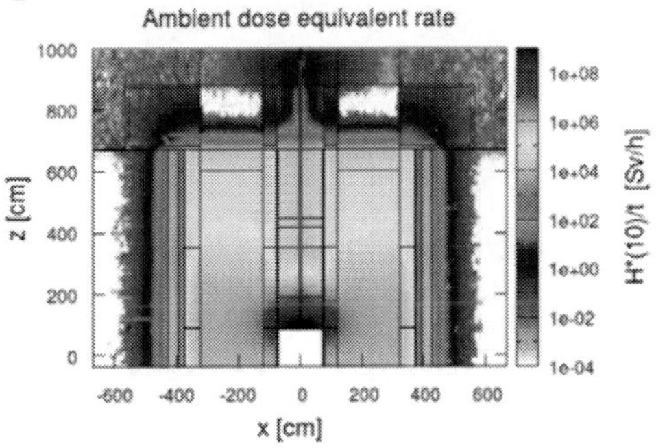

Figure 3. H*(10) rate [Sv/h] in the in-vessel structures, including all contributions to the radiation.

photons in the MeV range, which are not perfectly contained by the steel top cover plate (see Fig. 2). To fully contain this radiation and consolidate an already effective design, a thicker containment structure is planned.

3.2 Activation

An extensive analysis of the residual radiation has been performed. The operation of the system has been conservatively considered in a scheme of 3-months cycles in the sub-critical mode at the maximum beam intensity (I_{prot} = 4 mA), followed by 30 days stop. This irradiation profile has been implemented in the FLUKA radiation model and extended over 40 years, which is the planned MYRRHA life. The residual ambient dose equivalent rates in the system and 62 key regions selected for activation studies have been analysed for several points in time during the cooling, going from the end of the first cycle (EOC) until the end of the reactor life (EOL), which is assumed to be 40 years. An additional study of the residual radiation until a cooling of 10^4 years will be also presented.

In Fig. 4 the results for the residual dose rates in the structures around the core are reported for a

DOI: 10.1051/epjconf/201715305001

short/medium irradiation period (90 days sub-cycle), with a cooling time of 1 week. It can be observed the strong activation of the structures above the core, which contain high fractions of steel, especially in the hottest region above the core. The data also show an important residual dose rate in the nitrogen gas plenum, due to the activation of nitrogen from the reaction 14N(n,p)14C, which exhibits a very high neutron capture cross-section (1.83 b) in the thermal region. It has to be noticed that the total ^{14}C production would be in any case largely below the common regulatory limits, as reported by IAEA [6].

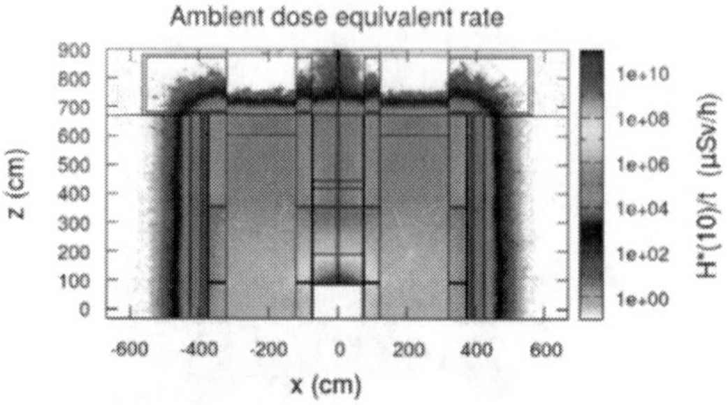

Figure 4. H*(10) residual dose rate [µSv/h] after 90 days of irradiation and 1 week cooling.

As example of the activation study after the end of the MYRRHA life, the residual activity in the diaphragm, the vessel and the liner are shown in Fig.5, for very long times (until 10^4 y) after the end of the reactor operation.

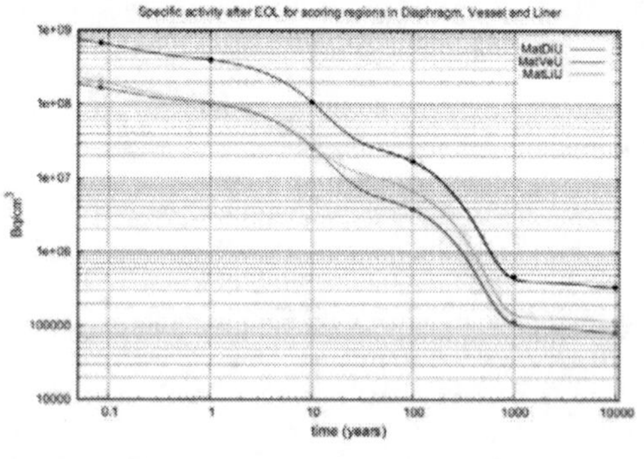

Figure 5. Residual activity [Bq/cm^3] in the Diaphragm, the Vessel and the Liner in the most irradiated region.

References

1. D. De Bruyn, H. Aït Abderrahim, P. Baeten and R. Fernandez, MYRRHA, the Multi-purpose Hybrid Research Reactor for High-tech Applications, *Proc. Int. Congress on Advances in NPPs (ICAPP '11)*, Nice, France, May 2-6, 2011, France Omnipress, 472-478 (2011).
2. MCNPX 2.6.0 Manual, LANL Report LA-CP-07-1473 (2008).
3. A. Ferrari, P. Sala, A. Fassò, J.Ranft, "FLUKA: A multi-particle transport code", CERN-2005-10 (2005), INFN/TC_05/11, SLAC-R-773.
4. G. Battistoni, S. Muraro, P.R. Sala, F. Cerutti, A. Ferrari, S. Rösler, A. Fassò, J. Ranft, "The FLUKA code: description and benchmarking", Proc. Hadronic Shower Simulation Workshop 2006, Fermilab 6-8 September 2006, AIP Conference Proceedings, 896, 31-49 (2007).
5. S.G. Mashnik, K.K. Gudima, R.E. Prael, A.J. Sierk, M.I. Baznat and N.V. Mokhov, "CEM03.03 and LAQGSM03.03 event generators for the MCNP6, MCNPX, and MARS15 transport codes", LANL Report LA-UR-08-2931 (2008).
6. IAEA Technical Reports Series No. 421, Management of waste containing tritium and carbon-14,Vienna (2004).

Radiological Shielding Design for the Neutron High-Resolution Backscattering Spectrometer EMU at the OPAL Reactor

Tunay Ersez[1,*], *Fernando* Esposto[1], and *Nicolas* R. de Souza[1]

[1]Australian Nuclear Science and Technology Organisation, Locked Bag 2001, Kirrawee DC NSW 2232, Australia

Abstract. The shielding for the neutron high-resolution backscattering spectrometer (EMU) located at the OPAL reactor (ANSTO) was designed using the Monte Carlo code MCNP 5-1.60. The proposed shielding design has produced compact shielding assemblies, such as the neutron pre-monochromator bunker with sliding cylindrical block shields to accommodate a range of neutron take-off angles, and in the experimental area - shielding of neutron focusing guides, choppers, flight tube, backscattering monochromator, and additional shielding elements inside the Scattering Tank. These shielding assemblies meet safety and engineering requirements and cost constraints. The neutron dose rates around the EMU instrument were reduced to < 0.5 μSv/h and the gamma dose rates to a safe working level of ≤ 3 μSv/h.

1 Introduction

In this paper we present modelling and design of compact radiation shielding required for operation of the EMU instrument, located at Australia's OPAL nuclear reactor. The acceptance criteria for the shielding design were (i) safety – dose rate ≤ 3 μSv/h at external surfaces of the instrument's experimental area, (ii) floor load limit < 20 tonnes/m^2, (iii) low cost, and (iv) for maintenance purposes "easy access" to components of the instruments [1, 2].

Prior to presenting our radiation shielding analysis a brief overview of the EMU instrument is given. It is worth noting that the design process integrated concurrent consideration of neutronic, engineering, and shielding requirements. An extensive description of the neutronic design and performance of this instrument will be presented elsewhere [3].

The EMU instrument is a cold-neutron backscattering spectrometer designed to measure inelastic neutron scattering. Its distinctive feature is that it has a tight elastic energy transfer resolution, of about 1 μeV. The instrument provides unique opportunities to study diffusion in complex materials, e.g. water molecules in confinement, polymers or biological molecules, particularly in the Asia-Oceania region [4].

EMU is the first instrument located on the CG3 neutron beam line of the OPAL Neutron Guide Hall [5]. The layout of the instrument is shown in Figure 1.

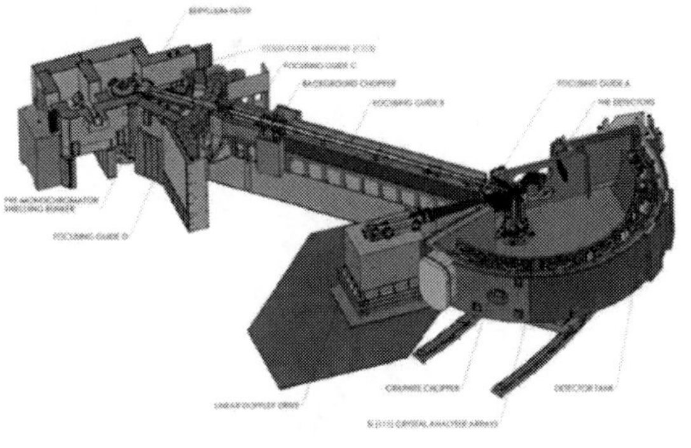

Figure 1. A sliced section view of the EMU instrument with the pre-monochromator bunker, focusing neutron guide sections, choppers, backscattering monochromator mounted onto the Doppler drive and Scattering Tank.

A pre-existing evacuated neutron guide of dimensions 200 mm (height) × 50 mm (width) feeds into the shared EMU/KOOKABURRA pre-monochromator bunker shielding assembly, and its bottom cross section will be intercepted ~200 mm further in air by the highly orientated pyrolytic graphite (HOPG) EMU pre-monochromator (take-off angles from 125.4° to 148°). The bunker shielding assembly for EMU also contains a beryllium filter and Focusing Guide Section D (FGD, entrance guide cross section 170 mm (height) × 50 mm (width)). To accommodate the range of take-off angles Pb cylindrical shielding blocks of radius 140 mm and 500 mm height have been designed. These blocks will be located at the section of the bunker where the neutron beam will exit towards the experimental area and they will reside on top of a hollow steel base (sides 10 mm thick), on which they move by sliding and are integrated

* Corresponding author: tunay.ersez@ansto.gov.au

with the tertiary shutter. The layout of the CG3 neutron guide shielding and shared EMU/KOOKABURRA pre-monochromator bunker is shown in Figure 2.

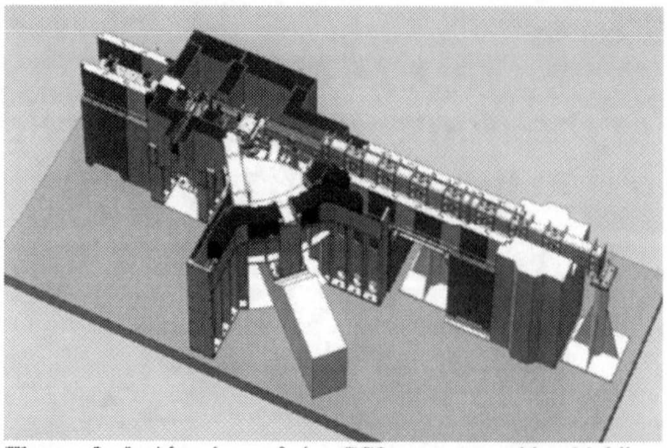

Figure 2. Inside view of the CG3 neutron guide shielding, shared EMU/KOOKABURRA pre-monochromator bunker walls (blue) and cylindrical blocks (blue).

The EMU instrument experimental area is enclosed by a pre-existing concrete wall shared with the KOOKABURRA instrument, which is of 2.5 m height and 300 mm thick up to the pre-monochromator bunker cylindrical blocks and 100 mm thick at the back of the cylindrical blocks. Also, enclosing the rest of the experimental area will be Gyprock Plasterboard panels with viewing windows, a wall from the pre-existing Out-of-Pile cold-neutron guide Bunker and the pre-existing CG3 shielding (Figure 3).

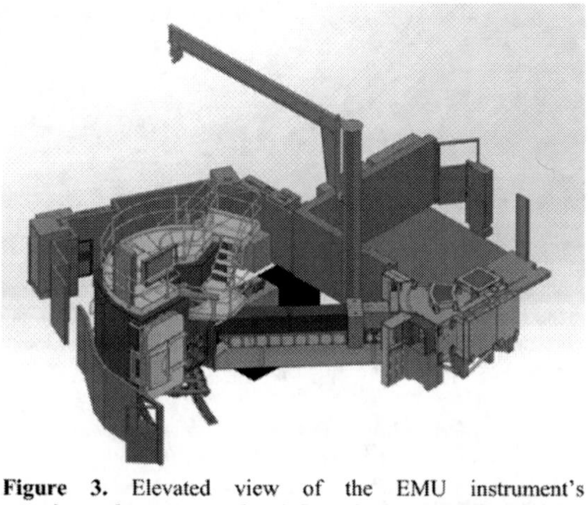

Figure 3. Elevated view of the EMU instrument's experimental area on the left and the KOOKABURRA instrument's experimental area on the right. The concrete wall and the Gyprock Plasterboard panels are shown.

The neutron beam will enter the EMU experimental area via the tertiary shutter, into the evacuated Focusing Guide Section C (FGC), which has an entrance guide cross section of 158 mm (height) × 79 mm (width) and exit guide cross section of 149 mm (height) × 85 mm (width) and is of length 816 mm. The guide surface is coated with m=2.5 supermirrors (SM) [6]. The vacuum chamber of FGC will be connected to the vacuum chamber of the Background Chopper (BC, spinning disc of outer radius 350 mm), to avoid the use of Al windows, which will minimise radiation levels. The BC disc consists of neutron absorbing material (^{10}B) and the backing support for the chopper and its vacuum chamber are composed of high purity AlMg4.5Mn. Focusing Guide Section B (FGB) is aligned with FGC and its vacuum shared with the BC vacuum, again to avoid use of Al windows. The FGB has an exit guide cross section of 50 mm (height) × 49 mm (width) and is of length 4046 mm, which will focus the neutron beam vertically and horizontally towards the N16 Al-window (0.82 mm thickness). The latter is the entrance to the evacuated Scattering Tank assembly. The FGB guide surface is coated with m=3.5 SM [6]. The Scattering Tank assembly will be constructed from 20 mm – 40 mm thickness stainless steel and will be under vacuum < 1 × 10^{-3} mbar.

Inside the Scattering Tank after the N16 Al window, there is another section of Focusing Guide (FGA), with an exit guide cross-section of 30 mm (height) × 30 mm (width) and of length 417 mm. The guide surface is coated with m=5 SM [6]. The neutron beam exiting FGA will be reflected by the HOPG crystal pieces (5 mm thickness) of the Graphite Chopper (GC) with 88 % efficiency towards the Backscattering Monochromator (BM) and the remaining 12 % will be transmitted to be fully absorbed by the ^{10}B-loaded epoxy layer set behind the HOPG crystals (Figure 4). The GC is also located within the Scattering Tank and serves as the second pre-monochromator (Figure 5). The reflected neutron beam from the GC is divergent and, before it reaches the Backscattering Monochromator (BM), will first pass through the N17 Al window (0.82 mm thickness) of the Scattering Tank and and then enter into an air flight tube constructed from steel of 10 mm thickness with exit dimensions of 400 mm (height) × 240 mm (width).

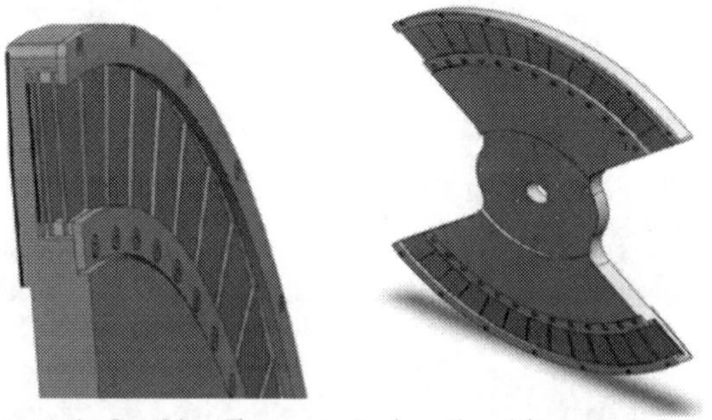

Figure 4. Graphite Chopper spinning disc (close-up at left, assembly at right) with HOPG crystal pieces at its periphery. See main text for a description of its interaction with the neutron beam.

The BM is formed in an 1800 mm spherical radius array of Si crystal elements (0.87 mm thickness) and is oscillated by the linear motor Doppler drive. The neutron beam will be reflected at 180° (i.e. back-scattered) from, and focussed by the BM to re-enter the Scattering Tank through the N17 Al window. It will then pass through the GC (HOPG crystals rotated out of the way) and will reach the scattering sample of interest to the instrument end user (Figure 5).

Figure 5. Elevated view inside the EMU Scattering Tank showing the FGA, GC, and cylindrical sample well insert locating the scattering sample.

Also, inside the Scattering Tank there will be ^{3}He linear position-sensitive neutron detectors, and Analyser Arrays formed in 1800 mm radius spherical arrays of Si crystal elements (0.87 mm thickness) glued on aluminium support frames (10 mm thickness) and spanning ~160° arc. The sample scatters neutrons into 4π sr. If a scattered neutron reaches the analyser and fulfils the backscattering Bragg condition, it is reflected back towards the sample. At which point, it will traverse the sample again and normally arrive at the detectors.

2 Method for computational modelling

A model of the EMU instrument shielding was constructed using MCNP 5-1.60 [7, 8], which took into consideration acceptance criteria (ii) – (iv). The effectiveness of the shielding design to meet acceptance criteria (i) was estimated by the Monte Carlo Method, considering neutrons and generated prompt γ-radiation. In the modelling the average surface flux tally (type F2) and the average cell flux tally (type F4) were used and the geometry splitting variance reduction method was implemented. The cross sections were based on ENDF/B-VI neutron data and MCPLIB04 photon data. Conversion from radiation fluxes to doses was in correspondence with NCRP-38, ANSI/ANS-6.1.1-1977 and ICRP-21 [9-11].

2.1 Geometry

The geometry of the EMU model is shown in Figures 6 – 10, which was constructed to match the neutronic component description given in the previous section. The pre-monochromator bunker shielding walls (top, bottom and sides) were Pb encased in steel, which is a combination of Fe (10 mm thick)/Pb (200 mm thick)/Fe (10 mm thick) and the internal surfaces of these walls, including the ceiling and base, were lined with MirroBor™ (5 mm thick) [12]. The gap size between the cylindrical block shields is 5 mm. These gaps have been covered with MirroBor™ of 5 mm thickness on the bunker internal face.

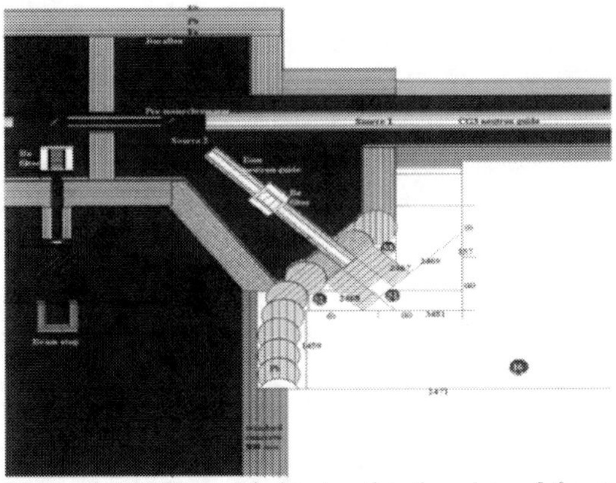

Figure 6. MCNP model showing the plan view of the pre-monochromators bunker and cylindrical block shields.

The experimental area bounding perimeter walls were treated as external surfaces at which dose rates were determined. This approach mirrors the operational requirement that personnel are prohibited from the experimental area when the tertiary shutter is open and the neutron beam is on; dose rates within the experimental area were nonetheless determined at a few locations as well. The bounding perimeter walls were sectioned appropriately up to a height of 5.5 m and each section of size 0.5 m (vertical) × 0.5 m (horizontal). The distances used were defined by the Neutron Guide Hall and instrument layout.

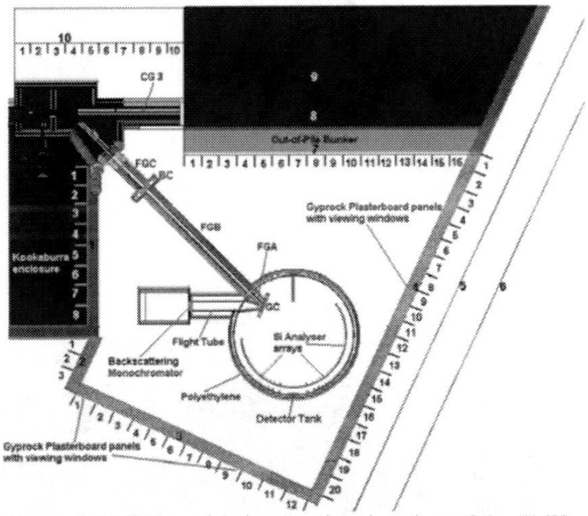

Figure 7. MCNP model showing the plan view of the EMU instrument at the 135° take-off angle configuration, with shielding.

Surface "1" is the common wall between the EMU and KOOKABURRA instruments. The wall is 300 mm thick standard concrete and height of 2.5 m. Surface "2" is the extension of Surface "1" but it will be made out of Gyprock Plasterboard panels with viewing window. Gyprock Plasterboard panels do not provide any radiation shielding per se and were thus modelled as air (or void). Also, Surfaces "3" and "4" will be made out of

Gyprock Plasterboard panels with viewing windows. Surfaces "5" and "6" are at 1 m and 2 m distances from Surface "4" respectively, and represent the elevated walkway and viewing area in the southern section of the Neutron Guide Hall. Surface "7" is the Out-of-Pile Bunker wall external surface and straight above it. Surfaces "8" and "9" are on top of the Out-of-Pile Bunker at 1 m and 2 m distances from Surface "7" respectively, and represent its accessible area. Surface "10" is at distance of 2 m from the EMU pre-monochromator and CG3 guide.

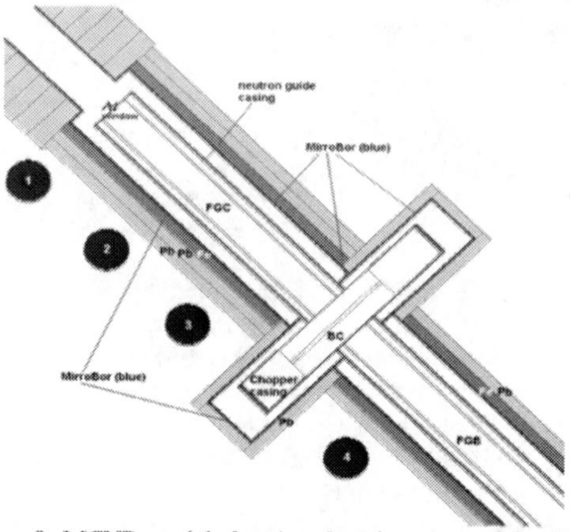

Figure 8. MCNP model showing the plan view of the FGC, FGB and BC surrounded by MirroBor™, Fe and Pb shielding.

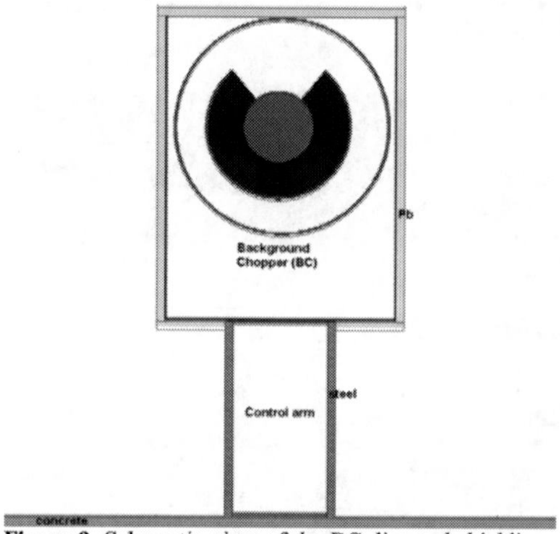

Figure 9. Schematic view of the BC disc and shielding above the control arm.

The instrument itself was modelled in the 135° take-off angle because, for most of its operation it will be very close to this configuration. The internal surface of the Scattering Tank is lined with Cd of 0.5 mm thickness and the exterior is covered with Polyethylene of 50 mm thickness, the latter mainly to slow down environmental neutrons and so reduce the neutron background noise in the detectors.

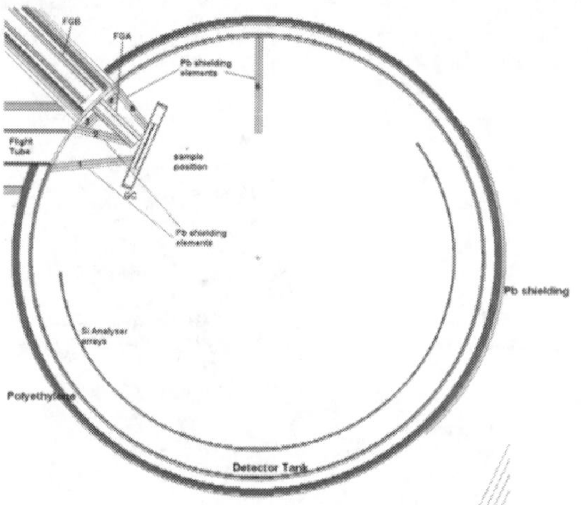

Figure 10. MCNP model showing the plan view of the EMU Scattering-Tank with external and internal shielding.

Also, all the void regions in the model were filled with air except the evacuated volumes of the focusing neutron guide (FGC, FGB, and FGA) and the Scattering Tank.

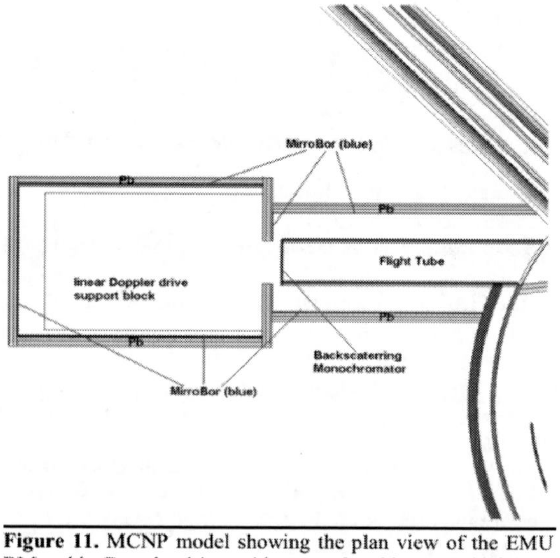

Figure 11. MCNP model showing the plan view of the EMU BM and its Doppler drive, with external and internal shielding.

2.2 Neutron sources

The simulations were performed using two sets of neutron sources for the pre-monochromator bunker and cylindrical block shields (Source 1 and Source 2) and four sets of neutron sources for the shielding design inside the experimental area and around its perimeter (Sources 3 to 6).

<u>Source 1:</u>

DOI: 10.1051/epjconf/201715305002

Maxwellian (T=20 K) mono-directional beam (along the CG3 neutron beam guide axis) and with a neutron current of 1×10^{12} neutrons.s^{-1}.

Sources 2 to 6 were derived from Vitess simulations [13].

Source 2:

Table 1. The energy spectrum from the EMU pre-monochromator.

Energy (meV)	Intensity (neutrons.s^{-1})
0 – 2.08	6.33×10^9
2.08 – 8.32	5.17×10^9
8.32 – 18.72	9.23×10^8
18.72 – 33.29	9.82×10^7

While the physical energy spectrum from the pre-monochromator consists in bands centred at 2.08, 8.32, 18.32, and 33.29 meV for most of its operation at 138° take-off angle, the above representation is a very good approximation for the purpose of our MCNP simulations. The total integrated reflected neutron current was 1.25×10^{10} neutrons.s^{-1}.

Sources 3 and 4:

Simulated as downstream mono-directional beams with energy 2.08 meV and neutron currents of 2.67×10^9 n.s^{-1} and 1.52×10^9 n.s^{-1}, respectively, these were used to determine dose rates resulting from the Al windows immediately before FGC and after FGB and at the Scattering Tank entrance and from neutrons absorbed by the Ni and Ti supermirror (SM) layers of the guides [14].

Source 5:

Represents the neutrons directed towards the BM with energy 2.08 meV and neutron current of 7.66×10^8 n.s^{-1}.

Source 6:

A mono-directional beam with a radius of 0 to 5 mm and directed towards the scattering sample, with energy of 2.08 meV and neutron current of 1.09×10^7 n.s^{-1}. To determine dose rates resulting from sample scattering simulations were carried out with typical samples of H_2O; Al_2O_3; SiO_2.

The MCNP code supplies a relative error associated with the number of events recorded in a tally cell. In the calculations performed for EMU, this relative error was lower than 5%, except those for which the dose rates were < 0.1 μSv/h. The total dose rates were evaluated as the sum of those resulting from each calculation performed with Source 1 and Source 2, and Sources 3 to 6, respectively.

3 Results and discussion

In the simulations performed, the neutrons were relatively easy to stop, the neutron dose rates around the EMU/KOOKABURRA pre-monochromator bunker and perimeter of the EMU experimental area were reduced to < 0.5 μSv/h by using 5 – 10 mm thick high density MirroBor™ on the internal surfaces of the pre-monochromator bunker, focusing guide and BC shielding, Flight Tube and BM shielding, covering gaps and lining the external surfaces of components, e.g., aluminium or steel.

Gamma radiation results from capture of neutrons by the selected shielding and various component materials. In the simulations performed, the gammas produced are from the capture of neutrons by the ^{27}Al (and also, from decay of ^{28}Al), found in the material of components, casings, and Al windows. As well, gamma radiation results from the ^{10}B in the GC, the Si of the BM, the steel material, and from neutrons absorbed by the ^{10}B, ^{nat}B, Si, Na and K contained in the neutron guide glass. The MirroBor™ used to attenuate the neutrons also contains ^{10}B. When irradiated with low energy neutrons ^{4}He (alpha particles) and ^{7}Li nuclei are emitted plus gammas of energy 0.48 MeV. ^{11}B does not activate on thermal neutron capture. For thermal neutron capture by ^{27}Al, the highest energy transition of gamma rays occurs at ~7.7 MeV. For Si it is ~3.5 MeV but there is another strong transition for Si occurring at ~4.9 MeV, which is equivalent to 90% of the strongest transition intensity. Therefore, relatively thick gamma absorbers were employed and in the models presented here Fe, Pb, and a combination of Fe/Pb/Fe was used.

3.1. MCNP simulations of shielding around the pre-monochromator bunker

When the sample shutter is closed the total (n + γ) dose rates around the cylindrical blocks and near the beam opening were ≤ 2.5 μSv/h. The total dose rate at the rear external surface of the pre-monochromators bunker shielding, near to where the pre-monochromator access doors will be located, was ≤ 6 μSv/h. On top of the bunker the maximum total dose rate was calculated to be 11.2 μSv/h. In order to meet the safety criteria time restrictions will be placed on personnel access to the bunker roof while the neutron beam is on; these restrictions have minimal operational impact. At other areas of the pre-monochromator bunker the total dose rates on the external surfaces were ≤ 3 μSv/h.

3.2 MCNP simulations inside the experimental area and around its perimeter

Following numerous simulations the shielding design inside the experimental area and around its perimeter was refined to include the following,

(i) The internal surfaces were lined with MirroBor™ of 5 mm thickness.

(ii) The FGC section's shielding walls (top, bottom and sides) – steel of 30 mm thickness and Pb of 60 mm thickness (Figure 8).
(iii) The BC shielding walls (top, bottom and sides) – Pb of 30 mm thickness (Figure 9).
(iv) The FGB section's shielding walls (top, bottom and sides) – steel of 30 mm thickness and Pb of 30 mm thickness (Figures 8 and 10).
(v) Shielding elements "1" to "8" inside the Scattering Tank – Pb of 50 mm thickness (Figure 10).
(vi) Pb of 25 mm thickness attached to the external rear surface of the Scattering Tank, with dimensions – arc length ~2570 mm × 1580 mm (height). The angle spread relative to the Scattering Tank centre is ~80° (Figure 10). Alternatively a Pb wall of 25 mm thickness can be installed at the projection of this Pb shielding element onto Surface "4".
(vii) Flight Tube, BM and linear Doppler drive and support block shielding walls (top, bottom and sides) – Pb of 50 mm thickness (Figure 11).

The steel layer in (ii) and (iv) above should prevent any high energy gammas that may eventuate from the SM coatings of neutron guide interacting with the Pb shielding to produce fast neutrons. Alternatively additional polyethylene-type shielding could be required on the outside of the Pb shielding.

For the combined neutron and gamma dose rates and BC in the open position the radiological doses were below the design limit of ≤ 3 μSv/h in most of the sections on the external surfaces of the experimental area's perimeter, for heights up to 5.5 m. The only exceptions were in Section 1 of Surface "3" where the dose rates were ~3.2 μSv/h (for height regions ≤ 2.5 m) and in Section 1 of Surface "7" the dose rates were in the range 3 μSv/h to 5.1 μSv/h (for height regions 3.5 m to 5.5 m). Similarly, when the BC is in the closed position Sections 1 to 4 of Surface "7" show that directly above the Out-of-Pile Bunker wall surface (i.e., > 3.5 m height) the gamma dose rate varies from 3 μSv/h to 11.4 μSv/h. However, there are existing barriers to keep persons standing on top of the Out-of-Pile Bunker away from its edge whereas persons may walk through or stand at areas located 1 m to 2 m away (Surfaces "8" and "9") and in these regions the dose rates are expected to be ≤ 3 μSv/h. Also, at Surface "10" the photon dose rates are expected to be ≤ 3 μSv/h.

When the tertiary shutter is open and the neutron beam is on, close to the BC shielding the dose rates were calculated to be in the range 15 μSv/h – 30 μSv/h and near the beginning of the FGC shielding and its Al window the dose rates were 100 μSv/h – 500 μSv/h (located at the positions of spheres labelled "1", "2", "3" and "4" shown in Figure 8). Therefore (as anticipated), while the instrument is operating persons will not be allowed to enter the experimental area.

The Pb shielding elements "1" to "8" in (v) were arranged with "1" to "5" and "8" positioned as vertical plates (Figure 10), while the other elements "6a" to "6b" are the top and bottom vertical plates above and below the FGA end section and "7" is the top cover. These Pb shielding elements were arranged such that the gammas produced from the Al window at the end of the FGB and the N16 and N17 Al windows were contained at the location which they were generated at, and such that those produced from the BM material were contained inside the Scattering Tank.

On the top of all shielding assemblies persons are not expected to be walking over, or spending any amount of time in the area while the neutron beam is on. Further, if the 3 μSv/h limit criterion is not required to be strictly adhered to in certain areas depending on the area occupancy factors for persons, then the above mentioned thicknesses for the shielding are conservative. Also, the results have shown that while the EMU instrument is operating the dose rates at the adjacent KOOKABURRA [15] and PELICAN [16] instruments and Polarised He-3 Station will be ≤ 3 μSv/h.

At the time the above simulations were performed MCNP 5-1.60 was the standard and validated code available, however, some of these results were checked with MCNP 6-1.0 [17, 18].

3.3 Radiation Surveys

Radiation surveys were conducted by ANSTO Health Physicists outside and inside of the instrument enclosure at 18 Radiation Base Points (RBP) E1 to I18, and permanent gamma radiation monitoring points GRM-1 and GRM-2 marked as γ, shown in Figure 12. The surveys were undertaken for the following configurations:

1. Tertiary shutter closed (baseline).
2. Tertiary shutter open, chopper system stopped (BC fully open, GC fully reflecting).
3. Tertiary shutter open, chopper system stopped (BC fully closed).
4. Tertiary shutter open, maintenance beam stop in place, chopper system stopped (BC and GC fully open).

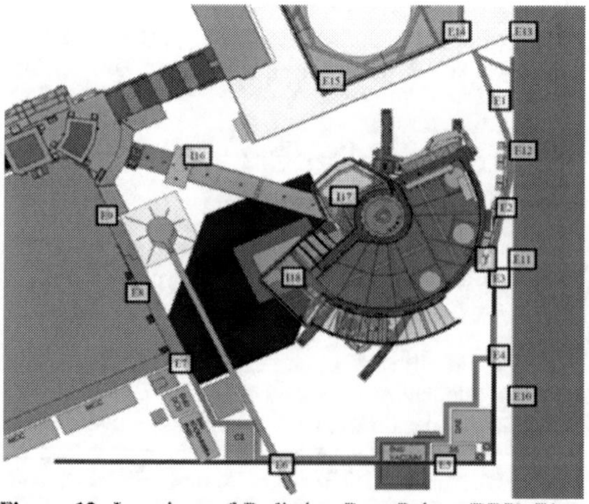

Figure 12. Locations of Radiation Base Points (RBP) E1 to I18, and permanent gamma radiation monitoring points GRM-1 and GRM-2 marked as γ, for radiation dose surveys.

Table 2. The radiation dose survey results (μSv/h) for configuration 4 with the tertiary shutter closed.

RBP	neutron	photon
E1	0.0	3.1
E2	0.0	0.3
E3	0.0	0.2
E4	0.0	0.2
E5	0.0	0.2
E6	0.0	0.1
E7	0.0	0.1
E8	0.0	0.2
E9	0.0	0.3
E10	0.0	0.5
E11	0.0	0.7
E12	0.0	0.6
E13	0.0	2.3
E14	0.0	2.0
E15	0.0	1.7
I16	0.0	0.6
I17	0.0	0.8
I18	0.0	0.2
GRM-1		0.6
GRM-2		0.5

The radiation surveys conducted for instrument configurations 1 to 3 have shown that the dose rates at the RBPs and permanent gamma radiation monitoring points were ≤ 3 μSv/h.

4 Conclusion

To meet the acceptance criteria and achieve radiological doses below the allowed limits, the shielding design for the EMU instrument has been refined by means of MCNP calculations. The latest shielding arrangement has confirmed the effectiveness of the design for the FGC; the BC; FGB; Al windows of the Scattering Tank; Flight Tube; BM and linear Doppler drive. Additional Pb shielding elements (vertical, top and bottom plates) of 50 mm thickness were arranged inside the Scattering Tank. These arrangements are consistent with safety requirements, more than meet the floor load limit (< 20 tonnes/m^2 limit), and cost constraints.

The current state of the instrument along with the shielding arrangement is shown in Figure 13. The instrument is in its initial stage of user operation at the time of this report.

Figure 13. Current state of the EMU instrument with the shielding arrangement.

This work was supported by the Neutron Beam Instruments Expansion Project (NBI2, ANSTO).

References

1. "Radiation Protection Plan for the OPAL Neutron Beam Facility", ANSTO Document Number: PO-OP-02 rev 1, May 2012.
2. *Reactor Beam Hall and Neutron Guide Hall Deflection Analysis Results Report*, ANSTO Document Number: RRRP-3220-EBECW-008-B, 31 July 2003.
3. N.R. de Souza et al., to be published (2017).
4. N.R. de Souza, A. Klapproth, G.N. Iles, Neutron News **27** (2016), pp. 20-21.
5. F. Klose, P. Constantine, S.J. Kennedy, R.A. Robinson, J. Phys.: Conf. Ser. **528** (2014), 012026.
6. "Focussing Guide", EMU – Backscattering Instrument, prepared by MirroTron for ANSTO, Document Number BRAGG-EMU-SPC-0041-00, January 2013.
7. X-5 Monte Carlo Team, *MCNP^{TM} – A General Monte Carlo N-Particle Transport Code, Version 5 1.40,* Report LA-UR-03-1987, LANL, U.S.A., April 2003 (Revised 10/3/05).
8. F. Brown, B. Kiedrowski, J. Bull, *MCNP 5 – 1.60 Release Notes,* Report LA-UR-10-06235, LANL, U.S.A., 2010.
9. NCRP Scientific Committee 4 on Heavy Particles, H.H. Rossi, chairman, "Protection Against Neutron Radiation," NCRP-38, National Council on Radiation Protection and Measurements (January 1971).
10. ANS-6.1.1 Working Group, M.E. Battat (Chairman), "American National Standard Neutron and Gamma-Ray Flux-to-Dose Rate Factors", ANSI/ANS-6.1.1-1977 (N666), American Nuclear Society, LaGrange Park, Illinois (1977).
11. ICRP Committee 3 Task Group, P. Grande and M.C. O'Riordan, chairmen, "Data for Protection Against Ionizing Radiation from External Sources: Supplement to ICRP Publication 15," ICRP-21,

International Commission on Radiological Protection, Pergamon Press (1971).
12. Neutron Shield: MirroBor™, Mirrotron Ltd., Konkoly-Thege ut 29-33, Budapest, Hungary, http://www.mirrotron.kfkipark.hu/shield.html
13. K. Lieutenant et al., Neutron instrument simulation and optimization using the software package VITESS, *Proc. SPIE 5536, Advances in Computational Methods for X-Ray and Neutron Optics,* 134 (October 21, 2004); http://www.helmholtz-berlin.de/forschung/oe/em/transport-phenomena/vitess/index_en.html
14. T. Ersez, G. Braoudakis, J.C. Osborn, Physica B **385-386** (2006), pp. 1268 -1270.
15. C. Rehm, A. Brule, A.K. Freund, S.J. Kennedy, J. Appl. Crystallogr. **46** (2013), pp. 1699-1704.
16. D.H. Yu, R.A. Mole, T. Noakes, S.J. Kennedy, R.A. Robinson, J. Phys. Soc. Japan **82** (2013), SA027.
17. D.B. Pelowitz, Editor, *MCNP6™ User's Manual Version 1.0,* Report LA–CP–13–00634 Rev. 0, LANL, U.S.A., 2013.
18. J.T. Goorley *et al., Initial MCNP6 Release Overview – MCNP6 Version 1.0,* Report LA– UR–13–22934, LANL, U.S.A., 2013 and in Nuclear Technology, 180, 298 (2012).

New infrastructure for studies of transmutation and fast systems concepts

Fabio Panza[1,a], *Gabriele* Firpo[2], *Guglielmo* Lomonaco[1,3], *Mikhail* Osipenko[1], *Giovanni* Ricco[1,4], *Marco* Ripani[1,4], *Paolo* Saracco[1] and *Carlo Maria* Viberti[2]

[1]Istituto Nazionale di Fisica Nucleare - Sezione di Genova, V. Dodecaneso, 33 - 16146 Genova (Italy)
[2]Ansaldo Nucleare C. so F.M. Perrone, 25 16152 Genova (Italy)
[3]GeNERG DIME/TEC - University of Genova, V. all'Opera Pia, 15/A - 16145 Genova (Italy)
[4]Centro Fermi – Compendio del Viminale – Piazza del Viminale 1 – 00184 Rome (Italy)

Abstract. In this work we report initial studies on a low power Accelerator-Driven System as a possible experimental facility for the measurement of relevant integral nuclear quantities. In particular, we performed Monte Carlo simulations of minor actinides and fission products irradiation and estimated the fission rate within fission chambers in the reactor core and the reflector, in order to evaluate the transmutation rates and the measurement sensitivity. We also performed a photo-peak analysis of available experimental data from a research reactor, in order to estimate the expected sensitivity of this analysis method on the irradiation of samples in the ADS considered.

1 Introduction

The scope of this work is the design optimization, via Monte Carlo simulations (with the MCNP6 [1] and MCB [2] codes), of a subcritical system to perform integral measurements on transmutation processes. Such a system may be an intermediate step between the zero-power ADS Guinevère [3] and the future Myrrha [4].
In this initial study we have considered:
- Actinide and FP pellets inserted into the core at two different positions, studying the burn-up in terms of variation of the initial isotopes and appearance of fission products
- Fission chambers with fissile or fissionable isotope depositions, studying the fission rate for a specific setup
- Photo-peak analysis of irradiated samples, as a method to determine the integral fission based on the appearance of specific FP and to determine the integral capture on U-238 based on the appearance of Np-239

2 ADS description

The geometry of the subcritical core is derived from [5], where the accelerator driver is a 70 MeV proton beam generated by a commercial cyclotron. With respect to the original studies on transmutation capabilities of above described machine [6], we increased the number of Fuel Assemblies (FAs) from 60 to 110 (increasing the Lead reflector radius accordingly) and changed the fuel from UO_2 with 20 % enrichment to the Superphenix MOX composition [7], obtaining a k_{eff} around 0.97 and a thermal power around 430 kW.
All the MCNP simulations reported here have been performed using a measured source spectrum obtained in a dedicated experiment [8]. We assumed a proton beam with 1 mA current, corresponding approximately to a total rate of neutron production from the beryllium target of $7.6 \cdot 10^{14}$ n/s. In Figure 1, the configuration and the selected irradiation positions in the xy plane are shown. The irradiation positions characteristics are reported in Table 1.

Table 1. Neutron fluxes in the 110 FAs configuration for the two considered positions (errors are statistical).

Position	Integral Flux ($n/cm^2/s$)
A	(1.53±0.01)E+13
B	(4.82±0.01)E+11

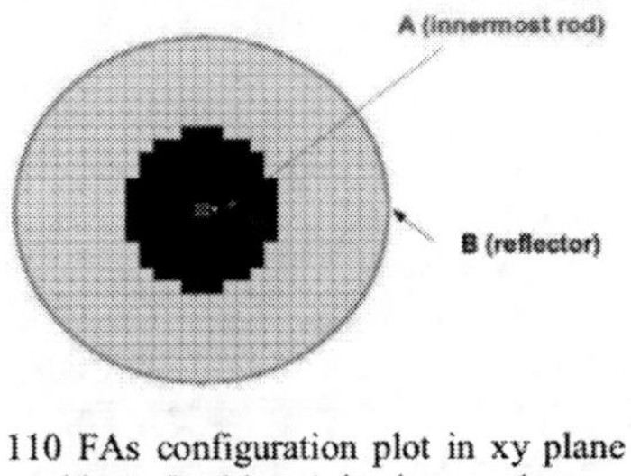

Figure 1: 110 FAs configuration plot in xy plane with the irradiation positions: Position A is close to the source and B at the reflector periphery. Purple: Beryllium target; black: reactor core; yellow: Lead reflector; green: Stainless Steel.

[a] Corresponding author: Fabio.Panza@ge.infn.it

The energy spectrum and axial distributions in the two positions close to the source (A) and in the reflector periphery (B) are plotted in Figure 2 and in Figure 3 respectively.

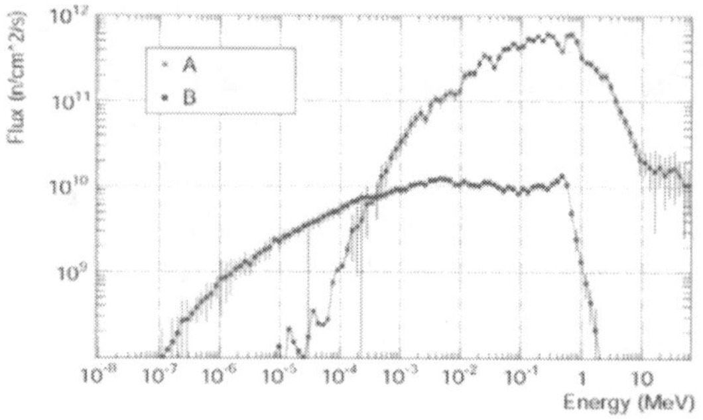

Figure 2: Energy flux distributions in the two positions near the source (A) and in the reflector (B).

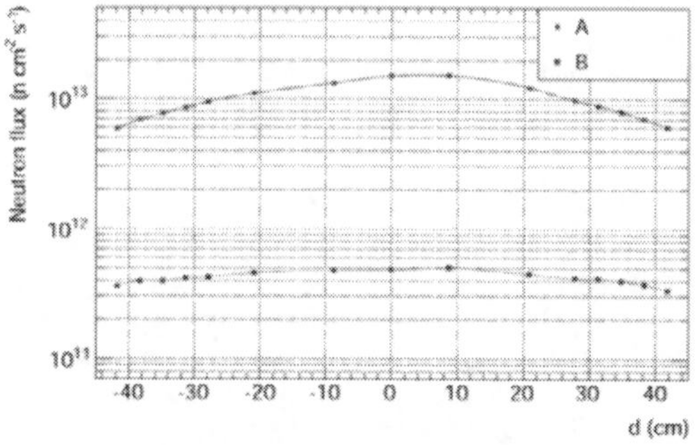

Figure 3: Axial flux distributions in the two positions near the source (A) and in the reflector (B): 0 represents the core mid-plane.

3 Minor actinides and FP irradiation

One of the purpose of these studies is to show how a low-power ADS can be used for integral measurements of nuclear properties relevant to future fast systems. Therefore, we studied the effect of specific irradiations by using the MCB code, in order to investigate the transmutation rate of selected nuclides. In particular, we simulated the irradiation of pellets of Minor Actinides (MA), Np-237, Am-241, Cm-244, as well as Medium Lived Fission Products (MLFP), Sr-90, Cs-137 and Long Lived Fission Products (LLFP), Se-79, Tc-99, Pd-107, Sn-126, I-129, Cs-135. The irradiations were performed by introducing in the simulations a dedicated irradiation channel near the neutron source in the equatorial position, in order to have the highest flux intensity (see Figure 1, position A). The dimensions of the pellet were 0.357 cm radius and 2 cm thickness. The evolution of the pellet composition was calculated by considering one year of irradiation (including the natural decay) in the ADS, or one year of pure decay, then comparing the two final compositions. The difference between the mass after 1 y of decay and 1 y of irradiation (*i.e.,* ***ΔM)*** , with respect to the initial pellet mass (*i.e,* ***M)*** are reported in the Tables 2 to 4.

Table 2. Results for the irradiation of MA pellets.

Nuclide	***ΔM/M (%)***
Np-237	6.35E-01
Am-241	6.90E-02
Cm-244	4.68E-02

Table 3. Results for the irradiation of MLFP pellets.

Nuclide	***ΔM/M (%)***
Sr-90	1.22E-03
Cs-137	1.14E-03

Table 4. Results for the irradiation of LLFP pellets.

Nuclide	***ΔM/M (%)***
Se-79	9.28E-03
Tc-99	1.53E-02
Pd-107	2.39E-02
I-129	9.25E-03
Cs-135	2.04E-03

As we can see from the tables, the contribution of the irradiation to the transmutation, obtained by subtracting the variation due to the decay, is about 10^{-1} % to 10^{-2} % for MA and about 10^{-2} % for LLFP.
The transmutation contribution for MLFP and for some LLFP is about 10^{-3} %, which can be easily understood if we consider their capture cross sections. In some of these cases, a softer neutron spectrum may be necessary, which may be the subject of specific optimizations of the system.

4 Fission chambers

We also performed an evaluation of the fission rate achievable in Fission Chambers (FC) with different depositions (U-235, U-238, Np-237, Pu-239, Pu-241, Am-241), for each of them assuming a typical mass m=10 μg and assuming 100% detection efficiency. In the simulation the FC were placed in either of the two above-mentioned positions (A and B in Figure. 1) within the core and in the Lead reflector respectively. The results are reported in Table 5.

Table 5. Fission rates R for fission chambers with different depositions. A and B are the measurement positions close to the source and in the reflector, respectively (see Figure 1).

Material	R (fiss/s) in A	R (fiss/s) in B
U-235	(7.15±0.20)E+05	(3.19±0.63)E+05
U-238	(2.51±0.03)E+04	(2.37±0.25)E+01
Np-237	(6.80±0.08)E+05	(1.47±0.12)E+05
Pu-239	(9.09±0.10)E+05	(2.45±0.12)E+05
Pu-241	(2.01±0.21)E+05	(1.62±0.42)E+03
Am-241	(1.60±0.11)E+05	(1.01±0.07)E+03

These simulations give us an idea of the integral measurements of fission rates of U-235, U-238, Np-237, Pu-239, Pu-241 and Am-241 in two different reactor positions, therefore with neutron fluxes that differ in both intensity and shape. In particular, with the assumed mass, which appears relatively modest, the counting rates obtained are high enough that high precision measurements can be performed within a few seconds to within several minutes (*e.g.*, in the case of U-238 in position B).

5 Gamma spectroscopy

As part of the integral measurements offered by the proposed ADS facility, we studied a possible method to experimentally estimate the MOX fuel burn-up. We explored the possibility to exploit gamma lines from FP appearing after irradiation. As a practical example, we have analyzed the gamma spectrum measured from a natural Uranium pellet irradiated for 6 hours at 250 W in the central channel of the TRIGA MARK II facility, a thermal research reactor of the LENA laboratory (University of Pavia).
An HPGe detector yielded the gamma spectrum that was analyzed with the Gamma Vision® code by ORTEC®. Our purpose was to evaluate the sensitivity and the systematic uncertainty of this measurement. To get a feeling of the systematic uncertainty, we compared the results of the Gamma Vision® code to manual fits performed by means of the ROOT analysis framework [9]. In Figure 4 we report the gamma spectrum of the natural Uranium pellet after the irradiation.

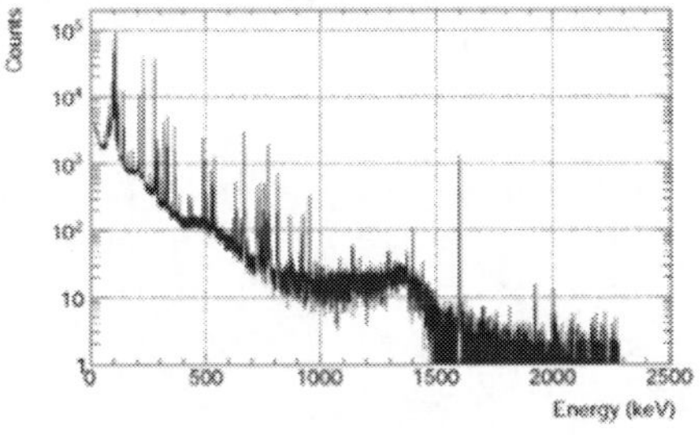

Figure 4: Gamma spectrum measured from a natural Uranium pellet irradiated for 6 hours at 250 W in the central channel of the TRIGA MARK II facility, a thermal research reactor of the LENA laboratory (University of Pavia).

We have considered some specific fission products featuring well-isolated and easily identifiable photo-peaks (La-140, Mo-99, Tc-99m) in the Gamma Vision® program and evaluated the activity of each single peak. To calculate the activity of each nuclide, we corrected the number of counts N in each peak according to the formula below, using five factors that take into account the decay during the irradiation and between the end of irradiation and the startup of counting (TDC), the total counting live time (LT), the branching ratio into that particular peak (BR), the detector efficiency (ε), and the self-absorption coefficient (A_c) [10], according to the following formula

$$A = \frac{N \cdot TDC}{LT \cdot BR \cdot \varepsilon \cdot A_c} \qquad (1)$$

The obtained numbers of counts for the selected peaks are reported in table Table 6.
Then, as an alternative analysis, we have considered the region around each peak and we have fitted the peak+background with a gaussian and a linear function, as shown in the following formula,

$$N(E) = a \cdot e^{-\frac{(E-b)^2}{2c^2}} + d \cdot E + e \qquad (2)$$

with five parameters, in order to obtain the counting rate in the signal region using the ROOT analysis package.
Obviously, in order to minimize the statistical uncertainties, we have considered the most populated (but well isolated) peaks for each isotope.
The fitted photo-peaks for La-140, Mo-99, Tc-99m, are shown in figures 5 to 7 respectively,

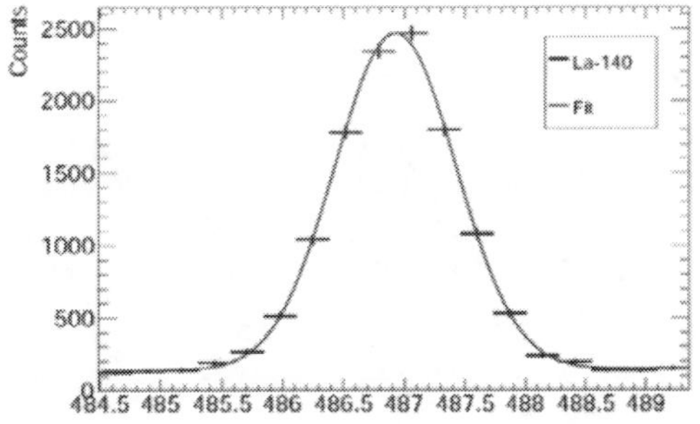

Figure. 5. The 487.03 keV photo-peak of La-140, along with the ROOT fit.

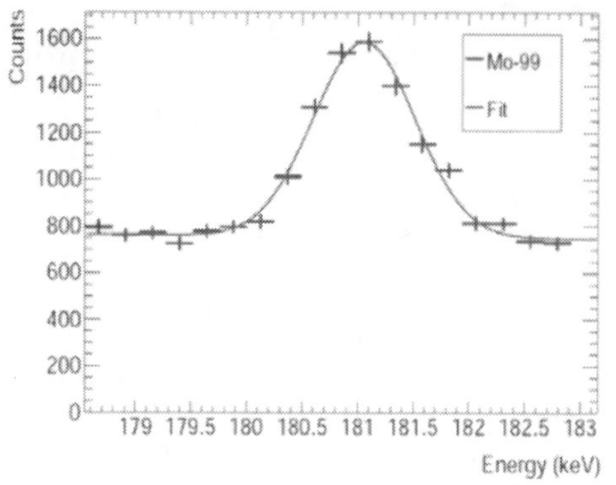

Figure. 6. The 181.09 keV photo-peak of Mo-99, along with the ROOT fit.

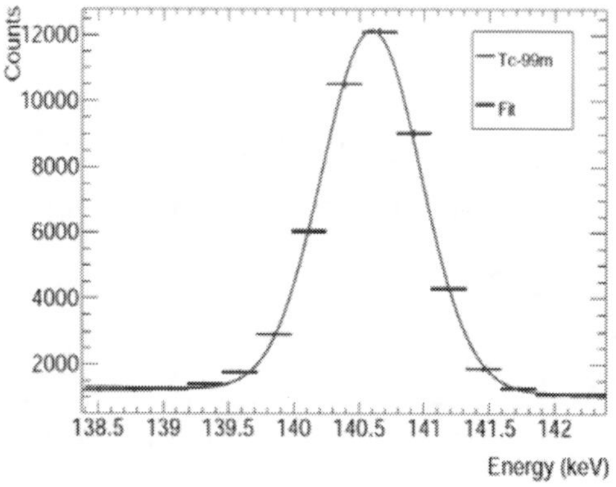

Figure. 7. The 140.51 keV photo-peak for Tc-99m, along with the ROOT fit.

Any difference between the peak area obtained with ROOT and Gamma Vision® beyond statistical uncertainties would be interpreted as systematic uncertainty. However, in the particular cases considered, we found the two independent results to be statistically compatible. The comparison between the two methods is reported in Table 6.

Table 6: Comparison of the ROOT fits and the GammaVision program for the areas of the photo-peaks of all three isotopes considered.

Nuclide	Counts. from ROOT	Counts from . GV
La-140	10667±17	10592±123
Mo-99	3921±129	3704±145
Tc-99m	39409±90	39001±256

Obviously, longer measurements would lead to smaller statistical errors, which could reveal a systematic difference between the two methods.

Once the activity has been calculated, it is possible to perform a fuel burn-up evaluation based on the appearance of the above-mentioned FP. In the present example where we analyzed the data from the TRIGA, we evaluated the burn-up of the U-235 contained in the natural uranium fuel. In the case of the ADS with MOX fuel, the method can be applied to evaluate the main U-235 and Pu-239 burn-up together (obviously Pu-239 is both created and destroyed; here we only measure the amount of Pu-239 that underwent fission). Considering the production of a specific fission product, knowing its corresponding yield in the fission process (Y_f) and its activity (A), the burn-up, or in other words the number of fissions, can be determined using the following formula:

$$n_F = \frac{A}{\lambda \cdot Y_f} \tag{3}$$

where λ is the nuclide decay constant. In Table 7, we report the fuel burn-up evaluated from different fission fragments.

Table 7: Calculated activities using ROOT fits for the three selected photo-peaks and corresponding evaluation of the U-235 burn-up in the data from the TRIGA reactor in Pavia.

Nuclide	Activity (Bq)	U-235 Fiss. mass (g)
La-140	(3.92±0.05)E+04	(5.18±0.04)E-11
Mo-99	(2.29±0.08)E+04	(5.06±0.19)E-11
Tc-99m	(2.30±0.06)E+05	(5.20±0.04)E-11

These results has been compared with the analytic formula which gave a U-235 fissioned mass of 5.14E-11 g by considering an effective fission cross section of 102 b for the central channel of the TRIGA MARK II reactor [10]. The results from the analysis of the three isotopes nicely agree with each other and with the expected value from the effective fission cross section. In order to find a possible application of this method to a MOX fuel (in which the fission fragments come mainly from U-238, U-235 and Pu-239), we propose to consider three different fission product activity measurements, thereby solving a system of three equations in three variables to distinguish the contributions from the different nuclides as reported below.

$$\begin{cases} N_{FP1} = Y_{U-235}^{FP1} \cdot N_{U-235} + Y_{U-238}^{FP1} \cdot N_{U-238} + Y_{Pu-239}^{FP1} \cdot N_{Pu-239} N_{FP1} \\ N_{FP2} = Y_{U-235}^{FP2} \cdot N_{U-235} + Y_{U-238}^{FP2} \cdot N_{U-238} + Y_{Pu-239}^{FP2} \cdot N_{Pu-239} N_{FP2} \\ N_{FP2} = Y_{U-235}^{FP3} \cdot N_{U-235} + Y_{U-238}^{FP3} \cdot N_{U-238} + Y_{Pu-239}^{FP3} \cdot N_{Pu-239} N_{FP3} \end{cases} \tag{4}$$

To estimate the irradiation time necessary to obtain the same quantity, for example, of La-140 from U-238, in the ADS as in the TRIGA, we have used a Bateman equation. The resulting irradiation time is about 1000 s. On the same line, we have also fitted the Np-239 photo-peak from the TRIGA data in order to evaluate the integral capture on U-238. In this case, the ROOT fit and the Gamma Vision® program turn out to be statistically incompatible, with a systematic difference of about 4 %: the results are shown in Table 8.

Table 8: Comparison of the ROOT fits and the Gamma Vision® program for the area of the photo-peak of Np-239.

Nuclide	Counts from. ROOT	Counts from . GV	Diff. (%)
Np-239	139551±141	144535±380	3.95±0.01

For the Neptunium-239 activity, the result is (3.72±0.01)E+05 Bq (from ROOT analysis), to be compared with the numerical calculation performed by using an effective cross section in [10], which gives 3.70E+05 Bq.
The fitted Np-239 photo-peak is shown in Figure 8.

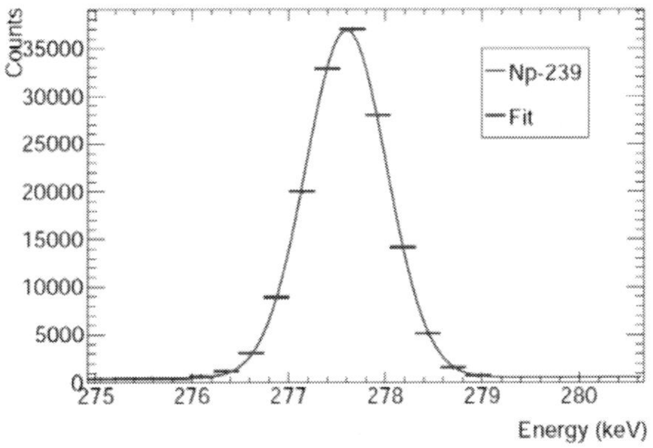

Figure. 8. The 277.60 keV photo-peak for Np-239, along with the ROOT fit.

In order to estimate the sensitivity of this method applied to the burn-up evaluation in a MOX fuel pellet placed in our reference ADS (same dimensions as the actinide pellets of section 4, irradiation time: 1y), it is useful to calculate the Minimum Detectable Activity (MDA) in the regions of the spectra where some MA photo-peaks will be present. The MDA at 95% CL is given by [5]:

$$MDA = \frac{2.71 + 4.65 \cdot \sigma_{bg}}{LT \cdot \epsilon \cdot BR} \qquad (5)$$

where:
- σ_{bg} is the poissonian background standard deviation
- LT is the live time of the data acquisition
- ε is the detector efficiency
- BR is the peak branching ratio

The results are reported in table 9, where we report the simulated activity for Am and Cm isotopes, along with the corresponding MDA. To calculate the MDA, it is necessary to make some assumption about the counts from the background in a specific measurement time. Therefore, as an indicative estimate, we assumed the same spectrum measured in the TRIGA of Pavia and the same total measurement live time. We see that Am-241, Am-243 and Cm-243 should all be detectable in a live-time of about 200 s. For the Cm-244, a much longer measurement time would be needed, in the order of a couple of days, since the MDA is significantly larger than the activity produced in the ADS in 1y. Finally, the Cm-245 is produced in a negligible and non-measurable amount.

Table 9: Sensitivity of the HPGe detector to MA photo-peaks

Nuclide	Gamma Energy (keV)	ε (%)	A (Bq) after 1y irradiation	MDA (Bq)
Am-241	59.54	3.9	6.91E+08	226.25
Am-243	74.76	5.4	7.18E+04	407.74
Cm-243	276.68	9.4	1.31E+03	133.91
Cm-244	42.82	2.06	6.05E+03	1.58E+05
Cm-245	130.05	15.39	9.97E-04	590.76

6 Conclusions

We simulated an Accelerator Driven System comprising a proton cyclotron with 70 MeV beam energy and 1 mA beam current. The core is made up by 110 fuel assemblies, each hosting 81 fuel rods with a Superphenix-type MOX. By simulating the insertion into the core (innermost rod) of various MA and FP pellets, we studied the variation of the initial isotopes, the appearance of fission products for the Actinides and the transmutation by capture of the FP. We observed that the percentage transmutation with respect to natural decay is about 10^{-2} to 10^{-1} % for MA, while for LLFP and MLFP is about 10^{-3} to 10^{-2} %.
We simulated the rate of fission chambers with fissile or fissionable isotope (U-235, U-238, Np-237, Pu-239, Pu-241, Am-241) depositions, considering two irradiation positions: close to the source and in the lead reflector periphery. Our results indicate that measurements of high precision can be performed within a few seconds to within several minutes.
We performed a photo-peak analysis of an irradiated natural Uranium sample (data taken at the TRIGA reactor of Pavia), as a method to determine the integral fission based on the appearance of specific FP and to determine the integral capture on U-238 based on the

DOI: 10.1051/epjconf/201715305003

appearance of Np-239. The analysis has been performed by evaluating the activity of the appeared fission product photo-peaks for La-140, Mo-99 and Tc-99m. We compared results for the peak area obtained from the Gamma Vision® program and from fits of the photo-peaks considered, performed with the ROOT analysis tool. The results are compatible within the statistical uncertainties, except for the U-238 capture where a 4 % discrepancy between the two analyses was observed. The methodology presented can be applied also to the study of the integral fission in the MOX fuel, assuming the fission products to come mainly from U-235, U-238 and Pu-239, by considering the activity of at least three different isotopes and by solving a system of three equations with three variables (the number of fissioned nuclides).

Acknowledgements

The research leading to these results has received funding from the European Atomic Energy Community's (Euratom) Seventh Framework Program FP7/2007-2011 under the Project CHANDA (Grant No. 605203)

References

1. T. Goorley, "MCNP6.1.1-Beta Release Notes", LA-UR-14-24680 (2014).
2. Cetnar J. et al., MCB - a continuous energy Monte Carlo Burnup code, OECD/NEA, Fifth international information exchange meeting, Mol (1998).
3. A. Kochetkov et al., Current progress and future plans of the FREYA Project, Proceedings of the Second International Workshop on Technology and Components of Accelerator-driven Systems, Nantes, France 21-23 May 2013
4. H. A. Abderrahim ., MYRRHA a flexible and fast spectrum irradiation facility, Proceedings of the 11th International Topical Meeting on Nuclear Applications of Accelerators (AccApp 2013), 5-8 August 2013, Bruges, Belgium
5. G. Ricco *et al.,* Eur. Phys. J. Plus, **129** (2014).
6. G. Lomonaco *et al.* Eur. Phys. J. Plus, **129(74)** (2014)
7. L. Mansani (Ansaldo Nucleare), private communication.
8. M. Osipenko *et al.* Eur. Phys. J. Plus, **129(68)** (2014)
9. B. Brun and F. Rademakers, ROOT, An Object Oriented Analysis Framework, Proceedings AIHENP'96 Workshop, Lausanne, Sep. 1996, Nucl. Inst. & Meth. in Phys. Res. A 389 (1997), https://root.cern.ch/
10. A. Borio di Tigliole, *et al., Preliminary TRIGA Fuel Burn-up evaluation by means Monte Carlo code and computation based on total energy released during reactor operation,* PHYSOR 2012, Knoxville, Tennessee, USA, April 15-20, 2012, on CD-ROM, American Nuclear Society, LaGrange Park, Illinois (2012).

DOI: 10.1051/epjconf/201715305004

Radiation Transport Calculation of the UGXR Collimators for the Jules Horowitz Reactor (JHR)

Yelko Chento[1,a], César Hueso[1], Imanol Zamora[2], Marco Fabbri[2], Cristina De La Fuente[2] and Asier Larringan[1]

[1]*IDOM Engineering, Consulting and Architecture S.A.U., Av. Zarandoa 23 48015 Bilbao, Spain, +34 944 797 600*
[2]*IDOM Engineering, Consulting and Architecture S.A.U., Gran Vía Carlos III 97 08028, Barcelona, Spain, +34 934 092 222*

Abstract. Jules Horowitz Reactor (JHR), a major infrastructure of European interest in the fission domain, will be built and operated in the framework of an international cooperation, including the development and qualification of materials and nuclear fuel used in nuclear industry. For this purpose UGXR Collimators, two multi slit gamma and X-ray collimation mechatronic systems, will be installed at the JHR pool and at the Irradiated Components Storage pool. Expected amounts of radiation produced by the spent fuel and X-ray accelerator implies diverse aspects need to be verified to ensure adequate radiological zoning and personnel radiation protection. A computational methodology was devised to validate the Collimators design by means of coupling different engineering codes. In summary, several assessments were performed by means of MCNP5v1.60 to fulfil all the radiological requirements in Nominal scenario (TEDE < 25µSv/h) and in Maintenance scenario (TEDE < 2mSv/h) among others, detailing the methodology, hypotheses and assumptions employed.

1 Introduction

Jules Horowitz Reactor (JHR), built on the Cadarache site, will be a major infrastructure of European interest in the fission domain, open to the international collaboration. The JHR will be built and operated in the framework of an international cooperation between several organizations bound by a Consortium Agreement, one of whose main aim is the development and qualification of materials and nuclear fuel used in nuclear industry.

The necessities to design and build new material testing reactors to support operation of the existing power reactors fleets and qualification of future technologies systems have led to the development of precise mechatronic systems compliant not only with the experimental requirements but also with the very demanding nuclear safety constraints, such as UGXR Collimators. These two multi slit gamma and X-ray collimation systems, located at the JHR pool (RER) and Irradiated Components Storage pool (EPI) Fig-1 and Fig-2, will allow to carry out the collimation of the hard X-rays towards the underwater part of the transmission imagining measurement system and the gamma beams between the immersed object to be measured and the experimental equipment located in the out of pile section, Fig-3.

These two Collimators, as well as support and hoisting structures for the gammametry detector, shall ensure proper shielding and shall also provide a second water barriers behind the pile liners.

Figure 1. JHR pool Collimator.

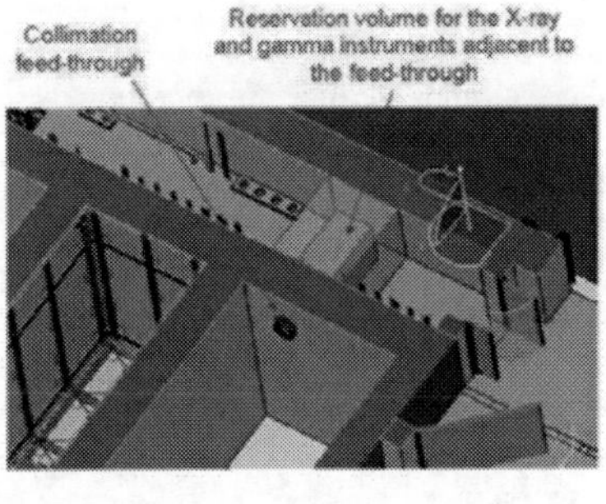

Figure 2. Irradiated Components Storage pool Collimator.

[a] Corresponding author: yelko.chento@idom.com

DOI: 10.1051/epjconf/201715305004

Several assessments are performed to fulfil all the radiological requirements, detailing the methodology employed as well as hypotheses and assumptions. Dose rate compliance is evaluated by mean of radiation transport calculations through MCNP5v1.60 [Ref. 1] and the use of SCALE6.1/ORIGEN-S [Ref. 2] for the source term definition, both intensity gamma and energy spectra.

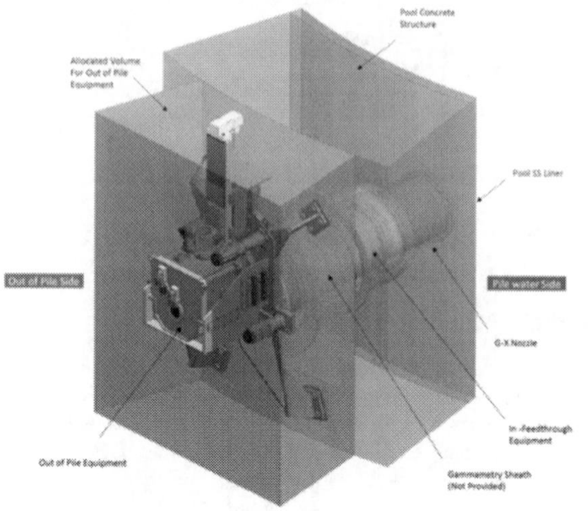

Figure 3. UGXR Collimators.

2 Requirements

Large amounts of radiation are expected to be produced by the spent fuel and X-ray accelerator, which implies that several aspects need to be verified to ensure adequate radiological zoning and radiation protection of the personnel. Three main requisites are defined that shall be taken into account during the design phase:

- A TEDE dose rate lower than 25 μSv/h in the technical gallery concerning UGXR equipment in the out of pile section under normal conditions.
- Existence, under normal conditions, of a significant difference between the shielded flux and the collimated flux, considering the Lanthanum-140 energy peak as a reference value. As well, a non-significant scattered component in the flux detected.
- In Maintenance situation, a TEDE dose rate lower than 2 mSv/h in the area concerning UGXR equipment in the out of pile section under normal conditions.

The correct compliance of the model with the Verification requisites is ensured, during the design phase, taking into account three requirements groups:

- Safety analyses (Structural Integrity & Biological Protection);
- Gamma viewing control (Geometrical viewing def. & Absence of leakage path);
- Mechanical performance (Movements accuracy & Drive dimensioning);

3 Baseline data and hypotheses

Multiple scenarios and hypotheses are defined that allowed to consider all the possible casuistry, so that even in the worst cases the Collimator adjacent galleries will be safe from the radiological point of view. The design shall ensure the dose rate requirements in:

- Nominal situation;
- Maintenance situation;
- Pre-collimator block loading/unloading situation;
- Other situations that shall be accomplished to satisfy the Nuclear Safety criteria (Leakage sensitivity analyses).

4 Model definition

The first step, avoiding the CAD generation by scratch, is to simplify and defeature UGXR Collimators models using SpaceClaim2015 [Ref. 3] package. The purpose of this is to allow successful conversion of the solid CAD bodies to MCNP5v1.60 [Ref. 1] through SuperMC [Ref. 4], obtaining a fast conversion process and also reducing the effort required to repair the resulting MCNP model.

It is convenient to remove details which are not important for radiation transport calculations, such as fillets, chamfers, bolt holes, etc. as shown in Fig-4. Whilst convertible to MCNP5v1.60 [Ref. 1], the presence of such details significantly increases calculation times.

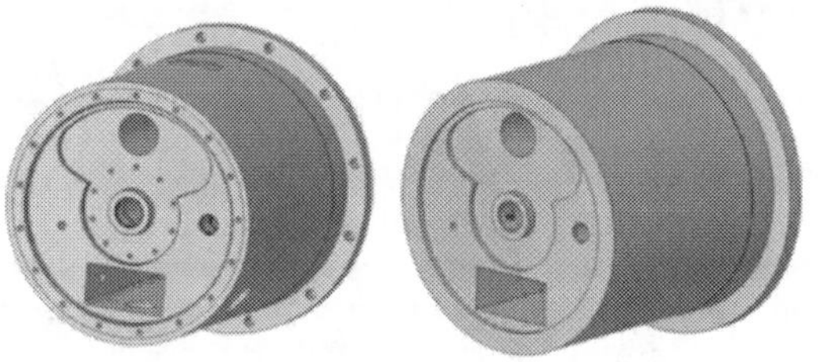

Figure 4. Comparison between original (left side) and simplified model.

The degree of simplification which is acceptable is a function of the photon importance of the features being considered. Generally, it is desirable to retain more detail of channels in the photon gradient direction, since photons would stream down them.

In order to maintain the correct mass of materials, such simplifications are always accompanied by homogenizations of materials and densities corrections. The geometry is modelled in a very detailed way, preserving the single components mass when possible.

Once the model is simplified and checked, the complex solids shall be decomposed into a combination of simple convex solids like box or sphere to allow the proper conversion by means of SuperMC [Ref. 4] to a Constructive Solid Geometry (CSG) required by MCNP5v1.60 [Ref. 1]. Therefore, manually assisted decomposition with SpaceClaim2015 [Ref. 3] is adopted and the whole model is decomposed into a number of convex solids which are then converted to MCNP files Fig-5.

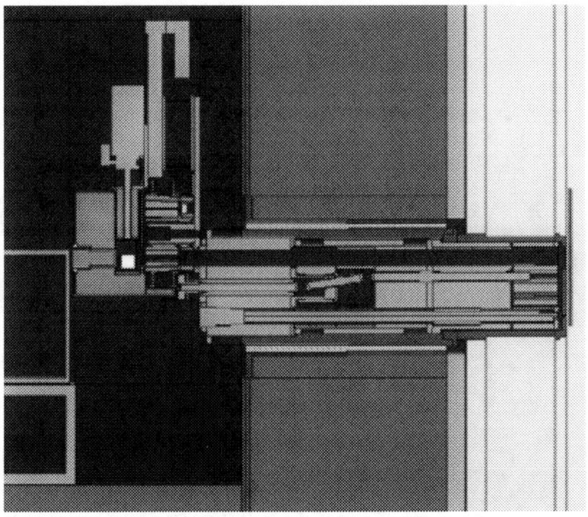

Figure 5. MCNP nominal model vertical cross section.

5 Source term & Radiation transport assessment

5.1 Source term

The source term considered is based on the maximum fission product activities from an experimental PWR MOX fuel cooled for 3 days. The intensity of photons per rod emitted by the irradiated fuel is 1.605E+15 photon/s with an average energy around 0.150 MeV. The source term is modelled as a cylindrical fuel pin facing UGXR Collimators.

It has to be remarked that all the mentioned cases are assessed assuming the same conservative source term defined through the fuel activity concentration, determining the energy spectra by means of SCALE6.1/ORIGEN-S [Ref. 2].

5.2 Radiation transport

Radiation transport calculations, which are very computational demanding, are fundamental to accomplish with the dose rate zoning requirements. The radiation transport assessments, which allowed performing the dose rate evaluation to the JHR personnel, are carried out by means of MCNP5v1.60 [Ref. 1]. AP ICRP-119 TEDE conversion coefficient [Ref. 5] has been employed because they are the most conservative ones.

The detailed evaluation for each mentioned scenario is one of the keys of this methodology, so a previous CAD of each scenario of the preliminary Collimators design is required to elaborate the geometric model. An iterative procedure allowed the optimization of the design for each of the scenarios under study.

The MCNP5v1.60 [Ref. 1] model contains the UGXR Collimator, the technical gallery, the pool and its concrete wall, the fuel pin source, the X-accelerator and its support. For conservatism no further objects are introduced in the technical gallery and the pin source is modelled without its support (see Fig-6). It has to be highlighted that from the radiological point of view, the internal equipment of the collimation feed-through plays an essential role in radiation protection in the out of pile section as it is the continuation of the biological protection formed by the concrete wall of the pool and thus it must ensure proper shielding.

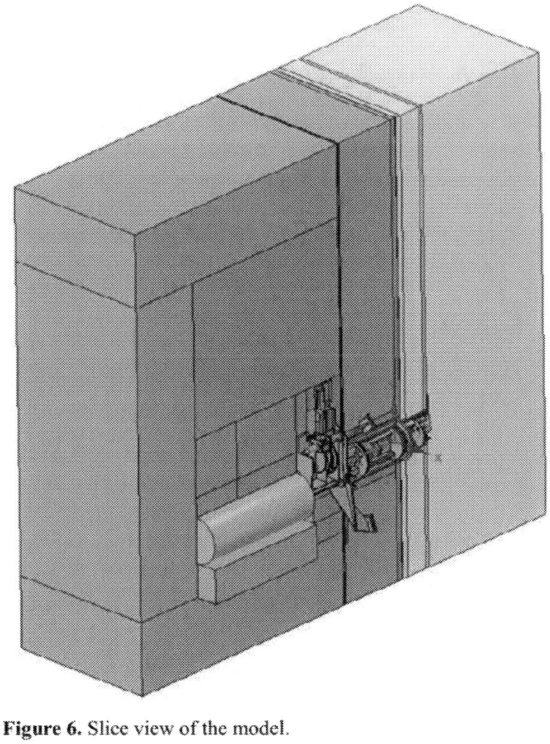

Figure 6. Slice view of the model.

Maintenance situation, Leakage situation and Pre-collimator block loading/unloading situation analyses introduce different changes with regarding Nominal situation model configuration, such as the modifications relative to the disposition of the fuel pin. These scenarios are complementary analyses to the Nominal situation where dose rate at technical gallery is calculated under different operational circumstances.

Considering different tallies outside and inside the Collimator allows to determine photon flux, thermal power and dose rate at different points of the Collimator but also in the technical gallery, where the JHR personnel is expected to be (see Fig-7). Furthermore, the radiological compliance has been assessed by the use of different MCNP5v1.60 [Ref. 1] tally typologies (e.g. F15/F25/F24) giving robustness to the analysis

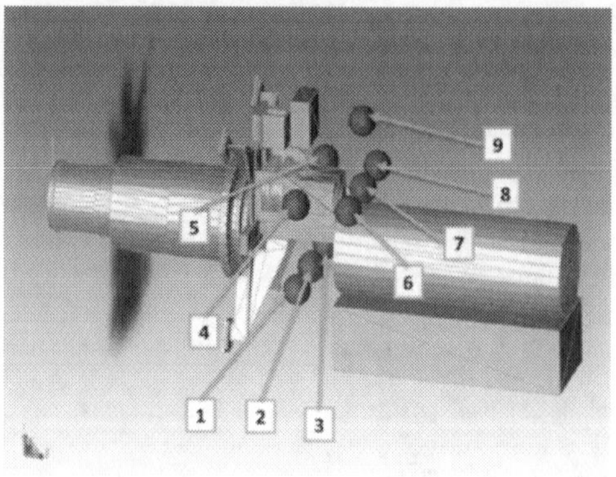

Figure 7. Some of the used tallies.

For the calculations, exponential transform variance reduction techniques are employed and optimized, as well as a mono-energetic weight window, by MCNP5v1.60 [Ref. 1] weight window generated previously, using a reduced density model. An iterative procedure allowed for the optimization of the variance reduction parameters, being lately checked via ADVANTG 3.0.3 code [Ref. 6] for each of the analyzed models.

6 Results

As expected, the dose rate decreases with the penetration inside the UGXR Collimators device due to the collimation effect and the shielding components. The primary photon dose contribution is found to be caused by the gamma direct coming from the source and they affect mostly the zone below the post-collimator wheel. Moreover, the secondary gamma in the technical gallery are generated almost all inside the lead detector shielding because of the large atomic number of the material.

For the "Nominal Situation" scenario, the dose rate estimated in the technical gallery is smaller than 25 μSv/h required, except for a small not-accessible area. The higher dose rate values are located in the UGXR Collimators lower area due to a downside primary photon streaming caused by the scattering of the collimated gamma flux against the post-collimator wheel and collimator as shown in Fig-8, Fig-9, Fig-10 and Fig-11.

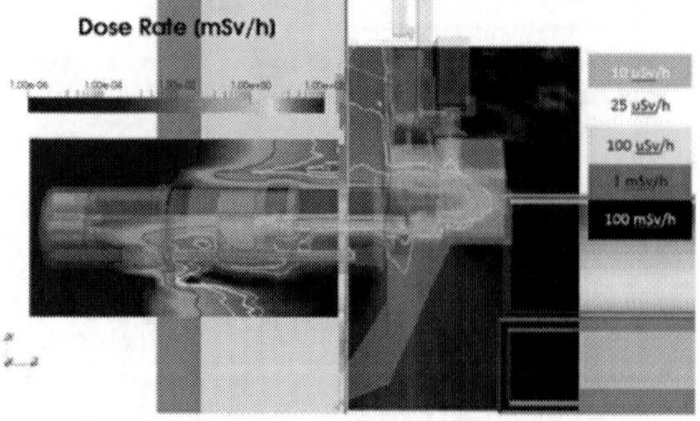

Figure 8. Dose Rate and contour plots for the Nominal situation.

Figure 9. Dose Rate field - 3D contour plots for Nominal situation.

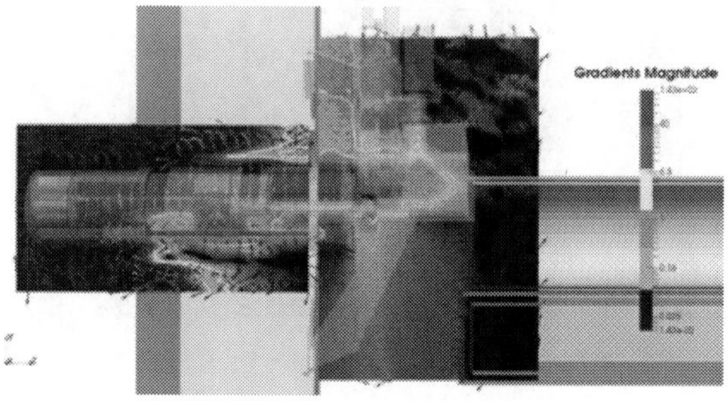

Figure 10. Dose Rate Gradient for Nominal situation.

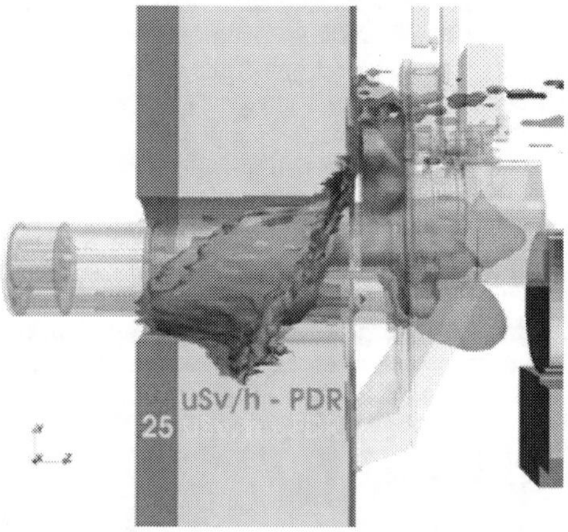

Figure 11. 25 μSv/h Dose Rate 3D contour plot comparison between Final Design (FDR) and Preliminary Design (PDR) for Nominal situation.

Regarding the "Maintenance Situation", thanks to the innovative design solution proposed, the radiological zone classification could be declassified from yellow (TEDE < 2 mSv/h) to green (TEDE < 25 μSv/h) with a positive impact on the overall occupational dose for the personnel as shown in Fig-12 and Fig-13. It has to be remarked that only for the upper zone of the technical gallery a dose rate larger than 10 μSv/h has been determined.

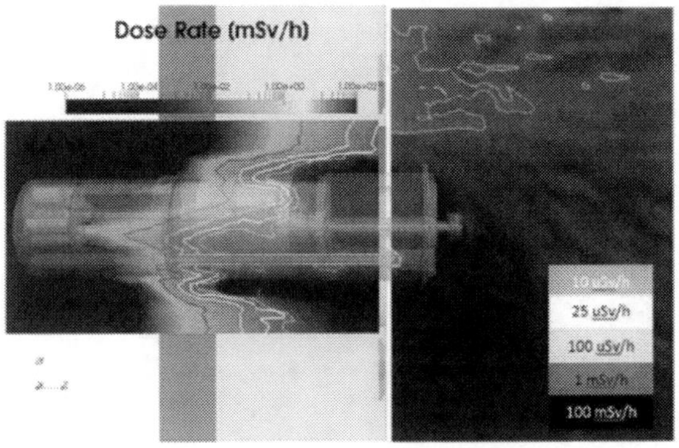

Figure 12. Dose Rate and contour plots for the Maintenance situation.

Figure 13. Dose Rate field - 3D contour plots for Maintenance situation.

From the leakage analyses it is noticeable the achieved design implicit safety shielding capacity of the UGXR Collimators and the concrete walls, not exceeding the 25 μSv/h required, even for the maintenance case as shown in Fig-14. The main contribution of the dose rate in the technical gallery is due to the secondary photons generated by Bremsstrahlung, whereas the lower streaming, which is caused mostly due to the leakage of primary that photons is not relevant.

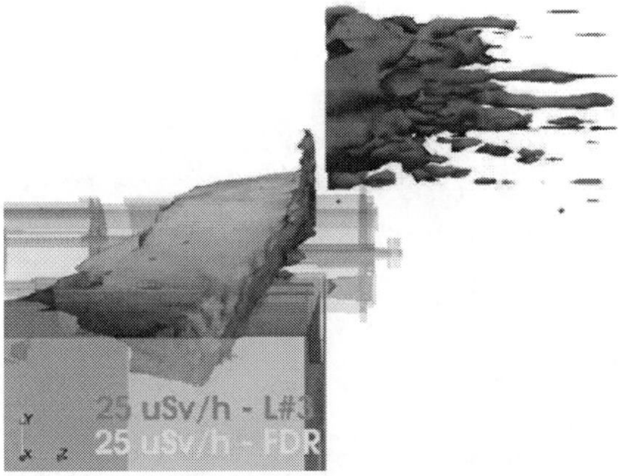

Figure 14. Dose Rate field - 3D contour plot comparison between Final Design (FDR) and Preliminary Design (PDR) for Leakage situation.

For the pre-collimator block loading/unloading operation, it has been confirmed the proper shielding capacity of the designed system as shown in Fig-15. The shielding action of both the pre-collimator utility head and the horizontal lead plug of the utility is evident, even when the pre-collimator is not present. Therefore, as expected, if the pre-collimator is present when the operation is started, the dose rates will be reduced by the extra-shielding.

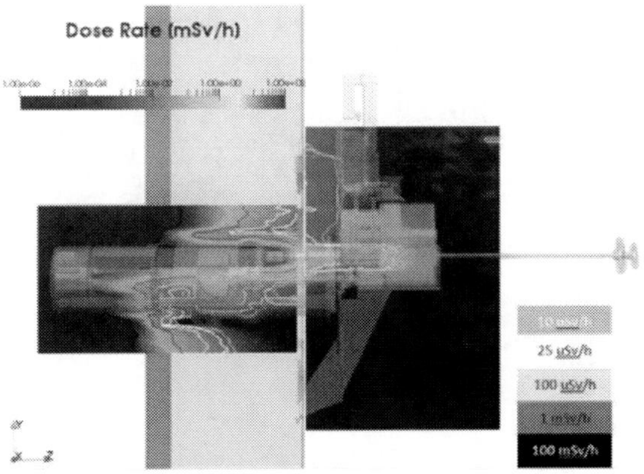

Figure 15. Dose Rate and contour plots for the Pre-collimator block loading/unloading situation.

Finally, a significant reduction of the photon flux of approximately 5 orders of magnitude is observed between the shielded flux and the collimated flux, in addition to a low scattered photon flux. Thus the collimation capacity of the UGXR Collimators is assured.

7 Conclusions

UGXR Collimators, located at the JHR pool and Irradiated Components Storage pool, dose rate compliance calculations are assessed by means of different engineering computational codes. The model Verification requirements has been satisfied, being able to verify the radiological design of the UGXR Collimators. Calculation codes selected for the development of this methodology are widely available, internationally used and validated in many studies.

It can be concluded that the UGXR Collimators design fulfils the dose requirement both for "Nominal and Maintenance Situation" including when applying conservatisms, as well as for the rest of the scenarios taken into consideration. Furthermore, the collimation capacity of the UGXR Collimators is checked through the significant difference between shielded and collimated flux, resulting in a flux incident to the crystal detector with a low scattered component and therefore less noise to signal will be observed.

It is worth to point out that several iterations are required until the design of the UGXR Collimators dose rate restrictions and optimal collimation capacity have been accomplished.

References

1. Los Alamos National Laboratory. MCNP – A General N-Particle Transport Code, Version 5 – Volume I: Overview and Theory, X-5 Monte Carlo Team, LA-UR-03-1987, April 2003.
2. Oak Ridge National Laboratory, SCALE: A Comprehensive Modeling and Simulation Suite for Nuclear Safety Analysis and Design, June 2011.
3. SpaceClaim Corporation, SpaceClaim2015.
4. FDS, SuperMC (Monte Carlo Automatic Modeling Program for Radiation) 5.1 Professional, 2015.
5. ICRP, International Commission on Radiological Protection. ICRP-119, Compendium of Dose Coefficients based on ICRP Publication 60.
6. Oak Ridge National Laboratory, ADVANTG 3.03 – Automated Variance Reduction Generator.

 DOI: 10.1051/epjconf/201715305005

Analysis and optimization of minor actinides transmutation blankets with regards to neutron and gamma sources

Timothée Kooyman[a,1], Laurent Buiron[1] and Gérald Rimpault[2]

[1]DEN/DER/SPRC/LEDC CEA Cadarache, 13108 Saint Paul lez Durance, France

[2]DEN/DER/SPRC/LEPh CEA Cadarache, 13108 Saint Paul lez Durance, France

Abstract. Heterogeneous loading of minor actinides in radial blankets is a potential solution to implement minor actinides transmutation in fast reactors. However, to compensate for the lower flux level experienced by the blankets, the fraction of minor actinides to be loaded in the blankets must be increased to maintain acceptable performances. This severely increases the decay heat and neutron source of the blanket assemblies, both before and after irradiation, by more than an order of magnitude in the case of neutron source for instance.
We propose here to implement an optimization methodology of the blankets design with regards to various parameters such as the local spectrum or the mass to be loaded, with the objective of minimizing the final neutron source of the spent assembly while maximizing the transmutation performances of the blankets. In a first stage, an analysis of the various contributors to long and short term neutron and gamma source is carried out while in a second stage, relevant estimators are designed for use in the effective optimization process, which is done in the last step. A comparison with core calculations is finally done for completeness and validation purposes. It is found that the use of a moderated spectrum in the blankets can be beneficial in terms of final neutron and gamma source without impacting minor actinides transmutation performances compared to more energetic spectrum that could be achieved using metallic fuel for instance. It is also confirmed that, if possible, the use of hydrides as moderating material in the blankets is a promising option to limit the total minor actinides inventory in the fuel cycle. If not, it appears that focus should be put upon an increased residence time for the blankets rather than an increase in the acceptable neutron source for handling and reprocessing.

1 Introduction

In the case of a closed nuclear fuel cycle, minor actinides transmutation is a potential solution to further decrease the radiotoxicity burden of the spent fuel, along with the footprint of the final geological repository, by decreasing the long-term activity and decay heat production of the spent nuclear fuel (Chabert, et al., 2011). This is achieved by removing minor actinides from the waste stream and submitting them to a neutron flux in order to obtain shorter-lived fission products.

This neutron flux can be obtained using various means, such as ADS (Tommasi, Bottollier-Curtet, Massara, Varaine, & Delpech, 2001) or critical fast reactors.. Only such kind of reactors will be considered in this work as successful implementation of minor actinides transmutation requires closure of the fuel cycle which can only be achieved using such reactors.. When considering critical reactors, two approaches can be distinguished. Minor actinides can either be incorporated in the reactor fuel, the so-called homogeneous approach or loaded in dedicated targets named Minor Actinides Bearing Blankets (MABB) located at the periphery of the reactor core. This last option is called heterogeneous transmutation. A detailed analysis of the advantages and drawbacks of each approach can be found in (NEA, 2012). In the homogeneous approach, the neutron spectrum hardening in the core leads to a negative impact on feedback coefficients and on core transient behavior, which means additional safety measures (power reduction, active systems) must be added. The entire fuel cycle is also "polluted" with minor actinides to some extent. However, once an equilibrium situation is reached, minor actinides production in the core is null. In the heterogeneous transmutation case, the "standard" fuel cycle and the transmutation fuel cycle are completely separated and the impacts on core operations are limited as the minor actinides are located in low flux level zones. However, minor actinides production continues in the core itself, which decreases the total transmutation performances of the whole.

To compensate for the low level of flux experienced at the periphery of the core by the transmutation blankets, it is necessary to increase the minor actinides content in the assemblies in order to maintain acceptable transmutation performances – namely in terms of mass consumed per unit of energy produced, usually expressed in kg/TWeh. This approach is limited by the subsequent increase in decay heat rate and neutron source of the irradiated blanket due to a higher curium production. This increase

[a] Corresponding author: timothee.kooyman@cea.fr

DOI: 10.1051/epjconf/201715305005

lengthens the required cooling time for the irradiated blankets, thus increasing the total minor actinides inventory in the fuel cycle. Additionally, the higher neutron emission increases the radioprotection requirements for handling and transportation of the blankets.

Depending on the corresponding limit for handling or reprocessing fast reactor spent fuel, either irradiated assembly decay heat or neutron emission can constitute a critical point for reprocessing. Considering the high uncertainty remaining on the effective limitations regarding reprocessing, it is currently uncertain which of this parameter will be dimensioning. Consequently, this paper will focus on the behavior of neutron source and associated dose rate with regards to minor actinides transmutation, decay heat considerations being treated separately.

We consider here an equilibrium situation for americium production and consumption in the fuel cycle, where the entire production of americium in the core is matched by consumption in minor actinides bearing blankets. Curium is discarded as a waste during the reprocessing step. Such a situation is for instance discussed in (Meyer et al, 2013).

In this case, the efficiency of the total transmutation process can be characterized by:

- The efficiency of americium destruction during irradiation, which is a measure of the number of reactor units to be equipped with blankets necessary to transmute the amount of americium produced in the cores.

- The total inventory of americium in the fuel cycle. This inventory depends on the irradiation time, the spent fuel cooling time and the manufacturing time of the new assemblies. The cooling time itself depends on the technological constraints associated with reprocessing. This inventory can be linked to the number of transports of radioactive material across a country, which should be as low as possible.

No explicit technological limit for handling or reprocessing spent fuel can be found as of now, considering that such a limit depends on the technological solutions used for assembly handling and transportation and on the reprocessing scheme available in the future, along with radioprotection considerations. However, it is possible to use the corresponding emission level of a standard fuel assembly as a reference point for comparison purposes and to work on a relative scale. This is detailed in the next part.

It has been shown that minor actinides transmutation performances and corresponding neutron source can be fully parametrized by the neutron spectrum and the amount of americium loaded in the blankets (Kooyman & Buiron, 2015). Considering the simple parametrization of the problem parameters and outputs, an optimization scheme under constraints was implemented in this work. Such a process is discussed here.

The physics of spent target assembly neutron source will be first characterized and compared to a standard fuel cycle assembly. In a second time, the general principle of an optimization methodology of minor actinides transmutation with regards to the fuel cycle constraints and more specifically to radioprotection constraints will be outlined. This methodology will then be applied and the results compared to complete core calculations.

2 Spent fuel neutron and gamma emissions analysis

Typical values for Sodium Fact Reactor (SFR) spent fuel after five years of cooling are given in Table 1. They were calculated using the SFR V2B core design as it can be found in (Sciora, et al., 2011). This core is a 3600 MWth homogeneous sodium fast reactor which was designed by CEA, EDF and Areva. Assembly total residence time is 2050 EFPD with a 5-batch management scheme. The ERANOS code system (Rimpault, 2002) was used for core calculations and the DARWIN code system for depletion calculations. (Tsilanizara, et al., 2000).

Table 1 : Spent fuel assembly characteristics after 5 years of cooling

Assembly mass (kg)	163
Decay heat (kW)	1,33
of which alpha (%)	55,4
of which beta (%)	23,7
of which gamma (%)	20,9
Neutron source (n/s)	1,22E+09
Mean gamma energy (MeV)	0,56
Maximum gamma energy (MeV)	3,40

The neutron source is dominated at 96 % by spontaneous fission of ^{244}Cm. Alpha decay heat is mainly coming from ^{244}Cm (40.7 %),^{238}Pu (37.6 %) and ^{241}Am (8.9%). Gamma and beta heating is distributed among various fission products. It can be inferred from this analysis that the addition of minor actinides in the fuel will have an impact on the decay heat and the neutron source by increasing the production of ^{244}Cm and ^{238}Pu and the amount of ^{241}Am in the fuel.

For comparison purposes, the same values are computed for a blanket assembly located in the 13th core ring loaded with 20 % in volume of americium oxide (AmO_2). The results are shown in Table 2. The americium isotopic vector used here is 75 % ^{241}Am and 25 % ^{243}Am. The blanket is irradiated for 4100 EPFD as it is considered in (Varaine, Buiron, Boucher, & Verrier, 2010).

Table 2 : Spent transmutation target characteristics after 5 years of cooling

Assembly mass (kg)	141
Decay heat (kW)	8,03
of which alpha (%)	97,7
of which beta (%)	1,4
of which gamma (%)	0,9
Neutron source (n/s)	1,54E+10
Mean gamma energy (MeV)	0,17
Maximum gamma energy (MeV)	3,40

A strong shift towards alpha heating in the target can be observed, due to the limited production of fission products compared to a standard fuel assembly. The total gamma power in the fuel assembly is 0.28 kW compared to 0.07 kW for the target. In both cases, the maximum energy for a gamma is due to ^{106}Rh decay, which is a fission product. As such, it can be reasonably assumed that the level of gamma shielding provided by handling devices and transportation casks for spent fuel assemblies is enough for target assemblies. For comparison purposes, the gamma spectrum after five years of cooling for an inner fuel assembly and a blanket assembly loaded with minor actinides is given below in Figure 1.

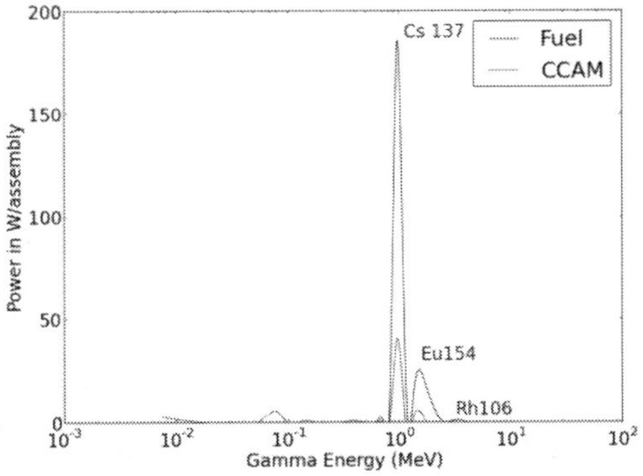

Figure 1 : Comparison of the gamma spectrum for a fuel and CCAM assembly

On the other hand, neutron source increases by a factor twelve between the two cases, which may severely hamper handling and transportation of the irradiated target assemblies if the cooling time is not prohibitively lengthened. Several options exist to make up for this increase, which include design of new transportations casks and handling machines or increased cooling times. However, as the half-live of ^{244}Cm is 18.1 years, a decrease by a factor twelve of the neutron source due to this isotope would require a prohibitive cooling time of 64.5 years. For an irradiation time of 4100 EFPD, this would mean nearly six times as many assemblies cooling down as being irradiated, or 14 tons of americium at various stages of cooling for 2.4 tons being irradiated in a SFR V2b.

Another option which will be investigated in the next part is to locally modify the neutron spectrum near the blankets to limit the production of ^{244}Cm and thus the total neutron source of the assembly. It should be pointed out here that ^{244}Cm production is highly sensitive to the isotopic composition of the americium vector used, as ^{244}Cm is almost only produced through $^{243}Am(n,\gamma)^{244}Cm$ reactions. This sensitivity will be characterized later on. Neutron spectrum modifications in the blankets have already been discussed, for instance in (De Saint Jean, 1998) or more recently in (Konashi, K.Ikeda, Itoh, Hirai, T.Koyama, & Kurosaki, 2015).

Considering that the neutron source is dominated by ^{244}Cm, the energy spectrum of the neutrons produced in the blankets can be considered constant during cooling and equal to the one of ^{244}Cm. This was verified by comparing the neutron spectrum at various cooling times, with mean variations in the neutron spectrum lower than 1.7 % between 5 and 100 years of cooling. Consequently, using dose coefficients taken from (ICRP 119) for antero-posterior neutron exposure of an anthropomorphic phantom and the spectrum shown in Figure 1, it is possible to evaluate the dose coefficient associated with the transmutation blankets neutron source at 317 pSv/cm². This value will be used for dose rate calculations in the following part. The contribution of (α,n) reactions is neglected as it is 10^3 lower than the one of spontaneous fission at any given time. The neutron dose rate of a standard fuel assembly after five years of cooling is 0.39 Sv/s while the corresponding value of a transmutation target is 4.88 Sv/s. The standard fuel neutron source (or dose rate) will be considered as the reference level in the next parts of this study.

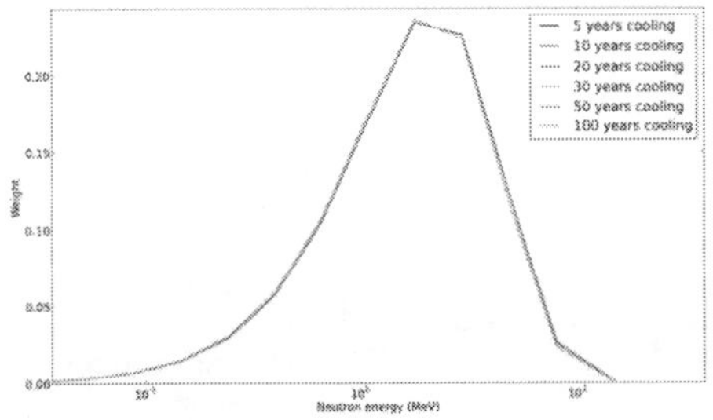

Figure 2 : Comparison of the normalized neutron spectrum at various cooling times

3 Outline of the optimization approach considered

We discuss in this part an optimization methodology of minor actinides transmutation blankets with regards to various parameters such as the local neutron spectrum in the blankets, the fraction of minor actinides loaded and the maximal acceptable limit for neutron emission at the end of cooling. This approach is based on the consideration that minor actinides transmutation can be characterized considering limited information on the neutron spectrum and the minor actinides loading, as discussed in (Kooyman & Buiron, 2015).

The r-factor, defined in Eq 1. as the inverse of the difference in neutron lethargy between creation (u_p) and absorption (u_d), was used here to parametrize the neutron spectrum. The higher this factor, the more energetic the spectrum is, with r factor around 0.35 in fuel assemblies and around 0.02-0.05 in hydride-moderated blankets.

$$r = \frac{1}{u_p - u_d} \quad (1)$$

The americium vector used contained 75 % of ^{241}Am and 25% of ^{243}Am. The americium concentration in the target was used as a second parameter and will be denominated *Am* in the next paragraphs.The maximal acceptable limit for neutron emission at the end of cooling was used as a third parameter and will be denominated S_{lim}.

The following approach was implemented: an initial calculation with a fixed core configuration with 40 % oxide fuel, 40 % coolant and 20 % ^{56}Fe as structures material was carried out, with 22.1 % of plutonium in the fuel. These values were chosen after considering various SFR designs. It was verified that the spectrum in the core did not influence the spectrum in the blankets. The neutron spectrum is computed using the ECCO cell code with a 33 groups energy mesh and the JEFF 3.1 nuclear data library (NEA/OECD, 2006). Then, this spectrum was used in source-based calculations of a blanket medium with a variable composition in terms of fuel, coolant and moderating material in order to cover as wide as possible a spectrum range. The data used for this approach are given below in Table 3. The cell calculations were carried out using the ECCO cell code (Rimpault, 2002).

Table 3 : Variation range of used parameters for cell calculations

Fuel type	Oxide, Nitride, Carbide, Metal (10 w% Zr)
Coolant type	Helium, Sodium, Lead-Bismuth Eutectic
Moderating material	MgO, Beryllium, ZrH_2
Moderating material variation range	0-10 vol%
AmO_2 volume fraction variation range	5-40 vol%

The americium bearing blanket medium is depleted for 4000 EPFD using a constant flux approximation with a flux level of 5e14 n/cm²/s representative of what can be found in radial blankets of a SFR V2b. As discussed in [13] for instance, this residence time is compatible with fuel and cladding swelling due to the lower neutron flux at the core periphery. For the core mentioned here above, this corresponds to 2375 kg of Americium loaded in 84 blankets assemblies. In such a configuration, the americium consumption in the blanket is roughly equal to twice the core production, which means only half of a given reactor fleet must loaded with MABB to achieve closure of the americium fuel cycle.

In the case of heterogeneous transmutation, the constant flux approximation is deemed realistic enough as blankets are exposed to an almost constant flux level from the core. Various quantities of interest are then computed, namely here transmutation rate and neutron source at various time steps. Other quantities can also be computed, such as decay heat or helium production. One thousand calculations were run to obtain a learning base for the construction of artificial neuron networks which are trained to evaluate the transmutation rate and the neutron source at various time steps (5,10,20,30,50 and 100 years) with the Am fraction and the r-factor as input data. This was done using the URANIE platform developed by CEA [26]. The transmutation rate was defined as the ratio of the americium mass consumed over the loaded americium and is expressed in % : $\tau = \frac{\Delta Am}{Am(t=0)} * 100$. The evolution of neutron source during cooling was approximated using the law described in Eq. 2.

$$S(t) = a\ln(t) + bt + c\sqrt{t} + d \quad (2)$$

Considering that the neutron spectrum in the blankets is also dependant on the americium fraction loaded into, artifical neural networks (ANNs) were trained to evaluate the various parameters of interest listed above depending on the r-factor and americium concentration in the fuel. An evaluation of the meta-modelling errors was done and is shown in Table 4. The ANNs were used to compute the neutron source levels at the calculated time steps and then the neutron source behaviour was fitted using the calculated points and Equation 2 as a fit function, as this approach was found to yield the most accurate results.

 DOI: 10.1051/epjconf/201715305005

Table 4 : Evaluation of some meta-modelling errors for transmutation rate and neutron source at 5 and 50 years. The cases annotated (Ann) correspond to artificial neuron networks calculations and the cases annotated (Reg) correspond to logarithmic regression of the neutron source.

	Transmutation rate	Neutron source after 5 years cooling (Ann)	Neutron source after 50 years cooling (Ann)	Neutron source after 5 years cooling (Reg)	Neutron source after 50 years cooling (Reg)
Mean error (%)	-0,34	0,06	-0,22	-0,09	1,88
Standard deviation (%)	4,14	3,50	3,68	3,49	3,82

Addition of minor actinides to the blankets has an hardening effect on the neutron spectrum by increasing capture rate in the epithermal energy range. Consequently, the r-factor of the spectrum in the blankets also depends on the Am concentration loaded and not all the combinations (r,*Am*) are physically achievable. Using the same approach as the one used to build the initial set, the allowable area in the (r, *Am*) plane for the algorithm to explore was computed. This area corresponds to realistic cases in terms of loaded mass and spectrum. Using hydrogenated material such as zirconium hydride (ZrH_2) as moderator highly increases the allowable area as it can be seen in Figure 3. However, this may lead to a safety concern in case of unprotected transients during which dissociation can occur (Terrani, M.Balooch, Wongsawaeng, S.Jaiyen, & Olander, 2010). For exhaustiveness and when necessary, we will consider in the following two cases : one with ZrH_2 use and one without. In the case without, the allowable area is much lower due to the lower moderating power of materials such as Be or MgO.

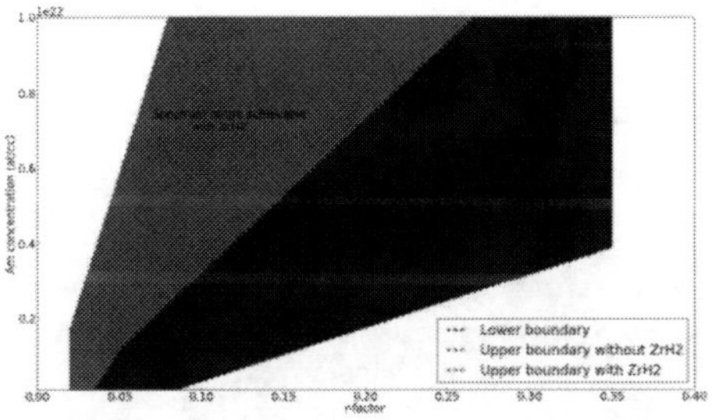

Figure 3 : (r,*Am*) diagram. The allowable range without ZrH_2 use is located between the two rightmost curves.

Two estimators were used to compare the solutions. The first one is representative of the total heavy nuclides inventory in the blankets. It is calculated based on the following assumptions :

- A minimal cooling time of five years
- A manufacturing time of two years at the end of cooling

We considered that an equilibrium was reached between minor actinides production in the core and consumption in the blankets over the complete fuel cycle. Consequently, if the neutron source after five years of cooling is below S_{lim}, the inventory estimator is calculated using Eq. 3, with T being the irradiation time in EFPD:

$$I(Am, R, T) = \left(1 + \frac{7*365}{T}\right) * Am \tag{3}$$

If not, the cooling time to reach a neutron source equal to S_{lim} is approximated by inverting the function used to compute the neutron source and the I estimator is calculated as shown in Eq 4. It can be directly seen that the total inventory depends both on the neutron source of the target and of the considered limit for reprocessing. A maximum cooling time of 100 years was considered here.

$$I(Am, R, T, S_{lim}) = \left(1 + \frac{T_{cooling}(S_{lim}) + 2 * 365}{T}\right) * Am \tag{4}$$

A second estimator based on the consumption of americium during irradiation is computed using Eq. 5 with T_r being the transmutation rate calculated as the percentage of americium having disappeared during irradiation. As the algorithm used minimizes a given objective function, the invert of the americium consumed was used.

$$C(Am, R) = \frac{1}{Am * Tr} \tag{5}$$

Using the meta-models of the various quantities of interest to compute the value of the estimators within the acceptable (r,*Am*) zone, the entire problem is fed to a genetic algorithm for optimization, which is carried with the objective of minimizing I and C, e.g. maximizing the consumption of minor actinides while minimizing the total inventory in the fuel cycle. All of the optimization calculations are carried out using the URANIE code from CEA (Gaudier, 2010) and the python module Scipy (Oliphant, 2007).

It should be mentioned here that an approximation is made here due to technological uncertainties. Indeed, the depletion calculations performed here yield the neutron source per gram of spent fuel. Now, consideration on fuel handling must be computed for an entire assembly, which means this value should be multiplied by the mass of the assembly, which itself depends on the type of fuel or coolant considered. As the information on the geometrical design of the target is not available, this methodology currently only takes into account neutron spectrum effects, regardless of the assembly design. Consequently, the integrated values which are discussed in the next parts are corresponding to a 'reference' target assembly of 141 kg of heavy metals.

Regarding actual technological feasibility of the designs, the domains shown in Figure 3 were compared with complete core calculations of the SFR V2b described in (Sciora et al, 2011) for various assembly designs using metal and oxide blankets along with various moderating material and were found to be consistent with these calculations.

4 Results

Results are shown in Figure 4, which represents the Pareto front and set for two cases, one where use of zirconium hydride is considered and one where is it not. This two sets corresponds to cases with are optimal in the Pareto-sense, e.g. for which no gain can be achieved for one objective without a loss in another one. In this case, they represent the cases which lead to the minimal cooling time (with regards to neutron source and dose) for a maximum minor actinides consumption. The Pareto front is the set of optimal cases in the initial parameters space while the Pareto zone is the set of optimal cases in the objectives space. It can be seen that, regardless of the use of zirconium hydride, the optimal solutions to the problem corresponds to the least energetic spectrum. However, no cooling time lower than 67 years was observed here, which is prohibitively long and consistent with the values calculated previously using only ^{244}Cm decay information. Consequently, the impact of lowering the neutron source limit for reprocessing was investigated.

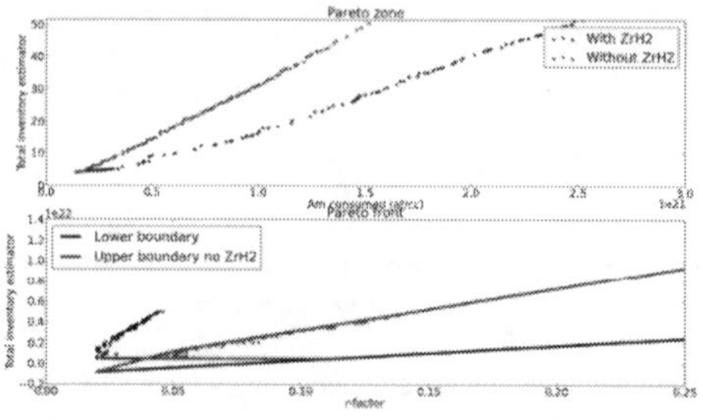

Figure 4 : Pareto front and zone for a considered limit equal to the neutron source of a standard fuel assembly after 5 years cooling (1.22 10^9 n/s/assembly or 0.39 Sv/s/assembly)

The impact of the considered limit on neutron dose rate is shown in Figure 5 . It can be seen that an increase in the allowable limit lead to better solutions in terms of inventories and minor actinides consumption. It can also be seen that increasing the reprocessing limit has an approximately twice bigger effect in the case without ZrH_2 compared to the case with ZrH_2. Nevertheless, the cases where zirconium hydride is used, even with the reference limit, remains the best option, except for a small subset of cases where the americium consumed is very low. The shape of the Pareto front is also not affected by the raising of the limit considered here.

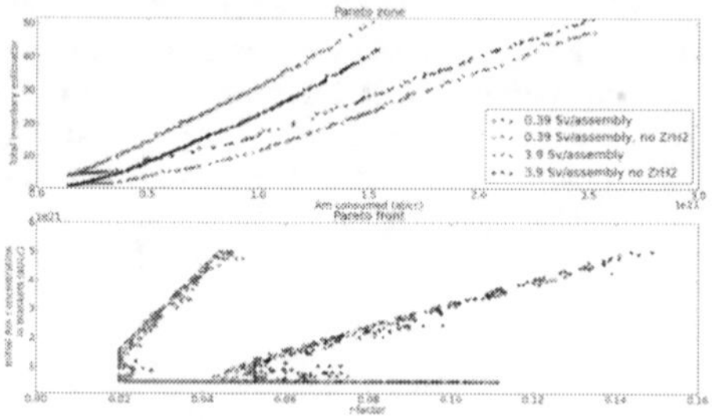

Figure 5 : Impact of the limit considered before reprocessing on the Pareto front and set.

Considering the high contribution of ^{244}Cm to neutron source, the sensitivity of the Pareto set and front to the isotopic vector of americium was also evaluated. Three calculations with varying ^{243}Am fraction were carried out and the results are shown below in Figure 6. For a limit value of 0.39 Sv/s, it can be seen that the isotopic vector considered has a very limited impact on the Pareto front. Indeed, for most cases, the cooling time required to reach the limit value is beyond 100 years, which is the limiting value considered here. Nevertheless, it can be observed that for high americium fraction, the cases with 10 % Am 243 exhibit a slightly lower inventory than the one with 40 %, which is consistent with the fact that Am 243 is the main precursor or 244Cm.

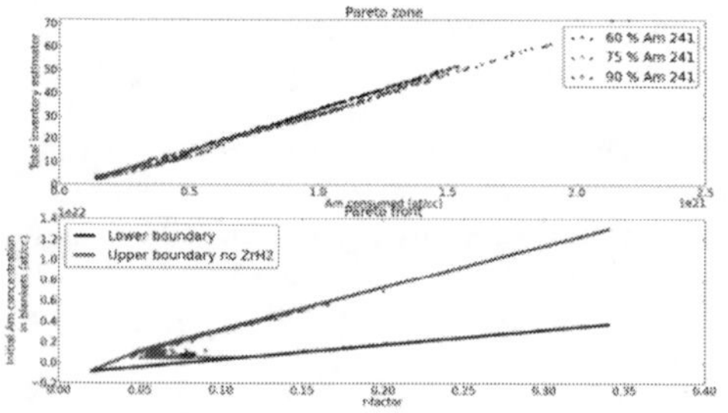

Figure 6 : Impact of the Americium isotopic vector on the Pareto front and set for cases without zirconium hydride. The results are similar when ZrH_2 is used. A limiting value of 0.39 Sv/s was considered here.

When the limit is increased to 3.9 Sv, as it is done in Figure 7, it can be seen that the impact of the americium vector is increased. As expected, when the ^{243}Am isotopic fraction is increased, the total inventory increases at similar performances. This is explained by the higher production of ^{244}Cm and subsequent increase in the neutron source of the irradiated fuel. Further investigations regarding the impact of the isotopic vector and the possible use of isotopic variations in limiting the total inventory are currently undergoing. This point illustrates the interest is of the optimization approach, as it allows to rapidly explore a wide range of design options.

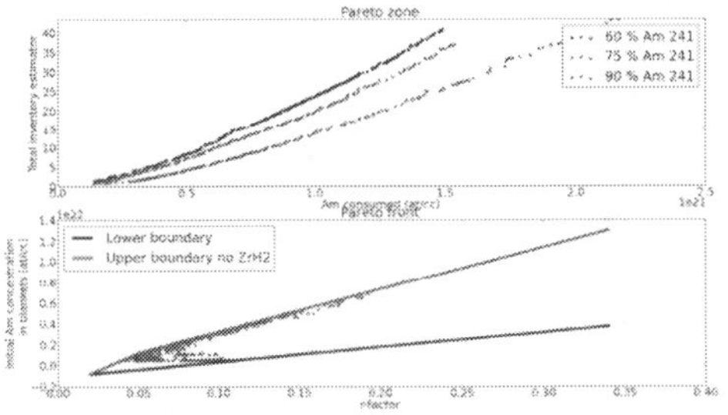

Figure 7 : Impact of the Americium isotopic vector on the Pareto front and set for cases with zirconium hydride. The results are similar when ZrH2 is used. A limiting value of 3.9 Sv/s was considered here.

For the irradiation time of 4000 EFPD considered here, it can be concluded that, aside from decreasing the minor actinides loaded in the blankets, the best option to limit the neutron source of the blankets and the cooling time is to use hydrides as moderating material in order to slow down the neutrons at the core periphery. If this option is not available, using moderating material such as beryllium or MgO appears to be the best option. However, it should be noted that, as it can be seen in Figure 5, the use of hydrides appears as a better solution than an increase in the reprocessing limit for given transmutation performances.

It is also possible to variate the irradiation time to evaluate the impact of this parameter. It can be seen below in Figure 8 that increasing the irradiation time to 6000 EFPD gives identical performances between cases without hydride and cases with hydride irradiated for 4000 EFPD. An increase of the irradiation thus appears as a potential replacement solution to the use of hydrides for neutron slowing-down in the blankets. However, this approach raises additional issues in terms of targets thermo-mechanical behavior at high fluence.

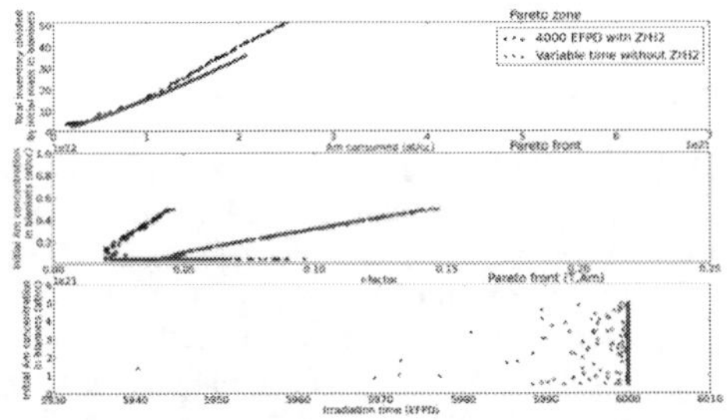

Figure 8 : Comparison of the Pareto front and set for a case with a 4000 EFPD irradiation time and zirconium hydride material vs a variable time without zirconium hydride.

It can also be observed that for the flux level considered here, increasing the irradiation time increases the overall performances. This conclusion may not stay true in the case of higher flux level or homogeneous transmutation due to the so-called 'curium peak' effect (Kooyman & Buiron, 2015) in which the curium concentration (and the neutron source) increases to a maximum and then decreases after a given fluence.

Finally, it is possible to use S_{lim} as an input parameter and to run the optimization process using r, Am and S_{lim} as variables. The comparison between a case without ZrH2 and a reprocessing limit allowed varying between 0.046 Sv/s and 4.6 Sv/s and a case with a fixed limit at 0.39 Sv/s with ZrH_2 is shown in Figure 10. First, it can be seen that the cases with the highest limit are optimal, which was expected. Additionally, it can be seen that for a small part of the americium consumption range (up to 7e20 at/cc), it is more interesting to increase the reprocessing limit than to use ZrH2 in terms of total inventory. However, above this threshold, use of ZrH2 yields the best results. It can be inferred from this and from Figure 5 that raising the maximal allowable neutron source is of interest only for small americium small consumption and that the use of ZrH2 becomes more interesting above a given consumption value which depends on the reprocessing limit.

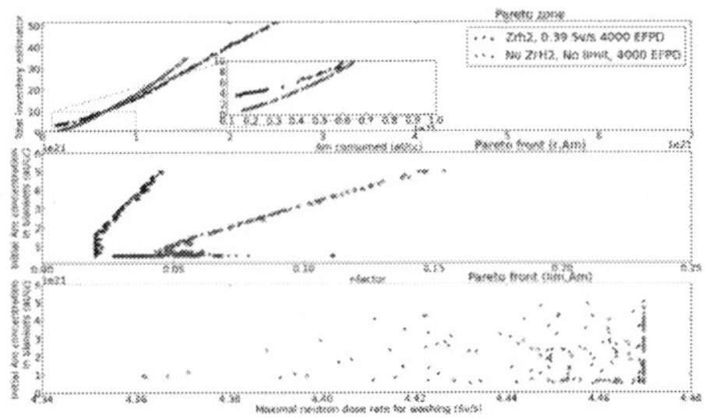

Figure 9 : Comparison of a case with the use of ZrH2 and a fixed neutron source limit at 0.39 Sv with a case without ZrH2 and a variable reprocessing limit

It appears from this analysis, that, apart from using hydrides as moderating material in the blankets, the best option to limit the neutron dose rate at a given time or the cooling time of the blankets is to increase the residence time of the blankets while using an available moderating material such as MgO or Be to slow down the neutron. Such an option may not be feasible due to material resistance constraint and especially pin pressurization due to helium release by decaying minor actinides, but the alternative of increasing the allowable limit for reprocessing may also present some technical difficulties due to prohibitive shielding thickness that may be required.

5 Comparison to core calculations

Core calculations were carried out to complete and verify the results of the optimization process. The SFR V2B design described in (Sciora, et al., 2011) was also used

 DOI: 10.1051/epjconf/201715305005

for this purpose. Four cases were compared with similar performances of 6 kg/TWeh, which corresponds to the one of a "reference" case with 20 % of americium oxide in volume and uranium oxide support matrix:

- The reference case
- A case with oxide and 10 vol% ZrH_2
- A case with oxide and 10 vol% MgO
- A case with metal fuel (as U10ZrAm with a smear density of 75 %)

The results are shown below in Table 5. It can be seen that the oxide and metal approach are roughly similar, mainly due to the fact that to compensate for the lower transmutation rate in the metal blankets due to the harder neutron spectrum, the amount of initial americium must be raised, thus increasing the overall production of curium. The case with ZrH_2 exhibits better performances than the two aforementioned cases, which can be explained by a "shift" towards heavier isotopes which consumes ^{244}Cm to yield ^{245}Cm (+118 % of ^{245}Cm in the moderated case). This shift slightly increases the long term neutron source due to the contribution of ^{246}Cm, which increases the cooling time compared to the reference case even though neutron source at 5 years is lower than the reference case. However as the required mass to achieve the same performances is only 71.5 % of the reference mass; the total impact on inventory is limited. The case with MgO exhibits a similar behavior than the case with ZrH_2 but to a lesser extent due to the limited moderating power of MgO. The Am inventory in the blankets necessary to obtain the same performances is slightly reduced while the cooling time is slightly increased due to an increase in the curium production. Consequently, the total americium inventory in the fuel cycle is slightly decreased, but to a lower extent than for the hydride case. These calculations are in good agreement with the results obtained using the methodology. This can also be seen looking at Figure 10, where the cores discussed in Table 5 have been plotted in the Pareto Front diagram from Figure 4. The cores with moderating material, which exhibit the best performances, are lying on the Pareto front, meaning they are optimal in this sense.

Table 5 : Comparison of the performances of oxide, metal and moderated oxide blankets with regards to neutron source

	Oxide	Metal	Oxide + MgO	Oxide + ZrH2
Assembly initial heavy metal mass	141	153	141	141
Am inventory in blankets	2375	2606	2316	1699
244Cm mass in the blankets at 5 years (kg)	113,1	110,8	114,3	105,6
Neutron source at 5 years (1e10n/s)	1,54	1,51	1,56	1,45
Cooling time to reach the level of a standard fuel assembly (days)	25335	24915	25455	25972
Estimated americium inventory (kg)	15476	16708	15161	11342

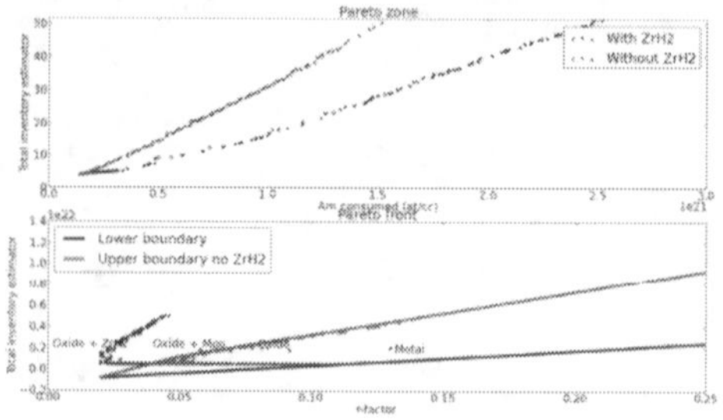

Figure 10 : Pareto Front with position of the cores discussed in Table 5 given

6 Conclusions

A new approach to consider transmutation issues related to fuel cycle parameters has been proposed and various heterogeneous transmutation cases have been compared using this methodology. When the spent fuel neutron dose is considered as the limiting factor for reprocessing, it appears that the optimal option in terms of cooling time and minor actinides transmutation performances is to use hydrides as moderating material in the blankets. However, the use of such a material may not be feasible due to potential dissociation issues. In this case, the best option is to use a less-effective moderating material such as beryllium or MgO and to increase the residence time of the blankets. These findings are consistent with the results of core calculations.

It is also shown here that no optimum can be found for minor actinides transmutation and spent blankets neutron dose rate in terms of neutron spectrum and mass to be loaded. Consequently, other factors such as technological feasibility of the required assembly design will be required to select a unique design option.

Further work will be carried out in the future with regards to the implementation of the optimization methodology, in order to take into account additional parameters linked to the fuel cycle and to core safety. Additionally, extensive analysis of the uncertainties and bias linked to the methodology use will be done.

References

C. Chabert, C.Coquelet-Pascal, A.Saturnin, G.Mathonniere, B.Boullis, D.Warin et al. : Technical and economic assessment of different options for minor actinides transmutation : the French case. (2011) *GLOBAL2011 Proceedings*

T.Kooyman, L.Buiron : Sensitivity analysis of minor actinides transmutation to physical and technological parameters. *EPJ-N*, **1** (15).

NEA Homogeneous versus heterogeneous recycling of transuranics in fast nuclear reactors (2012)

G. Rimpault : The ERANOS code and data system for fast reactor neutronic analyses. *PHYSOR Proceedings* (2002)

P.Sciora, D.Blanchet, L.Buiron, B.Fontaine, M.Vanier, F.Varaine et al. : Low void effect core design applied on 2400 MWth SFR reactor. *ICAPP 2011 Proceedings*

A.Tsilanizara, C.Diop, B.Nimal, M.Detoc, L.Luneville, M.Chiron : DARWIN: An Evolution Code System for a Large Range. *J. of Nuc. Sci & Tec., Supplement* **1**, 845-849 (2000)

F.Varaine, L.Buiron, L.Boucher, D.Verrier :Overview on homogeneous and heterogeneous transmutation in a new French SFR : reactor and fuel cycle impact. *11th IEPMT Proceedings* (2011).

F.Gaudier : URANIE : The CEA/DEN Uncertainty and Sensitivity platform. *Proc. Soc & Beha. Sci* **2** 7660-7661 (2010) (2010).

ICRP 119 : Compendium of dose coefficients based on ICRP publication 60. *Annals of the ICRP* **41** Supplement 1 (2012)

M.Meyer, et al : Scenarios for Minor Actinides Transmutation in the Frame of the French act for waste management *Proceedings of FR013* Conference, Paris. (2013)

K. Konashi et al : Enhancing minor actinides transmutation by irradiation of (MA,Zr)Hx in FBR blanket region, *Proceedings of GLOBAL2015* Paris (2015).

C. De Saint Jean : Americium once-through of moderated targets in a CAPRA core *Seminar Int. CAPRA conf*, Karlsruhe (1998) : Scenarios for Minor Actinides Transmutation in the Frame of the French act for waste management *Proceedings of FR013* Conference, Paris. (2013)

T.E. Oliphant : Python for scientific computing , *Computing in Science & Engineering* **9** (90) (2007).

K. Terrani, M. Balooch, D. Wongsawaeng, S. Jaiyen , D.Olander : The kinetics of hydrogen desorption from and adsorption on zirconium hydride *Journal of Nuclear Materials* **397** 61-68 (2010)

J. Tommasi, H. Bottollier-Curtet, S. Massara, F.Varaine, M.Delpech : Scenarios for waste management involving innovative systems (ADS) *Proceedings of GLOBAL2001* Paris (2001)

EPJ Web of Conferences 153, 05006 (2017) DOI: 10.1051/epjconf/201715305006
ICRS-13 & RPSD-2016

Reflector and Protections in a Sodium-cooled Fast Reactor: Modelling and Optimization

David Blanchet[1,a], Bruno Fontaine[1]

[1]*CEA, DEN/DER, F-13108 Saint-Paul-lez-Durance, France*

Abstract. The ASTRID project (Advanced Sodium Technological Reactor for Industrial Demonstration) is a Generation IV nuclear reactor concept under development in France [1]. In this frame, studies are underway to optimize radial reflectors and protections. Considering radial protections made in natural boron carbide, this study is conducted to assess the neutronic performances of the MgO as the reference choice for reflector material, in comparison with other possible materials including a more conventional stainless steel. The analysis is based upon a simplified 1-D and 2-D deterministic modelling of the reactor, providing simplified interfaces between core, reflector and protections. Such models allow examining detailed reaction rate distributions; they also provide physical insights into local spectral effects occurring at the Core-Reflector and at the Reflector-Protection interfaces.

1 Introduction

One of the main requirements for radial shields in a pool-type Sodium-cooled Fast Reactor (SFR) is to protect internal vessel structures from radiation damages. The Intermediate Heat Exchanger (IHX) also requires neutron shielding so as to minimize sodium activation in the secondary circuit. The natural choice for neutron absorber in radial shields is the boron carbide, and at a minimum cost, natural boron enrichment offers the longest lifetime. Between core and protections, reflectors can be seen as a transition zone. A good reflector material would preferably have high elastic scattering and low absorption cross sections to enhance the neutron flux level at the core periphery, hence preserving reactivity and flattening fuel burnup distributions. Reflectors are also foreseen to decrease the neutron flux level before protection assemblies in order to preserve their expected lifetime.

The present study is conducted to assess the neutronic performances of an MgO-based reflector material in comparison with a more conventional Stainless Steel reflector. The study also examines other promising candidates. Previous work has already been devoted to characterize the MgO reflector impact on the outer fuel region [2] and on the minor actinide transmutation in blankets [3]. The present study is revisiting the models and proposes a refined analysis, in particular on reactivity impacts, on power enhancement at the outer fuel region, and on the increased efficiency of the surrounding protections.

The present study is based on deterministic modelling. Monte-Carlo simulations are only used for punctual validation purposes.

Section §2 explains how reflector materials have been selected before investigation, which criteria are considered in the present study. Section §3 briefly describes the 1-D cylindrical model and presents the assessment of reflector materials based on this model. The ranking can be done by considering different criteria: based on albedo or on reactivity weights, but also in terms of shielding efficiency when they are associated with protections. Section §4 is revisiting the 1-D study, by upgrading a 2D R-Z model. Finally, section §5 presents some methodological aspects concerning the deterministic modelling and its validation by comparison with Monte-Carlo simulations.

2 Pre-Selection of Reflector Materials

The most desirable material characteristics for use in neutron reflectors can be listed below:

• High neutron albedos (ability to reflect neutrons into the core) to preserve reactivity,

• Power enhancement at the core periphery, for radial core power flattening and better fuel utilization,

• High resistance against thermal and mechanical loads (high conductivity, low thermal expansion, low Young modulus), in particular if the material is to be used without cladding,

• High resistance against irradiation damages (acceptable degradation of thermal and mechanical properties),

• Limited swelling under neutron irradiation,

• Low activation for dismantling purpose,

[a] Corresponding author: david.blanchet@cea.fr

- Industrial production perspectives,
- Sustainability of the supply chain,
- Acceptable cost (with considerations on the expected lifetime and on the recycling strategy).

Among the large number of potentially good reflector materials, the strategy retained in the present study was to pre-select those having the largest experimental feedbacks, and that could be used with much confidence in the start-up cores of ASTRID. In this regard, previous studies conducted in the past were of interest, like for instance those studies aiming at the development of target matrices for the transmutation of Minor Actinides in Phenix reactor [4] (MgO, $MgAl_2O_4$, $Y_3Al_5O_{12}$), but also some prospective studies that were conducted for the development of the Gas-cooled Fast Reactors (Zr_3Si_2, SiC) [5] or associated with fusion Reactors technologies [6] (V-4Ti-4Cr, Beryllium). Hydride materials, like CaH or ZrH (having a strong moderation impact), were directly disqualified from this pre-selection because of safety considerations (hydrogen risks in accidental conditions).

Hence, the resulting short list of reflector materials to be further examined in the present neutronic study are: *Magnesium Oxide* (MgO), *Spinel* ($MgAl_2O_4$), *Yttrium Aluminium Garnet* ($Y_3Al_5O_{12}$), *Zirconium Silicide* (Zr_3Si_2), *Silicium Carbide* (SiC), *Zirconium Carbide* (ZrC), *Titanium Carbide* (TiC), *Beryllium* (Be), *Vanadium alloy* (V-4Ti-4Cr), *depleted Uranium Oxide* (UO_2) as a fertile material, and *ferritic martensitic stainless steel* (denoted as "EM10"). These pre-selected materials are examined in a comparative study where the reference case is a standard reflector made of a stainless steel ("EM10"). The use of fertile material being excluded on grounds of proliferation concerns, it is considered here for exercise only.

3 A Simplified One-Dimensional Model for Material Assessment

3.1 Model Description

A simple one-dimensional cylindrical reactor model is setup with the aim of being almost representative of the ASTRID core. The geometry illustrated in Figure 1 is a radial cross-cut of the reactor model, comprising the fuel zone, the reflector zone (denoted as "R") and the radial protections or shielding (denoted as "P"), the rest of the geometry being filled with sodium from 350 cm up to 600 cm.

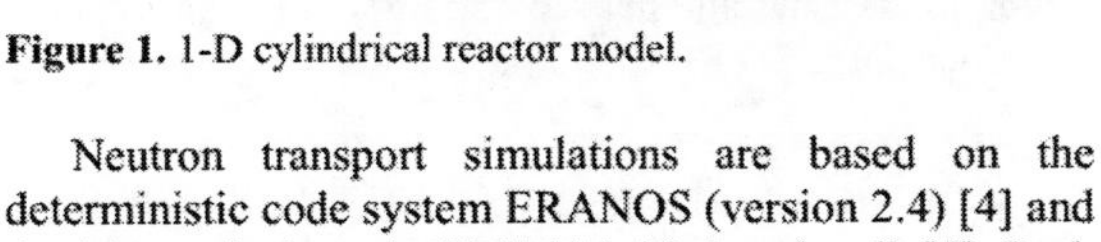

Figure 1. 1-D cylindrical reactor model.

Neutron transport simulations are based on the deterministic code system ERANOS (version 2.4) [4] and the Monte-Carlo code TRIPOLI-4® (version 9) [5]. Both codes are used in conjunction with the JEFF3.1 nuclear data library [6].

The Sn code BISTRO is used with a S4P1 angular quadrature and a spatial scheme based on finite differences. Two calculation routes are being considered: one is a "project route" based on a broad energy mesh in 33 groups; the other is a "reference route" based on a fine energy mesh in 1968 groups. For the rest of the study, they will be respectively denoted as "E2.33g" and "E2.1968g". The neutron cross sections are self-shielded (using the subgroup method) and homogenized at the sub-assembly level by the lattice code ECCO. "E2.33g" only differs from "E2.1968g" by the extra step of condensation in energy.

For simplicity reasons and to avoid any ambiguities when comparing with Monte-Carlo results, calculations are performed at room temperature (20°C). The geometry is discretized into mesh sizes of 1 cm, with "specular reflection" boundary conditions at R = 0 and "zero flux" boundary conditions at R = 600 cm. This simplified model is representative of radial flux gradients and local distributions at the various interfaces, but it is quite unrealistic for reactivity analysis. Indeed, the specular reflections imposed in axial directions are convenient for comparisons with Monte-Carlo simulations in which no leakage model is implemented, but the artificially high reactivity level induced by these conditions is preventing any comparative analysis on the reactivity-weight of reflectors. In deterministic modelling however, an axial buckling value can be easily derived to find a critical state, much more suitable for reactivity analysis. A second deterministic model is hence dedicated to reactivity analysis.

3.2 Material Ranking by Albedo

A priori, the best indicator for the assessment of reflector materials is a dimensionless quantity called the "albedo" denoted γ and defined as the ratio of the neutron flux reflected (or back scattered) by the material over the neutron flux entering the material:

$$\gamma = \frac{J-}{J+} \quad (1)$$

In the above equation, J stands for the neutron current at the interface between core and reflector, J- denoting the neutrons back-scattered into the core and J+ representing the neutron leakage from the core. Based on the 1-D model presented in section §3.1, Table 1 gives comparative results of albedos calculated for the different reflector materials.

Table 1. Reflector material ranking by albedo.

Rank	Material	Albedo
1	**MgO**	0,859
2	**$MgAl_2O_4$**	0,853
3	**$Y_3Al_5O_{12}$**	0,847
4	**Zr_3Si_2**	0,833
5	**SiC**	0,831

DOI: 10.1051/epjconf/201715305006

6	**ZrC**	0,827
7	**TiC**	0,823
8	**V-4Ti-4Cr**	0,821
9	**Be**	0,807
10	**EM10**	0,802
11	**UO_2**	0,744

According to Table 1, the ranking in terms of "*neutron reflection capabilities*" is indicated by decreasing albedos: MgO > $MgAl_2O_4$ > $Y_3Al_5O_{12}$ > Zr_3Si_2 > SiC > ZrC > TiC > V-4Ti-4Cr > Be > EM10 > UO_2.

3.3 Material Ranking by Reactivity weight

Another indicator providing a ranking of best reflector materials is the reactivity of the system. Reactivity results are given in Table 2.

Table 2. Reflector material ranking by reactivity.

Rank	Material	Reactivity (pcm)
1	**MgO**	30618
2	**$MgAl_2O_4$**	30607
3	**$Y_3Al_5O_{12}$**	30596
4	**Zr_3Si_2**	30579
5	**V-4Ti-4Cr**	30556
6	**SiC**	30553
7	**ZrC**	30550
8	**TiC**	30532
9	**EM10**	30517
10	**Be**	30494
11	**UO2**	30413

In the absence of axial leakage, such a 1-D model is only representative of the radial neutron leakages, and thereby exhibits an artificially very high reactivity level (because of neutron reflections). These reactivities are of course quite unrealistic and the differences found from one material to another are far from being representative of the results that could be obtained in a real 3D modelling with correct leakages and considering burnup distributions. However, it gives a fair idea of the potential ranking, since it almost confirms all the previous results based on albedos.

3.4 Material Ranking by Shielding Efficiency

The choice of a reflector material can also be guided by its efficiency in terms of neutron shielding when it is associated with radial protections made of neutron absorber. Let us give a fixed design for the radial protections in order to examine solely the impact of the reflector materials on the efficiency of the reflector – protection system. The indicator is the neutron flux attenuation. Neutron fluxes are calculated at R = 5 m from core center, which is the approximate location of Intermediate Heat Exchangers (IHX). The aim is not to obtain absolute values but to understand the trends. Neutron flux results are summarized in Table 3.

Table 3. Reflector material ranking by shielding efficiency.

Rank	Material	Φ at IHX ($n.cm^{-2}.s^{-1}$)
1	**Be**	126
2	**TiC**	449
3	**ZrC**	618
4	**MgO**	1801
5	**UO_2**	2161
6	**EM10**	2385
7	**$MgAl_2O_4$**	3095
8	**SiC**	3562
9	**$Y_3Al_5O_{12}$**	4483
10	**V-4Ti-4Cr**	5496
11	**Zr_3Si_2**	8997

According to Table 3, the best material ranking is: Be > TiC > ZrC > MgO > UO_2 > EM10 > $MgAl_2O_4$ > SiC > $Y_3Al_5O_{12}$ > V-4Ti-4Cr > Zr_3Si_2. This ranking is directly correlated with the moderation efficiency of the reflector material, since neutron moderation is improving the efficiency of the neutron absorbers in radial protections. These results are illustrated below in Figure 2.

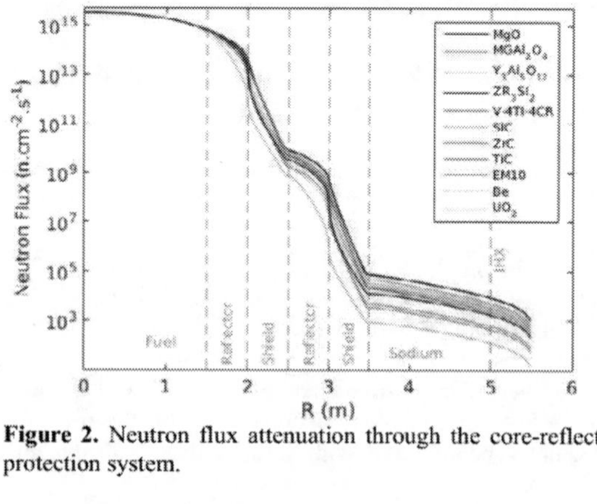

Figure 2. Neutron flux attenuation through the core-reflector-protection system.

One can however put in question the pertinence of such an indicator since the axial leakages are largely predominant in the ASTRID core concept. The next section is an attempt to address this problem by considering a 2D RZ modelling.

4 Considerations on Axial Leakage

DOI: 10.1051/epjconf/201715305006

This section examines a 2D-RZ modelling of the ASTRID reactor. This simple reactor model is presented in Figure 3 below.

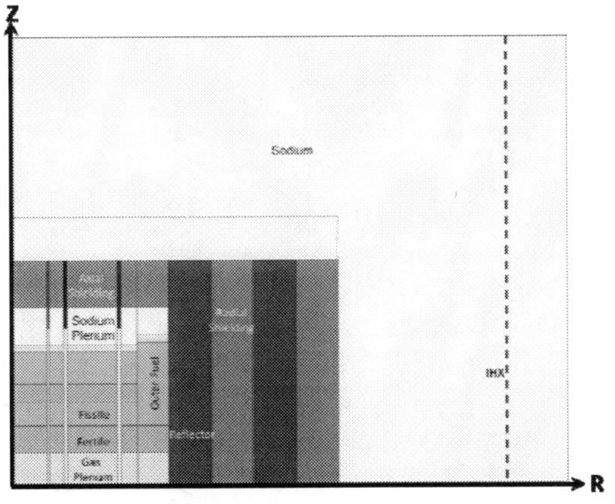

Figure 3. 2D-RZ model of the ASTRID reactor.

In a preliminary approach, most of design details concerning the core surroundings are voluntary ignored because information is lacking and there are still some unknowns in the specifications. For instance, the IHX is still in a design phase, with possible change in axial position, with regular updates in geometry or shielding options. Hence, the IHX is only represented here by its expected position (see the dashed line in Figure 3). Similarly, all vessel internals, sub-assembly head structures are simplified and represented by homogeneous sodium. As a consequence, the present model will not serve in determining absolute values, but will rather be useful to investigate shielding options in design studies, and to understand the main contributions.

Reactivity results are given in Table 4. These results are clearly confirming the trends already indicated by the one-dimensional study presented in §3.3.

Table 4. Reflector material ranking by reactivity (2D-RZ).

Rank	Material	Reactivity
-	-	pcm
1	MgO	7963
2	$MgAl_2O_4$	7829
3	$Y_3Al_5O_{12}$	7707
4	Be	7526
5	SiC	7514
6	ZrC	7471
7	TiC	7445
8	Zr_3Si_2	7408
9	V-4Ti-4Cr	7362
10	EM10	7139
11	UO_2	6645

The neutron flux is calculated and plotted in Figure 4.

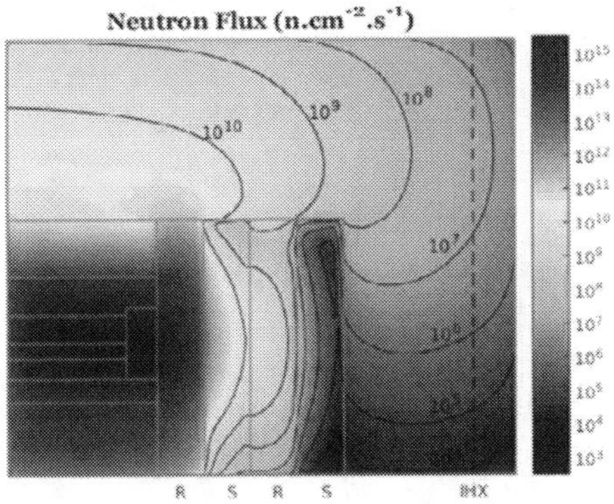

Figure 4. A 2D-RZ neutron flux map, illustrating the neutron attenuation in radial direction through the core-reflector-protection system, and the counterbalance by neutron bypass with axial leakages.

This illustration highlights a major contribution from the axial neutron leakages when determining the secondary sodium activation. According to this Figure, radial shielding looks already satisfactory and there is no need to further improve the efficiency of the radial reflector-protection systems. However, if one wants to decrease the level of secondary sodium activation in the IHX, most of the efforts should be devoted to develop better shielding against axial neutron leakages.

The heterogeneous axial arrangement of the ASTRID core is designed to minimize the sodium void reactivity effect during accident scenarios. This layout also results in significant axial neutron leakages under normal operation. Even though the upper shielding contains strong neutron absorbers, with highly enriched Boron carbide (up to 90% ^{10}B content) in the lower part and naturally enriched boron carbide in the upper part, the axial length available for protections is rather limited by severe constraints on total core height. In comparison, there is much space available for radial protections. In conclusion nothing serves to further optimize radial protections if neutrons are bypassing through axial leakages.

5 Validation studies by Comparison with Monte-Carlo Simulations

The validation studies presented in this section are restricted to the MgO reflector and to the more conventional Stainless Steel reflector. These studies were conducted to investigate local reaction rates at the interfaces (e.g. core-reflector), and also shielding aspects. Recommendations are issued for the improvement of deterministic modelling on these aspects.

5.1 Local interfaces

The core-reflector and the reflector-protection interfaces are very challenging for deterministic neutron transport modelling. The reflector is a scattering zone in sandwich between the fuel and the protections, which are two highly absorbing zones.

The MgO reflector material produces a slightly softer neutron spectrum than the stainless steel. Neutron slowing down is indeed favoured by lightest elements. This softening of the neutron spectrum has consequences on both sides: on the fuel side where the MgO is reflecting 20% more neutrons into the core than the stainless steel and on the protection side where the shielding is enhanced. The strength of the neutron flux attenuation can be observed with a log-scale in figure 5.

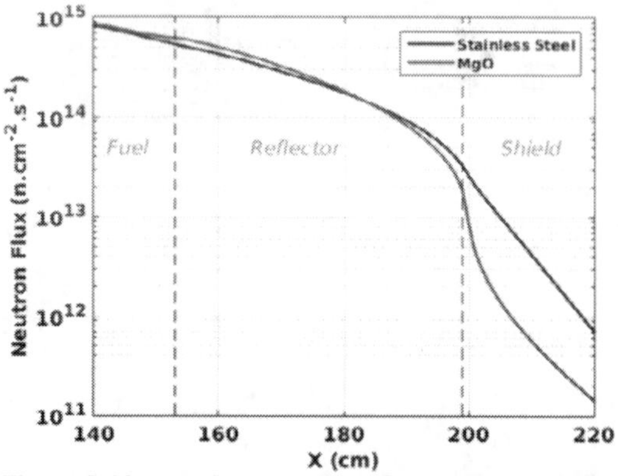

Figure 5. Neutron flux attenuation through the core-reflector-protection system, comparison between MgO and stainless steel reflectors.

The capture rate distributions are plotted in figure 6 and figure 7 for each of the two reflectors under study. The local spectral effects previously mentioned are increasing the capture rates, with characteristic peaking at the interfaces. In comparison to reference Monte-Carlo simulations, the project route calculations (defined in §3.1) are significantly overestimating the efficiency of the protections (+10% and +17% in capture rates at the reflector-protection interface respectively for stainless steel and MgO). The accurate prediction of these local effects would require better modelling of the cross sections; some improvements are suggested in the process of homogenizing neutron cross sections, both in radial protections and in fuel regions close to the reflectors.

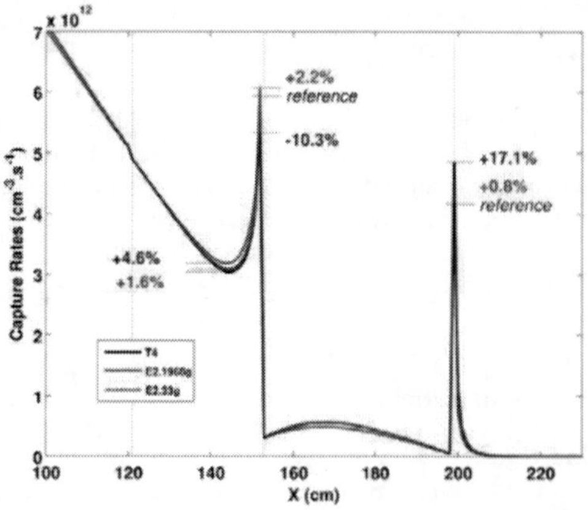

Figure 6. Capture rate distributions in the case of MgO reflector.

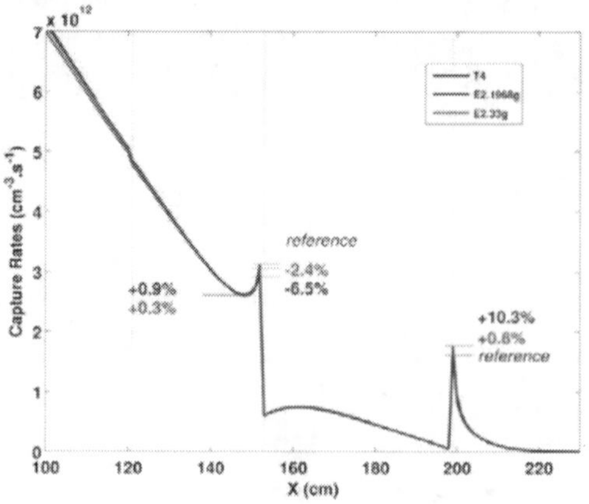

Figure 7. Capture rate distributions in the case of Stainless Steel reflector.

5.2 Shielding aspects

Monte-Carlo simulations are widely used in radiation shielding problems for the following reasons: it is easy-to-model any complex geometry, there is no discretization in energy/angle variables, the resolution method is naturally adapted to parallel machines and there are increasing computing resources available. Moreover, in the particular case of shielding calculations, for deep penetration problems where radiations are attenuated by several decades, or in bypass problems where very rare events are questioning the statistical significance of the results, some biasing games can be done. The variance reduction approach using biasing techniques however requires some verification because it introduces some deterministic modelling assumptions. These assumptions and the verification steps associated are briefly recalled hereafter for the specific case of the present study.

In a first step, analog simulations (i.e. simulations not using variance reduction, with the exception of implicit captures treatment…) are launched in critical mode. The fission source distributions are calculated. In a second step, a source calculation propagates the fission source distributions obtained at the previous step. Results obtained in step 1 and 2 are verified to be in very close agreement. Once verified, the source calculation can be improved in step 3, by using variance reduction technique. The biasing technique used in the TRIPOLI-4® code (with the INIPOND module) is based on the automatic generation of an importance map [10] [11]. The biasing energy mesh chosen for the present study is based on a 6 group structure defined by the following boundaries: 20, 4, 1, 10^{-2}, 10^{-4}, 10^{-6}, 10^{-11} MeV. The spatial mesh used for the ponderation is covering the whole cylindrical geometry of the reactor by ultra-thin steps of 1 cm length.

Step 1 and step 3 are simulated with 10^9 particles. The results are plotted together with the corresponding

ERANOS-2 results in Figure 8 and Figure 9 for the two cases of reflector. The variance reduction technique works perfectly to push neutrons through the reflectors and the protection rings. The very small statistical errors obtained in neutron flux close to the detector (at R = 500 cm) indicate a very good convergence of the simulations using the variance reduction technique; while natural simulations do not provide any result at all at this location.

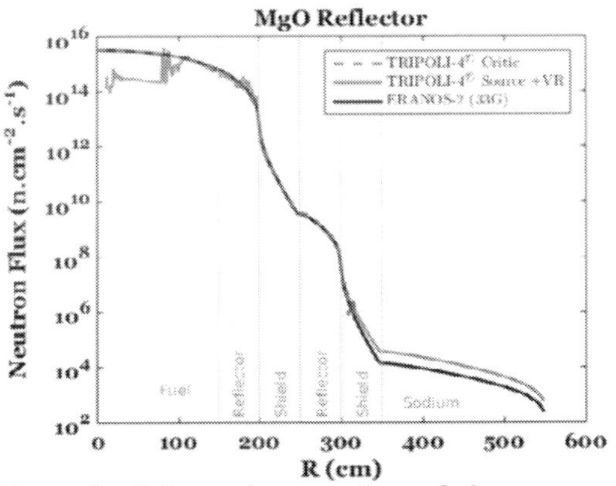

Figure 8. Code-to-code comparison of the neutron flux attenuation in the case of MgO reflector.

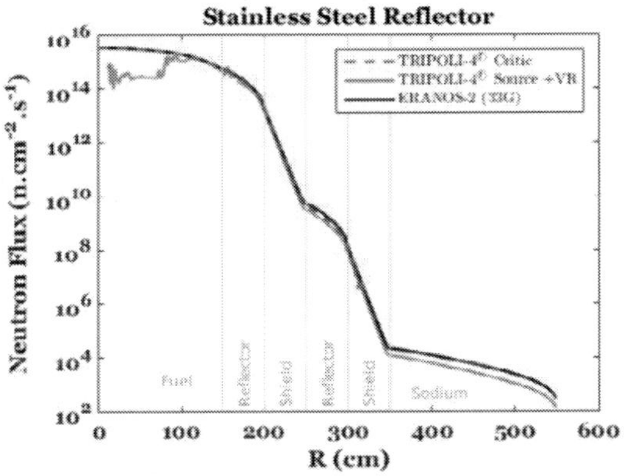

Figure 9. Code-to-code comparison of the neutron flux attenuation in the case of EM10 reflector.

The ERANOS-2 and TRIPOLI-4® neutron flux results are compared in terms of ratios in Table 5. Since we are dealing with simplified models, the absolute flux values are not of interest here, so only ratios are reported in Table 5. At R = 5 m, after 12 decades of attenuation, the neutron flux ratios are ~ 0.4 and ~1.9 respectively for the MgO and stainless steel reflectors.

Table 5. Ratio of ERANOS-2 and TRIPOLI-4® neutron flux results obtained at IHX location (R = 5 m).

	ΦE2 / ΦT4 (± 1σ)
MgO	0.38 (± 0.7%)
Stainless Steel	1.85 (± 2.7%)

This agreement seems quite reasonable for the purpose of doing parametric studies in a pre-design phase.

6 Conclusion

The Magnesium Oxide material is the reference choice for the ASTRID reflector. It has the required properties for a good reflector: radiation resistance and stability at high temperatures, low absorption and high scattering cross sections… It also benefits from a large experimental feedback since it was irradiated in the Phenix SFR as target matrix in the frame of Minor Actinide transmutation studies. The aim of the present study was to verify this choice of reflector material by comparing its neutronic performances with other material candidates, including a more conventional stainless steel reflector. The analysis is based upon the use of simplified 1-D and 2-D cylindrical models characterizing well the interfaces of interest: i.e. the core-reflector and the reflector-protection. The criteria retained for the ranking of reflector materials are purely based on neutronic considerations: reactivity of core-reflector systems, averaged albedo at core-reflector interfaces and, in association with protections, the efficiency in neutron attenuation, so as to minimize secondary sodium activation through the intermediate heat exchanger. The comparisons are clearly in favour of the MgO reflector, even though $MgAl_2O_4$ or $Y_3Al_5O_{12}$ are very close in terms of reactivity, and TiC or ZrC are slightly better in shielding. The flattening in radial core power distribution, which is important for better fuel utilization in the outer core region, was not explicitly considered as ranking criteria (but it is redundant with the reactivity one); however this aspect was investigated in details in [2].

Monte-Carlo simulations were conducted only for validation purposes, especially on local reaction rates at the interfaces (e.g. core-reflector), and also on shielding aspects. The deterministic results obtained with ERANOS-2 code system were found in reasonable agreement with the reference Monte-Carlo simulations using TRIPOLI-4®. Some improvements are suggested in the process of multi-group cross-section generation.

There is still a strong interest in developing 3D deterministic modelling, especially for design studies, since it provides useful indications for understanding trends and getting sensitivities; moreover, it allows propagation of uncertainties. This study is relying intentionally on very simplified reactor models (1-D and 2-D cylindrical) using deterministic codes. This approach is relevant for the design phase, because a lot of design specifications are still unknown or might be changing. For instance, the description of internal vessel structures, the position and geometry of the intermediate heat exchanger will have a direct impact on the determination of the secondary sodium activation. As long as the reactor description is incomplete, there is limited interest in determining only absolute values, because some of the most important modelling details could be forgotten.

References

1. J. Rouault, "The ASTRID Project: Status and Prospects at the Beginning of the Basic Design

DOI: 10.1051/epjconf/201715305006

Phase", in proceedings of ICAPP 2016, San Francisco, USA, 2016.
2. D. Blanchet, B. Fontaine, "On Core-Reflector Interface Effects in the ASTRID Sodium-Cooled Fast Reactor", in proceedings of ICAPP 2016, San Francisco, USA, 2016.
3. L. Buiron, B. Fontaine, "SFR Minor Actinide Bearing Blanket Fine Characterization Using Monte-Carlo", in proceedings of ICAPP 2016, San Francisco, USA, 2016.
4. J.M. Bonnerot et al., "Progress on inert Matrix Fuels for Minor Actinides Transmutation in Fast Reactor", in Proceedings of GLOBAL 2007, Bise, Idaho, USA, 2007.
5. M. Le Flem, J. Canel and S. Urvoy, "Processing and Characterization of Zr_3Si_2 for Nuclear Applications", Journal of Alloys and Compounds 465, 269-273, 2008.
6. D.L. Smith et al., "Development of Vanadium-Base Alloys for Fusion First-Wall-Blanket Applications", Fusion Engineering and Design 29, 388-410, 1995.
7. J.M. Ruggieri, J.F. Lebrat, J. Tommasi, P. Archier, "Status of ERANOS-2 code system validation for Sodium Fast Reactor Applications", in proceedings of FR09, Kyoto, Japan, 2009.
8. E. Brun, F. Damian, C.M. Diop, E. Dumonteil, F.-X. Hugo, C. Jouanne, "TRIPOLI-4®, CEA, EDF and AREVA reference Monte Carlo code", in Proceedings of SNA+MC2013, Paris, France, 2013.
9. A. Santamarina et al., "The JEFF-3.1.1 nuclear data library, JEFF report 22, validation results from JEF-2.2 to JEFF-3.1.1", vol. 6807. OECD, NEA, 2009.
10. J.-P. Both, J.-C. Nimal, T. Vergnaud, "Automated Importance Generation and Biasing Techniques for Monte-Carlo Shielding Techniques by the TRIPOLI-3 Code", Progress in Nuclear Energy, Vol 24, pp.273-281, 1990.
11. O. Petit, Y.-K. Lee, C.M. Diop, "Variance Reduction Adjustment in Monte-Carlo TRIPOLI-4® Neutron Gamma Coupled Calculations", Progress in Nuclear Science and Technology, Vol 4, pp.408-412, 2014.
12. N. Chapoutier, A. Janet, G. Nolin, M.-C. Ricol, A.-C. Scholer, D. Verrier, Ch. Venard, "ASTRID Core Shielding – Design Studies and Benchmark Analysis", in proceedings of ICAPP 2015, Nice, France, 2015.

Radiation transport out from the reactor core: to decouple or not to decouple

Kenneth W. Burn[1,*] and *Patrizio* Console Camprini[1]

[1]ENEA, Via M.M.Sole, 4, 40129 Bologna, Italy

Abstract. In the framework of the extension of the lifetime of currently operating reactors as well as of issues connected to decommissioning, accurate calculations of neutron and gamma responses outside the reactor core are increasingly being sought. Recently Monte Carlo calculations have been extended to provide a deep penetration capability incorporated within the eigenvalue calculation. This allows, in principle, neutron and gamma ray responses quite far outside the fissile region to be calculated within the same source-iteration scheme employed to define the neutronic responses in the fissile zone. In this paper, the new algorithm is compared to the classic decoupled approach - an eigenvalue calculation followed by a fixed source one - with the point of decoupling chosen as the fission sites. Two contrasting sample problems are discussed: a small fast research reactor and a large GEN-III Pressurized Water Reactor. The latter problem highlights the role of superhistories in maintaining the fundamental mode.

1 Introduction

In the last years work has been underway to extend the lifetime of currently operating reactors. In the case of Pressurized Water Reactors (PWR's), the radiation damage to the pressure vessel is a prime example of such evaluations. Furthermore there are an increasing number of requests to evaluate a variety of ex-core radiation responses linked to added safety devices or appraisals (ex-core neutron detectors, concrete ageing, detectors employed in severe accident scenarios, etc.). As far as decommissioning is concerned, accurate assessments of ex-core activation are required to develop dismantling and waste storage strategies, etc.

Normally the calculational approach to such problems, in particular when employing Monte Carlo, is to decouple the problem into an eigenvalue calculation of the fissile zone and a fixed source calculation going outside the core. The point of decoupling varies: it may be the flux leakage from the fissile zone or the fission sites within the fissile zone. Of these two options, employing the fission sites allows the number of phase space variables to be reduced to three: *viz.* x,y,z. (In particular the delicate question of discretization of the angle is avoided as fission is isotropic.) Notwithstanding, the fission site option still has several disadvantages:

- the spatial discretization still introduces some approximations that are difficult to quantify;
- for a core containing a variety of fuels and enrichments (and for a burnt core), it may not be feasible to tally the fission source at a level that is consistent with the material geometry description;
- to the authors' knowledge, current public Monte Carlo codes do not allow a complete pin-by-pin description of the fission source – any user-written modifications reduce the quality assurance;
- the whole process (eigenvalue calculation to generate the fission sites, interface code to create the source for the fixed source calculation, the fixed source calculation) is cumbersome and long, both in terms of computer and of human time.

Recently a new algorithm has been proposed to perform variance reduction (VR) within the source-iteration algorithm and therefore to allow in principle ex-core radiation responses to be estimated within a single eigenvalue calculation ([1–3]). In [1] and [3] the feasibility of the new approach was demonstrated in that localized in- and ex-core responses were calculated with a low error within the eigenvalue calculation. However the sample problems contained just a single ex-core problem (the rest were in-core).

Work is currently underway to test the new algorithm on ex-core problems in a more systematic fashion. In this paper the new algorithm is compared with the classic decoupled approach employing the fission sites at the point of decoupling for two contrasting sample problems: a small fast research reactor and a large GEN-III thermal power reactor.

2 Methodology

2.1 The Direct Statistical Approach

The Direct Statistical Approach (DSA) was originally developed for fixed source problems [4]. It can be distinguished from classical VR approaches such as the weight window generator [5] by the specific

* Corresponding author: kburn@enea.it

characteristic of estimating the second-moment and therefore the error.

The VR technique that the DSA aims to optimize is splitting and Russian roulette (RR) at surfaces in space and in energy (i.e. a 2-dimensional function), either independent of, or dependent on, the weight of the progenitor arriving at the surface [4]. Because the DSA uses the second-moment, in contrast to classical approaches that rely on the first-moment, it requires also the time. This is so as to be able to find the optimum trade-off between over-splitting that takes too much time (and therefore does not run a sufficient number of histories) and under-splitting that has too high a variance per source history.

2.2 Applying the DSA to the source-iteration algorithm in eigenvalue problems

The key aspect in applying the DSA to the source-iteration algorithm is the multi-response capability [4]. In applying VR to the eigenvalue problem, we are changing the sampling in the fissile zone and in the reflector and thereby distorting the fundamental mode. But the fundamental mode is the source for the next cycle in which again we are changing the sampling to improve the statistics of local detectors, be they in- or ex-core and logging the fundamental mode, the source for the next cycle, etc. Clearly there is a trade-off between the amount of distortion of the fundamental mode (to be minimized) and the alteration of the number of tracks at the local detector (to be maximized).

Thus we must add responses in the fissile zone (normally the fission neutron production rates in a sufficient number of spatial segments to mock-up the fundamental mode) which we name the global responses, to the local in- or ex-core responses for which VR is required. The DSA aims to optimize all these responses simultaneously.

It may be mentioned that first-moment approaches to optimizing VR can also contain a multi-response capability [6, 7]. In principle therefore such approaches could be employed in the same way in eigenvalue problems. Notwithstanding, comparisons between the DSA and classical approaches have only been made so far for fixed source problems [8] apart from a single assessment for an eigenvalue problem in [1].

2.2.1 Superhistories

Superhistories [9] lengthen the number of fission generations in each cycle, a cycle being defined as between normalization points, in other words an "iteration" in the source-iteration scheme. Note that the usual source-iteration scheme has one fission generation per normalization cycle. Although superhistories were introduced in [9] principally to mitigate the problem of underestimation of errors in the source-iteration scheme (and to a lesser extent the problem of response underestimation), in [1–3] they were reintroduced to improve the variances of local responses. Furthermore the VR parameters were varied according to the fission generation within the superhistory. This was successful for in-core responses, although the gain in efficiency was only a factor of 2 or so. Instead for ex-core responses there was virtually no gain. This was expected as it is the fissions in the core periphery that contribute to ex-core responses. Thus neutron transport within the fissile zone is unimportant for the local ex-core responses.

Notwithstanding a remark in the conclusion of [1] – "Allowing a number of fission generations between normalizations of the fission source (i.e. superhistories) may help in maintaining the fundamental mode ……" – the working hypothesis from that work was that superhistories of a single fission generation were appropriate for ex-core local responses whilst for in-core local responses, more than one fission generation per superhistory may improve somewhat the efficiency.

In the following sample problems, a single fission generation per superhistory was first chosen. This choice was satisfactory for the TAPIRO problem with a small, fast, core. The benchmark results of this problem provide the first main result of this paper. Instead for the large GEN-III thermal core, this choice produced results which were flawed. The second main result of this paper is the analysis of the cause of such flaws and subsequent indications of how to eliminate them.

3 TAPIRO fast reactor sample problem

TAPIRO is a helium-cooled 5 kW fast research reactor at the ENEA Casaccia centre, near Rome. It has a small, cylindrical (approximately 11 cm high, 12.5 cm diameter) highly-enriched core. Because of its relatively hard spectrum, it has recently been proposed to employ TAPIRO to irradiate actinides and measure their capture and fission cross-sections. To this end, a characterization was made of three experimental channels: diametral, tangential and radial, from the fissile region through the copper reflector, up to the concrete shield (see Figs. 1–3) [10]. In each mesh interval in the filled channels in Figs. 1–3, the total capture and fission reaction rates for the following isotopes were required: ^{235}U, ^{238}U, ^{237}Np, ^{239}Pu, ^{240}Pu, ^{242}Pu, ^{241}Am, ^{243}Am, ^{243}Cm, ^{244}Cm and ^{245}Cm.

Thus the problem is to calculate a substantial number of fluxes and actinide capture and fission rates at all the positions in the three channels shown in Figs. 1–3. Furthermore, the actinide capture cross-sections show a multitude of resonances (see for example Fig. 4) so that although the penetration is not large, the statistical difficulties are non-negligible.

Four calculational approaches were followed:

1) A single "brute force" analog eigenvalue calculation, using MCNP6.1 [11].
2) The classic decoupled approach employing the fission sites as the point of decoupling. For the second, fixed source, part of the calculation, analog VR was employed. Both parts of the calculation used MCNP6.1.
3) Like 2) but using VR in the second, fixed source, part of the calculation. For this, the DSA [4] that currently uses as vehicle MCNP5-1.40 [12], was

employed to generate the VR parameters. These parameters were then converted to a weight window and employed in MCNP6.1.

4) A single eigenvalue calculation including VR as described in [1–3]. This again uses as vehicle MCNP5-1.40 whilst the calculation with the optimized VR parameters was made with MCNP6.1.

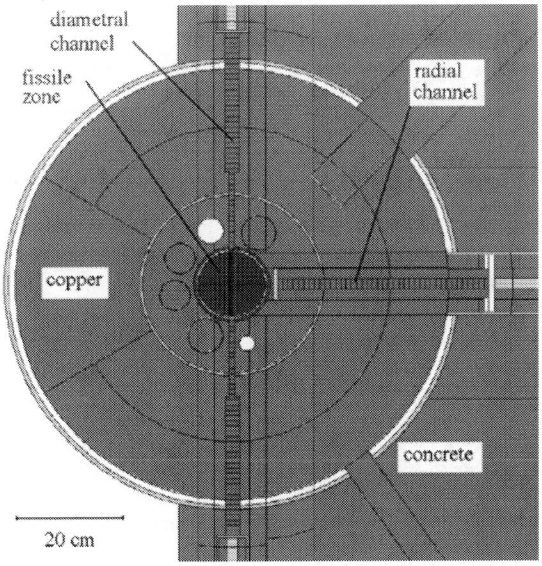

Fig. 1. TAPIRO: radial section at core mid-plane showing the diametral and radial channels.

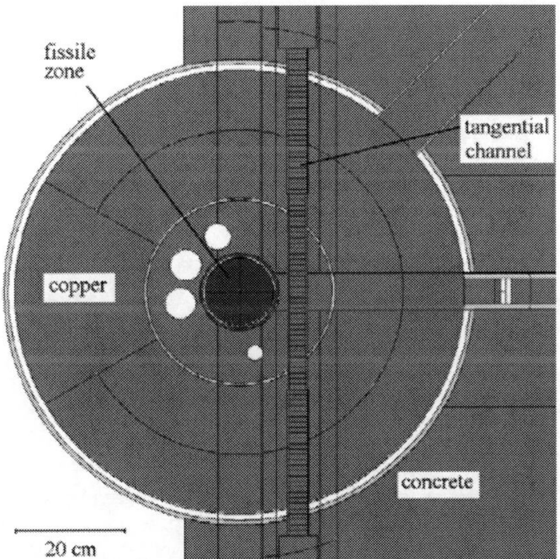

Fig. 2. TAPIRO: radial section 5 cm above core mid-plane showing the tangential channel.

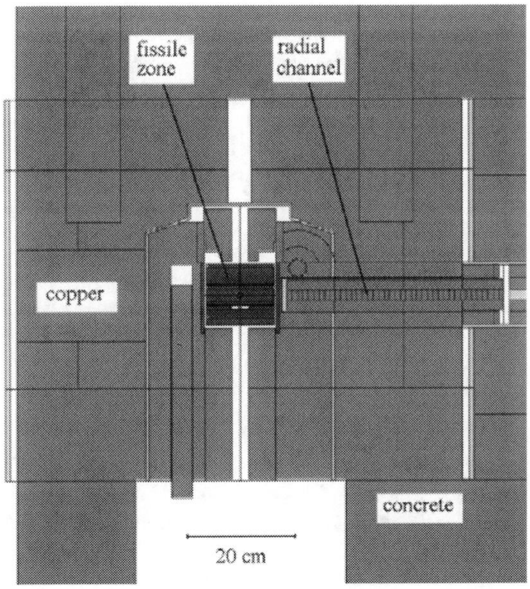

Fig. 3. TAPIRO: axial section showing the radial channel.

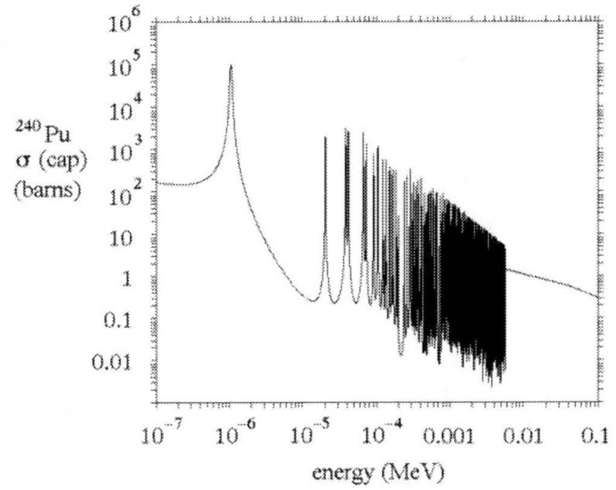

Fig. 4. ^{240}Pu neutron capture cross-section from JEFF 3.1.

Further aspects of the calculational approaches 2) – 4) are described as follows:

Approach 2): The spatial binning of the fissile zone was 2-dimensional, r-z, with 19 axial and 14 radial bins. This could be exactly modelled with the SDEF source card in MCNP.

Approach 3): The same fixed source as in approach 2) was used. The calculation that then read this source optimized the neutron splitting/RR parameters in 16 spatial regions and 10 energy groups with limits (MeV): 0.5, 0.025, 0.003, 9E-4, 2.5E-4, 2.5E-5, 2.5E-6, 2.5E-7 and 1E-8.

22 reaction rates (the capture and fission rates of the 11 isotopes) are requested in each of 226 meshes (96 meshes in the diametrical channel, 90 meshes in the tangential channel and 40 meshes in the radial channel). This is too large a number of responses for the DSA [4] which, in contrast to first moment approaches, must

maintain the identity of each adjoint flux estimate. The 22 cross-sections were reduced to a representative 5 (capture in ^{238}U, ^{239}Pu, ^{240}Pu and ^{244}Cm and fission in ^{243}Am) and the response meshes were merged to make a total of 12 macromeshes (4 in each channel).

Thus the problem was reduced to finding the optimum VR parameters in 16×10 phase space cells in the presence of 5×12 responses. As no weight variations were present other than those due to phase space surface splitting/RR, the weight-independent DSA was employed (see for example [13]). The resulting importances in space/energy were converted to a weight window and run with standard MCNP6.1.

Approach 4): In this case there is of course no source binning. The same 16 spatial regions and 10 neutron energy groups were employed as in the second part of the decoupled case in approach 3). Furthermore the same 5×12 responses were used as in the decoupled case. However now the fissile zone was divided into 8 segments and the fission rate was scored in each segment so as to maintain the fundamental mode. These 8 responses were added to the 60 ex-core responses. Again like the decoupled case, no weight variations were present. A superhistory of a single fission generation was employed.

In Figs. 5–7 are shown the results for the capture rates in ^{238}U, ^{242}Pu and ^{244}Cm respectively in the upper part of the diametral channel (see Fig. 1) for the four approaches: single eigenvalue analog, decoupled analog, decoupled VR and single eigenvalue VR. In Fig. 8 are shown the results for the capture rates in ^{240}Pu in the radial channel (see Figs. 1 and 3) again for the four approaches. Instead in Fig. 9 we see the fission rates in ^{245}Cm in the tangential channel (see Fig. 2).

Although not part of this comparison, it is of interest to note that the computer times were: 548.5 h, 638.9 h and 1071.6 h for the decoupled analog, decoupled VR and single eigenvalue VR calculations respectively. (The times with VR do not include the times required to generate the VR parameters which were however relatively small. The same does not apply to the human time of course.)

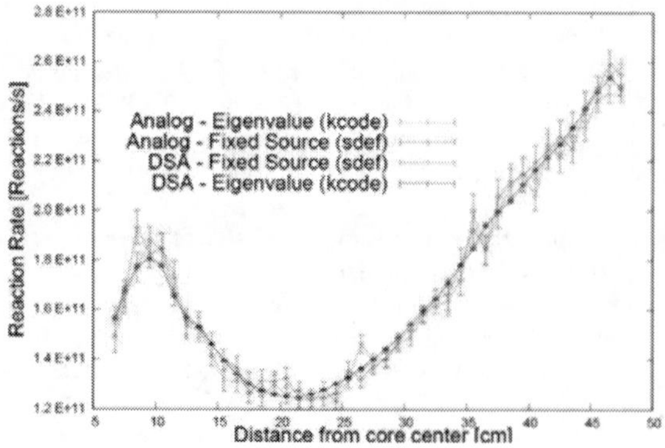

Fig. 5. ^{238}U capture rate in TAPIRO diametral channel.

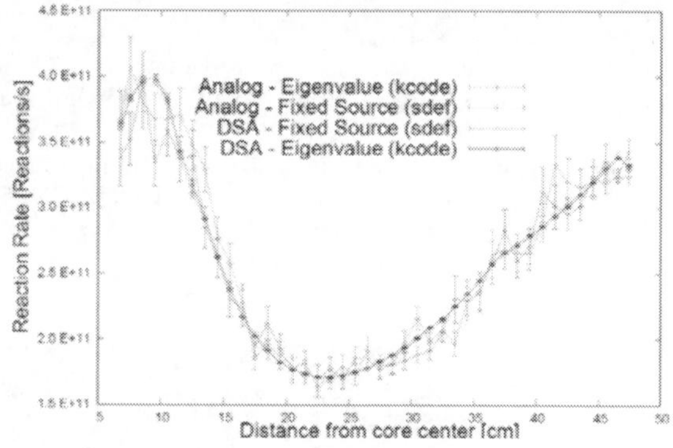

Fig. 6. ^{242}Pu capture rate in TAPIRO diametral channel.

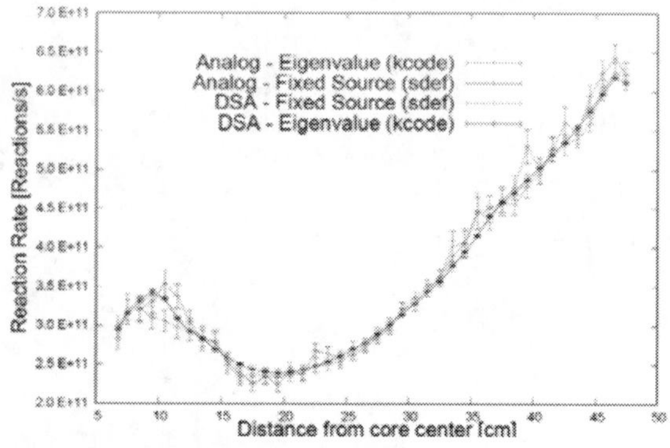

Fig. 7. ^{244}Cm capture rate in TAPIRO diametral channel.

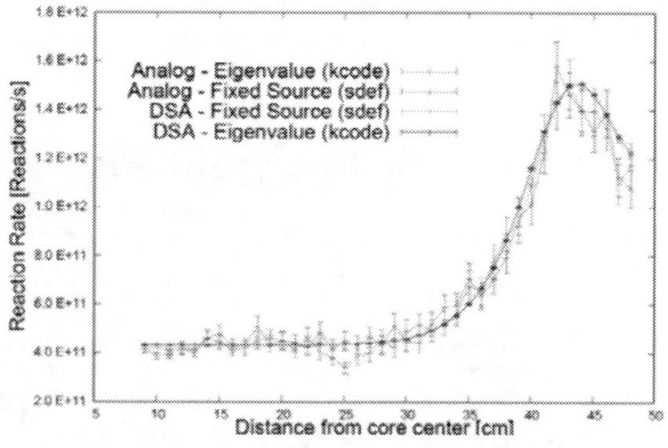

Fig. 8. ^{240}Pu capture rate in TAPIRO radial channel.

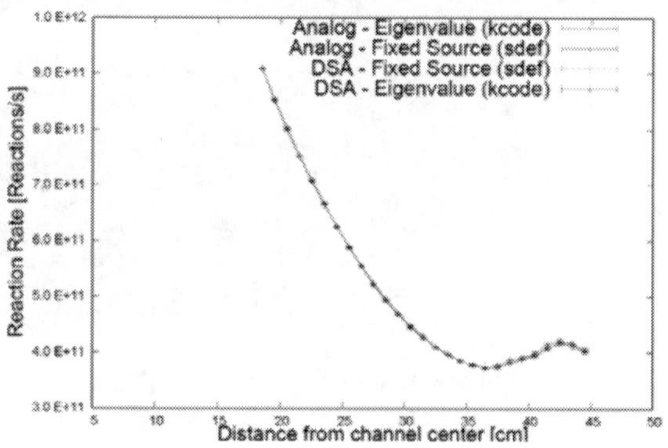

Fig. 9. ^{245}Cm fission rate in TAPIRO tangential channel.

For the small TAPIRO core and given the relatively fine source binning, the approximation involved in the decoupling can be neglected. Thus the results look to be at an appropriate benchmark level for the single calculation approach without decoupling for this kind of problem.

4 Gen-III power PWR

Axial and radial sections of the simplified PWR GEN3 model adopted with a heavy steel reflector are shown in Figs. 10–13. The detectors were situated outside the pressure vessel radially, between the vessel and the concrete and at various depths within the concrete, and axially, above the concrete at the bottom of the well and embedded in the concrete basemat. As well as neutron responses, gamma dose rates were also requested.

A pin-by-pin description of the fuel assemblies was adopted. Given that the core geometry is defined pin-by-pin, we wished also to define the fission source at the same level. Thus in contrast to the TAPIRO problem, the mechanism of decoupling of the calculation using the fission sites was relatively complex.

We did not attempt to run an analog calculation to calculate the fluxes and responses as the penetration was too great, especially for the responses embedded in the concrete basemat at the bottom of the well.

4.1 The decoupled calculation

4.1.1 The decoupled calculation: writing the source

The core contains 241 fuel assembles, each assembly with 17×17 fuel pin positions. After taking away the guide tubes (24 per assembly), we have a total of 63865 fuel pins (including Gd–doped ones). The fissile zone is approximately 4.2m high. Each pin must be divided into a sufficient number of axial segments to model reasonably the axial power profile. We used 37 axial segments, of variable length, but the same for all the pins.

Thus we had a total of 63865×37=2363005 tallies for the neutron source. Because of memory issues, it was necessary to divide the first part of the decoupled calculation that writes the neutron source into six parts, each part estimating the fission neutron source in a sub-set of the fuel pins. Each of these six calculations were identical (apart from the tallies), and starting from a converged fundamental mode fission distribution, ran 4000 cycles, each of 400000 source fission neutrons.

4.1.2 The decoupled calculation: reading the source

To read the resulting fission source distribution, MCNP does not allow a definition of the source at the fuel pin level when more than one fuel assembly is present. Thus a user-written source was required. The authors actually employed a dummy standard source and modified the code. Rather than keep all source date in memory, for each source history the spatial coordinates were read from a direct access file.

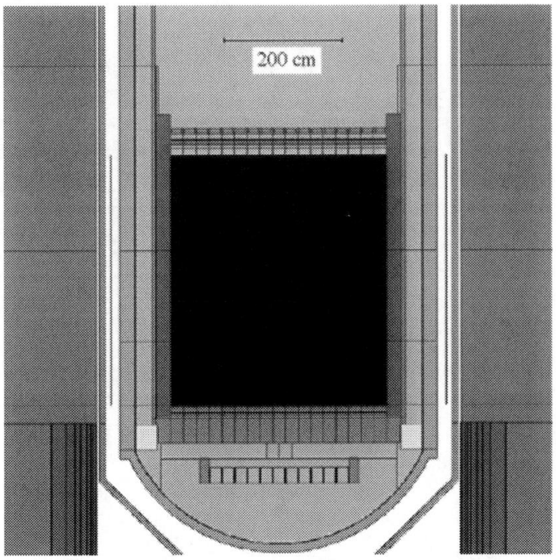

Fig. 10. Simplified model of large size PWR: axial section centred at the core mid-plane.

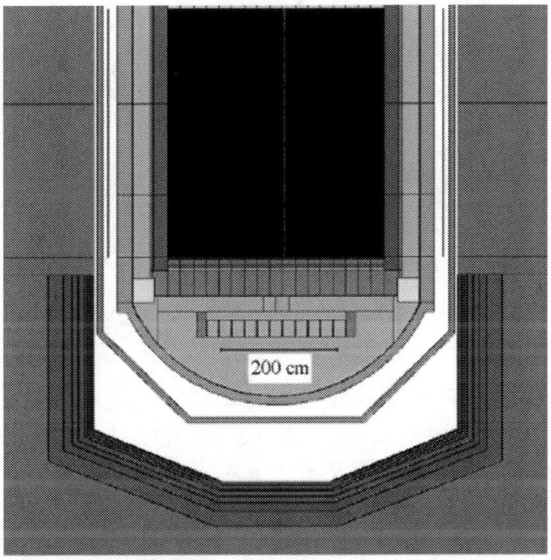

Fig. 11. Simplified model of large size PWR: axial section showing the bottom part of the well.

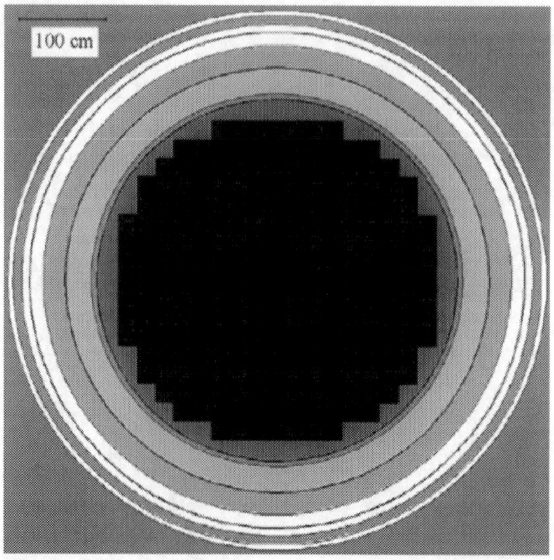

Fig. 12. Simplified model of large size PWR: radial section centred at the core mid-plane.

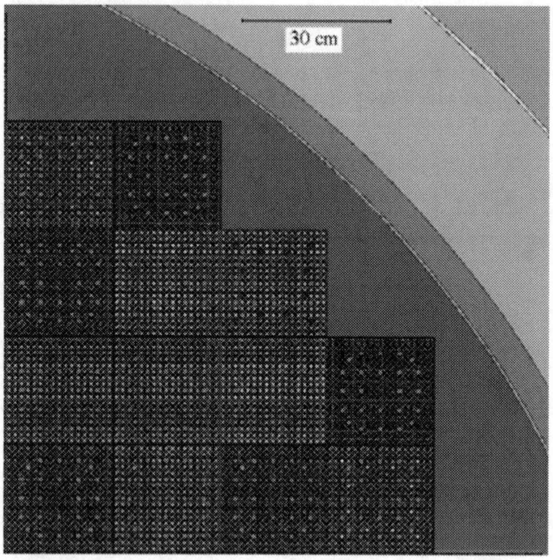

Fig. 13. Simplified model of large size PWR: radial section at the edge of the fissile zone.

Given that both neutron and gamma total fluxes, flux spectra and responses were requested at a variety of positions in the pressure vessel well and in the concrete surrounding the well, it was decided to divide the second part of the decoupled calculational suite into two: a neutron-only run and a n–gamma run with only gamma responses. In both cases the DSA [4] was employed. To generate the VR parameters, the number of neutron responses was reduced to 48 (with 44 spatial regions, 8 of which covered the fissile zone, and 7 energy groups with limits: 5, 2, 0.2, 0.02, 0.001 and 1E-6 MeV) and the number of gamma responses was reduced to 22 (with the same 44 spatial regions and 7 neutron energy groups plus 6 gamma energy groups with limits: 5, 3, 1.5, 0.7 and 0.2 MeV).

Having generated optimum or near-optimum VR parameters in the neutron-only and n–gamma cases, runs were made with these parameters, employing 390×10^6 source histories in the neutron-only case and 100×10^6 source histories in the n–gamma case. The results for all the requested responses, nearer or farther from the pressure vessel, were satisfactory and within the project constraints. At that point it was necessary to evaluate the uncertainties in the results arising from the decoupling.

4.2 The single eigenvalue calculation

It was desired to evaluate the approximations of a single radial bin for each fuel pin (which should underestimate the ex-core responses) and of the axial discretization of the source into 37 bins, by making two single eigenvalue calculations, one for the neutron-only case and the other for the n–gamma case.

The optimization of the DSA neutron and n–gamma calculations closely followed the fixed source cases and will not be described in detail. A key difference from the fixed source cases was the addition of 12 fission heating responses over the fissile zone making a total of 60 responses in the neutron-only case and 34 responses in the n–gamma case. This meant that the core spatial division into 12 segments to provide responses to maintain the fundamental mode was not the same as the core spatial division (8 segments) to control the neutron transport within the core. Although probably not critical, this is undesirable and was corrected in the later calculations. For both the neutron-only and the n–gamma cases the spatial regions and energy groups employed for VR were the same as in the decoupled cases. Again a superhistory of a single fission generation was employed.

Having reached the fundamental mode, cycle lengths of 350000 and 150000 fission neutrons were used in the neutron-only and n–gamma cases respectively.

4.3 Comparison of the results of the decoupled and single eigenvalue calculations

In Figs. 14 and 15 the axial profiles of the total neutron and total gamma fluxes radially just outside the pressure vessel are compared for the decoupled case and for two statistical samples (i.e. a different number of cycles of the same calculation) of the single eigenvalue case.

We see in these figures, especially in the neutron-only calculation, that there is a large distortion of the axial profile that is swamping the effect that we are looking for (the approximation of the decoupled calculation). In the rest of this section, we shall analyse the cause of such distortion. The analysis will focus on the single eigenvalue case.

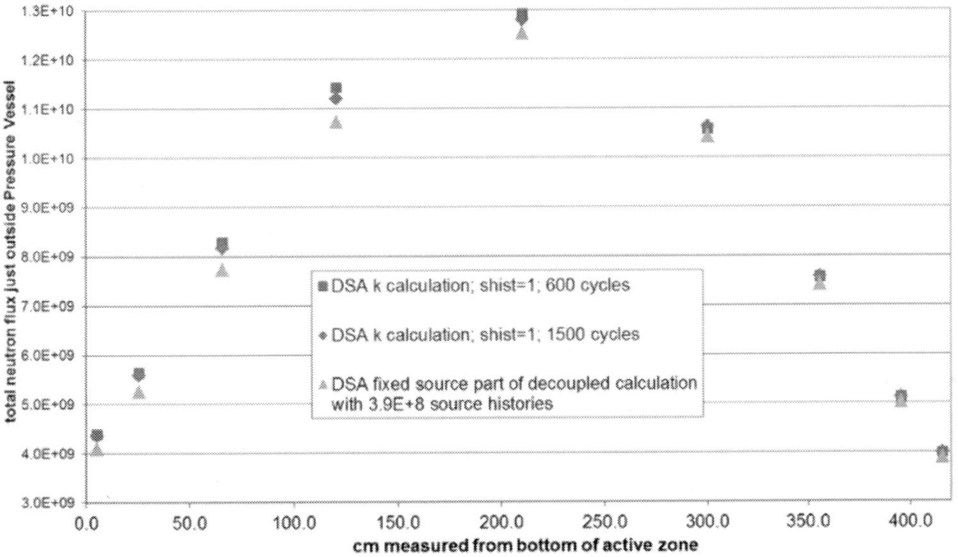

Fig. 14. Axial profiles of the total neutron flux just outside the pressure vessel of the Gen III PWR for the decoupled and single eigenvalue calculations.

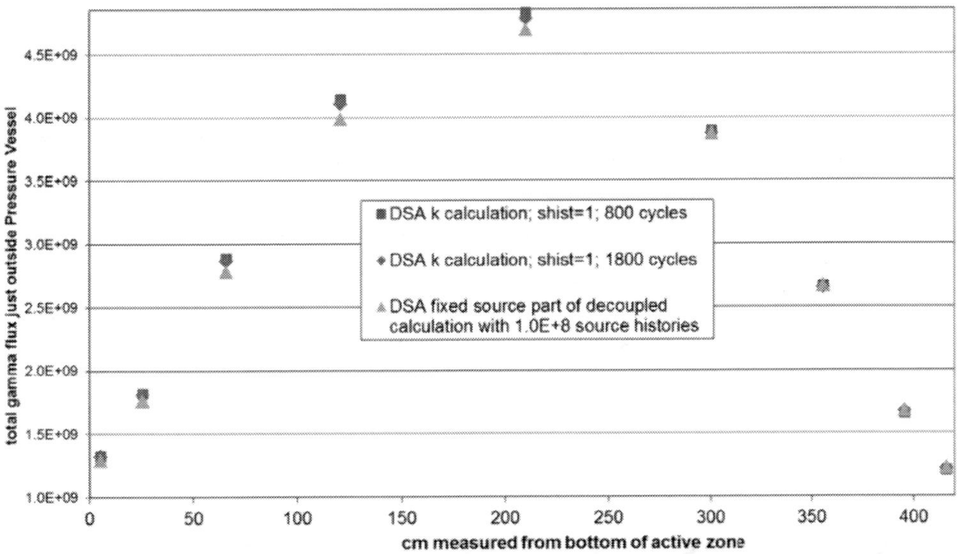

Fig. 15. Axial profiles of the total gamma flux just outside the pressure vessel of the Gen III PWR for the decoupled and single eigenvalue calculations.

4.4 Analysis of the single eigenvalue results

We suspected that the fundamental mode was not being properly maintained in the case of the single eigenvalue calculation. To examine this hypothesis, rather than look at the fluxes outside the vessel, we investigated the leakage neutron flux from the fissile zone into the radial reflector (see Figs. 12 and 13).

We firstly studied the evolution of the axial profile of the leakage neutron flux with the number of cycles run for two cases:

– a standard MCNP calculation with analog VR parameters and cycle size 350000 fission neutrons.

– a single eigenvalue calculation with a superhistory of one fission generation and a cycle size of 350000 fission neutrons. The same ex-core spatial cell configuration, energy group structure and ex-core responses were maintained from the previous calculations. Now instead the in-core spatial division and global responses were changed so that an identical spatial division of the fissile zone provided responses to maintain the fundamental mode as controlled the neutron transport within the core. This division was: three radial segments: the outer assembly ring, the next two assembly rings and the inner assemblies and five axial segments: < -147, -147 – -63, -63 – +63, +63 – +147 and > +147 (all cm with respect to the core mid-plane). Thus there were 15 core regions and 15 global, core responses making a total of 63 responses. There were correspondingly 51 spatial regions and, as previously, 7 neutron energy groups. As before the VR parameters were then optimized to the neutron local ex-core responses and global in-core responses simultaneously. A superhistory of a single fission generation was employed.

In Fig. 16 we see the evolution of the axial profile of the core neutron leakage flux for eight cases between 150 and 1500 cycles, with standard MCNP and analog VR. The axial profiles from the DSA run with exactly the same number of source fission neutrons are shown in Fig. 17. We see that unlike standard MCNP and analog VR, the DSA does not sufficiently maintain the fundamental mode.

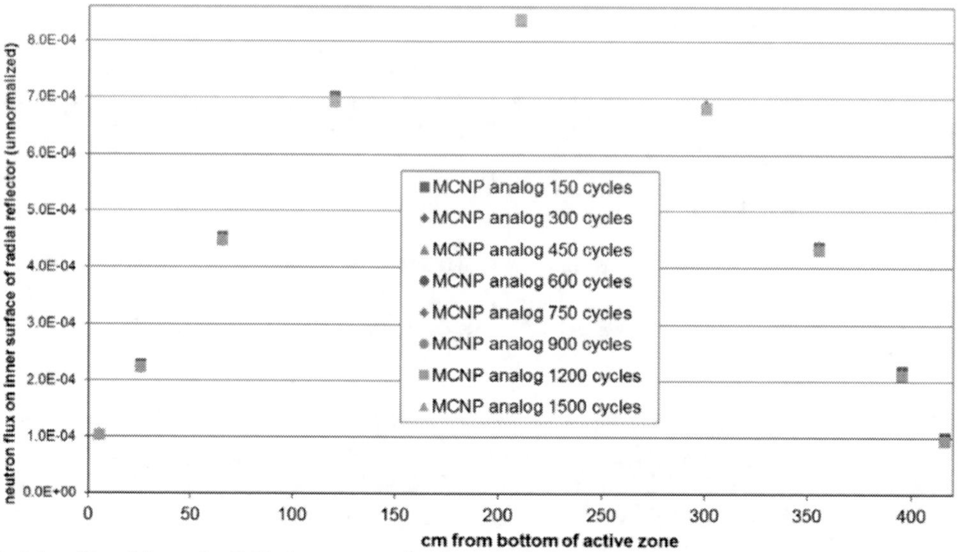

Fig. 16. Axial profiles of the total radial leakage neutron flux from the fissile zone of the Gen III PWR with standard analog VR MCNP eigenvalue calculations, increasing the number of cycles, each of 350000 neutrons, from 150 to 1500.

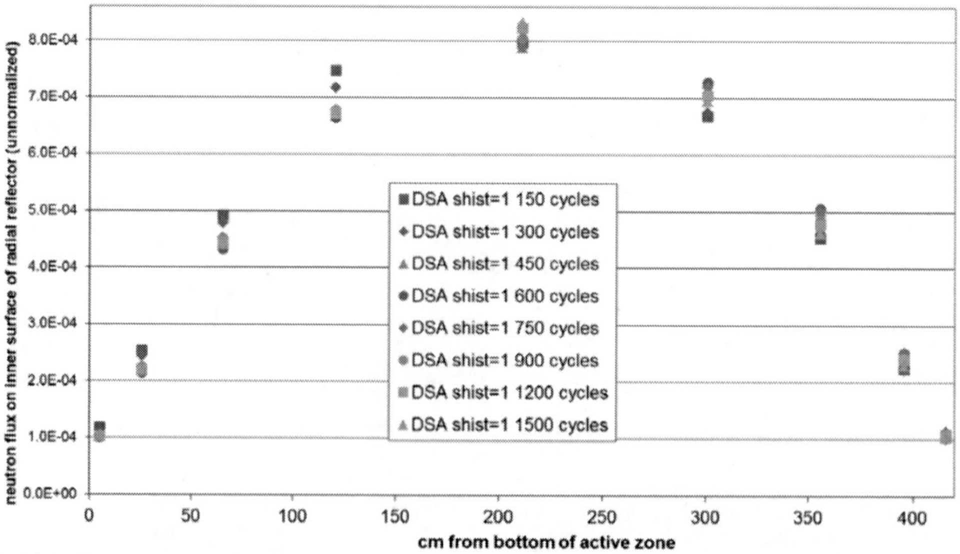

Fig. 17. Axial profiles of the total radial leakage neutron flux from the fissile zone of the Gen III PWR with DSA VR eigenvalue calculations with a superhistory of 1 fission generation, increasing the number of cycles, each of 350000 neutrons, from 150 to 1500.

Possible solutions to improve the convergence of the fundamental mode with the DSA were considered:

1) Increasing the number of in-core responses. Currently 15, the in-core responses might not have a sufficient weight in the minimization compared with the 48 ex-core responses.

2) Increasing the length of each superhistory: this should automatically raise the weight attached to the in-core responses due to transmission through the fissions of the response importances in future fission generations within a superhistory. It may also allow a better settling of the fission distribution due to longer in-core transport between normalization points.

The second option was adopted and superhistories of 1, 4, 12 and 25 fission generations were run. For each superhistory case, VR parameters were generated with the DSA in the 51×7 phase space regions including the 15×7 phase space regions in the core. Given the k_{eff} of the system of around 1.0132, fission generation numbers of: 150, 300, 450, 600, 750, 900, 1200 and 1500, were mocked up as normalization cycle numbers of:

- 36, 72, 108, 144, 180, 216, 288 and 361;
- 11, 22, 32, 43, 54, 65, 87 and 108;
- 4, 9, 13, 18, 22, 26, 35 and 44,

for the superhistory lengths of 4, 12 and 25 fission generations respectively.

For superhistory lengths > 1 these normalization cycle numbers would only approximate the required numbers of fission generations for analog importances. It was actually not possible to maintain the required numbers of fission generations because of splitting/RR between space/energy regions in the fissile zone. Furthermore there were many more ex-core collisions with the DSA compared to MCNP analog because of optimizing to the ex-core detectors.

There were 350000 fission neutrons starting each cycle in all four superhistory cases.

In Fig. 18 we see the evolution between 11 and 108 cycles of the axial profile of the core neutron leakage flux with the DSA for a superhistory of 12 fission generations, with again 350000 source fission neutrons per cycle. We see in Fig. 18 that the convergence of the fundamental mode has markedly improved. We now analyse further the effect of the length of the superhistory.

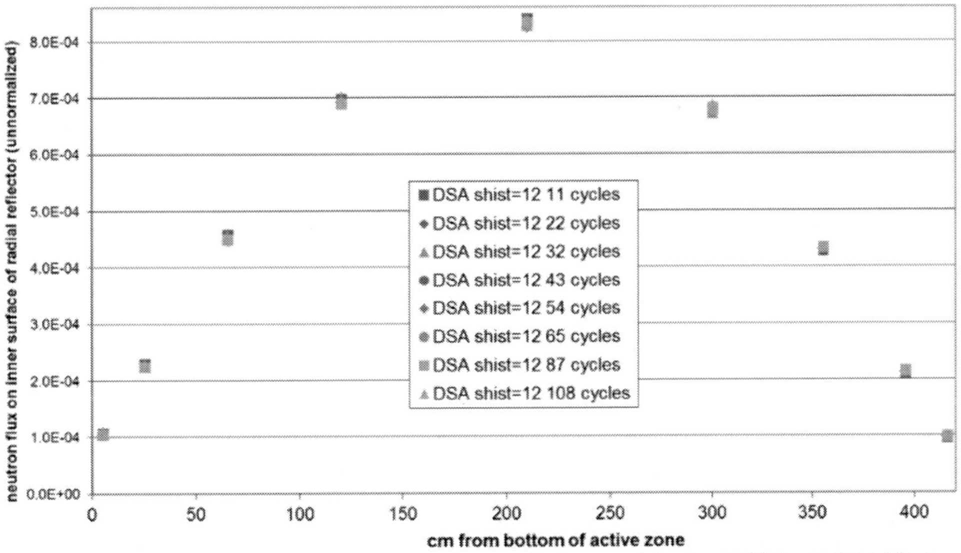

Fig. 18. Axial profiles of the total radial leakage neutron flux from the fissile zone of the Gen III PWR with DSA VR eigenvalue calculations with a superhistory of 12 fission generations, increasing the number of cycles, each of 350000 neutrons, from 11 to 108.

4.4.1 The impact of the length of the superhistories on the convergence of the fundamental mode

Figs. 19 show the axial profiles of the radial leakage fluxes of MCNP analog and the four DSA superhistory cases for cycle numbers of: 150, 300, 450, 600, 750, 900, 1200 and 1500 respectively, or their equivalent as discussed above for superhistory lengths > 1. The results are given as a ratio to the results of a separate, independent, analog MCNP run of 3500 cycles of 350000 neutron each. We see again that a superhistory of one fission generation does not sufficiently maintain the fundamental mode. From these figures we may infer that whilst a superhistory length of 4 is markedly better than 1, a length of 10 or so should be sufficient.

To try to establish a more level playing field for these results, we show in Table 1 a number of standard counters for the final calculations in Fig. 19h. The "loss to fission" row gives some measure of the statistical effort to maintain the fundamental mode. Although the DSA with a superhistory length of 1 has the lowest value (one quarter of analog MCNP), it is clear from Figs. 19 that four times the effort will still not bring these results near the quality of the others (and would anyway take four times the time of the other superhistory cases).

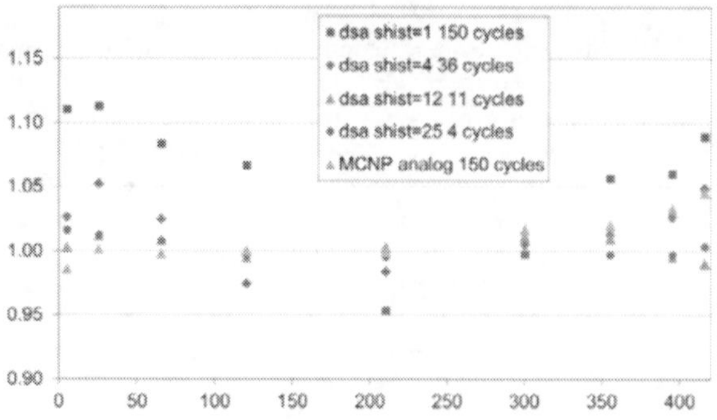

Fig. 19a. Analog MCNP run of 150 cycles.

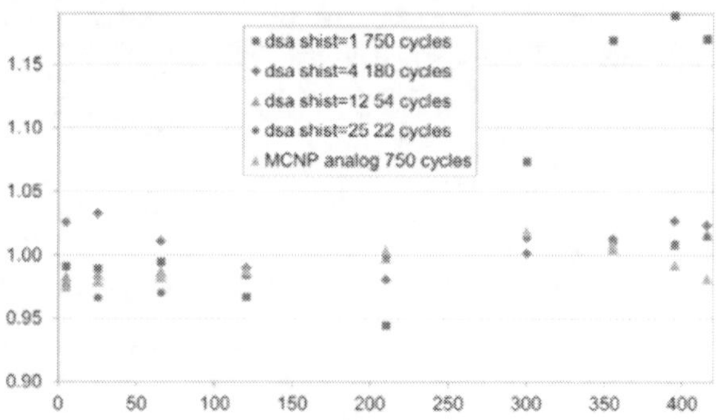

Fig. 19e. Analog MCNP run of 750 cycles.

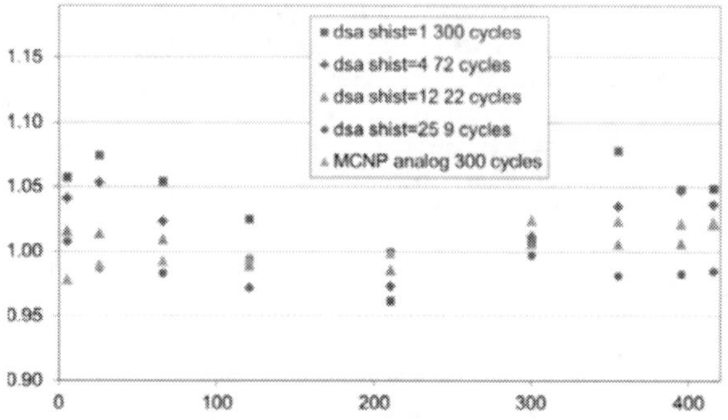

Fig. 19b. Analog MCNP run of 300 cycles.

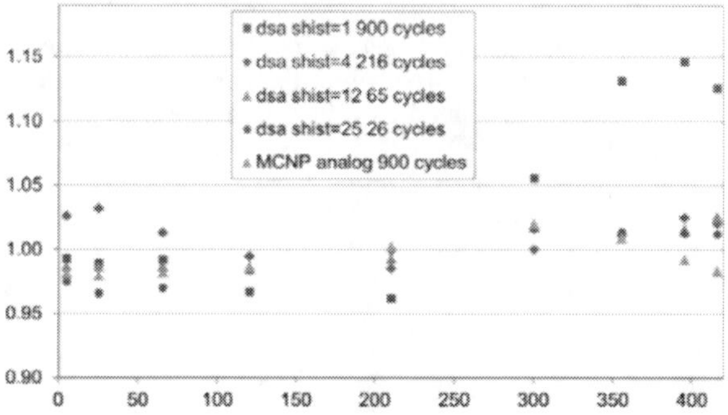

Fig. 19f. Analog MCNP run of 900 cycles.

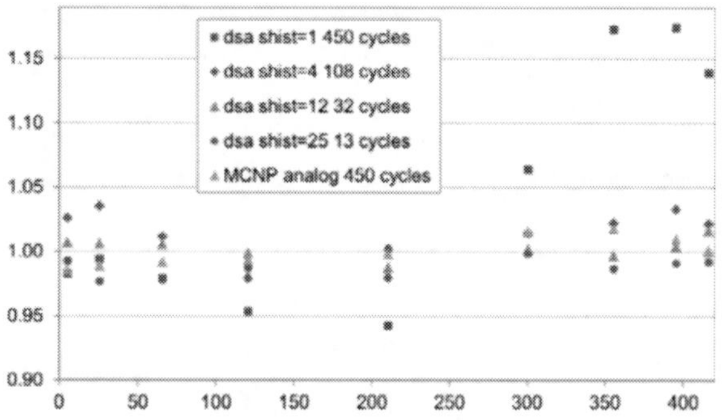

Fig. 19c. Analog MCNP run of 450 cycles.

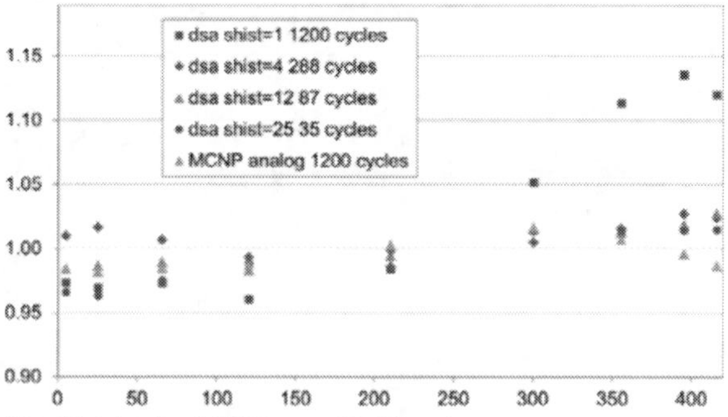

Fig. 19g. Analog MCNP run of 1200 cycles.

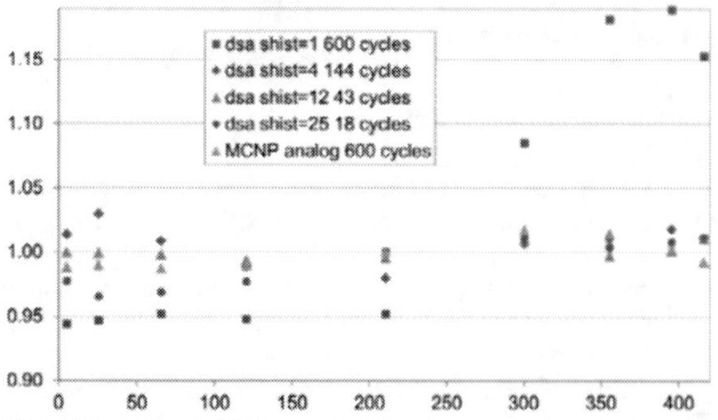

Fig. 19d. Analog MCNP run of 600 cycles.

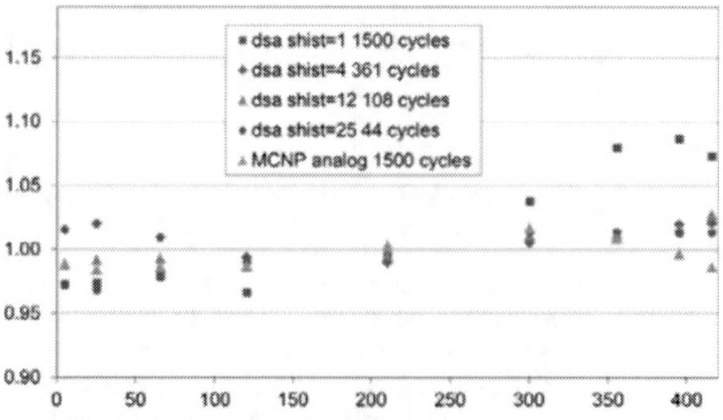

Fig. 19h. Analog MCNP run of 1500 cycles.

Figs. 19. Axial profiles of the radial leakage neutron flux of the Gen III PWR with an increasing number of cycles, each of 350000 neutrons, analog MCNP and DSA VR eigenvalue calculations with various "equivalent" length superhistories (ratios to an independent MCNP run of 3500 cycles, each of 350000 neutrons). The x axis denotes "cm from the bottom of the active zone".

Table 1. Some counters of interest in the MCNP 1500 cycle analog run and in the four DSA "equivalent" superhistory runs (G=billions, M=millions, K=thousands).

	MCNP analog 1500 cycles	DSA shist=1 1500 cycles	DSA shist=4 361 cycles	DSA shist=12 108 cycles	DSA shist=25 44 cycles
cycle source	525 M	525 M	126 M	38 M	15 M
loss to fission	199 M	50 M	161 M	285 M	346 M
total collisions	15 G	87 G	142 G	113 G	93 G
mins. computer time	11 K	40 K	46 K	38 K	31 K

To illustrate further the effect of superhistories in better maintaining the fundamental mode when variance reduction is present, we show in Figs. 20 the phase space cell importances (the z axis) found with the DSA for the four cases of superhistory lengths: 1, 4, 12 and 25 fission generations. The fifteen core regions, defined previously, have five axial segments ("z" in cm, going from bottom of core at NE to top of core at SW in Figs. 20) and three radial ones ("Inner" assemblies at NW, "Middle" assemblies and "Outer" assemblies at SE in Figs. 20). We remind ourselves that the seven neutron energy groups have limits: 5, 2, 0.2, 0.02, 0.001 and 1E-6 (MeV). Thus in Figs. 20 for each of the 5×3 core regions, there are 7 histograms, going from the highest energy red to the lowest energy blue.

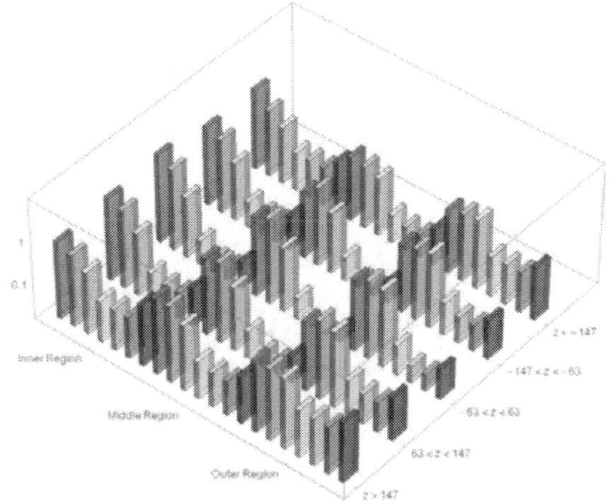

Fig. 20a. Superhistory of one fission generation.

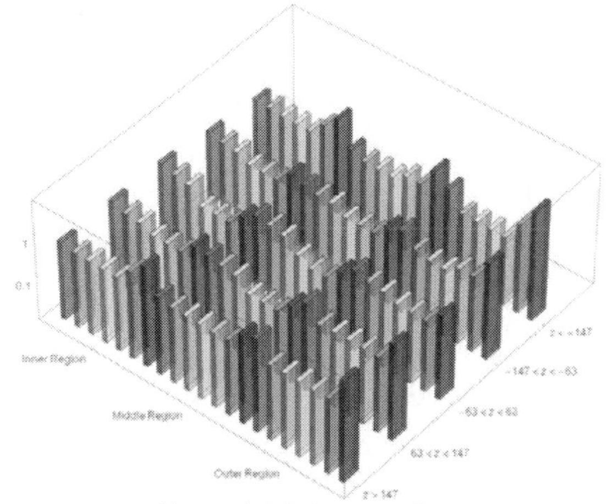

Fig. 20c. Superhistory of 12 fission generations.

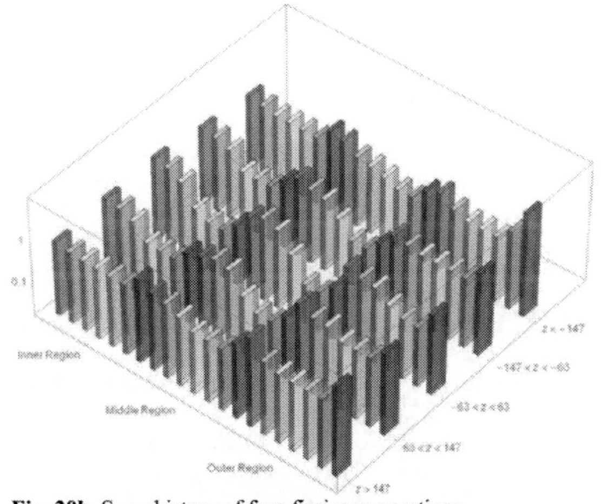

Fig. 20b. Superhistory of four fission generations.

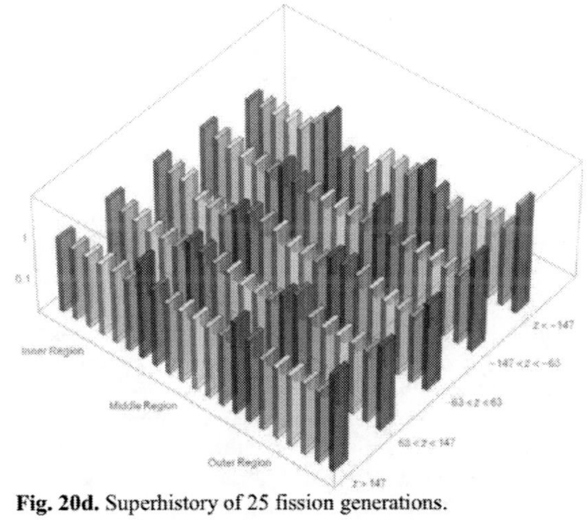

Fig. 20d. Superhistory of 25 fission generations.

Figs. 20. Importance topology in the fissile zone in space and energy for various length superhistories.

We see in Figs. 20 that the importance topology of the single fission generation per superhistory case is quite different compared with the three cases: 4, 12 and 25 fission generations per superhistory. For a single fission generation superhistory, the importance of the low energies is relatively low. Instead for 4, 12 and 25 fission generation superhistories it is relatively high, implying that fissions and their subsequent fission neutrons are important. Thus transport effects in the core are taken into account to a much greater degree than when all fissions are treated as absorptions. Also, although not obvious from Figs. 20, Figs. 20b–d in contrast to Fig. 20a have higher importances at low energies of the external assemblies at the bottom of the core compared with the top. This reflects the location of the deeper penetration ex-core local responses at the bottom of the well in Fig. 11.

5 Conclusion

The introduction of variance reduction within the Monte Carlo source iteration algorithm opened the path to the possibility of calculating ex-core responses at deep penetrations without decoupling the calculation with the inherent approximations associated with the decoupling.

Whilst for in-core local responses it had been shown that employing superhistories [9] and varying the VR parameters according to the fission generation within the superhistory can improve the multi-response variance, for ex-core responses no such gain had been found [1–3]. This was understandable because only peripheral fuel pins contribute to ex-core responses. Thus it had been imagined that superhistories need not be employed in ex-core problems and for in-core problems their only role is to improve the efficiency.

In this paper two sample problems involving ex-core responses are considered: the first a fast research reactor with a small highly enriched core, the second a simplified model of a large GEN-III PWR.

The first sample problem showed excellent agreement of the ex-core responses between the single eigenvalue run with VR (and a superhistory of a single fission generation), the decoupled run with VR and an analog decoupled run.

The second sample problem instead showed serious large-scale discrepancies between the single eigenvalue run with VR (and a superhistory of a single fission generation) and a decoupled run with VR. These discrepancies were not due to the approximations inherent in the decoupling but were caused by a flaw in the single eigenvalue calculation that did not maintain the fundamental mode sufficiently well. It was shown to a reasonably high probability that this was due to the use of the single fission generation per superhistory. Just four fission generations per superhistory improved greatly the stability of the fundamental mode, whilst 12 and 25 fission generations per superhistory improved it further. Probably 10 fission generations per superhistory would be sufficient (which is the same number recommended in [9] for other reasons).

Although only ex-core responses were considered here, the question of stability of the fundamental mode is of equal, if not greater, importance for in-core responses. Thus when employing variance reduction with the DSA within source-iteration eigenvalue calculations, superhistories with 10 or more fission generations should always be used for in- and for ex-core responses. With this *caveat*, a single calculation for ex-core responses should be more accurate and faster (in terms of human time) than a decoupled calculation, as well as involving a higher level of quality assurance.

Acknowledgements

Thanks to Mariya Brovchenko (IRSN, Fontenay-aux-Roses, France) for supplying a simplified generic PWR GEN-III model for calculation comparison and for defining the ex-core responses of interest.

Mario Carta and Georgios Glinatsis (ENEA, Casaccia, Italy) conceived the actinide measurement proposal and Valentina Fabrizio (ENEA, Casaccia, Italy) supplied the TAPIRO geometry model.

The computing resources and the related technical support used for this work have been provided by CRESCO/ENEAGRID High Performance Computing infrastructure and its staff*. CRESCO/ENEAGRID High Performance Computing infrastructure is funded by ENEA, the Italian National Agency for New Technologies, Energy and Sustainable Economic Development and by Italian and European research programmes, see http://www.cresco.enea.it/english for information.

*G. Ponti *et al.*, *Proc. of the 2014 International Conference on High Performance Computing and Simulation, HPCS 2014*, 1030

Dedication

This paper is dedicated to the memory of Herbert Rief (1930–2015), an excellent Monte Carlo developer and practitioner and a kind and gentle man.

References

1. K.W. Burn, Ann. Nucl. Energy, **73**, 218 (2014)
2. K.W. Burn, Ann. Nucl. Energy, **85**, 776 (2015)
3. K.W. Burn, *PHYSOR 2014*, JAERI-Conf 2014-003 (2014)
4. K.W. Burn, Nucl. Technol. **175-1** 138-145 (2011)
5. T.E. Booth, J.H. Hendricks, Nucl. technol./fus., **5** 90 (1984)
6. J.C. Wagner, E.D. Blakeman, D.E. Peplow, *Trans. Am. Nucl. Soc.* **97**, 630 (2007)
7. C.J. Solomon, A. Sood, T.E. Booth, *Proc. Int. Conf. Advances in Mathematics, Computational Methods, and Reactor Physics*, Saratoga Springs, NY, American Nuclear Society (2009)
8. T.E. Booth, K.W. Burn, Ann. Nucl. Energy **20(11)**, 733 (1993)
9. R.J. Brissenden, A.R. Garlick, Ann. Nucl. Energy **13(2)**, 63 (1986)
10. M. Carta *et al.*, ENEA internal report ADPFISS-LP2-083, rev.1 (2016)
11. J.T. Goorley *et al.*, LA-UR-13-22934 (2013)
12. MCNP Team, LA-UR-05-8617 (2005)
13. K.W. Burn, Nucl. Sci. Eng. **119**, 44 (1995)

Neutron-gamma flux and dose calculations in a Pressurized Water Reactor (PWR)

Mariya Brovchenko[1] [a], *Benjamin* Dechenaux[1], *Kenneth W.* Burn[2], *Patrizio* Console Camprini[2], *Isabelle* Duhamel[1], and *Arthur* Peron[3]

[1]Institut de Radioprotection et de Sûreté Nucléaire (IRSN), Fontenay-aux-Roses, 92260, France
[2]ENEA, Via M.M.Sole, 4, 40129 Bologna, Italy
[3]IRSN subcontractor

Abstract. The present work deals with Monte Carlo simulations, aiming to determine the neutron and gamma responses outside the vessel and in the basemat of a Pressurized Water Reactor (PWR). The model is based on the Tihange-I Belgian nuclear reactor. With a large set of information and measurements available, this reactor has the advantage to be easily modelled and allows validation based on the experimental measurements. Power distribution calculations were therefore performed with the MCNP code at IRSN and compared to the available in-core measurements. Results showed a good agreement between calculated and measured values over the whole core. In this paper, the methods and hypotheses used for the particle transport simulation from the fission distribution in the core to the detectors outside the vessel of the reactor are also summarized. The results of the simulations are presented including the neutron and gamma doses and flux energy spectra. MCNP6 computational results comparing JEFF3.1 and ENDF-B/VII.1 nuclear data evaluations and sensitivity of the results to some model parameters are presented.

1 Introduction

The Institute of Radiological Protection and Nuclear Safety (IRSN) acts as technical support to the French nuclear safety authorities. In this context, one of its missions is to improve nuclear safety through the enhancement of knowledge regarding operating nuclear power reactors. The issues concerning the aging and the characterisation of the reactor vessel require knowledge of the neutron and gamma radiation environment on the vessel and beyond.

Since few years, ENEA (Energia Nucleare e Energie Alternative, i.e. Nuclear Energy and Alternative Energy agency) and IRSN collaborate on Monte-Carlo simulations and Variance Reduction Techniques (VRT), which are used to determine neutron and gamma spectra in and out of the core.

Additionally, IRSN participates in the DISCOMS project [1], which stands for "DIstributed Sensing for COrium Monitoring and Safety", aiming at examining the status of the third barrier of confinement, in order to define possible mitigation strategies in case of a severe accident, namely: i) reactor pressure vessel breakthrough and corium relocation outside the vessel, ii) reactor pit concrete erosion and iii) corium cooling. One of the IRSN tasks in the frame of the DISCOMS project is to provide the radiation level in terms of neutron and gamma doses and flux energy spectra at various locations.

The work presented in this paper summarizes the studies performed on the radiation transport in the Tihange-I Belgian nuclear reactor and contributes to several issues requiring the knowledge of the radiation environment in and outside the reactor vessel.

2 Neutron flux distribution in the core

Tihange-I is a Pressurized Water Reactor (PWR), producing a nominal electric power of 900 MW. Following the first divergence of the core in 1975, start-up physics tests were performed. In this work, the authors used the data from the Hot Zero Power (HZP) tests, namely the flux distribution measurements that are freely available [2], and thus, were able to validate the calculations of the neutron fission distribution in the core. It is a tremendous advantage to have the possibility to validate the fission distribution in the core, since it is then used as the source term in the fixed source transport calculation to determine the neutron and gamma spectra outside the core.

2.1 Core model

The studied Tihange-I core is freshly fuelled with 3 different ^{235}U enrichments 1.95%, 2.55% and 3.10% (no burnup) and at HZP state (no thermal hydraulics; fixed

[a] Corresponding author: mariya.brovchenko@irsn.fr

conditions can be used). For this configuration, two separate measurement campaigns were executed, one by FRAMATOME[a] and another one by EDF, that resulted in two different fission distribution maps over a 1/8th slice of the core [2]. The moderator temperature and the boron concentration were measured, and these values were adopted for the modelling. The studied configuration corresponds to the case where no control rods are inserted into the core. Unfortunately, no information was available to the authors on the methods used to obtain the flux distribution maps, neither on the uncertainty of the values.

Power distribution calculations were made with the MCNP5 Monte Carlo code [3] that allows a detailed 3D geometry description of the core, as shown in figure 1.

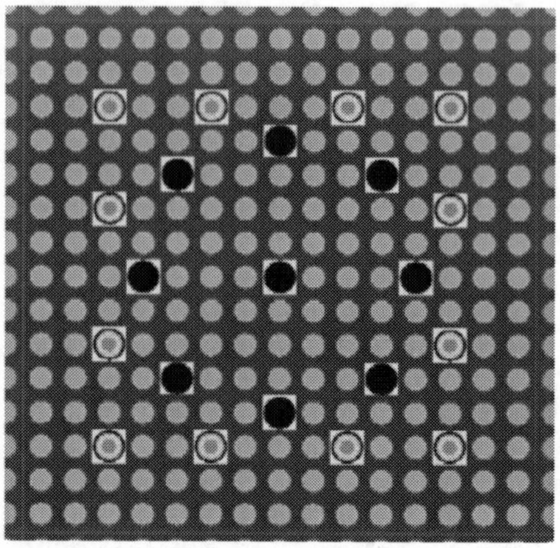

Fig. 1. Pin by pin modelling in the MCNP model, with 12 poisoned fuel rods and 9 guide tubes

2.2 Neutron flux distribution validation

An accurate neutron fission source distribution is a key ingredient to perform irradiation calculations outside the reactor primary vessel. The comparison of the flux distribution in the core allows core simulations on a realistic case to be validated. The measured and calculated flux distribution assembly by assembly is given in figure 2, corresponding to a 1/8th slice of the core. The statistical errors associated with these data are very low < 1%.

The relative errors between the calculated and the measured flux distribution are given in figure 3. Almost all discrepancies are lower than 2%. The major discrepancies are observed in the assemblies on the edge of the core. It should be noticed that the differences between the two sets of measurements and between measurements and calculations are of the same magnitude. Since no measurement uncertainty is provided, the calculated values can be considered in agreement with experimental ones.

[a] Areva Nuclear Power

	A	B	C	D	E
1	1.086				
	1.069				
	1.085				
2	1.087	1.081			
	1.089	1.069			
	1.092	**1.078**			
3	1.090	1.066	1.074		
	1.068	1.089	1.071		
	1.073	**1.088**	**1.081**		
4	1.066	1.078	1.076	1.099	
	1.085	1.066	1.094	1.088	
	1.082	**1.069**	**1.090**	**1.082**	
5	1.074	1.050	1.066	1.075	0.901
	1.059	1.078	1.059	1.089	0.901
	1.059	**1.068**	**1.056**	**1.064**	**0.900**
6	1.049	1.065	0.986	0.940	0.709
	1.073	1.045	1.004	0.954	0.706
	1.056	**1.036**	**0.985**	**0.936**	**0.711**
7	1.033	1.037	0.968	0.648	
	1.015	1.058	0.972	0.643	
	1.015	**1.052**	**0.977**	**0.652**	
8	0.921	0.675			
	0.885	0.671			
	0.912	**0.699**			

Framatome
EDF
MCNP

Fig. 2. Flux distribution in each assembly for a 1/8th slice of the core, experimental measurements and simulation results with MCNP5 and nuclear data library ENDF-B/VII.1. The assembly A1 corresponding to the central assembly of the core.

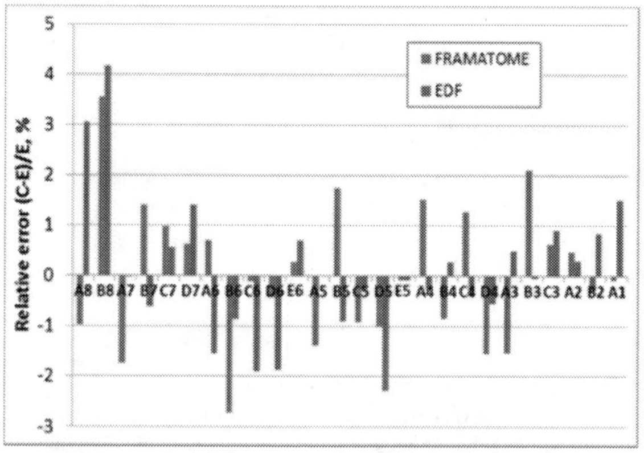

Fig. 3. Relative error (C-E)/E of the calculated flux distribution C compared to the experimental measurements E

3 Fixed source calculations

3.1 Model

To evaluate the neutron and gamma fluxes outside the vessel and in the basemat concrete, a detailed description of the internals in the vessel and the reactor pit were added to the model. The geometry outside the core was modeled as accurately as possible. As shown in figure 4, structures piercing the lower part of the vessel were modeled to take into account the instrumentation tubes.

Hypotheses maximizing the radiation transport through the lower part of the vessel were adopted for this model. The radial positions of the piercing tubes were based on the information from [2].

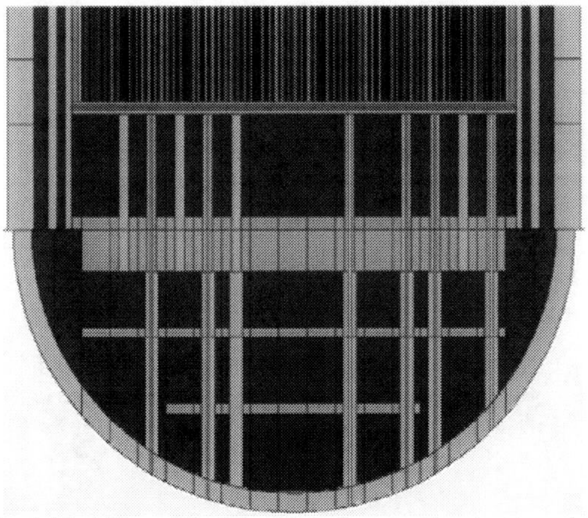

Fig. 4. Modelling of the internals and supporting structures in the bottom of the vessel

The responses of interest are defined on surfaces/volumes with symmetry around the Z-axis. The detectors were placed outside the vessel at the level of the active core. The responses were calculated with an axial segmentation, as detailed in figure 5. The radiation was also studied under the vessel bottom on the basemat surface and at different depths inside the basemat. As shown in figure 6, the reactor pit is divided into 3 rings.

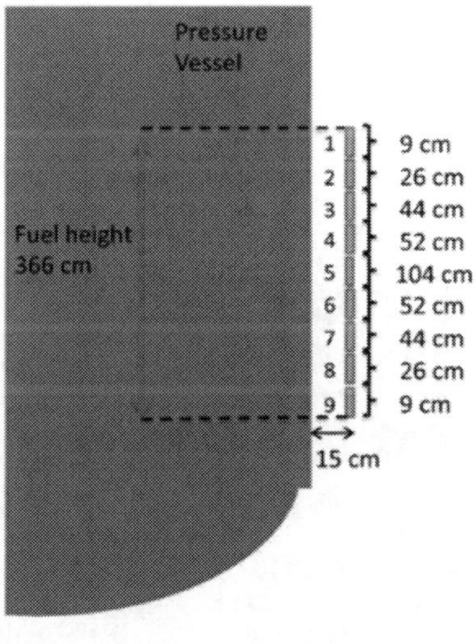

Fig. 5. Positions of the responses on the side of the vessel with the adopted axial segmentation

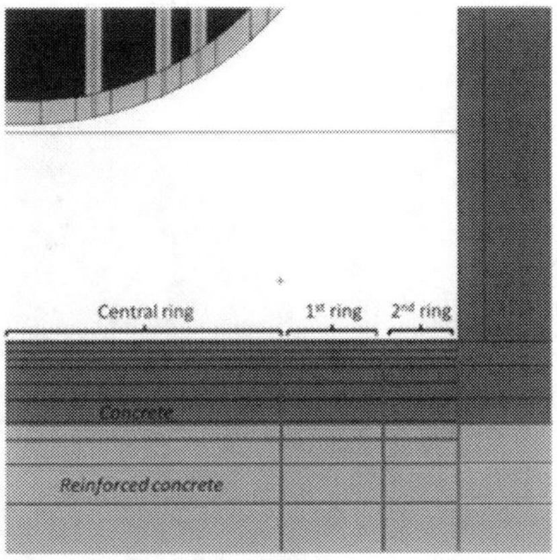

Fig. 6. Positions of the responses on the bottom of the basemat with a radial segmentation in 3 rings

3.2 Source term

The only source term considered in these calculations is the fission neutrons. The activation of structures outside the core is not taken into account in this study although it is, a priori, not negligible.

Previous studies showed that a flat radial distribution within an assembly leads to an overestimation of the responses outside the vessel [4]. For this reason, in this study the fission source distribution was defined in each pin.

The source term was calculated in two different ways: one by IRSN, the other by ENEA. The calculations were performed using MCNP6 [5]. In both sources, the radial fission distribution within a pin was considered as flat.

The fission source calculated by IRSN was based on 300 independent fundamental mode calculations with 5.10^7 simulated histories each, after 400 inactive cycles and using 100 000 particles per cycle. These calculations allowed a maximal standard deviation of the mean of 0.5% to be obtained in each pin. The axial distribution was taken as identical for all the pins in the core and 200 segments were used. This hypothesis was verified by considering the axial distribution in ten different pins. The nuclear data library used for the neutron source calculation was ENDF-B/VII.1.

In figure 7 is presented the fission rate distribution of the core evaluated at IRSN. As expected, the fission probability decreases rapidly in the assemblies on the edge of the core. For this core configuration, the assemblies with a higher enrichment are placed on the edge of the core. This leads to the highest fission rate at the interface between these assemblies and the assemblies closer to the center of the core.

DOI: 10.1051/epjconf/201715305008

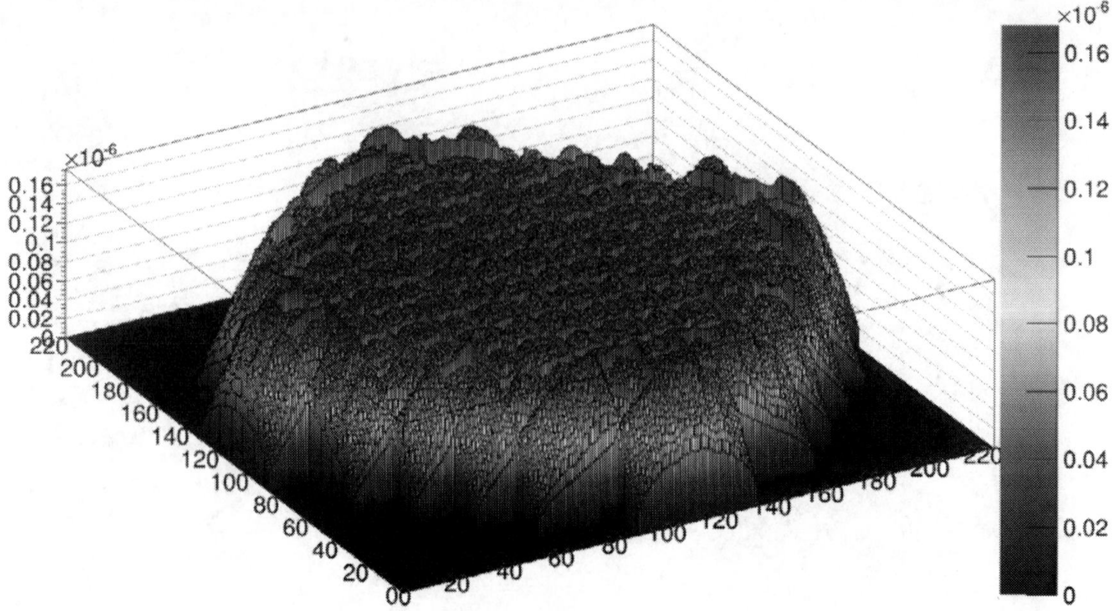

Fig. 7. Fission rate distribution in the core in arbitrary units evaluated by IRSN, each segment corresponding to a fuel pin

The fission source calculated by ENEA employed the nuclear data library JEFF-3.1 and was based on 512000 fission neutrons per cycle and 3600 cycles after the fundamental mode had been reached. An axial segmentation of 21 bins was employed. The lengths of the bins were not equal: the length of the first bin at the bottom of the fissile zone was 1 cm, then the length was doubled for each interval up to 16 cm, then 12 intervals each of length 24 cm, then the length was gradually reduced to finish at 8 cm for the top interval. The same axial segmentation was employed for all the bins. A separate axial variation of the neutron source for each pin was written. The resulting standard deviations on the source in each pin segment as given by MCNP (i.e. between source fission neutron histories and ignoring correlations) was typically 1% or less for the 24 cm central segments and 10-15% for the lowest 1 cm segment, apart from the external corner assemblies for which for the outer pins the deviations were 2% for the central segments and as much as 30% for the lowest segment.

It is not possible to define such a pin-by-pin source with a different axial variation for each pin in standard MCNP. Therefore the code had to be modified to accept the ENEA and the IRSN source.

The radiation source in this problem being the fissions occurring in the core of the reactor, the flux of neutral particles is strongly attenuated, especially in the concrete of the basemat. Therefore, such neutron-gamma shielding calculations require the use of variance reduction techniques.

3.3 Variance reduction methods

In such problems, the positions of the responses (beside the vessel, at the vessel bottom, within the basemat) and the physical quantities (gamma dose, fast and thermal neutrons) are numerous. Optimizing the set of importance maps (neutrons and gamma rays for different energy groups) to all of the responses is a challenge.

For this reason, the variance reduction (VR) parameters were calculated at ENEA using a detailed mathematical approach called the DSA (Direct Statistical Approach) that optimizes the VR parameters that govern the sampling [6]. The approach has been extended to include among other options, weight control and a multi-response capability [7].

The VR parameters calculated with the DSA can then be translated into classical Weight Window maps implemented in the MCNP code. The fixed source calculations were performed at IRSN using the Weight Window maps based on the VR parameters calculated with the DSA at ENEA.

The Weight Window parameters were calculated for 7 neutron energy groups and for 6 gamma energy groups. The parameters for neutrons with energy of 2-5 MeV are shown in figure 8. The white zones, imposed by the user, represent cells where no splitting or Russian roulette occurs. This is either because statistics were poor in the runs generating the VR parameters (for example because the region experienced both low fluxes and was unimportant to any of the responses), the region was too extensive to impose a single VR parameter, or there was a high probability of passing through the region without collision (for example in the case of air).

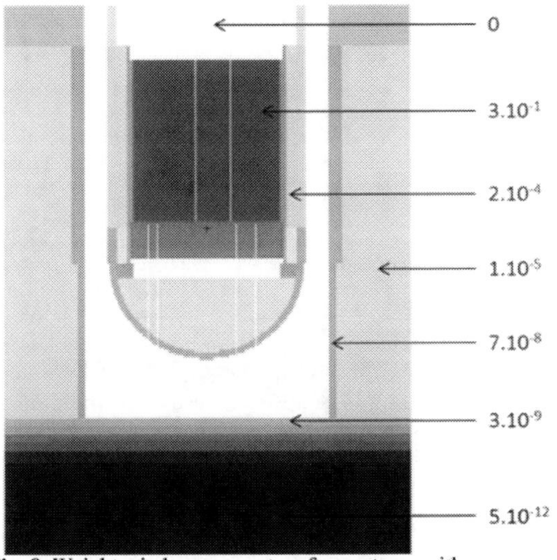

Fig. 8. Weight window parameters for neutrons with energy 2-5 MeV based on the DSA calculations

The Weight Window parameters for gamma rays with energy of 0.7-1.5 MeV are shown in figure 9. The white region in the core demonstrates that the gamma rays in this energy group created in the core by capture or fission do not contribute to any response outside the vessel (i.e. the importance is too low).

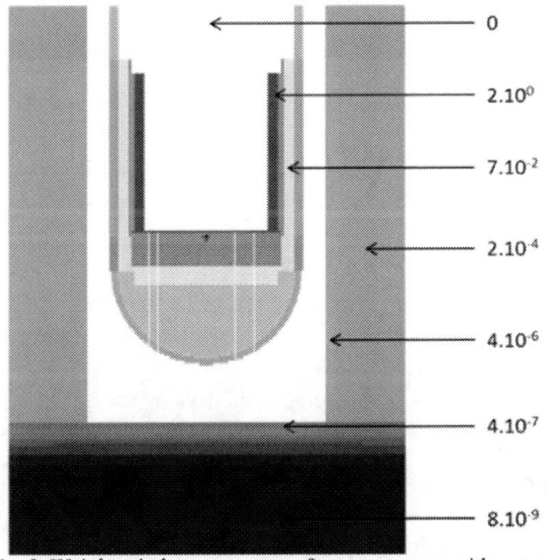

Fig. 9. Weight window parameters for gamma rays with energy 0.7-1.5 MeV based on the DSA calculations

4 Results

First, the results of the calculations and comparisons between the fixed source calculations performed at IRSN using ENDF-B/VII.1 library, and, at ENEA with JEFF-3.1 one, are presented in the following. Additional sensitivity calculations were performed at IRSN to see the impact of different physical parameters on the results and are detailed below.

4.1 Comparison of the two simulations

The axial distribution of the neutron flux on the side of the vessel (see figure 5) is presented in figure 10 for the simulations performed at IRSN and at ENEA. The neutron flux is divided into 3 groups: E_n < 0.1 MeV, 0.1 – 1 MeV, 1 – 20 MeV. Since irradiation effects on the vessel material is mainly due to rather fast neutrons, the neutrons with E_n > 0.1 MeV and E_n > 1 MeV were separated. The standard deviation of the calculations is presented in some figures, and is in most cases not visible since very low (< 1%).

The neutron flux behaviour in the figure 10 for the two simulations is very similar. Only small differences can be observed for the lower energy neutrons E_n < 0.1 MeV, that may be due either to the different nuclear data used or to the different source term definitions used in the simulations.

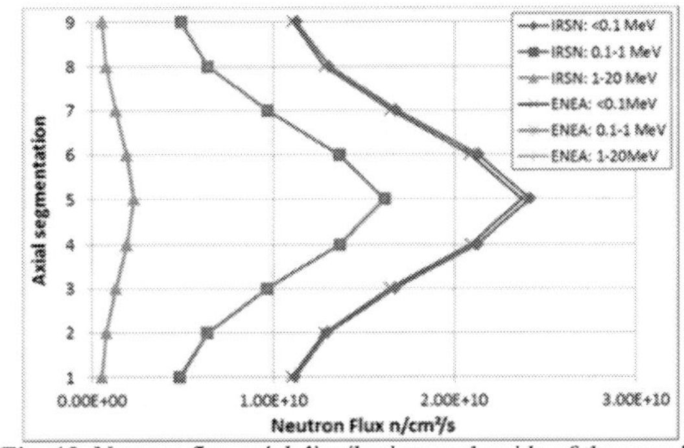

Fig. 10. Neutron flux axial distribution on the side of the vessel for 3 energy groups calculated by IRSN and ENEA

The energy distribution of the neutron flux is presented in figure 11. The neutron spectra for the central and the lower axial parts are very similar. On this logarithmic scale plot, the slight difference in the low energy neutrons observed previously cannot be seen anymore due to the scale of the figure. The slight differences between the two simulations for the neutrons with E_n around 10^{-3} eV at the lower segment between the two simulations is due to the low statistics, and is not of high interest here.

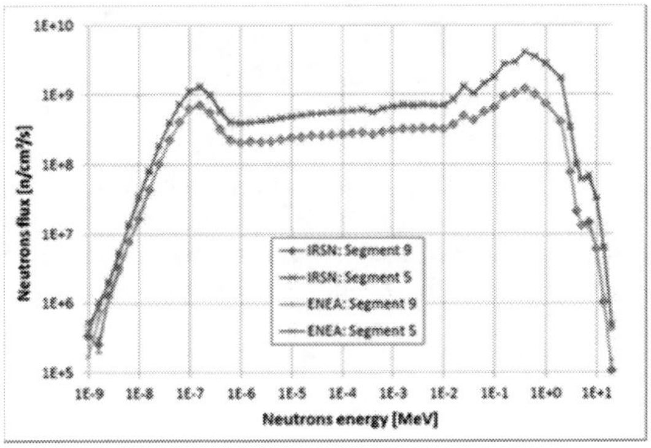

Fig. 11. Neutron spectra for the central part (segment 5) and the lower axial part (segment 9) on the side of the vessel calculated by IRSN and ENEA

Concerning the gamma rays, the energy spectra are presented in figure 12. As for neutrons, the total gamma ray flux in axial segment 5 is higher by a factor of 2 than the axial segment 9, even if the energy spectrum is in both cases very similar. Between the two simulations of IRSN and ENEA no difference can be observed, except for very low energy gamma rays E_g < 30 keV and very high energy rays E_g > 12 MeV, due to low sampling in these regions.

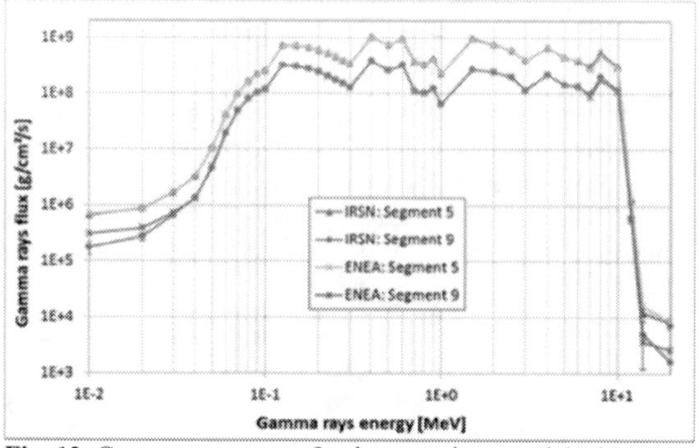

Fig. 12. Gamma ray spectra for the central part and the lower axial part on the side of the vessel calculated by IRSN and ENEA

Figure 13 illustrates the neutron spectra on the basemat for the three rings detailed in figure 6. The fast component of the neutron spectrum is much lower compared to the spectrum on the side of the vessel. The central ring in the basemat, directly under the vessel, has a slightly lower flux, especially for the faster component compared to the external rings. A good agreement can be observed between the two simulations.

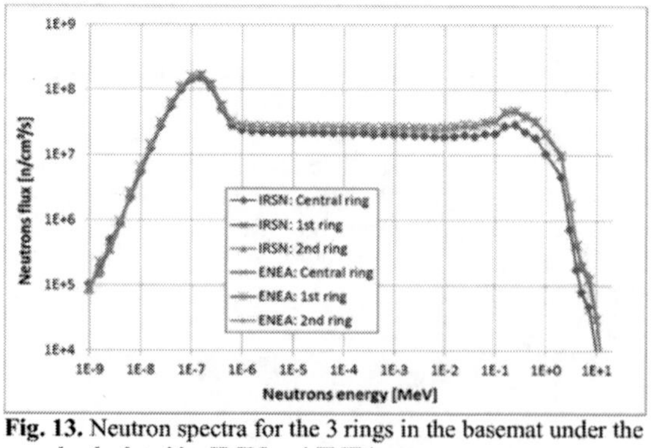

Fig. 13. Neutron spectra for the 3 rings in the basemat under the vessel calculated by IRSN and ENEA

Concerning the gamma ray flux on the basemat, the results are presented in figure 14. No significant difference can be observed between the different rings, nor between the two simulations, except for E_g > 12 MeV.

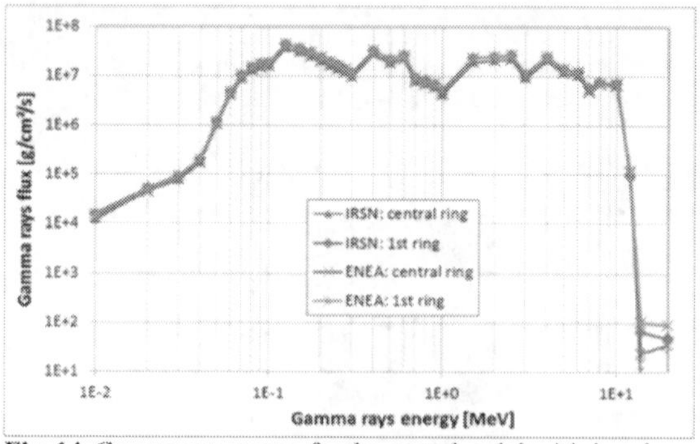

Fig. 14. Gamma ray spectra for the central and the 1st rings in the basemat under the vessel calculated by IRSN and ENEA

To give an idea of the neutron fluence in the basemat, considering the strong hypothesis that the presented reactor modelling is representative of the whole lifetime of the reactor (assumed to be 60 years), figure 15 shows the neutron fluence attenuation in the basemat for the 3 energy groups described previously based on calculations performed at IRSN. It is to be noticed that only the fastest component has an exponential attenuation in the basemat, since the two other groups have in addition to the attenuation also a "production" contribution due to the moderation of the neutrons in the concrete. The reinforcement in the concrete has no significant impact on the attenuation of the fast neutrons.

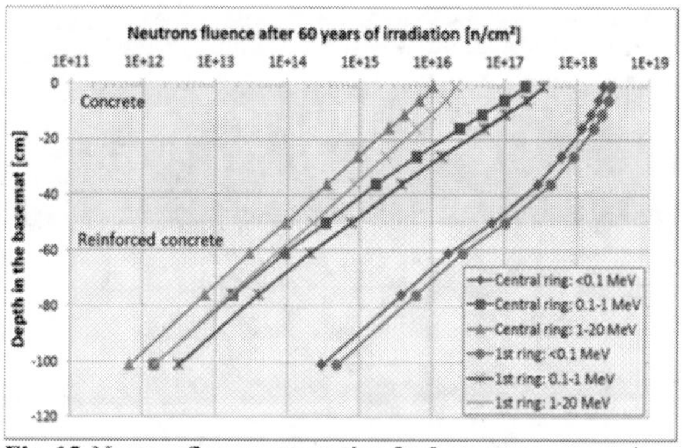

Fig. 15. Neutron fluence attenuation for 3 energy groups and different rings in the basemat

In a similar fashion to the neutron fluence, the gamma kerma and its attenuation in the basemat were calculated and the results calculated at IRSN are presented in figure 16. For the purpose of corium monitoring, studied within the DISCOMS project, optical fibers may be placed at different depths in the concrete. The kerma was calculated for an energy deposition in silica. The attenuation of the gamma kerma is not exponential since the neutron flux contributes to a gamma production in the concrete.

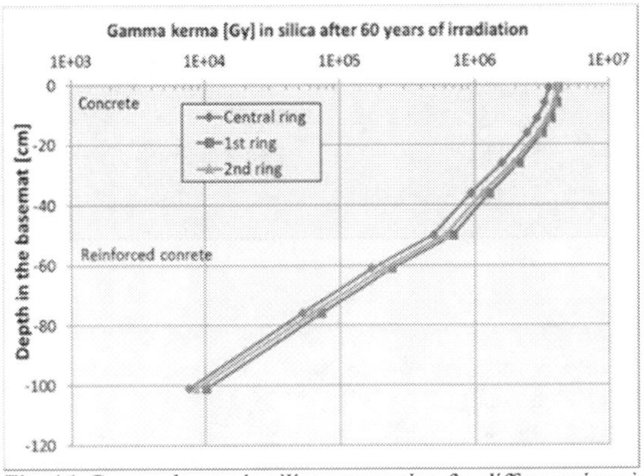

Fig. 16. Gamma kerma in silica attenuation for different rings in the basemat

4.2 Parameter sensitivity study

Varying different physical parameters of the nuclear reactor model gives an idea of the importance of these parameters on the calculated results. Instead of varying the geometrical dimensions, the densities/concentrations of different materials were modified. The variation of the parameter is not to be considered as real value since it was chosen to be high enough so as to be able to detect an impact on the calculated results.

To observe the impact of the water layer thickness inside the vessel (radial part) and in the bottom of the vessel, the water density was decreased by a factor of two. To observe the impact of the total steel thickness inside the vessel, composed of the thermal protection and the baffle, its density was decreased by a factor of two. Finally, the boron concentration was changed from 650 ppm to 0 ppm, to check if it has an impact on the radiation transport out of the vessel.

In figure 17 the impact of each parameter variation separately on the neutron flux spectrum on the side of the vessel in the central axial part is presented. As expected, almost no impact can be observed for the water density variation in the bottom of the vessel. Boron concentration variation has similarly no significant impact on the neutron flux except on the limits of the spectrum (low and high energy neutrons), which might be due to lower sampling. Steel thickness has an impact on the neutron flux, quietly proportional to the density variation, with a higher impact on the faster neutrons $E_n > 1$ MeV. The density of the water located in the vessel has an even higher impact. It can reach a factor of 4 to 5 on the neutron flux, except the fast neutrons with $E_n > 1$ MeV, where the impact is only of a factor 2 to 3.

Similar tendencies were observed for the gamma ray spectrum at the same location (see figure 18). Boron concentration modification has small impact on the overall gamma flux. The water density in the bottom of the vessel has no significant impact on the gamma flux. Decreasing steel density increases the gamma flux on the side of the vessel by a factor of 2 to 3, as for the neutrons, while the gamma flux increases by a factor of 3, with the water density reduction in the vessel.

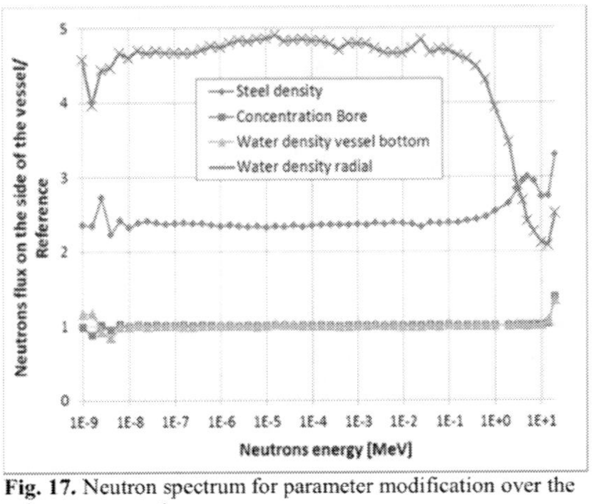

Fig. 17. Neutron spectrum for parameter modification over the reference case values

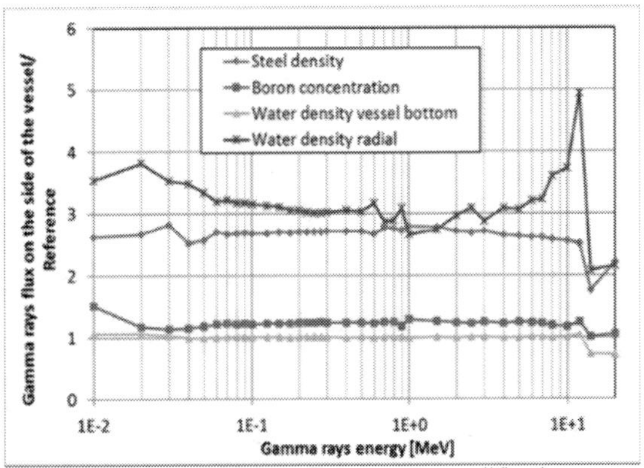

Fig. 18. Gamma ray spectrum for parameter modification over the reference case values

For these sensitivity calculations, other responses were also analysed, but are not presented here. It is nevertheless interesting to mention that for the neutrons and gamma rays on the basemat, the water density in the bottom of the vessel has no significant impact.

Concerning the variance reduction methods employed here, it is important to mention that the particle streaming through the piercing tubes (see figure 4) was not sampled in the runs that generated the VR parameters. This effect may not be negligible and will be examined separately.

5 Conclusion

In the framework of different R&D activities, IRSN aims to study the radiation environment in terms of neutron and gamma fluxes outside the vessel during the reactor operation.

To validate the neutron source distribution in the core, simulation results were compared to the available in-core measurements for the Tihange-I Belgian nuclear reactor.

The results of the fixed source ex-core calculations performed at IRSN and at ENEA using different nuclear data and different neutron-source spatial definition have shown a very good agreement. The variance reduction parameters optimized for the set of different neutron and gamma responses outside the vessel calculated by ENEA were used by both institutes. Then, the gamma ray and neutron spectra were evaluated outside the vessel. The neutron fluence and gamma kerma attenuation in the basemat concrete results were presented. From the parameter variation study, it was shown that the water thickness and the total steel thickness have a high impact on the neutron and gamma ray flux outside the vessel.

Acknowledgment

The IRSN work presented in this paper is co-funded by the French *"Programme Investissements d'Avenir"* referenced by n° ANR-11-RSNR-0007 and managed by ANR (National Research Agency).

References

1. P. Ferdinand et al., "DISCOMS: DIstributed Sensing for COrium Monitoring and Safety", *CANSMART 2015: International Conference on Smart Materials and Structures* - Vancouver, BC, Canada (2015)
2. H. Panek "Qualification du système NEPTUNE – Ébauche d'une nouvelle chaîne de calcul basée sur les codes APOLLO et Tortise", Note CEA-N-2092, (1979)
3. X-5 Monte Carlo Team, "MCNP - A general Monte Carlo n-particle transport code, version 5", Los Alamos National Laboratory, (2005)
4. M. Brovchenko et al., "Neutron-gamma flux and dose calculations for feasibility study of instrumentation implementation in the European Pressurized Reactor (EPR)", *submitted to ICRS-13 & RPSD-2016*
5. T. Goorley et al., "Initial MCNP6 Release Overview – MCNP6 version 1.0", LA-UR-13-22934, (2013)
6. A. Dubi et al., "General Statistical Model for Geometrical Splitting in Monte Carlo – parts I and II", *Transp. Theory Stat. Phys.* **14**(2), 167-193 and 195-221 (1985)
7. K.W. Burn, "Optimizing Monte Carlo to multiple responses: the Direct Statistical Approach, 10 years on", *Nuclear Technology* **175**, 138–145 (2011)

DOI: 10.1051/epjconf/201715305009

Characterization of Radiation Fields for Assessing Concrete Degradation in Biological Shields of NPPs

Igor Remec[1,*], Thomas M. Rosseel[1], Kevin G. Field[1] and Yann Le Pape[1]

[1]Oak Ridge National Laboratory, P.O. Box 2008, Oak Ridge, TN 37831-6170, USA

Abstract. Life extensions of nuclear power plants (NPPs) to 60 years of operation and the possibility of subsequent license renewal to 80 years have renewed interest in long-term material degradation in NPPs. Large irreplaceable sections of most nuclear generating stations are constructed from concrete, including safety-related structures such as biological shields and containment buildings; therefore, concrete degradation is being considered with particular focus on radiation-induced effects. Based on the projected neutron fluence values (E > 0.1 MeV) in the concrete biological shields of the US pressurized water reactor fleet and the currently available data on radiation effects on concrete, some decrease in mechanical properties of concrete cannot be ruled out during extended operation beyond 60 years. An expansion of the irradiated concrete database is desirable to ensure reliable risk assessment for extended operation of nuclear power plants.

1 Introduction

Numerous nuclear power plants (NPPs) in the USA have received approval to extend their licenses to operate up to 60 years and further extensions, to 80 years and beyond, are being considered to meet future national energy needs while reducing greenhouse gas emissions. NPPs lifetime extensions have renewed interest in long-term aging of materials in general and recently in concrete. Large structures of NPPs are built from concrete, including safety-related structures such as the biological shields and containment buildings. Concrete is a complex material and its properties change over time due to slow hydration, crystallization of amorphous components, reactions between cement paste and aggregates, and the effects of the local environment such as humidity, temperature, and for NPP concrete, radiation exposure.

Most of the legacy data on mechanical performance of irradiated concrete were analyzed by Hilsdorf et al. [1], (1978) who concluded that, for some concretes, neutron fluence of 1.0×10^{19} n/cm^2 may cause a reduction in compressive and tensile strength and a marked increase in volume. Although it was implied that the effects depend on the neutron fluence being "fast" or "thermal," the report did not quantify the differences and did not address the effect of the gamma rays. In addition, a recent review of Hillsdorf's data by Kontani et al.[2] (2013) revealed that most data points at high end of neutron fluence range were obtained at irradiation temperatures above 100°C and were using either aluminous cement or liquid glass aggregate, which is not representative of the environment and materials used in concrete for NPPs. Therefore it is not surprising that the recently completed Expanded Materials Degradation Analysis, jointly performed by the Department of Energy, the Nuclear Regulatory Commission and Nuclear Industry, identified the urgent need to develop a consistent knowledge base on irradiation effects in concrete [3]. A research program on concrete aging and degradation processes in NPPs was established within the Light Water Reactor Sustainability (LWRS) program under the US Department of Energy (DOE) and is being performed in partnership with the Electric Power Research Institute's Long-Term Operations (LTO) Program. This paper outlines recent progress in the irradiated concrete research program with a focus on the assessment of the radiation environment in concrete biological shields in NPPs.

2 Radiation Environments in Concrete Biological Shields

The radiation-induced concrete degradation has been historically correlated to neutron fluence; therefore, it is necessary to estimate the bounding fluence values for the US NPP fleet at the projected extended service. Due to the substantial differences in reactor designs and variations in plant operations, this effort is not trivial. There is no routine monitoring of radiation fields in biological shields of NPPs. However, every operating US NPP is required to conduct a reactor pressure vessel (RPV) surveillance program and a wealth of RPV fluence information is publicly available from the US NRC ADAMS database [4]. The RPV flux/fluence data need to be extrapolated in concrete biological shield (CBS). For this purpose coupled neutron and gamma ray transport

* Corresponding author: remeci@ornl.gov

calculations were performed for a three-loop and a two-loop pressurized water reactor (PWR), utilizing the slightly modified computer models previously generated for the RPV fluence investigations [5]. An example of the attenuation of the neutron and gamma flux through the pressure vessel and CBS is shown in Figure 1.

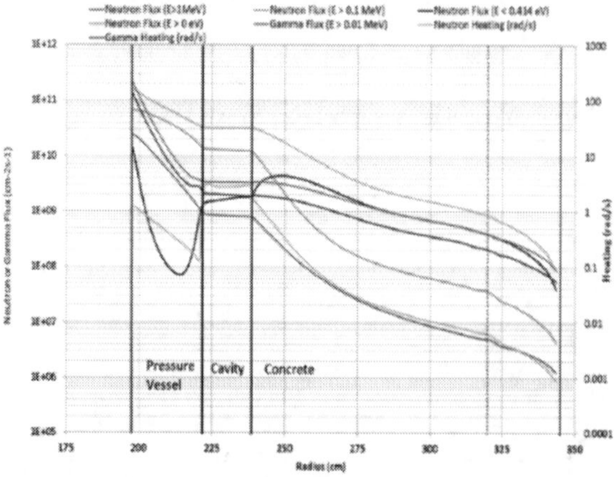

Figure 1. Radial variation of neutron flux at four energy ranges, gamma-ray flux, and heating rates for a three-loop pressurized water reactor.

With the data retrieved from ADAMS, and using the guidance obtained from transport calculations, Esselman and co-workers [6] estimated the neutron fluence values in the CBS, at 80 years of operation, for the current US NPP fleet. The results for the PWRs, which have higher expected neutron fluence values than the boiling water reactors, are shown in Fig. 2. The importance of the plant design is clearly observed: the four loop PWR reactors show significantly lower fluence values than the majority of both 2 and 3 loop plants.

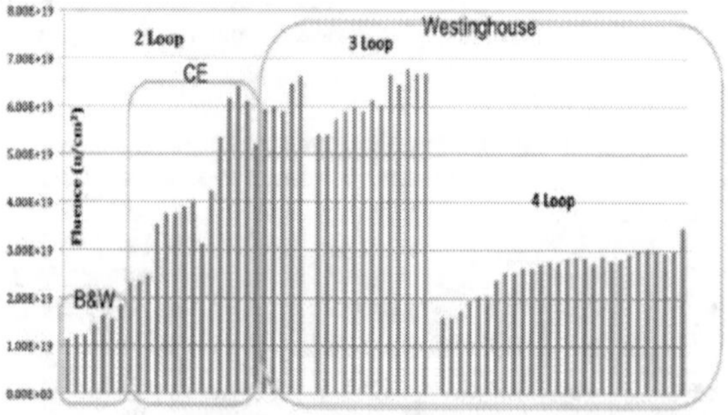

Figure 2. Summary of US PWR fleet neutron fluence (E > 0.1 MeV) at the biological shield extrapolated to 80 years and grouped by design and manufacturer [6].

3 Current irradiated concrete database

While a comprehensive review of the currently available results of irradiated concrete properties is presented in [7], a brief review of the parameter(s) used to characterize the field in which the concrete and its component samples were irradiated will be presented here.

Fig. 3 depicts the ratio of the compressive strength of irradiated concrete to compressive strength of unirradiated concrete versus neutron fluence. The data were taken from [7] and re-processed so that the symbols indicate the energy threshold of the neutron fluence depicted on the abscissa rather than the aggregate type. One notable feature in Fig. 3 is the large variation in residual compressive strength at any given neutron fluence value. The scattering of data, which are assembled from a large number of independent experiments, is caused by variations in numerous factors, such as different types of cement and aggregate, different cement-to-water ratios, different specimen sizes, variation in concrete curing and thermal treatment, and different temperatures of specimens during irradiation.

Of particular interest here is that the reported "neutron fluence" varies greatly between experiments. Some experiments gave only qualitative descriptions such as "slow" or "fast" neutron fluence, some used fluence of neutrons at energies above 0.1 MeV, and some above 1.0 MeV, some reported total (E > 0 MeV) fluence, and some experiments did not address the energy of neutrons at all. Moreover, the neutron spectrum was actually provided only for a few experiments. Due to the potentially large variations in neutron spectrum between experiments and even between different locations in the same irradiation, it is not possible to "re-normalize" the database to a common denominator, such as fluence of neutrons with energies E > 0.1 MeV. (This will be referred to as fluence with cutoff of 0.1 MeV.)

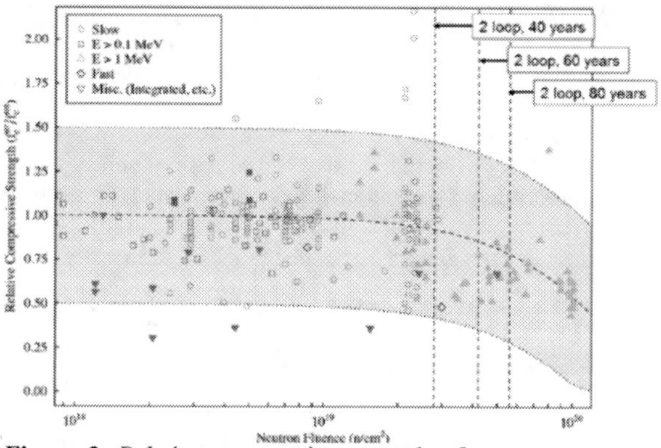

Figure 3. Relative compressive strength of concrete versus neutron fluence. The shapes of the symbols indicate neutron fluence energy cutoff. Solid symbols indicate specimen temperature during irradiation was above 100°C; hollow symbols correspond to temperature below 100°C. The fluence (E > 0.1 MeV) values for the two-loop PWR (from Table 1), are shown as dashed vertical lines.

Despite not being able to normalize the data to a unique cutoff energy, the neutron fluence cutoff value is crucial for the assessment of concrete degradation in nuclear power plants (NPPs). The values listed in Table 1, obtained in a transport analysis of a two-loop and three-loop PWR [5,8], indicate that both plants remain below the 1.0×10^{19} n/cm^2 fluence value for the 1 MeV cutoff, and therefore no radiation-induced degradation would be expected even for the 80 years of operation. However, for the 0.1 MeV cutoff, both plants exceed the

1.0 ×10^{19} n/cm^2 fluence level during 40 years of operation, and some concrete degradation may be expected. This is even more pronounced if the assessment is based on the integral fluence (E > 0 MeV).

Another observation related to Fig. 3 is that the data points at the high fluence end are dominated by the data reported for the fluence energy cutoff of 1 MeV. If the data are reported for the cutoff energy of 0.1 MeV, the data would move to significantly higher (approximately an order of magnitude) fluences, resulting in a reduced prediction of radiation degradation in the 1–10 × 10^{19} n/cm^2 fluence range.

Table 1. Predicted neutron fluence for the two PWRs analysed at 40, 60, and 80 years of operation assuming a 92% capacity factor.

Years Operation	PWR Type	Neutron Fluence (n/cm^2) E > 0.0 MeV	E>0.1 MeV	E > 1.0 MeV
40	2-loop	7.3×10^{19}	2.8×10^{19}	3.2×10^{18}
	3-loop	3.6×10^{19}	1.3×10^{19}	8.7×10^{17}
60	2-loop	1.1×10^{20}	4.2×10^{19}	4.9×10^{18}
	3-loop	5.4×10^{19}	2.0×10^{19}	1.3×10^{18}
80	2-loop	1.5×10^{20}	5.6×10^{19}	6.5×10^{18}
	3-loop	7.2×10^{19}	2.6×10^{19}	1.7×10^{18}

4 Atom displacements in aggregates

The radiation-induced swelling of aggregates, which leads to volumetric expansion of the concrete, is currently considered the dominant cause for loss of mechanical properties of irradiated concrete [7]. For this reason we investigated neutron induced atom displacements in aggregates. The displacement-per-atom (*dpa*) cross sections were generated with computer code Specomp [9] for several widespread aggregate minerals in concrete: Quartz (SiO_2), Calcite ($CaCO_3$), Anorthite ($CaAl_2Si_2O_8$), Microcline ($KAlSi_3O_8$), and Albite ($NaAlSi_3O_8$). Fig. 4 depicts the variation of *dpa* cross-sections with neutron energy; the shapes are very similar to the well-established *dpa* cross-section for silicon, which was taken from ASTM E 722 [10] and is shown for comparison.

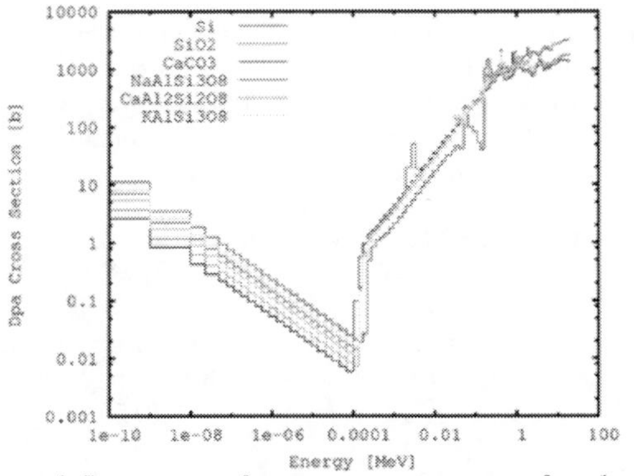

Figure 4. *Dpa* cross section versus neutron energy for selected minerals commonly used as aggregates in concrete.

These cross sections were folded with the neutron fluence rate spectrum in the cavity of a two-loop PWR and three-loop PWR, obtained from transport calculations, to calculate relative contributions to the *dpa* rate. The results plotted in Fig. 5 and Fig. 6 show that approximately 90% of the atom displacements are caused by the neutrons with energies between 0.1 MeV and 2 MeV while neutrons with energies above 1 MeV cause only about 20 % to 25 % of the total atom displacements. However, neutrons with energies above 0.1 MeV contribute more than 95% of all atom displacements. These statements hold for all materials considered and for the neutron spectrum in the cavity of a two-loop and three-loop PWR. This clearly indicates that from the perspective of creation of atom displacements in aggregates, neutron fluence with E > 0.1 MeV is more relevant correlation parameter than the fluence with E > 1MeV.

The values of the *dpa* rates for all aggregates considered agree to within ~ 10%, with the exception of calcite for which the *dpa* rate is about 20% smaller.

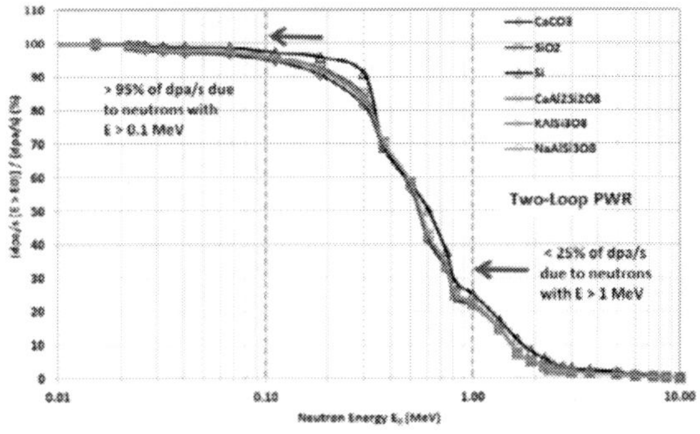

Figure 5. Relative contribution of neutrons with energies E > E_0 to the *dpa* rate for the neutron spectrum in the cavity of a two-loop PWR.

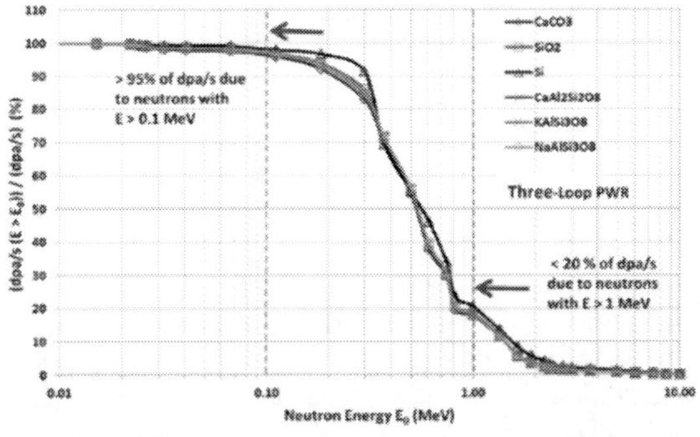

Figure 6. Relative contribution of neutrons with energies E > E_0 to the *dpa* rate for the neutron spectrum in the cavity of a three-loop PWR.

5 Neutron induced absorbed dose in aggregates

The absorbed dose was often used to assess and correlate the irradiation-induced degradation in particular in insulators [11]. Therefore the analysis described in Section 4 for *dpa* was repeated for the absorbed dose.

The results were very similar to those for *dpa* and will be just briefly summarized. Approximately 85 % of the absorbed dose rate is delivered by the neutrons with energies between 0.1 MeV and 2 MeV. The neutrons with energies above 1 MeV deliver ~ 25 % to 35 % of the absorbed dose rate. Neutrons with energies above 0.1 MeV are responsible for more than 95 % of the absorbed dose rate. These observations hold for all the aggregates considered and for the two-loop and the three-loop PWR plant analyzed. The results therefore indicate that, from the perspective of absorbed dose in aggregates, neutron fluence with E > 0.1 MeV is more relevant correlation parameter than the fluence with E > 1MeV.

6 Conclusions

In the neutron fluence rate spectra in the cavity of a two-loop and three-loop PWR about 90% of *dpa* and absorbed dose in several widespread components of aggregates in concrete: quartz, calcite, albite, anorthite, and microcline, is due to the neutrons in the energy range between 0.1 MeV and 2 MeV. The neutrons with energies above 1 MeV typically contribute only about 20% to 35% to the *dpa* and absorbed dose. The neutrons with energies above 0.1 MeV typically account for 95% or more of the *dpa* and absorbed dose in these materials. These observations, together with the experimental evidence that the radiation induced volumetric expansion of aggregates drives the degradation of concrete mechanical properties, indicate that fluence of neutrons with energy cutoff 0.1 MeV is better correlation parameter than fluence with threshold 1 MeV and should be used in the future. However, it is also recommended to report full neutron fluence rate energy spectrum and gamma-ray spectrum for all future concrete irradiation experiments.

Available database on irradiated concrete displays large variations in the level of degradation in concrete properties at any neutron fluence value and varying neutron fluence specifications. Besides, the concrete specimens used and the irradiation conditions may not have been representative of the condition in NPPs biological shields, which is particularly true at the fluence levels above $\sim 3.0\times10^{19}$ n/cm^2. Additional experiments are therefore needed; some of the undergoing efforts are described in [12].

The performance of concrete structures in the NPPs through the first 40 years of service has been very good and it is expected that the vast majority of these structures will continue to meet their functional and performance requirements during extended licensing periods. However, based on projected neutron fluence values (E > 0.1 MeV) in the concrete biological shields of the US PWR fleet and currently available irradiated concrete database, some decrease in the mechanical properties of concrete cannot be ruled out during extended operation beyond 40 years.

Acknowledgments

This manuscript has been authored by UT-Battelle, LLC, under contract DE-AC0500OR22725 with the US Department of Energy. The United States Government retains and the publisher, by accepting the article for publication, acknowledges that the United States Government retains a nonexclusive, paid-up, irrevocable, worldwide license to publish or reproduce the published form of this manuscript, or allow others to do so, for United States Government purposes.

References

1. H. Hilsdorf, J. Kropp, H. Koch, The effects of nuclear radiation on the mechanical properties of concrete, Special Publication of the American Concrete Institute 55, 223–254 (1978)
2. O. Kontani, Y. Ichikawa. A. Ishizawa, M. Takizawa, O. Sato, Irradiation effects on concrete structure, *International Symposium on the Ageing Management and Maintenance of Nuclear Power Plants*, 173–182 (2010)
3. H. Graves, Y. Le Pape, D. Naus, J. Rashid, V. Saouma, A. Sheikh, and J. Wall,. Expanded Material Degradation Assessment (EMDA), **Vol. 4**: Aging of Concrete, NUREG/CR-7153, ORNL/TM-2011/545. U.S. NRC, (2014)
4. U.S. NRC, Agency-wide Documents Access and Management System (ADAMS), http://www.nrc.gov/reading-rm/adams.html (2015)
5. I. Remec, Status Report on Defining a Unified Parameter for Characterization of Radiation Intended for Evaluation of Radiation-induced Degradation of Concrete, Technical Report, ORNL, (2014)
6. T. Esselman, et al., Expected Condition of Concrete at Age 80 Years of Reactor Operation, Technical Report, A13276-R-001, Lucius Pitkin, Inc., (2013)
7. K.G. Field, I. Remec, Y. Le Pape, Radiation effects in concrete for nuclear power plants – Part I: Quantification of radiation exposure and radiation effects, Nuclear Engineering and Design, **Vol. 282**, pp. 126-143, ISSN 0029-5493 (2015)
8. I. Remec, et al., Characterization of Radiation Fields in Biological Shields of NPPs for Assessing Concrete Degradation, EPJ Web of Conferences **106**, (2016)02002,http://dx.doi.org/10.1051/epjconf/2016 10602002
9. L.R. Greenwood, SPECOMP Calculations of Radiation Damage in Compounds, *ASTM-STP 1001*, pp. 598-602, (1989)
10. ASTM E722 Standard Practice for Characterizing Neutron Fluence Spectra in Terms of an Equivalent Monoenergetic Neutron Fluence for Radiation-Hardness Testing of Electronics, *Annual Book of ASTM Standards*, **Vol. 12.02** (2015)
11. P. Beynel, et al., Compilation of Radiation Damage Test Data, Part III: Materials used around high-energy accelerators, CERN 82-10, (1982)
12. T.M. Rosseel, et al., Recent Advances in Understanding Radiation Damage In Reactor Cavity Concrete, *Transactions, SMiRT-23, Manchester, United Kingdom - August 10-14*, (2015)

Quantitative Fissile Assay In Used Fuel Using LSDS System

Yong Deok Lee[1], *Ju Young* Jeon[1], and *Chang-Je* Park[2,*]

[1]Korea Atomic Energy Research Institute, P.O. Box105, Yusung, Daejeon, 305-600 Korea
[2]Sejong University, Gwangjin-gu, Seoul, 143-747 Korea

Abstract. A quantitative assay of isotopic fissile materials (U235, Pu239, Pu241) was done at Korea Atomic Energy Research Institute (KAERI), using lead slowing down spectrometer (LSDS). The optimum design of LSDS was performed based on economics, easy maintenance and assay effectiveness. LSDS system consists of spectrometer, neutron source, detection and control. LSDS system induces fissile fission and fast neutrons are collected at fission chamber. The detected signal has a direct relation to the mass of existing fissile isotopes. Many current commercial assay technologies have a limitation in direct application on isotopic fissile assay of spent fuel, except chemical analysis. In the designed system, the fissile assay model was setup and the correction factor for self-shield was obtained. The isotopic fissile content assay was performed by changing the content of Pu239. Based on the fuel rod, the isotopic content was consistent with ~2% uncertainty for Pu239. By applying the covering (neutron absorber), the effective shielding was obtained and the activation was calculated on the target. From the assay evaluation, LSDS technique is very powerful and direct to analyze the isotopic fissile content. LSDS is applicable for nuclear fuel cycle and spent fuel management for safety and economics. Additionally, an accurate fissile content will contribute to the international transparency and credibility on spent fuel.

1 Introduction

An isotopic fissile content assay is very important for the reuse of fissile materials through nuclear fuel cycle and the management of spent fuel. The accurate fissile content is a key point to be verified for increased safety and economics in the reuse and management. The pyro processing technology is under development in KAERI, as one option of nuclear fuel cycle, which increases the proliferation resistance. The pyro process produces the source material including trans-uranium to fabricate a fuel rod for sodium fast reactor (SFR). Therefore, the fissile content must be verified before the fabrication of fuel rod. Also, the fissile content is one of the factors to be given for the optimum design of spent fuel storage site.

A real time and non-destructive direct assay of fissile materials in spent fuel is currently big issue in Korea. The research on an interim and a permanent storage for final disposal has been performed. An accurate fissile material content can makes it possible for a maximum burnup credit on spent fuel storage site.

LSDS is under development to analyze the isotopic fissile content in spent fuel. The optimum design was done and the main parameters were investigated and determined for the system working. In the designed LSDS device[1], the source neutron is slowed down in the lead medium and produces continuously slowed down energy. The continuous energy spectrum enters into the fuel area and induces fissile fission with respect to the neutron energy. The fissile fast fission neutron is measured in the surrounding threshold fission chamber to discriminate the fast neutrons in the complex radiation field (intense gammas, source neutron, spontaneous fission neutron and (α, n) reaction).

The fissile assay model was setup and the fissile content assay was performed on the various plutonium contents. The self-shielding was calculated over the slowing down channel[2]. The mechanism for source neutron production was determined including the target. The energy deposition at each target layers by the incident electron was calculated and the produced radiations were evaluated. The effective shielding calculation was performed at the outside wall by applying the neutron absorber surrounding the spectrometer.

2 Assay Model

The detector measurement of fast fission neutron involves information on the fission of U235, Pu239, and Pu241 by the interrogation source neutron, with respect to the neutron energy. The assay model has a property that has linearity[3] between detection and fissile fission. The model is expressed like below,

$$y_i = k\epsilon[\nu_{1i}N_1 < \sigma_{f,1}\phi >_i + \nu_{2i}N_2 < \sigma_{f,2}\phi >_i + \nu_{3i}N_3 < \sigma_{f,3}\phi >_i]. \quad (1)$$

Where y_i is the detector signal at channel i, k is the normalization constant, and ϵ is the detector efficiency. 1, 2, and 3 represent the fissile materials, U235, Pu239, and

* Corresponding author: parkcj@sejong.ac.kr

Pu241 and N is the fissile mass. ν is the average neutron yield by the fission of each fissile material. σ_f is the fission cross section and ϕ is the source neutron intensity entering into the fuel rod. Therefore, $<\sigma_{f,1}\phi>_i$ represents the fissile fission reaction rate by the source neutron at channel i. In the system, the induced fissile fission neutron is detected by the surrounding threshold detectors. The fission neutron detection has a direct relation to the amount of fissile material. The configuration for the detector, fuel area and source is shown in figure 1.

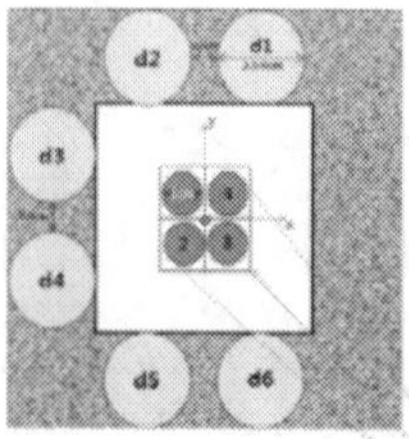

Fig. 1. Geometry of fuel rod and detector for fissile assay.

In the LSDS system, when the fuel is introduced into the assay area. The source neutron is captured in the fissile materials. The self-shielding factor must be applied for the correction of fission signals. The assay energy range is usually selected to involve the fission structures which represent the dominant properties of each fissile material. Figure 2 shows the fission signals before and after the application of the correction factor. In the simulation, the amounts of U235 and Pu241 were fixed to 1wt% and the content of Pu239 was varied from 1wt% to 10wt%. As shown in the figure, before the correction, the fission rate by increasing the amount of Pu239 does not match the amount increase. However, the corrected fission signal shows a well defined fission property with respect to the amount increase. Therefore, the correction in a fission signal is very important to get accurate assay results.

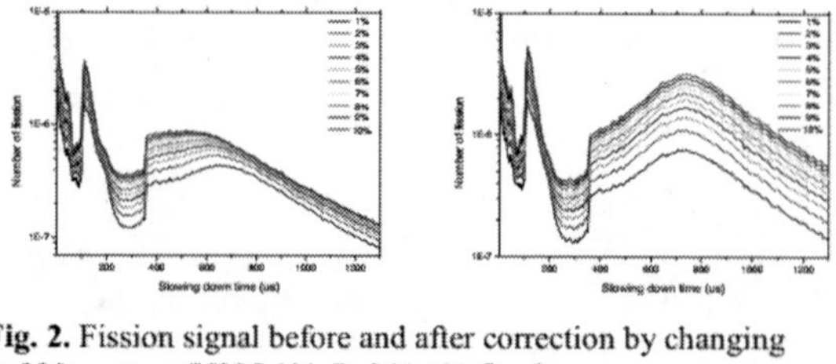

Fig. 2. Fission signal before and after correction by changing Pu239 content (U235 1%, Pu241 1% fixed).

3 Fissile assay

Based on the assay model in equation 1, an isotopic fissile assay was performed in the fuel rod for U235, Pu239 and Pu241. The detection signal in each channel is the sum of the fission by U235, Pu239 and Pu241. Therefore, from the fission signature, the assay range was selected to include each fissile dominant fission signature, 1keV through 1eV. Table 1 shows the summary of the assay results. In the content assay, the well agreed content of uranium and plutonium was obtained. For a change of 3 to 10% in the Pu239 content, the assay was obtained with 2~3% uncertainty compared to the actual value. However, for 1 and 2% content of Pu239, a relatively poor error was obtained. From the assay results, for highly enriched Pu239 cases, more accurate results were obtained. For the U235 assay, the results show good agreement to the actual mass. However, in the Pu241 case, the assay has a relatively low accuracy. The fission signature shows a similarity in the plutonium.

4 Source neutron

The intense source neutron is required in LSDS system to induce the continuous fissile fission. For effective neutron production, one section electron accelerator was decided with the target. The source neutron must be enough to get detection statistics for fission and overcome the radiation background. 30MeV electron accelerator and 500mA current were determined to generate ~10^{12} n's/sec at the target. 5 layered Ta plates were designed with Be at the end of plates[4]. The design was based on how to produce enough neutrons effectively and to cool down the plates without interference of the produced neutron spectrum. Figure 3 shows the total accelerating system to generate the source neutron. In the figure, short two columns were used to reduce the radiation leakage through the wall.

Because high energy electron enters into the target, the energy deposition on each plate was evaluated. In the target, not only electron but also gamma and neutron produced by interacting with electron deposit their energy on each plate. Therefore, the cooling system is considered not to melt down at the plate. The intense electron can penetrate Ta and Be layer and arrive at the lead. From the calculation, the electron is totally stopped at 19cm thickness of lead.

5 Shielding effect and activation

For effective and intense neutron production using electron accelerator, facility shielding is required. The optimum shielding design was done using concrete and borax combination[5]. The neutrons from the target, induced fissile fission and spontaneous fission by spent fuel leak out the spectrometer. The gamma rays are normally shielding in the lead spectrometer. The neutron interacts with the lead and shielding material. From the optimized facility shielding calculation[5], the shielding effectiveness was examined by applying the additional covering around the spectrometer to absorb leaked neutron. The covering (250cm in length and height) consists of HDPE-Borax with a 5cm thickness. Table 2 shows the calculated dose rate at the outside wall surface(cell #7 at figure 4) of the LSDS facility. When the HDPE-borax covers the spectrometer, the dose rate decreases 70% compared to that of the no covering case. Therefore, the application of covering around the spectrometer can reduce the facility wall thickness.

Table 1. Isotopic fissile content assay (U235:1%, Pu239:1%, Pu239: changed).

Mixed (%) U5-Pu9-Pu1	**1-1-1**	**1-2-1**	**1-3-1**	**1-4-1**	**1-5-1**	**1-6-1**	**1-7-1**	**1-8-1**	**1-9-1**	**1-10-1**
U-235	1.01	1.00	1.04	1.00	1.03	0.99	1.01	1.06	1.15	1.23
Error	0.09	0.10	0.12	0.10	0.11	0.11	0.11	0.09	0.12	0.12
Pu-239	1.05	1.82	3.10	4.01	5.09	5.94	7.04	8.00	9.26	10.01
Error	0.02	0.03	0.05	0.03	0.04	0.04	0.05	0.04	0.05	0.06
Pu-241	1.15	1.21	1.18	1.24	1.24	1.28	1.27	1.31	1.28	1.29
Error	0.02	0.02	0.03	0.02	0.02	0.02	0.02	0.02	0.02	0.02

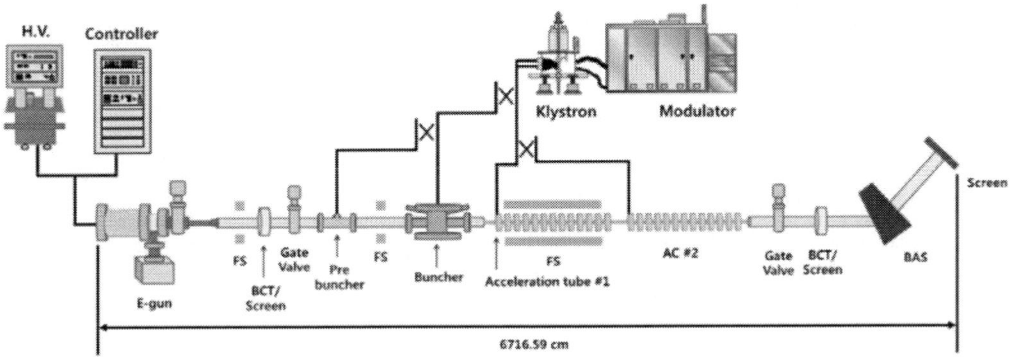

Fig. 3. Electron accelerator system.

Table 2. Dose rate at outside of facility wall by applying covering.

Material and Thickness	Dose rate(μSv/h)		
	Left wall	Right wall	Top wall
No covering	4.48E-02	3.49E-02	4.18E-02
HDPE-Borax (5cm)	1.32E-02	1.11E-02	1.08E-02

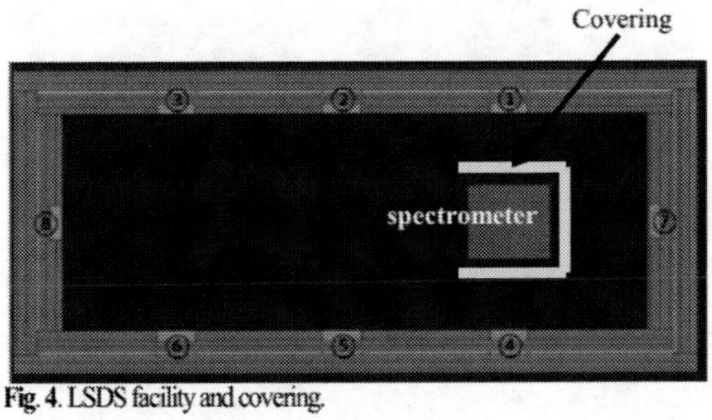

Fig. 4. LSDS facility and covering.

An activation evaluation was conducted on the Ta target. The target is located at the center of the spectrometer. The target should produce the neutron without degradation. The activation will affect the target life time. Table 3 shows the total activity by time after irradiation. After 1 day, the activity decreases drastically and after 90 days, it represents E-3Ci. Therefore, the activity decreases sufficiently based on the cooling time. After 90 days, the radiation can be shielded totally in the spectrometer.

6 Conclusion

In the designed LSDS, a mathematical model for an isotopic fissile assay was setup. The assay model involves all fissile fission signatures. The correction for self-shielding is a very important factor to represent the linear response in the fission. The correction improves the assay accuracy.

The isotopic fissile assay results show very good consistency with the actual mass for uranium and plutonium. In particular, the content for Pu239 was matched well to the actual content. It gives 2~3% uncertainty. The assay was in good agreement with the U235 content, with an ~2% error. However, the result for Pu241 was relatively poor compared to that of U235 and Pu239. In particular, by applying the additional covering shield around the spectrometer, a drastically decreased dose rate was obtained at the outside wall of the LSDS facility. This will influence the facility shielding size and economics.

The LSDS technique is very promising for an analysis of the isotopic fissile content in used fuel. In the LSDS, direct fissile fission is obtained in the detector. Therefore, the LSDS is very powerful and direct in an isotopic fissile assay. The LSDS is applicable for the nuclear fuel cycle and spent fuel management for safety and economics. Additionally, an accurate fissile material content will contribute to international transparency and credibility for nuclear material utilization and management.

Table 3. Activity change based on cooling time.

Material	Cooling time						
	1s	30m	1hr	3hr	1day	7 day	90day
Ta	Total activity (Ci)						
	3.05E+09	2.92E+09	2.80E+09	2.36E+09	3.96E+08	1.91E+03	1.11E-03

Acknowledgment

This work was supported by the Nuclear Research Foundation of Korea(NRF) grant funded by the Korea government(MSIP) (No. 2016904221).

References

1 Y.D. Lee, NET, 46, 6, 837 (2014)
2 Y.D. Lee, C.J. Park, G.I. Park, K.C. Song, "Self shielding in nuclear fissile assay using LSDS", *Trans. of the Korean Nuclear Society*, 345-346 (2012).
3 Y. Danon, *Experiments with a lead slowing-down spectrometer for assay of used nuclear fuel*, Rensselaer Polytechnic Institute, (2012
4 C.J. Park, M.K. Jaradat, L. Alawneh, Y.D. Lee, ANE, 49, 218 (2012)
5 J.D. Kim, NET, 47, 345 (2015)

A new shielding analysis method for used fuel transport and storage casks

Vincent Léger[1,*] and *Stavros* Kitsos[1]

[1]AREVA TN, Technical Department, 78180 Montigny-le-Bretonneux, France

Abstract. *To provide a cask with the largest possible loading capacity of spent fuel assemblies with the largest practicable burnup and shortest cooling time within all safety requirements, AREVA TN has adapted its design process and developed a more elaborated shielding analysis method. Taking advantage of the potential heterogeneities between sources of fuel assemblies to be loaded, and the self-shielding of assemblies loaded at the basket centre by the assemblies loaded at the basket periphery, the result of this method is expressed under the shape of a linear inequalities system allowing to optimize the cask capacity and performance.*

1 Introduction

Fuel assembly characteristics are evolving in order to improve performance of reactor operations. As a result enrichment is slightly increasing as well as burnup. In addition, in some case there is a need to unload reactor as soon as possible implying shorter cooling time.

To improve cask performance with the largest possible loading capacity of spent fuel assemblies, the largest practicable burnup and shortest cooling time while conforming to all safety requirements, AREVA TN has adapted its design process and has developed a more elaborated shielding analysis method to meet the latest regulatory recommendations.

This new method avoids to develop a safety demonstration based on a loading plan matrix with a limited number of homogeneous and non homogeneous loading options.

2 New regulatory practice

The 2009 IAEA recommendations (TS-R-1[1]) required that the maximum radioactive contents only be considered for shielding analyses under accident conditions of transport. Typical contents were used in order to demonstrate the compliance of the shielding thresholds under routine conditions of transport.

With the 2012 IAEA recommendations (SSR-6[2]), the maximum radioactive contents must now be considered under both conditions of transport and for all package types. As maximum radiation levels have not changed, routine conditions of transport are now the shielding relevant conditions in the licensing process. In parallel, in some countries, storage authorities have strengthened their recommendations and require to the definition of generic loading plans.

Thus, cask designers are facing the challenge of implementing a method to define maximum generic loading plans without restrictions on new cask design or cask license renewal.

Indeed, following the standard shielding analysis method leads to define a matrix of bounding loading plans based on limited heterogeneities from the maximum burnup and the minimum cooling time authorized in the case of a homogeneous loading plan. It allows a standard licensing process but this reaches a limited loading capacity and flexibility.

The new shielding analysis method developed by AREVA TN reconsiders the standard definition of the content in order to take advantage of potential heterogeneities between sources of the loaded fuel assemblies and of the self-shielding between the loaded fuel assemblies. The maximum authorized radioactive content is defined with only maximum neutron and gamma sources authorized in each basket lodgement.

* Corresponding author: vincent.leger1@areva.com

3 New shielding analysis method

3.1 Shielding relevant cask area

The key step of this method is based on a detailed shielding analysis in order to identify the shielding relevant cask areas. These areas are defined by the variations of shielding material thickness, by the specific basket geometry associated to the burnup, cooling time heterogeneities of loading plans, and by the radiation level thresholds. They correspond to zones on the cask surface or in the cask vicinity where the relative maximum radiation levels occur.

In this way, based on a 3D detailed geometry model[3,4] (Fig. 1), the geometric scope in which the radiation levels must be verified can be reduced.

Application to the TN®9-4 transport cask

The TN®9-4 cask is a transfer cask for the transport of 7 BWR UO_2 spent fuel assemblies (Fig. 2) from the Mühleberg NPP (KKM) to the Zwilag facility (Swiss centralized interim storage facility). These fuels assemblies are then loaded in the TN®24 BH which is designed for the transport and the interim storage of 69 BWR UO_2 spent fuel assemblies.

Regarding the maximum radiation levels generated at the contact of the TN®9-4 cask external surface, in the different axial and radial areas defined with an adapted external surface map (Fig. 1), the detailed shielding analysis performed allows reducing the fulfilment of the radiation level threshold, only

- in the bottom trunnion areas (Fig. 2),
- in the bottom radial areas following the angular directions 45°, 135°, 225° and 315° (Fig. 2).

To illustrate the strength of this geometrical scope reduction, a sensitivity analysis was performed on the radiation level distribution taking into account heterogeneous loading plans. Considering up to 7 fuel assemblies with a high average burn-up (60 GWd/t_{HM}) associated with fuel assemblies with a very low average burn-up (2 GWd/t_{HM}), the radiation level distributions of the 47 possible heterogeneous loading plans were analysed.

Regarding the fulfilment of the radiation level threshold (2 mSv/h at the contact[1,2]):

- the maximum radiation levels exceeding the radiation level threshold occur only in the radial directions 0° and 180°, i.e. at the contact of the bottom trunnions (Fig. 2);
- secondary maximum radiation levels always occur following the symmetrical radial directions 45°, 135°, 225° and 315° (Fig. 2);
- secondary minimum radiation levels always occur following the symmetrical radial directions 90° and 270° (Fig. 2).

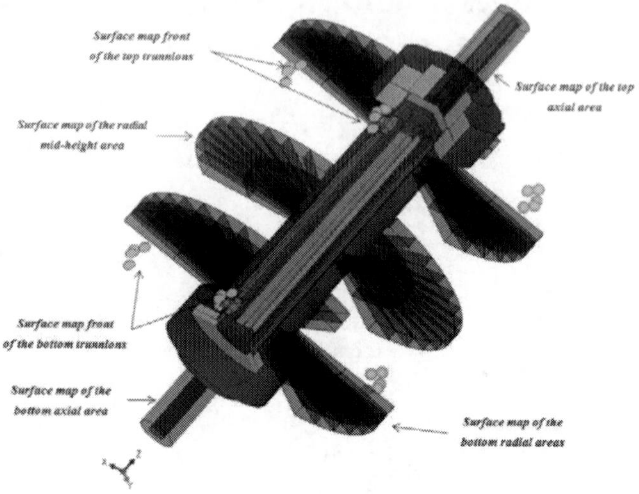

Figure 1: 3D detailed model of the TN®9-4 transport cask – External surface map

Thus, following the detailed shielding analysis performed, the reduction of the geometrical scope, in which the radiation levels must be calculated, is fully adapted to highly heterogeneous loading plans.

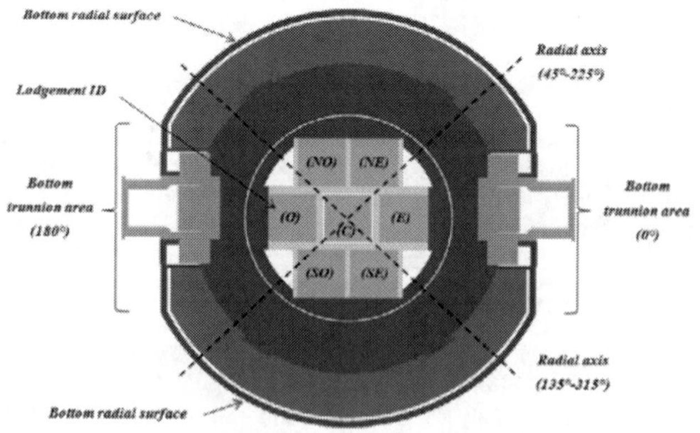

Figure 2: 3D detailed model of the TN®9-4 transport cask – Bottom trunnion section

3.2 Linear inequalities system

The new shielding analysis method followed by AREVA TN reconsiders the homogeneous model of the content in order to establish the maximal loading capacity of a cask. The result of this method is expressed under the shape of a linear inequalities system.

This method takes advantage of potential heterogeneities between sources of the fuel assemblies to be loaded through the comparison of the real fuel assemblies sources with cask specific reference source terms associated with the cask design. Futhermore the self-shielding of fuel assemblies loaded at the basket center by the fuel assemblies loaded at the basket periphery is integrated by the definition of the individual contribution of each source loaded in each basket lodgement to the radiation level generated in each shielding relevant cask area.

In this way, the first step is to define the maximum shielding capacity of the cask regarding radiation level threshold. To reach that goal, reference source terms (*RST*) are calculated, regarding each shielding relevant cask area, on the basis of a loading plan with uniformly distributed homogeneous sources for

- each neutron generation reactions: spontaneous fission and (α, n) reaction,
- primary and secondary gamma considering the shielding relevant gamma energy spectrum,
- gamma sources of ^{60}Co activation located in fuel rod end fittings and plenums.

Thus, a loading plan involving in each basket lodgement the same source, with a source strength equal to the reference source term, generates exactly the corresponding radiation level threshold.

The second step is to define the relative level of radioactivity authorized in each basket lodgement, taking into account the loading of the other lodgements. To reach that goal, basket lodgement factors (*f-factor*) are calculated on the basis of a loading plan with a homogeneous source uniformly distributed in one basket lodgement whereas other lodgements are loaded with fuel assemblies without source.

Associated with the same splitting between the nature of the emissions used for the definition of the reference source terms, this method leads to define the individual contribution of each basket lodgement to the dose rates in each shielding relevant cask area. The result of this method is expressed under the shape of a linear inequalities system allowing to verify the radiation level for any real loading plan.

For each shielding relevant cask area, the following linear inequality (*I*) is defined:

$$\sum_{lodg}\left(\sum_{j=1}^{2}\frac{f_j^{lodg}(n)\times S_j^{lodg}(n)}{RST_j\ (n)}+\sum_{i}\frac{f_i^{lodg}(\gamma)\times S_i^{lodg}(\gamma)}{RST_i\ (\gamma)}+\sum_{c}\frac{f_c^{lodg}(\gamma_{act})\times S_c^{lodg}(\gamma_{act})}{RST_c\ (\gamma_{act})}\right)\leq 1$$

where

- *j*: neutron source index, "j" = [1, 2] = [(α, n) reaction; spontaneous fissions],
- *i*: gamma energy group index,
- γ_{act}: activation gamma source,
- *c*: fuel axial zone index for activation gamma,
- *lodg*: basket lodgement ID,
- $S_j^{lodg}(n)$: neutron source strength "j" of the fuel assembly loaded in the basket lodgement "lodg",
- $S_i^{lodg}(\gamma)$: gamma source strength of energy group "i" of the fuel assembly loaded in the basket lodgement "lodg",
- $S_c^{lodg}(\gamma_{act})$: activation gamma source strength in the axial zone "c" of the fuel assembly loaded in the basket lodgement "lodg",
- $f_j^{lodg}(n)$: contribution factor of neutron source "j" loaded in the basket lodgement "lodg",
- $f_i^{lodg}(\gamma)$: contribution factor of gamma source of energy group "i" loaded in the basket lodgement "lodg",
- $f_c^{lodg}(\gamma_{act})$: contribution factor of activation gamma source loaded in the axial zone "c" of the basket lodgement "lodg",
- $RST_j\ (n)$: reference source term for neutron sources "j",
- $RST_i\ (\gamma)$: reference source term for gamma sources of the energy group "i",
- $RST_c\ (\gamma_{act})$: reference source term for activation gamma sources in the axial zone "c".

To optimize the cask capacity and to efficiently adapt the conservatism strength required, a normalization of the inequality results considering reference loading plans is performed.

In this way, neutron $\eta(n)$, gamma $\eta(\gamma)$ and activation gamma $\eta(\gamma_{act})$, effective normalization factors are calculated for each shielding relevant cask area.

Thus, the optimized lodgement loading capacity, in function of the emission and the shielding relevant cask area considered, is defined as follow:

- $EST_j^{lodg}(n) = \dfrac{\eta(n) \times RST_j(n)}{f_j^{lodg}(n)}$, neutron optimized loading capacity of the basket lodgement "lodg" regarding the shielding relevant cask area considered,
- $EST_i^{lodg}(\gamma) = \dfrac{\eta(\gamma) \times RST_i(\gamma)}{f_i^{lodg}(\gamma)}$, optimized loading capacity of the basket lodgement "lodg" for energy group "i" gamma and regarding the shielding relevant cask area considered,
- $EST_c^{lodg}(\gamma_{act}) = \dfrac{\eta(\gamma_{act}) \times RST_c(\gamma_{act})}{f_c^{lodg}(\gamma_{act})}$, optimized loading capacity of the basket lodgement "lodg" for activation gamma source loaded in the axial zone "c" and regarding the shielding relevant cask area considered.

Each linear inequality (*I*) is thus modified as follow:

$$\sum_{lodg}\left(\sum_{j=1}^{2}\frac{S_j^{lodg}(n)}{EST_j^{lodg}(n)} + \sum_i \frac{S_i^{lodg}(\gamma)}{EST_i^{lodg}(\gamma)} + \sum_c \frac{S_c^{lodg}(\gamma_{act})}{EST_c^{lodg}(\gamma_{act})}\right) \le 1$$

The linear inequalities system such defined could then be easily validating. Indeed any complete loading plan involving in one basket lodgement a source strength equal to the corresponding optimized loading capacity, generates at minimum a radiation level equal to the radiation level threshold in the associated shielding relevant cask area.

4 Cask loading performances and management

Using the linear inequalities system, the cask owner is able to directly assess the relevant radiation level of any loading plan. The major advantage of this approach is to allow the cask owner to take into account the real source strength heterogeneities of a real loading with a simple and efficient tool.

Indeed, through a simple spreadsheet, the use of the linear inequalities system only involves calculating the real sources of fuel assemblies to be loaded using a validated depletion code. In comparison to the time needed to perform a complete shielding calculation using a 3D Monte-Carlo code, using this method is decisive to give the cask user the maximum flexibility for the loading management.

Furthermore, a very detailed analysis of the cask loading performances could be obtained by converting the linear inequalities system in a numerical surrogate model (NSM). For this purpose, based on an adapted design space exploration software, an optimization workflow (Fig. 3) is defined in order to optimize the relevant cask loading parameters choosen ($\vec{x}$).

In this way, with a paralell coordinates plot (Fig. 4), the bounding loading plans regarding the level of heterogeneousness authorized are easily defined.

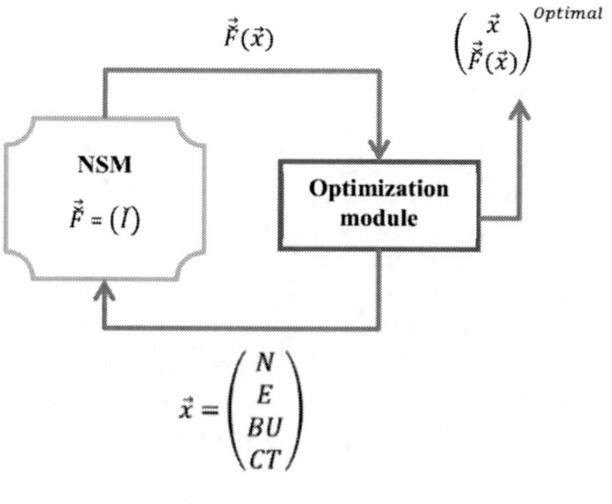

Figure 3: Optimization workflow using a NSM

In order to illustrate the strength of this numerical optimization workflow, in Figures 4 to 7 are plotted all the loading plans allowing to optimize the loading of a dual purpose cask considering maximum radiation level (I^{max}) and maximum cask component temperature (T^{max}). In this example, the cask basket, which presents more than 20 lodgements, is loaded in one or two different rows.

In Figure 4 are plotted all the possible loading plans if the cask basket is fully loaded with fuel assemblies (FA) presenting the same depletion characteristics (enrichiment (*E*), burnup (*BU*) and cooling time (*CT*)).

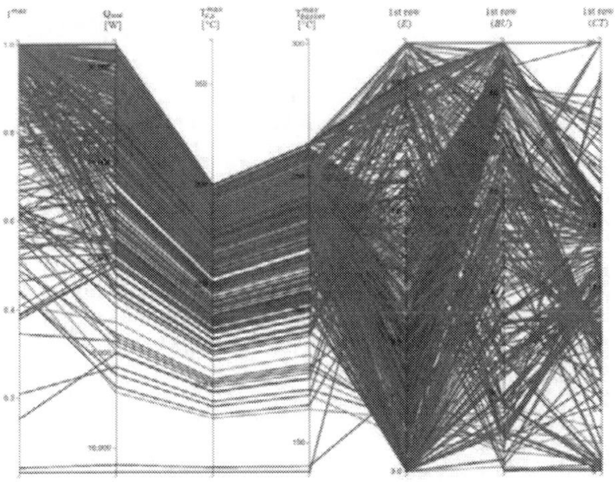

Figure 4: Paralell coordinates plot – Loading plans distribution – 1 loading row

For this cask loading strategy, it is interessant to see, that, whatever the FA depletion characteristics, the maximum FA and basket temperatures are always lower than respectively 300°C and 270°C, within a fulfilled inequality system. The maximum achievable total thermal power loaded (Q^{total}) is 30,3 kW.

In Figure 5, using numerical filters, the loading plan possibilities are reduced in order to respect design constrains. For this cask loading strategy, a minimum cooling time of 14 years is required and the radiation levels are always higher than 80% of the thresholds.

In the same way, in Figure 6 are plotted all the loading plans if the same cask basket is devided in two rows and if the spent fuel assemblies loaded in the same row have the same depletion characteristics (enrichiment (*E*), burnup (*BU*) and cooling time (*CT*)).

In comparison with the previous strategy, those allows to increase the maximum achievable total thermal power loaded within a fulfilled inequality system.

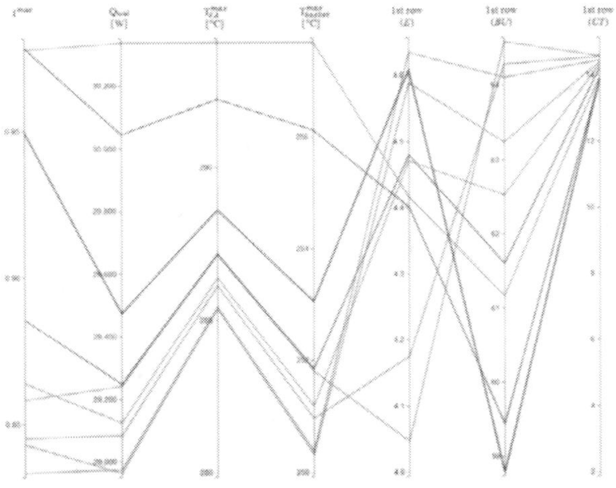

Figure 5: Paralell coordinates plot – Numerical filter application – 1 loading row

In Figure 7, using numerical filters, the loading plan possibilities are reduced in order to respect design constrains. It is interessant to see that, for this cask loading strategy, large ranges for FA depletion characteristics (enrichiment (*E*), burnup (*BU*) and cooling time (*CT*)) are available. The maximum achievable total thermal power loaded (Q^{total}) is 35 kW within a fulfilled inequality system.

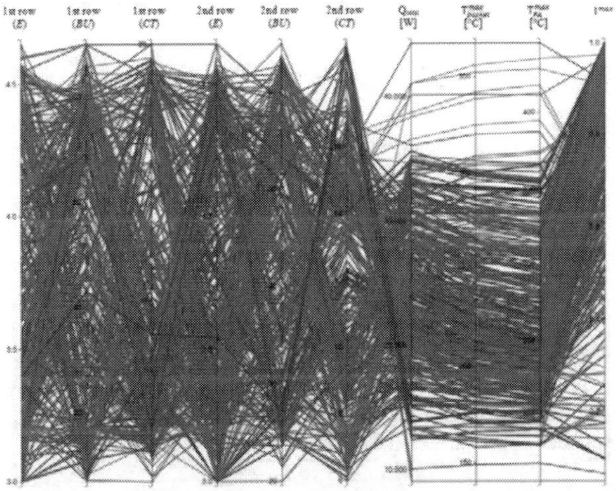

Figure 6: Paralell coordinates plot – Loading plans distribution – 2 loading rows

Thus, in the typical long term situation in which the burnup increases, the use of the linear inequalities system makes it easy to define an adapted long-term pool unloading strategy based on an optimized management of cold and hot fuel assemblies with a cooling time decrease.

 DOI: 10.1051/epjconf/201715305011

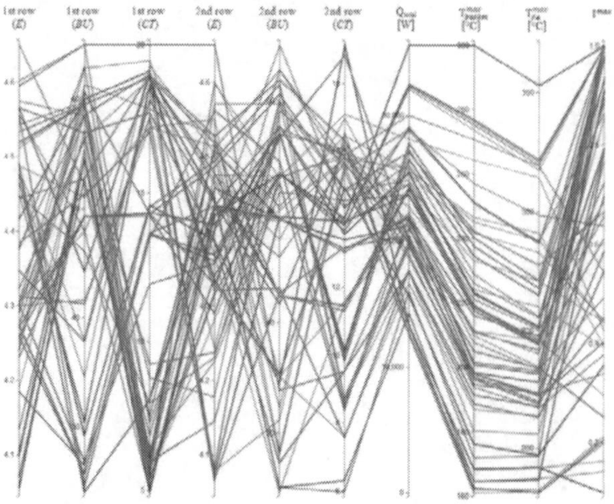

Figure 7: Paralell coordinates plot – Numerical filter application – 2 loading rows

Furthermore, the time at which cask can only be used with a reduced loading capacity, because of remaining spent fuels with high burnups and low cooling times, is deferred and the cask availability and performances are strengthened (Ref. 5).

This directly contributes to a decrease in the total number of required cask loadings and shipments and a significantly decrease of all radiation exposures according to ALARA objectives.

5 Conclusions

In order to avoid unnecessary restrictions of the authorized content characteristics while still meeting the new requirements, AREVA TN has adapted its licensing process and developed a more elaborated shielding analysis method.

The linear inequalities system, resulting from the new shielding analysis method developed by AREVA TN, presents a high implementation flexibility and allows taking advantage of the explicit characteristics of the fuel assembly inventory. The relevant technical advantages of this approach are the following:

- It avoids the necessity of defining in the license a maximal burnup and a minimum cooling time authorized in order to respect the radiation level thresholds.
- It avoids the necessity of defining a matrix of bounding loading plans in order to allow heterogeneous loading plans.
- It does not require any dose rate calculation for each loading.
- It provides a simple and efficient tool in the form of linear inequalities to verify that the radiation level thresholds are met.

This method has been validated by the German authorities (BfS, TÜV) for the design approval of the dual purpose TN®24 E package. This method has been also implemented for the design approval of the transport casks TN®G3 and TN®17 MAX, and for the license renewals of the TN®24 BH, TN®24 SH, TN®24 GB and TN®9-4 package.

Furthermore the determined linear inequalities system associated with the assessment of the thermal loading constrains allows the use of highly heterogeneous loading plans with the largest practicable burnup and the shortest cooling time.

The use of this validated method is decisive in giving the cask user the maximum flexibility for an appropriate cask loading strategy definition with an optimized management of cask loading plan in the long term.

References

1. International Atomic Energy Agency, IAEA Safety Standards for protecting people and the environment, *Regulations for the Safe Transport of Radioactive Material, Specific Safety Requirements*, SSR-6 (2012 Edition)

2. International Atomic Energy Agency, IAEA Safety Standards for protecting people and the environment, *Regulations for the Safe Transport of Radioactive Material, Safety Requirements*, TSR-1 (2009 Edition)

3. S. KITSOS, Proc. of the 15th Int. Symposium on the Packaging and Transportation of Radioactive Materials, *TN International Accurate Shielding Analysis for Casks*, PATRAM 2007, Miami, Florida, USA, (October 21-26, 2007)

4. C. NICOLETTI et al., Proc. of the 5th french-speaking scientific days, *Optimisation des Etudes en Radioprotection pour le Dimensionnement des Emballages de Transport et d'Entreposage de Matières Radioactives*, Journées SFRP, Paris, France, (March 25-26, 2014)

5. International Atomic Energy Agency, IAEA-TECDOC-1523, *Optimization Strategies for Cask Design and Container Loading in Long Term Spent Fuel Storage*, (December 2006)

Recalculation with SEACAB of the activation by spent fuel neutrons and residual dose originated in the racks replaced at Cofrentes NPP

Pedro Ortego[1,a], Alain Rodriguez[1], Candan Töre[1],
José Luis de Diego Compadre[2], Baltasar Rodriguez Quesada[2], Raul Orive Moreno[2]

[1]*SEA Ingeniería y Análisis de Blindajes S.L., Av. Atenas 75, Las Rozas 28232, Madrid, Spain*
[2]*IBERDROLA Generación Nuclear S.A.U., Cofrentes NPP, Valencia, Spain*

Abstract. In order to increase the storage capacity of the East Spent Fuel Pool at the Cofrentes NPP, located in Valencia province, Spain, the existing storage stainless steel racks were replaced by a new design of compact borated stainless steel racks allowing a 65% increase in fuel storing capacity. Calculation of the activation of the used racks was successfully performed with the use of MCNP4B code. Additionally the dose rate at contact with a row of racks in standing position and behind a wall of shielding material has been calculated using MCNP4B code as well. These results allowed a preliminary definition of the burnker required for the storage of racks. Recently the activity in the racks has been recalculated with SEACAB system which combines the mesh tally of MCNP codes with the activation code ACAB, applying the rigorous two-step method (R2S) developed at home, benchmarked with FNG irradiation experiments and usually applied in fusion calculations for ITER project.

1 Introduction

The reracking of a spent fuel pool is a frequent need in the effort to increase wet storage capacity of nuclear power plants. Cofrentes NPP, is a General Electric BWR-6 Mark III located in Valencia province, Spain, having two different wet storage ponds. In order to increase the storage capacity of the East Fuel Storage Pond (PACE in Spanish) the existing stainless steel racks were replaced by a new design of compact borated stainless steel racks allowing a 65% increase in fuel storing capacity.

More details about the rack characteristics, capacity improvement, extraction and installation procedure, etc, can be found in a previous paper [1]. At the removal of the old racks a decontamination process was performed eliminating completely the removable contamination but the intrinsic activity due to neutron irradiation needed to be estimated for designing the shielding of its interim storage at the plant.

The existing racks were irradiated by the residual neutron source originated in the (a,n) reactions and more important in the spontaneous fission of Cm242 and Cm244 isotopes.

The activation of the stainless steel rack channel had been calculated with the use of MCNP4B code using different libraries based on ENDF/B-IV and VI cross section evaluations. The neutron source was calculated with the use of the traditional SAS2H sequence inside the 4.4 version of SCALE [2].

Recently the activation has been recalculated with the use of SEACAB, a home developed system which couples the neutron flux map provided by the MCNP family of codes with the activation code ACAB. This system performs in an automated manner the activation and residual dose calculation according to the rigorous two-step method.

The final purpose of the activation calculation is the determination of the total Co60 photon source in the used racks to allow the design of a specific bunker inside the Almacén de Piezas de Baja Actividad o APBA (Low Activity Pieces Store), located at Cofrentes site.

Section 2 describes the estimation of a representative burnup and the activation calculation. Section 3 presents the dose estimates and Section 4 presents the conclusions.

2 CALCULATION PROCESS

The calculations of the spent fuel channel activation followed three different phases. First a representative and enveloping value of the fuel burnup was obtained by a detailed analysis of the management of the spent fuel pool positions during the previous 24 years of plant operation. Second the neutron source corresponding to such representative burnup was obtained assuming the typical irradiation cycle and the physical details of ATRIUM-10XP assembly design as enveloping of other fuel designs. Third the channel activation has been obtained with the use of MCNP4B [3] assuming a uniform neutron flux along the channel surface. Finally the activation has been recalculated with the use of SEACAB system and the results are compared with previous calculation results.

Additionally and in order to allow performing activation checks the photon dose in air at certain distances of the standing rack and of piled-up pieces of the rack structure have been calculated.

[a] Corresponding author: p.ortego@seaingenieria.es

DOI: 10.1051/epjconf/201715305012

2.1 Estimation of fuel assembly burnup

Cofrentes NPP keeps a complete record of the identity and characteristics of the fuel assembly located in each rack position. Nevertheless the calculation of the channel irradiation is a huge task due the large number of rack positions (1800), to the consideration of a detailed irradiation for each position and last but not least for the neutron cross talk between neighbour positions with different irradiation histories.

In general the rack positions have been occupied by the same assembly for long periods and the position has been modified occasionally at time periods associated to the refuelling outages, i.e. every 2 years as usual in the last few cycles. Therefore it is reasonable to assume that all occupied positions have reached the equilibrium in Co60 activity.

Concerning the photon source, only the peripheral positions of every rack are considered since the racks are planned to be stored in upright standing positions. The self-shielding provided by the 5mm stainless steel rack channels is considered large enough to rapidly reduce the relative importance of the inner channels.

Due to the high variation of the neutron source intensity according to the assembly burnup, the simple average of assembly burnup is not a correct procedure for determining a representative burnup value. Therefore the results of SAS2 calculations performed for different burnup values have been fitted to a polynomial expression of order 3.4.

The empty positions created a difficulty for estimating the average burnup since ignoring them meant to assume they had an assembly of the average burnup while giving a value of zero meant to neglect that they had been irradiated in previous time periods. Therefore the decay of the Co60 activity in the period when the rack position was empty was converted neutron intensity and further to a "reduced assembly burnup" by using the described expression, leading to the concept of "equivalent burnup".

The average values of equivalent burnup at the peripheral positions of the existing racks are presented in Table 1 where the left column corresponds to the assembly racks located at the North side of the PACE pond. It is seen that the range of burnups is not very wide and it can be stated that an enveloping value for the whole pond of 32 GWd/tU is reasonable. Using this enveloping value instead of the arithmetic average of about 24GWd/tU means multiplying the source by 2, given shape of the shape of the polynomial expression.

Table 1. Average value of equivalent burnup by rack.

Burnup in North-South direction (GWd/tU)					
N/A	N/A	25.3	27.9	27.7	28.1
20.6	22.7	22.6	22.2	22.4	22.2
30.5	30.8	31.3	30.6	30.5	30.0
25.8	25.9	25.4	20.1	26.7	25.8

2.2 Activation of the rack channel

As indicated above, the neutron source at the assembly rods has been obtained with SAS2 sequence of SCALE4.4. As representative fuel assembly the 10x10 ATRIUM-10XP has been used for having a similar uranium mass to the GE14 and higher than the SVEA Optima2 fuel design. The irradiation in core during cycles 13 to 15B has been modelled with 18 month cycles including 30 days of refuelling outage period. In these conditions the average neutron source per assembly is obtained as 1.34E+08 neutron/s.

As the rack channel completely contours the fuel assembly, a single value of material activation has been obtained. This has been done with the FM card capability of MCNP4B and using different cross sections, including LLLDOS and ENDF601 based on ENDF/B-IV and VI respectively. The neutrons have been born uniformly inside each pellet volume and uniformly in the vertical direction. The model considers each individual rod and distinguishes pellet, gap and clad for each rod.

The only available information on the cobalt impurity content corresponded to a semi-quantitative analysis performed on a sample of the stainless steel material used in the West pond (Piscina Almacenamiento Combustible Oeste) being 0.25% or 2500ppm. This value has been accepted for the PACE racks but as is obvious bears a large uncertainty.

The activation impact of an assembly occupying one position on the activation of the neighbour channels has been considered as well. The MCNP4B model is presented in Figure 1.

Figure 1 MCNP4B model of assembly and neighbour channels

2.3 Activation calculations with MCNP4B

The total specific activity per isotope in the enveloping rack channel is presented in Table 2 with the corresponding production reaction. It is seen that main activity corresponds to Co60 with 49.4kBq/cm3, Fe55 with 48.9kBq/cm3 and Ni63 with 22.1kBq/cm3.

Concerning the mid-term storage of the racks only the Co60 is important. The total Co60 activity per channel is calculated as follows:

Channel volume = 2 x 34 x 430 x 0.5 = 14525cm3

Including the overlaping part of the bended plates making the channel up. The activity of each rack channel is then calculated as:

Activity/channel = 14525x430x49400 = 7.18E+08Bq

and the total activity for a typical rack of 8x10 channels will be 5.7E+10 and for the largest rack, with 10x10 channels, will be 7.2E+10 Bq.

Table 2. Specific activity of different activation isotopes

Active isotope	Main reaction	Activity (Bq/cm3)
H3	Li6(n,a)H3	4.08E+01
Cl36	Cl35(n,g)Cl36	1.94E+03
Ca41	Ca40(,n,g)Ca41	5.71E+00
Mn54	Fe54(n,p)Mn54	1.45E-03
Ni59	Ni58(n,g)Ni59	1.44E+05
Co60	Co59(n,g)Co60	4.94E+04
Ni63	Ni62(n,g)Ni63	2.21E+04
Ag110m	Ag109(n,g)Ag110m	4.59E+01
Ag108m	Ag107(n,g)Ag108m	1.20E+01
Eu152	Eu151(n,g)Eu152	1.28E+01
Eu154	Eu153(n,g)Eu154	7.10E-01
Hf178m	Hf177(n,g)Hf178	1.40E+01
Fe55	Fe54(n,g)Fe55	4.89E+04
C14	N14(n,p)C14	3.83E+03
Mo93	Mo92(n,g)Mo93	5.31E+02

2.4 Recalculations with SEACAB

The SEACAB system has been benchmarked with the Frascati Neutron Generator irradiation and residual dose experiment [4].

The use of SEACAB requires the use of MCNPX or MCNP5 in order to obtain the detailed map of neutron flux in the whole energy spectrum. The so generated "mesh-tally" file is processed by SEACAB with an input file providing the fraction of material occupying each mesh volume. Additionally the irradiation and decay times are provided for their use in the activation-decay-activation cycle modelled by ACAB. Finally SEACAB gives the total activity in the channel and a distribution of activity point-by-point. This file is further used as SP card in the MCNP4B input to compute the dose rate at the position of interest as explained in next section.

2.5 Dose calculations

In order to advance the possible situations after racks extraction, it was necessary to provide the means to check the exact activation of each rack before its withdrawal from the pond. Therefore a set of photon dose calculations were performed by using the estimated activation to facilitate the comparison with the dose rate in air to be measured by Radiological Protection.

The dose in front of a row of standing racks has been calculated at short distance in conditions of direct view of a row of racks and behind different thicknesses of ordinary concrete and lead walls. The exact contribution of each row of channels has been calculated only for the first 3 rows, the dose contribution of the inner ones has been estimated with a row attenuation of 40% per row.

The dose rate obtained at contact is 4.58 mSv/h. This value is larger than the maximum dose rate allowed inside the APBA, hence certain thickness of concrete or lead was necessary to reach the maximum dose rate corresponding to controlled area.

By using an ordinary concrete wall of 40cm thickness the dose rate was calculated as 23uSv/h. By using a lead panel of 7cm thickness the dose rate was calculated as 21uSv/h.

The measurements performed indicated that the calculated activity was overestimated with high probability due to the conservatism of both assembly burnup and cobalt content in the stainless steel.

3 Summary and conclusions

The calculation of the activation of the used racks has been successfully performed with the use of MCNP4B codes and the cross sections based on ENDF/B-IV and VI evaluated libraries.

Dose calculations have been performed with the use of MCNP4B and the activity of Co60 calculated. The dose at contact with a row of racks in standing positions and the dose behind different thickness of concrete and lead as shielding materials has been calculated. This has allowed the preliminary definition of the characteristics of the bunker required for the storage of racks.

Recently the activation of the channel and the corresponding dose rate in air has been recalculated with SEACAB system of codes. The results are compared successfully with the results obtained previously, taking into account the differences in assembly burnup and in other calculation hypotheses.

References

1. Nuclear España, Sociedad Nuclear Española Magazine, **298**, July-August 2009
2. SCALE4.4
3. J.F. Briesmeister (Ed.), *MCNP General Monte Carlo N-Particle Transport Code, version 4B.*
4. Fusion Engineering and Design **98-99** October 2015 pp 2214-2218. C. Töre, A. Rodriguez & P. Ortego *SEACAB qualification with Frascati Neutron Generator residual dose measurements*

Test tasks for verification of program codes for calculation of neutron-physical characteristics of the BN series reactors

Georgy Tikhomirov[1a], Mikhail Ternovikh[1], Anton Smirnov[1], Ivan Saldikov[1], Rynat Bahdanovich[1] and Alexander Gerasimov[2]

[1]*National Research Nuclear University MEPhI (Moscow Engineering Physics Institute), Department of Theoretical and Experimental Physics of Nuclear Reactors, 31, Kashirskoe Shosse, 115409, Moscow, Russia.*
[2]*Institute for Theoretical and Experimental Physics, Department of safe nuclear reactors, 25, Bolshaya Cheremushkinskaya ul., 117218, Moscow, Russia.*

Abstract. System of test tasks is presented with the fast reactor BN-1200 with nitride fuel as prototype. The system of test tasks includes three test based on different geometric models. Model of fuel element in homogeneous and in heterogeneous form, model of fuel assembly in height-heterogeneous and full heterogeneous form, and modeling of the active core of BN-1200 reactor. Cross-verification of program codes was performed. Transition from simple geometry to more complex one allows to identify the causes of discrepancies in the results during the early stage of cross-verification of codes. This system of tests can be applied for certification of engineering programs based on the method of Monte Carlo to the calculation of full-scale models of the reactor core of the BN series. The developed tasks take into account the basic layout and structural features of the reactor BN-1200. They are intended for study of neutron-physical characteristics, estimation of influence of heterogeneous structure and influence of diffusion approximation. The development of system of test tasks allowed to perform independent testing of programs for calculation of neutron-physical characteristics: engineering programs JARFR and TRIGEX, and codes MCU, TDMCC, and MMK based on the method of Monte Carlo.

1 Introduction

The project "Proryv" was initiated in 2012. Its aim was development of new generation nuclear power technologies on the basis of closed nuclear fuel cycle with fast reactors including reactors of BN series. Specialized program codes should be developed and improved for this project. Validation of code complexes does not yet have a clearly defined methodology, mainly due to the inability to create for them a representative experimental base. We have to use the separate verification of major program codes included in such complexes. Test tasks are widely used for this purpose. Unfortunately, there are not so much test tasks dealing with fast reactors at the moment [1,2].

2 Classification of test tasks

Test tasks on neutron physical calculation can be divided into several categories, depending on the method of definition of the geometry, cross sections, and composition or applications and information sources:

1. Purely calculational (mathematical) test with specified coefficients for the solution of equations that has no real prototype (M - Mathematical test);

2. Calculation test, having a real reactor prototype with full initial data and all specified the necessary macro- and micro-cross sections and constants, which provides accurate solution (MP - Mathematical test with Prototype);

3. Test that has a real reactor prototype, however without required cross-sections. Concentrations of nuclides approximately correspond the real compositions. Some other characteristics can also be simplified in comparison with the prototype (PO - Prototype-Operational test);

4. Test, which simulates a real reactor without required cross sections. These tests are sometimes called operational tests. The concentrations of radionuclides and / or other data accurately correspond to any real state of the simulated reactor. As a rule, values of calculated functionals for these tests are based on the experimental data. They are used to validate the program codes. Tests of this type sometimes are not intended for public distribution, especially in the case of large efforts spent for evaluation of experimental data and in cases where there are potential buyers (O - Operational test).

Recently, calculation tests are often used instead of experimental tests because of high expenses of the lasts.

[a] Corresponding author: gvtikhomirov@mephi.ru

3 System of tests

The test system presented in this paper belongs to computing tests with BN-1200 as a prototype. The test system includes three consecutive tests that simulate different geometric models: fuel rods, fuel assemblies, and the core. Successive transition from a simple to a more complex geometry allows to identify the causes of discrepancies in the results on the early stage of code cross verification.

3.1 Test No. 1: fuel element

The purpose of this task is to calculate the burnup in a fuel rod as part of determining the characteristics of the reactor BN-1200. The task is solved both in a homogeneous and in heterogeneous geometry for a variety of types of plutonium isotopic compositions.

We consider the central fuel cell of fuel assembly of BN-1200 (Figure 1). Simple geometry allows to study in detail the calculation of burnup, compare models and approximations used in various calculation codes.

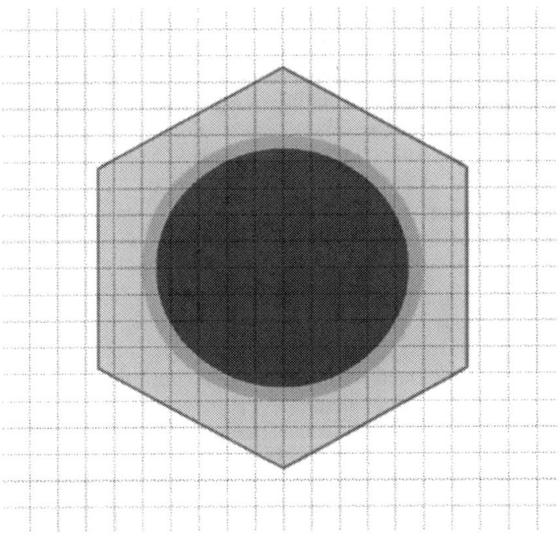

Fig. 1. Fuel element

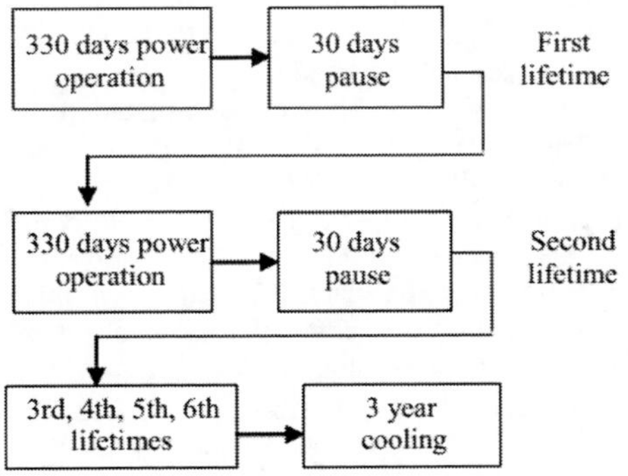

Fig.2. Structure of the calculated cycle

Calculations were performed for 6 fuel lifetimes with power operation guring 330 days and 30-days interval after power operatuin within the lifetime. So, the whole lifetime duration 360 days. The whole cycle considered for calculation (Figure 2) was about 9 years including 3 years cooling. Time steps $3\times10 + 6\times50$ days were used for calculations. Mixed nitride fuel was used in this calculation model. 28 isotopes were taken into account: ^{235}U, ^{236}U, ^{238}U; ^{237}Np, ^{239}Np; ^{238}Pu, ^{239}Pu, ^{240}Pu, ^{241}Pu, ^{242}Pu; ^{241}Am, ^{242m}Am, ^{243}Am; ^{242}Cm, ^{243}Cm, ^{244}Cm, ^{245}Cm; ^{14}N, ^{23}Na, ^{56}Fe, ^{52}Cr, ^{59}Ni, ^{96}Mo, ^{93}Nb, ^{55}Mn, ^{10}B, ^{11}B, ^{12}C. To account for the impact on the absorption of fission products, effective fission product can be used.

3.2 Test No. 2: fuel assembly

Calculation model of fuel assembly of the BN-1200 is considered in this test. We proposed to perform calculations consequently for the three models of fuel assemblies: heterogeneous height with 1 fuel layer, with 5 fuel layers and fully heterogeneous model (Figure 3). The first two models are the model of the "flat" core. Boundary conditions corresponding to reflection are assign only on the sides of the fuel assembly, while leakage of neutrons will be in the ends of fuel assembly. Scenario of irradiating wos the same as in test 1.

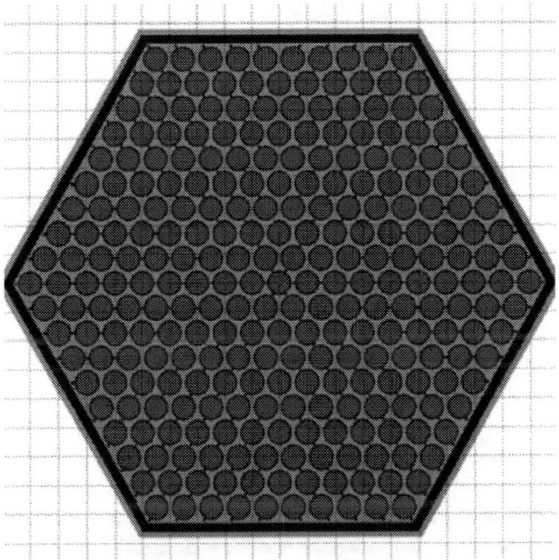

Fig. 3. Fuel assembly

3.3 Test No. 3: active core

This test was intended for simulation of reactor core of BN-1200. Conventionally, the test was divided into two sub-tasks. The first subtask was stationary one. Multiplication factor was the main calculated characteristic. Scenario of irradiation was the same as in the previous two tests. The calculation model was composed of hexagonal cells with compositions corresponding to the locatet in the cells fuel assemblies.

Seven types of fuel cells were selected radially to differ fuel compositions corresponding to the beginning of the cycle. Five axial zones with different fuel compositions were selected to consider different fuel burnup with the height of the core. The calculated model did not have an external zone of reproduction, raw material was admixed to the fuel assemblies of the core.

The second sub-task was intended for modeling of fuel recycle. Geometric model was the same as that described above. We supposed to calculate 60 fuel lifetimes in this test using the following algorithm of fuel load.

Duration of fuel lifetime was 360 days, including 330-days power operation and 30-days pause (Figure 2).

In the first three lifetimes, fuel assemblies with initial isotopic composition were loaded instead of discharged fuel assemblies. Discharged fuel assemblies were placed in 2-year intrareactor interim storage for cooling. Neutron fluxes in interim storage were assumed to be negligible. The duration of the external fuel cycle including reprocessing of fuel assemblies and the formation of a new load was assumed 1 year.

Beginning with the 4th lifetime, fuel loads were formed from the reprocessed fuel. Criticality for each lifetime provided by the choice of the mass fraction of plutonium. We considered 60-degree rotational symmetry of fuel assemblies, which was obtained from the real 360-degree schema of the load to reduce the computational expenses.

4 Calculated functionals

The following functionals were calculated in these tests: the effective multiplication factor, the isotopic composition in the end of fuel lifetime and after cooling, neutron induced reaction rates, temperature Doppler effect on fuel and steel, density factor for fuel and sodium, change of the reactivity with the fuel burnup, fuel reproduction parameters.

Analyzing the results of these calculations of test tasks, we can estimate an influence of the main sources of uncertainty on the error of neutron-physics calculations. Particular attention should be paid to the influence of the diffusion approximation, homogenization of fuel assemblies, and use of different constant libraries.

The developed system of test tasks can be used for verification of program codes for neutron-physical calculations that will improve the quality and reduce the time of certification of specialized code complexes.

5 Test results

5.1. Program codes for calculations

MMKKENO is program to calculate nuclear reactors by Monte Carlo method in multigroup approximation with scattering indicatrix in the P_n – approximation. The program includes modified geometric modules from the program MMKFK HRAN, OOBHEX and HEXBL. The program is designed to calculate the physical characteristics of nuclear reactor in the three-dimensional geometry. Program can calculate the following key characteristics: the effective multiplication factor; group fluxes; density of fissions; group absorption, fission, and leakage. The program has a number of possibilities, which require special skills and knowledge. One of these possibilities is introduction of areas with different value of neutrons. It allows to focus higher share of drawn neutron trajectories in areas with high value of neutrons. It also enables to use differential albedo, allowing to avoid the draw of neutron trajectories in the reflector. Use of these capabilities when solving criticality task allows to save computation time. However, the lack of experience may cause serious errors in the results.

MCU is a program aimed at modeling the transport of neutrons, gammas, electrons, and positrons in different three-dimensional systems by the Monte Carlo method. The program is under development during the period since 1982. Package MCU has a modular structure. Today the latest version of the package is MCU 5. This version contains the following modules: physical module, geometric module, source module, registration module, transport module, control module, burnup module, and others. Program uses each of the modules at certain calculation stage. The user creates an input file for calculation tasks, which looks like a set of text commands. In addition to single-processor mode, multiprocessor mode of calculation is possible. The paralleling is based on software Message Passing Interface (MPI). It should be noted that at the stage of calculation, each processor gets its own independent sequence of random numbers that does not overlap with sequences from other processors. An important step of calculation is the choice of libraries of evaluated nuclear data and the optimal scheme of operation of the submodules.

JARFR is a set of programs designed for three-dimensional calculation of neutron-physical characteristics of nuclear reactors in the multi-group diffusion approximation of neutron transport equation. The system of preparation of neutron-physical constants CONSYST/PRECONS based on libraries BNAB-93 and BNAB-RF is used as system of constants. The complex of programs JARFR is certified by Gosatomnadzor for neutron-physical calculations of fast reactors and implemented in a number of organizations. The program was developed by Scientific Research Center "Kurchatov Institute".

TRIGEX is a set of programs designed for serial (engineering) calculations of neutron-physical characteristics of fast reactors [5]. A set of programs uses internal program CONSYST to prepare constants. To use the obtained constants in the diffusion code TRIGEX, diffusion coefficient is calculated in this code by means of the transport approximation with the possibility of a convolution with the zero or first harmonic.

TDMCC (Time dependent Monte Carlo Code) is a program realizing a method of Monte-Carlo. This multi-purpose program can solve the tasks of evaluation of criticality and nuclear safety of both individual assemblies and active cores of the reactor plants, to evaluate radiation safety, to calculate the protection from radiation, to choose the parameters in order to obtain the specified value of the multiplication factor, and to perform calculations of fuel burnup and model fuel cycles of reactor installations. The program was developed in Russian Federal Nuclear Center – VNIIEF.

The multi-group calculation methods for fast reactors require the libraries of group cross sections covering practically the entire energy region from a few electron volts up to 15 MeV. The work on the preparation of group constants for fast reactors in our country was carried out in IPPE in Obninsk under the leadership of I. I. Bondarenko and M. N. Nikolaev. The system of constants BNAB is currently the main cross-section library, which is used in most calculations of fast reactors. The system of constants is periodically improved and updated with cross sections of new nuclides. Format of library of constants BNAB and its data are widely used abroad for calculation of reactors. The main advantage of this system is the ability of correct account of the resonance effects using the coefficients of self-shielding of cross sections.

5.2. The results of testing on model of fuel element (Test No. 1)

The development of the system of test tasks provided a possibility to begin cross-verification of programs for neutron-physical calculation used for the design of innovative reactor plants of new generation. This paragraph deals with the results of cross-verifications of engineering programs (JARFR and TRIGEX) and programs that provide the solution of neutron transport equation by Monte-Carlo method (MCU, SCALE – MMK, and TDMCC).

Test task No. 1 includes significant number of different variants of initial load. This paragraph presents the results of one of them. We considered the following functionals: Keff, the composition of unloaded fuel, reaction rates, etc. Values of these functionals calculated by each program were compared with average values over all programs. Difference between average value and separate value divided by average value was the respective deviation of this value. We will below refer to them simply as deviations having in view respective deviations.

Table 1 presents deviation of multiplication factor. Table 2 presents deviations of the compositions of unloaded fuel after 3240 days (9 years). Table 3 presents deviations of the rate of fission reaction, and table 4 presents deviations of reaction rate of capture.

According to the results of this work we can conclude about the influence of the diffusion approximation. An attention should be payed on the changes in deviations during a time. This is due to the difference of the chains of nuclide transformation incorporated and implemented in programs.

Table 1. Deviations of multiplication factor in Test No.1, % Δk/k

Moment of time	Progtams				
	MCU	SCALE - MMK	TDMCC	JARFR	TRIGEX
Beginning of cycle	-0,45	0,18	-0,53	0,3	0,49
End of 360-day lifetime	-0,51	-0,01	-	0,4	0,12
End of 9-yr cycle	-0,29	0,34	-	0,09	-0,14

Table 2. Deviations of fuel burnup inTest No. 1, %

Nuclide	Program			
	MCU	SCALE - MMK	JARFR	TRIGEX
^{238}U	-0,12	0,08	0,08	-0,05
^{238}Pu	-3,81	-0,35	6,77	-1,98
^{239}Pu	0,16	0,26	-0,32	-0,09
^{240}Pu	-0,53	-1,94	1,13	1,42
^{241}Pu	8,32	1,66	-11,54	3,87
^{242}Pu	-0,68	-1,21	0,87	1,05
Sum U	-0,12	0,14	0,01	-0,04
Sum Pu	0,1	-0,23	-0,23	0,37
Sum HM	-0,1	0,02	0,02	0,06

Table 3. Deviations of fission rate in Test No. 1, %

Nuclide:	Program			
	MCU	SCALE - MMK	JARFR	TRIGEX
^{235}U	1,33	-1,17	0	-0,2
^{238}U	-4,26	2,39	0,75	0,87
^{238}Pu	-2,48	1,85	0,05	0,48
^{239}Pu	0,67	-0,35	-0,05	-0,28
^{240}Pu	-1,38	2,89	-0,97	-0,65
^{241}Pu	1,14	-1,54	0,09	0,27
^{242}Pu	-3,11	5,04	-1,31	-1,01

Table 4. Deviations of capture rate in Test No. 1, %

Nuclide	Progran			
	MCU	SCALE - MMK	JARFR	TRIGEX
^{235}U	0,89	2,69	-1,71	-2,01
^{238}U	-0,27	-0,58	0,56	0,28
^{238}Pu	3,5	14,42	-11,5	-11,32
^{239}Pu	1,45	-0,93	0,01	-0,56
^{240}Pu	-4,11	0,71	1,82	1,35
^{241}Pu	4,17	3,97	-4,29	-4,58
^{242}Pu	0,04	-2,91	1,6	1,15
N	4,96	3,48	-4,94	-4,3
Na	5,15	-0,45	-2,29	-2,83
Fe	-0,33	-0,06	0,04	0,35
Ni	1,03	-5,23	1,83	2,02
Mo	1.28	-0,09	-0,57	-0,64
Nb	4,17	-12,35	3,41	3,02
Mn	-1,27	-10,87	5,7	4,78

5.3. The results of testing on model of fuel assembly (Test No. 2)

According to the developed system of test task, cross-verification of programs on the model of fuel element should be followed by their comparison on the model of fuel assembly. Test No. 2 deals with calculations on model of fuel assembly. This paragraph presents the results of the comparison calculations by the following programs: MCU, JARFR, SCALE-MMK and TRIGEX.

We performed comparison using following parameters: the dependence of the multiplication factor on time, the composition of unloaded fuel, reaction rates, etc. Table 5 presents the deviations of the multiplication factor. We see that deviation does not exceed 1%. For a given accuracy of calculations, this deviation demonstrates good agreement between the results of calculations obtained by different programs.

Table 5. Deviations of multiplication factor in Test No. 2 % Δk/k

Moment of time	Program			
	MCU	SCALE - MMK	JARFR	TRIGEX
Beginning of cycle	-0,59	-0,24	0,37	0,45
End of 360-day lifetime	-0,69	-	0,32	0,36
End of 9-yr cycle	-0,23	-0,12	0,21	0,15

Table 6 presents deviations of the rates of fission and table 7 presents deviations of the rates of capture.

Table 6. Deviations of fission rates in Test No. 2, %

Nuclide	Programs			
	MCU	SCALE - MMK	JARFR	TRIGEX
^{235}U	2,34	-0,99	-0,61	-0,8
^{238}U	-4,58	1,96	3,51	-1,27
^{238}Pu	-2,26	1,55	1,02	-0,4
^{239}Pu	1,07	-0,2	-0,2	-0,68
^{240}Pu	-1,53	2,33	0,92	-1,84
^{241}Pu	2,2	-0,93	-0,4	-0,93
^{242}Pu	-3,42	4,21	1,23	-2,39

Table 7. Deviations of capture rates in Test No. 2, %

Nuclide	Programs			
	MCU	SCALE - MMK	JARFR	TRIGEX
^{235}U	2,56	3,73	-3,4	-3,32
^{238}U	0,86	-0,23	-0,51	-0,12
^{238}Pu	5,62	15,33	-13,96	-13,39
^{239}Pu	3,62	0,72	-2,54	-2,05
^{240}Pu	-1,53	2	-0,33	-0,21
^{241}Pu	6,22	5,24	-6,27	-6,68
^{242}Pu	2,68	-1,61	-0,38	-0,8

6 Conclusions

System of test task is developed. Fast neutron BN-1200 reactor with nitride fuel was prototype for these tasks. The system of test task consists of three sequential tests based on different geometric models: fuel element, fuel assembly, and the active core. This system was used in independent testing of programs for calculation of neutron-physical characteristics: engineering programs JARFR and TRIGEX, and codes MCU, TDMCC, and MMK based on the method of Monte Carlo.

We selected following main functionals for cross-verification of codes: the effective multiplication factor, the change in isotopic composition of fuel during burnup, and rates of capture and fission reactions.

Analysis of the results of calculations of test tasks and their comparison allows to estimate the influence of the main sources of uncertainty on the accuracy of neutron-physical calculations.

The system of test task presented in this paper may be used in the verification of programs for neutron-physical calculations, which can be applied for the design and exploiting of new generation reactors.

References

1. IAEA-TECDOC-1623. February, 2010.
2. S. Belov, M. Farakshin, Working Material of the Research Coordination Meeting, Vienna, 20-24 November 2000. IAEA-RC-803.2. IWG-FR/103, 274-324.
3. J. Tommasi, Proposal for a benchmark on EFR calculations (SPRC/LEDC).
4. The comparaison of calculational results of efr-benchmark obtained by M. Semenov, J Tommasi, A. Zaetta, AIDA-MOX Theme 2. Utilization of Weapon Grade Plutonium in Reactor. Obninsk, Russia, April 24th-29th 1994.
5. A. Danilychev, A. Seregin, V. Stogov, Working Material of the Research Coordination Meeting, Vienna, 20—24 November 2000. IAEA-RC-803.2. IWG-FR/103, 342-345

DOI: 10.1051/epjconf/201715305014

Radiological environment within an NPP after a severe nuclear accident

Karin Andgren[1,a], Karin Fritioff[1], Anna Maria Blixt Buhr[1] and Tommi Huutoniemi[1]

[1]Vattenfall AB, SE-169 92 Stockholm, Sweden

Abstract. The radiological environment following a severe nuclear accident can be visualised on building layouts. The direct radiation in an area (or room) can be visualized on the layout by a colouring scheme depending on the dose rate level (for example orange for high gamma dose rate level and purple for an intermediate gamma dose rate level). Following the Fukushima accident, a need for update of these layouts has been identified at the Swedish nuclear power plant of Forsmark. Shielding calculations for areas where access is desired for severe accident management have been performed. Many different sources of radiation together with different types of shielding material contribute to the dose that would be received by a person entering the area. External radiation from radioactivity within e.g. pipes and components is considered and also external radiation from radioactivity in the air (originating from diffuse leakage of the containment atmosphere). Results are presented as dose rates for relevant dose points together with a method for estimating the dose rate levels for each of the rooms of the reactor building.

1 Introduction

In case of emergency and during drills, layouts describing the radiological environment within the reactor building are used by the emergency preparedness organisation or the first responders team. A mapping of the radiological environment can also provide support when planning accident management. Existing layouts for the three boiling water reactors of the Swedish nuclear power plant Forsmark, need to be updated. New equipment is installed following Fukushima, updated accident management strategies need to be considered and results from recently updated calculations of core inventory release fractions need to be taken into account. A project aiming to provide dose estimations for relevant dose points has therefore been carried out by Vattenfall Engineering on behalf of Forsmark. The project is on-going and it will be finalized during the spring 2016.

2 Severe accident scenario

Released fractions of the core inventory is based on an extremely unlikely worst case scenario, when no safety systems are functioning during 24 hours. Since no core cooling is assumed to be functioning during the scenario, a complete core melt will occur. The containment pressure can be very high in these types of scenarios and all operating reactors in Sweden therefore have passive filtered venting systems. This system is designed to filter and retain the volatile fission products (e.g. I and Cs) during depressurization of the containment. Noble gases will be released unfiltered to the environment. The selected scenario has a highly conservative approach, however, the assumptions regarding shielding, source geometries, densities, etc. aims to be as realistic as possible.

3 Source term

In NUREG-1465 [1], typical accident source terms for light-water nuclear power plants are listed. However, release fractions are (slightly) lower than results from plant-specific simulations using the modular accident analysis program, MAAP. Temporal resolution is also much better in the MAAP results. Therefore, the source term in the containment was decided using the plant-specific core inventories together with plant specific calculations of the release fractions. Radioactive nuclides of the following elements were included in the source term: Xe, Kr, Cs, I, Te and Sb. The source term was determined for various times after the accident initiation, namely 3 h, 6 h, 8 h, 21 h, 24 h, 7 d and 30 d.

The amount of gas and water is also calculated by MAAP. The source term in the gas volume as well as the water volume within the containment was therefore specified in Bq/m^3 for each of the time steps.

4 Contaminated systems

During the scenario several systems outside the containment will be contaminated at certain times depending on the possibility to start a system (e.g. a cooling system). It will then contain contaminated media

[a] Corresponding author: karin.andgren@vattenfall.com

and the activity concentration is assumed to be the same as that of the activity concentration inside the containment at that time.

An example is the containment spray system, used for depressurization and washing out aerosols from the containment gas. Ideally, the system will receive water from an external uncontaminated water tank. However, there is also a possibility to take water from the contaminated water of the condensation pool within the containment. Dose rates from the system containing contaminated water were determined.

5 Shielding calculations

Shielding calculations has been performed using MERCURAD for different pipe dimensions, components (e.g. heat exchangers) and also for the containment itself. For complex geometries, an Ecxel-based tool was developed.

5.1 Containment

The containment has a height of about 40 m and a diameter of about 20 m. Initial water level is about 10 m above the containment bottom. Radiation dose rates outside the containment (within the reactor building) were calculated. Core melt on the containment floor and containment penetrations were given special consideration. The figure below illustrates a pipe penetration and its isolation valve.

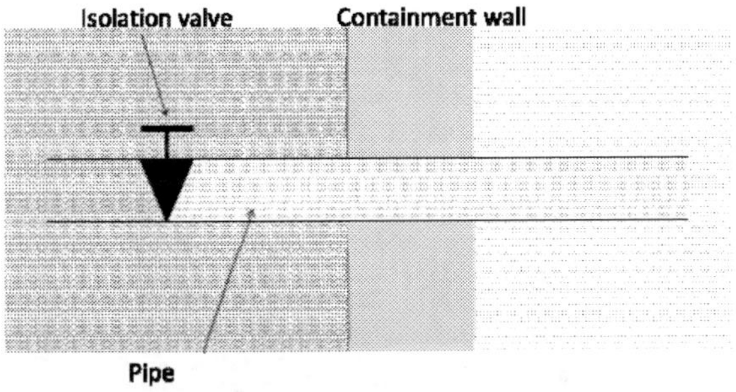

Figure 1. Pipe penetrating the containment

Other types of penetrations include personnel airlocks, electrical penetrations and equipment hatches. The shielding dimension and material varies depending on the type of containment penetration.

5.2 Pipes and components

Shielding calculations were done component wise for e.g. heat exchangers and pipes of different standardised geometries.

5.3 Complex geometries

Several rooms, interesting from accident management point of view, include several pipes of different dimensions and directions. In addition, the external radiation from the containment has to be considered (including the effect of relevant penetrations). There may also be radiation sources in adjacent rooms which has to be taken into consideration. An inventory of radiation sources for rooms of interest (using e.g. drawings and photographs) was made by engineers at Forsmark. Since there is a vast amount of complex geometries that needs to be modelled, the use of MERCURAD would be too time-consuming and a dedicated tool was therefore developed. A calculated unshielded flux was reduced using an Excel model that approximates attenuation and the build-up flux in different shielding materials and geometries.

6 Dose estimations

Resulting dose rates will be presented for each of the dose points that has been identified as relevant from an accident management perspective. This includes the access route to relevant areas,

In addition, a method will be developed for determining the dose in each room of the reactor building. There are about a hundred rooms in each of the three reactor buildings at Forsmark. The method for estimating the dose in all areas will be based on shielding calculations for single components and knowledge of attenuation for different thicknesses of shielding material (walls, doors, non-radioactive components, etc.)

In addition to the dose rate from external radiation from shielded radioactivity (within e.g. pipes and components), the dose rate from airborne radioactivity needs to be considered. Airborne radioactivity (dominated by the noble gases Xe and Kr) will leak via isolation valves, since the leakage rate is never zero, and into the reactor building. During a station blackout, the emergency ventilation system is unavailable and dose rates due to the diffuse leakage needs to be considered. However, if the power supply is working, the emergency ventilation system is used and the airborne radioactivity is then negligible.

7 Results

The project will be finalized during the spring 2016. Conclusion and results of the activity will be presented at the conference.

References

1. L. Soffer, S.B. Burson, C.M. Ferrell, R.Y. Lee and J. N. Ridgely NUREG-1465 (1995)

Update of the Nuclear Criticality Slide Rule for the Emergency Response to a Nuclear Criticality Accident

Matthieu Duluc[1,a], *Aurélie* Bardelay[1], *Cihangir* Celik[2], *Dave* Heinrichs[3], *Calvin* Hopper[2,b], *Richard* Jones[4], *Soon* Kim[3], *Thomas* Miller[2], *Marc* Troisne[5] and *Chris* Wilson[4]

[1]Institut de Radioprotection et de Sûreté Nucléaire (IRSN), B.P. 17, 92262 Fontenay-aux-Roses Cedex, France
[2]Oak Ridge National Laboratory (ORNL), P. O. Box 2008, MS-6170, Oak Ridge, Tennessee 37831, United States
[3]Lawrence Livermore National Laboratory (LLNL), 7000 East Avenue, Livermore, California 94551, United States
[4]Atomic Weapons Establishment (AWE), Aldermaston, Reading, RG7 4PR, United Kingdom
[5]Millennium, 16 avenue du Québec – SILIC 628, 91945 Courtabœuf, France

Abstract. AWE (UK), IRSN (France), LLNL (USA) and ORNL (USA) began a long term collaboration effort in 2015 to update the nuclear criticality Slide Rule for the emergency response to a nuclear criticality accident. This document, published almost 20 years ago, gives order of magnitude estimates of key parameters, such as number of fissions and doses (neutron and gamma), useful for emergency response teams and public authorities. This paper will present, firstly the motivation and the long term objectives for this update, then the overview of the initial configurations for updated calculations and preliminary results obtained with modern 3D codes.

1 Introduction

In 1997, Oak Ridge National Laboratory published the report "An Updated Nuclear Criticality Slide Rule" [1-2], as a tool for emergency response to a nuclear criticality accident. A similar document was produced by the Institut de Radioprotection et de Sûreté Nucléaire in 2000 [3].

This kind of document "permits continued updating of information during the evolution of emergency response, including exposure information about accident victims, estimates of potential exposures to emergency response re-entry personnel, estimates of future radiation field magnitudes, and number of fission ("fission yield") estimates, [1]" without precisely knowing the initial conditions leading to the criticality accident.

This document gives order of magnitude estimates of key parameters, useful for emergency response teams and public authorities. In practice, the "Slide Rule" provides estimates of the following information based upon variable times and distances from the accident:

- The magnitude of the number of fissions based on personnel or field radiation measurements,
- Neutron- and gamma-dose at variable unshielded distances from the accident,
- The skyshine component of the dose,
- Time-integrated radiation dose estimates,
- One-minute gamma radiation dose, and
- Dose-reduction factors for variable thicknesses of steel, concrete, and water.

The Slide Rule provides estimates for five unreflected spherical uranium systems that give general characteristics of operations typical of facilities licensed by the US Nuclear Regulatory Commission. An example of this Slide Rule is presented in Figure 1.

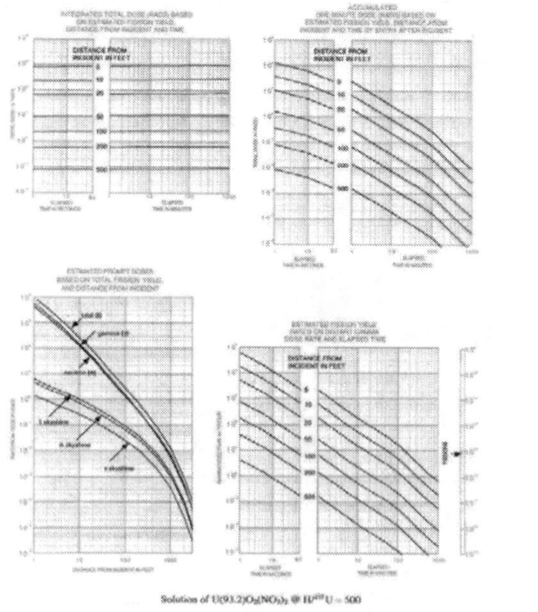

Fig. 1. "1997 Slide Rule" for the uranyl nitrate solution.

Several laboratories have determined the need to review, update, and expand the contents of the 1997

[a] Corresponding author: matthieu.duluc@irsn.fr
[b] Retired

Slide Rule. In particular, the only conversion factors used to provide doses (Henderson flux-to-dose factors [4]) are outdated, so additional conversion factors will be used. In addition, the results from old radiation transport tools and nuclear data used in originally developing the Slide Rule will be updated with modern tools and nuclear data and some simplifying assumptions will be relaxed.

2 Long term objectives of this update

As a result, a long term collaboration effort between AWE (UK), IRSN (France), LLNL (USA) and ORNL (USA), performed in the framework of the US DOE Nuclear Criticality Safety Program [5], began in order to update the Slide Rule with modern tools and add new configurations, taking into account the experience of several laboratories in using the 1997 Slide Rule.

The complete work, envisioned to spread over many years, will be divided into five steps:

1. Redo with modern radiation transport tools and nuclear data libraries, for the same configurations and assumptions, the calculations performed initially for the 1997 estimation of the doses;
2. Perform additional calculations to improve the quality/quantity of information given to the user of the Slide Rule in order to not only give a value but also the possible variations and the area of applicability of this value. These additional calculations might include:
   a. New configurations (impact of the geometry and composition of the source, new fissile media including plutonium systems, multiple layers of shielding, etc.);
   b. New flux-to-dose conversion factors (for dosimetry, radiological protection and instrumentation purposes);
   c. Impact of parameters on the result (sensitivity/uncertainty studies, such as thickness and composition of the ground, humidity and density of the air, etc.);
3. Review and improve the section regarding the estimation of the number of fissions;
4. Add other sections to the document like a section regarding actions to stop an on-going criticality accident (for example, standards with neutron poison).
5. Based on the previous work, the final task will be the development of a Slide Rule "application" for a handheld device (e.g. smartphone).

At the end, this work should improve the expertise for the real time response to a criticality accident in order to minimize the consequences of such an accident. In addition, this work will provide the opportunity to suggest experiments allowing the complete or partial validation of the tool results (benchmarking effort).

Another consequence of this collaborative effort might be the creation of "computer benchmarks" in order to test and validate the various variance reduction methods and to establish best practices when dealing with this kind of problem.

The present article is focused only on the first step with the presentation of preliminary results for the same configurations and assumptions used in the 1997 Slide Rule.

3 Initial configuration overview

3.1 Critical systems (source)

The critical uranium systems considered for the initial configuration of the 1997 Slide Rule were:

- Unreflected sphere of 4.95 wt% enriched aqueous uranyl fluoride, $U(4.95)O_2F_2$ and H_2O, solution having a hydrogen-to-^{235}U atom ratio of 410 (solution density = 2.16 g/cm^3),
- Unreflected sphere of damp 5 wt% enriched uranium dioxide, $U(5)O_2$, having a hydrogen-to-^{235}U atom ratio of 200 (uranium dioxide density = 2.5 g/cm^3),
- Unreflected sphere of 93.2 wt% enriched uranyl nitrate, $U(93.2)O_2(NO_3)_2$ and H_2O, solution having a hydrogen-to-^{235}U atom ratio of 500 (solution density = 1.075 g/cm^3),
- Unreflected sphere of 93.2 wt% enriched uranium sphere (metal density = 18.85 g/cm^3),
- Unreflected sphere of damp 93.2 wt% enriched uranium oxide, U_3O_8 plus water, having a hydrogen-to-^{235}U atom ratio of 10 (uranium oxide density = 4.15 g/cm^3).

Neutron and gamma doses were calculated as a function of distance from the surface of the critical sphere from 0.3048 to 914.4 m (1 to 3000 feet).

3.2 Model Description

The geometry for the 1997 Slide Rule models consisted of a simple 2-D air-over-ground configuration with the source located at the radial center of a right-circular cylinder. The radius and the height of the air cylinder was 1530 m. The center of the critical assemblies (spheres) were all 1 m above the ground. The ground is modelled as a 30.48 cm (1 ft) layer of concrete. The dimensions of the critical spheres and the composition of all materials can be found in references [1], [2] and [6].

Figures 2 and 3 present the model for these initial calculations. For more clarity, all the information needed to calculate the initial configuration was written in a specific document using a "benchmark format" [6]. Furthermore, a common file naming convention for the various cases has been adopted. The same principle will be used, in the future, for additional new configurations. An example is the following:

- SR-U-UN-G1-C1-d500-N.inp stands for:
  - SR: Slide Rule,
  - U: uranium,
  - UN: unreflected (no shielding),
  - G1: first case with ground,
  - C1: first case with uranium system (Uranyl fluoride (4.95%)),
  - d500: distance of 500 m from the critical sphere,
  - N: prompt neutron dose.

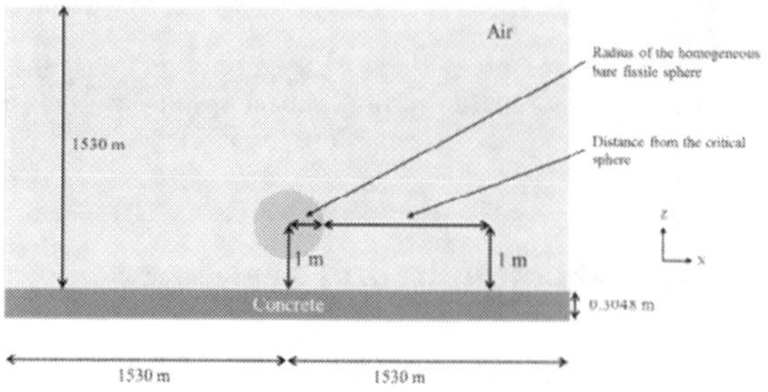

Fig. 2. X-Z elevation view of the initial configuration.

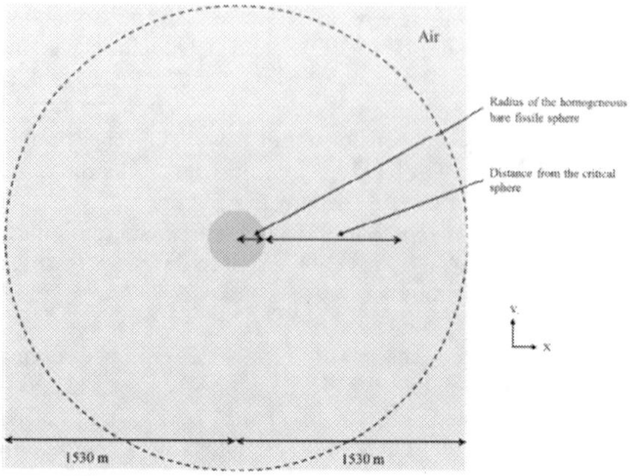

Fig. 3. X-Y plan view of the initial configuration.

3.3 Initial configuration calculations

The development of the 1997 Slide Rule information utilized the 2-D, discrete-ordinates, coupled neutron gamma-ray radiation transport code, DORT [17]. The source is modelled as a point at the source location with an energy spectrum equivalent to the leakage spectrum generated via a 1-D simulation with XSDRNPM [8]. GRTUNCL [18] was used to calculate the uncollided flux so DORT could use a distributed first collision source in order to mitigate ray effects.

The time-dependent sources were generated with the SCALE module ORIGEN-S which uses point-depletion methods to solve for the radioisotopic compositions after arbitrary irradiation/decay periods. This work assumed an instantaneous event and then tabulates the expected dose rates for various periods after the event for all five critical systems. Specifically, ORIGEN-S first estimates the production of radioisotopes instantaneously during the event, decays the concentrations to the selected times, and then computes the gamma-ray radiation source based on discrete gamma-ray line data. These sources are then utilized in 1-D discrete-ordinates calculations using the same methods, geometry, and codes as those of the prompt dose calculations to determine the leakage spectrum and then the 2-D methods are applied to perform delayed dose calculations.

4 Results and discussion for the initial configuration

This section presents and discusses the preliminary simulation results for the cases described in the previous section as well as the comparison with the 1997 Slide Rule. As specified in reference [6], without taking into account the skyshine calculations, at least 900 results are needed to cover all the cases. The laboratories used various codes and methods, presented hereafter. Every laboratory used the Henderson flux-to-dose conversion factors [4] in order to compare the results with the 1997 Slide Rule results.

4.1 Codes and methods used

4.1.1 MCNP

MCNP6.1 [7] was used with the ENDF/B-VII.1 cross sections library with a continuous energy representation of the cross sections. An F4 tally (i.e. track length estimate of cell flux) in the shape of a cylindrical shell was used to take advantage of the azimuthal symmetry of the problem. This cylindrical shell was 20 cm tall in the axial direction and 20 cm thick in the radial direction. For prompt doses, the total nu-bar was used. Weight windows, in space and energy, were used for distances beyond 500 m, using the MCNP Weight Windows Generator (with an iterative process, i.e. the weight windows used will provide a result and help creating a new and more effective weight windows that can be used for a new calculation).

For prompt doses, a two steps method was used. The first step is a static calculation (KCODE mode) to determine the distribution of fission neutron production inside the uranium sphere. Many possibilities were considered (and will be discussed later) but all the MCNP results presented were obtained with 20 spherical meshes (SMESH), having the same volume, in which the fission reaction rate were tallied. The second step used the results of the first step to describe a fixed source (SDEF mode) of fission neutrons. A Watt spectrum was used for the energy distribution. The prompt gamma and neutron doses were determined in the same calculation, the gammas being produced by the neutron interactions inside the uranium sphere. Indeed, in the second step, the fission neutron production is turned off (treated as absorption) but the gammas are produced (NONU = 0).

In order to produce the delayed gamma source, the coupling KENO/MONACO/COUPLE/ORIGEN from the SCALE package was used to calculate spectra and intensities. The criticality accident was assumed to occur in 1 μs. KENO calculates the distribution of fission neutron production inside the sphere. MONACO uses the KENO result and calculates the neutron flux inside the sphere. COUPLE uses the resulting MONACO neutron flux and creates problem dependent flux weighted cross sections to produce reaction rates. ORIGEN uses these reaction rates and performs the depletion and decay of the uranium systems. This method was used because it is very close to the method used to produce the 1997 Slide Rule. The MCNP calculations were performed on workstations. For prompt doses calculation, the simulation for each distance lasted about 15 hours on one processor (for less than 1E8 histories). The relative errors are below 5% and the 10 statistical checks from MCNP are passed except the Pareto slope check for some cases.

4.1.2 SCALE

SCALE 6.2 [8] was used with the ENDF/B-VII.1 cross section data. The libraries used with KENO and MAVRIC/MONACO both used a continuous energy representation of the cross sections. The CAAS analysis capability, coupling KENO and MAVRIC/MONACO, was used for this analysis, which is nearly an identical two-step methodology used for the MCNP analysis.

First, KENO was run with a cartesian mesh tally of the fission neutron production, which captured the asymmetry due to the ground 1 m below the center of each fissile assembly. Then MAVRIC/MONACO used the KENO tally as a fixed source, generated variance reduction parameters, and simulated the prompt doses. Region tallies were used in the model to calculate doses at the desired distances by introducing cells that were cylindrical shells in the actual problem geometry, as was done with MCNP. For the prompt dose calculations total nu-bar was used like the MCNP calculations.

The delayed gamma source was produced with the same coupling of KENO/MONACO/COUPLE/ORIGEN from the SCALE package that the MCNP calculations used. Once the delayed gamma intensities were obtained from the ORIGEN calculations, the fission mesh source was updated to represent delayed gammas instead of fission neutrons by changing particle type and energy spectrum. The spatial distribution of the fission neutron mesh source and delayed gamma mesh source are the same.

4.1.3 COG

COG 11.1 [9] was used with ENDF/B-VII.1 cross-section library data. COG is a general purpose, multi-particle, high-fidelity Monte Carlo code developed by Lawrence Livermore National Laboratory (LLNL). It provides accurate simulation results for complex 3-D shielding, criticality safety, and activation problems. A newly developed feature in COG 11.1 can generate, track and score delayed fission gamma (DFG) rays born between two given times [10]. Point-wise continuous cross-sections are used in COG and a full range of biasing options are available for speeding up solutions for deep penetration problems.

A direct one-step criticality/detector calculation method was applied for all neutron, and prompt and delayed gamma ray dose calculations. Each neutron and gamma particle is tracked from its birth in fission within the spherical fissile volume to its absorption in the system or score at the detectors at various distances in one single, massively parallel, COG supercomputer run with no variance reduction biasing applied. To activate the DFG option, DELAYEDPHOTONS (and associated time interval values) and DGLIB, are input in the BASIC and MIX blocks, respectively. Boundary-crossing detector option was activated to score the dose calculations.

4.2 Prompt doses results

4.2.1 Comparison

Prompt doses comparisons were performed between the 1997 Slide Rule and the modern codes. Figures 4 to 6 show this comparison for three cases and Figures 7 and 8 are focused on the code to code comparison for all cases.

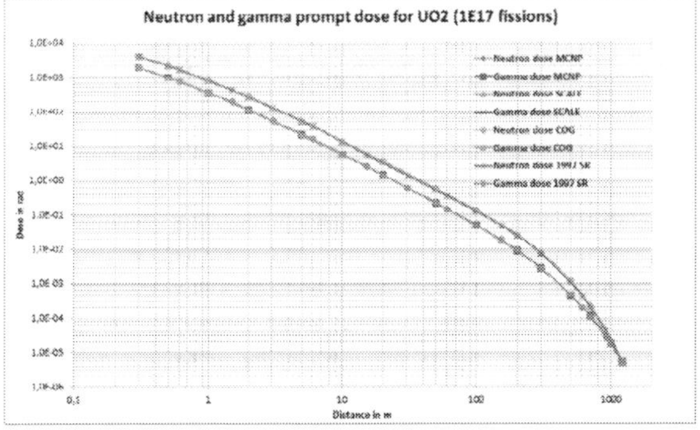

Fig. 4. Prompt doses comparison for uranium dioxide.

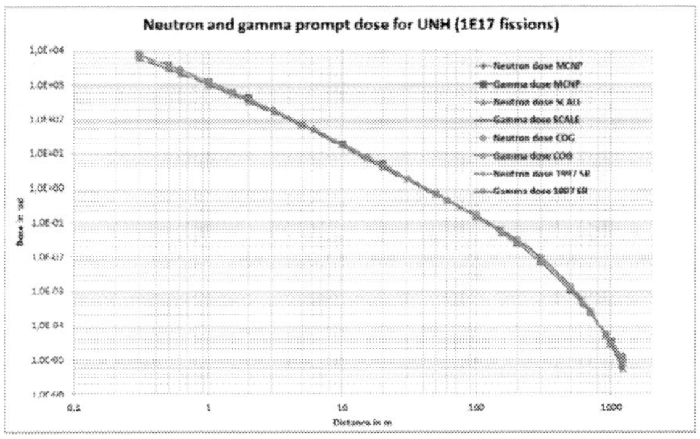

Fig. 5. Prompt doses comparison for enriched uranyl nitrate.

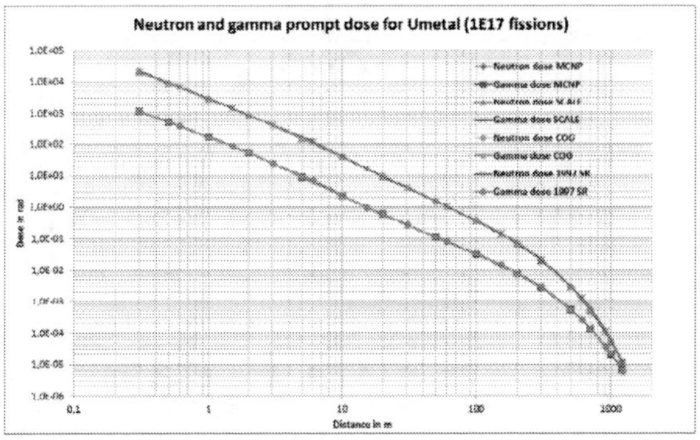

Fig. 6. Prompt doses comparison for enriched uranium metal.

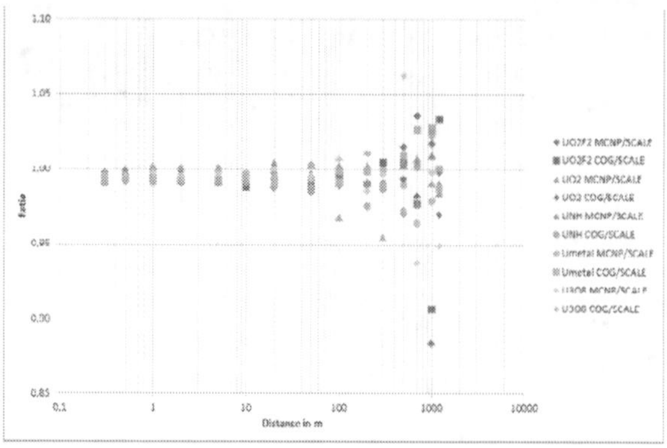

Fig. 7. Code to code comparison for all cases (neutron).

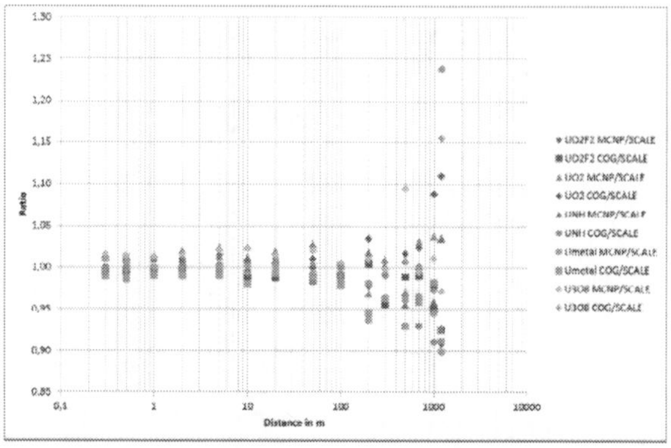

Fig. 8. Code to code comparison for all cases (prompt gamma).

A good agreement is observed between the 1997 Slide Rule results and the modern codes results for both prompt neutron and gamma doses for all cases. The discrepancies between codes are generally below 5%.

The corresponding relative error (2 σ) on the ratio between two codes is generally lower than 13% for neutrons (whatever the distance) and for gammas for distance lower than 300 m. After 300 m, the corresponding relative error on the ratio for gamma increases and can reach 17% for the MCNP/SCALE comparison and 53% for the COG/SCALE comparison.

This discrepancy between codes seems relatively independent of the distance for neutrons but increases for gamma doses with the distance. A possible explanation is that the prompt gamma contribution from the uranium sphere decreases with the distance increase. The gammas being produced by neutrons in the concrete and air, far from the uranium sphere, need to be correctly taken into account. This problem requires very good convergence not only for gammas created inside the uranium sphere, but also for neutrons that create gammas close to the detector. This problem seems more complicated than neutron dose determination and leads to higher discrepancies between codes.

One known difference between MCNP and SCALE gamma transport is MCNP's thick target bremsstrahlung model. This model accounts for the electromagnetic cascade of gammas and electrons that produce many low energy bremsstrahlung gammas. MCNP's thick target bremsstrahlung model accounts for these gammas, and allows users to not perform electron transport for geometries with thick shielding materials. All of SCALE's fixed-source radiation transport codes use gamma production data based on ENDF, which does not include this sort of bremsstrahlung. Regarding COG, whenever an electron-producing photon reaction occurs, COG checks whether the reaction occurred in a region enabled by the user for electron transport. If not (the case here), then the electron energy is immediately deposited.

4.2.2 Additional calculations

Many additional methods can be used to determine the doses, depending on the configurations. In particular, alternative ways to calculate the prompt doses were investigated with MCNP6.1 for some cases. These are divided into two categories.

The first category includes means to create the source and to use it (step one of the "two steps method"). One possibility is to use the Surface Source Write (SSW) card [11] to write the source points of a criticality calculation (KCODE mode), and then use it in a second calculation (via the Surface Source Read (SSR) card). It is an easy way to determine the source but, like the previous method (SMESH), one should pay attention to the convergence of the fission source location distribution, and run enough inactive batches. Besides, one thing to consider is that the continuum of the fission source is being approximated by a discrete set of points. It is then possible to run multiple instances of these same particles with different random number sequences. However it is not possible to generate any new points subsequently. Another limitation is that the sampling of the source variable can't be biased to improve the convergence rate of the problem.

The particle track (PTRAC) card can be used in order to write the starting energy of all energy neutrons and/or gammas produced by fission in order to be used in the second step. As an example, the Figure 9 illustrates the energy distribution of prompt neutron for uranyl nitrate.

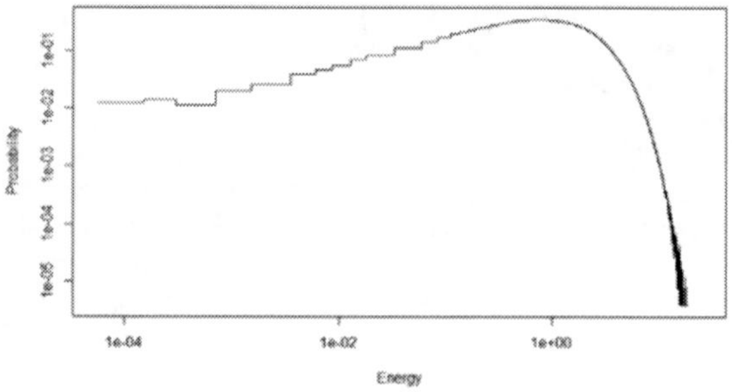

Fig. 9. Fission neutrons spectrum from PTRAC card for uranyl nitrate.

An additional possibility with the PTRAC card is to write the corresponding positions of the starting neutrons/gamma. In this case the resulting "method" looks like, in principle, the SSW/SSR method. This method, more accurate than taking the Watt spectrum, can have an impact on the results for some cases, in particular for highly "heterogeneous" systems (in terms of neutron spectrum and fissile material repartition).

Another possibility is to consider, for the second step, the source as a point source, which will save some computer time by not simulating particles that never leave the uranium sphere. This method needs to know the intensity, direction and spectrum of particles leaving the sphere. This method is presented in particular in the reference [12], using MCNP or SCALE. This also can be done with a deterministic code like XSDRNPM, which was used in the 1997 Slide Rule.

The second category includes variance reduction methods used to improve the calculation of the second step.

For example, the use of the F5 Tally (for a point or a ring) may improve the calculation results. It may also help the determination of the first weight windows, with the MCNP weight windows generation (WWG), that can be used later with an F4 Tally. On the contrary, the use of an F4 tally with a small volume (for example small sphere) is sometime necessary but will complicate the calculation. In this case, the use of DXTRAN sphere may be used to improve the particle sampling in a given region around the small volume. This technique and others like forced collision are discussed in [11].

Another possibility to improve the calculation is to use a deterministic code in order to obtain the adjoint flux and to generate the weight windows and the biased source for the Monte Carlo fixed source calculation, like the MAVRIC module of the SCALE package. For MCNP, this step can be done, for example, with ADVANTG [13] or ATTILA [14]. An example of adjoint flux is illustrated in the Figure 10.

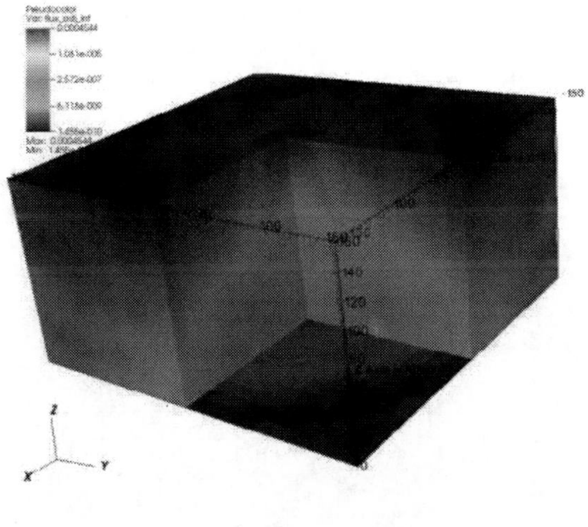

Fig. 10. Adjoint flux with ADVANTG for the case SR-U-UN-G1-C1-d700-G.

4.3 Delayed gamma doses

Delayed gamma doses comparisons were performed between the 1997 Slide Rule and the modern codes. Many decay times are considered in references [1] and [6] but only representative cases are presented hereafter. Figures 11 to 16 show these comparisons.

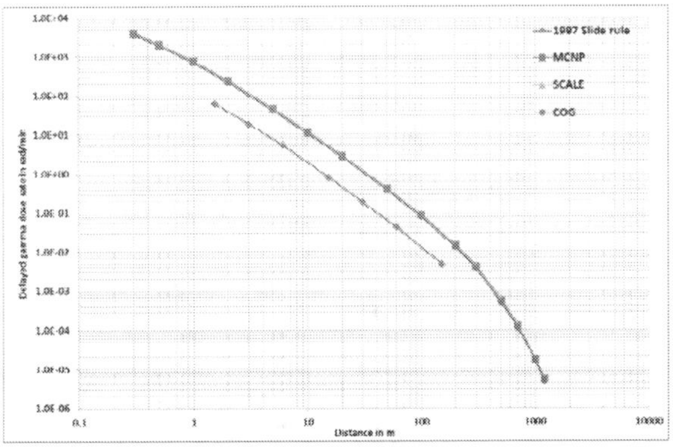

Fig. 11. Delayed gamma doses comparison for uranyl fluoride 1 s after the criticality accident.

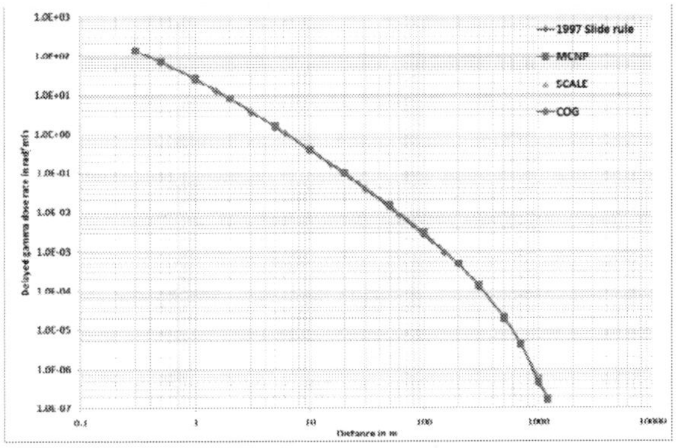

Fig. 12. Delayed gamma doses comparison for uranyl fluoride 1 min after the criticality accident.

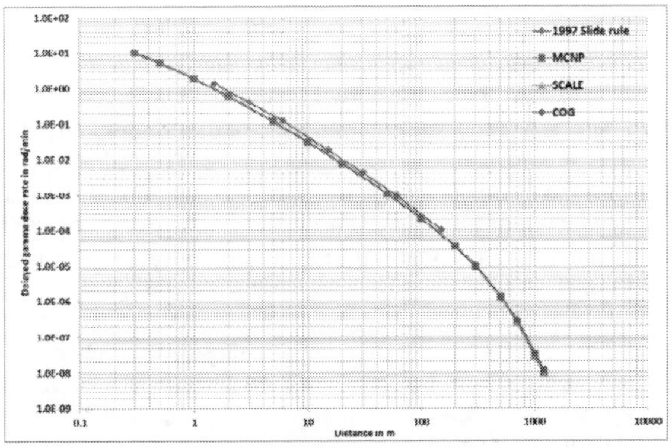

Fig. 13. Delayed gamma doses comparison for uranyl fluoride 10 min after the criticality accident.

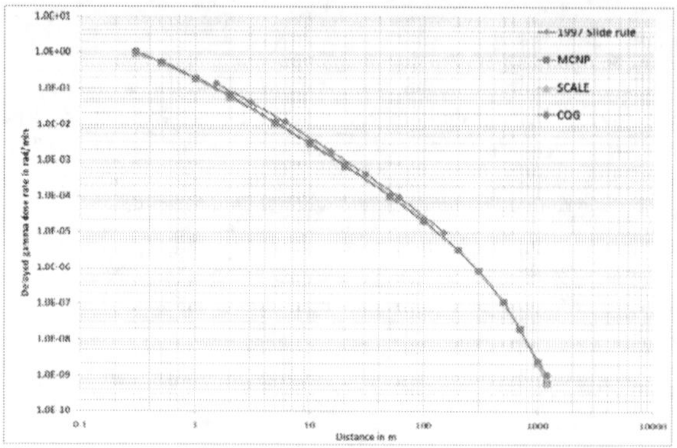

Fig. 14. Delayed gamma doses comparison for uranyl fluoride 100 min after the criticality accident.

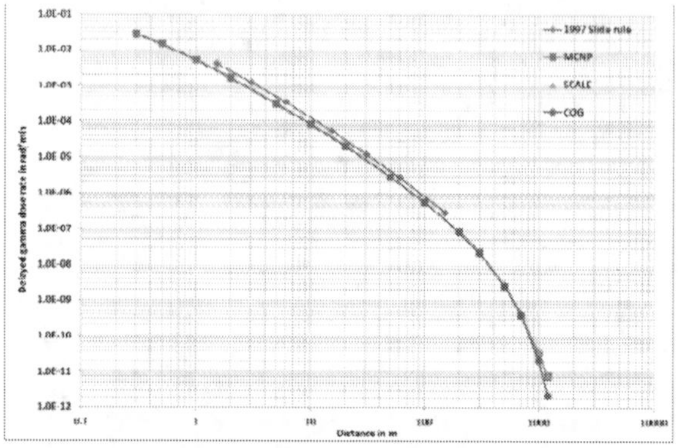

Fig. 15. Delayed gamma doses comparison for uranyl fluoride 1000 min after the criticality accident.

The previous figures present the results for the uranyl fluoride case but the following general observations are applicable for all cases:

- the modern code results are higher than the 1997 Slide Rule for times less than 1 minute, but as time increases (to 1 minute) this difference decreases,
- at 1 minute, the delayed modern code results agree very well with the 1997 Slide Rule,
- at time greater than 1 minute, the modern code results are slightly smaller than the 1997 Slide Rule, but this difference is fairly constant and is not a function of time above 1 minute.

These general observations might be slightly different for the different fissile media, because the characteristics of these media (in particular the enrichment and the moderation ratio) have an impact on the evolution of the isotopic composition, so on the delayed gamma source, which leads to some variations (see Figure 16 for example).

The complete explanation of these discrepancies is difficult to find because some assumptions made for the 1997 Slide Rule are unknown. However, the following elements can be suggested.

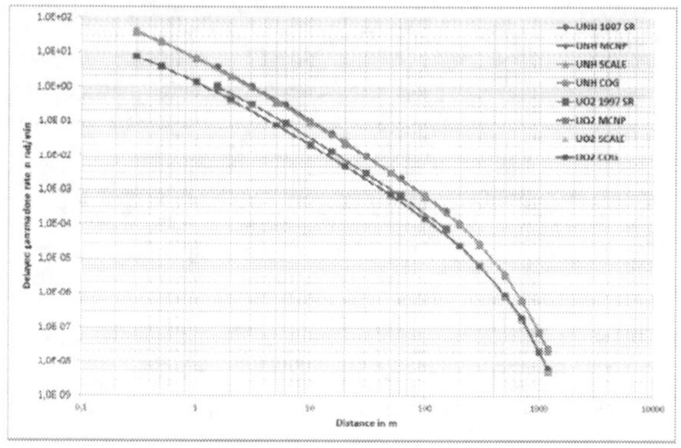

Fig. 16. Delayed gamma doses comparison for uranyl nitrate and UO_2 10 min after the criticality accident.

The first possibility to explain the discrepancies is the fact that ORIGEN data between 1997 and today has changed, especially with updated nuclear data. In particular for short decay times, the older gamma libraries were missing gamma spectral data for many of the short lived fission products [16]. This might be the best explanation for the discrepancies observed for results for times less than 1 minute. Use of 1997 nuclear data to run ORIGEN today was not tested here but could be done in the future in order to confirm and quantify the nuclear data effect.

It can also be noticed that, the modern code results are obtained by determining the instantaneous dose rate at a given time whereas the 1997 Slide Rule results considered an accumulated one minute dose, which take into account the intensity decrease of the source for short times. A part of the discrepancies between modern codes and 1997 Slide Rule may be explained because the intensity of delayed gamma greatly decreases for short decay time (in particular lower than 1 minute). As an illustration, the intensity of delayed gamma (given in gamma per second) as a function of the decay time after the accident is presented in Figure 17 for the uranyl fluoride case and shows that between 1 second and 1 minute, the intensity decreases by a factor ~30.

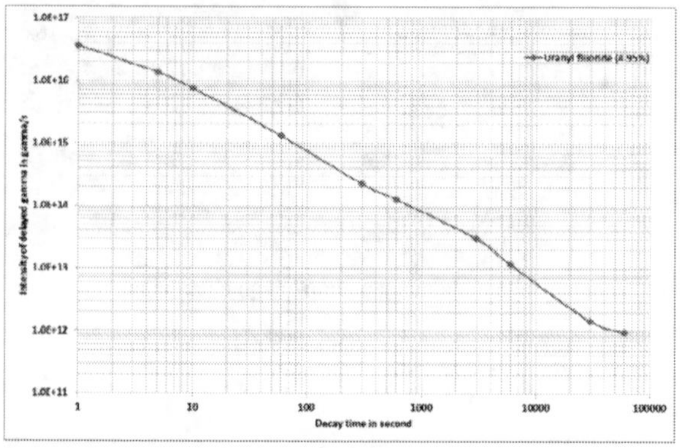

Fig. 17. Intensity of delayed gamma as a function of the decay time for the uranyl fluoride.

Finally, the 1997 Slide Rule states that "No delayed neutron contribution nor contributions from delayed gammas between 1 μs and 1 s were included in the dose curves". This assumption should have an impact on the results for the short times.

The code to code comparison is illustrated in Figures 18 to 22.

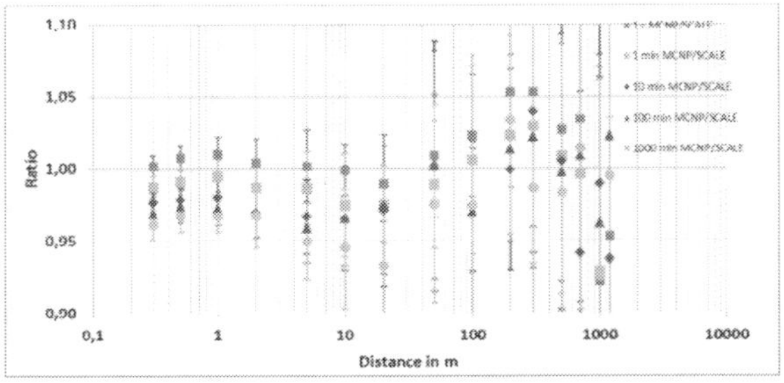

Fig. 18. Comparison between MCNP and SCALE for enriched uranium metal.

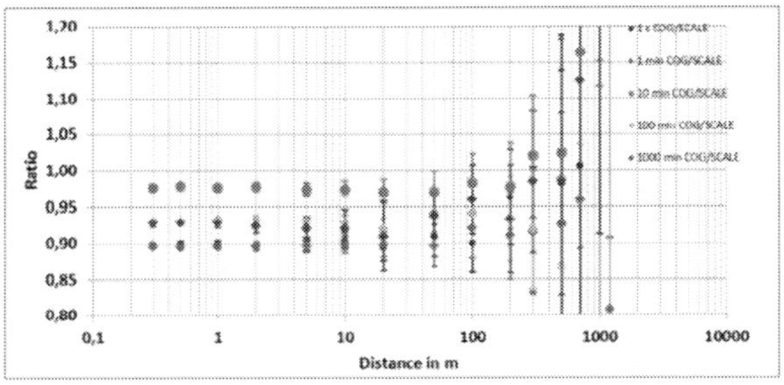

Fig. 19. Comparison between COG and SCALE for enriched uranium metal.

It can be seen that the discrepancies between codes increase with the distance after 500 meters. It may be explained by large uncertainties in the doses calculation results and not necessarily by a difference in the delayed gamma source. At 10 meters from the criticality accident (with no convergence issues), Figure 22 shows the comparison for all cases. Two behaviors can be observed. The comparison between MCNP and SCALE results shows a slight decrease trend with decay time. A possible explanation is that the delayed gamma sources were determined with the same "method" but with different options (in particular the energy binning) and code versions, which have an impact in particular on the sources intensity for long decay time.

The comparison between COG and SCALE shows an "up and down" behavior. The COG/SCALE parameter presents a slightly important discrepancy (0.82 – 0.95) for short time (less than 1 min), then a swing behavior (1.01 - 1.15) to finally come back to lower values (0.88 – 1.01). This behavior needs to be investigated.

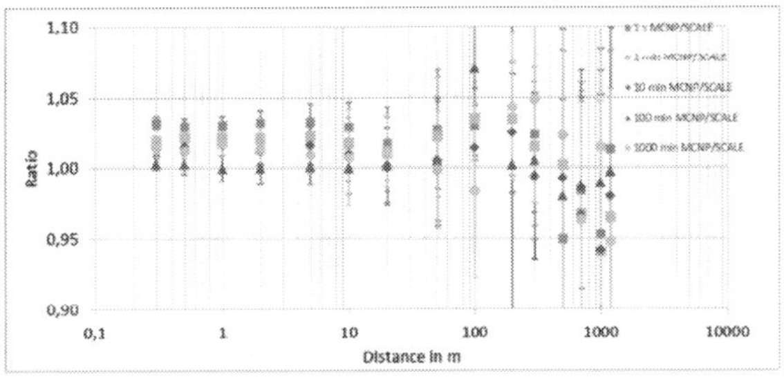

Fig. 20. Comparison between MCNP and SCALE for uranyl fluoride.

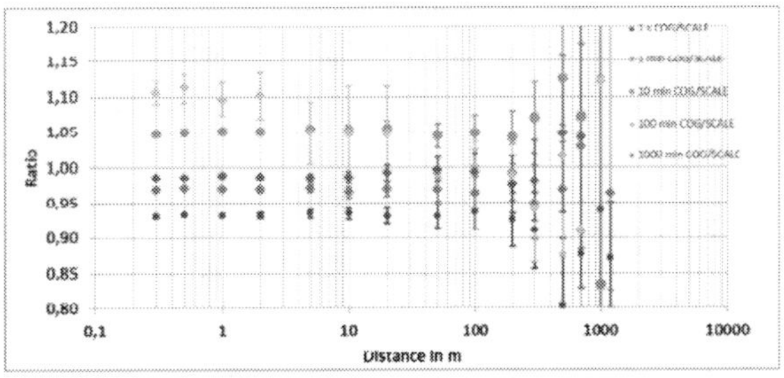

Fig. 21. Comparison between COG and SCALE for uranyl fluoride.

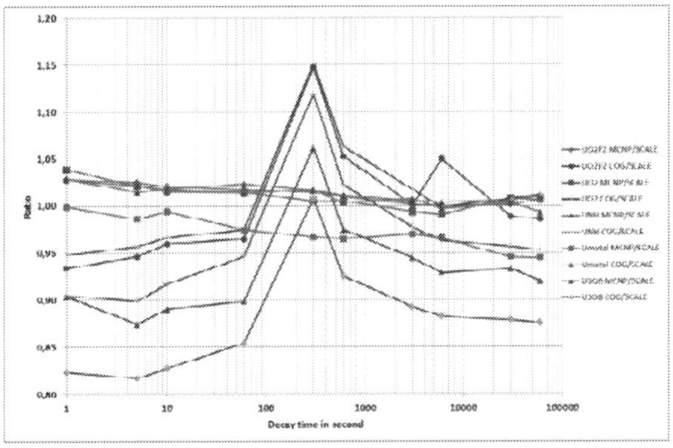

Fig. 22. Code to code comparison for all cases at 10 meters from the criticality accident.

5 Conclusions and perspectives

AWE (UK), IRSN (France), LLNL (USA) and ORNL (USA) began a long term collaboration effort in 2015 to update the nuclear criticality Slide Rule for the emergency response to a nuclear criticality accident. This document gives order of magnitude estimates of key parameters, useful for emergency response teams and public authorities. The first step of this update is to repeat with modern radiation transport tools, for the same configurations and assumptions, the calculations performed initially for the 1997 estimation of the doses.

Regarding the calculation of the prompt doses, the results from the modern tool used for this update are consistent. For the initial 2D configuration, the modern 3D tools results confirm the results obtained with the 2D tool used to create the 1997 Slide Rule. The impact of the use of new nuclear data is not visible on the update, with regard to the relative error (generally 5%). The interest of 3D capabilities will be developed for new configurations that will break the symmetry of the problem.

Regarding the determination of the delayed gamma source, some discrepancies occurred between codes. The detail analysis of these discrepancies was not the main purpose of this paper but additional effort should be done in order to understand the discrepancies. Other methods/codes should be investigated (like FISPACT or VESTA codes or the ACT card in MCNP) because only two methods were used for this paper.

In addition to the finalization of the calculations and the complete analysis of the results, effort will be devoted on the development and the calculation of new configurations (impact of the geometry and composition of the source, new fissile media including plutonium systems, impact of multiple layers of shielding, impact of the thickness and the composition of the ground, humidity of the air, etc.), using updated flux-to-dose conversion factors (for dosimetry, radiological protection and instrumentation purposes). The goal of this new work is to improve the quality/quantity of information given to the user of the "Slide Rule" in order to not only give a value but also the possible variations and the area of applicability of this value.

In parallel to this Slide Rule update effort (useful to limit the radiological consequences if a criticality accident should occur), this collaborative effort might be a good opportunity to create "computer benchmarks" in order to test and validate the various variance reduction methods and to establish best practices when dealing with this kind of problem (in particular fission source calculation).

Finally, this work will provide the opportunity to suggest experiments allowing the validation of the tool results (benchmarking effort). Indeed, the ICSBEP handbook for this subject is only composed of seven benchmarks, with only one case representative of a criticality accident [15].

References

1. B. L. Broadhead et al., "An Updated Nuclear Criticality Slide Rule: Technical Basis," Report NUREG/CR-6504 Vol. 1 ORNL/TM-13322/V1, (1997).
2. C. M. Hopper, and B. L. Broadhead, "An Updated Nuclear Criticality Slide Rule: Functional Slide Rule," Report NUREG/CR-6504, Vol. 2 ORNL/TM-13322/V2, (1998).
3. Fiches Accident-Type, IPSN internal report, (2000).
4. B. J. Henderson, "Conversion of Neutron or Gamma-Ray Flux to Absorbed Dose Rate," XDC 59-8-179, (1959).
5. NCSP website, http://ncsp.llnl.gov/
6. "Update of the nuclear criticality Slide Rule calculations - initial configuration," SR-U-UNREFLECTED-GROUND-001, IRSN internal report (2015).
7. J. T. Goorley et al., "Initial MCNP6 release overview – MCNP6 version 1.0," LA-UR-13-22934, (2013).
8. B. T. Rearden and M. A. Jessee, eds., "SCALE Code System," Oak Ridge National Laboratory report ORNL/TM-2005/39, Version 6.2. Available from Radiation Safety Information Computational Center as CCC-834, (2016).
9. "COG: A Multiparticle Monte Carlo Code Transport Code, Version 11.1," CCC-829, Released by Radiation Safety Information Computational Center (2015).
10. R. M. Buck et al., "COG11.1 Description, New Features, and Development Activities," pp. 910-922, International Conference on Nuclear Criticality Safety (ICNC 2015), Charlotte, NC, September 13-17, (2015).
11. B. C. Kiedrowski, "MCNP6 for criticality accident alarm systems – A Primer," LA-UR-12-25545 (2012).
12. T. M. Miller and D. E. Peplow, "Guide to performing computational analysis of criticality accident alarm systems," ORNL/TM-2013/211 (2013).
13. S. W. Mosher et al., "ADVANTG—An Automated Variance Reduction Parameter Generator," ORNL/TM-2013/416 Rev. 1 (2015).
14. "Attila User's Manual," Varian Security and Inspection Products, (2015).
15. T. M. Miller and al., "Neutron Activation Foil and thermoluminescent dosimeter responses to a bare pulse of the CEA Valduc SILENE critical assembly," ALARM-TRAN-AIR-SHIELD-001, International Criticality Safety Benchmark Evaluation Project, OECD/NEA/ICSBEP (2015).
16. I. Gauld, "Validation of ORIGEN-S decay heat predictions for LOCA analysis," PHYSOR 2006, Vancouver, BC, Canada, September 10-14, (2006).
17. M. B. Emmett and al., the DORT section of "A User's Manual for MASH 1.0 - A Monte Carlo Adjoint Shielding Code System," ORNL/TM-11778 (1992).
18. R. L. Childs, "GRTUNCL: First Collision Source Program," ORNL Informal Notes (1982).

Acknowledgements

This work was performed under the auspices of the U.S. Department of Energy by Oak Ridge National Laboratory and Lawrence Livermore National Laboratory under contracts DE-AC05-00OR22725 and DE-AC52-07NA27344, respectively, and was funded by the U.S. Department of Energy Nuclear Criticality Safety Program. IRSN authors would like to thank Sylvain BEYTOUT, from MILLENNIUM, for the preliminary calculations he did for this update.

Shielding analysis in the design phase of the new Emergency Operation Facility for Tihange Nuclear Power Plant

Gilles Genard[1,a], Romain Portal[1], Virginie Bouchat[1] and Serge Vanderperre[1]

[1]*Radwaste management – Decommissioning – Radiation protection, Tractebel, Ariane avenue 7, B1200 Brussels – Belgium*

Abstract. In the framework of the design studies for a new Emergency Operation Facility (EOF) for Tihange NPP, radiation protection analyses are needed to comply with effective dose rate criteria. In this aim, the shielding performed by, at the one hand, external walls and roof of the building and, on the other hand, internal walls, has been sized by means of MicroShield calculations. This paper explains how the calculations for external walls, doors, roof, floor and internal walls are made. The assumptions on the source terms and on the source geometry as well as the way the shielding is determined and the results of the sizing are presented.

1 Introduction

In the design phase of any new building on a nuclear site, such as a Nuclear Power Plant (NPP), considering the radiation protection aspects is mandatory to ensure workers and general population protection against ionizing radiations hazards. This is even more the case if radioactive contamination in the surrounding atmosphere of the building is anticipated or if radioactive sources are located in the building itself. As a consequence, designing shielding is necessary. At first, this shielding can be achieved by the building walls, the external and internal ones.

In the framework of the basic design studies for a new Emergency Operation Facility (EOF) for Tihange NPP, such studies are necessary. Indeed, this facility would be built on the site of the NPP and, due to its objective, which is to be used after a nuclear accident on one of the three units of the nuclear site, radioactive sources are expected around and inside the building. The basic design foresees that the inside of the EOF is subdivided in two zones: a Non Habitable Zone (NHZ) and a Habitable Zone (HZ). Consequently, appropriate shielding has to be put in place to ensure the safety of the workers managing the severe accident from the HZ of the EOF.

The source to consider for the shielding performed by external walls is the atmospheric contamination released around the building from the venting of one of the containment units through the filtered vent of this unit and the natural leak through the containment. The main source expected inside the building, to be considered to size internal walls, is due to the Heating, Ventilation and Air Conditioning (HVAC) filters that process external atmospheric contamination (and are therefore contaminated by the aforementioned atmospheric release) before supplying the air in the habitable zone of the building.

In the following, the paper tackles the study for the determination of minimum thicknesses of external walls, roof and door and of the internal walls, door and floor of the building.

2 Initial source term

The definition of the initial source term has been conducted by considering only one nuclear power unit affected by a severe accident. Note that, even if several units were affected by a severe accident, the wind cannot direct all the radioactive releases towards the EOF. To be conservative and knowing that the NPP units are similar enough (i.a. in terms of produced power) to present relatively comparable releases, the unit closest to the new EOF location has been taken into consideration.

Furthermore, in a conservative way, the wind is considered to be blowing the radioactive releases directly on the EOF. The wind speed is taken at 1.7 m/s, which is a relatively weak speed, limiting the atmospheric dispersion. No other aspect of the weather conditions (rain or dry atmosphere) is taken into account.

To determine the initial source term, several scenarios have been considered:

- The release of activity from the filtered vent,
- The release of activity through natural leaks of the containment unit,
- The release of activity from the filtered vent and through natural leaks of the containment unit.

For each scenario, activities of each radionuclide released in the atmosphere every 3 minutes during 10 days were calculated by means of the MELCOR code [1]. The period of 3 minutes is the time needed for the

[a] Corresponding author: gilles.genard@tractebel.engie.com

 DOI: 10.1051/epjconf/201715305016

contamination released from the vent chimney to reach the EOF with the considered wind speed.
The initial source term for each scenario, defined by the MELCOR code, features 1038 radionuclides. For each of these radionuclides, the released activities in function of time after initiation of the first venting, integrated during 3 minutes, are determined.

3 External shielding sizing

In the basic design, the EOF is 46 m long, 20 m wide and 4 m high and has a cuboid shape. Its external shielding will be assured by the four external walls and the roof, both in concrete, and, potentially, one external door in steel in one of those walls. The wall thickness will be identical for all four sides of the new building.

The dose rate calculations are performed by means of the MicroShield v5.05 code [2]. The latter is a calculation code, based on the "single point kernel" technique, which can determine gamma dose rate at a calculation point according to simple source and shielding geometries.

3.1 Sizing criterion

The shielding is meant to keep the effective dose rate below 1 mSv/h at 50 cm from the walls, which is considered to be the average walking distance from a wall, and at 1.30 m high, which is the average height of the personal dosimeters of the workers.

3.2 External walls and roof

3.2.1 Methodology

In order to ensure that the radiological criterion is complied with, three relevant locations in the EOF are considered (Points 1, 2 and 3, see Figure 1). The effective dose rate calculations are performed for each of these locations.

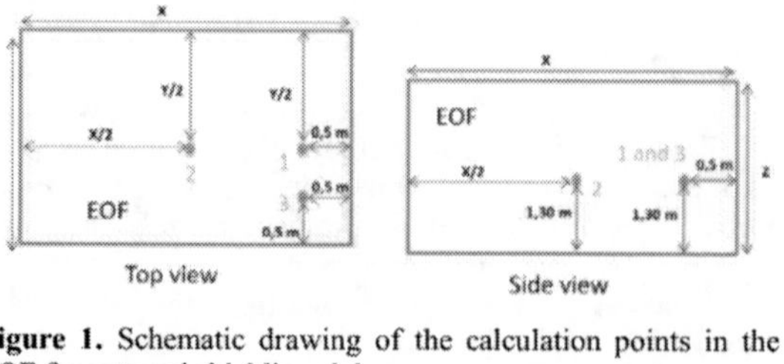

Figure 1. Schematic drawing of the calculation points in the EOF for external shielding sizing.

Point 1 is used to define the external wall thickness. To consider a conservative approach and keeping in mind that all walls will have the same thickness, this Point 1 is placed at 0.5 m of the nearest wall facing the potentially affected reactor building (RB) and in the middle of the same wall (to maximise dose rate). Point 2 is placed at the centre of the building and is used for the calculation of the roof thickness. Point 3 is placed in a corner at 0.5 m of two walls. This third point was chosen to check that workers located in this particular configuration are not exposed to a dose rate higher than the sizing criterion.

In order to comply with the sizing criterion of 1 mSv/h in the EOF, it is considered, for an initial stage, that half of the dose rate contribution inside the new building arises from the radiation through the lateral external walls and the other half of dose rate contribution arises from the radiation through the roof. Therefore, the calculations start by independently determining the minimal thicknesses for the external walls and roof by assuming that dose rates at Point 1 and Point 2 are below half of the sizing criterion (< 0.5 mSv/h). Afterwards, the total dose rate due to the contributions arising from all directions is calculated at Points 1 and 3 to verify the compliance with the 1 mSv/h criterion.

Therefore, the calculation steps are:

- Step 1: performing the sizing of the external wall (YZ plane in Figure 1) on the basis of the effective dose rate at Point 1 due to the contribution of the external radiation through this wall and by assuming that this dose rate contributes for half of the sizing criterion (< 0.5 mSv/h). The thickness obtained for this wall (side closest to the potentially impacted NPP unit) is then applied to the external walls of the three other building sides, conservatively,
- Step 2: performing the sizing of the roof (XY plane in Figure 1) on the basis of the effective dose rate at Point 2 due to the contribution of external radiation through the roof and by assuming that this dose rate contributes for half of the sizing criterion (< 0.5 mSv/h),
- Step 3: verification of the compliance with the sizing criterion (< 1 mSv/h) for Point 1 by calculating the total effective dose rate due to external radiation through the roof and the walls, except the wall on the side opposite to the releases (since this wall is the furthest from the potentially impacted NPP unit and the internal walls will also act as a shielding for the irradiation coming from there),
- Step 4: verification of the compliance with the sizing criterion (< 1 mSv/h) for Point 3 by calculating the total effective dose rate due to the external radiation through the same directions as Step 3.

To obtain the minimal thickness needed for the shielding, a curve representing the effective dose rate in function of the shielding thickness is obtained thanks to various calculations with different shielding thicknesses but the same source term. Then, the shielding thickness is determined as the first thickness value in centimetre beyond which the effective dose rate is below the considered sizing criterion (0.5 or 1 mSv/h depending on the step as explained previously).

3.2.2 Materials

The source term being an external one, it is composed of air with a density of 0.00122. The walls and roof are considered to be in concrete with a density of 2.35. The detailed compositions are the standard ones of MicroShield.

3.2.3 Source terms

For sizing external walls and roof, the retained activity for the MicroShield calculations is the maximum activity over time released from the filtered vent and through natural leaks of the containment unit (third scenario defined in Section 2). However, the release through natural leakage was proven to be negligible compared to the release from the filtered vent.

Due to the distance between the vent of the RB and the EOF, a dispersion of the released activity occurs. As illustrated in Figure 2, it is assumed that the total contamination vented during the time to reach the EOF is uniformly spread into two volumes: a cuboid volume and a pyramidal volume.

The cuboid volume, surrounding the EOF, is delimited by a base area which has the EOF dimensions increased by 5 m on all sides (see upper part of Figure 2) and a height identical to the one of the venting chimney (see lower part of Figure 2). This height is conservative since the radioactive releases, hotter than atmospheric air, should be dispersed at higher height than the vent of the RB and this decreasing effect on the source term is neglected here. The pyramidal volume joins the upper part of the venting chimney to the lateral side of the previously described cuboid volume. This volume is also very conservative considering the very low distribution angle of the releases.

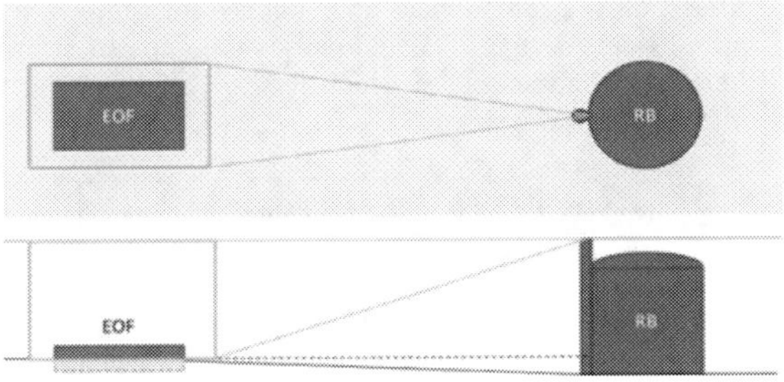

Figure 2. Schematic drawing of the source volume around the EOF coming from the vent of the RB.

For each of the 1038 radionuclides, the maximum activity selected among the activities calculated by the MELCOR code are then converted into activity concentration by taking into consideration the total volume obtained by adding the cuboid and pyramidal volumes.

Finally, the initial list of 1038 radionuclides is reduced down to 8 radionuclides on the basis of a general methodology developed by Tractebel [3] aiming at selecting the most relevant ones (in function of the magnitude of their activity, their fictive weight and their gamma ray energies). The selected radionuclides are radionuclides of krypton and xenon.

3.2.4 Models

The MicroShield code only enables using source volumes with simple geometries, which implies performing simplifications in a conservative way. To calculate the effective dose rates at the three calculation points, the cuboid and pyramidal source volumes are fractionated in five different volumes, as represented in Figure 3.

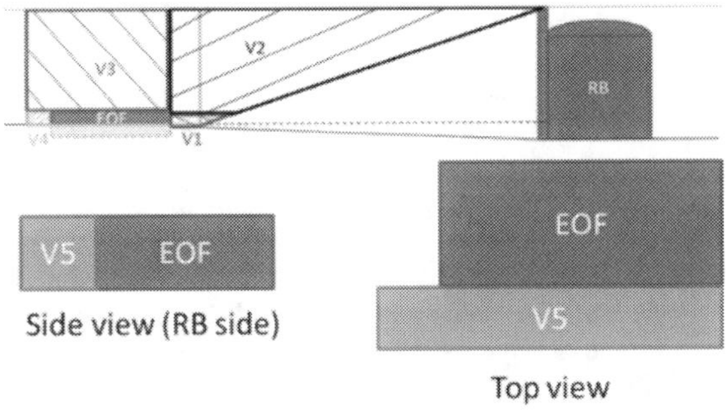

Figure 3. Definition of the volume fractions (V1, V2, V3, V4 and V5) of the source volumes.

Each model in the MicroShield code is defined by the shielding geometry and the volume fraction of the considered source term. However, MicroShield does not enable the modeling of a shielding whose height and width are smaller than those of the source volume. Furthermore, as it is not possible to model pyramidal source volume with this code, a cuboid source with an equivalent volume and having the same activity concentration must be used.

Consequently, the different models are:

<u>Model 1, used for dose rate calculations at Point 1 and Point 3, shown in Figure 4</u>:

• Source geometry is a cuboid with a width equal to the one of V1 and V2 (length of the EOF wall (20 m) with 5 m added on both sides), a height equal to the sum of the ones of V1 and V2 and a length determined to get a volume equivalent to V1+V2,

• Shielding is a concrete wall with a varying thickness,

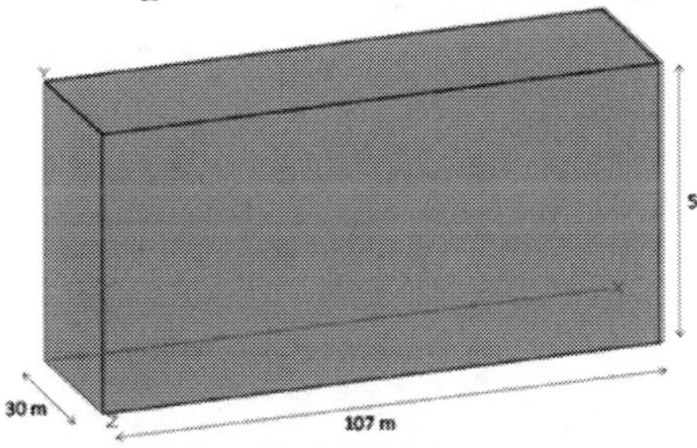

Figure 4. Representation of Model 1 as used in MicroShield.

<u>Model 2, used for dose rate calculation at Point 2</u>:

• Source geometry is a cuboid with a width equal to the one of V3 (length of the EOF roof (20 m) with 5 m added on both sides), a height equivalent to the one of V3 (the height of the vent chimney at which the EOF height was subtracted) and a length determined to get a volume equivalent to V2+V3,

• Shielding is a concrete roof with a varying thickness,

<u>Model 3, used for dose rate calculations at Point 1 and Point 2</u>:

• Source geometry is V3,

• Shielding is a concrete roof with a thickness defined by Step 2 of the methodology,

Model 4, used for dose rate calculations at Point 1 and Point 3:
• Source geometry is V5,
• Shielding is a concrete wall with a thickness defined by Step 1 of the methodology.

3.2.5 Results

Figure 5 presents the results of the calculation for Step 1 with dose rates calculated for several wall thicknesses. A curve is fitted on those calculated points and its intersection with the horizontal sizing criterion line (0.5 mSv/h) provides a minimal wall thickness of 49 cm. A similar curve is established for Step 2 in order to define a minimal roof thickness.

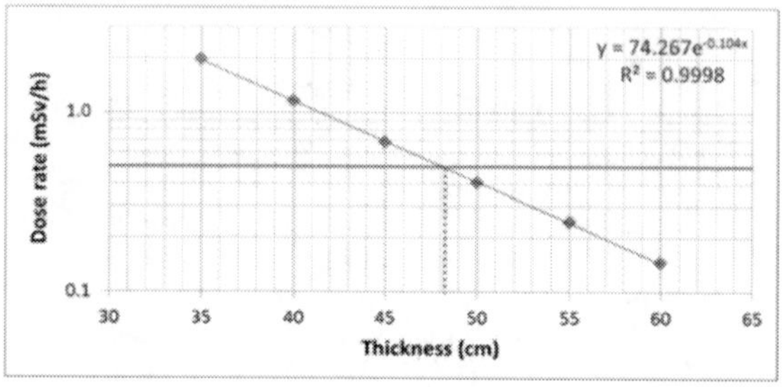

Figure 5. Dose rate evolution relative to the concrete thickness at Point 1 (the horizontal line being the 0.5 mSv/h sizing criterion).

The results of the application of the methodology mentioned in Section 3.2.1 can be summarized as follows:
• Step 1: minimal wall thickness is 49 cm,
• Step 2: minimal roof thickness is 50 cm,
• Step 3: effective dose rate at Point 1 is 0.74 mSv/h, which is well below 1 mSv/h,
• Step 4: effective dose rate at Point 3 is 0.68 mSv/h, which is well below 1 mSv/h.
Consequently, for this initial stage, the minimum thicknesses for the walls and roof are 49 cm and 50 cm, respectively.

3.3 External door

3.3.1 Methodology

The source term, the conditions and the models to be considered are the same as those considered for the walls thickness calculations. As a consequence, the necessary equivalent thickness of an external stainless steel door can be estimated based on the shielding thicknesses obtained for the concrete walls, by using the respective linear attenuation coefficients of concrete and steel.

As the way to size the wall shielding is conservative, in particular if the external door is foreseen on a building side that is not facing the venting chimney (see XZ plane in Figure 1), an optimisation process can be applied in order to reach a more realistic thickness for the door. It is considered that the contribution coming from external radiation through the concerned wall side (including through the door) could be increased up to an effective dose rate at Point 1 equal to the 1 mSv/h criterion instead of 0.74 mSv/h.

3.3.2 Results

The thickness of the wall where the external door will be integrated has been decreased from 49 cm to 18 cm of concrete. This induces a minimal stainless steel thickness estimated and rounded up to 5 cm.

4 Internal shielding sizing

The previous calculations have enabled to determine a thickness for the external walls, roof and external door of the EOF. However, they did not take into account the potential presence of radioactive sources inside the building.

The most important potential radioactive sources inside the EOF are the filters of the HVAC system. Two types of filters are featured in this system: charcoal filters for iodine elements and absolute filters for particulates. Both filter types are installed in two successive boxes (see Figure 6, the FA box to filter the Fresh Air from outdoors and the HZ box to filter the part of the filtrated fresh air used for the HZ ventilation). As represented in Figure 6, into each of these boxes, the incoming air passes successively through the charcoal filters and the absolute filters. The accumulation of activity on those filters only occurs when radioactive contamination is present outside the EOF and causes the apparition of an internal radiation source which has to be considered for internal shielding sizing.

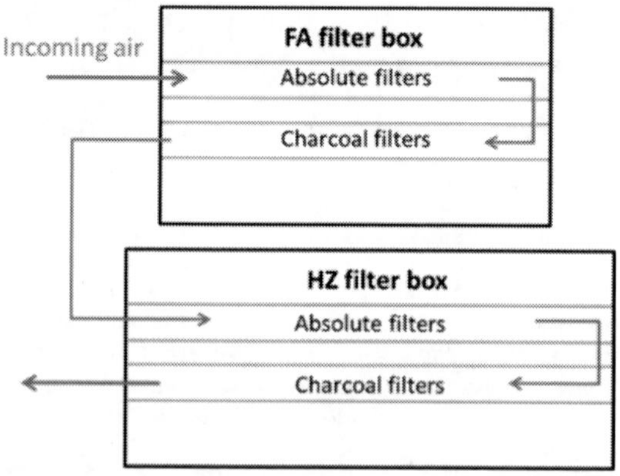

Figure 6. Schematic drawing of the FA and HZ filter boxes (side view).

The dose rate calculations for the internal shielding sizing are also performed by means of the MicroShield code, in its version 8.03 in this case [4].

4.1 Sizing criterion

1 mSv/h is still the maximum dose rate at which workers can be exposed in the HZ. However, a (maximum) contribution of 0.74 mSv/h is already to be taken into

account due to the contaminated air surrounding the EOF as detailed previously. Therefore, the criterion to use for the internal shielding sizing is 0.26 mSv/h at 0.5 m of the walls.

4.2 Methodology

The filtration boxes of the HVAC system are located in a room inside the NHZ of the EOF. As the main internal source terms arise from those filtration systems, the shielding has to be calculated for the wall separating this room from the habitable part of the new building. Therefore, the sizing calculation is related to the thickness of this wall and not to all other internal walls that do not have a shielding function. The floor of this room is also considered as a shielding, as a part of the HZ is located on the lower level, and its thickness has to be calculated.

Two FA and two HZ filter boxes are installed in parallel for redundancy purposes (see Figure 7). In a conservative approach, all the activity will be considered to be trapped in the filters located in the FA and HZ filter boxes closest from the wall of interest, i.e. no distribution between both redundant systems in parallel is taken into account.

For the wall sizing, the dose rate calculation is made at two calculation points as presented in Figure 7. Both points are at 0.50 m of the wall and at 1.30 m high for the reasons previously exposed in Section 3.1. Point 1 is placed at the centre of the considered HZ filter box while Point 2 is at the centre of the FA filter box of interest.

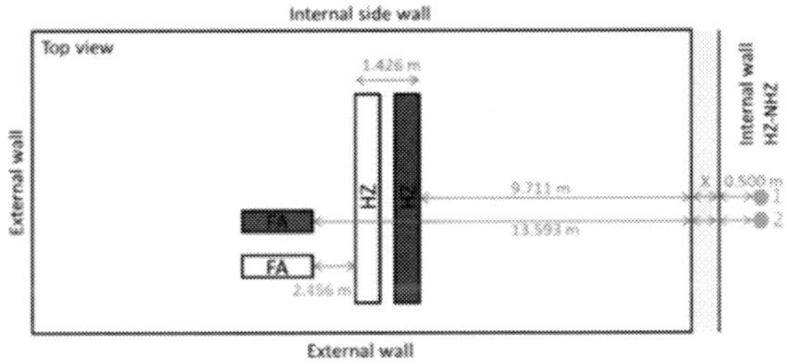

Figure 7. Schematic drawing of the calculation points in the EOF for internal wall sizing (top view). In red, the FA and HZ filter boxes used for dose rate calculations.

For the floor sizing, a similar approach is applied with Points 3 and 4 as shown in Figure 8. Point 3 is placed below the centre of the FA filter box while Point 4 is placed below the centre of the HZ filter box. Both are placed at 1.3 m high from the basement floor.

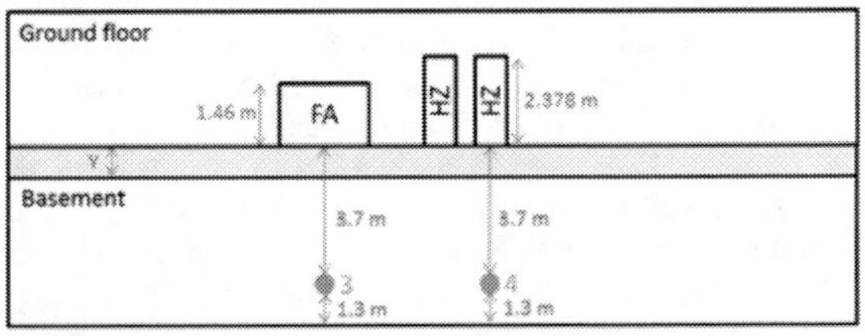

Figure 8. Schematic drawing of the calculation points in the EOF for floor shielding sizing (side view).

Note that these four calculation points are not related to those considered in Section 3.2. The shielding sizing will be established using the sum of the dose rates calculated at Points 1 and 2 for the wall and at Points 3 and 4 for the floor. Furthermore, for each of these calculation points, the dose rate results from the contribution of two source terms (a source term for absolute filters and a source term for charcoal filters).

4.3 Materials

The shielding is made of concrete (internal wall and floor) and of 304L stainless steel (external wall of the filter boxes). The material definition for the source terms is more complicated. For the absolute filters, it is a homogeneous mix of air (spaces between filters), 304L stainless steel (filter frame) and fiberglass paper, supposed to be 100% fiberglass for conservative purposes. For the charcoal filters, it is a homogeneous mix of air (spaces between filters), 304L stainless steel (filter frame) and carbon.

4.4 Source terms

The filters are supposed to be changed after 30 days of use following the occurrence of a severe accident. Note also that during the first venting, the ventilation system is kept closed to limit filter contamination. Consequently, to know the activity of the filters, the accumulation of particulates and iodine elements during this 30 day period needs to be determined. However, the temporal evolution of the volume activities around the EOF is established only for the first 10 days after severe accident (see Section 2). So, it was required to develop a methodology to calculate the accumulation beyond those 10 days. No radioactive decay of the filter accumulations is considered in order to be conservative and to simplify calculations. Furthermore, the release through natural leaks of the containment (second scenario in Section 2), which can not be neglected unlike for the external walls sizing, has a radically different behavior compared to the release from the filtered vent (first scenario in Section 2), meaning that two different approaches must be considered.

For the release through Natural Leak (NL), radionuclides with relatively long half-life show a volume activity around the EOF ($A^{NL}_{v,i}(t)$) which is approximately constant during the first ten days. Activity accumulation in the filters of the FA box is then calculated as follows [5]:

$$A^{FA,NL}_{acc,i}(\Delta t) = \frac{R_{max}}{\lambda_i}.(1 - e^{-\lambda_i.\Delta t}) \quad (1)$$

With:

- i, being representative of the studied radionuclide,
- R_{max}, a constant representing the accumulation of radionuclides on the filters per second, arbitrarily and conservatively defined by the maximum of the activity around the EOF released through the NL during the studied period, multiplied by the air flow rate D of the ventilation system ($R_{max} = Max(A^{NL}_{v,i}(t)).D$),
- Δt, the time after the severe accident,

• λ_i, the decay constant of the radionuclide i.

For the release through Filtered Vent (FV), radionuclides show a volume activity around the EOF ($A^{FV}_{v,i}(t)$) which is very high during the first venting and rapidly drops during the next ventings. The formula, used to determine the activity accumulation of each radionuclide after a time Δt in the filters of the FA box, is the following one:

$$A^{FA,FV}_{acc,i}(\Delta t) = \begin{cases} 0\ if\ \Delta t \leq t_i \\ \sum_{t=t_i}^{\Delta t} A^{FV}_{v,i}(t).D.180\ if\ t_i < \Delta t \leq t_f \\ A^{FA,FV}_{acc,i}(t_f)\ if\ \Delta t > t_f \end{cases} \quad (2)$$

With:

• i, being representative of the studied radionuclide,

• $A^{FV}_{v,i}(t)$, the volumetric activity outside the EOF at a time t, released from the filtered vent,

• D, the air flow rate of the ventilation system,

• Δt, the time after the severe accident,

• t_i, the time when the ventilation system will open permanently after the first venting,

• t_f, the time after which the activity has reach a constant value and is considered as equal to $A^{FA,V}_{acc,i}(t_f)$, this time being equal to 10 days.

The total activity accumulated on the FA filters is then the sum of the ones determined with formulas (1) and (2). Both previous formulas consider a maximum filtration efficiency of 100%. Usually, filtration efficiencies used for the Tihange NPP site are 99% for absolute filters and 94% for charcoal filters. Consequently, the activity in the filters of the HZ filter box, being after the FA filter box, will be not null but clearly inferior to activities accumulated in the FA filter box. Activity accumulation in the HZ filter box is calculated by applying the following formula:

$$A^{HZ}_{acc,i}(t) = \alpha.A^{FA}_{acc,i}(t) \quad (3)$$

With α being equal to 0.01 for absolute filters and 0.06 for charcoal filters.

The list of radionuclides to be considered is decreased following the aforementioned methodology [3]. From the initial list of 1038 radionuclides, 11 radionuclides are considered for the dose rate calculations, six of them being iodine isotopes. Furthermore, considering the fact that the two types of filters in each box are trapping different radionuclide types, two source terms are to be considered: radioactive particulates for the absolute filters and iodine isotopes for the charcoal filters. Both of those filters are integrated in the FA and HZ filter boxes. As there are two source terms (iodine and particulates) to take into account for each calculation point (Points 1 and 2 for the wall and Points 3 and 4 for the floor) and as the MicroShield code does not enable the modeling of more than one source term per calculation point, eight dose rate calculations have to be performed for shielding thicknesses determination:

• At Point 1, dose rates due to the iodine isotopes in the charcoal filters and the radioactive particulates in the absolute filters of the HZ filter box,

• At Point 2, dose rates due to the iodine isotopes in the charcoal filters and the radioactive particulates in the absolute filters of the FA filter box,

• At Point 3, dose rates due to the iodine isotopes in the charcoal filters and the radioactive particulates in the absolute filters of the FA filter box,

• At Point 4, dose rates due to the iodine isotopes in the charcoal filters and the radioactive particulates in the absolute filters of the HZ filter box.

As a consequence, eight specific models of calculations have to be developed and applied in order to determine, via dose rate calculations, the shielding thicknesses for the wall and the floor. These eight models are presented in the next section.

4.5 Models

A model includes the shielding (with all the elements between the source and the calculation point) and the source volume geometries. As previously mentioned, a total of eight models of calculation have to be established. Only a few of those models will be detailed here for illustration purposes.

Model 1 and Model 2 are used to calculate the dose rates at Point 2 due to iodine isotopes (as illustrated in Figure 9) and particulates in the FA filter box. Note that some volumes in Figure 9 are too thin to be identifiable on this drawing. Beside the source composition and density, the main difference between both models is the position of Point 2 compared to the source volume; the height of Point 2 being fixed and the height of the filters being different following the type (charcoal or absolute filters, see Figure 6). The source volume in Figure 9 (first volume on the left) has the dimensions of the charcoal filters for iodine isotopes but also of the absolute filters for the radioactive particulates, as found in the FA filter box.

Figure 9. Representation of Model 1 used in MicroShield.

Several shielding layers are considered between this source volume and the Point 2, with some simplifications made in a conservative way:

• The shielding elements present between the source and the Point 2 are the wall of the FA filter box, the four walls of the two HZ filter boxes, the internal wall to be sized and the air between those elements,

• The HZ filter boxes are supposed to contain only air (which is a conservative assumption since air has a lower shielding effect than the materials which are really present),

• The two walls of a HZ filter box are combined into one because the number of shielding in MicroShield is limited (6 mm thick instead of two 3 mm thick walls).

Model 3 and Model 4 are used to calculate the dose rates at Point 1 due to iodine isotopes and radioactive particulates (as illustrated in Figure 10) in the HZ filter box. As for previous models, the main difference between Model 3 and Model 4 is the position of Point 1 compared to the source volume (first volume on the left in Figure

10). For these models, the shielding elements present between the source and Point 1 are the wall of the HZ filter box, the internal wall to be sized and the air between those elements.

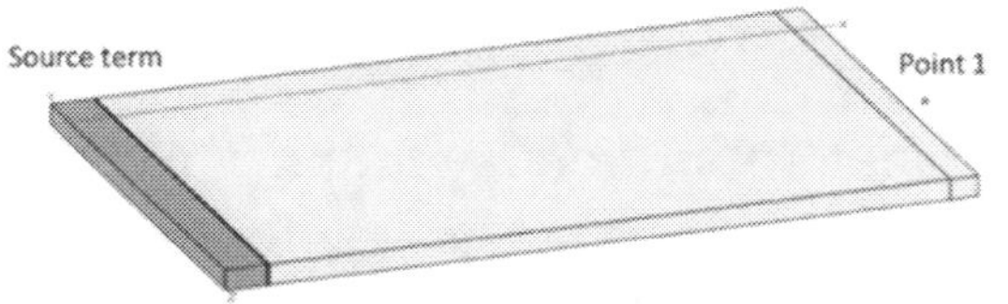

Figure 10. Representation of Model 4 used in MicroShield.

For the floor sizing, model definitions are similar to the ones for the wall. The source volume has the dimension of the FA or HZ filters. All shielding elements between the source and Point 3 or 4 must be considered. An example is shown in Figure 11, which presents the model used for calculation of the dose rate.

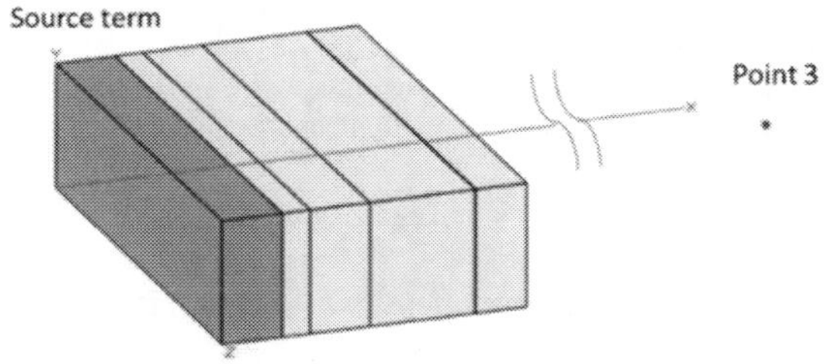

Figure 11. Drawing of the model used in MicroShield to define dose rate at Point 3 when the source term is the absolute filters of the FA filter box (the x axis being the vertical one).

In this example, the source volume (first volume on the left in Figure 11) has the dimensions of the absolute filters installed in the FA filter box. The different layers of shielding considered for this model are:

• The air between the absolute and charcoal filters inside the FA filter box (see Figure 6),
• The charcoal filters,
• The air between the charcoal filters and the bottom wall of the FA filter box,
• The steel for the bottom wall of the FA filter box (too small to be identified on the drawing),
• The concrete for the floor to be sized,
• The air down to Point 3 (localised at 1.3 m above the basement floor).

4.6 Results

For the internal wall sizing, calculations for the total dose rate provide a value of 0.0215 mSv/h without any shielding (i.e. wall thickness equal to 0 cm). Consequently, this dose rate is below the required 0.26 mSv/h which makes the shielding useless at this location. No consideration on door thickness is needed since the wall will not ensure any shielding purposes.

For the concrete floor sizing, Figure 12 shows total dose rates calculated for several floor thicknesses. A curve is fitted on those calculated points and its intersection with the sizing criterion line (0.26 mSv/h) provides a minimum thickness of 12.7 cm. However, the fitting function does not seem to be perfectly aligned with all the points. Therefore, to be conservative, a slightly higher concrete thickness of 13.5 cm might be more adequate.

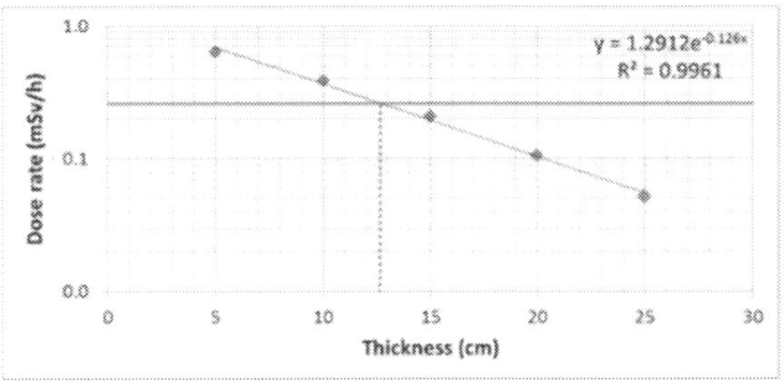

Figure 12. Total dose rate evolution relative to the floor concrete thickness (the horizontal line being the 0.26 mSv/h sizing criterion).

5 Conclusion

In this document, several methodologies are proposed for sizing the internal and external walls, floor, roof and doors of an Emergency Operation Facility (EOF) dedicated to manage a severe accident in a nuclear power plant. The particularity of such facilities is that the radioactive sources are expected both around and inside the building (considering the ventilation system).

All the proposed methodologies use the MicroShield code which is, usually, less adapted for complex geometries. The source term definitions and the models used for the shielding calculations with this code were developed and are illustrated, in this document, by calculations performed for the basic design of a new EOF for the Tihange nuclear power plant.

In that framework, it was considered that effective dose rate inside the habitable zone of the EOF had to be kept below 1 mSv/h after a severe accident, taking into account radioactive sources from inside and outside the building. Such a dose rate is achievable if a minimum thickness of about 50 cm for external walls and roof is considered. In addition, results have shown that internal shielding was not necessary, except for the floor where the HVAC filters are localised.

References

1. MELCOR Computer Code Manuals, Version 1.8.6, NUREG/CR-6119, Rev. 3, Sandia National Laboratories, Albuquerque, N.M. (2005).
2. Grove Engineering, MicroShield v5.05, Rockville, Maryland, USA (1999).
3. G. Genard, B. Gonze, D. Hulhoven, S. Vanderperre (Tractebel), *How to correctly choose the list of relevant radionuclides to assess dose uptake by workers?*, 2016 ISOE International Symposium (2016).
4. Grove Engineering, MicroShield v8.03, Rockville, Maryland, USA (2008).
5. Kenneth S. Krane, *Introductory Nuclear Physics* (John Willey & Sons, Sommerset, New Jersey, 1987).

DOI: 10.1051/epjconf/201715305017

A useful observable for estimating k_{eff} in fast subcritical systems

Paolo Saracco[1,2,*], *Walter* Borreani[1,2,**], *Davide* Chersola[1,2,***], *Guglielmo* Lomonaco[2,1,****], *Gianni* Ricco[1,3,†], and *Marco* Ripani[1,3,‡]

[1] *INFN - sez. Genova, Via Dodecaneso, 33 - 16146 Genova (Italy)*
[2] *GeNERG - DIME/TEC - University of Genova, Via all'Opera Pia, 15/A - 16145 Genova (Italy)*
[3] *CENTRO FERMI, Compendio del Viminale, Piazza del Viminale 1, 00184 Rome, Italy*

Abstract. The neutron multiplication factor k_{eff} is a key quantity to characterize subcritical neutron multiplying devices and for understanting their physical behaviour, being related to the fundamental eigenvalue of Boltzmann transport equation. Both the maximum available power - and all quantities related to it, like, e.g. the effectiveness in burning nuclear wastes - as well as reactor kinetics and dynamics depend on k_{eff}. Nevertheless, k_{eff} is not directly measurable and its determination results from the solution of an inverse problem: minimizing model dependence of the solution for k_{eff} then becomes a critical issue, relevant both for practical and theoretical reasons.

1 Introduction

As it is well known, the neutron multiplication factor k_{eff} is both a fundamental quantity for the evaluation of the behavior of subcritical reactors (i.e. Accelerator Driven Systems, ADS) and a paramount indicator to monitor its subcriticality level. However, it is not directly an observable: its determination is essentially an inverse problem and various ad-hoc methods [1–7] have been developed for its experimental determination from flux measurements; they have been tested against experimental results from MUSE [8], YALINA [9], DELPHI [10] and VENUS-F [11] with different findings, mainly depending on the subcriticality level. Additionally, it is important to highlight that their applicability is debatable when the value of k_{eff} is below 1.

The key point to understand the relevance of the parameter k_{eff} is that in an ADS it is mandatory to avoid neutron flux divergence in all conditions: as already underlined [12], the concept of *margin to criticality* is questionable, so that "on-line" monitoring of k_{eff} is, at present, an unavoidable safety requirement for all the facilities with a non-negligible power [8]. Among others, also the EU FP7 project FREYA has investigated this issue [13].

To exemplify the reason of the difficulties in the determination of k_{eff} from total neutron flux measurements $\Phi_T = \int \Phi(E)\,dE$, as they can be performed in small volumes by a fission chamber, we show in Figure 1 the MCNP simuation of a typical flux behaviour as a function of k_{eff} in different in-core positions; see Section 2 for a detailed description of this model (LEADS) and for the convention we use to identify tally positions, all in the same z-plane. LEADS model can be considered a typical Generation IV [15, 16] configuration. Two relevant facts can be inferred

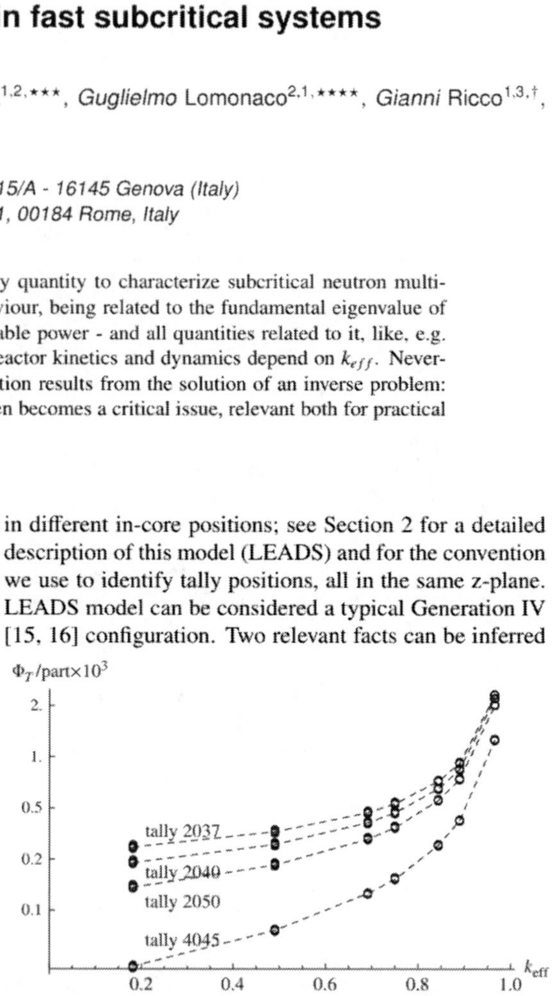

Figure 1. Logarithmic plot of the total neutron flux for 86 tallies at various distances from the source (see text).

from this plot: the first is that the supposed relation between a local quantity, the total neutron flux, and a global one, k_{eff}, is not functionally simple and moreover it is strongly position dependent; the second feature is not immediately evident from Figure 1 where actually a total of 86 different tallies are plotted together: however they are not randomly distributed in the core, but they have been chosen on purpose as the sets of tallies whose distances from the source differ from the one of 4 reference tallies (the one indicated in figure 1), by less than 5 mm. Then the second feature is that the flux position dependence effectively reduces with a very high confidence to a dependence

*e-mail: paolo.saracco@ge.infn.it
**e-mail: walter.borreani@unige.it
***e-mail: davide.chersola@edu.unige.it
****e-mail: guglielmo.lomonaco@unige.it
†e-mail: ricco@ge.infn.it
‡e-mail: ripani@ge.infn.it

only on the distance from the source: we think this to be a peculiar feature of a fast core, essentially because neutrons with longer mean free paths cannot distinguish fine details of the geometry.

What Figure 1 suggests is that any attempt to extract k_{eff} from total flux measurement [1–7] may be complicated and model dependent.

At least intuitively, being the fission neutron spectrum fast, one could expect that the "fast" component of the neutron flux could be more directly related to the ability of the core to multiply neutrons; however this is not sufficient because the fission rate is locally determined by the flux itself, so it becomes natural to normalize the fast flux component to the total one, by introducing a type of spectral index that we defined as [14]:

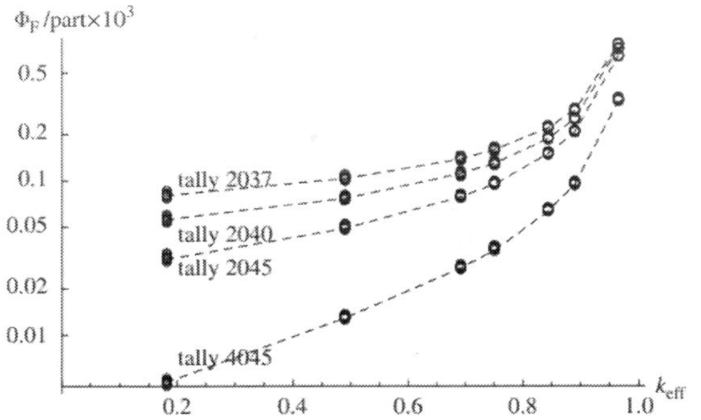

Figure 2. Logarithmic plot of the "fast" component of neutron flux for 86 tallies at various distances from the source (see text).

$$R(k_{eff}, \vec{x}; E_{th}) = \frac{\int_{E>E_{th}} \Phi(E)\, dE}{\int \Phi(E)\, dE} \quad (1)$$

where Φ is the (local) neutron flux, E is the neutron energy and E_{th} is some specific energy threshold. In (1) we explicitly indicated the possible functional dependences of the ratio R; however we expect this definition to be effective to determine k_{eff} when measured in positions far from the source, where fast flux component is determined mostly by fission. In particular one expects a relation between R and k_{eff} simpler than the one displayed in Figure 1, because the fast component of the flux behaves itself in a very similar way, see Figure 2.

In order to define in some way the "fast" component of the spectrum, we distinguished in [14] between neutrons with kinetic energy below and above 0.5 MeV. The choice of this (in some sense arbitrary cutoff) is related to the threshold energy for fast fission in some actinides; consequently the "fast" neutrons so defined could be able to produce a signal in suitably coated fission chambers, while "slow" neutrons could not. Starting from this definition, an alternative observable, possibly related to k_{eff}, is given by the measurement of the relative fast flux component on fuel rods, in an ADS sufficiently far from the central target. If we suppose that the primary fast neutrons from the source are almost completely absorbed in the innermost part of the core, the fast component of the flux in the peripheral rods should be mostly due to those secondary fission neutrons, which are only partially slowed down by interactions with fuel and coolant [14].

If these arguments are correct we expect R, for peripheral rods and for energy threshold equal to E_{th}, to depend essentially on the fission rate i.e. it should be represented by a proper function of k_{eff} alone. If that were the case, such an observable would offer the advantage of being measurable in all operational conditions. Moreover, it should be relatively independent from the type of neutron source, being essentially a characteristic of the specific core, and it could be derived from two quantities measured with similar devices. In particular, it could be used for monitoring the reactivity level periodically when the accelerator is running in continuous mode, thereby providing a real-time, source- and current- independent cross check of the power-to-current indicator without turning off the beam.

However, the extraction of k_{eff} from the direct measurement of R could in principle be affected and complicated by the dependence on other parameters such as geometry (rod dimensions, fuel to coolant volume ratio, etc.), composition (fuel and coolant materials) and the detector radial position into the core. The main result from [14] was that for peripheral rods the dependence of R on many core details, but material composition, is very mild, leaving a very simple k_{eff} dependence, which turns out to be essentially linear over a wide range of k_{eff} values (at least for $k_{eff} \gtrsim 0.7$).

This result confirms our previous guess grounded on physical arguments. In order to investigate to some extent this residual parameter dependence, we performed simulations of $R(k_{eff})$ using MCNP6 on the previously mentioned LEADS model (see section 2).

2 Model

In order to have a (relatively) realistic model, we assumed the design of a research fast ADS (LEADS), based on solid lead, described in [17] and [18] and already used also for transmutation purposes [19]. The reactor model has the following features:

- inner core (1.6 x 1.6 x 90 cm^3 hexahedral type lattice containing the 0.37 cm radius fuel rods surrounded by a 0.05 cm thick AISI steel cladding and embedded in solid lead) with 50 cm radius and 90 cm height, surrounded by a 120 cm radius and 160 cm high lead reflector, in turn contained within a 5 cm thick AISI steel vessel;
- 2 cm radius central axial beam pipe hosting the Be target;
- beam pipe filled with helium gas for the purpose of target cooling;
- a mixture of ^{238}U and ^{235}U as fuel.
- helium gas core cooling

A sketch of the used (LEADS) model is reported in Fig. 3.

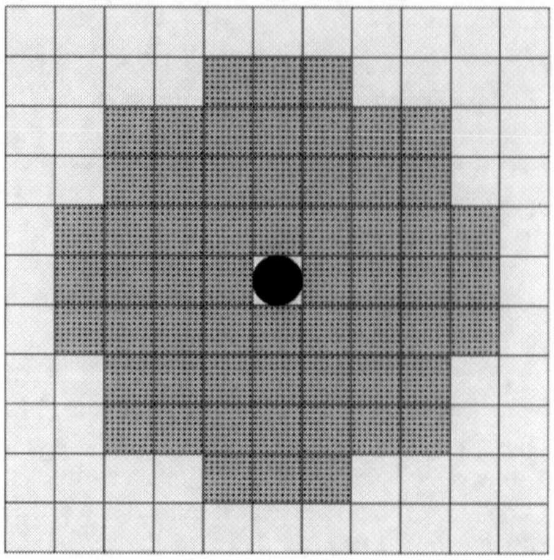

Figure 3. Core shape for LEADS model: in light grey the lead assemblies [14].

The present contribution aims to a more detailed investigation of the validity of the conclusion drawn in [14], in particular on the (residual) dependences of R on the choice of the cutoff energy E_{th} and on the position of the tally inside the core. To this purpose we evaluated the neutron flux in all the independent core positions for LEADS model: symmetry considerations restrict the number of different tallies to $81 \times 11 = 891$, see Figure 4. We identify each tally with four integer numbers, the first two indicating the x and y position of the fuels assembly (FA) with respect to the source in x=0, y=0, the second two indicating each individual tally inside a FA, from 01 to 81 from left to right and down to up (see fig. 4 for some examples of this convention). Flux has been averaged over a 2 cm high portion of the fuel pin at the mid plane in vertical direction.

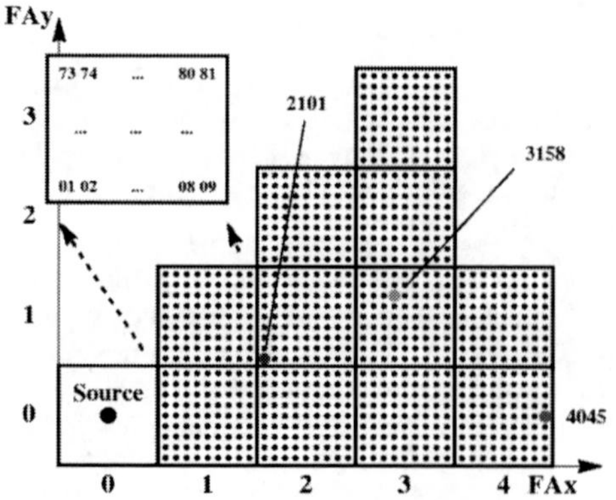

Figure 4. The convention used to identify tallies, In red tally 4045, the one used in [14], in green tally 3158 and in purple tally 2101. These tallies has been magnified with respect to real dimensions to make their identification easier.

3 Results

The most relevant conclusion drawn in [14] was that "*for $k_{eff} > 0.7$, the ratio calculated for LFR or GFR models turns out to show a quite similar trend well approximated by a straight line, while a different slope is observed for a sodium reactor. In our opinion, the results obtained show that such observable, after appropriate calibrations on the real systems under consideration, may be used for determining k_{eff} in all operational conditions, in particular for subcritical systems with different types of neutron source and different values of the beam current.*"

A first point to be investigated concerns the validity of a linear representation of the function $R\left(k_{eff}\right)$ at different positions in the core: we then considered all the tallies on the straight line from positions 1037 to 4045 and we performed two fits in the form $a + bk_{eff}$ and $c + dk_{eff} + ek_{eff}^2$, with E_{th} = 0.5 MeV, the same value assumed in [14]. Results for the coefficient of determination test [unfortunately usually indicated as R^2, here to not be confused with the function $R\left(k_{eff}\right)$ whose dependency of k_{eff} has been fitted] are shown in Figure 5: it is clear that for the

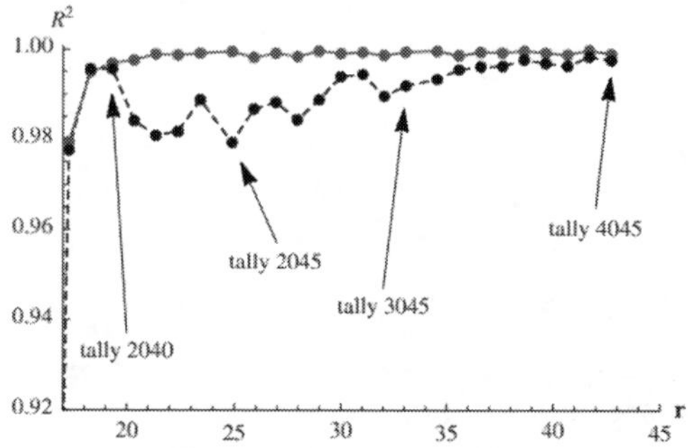

Figure 5. Coefficient of determination (R^2) tests for linear and parabolic fits to $R(k_{eff})$, for tallies from positions 1037 to 4045 as a function of their distance from the source. Upper (solid red) curve for the parabolic fit, lower (dashed black) curve for the linear fit. Tallies at lower distances have been omitted because they cannot be represented well by any of the two fits.

first few tallies near the reflector linear and parabolic fits to $R\left(k_{eff}\right)$ are equivalent, while as far as one considers tallies near to the source both become unable to represent the function. At intermediate distances parabolic fit is better than linear one. To better understand the meaning of the results in Figure 5 we show also the function $R\left(k_{eff}\right)$ for four different tallies 4045 (the original one used in [14]), 2045, 2040 and 2037, together with the corresponding linear (solid) and parabolic (dashed) fits. One immediately realizes that the increase in the coefficient of determination value observed for the linear fits in figure 5 around tally 2040, is accidentally due to a change of the sign in the second derivative of $R\left(k_{eff}\right)$ when moving towards the source.

We can observe another important feature in figure 6: the mean slope of $R\left(k_{eff}\right)$ decreases when moving towards the source: this confirms our original guess [14] that

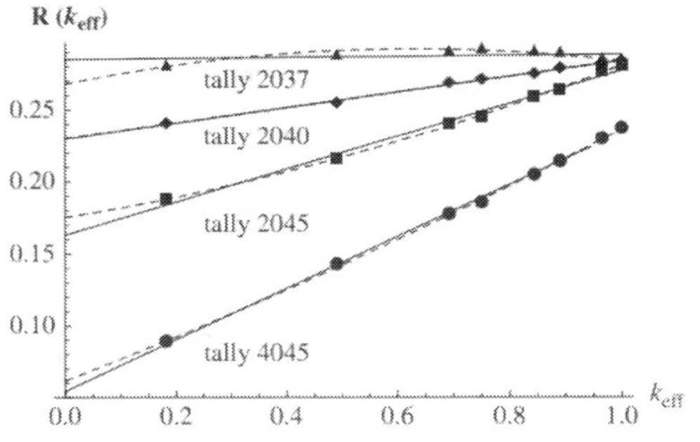

Figure 6. $R\left(k_{eff}\right)$ for 4 different tallies: 4045 (circles), 2045 (squares), 2040 (diamonds) and 2037 (triangles). Linear (solid) and parabolic (dashed) fits are also shown.

the best choice for the reference tally is at the border of the core (or near to it), because in this case the value of $R\left(k_{eff}\right)$ is more sensitive to k_{eff}.

3.1 E_{th} dependence

If we look at $R\left(k_{eff}\right)$ as an estimator for k_{eff}, sensitivity is larger when large variations in R are caused by small variations in k_{eff}, that is to say the optimal condition should be the maximum slope of $R\left(k_{eff}\right)$: in Figure 7 we show how the slope of $R\left(k_{eff}\right)$ depends on the choice of the energy cutoff parameter E_{th}. The maximum slope is attained

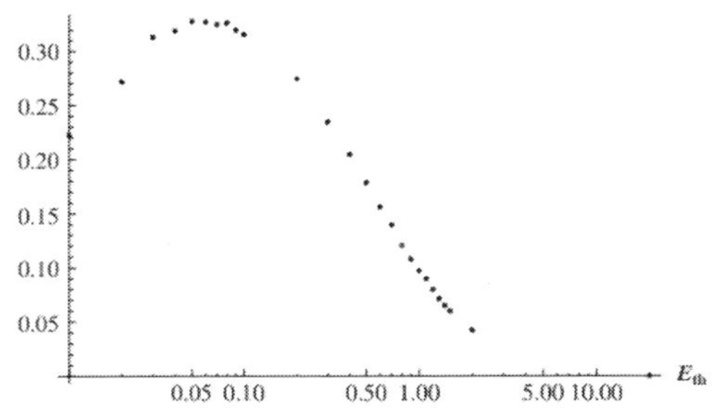

Figure 7. Variation of the slope of the linear fit to $R\left(k_{eff}\right)$ as a function of E_{th} for tally 4045.

for E_{th} in the range 0.03 – 0.1 MeV, which is lower then the value used naively in [14]. This may be due to the fact that fission spectrum is somewhat softened by the medium. However it may not be obvious to practically implement such a low thresold in an experiment.

3.2 Position dependence

We now examine the position dependence of R: this is not so important from a practical point of view, in that to use $R\left(k_{eff}\right)$ as en estimator for k_{eff} requires anyway to perform a (local) calibration. However, as we shall see, position dependence of R reveals some non trivial features, which in our opinion deserve some attention.

The neutron flux depends on energy and position, then in principle $R\left(k_{eff}\right)$ is really $R\left(k_{eff}; \vec{x}, E_{th}\right)$; in Figure 8 we show the countour plots of R, for various values of k_{eff}, as a function of xy-coordinates of pins.

From Figure 8 we see that an approximate dependence on the distance from the source is effectively realized. This is only a qualitative observation: to make it more quantitative we collect together all the results from tallies with (approximately) the same distance from the source. This is naturally not exactly possible: the best we can do is to collect together results from all tallies whose distances from the source are the same within a given tolerance (tol), or, that is the same, all tallies with a distance from the source $d \in [d_{ref} - tol/2, d_{ref} + tol/2]$. We give an example of such selection procedure in Figure 9. For evident

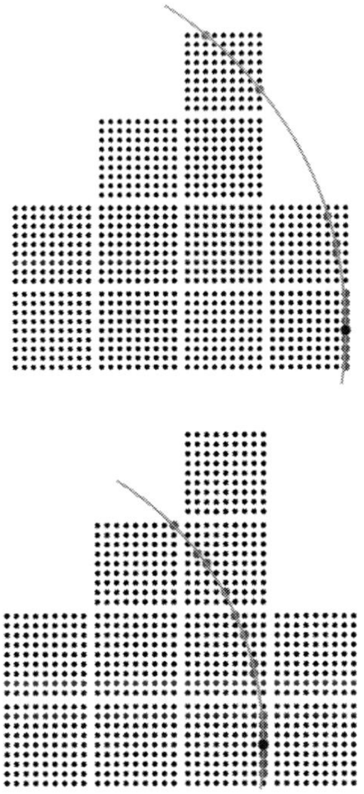

Figure 9. Two example of sets of tallies at the same distance from source with respect to a reference one (plotted in blue), with a tolerance d=2 mm

geometrical reasons the distribution of tallies selected by this procedure is more effective in testing different areas of the core at intermediate distances, as can be clearly seen in the figure. Then we show the same data as in Figure 6, but including all tallies within a tolerance of 2 mm. from the reference ones, together with the corresponding linear (solid) and parabolic (dashed) fits. Roughly speaking this means that $R\left(k_{eff}; \vec{x}, E_{th}\right)$ can be represented as $R\left(k_{eff}; \sqrt{x^2+y^2}, z, E_{th}\right)$; this (approximate) property holds better for a fast reactor, in our opinion, because of the longer mean free path of neutrons, which makes it more difficult to detect the specific details of geometry - including for instance pin diameter [14] - of the system

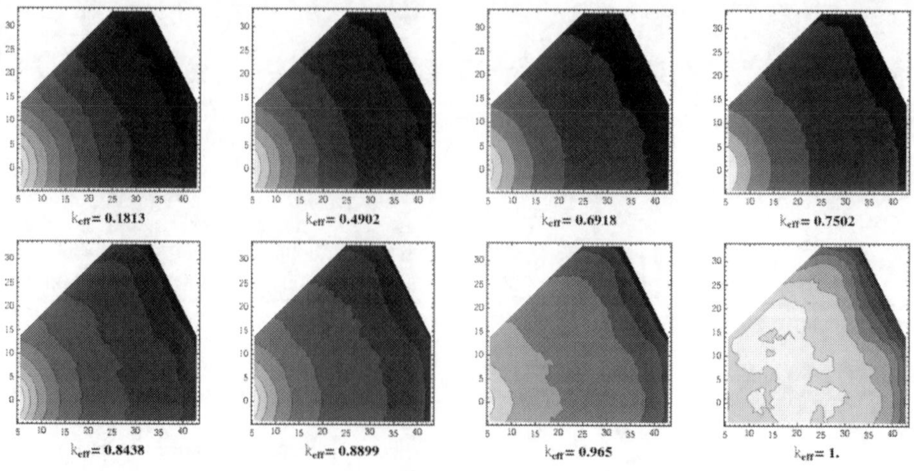

Figure 8. Countour plot of R as a function of tally position for various values of k_{eff} (E_{th} = 0.5).

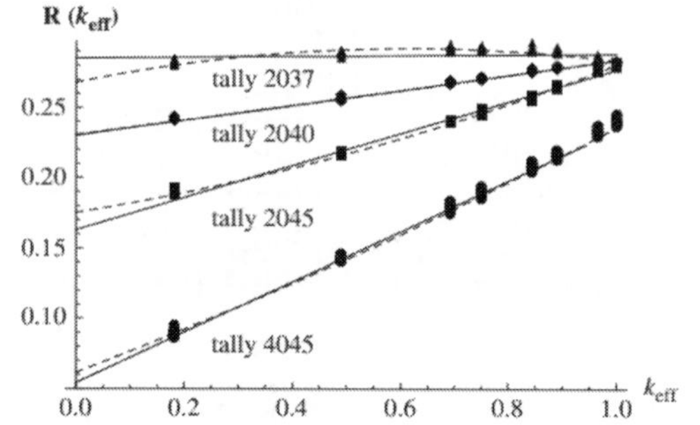

Figure 10. The same as Figure 6, but including all tallies at the "same distances" as reference ones, with a tolerance tol = 2 mm (see text).

as noted before in the introduction. We note also that the dispersions of values for different tallies at the same k_{eff} values are larger than for the case of the total neutron flux of Figure 1: there may be two reasons for this, the first being that since R is a ratio, absolute relative errors for numerator (the fast component of the flux) and the denominator (the total flux) sum up and the second being that the range of variability of R is reduced with respect to that of the total flux.

4 Conclusions

As a general conclusion we can confirm the results of [14]: R appears to be a promising and (relatively) robust estimator for k_{eff} in fast reactors, whose values can be determined by two measurements from the same or similar detector.

Moreover we observe that the maximum sensitivity of R to k_{eff} is obtained, for a fixed cutoff, at the boundary (but inside) the core. Therefore by appropriately choosing the energy threshold E_{th} as permitted by actual detectors, and by placing the detectors in that core region, the measurements can provide a good sensitivity to k_{eff} to be checked against calibrations.

We did not yet investigated how the efficiency of the detector for neutrons of different energies can influence the relation $R(k_{eff})$, a feature that can be studied by properly modifying the definition of R, and how this can interfere with the necessary calibration procedure.

References

[1] G.B. Bruna and J. Planchard. Computation of the flux in a slightly sub-critical reactor in the presence of external sources. Prog. Nucl. Eng. **25**, 1-6 (1991)

[2] A. Greenspan. A Generalized source multiplication method for determining reactivity. Nucl. Sci. Eng., **55**, 103-106 (1994)

[3] T. Chevret et al.. Reactivity measurement of the lead fast subcritical VENUS-F reactor using beam interruption experiments. In: Proceedings of PHYSOR 2014 - The Role of Reactor Physics Toward a Sustainable Future, Kyoto, Japan. ANS, n. 1087437, available on CD-ROM (2014)

[4] S. Chabod et al.. Reactivity measurement at Guinevere facility using the integral k_p method. In: Proceedings of PHYSOR 2014 - The Role of Reactor Physics Toward a Sustainable Future, Kyoto, Japan. ANS, n. 1087437, available on CD-ROM (2014)

[5] C. Berglof et al.. Spatial and source multiplication effects on the area ratio reactivity determination method in a strongly heterogeneous subcritical system. Nucl. Sci. Eng. **166**, 134-144 (2010)

[6] V. Becares. Monte Carlo assessment of spatial and energy effects in the VENUS-F subcritical configuration and application for reactivity determination. In: Proceedings of PHYSOR 2014 - The Role of Reactor Physics Toward a Sustainable Future, Kyoto, Japan. ANS, n. 1087437, available on CD-ROM (2014)

[7] G. Mila et al.. Pulsed neutron and source jerk experiments for reactivity assessment in deep subcritical configuration: a case study within the framework of the FREYA project. In: Proceedings of PHYSOR 2014 - The Role of Reactor Physics Toward a Sustainable Future, Kyoto, Japan. ANS, n. 1087437, available on CD-ROM (2014)

[8] R. Soule et al.. Neutronic studies in support of accelerator-driven systems: the MUSE experiments in the MASURCA facility. Nucl. Sci. Eng. **148**, 124-152 (2004)

[9] C. M. Persson et al.. Analysis of reactivity determination methods in the subcritical experiment Yalina. Nucl. Instr. Methods A, **554**, 374-383 (2005)

[10] J. L. Kloosterman. DELPHI: a new subcritical assembly at Delft University of Technology. In: Proceedings of PHYSOR 2004 - The Physics of Fuel Cycles and Advanced Nuclear Systems - Global Developments, Chicago, IL, USA. American Nuclear Society, IL, USA (2004)

[11] W. Uyttenhove et al.. The neutronic design of a critical lead reflected zero power reference core for on-line subcriticality measurements in accelerator driven systems. Ann. Nucl. Energy, **38**, 1519-1526 (2011)

[12] P. Saracco et al.. A preliminary study of an improved area method adapted to short time transients in subcritical systems. In: Proceedings of PHYSOR 2014 - The Role of Reactor Physics Toward a Sustainable Future, Kyoto, Japan. ANS, n. 1087437, available on CD-ROM (2014)

[13] W. Uyttenhove et al.. Static modal analysis of the current-to-flux subcriticality monitor for ADS. In: Proceedings of PHYSOR 2014 - The Role of Reactor Physics Toward a Sustainable Future, Kyoto, Japan. ANS, n. 1087437, available on CD-ROM (2014)

[14] D. Chersola, G. Ricco, M. Ripani and P. Saracco. An alternative observable to estimate keff in fast ADS. Ann. Nucl. Energy, **95**, 42-47 (2016)

[15] N. Cerullo and G. Lomonaco. Generation IV reactor designs, operation and fuel cycle. Chapter 13 in I. Crossland editor, Nuclear Fuel Cycle Science and Engineering, pag. 333-395, Woodhead Publishing (2012)

[16] D. Chersola, G. Lomonaco and R. Marotta. The VHTR and GFR and their use in innovative symbiotic fuel cycles. Progr. Nucl. Energy, **83**, 443-459 (2015)

[17] L. Mansani, M. Bruzzone, S. Frambati and M. Reale. An intrinsically safe facility for forefront research and training on nuclear technologies - General description of the system. Eur. Phys. J. Plus, **129**, 65 (2014)

[18] C. M. Viberti and G. Ricco. An intrinsically safe facility for forefront research and training on nuclear technologies - Core design. Eur. Phys. J. Plus, **129**, 66 (2014)

[19] G. Lomonaco, O. Frasciello, M. Osipenko, G. Ricco, M. Ripani. An intrinsically safe facility for forefront research and training on nuclear technologies - Burnup and transmutation. Eur. Phys. J. Plus, **129**, 74 (2014)

Methodology for the nuclear design validation of an Alternate Emergency Management Centre (CAGE)

César Hueso[1], Marco Fabbri [1,3], Cristina de la Fuente[1], Albert Janés[1], Joan Massuet[1], Imanol Zamora[1], Cristina Gasca[2], Héctor Hernández[2], J. Ángel Vega[2]

[1]Idom Ingeniería y Consultoría, Avda. Zarandoa, 23 Bilbao-Vizcaya-Spain 48015, +34 944797600
[2]ANAV Asociación Nuclear Ascó-Vandellòs II, L'Hospitalet de l'Infant-Tarragona-Spain 43890, +34 977818800
[3]Fusion for Energy, C/Josep Pla, 2, Torres Diagonal Litoral, Edif. B3 Barcelona-Spain 08019, +34 933201800

Abstract. The methodology is devised by coupling different codes. The study of weather conditions as part of the data of the site will determine the relative concentrations of radionuclides in the air using ARCON96. The activity in the air is characterized depending on the source and release sequence specified in NUREG-1465 by RADTRAD code, which provides results of the inner cloud source term contribution. Known activities, energy spectra are inferred using ORIGEN-S, which are used as input for the models of the outer cloud, filters and containment generated with MCNP5. The sum of the different contributions must meet the conditions of habitability specified by the CSN (Spanish Nuclear Regulatory Body) (TEDE <50 mSv and equivalent dose to the thyroid <500 mSv within 30 days following the accident doses) so that the dose is optimized by varying parameters such as CAGE location, flow filtering need for recirculation, thicknesses and compositions of the walls, etc. The results for the most penalizing area meet the established criteria, and therefore the CAGE building design based on the methodology presented is radiologically validated.

1 Introduction

After the earthquake and tsunami on March 11, 2011 in Fukushima Dai-chi, all nuclear plants in the European Union have been subjected to "stress tests". The Spanish nuclear sector has proposed, and the CSN has subsequently required, the creation of a centre to safely manage an emergency, called Alternate Emergency Management Centre (CAGE) located at the sites of the Nuclear Power Plants [Ref. 1]. Living conditions of the occupants of the CAGE in the event of a Severe Accident imply that TEDE must be <50 mSv and the equivalent dose to the thyroid is <500 mSv within 30 days following the accident [Ref. 2]. Given the weather conditions of each plant, the calculations are analogous to those supporting the Control Room and the different ways of radiation exposure or contamination are simulated [Fig. 1]. These paths that contributes to the dose are:

- Determination of dose due to inner radioactive cloud (within the CAGE).
- Determination of dose due to the presence of the radioactive cloud outside the CAGE.
- Determination of dose due to the accumulation of radionuclides in the filters.
- Determination of dose due to the proximity to the containment.

The variety of contributions to the dose has to be approached in an integral way. Each contribution is due to a different source term or a different interaction with the human body (i.e. external exposure, internal contamination, etc.) that have to be taken into account.
Considering that a radioactive cloud stands around the CAGE during the duration of the accident (720 hours), different situations arises.
Regarding the consequences of radioactive materials being incorporated in the inside CAGE atmosphere, external exposure and inhalation of radionuclides contributions have to be evaluated. This contribution requires the knowledge of the radiation transport mechanism and of the site meteorological data. To help solving this problem, the ARCON96, (Atmospheric Relative Concentration in building wakes) [Ref. 3] and RADTRAD 3.03 [Ref. 4] are applied.

* Corresponding author: cesar.hueso@idom.com

 DOI: 10.1051/epjconf/201715305018

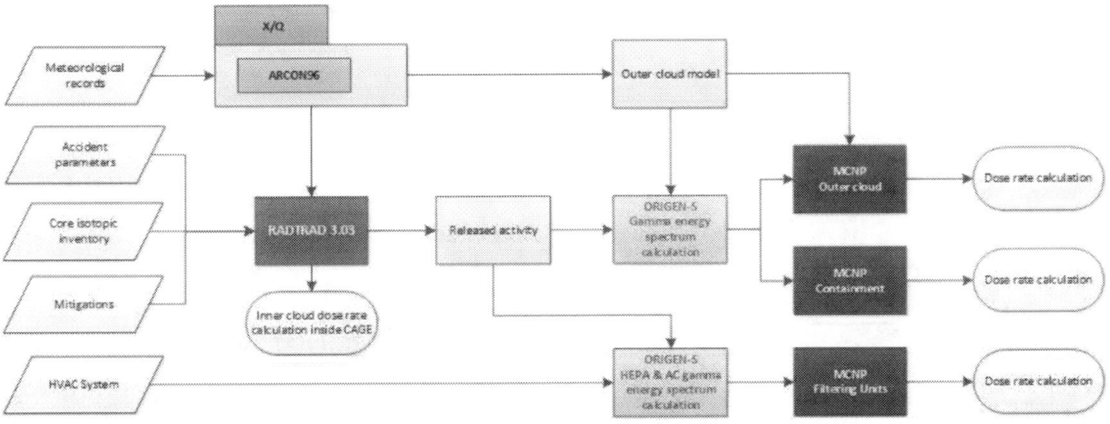

Fig. 1. Methodology applied to determine different dose contributions.

On the other hand, the fact of the radioactive cloud standing around the CAGE becomes a shielding problem where the source term is outside and the people to protect is inside. Therefore, a shielding has to be designed, mainly the concrete walls and doors. After assuming a geometry and applying the radionuclides activity released to the environment (RADTRAD), ORIGEN-S [Ref. 5] is used to "translate" the activities into gamma radiation energy spectra. These spectra are introduced as input data in a Monte Carlo radiation transport calculation by means of MCNP [Ref. 6]. This code delivers the outer cloud contribution.
In a similar way, the containment direct radiation is assessed. The only difference is that in this case, the inside containment activities (RADTRAD) are considered. This problem is highly demanding from a computational point of view because of the thicknesses of shielding (containment and CAGE concrete walls) and distance involved.
And last but not least, the outer radioactive cloud is being filtered by the filtering units. These HEPA and active carbon filters are not perfect and inner cloud contribution is due to their small inefficiency. Nevertheless the most of the radionuclides are accumulated as the filters work, resulting in a strong source term. To perform this calculation, the activities of filtered radionuclides, and their daughter´s activities, are taken into consideration. Again thanks to ORIGEN-S, these activities are "translated" into gamma energy spectra to be introduced as input data in a new radiation transport calculation that delivers the dose contribution due to the filtering units.

2 Assumptions and Input data

To carry out the necessary calculations by coupling the various codes that perform the methodology used to determine dose rates, there have to be considered situations and initial data that will determine the suitability of the resulting solutions. Then the input data and assumptions depending on the location of the CAGE are presented, as well as for each of the contributions to the final dose rate.

2.1 Diffusion factors

According to RG (Regulatory Guide) 1.23 Rev. 1 [Ref. 7], a Nuclear Power Plant should be able to get the weather information it requires to, among other objectives, determine the potential spread of radioactive material from an accident, so it can be deducted the amount of radionuclides resulting from the release into the environment of the considered source term. The ARCON96, it is a tool developed by the Nuclear Regulatory Commission to perform calculations of diffusion factors for habitability analysis of Control Rooms of Nuclear Power Plants in compliance with RG 1.194 [Ref. 8].
The following structure summarizes the different steps carried out to calculate the atmospheric relative concentrations (X/Q) of radioisotopes:
1.Obtaining meteorological data.
2.Process meteorological data.
- Calculation of the hourly averages.
- Calculation of atmospheric stability.

DOI: 10.1051/epjconf/201715305018

- Generation of meteorological files for ARCON96.

3.ARCON96 execution.

2.2 Obtain and process meteorological data

Weather information is provided by specific files, including the matrix of hourly frequencies, defined from the following time averages:
- Wind speed (in m / s) at different heights.
- Wind direction (in degrees) at different heights.
- Category stability (Pasquill, from A to G).

2.2.1 Hypothesis

- RG 1.194 considers representative hourly weather observations for more than 5 years.
- Height measurement data at 10 m and 29 m.
- An emission at ground level is assumed; conservative assumptions at the selected location distance.
- Conservatively, a height equal to intake 0 m is assumed.
- A 'terrain elevation difference' is taken equal to 0 m, since no data are available about it.
- Building line perpendicular to the direction of the release section.
- The angle between the CAGE and the emission source, considering not to locate the building in a predominant wind window.
- 90 degree wind window is taken.
- Distance from the emission point to CAGE: measured on the ground.
- Minimum wind speed 0.5 m/s.
- Surface roughness of 0.20 m.
- The initial values of σ_y and σ_z are equal to 0, as advised in RG 1.194.

2.3 Determination of dose

Once the diffusion factors have been obtained and therefore, the relative concentration of radionuclides known in the points to study, the analysis of the different routes of contribution to the dose within 30 days of accident principles is studied.

The aforementioned diffusion factors will be introduced as input data in the codes to be used for calculations of radiation transport.

2.3.1 Inner cloud contribution. Input data and assumptions

Determining dose inside the cloud will take place through software RADTRAD, as shown in Figure 1, since as specified in NUREG-1465 [Ref. 9] and the RG 1195 [Ref. 10] it is a code that incorporates adequate methodologies to meet dose determination.

Then the necessary data and hypotheses considered are as follows:
- Diffusion factors or X/Q obtained through the ARCON96 code.
- Flow diagram of HVAC system.
- Decontamination factor by natural deposition of elemental iodine.
- Chemical composition of radio-iodine, extracted value from NUREG-1465.
- Containment volume.
- Thermal power of the reactor.
- Reactor core inventory, it was assumed to be consistent with the inventory from the post-Fukushima stress test project.
- Overpressure flow calculation to define the radiological classification of CAGE.
- Net volumes of each of the areas of the CAGE.
- FGR 11: Limit Values of Radionuclide Intake and Air concentration and Dose Conversion Factors for Inhalation, Submersion, and Ingestion. 1988 [Ref. 11].
- FGR 12: External Exposure to Radionuclides in Air, Water and Soil [Ref. 12].

Furthermore, in regard to assumptions, the following are considered:
- Two different zones are considered in the CAGE. Zone A, which will be in excess of pressure compared to Zone B. Note that both of them are in overpressure from the atmospheric pressure.

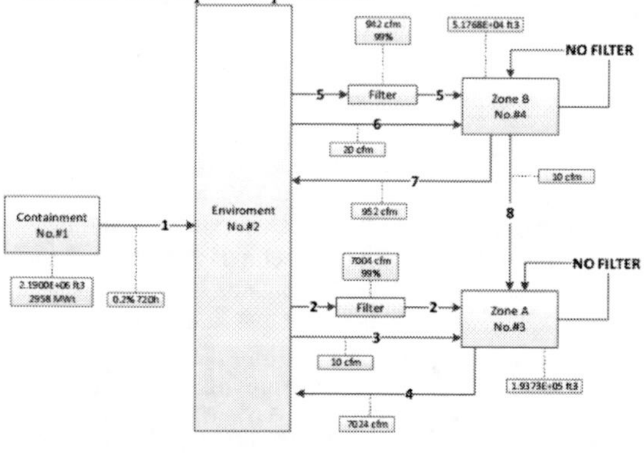

Fig. 2. Model RADTRAD.

- Release rates in the event of a severe accident in reactors PWR / BWR, are introduced into the RADTRAD code. The severe accident definition is in line with the post-Fukushima stress test accident definition. Note that the release fraction and timings for severe accident in RADTRAD are the ones from NUREG-1465, Tables 3.12 and 3.13.
- Isotopes of the source term are introduced by the corresponding external .nif file.
- Loading factors defined in RG 1.195 are used.
- Breathing rates according RG 1.195, being 3.5E^{-4} m^3/s at 720 hours.
- Credit is given to radioactive decay.
- Inflow of air and recirculation values established by the HVAC system are set for Zone A and B.
- In the compartment defined as containment credit is given to the natural deposition.
- A release of radionuclides to the environment is estimated corresponding to 0.2 % of the containment volume per day during the 30 days of the accident postulated, not only during the first 24 hours to add conservatism to the calculations.

- Outside contaminated air inlet is considered to be filtered by HEPA and active carbon filtering units.
- Infiltrations flow is introduced from the outside contaminated environment into Zone A 10 cfm / Zone B 20 cfm in line with the RG 1.78 [Ref. 13].
- The discharge rate compensates for infiltrations and the supply flow.
- Infiltrations flow from Zone B to Zone A 10 cfm.
- Several time steps are considered for evaluation according to the diffusion factor or X/ Q.

2.3.2 External cloud contribution. Input data and assumptions

Determining dose provided by the outer cloud to CAGE is carried out by coupling software RADTRAD 3.03, ORIGEN-S and MCNP as indicated in the flow diagram of Figure 1. RADTRAD 3.03 was used for estimating the release of radioactive materials into the environment in case of a severe accident. Then, using the diffusion factor determined by ARCON96, the average isotopic activity contained in the radioactive cloud in which is immersed the CAGE is obtained. Likewise, the corresponding gamma energy spectrum is determined by software ORIGEN-S, in which are entered as input data the activities obtained for each time interval. Finally, these gamma spectra are introduced in their respective MCNP5 simulations in order to characterize the cloud corresponding to the outer volume source. It is also needed:

- Simplified but representative model the geometry of the CAGE.
- Modeling the radioactive cloud as a semi-cylindrical source.
- Defining the parameters of interest of the simulation and the respective location where these values are extracted.

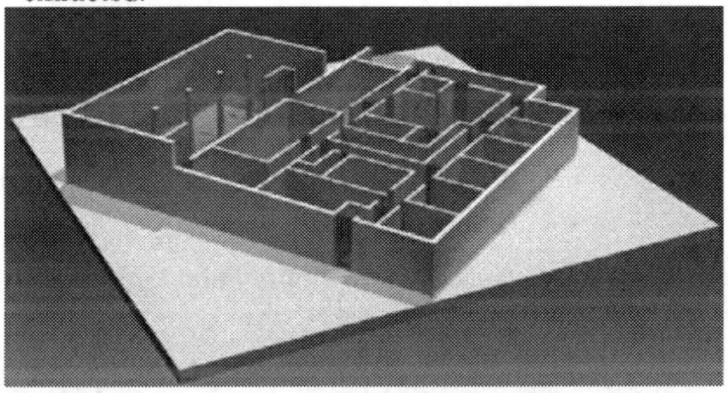

Fig. 3 Simplified geometry CAGE

Therefore, as input data are considered:

- Isotopic activity released to the outside environment.
- Diffusion factors or X/Q obtained through ARCON96 code.
- Figures 4 and 5 of RG 1.194 are used for the determination of the diffusion coefficients σ_z and σ_y. Note that once all the parameters from equation 1 form section 3.2 of the NUREG/CR6331 [Ref. 3] are known, the distance from the center of the plume (i.e. parameter y from the equation 1 of the present paper) can be calculated for each time step. This y parameter allows for the cloud volume definition.
- It is considered that the external environment is in the stability class of type G since thereby stability coefficients are minimized.
- "Still air" speed considered, being 0.5 m/s.
- Project drawings for the determination of the simplified geometry of CAGE.
- Gamma energy spectra for the characterization of the corresponding to the outer cloud over the CAGE volume source.
- Photon libraries MCPLIB84 included in the MCNP5 package.
- Radiological properties of the materials considered in the CAGE.

And the main assumptions are as follows:

- For simplistic effects it has been assumed that the cloud over the CAGE has a semi-cylindrical geometry. Therefore,

$$V_{cloud} = \tfrac{1}{2}\,\pi\, y^2 L \qquad (1)$$

V_{cloud} *(m3)*
y (m)
L (m)

- At each time step, a uniform concentration is assumed.
- Significantly, it is considered that the transport of the release to the CAGE is instantaneous.
- The most representative concrete walls of the CAGE are modeled.
- CAGE slabs and the outer ground to take into account the backscattering are modeled.
- No accumulation of radioisotope within the CAGE.
- The contribution of the neutron dose determination is neglected.
- In each time step, the intensity corresponding to the selected volume source spectrum previously calculated by ORIGEN-S is assumed.
- The weighting factor for the thyroid equivalent dose due to direct radiation is considered.
- Dose conversion factors are assumed according to ICRP119 [Ref. 14].

2.3.3 Contribution accumulation in filters. Input data and assumptions

All input data necessary for the definition of spectra by external cloud are necessary for determining activity and gamma energy spectra of radionuclides accumulated in the filters, noting that should generate new models to get the amount retained in filters.

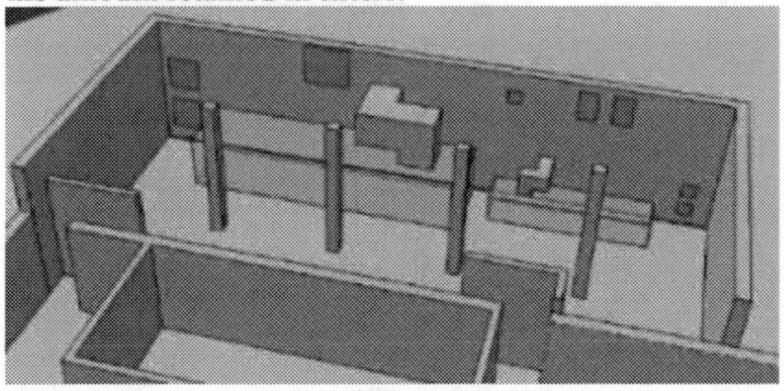

Fig. 4 Simplified Geometry Filtration Units

This activity retained in the filters thus becomes the source term for the Monte Carlo calculation, allowing the estimation of the thickness of shielding required or even the definition of the strategy for filter maintenance and management of the relevant waste.

The input data and particular hypothesis of this case would be:

- Volumetric flow HVAC system.
- Project drawings for the determination of the simplified geometry of the filters.
- Gamma energy spectra for characterization corresponding to the accumulation of radionuclides in the filtering units of CAGE volume source.
- All radionuclides will accumulate in the "virtual" filters, decaying in that filter.
- The activity generated by descendants is also considered in the accumulation of radionuclides in the filters.
- Inside the room HVAC, there are modeled only the two filter units, the walls corresponding to the labyrinths and walls defining this room.
- The outer dimensions of the filter units are modeled considering the geometric data of the drawings while the inner components are detailed by the typical constructive values.
- It has simplified model considering to assess the dose due to filters, are especially representative three zones: the interior of the room itself, protected areas through the mazes and inside the CAGE immediately behind the door in the adjoining area to HVAC room.
- Intensities resulting from the sum of the accumulation of isotopes in each time interval plus its corresponding decay daughters are considered. Thus, the fission products that are decaying in the filter at different time intervals are always considered.

2.3.4 Contribution by direct radiation from containment. Input data and assumptions

To carry out this simulation, it is proceeded in a similar way to that explained for the above cases.

Most of baseline data and hypotheses considered, coincide with those already set out throughout this document, so that only those that are particular to this model are mentioned:

- Drawings for determining the geometry of the simplified containment.
- It is considered conservatively that there is no leakage to the environment.
- Variance Reduction Techniques are used in the MCNP5 code because of the model dimensions and the shielding thicknesses that must be traversed.

3 Simulations

3.1. Location

Although the location of the buildings that house the CAGE obey multitude of conditions, among them is clearly identified, and so states the CSN in their design requirements, it should not be located in areas of predominant winds. If setting the building on one of these wind windows is mandatory, the design requirements in other areas will be influenced negatively, so each unit must locate the CAGE taking into account all factors, so that it can optimize the expenditures.

ARCON96 code utilization to determine the relative concentrations of radionuclides after a severe accident, allows to know the place where the concentration will be less. Especially sensitive to this situation would be the HVAC system, which may relax its demands in comparison to other place where concentrations were higher.

Once executed several cases, the X/Q are determined at different time intervals, providing the necessary data in the next phase.

3.1.1 Inner cloud

Once introduced the input data and the assumptions, the implementation of the necessary simulations proceeds.

The required results are TEDE and equivalent dose to the thyroid.

Throughout the project, there have been various adjustments that have enabled us to optimize the design of ventilation systems and sealing requirements of the building in general.

As an example may be mentioned that filtering recirculation is not required in the case of radiological accident, allowing cost optimization of the HVAC system.

3.1.2 Outer cloud

As previously stated, the radioactive cloud is a volumetric source term of gamma radiation, so to consider its contribution to the integral dose must be observed. The determination of this contribution can determine the thickness of the outer walls, in charge of providing the shielding necessary to maintain the habitability inside.

It is worth noting, that there may be cases where radiation limitation exceeds the limitation required from the seismic standpoint, prevailing over each other depending on the chosen location.

When it comes to characterize a source term, one must know its energy spectrum and the emission intensity (γ/s) as to equal activity, the contribution to the dose will depend on the isotopes considered.

This characterization of the source term, along with the model geometry, the definition of materials and measuring points (tallies), define the "input" of MCNP5. This code allows the determination of the direct radiation dose at different points defined by the user.

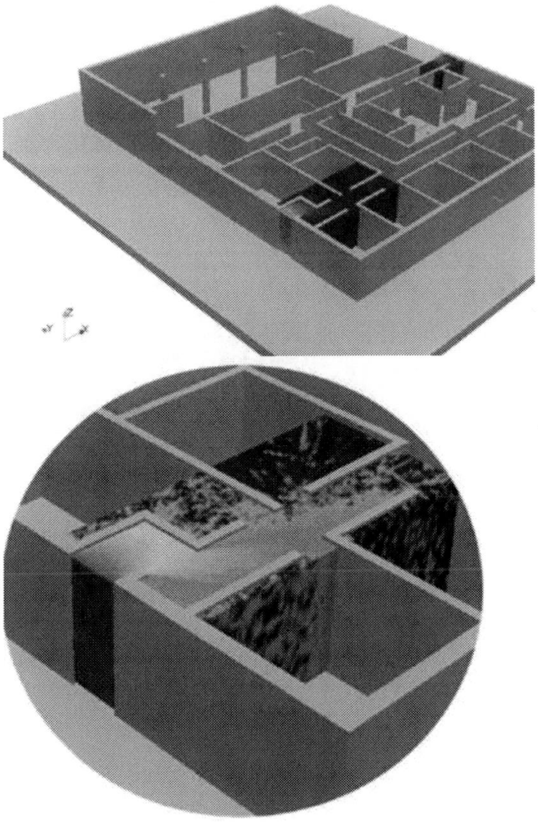

Fig. 5 Emergency exit detail

It should be noted that the outer cloud model has been paid to different hypotheses, finally opting for a semi-cylindrical representation of it that is considered conservative, as it homogenizes the limiting concentration at the selected location.

3.1.3 Filtering units

Similarly to the previous case, characterization of the source term is required, with the particularity that in this case, the concentration of radionuclides inside the filtration units increases over time, becoming a source term of great contribution to dose. The use of ORIGEN-S code allows taking into account the daughters; isotopes that are added to the retained radionuclides themselves.

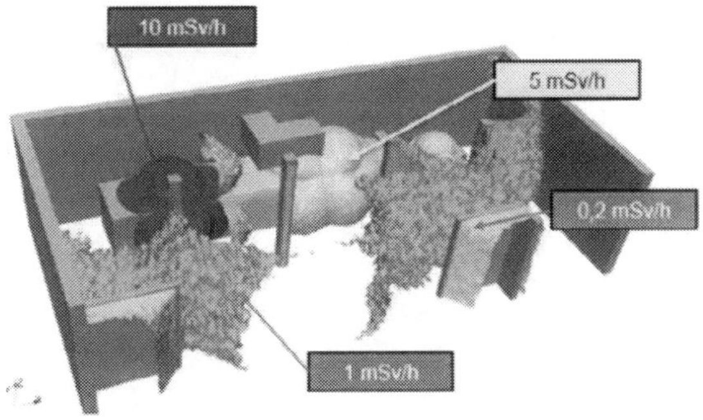

Fig. 6 Filtering Units iso-surface dose rates

Because of the magnitude of their contribution in this case the model is more detailed, simulating the materials of the filter housing and filter media. As shown in the previous figure, it has also been necessary to include labyrinthine accesses that reduce the dose.
Thanks to the results obtained by these simulations, the dimensions of the mazes and thickness of the separating wall of the room ventilation from rest of the building have been determined.

3.1.4 Containment

This simulation is similar to the outer cloud, with the proviso that the "cloud" is contained within the containment building.
Its contribution is found to be negligible after its evaluation.

4 Conclusions

Calculation codes selected for the development of this methodology are widely available, internationally used and validated in many studies. Therefore the robustness of the calculations depends primarily on the proper selection of input data and calculation assumptions.
The methodology disclosed herein allows for the modification of any of the parameters, so that it has developed a versatile method of radiological analysis.
Interaction with other areas' colleagues is key in the selection of (i.e. collaboration with HVAC specialists, civil design, etc.) the required valuable information.
On the other hand, knowledge of the design basis of the NPP where the CAGE is located it is essential to avoid incurring contradictions, being fundamental the existence of a fluid communication among all the interested parties.
It was required to characterize the HVAC room as an area of limited access to staff and managed directly by the Radiation Protection Department due to the dose rates obtained in the HVAC room.
After applying the methodology defined in this paper, the CAGE design is validated in compliance with the CSN requirements.
It is worth to point out that several iterations are required until the design of the building to imposed dose constraints has been tweaked.

References

1. Consejo de Seguridad Nuclear. Instrucciones Técnicas Complementarias. ITC´s a las Autorizaciones de Explotación. (2011).
2. Consejo de Seguridad Nuclear. Acta del Pleno 1.297, Anexo 4 *Criterios de Evaluación del CAGE.* (2013).
3. U.S. Nuclear Regulatory Commission. NUREG/CR-6331 *ARCON96 Atmospheric relative concentrations in building wakes*. (1997).
4. U.S. Nuclear Regulatory Commission. NUREG/CR-6604: RADTRAD *A simplified model for RADionuclide Transport and Removal and Dose estimation*. (1998).
5. OAK RIDGE NATIONAL LABORATORY. *SCALE: A Comprehensive Modeling and Simulation Suite for Nuclear Safety Analysis and Design*. (2011).
6. Los Alamos National Laboratory, *MCNP – A General N-Particle Transport Code, Version 5 –* **Volume I**: Overview and Theory, X-5 Monte Carlo Team. April 2003. LA-UR-03-1987.
7. RG 1.23 *Meteorological monitoring programs for nuclear power plants*. Rev. 1, U.S. Nuclear Regulatory Commission, (2007).
8. RG 1.194 *Atmospheric relative concentrations for control room radiological habitability assessments at Nuclear Power Plants*. (2003).
9. U.S. Nuclear Regulatory Commission. NUREG-1465: *Accident Source Terms for Light-Water Nuclear Power Plants*. (1995).
10. RG 1.195 *Methods and assumptions for evaluating radiological consequences of Design Basis Accidents at Light-Water Nuclear Power Reactors*. (2003).
11. EPA. FGR 11: *Limit Values of Radionuclide Intake and Air concentration and Dose Conversion Factors for Inhalation, Submersion, and Ingestion*. (1988).
12. EPA. FGR 12: *External Exposure to Radionuclides in Air, Water and Soil*.
13. RG 1.78: *Evaluating the Habitability of a Nuclear Power Plant Control Room during a Postulated Hazardous Chemical Release*. (2001).
14. ICRP. *Compendium of Dose Coefficients based on ICRP Publication 60. ICRP Publication 119. Ann. ICRP 41(Suppl.)*, ICRP. (2012).

AUTHOR INDEX

AUTHOR INDEX

AUTHOR INDEX

AUTHOR INDEX

AUTHOR INDEX

AUTHOR INDEX

AUTHOR INDEX

AUTHOR INDEX

EDP Sciences – Web of Conferences
17, Avenue du Hoggar
Parc d'Activité de Courtabouf
BP 112
F-91944 Les Ulis Cedex A
France

ISBN 978-1-5108-4812-2
9 781510 848122

13th International Conference on Radiation Shielding (ICRS-13) & 19th Topical Meeting of the Radiation Protection and Shielding Division of the American Nuclear Society (RPSD-2016)

EPJ Web of Conferences Volume 153 (2017)

Paris, France
3 - 6 October 2016

Part 2 of 2

edp sciences

13th International Conference on Radiation Shielding (ICRS-13) & 19th Topical Meeting of the Radiation Protection and Shielding Division of the American Nuclear Society (RPSD-2016)

EPJ Web of Conferences Volume 153 (2017)

Paris, France
3 - 6 October 2016

Part 2 of 2

Editors:

Fausto Malvagi
Fadhel Malouch
Cheikh M'Backe Diop

ISBN: 978-1-5108-4812-2

Printed from e-media with permission by:

Curran Associates, Inc.
57 Morehouse Lane
Red Hook, NY 12571

Some format issues inherent in the e-media version may also appear in this print version.

Printed by Curran Associates, Inc. (2017)

For additional information, please contact EDP Sciences – Web of Conferences
at the address below.

EDP Sciences – Web of Conferences
17, Avenue du Hoggar
Parc d'Activité de Courtabœuf
BP 112
F-91944 Les Ulis Cedex A
France

Phone: +33 (0) 1 69 18 75 75
Fax: +33 (0) 1 69 28 84 91

contact-edps@webofconferences.org

Additional copies of this publication are available from:

Curran Associates, Inc.
57 Morehouse Lane
Red Hook, NY 12571 USA
Phone: 845-758-0400
Fax: 845-758-2633
Email: curran@proceedings.com
Web: www.proceedings.com

TABLE OF CONTENTS

PART 1

PART 2

Development of a reliable estimation procedure of radioactivity inventory in a BWR plant due to neutron irradiation for decommissioning

Ken-ichi Tanaka[1,2,*] and Jun Ueno[3]

[1]The Institute of Applied Energy, 105-0003 Tokyo Japan
[2] Tokyo Institute of Technology, 152-8550 Tokyo Japan
[3]Genden Information System Company, 105-0004 Tokyo Japan

Abstract. Reliable information of radioactivity inventory resulted from the radiological characterization is important in order to plan decommissioning planning and is also crucial in order to promote decommissioning in effectiveness and in safe. The information is referred to by planning of decommissioning strategy and by an application to regulator. Reliable information of radioactivity inventory can be used to optimize the decommissioning processes. In order to perform the radiological characterization reliably, we improved a procedure of an evaluation of neutron-activated materials for a Boiling Water Reactor (BWR). Neutron-activated materials are calculated with calculation codes and their validity should be verified with measurements. The evaluation of neutron-activated materials can be divided into two processes. One is a distribution calculation of neutron-flux. Another is an activation calculation of materials. The distribution calculation of neutron-flux is performed with neutron transport calculation codes with appropriate cross section library to simulate neutron transport phenomena well. Using the distribution of neutron-flux, we perform distribution calculations of radioactivity concentration. We also estimate a time dependent distribution of radioactivity classification and a radioactive-waste classification. The information obtained from the evaluation is utilized by other tasks in the preparatory tasks to make the decommissioning plan and the activity safe and rational.

1 Introduction

Four commercial nuclear power reactors are undergoing decommissioning in Japan. Adding to the four reactors, five operating organizations declared to shut down permanently their six nuclear power reactors and to transit to decommissioning-stage in last fiscal year. Furthermore some reactors plan to start decommissioning in the next decade.

In Japan, a regulatory body requires an application whenever any organization of nuclear facility starts decommissioning. We call tasks for the application preparatory tasks for decommissioning. The preparatory tasks consist of five tasks as seen in Fig. 1; these are a radiological characterization, an investigation for plant characteristics, a planning for decommissioning activity, a safety assessment for decommissioning and a cost estimation for decommissioning. First two tasks, which are the radiological characterization and the investigation for plant characteristics, are called a plant characterization. These two should be performed in the earliest stage of the preparatory tasks. Outputs from them are referred to by other tasks. A reliability of an evaluation of radioactivity inventory, which is performed in the radiological characterization, is much important to plan optimized decommissioning planning and also to accomplish decommissioning activities safely and rationally [1].

The radiological characterization consists of three sub-tasks; these are an evaluation for neutron-activated materials, an evaluation for contaminated materials and an amount investigation of waste materials. Outputs from them are stored in database so as to be utilized by other tasks easily.

In order to perform the radiological characterization reliably, we improved a procedure of the evaluation of neutron-activated materials. The evaluation is composed of two calculation processes. One is a distribution calculation of neutron-flux [2, 3]. Another is a distribution calculation of radioactivity [4].

* Corresponding author: k-tanaka@iae.or.jp

To accomplish the calculations reliable, we need to verify them. Distributions of neutron-flux and radioactivity are calculated by using calculation codes and verified with measurements.

Fig. 1. Flow diagram of preparatory tasks

2 The procedure of the evaluation for neutron-activated materials.

2.1 Outline of the procedure

In order to perform the radiological characterization reliably, we improved a procedure of the evaluation for neutron-activated materials for a BWR. We perform four works in the procedure of the evaluation of neutron-activated materials as shown in Fig. 2. These are measurement of neutron-flux by using activation foils, distribution calculations of neutron-flux in a Primary Containment Vessel (PCV) by using calculation codes, distribution calculations of neutron-flux outside the PCV and time dependent distribution of radioactivity.

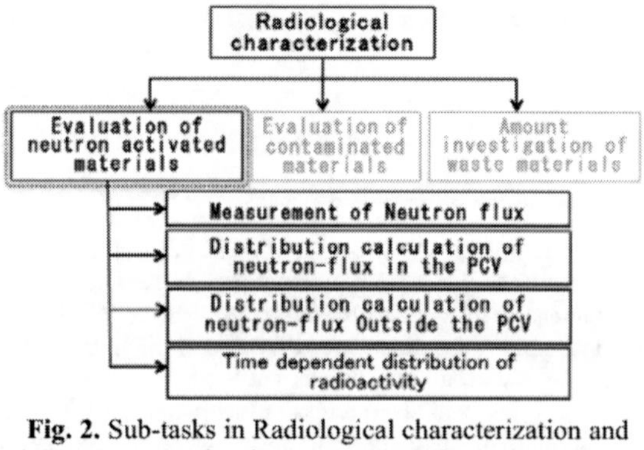

Fig. 2. Sub-tasks in Radiological characterization and activities in evaluation for neutron-activated material

Calculations are verified with measurements. By considering comparisons between calculations and measurements (C/Ms), we modified the calculation model repeatedly until our goal of reliability is achieved. When C/Ms are within one to ten, we judge that calculations are reliable and our goal of reliability is achieved [5].

2.2 Measurements of radioactivity due to neutron irradiation

We measured radioactivity in a range where neutron irradiation should be considered. The range is in the PCV and in a Main Steam line tunnel room (MS tunnel room). We performed measurements with activation foils. We implemented the measurements for the following two reasons. One was to obtain knowledge about both distribution of radioactivity and distribution of neutron-flux. Another was to verify validity of calculations.

We set activation foils at 30 locations where the characteristics of distribution of neutron-flux such as streaming, leakage, shielding, diffusion and others could be observed in the PCV as shown in Fig. 3. As seen in the right panel of the figure, six locations, these were from No.19 to No.24, were set to observe an effect of neutron streaming through a Feed Water (FW) piping hatch.

We used three kinds of foils as neutron activation detectors; these were Gold foil (Au), Gold foil cover by Cadmium (Au+Cd) and Nickel foil (Ni). Au and Au+Cd were used to measure neutron-flux in thermal energy range and in epi-thermal energy range [6]. We also use Ni foil that is threshold detector to measure neutron-flux in fast energy range [6].

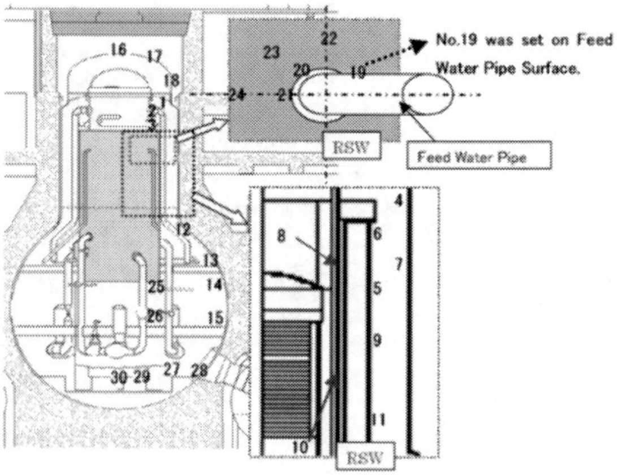

Fig. 3. Measurement locations of activation foils in the PCV

In order to measure neutron-fluxes that streams through the Biological Shielding Wall (BSW) both along a Main Steam line (MS line) and a Feed Water line (FW line) and that diffuses into the MS tunnel room, We also set ten foils in the MS tunnel room as seen in Fig. 4. In a simulation of neutron transport phenomena, a reliability of a downstream side in the phenomena depends on that of an upstream. In case of this measurement, the reliability of the estimation in the MS tunnel room would prove that in the BSW.

We set six sets of activation foils, which were No.1, 2, 3, 5, 6 and 10, around the MS line to observe neutron streaming through the MS line penetrating the BSW. One set, No.4, was located near the sidewall of the room. Other three sets, No.7, 8, and 9, were located on an aisle in an entrance of the MS tunnel room. These four foil sets were set to measure diffusion of neutron-flux in the MS tunnel room.

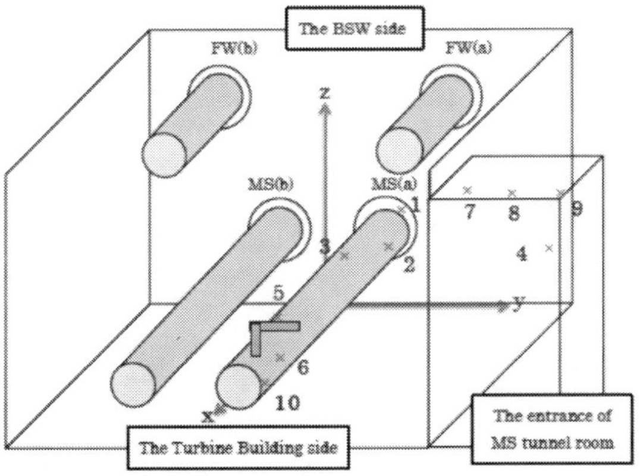

Fig. 4. Measurement locations of activation foils in the MS tunnel room

2.3 Distribution calculations of neutron-flux

We calculated three-dimensional (3D) distribution and two-dimensional (2D) distribution of neutron-flux around a reactor of the BWR in the PCV and inside the BSW. Furthermore, we performed 3D distribution calculation around the MS line and the FW line that penetrate the BSW and in the MS tunnel room. Areas where we set the ranges for these calculations are shown in Fig. 5.

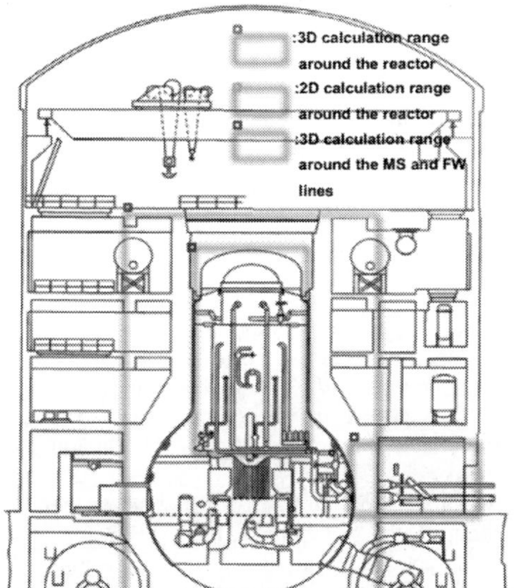

Fig. 5. Calculation ranges of 3D and 2D calculations

2.3.1 3D distribution calculation of neutron-flux around the reactor in the PCV

We performed 3D distribution calculation of neutron-flux around the reactor in the PCV by using 3D Discrete Ordinate method (Sn) calculation code TORT [2, 7]. A range of the calculation was an area surrounded with a blue square in Fig. 5. A purpose of the 3D calculation was to obtain knowledge of a characteristic of distribution of neutron flux. We set the range of the 3D calculation where important phenomena of neutron transport can be observed. We modeled the FW piping penetrating a Radiological Shielding Wall (RSW) that makes heterogeneity in neutron-flux distribution and so on. Both assessing the measurements and observing the 3D distribution of neutron-flux provided knowledge of the characteristics of distribution of neutron-flux in the PCV.

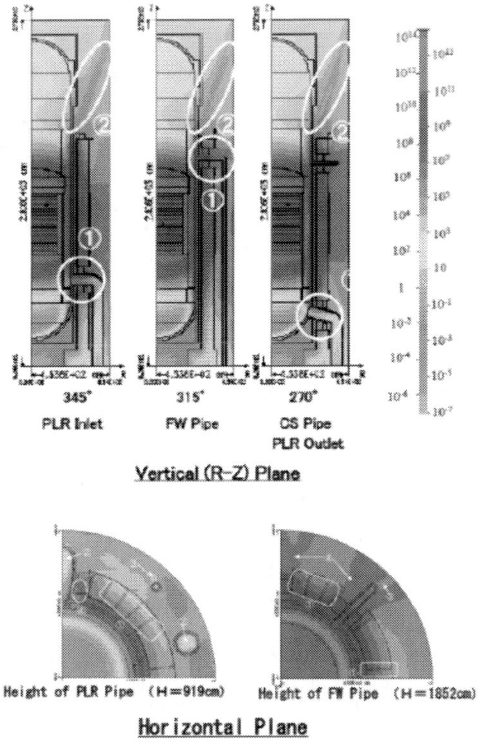

Fig. 6. Calculated 3D distribution of neutron-flux on vertical planes and on horizontal planes

We can find the following characteristics of neutron-flux distribution by observing contour maps of neutron-flux on r-z plane (the upper panel) in Fig. 6.

1) Although neutron leaks from hatches, water in the piping reduces amount of neutron leakage.
2) Although neutron streams through the gap between a Reactor Pressure Vessel (RPV) and a Radiological Shielding Wall (RSW), a stabilizer reduces amount of neutron streaming. The streaming neutron forms the area where flux level is rather high in the upper part of the PCV but the area is limited to a narrow range.

The lower panel in Fig. 6 also shows contour maps on r-θ plane. The heights of a plane in the left and the right panels are 919cm and 1852cm from the bottom of the PCV, respectively. Following characteristics are observed.

1) H-Beam in the RSW forms azimuthal heterogeneity in the distribution.
2) Water in the piping forms the area where flux level is rather low in the piping.
3) Shield effect of water in the piping forms the area where rather low flux behind the piping but the area is limited in a narrow range.
4) Although neutron leaks from hatches, water in the piping reduces amount of neutron leakage.

2.3.2 2D distribution calculation of neutron-flux in the PCV

We calculated 2D distribution of neutron-flux with 2D Sn code DORT [2, 7]. A range of the calculation was an area surrounded with a green square in Fig. 5. In order to perform the reliable 2D calculation of neutron-flux, we needed to generate the 2D model that could simulate neutron transport phenomena well in the PCV and to set optimized calculation conditions. At the onset, we generated a simple model, and causes of difference between calculated and measured fluxes were investigated. Then, we modified the 2D models by utilizing the knowledge obtained from measurements and 3D calculation.

The 2D distribution calculation started with a simple model. Only the RPV and concrete structures were modeled in the initial model. The 2D model was modified repeatedly until reliability was achieved. In order to simulate well the distribution of neutron-flux in the PCV, following phenomena were investigated and materials that would affect the 2D distribution of neutron-flux were taken into consideration in the final model.

1) Streaming through the gap between the RPV and the RSW.
2) Leakage from hatches and shielding effect of piping.
3) Upward streaming through the standpipes in the RPV, where steam is flowing through.
4) Shielding effect of H-Beam in the RSW.
5) Shielding (neutron absorption) effect of Control Rod Drive (CRD).
6) Shielding effect of the PCV internals in "Flask" area.

The best combination of mesh interval width and number of directional quadrature was also surveyed in the final model. Optimized calculation condition, which is the best combination of mesh interval width and numbers of directional quadrature, is as follows: 389 meshes in radial direction, 1104 meshes in axial direction and directional quadrature S16.

Fig. 7 shows that almost all of C/Ms of total neutron-flux at 30 locations meet the criteria of reliability that C/M is within one to ten.

Fig. 8 shows the 2D distributions of neutron-flux. The leftmost contour map shows the distribution of total neutron-flux. Contour maps of fast, epithermal and thermal neutron-flux are shown from the right neighbor side of that each in turn. The distribution characteristics of neutron-flux in the PCV, which were investigated and modeled, are observed in these maps: streaming and leakage are remarkable in fast and epithermal energy range (areas indicated "a" in the maps). Shielding effect by the PCV internals in "Flask" area affects thermal flux distribution.

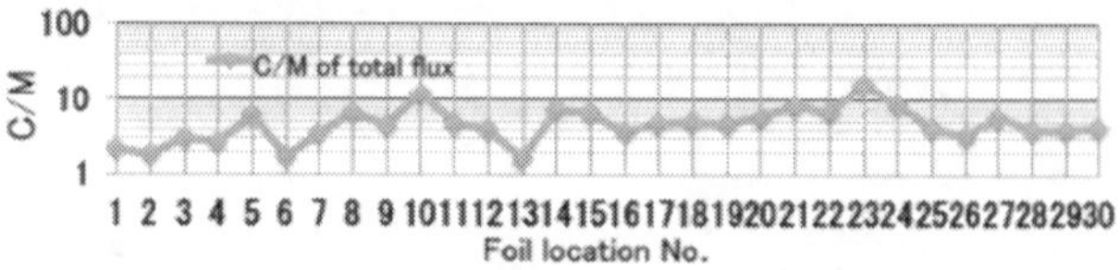

Fig. 7. Comparison between measured fluxes and 2D calculated fluxes (C/Ms)

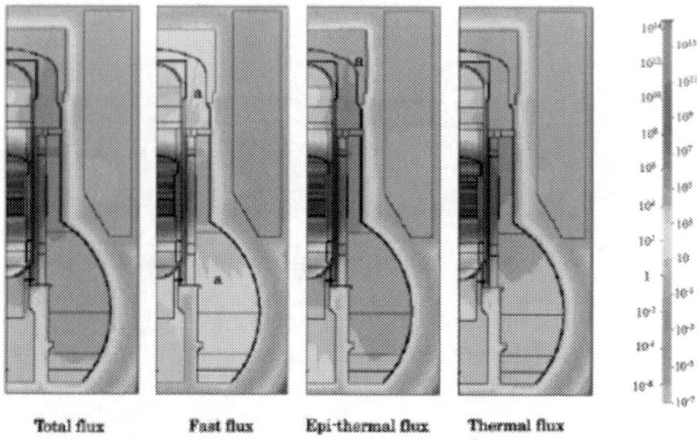

Fig. 8. Distribution maps of total flux, fast flux, epi-thermal flux and thermal flux

2.3.3 3D distribution calculation of neutron-flux outside the PCV

We also calculated 3D distribution of neutron-flux outside the PCV. The area includes a region where both the MS line and FW line penetrate the BSW and the MS tunnel room. The range is surrounded with a red square in Fig. 5.

According to a report by the regulatory body in Japan, materials, which are irradiated by neutrons to an effective-dose-rate less than 6.25 μ Sv/hr, can be treated as non-radioactive materials [8]. A purpose of this 3D calculation was an estimation of a boundary that distinguishes radioactive materials and non-radioactive materials in the BSW. In our work, we call the boundary on which materials are distinguished between radioactive materials and non-radioactive (NR) materials as NR boundary.

Reliable estimation of NR boundary is crucial in preparatory tasks for decommissioning of a nuclear power plant (NPP) to determine a decommissioning scenario, a dismantling design, a disposal planning and a decommissioning cost. Especially, the reliability would influence an accuracy of estimation for amount of very-low-level concrete waste because NR boundary would be located inside concrete of a BSW of a BWR. Amount of concrete waste is important since it is much larger than that of other materials from decommissioning activity of NPP.

The distribution calculation in the area required a separate estimation from other calculations because neutron transport phenomena in the area would be different from that of in the PCV. Neutron-streaming along the lines inside the BSW and neutron-diffusion in the MS tunnel room would occur in the area. Since these phenomena were hard to be simulated with 2D distribution calculation of neutron-flux, we performed 3D distribution calculation in the corresponding area. The calculation was performed with TORT by applying Sn-Sn coupling calculation technique [3,7]. We coupled the 2D calculation, which calculated neutron-flux distribution around the reactor, with the 3D calculation. Directional fluxes on an inner surface of the BSW given in the 2D calculation were set as Boundary Source that is one of calculation option in boundary conditions.

To simulate both neutron-streaming through the MS line and distribution of neutron-flux in the MS tunnel room well, we surveyed better combination between orders of scattering cross section and numbers of directional meshes. Because neutron-flux under epi-thermal energy range is dominant in the MS tunnel room, effect of order of scattering cross section (Pl) would be negligible. We judged that fine mesh set of the Level-symmetry directional mesh set such as S16 sets or biased directional mesh set was necessary so that calculation met the criteria for reliability.

According to C/Ms in the MS tunnel room, the calculation with S16 and P5 was the most effective to estimate NR boundary around the MS and the FW lines in our work. C/Ms in the MS tunnel room at ten locations are shown in Table 1. We judged that calculations were carried out with reliability by considering characteristics of distribution of neutron-flux in the MS tunnel room even if some C/Ms did not meet our criteria mentioned above. Nevertheless calculations resulted that NR boundary was located in the BSW concrete and that materials in the MS tunnel room were considered to be non-radioactive as seen in Fig. 9.

Table 1 C/Ms at foil locations in the MS tunnel room

Foil location	1	2	3	4	5	6	7	8	9
C/M	1.1	3.1	4.8	4.1	2.5	0.4	0.9	0.9	1.4

* Radioactivity at No.10 location was not detected.

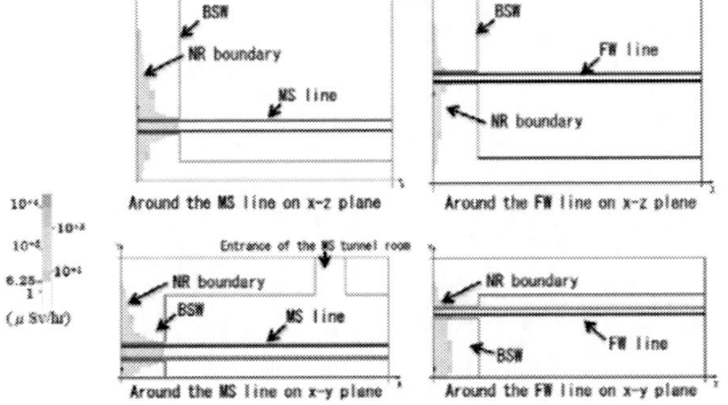

Fig. 9. NR boundary around the MS line and FW line

2.4 Distribution calculation of radioactivity due to neutron irradiation

We developed a new estimation procedure of time dependent distributions of radioactivity induced by neutron irradiation [9]. We performed activation calculations, which calculated generations and depletions of radioactive-nuclides due to neutron irradiation at whole spatial meshes of the 2D distribution calculation of neutron-flux. The calculations with the procedure could also provide time dependent 2D distributions of radioactivity of depleting with time after irradiation. The procedure was developed to estimate radioactivity distribution by considering characteristics of SCALE5.1/ORIGEN-S [10] calculation method and made it possible that we could estimate radioactivity distributions of various materials in much shorter computer time than the conventional method with the same level reliability.

We applied the procedure to the estimation of time dependent distributions of radioactivity inventory due to neutron irradiation around the PCV. Using the calculation results, we drew several kinds of time dependent radioactivity distribution maps. Other tasks referred to them to optimize the decommissioning strategy.

Fig. 10 shows one example of an application of the procedure. Distribution maps in the figure show time dependent distribution maps of a classification of radioactive waste for disposal and clearance from 20 years to 25 years after shut down. Area colored in green indicates ranges of very-low-level waste and areas colored in gray indicate ranges of clearance. Areas surrounded by red dotted square indicate a location of a Primary Line Recirculation Pump (PLR pump). These maps show that we will be able to treat the PLR pump as clearance when we will dismantle it 24 year after shut down.

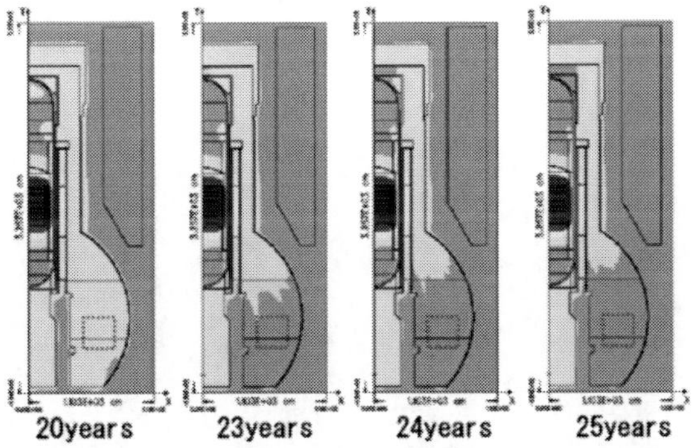

Fig. 10. Time dependent distribution maps of a classification of radioactive waste for disposal and clearance

3. Summary

We have developed the new procedure of evaluation for neutron-activated materials. We applied the procedure to an evaluation of radioactivity inventory of

BWRs that were declared to transit decommissioning-stage in the last fiscal year. The applications for decommissioning of BWRs have already been submitted and the regulatory body in Japan is reviewing the applications.

By applying the procedure to the BWRs, we obtained knowledge about the characteristics of distribution of neutron-flux in the PCV by measurements and by 3D distribution calculation of neutron-flux. 2D distribution calculation by referencing to the knowledge provided reliable 2D distribution of neutron-flux. The 2D distribution calculation accomplished our goal of reliability. By referencing to the 2D distribution of neutron-flux, we estimated time dependent distributions of radioactivity.

We also performed 3D distribution calculation of neutron-flux around the MS line and the FW line, which penetrate the BSW. We have been able to simulate neutron-streaming through the lines well. The distribution of neutron-flux around them provided that the boundary line is located in the BSW.

The information obtained from the evaluation would be utilized by other tasks in the preparatory tasks to make the decommissioning plan and the decommissioning activity safely and rationally.

The procedure of time dependent distribution of radioactivity due to neutron irradiation would play an important role in planning decommissioning strategy.

We performed this study as one of the decommissioning project promoted by JAPC. Authors thank the former director, Mr. Satoshi Karigome and the current director, Mr. Toyoaki Yamauchi for their understanding and the support for this study. We also thank members of the dismantling planning group of the decommissioning project department of JAPC for their useful information about the configuration of the internals in the PCV of Turuga NPP Unit 1 (TS-1). Members of Nuclear Service Corporation performed setting, collecting and radioactivity analysis of activation foils in the PCV of TS-1 by the contract. We thank them for their professional works and their useful advice in the planning of the measurement locations. A computer resource was provided from Genden Information System Corporation, a subsidiary of JAPC by a contract.

References

1. METI, *Roadmap for technology of LWR safety and human resource*, (1997) (in Japanese)
2. www.meti.go.jp/committee/sougouenergy/
3. Tanaka K, et.al., PNE 85 (2015) 254-270, (2015).
4. Tanaka K, et.al., PNE ,(to be published)
5. Tanaka K, et.al., proc. of 2013 AESJ annual meeting(2013) (in Japanese)
6. IAEA, Safety Series No.100, International Atomic Energy Agency, ISBN 9201240899, (1989).
7. Tsoulfanidis N, (Japanese translation by Sakai E), Japanese translation published by Gendaikougaku-sya, ISBN4874721281, (1986)
8. ORNL, CCC-650 DOORS3.2, Oak Ridge National Laboratory, (1998)
9. Advisory Committee for Natural Resources and Energy, Oct. 2007, "Report on Non-Radioactive waste from nuclear installation" (2007) (in Japanese)
10. Kurosawa M, et. al., Progress in Nuclear Science and Technology Volume 4 (2014) pp. 840 -843
11. Gauld C. I et. al., ORNL/TM-2005/39, Version 5.1, Vol. II, Book 1, Sect. F7, Oak Ridge National Laboratory, (2006)

Activation calculation for the dismantling and decommissioning of a light water reactor using MCNP™ with ADVANTG and ORIGEN-S

Luc Schlömer[1,a], Peter-W. Phlippen[1] and Bernard Lukas[2]

[1]*Wissenschaftlich-Technische Ingenieurberatung (WTI) GmbH, 52428 Jülich, Germany*
[2]*EnBW Kernkraft GmbH, 76661 Philippsburg, Germany*

Abstract. The decommissioning of a light water reactor (LWR), which is licensed under § 7 of the German Atomic Energy Act, following the post-operational phase requires a comprehensive licensing procedure including in particular radiation protection aspects and possible impacts to the environment. Decommissioning includes essential changes in requirements for the systems and components and will mainly lead to the direct dismantling. In this context, neutron induced activation calculations for the structural components have to be carried out to predict activities in structures and to estimate future costs for conditioning and packaging. To avoid an overestimation of the radioactive inventory and to calculate the expenses for decommissioning as accurate as possible, modern state-of-the-art Monte-Carlo-Techniques (MCNP™) are applied and coupled with present-day activation and decay codes (ORIGEN-S). In this context ADVANTG is used as weight window generator for MCNP™ i. e. as variance reduction tool to speed up the calculation in deep penetration problems. In this paper the calculation procedure is described and the obtained results are presented with a validation along with measured activities and photon dose rates measured in the post-operational phase. The validation shows that the applied calculation procedure is suitable for the determination of the radioactive inventory of a nuclear power plant. Even the measured gamma dose rates in the post-operational phase at different positions in the reactor building agree within a factor of 2 to 3 with the calculation results. The obtained results are accurate and suitable to support effectively the decommissioning planning process.

1 Introduction

The boiling water reactor (BWR) Philippsburg 1 (KKP1) and the pressurized water reactor (PWR) Neckarwestheim 1 (GKN I) were shut down and entered directly the post-operational phase which is covered by the operating licence held by the nuclear power plant owner. The decommissioning of a nuclear power plant (NPP) which is licensed under § 7 of the German Atomic Energy Act, following the post-operational phase requires a comprehensive licensing procedure including in particular radiation protection aspects and possible impacts to the environment. Decommissioning requires a decommissioning licence. This will include essential changes in requirements for the systems and components and will mainly lead to the immediate dismantling of the nuclear power plant which is one of the two fundamental decommissioning strategies in Germany.

In this context, neutron activation calculations for the structural components of KKP1 and GKN I have to be carried out to predict occurring activities and to estimate future costs for conditioning and packaging. To avoid an overestimation of the radioactive inventory and to calculate the expenses for decommissioning as accurate as possible, modern state-of-the-art Monte-Carlo-Techniques (MCNP6 [1]) combined with automated variance reduction tools (ADAVANTG [2]) are used and are coupled with the activation and decay module ORIGEN-S [3, 4]. This paper presents the calculation procedure and the obtained results with a validation by measured activations and gamma dose rates.

2 Calculation procedure

As pointed out in the introduction, the calculation of the activation inventory is done by a combination of two computational methods. First of all, a MCNP™ model has to be generated (cp. chapter 3) based on technical drawings of the reactor building. Additionally, the material composition and densities of all included components have to be identified. An essential part of the Monte-Carlo-simulation is the representation of the core (chapter 3.2) to map the appropriate neutron emission. Especially for the BWR model the choice of the correct water densities plays an important role because of occurring high gradients in axial direction inside the reactor pressure vessel (RPV).

According to the expected neutron flux distribution, the model is then divided into different segments, where

[a] Corresponding author: schloemer@wti-juelich.de

activation can be detected in relation to the local material composition.

For the first MCNP™ calculations, splitting and Russian roulette techniques were used as variance reduction tools and showed good results. Since ADVANTG was released in the second half of 2015, it is used by WTI as variance reduction tool for MCNP™ generating energy and space bias cards for the source and an adequate weight window mesh.

The results of the final MCNP™ calculation are reaction rates, neutron spectra and total neutron fluxes for the different segments. These results are used to execute an activation calculation with ORIGEN-S. The basis of this calculation are the burnup history of the reactor during its operation period and detailed material compositions including relevant trace elements to determine nuclide activities for several reference times. In a post-processing tool all information are merged to determine the specific activities of the components.

3 Calculation procedure

Exemplarily, Figure 1 shows an overview drawing and the derived MCNP™ model of KKP1. Not only neutron absorbing and moderating structures but also gaps which support neutron streaming are included. The structure is divided in some hundred segments for which activities are determined.

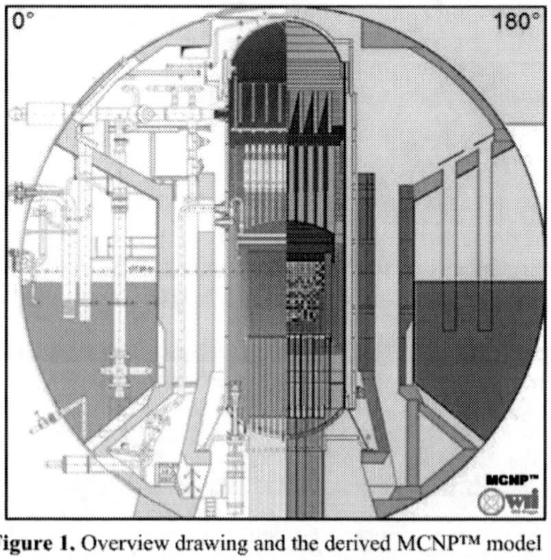

Figure 1. Overview drawing and the derived MCNP™ model

The calculation of neutron fluxes is based on the source distribution in MCNP™. An accurate core model must represent the neutron leakage and spectrum at the core boundary as far as they are relevant for activation reactions outside.

The local fission neutron source is derived from the axial burnup distribution of each fuel assembly in each cycle. Beside the burnup distribution, the water density influences the neutron leakage especially concerning the BWR-core model. Regarding the target to model an accurate representation of the core, transitions in fuel assembly shuffling influencing the neutron leakage have to be taken into account for both reactor types. Considering these aspects, steam-qualities with a local resolution are prepared for four operational phases of the BWR during full power reactor operation, where fuel shuffling schemes determine the neutron leakage from the core: (1) Start-up phase with radially flat power distribution, (2) introduction of higher enriched fuel and reduced neutron leakage, (3) introduction of spectral shift operation and transition to low-leakage fuel patterns and (4) introduction of additional orifices in second fuel assembly row with optimized flow rate in the outer fuel assembly rows. The relevant characteristics of the entire operational history of KKP1 are covered by these phases.

Regarding the PWR-core model of GKN I, the temperature distribution inside the vessel has been taken into account by varying the water density in axial direction. The boron concentration could be assumed as constant because no significant differences between the different cycles integrally appeared. Two different operational phases could be identified.

The local emission probability of neutrons is based on the local burnup-behaviour. The analysis of the provided distributions shows that fuel assemblies can be merged to different fuel regions in the case of the BWR. This simplification saves computing time because effectively only the outer fuel assembly regions influence the leaking neutron flux and thus the activation in the ex-core regions. In the case of the PWR each fuel assembly was modelled because of the asymmetric distributions during the operational phase.

The material composition of the active zone in both reactor types is based on a homogenized mixture of fuel, structural materials and water. The fissile material of the fuel consists of ^{235}U and ^{239}Pu. It is adjusted so that a k_{eff} of one in the MCNP™ model is achieved.

Because of the four operational different phases in KKP1, four separate MCNP™ runs with different core-models have to be set up. For GKN I two models resulted. These runs are reassembled in ORIGEN-S to calculate the activation considering the influences of all particular cycles.

4 Results

This chapter summarizes the results of the MCNP™ and coupled activation calculations and shows a validation of the calculated specific activations and the derived gamma dose rates.

4.1 General results of the calculation procedure

Exemplary the obtained neutron flux density distribution in the containment is shown in Figure 2 for the PWR GKN I. It can be seen that even in far distances from the neutron source, i. e. the modelled core, the total neutron flux density distribution is smooth indicating small relative errors. Additionally, the fulfilment of the ten statistical checks of the volume detector tallies in this region show that the MCNP™ calculation is converged.

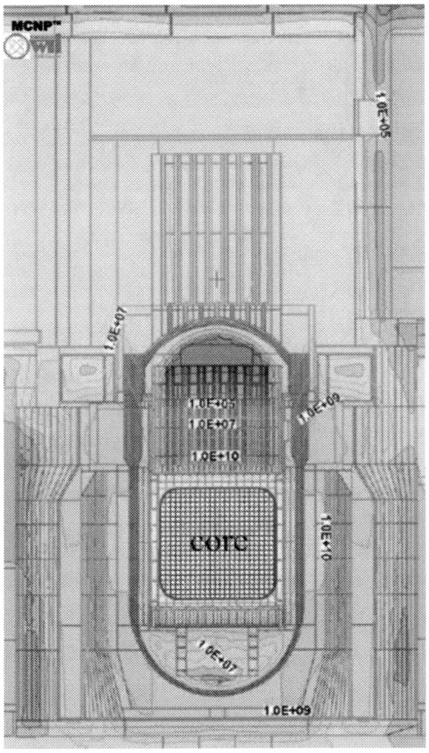

Figure 2. Neutron flux density distribution (1/(cm^2 s)) near the reactor pressure vessel of GKN I

To validate the calculated activities with ORIGEN-S, samples were taken inside the RPV, from the low activated regions inside the containment, and bore hole samples were taken from the biological shield. In this paper, only the results of the bore hole samples taken at the middle of the active zone and 4 m above the active zone are presented. The results are similar for PWR and BWR. The comparison of measured (M) and calculated (C) activities is presented as fraction (C/M) for the BWR KKP1.

Table 1 shows the obtained results using measured element concentrations as basis for the activation calculation. The bore hole samples contained not only the concrete and the armed concrete structure of the biological shield, but also a small probe of the RPV. Thus, the activities of both components could be measured and compared to the calculations. With regard to the publication of M. Pantelias and B. Volmert [5], where the validation of the neutron transport calculations with an in situ full-cycle foil activation for a PWR under exact-known boundary conditions is presented, about the same C/M-factors resulted as shown in Table 1. The activities of H-3 are overestimated by the calculation which can be explained by the following characteristic of H-3: After H-3 is formed it partially escapes from the components due to its high mobility. Generally, it can be seen that the activities calculated by the developed computational chain (cp. chapter 3) represent the measured activities very well.

Table 1. Comparison of measured (M) and calculated (C) activities for different bore hole samples (biological shield and RPV)

bore hole sample position	nuclide	concrete of biological shield			steel sample of RPV
		towards RPV	in the middle	towards annulus	
mid level of the active zone	H-3	7.1	6.7	-	*
	C-14	2.2	-	-	*
	Mn-54	*	*	*	2.3
	Co-60	0.9	5.6	-	1.2
	Cs-134	1.4	3.8	-	*
	Eu-152	4.1	3.0	-	*
	Eu-154	4.1	2.6	-	*
4 m above the active zone	H-3	19	2.9	2.8	*
	C-14	2.2	0.1	-	*
	Mn-54	*	*	*	6.8
	Co-60	0.9	-	4.6	1.8
	Cs-134	2.0	-	-	*
	Eu-152	5.7	-	2.0	*
	Eu-154	5.1	-	-	*

*: Not measured, -: Measured activity below detection limit

4.2 Results of the gamma dose rate calculations

In KKP1, the calculated activities are used to estimate the occurring dose rates in the post-operational phase. To validate this approach, the calculated nuclide activities are converted to photon spectra by ORIGEN-S and transferred to MCNP™ as source term. The occurring dose rate inside the containment can then be determined at different given receptor points. In the case of the validation, the gamma dose rate according to $\dot{H}^*(10)$ was measured at eleven receptor points (cp. Figure 3) after the decontamination of the primary circuit (Remark: The RPV was filled with water). Table 2 summarizes the results as C/M deviation. In the second row of Table 2 the results are shown without modification, i. e. a rotationally symmetric activation of the shroud. In this first calculation the deviations estimated for the receptor points at the level of the active zone (M4 to M8) are slightly different from the others. Including rotationally heterogeneous activation of the shroud, the results shown in the third row arise. This is due to the major contribution of the shroud to the total dose rate. Thus azimuthal varying heterogeneous activation has to be considered in dose rate calculations especially in the near region of high activated components.

DOI: 10.1051/epjconf/201715305020

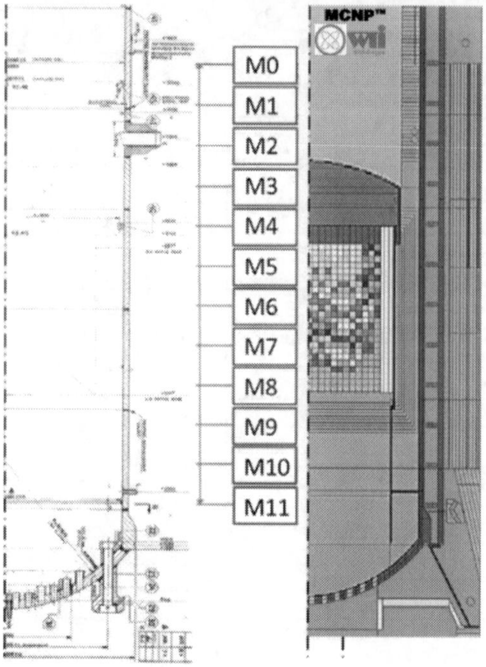

Figure 3. Receptor points M0 to M11 between reactor pressure vessel and biological shield for gamma dose rate measurements in KKP1 (BWR)

Table 2. Comparison of measured (M) and calculated (C) gamma dose rates at different receptor points in KKP1.

receptor point	C/M	C/M (corrected)
M0	1.2	1.2
M1	2.0	2.0
M2	3.0	2.7
M3	1.9	1.7
M4	3.1	2.3
M5	3.1	2.2
M6	3.2	2.1
M7	2.6	1.8
M8	4.0	3.0
M9	2.9	2.6
M10	2.9	2.9
M11	2.0	2.0

5 Conclusions and Outlook

The results presented in this paper clearly show that the developed procedure to determine activities can be applied for the calculation of the activation inventory of any type of nuclear power plants. The obtained results and even the estimated gamma dose rate predictions are very accurate and still conservative. In future, the calculated local activation inventories for a PWR will be used to calculate gamma dose rates occurring in this nuclear power plant type after shutdown. These results will be compared to measured dose rates. Based on the WTI experience, the developed calculation procedure can be used for the calculation of radioactive inventories of any type of nuclear facilities. The results are valuable contributions for the decommissioning planning and especially required for conditioning planning and the determination of package types and quantities.

References

1. T. Goorley, et al., *Initial MCNP6 Release Overview*, Nuclear Technology **180**, pp 298-315 (12/2012)
2. S. W. Mosher, et al., *ADVANTG-An Automated Variance Reduction Parameter Generator*, ORNL/TM-2013/416 (08/2015)
3. I. C. Gauld, *ORIGEN-S: Depletion Module to Calculate Neutron Activation, Actinide Transmutation, Fission Product Generation, and Radiation Source Terms*, in [4], Sect. F7 (06/2011)
4. *SCALE: A Comprehensive Modelling and Simulation Suite for Nuclear Safety Analyses and Design*, ORNL/TM-2005/39, Version 6.1 (06/2011), upgrade to Version 6.1.2 (02/2012)
5. M. Pantelias, B. Volmert, *Activation Neutronics for a Swiss Pressurized Water Reactor*, Nuclear Technology **192**, pp 278-285 (2015)

PANTHERE V2: MULTIPURPOSE SIMULATION SOFTWARE FOR 3D DOSE RATE CALCULATIONS

Gaël PENESSOT[1a], Éléonore BAVOIL[1], Laurent WERTZ[1], Fadhel MALOUCH[2], Thierry VISONNEAU[2] and Julien DUBOST[3]

[1]*EDF/SEPTEN/TE/RP, 12-14 avenue Antoine Dutriévoz, 69628 Villeurbanne, France*
[2]*CEA-Saclay DANS/DM2S/SERMA, 91191 Gif-sur-Yvette, France*
[3]*CS-SI, 45 rue Sainte Geneviève, 69006 Lyon, France*

Abstract. PANTHERE is a multipurpose radiation protection software developed by EDF to calculate gamma dose rates in complex 3D environments. PANTHERE takes a key role in the EDF ALARA process, enabling to predict dose rates and to organize and optimize operations in high radiation environments. PANTHERE is also used for nuclear waste characterization, transport of nuclear materials, etc. It is used in most of the EDF engineering units and their design service providers and industrial partners.

1 Introduction

PANTHERE is a radiation protection code used for 3D simulation of gamma dose rates and fluence for complex environment with distributed radioactive sources like nuclear facilities.

PANTHERE is intensively used by EDF engineering units and its industrial partners. It was originally used to prepare maintenance and modification operations in nuclear power plants, especially on the primary circuit like pressurizer heaters replacement and steam generator replacement. It is now extended to other fields like:
- shield design and zoning for EPR reactors
- dose calculations for the qualification of PWR equipments during severe accidents.
- decontamination and dismantling operations
- radiological characterization for applications associated to transport and nuclear waste management

PANTHERE is verified and validated by comparison with the reference Monte Carlo code TRIPOLI-4 of the CEA and gamma dose rates experimental measurements.

2 Physical models used

PANTHERE's computation kernel uses CEA solver NARMER, which is linked to the EDF R&D geometric library LIBCAD. This chapter focuses on the different physical models used by NARMER.

2.1 The point kernel integration method

The method used in NARMER is the point-kernel integration method with build-up factors. It is a macroscopic approach for gamma dose rate calculation, taking into account the interaction with matter. The dose equivalent rate (DER) for a source of activity A (gamma.cm^{-3}.s^{-1}) in a volume V_s is computed as in Eq. (1):

$$DER(E) = \sum_{0}^{E_{max}} F_D(E) \int_0^V \left[B(E) \frac{A(E)\, e^{-\sum_i \mu_i(E) x_i}}{4\pi (\sum_i x_i)^2} \right] dV \quad (1)$$

with:
- $F_D(E)$ the gamma fluence-to-dose conversion factor
- $\mu(E)$ the linear attenuation coefficient
- x the shield thickness encountered by a gamma ray along a straight line
- d the distance between the source and the calculation point
- $B(E)$ the build-up factor

2.2 The build-up factors

The build-up factors (B) are used to estimate the scattered radiation in matter. These factors are tabulated for all simple elements (from Z=1 to 98) and for 195 energy groups between 15 keV and 10 MeV. For multi-layer shields, the build-up factors are computed according to the "CEA/SERMA" formula [1].

3 PANTHERE architecture

PANTHERE is constituted of three main components: a computation kernel, a MySQL database for models management and sharing , and a 3D GUI to define the studies and both geometrical and radiological environment. It runs on a Linux operating system.

[a] Corresponding author: gael.penessot@edf.fr

3.1 Pre-treatment in the GUI

PANTHERE GUI allows the definition of geometrical and radiological models by the use of 9 parametrical primitives (e.g., cylinder, wall, etc.), frequently used to model nuclear facilities. They can be composed of 1 to 5 distinct media: 2 internal media (air/water) and 3 successive layers: "metal", "thermal insulation" and "biological protection".

Radioactive sources can be added to every layer of each object (both volume and surface sources). The spectrum is defined with energy rays, with nuclides or using energy groups. The nuclear data are taken from JEF2.2 [2].

Observation points are defined by their 3D coordinates. Measurement results can be added to each observation point to perform a readjustment of the predefined source activity (fitting).

Finally, results can be given according to several response functions: the equivalent dose rate H*(10) in $mSv.h^{-1}$, the "kerma in air" $mGy.h^{-1}$, the energetic fluence rate $MeV.cm^{-2}.s^{-1}$ and the uncollided flux $gamma.cm^{-2}.s^{-1}$.

Figure 1. Caption of the PANTHERE v2 GUI

3.2 Post-processing in the GUI

The results of the computation are displayed in form of source/point or nuclide/point matrix and can be exported in csv format or opened in a spreadsheet editor. Dedicated SALOME [3] modules allow for radiological zoning and isodose mapping.

PANTHERE results, combined with NPP maintenance program can be used as an input of CADOR, an optimization code for the installation of protection on NPP contaminated pipes. This code perform an optimization of the collective dosimetry by scheduling the task in order to reduce both painfulness of team in charge of installing protection and the dosimetry of team in charge of maintenance.

4 Physical model limitations

PANTHERE uses a simplified physical model to enhance calculation performance (10 to 25 times faster than stochastic methods like TRIPOLI-4). However, it restricts the usage of PANTHERE to domains where the approximations of Eq. (1) are pertinent.

The main restrictions are geometrical: the singularity of Eq. (1) imposes a minimum distance between source and observation point.

Finally, the build-up factors are validated for penetration lengths up to 50 MFP, and for energy ranging from 15 keV to 10 MeV. They should not be used outside this validation domain.

Fortunately, this domain covers most of studies for nuclear energy facilities.

5 Functionalities

5.1. CAD import

PANTHERE inherit SALOME's CAD modeler GEOM which allow s designing buildings, structures and piping. Moreover, PANTHERE allows the user to import CAD data from specific modelers like SolidWorks or PDMS. This functionality improves the efficiency of the geometrical modeling.

5.2 Source readjustment module

In most cases, calculations are compared to measures. PANTHERE features a module that adjusts theorical activity in model's sources to fit experimental measures. It allows the user to have a more realistic model and more relevant source contributions in order to design shelding.

5.3 Computational performance

The kernel of PANTHERE v2 improves computational performance by using an optimized geometrical representation and raytracer, and by implementing shared memory parallelization over the observation points, the sources and the nuclides.

Further acceleration could be obtained by distributed computing currently under development.

6 Conclusions

PANTHERE plays an important role in an ALARA regulatory process by providing simulation results for industrial radiation protection. Its use of a simplified computation method and its user-friendly GUI dedicated to radiation protection allow a reliable and fast industrial usage to maintain occupational radiation exposures as low as reasonably achievable in nuclear facilities.

Moreover, future releases will include beta (short term) and neutron (long-term) dose rates.

References

[1] C. Suteau, M. Chiron, G. Arnoud, Improvement of MERCURE-6's General Formalism for Calculating Gamma-Ray Build-up Factors in Multilayer Shields, *Nucl. Sci. & Eng.* 147 (2004), pp.419-424.
[2] JEF-2.2 Nuclear Data Library, NEA JEFF Report 17 (2000)
[3] http://www.salome-platform.org

Decommissioning Plan for European Spallation Source

Daniela Ene[1,a]

[1]*ESS ERIC, Environment Safety & Health Department, Lund, Sweden*

Abstract. This paper is a survey of the European Spallation Source initial decommissioning plan developed in compliance with Swedish Regulatory Authority requirements. The report outlines the decommissioning strategy selected and the baseline plan for decommissioning. Types and quantities of radioactive waste estimated to be generated at the final shut-down of the facility are further provided. The paper ends up with the analysis of the key elements of the decommissioning plan and the recommendations to the ESS management team..

1 Context

The European Spallation Source ERIC (ESS) is the European common effort in designing and building a next generation large-scale user facility for studies of the structure and dynamics of materials. The proposed schematic layout of the ESS facility is based on a linear driver (linac) directing the proton beam (5 MW of 2 GeV) of 2.8 ms long pulses with a 20 Hz on a tungsten target where neutrons are produced via spallation reactions. Further the neutrons will be moderated to thermal and subthermal energies in a couple of moderators placed around the target. The moderators feed 22 beamlines guiding the neutrons to the scattering instruments, mainly for neutron scattering research.

The envisaged period for the decommissioning of the ESS facility is 2065 - 2070 after an expected 40 years of lifetime. This paper is a survey of the ESS initial decommissioning plan developed in compliance with Swedish Regulatory Authority requirements defined for non-nuclear facilities, and following also the guidance provided by Swedish Nuclear Fuel and Waste Management Company (SKB).

2 Decommissioning strategy

The initial decommissioning plan is the first step, of the graded approach that should be applied to the development of the ESS decommissioning plan, in agreement with the international and national guidance. It is based on the status of the ESS-project, as reported in the Technical Design Report [1].

Based on comparative analysis of several parameters ESS selected as referenced decommissioning strategy: immediate dismantling option. The desirable final goal of decommissioning activities will be to return the site to green field status.

3 Baseline plan for decommissioning

The baseline plan for decommissioning consists of removing the components in a sequence starting with the most radioactive parts and ending with the least radioactive ones. Main tasks and time-scaling are roughly shown in Table 1.

Table 1. Main tasks of the ESS decommissioning project.

Schedule	Task
Phase #1	Remove highly activated components
Phase #2	Remove other ILW
Phase #3	Remove other LLW
Phase #4	Radiological clearance of buildings and site

The approach to be used involves the dismantling, segmenting and decontamination techniques that are expected to be effective for the ESS facility. A summary of the methodologies and types of equipment recommended for application in each of the area of the facility is furfer provided.

4 Radioactive waste expected at shut-down

All radioactive waste from ESS will be handled and disposed of within the Swedish system for management of radioactive waste. The waste classification system

[a]Daniela Ene: Daniela.ene@esss.se

used within this work is in accordance with Swedish policy and practice that define the following waste classes: i) free release material; ii) very low level waste (VLLW); iii) SFR-waste: short-lived; iv) SFL-waste: long-lived low-and intermediate level waste (LL-ILW); v) Heat generating waste. Swedish regulatory guidelines were used to classify ESS waste based on clearance index approach. The simplified scheme of ESS radwaste streams is provided in the Figure 1.

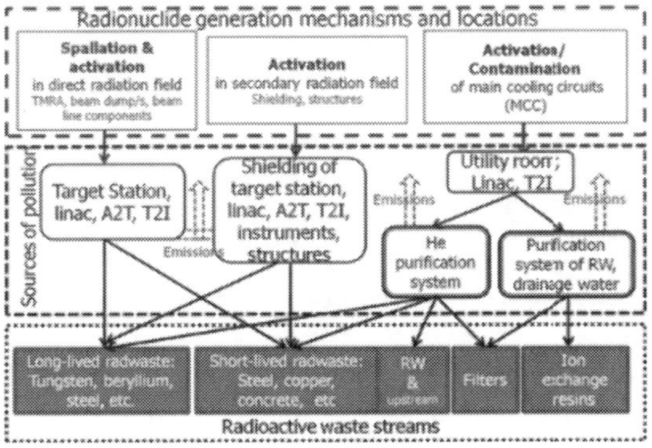

Figure 1. Simplified scheme of ESS radwaste streams

4.1 Waste radiological characterization

The amounts of ESS radioactive waste and classifications were derived using: i) precise calculations performed by means MCNPX2.6.0 [2] coupled with CINDER'90 [3] computer codes ii) scaling the activity from the operation experience of the existing spallation source installations for waste such it is difficult to predict level of activation or for components of the facility in stage of the pre-conceptual model.

For illustration purpose two examples are provided here. Firstly, the characterisation of the highest activated component, the tungsten target, that have a life time of five years. Table 2. gives the target classification in terms of waste based on the specific activity and its decay heat at various decay times after shut-down.

Table 2. Activity and decay heat as a function of decay time for tungsten target.

Decay time	Specific activity (TBq/m^3)	Decay Heat (kW/m^3)	Waste level
0	2.46E+06	2.17E+02	HLW
1week	5.86E+05	1.67E+01	
1y	6.32E+04	1.87E+00	
3y	2.28E+04	6.31E-01	
5y	1.56E+04	3.14E-01	
7y	1.19E+04	1.66E-01	ILW

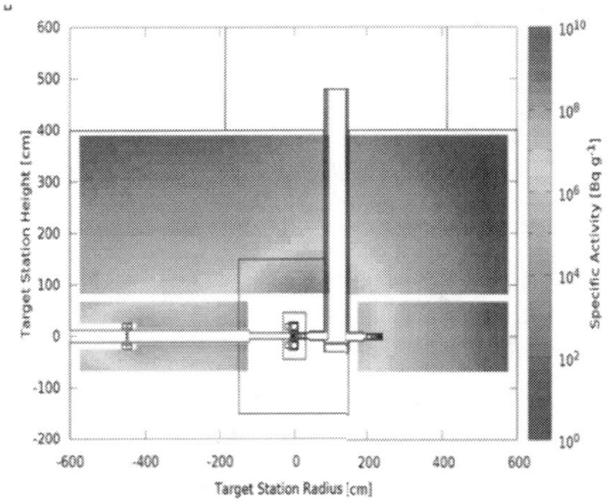

Figure 2. Specific activity of the target monolith at the final shut-down (40 years).

While the Figure 2 displays the map of the specific activity of a component having the lifetime of the facility, the shielding monolith.

Types and quantities of radioactive waste estimated to be generated at the final shut-down of the facility are further outlined.

Radwaste logistics developed at the ESS facility, during operations [4] was adapted to be used for decommissioning operations as well.

4.2 Summary of radwaste at shut-down

The sum-up of waste arising from entire ESS facility as decommissioning waste is provided. Decommissioning waste is breaking down in following categories:

- 11000 tons of metal waste**,** mainly Fe from the target monolith and the linac components dismantling likely to be SFR type waste;
- 1000 tons of waste may potentially be **SFL type** coming from parts of ILW from accelerator machine (10% from the total amount) and TS shielding monolith placed closest to the target (20% from the total);
- 3500 tons of concrete waste originating from the internal side of accelerator tunnel;
- 32000 tons of concrete VLLW from the dismantling of the shielding of the NBLs and instruments to be free released on site after to the storage for decay, if necessary;
- 800 tons of concrete VLLW from the dismantling of the Target monolith;
- 4000 tons of soil surrounding the concrete wall of the linac to be free released on site after to the storage for decay, if necessary.

Total generated decommissioning waste is about 58200 tons, from which 72% is expected to be free released. Note that the Clearance index of the soil around the concrete wall of the accelerator was also estimated and it was found that about one decay year after shut-down is enough to wait until the soil will be free released.

5 Key elements of decommissioning plan

The key elements of the decommissioning plan :
• Licensing conditions;
• Staffing and training;
• Organization and administrative control;
• Cost estimation;
• Waste management;
• Emergency management;
• Radiation and physical protection;
• On and off site monitoring;
• Quality assurance
are further presented and discussed, with emphasis upon the cost estimation.

The paper ends up with the recommendations to the ESS management team of actions that can be taken during the design phase for efficient dismantling and disposal and thus reducing any exposure levels of the personnel and decommissioning costs.

6 Conclusion

The main outcome of this report is the descriptions provided on all waste types arising during the decommissioning of ESS facility and the waste management considered. The waste amounts reported here are based upon the baseline design of the facility whose design now enters a more refined stage. The operation/maintenance plans of the facility shall be developed and optimized as well. Therefore the current data are estimations subjected to evolutions and regular updates

References

1. http://www.esss.se/ Technical Design Report, (2013)
2. D. B. Pelowitz, MCNPX User's Manual. Version 2.6.0. s.l.: Los Alamos National Laboratory, LA-CP-07-1473 (LANL, 2008).
3. Wilson, W. B. et al. "A manual for CINDER.90 version 07.4 and associated codes and data." 2007.
4. D. Ene, ESS Preliminary Waste Management Plan of European Spallation Source - 96154, Proceding of ICEM2013, Brusselles, Belgium, (2013).

Shielding Calculations on Waste Packages – The Limits and Possibilities of different Calculation Methods by the example of homogeneous and inhomogeneous Waste Packages

Mike Adams and Silva Smalian [a]

TÜV NORD EnSys Hannover GmbH & Co. KG, Hannover, Germany

Abstract. For nuclear waste packages the expected dose rates and nuclide inventory are beforehand calculated. Depending on the package of the nuclear waste deterministic programs like MicroShield® provide a range of results for each type of packaging. Stochastic programs like "Monte-Carlo N-Particle Transport Code System" (MCNP®) on the other hand provide reliable results for complex geometries. However this type of program requires a fully trained operator and calculations are time consuming. The problem here is to choose an appropriate program for a specific geometry. Therefore we compared the results of deterministic programs like MicroShield® and stochastic programs like MCNP®. These comparisons enable us to make a statement about the applicability of the various programs for chosen types of containers. As a conclusion we found that for thin-walled geometries deterministic programs like MicroShield® are well suited to calculate the dose rate. For cylindrical containers with inner shielding however, deterministic programs hit their limits. Furthermore we investigate the effect of an inhomogeneous material and activity distribution on the results. The calculations are still ongoing. Results will be presented in the final abstract.

1 Introduction

In medicine, in nuclear plants or in research accumulating radioactive waste has to be disposed of properly. In Germany the radioactive waste which cannot be reused or released has to temporarily be stored and finally be disposed of in a Federal final disposal site [1]. Before final disposal these wastes have to be submitted to a qualified conditioning process, during which the protection of staff members and the population from radioactive contamination is of highest priority.

To protect the population the dose rates and the nuclide inventory of radioactive waste ready for interim or final storage are strictly limited by both, the acceptance criteria of interim and final disposal sites [2] and by the transportation law. This has to be taken into consideration when conceiving a loading concept for radioactive waste campaigns and sizing radioactive waste containers.

To ensure these limitations the expected dose rates and nuclide inventory are beforehand calculated with suitable software. The user has to choose the right program for each type of packaging. Depending on the calculation parameters chosen deterministic programs with an easily accessible user interface such as MicroShield® already provide a range of results for simple geometries. These results, however, may differ by factor 4 from each other and therefore can´t be used efficiently for loading concepts and container design. Stochastic programs using the "Monte-Carlo N-Particle Transport Code System" (MCNP®) on the other hand provide reliable results for complex geometries, but require a fully trained operator. The question therefore is: Which program is the most efficient for which geometry? And how does an inhomogeneous material and activity distribution affect the results?

Comparing the results calculated with both, MicroShield® and MCNP®, we have examined the problem of the input-parameters with MicroShield® for some containers relevant for temporary and final disposal of waste. In addition, we limited the range of application of MicroShield® for these containers.

Our comparisons enable us to make a statement about the applicability of the various programs for chosen types of containers and distributions.

2 Calculation Method

To check the observance of the waste acceptance criteria for the dose rates and the nuclide inventory, conclusions must be drawn from dose rate measurements at containers to the activity inventory or, when conceiving loading concepts, from the known activity to

[a] Corresponding author: ssmalian@tuev-nord.de

the dose rate expected at the containers. For this purpose several shielding-programs are available of which we compared two.

2.1 MicroShield

MicroShield®, a product of Grove Software, Inc. was developed in the early 1990s in Virginia, USA, to perform shielding calculation. It is limited to and internationally recognized for a selection of/number of standard geometries. For multi-layer shields it can only be used under reservation and - as will be shown - with deeper understanding. The program is continuously being further developed and is now available in version 10.

2.2 MCNP

The program MCNP® is a trademark of Los Alamos National Security, Los Alamos National Laboratory and simulates the transport of particles through matter by using Monte Carlo algorithms. It thus allows the analysis of complex three-dimensional geometries with almost any material and source distribution (see fig. 12). Modelling effort and computing time are the only limiting factors. The program is constantly being further developed and is now available in version 6.1.

During the simulation of particle transport problems by means of the Monte Carlo method, each particle - from its generation until its disappearance (absorption or leaving the modelled area) - is monitored/examined. By using probability density functions and generated random numbers events for the particles are generated. The probability density functions are available for each material stored in tables.

The user provides all the relevant data as an input file, the drawing up of which is a complex task with MCNP®, therefore more error-prone than MicroShield and requiring a trained operator.

3 Comparison of the results

We have modeled different types of containers in constant conditions with both programs and compared the results. Our goal is to raise the awareness of the problem and make a statement about the range of application of the two differently convenient calculation methods, "MicroShield" and "MCNP", in regard to the reliability of the results provided by them.

In the considered volume sources, a homogeneous and an inhomogeneous distribution of the waste product and activity were assumed.

3.1. Thin walled geometries with homogeneous material and activity distribution

The MCNP® results are well in agreement with the MicroShield results for the build-up factor of the source. The MicroShield results differ by factor ~1,1 from each other. The build-up factor of the source clearly dominates these geometries, so that MicroShield® – provided the build-up factor is properly chosen - can be used to calculate the dose rate. Further marginal conditions, which are not modeled (e.g. inhomogeneous density or source distribution), lead to larger deviations than the choice of the calculation procedure.

3.2 Thin walled geometries with inhomogeneous material and activity distribution

The calculations are still ongoing. Results will be presented in the final abstract.

3.3 Thick walled geometries with homogeneous material and activity distribution

The results for a cylindrical container filled with iron (height 150 cm, diameter 106 cm, wall thickness 16 cm, top and bottom 18 cm, density 7 g/cm^3) and with 100 mm lead shielding inside are depending on the waste density modeled. The dose rate is calculated at a one-meter distance from the waste package at half height. The MicroShield® results for different build-up factors in otherwise constant conditions differ by factor 4. The MCNP® results are within this range and are either under-estimated by factor 0.5 or over-estimated by factor 2 by the MicroShield® results.

3.4 Thick walled geometries with inhomogeneous material and activity distribution

The calculations are still ongoing. Results will be presented in the final abstract.

4 Conclusion

Our comparisons make it possible to judge about the applicability of the various programs for chosen types of waste packages with homogenous material and activity distribution. For the thin-walled geometries the program MicroShield® is well suited to calculate the dose rate. For this purpose the source´s dominating build-up factor, which leads to the below mentioned deviations from MCNP®, has to be chosen. For some thick walled geometries, however, the program hits its limits. These examples clearly show the problem of choosing the right build-up factor at MicroShield®.

We expect to make a statement about the applicability of the various programs for chosen types of waste packages with inhomogeneous material and activity distribution. The calculations are still ongoing.

References

1. StrlSchV
Verordnung über den Schutz vor Schäden durch ionisierende Strahlen (Strahlenschutzverordnung-StrlSchV) vom 20. Juli
2. BfS
Anforderungen an endzulagernde radioaktive Abfälle

(Endlagerungsbedingungen) Stand: Dezember 1995 -
Schachtanlage Konrad-, ET-IB-79

Dose rate evaluation of workers on the operation floor in Fukushima-Daiichi Unit 3

Kaoru Matsushita[1,*], *Masahiko* Kurosawa[1] , *Keisuke* Shirai[2] , *Ippei* Matsuoka[3] and *Naoki* Mukaida[3]

[1]Toshiba Corporation, Yokohama, 235-8523, Japan
[2]Tokyo Nuclear Services Corporation, Tokyo, 110-0016, Japan
[3]Tokyo Electric Power Company Holdings, Inc. , Tokyo, 100-8560, Japan

Abstract. At Fukushima Daiichi Nuclear Power Plant Unit 3, installation of a fuel handling machine is planned to support the removal of spent fuel. The dose rates at the workplace were calculated based on the source distribution measured using a collimator in order to confirm that the dose rates on the operation floor were within a manageable range. It was confirmed that the accuracy of the source distribution was C/M = 1.0-2.4. These dose rates were then used to plan the work on the operation floor.

1 Introduction

In Unit3, spent fuel assemblies remain in the spent fuel pool (SFP). These will be removed from the pool and transported to a common pool. To do this, a new fuel handling machine (FHM) and workplace for machine maintenance will be installed over the SFP (Figure 1). Because people will be in the workplace, dose rates in the workplace needed to be evaluated to check the plan.

However, the reactor building of Unit 3 was severely damaged by a hydrogen explosion. The ceiling, floor and wall from an operation floor located in the top of Unit 3 were collapsed, the fuel handling machine and beams had also fallen onto the operation floor (Figure 2). Large pieces of debris from the wall, ceiling, machines and so on, were strewn all over the operation floor. Because of the debris, measurements could not be done to accurately evaluate the dose rate at the workplace. Therefore the situation on the operation floor was unknown.

From September 2011 until November 2013 a crane was used to remove the large debris on the operation floor. Once complete it was discovered that the operation floor was a high dose area and most of the floor was fragile.

To quantify the radiation field at the operation floor measurements were needed to calculate the source strength and distribution. To take these measurements we designed a collimator with a detector suited to the operation floor. This collimator and detector were used to quantify the source distribution. Finally, using the evaluated source distribution, the radiation field and dose rates at the workplace were calculated.

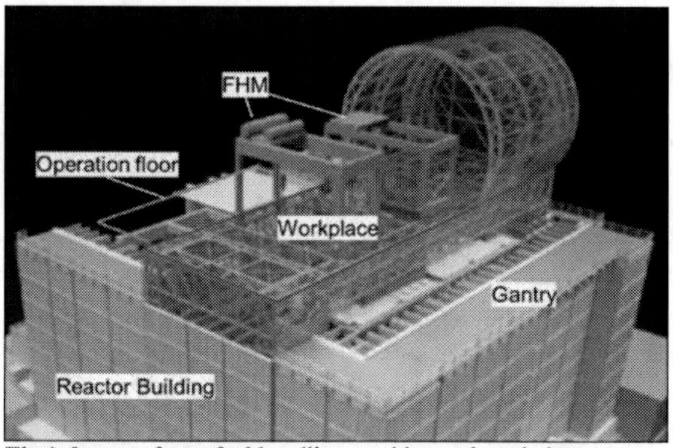

Fig.1. Image of new fuel handling machine and workplace

Fig.2. The operation floor of Unit 3 after removal large debris

2 Design of collimator

* Corresponding author: kaoru2.matsushita@toshiba.co.jp

2.1 Premise for designing collimator

There were high dose areas on the operation floor after a removal of the large debris. It was assumed that the small debris remained on the operation floor were radioactive and penetrated into the concrete floor. In addition, the operation floor was unstable and, in particular, the floor on the northwest side has mostly collapsed (see Figure 2). Thus neither people nor machines could come close to the operation floor; dose measurement needed to be done using a crane.

To mitigate the time needed to perform measurements using a crane, a 4-meter-by-4-meter mesh with measurements taken 50 cm above the floor was used to quantify the radiation source. 130 measurement points were needed to cover the area of interest using this mesh size. The strength of the source in each mesh was calculated using the relationship between unit radioactivity on a mesh and the calculated dose rates at a detector.

Because the measured dose rates from a target mesh will include dose rates from the other mesh, a collimator was used to attenuate the radiation from other mesh. Measurement data was collected using a wireless system and a Geiger-Muller (GM) counter. The GM has already been used at the Fukushima site with a wireless system.

Due to the project schedule using a new collimator was not an option. A collimator which had already been made for another measurement needed to be used. A vertical cross-section of the collimator is shown in Figure 3. It has a rectangular shape casing made of steel, filled with concrete with a cylindrical hole through the center of the concrete for gamma ray attenuation. This collimator was able to measure the dose rates from an area around and under the collimator by fixing GM counter in the hole.

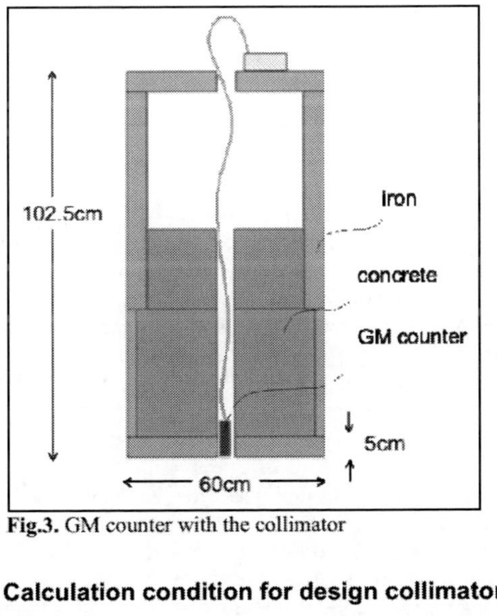

Fig.3. GM counter with the collimator

2.2 Calculation condition for design collimator

When measuring the dose rate from a target mesh, the results will include noise due to radiation from adjacent meshes. To understand the impact of noise from adjacent meshes, we examined the collimator configuration.

A disc-shape radiation source, with a radius of 400 cm and a uniform source strength, was assumed. The term D_{total} was defined as the dose rate from a radiation source in whole source region. The term $D_{0\text{-}200}$ was defined as the dose rate from the source assuming a 200 cm radius. The ratio of $D_{0\text{-}200}$ to D_{total} ($D_{0\text{-}200}/D_{total}$) was defined as a performance metric with a target value of over 0.99; a $D_{0\text{-}200}/D_{total}$ value of 0.99 means that the adjacent mesh provide < 1 % of the dose rate measured by the collimator. Therefore, if the radiation sources of adjacent mesh are two orders of magnitude (100x) higher than the target mesh, then no more than half of the dose rate at the target mesh would be noise from adjacent mesh. This level of conservatism was considered acceptable.

A radiation concentration was set based on the highest measured dose rate on the operation floor and the gamma spectrum was assumed to be from Cs134, Cs137, and Ba137m. "MCNP5" [1] was used for calculations with "MCPLIB04" [2] as the photoatomic data library and the flux to dose conversion factor is based on "ICRP Publication 74"[3].

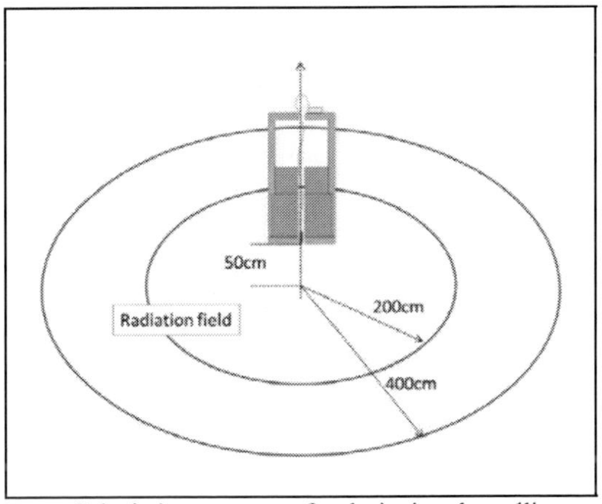

Fig.4. Calculation geometry for designing the collimator

2.3 $D_{0\text{-}200}/D_{total}$ of improved configuration collimator

The base case collimator shown in Figure 3, would have the GM counter set in the bottom of the hole since its $D_{0\text{-}200}/D_{total}$ was planned to be 0.99. However, it was found that a dose rate at the bottom of the hole was over the upper limit of the GM counter, 1000 mSv/h.

As a result, the GM counter was set in an upper region of the hole because the dose rate must be below the upper limit of the GM counter while maintaining the collimator configuration. However, $D_{0\text{-}200}/D_{total}$ was calculated to be 0.8, indicating the concrete region of the collimator was insufficient to attenuate the gamma rays. Additionally, moving the GM counter higher in the collimator reduced the source area from which gamma rays could come into the detector directly. Therefore, the detector was returned to the bottom of the collimator to expand the solid angle from which gamma rays could directly enter the detector. A 5 cm iron plate was installed on the bottom surface of the collimator to attenuate the dose rate at the detector and a 10 cm iron plate was added onto the outside surface to improve gamma ray attenuation in the concrete region. With these alterations, $D_{0\text{-}200}/D_{total}$ was 0.97. To increase the $D_{0\text{-}200}/D_{total}$ score, we extended the side plate down below the bottom plane of the detector to reduce dose rate from source between 200 cm and 400 cm scattered from the floor. With the extended side panel, it was found that $D_{0\text{-}200}/D_{total}$ was 0.99.

Relative dose rates maps are shown in Figure 5. The left map is a case of the base collimator and right map is an improved collimator. In both cases, the radius of the radiation source was set to 400 cm and the bottom of the collimator was located 50 cm above the floor. As the maps show, dose rates in the improved collimator were lower than in the basis collimator.

It was thought that the other configuration could effectuate the aim that $D_{0\text{-}200}/D_{total}$ was 0.99. However, we decided to prioritize a schedule of this project and to adopt the improved collimator and GM counter.

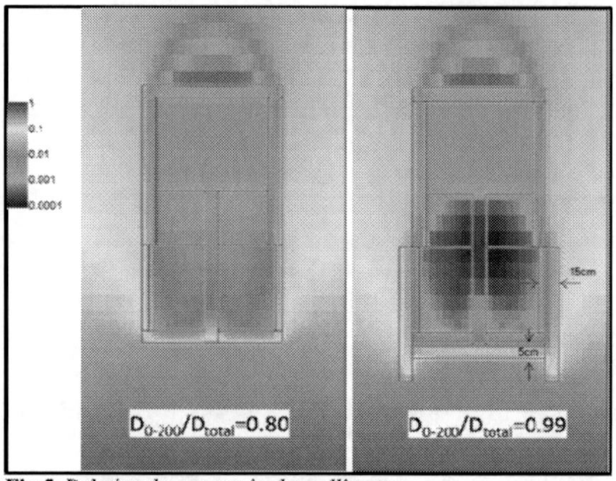

Fig.5. Relative dose maps in the collimator

This collimator and GM counter were hung 50 cm above the operation floor by a crane and a dose rate was measured at the each mesh. In addition, another unshielded GM counter was set 4.5 m above the GM counter in the collimator, i.e. at 5.0 m from the operation floor. Using this GM counter, unshielded dose rates at a distance of 5 m were obtained at the same time as the dose measurements at 50 cm with the collimator.

3 Evaluation of source distribution

3.1 Procedure of evaluation of source distribution

A flowchart showing how the evaluation of the source strength was calculated is shown in Figure 6. The source strength can be obtained using a relation between the target mesh which has a uniform source strength and a measured dose rate at the GM counter with the collimator. However the dose rate includes the dose rate from the adjacent mesh. Using the relationship between the adjacent mesh which have a uniform source strength and the dose rate at GM counter with the collimator, a dose rate from target mesh is obtained from the measurement dose rate. Therefore, the strength can be obtained using a relation of the target mesh and a dose rate at GM counter with the collimator.

Next, for variation of source strength, dose rates at 5 m from the operation floor wihout the collimator were calculated using the evaluated source strength. Ratios of caluculated dose rates at 5 m to measurement dose rates at 5 m obtained from GM counter (C/M) showed in 2.3. If ranges of C/Ms were over 2.0 or below 0.5, evaluated radiation concentration would be re-evaluated.

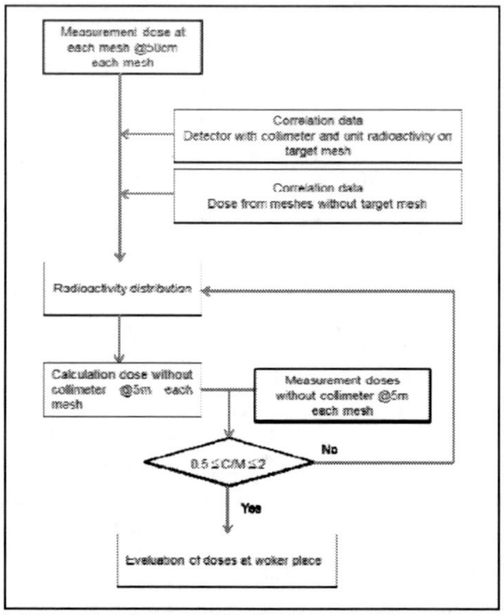

Fig.6. Flowchart of source distribution evaluation

3.2 Database of relation of radioactivity and location

It is necessary to evaluate the relationships between unit radioactivity of all 130 meshes and the calculated dose rates at the detector for all 130 meshes in advance to be able to quantify the source strength quickly. However, there were more cases needed to be calculated than could be accommodated by the schedule. In addition, the collimator had rectangular shape making it difficult to reproduce its orientation in the model geometry when measurements were being done by crane.

Accordingly, the relationship of the dose rate with the collimator and planar ring radiation source were evaluated by changing its radius from 0 m to 61 m and the results were used to create a database. This calculation geometry was similar to that shown in Figure 4.

Next, a small tool was created which can almost automatically evaluate the source strength of the target mesh using the procedure shown in 3.1. The tool extracts data depending on the distance between the detector within the collimator and the center of the target mesh.

Another challenge was that much of the operation floor was not flat or had collapsed. It was supposed that the source was mesh planar because the source distribution was used to evaluate dose rates at the workplace on the shield floor and therefore the planar source assumption was deemed acceptable.

For these calculations, the gamma ray spectrum was based on Cs134, Cs137, and Ba137m. "MCNP5" was used for the calculations with the "MCPLIB04" library and point detector tally F5. The flux to dose conversion factor from "ICRP Publication 74" was used.

3.3 Detector for Validation of evaluated source distribution

To validate the calculated source distribution, a detector without a collimator was set 5 m above the operation floor and the dose rates with the collimator were measured at the same time. Measured dose rates at 5m without the collimator were compared to calculated dose rates using the calculated source distribution.

The database mentioned in 3.2 and 3.3 is shown in Figure 7 for information. These are normalized by a dose rate at 0 cm of a distance from the center of GM counter to radiation source.

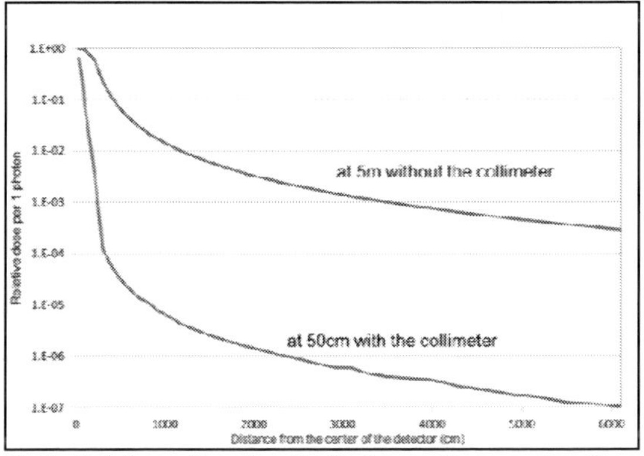

Fig.7. Relative radioactive distribution on the operation floor of Unit3

3.4 Evaluated source distribution

The first measurement after removal of the large debris was done on November, 2013. The source distribution was calculated using the procedures previous mentioned is shown in Figure 8. This map shows the ratio of the largest dose rate area ("Red" in the map) to the others. As a result of the map, the presence of a high radiation field in the meshes around the well plug. There are gaps between the plug and the floor with a couple of cm. The largest dose rate area is considered as the part of to which radioactive materials were strongly stuck by radioactive-gas-burst through a gap between the wellplug and the floor.

In addition, the dose rates at 5 m were calculated using the evaluated source distribution and C/Ms were calculated. The results are shown in Figure 9. The results show that the C/Ms were in the range of 1.0~2.4. In particular, C/Ms in the higher dose rate area around the shield plug were 1.0~1.3. Using this distribution, calculated dose rates would be conservative since almost all C/Ms were over 1.0. There were meshes which C/M were over 2.0 on outer edge and these C/Ms were attributed to higher settings of meshes around the wellplug. The high C/Ms had no impact on the dose rates at the workplace and this radiation distribution was available since the source strengths at the outer edges were relatively lower than other areas and the measured dose rates have a variation from measurement due to the crane operation.

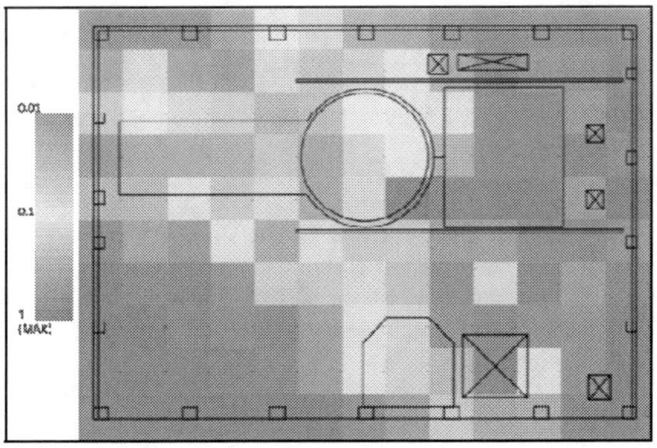

Fig.8. Relative radioactive distribution on the operation floor of Unit3

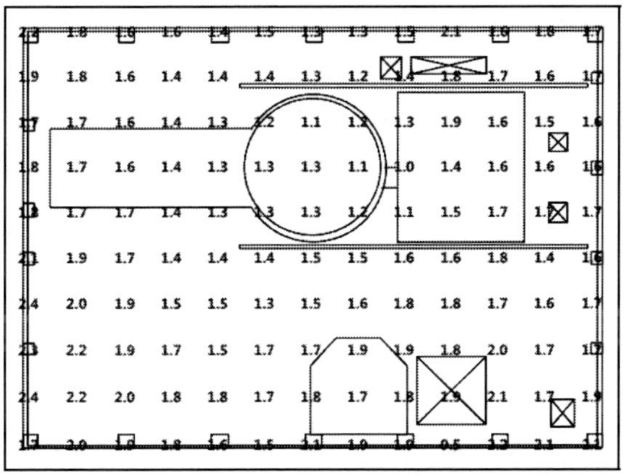

Fig.9. Ratios of calculated dose rates to measurement dose rates at 5m (C/M)

4 Evaluation of dose rates at workplace

4.1 Planned shields on the operation floor

Because the workers will be on the operation floor, the iron floors of 65 ~ 250 mm had been planned for use as shielding before the removal of the large debris and measurement. These shields were constituted of reed-shaped parts and the parts would be installed by crane. The planned shields are shown in Figure 10.

DOI: 10.1051/epjconf/201715305024

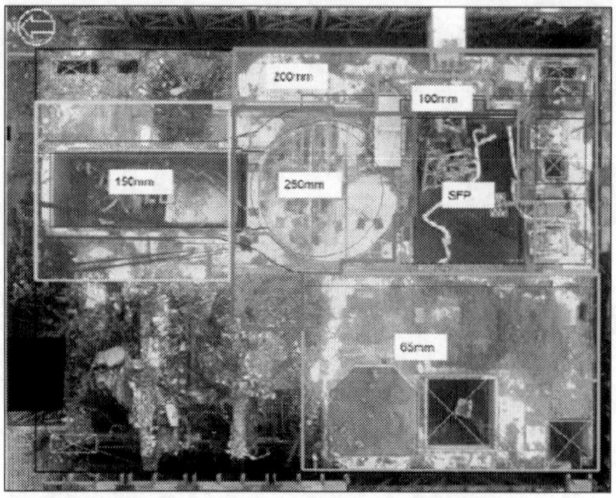

Fig.10. Planned shields on the operation floor with its thickness

As shown in Figure 2, the FHM and the workplace will be installed on the SFP. The floor of the workplace will be located at 6m above the operation floor while a installing a girder for the FHM. The dose rates at the workplace at 7 m above the operation floor considering the height of a person were calculated using "QAD-CGGP2R" [4] and the calculated source distribution.

4.2 Evaluation of the dose rate at workplace

It was found that the dose rates at some points were above the target dose rate of 1 mSv/h. The main causes were gaps between the shields and the operation floor and the inability to install more shielding due to the vulnerability of the building. Therefore, it was decided that additional shields should be designed to be installed in the structurally sound areas and that the smaller debris should be decontaminated.

In addition, debris outside of the reactor building and radioactive materials stuck on the walls would have to be considered following another measurement.

4.3 Examination of the dose rate from a gap between the shield floors

As shown Figure 10, shields in the northwest side were not planned before the workplace dose rate evaluation. However it was found that about 5 cm thick lead shields were needed for the northwest area despite the northwest area having relatively lower source strengths. It was impossible to directly install shields in the northwest area due to the structural integrity of the building. Therefore a shield structure which used the dryer and separator pit and the gantry for support was adopted.

The shields will consist of gutter shape parts the same as the other shield floors and the installation crane will allow gaps of about 45 mm between the parts on average. There is no workplace on these shield, however there are workplaces on the neighboring shields. The impact of how gaps between the shielding effect doses were calculated. Dose distributions at 7 m and 3.2 m from the operation floor and 1.2 m from the shields normalized by the doses at the center of the gap are shown. Increases, which width of increase was 40 cm at most were found around the gap, and the gap had little influence on the workplace.

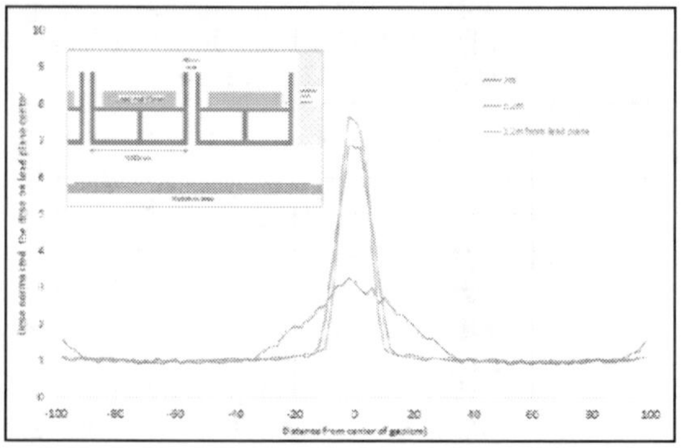

Fig.11. Dose rates normalized by dose rates on the center of the gap

Moreover, additional shields for use in the gaps between the shields and the operation floor were evaluated from the perspective of preventing gamma rays from reaching the workplace directly. In addition, a radiation distribution of the debris outside of the reactor building and the presence of radioactive materials on the wall were evaluated using the measured dose rate. Shields between the reactor building and the gantry were also examined.

The decontamination work on the operation floor and SFP also reduced the dose rate. To check a progress of the decontamination work and to decide whether to install the FHM and the shield floors quickly, another database of the dose rate at the workplace after installation of the shield floors per unit radioactivity (Bq/cm^2) was prepared. The dose rates could be calculated using only the change in the evaluated source distribution.

Measurements have been done several times and the dose rates have been used effectively for the planning of the work over the operation floor.

For all of this work, the dose rates in the workplace mentioned in 4.1 were confirmed to be below the 1 mSv/h dose rate target calculated. However there were over 1 mSv/h areas in some phases of the shield and FHM installation. Temporary shields and reduced work time would be examined for dose rate reduction of the workers.

As of August 2016, almost all of the shields shown in Figure 10 have been installed. Currently status of shields is shown in Figure 12.

Fig.12. The situation for the Shields as of August 2016

5 Conclusion

Using measurement data obtained with a collimator, reasonably accurate dose rates and source distributions could be calculated for the operation floor of Fukushima Daiichi Unit 3.

In addition, by developing dose rate databases based on unit radioactivity (Bq/cm^2) it is possible to calculate new dose rates using only the change in source distribution.

6 Summary

Using a collimator for dose rate measurement, radiation concentration in any site can be evaluated from the measured dose rates like in this case. However, designing a collimator suited for the situation will be needed.

Product names may be trademarks of their respective companies.

References

1. X-5 Monte Carlo Team, *MCNP - A General Monte Carlo N-Particle Transport Code, Version 5 Volume I: Overview and Theory*, LA-UR-03-1987, (Revised 2/1/2008)
2. M. C. White, *Photoatomic Data Library MCPLIB04: A New Photoatomic Library Based on Data from ENDF/B- VI Release 8*, Los Alamos Laboratory internal memorandum X-5:MCW-02-111(2002)
3. ICRP, *Conversion Coefficients for use in Radiological Protection against External Radiation*, ICRP Publication 74, Annals of the ICRP26(3/4), Pergamon Press, Oxford, (1996).
4. Y. Sakamoto and S. Tanaka, *QAD-CGGP2 and G33-GP2 : Revised Versions of QAD-CGGP and G33-GP Code with Conversion Factors from Exposure to Ambient and Maximum Dose Equivalents*, JAERI-M 90-110, (1990)

Evaluation of RAPID for a UNF cask benchmark problem

Valerio Mascolino[1,a], Alireza Haghighat[1,b], and Nathan J. Roskoff[1,c]

[1] *Nuclear Science & Engineering Lab (NSEL), Virginia Tech, 900 N Glebe Rd., Arlington, VA, 22203*

Abstract. This paper examines the accuracy and performance of the RAPID (Real-time Analysis for Particle transport and In-situ Detection) code system for the simulation of a used nuclear fuel (UNF) cask. RAPID is capable of determining eigenvalue, subcritical multiplication, and pin-wise, axially-dependent fission density throughout a UNF cask. We study the source convergence based on the analysis of the different parameters used in an eigenvalue calculation in the MCNP Monte Carlo code. For this study, we consider a single assembly surrounded by absorbing plates with reflective boundary conditions. Based on the best combination of eigenvalue parameters, a reference MCNP solution for the single assembly is obtained. RAPID results are in excellent agreement with the reference MCNP solutions, while requiring significantly less computation time (i.e., minutes vs. days). A similar set of eigenvalue parameters is used to obtain a reference MCNP solution for the whole UNF cask. Because of time limitation, the MCNP results near the cask boundaries have significant uncertainties. Except for these, the RAPID results are in excellent agreement with the MCNP predictions, and its computation time is significantly lower, 35 second on 1 core versus 9.5 days on 16 cores.

1 Introduction

To ensure criticality safety and maintain a high level of security of Used Nuclear Fuel (UNF), it is necessary to perform detailed Particle Transport simulations. The commonly used technique is the 3-D, continuous energy Monte Carlo method that is time-consuming and, for eigenvalue calculations, may have difficulties with the source convergence [1, 2]. Wenner and Haghighat [2] and Dufek [3] demonstrated that in criticality problems with high dominance ratio and/or under-sampling possibilities, the fission matrix (FM) methodology [1] should be used in order to be able to obtain unbiased solution.

The RAPID (Real-time Analysis for Particle transport and In-situ Detection) code system [4] is based on the Multi-stage Response-function Transport (MRT) methodology [5], and uses the FM formulation to determine eigenvalue, subcritical multiplication and fission density distribution of nuclear systems. This paper uses a benchmark problem similar to the GBC-32 Cask [6] to evaluate the accuracy and performance of RAPID against the results of a reference MCNP [7] solution. We performed a detailed sensitivity analysis on a single-assembly model in order to determine the MCNP eigenvalue parameters necessary to obtain a reliable reference solution. These results provided a guide for the selection of an appropriate set of parameters for the eigenvalue calculation for the whole cask.

The paper is organized as follows. Section 2 discusses the RAPID code system and its MRT approach. Section 3 introduces the benchmark problem. Section 4 introduces the methodology used to analyze results. Section 5 discuss the results of different cases to identify the best set of eigenvalue parameters, and compares the selected case with the RAPID calculation. Section 6 gives the conclusions and future work perspectives.

[a] e-mail: val@vt.edu
[b] e-mail: haghighat@vt.edu
[c] e-mail: roskofnj@vt.edu

2 RAPID Code System

To overcome the source convergence difficulty and computational effort of straightforward eigenvalue Monte Carlo calculations, the RAPID code system has been developed based on the MRT methodology. For a UNF pool or cask, RAPID is capable of quickly and accurately calculating eigenvalue, subcritical multiplication, pin-wise, axially dependent fission densities, and detector responses for monitoring a pool or cask. RAPID uses the Fission Matrix (FM) method for eigenvalue calculations, and the adjoint function methodology for the detector response calculations. RAPID partitions a problem into stages in which FM coefficients and importance functions are determined for different values of the key parameters such as enrichment, burnup, geometry, and cooling time. Then, it uses a system of equations to solve for eigenvalue and eigenfunction and the adjoint function methodology for determining a detector response. The methodology has been successfully demonstrated for spent fuel pools [4].

2.1 The Fission Matrix approach

The FM method can take two forms, depending on the type of problem. For a sub-critical multiplication problem, in which the fission source is driven by an independent source in the spent fuel (i.e., spontaneous fission and (α, n) reactions), the induced fission source in cell i is given by Equation 1.

$$F_i = \sum_{j=1}^{N} (a_{i,j} F_j + b_{i,j} S_j) \quad (1)$$

Where F_j is the induced fission source strength in fuel pin j, S_j is the intrinsic (or independent) source strength in fuel pin j, $a_{i,j}$ is the number of fission neutrons directly produced in fuel pin i due to a fission neutron born in fuel pin j, $b_{i,j}$ is the same as $a_{i,j}$ except for intrinsic source neutrons. These values are different because S and F have different spatial and spectral distributions within each cell. N is the total number of computational cells (which could be a whole assembly, a single fuel pin, or a fraction thereof).

We also consider the eigenvalue problem as in Equation 2.

$$F_i = \frac{1}{k} \sum_{j=1}^{N} a_{i,j} F_j \quad (2)$$

The method results in a set of N linear equations, which can be solved for $\vec{F}$ and k given the $a_{i,j}$ coefficients. The main task is to determine the coefficients $a_{i,j}$ and $b_{i,j}$ in the case of subcritical multiplication. Since not all the regions are coupled, i.e., have non-zero coefficients, we have developed a multi-layered approach to determine the necessary coefficients.

Coefficients $a_{i,j}$ are calculated through fixed-source Monte Carlo calculations, in which neutrons are originated in cell j and fission neutrons born in cell i's are tallied. The source neutrons are sampled over energy according to the Watt fission spectrum, isotropically. For $b_{i,j}$ the same applies except that neutrons are sampled according to the spontaneous fission and (α, n) reactions spectra.

3 Benchmark problem

The GBC-32 cask is loaded with 32 Westinghouse 17x17 OFA/V5 fuel assemblies (FA) of different burnups [6]. For the purpose of this work, the cask has been loaded with fresh fuel with 4% wt. enrichment only. The model consists of the 32 FAs contained in a stainless steel canister. The canister is equipped with Boral plates encased in Aluminum clads between the FAs, and it is placed in a Stainless Steel cylinder. The system is flooded with water. The model is depicted in Figures 1a and 1b. Figure 2 shows one assembly of the GBC-32 Cask.

For the purpose of analysis, two representative pins are highlighted (circled black) in Figure 2: pin (1,9) near the absorber panel on the left of the assembly, and pin (10,10) top-right of the central guide tube. These two pins have been chosen to plot fission density axial distributions because of being subject to the lowest and the highest values of fission density, respectively.

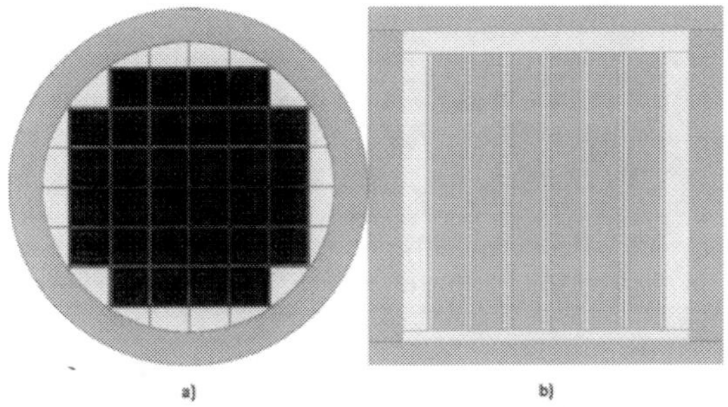

Figure 1. a) xy (at mid active height) and b) xz (at $y = 0$) cuts of the GBC-32 Cask model.

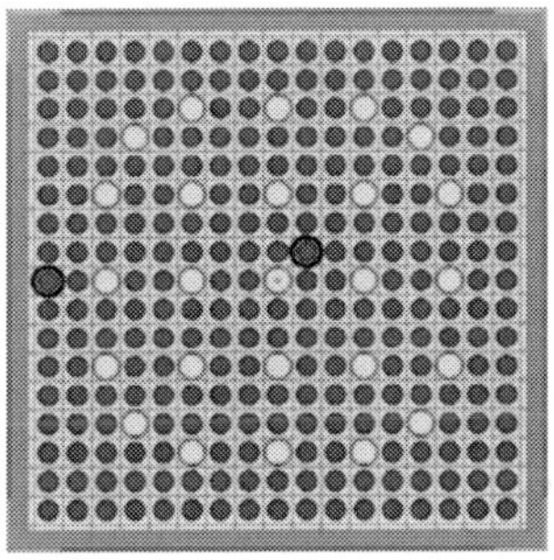

Figure 2. xy (at mid active height) cut of the unit cell of the GBC-32 Cask model.

4 Methodology

The aim of this work is to compare the RAPID calculated k_{eff} and pin-wise axial fission densities to those of a reference MCNP calculation. To obtain an accurate reference solution, an analysis is performed to obtain the best combination of the eigenvalue parameters used in the MCNP's KCODE calculations.

4.1 Determination of eigenvalue parameters in MCNP

In any Monte Carlo eigenvalue calculation, a major issue is achieving an accurate fission source distribution. This requires determination of three MCNP parameters, including number of skipped cycles (NSK), number of active cycles (NAC), and number of histories per cycle (NPS).

Additionally, the variance of the fission neutron source may be significantly biased by the presence of cycle-to-cycle correlation, which is difficult to detect and may cause MCNP reported variance to be underestimated [8]. This phenomenon arises from the fact that the power method [1] (used for eigenvalue Monte Carlo calculations) computes the fission source at iteration $k + 1$ based on the distribution at iteration k. The random numbers generated by the pseudo-random number generator (PRNG) for a certain iteration will be affected by the previous cycles, i.e., correlated. This effectively causes the fission density tallies to oscillate less than their actual statistical uncertainty, leading to an underprediction of the computed uncertainty. This effect usually arises in the fission density distribution tallies and not in the k_{eff} since the latter is an integral quantity.

In an attempt to detect the potential underprediction of the MCNP reported tally uncertainty, a replication study is performed for a fixed set of the eigenvalue parameters by changing the seed of the PRNG.

4.2 Analysis approach for the establishment of a reference MCNP single-assembly model

The Monte Carlo eigenvalue parameters, NPS, NSK, and NAC, have been investigated by analyzing the eigenvalue (k_{eff}), the normalized fission density(i.e., source distribution) and the associated uncertainties for the single-assembly model. The fission density is determined in 24 axial intervals for every fuel pin in the model, resulting in 6336 tally regions.

Convergence of the source is examined through the following techniques for different combinations of the eigenvalue parameters:

- The Shannon entropy stabilization as a function of generations [9]. It is worth noting that, as demonstrated by Wenner and Haghighat in [2], this method may fail to correctly diagnose the source convergence in loosely coupled problems.
- The L_∞, L_1, and L_2 norms of the relative differences of the fission neutron density tallies with respect to the most accurate calculation are computed. These norms are compared to the norms of the uncertainties associated with each tally for the particular calculation, and are defined as follows:

$$\begin{cases} L_\infty - norm: & \|X\|_\infty = \max_i |x_i| \\ L_1 - norm: & |X| = \frac{1}{N_t}\sum_{i=1}^{N_t} x_i \\ L_2 - norm: & \|X\| = \frac{1}{N_t}\sqrt{\sum_{i=1}^{N_t} x_i^2} \end{cases} \tag{3}$$

where X is a vector composed of N_t elements. X will be either the vector of relative differences or uncertainties. In principle, the norm of the relative difference vector should be lower than the norm of the statistical uncertainty if the source has converged. This would indicate that, on average, the variation of the results from case to case is within the statistical uncertainty. Relative differences are computed for each of the 6336 tally with respect to the most detailed calculation as follows:

$$R_{k,l} = \frac{S_{k,l} - S_{k,md}}{S_{k,md}} \qquad k = 1,6336 \tag{4}$$

where $R_{k,l}$ is the value of the relative difference of the k-th tally for the l-th case, $S_{k,l}$ is the value of the k-th tally for the l-th case, and $S_{k,md}$ is the value of the k-th tally for the most detailed (md) case.

- The Center of Mass (COM) [1] is calculated by:

$$\bar{r} = \frac{1}{M}\sum_{i=1}^{N_{tallies}} m_i r_i \tag{5}$$

where m_i is fission density at region i, r_i is the distance between region i and geometric center, M is the total fission density of the assembly, and $\bar{r}$ is the COM distance to the geometric center. The COM position evaluated in Equation 5 should oscillate around the geometric center of the model for a symmetric problem.

In addition to the aforementioned analyses, based on a select parameters combination we performed repetition of the same run with different PRNG seeds in order to identify the presence of possible cycle-to-cycle correlation through the approach showed in [8]. From the repetition of the run, the average MCNP calculated variance is calculated as follows:

$$\sigma_{MCNP} = \frac{1}{N_s}\sum_{i=1}^{N_s} \sigma_i \tag{6}$$

where N_s is the number of different seeds used and σ_i is the standard deviation obtained with MCNP for each of the runs.

Having N_s different values for the same quantity, the "real" standard deviation of the sample can be defined as:

$$\sigma_{real} = \sqrt{\frac{1}{N_s - 1}\sum_{i=1}^{N_s}(x_i - \bar{x})^2} \tag{7}$$

where x_i is the value of a certain quantity calculated for the i-th seed, and $\bar{x}$ is the average of all these values.

By calculating the value of these two standard deviations, an indication of the underestimation of the MCNP variance due to cycle-to-cycle correlation is provided:

$$f_s = \frac{\sigma_{real}}{\sigma_{MCNP}} \tag{8}$$

The f_s parameter is evaluated using Equation 8 for the system eigenvalue, k_{eff}, and for each of the axially dependent pin-wise fission density tallies.

Based on the above described evaluation process, a reference MCNP calculation is established for the single-assembly model to be compared with RAPID results.

4.3 Establishment of the full-cask MCNP reference model

Based on the results of Section 4.2, the set of parameters selected for the single-assembly are scaled to represent a full-cask model.

Since the assemblies are separated by absorbing racks, thus assemblies are loosely coupled, the whole-cask model is established by keeping NAC and NSK constant with respect to the single-assembly case.

The NPS parameter, in principle, should be scaled proportionally to the increase in the number of tally regions, resulting in a scale-up factor of 32. However, since a

NPS value of 10^6 (with 144 particles per tally region) has been used for the single assembly model, the direct-scaling approach would lead to prohibitive computation times. Therefore, we have used a reasonable NPS of 10^5 per assembly (demonstrated in Section 5.1), i.e., $3.2 \cdot 10^6$ for the full cask.

4.4 Comparison of MCNP reference models to RAPID calculations

The RAPID system calculation is compared to the MCNP reference calculation for both the single-assembly and the full-cask models. Note that the full-cask model is only partially converged because of time limitations. The quantities analyzed are the criticality eigenvalue parameter of the system k_{eff} and the pin-wise axial fission density distributions. The relative differences between the two methods are calculated and plotted, and maximum and average values are tabulated.

5 Results

This section is divided into four main parts: in Section 5.1 the parametric study of the single-assembly problem to assess adequate eigenvalue calculation parameters to establish a referene MCNP solution is presented; Section 5.2 addresses the cycle-to-cycle correlation issue; while in Sections 5.3 and 5.4 the results of the RAPID code are compared to the MCNP reference calculation for the single assembly and full cask models, respectively.

5.1 Eigenvalue Parameters

The parametric analyses are made by independently varying the three MCNP criticality parameters (NAC, NPS, and NSK) in order to isolate the effects of the variation of each of them.

The NAC study has been conducted fixing NSK to 1000 and NPS to 10^6, varying NAC from 500 to 3500 in steps of 500. The Shannon entropy behavior is presented in Figure 3 as a function of the cycle.

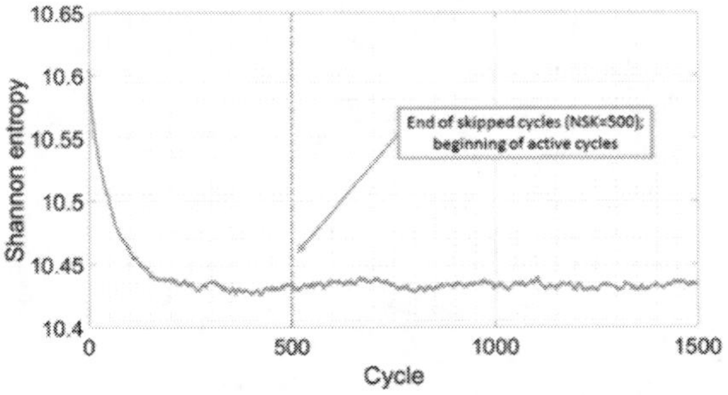

Figure 3. Shannon entropy as a function of cycles.

Figure 3 shows that the value of the Shannon entropy oscillates around a stable value for NAC greater than 300. Such a behavior is usually interpreted as a sign of source convergence [9].

Figure 4 presents the system eigenvalue, k_{eff}, as a function of NAC with the respective standard deviation.

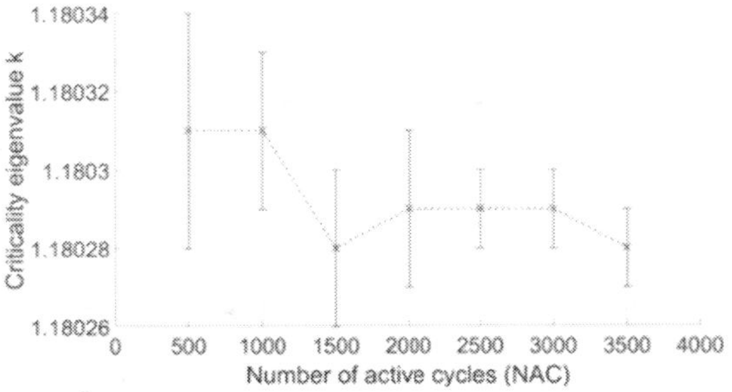

Figure 4. a) $k \pm \sigma_k$ for different NAC values.

Above figure shows that k_{eff} has converged, and the various runs differ by less than one standard deviation. Figure 5 instead shows the behavior of the statistical uncertainty σ_k as a function of NAC.

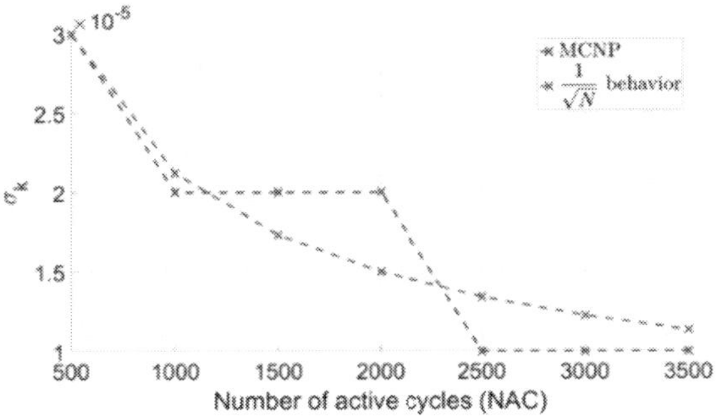

Figure 5. a) σ_k behavior for different NAC values.

The statistical uncertainty shown in Figure 5 correctly drops as $1/\sqrt{N}$, as expected based on the Central Limit Theorem [1].

Figure 6 shows the percentage distance of the COM from the expected geometric center, relative to half the active fuel height, for varying NAC:

$$\bar{r}_n(\%) = \frac{100}{H/2 \sum_{i=1}^{N_t} S_{i,n}} \sum_{i=1}^{N_t} r_i S_{i,n} \tag{9}$$

where r_i is the distance of the ith region from the center of the assembly, $S_{i,n}$ is the fission density value in the $i^t h$ region for the nth run, and H is the active fuel height.

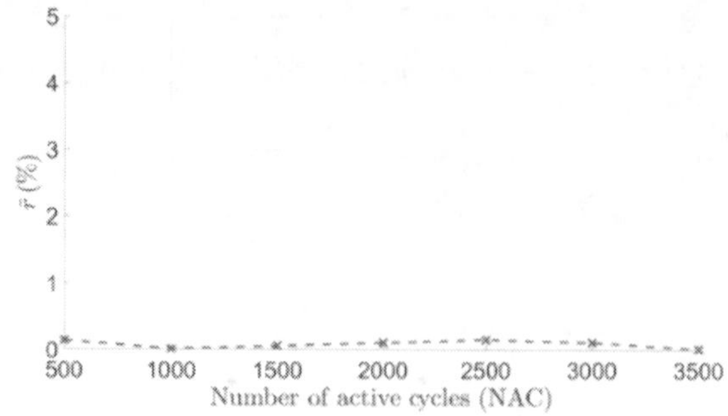

Figure 6. Center of Mass distance from geometrical center

Figure 6 shows that the COM oscillates of less than 0.5% (with respect to half the active fuel height along the z-direction). Such an oscillation is reasonable and indicates that the source has converged.

For the fission density distribution, the norms of the $R_{k,l}$ vector ($|R|_l$, $||R||_l$, and $||R||_{\infty,l}$) have been computed as described in Section 4.2 for every run l. Additionally, the norms of the vector $U_{k,l}$, containing the relative uncertainties of each tally k as output from MCNP ($|U|_l$, $||U||_l$, and $||U||_{\infty,l}$) have been calculated. Figures 7, 8, and 9 show the behavior of these norms as a function of NAC.

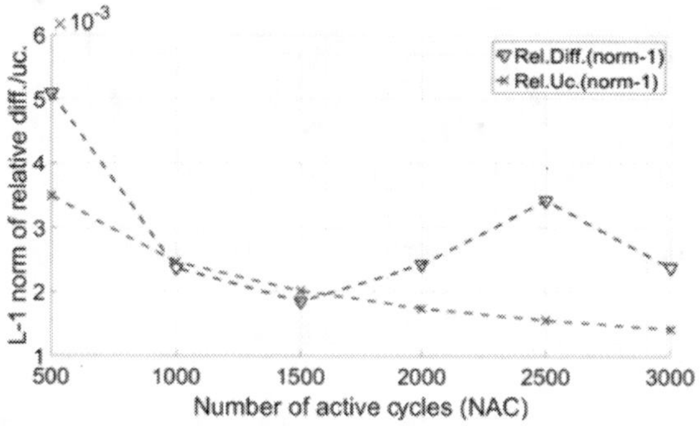

Figure 7. Comparison of relative differences and relative uncertainties L_1-norms, $|R|_l$ and $|U|_l$.

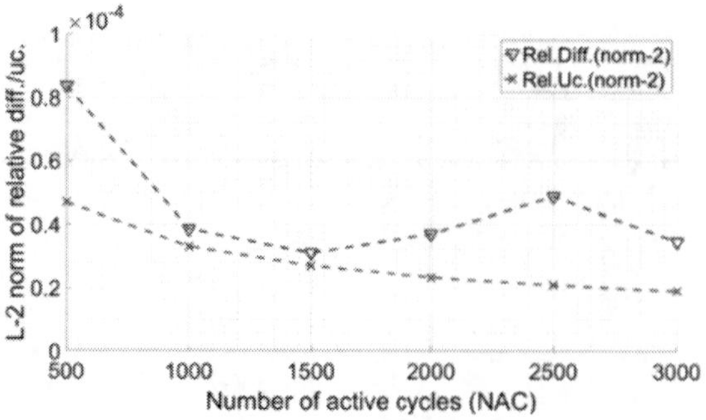

Figure 8. Comparison of relative differences and relative uncertainties L_2-norms, $||R||_l$ and $||U||_l$.

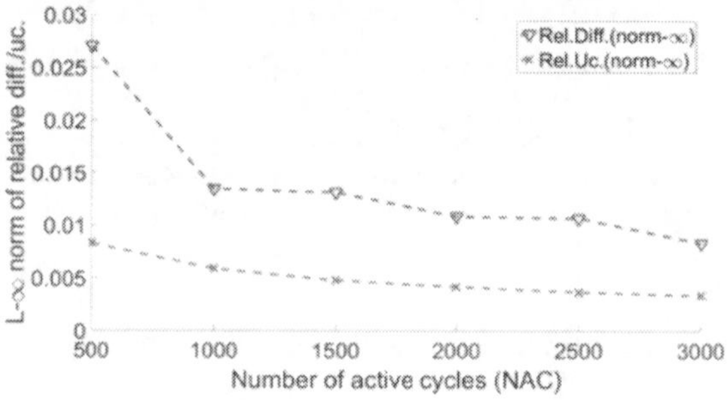

Figure 9. Comparison of relative differences and relative uncertainties L_∞-norms, $||R||_{\infty,l}$ and $||U||_{\infty,l}$.

Figures 7, 8, and 9 indicate that the relative differences are on average *higher* than the statistical uncertainty associated to the calculation.

Figures 10 and 11 show the fission density axial distribution in the representative pins (10,10) and (1,9) shown in Figure 2.

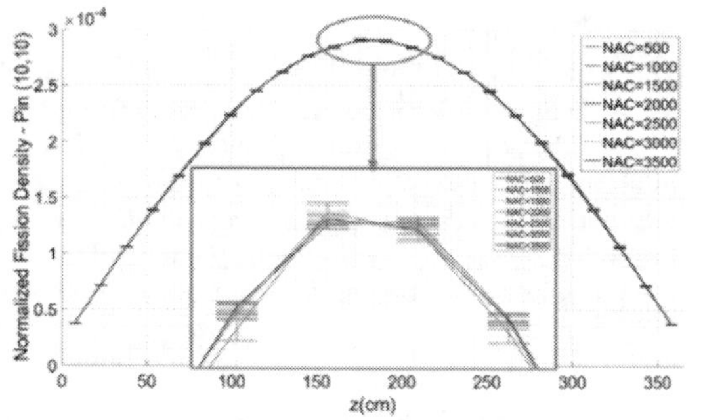

Figure 10. Fission density axial distribution in central pin - (10,10)

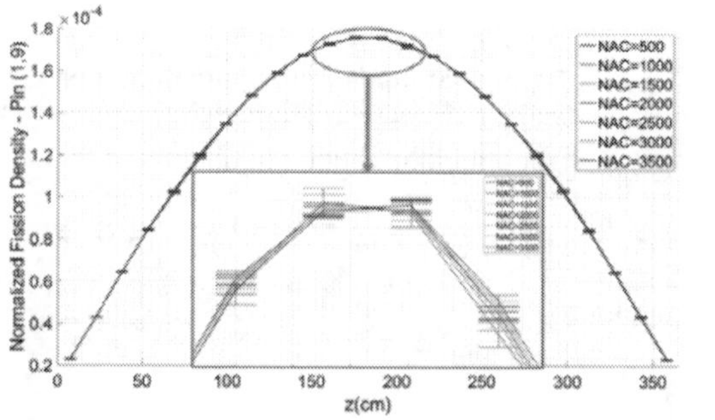

Figure 11. Fission density axial distribution in pin near absorber - (1,9)

Above figures show that the changes in the axially-dependent fission density distributions are on average higher than the MCNP predicted uncertainty. However, the shape is basically unaffected by the increasing number of active cycles, except for the NAC=500 (i.e., the lowest number considered) curve being slightly offset.

Similar analyses have been done for the NSK and NPS parameters independently, keeping fixed the other two. In all the cases, simlar results are obtained, leading to the following conclusions:

- The criticality eigenvalue, k_{eff} , converges and varies within the uncertainty for all the various runs;
- The shape of the fission density distribution is converged while on average the variation of the tallies is above the statistical uncertainty, suggesting that these could be underestimated;
- The Shannon entropy and COM behaviors are stabilized and would suggest that the fission source is not varying much and is symmetric as expected.
- The norms of the relative differences are higher than the statistical uncertainties ones.

The first three points suggest that the fission source has converged. However, the norms show an unexpected behavior: this can be caused either by a bias in the sample averages calculated by MCNP (i.e., a non-converged source, in contrast with the first three analyses) or an underestimation in the statistical uncertainties (e.g., due to cycle-to-cycle correlation). Section 5.2 analyzes the latter phenomenon.

5.2 Cycle-to-cycle correlation investigation

In order to assess the effect of the cycle-to-cycle correlation on the computed variance, we performed a replication study keeping the same set of MCNP criticality parameters and changing the PRNG seed. In this way, a "real" standard deviation is calculated and compared to the average MCNP uncertainty, as described in Section 4.2 and in [8]. This study was performed with smaller criticality parameters (NPS=1E6, NSK=300, and NAC=500) with respect to the study done in Section 5.1. This is necessary because significant computation is needed for running the 50 replication cases.

The approach is applied to both the criticality eigenvalue, k_{eff} , and each of the 6336 fission density tallies. Table 1 give k_{eff} and the associated standard deviations ("real" and MCNP calculated).

Table 1. Average k_{eff} , real and MCNP standard deviations, and their ratio.

$\bar{k}$	$\sigma_{MCNP}(pcm)$	$\sigma_{real}(pcm)$	f_σ
1.18028	3	2.59	0.8637

The analysis reveals that the "real" standard deviation, calculated through this replication study, is actually *lower* than the MCNP calculated one. This is mostly related to the fact that MCNP standard deviations for k_{eff} are provided with a resolution of 1 pcm. This result is in line with what is shown in Figure 4: the MCNP k_{eff} is converged and the estimation of the statistical uncertainty is accurate.

For the fission density values, the following quantities are analyzed:

- The maximum of the real to MCNP standard deviation ratios for all the k-th tallies:

$$f_{\sigma,max} = \max_k [f_{\sigma,k}] \quad (10)$$

- The average of the real to MCNP standard deviation ratios for all the k-th tallies:

$$f_{\sigma,avg} = \frac{1}{N_t} \sum_{k=1}^{N_t} f_{\sigma,k} \quad (11)$$

- The weighted average of the real to MCNP standard deviation ratios using the fission density source for all the k-th tallies:

$$f_{\sigma,wgt} = \frac{\sum_{k=1}^{N_t} f_{\sigma,k} \cdot S_k}{\sum_{k=1}^{N_t} S_k} \quad (12)$$

Table 2 summarizes the real and the MCNP standard deviations and these ratios. The statistical uncertainties for the fission density distribution are underestimated by the MCNP calculations by a factor greater than 2, identifying a strong cycle-to-cycle correlation.

Table 2. Real to MCNP standard deviation ratios for fission density tallies

	avg	*max*	*wgt*
σ_{real} (%)	0.72	1.60	0.67
σ_{MCNP} (%)	0.35	0.82	0.29
f_σ	2.29	3.45	2.28

A 3D plot of the f_σ ratios for these tallies is reported in Figure 12.

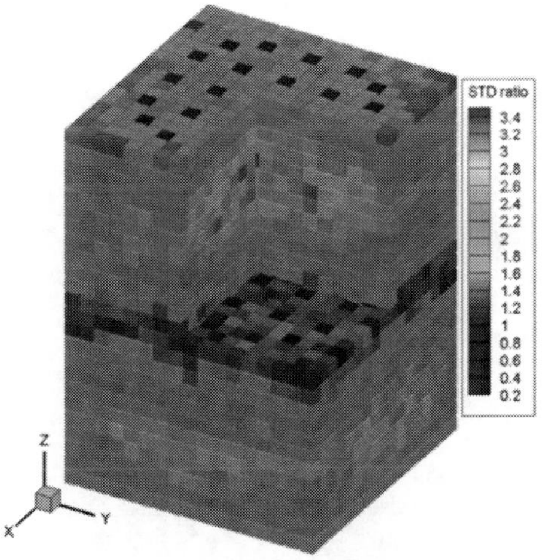

Figure 12. f_σ ratios for all the fission density tallies in the single-assembly model.

Figure 12 demonstrates that there is, *on average*, a significant underestimation of the uncertainties and that this underestimation changes significantly in different regions of the model.

As discussed in References [8, 9], cycle-to-cycle correlation is related to the method itself and it is not affected by the use of a "better" set of MCNP parameters. Therefore, we can compute the *adjusted* relative uncertainties for the NAC, NSK, and NPS parametric analyses (Section 5.1), multiplying each of the tally uncertainties by the weighted average ratios, $f_{\sigma,wgt}$. This would take into account the presence of cycle-to-cycle correlation in the comparison of the relative differences and relative uncertainties for the various cases. The relative differences L_1-norms, $|R|_l$, are plotted against the adjusted relative uncertainties L_1-norms, $|R|_{l,adj}$, for each of the parametric studies in Figures 13, 14, and 15.

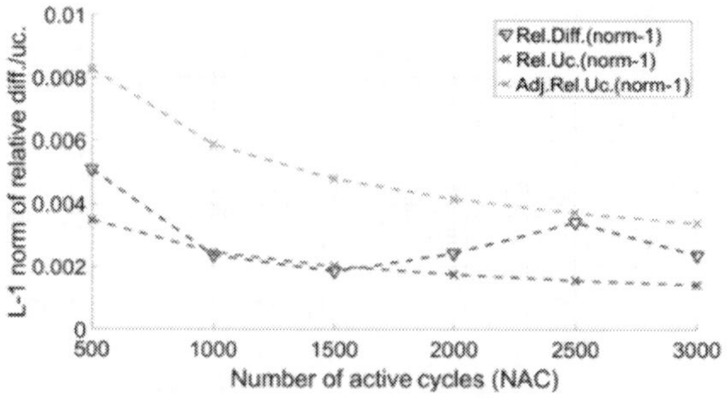

Figure 13. Comparison of relative differences and adjusted relative uncertainties Norm-1, $|R|_l$ and $|U|_{l,adj}$ as a function of NAC.

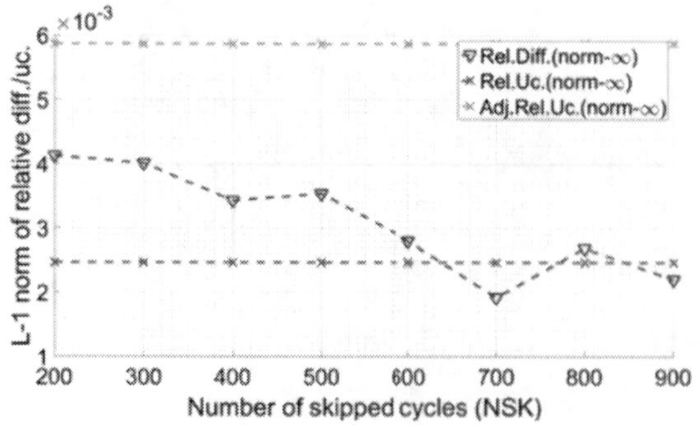

Figure 14. Comparison of relative differences and adjusted relative uncertainties Norm-1, $|R|_l$ and $|U|_{l,adj}$ as a function of NSK.

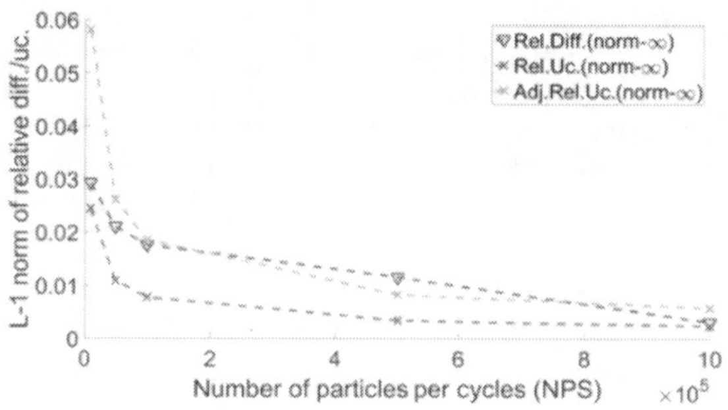

Figure 15. Comparison of relative differences and adjusted relative uncertainties Norm-1, $|R|_l$ and $|U|_{l,adj}$ as a function NPS.

The f_σ should hold similar results even for higher NAC, NSK, and NPS, since cycle-to-cycle correlation is independent of these parameters, as long as the source has converged, i.e., NSK is high enough [9]. Figures 13, 14, and 15 suggest that the tally variations are on average within the statistical uncertainty associated to the method for all the analyzed cases. Therefore, source convergence is achieved for reasonably lower values of the MCNP criticality parameters although the uncertainty calculated by the code is underestimated due to the cycle-to-cycle correlation. From this latter analysis we conclude that the NSK=500, NAC=1000, NPS=10^6 constitutes a suitable reference model for comparison with the RAPID code system.

5.3 Comparison of RAPID and MCNP results for a Single Assembly

The RAPID code was used to model the single assembly reflected in the $x-y$ directions. Table 3 presents the results of the eigenvalue calculation.

Table 3. k_{eff} comparison of RAPID compared to MCNP reference calculation.

Case	k_{eff}	Rel. Diff. [pcm]
MCNP Ref.	1.18030(2pcm[1])	-
RAPID	1.18091	52

[1] 1σ uncertainty

Above table shows that the RAPID k_{eff} calculation (NSK=500, NAC=1000, NPS=10^6) is in very good agreement with the MCNP reference calculation, with a relative difference of 52 pcm.

Table 4 presents the computational time required for both RAPID and MCNP.

DOI: 10.1051/epjconf/201715305025

Table 4. Computation of RAPID and MCNP required computation time.

Case	Num. Cores	Time [min]	Speedup
MCNP Ref.	16	666 (11.1 hours)	—
RAPID	1	0.102 (6.12 seconds)	6529

Table 4 indicates that the RAPID calculation on 1 computer core is 6500 times faster than the MCNP reference calculation on 16 computer cores.

Figure 16 shows the axial fission distribution of the entire assembly, i.e., integrated over the $x-y$ dimensions) for RAPID and the MCNP reference calculation. Note that all uncertainties on the MCNP reference calculation are $1-\sigma$. Uncertainties are within 0.05% for the $x-y$ integrated axially-dependent fission density and within 0.59% for the pin-wise axially-dependent one.

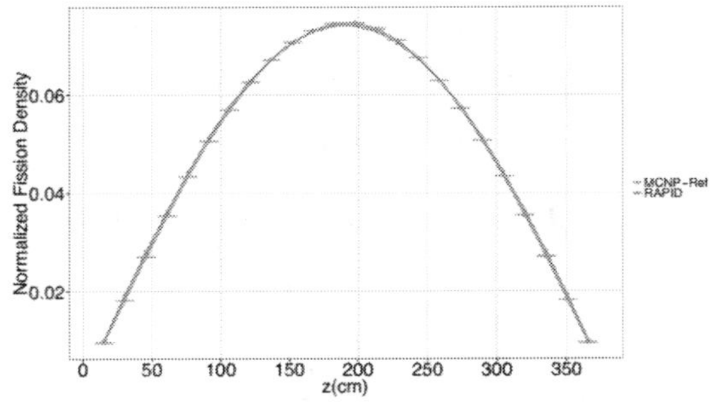

Figure 16. Axial fission density distribution (integrated over x-y dimensions). Error bars represent 1σ uncertainties.

Figure 17 shows RAPID to MCNP relative differences of pin-wise axially dependent fission density distributions.

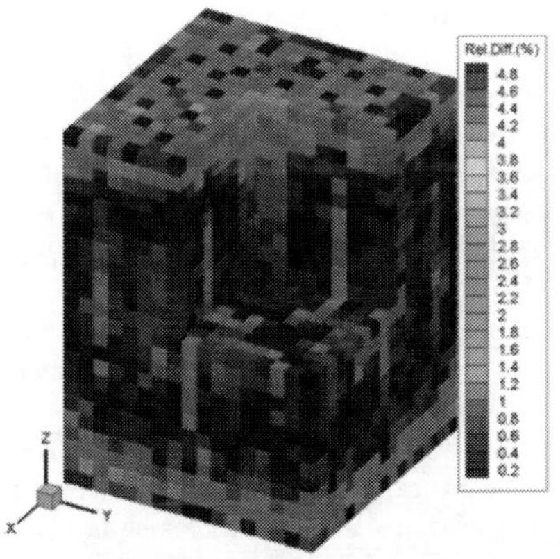

Figure 17. Relative differences of RAPID vs MCNP calculated pin-wise axially dependent fission density for the single-assembly.

The relative differences for all the regions are below 5%, with regions above 2% being those in the top and bottom layers. This behavior at the fuel rod tips is due to the assumed axial symmetry in the way FM coefficients are calculated. Maximum, average, fission-density weighted average and adjusted-weighted average relative it uncertainties for the MCNP single assembly calculation are given in Table 5. Table 6 gives maximum, average, and weighted-average relative differences of RAPID vs. MCNP.

Table 5. MCNP relative uncertainties for fission density distribution - Single-assembly

Relative uncertainties(%)			
Max.	**Avg.**	**Wgt. avg.**	**Wgt. avg. adj.**
0.59	0.25	0.21	0.48

Table 6. RAPID vs MCNP relative differences for fission density distribution - Single-assembly

Relative differences(%)		
Maximum	**Average**	**Weighted avg.**
5.00	0.92	0.69

Tables 5 and 6 demonstrate that RAPID and MCNP calculated fission densities are in excellent agreement.

5.4 Comparison of RAPID and MCNP results for the full-cask

As already discussed in Section 4.3, the set of parameters used for the full-cask reference MCNP calculation are NPS=$3.2 \cdot 10^6$, NSK=500, NAC=1000.

k_{eff} for RAPID and MCNP reference calculations are given in Table 7 for the full cask. These results show excellent agreement.

Table 7. k_{eff} comparison of RAPID compared to MCNP reference calculation - Full cask.

Case	k_{eff}	Rel. Diff. [pcm]
MCNP	1.14545(1pcm[a])	/
RAPID	1.14590	39

[a] 1σ statistical uncertainty

The maximum, average, and weighted average uncertainty values for the full-cask MCNP calculated fission density distribution tallies are reported in Table 8.

Table 8. Full-cask MCNP calculated fission density distribution uncertainties.

Relative uncertainties(%)		
Maximum	**Average**	**Weighted avg.**
9.04	1.62	1.15

DOI: 10.1051/epjconf/201715305025

Similar to the single-assembly case, the RAPID to MCNP relative differences for the pin-wise axially-dependent fission density distribution have been calculated. Figure 18 shows the 3D distribution of these relative differences.

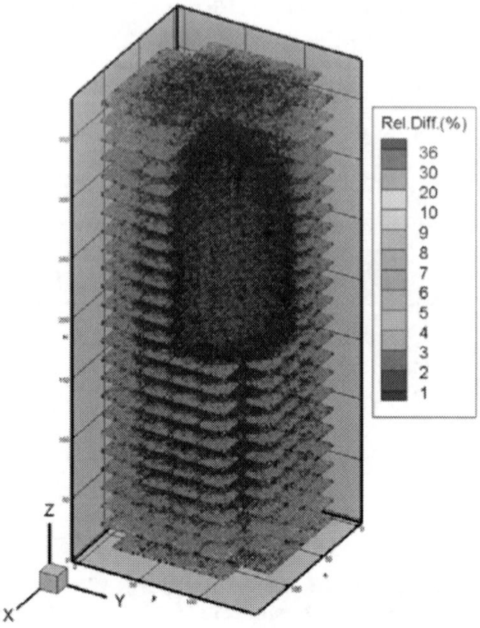

Figure 18. Relative differences of RAPID vs MCNP calculated pin-wise axially-dependent fission density for the full cask.

Figure 18 shows how the relative differences are lower than 2% the inner regions, where the MCNP statistical uncertainties are smaller. The highest relative differences are on the boundaries: it is important to note that, due to under-sampling [10], the peripheral fission densities calculated by MCNP are far from convergence and might be significantly biased. Accurate estimation of the fission density in these regions is required for detector responses estimation and dose calculations on the outside surface of the cask. Hence, the fission matrix approach should be used instead of straight-forward Monte Carlo calculations to achieve accurate unbiased detector responses. Cask dose calculation has been recently implemented in RAPID and demonstrated in [11]. The average relative difference, weighted on the fission density distribution, amounts to 1.56%, comparable to the statistical uncertainty reported in Table 8.

Table 9 shows the computational time requirements for both the methods, demonstrating how a large nuclear system as the full cask can be simulated in real time with RAPID, with a significant speedup with respect to MCNP.

Table 9. Computation time of RAPID and MCNP reference calculation - Single assembly.

Case	Num. Cores	Time [min]	Speedup
MCNP Ref.	16	13,767 (9.5 days)	—
RAPID	1	0.585 (35 seconds)	23,533

Figure 19 shows RAPID calculated 3-D fission density distribution.

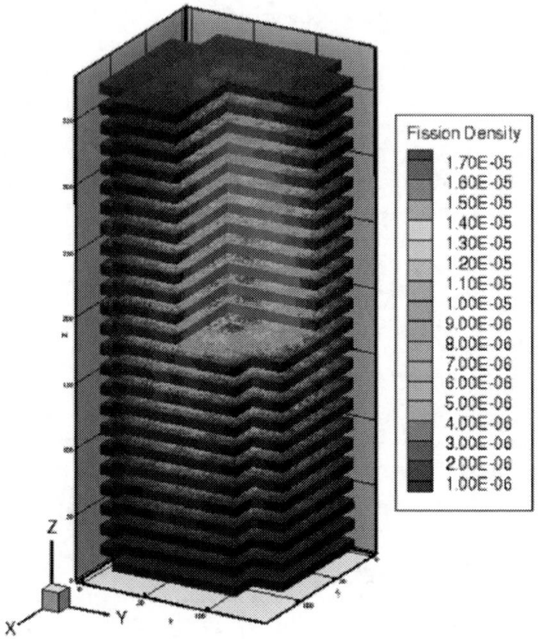

Figure 19. 3-D fission density distribution in the GBC-32 Cask.

6 Conclusions

This paper examines the performance of the RAPID code system for the simulation of a UNF cask benchmark, GBC-32. Because of the presence of absorbing plates in between the assemblies in a cask, any eigenvalue Monte Carlo simulation faces with the difficulty of particle under-sampling. Hence, it is necessary to examine carefully the source convergence. To do so, a detailed statistical analysis is performed using a single assembly model with reflective boundary conditions.

This analysis showed the importance of carefully selecting the criticality run parameters when using MCNP, and underlined that *substantial bias* in the standard deviation of the tallies might be caused by cycle-to-cycle correlation, related to the nature of criticality calculations and impossible to foresee. The RAPID code system, since it is based on a set of physically based transport coefficients that do not require any eigenvalue calculation, doesn't suffer from any of these difficulties.

Based on the parametric analysis, an appropriate set of eigenvalue parameters (NAC, NSK, NPS) have been selected to establish both a single-assembly and full-cask reference model for comparison with RAPID. It is demonstrated that the calculated eigenvalue and fission density distribution by the RAPID and MCNP code systems are in excellent agreement with the reference MCNP results for both the models, leading to average relative differences of the order of a few percents mostly related to boundary effects that will be addressed in future work.

In addition, under-sampling may cause biased boundary MCNP tallies even when large criticality parameters are used. This issue is avoided using the FM approach, since no Monte Carlo criticality calculations are necessary. The calculation of external doses and detector responses is highly affected by the accuracy of the outer regions, nearby the detector location, hence making the use of codes based on the FM methodology, such as RAPID, necessary to calculate these quantities.

Moreover, it is shown that the RAPID computation time is significantly (several orders of magnitude) smaller than the MCNP one, i.e., 35 seconds vs. 9.5 days on 16 cores. This speedup increases as the model becomes larger.

References

[1] A. Haghighat, *Monte Carlo Methods for Particle Transport* (CRC Press Taylor & Francis Group, 2015)

[2] M. Wenner, A. Haghighat, Progress in Nuclear Science and Technology **2**, 886 (2010)

[3] J. Dufek, W. Gudowski, Annals of Nuclear Energy **36**, 1270 (2009)

[4] W. Walters, N.J. Roskoff, A. Haghighat, *A Fission Matrix Approach to Calculate Pin-wise 3D Fission Density Distribution*, in *Proc. M&C 2015* (Nashville, Tennesse, 2015)

[5] A. Haghighat, K. Royston, W. Walters, Annals of Nuclear Energy **87**, 61 (2016)

[6] J.C. Wagner, *Computational Benchmark for Estimation of Reactivity Margin from Fission Products and Minor Actinides in PWR Burnup Credit Prepared by* (2000), ISBN 1800553684

[7] X-5 Monte Carlo Team, *MCNP-A General Monte Carlo N-Particle Transport Code, Version 5*, Los Alamos National Laboratory (2005)

[8] B.T. Mervin, S.W. Mosher, J.C. Wagner, G. Maldonado, Nuclear Science and Engineering **173**, 276 (2013)

[9] T. Ueki, F.B. Brown, *Stationarity diagnostics using Shannon entropy in monte carlo criticality calculation I: F test.*, in *American Nuclear Society 2002 Winter Meeting* (2002), pp. 17–21

[10] M. Wenner, A. Haghighat, *A combined diagnostic approach for Monte Carlo source convergence identification*, in *Proc. M&C 2009* (Saratoga Spring, NY, May 3-7, 2009)

[11] V. Mascolino, A. Haghighat, N.J. Roskoff, *External SNF Cask Dose Calculation Using RAPID*, in *Advances in Nuclear Nonproliferation Technology & Policy Conference* (Santa Fe, New Mexico, September 25-30, 2016)

DEMETERRES project: development of innovative technologies for removing radionuclides from contaminated solid and liquid matrices.

Pierre Chagvardieff[1], Yves Barré[2], Virginie Blin[3], Sylvain Faure[2], Anne Fornier[4], Didier Grange[5], Agnès Grandjean[2], Emmanuel Guiderdoni[6], Pascale Henner[7], Brice Siroux[3], Antoine Leybros[2], Marc Messalier[8], Hervé Paillard[5], Thierry Prévost[8], Malvina Rennesson[8], Catherine Sarrobert[1], Alain Vavasseur[1], Anne-Aliénor Véry[9]

[1]*CEA DRF, 13108 St Paul lez Durance, France*
[2]*CEA DEN, 30207 Bagnols sur Cèze, France*
[3]*CEA DEN, 91192 Gif sur Yvette, France*
[4]*CEA DRF, 92265 Fontenay aux Roses Cedex, France*
[5]*VEOLIA 1 rue Giovanni Battista Pirelli, 94410 Saint Maurice, France*
[6]*CIRAD Boulevard de la Lironde, 34398 Montpellier cedex 5, France*
[7]*IRSN, 13108 St Paul lez Durance, France*
[8]*AREVA 92084 Paris La Défense Cedex, France*
[9]*INRA, 2 place Pierre Viala, 34060 Montpellier Cedex 2, France*

Abstract. As part of the « post-accidental » management, the DEMETERRES project (RSNR PIA) proposes to develop innovative and environmentally friendly methods for removal of cesium and strontium from soils and liquid matrices in order to rehabilitate them for an agricultural use while minimizing the volume of generated wastes in accordance with the nuclear waste existing processes. Complementary approaches are used: they are based on physico-chemical technologies (such as foams flotation, supercritical CO_2 extraction, extractants in fluidized bed reactor …) and biological ones (bioextractants, phytoextraction) which concepts are described. These researches aim to design innovative and performing extractants in term of selectivity and to achieve the pilot reactor phase for each of them. These pilots will group in a network to provide a technological platform lasting the project, to which will be attached an available network of experts. The respective advances of these researches are presented, completed of tests initiated in Japan on contaminated soils through partnerships.

1 General context

The restoration of the environment following a nuclear accident is a question of debate. For instance, soil decontamination by removing top soil as initiated in Fukushima prefecture will generate a considerable volume of organic wastes which can't address the existing nuclear waste treatment process capacities: innovative process of decontamination are likely to be developed. Similarly, the first tests using phytoextraction are just under investigation: this technology needs additional basic knowledge on soil-plant Cs and Sr transfer to become more successful. The DEMETERRES project (Development of bio- and eco- technologies for effluents and soils remediation in support to a restauration strategy following a nuclear accident) intends to study the feasibility at the pre-industrial stage of a set of environmentally friendly complementary technologies for rehabilitation of solid (soils) and liquid (fresh and seawater) matrices using both physico-chemical and biological approaches.

Some are suitable for *ex-situ* treatment of high contaminated wastes, others for *in situ* treatment of large volumes of weakly contaminated soils. This project, involving CEA, AREVA, VEOLIA, IRSN, INRA and CIRAD as partners, has been selected in 2013 by the French PIA (Programme Investissement d'Avenir) as part of the "Recherche en Sûreté Nucléaire et Radioprotection" RSNR call and will end by 2018.

The project aims to provide the actors in the post-accidental phase – firstly, with a decision-support guide bringing together the performances of technologies developed in regard to French soils 30 km around nuclear plants (NPP) – secondly, with a pre-industrial platform allowing to make available the developed pilots and associated with an expertise network.

2 Modelling the adsorption behavior of Cs^+ and Sr^{2+} onto a soil

The main long-lasting radionuclides released into the environment as a result of a nuclear accident are Cs-137 and Sr-90. Prior to developing a decontamination strategy, one should understand the actual chemical state of radionuclides in the environment. This involves the understanding of the interactions of these radionuclides with the main compounds of soils: clay minerals, organic matter, bacteria, etc.

Cesium, as well as strontium, are strongly adsorbed in the upper horizon from a soil (0-10 cm), particularly on clay minerals [1]. This adsorption is controlled by different factors, such as the type of clay minerals (illite, smectite, vermiculite etc.) [2, 3], the presence of competitive ions (Ca^{2+}, Mg^{2+}, K^+, Na^+, etc.) [4-6], the physical and chemical properties of the soil (pH, ionic strength) [7], but also the presence of other compounds (carbonates, natural organic matter, etc.) [8, 9]. Taking into account these factors, the CEA has developed a multisite ion-exchange model. Its objective is simulating the uptake of an element in naturally occurring mineral phases and under various conditions [10, 11].

Considering general knowledge on natural environments, the study has firstly been focused on clay minerals, especially the three most representative from the French agricultural soils, *i.e.* smectite, illite and kaolinite. A literature review and specific additional adsorption tests have allowed developing a consistent thermodynamic database compiling the selectivity coefficients for both major cations (Ca^{2+}, Mg^{2+}, K^+, Na^+, H^+) and trace elements (Cs, Sr) towards these pure clay minerals. The reversibility of this adsorption has been verified.

The robustness and efficiency of the model is then tested by carrying out adsorption tests onto mixtures of clay minerals (see an example for cesium in Fig.1).

Fig. 1: Cesium adsorption isotherm on a mixture composed of 30% illite and 70% smectite (K_d: solid/solution distribution coefficient of cesium).

By comparing the experimental and simulated data, it is clear that cesium adsorption depends on its concentration. From the simulated curves, the contribution of illite to cesium adsorption is important at trace concentration and becomes less important at higher concentration. This is the direct consequence of illite having a smaller cationic exchange capacity than smectite, and higher affinity sites. However, an overestimation of the cesium adsorption is pointed out in the high concentration range, which should be further investigated. This kind of experiments allows verifying the model predictions and improves our understanding of the interactions involved in the adsorption processes of both cesium and strontium on clay minerals.

Parallel to these fundamental achievements, a physicochemical characterization of five French soils (representing 80% of the soils around the 20 target sites of the project: 19 Nuclear Power Plants and La Hague) was performed:

- CESTA (in the Bordeaux region).
- Étoile-sur-Rhône and Pusignan (region of the Rhône Valley).
- Digulleville and Herqueville (around La Hague).

It is then possible to verify and quantify the actual adsorption of cesium and strontium onto these soils in order to differentiate between adsorption mainly driven by clay minerals from the influence of other compounds, such as carbonates or natural organic matter.

The first results obtained using strontium and the Pusignan soil showed that experimental adsorption values and simulation are consistent. Though several approximations are used to simulate the strontium adsorption, the model succeeded in predicting the strontium uptake on Pusignan soil.

The results and understanding that will result from this study will be used in order to determine the most appropriate decontamination technologies regarding the type of contaminated soil.

3 *Ex situ* soil decontamination

For *ex-situ* soil remediation, two complementary physico-chemical technologies are developed: the first one is designed for clay soils and is based on the removal of clay microparticles (<25 microns) by a flotation foam process, the other one is devoted to sandy soils and rubles and is based on the extraction of the targeted radionuclide in supercritical CO_2 as an organometallic complex ("contaminant - extractant " complex).

3.1 Using particulate flotation foams

Particulate flotation foams are of great interest in mineral processing, water treatment and could be used for Cs radioactive soil decontamination [12]. For Cs extraction, the proof of concept has been established at the laboratory scale using commercial clays (illite). Clays particle fixing Cs are floated thanks a suitable foaming and collecting surfactant. In 2015, a laboratory scale unit has been implemented (Fig2, Fig3) in order to study and to optimize the process parameters with clays (% of surfactant, air injection rate, foam retention time...); measuring the overall extraction efficiency, selectivity and analyzing the size of the particles collected. Selectivity for the smallest clay particle size fraction (3 μm) in foams was observed and correlated with foam height and drainage time (see figure 4 below) [13].

DOI: 10.1051/epjconf/201715305026

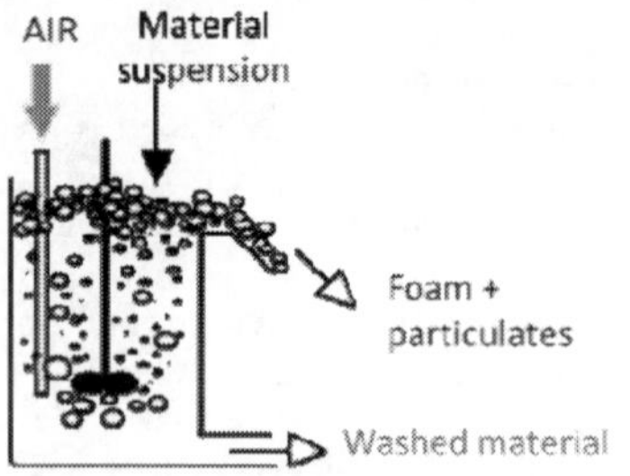

Fig 2: Principle of the particulate flotation foam process

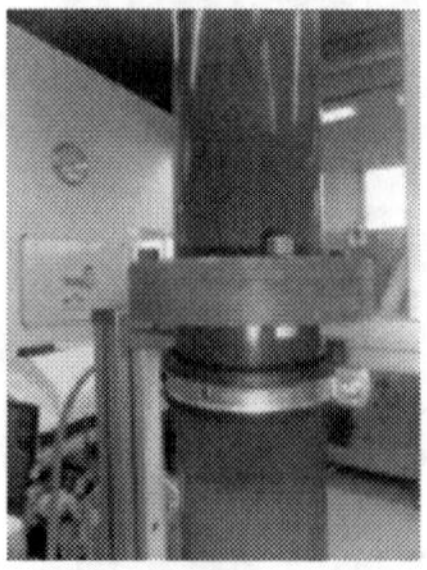

Fig 3: Foam building in a laboratory scale apparatus

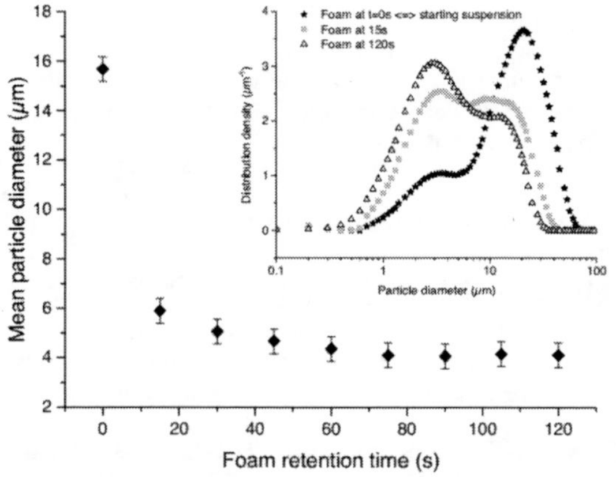

Fig 4: Mean particle diameter evolution with foam retention time in the foaming column. Size distributions at foam retention time of 0, 15 and 120 s

Tests with soils selected in the frame of the project have started and the first results are available. Preliminary pilot tests conducted in Japan on contaminated soils led to removal yields of the radioactive Cs close to 82% concentrated in 15 to 30% of the initial soil volume.

The fertility of treated soils will be tested after 3 months using a 50 % - 50 % w/w mixture prepared with an untreated soil (simulating a plowing). In parallel, a study of the bacterial biodiversity is being performed on treated and untreated soils to test the possible influence of the clay particles content decrease on the treated soil.

3.2 Using ionic flotation foam

The potential of an ionic flotation foam process has been explored for the removal of ionic Cs. A screening of extracting components in combination with suitable surfactants has been achieved. A foam composition including Prussian blue analogs in the form of highly specific nanoparticles Cs [14] [15] has been shown to be efficient at the laboratory scale

3.3 Using supercritical CO_2

Supercritical CO_2 (SC-CO_2) extraction process is one of the promising, smooth technologies which may be applied to soil remediation. SC-CO_2 exhibits gas-like mass transfer rates with liquid-like solvating abilities. Moreover, its high diffusivity and low viscosity enable it to penetrate and transport solutes from porous solid matrices to the SC-CO_2 phase. Furthermore, its flexibility in terms of ability to adjust solvating power by tuning operating pressure and temperature is quite interesting.

Direct extraction of metal ions by SC-CO_2 is inefficient due to charge neutralization requirement and weak ionic solute-solvent interactions. Hence, organic ligands are needed to chelate ions and make them soluble in SC-CO_2. Successful SC-CO_2 extraction of alkali metal ions was carried out using macrocyclic compounds with perfluorocarboxylic acids such as HPFOA [16]. Crown ethers and calixarenes have been, in particular, studied for their selectivity towards cesium. Synergistic effects between macrocycles and HPFOA improve solubility due to CO2-philic HPFOA. Solubilities of all these compounds have been measured (Fig. 5) in our operating range using high pressure cells coupled with gravimetric and/or analytical methods in CEA Marcoule.

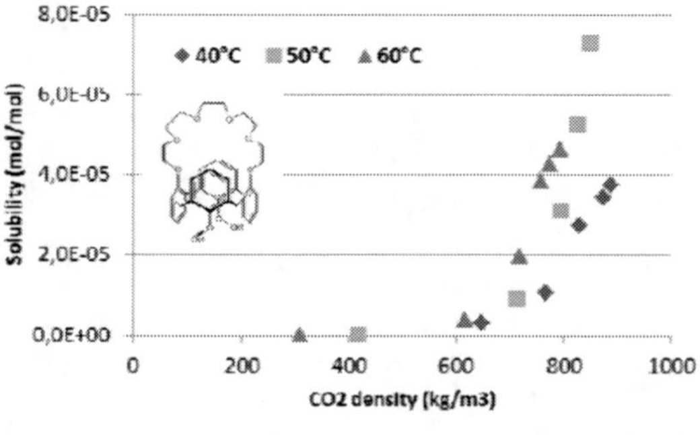

Fig 5: Example of solubility in supercritical CO_2 profile for used macrocycle: 25, 27-Bis (1-octyloxy)calix[4]arene-crown-6

Extraction set-up in batch dynamic mode is shown Fig.6. It includes a tank of CO_2, a pump to reach operating flow rate, a heat exchanger to maintain work temperature and a back pressure regulator to maintain operating pressure. CO_2 flows continuously through the

DOI: 10.1051/epjconf/201715305026

sample in extraction reactor, before being released in collectors downstream from regulation valve. Cs-organometallic complex is recovered in collectors.

Feasibility of SC-CO_2 extraction for decontamination of model silica soil has been studied with cesium contamination around 1mg.g^{-1} of soil using DB18C6/HPFOA extractant system. Process parameters influence such as operating pressure and temperature, extractant system amount, residual soil water content has been investigated in [17]. Low temperature (40°C) and high pressure (29 MPa) allow to optimize process ability for cesium removal. A compromise should be established, for a fixed CO_2 flow rate, between amounts of extractants and cationic exchanger. Moreover, residual moisture of silica soil plays a significant role concerning cesium extraction. By considering all these parameters, extraction yields up to 95.5% were obtained and demonstrated our concept.

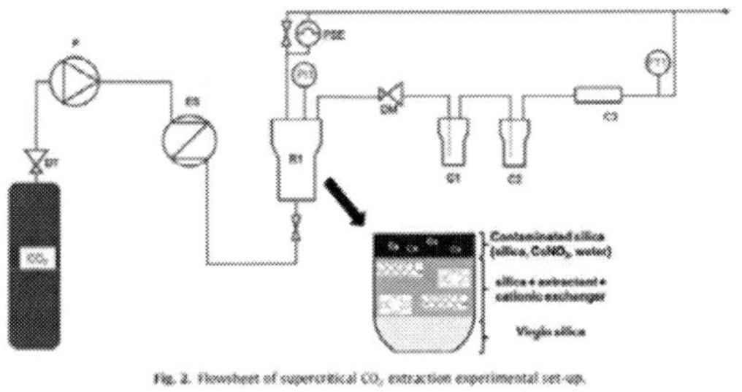

Fig 6: Principle of radionuclides extraction from sand and rubbles using supercritical CO_2

Further experimental runs with realistic cesium contamination content are in progress using ^{137}Cs in nuclear laboratory using gamma spectroscopy. In these runs, modified calixarene-crowns whose solubility was determined in our operating range will be used due to better selectivity towards radiocesium. These extractants have been functionalized to increase their solubility in SC-CO_2, while maintaining higher selectivity towards cesium.

4 *In situ* soil remediation

In situ restoration of contaminated soils can be achieved by phytoremediation. A drawback of this technic is a long term achievement and that it can only be performed on soils with low or medium levels of contamination. Yet, it appears complementary of *ex situ* remediation that stays the only way of treatment for heavy levels of contamination. Cesium (Cs) and strontium (Sr) are analogues of potassium (K) and calcium (Ca), respectively. The phytoextraction process is based on the transport of radionuclides *in planta* using plant transporters mechanisms usually dedicated to plant nutrition (Fig. 7). The first assays done in Japan have revealed a need for further understanding to enhance the capacity of plants to uptake and translocate Cs and Sr. Various strategies essentially based on genetic approaches are developed.

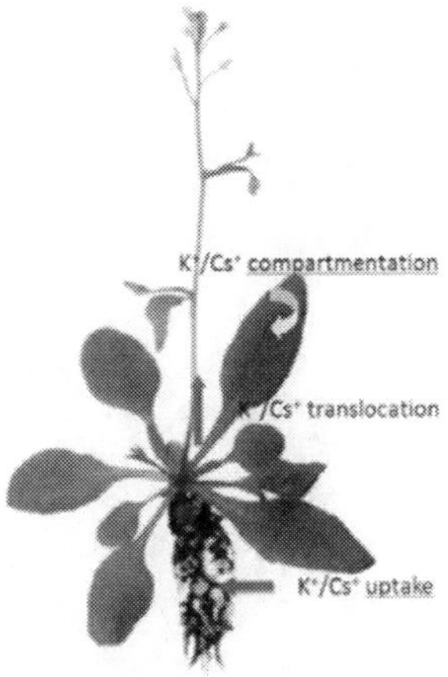

Fig 7: Phytoextraction principle

4.1 Fit the root architecture to the contamination profile

Adaptation of roots architecture to the contaminated zone, *e.g.* managing root architecture to locate the roots mainly in the first 10 cm, is a way to increase the yield of phytoextraction of radio-Cs. Conversely, deep rooting is a way to diminish the uptake of surface contaminant such as Cs. For that purpose, different genes have been targeted in Rice, which regulate the angle of development of secondary roots [18] (Fig 8). Work is now underway to characterize the impact of Rice roots architecture on Cs uptake and translocation using contaminated soils from Fukushima's area and artificial substrates based on perlite which mimics clay with a determined gradient of Cs contamination.

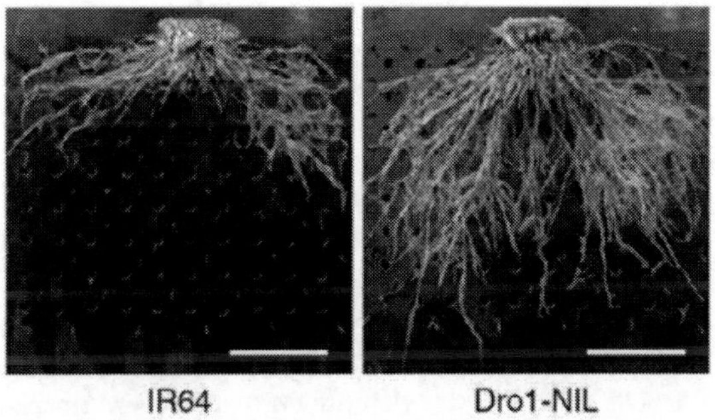

Fig 8: Rice lines with different roots architecture

4.2 Increase the radionuclides phyto-availability

Some substrates exuded by roots such as organic acids and protons are well-known to increase the phyto-availability of minerals such as Fe or Al by modifying their chemical speciation. We are studying a way to control organic acids and protons exudation by means of genetics, targeting genes that control these processes at the level of transporters and transcription factors in *Arabidopsis thaliana*. Validation of this concept is underway using contaminated soils from Fukushima's area.

4.3 Increase the uptake and translocation of radionuclides

Different plant transporters able to uptake K but also Cs have been identified in *Arabidopsis thaliana* and Rice. Depending on the genes considered, knock-out lines exhibited in hydroponic conditions a change in the transfer of radio-cesium to shoots from +200% to -50% when compared to the wild plant. These lines will be tested on contaminated soils and the corresponding genes targeted in agronomic crops to increase (phytoextraction) or decrease (safe-food) the transfer of radionuclides.

Due to chemical analogies and based on competition studies, Cs uptake in plants has been shown to proceed mainly by K transporters and to be potentially submitted to K homeostasis. In *Arabidopsis thaliana* as much as 90-100 genes are putative K transporters, some of them described for K uptake and a minority described as being able or not to also transport Cs. We tested candidates from different families of K transporters regarding their performances in K transport as detailed on Fig. 7, in particular for families in which at least one gene has been shown to respond to Cs. A screening of numerous knock-out lines of these potentially interesting candidates had been performed through Cs toxicity response assessment by germination test in liquid medium containing Cs and various levels of K. Lines whose response was different from reference plant (*e.g.* Fig 9.a, wild type ecotype ColumbiaO shows greener seedlings than mutant A after exposure to Cs)- tentatively because of different Cs accumulation level, were furtherly tested. If interest was confirmed at the end of the screening phase, Cs uptake and translocation experiments were performed (*e.g.* Fig 9.b., mutant A accumulated more Cs in leaves after 7 days than reference plant). In the following mutants are labelled by letters as description of their performance is still not totally published.

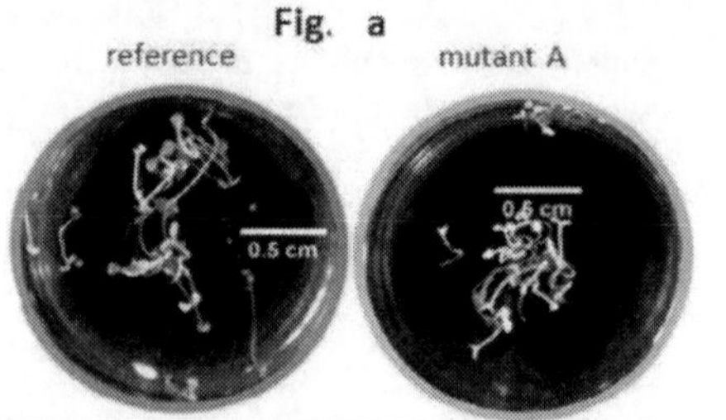

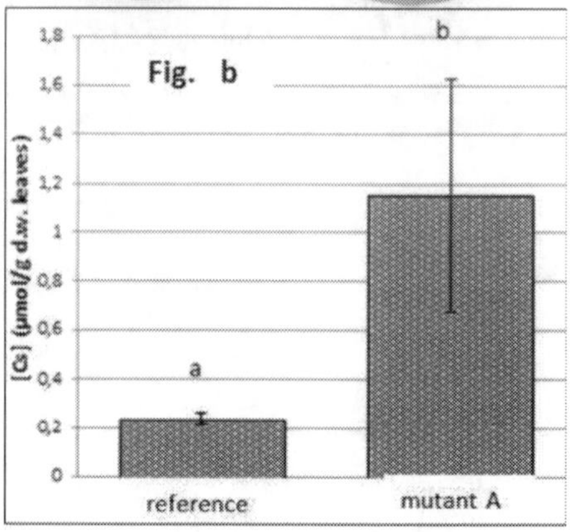

Fig 9: a) Cs toxicity in germinated seedlings and 7-days-Cs b) accumulation in 35 days old plants of mutant line A affected (knocked-out) in a K transporter compared to reference line Col0. Fig. 9a- 100µM K and 300 µM Cs; Fig. 9b. 100 µM K and 1 µM Cs.

Studies are under way for the full characterization of gene A and mutant performances regarding K and Cs uptake and translocation. Also are under way the first experiments testing these mutants using contaminated soils from Fukushima's area

4.4 The safe-food approach

When soils present a moderate contamination another strategy may be applied, it consists in growing plants with a limited capability to transfer Cs from the soils to the plant tissues. Such criterion depends on genetics determinants. This approach was started using Rice and targeting K transporters from KUP/HAK/KT and HKT families, most likely candidates to behave a high affinity transport of Cs. As previously mentioned, K transporters are numerous in plants and a first assay was heterologous expression in yeast in order to determine the selectivity of the encoded transporters concerning K and Cs, respectively. A mutant yeast line deficient in the main K transporters was used, *trk1 trk2 tok1*. Yeast growth was followed in the presence of various K and Cs concentrations coupled with the determination of Cs and K uptake by yeast transformed lines. This approach showed that several members exhibited a large permeability to Cs. However, when expressed in Xenopus oocytes, only one Rice transporter exhibited a high affinity transport of Cs. Work is now on the way to characterize the role of this transporter *in planta* using reverse genetics.

DOI: 10.1051/epjconf/201715305026

5 Liquid matrices decontamination

Decontamination of liquid matrices requires the development of extractants selective of Cs/Sr and economically acceptable as a way for developing technologies adapted to a « post-accidental » situation (mobility, maintenance…). Selective extraction on non-recycled support in columns (high performance selectivity) or fluidized-bed (multi-decontamination) pilot reactors is studied on two liquid matrices templates: freshwater and seawater. Two approaches are under progress to develop extractants selective of Cs and/or Sr.

5.1 Development of new chemical extractants

In this context a selective adsorbent material for Cs (SORBMATECH® 202, S202) was developed, consisting of potassium/copper ferrocyanide nanoparticles deposited in the mesoporosity of silica grains. This material is designed for column process due to its average grain size of 450 µm. Several batch experiments were carried out in order to obtain kinetic and thermodynamic data of the ion exchange. Kinetic experiment showed that the sorption equilibrium is reached in less than 5 min. Fast sorption kinetics is a significant asset for use in column.

Distribution coefficients $K_{d,Cs}$ were calculated for each of the isotherm experiments, as a function of Cs^+ remaining in solution. Isotherms performed with radioactive ^{137}Cs (pure water at pH 7, fresh water and seawater) were also used to calculate $K_{d,Cs}$ for very low Cs concentration. Each $K_{d,Cs}$ point is reported in 0. At low Cs^+ concentrations, the competitive ion effect is comparatively more effective: $K_{d,Cs}$ is about 2.5×10^6 mL/g at pH = 7 without competitive cation but decreases by one order of magnitude in seawater (K_d = 2×10^5 mL/g). The same behavior was observed for fresh water, although the lower concentration of competitive cations than in seawater leads to K_d close to the one in pure water (K_d = 10^6 mL/g). Despite these competitive effects, this material is undoubtedly a good candidate to decontaminate seawater effluent contaminated with traces of ^{137}Cs.

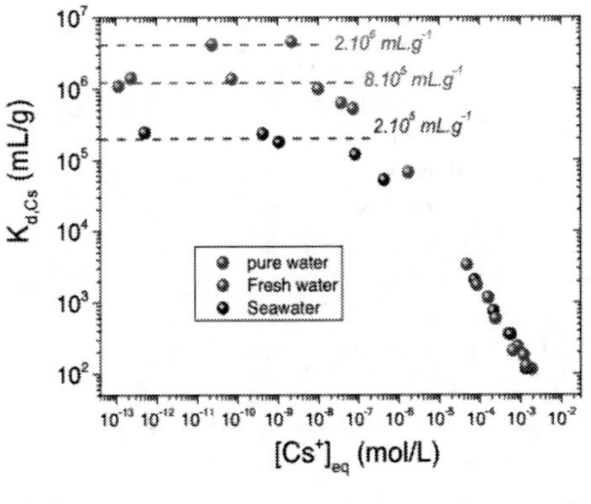

Fig.10: Variation of $K_{d,Cs}$ as a function of $[Cs^+]_{eq}$ for different type of aqueous solutions.

Column experiments were performed using a micro-column of 1 cm diameter and 10 cm high. The same amounts of material were put in the column and represented a bed height of 5 cm. A fresh water containing 0.5 mmol/L of $CsNO_3$ passed through this column at two different flow rates (20 BV/h and 200 BV/h). The breakthrough curves are reported on Figure 11, where no difference was observed between these two flow rates. Moreover, these breakthrough curves are typical of fast diffusion kinetic in the column with no detachment observed even at high flow rate.

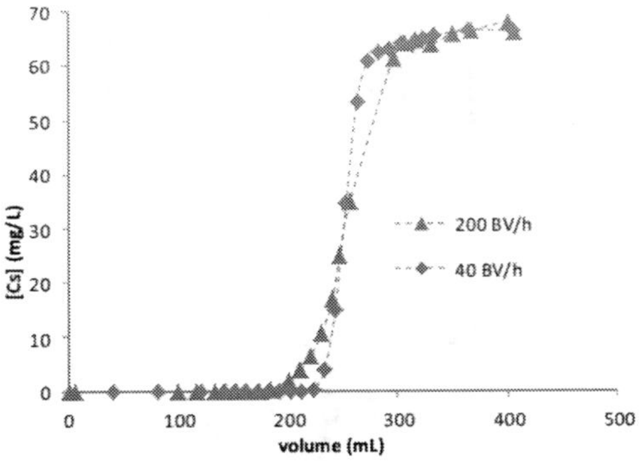

Fig.11: Breakthrough curves at 40 BV/h and 200 BV/h under same operational conditions (H_{bed}=5cm, freshwater, [Cs]=70 mg/L)

In Fukushima, the selective sorbents used to extract ^{90}Sr include sodium titanates and/or zeolites. However, ^{90}Sr removal efficiency by these ion exchange minerals is significantly reduced by high contents of competitive cations (Ca, Mg) and non-radioactive strontium isotopes in seawater. Traditionally, in the radioactive liquid effluent treatment plants, barium sulfate is used to remove radioactive strontium from effluents by a continuous coprecipitation process [19]. This process is selective but leads to large volumes of radioactive sludge that have to be managed. To address these concerns, our approach consists in coupling ion exchange and coprecipitation processes. A barium titanate material coupled with a precipitation with sulfate ions has proved to be an effective and selective sorbent for extraction of strontium from effluents of high salinity like seawater. This coupling demonstrates considerably high capacity and selectivity for strontium extraction with distribution coefficients K_d of 18200 mL/g, obtained from batch sorption tests with seawater enriched with ^{90}Sr (59355 Bq/L, m/V = 5 g/L). As evidenced by SEM-EDX pictures in Figure 12, the precipitation of $BaSO_4$ occurs at the surface of the titanate grains. However, Sr ions seem uniformly distributed in the $BaSO_4$ crystals and in the titanate grains. Therefore, the possible mechanism of Sr extraction involves the ion exchange with the Ba ions from the material and a coprecipitation of insoluble barium sulfate crystals formed at the surface of the titanate grains upon contact with the effluent containing sulfate ions.

 DOI: 10.1051/epjconf/201715305026

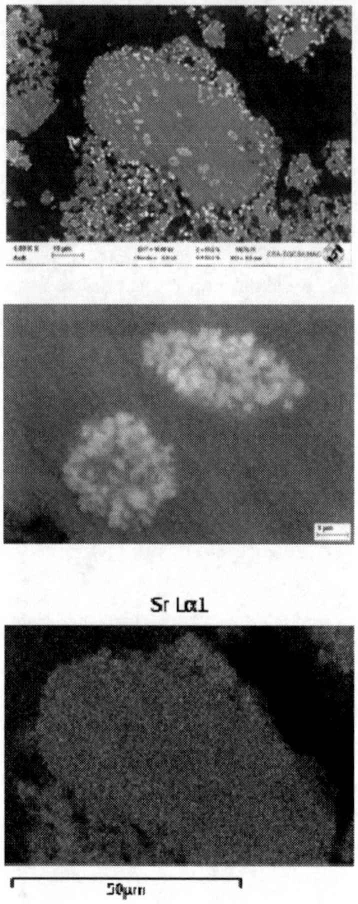

Fig. 12: SEM-EDX images of the barium titanate after a 24 h contact with a solution of $Sr(NO_3)_2$ ($[Sr^{2+}]$ = 8 mg/L) in Na_2SO_4 ($[SO_4^{2-}]$ = 2,7 g/L)

5.2 Drawing inspiration from living world to design innovative extractants

Some biological molecules, such as calmodulin, have a strong affinity for calcium, a strontium analog. Some peptides have been designed from the chelating domains (EF Hand) of calmodulin and are optimized after mutagenesis. Uranium-chelating peptides had previously been obtained using this approach [20]. Measurement of Sr affinity to screen the modified peptides (Fig 13) is made by fluorimetry. They will be further modified to increase their capacity to chelate Cs.

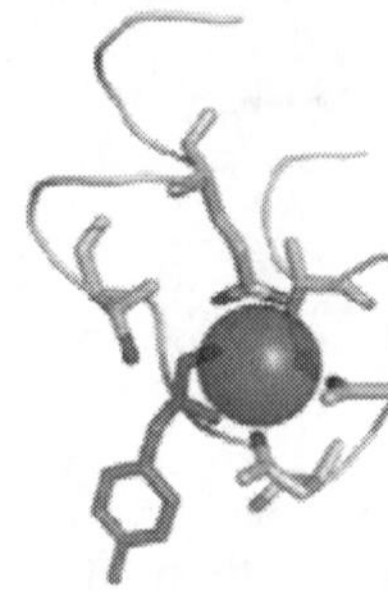

Fig 13: Modification (in red color) of a chelating domain of calmodulin to increase Sr affinity

Microorganisms (particularly soil bacteria) are also screened in presence of Sr or Cs to select strains with a high capacity of accumulation. From the first experiments, realized by incubating bacteria strains during 24h in freshwater containing 50 mg/L Sr, a strain accumulating 50 % of Sr has been isolated. These bacteria could be then used directly in bioreactors.

5.3 Mobile fluidized-bed pilot reactor

The fluidized-bed SAPHIRA™ technology developed by VEOLIA has been optimized and tested by AREVA / VEOLIA with water doped with Cs and Sr between November 2015 and March 2016.

Cesium is adsorbed with ferrocyanide salt. It is a 20 g/L suspension injected in the SAPHIRA™. During last week of trial, a specific adsorbant developpped by CEA / AREVA has been tested: SORBMATECH™. Basically, it is ferrocyanide salt linked to zeolithe which has the same adsorption properties as ferrocyanide salt but is easier to put in use (heavier: settling is better). Strontium is adsorbed by the precipitation of fresh $BaSO_4$ caused by the mixing in the SAPHIRA ™of $Ba(NO_3)_2$ and Na_2SO_4. This co-adsorption is currently performed in La Hague in a standard coagulation/flocculation facility. The main difference with SAPHIRA™ is the use of a single reactor, the absence of sludge recirculation and, therefore, the low sludge concentration within the equipment.

In DEMETERRES project the purpose is to develop a nuclearized, compact equipment which could be used in case of emergency after a nuclear accident.

5.3.1 SAPHIRA™ Pilot

The SAPHIRA™ pilot ran at 300 L/h. Hydraulic retention time was 60 minutes within the reactor.
Potable water was doped in line with 5ppm Cs and Sr (injection of concentrated solutions) and treated in the SAPHIRA™. Apart of ferrocyanide salt and $BaSO_4$, we also injected flocculant and sometimes coagulant in order to retain TSS (Total Suspended Solids). $BaSO_4$ was not problematic but ferrocyanide salt settled badly without the aid of coagulant and flocculant.
SAPHIRA™ was operated between 40 and 60 g/L TSS at the inlet of the reactor. The produced sludge was stored in cubitainers and incinerated: no dehydratation of sludge tested.

DOI: 10.1051/epjconf/201715305026

Fig 14: SAPHIRA™ pilot unit

5.3.2 Cesium adsorption by ferrocyanide salt

Decontamination factor ([Cs]initial/[Cs]out; known as "FD") is presented below (Fig 15). A decontamination factor of 100 means 99% removal yield. A dose of 50 ppm of ferrocyanide was kept for the tests

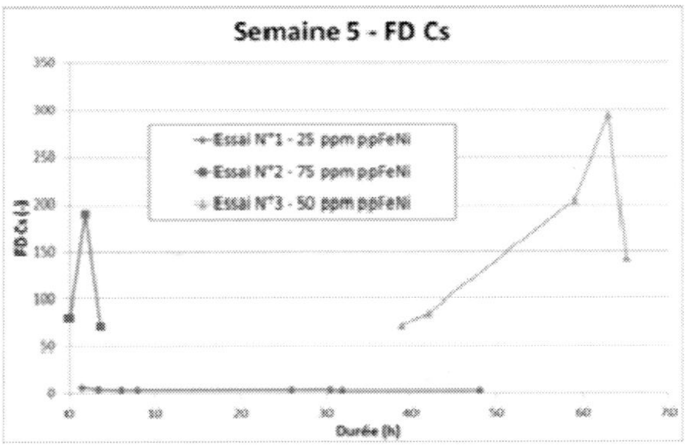

Fig 15: Cs Decontamination Factor according to the ferrocyanide concentration

Ferrocyanide salt is difficult to settle in the SAPHIRA™ because is very fine (around 10 microns). Small doses of coagulant and flocculant really help settling: At the outlet of the reactor the turbidity decreases from 60 NTU without, down to 10 NTU. With ferrocyanide and to achieve a FD_{Cs} of 70 it is necessary to inject a flocculant and set up a 20 µm filter. With the SORBMATECH™ FD_{Cs} is lower (20) but it is not necessary to inject flocculant.

The performances are limited due to the Cs concentration during inactive test. Higher DF could be expected with radioactive effluent.

5.3.3 Strontium adsorption by $BaSO_4$

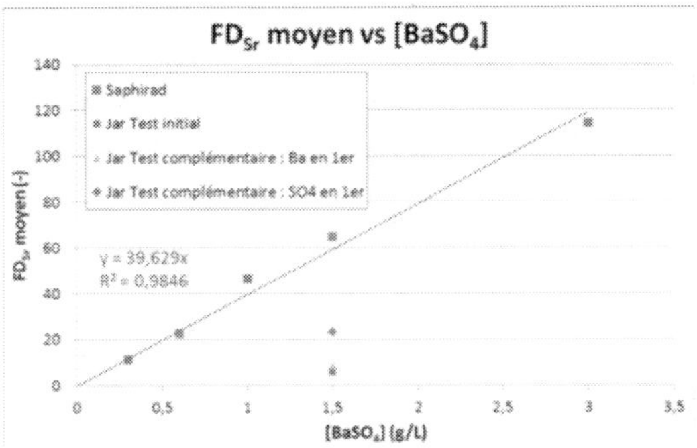

Fig 16: Sr Decontamination Factor according to the $BaSO_4$ concentration

The decontamination factor is around 60 with 1,5 g/L $BaSO_4$ precipitated. It is less efficient than cesium adsorption with ferrocyanide salt but this kind of result was expected. Results on the pilot are better than in jar-testing which would tend to prove that the sludge blanket has a positive effect. The mean FD_{Sr} at 1.5 g/L is 60 which means 98% removal yield.

$BaSO_4$ precipitates and settles well in the SAPHIRA™. The improvement of FD_{Sr} is also possible by increasing the concentration of the sludge bed in the reactor

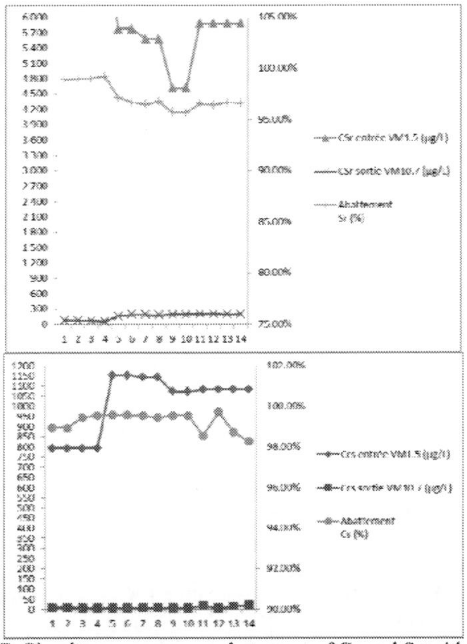

Fig 17: Simultaneous average abatement of Cs and Sr with $BaSO_4$

Average 97% for Sr (FD=51) and average 99% for Cs (FD= 166)

5.4 Mobile riser pilot reactor

Preconditioned or immersed ion-exchange columns have been developed. Pilot reactors of increasing capacity have been tested. SORBMATECH^{TM} feasibility assessment in large-capacity reactor required a previous production of several kg. A prototype of the NYMPHEA technology engineered by AREVA has been completed and tested. Its 1:1 scale-up is being evaluated.

6 Operational implementation of the project

On one part, a guidance for remediation of contaminated territories, leaning on the performances of the developed technologies, will be drafted. This guidance will integrate the soil analysis realized upstream to the project to define the typologies of grounds around the target sites. A further perspective is a decision-making tool for remediation strategies, integrating also REX of experimented technologies, and leaning on a multi-criteria methodology taking into account social and societal stakes in territories recovery.

On the other part, a preindustrial platform will be rigged, to make available access to the pilots developed within the framework of the project, both for the actors of the post-accidental phase and for new applications (extraction of other radionuclides, heavy metals…). To this platform will be associated an expertise network which could be mobilized in post-accidental situation and a set of trainings to perpetuate the implementation of processes and hardware.

Acknowledgements

The authors thank the French « Programme Investissement d'Avenir » (ANR-11-RSNR-0005 DEMETERRES) managed by the National Research Agency (ANR) for its financial support.

References

1 H. Amano, T. Matsunaga, S. Nagao, Y. Hanzawa, M. Watanabe, T. Ueno, Y. Onuma, Org. Geochem. **30**, 437 (1999)

2 M.H. Bradbury, B. Baeyens, J. Contam. Hydrol. **42**, 141 (2000)

3 X. Chen, S. Peng, J. Wang, J. Radioanal. Nucl. Chem. **303**, 509 (2015)

4 P. Jacquier, J. Ly, C. Beaucaire, Appl. Clay Sci. **26**, 163 (2004)

5 T Missana, M. García-Gutiérrez, Phys. Chem. Earth **32**, 559 (2007)

6 T. Missana, A. Benedicto, M. García-Gutiérrez, U. Alonso, Geochim. Cosmochim. Acta **128**, 266 (2014)

7 P.K. Verma, P. Pathak, P.K. Mohapatra, Radiochim. Acta **102**, 401 (2014)

8 C. Dumat, S. Staunton, J. Environ. Radioact. **46**, 187 (1999)

9 M. Samadfam, T. Jintoku, S. Sato, H. Ohashi, J. Nucl. Sci. Technol. **37**, 180 (2000)

10 E. Reinoso-Maset, J. Ly, J. Chem. Eng. Data **59**, 4000 (2014)

11 S. Savoye, C. Beaucaire, B. Grenut, A. Fayette, Appl. Geochem. **61**, 41 (2015)

12 Faure, S.; Messalier. M., Patent WO2013167728 A1. (2013)

13 Chapelain, J. C. M.; Faure, S.; Beneventi, D., Clay Flotation: Effect of TTAB Cationic Surfactant on Foaming and Stability of Illite Clay Microaggregates Foams. *Ind Eng Chem Res*, **55 (7)**, 2191–2201 (2016)

14 C. Micheau, P. Bauduin, O. Diat, Patent CEA WO2015063073 (2014)

15 D. Dedovets, P. Bauduin, J. Causse, L. Girard, O. Diat, Phys. Chem. Chem. Phys. **18**, 3188-3196 (2016)

16 S. Mochizuki, N. Wada, R. Smith Jr, H. Inomata, Anal. Commun. **36** :51-52 (1999)

17 A. Leybros, A. Grandjean, N. Segond, M. Messalier, O. Boutin, J. Environ. Chem. Eng. **4** :1076-1080 (2016)

18 N. Ahmadi, A. Audebert, MJ. Bennett A. Bishopp , A. Costa de Oliveira , B. Courtois , A. Diedhiou, A. Diévart , P. Gantet, A. Ghesquière , E. Guiderdoni, A. Henry , Y. Inukai, L. Kochian, L. Laplaze, M. Lucas, D. Trung Luu, B. Manneh, X. Mo, R. Muthurajan, C. Périn , A. Price, S. Robin, H. Sentenac, B. Sine, Y. Uga, AA. Véry, M. Wissuwa, Ping Wu and J. Xu Rice 7:29 (2014)

19 V. Pacary, Y. Barré, E. Plasari, *Method for the prediction of nuclear waste solution decontamination by coprecipitation of strontium ions with barium sulphate using the experimental data obtained in non-radioactive environment*, Chemical Engineering Research and Design, 88 **(2010)** 1142-1147.

20 MR. Beccia, S. Sauge-Merle, D. Lemaire, N. Bremond, R. Pardoux, S. Blangy, P. Guilbaud, C.Berthomieu J biol. Inorg. Chem., **20(5):** 905 – 19 (2015)

Continuous energy adjoint transport for photons in PHITS

Alex Malins[1,★], *Masahiko* Machida[1], and *Koji* Niita[2]

[1] *Center for Computational Science and e-Systems, Japan Atomic Energy Agency (JAEA), 178-4-4 Wakashiba, Kashiwa, Chiba 277-0871, Japan*
[2] *Research Organization for Information Science and Technology (RIST), 2-4 Shirane Shirakata, Tokai-mura, Naka-gun, Ibaraki 319-1195, Japan*

Abstract. Adjoint Monte Carlo can be an efficient algorithm for solving photon transport problems where the size of the tally is relatively small compared to the source. Such problems are typical in environmental radioactivity calculations, where natural or fallout radionuclides spread over a large area contribute to the air dose rate at a particular location. Moreover photon transport with continuous energy representation is vital for accurately calculating radiation protection quantities. Here we describe the incorporation of an adjoint Monte Carlo capability for continuous energy photon transport into the Particle and Heavy Ion Transport code System (PHITS). An adjoint cross section library for photon interactions was developed based on the JENDL-4.0 library, by adding cross sections for adjoint incoherent scattering and pair production. PHITS reads in the library and implements the adjoint transport algorithm by Hoogenboom. Adjoint pseudo-photons are spawned within the forward tally volume and transported through space. Currently pseudo-photons can undergo coherent and incoherent scattering within the PHITS adjoint function. Photoelectric absorption is treated implicitly. The calculation result is recovered from the pseudo-photon flux calculated over the true source volume. A new adjoint tally function facilitates this conversion. This paper gives an overview of the new function and discusses potential future developments.

1 Introduction

The Monte Carlo method can be inefficient for solving radiation transport problems with a small tally relative to the source. The inefficiency is that many particle histories during the course of a simulation fail to cross the tally region, and hence make no contribution to the calculation result. Large source, small tally geometries can require long computing times to converge the calculation result to a desired level of precision.

Adjoint Monte Carlo is one strategy for improving the efficiency of large source, small tally transport problems. In adjoint Monte Carlo, radiations are generated within the tally region and transported backwards through the geometry, undergoing any physical interactions in reverse. They contribute to the calculation result if they cross the source region. The efficiency benefit is conferred by solving a problem where the 'adjoint source region' is relatively much larger than the 'adjoint tally region', and thus a larger fraction of particle histories score [1].

The adjoint method is a promising strategy for environmental radioactivity transport problems. In these problems it is often necessary to calculate gamma ray dose rates at precise locations, where radioactivity dispersed widely within the environment contributes significantly to the result. For example, the source may be natural potassium, uranium and thorium series radioisotopes within soil, or radioactive fallout from a nuclear accident or test. In such cases decay photons emitted hundreds of metres from the detector are important [2].

Hoogenboom previously outlined a method of adjoint Monte Carlo for photons with continuous energy representation [3, 4]. One benefit of continuous energy representation over a multigroup scheme is that energy spectra can be evaluated with fine resolution, allowing the simulation of gamma spectrometer responses. Continuous energy representation is also important for accurately calculating radiation protection quantities, such as ambient dose equivalent ($H^{*}(10)$). However, there have only been a few implementations of adjoint Monte Carlo for photons with continuous energy representation in popular transport codes [5, 6].

This paper outlines the addition of a continuous energy photon adjoint function to the Particle and Heavy Ion Transport code System (PHITS) [7]. There are three main components to the code developments. First, an adjoint cross section library was developed adding adjoint incoherent scattering and pair production cross sections. Second, a new adjoint function was added to PHITS, allowing the transport and collision of pseudo-photons by coherent and incoherent scattering, and implicit treatment for photoelectric absorption. Finally, an adjoint tally function was developed to recover the forward calculation result from the pseudo-photon flux tallied over the real source volume.

★e-mail: malins.alex@jaea.go.jp

2 Overview of forward Monte Carlo photon transport

Monte Carlo radiation transport codes sample the Boltzmann transport equation

$$\chi(\mathbf{r}, E, \mathbf{\Omega}) = S(\mathbf{r}, E, \mathbf{\Omega}) + \widehat{C} \cdot \widehat{T} \cdot \chi(\mathbf{r}', E', \mathbf{\Omega}'). \quad (1)$$

The subject of the equation is the emission density of photons with energy E [eV], in the direction $\mathbf{\Omega}$, from point $\mathbf{r}$ in space, i.e. $\chi(\mathbf{r}, E, \mathbf{\Omega})$ [$\mathrm{cm^{-3}\,eV^{-1}\,sr^{-1}}$]. There are two contributions to $\chi(\mathbf{r}, E, \mathbf{\Omega})$: photons emitted by radiation sources, $S(\mathbf{r}, E, \mathbf{\Omega})$ [$\mathrm{cm^{-3}\,eV^{-1}\,sr^{-1}}$], and photons emerging from collisions, given by $\widehat{C} \cdot \widehat{T} \cdot \chi(\mathbf{r}', E', \mathbf{\Omega}')$.

Photons emerging from collisions at $\mathbf{r}$ first have to be transported to the collision site. This process is described by the transport operator, $\widehat{T}$, which operates on the emission density $\chi(\mathbf{r}', E', \mathbf{\Omega}')$. The result of the operation is the collision density $\psi(\mathbf{r}, E', \mathbf{\Omega}')$ [$\mathrm{cm^{-3}\,eV^{-1}\,sr^{-1}}$]:

$$\begin{aligned}\psi(\mathbf{r}, E', \mathbf{\Omega}') &= \widehat{T}(\mathbf{r}' \to \mathbf{r}, E', \mathbf{\Omega}') \cdot \chi(\mathbf{r}', E', \mathbf{\Omega}') \\ &= \int_0^\infty \Sigma_t(\mathbf{r}, E') e^{-\beta(\mathbf{r}, L, E', \mathbf{\Omega}')} \chi(\mathbf{r}', E', \mathbf{\Omega}') \mathrm{d}L. \quad (2)\end{aligned}$$

The integral here is over all path lengths $L = |\mathbf{r} - \mathbf{r}'|$ [cm] by which a photon can travel from $\mathbf{r}'$ to $\mathbf{r}$. The factor $\Sigma_t(\mathbf{r}, E')$ [$\mathrm{cm^{-1}}$] is the total macroscopic cross section for a photon with energy E' to undergo an interaction at $\mathbf{r}$. Assuming there is no directional dependence of the cross sections within the transport media, and that polarization effects can be discounted, the interaction cross sections are independent of a photon's direction of travel. The exponential term in Eq. (2) describes the attenuation faced by photons traversing towards the collision site, where

$$\beta(\mathbf{r}, L, E', \mathbf{\Omega}') = \int_0^L \Sigma_t(\mathbf{r} - l\mathbf{\Omega}', E') \mathrm{d}l. \quad (3)$$

The outcome of collisions is described by the collision operator

$$\begin{aligned}\widehat{C} \cdot \psi &= \widehat{C}(\mathbf{r}, E' \to E, \mathbf{\Omega}' \to \mathbf{\Omega}) \cdot \psi(\mathbf{r}, E', \mathbf{\Omega}') \\ &= \int_0^\infty \int_{4\pi} P_s(\mathbf{r}, E') \sum_A P_A(\mathbf{r}, E') \sum_j P_{j,A}(\mathbf{r}, E') \\ &\quad \cdot f_{j,A}(E' \to E, \mathbf{\Omega}' \to \mathbf{\Omega}) \psi(\mathbf{r}, E', \mathbf{\Omega}') \mathrm{d}E' \mathrm{d}\Omega'. \quad (4)\end{aligned}$$

The collision operator has intentionally been expended here into various factors, corresponding with steps or decisions taken in Monte Carlo sampling of the Boltzmann equation. The integral is over all possible incident photon energies and incident directions of travel to the collision site (note $\mathrm{d}\Omega'$ is a solid angle).

The first factor is the probability that a scattering rather than an absorption interaction occurs:

$$P_s(\mathbf{r}, E') = \frac{\Sigma_s(\mathbf{r}, E')}{\Sigma_t(\mathbf{r}, E')}. \quad (5)$$

Here $\Sigma_s(\mathbf{r}, E')$ [$\mathrm{cm^{-1}}$] is the total macroscopic cross section for all scattering interactions. Monte Carlo codes with implicit treatment of the photoelectric effect, i.e. photoelectric interactions are modelled as pure absorption processes into $E = 0$ eV, adjust the photon weight by $P_s(\mathbf{r}, E')$ upon each collision. Note PHITS offers more advanced treatments of the photoelectric effect in forward mode, including simulation of luminescence photons and ejected electrons [8], but the implicit treatment is outlined here as that is what is used in the adjoint mode.

The second factor in Eq. (4) is the probability that the scattering interaction occurs with an atom of type A amongst the various nuclides in the material:

$$P_A(\mathbf{r}, E') = \frac{\Sigma_{s,A}(\mathbf{r}, E')}{\Sigma_s(\mathbf{r}, E')}. \quad (6)$$

The sum of numerator over all atom types in the material is the denominator, i.e. $\Sigma_s(\mathbf{r}, E') = \sum_A \Sigma_{s,A}(\mathbf{r}, E')$.

The third factor is the probability for a type j scattering interaction to occur upon collision with the type A atom:

$$P_{j,A}(\mathbf{r}, E') = \frac{\Sigma_{j,A}(\mathbf{r}, E')}{\Sigma_{s,A}(\mathbf{r}, E')}. \quad (7)$$

Again the sum of the various possible numerators in this equation is the denominator, $\Sigma_{s,A}(\mathbf{r}, E') = \sum_j \Sigma_{j,A}(\mathbf{r}, E')$.

The macroscopic interaction cross sections are related to microscopic interaction cross sections, $\sigma_{j,A}(E')$ [b], by

$$\Sigma_{j,A}(\mathbf{r}, E') = n_A(\mathbf{r}) \sigma_{j,A}(E') \times 10^{-24}. \quad (8)$$

Here $n_A(\mathbf{r})$ [$\mathrm{cm^{-3}}$] is the number density of atoms of type A in the material at $\mathbf{r}$. The factor of 10^{-24} ensures $\Sigma_{j,A}(\mathbf{r}, E')$ has units of $\mathrm{cm^{-1}}$.

The final factor of the collision operator (Eq. (4)) is the normalized probability density function (pdf) for the possible outcomes of the scattering interaction:

$$f_{j,A}(E' \to E, \mathbf{\Omega}' \to \mathbf{\Omega}) = \frac{\mathrm{d}^2 \sigma_{j,A}(E')}{\mathrm{d}E \mathrm{d}\Omega} \cdot \frac{1}{\sigma_{j,A}(E')}. \quad (9)$$

The first factor on the right hand side of this equation is the differential microscopic cross section.

Monte Carlo codes sample Eq. (1) by simulating photon histories. A history is spawned by sampling the source term S. The photon is transported through the geometry to a collision site by sampling $\widehat{T}$. The weight of the photon history is multiplied by the factor P_s prior to sampling the collision itself. The collision nuclide and the type of interaction are selected according to P_A and $P_{j,A}$, respectively. The outcome of the interaction is determined by sampling $f_{j,A}$. The process of transport followed by collision repeats until either the photon leaves the simulation geometry, or its energy or weight fall outside the bounds of interest, upon which the history is terminated.

The result of a Monte Carlo simulation is typically calculated using a track length tally for the photon flux combined with a chosen response function. This corresponds to solving the integral:

$$R = \int \ldots \int \eta_\phi(\mathbf{r}, E', \mathbf{\Omega}') \phi(\mathbf{r}, E', \mathbf{\Omega}') \mathrm{d}^3 r \mathrm{d}E' \mathrm{d}\Omega'. \quad (10)$$

Here $\eta_\phi(\mathbf{r}, E', \mathbf{\Omega}')$ is the response function of choice, and $\phi(\mathbf{r}, E', \mathbf{\Omega}')$ [cm^{-2} eV^{-1} sr^{-1}] is the photon flux, related to the collision density by

$$\phi(\mathbf{r}, E', \mathbf{\Omega}') = \frac{\psi(\mathbf{r}, E', \mathbf{\Omega}')}{\Sigma_t(\mathbf{r}, E')}. \quad (11)$$

3 Adjoint Monte Carlo formalism

The PHITS adjoint function follows the formalism outlined by Hoogenboom [3, 4] and Gabler et al. [5]. The adjoint Boltzmann transport equation is:

$$\zeta^+(\mathbf{r}, E, \mathbf{\Omega}) = \eta_\phi(\mathbf{r}, E, \mathbf{\Omega}) + \widetilde{C} \cdot \widetilde{T} \cdot \zeta^+(\mathbf{r}', E', \mathbf{\Omega}'), \quad (12)$$

where superscript pluses and diacritic tildes denote adjoint quantities and operators, respectively. The first quantity, $\zeta^+(\mathbf{r}, E, \mathbf{\Omega})$, is the adjoint analogue of the forward emission density, i.e. the density of pseudo-photons emitted at $\mathbf{r}$. The operators $\widetilde{T}$ and $\widetilde{C}$ are analogues of the forward transport and collision operators, respectively.

Eq. (12) is sampled with the Monte Carlo method by simulating pseudo-photon histories. A pseudo-photon history is spawned by sampling an initial position, energy and direction from the 'adjoint source term', η_ϕ. This is in fact the tally response function for the corresponding forward transport problem. For a typical response function, such as a flux calculation, a pseudo-photon is spawned at a random location within the forward tally volume. The initial energy is chosen randomly from within the energy range of interest at the forward tally, and the initial direction of travel is also random.

The adjoint transport operator is basically the same as the forward transport operator:

$$\begin{aligned}\xi^+(\mathbf{r}, E', \mathbf{\Omega}') &= \widetilde{T}(\mathbf{r}' \to \mathbf{r}, E', \mathbf{\Omega}') \cdot \zeta^+(\mathbf{r}', E', \mathbf{\Omega}') \\ &= \widetilde{T}(\mathbf{r}' \to \mathbf{r}, E', -\mathbf{\Omega}') \cdot \zeta^+(\mathbf{r}', E', \mathbf{\Omega}') \\ &= \int_0^\infty \Sigma_t(\mathbf{r}, E') e^{-\beta(\mathbf{r}, L, E', -\mathbf{\Omega}')} \zeta^+(\mathbf{r}', E', \mathbf{\Omega}') \mathrm{d}L.\end{aligned} \quad (13)$$

The result of the transport operation is the pseudo-photon collision density, $\xi^+(\mathbf{r}, E', \mathbf{\Omega}')$. A pseudo-photon is transported to a collision site by randomly selecting a path length through the transport media from the exponential distribution in Eq. (13).

Although similar in form, the adjoint collision operator is different from the forward collision operator in a number of respects:

$$\begin{aligned}\widetilde{C} \cdot \xi^+ &= \widetilde{C}(\mathbf{r}, E' \to E, \mathbf{\Omega}' \to \mathbf{\Omega}) \cdot \xi^+(\mathbf{r}, E', \mathbf{\Omega}') \\ &= \int_0^\infty \int_{4\pi} P_s^+(\mathbf{r}, E') \sum_A P_A^+(\mathbf{r}, E') \sum_j P_{j,A}^+(\mathbf{r}, E') \\ &\quad \cdot f_{j,A}^+(E' \to E, \mathbf{\Omega}' \to \mathbf{\Omega}) \xi^+(\mathbf{r}, E', \mathbf{\Omega}') \mathrm{d}E' \mathrm{d}\Omega'.\end{aligned} \quad (14)$$

Each of the factors in this integral are described in turn.

The first factor, $P_s^+(\mathbf{r}, E')$, is the adjoint analogue of the forward scattering probability (P_s):

$$P_s^+(\mathbf{r}, E') = \frac{\Sigma_s^+(\mathbf{r}, E')}{\Sigma_t(\mathbf{r}, E')}. \quad (15)$$

The denominator here is the same as the forward case, i.e. the total macroscopic cross section for both absorption and scattering (Eq. (5)). However the numerator, $\Sigma_s^+(\mathbf{r}, E')$ [cm^{-1}], is a new quantity. It is the total adjoint macroscopic cross section for the scattering interactions. Note P_s^+ is not a probability per se, as it can take values greater than 1. The weight of a pseudo-photon is multiplied by $P_s^+(\mathbf{r}, E')$ upon each collision.

The second factor in Eq. (14) is the probability that the adjoint scattering interaction occurs with an atom of type A in the medium:

$$P_A^+(\mathbf{r}, E') = \frac{\Sigma_{s,A}^+(\mathbf{r}, E')}{\Sigma_s^+(\mathbf{r}, E')}. \quad (16)$$

This is a true probability as the sum of the possible numerators totals the denominator: $\Sigma_s^+(\mathbf{r}, E') = \sum_A \Sigma_{s,A}^+(\mathbf{r}, E')$.

The third factor is the probability for interaction type j to occur amongst the possible scattering interactions:

$$P_{j,A}^+(\mathbf{r}, E') = \frac{\Sigma_{j,A}^+(\mathbf{r}, E')}{\Sigma_{s,A}^+(\mathbf{r}, E')}. \quad (17)$$

Again the sum of the possible numerators, i.e. partial adjoint macroscopic cross sections for scattering, is the denominator: $\Sigma_{s,A}^+(\mathbf{r}, E') = \sum_j \Sigma_{j,A}^+(\mathbf{r}, E')$.

As for the forward case, the adjoint macroscopic cross sections are related to adjoint microscopic cross sections by the number density of atoms in the collision material:

$$\Sigma_{j,A}^+(\mathbf{r}, E') = n_A(\mathbf{r}) \sigma_{j,A}^+(E') \times 10^{-24}. \quad (18)$$

The adjoint microscopic cross sections, $\sigma_{j,A}^+(E')$, also have units of barns. They are calculated by integration of the differential cross section for the corresponding forward interaction. The integration range covers all incident photon energies that could yield an outgoing photon with energy equal to the colliding pseudo-photon:

$$\sigma_{j,A}^+(E') = \int_0^\infty \frac{\mathrm{d}\sigma_{j,A}(E)}{\mathrm{d}E'} \mathrm{d}E. \quad (19)$$

The final factor in Eq. (14) is the probability density function for the possible outcomes of adjoint scattering:

$$f_{j,A}^+(E' \to E, \mathbf{\Omega}' \to \mathbf{\Omega}) = \frac{\mathrm{d}^2\sigma_{j,A}^+(E')}{\mathrm{d}E\mathrm{d}\Omega} \cdot \frac{1}{\sigma_{j,A}^+(E')}. \quad (20)$$

The purpose of the adjoint simulation is to calculate the tally response of the corresponding transport forward problem. The response is given by:

$$\begin{aligned}R &= \int \ldots \int S(\mathbf{r}, E', \mathbf{\Omega}') \frac{\xi^+(\mathbf{r}, E', \mathbf{\Omega}')}{\Sigma_s^+(\mathbf{r}, E')} \mathrm{d}^3r \mathrm{d}E' \mathrm{d}\Omega' \\ &= \int \ldots \int S(\mathbf{r}, E', \mathbf{\Omega}') \phi^+(\mathbf{r}, E', \mathbf{\Omega}') \mathrm{d}^3r \mathrm{d}E' \mathrm{d}\Omega'.\end{aligned} \quad (21)$$

Here $\phi^+(\mathbf{r}, E', \mathbf{\Omega}') = \xi^+(\mathbf{r}, E', \mathbf{\Omega}')/\Sigma_s^+(\mathbf{r}, E')$ is the pseudo-photon flux.

4 Interactions

The PHITS adjoint mode currently supports coherent and incoherent scattering, and implicit photoelectric absorption. This section outlines calculation of the adjoint microscopic cross sections, and the Monte Carlo sampling strategies for the outcome of adjoint collisions.

4.1 Coherent scattering

4.1.1 Forward

Coherent scattering between a photon and an atom leads to a change in the direction of travel of the photon, but, to very good approximation, the energy of the photon is preserved. The differential cross section for the interaction is [9]:

$$\frac{\mathrm{d}\sigma_{\mathrm{coh}}(E')}{\mathrm{d}\mu} = \pi r_{\mathrm{e}}^2(1+\mu^2)F^2(Z,x') \times 10^{28}, \tag{22}$$

where $r_{\mathrm{e}} = 2.817 \times 10^{-15}$ m is the classical electron radius, and $\mu = \mathbf{\Omega}' \cdot \mathbf{\Omega}$ is the cosine of the scattering angle.

Eq. (22) consists of two parts: Thomson's cross section for scattering off a free electron, and the form factor correction to this, $F(Z,x')$, which accounts for atomic binding effects. Z is the atomic number of the collision atom, and x' [Å^{-1}] is the inverse wavelength of the recoiling atom:

$$x' = \frac{E'e(1-\mu)^{0.5}}{\sqrt{2}hc \times 10^{10}}. \tag{23}$$

Here $e = 1.602 \times 10^{-19}$ J eV^{-1}, $h = 6.626 \times 10^{-24}$ J s is Planck's constant, and $c = 2.998 \times 10^8$ m s^{-1} is the speed of light.

In a PHITS forward mode calculation, coherent scattering interactions are simulated as follows. The photon cross section library (based on JENDL-4.0 [10]) contains tabulated point data for the microscopic cross sections, $\sigma_{\mathrm{coh}}(E')$, calculated by integrating Eq. (22) numerically. The cross sections are stored against logarithmic values of the incoming energy E'. The library also contains tables for the form factor at point values of x', and for the integral of the form factor at point values of x'^2.

Upon a coherent scattering interaction, the scattering angle cosine is sampled from $f_{\mathrm{coh}}(E',\mu)$ using the rejection technique method from Carter and Cashwell [11]. A scattering azimuth is selected randomly from $[0, 2\pi]$.

4.1.2 Adjoint

The adjoint microscopic cross section for pseudo-photon coherent scattering from $(E',\mathbf{\Omega}')$ to $(E,\mathbf{\Omega})$ is:

$$\begin{aligned}\sigma^+_{\mathrm{coh}}(E') &= \int_0^\infty \frac{\mathrm{d}\sigma_{\mathrm{coh}}(E)}{\mathrm{d}E'}\mathrm{d}E \\ &= \int_{-1}^1 \frac{\mathrm{d}\sigma_{\mathrm{coh}}(E)}{\mathrm{d}\mu'}\frac{\mathrm{d}\mu'}{\mathrm{d}E'}\frac{\mathrm{d}E}{\mathrm{d}\mu}\mathrm{d}\mu \\ &= \int_{-1}^1 \frac{\mathrm{d}\sigma_{\mathrm{coh}}(E)}{\mathrm{d}\mu'}\mathrm{d}\mu. \end{aligned} \tag{24}$$

The steps above use the facts that $E' = E$, $\mu' = \mu$ and $x' = x$ for this interaction. The adjoint microscopic cross sections are thus identical to the forward microscopic cross sections at all energies, i.e. $\sigma^+_{\mathrm{coh}}(E') = \sigma_{\mathrm{coh}}(E')$. Likewise the probability density function for the scattering cosine is the same as the forward case:

$$\begin{aligned} f^+_{\mathrm{coh}}(E',\mu) &= \frac{\mathrm{d}\sigma^+_{\mathrm{coh}}(E')}{\mathrm{d}\mu}\cdot\frac{1}{\sigma^+_{\mathrm{coh}}(E')} \\ &= \frac{\mathrm{d}\sigma_{\mathrm{coh}}(E)}{\mathrm{d}\mu'}\cdot\frac{1}{\sigma_{\mathrm{coh}}(E)} \\ &= f_{\mathrm{coh}}(E,\mu'). \end{aligned} \tag{25}$$

PHITS thus samples a pseudo-photon coherent scattering event identically to a forward photon coherent scattering event. The code employs the same cross sections and rejection technique for the pdf as per the forward case.

4.2 Incoherent scattering

4.2.1 Forward

Incoherent scattering occurs between a photon and a quasi-free atomic electron. There is a correlated change in the energy and the direction of propagation of the photon after the interaction. The outgoing photon energy is:

$$E = \frac{E'}{1+k'(1-\mu)}, \tag{26}$$

where $k' = E'/E_{\mathrm{e}}$ is the incident photon energy relative to the rest energy of an electron, $E_{\mathrm{e}} = 0.511$ MeV. The differential cross section is [9]:

$$\frac{\mathrm{d}\sigma_{\mathrm{incoh}}(E')}{\mathrm{d}\mu} = \pi r_{\mathrm{e}}^2\left(\frac{E}{E'}\right)^2\left(\frac{E}{E'}+\frac{E'}{E}-1+\mu^2\right)S(Z,x') \times 10^{28}. \tag{27}$$

The factor $S(Z,x')$ is the incoherent scattering function, which corrects the Klein-Nishina cross section for atomic binding effects. The second argument is the wavelength of the recoiling target:

$$x' = \frac{E'e(1-\mu)^{0.5}(1+(k'^2+2k')(1-\mu))^{0.5}}{\sqrt{2}hc(1+2k'(1-\mu)) \times 10^{10}}. \tag{28}$$

For a forward incoherent scattering event, the PHITS cross section library contains tables for $\sigma_{\mathrm{incoh}}(E')$ against $\ln(E')$, obtained by numerical integration of Eq. (27). The library also contains $S(Z,x')$ values against x'.

To sample the pdf for the scattering cosine, a candidate is drawn in accordance to the Klein-Nishina component of Eq. (27) first. Kahn's rejection method is employed for photons below 1.5 MeV [12] and Koblinger's method for $E' > 1.5$ MeV [13], following Blomquist and Gelbard [14]. The candidate is then accepted or rejected against the incoherent scattering function [11]. Note for the calculation of $S(Z,x')$, Monte Carlo codes (e.g. MCNP [15]) commonly approximate x' using Eq. (23) instead of Eq. (28).

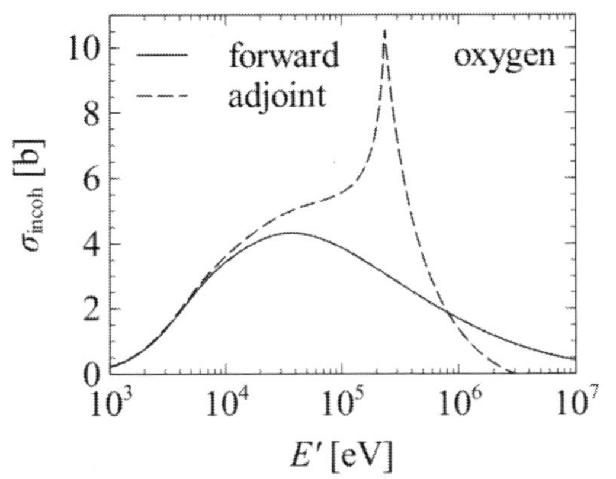

Figure 1. Forward and adjoint incoherent scattering cross sections for oxygen.

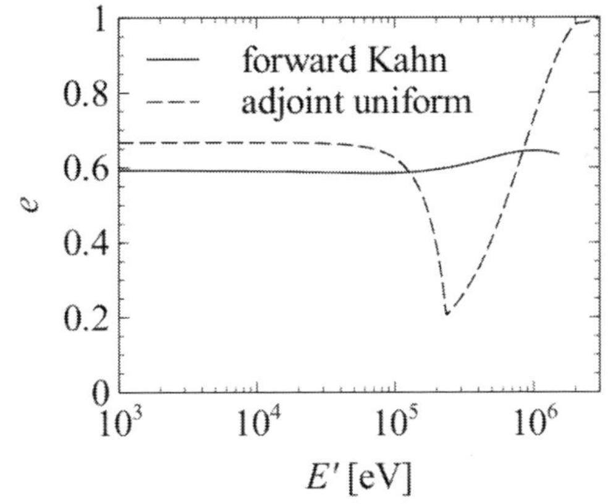

Figure 2. Acceptance efficiency of rejection methods for sampling the forward and adjoint Klein-Nishina distributions.

4.2.2 Adjoint

In adjoint mode, pseudo-photons gain (or preserve) energy upon incoherent scattering. The outgoing pseudo-photon energy is:

$$E = \frac{E'}{1 - k'(1-\mu)}. \tag{29}$$

The adjoint microscopic cross section is given by integrating numerically:

$$\begin{aligned}\sigma^+_{\text{incoh}}(E') &= \int_0^{E_{\max}} \frac{d\sigma_{\text{incoh}}(E)}{dE'} dE \\ &= \int_{\mu_{\min}}^1 \frac{d\sigma_{\text{incoh}}(E)}{d\mu'} \frac{d\mu'}{dE'} \frac{dE}{d\mu} d\mu \\ &= \int_{\mu_{\min}}^1 \pi r_e^2 \Big(\frac{E'}{E} + \frac{E}{E'} - 1 + \mu^2\Big) S(Z,x) \times 10^{28} d\mu \\ &= \int_{\mu_{\min}}^1 \pi r_e^2 \Big(\frac{1}{1-k'(1-\mu)} - k'(1-\mu) + \mu^2\Big) \\ &\quad \cdot S(Z,x) \times 10^{28} d\mu. \end{aligned} \tag{30}$$

During numerical integration, the scattered pseudo-photon energy E must be calculated using μ via Eq. (29) in order to evaluate x. It is necessary to set a minimum scattering cosine for when $E' \geq E_{\max}/(1 + 2E_{\max}/E_e)$, where $E_{\max}$ is the maximum energy of interest in adjoint photon calculations:

$$\mu_{\min} = \begin{cases} -1 & E' \leq E_{\max}/(1 + 2E_{\max}/E_e), \\ 1 - 1/k' + E_e/E_{\max} & \text{otherwise.} \end{cases} \tag{31}$$

We chose $E_{\max} = 3.0$ MeV for the PHITS adjoint mode, as this range covers the main radioactive decay photons. Without setting $E_{\max}$, the integral in Eq. (30) would diverge for $E' \geq E_e/2$.

An example of the adjoint microscopic incoherent scattering cross section is given in Fig. 1 for scattering off an oxygen atom. The adjoint cross section increases from low energies to peak at $E_{\max}/(1 + 2E_{\max}/E_e)$. At energies above this, the cross section decreases because the pseudo-photon scattering pathways to energies above $E_{\max}$ are truncated.

The shape of the adjoint cross section curve and the position of the peak in Fig. 1 necessarily depend on the value chosen for $E_{\max}$. As an example, the curves for germanium under two different choices of $E_{\max}$ are given in ref. [5].

The probability density function for the outcome of pseudo-photon incoherent scattering is:

$$\begin{aligned} f^+_{\text{incoh}}(E',\mu) &= \frac{d\sigma^+_{\text{incoh}}(E')}{d\mu} \cdot \frac{1}{\sigma^+_{\text{incoh}}(E')} \\ &= \pi r_e^2 \Big(\frac{1}{1-k'(1-\mu)} - k'(1-\mu) + \mu^2\Big) \\ &\quad \cdot \frac{S(Z,x) \times 10^{28}}{\sigma^+_{\text{incoh}}(E')}. \end{aligned} \tag{32}$$

The 'adjoint Klein-Nishina component' of this pdf can be sampled using the following rejection method. First a candidate μ is selected uniformly in $[\mu_{\min}, 1]$. This is accepted if:

$$\frac{1/(1-k'(1-\mu)) - k'(1-\mu) + \mu^2}{m} > r, \tag{33}$$

where r is a uniform random deviate in [0, 1] and m is the maximum value of the numerator in $[\mu_{\min}, 1]$:

$$m = \begin{cases} \frac{1}{1-k'(1-\mu_{\min})} - k'(1-\mu_{\min}) + \mu_{\min}^2 & E' \leq 2.020 \text{ MeV}, \\ 2 & \text{otherwise.} \end{cases} \tag{34}$$

Note the formula for m is conditional on the incoming pseudo-photon energy E'. This is because the maximum of the adjoint Klein-Nishina distribution switches from occurring at $\mu = \mu_{\min}$ to $\mu = 1$ around a crossover energy at $E' = 2.020$ MeV (for $E_{\max} = 3.0$ MeV). The crossover energy can be established for $E_{\max} \neq 3.0$ MeV by numerically solving

$$1/(1-k'(1-\mu_{\min})) - k'(1-\mu_{\min}) + \mu_{\min}^2 = 2. \tag{35}$$

Table 1. Data blocks in PHITS adjoint ACE cross section files. The block length N is defined in the ACE file header. Note: zero barn cross sections are stored as 0.0 in the tables.

#	Data	Unit	Length
1	Logarithms of energies for forward cross sections	ln(MeV)	N
2	Logarithms of forward incoherent cross sections	ln(b)	N
3	Logarithms of forward coherent cross sections	ln(b)	N
4	Logarithms of forward photoelectric cross sections	ln(b)	N
5	Logarithms of forward pair production cross sections	ln(b)	N
6	Incoherent scattering functions	-	21
7	Integrated coherent form factors	-	55
8	Coherent form factors	-	55
9	Average heating numbers	MeV per collision	N
10	Logarithms of energies for adjoint cross sections	ln(MeV)	$N+1$
11	Logarithms of adjoint incoherent cross sections	ln(b)	$N+1$

The acceptance efficiency of the adjoint Klein-Nishina distribution rejection routine per candidate scattering cosine is:

$$e = \frac{\int_{\mu_{\min}}^{1}(1/(1-k'(1-\mu)) - k'(1-\mu) + \mu^2)\mathrm{d}\mu}{m(1-\mu_{\min})} = \frac{\frac{1}{3} - \frac{k'}{2} - \frac{\mu_{\min}^3}{3} - \frac{k'\mu_{\min}^2}{2} + k'\mu_{\min} - \frac{\ln(1-k'(1-\mu_{\min}))}{k'}}{m(1-\mu_{\min})}. \quad (36)$$

The variation of e with the incident pseudo-photon energy E' is shown in Fig. 2. In forward mode, Kahn's method has a fairly flat acceptance efficiency of ≈ 60 % between 1 keV and 1.5 MeV. The rejection method developed for the adjoint mode generally has good acceptance efficiency, however there is a dip in efficiency between 150 and 650 keV.

Each candidate μ sampled from the adjoint Klein-Nishina distribution must then be accepted or rejected in line with the incoherent scattering function. Again μ must be used to evaluate x (approximated with Eq. (23), as per the forward case) in order to rejection sample $S(Z, x)$.

4.3 Photoelectric effect

PHITS adjoint mode implements an implicit treatment of the photoelectric effect. This marks a departure from forward mode, which offers more detailed physics simulation for the relaxations associated with photoelectric interactions.

In adjoint mode, the weight of a pseudo-photon is multiplied by P_s^+ upon each collision, then either a coherent or incoherent scattering event is forced. As it is possible for P_s^+ to be greater than one, pseudo-photon weights can become large and cause high variances in calculation results [5]. This risk should be monitored in adjoint calculation results.

5 Adjoint Cross Section Library

The original PHITS photon cross section library, which is based on JENDL-4.0 [10], was extended to facilitate the adjoint transport function. The adjoint specific library files are encoded in the ACE format. There is one file for each element, with filenames `XY000.j90p`. Characters `X` and `Y` are substituted with the element symbol, e.g. sodium is `Na000.j90p`. The second character is substituted with an underscore for elements with single character symbols, e.g. `O_000.j90p` for oxygen.

The first nine blocks of data in the new adjoint library files are identical to the original PHITS photon cross section files (Table 1). The blocks contain energies and microscopic cross sections for forward incoherent, coherent, photoelectric and pair production interactions, all stored logarithmically. They are followed by blocks containing incoherent scattering functions, integrated coherent form factors, coherent form factors, and average heating numbers.

Two new blocks follow to facilitate adjoint incoherent scattering. Block 10 contains the logarithms of energies as per block 1, but with an additional data point added at $E_{\max}/(1+2E_{\max}/E_e)$. The maximum in the microscopic adjoint incoherent scattering cross section occurs at this energy (Fig. 1). Block 11 contains logarithms of the microscopic cross sections for adjoint incoherent scattering, obtained by numerical integration of Eq. (30). PHITS interpolates between the adjoint incoherent scattering cross sections using log-log interpolation, as per the forward case.

Note separate data blocks are unnecessary for adjoint coherent scattering or photoelectric effect cross sections. The adjoint coherent scattering cross sections are equal to the forward cross section (section 4.1.2). As the adjoint photoelectric treatment is implicit, adjoint cross sections are not needed for this interaction. However, the forward photoelectric cross sections need to be read in to calculate $\Sigma_t(\mathbf{r}, E')$ for both adjoint transport and collision operations.

6 Pseudo-photon transport in PHITS

An adjoint simulation is called by the `iadjoint = 1` flag in the `[Parameters]` section of a PHITS input file. This instructs the code to simulate the pseudo-photon particle type. The user defines a source region in the input file that coincides geometrically to the tally region in the corresponding forward transport problem. Pseudo-photons are generated within the source with energies uniformly

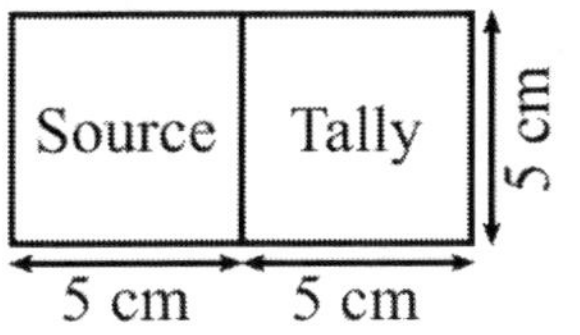

Figure 3. Geometry of the flux spectrum calculation.

between the minimum energy of interest for the forward tally and the highest photon energy emitted by the forward radiation source.

Each pseudo-photon is tracked through the geometry to a collision site, upon which its weight is adjusted by P_s^+ to account for non-absorption by the photoelectric effect. Then either a coherent or incoherent scattering interaction is forced. Coherent scattering is sampled identically to a forward mode calculation (section 4.1.1). A scattering cosine is sampled first from the Thomson distribution and then rejected against the coherent scattering form factors. A scattering azimuth is selected uniformly in $[0, 2\pi]$.

Upon incoherent scattering, the pseudo-photon scattering cosine is sampled according to the adjoint Klein-Nishina distribution and rejected against the incoherent scattering function (section 4.2.2). A scattering azimuth is selected uniformly in $[0, 2\pi]$.

7 PHITS adjoint tally

To obtain the calculation result the user defines the adjoint tally type in the input file (`[T-Adjoint]`). This tally type is the parallel of the forward track length tally, `[T-Track]` [8].

The geometry chosen for the `[T-Adjoint]` tally (`[mesh]` input) must correspond to the source geometry in the forward problem. The energy mesh of this tally (`[e-type]` input) should be the same as for the forward case. Finally the user defines the energy distribution of the photon source in the forward problem via input flags at the end of `[T-Adjoint]`.

PHITS tallies a pseudo-photon if it crosses the adjoint tally region with energy within the photon source energy range. The tally calculates the contribution of a pseudo-photon history to the calculation result using a track length estimator: WT_l/V. Here W is the pseudo-photon weight, T_l [cm] is the track length through the tally region, and V [cm^3] is the tally volume. Importantly, in a departure from the usual forward case, it is the energy of the pseudo-photon history upon generation in the adjoint source region that determines the energy bin that the history is tallied within for the `[T-Adjoint]` tally. This process recovers the calculation result of the corresponding forward problem (i.e. Eq. (21) is solved by `[T-Adjoint]`).

8 Examples

We present two example calculations to demonstrate the PHITS adjoint function. The first is a calculation for a

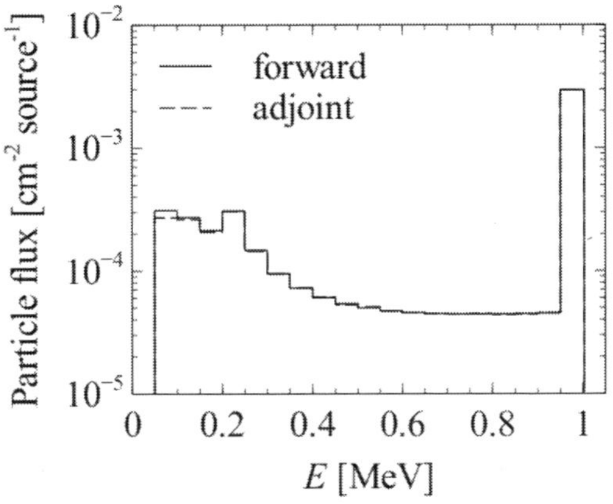

Figure 4. Photon flux spectrum calculated with forward and adjoint Monte Carlo for the geometry in Fig. 3.

photon spectrum demonstrating the accuracy of the adjoint method. The second is an application to a simple large source, small tally transport problem, to gauge the potential performance of the adjoint method for this class of problem.

8.1 Spectrum calculation

This calculation was for gamma ray spectrum inside an infinite oxygen medium, density $\rho = 1.0\,\mathrm{g\,cm^{-3}}$. The source and tally volumes were adjacent cubes, volume $125\,\mathrm{cm}^3$ (Fig. 3). Photons were emitted from the source with uniform probability between 0.95 and 1.0 MeV. The photon flux was tallied in 50 keV wide energy bins.

The calculated spectra are shown in Fig. 4. The adjoint method gives a basically identical result to the forward method at higher energies, to within statistical error. There is however a discernible difference between the fluxes calculated in forward and adjoint mode for the lowest energy bins (0.05–0.1 MeV). This is an artifact of the implicit treatment of the photoelectric effect in adjoint mode, which is an important interaction at these low energies. The forward mode calculation includes relaxation photons from photoelectric interactions, leading to the higher flux results in the lower energy bins.

8.2 Large source, small detector geometry

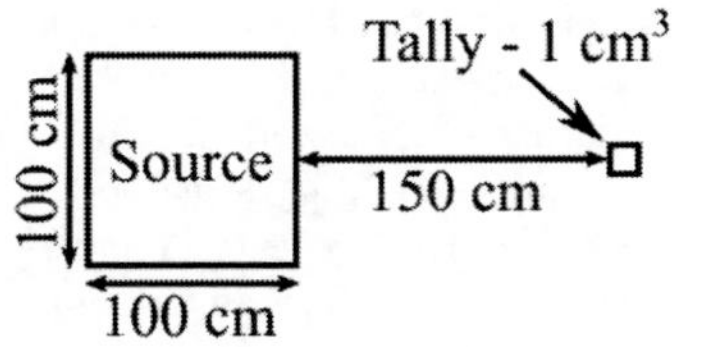

Figure 5. Geometry of simple large source, small detector test problem.

Table 2. Tally results, relative errors and CPU run times for flux calculation for the large source, small detector test problem. Forward mode results were calculated with the conventional `[T-Track]` tally (no acceleration), and with the point tally biasing method (`[T-Point]`). Adjoint mode results are in the `[T-Adjoint]` column.

	[T-Track]	[T-Point]	[T-Adjoint]
ϕ [$\times 10^{-6}$ cm^{-2} source^{-1}]	2.3603	2.4188	2.4056
Relative err. [%]	2.24	0.38	0.10
CPU time [min]	436	7.5	7.5

The second test problem consisted of a photon source that was relatively much larger than the tally (Fig. 5). The source and tally were within an infinite scattering medium consisting of oxygen, density $\rho = 0.01$ g cm^{-3}. A 1.0 m^3 source emitted photons uniformly in the energy range 0.05–1.0 MeV. A 1.0 cm^3 tally was located on the centre line from one of the source faces. The separation between the source and tally was 150 cm. The calculation result was the total flux over the 0.05–1.0 MeV energy range at the tally.

The calculation was performed in adjoint mode for 7.5 min run time on a 3.4 GHz four-core desktop computer. For comparison with an accelerated (biased) forward method, a calculation was executed with the PHITS point tally (`[T-Point]` [8]) for the same CPU run time. Finally the calculation was run in unbiased forward mode with the `[T-Track]` tally for 436 min, i.e. 58 times longer than the point and adjoint calculations.

The results of the fluxes and tally relative errors for the three calculation modes are given in Table 2. The relative error for the unbiased forward tally was an order of magnitude higher than for the point and adjoint calculation methods. This is despite the much longer run time and larger number of histories simulated. The adjoint and point tallies gave comparable flux results and calculation relative errors.

9 Conclusion and future developments

A new adjoint function has been added to PHITS for photon transport problems with continuous energy representation. Currently the function can treat coherent and incoherent scattering, and implicit photoelectric absorption. The function is based on three developments to PHITS: a new adjoint cross section library, a pseudo-photon simulation mode, and an adjoint tally function.

The new mode is a candidate for solving transport problems with small detector sizes relative to the source. In a simple test case for this class of transport problem, the adjoint mode offered orders of magnitude speed-up versus an unbiased forward calculation. The adjoint performance was comparable to a biased forward calculation using the point tally. A benefit of the adjoint method over the common point and ring tally forward biasing methods is it allows the tally volume to be any shape.

A number code developments are planned to improve the functionality of the PHITS adjoint mode. The developments are based on the point energy function [5], where pseudo-photons are forced to scatter to discrete energies. This will allow the incorporation of pair production interactions [4]. Pseudo-photon histories will be scattered to E_e to enable this interaction to occur.

The point energy function will also allow calculations for photon sources with discrete energies (e.g. line sources, such as radioactive decay gamma rays) [5]. Currently such calculations are not possible in the PHITS adjoint mode, as pseudo-photon histories are unlikely to cross the adjoint tally region with the requisite discrete energies to contribute to the calculation result. With a point energy function, pseudo-photons histories will be split off and forcibly scattered to the discrete energies after each collision. These sub-histories will then contribute to the calculation result if they cross the adjoint tally region.

References

[1] J.E. Hoogenboom, Ph.D. thesis, Delft University of Technology (1977)
[2] A. Malins, M. Okumura, M. Machida, H. Takemiya, K. Saito, *Fields of View for Environmental Radioactivity*, in *Proceeding of the International Symposium on Radiological Issues for Fukushima's Revitalized Future* (2015), pp. 28–34
[3] J.E. Hoogenboom, *Adjoint Monte Carlo Photon Transport in Continuous Energy Mode with Discrete Photons from Annihilation*, in *Proceedings of the PHYSOR 2000* (2000)
[4] J.E. Hoogenboom, Nucl. Sci. Eng. **143**, 99 (2003)
[5] D. Gabler, J. Henniger, U. Reichelt, Nucl. Instrum. Methods Phys. Res. Sect. B **251**, 326 (2006)
[6] L. Desorgher, F. Lei, G. Santin, Nucl. Instrum. Methods Phys. Res. Sect. A **621**, 247 (2010)
[7] T. Sato, K. Niita, N. Matsuda, S. Hashimoto, Y. Iwamoto, S. Noda, T. Ogawa, H. Iwase, H. Nakashima, T. Fukahori et al., J. Nucl. Sci. Technol. **50**, 913 (2013)
[8] K. Niita, N. Matsuda, S. Hashimoto, Y. Iwamoto, T. Sato, T. Furuta, S. Noda, T. Ogawa, H. Nakashima, T. Fukahori et al., *PHITS Ver. 2.88 User's Manual* (2016)
[9] J.H. Hubbell, H.A. Gimm, I. Øverbø, J. Phys. Chem. Ref. Data **9**, 1023 (1980)
[10] K. Shibata, O. Iwamoto, T. Nakagawa, N. Iwamoto, A. Ichihara, S. Kunieda, S. Chiba, K. Furutaka, N. Otuka, T. Ohasawa et al., J. Nucl. Sci. Technol. **48**, 1 (2011)

[11] L.L. Carter, E.D. Cashwell, *TID-26607: Particle-transport simulation with the Monte Carlo method* (1975)
[12] H. Kahn, *RM-1237-AEC: Applications of Monte Carlo* (1956)
[13] L. Koblinger, Nucl. Sci. Eng. **57**, 218 (1971)
[14] R.N. Blomquist, E.M. Gelbard, Nucl. Sci. Eng. **83**, 380 (1983)
[15] X-5 Monte Carlo Team, *MCNP - A General Monte Carlo N-Particle Transport Code, Version 5 Volume I: Overview and Theory*, Los Alamos National Laboratory (2008)

 DOI: 10.1051/epjconf/201715306002

Progress on the TRIPOLI-4®-Geant4 coupling

Davide Mancusi[1,*], Olivier Bringer[2], and Pascal Monot[3]

[1] *CEA, Centre de Saclay, DEN/DANS/DM2S/SERMA/LTSD, 91191 Gif-sur-Yvette CEDEX, France*
[2] *CEA, Centre de Saclay, DEN/DANS/DM2S/SERMA/LPEC, 91191 Gif-sur-Yvette CEDEX, France*
[3] *CEA, Centre de Saclay, DSM/IRAMIS/LIDYL/PHI, 91191 Gif-sur-Yvette CEDEX, France*

Abstract. The capability to simulate the transport of charged and/or high-energy hadrons (especially protons) is indispensable for a number of applications. This includes, among others, simulation studies concerned with radiation protection and decommissioning around accelerators and high-intensity laser facilities, as well as beam characterisation in spallation neutron sources. In the context of Monte-Carlo particle transport codes, solving these problems often requires the use of advanced variance-reduction techniques.
TRIPOLI-4® is a reference Monte-Carlo particle transport code for the simulation of low-energy ($\lesssim$ 20 MeV) neutrons and photons and offers a wide range of sophisticated variance-reduction schemes; however, it cannot be applied to the problems mentioned above because it lacks the capability to transport charged, high-energy hadrons.
This limitation can be circumvented by coupling TRIPOLI-4® with the Geant4 particle-transport toolkit. We present here the first results of this coupling.

1 Introduction

The TRIPOLI-4® Monte-Carlo particle-transport code [1] is capable of simulating the transport of neutrons, photons, electrons and positrons in matter. The code is accurately validated for neutron energies between thermal and approximately 20 MeV, which is the common upper limit on the availability of evaluated nuclear data. Electron and positron transport, on the other hand, is less reliant on the availability of nuclear data libraries. For this reason, the upper energy limit for TRIPOLI-4's electromagnetic cascade ($e^+/e^-/\gamma$) is of the order of 1 GeV [2], although systematic and extensive validation has been performed only up to about 10 MeV.

For some applications, the validity ranges indicated above are insufficient. For instance, radiation protection and commissioning/decommissioning of accelerator facilities may require the simulation of other particles (e.g. protons, deuterons, heavy ions...) and/or higher energies. This is the case, for instance, of proton accelerators devoted to cancer therapy, which typically employ proton energies of the order of 250 MeV [3]. Another example is given by facilities studying particle acceleration by high-intensity laser beams. At the time of writing, this promising technique can accelerate particles (mostly protons and electrons) up to energies of the order of 1 GeV on a micrometric length scale, which is orders of magnitude more efficient than any other existing acceleration technique. As far as radiation protection is concerned, particle energies below about 200 MeV are again often responsible for most of the dosimetry.

Therefore, it appears that a number of applications related to decommissioning and radiation protection require the ability to simulate the transport of electrons, positrons, photons, protons, neutrons and perhaps heavier ions up to some 250 MeV. This list can also be extended to include spallation neutron sources [4] and accelerator-driven systems (ADS) [5], both of which use spallation reactions induced by high-energy hadron beams (mostly protons) to abundantly produce energetic neutrons. The typical beam energies are slightly higher than the above and range from several hundred MeV to a few GeV. Finally, the ability to transport charged particles is also an essential stepping stone towards the simulation of many types of particle detectors, including (but not limited to) those involved in nuclear-reactor-core instrumentation. At present, the TRIPOLI-4 code is unable to meet the requirements specified by these applications.

One of the most noteworthy alternatives on the market of particle-transport codes is Geant4 [6–8], which was originally developed as a toolkit for the simulation of the response of LHC detectors and has been constantly extending its application field towards lower particle energies. An attractive aspect of Geant4 is that it offers a wide choice of physics models for the simulation of particle-matter interactions. Nowadays Geant4 is able to simulate transport of a very large number of particle types. It also includes a module for neutron transport using low-energy evaluated nuclear data libraries.

*e-mail: davide.mancusi@cea.fr

However, although a few variance-reduction schemes have recently been added to Geant4 [8], the toolkit remains mostly focused on analogue transport, which is an inefficient way of estimating e.g. the response of a heavily shielded detector or dosimeter. A large number of histories must be generated in order to obtain an accurate estimate of the desired response. TRIPOLI-4, on the other hand, offers a wide, well-established array of sophisticated variance-reduction solutions for neutron and photon transport that help mitigate this problem. Nevertheless, if one wishes to use TRIPOLI-4 and apply these techniques to the simulation of a Geant4 particle-detector model, one first needs to remodel the detector using TRIPOLI-4, which is inconvenient and error-prone.

In this work, we explore the possibility to couple TRIPOLI-4 to Geant4 in order to treat applications related to radiation protection, decommissioning around accelerators and high-intensity lasers, and spallation neutron sources. This coupled solution essentially allows users to benefit from "the best of both worlds". We briefly describe the implementation of the TRIPOLI-4/Geant4 coupling in Sec. 2. Section 3 presents the first verification results for coupled codes and serves as an illustration of the possible applications. Finally, Sec. 4 lays out the developments which are envisioned to upgrade the coupled code to production and collects our conclusions.

2 Description of the coupling

Geant4 is not a monolithic transport code, but rather a toolkit, i.e. a set of C++ shared libraries. Users are required to write their own C++ code to describe the problem geometry, the sources, the scores they are interested in, etc. From the developer's point of view, this situation is very practical. The toolkit approach offers a very large number of entry points for further development. This is an essential quality for the coupling because it frees developers from the history-loop straitjacket (source-transport-score-repeat).

The strategy of our development is to extend TRIPOLI-4 and provide new capabilities to TRIPOLI-4 users in a transparent way. For particles which are outside the native scope of TRIPOLI-4, the coupling should delegate the simulation task to Geant4. On the other hand, particles that are produced during the Geant4 transport phase but that can be handled by TRIPOLI-4 should be handed back to the latter.

In the framework of the standard Geant4 transport, one has the possibility to define *hooks* (*actions* in the Geant4 parlance), i.e. code snippets that are executed at specific stages of the transport simulation (e.g. before/after the loop, before/after each history, after each event, etc.). The TRIPOLI-4/Geant4 coupling relies heavily on actions for shuffling information back and forth between the two codes. One could say that the actions are actually the backbone of the coupling.

Additionally, the coupling requires carefully synchronised initialisation stages on both sides. This is because there are intra-code module dependencies (e.g. Geant4 primary generator actions must be instantiated after physics lists), as well as inter-code dependencies (e.g. we cannot set the Geant4 pseudo-random-number generator until we are finished with parsing the TRIPOLI-4 input file). In the following, we illustrate a few chosen technical aspects of the coupling.

2.1 Pseudo-random number generation

For the sake of reproducibility, it is important that Geant4 and TRIPOLI-4 share a common pseudo-random-number generator (PRNG). Both codes offer an abstract PRNG interface; we chose to write an adapter class that wraps TRIPOLI-4's PRNG within the Geant4 API. The reason for this is that the user directly interacts with TRIPOLI-4 through its input file, where they can modify the TRIPOLI-4 PRNG settings.

2.2 Geometry

In Geant4, the calculation geometry is typically described by a C++ class with a specific interface. The methods of this class define the volumes and materials used for transport. The source code for the class, along with the rest of the code describing the specific simulation problem the user is interested in (sources, scores, etc.), is compiled and linked with the Geant4 library.

In the context of the coupling, TRIPOLI-4 is able to read and interrogate a Geant4 geometry in the form of a compiled shared library. We have developed a small tool[1] to automate compilation and wrapping of the Geant4 geometry classes. Geometries parametrised by UI commands are also supported. Parallel geometries are currently not supported.

2.3 Sources (primary generator actions)

Particle sources (called *primary generator actions* in Geant4) can also be imported in a TRIPOLI-4/Geant4 simulation. Just like in the case of geometries, the source files describing the source are compiled into a shared library and imported by TRIPOLI-4. As for geometries, UI-parametrised Geant4 sources are supported.

2.4 Physics list

A Geant4 *physics list* is an organised collection of models that specifies which particles are to be transported and how transport should be simulated, as a function of particle type and energy. Although physics lists are outside the scope of the TRIPOLI-4/Geant4 coupling, they still play an important role for the user because they essentially determine the accuracy of the Geant4 transport.

The Geant4 user is in principle free to construct their own physics list from scratch. However, for the moment, the TRIPOLI-4/Geant4 coupling only supports the choice of one of the "stock" Geant4 physics lists, the default being `FTFP_INCLXX`.

[1] `g42so`: https://github.com/arekfu/g42so.

2.5 Physics

In its most basic use, the TRIPOLI-4/Geant4 coupling can perform particle transport with TRIPOLI-4 in a Geant4 geometry. In this case, only the Geant4 geometry package is involved. However, it is also possible to activate *physics-coupling* mode. In this mode, Geant4 will transport all particles that are not within the scope of TRIPOLI-4. This includes protons and heavy ions, but also neutrons, photons, electrons and positrons above and below the TRIPOLI-4 energy range. When Geant4 produces particles within the TRIPOLI-4 range, it hands them over to TRIPOLI-4 for further transport.

2.6 Scores

Users of the TRIPOLI-4/Geant4 coupling are not be expected to define separate scores for particles treated on either side of the coupling. Instead, they define scores in a normal TRIPOLI-4 input file, and the coupling ensures that TRIPOLI-4 scores are correctly and seamlessly fed from particles from either code.

Yet, it is impossible to feed some score types from the Geant4 side; this is the case, for instance, of point fluxes, which requires knowledge of the double-differential cross section for particle scattering, which is sometimes unavailable. In other cases, it is technically possible to feed the score, but care must be exercised. Consider energy deposition, for instance. Particles that are currently not transported by TRIPOLI-4 (such as nuclei) are assumed to locally depose their kinetic energy. If transport is enabled for these particles via the Geant4 coupling, their contribution must be properly subtracted from the local energy deposition, in order to avoid double counting.

3 Verification

We now turn to the verification of TRIPOLI-4/Geant4. A rather direct way to verify the coupling is to perform a coupled calculation with a minimal energy range for TRIPOLI-4 and compare the results with a pure Geant4 calculation. If low-energy particles never lead to the production of higher-energy particles, then the two calculation results are expected to be equal (up to statistical fluctuations) above TRIPOLI-4's high cutoff. This "trick" puts rather strict constraints on much of the coupling: for instance, one is led to conclude that scores must be correctly fed if the test passes. We have performed a number of checks using this approach and perfect statistical agreement was found in all cases.

As usual, it is useful to compare the results of the coupled code to some reference calculation code, which can be taken to be MCNP6 [9] or Geant4 itself. However, this is not necessarily a stringent test, because the physics implemented in different transport codes may be different. It is difficult to prove that any observed difference in the final result arises exclusively from the physics and are not due to some bug in the coupling. This is especially pertinent for calculations involving nuclear reactions without evaluated nuclear data. In this case, nuclear-reaction models must be employed to generate the final state of the reaction. The uncertainties here are much larger than for evaluated nuclear data; a factor of two between calculations performed with two different models is not uncommon.

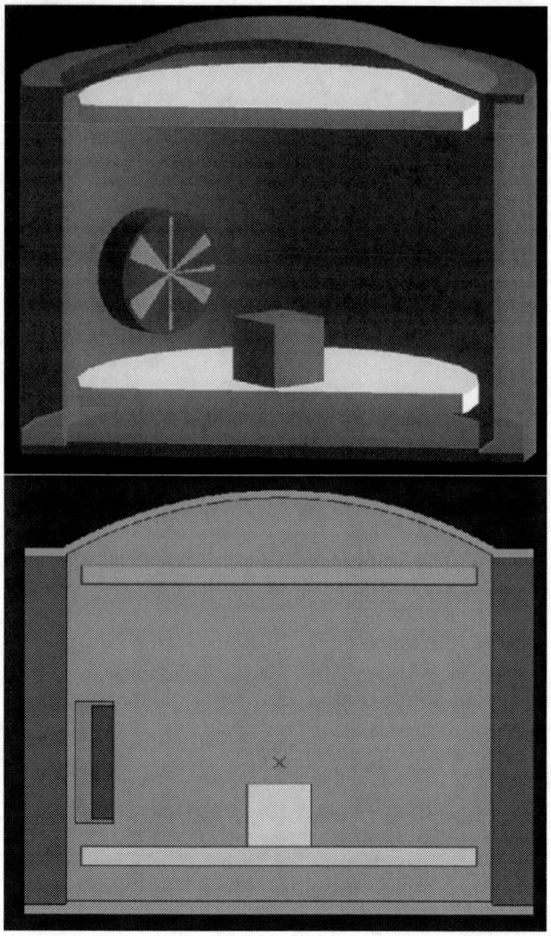

Figure 1. Views of the APOLLON enclosure geometry. The 3D view (top) was generated by the Geant4 OpenGL visualisation driver; the 2D view (bottom) was made with T4g, TRIPOLI-4's native visualisation tool. The cross on the 2D view indicates the source position.

3.1 Test case 1: APOLLON laser enclosure

Our first test case is a containment enclosure for the APOLLON laser facility [10]. APOLLON is a high-intensity laser (10 PW) which belongs to the Interdisciplinary Center on EXtreme Light (CILEX), located in Saclay, France. Figure 1 shows two views of the geometry. The point-wise mixed proton/electron source is placed at the centre, in the place indicated by the cross on the 2D view. The materials involved are aluminium alloys, with the exception of a glass parabola (the red region in the 2D view). The geometry was recreated by hand in Geant4 based on the original MCNPX input file. Note that the

Geant4 geometry was used for all the TRIPOLI-4 calculations.

The energy distributions of the source particles (protons and electrons) are most intense at low energy, with a long exponential tail that extends up to about 1 GeV. Both distributions are sharply anisotropic. As far as radiation protection outside the enclosure is concerned, low-energy protons and electrons are unimportant, because they are usually stopped without any secondary production. On the other hand, high-energy protons and electrons can yield large dose contributions, but are suppressed by the shape of the source spectrum. This argument suggests that the largest dose contributions come from intermediate energies; quantitative calculations show that energies up to ~ 250 MeV are responsible for most of the dosimetry. These energies are clearly outside TRIPOLI-4's native applicability range, at least as far as neutrons are concerned; therefore, the APOLLON enclosure is a good test case for the TRIPOLI-4/Geant4 coupling. For this paper, we only present calculations of neutron fluxes leaving the enclosure and generated by the proton source, but extensive calculations have actually been performed for in-enclosure and out-enclosure fluxes of photons and neutrons generated by both proton and electron sources.

An important remark is due at this point. Protons above a few MeV are quite efficient at inducing nuclear reactions in matter, which often result in the production of neutrons. Above about 150 MeV, evaluated nuclear data are typically not available and reaction events must be simulated using nuclear-reaction models (pre-equilibrium and intra-nuclear cascade models, coupled to statistical de-excitation models). Unfortunately, this comes at a price: large differences among models are to be expected. This makes it quite difficult to compare calculations performed with different transport codes, which do not necessarily employ the same physics models.

Fortunately, the high-energy transport codes we are considering here (Geant4 and MCNP6) do share one nuclear reaction model, namely INCL-ABLA. The version that is shipped with MCNP6 (INCL4.2), however, dates back to 2002 [11]; Geant4, on the other hand, contains the latest C++ version, INCL++ [12]. By default, the two model versions behave rather differently, but it is possible to adjust INCL++'s internal parameters to mimic INCL4.2 to some extent. This is the closest we can get to a comparison using the same model.

Figure 2 shows the result of calculations performed with TRIPOLI-4/Geant4, pure Geant4 and MCNP6. Geant4 uses the `FTFP_INCLXX_HP` physics list, with INCL++ in " INCL4.2 mode", both in stand-alone mode and within TRIPOLI-4. The MCNP6 calculations use the INCL4.2/ABLA model.

By comparing the pure Geant4 and the TRIPOLI-4/Geant4 calculations, an interesting observation can be made. The results are statistically compatible, despite the fact that the two codes use independent treatments of neutron transport below 20 MeV (handled by TRIPOLI-4 within the coupled calculation, and by NeutronHP within the Geant4 calculation). This can be interpreted as a verification of the NeutronHP sector against TRIPOLI-4's low-energy neutron transport. Note however that the cross sections used by the two codes are not identical: Geant4 uses the G4NDL library, which is essentially based on ENDF/B-VI, while the TRIPOLI-4 data library is based on JEFF-3.1.1. However, for the purpose of our fixed-source calculations, and for the precision we are seeking, the difference appears to be negligible.

Figure 2 also shows the result of a calculation performed with MCNP6, whose results are very similar to those obtained with TRIPOLI-4/Geant4 below 10 MeV. It is perhaps worth emphasising that MCNP6 cannot be considered as a reference code for these calculations insofar as the choice of the nuclear-reaction model induces an uncertainty on the calculated neutron flux that can be as large as a factor of two.

3.2 Test case 2: background estimation in Nucifer at Osiris

Nucifer [13] is a neutrino detector that was deployed near the compact Osiris research reactor core (70 MW). The Nucifer detector is well-suited for the detection of short-baseline neutrino oscillations due to its short distance (~ 7.2 m) from the reactor core. However, due to the extreme smallness of neutrino signal (about 1 neutrino event every 5 minutes), it is of capital importance to precisely estimate and possibly suppress any detection background. In the case at hand, one of the sources of background noise is represented by fast neutrons ($\gtrsim$ 2 MeV) coming from Osiris, which may mimic genuine neutrino events by inducing elastic scattering on nuclei contained in Nucifer's liquid scintillator.

These fast neutrons originate from fission events in Osiris's core. They can reach Nucifer after traversing about 3.5 m water (the reactor pool) and 2 m concrete. Figure 3 shows a schematic drawing of the geometry. Extremely few neutrons manage to traverse such enormous thicknesses without losing energy and dropping below the 2 MeV threshold energy, or without being absorbed. For this reason, the calculation of the background rate is presently impossible in Geant4.

It is however possible to tackle this problem using TRIPOLI-4's variance-reduction techniques. First, we performed a criticality simulation of the Osiris reactor core and stored neutrons leaving the core with energy larger than 2 MeV. Since the geometry of the Nucifer detector and its surroundings is modelled with Geant4, we imported the geometry using the TRIPOLI-4/Geant4 coupling and simulated the transport of the stored neutrons with a pure TRIPOLI-4 calculation. The exponential-transform method was applied, based on an importance map calculated by TRIPOLI-4's `INIPOND` module and subsequently adjusted by hand.

The calculation yields an attenuation factor for fast neutrons between Osiris and Nucifer of the order of 10^{-27}. This estimate, which is conservative, results in an expected elastic-scattering rate in the whole Nucifer detector volume of about 4×10^{-5} d^{-1}. This leads to the conclusion

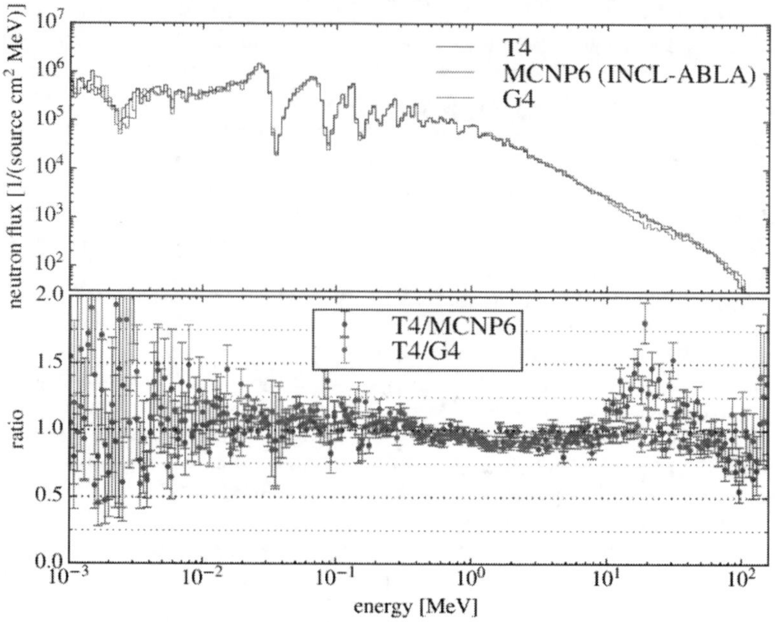

Figure 2. Top: energy-differential neutron flux generated by the proton source and leaving the enclosure, as calculated by TRIPOLI-4/Geant4, pure Geant4 and MCNP6.

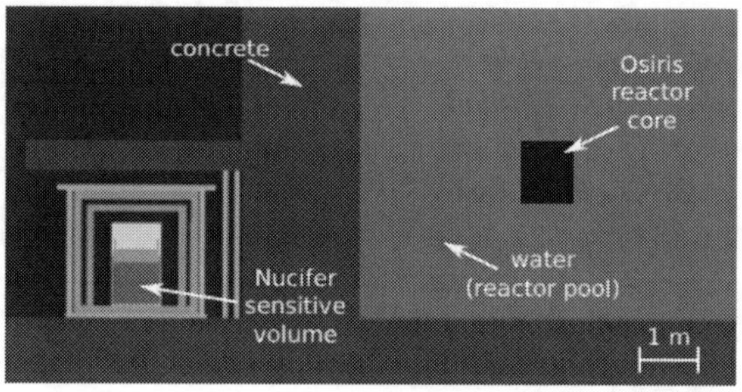

Figure 3. Cross section of the setup of the Nucifer-Osiris calculation.

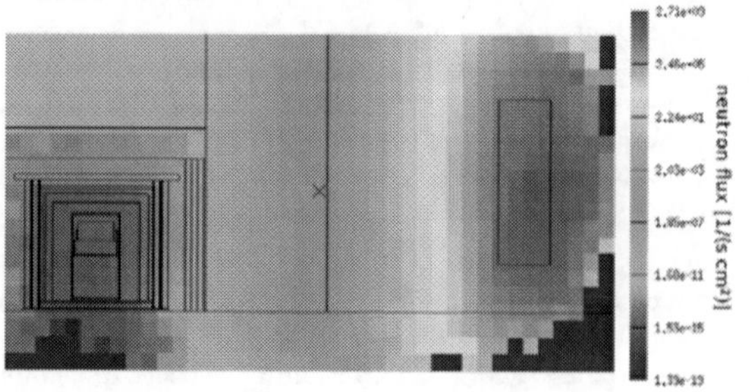

Figure 4. Cross-sectional colour map of neutron flux in the 8–20 MeV group, as calculated by TRIPOLI-4 in the Geant4 Nucifer geometry.

that the background noise induced by fast neutrons in Nucifer can safely be neglected.

3.3 Test case 3: n_TOF spallation source

The n_TOF facility [14] is a spallation neutron source located at CERN, which operates with a 20-GeV/c pulsed proton beam and which provides time-of-flight tagging of the neutron energy. One of the foremost advantages of n_TOF is that the energy of the produced neutrons spans more than twelve orders of magnitude, from thermal energies to the GeV range, allowing measurements for a wide range of applications. Accurate characterisation of the neutron source is crucial for these purposes, and some features of the neutron beam must be inevitably determined via numerical simulations.

Recently [15, 16], Geant4 simulations of the n_TOF spallation target were published. The studies focus on characterising the neutron beam (energy spectrum, resolution function, etc.) and the inevitable photon contamination. In order to show that the coupling is a suitable tool for this kind of studies, we have redone these simulations using TRIPOLI-4/Geant4.

Figure 5 shows the energy distribution per unit lethargy of neutrons leaving the spallation target in a 10°-cone directed towards the first experimental area (EAR1), as calculated by pure Geant4 and the TRIPOLI-4/Geant4 coupling. The black line represents a Geant4 calculation where molecular binding of hydrogen in water was neglected in scattering of thermal neutrons; this is the option that was used by Lo Meo *et al.* [15]. The blue line represents a Geant4 calculation where molecular binding of

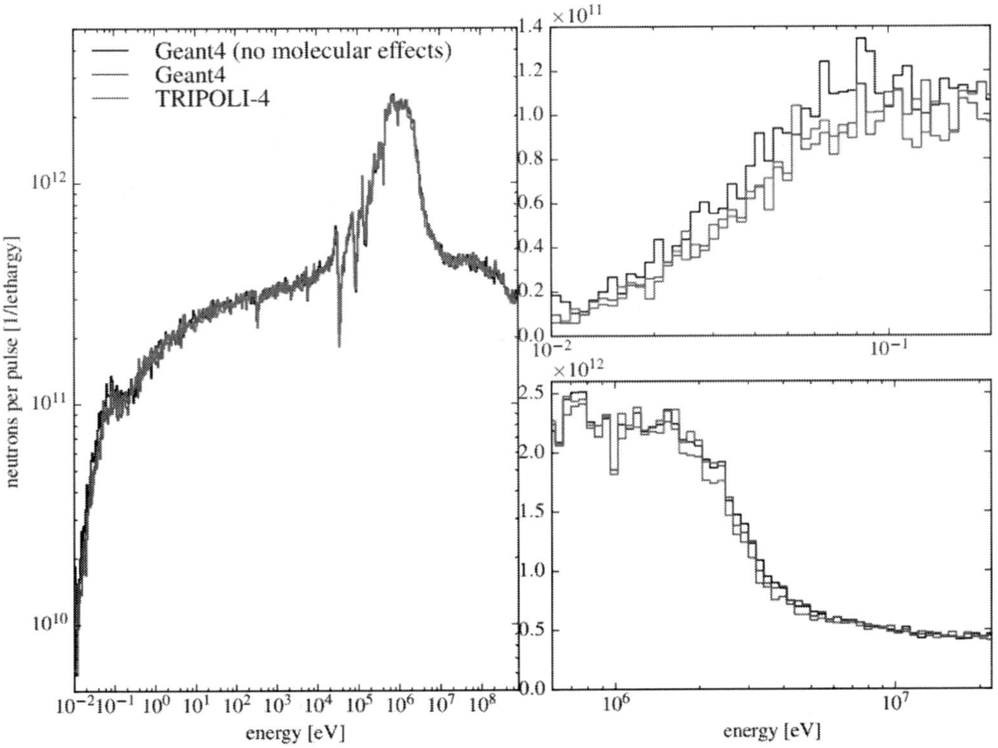

Figure 5. Energy spectrum of the spallation neutrons emerging from the n_TOF target in a 10°-cone directed towards the first experimental area (EAR1). Left: full spectrum in log-log scale; top right: zoom on the thermal region in log-linear scale; bottom right: zoom on the resolved-resonance region in log-linear scale.

hydrogen was correctly accounted for. Finally, the red line is the result of the TRIPOLI-4/Geant4 coupling. The same Geant4 physics list was used for all the calculations.

The zoom panels on the right of Fig. 5 highlight that the most relevant differences appear in the thermal and in the resolved-resonance ranges. As far as thermal scattering is concerned, we remark that the original Geant4 calculation overestimates the thermal flux by as much as 40%, while the other two calculations are statistically compatible. In the 1–10-MeV range, the shape of the spectrum in TRIPOLI-4/Geant4 is different from either Geant4 calculation. This might be due to differences in the nuclear data: as mentioned above, Geant4 uses the G4NDL library, which is essentially based on ENDF/B-VI, while the TRIPOLI-4 data library is based on JEFF-3.1.1.

Another interesting aspect that is not visible from Fig. 5 concerns CPU time. The TRIPOLI-4/Geant4 calculation is about 2.5 times faster than the Geant4 calculation. This is the ratio of CPU times for the whole calculation, and therefore includes the common high-energy transport phase. Moreover, the TRIPOLI-4/Geant4 calculation was run with a low-energy photon cutoff of 1 keV, while the Geant4 calculations used a photon production cut of 1 mm, which corresponds to 101.8 keV in lead. Therefore, we conclude that low-energy neutron transport is sensibly more efficient in TRIPOLI-4 than in Geant4's NeutronHP module. Part of this may be due to the use of implicit capture in TRIPOLI-4, which is unavailable in Geant4.

4 Perspectives and conclusions

We have described the progress made on the coupling between TRIPOLI-4 and the Geant4 particle-transport toolkit. The coupling allows TRIPOLI-4 to navigate inside Geant4 geometries. Geant4 particle sources can also be imported in TRIPOLI-4. Finally, and most importantly, Geant4 can be used to extend TRIPOLI-4's application scope by adding the capability to transport many additional particles. The coupling is driven from the TRIPOLI-4 side and it is designed in such a way that a TRIPOLI-4 user should be able to run an existing Geant4 calculation through the coupling with minimal additional effort.

The new capabilities of the coupled code have been illustrated by three specific test cases: radiation protection around a high-intensity laser beam, propagation of fast neutrons to a Geant4-modelled detector through severe attenuation, and characterisation of a neutron beam from a spallation source. When possible, TRIPOLI-4/Geant4 calculations have been compared to reference calculations performed with pure Geant4 and MCNP6, in order to verify a limited subset of the new features. The TRIPOLI-4/Geant4 results are found to be comparable to those of the other codes, and sensibly faster than equivalent calculations in Geant4.

Finally, we wish to trace the way for the future development of the TRIPOLI-4/Geant4 coupling. First, we need to consolidate the coupling. A sizeable body of V&V work remains to be done if one wished to cover all expected use cases. Further, minor features may need to be added to the code in order to satisfy particular user needs. Moreover, it is possible to boost the usefulness of the coupling by having TRIPOLI-4 feed particles to the Geant4 side of the coupling, which is almost never the case today. In particular, one could implement transport of charged particles produced in reactions which are of common interest for neutron detection, such as (n, f) and (n, α). This precious capability would pave the way to the detailed simulation of the response of several types of detectors which are outside TRIPOLI-4's scope today.

One limitation of the present version of the TRIPOLI-4/Geant4 coupling is that the user needs to be familiar with Geant4 to define the problem geometry in Geant4's native format. Alternatively, if a ROOT geometry is available, one may attempt conversion to the Geant4 format via GDML export/import. However, no simple solution exists for geometries described in TRIPOLI-4's native format, which is intrinsically different from ROOT's and Geant4's hierarchical geometry representations. One possible way to add support for native TRIPOLI-4 geometries would be to refactor Geant4's navigation behind an abstract interface, into which we could subsequently plug any one of TRIPOLI-4's existing geometry navigation systems. The bottleneck here is represented by the fact that the Geant4 geometry-navigation interface is quite dense. It is unclear to us whether the interface can be adapted to a non-hierarchical geometry engine.

The authors wish to acknowledge J. Gaffiot's collaboration for the simulation of the Nucifer/Osiris setup. We also wish to thank S. Lo Meo, C. Massimi and N. Colonna for kindly providing the geometry of the n_TOF spallation source and for contributing with discussions about the interpretation of the simulation results.

References

[1] E. Brun, F. Damian, C.M. Diop, E. Dumonteil, F.X. Hugot, C. Jouanne, Y.K. Lee, F. Malvagi, A. Mazzolo, O. Petit et al., Ann. Nucl. Energy **82**, 151 (2015)

[2] Y. Peneliau, DM2S report SERMA/LEPP/RT/02-3186/A, CEA-Saclay, France (2002)

[3] M. Goitein, A. Lomax, E. Pedroni, Phys. Today **55**, 45 (2002)

[4] D. Filges, F. Goldenbaum, *Handbook of spallation research. Theory, experiments and applications* (Wiley-VCH, Berlin, 2009)

[5] H. Aït Abderrahim, P. Baeten, D.D. Bruyn, J. Heyse, P. Schuurmans, J. Wagemans, Nucl. Phys. News **20**, 24 (2010)

[6] S. Agostinelli et al., Nucl. Instrum. Meth. A **506**, 250 (2003), official web site: http://geant4.cern.ch/.

[7] J. Allison et al., IEEE T. Nucl. Sci. **53**, 270 (2006)

[8] J. Allison, K. Amako, J. Apostolakis et al., Nucl. Instrum. Meth. Phys. Res. A (2016), in press

[9] T. Goorley, M. James, T. Booth et al., Nucl. Technol. **180**, 298 (2012)

[10] J. Zou, C. Le Blanc, D. Papadopoulos et al., High Power Laser Sci. Eng. **3**, e2 (2015)

[11] A. Boudard, J. Cugnon, S. Leray, C. Volant, Phys. Rev. C **66**, 044615 (2002)

[12] D. Mancusi, A. Boudard, J. Cugnon, J.C. David, P. Kaitaniemi, S. Leray, Phys. Rev. C (2014), submitted

[13] G. Boireau, L. Bouvet, A.P. Collin et al. (Nucifer Collaboration), Phys. Rev. D **93**, 112006 (2016)

[14] *n_TOF website*, https://ntof-exp.web.cern.ch/ntof-exp/

[15] S. Lo Meo, M.A. Cortès-Giraldo, C. Massimi, et al., Eur. Phys. J. A **51**, 160 (2015)

[16] Lerendegui-Marco, J., Lo Meo, S., Guerrero, C. et al., Eur. Phys. J. A **52**, 100 (2016)

FIFRELIN – TRIPOLI-4® coupling for Monte Carlo simulations with a fission model. Application to shielding calculations

Odile Petit[1,*], *Cédric* Jouanne[1], *Olivier* Litaize[2], *Olivier* Serot[2], *Abdelhazize* Chebboubi[2], and *Yannick* Péneliau[2]
[1]Commissariat à l'Energie Atomique et aux Energies Alternatives, DEN, SERMA, CEA, Université Paris-Saclay 91191 Gif-sur-Yvette, France
[2]Commissariat à l'Energie Atomique et aux Energies Alternatives, CEA/DEN/CAD/DER/SPRC CEA-Cadarache, 13108 Saint-Paul-Lez-Durance, France

Abstract. TRIPOLI-4® Monte Carlo transport code and FIFRELIN fission model have been coupled by means of external files so that neutron transport can take into account fission distributions (multiplicities and spectra) that are not averaged, as is the case when using evaluated nuclear data libraries. Spectral effects on responses in shielding configurations with fission sampling are then expected. In the present paper, the principle of this coupling is detailed and a comparison between TRIPOLI-4® fission distributions at the emission of fission neutrons is presented when using JEFF-3.1.1 evaluated data or FIFRELIN data generated either through a n/γ-uncoupled mode or through a n/γ-coupled mode. Finally, an application to a modified version of the ASPIS benchmark is performed and the impact of using FIFRELIN data on neutron transport is analyzed. Differences noticed on average reaction rates on the surfaces closest to the fission source are mainly due to the average prompt fission spectrum. Moreover, when working with the same average spectrum, a complementary analysis based on non-average reaction rates still shows significant differences that point out the real impact of using a fission model in neutron transport simulations.

1 Introduction

Monte Carlo transport calculations including fission sampling usually use nuclear data from international evaluation libraries such as JEFF-3.1.1 or ENDF/B-VII.1. Among those data, fission multiplicities for neutrons and gamma are provided only as mean values instead of a probability distribution function depending on the incident neutron energy, as observed physically. Moreover, the neutron emission spectrum may be restrictive since it is provided as a function of the incident neutron energy only and is thus the same for all fission neutrons produced by a given fission event. Consequently, when using these data in a Monte Carlo transport code, the energy conservation is preserved in mean value only, but not for each individual fission event.

On the contrary, fission models such as FIFRELIN [1] or FREYA [2] give access to a detailed simulation of the fission, including particle correlations and energy conservation. In the past, several works aimed at coupling various fission models with Monte Carlo transport codes or at integrating them, for example MCNP-DSP [3], MCNPX-PoliMi [4] or FREYA-TRIPOLI-4® [5]. Related calculations were mainly dealing with nuclear instrumentation applications for which individual events are collected and post-processed. However, in the context of shielding calculations in fission facilities, spectral effects of the use of a fission model for specific isotopes could be first analyzed in terms of average values provided by the Monte Carlo transport code. Secondly, an analysis based on non-average distributions could be performed in order to observe correlation effects between fission neutrons for example. These issues are addressed in the present paper by means of the TRIPOLI-4® [6] Monte Carlo transport code and the FIFRELIN fission model.

TRIPOLI-4® is a continuous-energy Monte Carlo transport code dedicated to shielding, reactor physics with or without depletion, criticality safety and nuclear instrumentation. A first coupling between TRIPOLI-4® and FREYA fission model has already been implemented [5] and consists in calling FREYA code by TRIPOLI-4® at each sampling of a fission event. In order to test another fission model, a new coupling with FIFRELIN code has recently been implemented starting from version 10 of TRIPOLI-4® code and works through external files between both codes.

FIFRELIN is a Monte Carlo code that simulates the de-excitation process of primary fission fragments leading to an estimation of prompt fission observables such as neutron/gamma/electron multiplicities and spectra in correlation with fission fragments. Primary fission fragments (after full acceleration and before prompt neutron emission) are completely determined by the knowledge of their mass A, charge Z, kinetic energy KE, excitation energy E* and spin/parity J^{π}. The three first characteristics (A, Z, KE) are sampled from experimental data or can be provided by external codes. The description of the initial (E*, J) entry zone is

* Corresponding author: odile.petit@cea.fr

performed through a mass dependent temperature ratio law $R_T(A)$ and a spin/parity distribution ruled by a spin cut-off parameter that can be modeled in different ways. Positive and negative parities are supposed to be equally likely. More details can be found in [7].

De-excitation cascades are simulated following the Becvar's notion of nuclear realization [8]. A nuclear realization is a random set of nuclear levels (energy, spin, parity) in association with partial widths for neutron, gamma or electron emission. Experimental data related to electromagnetic transitions in the discrete level region are taken from RIPL-3 database [9]. When nuclear level structure is completely unknown (in the continuum region), level density and strength function models are used. In between these regions, our partial knowledge of nuclear structure is completed by models up to a fixed maximum level density. In this way the whole available experimental information is accounted for.

FIFRELIN is ruled by five free input parameters driving the primary fission fragment excitation energy and spin-parity distributions (the (E*, J) entry zone discussed above). These five free parameters are determined to match a target observable such as the total average prompt neutron multiplicity (nu-bar). Once this procedure is completed, the whole set of fission observables can be compared with experimental results. The cascade can be modeled in two different modes, either *n/γ-uncoupled* or *n/γ-coupled*. In the first mode, neutron emission is modeled from Weisskopf statistical theory down to a spin dependent energy limit and gamma emission uses nuclear realization concept down to the ground state. The second mode is based on a Hauser-Feshbach formalism accounting for the competition between neutron and gamma emission at the very beginning of the cascade [10]. In this case the neutron emission is ruled from transmission coefficients tabulated on a specific energy-angular momentum grid. The nuclear realization concept is then extended for neutron emission up to the initial fragment excited state.

The purpose of this paper is to present the coupling details and give a preliminary application analyzing the impact of the use of FIFRELIN model on a TRIPOLI-4® shielding calculation including fission sampling. To meet this need, the ASPIS benchmark [11] was selected with several simple modifications for the purpose of this work. Section 2 of this paper is dedicated to the principle of the present coupling and Section 3 to its application to the modified ASPIS benchmark and to further investigations.

2 Coupling method

2.1 General principle

As already mentioned, the FIFRELIN-TRIPOLI-4® coupling was carried out through external files per fissile isotope of interest and therefore works in two consecutive steps: first, FIFRELIN code is run to produce a large amount of fission events for a given fissile isotope. For each fission event, specific data are stored in an output file that can be read by TRIPOLI-4® afterwards. This step can be repeated independently for several fissile isotopes of interest. Then, during the Monte Carlo neutron transport simulation, whenever a fission event needs to be sampled, TRIPOLI-4® reads the data stored in the appropriate FIFRELIN file, if it exists, at the current reading position for a new fission event (otherwise the evaluation files are used). Fission neutrons are directly produced accordingly in TRIPOLI-4® instead of sampling the number of fission neutrons and their energy and angle according to the evaluation distributions. The FIFRELIN files are read sequentially and need therefore to be as big as possible for each fissile isotope of interest, in order to provide as many statistically independent fission events as possible. Then, the fission neutrons carry on with their history through the usual neutron transport simulation of TRIPOLI-4®.

2.2 Validity and contents of a FIFRELIN file

The actual FIFRELIN list-mode (event-by-event) data file is simplified to contain an even number of lines (two lines for a fragment pair). Each line contains the nuclear charge, the mass number, the initial fragment kinetic energy, the characteristics of the cascade and the final (post-neutron) fragment kinetic energy. The characteristics of a cascade depend on the calculation scheme. In the n/γ-uncoupled mode it contains the neutron kinetic energy for every emitted neutron, while in the n/γ-coupled mode it contains the energy, spin and parity of every nuclear level and the kinetic energy in the laboratory frame of every neutron, gamma and conversion electron. In a very first step the emission is supposed to be isotropic in the laboratory frame. These events can be used to estimate a distribution like the prompt fission neutron spectrum allowing a comparison with estimators available in the code.

The nuclear realization concept is very time consuming. Because we were only interested in neutron emission, we started this work by using the n/γ-uncoupled mode and cut out gamma emission. Multi-threading was used to generate 2.10^8 fission fragment cascades on 100 threads for ^{235}U in the thermal incident neutron energy range. This prevented the use of a same history in the subsequent TRIPOLI-4® run.

Then, in a similar way, we used the n/γ-coupled mode to simulate around 4.10^7 fission fragment cascades. Even if gammas are not useful here, partial neutron and gamma widths have to be estimated to account for n/γ competition during the cascade. The major models used in this calculation scheme are:

- Koning-Delaroche optical model potential parametrization for neutron transmission coefficients
- Composite Gilbert-Cameron model for the level density taking into account an energy dependent Ignatyuk level density parameter
- Enhanced Generalized Lorentzian for E1 photon strength functions and Single Lorentzian for other multipolarities.

All these models and model parameters are described in RIPL-3 documentation [9].

2.3 Related distributions as retrieved by TRIPOLI-4®

As expected, using FIFRELIN instead of evaluated data induces modifications to the fission neutron multiplicity distribution and to the fission neutron spectrum. To make this clear, for both modes of FIFRELIN we compared several distributions related to fission neutron emission, obtained with two TRIPOLI-4® simulations (using either FIFRELIN data or JEFF-3.1.1 library) and involving thermal neutron induced fission on ^{235}U. Three quantities of interest were collected by specific TRIPOLI-4® tallies which do not take into account neutron transport simulation. Figure 1 compares prompt fission neutron multiplicities, Figures 2 and 3 present prompt fission neutron spectra and Figure 4 shows the distributions of the energies summed up for all neutrons emitted at a same fission event. Error bars are not presented since they are not visible in terms of orders of magnitude, compared to the displayed values. In the captions, "Fifrelin n" stands for the n/γ-uncoupled mode and "Fifrelin n/γ" for the n/γ-coupled mode of FIFRELIN.

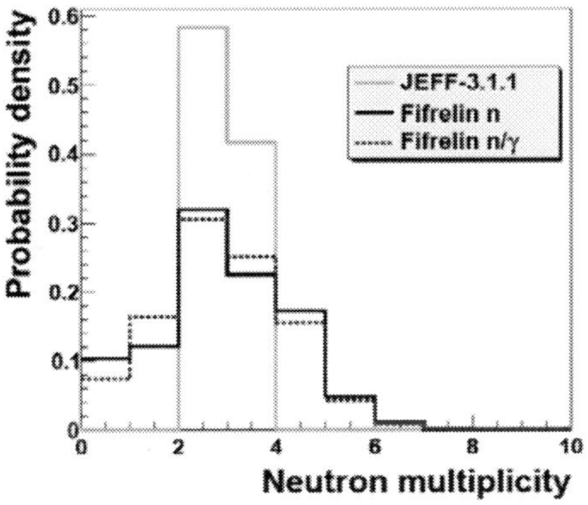

Fig. 1. Prompt fission neutron multiplicity distribution.

In Figure 1, the fission neutron multiplicity distribution obtained with FIFRELIN data spreads from 0 up to 9 neutrons emitted per fission event (with an average value of 2.431 for the n/γ-uncoupled mode and 2.413 for the n/γ-coupled mode) whereas only 2 or 3 fission neutrons can be emitted simultaneously when using the evaluated data (which corresponds to both integer numbers surrounding $\overline{\nu_p} = 2.417$ for ^{235}U at thermal incident neutron energies).

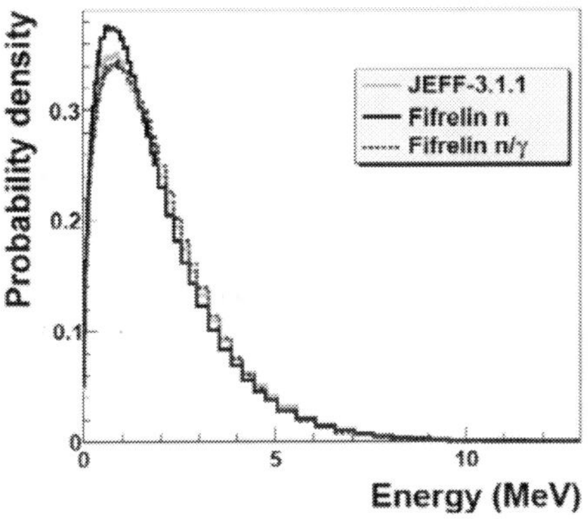

Fig. 2. Prompt fission neutron spectrum.

In Figure 2, the emission spectrum for fission neutrons is slightly shifted to lower energies when using FIFRELIN data based on the Weisskopf statistical theory, i.e. n/γ-uncoupled mode. A prompt spectrum with similar features would be obtained when using FREYA code [5], which also uses Weisskopf statistical theory for neutron emission. Moreover, the emission spectrum looks closer to the evaluated data in case of the Hauser-Feshbach formalism of FIFRELIN, i.e. n/γ-coupled mode. However, the lin-log representation of these spectra shows a slight discrepancy noticeable in Figure 3 at energies higher than approximately 2 MeV.

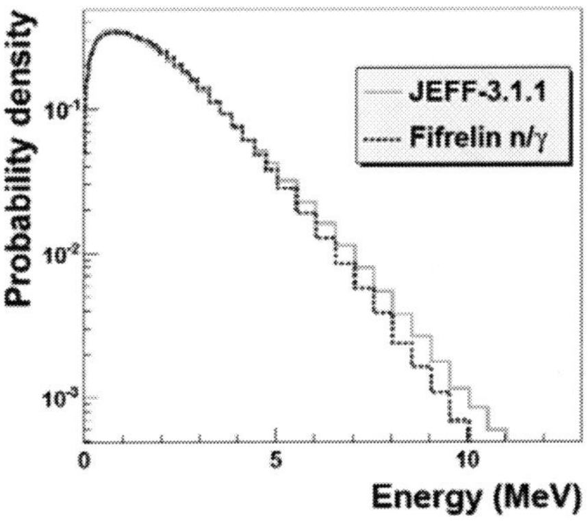

Fig. 3. Prompt fission neutron spectrum, lin-log representation.

Average values of these spectra are 2.035 MeV with JEFF-3.1.1 data and 1.943 MeV with the n/γ-uncoupled mode or 1.984 MeV with the n/γ-coupled mode of FIFRELIN.

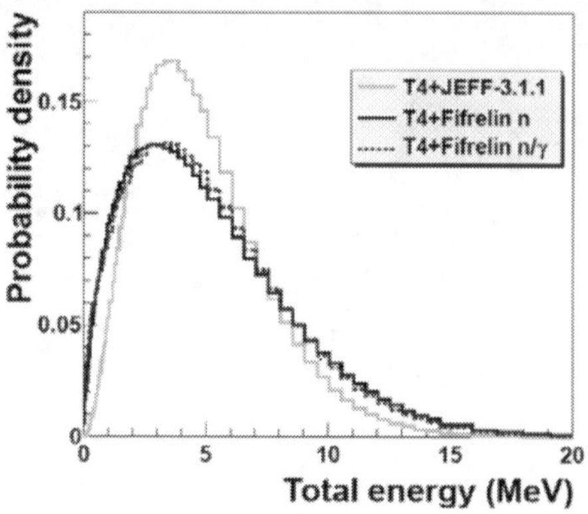

Fig. 4. Sum of neutron energies per fission.

The distributions plotted in Figure 4 are clearly not centered on the same energy: the distribution obtained with the evaluated data turns out to be more peaked and centered on a higher energy. Average values of these distributions are 4.907 MeV with JEFF-3.1.1 data and 4.720 MeV with the n/γ-uncoupled mode or 4.781 MeV with the n/γ-coupled mode of FIFRELIN. The impact of this difference after transport of the fission neutrons is investigated in the next Section.

3 Application to thermal ^{235}U in a fixed-source ASPIS calculation using FIFRELIN data

3.1 Description

We have selected the ASPIS benchmark from the SINBAD Database [11]. This configuration allows us to study the propagation of neutrons with energies higher than 40 keV from a fission plate into an iron bulk. Three different dosimeters are installed at various depths of penetration (from 5 cm to 1.2 m):

- ^{103}Rh(n,n'), threshold = 40 keV
- ^{115}In(n,n'), threshold = 339 keV
- ^{32}S(n,p), threshold ≈ 1.6 MeV.

Reaction rates are calculated for all these dosimeters.

Several modifications were made on this benchmark for the purpose of this study. With the aim of simulating fission events, a 10 eV-source was placed in the fission plate (instead of the usual Watt-spectrum together with fission sampling turned off).

When running TRIPOLI-4® with FIFRELIN data or with the evaluated data, the same following options were activated. Fission sampling was turned on, as well as the analog transport mode of the code [12]. No variance reduction was requested, in order to avoid an underestimation of the fission sampling. Finally, only prompt fission neutrons were sampled.

3.2 Results

No significant increase of the calculation time was noticed (less than 1%) when running TRIPOLI-4® with FIFRELIN data instead of the evaluated data for the fission sampling on ^{235}U, provided that the ahead calculation time of FIFRELIN is not taken into account. Tables 1 to 6 compare average surface neutron reaction rates in arbitrary units (a.u.) calculated by three simulations of the modified ASPIS benchmark, either using JEFF-3.1.1 library or n/γ-uncoupled mode of FIFRELIN (Tables 1, 3 and 5), or n/γ-coupled mode of FIFRELIN (Tables 2, 4 and 6), respectively on ^{103}Rh (Tables 1 and 2), ^{115}In (Tables 3 and 4) and ^{32}S (Tables 5 and 6) IRDF85 nuclei. Only results on the surfaces closest to the fission plate are shown. Relative standard deviations are given in percentages and relative errors of FIFRELIN results are presented in numbers of sigma, with reference to the JEFF-3.1.1 results. It should be noted that the convergence was better for the n/γ-uncoupled mode, due to a higher total number of fission events present in the related FIFRELIN files.

Table 1. Comparison of TRIPOLI-4® reaction rates on ^{103}Rh when using JEFF-3.1.1 or FIFRELIN n/γ-uncoupled mode.

Depth (cm)	R.Rate ± σ (%) with JEFF-3.1.1	R.Rate ± σ (%) with FIFRELIN n/γ-uncoupled	Rel. err. (nb. σ)
5.72	51.786 ± 0.101	50.525 ± 0.081	19.1
11.43	30.347 ± 0.095	29.731 ± 0.099	14.9
17.15	18.798 ± 0.128	18.395 ± 0.115	12.5
22.86	12.023 ± 0.133	11.807 ± 0.136	9.6
28.58	7.879 ± 0.156	7.756 ± 0.152	7.3
34.29	5.300 ± 0.180	5.217 ± 0.165	6.5

Table 2. Comparison of TRIPOLI-4® reaction rates on ^{103}Rh when using JEFF-3.1.1 or FIFRELIN n/γ-coupled mode.

Depth (cm)	R.Rate ± σ (%) with JEFF-3.1.1	R.Rate ± σ (%) with FIFRELIN n/γ-coupled	Rel. err. (nb. σ)
5.72	51.786 ± 0.101	51.743 ± 0.126	0.5
11.43	30.347 ± 0.095	30.335 ± 0.155	0.2
17.15	18.798 ± 0.128	18.722 ± 0.184	1.8
22.86	12.023 ± 0.133	12.023 ± 0.220	0.0
28.58	7.879 ± 0.156	7.870 ± 0.228	0.4
34.29	5.300 ± 0.180	5.311 ± 0.433	0.4

Table 3. Comparison of TRIPOLI-4® reaction rates on ^{115}In when using JEFF-3.1.1 or FIFRELIN n/γ-uncoupled mode.

Depth (cm)	R.Rate ± σ (%) with JEFF-3.1.1	R.Rate ± σ (%) with FIFRELIN n/γ-uncoupled	Rel. err. (nb. σ)
5.72	7.193 ± 0.162	6.842 ± 0.119	24.9
11.43	3.352 ± 0.155	3.202 ± 0.167	20.2
17.15	1.652 ± 0.218	1.574 ± 0.219	15.6
22.86	0.840 ± 0.265	0.805 ± 0.273	11.1
28.58	0.440 ± 0.354	0.425 ± 0.360	6.9
34.29	0.239 ± 0.435	0.231 ± 0.411	5.8

Table 4. Comparison of TRIPOLI-4® reaction rates on ^{115}In when using JEFF-3.1.1 or FIFRELIN n/γ-coupled mode.

Depth (cm)	R.Rate ± σ (%) with JEFF-3.1.1	R.Rate ± σ (%) with FIFRELIN n/γ-coupled	Rel. err. (nb. σ)
5.72	7.193 ± 0.162	7.171 ± 0.180	1.3
11.43	3.352 ± 0.155	3.347 ± 0.280	0.5
17.15	1.652 ± 0.218	1.637 ± 0.324	2.4
22.86	0.840 ± 0.265	0.839 ± 0.442	0.4
28.58	0.440 ± 0.354	0.438 ± 0.526	0.8
34.29	0.239 ± 0.435	0.237 ± 0.637	1.2

Table 5. Comparison of TRIPOLI-4® reaction rates on ^{32}S when using JEFF-3.1.1 or FIFRELIN n/γ-uncoupled mode.

Depth (cm)	R.Rate ± σ (%) with JEFF-3.1.1	R.Rate ± σ (%) with FIFRELIN n/γ-uncoupled	Rel. err. (nb. σ)
5.72	1.321 ± 0.232	1.219 ± 0.238	24.0
11.43	0.477 ± 0.352	0.440 ± 0.363	15.8
17.15	0.182 ± 0.563	0.169 ± 0.601	9.4
22.86	0.070 ± 0.804	0.066 ± 0.932	4.9
28.58	0.028 ± 1.345	0.027 ± 1.567	2.9
34.29	0.011 ± 1.892	0.010 ± 2.017	2.8

Table 6. Comparison of TRIPOLI-4® reaction rates on ^{32}S when using JEFF-3.1.1 or FIFRELIN n/γ-coupled mode.

Depth (cm)	R.Rate ± σ (%) with JEFF-3.1.1	R.Rate ± σ (%) with FIFRELIN n/γ-coupled	Rel. err. (nb. σ)
5.72	1.321 ± 0.232	1.256 ± 0.361	11.6
11.43	0.477 ± 0.352	0.455 ± 0.597	7.0
17.15	0.182 ± 0.563	0.170 ± 0.836	6.7
22.86	0.070 ± 0.804	0.067 ± 1.274	3.0
28.58	0.028 ± 1.345	0.027 ± 2.146	2.3
34.29	0.011 ± 1.892	0.010 ± 3.285	2.0

In the case of the n/γ-uncoupled mode of FIFRELIN, significant differences with JEFF-3.1.1 results are observed for all responses, particularly at the surface positions closest to the fission plate. This suggests that the spectral effect of FIFRELIN data noticed in the previous section, at the emission of fission neutrons, has still an impact on results after neutron transport to distances up to 34 centimeters.

As for the n/γ-coupled mode of FIFRELIN, Tables 2 and 4 show no difference for the first two responses: JEFF-3.1.1 and FIFRELIN results are within 3 sigma of uncertainties for ^{103}Rh and ^{115}In dosimeters. However, Table 6 shows discrepancies up to 11.6 sigma for the reaction rates on ^{32}S. When examining Figure 2 again, the prompt fission neutron spectrum of JEFF-3.1.1 is actually closer to FIFRELIN spectrum when using its n/γ-coupled mode, which can explain the lack of significant differences in Tables 2 and 4. But bigger discrepancies between JEFF-3.1.1 and n/γ-coupled mode spectra can be noticed in Figure 3 at energies higher than approximately 2 MeV, thus explaining that discrepancies can be seen on ^{32}S response only, which has the highest reaction rate threshold (1.6 MeV).

3.3 Further investigations

Further investigations were performed through specific modifications based on JEFF-3.1.1 nuclear data file, so as to better determine the origin of the differences noticed in Tables 1, 3, 5 and 6.

First, the same trend in results was observed when running TRIPOLI-4® with modified evaluated data of the prompt fission spectrum that consist in suppressing the dependence on the incident neutron energy by extending the evaluated data for thermal incident neutron energies to the whole incident energy range of the prompt ^{235}U fission spectrum. This means that the previously noticed differences are not due to the fact that FIFRELIN data, as used in this coupling work, do not depend on the incident neutron energy but are valid for a given energy range only.

Then, in another test, when replacing the prompt fission spectrum by the average prompt neutron fission spectrum, as illustrated in Figure 2, from respectively both modes of FIFRELIN, all TRIPOLI-4® results turned out to be very close to the results obtained with TRIPOLI-4®-FIFRELIN coupling (within 3 sigma of uncertainties). This suggests that the coupling results, which take advantage of the prompt fission neutron multiplicity distribution and of energy conservation per fission event, are not directly sensitive to the difference observed in Figure 4 (which remained unchanged after these respective modifications of the evaluated data), but only to the difference of average prompt fission spectrum.

In order to clarify the contribution of the fission model, apart from the sensitivity to a difference of average spectrum, we performed an analysis based on non-average reaction rates. Full distributions were built by means of simulations containing only one neutron. For that purpose, we modified the previous configuration in order to obtain a 4π-solid angle detection: this allowed us to take into account coincidences between neutrons emitted in a same fission event. The new configuration consisted in several concentric iron spheres and reaction rates were calculated at each spherical interface. We kept the previous modification of this Subsection for the simulation using evaluated data. As an example, Figure 5 shows very different distributions of ^{103}Rh reaction rate at the first interface (at a depth of 5.72 centimeters) obtained by simulating a single neutron per batch. In this figure, FIFRELIN events were produced with the n/γ-uncoupled mode of FIFRELIN.

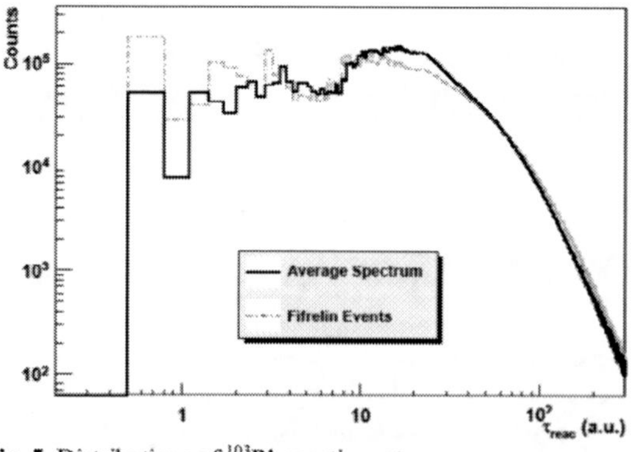

Fig. 5. Distributions of ^{103}Rh reaction rates.

In a similar way, different distributions would also be found with the n/γ-coupled mode of FIFRELIN when compared to the related average spectrum retrieved from evaluated data.

4 Conclusion

FIFRELIN fission model and TRIPOLI-4® Monte Carlo transport code have been coupled through external files, so as to simulate a detailed model of fission together with an accurate transport of the fission neutrons.

Calculations on a modified version of the ASPIS benchmark, either with JEFF-3.1.1 data or with the coupling with FIFRELIN code for the fission sampling on ^{235}U, showed significant differences on reaction rates which turned out to be mainly due to the slight difference of average prompt fission spectrum. However, the n/γ-coupled mode of FIFRELIN actually gave average reaction rates closer to those obtained with the evaluated data than its n/γ-uncoupled mode.

A complementary analysis was performed so as to clarify the real impact of the fission model after neutron transport, apart from the average trend: the distribution of reaction rates per fission event turned out to be very different with or without the use of FIFRELIN, highlighting the capability of the present coupling to take into account coincidences between neutrons emitted by a same fission event, which are still visible after neutron transport.

In a future work, other applications involving an analysis of events in coincidence could also be addressed, such as a time analysis of fission neutrons after transport.

TRIPOLI-4® is a registered trademark of CEA. We gratefully acknowledge EDF for their long term partnership and AREVA for their support.

References

1. O. Litaize, O. Serot, "Investigation of phenomenological models for the Monte Carlo simulation of the prompt fission neutron and gamma emission", Phys. Rev. C **82**, 054616 (2010)
2. J.M. Verbeke, J. Randrup, R. Vogt, "Fission Reaction Yield Algorithm FREYA for Event-by-Event Simulation of Fission", Comp. Phys. Comm. **191**, 178-202 (2015)
3. T.E. Valentine, "MCNP-DSP Users Manual", Oak Ridge National Laboratory, Oak Ridge, TN, ORNL/TM-13334 R2 (2001)
4. E. Padovani, S.A. Pozzi, S.D. Clarke, E.C. Miller, "MCNPX-PoliMi User's Manual", C00791 MNYCP, Radiation Safety Information Computational Center, Oak Ridge National Laboratory (2012)
5. J.M. Verbeke, O. Petit, "Stochastic analog neutron transport with TRIPOLI-4® and FREYA. Bayesian uncertainty quantification applied to neutron multiplicity counting", Nucl. Sci. Eng. **183**, 214-228 (2016)
6. TRIPOLI-4® Project Team, "TRIPOLI-4®, CEA, EDF and AREVA Reference Monte Carlo Code", Ann. Nucl. Energy, **82**, 151-160 (2015)
7. O. Litaize, O. Serot, L. Berge, "Fission modelling with FIFRELIN", Eur. Phys. J. A **51**, 177 (2015)
8. F. Becvar, "Simulation of gamma cascades in complex nuclei with emphasis on assessment of uncertainties of cascade-related quantities", Nucl. Instr. Meth. A **417**, 434-449 (1998)
9. R. Capote, M. Herman, P. Oblozinsky, P.G. Young, S. Goriely, T. Belgya, A.V. Ignatyuk, A.J. Koning,

S. Hilaire, V.A. Plujko, M. Avrigeanu, O. Bersillon, M.B. Chadwick, T. Fukahori, Zhigang Ge, Yinlu Han, S. Kailas, J. Kopecky, V.M. Maslov, G. Reffo, M. Sin, E.Sh. Soukhovitskii, P. Talou, "Reference Input Parameter Library (RIPL-3)", Nuclear Data Sheets **110**, Issue 12, 3107-3214 (2009)
10. D. Regnier, O. Litaize, O. Serot, "An improved numerical method to compute neutron/gamma deexcitation cascades starting from a high spin state", Comput. Phys. Commun. , **201**, 19-28 (2016)
11. I. Kodeli, H. Hunter, E. Sartori, "Radiation shielding and dosimetry experiments updates in the SINBAD database", Radiat. Prot. Dosim., **116**(1-4), 558-561 (2015)
12. O. Petit, E. Dumonteil, "Analog neutron transport for nuclear instrumentation applications with the Monte Carlo code TRIPOLI-4", Nucl. Technol., **192**, 259-263 (2015)

DOI: 10.1051/epjconf/201715306004

A versatile multi-objective FLUKA optimization using Genetic Algorithms

Vasilis Vlachoudis[1,a], Guido Arnau Antoniucci[1], Serge Mathot[1], Wioletta Sandra Kozlowska[1,2] and Maurizio Vretenar[3]

[1]Dep EN, CERN CH-1211, Switzerland

[2]Medical University of Vienna, Austria

[3]Accelerator and Technology Sector, CERN CH-1211, Switzerland

Abstract. Quite often Monte Carlo simulation studies require a multi phase-space optimization, a complicated task, heavily relying on the operator experience and judgment. Examples of such calculations are shielding calculations with stringent conditions in the cost, in residual dose, material properties and space available, or in the medical field optimizing the dose delivered to a patient under a hadron treatment. The present paper describes our implementation inside flair[1] the advanced user interface of FLUKA[2,3] of a multi-objective Genetic Algorithm[**Erreur ! Source du renvoi introuvable.**] to facilitate the search for the optimum solution.

1 Introduction

Monte Carlo programs like FLUKA[2,3] are often used to optimize complicated problems, with multiple free parameters and variable constraints. Some examples are:

- Shielding calculations, to find the optimal material configuration and geometry to reduce to some acceptable levels the dose at the same time reducing the cost and the space required;
- Target or dumps design, to find the best configuration of materials that can withstand the power of the beam, with reduce waste production and provide at the same time the best physics parameters;
- Optimize the radiotherapy treatment planning to deliver a prescribed dose to a volume of interest and minimizing the dose to nearby organs;
- and many others.

Such complicated simulation tasks, requires an experienced user to provide some initial educative guesses, and a large number of iterations until all the constraints are met and the optimum solution is found. In numerical analysis there are several optimization techniques that could be used to facilitate the task, each one with their advantages and drawbacks. The advantage of GA over the other algorithms that makes it suitable for Monte Carlo calculations is the no-dependence on gradients and derivatives, while every step is performed using random numbers. The present paper describes our implementation of a generic Genetic Algorithm [**Erreur ! Source du renvoi introuvable.**] solution that works seamlessly with the FLUKA Monte Carlo code.

[a] Corresponding author: Vasilis.Vlachoudis@cern.ch

2 Genetic Algorithm Implementation

In the field of artificial intelligence, Genetic Algorithm (GA) is a search that mimics the process of natural selection. It is commonly used to optimize problems using techniques inspired by natural evolution. The algorithm is based on a population of individuals, which is repeatedly modified/evolved in its environment until the target solution is reached. Our implementation of GA is written in Python as an extension to flair[1] the advanced user interface of FLUKA. Python[5] being an interpreted language is not as fast when compared to the compiled ones like C/C++, Fortran, however the CPU penalty that is added by using Python is minimal compared the time spent in FLUKA to evaluate each solution. However our implementation provides also a mechanism to fully parallelize the evaluation of each individual and takes care of the synchronization of the multiple threads. In the following sections we briefly described some of the terminology and our implementation.

2.1 Environment

The environment is the master class that contains the population, includes the basic parameters settings and provides the methods controlling the evolution of our population from generation to generation. The user can define the population size, maximum number of generations, as well as probabilities rates for crossover and mutation, and the goal for optimum solution. The class on initialization creates the initial population of random individuals, or it can be provided externally. Once everything is initialized the main loop is invoked.

At each iteration it is evaluating all individuals either in parallel or in synchronous order. Each individual evaluation requires one dedicated FLUKA job with the parameters defined in the chromosome of the individual. As soon as all the individuals are evaluated, if the desired goal is met the process stops.
Otherwise all individuals are sorted, based on their fitness function score, with higher preference the best individuals are recombined together (crossover) to produce new children individuals. In addition, randomly based on the mutation ration, some individuals are mutated, their chromosome are altered randomly according to a certain user defined pattern.
In our implementation and on user's choice we keep the best individuals untouched from one generation to another, the so called elitist selection.
The genetic algorithms have a tendency to converge towards the first local minimum that will find, unless if a high mutation rate is employed. However, with high mutation rates the converging process takes longer and can result in diverging from the solution. From lots of testing, we discovered that, as in the elitist approach, if we replace the worst individual with completely random one, it helped to converge faster, like in the case of smaller mutation rate but providing a better result.

2.2 Individual

Each individual is described by a chromosome which is represented as a string of alleles (genes).

2.2.1 *Allele*

Each allele is a python class that holds one or more parameters. In our implementation we have the following predefined alleles:

- Binary: holding a true/false value;
- Range: holding an arbitrary range of values from min to max;
- Integer: holding a integer value within a range
- Real-uniform: holding a floating point variable with a uniform distribution within a range;
- Real-Gaussian: holding a floating point number sampled from a normal distribution of values within a defined range;
- List: a list of predefined choices, can be anything of integers, reals or strings (e.g. material names)

However, the user can easily superclass the base allele class and create his own. Each allele must provide two methods, one for random assignment and one for mutation. In flair each allele is represented inside the FLUKA input file with a **#define** variable. A recent addition in flair is the extension of the pre-processor language giving the ability to perform complicated function evaluation for scalar, vectors or matrices and dynamically alter the settings inside the input file. The GA implementation is highly profiting from this addition to create parametric inputs for each individual.

2.2.2 *Crossover*

Crossover is the mating of two selected parent individuals for generating the child individuals of the next generation. There are numerous algorithms for that, currently we have implemented the single and two point crossover methods. Table 1 shows an example of a two point crossover method, where the two parent individuals P_1, P_2 mate to generate two new children C_1, C_2.

Table 1: Example of two-point crossover. Pivot points are represented with a double line.

P_1:	A	B	C	D	E	F	G	H	I
P_2:	a	b	c	d	e	f	g	h	I
C_1:	A	B	C	d	e	f	g	H	I
C_2:	a	b	c	D	E	F	G	h	I

Two pivot points are randomly sampled over the string of alleles and each child is composed by the sub-strings limited by the two pivot points alternating from the two parents.

2.2.3 *Mutation*

Mutation is a random alteration of an allele to a new random value. It is useful to enlarge the phase space of the search and to go out of local minima. Mutation typically has a very low probability rate typically 0.2-0.3%, if it is increased too much the search might fail to converge it is equivalent of increasing a lot the entropy of the system.

2.2.4 *Fitness function*

The evaluation or the cost of each individual is called "*fitness function*", which is the objective function we are trying to optimize (minimize or maximize) and it is user defined. To evaluate the fitness function in flair, each individual will create a dedicated FLUKA input file with the alleles appearing as **#define** cards in the beginning. Flair will launch the simulation with the input file and in the end collect and process the results. The output of the simulation is provided as input to the fitness function, which in turn evaluates the cost for the individual under study.
The fitness function is the most complicated part of the process, not from the technical point of view but from the conceptual. The user has to provide a fitness function that will summarize, classify within a single number the outcome of the simulation. For example, it is extremely difficult to relate the financial cost versus the physics optimization of the project or safety requirements. Inside flair we provide several methods of extracting the simulated results, in an easy way, however the user has to do the final python coding of the function. Multiple objectives can be described as the linear weighted sum of simpler quantities.

3 Test cases

Extensive tests have been carried out for the correct functioning of the algorithm as standalone or combined with FLUKA, for the fine tuning of the various parameters.

3.1 Parameter tuning

A simplified Monte Carlo optimization problem was devised to provide fast convergence and simulate the effect of each parameter. However, there is no generic guideline on having a unique set of parameters that works best for each problem. The individual tests where ranging from simple binary search problems, function fitting, travelling salesman problem etc..

The problem was consisting of a modified traveller salesman problem, with each node instead of having a fixed location, it includes also an uncertainty, fuzziness in the positioning. The program has to find the minimum path joining all nodes.

In the first run with 100 individuals we kept all parameters zero and with modified the cross-rate. Fig. 1 shows the spread of solutions from all individuals versus the generation number. As it is evident from the plot, the process quickly comes into convergence to the optimum solution based on the starting conditions. What is to remark here is that the higher the cross over parameter the steeper is the gradient of convergence. However, the final value stops on the first local minimum, and all solutions converge slowly to the same value. It is visible that there is no relation on the optimum reached and cross over rate.

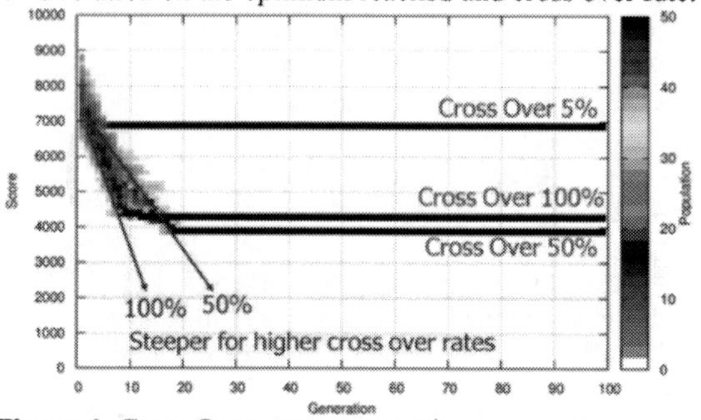

Figure 1. Cross Over parameter testing

In order to allow some individuals to escape the local minimum the mutation parameter is important, which provides extra perturbation to the system. In Fig. 2 is visible that high mutation rates of 10%, enlarges the spectrum of solutions however no good convergegnce can be achieved.

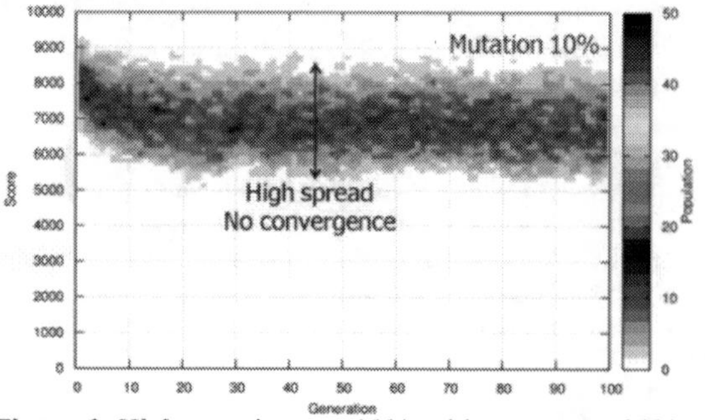

Figure 2. High mutation rate 10%, with cross-over 90%

Lowering the mutation rate, the spread becomes narrower. Also the smaller and slower rate of mutation leaves time to the cross over to minimize to the next local minimum Fig. 3. Lowering too much leads to a slower convergence.

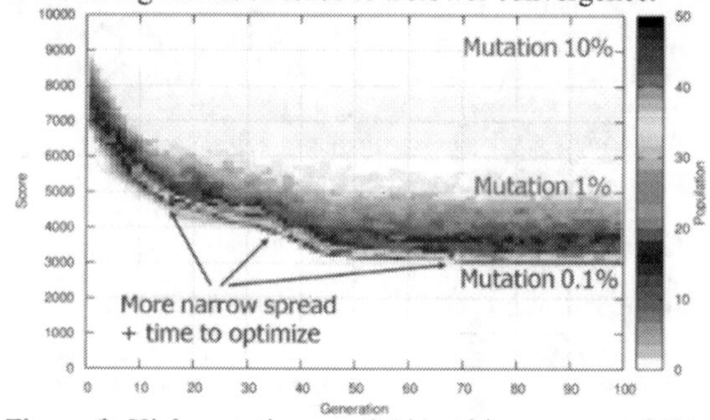

Figure 3. High mutation rate 0.1%, with cross-over 90%

The addition of elitism, where the most fitted solution is copied as such to the next generation ensure the best individual to be propagated from one generation to another. Furthermore, replacing the less fitted individuals with completely new purely random ones, helps in scanning better the phase space of solutions.

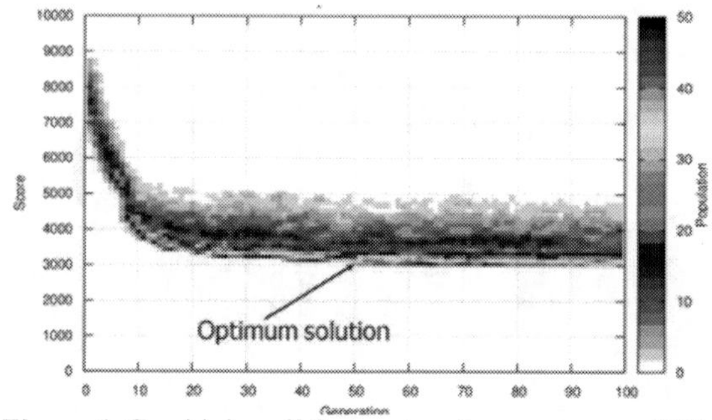

Figure 4. Combining all factors together, cross over 90%, mutation 0.1%, elitism 1, random 5.

Fig. 4 shows the convergence when all parameters are set close to the optimum, which an optimum solution reached already from the 50th generation.

Next we are going to describe two test cases in combination with FLUKA, where the code was used with success, as well as discussing the drawbacks and benefits.

3.2 HF-RFQ shielding

A possible RFQ-based accelerator[6] designed for the production of PET isotopes would be made of two directly coupled high frequency RFQs delivering a proton beam of 10 MeV in energy in a length of less than 5 meters (Fig.5). The main beam parameters are: proton energy: 10 MeV, average current: 20 μA, peak current: 500 μA, RF frequency: 750 MHz, duty cycle: 4 %.
This RFQ could be used as a compact accelerator for producing ^{18}F via the reaction $[^{18}O]H_2O(p,n)^{18}F$: 6.3 GBq/μA. Below 12 MeV, the neutron production rate is equivalent to the ^{18}F saturation yield [7], and for an average of 20 μA current we can consider a neutron

emission rate at saturation of 1.26 10^{11} n/s, with the spectrum as in Fig 6.

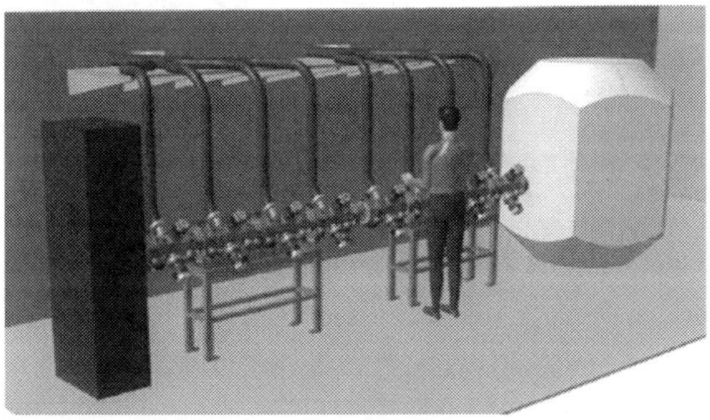

Figure 5. HF-RFQ conceptual design

The aim of the FLUKA simulation study was to find an adequate compact shielding, minimizing the dose from the emerging neutrons so the generator could be installed inside a hospital respecting the radioprotection requirements.

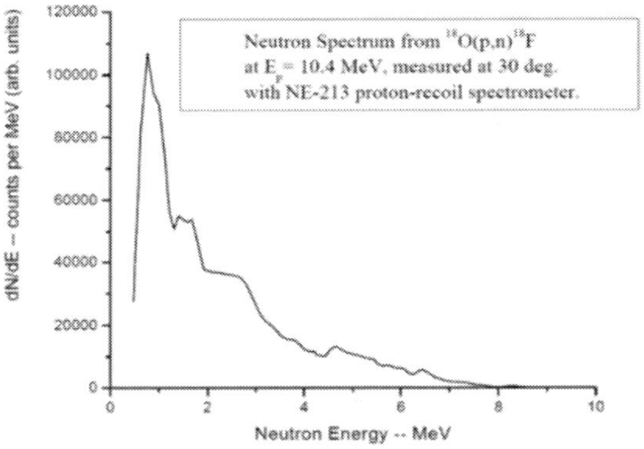

Figure 6. Neutron spectrum from $^{18}O(p,n)^{18}F$ reaction at E_p=10.4 MeV

3.2.1 *Setup*

A special source routine was written to model the outgoing neutron spectrum from Fig. 6 assuming isotropic angular distribution. The neutron target was placed in a shell structure, onion like spherical geometry like in Fig. 7, where the thickness of each layer as well the material was dynamically assigned via **#define** cards Fig. 8. Several materials were considered Polyethylene, Borated-Polyethylene, Iron, Copper, Tungsten, Concrete etc.

To save CPU time, the electromagnetic thresholds were raised, which leads to underestimation of the final dose. However it will not affect the concept of the shielding, since the neutrons are the ones responsible for the dose at the outer layers. Once a solution was attainted, the full simulation using also the electromagnetic part with normal thresholds was performed. Importance biasing was also used to enhance the results in the outer layers. Finally, a spherical scoring was used to enhance the statistics, profiting from the spherical symmetry.

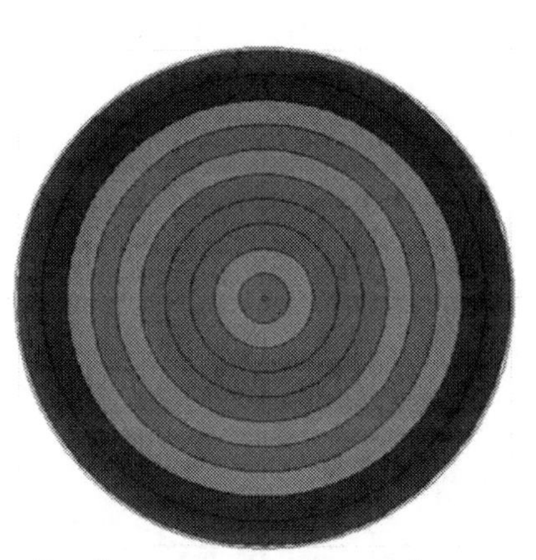

Figure 7. Geometry of the optimum solution. Cyan=Polyethylene, Brown=Iron, Green=Borated Polyethylene, every ring increases by 10cm the radius

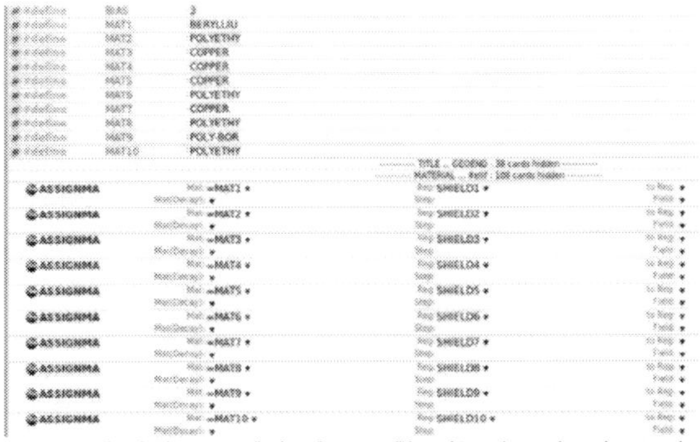

Figure 8. Snippet of the input file showing the dynamic definition of the materials to the various shells.

3.2.2 *Fitness function*

As a fitness function we used a weighted combination of two goals: *i)* minimizing the external diameter of the shielding, and *ii)* ensuring that the dose stays below the limit of 1μSv/h.

3.2.3 *Results*

A first run was performed with a population of 60 individuals and 40 generations, which resulted in a total of 2400 FLUKA runs. Using our cluster, each run lasted about 10min without the electromagnetic part. With 60 jobs running in parallel the total time for the first step of the optimization was about 8h, even though as it can be seen on Fig. 9, the fittest solution was already reached already from the 10th generation. The elitist selection guaranteed to keep this individual until the last generation. The first step provided a possible solution, which in later had to be refined with a complete run, including the electromagnetic transport, generating an initial population with individuals nearby the optimum solution of the first step (Fig 10). The total time of the second step was 10 times longer. The final solution resulted from the GA was a sandwich with the materials from inside to outside, polyethylene, copper and borated polyethylene (Fig. 7).

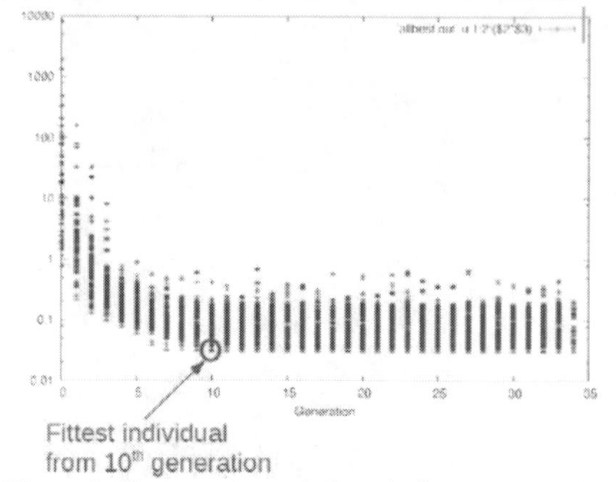

Figure 9. Fitness score evolution of 1st step transporting only neutrons versus the generation number

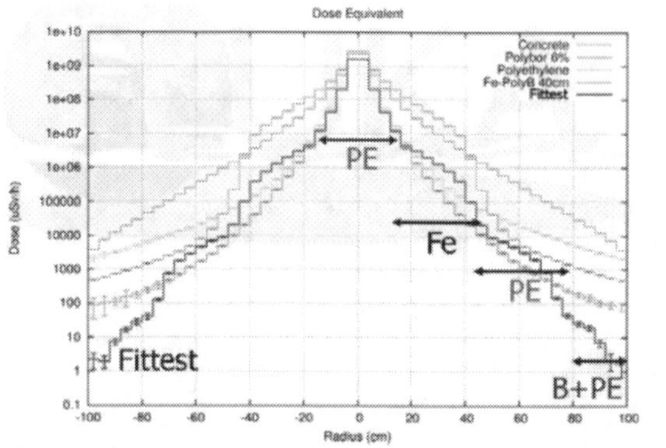

Figure 10. Dose equivalent versus radius for the optimum solution of the 2nd step, performing a full simulation with no cuts or simplifications.

3.3 Treatment Planning System optimization

Commercial Treatment Planning Systems for hadron therapy provide fast results using analytical models and parametric fits of the physical processes. On the other hand Monte Carlo provides superior results due to the detailed description of the physics, however it suffers from long execution times. In this project we are working in enhancing further the optimization result from the TPS using FLUKA based on the previous work [8]. The goal was to fine tune the particle intensity per spot to better approach the prescribed dose on the Planning Target Volume (PTV) and the same time minimizing the dose received by the Organs At Risk (OAR).

3.3.1 Simulation Setup

The simulation setup consisted on calculating the dose per each spot position on a 3D mesh binning, using the real voxel geometry generated from the DICOM-CT and the RT-PLAN. The output of the simulations was provided as input to the pre-optimizer [9]. The pre-optimizer is merging the 3D mesh binning's weighted by the individual intensity for each spot which we want to optimize.

3.3.2 Fitness function

The fitness function is the sum of the square of differences from the prescribed dose in the PTV plus the differences of the dose on the OAR times the Heaviside θ step function.

$$\chi(N) = \sum_{i \in PTV} \left(\widehat{D_{PTV}} - D_i\right)^2 + \sum_{j \in OAR} \left(\widehat{D_{OAR}} - D_j\right)^2 \times \theta\left(\widehat{D_{OAR}} - D_j\right) \quad (1)$$

3.3.3 Results

Two algorithms have been tested, a steep descent optimization and the genetic algorithm. The steep descent is calculating the maximum gradient direction of the fitness function, and using the golden-bisection algorithm the minimum is searched along this direction.
When a minimum is reached along the maximum gradient direction, the maximum gradient is recalculated at the minimum position, and the algorithm is repeated until a convergence is achieved.

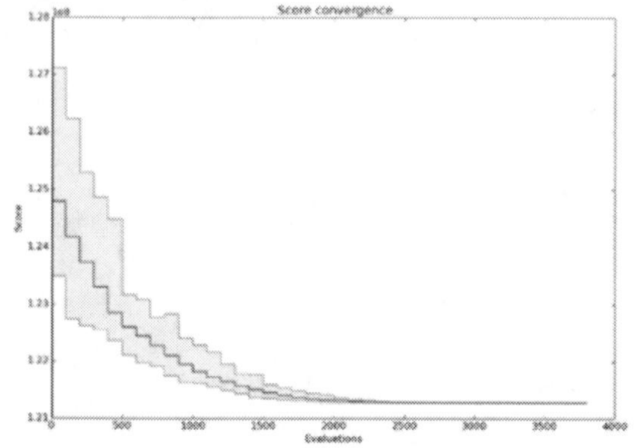

Figure 11. Solution convergence versus the number of evaluations

In the Genetic Algorithm each chromosome is represented as a vector of the beam intensities per spot. Running 60 individuals for 30 generations resulting in 1800 evaluations (Fig.11). The GA optimized the result much faster and gave a better convergence than the steep descent which stuck in a local minimum.

4 Conclusions

The present paper discusses the work in implementing a generic Genetic Algorithm approach directly coupled with the FLUKA Monte Carlo code via the flair interface. The implementation proved to be versatile enough to perform with ease optimization simulations. The main complications arises from the definition of the fitness function, which highly depends on the problem. It can range from either easily defined or being extremely complicated, especially when completely different requirements have to be merged together and represented by a single number.
The current implementation is able to work either in online mode like in the presented case of the HF-RFQ that the fitness function is calculated from the outcome of each simulation, or in offline mode like in the TPS optimization

case that first a large number of simulations is performed and then the GA is optimizing the combination of the simulations.

The main drawback of the GA is the requirement of a lot of iterations with each one requiring a full simulation with reasonable statistics. The CPU time needed can be considerable, unless if drastic optimizations in the simulation input are performed.

The benefit of using GA is that once the fitness function is defined the optimization procedure is almost autonomous. Since GA follows a random path approach, the optimization is converging as a Poisson distribution, with faster convergence when dealing with multiple free degrees of freedom to optimize.

References

1. V.Vlachoudis "*FLAIR: A Powerful But User Friendly Graphical Interface For FLUKA*" Proc. Int. Conf. on Math, Comp Meth & Reactor Physics, Saratoga Springs, NY 2009
2. G. Battistoni et al., "*The FLUKA code: Description and benchmarking*" Proc. Hadronic Shower Simulation Workshop, Fermilab 6-8 Sept 2006, IP Conf Proc 896, 31-49, (2007)
3. A. Fassò et al., "*FLUKA: a multi-particle transport code*" CERN-2005-10 (2005), INFN/TC_05/11, SLAC-R-773
4. Holland, John H. "*Adaptation in natural and artificial systems. An introductory analysis with application to biology, control, and artificial intelligence*" Ann Arbor, MI: University of Michigan Press (1975)
5. http://www.python.org
6. A.M.Lombardi et al. "Beam Dynamics in a High Frequency RFQ", Proc. of IPAC2015, Richmond, VA, USA (2015)
7. L.R.Carroll, "*Estimating the Radiation Source Term for PET Isotope Targets*", Workshop of Targetry and Target Chemistry, Triumf, CA, 2002
8. A.Mairani et al., "*A Monte Carlo-based treatment planning tool for proton therapy*", IOP Publishing 2013, Physics in Medicine and biology, Vol 58.
9. G.Arnau Antoniucci, "*FLUKA Pre-Optimizer for a Monte Carlo Treatment Planning System*", Openlab CERN Summer Student Report Sep 2015.

A computationally efficient moment-preserving Monte Carlo proton transport method in Geant4

D. A. Dixon[1], A. K. Prinja[2], A. P. McCartney[1], and H. G. Hughes[1]

[1]*Los Alamos National Laboratory, Los Alamos, NM 87545*
[2]*University of New Mexico, Dept. of Nuclear Engineering, Albuquerque, NM 87131*

The moment-preserving method, demonstrated as a viable alternative to condensed history for electrons, is extended to protons. Given the generality and the flexibility of the method, a discrete Coulomb scattering and discrete impact-ionization differential cross-section library for protons was readily developed and existing Geant4 electron discrete process and model classes were extended to make use of the new proton library. It is shown that levels of efficiency and accuracy similar to those demonstrated for electrons are obtainable for protons in straight-ahead, energy-loss problems. However, in problems with deflection, agreement is strongly dependent on the proton energy. That is, good agreement was observed in the few MeV range, while unsatisfactory agreement was observed in problems with proton energies above 100-MeV.

KEYWORDS: MCNP6, Monte Carlo, electron-photon transport, multiple-scattering, validation

I. Introduction

The interaction physics for charged particles such as electrons, protons, and heavy ions is characterized by extremely short distances to collisions typically followed by an infinitesimally small change in the particle direction and energy.[(1)] The associated single-scatter differential cross sections (DCS) are nearly singular about zero energy transfers and angular deflections rendering single-scatter Monte Carlo charged particle transport (CPT) simulations computationally impractical without significant computing resources. To overcome the computational burden, step-based Monte Carlo methods[(2–5)] are typically employed. These methods utilize infinite medium multiple scattering[(6)] and multiple excitation/impact-ionization energy-loss theories[(7)] condensing large numbers of collisions into individual collisions over a fixed spatial step. However, these schemes are known to suffer (with varying severity) from the inconsistency between the application of deterministic stepping to an inherently stochastic process. As a result, some step-based methods require ad hoc schemes for mitigating the impact of step-size dependencies and boundary crossing approximations.

In this paper, we extend the recently developed moment-preserving (MP) method demonstrated as a viable alternative to the condensed history method for electron transport,[(8)] to protons. Motivated by Lewis theory, analog DCSs are replaced in this approach by approximate DCSs that preserve a finite number of angular deflection and energy-loss moments of the analog DCS while approximating all remaining moments in terms of the exact lower order moments. This has been shown to yield longer mean free paths and less peaked cross sections,[(8)] which promotes the use of single-scatter Monte Carlo simulation. Moreover, accuracy scales with the number of moments preserved, while efficiency is achieved in most applications through the need to preserve only a limited number of moments.[(8)] Finally, the algorithms associated with the MP method are independent of the form of the underlying DCS, whether a theoretical result or tabulated data (numerical or experimental), and apply generically to all charged particle types. Therefore, significant reduction in code development and maintenance can be realized through use of a generic class that is extensible to a variety of charged particle types.

II. Method

Herein, we investigate the extension of the MP method to proton transport for three problems: the infinite medium problem (angular deflection only), the straight-ahead problem (energy-loss only), and a 1-D energy deposition problem (both angular deflection and energy-loss). We begin by introducing the analog cross section models used to generate the numerical benchmarks along with the corresponding DCS moments. We then define the discrete MP cross section models and summarize the procedure for computing discrete points and weights.

1. Differential cross-section models

The primary proton collision types explored herein are elastic Coulombic collisions with atoms (with cross sections denoted as Σ_a) and impact-ionization collisions with atomic electrons (with cross sections denoted as Σ_e) . For simplicity, we ignore delta-ray electrons, but accounting for delta-rays can be accommodated in the MP method as previously demonstrated.[(8)] We begin with a discussion of elastic Coulombic collisions with atoms as described by the Wentzel differential cross section.[(9,10)]

The Wentzel cross section describes collisions with atoms

DOI: 10.1051/epjconf/201715306005

where particles are deflected through some angle and impart energy to the atom depending on the angle through which the particle was deflected. For particles whose relativistic mass is not much smaller than the relativistic mass of the target, one must consider the cross section in the center-of-mass (COM) frame. The Wentzel differential cross section (units of cm^2) for protons in the COM frame is given by:

$$\Sigma_a(E_{lab}^{kin}, \mu) = 2\pi \left(\frac{r_e Z_p Z_t m_e c^2}{\beta_r p_{com}} \right)^2 [1 + 2\eta + \mu]^{-2}, \quad (1)$$

where

$$\eta = Z_t^{2/3} \left(\frac{m_e c^2 \alpha}{a_{TF} p_{com}} \right)^2 \left[1.13 + 3.76 \left(\frac{Z_p Z_t}{\beta_r} \right)^2 \right], \quad (2)$$

and E_{lab}^{kin} is the kinetic energy of the projectile in the lab frame, μ is the cosine of the angle between the initial and final momentum vector of the projectile, r_e is the classical electron radius, Z_p is the atomic number of the projectile, Z_t is the atomic number of the target atom, $m_e c^2$ is the electron rest mass energy, α is the fine structure constant, $a_{TF} = \frac{1}{2}\left(\frac{3\pi}{4}\right)^{2/3}$ is the Thomas-Fermi constant, β_r is the particle velocity per c in an "effective" frame, p_{com} is the COM momentum of the projectile, and η is the screening parameter. Definitions of the above terms can be found in the iliterature.[9,10]

Impact-ionization collisions are considered those in which the energy transferred by the projectile to a bound electron exceeds the binding energy of the target atom. A model for this collision type is the relativistic Rutherford DCS[1] for spin 1/2 particles, or:

$$\Sigma_e(E_{lab}^{kin}, Q) = N_e \frac{2\pi r_e^2 Z_p^2 Z_t m_e c^2}{\beta_{lab}^2 Q^2} \left[1 - \beta_{lab}^2 \frac{Q}{Q_{max}} + \frac{1}{2} \left(\frac{Q}{E_{lab}^{kin} + mc^2} \right) \right], \quad Q_{min} \le Q \le Q_{max}, \quad (3)$$

where Q_{min} is the mean ionization potential,

$$Q_{max} = (m_e c^2)^2 \beta_{lab}^2 \gamma_{lab}^2, \quad (4)$$

and $\gamma_{lab}^2 = 1/(1 - \beta_{lab}^2)$.

These cross sections will be used to establish the numerical benchmarks from which the accuracy and efficiency of the MP method will be evaluated. Furthermore, the MP DCS's are constructed from the moments of the analog DCS cross sections.

2. Differential cross section moments

As described in Dixon *et al.*,[8] differential cross section moments are central to the MP method so we now define the relevant DCS moments and then introduce the MP models used in this study.

Legendre moments of the Wentzel DCS in the COM are given by:

$$\Sigma_{n,\ell} = 2\pi \int_{-1}^{1} P_\ell(\mu) \Sigma_a(E_{lab}^{kin}, \mu) d\mu. \quad (5)$$

The moments contain important physical quantities such as the collision mean free path, $\lambda_a = 1/\Sigma_{a,0}$, and the transport mean free path, $\lambda_{tr} = 1/[\Sigma_{a,0} - \Sigma_{n,1}]$ which is related to the mean scattering cosine. In charged particle transport $\lambda_{tr}/\lambda_a >> 1$, which implies that particles undergo a staggering number of collisions before any significant change in the particle direction is observed. There is a well-known recurrence[2] for computing the moments in Eq. (5); however, we use a Gauss-Kronrod quadrature algorithm to compute the Legendre moments.

Now, we present the energy-loss moments of the relativistic Rutherford DCS given in Eq. (3):

$$Q_n(E) = \int_{Q_{min}}^{Q_{max}} Q^n \Sigma_e(E, Q) dQ, \; n = 0, 1, ... \quad (6)$$

Defining $C(E)$ as:

$$C(E) = \frac{4\pi r_e^2 Z_1^2 Z_2 m_e c^2}{\beta^2} \quad (7)$$

and inserting Eq. (3) into Eq. (6) for $n = 0$ gives:

$$\Sigma_{e,0} = C(E) \left[\left(\frac{1}{Q_{min}} - \frac{1}{Q_{max}} \right) - \frac{\beta^2}{Q_{max}} \ln \left(\frac{Q_{max}}{Q_{min}} \right) \right], \quad (8)$$

for $n = 1$:

$$\Sigma_{e,1} = C(E) \left[\ln \left(\frac{Q_{max}}{Q_{min}} \right) - \beta^2 \left(1 - \frac{Q_{min}}{Q_{max}} \right) \right], \quad (9)$$

and for for $n > 1$:

$$\Sigma_{e,n} = C(E) \left[Q_{max}^{n-1} \left(1 - \beta^2 \frac{n-1}{n} \right) - Q_{min}^{n-1} \left(1 - \beta^2 \frac{n-1}{n} \frac{Q_{min}}{Q_{max}} \right) \right]. \quad (10)$$

We note that at high energies, where $Q_{max}(\beta_{lab})$ can be significant, the higher moments increase with increasing order index n, while at lower energies they decrease with increasing n. Thus, while moments higher than two are often neglected when describing thick target energy spectra, they play a significant role in the transport of high energy protons through thin targets.

3. Discrete moment-preserving models

The form of the MP DCS used herein is referred to as the discrete model and for Coulomb elastic scattering is defined as:

$$\tilde{\Sigma}_a(E, \mu) = \sum_{\ell=1}^{L} \frac{\alpha_\ell(E)}{2\pi} \delta[\mu - \xi_\ell(E)], \quad (11)$$

where α_ℓ and ξ_ℓ are discrete points and weights respectively, and the total cross section is:

$$\tilde{\Sigma}_{a,0} = \sum_{\ell=1}^{L} \alpha_\ell(E). \quad (12)$$

Similarly, discrete impact-ionization DCS is defined as:

$$\tilde{\Sigma}_e(E, Q) = \sum_{k=1}^{K} \beta_k(E) \delta[Q - \gamma_k(E)], \quad (13)$$

where γ_k and β_k are the discrete points and weights respectively, and the total cross section is:

$$\tilde{\Sigma}_{e,0} = \sum_{k=1}^{K} \beta_k(E). \tag{14}$$

Construction of MP DCSs is somewhat generic in that one simply choses a preferred DCS representation using some number of parameters - in this case, discrete points and weights - and then sets up a system of equations for the parameters by enforcing moment preservation. The moment-preservation constraint is stated as follows:

$$\Sigma_{a,\ell} = \tilde{\Sigma}_{a,\ell}, \quad \ell = 1, 2, ..., L, \tag{15}$$

and

$$\Sigma_{e,n} = \tilde{\Sigma}_{e,n}, \quad n = 1, 2, ..., N, \tag{16}$$

for elastic Coulomb collisions and impact-ionization collisions respectively. By introducing Eqs. (12) and (13) into Eqs. (15) and (16) respectively, one can form a system of equations for the discrete points and weights. The resulting system of equations lends itself to a quadrature-based solution approach that is described in Dixon *et al.*[(8)]

As indicated in Eqs. (15) and (16), one enforces preservation of the higher-order moments and intentionally allows the zeroth moment to vary. In doing, the zeroth moment is self-determined by the DCS construction process and results in reduction of the zeroth moment (or the total cross section). This effect is seen clearly in Figs. (1) and (2) where the ratio of the approximate total cross section to the true total cross section is presented. Note that as the particle energy increases, the ratio grows because the DCS becomes more peaked at higher energies, thus, enhancing the impact of the MP DCS construction process. In contrast, increasing the number of points and weights reduces this effect because increasing the number of points and weights increases the level of moment preservation and the MP DCSs effectively go over to the analog DCSs.

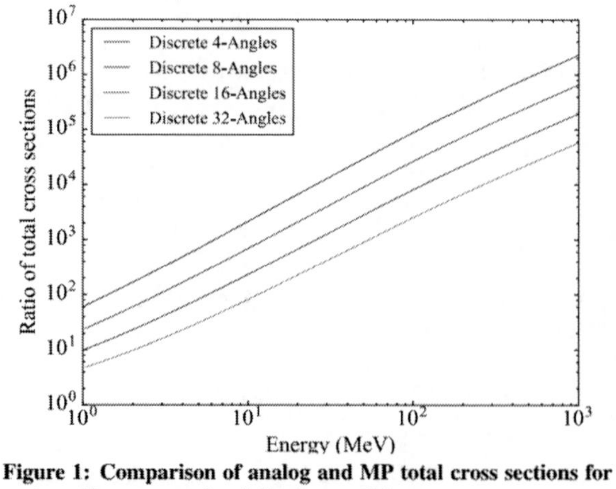

Figure 1: Comparison of analog and MP total cross sections for various discrete angular deflection models.

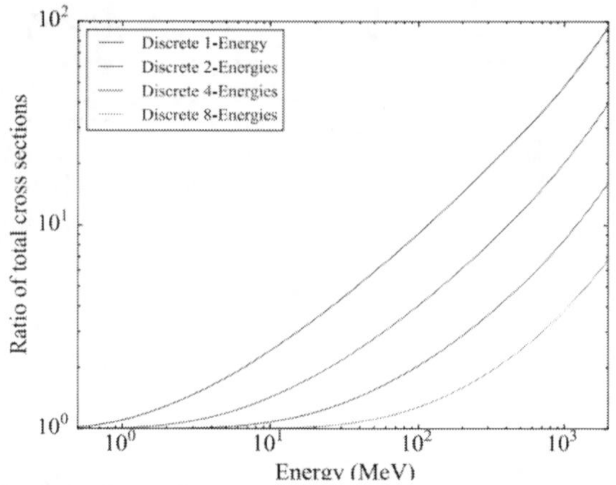

Figure 2: Comparison of analog and MP total cross sections for various discrete impact-ionization models.

III. Results

In the following sections, we present results from infinite medium transport models (angular distributions and so on), straight-ahead transport models (energy spectra), and a more complete transport model (energy deposition profiles).

1. Angular, longitudinal, and lateral distributions

Here, we study angular, longitudinal, and lateral distribution for protons (1-, 10-, and 100-MeV) in infinite media. Protons start at the origin and are followed until they reach a termination distance that is given by some fraction of the proton transport mean free path. Along the traversed path, protons may only undergo Coulombic collisions with target atoms and energy losses resulting during the collisions are neglected. When the proton reaches the end of the prescribed distance, a tally is made of the current direction, and the lateral and longitudinal displacement. Corresponding results are presented in Figs. 6 - 8 and Table 1.

In Fig. 6, angular distributions are presented for 1-, 10-, 100-MeV protons traversing distances of $\Sigma_{tr}/10$ and $\Sigma_{tr}/3$ in copper. Numerical benchmarks are compared to either 4- or 8-angle MP models. The angular distributions for 1-MeV and 10-MeV protons agree between the numerical benchmark and the MP models with exception of the 4-angle model for a traversed distance of $\Sigma_{tr}/10$. At 100-MeV, we observe significant disagreement between the numerical benchmark and the MP models (this disagreement is reflected in the energy deposition results).

Longitudinal and lateral displacement distributions are presented in Figs. 7 and 8 (note the difference in scale for these results). Keeping in mind the difference in scale, Figs. 7 and 8 show stronger disagreement as a function of increasing energy; particularly, in the tails of the distribution. Once again, the 8-angle model was the only model tested with consistent agreement and this agreement is only observed at for 1-MeV protons.

In general, the infinite medium results show that the discrete MP models tend to over-diffuse. This effect is exacerbated by increasing energy. Regarding efficiency, from Table 1 one might conclude that there is great potential for efficiency gains in the MP method. However, one will find that the MP method is very effective at extending the mean free path associated with Coulombic collisions with target atoms, but not as effective at extending the mean free path associated with impact-ionization. Therefore, in more realistic transport where MP models for Coulombic collisions with target atoms and electrons are combined, efficiency gains are limited by impact-ionization collisions, or the latter.

Table 1: Speed-up factors for computation of pathlength distributions in copper using 8-angle model. Speed-up factors are the ratio of the analog model simulation CPU time to that of a discrete model simulation CPU time .

Pathlength (cm)	1-MeV	10-MeV	100-MeV
$0.1\sigma_{tr}$	48	157	4033
$0.3\sigma_{tr}$	50	168	4445

2. Energy spectra

In this section, we study energy spectra of protons emerging from tungsten slabs with increasing thicknesses in a straight-ahead transport model where protons are restricted to energy-loss collisions. Fig. 9 presents a comparison of numerical benchmarks generated using the analog physics with various discrete models. For each of the spectra results, 10^6 particles were followed. Results indicating levels of accuracy and efficiency are compared in Fig. 9 and in Table 1 respectively.

For thinner slabs, the distributions are skewed towards higher energies and preservation of additional moments is required to capture this structure. As the slab thickness increases, the spectra broadens, indicating that higher-order moments are less important and a model with two discrete energies (preserving four moments) is sufficient. These results illustrate an advantage and a limitation of the MP method. That is, as the slab thickness is reduced to 0.1-cm, we observe discrete artifacts when using eight discrete points and weights. The artifacts arise in problems where an insufficient number of collisions occur resulting in preferred channels of energy-loss. This is easily resolved by increasing the number of discrete points and weights, or using alternative representations such as the hybrid model.[8] The opposite effect is observed in the 20-cm slab, where significant spreading in energy occurs via an increased number of collisions. Under these conditions, one can reduce the number of points and weights while maintaining agreement with the numerical benchmark. These two observables highlight one of the key features of the MP method - one can systematically control the accuracy of the method in varying scattering regimes. Table 1 shows that the efficiency of the MP method is model dependent and ranges from roughly a factor of 4 to 50 times faster than the computer time required to generate the numerical benchmark. This indicates that there is a clear connection between accuracy and efficiency in the MP method. As with most approximations, one must strike a balance between efficiency and accuracy.

Table 2: Speed-up factors for straight-ahead transport in tungsten. Speed-up factors are the ratio of the analog model simulation CPU time to that of a discrete model simulation CPU time.

	Discrete Model			
Slab Thickness (cm)	1-Energy	2-Energies	4-Energies	8-Energies
0.1	28	14	8	4
0.5	42	20	9	4
1.0	54	27	9	5
2.0	53	27	9	4
5.0	47	22	10	4
20.0	49	21	9	4

3. Energy deposition

In this section, energy deposition profiles for 7-MeV and 235-MeV protons in silicon and tungsten, respectively, are presented. In Figs. 3 - 5, energy deposition profiles for 7-MeV protons in silicon generated using various approximations are compared to a numerical benchmark. In Fig. 3 two different discrete MP models are compared to a numerical benchmark generated using the analog DCSs presented in Eqs. (1) and (3). As seen in the energy deposition profile, the two MP models (4-angles, 4-energies and 8-angles, 8-energies) are indistinguishable on visual inspection. However, the relative difference plot below the energy deposition profiles in Fig. 3 shows that there is disagreement in the first cell (closest to the source) and is between 3-4%. Agreement improves to fractions of a percent in the following cells until nearing the region of peak dose, where disagreement increases to a maximum of 2-3%. Finally, disagreement in the tail of the profile, or beyond the peak dose region, grows substantially. However, in this region, the energy deposited is rapidly falling to zero and uncertainty tends to overwhelm the solution.

In Fig. 4, three different approximations are compared to a numerical benchmark, where angular deflection is given by the analog DCS in Eq. (1) and energy-loss is given by the G4hIonisation model[11] (a multiple energy-loss straggling model). Similarly, the approximate models all combine the G4hIonisation model with various approximation representations of angular deflection (discrete MP representations or the Geant4 proton multiple-scattering model - i.e., Geant4 Protons). In contrast to the results in Fig. 3, there is no significant disagreement in the first cell. Like the results in Fig. 3, observable disagreement occurs in regions nearby the peak deposition. The Geant4 proton model tends to underestimate more significantly in the approach to the peak, while the 4-angle discrete MP model shows the greatest disagreement in the peak deposition region. In all cases, the greatest disagreement is with 4% of the numerical benchmark.

In Fig. 5, energy deposition profiles for 235-MeV protons in tungsten generated using various approximations are compared to a numerical benchmark. Again, the numerical benchmark was generated using the analog DCS in Eq. (1) for angular deflection and the G4hIonisation model for energy-loss. Two approximations are compared - an 8-angle discrete MP model and the Geant4 multiple-scatter model (each model was combined with the G4hIonisation model). Figure 5 shows clear disagreement between the numerical benchmark and the two approximations, where the most significant disagreement is attributed to the MP model. This disagreement is supported by the observations made in the discussion of the angular, longitudinal, and lateral distribution section, where it was noted that the discrete MP models tended to over-diffuse the solution - particularly, as energy increases.

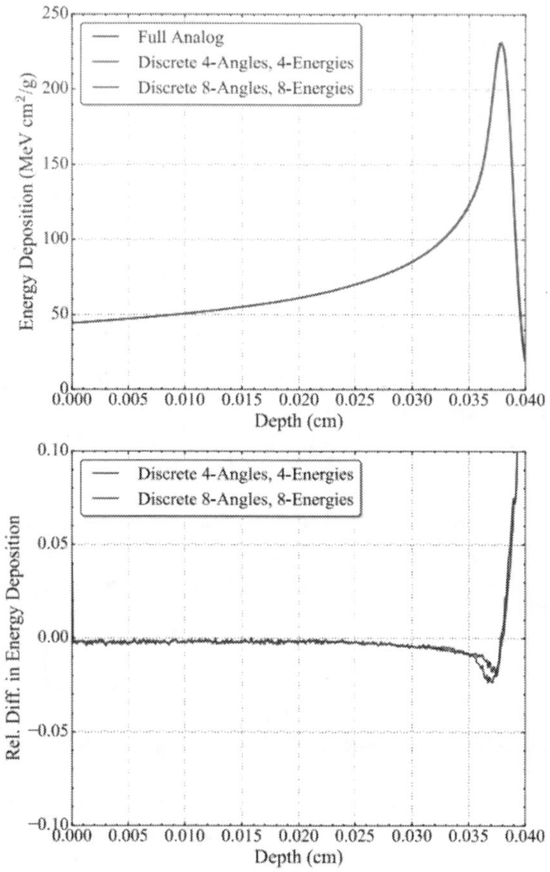

Figure 3: Comparison of proton energy deposition profiles for 7-MeV protons incident on a 1-D silicon slab. The benchmark was generated using the analog cross sections described in section 2, while the other profiles were generated using MP models. Relative differences are presented below the energy deposition profiles where differences are measured with respect to the benchmark.

In all cases, efficiency was strongly dependent on the range cut applied to electrons. That is, when applying a large range cut to electrons, the distance to a hard collision grows significantly. In effect, the G4hIonisation model is effectively an application of CSDA tables along the path of the proton between angular deflection collisions. Under these conditions, efficiency gains - measured with respect to the time required to compute the numerical benchmark - were on the order of 10^2. However, as the range cut is reduced, efficiency gains are reduced to less than a factor of ten or the time to completion using the approximate model is less than 10 times that of the analog model. From this, one can conclude that efficiency is dominated by impact-ionization collisions. In other words, one can develop models that are highly effective at extending the mean free path associated with angular deflection, but the effectiveness of the aforementioned model is limited by the shortest mean free path across all models considered.

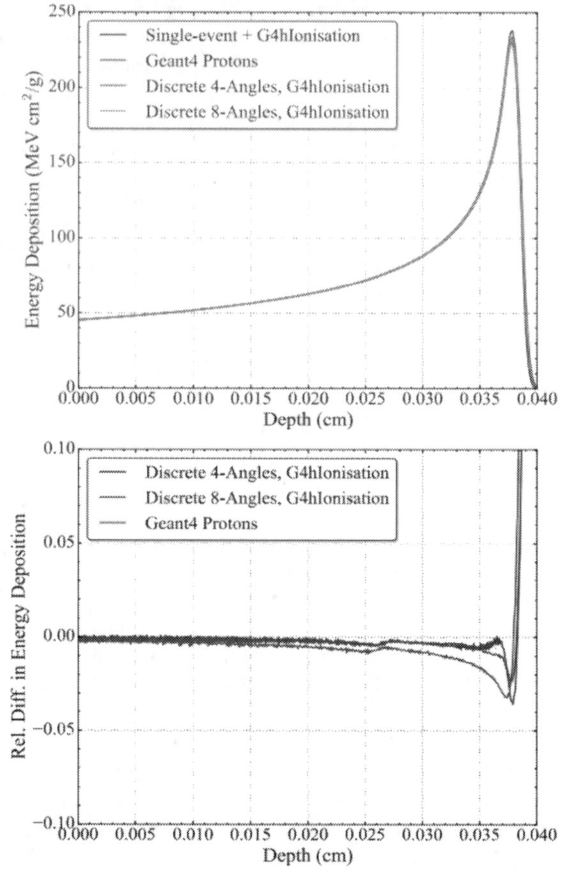

Figure 4: Comparison of proton energy deposition profiles for 7-MeV protons incident on a 1-D silicon slab. The numerical benchmark was generated using an analog Coulomb scattering model combined with the G4hIonization model (blue). The other profiles were generated using approximate methods including the standard Geant4 multiple-scattering treatment for protons (green), the discrete 4-angle MP model combined with the G4hIonization model (red), and the discrete 8-angle MP model combined with the G4hIonization model (teal). In addition, relative differences are presented below the energy deposition profiles where differences are measured with respect to the numerical benchmark.

IV. Conclusions

This paper served as an initial test of the applicability of the MP method for proton transport. A variety of discrete MP models were tested for use in computing angular, lateral, and longitudinal distributions, energy spectra, and energy deposition profiles. Consistent with previous work on electrons, accuracy and efficiency are both model and problem dependent. That is, efficient, low-order models can be used in thick problems where signif-

DOI: 10.1051/epjconf/201715306005

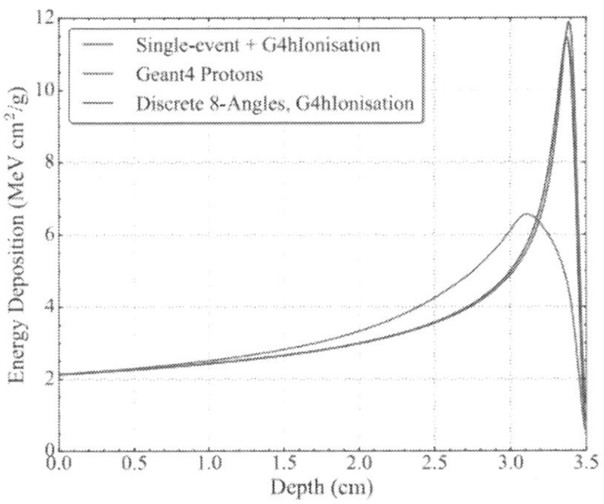

Figure 5: Comparison of proton energy deposition profiles for 235-MeV protons incident on a 1-D tungsten slab. The numerical benchmark was generated using an analog Coulomb scattering model combined with the G4hIonization model (blue). The other profiles were generated using approximate methods including the standard Geant4 multiple-scattering treatment for protons (green) and the discrete 8-angle MP model combined with the G4hIonization model (red). Relative differences are not presented for this case because the disagreement between the MP model and the numerical benchmark is .

icant spreading of the beam occurs. Alternatively, in peaked regimes, such as very thin slabs, less-efficient, higher-order models can be applied.

While there were similarities between observations made in this work and in previous work on electrons, systematic accuracy was not not observed where discrete representations of angular deflection were applied. In such problems, disagreement worsened with increasing energy. A somewhat speculative conclusion is that the discrete representation angular deflection is not appropriate for protons given the extreme peakedness - considerably more so than for electrons - of the analog DCS. Alternatively, it is possible that there is an inconsistency in the generation and the application of the discrete DCS for angular deflection, particularly, when boosting the deflection angle sampled in the COM to the lab frame. Moreover, it is possible that the expression for the accompanying energy-loss derived from relativistic kinematics should be considered when computing the discrete points and weights. That is, one might use Legendre/energy-loss moments with cross terms as input for the discrete cross section generation tool. Assuming that the discrete models are inappropriate for angular deflection, there are alternative MP formulations. Of particular interest is the hybrid model,[(8)] where discrete points and weights are applied over the small range corresponding to the peak of the DCS and the tail is captured exactly by the analog DCS.

Regarding the discrete impact-ionization models, we observed satisfactory efficiency and accuracy. That is, in the straight-ahead problems, application of various discrete impact-ionization models showed good agreement with the numerical benchmarks and efficiency gains ranged from 4 - 50 times faster than the analog model.

Finally, the Geant4 process and model classes used for electrons were easily extended for use in treating protons. This is because the process and model classes do not require modification, rather we simply supplied the new proton MP DCSs. So, generalization of the MP method to protons and heavy ions will mostly require identifying ideal analog DCSs and generating the corresponding MP DCSs that are appropriate for these particle types.

Acknowledgments

This work was funded by the US Department of Energy's National Nuclear Security Administration, under Los Alamos National Security, LLC, Contract No. DE-AC52-06NA25396.

References

1) R. Evans, *The Atomic Nucleus*, McGraw-Hill, New York, tenth printing edition, (1955).
2) M. Berger, "Monte Carlo Calculation of the Penetration and Diffusion of Fast Charged Particles," *Proc. Methods in Computational Physics*, volume 1 of *Statistical Physics*, p. 135–215, Advances in Research and Applications, Academic Press, 1963.
3) A. Bielajew and D. D. O. Rogers, "PRESTA: The Parameter Reduced Electron-Step Transport Algorithm for Electron Monte Carlo Transport," *Nucl. Instr. Meth. B*, **18**, 165-181 (1986).
4) A. Bielajew and F. Salvat, "Improved electron transport mechanics in the PENELOPE Monte-Carlo model," *Nucl. Instr. Meth. B*, **173**, 332-343 (2001).
5) F. Salvat, J. M. Fernández-Varea, and J. Sempau, *PENELOPE-2011: A Code System for Monte Carlo Simulation of Electron and Photon Transport*, Universitat de Barcelona, 2011.
6) G. Molière, "Theori der Streuung schneller geladener Teilchen I: Einzelstreuung am abgeschirmten Coulomb-Field," *Z. Naturforsch.*, **2a**, 133 (1947).
7) P. V. Vavilov, "Ionization Losses of High-Energy Heavy Particles," *Sov. Phys. JETP*, **5**, *4*, 920-923 (1957).
8) D. Dixon, A. Prinja, and B. Franke, "A computationally efficient moment-preserving Monte Carlo electron transport method with implementation in Geant4," *Nucl. Instr. Meth. B*, **359**, 20-35 (2015).
9) M. J. Boschini et al., "Nuclear and Non-Ionizing Energy-Loss for Coulomb Scattered Particles from Low Energy up to Relativistic Regime in Space Nuclear and Non-Ionizing Energy-Loss for Coulomb Scattered Particles from Low Energy up to Relativistic Regime in Space Radiation Environment," *Proceedings of the ICATPP Conference on Cosmic Rays for Particle and Astroparticle Physics* (2010).
10) *Principles of Radiation Interaction in Matter and Detection*, World Scientific, 4th edition, (2016).
11) S. Agostinelli et al., "Geant4—a simulation toolkit," *Nucl. Instr. Meth. A*, **506**, 250–303 (2003).

 DOI: 10.1051/epjconf/201715306005

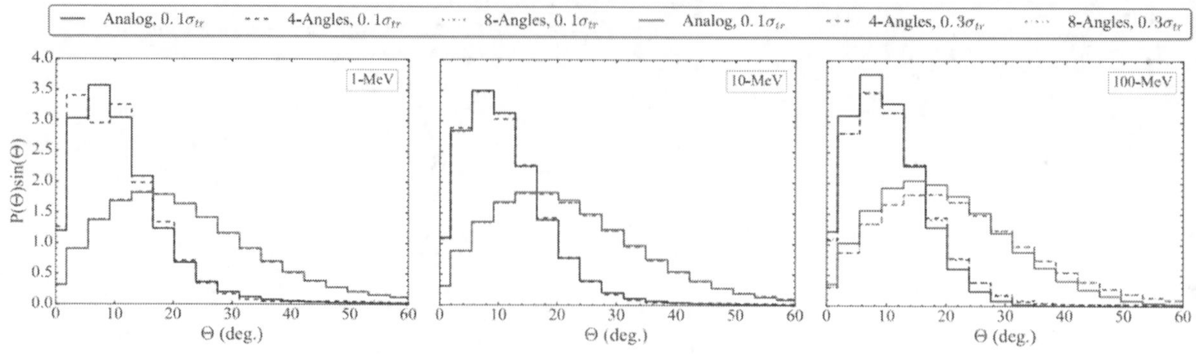

Figure 6: Comparison of angular distributions for 1-MeV, 10-MeV, and 100-MeV protons in an infinite medium of copper after traveling a distance of $\Sigma_{tr}/10$ and $\Sigma_{tr}/3$. The analog solutions (solid curve) were obtained using the Wentzel DCS and compared to a 4-angle (dash curve) and 8-angle (dash-dot curve) discrete model.

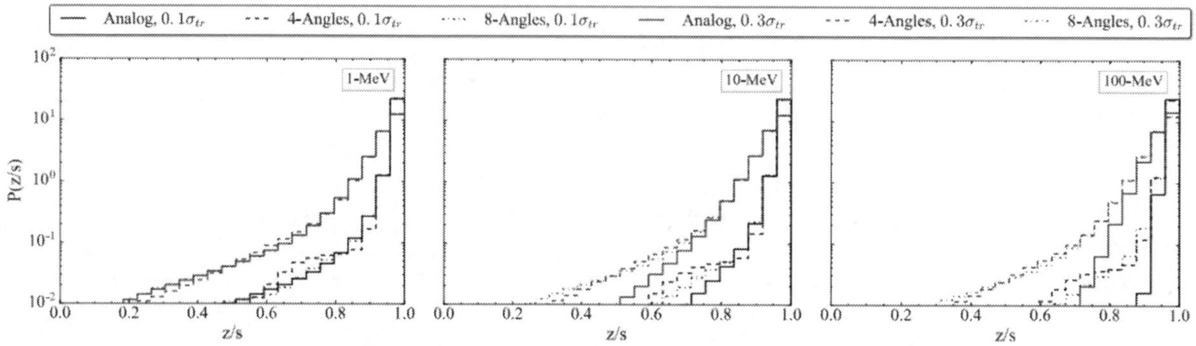

Figure 7: Comparison of longitudinal distributions for 1-MeV, 10-MeV, and 100-MeV protons in an infinite medium of copper after traveling a distance of $\Sigma_{tr}/10$ and $\Sigma_{tr}/3$. The analog solutions (solid curve) were obtained using the Wentzel DCS and compared to a 4-angle (dash curve) and 8-angle (dash-dot curve) discrete model.

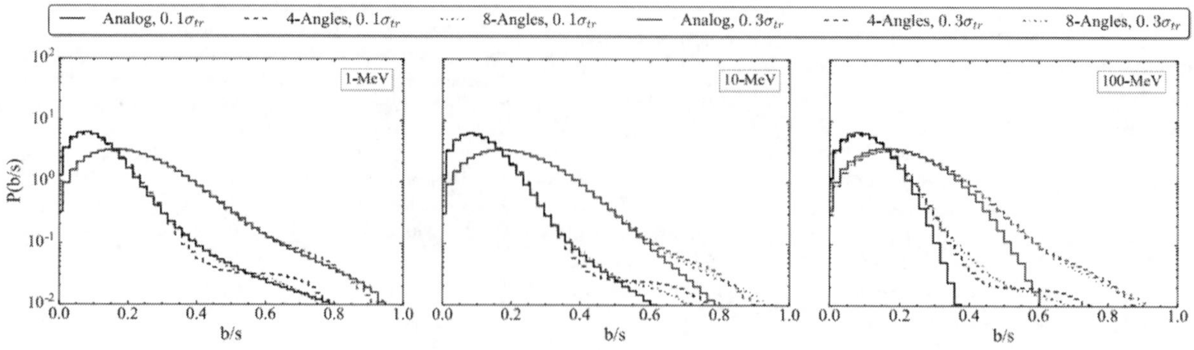

Figure 8: Comparison of lateral distributions for 1-MeV, 10-MeV, and 100-MeV protons in an infinite medium of copper after traveling a distance of $\Sigma_{tr}/10$ and $\Sigma_{tr}/3$. The analog solutions (solid curve) were obtained using the Wentzel DCS and compared to a 4-angle (dash curve) and 8-angle (dash-dot curve) discrete model.

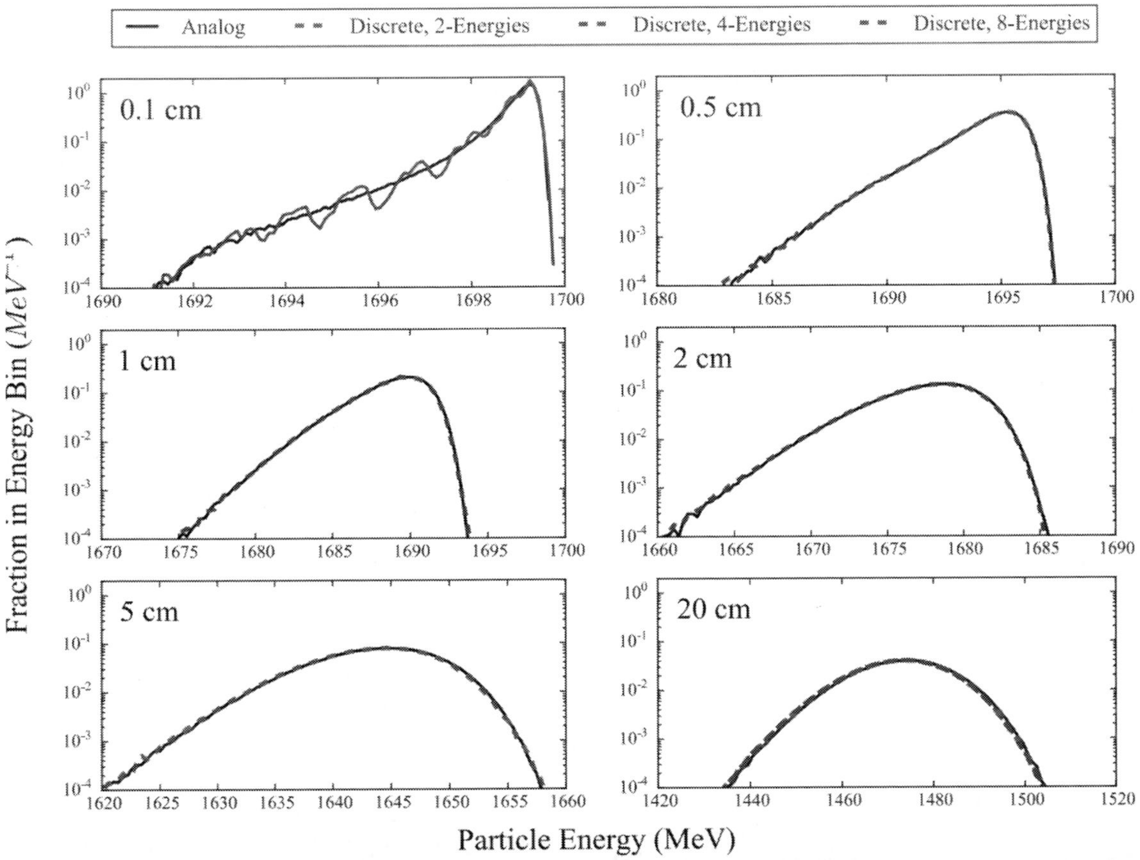

Figure 9: Comparison of energy spectra in 0.1-, 0.5-, 1-, 2-, 5-, and 20-cm tungsten slabs calculated using the analog model and various discrete models. In thicker slabs (bottom row), low-order models are sufficient, while in very thin (top row) slabs higher-order models are required. For the extreme case (0.1-cm) discrete artifacts are present with the most accurate model tested (eight discrete energies).

Adaptive Multilevel Splitting for Monte Carlo particle transport

Henri Louvin[1,*], Eric Dumonteil[2], Tony Lelièvre[3], Mathias Rousset[3], and Cheikh M. Diop[1]

[1] CEA Saclay, DEN/DM2S/SERMA/LTSD, F-91191 Gif-sur-Yvette
[2] IRSN, PSN-EXP/SNC, F-92262 Fontenay-aux-Roses
[3] Université Paris-Est, CERMICS (ENPC), INRIA, F-77455 Marne-la-Vallée

Abstract. In the Monte Carlo simulation of particle transport, and especially for shielding applications, variance reduction techniques are widely used to help simulate realisations of rare events and reduce the relative errors on the estimated scores for a given computation time. Adaptive Multilevel Splitting is one of these variance reduction techniques that has recently appeared in the literature. In the present paper, we propose an alternative version of the AMS algortihm, adapted for the first time to the field of particle tranport. Within this context, it can be used to build an unbiased estimator of any quantity associated with particle tracks, such as flux, reaction rates or even non-Boltzmann tallies. Furthermore, the efficiency of the AMS algorithm is shown not to be very sensitive to variations of its input parameters, which makes it capable of significant variance reduction without requiring extended user effort.

1 Introduction

The challenge in using Monte Carlo particle transport simulations for shielding applications is to minimize the computation time required to attain a reasonable variance on the quantity of interest, called *score*.

The basic approach of variance reduction techniques is to modify the simulation behaviour so as to increase rare events occurrence while keeping an unbiased estimator of the score.

In this view, *multilevel splitting techniques* were introduced to the field of particle transport by Kahn and Harris [1]. The principle of these techniques is to increase the number of simulated particles when approaching areas of interest of the geometry. Practically, the simulated space is divided into regions of importance, and the particles that pass from a less important to a more important region are duplicated. Each of the duplicated particles is given half the weight of the original to ensure that the simulation remains unbiased. Thus, more computation time is spent to simulate interesting particles rather than new particles from the source.

The downside of these techniques is that they require a fair knowledge of the system in order to accurately define the importance regions. More recently, a new method called Adaptive Multilevel Splitting (or AMS) has been proposed by Cérou and Guyader [2], and studied in a more general setting by Bréhier et al. [3]. This method also aims to duplicate the interesting particles of the simulation, but uses the knowledge accumulated during the simulation itself to determin on the fly which particle should be duplicated and at which point. One of the most interesting feature of AMS is that it requires little knowledge of the system to yield interesting results. The efficiency of the AMS in Monte Carlo simulations and its properties makes it attractive for computational physics problems that require precise rare event simulation. To this end, AMS was successfully extended to molecular dynamics simulations by Aristoff et al. for the resampling of reactive paths [4].

In this paper, we aim to apply the AMS algorithm to Monte Carlo particle transport, and demonstrate its efficiency for rare event simulations. In Section 2, we will describe a mathematical version of the AMS algorithm specifically designed to fit the requirements of particle transport. We introduce in Section 3.1 the context of the study, which is neutral particle transport with the Monte Carlo method. The core of this work is presented in Section 4, in which we introduce for the first time a pratical implementation of AMS within a Monte Carlo particle transport simulation. This version of the AMS algorithm was implemented in the development version of the Monte Carlo particle transport code TRIPOLI-4®. In the last Section of this paper, we present some of the results obtained using AMS within TRIPOLI-4®. These examples illustrate the validity of the algorithm as described in Section 4, as well as the efficiency of AMS as a variance reduction technique.

2 Mathematical setting

We present in this section a specific mathematical setting of the AMS algorithm, which will be afterwards fairly easy to adapt to the context of particle transport. This version

*e-mail: henri.louvin@cea.fr

of the AMS is a variant of the algorithm which perfectly fits the theoretical frame from [3], so that the estimator of the rare event occurence probability introduced in the following is unbiased.

2.1 Objective and setup

Let $X = (X_t)_{t\in\mathbb{N}}$ be a discrete-time Markov chain with values in $\mathbb{R}^d$, $d \in \mathbb{N}^*$. We define $\mathcal{P}$ as the path space, containing all possible realisations of the Markov chain X. Given T a subset of $\mathbb{R}^d$, we define the entrance time of the Markov chain X in T:

$$\tau_T = \inf\{t \in [0; \tau_f] : X_t \in T\},$$

where τ_f is the final stopping time of the Markov chain X, which we suppose finite.

Given an observable $\phi_T : \mathcal{P} \to \mathbb{R}$ such as $\phi_T = 0$ on the event $\tau_f < \tau_T$, we would like to estimate the average $\langle\phi_T\rangle$ of ϕ_T. Let us suppose now that the probability for (X_t) to enter T before τ_f is very small, i.e. $\mathbb{P}(\tau_T < \tau_f) \ll 1$. In such a case of rare event simulation, the AMS algorithm proposes to reduce the variance on the estimated average by increasing the number of Markov chains reaching the subset T. AMS is an iterative algorithm that consists of several steps. The first step is a basic Monte Carlo simulation of multiple replicas of the Markov chain X with a common initial state. Each of the following steps consists in the resampling of some Markov chains amongst the replicas, with an initial condition getting closer to T at each iteration of the algorithm.

2.2 AMS algorithm

2.2.1 Importance function

We define a so-called *importance function*, mapping $\mathbb{R}^d$ to $\mathbb{R}$:

$$\xi : \mathbb{R}^d \to \mathbb{R},$$

which is used to quantify the proximity of a point in $\mathbb{R}^d$ to the subset of interest T. The only requirement imposed on ξ is that there exists a constant $Z_T \in \mathbb{R}$ such that

$$\begin{aligned} \xi(x) &\geq Z_T \text{ if } x \in T \\ \xi(x) &< Z_T \text{ if } x \notin T. \end{aligned} \qquad (1)$$

We further define the *importance* of a Markov chain $X = (X_t)_{t\in\mathbb{N}}$ as the supremum of ξ along the chain:

$$\Xi(X) = \sup_{t\in\mathbb{N}} \xi(X_t). \qquad (2)$$

The ξ function is probably the most important ingredient of the AMS algorithm, since it is used to quantify the proximity of a path to the subset T. It is therefore important to choose a good function ξ with regards to the rare event we are trying to simulate. Even if the AMS algorithm is proven to yield an unbiased result *regardless of* ξ, the choice of an optimized importance function is expected to improve the variance reduction efficiency. We will discuss further the issue of the quality of the function ξ in Sect. 5.

2.2.2 Initialization

The AMS algorithm consists in an interacting system of weighted replicas. Given $n > 0$, we simulate n i.i.d. replicas of the Markov chain X, denoted by $X_0^j = (X_{0,t}^j)_{t\in\mathbb{N}}$, $j \in \{1,\ldots,n\}$. For all j, the initial state $X_{0,0}^j$ is a point x_0 located outside of T. We define Z_0 as

$$Z_0 = \xi(x_0).$$

Let us denote by $q \in \mathbb{N}$ the current iteration number, and by k the number of replicas resampled at each iteration.Within an iteration of the AMS algorithm, every replica has *the same weight*. The global weight at iteration q will be denoted W_q, with W_0 set to 1.

For the sake of simplicity, we present in the following the case where distinct paths cannot have the same importance. The general case is discussed in Section 2.3. Now we can start iterating on $q \geq 0$.

2.2.3 Iterations

(i) For each $j \in \{1,\ldots,n\}$, denote by $\tau_{q,f}^j$ the stopping time of the Markov chain X_q^j, and by S_q^j its importance (See (2):

$$S_q^j = \Xi(X_q^j). \qquad (3)$$

(ii) Sort the sample $(S_q^1,\ldots,S_q^n)$ in increasing order:

$$S_q^{(1)} < \ldots < S_q^{(k)} < \ldots < S_q^{(n)}.$$

(iii) Denote by Z_{q+1} the k-th order statistics $S_q^{(k)}$:

$$Z_{q+1} := S_q^{(k)}.$$

(iv) If $Z_{q+1} \geq Z_T$, stop iterating and go to the final step (See Sect. 2.2.5)

(v) For all $j \in \{1,\ldots,k\}$, resample replica $X_q^{(j)}$ according to the resampling kernel described in Sect. 2.2.4, and denote it by X_{q+1}^j

(vi) For all $j \in \{k+1\ldots,n\}$, define $X_{q+1}^j = X_q^{(j)}$

(vii) Update the global weight:

$$W_{q+1} = \frac{n-k}{n} W_q$$

(viii) Increment q and go to step (i)

2.2.4 Resampling process

At each iteration of the AMS algorithm, k replicas are resampled. These Markov chains undergo the following resampling procedure:

Let us denote the current iteration number by q and the associated resampling level by Z_{q+1}.

For each of the k replicas $X^{(j)}, j \in \{1,\ldots,k\}$ to be resampled, one of the remaining replica $X^{(i)}, i > k$ is *randomly* selected for duplication. We know for sure that the

Markov chain $(X_t^{(i)})_{t\in[0;\tau_f^{(i)}]}$ contains at least a state whose importance is greater than Z_{q+1} (otherwise it would have been resampled). We can therefore define

$$\tau_q^{(i)} = \inf\left\{t \in [0; \tau_{q,f}^{(i)}] : \xi(X_{q,t}^{(i)}) > Z_{q+1}\right\}$$

as the first time at which the replica $X^{(i)}$ has an importance greater than Z_{q+1}.

The resampled Markov chain $(Y_t)_{t\geq 0}$ is defined as a replica of the Markov chain X with initial condition

$$Y_0 = X^{(i)}_{q,\tau_q^{(i)}}.$$

2.2.5 Final step

Once the algorithm stops iterating, we denote by N the number of completed iterations. Given the bounded observable ϕ_T introduced in Sect. 2.1, we define

$$\hat{\phi}_T = \sum_{j=1}^{n} W_N^j \phi_T(X_N^j) \tag{4}$$

as an **unbiased** estimator of the average $\langle \phi_T \rangle$ [3].

2.3 Interpretation of the replicas weights

In this section we provide a practical interpretation of the AMS weights to give an intuition on the estimator (4). The mathematical proofs of the unbiasedness and consistency of the AMS estimator are not presented in this paper. We refer the reader to [2] and [3] for theoretical support.

At each iteration $q \geq 0$, the level Z_{q+1} is chosen in such a way that the probability for a path X_q^j to have an importance greater than Z_{q+1} (i.e. $\mathbb{P}(S_q^j \geq Z_{q+1} | S_q^j \geq Z_q)$) is estimated by

$$\hat{p}_q = 1 - \frac{k}{n}.$$

Keeping that in mind, we can see that the weights of the replicas at iteration $q + 1$ are nothing more than an estimate of the probability $\mathbb{P}(S_{q+1}^j \geq Z_{q+1})$.

In other words, the AMS algorithm provides us at each iteration q with a set of paths X_q^j, $j \in \{1, \ldots, n\}$, carrying a global weight, which can be interpreted as an estimate of the probability to have this particular set of paths instead of the paths sampled at iteration 0.

2.4 About the number of resampled replicas

The algorithm presented in Sect. 2.2 is an ideal case. In reality, it may occur that multiple replicas have the same importance. In that case, the number of replicas having an importance less or equal to the k-th lowest importance may very well be greater than k. When such a situation arises, **every** path which importance is less or equal to the level has to be resampled.

This modification has to be taken into account in the replicas weights. If the current iteration is the q-th and the number of particles to be resampled is $K' > k$, then the weight update at step (vii) of the algorithm has to be changed to:

$$W_{q+1} = \frac{n - K'}{n} W_q.$$

In some pathological configurations, all the replicas may happen to have the same importance, which would lead to extinction at the next iteration. In that case, the iterating process is interrupted and the algorithm goes straight to the final step (See Sect. 2.2.5), eventually yielding a null contribution.

3 Monte Carlo neutral particle transport

3.1 Neutron transport theory

The transport theory describes the mathematical framework needed to solve the equations governing the behavior of a gaz of non-interacting particles (e.g. neutrons, photons) in different materials, under different assumptions (spins are not taken into account, relativistic effects are neglected etc.). It has been discussed at length in [6] or [7]. In fissile materials, neutrons (and photons) can induce fissions, which results in the production of more particles. The modelling of such systems is achieved using the so-called critical linear Boltzmann equation. When the materials are not fissile, the standard linear Boltzmann equation can be used to describe the particle transport.

It is assumed that the neutral particles in homogeneous medium are transported along straight lines between collisions points and are randomly reoriented at each interaction (with eventual change of energy) [8]. As the flights lengths are exponentially ditributed, the underlying stochastic process governing the neutral particle transport is called *exponential flights*. We refer the interested reader to [9, 10] for recent developments on this topic.

3.2 Monte Carlo simulation of neutral particle transport

Numerically, transport problems are stochastically solved using Monte Carlo transport codes [11]. Those codes simulate directly the particles trajectories according to laws of probability provided by the so-called nuclear data (cross sections, energetic and angular transfers, secondary particles emission spectrums, etc.), which are available in international libraries. The flights lengths between interactions are randomly sampled, as well as the outgoing direction and energy of the particle after each collision with the medium.

Every Monte Carlo transport code relies on the particle tracking routine, which simulates the random life of the particles by transporting them from one interaction to another, until they are absorbed or leak out of the geometry.

Within a Monte Carlo simulation, the life of each transported particle is build step by step, which means that the state of the particle at a given time is determined exclusively from the knowledge of its state after the previous interaction. This makes the random process of particle transport a discrete-time markovian process. Therefore, the sequence of the phase-space coordinates of the particle (namely position, direction and energy) at the successive interactions is a Markov chain, so that the AMS algorithm described in Section 2 can be implemented.

4 Neutral particle transport with AMS

The goal of this section is to present a practical implementation of the AMS algorithm in the context of Monte Carlo particle transport.

4.1 Definitions and notations

Before going any further, let us introduce some definitions and notations which will be used throughout the following.

We define the position of a particle in the 6-dimensional *phase space* $\mathcal{S}$ as the set of coordinates $(\mathbf{X}, \mathbf{\Omega}, E)$, where $\mathbf{X}$ denotes the particle position in the 3-dimensional space, $\mathbf{\Omega}$ its direction, and E its energy.

A particle is considered *alive* while it is transported from an interaction to the next. When an interaction results in the capture of the particle, the transport stops and the particle is referred to as *killed*. We call *track* of a particle the sequence of its interaction points.

We call *geometry* the subspace of the phase space in which the simulation takes place. If a particle is transported to a point of the phase space that is not included in the geometry, this particle is considered leaking out of the geometry and is instantly killed.

The aim of the simulation is to estimate a *score* (typically a flux or a reaction rate) in a particular volume of the geometry we will refer to as *detector* or simply *target*.

4.2 Importance map

As we saw in Sect. 2.2.1, the geometry has to be provided with an importance function, so as to determine the regions of the system that are of interest to the simulation. Technically, we will use a function that associates any point of the phase space to an importance value, which is related to the probability for a particle located at this point to contribute to the final score. We will refer to this function as *importance function* or *importance map*, and denote it by

$$I(\mathbf{X}, \mathbf{\Omega}, E).$$

In order to satisfy the requirement (1), it is sufficient to define an importance I_{target} larger than any other value on the map, and to assign it to every point of the target volume. In TRIPOLI-4®, I_{target} is set to the numerical equivalent of positive infinity regardless of the importance values outside the target volume, so as to avoid mistakes.

4.3 The AMS algorithm

As described previously, the AMS algorithm is an iterative method. Each iteration includes the definition of a single splitting level alongside with the corresponding splitting process.

4.3.1 *Initialization (step $q = 0$)*

Let n be the initial number of simulated particles, and k the **user-defined** number of particles that are to be duplicated at each iteration of the algorithm. The initial particles are simulated using an analog Monte Carlo simulation (*i.e.* without variance reduction), and the track of each transported particle is kept in memory.

There are actually many ways to store a particle track. For the AMS algorithm to be the most performant, this track should only contain the emission point of the particle and the coordinates of the particle after each interacion with the geometry. The coordinates of a particle at an interaction point but before the interaction takes place should not be stored, as the importance function relate to the probability for a particle simulated from a given point of the phase space to contribute to the score, and that probability usually takes into account the fact that the particle is transported before interacting with the medium.

The global weight at iteration q is denoted by w_q, and w_0 is set to 1. We can now start iterating on $q \geq 0$.

4.3.2 *Iterations ($q \geq 0$)*

Each track T is given an importance value, defined as the maximum importance of the points of its track:

$$I(T) = \max_{i \in track} I(\mathbf{X}_i, \mathbf{\Omega}_i, E_i). \tag{5}$$

The n tracks are sorted according to their importance, and the k-th smallest track importance defines the splitting level I_{q+1}.

If I_{q+1} is greater than I_{target}, the iterating process stops. Otherwise, all the tracks that have an importance less or equal to the level I_{q+1} are suppressed, and the same number of tracks are randomly selected (with replacement) amongst the remaining tracks. The selected tracks are duplicated at their first collision point of importance greater than I_{q+1}, according to the resampling kernel presented in Sect. 2.2.4. The global weight is then updated:

$$\begin{aligned} w_{q+1} &= \frac{n - K_q}{n} w_q \\ &= \prod_{i=0}^{q} (1 - \frac{K_i}{n}), \end{aligned}$$

where K_q is the number of resampled particle tracks at iteration q (See Sect. 2.3).

4.3.3 *Final step and scoring*

As soon as a splitting level I_q is greater than I_{target}, the algorithm stops iterating. We define $N = q - 1$. The score $\hat{\phi}_{target}$ in the target is then computed using the standard estimators of Monte Carlo particle transport, as for an analog simulation, but using the last set of tracks.

An unbiased estimator of the flux in the target is built by weighting the estimated flux $\hat{\phi}_{target}$ by the probability of reaching the last splitting level (See Sect. 2.2.5):

$$\hat{\phi} = \hat{\phi}_{target}.w_N.$$

4.4 Strengths of the AMS

4.4.1 Number of contributions to the score

In order to reduce the variance when simulating rare events, it is interesting to have multiple small contributions to the score rather than a few large contributions and many zeroes.

The stopping criterion of the AMS algorithm offers a guarantee that *at least* $n - k + 1$ particles will reach the target and contribute to the score with a small weight w_N, except for pathological cases. However, the cases of extinction are rare and caused by the use of a very unadapted importance map.

4.4.2 Robustness with regard to the importance function

The estimator constructed as described in Sect. 4.3 is **unbiased** regardless of the parameter k [2]. The same holds true for any importance map under the single constraint [3]:

$$I_{target} \in]I_{source}, +\infty[. \quad (6)$$

In our implementation of the AMS algorithm, this condition is *always verified*, as we set the importance of the target to infinity regardless of the importance map (see Sect. 4.2). This allows us to use any function as importance. However, if the map does not properly describe the system, the result, though unbiased, will probably take a really long time to converge.

4.4.3 Usability

In addition to the usual parameters of any Monte Carlo particle transport simulation, AMS introduces only two free parameters: the number of duplicated particles k, and the importance map $I(\mathbf{X}, \mathbf{\Omega}, E)$.

As we already pointed out, any choice of k and I yields an unbiased result. The parametrization of the AMS algorithm is therefore quite simple. It is sufficient to define a purely geometric importance function, such as the inverse to the distance to the target, and to set any value for k less than n to reduce the variance. More efficiency can be achieved through the use of a precise importance map, taking into account the direction and energy of the particles.

4.4.4 Parallelization

A Monte Carlo simulation performed with AMS yields a single value as estimation of the quantity of interest. Therefore, multiple independant simulations are required to obtain a precise result. The arithmetic mean of the estimations over all simulations is computed, as it is the case in regular Monte Carlo. As the simulations remain totally independant until their output is gathered to compute the average result, a straightforward way of parallelizing AMS Monte Carlo is to run independant simulations on several processors simultaneously.

5 Validation and results

The AMS algorithm was implemented in the development version of the Monte Carlo particle transport code TRIPOLI-4® [5]. This 3D continuous-energy Monte Carlo code is developed by the Service d'Etudes des Réacteurs et de Mathématiques Appliquées (SERMA) at CEA Saclay.

In the following, we present some preliminary results obtained with TRIPOLI-4® using AMS and compare them to simulations with no variance reduction. The Monte Carlo simulations that do not use any variance reduction techniques are usually called *analog*, because they are directly analogous to the natural transport of the particles.

We aim to validate the AMS algorithm implementation and to estimate its efficiency as a variance reduction technique by comparing its results and performances to analog TRIPOLI-4® simulations.

5.1 Figure of merit

In a Monte Carlo simulation, the flux is estimated for several statistical independent groups of particles (or *batches*). The average flux and its associated variance are estimated from the series of batches. The purpose of variance reduction is to reduce the variance of the result for a given computation time. The efficiency of a variance reduction technique lies upon two factors. On one hand, the relative error σ of the average flux, and on the other hand the simulation time T. If we denote by N_B the total number of simulated batches, we have

$$\sigma \sim \frac{1}{\sqrt{N_B}} \quad \text{and} \quad T \sim N_B. \quad (7)$$

We use the figure of merit (FOM) to evaluate the efficiency of a Monte Carlo simulation. The FOM is defined as

$$FOM = \frac{1}{\sigma^2 T}, \quad (8)$$

and should remain relatively constant during the simulation according to (7). For different simulations of the same system, the simulation producing the largest FOM is either proportionally faster or more precise than the others. Therefore, the FOM can be used as a performance measurement for variance reduction.

5.2 Importance map calculation

The AMS algorithm implemented in TRIPOLI-4® can use several functions as importance maps.
The user can define a *purely geometrical* importance, such as the distance between a point and the particle source or the invert of the distance between a point and the AMS target.

Most of the time though, a more precise importance function is required to take into account the direction and energy of the simulated particles. To that end, the AMS has been linked to the so-called `INIPOND` module of

TRIPOLI-4®, which is used for other variance reduction techniques in this code and to automatically pre-computes an importance map (see [5]).

5.3 Bypass simulation

The first system consists of an extruded box filled with Helium. The dimensions of the box are $10cm \times 10cm \times +\infty$. A neutron flux is produced from an 15-MeV isotropic neutron source at one corner of the box and estimated inside an infinite cylinder of 1cm in diameter placed at the opposite corner. Inbetween is a massive infinite cylinder composed of highly concentrated Bore. The mean free path of neutrons in the Bore slab is less than 0.5 cm. The geometry for this problem is shown in Fig. 1.

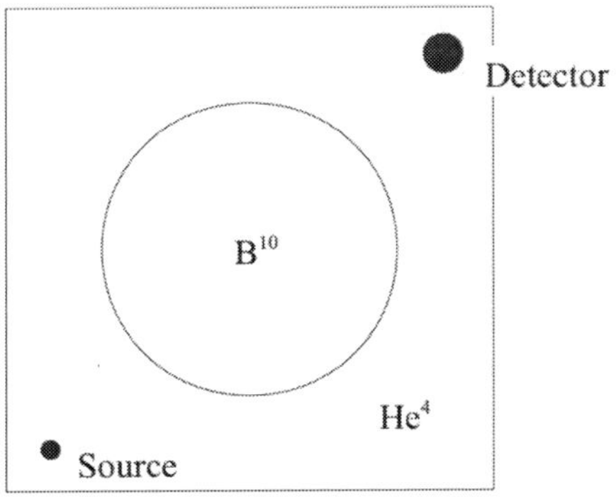

Figure 1. Bypass simulation configuration

The results obtained using TRIPOLI-4® without variance reduction and TRIPOLI-4® using the AMS algorithm are presented in Table 1. Several AMS simulations with 10^4 initial particles were performed, with different values for the parameter k, from 1 duplicated particle (0.01%) per iteration up to 5×10^3 (50%). In all simulations, the **INIPOND** importance map was used.

We observe that the AMS significantly improves the FOM regardless of the parameter k. The number of iterations required for the AMS algorithm to complete increases with k up to $\frac{k}{n} = 0.5\%$. Above this value, the computation time per simulation remains stable and the FOM gain relatively constant.

The smallest relative error for a given number of independent simulations is obtained with $k = 1$ (0.01%). However, this configuration requires a very large number of iterations for the AMS to complete, which has a strong impact on the computation time. It is therefore more interesting (regarding the FOM gain) to chose a greater k in order to speed up the simulation.

The study of the Bypass problem puts into light the apparent robustness of the AMS efficiency with regard to the parameter k. As long as it is not too small, its impact on the algorithm efficiency in reducing the variance remains stable. However, it is to be noted that the geometry of the Bypass configuration is fairly simple. We may find cases in which the number of splitting per iteration have a greater impact on the algorithm efficiency.

5.4 Neutron flux attenuation in water

The source of the second problem is a directional neutron point source, emitting neutrons according to a Watt spectrum and placed at the entry of a straight box filled with water. The detector is placed at 3 m from the source, and the neutron flux is estimated in 10-cm-thick slices between the source and the detector.

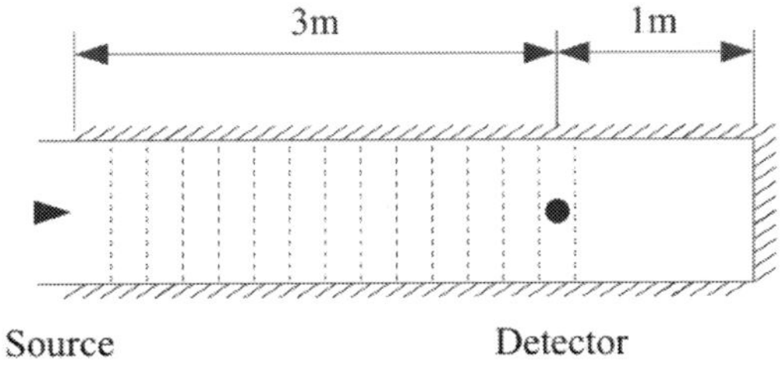

Figure 2. Water box simulation system

The neutron flux attenuation was estimated using both analog and AMS TRIPOLI-4® simulations. In a first realization, the importance function used for the AMS was the distance from the source point. The AMS was performed using 10^3 initial particles per batch and $k = 10^2$ splitting per iteration. Data were collected for both simulations during 65h. Fig. 3 presents the estimated flux with respect to the distance in the box, and the associated 99.7% confidence intervals. The FOM for these simulations are shown in Fig. 4.

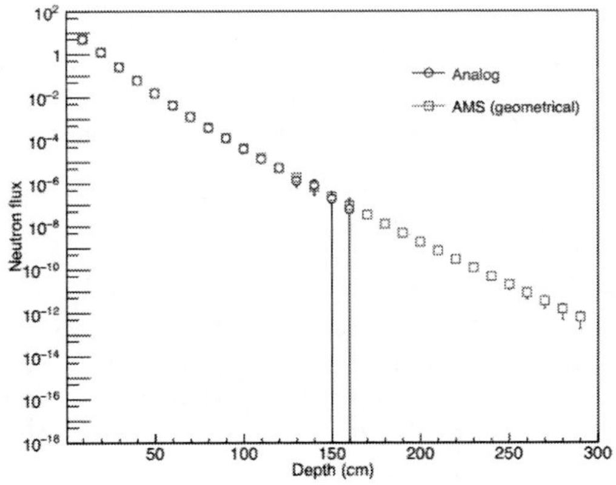

Figure 3. Flux attenuation in the Water box

Table 1. Comparison of TRIPOLI-4® with and without AMS for flux estimation in the Bypass geometry (48h calculations)

Simulation	Number of independent simulations	Mean	Relative error	FOM	Gain
Analog	14300	4.3042×10^{-10}	8.6838×10^{-2}	0.0008	**1**
AMS (0.01%)	210	4.0541×10^{-10}	1.5377×10^{-2}	0.0246	**31**
AMS (0.05%)	1705	4.2244×10^{-10}	7.1384×10^{-3}	0.1135	**147**
AMS (0.1%)	4038	4.1462×10^{-10}	3.8072×10^{-3}	0.3992	**517**
AMS (0.5%)	10170	4.1518×10^{-10}	2.3973×10^{-3}	1.0075	**1307**
AMS (1%)	10900	4.1547×10^{-10}	2.3948×10^{-3}	1.0123	**1313**
AMS (5%)	10884	4.1494×10^{-10}	2.4280×10^{-3}	0.9816	**1273**
AMS (10%)	11472	4.1451×10^{-10}	2.3604×10^{-3}	1.0387	**1347**
AMS (50%)	10829	4.1591×10^{-10}	2.3750×10^{-3}	1.0259	**1331**

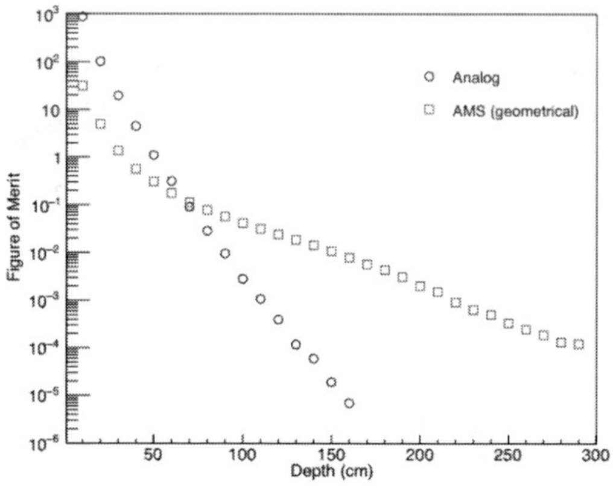

Figure 4. FOM comparison in the Water box

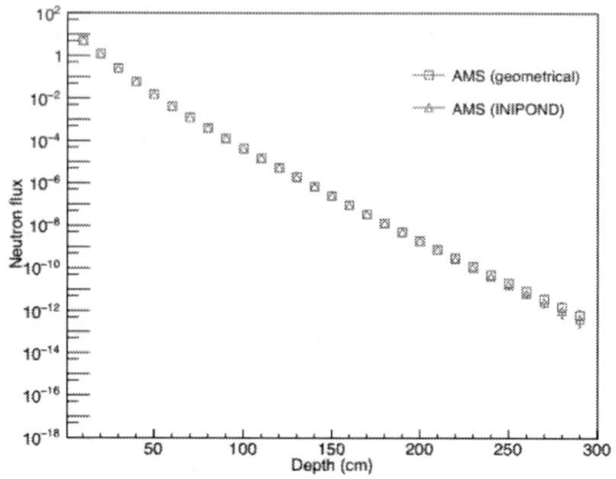

Figure 5. Flux attenuation in the Water box

It can be seen in Fig. 3 that the analog simulation in unable to estimate a flux farther than 160cm from the source, because of the strong attenuation of neutrons in water. However, the AMS simulation is able to accurately determine the neutron flux all the way from the source to the detector. Near the source point, the analog simulation is more efficient than the AMS, as shown in Fig. 4. The neutron flux in this part of the box is high enough for TRIPOLI-4® not to need variance reduction, which places the AMS at a disadvantage due to its impact on runtime.

The AMS mechanism starts to be interesting at a depth of 70cm. The gain in FOM increases up to 10^3 in favor of AMS at 160cm, from which point the analog simulation fails to estimate the flux.

In order to illustrate the role of the importance map within the AMS algorithm, the deep-penetration simulation was performed a second time, using the same parameters for AMS except for the geometrical importance function, which was replaced by the importance estimated using the `INIPOND` module. The fluxes obtained with both importance maps are shown in Fig. 5, and the associated FOM in Fig. 6.

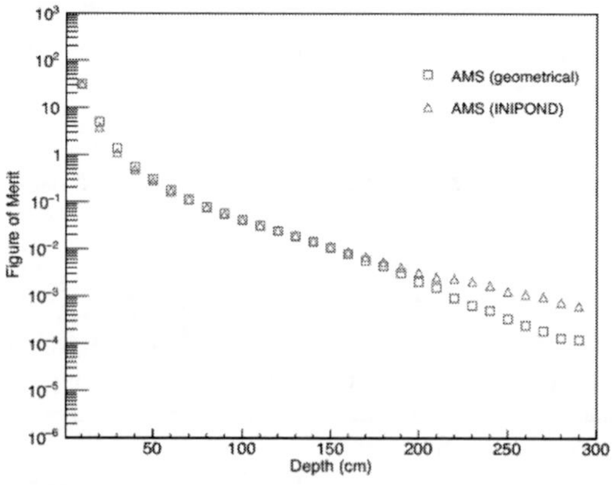

Figure 6. FOM comparison in the Water box

We observe that the use of the `INIPOND` map over the geometrical importance significantly improves the AMS efficiency up to a factor 4 at the target. We can notice that the FOM obtained with both maps are equivalent for depths below 160cm. However, we have to keep in mind that the parametrization of the `INIPOND` module required to get an accurate map can be far more complex than the input of a simple geometrical function.

5.5 Gamma spectroscopy

The AMS algorithm has the specificity to provide a group of particle tracks associated to a *global weight* (see Sect. 4). This feature let us hope to use AMS to reduce the variance on deposited scores, such as deposited spectrum (pulse height tally), energy or charge, for which all particles contributing to the score **must** have the same weight.

The use of other variance reduction techniques for quantities that depend on collective effect of a group of particles is a complex problem [12]. Currently, TRIPOLI-4® does not even support variance reduction with deposited scores. However, AMS offers a possibility to have an efficient variance reduction for deposited scores using the same algorithm than for other types of scores.

We present in Fig. 7 a deposited spectrum obtained by simulating the response of a high-purity Germanium (HPGe) detector placed at 50cm of a Sodium source. The gain in efficiency with AMS is shown in Fig. 8.

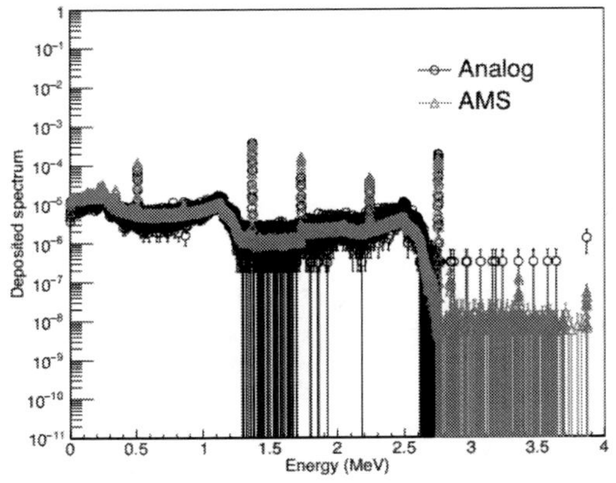

Figure 7. Deposited spectrum in the HPGe detector

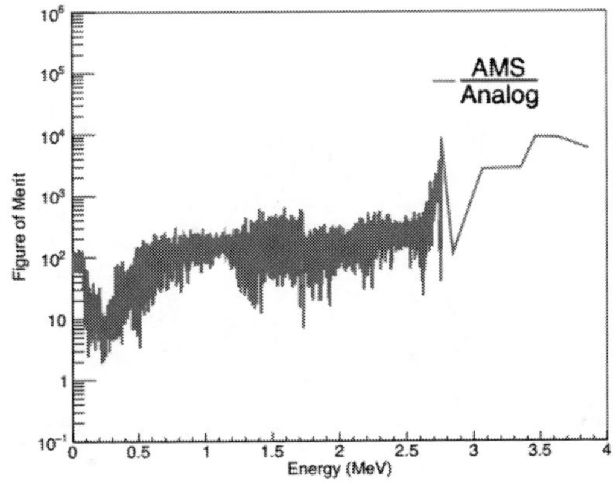

Figure 8. AMS gain in FOM over the deposited spectrum

These results are from photon-only simulations, as the coupled photons/electrons/positrons AMS is not yet fully operational.

6 Conclusion

These preliminary results show that the use of the Adaptive Multilevel Splitting algorithm in TRIPOLI-4® reduces significatively the variance. Its connection to the TRIPOLI-4® importance map calculation module enables us to use advanced parametrizations for the importance. The guarantee to have an **unbiased result** regardless of the importance map or the number of particles resampled at each iteration makes the AMS algorithm a viable alternative to existing variance reduction techniques in the most severe configurations. Furthermore, it can also be used for variance reduction on deposited scores without need to alterate the algorithm, as seen in Sect 5.5.

In the future, we plan to implement AMS variance reduction on deposited scores using coupled photons/electrons/positrons simulations as well as coupled neutrons/gamma problems.

Another concern is the comparison of the AMS efficiency to existing variance reduction techniques. A paper by the same authors is in preparation regarding this study [13].

Aknowledgement

TRIPOLI-4® is a registered trademark of CEA. We gratefully acknowledge EDF for their long term partnership and AREVA for their support.

The work of T. Lelièvre and M. Rousset is supported by the European Research Council under the European Union's Seventh Framework Programme (FP/2007-2013) / ERC Grant Agreement number 614492.

References

[1] H. Kahn, T.E. Harris, Nat. Bur. Stand. Appl. Math. Series, *Estimation of particle transmission by random sampling*, **12**:27-30 (1951)

[2] F. Cérou, A. Guyader, Stoch. Anal. Appl., *Adaptive multilevel splitting for rare event analysis*, **25(2)**:417-443 (2007)

[3] C.-E. Bréhier, M. Gazeau, L. Goudenège, T. Lelièvre, M. Rousset, Ann. App. Prob., *Unbiasedness of some generalized Adaptive Multilevel Splitting algorithms* (to be published)

[4] F. Cérou, A. Guyader, T. Lelièvre, D. Pommier, J. Chem. Phys., *A multiple replica approach to simulate reactive trajectories*, **134**:054108 (2011)

[5] E. Brun, F. Damian, C.M. Diop, E. Dumonteil, F.X. Hugot, C. Jouanne, Y.K. Lee, F. Malvagi, A. Mazzolo, O. Petit, J.C. Trama, T. Visionneau, A. Zoia, Ann. Nucl. Energ., *TRIPOLI-4®, CEA, EDF and AREVA reference Monte Carlo code*, **82**:151-160 (2015)

[6] J. J. Duderstadt, W. R. Martin, *Transport theory* (1979)

[7] G. I. Bell, S. Glasstone, *Nuclear reactor theory* (1970)

[8] K. Pearson, Nature, *The problem of the random walk*, **72**:294 (1905)

[9] A. Zoia, E. Dumonteil, A. Mazzolo, S. Mohamed, J. Phys. A: Math. Theor., *Branching exponential flights: travelled lengths and collision statistics*, **45(42)**:425002 (2012)

[10] A. Zoia, E. Dumonteil, A. Mazzolo, S. Mohamed, Phys. Rev. E, *Collision densities and mean residence times for d-dimensional exponential flights*, **83(4)**:041137 (2011)

[11] N. Metropolis, S. Ulam, J. Am. Stat. Assoc., *The Monte Carlo method*, **44(247)**:335-341 (1949)

[12] T. E. Booth, Nucl. Sci. Eng., *A Monte Carlo variance reduction approach for non-Boltzmann tallies*, **116**(2):113-124 (1994)

[13] E. Dumonteil, T. Lelièvre, H. Louvin, M. Rousset, *Adaptive Multilevel Splitting and Importance Sampling for Monte Carlo particle transport* (in preparation)

Recent developments in the TRIPOLI-4® Monte-Carlo code for shielding and radiation protection applications

Fadhel Malouch[1,a], Emeric Brun[1], Cheikh Diop[1], François-Xavier Hugot[1], Cédric Jouanne[1], Yi-Kang Lee[1], Fausto Malvagi[1], Davide Mancusi[1], Alain Mazzolo[1], Odile Petit[1], Jean-Christophe Trama[1], Thierry Visonneau[1], and Andrea Zoia[1]

[1] *Den-Service d'études des réacteurs et de mathématiques appliquées (SERMA), CEA, Université Paris-Saclay, F-91191 Gif-sur-Yvette, France*

Abstract. TRIPOLI-4® is a 3D continuous-energy Monte-Carlo particle transport code developed by CEA (SERMA) and devoted to shielding, reactor physics, criticality-safety and nuclear instrumentation. In this paper, we present the recent developments in the TRIPOLI-4® for shielding and radiation protection applications. Some of these additional features are already available in the TRIPOLI-4® version 10 released in December 2015. Other features are in development.

1 Introduction

TRIPOLI-4® is a three-dimensional and continuous-energy Monte-Carlo particle transport code developed by the "Service d'Etudes des Réacteurs et de Mathématiques Appliquées" (SERMA), at the CEA Paris-Saclay center. It is devoted to shielding, reactor physics, criticality safety and nuclear instrumentation [1].

TRIPOLI-4® has been developed starting from the mid of 1990s in C++. A new version of the code is typically released every two years. The TRIPOLI-4® version 10 has been released in December 2015 with in particular a Monte-Carlo depletion feature.

In this paper, we present the recent developments in the TRIPOLI-4® for shielding and radiation protection applications. Some of these additional features are already available in the TRIPOLI-4® version 10 released in December 2015. Other features are in development.

2 TRIPOLI-4® key features [1]

2.1 Tracked particles

TRIPOLI-4® is currently able to simulate four kinds of particles (with coupling):
- Neutrons from 20 MeV down to 10-5 eV,
- Photons from 50 MeV down to 1 keV,
- Electrons and positrons from 100 MeV down to 1 keV.

2.2 Nuclear Data

TRIPOLI-4® uses continuous energy cross sections processed with NJOY from any ENDF-6 format evaluation including JEFF-3, ENDF/B-VII, FENDL-2, JENDL-4. TRIPOLI-4® uses also probability tables from CALENDF in the unresolved resonance range.

2.3 Geometry

TRIPOLI-4® has its own geometry module allowing both surface-based and combinatorial representations. TRIPOLI-4® is also directly compatible with a geometry developed in the format of the ROOT software [2]. Thanks to its modularity, the code may be linked with any third party geometry with limited development effort.

2.4 Simulation modes

Three simulation modes are available in the TRIPOLI-4® code:

- "**Shielding**" mode, fixed-source simulation typically used for radiation protection and shielding analysis.
- "**Fixed_Sources_Criticality**" mode, fixed source simulation with treatment of fission events, typically used for subcritical fissile instrumentation problems.
- "**Criticality**" mode, used to find the fundamental mode and the associated fundamental eigenvalue of the critical Boltzmann equation.

2.5 Tallies

A non-exhaustive list of the tallies available in the TRIPOLI-4® code includes: volume, surface and point fluxes, surface currents, mesh tallies, reaction rates, deposited energy, built-in KERMA response functions, dose equivalent rate, dpa, gamma spectroscopy and effective multiplication factor k_{eff}.

[a] Corresponding author: fadhel.malouch@cea.fr

DOI: 10.1051/epjconf/201715306007

2.6 Variance reduction techniques

Standard variance reduction techniques are available in TRIPOLI-4®, such as implicit capture, particle splitting and Russian roulette. Moreover, TRIPOLI-4® has a special built-in variance reduction module, called INIPOND, based on the Exponential Transform Method, with an automatic pre-calculation of the importance map.

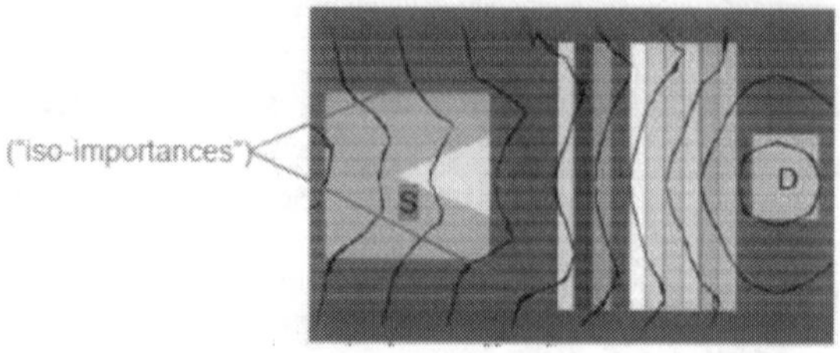

Fig. 1 Two-dimensional view with iso-importance lines of the photon importance map produced by TRIPOLI-4®. The source is on the left side and the detector on the right side [2].

2.7 Doppler broadening of the elastic scattering kernel

For the Doppler broadening of the elastic scattering kernel, the algorithm "Sampling of the Velocity of the Target nucleus" (SVT) is by default used in the TRIPOLI-4® code. The "Doppler Broadening Rejection Correction" (DBRC) and the "Weight Correction Method" (WCM) have been implemented in TRIPOLI-4® (version 9) to overcome the SVT limitations that affect resonant nuclei (such as ^{238}U) typically in the epithermal region.

2.8 Reactor period calculation

The asymptotic time behavior of neutron transport can be used in reactor start-up analysis or kinetics studies of nuclear systems. In TRIPOLI-4® version 10, the asymptotic reactor period is calculated as the inverse of the dominant eigenvalue. The algorithm is based on a modified alpha-k power iteration scheme [3].

2.9 Kinetics parameters computing

Kinetics parameters are key to the study of nuclear reactor dynamics and to safety issues in the context of transient or accidental reactor behavior. The Iterated Fission Probability (IFP) method implemented in TRIPOLI-4® version 10 allows computing the adjoint-weighted kinetics parameters: the delayed neutron fraction, the mean generation time and the Rossi Alpha [4].

2.10 Deposited charge

It is possible to calculate the spectrum of the charge deposited in a given volume by charged particles (electrons and positrons) using the DEPOSITED_CHARGE response recently implemented in TRIPOLI-4® version 10. The charge deposition can be used for nuclear instrumentation in the interpretation of signal of sensors irradiated in nuclear reactors, such as Self-Powered Neutron Detectors (SPNDs).

2.11 3D core fuel-depletion calculation

TRIPOLI-4® has been recently extended so as to cover depletion calculations by coupling the code to the MENDEL depletion solver developed by SERMA [5]. This coupling allows solving the Boltzmann-Bateman system of equations governing the evolution of materials under neutron irradiation. At a given time step, TRIPOLI-4® computes first the reaction rates in each depleted medium. Then, the MENDEL depletion solver calculates the end-of-step nuclide concentration for each region. TRIPOLI-4® version 10 depletion capability is based on C++ interfaces (accessible via C++ interpreter of ROOT [4]) that wrap the methods of both TRIPOLI-4® and MENDEL.

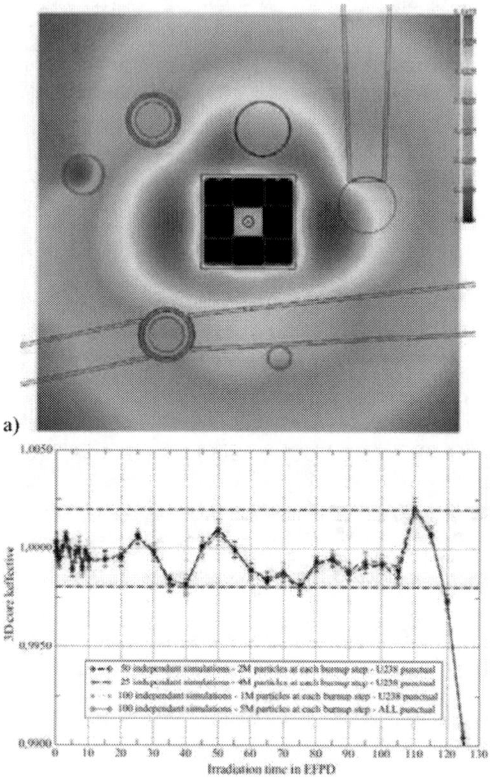

Fig. 2 a) Radial thermal flux distribution of the ORPHEE reactor (CEA, Saclay) calculated by TRIPOLI-4® for 3D core-depletion analysis. b) Evolution of the k_{eff} during irradiation. Time is expressed in effective full-power days (EFPD). Controls rods insertion is adjusted during irradiation.

2.12 Verification and Validation

The V&V test base is composed of several parts: elementary verification tests, criticality-safety benchmarks, shielding benchmarks, tests concerning parallel operations, tests covering the new features of the code. It includes several ICSBEP and SINBAD benchmarks, as well as proprietary benchmarks concerning CEA experimental facilities from CEA research centers [1].

We mention here an example recent shielding-V&V work that has been carried out on NAIADE 1 Water shielding benchmark [7]. For the nuclear decommissioning application, TRIPOLI-4® code capabilities on fission neutron deep penetration calculations were demonstrated. Both the shielding mode and the fixed-source sub-criticality mode of the code were validated against the NAIADE 1 water shielding benchmark. The variance reduction options of TRIPOLI-4® code are helpful to perform these calculations.

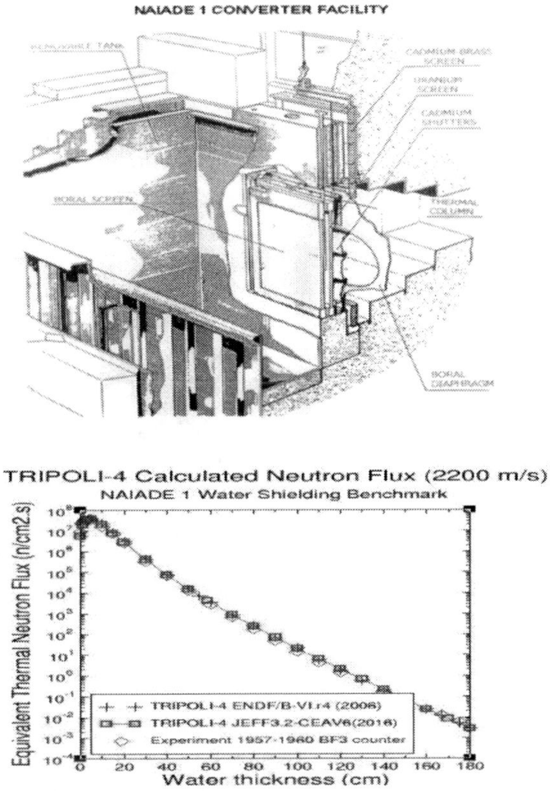

Fig. 3 Recent example of shielding-V&V work: NAIADE 1 Water shielding benchmark [7]

3 New features of TRIPOLI-4® version 10 for shielding and radiation protection applications

3.1 Thick-Target Bremsstrahlung for electromagnetic shower simulation

A simplified simulation mode for the electromagnetic shower, called Thick-Target Bremsstrahlung (TTB), has been implemented in TRIPOLI-4® version 10 to speed up coupled photon-electron-positron calculations. When this option is activated, secondary electrons and positrons produced by photon collisions are not transported, but part of their energy is converted into new bremsstrahlung photons. Fig. 4 shows a recent example of the use this feature for a benchmark of a sphere with a 20-MeV photon source in the center [8]. The simulation of the electromagnetic shower with TTB implicates a maximum difference of 30% in the deposited energy against a full coupled photon-electron-positron simulation, but the use of TTB speeds up calculations up 10 times.

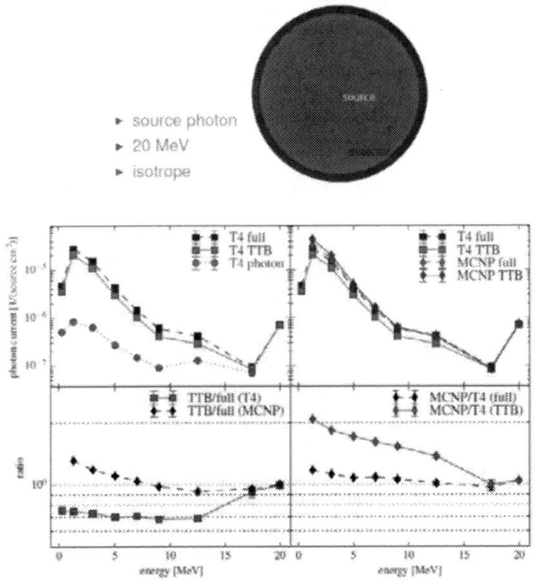

Fig. 4 Recent example of the use of the TTB feature. Here TTB vs full electromagnetic shower simulation: a maximum difference in deposited energy of 30% but acceleration up 10 times.

3.2 Analog neutron or photon transport with analog fission sampling

It is already possible with TRIPOLI-4® to perform a fully analog simulation (concerning both collisions and transport between collisions) for neutron and photon transport. TRIPOLI-4® version 10 enables analog fission simulation by sampling a full fission neutron multiplicity distribution using a coupling between TRIPOLI-4® and an external fission model providing fission sampling data (standard nuclear data libraries provide averaged neutron multiplicities only). It allows addressing time-dependent nuclear instrumentation applications needing detailed correlations between fission chains [9].

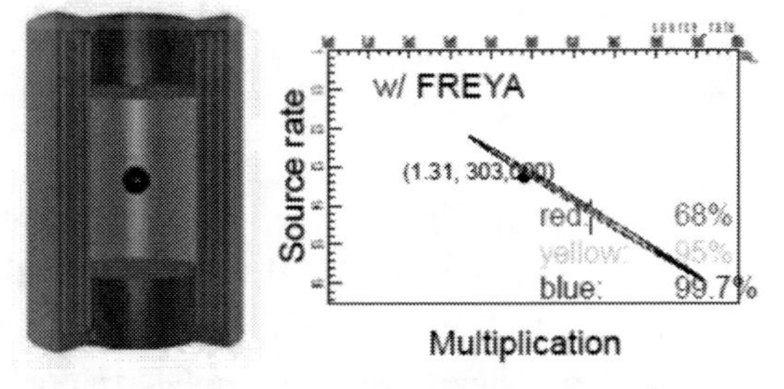

Fig. 5 Example of application: NMC (Neutron Multiplicity Counting) properly simulated by reconstructing the mass and multiplication of two objects by analyzing the measured signal from 3He tubes in a well counter [9].

3.3 "Replicate" option upgrading

With TRIPOLI-4®, we can perform a two-step calculation using first a global geometry to store the properties (energy, position, direction, weight) of particles crossing a given surface. Then, we use stored particles as surface sources for simulation on a local geometry. The REPLICATE option activates the particle splitting at the second-step simulation. This technique of variance reduction allows for an improved exploration of the whole phase space, especially when only a limited number of particles has been stored.

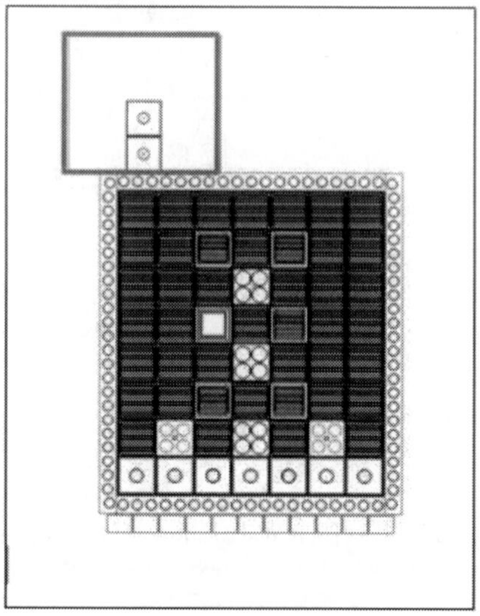

a) 1st step (global geometry)

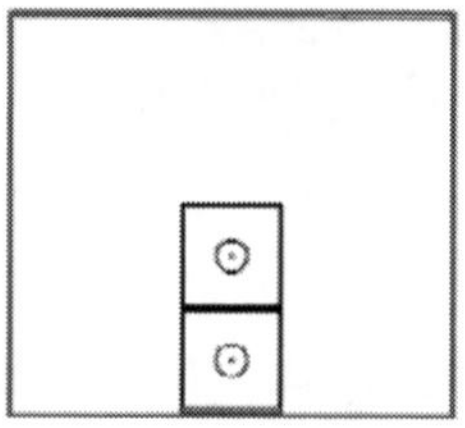

b) 2nd step (local geometry)

Fig. 6 Example of a two-step calculation in the case of the OSIRIS MTR reactor (CEA, Saclay). The blue rectangle (horizontal cross-section 30 cm x 30 cm) shows the boundary used for the storing of the particle characteristics. Inside the rectangle, we have two water boxes (horizontal cross-section 8 cm x 8 cm) with SPNDs monitors irradiated in the centers (about 3 mm of diameter) [8].

4 Features for shielding and radiation protection applications in development

4.1 Analog simulation with analog fission sampling

TRIPOLI-4® Monte Carlo transport code and FIFRELIN fission model (developed by CEA, Cadarache center, SPRC) have been coupled by means of external files so that neutron transport can then take into account fission distributions (multiplicities and spectra) that are not averaged, as is the case when using evaluated nuclear data libraries. Spectral effects on responses in shielding configurations with fission sampling are then expected [9].

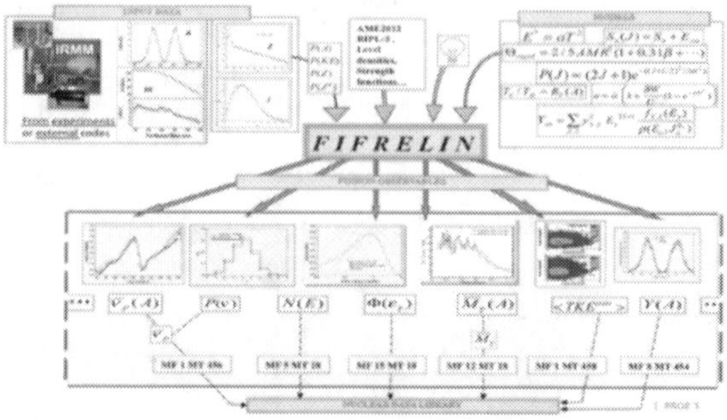

Fig. 7 FIFRELIN code (CEA, Cadarache) simulates the prompt part of the deexcitation process of binary fission.

4.2 TRIPOLI-4®/Geant4 coupling

The capability to simulate the transport of charged and/or high-energy hadrons (especially protons) is useful for, shielding, radiation protection and nuclear instrumentation applications. In the context of Monte-Carlo particle transport codes, solving these problems often requires the use of advanced variance-reduction techniques. TRIPOLI-4® is a reference Monte-Carlo particle transport code for the simulation of low-energy (< 20 MeV) neutrons and photons and offers a wide range of sophisticated variance-reduction schemes; however, it cannot be applied to the problems mentioned above because it lacks the capability to transport charged, high-energy hadrons. This limitation can be circumvented by coupling TRIPOLI-4® with the Geant4 particle transport toolkit (CERN) [10].

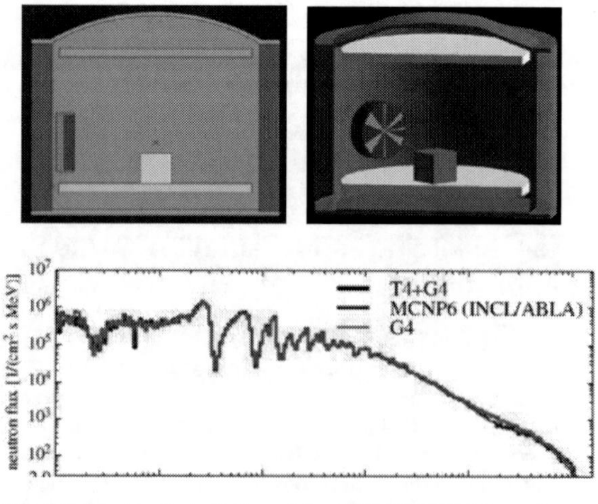

Fig. 8 Example of calculation of the APOLLON high intensity laser facility (CEA, Saclay) using TRIPOLI-4®/Geant4 coupling: proton-electron source; neutron-photon outgoing fluxes [13].

4.3 Rigorous two-step scheme for shutdown dose rate calculation

An activation scheme has been developed in TRIPOLI-4® for the shutdown dose rate calculations based on the Rigorous two-step approach (R2S). Two Monte Carlo transport calculations are used, the first for the neutrons to calculate the activation and second for photons emitted by activated materials. The production of the decay photons is performed by MENDEL inventory code (CEA) [14].

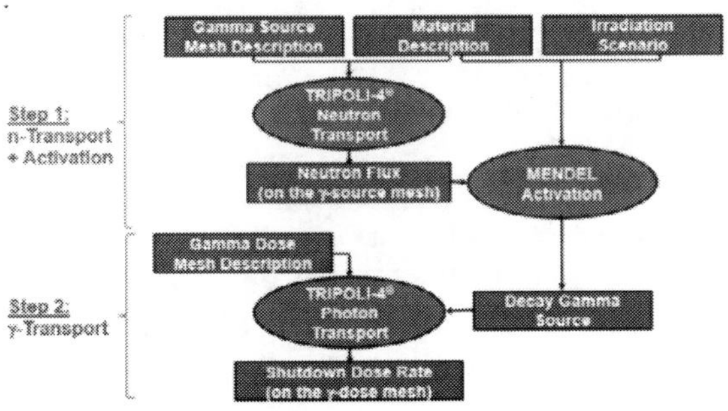

Fig. 9 the Rigorous two-step scheme for shutdown dose rate calculation using TRIPOLI-4® and MENDEL codes.

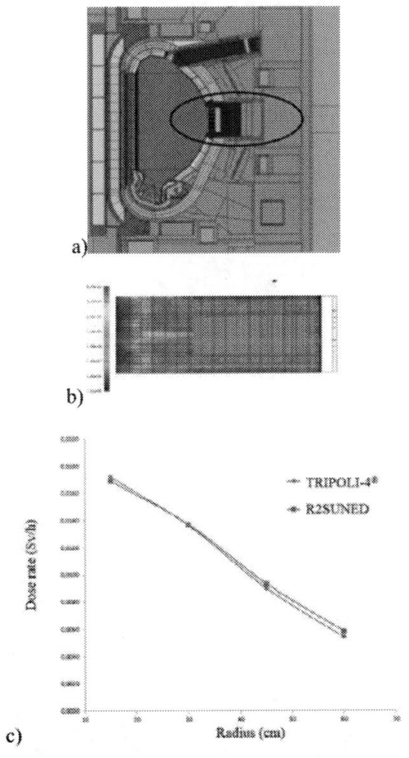

Fig. 10 Example of shutdown dose rate calculation using R2S TRIPOLI-4® in the case of a typical configuration of a port plug in ITER fusion facility [15]. a) ITER Equatorial Port Plug, b) Distribution of the total neutron flux calculated by TRIPOLI-4®, c) Shutdown dose rate at the rear face.

4.4 Variance Reduction using the method of Adaptive Multilevel Splitting (AMS)

A new technique of variance reduction based on the method of Adaptive Multilevel Splitting (AMS) has been developed in the TRIPOLI-4 code [14]. The AMS algorithm is an iterative method to help simulate rare events. Each iteration includes the definition of a single splitting level alongside with the corresponding splitting process:

- Classify simulated particle tracks and define a splitting level;
- Remove the particles that have not reach the threshold;
- Re-sample removed particles by splitting remaining ones.

These preliminary results show that the use of the Adaptive Multilevel Splitting algorithm in TRIPOLI-4® reduces significantly the variance. Furthermore, its connection to the TRIPOLI-4® importance map calculation module makes it quite simple to use [16].

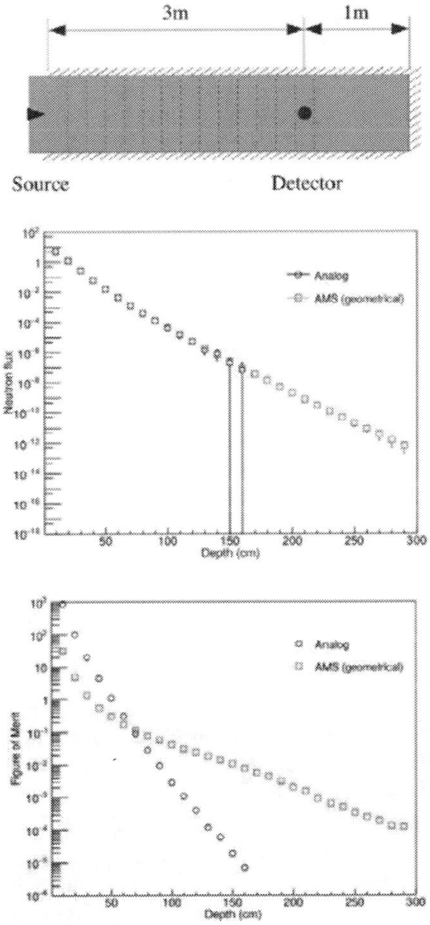

Fig. 11 Example of the use of the AMS method in TRIPOLI-4® in the case of a Deep penetration problem in water [16].

Acknowledgments

TRIPOLI-4® is a registered trademark of CEA. Authors gratefully acknowledge EDF for their long term support.

References

1. Brun, E., Damian, F., Diop, C.M., Dumonteil, E., Hugot, F.-X., Jouanne, C., Lee, Y.-K., Malvagi, F., Mazzolo, A., Petit, O., Trama, J.C., Visonneau, T., Zoia ,A., "TRIPOLI-4®, CEA, EDF and AREVA reference Monte Carlo code", Annals of Nuclear Energy, Volume 82, August 2015, Pages 151–160
2. Petit, O., Lee, Y.K., Diop, C., "Variance reduction adjustment in Monte Carlo TRIPOLI-4® neutron gamma coupled calculations", Progress in Nuclear Science and Technology Volume 4 (2014) pp. 408-412.
3. Zoia, A., Brun, E., Damian, F., Malvagi, F., "Monte Carlo methods for period calculations", Annals of Nuclear Energy, Volume 75, 2015, Pages 627–634.
4. Truchet, G., Leconte, P., Santamarina, A., Brun, E., Damian, F., Zoia, A.,"Computing adjoint-weighted kinetics parameters in TRIPOLI-4® by the Iterated Fission Probability method", Annals of Nuclear Energy, Volume 85, 2015, Pages 17–26.
5. Damian, F., Brun, E., "ORPHEE research reactor: 3D core depletion calculation using Monte-Carlo code TRIPOLI-4®", Annals of Nuclear Energy, Volume 82, 2015, Pages 203–216.
6. Brun, R., Rademakers, F., "ROOT - An Object Oriented Data Analysis Framework", Nuclear Instruments and Methods in Physics Research Section A: Accelerators, Spectrometers, Detectors and Associated Equipment, Volume 389, 1997, Pages 81-86. See also http://root.cern.ch
7. Lee, Y.K , "Neutron Deep Penetration Calculations in Light Water with Monte Carlo TRIPOLI-4® Variance Reduction Techniques", ICRS-13 & RPSD2016, Paris, France, October 3-6, 2016
8. Riz, D., "Calculation and use of multigroup cross sections including electron-photon cascade for a 3D Monte Carlo neutron-gamma transport code. Comparisons with MCNP-4B," in Proceedings of ANS PHYSOR, American Nuclear Society, 2000, CD-ROM..
9. Verbeke, J. M., Petit, O., "Stochastic Analog Neutron Transport with TRIPOLI-4 and FREYA: Bayesian Uncertainty Quantification for Neutron Multiplicity Counting", Nuclear Science and Engineering, Vol. 183, Nb 2, June 2016, p. 214-228.
10. Malouch, F., Lopez, F., Barbot, L., Fourmentel, D., "Calculation of neutron and gamma fluxes in support to the interpretation of measuring devices irradiated in the core periphery of the OSIRIS Material Testing Reactor", ANIMMA2015, Lisbon Portugal, April, 20-24, 2015.
11. Petit O., et al., "FIFRELIN-TRIPOLI-4® coupling for Monte-Carlo simulations with a fission model. Application to shielding calculations", ICRS-13 & RPSD2016, Paris, France, October 3-6, 2016.
12. "Geant4 is a toolkit for the simulation of the passage of particles through matter", http://geant4.cern.ch/
13. Mancusi, D., Bringer, O., Monot, P., "Progress on the TRIPOLI-4®-Geant4 coupling", ICRS-13 & RPSD2016, Paris, France, October 3-6, 2016.
14. Malouch, F, et al, "Recent developments in the TRIPOLI-4® Monte-Carlo code for fusion applications", 29th Symposium on Fusion Technology (SOFT 2016), Prague, Czech Republic, 5-9 September 2016.
15. Jaboulay, J.C., et al, "Rigorous-two-Step scheme of TRIPOLI-4® Monte Carlo code validation for shutdown dose rate calculation", ICRS-13 & RDSD2016, Paris, France, October 3-6, 2016.
16. Louvin, H., Diop C., Dumonteil, E., Lelièvre, T., Rousset, M., "Adaptive Multilevel Splitting for Monte Carlo particle transport", ICRS-13 & RPSD2016, Paris, France, October 3-6, 2016.

Recent Improvements of Particle and Heavy Ion Transport code System: PHITS

Tatsuhiko Sato[1,*], *Koji* Niita[2], *Yosuke* Iwamoto[1], *Shintaro* Hashimoto[1], *Tatsuhiko* Ogawa[1], *Takuya* Furuta[1], *Shin-ichiro* Abe[1], *Takeshi* Kai[1], *Norihiro* Matsuda[1], *Keisuke* Okumura[1], *Tetsuya* Kai[1], *Hiroshi* Iwase[3], and *Lembit* Sihver[4]

[1]Nuclear Science and Engineering Center, Japan Atomic Energy Agency, Shirakata, 2-4, Tokai, Ibaraki, 319-1195, Japan
[2]Research Organization for Information Science and Technology, Shirakata, 2-4, Tokai, Ibaraki, 319-1106, Japan
[3]High Energy Accelerator Research Organization, Oho 1-1, Tsukuba, Ibaraki, 305-0801, Japan
[4]Atominstitut, Technische Universität Wien, Stadionallee 2, 1020 Vienna, Austria

Abstract. The Particle and Heavy Ion Transport code System, PHITS, has been developed under the collaboration of several research institutes in Japan and Europe. This system can simulate the transport of most particles with energy levels up to 1 TeV (per nucleon for ion) using different nuclear reaction models and data libraries. More than 2,500 registered researchers and technicians have used this system for various applications such as accelerator design, radiation shielding and protection, medical physics, and space- and geo-sciences. This paper summarizes the physics models and functions recently implemented in PHITS, between versions 2.52 and 2.88, especially those related to source generation useful for simulating brachytherapy and internal exposures of radioisotopes.

1 Introduction

The Monte Carlo particle transport simulation code is an essential tool used in various fields of research such as radiation shielding, radiological protection, and medical physics. We are therefore developing Particle and Heavy Ion Transport code System, PHITS [1], that can simulate the transport of most particles with energy levels up to 1 TeV (per nucleon for ion) by using various nuclear reaction models and data libraries. The system is written in Fortran language and can be compiled using Intel Fortran 11.1 (or later versions) or GFortran (version 4.7 or 4.8). PHITS can be executed on the Windows, Mac, and Linux platforms. Distributed and shared memory parallelization techniques are made available using Message Passing Interface (MPI) protocols and open multi-processing (OpenMP) directives, respectively. Hybrid parallelization using both MPI and OpenMP is also feasible [2]. Various quantities, such as heat deposition, track length, and production yields, can be obtained via PHITS simulation using implemented "tally" estimator functions. Estimation of the time evolution of radioactivity is also feasible using DCHAIN-SP [3], which is also included in the PHITS package.

The latest version of PHITS available from the Organization for Economic Cooperation and

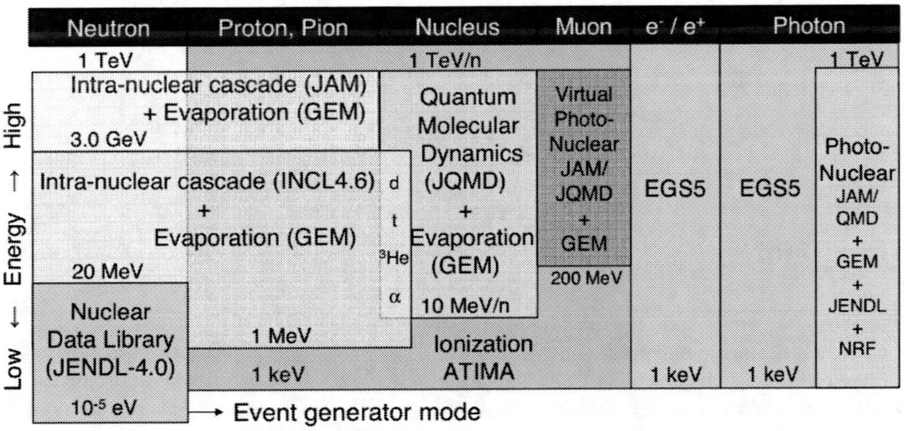

Figure 1. Map of the models and data libraries recommended for use in PHITS 2.88.

* Corresponding author: sato.tatsuhiko@jaea.go.jp

Development/Nuclear Energy Agency (OECD/NEA) databank [4] is 2.82 (Feb. 2016); however, a new version PHITS 2.88 will soon be released. In our previous report, we provided a detailed description of the features and functions of the models before version 2.52, implemented in PHITS [1]. In this paper, a brief summary of the physics models and functions implemented after the earlier mentioned version has been provided, especially those related to source generation for simulating brachytherapy and internal exposure of radioisotopes.

2 New Features

2.1 Upgrades for physics models and libraries

The map of the models and data libraries recommended for use in PHITS 2.88 is shown in Fig. 1. The major upgrades are listed below:

- The maximum energy of the particles that can be handled by PHITS has been extended from 100 GeV to 1 TeV (per nucleon for ions) because of the revisions made to the high energy nuclear reaction models JAM [5] and JQMD [6] as well as the implementation of the EGS5 algorithm [7]. This extension allowed PHITS to be more appropriately utilized in space applications. In addition, the implementation of the EGS5 algorithm improved the accuracy of electron transport simulation particularly for lower energies, which is very important for medical physics applications.
- The JQMD model was further improved to version 2.0, which enables more precise reproduction of the fragmentation cross sections by incorporating the reaction mechanisms that are particularly important for peripheral collisions [8]. Fig. 2 shows the charge changing cross section of 700 MeV/n ^{56}Fe ions incident to a natural carbon target calculated by JQMD and JQMD 2.0 in comparison with the measured data [9]. It is evident from the graph that JQMD 2.0 can reproduce the experimental data very well for high charge fragments, which are predominantly produced by peripheral collisions. This improvement is particularly important for cosmic-ray dosimetry and design of heavy ion therapy.
- The energy ranges of the photonuclear reactions simulated by PHITS were extended up to 1 TeV by implementing the models describing nuclear resonance florescence (NRF) [10] and high-energy photonuclear reactions [11]. The implementation of the NRF model enables the analysis of non-destructive assay of fissile nuclear materials using PHITS.
- The muon transport algorithm was improved by taking into consideration the mechanisms of virtual photonuclear reactions, pair production, bremsstrahlung, and negative muon capture reaction after the formation of a muonic atom [12]. As a benchmark of the newly implemented muon interaction models, underground muon intensities were calculated using PHITS 2.52 and 2.88, and compared with the experimental data [13,14]. The PARMA model [15,16], which was developed based on the airshower simulation performed by PHITS, was used for determining the vertical muon fluxes at the ground level. It can be seen in Fig. 3 that PHITS 2.88 can reproduce the experimental data very well, indicating the accuracy of newly implemented muon interaction models.
- The algorithm for de-excitation of a nucleus after an evaporation process was improved through the implementation of EBITEM [17], enabling precise estimation of γ-ray spectra and isomer production rates. This implementation allows PHITS to be utilized in the prompt γ-ray analysis, which is useful for various purposes such as design of on-line trajectory imaging for charged particle therapy based on gamma cameras.
- The models for calculating the total reaction cross sections were improved [18,19].
- The Kerma factors for some nuclei contained in JENDL-4.0 [20] were revised.
- A new reaction ejectile sampling algorithm to recover kinematic correlations from inclusive cross-section data, so-called event generator mode version 2.0, was

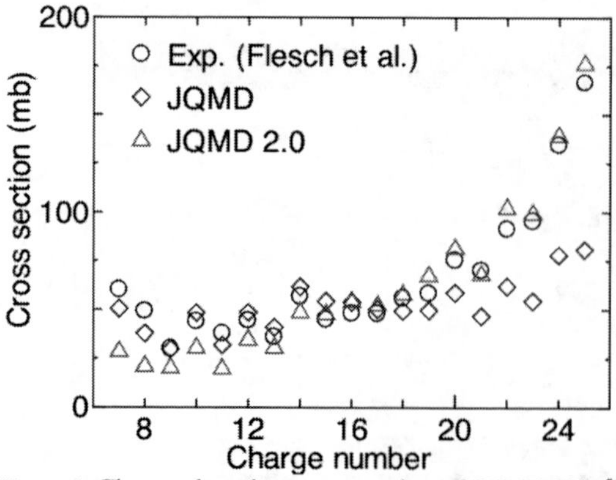

Figure 2. Charge changing cross section of 700 MeV/n ^{56}Fe ions incident to a natural carbon target calculated by JQMD and JQMD 2.0 in comparison with the measured data [9]

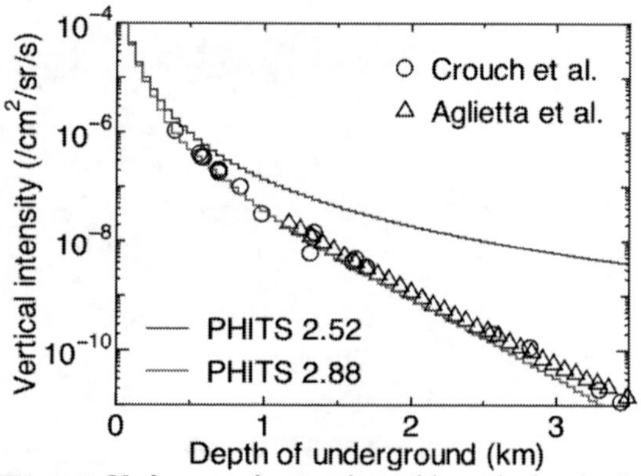

Figure 3. Underground muon intensities calculated using PHITS 2.52 and 2.88 in comparison with the experimental data [13,14]

implemented [21]. The energy spectrum of secondary neutrons and charge particles emitted from low-energy neutron interactions can be precisely determined on the basis of inclusive cross section data such as JENDL-4.0. As an example, the secondary neutron yields from ^{150}Nd(n,2n) reaction induced by 20 MeV neutrons calculated by the event generator mode versions 1 and 2 are shown in Fig. 4. The original inclusive cross section data contained in JENDL-4.0 are also plotted in the graph. The figure clearly indicates that the new version can reproduce the original data very well. This improvement enables more precise calculations of detector responses and dose equivalents due to neutron exposure below 20 MeV.

- DCHAIN-SP was improved by expanding the number of energy groups in its neutron-activation cross section libraries.
- A new approach for describing (d,xn) spectra at energies below 50 MeV was developed in combination with the intra-nuclear cascade model and the Distorted Wave Born Approximation (DWBA) [22]. Fig. 5 shows the double differential cross section of neutrons produced from natural lithium target bombarded by 40 MeV deuteron calculated by Intra-Nuclear Cascade of Liège (INCL) [23] coupled with and without DWBA. The corresponding experimental data [24] are also plotted in the graph. It is evident from the figure that the peaks observed over 40 MeV can be reproduced only by INCL with DWBA because such high-energy peaks are formed due to the discrete levels of nuclear structure, which can be considered only in DWBA. This improvement allows PHITS to be utilized in the design of neutron sources based on deuteron accelerators such as International Fusion Materials Irradiation Facility (IFMIF) project.
- The computational speed of ATIMA [25] in PHITS was improved to allow for the selection of ATIMA as the default stopping power calculation model.

2.2 Upgrades for support functions

In addition to the above-mentioned upgrades to the physics models and libraries, several user support functions were implemented for extending the capabilities of PHITS. The following list describes newly introduced functions in ver. 2.88, and the details of each are described in the PHITS manual [26]:

- The ability to read tetrahedral geometry (a kind of polygonal geometry) was implemented.
- The resolution of a detector can be taken into account when calculating the pulse-height tally.
- The statistical uncertainties in the results of tallies can be properly calculated even when the successive calculation mode using "dump" source.
- The absorbed dose in the unit of Gy can be calculated by considering the mass of the tally regions.
- A new tally for calculating the particle fluence at certain points, so-called the point estimator, was implemented.
- Particle fluxes can be calculated in the cylindrical co-ordinate as a function of the azimuth angle.
- Several user support functions were developed and introduced, including those for visualizing geometry errors, summing up two (or more) tally results, and defining a user's original tally.
- A conversion program from the DICOM image data to the PHITS input format, DICOM2PHITS, was developed.
- A function to automatically determine the appropriate weight windows (weight window generator) was implemented.
- A function that outputs PHITS tally results into a format usable by ParaView [27], which is an open-source, multi-platform data analysis and visualization application, was developed. For example, Fig. 6 shows the ParaView visualization of the absorbed doses in skeletal tissue of the International Commission on Radiological Protection (ICRP) adult male phantom [28] that has been irradiated by 1 GeV proton in the PHITS simulation.

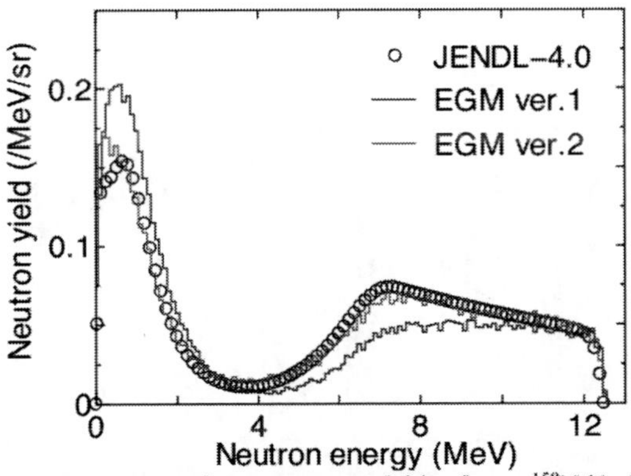

Figure 4. Secondary neutron yields from ^{150}Nd(n,2n) reaction induced by 20 MeV neutrons calculated by the event generator mode (EGM) versions 1 and 2, in comparison with the original inclusive cross section data contained in JENDL-4.0 [20]

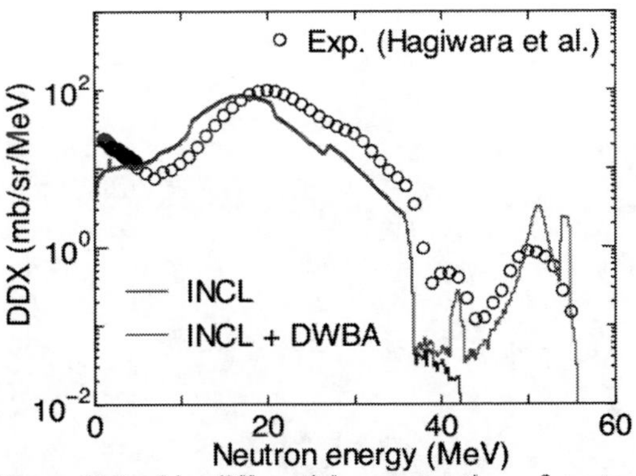

Figure 5. Double differential cross section of neutrons produced from natural lithium target bombarded by 40 MeV deuteron calculated by Intra-Nuclear Cascade of Liège (INCL) coupled with and without DWBA in comparison with the corresponding experimental data [24]

2.3 Upgrades for source generation

Recently, PHITS has been used for estimating the risk of internal exposure of radioisotopes [29] and therapeutic effects of brachytherapy on the basis of microdosimetric simulations. To conduct such simulations using the earlier versions of PHITS, the energy and emission probability of the radiations generated per nuclear transformation of radioisotopes were required to be specified by users by investigating the nuclear decay data such as level structure and branching ratios. However, it is sometimes difficult to evaluate the source term, particularly when the daughter nuclides of the radioisotopes are also radioactive.

Therefore, we developed a new source generation function that can automatically determine the discrete γ-ray spectra emitted from the decay of radioisotopes, including the contributions made by their daughter nuclides. The nuclear-decay data files for dose calculation, DECDC [30], which is equivalent to that in the database contained in ICRP Publication 107 [31], is employed as reference data for this new function. The input parameters to be specified in this function are the name and activity (Bq) of the initially existing radioisotopes and the time for the calculation of the activities after considering their decay chains. Fig. 7 shows the photon fluxes emitted one year after the decay of ^{144}Ce, which had had an initial activity of 100 Bq. Not only low-energy photons emitted from ^{144}Ce but also high-energy photons emitted from its daughter nuclides, such as ^{144}Pr and ^{144m}Pr, can be observed in the spectrum. Only the photon emission function can be dealt currently using this function; however, we intend to extend this functionality to enable handling of electron and alpha emissions as well.

It is occasionally required to consider the correlation of source particles for the exact reproduction of the internal exposure situations because some radioisotopes emit two (or more) radiations at once. Therefore, we have developed a function that generates multiple particles as an event while considering their angular correlation. This function is particularly important for simulating the microdosimetric profile of boron neutron capture therapy, because an α particle and a ^{7}Li ion are generated from the same location but opposite directions during a ^{10}B(n,α)^{7}Li reaction.

We also implemented source generation functions for reproduction of spontaneous fission neutrons [32] and reading of International Atomic Energy Agency (IAEA) phase space files [33]. The former is useful for simulations of radiological protection of internal exposure of spontaneous fission nuclides [34], whereas the latter is beneficial for medical physics simulations of X-ray therapy.

3 Conclusions

PHITS has been used in many countries for various applications, such as accelerator design, radiation shielding and protection, medical physics, and space- and geo-sciences. The tutorials of the code are organized several times per year in Japan, and sometimes in other countries such as France and Malaysia. Owing to the continuous improvement and promotion of the code, the number of registered users exceeded 2,500 in 2016, and has been rapidly increasing since then. A comprehensive benchmark calculation results of the new version of PHITS will be published elsewhere [35].

Acknowledgements

We are grateful for the support of Center for Computational Science & e-Systems (CCSE) of JAEA in promoting PHITS. We wish to thank Dr. Y. Namito and Dr. H. Hirayama of KEK for their advice in implementing the EGS5 algorithm. We would also like to thank Dr. Y. Mizuno of Kyoto Women's University for his advice in implementing the muon reaction models, Dr. L. Pong Hong of NAIS Co. Ltd. for his support in developing the function for generating fission neutron sources, and Mr. M. Adachi, Mr. T. Miura, and Mr. A. Wada of RIST for their assistance in the programing of PHITS. The function to read tetrahedral geometry was implemented under the support of Hanyang University Radiation Engineering Laboratory (HUREL), Korea. The function for tally results output in ParaView format was majorly programmed at

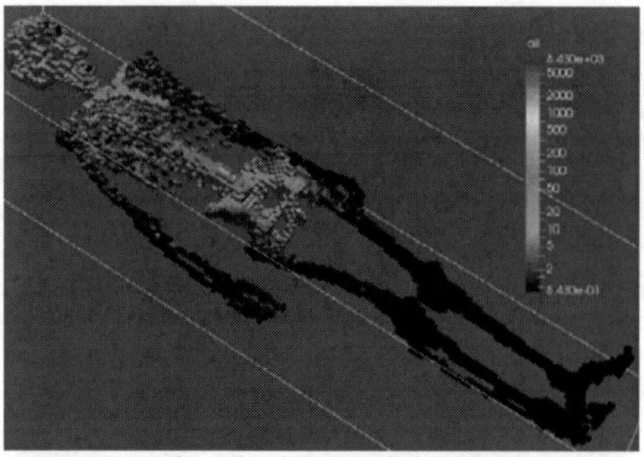

Figure 6. ParaView visualization of absorbed doses in the skeletal tissue of the ICRP adult male phantom that has been irradiated by 1 GeV proton in the PHITS simulation

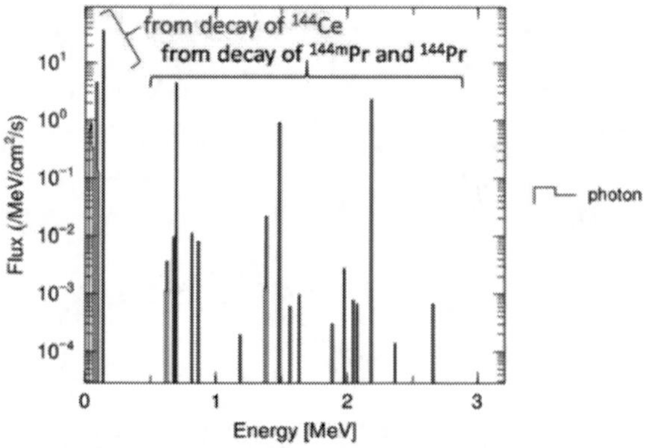

Figure 7. Photon fluxes emitted after one year of decay of ^{144}Ce with an initial activity of 100 Bq. The fluxes were obtained 1 cm away from the source location.

Visible Information Center, Inc. This work is partially supported by Japan Society for the Promotion of Science (JSPS), Grant-in-Aid for Scientific Research (C), Grant Number JP26461900.

References

1. T. Sato, K. Niita, N. Matsuda, S. Hashimoto, Y. Iwamoto, S. Noda, T. Ogawa, H. Iwase, H. Nakashima, T. Fukahori, K. Okumura, T. Kai, S. Chiba, T. Furuta and L. Sihver, Particle and Heavy Ion Transport Code System PHITS, Version 2.52, J. Nucl. Sci. Technol. **50**, 913-923 (2013).
2. T. Furuta, K.L. Ishikawa, N. Fukunishi, S. Noda, S. Takagi, T. Maeyama, K. Fukasaku, and R. Himeno, *Implementation of OpenMP and MPI hybrid parallelization to Monte Carlo dose simulation for particle therapy*, IFMBE Proceedings, **39**, 2099-2102. (2012).
3. T. Kai, F. Maekawa, K. Kosako, Y. Kasugai, H. Takada, and Y. Ikeda, *DCHAIN-SP 2001: High energy particle induced radioactivity calculation code*, JAERI-Data/Code 2001-016, (2001). [in Japanese]
4. http://www.oecd-nea.org/tools/abstract/detail/nea-1857/
5. Y. Nara, H. Otuka, A. Ohnishi, K. Niita, and S. Chiba, Relativistic nuclear collisions at 10A GeV energies from p+Be to Au+Au with the hadronic cascade model, Phys. Rev. C **61**, 024901 (2000).
6. K. Niita, S. Chiba, T. Maruyama, T. Maruyama, H. Takada, T. Fukahori, Y. Nakahara, and A. Iwamoto, Analysis of the (N,Xn') Reactions by Quantum Molecular-Dynamics Plus Statistical Decay Model, Phys. Rev. C **52**, 2620-2635 (1995).
7. H. Hirayama, Y. Namito, A.F. Bielajew, S.J. Wilderman, and W.R. Nelson, *The EGS5 code system*, SLAC-R-730 and KEK Report 2005-8, (2005).
8. T. Ogawa, T. Sato, S. Hashimoto, D. Satoh, S. Tsuda, and K. Niita, Energy-dependent fragmentation cross sections of relativistic ^{12}C, Phys. Rev. C **92**, 024614 (2015).
9. F. Flesch, S.E. Hirzebruch, G. Hüntrup, H. Röcher, T. Streibel, E. Winkel, W. Heinrich, Fragmentation cross section measurements of iron projectiles using CR-39 plastic nuclear track detectors, Radiat. Meas., **31**, 533-536 (1999)
10. T. Ogawa, S. Hashimoto, T. Sato, Development of general nuclear resonance fluorescence model, J. Nucl. Sci. Technol. **53**, 1766-1773 (2016).
11. S. Noda, S. Hashimoto, T. Sato, T. Fukahori, S. Chiba and K. Niita, Improvement of photonuclear reaction model below 140 MeV in the PHITS code, J. Nucl. Sci. Technol. **52**, 57-62 (2015).
12. S. Abe and T. Sato, Muon interaction models implemented into PHITS. J Nucl Sci Technol. **53**, 451-458 (2016).
13. M. Crouch, An improved world survey expression for cosmic ray vertical intensity vs. depth in standard rock, Proc. 20th Int. Cosmic Ray Conf.; 1987 Aug 2-15; Moscow (USSR). **6**, 165-168 (1987).
14. M. Aglietta, B. Alpat, E. D. Alyea et al., Muon "depth-intensity" relation measured by the LVD underground experiment and cosmic-ray muon spectrum at sea level, Phys. Rev. D. **58**, 092005 (1998).
15. T. Sato, Analytical Model for Estimating Terrestrial Cosmic Ray Fluxes Nearly Anytime and Anywhere in the World: Extension of PARMA/EXPACS, PLOS ONE **10**(12), e0144679 (2015).
16. T. Sato, Analytical Model for Estimating the Zenith Angle Dependence of Terrestrial Cosmic Ray Fluxes. PLOS ONE **11**(8): e0160390 (2016).
17. T. Ogawa, S. Hashimoto, T. Sato, K. Niita, Development of gamma de-excitation model for prediction of prompt gamma-rays and isomer production based on energy-dependent level structure treatment, Nucl. Instr. Meth. B 325, 35-42 (2014).
18. L. Sihver, A. Kohama, K. Iida, K. Oyamatsu, S. Hashimoto, H. Iwase, K. Niita, Current status of the "Hybrid Kurotama model" for total reaction cross sections, Nucl. Instr. Meth. B **334**, 34-39 (2014).
19. T. Sato, R. Kataoka, H. Yasuda, S. Yashiro, T. Kuwabara, D. Shiota and Y. Kubo, Air Shower Simulation for WASAVIES: Warning System for Aviation Exposure to Solar Energetic Particles, Radiat. Prot. Dosim. **161**, 274-278 (2014).
20. K. Shibata, O. Iwamoto, T. Nakagawa, N. Iwamoto, A. Ichihara, S. Kunieda, S. Chiba, K. Furutaka, N. Otuka, T. Ohsawa, T. Murata, H. Matsunobu, A. Zukeran, S. Kamada, and J. Katakura, JENDL-4.0: A New Library for Nuclear Science and Engineering, J. Nucl. Sci. Technol. **48**, 1-30 (2011).
21. T. Ogawa, T. Sato, S. Hashimoto, K. Niita, Development of a reaction ejectile sampling algorithm to recover kinematic correlations from inclusive cross-section data in Monte-Carlo particle transport simulations, Nucl. Instr. Meth. A **763**, 575-590 (2014).
22. S. Hashimoto, Y. Iwamoto, T. Sato, K. Niita, A. Boudard, J. Cugnon, J.C. David, S. Leray, D. Mancusi, New approach to description of (*d,xn*) spectra at energies below 50 MeV in Monte Carlo simulation by intra-nuclear cascade code with Distorted Wave Born Approximation, Nucl. Instr. Meth. B **333**, 27-41 (2014).
23. A. Boudard, J. Cugnon, J.C. David, S. Leray, and D. Mancusi, New potentialities of the Liège intranuclear cascade model for reactions induced by nucleons and light charged particles, Phys. Rev. C **87**, 014606 (2013).
24. M. Hagiwara, T. Itoga, N. Kawata, N. Hirabayashi, T. Oishi, T. Yamauchi, M. Baba, M. Sugimoto, and T. Muroga, Measurement of neutron emission spectra in Li(*d,xn*) reaction with thick and thin targets for 40-MeV deuterons, Fusion Sci. Technol. **48**, 1320-1328 (2005).
25. https://web-docs.gsi.de/~weick/atima/
26. http://phits.jaea.go.jp/manual/manualE-phits.pdf
27. http://www.paraview.org
28. International Commission on Radiological Protection, *Adult reference computational phantoms*, ICRP Publication 110, Ann ICRP **39**(2) (2009).

DOI: 10.1051/epjconf/201715306008

29. T. Sato, K. Manabe, N. Hamada, Microdosimetric analysis confirms similar biological effectiveness of external exposure to gamma-rays and internal exposure to ^{137}Cs, ^{134}Cs, and ^{131}I, PLOS ONE **9**(6), e99831 (2014).
30. A. Endo, Y. Yamaguchi and K.F. Eckerman, *Nuclear decay data for dosimetry calculation - Revised data of ICRP Publication 38*, JAERI 1347 (2005).
31. International Commission on Radiological Protection, *Nuclear decay data for dosimetric calculations*, ICRP Publication 107, Ann ICRP **38**(3) (2008).
32. J.M. Verbeke, C. Hagmnn, and D. Wright, *Simulation of neutron and gamma ray emission from fission and photofission*, UCRL-AR-228518 (2014).
33. https://www-nds.iaea.org/phsp/
34. A. Endo and T. Sato, Analysis of Linear Energy Transfers and Quality Factors of Charged Particles Produced by Spontaneous Fission Neutrons from ^{252}Cf and ^{244}Pu in the Human Body, Radiat. Prot. Dosim. **154**, 142-147 (2013).
35. Y. Iwamoto, T. Sato, S. Hashimoto, T. Ogawa, T. Furuta, S. Abe, T. Kai, N. Matsuda, R. Hosoyamada, and K. Niita, Comprehensive benchmark study for the recent version of the PHITS code, *submitted*.

Improvements in electron-photon-relaxation data for MCNP6

H. Grady Hughes[1,*]

[1]*Los Alamos National Laboratory, Los Alamos NM 87545 USA*

Abstract. After the initial release of the first production version of MCNP6 and its subsequent beta release, the extended electron-photon-relaxation capabilities of the code have continued to be developed. With the availability of newer data, several enhancements and corrections to the methods have been installed into the code. In this paper we discuss these improvements, including the development of three distinct approaches to sampling electron elastic scattering, the correction of an earlier data resolution problem, and a few formal corrections to the earlier ACE data file.

1 Introduction

The Los Alamos Monte Carlo transport code MCNP6 [1] has recently benefitted from significant work to improve the photon-electron transport and atomic relaxation capabilities of the code. The improvements [2-3] include extension of existing photon data to lower energies than previously supported; addition of new data such as subshell-specific photoelectric cross sections; completion of the incoherent and coherent form factor data to cover all scattering angles at all supported energies; new data for electron interactions including bremsstrahlung, elastic scattering, subshell-specific electroionization, and atomic excitation; a much more detailed representation for atomic relaxation; and an entirely new (to MCNP) approach to electron transport: the single-event method [4]. These improvements were made available with the initial production release of MCNP6 [1] along with an extended electron, photon, and relaxation data file eprdata12 [5] to support the new capabilities.

More recently new data [6] have become available and have now been incorporated into MCNP6, correcting a few issues with the previous work, and adding new sets of cross sections that were not included in eprdata12. The purpose of this paper is to introduce these newer changes and enhancements, which are expected to be released with the next public release of the MCNP6 code system.

2 Sources of the data

The database that supports these developments relies on an evolving set of libraries originally released by D. E. Cullen and collaborators [7-9] as separate data for photoatomic interactions (EPDL), atomic relaxation (EADL), and electroatomic interactions (EEDL). An updated version of this release was officially adopted in the ENDF/B VI.8 release of the Evaluated Nuclear Data File [10] and it was this source that was used to produce the ACE (A Compact ENDF) format library eprdata12 [5] for use in the initial production release of MCNP6 (and the subsequent beta release).

The latest improved version of this database was released by Cullen [6] in 2014. This is the EPICS2014 collection, which includes both an ENDF-formatted file and separate EPDL, EADL, and EEDL files. The newest enhancements to the developmental version of MCNP6 are primarily based on improvements and additions drawn from the EPICS2014 data.

3 Data and code enhancements

In this section we shall identify and discuss the additions and corrections to the single-event transport method at its current stage of development. There are two changes based on the new data that have potential importance for calculations and two that are primarily formal corrections.

3.1. Alternate elastic cross sections

The original eprdata12 presented for each element a tabulation of energy-dependent electron elastic scattering cross sections, together with a collection of secondary angular probability tables at a variety of electron energies. These data actually described the "large-angle" elastic scattering of the electron, specifying the cross section and subsequent deflection angle for electrons experiencing an elastic collision and scattering into a deflection angle in the range $-1 < \mu < 1 - 10^{-6}$, i.e. $\theta > \sim 1.4$ milliradians, with $\mu = \cos\theta$. Describing the elastic scattering with only these data is equivalent to making the approximation that scattering within the forward peak can be ignored, treating any scatter within the peak as a delta function in 1 - cos(θ). For many problems, this is a reasonable approximation, and this was all that was available in the initial releases of MCNP6 using eprdata12.

* Corresponding author: hgh@lanl.gov

With the new data of EPICS2014, two more possibilities become available. The first is the provision of a total electron elastic cross section. This is the sum of the large-angle cross section and the cross section for elastic events in which the electron deflection remains within the forward peak. To take advantage of the new data, one can sample for the occurrence of an elastic event using the total rather than the large-angle cross section. Then the ratio of the large-angle to the total cross section can be used in a rejection game to determine the angular sampling method. If the large-angle component is selected, the angular tables are used as in the released versions of MCNP6. If the in-peak component is chosen then the deflection angle is sampled from an analytic probability distribution

$$\frac{d\sigma}{d\mu} = \frac{A}{(\eta+1-\mu)^2}$$

recommended by the evaluators and given in the ENDF/B VI.8 release. Here $\mu = \cos\theta$, A is a normalization constant, and η is a screening angle used by Berger [11] and by Seltzer [12].

This two-component algorithm, combining both scattering ranges of the elastic interaction, should provide the most physically accurate simulation of the elastic scattering. It also provides the most expensive method in terms of computer time.

The second new possibility comes from the availability in the EPICS2014 data of a transport-corrected cross section for elastic scattering. This construct is often of interest in cases of large, highly forward peaked cross sections such as photon coherent scattering and electron elastic scattering, where its use can greatly reduce the effective cross section (reducing the computational burden) while still preserving the average scattering angle. The use of the transport-corrected cross section is an especially attractive option in highly scattering media, since the large number of elastic events tends to smooth out the details of the angular distribution, making this approach an increasingly good approximation precisely for the problems that need a more efficient computational method.

3.2. Large-angle near-peak elastic PDF

One of the issues that created some difficulties with the earlier eprdata12 release was insufficient precision in the tabulation of the angular probability distribution functions for large-angle electron elastic scattering, especially at high energies. Specifically, fewer digits than were needed to resolve the approach to the forward peak were provided in the ENDF/B VI.8 file. For example for 21.25 MeV electrons in iron, the last three points in the tabulation of scattering cosine *vs* probability density were

(0.999999, 4.60952E+05)
(0.999999, 6.01183E+05)
(0.999999, 8.16698E+05)

This situation clearly prevents the conversion of the angular distribution into a properly integrated, cumulative distribution that can be sampled in standard ways. To overcome this problem, a simple, but rather arbitrary adjustment was made to the data in eprdata12: each sequence of indistinguishable angles was collapsed to a single data point and the average probability density over the sequence was assigned to that point. The resulting cumulative angular distribution then had the expected form, and sampling of the elastic angular deflection could proceed as usual. In practice this was a serviceable expedient, but was hardly satisfactory on theoretical grounds, especially given the eventual availability of data lacking this flaw.

With the release of EPICS2014, this issue is now settled. Both the EEDL and ENDF formatted files of the new library present the tabulations of scattering angle *vs* probability density with sufficient numerical resolution to eliminate apparently coincident points in the distributions. Thus the preparation of the cumulative distributions and the Monte Carlo sampling for the scattering angle can now be dealt with in the standard way.

3.3. Average bremsstrahlung energy loss

It was the intention of the evaluators to provide a tabulation of the average energy of a bremsstrahlung photon as a function of the electron energy at the bremsstrahlung event. In the creation of ENDF/B VI.8, this tabulation was inadvertently reversed to present the average energy remaining to the electron after a bremsstrahlung event at the tabulated electron energy. For consistency with the ENDF release, this definition was retained in eprdata12, although MCNP6 made no actual use of this table. The EPICS2014 release restores the originally intended meaning of the table, so that the tabulated quantity is now the average bremsstrahlung photon energy. This form of the tabulation will also be maintained in the successor to the eprdata12 ACE-format file.

3.4. Compton Doppler broadening

The Compton Doppler broadening capability of MCNP allows the simulation of some of the effects of bound electrons on incoherent photon scattering. Data to support this capability have been available in photoatomic libraries including mcplib03 [13], mcplib04 [14] and eprdata12 [5]. Unfortunately, earlier versions of MCNP (MCNP5 and MCNPX) suffered from an inconsistency with the data: the probability that the incoherently scattered photon has interacted with a particular electron shell is presented in the data library as a tabulated probability density function, while the MCNP5 and MCNPX codes treated the table as a cumulative distribution. Thus the older codes produced incorrect results [15] for the distribution of shell interactions. To allow users of earlier versions of MCNP to avoid this bug, two new libraries, mcplib63, and mcplib84 [16] were produced with the probability in question already converted to cumulative form. These libraries can replace mcplib03 and mcplib04, respectively, enabling the older codes to sample Compton Doppler broadening as accurately as the current data can support.

This issue is moot for users of the official releases of MCNP6, since both of those versions automatically detect the form of the shell probability distribution, and convert the table to a cumulative distribution if needed.

Nevertheless, it seems appropriate to keep the eprdata library consistent with the newest of the traditional photoatomic libraries. Therefore the successor to eprdata12 will present the shell probabilities for Compton Doppler broadening as a cumulative distribution.

4 Ongoing developments

We are currently undertaking a broad review of benchmark calculations suitable for verification and validation of the electron-photon transport algorithms and data in MCNP6. This study will cover both the condensed-history and the single-event methods for electron transport, as well as the data and sampling methods for atomic relaxation. It is especially important that the newer data for electron elastic scattering be verified and validated, and that the practical computational aspects of the new methods be explored.

We are also conducting reviews of specific aspects of the transport, including the possibility of improvements in the details of the condensed-history algorithms, especially by increasing the use of the new eprdata libraries in several areas of condensed-history, where they have not previously been used.

5 Summary

We have provided a brief discussion of four recent enhancements and corrections to the electron, photon, and atomic relaxation data of the eprdata series of data libraries included with the MCNP6 distribution. These data and associated code improvements are undergoing verification and validation testing, and are projected to be a part of the next official release of the MCNP6 code system. A future extended version of this paper will present specific examples of the new capabilities and the results of practical calculations.

References

1. T. Goorley, M. James, T. Booth, F. Brown, J. Bull, L. J. Cox, J. Durkee, J. Elson, M. Fensin, R. A. Forster, J. Hendricks, H. G. Hughes, R. Johns, B. Kiedrowski, R. Martz, S. Mashnik, G. McKinney, D. Pelowitz, R. Prael, J. Sweezy, L. Waters, T. Wilcox, T. Zukaitis, Nuc. Tech. **180** 298-315 (2012).
2. G. Hughes, Prog. Nuc. Sci. and Tech. **4** 454-458 (2014)
3. H. G. Hughes, *SNA + MC 2013*, 03105 (2014)
4. H. G. Hughes, Los Alamos National Laboratory report LA-UR-12-21068, Rev. 3 (2013)
5. H. G. Hughes, Los Alamos National Laboratory report LA-UR-13-27377, Rev. 1 (2015)
6. D. E. Cullen, IAEA-NDS-218, IAEA Nuc. Data Section (2014)
7. D. E. Cullen, M. H. Chen, J. H. Hubbell, S. T. Perkins, E. F. Plechaty, J. A. Rathkopf, J. H. Scofield, Lawrence Livermore National Laboratory report UCRL-50400 6 Parts A-B, Rev. 4 (1989)
8. S. T. Perkins, D. E. Cullen, M. H. Chen, J. H. Hubbell, J. Rathkopf, J. Scofield, Lawrence Livermore National Laboratory report UCRL-50400 **30** (1991)
9. S. T. Perkins, D. E. Cullen, and S. M. Seltzer, Lawrence Livermore National Laboratory report UCRL-50400 **31** (1991)
10. Cross Section Evaluation Working Group, Brookhaven National Laboratory report BNL-NCS-17541, 8th Ed. (2000)
11. M. J. Berger, R. Wang, in *Monte Carlo Transport of Electrons and Photons* Ch. 2 (Plenum Press, New York, 1988)
12. S. M. Seltzer, Ibid. Ch. 7
13. M. C. White, Los Alamos National Laboratory report LA-UR-03-0787 (2003)
14. M. C. White, Los Alamos National Laboratory report LA-UR-03-1019 (2003)
15. B. C. Kiedrowski, F. B. Brown, M. C. White, D. K. Parsons, Los Alamos National Laboratory report LA-UR-12-00121 (2012)
16. M. C. White, Los Alamos National Laboratory report LA-UR-12-00018 (2012)

Release of RANKERN 16A

Adam Bird[1,*], *Christophe* Murphy[1], and *Geoff* Dobson[1]

[1]AMEC Foster Wheeler, King's Point House, Poundbury, Dorchester, Dorset DT1 3BW, UK

Abstract. RANKERN 16 is the latest version of the point-kernel gamma radiation transport Monte Carlo code from AMEC Foster Wheeler's ANSWERS Software Service. RANKERN is well established in the UK shielding community for radiation shielding and dosimetry assessments. Many important developments have been made available to users in this latest release of RANKERN. The existing general 3D geometry capability has been extended to include import of CAD files in the IGES format providing efficient full CAD modelling capability without geometric approximation. Import of tetrahedral mesh and polygon surface formats has also been provided. An efficient voxel geometry type has been added suitable for representing CT data. There have been numerous input syntax enhancements and an extended actinide gamma source library. This paper describes some of the new features and compares the performance of the new geometry capabilities.

1 Introduction

RANKERN is a well-established and powerful point-kernel Monte Carlo software tool for general gamma-ray radiation transport analysis for shielding and dosimetry applications. RANKERN is developed and licensed by AMEC Foster Wheeler's ANSWERS Software Service. The RANKERN package comprises not only the code itself but also nuclear data libraries, user documentation, productivity tools and user support services. Supporting geometry model visualisation and verification and results visualisation tools are also available.

The uncollided flux estimates produced by RANKERN are corrected for scattering using build-up factors; three build-up options are available. The first, known as the original method, expresses the build-up data in the form of cubic polynomials. The coefficients can be supplied directly or included from library data derived from a combination of reference's [1] and [2] and MCBEND[3] Monte Carlo simulations. In addition a method is available to estimate the importance of oblique build-up. The second method is a set of build-up libraries based on ANS-6.4.3 [4] recommendations. These extend the number of materials and penetration distances from 20 to 40 mfp and are the current default in RANKERN. The third option uses Diffusion Theory to calculate build-up factors. This is particularly useful for multi-layered shields without a dominant final shield material.

RANKERN is being continually developed to meet the needs of its users. RANKERN 16 is the latest version to be released in line with the ANSWERS vision to 'provide easy-to-use software that meets the current and emerging needs of the user community'. In the case of RANKERN 16 this vision focuses on the key areas of geometry modelling and user-friendliness. This paper details the developments incorporated into RANKERN 16 and some associated capabilities of the accompanying graphics software VisualWorkshop[5]. All the images in this paper were generated using VisualWorkshop.

2 Developments in RANKERN 16

2.1 Geometry Modelling

The geometry modelling system employed by RANKERN is known as Fractal Geometry (FG). FG is a well-established system of solid geometry modelling in which the problem geometry is subdivided into zones - defined as the intersections and differences of mathematical bodies. RANKERN has a large selection of body shapes ranging from simple bodies such as cuboids, cylinders and spheres to more complex bodies such as prisms, ellipses and tori. The bodies are assembled into structures called parts, which are self contained with their own local co-ordinate system to simplify the model construction. Parts may be included within other parts to any depth of nesting and a given part may be included more than once within the geometry. The ability to break down complex models into parts simplifies the preparation and checking of the input data. The FG geometry package is shared with the ANSWERS radiation transport Monte Carlo code MCBEND and the ANSWERS Criticality code MONK®[6]. RANKERN 16 benefits from improvements to geometry checking and the addition of two simple geometric bodies, a triangular prism and a rod with an elliptical cross section. The main enhancements come in the form of CAD geometry import capabilities which are detailed below.

* Corresponding author: adam.bird@amecfw.com

2.1.1 Import of a CAD file in the IGES format

This is a fully featured CAD import capability that provides particle tracking within the CAD model with no conversion or loss of accuracy of the geometry of the model. The interactive display within VisualWorkshop identifies the figures and sub-figure definitions of the IGES model making it simple to assign material compositions to regions of space within the CAD model. See Figure 1 for an example.

2.1.2 Import of a tetrahedral mesh format file

While there is little reason to convert a file to a tetrahedral mesh format for the sole purpose of importing into RANKERN it is quite common for these files to exist having been generated for a stress analysis or thermodynamics physics code for example. In this case gamma shielding assessments can be performed without the need to re-create the geometry model. Groups of tetrahedra that represent a named item within the mesh model are automatically assigned to a material of the same name in RANKERN. See Figure 2 for an example.

Figure 1. An example IGES model displayed by the 3D ray trace feature of VisualWorkshop .

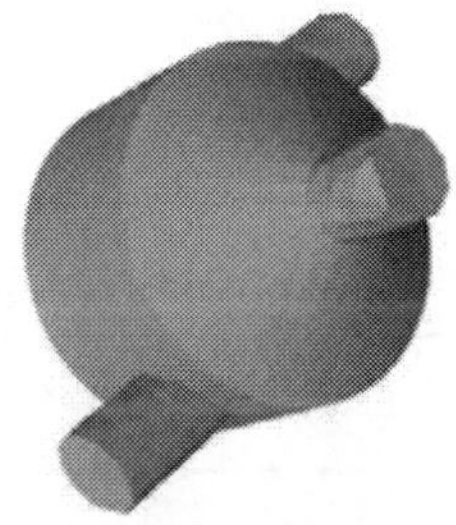

Figure 2. An example tetrahedral mesh model.

2.1.3 Import of a polygon surface format file

A triangular polygon surface geometry type can be imported or defined directly in the RANKERN input as a native FG body type.

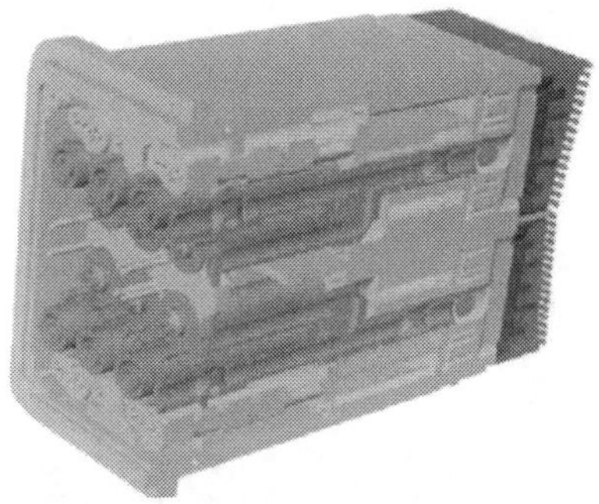

Figure 3. An example model that has been exported from a CAD package in a polygon surface format.

Many CAD packages have options for exporting a polygon surface description and this is usually faster to process than the IGES import and requires less memory than a tetrahedral mesh representation.

2.1.4 Voxel geometry representation

This geometry type can be imported or defined directly in the RANKERN input. A voxel geometry could be represented by many boxes. The voxel geometry type in RANKERN takes advantage of the regularity of the geometry to reduce memory requirements and provide efficient tracking.

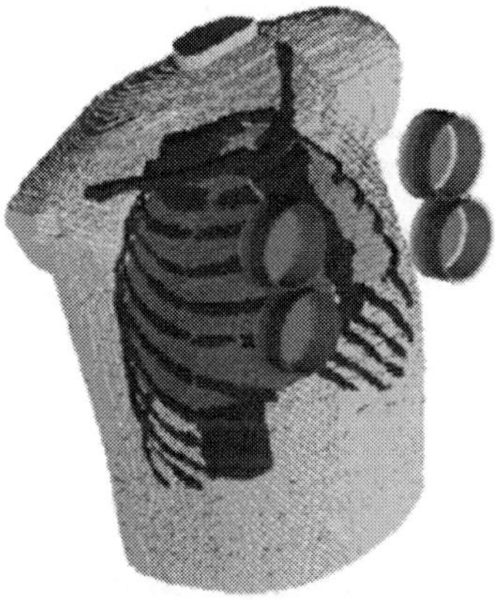

Figure 4. An example of a voxel geometry model from CT data with detectors defined using FG.

All of the geometry types described can be used in a mix and match way in a given model. Section 4 reports on the relative performance of the geometry methods.

2.2 Visualisation

VisualWorkshop version 3B, which accompanies RANKERN, includes some enhancements to contour plotting of results from RANKERN and can now display the particle tracks from a RANKERN calculation.

 DOI: 10.1051/epjconf/201715306010

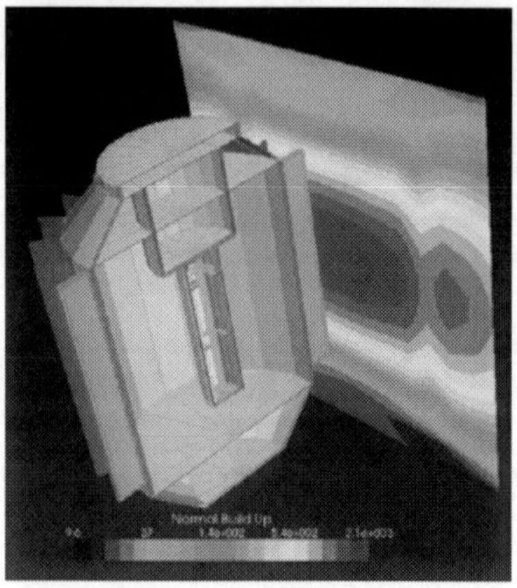

Figure 5. VisualWorkshop image showing a dose rate contour generated from a grid of RANKERN output points.

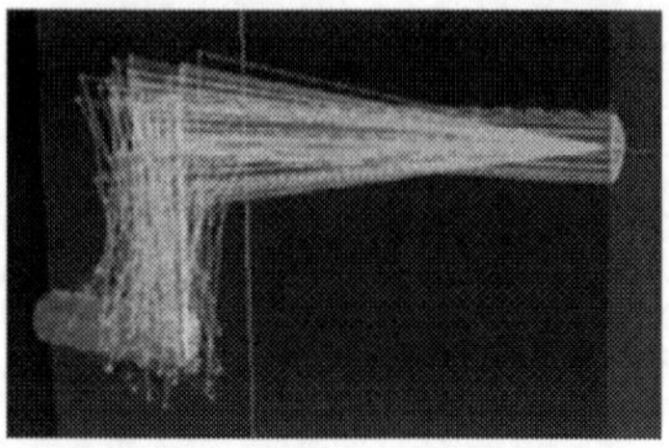

Figure 6. VisualWorkshop image showing RANKERN tracks through a duct.

3 Other enhancements

3.1 RKARD Package Enhancements

The RKARD system of data input provides a powerful method of using mathematical and logical functions to construct a flexible and efficient input model.

RANKERN 16 has enhanced RKARD facilities including:

- Double precision arithmetic to increase precision
- Exponentiation to any power
- Functions INT, ATAN, ACOS and ASIN
- Variables containing character string values
- Variables containing logical conditions
- Parameters used for looping containing variables
- Nested IF statement blocks
- ELSEIF statements
- Loop counters in conditions
- Extended condition statements
- Additional RKARD controls

3.2 Source Enhancements

Sources in RANKERN can be defined in terms of an energy group source spectrum or as an inventory of actinides. The latter can optionally be decayed before the calculation commences. For this release the source and decay libraries have been extended and now contain data for over 1800 actinides.

3.3 Looping

The input data for RANKERN can contain variables. Variables can be given lists of values and a calculation is run for each one. This is called a 'looping' calculation. A new command line argument specifying the loop index is available to select one of the calculations in a RANKERN looping case to run. This makes it possible to run RANKERN looping cases in parallel on a computing cluster.

3.4 New Random Number Generator

RANKERN now uses the same random number generator as the MCBEND and MONK codes. In previous versions of RANKERN if a seed was not provided then the same default seed was used each time. In RANKERN 16 if a seed is not provided it will be generated from the date and time of execution.

4 Performance implications of the geometry representation chosen

The native 3D Geometry capabilities of RANKERN are powerful and general and enable complex models to be constructed relatively easily. However, they need to be created by the modeller. Importing from 3rd party CAD software can save a user time. The relative performance of the imported geometry is important because the CAD geometry is not converted. As a result there are no geometric approximations and this comes with increased computational requirements.

Fractal Geometry, IGES, Polygon Surface and Tetrahedral Mesh versions of the model shown in Figure 5 were created and the relative computing time was compared for each one – see Table 1.

Table 1. Relative computing time for each geometry type.

Geometry Type	Relative Speed
Fractal Geometry	1.0
IGES	3.4
Polygon Surface	3.9
Tetrahedral Mesh	6.4

Thus there is a trade-off between ease of model preparation and run time. This is only likely to be an issue for very complex models with a lot of dose points or if a large number of calculations need to be performed.

5 Conclusions

This paper has summarised the developments incorporated into RANKERN 16. These cover improvements to the geometry modelling with the introduction of CAD import features and improvements to the RKARD input system amongst others. ANSWERS continues to be committed to meeting the current and future needs of its customers.

References

1. M Capo. *Ray build-up factors for point polynomial approximation of gamma isotropic sources*, APEX 510
2. J. N. Stone and H. H. Vav Tuyl, *PUSHLD - A code for calculating gamma dose rates for plutonium in various geometries*. HEDL-TME 73-89.
3. P. Cowan, G. Dobson, J. Martin, *Release of MCBEND11*, ICRS12-RPSD2012, Nara, Japan.
4. RSIC Data Library Collection. *ANS-6.4.3 Geometric progression gamma-ray build-up factor coefficients,* 1991. DLC-129
5. A. Bird and T. Fry, *Visual Workshop 2: A Model Viewer, Editor and Results Display Package for the Answers Shielding and Criticality Codes*, ICRS12-RPSD2012, Nara, Japan.
6. S. Richards, D. Long and P. Smith *MONK10: A Monte Carlo Code for Criticality Analysis.*, ICNC2015, Charlotte, NC, USA

Application of Inverse Gamma Transport to Material Thickness Identification with SGRD Code.

Philippe Humbert*

CEA DAM Ile-de-France, Bruyères-le-Châtel, 91297 Arpajon cedex, France

Abstract. SGRD (Spectroscopy, Gamma rays, Rapid, Deterministic) code is used to infer the dimensions of a one dimensional model of a shielded gamma ray source. The method is based on the simulation of the uncollided leakage current of discrete gamma lines that are produced by nuclear decay. Experimentally, the unscattered gamma lines leakage current is obtained by processing high precision gamma spectroscopy measurements. The material thicknesses are computed with SGRD using a fast ray-tracing algorithm embedded in a non-linear multidimensional iterative optimization procedure that minimizes the error metric between calculated and measured signatures. For verification, numerical results on a test problem are presented.

1 Introduction

Identification of nuclear radiation source characteristics is a subject of interest for nonproliferation and nuclear safeguard applications. Gamma spectroscopy is used for this purpose due to the sensitivity of these measurements to source isotopic composition and shielding materials properties. The determination of source characteristics using known signature measurements is an inverse transport problem. This subject has been studied by different authors (see for example [1-5]).

Here, we present the determination of the unknown material interface positions from the measured uncollided gamma line spectrum obtained by processing high precision gamma spectroscopy measurements. For this purpose, SGRD code [6], which was designed for fast computation of gamma ray leakage out of 1-D spherical assemblies, is used as a forward solver for iterative inverse transport calculations. The material thicknesses are computed using a non-linear multidimensional iterative optimization algorithm that minimizes the error metric between calculated and measured signatures. The optimization is performed using the gradient free Powell method [8,9].

For verification, numerical results on a test problem are presented. A synthetic gamma lines spectrum is used as input to the inverse transport solver and the obtained geometry is compared to the original one.

2 Uncollided Leakage Gamma Current Calculation

2.1. Gamma leakage

The observable is the uncollided gamma ray leakage line spectrum produced by the radioactive decay gamma source.

The number L of gamma particles leaving a sphere of radius R per unit time and solid angle without collision is calculated using the uncollided angular flux $\psi(R,\mu)$.

$$L = 2\pi R^2 \int_0^1 \mu\psi(R,\mu)d\mu \qquad (1)$$

In equation (1), μ is the cosine of the angle between the particle direction of flight and the spatial radial vector on the external surface.

2.2. Ray tracing

The direct flux is solution of the transport equation without scattering. It is solved by SGRD code using an accurate and fast ray-tracing method which gives the exact solution of the integral transport equation along discrete directions.

The shielded source is made of concentric spherical shells, each with a constant source intensity Q_i and total macroscopic cross section σ_i. As illustrated in figure 1, the angular discretization is such that a set of N_i characteristic rays is associated to each shell, from the most internal (i=1) to the most external shell with a non-null source (i=I).

* Corresponding author: philippe.humbert@cea.fr

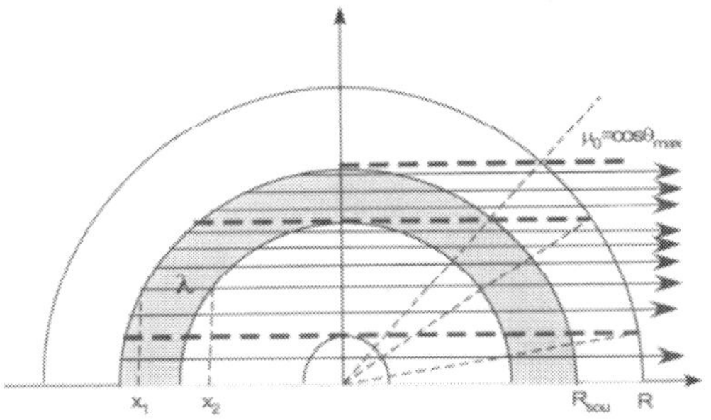

Fig. 1. 1-D spherical ray-tracing. The γ source is in the shaded shell.

The unscattered leakage is given by :

$$L = 2\pi R^2 \sum_{i=1}^{I} \frac{\Delta\mu_i}{N_i} \sum_{j=1}^{N_i} \mu_{ij} \psi(R, \mu_{ij}) \qquad (2)$$

In equation (2):

μ_{ij} characterizes the direction of flight associated to the jth ray crossing the ith shell,

μ_i is the cosine associated to the upper boundary of shell i and $\Delta\mu_i = \mu_i - \mu_{i-1}$.

The angular flux ψ is calculated using the following transmission equation:

$$\psi(x_{i+1}, \mu_j) = \psi(x_i, \mu_j) e^{-\sigma_i \lambda_i} + \frac{Q_i}{\sigma_i}\left(1 - e^{-\sigma_i \lambda_i}\right) \qquad (3)$$

The λ_i are the intersection lengths of the characteristics with the spherical shells.

$$\lambda_i = x_{i+1} - x_i \text{ with } x_i = \sqrt{R_i^2 - R^2(1-\mu^2)} \qquad (4)$$

2.3. Validation

The particle leakage from an homogeneous spherical source of radius R, opacity σ and intensity Q has an analytic expression [7].

$$L = \frac{Q}{8\sigma^3}\left[(1 + 2\sigma R)\exp(-2\sigma R) + \frac{(2\sigma R)^2}{2} - 1\right] \qquad (5)$$

The convergence of the ray tracing algorithm with the number of directions is illustrated in Table 1 for R=5cm, σ=1.484673 cm^{-1} and Q=1948.531 cm^{-3}.s^{-1}.

The ray tracing solver is faster and more accurate than S_N discrete ordinates for unscattered transport calculations (cf. figure 2).

Table 1. Convergence of the ray tracing and S_N solvers with the number of directions.

Number of directions	S_N Leakage (100 cells)	Ray-tracing Leakage
8	8188.1891	8120.3737
16	8142.4987	8125.7923
32	8131.6601	8127.6136
64	8129.0541	8128.1026
128	8128.4142	8128.2270
256	8128.2552	8128.2583
512	8128.2155	8128.2661
1024	8128.2056	8128.2680
Analytic	**8128.2679**	

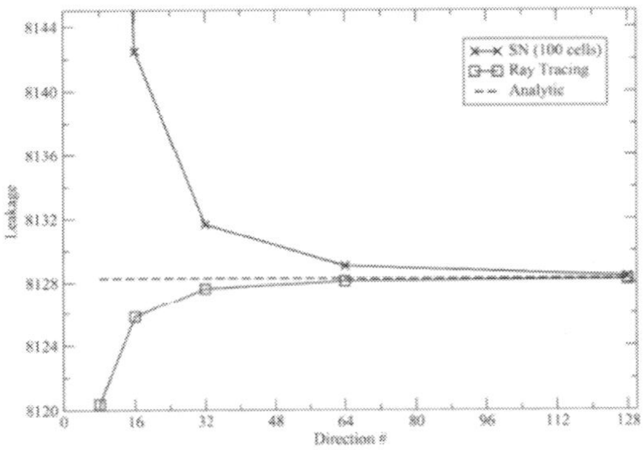

Fig. 2. Convergence to the analytic solution of the ray tracing and S_N leakage calculations with the number of directions.

3 Optimization - Inverse Transport

The goal of the inverse transport problem is to find the 1-D spherical assembly of source and shielding materials that produces the better agreement with the measured uncollided leakage gamma spectrum. Starting from a model composed of a given number of material atomic compositions and densities, the optimization algorithm automatically searches the shell thicknesses that give the best estimate of the measured line spectrum. The goal is then to minimize the error or distance between the measured and calculated spectrum. The quality of the estimation is quantified by the chosen error metric.

DOI: 10.1051/epjconf/201715306011

3.1. Error Metric

We use a least square method, the error is defined as :

$$E(x_1, x_2, \ldots, x_M) = 100\sqrt{\frac{1}{N}\sum_{i=1}^{N}\omega_i\left(\frac{L_i^{mes} - L_i^{calc}}{L_i^{mes}}\right)^2} \qquad (6)$$

L_i^{mes} : Measured leakage in line i,
L_i^{calc} : Calculated leakage in line i,
M : Number of material shells,
N : Number of spectral lines,
x_m : Thickness of shell m,
ω_i : Weight of line i.

Geometric or mass constraints can be incorporated in the error function. For instance one can favor the external radius around $R=R_0$ using the following error function where λ is the weight associated to the constraint.

$$E(\%) = 100\sqrt{\frac{1}{N}\sum_{i=1}^{N}\omega_i\left(\frac{L_i^{mes} - L_i^{calc}}{L_i^{mes}}\right)^2 + \lambda\left(\frac{R - R_0}{R_0}\right)^2} \qquad (7)$$

3.2. Stochastically Restarted Powell Optimization Method

The Powell method [8,9] is a multidimensional deterministic optimization algorithm which proceeds by a sequence of line minimizations along well chosen directions.

One advantage of the Powell method is that it needs to evaluate only the error function and does not necessitate the calculation of the gradient of the error with respect to the optimized variables for the choice of the successive minimization directions.

The Powell's method discarding the direction of largest decrease described in [8] is used to avoid the generation of linearly dependent set of directions.

The line minimizations are performed by successive bracketing of the minimum using Brent's method which switches between inverse parabolic interpolation and golden section.

If the error function has local minima, the Powell method can miss the best solution. In order to move from a local minimum the method can be stochastically restarted and another minimum is searched, the global minimum is estimated as the smallest local minimum after a given number of iterations.

We have implemented the following heuristic algorithm. The Powell optimization is embedded within a simple random optimization loop which is periodically restarted (cf. figure 3). The external radius of each shell is randomly sampled and the best configuration according to the error metric is retained and serves for a new Powell optimization. The best Powell's optimized geometry is retained.

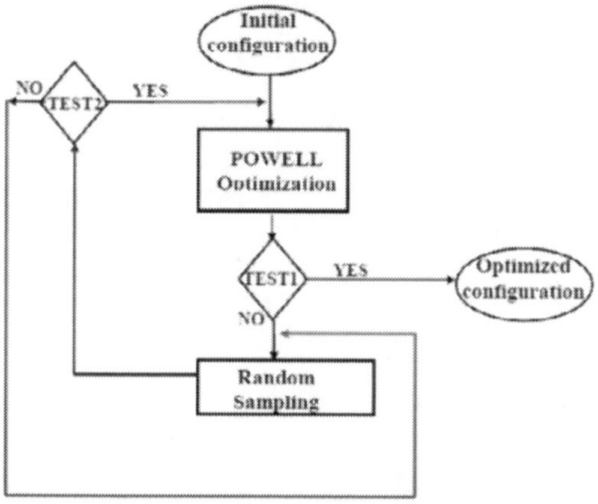

Fig. 3. Heuristic optimization algorithm : Powell optimization embedded within a simple random optimization loop.

In figure 3, TEST1 means that the total number of iterations has reached a maximum (200 for instance). The algorithm ends when TEST1 is true. TEST2 means that the number of random iterations is equal to a maximum (10 for instance) or that the random configuration is better than the previous ones. The Powell's optimization is triggered only when TEST2 is true. The loop in figure 3 is used to find a new random restart for the Powell's iterative algorithm which is not detailed in the figure.

4 SGRD Code

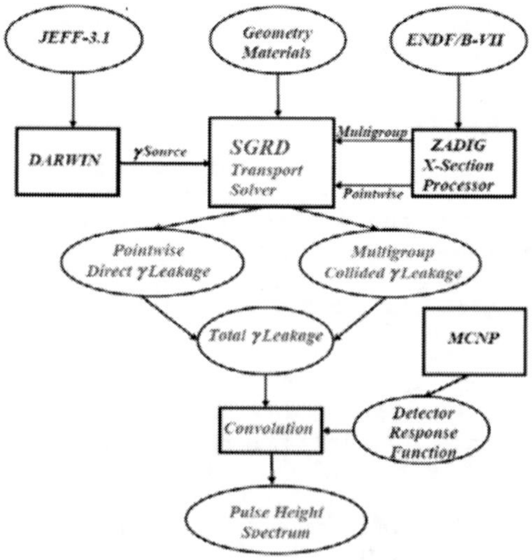

Fig. 4. SGRD input/output flow diagram for gamma spectra calculation.

SGRD code is primarily designed for fast computation of gamma pulse height spectra produced by spherical

 DOI: 10.1051/epjconf/201715306011

shielded sources [10]. The ray tracing solver is used for direct gamma lines transport and the collided flux is calculated using a discrete ordinates S_N solver, the gamma spectrum is obtained by convolution with the detector response function (cf. figure 4).

The code has been adapted for material interface position determination using the direct gamma lines intensity extracted from the measured gamma spectrum. For this purpose the Ray tracing solver is embedded in the stochastically restarted Powell minimization algorithm (cf. figure 5).

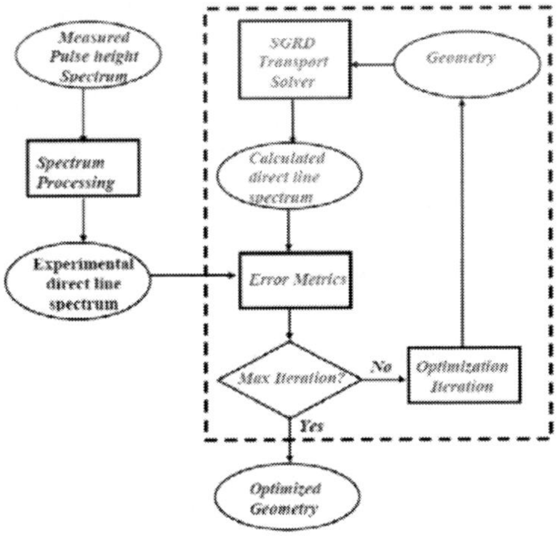

Fig. 5. Inverse transport for geometry search using SGRD ray-tracing solver.

A typical computation time for the optimization process is of the order of seconds or less on a personal computer (DELL® precision T5600).

5 Test Problem - Numerical Results

The test problem is a spherical natural uranium shell surrounded by polyethylene (CH_2) and an aluminum cover (cf. table 2).

Table 2. Test problem specification.

Materials	Air	Uranium	CH_2	Aluminum
Atomic Comp.	^{16}O 20% ^{14}N 80%	^{238}U 99.3% ^{235}U 0.7%	^{12}C 1/3 H 2/3	^{27}Al 100%
Density (g/cm^3)	0.0013	18.9	0.9	2.7
R_{EXT} (cm)	18	20	29	30

The gamma source is due to the uranium radioactive decay. Five characteristic lines are chosen for the optimization (cf. table 3).

Table 3. γ line characteristics and simulated leakage measurements.

Isotope	Energy (keV)	Intensity (γ/g/s)	Leakage (γ/sr/s)
^{235}U	143.76	8753	282.86
	185.71	45660.2	3259.60
	205.31	4004.4	386.95
^{238}U	766.36	39.7	12604
	1001.03	103.8	50599

The measurements are simulated using SGRD uncollided direct transport calculations.

A first optimized configuration is found after 10 Powell iterations. The external radius of the uranium and polyethylene shells are yet close to the solution, but the internal radius of the uranium shell is still wrong. In this case a correct solution is found after 159 iterations with several stochastic restarts as shown in figure 6 where the error value is plotted versus the number of iterations.

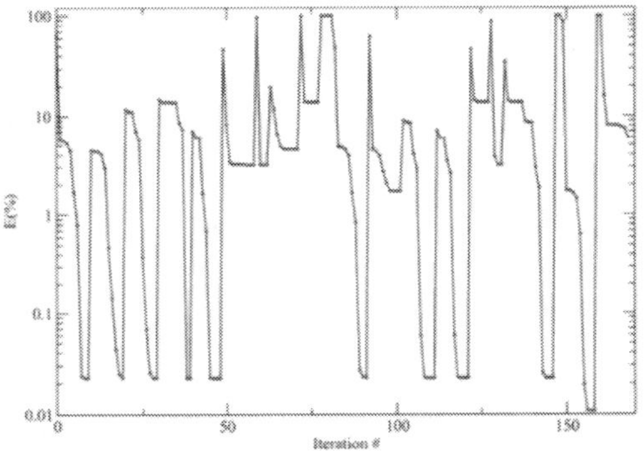

Fig. 6. Evolution of the error metric with the number of iterations.

The results of the optimization process using Powell method are presented in table 4 and 5.

Table 4. Results of the optimization, external radius.

Materials	Air	Uranium	CH_2	Aluminum
Reference R_{EXT}	18	20	29	30
Initialized R_{EXT} Iteration# 0 E=85.319%	9	16	23	30
Optimized R_{EXT} Iteration# 10 E=0.0226%	13.433	19.993	29.008	30.000
Optimized R_{EXT} Iteration# 159 E=0.0108%	17.967	19.998	29.002	30.000

DOI: 10.1051/epjconf/201715306011

Table 5. Results of the optimization, gamma lines intensities.

Energy (keV)	Initial (γ/sr/s)	Optimized (γ/sr/s)	"Measured' (γ/sr/s)
143.76	21.30	282.89	282.86
185.71	309.75	3259.80	3259.60
205.31	39.78	386.88	386.95
766.36	2815.44	12604.36	12604
1001.03	12807.21	50601.19	50599

The comparison between simulated (direct transport calculation), initial and optimized spectra (inverse transport calculation) is presented in figure 7.

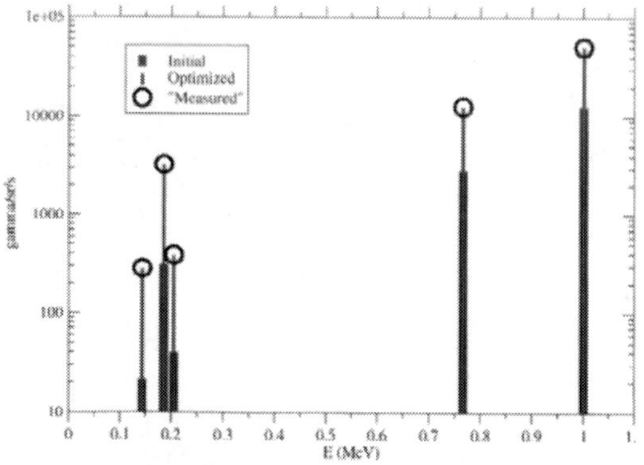

Fig. 7. Initial and optimized gamma line spectra compared to the measurement simulated by direct transport calculation.

6 Conclusion

Fast material thicknesses identification is performed using SGRD uncollided ray-tracing transport solver embedded in a gradient free stochastically restarted Powell iterative optimization loop.

We have presented the application of this methodology to a specific test problem. The code has been verified to work well using synthetic and measured data.

Future works will include the sensitivity analysis to measurement uncertainties, the use of other optimization parameters such as the source isotopic composition and the comparison with other optimization methods

References

1. J. A. Favorite, "Using the Schwinger Variational Functional for Solution of Inverse Transport Problems," *Nucl. Sci. Eng.*, **146**, 51-70 (2004)
2. K.C. Bledsoe, J.A. Favorite, and T. Aldemir, "Using the Levenberg-Marquardt Method for Solutions of Inverse Transport Problems in One- and Two-Dimensional Geometries," *Nucl. Tech.*, vol. 176, pp.106-126 (2011)
3. K.C. Bledsoe, J.A. Favorite, T. Aldemir, "Application of the Differential Evolution Method to Solving Inverse Transport Problems," *Nucl. Sci. Eng.*, **169**, pp. 208-221 (2011)
4. J.C. Armstrong, J.A. Favorite, "Identification of Unknown Interface Locations in Source/Shield System Using the Mesh Adaptive Direct Search Method," *Trans. of the Am. Nucl. Soc.*, **106**, pp. 375-377 (2012)
5. J. Mattingly, D. J. Mitchell, "A Framework for the Solution of Inverse Radiation Transport Problems," *IEEE Trans. Nucl. Sc.* **57** (6), 3734-3743 (2010)
6. Ph. Humbert, B. Méchitoua, "Fast Gamma Ray Leakage Spectra Simulation," *Proc. M&C 2009*, Saratoga Springs, New York, May 3–7, 2009, American Nuclear Society (2009) (CD-ROM)
7. K.M. Case, F. de Hoffmann, G. Placzek, "Introduction to the Theory of Neutron Diffusion," Vol.1, Los Alamos Scientific Laboratory report, USAEC (1953)
8. W. H. Press et al., *Numerical Recipes*, Cambridge University Press, Chapter 10 (1994)
9. R. Brent, "Algorithms for Minimization without Derivatives," Englewood Cliffs, N.J.: Prentice Hall, Chapter 7 (1973)
10. Ph. Humbert, B. Méchitoua, "Fast Computation of Pulse Height Spectra Using SGRD Code," Proc. ICRS-13 & RPSD-2016, Paris, France (2016)

Use of Lagrange Multipliers to Provide an Approximate Method for the Optimisation of a Shield Radius and Contents

Paul Warner[1,*]

[1]Rolls-Royce, Derby, UK

Abstract. For any appreciable radiation source, such as a nuclear reactor core or radiation physics accelerator, there will be the safety requirement to shield operators from the effects of the radiation from the source. Both the size and weight of the shield need to be minimised to reduce costs (and to increase the space available for the maintenance envelope on a plant). This needs to be balanced against legal radiation dose safety limits and the requirement to reduce the dose to operators As Low As Reasonably Practicable (ALARP). This paper describes a method that can be used, early in a shield design, to scope the design and provide a practical estimation of the size of the shield by optimising the shield internals. In particular, a theoretical model representative of a small reactor is used to demonstrate that the primary shielding radius, thickness of the primary shielding inner wall and the thicknesses of two steel inner walls, can be set using the Lagrange multiplier method with a constraint on the total flux on the outside of the shielding. The results from the optimisation are presented and an RZ finite element transport theory calculation is used to demonstrate that, using the optimised geometry, the constraint is achieved.

1 Introduction and Background

It is often required to design shielding around a radiation source in such a way that a radiation criterion is met on the outside of the shield. The shield itself may be composed of an array of different materials, each of which require optimisation in terms of thickness. A typical shield arrangement could be a steel water array composed of several steel shields placed at various distances from the source of radiation within a hydrogenous material. The question arises as to what is the overall radius of the shield and the optimum thicknesses of the steel shields given a constraint placed on the radiation level on the outside of the shield?

Although good neutron and gamma ray transport theory solvers are available to give predictions of the neutron flux or dose rate on the outside of a shield, the problem is still difficult since, if there are several shields to optimise, it is not clear how to decide on what the thickness of each shield should be. This is because the sensitivity of, for example, a steel shield in a water/steel array depends upon the distance from the source and also what steel shields are already included between it and a radiation source, such as a reactor core. In general, for this type of arrangement, the sensitivity of the dose rate on the outside of the shield is most sensitive to shields that are closest to the radiation source. This is due to the fact that, close to the reactor core, there will be a relatively high fast neutron population which is susceptible to inelastic scattering in steel. As this fast neutron population decreases by attenuation deeper into the shield, there will be relatively fewer fast neutrons and so this inelastic scattering process will be less efficient for outer steel shields. As a result, the outer steel shields become less effective.

The thicknesses of the shielding can be estimated by using a combination of engineering judgement and by running many design calculations using an appropriate neutron transport theory code to obtain an optimised design. However, in order to reduce the number of calculations it would be better to have a more systematic way of performing the optimisation.

This paper describes an approximate method for undertaking such an optimisation based upon minimisation of a function using Lagrange multipliers. The overall method is described and then applied to a theoretical small reactor model which is surrounded by a primary shield composed of a steel water array.

2 Description of the Theory

Since the major constraint is that placed on the total neutron flux (or dose rate) on the outside of the primary shield, it is required to represent the flux as a mathematical function of all the required variables:

$$\emptyset = F(x_1, x_2\, x_3 \ldots x_N) \tag{1}$$

The first variable x_1 is set equal to the shield radius. The other variables can be assigned arbitrarily. For example, x_2 could be the inner Primary Shield Tank (PST) steel wall thickness and x_3 could be a mid-wall and so forth.

The basic idea is to represent the change in flux on the outside of the PST by estimating the change as a function of radius assuming a starting guess. The other parameters are then represented as variations to this:

$$\emptyset = \varphi(x_1) + \delta\varphi(x_2) + \delta\varphi(x_3) \cdots \delta\varphi(x_n) \tag{2}$$

Where x_1 is the radius of the shield and x_2, x_3 etc are the steel shield thicknesses.

Each of the functions can be represented as a simple quadratic fit to a radiation transport theory code flux calculation:

* Corresponding author: paul.warner.nnppi@rolls-royce.com

$$\varphi(x_1) = \sum_{i=1}^{3} a_{i1} x_1^{i-1} \tag{3}$$

In order to supply constraints, a first order approximation is used by differentiating equation (1):

$$d\emptyset = \frac{\partial F}{\partial x_1} dx_1 + \frac{\partial F}{\partial x_2} dx_2 + \frac{\partial F}{\partial x_3} dx_3 + \cdots \tag{4}$$

In equation (4) the partial derivatives are the sensitivity of the flux to a given parameter and can be labelled as Gi for convenience. Using the geometry from a transport theory calculation as the starting guess for the PST flux $\emptyset_s$, the variation in the flux, to first order accuracy, can be given by:

$$\emptyset_s + d\emptyset = \emptyset_s + \sum_{i=1}^{N} G_i(x_i - x_s) \tag{5}$$

Where $x_s = 0$ for $i > 1$

If ϕ_c represents the user input constraint, eg the total neutron flux on the outside of the shield, then equations (2) and (5) may be combined using a Lagrange multiplier λ to give a general function L of all the variables:

$$L(x_1, x_2, x_3 \cdots x_N) = \sum_{j=1}^{N} \sum_{i=1}^{3} a_{ij} x_j^{i-1} - \lambda\left(\emptyset_s + \sum_{i=1}^{N} G_i(x_i - x_s) - \phi_c\right) \tag{6}$$

In the above equation, the coefficients a_{ij} represent the quadratic coefficients where the range of i is from 1 to 3 and that for j ranges from 1 to N (where N is the number of parameters).

The optimized solution is obtained by partial differentiation of the function in equation (6) with respect to each variable and equating to zero. This yields a set of simultaneous equations which can be solved for each variable. The solutions are:

If $x_{sj} = \mathit{Initial\ values\ for\ each\ of\ N\ parameters}$

Let

$$C = \emptyset_c - \emptyset_s + G_1 x_{s1}$$

$$\lambda = \sum_{i=1}^{N} \frac{\left(2C + G_i \frac{a_{2i}}{a_{3i}}\right)}{\frac{G_i^2}{a_{3i}}} \tag{7}$$

The basic shielding parameters are then given by:

$$x_i = \frac{\lambda G_i - a_{2i}}{2a_{3j}} \quad for\ i = 1, N \tag{8}$$

The rest of the paper describes the theoretical shield model, the calculation of the required constants and sensitivities and the results from the optimisation method. The result of the optimisation is confirmed using a transport theory calculation and suggestions are given for future improvements to the method.

3 Description of the Shielding Problem

The method is generally applicable to most shielding problems where a radiation source requires a shield design. Examples include shielding a fixed source, accelerator shielding and reactor shielding.

It is required to reduce the geometry to a relatively simple form such that it can be represented by a 2D calculation. In principle the method can be applied to more complex geometries but this would have to be balanced against the number of calculations required to generate the sensitivity data.

A further limitation of the approach, is that it is not applicable to any shielding calculations where there exists significant streaming paths. Such streaming paths could create a significant bypass around the steel arrays considered in this paper and would therefore need to be treated separately.

For this paper, a theoretical small reactor model was developed based upon a ^{235}U fission source. A two dimensional RZ schematic diagram of the model is shown in Figure 1.

The model contains a representation of a central reactor core with a uniformly distributed ^{235}U fission source equivalent to 500MW in strength. Outside the core, a number of typical steel components are modelled including steel baffles and a reactor pressure vessel. The shielding around the reactor is essentially a tank of water with a steel inner wall, two mid-wall shields and an outer steel shield.

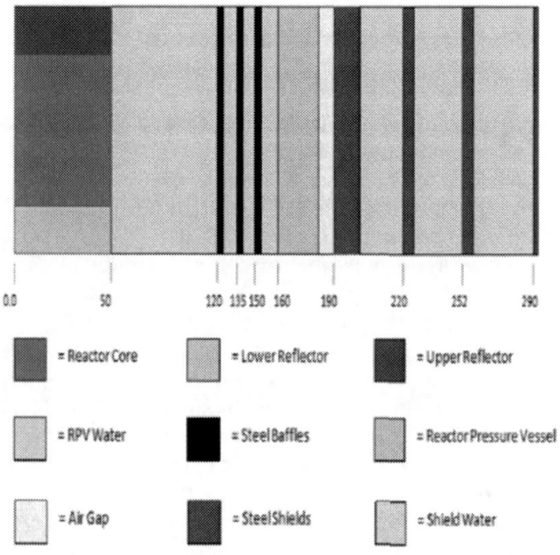

Fig. 1 Schematic Diagram of the Shielding Problem (All dimensions in cm).

The material geometry information for the model is given in Table 1 along with the relevant material thicknesses.

Table 1. Material Geometry Data.

Description	Number Densities (atoms cm^{-3})×10^{24}		Thickness (cm)
Upper Reflector	^{56}Fe $H_1(H_2O)$ ^{16}O	0.0339 0.0560 0.0280	-
Reactor Core	^{27}Al $H_1(H_2O)$ ^{16}O	0.0255 0.0560 0.0280	-
Lower Reflector	^{56}Fe $H_1(H_2O)$ ^{16}O	0.0339 0.0560 0.0280	-
Steel Baffles	^{56}Fe ^{52}Cr ^{58}Ni	0.0670 0.0090 0.0004	5.0
RPV	^{56}Fe ^{52}Cr ^{58}Ni	0.0670 0.0090 0.0004	20.0
Inner Primary Shield Wall	^{56}Fe ^{52}Cr ^{58}Ni	0.0670 0.0090 0.0004	4.0
Steel Shields	^{56}Fe ^{52}Cr ^{58}Ni	0.0670 0.0090 0.0004	2.0
Outer Primary Shield Wall	^{56}Fe ^{52}Cr ^{58}Ni	0.0670 0.0090 0.0004	2.0
Air Gap	^{27}Al	0.0022	-
Shield Water	$H_1(H_2O)$ ^{16}O	0.0600 0.0300	-
RPV Water	$H_1(H_2O)$ ^{16}O	0.0600 0.0300	-

A requirement was set to reduce the total neutron flux on the outside of the shield to fall just below 100 neutrons cm^{-2}s^{-1}. This value hence becomes the constraint for the Lagrange multiplier method. The method is then used to estimate the optimum thicknesses of the inner steel shield wall and also the two steel mid-walls shown in Figure 1.

4 Generating the Polynomial and Sensitivity Data for the Lagrange Method

The Lagrange multiplier method as described in this paper requires a set of polynomials and sensitivities to be generated in order to represent the variation in neutron flux on the outside of the shield as a function of the variables that require optimising.

In order to calculate the neutron fluxes at the outer edge of the shield, a simple two dimensional RZ finite element transport theory calculation was used. The model shown schematically in Fig 1 was meshed using a two dimensional RZ mesh made up of rectangular finite elements. The mesh contained 4408 elements and 4305 nodes. This mesh was input into the finite element even parity transport theory code EVENT [1]. The EVENT calculation was solved using a P_7 order for the spherical harmonic expansion and a P_3 order for the scatter cross sections. The BUGLE96 library was used to provide the basic cross section data in 47 neutron energy groups.

Using the model shown in Fig1 as a starting point for the EVENT calculation, the total neutron flux summed over all 47 energy groups was calculated on the outside of the shield at the core centreline. The value produced by EVENT was 171 neutrons cm^{-2} s^{-1} which was used as the starting flux for the Lagrange multiplier method given in equation (5).

The next step was to change the radius of the shield and then recalculate the neutron flux on the outside of the shield. This was repeated for several shield radii to give the flux as a function of radius. A simple quadratic curve fit within EXCEL was used to provide the fitting coefficients as shown in Fig 2.

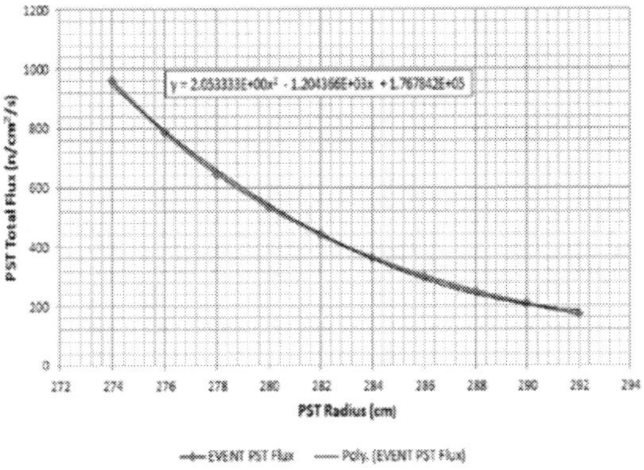

Fig. 2 Variation in Total Neutron Flux on the Outside of the Shield as a Function of Radius as Given by EVENT

A similar procedure was then adopted to derive the variation in the neutron flux on the outside of the shield while varying the primary shield inner steel wall. The results from the EVENT calculations are shown in Fig 3.

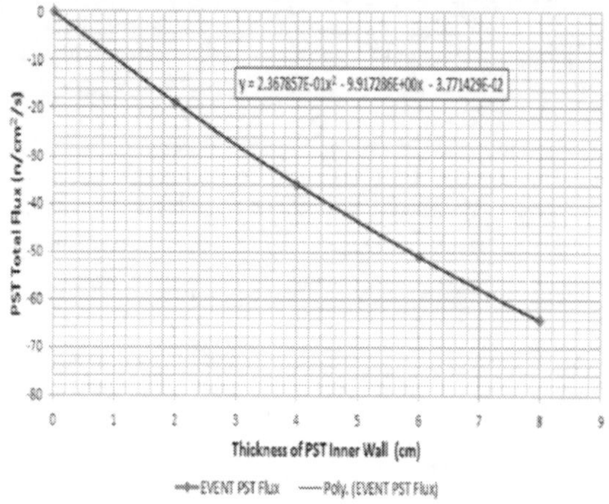

Fig. 3 Variation in Total Neutron Flux on the Outside of the Shield as a Function of Inner Steel Wall Thickness

One of the approximations inherent in using this method is that the variation in the thickness in each shield is independent of the thickness of other shields. This could be overcome using more complicated fitting techniques but this would detract from the overall ease of use of the approximate optimisation method. In order to ensure that this was not entirely lost a conservative approach to the two mid-wall shields was used. In this case, the variation in flux with thickness of the first mid-wall was calculated using the thickest primary shield inner wall. Similarly, the variation in primary shield flux with second mid-wall thickness was generated with both the thickest inner wall and first mid-wall. The results are shown in Figures 4 and 5.

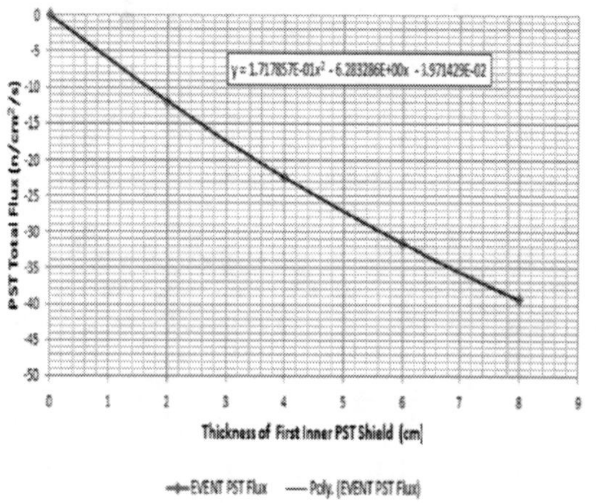

Fig. 4 Variation in Total Neutron Flux on the Outside of the Shield as a Function of First Steel Mid-Wall Thickness

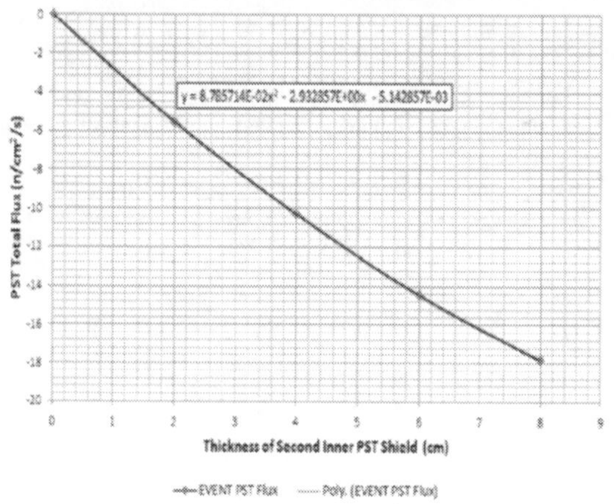

Fig. 5 Variation in Total Neutron Flux on the Outside of the Shield as a Function of Second Steel Mid-Wall Thickness

The sensitivities, G, in equation (5), are calculated by averaging the gradient of the respective plot over the range given in the figures. The quadratic fitting coefficients and the sensitivities are shown in Table 2.

Table 2. Fitting Parameters.

Item	Fitting Parameters and Sensitivities			
	a0	a1	a2	G
Shield Radius	1.76784E+5	-1.20436E+3	2.05333E+0	-21.180
Inner Shield Wall	-3.77142E-2	-9.91728E+0	2.36785E-1	-6.660
First Mid-Wall	-3.97142E-2	-6.28328E+0	1.71785E-1	-3.9187
Second Mid-Wall	-5.14285E-3	-2.93285E+0	8.78574E-2	-1.6900

5 Application of the Lagrange Method

It is evident from Figure 2 that the variation in flux as a function of radius is not a good approximation for a first order sensitivity. As a result it was necessary to run through first iteration of the Lagrange method in order to get a realistic estimate of the shield radius. This first iteration was then used to derive the sensitivities that are given in Table 2. Once the polynomial coefficients and sensitivities are known then equations (7) and (8) can be readily solved to obtain the required estimates for the shielding parameters using the flux constraint of 100 neutrons cm^{-2} s^{-1}.

As stated previously, applying the Lagrange method in the manner given above, gives a conservative result relative to the target flux of 100 neutrons cm^{-2} s^{-1}. A small adjustment was thus made to the final shield radius using the sensitivity data to give the required target.

6 Discussion of Results

The result of the optimisation is given in Table 3.The shield radius and the predicted steel shield thicknesses were modelled in a P_7 EVENT calculation (using a P_3 order for the scatter cross-sections) in order to find out how close the method gets to the constraint. The EVENT result is shown in Table 4. As can be seen, the results from the Lagrange multiplier method agree well with that produced from EVENT(close to the theoretical estimate of 100 neutrons cm^{-2} s^{-1}) while providing a reasonably practical optimisation of the shield thicknesses based upon the input sensitivities.

The results in Table 3 also show the expected behaviour in that the sensitivity to steel thickness increases as the shield is moved closer to the core. This is due to the effect caused by inelastic scattering of neutrons in steel which slows down fast neutrons to intermediate level. Once at these energies, the elastic scattering due to hydrogen in water becomes more important. The more steel placed between the core and the edge of the shield the less effective it becomes, since there will be fewer fast neutrons to slow down. The result is that the outer steel shields become increasingly less effective and hence the sensitivity of the flux on the outside of the shield becomes less. This is also shown in Figures 3 to 5. The effect in the Lagrange multiplier method is to make inner shields thicker than the outer shields. As can be seen the results exhibit this effect.

Table 3. Shield Parameters predicted by the Method

Item	Predicted Value for Parameter (cm)
Shield Radius	286.8
Inner Shield Wall	10.5
First Mid-Wall	9.8
Second Mid-Wall	9.5

Table 4. Comparison of the Estimated Fluxes using the Lagrange Fitting Parameters with an Event Calculation

Method	Total Neutron Flux at the Edge of the Shield (neutrons cm^{-2} s^{-1})
Lagrange Prediction	100
EVENT	90

7 Overall Conclusions and Future Development

In this paper an approximate method has been described for the purpose of optimisation of a shield array of materials. The method, based upon Lagrange multipliers, has been shown to produce practical results for a theoretical reactor model surrounded by a steel water array. The method has been applied to the array and the results shown to agree with the validated transport theory code EVENT.

A major approximation used in the method is that of first order sensitivities. This has been shown to be mitigated by running an initial iteration of the algorithm in order to get close to the final shield radius. The sensitivities can then be re-calculated about the corrected radius and these can be used to improve the final results.

A second approximation associated with the method is that the sensitivities and fitting are calculated for each shielding parameter independently. This means that the interplay in the energy spectrum of the neutrons resulting from the variation in the thickness of the shields is not accurately modelled. The approach adopted in this paper was to calculate the sensitivity of each shield while the thickest previous shield was in place. This approach results in an overall conservative result with respect to any radiation criteria or constraint placed on the flux or dose rate on the outside of the shield. It is likely that as more shields are included in the array, this approximation may become too conservative and another approach may be required.

To overcome the aforementioned issue, a future improvement to the method would be to calculate several sets of sensitivities for two thicknesses of steel shields within the array corresponding to the thickest steel shield and also the thinnest. By iterating the method and using interpolation to the most appropriate value the method could be made more accurate. This would give a much more accurate estimate of the effect of energy spectrum on sensitivities for the appropriate shield thickness. The disadvantage of this approach is that twice as many transport theory calculations would be required to calculate the sensitivities.

Currently, it takes each EVENT calculation approximately 10 minutes on a single processor machine and a minimum of 3 calculations are required to fit the quadratic data for each shield. Hence, as the number of shields is increased it would mean doubling the number of transport theory calculations which may become impractical. This would need to be considered as part of a future study.

The method has not been applied to gamma ray transport. There is no reason why the method should not be successful for this case, but, often it is required to optimise a shield for both neutron and the gamma rays simultaneously. A future improvement would be to develop a fitting function based upon the sum of neutron and gamma dose rate on the outside of the shield. This would enable an approximate optimisation of a shield for both neutrons and capture gamma rays that result from neutron capture within the shield material.

Finally, it is intended to test the method more robustly in the future using a genetic algorithm in conjunction with a fast transport theory solver.

References

1. P. Warner, C. de Oliveira, J. Nuc. Sci. **37** 466 (2000)

Rapid evaluation of particle properties using inverse SEM simulations

Kursat B. Bekar[1], *Thomas M. Miller*[1,*], *Bruce W. Patton*[1], *and Charles F. Weber*[1]

[1]*Oak Ridge National Laboratory, Reactor and Nuclear Systems Division, P.O. Box 2008, Oak Ridge, Tennessee, 37831, USA*

Abstract. The characteristic X-rays produced by the interactions of the electron beam with the sample in a scanning electron microscope (SEM) are usually captured with a variable-energy detector, a process termed energy dispersive spectrometry (EDS). The purpose of this work is to exploit inverse simulations of SEM-EDS spectra to enable rapid determination of sample properties, particularly elemental composition. This is accomplished using penORNL, a modified version of PENELOPE, and a modified version of the traditional Levenberg–Marquardt nonlinear optimization algorithm, which together is referred to as MOZAIK-SEM. The overall conclusion of this work is that MOZAIK-SEM is a promising method for performing inverse analysis of X-ray spectra generated within a SEM. As this methodology exists now, MOZAIK-SEM has been shown to calculate the elemental composition of an unknown sample within a few percent of the actual composition.

1 Introduction

Analysis of small particles and material surfaces has progressed rapidly with the development of the scanning electron microscope (SEM) over the past half century. The characteristic X-rays produced by the interactions of the electron beam with the sample are usually captured with a variable-energy detector, a process termed energy dispersive spectrometry (EDS). Measuring the chemical composition and crystallographic phase of small samples using electrons and X-rays in SEM instruments is termed electron probe microanalysis (EPMA). By focusing an electron beam to a diameter of a few nanometers and measuring X-rays fluoresced from the sample when it is struck by fast electrons, an analyst can determine the elements present (qualitative analysis) with a spatial resolution well below 1 μm. The problem of accurately measuring the relative elemental abundances (quantitative analysis) is more difficult and has only been conducted acceptably on samples that are flat, highly polished, and chemically uniform. When these conditions are met, the chemical composition can be measured with great precision by comparison with known standards that have been carefully prepared. However, most real-world environmental samples do not conform to these restrictions. Hence, it is desired to obtain both qualitative and quantitative analyses without using standards and without meticulous sample preparation.

The purpose of this work is to exploit inverse simulations of SEM-EDS spectra to enable rapid determination of sample properties, including both qualitative and quantitative chemical analyses. This requires a method to model the electron interactions with small samples that occur in an SEM and the subsequent generation of X-rays and an algorithm to minimize the difference between the measured and calculated SEM spectra by iteratively updating the elemental composition.

2 penORNL

The first task of this work was to evaluate the radiation transport codes available for modeling coupled electron / photon physics [1]. This survey showed that the Monte Carlo (MC) radiation transport code PENELOPE [2,3] provided the best combination of accurate physics, variance reduction specifically tailored to the generation of characteristic X-rays, and speed on a single processor. A second code, MCNP6 [4], had the benefits of being a parallel code and including the ability to tally pulse heights, but was slower in execution on a single processor.

PENELOPE has only been developed as a serial code, since the primary focus of the developers is the electron/photon physics modeling. They assumed that most users would write their own main program or driver program. Because MC codes can achieve massive speedups using multiple processors, a new main program was written to create a parallel version of PENELOPE. This new parallel version of the code is named penORNL [5]. The development of penORNL created a code that had all the most important features needed: accurate

*Corresponding author: millertm@ornl.gov

This manuscript has been authored by UT-Battelle, LLC, under Contract No. DE-AC05-00OR22725 with the U.S. Department of Energy. The United States Government retains and the publisher, by accepting the article for publication, acknowledges that the United States Government retains a nonexclusive, paid-up, irrevocable, worldwide license to publish or reproduce the published form of this manuscript, or allow others to do so, for United States Government purposes. The Department of Energy will provide public access to these results of federally sponsored research in accordance with the DOE Public Access Plan (http://energy.gov/downloads/doe-public-access-plan).

DOI: 10.1051/epjconf/201715306013

physics, an efficient algorithm, variance reduction, parallel capability, and a preliminary implementation of pulse-height tallies.

3 Detector response function

To compare MC simulation results to experimental data from the SEM-EDS process, a detector response function is used to generate the detector pulse-height spectrum from the photon line spectrum calculated by the MC simulation of the X-rays incident on the detector. The detector response function is defined as the pulse-height distribution for any incident monoenergetic X-ray, indicated by $R(E',E)$, where E' is the incident photon energy and E is the pulse-height energy. All our calculations used a typical Si(Li) detector response which was imported from the PENELOPE code package:

$$R(E,E') = \frac{1}{\sigma\sqrt{2\pi}} \exp\left[-\frac{(E-E')^2}{2\sigma^2}\right], \quad (1)$$

where σ is the standard deviation of the photopeak and its relation to the full-width-at-half-maximum is given as $\text{FWHM} = 2.35\sigma$. To get the relationship of detector photopeak standard deviation with respect to X-ray energy, a non-linear model is used:

$$\sigma(E') = \sqrt{a_1 + a_2 E'}\,. \quad (2)$$

A set of calculations was performed to optimize the coefficients a_1 and a_2 in the response function, to obtain better agreement between the spectra computed by penORNL and the measured spectra. Table 1 presents the optimal values of these coefficients.

Table 1. Coefficients used in the detector response function definition

Coefficient	Optimal value
a_1	0.006179
a_2	0.00255/keV

If $M(E)$ represents the MC line spectrum computed by penORNL, a realistic detector spectrum $C(E)$ can be calculated by convolution with the response function:

$$C(E) = \int_0^{E_{max}} R(E,E')\, M(E')\, dE'\,. \quad (3)$$

In reality, discrete energy bins are used to describe detectors responses, and the continuous response function in Eq. (1) is replaced by the response matrix $\mathbf{R} = (r_{ij})$, with elements $r_{ij} = R(E_i, E_j)$. The convolution integral in Eq. (3) is likewise replaced by the discretized form:

$$C_i = C(E_i) = \sum_{j=1}^{N} R(E_i, E_j) \times M(E_j)\,, \quad (4a)$$

which can be expressed in matrix form as

$$C = \mathbf{R}M, \quad (4b)$$

where $M_j = M(E_j)$, and $M = (M_1, M_2, \ldots, M_N)^T$ and $C = (C_1, C_2, \ldots, C_N)^T$ are vectors of computed line spectra and detector response spectra, respectively.

A measured SEM-EDS spectrum and a typical MC line spectrum computed by penORNL for the same standard (apatite) are shown in Figure 1(a). After the convolution operation was performed, the computed spectrum was represented in a more realistic shape, as presented in Figure 1(b). The difference (%) shown in Figure 1(b) calculated between the measured spectrum and the convoluted spectrum is mostly due to improperly estimated bremsstrahlung background. Differences between the actual measurement and simulation can cause differences in the bremsstrahlung background; therefore, we observe such a difference after the convolution process.

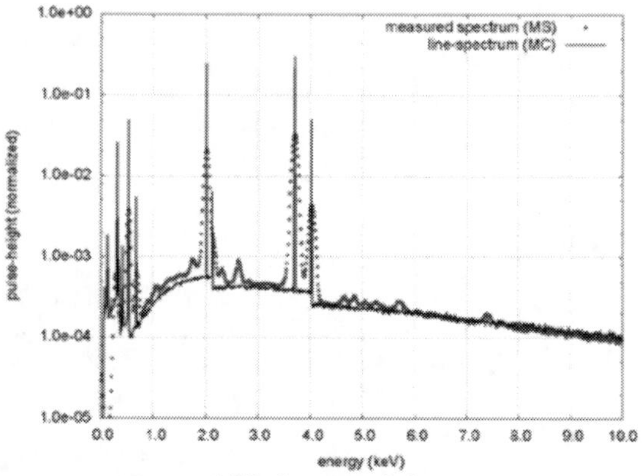

(a) Before convolution

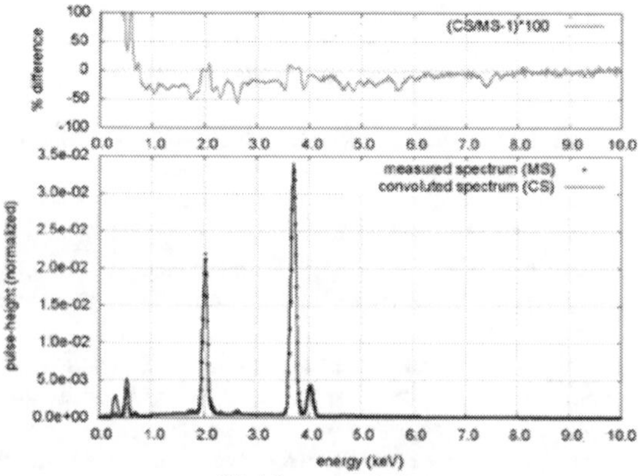

(b) After convolution

Figure 1. Results before and after the convolution process on the apatite (O:F:P:Ca) sample.

4 Inverse analysis of SEM-EDS spectra

Traditional SEM-EDS analysis uses properly calibrated standards to identify the elements in an unknown sample. Usually, X-ray energies from the unknown sample shown in the EDS spectrum are compared with known characteristic X-ray energy values to determine the presence of an element in a sample. Quantitative results can be obtained from peak heights, i.e., the relative X-ray counts at the characteristic energy levels for the sample constituents. X-ray intensity ratios are calculated by using the sample intensity and the standard specimen intensity for each element present in the unknown sample. The measured intensity ratios are equivalent to the ratios of mass or weight fractions of the constituents in the sample.

A complete MC simulation of sample interactions covers the scattering of the primary radiation and includes second- and higher-order effects such as the enhancements of fluorescence lines by higher-energy fluorescence or scattered radiation. As a result, it is possible to predict the complete spectral response of an SEM-EDS spectrometer. A significant advantage that the quantification scheme based on MC simulation has over traditional analysis methods is that the simulated spectrum can be compared directly to the experimental data in its entirety, taking into account not only the fluorescence line intensities but also the scattered background of the EDS spectrum. This is coupled with the fact that MC simulations are not limited to first- or second-order approximations or to ideal geometries.

4.1 Theory

Our inverse methodology is a multilevel iterative search algorithm that seeks a solution for the parameter vector to minimize the differences between the measured SEM-EDS X-ray spectrum and the computed spectrum of the unknown sample. The elements of the parameter vector correspond to the elemental inventories of the sample. This minimization operation is performed over the entire spectrum rather than using only characteristic X-ray peaks. At a high level, the inverse analysis of a measured SEM-EDS spectrum proceeds via the following steps.

1. An initial guess of the elemental composition must be provided and must include all the possible elements within the unknown sample. Effectively, this step constitutes a qualitative analysis of the spectrum.
2. The initial elemental composition is used as input to penORNL to calculate an SEM-EDS spectrum. If the difference between the calculated and measured spectra is less than a prescribed tolerance, the elemental composition is assumed to be optimal and the analysis stops. Since specific elemental inventories have been provided, this step is a quantitative analysis. This step is referred to as an outer iteration.
3. If the convergence criterion in Step 2 is not met, then the elemental inventories are modified via an iterative algorithm to minimize the difference between the calculated and measured spectra. When some prescribed convergence criterion is met, then an updated elemental composition has been determined. Each of these iterations is referred to as an inner iteration.
4. To ensure that the new elemental composition is indeed the best result, this procedure returns to Step 2 to repeat the MC simulation using the elemental composition produced by Step 3 as the input in Step 2. In other words, a new outer iteration is started.

4.2 Levenberg–Marquardt

The minimization problem for the SEM-EDS inverse calculations can be expressed as a sum-of-squared-error minimization with the following objective function:

$$Y(\alpha) = \sum_{i=1}^{N} [X_i - C_i(\alpha)]^2, \qquad (5)$$

which is minimized with respect to the parameter vector α. X_i is the measured spectrum at channel i, and C_i is the computed spectrum for the given parameter vector α at channel i [see Eq. (4a)]. The problem described in Eq. (5) is a nonlinear least-squares problem and can be solved by the Levenberg–Marquardt nonlinear optimization algorithm [6,7]. This algorithm requires a penORNL calculation for each perturbation made to the elemental weight fractions of the parameter vector and additional calculations for the derivatives (see details in reference 8).

4.3 Modified algorithm for SEM-EDS

Although the traditional Levenberg–Marquardt algorithm is a reliable approach, it is not feasible for the scope of this project because it requires multiple penORNL calculations at each inner iteration. Hence a modified version of the search procedure has been developed that utilizes a linearization of the computed solution. In this methodology, the MC line spectrum is decomposed into two components and linearized:

$$M_j(\alpha^k) = B_j(\alpha^0) + \sum_l \frac{\alpha_l^k}{\alpha_l^0} P_{jl}(\alpha^0), \qquad (6)$$

where B_j is the background spectrum, P_{jl} is the line spectrum in channel j due to element l, and the summation is taken over all elements considered. As noted by the dependence on α^0, the background spectrum B_j and the line spectrum P_{jl} remain constant through all inner iterations. The only quantity that changes is α_l^k, which is the updated inventory of element l at inner iteration k. Inherent in this formulation are the following assumptions.

1. The characteristic line intensities of the spectrum are approximately proportional to the amount of each element in the sample (parameter vector). As a result, the characteristic line spectrum can be expressed with a linear dependence on the parameter vector.
2. The background component does not change with the parameter vector.

As a result of the linearization, a modified version of the Levenberg–Marquardt iteration involves a single penORNL calculation at the beginning of each outer iteration. However, the linearization also implies that a globally optimal solution may not be achieved at the end of the search process. Rather, it only provides a best estimate for a given penORNL simulation because the background is assumed constant. To correct this problem, we perform another penORNL calculation using the optimal parameter vector from the previous set of inner iterations. This allows updating of the background and characteristic line spectra, after which a new optimal vector is determined using the linearized search process. This multilevel search process is repeated until no further improvement is obtained.

An outline of the multilevel search algorithm is depicted in Figure 2. Starting with the fixed sample, detector/electron beam geometry, and an initial parameter vector (discussed below), the search algorithm performs a penORNL calculation to simulate the SEM-EDS process. Then, the algorithm computes the sum of the squares of the difference between the computed spectrum and the experimental data. If the desired convergence criterion is satisfied, then the calculation is terminated. Otherwise, the current parameter vector and corresponding computed spectrum are used as input for the modified Levenberg–Marquardt search process to seek a better solution vector. Another penORNL simulation is initiated using the updated elemental compositions, and the above procedures are repeated until the best solution vector is found.

In this methodology, there are two iteration loops: (1) outer iteration (or *step*), in which the penORNL calculation is performed, and (2) inner iteration (or just iteration), which is the iteration of the modified Levenberg–Marquardt search process. The characteristic line contribution to the spectrum is updated during Step 2 through the linear dependence on α^k, whereas the background component is only updated after a penORNL calculation is performed at the start of Step 1. This arrangement might degrade the overall performance for some cases in which the background contribution is significantly higher than the contribution of the characteristic lines. Another problem arises if the characteristic lines of two or more elements are scored in the same channel of the MC line spectrum. To minimize the effects of this issue, the line emission probabilities from the penORNL simulation are provided to the search algorithm in order to make a first order estimate of the intensity of the characteristic lines within the same channel of the MC line spectrum.

The modular code framework MOZAIK-SEM was developed to perform the inverse calculations described above. It is patterned after an existing code MOZAIK, which is a code package that enables automatic geometry optimization for nuclear applications [9]. One of the benefits of this framework is its modularity. For example, penORNL and its auxiliary modules (preprocessor and postprocessor modules) can be replaced with another simulation code without modifying the MOZAIK-SEM framework. Similarly, the modified Levenberg–Marquardt methodology could be replaced with another optimization approach. This flexibility allows one to test many options within the same analysis framework.

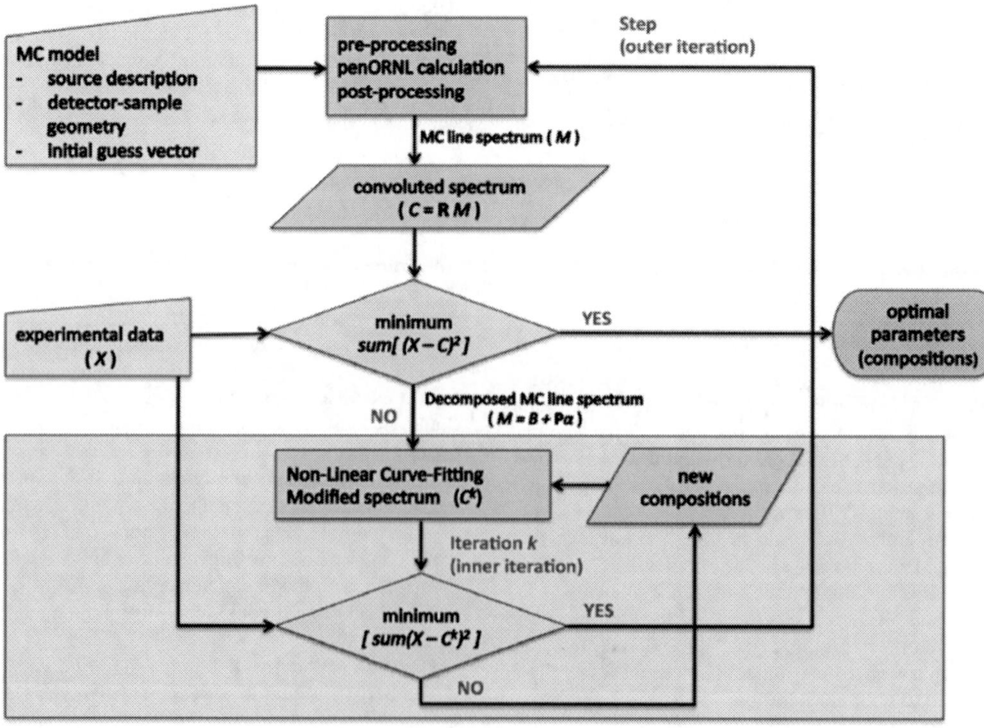

Figure 2. Outline of the iterative search process.

5 Initial guess estimation

Inverse calculations require specification of an initial parameter vector (the elements and their inventories). MOZAIK-SEM does support a user-defined initial parameter vector, although an automatic specification is more useful if no elemental knowledge is available. For this purpose, an initial guess-generation module (MOZAIK-IGM) was developed and integrated into MOZAIK-SEM. This module enables a fully automated search process to construct an initial guess of the parameter vector, which represents both qualitative and quantitative analysis without user insight. It is divided into three submodules, described in the following subsections. The tasks described in the subsections are performed sequentially in the order described.

5.1 Peak analysis

The Peak Analysis submodule involves the following tasks:
1. Performs smoothing in the reference spectrum (measured spectrum) using one of several widely used algorithms
2. Analyzes the reference spectrum and marks the possible peak locations
3. Obtains (from the penORNL element database) X-ray line energies close to the peak locations that have been marked
4. Confirms the peak locations if they match any X-ray line energy in the penORNL element database; and
5. Removes from the element list any element whose characteristic X-ray lines do not match any peaks in the reference spectrum

The Peak Analysis submodule uses either exact X-ray line energies or the X-ray line energies within an energy interval (energy of the marked peak ± 50 eV) to confirm the marked peak as a characteristic X-ray.

Figure 3 demonstrates the peak identification performed by the Peak Analysis submodule for the reference spectrum of an Inconel sample. In this case, the reference spectrum was computed by penORNL with 10^9 source particles, which allows the Peak Analysis submodule to skip the smoothing step since noise on the reference spectrum is minimal. The methodology was able to confirm all identified peak locations because each peak matches at least one characteristic X-ray line in the element database. The Peak Analysis submodule provides these results (identified peaks and elements associated with these peaks) as an input to the next submodule within the MOZAIK-IGM module.

5.2 Spectrum reconstruction

An algorithm was developed to construct a spectrum from the individual spectra of all the elements identified by the Peak Analysis submodule. A simulated spectrum database has been generated by performing a penORNL calculation for most naturally occurring elements in the periodic table (Z = 5–92) and convoluting with the generic detector response function described in section 3.

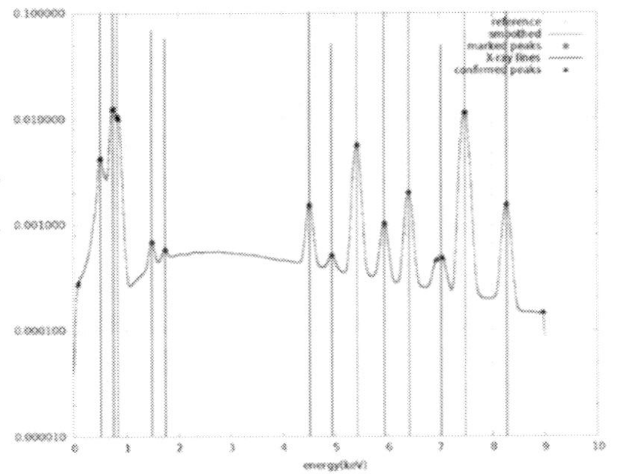

Figure 3. Characteristic X-ray peaks identified and marked by the Peak Analysis submodule for Inconel.

After obtaining the element list from the Peak Analysis submodule, the Spectrum Reconstruction submodule constructs the sample spectrum by multiplying each elemental spectrum from the database by the corresponding elemental weight fraction and adding all these together. Then it passes this constructed spectrum to the next submodule.

One of the deficiencies in this approach is that it does not include the matrix effect whereby one element's characteristic X-ray can excite another element's electrons, which can subsequently produce a characteristic X-ray from the other element. Therefore, this constructed spectrum is preliminary and must be refined by further calculation.

5.3 Nonlinear least-squares search

This submodule uses a simple optimization scheme to find optimal weights for the spectrum construction process. The purpose of this step is to minimize the size of the parameter vector (by removing elements whose contribution to the composite spectrum is insignificant) rather than to obtain correct weights. The elemental weights are preliminary at this point (matrix effects have not been considered), and final values are only assigned after a full physics calculation is performed. This optimization in the Initial-Guess Module is necessary because the Peak Analysis submodule might add two or more elements that have characteristic lines matching the same identified peak in the reference spectrum.

6 Verification and validation

In this paper, the result of two verification simulations and one validation simulation are presented. In the verification simulations the "experimental data" are actually the results of penORNL simulations using the exact weight fractions of a sample with different random number seeds and smaller MC uncertainties than the MOZAIK-SEM inverse simulations. In other words, the verification simulations are primarily a test of our inverse analysis methodology implemented in MOZAIK-SEM. The validation simulation uses actual experimentally measured data provide by NIST, which tests our inverse analysis methodology in MOZAIK-SEM and the PENELOPE physics in penORNL.

6.1 Verification with U_3O_8

For this verification simulation, the sample consists of a cube of U_3O_8 (6.83 g/cm³) measuring 50 μm on all sides. It was placed 2 cm away from the 25 keV electron beam source, and the X-ray spectrum was tallied for a generic Si(Li) detector. Energy bin widths were set to 10 eV to match the typical channel width in SEM-EDS detectors. Since this is a verification calculation, a reference spectrum was created by penORNL simulating 1 billion source particles (electrons) to reduce the statistical noise in the X-ray spectra.

One difficult aspect of this spectrum is that the sample contains a light element, oxygen. Traditional quantitative analysis has had difficulty correctly identifying the weight fractions of light elements. On the other hand, there are few overlapping peaks in the X-ray spectrum and only U and O are included in the initial guess of the elemental composition (parameter vector).

The initial parameter vector for this verification simulation contained a guess that half of the unknown sample was U and the other half O, by weight. Figure 4 compares the reference U_3O_8 spectrum and the spectrum of the composition MOZAIK-SEM calculated. Also included in Figure 4 is a plot of the percent difference between the reference and MOZAIK-SEM spectra. There is very little observable difference between the two spectra in Figure 4, but the largest percent difference in the spectra occurs in the tails of photopeak at 3 keV. The actual composition of the U_3O_8 sample (weight fractions) is shown in Table 2, which also includes the elemental composition calculated by MOZAIK-SEM along with the percent difference between the calculated and reference elemental composition. The percent difference between the reference and MOZAIK-SEM O weight fraction is nearly 1.2%, but the absolute difference for both elements' weight fraction is less than 0.01.

6.2 Verification with K497

This verification simulation is nearly identical to the previous simulation. However, the composition of the sample has been changed from U_3O_8 to that of a NIST standard glass, K497 (1.79 g/cm³).

This verification simulation is interesting because K497 is composed of 12 elements, and some of these elements have weight fractions significantly less than others. Moreover, several characteristic X-ray lines of the 12 elements overlap, contributing to the difficulty of this problem.

Table 3 presents the weight fractions of each element estimated by MOZAIK-SEM. The results of this verification test are a mix of positive and negative observations:

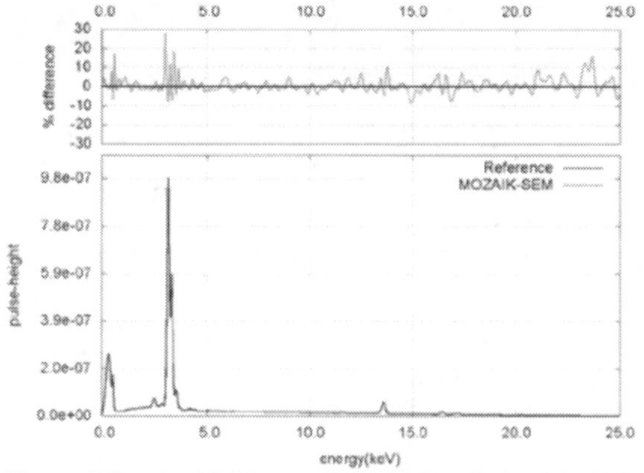

Figure 4. Results from inverse SEM-EDS analysis of the U_3O_8 sample. Inverse analysis was accomplished in 4 outer iterations.

Table 2. Elemental composition of the U_3O_8 sample (10^8 source particles in each penORNL simulation within MOZAIK-SEM).

Element	Reference	MOZAIK-SEM	% diff
U	0.848001	0.84980	0.2122
O	0.151999	0.15020	-1.1836

1. All elements with weight fractions greater than 0.01 (O, Mg, Al, P) were estimated with relative error less than 3%; Fe also was estimated very well. The oxygen prediction with less than 1% error is especially significant.
2. For trace elements (weight fractions < 0.01), the estimated weight fractions contained larger relative error (even double or triple digits). The element B was completely discarded. These errors are at least partly related to MC uncertainties, as the characteristic X-ray peaks of these elements were statistically indistinguishable from the background portion of the simulated spectrum.
3. Unusual behavior was exhibited by P and Zr. These elements have characteristic X-ray line energies that are very close to each other and their peaks overlap. The search process estimated the total weight fraction for these two elements with a 0.65% difference (reference, 0.3318 + 0.004 = 0.3358; MOZAIK-SEM, 0.3229 + 0.01507 = 0.33797) but could not resolve the contribution of each element individually—the Zr weight fraction is overestimated, and the P weight fraction is (slightly) underestimated.
4. The absolute errors for all elements except Zr are less than 1% of the total assay. Zr is actually present as a trace quantity (fraction 0.004) but is predicted to have small, but non-trace, assay fraction (1.5%).

Figure 5 presents the comparison of the reference spectrum and the calculated inverse spectrum. The plot shown at the top of Figure 5 compares the spectra over the full energy range, and the plot at the bottom presents the same comparison over the energy range 0.0 to 5.0 keV. Discrepancies in the spectrum plots themselves are indistinguishable, although the difference plots indicate slight errors, especially around peak energies.

DOI: 10.1051/epjconf/201715306013

Table 3. Elemental composition of the K497 sample (10^8 source particles in each penORNL simulation within MOZAIK-SEM).

Element	Reference	MOZAIK-SEM (± MC error)	% diff
B	0.0005	0.00000 (± 0.00000)	---
O	0.5245	0.52648 (± 0.00035)	0.38
Mg	0.0521	0.05129 (± 0.00042)	-1.56
Al	0.0578	0.05679 (± 0.00036)	-1.74
Si	0.0013	0.00003 (± 0.01154)	-97.66
P	0.3318	0.32290 (± 0.00035)	-2.68
Ti	0.0021	0.00230 (± 0.00119)	9.51
Fe	0.0024	0.00243 (± 0.00112)	1.22
Zr	0.0040	0.01507 (± 0.00007)	276.74
Ce	0.0062	0.00549 (± 0.00067)	-11.43
Ta	0.0080	0.00933 (± 0.00101)	16.56
Pb	0.0092	0.00789 (± 0.00142)	-14.19

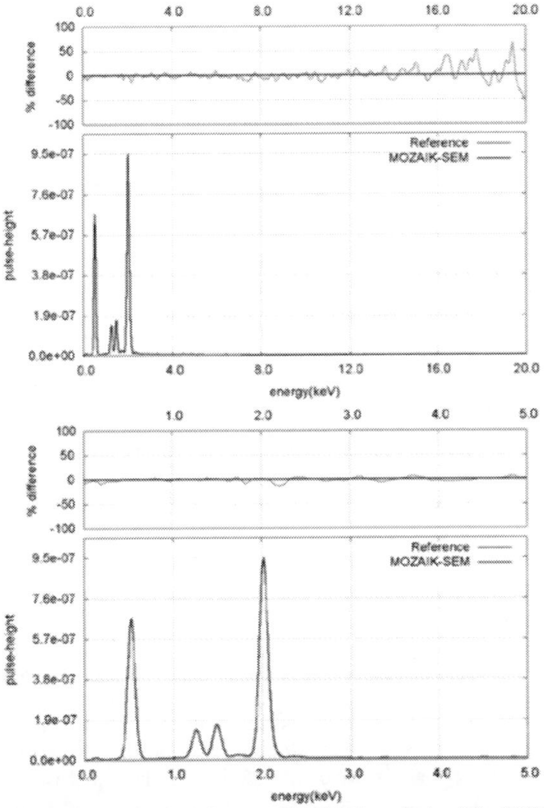

Figure 5. Results from inverse SEM-EDS analysis of the K497 glass sample. Inverse analysis was accomplished in 22 outer iterations.

6.3 Summary of Verification

All the results from the verification study are summarized below.

1. The methodology within MOZAIK-SEM is a feasible tool for SEM-EDS inverse calculations.
2. The MC errors in the results from penORNL simulations limit the performance of MOZAIK-SEM for trace elements. Better results can be obtained for elements with very low weight fractions by performing longer penORNL calculations that produce smaller MC errors for the background portion of the spectrum and legitimate small photopeaks within the spectrum.
3. In these examples, MOZAIK-SEM estimated the weight fraction of the elements correctly if the fraction was larger than 0.01. The MC errors for large photopeaks were usually on the order of 0.01 or less. In almost every case, the predicted elemental composition was within 2% of the actual assay value.
4. MOZAIK-SEM resolved many overlaps within the spectrum; however, it demonstrated some problems with the following challenging cases.

A. *Overlaps in the MC line spectrum.* This occurs when two or more elements' characteristic X-ray line energies are so close that the MC code scores these lines in the same energy bin of the tally.

B. *Overlaps in the convoluted spectrum.* This occurs when two or more elements' characteristic X-ray line energies are different enough that they appear in different energy bins of the MC line spectrum. However, when the MC line spectrum is convoluted with the detector response function, the characteristic X-ray energies are close enough that only a single photopeak appears in the convoluted spectrum.

6.4 Validaton with fluorapatite

In this validation calculation a measured spectrum provided by NIST for a material standard was used to evaluate the performance of MOZAIK-SEM. The material to be considered is the mineral fluorapatite $Ca_{10}(PO_4)_6F_2$, whose elemental weight fractions are given in the second column of Table 4 (3.15 g/cm^3). Using an initial guess distributed uniformly among the elements produces the estimated weight fractions given in the third column, which contain noticeable discrepancies. However, all elements are estimated to within 5% absolute error.

Table 4. Elemental composition of the apatite sample (10^8 source particles in each penORNL simulation within MOZAIK-SEM).

Element	Reference	MOZAIK-SEM	% diff
O	0.38071	0.30041	-21.092
F	0.03767	0.03506	-6.928
P	0.18426	0.21519	16.786
Ca	0.39736	0.44937	13.089

The following are several possible issues that could have contributed to the error in predicted weight fractions.

1. As shown in Figure 6, the peak centroid energies for the two spectra are slightly different, which means the energy bin structure in the penORNL simulation may be different from that of the reference spectrum. The energy bin structure of the simulation is based on data provided in the NIST data file. It is possible that the detector was not calibrated immediately before the measurement was taken, so the recorded bin structure may not be correct. Additionally, the relationship between channel number

and X-ray energy may not be linear, which appears to be the assumption applied by NIST.

2. All our calculations used a generic detector response function that works well for the verification calculations (see section 3). A response function tailored to the particular detector, and possibly the SEM itself, might improve the results.

3. It is possible that the electron source specification in our computational model is not consistent with the actual SEM used because the backgrounds from both spectra have different slopes (Figure 7). The electron spectrum was reported as a single energy, which is likely not correct. That energy is probably the mean electron energy, but no information about the energy distribution is available.

4. Figure 7 compares the measured spectrum with a computed spectrum using the known elemental composition. It indicates that the measured spectrum exhibits some peaks that are not shown in the computed spectrum. The energies of these peaks do not match the characteristic X-ray line energies of the elements in the apatite sample, so they do not appear in the computed penORNL spectrum.

A. If these "extra" peaks do not belong to the elements in the sample, they might be sum peaks, which our generic response function does not account for.

B. Another possibility is that these are characteristic X-rays of some contaminant element (in the sample itself or in the sample environment) that is not in our computational model.

C. The final possibility is that there is an unidentified error (or errors) in the penORNL physics model or in the atomic data.

For all these possible reasons, MOZAIK-SEM could not estimate the weight fraction for this sample more accurately. At this time, not enough information is available to rule out any of these explanations. Because the sample represents an archived spectrum made several years ago, it is not likely that additional information about the measured spectrum can be obtained.

7 Conclusions and future work

The overall conclusion of this work is that MOZAIK-SEM is a promising method for performing inverse analysis of X-ray spectra generated within a SEM, but additional work is needed to refine and validate this methodology. The methodology implements a two-tiered strategy: an outer iteration, which requires a full MC calculation, and an inner iteration, which refines elemental estimates quickly. The latter uses a modified Levenberg–Marquardt iteration to solve a linearized problem, which reduces the inverse analysis into two pieces, the photopeaks and the background. This approach has the benefit of reducing the number of MC simulations required, although the background portion of the spectrum and the matrix effects on the spectrum are only updated after each outer iteration.

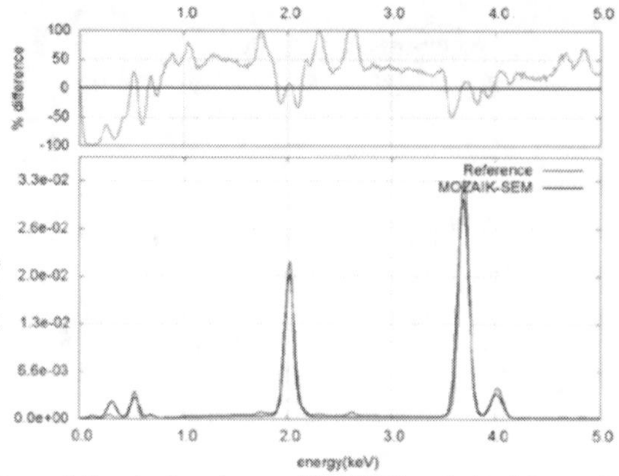

Figure 6. Results from inverse SEM-EDS analysis of the apatite sample. Reference is the standard's spectrum from NIST.

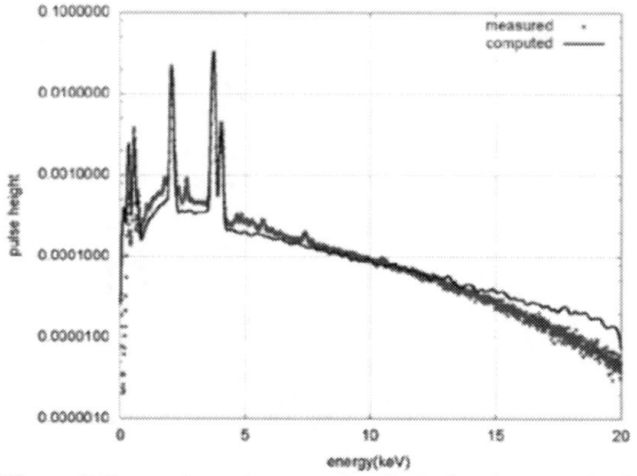

Figure 7. Comparison of measured and calculated spectra for the apatite sample. The measurement was supplied by NIST and the computed spectrum by penORNL using the actual elemental composition.

Verification simulations with MOZAIK-SEM revealed that this methodology can successfully identify the weight fractions of low-Z elements such as oxygen and fluorine. However, like other quantitative analysis methods, this method can also suffer when X-ray line energies for multiple elements are very close together. A problem unique to this method is in distinguishing small photopeaks due to trace elements from MC statistical noise in the X-ray background; the problem occurs when the background MC uncertainty is of the same order of magnitude as the weight fraction of a trace element. Simulating more particle histories during the inverse analysis helps reduce this issue, but that increases the computational burden in execution time or additional processors. Finally, the validation simulation with MOZAIK-SEM was less successful than the verification simulations, at least in part due to vaguely defined conditions of the experimental measurement. A number of prominent differences exist between the spectrum measured by NIST and the spectrum simulated by penORNL. In the end, no source of these differences can be absolutely identified, so it is impossible to state if these differences are due to the measurements, MOZAIK-SEM methodology, penORNL physics, or the interaction data it uses.

For the MOZAIK-SEM methodology to be improved, two issues need to be addressed. First, a better method to deal with X-ray line energies that are very close together should be investigated. The method as it exists now works fairly well, but there is clearly room for improvement. Second, and more importantly, additional validation simulations should be performed, but this should be done as part of a much closer collaboration between the experimentalist and modeler. For the validation effort to be successful, future measurements should include (1) an energy calibration of the SEM-EDS detector, (2) an energy calibration of the SEM electron source, and (3) greater details of the geometry, including the sample, detector, and, of lesser importance, the internal configuration of the SEM sample chamber.

Acknowledgments

This work was performed with funds provided by the US DOE through the Nonproliferation Enabling Capabilities program within the Office of Defense Nuclear Nonproliferation Research and Development.

The authors of this report gratefully acknowledge the assistance provided by NIST, namely, John Henry Scott, Nicholas W. M. Ritchie, and Dale E. Newbury.

Finally, the authors would also like to thank Francesc Salvat for his assistance and insight into PENELOPE.

References

1. T. M. Miller, B. W. Patton, and C. F. Weber, Trans. of the Amer. Nuc. Soc. **110**, 497 (2014)
2. F. Salvat, J. M. Fernandez-Varea, and J. Sempau, *PENELOPE, a code system for Monte Carlo simulation of electron and photon transport*, Universitat de Barcelona (2011)
3. F. Salvat et al., *Microchimica Acta* **155**, 67 (2006)
4. MCNP6 Development Team, *MCNP6 User's Manual Version 1.0*, Los Alamos National Laboratory, LA-CP-13-00634, Rev. 0 (2013)
5. K. B. Bekar et al., *ANS MC2015 – Joint International Conference on Mathematics and Computation (M&C), Supercomputing in Nuclear Applications (SNA) and the Monte Carlo (MC) Method* (2015)
6. K. Levenberg, Quart. of App. Math. **2**, 164 (1944)
7. D. W. Marquardt, J. Soc. Ind. and App. Math. **11**, 431 (1963)
8. K. B. Bekar et al., *Rapid Evaluation of Particle Properties Using Inverse SEM Simulations*, Oak Ridge National Laboratory, ORNL/TM-2015/779 (2016)
9. K. B. Bekar, *Modular Optimization Code Package: MOZAIK*, PhD dissertation, The Pennsylvania State University, University Park, Pennsylvania (2008)

EPJ Web of Conferences 153, 06014 (2017) DOI: 10.1051/epjconf/201715306014
ICRS-13 & RPSD-2016

Radiation Transport in Random Media With Large Fluctuations

Aaron Olson[1,2,⋆], *Anil* Prinja[1,⋆⋆], and *Brian* Franke[2,⋆⋆⋆]

[1] *University of New Mexico (UNM), Albuquerque (USA)*
[2] *Sandia National Laboratories (SNL), Albuquerque (USA)*

Abstract. Neutral particle transport in media exhibiting large and complex material property spatial variation is modeled by representing cross sections as lognormal random functions of space and generated through a nonlinear memory-less transformation of a Gaussian process with covariance uniquely determined by the covariance of the cross section. A Karhunen-Loève decomposition of the Gaussian process is implemented to efficiently generate realizations of the random cross sections and Woodcock Monte Carlo used to transport particles on each realization and generate benchmark solutions for the mean and variance of the particle flux as well as probability densities of the particle reflectance and transmittance. A computationally efficient stochastic collocation method is implemented to directly compute the statistical moments such as the mean and variance, while a polynomial chaos expansion in conjunction with stochastic collocation provides a convenient surrogate model that also produces probability densities of output quantities of interest. Extensive numerical testing demonstrates that use of stochastic reduced-order modeling provides an accurate and cost-effective alternative to random sampling for particle transport in random media.

1 Introduction

Strongly heterogeneous media arise in several applications that include radiation shields, nuclear fuel, BWR moderators, clouds, planetary and stellar atmospheres, turbulent gases and plasmas. Neutral and charged particle transport computations in such media have relied heavily on formulating transport equations with spatially random coefficients (physical data) and developing solution methods to deal with the additional stochastic dimensions. Attempts at developing approximate closures that yield only low order statistical information (e.g., mean and variance of the flux) have proved to be highly restrictive under real physics conditions [1, 2] or rely on techniques that require fluctuation amplitudes to be small for robustness [3]. Recently, stochastic spectral methods such as polynomial chaos and stochastic collocation [4] have been developed for aleatoric uncertainty quantification and sensitivity analysis, and successfully applied in radiation transport work [5, 6]. Advances in UQ techniques have tended to focus on efficiently handling large numbers of uncertain variables but the rigorous stochastic basis of the approach also promotes its use in situations where stochasticity is due to spatial heterogeneity and the associated uncertainty is large. Here we apply these techniques to radiation transport in media with spatially randomly varying cross sections without restriction on fluctuation amplitudes. Specifically, we represent the cross sections as a lognormal spatial random process with specified mean, variance and covariance function and use a KL-decomposition to efficiently generate cross section realizations that are strictly positive. Woodcock Monte Carlo [7, 8] is then used to simulate transport using random sampling of cross sections and deterministic sampling based on a stochastic collocation technique. A polynomial chaos expansion (PCE) model is built for the stochastic response, which enables construction of probability density functions of physical quantities of interest. Numerical results for the mean and variance in the scalar flux and leakage currents are obtained. We focus in this work on demonstrating our approach to solving random medium transport problems and defer application to specific problems to a future investigation.

⋆e-mail: aolson@sandia.gov
⋆⋆e-mail: prinja@unm.edu
⋆⋆⋆e-mail: bcfrank@sandia.gov

2 Problem Statement and Stochastic Model

The random transport equation of interest is given by

$$\mu\frac{\partial\psi(x,\mu,\omega)}{\partial x}+\Sigma_t(x,\omega)\psi(x,\mu,\omega)=\frac{\Sigma_s(x,\omega)}{2}\int_{-1}^{1}d\mu'\psi(x,\mu',\omega),$$
$$0\le x\le L;\qquad -1\le\mu\le 1$$
$$\psi(0,\mu)=\delta(1-\mu),\quad \mu>0;\qquad \psi(L,\mu)=0,\quad \mu<0,\tag{1}$$

where x and μ are the particle spatial and angular variables and the label ω denotes an element in the sample space, i.e., a particular realization. Macroscopic total or scattering cross section r, $r=\{t,s\}$, is represented as the product of microscopic cross section r and the stochastically dependent material atom density N_{at}:

$$\Sigma_r(x,\omega) = \sigma_r N_{at}(x,\omega). \quad (2)$$

The atomic density $N_{at}(x,\omega)$ is modeled as a lognormal random field derived from Gaussian random field $g(x,\omega)$,

$$N_{at}(x,\omega) = \exp\left[g(x,\omega)\right], \quad (3)$$

yielding all macroscopic cross sections for the material and ensuring that these values are always positive.

2.1 The Karhunen-Loève Expansion

The optimal spectral expansion that maintains second-order statistics of the Gaussian random field $g(x,\omega)$ is given by the Karhunen-Loève (KL) expansion,

$$g(x,\omega) = \langle g\rangle + \sqrt{v_g}\sum_{k=1}^{\infty}\sqrt{\gamma_k}u_k(x)\xi_k(\omega), \quad (4)$$

where $\langle g\rangle$ and v_g are the average and variance of the Gaussian process and γ_k, u_k, and ξ_k are the eigenvalue, eigenfunction, and random variables associated with the KL expansion. The KL eigenvalues and eigenfunctions are given by the solution of the following Fredholm integral equation of the second kind with the normalized covariance as the kernel [9]:

$$\int_0^L c_g(x,x')u(x')dx' = \gamma u(x), \quad (5)$$

where $c_g(x,x') = C_g(x,x')/v_g$, and $C_g(x,x')$ is the process covariance. The eigenvalues are all positive and ordered with decreasing magnitude, which makes computation feasible by truncation of the KL expansion at an order that captures most of the variability, depending on the strength of correlation [5, 9]. The eigenspectrum can be obtained analytically for several covariance functions including an exponential covariance, characterized by a correlation length λ_c, and linear covariance [9].

The memory-less nonlinear transformation of the Gaussian process requires solving for the mean, variance, and covariance of the underlying Gaussian process in terms of the statistics of the cross section that is modeled as a lognormal process. It is not hard to relate the mean and variance of the processes,

$$\langle g\rangle = \ln\left(\frac{\langle N_{at}\rangle^2}{\sqrt{v_{N_{at}} + \langle N_{at}\rangle^2}}\right); \quad v_g = \ln\left(\frac{v_{N_{at}}}{\langle N_{at}\rangle^2} + 1\right), \quad (6)$$

and the normalized covariances are related [5] by

$$c_g(x,x') = \frac{\ln\left(c_{N_{at}}(x,x')\frac{v_{N_{at}}}{\langle N_{at}\rangle} + 1\right)}{\ln\left(\frac{v_{N_{at}}}{\langle N_{at}\rangle^2} + 1\right)}. \quad (7)$$

The eigenspectrum of the Gaussian KL process must be solved numerically since the Fredholm equation, Eq. (5), is not solvable analytically for the covariance of the Gaussian process given by Eq. (7).

2.2 Numerical Solution of the KL Eigenspectrum

The Nyström method numerically solves the Fredholm integral equation for the Karhunen-Loève eigenspectrum by discretizing over the domain using nodes x_j and weights w_j,

$$\sum_{j=1}^{N_{Ny}} w_j c_g(x_i,x_j)u_k(x_j) = \gamma_k u_k(x_i), \quad k = 1,\ldots,N_{Ny}, \quad (8)$$

where N_{Ny} is the number of nodes in the discretization, and eigenvector values are solved at nodes x_i. The numerical solve of the Fredholm equation is written in matrix notation as

$$\mathbf{CWu}_k = \gamma_k\mathbf{u}_k, \quad (9)$$

where matrix $\mathbf{C}$ is a symmetric positive semi-definite matrix with elements $C_{ij} = c_g(x_i,x_j)$, matrix $\mathbf{W}$ is a diagonal matrix with values w_j, and vector $\mathbf{u}_k$ spans the physical domain for each KL eigenvalue k. Eq. (9) is reformed as

$$\mathbf{B}\mathbf{u}_k^* = \gamma_i\mathbf{u}_k^*, \quad (10)$$

where the matrix $\mathbf{B}$ and vector $\mathbf{u}_k^*$ are given by:

$$\mathbf{B} = \mathbf{W}^{\frac{1}{2}}\mathbf{C}\mathbf{W}^{\frac{1}{2}}; \quad \mathbf{u}_k^* = \mathbf{W}^{\frac{1}{2}}\mathbf{u}_k, \quad (11)$$

and $\mathbf{W}^{-\frac{1}{2}}$ is the diagonal matrix with values $\sqrt{w_j}$. The eigevector can then be evaluated as:

$$\mathbf{u}_k = \mathbf{W}^{-\frac{1}{2}}\mathbf{u}_k^*. \quad (12)$$

The simplest way to evaluate eigenfunction $u_k(x)$ using eigenvector $\mathbf{u}_k$ is to return the eigenvector value which corresponds to the eigenvector node x_i closest to x. One evaluation of $u_k(x)$ requires sampling only one value of $\mathbf{u}_k$ and the resulting eigenfunction is a discontinuous step function in physical space. The accuracy of $u_k(x)$ is improved by evaluation using an interpolating function provided in the Nyström method theory [10, 11]:

$$u_k(x) = \frac{1}{\gamma_k}\sum_{j=1}^{N_{Ny}}\sqrt{w_j}\mathbf{u}_{k,j}^* c_g(x,x_j). \quad (13)$$

The resulting eigenfunctions $u_k(x)$ are more accurate than when using the first interpolation scheme, continuous, and differentiable, though sampling a value of x requires N_{Ny} evaluations of $\mathbf{u}_k$, making this interpolation scheme computationally expensive. A third interpolation scheme linearly interpolates between values of the eigenvector $\mathbf{u}_k$ and defaults to use of the interpolating function in Eq. (13) when x is near the edge of the slab and therefore not bounded by values of $\mathbf{u}_k$. This interpolation scheme appears to be nearly as accurate as the scheme given in the Nyström theory, is continuous but not differentiable, and assuming that each value of x is equally likely to be sampled has a cost of $1 + 2\frac{N_{Ny}-1}{N_{Ny}}$–roughly three–evaluations of $\mathbf{u}_k$ to sample a value of $u_k(x)$. This linear interpolation scheme is used in this work.

With the Gaussian process completely characterized and realizations of independent standard normal random

 DOI: 10.1051/epjconf/201715306014

variables readily sampled, realizations of cross sections are obtained from the truncated KL expression:

$$\Sigma_r(x,\omega) = \sigma_r \exp\left[\langle g\rangle + \sqrt{v_g}\sum_{k=1}^{K}\sqrt{\gamma_k}u_k(x)\xi_k(\omega)\right]. \quad (14)$$

To demonstrate the positivity and moment preserving properties of these lognormal processes, we plot fifty Gaussian random process realizations and fifty lognormal random process realizations each with $\langle g\rangle = 0.75$ cm and $v_g = 0.225$ cm^2 over a slab of length $L = 5$ cm in Fig. 1 and Fig. 2 respectively. Where realizations of the Gaussian

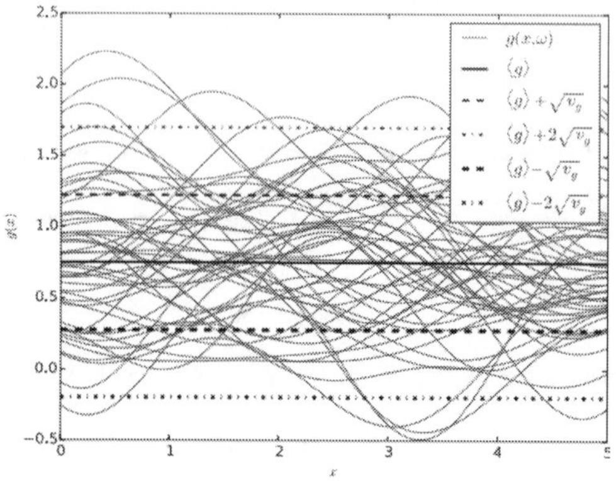

Figure 1. Fifty Gaussian Realizations

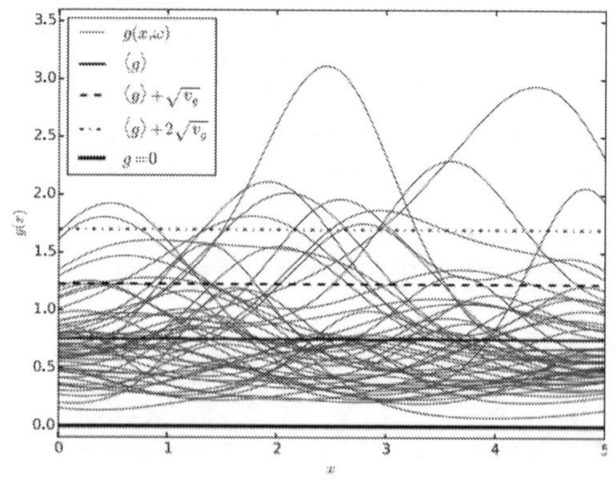

Figure 2. Fifty Lognormal Realizations

process contain negative values, realizations of the lognormal process do not. In order to recover the input variance, the lognormal process contains more values near zero and fewer values which may be significantly larger than the process mean.

As an implementation verification test and to demonstrate the increased accuracy of the Karhunen-Loève expansion with larger truncation orders K, the mean and standard deviation of an ensemble of 50,000 realizations were sampled at 20 different points in a slab using KL truncation orders of 3, 7, and 15. The observed sample mean and standard deviation values are plotted in Fig. 3 along with the input mean ($\langle\Sigma_r\rangle = 0.75$ cm) and standard deviation ($\sqrt{v_{\Sigma_r}} = \sqrt{0.225} \approx 0.474$ cm). Artifacts of the spatial

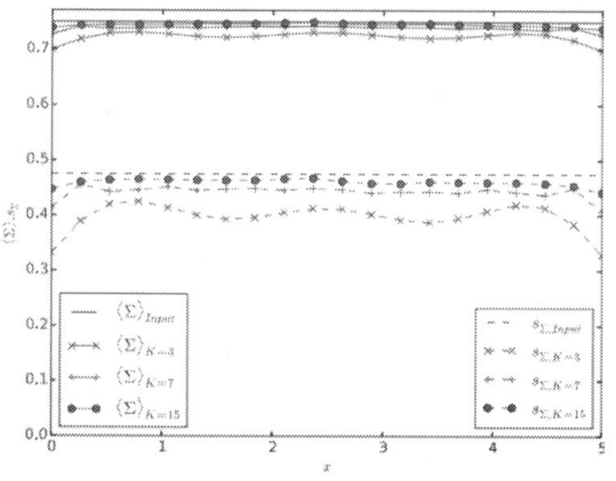

Figure 3. Observed Mean and Standard Deviation in Ensemble of Lognormal Random Cross Section

shapes of the eigenfunctions can be seen in the oscillatory shape of the sampled moments, which are mitigated by increasing number of terms retained in the KL-expansion. The moments of the cross sections are preserved over the ensemble with more accuracy for larger truncation orders K.

3 Monte Carlo Transport Solver for Spatially Continuous Cross Sections

Numerical simulation of particle transport on different realizations of the stochastic transport equation can be performed using a variety of methods. Deterministic solution methods require discretization of the geometry and, unless finite elements are used with mesh boundaries and basis functions built to match the Nyström interpolating scheme, introduce an error source. The traditional approach to Monte Carlo transport requires spatially discrete cross section values which could be achieved through a discretization in physical space, again introducing an error. Monte Carlo transport could be effected exactly by computing distance to collision based on integrals of the total cross section, the optical thickness, though the cross section profiles of interest here are not analytically integrable requiring numerical integration which introduces error and is likely to be expensive. Transport is solved here introducing only resolvable statistical error using Monte Carlo transport with Woodcock sampling.

Monte Carlo transport with Woodcock sampling [7], Woodcock Monte Carlo (WMC) for short, samples a distance to potential collision based on a fictitious total cross section equal to or larger than the maximum total cross section over a particle flight. The value of the total cross section is compared to the value of the fictitious total cross section using a pseudo-random number to determine if the particle undergoes a collision or continues streaming. As long as WMC computations are not dominated by rejected potential interactions, WMC efficiently and exactly, within statistical noise, computes transport results. We note that WMC is used regularly in at least one production Monte Carlo transport code [7]. The WMC solver used in this work has been utilized previously [8] and benchmarked against a suite of problems for which an analytic or near analytic solution is available.

4 Stochastic Solution Methods

The Karhunen-Loève expansion with a truncation of order K is used to characterize a random field as a function of K Gaussian random variables. Stochastic solution methods introduced here resolve the effects of these remaining random variables yielding statistical averages of particle flux and current and probability density function of leakage currents.

Statistical averages of a quantity of interest φ, here either particle flux in a cell or a leakage current, are defined with respect to the joint probability density $P(\xi_1, \ldots, \xi_N)$ over all random variables retained in the KL-expansion as

$$\langle \phi^m \rangle = \int_{\xi_N} \cdots \int_{\xi_1} \phi^m(x, \xi_1, \ldots, \xi_N) P(\xi_1, \ldots, \xi_N) d\xi_1 \cdots \xi_N. \quad (15)$$

Standard normal random variables in the KL expansion are known to be independent [9] such that the joint probability density factorizes into a product of standard normals for each random variable, vis.,

$$P(\xi_1, \ldots, \xi_K) = \prod_{k=1}^{K} P(\xi_k), \; \forall \, \xi_k \in N(0, 1). \quad (16)$$

The integral in Eq. (15) is solved in one of two ways here, either through random or deterministic sampling of the random variables. Random sampling solutions provide probability density functions of the response built from a number of values equal to the number of transport evaluations. Solutions from the deterministic sampling approach are used to create a surrogate model of the response from which probability density functions are constructed using an arbitrary number of samples. Once a surrogate model is built, generation of probability density functions is much cheaper than generation using random sampling.

4.1 Random Sampling

Moments of the response can be solved by numerically evaluating the integral on the right-hand-side of Eq. (15) using random sampling (RS):

$$\langle \varphi^m \rangle \approx \frac{1}{R} \sum_{i=1}^{R} \varphi^m(x, \xi_1^{(i)}, \ldots, \xi_K^{(i)}), \quad (17)$$

where R is the total number of sampled realizations and $\xi_k^{(i)}$ are the randomly sampled values of the standard normal random variables. Transport results are obtained for each of R realizations to yield ensemble averaged response moments which converge towards the solution as $R^{-0.5}$. The approach is dimensionally agnostic, meaning the convergence rate is not affected by the number of random dimensions. This method is appropriately called a Monte Carlo method, but we reserve that title in this work for Monte Carlo particle transport to avoid confusion.

4.2 Stochastic Collocation

Moments of the response can also be solved by numerically evaluating the integral on the left-hand-side of Eq. (15) using multi-dimensional quadrature:

$$\langle \varphi^m \rangle \approx \sum_{q_K}^{Q_K} \cdots \sum_{q_1}^{Q_1} w_1^{(q_1)} \cdots w_K^{(q_K)} \varphi^m(x, \xi_1^{(q_1)}, \ldots, \xi_K^{(q_K)}), \quad (18)$$

where Q_k is the order of quadrature used on the k-th random variable and $w_k^{(q_k)}$ and $\xi_k^{(q_k)}$ are the weights and nodes of quadrature points. Such application of quadrature over the stochastic domain is called stochastic collocation (SC). Gauss-Hermite quadrature, being the optimal non-nested quadrature for integration of polynomials over Gaussian probability basis functions, is used, and if the response surface is sufficiently smooth, such Gauss-Hermite quadrature is effective. The stochastic collocation method, for isotropic collocation grids, $Q_k = Q_{k'} \forall k, k'$, convergences exponentially in log-log space. It does, however, suffer from the curse of dimensionality such that its convergence is slower with an increased number of random variables. The number of random variables below which SC is more efficient than RS depends on the smoothness of the response surface and the error tolerance desired in the response–smoother surfaces and tighter error tolerances favor SC over RS. For reasonably smooth problems and typical error tolerances, SC tends to be more efficient than RS for roughly 10 to 20 random variables. For the tensor product collocation grids used in this work the total number of realizations over which transport must be solved is equal to the product of the quadrature order in each dimension:

$$R_{SC} = \prod_{k=1}^{K} Q_k. \quad (19)$$

Anisotropic stochastic collocation grids, for which $\exists k, k' : Q_k \neq Q_{k'}$, are especially effective when integrating over KL random variables since KL terms are known to decrease monotonically in magnitude [12], though it is difficult to a priori know the optimal quadrature orders Q_k. Adaptive methods which estimate the error contribution of each dimension may be useful in choosing the optimal order. Additional efficiency gains may be achievable using sparse collocation grids [5, 6, 12].

4.3 The Polynomial Chaos Expansion

While the stochastic collocation method does not provide full distributions of response values, the SC method can be used to build a surrogate model of the response using a polynomial chaos expansion (PCE). The polynomial chaos expansion using Hermite polynomials can be expressed as:

$$\begin{aligned}\varphi(\xi_{i_1},\xi_{i_2},\ldots) =&\widehat{u}_0 H_0 + \sum_{i_1=1}^{\infty}\widehat{u}_{i_1}H_1(\xi_{i_1})\\ &+\sum_{i_1=1}^{\infty}\sum_{i_2=1}^{i_1}\widehat{u}_{i_1 i_2}H_2(\xi_{i_1},\xi_{i_2})\\ &+\sum_{i_1=1}^{\infty}\sum_{i_2=1}^{i_1}\sum_{i_3=1}^{i_2}\widehat{u}_{i_1 i_2 i_3}H_3(\xi_{i_1},\xi_{i_2},\xi_{i_3})+\cdots,\end{aligned} \tag{20}$$

where instances of $\widehat{u}$ represent coefficients to be determined and H_n represent n-order multivariate probabilists' Hermite polynomials. The K-dimensional PCE truncated at a total polynomial order of I can be written more succinctly using multi-indices $\mathbf{i}$ and $\vec{\xi}$ as

$$\varphi(\vec{\xi}) = \sum_{\iota=0}^{I_t}\widehat{u}_{\mathbf{i}(\iota)}H_{\mathbf{i}(\iota)}(\vec{\xi}), \tag{21}$$

where the total number of PCE terms is equal to

$$I_t + 1 = \frac{(K+I)!}{K!I!}. \tag{22}$$

Since the probability densities in each dimension are independent, the multidimensional Hermite polynomials factor into the product of univariate Hermite polynomials:

$$H_{\mathbf{i}(\iota)} = \prod_{k=1}^{K} H_{n_k}(\xi_k),$$
$$n_k \in \{0,\ldots,I\}\forall k \in\{0,\ldots,K\},\quad \sum_{k=1}^{K} n_k \le I. \tag{23}$$

The coefficients of the PCE are solved by projecting the expansion in Eq. (21) over the polynomial chaos and using stochastic collocation to compute the resulting inner products:

$$\left\langle \varphi(\vec{\xi}), H_{\mathbf{j}}(\vec{\xi})\right\rangle = \left\langle \sum_{\iota=0}^{I_t}\widehat{u}_{\mathbf{i}(\iota)}H_{\mathbf{i}(\iota)}(\vec{\xi}), H_{\mathbf{j}}(\vec{\xi})\right\rangle, \tag{24}$$

where the angular brackets represent integration over the phase space with weight function given by the probability density $p(\vec{\xi})$:

$$\langle f,g\rangle = \int_{\xi_I}\cdots\int_{\xi_1} f(\vec{\xi})g(\vec{\xi})p(\vec{\xi})d\vec{\xi}. \tag{25}$$

The right-hand-side of Eq. (24) is simplified using orthogonality of the Hermite polynomial:

$$\int_{-\infty}^{\infty} H_i(\xi)H_j(\xi)\frac{1}{\sqrt{2\pi}}\exp\left[-\frac{\xi^2}{2}\right]d\xi = j!\,\delta_{ij} = \frac{1}{a}\delta_{ij} \tag{26}$$

to obtain for the expansion coefficients:

$$\widehat{u}_{\mathbf{i}} = a_{\mathbf{i}}\left\langle \varphi(\vec{\xi}), H_{\mathbf{i}}(\vec{\xi})\right\rangle, \tag{27}$$

where $a_{\mathbf{i}}$ is defined for one dimension in Eq. (26). The integral on the right-hand-side of Eq. (27) is then numerically evaluated using the stochastic collocation weights $w_k^{(q_k)}$, nodes $\xi_k^{(q_k)}$, and solutions $\varphi(\vec{\xi}^{(\mathbf{q})})$ generated when solving response moments in Eq. (18):

$$\widehat{u}_{\mathbf{i}} \approx a_{\mathbf{i}}\sum_{q_K=1}^{Q_K}\cdots\sum_{q_1=1}^{Q_1} w_1^{(q_1)}\ldots w_K^{(q_K)}\varphi(\vec{\xi}^{(\mathbf{q})})$$
$$H_{\mathbf{i}_1}(\xi_1^{(q_1)})\ldots H_{\mathbf{i}_K}(\xi_K^{(q_K)})p(\vec{\xi}^{(\mathbf{q})}), \tag{28}$$

where $\mathbf{q}$ is a multi-index of values q_k.

The polynomial order in any stochastic dimension should not be larger than the quadrature order used in that dimension. In this work we set the total PCE polynomial order equal to the largest quadrature order used, vis.,

$$I := max(Q_k),\ \forall k \in \{0,\ldots,K\}, \tag{29}$$

but additionally truncate any terms for which the polynomial order in a given dimension is greater than the quadrature order used in that dimension. The resulting number of PCE terms is thus

$$I_t + 1 \le \frac{(K+I)!}{K!I!}. \tag{30}$$

The model constructed for a KL truncation order K will be more accurate for higher polynomial orders I and collocation orders Q_k. Once constructed the surrogate model can be used to approximate the response corresponding to a new set of random variable values without solving transport over an additional realization. The surface can be cheaply sampled to, for instance, produce a visual of the response surface, estimate the response for a new set of input parameters, or create a probability density function of the response.

A polynomial fit of the system response does not guarantee purely positive values, especially when the response is near zero over much of stochastic domain. We introduce a positive-preserving transformation of the PCE by fitting the response to

$$\varphi(\vec{\xi}) = \left(\left(\sum_{\iota=0}^{I_t}\widehat{u}_{\mathbf{i}(\iota)}H_{\mathbf{i}(\iota)}(\vec{\xi})\right)^2\right)^n, \tag{31}$$

where n is a free parameter. Coefficients are then solved by applying stochastic collocation over the integral of

$$\widehat{u}_{\mathbf{i}} = a_{\mathbf{i}}\left\langle \left(\left(\varphi(\vec{\xi})\right)^{\frac{1}{n}}\right)^{\frac{1}{2}}, H_{\mathbf{i}}(\vec{\xi})\right\rangle. \tag{32}$$

A value of $n = 0.5$ has no effect on evaluation of coefficients for transport problems since transport response is always positive, and simply takes the absolute value of PCE values. Values of n smaller than 0.5 map the response values less than 1 to smaller positive values and values greater than 1 to larger values, creating a more sharply varying surface, whereas values of n greater than 0.5 map all values closer to 1. We hypothesize that mapping values which the PCE will fit closer to 1 increases the accuracy of the fit, as long as values of n are not chosen to be so large that numerical roundoff of exponential computations introduces significant error. As the value of n approaches infinity, all values map to 1 within machine precision and all information defining the surface is lost. We somewhat arbitrarily have chosen a value of $n = 1$ for the results in this paper. Other positive-preserving transformations are certainly conceivable and are a topic of future work.

With or without a positive-preserving transformation, finite polynomial fits of a response which spans an unbounded domain will, in regions of low probability, include values which are very high. The simplest method for handling such very large sampled values when constructing a probability density function is to ignore them since they are sampled infrequently. Perhaps a screening method which rejects samples taken from regions of sufficiently low probability would treat this problem.

5 Numerical Results

Solution of the stochastic transport equation is demonstrated on a problem described by microscopic total and scattering cross sections $\sigma_t = 2.0\,\mathrm{cm}^2$ and $\sigma_s = 1.0\,\mathrm{cm}^2$, atomic density average and variance of $\langle N_{at}\rangle = 1.0\,\mathrm{cm}^{-3}$ and $\upsilon_{N_{at}} = 1.0\,\mathrm{cm}^{-6}$, and an exponential covariance function with correlation length $\lambda_c = 1.5$ cm.

Realizations of the random cross sections are generated using a lognormal transformation of the Karhunen-Loéve (KL) expansion truncated at expansion order $K = 7$ in which the eigenspectrum is solved using the Nystöm method (NM) with an $N_{Ny} = 100$ node discretization. Transport results are obtained using Woodcock Monte Carlo (WMC) with $N = 1.6 \times 10^5$ particle histories. KL random variable values are chosen either using random sampling (RS) with $R = 2 \times 10^5$ samples, isotropic stochastic collocation (iSC) with quadrature orders $\mathbf{Q} = \{5, 5, 5, 5, 5, 5, 5\}$ for a total of $R_{SC} = 78,125$ realizations, or anisotropic stochastic collocation (aSC) with quadrature orders $\mathbf{Q} = \{6, 5, 5, 4, 4, 3, 3\}$ for a total of $R_{SC} = 21,600$ realizations. Probability density functions (PDFs) of the reflectance and transmittance are generated using $R = 2 \times 10^5$ samples when using the RS method, and by sampling from a constructed polynomial chaos expansion (PCE) model, both with and without the positive-preserving transformation, 1×10^6 times when solving using SC.

These solver parameters were chosen as the result of convergence studies and the transmittance standard deviation relative error is estimated using these parameters to be less than 3 %. Perhaps the most noteworthy convergence study was the one performed on the choice of Karhunen-Loève truncation order K, using the transmittance standard deviation as the output of interest. A KL truncation order of $K = 10$ was used to estimate the correct solution, holding all other solver parameters the same, and the relative difference of the solution given by values of $K \leq 10$ and $K = 10$ is plotted in Figure 4. The estimated relative error contribution to the transmittance standard deviation based on the truncation order of the KL expansion is therefore 0.8 %. A similar study was performed for choosing

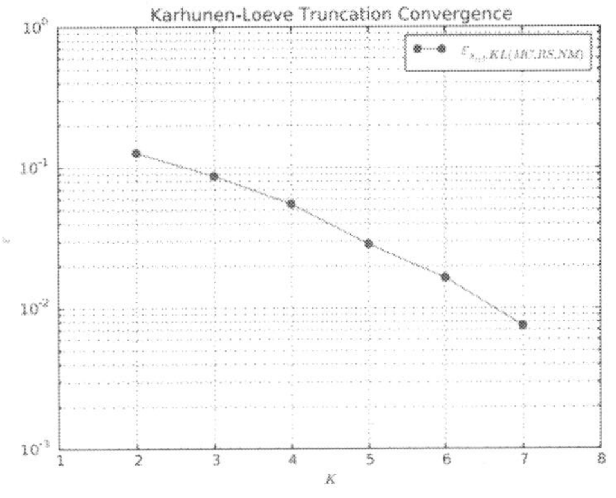

Figure 4. Standard Deviation Relative Error of KL Truncation

Table 1. Leakage Values and Cost

Transmittance	mean	deviation	R or R_{SC}
RS	0.01289	0.03376	200,000
iSC	0.01292	0.03335	78,125
aSC	0.01283	0.03388	21,600

Reflectance	mean	deviation	R or R_{SC}
RS	0.115138	0.000983	200,000
iSC	0.115162	0.001018	78,125
aSC	0.115151	0.000964	21,600

the number of particle histories N, the number of random samples R when using random sampling, and the quadrature orders $\mathbf{Q}$ when using stochastic collocation. The SC convergence study was performed using isotropic stochastic collocation such that the quadrature order in each dimension was increased at that same rate. Error was estimated using several anisotropic SC grids chosen based on intuition, and one of them was chosen for this problem. Problem parameters were chosen for each solver such that the estimated bound on the total error of the transmittance standard deviation was about 3 %.

Table 1 records the transmittance and reflectance values yielded by the RS, iSC, and aSC methods and the number of realizations used to produce these results. The greatest difference in transmittance standard deviation values between these solutions is about 1.6 %, less than the estimated bound on the error of 3 %. The largest relative difference between results for any quantity is a relative difference of about 5.6 % for the reflectance standard deviation. Error in these final quantities can be reduced by refining solver parameters such as choosing a larger KL truncation order K, using more particle histories, or increasing the number of realizations either directly in random sampling or by using higher-order quadrature in the SC method. Data from convergence studies, such as the plot in Figure 4, can be used to estimate the change of system parameters that will reduce the total error for the lowest computational cost.

Having solved the problem using each method to roughly the same accuracy, it is clear that for this problem the anisotropic stochastic collocation implementation solves the problem most efficiently, requiring about 10 times less realizations than random sampling. There is likely a more efficient set of anisotropic quadrature orders than the somewhat arbitrarily chosen set used here and an appropriate adaptive method may be able to find that set of quadrature orders efficiently. Isotropic stochastic collocation is about 2.5 times more efficient than random sampling for this problem. The SC methods are expected to be yet more efficient for more highly correlated problems in which the correlation length λ_c is larger, since the magnitude of successive KL terms falls off more quickly for more correlated processes allowing more liberal truncation of the expansion and a higher degree of anisotropy.

Scalar flux mean and relative standard deviation solved in 100 cells throughout the slab are given in Figures 5 and 6. Scalar flux values yielded by each of the three

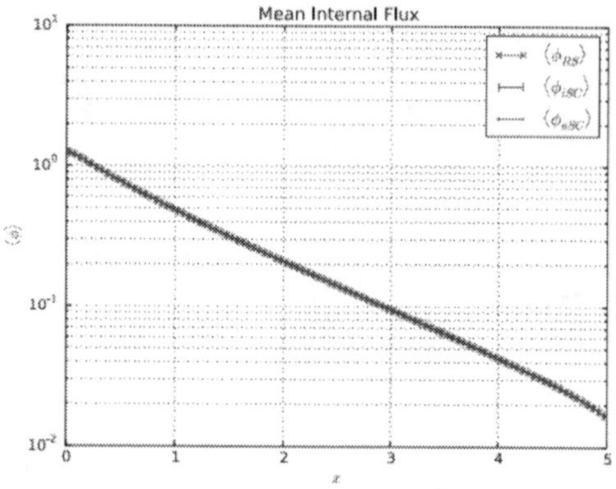

Figure 5. Mean Scalar Flux in Slab

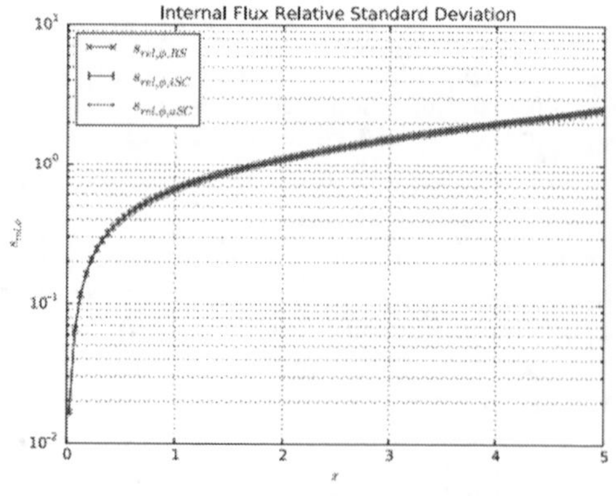

Figure 6. Relative Standard Deviation of Scalar Flux in Slab

methods match up well with one another. The greatest deviation between the methods appears to be near the transmissive boundary at which the mean flux is a little higher and relative standard deviation is a little lower for the isotropic stochastic collocation method than the others. This observation is consistent with the transmittance values reported in Table 1.

Whereas the mean and standard deviation of leakage values provide information often useful in shielding applications, information beyond the first two moments may be of value. For example, the median expected transmittance or the largest credible transmittance may be of more interest. Probability density functions are therefore generated which provide a variety of information beyond that given by the mean and standard deviation of leakage values.

Probability density functions of the transmittance generated with random sampling and isotropic stochastic collocation are given in Figure 7 and with random sampling and anisotropic stochastic collocation in Figure 8. Each plot contains results generated by constructing a PCE model in a straightforward way and using the positive-preserving transformation introduced here with free parameter $n = 1$. PDFs generated using the transformation with a free parameter value of $n = 0.5$ would be the same as those generated by the PCE model without transformation but in which all negative values are tallied as positive values of the same magnitude. Five times as many sam-

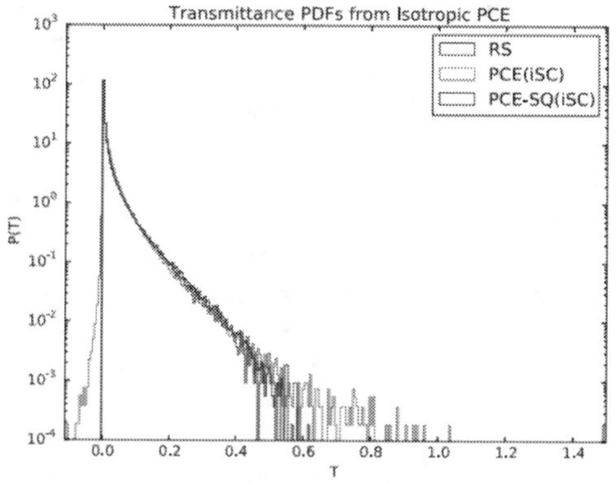

Figure 7. Transmittance Probability Density Functions

ples were taken to form the PDFs constructed from PCE models than those constructed using RS since it is much cheaper to sample a PCE model of the response once constructed than to solve transport results over additional realizations using RS. The PDF generated from the naive implementation of the PCE contains negative values due to the truncation of the PCE, making the positive-preserving transformation useful to maintain physically meaningful results. With or without this transformation, the polynomial fit over an unbounded support given by the PCE contains a low probability of sampling artificially large values. A method to screen out such samples may be worth investigating. The tapering off of high values in the PDF produced using iSC is also likely due to use of a polynomial fit over an unbounded support. It appears that having a quadrature order of 6, as in the anisotropic case, as opposed to 5, as in the isotropic case, in the first KL term re-

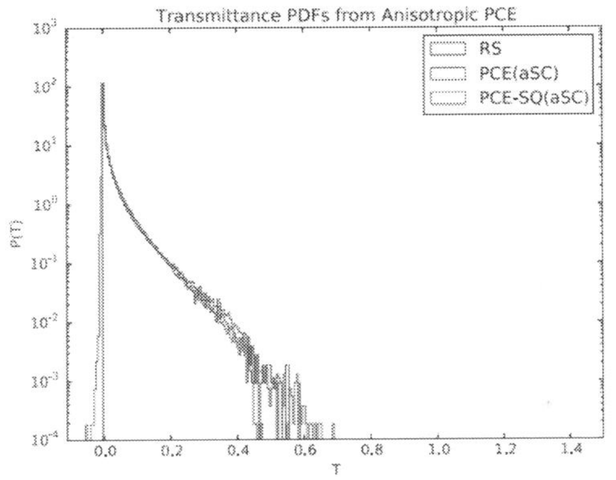

Figure 8. Transmittance Probability Density Functions

duces much of this effect. The transmittance PDFs generated from PCE models using the PCE transformation presented here contain less larger values. Comparing against the PDF produced using RS, this appears to be a good trait in the iSC case for this problem and a bad trait in the aSC case for this problem.

The PDFs produced using the PCE model closely match the PDFs generated with random sampling, and provide more information than a mean and variance. It is unclear whether using the positive-preserving transformation produced PDFs that were more accurate than the PDFs produced without the transformation on the positive portion of the domain, but the positive-preserving transformation did a good job a resolving the steep drop-off in transmittance values near zero. It is possible that other positive-preserving transformations, such as a maximum entropy representation, may prove to be a better choice than the one presented here; this is a topic of future work.

Likewise, PDFs of the reflectance generated using random sampling and isotropic stochastic collocation are given in Figure 9 and PDFs of the reflectance generated using random sampling and anisotropic stochastic collocation are given in Figure 10. While maintaining positive values did not turn out to be as important in modeling the reflectance response as modeling the transmittance response, the results generated using the positive-preserving transformation do not appear less accurate than when not using the transformation. PDFs of the response generated from PCE models deviate somewhat more from the PDF generated using random sampling for the reflectance than the transmittance. This is not an unexpected result for the problem parameters chosen since the standard deviation of the transmittance was the quantity that solver parameters were chosen to reduce the error of below a certain tolerance and the relative difference in the reflectance values were larger than the relative difference in transmittance values (Table 1). The PDFs generated from the PCE models could be made more accurate by increasing one or more of the solver parameters such as using more KL terms, more particle histories, more random samples, or a higher PCE truncation yielding more higher-order polynomial cross terms in the expansion.

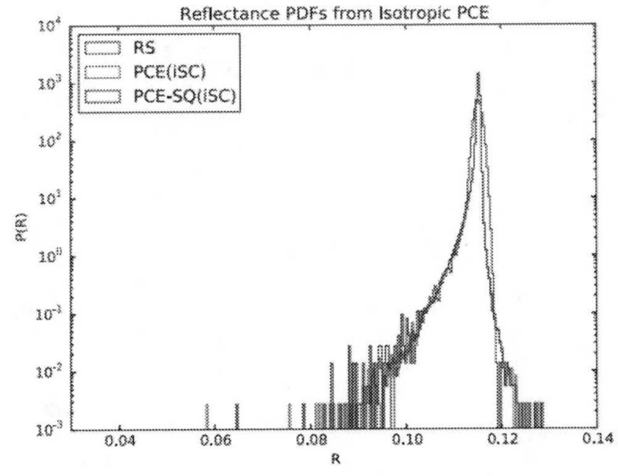

Figure 9. Reflectance Probability Density Functions

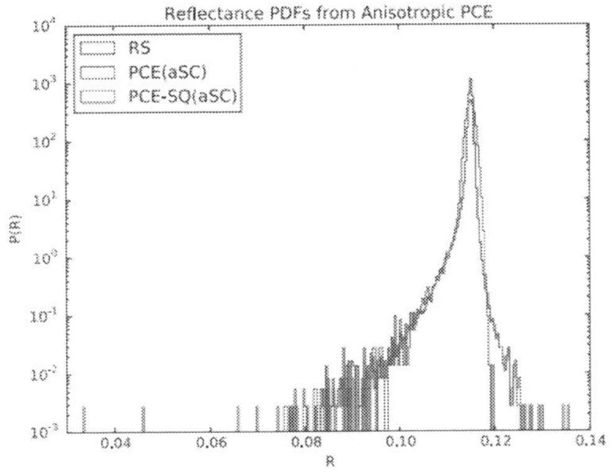

Figure 10. Reflectance Probability Density Functions

6 Conclusions

We have demonstrated that a combination of stochastic spectral representation and Woodcock Monte Carlo simulation allows efficient computation of radiation transport in strongly random media. The use of a lognormal distribution to represent the fluctuations in medium properties facilitates enforcement of strict positivity of cross section realizations, a condition that has challenged standard approaches that assume Gaussian fluctuations [8]. The fact that the lognormal random process is a memory-less nonlinear transformation of a Gaussian process further enables efficient and accurate reconstruction of the random cross section realizations using a Karhunen-Loève representation of the cross section. The robustness of this model enables fluctuations of arbitrarily large amplitude to be studied, using both random sampling of Karhunen-Loève realizations and more efficient stochastic collocation techniques. The polynomial chaos expansion is shown to be a useful surrogate model of the response populated using stochastic collocation results which allows for efficient construction of probability density functions of out-

 DOI: 10.1051/epjconf/201715306014

put variables. The use of a positive-preserving transformation of the polynomial chaos expansion further increases its usability by enforcing that all output variable values are positive with no apparent loss of accuracy in the representation. Further work on these techniques may include examination of covariance functions other than the exponential covariance, advanced stochastic collocation grids such as adaptive or sparse grids, use of a different positive-preserving transformation for the surrogate model of the response such as a lognormal transformation, or more judiciously selected ceiling cross sections to enhance the efficiency of the Woodcock Monte Carlo solver.

Acknowledgments

The first author was supported by the Department of Energy Nuclear Energy University Programs Graduate Fellowship.

Sandia National Laboratories is a multi-mission laboratory managed and operated by Sandia Corporation, a wholly owned subsidiary of Lockheed Martin Corporation, for the U.S. Department of Energy's National Nuclear Security Administration under contract DE-AC04-94AL85000.

References

[1] Adams, M. L. and Larsen, E. W. and Pomraning, G. C., Journal of Quantitative Spectroscopy and Radiative Transfer **42**, 253-266 (1989)

[2] Pomraning, G. C., *Linear kinetic theory and particle transport in stochastic mixtures* (World Scientific Publishing Co. Pte. Ltd., River Edge, New Jersey USA, 1991)

[3] Prinja, A. K. and Gonzalez-Aller, A., Nuclear Science and Engineering **124**, 89-96 (1996)

[4] Le Maître, O. P. and Knio, O. M., *Stochastic finite elements: A spectral approach* (Springer-Verlag New York, Inc., 2003)

[5] Fichtl, E. D. and Prinja, A. K., Journal of Quantitative Spectroscopy and Radiative Transfer **112**, 646-659 (2011)

[6] Ayres, D. and Eaton, M. D., Annals of Nuclear Energy **80**, 379-402 (2015)

[7] Leppänen, J., Nuclear Science and Engineering **174**, 318-325 (2013)

[8] Olson, A. J. and Franke, B. C. and Prinja, A. K., *Woodcock Monte Carlo transport through binary stochastic media*, Transactions of the American Nuclear Society, Anaheim, CA (ANS, November 2014)

[9] Ghanem, R. G. and Spanos, P. D., *Spectral methods for uncertainty quantification with applications to computational fluid dynamics* (Springer, Springer Dordrecht Heidelberg London New York, 2010)

[10] Sun, S. and Zhao, J. and Zhu J., Information Fusion **26**, 36-48 (2015)

[11] Betz, W. and Papaioannou, I. and Straub, D., *Assessment of methods for the numerical solution of the Fredholm integral eigenvalue problem*, 11th International Conference on Structural Safety, & Risk 2013, New York, NY (IASSAR, June 2013)

[12] Nobile, F. and Tempone, R. and Webster, C. G., SIAM Journal of Numerical Analysis **5**, 2411-2442 (2008)

Research on Primary Shielding Calculation Source Generation Codes

Zheng Zheng[1,*], *Qiliang* Mei[1], *Hui* Li[1], *Danhua* Shangguan[2], *Guangchun* Zhang[2]
[1]Shanghai Nuclear Engineering Research and Design Institute Hongcao Road No. 29, Shanghai, China
[2]Institute of Applied Physics and Computational Mathematics Fenghao East Road No. 2, Beijing, China

Abstract. Primary Shielding Calculation (PSC) plays an important role in reactor shielding design and analysis. In order to facilitate PSC, a source generation code is developed to generate cumulative distribution functions (CDF) for the source particle sample code of the J Monte Carlo Transport (JMCT) code, and a source particle sample code is deveoped to sample source particle directions, types, coordinates, energy and weights from the CDFs. A source generation code is developed to transform three dimensional (3D) power distributions in xyz geometry to source distributions in $r\theta z$ geometry for the J Discrete Ordinate Transport (JSNT) code. Validation on PSC model of Qinshan No.1 nuclear power plant (NPP), CAP1400 and CAP1700 reactors are performed. Numerical results show that the theoretical model and the codes are both correct.

1 Introduction

Primary shielding is composed of in-core internals surrounding the core (including baffle and barrel), water (including water between baffle and barrel, and water between barrel and Reactor Pressure Vessel (RPV)), and RPV and concrete surrounding RPV. PSC plays an important role in evaluating the radiation dose of personnel, insuring the safety of RPV and reactor internals during the whole lifetime, and preventing the concrete shielding as well as the outer components and structures from an excess of activation. PSC mainly consists of the following steps: (1) determine the core neutron and photon sources, (2) perform neutron and photon transport calculations, (3) derive the dose rate, heating rate and displacements per atom (DPA) etc. by multiplying response functions with fluence rates.

Both the Monte Carlo method and the Discrete Ordinate method are usually adopted to perform neutron and photon transport calculations. JMCT[1] is a 3D neutron-photon transport Monte Carlo code developed by Institute of Applied Physics and Computational Mathematics (IAPCM). JMCT provides a source particle sample code for user to define space, energy, direction and weight for source particles. JSNT[2] is a 3D neutron-photon transport Discrete Ordinate code developed by IAPCM.

In order to facilitate PSC, we develop a source generation code and a source sample code for JMCT, as well as a source generation code for JSNT. Both the CDFs of source particles for JMCT and 3D source distributions of each mesh for JSNT are calculated using 3D power distribution, fission fraction, fission spectrum, number of particles and energy released per fission etc. We apply the source codes on PSC model of Qinshan No.1 NPP, CAP1400 and CAP1700 reactors, with the aim to validate the source codes, and analyse the effects of the variations of fuel assembly burnup and enrichment on the fluence rates.

2 Theoretical Model

2.1 JMCT source generation and sample methods

Firstly, the probability density functions (PDF) and CDF of source particles are calculated using 3D power distribution, fission fraction, fission spectrum, number of particles and energy released per fission etc. Secondly, source particle directions, types, coordinates, energy and weights are sampled from the CDFs.

2.1.1 Cumulative distribution functions calculation

The energy PDF of the mth assembly is:

$$\chi'(m,g',p) = \frac{\sum_{n=1}^{N'} f'(m,n)\cdot v(n,p)\cdot \chi(n,g',p)\cdot T(g',p)}{\sum_{n=1}^{N'}\left\{ f'(m,n)\cdot v(n,p)\cdot \sum_{g=1}^{G(p)} \chi(n,g,p)\cdot T(g,p)\right\}} \quad (1)$$

Where g,g' is energy index, m is assembly index, n is the fission nuclide index, p is the particle type index, p=1 is neutron, p=2 is photon, N' is the number of nuclides, $\chi'(m,g',p)$ is the PDF of particle p, group g' and assembly m, $f'(m,n)$ is the fission fraction of nuclide n and assembly m. $v(n,p)$ is number of

* Corresponding author: zhengzheng@snerdi.com.cn

particles released per fission for particle p and nuclide n, particles/s, $\chi(n,g,p)$ is the probability of particle p, group g and nuclide n, $G(p)$ is the number of groups for particle p, $T(g,p)$ is the energy bias factor for particle p and group g.

The energy CDF of the mth assembly is:

$$E(m,g",p)=\sum_{g'=1}^{g"}\frac{\chi'(m,g',p)}{\sum_{g=1}^{G(p)}\chi'(m,g,p)},g"=1,...,G(p) \quad (2)$$

Where $E(m,g",p)$ is the CDF of particle p, group $g"$ and assembly m.

The space PDF of the qth mesh is:

$$S(i,j,k,p)=$$
$$P(i,j,k)\cdot\frac{C'\cdot B'(m)}{K'(m)} \quad (3)$$
$$\cdot\sum_{n=1}^{N'}\left\{f'(m,n)\cdot v(n,p)\cdot\sum_{g=1}^{G(p)}\left[\chi(n,g,p)\cdot T(g,p)\right]\right\}$$

Where $S(i,j,k,p)$ is the source of particle p and mesh (i,j,k), particles/(cm^3·s), $P(i,j,k)$ is the power of mesh (i,j,k), W/cm^3, C' is unit conversion factor, 6.241508×10^{12}MeV/(s·W), $B'(m)$ is the space bias factor of assembly m, $K'(m)$ is the energy released per fission of assembly m, MeV.

$f'(m,n)$ and $K'(m)$ can be calculated by interpolation from library by assembly burnup and enrichment. The library includes fission fraction and energy released per fission of 6 fission nuclides (U-235, U-238, Pu-239, Pu-240, Pu-241 and Pu-242) with 25 burnups and 5 enrichments.

The space CDF of the qth mesh is:

$$D(q,p)=\sum_{q'=1}^{q}\frac{S'(q',p)\cdot V'(q')}{\sum_{q"=1}^{Q}S'(q",p)\cdot V'(q")},q=1,...,Q \quad (4)$$

Where $D(q,p)$ is the CDF of particle p and mesh q, $S'(q',p)$ is the source of particle p and mesh q', particles/(cm^3·s), $V'(q')$ is the volume of mesh q', cm^3, Q is the number of mesh.

The number of the pth particle is:

$$N(p)=\sum_{i,j,k}S(i,j,k,p)\cdot V'(i,j,k)\cdot R(p) \quad (5)$$

Where $N(p)$ is the number of particle p, $V'(i,j,k)$ is the volume of mesh (i,j,k), cm^3, $R(p)$ is the particle type bias factor of particle p.

The particle type CDF of the pth particle is:

$$F(p)=\sum_{p'=1}^{p}\frac{N(p')}{\sum_{p"=1}^{P}N(p")},p=1,...,P \quad (6)$$

Where $F(p)$ is the particle type CDF of particle p, P is the number of particle types.

Directions of neutrons and photons released in fission are isotropic, which can be determined using isotropic sample subroutine.

Before the biased PDFs and CDFs are calculated, the bias factors should be normalized as follows:

$$T(g,p)=\frac{T'(g,p)}{\sum_{g'=1}^{G(p)}T'(g',p)}\cdot G(p),\sum_{g=1}^{G(p)}T(g,p)=G(p) \quad (7)$$

$$B'(m)=\frac{B"(m)}{\sum_{m=1}^{M}B"(m)}\cdot M,\sum_{m=1}^{M}B'(m)=M \quad (8)$$

$$R(p)=\frac{R'(p)}{\sum_{p=1}^{P}R'(p)}\cdot P,\sum_{p=1}^{P}R(p)=P \quad (9)$$

Where $T'(g,p)$ is the un-normalized energy bias factor for particle p and group g, $B"(m)$ is the un-normalized space bias factor of assembly m, $R'(p)$ is the un-normalized particle type bias factor of particle p, M is the number of assembly.

2.1.2 Source particle sample method

Considering the dependent relationship of variables, the source particle sample sequence is the reverse of the CDF calculation sequence, which is: direction, particle type, space and energy.

Directions are determined by isotropic sample.

For particle type, select random number $\xi_p\in(0,1]$, determine index p which satisfies $F(p-1)<\xi_p\le F(p)$.

For space, select random number $\xi_q\in(0,1]$, determine index q which satisfies $D(q-1,p)<\xi_q\le D(q,p)$. Select random number $\xi_{q1},\xi_{q2},\xi_{q3}\in(0,1]$, calculate particle coordinates as follows:

$$\begin{cases}x=x_q^1+\xi_{q1}\cdot(x_q^2-x_q^1)\\y=y_q^1+\xi_{q2}\cdot(y_q^2-y_q^1)\\z=z_q^1+\xi_{q3}\cdot(z_q^2-z_q^1)\end{cases} \quad (10)$$

Where x,y,z are particle coordinates, cm, x_q^1,x_q^2 are lower and upper boundary of mesh q along axis x, cm, y_q^1,y_q^2 are lower and upper boundary of mesh q along axis y, cm, z_q^1,z_q^2 are lower and upper boundary of mesh q along axis z, cm.

For energy, select random number $\xi_g\in(0,1]$, determine index g which satisfies $E(g-1,q,p)<\xi_g\le E(g,q,p)$. Select random number $\xi_{g1}\in(0,1]$, calculate particle energy as follows:

$$E'=E_g^1+\xi_{g1}\cdot(E_g^2-E_g^1) \quad (11)$$

Where E' is particle energy, MeV, E_g^1,E_g^2 are lower and upper boundary of group g, MeV.

When the bias factors are used, in order to get unbiased results, source particle weight should be adjusted as follows:

$$w' = \frac{w}{R(p)\cdot B'(m)\cdot T(g,p)}\cdot\frac{N^{\dagger}}{N''} \quad (12)$$

Where w' is biased weight, w is unbiased weight, $N^{\dagger}$ is the number of biased particles, N'' is the number of unbiased particles.

Based on the CDF calculation method and source particle sample method above, a source generation code and a source sample code for JMCT are developed. Fig. 1 shows the flowchart of the source codes.

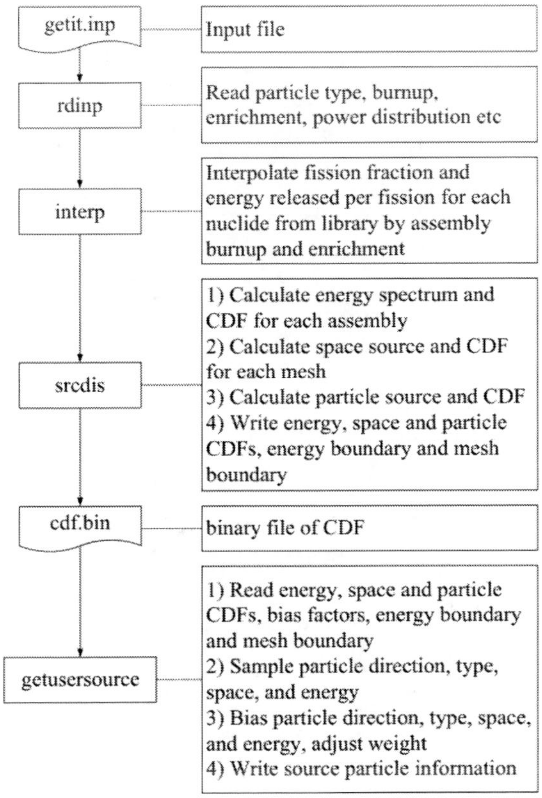

Fig. 1. Flowchart of CDF calculation and source particle sample for JMCT

2.2 JSNT source generation method

Firstly, the pin mesh source is calculated using 3D power distribution, fission fraction, fission spectrum, number of particles and energy released per fission. Secondly, the geometry mesh source is calculated using the pin mesh source by the area or volume weighted method.

The pin mesh source is calculated as follows:

$$S''(i,j,k,g) = P(i,j,k)\cdot C^{\dagger}(m,g)$$
$$= P(i,j,k)\cdot\frac{C'\cdot B'(m)}{K'(m)}\cdot\sum_{n=1}^{N'}\left[f'(m,n)\cdot v(n)\cdot\chi''(n,g)\right] \quad (13)$$

Where $S''(i,j,k,g)$ is the source of group g and pin mesh (i,j,k), particles/(cm^3·s), $C^{\dagger}(m,g)$ is power to source transform factor, particles/s/MW, $\chi''(n,g)$ is the probability of group g and nuclide n.

The geometry mesh source is calculated as follows:

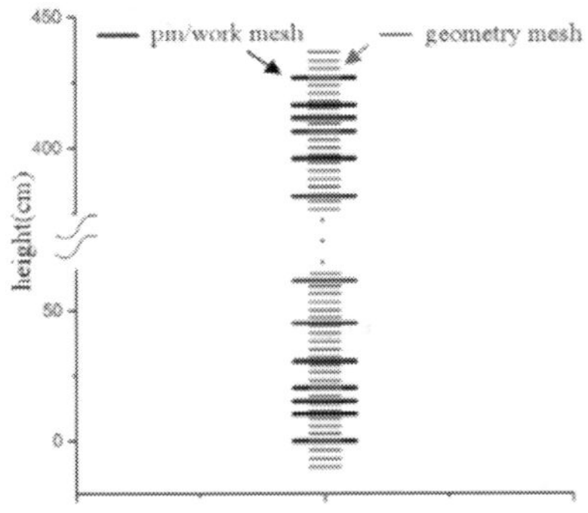

(a) axial mesh

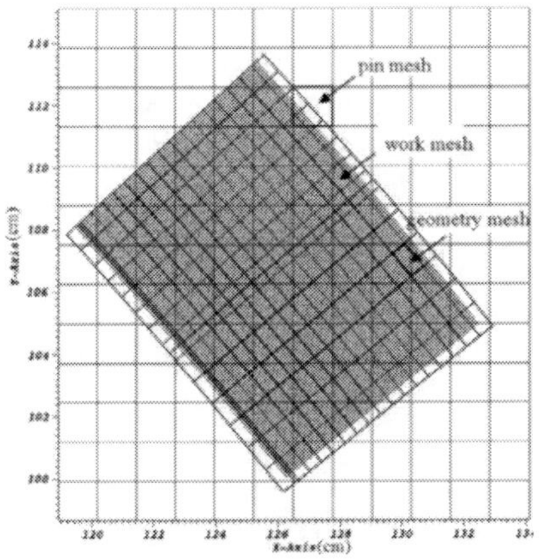

(b) radial mesh

Fig. 2. Geometry transformation sketch map

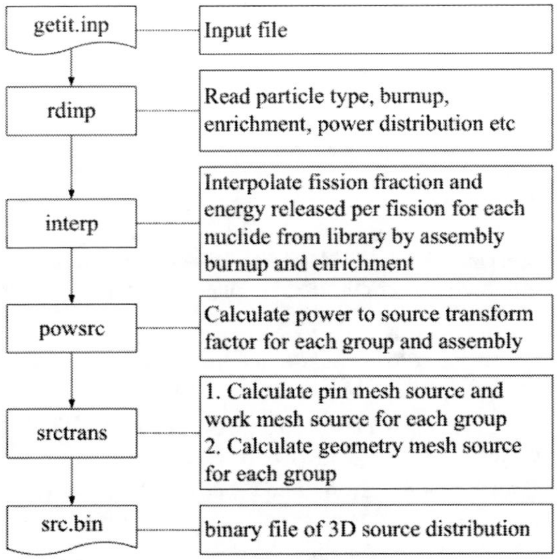

Fig. 3. Flowchart of the source generation code for JSNT

1) Define a group of $r\theta z$ work mesh, for each work mesh, set the source of the work mesh equals to the source of pin mesh if the center of the work mesh lies in the pin mesh.

2) As Fig. 2 and Formula (14) show, calculate the source of each geometry mesh by the area or volume weighted method.

$$S^{\vee}(q',g)=\sum_{q\in q'} S^{\wedge}(q,g)\cdot V'(q) \Big/ \sum_{q\in q'} V'(q) \quad (14)$$

Where $S^{\vee}(q',g)$ is the source of group g and geometry mesh q', particles/(cm^3·s), $S^{\wedge}(q,g)$ is the source of group g and work mesh q, particles/(cm^3·s).

Fig. 3 shows the flowchart of the source generation code for JSNT.

3 Numerical Results

We perform validation on PSC model of Qinshan No.1 Reactor, and apply theses codes on CAP1400 and CAP1700 reactors.

3.1 Qinshan No.1 reactor surveillance capsule Fast Neutron Fluence Rate (FNFR) calculation

Qinshan No.1 Reactor is a 966MW (thermal power) pressurized light-water reactor (PWR) designed by Shanghai Nuclear Engineering Research and Design Institute (SNERDI) of China and placed in operation in December of 1991. It is owned by China National Nuclear Corporation Company (CNNCC) and operated by Qinshan Nuclear Power Limited Company (QNPLC). The core of the Qinshan No.1 Reactor consists of 121 assemblies and is surrounded by the core baffle, barrel, pressure vessel, and biological shield. According to the requirement of RPV material radiation surveillance program, Qinshan NPP shutdowns the reactor on June 6 of 2014 (end of the 15th fuel cycle), and removes the 6th radiation surveillance capsule (8 in total). The FNFR of this capsule is important for the RPV radiation evaluation. China Institute of Atomic Energy (CIAE) extracts samples from dose box of surveillance capsule, measures the activation, and calculates the neutron fluence rate for energy greater than 0.1MeV, 0.5MeV and 1.0MeV by using a neutron spectrum calculation code. SNERDI calculates the FNFR by using JMCT and JSNT codes with newly developed source codes.

The geometry models of Qinshan are shown schematically in Fig. 4, which also show the locations of the capsule. Both radial and axial core power distributions are processed as follows:

$$\bar{P}(N)=\sum_{n=1}^{N}\left[P(n)\cdot T(n)\right] \Big/ \sum_{n=1}^{N} T(n) \quad (15)$$

Where $\bar{P}(N)$ is the average relative power from cycle 1 to cycle N, $P(n)$ is relative power for cycle n, $T(n)$ is the effective full power day (EFPD) for cycle n, days.

In JMCT calculation, 2E+9 neutrons are simulated. Geometry splitting with Russian Roulette and space bias factors are used to improve the efficiency. Point detectors are used to tally the FNFRs. In JSNT calculation, the number of meshes is about 200 million for 1/8 $r\theta z$ geometry, multi-group library is BUGLE-96[3] library with 47 group neutron and 20 group photon, the order of the Legendre expansion of the scattering cross-section is 3, the order of the fully symmetric angular quadrature set is S_8, and the point-wise flux convergence criterion is 0.001. The source energy spectrum is taken as U-235 fission spectrum.

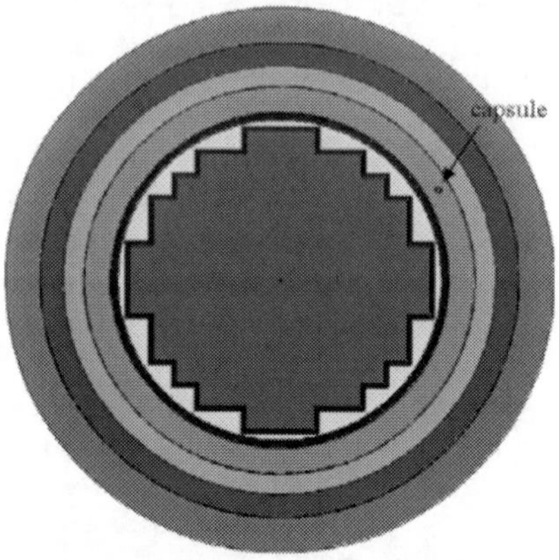

(a) xy model

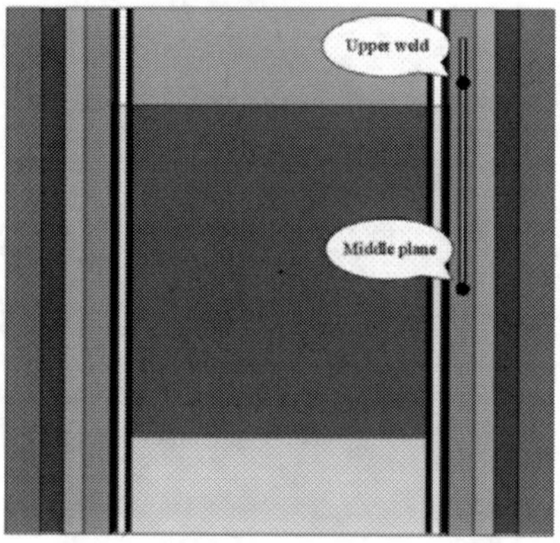

(b) xz model

Fig. 4. Geometry model of Qinshan reactor

Table 1 gives FNFRs of Qinshan reactor for nominal power. As Table 1 shows, the relative errors are within 10% and 30% for the middle plane and upper weld, respectively. The measurement uncertainty is about 15%, which comes from target nuclides weights of detector samples, nuclear cross-sections, and core power etc. The calculation uncertainties are about 10% and 20% for the middle plane and upper weld, respectively, which come from power distribution, geometry and material etc. Considering the uncertainties above, the results of JMCT and JSNT are comparable with measurements.

Table 1. FNFRs of Qinshan reactor

Position	Energy (MeV)	Codes	FNFR (n·cm^{-2}·s^{-1})	Relative error

				(%)
Middle plane	≥0.1	Measure	1.404E+11	-
		JMCT	1.449E+11	3.24
		JSNT	1.545E+11	10.05
	≥0.5	Measure	9.725E+10	-
		JMCT	1.007E+11	3.56
		JSNT	1.068E+11	9.85
	≥1.0	Measure	6.552E+10	-
		JMCT	6.696E+10	2.21
		JSNT	7.179E+10	9.58
Upper weld	≥0.1	Measure	1.594E+10	-
		JMCT	1.133E+10	-28.90
		JSNT	1.216E+10	-23.68
	≥0.5	Measure	1.103E+10	-
		JMCT	8.040E+09	-27.08
		JSNT	8.722E+09	-20.89
	≥1.0	Measure	7.415E+09	-
		JMCT	5.662E+09	-23.63
		JSNT	6.098E+09	-17.75

3.2 CAP1400 reactor RPV neutron and photon fluence rate calculation

CAP1400 is an advanced passive generation III PWR designed by SNERDI of State Power Technology Corporation (SNPTC) of China, which is belonged to State Power Investment Corporation (SPIC). The thermal power of CAP1400 is 4040MW with 193 assemblies and a low-leakage refueling sheme. Different from previous reactors, the core shroud of CAP1400 is a highly heterogeneous structure composed of baffle and 6 half ring segments. Since it is subject to high neutron and photon fluence rates, both thermal stresses and iron DPA become determining parameters for its sizing. The present work aims to calculate neutron and photon fluence rates from barrel to RPV, which can be used to derive heat generation rate and DPA for the core shroud.

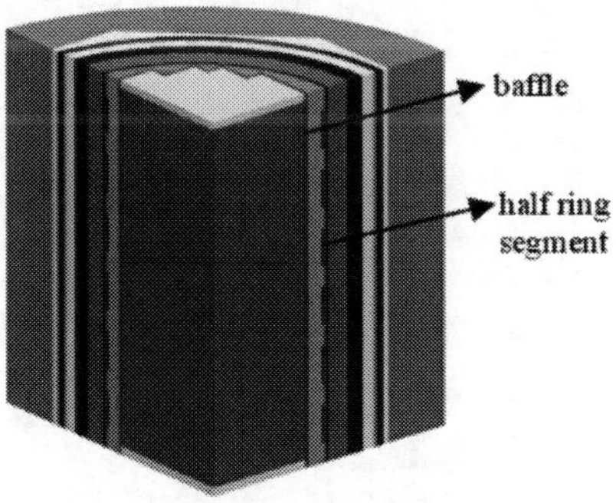

Fig. 5. Geometry model of CAP1400

Fig. 5 shows the geometry model of CAP1400. The cycle-average power distribution of equilibrium cycle (Cycle 6) is used to determine the core source. Fig. 6 and Fig. 7 are the radial and 3D neutron and photon source distributions of JMCT and JSNT, respectively. In JMCT calculation, number of particles (NPS) is 5E+9, superimposed mesh tallys are used to give fluence rate distributions. In JSNT calculation, almost the same parameters are used as in the previous JSNT calculation for Qinshan No.1 reactor surveillance capsule FNFR calculation. The source energy spectrum of each assembly is a mixed fission spectrum determined by the fission fractions of U and Pu.

Fig. 6. Radial neutron and photon source of JMCT

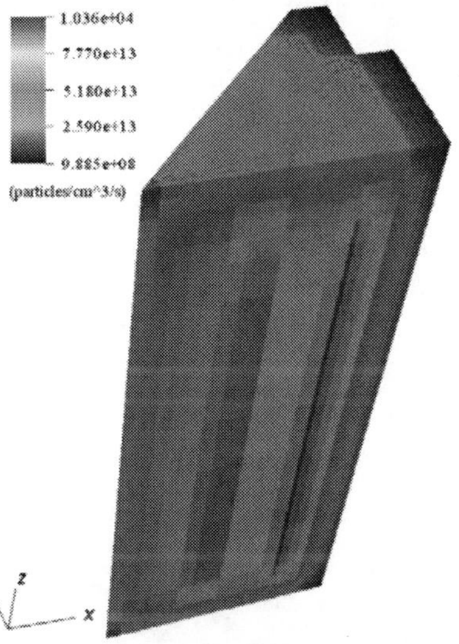

Fig. 7. 3D neutron and photon source of JSNT

Fig. 8 gives the radial neutron and photon fluence rates and error distributions from barrel to RPV. As Fig. 8 shows, most of the relative errors of neutron fluence rates of JMCT and JSNT are within 20%, which satisfy the requirements for different radiation shielding calculation methods of U.S. Nuclear Regulatory Commission (NRC) RG 1.190[4]. While most of the relative errors of photon fluence rates with photon energy greater than 1MeV and between 0.1MeV and 1.0MeV of JMCT and JSNT are within 20%, the relative errors with photon energy less than 0.1MeV are about

40%. This is probably attributed to the difference from multigroup and continuous libraries. We will investigate this phenomenon in the future.

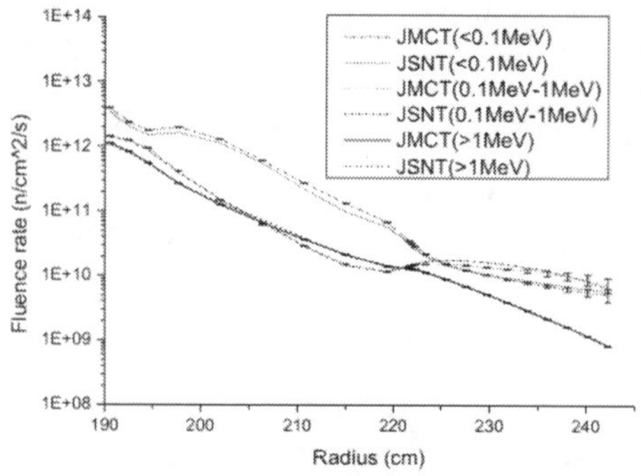

(a) neutron fluence rate

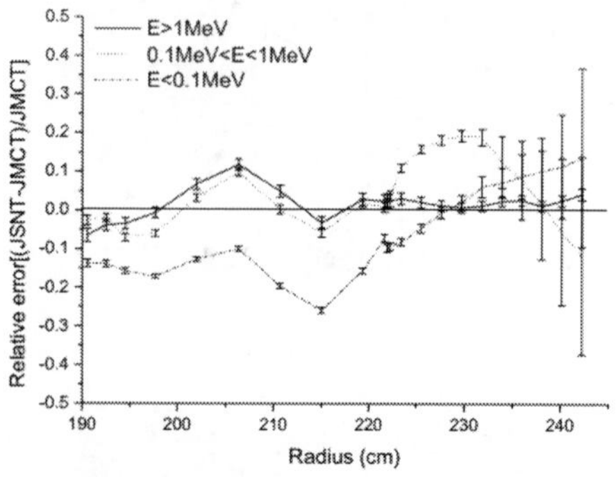

(b) neutron fluence rate error

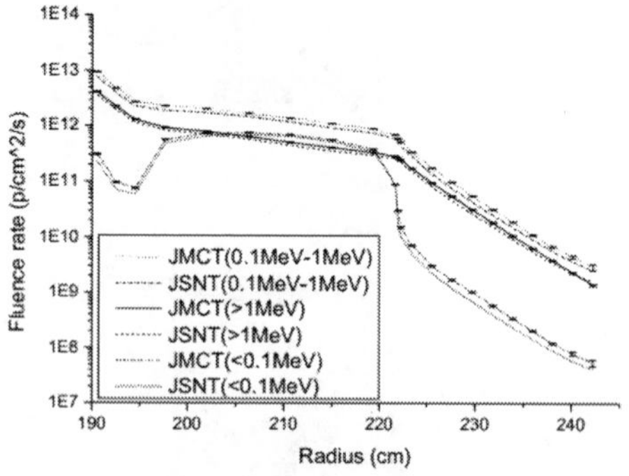

(c) photon fluence rate

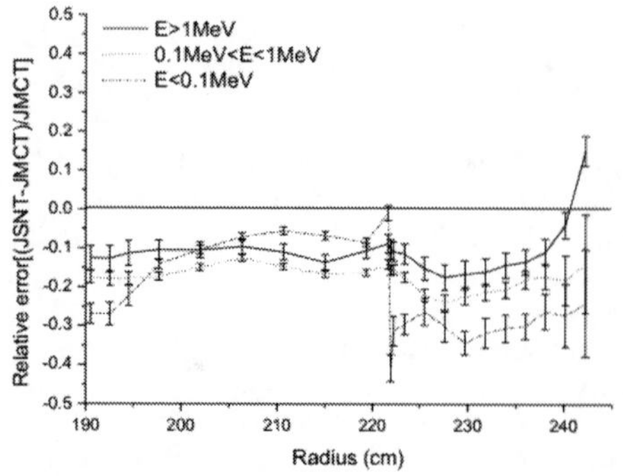

(d) photon fluence rate error

Fig. 8. Neutron and photon fluence rates and error

3.3 CAP1700 reactor RPV FNFR calculation

CAP1700 is a conceptual innovation advanced passive generation III PWR designed by SNERDI, which is aimed to improve the safety and economics of PWRs. Compared with CAP1400, CAP1700 has a new-style baffle with width increasing from 3cm to greater than 10cm, leading to a lower FNFR on RPV. The objective of this work is to analyse the effects of the fuel assembly burnup and enrichment variations on the FNFR of RPV.

Fig. 9 shows the geometry model of CAP1700. Table 2 gives the maximum FNFR (greater than 1.0MeV) on the inner surface of RPV. As Table 2 shows, the FNFR using U an Pu mixed spectrum is greater than that using U spectrum by 10% to 15%. Compared with U, Pu releases more neutrons and energy for each fission, and the fission spectrum is harder. These factors result in the increase of the FNFR using U and Pu mixed spectrum. Since Pu builds up with the increase of burnup, this effect should be considered in core source determination.

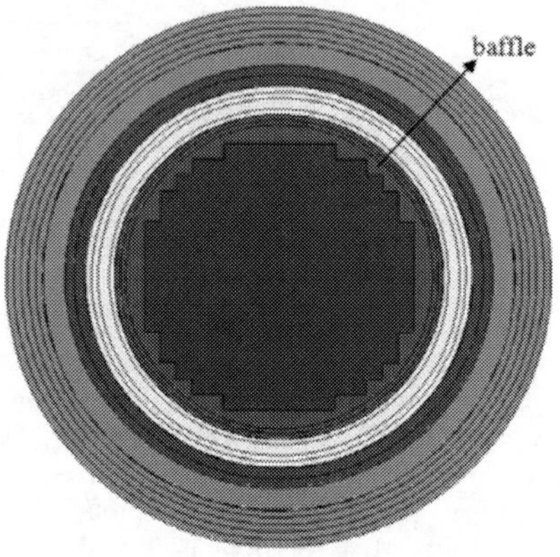

(a) xy model

(b) xz model

Fig. 9. Geometry model of CAP1700

Table 2. Maximum FNFR of CAP1700

Codes	Fission spectrum	FNFR ($n \cdot cm^{-2} \cdot s^{-1}$)	Relative error (%)
JMCT	U-235	1.07E+08	-
JMCT	U and Pu	1.20E+08	11.68
JSNT	U-235	1.09E+08	1.78
JSNT	U and Pu	1.23E+08	15.33

4 Conclusions

As core source is the precondition for reactor shield calculation, the objective of this work is to investigate source generation methods and develop source codes for a Monte Carlo code JMCT and a Discrete Ordinate code JSNT. The source codes consider burnup and enrichment of fuel assembly, and supply particle type, energy and space bias factors for users to improve calculation efficiency. The performed PSCs shows that the source codes developed are correct. The source codes developed can be used to support other Monte Carlo codes and the Discrete Ordinate codes without large modification. We will extend the sources codes for other reactor types in the future.

References

1. JMCT User's Manual, Institute of Applied Physics and Computational Mathematics, 2015
2. JSNT User's Manual, Institute of Applied Physics and Computational Mathematics, 2015
3. BUGLE-96, Coupled 47 Neutron, 20 Gamma-Ray Group Cross Section Library Derived from ENDF/B-VI for LWR Shielding and Pressure Vessel Dosimetry Applications, DLC-185, Oak Ridge National Laboratory, 1999
4. Regulatory Guide 1.190 , Calculational and Dosimetry Methods for Determining Pressure Vessel Neutron Fluence, Nuclear Regulatory Commission, 2001

AREVA Developments for an Efficient and Reliable use of Monte Carlo codes for Radiation Transport Applications

Nicolas Chapoutier[1,*], *François* Mollier[2], *Guillaume* Nolin[2], *Matthieu* Culioli[2], *Jean-Reynald* Mace[3]

[1]AREVA NP Inc, Solomon Pond Park, Marlborough, MA 01752
[2]AREVA NP, 10-12 Rue Juliette Recamier, 69456 LYON CEDEX 06
[3]AREVA, 1 Place Jean MILLIER, 92084 PARIS LA DEFENSE CEDEX

Abstract. In the context of the rising of Monte Carlo transport calculations for any kind of application, AREVA recently improved its suite of engineering tools in order to produce efficient Monte Carlo workflow. Monte Carlo codes, such as MCNP or TRIPOLI, are recognized as reference codes to deal with a large range of radiation transport problems. However the inherent drawbacks of theses codes - laboring input file creation and long computation time - contrast with the maturity of the treatment of the physical phenomena. The goals of the recent AREVA developments were to reach similar efficiency as other mature engineering sciences such as finite elements analyses (e.g. structural or fluid dynamics). Among the main objectives, the creation of a graphical user interface offering CAD tools for geometry creation and other graphical features dedicated to the radiation field (source definition, tally definition) has been reached. The computations times are drastically reduced compared to few years ago thanks to the use of massive parallel runs, and above all, the implementation of hybrid variance reduction technics. From now engineering teams are capable to deliver much more prompt support to any nuclear projects dealing with reactors or fuel cycle facilities from conceptual phase to decommissioning.

1 Introduction

In the context of the rising of Monte Carlo transport calculations for any kind of application, AREVA recently improved its suite of engineering tools in order to produce efficient Monte Carlo workflow. Monte Carlo codes, such as MCNP [1] or TRIPOLI [2], are recognized as reference codes to deal with a large range of radiation transport problems. However the inherent drawbacks of theses codes - laboring input file creation and long computation time - contrast with the maturity of the treatment of the physical phenomena. The goals of the recent AREVA developments were to reach similar efficiency as other mature engineering sciences such as finite elements analyses (structural or fluid dynamics for example). 3 axes of development are presented in this paper:

- VICTORIA: a versatile Graphical User Interface based on CAD (Computer-Aided Design) tools to address radiation protection solvers,
- Calculation runs speed up,
- Chaining of Monte Carlo results to other solvers and engineering fields.

2 VICTORIA

VICTORIA (for Various Instances CAD calculation TOol and Radiation-propagation Interface by AREVA) is a software dedicated to AREVA engineering teams to address input decks and the post-processing of radiation transport simulations (notably MCNP and TRIPOLI).

2.1 Challenges of the VICTORIA software

The original idea of the project came from the observation that radiation solvers do not embed enough efficient environments for engineering workflow. Indeed geometry creation and the definition of other simulation parameters have to be defined by an ASCII file based on keywords. Compared to other engineering sciences no aided engineering software permits to assist the user to efficiently perform analyses. Especially no general CAD integration is natively available. The modification of existing models can be complex and source of human errors. The primary author of a Monte Carlo model is often the only one capable to adapt it for other purpose or design modifications. These drawbacks do not meet the profitable requirements of the engineering projects. Indeed prompt iterations are expected to progress efficiently in interface with other disciplines.

Also the engineering teams are sometimes using different radiation protection solvers as point-kernel or Monte Carlo calculation codes (MCNP or TRIPOLI according to the project context). Despites their similarities each solver has its own syntax and specificities and a model built for one solver cannot be directly reused to another. In term of skill management,

* Corresponding author: nicolas.chapoutier@areva.com

keeping trained engineers for several solvers is also a challenge.

2.2 Technology choices

To fulfil the objectives of the VICTORIA software, AREVA based its developments on the ANSYS WORKBENCH platform. ANSYS, world leader for Computer Aided Engineering software, has a reliable experience of integrated simulation system. The ANSYS WORKBENCH platform gathers finite elements solvers (structural analyses, computational fluid dynamics (CFD) for example) with a common and compatible data models management. Also 3D modelling software is included to create geometries, edit, clean up or repair CAD imports.

The ACT (ANSYS Customization Toolkit) product recently released by ANSYS enables to add customization to the WORKBENCH platform. This application programming interface permits to redesign the purpose of the traditional ANSYS analyses for radiation protection solvers. Existing features: CAD modeling, meshing algorithm, 3D post-processing are directly inherited from the platform for customization. The specific developments in ACT can follow the future new releases of the ANSYS Workbench platform, meaning that all new improvements from ANSYS in the standard platform will be accessible by the VICTORIA users. In other words VICTORIA is not a standalone application and its sustainability can be ensured.

2.3 Overview of VICTORIA

2.3.1 VICTORIA Workflow

The creation of a radiation protection analyses is based on a sequence of several steps:

- Material database creation,
- 3D geometry importation and / or creation,
- Model: material assignment to bodies,
- Setup: simulations parameters (particles type, sources and tally creation),
- Solution and results: including post-processing features: import of output files for 2D/3D rendering of tallies.

Each workflow defined in a project is pictured on the project schematic front page. The project schematic offers a comprehensive and visual overview of the cases dealt. Duplication and modification of whole workflow can be done. The saved projects stored all the needed information ensuring traceability and reliable management of the models.

As an overview, in the figure 1, 2 MCNP analyses are run. Both analyses shared the same material libraries and 2D geometry. The review of such studies where identical portions of the model are used for different cases is speeded up.

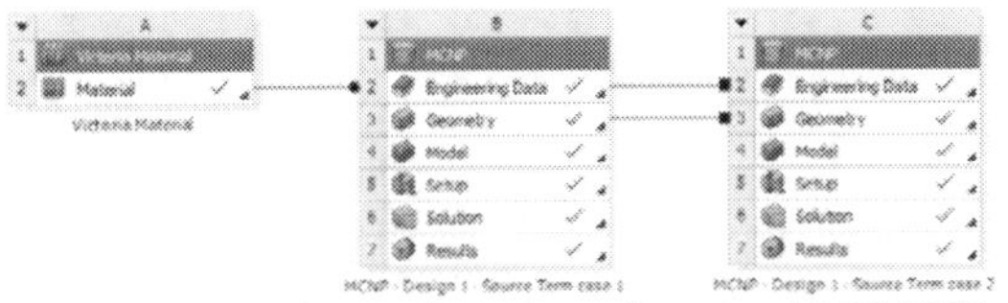

Figure 1. Screen shot on the project schematic with 2 MCNP analyses

At any steps of the workflow definition the user can modify upstream parameters. For example, the modification of geometrical property (modification of a thickness, addition of parts…) will be automatically taken into account into the downstream parameters (update of material assignment, and source or tally properties for example). This feature is perfectly adapted for design phase. Indeed sensitivity analyses, or design modifications can be easily done. All the unchanged parameters will be inherited from the previous analysis. So the engineer on the basis of a workflow can efficiently study alternative designs.

Also as pictured on figure 2, the user can convert workflow from one solver to another. This means that all objects (a material, a geometry, a source or a tally defined on a part of the model) can be addressed either in the MCNP or TRIPOLI syntax.

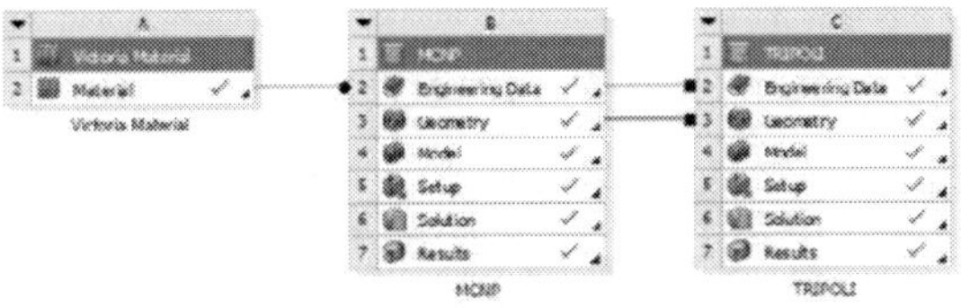

Figure 2. Screen shot on the project schematic – MCNP and TRIPOLI cases

Figure 2 shows a project with 2 analyses. Here the MCNP and TRIPOLI cases shared the same 3D geometry meaning that the geometry built is translated both in MCNP and TRIPOLI syntax.

In order to convert each objet of the workflow either in MCNP or TRIPOLI syntaxes each attributes have the capability to be addressed to the dedicated syntax. This bidirectional expression is doable because of the relative similarities in the two solvers. For a large majority of MCNP keywords there is its equivalent in the TRIPOLI syntax. If a specific attribute exists in solver without any equivalent in the other, the attributes is exposed in the Graphical User Interface (GUI) only in the frame of the adequate workflow and must be filled by the user. This is particularly notable for the simulation parameters where each solver has its own logic. In order to smooth the experience of the user default values are set corresponding to typical parameters for transport calculations. If needed these values can be redefined by the user thanks to the GUI.

2.3.2 Material Database

A dedicated GUI has been developed in order to create material databases useable for the radiation

 DOI: 10.1051/epjconf/201715306016

solvers. Composition (chemical distribution or Plutonium vectors for example) and density are defined by the user for each material. Routines are implemented in order to compute mass density from atomic concentration or vice versa. The mixture of material is also allowed with the automatic computation of the resulting density and composition. Temperature dependent cross section and thermal treatment options are also callable. Natural abundance of elements are stored to aid the material definition for neutron problems.

The dataset structure built to manage these databases allows the storage and the sharing between several users. This ensures traceability and quality.

2.3.3 Geometry and CAD integration

The surfaces allowed by Monte-Carlo solvers are reduced to basics shapes: planes, cylinders, sphere, cones, torus, ellipsoids, hyperboloid and paraboloid. Other surfaces as splines are discarded. The algorithm embedded within VICTORIA converting 3D bodies to Monte-Carlo geometry syntaxes deals with the main useful shapes: planes, sphere, cylinder, cone and torus. Each of these surfaces is also correctly written when coordinates transformations are applied (translation and/or rotation). In the case of torus expression for MCNP the transformations are limited, as allowed by the solver, to transformations parallels to the axis of the global system.

The geometry creation and preparation are the main time consuming step. As described above the geometry creation can be done by direct modeling and / or CAD or MCNP/TRIPOLI file import.

The challenge of the geometry preparation is to get a set of bodies whom the shapes comply with the surfaces allowed and the limitations of the conversion algorithm. This is particularly true when CAD files are used. Generally CAD files are not built for radiation purposes and may contains spline surfaces, complex bodies and irrelevant details for radiation aspect. It is required to limit the complexity of the bodies in terms of the number of surfaces and ambiguity surfaces involved (ambiguity surfaces are surfaces which are not part of the external boundaries of a solid but are needed to decompose spatially the body in a set of non-overlapping shapes). To achieve this difficulty, the user is assisted by several tools added to the 3D modeler such as:

- detection of unsupported surfaces,
- capability to replace a surface by a supported one,
- detection of unsupported bodies,
- spatial decomposition of complex bodies.

Also because the Monte Carlo geometry must cover any coordinate points of the system, the addition of internal or enclosure volumes inside and around equipment can be built in the 3D modeler. The same approach in term of preparation (spatial decomposition) is also required for these volumes.

Figure 3 presents how the CAD model of a steam generator is prepared by deleting useless details and unsupported surfaces (e.g. blended surfaces).

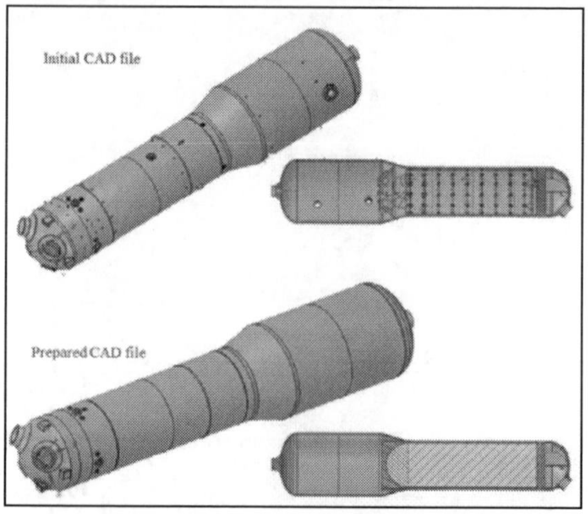

Figure 3. Example of geometry preparation

Tools have also been integrated to deal with lattice structures. A common add-in to the 3D modeler GUI is used to generate both MCNP and TRIPOLI lattices for Cartesian and hexagonal pattern.

VICTORIA developments also address the latest features of MCNP6. Notably MCNP6 permits to describe geometries with Unstructured Meshes. Instead of building the geometry based on traditional surfaces (or macro bodies), the bodies are defined as a set of tetrahedrons. This offers the capability to convert any kind of CAD shapes to MCNP6.

Figure 4 presents an extensive test case of this feature applied to model the entire quarter of a reactor building. More than 400 000 tetrahedrons are used in this model. However such technic has inherent drawbacks as an increase of the computer runs of a factor of ~10 for equivalent configuration modeled with traditional surfaces. However this functionality prefigures what could be in a next future a more global approach to deal with CAD integration for Monte Carlo simulation and demonstrates the versatility of the software using mesh algorithms available in the ANSYS platform.

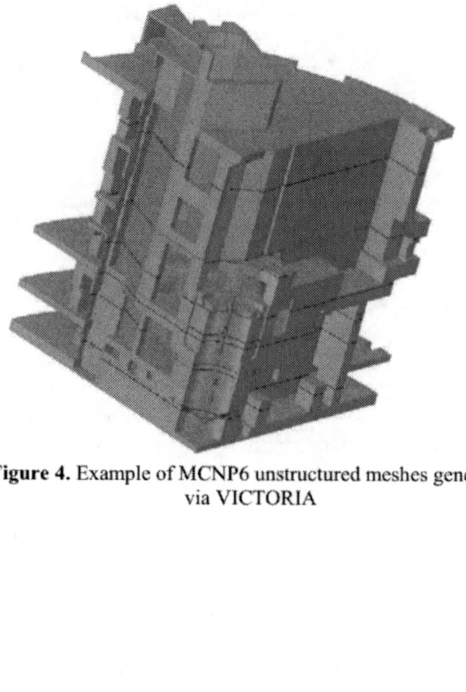

Figure 4. Example of MCNP6 unstructured meshes generates via VICTORIA

2.3.4 Model and setup definition

In this step of the workflow the user can:

- assign the materials to volumes,
- set the simulation parameters (mode, termination criterion…),
- define the sources,
- define the tallies.

Figure 5 presents a view of a PWR model with material assignment.

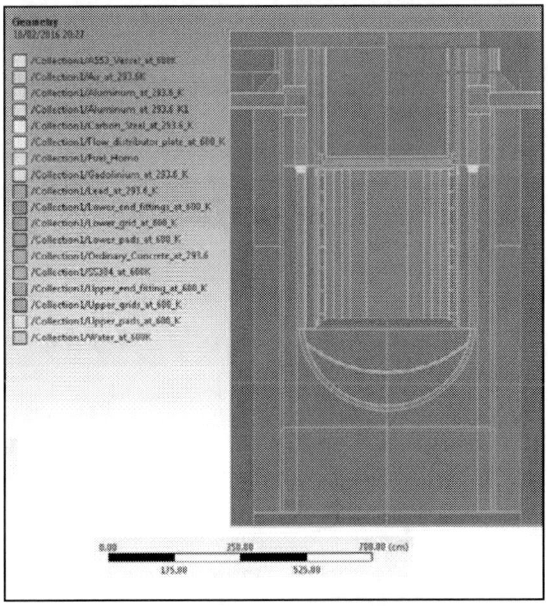

Figure 5. Example of a PWR model generated with VICTORIA

The sources can be defined for both neutron and gamma problems. The shape of sources can be a set of points, on surfaces or on volumes (cells rejection).

The surface sources definition is particularly used to study problems with contamination on systems. The locations of the sampling of source particles on the surfaces are determined by a specific routine.

For volume sources, the syntax generated is a combination of sampling volumes associated with cell rejection. The algorithm embedded within VICTORIA determines the shape and the size of these elementary volumes (box, cylinder or sphere) in order to limit the waste of computer time for cell rejection.

All types of tallies can be defined in the GUI: point, surface, cell tally and also meshed tallies: structured tallies (FMESH and TMESH cards of MCNP and extended mesh of TRIPOLI) and Unstructured Meshes (EMBEE card of MCNP6). Response functions (flux, dose, reaction rate) and energetic bins are also callable.

At the end of this step the solver input file is exported in a user defined directory. The solver run is not part of the VICTORIA workflow. The run is done according to the user preference (linux or windows) and the cluster availabilities.

2.3.5 Post processing

Post-processing features allows the parsing of solver output files to display 3D rendering and edit Excel sheets of results. Figure 6 is such an example of neutron mesh tally superimposed on the lower part of the reactor cavity pit. All the meshed tallies of the solvers (EXTENDED mesh for TRIPOLI and FMESH, TMESH and EMBEE for MCNP) can be displayed, 3D rendering capabilities are inherited from the mature finites elements tools (iso-lines, clipping, probe, etc…).

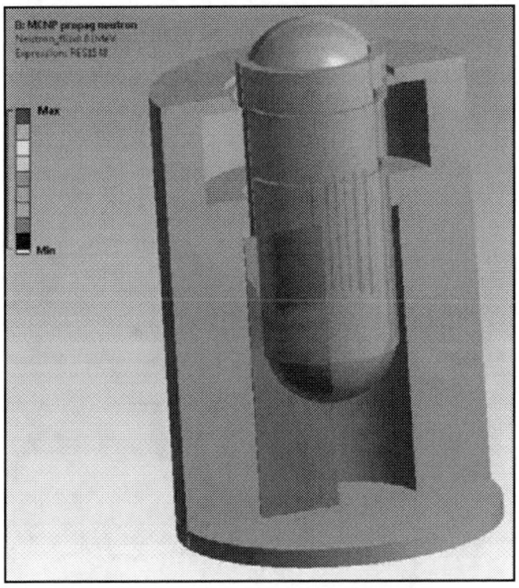

Figure 6. Example of 3D mesh tally rendering

The generation of CSV files or mesh based file format (VTK) is also commonly done to simplify the interface with other tools or engineering fields. This is discussed in section 4.

3 Acceleration of calculations

Efforts have been done to decrease the computer running time of the calculations. In addition to the use of parallel runs, the use of hybrid variance reduction technics is deployed for MCNP usage. The term hybrid is used because of the methods imply the use of deterministic code prior to the Monte Carlo run. For transport calculation the CADIS method [3] is used to generate MCNP Weight Windows input files.

More precisely CADIS method is used to improve the streaming of particles at one specific location. In that case an adjoint calculation is run with a SN 3D deterministic code [4] on an equivalent model of the MCNP case. The Weight Windows values of the user defined mesh are inversely proportional to the adjoint flux distribution of a point source located at the target coordinate.

For problems requiring to reach convergence criteria in several locations the Forward Weighted CADIS method is used. In that case a forward calculation is firstly run to assess the adequate strengths of the adjoint sources.

The FW-CADIS method has demonstrated its efficiently on a large set of study case both from neutron and gamma for several applications (shielding, fluence, energy deposition, activation). The figure below shows an example of fast fluence calculation on a PWR model. In that case, the suggested area of convergence is the entire model (from the fuel active part to the concrete cavity pit). Both traditional and extended beltline area are calculated simultaneously. For this demonstrative case the size of the mesh tally voxel is 8x8x19cm^3. On that example the fast fluence in the extended area (nozzle region) is calculated with a relative error of 2% after 2 hours of calculation and 0.7% after 18 hours of calculation with 7 tasks at 2.93GHz.

Figure 7. Example of MCNP fluence calculations on extended beltline area

Developments have also been done in order to run CADIS methods on Unstructured Mesh syntax of MCNP6. On figure 8, the FW-CADIS method has been applied to the model presented on figure 4 in order to calculate the neutron flux distribution in the concrete civil work. Reaching simultaneously convergence criteria in several locations is particularly useful for activation analyses.

Hybrid technics offer a global approach and ready to use for the engineering teams. The main difficulties remain on the choice of an appropriate mesh especially while dealing with small penetrations.

Figure 8. Example of the calculation of neutron flux in the civil work of a whole reactor

4 Chaining Monte-Carlo simulations with external software

As the result of the easy creation of large set of tallies on 3D structures with VICTORIA, the Monte Carlo results can be easily given as input data to external software. This section illustrates two applications of chaining calculations.

4.1 Chaining Monte Carlo simulation results with finite elements analyses

In this application VICTORIA is used to generate a MCNP input files for energy deposition (neutron and gamma) calculations on the reactor internals. This study case is the heavy neutron reflector surrounding the EPR^{TM} core. The heavy neutron reflector (HNR) is a thick steel equipment drilled by cylindrical holes to allow the flow of water coolant. The challenge of this analysis is to calculate the energy deposition on the steel part and on a 3D user defined mesh. The geometry of the heavy neutron reflector is so built by a CAD import, then the part is sliced along a Cartesian pattern in order to define an energy deposition tally on each resulting piece of reflector. Figures 9 and 10 show this process until the integration within the whole MCNP model of the core.

Figure 9. Spatial decomposition of equipment for energy deposition calculations

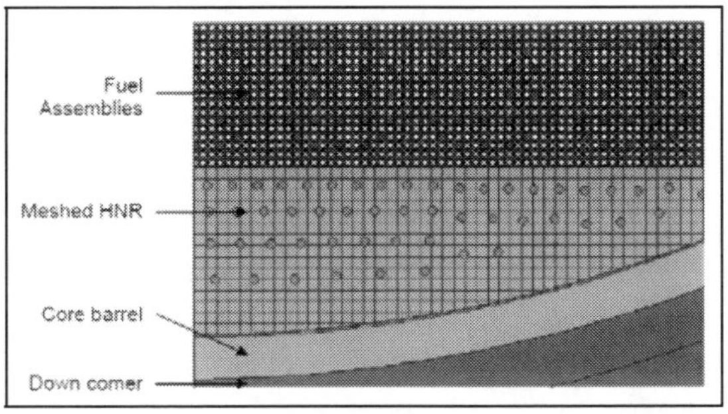

Figure 10. Integration of the meshed equipment in the MCNP model

The MCNP results of each tallies are then post-processed to a CSV file containing the coordinates values of the center of each cells/tallies and the corresponding energy deposition value.

Such file format is readable by the main finite elements software packages to run thermomechanical analyses.

Such workflow is presented in figure 11 for the ANSYS Workbench software where the energy disposition is loaded as an external data of heat generation (W/cm^3).

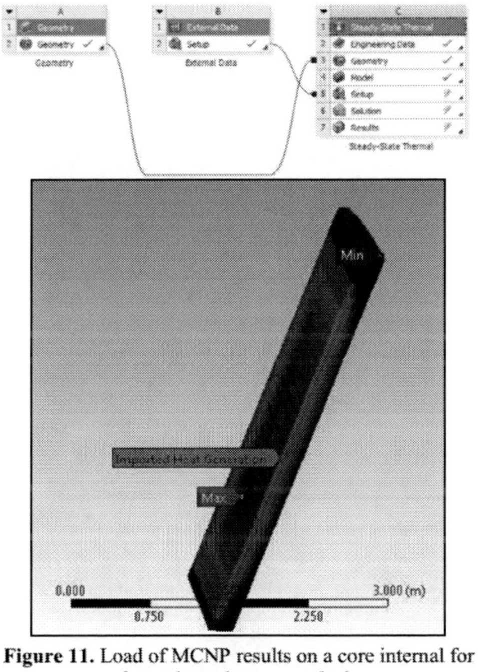

Figure 11. Load of MCNP results on a core internal for thermal steady state analysis

4.2 Chaining with depletion codes for activation analyses

Another kind of application requiring an efficient management of MCNP data results is the use of neutron spectrum calculations for activation assessment. In the context of the dismantling of reactors the prediction of activities inventory of structures surrounding the core is one of the key driver of the cost of the dismantling. Indeed the activities inventory is the input for waste classification assessment, doses evaluation leading to define the decommissioning scenario and schedule and related safety analyses. The AREVA NP methodology is based on Monte-Carlo 3D transport and is being used to support the dismantling of the French reactor fleet [7].

Once the neutron spectrum is calculated during power operation on a user defined mesh the data are post-processed to a VTK file format structure permitting storage and 3D visualization.

Based on the detailed material composition and power history an activation and depletion calculation (DARWIN or SCALE) is executed on each voxel. The results scalars fields are nuclide inventories, waste classification indicator, gamma source terms on user specified decay time steps.

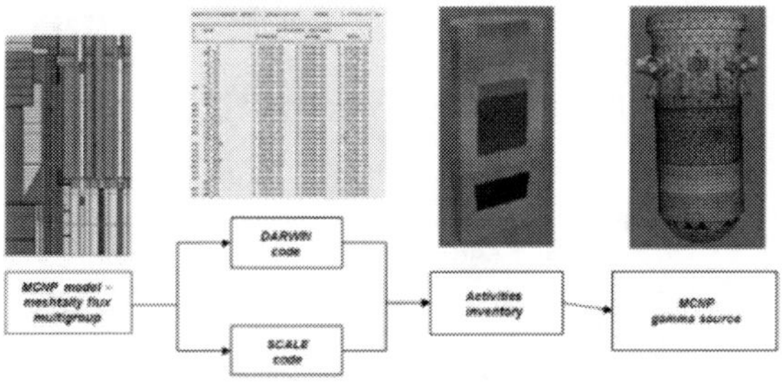

Figure 12. Illustration of the coupling between MCNP flux calculations and activation/depletion calculation codes

This chaining avoids human errors in the management of these huge quantities of data (storage of neutron spectrum and history, creation of depletion calculation code inputs, post-treatment of extraction of inventories quantities). The activation gradient can so be calculated on an adequate 3D mesh avoiding considering penalizing values for all the equipment/concrete solid. More generally having an automate workflow expedites modeling reliable and high fidelity evaluations.

5 Conclusions

This paper presents recent progress of AREVA developments. The suite of engineering tools has been drastically upgraded in order to offer Monte Carlo calculations as a reliable and efficient solution for radiation propagation projects. The main development is the VICTORIA software (Various Instances CAD calculation TOol and Radiation-propagation Interface by AREVA) offering CAD integration for both MCNP and TRIPOLI solvers. The calculations running times have been decreased by using modern technics. The use of Monte Carlo solvers is widespread from conceptual design phase [5, 6] to decommissioning [7] for any kind of reactor fleet or nuclear facility.

The versatility of the tools developed enables prompt answer to the needs as:

- New plant design and life extension analyses (Fluence, heating, sensors responses and

embrittlement of equipment surrounding the core),
- Radiation protection studies and unpredicted radiological hazards,
- Support to Dismantling and Decommissioning for waste characterization, inventories and dose to operators calculations.

Monte Carlo methods are now much more affordable for projects avoiding simplifications and conservatisms of simpler methods. Monte-Carlo 3D with continuous energy analysis are superseding previous approaches where 2D models with multi group energy bins were applied.

Future developments are planned for the VICTORIA software to improve the user experience with new features.

References

1. F. Brown, B. Kiedrowski, J. Bull, "MCNP5-1.60 Release Notes", LA-UR-10-06235, (2010)
2. E. Brun, F. Damian, Eric Dumonteil, "TRIPOLI-4 version 8 User Guide, CEA-R-6316", (2013)
3. J. C. Wagner, E. D. Blakeman, D. E. Peplow, "Forward-weighted CADIS method for variance reduction of Monte Carlo calculations of distributions and multiple localized quantities", *in International Conference on Mathematics, Computational Methods & Reactor Physics – M&C 2009*, Saratoga Springs, N.Y., (2009)
4. Th. M. Evans et al. "Denovo: a new Three-dimensional parallel discrete ordinates code in scale", *Nuclear Technology vol. 171,* (2010)
5. N. Chapoutier, A. Janet, G. Nolin, "ASTRID Core Shielding – Design Studies and Benchmark Analysis", *Proceedings of ICAPP 2015, Paper 15305*, (2015)
6. C. Jammes, N. Chapoutier, P. Filiatre, "Neutron flux monitoring system of the French GEN-IV SFR: assessment of diverse solutions for in-vessel detector installation", Nuclear Engineering and Design; **v. 270**, (2014)
7. M. Culioli, N. Chapoutier, S. Barbier, S. Janski, "State of the art of Monte Carlo technics for reliable activated waste evaluations", *International Symposium on PREparation for DECommissionning*, Lyon (2016)

JMCT Monte Carlo Simulation Analysis of BEAVRS and SG-III Shielding

Deng Li [1,2*], *Li Gang* [1,2], *Zhang Baoyin* [1,2], *Shangguan Danhua* [1,2], *Ma Yan* [1,2], *Hu Zehua* [1,2], *Fu Yuanguang* [2], *Li Rui* [2], *Shi Dunfu* [2], *Hu Xiaoli* [2], *Wang Wei* [1,2]

1 Institute of Applied Physics and Computational Mathematics (IAPCM), Beijing, China

2 CAEP Software Centre for High Performance Numerical Simulation(CAEP-SCNS), Beijing, China

Abstract. JMCT is a general purpose Mont Carlo neutron-photon-electron or coupled neutron/photon/electron transport code with a continuous energy and multigroup. The code has almost all functions of a general Monte Carlo code which include the various variance reduction techniques, the multi-level parallel computation of MPI and OpenMP, the domain decomposition and on-fly Doppler broadening, etc. Especially, JMCT supports the depletion calculation with TTA and CRAM methods. The input uses the CAD modelling and the calculated results use the visual output. The geometry zones, materials, tallies, depletion zones, memories and the period of random number are enough big for suit of various problems. This paper describes the application of the JMCT Monte Carlo code to the simulation of BEAVRS and SG-III shielding model. For BEAVRS model, the JMCT results of HZP status are almost the same with MC21, OpenMC and experiment. Also, we performed the coupled calculation of neutron transport and depletion in full power. The results of ten depletion steps are obtained, where the depletion regions exceed 1.5 million and 120 thousand processors to be used. Due to no coupled with thermal hydraulics, the result is only for reference. Finally, we performed the detail modelling for Chinese SG-III laser facility, where the anomalistic geometry bodies exceed 10 thousands. The flux distribution of the radiation shielding is obtain based on the mesh tally in case of Deuterium-Tritium fusion reaction. The high fidelity of JMCT has been shown.

1 INTRODUCTION

The high fidelity particle transport system JPTS(J Particle Transport System) has been developed for simulation of reactor full-core and shielding. This package is developed based on the three support framework JASMIN[1], JAUMIN and JCOGIN[2], where JASMIN is an adaptive structured mesh infrastructure, JAUMIN is an adaptive unstructured mesh infrastructure and JCOGIN is a parallel combinatorial geometry infrastructure. The JPTS package can do large scale parallel computation. Here, we mainly introduce a general purpose 3-D Monte Carlo transport code JMCT(J Monte Carlo Transport)[3]. Two models are chosen as test examples, where model one is the BEAVRS which is permitted by MIT Computational Reactor Physics Group in M&C2013 conference[4]. Another model is from the Chinese SG-III laser facility.

For BEAVRS model, we finish the detail modelling and simulation for the full core in hot zero power(HZP) status. 95% pins have less 1% standard deviation. The detailed pin-power density distribution, standard deviation etc. are shown. Then, we make the coupled calculation of neutron transport and depletion in full power status. Due to the memory resume too large, the simulation is done in case of 30/398 axial plane, where the depletion regions exceed 1.5 million. The simulation uses 120 thousand processors in Chinese TianHe-II supercomputer. Since the BEAVRS model involves the coupled with thermal hydraulics. At present, JMCT isn't coupled with thermal dynamic (being done), so the result is only for reference.

For SG-III model, the geometry is very complicated and irregular; the total geometry bodies exceed ten thousand. We order built some special body, such as optical instrument. The neutron and photon flux distributions of all building are given in case of the deuterium-tritium fusion reaction. The result has been used for the theory evidence of shielding design.

2 INTRODUCE OF JPTS PACKAGE AND JMCT MONTE CARLO CODE

2. 1 JPTS Package

JPTS package is developed by IAPCM. It contains the four applied codes JNuDa, JSNT, JMCT, JBURN and a suit of data libratory NuDa. Furthermore, the CAD pre-processor JLAMT and view post-processor TeraVAP are equipped(see Figure1). Where:

* Corresponding author: deng_li@iapcm.ac.cn
This work was performed under the National Energy Administration of China (2015ZX06002008) and the Technology and Industry for National Defense (C1520110002).

DOI: 10.1051/epjconf/201715306017

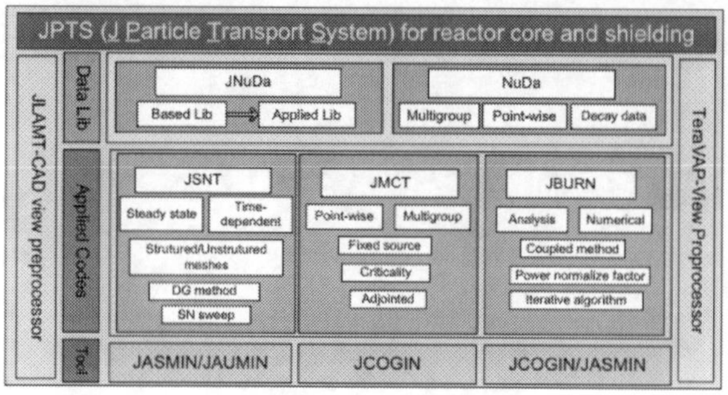

Fig. 1. JPTS package flow

Currently, JSNT S_N Code Support [5,6]

1) **Particle type**: neutron, photon or coupled of neutron and photon.
2) **Mode**: forward/adjoint.
3) **Problem**: fixed-source/criticality.
4) **Space**: 3D/2D, Cartesian /cylindrical geometry.
5) **Mesh**: non-uniform structured (JSNT-S)/unstructured mesh (JSNT-U).
6) **Energy**: multigroup, anisotropic P_N scattering (N=1, 3, 5).
7) **Parallelization**: massive parallel computing (space-angle parallelization).
8) **Algorithm**: acceleration algorithms (rebalance methods, multigrid methods).
9) **Input**: visualization modelling and automatic mesh generation, multiple choice of spatial discretion (TWD, DFEM ...).
10) **Output**: visualization analysis.
11) **Support Framework**: JASMIN[1].

Currently, JBURN Burnup Code Support

1) **Analysis**: Transmutation Trajectory Analysis (TTA).
2) **Numerical**: Chebyshev Rational Approximation Method (CRAM).
3) **Mode:** inner coupled with JMCT.
4) **Depletion regions:** >1 millions.
5) **Support Framework**: JCOGIN[2].

Currently, JNuDa Cross-Section Code Support

Continuous point-wise/Multigroup/Decay Data, where

1) Point-wise cross-section about 450 nucleus.
2) Multigroup library of 47/172 group for neutron and 20 groups for gamma.
3) Decay data (>1500 elements).

Currently, JLAMT Pre-processor Support[7]

1) **Geometry:** sphere, cylinder, rectangle, et al., some special geometrical body can be ordered.
2) **Repeat structure:** especial supporting the same geometry with the different material.

Currently, TeraVAP Post-processor Support

Scale: TB scale data and parallel visualization output.

2.2 JMCT Monte Carlo Code[3]

JMCT is a general purpose 3-D Monte Carlo transport code of neutron, photon, electron or coupled neutron/photon/electron with the combinatorial geometry. Currently, JMCT support:

1) **Particle type**: neutron, photon, electron and coupled neutron/photon/electron.
2) **Mode**: forward/adjoint/burnup.
3) **Problem**: fixed-source/criticality.
4) **Space**: 3D combinatorial geometry.
5) **Energy**: continuous/multigroup (P_5).
6) **Source**: standard source/pin-by-pin source/user defined source.
7) **Tally**: point/surface/cell/mesh.
8) **Algorithms**: domain decomposition[8]/uniform tally density[9] /mesh tally and mesh windows. etc.
9) **Parallelization**: MPI (particle)+OpenMP (domain).
10) **Input/Output**: CAD modelling and visualization.
11) **Temperture**: on-fly Doppler broadening[10].
12) **Fast critical search of boron concentration:** only one step.
13) **Tally types**: keff, point/surface/volume flux, energy deposition, power and reactivity etc.
14) **Support Framework**: JCOGIN[2].

3 TESTS

3.1 BEAVRS Model

3.1.1 Introduction of Model

The BEAVRS model was released by MIT Computational Reactor Physics Group in July 7, 2013 (www.crpg.mit.edu). It includes detailed specification of operating 4-loop Westinghouse PWR (3411MW), two cycles of measured data, HZP/full power data, fuel loads by assembly as built, three enriched fuels(1.6%, 2.4% and 3.1%). The detailed data is in reference[4]. Two cycles of measured data can be used to validate high-fidelity core analysis codes. The basic data is as following:
Fuel assemblies: 193.
Axial planes: 398.
Pins/assembly: 289(17×17, where 264 fuel pins and 25 guide tubes).
Total tally regions: 22,199,246 (193×17×17×1×398).
Total regions: 44,398,492 (193×17×17×2×398).
Requirement: ≤1% standard deviation for 95% fuel pin-powers.
The part results of MC21[11] and OpenMC[12] were presented in PHYSOR2014[13].

3.1.2 Simulated Results

(1) Results in HZP status

HZP status is simulated in 398 layers in axial direction. Due to the memory exceeding the limit of single core, the space domain is decomposed into 8 parts(figure 2(f)). Eight pin types(figure 2(a)) and nine types of assembles(figure 2(b)). The tally was for all pin fuel regions, the simulation tracks 4 million neutrons each cycle, it discards 400 cycles of 1000 cycles. Table 1 shows the standard deviation distribution in 95% confidence level for all of pins. Table 2 shows the keff comparison of JMCT, OpenMC and MC21 in different location of control rods and boron concentrations. Table 3 shows the reactivity worth of control rods in 556K. Figure 3 shows the comparison of pin-power distribution, difference at axial elevation of peak power and the comparisons of the MC21 and JMCT powers in axial. The maximal difference is 3.17%. Figure 4 shows the detectors tallies in meter pipes between JMCT and experiment. The maximal difference is -14.77% in B13 assemble and minimum power assemble is -5.648% in L15 assemble. Figure 5 shows the axial power shape of the B13(maximal difference) and L15(minimum power assembles) between MC21 and JMCT as compared to experiment.

(a) Pins (eight types)

(b)assembles (nine types)

(c)grid

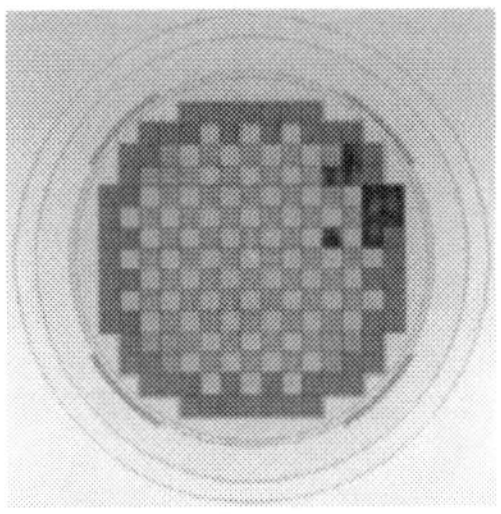

(d)core in radial

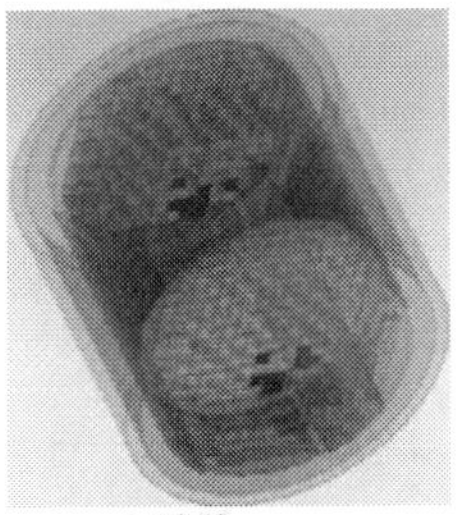

(e)full core

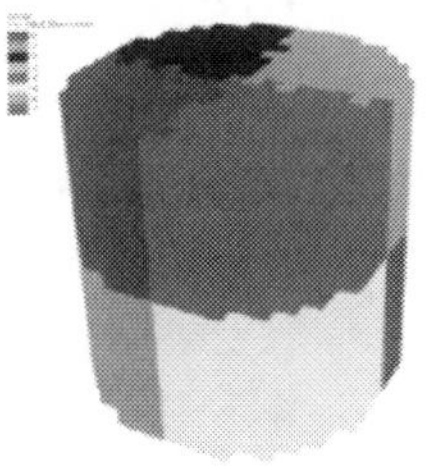

(f)2×2×2 domain decomposition

Fig.2. BEAVRS modelling by JLAMT

Table 1. Max and min pin error of flux and energy deposition

Count	MAX	MIN	95%	99%
Flux	0.0091	0.00118	<0.00332	<0.00423
Energy deposition	0.01933	0.00254	<0.0075	<0.00955

Table 2. keff comparison in different control rod statuses and boron concentration

HZP Critical Boron Evaluation	Boron Concentration	JMCT (95% confidence leave)	OpenMC (95% confidence leave)	MC21 (95% confidence leave)
ARO	975	1.000479±0.000030	0.99920±0.00004	0.9992614±0.000004
D in	902	1.002174±0.000030	1.00080±0.00004	
C,D in	810	1.001419±0.000032	1.00023±0.00005	
A,B,C,D in	686	0.9999172±0.000032	0.99884±0.00004	
A,B,C,D,SE,SD,SC in	508	0.9983806±0.000032	0.99725±0.00004	

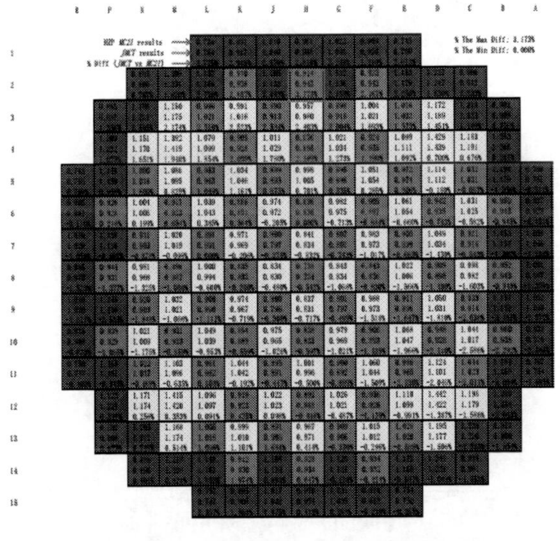

Fig. 3. Comparison of pin power distribution and difference at axial elevation of peak power between MC21 and JMCT

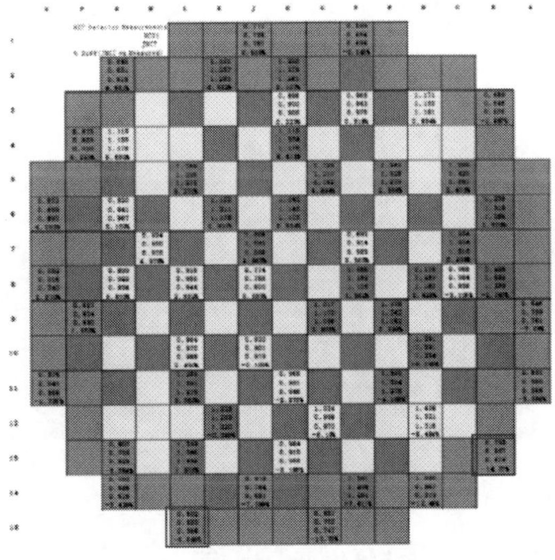

Fig. 4. Comparison of the detectors tallies in meter pipes

Table 3. Comparison of reactivity worth of control rod in different statuses and boron concentration

HZP Bank worth	Boron	Measure	MC21	OpenMC	JMCT
D	938.5	788	773	771±6	770±6
C with D in	856	1203	1260	1234±7	1258±6
B with D,C in	748	1171	1172	1197±7	1162±6
A with D,C,B in	748	548	574	556±6	578±6
SE with D,C,B,A in	597	461	544	501±6	543±6
SD with D,C,B,A,SE in	597	772	786	844±6	781±6
SC with D,C,B,A,SE,SD in	597	1099	1122	1049±6	1107±6

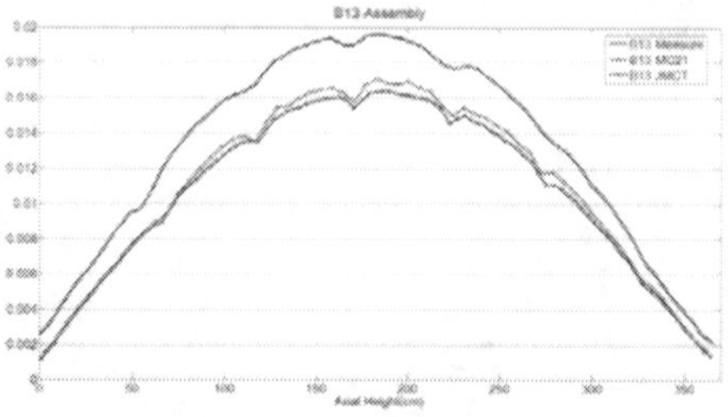

(a)axial power shape in B13 assembly
(with maximal difference)

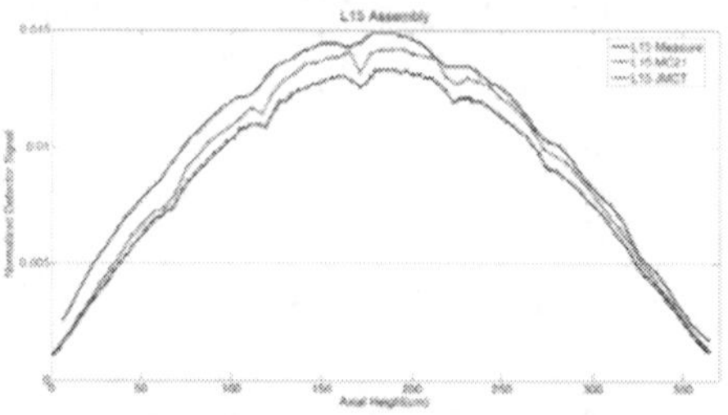

(b)axial power shape in L15 assembly
(with minimum power)

Fig. 5. The axial power shape comparison of MC21, JMCT and experiment

(2)Result in Full Power Status

The coupled neutron transport and depletion is run in 30/398 axial planes, where the depletion region is up to 1528560 (193×264×30). The space domain is also decomposed into 8 parts. It takes about one hour with 120,000 cores on Chinese TianHe-II computer. The result of the tenth steps is obtain. Figure 6 shows the power distribution of some status. Due to the JMCT no coupled with thermal hydraulics, the result is only for reference. For full simulation of BEAVRS model, due to the large memory resume, we predict that the goal need a long time and impossible before 2018.

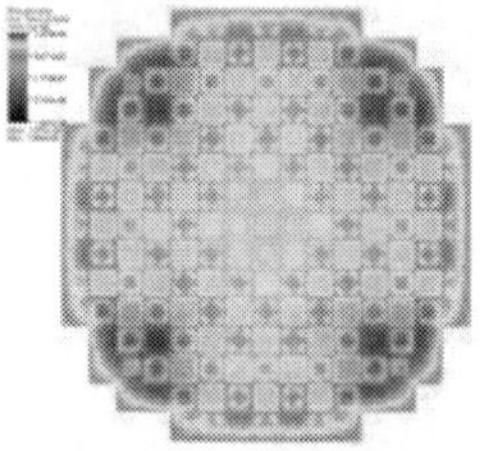

(a)pin power in radial

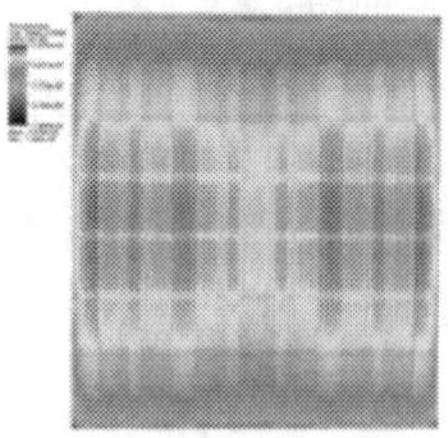

(b)pin power in axial

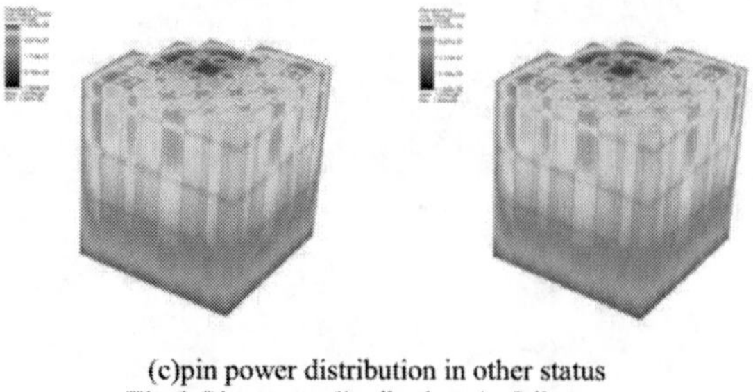

(c)pin power distribution in other status

Fig.6. Pin power distributions in full power

3.2 Chinese SG-III Laser Model

Chinese SG-III laser device is with 48 laser beams and size in 45m×45m×53m. Diameter of target chamber is 6 m. Power is 50 TW. It is applied to drive the nuclear fusion reaction by the laser energy. Figure 7 shows the modelling by JMCT pro-processor JLAMT, where figure 7(a)(b) show the locations of tally floors and figure 7(c) shows the building (six floors in ground). The tally is for all floors (seven floors in total). Mesh tally does and it has about 0.63 million meshes. The 0.4 billion neutron histories are simulated by 1024 cores. The 3.1 CPU hours are taken. Where the source is a 14.1 MeV deuterium and tritium(D-T) neutron point source. Figure 8(a)-(b) gives the neutron and photon flux distributions in the base of the fourth floor. Figure 8(c) gives the energy distributions of each floor. Figure 8(d) gives a part of flux distribution in building.

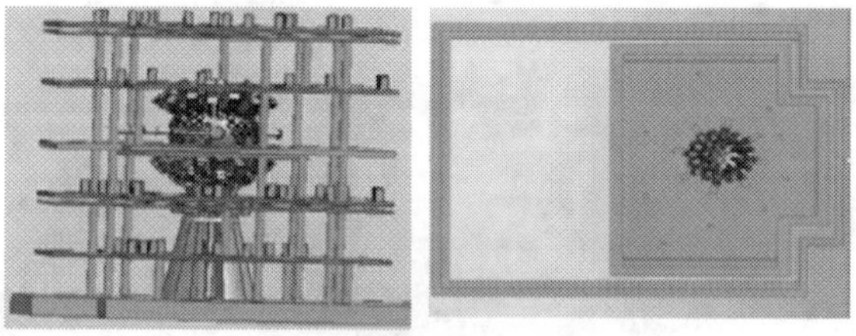

(a)location of the tally floors

(b)building of SG-III device

Fig.7. Section of SG-III laser devic

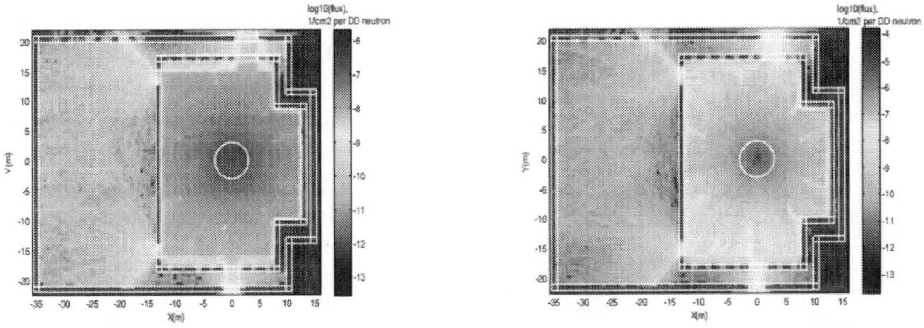

(a)neutron flux in horizontal plane (b)photon flux in horizontal plane

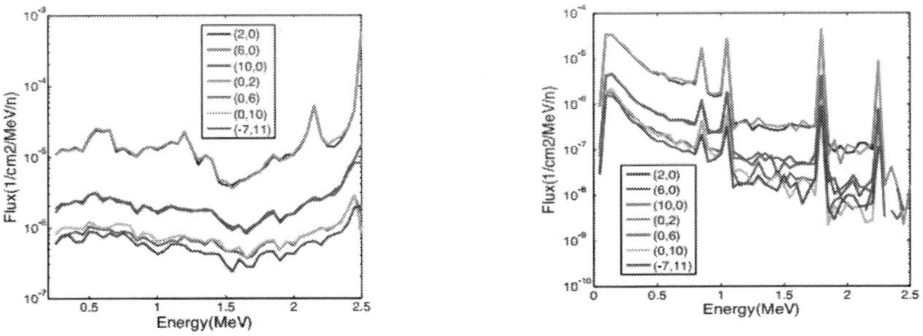

(c)neutron energy spectrum in each floor (d)photon energy spectrum in each floor

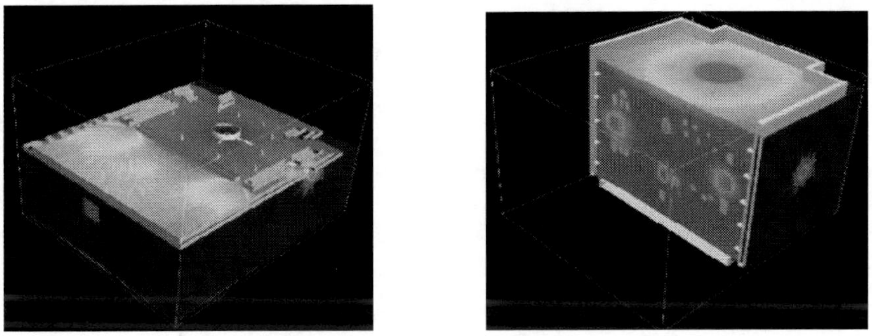

(e)flux distribution of full building (moving)

Fig.8. Flux and energy spectrums in D-T reaction for SG-III model

4 Conclusion

JMCT Monte Carlo code is developed and with the capability of the full-core pin-by-pin simulation. It well suits to simulate the large nuclear power reactors and radiation shielding of some large facilities. At present, JMCT still exist no symmetry in power distribution after several burnup steps. We are analysing the reason. On the other hand, the depletion complicating uncertainty quantification and propagation of error will be considered in our next work. Furthermore, it needs to find some new methods to reduce the computational fee. At present, some challenges still exist in simulation of the BEAVRS model. Some new algorithms are being developed.

References

1. Z. Mo, A. Zhang, X. Cao, et al. "JASMIN: a parallel software infrastructure for scientific computing". Front. Compute. Sci. China, 2010, 4(4): 480-488.
2. Baoyin Zhang, Gang Li, Li Deng, et al., "JCOGIN: A PARALLEL PROGRAMMING INFRASTRUCTURE FOR MONTE CARLO PARTICLE TRANSPORT," PHYSOR 2014 – The Role of Reactor Physics Toward a Sustainable Future The Westin Miyako, Kyoto, Japan, September 28 – October 3, 2014.

3. Li Deng, Tao Ye, Gang Li et al., "3-D Monte Carlo Neutron-Photon Transport Code JMCT and Its Algorithms", PHYSOR 2014, Kyoto, Japan, September 28 – October 3, 2014.
4. "MIT Benchmark for Evaluation And Validation of Reactor Simulations," RELEASE rev. 1.0.1, MIT Computational Reactor Physics Group, July 7, 2013.
5. Tangpei Cheng, Junxia Wei, Guangchun Zhang, Huayun Shen, Li Deng. Acceleration of Discrete Ordinates Calculations using Parallel Partial Current Rebalance Algorithm and Algebraic Multigrid Solver. The Physics of Reactors conferences (PHYSOR 2016), American Nuclear Society, 2016.
6. Tangpei Cheng, Junxia Wei, Huayun Shen, Bin Zhong, Li Deng. Development of Parallel 3D Discrete Ordinates Transport Program on JASMIN Framework. The 7th International Conference on Modelling and Simulation in Nuclear Science and Engineering, Canadian Nuclear Society, 2015.
7. Yan Ma, YuanGuang Fu, GuiMing Qin, et al, "The large-scale auto modelling tool for Monte Carlo simulations," 7th International Conference on Modelling and Simulation in National Science and Engineering, 2015, October 18-21, Ottawa, Canada.
8. G. Li, B. Zhang, L. Deng et al. "Combinatorial Geometry Domain Decomposition Strategies for Monte Carlo Simulations," M&C 2013, Sun Valley, Idaho, May 5-9, pp.434-443(2013).
9. Danhua ShangGuan, Gang Li, Baoyin Zhang, Li Deng, et al., "Uniform Tally Density–Based Strategy for Efficient Global Tallying in Monte Carlo Criticality Calculation," NUCLEAR SCIENCE AND ENGINEERING, 182(4), 555-562, 2016.
10. Liu Xiong-Guo, Deng Li, Hu Ze-Hua et al., "Study of on-the-fly Doppler broadening in JMCT Program," Acta Phys. Sin., 65(9), 092501-092505, 2016.
11. D. P. Griesheimer, et al., "MC21 v.6.0 - A Continuous-Energy Monte Carlo Transport Code with Integrated Reactor Feedback Capabilities," Ann. Nucl. Energy, in press, 2015.
12. P.K. Romano, et al., "OpenMC: A state-of-the-art Monte Carlo code for research and development," Ann. Nucl. Energy, in press, 2015.
13. Daniel J. Kelly, Bryan R. Herman, et al., "Analysis of Select BEAVRS PWR Benchmark Cycle 1 Results Using MC21 and OpenMC," PHYSOR 2014, Kyoto, Japan, September 28 – October 3, 2014.

Simplification of an MCNP model designed for dose rate estimation

Alexander Laptev[1] and *Robert* Perry[1,2,*]

[1]Los Alamos National Laboratory, P.O. Box 1663, Los Alamos, NM 87545
[2]Present: GAEA Scientific, 394 Catherine Ave., Los Alamos, NM 87544

Abstract. A study was made to investigate the methods of building a simplified MCNP model for radiological dose estimation. The research was done using an example of a complicated glovebox with extra shielding. The paper presents several different calculations for neutron and photon dose evaluations where glovebox elements were consecutively excluded from the MCNP model. The analysis indicated that to obtain a fast and reasonable estimation of dose, the model should be realistic in details that are close to the tally. Other details may be omitted.

1 Introduction

Gloveboxes are often the subject of radiological engineering design reviews. The Monte Carlo N-Particle transport code, MCNP, [1] is the primary tool used to estimate a worker's external dose from radiological activities performed in a glovebox. Coding a complete glovebox model is very time consuming. This research provides a technical analysis of the changes in the results following several simplifications of a complete glovebox MCNP model. It provides insight on how to efficiently build a simple glovebox model that adequately describes the problem resulting in considerable savings of time.

2 Original MCNP Model

The glovebox model under discussion is a complex glovebox with several layers of protective shielding (lead and polymethyl methacrylate, PMMA) and water doors underneath it. Both the front and back of the glovebox had three upper and lower workstations with a total of six workstations. Each work station had a single window and two glove ports. Three larger windows were located between the three upper and lower workstations. All the windows were made from leaded glass. The gloves used in the model were leaded. An airlock port was located on both ends of the glovebox.

The model contained four 200 gram plutonium-238 dioxide sources. It was assumed that two sources would be bare plutonium-238 dioxide and these sources were centered in each half of the glovebox. The remaining two sources were encased in steel containers and placed in opposing front corners of the glovebox.

To characterize the radiation field, three torso-sized tallies were used. The three tallies are hereby referred to as Tally #1, Tally #2, and Tally #3. Tally #1 was placed against the center of the bottom row of the middle workstation at a distance of 1 mm from the additional shielding. Tally #2 was placed in an identical location as Tally #1 with the only difference being Tally #2 was placed 30 cm away instead of 1 mm. Tally #3 was placed against the center of the bottom row of the right workstation at a distance of 1 mm from the additional shielding. Tally #1 matches the most probable worker's position. The position of Tally #3 was selected to characterize the dose rate at the right workstation (it is the also the same for the left workstation). The dose rate at the right and left workstations is expected to be higher because of the spare source is placed in each front corner of the glovebox.

The original model of the glovebox is presented in Fig. 1. The yellow box represents the concrete floor of the room. The steel walls and waters doors of the glovebox are shown in blue with red glove ports (leaded gloves) and green windows. The outer layers of additional shielding (lead, stainless steel, and PMMA) are shown in pink. One of the two bare plutonium-238 oxide sources, which is visible inside of the glovebox, is shown as a small, yellow, horizontal cylinder while one of the two contained sources is shown as a small, grey, vertical cylinder. The numbered, translucent squares are the tallies, and the wire frame represents the universe recognized by MCNP.

As mentioned, the original glovebox model was very complicated. The MCNP input geometry (cell, surface, and transformation cards) consisted of 1071 rows, and the number of lines is roughly proportional to the time spent by a developer programming. As much as $5 \cdot 10^8$ neutrons were simulated for the neutron dose rate calculation and $8.4 \cdot 10^9$ photons were simulated for the photon dose rate calculation. The CPU time required to process that many neutrons and photons, respectively, is 36,506 and 44,234 min. The estimated photon and neutron effective doses for Tallies #1 through #3 are

* Corresponding author: RTP@gaea-scientific.com

shown in Table 1. The results have a statistical uncertainty of $\leq 0.1\%$ for neutrons and $\leq 0.5\%$ for photons.

Fig. 1. Original MCNP model. The model includes the additional shielding, the tallies, the concrete floor, the four sources, all of the glovebox, and the water doors. The tallies are numbered. The left side of the glovebox and the additional shielding were removed in the figure to expose the two of the sources contained in the glovebox.

Table 1. Estimated Effective Dose Rates for Iteration 0.

Tally #	Estimated Dose Rates, mrem/hr	
	Photon	Neutron
1	0.302	4.76
2	0.226	2.75
3	0.577	9.97

3 Simplifying Iterations

The simplification of the original model was made in a series of consecutive calculations. The list of calculations along with the number of lines in the MCNP input file are presented in Table 2. At each step in the simplification process the estimated photon and neutron doses are compared to both the original model and the previous calculation.

3.1 Iteration 1: Reduced Outside World Boundaries

For this iteration, the size of the modelled air universe that surrounded the glovebox was reduced (see Fig. 2). The advantage of limiting the universe was that the CPU time was decreased. For example, for the neutrons, CPU time was reduced by 24%. The reduction in CPU time was attributed to particles that escaped the glovebox having fewer interactions outside of the glovebox.

Table 2. Simplifying iterations.

#	Description	# of MCNP lines	CPU Time, min	
			Photon	Neutron
0	Original model	1071	44,234	36,506
1	Reduced Outside World Boundaries	1064	36,234	27,850
2	Single Wall and Bottom	518	14,359	7,132
3	Single Wall	513	15,874	6,525
4	Simple Single Wall	473	11,612	5,489
5	Removed Upper Workstations	313	10,937	4,427
6	Removed Gloves	295	9,012	4,586
7	Removed Concrete Floor	290	8,285	3,221
8	Removed Water Door	278	9,384	2,972
8*	Additional Iteration: Removed Water Doors with Floor (Comparison with Iteration 6)	283	9,433	3,283

Also the MCNP input file was reduced slightly to 1064 lines by ignoring unnecessary details such as the filter case on the top of the glovebox.

The observed change in the dose rates was less than 0.3% in absolute value (Table 3). The decrease reflected less of a contribution from radiation being scattered off the air surrounding the glovebox towards the tallies. Fig. 3 presents the change in the neutron dose rates, the MCNP input file size, and CPU time observed during current iteration.

Fig. 2. MCNP glovebox model for Iteration 1: Reduced outside world boundaries.

3.2 Iteration 2: Single Wall and Bottom

This iteration simplified the glovebox model to only the front (with additional shielding) and bottom of the glovebox, the front of the water door, the concrete floor, and the sources (see Fig. 4). This iteration decreased the size of input file down to 518 lines and the CPU time to 7,132 and 14,359 minutes, for neutron and photon doses respectively.

Table 3. Estimated Effective Dose Rates for Iteration 1: Reduced outside world boundaries.

Tally #	Estimated Dose Rates, mrem/hr		Percent Change vs Iteration 0	
	Photon	Neutron	Photon	Neutron
1	0.302	4.75	-0.0	-0.1
2	0.226	2.74	-0.2	-0.3
3	0.576	9.96	-0.1	-0.1

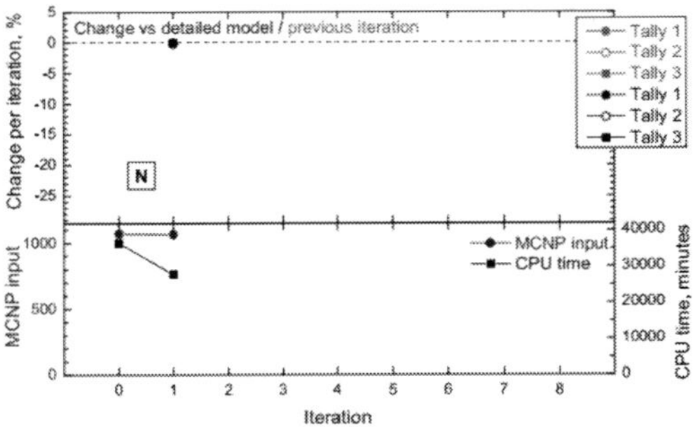

Fig. 3. The change in the neutron dose rates (upper plot) as well as the MCNP input file size and CPU time (lower plot) observed during Iteration #1.

The estimated photon and neutron effective doses and their relative change from both Iteration 0 and Iteration 1 are shown in Table 4. The change in the dose rates was noticeable, especially in the neutron dose rate, which changed approximately 10%. The 10% decrease in the neutron dose rate reflected a missing contribution from neutrons scattering off the thick additional shielding external of the glovebox towards the tallies. Although it was not as prominent, a similar effect was seen for photons. Fig. 5 presents the change in the neutron dose rates, the MCNP input file size, and CPU time observed during current iteration.

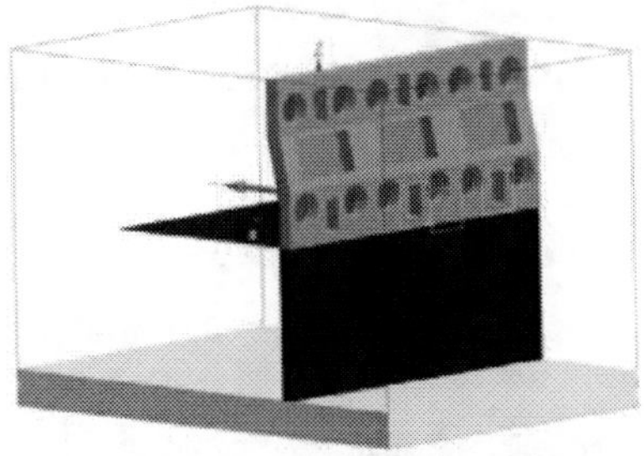

Fig. 4. MCNP glovebox model for Iteration 2: Single Wall and Bottom.

Table 4. Estimated Effective Dose Rates for Iteration 2: Single Wall and Bottom.

Tally #	Estimated Dose Rates, mrem/hr		Percent Change vs Iteration 0 / 1	
	Photon	Neutron	Photon	Neutron
1	0.297	4.30	-1.6 / -1.5	-9.7 / -9.6
2	0.220	2.38	-2.5 / -2.3	-13.5/-13.2
3	0.569	9.11	-1.3 / -1.3	-8.6 / -8.5

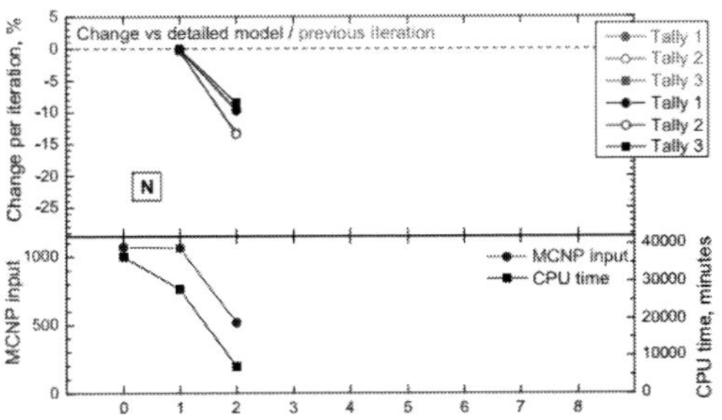

Fig. 5. The change in the neutron dose rates (upper plot) as well as the MCNP input file size and CPU time (lower plot) observed during Iteration #2.

3.3 Iteration 3: Single Wall

In this iteration, the glovebox bottom was removed from the model, leaving only the front wall (with additional shielding), the front of the water door, the sources, and the concrete floor (see Fig. 6). The size of input file was reduced to 513 lines. The CPU time was decreased by 9% for neutrons but increased by 11% for photons (compare with Iteration 2). The increase in the photons CPU time indicated that some photons, which were attenuated by the bottom of the glovebox in Iteration 2, were traced for a larger number of interactions and, therefore, more CPU time was used.

The estimated photon and neutron effective doses and their relative change from both Iteration 0 and Iteration 2 are shown in Table 5. Once again, the change in the neutron dose rate was nearly 10%. The noticeable change in the neutron dose rate reflected a missing contribution from neutrons being scattered off the bottom of the glovebox. This effect was similar to the removal of the thick additional shielding in Iteration 2.

The photon dose rate profile was altered by this simplification. The dose rate decreased in the tallies located close to glovebox (Tallies #1 and #3) and increased by more than 6% in Tally #2. The photon dose rate decrease observed in Tallies #1 and #3 was identical to the mechanism that decreased the neutron dose rate: there were no photons being scattering off the bottom of the glovebox. The photon dose rate increase observed in Tally #2 can be explained by more photons being free to scatter off the air and concrete floor since they were not attenuated by the bottom of the glovebox. Fig. 7 presents the change in the neutron dose rates, the MCNP input file size, and CPU time observed during current iteration.

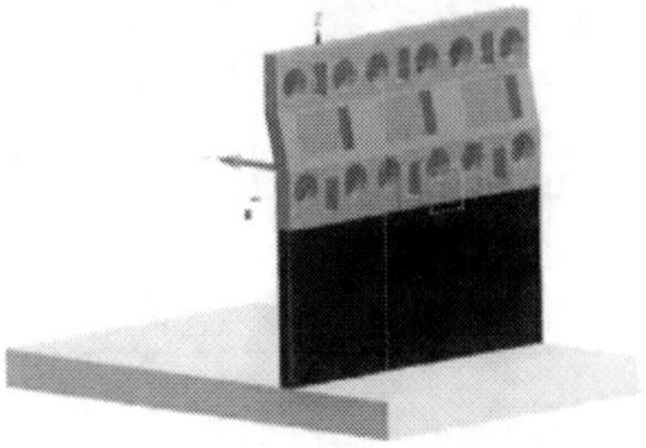

Fig. 6. MCNP glovebox model for Iteration 3: Single Wall.

Table 5. Estimated Effective Dose Rates for Iteration 3: Single Wall.

Tally #	Estimated Dose Rates, mrem/hr		Percent Change vs Iteration 0 / 2	
	Photon	Neutron	Photon	Neutron
1	0.287	3.85	-5.2 / -3.7	-19.1/-10.3
2	0.234	2.31	3.7 / 6.4	-16.1/-3.1
3	0.558	8.31	-3.3 / -2.0	-16.7/-8.8

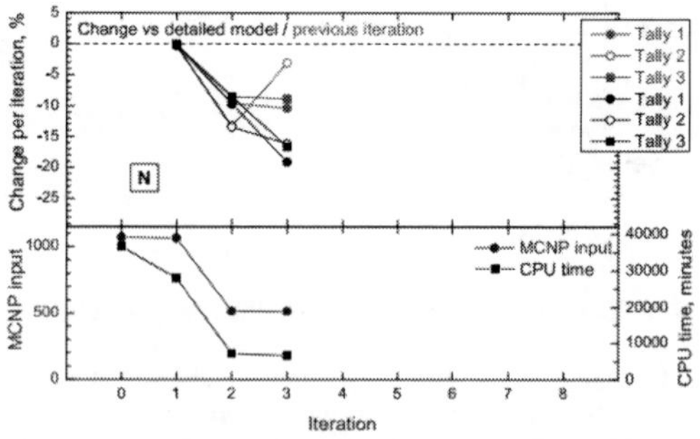

Fig. 7. The change in the neutron dose rates (upper plot) as well as the MCNP input file size and CPU time (lower plot) observed during Iteration #3.

3.4 Iteration 4: Simple Single Wall

In this iteration, the front wall was simplified by replacing the complex, angled wall with a straight, simple wall (see Fig. 8). This iteration reduced the size of input file to 473 lines, and decreased the CPU time for neutrons by 16% and for photons by 27% (compare with Iteration 3).

The estimated photon and neutron effective doses and their relative change from both Iteration 0 and Iteration 3 are shown in Table 6. The observed changes in dose rates were negligible compared to the previous iteration, allowing a conclusion to be drawn that geometrical particularities, such as a curvature, are not essential to the model. Fig. 9 presents the change in the neutron dose rates, the MCNP input file size, and CPU time observed during current iteration.

Fig. 8. MCNP glovebox model for Iteration 4: Simple Single Wall.

Table 6. Estimated Effective Dose Rates for Iteration 4: Simple Single Wall.

Tally #	Estimated Dose Rates, mrem/hr		Percent Change vs Iteration 0 / 3	
	Photon	Neutron	Photon	Neutron
1	0.286	3.86	-5.3 / -0.1	-19.1 / 0.1
2	0.233	2.32	3.3 / -0.4	-15.6 / 0.6
3	0.558	8.31	-3.3 / 0.0	-16.7 / 0.0

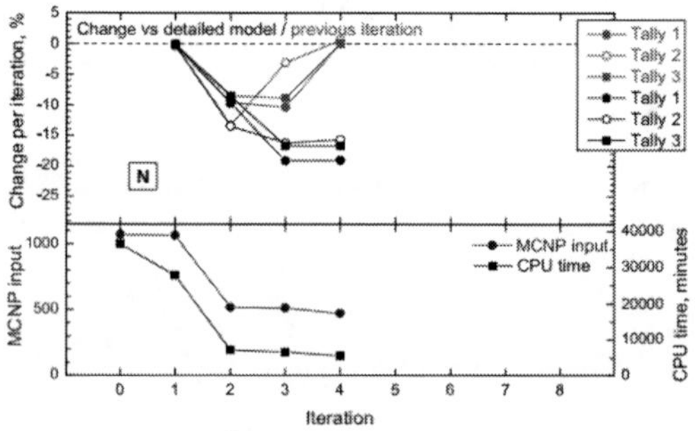

Fig. 9. The change in the neutron dose rates (upper plot) as well as the MCNP input file size and CPU time (lower plot) observed during Iteration #4.

3.5 Iteration 5: Removed Upper Workstations

This iteration removed the upper three workstations from the model (see Fig. 10). This iteration simplified the input file down to 313 lines and subsequently decreased the CPU time for neutrons by 19% (compare with Iteration 4).

The estimated photon and neutron effective doses and their relative change from both Iteration 0 and Iteration 4 are shown in Table 7. Like the previous iteration, there was little change in either the photon or neutron dose rates. This proved that the location of the tallies was sufficiently far from the upper work stations to minimize the contribution from radiation that penetrated the top of the glovebox or that passed through the upper workstations. Fig. 11 presents the change in the neutron dose rates, the MCNP input file size, and CPU time observed during current iteration.

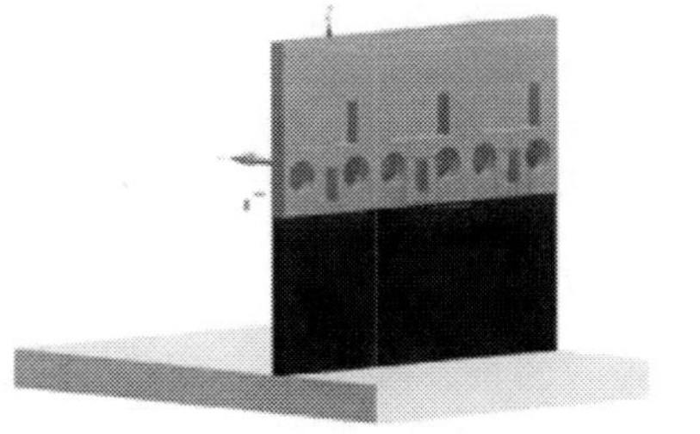

Fig. 10. MCNP glovebox model for Iteration 5: Removed Upper Workstations.

Table 7. Estimated Effective Dose Rates for Iteration 5: Removed Upper Workstations.

Tally #	Estimated Dose Rates, mrem/hr		Percent Change vs Iteration 0 / 4	
	Photon	Neutron	Photon	Neutron
1	0.286	3.85	-5.3 / -0.0	-19.0 / -0.0
2	0.232	2.31	2.9 / -0.4	-16.2 / -0.7
3	0.558	8.31	-3.3 / -0.0	-16.7/-0.0

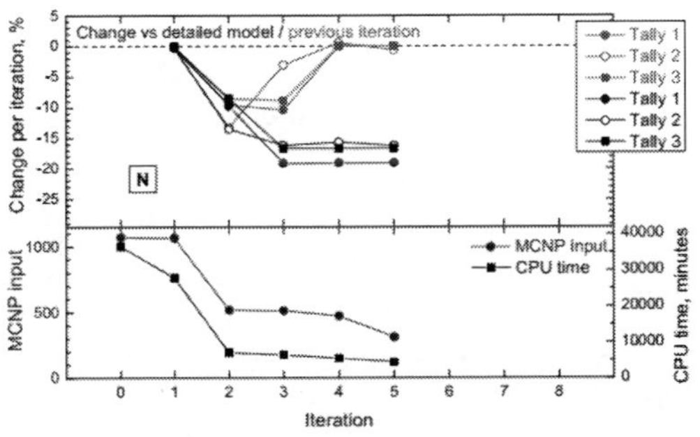

Fig. 11. The change in the neutron dose rates (upper plot) as well as the MCNP input file size and CPU time (lower plot) observed during Iteration #5.

3.6 Iteration 6: Removed Gloves

This iteration removed the leaded gloves from the lower glove ports (see Fig. 12). This simplification reduced the input file to 295 lines. The CPU time for neutrons was increased by 4% while for photons it reduced by 18% (compare with Iteration 5).

The estimated photon and neutron effective doses and their relative change from both Iteration 0 and Iteration 5 are shown in Table 8. Observations showed that the change in the neutron dose rate was minimal but the photon dose rate increased significantly. The photon dose rate increase was the result of a significant contribution of low-energy photons (energy less than 25 keV) emitted by the unshielded sources. Although the obtained result for photon dose is technically correct for the current MCNP model, it is not practical. In a real situation, those low-energy photons are easily attenuated by items like a source container, gloves, etc. Fig. 13 presents the change in the neutron dose rates, the MCNP input file size, and CPU time observed during current iteration.

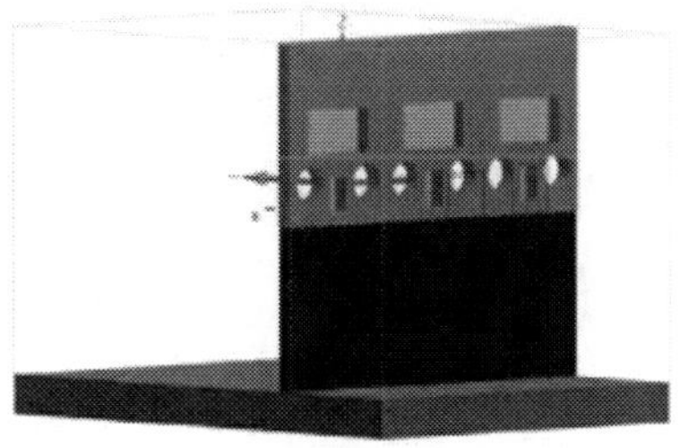

Fig. 12. MCNP glovebox model for Iteration 6: Removed Gloves.

Table 8. Estimated Effective Dose Rates for Iteration 6: Removed Gloves.

Tally #	Estimated Dose Rates, mrem/hr		Percent Change vs Iteration 0 / 5	
	Photon	Neutron	Photon	Neutron
1	0.495	3.89	64 / 73	-18.4 / 0.8
2	0.926	2.32	310 / 299	-15.8 / 0.5
3	0.759	8.31	32 / 36	-16.6 / 0.1

3.7 Iteration 7: Removed Floor

The concrete floor was removed in this iteration (see Fig. 14). After the concrete floor was removed the size of the input file was reduced to 290 lines. Without the floor, there was less neutron and photon scattering. Thus, there were less interaction events and particles escaped the universe faster. Therefore, the CPU time reduced (compare with Iteration 6). That reduction was especially significant for neutrons, by 30%.

DOI: 10.1051/epjconf/201715306018

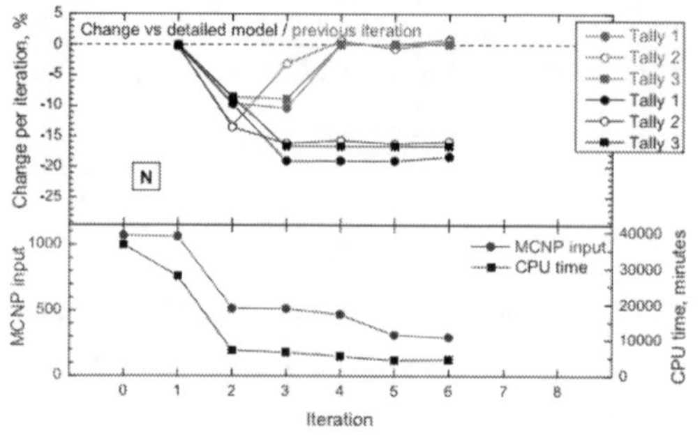

Fig. 13. The change in the neutron dose rates (upper plot) as well as the MCNP input file size and CPU time (lower plot) observed during Iteration #6.

The estimated photon and neutron effective doses and their relative change from both Iteration 0 and Iteration 6 are shown in Table 9. Removing the concrete floor produced an effect similar to the removal of the back, sides, and top of the glovebox and additional shielding in Iteration 2. The decrease in the dose rates was less than that observed in Iteration 2 because the distance between the tallies and the floor was greater than the distance between the tallies and the back, sides, and top of the glovebox and the additional shielding that was removed in Iteration 2. In addition, the magnitude of the decrease in the neutron dose rate was not as pronounced because concrete is not as effective moderator as PMMA. Fig. 15 presents the change in the neutron dose rates, the MCNP input file size, and CPU time observed during current iteration.

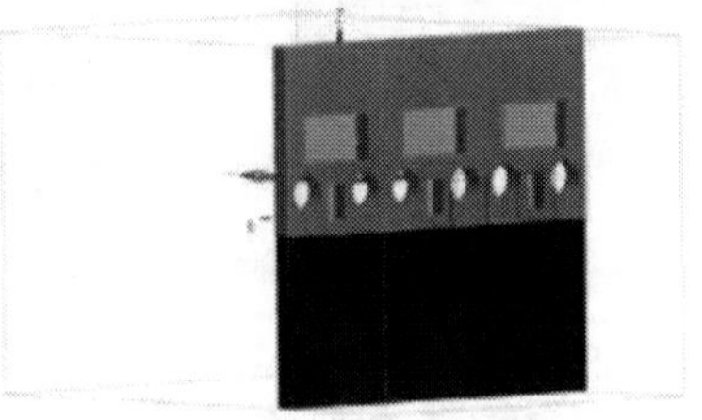

Fig. 14. MCNP glovebox model for Iteration 7: Removed Floor.

Table 9. Estimated Effective Dose Rates for Iteration 7: Removed Floor.

Tally #	Estimated Dose Rates, mrem/hr		Percent Change vs Iteration 0 / 6	
	Photon	Neutron	Photon	Neutron
1	0.491	3.75	63 / -0.7	-21.3 / -3.6
2	0.923	2.23	309 / -0.3	-19.0 / -3.8
3	0.756	8.20	31 / -0.4	-17.7 / -1.3

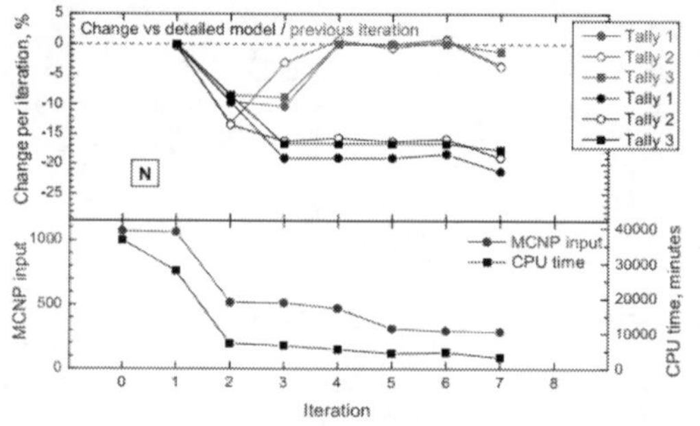

Fig. 15. The change in the neutron dose rates (upper plot) as well as the MCNP input file size and CPU time (lower plot) observed during Iteration #7.

3.8 Iteration 8: Removed Water Door

The water door was removed in this iteration (see Fig. 16). This reduced the size of the input file to 278 lines. The CPU time for neutrons was increased by 10% (compare with Iteration 7).

The estimated photon and neutron effective doses and their relative change from both Iteration 0 and Iteration 7 are shown in Table 10. There was little change in the dose rates compared with the previous iteration, except the neutron dose rate was 12.5% less in Tally #2. The decrease in the neutron dose rate in Tally #2 upon the removal of the water door was an indicator that the geometry had been over-simplified. It should be noted that although the MCNP results for this iteration are correct, they are not realistic. Both the concrete floor and the water door scattered neutrons, allowing more neutrons to be sent in the direction of the tallies. Tallies #1 and 3 were too close to have additional neutrons scattered in their direction but Tally #2 was sufficiently far enough from the glovebox to have additional neutrons scattered into it from both the concrete floor and the water door. Fig. 17 presents the change in the neutron dose rates, the MCNP input file size, and CPU time observed during current iteration.

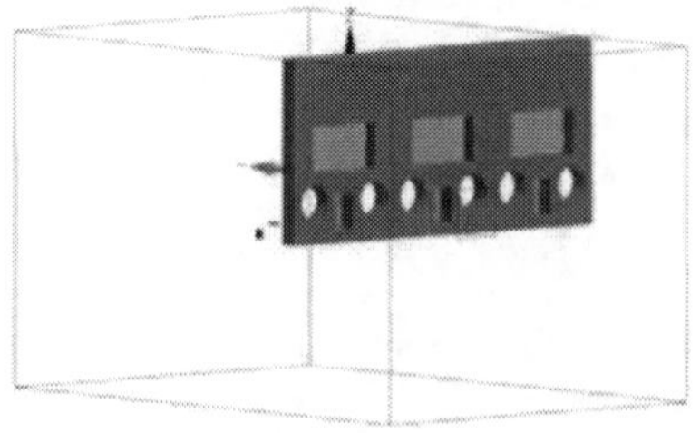

Fig. 16. MCNP glovebox model for Iteration 8: Removed Water Door.

3.9 Additional Iteration: Removed Water Doors with Floor (Comparison with Iteration 6)

In order to determine the level of influence the water door had on the estimated dose rates when there is radiation scattered on a concrete floor, the model from

Iteration 6 was taken as a starting point and the water door was removed from it. The computational model for this additional iteration is presented in Fig. 18. The size of the input file was 283 lines. The CPU time for neutrons was decreased by 28% and for photons was increased by 5% (compare with Iteration 6).

Table 10. Estimated Effective Dose Rates for Iteration 8: Removed Water Door.

Tally #	Estimated Dose Rates, mrem/hr		Percent Change vs Iteration 0 / 7	
	Photon	Neutron	Photon	Neutron
1	0.493	3.69	63 / 0.5	-22.4 / -1.5
2	0.931	1.95	312 / 0.8	-29.0/-12.5
3	0.741	8.05	29 / -2.0	-19.1 / -1.8

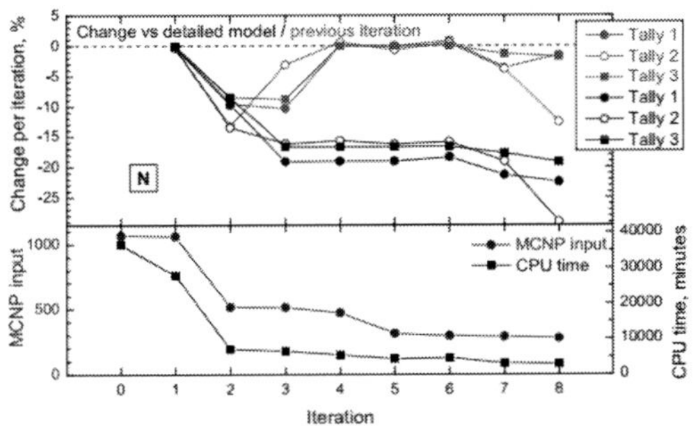

Fig. 17. The change in the neutron dose rates (upper plot) as well as the MCNP input file size and CPU time (lower plot) observed during Iteration #8.

The estimated photon and neutron effective doses and their relative change from both Iteration 0 and Iteration 6 are shown in Table 11. As expected, both the photon and neutron dose rates increased when the water door was removed. This increase can be explained by an extra contribution of particles being scattered off the floor and being collected by the tallies. The most significant change was induced in Tally #2. Which was the expected result considering that this Tally is more exposed to the radiation scattered off the floor than other tallies.

4 Summary

A summary of the effects on the dose rates from the various simplifications is presented in Fig. 19 for neutrons and Fig. 20 for photons. The summary of the MCNP input file size and the CPU time is shown in Table 1.

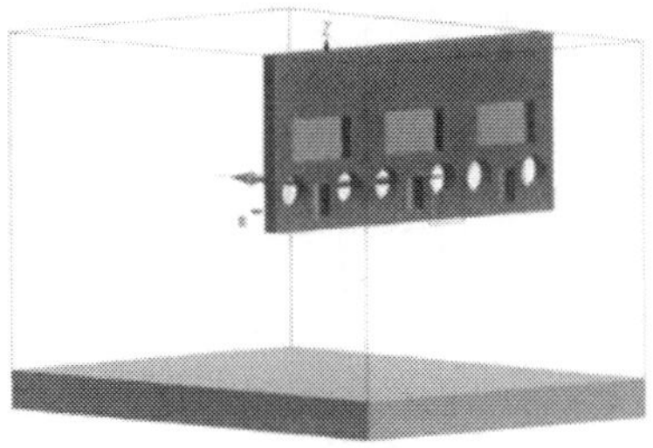

Fig. 18. MCNP glovebox model for the Additional Iteration: Removed Water Doors with Floor (Comparison with the Iteration 6).

Table 11. Estimated Effective Dose Rates for the Additional Iteration: Removed Water Doors with Floor (Comparison with the Iteration 6).

Tally #	Estimated Dose Rates, mrem/hr		Percent Change vs Iteration 0 / 6	
	Photon	Neutron	Photon	Neutron
1	0.529	4.22	75 / 6.9	-11.3 / 8.7
2	0.990	2.78	338 / 6.9	1.2 / 20.2
3	0.775	8.50	34 / 2.1	-14.8 / 2.2

The analysis indicated that even the most simplified glovebox model doesn't result in a crucial deviation from the results obtained with a complex model. It was found that the iterations 2, 3, and the additional iteration from Section 3.9 produced the largest change in the estimated doses (excluding iteration 6 for photons when there was nothing between sources and tallies but atmospheric air). That change was about 10% per iteration for neutrons and up to 5% for photons.

The total change in neutron dose rates between the simple and complex model ranged from 20 to 30% and for photons was about 5% (excluding 6 and subsequent iterations). Also the additional iteration (Section 3.9) demonstrated that some changes may compensate changes accumulated over other simplifying iterations. The main conclusion from the analysis is that it is possible to get a reasonable estimate of the anticipated dose rates using a simplified MCNP model. The level of the simplification can be selected based on the acceptable uncertainties in calculated dose rates.

The list below summarizes the findings of this exercise:

- A reduced outside world boundaries do not change the dose rates.
- Removing any additional bulky shielding placed on a glovebox or a structure element like a glovebox wall, which is close to a source, can result in underestimating the neutron dose rate by approximately 10% and the photon dose rate by approximately 5%.

- Adding complex geometry, such as curvatures, to windows, walls, etc. does not sufficiently influence either the photon or neutron dose rates.
- Adding structural details, which are sufficiently far from tally (like upper workstations in the model under consideration), does not influence either the photon or neutron dose rates.
- To avoid the significant contribution to the photon dose rate from low-energy photons, always place a material between the source and tally.

A good rule of thumb is start with the simplest approach. Complexity may be added to increase the accuracy of the model if the dose rates exceed or are close to the design limits.

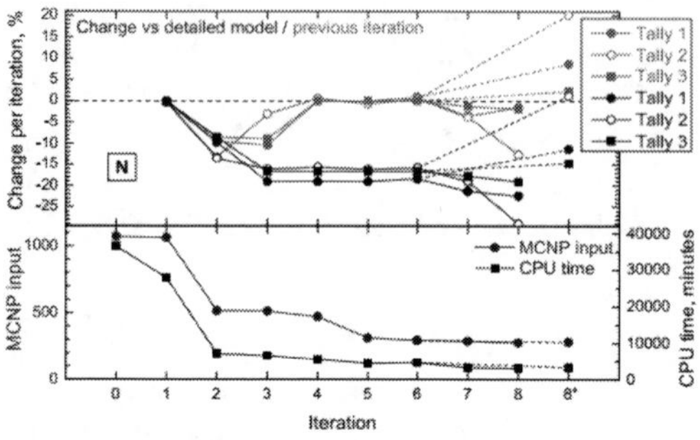

Fig. 19. Summary of the change in the neutron dose rates from the course of consequent iterations on the glovebox model simplification (upper plot) as well as a summary of the MCNP input file size and CPU time (lower plot). The additional iteration (Section 3.9) is marked as "8*". Dashed lines represent a tendency of the change in the dose rates of this iteration after Iteration #6.

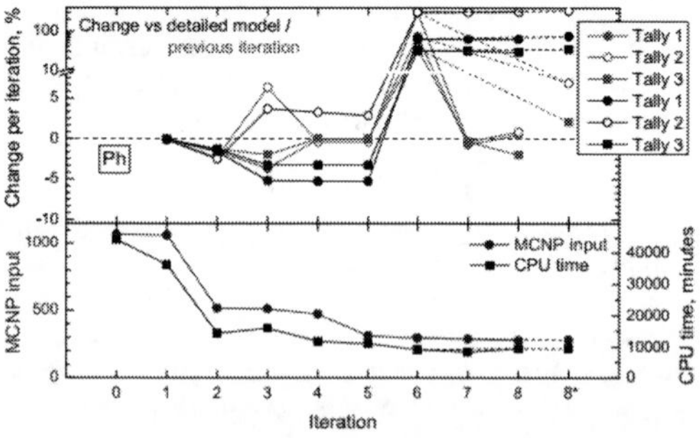

Fig. 20. Summary of the change in the photon dose rates from the course of consequent iterations on the glovebox model simplification (upper plot) as well as a summary of the MCNP input file size and CPU time (lower plot). The additional iteration (Section 3.9) is marked as "8*". Dashed lines represent a tendency of the change in the dose rates of this iteration after Iteration #6.

5 Conclusion

The analysis indicated that most of the particle fluence across the tallies penetrated the front wall of the glovebox. Thus to obtain a fast and reasonable estimation of dose, the model should be realistic in details that are close to the tally. Other details may be omitted.

Acknowledgment

This work was conducted while in the LANL Group RP-PROG. We wish to acknowledge the contributions made by the group members: J.M. Bean, C.J. Bianconi, M.A. Griffin, L.D. Hetrick, P.S. Hoover, B.T. Keller, G.R. Murrell, and R.C. Rangel.

Reference

1. X-5 Monte Carlo Team, *MCNP – A General Monte Carlo N-Particle Transport Code, Version 5* (LA-UR-03-1987, LANL, Los Alamos, 2003)

CAD-Based Monte Carlo Neutron Transport Analysis for KSTAR

Geon Ho Seo, Sung Hoon Choi and Hyung Jin Shim*

Nuclear Engineering Department, Seoul National University, 08826 Seoul, Korea

Abstract. The Monte Carlo (MC) neutron transport analysis for a complex nuclear system such as fusion facility may require accurate modeling of its complicated geometry. In order to take advantage of modeling capability of the computer aided design (CAD) system for the MC neutronics analysis, the Seoul National University MC code, McCARD, has been augmented with a CAD-based geometry processing module by imbedding the OpenCASCADE CAD kernel. In the developed module, the CAD geometry data are internally converted to the constructive solid geometry model with help of the CAD kernel. An efficient cell-searching algorithm is devised for the void space treatment. The performance of the CAD-based McCARD calculations are tested for the Korea Superconducting Tokamak Advanced Research device by comparing with results of the conventional MC calculations using a text-based geometry input.

1 Introduction

The Monte Carlo (MC) particle transport analysis for a complex system such as an accelerator or a fusion facility may require accurate modeling of the complicated geometry. However its manual modeling of all the geometrical objects in a text input file is lengthy and error-prone. In order to overcome this problem, techniques for automatic coupling of MC codes with the computer aided design (CAD) system have been widely studied and applied for various transport analyses [1-6].

There have been two kinds of approaches to develop MC code systems utilizing the CAD data: the external format conversion and the CAD kernel imbedded MC simulation. The first approach includes several interfacing programs such as McCad [1], MCAM [2], GEOMIT [3] etc. which were developed to automatically convert the CAD data into the MCNP [7] geometrical input data. This approach makes the most of the existing MC codes without any modifications, but implies latent data inconsistency due to the difference of the geometry modeling system. In the other approach, a MC code directly utilizes the CAD data for the MC particle tracking or an internal conversion to an internal data structure of the constructive solid geometry (CSG) and/or boundary representation (B-rep) modeling with help of an imbedded CAD kernel. MCNP-BRL [4], DAGMC [5], and OiNC [6] have demonstrated their capabilities of the CAD-based MC simulations.

The Seoul National University MC code, McCARD [8], has demonstrated [9] its CAD-based MC neutronics analysis capability for the HANARO research reactor [10]. In McCARD, the CAD geometrical data are automatically converted into the internal geometrical data structure with help of an imbedded CAD kernel, OpenCASCADE (OCC) [11]. Recently authors have augmented a McCARD version with the CAD-based geometry processing module (or McCARD/CAD) by an efficient algorithm for the void space treatment required for the fusion facility analysis. In this paper, we examine the performance of McCARD/CAD for the neutronics analysis of Korea Superconducting Tokamak Advanced Research (KSTAR) device [12].

2 Geometry Processing in McCARD/CAD

2.1. Distance to Surface Calculation Module

In a course of the MC particle tracking, geometrical modeling data are required to calculate the minimum distance to surface (DTS) from a particle at a given location and direction in a cell. For efficient DTS computations, cells bounded by elementary surfaces such as planar, cylindrical, spherical, conical, and toroidal surfaces in the CAD geometrical data file are converted into the internal geometrical data structure in McCARD. When a solid surface is not the elementary such as a B-spline or a rotational one, then OCC library functions are directly utilized to calculate the DTS. Figure 1 shows an algorithm of the McCARD/CAD DTS calculation.

* Corresponding author: shimhj@snu.ac.kr

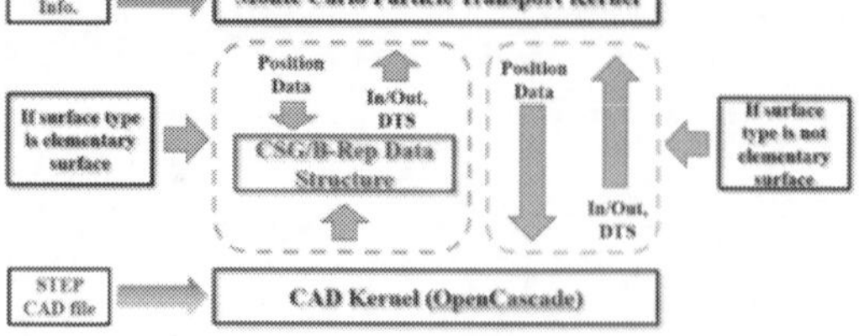

Fig. 1. Algorithm of McCARD/CAD DTS calculation

2.2. Cell Conversion

In the McCARD CAD-based geometry processing module, a solid in the CAD model whose boundaries are represented by elementary surfaces such as plane, cylindrical, spherical, conical, toroidal surfaces is converted into the McCARD cell model in which a cell is defined by Boolean operations such as 'AND', 'OR', 'NOT' of the boundary surfaces. For this cell conversion, the following geometrical information for each solid is required.

- surface equations of solid boundaries
- direction (+, –) of each boundary surface including the solid
- Boolean operators ('AND' or 'OR') between surfaces

The surface equations and directions of bounding surfaces can be readily obtained by the OCC library functions. In order to represent a concave-type solid into a union of convex-type cells in which the surfaces are connected by the 'AND' operator, the following algorithms are used.

① Find all vertices which make up a solid.
② If the vertices are located only in one side of a plane which bounds the solid, then the plane surface is assigned to convex-type. If not, the plain becomes concave-type.
③ If a plane is concave-type, divide the solid by the plane and connect the divided solids by the union ('OR') operator.
④ Repeat the process step ①~③ until all the divided solids becomes convex.
⑤ Define the divided convex solid by the intersections of its boundary surfaces using the 'AND' operators and combine all the divided convex solids with the 'OR' operators

2.3. Void Space Treatments

An artificial void cell per each subsystem which can access all the surface data of the subsystem is automatically created in McCARD/CAD to simulate particle tracks in its void spaces. When a particle enters the void cell, the minimum distance to surface is selected among distances to all the surfaces with checking their validities. In order to enhance the efficiency, a cell which the particle enters crossing the selected surface is searched by priorities of neighbor cells which are updated during the MC particle tracking.

3 Application for KSTAR

The neutron flux of the KSTAR fusion research device [11] is calculated by McCARD/CAD. Figure 2 shows the CAD model of KSTAR used in the McCARD/CAD calculations. The KSTAR CAD model consists of plasma facing components, a vacuum vessel, poloidal field coils, and toroidal field coils. For the comparison, the McCARD calculations using text geometry inputs are also conducted with or without the void space treatments. The McCARD calculations are performed on one billion neutron sources with the Gaussian fusion energy spectrum using the ENDF/B-VII.0 continuous energy cross section libraries.

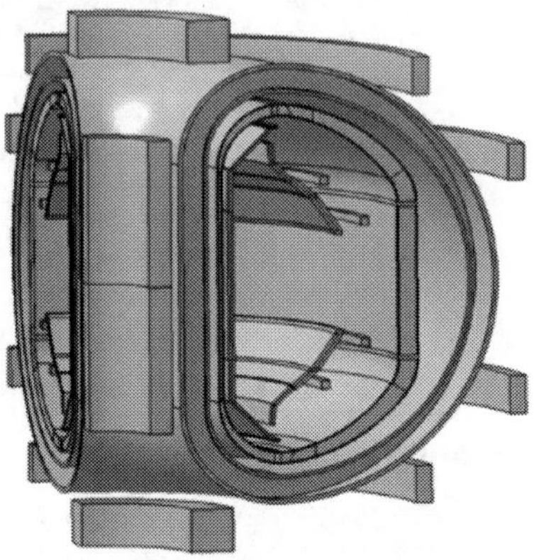

Fig. 2. CAD model of KSTAR

Figures 3 and 4 show the comparisons of neutron fluxes in the vacuum vessel and the toroidal field coil, respectively, calculated by McCARD/CAD and McCARD using a text input file. In the figures, the x axis denotes the angular location of the vessel or coil. From the figures, we can see that the flux results are identical within their statistical confidence intervals. Table 1 shows the comparison of elapsed times of three McCARD calculations. From the table, we can observe that the McCARD/CAD with the void space treatment takes about 57% additional times comparing to the McCARD without the void space treatment.

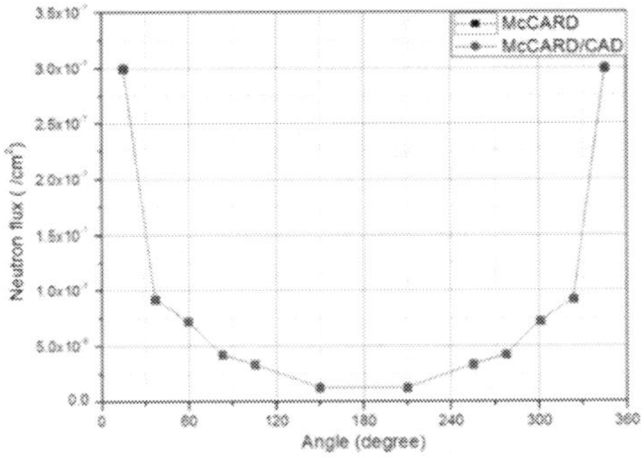

Fig. 3. Comparison of neutron flux in the vacuum vessel calculated by McCARD/CAD and McCARD

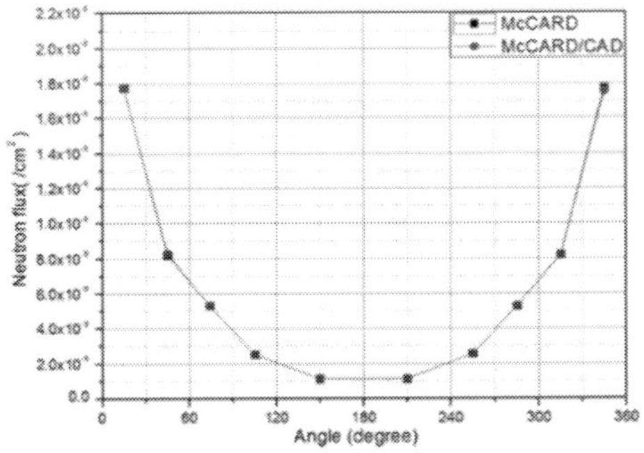

Fig. 4. Comparison of neutron flux in the toroidal field coil calculated by McCARD/CAD and McCARD

Table 1. Comparison of McCARD calculation times

	McCARD /CAD	McCARD with a text input	
Void space treatment	on	on	off
CPU time [hour]	25.31	24.73	16.07

References

1. H. Tsige-Tamirat, U. Fischer, A. Serikov, S. Stickel, *Fusion Eng. Des.*, **83**, 1771 (2008)
2. Y. Wu and FDS Team, *Fusion Eng. Des.*, **84**, 1987 (2009)
3. S. Sato, et al., *Nucl. Technol.*, **168**, 843 (2009)
4. K.B. Bekar, T.M. Evans, *Trans. Am. Nucl. Soc.*, **101**, 623 (2009)
5. P.P.H. Wilson, et al., *Fusion Eng. Des.*, **85**, 1759 (2010)
6. F. Fleurot, K. Searson, "OiNC, Criticality Benchmark Models in CAD," *ICNC2011-International Conference on Nuclear Criticality*, Edinburgh (2011)
7. J. F. Briesmeister, *MCNP-A General Monte Carlo N-Particle Transport Code, Version 4C*, LA-13709, Los Alamos National Laboratory (2000)
8. H.J. Shim, et al., *Nucl. Eng. Technol.*, **44**, 161 (2012)
9. H.J. Shim, et al., *Ann. Nucl. Energy*, **82**, 48 (2015)
10. H. Kim, et al., *J. Nucl. Sci. Technol.*, **33**, 527-538 (1996)
11. OpenCASCADE™, www.opencascade.org
12. G.S. Lee, et al., *Nucl. Fusion*, **40**, 575 (2000).

An Augmented γ-Spray System to Visualize Biological Effects for Human Body

Seiya Manabe[1,*], *Hideki* Tenzou[2], *Takaaki* Kasuga[1], *Yukiko* Iwakura[1] and *Robert* Johnston[2]

[1]National Institute of Technology, Kagawa College, Advanced Course in Electronics, Information and Communication Engineering, Electronic Systems Course, 551 Kohda, Takuma-cho, Mitoyo-shi, Kagawa, 769-1192, Japan
[2] National Institute of Technology, Kagawa College, Department of Electronic Systems Engineering, 551 Kohda, Takuma-cho, Mitoyo-shi, Kagawa, 769-1192, Japan

Abstract. The purpose of this study was to develop a new educational system with an easy-to-use interface in order to support comprehension of the biological effects of radiation on the human body within a short period of time. A paint spray-gun was used as a gamma rays source mock-up for the system. The application screen shows the figure of a human body for radiation deposition using the γ-Sprayer, a virtual radiation source, as well as equivalent dosage and a panel for setting the irradiation conditions. While the learner stands in front of the PC monitor, the virtual radiation source is used to deposit radiation on the graphic of the human body that is displayed. Tissue damage is calculated using an interpolation method from the data calculated by the PHITS simulation code in advance while the learner is pulling the trigger with respect to the irradiation time, incident position, and distance from the screen. It was confirmed that the damage was well represented by the interpolation method. The augmented γ-Spray system was assessed by questionnaire. Pre-post questionnaire was taken for our 41 students in National Institute of Technology, Kagawa College. It was also confirmed that the system has a capability of teaching the basic radiation protection concept, quantitative feeling of the radiation dose, and the biological effects

1 Introduction

After the accident at Tokyo Electric Power Company's (TEPCO) Fukushima Dai-Ichi Nuclear Power Plant, not only specialists in radiation but also a lot of people in general have a growing interest in radiation protection. It's important to learn how to protect yourself from radiation through quantitative discussions for the safe use of radiation or decontamination work aimed at reducing radiation levels. Some handbooks and e-Learning tools have been prepared for use in schools for relatively-long courses to teach deterministic effects and probabilistic effects with the units themselves. Some learners, however, often get confused between the absorbed dose D [Gy] and the effective dose H [Sv]. It's difficult for the lay public to understand both the physical definitions and the calculation methods of each intuitively [1]. Therefore, the purpose of this study was to develop a new short interactive educational demo system with an easy-to-use interface to support comprehension of the biological effects that radiation can have on the human body.

2 System overview

Figure 2 shows an overview of the new system. It has a spray-type user interface. The learner stands in front of a PC monitor and aims the virtual γ-Sprayer at the monitor where an image of the human body is displayed. The time elapsed pulling the trigger is proportional to the total number of virtual γ-rays emitted from the nozzle. The position and orientation of the gun is tracked by a "Leap Motion" non-contact sensor. Irradiation points are calculated from the sensor data. The learner can control the irradiated centre position by handling the γ-Sprayer and can focus on the centre by changing the distance from the monitor. The learners can intuitively understand that there are three ways to protect themselves from radiation. One of them is to keep your distance from radioactive materials. The second point is to reduce the exposed time to radiation. The learners can observe the change of effective dose depending on irradiation

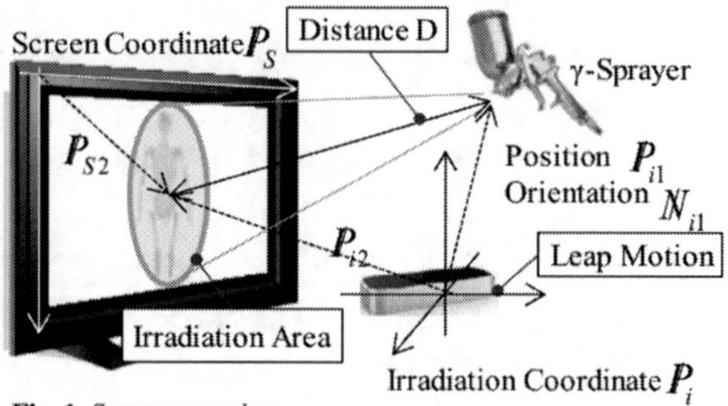

Fig. 1. System overview.

* Corresponding author: manabe@aees.kyushu-u.ac.jp

positions. Thus, the third point is to understand which organs or tissues have higher radiation sensitivity. The radioactivity is selectable through the GUI.

The educational system was developed on a 64-bit Windows 8 platform in the Processing programming language. The PC was equipped with an Intel Core i7-4820K (3.70GHz), 16 GB of RAM and a NVIDIA GeForce GTX 770 graphic card.

2.1 Data flow diagram

Figure 2 shows the data flow diagram of the system. The educational system is required to display the radiation biological effects and effective dose in real-time. The system was designed to estimate the dose rate for any incident position on the human body by using an interpolation approximation to save time calculating radiation damage. The PHITS (Particle and Heavy Ion Transport code System) Ver. 2.70 simulation code [2] was used to calculate the dose data per number of incident γ-rays hitting the organs and tissue for each incident position at 10 cm intervals. The dose data for each incident position on the human body for the pre-calculations was called the "Response functions". The MIRD (Medical Internal Radiation Dose) phantom [3] was used as the physical model of the human body.

The "Leap Motion" non-contact sensor acquires the position of the virtual γ-Sprayer. Irradiation points on the monitor are calculated from the position and orientation of the γ-Sprayer. The effective dose and equivalent dose of organs and tissues are calculated by multiplying the interpolation values based on response functions of both the duration the trigger is engaged and the pressure applied to the trigger.

The damage for each organ in the human body is rendered at an interval of 0.5 seconds. Learners can see the effects that the radiation has on each organ in the textboxes on the monitor in real time.

2.2 Mechanical System

2.2.1 Tracking

A Leap Motion sensor was used to track the position and orientation of the γ-Sprayer. The gesture sensor has a view field of approximately 150° and uses a depth sensor to track hand features down to 1/100th of a millimetre. The Leap Motion works with two infrared (IR) cameras and three IR LEDs as a depth sensor in a limited field of approximately 61 cubic centimetres. Using the stereoscopy provided by using both cameras, the device can minimize position errors of the γ-Sprayer. The sensor was set on the same stage as the monitor. The relative position $P_0(x_0, y_0)$ between the sensor and the monitor must be calibrated at start-up. The γ-Sprayer is identified through acquiring the shape of the nozzle parts by using the Leap Motion V2 Beta Software Development Kit (SDK).

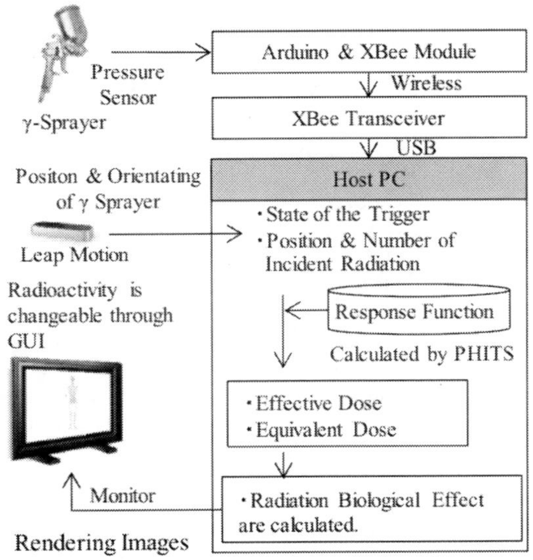

Fig. 2. Data flow diagram.

2.2.2 γ-Sprayer Controller

Figure 3 is a structure of γ-Sprayer. EARTH MAN Takagi suction-feed spray gun was manufactured by Takagi CORPORATION and remodelled to γ-Sprayer Controller. A pressure sensor was attached to the trigger to measure the position of the trigger. The data is sent while the trigger is pulled. The analogue data is converted to digital data by an Arduino Uno R3 and the digital data is sent via an "XBee" wireless module to the main program on a host PC. In the main program, the emission number (intensity) N of the γ-rays is calculated by the following equation (1).

$$N = N_0 \cdot dT_{\mathrm{trig}} \cdot F_{\mathrm{trig}} \quad (1)$$

Where, N_0 is a constant value and dT_{trig} is the elapsed time of the trigger at the trigger pressure force F_{trig}. The spray spreads out γ-rays in a solid angle determined by $2\pi\,(1-\cos 5°)$.

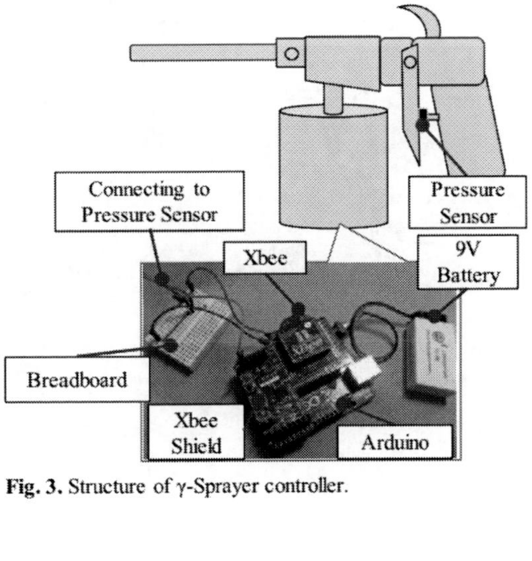

Fig. 3. Structure of γ-Sprayer controller.

2.3 Physical model

2.3.1 Response Functions calculated by PHITS

Response functions were calculated by using the PHITS Monte Carlo simulation code and a MIRD mathematical human phantom placed in a vacuum. PHITS can simulate the transport of all particles such as nucleons, nuclei, mesons, photons, and electrons in arbitrary geometries composed of any element, compound or mixture. The phantom is a heterogeneous mathematical representation of the human body to estimate absorbed doses from internal and external exposure. In order to estimate an accurate effective dose, it was necessary to distinguish the absorbed dose in the red bone marrow from that of the bone surface. Both of them are, however, defined as one material, "Bone", in the MIRD phantom. Therefore, the absorbed dose of "Bone" was divided into that of the red bone marrow and that of the bone surface according to their masses.

The height and weight of the phantom body is based on the national mean Japanese male at 174 cm and 73 kg respectively. Table 1 shows the elemental compositions and densities of each organ and tissue for the MIRD phantom model. 1.0 MeV γ-rays from Co60 were selected as the incident radiation for this system. Response functions for each incident position and angle were calculated at a 10 cm interval. The irradiation system in this system is showed in Figure 4. Response functions have relationship between irradiation position on human body (x, y), distance from human body to radiation source L, horizontal angle φ and vertical angle θ. The parameter L was set from 200 cm to 700 cm by 100 cm. The φ or θ was set from 0° to 180° by 30°. Response functions are saved as text files in a direct folder of the running program. Absorbed dose is estimated by using interpolation method in real time.

Table 1. Elemental compositions and densities in each organs and tissue for the MIRD phantom model.

Organ	Density [g/cm³]	H	C	N	O	Ca	Na	P	S	Cl	K
Lung	0.30	10	10	2.9	76	-	0.2	0.1	0.2	0.3	0.2
Bone	1.49	7.3	26	3.1	48	10	0.3	5.1	0.2	0.1	0.1
Soft Tissue	0.99	11	23	2.5	64	-	0.1	0.1	0.2	0.1	0.2

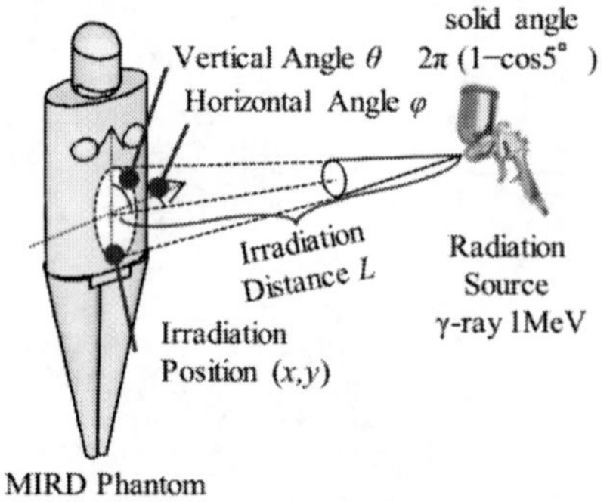

Fig. 4. The target of human body is being irradiated by using spray-type interface.

2) Radiation Dose Estimation Method

Equivalent dose H_T is calculated by the following equation (2) to determine the biological effects of the radiation.

$$H_T = \Sigma_R W_R \cdot D_{T,R} \quad (2)$$

Where, $D_{T,R}$ is the absorbed dose deposited in an organ or tissue T by radiation R, and W_R is the radiation weighting factor. W_R is dependent on the type and energy of radiation R. Our current system selects γ-rays as the incident radiation. The radiation weighting factor W_R is '1'.

Effective dose E is calculated by the following equation (3).

$$E = \Sigma_T W_T \cdot H_T \quad (3)$$

Where, W_T is the tissue weighting factor defined by regulation as showed in Table 2. E is the tissue-weighted sum of the equivalent doses in all specified organs and tissues. The W_T is referred to the ICRP Publication 60 [4]. The E represents the stochastic health risk, which the probability of cancer induction and genetic effects of ionizing radiation delivered to those body parts.

Each damage or symptom for human body is estimated based on the absorbed doses, the equivalent doses or the effective doses given by above equation (2) and (3). Deterministic effects considered in this system are listed in Table 3. Deterministic effects have thresholds below which the effects do not occur. The threshold for each symptom is also listed in Table 3. The symptoms for each of the organs or tissues were calculated based on the values of the absorbed doses or the effective doses in real time. Probabilistic effects considered in this system are fatality cancer incidences in each parts of human body. Fatality cancer incidences are propositional to equivalent doses of each part. Fatality probability coefficients in each part are also refed to ICRP Pub. 60 and listed in Table 4. The irradiated effective dose is compared with the general exposure. Considered general exposures are listed in Table 5.

Table 2. Tissue weighting factor for each organs and tissue.

Organ or Tissue	Tissue Weighting Factor
Gonands	0.20
Bone Marrow (Red)	0.12
Colon	0.12
Lung	0.12
Stomach	0.12
Bladder	0.05
Breast	0.05
Liver	0.05
Oesophagus	0.05
Thyroid	0.05
Skin	0.01
Bone Surfaces	0.01
Remainder	0.05

Table 3. List of biological effects from ICRP Pub. 60.

Absorbed Dose[Gy]	**Part**	**Biological Effect**
0.15	Gonand	Temporary Sterilization
3.5		Permanent Sterilization
0.5	Bone Marrow	Disturbance of Hematopoiesis
Effective Dose[Sv]	**Biological Effect**	
0.1	Mutation in Chromosome Banding	
0.5	Lymphocytic Leukopenia	
1	10% People have Nausea	
3	Dead in 30 - 60 Days due to Injury to Bone Marrow	
5	Dead in 10 - 20 Days due to Injury to GI Tract and	
15	Dead in 1- 5 Days due to Injury to Nervous System	

Table 4. List of fatality cancer incidences from ICRP Pub. 60.

Organ or Tissue	**Fatality Cancer Incidences[$10^{-4}Sv^{-1}$]**
Bladder	30
Bone Marrow (Red)	50
Bone Surfaces	5
Breast	20
Colon	85
Liver	15
Lung	85
Esophagus	30
Stomach	110
Thyroid	8
Whole Rate	500

Table 5. List of general exposure considered in the system.

General Exposure	**Equivalent Dose[mSv]**
CT	6.9
Chest Roentgenography	0.05
PET	2.2
Trip (Tokyo to NewYork)	0.2
Background Radiation	2.4

3 Demonstration

The demonstration of the system is shown in Fig. 5. The small intestine around the left side of the body was irradiated for about 3 minutes under the conditions of having 10 GBq of γ-rays emitted from the nozzle with 5° beam divergence. Irradiated areas can be seen as small squares drawn around the small intestine on the target human body as shown in Fig. 6(A). The colour has changed from blue to red proportionally to the sum of the effective dose irradiated from each point. The equivalent dose for each organ or tissue on the body on the left is shown in Fig. 6(B). The learner can see the different damage sensitivity for each tissue. The name of the symptom caused by exceeding the dose threshold is shown in the Text Box on the area (C) in Fig. 6 with pointing the parts. Probabilistic effects are shown in the

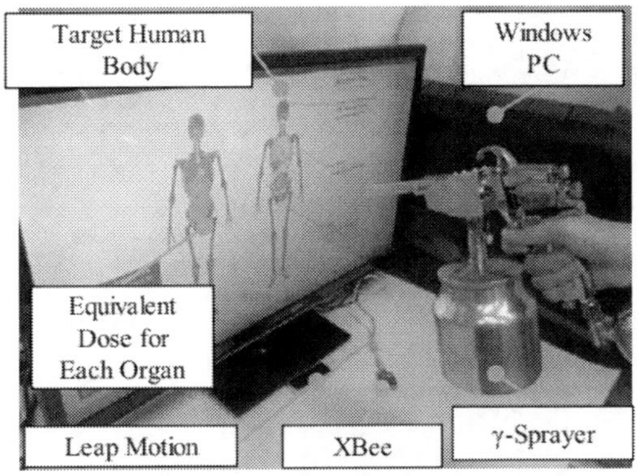

Fig. 5. The target of human body is being irradiated by using spray-type interface.

Text Box on the area (D) in Fig. 6. Learners can observe two kinds of effects of radiation in real time. Thus, learners can understand a basic concept of radiation protection. Comparison between the general exposures and the irradiated effective dose is shown in the Text Box on the area (E) in Fig. 6. Learners can have experience a quantitative feeling about the relationship between doses of general exposure and doses which cause the symptom intuitively. The irradiation conditions such as the radioactivity and the scale factor of the lapse time are changeable through a setting panel as shown in Fig. 6(F).

4 Evaluation

4.1 Confirmation of the accuracy of the dose estimation

Figure 7 and 8 illustrates the interpolated effective doses and the effective doses calculated by PHITS per incident γ-rays corresponding to each vertical angle θ and horizontal angle φ. A body around heart was aimed. Effective doses in 15°, 45°, 50°, 65°, 75°, 105° and 135° were estimated based on response functions in real time. Effective doses are well-reproduced within the error of approximately 20 %.

Figure 9 is the images for equivalent dose to show the damage for each organ. Figure 9(A) shows the non-interpolated calculation. Figure 9(B), (C) and (D) are the images for 5 cm, 10 cm, and 15 cm of each interval distance, respectively. Figure 10 shows the error of hue depending on interval distance in two irradiation points. The difference of hue contrast between the non-interpolated calculation and the interpolated calculation was over 20 % more than 15 cm interval. Therefore, the interval was decided to be 10 cm in the system.

4.2 Educational assessment

The augmented γ-Spray system was assessed by questionnaire listed in Table 6. Pre-post questionnaire was taken for our 41 students in National Institute of Technology, Kagawa College. Their major is electronics and they hadn't taken any classes about radiological

DOI: 10.1051/epjconf/201715306020

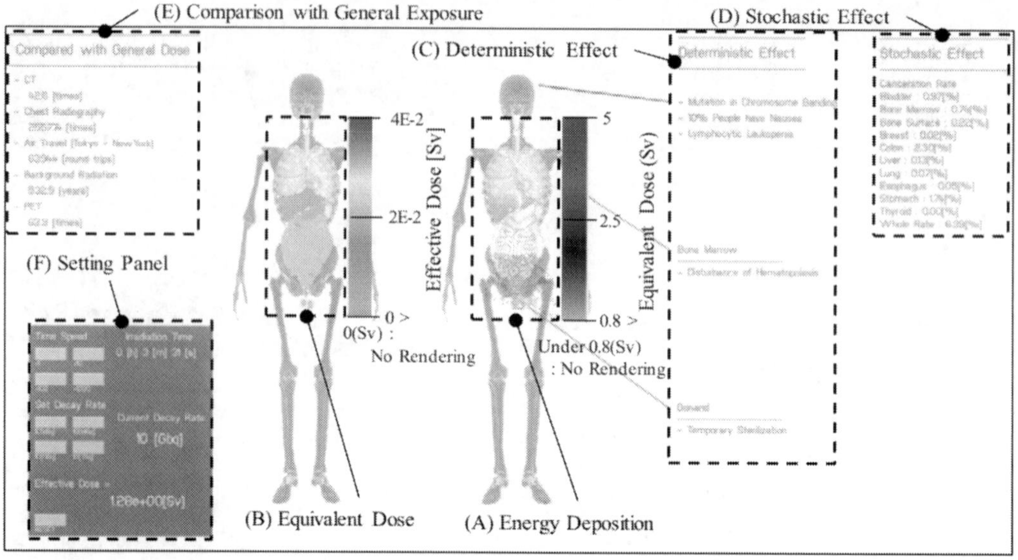

Fig. 6. Screenshot of projection images

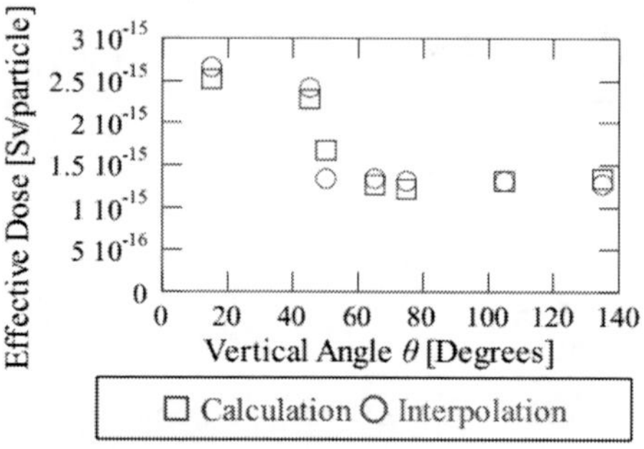

Fig. 7. Comparison of the calculated effective dose and the interpolated effective dose per unit γ-ray irradiated from each vertical angles.

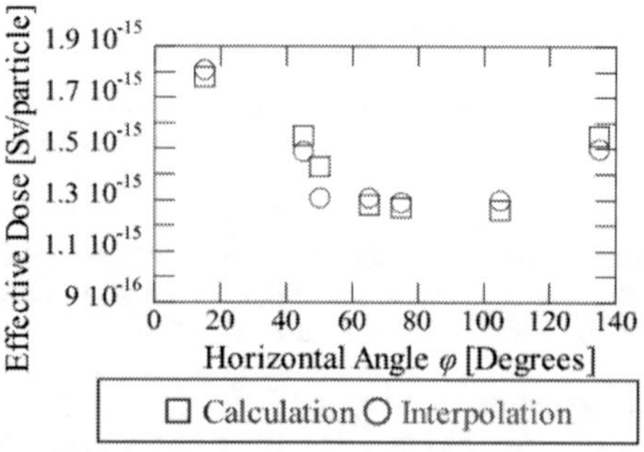

Fig. 8. Comparison of the calculated effective dose and the interpolated effective dose per unit γ-ray irradiated from each horizontal angles.

engineering or radiation medical physics. Questions of Q1 and Q5 were created learners' basic physical knowledge to protect from radiations. The aim of Q2, Q6, and Q7 was to survey basic understand the difference between definition of [Gy], [Bq], and [Sv]. Q3, Q4, Q8, and Q9 were created to assess basic comprehension about biological effects on human body. The augmented γ-Spray system can show the both effects in the monitor

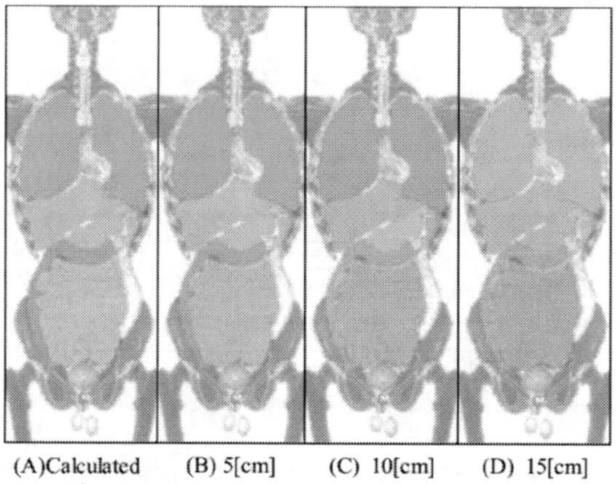

Fig. 9. Visual evaluation of the equivalent dose depending on interval distance.

in real time. It's important to confirm learners' understand about the differences between deterministic effect and stochastic effect.

Figure 10 is a result of the questionnaire survey for the 41 learners. Figure 10 illustrates the changes of percentage of correct answer for both pre and post questionnaires. Before brief demonstration to explain how to use the augmented γ-Spray system, the learners were taken pre-questionnaire. After the pre-questionnaire, each learner touched the trigger of γ-Spray for about 5 minutes. We can see that the average percentage of the correct answer has grown up over 40 % on the average after using the system. It was confirmed that the system has a capability of teaching the basic radiation protection concept, quantitative feeling of the radiation dose, and the biological effects.

Table 6. Questionnaires for learners.

No.	Question	Choice
Q1	Are there any relationships between the radiation damage and irradiation time?	Yes, No
Q2	Which unit means the activity of a quantity of radioactive material in which one nucleus decays per second?	Bq, Sv, Gy
Q3	How many times can you accept when a doctor ask you to undergo an X-ray inspection in one year?	Zero, a few times, more, couple dozen times
Q4	Dose a radiation associated lesion of agenesis, an hematopoiesis or reduction of lymph have a character that once the dose threshold has been exceeded, the severity of an effect increases with dose?	Yes, No
Q5	A radiation intensity is changed by	hardness of a RI, distance from a RI , temperature of a RI
Q6	Which unit means absorbed energy per kilogram of the matter?	Bq, Sv, Gy
Q7	Which unit indicates the equivalent biological effect of the deposit of a joule of radiation energy in a kilogram of human tissue?	Bq, Sv, Gy
Q8	Dose a probability of cancer occurrence increases with effective radiation dose?	Yes, No
Q9	Are there any differences of damage sensitivity between each body organ for irradiations?	Yes, No

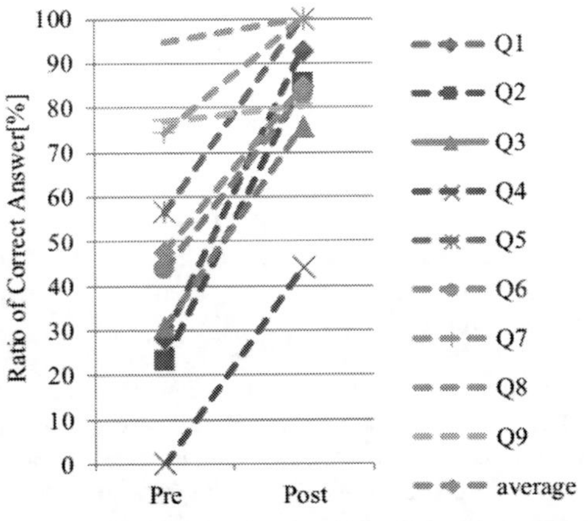

Fig. 10. Result of the questionnaire survey about teaching effectiveness of the system.

5 Summary

A new visualization system with a hand-held interface for education in radiation protection was proposed and demonstrated in this paper. The demonstration highlights some of the features used to help teach the concepts that the severity of radiation damage is related to exposure time and distance from the source and that different organs and tissues in the body have varying degrees of sensitivity to radiation. The learners can see radiation damage occurring in real time while the learner is pulling the trigger. Interpolation method was used to calculate equivalent dose in real time on each organ. The response functions for the method were calculated by using the PHITS Monte Carlo simulation code and a MIRD mathematical human phantom placed in a vacuum. It was confirmed that effective doses by interpolation method are well-reproduced within the error of approximately 20 % at 10cm interpolation interval. The augmented γ-Spray system was assessed by pre-post questionnaire for college students. The learners could understand a quantitative feeling about the relationship between the three units of radioactivity [Bq], absorbed dose [Gy], and effective dose [Sv]. It was also confirmed that the system has a capability of teaching the basic radiation protection concept, quantitative feeling of the radiation dose, and the biological effects.

References

1. J. Hudzietzová, J. Sabol, Int. J. Environ. Res. Public Health, **1**, 132-139, (2014)
2. T. Sato et al., J. Nucl. Sci. Technol, **50**, 913-923 (2013)
3. ICRP, *ICRP Publication 74. Ann. ICRP 26 (3-4)*, 15-33 (1996)
4. ICRP, *ICRP Publication 60. Ann. ICRP 21 (1-3)*, 6-153 (1991)

Validation of the MCNP6 electron-photon transport algorithm: multiple-scattering of 13- and 20-MeV electrons in thin foils

David A. Dixon[1] and H. Grady Hughes[1]

[1]Los Alamos National Laboratory, Los Alamos, NM 87545

This paper presents a validation test comparing angular distributions from an electron multiple-scattering experiment with those generated using the MCNP6 Monte Carlo code system. In this experiment, a 13- and 20-MeV electron pencil beam is deflected by thin foils with atomic numbers from 4 to 79. To determine the angular distribution, the fluence is measured down range of the scattering foil at various radii orthogonal to the beam line. The characteristic angle (the angle for which the max of the distribution is reduced by $1/e$) is then determined from the angular distribution and compared with experiment. Multiple scattering foils tested herein include beryllium, carbon, aluminum, copper, and gold. For the default electron-photon transport settings, the calculated characteristic angle was statistically distinguishable from measurement and generally broader than the measured distributions. The average relative difference ranged from 5.8% to 12.2% over all of the foils, source energies, and physics settings tested. This validation illuminated a deficiency in the computation of the underlying angular distributions that is well understood. As a result, code enhancements were made to stabilize the angular distributions in the presence of very small substeps. However, the enhancement only marginally improved results indicating that additional algorithmic details should be studied.

KEYWORDS: MCNP6, Monte Carlo, electron-photon transport, multiple-scattering, validation

I. Introduction

It is of interest to establish a comprehensive electron-photon validation suite for the MCNP ® version 6.2 code system.(1) Though previous efforts to validate the electron-photon transport in earlier versions of MCNP6 were made,(2–4) these validations correspond to unsupported versions of the code and are somewhat limited in scope. New validation efforts are underway(5) that test the current version of the code containing improvements to the electron-photon transport capabilities.(6,7)

The purpose of this paper is to benchmark MCNP6 against the measurements of Ross *et al.*(8) Ross *et al* measured angular distributions corresponding to electron pencil beams of 13- and 20-MeV normally incident on single-element (with exception of a titanium alloy), thin foils with atomic numbers from 4 to 79. Electron fluence was measured roughly 120 cm from the beam exit window at various radii orthogonal to the beam line. While a number of similar benchmarks exist, uncertainties in the accelerator descriptions associated with these similar experiments somewhat limit their utility. Whereas, Ross *et al* determined angular distributions where the beam characteristics are well known, with no free parameters, making this benchmark ideal for validation of electron transport codes. Though this benchmark is valuable to codes intended for medical physics applications, it is also valuable for general validation of multiple-scattering models. As such, this experimental benchmark will be added to the MCNP6 electron-photon validation suite.

As the emphasis of this paper is the validation test rather than MCNP6 electron-photon transport algorithm, we briefly describe the relevant physics and identify additional references that provide greater detail. The MCNP6 electron-photon transport algorithm is considered a class I condensed history method. In MCNP6, electrons traverse a substep distance where both angular deflection and energy-loss are sampled (i.e. the primary processes impacting the angular distributions studied herein). In MCNP6, angular deflection resulting from multiple elastic scattering of electrons is given by the Goudsmit-Saunderson distribution,(9,10) and energy-loss resulting from multiple inelastic scattering is given by the Landau distribution.(11)Additionally, secondary electrons and bremsstrahlung photons are sampled along the substep, and the necessary atomic relaxation processes are considered. A substep is determined by dividing an energy step, or the range corresponding to some ΔE_k, into M smaller steps, where M is an increasing function of the target atomic number. Calculation of the energy steps, substeps, and the necessary electron transport data are completed during a preprocessing stage for a fixed grid of k energies. While electron and photon transport is coupled, the impact of photons on the angular distribution is nominal, and therefore, not discussed herein. More detail on the electron-photon transport algorithm is given by Hughes.(12)

The accuracy of the aforementioned processes depends on both algorithmic limitations and the underlying differential cross sections. In essence, this validation tests both potential sources of error. However, assume for a moment that the underlying differential cross sections are sufficiently accurate. Under these conditions, algorithmic error would account for any disagreement between simulation and experiment. Specific examples of algorithmic error relevant to this study include the stability of the underlying angular distributions, the boundary crossing approximation, the solution dependence on the step-size, and the treatment of angular deflection attributed to impact-ionization collisions for both the primary and the knock-on electron (again, details on each of these can be found in the literature[12,13]).

II. Method

1. Experiment

The characteristics of the experiment are detailed in Ross *et al.*[8] Below we provide sufficient detail to understand the simulation setup and the validation results. For the purposes of this paper, we consider a beam of electrons emerging from an accelerator exit window (see Fig. 1). The beam is effectively a pencil beam with some spread in space, angle, and energy. However, only the spread in space is considered. The spatial distribution of the experimental beam in transverse directions did not have cylindrical symmetry, but could be approximated by a gaussian with a full-width at half-max of approximately 1-mm.[8] Immediately upstream of the exit window is the scattering foil that is followed by the remainder of the apparatus including a transmission monitor chamber, a helium bag, and then the detector.

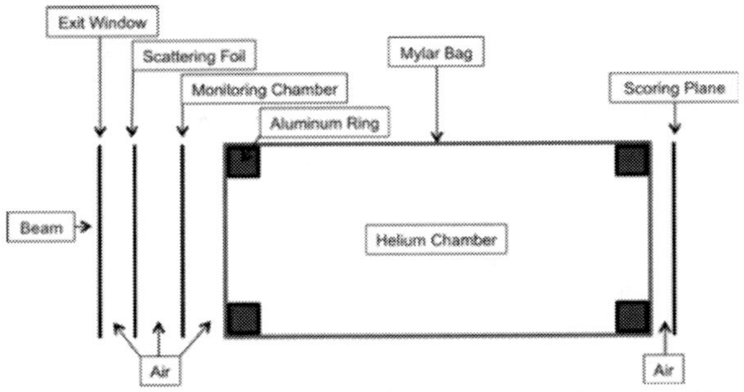

Figure 1: Schematic of experimental apparatus (not to scale).

To sufficiently resolve the scattered electrons a significant distance between the foil and the detector was required. What would be an air column was replaced by a helium column to reduce the scattering power of the region between the foil and the detector. For this reason, a Mylar bag was constructed to contain the helium.

The scattered electron distributions were measured using both an ionization chamber and diode detectors. The point of measurement was assumed to be the center of the detector. The detectors were mounted on a linear translator positioned orthogonal to the beam line. The detectors were then moved along the translator to determine the fluence at a given radial distance from the beam line that is then converted to angle.

2. Simulation

In this work, the experimental apparatus was approximated by several cylinders. The position and thickness of the cylinders were taken from the description of the key experimental components in Ross *et al.*[8] The electron source was placed at the origin and directed along the positive x-axis. The beam was assumed to be Gaussian in the transverse directions with a FWHM=1-mm. One billion source particles were followed in each simulation. The scoring plane was taken to be the plane orthogonal to the beam located at a point furthest from the source. The F2 surface flux tally was applied to this plane, which was subdivided into rings with increasing radii at 0.1 cm spacings from 0.1 cm out to 21 cm (this is accomplished using the *FS* or tally segment card).

The default photon physics was used to transport photons; however, it is of interest to note that the new EPRDATA library[7] (i.e. the ".12p" extension) was used for photon data.

For electrons, three physics settings were studied including the default physics, and two other settings where the impact of modifying *ESTEP* and *EFAC* are studied in isolation. *ESTEP* controls the number of substeps per energy step and presumably improves accuracy in problems with optically thin regions where electrons stream through these regions without colliding. Under such conditions, truncation of the electron substep is likely and results in an approximation of the angular deflection distribution. Therefore, one can increase *ESTEP* to ensure at least a few of collisions occur before reaching a boundary. *ESTEP* is applied to specific materials, so one can modify the substep size only for the material of interest that is contained in the thin cells. One should always keep in mind that reducing the substep size increases runtime. The following two tables present the *ESTEP* settings necessary to ensure that roughly 10 condensed collisions occur in each foil. These settings correspond to results labeled as *ESTEP*. Default values of *ESTEP* are for the materials used herein are: Be = 2, C= 3, Al = 5, Cu= 7, Ta = 12, and Au= 13.

Table 1: *ESTEP* parameter analysis for 13-MeV simulations. *ESTEP* was determined such that there were approximately 10 substeps per foil thickness (that is, electrons will suffer roughly 10 condensed collisions when traversing the foil).

Material	Thickness (mg/cm2)	Step size (g/cm^2)	*ESTEP*
Be	926	6.29×10^{-1}	8
Al	70.1 140	5.33×10^{-1}	80 40
Cu	43.0 86.4 129.6 174.5	4.95×10^{-1}	119 63 42 35
Ta	44.3	4.13×10^{-1}	24
Au	31.2 54.8 93.7	3.96×10^{-1}	130 78 52

In the context of this work, *EFAC* has a similar effect, but *EFAC* is a global parameter in the sense that it is used to adjust the coarseness, or fineness, of the energy grid that is material independent. So, rather than selectively reducing the substep size in thin cells, the substep size is reduced in all cells. By default *EFAC* =0.917. Increasing it to 0.99 (maximum permissible value) the substep size decreases, and runtime increases.

Table 2: *ESTEP* parameter analysis for 20-MeV simulations. *ESTEP* was determined such that there were approximately 10 substeps per foil thickness (that is, electrons will suffer roughly 10 condensed collisions when traversing the foil).

Material	Thickness (mg/cm2)	Step size (g/cm^2)	*ESTEP*
Be	926	9.078×10^{-1}	10
C	546	7.83×10^{-1}	15
Al	70.1 140 274	7.22×10^{-1}	105 55 30
Cu	43.0 86.4 129.6 174.5	6.20×10^{-1}	147 77 49 42
Ta	44.3	4.76×10^{-1}	24
Au	54.8 164.2	4.53×10^{-1}	91 39

3. Results Processing

Characteristic angles are determined by fitting a Gaussian to the measured and calculated angular distributions, $R(\theta)$, over a partial range. The partial range corresponds to all values where the distributions are greater than approximately R_{max}/e or $1/e$ if the distribution is nomalized such that the amplitude is unity. The fit is then used to obtain the characteristic angle by setting the Gaussian equal to $1/e$ and solving for $\theta = \theta_{1/e}$. That is,

$$R(\theta_{1/e}) = \frac{R_{max}}{e} = a\exp(-\theta^2_{1/e}/b), \quad (1)$$

where a and b are the Gaussian fit parameters obtained in the first step. Therefore,

$$\theta_{1/e} = \sqrt{-b\ln\left(\frac{R_{max}}{ea}\right)}, \quad (2)$$

or if $R_{max} = a = 1$, $\theta_{1/e} = \sqrt{b}$.

Ross *et al* provide the experimental results where the independent variable is given in either the raw form, or shifted. The shifted results are used herein, as they correspond to a symmetric distribution necessary for a consistent comparison.

III. Results

The characteristic angles of the measured and calculated angular distributions are presented in Tables 4 and 5. In these tables there are several characteristic angles reported as described below:

- Ross *et al.*: Characteristic angles reported by Ross *et al.*(8) using the fitting technique described therein.
- Dixon and Hughes: Characteristic angles determined by applying the fitting technique described herein to the data from Ross.(8)
- Calc.: Characteristic angles determined by applying the fitting technique described herein to the angular distributions calculated using MCNP6.

Furthermore, there are three different calculated results corresponding to different physics settings: the default physics; the default physics with modified *ESTEP* parameters, and the default physics modified *EFAC* parameter of 0.99.

The average difference (i.e. an arithmetic average of the rel. diff. over each of the foils) in the measured and calculated characteristic angle for 13-MeV and 20-MeV electrons for the physics settings tested are presented in Table 3. According to Faddegon(14) *et al.*, for 13-MeV and 20-MeV electrons the average difference in the measured and calculated characteristic angle for EGSnrc was 1.3% and 1.1% respectively; for PENELOPE the average difference was 1.1% and 1.1% respectively; and for Geant4 the average difference was 0.7% and 0.9% respectively.

Table 3: Average of the relative differences between measured and calculated characteristic angles for 13-MeV and 20-MeV electrons for three different physics configurations including: default, *ESTEP*, and *EFAC*.

	Average Relative Difference	
Physics Setting	13-MeV	20-MeV
Default	6.4%	5.8%
ESTEP	10.1%	12.2%
EFAC	7.3%	9.5%

The average values presented above were determined from the relative difference results in Tables 4 and 5. The values in Tables 4 and 5 compare characteristic angles corresponding to MCNP6 computed angular distributions (i.e. Calc.) with the Dixon and Hughes column such that a consistent fitting technique is used between the compared results. The characteristic angle reported by Ross(8) was included to demonstrate that the fitting technique used by Dixon and Hughes gives characteristic angles that are in agreement with those reported by Ross.(8)

In general, the characteristic angles are overestimated as seen in Tables 4 and 5 and in Figs. 2 and 3. In Figs. 2 and 3, the angular distribution for 13- and 20-MeV electrons on several foils with the default physics are presented.

IV. Conclusions

The multiple-scattering validation test of the MCNP6 Monte Carlo code system reveals deficiencies in the treatment of angular deflection in MCNP6 for the default electron-photon transport settings. Moreover, refinement of the electron transport physics via *ESTEP* and *EFAC* only exacerbate the disagreement between the measured and calculated results. The cause of the disagreement remains an open question, but it is widely known that the class I condensed algorithm for electron transport contained in MCNP6 has inherent limitations. Of relevance to this work, these limitations include both the boundary crossing approximation and stepsize artifacts. Boundary crossing approximations are particularly sensitive when transporting electrons between high and low density regions. The Fano cavity test should be considered to quantify the impact of this effect. Stepwise artifacts result from the application of angular

Table 4: Comparison of measured and calculated characteristic angles for 13-MeV electrons on Be, Al, Cu, Ta, and Au. Three different physics configurations are studied: default, *ESTEP*, and *EFAC*=0.99.

				Default		*ESTEP*		*EFAC*	
Material	Thickness (mg/cm2)	Ross *et al.* (deg.)	Dixon & Hughes (deg.)	Calc. (deg.)	Rel. Diff. (%)	Calc. (deg.)	Rel. Diff. (%)	Calc. (deg.)	Rel. Diff. (%)
Be	926	8.143	8.089	8.363	3.382	8.846	9.353	9.081	12.26
Al	70.1	4.003	3.981	4.308	8.223	4.558	14.48	4.289	7.745
	140	5.268	5.226	5.602	7.196	5.798	10.93	5.652	8.150
Cu	43.0	4.219	4.167	4.541	8.968	4.745	13.86	4.450	6.786
	86.4	5.630	5.562	5.970	7.310	6.161	10.74	5.956	7.056
	129.6	6.861	6.803	7.115	4.593	7.323	7.645	7.210	5.989
	174.5	7.956	7.911	8.521	7.704	8.681	9.722	8.619	8.938
Ta	44.3	5.558	5.503	5.812	5.608	5.787	5.158	5.799	5.385
Au	31.2	4.878	4.798	5.061	5.473	5.346	11.43	5.164	7.626
	54.8	6.329	6.260	6.568	4.911	6.769	8.124	6.650	6.216
	93.7	8.243	8.231	8.812	7.052	9.013	9.495	8.986	9.163

Table 5: Comparison of measured and calculated characteristic angles for 20-MeV electrons on Be, C, Al, Cu, Ta, and Au. Three different electron physics configurations are studied: default, *ESTEP*, and *EFAC*=0.99.

				Default		*ESTEP*		*EFAC*	
Material	Thickness (mg/cm2)	Ross *et al.* (deg.)	Dixon & Hughes (deg.)	Calc. (deg.)	Rel. Diff. (%)	Calc. (deg.)	Rel. Diff. (%)	Calc. (deg.)	Rel. Diff. (%)
Be	926	5.238	5.214	5.434	4.234	5.694	9.208	5.839	11.99
C	546	5.132	5.108	5.198	1.764	5.456	6.825	5.561	8.873
Al	70.1	2.653	2.634	2.873	9.094	3.148	19.51	2.877	9.221
	140	3.484	3.463	3.657	5.611	4.018	16.02	3.823	10.39
	274	4.777	4.750	4.978	4.798	5.269	10.94	5.268	10.91
Cu	43.0	2.790	2.768	3.025	9.284	3.279	18.49	2.999	8.375
	86.4	3.714	3.685	3.891	5.605	4.245	15.21	4.041	9.666
	129.6	4.493	4.454	4.758	6.843	4.979	11.81	4.900	10.02
	174.5	5.198	5.147	5.429	5.496	5.724	11.22	5.688	10.51
Ta	206.3	7.913	7.809	8.207	5.099	8.306	6.364	8.793	12.60
Au	54.8	4.127	4.111	4.382	6.590	4.652	13.17	4.563	10.99
	164.2	7.278	7.258	7.593	4.605	7.775	7.113	8.025	10.57

deflection data that is not fully converged. That is, MCNP6 currently allows for no more than 240 series solution terms when computing the angular deflection distributions. In many cases, additional terms are required to ensure that the solution has converged. The code was modified to determine if allowing more terms would improve the angular distributions. We found that allowing for several thousand terms in the computation of the underlying angular deflection distributions, the agreement in the characteristic angle improved by roughly 1% for each of the different physics settings.

Acknowledgments

This work was funded by the US Department of Energy's National Nuclear Security Administration, under Los Alamos National Security, LLC, Contract No. DE-AC52-06NA25396.

The authors would like to give many thanks to Daren Sawkey for his careful explanation of the validation process for the multiple-scattering experiment.

References

1) T. Goorley et al., "Initial MCNP6 Release Overview," *Nuclear Technology*, **180**, 298 (2012).
2) D. J. Whalen, D. E. Hollowell, and J. S. Hendricks, "MCNP: Photon Benchmark Problems," LA-12196, Los Alamos National Laboratory (1991).
3) D. P. Gierga and K. J. Adams, "Electron/Photon Verification Calculations Using MCNP4B," LA-13440, Los Alamos National Laboratory (1999).
4) P. Libby, G. Hughes III, and J. Goorley, "Electron Transmission and Backscatter Verification Calculations Using MCNP5," LA-UR-03-5751, LANL (2003).
5) D. Dixon and H. Hughes, "Validation of MCNP6 for Electron Energy Deposition in Extened Media," *Trans. Am. Nucl. Soc.* (2015).
6) H. G. Hughes, "Improved Logic for Sampling Landau Straggling in MCNP5," LA-UR-05-4404, Los Alamos National Laboratory (2005).
7) H. G. Hughes, "Recent Developments in Low-Energy Electron/Photon Transport for MCNP6 - Revision 4," LA-UR-12-24213 rev 4, Los Alamos National Laboratory (2013).
8) C. K. Ross et al., "Measurement of Multiple Scattering of 13

and 20 MeV electrons by thin foils," *Med. Phys.*, **35**, 4121–4131 (2008).
9) S. Goudsmit and J. L. Saunderson, "Multiple Scattering of Electrons," *Phys. Rev.*, **57**, 24–29 (1940).
10) S. Goudsmit and J. L. Saunderson, "Multiple Scattering of Electrons. II," *Phys. Rev.*, **58**, 36–42 (1940).
11) L. Landau, "On the Energy Loss of Fast Particles by Ionization," *J. Phys.*, **8**, 201 (1944).
12) H. G. Hughes, "Treating Electron Transport in MCNP," LA-UR-96-4583, Los Alamos National Laboratory (1996).
13) S. M. Seltzer, "An Overview of ETRAN Monte Carlo Methods," *Proc. Monte Carlo Transport of Electrons and Photons*, volume 38 of *Ettore Majorana International Science*, p. 153–181, Physical Sciences, Plenum Press, 1988.
14) B. A. Faddegon et al., "Accuracy of EGSnrc, Geant4, and PENELOPE Monte Carlo Systems for Simulation of Electron Scattering in External Beam Radiotherapy," *Phys. Med. Biol.*, **54**, 6151-6163 (2009).

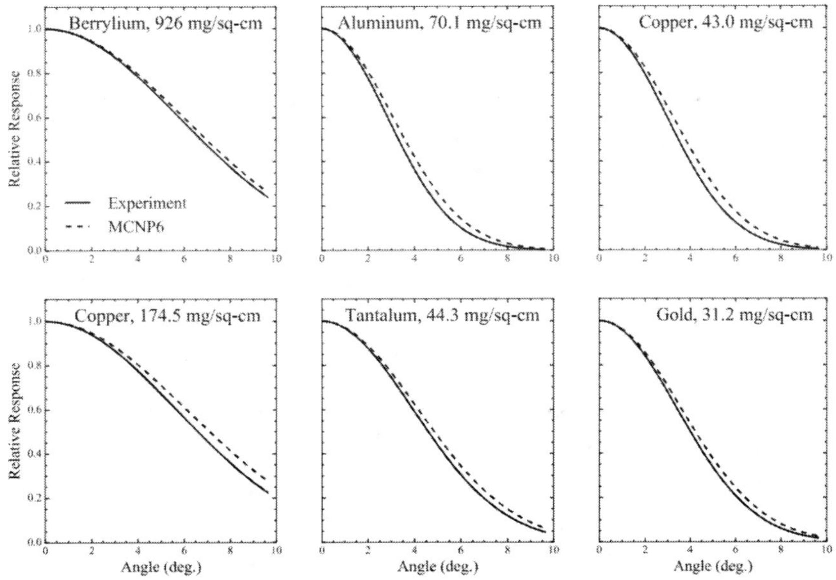

Figure 2: Comparison of Gaussian fits to experiment (solid curve) and calculated (dashed curve) angular distributions for 13-MeV electrons on various foils.

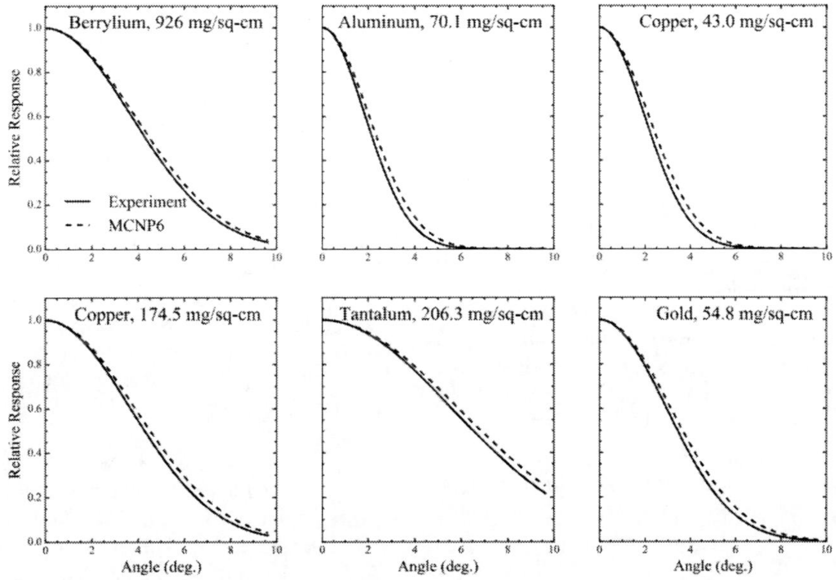

Figure 3: Comparison of Gaussian fits to experiment (solid curve) and calculated (dashed curve) angular distributions for 20-MeV electrons on various foils.

Performance Study of Monte Carlo Codes on Xeon Phi Coprocessors — Testing MCNP 6.1 and Profiling ARCHER Geometry Module on the FS7ONNi Problem

Tianyu Liu[1], *Noah* Wolfe[1], *Hui* Lin[1], *Kris* Zieb[1], *Wei* Ji[1], *Peter* Caracappa[1], *Christopher* Carothers[1], and *X. George* Xu[1,*]

[1]*Nuclear Engineering, Rensselaer Polytechnic Institute (RPI), Troy, New York, USA, 12180*

Abstract. This paper contains two parts revolving around Monte Carlo transport simulation on Intel Many Integrated Core coprocessors (MIC, also known as Xeon Phi). (1) MCNP 6.1 was recompiled into multithreading (OpenMP) and multiprocessing (MPI) forms respectively without modification to the source code. The new codes were tested on a 60-core 5110P MIC. The test case was FS7ONNi, a radiation shielding problem used in MCNP's verification and validation suite. It was observed that both codes became slower on the MIC than on a 6-core X5650 CPU, by a factor of ~4 for the MPI code and, abnormally, ~20 for the OpenMP code, and both exhibited limited capability of strong scaling. (2) We have recently added a Constructive Solid Geometry (CSG) module to our ARCHER code to provide better support for geometry modelling in radiation shielding simulation. The functions of this module are frequently called in the particle random walk process. To identify the performance bottleneck we developed a CSG proxy application and profiled the code using the geometry data from FS7ONNi. The profiling data showed that the code was primarily memory latency bound on the MIC. This study suggests that despite low initial porting effort, Monte Carlo codes do not naturally lend themselves to the MIC platform — just like to the GPUs, and that the memory latency problem needs to be addressed in order to achieve decent performance gain.

1 Introduction

In recent years hardware acceleration using Many Integrated Core coprocessors (MICs) made by Intel or Graphics Processing Units (GPUs) by Nvidia has become increasingly common in scientific computing. As of June 2016, these two types of accelerators have been employed in 89 supercomputers on the Top-500 list [1] and, most impressively, in 19 out of the top 20 supercomputers on the Green-500 list [2] which ranks the systems in energy efficiency (floating-point operations per Joule). Several national laboratories in the U.S. have been building their next-generation supercomputers based on accelerators, notably "Trinity" at LANL, "Cori" at NERSC, "Aurora" at ANL, "Summit" at ORNL and "Sierra" at LLNL.

With the advent of these new platforms, of our particular interest in nuclear engineering and science community is how to effectively speed up some routinely used but extremely slow applications such as Monte Carlo simulation of radiation transport. Two specific questions from developers are: ① *how hard is it to port existing codes to accelerators, how good is the performance and what is the bottleneck*, ② *how hard is it to perform accelerator-specific optimization?*.

*e-mail: xug2@rpi.edu

1.1 Easiness of porting

Currently the MICs (Knights Corner generation) and GPUs (Kepler and Maxwell generations) are not binary compatible with the CPUs, which means existing programs cannot directly run on accelerators.

For GPUs, the codes need to be rewritten in Nvidia's GPU-specific Application Programming Interface (API) called CUDA [3]. Users' programming responsibility typically includes determining parts of codes to be run on the GPUs in parallel, devising appropriate multithreading strategy, modifying those parts of codes as GPU kernel or device functions, managing GPU memory whose address space is separate from the host system memory, and controlling data transfer between the host and GPUs. At present, porting large-scale production or legacy codes completely to CUDA could be a onerous task despite not being entirely impossible. None the less, over years we have observed a steady evolution in CUDA with respect to programmability. Examples include the development of "CUDA runtime API" built on the original low-level driver API, which significantly reduces the amount of boilerplate codes and improves readability, "unified memory" which carries the burden of memory management to some extent by eliminating the need for explicit data copy. The cost of porting will likely continue to drop in the years to come. In addition, alternative languages do exist, such as

 DOI: 10.1051/epjconf/201715306022

the compiler directive based OpenACC [4] and new version of OpenMP (> 4.0) [5], to facilitate code porting at the cost of less functionality and lower performance than CUDA.

For MICs, code porting is usually easier. The fastest solution is a straightforward "recompile and run". Using special compiler flags MIC-compatible programs can be directly generated, and the MIC coprocessor is simply used as a separate compute node. Production codes can be ported this way effortlessly to enable a quick test drive. The drawback, however, is that the entire program has to run on the MICs and the serial parts can be very slow due to low single-core performance. Besides, the entire data to be used by the serial and parallel parts need to fit the memory on the MIC node and be copied to it beforehand. Solutions without this limitation are available but require code changes. Examples are (1) Intel MPI [6]. It allows communications between MPI processes on the host node and MIC node, thus making CUDA-like, heterogeneous computing possible. (2) Compiler directive based languages such as OpenACC, OpenMP and Intel offload pragma [7]; (3) Intel APIs such as "Cilk Plus" [8] which uses a shared memory model analogous to CUDA's unified memory, "Coprocessor Offload Infrastructure (COI)" [9] which is a low-level, versatile API that permits fine control of memory allocation and copy.

1.2 Potential problems of directly ported code

While MIC's "recompile and run" significantly reduces the initial porting cost, we have some *theoretical* concern over the performance of Monte Carlo codes generated this way. (1) The MICs are most suited to vectorized computations for compute-bound programs. The unique Single-Instruction Multiple Data (SIMD) 512-bit wide registers on the MICs allow simultaneous operations on 8 double-precision data at a single instruction. However, the history-based Monte Carlo codes cannot directly benefit from this feature: Each compute thread still tracks a single particle at a time instead of 8; the prevalent `do-while` loops that require indefinite steps to exit, and the data dependency across different iterations within a loop, essentially inhibit compiler's capability to perform automatic vectorization. (2) The Knights Corner (KNC) generation of MICs use the legacy in-order execution processors. Such processors will stall when waiting for the data to be loaded from the memory to the cache then to the registers. Hardware threading on the MICs mitigates this problem but to a limited degree. Monte Carlo codes usually require intensive memory access in a random, scattered pattern, and there is a fair chance that data are not available in the cache when needed for computation and must be loaded from the memory, leading to processor stall and long latency.

This paper uses *experimental* approach to answer question ① by examining two Monte Carlo codes on the MIC platform. In the first part of the paper, MCNP 6.1 was recompiled into MIC-compatible parallel codes with multithreading and multiprocessing capacity, respectively. The performance was studied using a strong scaling test. In the second part, the performance of a Constructive Solid Geometry (CSG) proxy application in ARCHER was characterized. Both experiments used the radiation shielding benchmark problem "FS7ONNi" in MCNP's verification and validation suite.

2 Part 1: Testing of MCNP 6.1

2.1 Method

The MCNP 6.1 source code and build system (using the GNU make tool set) support build configurations for both OpenMP (shared memory model) and MPI (distributed memory model) parallel computing approaches. Cross compilation of the large MCNP 6.1 code base was performed on the host x86/x86_64 machine in order to generate MIC compatible binary executables to run on the MIC device. The Intel fortran ("ifort") compiler was used to compile the fortran code, and the Intel C ("icc") and MPI ("mpiicc") compilers were used to compile the C code for OpenMP and MPI executables respectively. The version of these tools is 16.0.2. During cross compilation, the `-mmic` flag was added to both the compiler and linker to reference the MIC compatible libraries so that the emitted binary was able to execute the Initial Many Core Instructions (IMCI) on the MIC.

The test case FS7ONNi is a neutron transport problem, where a disc source is placed in a room ($9.0\times6.9\times6.8\ m^3$) with multiple shielding objects in it. There are 178 cells and 61 surfaces in total. The flux at a certain point in the room (point detector tally) is calculated.

The MIC coprocessor is 5110P with KNC architecture, which has 60 cores at 1.052 GHz and 7.75 GB memory (ECC enabled). Two CPUs were used to compare against the MIC. They are X5650 CPU with Westmere architecture and 6 cores at 2.66 GHz, and E5-2697 v3 CPU with Haswell architecture and 14 cores at 2.6 GHz. Each core of the MIC supports four hardware threads (also commonly known as hyperthreads), whereas that of the CPUs support two.

2.2 Result

2.2.1 OpenMP/MPI on a single MIC

Both the OpenMP and MPI versions of MCNP were run in the "native" execution mode. In this mode, the simulation was entirely performed on the MIC alone. The executable files, cross-section data, user-input, and dynamic libraries were uploaded to the MIC in advance.

In the first test one MIC coprocessor was used. The threads or processes were evenly distributed among 60 physical cores. Simulation with only 1 particle was performed at first to determine the time of initialization and finalization. This process was found to be ~22 seconds on the KNC MICs (due to low single-core performance) and ~2 seconds on the Westmere and Haswell CPUs. This value was subtracted from the time of complete simulation to derive the time of parallel particle transport.

DOI: 10.1051/epjconf/201715306022

The scalability of OpenMP and MPI on the MIC is compared in Figure 1. The sweet spot was found to be 60 processes for the MPI version with 1 process pinned to each physical core, and 40 threads for the OpenMP version with one thread on each physical core where some cores were unused. At 60 or less MPI processes or OpenMP threads, one process or thread possesses full utilization of its corresponding core's cache and memory access system. The particle transport process consists of a large amount of irregular memory accesses. Therefore, memory latency is vital in order to supply the execution pipelines of each core. Moving past 60 MPI processes, or OpenMP threads, increases the demand on each core's cache and memory access system. As a result, performance is diminished as execution pipelines are starved waiting for memory. Furthermore, it was observed that over subscription of MIC resulted in serious performance degradation. For example, when 480 threads were launched, the computation time almost quadrupled that of 240 threads. Such cases always have remarkably high "system CPU time", which is presumably caused by a disproportionate increase in the parallel overhead. It is worth mentioning that despite better performance, MPI with distributed memory model is not intended for use on a single node such as one MIC card, as the memory is soon to be consumed by the processes with separate address space. In fact, for our test case one MIC can only run 120 processes at a time.

According to [10], the scalability of multithreaded OpenMP was largely reduced when the compiler flag `-heap-arrays` was applied to an earlier version of compiler. We compiled the program with `-heap-arrays` and `-no-heap-arrays` (this is also compiler's default flag) flags respectively and compared their performance in Figure 1. In general, `-no-heap-arrays` resulted in faster code, since this flag places statically allocated arrays on the stack which has faster access than the heap memory. However, it did not improve scalability appreciably.

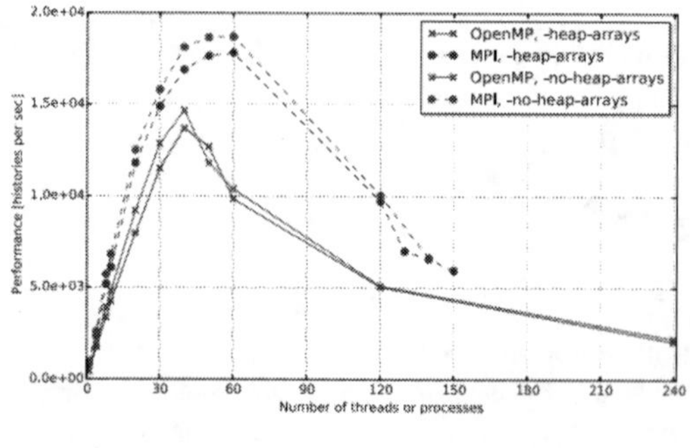

Figure 1: Comparison of MCNP 6.1 OpenMP and MPI performance when the program is compiled with `-heap-arrays` and `-no-heap-arrays` flags respectively. Only the time of parallel particle transport is considered. The performance is normalized to particle histories simulated per second.

The OpenMP scalability of MCNP 6.1 on the MIC and CPUs are compared in Figure 2. On both CPUs the scalability are much better, as linearity holds until all the physical cores are used, beyond which hardware threads help to hide memory access latency and improve the performance to some extent. Compared to Westmere and Haswell CPUs, the peak performance on the MIC was found to be lower by a factor of 4.2× and 14×, respectively, shown in Table 1. This is believed to be a result of the underlying interconnection network of cores and memory controllers within the MIC architecture. The current generation MIC based on KNC architecture uses a bi-directional multi-layered ring interconnect for data transfer between cores and memory controllers and can quickly get saturated with the irregular memory access of MCNP. The future Knights Landing (KNL) and Knights Hill (KNH) architectures with their proposed 2D mesh interconnect may benefit memory performance.

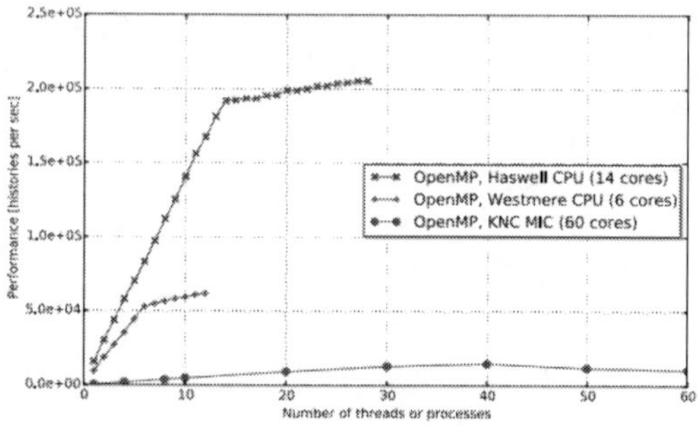

Figure 2: Strong scaling of MCNP 6.1 on the MIC and CPUs. Only the time of parallel particle transport is considered. The performance is normalized to particle histories simulated per second. The Haswell CPU outperforms KNC MIC by a wide margin.

Table 1: Peak performance comparison of MCNP 6.1 on the CPUs and MIC. The compute time of the parallel particle transport is normalized to per 1 million particles.

Processor	Number of execution unit	Compute time[sec]
6-core Westmere CPU	12 threads	16
14-core Haswell CPU	28 thread	4.9
60-core KNC MIC	40 threads	85

3 Part 2: Development and Profiling of CSG Module in ARCHER

3.1 Method

3.1.1 CSG Module and Data Structure

The CSG module in ARCHER is based on OpenMC [11], an open source neutronics Monte Carlo transport code in Fortran. Several basic features have been inherited from it, such as the capability to handle universe-based geometries

DOI: 10.1051/epjconf/201715306022

and lattice structures. Our modifications include porting it to ARCHER's CPU-GPU-MIC framework, rewriting in C++11, adding the functionality of reading and parsing MCNP input files.

The CSG data consists of cell and surface arrays, shown in Figure 3. Both arrays are a list of pointers, with each element pointing to an object. The objects are not contiguous in memory since they are created at different points when the MCNP input file is parsed. The cell objects are instantiated from a single class, whose data members include cell ID, universe ID, material ID and a vector of indices of the bounding surfaces. The surface objects vary in size, as they are instantiated from different classes (planes, cylinders, spheres, etc). These surface classes are derived from a common base class whose data members include surface ID and vectors of indices of the cells in the positive or negative sense of the surface. The difference of these surface classes is the number of geometry parameters. For instance, a plane perpendicular to x axis contains 1 parameter ($x0$), while a sphere contains 4 ($x0, y0, z0, r$). In order to obtain all data in a cell or surface object from the memory, the program needs to undergo a few levels of indirection — the pointer of the object, the object itself, and the index vectors. This data layout indicates that memory access can be subject to long latency in case of cache misses on the MIC, whereby the data are not readily available in the cache. Cache misses are simulated in our proxy application described in the next section.

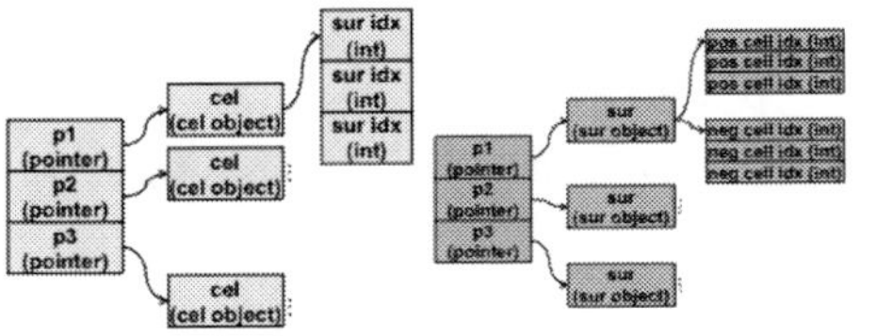

Figure 3: Memory layout of CSG cell (yellow) and surface (green) data structure. Separate boxes indicate that the data are not contiguous in memory. Accessing all data in a cell or surface requires several levels of indirection.

3.1.2 CSG Proxy Application

The functions in the CSG module are intensively used in the particle tracking process. In order to study the properties of this module we developed a proxy application for it. A proxy application [12] is *a catchall term for the simplification of characteristics of real applications that are of interest to DOE*. Our CSG proxy application falls into the "mini app" category where the CSG module combined with an artificial, simplified interaction model forms a stand-alone application. The interaction model assumes one-group transport and only simulates absorption ($\Sigma_a = 0.001 cm^{-1}$) and isotropic scattering ($\Sigma_a = 0.099 cm^{-1}$). The particle's path-length in the entire spatial region is scored as the program output.

The workflow of the CSG proxy application is shown in algorithm 1. In a full Monte Carlo transport code, some events such as constructing macroscopic cross-sections, interpolating pretabulated data, and accumulating tallies, entail memory read or write operations to other memory and can cause the CSG data to be evicted from the cache. This effect has been taken into account in the proxy application. Specifically, all the cell and surface objects were manually evicted from both L1 and L2 cache on the MIC. The eviction was implemented using compiler intrinsic function `_mm_clevict(p, hint)`, where `p` is the memory address and `hint` specifies eviction mode. Allowed options include eviction from L1 only and from both L1 and L2.

Table 2 lists three most time-consuming functions in the CSG proxy application. The instrumented function `EvictCSGCache` appears on the top mainly because the overhead of artificial L2 cache eviction is very large. Our current focus is to study the 2nd and 3rd hotspots.

Table 2: Top 3 hotspots in the CSG proxy application.

Function	Time percentage [%]
EvictCSGCache	56%
FindCell (FC)	13%
FindDistanceToBoundary (FD)	12%

To characterize the performance of CSG proxy application, we derived the following 5 performance metrics from the profiling results. The exact definition of these metrics can be found in Intel's reference [13]. The profiler "Intel VTune Amplifier XE 2016" was used to collect the metrics.

- *Vectorization Intensity (VI).*
  A measure of how effectively the 512-bit wide SIMD register is utilized overall. Higher is better. On MICs both scalar and vector operations use vector registers via masks. For pure scalar operations, one bit of the mask is set so that only one 64-bit double-precision data element is applied (VI=1), while for full vector operations, all bits of the mask are set, allowing all 8 data elements (VI=8).
- *L1 compute to data access ratio (L1CD).*
  A measure of how many computations are performed per L1 cache access. Higher is better. When L1CD is smaller than VI, the program is considered not computationally dense.
- *L2 compute to data access ratio (L2CD).*
  A measure of how many computations are performed per L2 cache access. Higher is better. When L2CD is smaller than $100 \times$ L1CD, the program is considered to have too many L1 cache misses [13].
- *L1 hit rate (L1H).*
  A measure of the probability that the data hits L1 cache per memory access. Higher is better. This metric is a conservative estimate of L1 hits, as it takes into account a special type of L1 cache misses, where the data are not in L1 cache but are being prefetched into L1. When L1H is smaller than 95%, the program is considered to have too many L1 cache misses [13].

 DOI: 10.1051/epjconf/201715306022

- *Estimated Latency Impact (ELI).*
  An approximation of the clock cycles spent on L2 cache access and memory access per L1 cache miss. Lower is better. On MICs, a conservative estimate of L2 hit rate is unavailable due to the limitation that the special type of L2 cache misses, where the data are not in L2 cache but are being prefetched into L2, cannot be properly determined. Intel recommend using ELI as a workaround. If all data not in L1 are found in L2, then ELI=21 (lower bound), which is the clock cycles of a L2 cache access after a L1 cache miss. If there are L2 cache misses, it will take extra hundreds of clock cycles to load the data from the system memory. When ELI is larger than 145 clock cycles, the program is considered to suffer from long access latency due to L2 cache misses [13].

3.2 Results

3.2.1 Performance Characterization

The profiling results for the two time-consuming functions `FindCell` and `FindDistanceToBoundary` are listed in Table 3. The VI values of both functions are 2, which means the SIMD feature of MIC is almost completely untapped. In fact, by checking compiler's vectorization report, no automatic vectorization was made for these functions at all, and VI values should be 1. It is speculated that the mask operations in scalar calculations also count as vector operations and hence contribute to VI. The fact that L1CD values are lower than VI is an indicator that the code is not computationally dense but instead memory access demanding. Specifically, the cache utilization is inefficient, confirmed by the borderline low L1H, the significantly low L2CD, and the high ELI value. These values suggest that due to L1 cache (only costing 1 clock cycle) misses, access to L2 cache (costing 21 clock cycles) is frequent. To exacerbate the situation, those data that miss L1 cache mostly do not hit L2 cache, and thus can only be loaded from the main memory. The resulting performance penalty is quantified by the ELI values, which exceed the 145 clock cycle threshold. These data indicate that the two functions are memory latency bound.

Table 3: Profiling result of ARCHER's CSG proxy application on the MIC. Refer to subsubsection 3.1.2 for the metrics' full name. FC: FindCell, FD: FindDistanceToBoundary.

Metrics	FC	FD	Investigate if
VI	2	2	< 8
L1CD	1.3	1.4	< VI
L1H	96%	92%	< 95%
L2CD	34	18	< 100 × L1CD
ELI	1135	972	> 145

It should be pointed out that that the geometry data in the FS7ONNi test case are not sufficiently large in size. Therefore two important metrics — L1 and L2 translation lookaside buffer (TLB) miss ratios [13] were found trivial in this performance study. They are, however, useful when data significantly exceed the memory page size, i.e. 2 MB on the MIC.

The result of the strong scaling test is shown in Figure 4. Again, only the parallel particle transport was timed. The OpenMP thread affinity was set to "balanced" whereby threads are evenly spread across 60 cores. Scalability observed was not satisfactory as the performance increase started to become sublinear before 60 physical cores are used. However, what is better than part 1 is that the increase continued and reached its peak when all 240 hardware threads were used. Here the compute performance of MIC and CPU codes was not compared. This is because on the CPU, there is no intrinsic function equivalent to `_mm_clevict(p, hint)` that simply performs cache eviction. The most similar, CPU-specific intrinsic function `_mm_clflush(p)` invalidates the cache line but also writes data back to the memory. Therefore it has large overhead and the performance result would be heavily biased against the CPU.

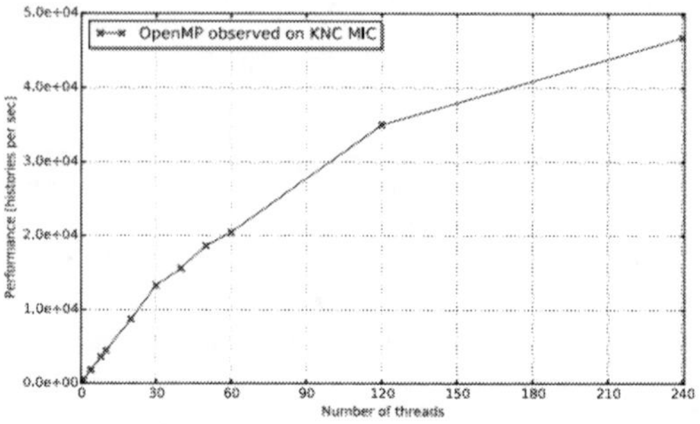

Figure 4: Strong scaling of ARCHER's CSG proxy application on the MIC. Only the time of parallel particle transport is considered. The performance is normalized to particle histories simulated per second.

4 Discussion

The bottleneck of Monte Carlo codes on the MICs are the lack of vectorization and excess of memory access latency. Solutions to these problems exist, in spite of being only effective to some specific parts of Monte Carlo codes. For example, the macroscopic cross-section construction subroutine is usually one of the hotspots in neutronics transport simulation. This subroutine has a relatively simple structure and offer good opportunity for optimization. In our recent study [14], we optimized the XSBench code [15] to the accelerators. XSBench abstracts the macroscopic cross-section construction process from a full Monte Carlo code OpenMC [11]. Compared to the original XSBench run on a 6-core CPU, our directly ported codes were found to have a speedup factor of only 2.7× on a MIC (60 cores) and 1.4× on a GPU (15 streaming processors), whereas these numbers rose to 6.0× and 8.1×, respectively, as a result of several optimization techniques that focuses on vectorizing the computation and hiding memory latency. Another example is the study by

[16], where the OpenMC code was tested on the MICs and OpenMP-specific optimizations were made to reduce tally overhead, such as replacing OpenMP's `critical` sections with `atomic` operations.

5 Conclusions

In this study we tested two Monte Carlo codes on the Knights Corner generation of MIC coprocessor: the production code MCNP 6.1, and the CSG module of our ARCHER code developed specifically for the heterogeneous architecture. The experiments have confirmed our concern that the Monte Carlo codes do not naturally lend themselves to the KNC MIC coprocessors, because the 512-bit wide SIMD registers are underutilized, and the random, scattered memory access pattern is not cache-friendly and causes long memory latency. Such problems also occur to the GPUs which implement Single instruction, multiple thread (SIMT) model and have even less amount of cache per thread. Software optimizations and algorithm improvements must be conducted in order for Monte Carlo codes to achieve decent performance on accelerators.

Acknowledgments

We thank OpenMC community for the high quality code and Intel Corp. for the hardware donation.

References

[1] *Top 500 list* (2016), `https://www.top500.org`

[2] *Green 500 list* (2016), `http://www.green500.org/`

[3] Nvidia, *CUDA C Programming Guide* (2016)

[4] OpenACC committee, *The OpenACC application programming interface, version 2.5* (2015)

[5] OpenMP committee, *The OpenMP application programming interface, version 4.5* (2015)

[6] Intel, *Intel MPI library reference manual for Linux OS* (2016)

[7] Intel, *Intel C++ compiler 16.0 user and reference guide* (2016)

[8] Intel, *Intel Cilk Plus language extension specification, version 1.2* (2013)

[9] C.J. Newburn, S. Dmitriev, R. Narayanaswamy, J. Wiegert, R. Murty, F. Chinchilla, R. Deodhar, R. McGuire, *Offload compiler runtime for the Intel Xeon Phi coprocessor*, in *Parallel and Distributed Processing Symposium Workshops & PhD Forum (IPDPSW), 2013 IEEE 27th International* (IEEE, 2013), pp. 1213–1225

[10] O. Mikhail, *Openmp and -heap-arrays not compatible since ifort 13?* (2016)

[11] P.K. Romano, B. Forget, Annals of Nuclear Energy **51**, 274 (2013)

[12] M. Heroux, R. Neely, S. Swaminarayan, Tech. Rep. LA-UR-13-20460 and LLNL-TR-592878, Sandia, LLNL, LANL (2013)

[13] S. Cepeda, *Optimization and Performance Tuning for Intel Xeon Phi Co-processors, Part 2: Understanding and Using Hardware Events* (2012)

[14] T. Liu, N. Wolfe, C.D. Carothers, W. Ji, X.G. Xu, *Optimizing the Monte Carlo neutron cross-section construction code, XSBench, for MIC and GPU platforms (in press)* (2016)

[15] J.R. Tramm, A.R. Siegel, T. Islam, M. Schulz, *XSBench - The development and verification of a performance abstraction for Monte Carlo reactor analysis*, in *PHYSOR 2014 - The Role of Reactor Physics toward a Sustainable Future* (Kyoto, 2014)

[16] D. Ozog, A.D. Malony, A. Siegel, *Full-core PWR transport simulations on Xeon Phi clusters*, in *M&C + SNA + MC 2015 — Joint International Conference on Mathematics and Computation (M&C), Supercomputing in Nuclear Applications (SNA) and the Monte Carlo (MC) Method* (American Nuclear Society, 2015)

Appendix

```
input : Geometry data
input : n: number of histories
output: Particle path-length in all cells

for i ← 0 to n − 1 do
    InitializeParticle()
    while true do
        if cell not found then
            FindCell()
        end
        d = FindDistanceToBoundary()
        s = SampleDistanceToCollisionSite()
        UpdatePosition()
        ScorePathLength()
        EvictCSGCache() // manually evict data
        if d < s then
            BoundaryCrossing() // contains calls of
            FindCell()
        else
            Collision()
        end
        if particle is killed then
            break
        end
    end
end
```

Algorithm 1: Pseudocode of the CSG proxy application

DOI: 10.1051/epjconf/201715306023

Quantification of the validity of simulations based on Geant4 and FLUKA for photo-nuclear interactions in the high energy range

Lina Quintieri[1,*], *Maria Grazia* Pia[2], *Mauro* Augelli[3], *Paolo* Saracco[2], *Marco* Capogni[1], and *Guido* Guarnieri[4]

[1] *ENEA Centro Ricerche della Casaccia, Roma-Italy*
[2] *INFN Sezione di Genova-Italy*
[3] *CNES, Toulouse-France*
[4] *ENEA Centro Ricerche di Portici, Napoli-Italy*

Abstract. Photo-nuclear interactions are relevant in many research fields of both fundamental and applied physics and, for this reason, accurate Monte Carlo simulations of photo-nuclear interactions can provide a valuable and indispensable support in a wide range of applications (i.e from the optimisation of photo-neutron source target to the dosimetric estimation in high energy accelerator, etc). Unfortunately, few experimental photo-nuclear data are available above 100 MeV, so that, in the high energy range (from hundreds of MeV up to GeV scale), the code predictions are based on physical models. The aim of this work is to compare the predictions of relevant observables involving photon-nuclear interaction modelling, obtained with *GEANT4* and *FLUKA*, to experimental data (if available), in order to assess the code estimation reliability, over a wide energy range. In particular, the comparison of the estimated photo-neutron yields and energy spectra with the experimental results of the n@BTF experiment (carried out at the Beam Test Facility of DaΦne collider, in Frascati, Italy) is here reported and discussed. Moreover, the preliminary results of the comparison of the cross sections used in the codes with the"evaluated' data recommended by the IAEA are also presented for some selected cases (W, Pb, Zn).

1 Introduction

The interest in a reliable and accurate implementation of photo-nuclear physics in Monte Carlo codes spans over a wide range of applications and is ever more increasing. Photonuclear physics plays a relevant role in many fundamental research and applicative fields: from the design and optimisation of targets for photo-hadronic sources (i.e accelerator driven neutron and muon source, radionuclide production for pharmaceutical applications, etc.), to dosimetry and radioprotection concerns in high energy physics (i.e high intensity and high brilliance gamma sources), from theoretical astrophysics, where the photonuclear could have a fundamental role in stellar nucleosynthetic, to the plasma physics of fusion reactors. In this frame, the goal of this work is to investigate quantitatively the reliability of widely used Monte Carlo codes to simulate experimentally relevant observables involving photon-nuclear interaction modelling. We have started an assessment of the reliability of predictions of GEANT4[1] and FLUKA [2] and some preliminary results of this complex work (still in progress) are presented and discussed in this report. The comparison of measured observables with simulated ones, in a high energy scenario for which experimental data are available, is also presented. The comparison relies on suitable quantitative validation tests, based on statistical data analysis methods (i.e.χ^2 test). The difference in prediction capabilities of different Monte Carlo codes can originate mainly at two different levels, that represent the sequential steps of the physics implementation inside the code itself:

- first level: different used data library cross sections
- second level: different reconstruction of the secondary final state (energy spectra, angular distribution, etc)

As first step, the investigative analysis (to find the reason of the possible differences) foresees to compare the cross sections used in the specified codes with the IAEA recommended photonuclear cross sections[3], mainly focusing on selected materials, widely used in high energy accelerator context (lead, tungsten, copper, etc). The IAEA cross sections have been chosen as reference term for the comparison, since this data collection has been obtained by an accurate selection of the worldwide experimental measurements, in order to provide the scientific community with a "recommended" and "selected" evaluated photonuclear data. This collection contains a relatively complete photo-nuclear data for 164 isotopes and has been made available in 2000 as ENDF-6 formatted files. In case of lack of experimental data, the comparison among the results obtained with independent codes, even if not appropriate to validate the simulation observables, could highlight special needs of adopting conservative approach

*e-mail: lina.quintieri@enea.it

for those cases potentially found not in agreement or, in addition, address experiments for benchmarking purposes in the future high-energy neutron calibration fields or high brilliance and high intensity gamma source (i.e, ELI [4]).

2 Physics Overview

Photo-nuclear interactions consist of absorption of a photon by a nucleus, that is brought in an excited state. The excited nucleus then undergoes the de-excitation process, emitting secondary particles (neutron, protons, pion, etc) and possibly undergoing fission, depending on the photon energy and target nuclei. The particle photo-production is a threshold reaction (a nucleon can be delivered from a nucleus only if the energy transferred by the photon overcomes the average binding energy of its nucleons). This energy threshold is lower for heavy nuclei for which photo-nuclear reactions are almost equivalent to photo-neutron ones, since the emission of protons is strongly repressed by the large Coulomb barrier (whose height is proportional to the nucleus atomic number). The reverse happens below atomic number Z=20, for which proton yield is, in general, larger than the neutron one. The energy threshold for heavy nuclei goes from 5 to 7 MeV and for medium or light nuclei it is around 15 -20 MeV. Moreover, the photo-neutron cross sections exhibit a maximum at photon energy between 13-18 MeV for heavy nuclei and 20-23 MeV for light nuclei (A < 40). nucleons).

For modelisation purpose, five energy regions generally can be distinguished:

- Nuclear Resonance Fluorescence (E_γ < 2MeV). In this process the photon induces a magnetic dipole vibrational resonance of the nucleus. It is about a hundred times weaker than the giant dipole resonance and is extremely narrow, on the order a few keV. The key feature is that it produces characteristic, i.e. isotope dependent, gamma-ray emission lines
- Giant dipole resonance ($5 < E_\gamma < 30$ MeV). The electric field of the photon transfers its energy to the whole nucleus by inducing an oscillation (known as giant resonance oscillation), which leads to a relative displacement of tightly bound neutrons and protons inside the nucleus. Absorption of the incident photons excites the nucleus to a higher discrete energy state and the extra energy is emitted in the form of nucleons (neutron, proton, etc).
- Quasi deuteron absorption ($30 < E_\gamma < 140$ MeV). When the photon energy is greater than 35 MeV, the cross section for the Giant Resonance neutron production decreases rapidly. For $35 < E_\gamma < 140$ MeV, the photo-neutron production is due to Quasi Deuteron effect. In this process, the incident photon interacts with the dipole moment of a neutron-proton pair inside the nucleus rather than with the nucleus as a whole.
- Delta resonance ($E_\gamma > 140$ MeV). Above 140 MeV photo-neutrons are produced via photo-pion production
- Photo-meson region ($E_\gamma > 350$ MeV)

3 n@BTF: experimental results and Monte Carlo code benchmarking

High energy electrons interact with heavy target (i.e. material with high atomic number), producing mainly bremsstrahlung radiation with a continuous energy spectrum, whose end point is equal to the maximum primary electron energy. These secondary photons can excite the nuclei of the target, which decay back into the fundamental state by evaporating one or more neutrons or expelling other charged or neutral hadrons (pions, protons, etc.), depending on the primary electron energy and the target nucleus. A photo-neutron source by an electron or positron pulsed beam has been realised in the frame of the experiment "n@BTF"[5], in 2010, at the Beam Test Facility (BTF) [6] of the National Laboratory of Frascati, near Rome (Italy). The BTF is a transfer line, driven by a pulsed magnet that allows diverting electrons or positrons, normally released to the DaΦne [7] damping ring, towards a $100m^2$ experimental hall, where the n@BTF experiment has been carried out (Figure 1 shows the experimental set-up). In the n@BTF experiment, neutrons are produced sending 510 MeV electrons to impinge on an optimised tungsten target (7cm diameter, 6 cm height). The photons from the electromagnetic cascade may excite the W nuclei, resulting mainly in neutron production in the MeV region related to the Giant Resonance mechanisms, even if the energy spectrum spans over more than 9 decades of energy (from few meV up to hundred of MeV). Neutrons with maximum energy equal to that of the primary beam could be generated.

Figure 1. The n@BTF experimental set-up

The intensity of the neutron beam and its fluence rate energy distribution at a well established point of reference in the irradiation room were predicted by Monte Carlo simulations and measured with a Bonner Sphere Spectrometer (BSS). Due to the large photon contribution and the pulsed time structure of the beam, passive photon-insensitive thermal neutron detectors were used as sensitive elements of the BSS. For this purpose, a set of Dy activation foils was used. In the performed Monte Carlo simulations with GEANT4 and FLUKA all the characteristics of the available electron beam, as reported in Table 1 (primary beam energy spread, spot size, energy distribution, etc.), have been taken properly into account.

In Figure 2 the lethargic (E dΦ/dE) spectra, normalised to the total neutron flux, estimated by FLUKA (with version 2006.3 and confirmed recently with version

Table 1. BTF main electron beam parameters

Parameters	Value
Energy range	25-750 MeV
Transverse emittance (both planes) @510 MeV	1 mm mrad
Energy spread	1 %
Repetition rate	1-50 Hz
Number of particles per pulse	$1\text{-}10^{10}$
Macro bunch duration	1 or 10 ns
Spot size (at the line exit)	2mm (single particle) 2cm (high multiplicity)

Table 2. Total neutron flux per pimary particle [cm^{-2}]

Exp.measurement	$8.04 \cdot 10^{-7} \pm 3\%$
FLUKA	$8.10 \cdot 10^{-7} \pm 4\%$
Geant4	$4.81 \cdot 10^{-7} \pm 3\%$

2011.2c.3) and GEANT4 (version 10.1.0), respectively, are shown together with the normalised lethargic spectrum measured by the Bonner Spheres spectrometer. A good agreement between measurements and simulation is found for FLUKA, that provides an accurate reconstruction of the experimental resonance, both in energy position and in amplitude, as assessed by χ^2 test. On the contrary, the GEANT4 simulation of the "n@BTF " experiment gives results that underestimate of more than 30% the photo-neutron integrated yield. The experimental and calculated neutron fluxes, collected over a sphere of 10 cm diameter, whose centre is 1.49 m far away from the target along one of the two extraction lines (red arrows in Figure 3), are reported in Table 2.

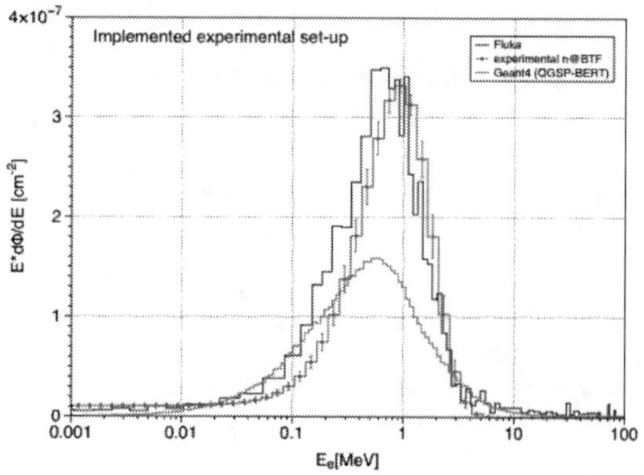

Figure 2. n@BTF fluence energy spectrum (normalized to the total flux) at the reference point (1.49 form the target center)

In order to assess and check what could be the cause of the different GEANT4 predictions, a more simplified geometry has been simulated in order to focus only on the photo-nuclear physics (reducing the spurious dependencies on other non relevant physics and geometric aspects). A "unit test" (identified as "n@BTF unit case") has been built to compare the different code estimations of photo-neutron escaping from a simpler geometry, made only of the n@BTF target in the vacuum and neglecting the complete experimental apparatus (beam dump, layered shielding box around the target itself, as shown in Figure 3).

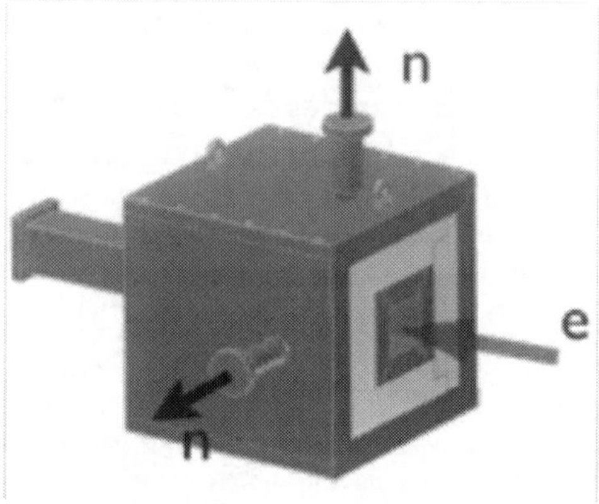

Figure 3. n@BTF: layered (polyethylene-lead) shield box around target and beam dump

In Figure 4 the energy spectra of the photo-neutrons produced in the target, estimated, respectively, with GEANT4 and FLUKA for the unit case, are compared. The results are reported in term of neutron per unit primary electron, integrated over all the solid angle and shown in lethargic representation. The estimations of the source term with the "n@BTF unit test case" give a definitive confirmation that the difference are due to the mechanism of production of neutrons in the target rather than in the neutron transportation and interaction with other materials.

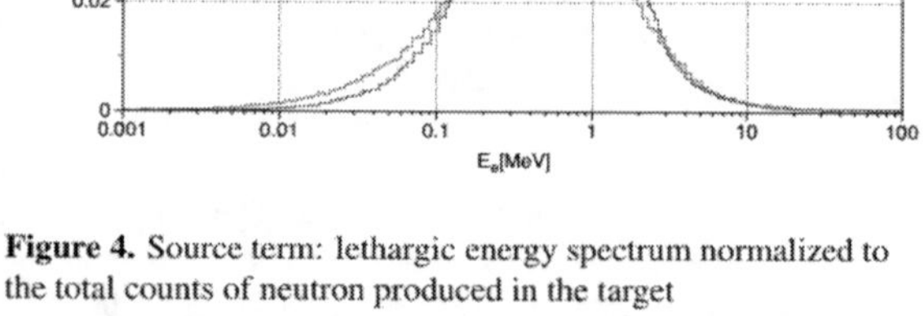

Figure 4. Source term: lethargic energy spectrum normalized to the total counts of neutron produced in the target

Furthermore, it has also been checked if the differences between the GEANT4 predictions and those of the other

Table 3. GEANT4 and FLUKA photo-production yields for 510 MeV electron on W-bulk

510 MeV Electrons	Neutron per primary	Proton per primary
Geant4 n@BTF unit test	0.12 ± 3%	1.74 10^{-3} ± 6%
Geant4 Hadr01	0.13 ± 3%	1.76 10^{-3} ± 6%
Fluka	0.22 ± 0.45%	1.44 10^{-3} ± 0.9%

Table 4. GEANT4 and FLUKA photo-production yield for 510 MeV gamma on W-bulk

510 MeV Gammas	Neutron per primary	Proton per primary
Geant4 n@BTF unit test	0.14 ± 3%	3.53 10^{-3} ± 6%
Geant4 Hadr01	0.15 ± 3%	3.55 10^{-3} ± 6%
Fluka	0.26 ± 0.12%	2.77 10^{-3} ± 2%

codes could be attributed to the different efficiency in conversion from high energy electrons to gamma showers. For this reason, several simulations have been performed (with GEANT4 and FLUKA) to analyse the results obtained respectively with 510 MeV mono-energetic gammas as primary beam and comparing the results with the photo-neutrons estimated with 510 MeV primary electrons. The results of these simulations allow to conclude that no significant difference is related to the electromagnetic conversion from high energy electrons to photons, since, in both cases, the photo-neutron yield predicted by GEANT4 underestimates of the same amount the value predicted by FLUKA, as documented in Table 3 and 4. Three "reference" physics lists (including hadronic and photo-nuclear physics) for GEANT4 have been alternatively used: QGS-BERT, QGS-BERT- HP and Shielding. The results obtained from these simulations are in excellent agreement. Finally, in order to check that the obtained results are not dependent by the specific built application, another independent application has been used: the 'Hadr01' from the extended examples provided in the GEANT4 packages (simply specifying the geometry and material of the target and the energy and type of the primary particles). The differences in the results from the "n@BTF unit case" and the "adapted Hadr01" case, could come mainly from the different scoring solutions implemented respectively in each application to estimate the neutron yield and energy spectrum. As shown in Table 3 and 4, the estimations of the neutron yields obtained with the two independent applications, in both cases of 510 MeV primary electrons and 510 MeV primary gammas, agree within few percent difference (<8%).

At this stage of the analysis it is not really clear why the prediction of GEANT4 are underestimating the experimental values, and the preliminary checks seem to demonstrate that the underestimation is not depending on the failure of the specific application used, even if further investigation in this sense can been pursued. More reasonably, differences can originate from the implementation of the photo-nuclear physics at some (still not well identified) stage. We want to individuate and possibly address the solution of some critical implementation in GENAT4 (if any, after careful investigation) or eventually to promote an improvement of the photo-nuclear implementation inside the code itself. The first step of the analysis is the comparison of the photo-nuclear cross sections used in each code with the evaluated IAEA photo-nuclear data, starting from those cases in which we are more directly interested. Anyway, the complete analysis will foresee to compare as exhaustively as possible all the photo-nuclear cross sections and finally to focus on the fitting procedure characteristic of each code to extend the cross sections in the energy range, when experimental data are missing or to evaluate (according the different parameterisation applied in each code) the cross sections for those cases non included in the data library. Successively, in a longest time scale, the models implemented to obtain the secondary particles and their final states (particle typology, energy dependence, angular cross section, etc.) will be also analysed and quantitatively compared.

4 Implementation of Photonuclear physics in Geant4

Photo-nuclear interaction cross sections in GEANT4 database are approximated for all the nuclei and all the energies (from hadron production threshold to about 50 GeV). The GEANT4 database is made of parameterised cross sections of 51 nuclei (see Table 5 and Table 6). The cross section parameterisation in the different energy range is expressed as function of: the Atomic mass "A", the energy "e" and several other parameters that are calculated for a list of selected 14 nuclei and interpolated for all the other nuclei. The photonuclear cross sections are parameterised in the *G4PhotoNuclearCrossSection*: the evaluated data of the 51 nuclei are inserted as G4double arrays and the fit algorithms for the parameterisation of all the rest of nuclei (not included in the GEANT4 data base) are implemented according five energy regions, each corresponding to the physical process that dominates.

- The Giant Dipole Resonance (GDR) region, depending on the nucleus, extends from 10 MeV up to 30 MeV. It usually consists of one large peak, though for some nuclei several peaks appear.
- The "quasi-deuteron" region extends from around 30 MeV up to the pion threshold and is characterised by small cross sections and a broad, low peak.
- The Δ region is characterised by the dominant peak in the cross section which extends from the pion threshold to 450 MeV.
- The Roper resonance region extends from roughly 450 MeV to 1.2 GeV. The cross section in this region is not strictly identified with the real Roper resonance because other processes also occur in this region.

 DOI: 10.1051/epjconf/201715306023

Table 5. Evaluated photonuclear cross sections available in high energy range (up to 50 GeV)

1H	2H	3He	4He	6Li	7Li	9Be	12C
16O	27A	Cu	Sn	Pb	238U		

Table 6. Evaluated photonuclear cross sections available in low energy range (up to 105 MeV).

2H	4He	6Li	7Li	9Be
12C	14N	15N	16O	19F
23Na	24Mg	27Al	28Si	32S
34S	40Ca	54Fe	55Mn	56Fe
58Ni	58.9Co**	63.5Cu**	65.4Zn**	76Se
82Se	107.9Ag	112.4Cd**	118.7Sn**	126.9I**
154Sm	156Gd	159Tb	165Ho	168Er
174Yb	178Hf	180Hf	181Ta	184W
186W	197Au	204.4Tl	207.2Pb**	209Bi
232Th	235U	238U	239Pu	

** Natural element (mixture of isotopes as happen in nature)

- The Reggeon-Pomeron region extends upward from 1.2 GeV.

Currently the 14 nuclei reported in Table 5 are used in the parameterisation to extrapolate toward high energy. Their evaluated cross sections sections are available up to 50 GeV[1]. The list of the the elements whose evaluated cross sections is available in GEANT4 from 2 MeV up to 105 MeV is reported in Table 5.

The photo-nuclear cross sections cover incident photon energies from the hadron production threshold upward. Final states in gamma-nuclear and electro-nuclear physics are described using chiral invariant phase-space decay at low energies, and quark gluon string model at high energies. The photo-nuclear physics is activated by the *G4PhotoNuclearProcess* class (normally included in the GammaPhysics list), through the method *GetCrossSectionDataStore* that allows to instantiate the *G4PhotoNuclearCrossSection*. In the GammaPhysics list both the photo-nuclear process and the hadronic model for determining the final state of products (energy and angular distribution to be used for the ensuing transport) are introduced.

5 Implementation of Photo-nuclear physics in FLUKA

An important upgrade for the photo-nuclear physics in FLUKA was done in 2005, when the data library was completed and extended in such a way that cross section data for 190 nuclides were made available. If experimental cross sections were not available, then Lorentz fits of the existing data could be used to derive them: for Z> 29 typically the parameters of the fit (as peak energy, peak height and width) are taken from the Atlas of Dietrich and Berman [8] except for Pr, Au and Pb for which the parameters used are those published by Berman et al [9]. For Z< 29,"parameterised" Lorentz parameters are used, according general formulas giving the Lorentz parameters (peak energy, peak height, width) as function of A and Z (as reported in Berman and Fultz[10]). All cross sections are extrapolated to zero at threshold and forced to join smoothly the Quasi Deuteron cross section curve at the upper end of the Giant Dipole Resonance range.

FLUKA models photo-nuclear interactions over the energy range from MeV to TeV and four regions are distinguished for modelling purpose[11]:

1. Giant Dipole Resonance (GDR): $7 < E < 30MeV$
2. Quasi Deuteron Resonance (QD):30 <E <$140MeV$
3. Delta Resonance: $E > 140MeV$
4. High energy region:$E > 720MeV$

For each of these four models, two algorithms have been implemented to provide the value of the total interaction cross sections as a function of photon energy and target nucleus, and the initial energy and momentum transfer between the photon and one or more particles inside the nucleus. Once the photon energy has been transferred, further interactions of the particles involved are described by the normal intra-nuclear cascade, pre-equilibrium and evaporation models implemented in FLUKA.

6 Preliminary results: comparison between the MC cross sections and the IAEA recommended ones

For a long time photo-nuclear processes were neglected in Monte Carlo codes mainly due to the lack of complete evaluated data for applications. The need for evaluation methods for photo-nuclear data arises because it is difficult to develop a complete photo-nuclear data file on the basis of measured cross section alone. These data were often obtained form different kind of photon sources, causing significant discrepancies and, still today, there is a lack of data in a number of important cases. The reasons of the difficulty to produce evaluated photo-nuclear data can be briefly summarised in the following itemize:

- Experimental photo-nuclear data from different laboratories (Lawrence Livermore National Laboratory, CEA, Saclay, Argonne National Laboratory, Illinois etc.) often show discrepancies that must be resolved in the evaluation process
- Most of the existing spectral measurements are for bremsstrahlung, while only the measurements from mono-energetic sources give emission spectra directly useful for cross-section evaluations

[1]Note that all these elements are used also in the fitting procedure to reconstruct the GDR in the low energy region (together with the elements of Table 6) with the exception of 2H and 3He that are used only for the high energy cross-section interpolation

- Photo-nuclear data are isotopic in nature, the cross sections showing irregular dependence on atomic number (Z) and atomic mass (A). Thus while photo-atomic data are readily tabulated by element, photo-nuclear data must be tabulated for each isotope of an element

This explains why a relatively complete photo-nuclear data file became available only in 2000 [3] in order to be promptly distributed to the whole scientific community. The IAEA Compilation and Evaluation of Photonuclear Data for Applications, released in 2000 in ENDF format for 164 isotopes, represents still today the internationally recognised data file of evaluated photo nuclear cross-sections for use in transport simulation codes. The term evaluated means that cross-sections or other data have been optimised: 1) assessing the most probable value from a collection of experimental data; 2) performing theoretical calculations with models and parameters that have been optimised to experimental data or 3) using a combination of experimental data and theoretical calculation to ascertain best or recommended values. Taking into account the previous considerations, the IAEA recommended evaluated photo-nuclear data have been assumed, in this work, as the reference respect to which evaluate the conformity or eventual remarkable differences of the implemented photo-nuclear cross section of GEANT4 and FLUKA. The photo-neutron cross sections to which we refer in the tests are the "total photo-absorption"ones. For sake of clarity, the photo-neutron cross section definition is reported in the following formula [3]:

$$\sigma(\gamma, sn) = \sigma(\gamma, n) + \sigma(\gamma, np) + \sigma(\gamma, 2n) + \sigma(\gamma, 2np) \\ +\sigma(\gamma, n2p) + \sigma(\gamma, 3n)....... + \sigma(\gamma, F)$$

that expresses the photo-neutron cross-section as the sum of all cross sections for neutron production reactions, taking into account the photo-neutron multiplicity. The sum of the photo-neutron cross section as defined above with the photo-charged reaction cross section gives the' *total photo-absorption'* cross-section:

$$\sigma(\gamma, abs) = \sigma(\gamma, sn) + \sigma(\gamma, p) + \sigma(\gamma, 2p) \\ +\sigma(\gamma, d) + \sigma(\gamma, dp) + \sigma(\gamma, \alpha) + ...$$

For heavy nuclei the photo-absorption cross-section can be approximated with the first term of the above two equations

$$\sigma(\gamma, abs) = \sigma(\gamma, sn)$$

Tungsten and lead are among the most important materials for which accurate photo-nuclear data are needed, since they are extensively used as structural, shielding and bremsstrahlung target materials (together with Be, Al Si, Ti, V, Cr, Fe, Co, Ni, Cu, Zn, Zr, Mo, Sn, Ta). For this reason and since these materials are used in the n@BTF experiment (in the target and shied, respectively), they have taken priority in the analysis hereafter presented. The comparisons of the total photo-absorption cross sections

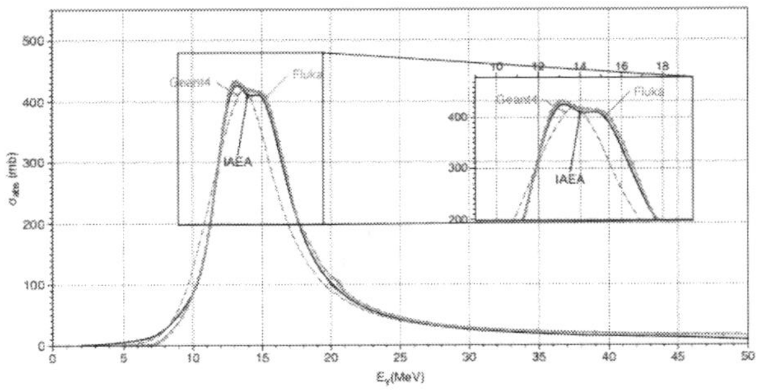

Figure 5. Photoneutron cross sections for natural tungsten: GEANT4, FLUKA and IAEA

for natural tungsten, lead and zinc (up to 50 GeV) with the IAEA evaluated values are reported, respectively in Figure 5, 6 and 7.

Validation of predictions with accurate statistical tests (when experimental values are available) also on the secondary particles produced (energy spectra, angular distribution) are envisaged. The main goal of this work is to provide, at the end of the analysis still in progress, an objective guidance to the scientific community regarding the use of Monte Carlo codes as instruments for the design and optimisation of facilities involving photo-nuclear interactions at high energies.

6.1 Natural tungsten photo-absorption cross-sections

Geant4 cross section library includes data for the following two tungsten isotopes: 184W and 186 W up to 105 MeV, while for higher energy the cross section values are obtained by fitting the data of 14 listed nuclides in Table 5 for which the cross section is reported up to 50 GeV, by means of parameters that are obtained as a function of 'A', 'Z' and 'energy' according suitable models in different energy range. In order to extract the cross section for the natural tungsten, a proper application has been developed, in which only the photo-nuclear physics is activated.

The Kolmogorov-Smirnov test with null hypothesis affirming that data vectors GEANT4 and IAEA are from populations with the same distribution, at the 1% significance level, produced the rejection of the null hypothesis.

6.2 Natural lead photo-absorption cross sections

The Kolmogorov-Smirnov test with null Hypothesis: <<"GEANT4 estimates the recommended IAEA distribution">>, at 1% significance level, produces the rejection of the null Hypothesis (p=2.3E-3). On the contrary, the null Hypothesis: IAEA (and FLUKA that complies) estimates the experimental distribution (Ahrens 1985), with significance level α=0.01 is not rejected

 DOI: 10.1051/epjconf/201715306023

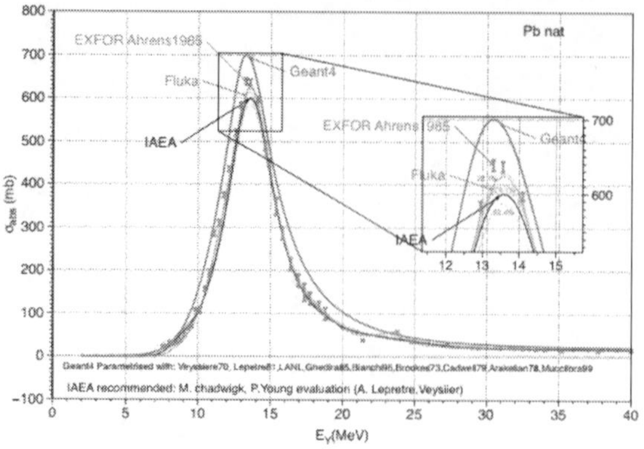

Figure 6. Photo-neutron cross-sections for natural lead: GEANT4, FLUKA and IAEA recommended

6.3 Natural zinc photo-absorption cross-sections

In this preliminary analysis, the natural Zn total photo-absorption cross sections, used in GEANT4 and FLUKA codes, have been also compared to the corresponding IAEA evaluated values. In the documentation[12]showing the approximation of the photo-nuclear interaction cross section for all the elements whose nuclear data are tabulated in the *G4PhotoNuclearCrossSection* class, the nat-Zn is the only missing case. Finally, Kolmogorov-Smirnov Test with significance level alpha=0.01, and null Hypothesis affirming that GEANT4 estimates the recommended IAEA distribution is rejected.

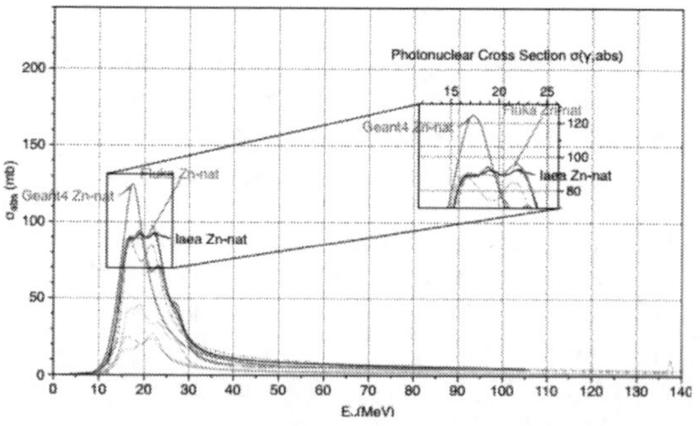

Figure 7. Photo-neutron cross-sections for natural zinc: GEANT4, FLuka and IAEA recommended

7 Conclusion and future plan

Statistical methods are used both to quantify the consistency of simulations based on individual Monte Carlo codes with experimental data and to establish the significance of differences. At this stage of the analysis and for the specific case of measurements carried in high energy range at DaΦne Beam Test Facility, FLUKA predictions show a better agreement with experimental data, respect to GEANT4. Anyway in FLUKA the original cross section database, contrarily to GEANT4, are not accessible directly by the user. In order to investigate the reason of why GEANT4 underestimates the neutron yield for the n@BTF experiment, a rigorous analysis of comparison of the implemented photonuclear data for the concerned materials with the IAEA evaluated data has been started. The Kolmogorov-Smirnov tests performed for W, Pb an Zn to check the compliance of the GEANT4 with the evaluated and recommended IAEA cross sections, produced negative results (contrarily to FLUKA).

Acknowledgements

Photo-nuclear calculations are typically complex and computationally intensive, since they require extensive libraries of isotopic data, high statistics simulations and, often the application of adequate biasing techniques, due to the low photo-neutron cross sections. In order to produce accurate predictions for the photo- neutron yields, large number of events are simulated and variance reduction methods are applied. The ability to exploit techniques for parallel calculations is fundamental to achieve statistically significant results within a reasonable time. The computing resources and the related technical support used for this work are provided by the CRESCO/ENEAGRID High Performance Computing infrastructure [13]. The CRESCO/ENEAGRID infrastructure is funded by ENEA, the Italian National Agency for New Technologies, Energy and Sustainable Economic Development and by national and European research programs.

References

[1] S.Agostinelli et al (2003),'Geant4-a simulation toolkit', Nuclear Instruments and Methods in Physics Research A5 06, pp. 250-303

[2] A. Fasso et al. (2005), 'A multi-article transport code', 'CERN-2005-10INFN/TC-05/11

[3] Handbook on Photonuclear Data for Applications Cross-sections and Spectra. IAEA-TECDOC1178, 2000

[4] https://eli-laser.eu.

[5] R. Bedogni et al. (2011), 'Experimental and numerical characterisation of the neutron field produced in the n@BTF Frascati photo-neutron source', Nuclear Instruments and Methods in Physics Research A, 659, pp. 373-377

[6] B Buonomo,et a. (2008), 'A wide range electrons, photons, neutrons beam facility', Proceedings of 11th European Particle Accelerator Conference EPAC-2008, Giugno 23-27, Genoa,Italy

[7] G. Vignola, The DAFNE Project Team, 'DAFNE: the Frascati Phi- Factory', 1991 Particle Accelerator Conference, San Francisco, May 1991, p. 68..

[8] S. Dietrich, er al(1988), 'Atlas of Photoneutron Cross Sections Obtained with Monoenergetic Photons', Atomic Data and Nuclear Data Tables, Vol. 38, p.199.

[9] B.Berman et al. (1987), '..',Phys. Rev. C 36, 128

DOI: 10.1051/epjconf/201715306023

[10] B.Berman,et al. (1975), 'Measurements of the giant dipole resonance with monoenergetic photons', Rev. Mod. Phys. 47, 713

[11] A.Fasso et al. (1994), Proc. 8th Int. Conf on Radiation Shielding, Arlington (Texas) Apr. 24-28, pp 643-649

[12] M.V. Kssov (2002), "Approximation of photo nuclear interaction cross-sections", Eur.Phys.Journal A. 14, pp. 377-392

[13] G. Ponti et al. (2014), "The role of medium size facilities in the HPC ecosystem: the case of the new CRESCO4 cluster integrated in the ENEAGRID infrastructure", Proceedings of the International Conference on High Performance Computing and Simulation, HPCS 2014, art. no. 6903807, 1030-103.

DOI: 10.1051/epjconf/201715306024

Evaluation of RayXpert® for shielding design of medical facilities

Sylvie Derreumaux[1,*], *Sophie* Vecchiola[1], *Thomas* Geoffray[2], and *Cécile* Etard[1]

[1]Institut for radiation protection and nuclear safety, Radiation protection expertise department, BP17 92260 Fontenay aux Roses, France
[2]D&S, 573 Avenue de l'Hermitage, 30200 Bagnols-sur-Cèze, France

Abstract. In a context of growing demands for expert evaluation concerning medical, industrial and research facilities, the French Institute for radiation protection and nuclear safety (IRSN) considered necessary to acquire new software for efficient dimensioning calculations. The selected software is RayXpert®. Before using this software in routine, exposure and transmission calculations for some basic configurations were validated. The validation was performed by the calculation of gamma dose constants and tenth value layers (TVL) for usual shielding materials and for radioisotopes most used in therapy (Ir-192, Co-60 and I-131). Calculated values were compared with results obtained using MCNPX as a reference code and with published values. The impact of different calculation parameters, such as the source emission rays considered for calculation and the use of biasing techniques, was evaluated.

1 Introduction

For the past few years, demands for expert evaluation of medicine, industrial and research facilities dimensioning have been increasing. To respond to those requests, an efficient simulation tool is necessary. This tool should be easy to use and should offer a good compromise between calculation time and result accuracy. The software RayXpert® has been selected for this purpose.

Before using this software in routine, the impact of different calculation parameters was evaluated and basic configurations were validated.

This article describes the validation of RayXpert® for exposure and transmission calculations with the radionuclides most used in medicine for therapeutic purposes: Ir-192 and Co-60 in brachytherapy and I-131 in nuclear medicine. Preliminary results are presented.

2 Materials and methods

In this study, calculations and comparisons of gamma constants and tenth value layers (TVL) have been performed. Results with RayXpert® have been compared with those obtained by using MCNPX 2.7.0, that has been considered as a reference code, and with values published in the literature.

2.1 RayXpert® version 1.3.4.0

The software RayXpert® is developed by Tests and RADiations (TRAD, France). It uses Monte Carlo method implemented in GEANT 4 V10.0 to simulate photon and electron transport for energies between 1 keV and 100 MeV. It is implemented on PC Windows and has the advantage of having a user-friendly interface that allows easy and rapid building of complex 3D geometries (Computer-Aided Design).

RayXpert® offers the possibility of considering radioactive sources of continuous electron or photon spectra (ie: for modelling conventional radiology or radiotherapy X-ray sources).

Some important calculation parameters are automatically assessed by the software, such as the definition of the energy spectra of radionuclides' emission (database and selection of the rays), materials' composition, conversion coefficients, and the weights of the volumes for the Weight Windows variance reduction method. This enables a straightforward use of the software.

In this study, RayXpert® version 1.3.4.0 was used.

Energy spectra of radionuclides' emission are extracted from the EAF-2010 decay data library (EASY-2010 system) or from the JEFF-3.1/3.1.1 radioactive decay data sub-library [1]. For this study, the EAF database was chosen as it is the default option.

Calculations take into account the emission photon rays contributing to 99% of the emitted energy, with an automatic selection by the software ("RayXpert® criterion"):

$$\sum_{i=1}^{n\_simul} P(E_i)E_i = 0.99 \sum_{i=1}^{n\_spectrum} P(E_i)E_i \quad (1)$$

* Corresponding author: sylvie.derreumaux@irsn.fr

Where $P(E_i)$ is the radiation intensity for a given ray i of energy E_i, n_simul is the number of rays that are considered for the simulation and n_spectrum is the total number of rays of the spectrum extracted from the decay data library.

The results are expressed in terms of ambient dose equivalent H*(10). The conversion coefficients from fluence to ambient dose equivalent, H*(10), are automatically applied by the software and are extracted from the publication 74 of ICRP [2].

RayXpert® proposes a database of materials, whose compositions correspond to the one given in the PNNL-15870 rev 1 [3]. The choice of a given material for a given volume automatically attributes that composition to the volume.

In order to improve the convergence, different variance reduction methods are proposed as options: the Weight Windows, the thick target bremsstrahlung approximation (TTB) and distance or energy cut-off [1]. Moreover, calculation time can be reduced by the parallelization of calculations.

2.2 MCNPX 2.7.0

MCNPX (Monte Carlo N-Particle eXtended) is developed and maintained by the Los Alamos National Laboratory [4]. The code deals with transport of neutrons, gamma rays, and coupled transport, i.e., transport of secondary gamma rays resulting from neutron interactions. The MCNPX code can also treat the transport of electrons, both primary source electrons and secondary electrons created in gamma-ray interactions.

In this study, MCNPX version 2.7.0 was used.

Energy spectrum of the source's radiation is defined by the user. For this study, energy spectra have been extracted from ICRP 107 [5] and a selection of the most contributive rays was performed to improve calculation times, according to a home defined reference method.

The selection rule for a given ray i is as follows ("IRSN criterion"):

$$\frac{P(E_i)}{\sum_i P(E_i)} \geq 10^{-4} \qquad (2)$$

Where P(Ei) is the radiation intensity for a given ray i (energy Ei) and Σ_i P(Ei) is the sum of all intensities.

Furthermore, radiations with energy below 5 keV were not considered because their contribution to the dose rate at a distance of 1 m in air is not significant.

In order to be consistent with the calculations made with RayXpert® (see 2.1), the composition of materials was also extracted from the PNNL-15870 rev 1 [3].

The photon fluence in the detector was calculated. The conversion coefficients from fluence to H*(10) were extracted from the ICRP 74 [2].

For this study, the convergence has been improved by the use of geometry splitting variance reduction method [4].

2.3 Comparison between RayXpert® and MCNPX

RayXpert® modelling validation was performed by calculating the gamma constant and transmission through different shielding materials for Ir-192, Co-60 and I-131, the most used radioisotopes in brachytherapy and in nuclear medicine treatments, and by comparing the results obtained with RayXpert® and with MCNPX.

Calculations were performed using MCNPX with the "IRSN criterion" for the selection of the emission rays (see 2.2) and RayXpert® without biasing method.

The source was quasi punctual (sphere with a radius of 0.5 mm) and in order to throw off self-attenuation, its material was set to air with a density of 0.001. Indeed, the activity of the medical radioactive sources used is the apparent activity.

The ambient dose equivalent rate, $\dot{H}$*(10), was calculated in air at a distance of 1 m from the source, for a source activity of 18.5 GBq for Ir-192 (typical activity of a new source in pulsed dose rate (PDR) brachytherapy), 3.7 GBq for I-131 (typical activity for the treatment of thyroid cancer) and 74 GBq for Co-60 (typical activity of a new source in high dose rate (HDR) brachytherapy). The detector was a sphere of air with a radius of 2.5 cm.

For calculation of the gamma dose constant, $\dot{H}$*(10) was normalized to a source activity of 1 GBq.

For the calculation of transmission, the shield was a quasi-infinite wall (10 m x 10 m x thickness) centred at 50 cm from the source. Calculations were performed for several typical materials: ordinary concrete, steel (HT9), lead, brick (common silica) and leaded glass, with densities of 2.3, 7.87, 11.35, 1.8 and 6.62 $g.cm^{-3}$, respectively. For each material, calculations were made for thicknesses of approximately n tenth-value layers: B_n = n x B_1 where B_1 ~ TVL (n = 1, 2 and 3 using MCNPX and n = 1 and 2 using RayXpert®).

For Ir-192 and Co-60, the B_1 values for concrete, steel and lead were obtained from TVL values for large attenuation published in the IAEA Safety Reports Series Report No. 47 [6]. For brick and leaded glass, the B_1 value for each radionuclide was determined applying an inverse density relationship to the concrete B_1 value. For leaded glass, preliminary transmission calculations allowed to get a better estimation of TVL and to adjust the value of B_1 (the inverse density relationship leaded to an overestimation of TVL).

For I-131, the Ir-192 B_1 values were considered, as the average emission photon energies are very close (0.36 MeV and 0.38 MeV for Ir-192 and I-131, respectively).

For each radionuclide and each barrier material, the cumulative tenth-value layer, TVL_c, and the successive tenth-value layers, TVL_n (for n=1,2...), were calculated.

TVLc is the approximate TVL value for large attenuation. It was estimated as follows: for a given barrier thickness B_n:

$$TVL_c = -B_n/log_{10}(T_n) \qquad (3)$$

Where T_n is the transmission factor for barrier B_n: T_n = $H^*(10)_{with}/H^*(10)_{without}$, where $H^*(10)_{with}$ and

$H^*(10)_{without}$ are the ambient dose equivalent at calculation point with or without barrier, respectively.

TVL_n is the nth TVL: TVL_1 is the 1[rst] TVL, TVL_2 is the 2nd TVL... It was estimated as follows: for two successive barrier thicknesses B_{n-1} and B_n:

$$TVL_n = -(B_n - B_{n-1})/log_{10}(T_n/T_{n-1}) \quad (4)$$

2.4 Influence of the ray selection rule

In order to evaluate the influence of the RayXpert® selection rule for the photon emission rays, ambient dose equivalent rate, $\dot{H}^*(10)$, was calculated using MCNPX and energy spectra obtained with the RayXpert® criterion (paragraph 2.1) on the one hand, and with the IRSN criterion on the other hand (paragraph 2.2).

Calculations were performed in air at a distance of 1 m from the source, without shield and with a shield of different materials and thicknesses, for an Ir-192 source of 18.5 GBq and for an I-131 source of 3.7 GBq.

2.5 Impact of the use of biasing technics

In order to evaluate the impact of biasing technics proposed by RayXpert®, ambient dose equivalent rate $\dot{H}^*(10)$ was calculated for Ir-192 using TTB approximation and energy cut-off, either separately or jointly, for different shielding materials and different thicknesses. When applied, the energy cut-off values were 10 keV for photons and 100 keV for electrons.

3 Results and discussion

3.1 Gamma dose constant

For each radionuclide, the gamma dose constant was obtained by the calculation of $\dot{H}^*(10)$ at a distance of 1 meter from the source in air and without barrier. Results are presented in table 1, together with published data.

There is an excellent agreement between RayXpert® and MCNPX for Ir-192 and Co-60, with calculated values of 138 and 139 $\mu Sv.m^2.h^{-1}.GBq^{-1}$ for Ir-192 and 340 and 341 $\mu Sv.m^2.h^{-1}.GBq^{-1}$ for Co-60, with RayXpert® and MCNPX, respectively. However, a significant difference (~ 9%) appears for I-131, with values of 60 and 66 $\mu Sv.m^2.h^{-1}.GBq^{-1}$, using respectively RayXpert® and MCNPX.

Considering the statistical uncertainties, the values obtained by MCNPX are in good agreement with the exposure values published in 2004 by Delacroix et al. for a distance of 1 meter from a punctual source [7]: 135, 66 and 351 $\mu Sv.m^2.h^{-1}.GBq^{-1}$, respectively for Ir-192, I-131 and Co-60. For I-131, the gamma dose constant calculated by MCNPX corresponds to the value recommended by the DIN 6844-3 standard [8]. On the contrary, the results are quite different from the values published in 1981 by the Oak Ridge National Laboratory [9]: 159.9, 76.5 and 370.3 $\mu Sv.m^2.h^{-1}.GBq^{-1}$, respectively for Ir-192, I-131 and Co-60.

Table 1. Gamma dose constant calculated with MCNPX and RayXpert® for Ir-192, I-131 and Co-60, and data published by Delacroix [7], DIN [8] and ORNL [9]. Relative statistical uncertainties are shown in brackets as percentages.

Radio-nuclide	Gamma dose constant ($\mu Sv.m^2.h^{-1}.GBq^{-1}$)				
	Ray-Xpert®	MCNPX	Delacroix	DIN 6844-3	ORNL
Ir-192	138 (1.3)	139 (2.0)	135		159.9
I-131	60 (0.7)	66 (0.9)	66	66	76.5
Co-60	340 (1.9)	341 (1.9)	351		370.3

3.2 Transmission and TVL

Transmission was calculated for a sample of shields (materials and thicknesses). For each radionuclide, $\dot{H}^*(10)$ was calculated at a distance of 1 meter from the source in air, with and without barrier. Results are presented in tables 2, 3 and 4 for Ir-192, I-131 and Co-60, respectively.

There is an overall good agreement between RayXpert® and MCNPX regarding transmission calculation: the relative difference between transmitted $\dot{H}^*(10)$ is lower than 10% for the three radionuclides and all the barriers. The discrepancies are included in the calculation uncertainty (2σ), except perhaps for one configuration (I-131, 40 cm brick).

Using the results obtained for transmission calculation, TVL_c and TVL_n were estimated according to formulae (3) and (4). Results are presented in tables 5, 6 and 7 for Ir-192, I-131 and Co-60, respectively.

For each radionuclide and each barrier material, TVL_c was calculated using transmission factor for the largest barrier that could be simulated given the calculation time: with MCNPX: B_3 for I-192 and I-131 and B_2 for Co-60; with RayXpert®: B_2. In order to compare the results, TVLc was also calculated using transmission factor for thickness B_2 with MCNPX for I-192 and I-131.

The successive tenth-value layers, TVL_n, were evaluated using the transmission factors for successive barrier thicknesses: TVL_1 and TVL_2 with RayXpert®, TVL_1, TVL_2 and TVL_3 with MCNPX.

The TVL_c and TVL_n values calculated with RayXpert® are in very good agreement with those obtained with MCNPX: the difference is globally lower than 5%.

As expected, TVL_2 is lower than TVL_1 for concrete, steel and brick, for the three radionuclides, due to the build-up effect. On the contrary, TVL_2 is higher than TVL_1 for lead and leaded glass (except for Co-60), due to radiation hardening while penetrating the barrier (photoelectric effect for low energy photons).

Regarding Ir-192, the TVL_c values obtained with MCNPX for the largest barrier thickness (B_3) are very close or equal to the TVL published by the IAEA SRS Report No. 47 for large attenuation (15.2 cm, 1.6 cm and 4.3 cm, respectively for ordinary concrete, lead and steel) [6]. Also, TVL_1 for steel matches the value recommended by IAEA (4.9 cm).

Table 2. Transmission calculated with MCNPX and RayXpert® for Ir-192 (18.5 GBq), and relative difference (Δ). Relative statistical errors (1σ) are shown in brackets (%).

Barrier		Ḣ*(10) (Sv/h)		Δ (%)
Material	B_n (cm)	RayXpert®	MCNPX	
Air	-	2.55E-03 (1.3)	2.56E-03 (2.0)	-0.5
Concrete	15	4.21E-04 (3.1)	4.45E-04 (2.4)	-5.4
	30	3.77E-05 (3.9)	4.03E-05 (1.1)	-6.5
	45	-	3.20E-06 (1.3)	-
Lead	1.5	1.27E-04 (3.5)	1.38E-04 (3.1)	-7.7
	3	1.52E-05 (5.1)	1.54E-05 (8.4)	-1.6
	4.5	-	2.31E-06 (2.8)	-
Steel	5	2.51E-04 (2.1)	2.48E-04 (2.2)	1.3
	10	1.54E-05 (5.4)	1.52E-05 (1.6)	1.3
	15	-	8.99E-07 (0.8)	-
Brick	20	4.41E-04 (1.0)	4.40E-04 (2.5)	0.1
	40	3.86E-05 (2 .5)	3.78E-05 (0.8)	2.2
	60	-	3.08E-06 (1.8)	-
Leaded glass	3	1.45E-04 (2.2)	1.56E-04 (2.6)	-7.3
	6	1.77E-05 (5.0)	1.78E-05 (4.3)	-0.4
	12	-	3.66E-07 (5 .1)	-

Table 3. Transmission calculated with MCNPX and RayXpert® for I-131 (3.7 GBq), and relative difference (Δ). Relative statistical errors (1σ) are shown in brackets (%).

Barrier		Ḣ*(10) (Sv/h)		Δ (%)
Material	B_n (cm)	RayXpert®	MCNPX	
Air	-	2.23E-04 (0.7)	2.45E-04 (0.9)	-9.0
Concrete	15	4.13E-05 (2.1)	4.16E-05 (1.9)	-0.7
	30	3.89E-06 (2.8)	3.89E-06 (0.3)	0.0
	45	-	3.16E-07 (0.3)	-
Lead	1.5	1.17E-05 (2.9)	1.16E-05 (2.4)	0.6
	3	1.73E-06 (2.9)	1.77E-06 (2.1)	-2.5
	4.5	-	3.27E-07 (2.0)	-
Steel	5	2.37E-05 (2.0)	2.40E-05 (0.9)	-1.3
	10	1.33E-06 (4.0)	1.45E-06 (0.6)	-8.4
	15	-	8.93E-08 (0.6)	-
Brick	20	4.08E-05 (2.0)	4.10E-05 (0.2)	-0.5
	40	3.95E-06 (3.1)	3.64E-06 (0.3)	8.5
	60	-	2.93E-07 (0.3)	-
Leaded glass	3	1.34E-05 (2.1)	1.40E-05 (2.8)	-4.6
	6	1.86E-06 (3.9)	2.03E-06 (4.3)	-8.6
	12	-	6.36E-08 (3.7)	-

Table 4. Transmission calculated with MCNPX and RayXpert® for Co-60 (74 GBq), and relative difference (Δ). Relative statistical errors (1σ) are shown in brackets (%).

Barrier		Ḣ*(10) (Sv/h)		Δ (%)
Material	B_n (cm)	RayXpert®	MCNPX	
Air	-	2.52E-02 (1.9)	2.53E-02 (1.9)	-0.4
Concrete	25	3.67E-03 (1.4)	3.55E-03 (2.0)	3.4
	50	2.83E-04 (4.7)	2.70E-04 (1.2)	4.7
Lead	4	3.61E-03 (1.5)	3.58E-03 (1.0)	0.8
	8	4.12E-04 (4.5)	3.77E-04 (0.6)	9.3
Steel	8	3.13E-03 (1.7)	3.16E-03 (1.8)	-0.9
	16	2.09E-04 (5.8)	2.07E-04 (1.0)	0.8
Brick	30	4.51E-03 (1.2)	4.40E-03 (1.4)	2.5
	60	4.44E-04 (3.7)	4.33E-04 (0.3)	2.5
Leaded glass	8	3.14E-03 (2.4)	3.20E-03 (1.0)	-1.9
	16	2.70E-04 (3.9)	2.76E-04 (0.6)	-2.3

Table 5. TVL_c and TVL_n calculated using the transmission factors obtained with MCNPX and RayXpert® for Ir-192.

Barrier		TVL_c (cm)		TVL_n (cm)	
Material	B_n (cm)	Ray-Xpert®	MCNP-X	Ray-Xpert®	MCNP-X
Concrete	15	-	-	19.2	19.7
	30	16.4	16.6	14.3	14.4
	45	-	15.5	-	13.6
Lead	1.5	-	-	1.2	1.2
	3	1.3	1.4	1.6	1.6
	4.5	-	1.5	-	1.8
Steel	5	-	-	5.0	4.9
	10	4.5	4.5	4.1	4.1
	15	-	4.3	-	4.1
Brick	20	-	-	26.2	26.1
	40	22.0	21.8	18.9	18.7
	60	-	20.5	-	18.4
Leaded glass	3	-	-	2.4	2.5
	6	2.8	2.8	3.3	3.2
	12	-	3.1	-	3.6

Table 6. TVL_c and TVL_n calculated using the transmission factors obtained with MCNPX and RayXpert® for I-131.

Barrier		TVL_c (cm)		TVL_n (cm)	
Material	B_n (cm)	Ray-Xpert®	MCNP-X	Ray-Xpert®	MCNP-X
Concrete	15	-	-	20.5	19.5
	30	17.1	16.7	14.6	14.6
	45	-	15.6	-	13.8
Lead	1.5	-	-	1.2	1.1
	3	1.4	1.4	1.8	1.8
	4.5	-	1.6	-	2.0
Steel	5	-	-	5.1	5.0
	10	4.5	4.5	4.0	4.1
	15	-	4.4	-	4.1
Brick	20	-	-	27.1	25.8
	40	22.8	21.9	19.7	19.0
	60	-	20.5	-	18.3
Leaded glass	3	-	-	2.5	2.4
	6	2.9	2.9	3.5	3.6
	12	3.4	3.3	4.1	4.0

Table 7. TVL_c and TVL_n calculated using the transmission factors obtained with MCNPX and RayXpert® for Co-60.

Barrier		TVL_c (cm)		TVL_n (cm)	
Material	B_n (cm)	Ray-Xpert®	MCNP-X	Ray-Xpert®	MCNP-X
Concrete	25	-	-	29.9	29.3
	50	25.6	25.4	22.5	22.3
Lead	4	-	-	4.7	4.7
	8	4.5	4.4	4.2	4.1
Steel	8	-	-	8.8	8.9
	16	7.7	7.7	6.8	6.8
Brick	30	-	-	40.2	39.5
	60	34.2	34.0	29.8	29.8
Leaded glass	8	-	-	8.9	8.9
	16	8.1	8.2	7.5	7.5

Concerning Co-60, the TVLc values obtained with RayXpert® and MCNPX for the barrier thickness B2 are consistent with the IAEA TVL (21.8 cm, 4.1 cm and 7.1 cm, respectively for ordinary concrete, lead and steel); however they are higher than the IAEA values, suggesting that these correspond to thicknesses larger than B2. TVL1 for steel match the value recommended by IAEA (8.7 cm), while TVL1 for concrete are significantly higher (29.9 cm and 29.3 cm versus 24.5 cm).

Significant discrepancies appear between our results and the TVL1 and TVL_e values published by Delacroix for lead and steel (TVL_e is the equilibrium TVL; it can be approximated by TVL_3). The Delacroix et al. data are systematically higher, for the three radionuclides, especially the TVL_1 values: 2.6 cm, 2.4 cm and 6.6 cm for Ir-192, I-131 and Co-60, respectively, for lead; 7.8 cm, 7.7 cm and 10.3 cm for Ir-192, I-131 and Co-60, respectively, for steel [7].

3.3 Influence of the ray selection rule and of the automatic energy distribution of the results

Ambient dose equivalent rate, Ḣ*(10), was calculated using MCNPX and emission energy spectra obtained with the RayXpert® selection criterion on the one hand, and with the IRSN criterion on the other hand, for Ir-192 and I-131 and for different barrier materials and thicknesses. Results are presented in tables 8 and 9 for Ir-192 and I-131, respectively.

The relative difference between Ḣ*(10) calculated with the IRSN criterion and with the RayXpert® criterion is lower than 5% and is included in the calculation uncertainty (2σ), for Ir-192 and I-131. The average relative difference is lower than 1% (0.0% for Ir-192 and -0.4% for I-131).

It appears that the RayXpert® automatic selection rule for the emission photon rays does not impact significantly the calculated ambient dose equivalent rate for Ir-192 and I-131.

For I-131, the relative difference between H*(10) calculated with both criteria and without barrier is only -2.1%. This cannot explain the significant difference obtained for the gamma dose constant calculated with RayXpert® and MCNPX (see 3.1). As calculations with MCNPX were performed using the ICRP 107 emission spectrum, calculations were also made using the EAF-2010 spectrum with MCNPX, with both selection criteria. This gave the same results as using the ICRP 107 emission spectrum (2.46E-04 Sv/h with IRSN criterion and 2.39E-04 Sv/h with RayXpert criterion). However, again a small difference (-2.8%) appears between the results obtained with both criterion, suggesting that the RayXpert® selection criteria could explain partially the difference between the gamma dose constant calculated with RayXpert® and MCNPX.

The emission photon rays that were considered for simulation using the different approaches are shown in figure 1. It can be seen that the RayXpert® criterion (RXT) strongly reduces the number of rays considered for the simulation, irrespective of the emission spectrum (ICRP 107 or EAF-2010) or of the software (RayXpert® or MCNPX). The removed rays correspond to the lowest energies or to the lowest intensities with intermediate energies.

Table 8. Ḣ*(10) calculated with MCNPX for Ir-192 with IRSN and RayXpert® criteria for selection of the photon emission rays, and relative difference (Δ). Relative statistical uncertainties (1σ) are shown in brackets as percentages.

Barrier		Ḣ*(10) (Sv/h)		Δ (%)
Material	B_n (cm)	IRSN criterion	RayXpert® criterion	
Air	-	2.56E-03 (2.0)	2.56E-03 (2.0)	0.1
Concrete	15	4.45E-04 (2.4)	4.40E-04 (6.1)	-1.1
	30	4.03E-05 (1.1)	4.05E-05 (1.1)	0.4
	45	3.20E-06 (1.3)	3.25E-06 (0.9)	1.7
Lead	1.5	1.38E-04 (3.1)	1.32E-04 (3.1)	-4.1
	3	1.54E-05 (8.4)	1.54E-05 (2.0)	0.0
	4.5	2.31E-06 (2.8)	2.32E-06 (3.2)	0.5
Steel	5	2.48E-04 (2.2)	2.56E-04 (3.0)	3.4
	10	1.52E-05 (1.6)	1.51E-05 (1.5)	-0.6
	15	8.99E-07 (0.8)	8.99E-07 (0.8)	0.0

Table 9. Ḣ*(10) calculated with MCNPX for I-131 with IRSN and RayXpert® criteria for selection of the photon emission rays, and relative difference (Δ). Relative statistical uncertainties (1σ) are shown in brackets as percentages.

Barrier		Ḣ*(10) (Sv/h)		Δ (%)
Material	B_n (cm)	IRSN criterion	RayXpert® criterion	
Air	-	2.45E-04 (0.9)	2.40E-04 (0.9)	-2.1
Concrete	15	4.16E-05 (1.9)	4.19E-05 (1.8)	0.7
	30	3.89E-06 (0.3)	3.88E-06 (0.3)	-0.3
	45	3.16E-07 (0.3)	3.14E-07 (0.3)	-0.6
Lead	1.5	1.16E-05 (2.4)	1.21E-05 (2.3)	4.5
	3	1.77E-06 (2.1)	1.74E-06 (2 .1)	-1.8
	4.5	3.27E-07 (2.0)	3.27E-07 (2.0)	0.0
Steel	5	2.40E-05 (0.9)	2.35E-05 (0.9)	-2.0
	10	1.45E-06 (0.6)	1.44E-06 (0.6)	-0.7
	15	8.93E-08 (0.6)	8.81E-08 (0.6)	-1.3

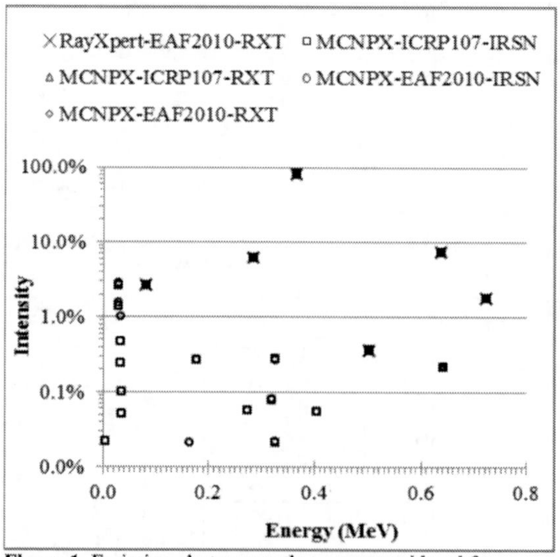

Figure 1. Emission photon rays that were considered for simulation using the different approaches.

Further investigations have been performed with RayXpert®, regarding the automatic energy distribution of the results. Indeed, the software distributes the calculated fluences as a histogram with energy bins automatically determined: the energy binning is logarithmic, with 10 steps per decade (default option). Fluences are converted to H*(10) using the conversion coefficients from the publication 74 of ICRP [2] (paragraph 2.1). As the software offers to the user the possibility to define any energy distribution (input file), simulations were also made (calculations in air without barrier) using the energy steps of the ICRP 74 A.21 table. For I-131, the calculated value for H*(10) was 2.38E-04 Sv/h (± 0.7%).

The significant increase of the H*(10) value (+ 7%) using the ICRP 74 energy steps instead of the default option for the energy distribution (see table 3) can explain the difference observed for the gamma dose constant calculated with RayXpert® and MCNPX, if this effect is added to the small impact of the ray selection rule (~ 2-3%). Note that the simulations with MCNPX were made using the ICRP 74 energy steps for the conversion coefficients (calculation of H*(10) with tally F4).

3.4 Impact of the use of biasing technics

$\dot{H}$*(10) was calculated for Ir-192 using TTB approximation and energy cut-off, either separately or jointly, with concrete, lead and steel barrier. Results are presented in table 10 for barrier thickness B_1 (15 cm, 1.5 cm and 5 cm for concrete, lead and steel, respectively).

The use of biasing methods in RayXpert® reduces significantly the calculation time and/or the statistical error, with a relative difference in $\dot{H}$*(10) (between results obtained with and without biasing method) lower than 10%, which is acceptable for radiation protection purposes.

Table 10. $\dot{H}$*(10) calculated with RayXpert® for Ir-192 and a barrier of thickness B_1, with and without biasing methods (TTB approximation and energy cut-off), statistical errors, relative difference (Δ) (with and without biasing) and calculation times.

Barrier material	Biasing method	$\dot{H}$*(10) (Sv/h)	Error (%)	Δ (%)	Time (h)
Concrete	No bias	4.21E-04	3.1	-	6.4
	TTB	4.55E-04	2.9	8.0	1.1
	E_{cut}	4.46E-04	1.7	5.9	3.3
	TTB + E_{cut}	4.56E-04	1.6	8.3	1.0
Lead	No bias	1.27E-04	3.5	-	23.2
	TTB	1.30E-04	2.6	1.9	3.1
	E_{cut}	1.25E-04	2.4	-1.9	4.2
	TTB + E_{cut}	1.30E-04	1.7	2.3	3.4
Steel	No bias	2.51E-04	2.1	-	40.8
	TTB	2.50E-04	1.1	-0.3	15.5
	E_{cut}	2.53E-04	1.7	0.7	8.2
	TTB + E_{cut}	-	3.1	-	-

4 Conclusion

The influence of the parameters implemented in RayXpert® is better known and transmission calculation for some basic configurations has been validated for the radionuclides most used in medical therapy facilities.

Regarding the gamma dose constant, there is an excellent agreement between RayXpert® and MCNPX for Ir-192 and Co-60. However, a significant difference (~ 9%) appears for I-131, the value obtained with MCNPX corresponding to the one recommended by the DIN 6844-3 standard. This difference can be explained by the automatic definition by RayXpert® (default option) of the energy distribution of the fluence results (~ 7%) together with a small impact of the RayXpert® selection rule for the emission photon rays (~ 2-3%). For all radionuclides, the results are quite different from the values published in 1981 by the Oak Ridge National Laboratory.

There is an overall good agreement between RayXpert® and MCNPX regarding transmission calculation: the relative difference between transmitted $\dot{H}$*(10) is lower than 10% for the three radionuclides and all the investigated barrier materials and thicknesses.

Using the results obtained for transmission calculation, the cumulative tenth-value layer, TVL_c, and the successive tenth-value layers, TVL_n, were calculated for different typical shielding materials (ordinary concrete, lead, steel, brick and leaded glass). The values calculated with RayXpert® and MCNPX are in very good agreement, and are consistent with the data recommended by the IAEA SRS Report No. 47 (except TVL_1 for concrete). On the other hand, significant discrepancies appear between our results and the values published by Delacroix for lead and steel.

The RayXpert® selection rule for the emission photon rays does not impact significantly the calculated ambient dose equivalent rate if it is applied in MCNPX.

Finally, biasing methods proposed in RayXpert® were checked (TTB approximation and energy-cutoff). They reduce significantly the calculation time and/or the statistical error, while having an impact on the results lower than 10%, which is acceptable for radiation protection purposes.

This work should now be pursued along three major lines. Firstly, the calculations should be extended to other radionuclides encountered in medical facilities. Then the efficiency of RayXpert® should be tested for routine use and realistic configurations found in brachytherapy and nuclear medicine facilities. Finally, the possibility to use this software for radiology facilities should be studied.

References

1. RayXpert, *Manuel d'utilisation* (2014).
2. ICRP Publication 74, *Conversion coefficients for use in radiological protection against external radiation* (1996).
3. R.J. McConn Jr, C.J. Gesh, R.T. Pagh, R.A. Rucker, R.G. Williams III, *Compendium of Material*

DOI: 10.1051/epjconf/201715306024

Composition Data for Radiation Transport Modeling, PNNL-15870 rev1 (2011).
4. MCNPX™ *User's Manual* Version 2.7.0 (2011).
5. ICRP Publication 107, *Nuclear Decay Data for Dosimetric Calculations* (2007).
6. IAEA Safety Reports Series Report No. 47, *Radiation protection in the Design of Radiotherapy Facilities* (2006).
7. Delacroix et al., *Guide pratique Radionucléides et Radioprotection*, 4ème édition, EDP Sciences (2004).
8. DIN 6844-3: 2006-12, *Nuclear medicine departments - Part 3: Radiation protection calculations - Nuklearmedizinische Betriebe - Teil 3: Strahlenschutzberechnungen* (2006).
9. L. M. Unger and D. K. Trubey, *Specific Gamma-Ray Dose Constants for Nuclides Important to Dosimetry and Radiological Assessment*, ORNL/RSIC-45 (1981).

DOI: 10.1051/epjconf/201715306025

A Deep Penetration Problem Calculation Using AETIUS: An Easy Modeling Discrete Ordinates Transport Code UsIng Unstructured Tetrahedral Mesh, Shared Memory Parallel

Jong Woon KIM [1*] and Young-Ouk LEE [1]

[1]*Korea Atomic Energy Research Institute, 989 Daeduck-daero, Yuseong-gu, Daejeon 305-353, Korea*
*Corresponding Author, E-mail: jwkim@kaeri.re.kr

Abstract

As computing power gets better and better, computer codes that use a deterministic method seem to be less useful than those using the Monte Carlo method. In addition, users do not like to think about space, angles, and energy discretization for deterministic codes.

However, a deterministic method is still powerful in that we can obtain a solution of the flux throughout the problem, particularly as when particles can barely penetrate, such as in a deep penetration problem with small detection volumes.

Recently, a new state-of-the-art discrete-ordinates code, ATTILA, was developed and has been widely used in several applications. ATTILA provides the capabilities to solve geometrically complex 3-D transport problems by using an unstructured tetrahedral mesh.

Since 2009, we have been developing our own code by benchmarking ATTILA. AETIUS is a discrete ordinates code that uses an unstructured tetrahedral mesh such as ATTILA. For pre- and post- processing, Gmsh is used to generate an unstructured tetrahedral mesh by importing a CAD file (*.step) and visualizing the calculation results of AETIUS. Using a CAD tool, the geometry can be modeled very easily.

In this paper, we describe a brief overview of AETIUS and provide numerical results from both AETIUS and a Monte Carlo code, MCNP5, in a deep penetration problem with small detection volumes. The results demonstrate the effectiveness and efficiency of AETIUS for such calculations.

KEYWORDS: deterministic code, discrete ordinates method, unstructured tetrahedral mesh, deep penetration, high S_N order, comparison

I. Introduction

These days, when we develop or use a deterministic code, a lot of people ask what is its advantage compared to computer codes that use a Monte Carlo method such as MCNP5[1].

The reason for its use is a continual increase in computing power and parallel processing capabilities. Users prefer to use a method that has less approximation such as a Monte Carlo method.

Even for the deterministic codes, the results of a Monte Carlo codes are used as a reference calculation for cross checking.

Then, does a deterministic method have any better features than a Monte Carlo method? This is the main motivation of this paper and why we chose a deep penetration problem with small detection volumes.

In this paper, we describe a brief overview of AETIUS and provide numerical results from both AETIUS and a Monte Carlo code, MCNP5, in a deep penetration problem with small detection volumes.

II. Method

Before naming our code as AETIUS (An Easy modeling Transport code usIng Unstructured tetrahedral mesh, Shared memory parallel), it was tested using several applications[2-4]. MUST (Multi-group Unstructured geometry S_N Transport) is a twin code that uses C++[5,6].

AETIUS is programed using f90 and uses Gmsh[7] as a pre- and post- processing progam. A white boundary condition on an arbitrary surface, the first collision source on the volume source, and shared memory parallel capabilities have been added to AETIUS.

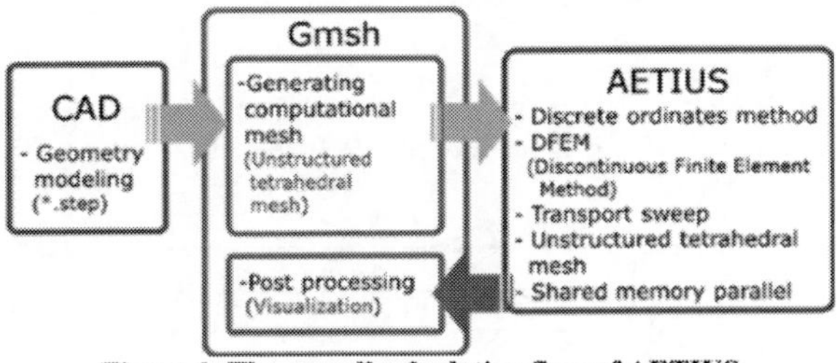

Figure 1: The overall calculation flow of AETIUS

Compare to the determinisitc codes use a regular mesh, one of the merits of MCNP5 is its capability to deal with a complicated geometry without any computational burden.

However, with the help of a CAD tools (e.g., CATIA) and Gmsh, we can finish complicated geometry modeling for AETIUS very easiliy even though the geometry is a helical tube in a cube as shown in Figure 2. This modeling is difficult to accomplish without using a CAD tool and unstructured tetrahedral mesh.

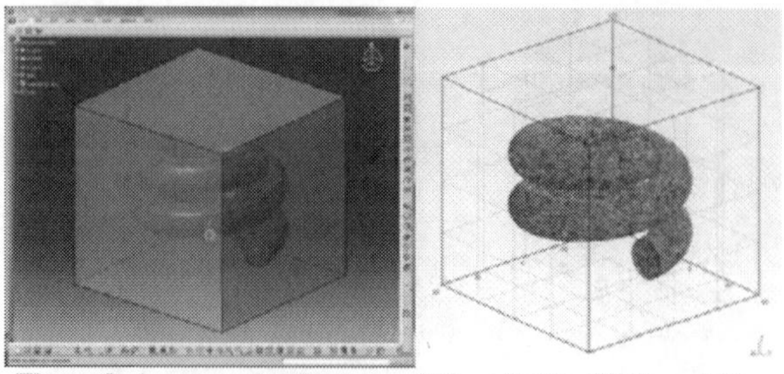

Figure 2: An example of easy modeling: (left) a CAD modeling with CATIA and (right) an unstructured tetrahedral mesh with Gmsh

1. Space

Every determinsitic code requires spatial discretization to solve the Boltzmann transport equation. An unstructured tetrahedral mesh has been used for structural mechanics, fluid dynamics, and so on. However, it has not been widely used for nuclear reactor designs or radiation shielding problems.

To deal with an unstructured tetrahedral mesh in the discrete ordinates transport code, Wareing T.A. et al. proposed a discontinuous finite element method (DFEM)[(8)] in 2001 and developed the ATTILA[(9)] code.

AETIUS also uses DFEM for space discretization. DFEM is used for solving the balance equation between adjacent tetrahedral elements according to the direction of angle in a discrete set.

2. Angle

We recently implemented a triangular and rectangular Chebyshev-Legendre[(10)] quadrature to AETIUS so that we can increase the angular quadrature up to S_{100}.

If the problem is ray-effect dominant, we may obtain better results using the first collision source method. However, if the problem consists of a ray-effect dominant zone (e.g., air) and a scattering dominant zone (e.g., concrete) together, the capability to increase S_N order over S_{20} may be useful because the first collision source method alone may not be enough to deal with this kind of problem.

3. Energy

We also use a multi-group library. The libraries currently in use[(11)] were generated based on ENDF/B-VII.0 and listed in Table 1.

Name	Group structure	
	Neutron	Gamma
VITAMIN-B6	199	42
BUGLE-96	47	20
VITAMIN-J	175	42
LANL-30	30	12
LANL-80	80	24
SCALE-44	44	18

Table 1: A list of available multi-group libraries for AETIUS

4. Shared memory parallel

AETIUS is a three-dimensional discrete ordinates code, which solves the Boltzmann transport equation on an unstructured tetrahedral mesh with DFEM. DFEM requires an unknown angular flux at four vertexes per tetrahedral element. The angle and energy are discretized through the angular quadrature and multi-group structure, respectively.

For example, in the source code, the angular flux is given as Eq. (1), where 4 indicates four vertexes per element, elem is the index of tetrahedral element, n is the index of discretized angle, and g is the index of energy group.

$$\psi\ (4,\ \mathrm{elem},\ n,\ g) \qquad (1)$$

If we have to solve a large 3D problem so that the number of elements, angular quadrature, or energy group need to be increased, then it requres more memory and a longer calculation time according to a variable size increase.

OpenMP[(12)] is implemented in AETIUS, and is easy to be implemented in an existing serial code. Besides, AETIUS requires a large amount of memory when solving a large problem, shaered memory parallel might be more suitable.

Currently, we run AETIUS on a server that has four Intel Xeon E7-4890v2 2.8Ghz CPUs using 120 threads with 512 Gb of memory.

III. Application

1. Description

To see the effectiveness and efficiency of AETIUS, we chose a deep penetration problem with small detection volumes, as shown in Figure 3.

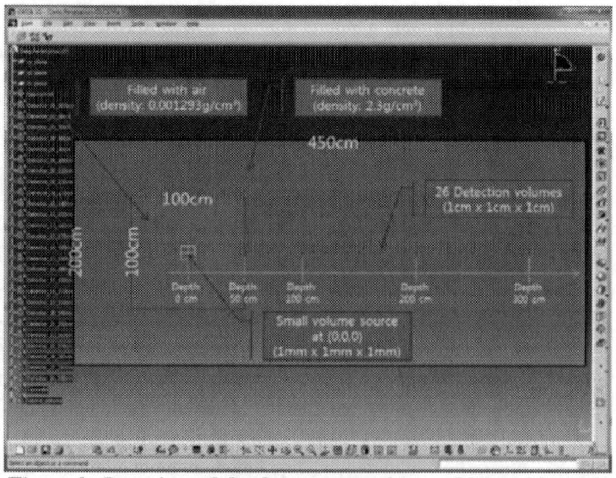

Figure 3: Overview of the deep penetration problem with small detection volumes

A 1m×1m×1m cube is located at the center of origin (0,0,0) and filled with air. Concrete covers the outside of the cube, as shown in Figure 3. The thickness in the x-direction is 3m and in the other directions is 50cm.

A small volume source zone ($1mm^3$) is prepared for one of two AETIUS calculations. 26 detection volumes (1cm×1cm ×1cm) are located along the x-axis from 50cm to 300cm with 10cm spacing.

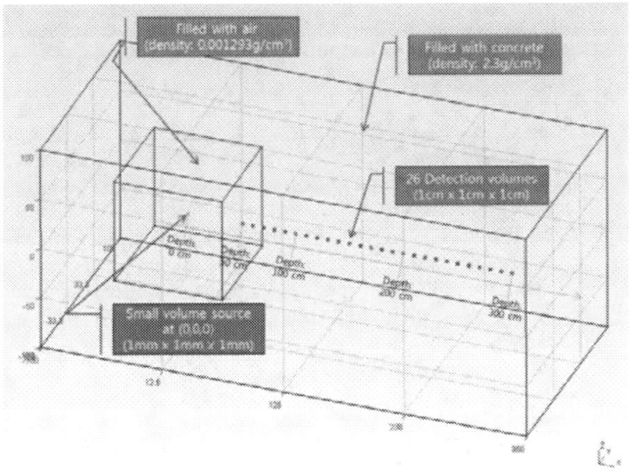

Figure 4: Three-dimensional view of a deep penetration problem with small detection volumes

The computational mesh is generated by Gmsh and shown in Figure 5.

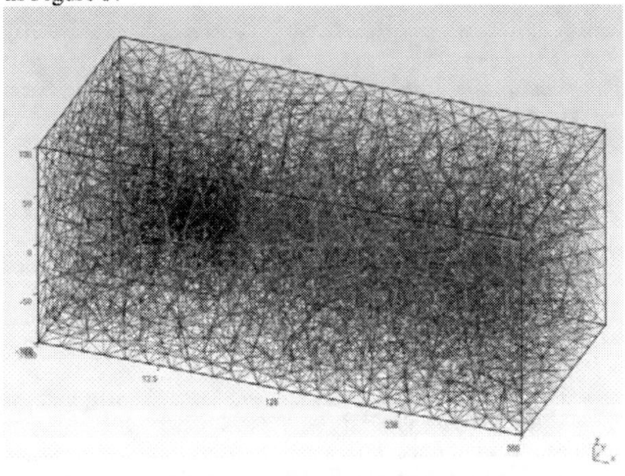

Figure 5: The computational mesh (generated by Gmsh, 61,562 unstructured tetrahedrons)

Transport calculations are performed on a deep penetration problem with small detection volumes. This problem seems very simple and easy. However, in view of the Monte Carlo transport code, it may be simple but not easy to obtain solutions with a reliable confidence interval (the relative error is less than 0.1 for all other tallies)(1).

It is hard to send particles deep into the concrete block and the tally zones (detection volumes) are small ($1cm^3$).

To obtain a satisfactory solution with MCNP5, variance reduction techniques such as splitting or weight window should be used.

In MCNP5 calculation, the neutron source is evenly sampled between 15MeV and 17MeV, which is the 1st group of LANL-30 group structure. Some other calculation parameters are listed in Table 2.

	MCNP5	AETIUS	
		with FCS	without FCS
Source strength	Point source: 1 source particle/sec at origin (0,0,0)	Point source: 1 source particle/sec at origin (0,0,0)	Volume source: 10^3 source particle/cm^3-sec, volume source zone ($10^{-3}cm^3$)
Source spectrum	Source is given in the 1st group of LANL-30		
Library	ENDF-B/VII.0		
Energy group structure	Continuous energy	LANL-30	
Material density	Air : 0.001293 g/cm^3 Concrete : 2.3 g/cm^3		
P_N order	n/a	3	
S_N order	n/a	Triangular Chebyshev-Legendre S_{40}	
Calc. options	weight window	FCS with point source	Volume source only
Error criterion	nps :5×10^9	1×10^{-4}	
Parallel	MPI (50 cores)	OpenMP (120 cores)	

FCS : first collision source method

Table 2: Calculation parameters

2. Results

The calculated total neutron flux from AETIUS is visualized through Gmsh and shown in Figure 6. This is a sectional view of the lower part of the total neutron flux isosurface distribution.

One of the key advantages of using a discrete ordinates code such as AETIUS is that we can obtain a solution of the flux throughout the problem. This is particularly useful when we wish to understand how the flux changes throughout the modelled geometry.

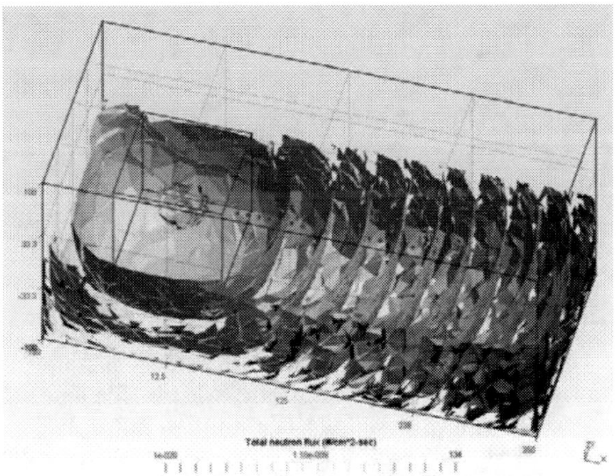

Figure 6: Visualized total neutron flux isosurface distribution (sectional view of the lower part)

Volume averaged fluxes on the 26 detector volumes are compared with those of MCNP5 and are shown in Figure 7. These fluxes of AETIUS were obtained through a single run. However, for MCNP5, we have to run it several times to obtain the weight window and re-run it again. Although we used weight window in the MCNP5 calculation, the average fluxes in the last two detection volumes could not be obtained.

In view of relative error, volume-averaged fluxes for the 17 detection volumes near source area (from 50cm to 210cm) are within a reliable confidence interval and are well matched with those of AETIUS.

However, for last 9 detection volumes, the relative errors of MCNP5 result are getting increased as depth is increased. Moreover, relative errors of last two detection volumes are zero since no particles are tallied.

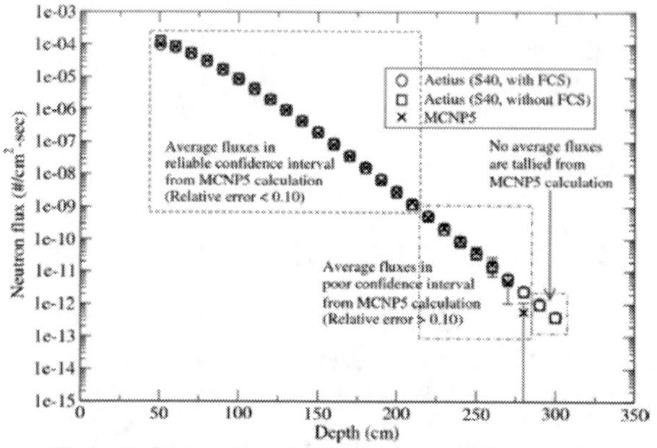

Figure 7: Comparison of volume averaged fluxes on 26 detection volumes

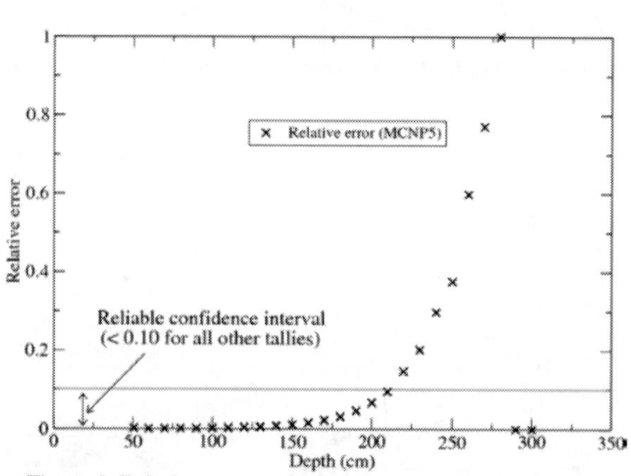

Figure 8: Relative error of MCNP5 on 26 detection volumes

IV. Conclusion

We discribed a brief overview of AETIUS and provided numerical results from both AETIUS and MCNP5 for a deep penetration problem with small detection volumes.

Although the results of the two codes are obtained using completely different calculation methods, they show a good agreement.

Deterministic codes (e.g., AETIUS) require many spatial elements, high S_N/P_N orders, and many energy groups to obtain a good solution, which could require many iterations and longer calculation time.

Similarly, Monte Carlo codes (e.g., MCNP5) also require longer running time and many iterations to optimize the variance reduction techniques (e.g., splitting and weight window).

However, deterministic codes can provide a rough estimate of the flux throughout a problem relatively quickly. This can help to identify areas of a problem where shielding may or may not be performing as well as expected.

References

1) X-5 Monte Carlo Team, *MCNP - A General Monte Carlo N-Particle Transport Code, Version 5, Volume II: User's Guide*, LA-CP-03-0245, Los Alamos National Laboratory (2008).
2) J.W. Kim, S.G. Hong, and Y. Lee, "Computational efficiency of a modified scattering kernel for full-coupled photon-electron transport parallel computing with unstructured tetrahedral meshes," *Nuclear Engineering and Technology*, vol. 46, no. 2, 263-272 (2014).
3) J.W. Kim, C.W. Lee, Y. Lee, D. Lee, and S. Cho, "Development of discrete ordinates code supporting unstructured tetrahedral mesh and applied in neutronics analysis for the Korea Helium Cooled Ceramic Reflector Test Blanket Module," *Fusion Engineering and Design*, 89, 1172-1176 (2014).
4) J.W. Kim, C.W. Lee, Y. Lee, D. Lee, and S. Cho, "Preliminary study on applying discrete ordinates code supporting unstructured tetrahedral mesh to the 40-degree toroidal segment ITER model," *Fusion Science and Technology*, 68, 652-656 (2015).
5) S.G. Hong, J.W. Kim, and Y. Lee, "Development of MUST (Multi-group Unstructured geometry S_N Transport) Code," *Transaction of the Korean Nuclear Society Autumn Meeting*, Gyeongju, Korea (2009).
6) J.W. Kim, S.G. Hong, and Y. Lee, "MUST code verification and validation on the shielding test problems," *ICRS-12 & RPSD-2012*, Nara, Japan (2012).
7) C. Geuzaine, J.-F. Remacle, "Gmsh: a three-dimensional finite element mesh generator with built-in pre- and post-processing facilities," *Int. J. Numer. Meth. Eng.* 79 (11) 1309-1331 (2009).
8) T.A. Wareing, J.M. McGhee, J.E. Morel, and S.D. Pautz, "Discontinuous finite element S_N methods on three-dimensional unstructured grids," *Nuclear Science and Engineering*, 138, 256-268 (2001).
9) J.M. McGhee, T.A. Wareing, and D.A. Barnett Jr., *Attila User's Manual*, Transpire Inc. (2007).
10) K.D. Lathrop and B.G. Carlson, *Discrete ordinates angular quadrature of the neutron transport equation*, LA-3186, Los Alamos National Laboratory (1965).
11) D.H. Kim, C. Gil, and Y. Lee, "Validation of an ENDF/B-VII.0-based neutron and photon shielding library in MATXS-format," *Journal of the Korean Physical Society*, vol. 59, no. 2, 1199-1202 (2011).
12) Barbara Chapman et al., *Using OpenMP*, The MIT Press, Cambridge, Massachusetts, London, England (2008).

Adaptive angle and parallel multigrid for deterministic shielding problems

Steven Dargaville[1,a], Andrew Buchan[1,b], Richard Smedley-Stevenson[1,2,c], Paul Smith[1,3,d], and Chris Pain[1,e]

[1] *Applied Modelling and Computation Group (AMCG), Imperial College London, UK*
[2] *AWE, Aldermaston, UK*
[3] *ANSWERS Software Services, AMEC-FW, Dorset, UK*

Abstract. Traditionally, solver technology and the space/angle discretisations are intimately linked; sweep-based (wavefront) methods are typically used with DG FEM in space and Sn in angle to solve the Boltzmann-transport equation. These parallelise well (scaling to >100,000 cores) on structured grids, however achieving good scaling on unstructured grids is still an open problem. This talk will focus on alternate space/angle discretisations and solver technology we have been developing within the Applied Modelling and Computation Group (AMCG) at Imperial College. These approaches enables the use of traditional angular discretisations like Pn, Sn, along with new approaches based on linear and haar wavelets. We can use these angular discretisations to perform regular and goal-based anisotropic adaptivity in angle, focusing resolution in important directions. We have also been developing multigrid solver technology which does not require sweep-based methods, allowing the possibility of excellent scaling on unstructured grids.

1 Introduction

In this work, we focus on the steady-state, first-order integro-differential form of the BTE, given by

$$\mathbf{\Omega}\cdot\nabla\psi(\boldsymbol{r},\mathbf{\Omega},E)+\Sigma_{\mathrm{t}}\psi(\boldsymbol{r},\mathbf{\Omega},E)=\int_{\mathbf{\Omega}'}\int_{E'}\Sigma_{\mathrm{s}}(\boldsymbol{r},\mathbf{\Omega}'\to\mathbf{\Omega},E'\to E)\psi(\boldsymbol{r},\mathbf{\Omega}',E')\mathrm{d}E'\mathrm{d}\mathbf{\Omega}'+S_{\mathrm{e}}(\boldsymbol{r},\mathbf{\Omega},E),\quad(1)$$

where $\psi(\boldsymbol{r},\mathbf{\Omega},E)$ is the angular flux at spatial position $\boldsymbol{r}$, in direction $\mathbf{\Omega}$ and at energy E. The total and scatter macroscopic cross sections are given by Σ_{t} and Σ_{s}, respectively, with S_{e} the source term.

We provide a brief outline of our discretisation of (1), below, taken from [1]. In short, we use a sub-grid scale FEM on an unstructured grid to provide a high-order, stable discretisation. For the sub-grid scale formulation, we decompose the solution as $\psi=\phi+\theta$, where ϕ and θ are the solutions on the "coarse" and "fine" scales, respectively. We approximate the solution on the coarse scale with a

[a] e-mail: s.dargaville@imperial.ac.uk
[b] e-mail: andrew.buchan@imperial.ac.uk
[c] e-mail: r.smedley-stevenson@imperial.ac.uk
[d] e-mail: Paul.Smith3@amecfw.com
[e] e-mail: c.pain@imperial.ac.uk

DOI: 10.1051/epjconf/201715306026

continuous finite element representation spanned by η_N basis functions, while on the fine scale we use a discontinuous representation spanned by η_Q basis functions (like [2]). Doing this, we can recover the system

$$\begin{pmatrix} \mathbf{A} & \tilde{\mathbf{B}} \\ \mathbf{C} & \tilde{\mathbf{D}} \end{pmatrix} \begin{pmatrix} \tilde{\mathbf{\Phi}} \\ \tilde{\mathbf{\Theta}} \end{pmatrix} = \begin{pmatrix} \mathbf{S_\Phi} \\ \mathbf{S_\Theta} \end{pmatrix}, \tag{2}$$

where $\mathbf{S_\Phi}$ and $\mathbf{S_\Theta}$ are the discretised source and $\tilde{\mathbf{\Phi}}$ and $\tilde{\mathbf{\Theta}}$ are vectors containing the coefficients of the coarse and fine discretised solutions, $\tilde{\phi}_{i,j}$ and $\tilde{\theta}_{i,j}$, respectively. This is a Schur complement of the block $\mathbf{D}$, or

$$(\mathbf{A} - \mathbf{B}\mathbf{D}^{-1}\mathbf{C})\tilde{\mathbf{\Phi}} = \mathbf{S_\Phi} - \mathbf{B}\mathbf{D}^{-1}\mathbf{S_\Theta}. \tag{3}$$

Equation 3 has the same number of DOF as the continuous problem, but now involves the inversion of $\mathbf{D}$. As is typical in Schur complements, we apply approximations to $\mathbf{D}$ to increase its sparsity. The inversion of $\mathbf{D}$ can be performed element by element (in practice, given the approximations applied, $\mathbf{D}^{-1}$ is often small enough to be stored in memory). Solving (3) can therefore be considered equivalent to solving the continuous problem modified by a local, element-based stabilisation term. The general nature of this discretisation makes it easy to apply different (adaptive) angular discretisations.

2 Anisotropic angular adaptivity

Goal-based adaptive technology requires solving an adjoint problem, which allows the targeting of resolution where necessary to reduce the error in a goal, like the average flux over a region/surface (see [3–5], also see [6–9] for more examples of the use of angular adaptivity). Figures 1 and 2 show the results of using goal-based angular adaptivity using linear wavelets to discretise in angle. Figure 1b shows that a given error in the average flux of the "green" region show in Figure 1a, is achieved with around an order of magnitude less unknowns than a uniform S_n calculation.

This is because the goal-based angular adaptivity has focused angular resolution only in the duct region, as shown in Figure 2a. Figure 2b shows that this angular resolution is applied anisotropically on the sphere.

3 Multigrid solver

Multigrid has been used with great success in radiation transport, applied to the spatial, angular and energy domains, along with combinations of the three. This started with S_n discretisations in angle, with multigrid applied both in angle [10, 11] and space [12–15]. Spherical harmonics (P_n) have also been used, again with multigrid performed in both space [16–19] and angle [20]. Stepping away from purely spatial or angular multigrid, mixed space/angle [21, 22] and space/angle/energy [23] multigrid schemes with S_n have been explored recently. Multigrid in energy has also been used to improve the parallel performance of DENOVO [24].

We have been developing matrix-free multigrid solvers for the adapted discretisations described above. With this solver, we can precondition with spatial, angular or mixed space/angle multigrid. This is with the goal of improving the scaling on unstructured grids, when compared with purely sweep-based solvers. Table 1 shows that applying uniform angular resolution with level 4 Haar wavelets (corresponds to 256 angles, see [5]), results in 80% strong scaling efficiency on 6144 cores.

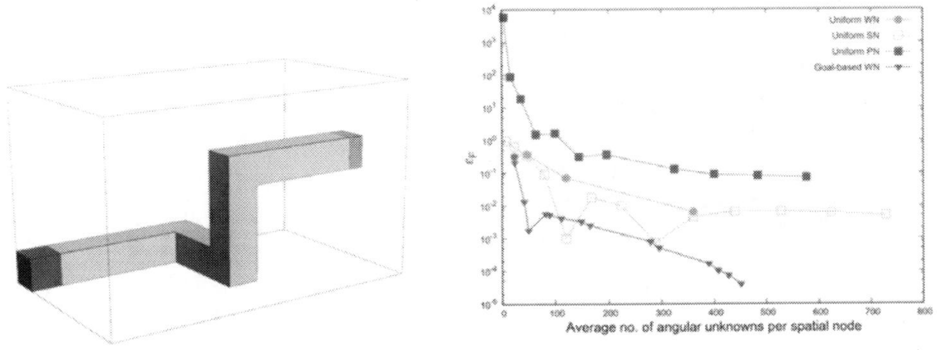

(a) Geometry of the 3D "duct" problem. Red region is a source, with the "goal" set as the average flux over the green region. The duct is a vacuum, surrounded by a scattering region with the bulk a heavy absorber.

(b) Convergence of the error in the average flux in the goal region using adaptive linear wavelets.

Figure 1: Problem geometry and convergence results for a 3D "duct" shielding problem [4, 5]

Table 1: Strong scaling with H_{44} (256 angles)

Cores	Time(s)	Elements	DOFs	DOFs/Core	Scaling
72	7451.3	1,290,000	41,034,240	2,279,680	-
3072	197.6	1,290,000	41,034,240	53,430	88%
6144	109.7	1,290,000	41,034,240	26,715	79.6%

References

[1] S. Dargaville, M.A. Goffin, A.G. Buchan, C.C. Pain, R.P. Smedley-Stevenson, P.N. Smith, G. Gorman, Annals of Nuclear Energy **86**, 99 (2015)

[2] T.J.R. Hughes, G. Scovazzi, P.B. Bochev, A. Buffa, Computer Methods in Applied Mechanics and Engineering **195**, 2761 (2006)

[3] M.A. Goffin, C.M.J. Baker, A.G. Buchan, C.C. Pain, M.D. Eaton, P.N. Smith, Journal of Computational Physics **242**, 726 (2013)

[4] M.A. Goffin, A.G. Buchan, A.C. Belme, C.C. Pain, M.D. Eaton, P.N. Smith, R.P. Smedley-Stevenson, Annals of Nuclear Energy **71**, 60 (2014)

[5] M.A. Goffin, A.G. Buchan, S. Dargaville, C.C. Pain, P.N. Smith, R.P. Smedley-Stevenson, Journal of Computational Physics **281**, 1032 (2015)

[6] A.M. Watson, Ph.D. thesis, Texas A&M University (2005)

[7] J.C. Stone, Ph.D. thesis, Texas A&M University (2007)

[8] H. Park, C.R.E. de Oliveira, Nuclear Science and Engineering **161**, 216 (2009)

[9] K. Rupp, T. Grasser, A. Jungel, *Adaptive variable-order spherical harmonics expansion of the Boltzmann Transport Equation*, in *2011 International Conference on Simulation of Semiconductor Processes and Devices (SISPAD)* (2011), pp. 151–154

[10] J.E. Morel, T.A. Manteuffel, Nuclear Science and Engineering **107**, 330 (1991)

[11] B. Turcksin, J.C. Ragusa, J.E. Morel, Transport Theory and Statistical Physics **41**, 1 (2012)

[12] P.F. Nowak, Ph.D. thesis, University of Michigan (1988)

 DOI: 10.1051/epjconf/201715306026

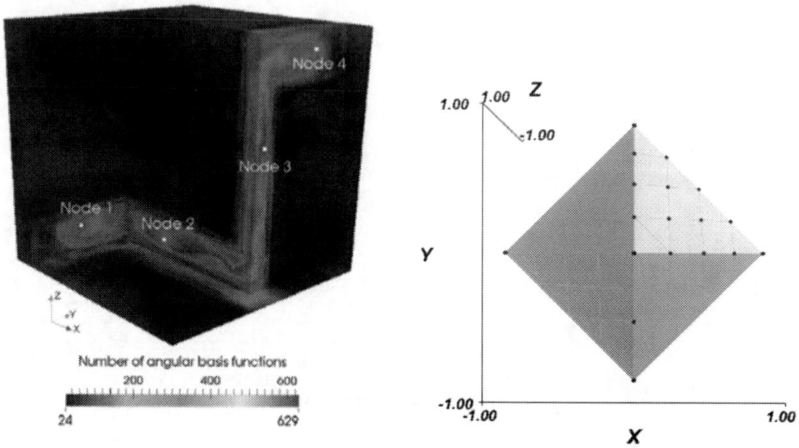

(a) Number of angular unknowns

(b) Position of wavelets at Node 2 shown in Figure 2a. Most of the wavelets are focused in the direction pointing down the duct region.

Figure 2: Anisotropic goal-based adaptivity with linear wavelets applied to 3D "duct" shielding problem [4, 5]

[13] T. Manteuffel, S. Mccormick, J. Morel, S. Oliveira, G. Yang, SIAM Journal on Scientific Computing **16**, 601 (1995)

[14] T. Manteuffel, S. McCormick, J. Morel, G. Yang, SIAM Journal on Scientific Computing **17**, 1449 (1996)

[15] G. Kanschat, J. Ragusa, SIAM Journal on Scientific Computing **36**, A2326 (2014)

[16] S. Oliveira, Y. Deng, Progress in Nuclear Energy **33**, 155 (1998)

[17] B. Chang, B. Lee, ETNA. Electronic Transactions on Numerical Analysis [electronic only] **15**, 132 (2003)

[18] P. Brown, B. Lee, T. Manteuffel, SIAM Journal on Scientific Computing **25**, 513 (2003)

[19] B. Chang, T. Manteuffel, S. McCormick, J. Ruge, B. Sheehan, SIAM Journal on Scientific Computing **29**, 1900 (2007)

[20] C.R.E. de Oliveira, C.C. Pain, M. Eaton, *Hierarchical angular preconditioning for the finite element-spherical harmonics radiation transport method*, in *Proceedings of PHYSOR 2000 ANS International Topical Meeting on Advances in Reactor Physics and Mathematics and Computation into the Next Millenium, Pittsburgh, USA* (2000)

[21] B. Lee, SIAM Journal on Scientific Computing **31**, 4744 (2010)

[22] B. Lee, SIAM Journal on Scientific Computing **32**, 2497 (2010)

[23] B. Lee, Numerical Linear Algebra with Applications **19**, 773 (2012)

[24] R.N. Slaybaugh, T.M. Evans, G.G. Davidson, P.P.H. Wilson, Journal of Computational Physics **242**, 405 (2013)

Polynomial axial expansion in the Method of Characteristics for neutron transport in 3D extruded geometries

Laurent Graziano[1,⋆], *Simone* Santandrea[1,⋆⋆], and *Daniele* Sciannandrone[1,⋆⋆⋆]

[1] *CEA, Saclay (France)*

Abstract. Recently a solver based on the Method Of Characteristics (MOC) for 3D extruded geometries has been developed in the APOLLO3® project [1]. In this method the domain is divided in regions and the Step Characteristics (SC) approximation is used in each of them to represent the solution of the transport problem. Since the biggest degree of heterogeneities is found along the radial direction, the idea proposed in this paper is to keep the SC approximation to compute the solution over the radial plane and to implement a polynomial expansion of the flux along the vertical direction. In fact the results of the previous works [1] show that the flux gradient along the z axis are likely to be represented by a classical polynomial basis. In this paper we weigh up the benefits of using a less refined mesh.

1 Introduction

In the recent years a solver based on the Method of Characteristics (MOC) allowing the treatment of 3D extruded geometries has been developed inside the TDT module of APOLLO3® [1][2] [3]. The standard Step Characteristics (SC) approximation is used and results show an excellent agreement with Monte-Carlo simulations. However a fine mesh refinement is needed to converge, due to strong flux gradients. In this paper an improvement of this method is proposed: the results of the previous work show that the flux gradients are likely to be represented by a polynomial basis along the vertical direction. Since most of the geometrical and physical heterogeneities are radially located, the SC approach is preserved to represent the solution over the radial plane. As a matter of fact the strong irregularities in the geometrical meshes prevent from an efficient use of a polynomial expansion. On the contrary along the axial direction the computational meshes assume a Cartesian shape, well suited for a polynomial representation of sources and fluxes, as shown in Figure 1. A suitable polynomial development in this direction allows us to approximate the strong flux gradients without the help of a large number of axial meshes. The seminal ideas of this method are already given in [3]. They are here developed in a different form.

2 Equation derivation

2.1 Polynomial basis definition

The starting point is the mono-group, steady state neutron transport equation is a geometrical domain D and boundary Γ:

$$\left[\vec{\Omega}\nabla + \Sigma_t(\vec{r})\right]\Psi(\vec{r},\vec{\Omega}) = q(\vec{r},\vec{\Omega}) \quad \vec{r}\in D,\ \vec{\Omega}\in S_{4\pi}$$
$$\Psi(\vec{r},\vec{\Omega}) = \beta\Psi(\vec{r},\vec{\Omega}) \quad r\in\Gamma,\ \vec{\Omega}\in S^+_{4\pi}(\vec{r}) \tag{1}$$

where $q(\vec{r},\vec{\Omega})$ is the source term, $S_{4\pi}$ stands for the unit sphere directions, $S^+_{4\pi}(\vec{r})$ is the set of exiting directions from D in the point $\vec{r}$ and β is a function relating entering and exiting direction.

A local polynomial basis can be used to express sources and flux moments:

$$\vec{P}(\bar{z}_r) = \{\bar{z}_r^p = \left(\frac{z_r - \bar{z}_r}{\Delta z_r/2}\right)^p,\quad 0\le p\le N_p\} \tag{2}$$

where Δz_r is the height of the given region, $z_r \in [-\Delta z_r/2, \Delta z_r/2]$ is the local coordinate for the r_{th} region,

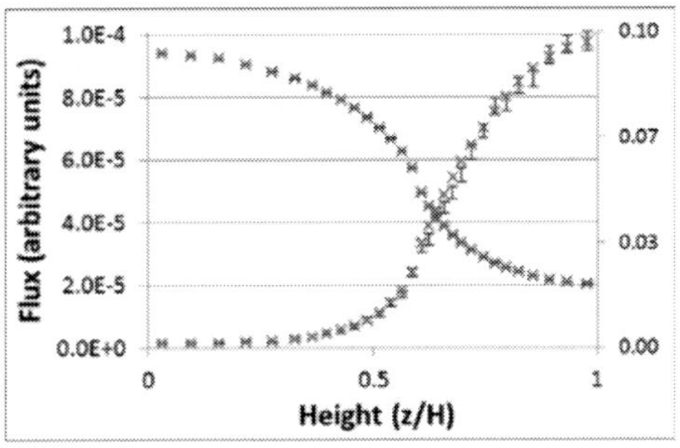

Figure 1: Axial distribution of the flux for a thermal (red) and a fast (blue) group inside the fuel pin. The figure also shows the reference Monte Carlo solution with associated error bars [3].

⋆e-mail: laurent.graziano@cea.fr
⋆⋆e-mail: simone.santandrea@cea.fr
⋆⋆⋆e-mail: daniele.sciannandrone@cea.fr

$\bar{z}_r$ is the value of the axial coordinate at the region center and N_p is the chosen order for the polynomial expansion.

Thanks to this definition $\bar{z}_r \in [-1, 1]$ for every region, independently on the region height. In this way passing from a region to the next one through an horizontal border the polynomial value only change is sign, switching from $+1$ to -1. This property allows to avoid a certain number of operations during the trajectory sweep, as it will be shown later.

From the local polynomial basis $\vec{P}(\bar{z})$ is possible to retrieve an orthonormalized basis $\vec{\xi}(\bar{z})$ through the Gram-Schmidt orthonormalization algorithm. The coefficients of this algorithm can be used to form a lower triangular matrix that allows to switch from the orthonormal to the polynomial basis:

$$\boxed{\vec{\xi}(\bar{z}) = \bar{\bar{\mathbf{M}}}^{-1}\vec{P}(\bar{z})} \tag{3}$$

where:

$$\bar{\bar{\mathbf{M}}} = \begin{bmatrix} 1 & 0 & 0 & 0 \\ m_{10} & m_{11} & 0 & 0 \\ m_{20} & m_{21} & m_{22} & 0 \\ m_{30} & m_{31} & m_{32} & m_{33} \\ \vdots & \vdots & \vdots & \vdots\ddots \end{bmatrix} \quad with \quad \begin{aligned} m_{pj} &= \frac{<P_p, \xi_j>}{\|\tilde{\xi}_p(\bar{z})\|} \\ m_{pp} &= \|\tilde{\xi}_p(\bar{z})\| \end{aligned}$$

The terms $< P_p, \xi_j >$ are computed using the geometrical matrix $\bar{\bar{\mathbf{P}}}$:

$$\bar{\bar{\mathbf{P}}} = \frac{1}{V_r}\int d\vec{r}\, \vec{P}(\bar{z}) \otimes \vec{P}(\bar{z}) = \int_{-1}^{+1} d\bar{z}\, \vec{P}(\bar{z}) \otimes \vec{P}(\bar{z}) \tag{4}$$

which elements read:

$$\bar{\bar{\mathbf{P}}}_{pp'} =< P_p, P_{p'} >= \begin{cases} \frac{1}{p+p'+1} & for \quad p + p' \ even \\ 0 & for \quad p + p' \ odd \end{cases}$$

and V_r denotes the volume of the r region.

2.2 Source expansion

The source term can now be expanded over the orthonormalized basis:

$$q(\vec{r}, \vec{\Omega}) = \sum_r \theta_r(\vec{r})\vec{\xi}(\bar{z}_r) \cdot \vec{q}_r(\vec{\Omega}) \tag{5}$$

where θ_r is the characteristic function for the r_{th} region and $\vec{q}_r$ are the orthonormal coefficients of the source moments:

$$\vec{q}_r(\vec{\Omega}) = \sum_n A_n(\vec{\Omega}) \cdot \vec{q}_r^{\,n}(\vec{\Omega}) = \sum_n A_n(\vec{\Omega}) \cdot \left[\Sigma_r^n \vec{\phi}_r^n + \vec{S}_r^n\right] \tag{6}$$

Σ_r^n is the macroscopic in-group transfer cross section, A_n the real spherical harmonics, $\vec{\phi}_r^{\,n}$ the coefficients of the flux moments expansion and $\vec{S}_r^{\,n}$ is the external source, accounting for fissions and transfer from other groups. The sum over n stands for:

$$\sum_n A_n(\vec{\Omega})\vec{q}_r^{\,n}(\vec{\Omega}) = \sum_{l=0}^{l=K} \sum_{m=-l}^{m=l} A_l^m(\vec{\Omega})\vec{q}_{r,l}^{\,m}(\vec{\Omega})$$

Where K is the anisotropy order.

Thanks to the orthonormality property, by projecting $q(\vec{r}, \vec{\Omega})$ over the polynomial basis $\vec{\xi}(\bar{z})$ we get:

$$\begin{aligned} \int_{-1}^{+1} d\bar{z}\, \vec{\xi}(\bar{z})\, q(\vec{r}, \vec{\Omega}) &= \int_{-1}^{+1} d\bar{z}\, \vec{\xi}(\bar{z}) \otimes \vec{\xi}(\bar{z})\, \vec{q}_r(\vec{\Omega}) \\ &= \begin{cases} 0 & if \quad p \neq p' \\ \vec{q}_r(\vec{\Omega}) & if \quad p = p' \end{cases} \end{aligned}$$

This important property is very useful, as shown later, and justify the choice of a orthonormalized basis.

To switch from the orthonormal source coefficients to the polynomial one the following relation is used, retrieved from the definition (3):

$${}^t\left(\bar{\bar{\mathbf{M}}}^{-1}\, \vec{P}(\bar{z})\right)\vec{q}_r(\vec{\Omega}) = {}^t\left(\vec{P}(\bar{z})\right) \cdot \vec{q}_{r,pol}(\vec{\Omega})$$

From which we get:

$$\boxed{\vec{q}_{r,pol} = {}^t\bar{\bar{\mathbf{M}}}^{-1} \cdot \vec{q}_r(\vec{\Omega})} \quad and \quad \boxed{\vec{q}_r = {}^t\bar{\bar{\mathbf{M}}} \cdot \vec{q}_{r,pol}(\vec{\Omega})} \tag{7}$$

where $\vec{q}_{r,pol}(\vec{\Omega})$ is the coefficient vector referring to the polynomial basis $\vec{P}(\bar{z})$.

2.3 Transmission equation

The SC numerical MOC transmission equation obtained from the integral transport equation reads:

$$\Psi_r(t, \vec{\Omega}) = \Psi_r(0, \vec{\Omega})e^{-\tau(0,t)} + \beta(t, \vec{\Omega})\, q_r(t, \vec{\Omega})$$

where $\Psi_r(0, \vec{\Omega})$ is the entering flux at the trajectory beginning, $\Psi_r(t, \vec{\Omega})$ is the flux in a generic point t along the trajectory and:

$$\beta_r(t, \vec{\Omega}) = \left[\frac{1 - e^{-\tau(0,t)}}{\Sigma_r}\right] \qquad \tau(0, t) = \int_0^t dt'\, \Sigma_r$$

are called respectively the *transmission coefficient* and the *optical length*.

The source expansion of Eq.(5) is now used. Eq.(3) allows us to switch to the standard polynomial basis and a local system of coordinates along the trajectory is used to express the $\bar{z}$ component:

$$\bar{z}_r = z_r^{in} - \bar{z}_r + \mu\, t \qquad \mu = cos(\theta) \qquad t \in (0, l_r)$$

z_r^{in} is the z coordinate of the entering point of the trajectory in the r region, θ is the polar component of the $\vec{\Omega}$ direction. The polynomial transmission equation obtained in this way reads:

$$\Psi_r(t, \vec{\Omega}) = \Psi_r(0, \vec{\Omega}) \cdot e^{-\tau(0,t)} + \tag{8}$$
$$\sum_{p=0}^{N_p} \left\{ \left[\sum_{k=0}^{p} c_{p,k}\, P_k(z_r^{in})\, \mu^{p-k} \left(\frac{2}{\Delta z_r}\right)^{p-k} E_{p-k}(\tau(t))\right] \cdot \frac{\left(\vec{q}_{r,pol}(\vec{\Omega})\right)_p}{\Sigma_r} \right\}$$

where:

$$c_{pk} = \frac{p!}{k!(p-k)!}$$

$$E_{p-k}(\tau(t)) = \frac{1}{\Sigma_r^{(p-k)}} \int_0^{\tau(t)} d\tau' \tau'^{p-k} e^{(\tau'-\tau(t))} \quad (9)$$

and $P_k(\bar{z}_r^{in})$ is defined in (2).
For numerical reasons it is useful to rearrange the transmission equation as follows:

$$\Psi_r(t, \vec{\Omega}) = \Psi_r(0, \vec{\Omega}) \cdot e^{-\tau(0,t)} + \vec{P}(\bar{z}_r^{in}) \cdot \vec{T} \quad (10)$$

where:

$$T_k = \sum_{p=max(k,1)}^{N_p} F_{p,k} \frac{\left(\vec{q}_{r,pol}(\vec{\Omega})\right)_p}{\Sigma_r}$$

$$F_{p,k} = c_{p,k}\, \mu^{p-k} \left(\frac{2}{\Delta z}\right)^{p-k} E_{p-k}(\tau(t))$$

This final form of the transmission equation allows to minimize the number of floating point operations to be executed on-the-fly during the sweep, for each chord. The term $\vec{T}$, on the other hand, can be pre-computed during the classification phase. The classification process is widely explained in [3]. In a few words, the chord classification is a procedure that exploits the regularity of extruded geometries to recognize chords with the same length. Starting for a given 2D chord, a given polar angle and moving vertically, every chord that lies on that basic 2D chord and intersect two successive vertical surface will have the same length. Fig. 2 gives a visual interpretation of the concept. These chords intersecting two successive vertical surfaces are referred to as the *V-chords* in [3] . In the same way *H-chords* are defined as chords intersecting two successive horizontal surfaces, and *M-chords*, as chords intersecting a horizontal and then a vertical surface (or vice versa). The advantage of the classification process is that when the same chords is found several times, its coefficient can be computed just once. Since the $\vec{T}$ term is computationally expensive, the possibility to strongly reduce the number of times that it must be computed represent a big advantage.

Computational cost

The transmission sweep is one of the most expensive part of the iterative scheme in computational terms. The transmission equation adapted to the polynomial case derived in this section is clearly more expensive than the classical SC transmission, which asks for one floating point operation per chord (for classified chords). On the other hand the polynomial equation needs $1+N_p$ operations (N_p being the polynomial degree). Moreover, it requires to know at each point during the trajectory sweep the value of the axial coordinate and all its powers. If this information is computed on-the-fly it requires $1+N_p$ additional operations. In addition, even the coefficients that ca be pre-computed during the classification phase are much more expensive. For these reason for an equivalent tracking (so the same amount and type of chords) the polynomial solution will be more expensive in computational terms.

2.4 Escape coefficient computation

The escape coefficients $E_{p-k}(\tau)$ can be computed integrating Eq. (9). However, by integrating by parts this equation, a useful recursive relation can be obtained. This allows to compute only the 0_{th} order coefficient and to retrieve the others:

$$E_0(\tau) = 1 - e^{-\tau}$$

$$E_b(\tau) = l^b - \frac{b}{\Sigma_r} E_{b-1}(\tau) \qquad 1 \leq b \leq N_p \quad (11)$$

The generic escape coefficient E_b can be expressed as a function of the first one through the relation:

$$E_0(\tau) = 1 - e^{-\tau}$$

$$E_1(\tau) = l - \frac{1}{\Sigma_r} + \frac{1}{\Sigma_r} e^{-\tau}$$

$$E_2(\tau) = l^2 - \frac{2}{\Sigma_r} l + \frac{2}{\Sigma_r^2} - \frac{2}{\Sigma_r^2} e^{-\tau}$$

$$E_3(\tau) = l^3 - \frac{3}{\Sigma_r} l^2 + \frac{3 \cdot 2}{\Sigma_r^2} l - \frac{3 \cdot 2}{\Sigma_r^3} + \frac{3 \cdot 2}{\Sigma_r^3} e^{-\tau}$$

$$E_b(\tau) = \sum_{i=0}^{b} \left[\frac{b!}{i!} \frac{l^i}{(-\Sigma_r)^{b-i}} \right] - \frac{b!}{(-\Sigma_r)^b} e^{-\tau}$$

Unfortunately the recursive relation Eq. (11) is ill-conditioned for small values of the total cross section. Another possible approach is to start with the highest order coefficient and retrieve the others with the corresponding backward recursive relation. This is a more stable approach for small τ values. In this case the highest order coefficient must be computed. The more efficient way to do this is through the use of tabulated values. Again, small values of the optical length cause big instabilities not only in the use of the recursive relation, but also in the initial computation of the highest order coefficient.

To avoid this problem the strategy adopted is to tabulate a different function. The escape coefficient Eq. (9) is computed as:

$$E_b(\tau) = l^b E_b^T(\tau)$$

$$E_b^T(\tau) = \frac{1}{\tau^b} \int_0^{\tau(t)} d\tau' \tau'^b e^{(\tau'-\tau(t))} \quad (12)$$

The function $E_b^T(\tau)$ happens to be well-conditioned even for small τ values. Integrating (12) by parts a forward and a backward relation can be retrieved:

$$forward \rightarrow \quad E_b^T(\tau) = 1 - \frac{b}{\tau} E_{b-1}^T(\tau)$$

$$backward \rightarrow \quad E_b^T(\tau) = \frac{\tau}{b+1} \left[1 - E_{b+1}^T(\tau)\right]$$

Also in this case the backward relation is the more stable for small values of the optical length. A table must be pre-computed for the highest order escape coefficient. Instead of integrating the relation (12) a more efficient approach can be the following: the forward recursive relation is used starting from the 0_{th} order value for large values of the optical length and a Maclaurin series is used to expand the function $e^{-\tau}$ for small τ values.

 DOI: 10.1051/epjconf/201715306027

For large τ:

$$E_0^T(\tau) = 1 - e^{-\tau}$$
$$E_b^T(\tau) = 1 - \frac{b}{\tau}E_{b-1}^T(\tau)$$

For small τ:

$$E_0^T(\tau) = 1 - e^{-\tau} = \tau - \frac{\tau^2}{2} + \frac{\tau^3}{3!} - ...$$
$$E_1^T(\tau) = 1 - \frac{1}{\tau}E_0^T(\tau) = \frac{\tau}{2} - \frac{\tau^2}{3!} + \frac{\tau^3}{4!} - ...$$
$$E_2^T(\tau) = 1 - \frac{2}{\tau}E_1^T(\tau) = \frac{2}{3!}\tau - \frac{2}{4!}\tau^2 + \frac{2}{5!}\tau^3 - ...$$
$$E_3^T(\tau) = 1 - \frac{3}{\tau}E_2^T(\tau) = \frac{3!}{4!}\tau - \frac{3!}{5!}\tau^2 + \frac{3!}{6!}\tau^3 - ...$$
$$E_b^T(\tau) = \sum_{i=1}^{\infty}(-)^{i+1}\frac{b!}{(b+i)!}\tau^i$$
$$= \frac{\tau}{b+1}\left[1 - \frac{\tau}{b+2}\left[1 - \frac{\tau}{b+3}\left[1 - \frac{\tau}{4}\right]\right]\right]$$

2.5 Balance equation

A balance equation can be obtained by projecting the transport equation (1) over $\vec{\xi}(\bar{z})$:

$$\frac{1}{V_r}\int_r d\vec{r}\vec{\xi}(\bar{z})\left(\vec{\Omega}\nabla\Psi(\vec{r},\vec{\Omega})\right) + \frac{\Sigma_r}{V_r}\int_r d\vec{r}\vec{\xi}(\bar{z})\Psi(\vec{r},\vec{\Omega}) = \vec{q}(\vec{r},\vec{\Omega})$$

where the integral is performed over the r region.

The following definition is given for the orthonormal coefficients of the angular flux:

$$\vec{\Psi}(\vec{r},\vec{\Omega}) = \frac{1}{V_r}\int_r d\vec{r}\vec{\xi}(\bar{z})\Psi(\vec{r},\vec{\Omega}) \quad (13)$$

The equation is rearranged, the basis switch is done again with Eq. (3) and the following useful property of the polynomial basis $\vec{P}(\bar{z})$ is used:

$$\frac{\partial P_p}{\partial z} = \frac{p}{(\Delta z/2)}P_{p-1}$$

The final balance equation obtained is:

$$\boxed{\Sigma_r\vec{\Psi}(\vec{r},\vec{\Omega}) = \vec{q}_r(\vec{r},\vec{\Omega}) - \bar{\bar{\mathbf{M}}}^{-1}\cdot\Delta\vec{J}(\vec{r},\vec{\Omega}) + \mu\,\bar{\bar{\mathbf{C}}}\,\vec{\Psi}(\vec{r},\vec{\Omega})} \quad (14)$$

where the current term is:

$$\Delta\vec{J}(\vec{r},\vec{\Omega}) = \frac{1}{V_r}\sum_{\substack{t\|\vec{\Omega}_j \\ t\cap r}}\Delta_t\left[\vec{P}\left(\bar{z}_r^{out}\right)\Psi_r^{out} - \vec{P}\left(\bar{z}_r^{in}\right)\Psi_r^{in}\right]$$

Here, $\substack{t\|\vec{\Omega}_j \\ t\cap r}$ means that the sum is performed for all the trajectories intersecting the r region parallel to the direction $\vec{\Omega}_j$.

The $\bar{\bar{\mathbf{C}}}$ matrix reads:

$$\bar{\bar{\mathbf{C}}} = \bar{\bar{\mathbf{M}}}^{-1}\cdot\left[\frac{p}{\Delta z/2}\right]\cdot\bar{\bar{\mathbf{S}}}\cdot\bar{\bar{\mathbf{M}}}$$

and the shift matrix $\bar{\bar{\mathbf{S}}}$ is:

$$\bar{\bar{\mathbf{S}}} = \begin{bmatrix} 0 & 0 & 0 & 0 \\ 1 & 0 & 0 & 0 \\ 0 & 1 & 0 & 0 \\ & & & \\ 0 & 0 & \ddots & \ddots \end{bmatrix}$$

Since the profile of the $\bar{\bar{\mathbf{C}}}$ matrix is lower triangular with zeros on the main diagonal the components of $\vec{\Psi}(\vec{r},\vec{\Omega})$ can be computed treating the system as a standard lower triangular system of equations, solving starting from the *0* order polynomial component.

2.6 Rebalancing

The tracking strategies used in the Method of Characteristics for three-dimensional geometries are widely explained in [1]. The same strategies apply to the polynomial version, since the same type of geometries are treated.
As already said, the use of a polynomial development leads to a lower numbers of axial meshes. This means regions with bigger volumes that could be integrated, with similar relative numerical errors, using a bigger axial integration parameter. Fig.2 represents the radial and axial integration parameters, respectively Δr and Δs. As supposed, to a bigger Δs value corresponded an equivalently precise volume integration. Unfortunately, the same thing can not be said about the precision of the solution. Releasing the Δs value caused big instability issues, that lead in most of the cases to a divergent solution.
The problem has been identified in a difference between the balance used to compute the flux moments and the one coming from the transmission equation. A *rebalancing* technique has been implemented to force the two balances to coincide. The equations for this technique are retrieved starting by integrating the transmission equation along a chord of length L:

$$L\overline{\Psi}(L,\vec{\Omega}) = \int_0^L dt\,\Psi_r(t,\vec{\Omega}) = \int_0^L dt\,\Psi_r(0,\vec{\Omega})\,e^{-\tau(0,t)} +$$
$$\int_0^L dt\sum_{p=0}^{N_p}\left\{\left[\sum_{k=0}^{p}c_{p,k}\,P_k(z^{in})\,\mu^{p-k}\left(\frac{2}{\Delta z}\right)^{p-k}E_{p-k}(\tau(t))\right]\cdot\frac{\left(\vec{q}_{r,pol}(\vec{\Omega})\right)_p}{\Sigma_r}\right\}$$

The only terms depending on t are:

$$\int_0^L dt\,\Psi_r(0,\vec{\Omega})\,e^{-\tau(0,t)} = \Psi_r(0,\vec{\Omega})\left[\frac{1-e^{-\Sigma L}}{\Sigma}\right]$$
$$\int_0^L dt\,E_{p-k}(\tau(t)) = \frac{L^{p-k+1}}{p-k+1} - \frac{E_{p-k}(\tau(L))}{\Sigma}$$

DOI: 10.1051/epjconf/201715306027

So we obtain:

$$
\begin{aligned}
&\Sigma L\overline{\Psi}(L,\vec{\Omega}) = \Psi_r(0,\vec{\Omega}) - \Psi_r(0,\vec{\Omega})e^{-\Sigma L} + \\
&-\sum_{p=0}^{N_p}\left\{\left[\sum_{k=0}^{p} c_{p,k}\, P_k(z^{in})\,\mu^{p-k}\left(\frac{2}{\Delta z}\right)^{p-k} E_{p-k}(\tau(t))\right]\cdot\frac{\left(\vec{q}_{r,pol}(\vec{\Omega})\right)_p}{\Sigma_r}\right\} \\
&+\sum_{p=0}^{N_p}\left\{\left[\sum_{k=0}^{p} c_{p,k}\, P_k(z^{in})\,\mu^{p-k}\left(\frac{2}{\Delta z}\right)^{p-k} \frac{L^{p-k+1}}{p-k+1}\right]\cdot\frac{\left(\vec{q}_{r,pol}(\vec{\Omega})\right)_p}{\Sigma_r}\right\} \\
&= \Psi_r(0,\vec{\Omega}) - \Psi_r(L,\vec{\Omega}) + \sum_{p=0}^{N_p}\left[\gamma_p(t)\cdot\left(\vec{q}_{r,pol}(\vec{\Omega})\right)_p\right]
\end{aligned}
$$

where:

$$
\gamma_p(t) = \sum_{k=0}^{p} c_{p,k}\, P_k(z^{in})\,\mu^{p-k}\left(\frac{2}{\Delta z}\right)^{p-k}\frac{L^{p-k+1}}{p-k+1}
$$

Integrating now the equation over the volume of each region we obtain an average angular flux. Multiplying by the real spherical harmonic and integrating over the angular variable we obtain a balance equation for the angular flux moments, only for the constant polynomial value:

$$
\begin{aligned}
&\Sigma V_r \phi^n_{0,r} = \Sigma\frac{1}{4\pi}\oint d\vec{\Omega}\, A_n(\vec{\Omega})\int_{\delta r} d_{2r\perp}\, L\overline{\Psi}_r(L,\vec{\Omega}) = \\
&\frac{1}{4\pi}\oint d\vec{\Omega}\, A_n(\vec{\Omega})\int_{\delta r} d_{2r\perp}\left(\Psi_r(0,\vec{\Omega}) - \Psi_r(L,\vec{\Omega})\right) + \\
&\frac{1}{4\pi}\oint d\vec{\Omega}\, A_n(\vec{\Omega})\int_{\delta r} d_{2r\perp}\sum_{p=0}^{N_p}\left[\gamma_p(t)\cdot\left(\vec{q}_{r,pol}(\vec{\Omega})\right)_p\right]
\end{aligned}
$$

where the volume integral has been decomposed coherently with the trajectory-based spatial discretization, and computed numerically as:

$$
\int_r d\vec{r} = \int_{\delta r} d_{2r\perp}\int_0^L dt \simeq \sum_{\substack{t\parallel\vec{\Omega}_j\\ t\cap r}}\int_0^L dt
$$

Eq.(6) and Eq.(7) are now used to expand the source term and to switch to the orthonormal basis, obtaining (written as the numerical equivalent form):

$$
\begin{aligned}
\Sigma V_r\phi^n_{0,r} &= \Sigma\sum_{\vec{\Omega}_j\in S_N}\omega(\vec{\Omega}_j)A_n(\vec{\Omega}_j)\left(\Psi_r(0,\vec{\Omega}) - \Psi_r(L,\vec{\Omega})\right) + \\
&+\sum_{p=0}^{N_p}\sum_{n'}\left[\sum_{p'=0}^{N_p}\bar{\bar{M}}^{-1}_{pp'}\sum_{\vec{\Omega}_j\in S_N}\omega(\vec{\Omega}_j)A_n(\vec{\Omega}_j)A_{n'}(\vec{\Omega}_j)\cdot\right. \\
&\left.\cdot\sum_{\substack{t\parallel\vec{\Omega}_j\\ t\cap r}}\Delta_t\cdot\gamma_{p'}(t)\right]\cdot q^{n'}_{p,r}
\end{aligned}
$$

The last term can be expressed in a tensorial form, defining the operator Γ with dimension $N_p \leftarrow N_p * N_{mom}$:

$$
\begin{aligned}
\left(\bar{\bar{\Gamma}}:\bar{\vec{q}}_r\right)_n &= \sum_{p=0}^{N_p}\sum_{n'}\left[\sum_{p'=0}^{N_p}\bar{\bar{M}}^{-1}_{pp'}\cdot\right. \\
&\left.\cdot\sum_{\vec{\Omega}_j\in S_N}\omega(\vec{\Omega}_j)A_n(\vec{\Omega}_j)A_{n'}(\vec{\Omega}_j)\sum_{\substack{t\parallel\vec{\Omega}_j\\ t\cap r}}\Delta_t\cdot\gamma_{p'}(t)\right]\cdot q^{n'}_{p,r}
\end{aligned}
$$

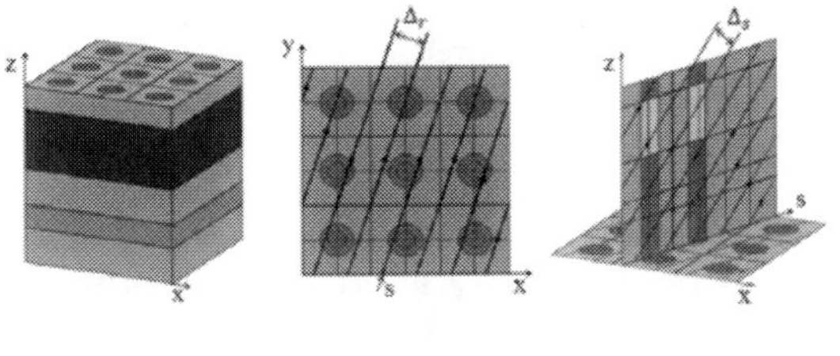

Figure 2: Tracking strategy for 3D axial geometries. On the left the whole geometry, composed by a 2D unstructured geometry extruded along z. In the middle are shown the 2D trajectories, with spacing Δr. Finally on the right each 2D trajectory generate a *sz-plane* on which the actual 3D trajectory is traced, with axial spacing Δs[1].

where ":" refers to the double tensorial reduction.

This new balance equation can be compared to the standard balance equation, that reads:

$$
\begin{aligned}
\Sigma V_r\phi^n_{0,r} =& \Sigma\sum_{\vec{\Omega}_j\in S_N}\omega(\vec{\Omega}_j)A_n(\vec{\Omega}_j)\left(\Psi_r(0,\vec{\Omega}) - \Psi_r(L,\vec{\Omega})\right) + \\
&+ V_r\left(\bar{\bar{S}}_p\cdot\vec{q}_{0,r}\right)_n
\end{aligned}
$$

where the *Spurious* matrix $\bar{\bar{S}}_p$ reads:

$$
\bar{\bar{S}}_p = \frac{1}{4\pi}\oint d\vec{\Omega}\, A_n(\vec{\Omega})\otimes A_n(\vec{\Omega})
$$

A rebalancing equation can be obtained forcing the two equations to give the same result. In particular, since only the last term is different in the two equations, a correction to be applied to the 0_{th} polynomial order transport source can be obtained, in order to let this two terms to coincide:

$$
V_r\,\bar{\bar{S}}_p\cdot\vec{q}_{0,r} = \bar{\bar{\Gamma}}:\left[\bar{\vec{q}}_r + \delta\vec{q}_{0,r}\right]
$$

from which we obtain:

$$
\boxed{\delta\vec{q}_{0,r} = \bar{\bar{\Gamma}}_0^{-1}\cdot\left[V_r\,\bar{\bar{S}}_p\cdot\vec{q}_{0,r} - \bar{\bar{\Gamma}}:\bar{\vec{q}}_r\right]}
$$

where $\bar{\bar{\Gamma}}_0$ is:

$$
\begin{aligned}
&\left(\bar{\bar{\Gamma}}_0\cdot\delta\vec{q}_{0,r}\right)_n = \\
&\sum_{n'}\left[\sum_{\vec{\Omega}_j\in S_N}\omega(\vec{\Omega}_j)A_n(\vec{\Omega}_j)A_{n'}(\vec{\Omega}_j)\sum_{\substack{t\parallel\vec{\Omega}_j\\ t\cap r}}\Delta_t\cdot\gamma_0(t)\right]\cdot\delta q^{n'}_{0,r}
\end{aligned}
$$

where the $\bar{\bar{M}}^{-1}$ and his related sum over p' disappeared since the matrix is lower triangular.

3 Results

The polynomial axial MOC method has been validated using the same calculation scheme and the same case of a previous work[3]. This case is a heterogeneous three-dimensional assembly of the innovative Sodium-cooled

Fast Breeder Reactor ASTRID, developed at the CEA in France. The accuracy of the solution is analyzed comparing the results with a reference Monte Carlo simulation obtained with the TRIPOLI4® code, also developed at the CEA.

The two-dimensional section of the ASTRID assembly used in the calculation is shown in Fig. 3a, while Fig. 3b shows the actual computational domain. Axially, the geometry is composed by a fertile part superposed to a fissile section.

The multi-group cross sections have been computed with the Subgroup method of the AUTOP module of APOLLO3®. The self-shielding procedure and the 2D-3D equivalence applied is explained in [2][3]. A 1968 groups discretization of the energy is used for the multi-group calculation.

The presence of two parts with different materials and composition leads to the strong flux gradient, shown in Fig. 1. For this reason the Step approximation needs a big number of axial meshes, in particular in the interface section, and justifies the idea of a polynomial development. Fig. 4 shows the amount of meshes needed both in the Step and in the polynomial case to properly describe the flux gradients. The use of a polynomial development allowed to decrease the meshes number, from 30 to 2.

Although the number of axial meshes decreased of about one order of magnitude, the number of chords did not decreased proportionally (for the same values of the integration parameters described in section 2.6). In fact, the most abundant chords even with 30 planes in this geometry (and probably in most of them) are the *V-chord*, connecting two vertical surfaces. The reason lies in the length of the average 2D chord, smaller than the average plane height. Reducing the number of meshes the number of *M-chords* decreased, while the *H-chords* basically disappeared, leading to a percentage of *V-chords* > 95%. Nevertheless the total number of chords slightly decreased. The *V-chords* became almost the only type. The bigger percentage of *V-chords* in comparison with the step case represent an advantage, since it is possible to pre-compute the transmission coefficients for classified chords. However, since the amount of *V-chords* was already high in the step case and since the number of floating operation per chord in the polynomial case is much higher than in the SC approximation, as explained is section 2.3, the hoped advantage in terms of computational time has not been achieved. Calculations have been performed on a Xeon E5-2680@2.8 GHz with two CPUs sharing 256GB of main memory. Each CPU is composed of 10 cores, with 12GB of memory each.

Table 1 shows a precise comparison of chords type and amount between Step and Polynomial method as well as the reactivity errors and computational times for both methods.

Figs. 5, 6, 7 shows the absolute errors on the absorption and fission reaction rates of the Pu_{94}^{239}, U_{92}^{235} and U_{92}^{238} isotopes in the four fuel pins homogenization zones (external, internal, fissile, fertile), over the 33 homogenization groups. Comparison with TRIPOLI4® are done normalizing each rate to the total absorption reaction rate.

Table 1: Comparison between the Step and the Polynomial method regarding the chords amount, types and computational time, using the same integration parameters. Since the DP_N acceleration is available for the Step method, but not yet for the Polynomial development, both calculation are performed without acceleration. M stands for million.

	Step	**Polynomial**
Δr (cm)	0.05	0.05
Δs (cm)	0.2	0.2
Total number of chords	44.79 M	38.06 M
• number of V chords	31.75 M	37,10 M
• number of H chords	1.38 M	0
• number of M chords	11.66 M	0.96 M
Classified chords	73.97 %	97.47 %
Chord reduction	-	-15.03 %
Computational time	14 393 s	20 069 s
Time overhead	-	+39.44 %
Reactivity error (pcm)	-1.38	+9.39

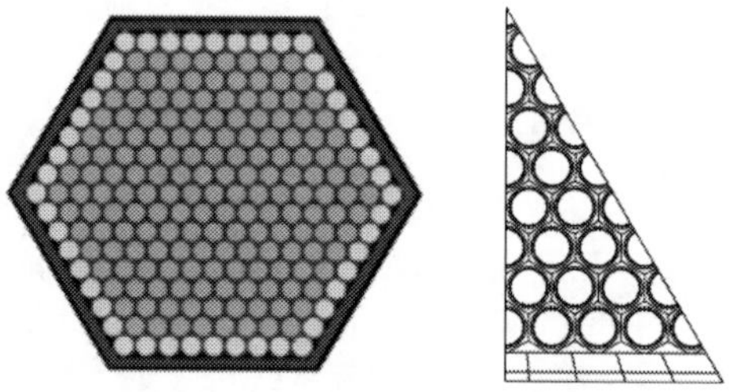

(a) (b)

Figure 3: On the left the two dimensional section of the ASTRID assembly with $2\pi/12$ symmetry. The two different colors of the fuel pins represents the two self-shielding areas. On the right the actual computational domain, that corresponds to one twelve of the full assembly.

4 Conclusions and perspectives

The polynomial axial development has been implemented for the 3D Method of Characteristics. Results show a good agreement with the Monte Carlo simulation and with the results obtained in the previous work, where the 3D SC method has been developed[2][3].

The tracking size for the Polynomial case, thanks to a lower amount of meshes, is smaller and the percentage of classified chords is bigger. Nevertheless the computational cost is still higher for the polynomial case. Two factors may come in use to increase the computational performances of the Polynomial method, that we will try to

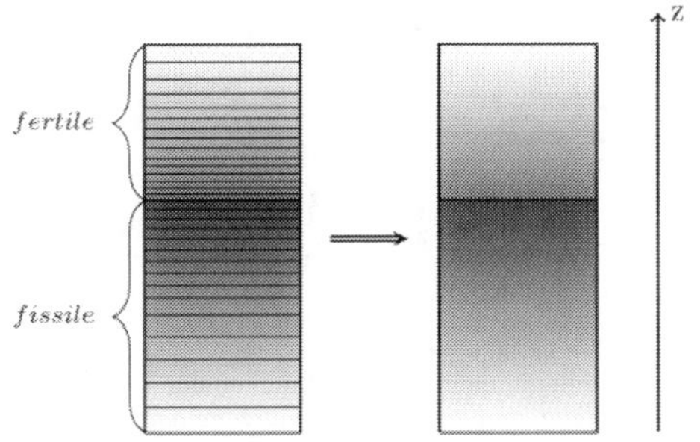

Figure 4: Difference between the Step and the Polynomial axial discretization.

develop. First, the DP_N acceleration technique that will be developed should take advantage of the lower amount of meshes both for the important memory occupation that this method requires, and for the computational cost that strongly depends on the number of unknowns. Secondly, thanks to the bigger axial size of the meshes, an equivalent precise solution should be obtained using a bigger axial integration parameter (Δs) and a significant lower number of chords. This analysis is already in progress, that is why the *rebalancing* technique explained in Sec. 2.6 has been implemented. Tests carried out with bigger Δs seem to show equally precise values in term of reactivity. However we cannot consider these results conclusive, since the reactivity and the integrated reactions rates proved to be well calculated, even if the axial profile of the flux was not. This phenomena observed while using the Step method is not yet well understood, however it may cause errors for example in an evolution calculation. For this reason we leave the discussion of the possibility to increment the value of the axial integration parameters for the future, when conclusive results will be available.

References

[1] D. Sciannandrone, S. Santandrea, R. Sanchez, *Optimized tracking strategies for step MOC calculations in extruded 3D axial geometries, Annals of Nuclear Energy* (2015)

[2] D. Sciannandrone, S. Santandrea, R. Sanchez, J. F. Vidal et al., *Coupled fine-group three-dimensional flux calculation and subgroups method for a FBR hexagonal assembly, Proc. Int. Conf. ANS MC2015, Nashville, Tennesse* (April 19-23, 2015)

[3] D. Sciannandrone, *Acceleration and higher order schemes of a characteristics solver for the neutron transport equation in 3D axial geometries*, PHD thesis, Université Paris Sud-Orsay (2015)

[4] P. Archier, J-M Palau, J-F Vidal, S. Santandrea, D. Sciannandrone, *Validation of the newly implemented 3D TDT-MOC solver of APOLLO3® code on a whole 3D SFR heterogeneous assembly*, Physor 2016, American Nuclear Society, Sun Valley Idaho, USA May 1-5 (2016)

 DOI: 10.1051/epjconf/201715306027

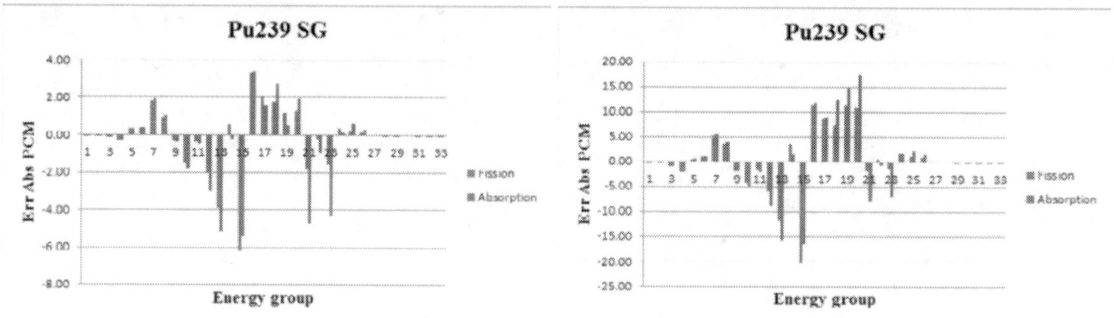

(a) External radial zone, fissile layer

(b) Internal radial zone, fissile layer

Figure 5: Absolute error over the 33 homogenization groups of the absorption and fission reactions rates for the isotope Pu_{94}^{239}.

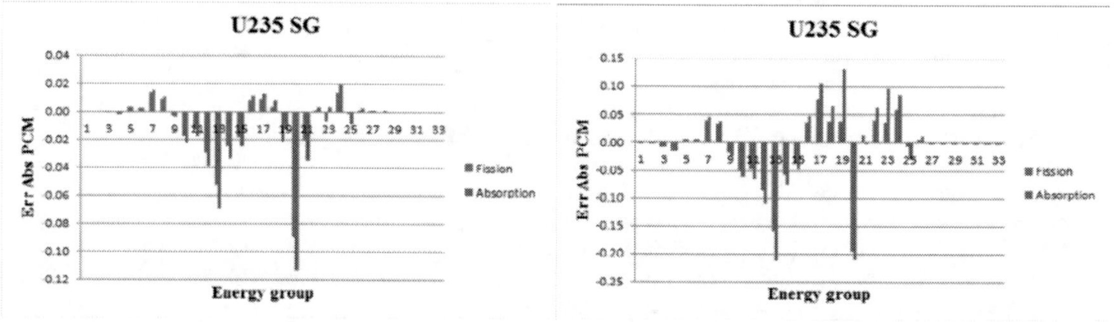

(a) External radial zone, fissile layer

(b) Internal radial zone, fissile layer

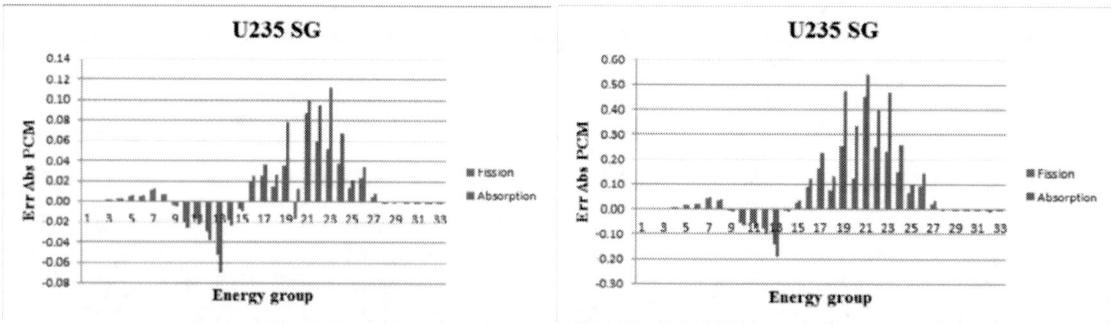

(c) External radial zone, fertile layer

(d) Internal radial zone, fertile layer

Figure 6: Absolute error over the 33 homogenization groups of the absorption and fission reactions rates for the isotope U_{92}^{235}.

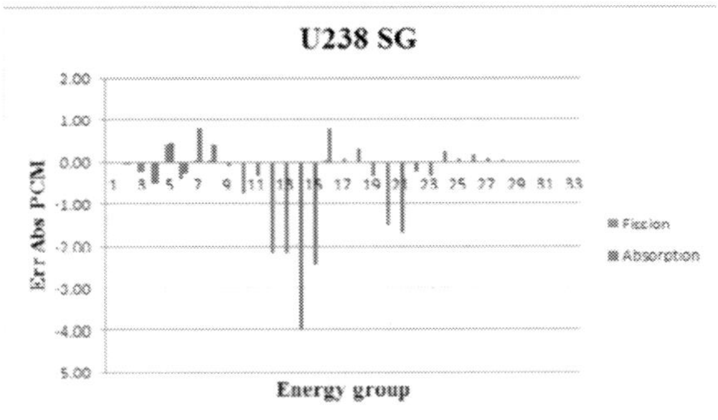

(a) External radial zone, fissile layer

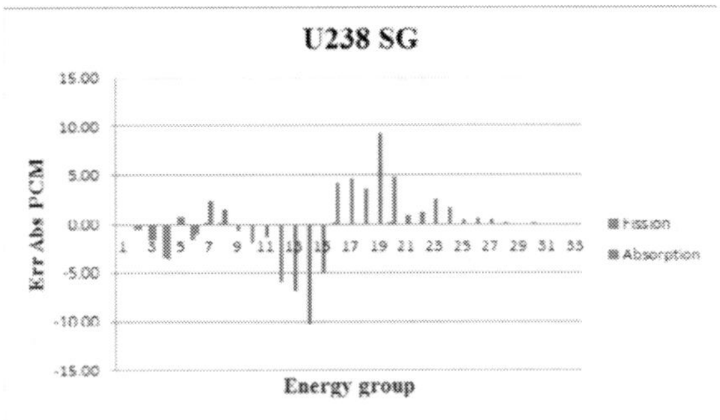

(b) Internal radial zone, fissile layer

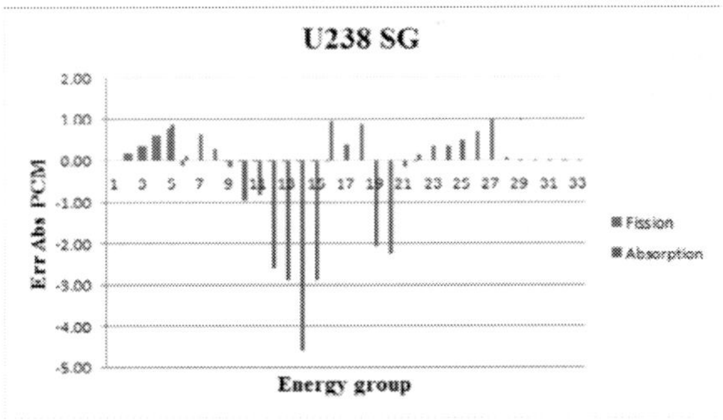

(c) External radial zone, fertile layer

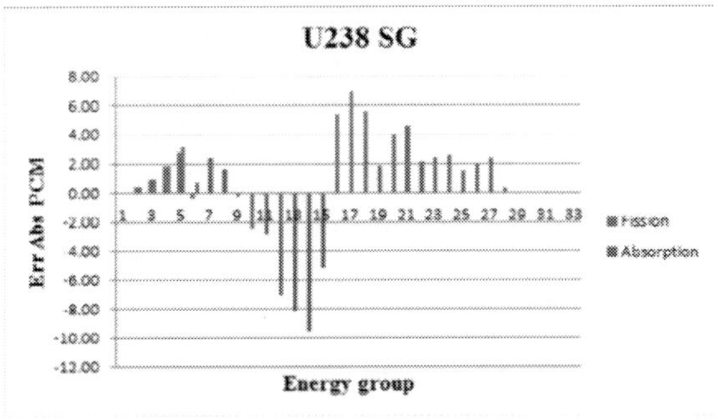

(d) Internal radial zone, fertile layer

Figure 7: Absolute error over the 33 homogenization groups of the absorption and fission reactions rates for the isotope U_{92}^{238}.

NARMER-1: a photon point-kernel code with build-up factors

Thierry VISONNEAU[1,a], Laurence PANGAULT[1], Fadhel MALOUCH[1], Fausto MALVAGI[1], Florence DOLCI[1]

[1] *Den-Service d'étude des réacteurs et de mathématiques appliqués (SERMA), CEA, Université Paris-Saclay, F-91191, Gif-sur-Yvette, France*

Abstract. This paper presents an overview of NARMER-1, the new generation of photon point-kernel code developed by the Reactor Studies and Applied Mathematics Unit (SERMA) at CEA Saclay Center. After a short introduction giving some history points and the current context of development of the code, the paper exposes the principles implemented in the calculation, the physical quantities computed and surveys the generic features: programming language, computer platforms, geometry package, sources description, etc. Moreover, specific and recent features are also detailed: exclusion sphere, tetrahedral meshes, parallel operations. Then some points about verification and validation are presented. Finally we present some tools that can help the user for operations like visualization and pre-treatment.

1 Introduction

Starting from the 60s, CEA launched the development of MERCURE, a gamma-ray shielding code using the point-kernel integration formalism. MERCURE-3 was published in 1967 [1], and MERCURE-4 in 1974 [2]. Later in the 80s, EDF and COGEMA (AREVA) became sponsors of MERCURE for its development and validation. MERCURE-5v3 released in 1995 [3] followed by MERCURE-6 [4]. MERCURE uses a Monte Carlo method for the point-kernel integration and build-up factors to take scattered photons into account. MERCURE-6 has enabled the build-up factor calculation for composed shields by using new algorithms developed by SERMA (Reactor Studies and Applied Mathematics Unit) [5]. Then, the development of NARMER-1 has been launched to integrate both gamma-ray attenuation and reflection features (implemented by MERCURE-6 and NARCISSE-3 [6] respectively). For this purpose, the code has been re-written in C++ allowing a modern programming based on object oriented design, with a great care to some extra-functional properties such as modularity, extensibility, maintainability and testability. Ongoing works are related to the improvement of reflection calculations, parallel computing, interoperability with CAD geometries, verification and validation, and integrability in the EDF, CEA software platforms. The development is also based on a components sharing strategy with other SERMA codes for example for the base libraries (geometry, composition, source term, etc.)

NARMER-1 is part of a set of software developed by SERMA in the broad field of reactor studies. It is closely related to TRIPOLI-4® [7] as it uses the same geometry package allowing easier comparison to this reference Monte Carlo code for validation purpose. Depending on the period, up to 5 people have been contributing to MERCURE and now to NARMER-1, their activities covering development, integration & architecture, V&V, documentation, user support, distribution and licensing.

2 Principles

2.1 Straight-line attenuation

Considering an isotropic source represented by a Dirac in space and energy, the point-kernel of attenuation has the following classical expression:

$$K_0(\vec{r},\vec{r_s},E) = \frac{e^{-\sum_i \mu_i(E).t_i}}{4\pi||\vec{r}-\vec{r_s}||^2} \qquad (1)$$

where $\mu_i(E)$ is the attenuation coefficient while t_i is the track length in the homogeneous medium i for the energy E.

The point-kernel represents the direct photon flow, without scattering. Diffusion contributions are taken into account by a build-up factor which depends on the computed physical quantity, the material and the track length in the medium. Indeed, the build-up factor is defined by the quotient of the response taking into account the diffusion contributions by the response without scattering.

[a] Corresponding author: thierry.visonneau@cea.fr (T. Visonneau)

2.2 Single-layer shield build-up factors

Single-layer shield build-up factors are calculated in a homogenous and infinite medium for a source defined as punctual, isotropic, mono-kinetic, placed in a spherical geometry. They have been calculated by running a discrete ordinates transport code, TWODANT [8], and following a method described by KITSOS, et al. [9]. Energy groups, angular mesh and space mesh have met the following features: 218 energy groups from 1 keV to 10 MeV, P7 for the Legendre expansion of the scattering cross sections, S96 for the Gauss-Legendre angular quadrature and at least 28 spatial meshes per mean free path. The build-up factors take the following phenomena into account: photoelectric effect, coherent and incoherent scattering, pair production, secondary sources of bremsstrahlung and fluorescence. Cross sections come from JEF2.2 evaluation files [10]. The build-up factors are available for each single chemical element from hydrogen (Z=1) to californium (Z=98), and are tabulated in 22 thicknesses (from 0.5 to 50 mean free path) and 195 energy groups (from 15keV to10MeV). Energy groups are identical to energy groups of total microscopic cross sections.

For a homogeneous material composed of many chemical elements, the build-up factors are computed on the fly by the code which determines an equivalent atomic number of the mixture (Z_{eq}). For a given energy group g, the total macroscopic cross-section per electron of the mixture is calculated in equation (2) while equation (3) gives the total cross section per electron.

$$\tilde{\sigma}_g^m = \frac{\sum_i N_i \tilde{\sigma}_g^i}{\sum_i N_i} \quad (2)$$

$$\tilde{\sigma} = \frac{\sigma}{Z} \quad (3)$$

The equivalent atomic number of the mixture ($Z_{eq,g}$) is obtained with respects to the corresponding computed cross-section ($\tilde{\sigma}_g^m$). Thus, the build-up factors of the mixture correspond to the build-up for (Z_{eq}) obtained by a log-log interpolation of the $B(Z)$ function, by energy group, and as a function of the number of mean free path.

2.3 Multi-layer shield build-up factors

In this section layers are noted L_i numbered increasingly from the source, Z_i, X_i (in unit of mfp) and B_i are the corresponding atomic number, thickness and build-up, respectively.

Considering a double layer shield, NARMER-1 implements the SERMA formula (4) to compute the corresponding build-up factor.

$$B_{1,2}^{Serma}(X_1, X_2) = \frac{B_1(X_1)}{B_2(X_1)} B_2(X_1 + X_2) F_{1,2}(X_1, X_2) \quad (4)$$

The function $F_{1,2}$ is defined by equation (5).

$$F_{1,2}(X_1, X_2) = \left[1 + \frac{X_2^{\beta_{1,2}}}{\alpha_{1,2}}\right]^p \quad (5)$$

$\alpha_{1,2}$, and $\beta_{1,2}$ are defined by equation (6) and (7).

$$\log(\alpha_{1,2}) = a \log^3(X_1) + b \log^2(X_1) + c \log(X_1) + d \quad (6)$$

$$\log(\beta_{1,2}) = e \log^3(X_1) + f \log^2(X_1) + g \log(X_1) + h \quad (7)$$

The nine coefficients a, b, c, d, e, f, g, h and p are determined empirically for nine materials (nitrogen, water, aluminium, iron, zirconium, barium, tungsten, lead and uranium) and for 195 energy groups from 15 keV to 10 MeV. For the materials that are not part of that library, coefficients of materials whose atomic numbers are the closest to the ones of the double layer shield materials are used.

For a multi-layer (more than two) shield, NARMER-1 uses an iterative process that replaces, from source to calculation point, two layers by an equivalent layer. First, the equivalent atomic number Z_{eq} is determined by using a neural network trained on large training set. This equivalent atomic number Z_{eq} is the closest to a solution to equation (8), where Z is the unknown quantity.

$$B(E, Z, X_1 + X_2) = B_{1,2}^{Serma}(E, X_1, X_2) \quad (8)$$

As shown by Table 1, the neural network is divided into five regions that are delimited with respect to physical considerations about sharp variations of Z_{eq} induced by fluorescence [5].

Table 1. Description of the five neural networks.

network	energy domain	material domain
1	300keV to 10MeV	$Z_1 < Z_2$
2		$Z_1 > Z_2$
3	15keV to 300keV	$Z_1 < Z_2$ and $Z_1 \geq 27$
4		$Z_1 > Z_2$ and $Z_2 \geq 27$
5		$Z_1 < 27$ and $Z_2 < 27$

Then, the equivalent thickness (X_{eq}) of the equivalent material is calculated to reproduce, with the Z_{eq} material, the double-layer build-up, by solving equations (9) where X is the unknown quantity.

$$B(E, Z_{eq}, X) = B_{1,2}^{Serma}(E, X_1, X_2) \quad (9)$$

The above requirements are summarized in equations (10).

$$\begin{cases} B(E, Z_{eq}, X_{eq}) = B_{1,2}(E, X_1, X_2) \\ X_{eq} \simeq X_1 + X_2 \end{cases} \quad (10)$$

If Z_{eq} is a good approximation, X_{eq} is close to $X_1 + X_2$.

Hence, considering a multi-layer shield composed of N layers, the iterative process starts with the first steps shown by Fig. 1.

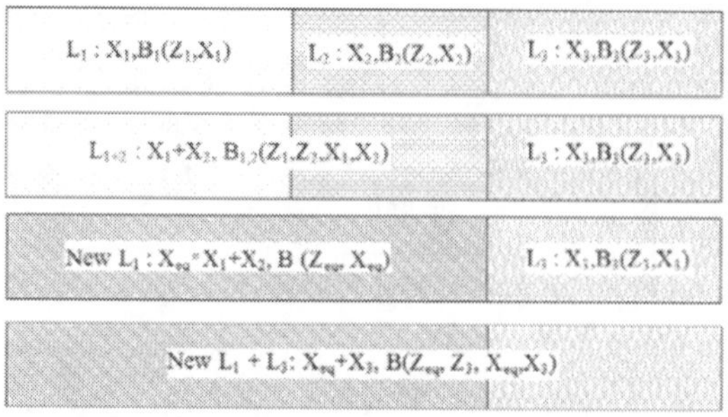

Fig. 1. Determination of a three layer shield build-up factor.

2.4 Source integration

For a source S distributed over energy and space, the response R_α described by a response function $f_\alpha(E)$ to be calculated follows the equation (11).

$$R_\alpha(\vec{r}) = \int_{E_{min}}^{E_{max}} dE_s \int_{V_s} d\vec{r_s}\, S(E_s, \vec{r_s})\, f_\alpha(E)\, K_0[(\vec{r_s}, E_s) \to \vec{r})]\, B_\alpha[(\vec{r_s}, E_s) \to \vec{r})] \quad (11)$$

NARMER-1 uses a Monte Carlo method to integrate points-kernel with build-up factor over the source domain. Considering the spatial and energy meshes defining the source, the code performs a stratified sampling as variance reduction technique. New sampling points are added in each cell proportionally to the variance of the cell, until the variance on the global result reaches the asked precision.

2.5 Computed physical quantities

The following response functions are available in NARMER-1 calculations: uncollided flux, ambient dose equivalent rate H*(10), energy fluence, and air KERMA.

Continuous energy and multi-group cross sections come from JEF2.2. Conversion coefficients from fluence to dose equivalent rate and air KERMA come from ICRP74 [11].

3 Generic features

3.1 Programming language, computer platforms

NARMER-1 has been developed in C++. It takes advantage of the principles of object oriented design, especially the open/closed principle making future developments easier. It uses python as the user interface. It is currently developed on Linux system but is maintained and available on both Linux and Windows.

3.2 Geometry package

NARMER-1 comes with the TRIPOLI-4® native geometry package, allowing for both a pure surface-based representation, and a combinatorial representation with predefined shapes and boolean operators (any combination of these two kinds of representations can be adopted).
As well as TRIPOLI-4®, NARMER-1 has been also made compatible with any geometry developed in the format of the ROOT software (Brun and Rademakers, 1997). This allows the user to directly use (without recompiling) an already built ROOT geometry.
More generally, NARMER-1 may be linked to any third-party geometry module providing an Application Programming Interface for the current particle position and the next intersection in the direction of flight.
Details concerning visualization tools for building and checking the geometry will be provided in Section 4.

3.3 Sources description

Sources are defined with an energetic spectrum and an emitting volume according to a mesh at a given position.

3.3.1 Energy Distribution

Line spectrum can be used as well as group spectrum based on a 195 bounds pre-defined mesh.

3.3.2 Spatial Distribution

Volume based sources can be described using the following type of meshes: Cartesian, Spherical, Cylindrical, Conic, Toroidal, General Conic, Constant Width Conic, and General Cylindrical.

Surface based sources can be described using the following type of meshes: Cartesian, Spherical Shell, Cylindrical Shell, Disk, Conic Shell, and Toroidal Shell.

4 Specific or recent capabilities

4.1. Exclusion sphere

In order to perform computations for some detectors that can be very close to the emitting volume or inside the emitting volume, it is necessary to avoid a singularity due to the $1/r^2$ factor that can lead to infinite variance. Therefore the code uses an exclusion sphere. The radius of the sphere can be set by the user.

4.2 Tetrahedral meshes

For specific uses, geometry and emitting volumes are modelled using computer-aided design software allowing easier definition of very complex 3D scenes. In this context the code has the possibility to define tetrahedral meshes for the definition of emitting volumes. Observed performances of the sampling method allow computation in good conditions.

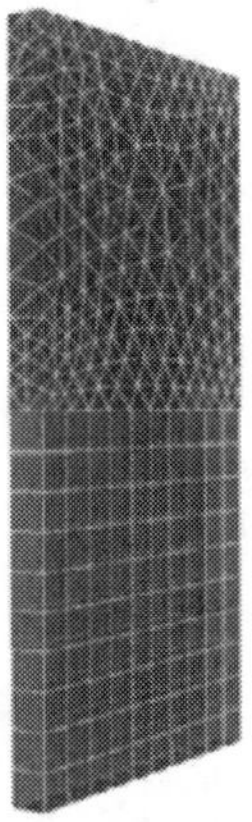

Fig. 2. A detector point at z=0 in front two emitting meshes, a tetrahedral mesh in the region z>0 and a cartesian mesh in the region z<0

As an illustration also allowing some considerations about verification and validation, the simple test case Fig. 2 shows a detector point belonging to a median plan defined by z=0 separating two identical emitting volumes having the form of a parallelepiped. The emitting volume in the region z>0 is described by a tetrahedral mesh and the one in the region z<0 is described by a cartesian mesh. NARMER-1 detailed results show that the geometrical intensity is identical for both emitting volumes, and also that the contribution to the global result is close to 50% for each source.

4.3 Double Energy Angle Differential Albedos

Lacunar media shielding studies can be dealt with albedo concept which is an alternative to the "exact" transport method calculations (SN, Monte Carlo). For multiple reflections on lacunar medium walls, one needs double energy-angle differential albedos because they are function of the incident energy. Reflection feature is a point of interest for EDF sponsor. R&D development and validation studies in this domain are still under progress at SERMA.

4.4 Parallel operations

Complex 3D scenes with a high number of gamma sources and a lot of geometrical volumes to cross can lead to substantial simulation times. Therefore, some developments and validation studies are currently under progress to use multi-threads algorithms.

5 Verification and Validation

Since the NARMER-1 code is attended to be used for radiation protection purposes, verification and validation are crucial issues. Verification is supported in particular by non-regression testing. Excellent results are observed about non-regression results with respect to MERCURE-6. Validation is based on comparison with calculations performed with the TRIPOLI-4® Monte Carlo code as well as with experimental benchmarks.

Here are presented four cases illustrating the validation aspect: two basic tests and two realistic cases in which the ambient dose equivalent rate H*(10) is calculated with NARMER-1 and confronted with TRIPOLI-4® results.

The definition of the geometry is common for both types of calculation.

The relative difference of H*(10) of NARMER-1 from TRIPOLI-4® is defined by equation (12):

$$Rel.Diff._{H*(10)} = \frac{Result_{NARMER-1} - Result_{TRIPOLI-4^{®}}}{Result_{TRIPOLI-4^{®}}} \quad (12)$$

5.1 Basic configurations

5.1.1 Multi-layers configuration

In this case, a point-like γ source is surrounded by six concentric spherical screens (from the centre to the outer blanket: Al - Cu - Al - Zr - Cu - Zr). Calculations were done for the following discrete values of γ energy: 1 MeV, 0.4 MeV, 0.32 MeV, 0.3 MeV, 0.295 MeV, 0.2 MeV and 0.1 MeV.

The ambient dose equivalent rate H*(10) is calculated at six points located at 7.5 cm, 12.5 cm, 17.5 cm, 22.5 cm, 27.5 cm and 32.5 cm from the source represented by dots in Fig. 3.

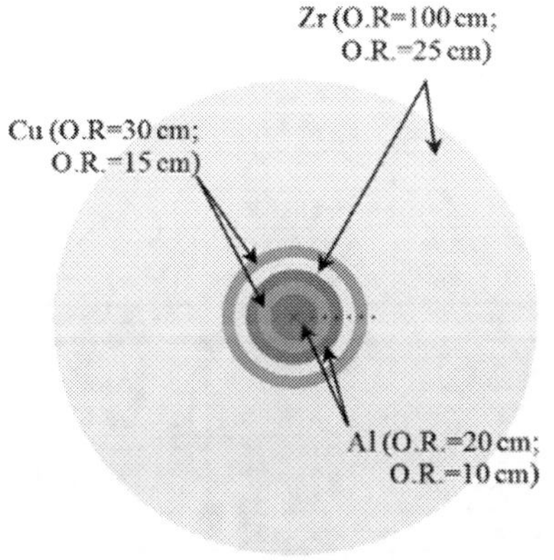

Fig. 3. Radial cut view of the multi-layers configuration.

The TRIPOLI-4® calculations concerned the transport of γ rays only. Variance reduction technique was required for the 0.1 MeV source in order to converge.

The results of the relative difference of H*(10) of NARMER-1 from TRIPOLI-4® are illustrated for each energy as a function of the distance from the source on

Fig. 4 whereas H*(10) numerical values of NARMER-1 calculations are shown in Table 2.

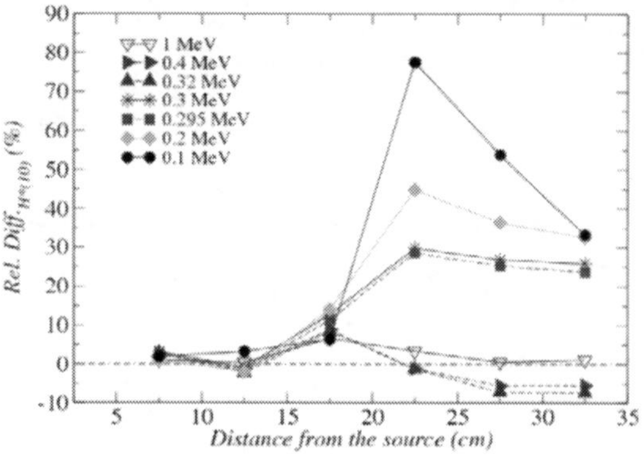

Fig. 4. Relative difference of H*(10) NARMER-1 calculations with respect to TRIPOLI-4® calculations for the multi-layers configuration.

Above 300 keV of γ energy, the relative difference does not exceed 10%. However, below 300 keV, NARMER-1 results overestimate TRIPOLI-4® significantly. We note that 300 keV represents the cut energy in the definition of the neural networks domains used to determine the equivalent atomic number of the fictive layer for build-up calculations (see Table 1). Cu is replaced by Fe in the material list of the double-layer build-up formula. The back-scattering flux is overestimated in the third layer (Al) due the heavier material (Zr) in the fourth layer. This test case uses multiple times the iterative process (Fig. 1).

Table 2. Numerical results of NARMER-1 for H*(10) calculations (μSv/h) of a multi-layers configuration (convergence criteria is 1%).

(cm)	1 MeV	0.4 MeV	0.32 MeV	0.3 MeV
7.5	9.778E-05	4.867E-05	4.025E-05	3.821E-05
12.5	2.694E-05	1.001E-05	7.209E-06	6.598E-06
17.5	1.112E-05	3.658E-06	2.482E-06	2.276E-06
22.5	4.643E-06	9.907E-07	5.715E-07	6.250E-07
27.5	2.187E-06	3.419E-07	1.677E-07	1.803E-07
32.5	1.030E-06	1.105E-07	4.565E-08	4.611E-08

	0.295 MeV	0.295 MeV	0.1 MeV
7.5	3.747E-05	2.561E-05	1.072E-05
12.5	6.359E-06	2.916E-06	1.555E-07
17.5	2.174E-06	8.218E-07	8.457E-09
22.5	5.893E-07	1.455E-07	9.245E-11
27.5	1.667E-07	2.270E-08	1.933E-13
32.5	4.180E-08	3.199E-09	3.827E-16

5.1.2 Simplified water pipe

This case corresponds to a 0.1 cm thick, 8.1 cm outer radius, 200 cm high, steel pipe filled with water.

A point-like γ source is located at the pipe's centre with three different energies: 10 MeV, 1 MeV and 0.1 MeV.

The ambient dose equivalent rate H*(10) is calculated at three locations distant from the source of 5 cm (in the water), 8.05 cm (in the steel) and 9 cm (in the air). These locations are represented by dots in Fig. 5.

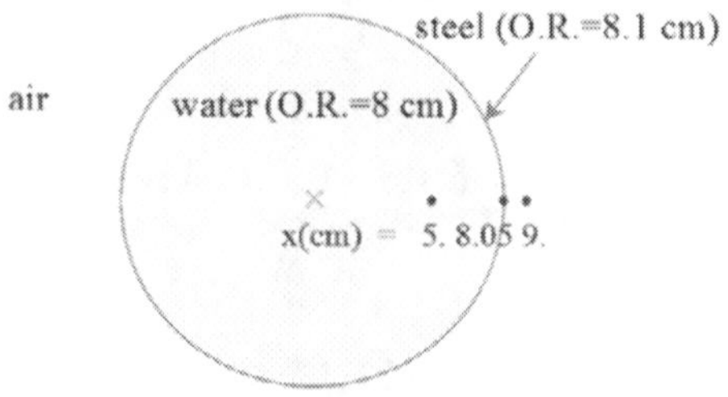

Fig. 5. Radial cut view of a simplified water pipe.

The whole electromagnetic cascade has been taken into account for the 10 MeV while only γ rays have been transported for the 1 MeV and 0.1 MeV sources in the TRIPOLI-4® calculations.

Results of the H*(10) relative difference as defined by equation (12) are shown for different energy source as a function of the distance from the source in Fig. 6 whereas H*(10) numerical values of NARMER-1 calculations are shown in Table 3.

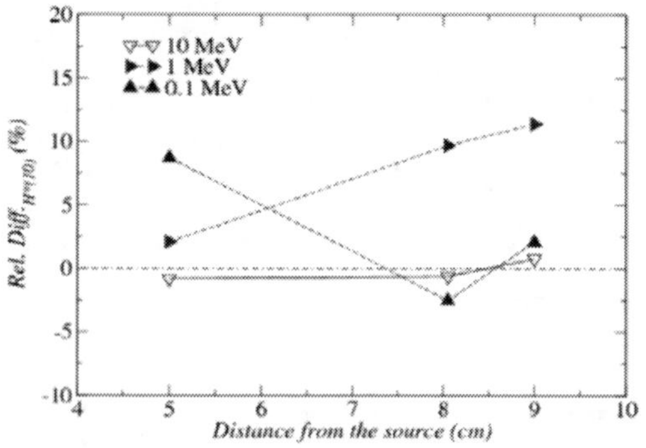

Fig. 6. Relative difference of H*(10) from NARMER-1 to TRIPOLI-4® for the simplified water pipe configuration.

It can be seen that the relative difference remains below about 10% in all cases which reflects the good agreement between the two codes.

Table 3. Numerical results of NARMER-1 for H*(10) calculations (μSv/h) of a simplified water pipe configuration (convergence criteria is 1%).

	5.	8.05	9.
10 MeV	1.158E-02	4.296E-03	3.418E-03
1 MeV	2.383E-03	8.771E-04	6.949E-04
100 keV	4.824E-04	8.764E-05	5.610E-05

5.2 Realistic configurations

5.2.1 Water pipe with protection

This case corresponds to a realistic water pipe which is made of a hollow steel cylinder (inner radius 30 cm, 2 cm thick) full of water which is covered by a 4 cm thick lead layer.

The source is distributed on the volume of the water and takes several discrete values (7 MeV, 3 MeV, 1 MeV and 0.6 MeV).

The ambient equivalent dose is calculated at different distances from the centre of the pipe: 30 cm (interface water/steel), 32 cm (interface steel/lead), 36 cm (interface lead/air), and in the air at 40 cm, 45 cm, 50 cm, 55 cm, 60 cm, 80 cm and 90 cm represented by dots in Fig. 7.

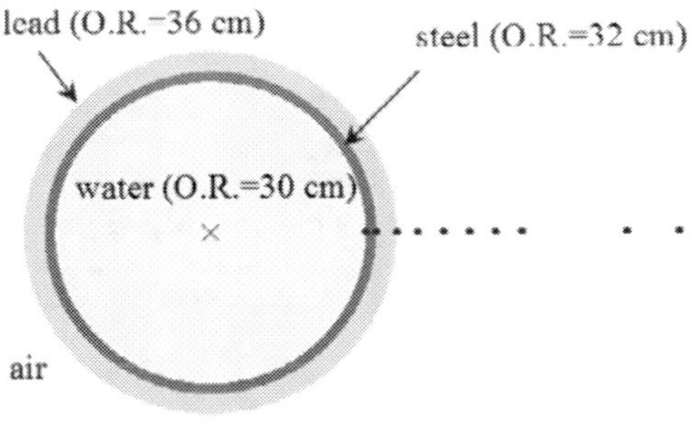

Fig. 7. Radial cut view of the water pipe with protection.

The full electromagnetic cascade has been taken into account for the TRIPOLI-4® calculations for the 7 MeV and 3 MeV sources. Only γ rays have been transported for the two others energy source.

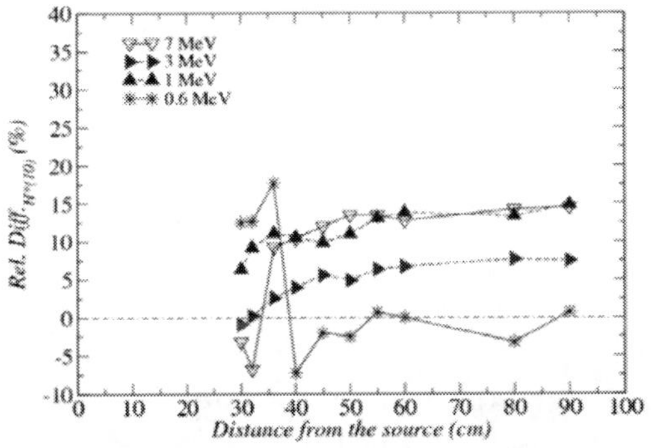

Fig. 8. Relative difference of H*(10) NARMER-1 calculations with respect to TRIPOLI-4® calculations for the configuration of a water pipe with protection.

The difference between NARMER-1 and TRIPOLI-4® is between -7% and +18%. This difference does not vary linearly with the source energy because of different physical effects: Compton effect is dominant at low energy whereas at high energy pair creation can occur.

Table 4. Numerical results of NARMER-1 for H*(10) calculations (μSv/h) of a water pipe with protection (convergence criteria is 1%).

cm	7 MeV	3 MeV	1 MeV	0.6 MeV
30	1.183E+00	6.066E-01	2.490E-01	1.651E-01
32	5.579E-01	2.581E-01	6.812E-02	3.112E-02
36	4.942E-02	2.288E-02	1.123E-03	2.773E-05
40	4.334E-02	2.020E-02	9.948E-04	2.469E-05
45	3.829E-02	1.771E-02	8.757E-04	2.187E-05
50	3.416E-02	1.577E-02	7.812E-04	1.954E-05
55	3.068E-02	1.425E-02	7.048E-04	1.767E-05
60	2.788E-02	1.294E-02	6.467E-07	1.617E-05
80	2.028E-02	9.455E-03	4.780E-04	1.199E-05
90	1.761E-02	8.241E-03	4.222E-04	1.069E-05

5.2.2 UO_2 - PuO_2 source

A cylinder of PuO_2-UO_2, in which the γ source is equally distributed, is located in front of a series of screens as drawn on Fig. 9.

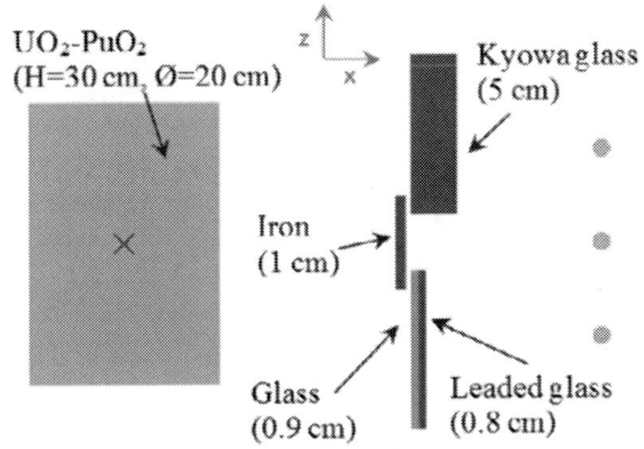

Fig. 9. Radial cut view of the UO2-PuO2 configuration.

The configuration of the series of 40 cm long screens is defined as following:

- A 1 cm thick, 10 cm tall, iron screen is located at 28.5 cm from the center of the source,
- A 5 cm thick, 17 cm tall, Kyowa-glass screen is located at 30 cm from the center of the source
- A 0.9 cm thick, 17 cm tall, glass screen is located at 30 cm from the center of the source close to a 0.8 cm thick, 17 cm tall, leaded glass screen situated at 30.9 cm from the center of the source.

The source energy varies from 15 keV to 3.67 MeV and its distribution is illustrated in Fig. 10.

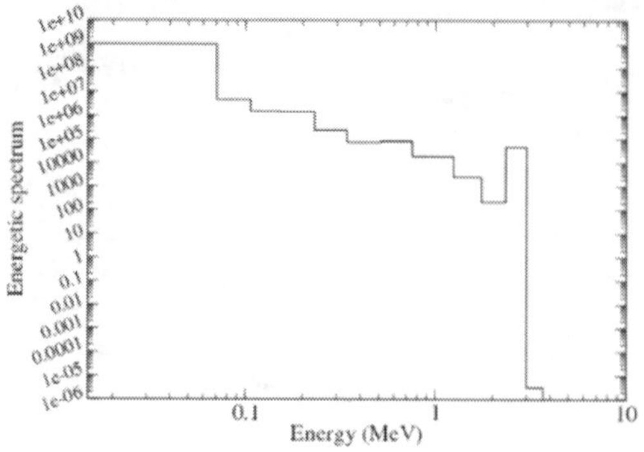

Fig. 10. Energy distribution of the γ source.

The atomic compositions are noted in the Table 5.

Table 5. Atomic compositions of the UO_2-PuO_2 test case

Material	Density (g/cm^3)	Element	Composition (10^{24} atoms/cm^3)
Air	0.0013	N	4.34e-05
		O	1.07e-05
Kyowa – glass	1.6	C	3.515e-02
		H	5.554e-02
		O	1.227e-02
		Pb	1.395e-03
Glass	2.52	Si	1.714e-2
		B	1.073e-03
		Ba	1.989e-05
		Na	4.654e-03
		O	4.636e-02
		Ca	2.674e-03
Leaded glass	4.2	Si	1.513e-02
		Pb	6.654e-03
		O	3.842e-02
		Ca	2.685e-03
UO_2-PuO_2	2.5	U	5.018e-03
		Pu	5.540e-04
		O	1.114e-02

The ambient dose equivalent rate H*(10) is calculated at three points, respectively at coordinates (50.,0.,-10.), (50.,0.,0.) and (50.,0.,10.), symbolized by dots in Fig. 9. The Table 6 presents the results of the calculations made with NARMER-1 and TRIPOLI-4® and their comparison.

Table 6. H*(10) calculated with NARMER-1 and TRIPOLI-4®

z	NARMER-1 H*(10) (μSv/h)	NARMER-1 σ (%)	TRIPOLI-4® H*(10) (μSv/h)	TRIPOLI-4® σ (%)	$Dev_{H*(10)}$ (N-T)/T (%)
10.	1.55e+03	0.12	1.29e+03	0.04	19.7
0.	2.10e+03	0.19	1.66e+03	0.07	26.8
-10.	1.93e+03	0.14	1.54e+03	0.05	25.6

NARMER-1 values overestimate TRIPOLI-4® values of about 20%.

6 Visualization and pre-treatment

Sharing the same geometry package as TRIPOLI-4®, NARMER-1 can be used with the pre-and-post treatment tool distributed with TRIPOLI-4®, known as "Visualizer T4G". The user also has possibility to check the definition of the geometry by using the last verification tools implemented in TRIPOLI-4®, including "Check Geom" and "Check Tracking" functionalities.

7 Summary and conclusion

NARMER-1 is a simplified code that allows to perform fast calculations to evaluate ambient dose equivalent H*(10) and KERMA for radiation protection studies.

A validation study has been done with comparison to TRIPOLI-4® considered as a reference code.

Basic configurations and sophisticated configurations show overall a good agreement between the two codes. In some cases, we observe an overestimation of NARMER-1 with respect to TRIPOLI-4®. The results are conservative for radiation protection studies.

NARMER-1 calculations are much faster than TRIPOLI-4® ones: NARMER-1 calculations last less than one minute and can reach several minutes for complicated configurations. These calculations times remain orders of magnitude below those met for TRIPOLI-4®.

Acknowledgments

The authors gratefully acknowledge EDF for its support to the NARMER-1 project.

References

[1] C. Devillers, *"Programme MERCURE-3, Atténuation en ligne droite dans une géométrie à trois dimensions"*, Internal Report, CEA-R 3264 (1967)

[2] C. Devillers, C. Dupont, "*MERCURE-4 – Three Dimensional Monte-Carlo Program for the integration of line of sight point attenuation kernels*", Internal Report, CEA-N-1726 (1974)

[3] C. Dupont, *"MERCURE-5.3 : Un Programme de Monte-Carlo à trois dimensions pour l'intégration de noyaux ponctuels d'atténuation en ligne droite."*, Internal Report, Rapport DMT 95/552 SERMA/LEPP 95/1817 (1995)

[4] A. Assad, M. Chiron, J.-C. Nimal, C.M. Diop, and P. Ridoux, "*General Formalism for Calculating Gamma-Ray Buildup Factors in Multilayer Shields into MERCURE-6 Code*", Jour. Nuc. Sci. and Tech., Supplement 1, p. 493- 497 (March 2000)

[5] C. Suteau, M. Chiron, "*Improvement of MERCURE-6 General Formalism for calculating gamma-ray build-up factors in multilayer shields*", Nucl. Sci. Eng., 147, 43-55, (2004)

[6] L. Bourdet, Internal Report, Rapport DMT 95/482 SERMA/LEPP 95/1796 (1995)

[7] E. Brun, F. Damian, C.M. Diop, E. Dumonteil, F.X. Hugot, C. Jouanne, Y.K. Lee, F. Malvagi, A. Mazzolo, O.

Petit, J.C. Trama, T. Visonneau, A. Zoia, "*TRIPOLI-4®, CEA, EDF and AREVA reference Monte Carlo code*", Annals of Nuclear Energy, 82 (2015) 151–160.
[8] R. E. Alcouffe, F. W. Brinkley, D. R. Marr, and R. D. O'Dell, "*User's Guide for Twodant: A Code Package for Two-Dimensional, Diffusion-Accelerated, Neutral-Particle, Transport*", LA-10049-M, Revised, Vol. 1, p. 1, Group X-6, Los Alamos National Laboratory (Nov. 1989).
[9] S. Kitsos, C. M. Diop, A. Assad, J. C. Nimal, and P. Ridoux, "*Improvement of Gamma-Ray Sn Transport Calculations Including Coherent and Incoherent Scatterings and Secondary Sources of Bremsstrahlung and Fluorescence: Determination of Gamma-Ray Buildup Factors*", Nucl. Sci. Eng, 123, 215-227 (1996)
[10] The JEF-2.2 Nuclear Data Library, NEA JEFF Report 17 (2000).
[11] ICRP (International Commission on Radiological Protection) 74 (1995); Conversion Coefficients for use in Radiological Protection against External Radiation.

A New Point Kernel Sampling Scheme and Monte Carlo Transport

Benoît Avez[1,⋆] and *Laurent* Bindel[1,⋆⋆]

[1] *Millennium, P.A. Courtabœuf, 16 avenue du Québec, 91 940 Villebon-sur-Yvette, France.*

Abstract. Starting from an existing point kernel intergrated over the bounding surface of a volume source [1], the present contribution proposes a new point kernel sampling scheme. This sampling scheme is shown to be more efficient than the standard one because the integration proceeds from line segment sources instead of point sources. A new kind of Monte Carlo transport process is proposed. It is called flux transport from a line source and is expected to be more efficient than the standard source transport.

1 Context

For some nuclear engineering purposes, the calculation speed can a parameter of importance. As examples of such applications, one can cite radiation protection, with virtual reality simulators where quick and yet accurate dose rate calculations are needed, or nuclear measurement softwares, that often rely on the calculation of transfer functions in order to process a given spectrum. For basically all softwares that need basically real-time results, the calculation time is crucial. As a consequence, researches aiming at finding new methodologies for point kernel calculation or Monte-Carlo transport are still topical to face the near-future of real-time embedded systems. In this paper, we present a new sampling scheme for the point kernel formula, discuss its advantages over the clasical one, and propose a new Monte-Carlo transport process.

2 New sampling procedure for the point kernel formula

The classical point kernel formula giving the collisionless scalar flux at a point detector of an isotropic, monoenergetic and volumetric source of spatial pdf $p(\mathbf{r})$ and total emission rate A is the well known formula:

$$\Phi(\mathbf{r}) = A \iiint \frac{p(\mathbf{r}')\, e^{-\int_\gamma \mu(s)ds}}{4\pi\|\mathbf{r}-\mathbf{r}'\|^2}\, d\mathbf{r}', \qquad (1)$$

where γ is the straight line path between the detector point $\mathbf{r}$ and the source point $\mathbf{r}'$ (hereafter referred to as the "straight line path", with $\mathbf{r}' = \mathbf{r} - s\,\mathbf{\Omega}$ and $\mathbf{\Omega} = \frac{\mathbf{r}-\mathbf{r}'}{\|\mathbf{r}-\mathbf{r}'\|}$, and where $\mu(s)$ is the total macroscopic photon cross-section at coordinate s of the straight line path γ. For a spatially homogeneous source distribution within the volume V_S, $p(\mathbf{r}) = \frac{1}{V_S}$ is constant. Taking $\mathbf{r}$ as the origin of coordinates, and expressing Eq. (1) in spherical coordinates leads to:

$$\Phi(\mathbf{r}) = \frac{A}{V_S} \iiint_{V_S} \frac{e^{-\int_\gamma \mu(s')ds'}}{4\pi s^2}\, s^2\, ds\, d\mathbf{\Omega}. \qquad (2)$$

For such a simple source distribution, it is interesting to see from Eq. (2) that, for a given solid angle, the integration over s is analytical. As a consequence, the partial contribution to the scalar flux of all possible source points of direction $\mathbf{\Omega}$ towards the detection point $\mathbf{r}$ reads[1]:

$$\varphi(\mathbf{r},\mathbf{\Omega}) = \frac{A}{V_S} \int_\Gamma \frac{e^{-\int_\gamma \mu(s')ds'}}{4\pi s^2}\, s^2\, ds,$$
$$= \frac{A}{4\pi\, V_S}\, e^{-\sum_{i\in[0;s_{in}]} \mu_i d_i} \times \frac{1 - e^{-\mu_s(s_{out}-s_{in})}}{\mu_s}, \qquad (3)$$

where Γ is the source segment that generates with respect to the detection point $\mathbf{r}$ all possible straight line paths γ of direction $\mathbf{\Omega}$. In the above equation Eq. (3), s_{in} and s_{out} stands for the abscissa of the entrance and exit points of the source volume along the line generated by $\mathbf{\Omega}$, μ_i stands for the total macroscopic photon cross-section of material i that is crossed along that line, and d_i the track length in material i (the sum over the indices i run over all materials that are crossed from the detector point up to the entrance of the source entrance, and index s is specific to the source volume). The latter Eq. (3) was used as an intermediate step to derive Eq. (5) of reference [1].

In a Monte-Carlo evaluation of the integral Eq. (2), this makes it possible to reduce the dimensionality of the problem. One can compute the associated marginal probability density $\tilde{p}(\mathbf{\Omega})$, integration of $p(\mathbf{r}) = \frac{1}{V_S}$ on the line of all possible sources points for a given solid angle $\mathbf{\Omega}$, which

⋆e-mail: benoit.avez@millennium.fr
⋆⋆e-mail: laurent.bindel@millennium.fr

[1] Here, we implicitely assume that the homogeneous source is distributed within a convex volume. A possible way to circumvent this assumption will be adressed further.

reads [2]:

$$\tilde{p}(\mathbf{\Omega}) = \int_{\Gamma} \frac{1}{V_S} s^2 \, ds, \\ = \frac{s_{out}^3 - s_{in}^3}{3 V_S}. \quad (4)$$

The solution of Eq.(2) can be recast under the following lines:

$$\Phi(\mathbf{r}) = \iint \left[\frac{\varphi(\mathbf{r}, \mathbf{\Omega})}{\tilde{p}(\mathbf{\Omega})} \right] \tilde{p}(\mathbf{\Omega}) \, d\mathbf{\Omega}. \quad (5)$$

Sampling $\mathbf{\Omega}$ from the marginal pdf Eq. (4), which is exactly the probability density function obtained by uniformly sampling the source volume and considering only $\mathbf{\Omega}$ as a relevant variable, we are left with the following Monte-Carlo formula:

$$\Phi(\mathbf{r}) = \lim_{N\to\infty} \left[\frac{1}{N} \sum_{i=1}^{N} \frac{\varphi(\mathbf{r}, \mathbf{\Omega_i})}{\tilde{p}(\mathbf{\Omega_i})} \right]; \quad (6)$$

with numerator and denominator obtained respectively from Eq. (3) and Eq. (4). This formula Eq. (6) is the main result of this paper. It is based on the so-called *use of expected values* (see, e.g., section 4.2 of ref. [2]) variance reduction technique.

3 Practical aspects of the new sampling procedure

3.1 Implementation of the formula

The implementation of the Monte-Carlo evaluation of the point-kernel formula using this sampling scheme is illustrated in Fig. 1 and follows Eqns. (3-4;6):

- Select a random point within the source volume - the choice of a given point is just a choice of a given solid angle;
- Compute the contribution to the scalar flux of this given solid angle Eq. (4) and its marginal probability density Eq. (3);
- Add the ratio of the two preceeding quantities to the current scalar flux evaluation, and iterate in order to reach the desired statistical precision.

One can notice that the implementation of this scheme is easy to achieve in existing point-kernel codes: only one additional surface has to be tracked (back) together with the use of Eq. (6).

3.2 Efficiency of the proposed point-kernel formula

In this section, the efficiency of the proposed point-kernel formula is illustrated through two test cases:

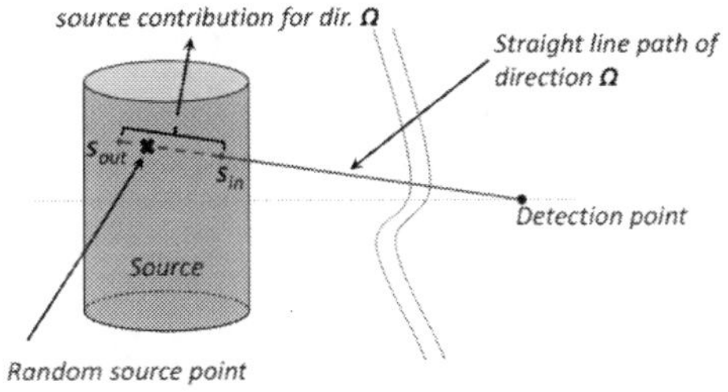

Figure 1. Illustration of the proposed point-kernel formula: for a given random source, contribution to the detection point comes from the full source segment between s_{in} and s_{out}.

- a 1st test case with a constant geometry with a source of varying density;
- a 2nd test case with a constant source material with several geometric extensions.

The efficiency of the method is evaluated by comparing the relative (statistical) error R of the proposed point-kernel formulation and the standard one for the same number of Monte-Carlo samples (100000).

3.2.1 1 m^3 cubic container filled with iron at several densities

The model, a cubic meter filled with iron at several densities is presented in Fig. 2. First, the proposed method

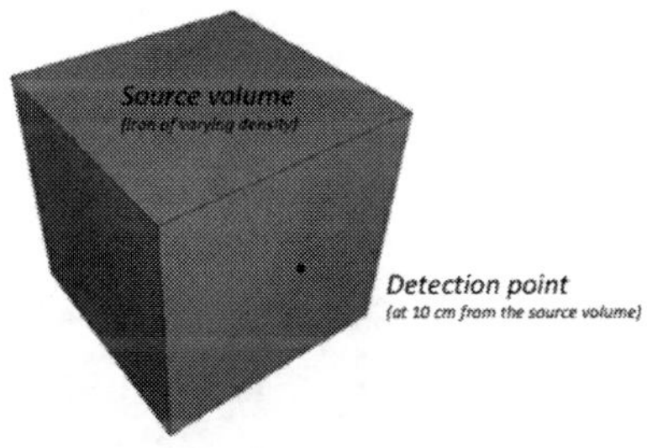

Figure 2. Test case #1: 1 m^3 cube of varying iron density. The detection point is located at 10 cm from the source volume.

is benchmarked for this test-case with iron at 8 $g.cm^{-3}$ against the standard point-kernel formula implemented in the same program, and is also benchmarked against MCNP6 [3] in similar conditions (i.e. uncollided photon flux of a F5 tally with simple physics treatment and energetic transport cut-off just below the considered energies). The results of the method comparison with respect

[2]Being a probability density, $\tilde{p}(\mathbf{\Omega})$ has the property $\iint \tilde{p}(\mathbf{\Omega})\, d\mathbf{\Omega} = 1$. One can also notice that Eq. (4) could be used (up to the $\frac{1}{V_S}$ factor) to estimate the volume of a given cell by isotropically sampling $\mathbf{\Omega}$ from a given point (possibly within the volume for more efficiency, in which case with $s_{in} = 0$).

to standard point-kernel formulation (including the MCNP uncollided F5 evaluation) are presented in Table 1. A good

Table 1. Comparison of fluxes and relative errors of the improved point-kernel formulation with respect to the standard point-kernel method and MCNP. Number of Monte-Carlo samples is 10 millions.

E_γ	MCNP		Standard		Improved	
MeV	Φ	err.	Φ	err.	Φ	err.
0.1	1.57	1.0%	1.53	1.0%	1.54	0.04%
0.15	2.91	0.7%	2.87	0.7%	2.87	0.04%
0.2	3.82	0.6%	3.77	0.6%	3.78	0.04%
0.3	4.92	0.6%	4.87	0.6%	4.87	0.04%
0.4	5.66	0.5%	5.60	0.5%	5.60	0.04%
0.5	6.27	0.5%	6.21	0.5%	6.21	0.04%
0.6	6.81	0.5%	6.74	0.5%	6.74	0.04%
0.8	7.77	0.4%	7.70	0.4%	7.70	0.04%
1	8.64	0.4%	8.58	0.4%	8.57	0.04%
1.5	10.6	0.4%	10.5	0.4%	10.5	0.04%
2	12.0	0.3%	12.0	0.3%	11.9	0.04%

matching between the three methods is achieved in terms of fluxes (mean fluxes), and the small differences between MCNP and our implementations are likely due to different cross-sections and cross-section treatments. However, whereas standard point-kernel evaluations (MCNP and our implementation) have comparable relative errors, the relative error of the improved point-kernel formulation has, for this test case, an improvement of approximately a factor of ten.

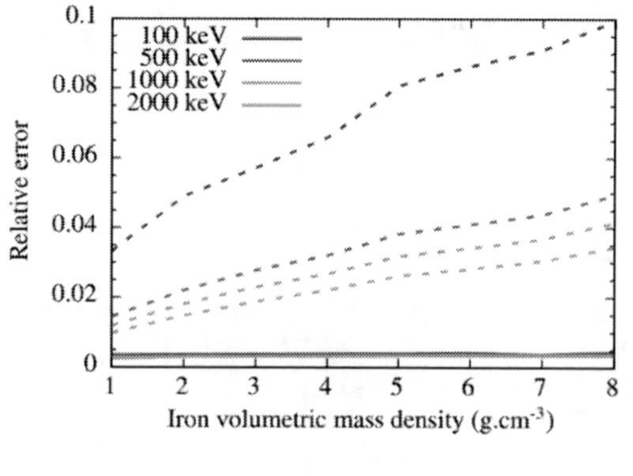

Figure 3. Evolution of the relative error with respect to the iron density - dashed: standard formula, solid: improved formula.

We now focus on test case #1, with iron densities ranging from 1 g.cm^{-3} to 8 g.cm^{-3}. Relative errors obtained at several energies for the standard point-kernel formulation and for the improved one are shown in Fig. 3. This figure shows that with the improved method :

i. the relative error is almost constant as a function of the iron density whereas it increases in the standard formulation case,

ii. the relative error is approximately the same for all considered energies, ranging from 100keV to 2MeV, whereas it is not the case for the standard formulation,

iii. the relative error is always smaller than the one obtained with the standard point-kernel formula, and

iv. best improvement of the proposed point-kernel formulation occurs at low energies, where the standard formulation is the less efficient.

3.2.2 $n \times n$ array of 1 m^3 cubic containers filled with concrete of density 1.5 g.cm^{-3}.

The model is presented in Fig. 4. The results of arrays

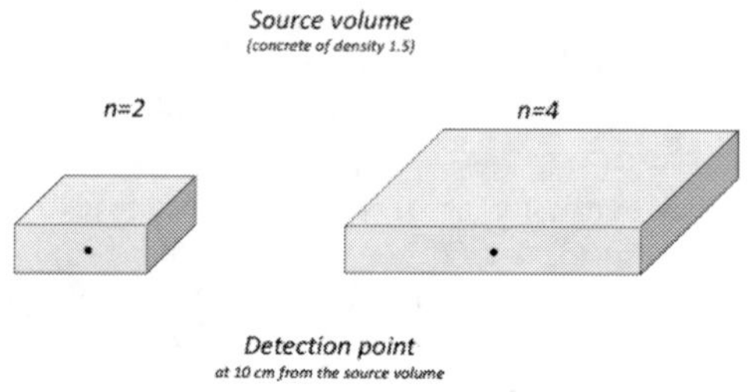

Figure 4. Test case #2: $n \times n$ array of 1 m^3 cubic containers filled with concrete of density 1.5 g.cm^{-3}. The detection point is located at 10 cm from the source volume.

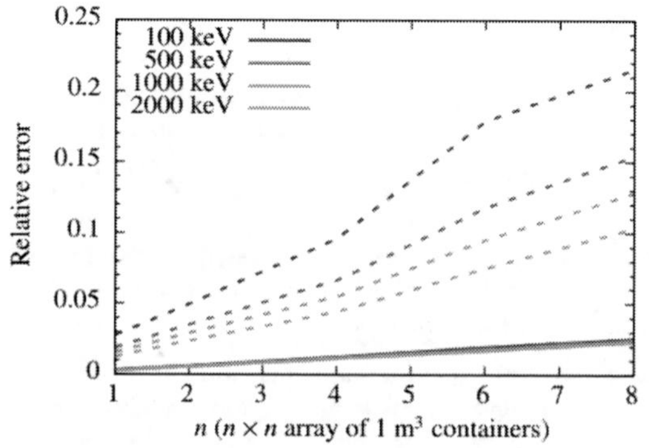

Figure 5. Evolution of the relative error with respect to the number of containers - dashed: standard formula, solid: improved formula.

of $n = 1$ to $n = 8$ are illustrated in Fig. 5. The latter shows a similar trends for the relative error as compared to the first test case. One difference, as compared to the preceding test-case, is the fact that the relative error is not constant with respect to the size of the source, but slightly increases. This dispersion increase is due to the geometrical extension of the source.

4 Conclusion for the new sampling procedure

To summarize our main findings:

- Based on the use of expected values, we derived a new point-kernel formula by directly integrating all source contributions of a given solid angle to the scalar flux;
- This formulation can easily be implemented into an existing point-kernel based code: as compared to standard Monte-Carlo point-kernel evaluations, a *single* additional surface crossing has to be computed in the proposed formulation;
- The improvements, in terms of variance reduction, are particularly interesting for "optically thick" sources, i.e. for very absorbing source materials or for extended sources.

This technique has also the advantage not to depend on biaising parameters or on a specific mesh that are often also energy dependent, like in importance sampling. Finally, as already stated previously, the current derivation has been performed assuming a convex source volume. However, When the source volume is not convex, the same formula, without additional complexification, can still be applied. Indeed, in such cases, it is sufficient, for a drawn random source point, to compute the marginal probability density and the contribution to the scalar flux not on the full intersection between the generated straight line and the source volume (i.e. several disconnected segments), but only of the (connected) segment that contains the drawn point.

5 Towards a new Monte-Carlo transport process

The transport equation of an isotropic source reads:

$$[\mathbf{\Omega}.\nabla_{\mathbf{r}} + \mu_t(\mathbf{r},E)]\,\phi(\mathbf{r},E,\mathbf{\Omega}) = \iiint \mu_s(\mathbf{r},E'\to E,\mathbf{\Omega}'\to\mathbf{\Omega})\,\phi(\mathbf{r},E',\mathbf{\Omega}')\,dE'd\mathbf{\Omega}' + \frac{S(\mathbf{r},E)}{4\pi}, \quad (7)$$

where μ_t and μ_s are the total and scattering macroscopic cross-sections, S the source term, and whose left hand side inversion can be done (for instance, on a bounded domain with zero entering flux) using the the integral operator $\hat{T}_{\mathbf{r},E,\mathbf{\Omega}}[\,]$:

$$\hat{T}_{\mathbf{r},E,\mathbf{\Omega}}[\,] \equiv \int_\Gamma ds\, e^{-\int_\gamma \mu_t(s',E)ds'}\,[\,] \quad (8)$$

where the integral domain Γ of s is ranging from 0 to $|\mathbf{r}_b - \mathbf{r}|$, with $\mathbf{r}_b$ the zero incoming flux boundary in direction $-\mathbf{\Omega}$ with respect to $\mathbf{r}$ (and where s' is ranging from 0 to s). In order to simplify the notations, we now condense the notations. One defines the scattering operator $\hat{C}$:

$$\hat{C}_{\mathbf{r},E,\mathbf{\Omega}}[\,] \equiv \int dE' \iint d\mathbf{\Omega}'\,\mu_s(\mathbf{r},E'\to E,\mathbf{\Omega}'\to\mathbf{\Omega})\,[\,] \quad (9)$$

and, taking $P \equiv \mathbf{r}, E, \mathbf{\Omega}$, the flux takes the following Neumann series formulation:

$$\begin{aligned}\phi(P) = &\quad \hat{T}_P\left[\frac{S}{4\pi}\right]\\ &+ \hat{T}_P\hat{C}_{P'\to P}\hat{T}_{P'}\left[\frac{S}{4\pi}\right]\\ &+ \hat{T}_P\hat{C}_{P'\to P}\hat{T}_{P'}\hat{C}_{P''\to P'}\hat{T}_{P''}\left[\frac{S}{4\pi}\right]\\ &+ \ldots \end{aligned} \quad (10)$$

Since $\hat{T}_P\left[\frac{S}{4\pi}\right]$ is the (uncollided) flux, the preceeding form Eq. (10) suggests that (i) the uncollided (directed) flux $\varphi(P) = \hat{T}_P\left[\frac{S}{4\pi}\right]$ can be taken, from Eq. (3), as the 0th order quantity, and (ii) higher orders are built of successive applications of $\hat{T}_P\hat{C}_{P'\to P}$ applied to $\varphi(P)$. This makes it possible to compute the total scalar flux from Eq. (10):

$$\begin{aligned}\Phi(\mathbf{r},E) = \iint d\mathbf{\Omega}\ \{&\varphi(P)\\ &+ \hat{T}_P\hat{C}_{P'\to P}\varphi(P')\\ &+ \hat{T}_P\hat{C}_{P'\to P}\hat{T}_{P'}\hat{C}_{P''\to P'}\varphi(P'')\\ &+ \ldots\} \end{aligned} \quad (11)$$

This formulation can serve as the basis of a new Monte-Carlo sampling scheme in which a flux is transported instead of a particle, and in which a contribution to the tally is done at each collision, in the spirit of next-event estimators (see e.g. Eq. (5.80) of [4]). The latter can take advantage of the source line flux evaluation Eq. (3) for the evaluation of the (directed) uncollided flux φ. Efforts are currently being done in that direction.

References

[1] L. Bindel, A. Gamess, E. Lejeune, J. Nucl. Sci. Technol. **37** suppl. 1, p. 512-514 (2000)

[2] M. H. Kalos and P. A. Whitlock, *Monte Carlo methods. Second edition* (WILEY-VCH, 2008)

[3] T. Goorley, et al., *Initial MCNP6 Release Overview*, Nuclear Technology **180**, p 298-315 (2012)

[4] M. H. Kalos, F. R. Nakache and J. Celnik, *Computing Methods in Reactor Physics*, Chapter 5 (GORDON AND BREACH, 1968)

Using the MCNP Taylor series perturbation feature (efficiently) for shielding problems

Jeffrey Favorite*

X-Computational Physics Division, XCP-3, P.O. Box 1663 MS F663, Los Alamos National Laboratory, Los Alamos, NM 87545 USA

Abstract. The Taylor series or differential operator perturbation method, implemented in MCNP and invoked using the PERT card, can be used for efficient parameter studies in shielding problems. This paper shows how only two PERT cards are needed to generate an entire parameter study, including statistical uncertainty estimates (an additional three PERT cards can be used to give exact statistical uncertainties). One realistic example problem involves a detailed helium-3 neutron detector model and its efficiency as a function of the density of its high-density polyethylene moderator. The MCNP differential operator perturbation capability is extremely accurate for this problem. A second problem involves the density of the polyethylene reflector of the BeRP ball and is an example of first-order sensitivity analysis using the PERT capability. A third problem is an analytic verification of the PERT capability.

1 Introduction

The Taylor series or differential operator perturbation feature [1, 2] was introduced into MCNP [3] in version 4B in the mid-1990s. The perturbation feature has been shown to be unreliable for estimating perturbations in eigenvalue (k_{eff}) problems [4, 5]. However, for fixed-source (shielding) problems, the perturbation feature can be extremely accurate and deserves to be more widely used. The Taylor series perturbation feature is invoked with the PERT card.

In publications until now [e.g., 2], a separate MCNP PERT card was used for each perturbed point in a parameter study. In fact, the MCNP manual [3] recommends two PERT cards for each perturbation so that the first- and second-order Taylor terms can be examined independently. The manual says that "each perturbation increases running time by 10%-20%" [3]. Limiting the number of PERT cards is therefore desirable.

This paper shows how to compute the two coefficients of a two-term Taylor expansion using only two PERT cards. These coefficients can then be used to estimate the perturbed response for *any* point in a parameter study. The statistical uncertainty can be well estimated from these two PERT cards; it can be computed exactly using three more PERT cards.

In addition, this paper shows how to use the PERT card to perform first-order sensitivity analysis such as may be used in uncertainty quantification.

We emphasize that the results of this paper apply only to fixed-source problems, *not* eigenvalue problems.

2 Taylor series coefficients

A Taylor series expansion of a response c that is a function of some reaction cross section σ_x is

$$c(\sigma_x) = c(\sigma_{x,0}) + \left.\frac{dc}{d\sigma_x}\right|_{\sigma_{x,0}} \Delta\sigma_x + \frac{1}{2}\left.\frac{d^2c}{d\sigma_x^2}\right|_{\sigma_{x,0}} (\Delta\sigma_x)^2 + \cdots, \quad (1)$$

where $\sigma_{x,0}$ is the reference value of the cross section and

$$\Delta\sigma_x \equiv \sigma_x - \sigma_{x,0}. \quad (2)$$

Define the first- and second-order Taylor series terms as

$$\Delta c_1 \equiv \left.\frac{dc}{d\sigma_x}\right|_{\sigma_{x,0}} \Delta\sigma_x \quad (3)$$

and

$$\Delta c_2 \equiv \frac{1}{2}\left.\frac{d^2c}{d\sigma_x^2}\right|_{\sigma_{x,0}} (\Delta\sigma_x)^2, \quad (4)$$

respectively. Define p_x as the relative cross-section change,

$$p_x \equiv \frac{\Delta\sigma_x}{\sigma_{x,0}}. \quad (5)$$

Then, using the chain rule, the Taylor series terms can be written conveniently as

$$\Delta c_1 = \left.\frac{dc}{dp_x}\right|_{p_x=0} p_x \quad (6)$$

and

$$\Delta c_2 = \frac{1}{2}\left.\frac{d^2c}{dp_x^2}\right|_{p_x=0} p_x^2. \quad (7)$$

For notational convenience, define $c_0 \equiv c(\sigma_{x,0})$.

* Corresponding author: fave@lanl.gov

3 Taylor series coefficients and the MCNP PERT card

3.1 Computing the coefficients

At present, the MCNP perturbation capability, invoked with the PERT card, uses a two-term Taylor expansion with no cross terms [6, 3]. The perturbation feature estimates the derivatives in Eqs. (6) and (7) and multiplies them by the relative perturbation p_x (or p_x^2) computed from the user input.

Define coefficients

$$c_1 \equiv \left.\frac{dc}{dp_x}\right|_{p_x=0} \tag{8}$$

and

$$c_2 \equiv \left.\frac{1}{2}\frac{d^2c}{dp_x^2}\right|_{p_x=0}. \tag{9}$$

These coefficients can be computed from any single arbitrary reference perturbation amount $p_{x,r}$ using, from Eqs. (6) and (7),

$$c_1 = \frac{\Delta c_1(p_{x,r})}{p_{x,r}} \tag{10}$$

and

$$c_2 = \frac{\Delta c_2(p_{x,r})}{p_{x,r}^2}. \tag{11}$$

Now the perturbed response due to an arbitrary perturbation p_x can be computed using

$$c_{\text{PERT}}(p_x) = c_0 + c_1 p_x + c_2 p_x^2. \tag{12}$$

Equation (12) [with Eqs. (10) and (11)] is the main conclusion of this paper. With this equation, a parameter study can be done for any number of perturbed values of the parameter with only two MCNP PERT cards, rather than the two PERT cards per perturbation that are recommended in the manual. A detailed prescription is given in Sec. 4.

3.2 Approximate uncertainty

The standard deviations of the Monte Carlo estimates of c_1 and c_2, s_{c_1} and s_{c_2}, are

$$s_{c_1} = \frac{s_{\Delta c_1}}{|p_{x,r}|} \tag{13}$$

and

$$s_{c_2} = \frac{s_{\Delta c_2}}{p_{x,r}^2}, \tag{14}$$

where $s_{\Delta c_1}$ and $s_{\Delta c_2}$ are the standard deviations of the Monte Carlo estimates of Δc_1 and Δc_2 [meaning $\Delta c_1(p_{x,r})$ and $\Delta c_2(p_{x,r})$ here and in the rest of this paper]. The standard deviation of the unperturbed tally c_0 is s_{c_0}. With these quantities, the variance of the estimated perturbed tally of Eq. (12), estimated by assuming that c_0, Δc_1, and Δc_2 are uncorrelated (subscript "*unc.*"), is

$$s^2_{c_{\text{PERT},unc.}} = s^2_{c_0} + s^2_{c_1} p_x^2 + s^2_{c_2} p_x^4. \tag{15}$$

3.3 Exact uncertainty

Typically, and in the prescription given in Sec. 4, c_0, Δc_1, and Δc_2 are all computed from the same set of histories in the same cells. They are likely to be highly correlated. The effect of the correlations among c_0, Δc_1, and Δc_2 on the variance $s^2_{c_{\text{PERT}}}$ can be computed exactly.

The covariance matrix for c_{PERT} is

$$\underline{\underline{C}} = \begin{bmatrix} s^2_{c_0} & \rho_{c_0\Delta c_1} & \rho_{c_0\Delta c_2} \\ \rho_{c_0\Delta c_1} & s^2_{\Delta c_1} & \rho_{\Delta c_1\Delta c_2} \\ \rho_{c_0\Delta c_2} & \rho_{\Delta c_1\Delta c_2} & s^2_{\Delta c_1} \end{bmatrix}, \tag{16}$$

where the covariances are

$$\rho_{xy} \equiv r_{xy} s_x s_y, \tag{17}$$

where the correlation coefficients are

$$r_{xy} \equiv \frac{N\sum(x_i y_i) - \left(\sum x_i\right)\left(\sum y_i\right)}{\sqrt{\left[N\sum x_i^2 - \left(\sum x_i\right)^2\right]\left[N\sum y_i^2 - \left(\sum y_i\right)^2\right]}}, \tag{18}$$

where N is the number of histories sampled and x_i and y_i are the x and y scores for the ith history.

We now find the correlation coefficient r_{xy} in terms of x, y, and z, where $z = x + y$. The variance in a Monte Carlo estimate of z, implicitly accounting for correlations in x and y, is

$$\begin{aligned} s_z^2 &= \frac{1}{N^2}\sum z_i^2 - \frac{1}{N^3}\left(\sum z_i\right)^2 \\ &= \frac{1}{N^2}\sum (x_i + y_i)^2 - \frac{1}{N^3}\left(\sum (x_i + y_i)\right)^2 \\ &= \frac{1}{N^2}\left(\sum x_i^2 + \sum y_i^2 + 2\sum (x_i y_i)\right) \\ &\quad - \frac{1}{N^3}\left[\left(\sum x_i\right)^2 + \left(\sum y_i\right)^2 + 2\left(\sum x_i\right)\left(\sum y_i\right)\right]. \end{aligned} \tag{19}$$

The variance in a Monte Carlo estimate of the sum $x + y$, *not* accounting for correlations in x and y, is

$$\begin{aligned} s_{x+y}^2 &= s_x^2 + s_y^2 \\ &= \frac{1}{N^2}\left(\sum x_i^2 + \sum y_i^2\right) - \frac{1}{N^3}\left[\left(\sum x_i\right)^2 + \left(\sum y_i\right)^2\right] \end{aligned} \tag{20}$$

Using Eq. (20) in Eq. (19) yields

$$s_z^2 = s_{x+y}^2 + \frac{2}{N^3}\left[N\sum (x_i y_i) - \left(\sum x_i\right)\left(\sum y_i\right)\right]. \tag{21}$$

Rearranging Eq. (21) and using the result in Eq. (18) [with the first line of Eq. (19), replacing z with x and y] yields

$$\begin{aligned} r_{xy} &= \frac{\frac{N^3}{2}\left(s_z^2 - s_{x+y}^2\right)}{\sqrt{\left(N^3 s_x^2\right)\left(N^3 s_y^2\right)}} \\ &= \frac{s_z^2 - s_{x+y}^2}{2 s_x s_y}, \end{aligned} \tag{22}$$

and using the first line of Eq. (20) yields

$$r_{xy} = \frac{s_z^2 - s_x^2 - s_y^2}{2 s_x s_y}. \tag{23}$$

Equation (23) will be applied for $(x,y) \in (c_0, \Delta c_1, \Delta c_2)$ and $z \in (c_0 + \Delta c_1, c_0 + \Delta c_2, \Delta c_1 + \Delta c_2)$. Using Eq. (23) in Eq. (17) yields

$$\rho_{xy} = \tfrac{1}{2}(s_z^2 - s_x^2 - s_y^2). \tag{24}$$

The sensitivity vector for c_{PERT} [from Eq. (12) with Eqs. (10) and (11)] is

$$\underline{S} = \left[\frac{\partial c_{\text{PERT}}}{\partial c_0}, \frac{\partial c_{\text{PERT}}}{\partial \Delta c_1}, \frac{\partial c_{\text{PERT}}}{\partial \Delta c_2} \right]^T = \left[1, \frac{p_x}{p_{x,r}}, \left(\frac{p_x}{p_{x,r}} \right)^2 \right]^T. \tag{25}$$

where superscript T indicates the transpose.

Using the "sandwich rule," the total variance in c_{PERT} is

$$s_{c_{\text{PERT}}}^2 = (\underline{S})^T \underline{\underline{C}} (\underline{S}). \tag{26}$$

Using Eqs. (15), (16), (24), and (25), Eq. (26) becomes

$$\begin{aligned} s_{c_{\text{PERT}}}^2 &= s_{c_{\text{PERT},unc.}}^2 + \left(s_{c_0+\Delta c_1}^2 - s_{c_0}^2 - s_{\Delta c_1}^2 \right) \left(\frac{p_x}{p_{x,r}} \right) \\ &+ \left(s_{c_0+\Delta c_2}^2 - s_{c_0}^2 - s_{\Delta c_2}^2 \right) \left(\frac{p_x}{p_{x,r}} \right)^2 \\ &+ \left(s_{\Delta c_1+\Delta c_2}^2 - s_{\Delta c_1}^2 - s_{\Delta c_2}^2 \right) \left(\frac{p_x}{p_{x,r}} \right)^3. \end{aligned} \tag{27}$$

4 Prescription for perturbations

4.1 Basic

The prescription to estimate a response c for any number of perturbations of a parameter p_x is as follows:

Choose a reference perturbation $p_{x,r}$; a convenient choice is $p_{x,r} = 1$.

Set up two PERT cards with the density equal to [from Eqs. (2) and (5)] $\sigma_{x,r} = \sigma_{x,0}(1 + p_{x,r})$. Obviously, if $p_{x,r} = 1$, then $\sigma_{x,r} = 2\sigma_{x,0}$. One PERT card uses METHOD = 2 and the other uses METHOD = 3.

Run the problem using MCNP.

The METHOD = 2 perturbation result is $\Delta c_1(p_{x,r}) \pm s_{\Delta c_1}$; use Eq. (10) to compute c_1. The METHOD = 3 perturbation result is $\Delta c_2(p_{x,r}) \pm s_{\Delta c_2}$; use Eq. (11) to compute c_2. Compute the standard deviations using Eqs. (13) and (14).

A practically continuous curve of $c_{\text{PERT}}(p_x)$ can now be created using Eq. (12) with uncertainties estimated using Eq. (15).

4.2 Checking accuracy

As stated in the manual, it is a good idea to test whether a second-order Taylor expansion is accurate for the largest perturbation that is expected. One way to do this is to compare the first- and second-order terms $c_1 p_x$ and $c_2 p_x^2$ with the total $c_{\text{PERT}}(p_x)$, as suggested in the manual. This can be done for all values of p_x used in the analysis without additional PERT cards. From Eq. (12), the ratio of the second-order term to the first-order term is

$$\frac{\text{2nd}}{\text{1st}} = \frac{c_2 p_x^2}{c_1 p_x} = \frac{c_2}{c_1} p_x, \tag{28}$$

which is linear with the relative perturbation.

However, contrary to statements in the manual, it is not always true that the second-order term should contribute much less to the total than the first-order term. For example, consider the notional curve shown in Fig. 1. The base case of $\sigma_0 = 2$ is at or near the optimum for response R. The first-order Taylor term is zero or close to zero, and the second-order term dominates. Higher-order terms may be zero. Caution is advised.

Where possible, it is wise to use a direct calculation of the perturbed tally to test $c_{\text{PERT}}(p_x)$ for the maximum values of $\pm|p_x|$ that are of interest.

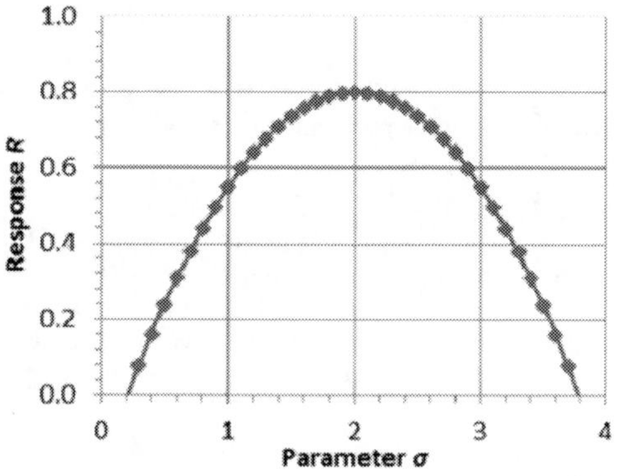

Fig. 1. Notional graph of a response R as a function of a parameter σ; $\sigma_0 = 2$.

4.3 Exact uncertainty

Exact uncertainties that account for the correlations among c_0, Δc_1, and Δc_2 can be computed as follows:

Set up three more PERT cards in addition to those set up in Sec. 4.1 (with the same density as in Sec. 4.1). One of the new PERT cards uses METHOD = –2, one uses METHOD = –3, and one uses METHOD = 1.

Run the problem using MCNP.

The METHOD = –2 perturbation result is $c_0 + \Delta c_1(p_{x,r}) \pm s_{c_0+\Delta c_1}$. The METHOD = –3 perturbation result is $c_0 + \Delta c_2(p_{x,r}) \pm s_{c_0+\Delta c_2}$. The METHOD = 1 perturbation result is $\Delta c_1(p_{x,r}) + \Delta c_2(p_{x,r}) \pm s_{\Delta c_1+\Delta c_2}$.

Exact uncertainties may now be computed using Eqs. (15) and (27). Note that the perturbation results for METHOD = −2, −3, and 1 are not used; only the corresponding uncertainties are used.

5 Prescription for first-order sensitivity analysis

5.1 Basic

The first-order relative sensitivity coefficient of a response c to some reaction cross section σ_x is defined as

$$S_{c,\sigma_x} \equiv \frac{\sigma_{x,0}}{c_0} \frac{dc}{d\sigma_x}\bigg|_{\sigma_{x,0}} . \tag{29}$$

Using $\sigma_x = \sigma_{x,0}(1+p_x)$ in Eq. (29) with the chain rule yields

$$S_{c,\sigma_x} = \frac{1}{c_0} \frac{dc}{dp_x}\bigg|_{p_x=0} , \tag{30}$$

or, from Eq. (8),

$$S_{c,\sigma_x} = \frac{c_1}{c_0}. \tag{31}$$

The prescription to estimate the first-order sensitivity coefficient is now clear: Follow the prescription of Sec. 4.1 to compute c_1, and divide the result by the unperturbed response c_0. The estimated variance of the sensitivity coefficient, assuming c_0 and c_1 are uncorrelated (subscript "*unc.*"), is

$$s_{S_{unc.}}^2 = S_{c,\sigma_x}^2 \left[\left(\frac{s_{c_0}}{c_0}\right)^2 + \left(\frac{s_{c_1}}{c_1}\right)^2 \right]. \tag{32}$$

Equation (13) gives s_{c_1}. In this approximation, only METHOD = 2 is required.

5.2 Exact uncertainty

The sensitivity vector for S_{c,σ_x} [from Eq. (31) with Eq. (10)] is

$$\underline{S} = \left[\frac{\partial S_{c,\sigma_x}}{\partial c_0}, \frac{\partial S_{c,\sigma_x}}{\partial \Delta c_1} \right]^T = \left[\frac{-\Delta c_1}{p_{x,r} c_0^2}, \frac{1}{p_{x,r} c_0} \right]^T . \tag{33}$$

The covariance matrix for S_{c,σ_x} is

$$\underline{\underline{C}} = \begin{bmatrix} s_{c_0}^2 & \rho_{c_0 \Delta c_1} \\ \rho_{c_0 \Delta c_1} & s_{\Delta c_1}^2 \end{bmatrix}, \tag{34}$$

with

$$\rho_{c_0 \Delta c_1} = \tfrac{1}{2}(s_{c_0+\Delta c_1}^2 - s_{c_0}^2 - s_{\Delta c_1}^2), \tag{35}$$

as in Sec. 3.3. Using Eq. (26) (the sandwich rule) and Eqs. (31) and (32), the total variance of the sensitivity coefficient is

$$s_S^2 = s_{S_{unc.}}^2 - \left(s_{c_0+\Delta c_1}^2 - s_{c_0}^2 - s_{\Delta c_1}^2\right)\left(\frac{S_{c,\sigma_x}^2}{c_0 \Delta c_1}\right). \tag{36}$$

In this case, as in Sec. 4.3, the METHOD = 2 result gives $s_{\Delta c_1}$, and a PERT card with METHOD = −2 is needed to give $s_{c_0+\Delta c_1}$.

6 Numerical examples

6.1 Neutron detector polyethylene density

As an example, consider a detailed model of a neutron detector consisting of 15 helium-3 tubes with an active height of 15 in. (Fig. 2) and pressure of 10 atm. arranged in three rows in a high-density polyethylene (HDPE) moderator (Fig. 3). The nominal density of the HDPE is

Fig. 2. Three-dimensional rendering of the neutron detector. The light blue material is HDPE, within which are arranged the helium-3 tubes. The height of the HDPE is 16.59 in.

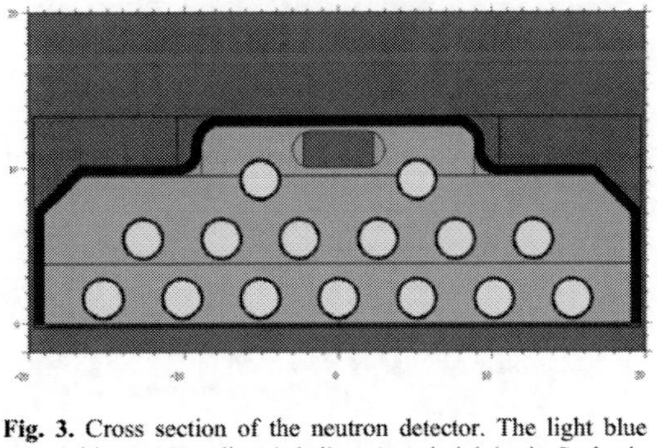

Fig. 3. Cross section of the neutron detector. The light blue material is HDPE, yellow is helium-3, and pink is air. Scales in centimeters.

taken to be 0.962 g/cm^3. How does the response of the detector change as the HDPE density changes?

An Am-Be source was modeled 30 cm from the front face of the detector and centered vertically with respect to the active height of the tubes. Detector counts were modeled as captures in helium-3 in the active part of the tubes.

The total efficiency as a function of HDPE density is plotted in Fig. 4. (The total efficiency is the number of counts per source particle emitted, and for this calculation it is not necessary to specify the source strength.) Seven direct calculations (using perturbed densities) as well as the unperturbed (reference) direct result are compared with a second-order Taylor series estimate computed using coefficients calculated during the run of the unperturbed reference case. Uncertainties for the Taylor series estimates are the uncorrelated estimates of Eq. (15). Figure 4 shows that the Taylor series estimate of the MCNP PERT card provides an extremely accurate and fast method for performing this sensitivity study.

Adding the two PERT cards increased the run time by 6%.

Including all correlations among the Taylor series coefficients changes the standard deviation from 4.67 × 10^{-6} to 4.43 × 10^{-6} for the largest negative perturbation and from 4.42 × 10^{-6} to 4.49 × 10^{-6} for the largest

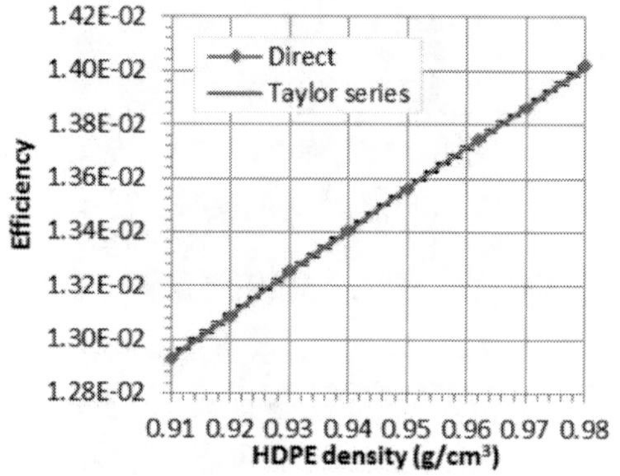

Fig. 4. Efficiency for the neutron detector.

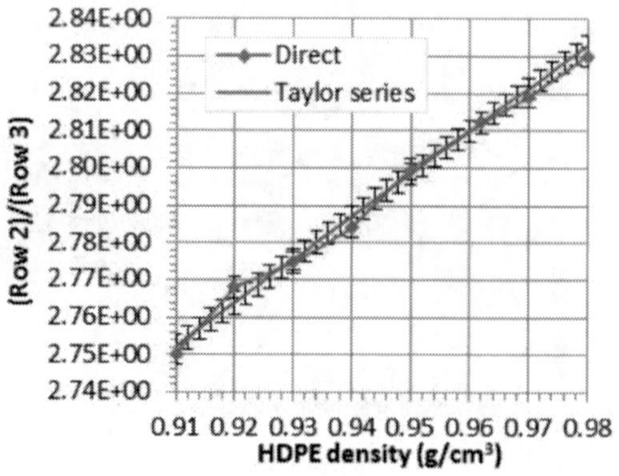

Fig. 5. Row ratio for the neutron detector.

positive perturbation. Adding three more PERT cards increased the run time (from the case with two PERT cards) by only 5%.

The advantage of a continuous parameter study is not apparent in Fig. 4 since the direct results are relatively easy to acquire. However, another quantity of interest for the neutron detector is the "row ratio," or the ratio of the sum of the counts in the second, third, fourth, and fifth tubes in the middle row (called row 2) to the sum of the counts in the two tubes in the back row (called row 3, the top row in Fig. 3).

Figure 5 shows the row ratio for the seven perturbed configurations and the unperturbed reference case. Statistical uncertainty causes "wiggle" in the curve through the points.

The Taylor series estimate, also shown on Fig. 5, was obtained as the ratio of the individual Taylor series estimates for the count rates for rows 2 and 3. Uncertainties for the Taylor series estimates use the uncorrelated estimate of Eq. (15) for each row.

Including all correlations among the Taylor series coefficients changes the standard deviation from 3.10 × 10^{-3} to 2.99 × 10^{-3} for the largest negative perturbation and from 2.85 × 10^{-3} to 2.88 × 10^{-3} for the largest positive perturbation.

Comparing the contributions of the first- and second-order terms, as suggested in the MCNP manual [3], can easily be done for each perturbed point using Eq. (12).

6.2 Neutron reflector polyethylene density

This problem illustrates the use of the PERT card for the sensitivity of a response to the mass density of a neutron reflector. The response is the total neutron leakage from a spherical system. The system is the polyethylene-reflected BeRP ball [7, 8] described in Table 1. The density of the polyethylene is perturbed.

Table 1. BeRP ball geometry and materials.

Material	Outer Radius (cm)	Density (g/cm^3)
α-Pu	3.794	19.6
NO Gas Fill	3.82829	0.00129
Steel	3.85879	7.62
Polyethylene	11.47879	0.95

The exact leakage for each perturbation is compared with the first- and second-order Taylor series perturbation estimates in Fig. 6. Error bars of one standard deviation are present but not visible. The second-order estimate of Eq. (12) produces the correct curvature near the unperturbed density (0.95 g/cm^3), but it is too large for negative density perturbations and too small for positive density perturbations. A higher-order

DOI: 10.1051/epjconf/201715306030

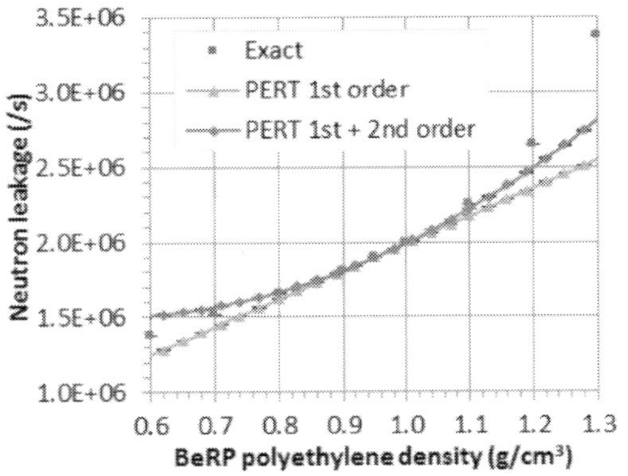

Fig. 6. Neutron leakage as a function of BeRP polyethylene density.

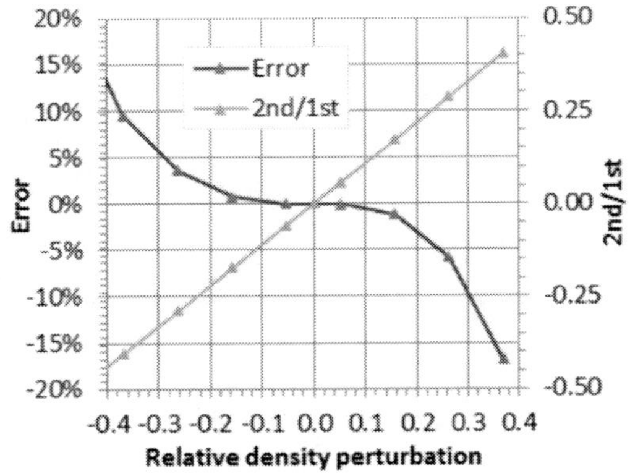

Fig. 7. Error in the second-order perturbation estimate and ratio of second-order term to first-order term for the BeRP ball problem.

Taylor expansion is needed for more accuracy in the range of perturbations shown.

The error is quantified in Fig. 7 and presented with the ratio of the second-order term to the first-order term, as given by Eq. (28). For this problem, the error is within ±10% when the ratio is –0.40 to 0.33.

We turn now to the first-order sensitivity of the leakage to the polyethylene density, computed using Eq. (31). The first-order PERT method is compared with central-difference estimates computed using

$$S_{c,\rho}^{CD} \approx \frac{\rho_0}{c_0}\left(\frac{c(\rho_0+h)-c(\rho_0-h)}{2h}\right), \tag{37}$$

where h is the change made in the density to compute the central difference. It is important to choose the perturbation h small enough that the points $c(\rho_0-h)$, c_0, and $c(\rho_0+h)$ are on a line but large enough that the difference $c(\rho_0+h)-c(\rho_0-h)$ can be calculated accurately [9], and, if a Monte Carlo code is used, with a small uncertainty. Central-difference estimates using Eq. (37) with different values of h and MCNP results for c are shown in Table 2 with Monte Carlo relative uncertainties (1s). The difference between the PERT estimate and the central-difference estimate is shown in terms of the number of standard deviations of difference.

Table 2. Sensitivity of neutron leakage to polyethylene density.

Method	h (g/cm³)	Sensitivity (%/%)	Diff. w.r.t. PERT (Ns)[(a)]
PERT	N/A	0.9270 ± 0.23%	N/A
Central Diff.	0.05	0.9370 ± 0.32%	1.94
Central Diff.	0.15	0.9949 ± 0.11%	20.7
Central Diff.	0.25	1.136 ± 0.07%	69.4

(a) Number of standard deviations of difference.

The central-difference estimate using the smallest value of h (0.05 g/cm³) is just within 2s of the first-order PERT value of the sensitivity. Which is the more accurate value? The three points of the sensitivity are not exactly on a line: The Pearson correlation coefficient for the points is 0.9994. In our judgment, the PERT value is the most accurate.

6.3 Analytic monodirectional one-group slab

This problem applies the PERT card to the reaction rate inside a slab when there is a monodirectional source and no scattering. The solution is analytic, so this is a good verification problem for the PERT capability.

A boundary source of strength q impinges on the left, at x = 0 cm, and the width of the slab is X. The macroscopic cross section is Σ. The flux $\phi(x)$ at any point x within the slab is

$$\phi(x) = qe^{-\Sigma x} \tag{38}$$

and the total reaction rate R within the slab is

$$\begin{aligned} R &= \int_0^X dx\Sigma\phi(x) \\ &= \int_0^X dx\Sigma q e^{-\Sigma x} \\ &= q\left(1-e^{-\Sigma X}\right). \end{aligned} \tag{39}$$

The cross section Σ is perturbed a relative amount p; in accordance with Eq. (5), we write the cross section as

$$\Sigma = \Sigma_0(1+p). \tag{40}$$

Using Eq. (40) in Eq. (39), the reaction rate is

$$R = q\left(1-e^{-\Sigma_0(1+p)X}\right). \tag{41}$$

From Eq. (41), the first derivative of R with respect to p is

$$\frac{\partial R}{\partial p} = q\Sigma_0 X e^{-\Sigma_0(1+p)X}. \tag{42}$$

The second derivative is

$$\frac{\partial^2 R}{\partial p^2} = -q(\Sigma_0 X)^2 e^{-\Sigma_0(1+p)X}. \tag{43}$$

The derivatives evaluated at $p = 0$ are

$$\left.\frac{\partial R}{\partial p}\right|_{p=0} = q\Sigma_0 X e^{-\Sigma_0 X} \tag{44}$$

and

$$\left.\frac{\partial^2 R}{\partial p^2}\right|_{p=0} = -q(\Sigma_0 X)^2 e^{-\Sigma_0 X}. \quad (45)$$

The two-term Taylor series expansion is

$$R_{TS}(p) = R_0 + \left.\frac{\partial R}{\partial p}\right|_{p=0} p + \frac{1}{2}\left.\frac{\partial^2 R}{\partial p^2}\right|_{p=0} p^2$$
$$= R_0 + R_1 p + R_2 p^2, \quad (46)$$

where R_1 and R_2 are c_1 and c_2 of Eqs. (8) and (9).

We now apply the parameters of Table 3 to this problem. Analytic values of R_0, R_1, and R_2 [from Eqs. (41), (44), and (45), respectively] are shown in Table 4 and compared with values computed using MCNP. R_0 is, of course, a regular reaction-rate cell tally. R_1 and R_2 were computed using the PERT card as discussed in Sec. 4.1. The differences between the MCNP results and the analytic values are well within one standard deviation.

A parameter study is shown in Fig. 8. The exact reaction rate, from Eq. (41), is compared with the two-term Taylor series of Eq. (46) when the coefficients are computed analytically and when the coefficients are from the MCNP PERT capability. The two curves for the Taylor series are indistinguishable. Figure 8 shows that a two-term Taylor expansion is extremely accurate for cross-section perturbations of ±10% for this problem; beyond that, higher-order terms may be needed.

Figure 9 quantifies these observations, showing the error in the second-order PERT estimate as well as the ratio of the second- to the first-order Taylor term. For this problem, even when the second-order term is about the same size as the first-order term, the error in the Taylor series estimate is less than 0.6%.

Table 3. Parameters of the analytic slab problem.

Parameter	Value
q	1.0 cm^{-2}s^{-1}
X	1.0 cm
Σ_0	5.2 cm^{-1}

Table 4. Coefficients of the Taylor expansion.

Coefficient[(a)]	Analytic	MCNP PERT	N_S[(b)]
R_0 (s^{-1})	0.994483	0.994428 ± 0.01%	0.57
R_1 (s^{-1})	0.0286861	0.0287011 ± 0.83%	0.06
R_2 (s^{-1})	−0.0745840	−0.0746330 ± 0.48%	0.14

(a) The right side of Eq. (39) is multiplied by a unit area, making the units work out to s^{-1}.
(b) Number of standard deviations of difference.

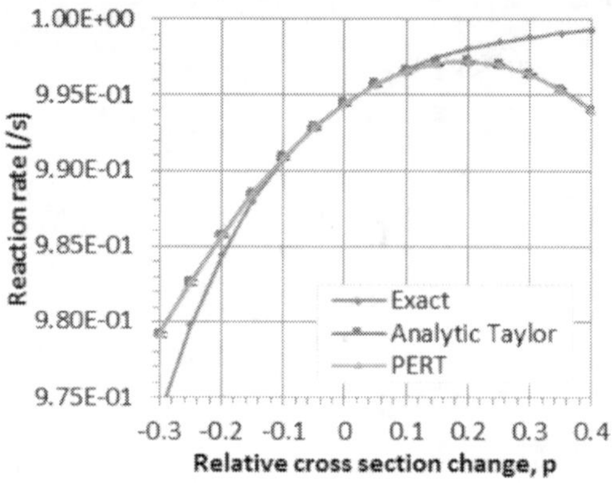

Fig. 8. Exact reaction rate and Taylor series estimates for the analytic problem.

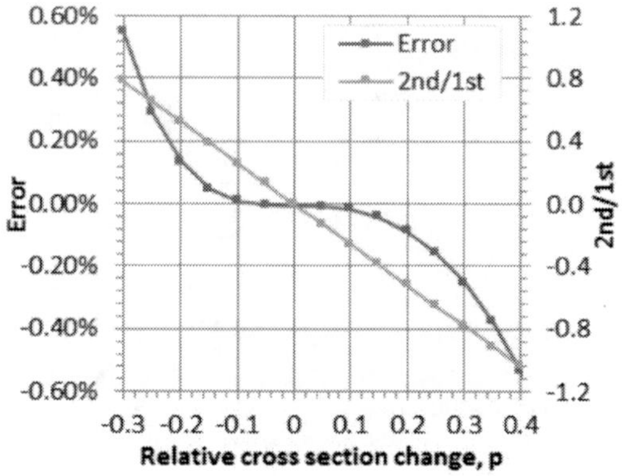

Fig. 9. Error in the second-order perturbation estimate and ratio of second-order term to first-order term for the analytic problem.

7 Summary and conclusions

The MCNP perturbation capability can be extremely accurate for shielding problems. This paper has demonstrated that it is not necessary to use a PERT card (or two PERT cards) for every perturbed point in a parameter study. Just two PERT cards suffice to obtain the Taylor series coefficients, generate a near-continuous curve of the resulting perturbed response, and estimate the statistical uncertainty in the perturbed response.

This paper has focused on cross-section and density perturbations. Material-substitution perturbations require special treatment [6], but it might be possible to apply the methods of this paper. On the other hand, the first-order sensitivity of a response to material composition changes can be computed using the first-order sensitivity to each of the individual nuclide densities, computed as described in Sec. 5. This topic is receiving new attention [10, 11].

The run times for the test problem of Sec. 6.1 increased by only ~20% of the rule of thumb quoted in the manual [3].

The PERT card must not be used to perturb problem parameters that would cause the source spatial, spectral, or angular distribution to be perturbed.

References

1. G. McKinney and E.T. Cheng, "MCNP Sensitivity Method Development and Applications," Trans. Am. Nucl. Soc., **46**, 278-279 (1984)
2. G.W. McKinney and J.L. Iverson, "Verification of the MCNP Perturbation Technique," *Proc. 1996 Top. Mtg. Rad. Prot. Shielding*, Vol. 2, 959-966, No. Falmouth, Mass., April 21-25 (1996)
3. T. Goorley et al., "Initial MCNP6 Release Overview," Nucl. Technol., **180**, 298-315 (2012)
4. J.A. Favorite, "An Alternative Implementation of the Differential Operator (Taylor Series) Perturbation Method for Monte Carlo Criticality Problems," Nucl. Sci. Eng., **142**, 327-341 (2002)
5. J.A. Favorite, "On the Accuracy of the Differential Operator Monte Carlo Perturbation Method for Eigenvalue Problems," Trans. Am. Nucl. Soc., **101**, 460-462 (2009)
6. J.A. Favorite and D.K. Parsons, "Second-Order Cross Terms in Monte Carlo Differential Operator Perturbation Estimates," *Proc. Int. Conf. Mathematical Methods for Nuclear Applications*, Salt Lake City, Utah, Sept. 9-13, CD-ROM (2001)
7. J. Mattingly, "Polyethylene-Reflected Plutonium Metal Sphere: Subcritical Neutron and Gamma Measurements," Sandia National Laboratories Report SAND2009-5804 Revision 3 (Rev. July 2012)
8. E.C. Miller et al., "Computational Evaluation of Neutron Multiplicity Measurements of Polyethylene-Reflected Plutonium Metal," Nucl. Sci. Eng., **176**, 167-185 (2014)
9. W.H. Press et al., *Numerical Recipes in FORTRAN: The Art of Scientific Computing*, 2nd ed., Chap. 5.7 (reprinted with corrections) (Cambridge University Press, 1994)
10. J.A. Favorite et al., "Adjoint-Based Sensitivity and Uncertainty Analysis for Density and Composition: A User's Guide," Trans. Am. Nucl. Soc., **115**, 669-672 (2016)
11. J.A. Favorite et al., "Adjoint-Based Sensitivity and Uncertainty Analysis for Density and Composition: A User's Guide," Nucl. Sci. Eng., **185**, article in press (2017)

EPJ Web of Conferences 153, 06031 (2017) DOI: 10.1051/epjconf/201715306031
ICRS-13 & RPSD-2016

Acceleration of MCNP calculations for small pipes configurations by using Weigth Windows Importance cards created by the SN-3D ATTILA

Eric Castanier[1,*], *Loic* Paterne[1], and *Céline* Louis[1]

[1]AREVA PROJETS, 1 rue des hérons, 78182 Montigny le Bretonneux, France

Abstract. In the nuclear engineering, you have to manage time and precision. Especially in shielding design, you have to be more accurate and efficient to reduce cost (shielding thickness optimization), and for this, you use 3D codes. In this paper, we want to see if we can easily applicate the CADIS methods for design shielding of small pipes which go through large concrete walls. We assess the impact of the WW generated by the 3D-deterministic code ATTILA versus WW directly generated by MCNP (iterative and manual process). The comparison is based on the quality of the convergence (estimated relative error (σ), Variance of Variance (VOV) and Figure of Merit (FOM)), on time (computer time + modelling) and on the implement for the engineer.

1 Introduction

In the nuclear engineering, you have to manage time and precision. Especially in shielding design, you have to be more accurate and efficient to reduce cost (shielding thickness optimization), and for this, you use 3D codes. On the other hand, run 3D Monte Carlo or deterministic codes requires time which is not always compatible with the projects dead line. So, to solve this dilemma, the Oak Ridge National Laboratory (ORNL) has developed two methods for increasing the efficiency of fixed-source MC simulations: The Consistent Adjoint Driven Importance Sampling (CADIS) method [1] and the Forward Weighted CADIS (FW-CADIS) method [2]. These methods utilize the results of approximate 3-D discrete ordinates transport calculations to generate consistent space- and energy-dependent source and transport (weight windows - WW) biasing parameters for MCNP code [3].

In this paper, we want to see if we can easily applicate these methods for design shielding for typical penetrations met in the French Fuel Cycle Re-processing plant located in La Hague (see figure 1).

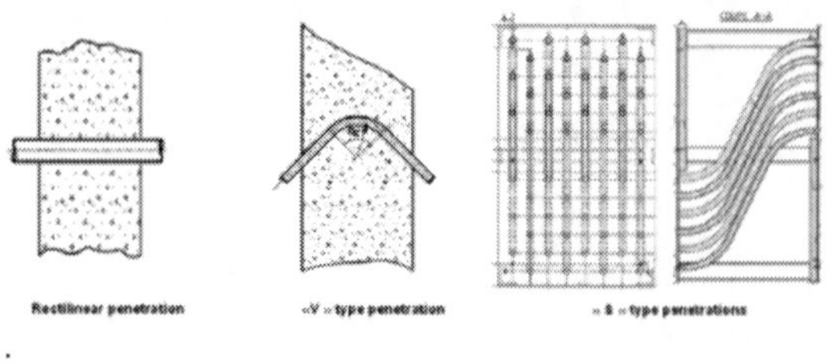

Fig. 1. View of the typical penetrations.

For these penetrations, we need sometimes to design shielding to reduce the shielding defect in the wall due to these penetrations.

2 Methods

The goal of this study is to evaluate the impact of the WW generated by the 3D-deterministic code ATTILA [4] versus WW directly generated by MCNP (iterative and manual process – see figure 2).
In the iterative method; we chain several MCNP calculations where we use the weight window card created by the previous calculation.

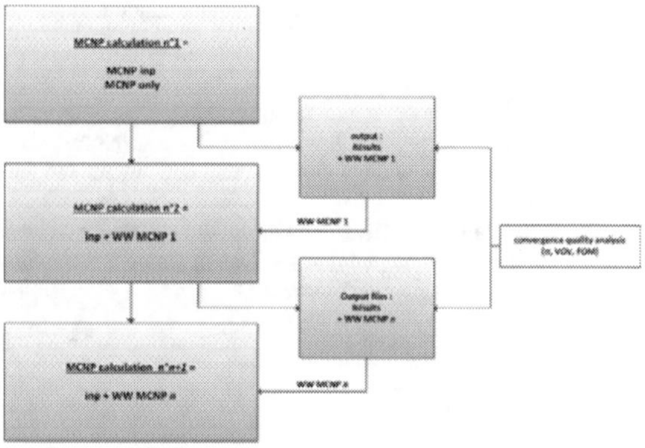

Fig. 2. Iterative method.

In the step 1, we create our first weight windows card. In the same time, we analyze the convergence quality and if it is not acceptable, we have to run a second calculation.

* Corresponding author: eric.castanier@areva.com

For this one, we use the weight window card created in the first calculation. We analyze the convergence quality of the result and so on to reach a good convergence.

With the CADIS method, we create the first weight window card by a 3-D discrete ordinates transport calculations (see figure 3).

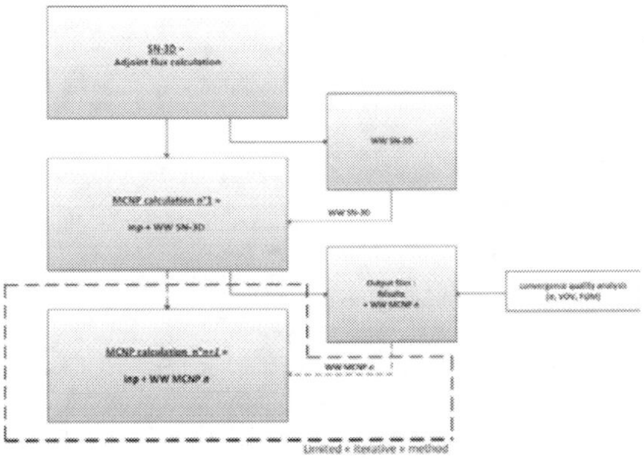

Fig. 3. CADIS method.

It allows to reduce significantly the MCNP iterative calculations.

The comparison is based on the quality of the convergence (estimated relative error (σ), Variance of Variance (VOV) and Figure of Merit (FOM)), on time (computer time + modelling) and on the implement for the engineer.

3 Calculations

3.1 Geometry

The configuration is very simple (see fig. 2): an homogeneous cubic source of caesium-137 (100 cm x 100 cm x 100 cm) faces a pipe crossing a concrete 1m-thick wall.

To discern the importance of the different parameters of the deterministic code (SN order, mesh…), 3 diameters for pipes are considered: 1 cm, 5 cm and 10 cm.

Table 1. Source characteristics.

Iso	E (MeV)	Gamma/s
Cs137	0.662	0.85

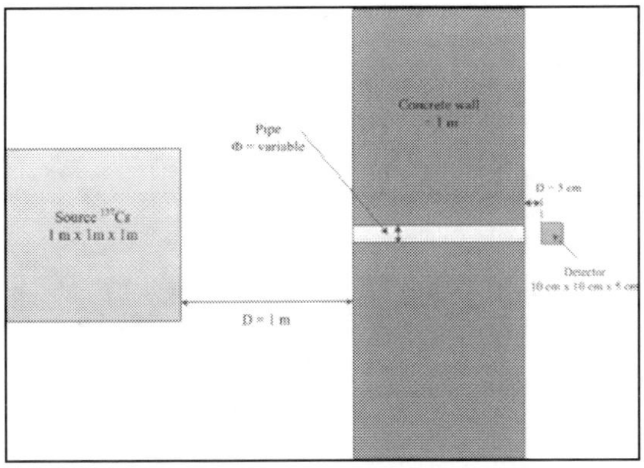

Fig. 4. View of the geometric configuration.

3.2 Codes

Dose Equivalent Rates (DER) due to gamma photons are calculated with the MCNP code, version 4C2 [3]. MCNP is a continuous-energy transport code for simulating nuclear processes based on the Monte Carlo method (stochastic method). It enables the transport of particles (neutrons, photons, electrons…) through the matter and simulates their interactions by means of probability laws. As the method is based on a stochastic method, the results go along with a statistical error. The upper limit criterion of this error is set to 10 %.

The DER is obtained with the "Track length estimate of cell flux", called F4 tally, in a parallelepiped cell (10 cm x 10 cm x 5 cm).

The 3-D discrete ordinates transport code used to generate consistent space- and energy-dependent source and transport (weight windows) is ATTILA 8.0 code [4]. ATTILA 8.0 is a three dimensional (3-D) S_n transport theory code. Attila solves the linearized Boltzmann transport equation in first order form using a tri-linear discontinuous spatial differencing on an arbitrary tetrahedral mesh. Attila comes with a graphical user interface (GUI) which allows direct integration with CAD software programs such as Solidworks®.

Applied flux to dose conversion coefficients are extracted from ICRP 74 [5].

3.3 WW generation

The mesh parameters fixed to realize the WW generation by ATTILA are shown on table 2.

Table 2. Mesh dimensions.

Pipe diameter	Mesh sizing (cm) Wall	Source	Global mesh
1 cm		5	
5 cm	6	6	40
10 cm		7	

Calculations are performed with a triangular Chebychev Legendre quadrature and an Sn order of 24 to 42.
A gamma cross section library based on RADION15 XS library with a specific group structure (34 groups, 11 MeV – 1E-3 MeV) is used.

Unfortunately, we have no choice but to define a uniform WW mesh with ATTILA, while we can optimize it with the WW generator from MCNP (see figure 5).

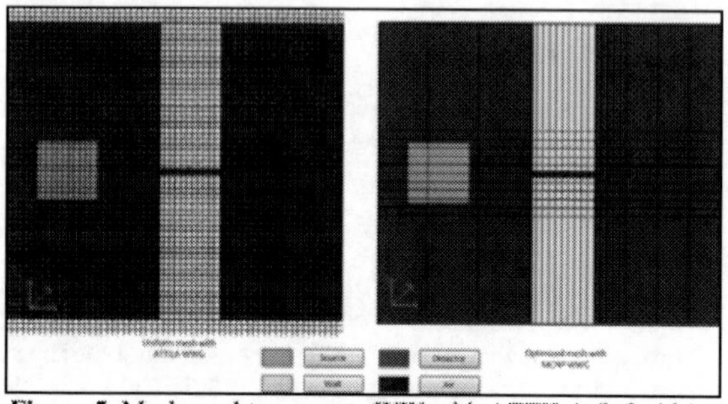

Figure 5. Mesh used to generate WW with ATTILA (left side) and MCNP (right side).

4 Results

4.1 Adjoint flux

One example of the adjoint flux is shown on the figure 6.

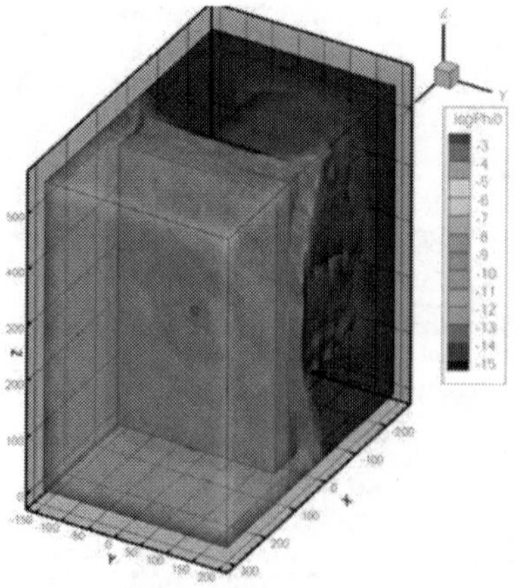

Figure 6. View of the adjoint flux

4.2 WW card

One example of the WW card generated either by ATTILA or by MCNP is shown on the figure 7.

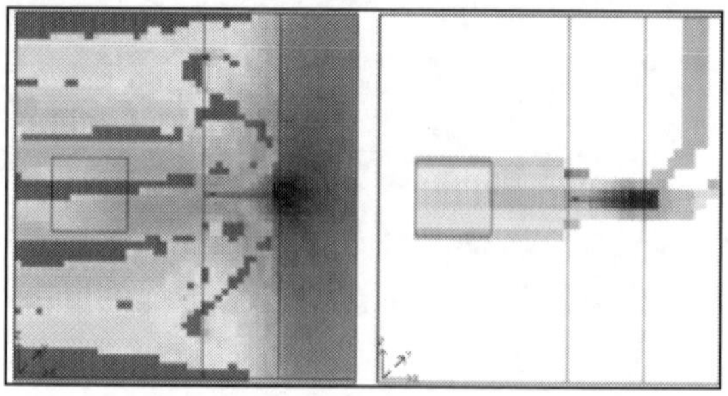

Figure 7. WW card generated by ATTILA (left side) and MCNP (right side)

The particles which go from yellow/green region to dark blue region will be facilitated (and so well sampled) while the ones which go to the orange will be killed rapidly. We can see on the map created by ATTILA some non-well defined regions (in red) due to negative flux.

On the map created by ATTILA, we can see in red some non-well defined regions due to negative flux. If there are too many regions with negative flux, especially around the regions of interest, you will generate a bad weight window and you will have many problems with the MCNP calculation.

4.3 Results - discussion

For rectilinear pipes with large diameter ($\Phi > 10$ cm), the WW generated by ATTILA is not efficient: it consumes time and provides no more information than a simple MCNP calculation with using usual convergence technics.

For rectilinear pipes with smaller diameter ($\Phi \leq 5$ cm), we had to increase the SN order (24 to 42 due to negative flux). The ATTILA's meshing and the normalization are significant on the final MCNP DER's value. It is important to be very careful on the ATTILA calculation to be sure that it generates a good WW card. In any case, we found it necessary to create a new WW card with MCNP to make sure to get a consistent result.

4.4 Future

The ATTILA 9.0 code was released at the beginning of spring. It offers new features like direct MCNP6 calculations. We could also define region specific mesh like in the MCNP weight windows generator. The next step will be to see if our results are more consistent with these new features.

References

1. J.C. Wagner, A. Haghighat, Nucl. Sci. **128**, 186 (1998)
2. J.C. Wagner, D.E. Peplow, S.W. Mosher, Nucl. Sci. **176**, 37 (2014)
3. MONTE CARLO TEAM, ''MCNP—A General Monte Carlo N-Particle Transport Code, Version 4c'' LA -13709-M, Los Alamos National Laboratory (2000)
4. J.M. McGhee, T.A. Wareing and D.A. Barnett Jr, Attila User's Manual, Transpire Inc. (2007)
5. ICRP publication 74, Conversion Coefficients for use in Radiological Protection against External Radiation, Annals of the ICRP 26(3/4) (1996)

Using MCBEND for neutron or gamma-ray deterministic calculations

Dobson Geoff[1,*], *Bird* Adam[1] , *Tollit* Brendan[1] and *Smith* Paul[1]

[1]Amec Foster Wheeler, Kings Point House, Queen Mother Square, Poundbury, Dorchester, Dorset, DT1 3BW, United Kingdom

Abstract. MCBEND 11 is the latest version of the general radiation transport Monte Carlo code from AMEC Foster Wheeler's ANSWERS® Software Service. MCBEND is well established in the UK shielding community for radiation shielding and dosimetry assessments. MCBEND supports a number of acceleration techniques, for example the use of an importance map in conjunction with Splitting/Russian Roulette. MCBEND has a well established automated tool to generate this importance map, commonly referred to as the MAGIC module using a diffusion adjoint solution. This method is fully integrated with the MCBEND geometry and material specification, and can easily be run as part of a normal MCBEND calculation. An often overlooked feature of MCBEND is the ability to use this method for forward scoping calculations, which can be run as a very quick deterministic method. Additionally, the development of the Visual Workshop environment for results display provides new capabilities for the use of the forward calculation as a productivity tool. In this paper, we illustrate the use of the combination of the old and new in order to provide an enhanced analysis capability. We also explore the use of more advanced deterministic methods for scoping calculations used in conjunction with MCBEND, with a view to providing a suite of methods to accompany the main Monte Carlo solver.

1 Introduction

MCBEND 11 [1] is the latest version of the general radiation transport Monte Carlo code from AMEC Foster Wheeler's ANSWERS® Software Service. MCBEND is well established in the UK shielding community for radiation shielding and dosimetry assessments. The MCBEND package comprises the code, a range of nuclear data libraries, user and application guides, and geometry visualisation and results tools provided by VisualWorkshop.

Some Monte Carlo radiation transport codes have a pedigree of several decades, and the MCBEND code falls into this category. As such, there have been many model developments, including some that have stood the test of time and others that have not. In this paper we explore the history of some of these methods, and show how some of the older methods might have new relevance, in particular when mixed with newer features. In particular we look at two strands of development involving deterministic methods to show how this mix of the old and the new might provide additional tools for shielding applications.

The first strand involves the use of an often overlooked feature in MCBEND that can be used to provide a diffusion solution in a matter of seconds, using the same model input as the full Monte Carlo solution. The use of modern tools for processing and displaying results makes this an attractive option for performing scoping calculations. By revisiting the method to make it compatible with more MCBEND features (eg newer source modules) this becomes relevant to the current user.

The second route we explore is an evolution of the way a tetrahedral based geometry description can be utilised in MCBEND. Originally this feature was developed as a method for representing CAD models in MCBEND, but it has now been extended to provide a toolkit to switch between a modern, fully featured, deterministic methodology based on a tetrahedral mesh, and the MCBEND Monte Carlo method. This is illustrated by use with the FETCH2 code developed at Imperial College, London [2].

2 Scoping calculations with MCBEND

The Scoping Tool is an often overlooked feature of MCBEND. This mature feature provides a fast scoping capability based on the diffusion solver used by the MCBEND 'MAGIC' option for generating importance maps. Usually this method is used to accelerate the main MCBEND Monte Carlo solution, and is used routinely for this purpose in adjoint mode. However it can also be run in forward mode to provide scoping calculations.

When used to support acceleration, the importance map generated is used in conjunction with Splitting/Russian Roulette. The source for the adjoint is

* Corresponding author: geoff.dobson@amecfw.com

 DOI: 10.1051/epjconf/201715306032

taken from a response function appropriate to the forward calculation. Part of the attraction of the method for acceleration purposes is the integration of MAGIC with the MCBEND geometry and material specification units; this means the method can be run easily as part of a normal MCBEND calculation.

The method is quick and easy to use, with the MCBEND geometry mapped onto either a Cartesian or a Cylindrical 3D mesh automatically.

This ease of use also extends to the forward mode. The integration of the scoping tool with MCBEND means that results can be generated everywhere in the MCBEND model and displayed in the VisualWorkshop environment. An example results display is shown in the Figure below for a PWR case.

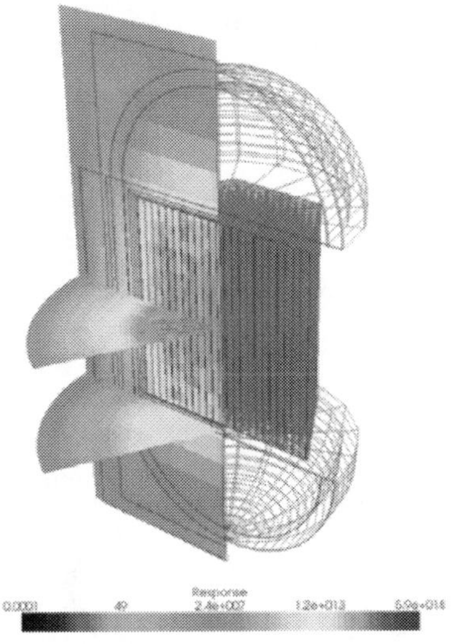

Neutron Dose in PWR, < 5 second run time.

Fig. 1. Visual Workshop image of Dose in a PWR.

2.1 How the scoping tool works

A diffusion calculation is carried out by the MAGIC module, using the splitting mesh and standard 33, 28 or 22 group schemes, depending on the calculation method. Sources and materials defined in the MCBEND input are mapped to the splitting mesh used by the diffusion calculation.

The results of the diffusion calculation are available in the computational mesh for the diffusion solver (the splitting mesh), but there is also an option to provide results in the material mesh, in which case the results are mapped back to the zones that define the material, facilitating easy direct comparison of results with a Monte-Carlo run. If scoring in the splitting mesh is requested the results are available for VisualWorkshop to display.

The two figures above show the display of dose in a flask model and also the display of an isosurface for flux. We return to this flask model in Section 2.3.

Fig. 2. Display of Dose for a flask model.

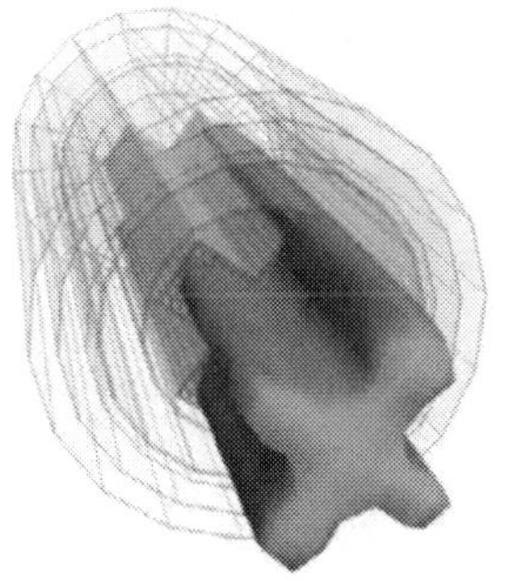

Fig. 3. Isosurface flux for a flask model.

2.2 Historical Restrictions

The released version of MCBEND, MCBEND11A Release Update 1, provides for this scoping style calculation for use only in neutron calculations. It also requires the source to be specified in an older, deprecated, set of modules referred to as "Simple Source". This source module, together with an alternative "Complex Source" module have been largely superseded by a "Unified Source" module in MCBEND, which combines the functionality of the two old modules, and is written in modern FORTRAN. There are still however a few features in MCBEND that require the older source modules, and the scoping tool is one of them. Additionally, the source for scoping calculations was restricted to being a histogram source (as a function of energy), with the source separable in space and energy.

These restrictions have now been relaxed, and development versions of MCBEND now allow scoping calculations to be performed using the Unified Source module, and also no longer require the source description to be separable in energy and space. Currently this is done by a Monte Carlo sampling of the source description. Additionally the restriction to neutrons has been removed, so that gamma calculations can also be performed with the scoping tool.

2.3 A case study in the use of the scoping tool

In this example, the scoping tool is illustrated by use of a transport flask case study. This example models UO2 fuel rods in a flask, partially flooded. It consists of a set of five fuel baskets in a partially-flooded flask as shown below.

Fig. 4. Cross-section of a flask model.

Initially the splitting mesh from the calculation of the importance map in the original case was used. This was a simple RZ splitting mesh, with no azimuthal discretisation. A few boundaries were added in the Z dimension to coincide with the plane surfaces of external scoring regions. However, the simple radial subdivision within the flask, with no angular subdivision, effectively homogenized the materials within this region, causing the doses in the flask components to be underpredicted by a factor of two or three.

An improved mesh was then specified in the R and θ dimensions, which is illustrated below.

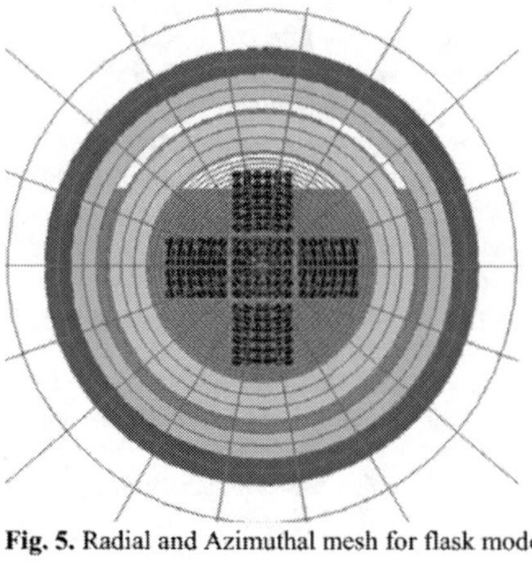

Fig. 5. Radial and Azimuthal mesh for flask model.

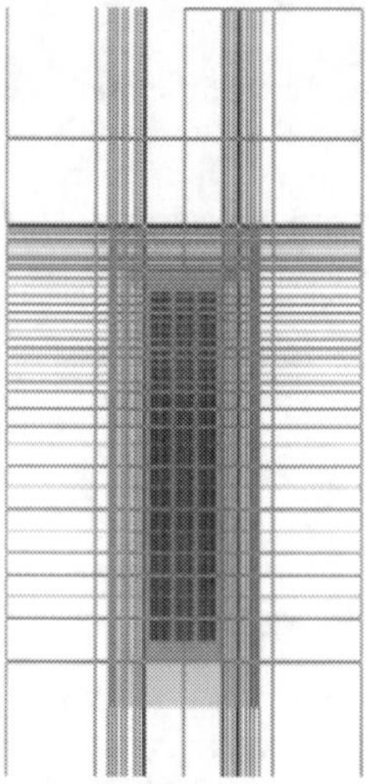

Fig. 6. Axial mesh for flask model.

The revised results, give reasonable agreement between the diffusion and Monte Carlo calculations for the flask components. However, the doses at the lid and 1m above the lid were grossly exaggerated. It was judged that this problem was due to a pathological effect – the diffusion equations predict that scattering can occur in a void, permitting neutrons to reach the lid after they have passed through the (thinner) flask wall.

The problem was solved by specifying black absorber round the side of the flask and beyond a radius of 1m, as shown below.

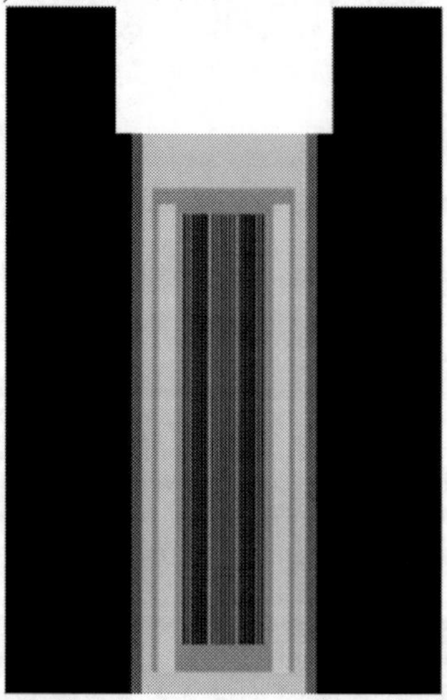

Fig. 7. Absorber boundary conditions for flask model.

In a range of neutron case calculations it has proved possible to achieve an accuracy of about 20 or 30% in most scoring regions, including at the lid and 1m above the lid. However it can be important to understand the limitations of diffusion solutions, as illustrated by this example. In this case the pathological effects, due to neutrons from the side being scattered to the lid could be avoided through the use of suitably positioned black absorbers.

The automatic nature of the scoping tool, with minimal user intervention, coupled with the availability of graphical output processing and graphical representations of the geometry make it a very flexible scoping tool. This is further improved by upgrades to provide compatibility with the most up-to-date modules in MCBEND (including the Unified Source module) and a relaxation of historical restrictions on its use.

3 Tetrahedral mesh capabilities

MCBEND has the capability to track in a tetrahedral mesh. This capability was originally developed over ten years ago in order to allow import of a CAD mesh. CAD import for ANSWERS codes has subsequently been developed further in a number of ways, see for example [3], but the tetrahedral tracking capability has opened up a route to partnering MCBEND with state-of-the-art deterministic codes, such as Imperial College's FETCH2 code which is currently being evaluated for possible use with MCBEND.

MCBEND provides two implementations of a tetrahedral mesh, the first uses Woodcock tracking in Hole geometries to perform the Monte Carlo calculation. Woodcock tracking is an alternative to conventional tracking and avoids the need to calculate the distance to surfaces by instead using the minimum mean free path of all the materials in the region of interest, and using this to sample for the distance between collisions. To compensate for the modification to the mean free path, the idea of pseudo-collisions is introduced in which a sampling process takes place following a collision to decide whether the collision is a real collision, or a "do nothing" pseudo collision. The sampling process for real collisions is based on the ratio of the minimum mean free path to the actual mean free path. This form of tracking provides an exact representation of the transport equation.

The second method is explicit (conventional) tracking to the tetrahedral surfaces. This is still fast as only plane surfaces are involved. One potential advantage of the explicit tracking approach is that it facilitates the use of track length scoring in the tetrahedral mesh. Although we now have an implementation of this form of scoring, it may not be such an advantage in practice because it is not necessary for the scoring mesh to be based on the tracking mesh. In MCBEND we have a "Unified Tally" module where scoring bodies can be defined independently of the geometry that defines the materials, with tracking to scoring body surfaces only being necessary at a change of direction of the particle. This may be more useful than scoring in the tetrahedral mesh in practice.

Although originally intended for modelling CAD geometries in MCBEND, the application we are concerned with here is the potential for sharing a mesh/geometry with deterministic codes, including state-of-the-art methodologies. This is explored below using the FETCH2 code.

3.1 FETCH2

FETCH2 is a successor to Imperial College's EVENT code which has been widely used for many years for radiation transport calculations. It is a general purpose finite element neutral particle transport solver for neutrons and photons in critical and sub-critical systems. It is written to be highly parallelisable and scales well up to many thousands of processors.

FETCH2 is built around a novel finite element method which combines the advantages of continuous and discontinuous finite element methods. The solver is based on matrix-vector multiplications which ensures that it is future proofed against advances in computer technology, such as use on GPGPUs and Xeon Phi coprocessors. FETCH2 uses domain decomposition to distribute the spatial mesh over multi-processor machines together with a space-angle multigrid scheme to increase the efficiency of the solvers.

FETCH2 can solve the transport equation using the following angular discretisation methods:

- Spherical Harmonics (P_n)
- Discrete Ordinates (S_n)
- Linear Octahedral Wavelets
- Haar Wavelets

making it suitable for shielding, criticality and reactor physics applications.

To reduce the burden on the user, FETCH2 has the unique ability to anisotropically adapt both the spatial and angular resolutions, either globally or to some goal (such as a detector response).

To illustrate the use of FETCH2 in conjunction with MCBEND, we apply the codes to a waste store model, which can be represented in MCBEND either by using the usual simple body representation, or using a tetrahedral representation.

3.2 Waste Store Model

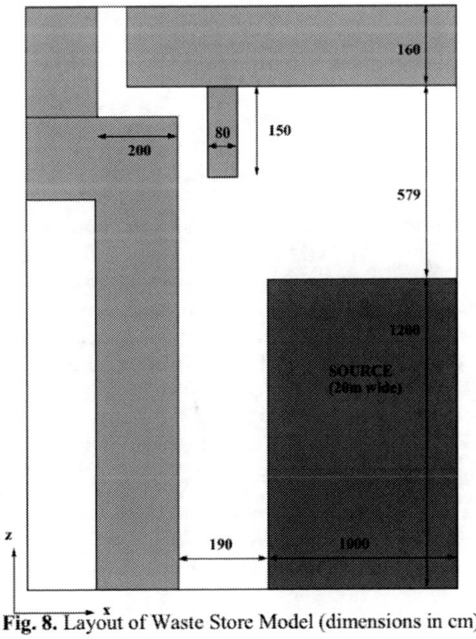

Fig. 8. Layout of Waste Store Model (dimensions in cm).

In order to test the capabilities for sharing a common tetrahedral mesh between FETCH2 and MCBEND, the waste store model shown above was used as a representative shielding problem.

Here the material in green is concrete, and the duct has dimensions 40 cm in the x direction, 60 cm in the y direction. We are interested in the gamma-ray dose-rate both at the exit of the duct, but also in the variation along the duct. The dominant contribution to the dose-rate arises from particles which enter the mouth of the duct and scatter at the corners before reaching the duct exit.

The first FETCH2 model used for the waste store example had a fixed 3D mesh (ie with no mesh adaptivity). The open source GMSH tool [4] was used to generate the mesh which incorporated approximately 423000 elements, with the tetrahedral element size mostly of order 65 cm, but refined to about 5 cm in the vicinity of the duct. Eleven gamma groups were used from the Bugle96 library. The methods used were linear continuous plus piecewise constant Subgrid Scale. The scatter model was Spherical Harmonics anisotropic scatter (using P_0 up to P_5).

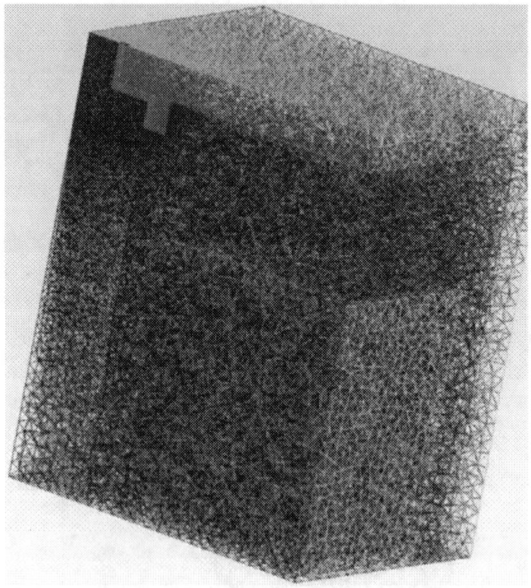

Fig. 9. Tetrahedral mesh for Waste Store.

Matrix free spatial multigrid solvers were applied using PETSc [5] FGMRES smoothing. For the 3D model, the calculations were run on 64 parallel processors. Doses were scored at duct centre points (linear interpolation within bounding element). Doses were calculated using ICRP-51 (AP and LAT). The angular discretisation was Equal Weight S_n quadrature. A comparison of the FETCH2 calculations with MCBEND using the tetrahedral mesh and explicit tracking is given in the table below.

Table 1. Comparison of FETCH2 and MCBEND Gamma Dose Results (micro Sv/hr) for Waste Store model.

Model	Entrance	First Corner	Second Corner	Exit
MCBEND	7.245E-02 (0.6%)	2.355E-03 (1.8%)	3.017E-06 (5.3%)	4.393E-09 (12.5%)
FETCH2 (S10 AP Dose)				
P_0 scatter	8.548E-02	5.406E-03	5.704E-06	2.398E-08
P_1 scatter	8.379E-02	5.764E-03	6.493E-06	2.373E-08
P_3 scatter	7.679E-02	5.426E-03	5.369E-06	1.112E-08
P_5 scatter	7.369E-02	5.181E-03	7.165E-06	1.553E-08

For the P_5 case, the FETCH2 calculations are generally within a factor two of the MCBEND calculations, and so FETCH2 is in reasonable agreement with MCBEND. The MCBEND calculations used the Forced Flight acceleration method in addition to other acceleration methods. Forced Flight causes a particle to be transported directly to a defined geometrical interface at every collision, with the particle weight suitably adjusted to avoid any bias. A feature of forced flight that was new in MCBEND 11 was the ability to have multiple forced flight interfaces acting in series, and this facility was used here.

There are still some known model differences between the FETCH2 calculations and the MCBEND calculations in the area of source definition, and nuclear data. Additionally there is ongoing analysis to assess a range of methods available in FETCH2, including:

- Wavelets for angular discretisation
- The available Multi-Grid Solvers
- The Adaptive resolution capability

FETCH2's powerful adaptive resolution capability is briefly illustrated below, as it incorporates features of interest in their own right.

Two approaches to adaptive resolution are available in FETCH2: Regular adaptivity which aims to reduce error everywhere; and Goal based adaptivity which aims to reduce the error for a specified goal. The Goal based method relies on solving the forward and adjoint transport equation – the latter providing the "importance" similar to MAGIC.

The adaptive resolution is fully anisotropic (anisotropic in space and angle). However the illustration below uses spatial adaptivity alone for a two dimensional slice of the waste store model, using:

- S_4 angular quadrature
- Solution adapted 10 times
- Two adaptivity tolerances

The modelling choices allow fast turnaround of calculations with modest computing resources.

The Figure below illustrates the use of regular spatial adaptivity. The plot on the right has the finer adaptive tolerance. The figure illustrates:

- Resolution highlighting ray effects (S4)
- Little resolution in the duct (flux varies from order 10 to 10E-9 making defining a regular adaptive tolerance challenging)
- Decreasing the tolerance begins to add more resolution to the duct as well as substantially more resolution elsewhere.

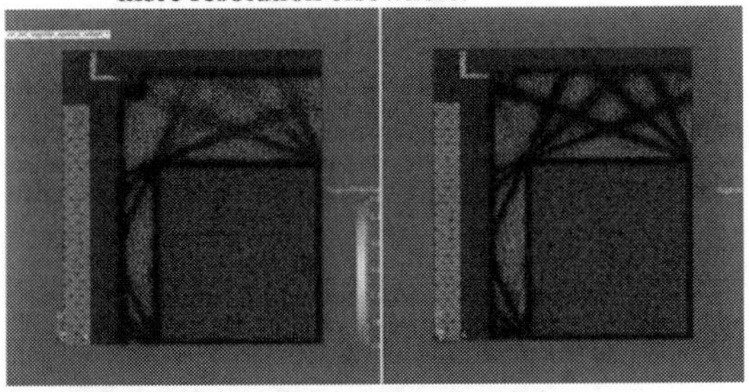

Fig. 10. FETCH2 Spatial Adaptivity (Regular).

The next figure now uses Goal based spatial adaptivity, with the goal set as average flux in a thin region near the duct exit. The goal based adaptivity clearly out-performs the regular adaptivity in this case, in the sense of achieving much more resolution in and around the duct, and also in the dominant pathway to the duct.

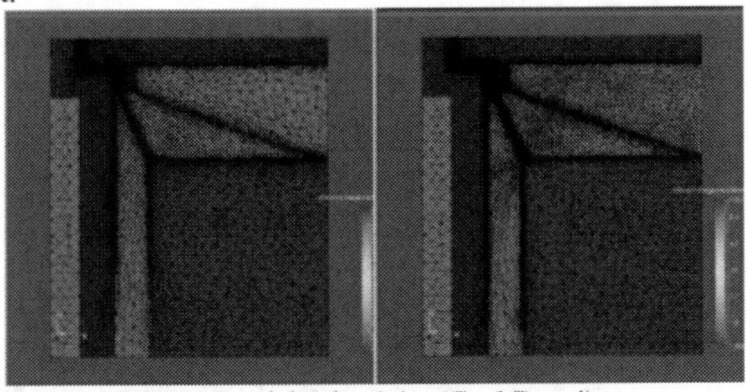

Fig. 11. FETCH2 Spatial Adaptivity (Goal Based).

The figure below shows a close up of the duct region, where the effects of the spatial adaptivity can be seen more clearly. The darker regions where the mesh is of finer resolution will likely be a mix of ray effects and real physical effects.

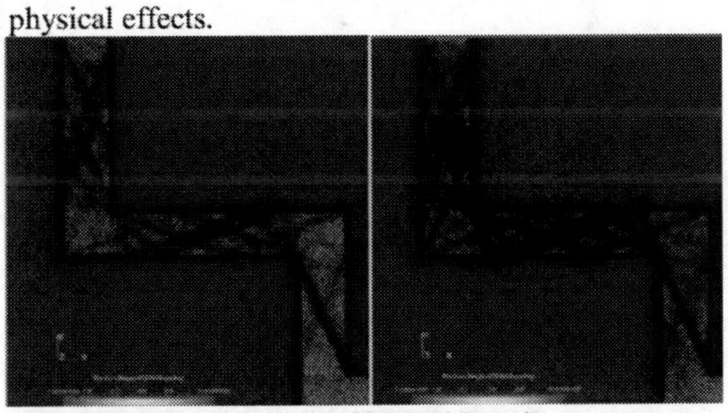

Fig. 12. FETCH2 Spatial Adaptivity (Goal Based).

4 Conclusions

This paper describes how a number of features in MCBEND of various vintages have been combined to allow additional tools for shielding applications. The adjoint diffusion capability used for generating an importance map for the main Monte Carlo calculation also provides for a very quick running forward scoping tool for neutron calculations. This has now been extended to include gamma calculations, and also to be compatible with the newer source module, Unified Source, which was released with MCBEND 10. Additionally the improved facilities for results display in Visual Workshop, where results can be overlaid on the geometry gives this old feature a new lease of life which can allow calculations and results to be displayed in seconds.

At the other end of the deterministic scale is the ability to track and score on a tetrahedral mesh, which allows MCBEND to partner state-of-the-art deterministic codes if these codes use a tetrahedral spatial discretisation. One such code is the FETCH2 code developed at Imperial College, London. This makes it possible to run MCBEND calculations on the same mesh as is used for the deterministic calculation, with benefits for porting models between codes. Although MCBEND can score in the tetrahedral mesh, it can also score in bodies dedicated to scoring, such as subdivided cylinders and cuboids, using the latest scoring module, Unified Tally, which was released with MCBEND 11. This module can be used in conjunction with a model based on a tetrahedral mesh for tracking, providing benefits for sharing model between codes.

The dusting off and extension of older methods, combined with some more recent capabilities has allowed the old and the new to come together to create new tools for the shielding practitioner, and illustrates a benefit of keeping older methods current, and also as testing methods advance to much more automation, there is a benefit to ensuring that older features are comprehensively covered in any test suite as far as practical.

The authors gratefully acknowledge the work done by David Picton on which much of the description of the scoping tool was based. Thanks also to members of the Imperial College AMCG Department: Christopher Pain, Steven Dargaville, Andrew Buchan and Richard Smedley-Stevenson.

References

1. P. Cowan, G. Dobson, J. Martin, *Release of MCBEND11*, ICRS12-RPSD2012, Nara, Japan.
2. S. Dargaville, A Buchan, R Smedley-Stevenson, P Smith, C Pain, *Adaptive angle and parallel multigrid for deterministic shielding*, ICRS13-RPSD2016, October 2016, Paris, France.
3. A. J. Bird, C Murphy, G. P. Dobson, *Release of RANKERN16A*, ICRS13-RPSD2016, October 2016, Paris, France.
4. C. Geuzaine and J.-F. Remacle. *Gmsh: a three-dimensional finite element mesh generator with built-in pre- and post-processing facilities.* International Journal for Numerical Methods in Engineering 79(11), pp. 1309-1331, 2009.
5. Satish Balay et al, PETSc Web page, https://www.mcs.anl.gov/petsc, 2016

EPJ Web of Conferences **153**, 06033 (2017) DOI: 10.1051/epjconf/201715306033
ICRS-13 & RPSD-2016

Monte Carlo simulation of particle-induced bit upsets

Frédéric Wrobel[1,*], Antoine Touboul[1], Jean-Roch Vaillé[1,2], Jérôme Boch[1], and Frédéric Saigné[1]

[1]*Université de Montpellier, Institut d'Electronique et des Systèmes, UMR-CNRS 5214, France*

[2]*Université de Nîmes, France*

Abstract. We investigate the issue of radiation-induced failures in electronic devices by developing a Monte Carlo tool called MC-Oracle. It is able to transport the particles in device, to calculate the energy deposited in the sensitive region of the device and to calculate the transient current induced by the primary particle and the secondary particles produced during nuclear reactions. We compare our simulation results with SRAM experiments irradiated with neutrons, protons and ions. The agreement is very good and shows that it is possible to predict the soft error rate (SER) for a given device in a given environment.

1 Introduction

Radiation-induced failures in electronic devices represent a major concern for microelectronic reliability, especially for aerospace applications [1-3]. In satellites, devices are exposed to protons, electrons and heavy ions leading to malfunctions. In the following, we will consider a specific kind of malfunction that is called single event effect (SEE), meaning that a unique particle is able to trigger a failure. The cosmic particles that are at play are protons and heavy ions. Moreover, cosmic rays are able to interact with molecule nucleus of the atmosphere and produce secondary particles whose flux depends on the altitude. Fig. 1 gives the typical particle flux calculated in the atmosphere as a function of altitude with QARM [4]. Even if different kinds of particles exist, the main contribution to the SEE rate in atmosphere is actually attributed to neutrons.

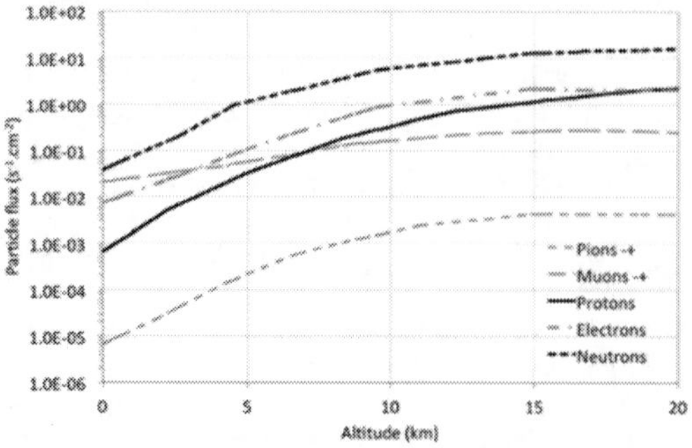

Figure 1. Particles fluxes in the atmosphere as a function of altitude (simulated with QARM).

* Corresponding author: frederic.wrobel@umontpellier.fr

One major issue for memory devices is when a unique particle is able to upset a bit from its value to its opposite. These Single Event Upsets (SEU) can occur in space and also in atmosphere and it is useful to be able to predict the number of failures induced in a given memory at a given location (i.e. as a function of altitude and latitude). To do so, we developed the Monte Carlo code MC-ORACLE. This code simulates the passage of particles in a complex structure composed of multiple volumes with various materials (silicon, silicon dioxide, metals…). Secondary ions produced during nuclear reactions are treated in the energy range of 1MeV-200MeV, which is representative of the atmospheric spectrum. The ionizing particles are tracked in the structure and the released energy is determined in the region of the device that is known to be sensitive. Using an SEU criterion it is then possible to determine whether an error occurs or not. Bit flip cross sections are determined as a function of incident energy or as a function of ion stopping power. The Soft Error Rate can also be calculated for a specific location, which is defined by the altitude and the latitude in atmosphere or for a given mission in Space.

2 The Monte Carlo approach

We developed the MC-Oracle code, for which a first simplified version has been presented in [5]. MC-Oracle is a Monte Carlo tool that allows investigating Single Event Upsets (SEU), Multiple Cell Upsets (MCU), and Single Event Transients (SET). The bit flip cross sections as well as the soft error rate (SER) are the main quantities that are provided by the code.

Fig. 2 represents the flowchart of the code, which

shows three main parts: the structure definition, the particle-structure interaction and the effect on the electronic device.

First, we need to define the geometry and the materials that compose the device. We also need to give the location of the electrodes transistors, as they will be able to collect the charge deposited in the structure by the ionizing particles. Moreover, it is necessary to define the kind of particles that interact with the device (neutrons from atmosphere, or protons or ions from space).

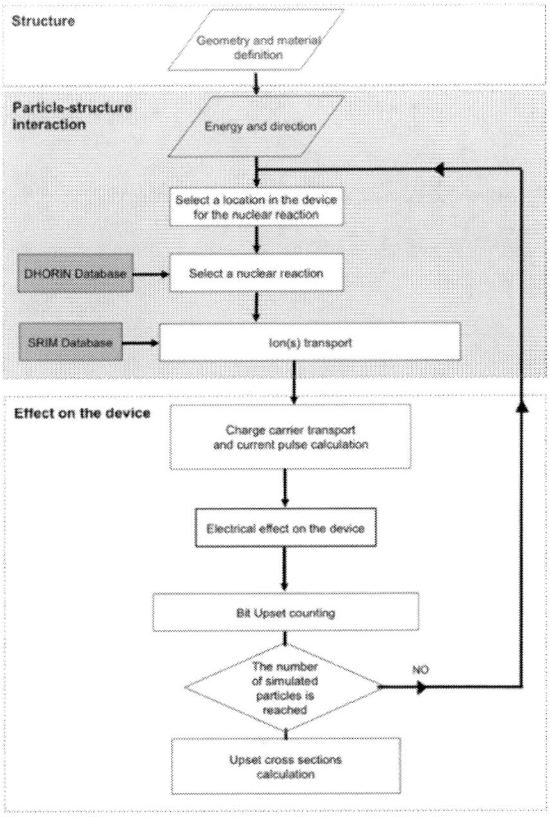

Figure 2. Flowchart of the Monte Carlo tool (MC-Oracle)

Once the definition of the simulation is done, MC-Oracle starts the simulation with a random choice of an incident particle, its position, energy and direction. Obviously, this is done with respect to the distributions of energy and direction. The initial position is selected outside the device. For incident ions only ionization is accounted for as it has already been established that nuclear reactions induced by ions have a very low contribution to upset bit rate.

For incident neutron and proton, nuclear reactions have been pre-calculated with the DHORIN code [6]. This latest uses the model of 2D-exciton model for pre-equilibrium step, the formulation of Hauser-Feshbach and the generalized evaporation model. Elastic reactions are treated with the ECIS code that uses spherical optical models from the RIPL2 database. The DHORIN database is composed of several monoenergetic simulation files that contain millions of nuclear reactions details about the secondary ions (nature, energy and direction). These ionizing particles are transported within the device and the deposition of energy is calculated in the sensitive regions using the SRIM database made of the stopping power of various ions as a function of energy [7].

Next, the effect of the energy deposition in the device is treated. To do so, the tool divides each ion track into small fragments that spherically diffuse their charge carriers and can be collected by the drain electrode that is itself divided into elemental sections. Then, integrating the charge diffusion along the whole track and over the whole drain surface gives the current pulse that is generated by the nuclear reaction at the electrode of each transistor. A basic electrical model of the memory cell is finally used to state whether the calculated transient current is able to trigger a cell upset or not.

In our simple study, the 3D-simulated device structure is a bulk made of a silicon layer with a surface of 30 μm x 30 μm and a thickness of 20 μm. Above this bulk, we added a layer of 10 μm of silicon dioxide to account for the Back End Of Line (BEOL). At the Si/SiO_2 interface, we simulate 20 x 20 SRAM cells, each bit being modeled by 4 transistors (we did not account for the pass gate transistors that are less sensitive). For each cell we have two OFF transistors (one NMOS and one PMOS), which are likely to collect the charge.

The SEU cross section is finally evaluated by calculating the ratio of the number of bit flips and the simulated particle fluence. In the case of non-monoenergetic incident particles, the spectrum can be provided to the code. In that case, it is more convenient to calculate the soft error rate (SER), which directly account for the real environment.

3 Results

In this section we present some results obtained with MC-ORACLE for devices irradiated with heavy ion, proton and neutron. Figure 3 shows the comparison between the Monte Carlo approach and the experimental results for a 65nm SRAM irradiated with heavy ions [8]. The calculated Single Event Bit upset cross section is in good agreement with experimental results above 0.5 $MeV.cm^2/mg$ which shows that the modeling is efficient. At lower LET a singular peak is observed and is predicted by the Monte Carlo tool, though the order of magnitude is not correct. This peak corresponds to irradiation with protons for which the ionization is efficient to trigger SEU very near the Bragg's peak.

 DOI: 10.1051/epjconf/201715306033

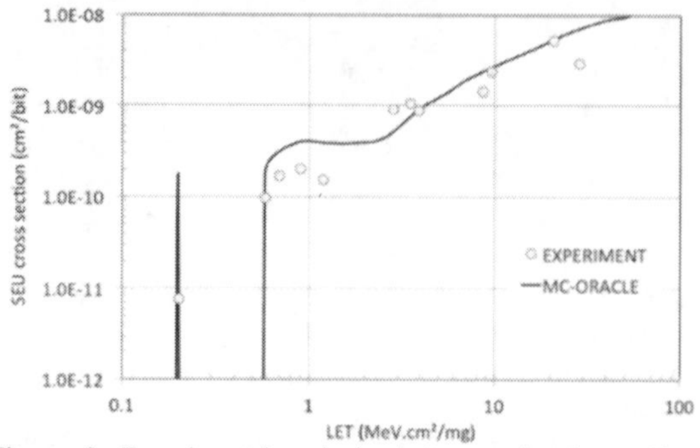

Figure 3. Experimental and simulation results for a 65nm SRAM irradiated with ions.

Figure 4 shows the comparison between the Monte Carlo approach and the experimental results for a 90nm SRAM irradiated with protons [9] and as a function of proton energy. The calculated cross section is in good agreement with the experimental data. In figure 5, we plotted the results obtained with the 90nm SRAM irradiated with neutrons. Neutrons are here not monoenergetic and the energy distribution has actually the same shape than the atmospheric spectrum. The results are then given in FIT/Mbit, one Failure In Time being the number of failures during one billion of hours. The soft error rate is calculated for different supply voltage V_{DD} that are characteristic of the normal functioning of the device. Here again the simulation results are in good agreement with experimental data.

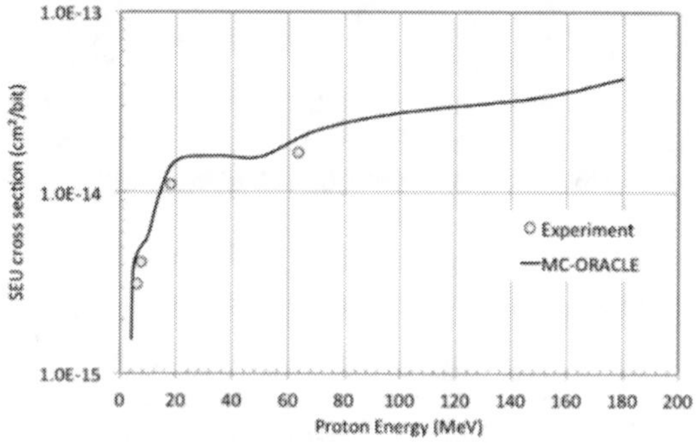

Figure 4. Experimental and simulation results for a 90nm SRAM irradiated with protons.

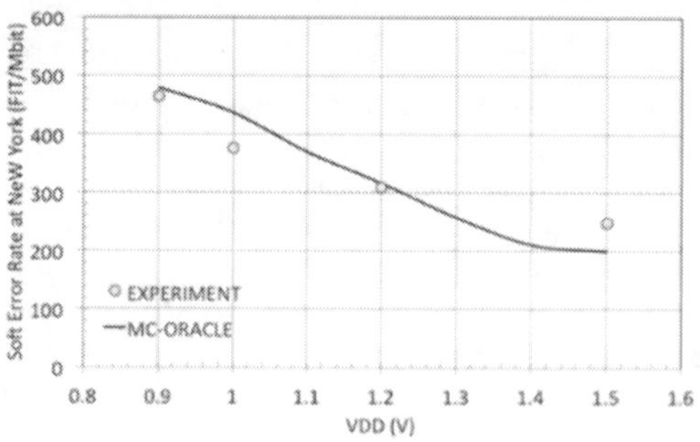

Figure 5. Experimental and simulation results for a 90nm SRAM irradiated with neutrons for various voltages.

4 Summary and conclusions

In order to address the issue of radiation-induced failures in electronic devices we developed the MC-Oracle tool. This code is able to transport the particles within the device and, based on the energy deposition, is able to calculate the transient current induced by the primary particle and the secondary particles produced during nuclear reactions.

We show that our approach is consistent by comparing SRAM experiments under neutron, proton and ion with our simulations results. The agreement is very good and shows that it is possible to predict the soft error rate (SER) for a given device in a given environment.

References

1. R. Baumann, "Soft errors in advanced computer systems," IEEE Trans. Device Mater. Rel., vol. 22, no. 3, pp. 258–266, May–Jun. 2005.
2. T. Karnik and P. Hazucha, "Characterization of soft errors caused by single event upsets in CMOS processes," IEEE Trans. Depend. Secure Comput., vol. 1, no. 2, pp. 128–143, Apr.–Jun. 2004.
3. R. D. Schrimpf, K. M. Warren, R. A. Weller, R. A. Reed, L. W. Massengill, M. L. Alles, D. M. Fleetwood, X. J. Zhou, L. Tsetseris, and S.T. Pantelides, "Reliability and radiation effects in IC technologies," in Proc. IEEE Int. Reliability Physics Symp., Apr. 27–May 1 2008, pp. 97–106.
4. F.Lei,A.Hands,S.Clucas,C.Dyer,andP.Truscott,"Improvemen tsto and validations of the QinetiQ atmospheric radiation model (QARM)," in Proc. 8th Eur. Conf. Radiation and Its Effects on Components and Systems, 2005, pp. D3-1–D3-8.
5. F. Wrobel, F. Saigné, « MC-ORACLE: A tool for predicting Soft Error Rate », Computer Physics Communications 182 (2011), 317-321
6. F. Wrobel, "Detailed history of recoiling ions induced by nucleons," Computer Physics Communications, vol. 178, no. 2, pp. 88–104, Jan. 2008.
7. Available: http://www.srim.org/SRIM
8. B. D. Sierawski, K. M. Warren, R. A. Reed, R. A. Weller, M. M. Mendenhall, R. D. Schrimpf, R. C. Baumann, and V. Zhu, "Contri- bution of low-energy (MeV) neutrons to upset rate in a 65 nm SRAM," in Proc. IEEE Reliability Physics Symp., 2010, pp. 395–399.
9. G. M. Swift, G. R. Allen, C. W. Tseng, C. Carmichael, G. Miller, and J. S. George, "Static upset characteristics of the 90 nm virtex-4QV FPGAs," in Proc. IEEE Radiation Effects Data Workshop, 2008, pp. 98–105.

Reduced Variance using ADVANTG in Monte Carlo Calculations of Dose Coefficients to Stylized Phantoms

Mauritius Hiller[1,a], Michael Bellamy[1], Keith Eckerman[1], Nolan Hertel[1,2]

[1]Oak Ridge National Laboratory, Center for Radiation Protection Knowledge, P.O. Box 2008 MS 6335, Oak Ridge, TN 37831-6335
[2]Georgia Institute of Technology, 770 State Street, Atlanta, GA 30332-0745

Abstract. The estimation of dose coefficients of external radiation sources to the organs in phantoms becomes increasingly difficult for lower photon source energies. This study focus on the estimation of photon emitters around the phantom. The computer time needed to calculate a result within a certain precision can be lowered by several orders of magnitude using ADVANTG compared to a standard run. Using ADVANTG which employs the DENOVO adjoint calculation package enables the user to create a fully populated set of weight windows and source biasing instructions for an MCNP calculation.

1 Introduction

The calculation of dose coefficients to human phantoms for Monte Carlo calculations sets a huge demand on available computing power. Efforts that help reduce the variance of the calculation and increase the efficiency greatly facilitate the solution of such problems.

While the applicability of standard variance reduction methods implemented in MCNP is limited and, especially at low energies, the time and effort required to create appropriate weight windows is excessive. However, the ADVANTG (1) code can create weight windows in a short time frame.

In this paper, an example the use of the ADVANTAG code to increase both the speed and efficiency of an MCNP6 computation of organ doses for a stylized phantom fully emerged in contaminated water is presented.

2 Methods

The calculations shown herein focus on the determination of the dose coefficient for the adrenals, a small organ, inside the body that presents a particular challenge to generate statistically acceptable uncertainties when it is used as a tally cell. Computations were performed and presented for photons in an energy range from 10 keV to 500 keV. The calculations were terminated at an error of 3 % in the adrenals.

The phantom used for the calculations is the ORNL stylized adult phantom.

The calculations were performed on an Intel core i7 CPU with 4 cores. MCNP 6.1.1 (2) was used with the OMP multi core support using the "task" setting.

2.1 Situation Prior to Optimization

The stylized phantom is fully immersed in water. The water is in a cubical box with 4 m side length, whereas the adult male is about 80 cm wide and 180 high and has been centred in the box. All water is considered contaminated with photon emitters so the phantom is irradiated from all sides. The simulated photon energy range is from 10 – 500 keV.

2.2 ADVANTG Setup

The ADVANTG code (Mosher et al. 2015) which creates a weight window input file for MCNP using the Denovo code was used in this problem. Denovo uses a multigroup adjoint calculation to determine the weight windows. This makes it possible to create completely assigned weight windows even for a fine mesh at low energies in a short time. Figure 1 shows a set of weight windows generated by ADVANTG, were regions of highest interest are shown in blue while regions of least interest are shown in red.

Besides the weight windows, ADVANTG also provides a source bias setting.

An MCNP input file was produced with ADVANTAG determined weight windows. Then the run of the optimized input file was stopped when an error of 3 % in the adrenals was achieved.

3 Results

The CPU time needed to run ADVANTG to create the modified MCNP input file and the CPU time then needed to run the MCNP file were summed and are compared to the results from MCNP runs which did not use any variance reduction techniques in Table 1.

[a] Corresponding author: hillermm@ornl.gov

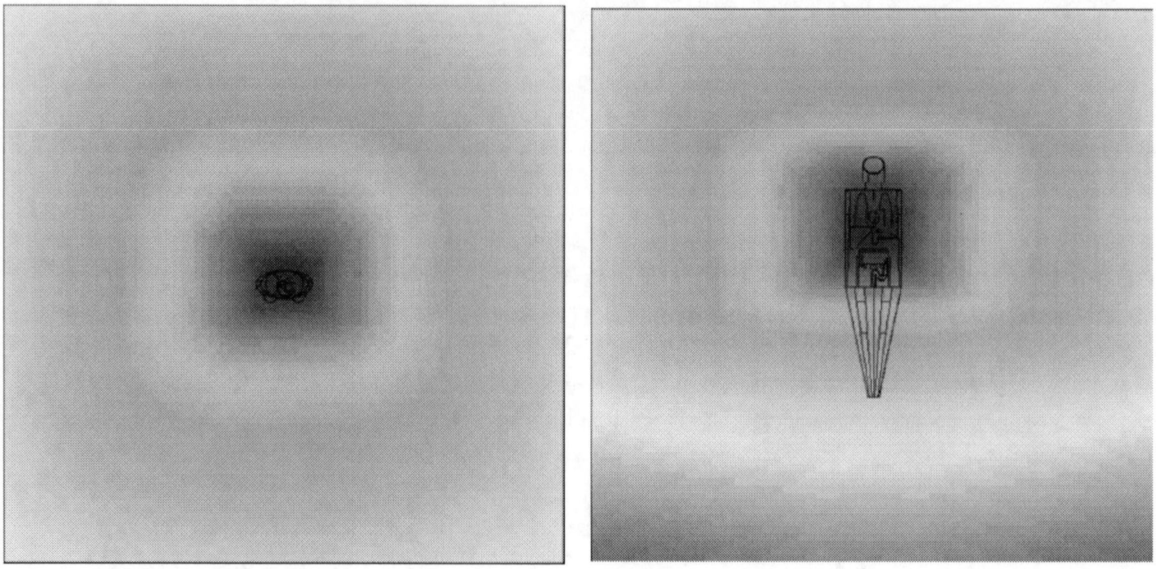

Figure 1: Front and top view of the phantom with the weight windows assigned by ADVANTG. Areas of high importance are shown in blue areas of low importance in red.

Figure 2 shows the run time of the unmodified run against the run time of the ADVANTG code plus the modified run. The time gained using ADVANTG can be seen as the distance from the top of the bar of the ADVANTG run to the line of the unmodified runs.

It can be see seen that ADVANTG helps to significantly speed up the runs by for lower energies. The time needed to run the Denovo code is relatively constant with the energy. Therefore, the effect on the speed of the computation of ADVANTG is less towards higher energies, in fact the direct runs are faster for energies above 100 keV.

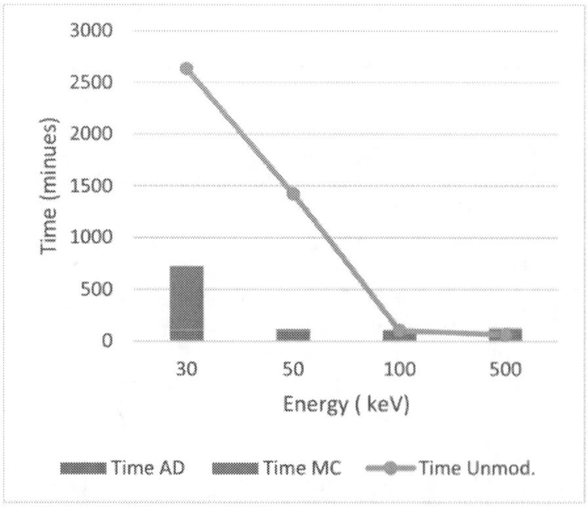

Figure 2: Run-times of unmodified MCNP runs against ADVANTG plus modified MCNP runs.

4 Discussion

The reduced efficiency towards higher energy calculations can be explained with the time needed to create the optimizations in ADVANTG. For higher energies, the runs without optimization finish in a time, shorter than the ADVANTG run, thus it becomes unfavourable to optimize the problem with ADVANTG prior to its MCNP run. Photons with energies above 100 keV more easily penetrate the phantom and reach the inner organs.

References

1. W. Mosher, S. R. Johnson, A. M. Bevill, A. M. Ibrahim, C. R. Daily, T. M. Evans, J. C. Wagner, J. O. Johnson, R. E. Grove. ADVANTG - An Automated Variance Reduction Parameter Generator. ORNL/TM-2013/416 Rev. 1, Oak Ridge National Laboratory, Oak Ridge, TN, 2015.
2. D. B. Pelowitz, A. J. Fallgren, G. E. McMath. MCNP6 User's Manual, Code Version 6.1.1beta. LA-CP-14-00745, Rev. 0, 2014.

Table 1: Run-times for Monte Carlo runs with and without prior optimization in ADVANTG.

	Unmodified		ADVANTG			Unmodified / ADVANTG	
E (keV)	Time MC (min)	Mean (MeV/g)	Time AD (min)	Time MC (min)	Mean (MeV/g)	Time	Mean
500	65.9	3.947E-09	120	8	3.92E-09	0.51	1.01
100	105.0	5.543E-10	105	4	5.67E-10	0.96	0.98
50	1425.4	1.44E-10	101	18	1.48E-10	12.02	0.97
30	2633.3	2.15E-11	108	617	2.15E-11	3.63	1.00
10	No convergence in reasonable time frame of several days of calculation						

Measurements and analyses of the distribution of the radioactivity induced by the secondary neutrons produced by 17-MeV protons in compact cyclotron facility

Norihiro Matsuda[1,*], *Yuichi* Izumi[2], *Yoshiyuki* Yamanaka[2], *Toshiyuki* Gandou[2], *Masaaki* Yamada[2], and *Koji* Oishi[2]

[1]Japan Atomic Energy Agency, 2-4 Shirakata, Tokai-mura, Naka-gun, Ibaraki, Japan
[2]Japan Environment Research Co., Ltd, 6-24-1 Nishishinjyuku, Shinjyuku, Tokyo, Japan

Abstract. Measurements of reaction rates by secondary neutrons produced from beam losses by 17-MeV protons are conducted at a compact cyclotron facility with the foil activation method. The experimentally obtained distribution of the reaction rates of ^{197}Au (n, γ) ^{198}Au on the concrete walls suggests that a target and an electrostatic deflector as machine components for beam extraction of the compact cyclotron are principal beam loss points. The measurements are compared with calculations by the Monte Carlo code: PHITS. The calculated results based on the beam losses are good agreements with the measured ones within 21%. In this compact cyclotron facility, exponential attenuations with the distance from the electrostatic deflector in the distributions of the measured reaction rates were observed, which was looser than that by the inverse square of distance.

1 Introduction

Recently, high-energy electron linear accelerators for external beam therapy (EBT) for tumor treatment and compact cyclotrons to produce radiotracers for the positron emission tomography (PET) have been increasing in the medical field. There are over 1,200 accelerators in Japan as of March, 2015 according to statistics [1]. Lifetime of the accelerators is expected about 20 years [2]. There will surely come a day to replace the existing accelerator with newer models or decommission the facility itself. When the accelerator is replaced or the facility is decommissioned, we need to evaluate the total amount of radioactivity (inventory) induced by the day-to-day accelerator operations. Wastes arising during the replacing and decommissioning process aren't always radioactive wastes because clearance criteria have been introduced in regulations of Japan since 2012. We can substantially reduce the amount of radioactive waste, especially for several thousands of m^3 of concrete, by release from regulatory control based on the clearance criteria. Thus, recognizing and separating the radioactive wastes depending on the radioactivity or concentration of radioactivity is of the upmost importance to reduce the amount of those and to save the cost. There are only several reports before 2012 on the decommissioning for those accelerator facilities in Japan [3]-[7].

In planning process of the decommissioning, we also need to estimate exposure dose by gamma-rays from the radioactive nuclides induced by accelerator operations for workers engaged in the decommissioning operations in advance. Spatial distributions of ambient dose equivalent rates are measured on a periodic basis at some representative points. The information based on measurements will be available for the dose estimations. However, the spatial distributions of ambient dose equivalent rates are easily and drastically changed with time by the progress of decommissioning operations and physical decay of radioactive nuclides. The relatively high-activated accelerator and the components should be transported to off-site for reducing radiation exposure and keeping them as low as reasonable achievement. On the other hand, the radioactivity in their surrounding structure such as concrete walls is relatively low compared to that in the accelerator. For the concrete, long-half-life radioactive nuclides such as ^{152}Eu, ^{154}Eu, ^{60}Co, produced from minute impurities may become a problem for the dose estimations [2]. It can't be expected to reduce the ambient dose equivalent rates by physical decay during the decommissioning process. And it is hard to remove them to off-site without dismantling them. Therefore, we must estimate the exposure dose for workers in different situations throughout the entire period.

Thus, useful tools are required to evaluate the radioactivity and estimate the ambient dose equivalent rates in accelerator facility explained above. Particle and Heavy Ion Transport code System, PHITS [8], coupled with high energy particle induced radioactivity calculation code, DCHAIN-SP 2001 [9] (hereinafter called PHITS-DCHAIN), is prepared in the package of the PHITS which is available from the Organization for Economic Cooperation and Development/Nuclear Energy Agency (OECD/NEA) databank [10]. The

* Corresponding author: matsuda.norihiro@jaea.go.jp

DOI: 10.1051/epjconf/201715307001

evaluation procedures for the accelerator facility with the PHITS-DCHAIN are as follows:

(i) Evaluation of average fluxes of secondary neutrons generated from beam losses during beam operation at targets, injection/extraction apparatus, dumps etc.,

(ii) Evaluation of induced radioactivity in materials after the beam operation based on the average neutron fluxes obtained above, and

(iii) Evaluation of ambient dose equivalent rates by gamma-rays from the radionuclides obtained above.

Hence, these evaluations depend on the assumption of the beam losses. These calculations are started from the beam losses. The principal beam losses in the accelerator facility, such as targets, collimators, and beam dumps are well-known as beam current to keep radiation safety and to protect accelerators. The beam current is measured from the electric and magnetic fields created by the beam. For large-scale facilities, beam loss monitors and radiation monitors are installed throughout the facility, and are used for detecting uncontrolled losses. On the other hand, most small-scale facilities for medical purposes don't have such monitors. Therefore, additional measurements are needed to understand more about the beam losses and the positions for such facilities. The PHITS-DCHAIN would be available for deducing the beam losses from the additional measurements, such as ambient dose equivalent rates, induced radioactivity, or neutron fluxes.

In this study, measurements of secondary neutrons on the concrete walls generated from beam losses are carried out with the foil activation method in a compact cyclotron facility. The measured results are compared with calculations using PHITS-DCHAIN. The characteristics of distributions of the secondary neutron flux in the facility are identified through the measurements and the analyses.

2 Methods

2.1. Arrangement of compact cyclotron facility

Figure 1 shows a top cross-sectional view of the compact cyclotron facility. There is a cyclotron (BC-1710: The Japan Steel Works, Ltd.) and it has two beam transport lines (X- and Y-course). This compact cyclotron can accelerate protons up to 17 MeV, deuterons up to 10 MeV/u, Helium-3 up to 26 MeV/u, and Helium-4 up to 20 MeV/u. The X-course is a straight-line which lies on the middle of the facility, 7.2 m extended from the cyclotron. The Y-course is a branched line by switching magnet (SM) at a 35 degrees angle in the middle of the straight-line, of which length is 4.5 m. A target, a collimator and a Faraday cup are put in a target chamber connected at the end of both beam transport lines.

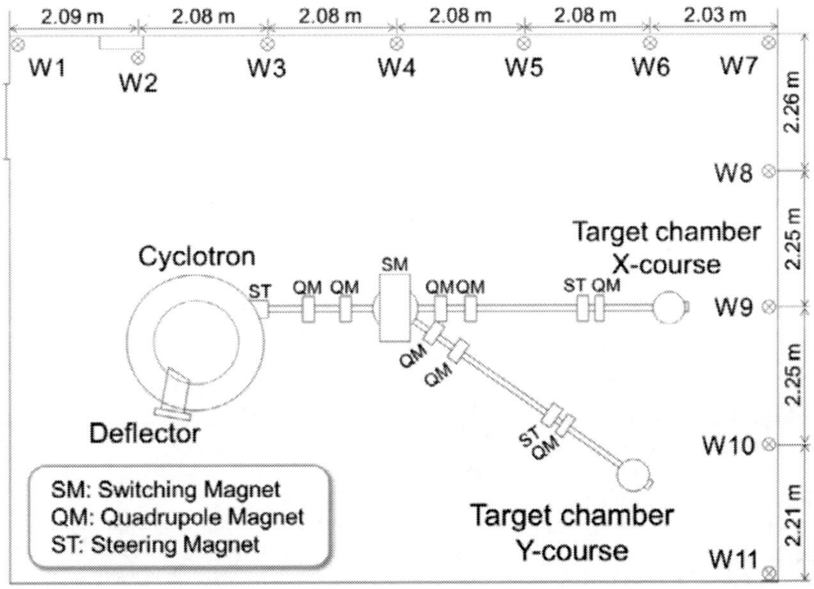

Figure 1. Top cross-sectional view of a compact cyclotron facility, which have a cyclotron, two beam transport lines and target chambers (X- and Y-course), and some magnets.

2.2. Irradiation target on X-course

During beam operation in this study, 17-MeV protons were transported only to the X-course (not transported to the Y-course). The proton beam was shaped by passing through the beam slit (collimator) of graphite (3 mm thick) with a slit aperture in the central, of which height was 10 mm and width was 5 mm. The shaped beam was impinged on the aluminum filter (0.05 mm thick), the irradiation target of SS316 (10 mm height × 5 mm width × 0.15 mm thick), and the Faraday cup (dump) of graphite (3 mm thick). These positions of beam losses from the beam slit to the Faraday cup in the target chamber, shown in Fig. 2 are hereinafter collectively called the X-target.

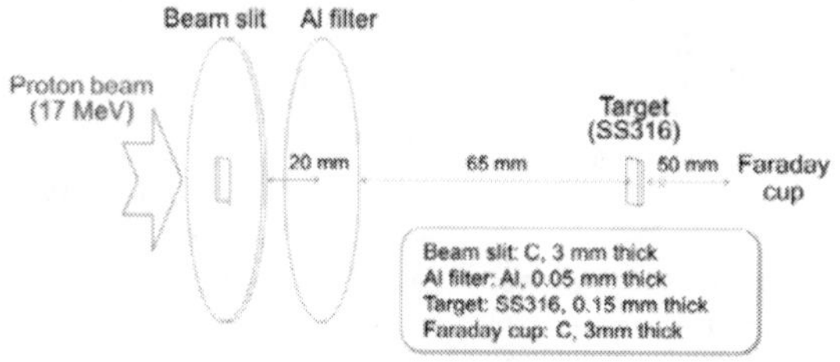

Figure 2. Arrangement of a beam slit (3 mm), an aluminum filter (0.05 mm), a SS316 target (0.15 mm) and a Faraday cup (3 mm), which are collectively called the X-target at the end of beam transport line (X-course).

2.3. Beam current to the X-target

Beam operating time of the cyclotron was 3 hours in this study. The average beam current transported to the X-target was 14.5 μA. And, the average beam current lost by the SS316 target was 1.9 μA. The beam losses at the beam slit, the aluminum filter, the SS316 target, and the Faraday cup were calculated with PHITS-DCHAIN. The calculated beam loss ratios were 77%, 0.01%, 12%, and 12%, respectively. Thus, the calculated beam losses at the SS316 target corresponded to 1.7 μA.

2.4. Beam losses at an electrostatic deflector

In a typical compact cyclotron with an electrostatic deflector for beam extraction, about 50% of accelerating protons would be lost by the electrostatic deflector [11]. The electrostatic deflector, which made from phosphorus deoxidation copper (Cu, > 99.90%) should be considered as beam loss point in this study. Since we can't directly obtain the beam current at the electrostatic deflector, the average one lost by the electrostatic deflector was estimated by beam current at beam probes A and B (shown in Fig. 3), and to the X-target. During a beam commissioning operation, the beam current at beam probes A and B was 4.8 and 3.0 μA, respectively, and that to the X-target was 2.3 μA. In this compact cyclotron, it was found that 38% of accelerating beam was lost in the vacuum chamber (between the beam probes A and B). Normalization coefficient was about 0.3 for estimating the beam current lost by the electrostatic deflector from that to the X-target. The estimated average beam current lost by the electrostatic deflector was 4.4 μAv (= 14.5 μA × 0.3). The beam losses at the electrostatic deflector were about 2.6 times larger than those at the SS316 target.

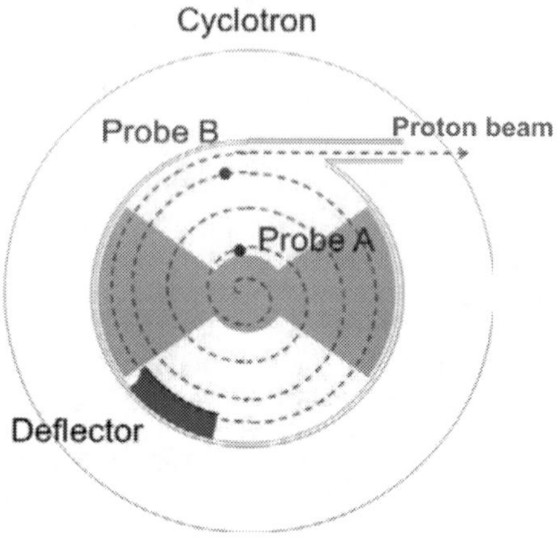

Figure 3. Arrangement of an electrostatic deflector, and beam probes (A and B) in a vacuum chamber of the compact cyclotron.

2.5. Uncontrolled losses

Uncontrolled losses may be occurred by operation failure around the beam lines such as quadrupole magnets (QM), steering magnets (ST) and switching magnet (SW) in Fig. 1. These uncontrolled beam losses kept to low level took no thought in this study. The accelerating beam loss in the vacuum chamber didn't also consider because of low-energy.

2.6. Activation measurements

Activation foils (0.02 mm thick Gold) were installed on concrete walls at the same level as the beam line (1.3 m height above the floor) shown as W1 to W11 in Fig. 1. The W1 to W11 were divided into two sides to the beam direction on X-course: left side wall (from W1 to W7) and front side wall (from W7 to W11).

After the beam operation to the X-target, the gamma-ray spectra from the activation foils were measured with the high-purity germanium (HPGe) detector (ORTEC: GMX-20195-S). The reaction rates of ^{197}Au (n, γ) ^{198}Au, which could be expressed in produced ^{198}Au atoms per target atom per incident proton, are determined after being corrected to take into account the beam current fluctuation during the irradiation. The peak efficiency of the HPGe detectors was determined by using the Gamma

Studio software of ORTEC. The components of estimated errors in reaction rates were the counting statistics and the detector efficiency.

3 Results and discussion

3.1. Experimental results

Measured reaction rates of ^{197}Au (n, γ) ^{198}Au on the concrete walls (from W1 to W11 in Fig. 1) are listed in Table 1 and showed in Fig. 4. In the figure, two peaks are observed at W2 (broad) and W7 (sharp). The reaction rate at W2 shows the largest value among all measurements. The value at W2 is 1.5 times larger than that at W7, which is the least value. On the other hand, the value at W9 clearly shows a peak on the front side wall (from W7 to W11), which is 1.3 times larger than two ends of that (W7 and W11). This result indicates that there are one or more principal beam loss points (at least two beam loss points from the number of peaks in the figure, such as around the compact cyclotron and the target chamber of the X-course).

Table 1. Measured reaction rates of ^{197}Au (n, γ) ^{198}Au on the concrete walls in the cyclotron facility.

Position	Reaction rate (atom^{-1} proton^{-1})
W1	$9.03\times10^{-18} \pm 7.10\times10^{-19}$
W2	$9.50\times10^{-18} \pm 7.11\times10^{-19}$
W3	$8.30\times10^{-18} \pm 7.01\times10^{-19}$
W4	$7.89\times10^{-18} \pm 6.55\times10^{-19}$
W5	$6.69\times10^{-18} \pm 5.59\times10^{-19}$
W6	$5.57\times10^{-18} \pm 5.25\times10^{-19}$
W7	$5.60\times10^{-18} \pm 5.30\times10^{-19}$
W8	$6.24\times10^{-18} \pm 5.87\times10^{-19}$
W9	$8.76\times10^{-18} \pm 6.90\times10^{-19}$
W10	$7.00\times10^{-18} \pm 6.10\times10^{-19}$
W11	$5.79\times10^{-18} \pm 5.72\times10^{-19}$

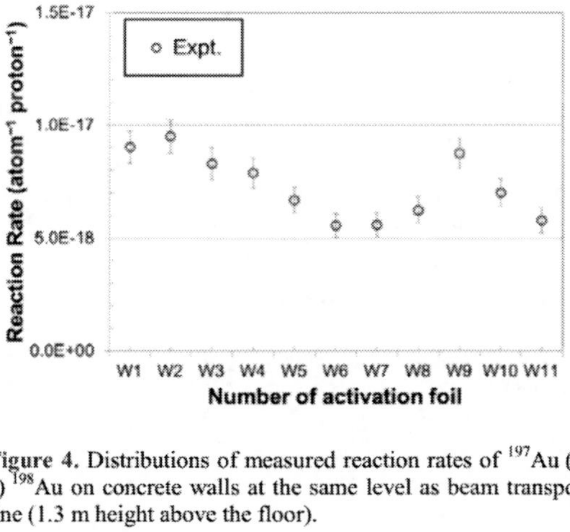

Figure 4. Distributions of measured reaction rates of ^{197}Au (n, γ) ^{198}Au on concrete walls at the same level as beam transport line (1.3 m height above the floor).

3.2 Calculation results

The experimental data on the concrete walls were compared with calculation results by using PHITS-DCHAIN adopting the Japanese Evaluated Nuclear Data Library (JENDL) -4.0 cross-section library [12]. In the calculation, the cyclotron facility including material compositions, beam energy, beam current (beam losses), and beam radius (1.1 cm) on the X-course, were precisely modelled. Absorption cross-section data of ^{197}Au (n, γ) ^{198}Au based on the JENDL dosimetry file [13] was used to convert from the calculated neutron fluxes with PHITS-DCHAIN to the reaction rates, listed in Table 2. The statistical errors (uncertainties) of calculation results for neutron transport calculations were far lower than 10%.

Table 2. Absorption cross-section data of ^{197}Au (n, γ) ^{198}Au used as conversion coefficient to the reaction rates

Reaction rate			
Energy (MeV)	Cross-section (cm^{-2})	Energy (MeV)	Cross-section (cm^{-2})
1.00×10^{-10}	8.756×10^{-23}	7.43×10^{-1}	8.498×10^{-26}
4.14×10^{-7}	2.583×10^{-23}	9.07×10^{-1}	8.036×10^{-26}
1.12×10^{-6}	2.563×10^{-23}	1.11×10^{0}	7.807×10^{-26}
2.38×10^{-6}	1.295×10^{-22}	1.35×10^{0}	6.973×10^{-26}
5.04×10^{-6}	3.058×10^{-22}	1.65×10^{0}	6.089×10^{-26}
1.07×10^{-5}	1.813×10^{-24}	2.02×10^{0}	4.954×10^{-26}
2.26×10^{-5}	1.711×10^{-23}	2.46×10^{0}	3.631×10^{-26}
1.01×10^{-4}	1.115×10^{-23}	3.01×10^{0}	2.513×10^{-26}
4.54×10^{-4}	1.126×10^{-23}	3.68×10^{0}	1.423×10^{-26}
1.58×10^{-3}	3.676×10^{-24}	4.49×10^{0}	6.204×10^{-27}
3.35×10^{-3}	1.741×10^{-24}	5.49×10^{0}	2.502×10^{-27}
7.10×10^{-3}	1.060×10^{-24}	6.70×10^{0}	8.175×10^{-28}
1.50×10^{-2}	7.005×10^{-25}	8.19×10^{0}	1.624×10^{-28}
3.18×10^{-2}	4.647×10^{-25}	1.00×10^{1}	2.756×10^{-29}
8.65×10^{-2}	3.216×10^{-25}	1.22×10^{1}	2.862×10^{-30}
1.50×10^{-1}	2.659×10^{-25}	1.35×10^{1}	8.036×10^{-31}
2.24×10^{-1}	2.152×10^{-25}	1.49×10^{1}	3.194×10^{-31}
3.34×10^{-1}	1.568×10^{-25}	1.75×10^{1}	1.060×10^{-31}
4.98×10^{-1}	1.122×10^{-25}	1.96×10^{1}	---

Calculated results of the reaction rates on the concrete walls were shown in Fig. 5 with the measured ones. Contributions for the reaction rates from the beam losses at the electrostatic deflector (blue dashed line) were larger than those at the X-target (red dotted line) on the whole, and accounted for nearly 90% of the measured ones except for ones at W9 and W10. The calculated total reaction rates (summation of the both, black solid line) were good agreements with the measured ones within 21% (C/Es: 0.91~1.21).

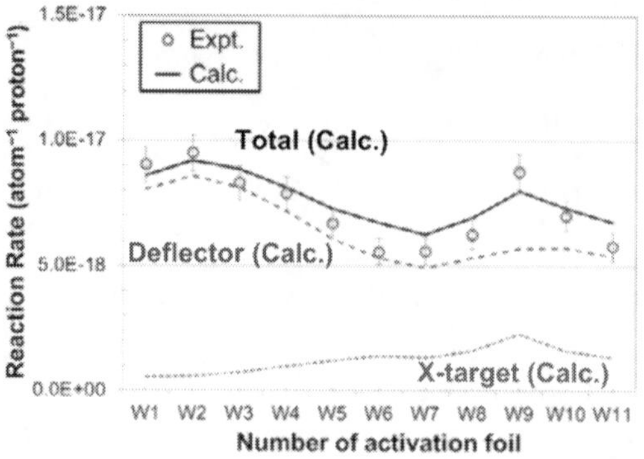

Figure 5. Distributions of measured reaction rates of ^{197}Au (n, γ) ^{198}Au on concrete walls, and calculated ones from beam losses at the X-target (red dotted line), the electrostatic deflector (blue dashed line), and the total (summation of the both, black solid line).

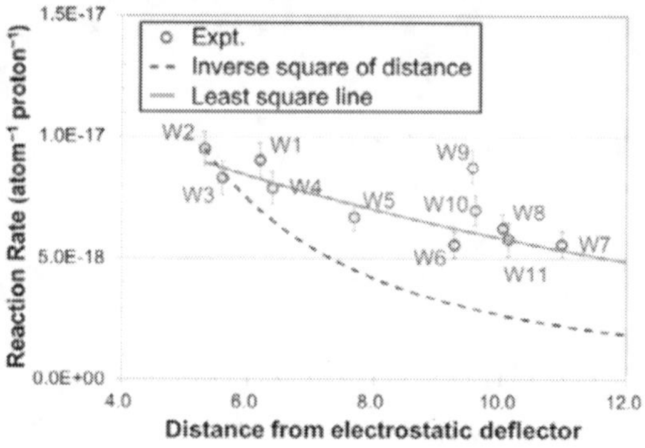

Figure 6. Distributions of measured reaction rates of ^{197}Au (n, γ) ^{198}Au on concrete walls with distance from the electrostatic deflector and the fitting line obtained by inverse square of distance (blue dashed line) and least-square method, except for that at W9 and W10 (green solid line).

3.3 Distributions of measured reaction rates in the facility

According to the calculations results of the reaction rates by PHITS-DCHAIN, secondary neutrons produced from the beam losses at the electrostatic deflector are dominant in this facility, except for ones at W9 and W10. We assume the electrostatic deflector as principal isotropic source of the secondary neutrons. Figure 6 shows distribution of the measured reaction rates with the distance from the electrostatic deflector. If there is an isotropic source, distributions of measured reaction rates will be inversely proportional to the square of the distance from the electrostatic deflector. However the distributions of measured reaction rates disagreed with the attenuation curve by the inverse square of distance from the electrostatic deflector (blue dashed line), exponential attenuation with the distance was observed (green solid line). The attenuation curve was obtained by the method of least squares, except for the measured reaction rates at W9 and W10. The attenuation curve of the measured reaction rates was looser than that by the inverse square of distance. This tendency about the exponential attenuations with the distance may make the evaluation for distributions of radioactivity and ambient dose equivalent rates in the compact cyclotron facility much easier.

4 Conclusion

The measurements on the concrete walls of the reaction rates of ^{197}Au (n, γ) ^{198}Au with the foil activation method have been carried out in the facility having the compact cyclotron with the electrostatic deflector. In this study, secondary neutrons in the facility were produced by 17-MeV protons transported to the target chamber of the X-course and from the beam losses at the electrostatic deflector during the accelerator operation. The measured data indicated that at least two beam loss points existed not only at the irradiation target but also around the compact cyclotron. The distributions of the measured reaction rates were calculated by using the PHITS-DCHAIN adopting the JENDL-4.0 cross-section library. The calculated results were good agreement with the measured ones within 21% (0.91~1.21). Moreover, it was found that the distributions of the measured reaction rates except for W9 and W10 showed exponential attenuation with the distance from the electrostatic deflector. The attenuation curve obtained by the method of least squares was looser than that by the inverse square of distance. This tendency about the distributions may make the evaluation of radioactivity and ambient dose equivalent rates in the compact cyclotron facility much easier. We need to investigate further identifying the distribution of the measured reaction rates, beam losses in the compact cyclotron facility.

Acknowledgements

We wish to thank Dr. K. Suyama for his support and advice on this study, and Dr. H. Nakashima for his help on establishing this study.

References

1. Japan radioisotope association, Statistics on the use of radiation in Japan 2015 (2016), http://www.jrias.or.jp/report/pdf/riyoutoukei2015.pdf
2. International atomic energy agency, Cyclotron produced radionuclides: guidelines for setting up a facility, International atomic energy agency technical reports series No. 471, (2009).
3. A. Endo, Dismantling work of the JAERI electron linac and measurement of induced radionuclides, J. Health Phys., 34, 161-165 (1999) in Japanese.
4. A. Yamadera, Dismantling work of cyclotron of Tohoku University and measurement of induced radionuclides, J. Health Phys., 34, 166-170 (1999) in Japanese.
5. T. Ishimoto, H. Kubota, T. Mori, K. Terakawa, K. Tani and K. Ishii, Decommissioning of the Medical Compact cyclotron, Journal of the RANDEC, 31, 21-31 (2005) in Japanese with English summary.
6. K. Masumoto, Decommissioning of accelerator facility – Example of the Tanashi branch, High Energy Accelerator Research Organization, 39, 30-43 (2009) in Japanese with English summary.
7. K. Ito, Y. Nakata, H. Matsuda and N, Sato, Decommissioning of the Medical cyclotron in National Center of Neurology and Psychiatry, Kakuigaku, 48, 109-119 (2011) in Japanese with English summary.
8. T. Sato, K. Niita, N. Matsuda, S. Hashimoto, Y. Iwamoto, S. Noda, T. Ogawa, H. Iwase, H. Nakashima, T. Fukahori, K. Okumura, T. Kai, S. Chiba, T. Furuta and L. Sihver, Particle and Heavy Ion Transport Code System PHITS, Version 2.52, J. Nucl. Sci. Technol. **50**, 913-923 (2013).
9. T. Kai, F. Maekawa, K. Kosako, Y. Kasugai, H. Takada, and Y. Ikeda, *DCHAIN-SP 2001: High energy particle induced radioactivity calculation code*, JAERI-Data/Code 2001-016, (2001). [in Japanese]
10. http://www.oecd-nea.org/tools/abstract/detail/nea-1857/
11. Y. Kumada, *Challenges to compact cyclotron*, Radiation & Industries 102, 30-35 (2004). [in Japanese]
12. K. Shibata, O. Iwamoto, T. Nakagawa, N. Iwamoto, A. Ichihara , S. Kunieda, S. Chiba, K. Furutaka, N. Otuka, T. Ohsawa, T. Murata, H. Matsunobu, A. Zukeran, S. Kamada, and J. Katakura, JENDL-4.0: A New Library for Nuclear Science and Engineering, J. Nucl. Sci. Technol. **48**, 1-30 (2011).
13. K. Kobayashi, T. Iguchi, S. Iwasaki, T. Aoyama, S. Shimakawa, Y. Ikeda, N. Odano K. Sakurai, K. Shibata, T. Nakagawa and M. Nakazawa, *JENDL Dosimetry file 99 (JENDL/D-99)*, JAERI 1344, (2002).

ANITA-IEAF activation code package – updating of the decay and cross section data libraries and validation on the experimental data from the Karlsruhe Isochronous Cyclotron

Manuela Frisoni*

ENEA, CR "E.Clementel", Via Martiri di Monte Sole 4, 40129 Bologna, Italy

Abstract. ANITA-IEAF is an activation package (code and libraries) developed in the past in ENEA-Bologna in order to assess the activation of materials exposed to neutrons with energies greater than 20 MeV. An updated version of the ANITA-IEAF activation code package has been developed. It is suitable to be applied to the study of the irradiation effects on materials in facilities like the International Fusion Materials Irradiation Facility (IFMIF) and the DEMO Oriented Neutron Source (DONES), in which a considerable amount of neutrons with energies above 20 MeV is produced. The present paper summarizes the main characteristics of the updated version of ANITA-IEAF, able to use decay and cross section data based on more recent evaluated nuclear data libraries, i.e. the JEFF-3.1.1 Radioactive Decay Data Library and the EAF-2010 neutron activation cross section library. In this paper the validation effort related to the comparison between the code predictions and the activity measurements obtained from the Karlsruhe Isochronous Cyclotron is presented. In this integral experiment samples of two different steels, SS-316 and F82H, pure vanadium and a vanadium alloy, structural materials of interest in fusion technology, were activated in a neutron spectrum similar to the IFMIF neutron field.

1 Introduction

The evaluation of the neutron induced material activation plays an important role for the development of future fusion power plants for issues related to safety, engineering design and radioactive waste management.

For these devices the activation codes and cross section libraries handling neutron energies up to 20 MeV are quite adequate. Besides, in order to study the irradiation effects on fusion materials, some facilities have been proposed to produce accelerator-based neutron sources at sufficient intensity to test samples of candidate materials to be used in future fusion reactors.

The International Fusion Materials Irradiation Facility (IFMIF) and, more recently, the DEMO Oriented Neutron Source (DONES), more tightly focused on DEMO needs, have been proposed to be such dedicated facilities. In both these plants the neutron source is produced through the reaction of 40 MeV deuterons impinging on a liquid lithium target and a considerable amount of neutrons with energies above 20 MeV is produced.

The availability of reliable nuclear data and activation codes in the neutron energy range extended over 20 MeV is required in order to perform activation calculations for these devices.

To this end, the ANITA-IEAF activation package (code and libraries) was developed in the past in ENEA-Bologna [1], [2] for the activation characterization of materials exposed to neutrons with energies up to 150 MeV. It traces back to the ANITA-2000 code package released by ENEA and freely distributed at OECD-NEADB [3] and ORNL-RSICC [4].

ANITA-IEAF provides the radioactive inventories of materials exposed to neutron irradiation, continuous or stepwise. The activity, isotopic nuclide density, decay heat, biological hazard, clearance index and gamma ray source spectra, in the standard 42 groups VITAMIN-J energy structure, are calculated at shutdown and at different cooling times.

The original package contains: a) the activation code ANITA-IEAF, able to manage the large amount of reactions with threshold higher than 20 MeV and up to 150 MeV, b) the activation cross section library based on the Intermediate Energy Activation File IEAF-2001 [5], c) the decay data library (file fl1) containing the quantities describing the decay properties of the unstable nuclides and d) the library (file fl2) containing the gamma ray spectra emitted by the radioactive nuclei.

The fl1 and fl2 files of the ANITA-IEAF code package, originally based on the evaluated nuclear data library FENDL/D-2.0 [6], were recently updated on the basis of the JEFF-3.1.1 Radioactive Decay Data Library [7]. The ANITA-IEAF code actually uses a neutron activation cross section library based on the data of the European Activation File EAF-2010 [8] provided in the package EASY-2010 [9].

The present paper summarizes the main characteristics of the updated version of the ANITA-IEAF activation code package.

The validation of ANITA-IEAF is an essential element for its quality assurance in view of its

* Corresponding author: manuela.frisoni@enea.it

applicability to activation calculations (radioactive inventories and decay gamma sources) in facilities like IFMIF and DONES.

In this paper, the validation effort related to the comparison between the code predictions and the activity measurements obtained from the Karlsruhe Isochronous Cyclotron is presented [10], 11]. In this experiment four structural materials of interest in fusion technology were investigated. Samples of standard steel SS-316, low activation steel F82H, pure vanadium and vanadium alloy (V-4Ti-4Cr) were irradiated in a neutron spectrum resembling that of the IFMIF facility and the specific activities of many radionuclides were measured at various cooling times.

The ratios C/E between calculated (C) and experimental (E) values are shown and discussed in this paper.

The ANITA-IEAF results are also compared with those obtained through the FISPACT-II code [12], contained in the European Activation System-II (EASY-II) [13], using the same EAF-2010 neutron activation cross section library, in order to evaluate the responses of the two codes.

Moreover, due to the fact that the specific activities are mainly sensitive to the activation cross sections, a comparison between the experimental results and the calculations performed with different activation cross section libraries is presented (i.e. ANITA-IEAF/EAF-2010, ANITA-IEAF/IEAF-2001 and FISPACT-II/TENDL-2014 calculations).

2 Main features of the updated ANITA-IEAF activation code package

The schematic block diagram of the data/libraries required by the ANITA-IEAF code is shown in Figure 1.

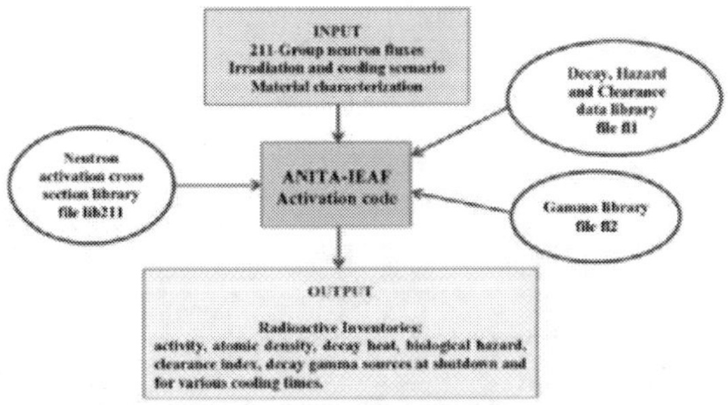

Fig. 1. ANITA-IEAF activation code block diagram.

The Decay, Hazard and Clearance data library (file fl1) provides the information describing the decay properties of the unstable nuclides.

It contains, for each nuclide, the decay mode, the branching ratios, the decay constant (s^{-1}), the total energy (MeV) released in the decay and the energy (MeV) released in the form of gamma or X-rays. The decay data of the new fl1 file were taken from the JEFF-3.1.1 Radioactive Decay Data library (MF=8, MT=457 section in the ENDF-6 [14] standard notation) and properly converted to the format required by the ANITA-IEAF code.

The fl1 file provides also for each radionuclide the clearance level C_L (Bq/g). This quantity allows to establish if a radioactive material can be potentially moved out of the originating facility and recycled. The safe handling of radioactive waste is recognized as crucial to ensure protection of human health and the environment. IAEA publish regulations on these issues and [15] provides information on suggested clearance level values for a set of important radionuclides. The clearance level data contained in the new fl1 file have been updated by including the information from [15] and [16]. The updated fl1 file actually provides data for 3433 nuclides.

The gamma library (file fl2) contains the gamma ray spectra emitted by the radioactive nuclei in the VITAMIN-J 42-γ energy group structure. It was updated on the basis of the gamma radiation spectra contained in the JEFF-3.1.1 Radioactive Decay Data library. The data given in the fl2 gamma library are used in ANITA-IEAF in order to compute the intensity and the energy distribution of the gamma-rays emitted by the irradiated material. The decay gamma-ray source (Photons/cm^3s) calculated by ANITA-IEAF may be given as input to a radiation transport code in order to define and establish shielding design criteria related to beam-off operational phases and to evaluate the dose equivalent rates.

The original ANITA-IEAF activation code used a 256-energy group structure up to 150 MeV and an activation cross section library based on IEAF-2001.

The updated ANITA-IEAF code actually uses an activation library based on the EAF-2010/VITAMIN-J+ (211-energy group structure, up to 55 MeV) neutron-induced cross section library contained in the package EASY-2010, named "eaf_n_gxs_211_flt_20010".

In the ANITA-IEAF code new MT numbers have been introduced, from 151 to 200, the same as in FISPACT code [12], that allow describing reactions with up to 8 emitted particles corresponding to a set of unallocated numbers in the standard ENDF-6 format.

Each reaction corresponds to a well-defined ZA (ZA=Z*1000+A) difference with respect to the target nuclide and this ZA is used in the activation code to create and follow the decay chains.

The MODBIN module must be used for the conversion of the EAF-2010 card image file to the "lib211" file in binary format as required by the ANITA-IEAF code (see Fig.1). This module was modified accordingly with the changes introduced in the updated ANITA-IEAF code.

3 ANITA-IEAF calculations

The ANITA-IEAF predictions were compared with the activity measurements obtained from the Karlsruhe Isochronous Cyclotron. In this experiment, a saturation thick target of natural lithium was irradiated with 40 MeV deuterons. The resulting neutron spectrum and yield were measured by multi-foil activation. Samples of two different steels, SS-316 and F82H, pure vanadium and a vanadium alloy were activated in the neutron field. These are structural materials of interest in fusion technology.

For each sample, the specific activities of many radionuclides, in Becquerel per kilogram, were determined by gamma spectrometry. The details of the experimental set-up, taken from [10], (i.e. material composition, irradiation time, neutron flux, etc.) used in the ANITA-IEAF input in order to model the experiment, are given in the following.

3.1 Neutron Flux

The Karlsruhe Isochronous Cyclotron produced a neutron spectrum almost identical to the IFMIF high-flux test region spectrum but at a lower intensity.

The Centre position neutron flux [10], in the 211-energy group structure, was used in the ANITA-IEAF calculations (see Fig. 2).

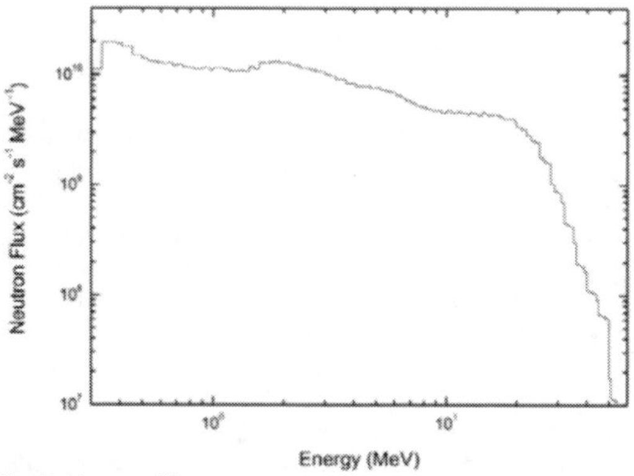

Fig. 2. Neutron Flux.

3.2 Activation parameters

The activation parameters used in the calculations, i.e. the neutron flux Φ ($cm^{-2}s^{-1}$) and the irradiation time (s) for each sample, are reported in Table 1.

Table 1. Activation parameters

Material	Neutron Flux density Φ(cm^{-2} s^{-1})	Irradiation time (s)
SS-316	4.10 E11	7525
F82H	4.14 E11	7525
V- alloy	4.22 E11	7525
V- pure	4.27 E 11	7525

3.3 Samples compositions

The elemental compositions in weight % used in the calculations are given in Table 2.

3.4 Cooling times

The measurements on the activated samples were performed at different cooling times ranging from about 1 hour up to 150 days. The cooling times are given in Table 3.

Table 2. Compositions of samples in wt.%

Element	SS-316	F82H	V-alloy	V-pure
B	0.0002	0.0002	0.0005	0.0005
C		0.09	0.0045	0.0027
N		0.008	0.011	0.013
O			0.037	0.041
Mg			0.001	0.001
Al	0.05	0.003	0.020	0.025
Si	0.4	0.11	0.034	0.017
P		0.002	0.003	0.003
S		0.002	0.001	0.001
Ti	0.15	0.01	4.1	
V		0.16	91.8483	99.8658
Cr	17.5	7.70	3.9	
Mn	1.8	0.16		
Fe	65.1648	89.7467	0.021	0.016
Co	0.03	0.005		
Ni	12.3	0.02		
Cu	0.1	0.01	0.005	0.005
Nb	0.005	0.0001	0.0087	0.004
Mo	2.5	0.003	0.005	0.005
Ta		0.02		
W		1.95		

Table 3. Cooling times

SS-316	F82H
5965 s ≈ 1.7 h	4173 s ≈ 1.2 h
18848 s ≈ 5.2 h	18011 s ≈ 5 h
93333 s ≈ 1.1 d	92043 s ≈ 1.1 d
178860 s ≈ 2.1 d	158665 s ≈ 1.8 d
432539 s ≈ 5 d	424416 s ≈ 4.9 d
2520240 s ≈ 29.2 d	2922060 s ≈ 33.8 d
12962640 s ≈150 d	13036200 s ≈ 150.9 d

V-alloy	V-pure
5477 s ≈ 1.5 h	4983 s ≈ 1.4 h
17189 s ≈ 4.8 h	16396 s ≈ 4.6 h
90710 s ≈ 1 d	90001 s ≈ 1 d
417962 s ≈ 4.8 d	412474 s ≈ 4.8 d
2254080 s ≈ 26.1 d	3100500 s ≈ 35.9 d
13341780 s ≈154.4 d	13146900 s ≈ 152.2 d

4 Comparison between ANITA-IEAF calculations and experimental results

The results of the comparison between the measured activities and the corresponding ANITA-IEAF predictions are given in Fig.3 ÷ Fig.9.

The comparison is shown in terms of calculated-to-experimental activities ratios (C/E) estimated as average over the cooling times for each isotope of the measured material sample. The error bars (1σ) account for the experimental uncertainties only taken from [10].

For SS-316, F82H and V-alloy samples the C/E ratios of the main contributors to the total activity are shown separately, whereas for V-pure the C/E values of all the isotopes are given together.

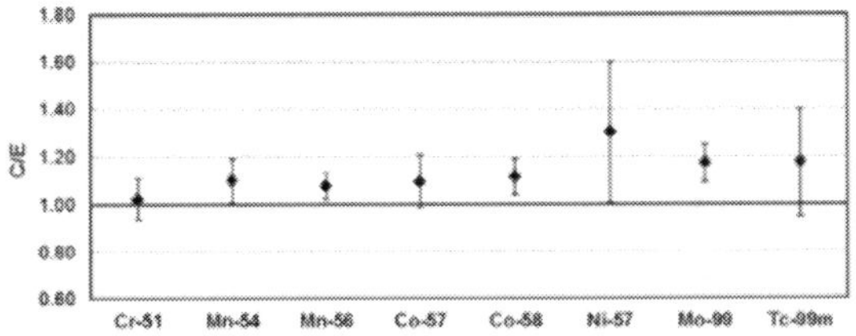

Fig. 3. SS-316 specific activity: calculation to experiment ratios (C/E). Main isotope contributors.

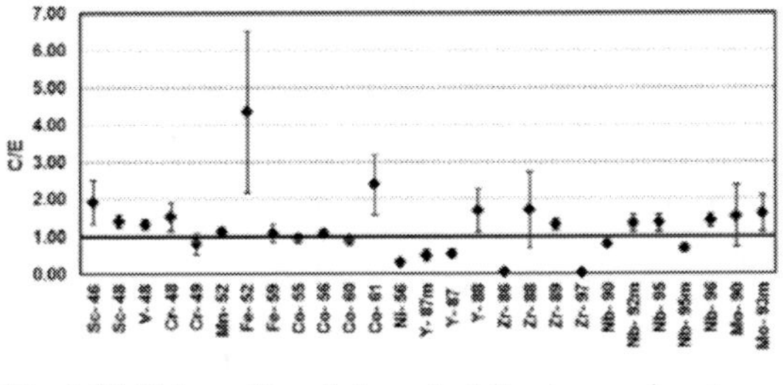

Fig. 4. SS-316 specific activity: calculation to experiment ratios (C/E). Other isotopes.

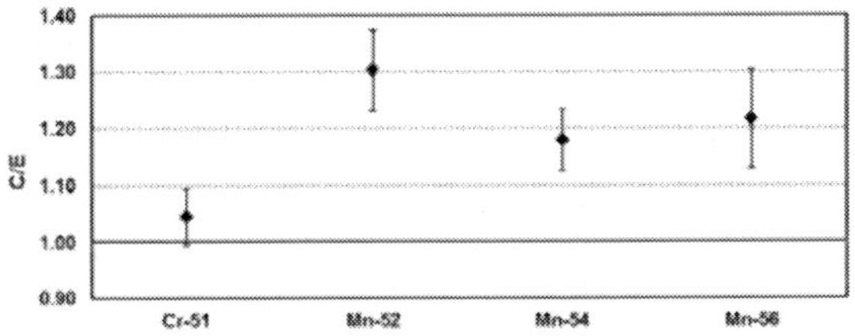

Fig. 5. F82H specific activity: calculation to experiment ratios (C/E). Main isotope contributors.

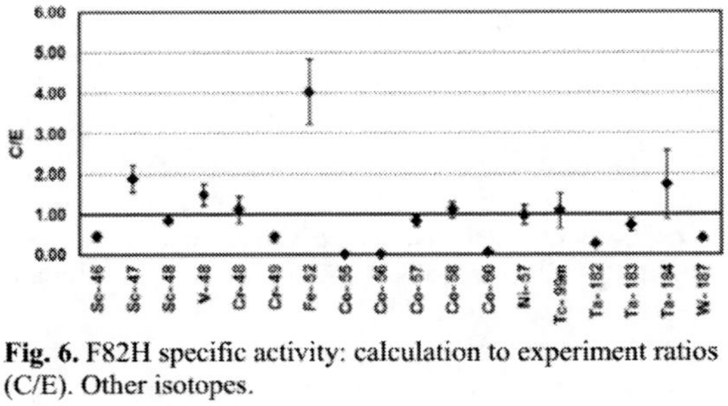

Fig. 6. F82H specific activity: calculation to experiment ratios (C/E). Other isotopes.

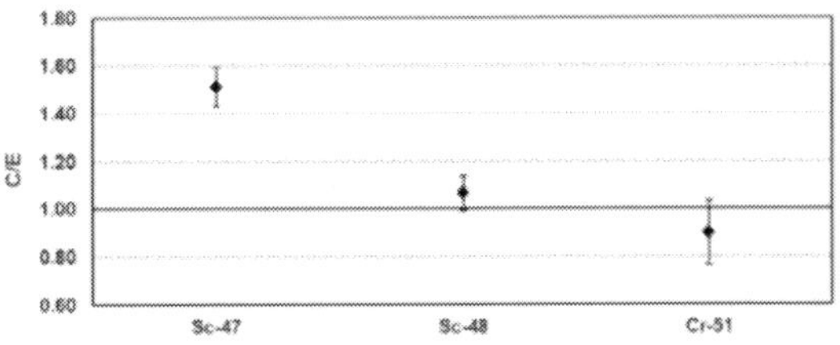

Fig. 7. V-alloy specific activity: calculation to experiment ratios (C/E). Main isotope contributors.

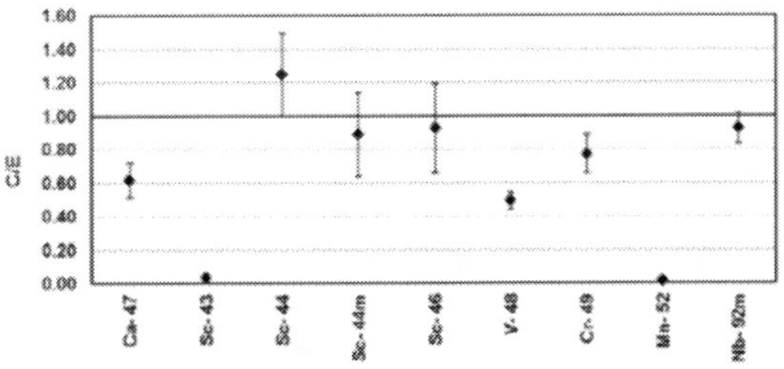

Fig. 8. V-alloy specific activity: calculation to experiment ratios (C/E). Other isotopes.

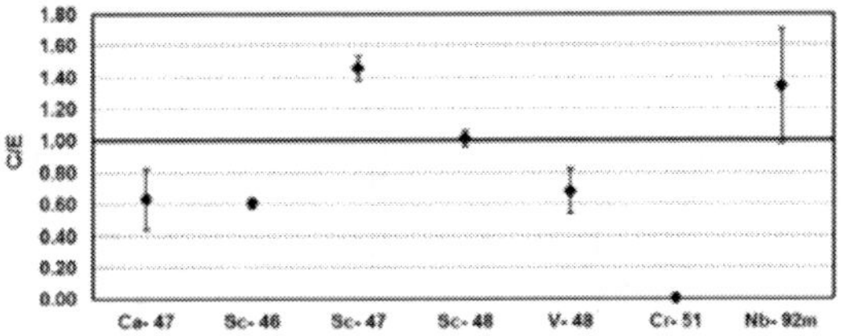

Fig. 9. V-pure specific activity: calculation to experiment ratios C/E). All isotopes.

The comments listed in the following are drawn from the results analysis.

4.1 SS-316 steel sample

- For the irradiated SS-316 steel sample, 35 isotopes have been analyzed. Among the measured nuclides the main contributors to the total activity are: Cr-51, Mn-54, Mn-56, Co-57, Co-58, Ni-57, Mo-99 and Tc-99m. The corresponding C/E values are shown in Fig.3.
- Cr-51: is an important contributor, up to 50% , for all the cooling times considered. The agreement between experimental and calculated activities is quite good (discrepancies within 6%).
- Mn-54: its contribution is not relevant, 0.1%, in the first two hours of cooling time where the discrepancy is large (24%). The importance of Mn-54 increases up to 16.1% at the last cooling time considered. The corresponding C/E values show discrepancies within 20%.
- Mn-56: is the most important contributor to the total activity, from 65% to 78%, up to 6 hours of cooling time. The measured values and the ANITA-IEAF predictions, in terms of C/E, show a very good

agreement. For all the cooling times the discrepancies are lower than 10%.

- Co-57: is an important activation contributor, up to 17.7% at 150 days of cooling time. The measured values and the ANITA-IEAF predictions, in terms of C/E, show a good agreement. For all the cooling times the discrepancies are lower than 15%.
- Co-58: this nuclide becomes significant after 1 day of cooling time; its contribution grows up to 22% after 29 days of cooling time. The C/E discrepancies are always within 15%.
- Ni-57: is an important product in the first 5 days (half-life≈36h); its contribution increases from 1% at the first cooling time up to about 16% at 2.1 days. The discrepancies range from 24% to 42%.
- Mo-99: its contribution is about 4% from 1 to 5 days of cooling time. The C/E values show discrepancies up to 24%.
- Tc-99m: its contribution is about 4% in the range between 1 and 5 days of cooling time. For these times the C/E values show discrepancies within 25%. The largest discrepancies are found at the cooling times where its contribution is negligible.
- The average C/E values of Cr-51, Co-57, Tc-99m show a good agreement within the error bars. The discrepancies are within 20% for all the isotopes except for Ni-57. In general the calculations predict an overestimation of the experimental results.
- The other less important nuclides show larger discrepancies as, mainly, Sc-46 (C/E=1.91), Co-61 (C/E=2.39), Fe-52 (C/E=4.34), Zr-86 (C/E=0.03) and Zr-97 (C/E=0.02) (see Fig. 4).

4.2 F82H steel sample

- For the irradiated F82H steel sample, 22 isotopes have been analysed. The main isotope contributors to the total activity are: Cr-51, Mn-52, Mn-54 and Mn-56. The corresponding C/E values are reported in Fig.5.
- Cr-51: is an important contributor, up to 60%, for all the cooling times. The discrepancies between experimental and calculated activities are within 7%, showing a very good agreement.
- Mn-52: its contribution is relevant up to 5 days of cooling time; the discrepancies between experimental and calculated activities are large, up to 43%, showing an overestimation of the experimental results.
- Mn-54: its contribution increases from about 7% at the first cooling time up to 23.7% at 150 days. The corresponding C/E values show discrepancies up to 19%.
- Mn-56: is the most important contributor to the total activity, 98%, up to 5 hours of cooling time. For these times the C/E values show discrepancies within 10%. The largest discrepancy, about 60%, is found at the cooling time where its contribution is negligible (lower than 0.2%).
- the average C/E value of Cr-51 only shows a good agreement within the error bars with a discrepancy lower than 10%. The biggest discrepancy is found for Mn-52 (30%). In general the calculations predict an overestimation of the experimental results.
- The other less important nuclides show larger discrepancies as, mainly, Fe-52 (C/E=4.0), Co-55 (C/E=2.4E-03), Co-56 (C/E=0.01) and Co-60 (C/E=0.05).

4.3 Vanadium alloy sample

- For the irradiated V-alloy sample, 12 isotopes have been analyzed. The main isotope contributors to the total activity are: Sc-48, Cr-51 and Sc-47. The corresponding C/E values are reported in Fig.7.
- Sc-48: is the most important contributor to the total activity up to 5 days of cooling time. At these times the C/E values show a good agreement, discrepancies within 11%, between experimental and calculated results. The largest discrepancy, greater than 50%, is found at 26 days of cooling time where its contribution is negligible (lower than 0.1%).
- Cr-51: is an important contributor, up to 56%, for all the cooling times. The biggest discrepancy between experimental and calculated activities is found at 1 day cooling time where the discrepancy grows up to 27%. The calculated values are in general underestimate with respect to the experimental data.
- Sc-47: is an important contributor, up to about 54%, for all the cooling times considered. The discrepancies between experimental and calculated activities are up to 86%, showing an overestimation of the calculated results with respect to the experimental ones.
- The average C/E values of the previous nuclides show discrepancies within 20% except for Sc-47.
- For the other less important nuclides the major discrepancies are found for Sc-43 and Mn-52. All the experimental results are underestimated except for Sc-44 (see Fig.8).

4.4 Pure-Vanadium sample

- For the irradiated V-pure sample, 7 isotopes have been analyzed. The main isotope contributors to the total activity are: Sc-48, Sc-47 and Sc-46. The pure-Vanadium C/E values for all the isotopes are reported in Fig.9.
- Sc-48: is the most important contributor, up to 56%, to the total activity for all the cooling times considered. The C/E values show a very good agreement, discrepancies within 4%, between experimental and calculated results.
- Sc-47: is an important contributor, up to 57%, up to 5 days of cooling time. The agreement between experimental and calculated activities is rather poor. The calculated results show a general overestimation of the experimental values.
- Sc-46: contributes up to 6% to the total activity. The calculated activities show an underestimation, up to 40%, of the experimental ones.
- For the other nuclides the agreement between experimental and calculated activities is rather poor. In

DOI: 10.1051/epjconf/201715307002

particular, for Cr-51, the calculated results show a huge underestimation of the experimental values.

5 Comparison between ANITA-IEAF and FISPACT-II calculations

The ANITA-IEAF results were compared with those obtained by the FISPACT-II code, a completely new inventory code that replaces FISPACT-2007 [17].

In this section the results obtained by the two codes, by using the same EAF-2010/VITAMIN-J+ (211-energy group structure) neutron activation cross section library, are shown in Fig.10 ÷ Fig16.

The ANITA-IEAF and FISPACT-II codes give quite similar results with differences within 2% for all the isotopes, except for Mn-52 (≈7%) for the SS-316 and F82H samples.

The discrepancies between the ANITA-IEAF and FISPACT-II calculated values can be explained as due to differences in the decay data (as life-times, branching ratios) and/or a different numerical treatment of the nuclear chains.

It has to be underlined that ANITA-IEAF uses decay data based on the JEFF-3.1.1 RDD library, whereas FISPACT-II uses the data contained in the libraries of the EASY-2010 system.

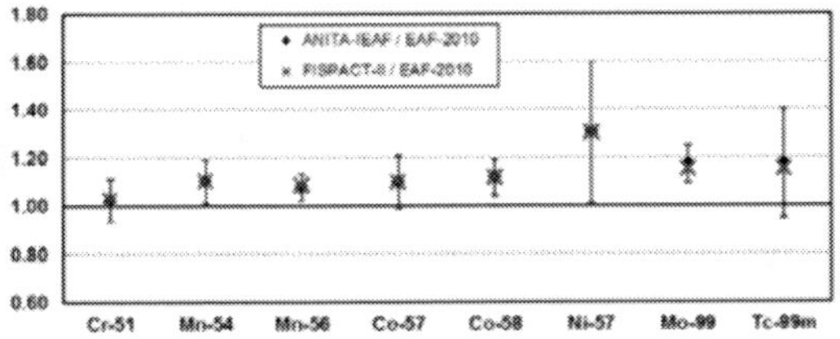

Fig. 10. SS-316 specific activity: comparison between ANITA-IEAF and FISPACT-II calculations. Main isotope contributors.

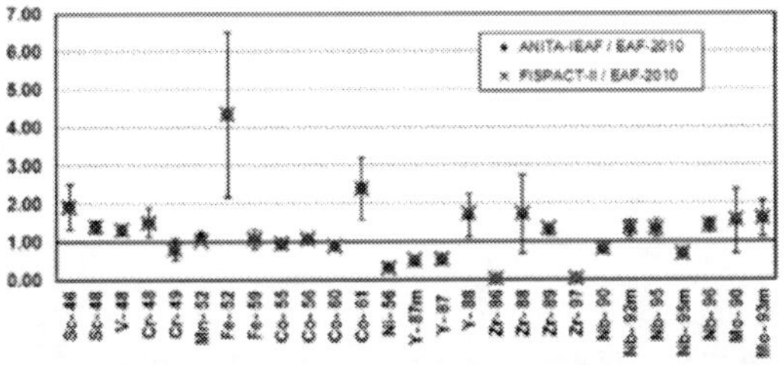

Fig. 11. SS-316 specific activity: comparison between ANITA-IEAF and FISPACT-II calculations. Other isotopes.

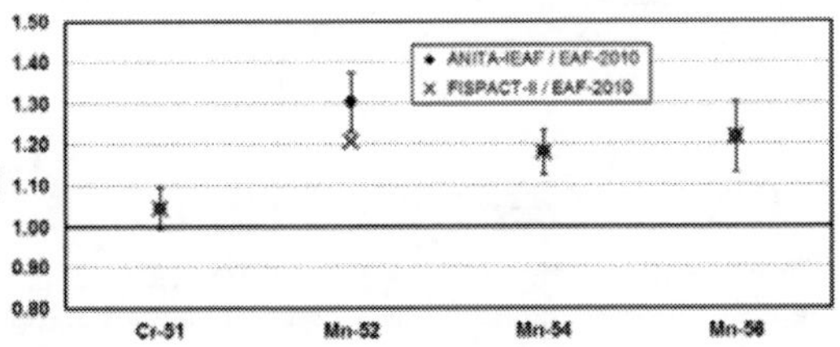

Fig. 12. F82H specific activity: comparison between ANITA-IEAF and FISPACT-II calculations. Main isotope contributors.

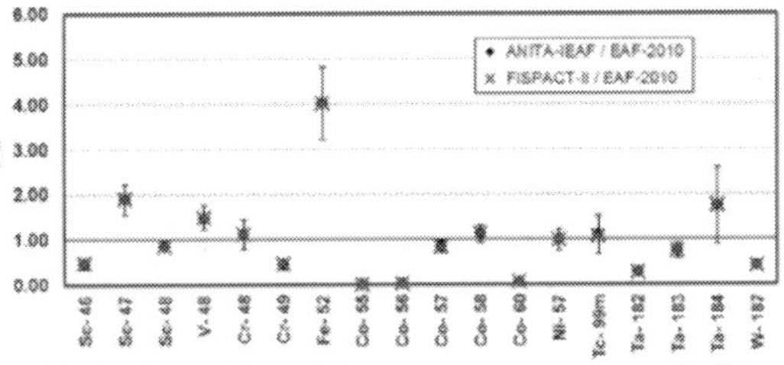

Fig. 13. F82H specific activity: comparison between ANITA-IEAF and FISPACT-II calculations. Other isotopes.

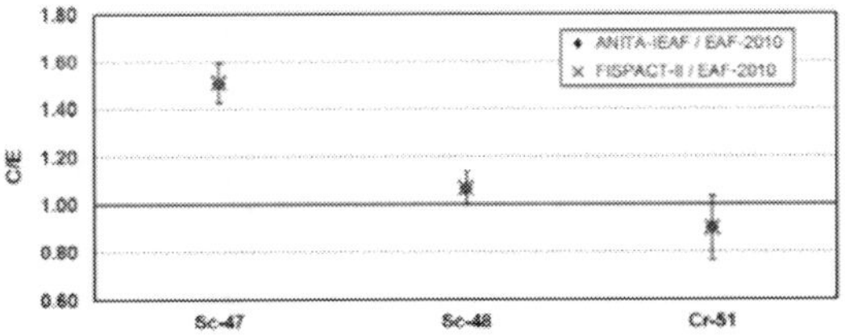

Fig. 14. V-alloy specific activity: comparison between ANITA-IEAF and FISPACT-II calculations. Main isotope contributors.

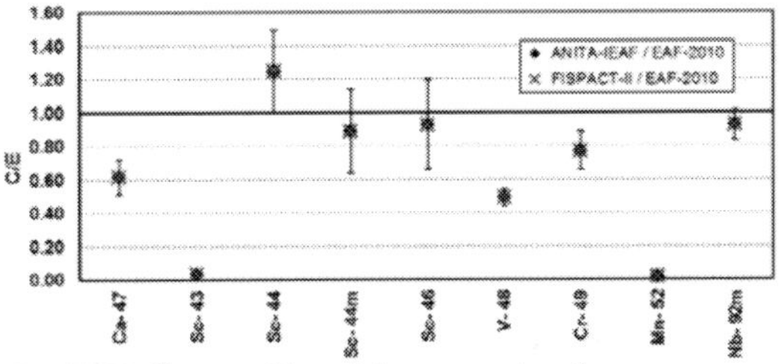

Fig. 15. V-alloy specific activity: comparison between ANITA-IEAF and FISPACT-II calculations. Other isotopes.

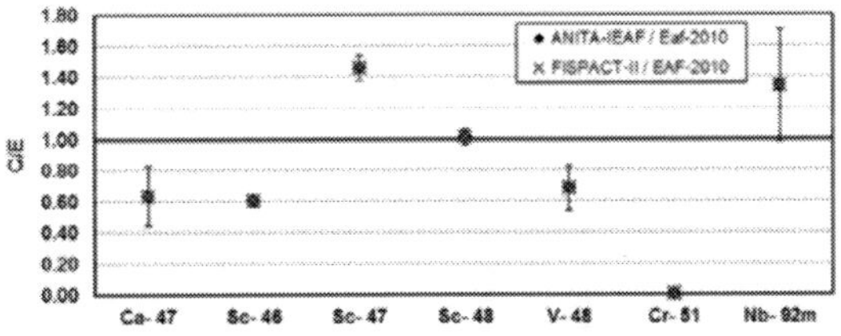

Fig. 16. V-pure specific activity: comparison between ANITA-IEAF and FISPACT-II calculations. All isotopes.

6 Activation cross section libraries comparison

The results reported in the previous section confirmed that the ANITA-IEAF and FISPACT-II codes give quite comparable results when using the same neutron activation cross section library.

Their predictions are strictly dependent on the data contained in the libraries used.

The data contained in these libraries represent the most critical point of the activation calculations more than the activation codes themselves.

Nuclear data have been continuously updating and different libraries (both for decay data and activation cross sections) have become available over time.

In particular, the specific activities, under analysis in this benchmark, are mainly sensitive to the neutron activation cross sections.

A comparison between the experimental results and calculations performed using different activation cross section libraries is presented in this section.

The FISPACT-II code uses as principal sources of cross section data the different generations of the TALYS-based Evaluated Nuclear Data Libraries. In the present calculations the latest TENDL-2014 [18] evaluated cross section data library, in the 709-energy group structure contained in the package released by OECD-NEADB [19], was used. These data are provided in ENDF-6 format.

Three different calculations are compared, namely: a) "updated" ANITA-IEAF code and EAF-2010 library, b) "original" ANITA-IEAF code and IEAF-2001 library from [1], c) FISPACT-II code and TENDL-2014 library.

The results are shown in Fig.17 ÷ Fig.23. The error bars (1σ) are associated to the FISPACT-II results and take into account both the experimental and the calculation uncertainties as provided by the FISPACT-II code.

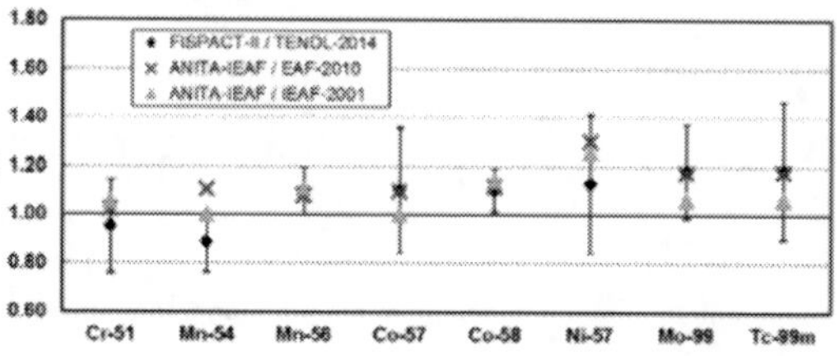

Fig. 17. SS-316 specific activity. Main isotope contributors.

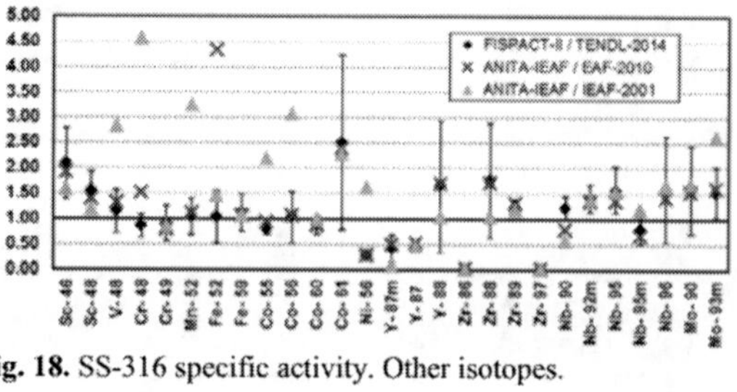

Fig. 18. SS-316 specific activity. Other isotopes.

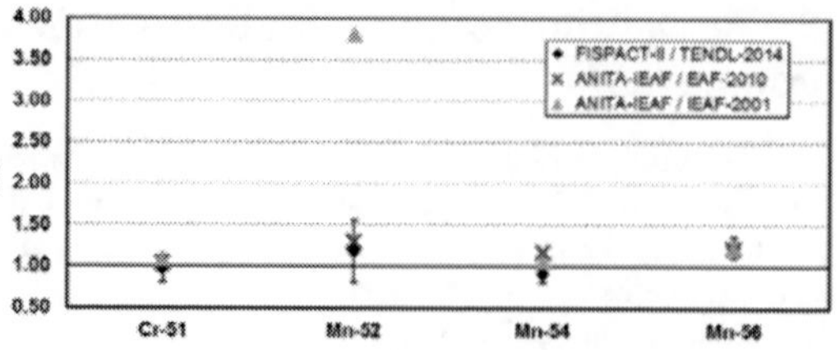

Fig. 19. F82H specific activity. Main isotope contributors.

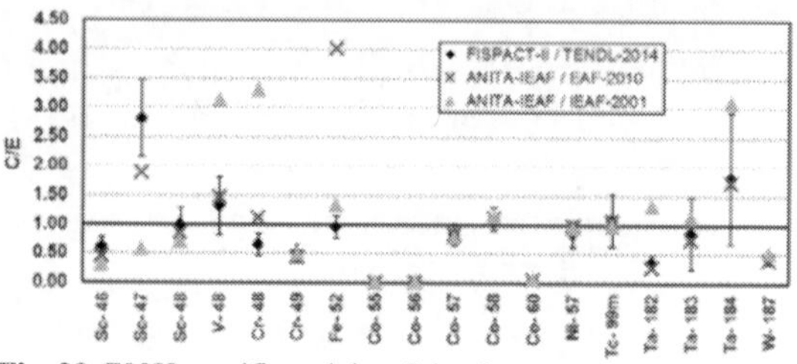

Fig. 20. F82H specific activity. Other isotopes.

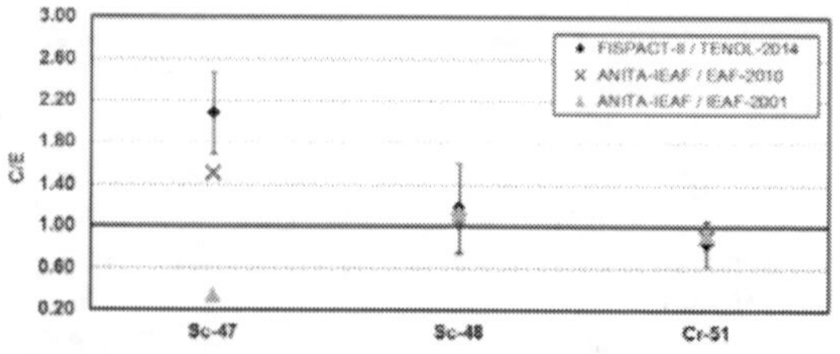

Fig. 21. V-alloy specific activity. Main isotope contributors.

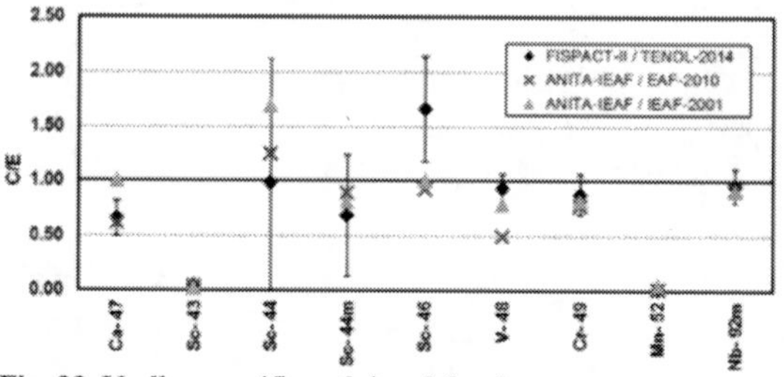

Fig. 22. V-alloy specific activity. Other isotopes.

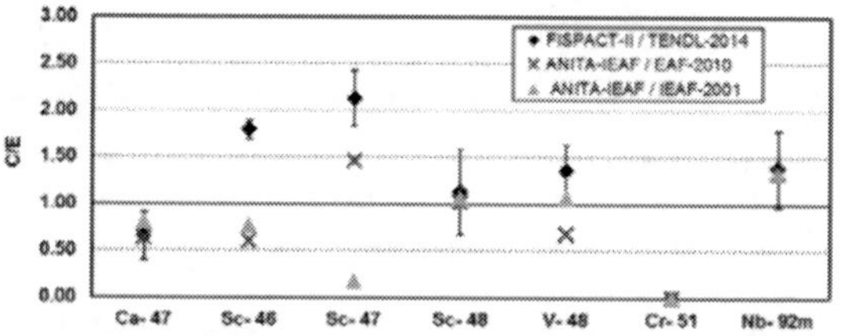

Fig.23. V-pure specific activity. All isotopes.

The analysis of the results confirms once more what previously stated, that is the relevance of the cross section data on the responses of the activation calculations. The most recent libraries don't always provide better results when compared with the experimental data.

Some minor discrepancies among the calculations can be attributed to differences in the decay data.

More specific conclusions for each sample are given in the following.

6.1 SS-316 steel sample

Concerning the main isotope contributors (see Fig.17) the results obtained by the three calculations show discrepancies within 15% among them. In some cases, as

for Ni-57, the more recent TENDL-2014 library gives better results compared with the experimental value.
For some isotopes, as Mn-54, Co-57, Mo-99 and Tc-99m, the "more dated" IEAF-2001 library shows a better behaviour.

As a general trend, the results obtained by FISPACT-II for all the isotopes show an agreement with the experimental data within the error bars comprehensive also of the calculation uncertainties.

For the other less important isotopes (see Fig.18), the huge discrepancies shown by the IEAF-2001 library for V-48, Cr-48, Mn-52, Co-55, Co-56 and Mo-93m are strongly reduced by both EAF-2010 and TENDL-2014 libraries.

All the calculations show a large underestimation for Zr-86 and Zr-97. For Y-88 and Zr-88 the IEAF-2001 results show a better agreement with the experimental data.

6.2 F82H steel sample

Regarding the main isotope contributors (see Fig.19) the three calculations give quite comparable results for Cr-51, Mn-54 and Mn-56. For Mn-52 the IEAF-2001 value shows a big discrepancy (C/E=3.8) strongly reduced by EAF-2010 and TENDL-2014 calculations.

For the other less important isotopes (see Fig.20) a different behaviour is found for each nuclide. The huge discrepancies shown by the IEAF-2001 library for V-48, Cr-48 and Ta-184 are reduced in both EAF-2010 and TENDL-2014 calculations.

For some isotopes all the calculations show a large underestimation, as, for example, Co-55, Co-56 and Co-60. For Sc-47, IEAF-2001 gives an underestimation of the experimental value whereas TENDL-2014 and EAF-2010 give an overestimation (with a slight better agreement for EAF-2010).

6.3 Vanadium alloy sample

Concerning the main isotope contributors (see Fig.21) for Sc-48 and Cr-51 the three calculations give comparable results. For Sc-47 all the calculations show big discrepancies with respect to the experimental data but a different behaviour. The IEAF-2001 library predicts an underestimation whereas both EAF-2010 and TENDL-2014 give an overestimation.

For the other less important isotopes (see Fig.22) different conclusions can be derived depending by the nuclide. For Ca-47, for example, IEAF-2001 shows a better agreement with the experimental value whereas, for Sc-44, TENDL-2014 gives the best evaluation.

All the calculations show a large underestimation for Sc-43 and Mn-52.

6.4 Pure-Vanadium sample

For Ca-47, Sc-48 and Nb-92m the three calculations give similar results (see Fig.23). For Sc-46 better results are provided by EAF-2010 and IEAF-2001 calculations.

For Sc-47 all the calculations show big discrepancies with respect to the experimental data but a different behaviour. The IEAF-2001 calculations predict an underestimation whereas both EAF-2010 and TENDL-2014 give an overestimation.

All the calculations show a huge underestimation for Cr-51.

7 Conclusion

ANITA-IEAF is an activation package (code and libraries) developed in the past in ENEA-Bologna in order to assess the activation of materials exposed to neutrons with energies greater than 20 MeV.

An updated version of the ANITA-IEAF activation code package has been recently developed. It is suitable to be applied to the study of the irradiation effects on materials in facilities like the International Fusion Materials Irradiation Facility (IFMIF) and the DEMO Oriented Neutron Source (DONES), in which a considerable amount of neutrons with energies above 20 MeV is produced through the reaction of 40 MeV deuterons impinging on a liquid lithium target.

In this paper the main characteristics of the updated version of ANITA-IEAF, able to use decay and cross section data based on the most recent nuclear evaluated data libraries, are described.

The validation effort related to the comparison between the code predictions and the activity measurements obtained from the Karlsruhe Isochronous Cyclotron, for SS-316, F82H, pure vanadium and a vanadium alloy samples, is presented.

The ANITA-IEAF results are compared with those obtained by the FISPACT-II code by using the same EAF-2010 neutron activation cross section library. The results obtained by the two codes are quite comparable confirming that ANITA-IEAF is a reliable tool for the activation calculations in the frame of the fusion field.

In order to check the relevance of the cross section data on the activation responses, a comparison between the experimental results and the calculations performed with different activation cross section libraries is presented (i.e. ANITA-IEAF/EAF-2010, ANITA-IEAF/IEAF-2001 and FISPACT-II/TENDL-2014 calculations).

The calculation/experiment comparison shows that the neutron activation cross sections also of the latest evaluated libraries, at least for some nuclides, need a further assessment and improvement.

References

1. D.G. Cepraga, M. Frisoni, G. Cambi, IFMIF Shielding calculations: ANITA-IEAF activation code package validation against the Karlsruhe Isochronous Cyclotron experiment, ENEA Report FUS-TN-SA-SE-R-070 (2003)
2. D.G. Cepraga, M. Frisoni, G. Cambi, ANITA-IEAF: a code package for performing fusion material transmutation and activation analysis induced by

intermediate energy neutrons, Fusion Engineering and Design **69**, 719 (2003)
3. D.G. Cepraga, G. Cambi, M. Frisoni, G.C. Panini, ANITA-2000, OECD-NEA Data Bank NEA-1638 (2000)
4. D.G. Cepraga, G. Cambi, M. Frisoni, G.C. Panini, ANITA-2000, RSICC CCC-693 (2002)
5. U. Fischer, D. Leichtle, U.von Mollendorff, I. Schmuck, ZZ-IEAF-2001, Intermediate Energy Activation File, OECD-NEA Data Bank NEA-1656/01
6. R.A. Forrest, F.M. Mann, FENDL/D-2.0, Decay Data Library for Fusion Applications, summary documentation by A.B. Pashchenko and H.Wienke, report IAEA-NDS-178 (1997)
7. M.A. Kellet, The JEFF-3.1.1 Decay Data Library, JEFDOC-1188 (2007)
8. J.-Ch. Sublet, L.W. Packer, J. Kopecky, R. A. Forrest, A. J. Koning, D.A. Rochman, The European Activation File: EAF-2010 neutron-induced cross section library, CCFE-R (10) 05
9. EASY-2010 OECD NEA Data Bank NEA-1564/13 (2011)
10. U.von Mollendorff, F. Maekawa, H. Giese, H. Feuerstein, A nuclear simulation experiment for the International Fusion Materials Irradiation Facility (IFMIF), FZKA 6764 (2002)
11. U. von Mollendorff, F. Maekawa, H. Giese, P.P.H. Wilson, Experimental test of structural materials activation in the IFMIF neutron spectrum, Fusion Engineering and Design **51-52** 919 (2000)
12. J-Ch. Sublet, J.W. Eastwood, J.G. Morgan, M. Fleming, M.R. Gilbert, The FISPACT-II User Manual UKAEA-R(11)11 Issue 7, (2015)
13. J-Ch. Sublet, J. Eastwood , G. Morgan, A. Koning, D. Rochman, EASY-II: a system for modelling of n, d, p, γ and α activation and transmutation processes, Joint International Conference on Supercomputing in Nuclear Applications and Monte Carlo 2013 (SNA + MC 2013*)*, La Cité des Sciences et de l'Industrie, Paris, France, October 27-31, 2013
14. ENDF-6 Formats Manual Data Formats and Procedures for the Evaluated Nuclear Data Files ENDF/B-VI and ENDF/B-VII Written by the Members of the Cross Sections Evaluation Working Group, Edited by M. Herman and A. Trkov (2010).
15. Application of the Concepts of Exclusion, Exemption and Clearance, IAEA Safety Standard Series No. RS-G-1.7, IAEA Vienna (2004)
16. IAEA, Clearance levels for radionuclides in solid materials – Application of exemption principles, IAEA-TECDOC-855, IAEA, Vienna (1996)
17. R.A. Forrest. FISPACT-2007: User Manual, Technical Report UKAEA FUS 534, EURATOM/UKAEA Fusion Association (2007)
18. A.J Koning, D.Rochman, TENDL-2014, TALYS-based Evaluated Nuclear Data Library, http://www.talys.eu/tendl-2014 (2014)
19. J.-Ch. Sublet, J.W. Eastwood, J. G. Morgan, M. Fleming, M. R. Gilbert, FISPACT-II, OECD-NEA Data Bank NEA-1890 (2016)

Production and testing of the ENEA-Bologna VITJEFF32.BOLIB (JEFF-3.2) multi-group (199 n + 42 γ) cross section library in AMPX format for nuclear fission applications

Massimo Pescarini*, *Roberto* Orsi, and *Manuela* Frisoni

ENEA, Italian National Agency for New Technologies, Energy and Sustainable Economic Development, "E. Clementel" Research Centre, I-40129 Bologna, Italy

Abstract. The ENEA-Bologna Nuclear Data Group produced the VITJEFF32.BOLIB multi-group coupled neutron/photon (199 n + 42 γ) cross section library in AMPX format, based on the OECD-NEA Data Bank JEFF-3.2 evaluated nuclear data library. VITJEFF32.BOLIB was conceived for nuclear fission applications as European counterpart of the ORNL VITAMIN-B7 similar library (ENDF/B-VII.0 data). VITJEFF32.BOLIB has the same neutron and photon energy group structure as the former ORNL VITAMIN-B6 reference library (ENDF/B-VI.3 data) and was produced using similar data processing methodologies, based on the LANL NJOY-2012.53 nuclear data processing system for the generation of the nuclide cross section data files in GENDF format. Then the ENEA-Bologna 2007 Revision of the ORNL SCAMPI nuclear data processing system was used for the conversion into the AMPX format. VITJEFF32.BOLIB contains processed cross section data files for 190 nuclides, obtained through the Bondarenko (f-factor) method for the treatment of neutron resonance self-shielding and temperature effects. Collapsed working libraries of self-shielded cross sections in FIDO-ANISN format, used by the deterministic transport codes of the ORNL DOORS system, can be generated from VITJEFF32.BOLIB through the cited SCAMPI version. This paper describes the methodology and specifications of the data processing performed and presents some results of the VITJEFF32.BOLIB validation.

1 Introduction

In the last two decades, the three-dimensional (3D) deterministic discrete ordinates (S_N) transport codes using group-wise cross section libraries extended their practical simulation capability to an enlarged variety of nuclear systems with different neutron spectra and high geometrical complexity. In particular they increased their competitiveness with respect to the 3D Monte Carlo codes in radiation shielding and damage safety analyses in nuclear fission reactor applications. Comparable or even more convenient performances in terms of CPU times were in fact obtained with the same calculation precision, similar description capability of complex geometries, suitable simulation of different neutron and photon spectral conditions. Various factors concurred to determine specifically the increased flexibility and convenience of the 3D SN codes: 1) the impressive increased performance of modern computers, 2) innovative fine-group pseudo-problem-independent cross section libraries with upscatter cross sections, 3) new ancillary pre/post-processor systems of programs dedicated to 2D and 3D deterministic transport codes and 4) no need to justify, as in the case of the Monte Carlo codes, the statistics adopted with respect to the requirements of the nuclear safety authorities. For example the ORNL VITAMIN-B6 [1] (ENDF/B-VI.3 [2] data; ORNL, 1996), the ORNL VITAMIN-B7 [3] (ENDF/B-VII.0 [4] data; ORNL, 2011) and the ENEA-Bologna VITJEFF311.BOLIB [5] (JEFF-3.1.1 [6] [7] data; ENEA-Bologna, 2011) are examples of libraries of the cited type, based on the Bondarenko [8] method for neutron resonance self-shielding and temperature effects. Concerning the ancillary systems to the S_N transport codes, the free availability, since 2001 of the ENEA-Bologna BOT3P system, based on combinatorial geometry algorithms, it is underlined here. BOT3P, now available in the BOT3P-5.3 [9] version, can be used in particular for the automatic generation and the graphical verification of the spatial mesh grids of the geometrical model of the following S_N 3D transport codes: TORT-3.2 [10] (ORNL DOORS [11] system], LANL PARTISN [12], KATRIN (KIAM CNCSN 2009 [13] system) and the commercial code RAPTOR-M3G [14], currently used [15] by Westinghouse together with BOT3P. Up to recent times, only the 3D Monte Carlo codes could obtain satisfactory results with high calculation precision in the description of complex geometries. At present, with the recent availability of the cited ancillary pre/post-processor systems, the 3D S_N codes increased dramatically their geometrical simulation capability and this induced an increased interest to give more availability of updated group-wise cross section libraries. This is justified both in terms of the previously cited fine-group coupled n/γ pseudo-problem-independent mother libraries and broad-group

* Corresponding author: massimo.pescarini@enea.it

coupled n/γ working libraries with parameterized sets of self-shielded cross sections for a specific type of fission reactor, obtained through problem dependent cross section collapsing from the previously cited fine-group mother libraries. Examples of broad-group working libraries for LWR shielding and pressure vessel dosimetry applications are: BUGLE-96 [1] derived from VITAMIN-B6, BUGLE-B7 [3] from VITAMIN-B7 and BUGJEFF311.BOLIB [16] from VITJEFF311.BOLIB. BUGJEFF311.BOLIB is in particular the first BUGLE-96-type freely released working library based on JEFF data. This is the background which encouraged the ENEA-Bologna Nuclear Data Group to continue the generation of updated fine-group cross section libraries of VITAMIN-B6-type and derived collapsed working libraries of BUGLE-96-type in the group structures and data processing methodology presently recommended by the US ANSI/ANS-6.1.2-1999 (R2009) [17] standard. In the present paper in particular the generation of the VITJEFF32.BOLIB [18] fine-group coupled neutron and photon (199 n + 42 γ) cross section library in AMPX format for nuclear fission applications, based on the OECD-NEA Data Bank JEFF-3.2 most updated evaluated nuclear data library, is described.

2 Processing codes and procedures

The VITJEFF32.BOLIB library was derived (see Figs. 1 and 2 and Table 1) from cross section data files in GENDF format, generated through the LANL NJOY-2012.53 [19] nuclear data processing system. The following modules of NJOY were used to process neutron interaction (n-n), gamma-ray production (n-γ) and gamma-ray interaction (γ-γ) data, from the JEFF-3.2 incident neutron and photo-atomic data in ENDF-6 format into the GENDF format. Specifically, the MODER, RECONR, BROADR, THERMR, HEATR, GASPR, PURR, GROUPR modules were used for the incident neutron data while the MODER, RECONR and GAMINR modules were used for the photo-atomic data. Then the revised version of the SMILER module of the ENEA-Bologna 2007 Revision [20] of the ORNL SCAMPI [21] nuclear data processing system was used to translate the multi-group data from the GENDF format into the AMPX master library format of VITJEFF32.BOLIB. It is underlined that, in particular, this ENEA revised version of SMILER permits to obtain separately the prompt fission spectrum (MF=6 and MT=18), the delayed fission spectrum (MF=5 and MT=455) or the total fission spectrum needed for fixed source transport calculations, differently from the original ORNL SMILER version, used to generate VITAMIN-B6, which permits to obtain only the prompt component. Then the module AIM of SCAMPI was used to convert the master cross section library from binary to BCD format. The availability of VITJEFF32.BOLIB permits now to collapse new updated self-shielded broad-group working cross section libraries which, e.g., can be produced in the FIDO-ANISN format with the previously cited ENEA-Bologna 2007 Revision of SCAMPI from VITJEFF32.BOLIB. More specifically, the cross sections in AMPX format can be collapsed through the MALOCS module, can be self-shielded through the BONAMI module and finally can be produced into the AMPX or FIDO-ANISN formats. The cross sections in AMPX format can be used by the XSDRNPM code (1D S_N), included in the SCAMPI or in the SCALE [22] systems. Moreover the cross sections in FIDO-ANISN format can be used for example by the discrete ordinates (S_N) transport codes ANISN-ORNL (1D), DORT (2D) and TORT (3D) of the DOORS system of deterministic codes. The present calculations were performed on a Personal Computer (CPU INTEL Pentium III, 448 MB of RAM; FSF-g77 version 0.5.26 FORTRAN compiler for NJOY and f77 Absoft version 5.0 FORTRAN 77 compiler for SCAMPI) with the Linux Red Hat 7.1 operative system.

3 Library specifications

A set of 190 JEFF-3.2 cross section data files was processed for the VITJEFF32.BOLIB library. The increased number of materials with respect to the VITAMIN-B6 (120 materials) library is mainly caused by the almost complete availability of isotopic data files in the JEFF-3.2 evaluated nuclear data library in substitution of natural element data files. The Bondarenko (f-factor) method was used for handling resonance self-shielding and temperature effects. As for VITAMIN-B6, all the 184 standard (not bound) nuclides were processed at 4 temperatures (300 K, 600 K, 1000 K and 2100 K). Concerning the background cross section σ_0 values adopted, a conservative choice was performed with respect to both ORNL VITAMIN-B7 and VITAMIN-B6 libraries. Differently from VITAMIN-B6 and similarly to VITAMIN-B7, each standard nuclide was processed with the same set of twelve values (1.0E-06, 0.01, 1, 10, 50, 100, 300, 1000, 1.0E+04, 1.0E+05, 1.0E+06 and 1.0E+10 barns) for σ_0. In particular the 1.0E+10 value corresponds to the infinite dilution background cross section whereas 1.0E-06 represents the total self-shielding value. This approach, i.e. the same set of twelve background cross sections for each standard nuclide, was more similar to that employed in the VITAMIN-B7 generation but, in addition to the values used for the data processing of VITAMIN-B7, the 1.0E-06 and 0.01 values were included. The 1.0E-06 value was chosen as near as possible to 0.0 (in line with the values used in the past for getting close to 0.0), according to the previously generated OECD-NEADB VITAMIN-J [23] library. Concerning then the 0.01 value, it is underlined that this additional σ_0 numerical value improves further (see [24]), for example, the neutron cross section self-shielding of Fe-56 in natural iron and of Co-59 (only isotope in natural cobalt), through a more precise self-shielding factor interpolation. Thermal scattering cross sections were produced for six additional bound nuclides which were processed at all the temperatures available in the JEFF-3.2 thermal scattering law data file which is the same as the ones included in the JEFF-3.1.2, JEFF-3.1.1 and JEFF-3.1 evaluated nuclear data libraries: H-1 in light

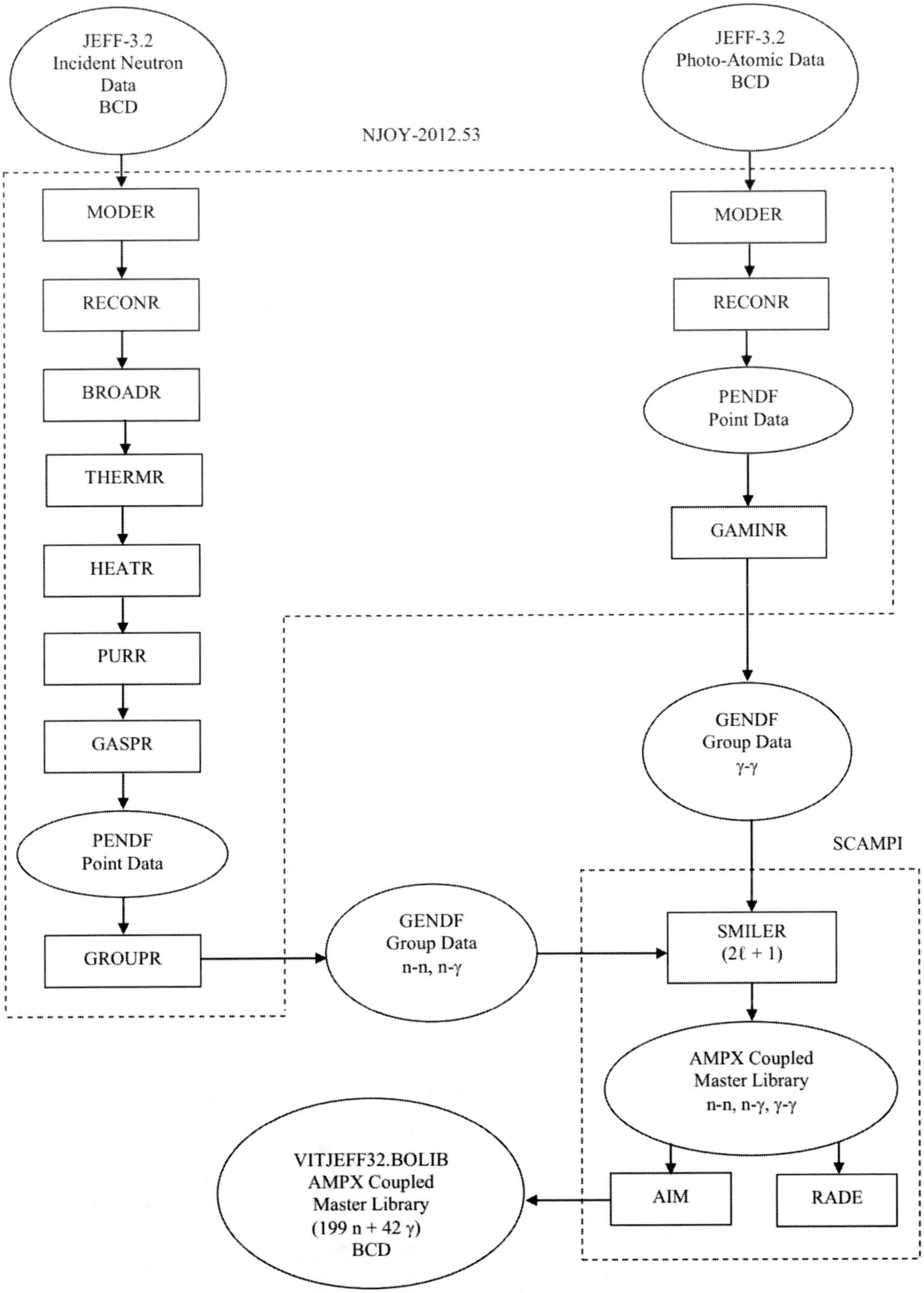

Fig. 1. Procedure for generating the VITJEFF32.BOLIB standard nuclides in AMPX format from JEFF-3.2.

DOI: 10.1051/epjconf/201715307003

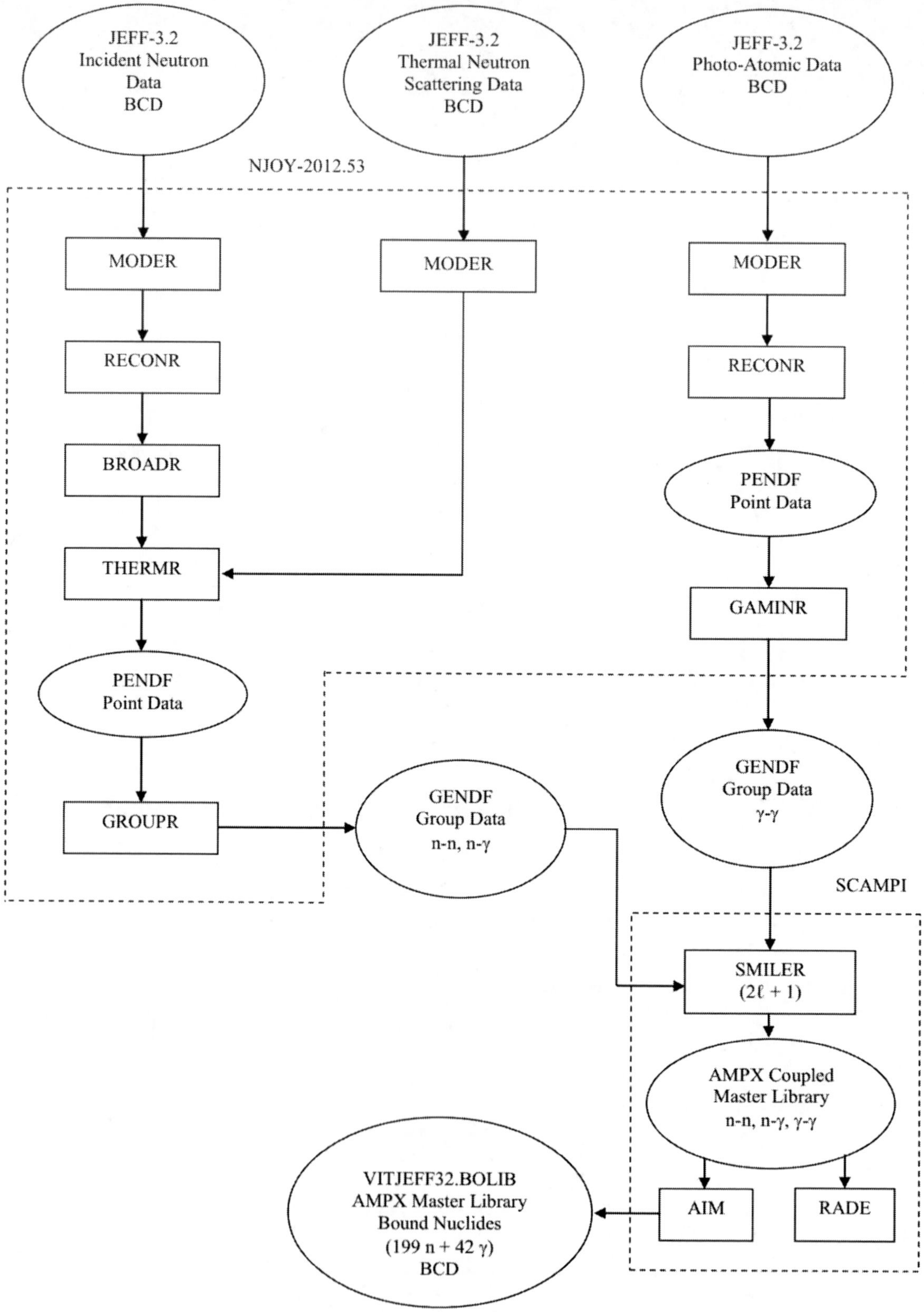

Fig. 2. Procedure for generating the VITJEFF32.BOLIB bound nuclides in AMPX format from JEFF-3.2.

 DOI: 10.1051/epjconf/201715307003

Table 1. Modules from the NJOY-2012.53 and SCAMPI nuclear data processing systems used to process VITJEFF32.BOLIB.

NJOY-2012.53 System

Module	Function
MODER	Converts between ENDF standard coded mode and the NJOY blocked binary mode.
RECONR	Reconstructs point-wise cross sections from JEFF resonance parameters and interpolation schemes.
BROADR	Doppler-broadens and thins point-wise cross sections.
THERMR	Produces cross sections and energy-to-energy matrices for free or bound scatterers in the thermal energy range.
HEATR	Generates heat production cross sections (KERMA factors) and damage-energy production.
GASPR	Computes gas production cross sections.
PURR	Computes probability tables and effective point-wise self-shielded cross sections in the unresolved energy range.
GROUPR	Generates self-shielded multi-group cross sections and group-to-group scattering and photon production matrices in GENDF format.
GAMINR	Computes multi-group photo-atomic cross sections, KERMA factors, group-to-group photon scattering matrices.

SCAMPI System (ENEA-Bologna 2007 Revision)

Module	Function
SMILER	Translates GENDF files produced by NJOY into AMPX master interface format.
RADE	Performs sanity and consistency tests on multi-group libraries.
AIM	Converts master cross section libraries from binary format to BCD (or vice-versa).

water, H-1 in polyethylene, H-1 in zirconium hydride (not contained in the VITAMIN-B6 library), H-2 in heavy water, C in graphite and Be in beryllium metal. VITJEFF32.BOLIB adopts the same problem-independent group structure of the VITAMIN-B6 library with 199 neutron and 42 photon energy groups, able to treat not only "fast" but also "thermal" neutron (water or graphite-moderated) reactor systems, through the adequate group structure at lower energies. It is underlined that the thermal neutron energy range below 5.043 eV, i.e. the range of the groups which include up-scatter, contains 36 groups. The photon energy group structure includes the top energy group with uppermost limit extended to 30 MeV. The neutron and photon weighting functions are the same as in VITAMIN-B6. The neutron weighting function is of the form typically chosen for fission reactor shielding problems, i.e. it consists of a smoothly varying combination of the following neutron spectra:

1. Maxwellian thermal spectrum (kT = 0.025 eV)
   $W_1(E) = C_1Ee^{-E/kT}$ from 1.0E-5 eV to 0.125 eV;
2. l/E slowing down spectrum
   $W_2(E) = C_2/E$ from 0.125 eV to 820.8 keV;
3. Fission spectrum (θ = 1.273 MeV)
   $W_3(E) = C_3E^{1/2}e^{E/\theta}$ from 820.8 keV to 20 MeV.

This corresponds to an IWT=4 option in GROUPR with the C_1, C_2 and C_3 constants assuring the continuity of the spectrum function. The fission temperature θ has been adjusted to better reflect the neutron spectrum in a thermal reactor (θ = 1.273 MeV versus θ = 1.41 MeV for the VITAMIN-E [25] library dedicated to fast fission and fusion applications). The photon weighting spectrum consists of a l/E spectrum with a "roll-off" of the spectrum at lower energies to represent photoelectric absorption and a similar drop-off of the spectrum at higher energies corresponding to the Q-value for neutron capture. This corresponds to the IWT=3 option in GAMINR. In VITJEFF32.BOLIB the order of scattering used for both neutrons and photons is P7, for nuclides with Z=l through Z=30 (zinc) and P5 for the remainder of the nuclides. The numerical values of the fractional error tolerances in NJOY for both the libraries, 0.1% for resolved resonance reconstruction and for linearization in RECONR and 0.1% for thinning in BROADR, were reduced with respect to the 0.2% corresponding values in VITAMIN-B6. About the available neutron fission spectra, the total (prompt + delayed) neutron fission spectra χ for U-235, U-238 and Pu-239, obtained from the specific GENDF files through the previously cited ENEA-Bologna version of SMILER and the ICE modules of SCAMPI, were included in tabulated form into each package of the two libraries. It is underlined that the corresponding data in VITAMIN-B6 include only the prompt spectrum components.

4 Examples of library validation

VITJEFF32.BOLIB was extensively tested on about 80 thermal, intermediate and fast spectrum neutron criticality benchmark experiments. The experimental details (geometry, composition and k-effective) were taken from the ICSBEP Handbook [26]. The data processing of the problem-dependent working cross section libraries and the transport calculations were performed using the ENEA-Bologna 2007 Revision of SCAMPI. In particular, the BONAMI module generated the neutron self-shielded cross section working libraries and the XSDRNPM 1D discrete ordinates (S_N) transport module calculated the k-effective (k_{eff}) results. All the calculations were performed in spherical geometry, in the P_5-S_{16} approximation. In particular, P_5 corresponds to L=ℓ-max, the maximum order of the Legendre polynomial (P_ℓ) expansion of the scattering cross section matrices and S_{16} represents the order of the flux angular discretization.

The k_{eff} results obtained with VITJEFF32.BOLIB were compared with the corresponding results calculated using the VITJEFF311.BOLIB library, previously processed in ENEA-Bologna through NJOY-99.259 [27].

A simplified data processing and validation flow-chart of the VITJEFF32.BOLIB library is shown in Fig. 3.

A partial set of k-effective results for metallic fast neutron spectrum criticality benchmark experiments is presented in Table 2, where the most relevant differences between the results obtained from the VITJEFF32.BOLIB and VITJEFF311.BOLIB libraries were found.

Following the denominations reported in the ICSBEP handbook, the k-effective results, obtained in the 1D transport calculations using the VITJEFF32.BOLIB library and reported in Table 2, refer to HEU (Highly Enriched Uranium) benchmark experiments, to U-233 benchmark experiments and to Pu-239 benchmark experiments. The following acronyms are used in the cited table: NU (natural uranium), DU (depleted uranium) and WC (wolfram carbide or tungsten carbide).

It is underlined that in the ICSBEP Handbook the assumption of a normal probability distribution for the k_{eff} values of the "experimental or benchmark-model" configurations is considered usually reasonable. Consequently the k_{eff} values reported in Table 2 are the mean values of the assumed normal distributions and their associated uncertainties $\pm\Delta k_{eff}$ correspond to one standard deviation (1σ) which represents a confidence level with 68% probability that the measurement result is contained within $\pm1\sigma$.

The VITJEFF32.BOLIB HEU-MET-FAST k_{eff} results present a systematic increase ranging from about 100 to about 700 pcm and are in general less adherent to the experimental results with respect to the corresponding results obtained with VITJEFF311.BOLIB. On the other hand the results obtained using VITJEFF32.BOLIB are almost always overestimated with respect to the corresponding k_{eff} experimental results.

The VITJEFF32.BOLIB U-233-MET-FAST k_{eff} results appear systematically lower of about 200÷400 pcm and more adherent to the experimental results with respect to the corresponding results obtained with VITJEFF311.BOLIB. On the other hand the results obtained using VITJEFF32.BOLIB are systematically overestimated with respect to the corresponding k_{eff} experimental results.

The VITJEFF32.BOLIB PU-MET-FAST k_{eff} results appear systematically higher of about 100÷300 pcm and in general less adherent to the experimental results with respect to the corresponding results obtained with VITJEFF311.BOLIB. The results obtained using VITJEFF32.BOLIB are in general overestimated with respect to the corresponding k_{eff} experimental results.

It is underlined that all the cited results are always contained within a confidence level of the total experimental error of $\pm3\sigma$ (99.7% probability that the measurement result is between ±3 standard deviations).

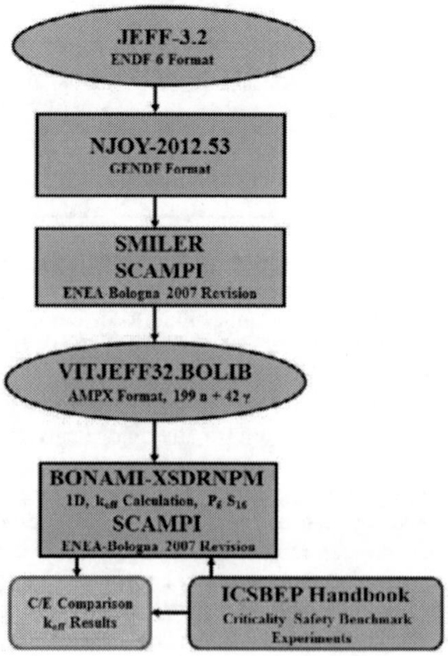

Fig. 3. Simplified data processing and validation flow-chart of the VITJEFF32.BOLIB Library.

Table 2. k-effective calculated results for metallic criticality safety benchmark experiments with fast neutron spectrum.

ICSBEP Handbook Benchmark Name	Reflector	Experimental(*) or Benchmark-Model $k_{eff} \pm \Delta k_{eff}$(pcm)	Geometry/ P_L - S_N	VITJEFF32. BOLIB k_{eff}	VITJEFF311. BOLIB k_{eff}
Highly Enriched Metallic Uranium with Fast Neutron Spectrum					
HEU-MET-FAST-001-001 (GODIVA)	--	1.00000 ± (100)	1D Sph./ P5-S16	1.00051	0.99533
HEU-MET-FAST-002-001 (TOPSY)	NU	1.00000 ± (300)	1D Sph./ P5-S16	1.00779	1.00095
HEU-MET-FAST-003-001	NU (2in)	1.00000 ± (500)	1D Sph./ P5-S16	0.99865	0.99373
HEU-MET-FAST-003-002	NU (3in)	1.00000 ± (500)	1D Sph./ P5-S16	0.99887	0.99345
HEU-MET-FAST-003-003	NU (4in)	1.00000 ± (500)	1D Sph./ P5-S16	1.00415	0.99825
HEU-MET-FAST-003-004	NU (5in)	1.00000 ± (300)	1D Sph./ P5-S16	1.00269	0.99643
HEU-MET-FAST-003-005	NU (7in)	1.00000 ± (300)	1D Sph./ P5-S16	1.00742	1.00071
HEU-MET-FAST-003-006	NU (8in)	1.00000 ± (300)	1D Sph./ P5-S16	1.00776	1.00092
HEU-MET-FAST-003-007	NU (11in)	1.00000 ± (300)	1D Sph./ P5-S16	1.00849	1.00148
HEU-MET-FAST-003-008	WC (1.9in)	1.00000 ± (500)	1D Sph./ P5-S16	1.00461	1.00146
HEU-MET-FAST-003-009	WC (2.9in)	1.00000 ± (500)	1D Sph./ P5-S16	1.00613	1.00364
HEU-MET-FAST-003-010	WC (4.5in)	1.00000 ± (500)	1D Sph./ P5-S16	1.01029	1.00897
HEU-MET-FAST-003-011	WC (6.5in)	1.00000 ± (500)	1D Sph./ P5-S16	1.01479	1.01416
HEU-MET-FAST-003-012	Ni (8in)	1.00000 ± (500)	1D Sph./ P5-S16	1.01172	1.00922
HEU-MET-FAST-022-001	Al	1.00000 ± (190)	1D Sph./ P5-S16	1.00005	0.99527
HEU-MET-FAST-027-001	Pb	1.00000 ± (250)	1D Sph./ P5-S16	1.00458	1.00052
HEU-MET-FAST-028-001 (FLATTOP-25)	NU	1.00000 ± (300)	1D Sph./ P5-S16	1.00895	1.00218
HEU-MET-FAST-057-001	Pb	1.00000 ± (200)	1D Sph./ P5-S16	0.99684	0.99425
HEU-MET-FAST-057-002	Pb	1.00000 ± (230)	1D Sph./ P5-S16	1.00437	1.00122
Metallic U-233 with Fast Neutron Spectrum					
U233-MET-FAST-001-001 (JEZEBEL-233)	--	1.00000 ± (100)	1D Sph./.P5-S16	1.00083	1.00383
U233-MET-FAST-002-001	HEU	1.00000 ± (100)	1D Sph./.P5-S16	1.00011	1.00257
U233-MET-FAST-002-002	HEU	1.00000 ± (110)	1D Sph./.P5-S16	1.00166	1.00342
U233-MET-FAST-003-001	NU	1.00000 ± (100)	1D Sph./.P5-S16	1.00189	1.00575
U233-MET-FAST-003-002	NU	1.00000 ± (100)	1D Sph./.P5-S16	1.00363	1.00719
U233-MET-FAST-006-001 (FLATTOP-23)	NU	1.00000 ± (140)	1D Sph./.P5-S16	1.00472	1.00665
Metallic Pu-239 with Fast Neutron Spectrum					
PU-MET-FAST-001-001 (JEZEBEL)	--	1.00000 ± (200)	1D Sph./ P5-S16	1.00044	0.99889
PU-MET-FAST-002-001 (JEZEBEL-240)	--	1.00000 ± (200)	1D Sph./ P5-S16	1.00267	1.00277
PU-MET-FAST-006-001 (FLATTOP-PU)	NU	1.00000 ± (300)	1D Sph./ P5-S16	1.00642	1.00305
PU-MET-FAST-008-001 (THOR)	Th	1.00000 ± (60)	1D Sph./ P5-S16	0.99861	1.00124
PU-MET-FAST-009-001	Al	1.00000 ± (270)	1D Sph./ P5-S16	1.00468	1.00401
PU-MET-FAST-010-001	NU	1.00000 ± (180)	1D Sph./ P5-S16	1.00257	1.00091
PU-MET-FAST-011-001	H2O	1.00000 ± (100)	1D Sph./ P5-S16	1.00242	0.99943
PU-MET-FAST-018-001	Be	1.00000 ± (300)	1D Sph./ P5-S16	1.00290	1.00200
PU-MET-FAST-023-001	Graphite	1.00000 ± (230)	1D Sph./ P5-S16	1.00027	0.99850
PU-MET-FAST-024-001	Polyethylene	1.00000 ± (200)	1D Sph./ P5-S16	1.00227	0.99948
PU-MET-FAST-030-001	Graphite	1.00000 ± (210)	1D Sph./ P5-S16	1.00439	1.00395
PU-MET-FAST-031-001	Polyethylene	1.00000 ± (210)	1D Sph./ P5-S16	1.00607	1.00385

(*) The ± Δk_{eff} uncertainties correspond to 1 standard deviation (1σ).

5 Conclusion

The present library was transferred to OECD-NEADB where it received the denomination NEA-1891 ZZ VITJEFF32.BOLIB and is presently in free distribution. The generation of a collapsed working library in FIDO-ANISN format of self-shielded cross sections into the recommended BUGLE-96 (47 n + 20 γ energy groups) group structure, dedicated to neutron/photon shielding and LWR pressure vessel dosimetry in nuclear safety transport analyses, can be now performed using the updated JEFF-3.2 data and the ENEA-Bologna 2007 Revision of the SCAMPI system. It is considered important to extend further the availability of group-wise cross section libraries to offer, in particular, to the users of 3D deterministic codes a flexible possibility to generate working cross section libraries based on updated data files. Finally it is noted that the use of the deterministic transport codes together with the necessary multi-group working cross section libraries is often preferred by the industrial organizations that must fulfil quality assurance procedures in reactor safety calculations. In fact there is no need, as in the case of the Monte Carlo codes, to justify the validity of the statistics adopted since the deterministic codes are based on rigorous analytical solutions of the neutral particle transport equations.

References

1. J.E. White, D.T. Ingersoll, R.Q. Wright, H.T. Hunter, C.O. Slater, N.M. Greene, R.E. MacFarlane, R.W. Roussin, Production and Testing of the Revised VITAMIN-B6 Fine-Group and the BUGLE-96 Broad-Group Neutron/Photon Cross-Section Libraries Derived from ENDF/B-VI.3 Nuclear Data, ORNL, Oak Ridge, ORNL-6795/R1, NUREG/CR-6214, Revision 1 (January 1995)
2. P.F. Rose, ENDF/B-VI Summary Documentation, BNL, Brookhaven, BNL-NCS-17541 (ENDF-201) 4th Edition (October 1991)
3. J.M. Risner, D. Wiarda, M.E. Dunn, T.M. Miller, D.E. Peplow, and B.W. Patton, Production and Testing of the VITAMIN-B7 Fine-Group and BUGLE-B7 Broad-Group Coupled Neutron/Gamma Cross-Section Libraries Derived from ENDF/B-VII.0 Nuclear Data, ORNL, Oak Ridge, ORNL/TM-2011/12, NUREG/CR-7045 (September 2011)
4. M.B. Chadwick et al., ENDF/B-VII.0: Next Generation Evaluated Nuclear Data Library for Nuclear Science and Technology, Nuclear Data Sheets, Volume **107**, Number 12, pp. 2931-3060 (December 2006)
5. M. Pescarini, V. Sinitsa, R. Orsi, VITJEFF311.BOLIB - A JEFF-3.1.1 Multigroup Coupled (199 n + 42 γ) Cross Section Library in AMPX Format for Nuclear Fission Applications, ENEA-Bologna Technical Report UTFISSM-P9H6-003 (November 10, 2011); ENEA-Bologna Technical Report UTFISSM-P9H6-003 Revision 1 (March 14, 2013)
6. *The JEFF-3.1.1 Nuclear Data Library*, JEFF Report 22, OECD-NEA Data Bank (2009)
7. *The JEFF-3.1 Nuclear Data Library*, JEFF Report 21, OECD-NEA Data Bank (2006)
8. I.I. Bondarenko, M.N. Nikolaev, L.P. Abagyan, N.O. Bazaziants, *Group Constants for Nuclear Reactors Calculations* (Consultants Bureau, New York, 1964)
9. R. Orsi, BOT3P Version 5.3: A Pre/Post-Processor System for Transport Analysis, ENEA-Bologna Technical Report FPN-P9H6-011 (October 22, 2008)
10. W.A. Rhoades, D.B. Simpson, The TORT Three-Dimensional Discrete Ordinates Neutron/Photon Transport Code (TORT Version 3), ORNL, Oak Ridge, ORNL/TM-13221 (October 1997)
11. DOORS3.1: One-, Two- and Three-Dimensional Discrete Ordinates Neutron/Photon Transport Code System, ORNL, Oak Ridge, RSIC Computer Code Collection CCC-650 (August 1996)
12. R.E. Alcouffe, R.S. Baker, J.A. Dahl, S.A. Turner, R. Ward, PARTISN: A Time-Dependent, Parallel Neutral Particle Transport Code System, LANL, Los Alamos, LA-UR-08-07258 (Revised November 2008)
13. A.M. Voloschenko, S.V. Gukov, V.P. Kryuchkov, A.A. Dubinin, O.V. Sumaneev, The CNCSN One-, Two- and Three-Dimensional Coupled Neutral and Charged Particle Discrete Ordinates Code Package, M&C-2005, Avignon, France (2005)
14. G. Longoni, S.L. Anderson, Reactor Dosimetry Applications Using RAPTOR-M3G: A New Parallel 3-D Radiation Transport Code, ISRD-13, Akersloot, The Netherlands (2008)
15. F.A. Alpan, J.A. Kulesza, Comparison of Standard Light Water Reactor Cross-Section Libraries Using the United States Nuclear Regulatory Commission Pressurized Water Reactor Standard Core Loading Benchmark Problem, ISRD-15, Aix-en-Provence, France (2014)
16. M. Pescarini, V. Sinitsa, R. Orsi, M. Frisoni, BUGJEFF311.BOLIB - A JEFF-3.1.1 Broad-Group Coupled (47 n + 20 γ) Cross Section Library in FIDO-ANISN Format for LWR Shielding and Pressure Vessel Dosimetry Applications, ENEA-Bologna Technical Report UTFISSM-P9H6-002, (May 12, 2011); ENEA-Bologna Technical Report UTFISSM-P9H6-002 Revision 1 (March 14, 2013)
17. American National Standard, American Nuclear Society, Neutron and Gamma-Ray Cross Sections for Nuclear Radiation Protection Calculations for Nuclear Power Plants, ANSI/ANS-6.1.2-1999 (Reapproved 2009)
18. M. Pescarini, R. Orsi, M. Frisoni, VITJEFF32.BOLIB - A JEFF-3.2 Multi-Group Coupled (199 n + 42 γ) Cross Section Library in AMPX Format for Nuclear Fission Applications ENEA-Bologna Technical Report SICNUC-P9H6-002 (February 2016)
19. R.E. MacFarlane, D.W. Muir, R.M. Boicourt, A.C. Kahler, The NJOY Nuclear Data Processing System, Version 2012, LANL, Los Alamos, LA-UR-12-

27079 Rev, Updated for NJOY2012.50 (February 2015).
20. V. Sinitsa, M. Pescarini, ENEA-Bologna 2007 Revision of the SCAMPI (ORNL) Nuclear Data Processing System, ENEA-Bologna Technical Report FPN-P9H6-006 (September 13, 2007)
21. SCAMPI, ORNL, Oak Ridge, RSIC PSR-352, (September 1995). Available from OECD-NEA Data Bank as PSR-0352/06 SCAMPI, last version of the ENEA-Bologna 2007 Revision (see [20]) (July 2015)
22. SCALE: A Comprehensive Modeling and Simulation Suite for Nuclear Safety Analysis and Design, ORNL, Oak Ridge, ORNL/TM-2005/39, Version 6.1 (June 2011)
23. E. Sartori, VITAMIN-J, a 175 Group Neutron Cross Section Library Based on JEF-1 for Shielding Benchmark Calculations, OECD-NEA Data Bank, JEF/DOC-100 Preliminary (October 1985)
24. C. Konno, K. Takakura, K. Kondo, S. Ohnishi, K. Ochiai and S. Sato, Important Remarks on Latest Multigroup Libraries, Progress in Nuclear Science and Technology, Vol. **2**, pp. 341-345 (2011)
25. C.R. Weisbin, R.W. Roussin, J.J. Wagschal, J.E. White, R.Q. Wright, VITAMIN-E: An ENDF/B-V Multigroup Cross-Section Library for LMFBR Core and Shield, LWR Shield, Dosimetry and Fusion Blanket Technology, ORNL, Oak Ridge, ORNL-5505 (ENDF-274) (February 1979)
26. ICSBEP, OECD-NEA NSC, NEA/NSC/DOC(95)03 (September 2004 Edition)
27. R.E. MacFarlane, NJOY-99, "README0", ORNL, RSIC PSR-0480/02 (December 1999)

Monitoring System for the Gold Target by Radiation Detectors in Hadron Experimental Facility at J-PARC

Ryotaro Muto[1,*], *Keizo* Agari[1], *Kazuya* Aoki[1], *Kotaro* Bessho[1], *Masayuki* Hagiwara[1], *Erina* Hirose[1], *Masaharu* Ieiri[1], *Ruri* Iwasaki[1], *Yohji* Katoh[1], *Jun-ichi* Kitagawa[1], *Michifumi* Minakawa[1], *Yuhei* Morino[1], *Kiwamu* Saito[1], *Yoshinori* Sato[1], *Shin'ya* Sawada[1], *Yoshihisa* Shirakabe[1], *Yoshihiro* Suzuki[1], *Hitoshi* Takahashi[1], *Kazuhiro* Tanaka[1], *Akihisa* Toyoda[1], *Hiroaki* Watanabe[1], and *Yutaka* Yamanoi[1]

[1]High Energy Accelerator Research Organization, 1-1 Oho, Tsukuba, Ibaraki 3050801, JAPAN

Abstract. At the Hadron Experimental Facility in J-PARC, we inject a 30-GeV proton beam into a gold target to produce secondary particle beams required for various particle and nuclear physics experiments. The gold target is placed in a hermetic chamber, and helium gas is circulated in the chamber to monitor the soundness of the target. The radioactivity in helium gas is continuously monitored by gamma-ray detectors such as a germanium detector and a NaI(Tl) detector. Beam operations with those target-monitoring systems were successfully performed from April to June and October to December 2015, and from May to June 2016. In this paper, the details of the helium gas circulation system and gamma-ray detectors and the analysis results of the obtained gamma-ray spectra are reported.

1 Introduction

The Hadron Experimental Facility (HEF) is one of the experimental facilities in Japan Proton Accelerator Research Complex (J-PARC) [1] located in Tokai, Ibaraki, Japan. In the HEF, we inject the proton beam into a gold target to produce secondary particles such as kaons and pions and use them for various particle and nuclear experiments [2]. Figure 1 shows the accelerator configuration of the J-PARC and Figure 2 shows the schematic view of the HEF.

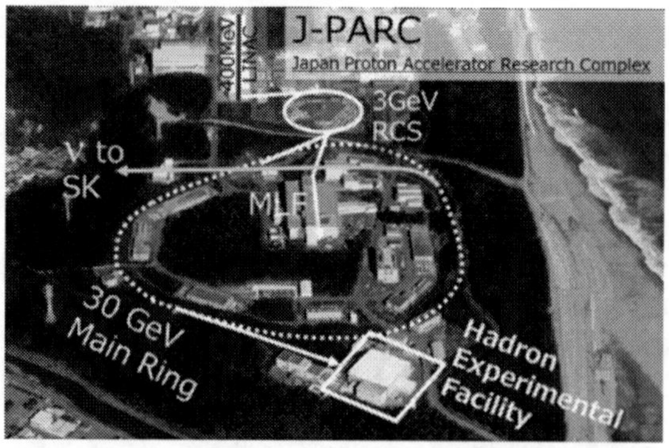

Fig. 1. Accelerator configuration of the J-PARC. Protons are accelerated up to 30 GeV through the 400-MeV linac, 3-GeV Rapid Cycle Synchrotron, and 30-GeV Main Ring and slowly extracted to the Hadron Experimental Facility (HEF).

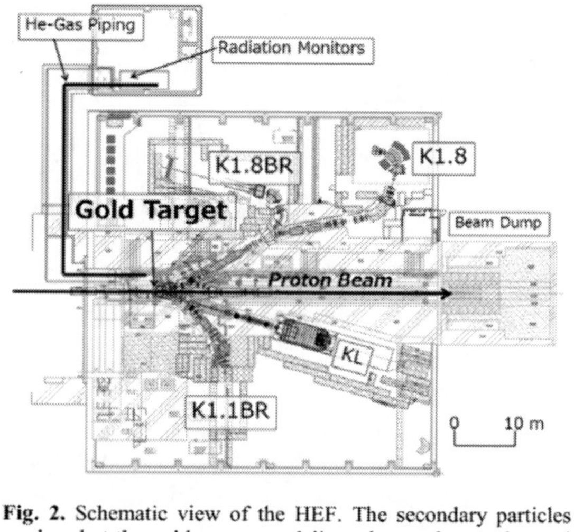

Fig. 2. Schematic view of the HEF. The secondary particles produced at the gold target are delivered to each experimental area. The locations of He-gas piping and radiation monitors for the monitoring system are also shown.

The proton beam accelerated up to 30 GeV is slowly extracted from the Main Ring for 2 seconds within 6 or 5.52 seconds-accelerator cycles and injected into the gold target [3]. The gold target is bonded onto a water-cooled copper base by the heat isostatic pressing (HIP) method. Figure 3 shows a photograph and a schematic cross section of the gold target.

* Corresponding author: ryotaro.muto@kek.jp

DOI: 10.1051/epjconf/201715307004

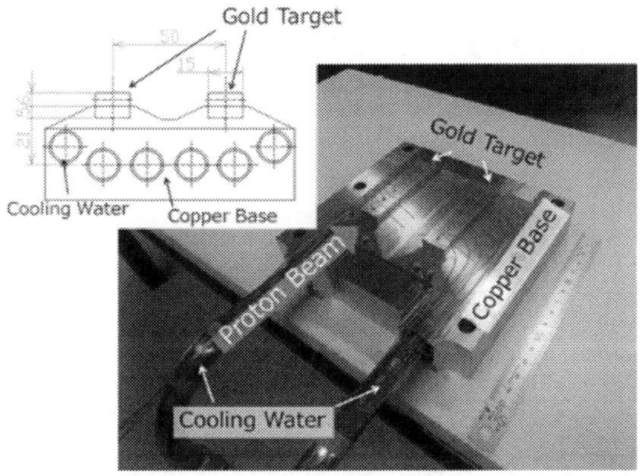

Fig. 3. Photograph and schematic cross section of the gold target. Two gold targets are mounted on a water-cooled copper base: one is for the beam operation and the other is a spare.

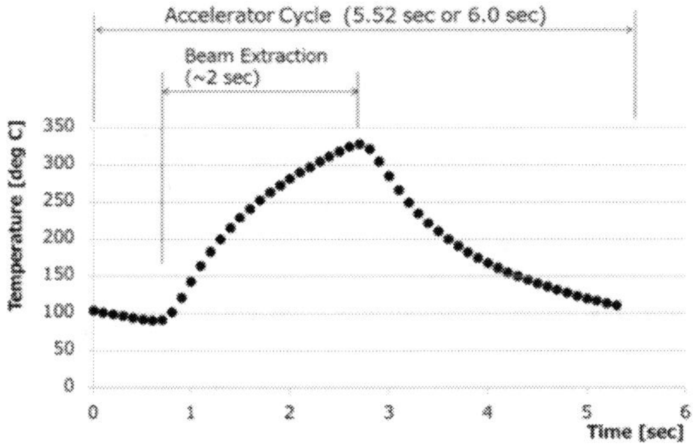

Fig. 4. Typical target temperature variations during an accelerator cycle. The temperature is measured at the top of the target by a thermocouple. The target temperature rises during the beam extraction, which lasts approximately 2 seconds, and then decreases. The maximum temperature is approximately 320 °C with the 42-kW beam.

The temperature of the target is measured by thermocouples, and it reaches a maximum of 350 °C with the normal 50-kW beam. Figure 4 shows the typical temperature variation of the gold target measured with a 42-kW proton beam. The target temperature depends not only on the beam power but also on the position, profile and instantaneous rate of the extracted beam; thus, these beam properties are continuously monitored. When a large deviation from the normal operation is detected, the beam extraction stops within several milliseconds. In addition to these measures to protect the target, a hermetic chamber for the gold target and a helium gas circulation system with gamma ray detectors are installed. In case of target failure, the temperature of the target rises abnormally and the accumulated radioactive materials are released from the damaged target [4]. The hermetic chamber prevents the radioactive materials from spreading. To detect such events the chamber is filled with circulating helium gas and the rate and energy spectrum of the gamma rays from radionuclides in helium gas are continuously monitored by a germanium detector and a NaI(Tl) detector.

2 Helium gas circulation system

Figure 5 shows a schematic view of the helium gas circulation system.

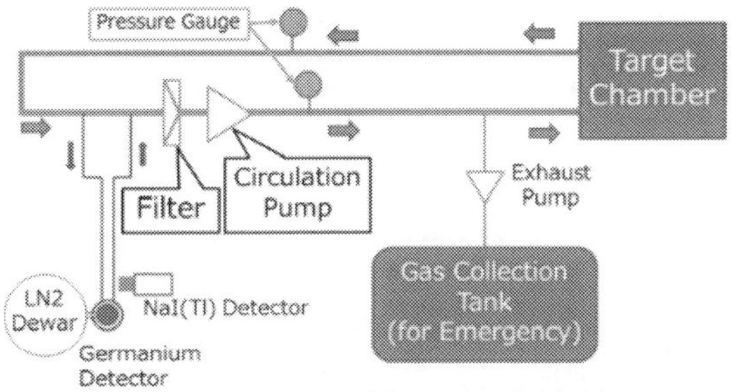

Fig. 5. Schematic view of the helium gas circulation system.

The helium gas circulation system is placed in the machine building located next to the Hadron Experimental Hall. The target chamber and circulation system are connected using gas pipes, which are 168-m long and have a 1 1/4-inch nominal diameter for the most part. The total volume of helium gas in the target chamber and gas pipes is 0.35 m^3 with a pressure of 80 kPa. The gas is circulated by a metal bellows pump and its flow rate is 3.6 m^3/hour. A filter unit composed of a HEPA filter and a charcoal filter is mounted to reduce contaminations in the gas. The status of the circulation system is monitored by pressure gauges, flowmeters, a dew-point meter and a helium purity monitor. Approximately 10 % of the gas branches off to be monitored by gamma-ray detectors. The flow rate in the branch line is 0.35 m^3/hour. The radioactive materials generated in the target chamber are estimated to reach the detectors in approximately 90 s. In case the gamma-ray detectors observe high-rate gamma rays in helium gas, the gas is collected in a gas collection tank to localize the radioactive materials. The 99 % gas can be collected less than 10 minutes in the tank.

The hermetic target chamber has two beam windows made of titanium alloy (Ti-6Al-4V). The thicknesses of the beam-entry and beam-exit windows are 2 and 4 mm, respectively. Further details of the target chamber could be found in [3].

3 Germanium detector for monitoring radioactivity in the circulating helium gas

The spectra and rate of the gamma rays from helium gas are acquired and monitored by a high-purity germanium detector system made by Mirion Technologies (Canberra), Inc. [5]. We referred to the concepts of the previous monitoring system built for the neutron production target at JSNS [6] for developing the system. The energy resolution of the germanium detector is less than 1.8 keV in FWHM at 1332 keV and the energy range covered is from 40–4000 keV. Figure 6 shows photographs of the germanium detector system. The dimensions of the detector are shown in Figure 7. The helium gas pipe winds twice around the detector. The total efficiency for the detection of the gamma rays in the gas system is estimated to be 8×10^{-7} at 1332 keV. The detector and gas pipe are covered by 100-mm-thick lead shields to suppress backgrounds, and 3-mm-thick copper plates are placed between the lead shields and the detector to reduce the effect of X-rays from lead shields. The thickness of the copper plates was determined according to [7]. Figure 8 shows the gamma-ray spectra with and without the copper plates. The copper plates effectively reduce X-rays in the energy range below 100 keV. The detector is cooled by the liquid nitrogen stored in a dewar. The dewar has a pulse-tube cryocooler and there is no need to add liquid nitrogen during a beam operation which typically lasts several months.

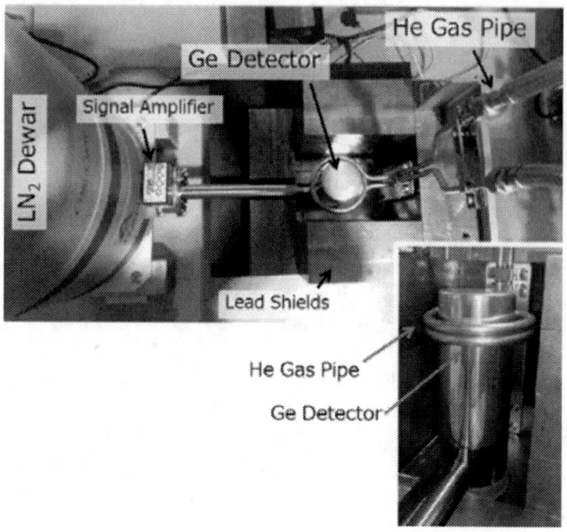

Fig. 6. Photographs of the germanium detector system without copper plates and a part of lead shields. The lower right photograph shows the zoomed-in view of the germanium detector and helium gas pipe winding around the detector.

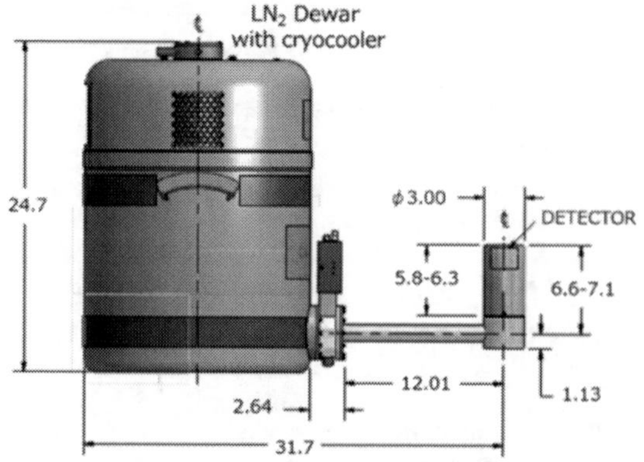

Fig. 7. Dimensions of the germanium detector system in inches.

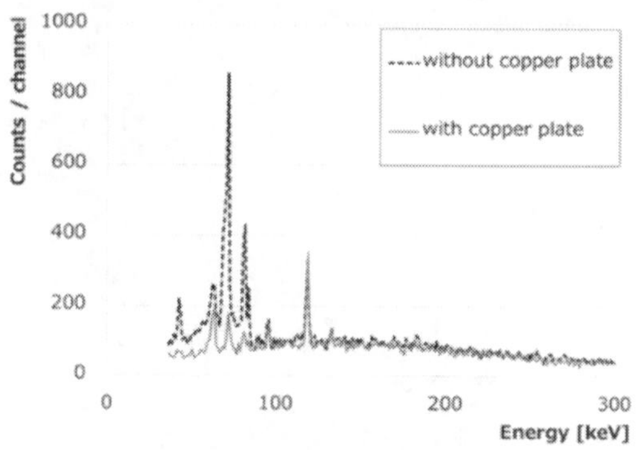

Fig. 8. Gamma-ray spectra with and without the copper plates between the lead shields and the germanium detector.

3.1 Data processing of the germanium detector

Figure 9 shows the schematic configuration of the data processing system of the germanium detector. The pulse height of the signals from the germanium detector are analyzed and digitized by a multichannel analyzer (MCA), and collected by a desktop computer located near the germanium detector. The desktop computer has a contact output and produces a beam abort signal when the following parameters exceed the predetermined thresholds:

- Total counting rate in the energy range from 40 keV to 4 MeV
- One hour accumulations of the gamma-ray counts in ten energy ranges of interest for focused monitoring of the nuclides expected to be produced in the target and the beam windows
- The dead time of the measurement (to detect the saturation of the detector due to a large number of gamma rays)
- Network response time between the desktop computer and the database server

The beam operation stops within several milliseconds after the beam abort signal is detected.

The energy spectra accumulated in an hour and the total counting rate averaged in a minute are transferred via the network to a database server (Microsoft SQL server) located in a control room. The acquired data are stored in a RAID 5 disk array and backed up in a network attached storage (NAS) once a day.

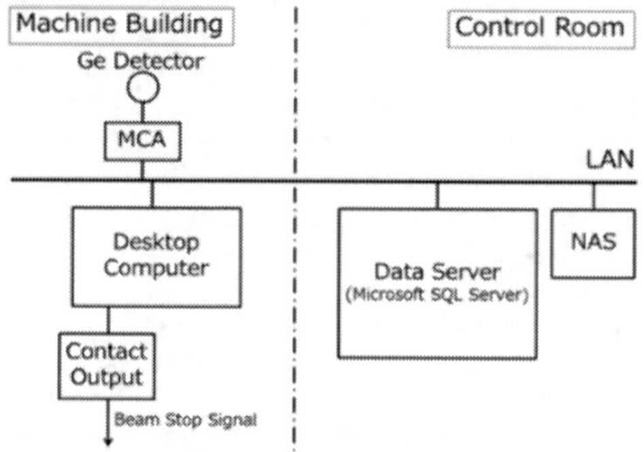

Fig. 9. Schematic diagram of the system used to process the data from the germanium detector.

4 NaI(Tl) detector for monitoring the rate of gamma rays from the circulating helium gas

The rate of the gamma rays is also monitored by a NaI(Tl) detector, which has a higher rate tolerance, made by Mirion Technologies (Canberra), Inc. [5]. It is mounted just above the lead shields of the germanium detector. Figure 10 shows a photograph of the NaI(Tl) detector. The detector is composed of a 2 x 2 inch NaI(Tl) crystal and a photomultiplier and covered by 50-mm-thick lead shields. The detector is placed at a distance of 10 mm from the helium gas pipe and the total detection efficiency is estimated as 6×10^{-7} at 1332 keV. The counting rate of the gamma rays whose energies are larger than 250 keV is converted into a voltage by an ORTEC Model 661 Ratemeter [8]. If the Ratemeter output exceeds the predetermined threshold level, the beam extraction is immediately aborted.

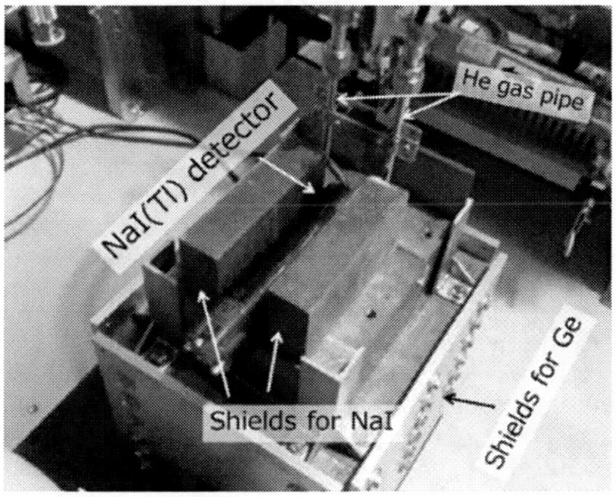

Fig. 10. Photograph of the NaI(Tl) detector without lead shields at the top and in front of the detector.

5 Beam operations with the gamma-ray detectors and helium gas circulation system

The installation of the system described above was completed by March 2015 and beam operations were performed from April to June and October to December 2015, and from May to June 2016. The beam power in the operations was 3 kW initially and gradually increased up to 42 kW. The data from germanium and NaI(Tl) detectors were successfully acquired through the beam operations.

5.1 Identified radioactive nuclides

Figure 11 shows an example of the gamma-ray spectrum in the 42-kW normal beam operation accumulated in an hour. From the spectrum, Hg-192, which is presumed to be produced in the gold target and transported to the detector in the gaseous phase, and its daughter nuclide Au-192 were identified by the peak positions and half-lives. C-10, O-14, O-19, Ne-23, Ne-24, Na-24, Ar-41, Sc-44m, I-119, I-121, I-123, Xe-123, Xe-125, Hg-193m, and Hg-195m were also identified.

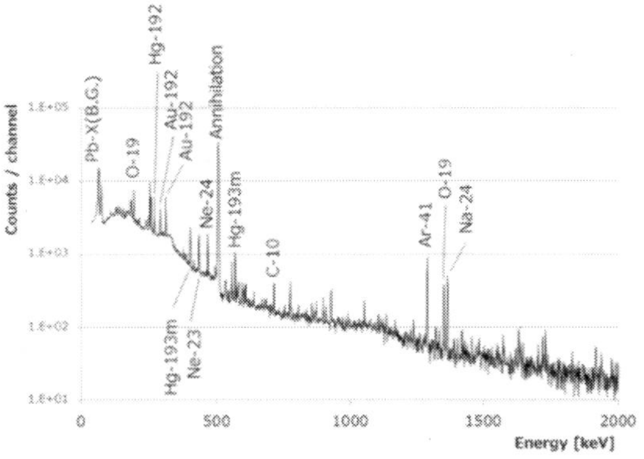

Fig. 11. Example of the gamma-ray spectrum obtained by the germanium detector. The spectrum was accumulated in an hour with the 42-kW beam. The major radionuclides identified by the energy and half-lives are indicated.

Figure 12 shows the trend graphs of the counting rates of the germanium detector corresponding to the radioactive nuclides identified with the 42-kW beam. The amount of each radioactive nuclide is stably monitored during the beam operation.

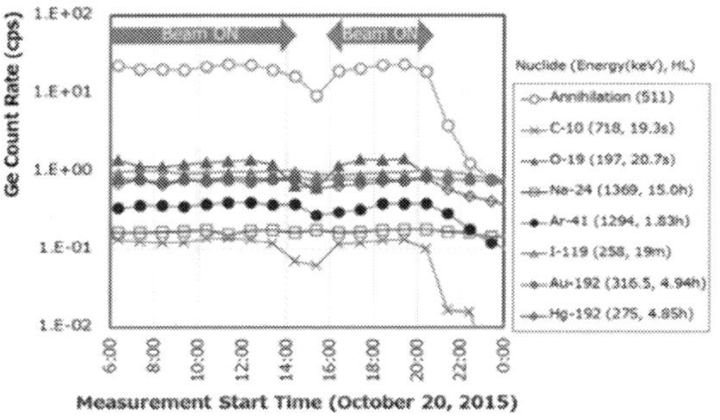

Fig. 12. Counting rates of the germanium detector corresponding to the identified radioactive nuclides as a function of measurement start time. Each counting rate is derived from the gamma ray spectrum accumulated in an hour.

5.2 Response time of the germanium detector for the radioactive materials produced in the target chamber

The target chamber and gamma-ray detectors are connected using 168-m long helium gas pipes. Thus it takes non-negligible time for the radioactive nuclides produced in the target chamber to reach the gamma-ray detectors. We measured this response time at the start of the beam operation. Figure 13 shows the results. The count rate of the germanium detector increases approximately 50 seconds after the start of the beam operation in both the first and second measurements. With this result, we establish a regulation for the beam operation: when an abnormal beam condition is detected, we watch the counting rate and the spectrum of the germanium detector carefully for 5 min to determine the soundness of the gold target.

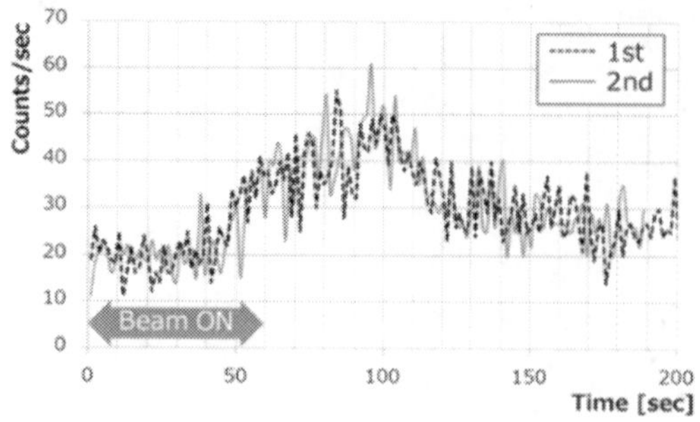

Fig. 13. Results of the response time measurements for the germanium detector. The beam operation started at time 0 and lasted for 60 seconds in this measurement.

5.3 Dependence of the amount of radioactive nuclides on beam power and target temperature

The radioactive nuclides detected by the gamma-ray detectors are mainly produced in the gold target and the titanium-alloy beam windows of the target chamber, and the amounts of radioactive nuclides depend on the production rates and release rates of the nuclides in the bombarded materials. Thus, we investigated the dependences of the count rates for the identified radioactive nuclides on the beam power and the target temperature. Figure 14 shows the measured maximum target temperature as a function of the beam power, which increases approximately linearly with the beam power. We performed two beam operations with the same beam power of 42 kW in the winter of 2015 and spring of 2016, and the maximum target temperature in the latter beam period (302 K) was higher than that of former (277 K) because of the shorter beam extraction duration and the beam position difference.

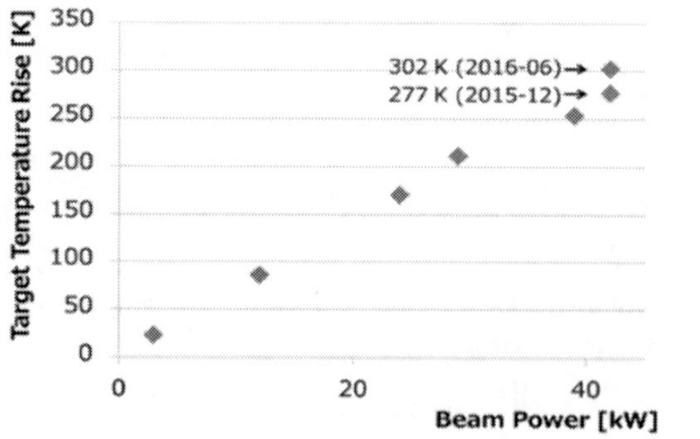

Fig. 14. Maximum target temperature rise as a function of the beam power.

Figure 15 shows the dependences of the counting rates for the identified radionuclides measured by the germanium detector. We used the data from the winter run of 2015 (target temperature rise was 277 K) for the 42 kW beam power. Lighter nuclides such as O-19, Ne-23, Na-24, and Ar-41 have approximately linear dependence on the target temperature. On the other hand, heavier nuclides such as Au-192 (daughter of Hg-192), Hg-192, Hg-193m, and Hg-195m are observed only with the beam powers higher than 29 kW. This indicates that there are factors other than the beam power to determine the amount of detected radionuclides.

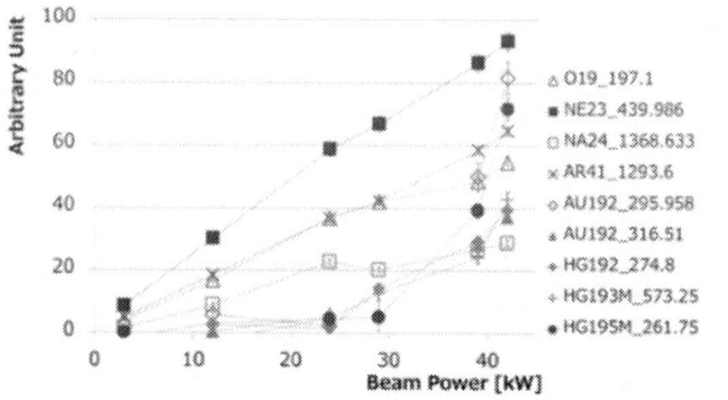

Fig. 15. Beam power dependence of the amount of the radionuclides identified by the germanium detector.

Thus, we investigated the target temperature dependence of the counting rates of the identified radionuclides using two beam operations with the same beam power of 42 kW and different target temperature rises (277 K and 302 K). Figure 16 shows the ratios of the counting rates of the radionuclides in 302 K data to 277 K data. In the ratios for nuclides lighter than Ar-41, the dependence on the target temperature is not clearly seen, which implies that those lighter nuclides were mainly produced in the titanium-alloy beam windows. By contrast, nuclides heavier than I-121, which are considered to be produced in the gold target, have ratios ranging from 2 to 4. The radionuclides produced in the gold target diffuse to its surface and are released into the circulating helium gas, thus the temperature dependence of the counting rates of radionuclides is possibly explained by the combination of the temperature dependence of the diffusion and release processes. First, the diffusion in a solid state material is governed by Fick's first law, and the temperature dependence of the diffusion coefficient D, which is an essential parameter characterizing the diffusion process, is described by the Arrhenius equation $D(T) = D_0 \exp(-E_a/kT)$, where E_a is the activation energy for diffusion [9]. Second, the release rate also depends on the temperature. For example, the vapor pressure of mercury at 600 K is approximately twice that at 570 K [10]. Those factors certainly depend on the nuclide, but the measurement accuracy is insufficient for the discussion on nuclide dependence. In future beam operations, we will collect data with different beam powers and temperatures and perform further studies.

 DOI: 10.1051/epjconf/201715307004

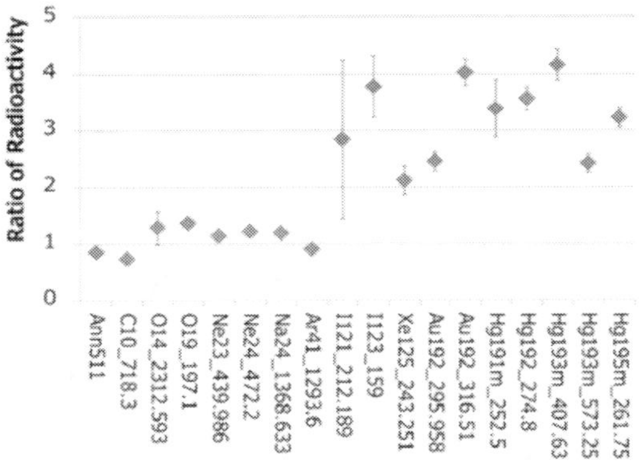

Fig. 16. Ratios of the counting rates of the identified radionuclides in 302-K data to those in 277-K data.

6 Summary

We developed a target monitoring system composed of a hermetic target chamber, helium gas circulation system, and gamma-ray detectors and successfully operated the system for beam operations. Several radioactive nuclides produced by the beam were identified, and it was shown that the amounts of radioactive nuclides depend on the target temperature as well as beam power. The next beam operation is scheduled for March 2017 and further studies will be done.

References

1. Joint Project Team of JAERI and KEK, KEK Report **99-4** (1999) and JAERI-Tech **99-056** (1999)
2. K. Agari et al., Prog. Theor. Exp. Phys., 02B009 (2012)
3. H. Takahashi et al., Journal of Radioanalytical and Nuclear Chemistry **305-3** (2015) 803-809
4. M. Hagiwara et al., JPS Conf. Proc., **8** (2015) pp. 051007-1 - 051007-6
5. http://www.canberra.com/
6. Y. Kasugai, K. Otsu and T. Kai, Proceedings of the ICANS-XIX, 19th meeting on Collaboration of Advance Neutron Source (2010)
7. K. Komura, RADIOISOTOPES, **55** (2006) 233-243 (in Japanese)
8. http://www.ortec-online.com/download/661.pdf
9. R. Kirchner, Nucl. Instr. Meth. **B 70** (1992) 186
10. M. L. Huber, A. Laesecke, and D. G. Friend, NISTIR 6643 (2006)

Study on the Characteristics of a Scintillator for Beta-ray Detection using Epoxy Resin

Jong Soo Nam*, *Yong Seok* Choi, *Sang Bum* Hong, *Bum Kyung* Seo, *Jei Kwon* Moon, and *Jong Won* Choi

Korea Atomic Energy Research Institute, Decontamination and Decommissioning Technology Development Division, 34057 Daejeon, Republic of Korea

Abstract. A thin plate of a plastic scintillator for detecting a beta-ray was developed. The plastic scintillator was made using epoxy resin and organic scintillators such as 2.5-diphenyloxazole (PPO) and 1,4-bis [5-phenyl-2-oxazole] benzene (POPOP). The mixture ratio of epoxy resin and the organic scintillators was determined using their absorbance, transmittance, emission spectra, and transparency. Their optimal weight percentage of PPO and POPOP in the organic scintillators was adjusted to 0.2 wt%:0.01 wt%. The prepared plastic scintillator was used to measure the standard source of Sr-90. The pulse height spectra and total counts of the prepared plastic scintillator were similar to a commercial plastic scintillator. Based on the above results, a large-area plastic scintillator was prepared for rapid investigation of a site contaminated with Sr-90. The prepared large-area plastic scintillator was evaluated for the characteristics in the laboratory. The evaluation results are expected to be usefully utilized in the development of a large-area plastic scintillation detector. The large-area plastic scintillation detector developed on the basis of the evaluation results is expected to be utilized to quickly measure the contamination of Sr-90 in the grounds used as a nuclear power facility.

1 Introduction

The methods for detecting radiation are ionization, excitation, scintillation, extinction and a chemical reaction [1]. Radiation detection using scintillator light produced in a material is one of the oldest and most useful techniques for the detection of a variety of radiation [2, 3]. Scintillation detectors are widely used for radiation measurements in nuclear and high-energy physics. In addition, new applications in the medical field were recently found [4-6]. A detector using plastic scintillators is well known to have an easy operation because it consists of a chemically stable material [7]. Several studies have been conducted to include the features, such as neutron radiation hardness [8] and sensitivity [9-11], as well as reduced manufacturing costs [12, 13]. A large-area plastic scintillator of an affordable price has been required by many researchers of nuclear and high-energy physics fields [14-17].

In general, a plastic scintillator using a polymer such as polymethylmethacrylate (PMMA), polyvinyltoluene (PVT) or polystyrene (PS) is added to an organic scintillator. As an organic scintillator, the first solute is p-terphenyl or 2.5-diphenyloxazole (PPO), and the second solute is 1,4-bis [5-phenyl-2-oxazol] benzene (POPOP) [18]. A method for preparing a plastic scintillator is a mixture of a polymer and organic scintillators used for thermal polymerization. Recently, the light generated in the scintillator owing to the external radiation is able to transmit the long distance by combining the optical fiber and the plastic scintillator [19]. However, this method may result in a loss of light in a combination of the optical fiber and the plastic scintillator.

In this study, we prepared a plastic scintillator whose manufacturing process is simple and can be freely shaped. A thin plate of a plastic scintillator was manufactured using epoxy resin as a polymer. The plastic scintillator was made by mixing epoxy resin and organic scintillators under various conditions. We investigated the properties such as the absorbance, transmittance, emission spectra, and transparency of the prepared plastic scintillator. The optimal mixture ratio to prepare the plastic scintillator was derived from the above results. Using the derived results, we made the large-area plastic scintillator, which can quickly measure the contamination site and evaluated characteristics of the large-area plastic scintillator in the laboratory.

2 Method

The polymer materials for the preparation of a plastic scintillator were mainly PMMA, PVT or PS. However, PMMA, PVT and PS generate a bad smell in the thermal polymerization process and require a relatively long drying time. A polymer was used as an epoxy resin (WE-300A, Won chemical co. Korea) because no odor

* Corresponding author: namjs@kaeri.re.kr

resulted from thermal polymerization, and the drying time is short. Organic scintillators were added to the PPO as the first solute and the POPOP as the second solute.

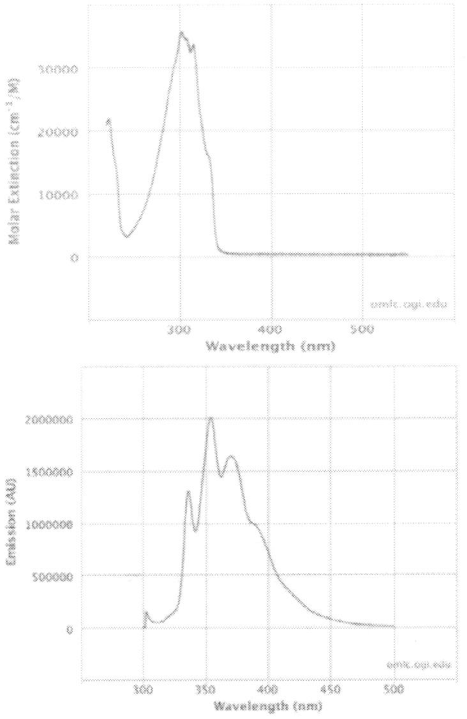

Figure 1. Absorbance (up) and emission (down) spectra of PPO

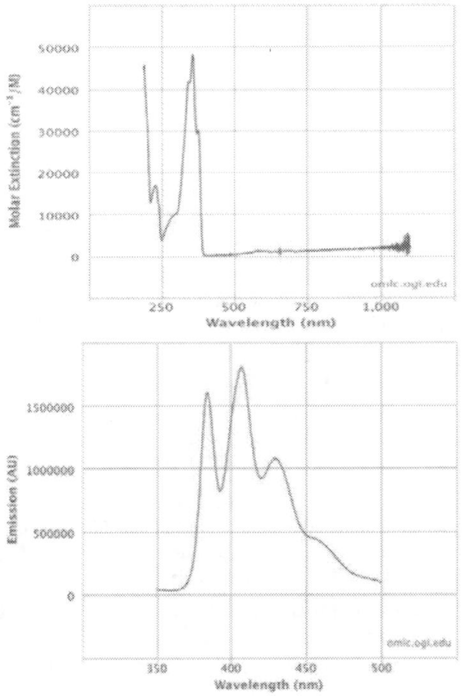

Figure 2. Absorbance (up) and emission (down) spectra of POPOP

Absorbance and emission spectra of PPO and POPOP are shown in Figures 1 and 2 [20, 21]. A polymer material such as an epoxy resin absorbs the energy of the incident radiation. The first solute (PPO) emits scintillation of light in an ultraviolet range (wavelength of 200 to 400 nm) by receiving the absorbed energy. However, the wavelength of the ultraviolet range can be easily absorbed in a polymer and is not suitable for the scintillation response of a photomultiplier tube (PMT). The second solute (POPOP) can reduce the absorbance caused in the medium by emitting a blue wave (wavelength of 450 to 490 nm) of a long wavelength through the absorbance of scintillation caused from the PPO. In addition, it is a wave shifter for moving the wavelength responsible for the PMT.

The plastic scintillator is made using epoxy resin, PPO and POPOP. The preparation procedure of the plastic scintillator is shown Figure 3. The epoxy resin of a polymer was added in the PPO and POPOP in accordance with the mixture ratio. The scintillation solution into the hardener (WE-300B, Won chemical co. Korea) was mixed by a stirrer for 2 hours at room temperature. After identifying no air bubbles of a polymer solution, the thickness of the plastic scintillator using the polyethylene film to facilitate the removal from the mould was placed in a scintillator volume of approximately 2 to 3 mm. The polymer solution is spread evenly above the polyethylene film. The temperature of the thermal polymerization was successively adjusted at 90 degrees for 12 hours, 110 degrees for 5 hours, and 130 degrees for 2 hours [22].

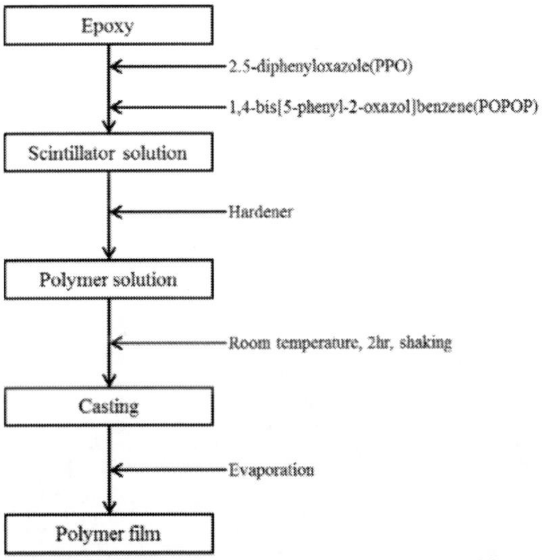

Figure 3. Preparation procedure of the plastic scintillator

The mixture ratio used to make the plastic scintillator is shown in Table 1. The plastic scintillator made a 50 mm diameter to evaluate the characteristics according to the mixture ratio. The substance of the plastic scintillator regarding the total weight of epoxy resin was 0.1 to 1 wt% and 0.01 to 0.05 wt% for PPO and POPOP, respectively. The optical properties such as the absorbance, transmittance, and emission spectra in the visible light of the prepared plastic scintillator measured using a spectrofluorometer. PMT (H5211A-MOD, Hamamatsu) as a light measuring device was used. The amplification of the scintillation light using PMT was achieved.

Table 1. The mixture ratio of organic scintillators by PPO and POPOP

Sample No.	PPO (wt%)	POPOP (wt%)	Sample No.	PPO (wt%)	POPOP (wt%)
1	0.1	0.01	7	0.1	0.0025
2	0.2	0.01	8	0.1	0.005
3	0.4	0.01	9	0.1	0.01
4	0.6	0.01	10	0.2	0.01
5	0.8	0.01	11	0.2	0.025
6	1	0.01	12	0.2	0.05

3 Results

The prepared plastic scintillator was observed with the naked eye. The yellow color phenomenon in the plastic scintillator increased as the amount of PPO occurred. It was founded to be present in the plastic scintillator without melting according to the amount of POPOP. To the naked eye, we could see the transparency of the plastic scintillator by adding the least amount of PPO and POPOP.

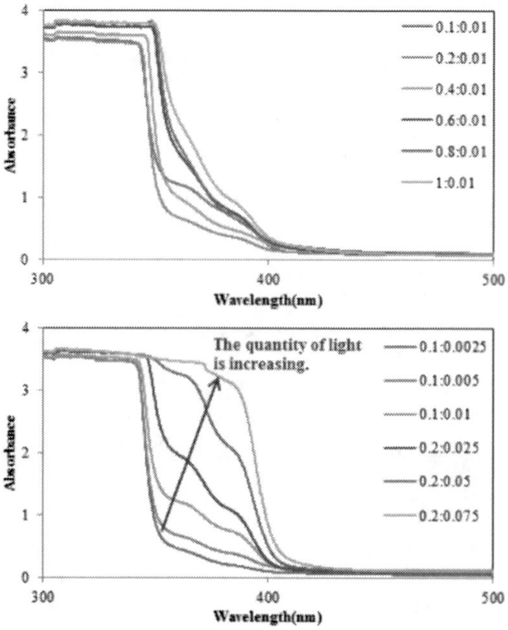

Figure 4. Absorbance spectra according to the amount of PPO (up) and POPOP (down)

The optical properties such as the absorbance, transmittance, and emission spectra in the visible light (wavelength of 380 to 800 nm) of the prepared plastic scintillator were measured using a spectrofluorometer.

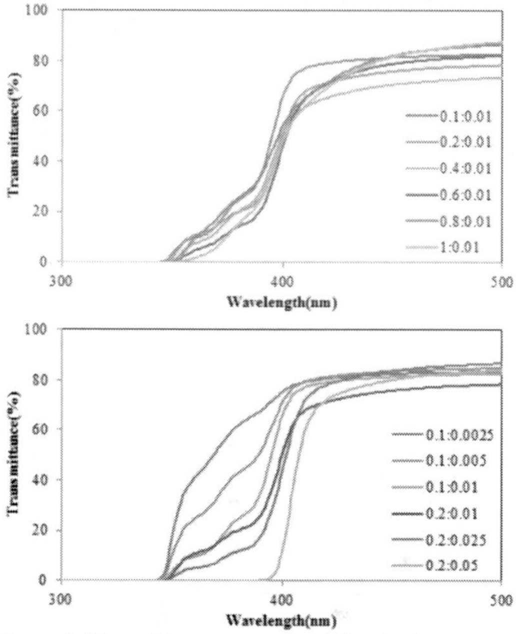

Figure 5. Transmittance spectra according to the amount of PPO (up) and POPOP (down)

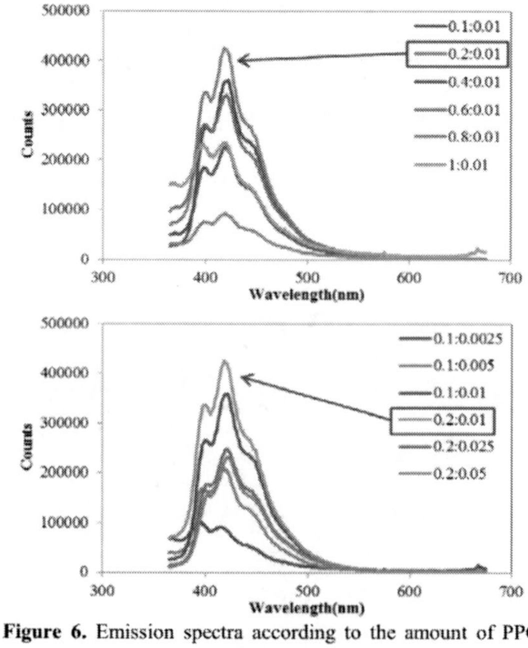

Figure 6. Emission spectra according to the amount of PPO (up) and POPOP (down)

The absorbance, transmittance, and emission spectra were observed to derive the optimal mixture ratio of the plastic scintillator. Figures 4, 5, and 6 show the light absorbance, transmittance, and emission spectra according to the amount of PPO and POPOP, respectively. Even if the amount of PPO increases, the absorbance and emission spectra did not show a consistent trend. The transmittance of all plastic scintillators has confirmed to be excellent in the 70 % range. However, if the amount of POPOP increases, the absorbance and emission spectra did show a growing trend. These results indicate that the amount of PPO is independent, and the greater the amount of increase in POPOP, the more the quantity of light that is increasing. However, the plastic scintillator increased as the amount of PPO was remarkably changed to yellow. In addition, there is a problem without melting to POPOP.

Based on the above measurement result, the best plastic scintillator is slightly yellowing, and excellent in absorbance and transmittance, and its amount of emission of light is the best. The weight percentage of PPO and POPOP in an organic scintillator was adjusted to 0.2 wt%:0.01 wt%.

The detection efficiency of a plastic scintillator was evaluated based on the content prepared to the optimal mixture ratio. A plastic scintillator was optically connected to the PMT with optical grease and wrapped with a black tape after placing a white paper as a reflector [23]. Figure 7 shows the plastic scintillator measurement setup. The measurement system was placed in a light shield box before loading in the darkroom. The distance of the plastic scintillator and standard source was maintained at 20 mm.

The standard source is Sr-90 with a radioactivity of 0.1 μCi. Figure 8 shows the decay scheme of Sr/Y-90, which emits beta-rays without emitting a gamma-ray. The contamination of Sr-90 is measured to have a high-energy beta radiation of 2.3 MeV emitted by beta decay of Y-90.

Figure 9 shows the spectra of the commercial and prepared plastic scintillator. As shown in the figure, the pulse height spectra and total counts of the prepared plastic scintillator were similar to a commercial plastic scintillator.

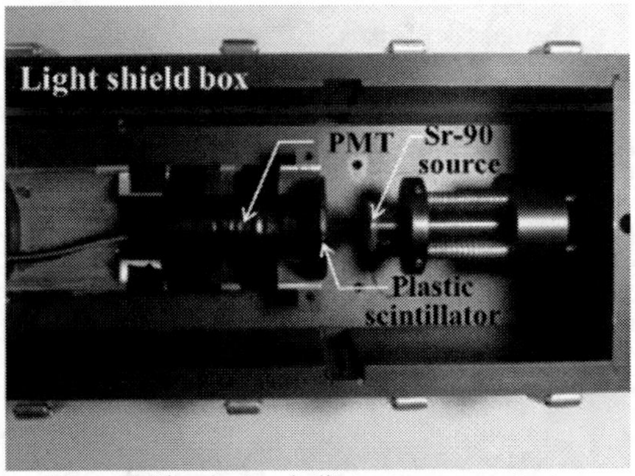

Figure 7. The plastic scintillator measurement setup

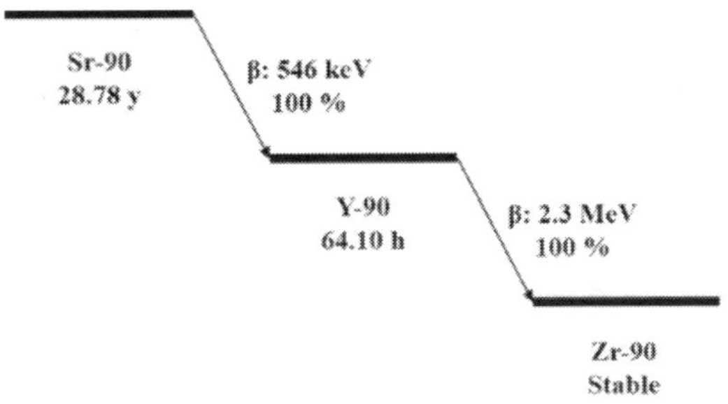

Figure 8. Decay scheme of Sr-90

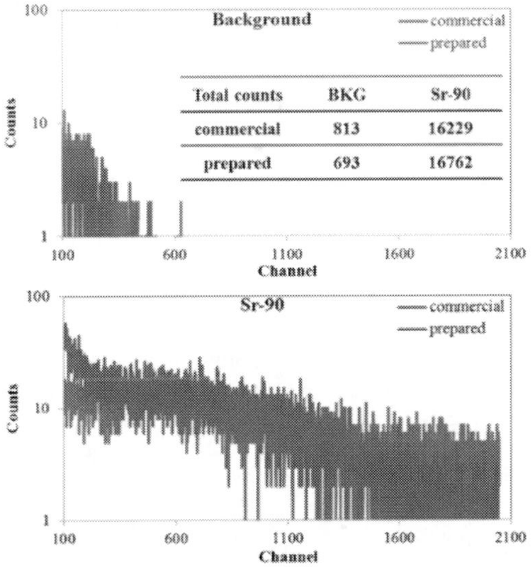

Figure 9. Radiation spectra of the plastic scintillator

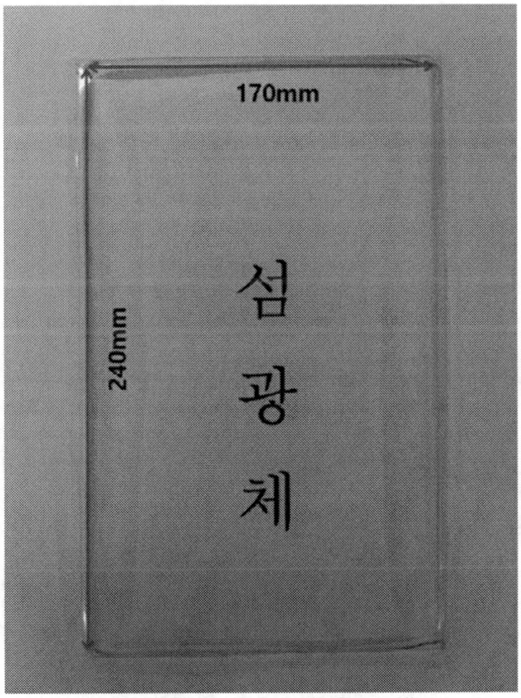

Figure 10. The large-area plastic scintillator

DOI: 10.1051/epjconf/201715307005

Based on the characteristic results, a large-area plastic scintillator was prepared. The size of the large-area plastic scintillator was similar to the window size of a typical pancake-type αβ surface contamination counter. The size of the large-area plastic scintillator is 240 mm (length) x 170 mm (width) x 3 mm (thickness) (Figure 10). In order to make the large-area plastic scintillator, it was produced above a metal mould of the same size. Polyethylene film was used as a releasing agent to ease the removal of the large-area plastic scintillator from the metal mould. The large-area plastic scintillator was pressed by a metal lid to prevent twisting.

We wanted to evaluate the characteristics of the large-area plastic scintillator. It was difficult to connect them because of the difference in size between the large-area plastic scintillator and PMT.

4 Conclusion

A thin plate of a plastic scintillator with a simple preparation process can be freely shaped using epoxy resin and organic scintillators such as PPO and POPOP. PPO emits a scintillation of light in the ultraviolet range, and POPOP is a wave shifter for moving the wavelength responsible for the PMT. When preparing the plastic scintillator, the temperature of the thermal polymerization rose sequentially. Otherwise, a thermal polymerization at high temperature (for example 130 degrees for 2 hours) was crushed, or at too low a temperature (for example, less than 80 degrees) was weak to be easily broken down. The mixture ratio of PPO and POPOP was determined using their absorbance, trasmittance, emission spectra, and transparency. The optimal weight percentage of PPO and POPOP in an organic scintillator was determined to be 0.2 wt%:0.01 wt%. The pulse height spectra of a plastic scintillator were evaluated based on the content prepared to the optimal mixture ratio, and the prepared plastic scintillator was similar to a commercial plastic scintillator.

Based on the above results, the large-area plastic scintillator of the window size of a typical pancake-type αβ surface contamination counter was prepared. We want to evaluate the characteristics of the large-area plastic scintillator. However, there were the difficulties in evaluating the characteristics of the large-area plastic scintillator. The cross-sectional area of the large-area plastic scintillator is significantly different to PMT. The tool will be necessary to connect the large-area plastic scintillator and PMT. The tool for connecting them generally utilizes a light guide. The material of the light guide is used in a highly transparent acrylic. Acrylic of a pyramid shape to connect them will be made. Then, a characteristic evaluation of the large-area plastic scintillator needs to be carried out. In addition, the light guide serves to enhance the focusing efficiency of the light.

It will be utilized for the preparation of large-area plastic scintillators for a rapid investigation of the decommissioning site of nuclear power plants contaminated with Sr-90.

Acknowledgement

This work has been carried out under the long-term nuclear research and development program (2012M2A8A5025996) sponsored by Ministry of Science, ICT and Future Planning of Korea.

References

1. J.H.Kim, K.S.Joo, Journal of IKEEE **18**, 456 (2014)
2. G.F.Knoll, *Radiation Detection & Measurement*, 219 (2010)
3. S.N.Ahmed, *Physics & Engineering of Radiation Detection*, 319 (2007)
4. T.K.Lewellen, Phy Med Biol **53**, R287 (2008)
5. L.Archambault, T.M. Briere, F.Pönish, L.Beaulieu, D.A.Kuban, A.Lee, S.Beddar, Int J Radiat Oncol **78**, 280 (2010)
6. A.Vandenbroucke, A.M.K.Foudray, P.D.Olcott, C.S.Levin, Phy Med Biol 55, 5895 (2010)
7. M.Watanabe, M.Katsumata, H.Ono, T.Suzuki, H.Miyata, Y.Itoh, K.Ishiba, M.Tamura, Y.Yamaguchi, Nucl Instrum Meth A **770**, 197 (2015)
8. A.Quaranta, S.Carturan, T.Marchi, M.Cinausero, C.Scian, V.L.Kravchuk, M.Degerlier, F.Gramegna, M.Poggi, G.Maggioni, Opt Mater **32**, 1317 (2010)
9. V.B.Brudanin, V.I.Bregadze, N.A.Gundorin, D.V.Filossofov, O.I.Kochetov, I.B.Nemtchenok, A.A.Smolnikov, S.I.Vasiliev, Particles and Nuclei Letters **109**, 69(2001)
10. L.Ovechkina, K.Riley, S.Miller, Z.Bell, V.Nagarkar, Physics Procedia **2**, 161 (2009)
11. A.Quaranta, S.Carturan, T.Marchi, M.Buffa, M.Degerlier, M.Cinausero, G.Guastalla, F.Gramegna, G.Valotto, G.Maggioni, V.L.Kravchuk, J Non-Cryst Solids **357**, 1921 (2011)
12. A.Pla-Dalmau, A.D.Bross, K.L.Mellott, Nucl Instrum Meth A **466**, 482 (2001)
13. H.Nakamura, Y.Shirakawa, S.Takahashi, H.Shimizu, Europhysics Letters **95**, 22001 (2011)
14. The CMS collaboration, *The Compact Muon Solenoid Technical Proposal*, CERN/LHCC 94-38 (1994)
15. The CDF Ⅱ collaboration, *The CDF-Ⅱ Detector Technical Design Report*, Fermilab-Pub-96/390-E (1996)
16. A.L.C.Sanchez, H.Miyata, N.Nakajima, H.Ono, Y.Fujii, S.Itoh, F.Kajino, J.Kanzaki, K.Kawagoe, S.Kim, S.Kishimoto, T.Matsumoto, H.Matsunaga, A.Nagano, R.Nakamura, K.Sekiguchi, T.Takeshita, N.Uchida, Y.Yamada, S.Yamamoto, S.Yamauchi, Nucl Instrum Meth A **546**, 535 (2005)
17. H.Ono, H.Miyata, S.Iba, N.Nakajima, A.L.C.Sanchez, Y.Fujii, S.Itoh, F.Kajino, J.Kanzaki, K.Kawagoe,

S.Kim, S.Kishimoto, T.Matsumoto, H.Matsunaga, A.Nagano, R.Nakamura, K.Sekiguchi, T.Takeshita, Y.Tamura, S.Yamauchi, Nucl Instrum Meth A **600**, 398 (2009)
18. B.K.Seo, G.H.Kim, Z.H.Woo, W.J.Oh, K.W.Lee, M.J.Han, Analytical Science & Technology **18**, 495 (2005)
19. C.H.Park, J.H.Moon, B.K.Seo, Radiat Meas **46**, 687 (2011)
20. http://omlc.org/spectra/PhotochemCAD/html/0.20.html
21. http://omlc.org/spectra/PhotochemCAD/html/0.77.html
22. J.S.Nam, Y.S.Choi, S.B.Hong, B.K.Seo, J.K.Moon, J.W.Choi, *Abstracts of Proceedings of the Korean Radioactive Waste Society AUTUMN 2015*, **13**, 287 (2015)
23. E.P.Jacosalem, S.Iba, N.Nakajima, H.Ono, A.L.C.Sanchez, A.M.Bacala, H.Miyata, GLD Calorimeter Group, Pramana **69**, 1051 (2007)

Photoneutron Flux Measurement via Neutron Activation Analysis in a Radiotherapy Bunker with an 18 MV Linear Accelerator

Yiğit Çeçen[1,2,*], *Tuğçe* Gülümser[1,2], *Çağrı* Yazgan[1,2], *Haris* Dapo[1], *Mahmut* Üstün[1], *Ismail* Boztosun [1]

[1] Akdeniz University, Nuclear Sciences Application and Research Center, 07070, Antalya, Turkey
[2] Akdeniz University, School of Medicine, Department of Radiation Oncology, 07070, Antalya, Turkey

Abstract. In cancer treatment, high energy X-rays are used which are produced by linear accelerators (LINACs). If the energy of these beams is over 8 MeV, photonuclear reactions occur between the bremsstrahlung photons and the metallic parts of the LINAC. As a result of these interactions, neutrons are also produced as secondary radiation products (γ,n) which are called photoneutrons. The study aims to map the photoneutron flux distribution within the LINAC bunker via neutron activation analysis (NAA) using indium-cadmium foils. Irradiations made at different gantry angles (0°, 90°, 180° and 270°) with a total of 91 positions in the Philips SLI-25 linear accelerator treatment room and location-based distribution of thermal neutron flux was obtained. Gamma spectrum analysis was carried out with high purity germanium (HPGe) detector. Results of the analysis showed that the maximum neutron flux in the room occurred at just above of the LINAC head (*$1.2x10^5$ neutrons/cm^2.s*) which is compatible with an americium-beryllium (Am-Be) neutron source. There was a 90% decrease of flux at the walls and at the start of the maze with respect to the maximum neutron flux. And, just in front of the LINAC door, inside the room, neutron flux was measured less than 1% of the maximum.

1 Introduction

Cancer is one of the most common health problem known and the number of cancer patients increase every year all over the world [1]. New methods for cancer treatment are developed with developing technology [2]. One of the methods developed for the treatment of cancer patients is the treatment by radiation, known as radiotherapy.

Radiotherapy centers use high energy X-rays produced by LINACs to treat patients. Electrons are accelerated into a tungsten target to produce bremsstrahlung photons [3]. These photons are produced by electrons at different energies ranging from 4 to 25 MeVs.

By design, there are high atomic number materials inside the accelerator head (such as target, smoothing filter, collimators). If the bremsstrahlung X-rays' energy exceeds the threshold energy of these materials, photonuclear reactions such as neutron break off may occur. These neutrons are referred to as photoneutrons. Produced neutrons may cause a nuclear reaction interacting with the surrounding material. These materials, which become active with (n,γ) reactions, may release gamma radiation to the environment [4].

Resulting photoneutrons spread out easily from the accelerator head and scatter into the room. As a result of these neutrons, an additional dose occurs in the patient's body [5]. In addition to unwanted additional dose in the patient; due to high radiobiological effects of the neutrons, significant amount of side effects may develop for the patient [6]. These unwanted doses and radiobiological activity also increase the risk of secondary cancer [7]. Products resulting from the photoactivation event also threaten the health of the employees and the people who are inside or outside facility [5, 8, 9]. Electronic devices in the treatment room are also adversely affected by the production of photoneutrons [9, 10]. All of the materials found in the LINAC room are capable of doing neutron absorption reactions [11].

Flux distribution of the photoneutrons in the treatment room depends on many variables. The room design hosting the accelerator (the room's geometry and consisting materials of the shielding), the type of accelerator (design and materials it contains), the accelerator's radiation beam type, the planned treatment area (wedge, multi-leaf collimator, etc.) and the resulting neutron spectrum are the factors that affect photoneutron distribution in the treatment room [5].

* Corresponding author: ycecen@akdeniz.edu.tr

DOI: 10.1051/epjconf/201715307006

To create a suitable radiation protection protocol for the device, deposited neutron dose by workers and patients should be calculated. To calculate the dose of the occurring neutrons, characteristics of neutrons in the therapy room (flux, distribution, etc.) must be known [12].

2 Materials & Methods

To obtain the neutron distribution in the radiotherapy room, thermal neutron flux was measured in different locations. Thermal neutrons will generate more radiation damage in living tissue and in circuit materials due to high thermal absorption cross sections of water, carbon and silicon [13].

In this study, Philips SLI-25 linear accelerator was used as a photo-nuclear source. For the analysis AMATEK-ORTEC brand GMX-20195-P model n-type high purity germanium (HPGe) semi-conductor detector was used.

Indium and cadmium metal foils were used for neutron activation analysis (NAA), which both have high neutron absorption cross sections. These foils are activated as a result of neutron absorption. As shown in figure 2.1, indium cross section is high for thermal and epithermal energy regions [14]. Measurements with bare indium foils gave information about the neutrons in these energy ranges.

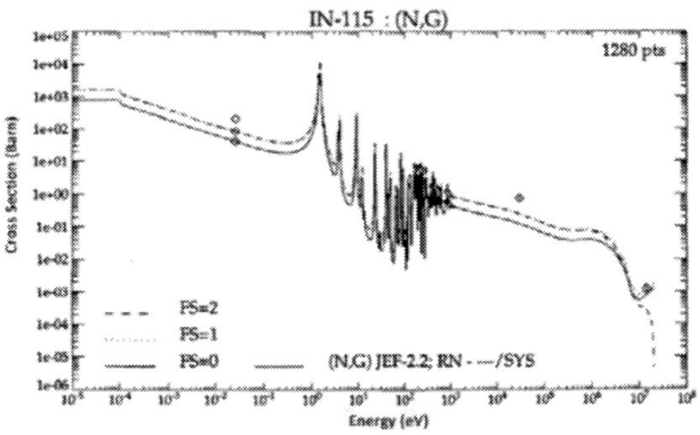

Figure 2.1. Indium-115 Neutron Absorption Cross Section [15]

The cadmium cross section is high for neutrons in thermal and high energy regions (Figure 2.2) [14]. For this reason, indium foils were covered with the cadmium foils and same experimental setup was prepared and measurements were taken. Thus, thermal neutrons were absorbed by cadmium before they reach the indium and only epithermal neutrons were absorbed by the indium foils.

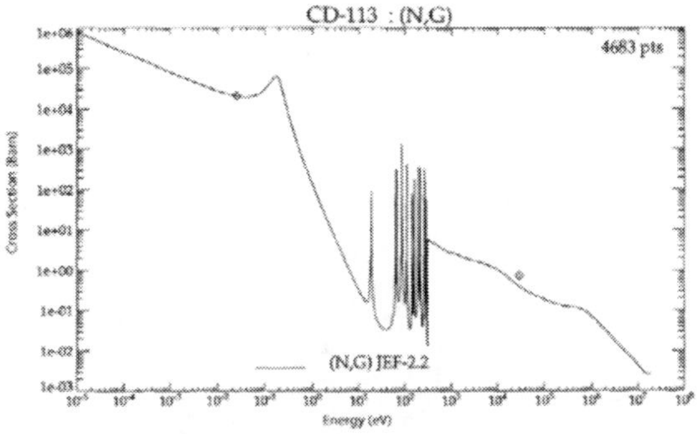

Figure 2.2. Cadmium-113 Neutron Absorption Cross Section [15]

To obtain the information for the thermal neutron flux, differences between the measured data must be studied between bare indium and cadmium covered indium foils.

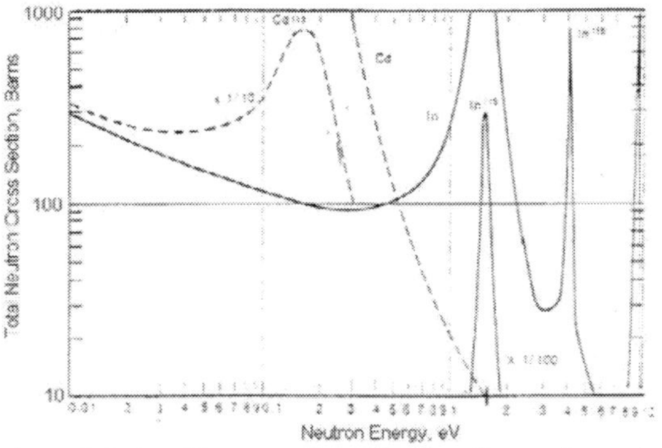

Figure 2.3. Total Neutron Cross Section Comparison of Cadmium-113 and Indium-115

$$ {}^{1}_{0}n + {}^{115}_{49}In \xrightarrow{T_{1/2}=54.2\,min} {}^{116}_{49}In \xrightarrow{T_{1/2}=14s} $$
$$ {}^{116}_{50}Sn^{*} + \beta^{-} + \overline{\nu} \xrightarrow{T_{1/2}=14.10\,h} {}^{116}_{50}Sn + \gamma \quad (2.1) $$

As shown in eq. 2.1, ^{115}In absorbs one neutron and then it becomes ^{116}In in 14 seconds. Half-life of ^{116}In is 54.2 minutes. ^{116}In isotope at excited state emits β^- radiation because of neutron excess. After the decay, ^{116}In becomes ^{116}Sn which has a half-life of 4.1 hours. During this process an anti-neutrino is emitted by ^{116}In for energy and momentum protection [16]. Experiments were carried out for different locations by resetting the activity of indium foils. Due to 4.1 hours half-life of ^{116}Sn, two days of cooling period was planned between experiments which correspond to about ten half-lives.

For the analysis of photoneutron flux in the radiotherapy room, 18 MeV bremsstrahlung photon energy was used. 7 different experimental setups and a total of 91 separate locations were prepared. Indium and cadmium-coated indium foils were irradiated at 10000 MU (Monitor Units - irradiation time) with 0°, 90°, 180° and 270° gantry angle (irradiation angle). In this regard,

13 indium and then cadmium plated indium irradiations were made by placing in water used as a moderator, as shown in Figure 2.4. Water is a fine moderator which has a high neutron scattering cross section and thus allows thermalization [17].

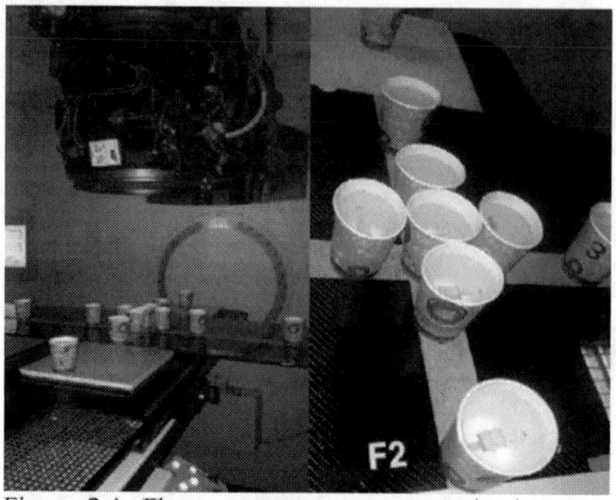

Figure 2.4. Flux measurement setup at the isocenter plane.

The cups that were used in the experiments have the same water height and the foils were placed at the same positions. Products of indium decay are shown in the diagram in Figure 2.5. ^{116m}In ($T_{1/2}$ = 54.2 min) with a 99.94% probability at 1293.4 keV peak energy was counted and thus neutron flux information was obtained. The counting process was performed with the HPGe detector system and the data was analyzed by the GF3 gamma spectroscopy analysis software.

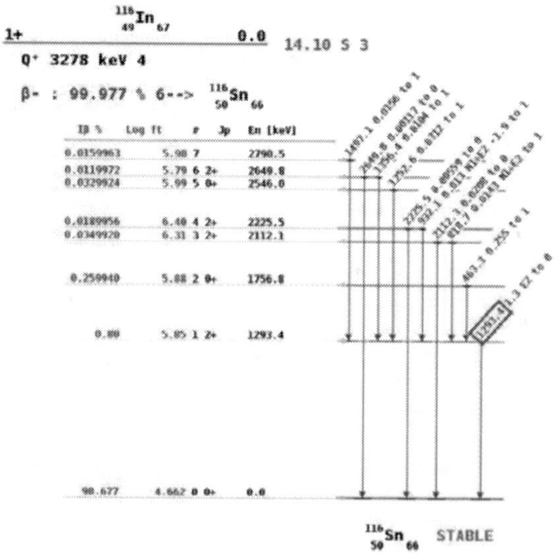

Figure 2.5. ^{116}In Decay scheme [18]

Firstly, bare indium foils were irradiated and measurements were taken. The resulting count (C_0 (bare)) was recorded. Then indium foils coated with cadmium foils were irradiated, which are permeable to epithermal neutrons. Similarly, (C_0 (Cd)) was measured and recorded. After the measurements, the thermal neutron flux was obtained using the following equations.

$$C_0 = \frac{I\lambda}{\varepsilon(1 - e^{-\lambda t_1})(e^{-\lambda t_2} - e^{-\lambda t_3})} \quad (2.2)$$

C_0; irradiated foils saturation activity, I; # of counts between t_3-t_2, λ; radioactive decay constant of ^{116}In foils (**λ**=0,0002130), ε; the internal and geometric efficiency of the detector used in measurement, t_1; foil irradiation time, t_2; remaining time between the start of the measurement and end of the irradiation, t_3; refers to the time period between end of the irradiation and end of the measurement.

$$C_0(\text{bare}) = N(\phi_{th}\sigma_{th} + \phi_r\sigma_r) \quad (2.3)$$

$$C_0(\text{Cd}) = N(\phi_r\sigma_r) \quad (2.4)$$

The terms used in the equations; N; atom density of the foil (atoms/cm^3), ϕ_{th}; thermal neutron flux (neutron/cm^2·s), σ_{th}; thermal neutron absorption cross section of indium foils (cm^2), ϕ_r; resonance neutron flux (neutron/cm^2·s), σ_r: resonance neutron absorption cross section of indium foils (cm^2).

The difference of counts taken by bare indium foils and indium foils coated by cadmium foils equals to the number of thermal neutrons.

$$C_0(\text{bare}) - C_0(\text{Cd}) = N\phi_{th}\sigma_{th} \quad (2.5)$$

To calculate the atom density of foils equation (2.6) was used.

$$N = \frac{m}{M}N_A \quad (2.6)$$

To obtain thermal neutron flux, following procedures were performed.

$$R = \frac{C_0(\text{bare})}{C_0(\text{Cd})} \quad (2.7)$$

$$R - 1 = \frac{N\phi_{th}\sigma_{th}}{C_0(\text{Cd})} \quad (2.8)$$

$$\phi_{th} = \frac{C_0(\text{Cd})}{N\phi_{th}}(R - 1) \quad (2.9)$$

3 Results

Results of the analysis show that the maximum neutron flux in the room occurs at the top of the LINAC head towards the gun direction. This was expected since most of the neutrons are produced when the electron beam hits the tungsten target and at the primary collimator. Both the target and the primary collimator are located at the top of the gantry head. The maximum thermal neutron flux obtained was $1.2x10^5$ neutrons/cm^2.s which is higher than a standard americium-beryllium (Am-Be) neutron source.

At the isocenter plane (SSD=100 cm), the fluxes were $5.4x10^4$ at the center, $1.5x10^4$ at 2.5 m away and $9.9x10^3$ n/cm^2.s at the room wall which is 3.8 m away from isocenter. The flux at the maze entrance was measured nearly six in a ten thousand less (81 n/cm^2.s).

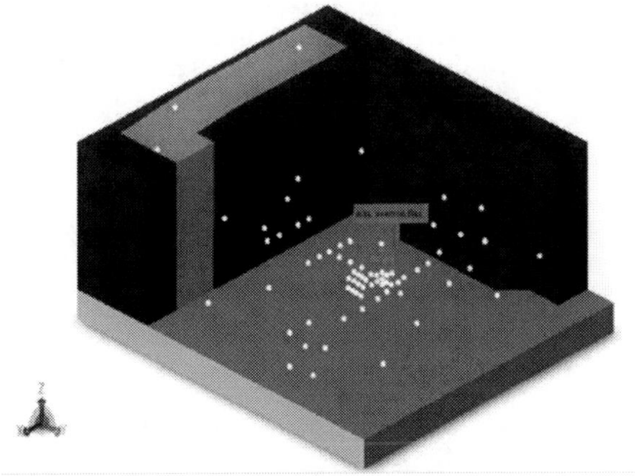

Figure 3.1. Positions of measurement (91 points) in the linear accelerator room

To set up an example, thermal neutron flux distributions are shown for 2 experimental setups of the total of 7 experimental setups.

Experimental Parameters
X-Ray
Gantry=0^0
10000 MU (avg. irradiation time 30 min)
18 MV (18 MeV brems spectrum)
40x40 cm^2 area (collimation)
SSD =100 cm (Distance to target)

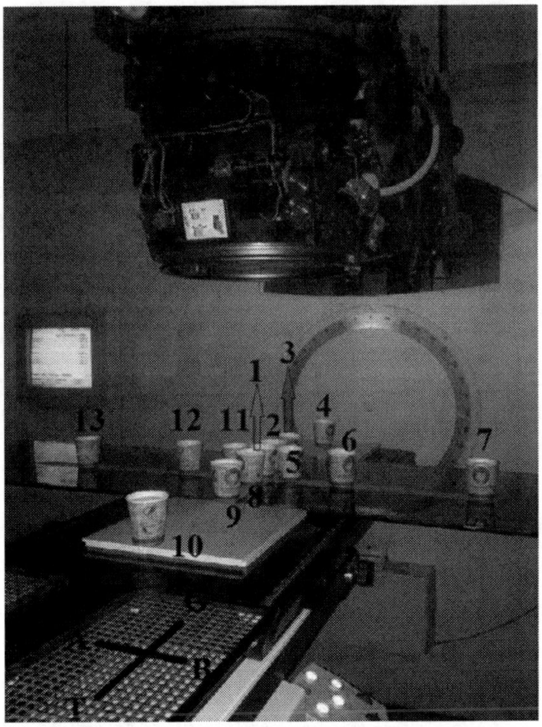

Figure 3.2. Location of indium and cadmium foils for experiment #1

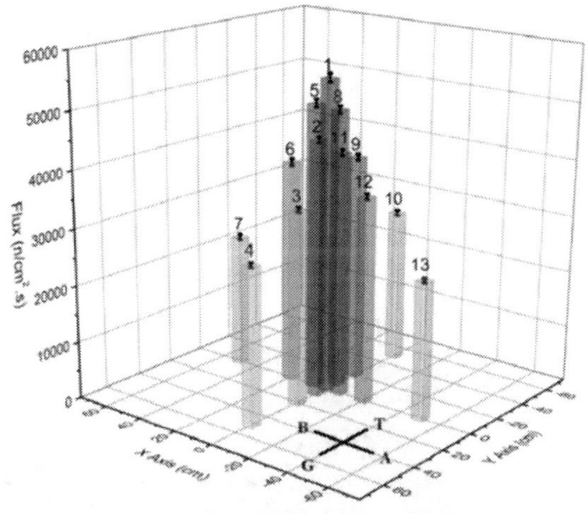

Figure 3.3. Space dependent photoneutron flux distribution graph obtained from experiment #1

Figure 3.4. Location of indium and cadmium foils for experiment #2

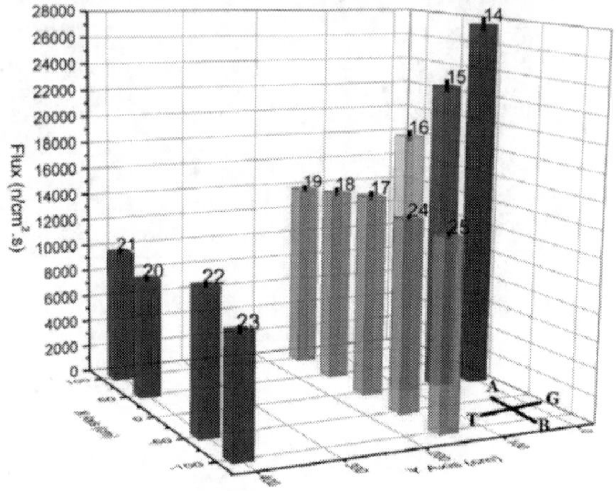

Figure 3.5. Space dependent photoneutron flux distribution graph obtained from experiment #2

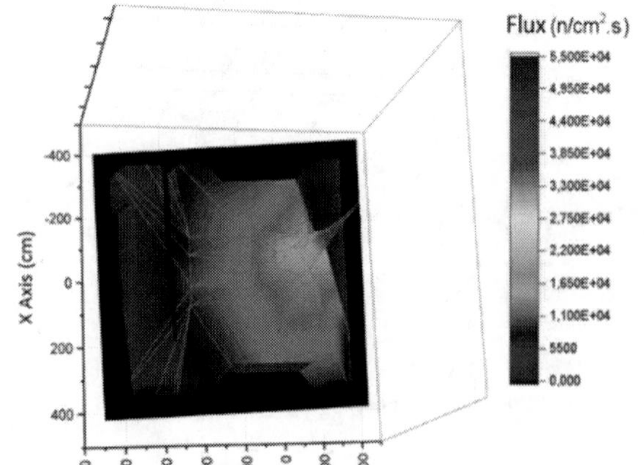

Figure 3.6. Thermal neutron flux distrubution in LINAC room (SSD=100cm, 10000 MU)

4 Discussion

When obtained results were analyzed, maximum photoneutron flux occurred at the center of the LINAC and showed a descending gaussian distribution to the walls. Thermal photoneutron flux at the field center was $1.2\text{x}10^5 \pm 740$ n/cm².s. At the doorway just after the door, thermal neutron flux was 81 ± 46 n / cm².s.

A similar study was performed by Yücel et al. A Varian Clinac DHX device and RW3 solid phantom were used. For the NAA, gold and cadmium-coated gold foils were used. The measurements were performed at the field center, field edges and different positions in the room at depths of 0, 5 and 10 cm water. During the experiment; gantry angle was zero, irradiation time was 5000 MU, the field size was 20x20 cm² and SSD was 100 cm. Thermal neutron flux measurement results were at surface (d=0 cm) $6.65 \pm 0.20 \times 10^5$ n/cm².s, at depth 5 cm $3.73 \pm 0.11 \times 10^6$ n/cm².s and depth 10 cm of water $2.18 \pm 0.06 \times 10^6$ n/cm².s. At the door, flux was $2.36 \pm 1.30 \times 10^2$ n/cm².s [19]. These results are comparable with our findings. The differences occur from different design-brand of the accelerator and other measurement parameters.

Esposito et al, studied NAA with Elekta Precise linear accelerator device which is nearly the same device that we used in the experiment. The photon energy was chosen as 18 MV. The field size was 15x15 cm² and gold foils were used for NAA. After the measurement, the maximum neutron flux was observed at 1 cm away from the center and at SSD 100 cm flux was 48% of the maximum flux. At the maze entry, the flux was 15% of the maximum flux [20]. The results were given in terms of dose values so it would be appropriate to compare the results relatively with our study. When the results were compared with respect to the maximum flux, they appeared to be quite similar. In our study, 100 cm away from the center, the flux equals to 45% of the maximum neutron flux, and at 1.5m away from center it is 42% of the maximum flux. At the maze entry it was 13% of maximum flux.

Konefal et al, studied with a 15 MV Siemens Primus device. They used indium and cadmium-coated indium foils for the NAA. They took measurements at various points in treatment room. As a result, the maximum thermal neutron flux was obtained at the center of field at SSD 100 cm. Thermal neutron flux at the wall was 67% of the measurement taken at the field center. 1.5 m away from the center, the measurement showed that the thermal neutron flux was 91% of the maximum thermal neutron flux. The measurement that was taken at the intersection of the maze was 21% of the maximum thermal neutron flux [14]. In our study, the thermal neutron flux values were 33%, 42%, and 13% respectively.

Liu et al, used a Varian Clinac 21EX linear accelerator. Photon energy was chosen as 15 MV and indium foils were used in this study. The thermal neutron flux change was investigated from field center to the field edge. The results showed that the thermal flux ratio to the maximum flux at 20 cm away from center was 80% for direction A (Left side of the LINAC), and for direction B (right side) at same distance the result was 50% of the maximum thermal neutron flux [21]. The thermal neutron fluxes at the same points in our study were 65% and 71% respectively. One should expect similar results due to symmetry.

Sanchez et al, used indium foils in the linear accelerator room during their experiment. The thermal neutron flux at SSD 100 cm and field size of 30x30 cm^2 was 7.28×10^4 $n/cm^2.s$ [22]. In our study the thermal neutron flux at same condition was obtained as $5.4 \times 10^4 \pm 744$ $n/cm^2.s$. The order of results is the same for the two studies. However, the model of linear accelerator was not specified.

In this study, we used indium and cadmium foils for the measurement of the thermal neutron flux at various locations throughout the room. The maximum thermal neutron flux was detected just above the head of the LINAC towards the gun direction ($G=0^0$). The neutron flux decreased from the field center towards the edge of the room which followed a gaussian distribution. The studies that were compared with our study included some different parameters which caused some deviations in the results. Different parameters were; different types of activation foils, amount of given MU, model and brand of LINACs and types of detectors. Nedaei et al, showed that the Varian brand devices caused three times more photoneutron than the Elekta brand devices [23]. Liu et al, discussed that the experiments that were carried out with activation foils caused a standard deviation between 12.2%-13.9% because of normalization of foils and the self-absorption phenomena [21].

After an extensive literature research, we can say that measurement of thermal neutron flux at 91 different locations inside a radiotherapy room was the most comprehensive study that was carried out up to date. The experimental parameters varied extensively; such as, different gantry angles, different MU values, inside and outside field measurements, etc. Thus, we could map the thermal neutron flux distribution throughout the room.

5 Conclusions

Recently, radiotherapy in the treatment of cancer patients has become an important modality. The radiation energy used is increasing with the development of new technologies. But if the necessary measures are not taken, increasing the radiation energy can give damage to patients, employees and environment. One should understand the consequences of photoneutron production in linear accelerators and should analyze the flux and energy of neutrons which were formed in the treatment rooms.

In our study, the neutron flux distribution within the bunker was measured with detail using 91 points. The flux decreases as we move away from the isocenter, which is compatible with the literature. The magnitude of the neutron fluxes shows that there is a significant amount of neutron dose within the room. The corresponding neutron dose to the patient however is only 0.1-0.3 % of the total dose. However, neutrons have a high RBE and this unwanted dose is not calculated with the treatment planning system (TPS).

Photoneutron production is an interesting research topic which has both experimental and simulation legs. The future work would be to compare the results with the Monte Carlo simulations.

References

1. A. Jemal, T. Murray, A. Samuels, A. Ghafoor, E. Ward, M.J. Thun, Cancer Statistics. American Cancer Society. **53**: 5-26 (2003)
2. J.I. Spicka, H.S. Kim, D.W. Oh, V. Marable, K. Fleury, Equal surface dose compensation. American Association of Medical Dosimetrists. **14**: 287-290 (1989)
3. F. Biltekin, M. Yeğiner, G. Özyiğit, Investigating in-field and out-of-field neutron contamination in high-energy medical linear accelerators based on the treatment factors of field size, depth, beam modifiers, and beam type. Physica Medica. **31**: 517-523 (2015)
4. V. Vylet, J.C. Liu, Radiation Protection at High Energy Electron Accelerators. Oxford Journals., **96**: 333-343 (2011)
5. B. Juste, S. Morató, R. Miró, G. Verdú, S. Díez, MCNP6 unstructured mesh application to estimate the photoneutron distribution and induced activity inside a linac bunker. Radiation Physics and Chemistry. (2016)
6. International Commission on Radiological Protection. ICRP Publication 60, Recommendations of the International Commission on Radiological Protection (1990) (http://www.icrp.org/publication.asp?id=icrp%20publication%2060)
7. R.M. Howell, M.S. Ferenci, Investigation of secondary neutron dose for 18 MV dynamic MLC IMRT delivery. American Association of Physicists in Medicine **32**: 786-793 (2005)
8. J.P. Lin, W.C. Liu, C.C. Lin CC, Investigation of photoneutron dose equivalent from high-energy photons in radiotherapy. Applied Radiation and Isotopes, **65**: 599–604 (2007)
9. L. Sajó-Bohus, H.R. Vega-Carrillo, H.S. Virk, SSNTD technique in photo-neutron applications. Solid State Phenomena. **239**: 180-214 (2015)
10. E.D. Conrad, Transient Radiation Effects In Electronics. Advances in Nuclear Science and Technology. **3**: 143-203 (1966)
11. A. Naseria, A. Mesbahia, A review on photoneutrons characteristics in radiation therapy with high-energy

photon beams. Reports of practical oncology and radiotherapy. **15**: 138-144 (2010)

12. M. Králik, K. Turek, Characterisation of neutron fields around high energy X-ray radiotherapy machines. Rad. Prot. Dosim. **110**: 503-507 (2004)
13. C.L. Allred, Effects of Radition On Silicon and Borosilicate Glass. MIT Materials Science and Engineering, Massachusetts (2003)
14. A. Konefał, A. Orlefb, M. Dybekc, Z. Maniakowskib, K. Polaczek-Grelika, W. Zippera, Correlation between radioactivity induced inside the treatment room and the undesirable thermal/resonance neutron radiation produced by linac. Physica Medica. **24**: 212-218 (2008)
15. International Nuclear Data Committee, IAEA Annual Report, Atlas of Neutron Capture Cross Sections (1997) (http://www.iaea.org/inis/collection/NCLCollectionStore/_Public/28/060/28060364.pdf)
16. J.L. Lawson, J.M. Cork, The radioactive isotope indium. Physical Review. **52**: 531-535 (1937)
17. J.R. Lamarsh, A.J. Baratta, Radiation Shielding. In: Horton MJ ed. Introduction to Nuclear Engineering. 3rd ed. New Jersey: Prentice Hall p: 548-604 (2001)
18. International Atomic Energy Agency (IAEA), Nuclear Data Services, ^{116}In Decay Radiation Scheme (https://www-nds.iaea.org/relnsd/vcharthtml/VChartHTML.html)
19. H. Yücel, İ. Çobanbaş, A. Kolbaşı, A.Ö. Yüksel, V. Kaya, Measurement of Photo-Neutron Dose from an 18-MV Medical Linac Using a Foil Activation Method in View of Radiation Protection of Patients. Nuclear Engineering and Technology. **48**: 525-532 (2016)
20. A. Esposito R. Bedogni, L. Lembo, M. Morelli, Determination of the neutron spectra around an 18MVmedical LINAC with a passive Bonner sphere spectrometer based on gold foils and TLD pairs, Radiation Measurements. **43**: 1038-1043 (2008)
21. M. Liu, S.S. Huang, W.S. ShanLiu, D.M. Yea, L.K. Pan LK, C.Y. Chen CY, Distribution of spatial photo neutrons inside a 70kg water phantom via neutron activation analysis. Applied Radiation and Isotopes. **68** : 1816–1821 (2010)
22. F. Sanchez, G. Madurga, R. Arrans, Neutron measurements around an 18 MV linac. Radiotherapy and Oncology. **15**: 259-265 (1989)
23. H.A. Nedaie, H. Darestani, N. Banaee, N. Shagholi, K. Mohammadi, A. Shahvar, E. Bayat, Neutron dose measurements of Varian and Elekta linacs by TLD600 and TLD700 dosimeters and comparison with MCNP calculations. Journal of Medical Physics. **39**: 10-17 (2014)

Precise calculation of neutron-capture reactions contribution in energy release for different types of VVER-1000 fuel assemblies

Georgy Tikhomirov*, *Rynat* Bahdanovich, and *Phu* Pham

National Nuclear Research University MEPhI (Moscow Engineering Physics Institute), Department of Theoretical and Experimental Physics of Nuclear Reactors, 115409 Kashirskoe shosse 31, Moscow, Russia

Abstract. Precise calculation of energy release in a nuclear reactor is necessary to obtain the correct spatial power distribution and predict characteristics of burned nuclear fuel. In this work, previously developed method for calculation neutron-capture reactions – capture component – contribution in effective energy release in a fuel core of nuclear reactor is discussed. The method was improved and implemented to the different models of VVER-1000 reactor developed for MCU 5 and MCNP 4 computer codes. Different models of equivalent cell and fuel assembly in the beginning of fuel cycle were calculated. These models differ by the geometry, fuel enrichment and presence of burnable absorbers. It is shown, that capture component depends on fuel enrichment and presence of burnable absorbers. Its value varies for different types of hot fuel assemblies from 3.35% to 3.85% of effective energy release. Average capture component contribution in effective energy release for typical serial fresh fuel of VVER-1000 is 3.5%, which is 7 MeV/fission. The method will be used in future to estimate the dependency of capture energy on fuel density, burn-up, etc.

1 Introduction

Calculation of energy release in fuel core of nuclear reactor plays an important role both at the stage of design and during the operation. As of today the majority of energy release estimations leave out of account capture component of energy release, which is connected with neutron-capture reactions ((n,γ), (n,α), etc.). This component is defined by the nuclide composition of fuel core and neutron spectrum. That is why it varies for different burn-ups, fuel designs and is hard to calculate.

Use of underestimated value of energy release by engineering programs could produce considerable error in distribution of energy release in a fuel core and in characteristics of spent nuclear fuel. The precision of energy release calculation is important for prediction of reactor's antineutrino spectrum [1].

Different fuel assembly models of VVER-1000 were calculated in order to estimate the influence of several factors on capture component. An equivalent cell model was considered in order to test the developed method. After that, different types of fuel assembly models were calculated to confirm the obtained results and to examine the influence of fuel enrichment and presence of burnable absorbers on the capture component of energy release.

On the first stage of effective energy release calculation, we used MCNP 4 and MCU 5 codes to obtain nuclear reaction rates for each isotope in the developed model (equivalent cell model, fuel assemblies 13ZS and 30ZSV). Results obtained by two codes showed good agreement. Further, we used only MCU code to calculate other types of fuel assemblies (16ZS, 24ZS, 24ZSV, 24ZSW, 35ZS, 35ZSU).

It is important to note, that all the calculations using MCNP code were performed in the past and discussed in [2, 3]. The temperatures of materials were specified by accessible cross-section libraries. In this work, we re-examined the results using improved method for the effective energy release calculation.

2 Calculation of effective energy release components

Average energy released after one fission (fission energy yield, E_f) could be represented as:

$$E_f = E_k + E_n + E_{\gamma\,inst} + E_{\gamma\,del} + E_\beta + E_\nu \qquad (1)$$

E_k – kinetic energy of fission products, E_n – kinetic energy of instant and delayed fission neutrons, $E_{\gamma\,inst}$ and $E_{\gamma\,del}$ – energy of instant and delayed gamma rays, E_β – kinetic energy of delayed beta particles, E_ν – kinetic energy of antineutrinos.

Part of a fission energy, which dissipates in a fuel core, is called effective fission energy (E_{eff}):

$$E_{eff} = E_f - E_\nu - \Delta E_n - \Delta E_{\beta\gamma} \qquad (2)$$

* Corresponding author: GVTikhomirov@mephi.ru

ΔE_n – adjustment, which takes into account loss of fission neutrons kinetic energy after absorption, which leads to fission, $\Delta E_{\beta\gamma}$ – adjustment for delayed energy, which considers that beta decays of fission products are incomplete and shows how much energy will be radiated in future.

Considering that some energy is also released in neutron-capture reactions (capture energy), effective energy release (E_{efe}) is:

$$E_{efe} = E_f - E_v - \Delta E_n - \Delta E_{\beta\gamma} + E_{capt} \quad (3)$$

$$E_{capt} = E_{capt\ inst} + E_{capt.\ del} \quad (4)$$

E_{capt} – energy released in neutron-capture reactions, sum of instant capture energy ($E_{capt\ inst}$) and delayed capture energy ($E_{capt.\ del}$), released during decay of products of neutron-capture reactions. The main contributors to capture energy in fuel core are (n,γ) and (n,α) reactions [1, 4-7].

Developed method for calculation of effective energy release is based on the use of high-precision neutron codes (MCNP, MCU, SCALE, SERPENT, etc.) for reaction rates estimations in a fuel core. Using these codes allow conducting precise calculations and obtaining results close to the experimental ones. The precision of such codes is defined by the used evaluated nuclear data libraries.

Calculation of energy release in a nuclear reactor was divided into three stages:

1. Development of reactor's mathematical model and selection of important nuclear reactions and isotopes. Complexity of the model and its detailing could considerably influence the results.
2. Computation of fission and capture reaction rates in developed model with the use of high-precision neutron codes.
3. Calculation of effective energy release components with the use of obtained reaction rates and fission, neutron-capture ((n,γ), (n,α), etc.) and decay energy yields table values based on data from Nubase2012 and ENDF/B-VII.I [8, 9].

In order to obtain the correct results the neutron multiplication factor (K_{eff}) in the developed model should be equal unity. Otherwise, one could get the wrong neutron balance.

3 Calculated models and results

3.1 Equivalent cell model

In order to test the developed method for calculation of capture reactions contribution in effective energy release we calculated an equivalent cell model of VVER-1000 with neutron reflection on the boarders (Figure 1, Table 1).

As in a nuclear reactor, the multiplication factor in this model is equal unity (^{10}B put into the moderator, equal to 21.7 and 17.8 g (H_3BO_3) / kg (H_2O) for cold and hot states). Calculation of energy release was conducted only for cold and hot fresh fuel. The concentration of fission products was assumed much lower than their equilibrium concentration (negligible).

In order to calculate the delayed beta and gamma energy we used an adjustment $\Delta E_{\beta\gamma}$. After 0.5 day from the reactor start-up $\Delta E_{\beta\gamma}$=1.38 MeV/fission (for uranium fuel, according to formulas from [1]). The same value of $\Delta E_{\beta\gamma}$ we obtatined for all models presented in the paper. This amount of energy should be substracted from the table fission energy yield, the same as the energy of antineutrinos.

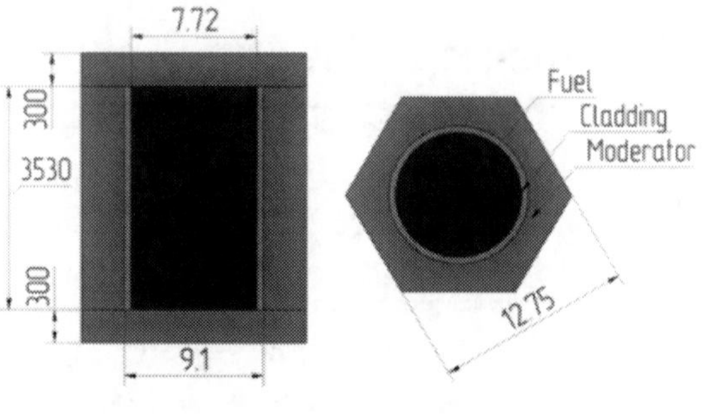

Fig. 1. Dimensions of the equivalent cell model (mm).

Capture energy which is given in the results contains instant capture energy and delayed capture energy calculated for 0.5 days from the reactor start-up (Table 2). We should note, that the delayed capture energy is approximately 5.5% of capture energy and it is defined by the decay of ^{239}U. This contribution could increase for burned fuel.

Table 1. Material characteristics of the equivalent cell model.

Material	Composition	Density g/cm^3
Fuel	UO_2, 3% ^{235}U	10.0
Cladding	Zr + 1% ^{93}Nb	6.6
Moderator	H_2O + ^{10}B	0.715

The difference in the functionals calculated using MCU and MCNP codes is small. It exists mostly because of the difference in the evaluated nuclear data libraries.

Code MCU uses continuous temperature dependent cross-section library VESTA in the thermal energy range and 26-group BNAB data library in the intermediate energy range. However, resolved resonances are calculated using Breit-Wigner or Adler-Adler formalism. For the fast energy range (higher than 100 KeV) we used continuous cross-section library at 300 K based on ENDF/B-VII.0. MCNP uses continuous cross-section libraries for all energy ranges and temperatures. That could be the reason why the difference for hot state is more that for cold state.

In addition, there is a simplification in the capture energy release calculation using MCU code. MCU can calculate (n,γ), (n,α), etc. reaction rates only using continuous cross-sections (which we had only for 300 K temperature). That is why we made a simplification based on the results of MCNP calculation. We consider neutron capture (n,c) as (n,γ) reaction for all nuclides except ^{10}B and ^{16}O for which (n,α) is dominant. According to our estimations, such simplification could cause 0.3% decrease in capture energy value for the hot state and 0.4% for the cold state.

Table 2. Results of the equivalent cell model calculation (energy in MeV/fiss.).

Functional	MCNP	MCU	$\frac{MCNP - MCU}{MCNP}$
Cold state: T_{fuel}=293.6 K, $T_{cladding}$=293.6 K, $T_{moderator}$=293.6K			
K_{eff}	1.0161	1.0171	-0.10%
ΔE_n	0.07	0.07	0.00%
$E_{capt.}$	6.15	6.19	-0.65%
E_{efe}	198.4	198.4	0.00%
E_{capt} / E_{efe}	3.10%	3.12%	-0.65%
Hot state: T_{fuel}=1200 K, $T_{cladding}$=900 K, $T_{moderator}$=600 K			
K_{eff}	1.0075	1.0027	0.48%
ΔE_n	0.08	0.08	0.00%
$E_{capt.}$	6.69	6.77	-1.20%
E_{efe}	199.0	199.0	0.00%
E_{capt} / E_{efe}	3.36%	3.40%	-1.22%

Capture energy for the hot fresh fuel is 6.7 MeV/fiss., that is 3.4% of the effective energy release; for the cold fresh fuel is 6.2 MeV/fiss., that is 3.1% of the effective energy release. The difference is caused by the Doppler resonance broadening of ^{238}U after heating of the fuel. Neutron capture on ^{238}U increases by 12%, on ^{10}B decreases by 10% and on hydrogen decreases by 3%, effect for other isotopes is small. Energy release in (n,γ) reaction on ^{238}U is higher than in (n,α) on ^{10}B that is why the capture energy increases. Boron concentration is lower in hot state because of the Doppler resonance broadening (^{238}U).

3.2 Modified serial fuel assemblies

Models of modified 13ZS and 30ZSV fuel assemblies with neutron reflection on the boarders were calculated in order to estimate influence of complex geometry and presence of burnable absorbers on capture energy component (Figure 2). Dimensions of the fuel rods are the same as for equivalent cell model (Figure 3, 4). Material compositions are in Table 3. Fuel assembly's pitch is 236 mm.

The results are presented in Table 4. Difference in the results obtained by MCU and MCNP is small and it is the same order of magnitude as for equivalent cell model.

Capture energy component contribution in effective energy release for hot state of fuel assembly without gadolinium (type 13ZS) is 3.4%, and for fuel assembly with gadolinium (type 30ZSV) is 3.9%. The results demonstrate that effective energy release increases by 0.9 MeV/fiss. (or 0.5%) for fuel with gadolinium. It is because the energy release in (n,γ) reaction on gadolinium is three times higher than in (n,α) reaction on boron.

Table 3. Characteristics of modified serial fuel assemblies' materials.

Material	Composition	Density g/cm^3
Fuel	UO_2, 3% ^{235}U	9.5
Fuel with gadolinium	UO_2 (2.4% ^{235}U) + 6.4% $^{157}Gd_2O_3$	9.34
Cladding	93.33% Zr + 1% ^{93}Nb + 4% Fe + 0.6% Ni + 1% Cr + 0.06% Ti + 0.01% C	6.6
Steel (08X18H10T)	69.9% Fe+2% + 18.5 % ^{52}Cr+10.5 % Ni + 1% Ti + 0.1% C	7.7
Moderator	H_2O + x% H_3BO_3	0.715

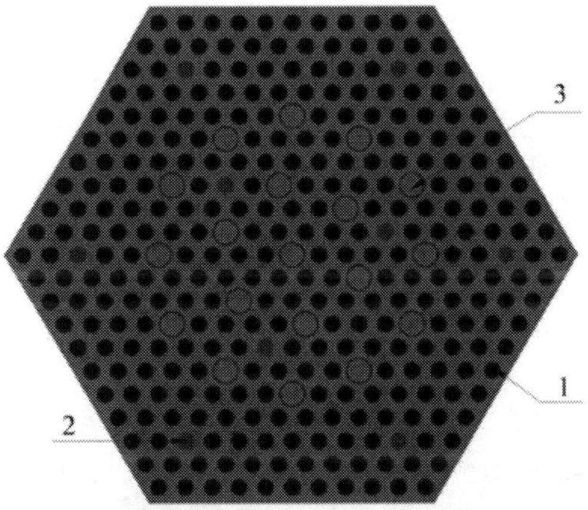

Fig. 2. Fuel assembly 13ZS (considering all fuel rods the same) and fuel assembly 30ZSV with gadolinium. 1 – fuel rod (3% ^{235}U), 2 – fuel rod with gadolinium (2.4% ^{235}U, 6.4% $^{157}Gd_2O_3$), 3 – guide channel.

Capture energy increases in the hot state compared with the cold state. However, tDoppler resonance broadening causes smaller effect than for equivalent cell model, probably because of the presence of guide channels with the moderator. Fraction of neutron capture on boron did not decreases too much with the heating of the materials, only by 3.2% for 13ZS and by 3.6% for 30ZSV. Neutron capture on ^{238}U increases by 3.5% for 13ZS and by 3.4% for 30ZSV, effect for other isotopes is negligible.

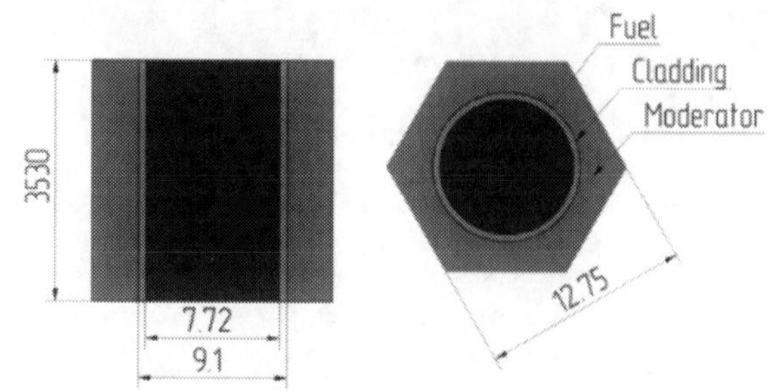

Fig. 3. Characteristics of a fuel rod (mm).

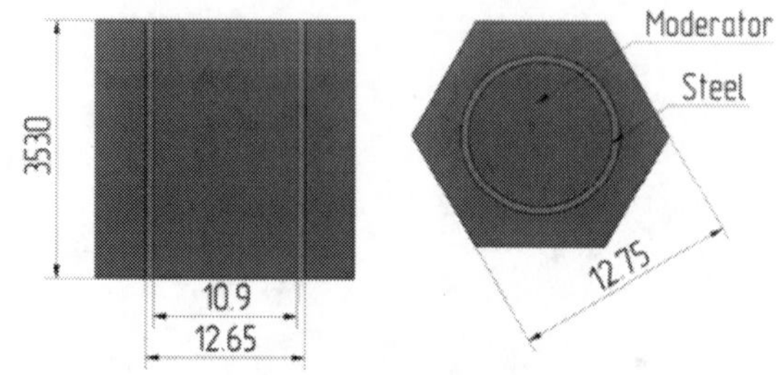

Fig. 4. Characteristics of a guide channel (mm).

Table 4. Results of 13ZS and 30ZSV fuel assemblies' calculation (energy in MeV/fiss.).

Functional	MCNP[1]	MCU[1]	MCU[2]	$\frac{MCNP^1 - MCU^1}{MCNP^1}$
13ZS				
H_3BO_3, g/kg	15.7	15.7	14.17	-
K_{eff}	1.008	1.0072	9.9975	0.08%
ΔE_n	0.076	0.074	0.075	2.63%
$E_{capt.}$	6.64	6.72	6.83	-1.20%
E_{efe}	198.9	199.0	199.1	-0.05%
E_{capt} / E_{efe}	3.34%	3.38%	3.43%	-1.15%
30ZSV				
H_3BO_3, g/kg	9.5	9.5	7.7	-
K_{eff}	1.0027	1.0038	1.0007	-0.11%
ΔE_n	0.076	0.074	0.073	2.63%
$E_{capt.}$	7.60	7.64	7.73	-0.53%
E_{efe}	199.8	199.9	200.0	-0.05%
E_{capt} / E_{efe}	3.80%	3.82%	3.86%	-0.48%

[1] T_{fuel}=293.6 K, $T_{cladding\ and\ steel}$=293.6 K, $T_{moderator}$=600 K
[2] T_{fuel}=1200 K, $T_{cladding}$=600 K, $T_{moderator\ and\ steel}$=578 K, black border on the ends

3.3 Serial fuel assemblies

Models of serial fuel assemblies (16ZS, 24ZS, 24ZSV, 24ZSW, 35ZS, 35ZSU) with mirror reflection condition on the lateral border and black condition on the ends were calculated using MCU code in order to obtain the precise values of capture energy component (Figure 5). Dimensions of fuel rods are the same as for 13ZS and 30ZSV models (Figure 3, 4). Characteristics of materials are in Table 5 [10, 11]. Calculated K_{eff} is 0.999 for all the models.

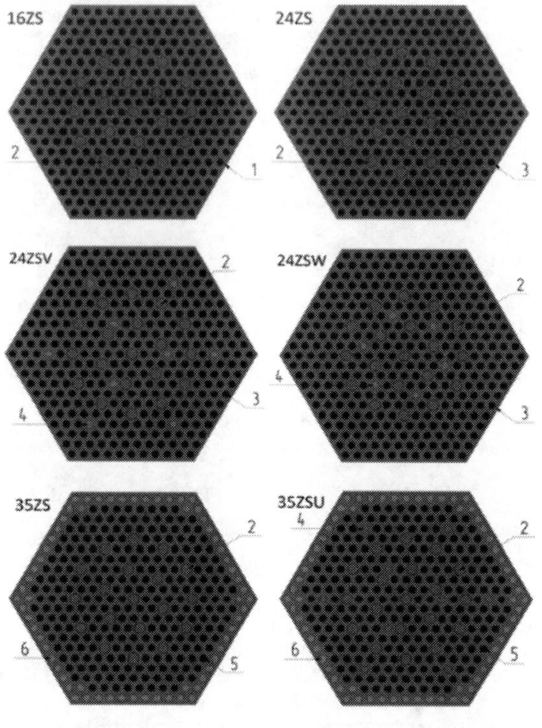

Fig. 5. Models of VVER-1000 serial fuel assemblies (16ZS, 24ZS, 24ZSV, 24ZSW, 35ZS, 35ZSU). 1 – fuel rod (1.6% ^{235}U), 2 – guide channel, 3 – fuel rod (2.4% ^{235}U), 4 – fuel rod with gadolinium (2.4% ^{235}U, 8% Gd), 5 – fuel rod (3.6% ^{235}U), 6 – fuel rod (3.3% ^{235}U).

According to the results (Table 6) capture energy contribution in effective energy release depends on the fuel assembly's type and varies starting from 3.35% to 3.79% for serial VVER-1000 fuel assemblies. There is a tendency of decreasing of capture energy with the increase of boric acid concentration. It is because energy release in (n,α) reaction on ^{10}B is much lower than energy release in (n,γ) reaction on other isotopes.

Capture energy decreases with the increase of fuel enrichment (see results for 16ZS, 13ZS, 24ZS, 35ZS). The more fuel enrichment the more boric acid should be add to achieve unity of effective multiplication factor.

The dependency of capture energy on fuel enrichment is non-linear (Figure 6), the following formula could be used to calculate capture energy:

$$E_{capt}(x) = 0.030x^4 - 0.342x^3 + 1.547x^2 - 3.408x + 9.865 \quad (5)$$

Table 5. Characteristics of serial fuel assemblies' materials.

Material	Composition	Density g/cm^3	Temperature, K
Fuel	UO_2, 3% ^{235}U	9.5	1200
Fuel with gadolinium	UO_2 (2.4% ^{235}U) + 8% Gd_2O_3	9.5	1200
Cladding	Zr + 1% ^{93}Nb	6.6	600
Steel (08X18H10T)	69.9% Fe + 2% + 18.5 % ^{52}Cr +10.5 % Ni + 1% Ti + 0.1% C	7.7	578
Moderator	H_2O + x% H_3BO_3	0.715	578

Table 6. Results of serial fuel assemblies calculation (energy in MeV/fiss.).

Parameter	16ZS	24ZS	35ZS
Max x% ^{235}U	1.6	2.4	3.6
H_3BO_3, g/kg	3.63	10.06	19.10
ΔE_n	0.08	0.07	0.07
$E_{capt.}$	7.16	6.85	6.65
E_{efe}	199.4	199.1	198.9
E_{capt} / E_{efe}	3.59%	3.44%	3.35%
Parameter	24ZSV	24ZSW	35ZSU
Max x% ^{235}U	2.4	2.4	3.6
H_3BO_3, g/kg	4.75	6.70	13.29
ΔE_n	0.07	0.07	0.07
$E_{capt.}$	7.57	7.30	7.17
E_{efe}	199.8	199.5	199.4
E_{capt} / E_{efe}	3.79%	3.66%	3.60%

where x – fuel enrichment in weight percent, x=[1.6%, 3.4%]. However, dependency of critical boric acid concentration on fuel enrichment is linear:

$$Cboric\ (x) = 8.000x - 9.149 \quad (6)$$

This two formulas were obtained from the modeling of 16ZS fuel assembly with different fuel enrichments.

Concerning boric acid concentration, for assemblies 13ZS, 24ZS, 35ZS, 35ZSU it exceeds concentration allowed during normal operation on power. However, one should take into account that modeled fuel assemblies are put into fuel core of VVER-1000 together. Exceeding of normal boric acid concentration shows that it is not possible to assemble fuel core from that types of fuel assemblies alone. In addition, one should consider that there is a B_4C rod regulation system with one group of rods always dropped into fuel core. That is why the estimations are reasonable.

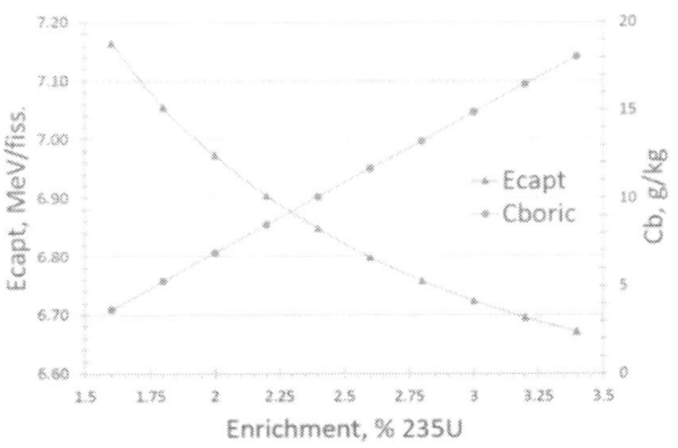

Fig. 6. Capture energy (Ecapt) and boric acid concentration (Cboric) on fuel enrichment.

In the case of fuel core assembled of all modeled serial fuel assemblies, average capture energy is 7 MeV (3.5% of effective energy release). It is recommended to use this value for calculation of power distribution and fuel burn-up.

Interesting to note that difference in effective energy release in Table 6 is caused by the difference in capture energy. Effective energy release less capture energy (E_{eff}) is 192.25 MeV/fiss. for all fuel assembly models.

It is evident that energy from (n,α) reactions releases almost at the same point where the reaction occurs (mostly in the moderator). If we consider that the effective fission energy and capture energy from (n,γ) reactions is allocated in the fuel rods (absorption of gamma rays is much higher for elements with higher number of electrons (Z)), than we could estimate capture energy distribution in the assembly. For fuel assembly 24ZS 2.8% of E_{efe} (part of the capture energy) releases in the fuel rods and 0.6% of E_{efe} in the moderator. In comparison, for the fuel assembly 24ZSV with gadolinium 3.4% of E_{efe} releases in the fuel rods and 0.4% of E_{efe} in the moderator. Thus, presence of gadolinium could increase effective energy release in fuel rods by 0.6%.

Capture energy and effective energy obtained previously for fuel assemblies (13ZS and 30ZSV), using old method [2, 3], differs from the values presented in this paper by 4% and 0.6%, correspondingly. The difference is caused by considering the delayed capture energy, ΔE_n and $\Delta E_{\beta\gamma}$ adjustments in the effective energy release formula (3).

4 Conclusions

Method for calculation of capture energy contribution in effective energy release in fuel core of a nuclear reactor was improved. It was tested on a simple model of

VVER-1000 equivalent cell and eight models of a fuel assembly (13ZS, 30ZSV, 16ZS, 24ZS, 24ZSV, 24ZSW, 35ZS, 35ZSU). Results were obtained using MCU and MCNP computer codes.

It is shown, that capture energy release decreases with the growth of fuel enrichment for VVER-1000 and neutron-capture reactions contribution in energy release depends on fuel assembly's type. Formula (5) was produced in order to estimate dependency of capture energy on fuel enrichment.

Results of the equivalent cell modelling are very close to the results of the fuel assembly modelling. Discrepancy in capture energy is less than 1% (for hot states). Thus, one could use a simple equivalent cell model for the estimation of the effective energy release and its dependency on different factors (burn-up, enrichment, etc.).

Results obtained by MCU and MCNP codes insignificantly differ from each other both for cold and hot states of the models. Implemented simplification concerning calculation of capture energy release using MCU code is valid. It could produce 0.3% error in the capture energy value, which leads to 0.01% error in the effective energy release.

Although boric acid concentration for some fuel assemblies exceeds normal value for reactor's operation, the obtained results are justified, because the fuel core of VVER-1000 is assembled using different types of fuel assemblies.

Average energy release for hot fresh fuel of VVER-1000 is 199 MeV/fiss., contribution of capture energy is 3.5%.

The obtained results show that energy release for fuel without gadolinium is less by 0.5% than for fuel with gadolinium. Presence of gadolinium could also increase effective energy release in fuel rods by approximately 0.6%, because of capture energy component.

The improved method enables to calculate precise value of capture energy contribution in effective energy release for different types of nuclear reactor, and to estimate its dependency on the fuel burn-up, fuel enrichment and density, composition of fuel core, etc. Difference in the effective energy release between old method and improved method is 0.6%. It is caused by the difference in effective energy release calculation method.

This work was performed at the Department of Theoretical and Experimental Physics of Nuclear Reactors in the framework of 5-100 project. BASOV and CHERENKOV clusters of MEPhI were used for the preliminary calculations.

References

1. V.I. Kopeikin, L.A. Mikaelyan, V.V. Sinev Phys. of At. Nuc., **T. 67. № 10.**, p. 1892-1899 (2004)
2. R.B. Bahdanovich, G.V. Tikhomirov, I.S. Saldikov, E.F. Kryuchkov *PHYSOR 2014* (Kyoto, Japan, 2014)
3. R.B. Bahdanovich, G. V. Tikhomirov, I. S. Saldikov, M.Yu.Ternovykh, A. S. Gerasimov *GLOBAL 2015* (Paris, France, 2015)
4. A.Y. Kramerov, Y.V. Shevelev *Engineering calculations of nuclear reactors* (Energoatomizdat, Moscow, Russia, 1984)
5. X. B. Ma, W. L. Zhong, L. Z. Wang, Y. X. Chen, and J. Cao Physical Review C - Nuclear Physics, **Vol. 88, Issue 1** (2013)
6. M.F. James Journal of Nuclear Energy, **Vol. 23, Issue 9**, Pages 517-536 (1969)
7. A. Persic, A. Trkov *Nuclear Energy in Central Europe '99* (Portoroz, Slovenia, 1999)
8. B. Pritychenko, A. Sonzogni, Q-value Calculator (QCalc) NNDC, Brookhaven National Laboratory URL: http://www.nndc.bnl.gov/qcalc/
9. Nuclear Energy Agency, Java-based Nuclear Data Information System (JANIS), available online. URL: http://www.oecd-nea.org/janis/
10. A.K. Gorohov, Yu.G. Dragunov, G.L. Lunin, A.N. Novikov, V.I. Tsofin, Yu.A. Anan'ev, *Justification of neutron-physical and radiation characteristics of VVER designs* (IKC Academkniga, Moscow, 2004)
11. B.R. Bergelson, V.V. Belonog, A.S. Gerasimov, G.V. Tikhomirov Atomic Energy **Vol. 109, Issue 4**, Pages 240-245 (2011)

^{10}B+ZnS(Ag) as an alternative to ^{3}He-based detectors for Radiation Portal Monitors

Karen Arlet Guzman-Garcia[1*], *Hector Rene* Vega-Carrillo[2], *Eduardo* Gallego[1], *Juan Antonio* Gonzalez-Gonzalez[3], *Alfredo* Lorente[1] and *Sviatoslav* Ibañez-Fernandez[1]

[1]Universidad Politécnica de Madrid, Departamento de Ingeniería Energética, José Gutierrez Abascal, 2, 28006, Madrid, Spain
[2]Universidad Autónoma de Zacatecas, Unidad Académica de Estudios Nucleares, Ciprés 10, 98068, Zacatecas, Zac., Mexico
[3]Universidad Politécnica de Madrid, Laboratorio de Ingeniería Nuclear, C. Prof. Aranguren 3, 28040, Madrid, Spain

Abstract. Typical radiation portal monitor systems, RPM, deployed to detect illicit trafficking of radioactive materials include a set of gamma-ray detectors and neutron detectors. Usually the employed neutron detectors are pressurized ^{3}He-based neutron detectors tubes. Due the shortage of ^{3}He reported since 2009, the amount of ^{3}He available for use in gas proportional counter neutron detectors has become limited, while the demand has significantly increased, especially for homeland security applications. For this reason, many different alternatives are being investigated for its use in RPM systems. The aim of this work is to study a scintillation detector ZnS(Ag) mixed with highly enriched ^{10}B, ^{10}B+ZnS(Ag). Using Monte Carlo methods, MCNPX code, the response of two neutron detectors based on ^{10}B+ZnS(Ag), manufactured by BridgePort Instruments LLC with different geometries, were estimated by calculating the number of $^{10}B(n,\alpha)^{7}Li$ reactions for 29 monoenergetic neutron sources. Measurements and models were made, and both detectors were compared. The importance of the distance with respect to the ground was studied. The response with a ^{252}Cf moderated neutron source (0.5 cm lead and 2.5 cm polyethylene) was calculated in order to compare with other studied alternatives in the USA by Pacific National Northwest Laboratory, PNNL. With these results we conclude that neutron detectors using ^{10}B+ZnS(Ag) are an interesting alternative for replacing ^{3}He detectors. From the analysis with MCNPX we propose an improvement in the detector design.

1 Introduction

Since September 11th, 2001 the security systems in the United States of America improved significantly, adding radioactivity detection systems with neutron detectors, which until then only had gamma-ray detectors [1].

Currently the Radiation Portal Monitors (RPM) used for interception of illicit materials at borders are a set of gamma-ray detectors also including highly sensitive ^{3}He-based neutron detection systems. The detection of neutrons in the international trade of commodities is a serious concern, since a neutron signature from a vehicle may indicate the presence of SNM [2].

The main reason for having neutron detection capability is the need to detect fission neutrons from Special Nuclear Materials, SNM, like ^{239}Pu. Plutonium itself is a significant neutron source, while uranium in large quantities can be detected by its neutron signature. Since the shielding of neutrons can be difficult, neutron detection is an important means to finding SNM.

There are three basic requirements for neutron detectors for safeguards applications involving detection and measurements of SNM: 1) High absolute detection efficiency; 2) detector's low intrinsic gamma ray sensitivity; and 3) maintaining neutron detection efficiency when simultaneously is exposed to high gamma ray exposure rate. ^{3}He proportional counters clearly fulfil these requirements [3].

The use of ^{3}He as a neutron detector material has the great advantage that ^{3}He is only sensitive to neutrons and its sensitivity in proportional counters to gamma ray is negligible (pileup effects only become a problem in radiation fields of ~1 R/h). The proportional counters tubes containing ^{3}He are very simple in design, mechanically robust over a wide range of environmental conditions, and do not degrade over years of operation [4].

Sources of most concern include: complete weapons of mass destruction (WMD), improvised nuclear devices (IND); Special Nuclear Material (SNM) for weapons

* Corresponding author: ingkarenguzman@gmail.com

production, including plutonium and highly enriched uranium (HEU); and material or assemblies for radiological dispersal devices (RDD), usually known as "dirty bombs" [5].

The amount of nuclear material needed to manufacture a weapon of mass destruction is relatively small. In Figure 1 the radius of a sphere of bare fissile materials (Godiva) necessary to have a critical system are shown. Above 6.2 cm-radius of pure ^{239}Pu or war grade ^{239}Pu (Pu WG) becomes critical.

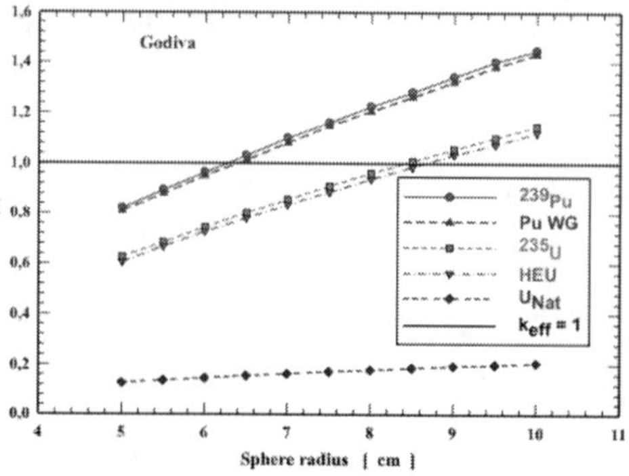

Figure 1. Effective multiplication factor of spheres of different radius with different nuclear materials (PuWG, HEU and ^{235}U).

Due the shortage of ^{3}He, reported since 2009, several efforts have been carried out to find alternative detectors for RPM, seeking similarity to those already installed with ^{3}He features [6].

The aim of this work was to study the properties of scintillation ^{10}B+ZnS(Ag) detectors to detect neutrons in RPMs. This type of detector has a scintillator ZnS(Ag), mixed with enriched ^{10}B. The neutron response of two ^{10}B+ZnS(Ag) detectors, with different geometries, were calculated using Monte Carlo methods with the MCNPX 2.7.0 code [7]. The neutron response was assessed from the number of $^{10}B(n,a)^{7}Li$ reactions for 29 monoenergetic neutron sources whose energies were between 10^{-9} to 20 MeV. Also, the actual neutron responses were evaluated through experimental measurements with small ^{252}Cf sources.

2 Materials and methods

2.1 Description of the N-15 and N-48 neutron detectors of ^{10}B+ZnS(Ag)

The N-48 and N-15 were manufactured by BridgePort Instruments LLC [8, 9]. Both use a mixture of ZnS(Ag) with ^{10}B, as neutron detection and screens on polymethyl methacrylate (PMMA) that guide the light pulses to a photomultiplier tube (PTM). The PTM has an embedded high voltage supply and multichannel analyzer eMorpho® digital electronics. The external size of the N-15 detector is 23 x 36 x 4 cm, and the external dimensions of the N-48 detector are 141.5 x 16.7 x 6.35 cm. Figure 2 shows the overall geometry of both detectors.

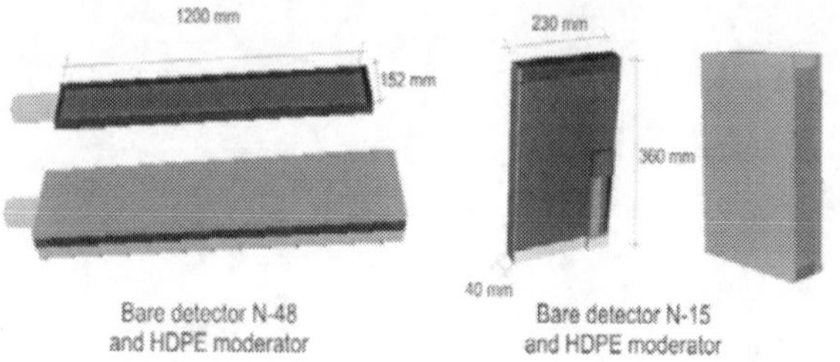

Figure 2. N-15 and N-48 geometries (MCNPX model)

In both cases, the detector's sensitive area is composed of five transparent ~ 0.017 cm-thick layers of a mixture of ZnS(Ag) and 95% ^{10}B enriched boron. These layers are arranged in four plates of PMMA sizing 23 x 36 x 0.635 cm for the N-15 and 120 x 15.2 x 0.635 cm for the N-48 detector. The PMMA function is twofold, as light guide and as neutron moderator. All is surrounded by ~8μm thick aluminium mylar as light reflector. The detector configuration is displayed in Figure 3.

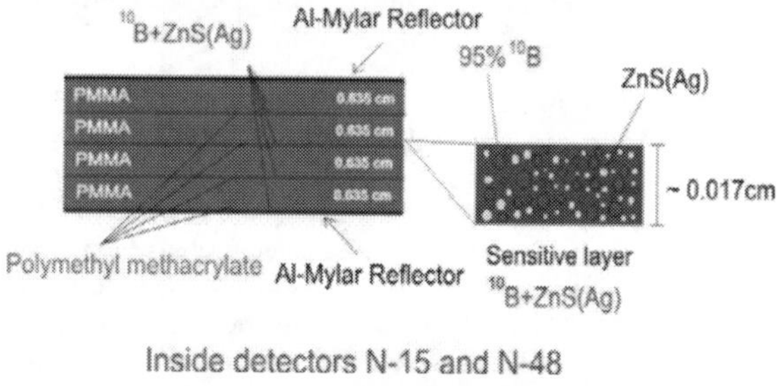

Figure 3. Internal configuration of the detectors.

Each detector has an outer moderator made of high density polyethylene (0.94 g cm^{-3}), HDPE.

For the N-15 detector the moderator thickness is 24 mm in the front, lateral faces, and top, bottom, 36 mm while 48 mm-thick in the back (24+36+48 mm). In the N-48A detector the moderator is 25 mm-thick in front, top, bottom, and lateral faces, while 50 mm-thick in the back (25+50 mm). The HDPE moderator on both detectors can be seen in Figure 4.

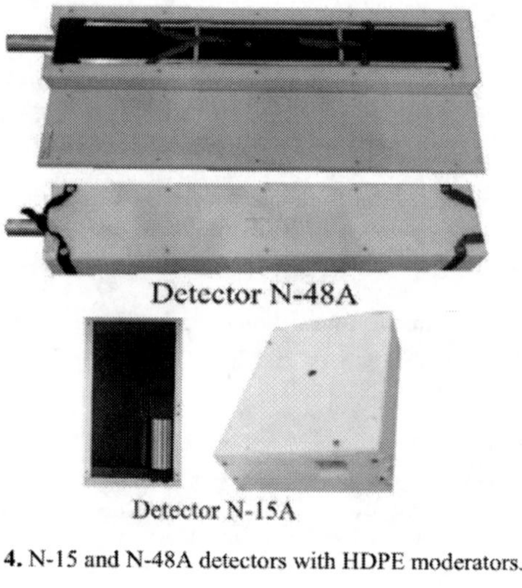

Figure 4. N-15 and N-48A detectors with HDPE moderators.

2.2 Monte Carlo calculations

2.2.1 Detectors response

Using the MCNPX 2.7.0 code, models of both detectors were built including all the detectors details such as: the sensitive layers, the PMMA, the PMT and the moderators. Here, the PMT was modelled as an empty cylinder of glass. For each detector the response was calculated using 29 monoenergetic neutrons with energies from 10^{-9} to 20 MeV. The response was estimated for the bare detector (without moderator) and with the HDPE moderator. The response was estimated using tally f4 [10, 11] where the number of $^{10}B(n,\alpha)^{7}Li$ reactions were calculated.

2.2.2 Indoor response

The detector models were also coupled to the model of the Neutron Measurements Laboratory of the Universidad Politécnica de Madrid (LMN-UPM) [12] and the response against $^{252}Cf_{UPM}$ neutron sources to 200 cm were calculated in similar conditions to those in the experimental measurements, Figure 5.

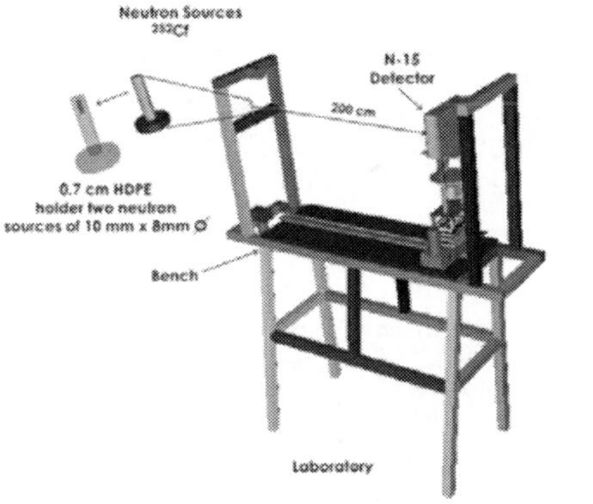

Figure 5. Configuration of the in-laboratory model (with N-15detector).

2.2.3 Outdoor response

A model was made with two N-15 detectors, in similar conditions to the outdoors measurements, without walls from a laboratory, representing normal operation conditions. Here, the detector response to neutrons produced by $^{252}Cf_{UPM}$ located 220 cm from the detector and 115 cm above the floor was calculated. In Figure 6 an scheme of the model is shown.

2.2.4 Effect on the response due to floor-to-detector distance.

Both, N-15 and the N-48 detectors, were positioned 200 cm from the $^{252}Cf_{UPM}$ source and the response was calculated varying the distance between the detector's centres and the floor, as it is shown in Figure 7. Here, the $^{252}Cf_{UPM}$ source was fixed at 100 cm above the floor. This response was in the aim to study the room-return effect.

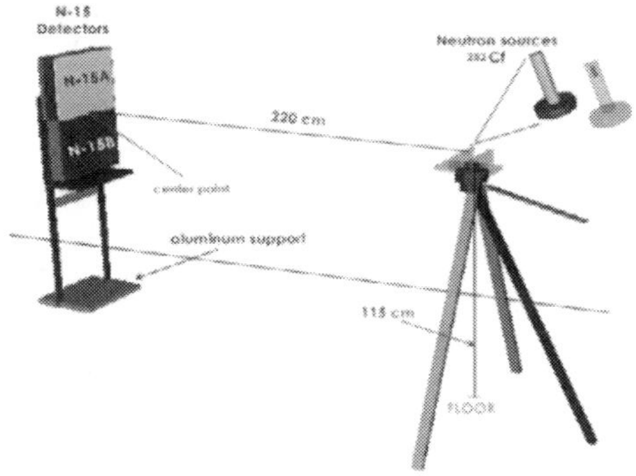

Figure 6. Configuration of the outdoors model (N-15 detectors).

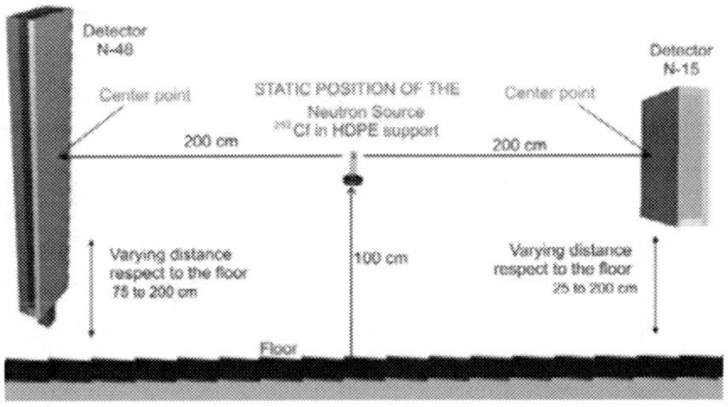

Figure 7. N-15 and N-48 outdoors (MNCPX model) (varying distances of detectors with respect to the ground).

2.2.5 Response with a moderated neutron source of ^{252}Cf

Different studies of alternatives to ^{3}He neutron detectors have been made in the USA by the Pacific Northwest National Laboratory, PNNL [13]. A ^{252}Cf neutron source was used for the tests of neutron sensors; in order to reduce the gamma-ray flux from the ^{252}Cf, the source is surrounded by at least 0.5 cm of lead, and to moderate the neutron spectrum the neutron source is surrounded with 2.5 cm of Polyethylene.

They recommended that the absolute detection efficiency for such the ^{252}Cf source, located 200 cm perpendicular to the geometric midpoint of the neutron detector sensor, shall be greater than 2.5 cps/ng of ^{252}Cf [13] and in accordance with the ANSI requirements [14].

Three different models for the N-15 neutron detector were made, using the neutron source described above (0.5 cm lead and 2.5 cm polyethylene) at a distance of 200 cm, in order to compare some previous results with others alternatives studied by PNNL.

The three different models were as shown in Figure 8, for the N-15 detector; A) an array like in Figure 6, with the only difference in the neutron source $^{252}Cf_{PNNLL}$ and at 200 cm respect to the neutron source; B) only one neutron detector, N-15 at 200 cm respect to the geometric midpoint of the detector in horizontal position, and C) one N-15 detector at 200 cm in vertical position.

DOI: 10.1051/epjconf/201715307008

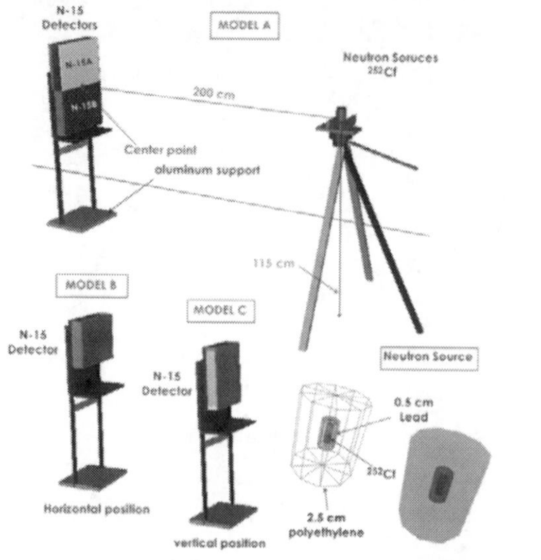

Figure 8. Configuration of the three models with different moderated neutron source conditions (2.5 cm polyethylene and 0.5 cm lead)

An extra model was made to determine the response in cps/ng of $^{252}Cf_{PNNL}$ at 200 cm in this conditions for the N-48 neutron.

In all the Monte Carlo calculations the number of histories was large enough to obtain uncertainties less than 5%. For the calculations, the cross sections were taken from the ENDF/B-VI library, where the S(α, β) treatment was included to take into account the thermalized neutron interactions [10].

2.3 Measurements

Measurements were carried out using a $^{252}Cf_{UPM}$, the experimental array was the same used in the Monte Carlo modelling. Here, the count rates per ng of ^{252}Cf was determined.

Also, measurements were carried out outside the laboratory, emulating the actual conditions of detectors in the checking points at borders, where the detectors were outside the laboratory.

The measurements were carried with two small $^{252}Cf_{UPM}$ neutron sources; these sources are double encapsulated within two steel cylinders of 1mm thickness with dimensions 10 mm x 8 mm Ø, which are moderated with 0.7 cm thickness of HDPE, 0.94 gr/cm^3. The HDPE is also the support for positioning the sources in the bench or in the irradiation point as shown in Figure 9.

For the laboratory measurements, the $^{252}Cf_{UPM}$ source was located at 200 cm from the detectors, as shown in Figure 10.

For the outdoor measurements, the $^{252}Cf_{UPM}$ source was located at 220 cm from the N-15 neutron detectors, and these were 115 cm above the floor respect to the centre point, Figure 11.

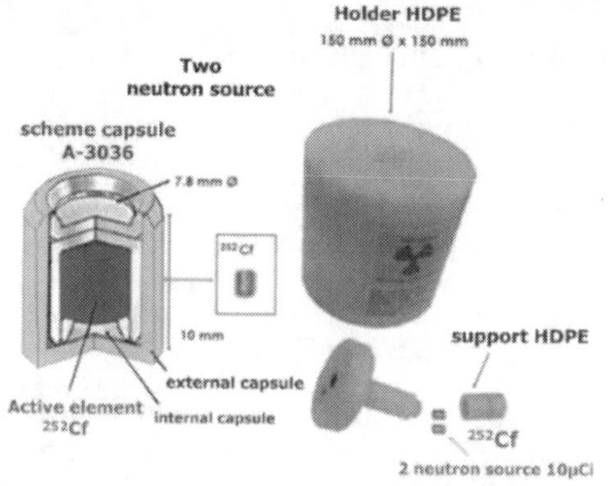

Figure 9. Neutron sources of Laboratory of the Universidad Poltecnica of Madrid, UPM.

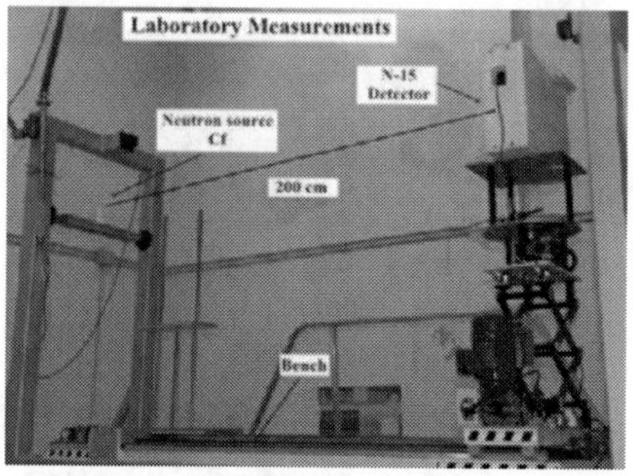

Figure 10. Indoor measurements with the N-15 detector at the LMN-UPM facility.

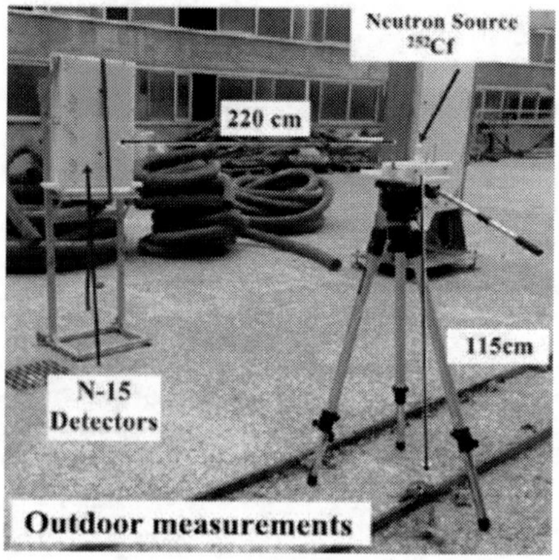

Figure 11. outdoor measurements with the N-15 detector

3 Results

3.1. Calculated responses

Figure 12 shows the response of both detectors, bare and with the HDPE moderators. Responses are the amount of $^{10}B(n,\alpha)^7Li$ reactions per neutron from the source, in a range of monoenergetic neutrons from 10^{-9} to 20MeV.

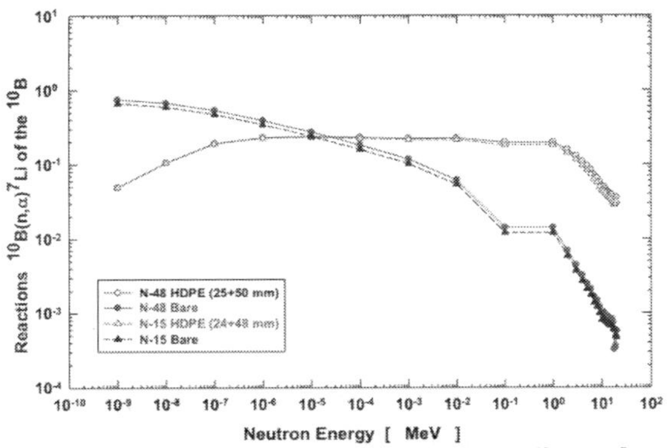

Figure 12. N-48 and N-15 detectors response, $^{10}B(n,\alpha)^{7}Li$ reactions per neutron emitted from the source

Both bare detectors have approximately the same response, being slightly larger for the N-48 detector. Without moderator, both detectors have a large response for low energy neutrons decreasing for larger energy neutrons. When the moderators are included both detectors have roughly the same response, being smaller for low energy neutrons (<0.1 eV) and when the energy is > 1 MeV. From 0.1 eV to 1 MeV the response is almost flat, being almost independent from neutron energy.

3.2. Detection capability in cps/ng ^{252}Cf to 200 cm and 220 cm

Table 1 shows the calculated and measured count rates per ng of $^{252}Cf_{UPM}$ at 200 cm indoors for both neutron detectors, N-15 and N-48, and outdoors at 220 cm, N-15 detectors. The calculated values were scaled up to neutron source strength, where 1 ng of ^{252}Cf produces 2340 n/s [15].

Table 1. Count rate per ng of ^{252}Cf.

	Indoors	
Detector	**Measurements [cps/ng ^{252}Cf]**	**MCNPX [cps/ng ^{252}Cf]**
N-48 **[200 cm]**	1.77± 0.10 [7]	1.76 ± 0.01
N-15 **[200 cm]**	0.76 ± 0.20 [15]	0.73 ± 0.01
	Outdoor	
N-15 **[220 cm]**	0.61 ± 0.20	0.68 ± 0.02

Measured and calculated count rates were corrected by the detector efficiencies that were previously determined [7, 16]. For both conditions, the measured results are in good agreement with the MCNPX calculations. By comparing the indoor and outdoor results for the N-15 detector, the differences are not relevant. Comparing both detectors under indoor conditions the detector N-48 show a better response per ng of ^{252}Cf, probably due to the detector size.

3.3. Detection capability in cps/ng ^{252}Cf to 200 cm, varying the position respect to the ground

Table 2 shows the calculated detectors response, in cps per ng of ^{252}Cf, under outdoor conditions with the centre of both detectors located at varying distance from the floor ($^{252}Cf_{UPM}$). Uncertainties of these results are only those from the Monte Carlo calculations.

Table 2. Importance of the position of detectors above ground for outdoor measurements at 200 cm from the ^{252}Cf source (MCNPX calculations).

Distance above ground (cm)	**Detector N-48 cps/ng $^{252}Cf_{UPM}$**	**Detector N-15 cps/ng $^{252}Cf_{UPM}$**
200	1.16 ± 0.06	0.57 ± 0.04
175	1.33 ± 0.06	0.66 ± 0.04
150	1.48 ± 0.07	0.70 ± 0.04
125	1.60 ± 0.07	0.81 ± 0.05
100	1.67 ± 0.07	0.85 ± 0.05
75	1.70 ± 0.07	0.86 ± 0.05
50	-	0.83 ± 0.05
25	-	0.78 ± 0.04

The detectors' efficiency varies depending on the position with respect to the floor. Due to the size of N-48 detector, the minimum allowable distance between the centre of the detector and the floor was 75 cm. In all cases, the count rate of N-48 detector is larger than the count rate of N-15, due to the detector size. Nevertheless, the source-to-detector's centres distance is the same and both detectors reach the largest count rate when their centres are allocated at 75 cm above the floor, probably due the amount of scattered neutrons from the floor.

The efficiency of both detectors changes with their position with respect to the floor. Therefore, it is concluded that the use of these detectors in actual RPM should be characterized in situ according to their position above ground, thereby selecting the best efficiency in each case in situ; that is, to calibrate the detector.

3.4. Detection capability in cps/ng ^{252}Cf to 200 cm with a moderated neutron source

Table 3 shows the calculated count rates per ng of ^{252}Cf at 200 cm for outdoors conditions, in three different models for the N-15 (Figure 8) neutron detector and one for the N-48, with a moderated neutron source with 2.5 cm of HDPE and gamma-shielded with 0.5 cm thickness of lead, $^{252}Cf_{PNNL}$. The calculated values were scaled up to neutron source strength, where 1 ng of ^{252}Cf produces 2,340 n/s [15].

The N-15 detectors response with the UPM sources is ~ 0.81± 0.05 cps/ng $^{252}Cf_{UPM}$ and for the model neutron source similar PNNL sources is ~ 0.93 ± 0 .04 cps/ng $^{252}Cf_{PNNL}$

Table 3. Count rate per ng of ^{252}Cf at 200 cm (MCNPX calculations, with the $^{252}Cf_{PNLL}$).

	Two detectors	
Model	**N-15A [cps/ng ^{252}Cf] ^{252}Cf$_{PNLL}$**	**N-15B [cps/ng ^{252}Cf] ^{252}Cf$_{PNLL}$**
Model A	0.90 ± 0.05	0.94 ± 0.05
	N-48A [cps-ng^{252}Cf] ^{252}Cf$_{PNLL}$	
Model N-48	1.71 ± 0.07	

The N-48 detector response with UPM sources is ~ 1.60 ± 0.07 cps/ng $^{252}Cf_{UPM}$ and for the model neutron source similar PNNL sources is ~ 1.71 ± 0 .04 cps/ng $^{252}Cf_{PNNL}$. These results correspond to the neutron detectors in outdoor conditions and the same distance above floor in order to compare with other alternatives previously studied by the PNNL.

4 Conclusions

Using Monte Carlo methods the response functions of two different ^{10}B+ZnS(Ag) detectors were calculated, as well their performance against ^{252}Cf neutrons. Measurements were carried out indoors under laboratory conditions and outdoors emulating their use in border checking points.

The response increases significantly in efficiency for both detectors, when calculation is based on the PNNL neutron source $^{252}Cf_{PNNL}$ configuration, probably due to the larger moderation of the source which increases the thermal neutron flux, thus obtaining a better efficiency than that reported with the sources used in UPM.

The ^{10}B+ZnS(Ag) detectors are an alternative to replace ^{3}He detectors in RPMs. The N-15 detectors are considered suitable for portable backpack detectors. The N-48 detector is close to be considered a replacement for ^{3}He detectors. An improvement in the geometry of the detector increasing the amount of ^{10}B could increase the detector efficiency aiming to reach 2.5 cps/ng ^{252}Cf, defined as a goal to use this type of detectors as an alternative in RPMs [13].

Regardless the type of detector, the use of moderator allows to have a flat response in a wider energy range.

The position of the detector with respect to the ground is an important feature due to the detector response to scattered neutrons [17].

References

1. J.H. Ely, R.T. Kouzes, J. Schweppe, E. Siciliano, D. Strachan, D. Weier, Nucl. Instrum. Meth. Phys. Res A, **560**, 373 (2006).
2. R. T. Kouzes, J. H. Ely, L. E. Erickson, W. J. Kernan, A. T. Lintereur, E. R. Siciliano, D. C. Stromswold, M. I. Woordring, Report PNNL-**19311** Pacific Northwest National Laboratory, USA. (2010).
3. A. P. Simposon, S. Jones, M. J. Clapham, S. A. McElhaney, INMM5 2nd Annual Meeting, July 17-21, Palm Desert, California, (2011).
4. R. T. Kouzes, E. R. Siciliano, H. J. Ely, P. E. Keller, R. J. McConn, Nucl. Instrum. Meth. Phys. Res A, **584**, 383 (2008).
5. R. T. Kouzes, H. J. Ely, E. L. Erikson, W.J. Kernan, A.T. Lintereur, E. R. Siciliano, D.L. Stephens, D.S. Stromswold, R. M Van Ginhoven, M. L. Woodring. Nucl. Instrum. Meth. Phys. Res A, **623**, 1035 (2010).
6. K. Zeitelhack, R. Cooper, B. Guerard, K. Soyama, D. Greenfield, G. Kemmerling, T. Wilpert, N. Rhodes, O. Kiselev, M. Klein, R. Engels, G. Smith, L. I. Defendi, Report on the meeting of detector Experts, FRM II, Grenoble France on July 7-8, (2009).
7. D. B. Pelowitz, MCNPX™ User's Manual, Version 2.7.0, Report LA-Cp-11-00438. Los Alamos National Laboratory, USA. (2011)
8. BridgePort Instruments, LLC, (2012).
9. BridgePort Instruments, LLC, (2013).
10. H. R. Vega-Carrillo, R. Barquero, G. A. Mercado, Int. J. Radiat. Res., **11**, 148 (2014).
11. H. R. Vega-Carrillo, K. A. Guzmán-García, E. Gallego, A. Lorente, Radiat. Meas., **69**, 30 (2014).
12. H. R. Vega-Carrillo, E. Gallego, A. Lorente, I. P. Rubio, R. Mendez, Appl. Radiat. Isot., **70**, 1603 (2012)
13. M. L. Woondring, J. H. Ely, R. T. Kouzes, D. C. Stromswold, Report PNNL-19726 Pacific Northwest National Laboratory, USA. (2010).
14. ANSI N42.35-2006 standard (2007).
15. H. R. Vega-Carrillo, Rev. Mex. Fis., **34**, 25 (1988)
16. K. A. Guzman-Garcia, H. R. Vega-Carrillo, E. Gallego, A. Lorente, R. Mendez, J. A. Gonzalez, S. Ibañez-Fernandez, ISSSD 2015, Leon, Gto., Mexico Sept 26-30 (2015).
17. H. R. Vega-Carrillo, E. Gallego, E. Manzanares-Acuña, A. Lorente, R. Barquero, A. Martin-Martin, J. L. Gutierrez Villanueva, Rev. Mex. Fis., **54**, 63 (2008).

DOI: 10.1051/epjconf/201715307009

Improvement of ^{93m}Nb and ^{103m}Rh activity measurement methodology for reactor dosimetry

J. Riffaud[1,a], M.-C. Lépy[1], C. Domergue[2], H. Philibert[2], C. Destouches[2], N. Thiollay[2], J.M. Girard[2]

[1]CEA, LIST, Laboratoire National Henri Becquerel (LNE-LNHB), CEA-Saclay, Bât 602, PC 111, 91191 Gif-sur-Yvette CEDEX, France
[2]CEA, DEN, DER, Laboratoire de Dosimétrie, Capteurs et Instrumentation (LDCI), Cadarache, 13108 Saint Paul-lez-Durance, France

Abstract. Reactor dosimetry is based on the analysis of the activity of irradiated dosimeters, such as ^{93m}Nb and ^{103m}Rh. The activity measurement of these dosimeters is conventionally performed by X-ray spectrometry, but the low-energy of emitted photons makes it difficult to derive reliable results with low uncertainties. Approaches to improve these characterisations are presented: they include high accuracy efficiency calibration of a HPGe detector using both experiments and Monte Carlo simulation, calculation of corrective factors for the geometry (self-absorption) and self-fluorescence effects. Improvement of the knowledge of the ^{103m}Rh decay scheme is also required and a specific experiment is proposed, including activity measurement of a ^{103m}Rh solution by liquid scintillation, and measurement of the photon emission intensities by X-ray spectrometry. A method for calculating coefficients to take into account the self-fluorescence effects in dosimeters is also suggested to improve the uncertainties on activity measurements.

1. Introduction

Reactor dosimetry aims at determining the neutron flux density received during an irradiation and to characterise the energy distribution of neutrons. This technique is based on the analysis of the activity of irradiated dosimeters including isotopes which are subject to activation or fission reactions under the neutron flux. For example, inelastic scattering reactions (n, n') are useful to characterise neutron beams with energies around 1 MeV [1-4]. Both reactions ^{93}Nb(n,n')^{93m}Nb and ^{103}Rh(n,n')^{103m}Rh are of particular importance for dosimetry in reactor. The measurement of the activity of Nb and Rh dosimeters provides the basic data that can be traced back to the reactor operating information.
The activity measurement of such dosimeters is conventionally performed by X-ray spectrometry. Unfortunately, the low-energy of emitted photons makes it difficult to derive results with sufficiently low uncertainties. Indeed, the efficiency calibration of the detector must be accurately performed and self-attenuation in the dosimeter must be taken into account. In addition, niobium and rhodium are not 100 % pure, other elements are present. These impurities and other isotopes of niobium and rhodium created during irradiation are activated by neutrons and induce fluorescence in the dosimeters which increase their activity. In the present study, different aspects of the methodology used to measure the dosimeters activity are examined and improvements of the usual measurement procedure are provided.

2. ^{93m}Nb and ^{103m}Rh decay schemes

Activation of niobium dosimeters produces ^{93m}Nb, characterised with a half-life of 16.12 (15) years, which decays by gamma transition to the ground state of the stable nuclide ^{93}Nb (Figure1 – left panel). This transition is highly converted thus leading to the emission of characteristic X-rays of niobium, with energies for K X-rays between 16.5 keV and 19 keV. Their emission intensities are quoted with relative standard uncertainties of the order of 2% [5].

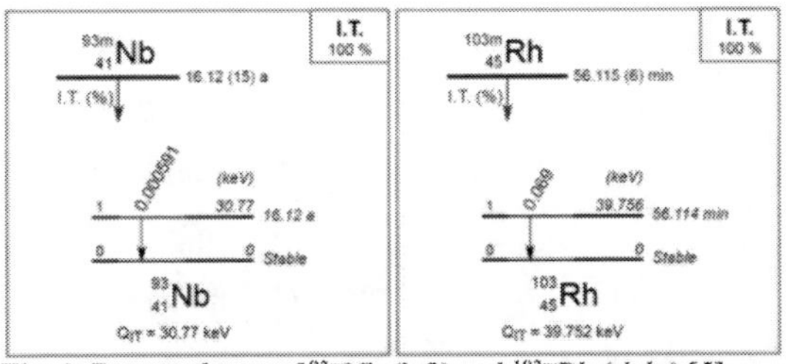

Fig. 1. Decay scheme of ^{93m}Nb (left) and ^{103m}Rh (right) [5]

The decay scheme of ^{103m}Rh is similar to that of ^{93m}Nb (Figure1 – right panel). Its half-life is short (56.115 (6) minutes) and the transition from the isomeric state to the ground state of ^{103}Rh is highly converted, resulting in X-ray emissions, with energies for K transitions between 20 keV and 23.2 keV. The emission intensities of the K X-rays are specified with relative standard uncertainties between 5.5 and 7% [6].

[a] Corresponding author: jonathan.riffaud@cea.fr

3. X-ray spectrometry

3.1 Efficiency calibration

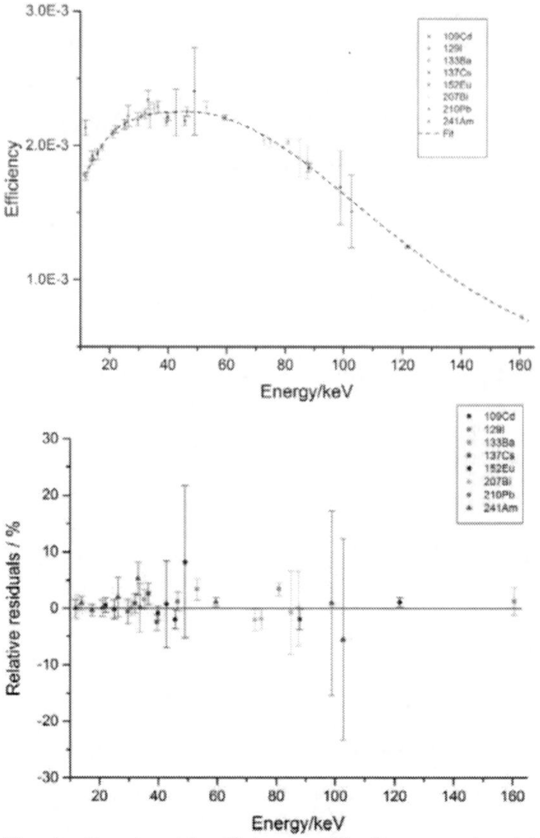

Fig. 2. Experimental efficiency calibration: experimental values and fitted curve. The lower panel displays relative residuals.

X-ray spectrometry is performed using a high-purity germanium (HPGe) detector. The experimental efficiency calibration is defined for a reference geometry, using radioactive standard point sources for a given source-detector distance (8 cm). The radionuclides used are ^{88}Y, ^{109}Cd, ^{133}Ba, ^{137}Cs, ^{152}Eu, ^{207}Bi, ^{210}Pb and ^{241}Am. The radioactive solutions are standardised at the Laboratoire National Henri Becquerel by absolute methods with relative combined uncertainties less than 1%. Then a weighted drop of solution is deposited between two sheets of Mylar® to prepare point sources for γ- and X-ray spectrometry. These are used to establish the full-energy peak efficiency calibration of HPGe detectors, according to:

$$\varepsilon(E) = \frac{N(E)}{A\,I(E)\,t} \prod_i C_i \qquad (1)$$

Where ε(E) is the detection efficiency at the energy "*E*", *N(E)* is the number of counts under the full-energy peak with energy "*E*", *A* the activity of the radionuclide included in the standard point source, *I(E)* the emission intensity line with energy line "*E*", *t* the acquisition time and C_i stands for different correction factors.
The experimental values are fitted using least-square method, to have an efficiency value for any energy, with the use of a poly-logarithmic function (Figure 2). Both experimental and efficiency values are obtained with 1.5 % relative combined standard uncertainties.

3.2 Efficiency calculation with Monte Carlo method

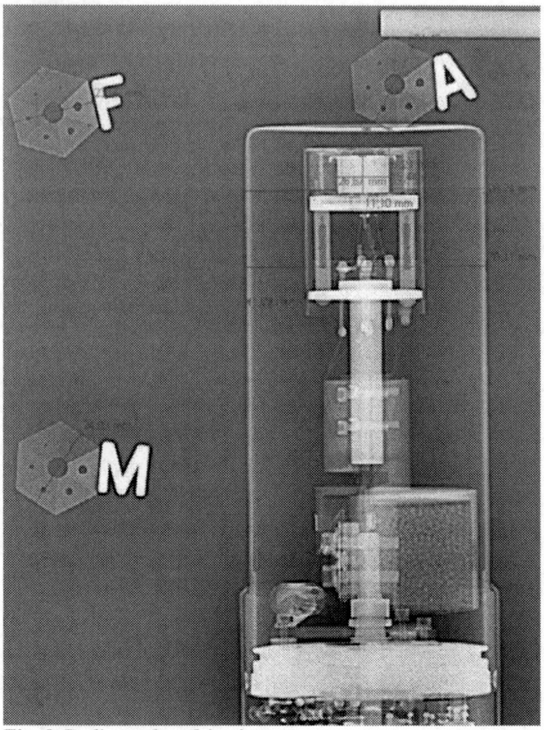

Fig. 3. Radiography of the detector

Due to the poor knowledge of the photon emission intensities of radionuclides in the low energy range, the efficiency calibration curve was refined using Monte Carlo simulation. PENELOPE [7] and GEANT4 [8] code were used and optimisation of physical parameters of the germanium crystal was necessary to adjust the efficiency curve obtained with the simulation with the experimental fit. We started with the dimensions provided by the manufacturer and those which can be determined from the radiography of the detector (Figure 3). The result (Figure 4) was obtained after many changes in the crystal dimensions with relative differences between experimental and simulations lower than 1%. The results obtain with GEANT4 are very close to those of PENELOPE, the low-energy models used with GEANT4 is the PENELOPE physics package.

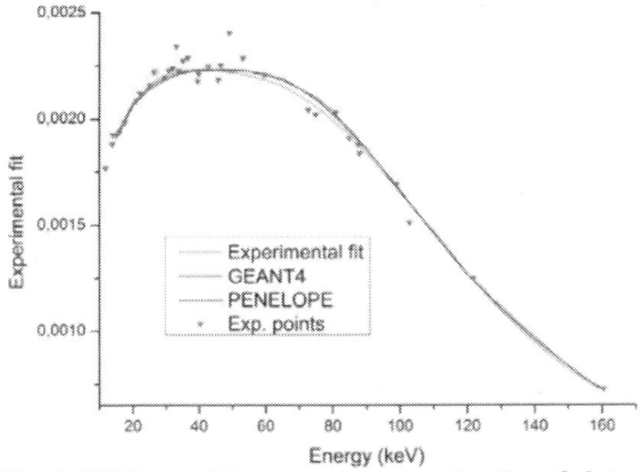

Fig. 4. Efficiency of the spectrometer as a function of photon energy. The experimental efficiency curve is shown by the green curve and the Monte Carlo computed efficiency by the red curve with the use of GEANT4 code and the blue curve with PENELOPE code.

4 Measurement procedures

4.1 ^{93m}Nb

The dosimeter activity measurement is carried out using the calibrated HPGe detector, at the same distance than the reference point sources. The activity measurement, *A*, is obtained from the number of counts under the full-energy peaks of K X-rays, and calculated with the following formula:

$$A = \frac{N(E)}{I(E)\,\varepsilon(E)\,t}\prod_i C_i \qquad (2)$$

In addition to corrections for radionuclide decay and coincidence summing, such as used for the efficiency calibration, further specific corrections must be taken into account to obtain the activity of the dosimeters. The sample is in the shape of a metal strip, 5 mm long, 1 mm wide and 19.89 µm thick. Corrective factors must be applied to take into account the change in geometry and self-absorption of X-rays in the material thickness.

Self-absorption and geometry corrections are computed by Monte Carlo simulation. In spite of the thinness of the dosimeter (20 µm) the self-absorption correction is around 20%. Currently, the relative standard uncertainty on activity measurement is about 6%. Correction coefficients for fluorescence caused by each impurity were calculated in 1972. The value of these corrections depends on the activity of the impurities compared to the activity of the dosimeter and can vary from 0.5% to 7%. The activity of the impurities is measuresd by classical γ-ray spectrometry.

4.2 ^{103m}Rh

The measurement activity procedure for reactor dosimetry is the same as in the case of ^{93m}Nb; however, these are pellets of pure ^{103}Rh which are irradiated. But uncertainties on the X-ray intensities lead to high uncertainties in measurement results, of the order of 10%. Self-absorption and geometry corrections are taken into account and computed as for ^{93m}Nb. With a thickness of 50 µm, the self-absorption in rhodium dosimeters is about 50%. The correction coefficients for fluorescence have never been calculated for rhodium impurities. Because, unlike the niobium, only the fluorescence induced by iridium impurities can disturb the measurements, and should be considered.

5 Improvement of decay data

Emission intensities of X-rays in the recommended data for isomeric transition are determined from the γ-ray transition probabilities, internal conversion coefficients and fluorescence yield values [9]. The knowledge of the fluorescence yields for ^{93m}Nb is rather poor and based on old experiments [10]. All data for ^{103m}Rh, are not well known and leading to high uncertainties on emission intensities [10]. Two experiments were performed to measure the K fluorescence yields of Nb and Rh and the emission intensities of K X-rays of ^{103m}Rh, in order to improve the knowledge of the decay data.

5.1 K fluorescence yields measurements

New measurements of the K fluorescence yields of niobium and rhodium were performed, taking advantage of optimized experimental facilities, using tunable monochromatic X-rays [10]. The experiment included two steps:

5.1.1 Mass attenuation coefficients

First, accurate values of the mass attenuation coefficients ($\frac{\mu}{\rho}(E)$) of niobium and rhodium were measured at the SOLEIL synchrotron [11]. These coefficients were obtained following an experimental procedure previously established and validated for similar measurements on different metallic targets [12,13]. Mass attenuation coefficients, were measured in transmission mode using a monochromatic parallel photon beam with energy *E*, under normal incidence to the sample with thickness *x*. The measurement consists in measuring the intensity of a monochromatic photon flux in front the sample, $I_0(E)$, and immediately behind it $I(E)$. For each energy step, the mass attenuation coefficient can be calculated assuming the Beer-Lambert law:

$$I(E) = I_0(E)\, e^{-\frac{\mu}{\rho}(E)\rho x} \qquad (3)$$

where μ is the linear attenuation coefficient of the target material and ρ is the material density.

5.1.2 Fluorescence yields

In the second step, the fluorescence yields are measured in a reflection mode (Figure 5) where the target (niobium or rhodium) is placed at the angle of 45° relative to the incident monochromatic photon beam, with energy E_0 and intensity I_0. Electronic rearrangement following photoelectric effect induces X-ray emission with energy *Ei* according to the partial K fluorescence yield ω_{Ki}. These are recorded with an energy-dispersive detector (HPGe) installed at 45° from the target. Thus, the full-energy peak areas of each line, Ni, allow deriving the partial fluorescence yield, according to:

$$\omega_{Ki} = \frac{4\pi}{\Omega}\frac{N_i}{\varepsilon_i}\frac{1}{I_0\tau_K}\frac{\mu_0+\mu_i}{1-e^{-\frac{\mu_0+\mu_i}{\sin\frac{\pi}{4}}l}} \quad (4)$$

Where:

Ω : detection solid angle,

ε_i: full-energy peak (FEP) efficiency of the HPGe detector,

τ_K : photoelectric interaction cross section in the K shell

μ_0 and μ_i : mass attenuation coefficients of the target material for the incident and the emitted photon beams, respectively.

Using nine incident energies ranging from 20 keV to 25 keV, the total K fluorescence yield for niobium was obtained as ω_K= 0.724 (14). For rhodium, results were derived with an average of eight measurements with incident photons in the 26 keV to 30 keV energy range. The total K fluorescence yield of rhodium was computed as ω_K= 0.814 (41).

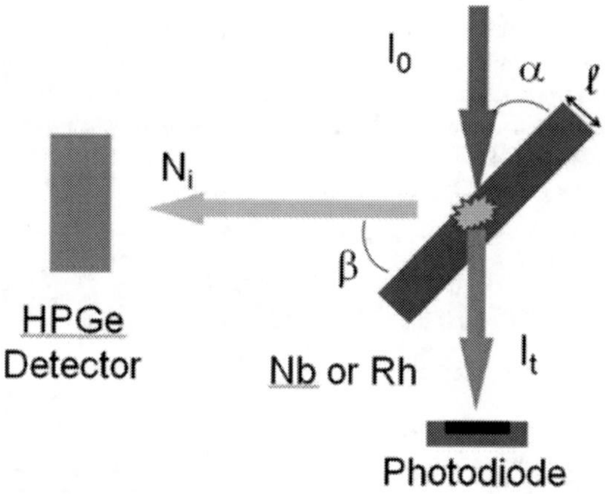

Fig. 5. Geometry of the experimental setup for fluorescence yields measurements

5.2 Photon emission intensities of ^{103m}Rh

A specific experiment was planned in order to measure the absolute photon emission intensities of ^{103m}Rh. The procedure includes activity measurement using a primary method, liquid scintillation, and photon emission intensity measurement using X-ray spectrometry, on the same sample.

5.2.1 Source preparation

Due to the short half-life of ^{103m}Rh, a specific irradiation was conducted at the nuclear reactor ISIS (CEA Saclay), immediately before chemical preparation of the samples for measurements. The dissolution of pure rhodium is very complex and difficult to obtain in the laboratory. For this experiment the sample consist in a rhodium(III) chloride powder, with the formula $RhCl_3(H_2O)_n$. The $RhCl_3$ was irradiated in the reactor core, placed in a thermal neutron filter to avoid activating chlorine. The powder is soluble in water and allowed having samples quickly ready for measurements in liquid scintillation and spectrometry.

5.2.2 Activity measurement

The activity measurement was performed by liquid scintillation (LS), using the triple-to-double coincidence ratio method (TDCR). After dissolution, the rhodium source is mixed with a commercial liquid scintillator and the scintillating sources are measured using a triple photomultiplier LS counter. The detection efficiency is deduced from the triple to double coincidence ratio, using a model describing the light emitted by the source, from the radiation-matter interaction to the statistics of the light emission. This method is described in detail in reference [14]. The uncertainty of the activity is evaluated with a Monte Carlo method, following the recommendations of [15].

5.2.3 Photon emission intensity measurement

In parallel, X-ray spectrometry allows to determine photon emission intensities. A specific container was optimized for measuring low-energy X-rays. It was filled with the radioactive solution and placed at the reference distance. The efficiency calibration curve required corrective factors to take into account the geometry difference between a point source and the volume source. These corrections can be obtained by Monte Carlo method. Accurate processing of the X-ray spectra requires deconvolution of the K X-ray lines with peak shape taking into account scattering effect (low-energy tailing). The results are under processing but it is expected to derive photon emission intensities with relative combined standard uncertainties around 3%.

6 Self-fluorescence corrections

In this section we focus on the additional fluorescence induced in niobium dosimeters. The typical composition of niobium foils used for dosimetry includes other elements (Table 1). The quantities were determined by mass spectrometry or by neutron activation method (Ta). These impurities are activated during irradiation in the reactor. For example, ^{182}Ta is activated by radiative capture from ^{181}Ta, ^{181}Ta(n,γ)^{182}Ta. In addition, other radioactive isotopes of Nb are created, such as ^{92}Nb and ^{94}Nb by ^{93}Nb(n,2n)^{92}Nb and ^{93}Nb(n,γ)^{94}Nb reactions. After irradiation, some of these impurities are still present during measurements, because of their half-lives, and can be troublesome: ^{94}Nb, ^{95}Nb, ^{92m}Nb and ^{182}Ta.

Table 1. Impurities present in niobium dosimeters

Elements	Quantities (ppm)
Mo	14
W	12
Ta	4.6
Zr	2
Fe	0.06
Ni	0.02
Cr	0.004
Co	0.0003

Moreover, the presence of ^{60}Co has been occasionally observed during measurement. Photons and electrons originating from the decay of these impurities (blue arrow on Figure 6) can interact with atoms of the dosimeter itself by photoelectric effect or other electromagnetic interactions. The dosimeter material is ionised, and photo-electrons are thus ejected from atoms of niobium. This is followed by electron rearrangement, with X-rays emission of niobium. This fluorescence effect generates an extra niobium X-ray emission (red arrows on Figure 6) which is added to the one due to the decay of ^{93m}Nb (green arrows on Figure 6). Consequently, this increase of the number of counts under the full-energy peaks of ^{93m}Nb leads to an overestimation of the dosimeter activity and corrective factors and must be taken into account to derive the activity of ^{93m}Nb only.
The self-fluorescence coefficients were calculated analytically in 1972 (Table 2) for ^{182}Ta, ^{94}Nb, ^{95}Nb and ^{92m}Nb. The coefficients are expressed in $s^{-1}.Bq^{-1}$, indicating the number of fluorescence X-rays emitted per second and per decay of the impurity.

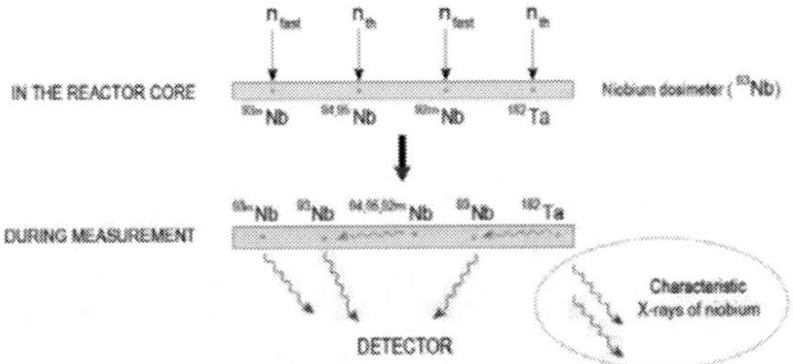

Fig. 6. Scheme of the self-fluorescence in a niobium dosimeter

However, they were calculated only for main energy lines and did not take into account the electrons (or positron) interactions for ^{182}Ta and ^{92m}Nb and the self-absorption of Nb X-rays in the dosimeter following the electronic rearrangement [16]. It was thus decided to actualize the contributions of the impurities in the dosimeters spectra using Monte Carlo method.

7 Monte Carlo method and results

7.1 GEANT4 simulation

The assessment of the coefficients for all radionuclides mentioned above requires simulating radioactive decay according to the decay scheme for each radionuclide. This means taking into account the branching ratios, electronic and photonic emission intensities and their interactions with matter and atomic relaxation processes. GEANT4 enables to simulate radioactive decay thanks to the module "*G4RadioactiveDecay*" and its associated classes. Instead of decay data taken from ENSDF files [17], we preferred to use, when they were available, decay data recommended by the Decay Data Evaluation Project (DDEP) and published by the LNHB [18]. For ^{94}Nb and ^{92m}Nb, we took ENSDF data. Thus we used PENNUC code, which simulate random decay pathways of radioactive nuclides, and associated decay data files [19]. PENNUC generates file a called "CASCADE" which is then read by GEANT4, simulating radioactive decay while taking into account the branching ratios, emission intensities (for electrons, photons and positrons) and internal conversion with recommended data. The electromagnetic physics used in the GEANT4 simulation is the physics models of PENELOPE for electrons, positrons and photons, which is dedicated to the low-energy interactions (Atomic effects, fluorescence, Doppler broadening …).
The geometry of the simulation is the same as for efficiency calculation, but the source is not a point but a niobium volume with the same dimensions as those of dosimeters under study. The particles (photons, electrons and positrons) are emitted inside the niobium volume with a random starting position and a random direction within a solid angle of 4π steradian.
The energy of any particles deposited in the detector volume, is recorded in a root file. The result is an energy spectrum from which the self-fluorescence coefficient is calculated for a given impurity. This coefficient is obtain with the following formula

$$C_{N,X} = \frac{\sum_i N_{X_i}}{N_D} \frac{1}{\varepsilon_X} \qquad (5)$$

Where "$C_{N,\ X}$" is self-fluorescence coefficient for the radionuclide "N" (N = ^{182}Ta, ^{95}Nb…) and the specific characteristic X-ray "X" (X = K_α or K_β), "N_{Xi}" is the number of counts under the full-energy peak for the line "X_i" ($i = [1, 2]$ for K_α or $i = [1, 5]$ for K_β), "N_D" is the number of simulated decay and "ε_X" is the efficiency for the mean energy of the line "X". The total coefficient "C_N" for a radionuclide is simply the addition of the "$C_{N,\ K\alpha}$" and "$C_{N,\ K\beta}$" coefficient values.
Simultaneously, we count the number of photons exiting the niobium volume by discriminating the type of the particle, its energy and the process responsible of the ionisation of the niobium atom: "photo-electric effect", "electron ionisation" and "Compton scattering". Thus, it is possible to determine the particles at the origin of the atom ionization and subsequent fluorescence emission:

electrons for "electron ionisation" or photons for "photo-electric effect" or "Compton scattering".

7.2 Results and discussion

The coefficients were calculated from the spectra obtained by the simulations for the niobium energy lines. The treatment of the spectra is quite simple since full-energy peaks appear as Dirac functions from which the background was subtracted. The background comes from the energy deposited indirectly by electrons (Bremsstrahlung) and Compton scattering. The coefficients values were obtained from simulations with several decays for each radionuclide. Table 2 shows the results obtained, with standard uncertainties calculated according to the GUM [15], and compares them with the coefficients assessed in 1972. These results are valid only for a solid angle less or equal to the one of the geometry in the simulation.

For ^{182}Ta, both recent and former values are rather close, although values in 1972 did not take into account electron interactions and self-absorption of Nb fluorescence X-rays in the dosimeter. By tracking the origin of the fluorescence of niobium as explained above, we can determine the ratio of X-rays number resulting from the electron interactions in the dosimeter on the total number of fluorescence X-rays. For this impurity, twenty percent of the X-ray fluorescence is due to electrons. The other part is due to the photoelectric and Compton effects.

Table 2. Comparison between coefficients obtained in this work and coefficients calculated in 1972.

Impurity	Our work ($s^{-1}.Bq^{-1}$)	In 1972	Relative difference
^{92m}Nb	0.00067 (4)	0.021	> 3000%
^{94}Nb	0.0119 (2)	0.0055	54%
^{95}Nb	0.00310 (6)	0.0033	11.1%
^{182}Ta	0.0645 (6)	0.077	19.4%
^{60}Co	0.00821 (11)	-	-

As for ^{182}Ta, both recent and former values for ^{95}Nb agree quite well. In 1972 they tried to take into account interactions of electrons coming from the β^- decay [16]. They checked their calculations experimentally, and were able to find a value reasonably good. In addition, knowledge on the decay scheme and data has not changed significantly since 1972, whether on the beta spectrum, the half-life, etc.

For ^{94}Nb, both recent and former values differ by a factor two. As for ^{95}Nb, they tried to take into account in 1972 interactions of electrons and evaluated experimentally their calculations. But, knowledge on the decay scheme and data changed significantly since 1972, especially data on the beta spectrum. For example, the maximum energy of the spectrum has decreased from 610 keV in 1972 to 470 keV today. For these two radionuclides, the contribution of the electrons to the fluorescence emission is higher than eighty percent. The consideration of their interactions is essential, however the theoretical calculation and experimental validation is very complex especially with the calculation means available at the time.

For ^{92m}Nb, both recent and former values disagree by a factor upper than thirty. In 1972, they deduced the coefficient by observing the evolution of the ratio of Zr X-rays Nb peaks [16]: ^{92m}Nb disintegrates to ^{92}Zr and emits Zr X-rays. They have probably considered that the radionuclide is only emitting Zr X-rays in the dosimeter, although this may not be the case.

The magnitude of corrections on the dosimeter activity for each impurity depends on the activity of this impurity. However we notice that the ^{182}Ta and ^{94}Nb can be troublesome if a significant activity is observed relatively to ^{93m}Nb activity. For ^{92m}Nb, the impact is quite negligible, especially since it is generally observed with low activities.

Conclusion

Low-energy X-ray spectrometry applied to activity measurement of solid dosimeters is challenging. Different aspects have been examined in the present study and different experiments have been conducted to improve the atomic data of rhodium and niobium, as well as the decay scheme of ^{103m}Rh. The importance of corrective factors has been underlined and Monte Carlo simulation has been used to compute self-absorption and fluorescence effects. The use of Monte Carlo method enables more accurate calculations of the self-fluorescence corrections than previously. Moreover, modern particle transport codes make possible to take into account all the effects occurring at low energies and to clearly identify the main ones. Due to the present study, significant improvement in the measurement methodology for solid dosimeters is expected.

The authors would like to thanks H. Carcreff and O. Vigneau (CEA-DEN) for their valuable assistance and support in the Rh irradiation experiment.

References

1. A. Ballesteros, L. Debarberis, W. Voorbraak, J. Wagemans, P. D'hondt, Prog. Nucl. Energy **52**, 615-619 (2010)
2. W. P. Voorbraak, T. Kekki, T. Serén, M. van Boxstaele, J. Wagemans, J.R.W. Woittiez, *Proceeding of the FISA-2003 Conference* (2003), https://cordis.europa.eu/fp5-euratom/src/ev-fisa2003.htm
3. V. Sergeyeva, C. Domergue, C. Destouches, J.-M. Girard, H. Philibert, D. Beretz, J. Bonora, N. Thiollay, A. Lyoussi, EPJ Web Conf. **106**, (2016) DOI: http://dx.doi.org/10.1051/epjconf/2016106050 11
4. C. Domergue, D. Beretz, C. Destouches, J.-M. Girard, H. Philibert, J. Plagnard, *Advancements in*

Nuclear Instrumentation Measurement Methods and their Applications (ANIMMA), 2009 First International Conference on, 1-6 (2009) DOI: 10.1109/ANIMMA.2009.5503820

5. M.-M. Bé, V. Chisté, C. Dulieu, M.A. Kellett, X. Mougeot, A. Arinc, V.P. Chechev, N.K. Kuzmenko, T. Kibédi, A. Luca, A.L. Nichols. *Table of Radionuclides, Monographie BIPM-5* **8** (BIPM, Pavillon de Breteuil, F-92310 Sèvres, France, 2016)
6. *Private Communication*. M.-M. Bé, M.A. Kellett (2016)
7. F. Salvat, *PENELOPE-2014: A Code System for Monte Carlo Simulation of Electron and Photon Transport* (OECD/NEA Data Bank, Issy-les-Moulineaux, France, 2015)
8. S. Agostinelli et al., Nucl. Instrum. Methods Phys. Res., Sect. A **506**, 250-303 (2003)
9. M.-M. Bé, *Table of Radionuclides – Introduction* (Technical report, CEA, LIST, LNE-LNHB, 2011)
10. J. Riffaud, M.-C. Lépy, Y. Ménesguen, A. Novikova, X-Ray Spectrom. (to be published)
11. Soleil synchrotron (2007), http://www.synchrotron-soleil.fr/portal/page/portal/Accueil
12. Y. Ménesguen, M.-C. Lépy, Nuc. Instrum. Methods Phys. Res., Sect. B **268**, 2477-2486 (2010)
13. Y. Ménesguen, M. Gerlach, B. Pollakowski, R. Unterumsberger, M. Haschke, B. Beckhoff, M.-C. Lépy, Metrologia **53**, 7-17 (2016)
14. P. Cassette, M.-M. Bé, F. Jaubert, M.-C. Lépy, Appl. Radiat. Isot. **60**, 439-445 (2004)
15. JCGM, *Evaluation of measurement data - Supplement 1 to the "Guide to the expression of uncertainty in measurement" - Propagation of distributions using a Monte Carlo method* (BIPM, 2008)
16. R. Lloret, *Utilisation du niobium comme intégrateur de fluences élevées de neutrons rapides* (Technical report, CEA-CEN/G-Service des piles, 1973)
17. Chapter 44, *Geant4 Physics Reference Manual* (2015)
18. M.A. Kellett, EPJ Web Conf. (to be published)
19. E. Garcia-Torano, V. Peyres, M.-M. Bé, C. Dulieu, M.-C. Lépy, F. Salvat, Nuc. Instrum. Methods Phys. Res., Sect. B (to be published)

Measurement of neutron energy spectra for E_g=23.1 and 26.6 MeV mono-energetic photon induced reaction on ^{nat}C using laser electron photon beam at NewSUBARU

Toshiro Itoga[1,*], *Hiroshi* Nakashima[2,4], *Toshiya* Sanami[2,3,6], *Yoshihito* Namito[3,6], *Yoichi* Kirihara[3], *Shuji* Miyamoto[5], *Akinori* Takemoto[5], *Masashi* Yamaguchi[5] and *Yoshihiro* Asano[2,5]

[1]JASRI, 1-1-1, Kouto, Sayo-cho, Sayo-gun, Hyogo 679-5198, Japan
[2]RIKEN SPring-8 Center, 1-1-1, Kouto, Sayo-cho, Sayo-gun, Hyogo 679-5198, Japan
[3]KEK, 1-1, Oho, Tsukuba-shi, Ibaraki 305-0801, Japan
[4]JAEA, 2-4 Shirakata, Tokai-mura, Naka-gun, Ibaraki 319-1195, Japan
[5]LASTI University of Hyogo, 3-1-2 Koto, Kamigori-cho, Ako-gun, Hyogo 678-1205, Japan
[6]The Graduate University for Advanced Studies (SOKENDAI), 1-1 Oho, Tsukuba, Ibaraki 305-0801, Japan

Abstract. Photo-neutron energy spectra for E_g=23.1 and 26.6 MeV mono-energetic photons on ^{nat}C were measured using laser Compton scattering facility at NewSUBARU BL01. The photon energy spectra were evaluated through measurements and simulations with collimator sizes and arrangements for the laser electron photon. The neutron energy spectra for the ^{nat}C(g,xn) reaction were measured at 60 degrees in horizontal and 90 degrees in horizontal and vertical with respect to incident photon. The spectra show almost isotropic angular distribution and flat energy distribution from detection threshold to upper limit defined by reaction Q-value.

1 Introduction

In the shielding design of electron accelerator facilities, the double differential cross-section (DDX) data, which fully describe energy and angle of emitted particles, of the photo-neutron production reaction are necessary in order to determine thicknesses and layouts of shielding wall with considering neutron transport and penetration. Especially, the data for materials consisting of collimator, target and beam dump are quite important because of continuous beam loss during operation. Graphite is one of the materials due to relatively low neutron yield, long radiation length and high melting point.

Until now, there are few experimental data of photo-neutron production DDX since there were no facilities providing intense mono-energetic photons for the experiment. Only the total photo-neutron production cross-section data have been provided so far even in giant resonance energy region, mainly for a physical interest on nuclear structure [1].

Thus, neutrons having Maxwellian energy and isotropic angular distribution are assumed in the shielding design. This assumption will be adequate for relatively heavy nuclei [2], however should be confirmed experimentally for light nuclei due to less population of nucleus. For carbon photo neutron reaction, Noda et al. pointed out that an isospin selection rule affects energy dependence of total cross section [3]. Experimental DDX data on carbon are highly desired to confirm these points.

Recently, the laser electron photon beam facilities are in operation using a laser Compton backscattering (LCS) technique in the world. At NewSUBARU BL01 [4], mono-energetic photon up to 75 MeV is obtained by appropriate combinations of the electron beam energy, laser wave-length and collimators. The variable energy, intense monochromatic photon source is useful to measure the DDX data.

In this paper, we present the spectral measurements and simulations for the circular polarized laser electron photon beam and photo-neutron at NewSUBARU BL01. Measurements and simulations were performed to evaluate energy spectrum of LCS photon. Measurements were performed to obtain the photo-neutron energy spectrum and the angular distribution for 23.1 and 26.6 MeV gamma induced reaction on graphite. In addition, the target activity was also measured in order to confirm the absolute yields of the neutron.

2 Experiments

Figure 1 shows the experimental set-up at NewSUBARU BL01. The laser electron photon beam is generated by backward Compton scattering of laser photon. The maximum photon energies of 23.1 and 26.6 MeV are obtained for 1149 and 1223 MeV electrons, respectively, with Nd:YVO_4 laser (1064 nm). Circular polarization was chosen by using a λ/4 wave plate to avoid influence

* Corresponding author: itoga@spring8.or.jp

from polarization [5]. The two 100 mm thick lead collimators with the aperture diameters of 3 and 2 mm were set at 15.6 and 18.5 m from scattering point. By defining scattering angle using the collimators the nominal energy spreads of photon were estimated to be 1.2 and 1.6 MeV. A cylindrical graphite target, 100 mm long × 10 mm in diameter (1.92 g/cm^3) was set at 26 m away from the laser scattering point.

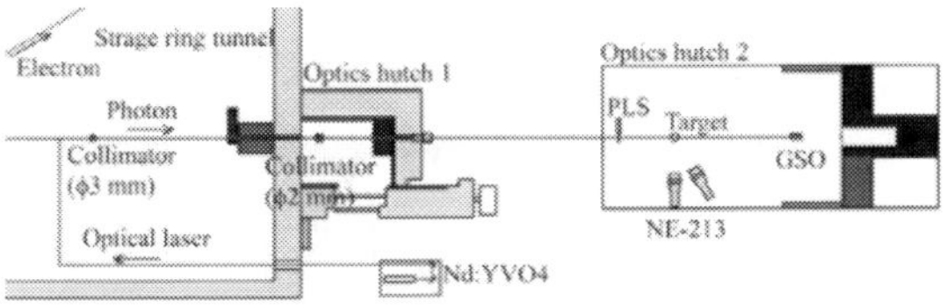

Figure 1. Schematic view of NewSUBARU BL01. PLS indicate the plastic scintillation detector.

The photon and neutron measurements were carried out in the optics hutch 2. The absolute intensity and energy width of the photon were measured using a Ce:Gd_2SiO_5 (GSO) scintillator with the size of H76 × W76 × L180 mm [6]. Photon intensity was monitored by a thin plastic scintillation detector (5 mm in thickness) during the neutron measurements.

Three NE-213 organic liquid scintillators with the size of 127 mm in diameter × 127 mm long were set at the 709, 647 and 659 mm from the target, The detector position with respect to the photon beam direction were 60 and 90 degrees in horizontal direction and 90 degrees in vertical direction. Neutron energy was determined by the time-of-flight method using a single electron bunch ring operation and pulsed laser (25 kHz) in order to avoid the frame overlap and mitigate Bremsstrahlung photons.

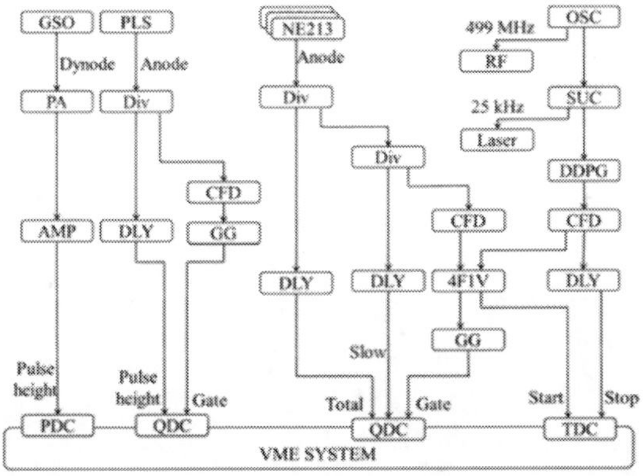

Figure 2. Block diagram of the photon and neutron measurement circuit. PLS: plastic scintillation detector, PA: preamplifier, AMP: amplifier, Div: signal divider, DLY: nano second delay, CFD: constant fraction discriminator, GG: gate generator, OSC: oscillator, SUC: synchronous universal counter, DDPG: digital delay pulse generator, 4F1V: 4 fold 1 veto, PDC: peak sensing analogue to digital converter, QDC: charge integrating analogue to digital converter, TDC: time to digital converter.

The electronic circuits show in fig. 2 was used for the photon and neutron measurements. By using QDC, the charge of the total and slow (decay) components of the anode signals were integrated for the discrimination between neutron and photon events. The neutron flight time was measured by TDC. The pulse height information of the PLS was also digitized by using QDC. These digital data were stored in a personal computer via VME system. The history of photon beam intensity for activation analysis was also recorded using PLS signals.

After the neutron measurements, the activities of ^{11}C accumulated in each targets were measured by detecting 511 keV annihilation gamma-rays with a high-pure Ge detector (ORTEC GEM20P4-70) and a multi-channel analyzer in order to confirm the absolute yields of the neutron. The dead times during the gamma-ray counting were less than 1 %.

3 Data analysis

Time of flight spectra were deduced for the neutron events identified by the pulse shape discrimination technique and converted into energy spectra. The lower energy threshold was set based on pulse height of ^{137}Cs, ^{60}Co, ^{22}Na Compton edge [7]. Figure 3 shows the neutron detection efficiency used in this data analysis. The neutron detection efficiency of the NE-213 scintillator was obtained by the calculation with SCINFUL-QMD code [8] for the energy region above 3 MeV. For the energy region up to 3 MeV, the neutron detection efficiency obtained experimentally by using a ^{252}Cf neutron source with the NBS spectrum [5]. The difference of detection efficiency was caused by the characteristics of the detectors and the distances between the target and the detectors. Finally, the energy spectra were normalized by the solid angle and the number of incident photons. In the neutron measurements, the time average of the photon intensity was 2.4 x 10^5 photons/s on the target with the laser power of 26.9 W.

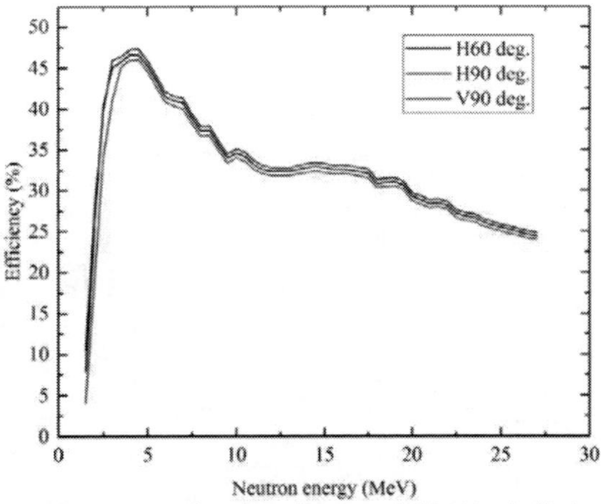

Figure 3. Neutron detection efficiency of NE-213 scintillation detector.

The activation cross sections were esitmated by considering the peak counts of gamma-ray spectra, the peak efficiency of the high-pure Ge detector calculated by the EGS5 code [9] and the beam fluctuation during the irradiation. The activation cross sections corrected for the beam fluctuation are given as following equation.

$$\sigma = \frac{\lambda C}{N\varepsilon\gamma}\left(\sum_{i=1}^{n} N_i\right)^{-1}\left\{e^{-\lambda t_c}\left(1 - e^{-\lambda t_m}\right)\right\}^{-1} \quad (1)$$

where,

$$N_i = \frac{\varphi_i}{\Delta t}\left\{\left(1 - e^{-\lambda\,\Delta t}\right)e^{-\lambda(n-i)\Delta t}\right\}$$

λ is a decay constant, C is a net counts of gamma-ray peak area, N is a number of atoms in the target, ε is a peak efficiency, γ is a branching ration of gamma-ray, t_c is a cooling time, t_m is a measurement time, ϕ_i is a number of photons for irradiation time interval Δt.

In the simulation using EGS5 code, the energy of the laser electron photon E_γ was defined by the flowing equations [10].

$$E_\gamma = \frac{E_L(\beta\cos\alpha + 1)}{1 - \beta\cos\theta + \frac{E_L}{\gamma m_e c^2}\{1 + \cos(\alpha+\theta)\}}\ , \quad (2)$$

where γ is the Lorentz factor of the electron, E_L is the energy of the laser photons, $\beta = v/c$ (v is the electron velocity and c is the speed of light), α is the angle between the electron beam and laser photon direction, θ is the photon emission angle respect to the direction of the electron beam and $m_e c^2$ is the electron rest mass. The angler distribution of the laser electron photon was obtained as

$$\frac{d\sigma}{d\Omega} = \frac{\beta(\sin\theta - \sin\alpha) + \sin(\alpha+\theta)}{\gamma^2(1+\beta\cos\alpha)(1-\beta\cos\theta)^2\sin\theta}\frac{r_0^2}{2}R^2\left(R + \frac{1}{R} - 1 + \cos^2\Theta_{ER}\right), \quad (3)$$

where

$$\frac{1}{R} = 1 + \frac{\gamma(1+\beta\cos\alpha)E_L}{m_e c^2}(1 + \cos\Theta_{ER}),$$

$$\cos\Theta_{ER} = \frac{1 + \cos(\alpha+\theta)}{\gamma^2(1+\beta\cos\alpha)(1-\beta\cos\theta)} - 1,$$

r_0 is the classical electron radius [10]. The simulation results were normalized by the peak count of the experimental results. The collimator hole sizes and positions, and energy spread, beam size and divergence of electron beam were taken into account in the simulation. Electron beam parameters considering in the simulation were listed in table 1.

Table 1. Electron beam parameters in the simulation.

Maximum photon energy	23.1 MeV	26.6 MeV
Beam energy	1149 MeV	1233 MeV
Holizontal beam size (σ)	283 μm	304 μm
Vertical beam size (σ)	58.4 μm	62.68 μm
Holizontal angular divergence (σ)	277 μrad	298 μrad
Vertical angular divergence (σ)	9.1 μrad	9.8 μrad
Energy spread	0.047%	0.047%

4 Results and discussion

4.1 Photon spectrum and pulse height spectrum of GSO detector

Figure 4 shows the photon energy spectrum on the target obtained by the simulation. According to this calculation, we estimated that the energy spreads for the 23.1 and 26.6 MeV photons were 1.29 and 1.69 MeV in FWHM, if the collimators were placed at appropriate positions.

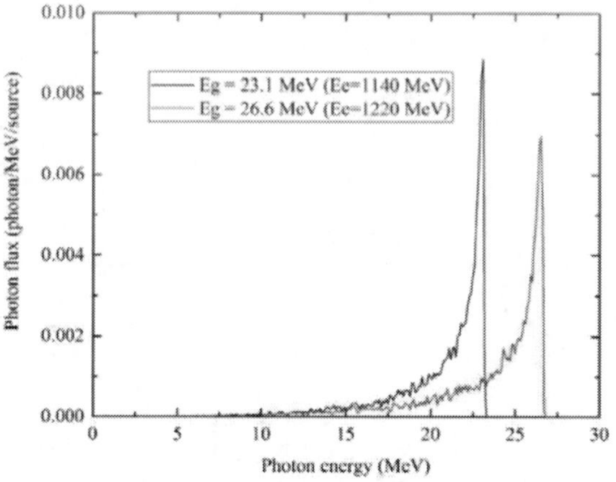

Figure 4. Simulation results of energy spectrum of the laser electron photon induced on the target for 1149 and 1233 MeV electron with 1064 nm laser photon, 3 mm and 2 mm collimators.

Figure 5 shows the measured pulse height spectrum of the GSO detector in comparison with the simulation result obtained by EGS5 code. Energy calibration of the GSO detector was determined using peak energy of the laser electron photon of the simulation result. The energy resolution of 5.0% and 6.0% at 23.1 and 26.6 MeV respectively for the GSO detector were estimated from experimental results. In the lower energy region, the simulation results show high intensity rather than the experimental data. One of the reasons of this difference is an overestimation of angular divergence of the electron beam.

DOI: 10.1051/epjconf/201715307010

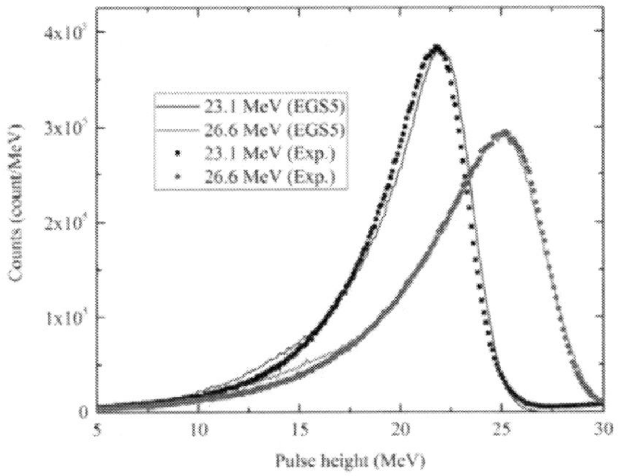

Figure 5. Pulse height spectra of GSO detector at 23.1 and 26.6 MeV in comparison with simulation result by ES5 code.

4.2 Neutron energy spectrum

Figure 6 shows the neutron energy spectra at E_g=23.1 and 26.6 MeV from the ^{nat}C(g, xn) reaction above 1.75 MeV with the simulation results of the PHITS2.82 [11]. In the fig. 6, the error bars are considered with statistical errors, the errors of the neutron detection efficiency (5%) [12] and the errors of the number of photon (5%). The photon energy spectra in fig. 5, the graphite target and NE-213 detector were considered in the simulation.

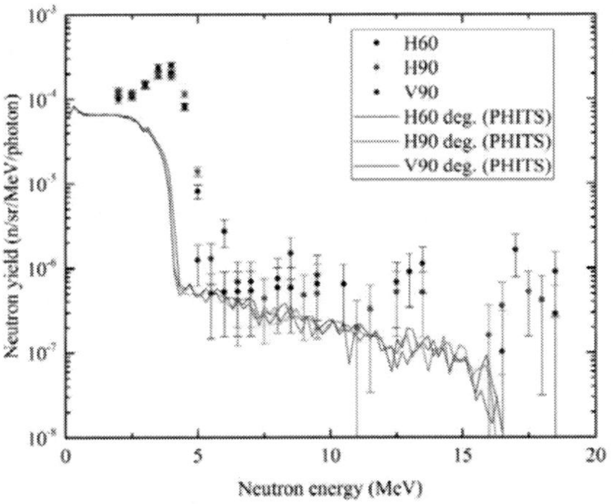

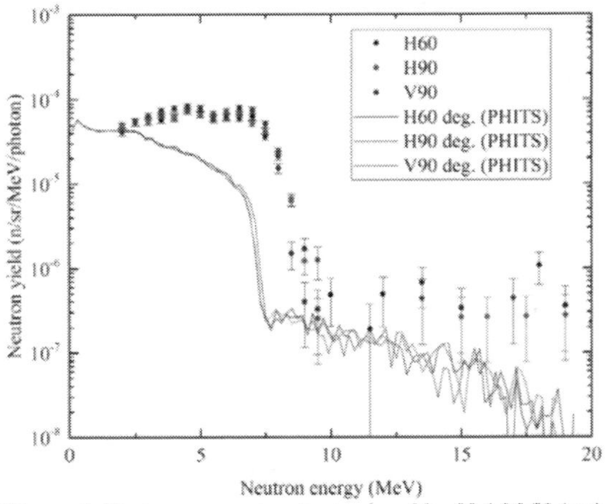

Figure 6. Neutron energy spectra produced by 23.1 MeV (top) and 26.6 MeV (bottom) quasi-mono-energetic photons in comparison with the simulation results by PHITS code. H indicates the horizontal direction and V indicates the vertical direction with respect to the direction of laser electron photons.

In the case of the detector at H60 degree, the energy resolution is less than 17.4% because of relatively short flight path. The neutrons from the ^{12}C(g, n) reaction are dominant even for low reaction Q value (-18.7 MeV). Above 4 and 8 MeV, a few events which seem to the neutrons from the ^{13}C(g, n) reaction (Q-value = -4.95 MeV) were observed. The spectra show similar shapes each other for all the angles as expected, however, their energy dependencies are different from Maxwelian distribution. The simulation results are smaller than the experimental data especially around the maximum neutron energy from the ^{12}C(g, n) reaction.

The neutron yields were compensated for the neutron attenuation in the graphite target in order to clarify neutron angular distribution. The attenuation coefficient f was obtained by following equation.

$$f = \exp\left(-\frac{\rho N_A}{m}\sigma t\right), \qquad (4)$$

where ρ is the target density, N_A is the Avogadro constant, m is the atomic weight, σ is the neutron total cross section and t is the neutron path length in the target. The JENDL-4.0 library [13] was referred to as obtain the total neutron cross section.

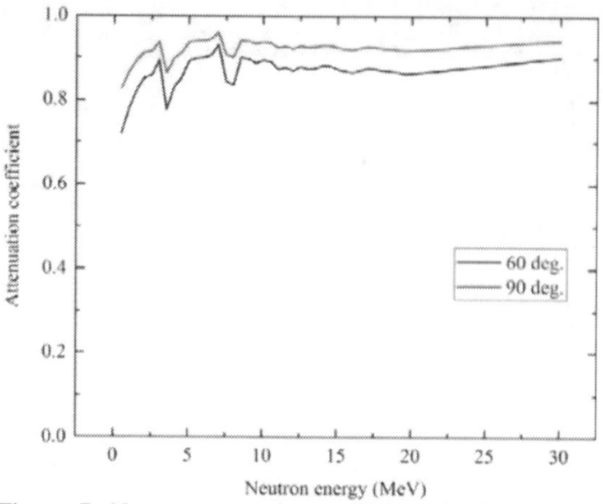

Figure 7. Neutron attenuation coefficients in the graphite target.

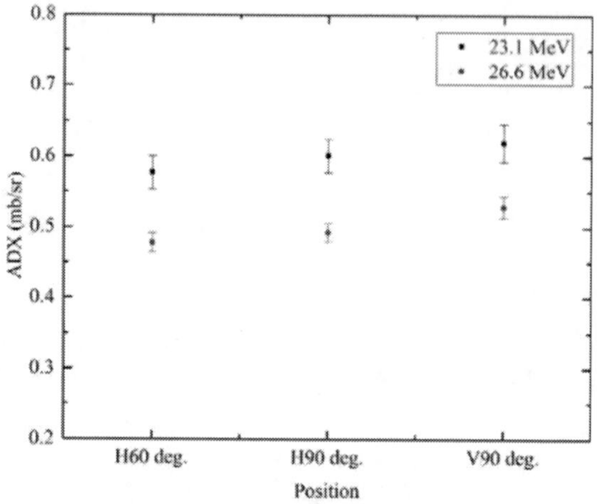

Figure 8. Neutron angular differential cross section (ADX) for $^{nat}C(g,xn)$ reaction.

Figure 7 shows the calculated neutron attenuation coefficients in the graphite target. The path lengths are 5.8 and 5.0 mm for 60 degree and 90 degree respectively assuming that the neutrons were produced at centre of the target. After the compensation, the neutron energy spectra were integrated above 1.75 MeV with dividing number of carbon nuclei in the target. Figure 8 shows neutron angular differential cross section (ADX) for the $^{nat}C(g, xn)$ reaction.

The neutron yield distribution shows nearly isotropic above 1.75 MeV. Production cross sections for the neutrons above the detection threshold ($E_n > 1.75$ MeV) were deduced from an average of three positions. Figure 9 shows the neutron production cross section for the $^{nat}C(g, xn)$ reaction in comparison with the result of PHITS code. The simulation with the mono-energetic photon also counted number of neutrons the energy of which is above 1.75 MeV. The simulation results are smaller than the experimental results.

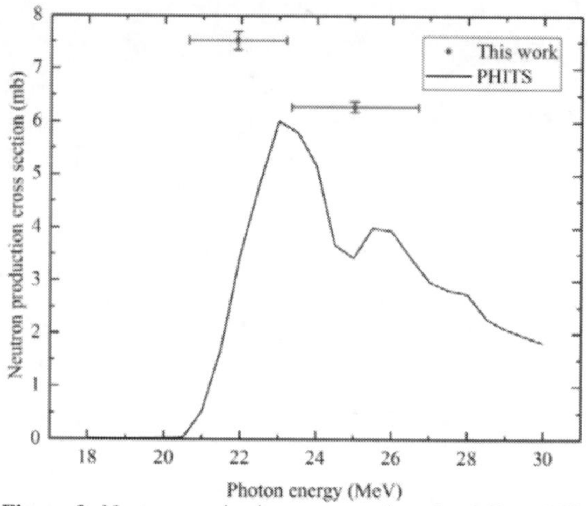

Figure 9. Neutron production cross section above $E_n > 1.75$ MeV with the results of PHITS calculation. The energy of this work means the centroid of the laser electron photon spectrum. Error bars of energy indicate the energy spread of the laser electron photon.

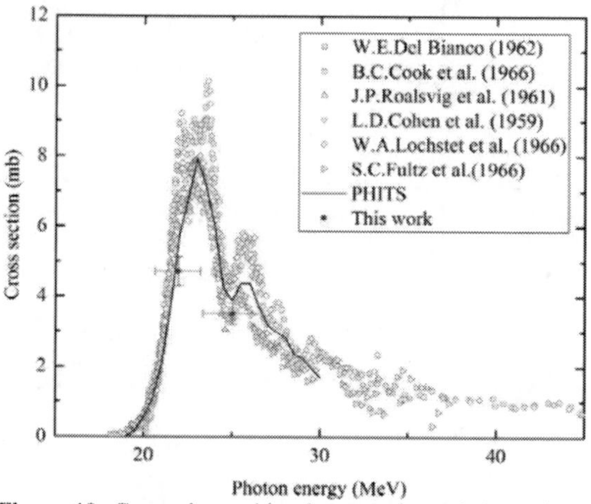

Figure 10. Comparison with other experimental data of the $^{12}C(g, n)^{11}C$ reaction. The energy of this work means the centroid of the laser electron photon spectrum. Error bars of energy indicate the energy spread of the laser electron photon.

4.3 Activation cross section

Figure 10 shows the activation cross sections for the $^{12}C(g,n)^{11}C$ reaction at E_g = 23.1 and 26.6 MeV in comparison with other experimental data [1, 14-18] and the simulation results of the PHITS code with mono-energetic photon. The produced ^{11}C nuclei in the target were counted in the simulation. The present data are close to other experimental data and simulation results.

DOI: 10.1051/epjconf/201715307010

5 Conclusions

The spectral measurements and simulations have been carried out for the circular polarized laser electron photon beam and photo-neutron at NewSUBARU BL01.

From experimental results, the energy resolutions of GSO detector were estimated 5% and 6% for 23.1 and 26.6 MeV quasi-mono-energetic photon respectively. The simulation results for the pulse height distribution of GSO scintillation detector with these energy resolutions were in good agreement with the experimental data.

Using this quasi-mono-energetic photon source, the neutron energy spectra for the $^{nat}C(g,xn)$ reaction with respect to the neutron emission angle were measured with the time-of-flight technique above 1.75 MeV. The spectra show nearly isotropic distribution and the present data were larger than the simulation results using PHITS code. On the contrary, the results of activation cross section for the $^{12}C(g,n)^{11}C$ reaction were smaller than the simulation result. Further analysis is necessary to reversal reason of this discrepancy for both experimental and calculation.

References

1. S. C. Fultz, J. T. Caldwell, B. L. Berman, R. L. Bramblett and R. R. Harvey, Phys. Rev., **143**, 790 (1965)
2. V. V. Verbinski and W. R. Burrus, Phys. Rev., **177**, 1661 (1969)
3. S. Noda, S. Hashimoto, T. Sato, T. Fukahori, S. Chiba and K. Niita, Nucl. Sci. and Tech., **52**, 57 (2015)
4. Y. Asano, S. Miyamoto and LEPS-II collaboration, Prog. Nucl. Sci. Tech., **4** 252 (2014)
5. Y. Kirihara, T. Itoga, T. Sanami, H. Nakashima, Y Namito, S. Miyamoto, A. Takemoto, M. Yamaguchi and Y. Asano, in this proceedings
6. K. Horikawa, S. Miyamoto, S. Amano and T. Mochizuki, Nucl. Instr. and Meth. A, **618** 209 (2010)
7. K. Shin, H. Tokumaru, M. Yoshida and T. Hyodo, Mem. Fac. Eng. Kyoto Univ., **41**, 116 (1979)
8. D. Satoh, S. Kunieda, Y. Iwamoto, N. Shigyo and K. Ishibashi, Nucl. Sci. Tech., **2** 657 (2002)
9. H. Hirayama, Y. Namito, A.F. Bielajew, S.J. Wilderman and W.R. Nelson, SLAC-R-730, KEK Report 2005-8 (2005)
10. Y. Taira, M. Adachi, H. Zen, T. Tanikawa, M. Hosokawa, Y. Takashima, N. Yamamoto, K. Soda and M. Katoh, Nucl. Instr. and Meth. A, **637** S116 (2011)
11. T. Sato, K. Niita, N. Matsuda, S. Hashimoto, Y. Iwamoto, S. Noda, T. Ogawa, H. Iwase, H. Nakashima, T. Fukahori, K. Okumura, T. Kai, S. Chiba, T. Furuta and L. Sihver, J.Nucl.Sci.Technol. **50**, 913 (2013).
12. S. Meigo, Nucl. Instr. and Meth. A, **401**, 365 (1997)
13. K. Shibata, O. Iwamoto, T. Nakagawa, N. Iwamoto, A. Ichihara, S. Kunieda, S. Chiba, K. Furutaka, N. Otuka, T. Ohsawa, T. Murata, H. Matsunobu, A. Zukeran, S. Kamada, and J. Katakura, J. Nucl. Sci. Technol. **48**, 1 (2011).
14. W. E. Del Bianco and W. E. Stephens, Phys. Rev., **126**, 709 (1962)
15. B. C. Cook, J. E. E. Baglin, J. N. Bradford and J. E. Griffin, Phys. Rev. **143**, 724 (1966)
16. J. P. Roalsvig, I. C. Gupta and R. H. H. Haslam, Can. J. Phys., **39**, 643 (1961)
17. L. D. Cohen and W. E. Stephens, Phys. Rev. Lett., **2**, 263 (1959)
18. W. A. Lochstet and W. E. Stephens, Phys. Rev., **141**, 1002 (1966)

Indirect measurement of field emission electron current from the main superconducting cavities of compact ERL at KEK

Hiroshi Matsumura[1,*], *Hajime* Nakamura[1], *Akihiro* Toyoda[1], *Ken-ichi* Hozumi[1], *Hiroshi* Sakai[1], *Kazuhiro* Enami[1], *Takaaki* Furuya[1], *Kenji* Shinoe[1], *Kensei* Umemori[1], *Kaiichi* Haga[1], *Shogo* Sakanaka[1], *Masaru* Sawamura[2], and *Enrico* Cenni[3]

[1]High Energy Accelerator Research Organization (KEK), Tsukuba, Ibaraki 305-0801, Japan
[2]Japan Atomic Energy Agency (JAEA), Tokai-mura, Ibaraki 319-1195, Japan
[3]The Graduate University for Advanced Studies, Tsukuba, Ibaraki 305-0801, Japan

Abstract. The field emission electron currents from the main superconducting cavities (Cavities #3 and #4) of compact ERL at KEK, Japan, were estimated indirectly from photon dose rates measured around the cavities and on the roof of the compact ERL room. The field emission electron currents estimated from the photon dose rates measured around the cavities are in good agreement with those on the roof of the compact ERL room. The field emission electron currents increased steeply with the applied voltage. The field emission electron currents corresponding to the applied voltages were different between Cavity #3 and Cavity #4. We found that the field emission electron current exceeded 1 μA at 13.5 MV for Cavity #3 and at 15.5 MV for Cavity #4. This result was used in considering unexpected loss of field emission electrons.

1 Introduction

The compact Energy Recovery Linac (compact ERL)—an electron accelerator with a maximum energy of 26 MeV—has been under development at the compact ERL room at KEK, Tsukuba, Japan [1]. The compact ERL consists partly of the injector and the recirculating loop, as shown in Figure 1. Electrons are accelerated to a maximum energy of 6 MeV in the injector. The accelerated electrons are transferred to the recirculating loop and accelerated to a maximum energy of 26 MeV by a set of main superconducting cavities (Cavities #3 and #4) [2]. The 26 MeV electrons are transported round the recirculating loop until they reach the main superconducting cavities again. Subsequently, the electrons are decelerated to 6 MeV in the main superconducting cavities, and finally, the beam is terminated at the beam dump. In the compact ERL, the main superconducting cavities play an important role for acceleration and deceleration of electrons.

When a voltage is applied to the main superconducting cavities, field emission electrons are generated inside those cavities [3]. A portion of the generated electrons is accelerated inside the cavities in the same way as operated electron beam and consequently emitted from both ends of the superconducting cavities. Loss of field emission electrons in the recirculating loop is considered an unexpected radiation source. Before the main superconducting cavities (Cavities #3 and #4) were

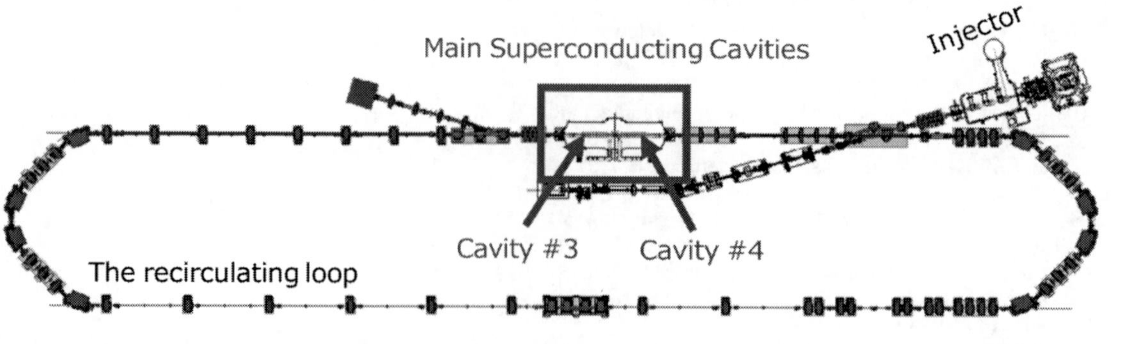

Fig. 1. Location of the main superconducting cavities in compact ERL installed in the compact ERL room in KEK. The main superconducting cavities consist of Cavities #3 and #4.

* Corresponding author: hiroshi.matsumura@kek.jp

installed in the compact ERL in 2014, we had to know whether the field emission electron current was acceptable for machine safety and radiation safety.

In this study, therefore, we indirectly measured the field emission electron current from each main superconducting cavity. In particular, the field emission electron currents were obtained through measurements and simulations of secondary radiation photons and subsequently compared. The secondary photons were produced by bombarding stainless steel (SUS) blind flanges, used for closing both ends of the superconducting cavities, with field emission electrons.

2 Measurements

2.1 Outline of measurements

During measurements of the field emission electron currents from the main superconducting cavities, the assembly of the main superconducting cavities were closed using 1-cm-thick SUS blind flanges. Furthermore, 20-cm-thick lead and 50-cm-thick concrete shield blocks were added to each side of the main superconducting cavities, as shown in Figures 2a and 2b.

While a voltage was applied to each main superconducting cavity, the field emission electrons are generated and accelerated inside their own cavity. The SUS blind flanges are bombarded with the accelerated electrons and bremsstrahlung was consequently generated. We can indirectly estimate the field emission electron current from the flux of the photons generated at the SUS blind flanges. In this study, the photon flux was measured as photon dose rate.

The field emission electrons can be accelerated toward both ends of a superconducting cavity. However, a previous study [3] found that the field emission electron currents in Cavities #3 and #4 were quite different in opposing directions. In the case of Cavity #4, the intensity of the electrons emitted to the right in Figure 1 was much higher than those emitted to the left. However, in the case of Cavity #3, the intensity of the electrons emitted to the left in Figure 1 was much higher than those emitted to the right. Therefore, in this study, the field emission electron currents were measured in only the higher intensity directions (to the left in Cavity #3 and to the right in Cavity #4 in Figure 1).

The photon dose rates around the main superconducting cavities in the compact ERL room were measured using ThermoLuminescence Dosimeters (TLDs). This method is described in detail in Section 2.2. Furthermore, the photon dose rates on the roof of the compact ERL room were measured using survey meters. This method is described in Section 2.3.

The measured photon dose rates were compared with photon dose rates calculated by MARS15 [4] and converted into field emission electron current. The details of the MARS15 calculation and the method of converting a photon dose rate to a field emission electron current are described in Sections 3 and 4, respectively.

2.2 Measurement of photon dose rates around Cavities #3 and #4 using TLDs

2.2.1 Measurement procedures

TLDs (Panasonic, UD-813PQ4) were used to measure of the photon dose rates around the cavities. First, ten TLDs were installed around Cavity #4 on the beam height plane. The locations of the installed TLDs are shown in Figure 2a. To measure the photon dose rates from Cavity #4, 14.6 MV was applied to Cavity #4 for 22.49 min. (duration over 14.1 MV). After the operation, the TLDs were removed from the compact ERL room.

On a different day, 16 TLDs were installed on the beam height plane around Cavity #3 (exception: #11, #12, and #13 were +63 cm above the beam height). The locations of the installed TLDs are shown in Figure 2b. To measure the photon dose rates from Cavity #3, 13.5 MV of voltage was applied to Cavity #3 for 10.67 min. (duration over 13.0 MV). After the operation, the TLDs were removed from the compact ERL room.

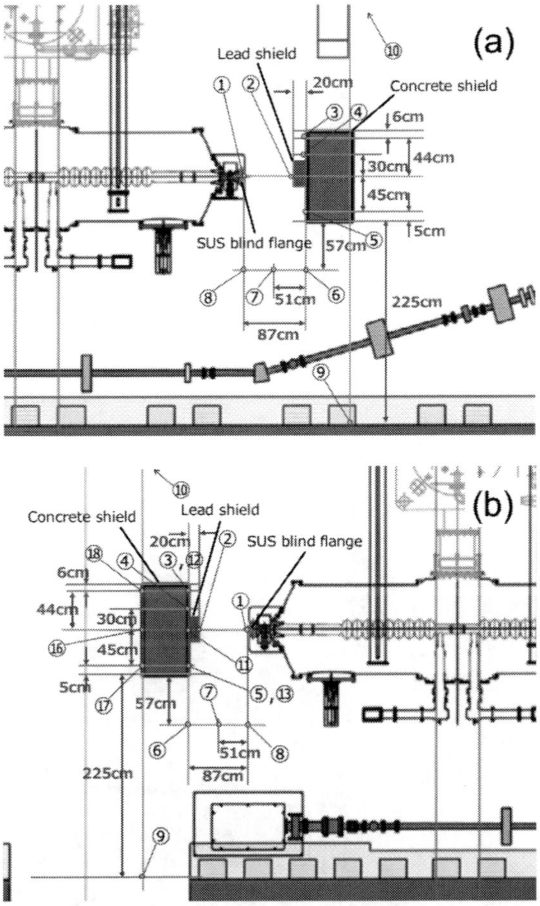

Fig. 2 a. Locations of the TLDs (#1 to #10) installed around Cavity #4 in the compact ERL room. **b.** Locations of the TLDs (#1 to #18, no #14 and #15) installed around Cavity #3 in the compact ERL room.

The photon dose could be read from each exposed TLD. Photon dose rates were obtained from the measured photon doses and the duration of applied voltage.

2.2.2 Measurement results

The measured photon dose rates around Cavities #4 and #3 are shown in Figures 3a and 3b, respectively. The photon dose rates were distributed from 0.011 to 382 Sv/h, depending on the location of the installed TLD. Though the voltage applied to Cavity #3 (13.5 MV) was lower than that applied to Cavity #4 (14.6 MV), the photon dose rate of Cavity #3 was much higher than Cavity #4. The field emission electron currents from Cavity #3 were found to be much higher than those of Cavity #4.

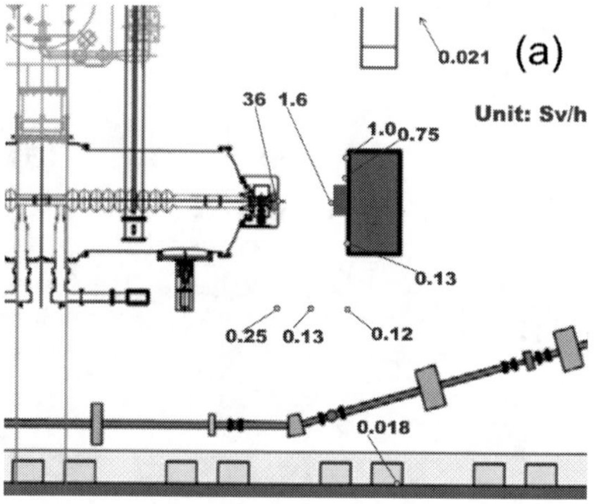

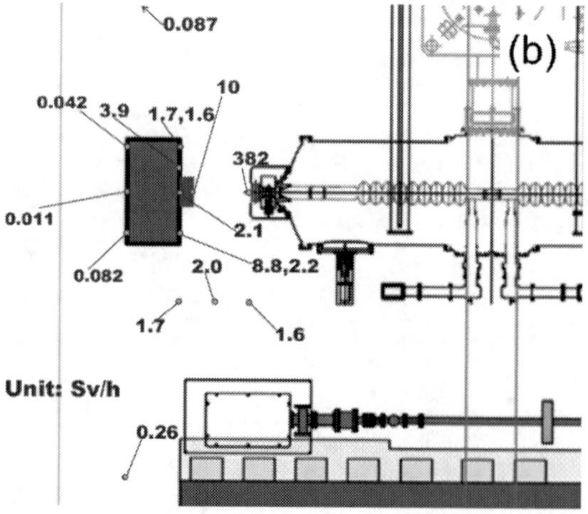

Fig. 3 a. Measured photon dose rates (Sv/h) from Cavity #4. **b.** Measured photon dose rates (Sv/h) from Cavity #3.

2.3 Photon dose rate measurement on the roof of the compact ERL room using survey meters

2.3.1 Measurement procedures

The photon dose rates on the roof of the compact ERL room during application of voltage to each cavity were measured using a NaI (Tl) scintillation survey meter (Aloka, Model: TCS-161) and an ionization chamber type survey meter (Aloka, Model: ICS-311). The shield structure of the compact ERL room is shown in Figure 4. The roof of the compact ERL room is concrete (thickness: 1 m, density: 2.32-2.36 g/cm^3). Field emission electrons travel along the beamline and hit the SUS blind flange at the red point in Figure 4. The distance between the ceiling of the room and beamline height is 2.8 m. The applied voltage was increased up to 13.5 MV for Cavity #3 and 15.5 MV for Cavity #4.

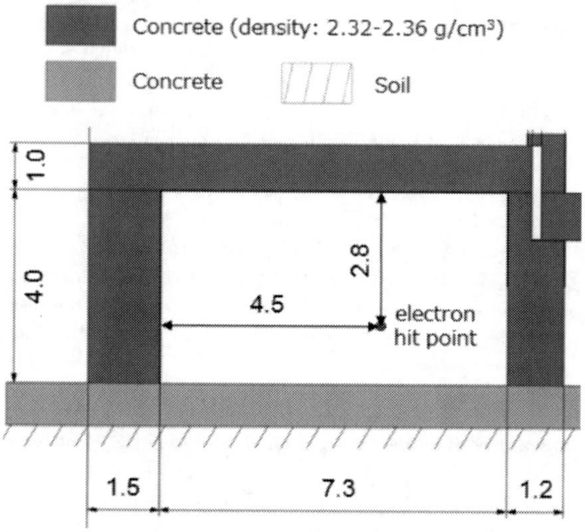

Fig. 4. Longitudinal-sectional view of the compact ERL room perpendicular to the flight direction of the field emission electron and including the electron hit point.

2.3.2 Measurement results

The maximum photon dose rate on the roof of the compact ERL room occurs just above the additional 50-cm-thick concrete shield. The photon dose rates obtained by the NaI (Tl) scintillation survey meter were comparable with those obtained by the ionization chamber type survey meter. The maximum photon dose rates measured by the NaI (Tl) scintillation survey meter are plotted as a function of applied voltage in Figure 5.

The field emission electron currents were found to increase steeply with applied voltage. Field emission from Cavity #3 began to increasing at an applied voltage ~2.5 MV lower voltage than for Cavity #4.

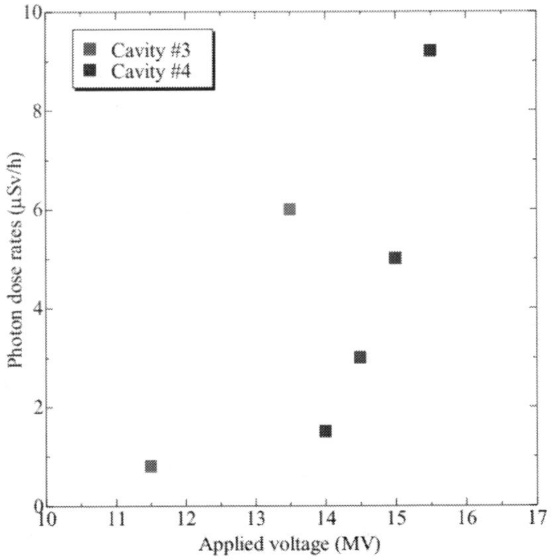

Fig. 5. Maximum photon dose rates measured on the roof of compact ERL room as a function of voltage applied to Cavities #3 and #4.

3 Calculation by MARS15

3.1. Calculation of photon dose rates by monochromatic energy electron

The photon dose rates (mSv/h) per μA of electron current were calculated using MARS15 for comparison with the measured dose rates in order to estimate field emission electron currents. In the MARS15 calculation, a SUS blind flange was bombarded with electrons at energies of 2.5, 5.0, 7.5, 10.0, 12.5, 15.0, 17.5, and 20.0 MeV. The density of the concrete roof was input as 2.34 g/cm^3. Then, the photon dose rates for each incident electron energy were obtained at the measurement locations.

As an example, a distribution of the calculated photon dose rates induced by 15.0 MeV electrons at 0.5 μA on the horizontal plane at incidence height is shown in Figure 6. In this calculation, the incident electrons struck the SUS blind flange perpendicular to its plane and at its center, although it has been observed that flight direction of the field emission electrons deviated a little from the center [6]. The strong photon dose rate was estimated to be in the forward direction.

Another example of a distribution of the calculated photon dose rates on the roof of the compact ERL room is shown in Figure 7. Here, energy and current of the incident electrons were also 15 MeV and 0.5 μA, respectively. As observed by the measurement, the maximum photon dose rate on the roof of the compact ERL room occurs just above the additional 50-cm-thick concrete shield.

The photon dose rates per μA electron current at the measurement locations calculated by MARS15 are plotted as a function of incident electron energy in Figure 8. The lines were obtained by least-square fits of a function to the plotted values. The photon dose rates increased with increasing incident electron energy.

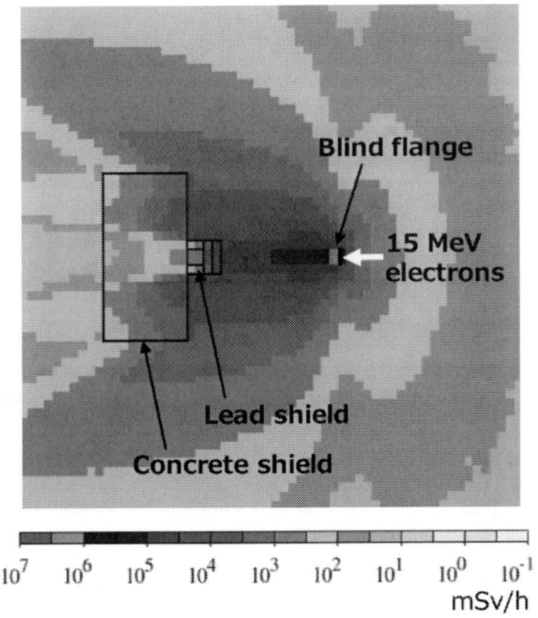

Fig. 6 Distribution of calculated photon dose rates induced by 15 MeV electrons at 0.5 μA on a horizontal plane at incidence height. The photon dose rate distribution was calculated by MARS15.

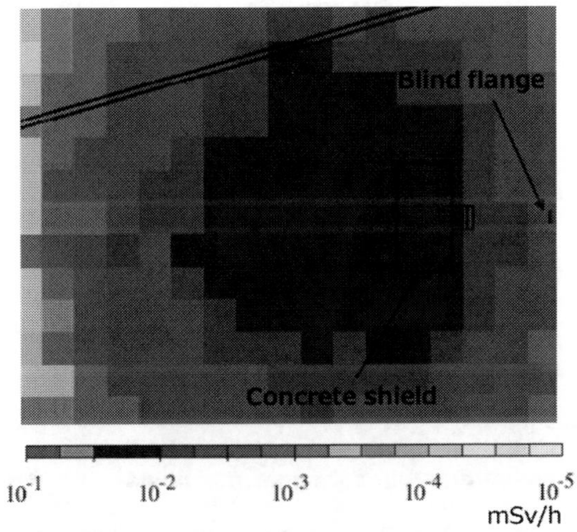

Fig. 7 Distribution of calculated photon dose rates induced by 15 MeV electrons at 0.5 μA on the roof of the compact ERL room. The photon dose rate distribution was calculated by MARS15.

3.2 Calculation of photon dose rates corresponding to the measured dose rates

The energies of the field emission electrons from the main superconducting cavities depend on the location at which the electrons are generated [3]. Here, a uniform energy spectrum model can be applied to the field emission electrons to calculate the photon dose rate. Because the dose rates in Figure 8 were obtained by monochromatic energy electron, the calculated dose rate corresponding to the measured dose rate was obtained from the integral value of dose rate for monochromatic energy in Figure 8 from zero to a maximum energy. The field emission electron current was estimated by comparing the measured and calculated photon dose rates.

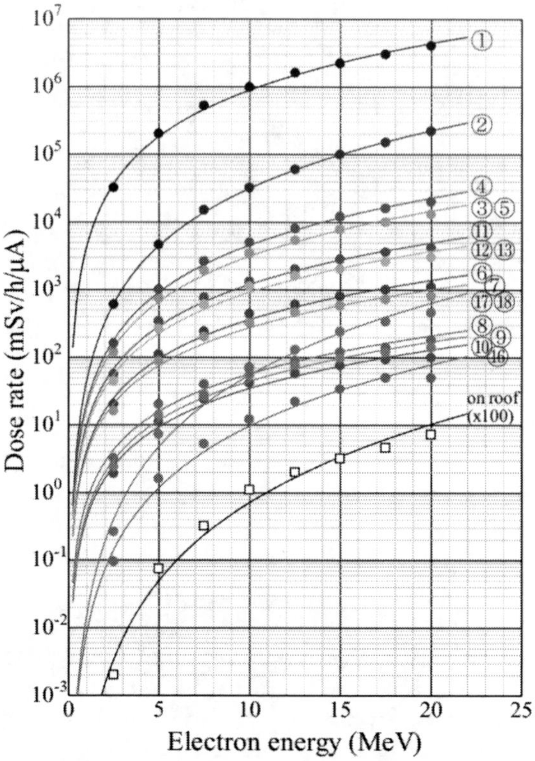

Fig. 8. Incident electron energy dependences of photon dose rates per μA electron current at the measurement locations calculated by MARS15. Numbers in circles correspond to the measurement locations shown in Figures 2a and 2b.

4 Field emission electron currents

4.1 Estimation from the photon dose rates measured around Cavities #3 and #4

The electron energy spectra were assumed to be distributed uniformly up to 12 MeV for Cavity #4 and up to 13 MeV for Cavity #3 at the SUS blind flange. Field emission electron currents arriving at the SUS blind flange from Cavities #3 and #4 were estimated by comparing the measured photon dose rate with the calculated photon dose rate. In Figure 9a and 9b, the field emission electron currents arriving at the SUS blind flange from Cavities #4 and #3, respectively, are indicated. These currents were estimated from photon dose rates for each location. The estimated field emission electron currents at locations #1, #2, and #8 were lower than others. This is due to the fact that the field emission trajectories are anisotropic, that was observed previously in [6]. The estimation at sharp and large angle directions is susceptible to the anisotropic trajectory. The estimated field emission electron currents agreed sufficiently for order estimation, except for the values at locations #1, #2, and #8 in Figures 2a and 2b.

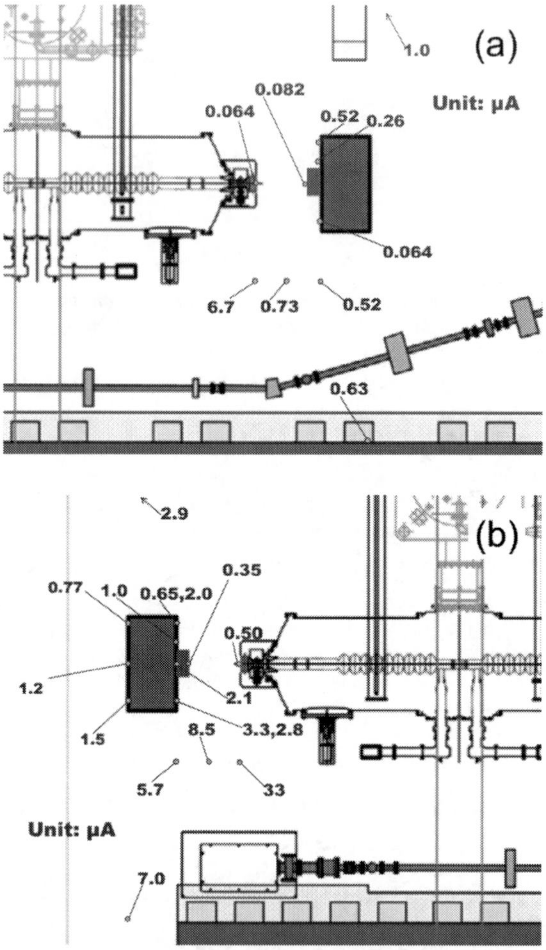

Fig. 9 a. Field emission electron currents (in units of μA) arriving at the SUS blind flange from Cavity #4. **b.** Field emission electron currents (in units of μA) arriving at the SUS blind flange from Cavity #3. The field emission electron currents were estimated from photon dose rates for respective locations.

The field emission electron current from Cavity #4 at 14.6 MV was estimated to be 0.53 ±0.30 μA. The field emission electron current from Cavity #3 at 13.5 MV was estimated to be 3.0 ±2.5 μA. These values were obtained as averages of individual estimated field emission electron currents (values at the locations #1, #2, and #8 in Figures 2a and 2b were omitted).

4.2 Estimation from the photon dose rates measured on the roof of the compact ERL room

The field emission electron currents arriving at the SUS blind flange from Cavities #3 and #4, estimated from the photon dose rates on the roof of the compact ERL room at various applied voltages, are plotted as a function of applied voltage in Figure 10. Here, we assumed that the electron spectra were distributed uniformly for energies up to 11 MeV (for 11.5 MV), 13 MeV (for 13.5, 14.0, and 14.5 MV), and 14 MeV (for 15.0 and 15.5 MV).

In the case of Cavity #3, the field emission electron current of 1.2 μA at 13.5 MV is consistent with 3.0 ±2.5 μA from the photon dose rates measured around Cavity #3 in Section 4.1. In the case of Cavity #4, the field emission electron current of 0.6 μA at 14.5 MV also agreed with 0.53 ±0.30 μA obtained previously in Section 4.1. Therefore, the photon dose rate on the roof is a good indicator of the field emission electron current. Field emission electron current increased exponentially as a function of applied voltage. Then, field emission electron current exceeding 1 μA was observed at 13.5 MV for Cavity #3 and at 15.5 MV for Cavity #4.

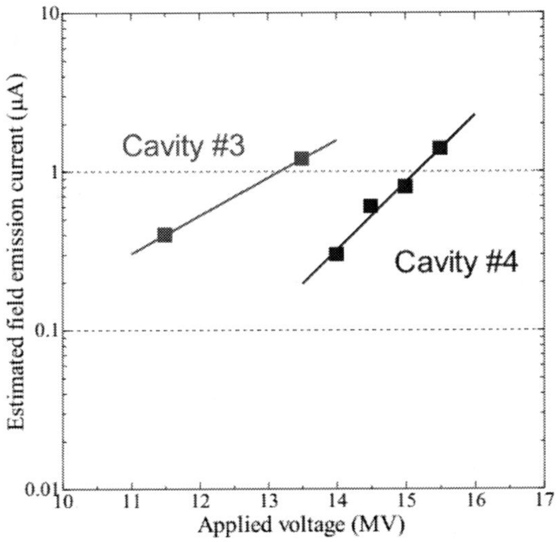

Fig. 10. Estimated field emission electron current arriving at the SUS blind flange fom Cavity #3 and Cavity #4 as a function of applied voltage. The values were estimated from the photon dose rates on the roof of the compact ERL room.

5 Conclusions

The field emission electron currents from the main superconducting cavities of compact ERL at KEK were estimated indirectly from photon dose rates measured around the cavities and on the roof of the compact ERL room. The field emission electron currents increased steeply with applied voltage. The field emission electron currents corresponding to the applied voltages were different between Cavities #3 and #4. The field emission electron current exceeding 1 μA was observed at 13.5 MV for Cavity #3 and at 15.5 MV for Cavity #4. In order to suppress the field emission electron current to sufficiently low level, 8.58 MV was applied to both Cavities #3 and #4 for the first recirculating loop operation in 2014.

References

1. T. Obina, M. Adachi, S. Adachi, T. Akagi, M. Akemoto, D. Arakawa, S. Araki, S. Asaoka, M. Egi, K. Enami, K. Endo, S. Fukuda, T. Furuya, K. Haga, K. Hara, K. Harada, T. Honda, Y. Honda, H. Honma, T. Honma, K. Hosoyama, K. Hozumi, A. Ishii, X. Jin, E. Kako, Y. Kamiya, H. Katagiri, H. Kawata, Y. Kobayashi, Y. Kojima, Y. Kondou, T. Konomi, A. Kosuge, T. Kume, T. Matsumoto, H. Matsumura, H. Matsushita, S. Michizono, T. Miura, T. Miyajima, H. Miyauchi, S. Nagahashi, H. Nakai, H. Nakajima, N. Nakamura, K. Nakanishi, K. Nakao, K. Nigorikawa, T. Nogami, S. Noguchi, S. Nozawa, T. Ozaki, F. Qiu, H. Sagehashi, H. Sakai, S. Sakanaka, S. Sasaki, K. Satoh, Y. Seimiya, T. Shidara, M. Shimada, K. Shinoe, T. Shioya, T. Shishido, M. Tadano, T. Tahara, T. Takahashi, R. Takai, H. Takaki, T. Takenaka, O. Tanaka, Y. Tanimoto, N. Terunuma, M. Tobiyama, K. Tsuchiya, T. Uchiyama, A. Ueda, K. Umemori, J. Urakawa, K. Watanabe, M. Yamamoto, N. Yamamoto, Y. Yamamoto, Y. Yano, M. Yoshida, R. Hajima, M. Mori, R. Nagai, N. Nishimori, M. Sawamura, T. Shizuma, M. Kuriki, *Recent developments and operational status of the compact ERL at KEK*, (Proceedings of IPAC2016, Busan, Korea, May 8-13, 2016)
2. H. Sakai, K. Enami, T. Furuya, M. Satoh, K. Shinoe, K. Umemori, M. Sawamura, E. Cenni, *High power CW tests of cERL main-linac cryomodule* (Proceedings of SRF2013, Paris, France, 2013).
3. E. Cenni, T. Furuya, H. Sakai, K. Umemori, K. Shinoe, M. Satoh, M. Sawamura, *Field emission simulation for KEK-ERL 9-cell superconducting cavity* (Proceedings of IPAC2012, New Orleans, USA, 2012)
4. H. Matsumura Y. Kishimoto, T. Miura, K. Hozumi, K. Haga, S. Sakanaka, K. Nigorikawa, S. Nagahashi, T. Obina and N. Nakamura, KEK-Internal **2015-6** (2015)
5. N. V. Mokhov, S. I. Striganov, *MARS15 overview*, (Proceedings of Hadronic Shower Simulation Workshop Fermilab September 2006), AIP Conf. Proc. **896**, 50 (2007)
6. E. Cenni, K. Enami, T. Furuya, H. Sakai, M. Satoh, K. Shinoe, K. Umemori, *Field emission measure during cERL main LINAC cryomodule high power test in KEK* (Proceedings of SRF2013, Paris, France, 2013)

Study of pulse shape discrimination for a neutron phoswich detector

Jessica Hartman[1] and *Alexander* Barzilov[1]

[1]Department of Mechanical Engineering, University of Nevada, Las Vegas, 4505 S. Maryland Parkway, Las Vegas, NV, 89154, USA

Abstract. A portable phoswich detector capable of differentiating between fast neutrons and thermal neutrons, and photons was developed. The detector design is based on the use of two solid-state scintillators with dissimilar scintillation time properties coupled with a single optical sensor: a ^{6}Li loaded glass and EJ-299-33A plastic. The on-the-fly digital pulse shape discrimination and the wavelet treatment of measured waveforms were employed in the data analysis. The instrument enabled neutron spectrum evaluation.

1 Introduction

Nuclear technologies are utilized in power industry, medical isotope production, food irradiation, scanning of cargo, and other applications. The worldwide expansion of nuclear technologies calls for portable instruments that enable detection and quantification of radiation sources and materials. The development of instruments capable of fast neutron, thermal neutron, and photon detection is an ongoing effort [1]. The ability for simultaneous analysis of these three signatures provides an additional level of certainty in the radiation assay. Besides, neutron energy is a unique signature for nuclear material or radiation source identification. Previous development efforts were focused primarily on the application of the multi-scintillator design to gamma-ray, alpha- and beta-particle measurements [2, 3].

To meet the goal of the neutron source detection and its identification, this research project was focused on the development of a portable phosphor sandwich (phoswich) detector capable of differentiating between fast neutrons, thermal neutrons, and photons. A neutron phoswich design is based on the use of two scintillators with dissimilar times of scintillation coupled with a single optical sensor – a photomultiplier tube (PMT).

2 Phoswich detector

The instrument employs two solid-state scintillators with the dissimilar properties of the scintillation light decay. The first scintillator is a glass disc (5-cm in diameter and 1-mm thick) loaded with 1.58×10^{22} of ^{6}Li atoms/cm^3 enabling high sensitivity to thermal and epithermal neutrons, and a low sensitivity to high energy photons and fast neutrons [4]. The thermal-neutron-induced scintillation decay time for this cell is about 60 ns. The second scintillator is a cylindrical cell (5-cm in diameter and 5-cm long) of the EJ-299-33A plastic.

This plastic offers pulse shape discrimination (PSD) properties which enable the differentiation between neutrons and gamma rays [5]. It has a scintillation time constant of about 4 ns for photons, and about 30 ns for the neutrons through the proton recoil. Two scintillator cells were coupled to a single PMT. The computer controlled high-voltage (HV) base was used to supply power to the PMT. The photomultiplier's anodet signals were processed with a DAQ chip – a miniature digitizer eMorpho (Bridgeport Instruments). The digitizer was connected to a small Linux computer (BeagleBone) equipped with a LCD touchscreen. A WiFi module was used for the wireless communication. A rechargeable Li-ion battery was utilized for portability of the system. The layout of the instrument is shown in Fig. 1. The dimensions of the detector are 9 cm × 25 cm × 17 cm. The portable detector module is shown in Fig. 2.

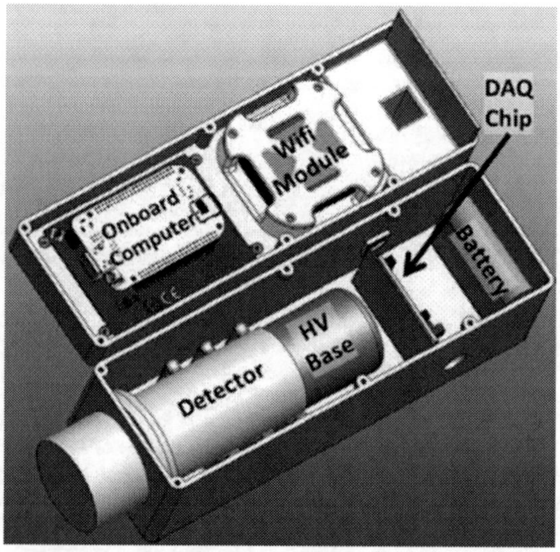

Fig. 1. Internal layout of the instrument.

 DOI: 10.1051/epjconf/201715307012

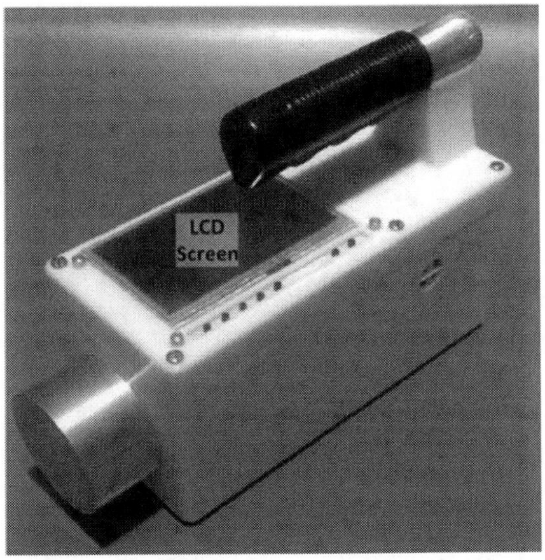

Fig. 2. Portable detector module.

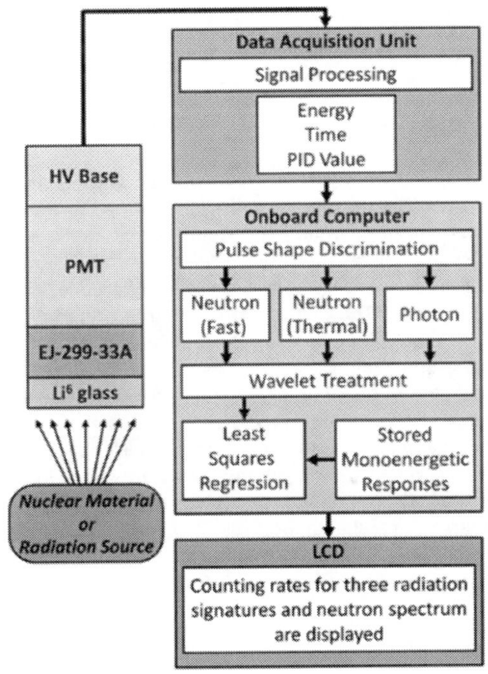

Fig. 3. Scheme of operation of a phoswich detector.

The scheme of the phoswich operation is shown in Fig. 3. Incident radiation emitted from a source enters the phoswich assembly, reacting with the scintillator cells to produce scintillation photons. The PMT anode signals generated due to these reactions are sent to the DAQ chip. This digitizer processes the data using the list mode and assigns the energy, time, and particle identification (PID) values to each of the incoming signals.

The PID value is a ratio of the integral under the tail of the signal to the integral of the front part of the pulse. The PID enables segregating waveforms generated by photons, thermal neutron, and fast neutrons. The IT and PIT integration time values (see Fig. 4) were set up to achieve the optimal figure of merit (FOM) of the PSD.

Processed signals are then sent to the onboard computer, where the PID value is used to carry out the on-the-fly PSD analysis thus separating the signals caused by different incident radiations.

Once the digital waveforms were segregated, a wavelet treatment (Eq. 1) was applied to the waveforms $R(E)$ using the scaling function φ with unique scale (a) and shift (b) parameters, generating a wavelet Ψ_n which accurately plots the shape of the response while smoothing the data:

$$\Psi_n = \int_{-\infty}^{\infty} R(E)\left(\frac{1}{\sqrt{a}}\right)\varphi\left(\frac{E-a}{b}\right)dE \qquad (1)$$

The inherent ability of the wavelets to maintain key signal features while reducing noise sensitivity during fitting procedures makes it ideal for use in the spectral unfolding.

A least squares regression technique was utilized for decomposition of a polyenergetic neutron detector response as the sum of monoenergetic neutron responses. The discrete form of the Fredholm integral was utilized (Eq. 2).

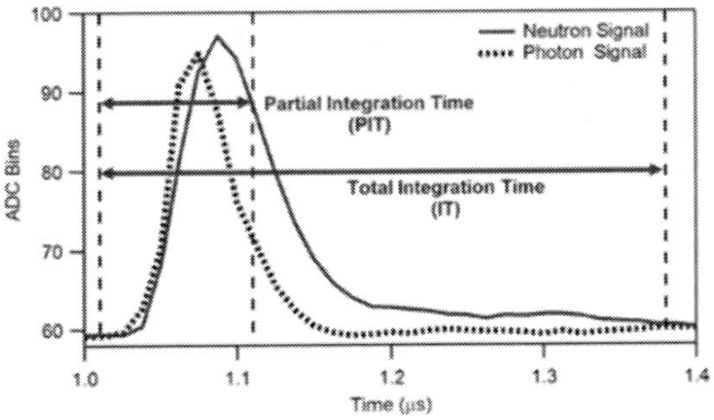

Fig. 4. Digitized PMT signal waveforms for the EJ-299-33A scintillator cell.

$$N_i + e_i = \sum_j R_{ij} S_j \qquad (2)$$

where N_i is the data for a channel i, e_i is a discrepancy between the reconstructed and measured values, R_{ij} is the monoenergetic response for neutron energy j, and S_j is a weight.

The responses R_{ij} were measured using accelerator-based monoenergetic neutron sources and stored as a data library. Gaseous tritium and deuterium targets with incident proton and deuteron beams were used. $T(p,n)^3He$, $D(d,n)^3He$, and $T(d,n)^4He$ reactions were employed to generate mono-energetic neutrons. The angular dependence of the neutron yield for these reactions [6] allowed measurements of response functions for neutron energies ranging from 0.1 MeV to 8 MeV and from 12 MeV to 20 MeV.

DOI: 10.1051/epjconf/201715307012

The code supplied with the math libraries was used to solve Eq. 2 for the weights S_j by minimization of the error e_i. The weights represent the unfolded neutron spectrum.

3 Results

PSD capabilities of the phoswich system were experimentally evaluated. Fig. 5 shows segregated fast neutron and gamma-ray signals measured using the plastic cell and a 2-Ci PuBe source. The slower scintillation decay time for the ^{6}Li-loaded cell also allowed measuring the thermal-neutron count rate.

The PSD FOM was determined using full width half maximum values and the separation between neutron and photon distributions shown in the plot of count rates versus the particle ID values (Fig. 6). The FOM was measured as 1.3 for the plastic scintillator cell.

The unfolded neutron spectra for deuterium-deuterium (DD) and deuterium-tritium (DT) neutron sources are shown in Fig. 7. The spectral unfolding technique did show sensitivity to the stored detector response functions.

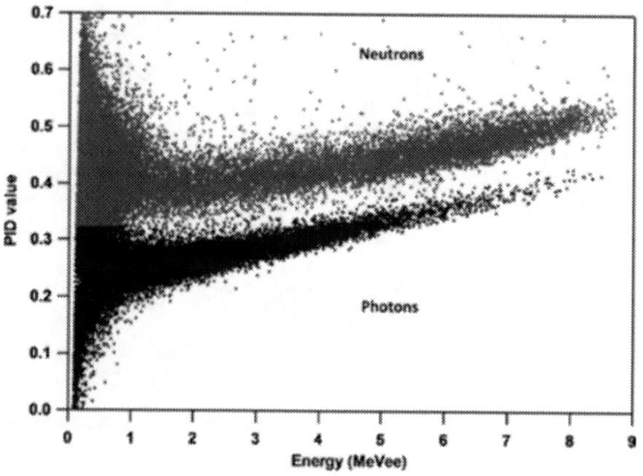

Fig. 5. PSD plot for the EJ-299-33A cell.

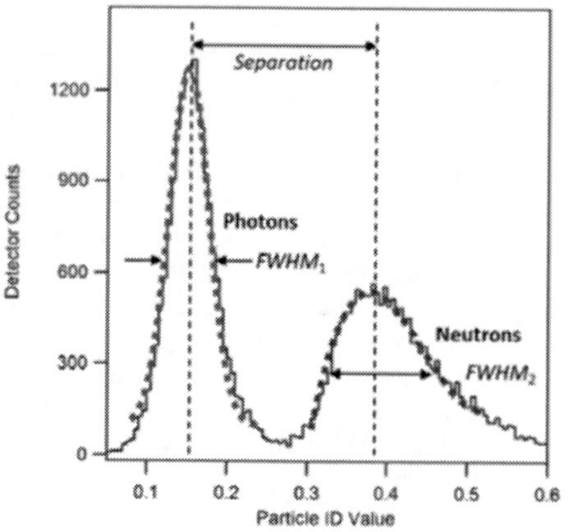

Fig. 6. Neutron/photon FOM.

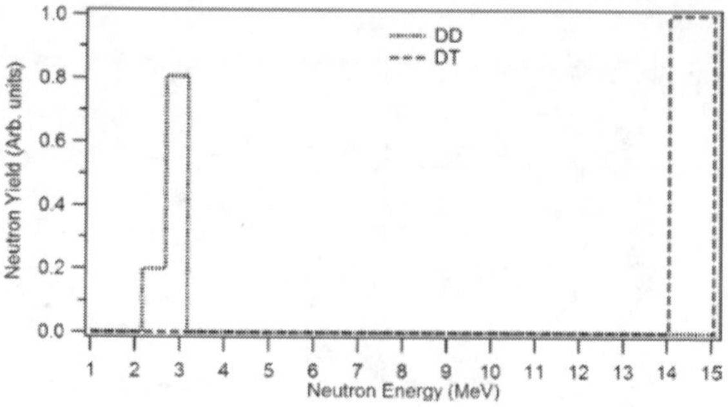

Fig. 7. Unfolded spectra for DD and DT neutron sources.

4 Conclusions

A phoswich neutron detector of the small form factor was designed. The detector enabled onboard data analysis with a wireless communication of the measured data. The phoswich was capable of simultaneously measuring gamma-ray, thermal neutron, and fast neutron count rates, and the fast-neutron spectrum evaluation. The results were reported on the LCD screen, displaying count rates for radiation signatures and the unfolded neutron spectrum.

References

1. N. Zaitseva, A. Glenn, H.P. Martinex, L. Carman, I. Pawelczak, M. Faust, S. Payne, "Pulse shape discrimination with lithium-containing organic scintillators," Nucl. Instrum. Meth., 729, 747 (2013).
2. S. Usuda, S. Sakurai, K. Yasuda, "Phoswich detectors for simultaneious counting of α-, β(γ)-rays and neutrons," Nucl. Instrum. Meth., 388, 193 (1997).
3. S. Yamamoto, H. Ishibashi, "Development of a three-layer phoswich alpha-beta-gamma imaging detector," Nucl. Instrum. Meth., 785, 129 (2015).
4. L. Greenwood, N. Chellew, G. Zarwell, "Comparative analysis of pulse shape discrimination methods in a ^{6}Li loaded plastic scintillator," Nucl. Instrum. Meth., 788, 146 (2015).
5. Eljen Tech: http://www.eljentechnology.com.
6. H. Liskien, A. Paulsen, "Neutron production cross sections and energies for the reactions T(p,n)^{3}He, D(d,n)^{3}He, and T(d,n)^{4}He, " Atomic Data and Nuclear Data Tables (1973).

Intrinsic noise of a superheated droplet detector for neutron background measurements in massively shielded facilities

Ana C. Fernandes[1,*], *Tomoko A.* Morlat[1], *Miguel* Felizardo[1], *Andreas* Kling[1], *José* G. Marques[1], *Maria I.* Prudêncio[1], *Rosa* Marques[1], *Fernando P.* Carvalho[1], *Ignácio* Lázaro Roche[2] and *Thomas A.* Girard[3]

[1]Centro de Ciências e Tecnologias Nucleares, Instituto Superior Técnico, Universidade de Lisboa, 2695-066 Bobadela, Portugal
[2]Laboratoire Souterrain à Bas Bruit, University of Nice, University of Avignon, Centre National de la Reserche Scientifique, Aix-Marseille University, Observatoire de la Côte d'Azur, 84400 Rustrel, France
[3]Departamento de Física, Faculdade de Ciências, Universidade de Lisboa, 1749-016 Lisboa, Portugal

Abstract. Superheated droplet detectors are a promising technique to the measurement of low-intensity neutron fields, as detectors can be rendered insensitive to minimum ionizing radiations. We report on the intrinsic neutron-induced signal of C_2ClF_5 devices fabricated by our group that originate from neutron- and alpha-emitting impurities in the detector constituents. The neutron background was calculated via Monte Carlo simulations using the MCNPX-PoliMi code in order to extract the recoil distributions following neutron interaction with the atoms of the superheated liquid. Various nuclear techniques were employed to characterise the detector materials with respect to source isotopes (^{238}U, ^{232}Th and ^{147}Sm) for the normalisation of the simulations and also light elements (B, Li) having high (α,n) neutron production yields. We derived a background signal of ~10^{-3} cts/day in a 1 liter detector of 1-3 wt.% C_2ClF_5, corresponding to a detection limit in the order of 10^{-8} n cm^{-2}s^{-1}. Direct measurements in a massively shielded underground facility for dark matter search have confirmed this result. With the borosilicate detector containers found to be the dominant background source in current detectors, possibilities for further noise reduction by ~2 orders of magnitude based on selected container materials are discussed.

1 Introduction

A superheated droplet detector (SDD) consists of a homogeneous suspension of micrometric superheated liquid droplets inside a viscous gel of matched density, each droplet functioning as a small bubble chamber [1, 2]. These detectors are traditionally used for personal neutron dosimetry [3, 4], but have also been employed in high energy neutron spectrometry [5, 6], registration of high and intermediate heavy ions [7] and direct search for dark matter [8, 9]. At C^{2}TN superheated droplet detectors (SDD) of 1-3 wt.% C_2ClF_5 in a food gel-based matrix are developed and operated with an acoustic instrumentation for signal detection. The detectors have been used for the detection of rare events in the context of astrophysics [9] and more recently for alpha particle (α) measurements [10].

The response of the device is based on the Seitz "thermal spike" model [3, 11]: if the energy deposition is sufficiently large within a sufficiently short distance in the droplet, the droplet vaporises. These limits depend on the liquid and its thermodynamic degree of superheating: typically the higher the superheat, the lower the energy required for bubble nucleation. A selection of the thermodynamic operation conditions can render the SDD insensitive to energetic muons, electrons and photons < 6 MeV that plague most rare event detection experiments targeting heavy ionising radiations. We therefore obtain a low-noise instrument for the characterisation of neutron environments at extremely low intensities where the reduced radiation intensity hampers the use of most experimental methods. The application is particularly interesting for shielded facilities employed in low-noise gamma-spectroscopy [12], intrinsic soft error rate measurements in electronic devices [13], investigation of neutrino and neutrino-less physical processes and direct search for dark matter [14]. The intrinsic detector background becomes critical, for it determines the ultimate detection limit thereby affecting the measurement accuracy and precision.

The SDD signal was measured at an underground facility in LSBB [15] following various efforts to reduce the environmental neutron radiation that originated from the cavern structural materials [16, 17] and the α background induced by atmospheric radon that might diffuse into the detector gel [18]. The neutron signal corresponding to the experimental set-up was calculated independently, using the stochastic MCNP-5 code [19] to model neutron transport from and within the entire facility [20]. The calculations showed that >85% of the SDD signal originated from its own materials, namely the borosilicate glass container. It was therefore concluded that the measurement at this massively

* Corresponding author: anafer@ctn.tecnico.ulisboa.pt

shielded facility provided a direct determination of the detector intrinsic noise.

For the simulations, additional codes were employed to extract the reaction rates from the neutron spectra. These included FLXPRO [21] to manage reaction cross sections and uncertainties, and SPECTER (based on the ENDF-5 cross section data from the 1980's) to derive the distribution of nuclear recoils that ultimately produce the detector signal [22]. Neutron production yields from (α,n) reactions were obtained from the USD code [23] which was the only publicly-available software providing (α,n) spectra for α energies covering the ^{238}U and ^{232}Th decay chains. The total neutron yields from USD have the same order of magnitude (discrepancy of 25-45%) as those from other (α,n) codes and experimental data; however, unrealistic sharp peaks are observed at the lower end of the energy distribution. Although these features spread out in the hydrogenous gel of the SDD [20] they are expected to affect the characterisation of the bulk superheated liquid devices under development [24].

In this paper we investigate the intrinsic background of the SDD using the MCNPX-PoliMi-2.0 code [25] to extract directly the recoil distribution using the most recent cross section and angular data. The (α,n) data is mostly obtained from the SOURCES-4A code extended to 10 MeV α's [26]. New experimental data regarding emitters and relevant (α,n) isotopes in the gel is now included in the model. The revised detection limit is determined for the current detector design, and further developments towards improving its performance are identified.

2 The SDD

2.1 General description

As described in Section 1, the threshold conditions required for the phase transition of a droplet set thermodynamically-defined minimum values for the radiation energy and linear energy transfer. Our SDDs are usually operated at 9 °C and 2 bar. Pressurization improves the detector life and reduces Rn diffusion into the gel. The SDD sensitivity to radiations simultaneously depositing less than 8 keV at less than 160 keV μm^{-1} is inhibited, rendering the detector sensitive only to nuclei recoiling after neutron interactions, and α's. In the case of neutrons, the threshold conditions define a critical recoil energy (E_{rec}) hence a minimum neutron energy (E_{min}) for the induction of a nucleation - the relation between E_{rec} and E_{min} being determined by the kinematics of the nuclear interaction [27].

In order to gain sensitivity the concentration of superheated liquid is larger than that of commercial devices (<0.1 wt.%). The gel becomes opaque inhibiting the visual detection of the events; a low-frequency acoustic instrumentation is employed to detect the shock wave formed by the bubble expansion in the gel [28].

Figure 1 shows a 150 ml SDD with 2 wt.% C_2ClF_5. The container is made of laboratory-grade borosilicate glass (Schott Duran, Germany). The cap is modified with a feedthrough that allows detector pressurization, pressure monitoring and signal acquisition with a microphone embedded in a 1 cm glycerine layer that covers the gel.

Fig. 1. A Sm-doped 150 ml SDD equipped with microphone and valve for pressurisation. The raw acoustic signal is displayed in the background. A discard microphone is shown in front of the detector.

The detector lifetime is limited by the fracture of the gel which in turn is promoted by the growth of gas bubbles in an exposed detector. Underground testing of an SDD with improved fabrication protocols towards lifetime extension [29] yielded an intrinsic lifetime of 100 days.

Signal treatment and analysis allowed to determine the characteristic frequency signature of events originated by radiation and identify them among other acoustic, seismic, or gel-related noises [30]. In addition, recoil and α- induced events can be discriminated based on signal amplitudes, with α's producing larger amplitudes as a result of the production of various proto-bubbles along the Bragg peak [31, 32].

Further details on SDD fabrication, instrumentation and particle discrimination can be found in Ref. 9 and references therein. A modification was recently introduced relatively to the standard fabrication protocol, with the droplet shearing step occurring at a higher temperature (60°C) than standard (45°C) in order to decrease the gel stiffness.

The particle discrimination capability (due to the existence of a gap between recoil and α-amplitude distributions) is linked to the droplet size distribution [33]. The size distribution was recently measured by optical microscopy for 150 ml devices (Fig. 2); a gaussian fit yields an average radius of 4.6 μm. The average droplet size is significantly smaller than previously reported for 1 liter standard devices (30 μm); a possible cause for the discrepancy is the increased efficiency of the droplet shearing due to the smaller device volume and gel stiffness.

DOI: 10.1051/epjconf/201715307013

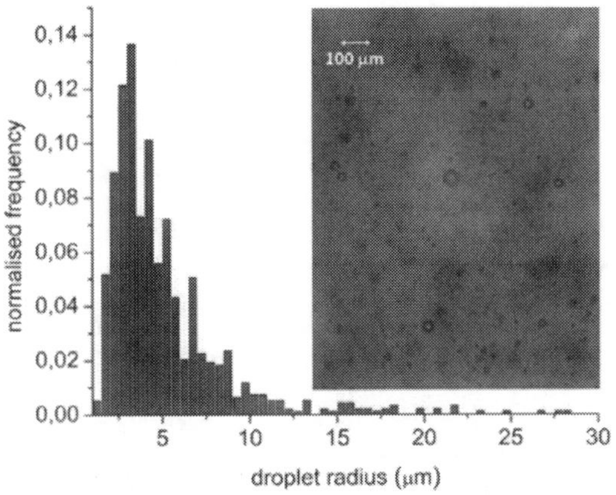

Fig. 2. Droplets (magnification 100x) and droplet size distribution in a 150 ml fresh SDD (sample size: 930 droplets).

The size distribution is found to be preserved after a recompression stage that reliquifies the gas bubbles allowing the regeneration of an exposed detector. This facilitates neutron spectrometry, in which the temperature-varying method at fixed pressure is employed to modify E_{min}. As the temperature increases, neutrons of lower energies are gradually included in the measurement (Fig. 3). From the difference between the detector signals at different temperatures the fluence rate of neutrons between the corresponding E_{min} can be derived [34]. However, as measurements at moderate temperatures are expected to decrease the SDD lifetime, the detector regeneration possibility is an interesting alternative to the application of fresh devices with potentially different sensitivities.

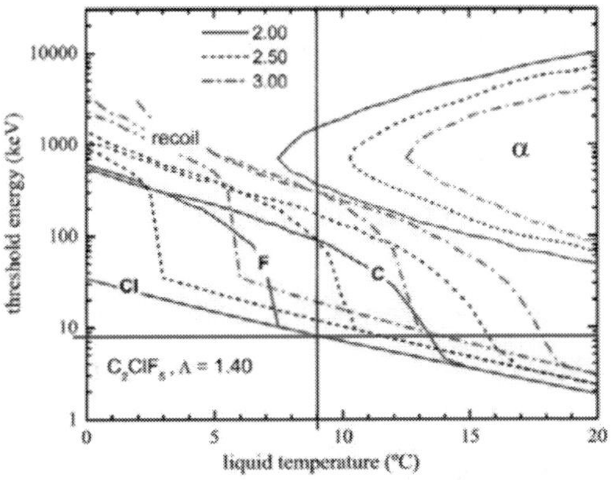

Fig. 3. Threshold energy of nuclear recoils and alpha particles for C_2ClF_5 as a function of temperature, at different pressures. The vertical and horizontal lines denote the standard operating conditions (9°C and 2 bar).

2.2 Materials

The choice and characterisation of materials is crucial for the development of a low-noise neutron detector, as most materials contain traces of radioisotopes that produce neutrons either by spontaneous fission or via (α,n) reactions. We are mostly concerned with ^{238}U and ^{232}Th due to the emission of α's in the course of the decay chain and neutron emission following the spontaneous fission of ^{238}U. Samarium is a natural emitter of low energy α's (from ^{147}Sm, with a natural abundance of 14.6%) whose presence in the gel justifies evaluation [23]. The presence of light elements with high (α,n) production yields – namely Boron – in material compositions was also considered.

Neutron activation analysis, gamma- and α-spectrometry and ion-coupled plasma spectrometry (ICP-MS) were used to quantify the emitters, whereas the determination of B and Li employed ion beam analysis techniques and ICP-MS. The results are summarised in Table 1. When concentrations are below the detection limits, the latter are quoted. Materials' composition and density were extracted from Ref. 20; that from the gel was modified to include the superheated liquid as required for MCNPX-PoliMi. For neutron transport and recoil calculation, the revised composition (in weight) for an SDD with 2 wt.% superheated liquid is: 8.92% H, 33.43% C, 0.74% N, 55.22 O, 1.23% F and 0.46% Cl. The possible traces of B and Li in the gel were considered only to derive the (α,n) source term (Section 3.2.1).

Table 1. Measured concentrations (in μg/g) of neutron and α-emitters and selected light elements in the detector materials.

Isotope or Element	Borosilicate	Gel	Microphone
^{238}U	0.220±0.033	<0.001	<1.09
^{232}Th	0.310±0.027	<0.01	-
^{147}Sm	-	<0.001	-
B	(4.1±0.2) x10^4	<0.03	~0
Li	-	<0.005	-

3 Intrinsic neutron-induced noise

3.1 Measurement

3.1.1 Brief description of the experimental set-up

The SDD background noise was measured in the GESA room of LSBB, equipped with a water-based neutron shield. The facility is located at 500 m depth in calcite rock, corresponding to 1500 m water-equivalent (mwe). A schematic view of GESA with the experimental set-up is shown in Fig. 4. A detailed description of the facility including dimensions can be found in Refs. 9, 20. The cavern walls and ceiling are finished with concrete and a steel lining. Fifteen SDDs of 1 liter, fabricated in an adjacent clean room, sat on a thermostatic water bath of 750 liter for temperature control, with the water level covering them up to the glycerine level. Each detector glass container weighted 740 g.

A water shield was installed around the tank for protection against environmental neutrons from the cavern structural materials (mostly from the concrete). The temperature of the water bath and individual detectors pressure were monitored in order to evaluate the stability of the thermodynamic operation conditions. This information, together with the signal from each detector was acquired continuously and available on-line and remotely.

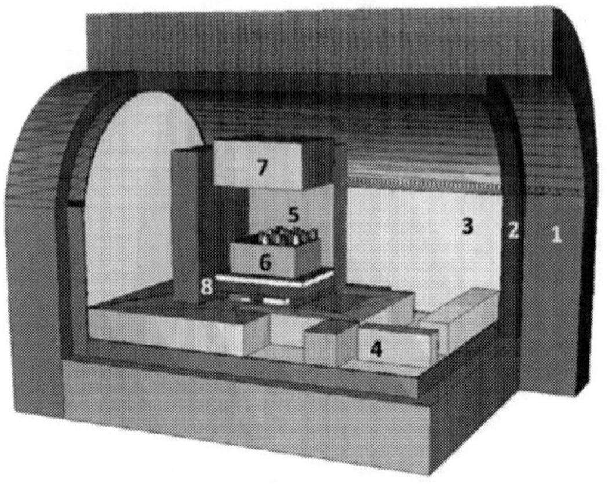

Fig. 4. Schematic view of the GESA room with fifteen SDDs in a water bath surrounded by a neutron shield (some parts were removed from the figure for clarity). 1: rock; 2: concrete; 3: steel lining; 4: concrete floor structures; 5: detectors; 6: tank (water below the detectors); 7: water shield; 8: wooden tank support with various polyethylene and paraffin layer/inserts.

3.1.2 Radiation backgrounds

The water-based shielding, with a thickness of 50-75 cm was designed using a detailed MCNP model of the facility (also displayed in Fig. 4). MCNP outputs were normalised to the measured ^{238}U and ^{232}Th concentrations in the various structural, shield and detector materials. A reduction of the neutron field in the facility by 2 orders of magnitude was achieved, with on-detector neutron fluence rate values in the order of 10^{-8} cm^{-2} s^{-1}. The overall calculated event rate is 0.372±0.097 evt/kgd, of which 86% originate in the detector glass [20].

Efforts to reduce the α-induced signal due to the diffusion of atmospheric radon into the detector walls include purging the air 10 times per day (reducing the maximum Rn concentration from 1000 to 60 Bq m^{-3}) and circulating the water bath in order to replace the top 1 cm layer each minute [9, 18, 31].

The influence of Rn adsorption on the glass container stemming from the water bath and consequent implantation of the decay daughters originating (α,n) neutrons is evaluated for the first time. For equilibrium concentrations of Rn in air and water [35], the amount of Rn available for adsorption is 22 mBq/liter. This is actually an upper limit, since water circulation is neglected. The total Rn activity in the pool water is ~16 Bq. Due to the geometrical arrangement the detectors occupy only about 15% of the pool volume reducing further the amount of Rn available for adsorption at the glass to 3 Bq (for the remaining volume adsorption at the pool walls - which have a surface area ca. five times larger than the detectors – exhibits an alternative sink). In addition, the amount of Rn descendants implanting in the glass is less than 50% of the adsorbed. Comparing the <2 Bq of implanted activity from Rn in the surrounding water to the 44.5 Bq Rn contained in the 11.1 kg of glass in equilibrium (2.74 Bq kg^{-1} from ^{238}U and 1.27 Bq kg^{-1} from ^{232}Th decay chains) shows that any additional Rn activity due to an eventual adsorption of Rn and its subsequent recoil implantation at the water/glass interface has a negligible influence. Finally, the SDDs are fabricated in the clean room with similar Rn levels for various days before installation in the experiment, thereby "losing memory" of previous Rn exposure (half-life of ^{222}Rn from ^{238}U is 3.8 d; of ^{220}Rn from ^{232}Th is 56 s).

Cosmic muons interacting with rock originate high energy neutrons that can contribute to the detector signal. Estimates of the corresponding neutron field were based on data from Refs. 36, 37, and are reported in detail in Ref. 20. Data from other underground facilities were interpolated in order to derive the neutron energy distribution up to 3 GeV for GESA's rock properties (namely density: 2.61 g cm^{-3} and average atomic number: 23.4). MCNPX [38] was employed to estimate the neutron-induced signal in the SDDs, yielding an event rate that is 0.6% of total.

3.1.3 Experimental results

A measurement stage of 75 days was performed using fifteen SDDs of 1 liter comprising a total of 215 g superheated liquid. Three detectors failed within days as a result of overpressure and were excluded from the measurement data resulting in a liquid loss of 45 g. Another detector of 11.5 g liquid failed after 48 days of measurements; its results have been accounted for. There is a trend for the increase of detector pressure with time as bubbles are formed and the gel swells. Data at >2.2 bar has been excluded in order to maintain the recoil energy threshold of 8 keV. This yield an exposure loss down to 6.71 kgd.

A total of 1982 events were registered. Of these, 1171 are correlated among the various SDDs indicating an origin diverse from radiation. Signal analysis indicated that 95% of the uncorrelated events corresponded to environmental noise events, gel fractures and gas microleaks. The characteristic frequency signature allowed to identify 30 nucleations, of which 29 were induced by α's. An overall result of 1 recoil event is derived corresponding to 0.15 evt/kgd that originate from the detector container, as described in Section 3.1.2.

3.2 Calculation

3.2.1 Neutron production yields

The intrinsic neutron-induced noise of a 1 liter SDD at 2 wt.% liquid originating from its container, gel and microphone was calculated using MCNPX-PoliMi. A simple geometry model was used: the SDD glass container is represented as a box of square cross section (8 cm internal size) of 5 mm thickness having gel up to 12 cm height and a 2 cm glycerine layer embedding a punctual microphone at the geometric centre.

The neutron source description included the Watt spontaneous fission spectrum with tabulated parameters (the intensity being derived from the measured ^{238}U concentrations and decay data), as well as (α,n) yields and spectra calculated using the modified SOURCES-4A code. Isotropic emission was considered. Secular equilibrium in the decay chains was assumed. This is the case for ^{238}U in the borosilicate glass used, for which there is no evidence of disequilibrium in the ^{226}Ra sub-chain [39]. With respect to the microphone material, SOURCES lacks neutron production data in Fe and the USD code was used instead.

Production yields and energy distributions are shown in Table 2 and Fig. 5. For the gel, upper limits were derived for the detection thresholds of emitters and light elements (Table 1). The latter were found to have no impact on the calculated neutron yields.

The complex structure of the USD data for Fe contrasts with the smooth distribution retrieved by SOURCES for Th (and in general); a possible cause for this effect could be a deficient calculation of the continuous α energy loss in the material by the USD code [40]. The Sm-induced distribution is a particular case where the sharp features retrieved by SOURCES are due to the fact that at low α energies only a very limited number of target isotopes and states contributes to the production of neutrons and extends up to low energies. The maxima shown in Fig. 5 represent (by increasing energy) the contribution of the first excited and ground states of $^{20/21}Ne$ isotopes and of ^{16}O [40, 41].

Table 2. Neutron yields (n $\mu g^{-1} y^{-1}$) from spontaneous fission and decay-induced (α,n) reactions due to the presence of emitters in the SDD materials. Secular equilibrium is assumed for ^{238}U and ^{232}Th decay chains.

Reaction / Material	^{238}U	^{232}Th	^{nat}Sm
Spontaneous fission	0.430	3.84×10^{-6}	0
(α,n) in borosilicate	3.185	1.142	4.86×10^{-5}
(α,n) in gel	0.633	0.267	2.98×10^{-5}
(α,n) in microphone	0.165	0.173	0

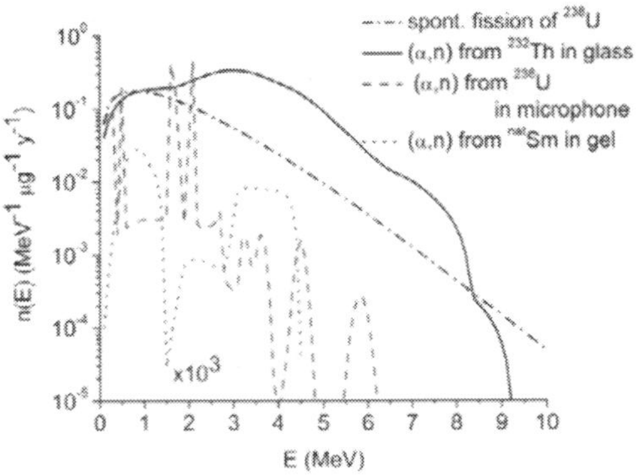

Fig. 5. Energy distribution of (α,n) neutrons due to the presence of ^{238}U and ^{232}Th (in secular equilibrium) and ^{nat}Sm for various SDD materials. The Watt distribution representing the spontaneous fission of ^{238}U is also represented.

3.2.2 Event-producing reactions

The various neutron interactions with the liquid atoms were evaluated with respect to the threshold conditions referred in Section 2.1. We considered elastic and inelastic scattering in F, Cl and C, and the transmutation reactions with positive Q-value: (n,α) and (n,p) in ^{35}Cl. Previous investigations [20] allow to reduce the number of the relevant contributing reactions; their E_{rec} and corresponding E_{min} are shown in Table 3. E_{min} is in the order of 10-100 keV for scattering and 0 for the transmutation reactions.

Table 3. Threshold recoil and minimum neutron energy for the event-producing reactions in C_2ClF_5 at 9°C and 2 bar (elastic scattering: (n,n); inelastic scattering: (n,n')).

Reaction	E_{rec} (keV)	E_{min} (keV)
$^{nat}C(n,n)^{nat}C$	115	402
$^{nat}F(n,n)^{nat}F$	8	43
$^{nat}Cl(n,n)^{nat}Cl$	8	75
$^{nat}F(n,n')^{nat}F$	8	120

3.2.3 Neutron fluence and event rates

The detection efficiency curve of the SDD deviates from a perfect step function due to statistical fluctuations in the energy deposition process. Experimental results yield a best fit to the function $1-\exp[-\Gamma[E/E_{rec}-1)]$ for $E \geq E_{rec}$ with $\Gamma=4.2\pm0.3$, and 0 otherwise [9]. The calculated recoil distribution was convoluted with the efficiency curve when calculating the detector signal.

The on-detector calculated neutron energy distribution calculated with MCNPX is shown in Fig. 6 in the lethargy representation. It has the characteristic shape of a moderated neutron spectrum as can be found in nuclear fission reactors, with thermal (<0.5eV) and fast (>1MeV) fluence rates of $2.7x10^{-8}$ and $4.6x10^{-8}$ $cm^{-2}s^{-1}$, respectively. The contribution of the (α,n) from ^{238}U in the microphone is included in Fig. 6 in order to evidence the spreading of the source spectra features as neutrons are moderated in the hydrogenous gel.

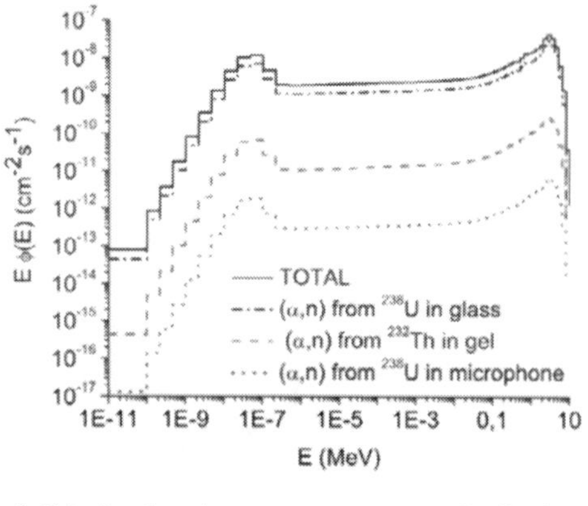

Fig. 6. Calculated on-detector neutron energy distribution, with discrimination of some contributing reactions.

The fluence rate beyond 43keV (the smallest E_{min}) is $7.0x10^{-8}cm^{-2}s^{-1}$, of which 99% originate from the borosilicate glass container. The contribution of each reaction in borosilicate is the following: 7% from ^{238}U spontaneous fission, 62% from ^{238}U decay-induced (α,n) production and 31% from ^{232}Th decay-induced (α,n) production.

The energy distribution of recoils was extracted with MCNPX/PoliMI, with neutron and recoil energy cuts at 40 and 8 keV, respectively. Table 4 summarises the event rate per unit of liquid mass due to the various reactions and materials. The total event rate is 20% higher than previous calculations [20] mostly due to the application of increased (α,n) yields, thereby confirming the uncertainty estimate of ~20% that was made then.

Table 4. Event rate (count $kg^{-1}d^{-1}$) from spontaneous fission and decay-induced (α,n) reactions due to the presence of emitters in the SDD materials. Secular equilibrium is assumed for ^{238}U and ^{232}Th decay chains.

Material	^{238}U sp.fiss.	^{238}U (α,n)	^{232}Th (α,n)	^{nat}Sm (α,n)
Borosilicate	3.75 $x10^{-2}$	0.288	0.144	-
Gel	<3.81 $x10^{-4}$	<5.53 $x10^{-4}$	<3.23 $x10^{-4}$	<3.74 $x10^{-8}$
Microphone	<1.91 $x10^{-6}$	<7.49 $x10^{-5}$	-	-
TOTAL	0.41±20%			

We thereby calculate a detector signal originating from its own emitters of $5.9x10^{-3}$ counts per day for a 1 liter device with 14.3 g of C_2ClF_5 (corresponding to the average mass in the experiment). The calculated neutron detection efficiency, i.e., counts per neutron entering the gel volume) is 0.6%. With a detector surface of ~667 cm^2 a detection limit of $\sim 2x10^{-8}$ n $cm^{-2}s^{-1}$ is derived. This is two orders of magnitude lower than the neutron fluence rate in underground facilities at moderate (1-2 km water-equivalent) depth [20, 42]. In spite of the reduced detection efficiency, the time required for the characterization of such neutron environments (~1 count/day) in 10 energy groups with 10 detectors operating in the varying temperature method is ~2 weeks, which is easily manageable with remote monitoring and control such as that employed e.g. in the SIMPLE experiments that run continuously over ~3 months.

Possibilities for future reduction of the background noise should focus on detector container materials at decreased emitter and Boron concentrations. Polymethylmethacrylate samples from various producers have been evaluated by neutron activation analysis yielding less than 0.05 ppm ^{238}U. With a (α,n) yield ~5% than that of borosilicate glass, an overall reduction by two orders of magnitude is in principle achievable - this corresponds to the upper limit for the signal induced by the detector gel itself.

4 Conclusions

The intrinsic neutron background signal of C_2ClF_5 SDDs was measured in an underground facility and simulated using the MCNPX/PoliMI code. The measured and calculated event rates were 0.15 and 0.41 count kg^{-1} d^{-1}, respectively. This level of agreement is satisfactory for rare event measurements: should there be one more neutron detected and the agreement would be within the uncertainty of the calculation (dominated by that in (α,n) yields). The detection limit in the order of 10^{-8} $cm^{-2}s^{-1}$ is is 2 orders of magnitude higher than the neutron environment in underground facilities at moderate depth. The device is therefore adequate to the characterisation of various facilities for low radiation background applications. If necessary, the current device can be upgraded for a significant background reduction simply by modifying the container material.

Acknowledgements

We thank Vitaly Kudryavtsev (University of Sheffield) for offering the improved version of SOURCES-4A and Gennadiy Vlaskin (ICTP, Moscow) for the interpretation of the (α,n) distributions. Luís Alves (C^2TN) is acknowledged for the microscope usage.
This work was supported by Fundação para a Ciência e Tecnologia (Portuguese Ministry of Science) via grants IF/00628/2012/CP0171/CT0008 and SFRH/BPD/94028/2013 and projects UID/Multi/04349/2013 and PTDC/EEI-ELC/2468/2014.

References

1. R.E. Apfel, Nucl. Instrum. Methods 162, 603 (1979)

2. H. Ing, H.C. Birnboim, Nucl. Tracks Radiat. Meas. **8**, 285 (1984)
3. F. d'Errico, Nucl. Instrum. Methods **B184**, 229 (2001)
4. H.W. Bonin, G.R. Desnoyers, T. Cousins, Radiat. Prot. Dosim. **46**, 265 (2001)
5. F. d'Errico, W.G. Alberts, G. Curzio, S. Guldbakke, H. Kluge, M. Matzke, Radiat. Prot. Dosim. **61**, 159 (1995)
6. F. d'Errico, A. Prokofiev, A. Sanniko, H. Schuhmacher, Nucl. Instrum. Methods **A505**, 50 (2003)
7. S.L. Guo, et al., Radiat. Meas. **31**, 167 (1999)
8. S. Archambault, et al., Phys. Lett. B **711**, 153 (2012)
9. M. Felizardo, et al., Phys. Rev. D **89**, 072013 (2014)
10. A.C. Fernandes, et al., E3S Web of Conferences: i-DUST 2016 (in press)
11. F. Seitz, Phys. Fluids **1**, 2 (1958)
12. F.J. Maringer, D. Arnold, U. Wätjen, Appl. Radiat. Isot. **109** (2016)
13. A. Lesea, K. Castellani-Coulié, G. Waysand, J. Le Mauff, C. Sudre, IEEE Trans. Nucl. Sci. **55**, 2148 (2008)
14. N. Fornengo, M. Regis, H.-S. Zechlin, J. Phys. Conf. Ser. **718** (2016)
15. http://lsbb.oca.eu
16. A.C. Fernandes, M. Felizardo, A. Kling, J.G. Marques, T. Morlat, Nucl. Instr. Meth. **A 623**, 960 (2010).
17. A.C. Fernandes, et al., E3S Web of Conferences **4**, 03003 (2014)
18. T.A Girard, M. Felizardo, A.C. Fernandes, T.A. Morlat, A.R. Ramos, Proc. i-DUST 2010, 03003 (2011). http://dx.doi.org/10.1051/idust/201103003
19. X-5 Monte Carlo Team, LA-UR-03-1987, Los Alamos National Laboratory, 2003
20. A.C. Fernandes, et al., Astrop. Phys. **76**, 43 (2016)
21. F.W. Stallmann, NUREG/CR-4349, ORNL/TM-9333, Oak Ridge National Laboratory, 1985.
22. L.R. Greenwood, R.K. Smither, ANL/FPP/TM-197, Argonne National Laboratory, 1985
23. D.M. Mei, C.F. Zhang, A. Hime, Nucl. Instrum. Methods **A 606**, 651 (2009). Data available at http://neutronyield.usd.edu
24. M. Felizardo, et al., E3S Web of Conferences: i-DUST 2016 (in press)
25. S.A. Pozzi, S.D. Clarke, W.J. Walsh, et al., Nucl. Instrum. Methods A 694, 119 (2012)
26. V. Tomasello, V.A. Kudryavtsev, M. Robinson, Nucl. Instrum. Methods **A 595**, 431 (2008)
27. T. Mayer-Kuckuk, *Kernphysik* (Teybner, Stuttgart, 1984)
28. M. Felizardo, et al., Nucl. Instrum. Methods **A 585**, 61 (2008)
29. T. Morlat, D. Limagne, G. Waysand, J.I. Collar, Nucl. Instrum. Meth. **A 560**, 339 (2006)
30. M. Felizardo, et al., Nucl. Instrum. Methods **A 589**, 72 (2008)
31. M. Felizardo et al., Phys. Rev. Lett. **105**, 211301 (2010)
32. T.A. Girard, et al., Phys. Rev. Lett. **108**, 259002 (2012)
33. M. Felizardo, et al., Astrop. Phys. (to be published)
34. A.R. Ramos et al., Radiat. Prot. Dosim. **115**, 398 (2005)
35. V.F. Weigel, Chem. Ztg. **102**, 287 (1978)
36. D.M. Mei, A. Hime, Phys. Rev. D **83**, 053004 (2006)
37. Y.F. Wang, V. Balic, G. Gratta, A. Fassó, S. Roesler, A. Ferrari, Phys Rev D **64**, 013012 (2001)
38. L.S. Waters, G.W. McKinney, J.W. Durkee, M.L. Fensin, J.S. Hendricks, M.R. James, R.C. Johns, D.B. Pelowitz, AIP Conf. Proc. **896**, 81 (2007)
39. P. Jagam, J.J. Simpson, Nucl. Instrum Methods **A 324**, 389 (1993). *Erratum:* Nucl. Instrum. Methods **A 334**, 657 (1993)
40. A.C. Fernandes, A. Kling, G.N. Vlaskin (this volume)
41. G.N. Vlaskin, Yu.S. Khomyakov, V.I. Bulanenko, Atomic Energy 117, 357 (2015)
42. S. Cebrián, B. Beltrán, J. M. Carmona, et al., Nuclear Physics B (Proc. Suppl.) 138 (2005) 65–67

Feasibility study on real-time γ -ray spectrum / dose measurement system

Mina Kobayashi[1,*], *Fuminobu* Sato[1], *Sachie* Kusaka[1] and *Isao* Murata[1]

[1] Division of Sustainable Energy and Environmental Engineering, Graduate School of Engineering, Osaka University, Yamada-oka 2-1, Suita, Osaka, 565-0871 Japan

Abstract. Recently, medical applications of radiation have been widely spread. However, exposure of medical staffs is sometimes not focused on because treatment of patients is the first priority. It is thus important to decrease exposure for them as much as possible. The purpose of this study is to develop a system which can measure energy spectrum and dose of gamma-rays at the same time in real time in medical application spots. As a result, the medical staff could be guided to aware the risk of radiation and finally the exposure dose to them could be suppressed substantially. We first decided to use a CsI(Tl) scintillator as the gamma-ray detection device. A Multi-Pixel Photon Counter (MPPC) was attached to the scintillator to detect gamma-ray signals. Pulse height spectra were measured with several standard gamma-ray sources. The detection efficiency and energy resolution were deduced from the measured results and the detection efficiency was compared with the calculation result by MCNP5. After evaluating the response function, the energy spectrum was derived with the spectrum type Bayesian estimation and the sequential Bayesian estimation procedure. From the result, it was confirmed that the sequential Bayesian estimation could be applied to real time measurement of gamma-ray energy spectrum and dose.

1 Introduction

In recent years, radiations have been used very often in medical field, *i.e.*, radiation therapy, production of radioactive drugs and so on. There are risks of radiation exposure to medical staffs in the field by secondary produced gamma-rays. However, the exposure dose of medical staffs is not so seriously focused on so far, because treatment of patients is the first priority in hospitals. The medical staffs notice their radiation exposure, however, work for patients devotedly. The purpose of this study is to develop a spectrometer (monitor) which can show gamma-ray energy spectrum and the dose simultaneously in real time to make them understand the risk of radiation appropriately and to finally decrease the exposure dose of the medical staffs. The staffs will be able to engage in medical treatments with radiations, recognizing their exposure dose and energy information (gamma-ray energy spectrum) at the same time for their own safety. The monitor thus has another advantage as in the following: In the past study, there were monitors which can indicate the pulse height spectrum and/or dose [1]. However, there were no monitors which can show the gamma-ray energy spectrum and dose at the same time. The monitor which we seek estimate the energy spectrum of gamma-rays, with which the dose can be evaluated accurately.

In this paper, the result of basic performance evaluation of a prototype gamma-ray spectrum and dose measuring system and conversion procedure from the pulse height spectrum to energy spectrum are described.

2 Measuring system

2.1 Design criteria of the system

The essential conditions for designing the spectrometer are as follows.

(1) The spectrometer is small and light for practical use by medical staffs.
(2) The elemental measuring device has high detection efficiency and good energy resolution so as to reproduce even energy spectrum of gamma-rays.
(3) In regard to the dynamic range, the upper energy of gamma-rays was set to 3 MeV, because radioactive nuclides, which are produced in the operation of accelerators for particle radiation therapy or making radioisotopes for medicine, can emit gamma-rays having an upper energy of around 3 MeV.

To meet these conditions, scintillators are clearly preferable. Candidate scintillators must have a similar performance to NaI(Tl) because NaI(Tl) is regarded as a convenient and most popular detector for measuring gamma-rays among various scintillators. But NaI(Tl) has deliquescency, hence the measurement system tends to be large and heavy, because a metal case is required to keep the NaI(Tl) crystal. Therefore, instead of NaI(Tl), CsI(Tl) is used in this study, because CsI(Tl) has almost the same performance as NaI(Tl). Moreover CsI(Tl) has

* Corresponding author: mkobayashi@qr.see.eng.osaka-u.ac.jp

no deliquescency. Then, we used a Multi-Pixel Photon Counter (MPPC) as a photon counting device, because the MPPC is really smaller and lighter than a photomultiplier tube (PMT).

In this study, as a prototype of the present measuring system of gamma-rays, we chose a CsI(Tl) crystal of 3 x 3 x 3 cm^3 and measured the basic performance of the CsI(Tl) attached to the MPPC.

2.2 Derivation of energy spectrum by Bayesian estimation

The present measuring system utilizes a CsI(Tl) detector to measure the pulse height spectrum of gamma-rays. After the pulse height spectrum is obtained, it is converted into the energy spectrum and furthermore from the energy spectrum the dose rate is estimated with the flux-dose conversion factor. The point in this study is the energy spectrum and dose can be evaluated at the same time by conducting conversion from the pulse height spectrum to the energy spectrum in real time. This is the goal of the present study.

For this purpose, we employ the Bayesian estimation process to convert from the measured pulse height spectrum to the energy spectrum continuously in real time. In our laboratory, the spectrum type Bayesian estimation has been used for years to estimate various energy spectra, in which as the first step the pulse height spectrum is measured and thereafter it is converted into the energy spectrum as the post-process. In this study, this flow is continuously conducted for each count during measurement by the sequential Bayesian estimation [2], the detail of which is described in the following.

2.2.1 Bayesian estimation

The Bayes theorem is known as a conditional probability in mathematics. There are two events A and B. The probability of B is $P(B)$ and the conditional probability of A for given B is $P(A|B)$. When $P(B)$ is not zero, $P(A|B)$ is written by following equation.

$$P(A|B) = \frac{P(B \cap A)}{P(B)} \quad (1)$$

From Eq. (1), the next equation can be derived by considering $P(B \cap A) = P(A \cap B) = P(B \mid A) \cdot P(A)$.

$$P(A|B) = \frac{P(B|A) \cdot P(A)}{P(B)} \quad (2)$$

In Eq. (2), when A_j (j = 1~n) instead of A is a mutually exclusive event and they include the whole events in the sample space S, an arbitrary event B is written by Eq. (3).

$$\begin{aligned} P(B) = {} & P(A_1) \cdot P(B|A_1) \\ & + P(A_2) \cdot P(B|A_2) \\ & + \cdots + P(A_n) \cdot P(B|A_n) \\ = {} & \sum_{j=1}^{n} P(A_j) \cdot P(B|A_j) \end{aligned} \quad (3)$$

From the above, a conditional probability of A_j for given B is written by Eq. (4).

$$P(A_j|B) = \frac{P(B|A_j) \cdot P(A_j)}{\sum_{j=1}^{n} P(B|A_j) \cdot P(A_j)} \quad (4)$$

This is the basic expression of the Bayesian estimation. In Eq. (4), $P(A_j)$ and $P(A_j|B)$ are called prior probability and posterior probability, respectively. In the general way of Bayesian estimation, first of all, a prior probability $P(A_j)$ is assigned an arbitrary probability value, and then $P(B|A_j)$ evaluated beforehand is substituted in Eq. (4). Finally a posterior probability $P(A_j|B)$ is obtained. By repeating this process, we can obtain more reliable probability.

In this study, by applying this essential process one after another at a time of count, we can estimate the energy spectrum in real time. Eq. (5) shows the relation of the response function, pulse height spectrum and energy spectrum in radiation measurement.

$$\begin{pmatrix} y_1 \\ \vdots \\ y_i \\ \vdots \\ y_n \end{pmatrix} = \begin{pmatrix} R_{11} & R_{12} & \cdots & R_{1j} & \cdots & R_{1n} \\ \vdots & & & \vdots & & \vdots \\ R_{i1} & & & R_{ij} & & R_{in} \\ \vdots & & & \vdots & & \vdots \\ R_{n1} & \cdots & & R_{nj} & \cdots & R_{nm} \end{pmatrix} \cdot \begin{pmatrix} \varphi_1 \\ \vdots \\ \varphi_j \\ \vdots \\ \varphi_n \end{pmatrix} \quad (5)$$

In Eq. (5), y_i is the pulse height spectrum called posterior probability in the Bayesian estimation. It means, when a spectrometer detects a pulse height signal at channel i, the signal would come from an event originated from energy E_j in the energy spectrum φ, which is called prior probability. R_{ij} means a likelihood, which is defined as the response function of the detector for each energy, E_j, in radiation measurements. We can derive the probability of the energy spectrum at j from a measured pulse height signal at i with the estimated response function of the detector, R_{ij}, in the following equation rewritten from Eq. (4).

$$P(j|i) = \frac{R_{ij}\varphi_j}{\sum_{j=1}^{n} R_{ij}\varphi_j} \quad (6)$$

The numerator of the right side means the conditional relative probability of a detected signal at i to be assigned to φ_j as understood from Eq. (5).

2.2.2 Spectrum type Bayesian estimation

A pulse height spectrum is measured in advance in the spectrum type Bayesian estimation. As in Eq. (5), it means after the measurement y_i has a certain count. With Eq. (6), it can be expected that the posterior probability would be estimated by Eq. (7).

$$P(j|i) = y_i \frac{R_{ij}\varphi_j}{\sum_{j=1}^{n} R_{ij}\varphi_j} \quad (7)$$

By summing up y_i for pulse height i, the following Eq. (8) can be obtained.

$$P(j|i) = \sum_{i=1} y_i \frac{R_{ij}\varphi_j}{\sum_{j=1}^{n} R_{ij}\varphi_j} \quad (8)$$

This is known to be an extended interpretation of the Bayesian estimation. By using the posterior probability as the prior probability, estimation can be carried out repeatedly. After an appropriate number of iterations, the converged estimation would be obtained. The spectrum type Bayesian estimation is usually applied to radiation measurements. However, in the present study, estimation should be carried out continuously during the measurement. For that purpose, the sequential Bayesian estimation was adopted instead as detailed in the next section.

2.2.3 Sequential Bayesian estimation

The first step of the sequential Bayesian estimation is to set an arbitrary initial spectrum to φ_j. We then calculate the porior probability of φ at energy j from a measured signal y at channel i with the response function R_{ij} using Eq. (6). As a result, the φ_j is adjusted by adding one contribution of the detected signal at channel i. By repeating this process for every count, we could obtain the converged energy spectrum, φ_j. However, in reality this process is known to be not properly working as detailed in Ref. 2 by Iwasaki. Then as pointed out by Iwasaki, we introduce a factor α which adjusts the rate of correction to the prior probability. It can control how strongly the prior can affect compared to the next posterior probability [2]. The estimation is carried out by Eq. (9).

$$\varphi_j = (1-\alpha)\varphi_j + \alpha \cdot \frac{R_{ij}\varphi_j}{\sum_{j=1}^{n} R_{ij}\varphi_j} \quad (9)$$

There is no precedent which showed the energy spectrum could be estimated accurately by the sequential Bayesian estimation. Generally, it is known that the convergence of the sequential Bayesian estimation is very slow. However we think even if the spectrum estimation is not completely converged, we can estimate the dose, because the dose is theoretically an integral quantity of the energy spectrum and it has to be converged rather faster.

3 Experimental methods

We made up of a prototype measuring device with a CsI(Tl) crystal (3 cm x 3 cm x 3 cm) and MPPC (C11206-4689) as shown in Fig. 1.

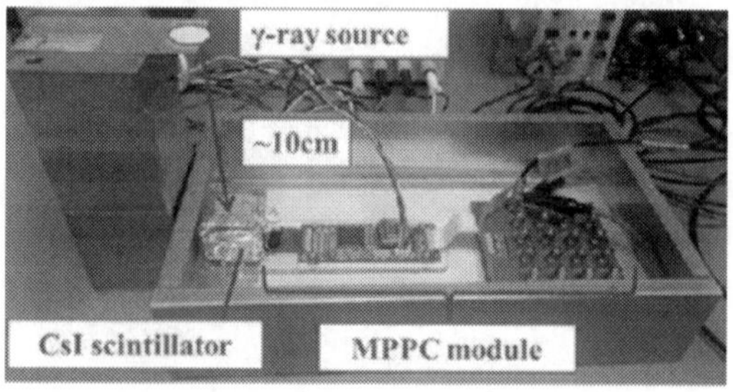

Fig. 1. Prototype system

The pulse height spectra were measured with several standard gamma-ray sources (^{22}Na, ^{60}Co, ^{133}Ba, ^{137}Cs and ^{241}Am) to check the detection efficiency and energy resolution. The details of the standard gamma-ray sources are summarized in Table 1.

Table 1. Detail of the standard gamma-ray sources

Nuclide	Energy (MeV)	Half-life	Radioactivity (Bq)
^{22}Na	0.551	2.601y	3.57E+05
	1.274		
^{60}Co	1.173	5.271y	6.84E+05
	1.332		
^{133}Ba	0.356	10.52y	6.84E+05
^{137}Cs	0.662	30.07y	2.50E+05
^{241}Am	0.059	432.2y	3.77E+05

The CsI(Tl) crystal was attached to a MPPC and was covered with Teflon tape and aluminum foil to shield light. They were placed in an aluminum box together with a power control module. A standard gamma-ray source was positioned at 10 cm from the CsI(Tl) scintillator surface. Other equipment like a power supply, multi-channel analyzer (MCA) and so on are outside the box. Fig. 2 shows the block diagram of the electronic circuit of the prototype detector. Pulse height spectra were measured by the MCA controlled by a laptop computer.

The measurements were completed when the net area count of the main peak exceeds 10000 counts. From the measured pulse height spectrum, detection efficiency and energy resolution were determined. The measured detection efficiencies were compared with simulations by MCNP5 (Monte Carlo N-Particle Transport Code) [3] to verify the validity of the code and the performance to apply to simulations of other size crystals.

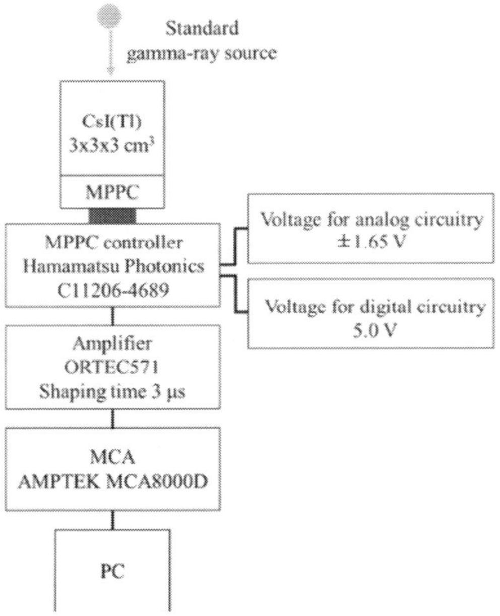

Fig. 2. Block diagram of the electronic circuit of the prototype detector.

4 Result and discussion

4.1 Basic performance of CsI(Tl)

We measured pulse height spectra for ^{22}Na, ^{60}Co, ^{133}Ba, ^{137}Cs and ^{241}Am, and compared the results with NaI(Tl). The comparison result is described in Figs. 3 and 4. These two figures indicate the performance of the CsI(Tl) scintillator, *i.e.*, the detection efficiency and energy resolution, was better than NaI(Tl), which is a normally utilized scintillator for gamma-ray dose measurement.

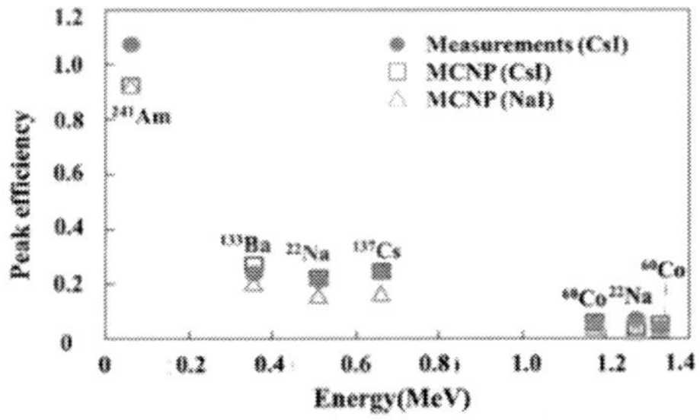

Fig. 3. Measured detection efficiency of CsI compared with the simulation result of NaI and CsI by MCNP5. The agreement is excellent except ^{241}Am between measurement and calculation for CsI. Also the performance is better than NaI having the same size.

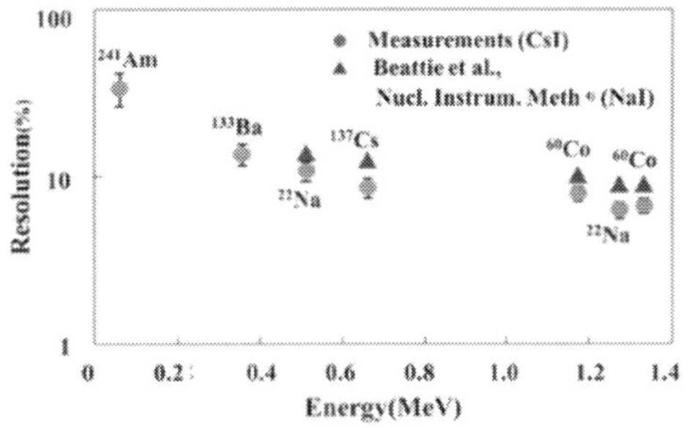

Fig. 4. Measured energy resolution of CsI compared with calculated values of NaI by Beattie et al.[4]. The energy resolution of CsI is found to be better than NaI.

Fig. 5 also shows comparison of the measured pulse height spectrum for ^{22}Na with the calculation result by MCNP5, in which two strong photo peaks of 0.511 MeV and 1.27 MeV are clearly observed. From the figure, an excellent agreement was obtained. Also, in the measured spectrum, a clear backscattering peak is observed. This should be removed in the real detector response function. In the present case, the backscattering peak was correctly reproduced by appropriately considering surrounding materials of the scintillator like a desk and other massive items. It means it would be possible to evaluate the detector response function properly by considering the precise structure of the present detector including the casing and inside. From the present comparison, the validity of the calculation code, MCNP5, was finally confirmed.

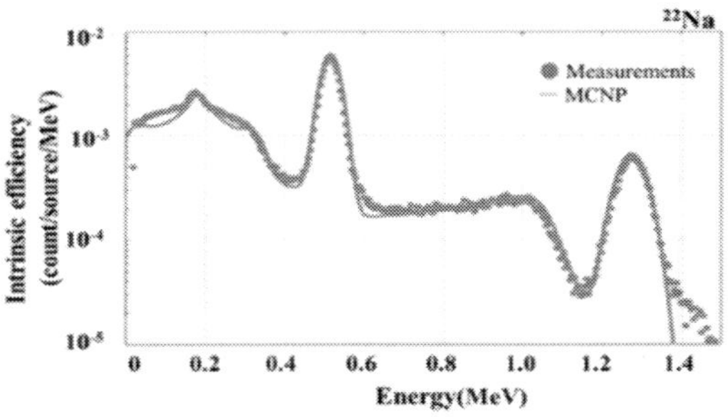

Fig. 5. Measured pulse height spectrum for ^{22}Na compared with MCNP5 calculation.

4.2 Evaluation of response function

To carry out the Bayesian estimation, the response function of the prototype CsI(Tl) spectrometer was evaluated. Fig. 6 shows examples of pulse height spectra for monoenergetic gamma-rays up to 3 MeV. Pulse height spectra evaluated for every 10 keV from 100 keV to 3 MeV are compiled to make the response function of the present prototype spectrometer. In Fig. 6, we can find that a backscatter peak appears for each energy, and single or double escape peak appear for energies larger than around 2 MeV. Moreover especially at 3 MeV, the double escape peak is larger than the photoelectric peak.

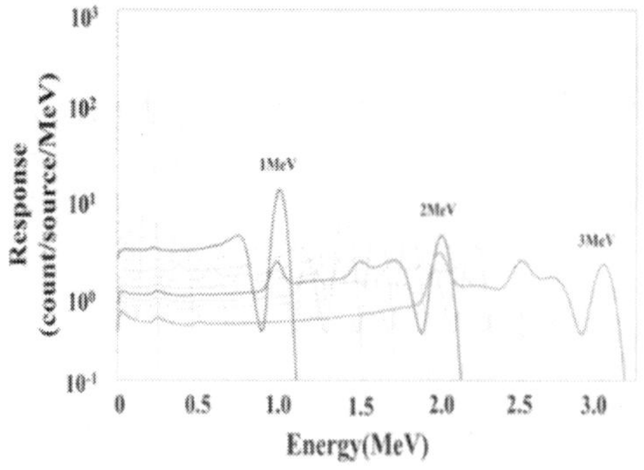

Fig. 6. The response function of a 3 x 3 x 3 cm^3 CsI scintillator to monoenergetic gamma-rays up to 3 MeV.

From the above result, the energy spectrum is clearly different from the pulse height spectrum, and cannot be estimated directly from the pulse height spectrum. Therefore the unfolding process is needed to derive the energy spectrum.

4.3 Unfolding by Bayesian estimation

The measured pulse height spectra were unfolded by the Bayesian estimation process to estimate the gamma-ray energy spectra with the evaluated response function described in Sec. 4.2. In this study, we carried out comparison between the result of the spectrum type Bayesian estimation and the sequential Bayesian estimation to test and confirm the performance of the latter estimation method, because the spectrum type Bayesian estimation is commonly used in radiation measurements.

Fig. 7 shows the unfolding result of the pulse height spectrum for ^{22}Na shown in Fig. 5 with the spectrum type Bayesian estimation.

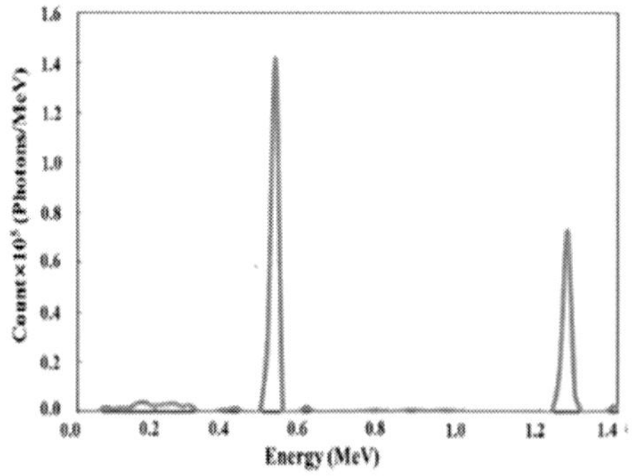

Fig. 7. The energy spectrum for ^{22}Na calculated by the spectrum type Bayesian estimation.

In Fig. 7 clear two peaks of 0.511 MeV and 1.27 MeV are observed, which correspond to emitted two gamma-rays from ^{22}Na. There are no any other significant peaks seen except the two peaks. Moreover the ratio of the number of emitted gamma-rays is estimated to be 0.511 MeV :1.27 MeV = 1.73:1, which was calculated by making the ratio of the number of the peak net counts for the two peaks. The measured ratio is in good agreement with the literature value of 0.511 MeV : 1.27 MeV = 1.8:1.

In the sequential Bayesian estimation, at first we need to determine α value in Eq. (9) so as to give the most accurate result. At the same time, the minimum required number of counts should be evaluated for real applications. They were finally determined by considering the following points.

(i). Deviation between measured spectra (M_i) and folding spectra (y_i) of unfolding result with Eq. (5)

(ii). Reproducibility of the emission rate of gamma-rays for spectra after unfolding

In regard to point (i), the deviation for α value and various number of counts for ^{133}Ba is shown in Fig. 8 as an example. Here, the deviation is calculated as $1/N\sqrt{\sum_i (M_i - y_i)^2}$. N means the total number of counts of gamma-rays in the experiment. As Fig. 8 shows, the deviation becomes less than 1% in 50000 counts, moreover, the deviation becomes large for α > 0.1. A similar tendency was seen for other nuclides as well.

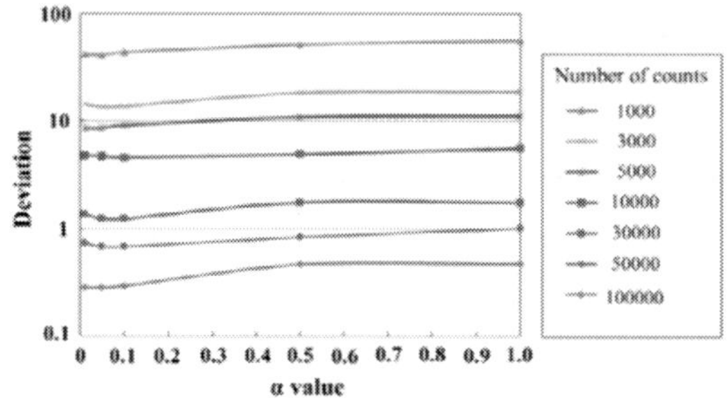

Fig. 8. Deviation for ^{133}Ba for various numbers of counts as a function of α.

Therefore we examined point (ii) to determine the best α value for α ≤ 0.1. Table 2 shows C/E indicating the reproducibility of the emission rate of gamma-rays. "C" is the evaluated number of emitted gamma-rays in the literature. "E" is the number of emitted gamma-rays summed up in each peak of the unfolded spectrum when the iteration number is 1000 in the case of the spectrum type Bayesian estimation and when the number of measured signals is 10000, 50000, 100000 with the sequential Bayesian estimation.

From Table 2, the precision of the sequential Bayesian estimation is confirmed to be the same level as the spectrum type Bayesian estimation. By considering the results for the three α values of 0.01, 0.05 and 0.1, the best value is 0.05 over 50000 counts. On the other hand, if examining the variance of C/E, the variance is converged for the number of counts over 50000 as shown in the bottom of Table 2, where the variance is calculated as $1/N'\sum(C/E-1)^2$. N' means the number of data of C/E. In conclusion, in the present detector setup, 50000 counts are consequently thought to be enough for real applications for α = 0.05.

Table 2. Comparison of the number of emitted gamma-rays (C/E) estimated by both Bayesian methods at 1000 times iterations or 10000, 50000 and 100000 counts and variance of C/E.

Nuclide	Energy (MeV)	The spectrum type Bayesian estimation (1000 times iterations)	The sequential Bayesian estimation (10000 counts)			The sequential Bayesian estimation (50000 counts)			The sequential Bayesian estimation (100000 counts)		
			α=0.01	α=0.05	α=0.1	α=0.01	α=0.05	α=0.1	α=0.01	α=0.05	α=0.1
^{22}Na	0.511	1.41	1.25	1.09	0.98	1.28	1.16	1.07	1.51	1.16	1.10
	1.274	1.36	1.55	1.61	1.74	1.46	1.28	1.27	1.10	1.06	1.10
^{60}Co	1.173	1.31	1.38	1.29	1.23	1.20	1.38	1.54	1.19	1.03	0.92
	1.332	1.28	1.32	1.36	1.25	1.40	1.05	0.87	1.27	1.15	1.20
^{133}Ba	0.356	0.72	0.80	0.77	0.60	0.76	0.68	0.48	0.65	0.54	0.49
^{137}Cs	0.662	0.81	0.82	0.75	0.73	0.80	0.71	0.68	0.77	0.69	0.68
Variance of C/E		0.10	0.11	0.12	0.15	0.10	0.07	0.13	0.09	0.06	0.07

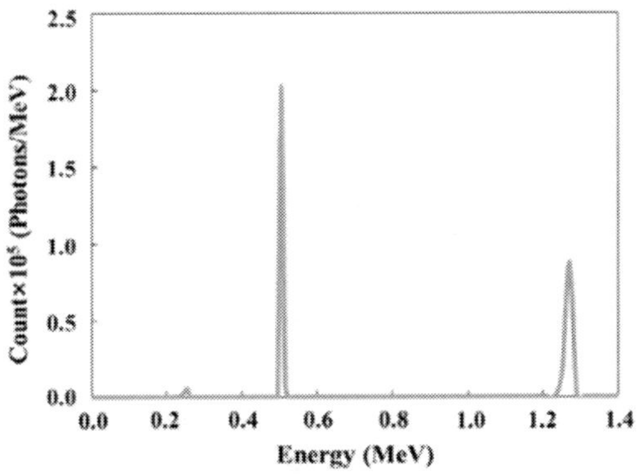

Fig. 9. The energy spectrum for ^{22}Na obtained by unfolding by the sequential Bayesian estimation when the number of measured signals is 50000 counts and α = 0.05.

Fig. 9 shows an example of the unfolding result of the pulse height spectrum for ^{22}Na with the sequential Bayesian estimation in 50000 counts, α = 0.05. In the figure, two peaks of 0.511 MeV and 1.27 MeV are observed, which correspond to the two emitted gamma-ray energies for ^{22}Na. The obtained ratio of the number of emitted gamma-rays for the two energies is 0.511 MeV: 1.27 MeV = 1.99:1. Though the ratio is a little larger than the literature value, the agreement is acceptable. Also, it was found that the result of the sequential Bayesian estimation fluctuates depending on the number of counts, that is, the estimation is a little unstable for the detector count. We should carefully examine this point in the next step. However, from the examination carried out so far in such as the result of Fig. 9, it was confirmed that the sequential Bayesian estimation would have a potential ability to apply to real-time and simultaneous gamma-ray dose and spectrum estimation.

5 Conclusions

We have been developing a new spectrometer which can display the energy spectrum and dose of gamma-rays at the same time in real time. As a detection device of gamma-rays, CsI(Tl) scintillator (3 x 3 x 3 cm^3) was chosen because the basic performance of CsI(Tl), *i.e.*, the detection efficiency and energy resolution, was similar to commonly used NaI(Tl) scintillator. The CsI(Tl) crystal was attached to a MPPC in an aluminum box to constitute a prototype detector. By using this detector, measurements were carried out to test the basic performance. After that, the possibility of data processing in real time was examined by application of the Bayesian estimation. As the result, we found that the sequential Bayesian estimation could realize measurement of gamma-ray energy spectrum and dose in real time. As a future work, we will carry out a more detailed measurement and thereafter, develop an actual spectrometer.

References

1. T. Nakamori, et al. *Development of a gamma-ray imager using a large area monolithic 4 x 4 MPPC array for a future PET scanner*, JINST **7** C01083(2012)
2. S. Iwasaki, *A new approach for unfolding problems based only on the Bayes' Theorem, Proceedings of the 9th International Symposium on Reactor Dosimetry*, pp. 245-252 (1996)
3. X-5 Monte Carlo Team, *MCNP – A General N-Particle Transport Code, Version 5,* Los Alamos National Laboratory, LA-UR-03-1987 (2003)
4. R.J.D.Beattie, J.Byrne., *A Monte Carlo program for evaluating the response of a scintillation counter to monoenergetic gamma rays*, Nuclear Instrument and Method **104**, 163-168, (1972)

Thought Experiment to Examine Benchmark Performance for Fusion Nuclear Data

Isao Murata[1,a], Masayuki Ohta[2], Sachie Kusaka[1], Fuminobu Sato[1], Hiroyuki Miyamaru[3]

[1]*Division of Electrical, Electronic and Information Engineering, Osaka University, Yamada-oka 2-1, Suita, Osaka, 565-0871 Japan*
[2]*Center for Atomic and Molecular Technology, Osaka University, Yamada-oka 2-1, Suita, Osaka, 565-0871 Japan*
[3]*Organization for Industry, University and Government Cooperation, Osaka Prefecture University, Sakai, Osaka 599-8570, Japan*

Abstract. There are many benchmark experiments carried out so far with DT neutrons especially aiming at fusion reactor development. These integral experiments seemed vaguely to validate the nuclear data below 14 MeV. However, no precise studies exist now. The author's group thus started to examine how well benchmark experiments with DT neutrons can play a benchmarking role for energies below 14 MeV. Recently, as a next phase, to generalize the above discussion, the energy range was expanded to the entire region. In this study, thought experiments with finer energy bins have thus been conducted to discuss how to generally estimate performance of benchmark experiments. As a result of thought experiments with a point detector, the sensitivity for a discrepancy appearing in the benchmark analysis is "equally" due not only to contribution directly conveyed to the deterctor, but also due to indirect contribution of neutrons (named (A)) making neutrons conveying the contribution, indirect controbution of neutrons (B) making the neutrons (A) and so on. From this concept, it would become clear from a sensitivity analysis in advance how well and which energy nuclear data could be benchmarked with a benchmark experiment.

1 Introduction

There are a lot of nuclear data measurements, named differential experiments, carried out so far with DT neutron sources. Also there exist many benchmark experiments, i.e., integral experiments, to verify the nuclear data validity. They are of course aiming at fusion reactor development finally. Differential experiments can check the accuracy of evaluated nuclear data at 14 MeV directly. Integral experiments are vaguely expected to validate the nuclear data also below 14 MeV as well as at 14 MeV. However, no precise studies concerning this matter were performed systematically so far. On the other hand, as for numerical calculations with transport codes, it is quite common to analyze benchmark experiments using Monte Carlo codes. It would become very useful if benchmark performance for nuclear data could directly be analyzed beforehand from the Monte Carlo calculations. In the author's group, various fusion neutronics experiments have been carried out so far with a 14 MeV neutron source, OKAVIAN. We thus started to examine how well integral benchmark experiments with DT neutrons can play a benchmarking role for energies below 14 MeV using a general purpose Monte Carlo code MCNP.[1)]

We calculated an energy spectrum of neutrons making neutrons or gamma-rays detected by a detector in a leakage neutron or gamma-ray spectrum measurement. This spectrum was named "neutron spectrum before last collision", φ_{last}, in this study. Calculation of φ_{last} was done with a point detector as in the following procedure. For each detected contribution, there exists a neutron or gamma-ray, named (A), conveying the contribution to the detector. The information, i.e., the contribution and energy of (A) is summed up to make a measured spectrum in the calculation. Now, we think of another neutron (B), making the neutron or gamma-ray, (A), conveying the contribution. The spectrum of φ_{last} can be evaluated by contribution conveyed by particle (A) with the energy of particle (B) instead of (A).

As a result of the series study[2,3)], especially for DT neutron benchmark experiments, it was found that for secondary gamma-ray spectrum measurements nuclear data at around 14 MeV were dominantly benchmarked. In return, for neutron spectrum measurements those below 14 MeV were also benchmarked fairly well in addition to 14 MeV, because neutrons could rapidly be moderated in a sample material. In conclusion, to make benchmark experiments efficient, a spectrum shifter made with beryllium should be used to make the incident neutron spectrum softer especially for gamma-ray spectrum measurements. For neutron spectrum measurements, it was pointed out that reduction of experimental time could be expected.

a Corresponding author: murata@see.eng.osaka-u.ac.jp

In benchmark analyses carried out so far[2,3], we set only one energy boundary at 10 MeV to examine φ_{last}, meaning the number of energy groups is just two (14 MeV and other than 14 MeV), because the incident particle is a mono-energetic 14 MeV neutron and we wanted to know simply if the measured neutron spectrum below 14 MeV could validate the nuclear data not only for 14 MeV but also below 14 MeV. This concept worked well for performance analysis of the benchmark with DT neutrons. But in general cases it seemed to be insufficient to employ one boundary. The key point is how we should make a discussion in the general cases in the same manner as the case of two energy groups. As shown in **Fig. 1** which is the case of a leakage neutron spectrum by DT neutrons, it is found that neutrons (B) are already staying in a bin close to the energy bin of neutrons (A). In this sample size (about several tens cm thick), it can be found that the neutron spectrum mostly reaches its equilibrium one. For gamm-ray spectrum case in **Fig. 2**, in contrary, gamma-rays produced by 14 MeV neutrons are found to be dominant. It means especially for neutrons that finer energy meshes are essential and indispensable in general cases. Consequently, two problems were pointed out to discuss the benchmark performance in Ref. 3).

Here, one of the two is introduced, because it is directly related to the present objectives. In our previous studies the problem did not explicitly appear, because the energy boundary is only at 10 MeV, meaning that only 14 MeV neutrons and neutrons below 14 MeV exist. However, if using finer energy bins, there must exist various energetic neutrons, i.e., neutrons (A) conveying contribution to a certain energy mesh of the spectrum, neutrons (B) making neutrons (A), neutrons (C) making neutrons (B) and so on. The key issue is that one has to think of not only neutrons (A) but also other neutrons created during the whole particle history from the source.

In the present study, we carry out a thought experiment to precisely examine which neutron, i.e., nuclear data for the neutron energy, created during a transport history is benchmarked by the measured detector contribution. The change appearing in the measured spectrum is examined, if the cross section value were artificially changed to be a little over- or underestimated. From the result, the benchmark performance sensitivity due to discrepancy between measurement and calculation is discussed. Investigating the whole thought experiment, we discuss how well neutrons created during the transport are benchmarked. And finally we examine that we reach the conclusion that there would be no need to calculate adjoint function for benchmark performance (sensitivity) analysis.

2 Thought Experiment

As a thought experiment, we think of a DT neutron incidence to a sample as shown in **Fig. 3**. The sample is so thin that it can be assumed that scattering occurs just twice at maximum. Also, assuming that emitted secondary neutrons via first and second scattering can be measured separately. It means that the spectrum from the first scattering corresponds to double differential cross section (DDX) and the second one is a distortion term by scattering. The mixture of the two is a neutron spectrum to be measured.

In the separate measuremtn, we think of how effectively each spectrum plays a benchmarking role. Naturally, the neutron spectrum made directly from the first scattering plays a benchmarking role for the incident 14 MeV neutrons (n_0). In this case the contribution is conveyed by neutrons (n_1) to the detector. The second spectrum is made up of contributions created by neutrons (n_1) emitted via nuclear reactions of the incident neutrons (n_0). The contributions are delivered by neutrons (n_2). Therefore, the spectrum can play a benchmarking role for n_1. The point is 'whether the second one could have a benchmarking role also against the incident neutrons (n_0)'. Seemingly, it is quite likely, because neutrons (n_1) producing the second spectrum are originally created by neutrons (n_0). Now, the problem is, if so, 'whether the benchmark performance degree for both of n_0 and n_1 is the same'.

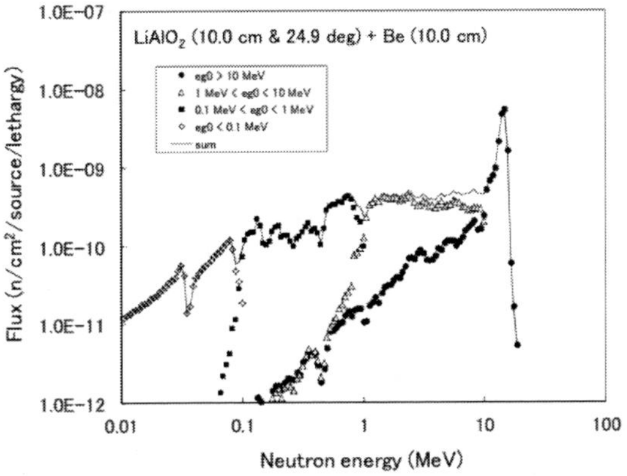

Fig. 1. Calculated leakage neutron spectrum for Li_2TiO_3 by DT neutrons. 'eg0' is the energy of neutron spectrum before last collision, φ_{last}, and 'sum' is the calculated neutron spectrum. ●,△,■ and ◇ indicate to partial spectra made by each φ_{last}.

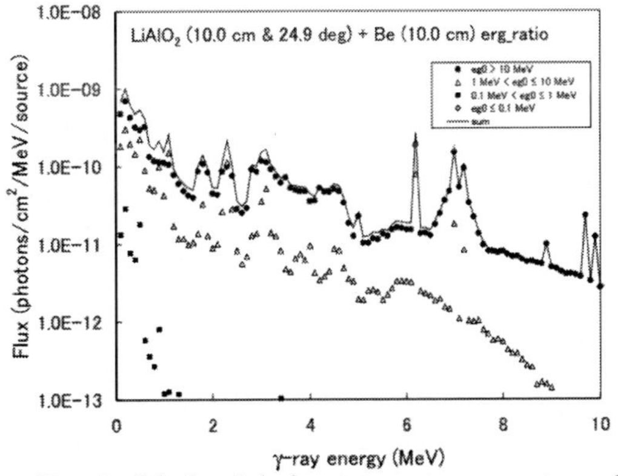

Fig. 2. Calculated leakage gamma-ray spectrum for Li_2TiO_3 by DT neutrons. Other condition is the same as Fig. 1.

Table 1 Error sensitivity analysis result in a simple benchmark model.

	Cross section assumption		Comparison of experiment and calculation		
	14 MeV	Below 14 MeV	n_1 spectrum	n_2 spectrum	Total
(1)	−10 %*1	Correct	−10 %*2	−10 %	−10%
(2)	Correct	−10 %	Correct	−10 %	Underestimated a little*3
(3)	−5 %	−5 %	−5 %	−5 ~ −10%	−5 ~ −10%*4

*1 Cross section is assumed to be underestimated. *2 Underestimation of calculation
*3 Because n_1 spectrum is dominant. *4 The value is expected to be close to -5%.

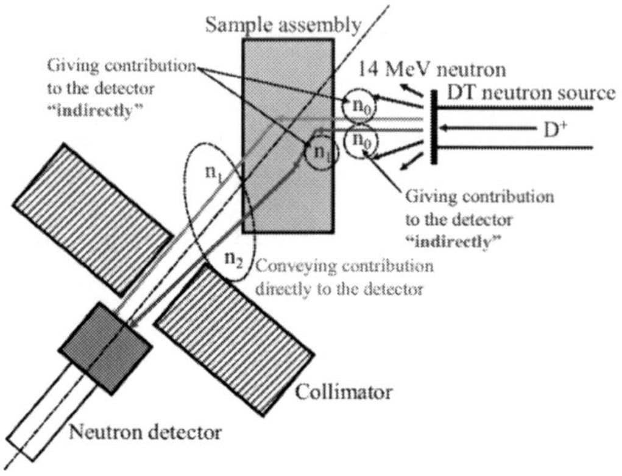

Fig. 3 Schematic arrangement of neutron benchmark experiment for the thought experiment.

2.1 Simple benchmark sensitivity

Now thinking of a similar benchmark experiment to Fig. 2 with a relatively thin slab sample, the benchmark sensitivity is examined with a simple condition as follows: Assuming cross sections except 14 MeV are correct, and that of 14 MeV is underestimated by 10 % (By this, the produced DDX by 14 MeV neutron shows a little smaller spectrum by 10 %.) In this case, if comparing the measured spectrum with the calculation, the calculated n_1 spectrum shows 10 % smaller in the whole energy region. This means that the number of neutrons making the second spectrum is decreased by 10 %. Thus, the spectrum of n_2 neutrons produced by n_1 shows 10 % smaller also. As a total of n_1 and n_2, the estimated measured spectrum shows 10 % smaller. The result is summarized in **Table 1**. The above result indicates that both spectra due to the first and second scatterings play equally the benchmarking role at the energy of 14 MeV.

Inversely, assuming the cross section of 14 MeV is correct and others below 14 MeV are smaller by 10 %, a similar simple thought experiment is done as follows: The n_1 spectrum calculated is clearly correct (showing the same as the measurement). As for the n_2 spectrum, the absolute value becomes smaller by 10 % because of 10% smaller cross sections for neutrons n_1, although the intensity of neutrons n_1 (produced as a result of the first scattering by n_0) making the n_2 spectrum is correct. The summed-up spectrum to be measured shows a smaller value by less than 10 %, probably the absolute value is smaller, because in reality the spectrum of n_1 is dominant.

Then, in case of the cross section being 5 % smaller in both energy ranges, meaning 5 % smaller in the whole energy range, the n_1 spectrum shows 5 % smaller. Because the spectrum of neutrons making the n_2 spectrum is also 5 % smaller, the n_2 spectrum has to be smaller by 5 %. The total result shows a decrease of between 5 and 10 %, probably close to -5%.

In summary, when the calculated spectrum shows a 10 % smaller value in energies below the elastic peak of 14 MeV compared to an experiment, the benchmark activity is to find out the reason why the discrepancy happens. From the above sensitivity analysis, it would be the most likely that the cross section of 14 MeV would be wrong (smaller), then second-likely that both would be underestimated, and the probability is the smallest that the cross sections only below 14 MeV are incorrect.

The important conclusion is that this kind of evaluation is of course not possible if only neutron spectrum before last collision is taken into consideration.

2.2 Precise benchmark sensitivity

Based on the results obtained above, a more precise benchmark sensitivity is next examined with a practical model, that is, a leakage spectrum measurement with a slab assembly by 14 MeV neutron incidence. Here, we focus on an energy bin of 5 to 6 MeV, for example, in the measured spectrum. Similar to the previous section, the number of scatterings is assumed to be up to twice. In this case, some measured neutrons may be produced by 14 MeV neutron's nuclear reactions, and some others may be via nuclear reactions by neutrons below 14 MeV, which are created by the 14 MeV neutron's nuclear reactions. The former measured neutrons are named n_1, and the latter is n_2, in the same manner as the previous section. Now, we employ the following assumption additionally that the ratio of the detector contribution in the energy bin conveyed by n_1 is $1-a$ ($0 < a < 1$), and that by n_2 is a as shown in **Fig. 4**. So, if the C/E is 0.9 and the error contribution is simply proportional to contribution of the whole spectrum, it could be expected

DOI: 10.1051/epjconf/201715307015

Table 2 Error sensitivity analysis result in a precise benchmark model.

Measured	n_1 spectrum	n_2 spectrum	Error sensitivity to the energy bin	Total	Ratio	Expectation
10 % underestimated nuclear data at 14 MeV and correct for others to examine n_0 sensitivity						
n_1	−10 %	–	$-10(1-a)$ %	−10 %[*2]	1	$1-a$
n_2	−10 %[*1]	−10 %	$-10a$ %			
Correct nuclear data for 14 MeV and 10 % underestimated for others to examine n_1 sensitivity						
n_1	0	–	0	$-10a$ %[*3]	a	a
n_2	0[*1]	−10 %	$-10a$ %			

*1 As a source to make n_2. *2 n_0 sensitivity *3 n_1 sensitivity

that the origin of the error, 10%, could be divided into $-10(1-a)$ % for 14 MeV neutrons and $-10a$ % for n_1, because n_1 is produced by a 14 MeV neutron and n_2 is produced by n_1. However, as detailed in the following, it will be made clear that this is not the case.

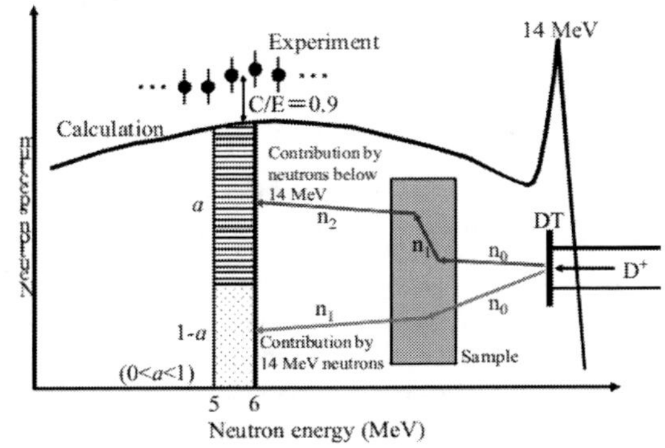

Fig. 4 Detector contribution ratio a in neutron spectrum.

In the condition above, from the result in Sec. 2.1, it can be expected that the probability is larger that the cross section of 14 MeV is incorrect. In other words, the crucial point is that n_1 is the source term to make n_2. Again we think of the condition, C/E=0.9.

At first it is assumed that the cross section at 14 MeV is underestimated by 10 % and others are correct. In this case, as shown in **Table 2**, the calculated n_1 spectrum shows an underestimated result by −10 %, and the error sensitivity of bin 5~6 MeV due to n_1 is estimated to be $-10(1-a)$ %. Though cross sections of scattering induced by n_1 are correct, the spectrum of n_1 as a source term for the following nuclear reaction is smaller by −10 %. The calculated spectrum of n_2 thus becomes −10 %, because other cross sections than 14 MeV are assumed to be correct. Consequently, The error sensitivity of bin 5~6 MeV due to n_2 is $-10a$ %. As a result, the total sensitivity of the bin is calculated by summing up n_1 and n_2 error sensitivities, i.e., −10 %.

Second, it is assumed that the cross section at 14 MeV is correct and others are underestimated by 10 %. In this case, the calculated spectrum of n_1 is correct, and the error sensitivity of bin 5~6 MeV due to n_1 is estimated to be zero. The spectrum of n_1 as a source term for the following nuclear reaction is also correct. However, the spectrum of n_2 possesses −10 % error, because other cross sections than 14MeV are assumed to be underestimated by 10%. The error sensitivity of bin 5~6 MeV due to n_2 is thus $-10a$ % and the total sensitivity of the bin is thus $-10a$ %.

As a sensitivity ratio, 14 MeV : Others = 1 : a, meaning the probability is higher that the cross section at 14 MeV is not correct, as expected also in Sec. 2.1. However, as described earlier it was expected the ratio is $1-a$: a. Both conclusions do not agree with each other. In the next section, the reason why the disagreement appears will be made clear.

2.3 Benchmark sensitivity contribution of each scattering neutron

In the beginning of Sec. 2.2, out of the contribution in the bin 5~6 MeV, it was seemingly expected that the ratio of $1-a$ came from n_1 and ratio, a, from n_2. However, this was available in case that we focused just on neutrons before last collision. In the conclusion of Sec. 2.2, considering another nuclear reaction forming spectrum n_2, the error sensitivity became

$$14\ \text{MeV} : \text{Others} = 1 : a \qquad (1).$$

In fact, this sensitivity is deeply related to the place where detector contribution comes from. Practically, n_2 surely conveys contribution a in the energy bin 5~6 as shown in Fig. 4. However, n_2 is created by n_1 via nuclear reaction. It means that it can be said that n_1 would convey the contribution indirectly to the detector. More precisely, it can be concluded from a physical consideration with 'importance' proposed by Lewins[4] with the result in Table 2 that "Error sensitivity for a certain neutron is proportional to the total detector contribution neutrons give directly and indirectly." Concretely, the detector contribution 14 MeV neutrons, n_0, give includes partly contribution, $1-a$, through neutron n_1 as shown in **Table 3**. This is a direct contribution made by a 14 MeV neutron. And there is another contribution of n_1, a, conveyed by n_2. n_2 is made by n_1's nuclear reaction. This is also a direct contribution of n_1 conveyed by n_2. Since n_1 is created by n_0, it can be thought that n_0 has an indirect contribution a. This is the same concept as 'importance' proposed by Lewins. As a result, n_0 has direct contribution, $1-a$, and indirect contribution, a, leading to the conclusion that n_0 has the total contribution,

$$(1-a)+a=1 \quad (2).$$

On the other hand, the contribution of other neutrons, n_1, is just a, directly made by themselves. Consequently, the detector contribution ratio can be estimated to be

$$14 \text{ MeV} : \text{Others} = 1 : a \quad (3),$$

as summarized in Table 3. This result is surprisingly the same as the error sensitivity ratio discussed in Sec. 2.2 and shown in Table 2.

Now, the problem is why the above expectation holds. The error sensitivity of 14 MeV neutrons in the case of assuming cross sections except 14 MeV are correct and that of 14 MeV is underestimated by 10 %, described in the previous section, shows

$$-10(1-a)\,\% + (-10a\,\%) = -10\,\% \quad (4).$$

It seems that a sensitivity of neutrons other than 14 MeV, that is, $-10a$ % is added. This is just caused by the fact that 14 MeV neutrons produce n_1. More practically, underestimation of cross section at 14MeV propagates in the lower energy neutrons, n_1, created by n_0. This is the indirect contribution. If removing an initially given error of −10 %, the total ratio unity in Table 2 is exactly the same as the contribution summation calculation just above. On the other hand, the error sensitivity of other neutrons n_1 in Sec. 2.2 is $-10a$ %. This just means if removing −10 %, the term a is left, which is similarly the same as the contribution in Table 3.

Table 3 Direct/indirect contribution ratio to the detector.

Contribution	Direct	Indirect	Total
14 MeV (n_0)	$1-a$	a	1
Others (n_1)	a	–	a

The discussion carried out so far concludes as follows: In case of benchmarking the nuclear data with a measured neutron or gamma-ray spectrum, if any discrepancy appears, one wants to know which energy neutrons are concerned for this discrepancy. Or before benchmark experiments, one wants to know in which energetic neutrons the expeiments benchmark. According to the present thought experiments, the benchmark sensitivity for the experiment is "equally" due not only to neutrons (named (A)) making neutrons conveying contribution directly to the deterctor, but also due to neutrons (B) making neutrons (A). In this case, it can be recognized that neutrons (B) convey the same contribution indirectly through (A), which is called indirect contribution. Similarly neutrons (C) making (B) have the same indirect contribution as (B) and so on, i.e., all the ancestor neutrons have the same contribution (error sensitivity). This gives us an very important hint for the benchmark error sensitivity as in the following section:

2.4 Benchmark sensitivity with Lewins importance in forward Monte Carlo calculations

As Lewins proposed[4)], in particle transport calculations, importance is regarded as contribution each particle has. The Lewins' contribution is defined as all the contribution which will be made by all the progenies of the particle in the whole transport calculation. According to the present thought experiments, it was found that the contribution governing the error sensitivity in Table 3 was exactly the same as the Lewins's contribution (importance). It means that benchmark error sensitivity could be evaluated by importance which can be estimated by forward Monte Carlo calculations. It also means error sensitivity could be predicted in forward Monte Carlo calculations, i.e., without adjoint function.

As a next phase, a more theoretical approach to form the mathematical expression with the present thought experiment and Lewins' style importance. And practical Monte Carlo calculations are planned to carry out to confirm the validity.

3. Conclusion

Thought experiments have been conducted to discuss how to estimate performance of benchmark experiments for neutron nuclear data for fusion reactor design. We carried out simple and precise thought experiments to examine which energy neutron is deeply concerned to the discrepancy, in case that there appears discrepancy between an experiment and its analysis with a transport code. As a result of the thought experiments, the benchmark sensitivity is "equally" due not only to neutrons (named (A)) making neutrons conveying contribution directly to the deterctor, but also due to neutrons (B) making neutrons (A). The neutrons (B) convey the same contribution indirectly through (A), which is called indirect contribution. Similarly all the ancestor neutrons have the same contribution (error sensitivity), which is an indirect contribution.

The above result leads the following conclusion: Because the Lewins's contribution (importance) is exactly the same as contribution governing the error sensitivity, benchmark error sensitivity could be evaluated by importance which can be estimated by forward Monte Carlo calculations. It also means error sensitivity could be predicted in forward Monte Carlo calculations, i.e., without adjoint function. With this procedure, it would become possible that from a sensitivity analysis in advance we could know how well and which energy nuclear data could be benchmarked with a benchmark experiment.

References

1) MCNP Team, "MCNP5-1.40 RSICC Release Notes", *LA-UR-05-8617* (2005).
2) M. Ohta et al., "Preliminary spectrum shifter design for

intermediate energy nuclear data benchmark experiments with DT neutrons," *Fusion Eng. Des.*, **84**, 1446-1449 (2009).
3) I. Murata, M. Ohta, H. Miyamaru, K. Kondo, S. Yoshida, T. Iida, K. Ochiai, C. Konno, "Performance Analysis of Fusion Nuclear-data Benchmark Experiments for Light to Heavy Materials in MeV Energy Region with A Neutron Spectrum Shifter," *J. Nucl. Mater.*, **417**, 1127-1130(2011).
4) J. Lewins, "IMPORTANCE The Adjoint Function", PERMAMON PRESS (1965).

Comparison with simulations to experimental data for photo-neutron reactions using SPring-8 Injector

Yoshihiro Asano[1,*]

[1]XFEL/SPring-8 Center, RIKEN 1-1 Koto Sayo Hyogo 679-5148, Japan

Abstract. Simulations of photo-nuclear reactions by using Monte Carlo codes PHITS and FLUKA have been performed to compare to the measured data at the SPring-8 injector with 250MeV and 961MeV electrons. Measurement data of Bismuth-206 productions due to photo-nuclear reactions of ^{209}Bi(γ,3n) ^{206}Bi and high energy neutron reactions of ^{209}Bi(n,4n)^{206}Bi at the beam dumps have been compared with the simulations. Neutron leakage spectra outside the shield wall are also compared between experiments and simulations.

1 Introduction

Since 1997, the 8GeV class synchrotron radiation facility, SPring-8, has been operated successfully for the users. The shielding designs of the SPring-8 accelerators were performed by using Jenkins' formula [1] for the bulk shielding in the lateral direction and modified Swanson's formula [2] in the forward direction. To verify the verification of the shielding design, benchmark experiments were performed by using Helium-3 high sensitive neutron counter with polyethylene moderators, TLD dose meters and Polymeric track detectors. The dose distributions within the shield tunnel were also measured by using LiF Thermo-Luminesence detectors and Polymetric neutron track detectors (CR-39) and the data were presented at ICRS9 [3]. The high energy neutron distributions around the beam dump of the accelerated electron are also measured by using bismuth detectors. Those experimental data were presented and compared with the empirical formula and EGS4 [4] with photo-neutron production cross section data [5].

In addition to FLUKA [6,7], the Monte Carlo code PHITS2.82 [8,9] have been improved recently to simulate the photo-nuclear reaction calculation. These simulation codes are much useful for radiation shielding design of electron accelerator machines as well as proton accelators with good usability. The neutrons and gamma dose distributions around the beam dumps of the SPring-8 injector have been simulated by using both Monte Carlo codes. However experimental data, especially photonuclear reaction data are quite insufficient to compare to the simulations.

Bismus is a useful metal to measure high energy neutrons and gamma-rays because of the neutron and gamma energy dependence of the nuclear production of the bithmus radioisotopes. At ICRS9, the data of bismus-206 due to 250MeV electrons were presented. In addition to the data produced by 250 MeV electrons, the experimental data produced by 961 MeV electrons are presented and both data have been compared to simulations. The neutron spectra outside the shield tunnel are compared between the PHITS simulations and experimental data.

2 Experiments

The measurements were performed in 1998. At that time, the injector of SPring-8 had two dumps, one was 250MeV (L1) dump and the other was 1.15GeV (L2) dump. The geometrical configuration is shown in Fig.1. At the experiments, the 250MeV and 961 MeV electron beam that lead to L1 dump and L2 dump were employed and the data are indicated in Table 1, respectively. The beam dumps have double cylindrical structures, made of graphite for the inner layer and iron for the outer layer to reduce photo-neutron production and shielding.

The neutron leakage spectra outside the shield tunnel were measured by using extremely high sensitive helium-3 counter with polyethylene moderators, and the data were unfolded with SAND-II code [10]. Two types (small and large) of bismuth ingot were set around the beam dumps to measure the high energy neutrons and gamma-rays.

Table 1. Electron beam conditions of the linac injector during the experiments.

Dump	250MeV (L1dump)	1.15GeV (L2dump)
Electron energy	250MeV	961MeV
Frequency	60pps	60pps
Peak current (40ns)	143.0mA	134.3mA
Total electrons	3.48 x10^{6}	7.75x10^{15}

DOI: 10.1051/epjconf/201715307016

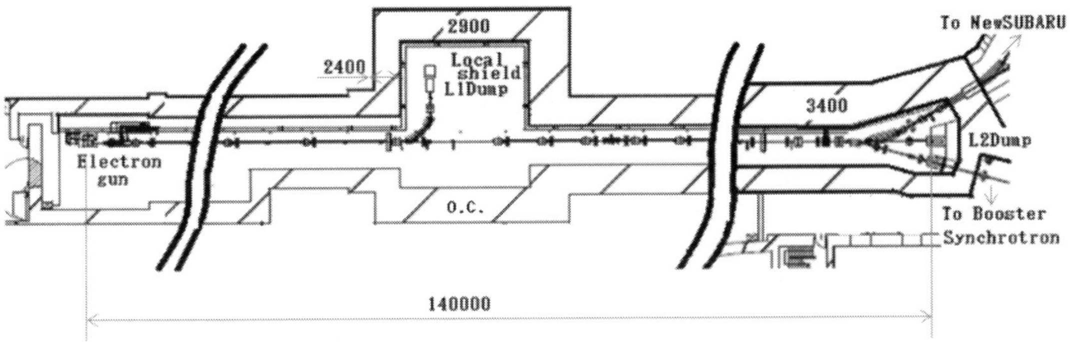

Fig. 1. Conceptual plane view of SPring-8 injector. (L1Dump and L2Dump are 250MeV dump and 1.15GeV dump, respectively. Unit is mm) At the end station of the linac, accelerated electrons can be extracted to 3 ways. One is to L2 dump, one is to booster synchrotron, and the other is to small SR ring, NewSUBARU.

3 Simulations

Neutron and gamma effective dose (Worst geometry) simulations around the L1 dump by using FLUKA are shown in Figs.2 and 3, respectively. PHITS simulations for effective dose distribution (AP geometry) due to 961 MeV are shown in Figs.4 and 5 for neutron and gamma doses, respectively.

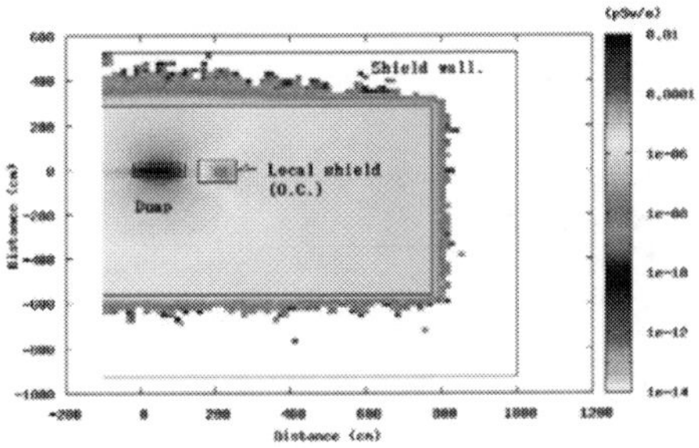

Fig.2. Neutron dose distribution at L1 dump (Horizontal plan at electron beam level, unit:pSv/e)

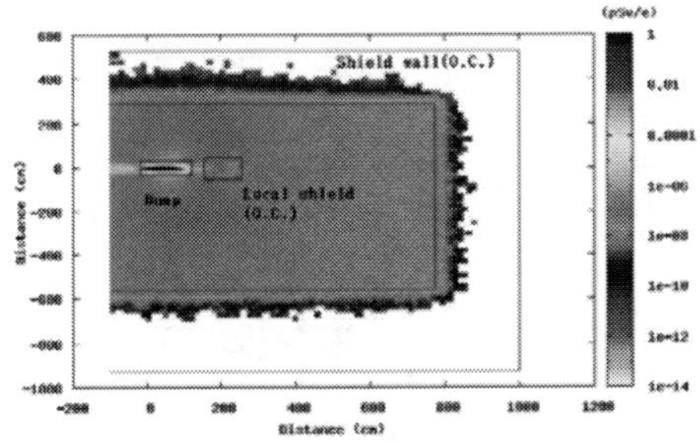

Fig.3. Gamma dose distribution at L1 dump. (Horizontal plan at electron beam level, unt;pSv/e)

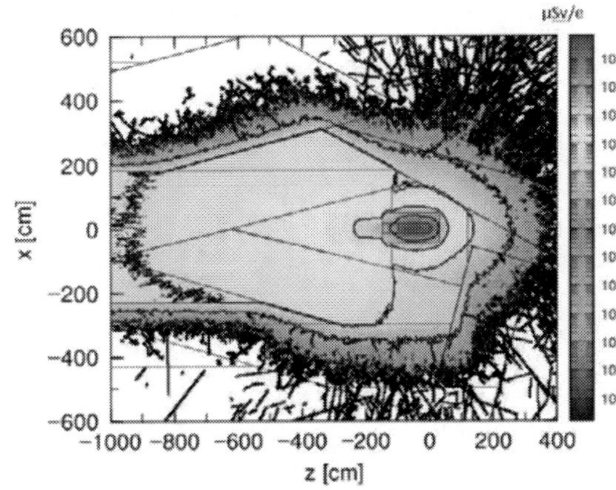

Fig.4. Neutron dose distribution due to 961MeV electrons. (Horizontal plane at electron beam level, unit:μSv/e)

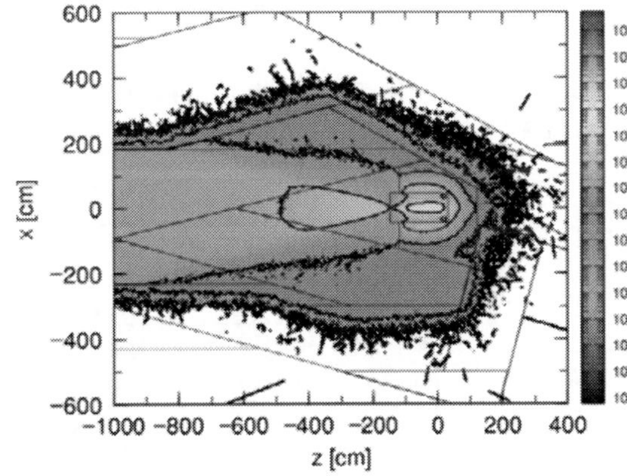

Fig.5. Gamma dose distribution due to 961MeV electrons. (Horizontal plane at electron beam level, unit:μSv/e)

4 Comparison between simulations and experiments

4.1 Bismuth isotopes production around the beam dumps

Small and large size bismuth ingots were employed for activation detectors and gamma-rays from bismuth detectors were measured after irradiation. The size of small detectors is 1.12 cm and 2.9 cm for the thickness and the diameter, respectively. The large detectors are 2.3 cm thick and 8 cm in diameter. In those cases, the self-shielding effect of detectors cannot be ignored to measure the gamma-rays. Figure 6 shows the self-shielding factors depending on gamma-ray energy and the size of bismuth detectors. These are PHITS simulations on the assumption of the isotropic distribution of the radioisotopes within the detectors. The assumption of the isotropic distribution is reasonable because of not so much thick detectors for high energy photons and neutrons and it's confirmed by using PHITS simulations. The previous data [3] were without

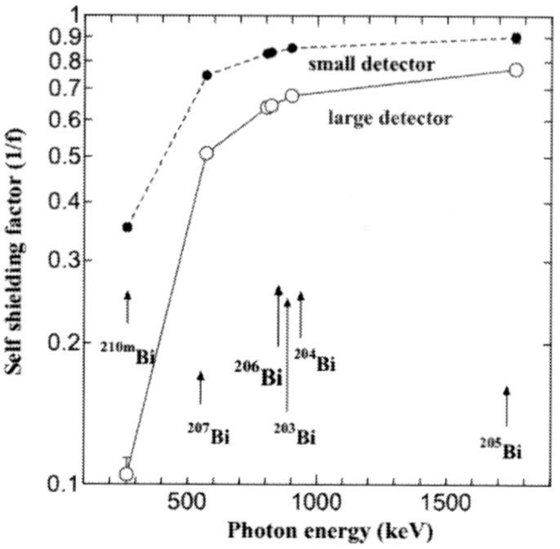

Fig.6. Self-shielding effect for the gamma-ray measurements of the bismuth radioisotopes depending on the photon energy and size of the detectors. The arrows of bismuth radioisotopes indicate the photon energy of main emissions. Full and open circles are for small and large detectors, respectively.

considering these self-shielding effects so that these data must be corrected, especially for large detectors. The PHITS simulations of bismuth isotopes productions have been performed using [T-Dchain] command and DCHAIN-SP [11] with weighting factor pnimul=100 of photo-nuclear reaction.

4.1.1 250MeV electrons (L1 dump)

The bismuth-206 production distribution around the L1 dump was measured and compared to the Monte Carlo simulations. The bismuth detectors were set up as shown in Fig.7. In the experimental paper [3], the bismuth-206 production calculations were performed by using EGS4 with the library for $^{209}Bi(\gamma,3n)^{206}Bi$ reaction created by themselves using MCPHOTO [12] + PICA [13]. The reference paper also considered for $^{209}Bi(n, 4n)^{206}Bi$ reaction by using ENDF-B/VI library.

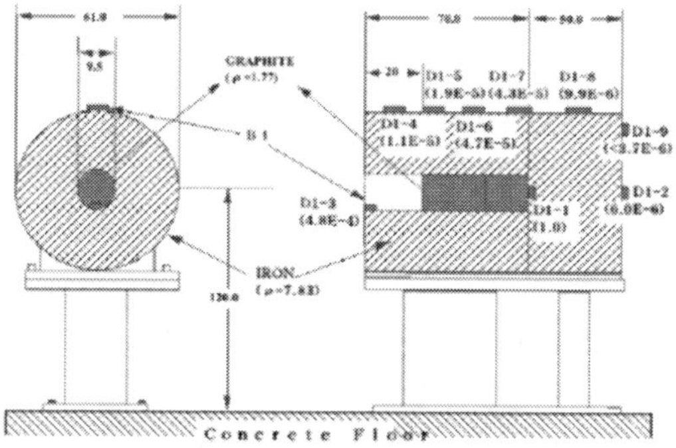

Fig.7. Illustration of the 250 MeV electron beam dump (L1) and the location of the bismuth detectors. D1-1 to D1-9 are bismuth detectors and the figures in parenthesis under the detector number are the relative values of produced bismuth-206, which are normalized to the D1-1 detector values [3].

Table 2. Bismuth-206 production distribution due to 250MeV electron injection into dump. The errors are considered only for statistical errors. The figures in the parentheses are pro forma amounts because of large statistical errors. (*; without considering self-shielding, *2; with considering self-shielding, *3;unit is converted from (g^{-1} $electron^{-1}$) [3] to (Bq/g)/(e/s))

Bi detector	Measured data* (Bq/g)/(e/s)	Measured data*2 (Bq/g)/(e/s)	EGS4 cal. *3 (Bq/g)/(e/s)	FLUKA cal. (Bq/g)/(e/s)	PHITS cal. (Bq/g)/(e/s)
D1-1(Small)	$1.45x10^{-7}\mp0.36\%$	$1.74x10^{-7}\mp0.36\%$	$3.7x10^{-7}$	$2.27x10^{-7}\mp0.50\%$	$2.21x10^{-7}\mp1.2\%$
D1-2(Small)	$8.47x10^{-13}\mp25\%$	$1.02x10^{-12}\mp25\%$	$6.0x10^{-12}$	$(6.65x10^{-12}\mp14\%)$	-
D1-3(Small)	$6.92x10^{-11}\mp2.3\%$	$8.34x10^{-11}\mp2.3\%$	$2.8x10^{-10}$	$(1.40x10^{-11}\mp9.6\%)$	$(1.70x10^{-10}\mp40\%)$
D1-4(Large)	$1.61x10^{-12}\mp2.9\%$	$2.52x10^{-12}\mp2.9\%$	$1.2x10^{-12}$	$(2.60x10^{-12}\mp14\%)$	-
D1-5(Large)	$2.75x10^{-12}\mp2.2\%$	$4.31x10^{-12}\mp2.2\%$	$6.2x10^{-12}$	$5.03x10^{-12}\mp3.7\%$	$(2.38x10^{-10}\mp23\%)$
D1-6(Large)	$6.79x10^{-12}\mp1.4\%$	$1.06x10^{-11}\mp1.4\%$	$1.4x10^{-12}$	$1.30x10^{-11}\mp2.7\%$	-
D1-7(Large)	$6.21x10^{-12}\mp1.5\%$	$9.74x10^{-12}\mp1.5\%$	$7.5x10^{-12}$	$1.30x10^{-11}\mp4.5\%$	$(1.36x10^{-10}\mp41\%)$
D1-8(Large)	$1.40x10^{-12}\mp3.1\%$	$2.19x10^{-12}\mp3.1\%$	$2.9x10^{-12}$	$(1.78x10^{-12}\mp14\%)$	-
D1-9(Small)	$<5.35x10^{-13}$	$<6.45x10^{-13}$	$5.2x10^{-14}$	-	-

Table 2 indicates the results of the FLUKA and PHITS simulations together with the measurement data and the EGS4 calculations. In the simulations, FLUKA indicates about 1.30 times higher than that of the measurement data of D1-1 detector. The FLUKA simulation shows close agreement with the PHITS results for the bismuth-206 production within the D1-1 detector. For another detecors except D1-2 and D1-3, the FLUKA simulations show good agreement within about 30% discrepancies with the measured data. Large stastistical errors are indicated for the D1-2, D1-3 detectors, and the PHITS simulations have also large stastical errors for another detectors. Experimental data of another bismuth radioisotopes with the self shielding corrections are shown in Table 3 including PHITS simulations for D1-1 detector. In Table 3, «unknown » means that a long decay time was needed to measure the gamma-rays because of very high activities. Therefore, the information of bismuth-204 and 203 was cleared.

Table 3. Bismuth radioisotopes production within D1-1 and D1-6 detecors. (* ; measurement data)

Bi-det. No	Exp.(D1-1)* (Bq/g)/(e/s)	PHITS Cal (D1-1) (Bq/g)/(e/s)	EXP.(D1-6)* (Bq/g)/(e/s)
^{207}Bi	2.68×10^{-6} ∓6.4%	3.66×10^{-6} ∓0.22%	$<2.37 \times 10^{-10}$
^{205}Bi	1.01×10^{-7} ∓1.0%	2.52×10^{-8} ∓1.4%	9.95×10^{-12} ∓ 4.8%
^{204}Bi	unknown	-	8.94×10^{-13} ∓ 16%
^{203}Bi	unknown	-	1.24×10^{-12} ∓ 31%

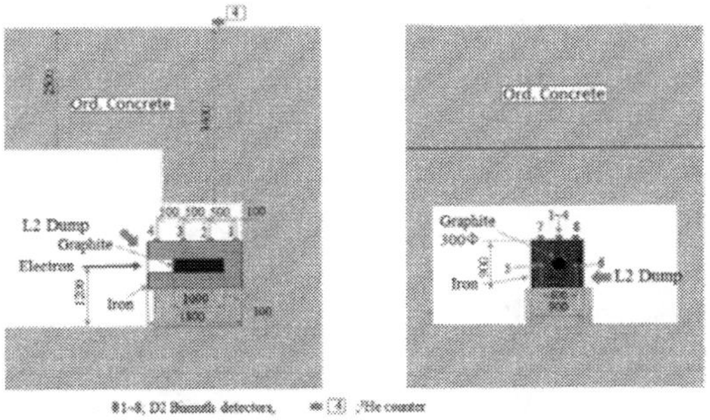

Fig.8. Configuration of bismuth detectors around the L2 beam dump (Vertical plane, left figure is parallel and right is perpendicular against the electron beam axis)

4.1.2 961MeV electrons (L2 dump)

The bismuth-206 production distribution around the L2 dump using 961 MeV electrons was measured as well as 250MeV electrons, and the configuration of bismuth detectors are as shown in Fig.8. The measurement data and simulation results are listed in Table 4. Both simulations show agreement within about ∓50 % as shown in Fig.9 even though simulations have large stastical errors. Another bismuth radioisotopes which were produced in the detectors of D2-2, D2-3,and D2-4 are listed in Table 5 including the PHITS simulations for D2-3 detector. The measurement data of Tables 2 to 5 are useflu to estimate the photon and neutron energy distributions because of the threshold energy dependence on the photon and neutron energy for each reactions such as 22.44MeV, 29.48MeV, 37.97MeV, and 45.16MeV for ^{209}Bi(γ ,3n)^{206}Bi, ^{209}Bi(γ ,4n)^{205}Bi, ^{209}Bi(γ ,5n)^{204}Bi, and ^{209}Bi(γ ,6n)^{203}Bi, respectively. It is difficult to estimate the amount of ^{210}Bi experimentally and directly, however it is easy to compare between the simulations. In comparison with the simulations, ^{210}Bi productions are listed as indicated in Table 6, and shown in good agreements.

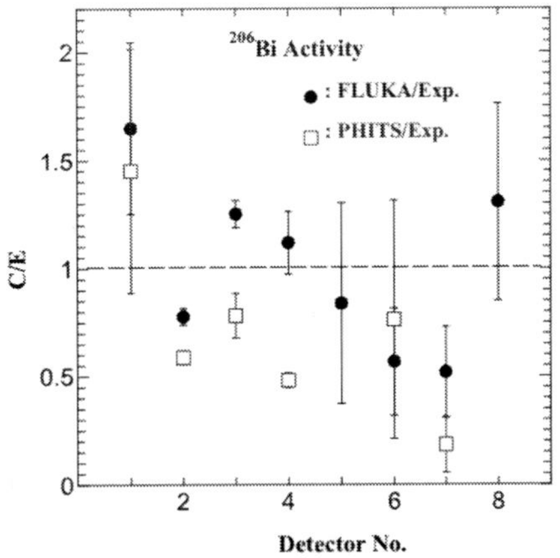

Fig.9. Ratios of simulations to experimental data (Full circles and open squres are the data of FLUKA and PHITS, respectively. The errors are only considering stastical errors.)

4.2 Leakage neutron spectrum outside the roof of shield tunnel

Leakage neutron spectra have been simulated by using PHITS with the [T-Point] tally of which algorism is same as the Point Detector of MCNP5 [14] because of strong attenuation through thick shield walls. In these simulations, two points have been performed to compare the experimental data [3]. One is located directly above the L1 dump outside the roof of the tunnel with the shield thickness of 190 cm ordinary concrete, and the other is located directly above the L2 dump outside the roof of the tunnel with the shield thickness of 360 cm ordinary concrete as indicated in Figs. 1 and 8. These points are under air conditioning so that it can be avoided the change of the environmental conditions, relatively. In the simulations, the density of 2.26 g/cm^3 has been employed for ordinary concrete. Figs. 10 and 11 show the simulation results in comparison with the experimental data that were measured by using extremely high sensitive helium-3 counter with multi-polyethylene moderators and unfolded the measurement data. These simulations show reasonably agree with the experimental data. However, the results of the simulation for the 250MeV electron case indicate slightly high in the region of high energy in comparison with the experimental data as shown in Fig.10. On the other hand, Fig.11 shows that the simulation results for the 961 MeV electron case indicate the slightly lower spectrum than that of experimental data in the region of intermediate

Table 4 Bismuth-206 production distribution due to 961MeV electron injection into L2 dump. The errors are considered only for statistical errors. The figures in the parentheses are pro forma amounts because of large statistical errors. (*; without considering self-shielding, *2; with considering self-shielding, *3;unit is converted from (g^{-1} $electron^{-1}$) [3] to (Bq/g)/(e/s)

Bi detector No.	Measured data* (Bq/g)/(e/s)	Measured data*2 (Bq/g)/(e/s)	FLUKA cal. (Bq/g)/(e/s)	PHITS cal. (Bq/g)/(e/s)	C/E (FLUKA)	C/E (PHITS)
D2-1(Small)	$1.08\times10^{-11}\mp17\%$	$1.31\times10^{-11}\mp17\%$	$(2.16\times10^{-11}\mp17\%)$	$(1.90\times10^{-11}\mp35\%)$	$(1.7\mp24\%)$	$(1.5\mp39\%)$
D2-2(Large)	$2.01\times10^{-11}\mp1.8\%$	$3.15\times10^{-11}\mp1.8\%$	$2.45\times10^{-11}\mp4.6\%$	$1.85\times10^{-11}\mp5.0\%$	$0.78\mp4.9\%$	$0.59\mp5.3\%$
D2-3(Large)	$1.14\times10^{-11}\mp2.3\%$	$1.79\times10^{-11}\mp2.3\%$	$2.24\times10^{-11}\mp4.5\%$	$(1.40\times10^{-11}\mp13\%)$	$1.3\mp5.1\%$	$(0.78\mp13\%)$
D2-4(Large)	$1.56\times10^{-12}\mp6.8\%$	$2.45\times10^{-12}\mp6.8\%$	$2.74\times10^{-12}\mp11\%$	$1.18\times10^{-12}\mp2.2\%$	$1.1\mp13\%$	$0.48\mp7.1\%$
D2-5(Small)	$3.50\times10^{-12}\mp24\%$	$4.26\times10^{-12}\mp24\%$	$(3.57\times10^{-12}\mp50\%)$	-	$(0.84\mp55\%)$	-
D2-6(Small)	$5.22\times10^{-12}\mp18\%$	$6.28\times10^{-12}\mp18\%$	$(3.56\times10^{-12}\mp40\%)$	$(4.79\times10^{-12}\mp70\%)$	$(0.57\mp44\%)$	$(0.76\mp72\%)$
D2-7(Large)	$1.09\times10^{-12}\mp7.8\%$	$1.71\times10^{-12}\mp7.8\%$	$(8.87\times10^{-13}\mp40\%)$	$(3.13\times10^{-13}\mp13\%)$	$(0.52\mp41\%$	$(0.18\mp70\%)$
D2-8(Large)	$1.02\times10^{-12}\mp8.0\%$	$1.60\times10^{-21}\mp8.0\%$	$2.09\times10^{-12}\mp34\%$	$(1.40\times10^{-11}\mp13\%)$	$(1.31\mp35\%)$	-

Table 5 Bismuth radioisotopes production within D2-2, 3, and 4 detecors. (uni ;(Bq/g)/(e/s), * ; measurement data)

Bi	(D2-2)*	(D2-3)*	PHITS Cal. (D2-3)	(D2-4)*
^{205}Bi	3.22×10^{-11} $\mp6.6\%$	1.86×10^{-11} $\mp8.2\%$	1.52×10^{-11} $\mp36\%$	3.44×10^{-12} $\mp18\%$
^{204}Bi	1.89×10^{-11} $\mp3.1\%$	9.13×10^{-12} $\mp4.0\%$	1.16×10^{-11} $\mp46\%$	1.52×10^{-12} $\mp13\%$
^{203}Bi	1.65×10^{-11} $\mp6.1\%$	9.51×10^{-12} $\mp8.4\%$	3.48×10^{-12} $\mp17\%$	6.42×10^{-13} $\mp55\%$

Table 6 Simulation of bismuth-210 production

Det. No	FLUKA cal. (Bq/g)/(e/s)	PHITS cal. (Bq/g)/(e/s)	$\frac{\text{FLUKA cal}}{\text{PHITS cal.}}$
No.1	$5.64\times10^{-11}\mp1.7\%$	$4.22\times10^{-11}\mp1.9\%$	$1.3_4\mp2.6\%$
No.2	$3.34\times10^{-11}\mp1.6\%$	$4.34\times10^{-11}\mp1.0\%$	$0.77_0\mp1.9\%$
No.3	$3.04\times10^{-11}\mp0.88\%$	$2.97\times10^{-11}\mp0.85\%$	$1.0_2\mp1.2\%$
No.4	$9.99\times10^{-12}\mp1.9\%$	$1.05\times10^{-11}\mp3.2\%$	$0.95_1\mp3.7\%$
No.5	$2.64\times10^{-12}\mp6.4\%$	$2.67\times10^{-12}\mp1.7\%$	$0.98_9\mp6.6\%$
No.6	$2.74\times10^{-12}\mp9.6\%$	$3.24\times10^{-12}\mp2.3\%$	$0.84_6\mp9.9\%$
No.7	$8.60\times10^{-12}\mp2.5\%$	$8.11\times10^{-12}\mp6.1\%$	$1.0_6\mp6.6\%$
No.8	$8.24\times10^{-12}\mp2.0\%$	$7.70\times10^{-12}\mp4.1\%$	$1.0_7\mp4.6\%$

energy region. These discrepancies need future investigations as well as the improvements of the statistical errors.

5 Summary

The experimental data of bismuth radioisotopes distribution at around the beam dumps produced by 961 MeV electrons are presented as well as that of due to 250 MeV electrons. These data are useful to compare to the simulations. FLUKA and PHITS Monte Carlo simulations have been performed to compare to these data and the both simulations show reasonably agreements in a range of within about ∓50%. The PHITS simulations are performed for the leakage neutron energy spectra to compare to the experimental data, and these show slightly different shapes. These discrepancies need future investigations.

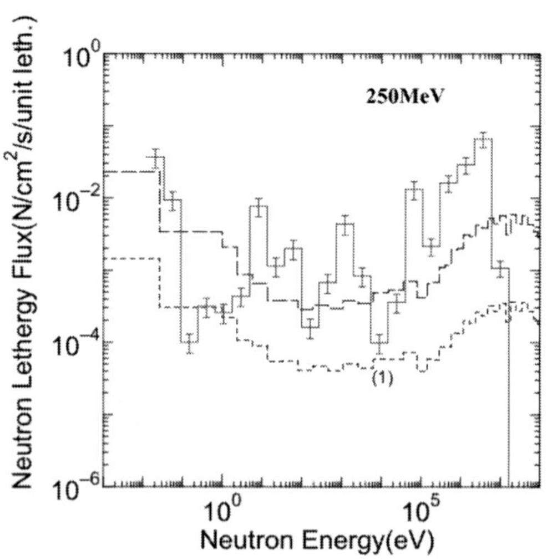

Fig.10 Comparison of the leakage neutron energy spectrum at the roof outside the shield tunnel and directly above of the L1 dump. (Solid red line indicates the simulation results, and dashed line is the experimental data. Dotted line indicates the neutron back ground spectrum)

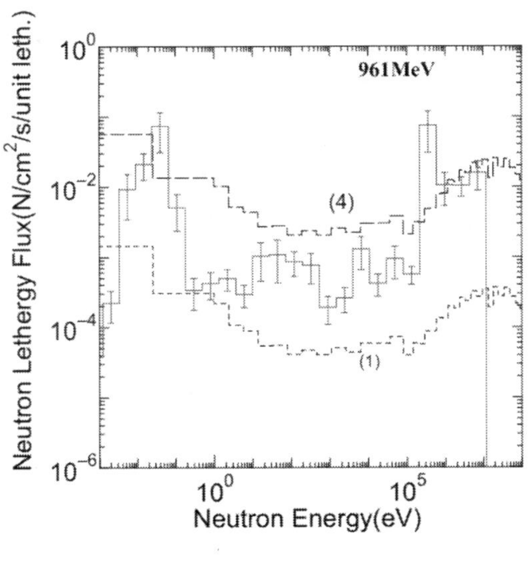

Fig.11 Comparison of the leakage neutron energy spectrum at the roof outside the shield tunnel and directly above of the L2 dump. Lines indicate the same as Fig.10.

The author wishes to thank Dr. Tatsuhiko.Sato of JAEA and the developer group members for thier help and valuable discussions to use PHITS.

References

1. T.M. Jenkins, *Neutron and photon measurements through concrete from a 15 GeV electron beam on a target-Comparison with models and calculations,* Nuclear Instruments and Methods 159 265 (1979)
2. E.Braeuer, *Radiation shielding for the 6 GeV, ESRF,* ESRF/SHILED/88-04 (1988).
3. Y.Asano & N.Sasamoto, *Benchmark Experiment for Radiation Shielding around the Beam Dump of the SPring-8 Injector,* J.of Nuclear Science and Technology Supple. 1 535-539 (2000)
4. W.R.Nelson et al. *The EGS4 code system,* SLAC-265 UC-32 (1985)
5. S.S.Dietrich et al. *Atlas of photoneutron cross section obtained with monoenergetic photons,* Atomic Data and Nucl. Data tables 38 (1988)
6. G.Battistoni et al. *The FLUKA code : Discription and benchmarking,* Proc. Of the Hadronic Shower Simulation Workshop 2006, AIP conf. proceeding 896,31-49 (2007)
7. A.Fasso et al. *FLUKA a multi-particle transport code,* CERN-2005-10 INFN/TC/05/11 SLAC-R-773 (2005)
8. T.Sato et al. *Particle and heavy ion transport code system PHITS, version 2.52,* J. Nucle. Sci. Technol. 50:9 913-923 (2013)
9. S.Noda et al. *Improvement of photonuclear reaction model below 140MeV in the PHITS code,* J. Nucle. Sci. and Technol,00223131.2014.923349, (2014)
10. P.J.Griffin et al. *User's Manual for SNL-SAND-II Code,* AFWL-TR 67-41, CCC-112 SAND93-3957 UC-713 (1969)
11. T.Kai et al. *DCHAIN-SP 2001 :High energy particle induced radioactivity calculation code,* JAERI-Data-Code-2001-016 (2001) in Japanese
12. N.Kishida, *Methods used in photonuclear data evaluation at JNDC,* Proc. Int. Synposium on Nuclear Data Eval BNL 598 (1992)
13. T.A.Gabriel et al., *PICA Monte Carlo Medium-Energy photon-induced intranuclear Cascade Analysis Code System,* ORNL CCC-160 (1972)
14. X-5 Monte Carlo Team. *MCNP-A General Monte Carlo N-Particle Transport Code, Version 5,* LA-UR-03-1987 (1987)

Shielding calculations with SCALE/MAVRIC and comparison with measurements for the TN85 cask with vitrified high level radioactive waste

Holger Thiele[a] and Frank-Michael Börst

Bundesamt für Kerntechnische Entsorgungssicherheit (BfE), Willy-Brandt-Straße 5, 38226 Salzgitter, Germany

Abstract. A series of dose rate/spectra measurements in the German interim storage facility Gorleben was carried out at a TN85 cask in April 2009. This type of cask is used for the transport and interim storage of vitrified high level radioactive waste (HAW) from reprocessing. The aim of this work is to assess the shielding component MAVRIC of the SCALE code system with these measurements for the use in the German Bundesamt für Kerntechnische Entsorgungssicherheit (BfE).

1 Introduction

Aim of this paper is the presentation of the comparison of dose rates measured at a TN85 cask at the interim storage facility Gorleben with calculations.

The Bundesamt für Kerntechnische Entsorgungssicherheit (BfE) as the German competent authority for approval of design of packages for the transport of radioactive material uses the Standard Computer Analysis for Licensing Evaluation (SCALE) [1] code system developed by Oak Ridge National Laboratory (ORNL) as a calculation tool for shielding evaluation and for decision whether a new package design meets all applicable dose rate limits of the transport regulations. The main field of appclication of shielding calculations at BfE are thick walled metal caks.

In April 2009 radiation measurements were carried out at a TN85 cask for transport and interim storage of high level radioactive waste from reprocessing (HAW). These measurements were conducted in the interim storage facility Gorleben and included neutron and gamma dose rate measurements and spectrometric measurements for neutrons.

First results exclusively for neutron measurements were presented by Rimpler et al. at the NEUDOS-11 symposium [2].

In this paper shielding calculations for the TN85 cask are presented that were performed using the MAVRIC sequence from the SCALE system. MAVRIC calculates dose rates by performing a three-dimensional Monte-Carlo analysis (MONACO) using an automated biasing procedure. The biasing parameters are generated from results of a three-dimensional adjoint discrete ordinates calculation.

For these calculations the necessary source terms were generated with ORIGEN-ARP from the SCALE system. The HAW canisters as well as the basket and the cask body were modelled in detail using SCALE General Geometry Package (SGGP) geometry.

2 Description of cask and its contents

The TN85 casks consists of a thick-walled finned cylindrical cask body made of forged steel. In storage configuration, i.e. the configuration in which the cask has been during the measurements, the cask is closed with a primary and a secondary lid, each made of stainless steel, and their respective screws and seals. For more effective neutron shielding, chambers filled with borated resin are attached to the outside of the cask body. The same resin is used for the axial neutron shielding at the cask bottom and on the primary lid. The outer dimensions of this cask in the storage configuration as the basis for the measurements and calculations are ~ 2.3 m diameter and ~ 6 m in height; the loaded cask weight is ~ 107 t. The cask is loaded with 28 HAW canisters in four levels with six canisters arranged concentrically and one central canister in each level. The general configuration of the TN85 cask is shown in Fig. 1.

The HAW canisters contain actinides and fission products in a matrix of borosilicate glass. The gamma radiation around this cask mainly results from the fission products ^{134}Cs, ^{137}Cs, and ^{154}Eu. Neutron radiation is mainly caused by spontaneous fission of ^{244}Cm and (α,n) reactions of ^{244}Cm and ^{241}Am with oxygen or boron. The average source strength ratio of spontaneous fission to (α,n) reactions for the calculations presented in this paper was determined with ORIGEN-ARP to be 1.3 (single standard deviation ± 10%). The decay of radionuclides between time of loading and time of measurement has been taken into account.

[a] Corresponding author: holger.thiele@bfe.bund.de

 DOI: 10.1051/epjconf/201715307017

3 Results of the dose rate measurements

The dose rate measurements were carried out at a stand-alone TN85 cask in storage configuration at the Gorleben interim storage facility.

Commercial devices were used for measuring gamma and neutron dose rates and were calibrated according to International Commission on Radiological Protection (ICRP) Publication 60 (ICRP 60) [3]. The gamma dose rates were measured using the dosemeter AD 6 (Automess), the neutron dose rates were measured with an LB 6411 monitor (Berthold).

The results of the measurements are plotted in Fig. 3 for gamma dose rates and in Fig. 4 for neutron dose rates.

According to manufacturer's data, the uncertainty of the gamma dose rate is ± 30% at maximum (relative to ^{137}Cs within an energy range from 30 keV to 2 MeV). Concerning the dose rate $\dot{H}^*(10)$ for neutrons, the LB 6411 has an energy depending response of ± 40% at maximum (within an energy range from 50 keV to 10 MeV, relating to 1 MeV).

At one reference position the neutron dose rate was derived from neutron spectra measured by means of a Bonner multisphere spectrometer as reference value for the evaluation of the neutron dose rates measured by commercial devices and is also shown in Fig. 4. The uncertainty of the dose rate derived with the Bonner multisphere is assessed with ± 15% at maximum.

4 Modelling cask and contents

The creation of the source terms, the modelling of the cask and its contents and finally the calculations were carried out with SCALE 6.1.3. In the MAVRIC calculation file, the geometry of the 28 HAW canisters as well as the basket and the cask body were modelled using SGGP geometry (Fig. 2). Additionally to the Safety Analysis Report [5] the "QA Life Time Records" of the TN85 cask [6] were taken into account to adapt the dimensions of the parts of the cask and its material compositions individually.

For the shielding calculations the coupled 200 neutron group - 47 gamma group library was chosen (v7-200n47g). The library is based on ENDF/B-VII data.

To obtain the gamma and neutron "ambient dose rate equivalent" $\dot{H}^*(10)$ according to ICRP 60, built-in conversion coefficients based on ICRU 57 were used. They are identical to the coefficients published in ICRP Publication 74 [4].

The dose rates were calculated by placing point detectors along the 90°(trunnion) and 180° axis of the cask at a distance of ~ 28 cm respectively ~ 16 cm from the cask surface for both the gamma dose rate calculations and the neutron dose rate calculations. The chosen distance represents the distance between the measurement device and the top of the radial cooling fins of the cask.

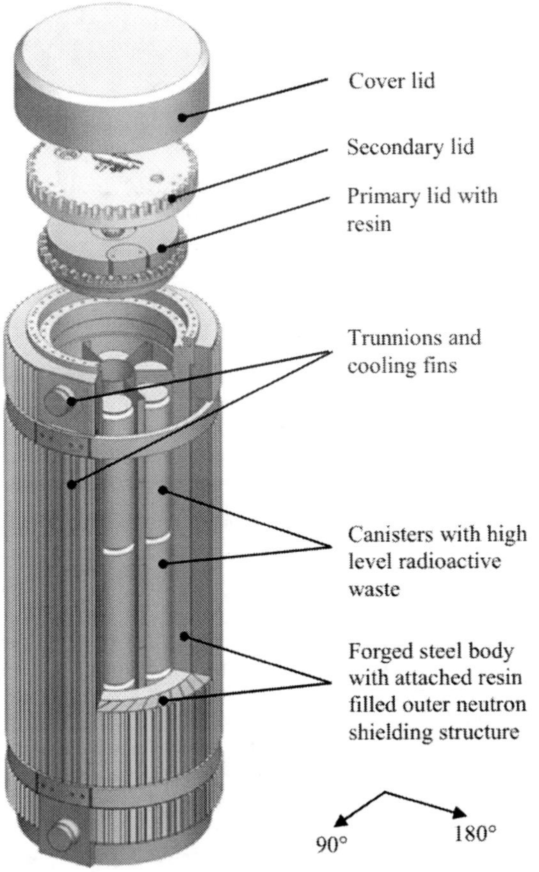

Figure 1. TN85 cask in storage configuration

The loading plan of the cask describes the content of fission products and actinides in each HAW canister with their activities and masses, respectively. For the calculations these data had to be converted into energy-dependent source strengths. These source strengths were obtained by decay calculations using ORIGEN-ARP software. Then, the source region was modelled with separate source strengths and spectra for each of the 28 HAW canister. The inventory that has to be considered for radiation purposes consists of six nuclides for the gamma emission (^{90}Sr, ^{106}Ru, ^{134}Cs, ^{137}Cs, ^{144}Ce, and ^{154}Eu) and three nuclides for the neutron emission (^{237}Np, ^{241}Am, and ^{244}Cm).

The neutron sources induced by (α,n)-reactions have been calculated by the use of the capabilities of ORIGEN-ARP, taking into account several target elements of the vitrified glass taken from [7].

5 Results of calculation

The results of the calculations and the measurements are shown in Fig. 3 for the gamma dose rates and in Fig. 4 for the neutron dose rates.

The standard deviations of the calculations for the gamma and neutron dose rates were about 4 % and 1.5 %, respectively.

At the 90°-axis where the trunnions are located the calculated gamma dose rates are lower than the measured values, at the middle region of the cask by 49 % on average (Fig. 3). For the height at 30 cm over ground the calculated gamma dose rate is about 22 % higher than measured. One reason might be the location of the measurement devices that were not placed directly in front of the trunnion like the point tally of the calculation. Since two AD 6 were used simultaneously the positions of both devices were slightly shifted from the trunnion centre.

The neutron dose rates calculated for the middle region of the cask at the 90°-axis show slight deviation from the measured values of about -19 % on average (Fig. 4). Deviations that are above of the measured values for neutrons appear especially in the regions around the trunnions near the ends of the cask with up to 33 % and a shifted maximum.

Compared to the dose rate measured by the Bonner sphere system at the 90°-axis, the results of the LB 6411 measurement and the calculations show deviations of about -10 % and -25 %, respectively.

At the 180°-axis the differences of the measured and calculated values of the gamma and neutron dose rates for the middle region of the cask are equal to those at the 90°-axis. Solely at two regions around the circumferential rings which replace the radial shock absorbers during the interim storage (at heights of about 90 cm and 510 cm) much lower neutron dose rates (by about 50 %) were calculated than measured. In opposite, at the 600 cm level, higher neutron dose rates have been calculated than measured.

The deviations in the calculated dose rates for regions in the vicinity of heterogeneities in the cask structure and especially in the top region may be caused by various reasons, e.g. the filling heights and composition of the contents of the canisters, uncertainties in the description of the nuclide inventory, modelling the cask structures and materials and cask surrounding structures.

Some potential sources of deviations have been quantified and examined:

The TN85 cask has not been placed in empty space. Hence, some influences of the surrounding material have to be taken into account. A plate of concrete (radius=400 cm, height=20 cm, material: "regular concrete" from the SCALE material library) has been modelled as floor structure below the cask. The possible influence has been checked by some additional calculations which do not show significant influence on the results.

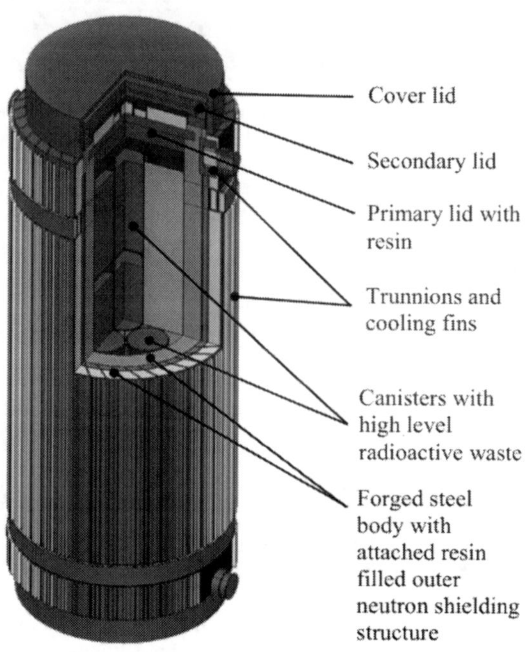

Figure 2: SGGP-Model of the TN85 cask in storage configuration

The background dose rate level during the measurements has been quantified and can also be excluded as possible source of deviations.Variations within possible parameter ranges for the density of the forged steel body (reduction of the density from 7,85 g/cm^3 to 7,75 g/cm^3) as well as the composition of the resin (variation of the composition) do not show any significant effect on the results.

As a further potential source of deviations a special feature of the design of the cask has been analysed. The neutron shield is the outermost shell and this causes secondary gamma radiation induced by (n,γ)-reactions with the constituents (H, B, C, O) of the resin. There is no significant gamma shield beyond the neutron shield. The contribution of secondary gamma radiation to the gamma dose rate is in the same order of magnitude as the primary gamma radiation therefore. Effects are discussed in section 6.

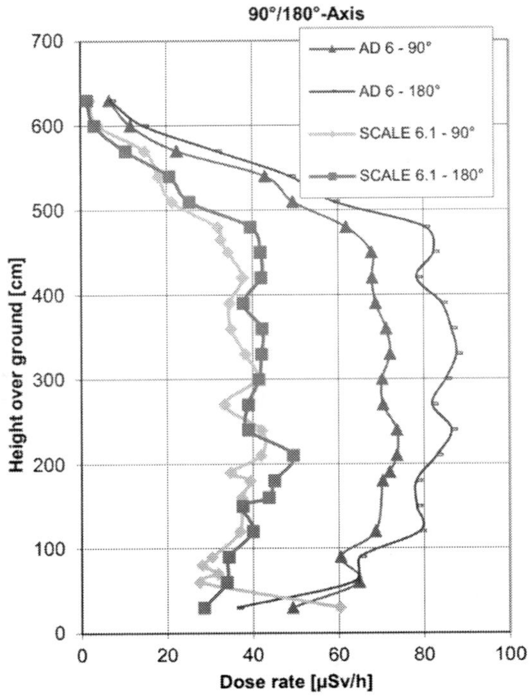

Figure 3. Gamma dose rates

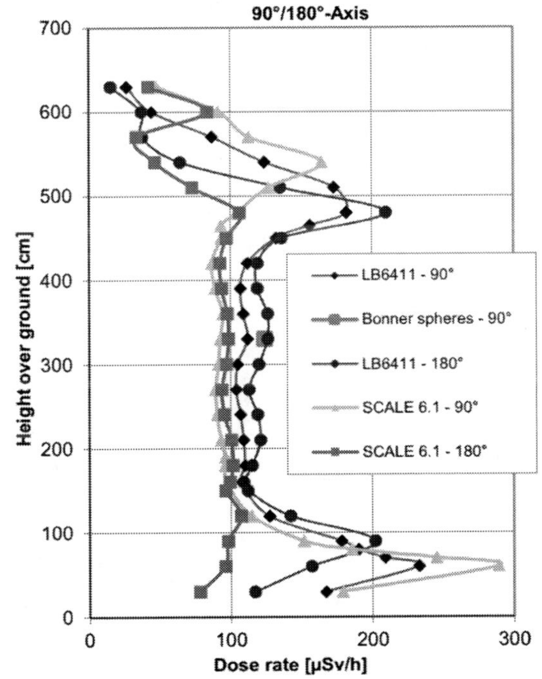

Figure 4. Neutron dose rates

6 Discussion and Conclusions

The aim of this work has been to show the applicability of the shielding sequence MAVRIC of the SCALE 6.1.3 package to assess the shielding capacity of thick walled casks for HAW canisters.

The calculated dose rates are generally lower than the measured values with some exceptions. Variations in the cask model (materials, dimensions of the components) and the consideration of surrounding structures seem not to be the source of the deviations as mentioned in section 5.

Another typical reason responsible for deviations between measured and calculated dose rates are energy dependencies of the dose rates displayed by the measurement devices, here AD 6 for gamma radiation and LB 6411 for neutron radiation.

The AD 6 has an energy dependency that shows overestimation of the dose rate for all energies above 662 keV, the energy of calibration by ^{137}Cs (Fig. 5). The energy dependence is specified up to an energy of 1.25 MeV, with a maximum overestimation by a factor of 1.35 [8]. For energies below 662 keV the dose rate is underestimated, but in the energy range with significant photon flux at the TN85 cask, i.e. above 65 keV, the underestimation is less than 13 % (Figs. 5 and 6) [8].

In the case of the TN85 cask, the energy group resolved photon flux calculated with SCALE by BfS (Fig. 6) combined with the energy dependency for the AD 6 (Fig. 5) and the fluence-to-dose conversion factors for $H^*(10)$ leads to a dose rate that would be displayed by an AD 6 detector that is about 75 % higher than the gamma dose rate without the energy dependency. This is an important outcome to explain the pronounced lower calculated dose rates compared to the gamma dose rates displayed by the AD 6 dosemeters. After applying an appropriate correction factor to the gamma dose rates the differences between the corrected measurement results and the results of the calculations for the majority of the calculated dose rates are about 20 % and less.

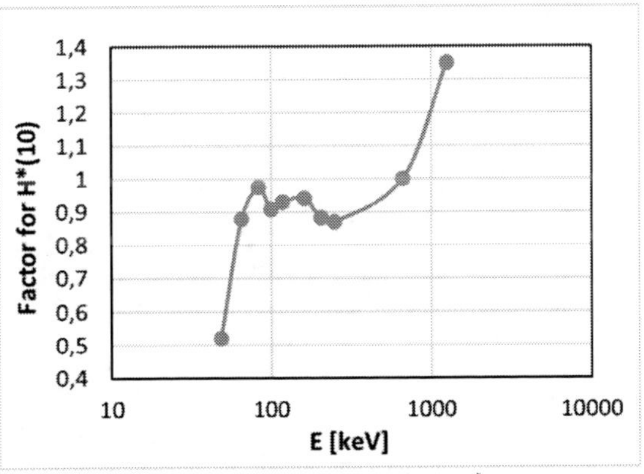

Figure 5: Energy dependency of the displayed $\dot{H}^*(10)$ values of the AD 6

This leads to the conclusion that even for a measurement device known to provide appropriate covering values for radiation protection purposes an analysis of the influence of the energy dependence needs to be performed when used for benchmark measurements.

The LB 6411 neutron probe has a given deviation of ± 40 % for $\dot{H}^*(10)$ in the energy range between 50 keV and 10 MeV [9]. The majority of the calculated results is within this range of deviation.

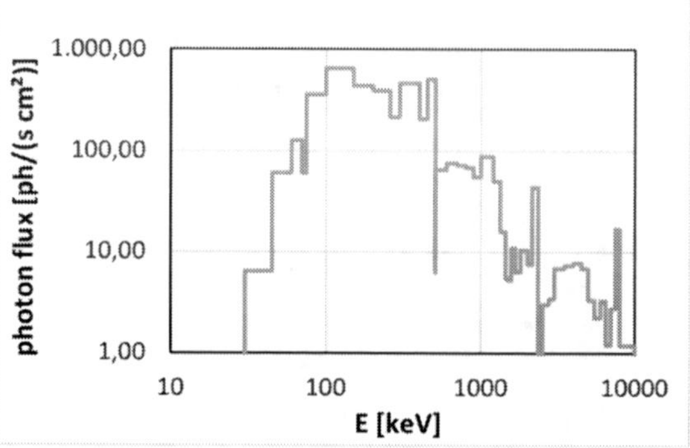

Figure 6: Energy group resolved photon flux at the cask

Furthermore, the point tallies of SCALE/MAVRIC are the standard tallies for all shielding calculations. In opposite to the point detectors of the calculations, real neutron probes have dimensions of spheres with diameters of about 30 cm. In these cases, questions concerning geometrical influences of the probe spheres compared to point detectors and averaging effects need to be quantified.

With this in mind, the shielding sequence of the SCALE package is a suitable tool for assessing the shielding capacities of thick walled casks like the TN85.

It is planned furthermore to use the continuous energy features of the newly released SCALE 6.2 for an assessment of the included MAVRIC shielding sequence.

References

1. Scale: A Comprehensive Modeling and Simulation Suite for Nuclear Safety Analysis and Design, ORNL/TM-2005/39, Version 6.1, June 2011. Available from Radiation Safety Information Computational Center at Oak Ridge National Laboratory as CCC-785.
2. A. Rimpler, F.-M. Börst, D. Seifarth, Neutron measurements around a TN85-type storage cask with high-active waste, Radiation Measurements 45 (2010) 1290-1292.
3. "ICRP Publication 60: 1990 Recommendations of the International Commission on Radiological Protection", Ann. ICRP, Vol. 21 (No. 1-3), International Commission on Radiological Protection (1991)
4. "ICRP Publication 74: Conversion Coefficients for Use in Radiological Protection against External Radiation", Ann. ICRP, Vol. 26 (No. 3/4), International Commission on Radiological Protection (1996).
5. Safety Analysis Report for the TN85, 3644-2-Z (Rev. 8)
6. QA Life Time Records for the TN85-03 CASK, 30.10.2007
7. COGEMA Branche Retraitement (300 AQ 016), Specifications of Vitrified Residues Produced from Reprocessing at UP 2 or UP 3-A La Hague Plants, Second Series, July 1986
8. Product manual for the 6150AD® detector series, Automation und Messtechnik GmbH (automess)
9. Product manual for the LB 6411 neutron probe, BERTHOLD TECHNOLOGIES GmbH & Co. KG

 DOI: 10.1051/epjconf/201715307018

Optimal Shape of a Gamma-ray Collimator: single *vs* double knife edge

Albert Metz [1*], and *Alfred* Hogenbirk [2]

[1] NRG, Consultancy & Services, 6800 ES Arnhem , The Netherlands
[2] NRG, Nuclear Operations, 1755ZG Petten , The Netherlands

Abstract. Gamma-ray collimators in nuclear waste scanners are used for selecting a narrow vertical segment in activity measurements of waste vessels. The system that is used by NRG uses tapered slit collimators of both the single and double knife edge type. The properties of these collimators were investigated by means of Monte Carlo simulations. We found that single knife edge collimators are highly preferable for a conservative estimate of the activity of the waste vessels. These collimators show much less dependence on the angle of incidence of the radiation than double knife edge collimators. This conclusion also applies to cylindrical collimators of the single knife edge type, that are generally used in medical imaging spectroscopy.

1 Introduction

For the characterization of radioactive waste by means of γ-radiation two types of scanners are available. The Tomographic Gamma Scanner (TGS) provides a two dimensional image of slabs from a waste vessel, with a typical height of a few centimetres [1]. The Segmented Gamma Scanner (SGS) also makes use of measurements from different vertical segments, but can only determine the overall activity of a segment [2]. In both types of scanners the waste vessel is usually rotated about the vertical axis as shown in Figure 1.

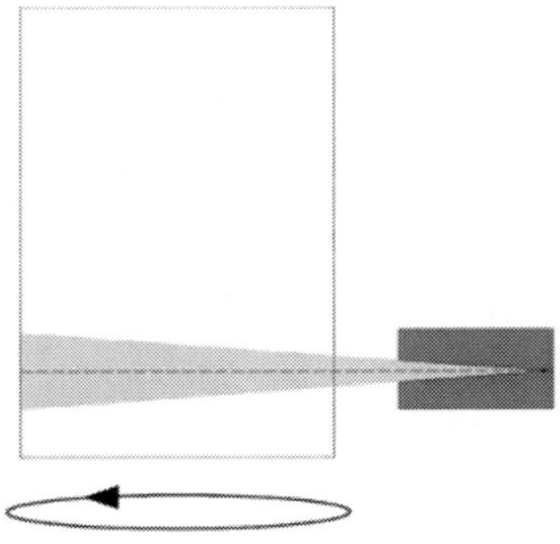

Fig. 1. Schematic view of a waste scanner. Radiation from a segment of the rotating waste vessel (left) is collected in a collimated beam.

In order to obtain a spatial resolution in the axial direction (*i.e.* to be able to differentiate between signals from different vertical segments), it is necessary to limit the angular acceptance of the detection system. This can be arranged by means of a collimating system, which ideally should provide a response that is independent of the angle of incidence within the measured axial segment. Because of the finite angle of acceptance, and also because the vessel is rotated, it is unavoidable that measurements from different segments have some overlap.

Conditions for collimation in the horizontal direction are less severe. When tomographic or imaging techniques can be applied, the horizontal location of the waste within the segment can be determined, so that an efficiency correction can be applied.

For high-level radioactive waste the SGS technique is usually applied, as is done in the VINISH setup at NRG [3]. In this setup the collimating system consists of separate horizontal and vertical collimators, each of which collimates only in one direction. These are mounted on a sliding system, so that collimators can be selected with different apertures. By reducing the collimator aperture, the count rate in the detector can be reduced, so that waste vessels with both high and low activities can be analysed. For the highest activities the collimator aperture should be reduced as much as possible. In that case a considerable fraction of the transmitted beam passes through the collimator body, which is usually made from tungsten.

In this paper we use Monte Carlo simulations to investigate the behaviour of collimating systems with different geometries. These geometries are described in section 2 of this paper. The setup of the simulations is described in section 3. Results on the effect of the acceptance angle, γ-energy, split width and collimator thickness are reported in section 4. In section 5 results are reported for cylindrical collimators, that are

* Corresponding author: metz@nrg.eu

DOI: 10.1051/epjconf/201715307018

commonly used in medical imaging applications, and a comparison with published theoretical results is made. Discussion and conclusions are presented in sections 6 and 7.

2 Collimator types

In this paper we consider two collimator types that have an (apparently) well-defined angle of acceptance. One type has a so-called double knife edge (DKE) and is shown in Figure 2. The other is of the single knife edge type (SKE) and is shown in Figure 3. The angles α and ϕ denote, respectively, the acceptance angle of the collimator and the angle between the – supposedly parallel – beam and the surface normal.

Initially we will consider collimators that have a 1-dimensional slit, *i.e.* they are tapered in the plane of the drawings, but their cross sections do not change in the direction parallel to this plane.

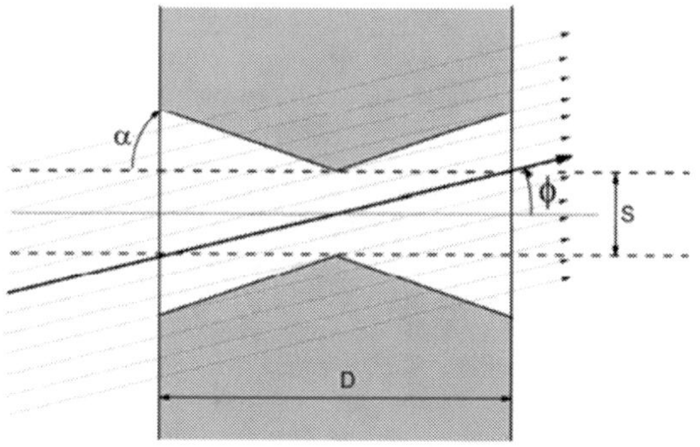

Fig. 2. Cross section of the double knife edge collimator. The symbol S denotes the collimator split, D is the thickness of the collimator. The other symbols are described in the text.

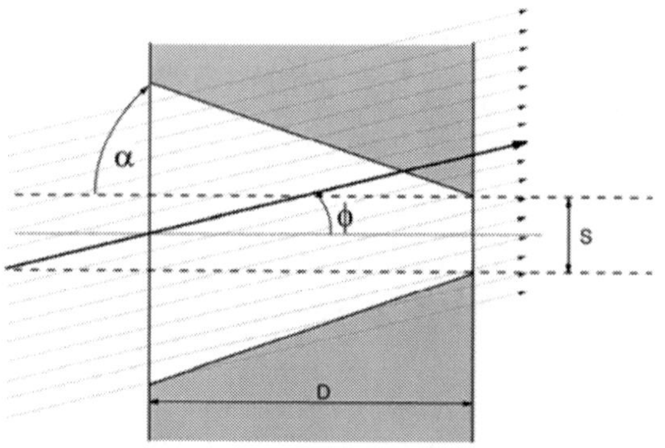

Fig. 3. Cross section of the single knife edge collimator. The meaning of the symbols is as in Fig.2.

3 Monte Carlo Simulations

The imaging properties of cylindrical collimators have been investigated by various authors. For some geometries analytical results have been obtained. These will be discussed in section 5. For the 1D knife edge collimators that form the main subject of this paper no analytical results have been published to the authors' knowledge.

Therefore Monte Carlo simulations have been carried out with MCNPX [4]. For the 1D-cases a line source was used, producing a parallel beam of mono-energetic γ-rays that impinges on the left side of the collimator, as shown in Figure 4. The resulting parallel beam closely approximates the actual situation in a waste scanning setup, where the source detector distance is usually much larger than the opening of the collimator. Transmitted particles were tallied using a flux tally (F2) for a circular surface directly behind the collimators. This is a model for the gamma detector, which will be placed on this location in the case of an actual measurement. The length of the line source (10 cm) was chosen sufficiently large to ensure that both the collimator and the tally surface (diameter 5 cm) are fully covered by the γ-beam for all relevant angles of incidence ϕ. An energy filter of ± 5 keV around the initial γ-energy was applied, so that only unscattered photons were selected. For these photons the transmission of the collimator was calculated by comparison with the photon flux for the case where the collimator is replaced by an air volume in an otherwise unchanged geometry. For the 1D-cases a total of 10^6 particle histories was sufficient to obtain good statistics, with relative errors of the order of 1 % for a transmission of 10^{-2}, and 5 % for a transmission of 10^{-3}.

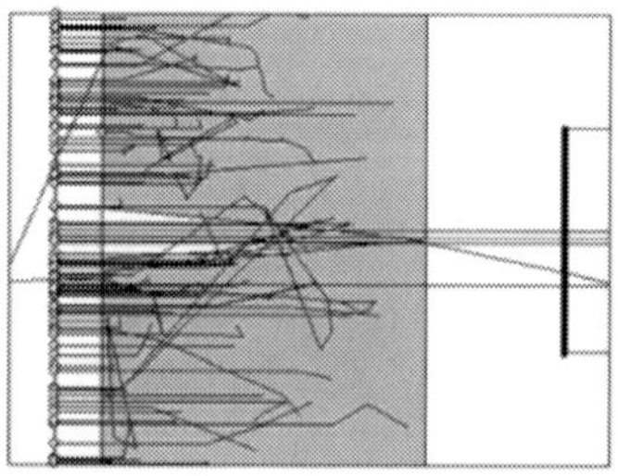

Fig. 4. Cross section of the MCNP model. A parallel beam of γ-rays emerges from the line source on the left. The thick vertical line on the right marks the tally plane.

DOI: 10.1051/epjconf/201715307018

4 Results for 1D-collimators

In all our simulations we find that DKE collimators show a strong dependence of the transmission on the angle of incidence ϕ, whereas SKE collimators have a nearly ϕ-independent response within the angle of acceptance α. As an example we show the results for a typical case, using a photon-energy of 1 MeV, a tungsten collimator with a thickness D of 7 cm, zero split S and an acceptance angle α of 5.7° (corresponding to $\tan(\alpha) = 0.1$). The input parameters are summarized in Table 1.

Table 1. MCNP Input parameters for 1D-collimators.

Material	Tungsten
D (cm)	7.0
S (cm)	0.0
tan (α)	0.1
E (MeV)	1.0
nps	10^6

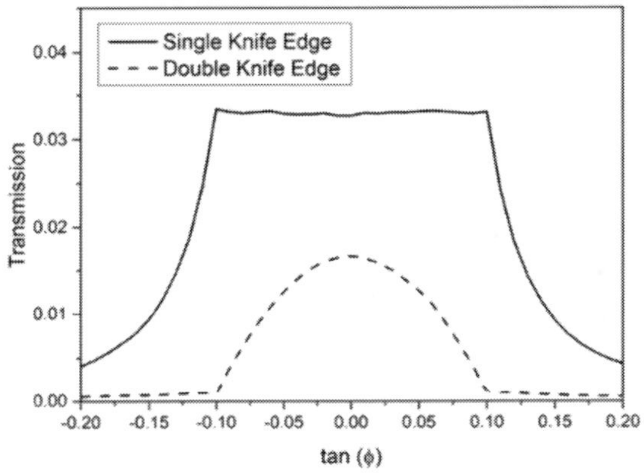

Fig. 5. Transmission curves for single and double knife edge collimators. The relevant parameters are given in Table 1.

For angles of incidence within the acceptance angle of the collimator, ranging from $\tan(\phi) = -0.1$ to $\tan(\phi) = 0.1$, the SKE transmission curve is practically independent of the angle of incidence ϕ, but it shows considerable 'leakage' outside of this region. The DKE collimator, on the other hand, has a strong ϕ-dependence inside the acceptance angle, but outside of this region it drops to zero much more rapidly.

4.1 Effect of acceptance angle

For practical applications it is of interest to investigate the dependence on the acceptance angle. Results for $\tan(\alpha) = 0.01$ to 0.1 (with α ranging from 0.57° to 5.7°) are shown in Figures 6 and 7.

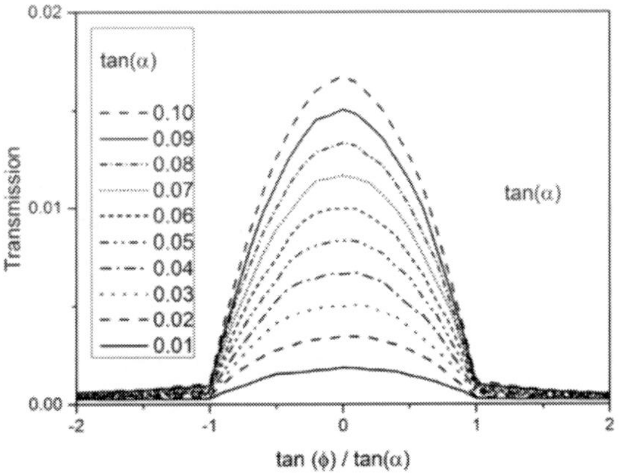

Fig. 6. Transmission curves for DKE collimators for various acceptance angles α. The horizontal axis is scaled to $\tan\alpha$, to allow for comparison of the different cases. The relevant parameters are given in Table 1.

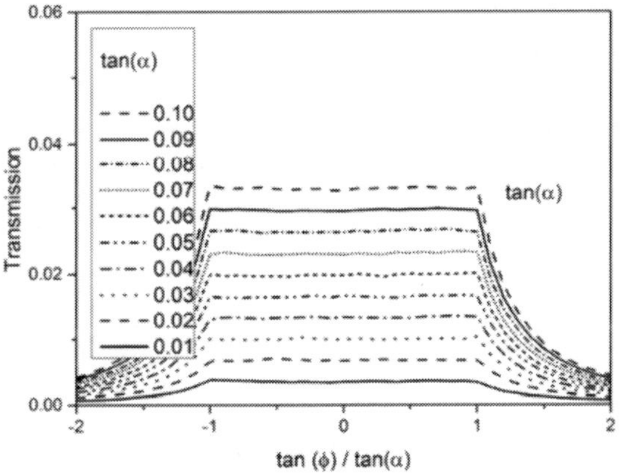

Fig. 7. Transmission curves for SKE collimators for various acceptance angles α.

From these curves it is clear that the qualitative behaviour of the two collimator types (flat response for SKE, versus strongly varying response for DKE) is independent of the acceptance angle α.

4.2 Effect of γ-Energy

A further point of interest is the dependence of the transmission upon the γ-energy. In Figures 8 and 9 transmission curves for various energies are shown, ranging from 0.10 to 1.50 MeV.

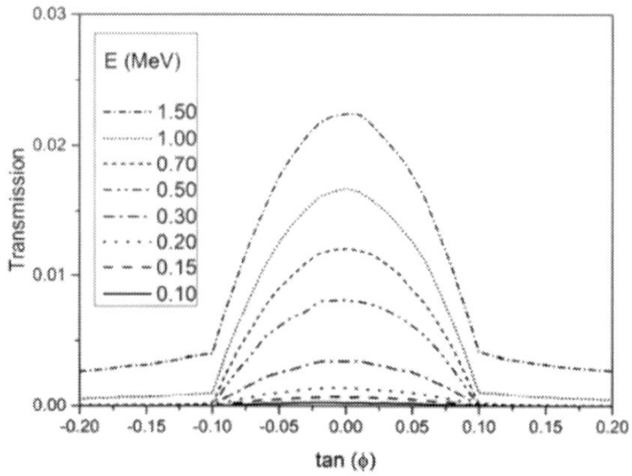

Fig. 8. Transmission curves for DKE collimators for various opening γ-energies E. The relevant parameters are given in Table 1.

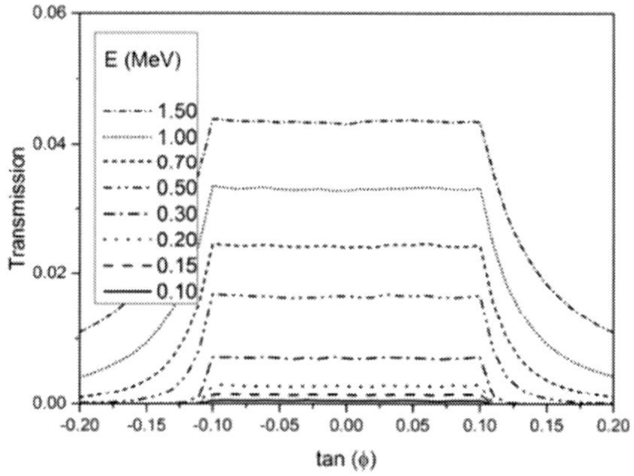

Fig. 9. Transmission curves for SKE collimators for various opening γ-energies E. The relevant parameters are given in Table 1.

From the transmission curves in Figure 9 it can be observed that the amount of leakage outside of the acceptance angle α, which is typical for the SKE collimator, increases with γ-energy. Some leakage is also observed for the DKE collimator (Figure 10), but in this case the effect is much smaller.

4.3 Effect of Collimator Split

When a moderate reduction of the flux is desired, e.g. when measuring a waste vessel of medium or low activity, it is desirable to increase the transmission of the collimator system. This can in principle be achieved by using a collimator split *S* (see Figures 2 and 3) different from zero. In Figures 10 and 11 transmission curves for various collimator splits are shown, ranging from 0.1 to 1.0 cm.

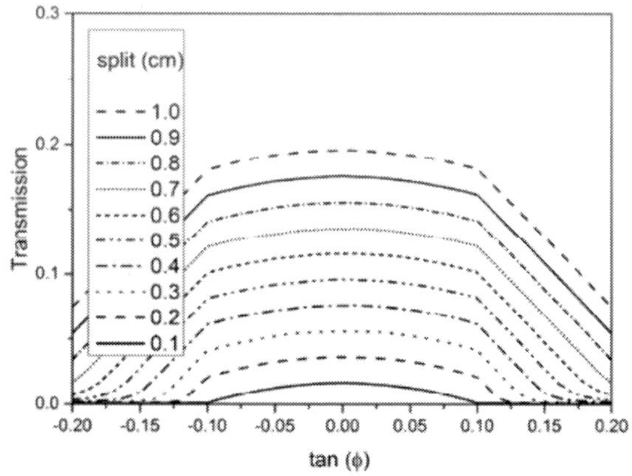

Fig. 10. Transmission curves for DKE collimators for various values of the collimator split width *S*. The other relevant parameters are given in Table 1.

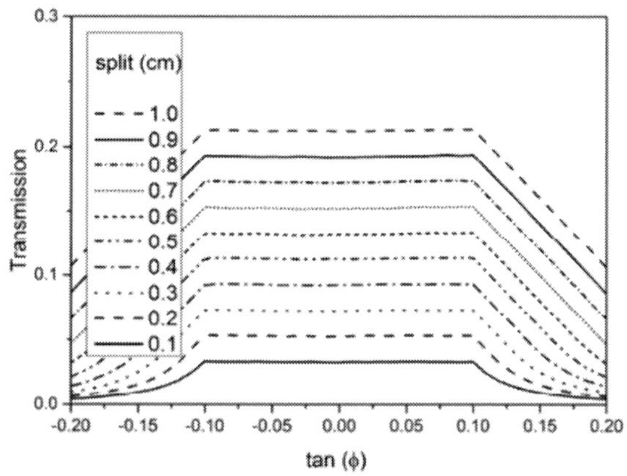

Fig. 11. Transmission curves for SKE collimators for various values of the collimator split width *S*. The other relevant parameters are given in Table 1.

From the transmission curves in Figures 10 and 11 it can be observed, that the transmission increases with increasing split width. The SKE curves (Figure 11) remain flat within the acceptance angle. The DKE curves (Figure 10) become relatively less curved with increasing split width.

4.4 Effect of Collimator Thickness

In this section the dependence of the transmission upon the collimator thickness D is studied. In particular it is interesting to investigate to what extent the leakage of γ-flux outside of the acceptance angle α can be reduced by increasing the collimator thickness. Within a practical thickness range from D = 3 cm up to 9 cm such effects can be well observed for a γ-energy of 0.5 MeV. As in the other simulations an acceptance angle with tanα=0.1 and zero split were used. The results are shown in Figures 12 and 13.

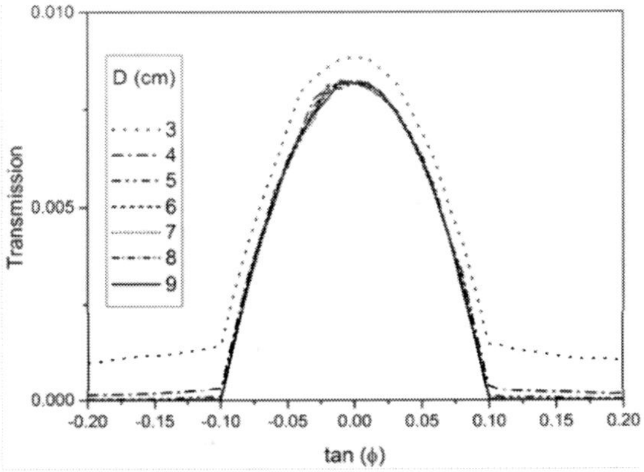

Fig. 12. Transmission curves for DKE collimators for various values of the collimator thickness D. The other relevant parameters are given in Table 1 with the exception of the γ-energy E which is set to 0.5 MeV.

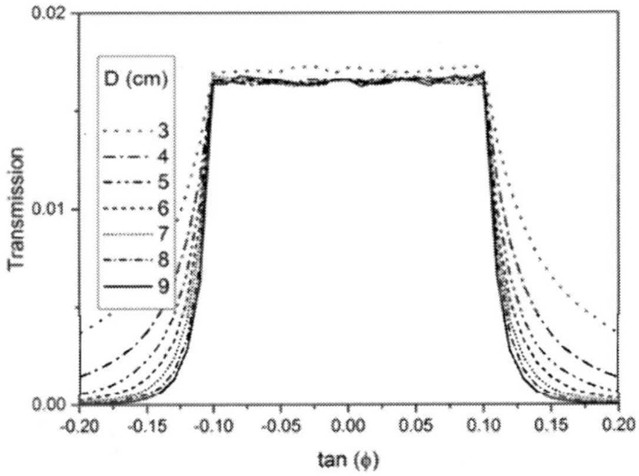

Fig. 13. Transmission curves for SKE collimators for various values of the collimator thickness D. The other relevant parameters are given in Table 1 with the exception of the γ-energy E which is set to 0.5 MeV.

From the results in Figures 13 it can be observed that the transmission within the angle of acceptance is largely independent of the collimator thickness, but that the amount of leakage for the SKE collimator can be significantly reduced by increasing the collimator thickness. For the DKE collimator both the transmission and the leakage can be seen to be largely independent of the collimator thickness, with the exception of the case D=3 cm, which is insufficient to complete block 0.5 MeV γ-rays outside of the acceptance angle of the collimator.

5 Results for cylindrical collimators

Cylindrical collimators are widely used in (medical) imaging applications. In those applications the photon energy is lower and the collimator dimensions are smaller than those commonly used in waste scanners. Moreover the distance from the source to the collimator is chosen relatively small, to achieve a large magnification. In some cases researchers have even used an 'inverted' single knife edge collimator (which diverges from a narrow opening near the source to a wider opening near the detector), to make the distance to the source even smaller [5]. From a qualitative understanding of the physical principles – as discussed in section 6 – it seems likely that the differences that we have observed between one dimensional SKE and DKE collimators, will also occur for cylindrical collimators, with a 3D conical opening.

The behaviour of cylindrical collimators for imaging purposes has been extensively studied, with special attention for imaging properties, in particular the spatial resolution. In a number of publications the so called sensitivity has also been studied, a quantity that is equivalent to the transmission reported in this paper. Analytical results for sensitivity of the DKE collimator have been reported by Metzler *et al.* [6] and for the SKE case by Alnaghy *et al.* [7]. These authors make a distinction between the geometric sensitivity, which accounts for the photons that pass through the collimator aperture, and the penetration sensitivity, that accounts for the photons that pass through the collimator walls. By choosing a split of zero width, the geometric sensitivity can be reduced to zero, so that the transmission is completely described by the penetration sensitivity S_{pen}.

For the cylindrical DKE collimator with zero split (No-Hole Approximation) Metzler *et al.* [6] present a formula, which can be rewritten as:

$$S_{\text{pen;DKE}}(\phi) \approx C \times \cos^5\phi \times (\tan^2\alpha - \tan^2\phi)^{3/2} \quad (1)$$

where C is a constant, our ϕ corresponds to $\pi/2 - \theta$ and our α to $\alpha/2$ in [6].

The formula that Alnaghy *et al.* [7] present for the cylindrical SKE can be rewritten as:

$$S_{\text{pen;SKE}}(\phi) \approx C \times \cos^5\phi \times \left(1 - \frac{\cos^2\alpha}{\cos^2\phi}\right)^{\frac{3}{2}} \quad (2)$$

where C is a constant, our ϕ corresponds to $\pi/2 - \theta$ and our α to $\alpha/2$ in [7].

DOI: 10.1051/epjconf/201715307018

5.1 Effect of γ-Energy for cylindrical collimators

We have investigated the transmission curves for cylindrical collimators for various energies, with the parameters given in Table 2. Because the transmission of the cylindrical collimator is much lower, a higher number of particle histories (nps = 10^8) was required to achieve sufficient statistics , with relative errors of the order of 0.3 % for a transmission of 10^{-3}, and 2 % for a transmission of 10^{-4}.. Instead of the line source that was used for the 1D simulations, a circular surface source was used. The lateral extension of this source (10 cm)was chosen sufficiently large to make sure that both the collimator and the tally surface (diameter 5 cm) were fully covered by the γ-beam for all relevant angles of incidence ϕ.

Table 2. MCNP Input parameters for cylindrical collimators.

Material	Tungsten
D (cm)	7.0
S (cm)	0.0
tan α	0.1
nps	10^8

The results for cylindrical collimators are shown in Figures 14 and 15. From these figures it can be observed that the transmission of a beam of parallel γ-rays behaves in a very similar way as for one dimensional collimators. For the DKE case the transmission curves show a curved profile within the acceptance range, whereas for SKE collimators the transmission profile is practically flat.

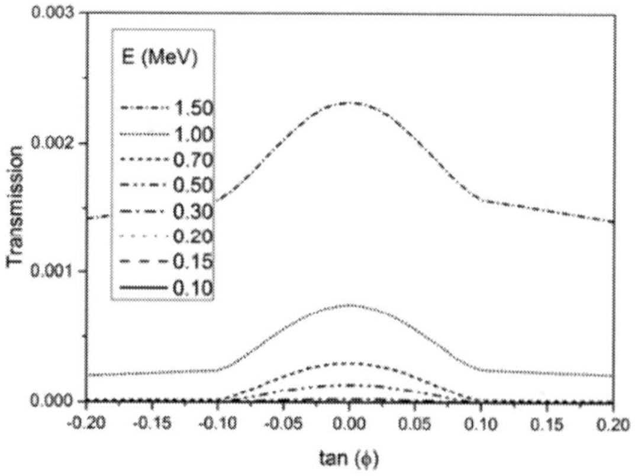

Fig. 14. Transmission curves for cylindrical DKE collimators for various γ-energies. The other relevant parameters are given in Table 2.

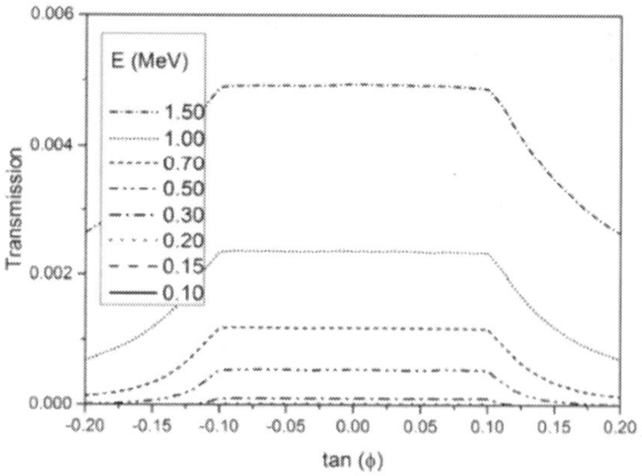

Fig. 15. Transmission curves for cylindrical SKE collimators for various γ-energies. The other relevant parameters are given in Table 2.

5.2 Reversed cylindrical collimators

Because reversed cylindrical SKE collimators are occasionally used in imaging studies [5], these were also investigated. In a reversed SKE collimator, the conical opening diverges towards the detector. The results are shown in Figure 16. Apart from small differences due to the statistical nature of the simulations, the results shown in Figure 16 are identical with those from Figure 15.

This result shows that for this particular case (zero split and parallel γ-beam) the forward *vs.* backward orientation of the collimator does not influence its transmission characteristics.

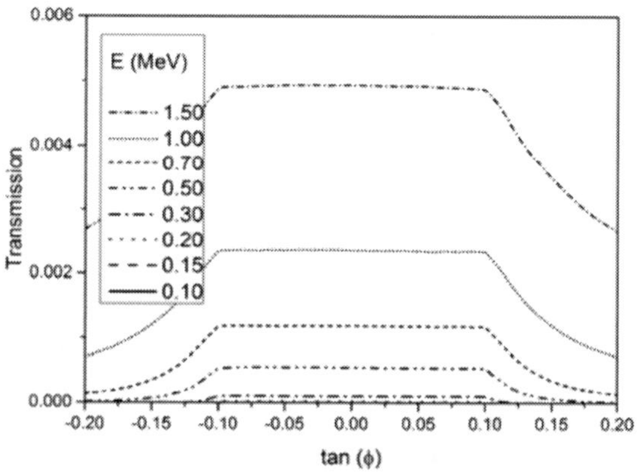

Fig. 16. Transmission curves for reversed cylindrical SKE collimators for various γ-energies. The other relevant parameters are given in Table 2.

6 Discussion

6.1 Comparison with theoretical results

6.1.1 Cylindrical DKE collimator

In Figure 17 we show a comparison between the results from our simulation, and the theoretical formula by Metzler *et al.* [6]. The constant C in formula 1 was adapted to match the maxima of both curves. The parameters used for the simulation are as in Table 2, and the γ-energy was set to 0.5 MeV.

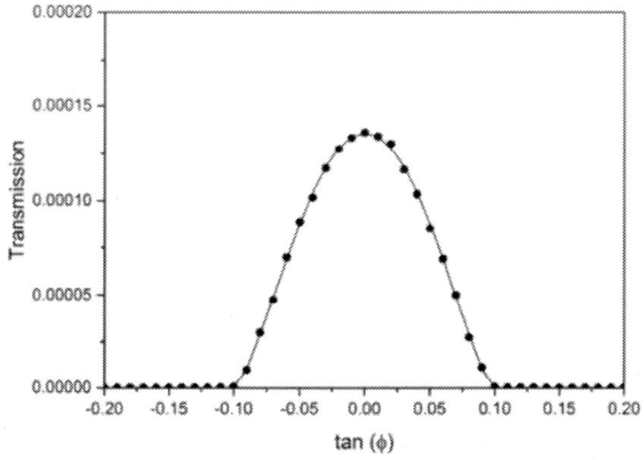

Fig. 17. Comparison between MCNP simulations for a 3D DKE collimator and theoretical results by Metzler *et al.* [6].

From this figure it can be seen that there is a very good agreement between theory and simulation. Note, however, that for higher γ-energies the simulations predict a distinct tail for angles of incidence outside acceptance angle (as shown e.g. in Fig. 14), whereas Equation (1) predicts zero transmission for these conditions.

6.1.2 One dimensional DKE collimator

In Figure 19 we show a comparison between the results from our simulation and the theoretical formula by Metzler *et al.* [6]. The constant C in formula 1 was adapted to match the maxima of both curves. The parameters used for the simulation are as in Table 2, and the γ-energy was set to 0.5 MeV.

From Figure 19 it can be seen that the theoretical formula describes a somewhat narrower profile than the simulations. However, the theoretical formula that was derived for a 3D cylindrical collimator and therefore does not necessarily apply to the 1D case for which the simulations were carried out.

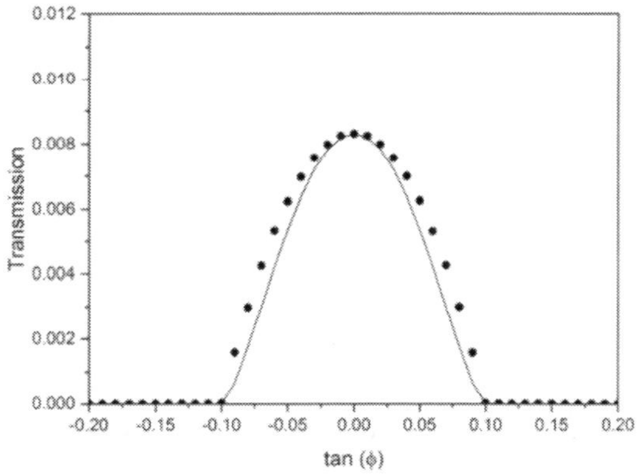

Fig. 18. Comparison between MCNP simulations for a 1D DKE collimator and theoretical results by Metzler *et al.* [6] for a 3D cylindrical collimator.

6.1.3 Cylindrical SKE collimator

In Figure 18 we show a comparison between the results from our simulation, and the theoretical formula by Alnaghy *et al.* [7]. The constant C in formula 2 was adapted to match the maxima of both curves. The parameters used for the simulation are as in Table 2, and the γ-energy was set to 0.5 MeV.

From this figure it can be seen that there is little agreement between theory and simulation. The reason for this discrepancy is unclear, but is must be noted that the theoretical formula is an approximation for the case of a point source, located closely to the collimator, whereas the simulations were carried out for a source at infinite distance.

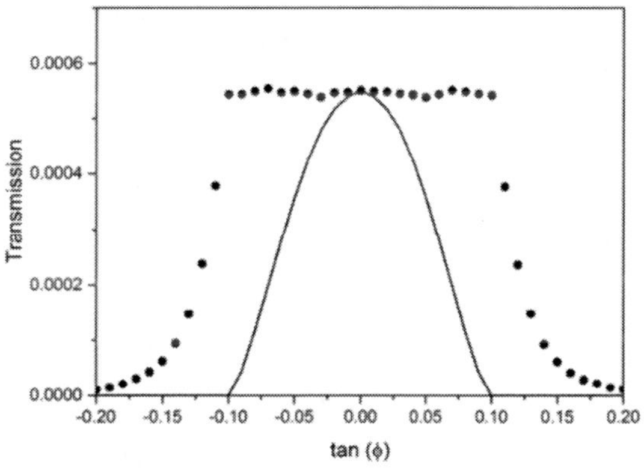

Fig. 19. Comparison between MCNP simulations and theoretical results by Alnaghy *et al.* [7].

6.2 Qualitative explanation

Since there is a marked discrepancy between the theoretical formula by Alnaghy *et al.* [7] and the MCNP simulations reported in this paper, a simple geometrical argument will be presented, that qualitatively explains the observed behaviour of the different collimator types. It is assumed that γ-rays, which have to transverse less than half the thickness *D* of the collimator, pass without attenuation while γ-rays, that have to transverse a greater length, are completely absorbed. This is illustrated in Figure 20 for the DKE collimator.

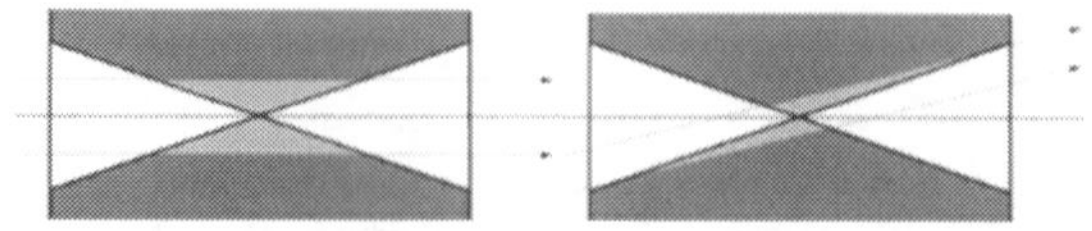

Fig. 20. Qualitative explanation of the angular dependence of the transmission of a DKE collimator. The transparent region is coloured light grey, the opaque region dark grey.

From this figure it can be observed that the effective open area of the collimator rapidly diminishes with increasing angle of incidence, and will completely be closed when the incoming beam grazes the collimator surface, i.e. when ϕ equals α. The case of the SKE collimator is illustrated in Figure 21.

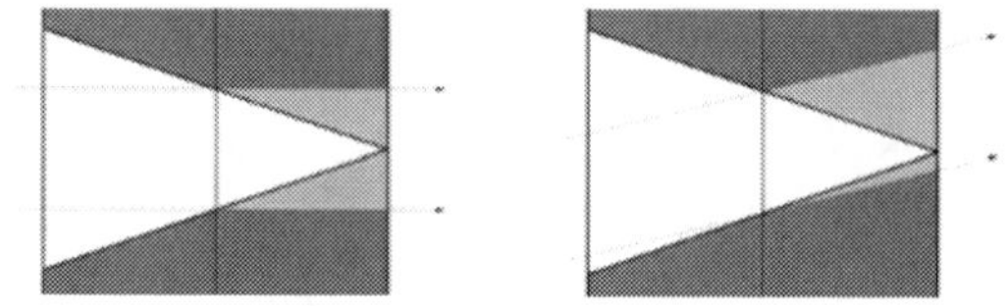

Fig. 21. Qualitative explanation of the angular dependence of the transmission of a SKE collimator. The transparent region is coloured light grey, the opaque region dark grey.

From this figure it can be observed that the effective open area of the collimator does not change with increasing angle of incidence. Beyond grazing incidence, i.e. when $\phi > \alpha$, the transmitted fraction of the beam will be gradually reduced. This qualitative model describes the results from the MCNP simulations quite well.

7 Conclusions

In this paper we have reported results on MCNP calculations used to determine the optimal type of collimator to be used for vertical collimation in a waste scanning system. From the two collimator types that were studied, the SKE collimator was found to have a transmission curve that is practically independent of the angle of incidence, whereas the DKE collimator shows strong variations in transmission. Therefore the SKE collimator is to be preferred when a conservative estimate of the activity inside the waste vessels has to be reported.

A disadvantage of the SKE collimator is that it shows considerable flux leakage outside of the angle of acceptance, especially for high γ-energies. This can be remedied to some extent by increasing the thickness of the collimator.

For waste vessels of lower activity the transmission can be increased by increasing the width of the collimator split. In that case the flat transmission profile of the SKE collimator is preserved, but at larger split widths the difference between SKE and DKE collimators gradually disappears.

Theoretical results for knife edge collimators are only available for cylindrical collimators. They agree well with our MCNP results for DKE collimators, but for SKE collimators marked discrepancies were found. Further theoretical work on SKE collimators is required to clarify these discrepancies.

Acknowledgements

We would like to thank Pieter van Mullekom and Jürgen Willems at Nuclear Fields, for providing insight into possible applications of collimating systems, and Marlies Goorden at Delft University for fruitful discussions about the use of collimators in medical imaging. This work was supported by the Dutch Ministry for Economic Affairs.

DOI: 10.1051/epjconf/201715307018

References

1. J Steven Hansen, *Application Guide To Tomographic Gamma Scanning Of Uranium And Plutonium*, **LA-UR-04-7014**, Los Alamos National Laboratory (2004) TGS
2. T. Bücherl, E. Kaciniel, Ch. Lierse, *Synopsis of Gamma scanning systems*, Report **WG-A-01**, European Commission (1998)
3. https://www.nrg.eu/about-nrg/radioactive-waste-project/technical-equipment/gamma-scanner-for-characterisation-vinish.html?L=1
4. D.B. Pelowitz (ed.), *MCNPX user's manual, version 2.6.0*, Report **LA-CP-07-1473**, Los Alamos National Laboratory (2008)
5. Oleksandra Ivashchenko, Frans van der Have, Marlies C. Goorden, RuudM. Ramakers, and Freek J. Beekman, J. Nucl. Med. **56**, 470 (2015)
6. S.D. Metzler, J.E. Bowsher, M.F. Smith, R.J/ Jaszczak, IEEE Trans. Med. Imaging, **20**, 730 (2001)
7. Saree Alnaghy, Mitra Safavi-Naeini, Daniel R. Franklin, Zhangbo Han, Dean L. Cutajar, Marco Petasecca, Michael Lerch, Anatoly B. Rosenfeld, IEEE Trans. Nucl. Sci., **63**, 1375 (2016)

 DOI: 10.1051/epjconf/201715307019

Estimation of material degradation of VVER-1000 baffle

Davit Harutyunyan[1,2,*], *Michal* Košťál[2], *Stanislav* Vandlík[2], *Anna* Hojná[2], *Martin* Schulc[2], *and Stanislav* Flibor[2]

[1]Department of Nuclear Reactors, FNSFE, CTU Prague, V Holešovičkách 2, Prague 8, Czech Republic
[2] Research Center Rez Ltd., 250 68 Husinec-Rez 130, Czech Republic

Abstract. The planned lifetime of the first commercial VVER-1000 units were designed for 30 to 35 years. Most of the early VVER plants are now reaching and/or passing the 35-year mark. Service life extension for another 10 to 30 years is now under investigation. Life extension requires the evaluation of pressure vessel internals degradation under long-term irradiation. One of the possible limiting factors for the service life of VVERs is a void swelling of the Russian type titanium stabilized stainless 08Ch18N10T steel used to construct the baffle surrounding the core. This article aims to show first steps towards deeper analysis of the baffle degradation process and to demonstrate the possibilities of precise calculation and measurements on the VVER-1000 mock-up in LR-0 reactor.

1 Introduction

At present, more knowledge of void swelling formation and development in the thermal reactor conditions is needed. Reliable data of voids swelling have been obtained in conditions of different neutron spectrum - in fast reactors or accelerators [1], [2]. Data shows 20 % and higher void swelling presence in the steels results in a drastic and total loss of strength and plasticity, but the findings were obtained for the steels irradiated up to 72 dpa at fast neutron spectrum (BOR-60 reactor) at about 400-550°C.

Certainly, the impact of voids swelling on material behavior, especially on mechanical strength or fracture resistance is very devastating, but it is a question if these warnings are applicable to Light Water Reactors LWRs, because total void swelling observed to date in Flux Thimble Tube of Pressure Water Reactors PWRs, up to 80 displacement per atom dpa, has been relatively low, under 1 % [3]. One dpa means that every atom has been displaced once on average. It is material dependent measure of an exposure.

2 Void swelling in VVER -1000 baffle

The degradation effect from gamma rays in the case of VVER-1000 reactor pressure vessel (RPV) does not exceed 4 % of total dpa. In case of baffle the source of material degradation is mainly by high energy gamma origin by neutron capture.

Our calculations for VVER-1000 mock-up assembled at LR-0 zero power experimental reactor (Research Centre Rez) show that neutron share on radiation heating in thicker parts of baffle is less than 3 %. Deposited energy from radiation heating is transferred to the steel lattice causing atomic displacements – radiation damage. The damage causes changes of material behavior including some void swelling which results an increase in the external dimension of solid materials after irradiation.

Void swelling of the reactor internal parts is closely related to the temperature at the site. Increasing temperature causes higher probability of gas formation and trapping it in the barrel material. This temperature may be significantly higher than the temperature of the coolant in internal parts of reactor because there is significant heating due to gamma radiation. Gamma caused heating got origin by fission, decay of fission products in fuel or neutron interactions. Mainly in thicker parts of the baffle, an important role is played by the gamma originated by thermal capture on steel (7.63MeV and 7.65 MeV) or on $_1^1H$ (2.23 MeV).

Impact of gamma irradiation from capture on $_1^1H$ is notable in vicinity of cooling channels (1-13 on the Fig.1). Since, these reactions are caused mainly by thermal neutrons, theirs correct description is fundamental for accurate gamma field distribution description. Temperature shift in baffle material is caused mainly by gamma heating in contrast to the surrounding water. Eq. 1 estimates void swelling originated by radiation impact [4]:

$$S(k,D,T) = R(k)\varphi(D - D_0(k,T))\exp\left\{-\frac{(T - T_{\max}(k))^2}{2\sigma_T^2(k)}\right\} \quad (1)$$

where $S(k,D,T)$ is the swelling (%), D is the damage dose (dpa), T is the irradiation temperature (°C), k is the dose rate (dpa/s), $\varphi(x)=x\cdot\theta(x)$, $\theta(x)$ is Heaviside unit step function, $\theta(x)=1$, $x > 0$ or 0, $x \leq 0$,

* Corresponding author: davit.harutyunyan@cvrez.cz

$R(k)=r_0-r_k$ ln k is the swelling rate at steady state (%/dpa), $D_0(k,T)=d_0-d_TT+dk$ ln k is the incubation period (dpa), $T_{max}(k)=T_0+T_k$ ln k is the peak swelling temperature (°C), $\sigma_t(k)=\sigma_0-\sigma_k$ ln k is the temperature dispersion (°C).

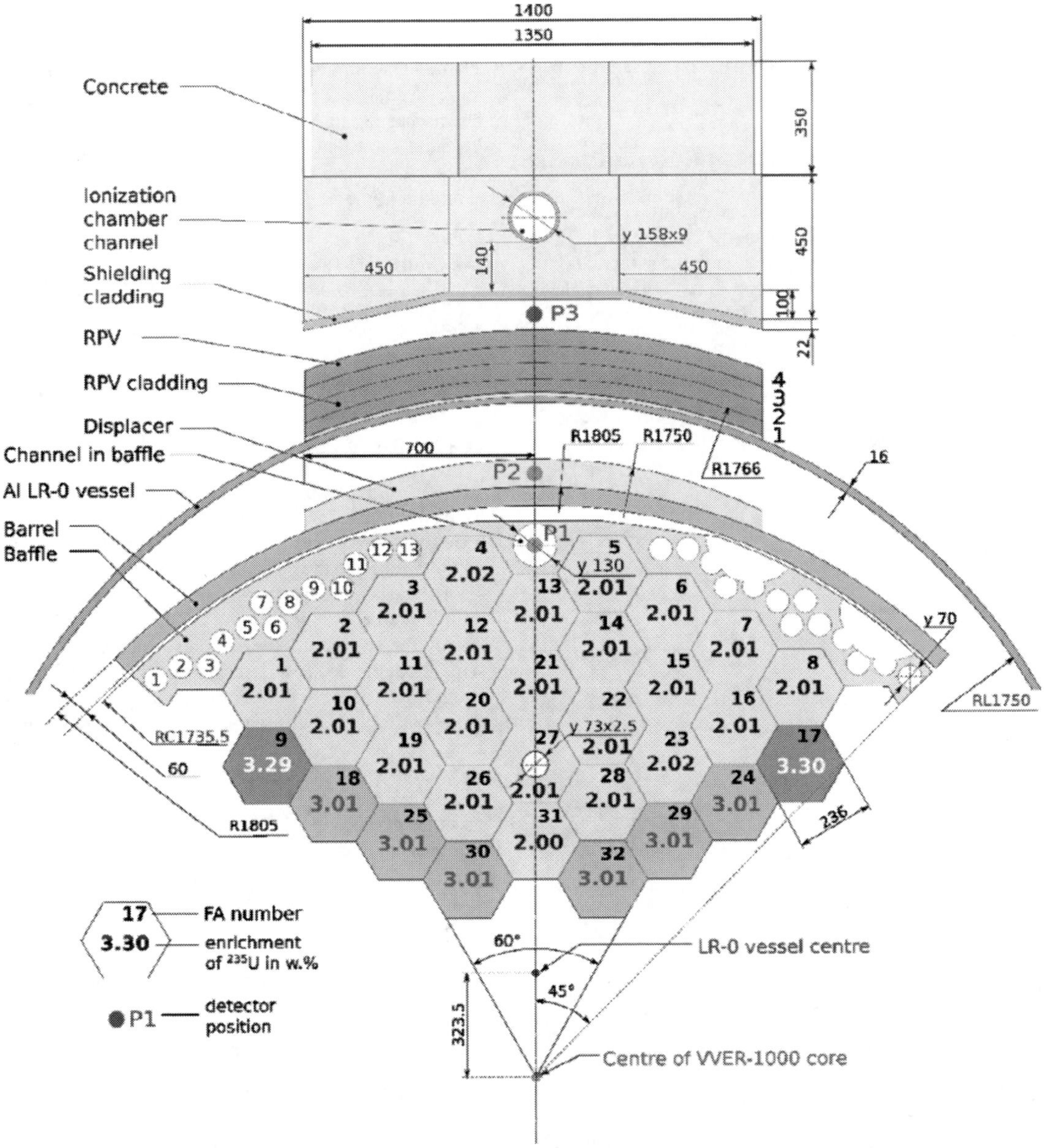

Fig. 1 Section of the VVER-1000 mock-up in LR-0 reactor in the x-y plane (Dimensions in mm).

Empirical equation with dependence on dpa rate was employed in computations. The creation of Frenkel pairs – dpa according to widely used NRT theory [5] is given by the Kinchin–Pease see formula 2.

$$DPA = k\frac{E_a}{2E_d} \quad (2)$$

where E_a stands for available energy, E_d for the Lindhard cut-off energy required for displacing one atom from its normal site and k for displacement efficiency. However, this formula does not account for any thermally activated recombination of damage, nor the well-known fact that in metals the damage production is for high energies only approximately 20 % of the Kinchin-Pease prediction [6]. Formula also does not take into account ion channelling, which may in crystalline or polycrystalline materials reduce the nuclear deposited energy and thus the damage production dramatically for some ion-target combinations. This suggests that calculated value is upper limit of dpa estimate. Incident particles of different types and energies produce different types of primary damage (neutrons, heavy charged particles, electrons, photons), they produce differences in secondary damage accumulation. To estimate radiation damage we need to know both energy spectrum and absolute flux level. Fission fragments have limited range, primarily impacts the fuel. Most important source of dislocations in the baffle is caused by fast neutrons.

3 Experimental and calculation methods

Displacement per atom and gamma heating calculations were done with MCNP6 model of LR-0 VVER-1000 mock-up. This model is validated with experimental data gathered on VVER-1000 mock-up measurements [7] [8] [9]. The MCNP calculations provide data which are impossible to measure, such as deposited energy of particles. Completing this data with thermo-hydraulic computation, it is possible to obtain all values in Eq. 1 for the void swelling determination. To obtain required data for void swelling determination approximately 27×23 cm part of reactor baffle was chosen, see Fig.2.

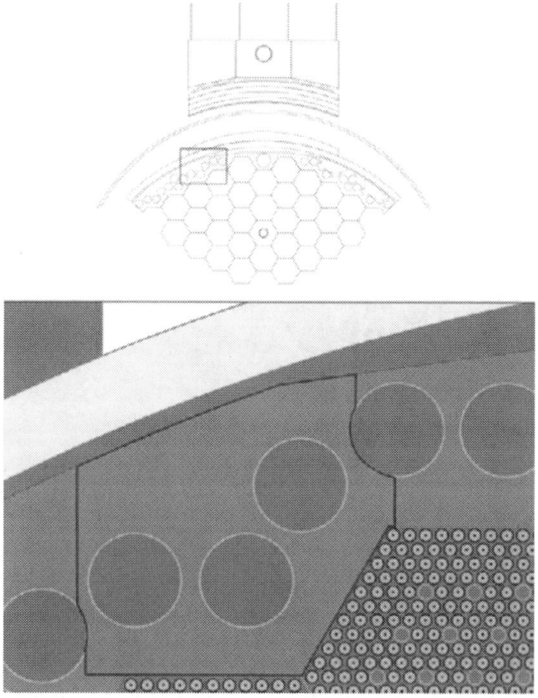

Fig. 2 Selected part of baffle.

3.1. Experimental methods

The mock-up core consists of 32 dismountable assemblies with enrichment 2 %, 3 % and 3.3 % of 235U. Demineralized water with dissolved boric acid with 4.6 ± 0.1 g/l is used as a moderator. The fuel assemblies are of shortened VVER type, (sintered UO2 pellets, outer diameter 7.53 mm, central inner hole 1.4 mm, active length 125 cm, total length 135.7 cm, Zr cladding tube with outer diameter 9.15 mm, wall thickness 0.72 mm). Mock-up includes radially full scale VVER-1000 barrel, baffle simulators and concrete shielding. For computational models validation the neutron spectra measurements were used in points P2 and P3 (see Fig.1). Neutron spectra below 1 MeV were measured using proton recoiled method with a hydrogen chamber, which was operated at 400 kPa over an energy range from 0.111 MeV to 0.368 MeV and at 1000 kPa in the higher range. For energies from 1 MeV to 10 MeV a stilbene scintillator was used with neutron and gamma pulse shape discrimination. Details of this measurement could be found in [10]

3.2. Calculation methods

Gamma heating and dpa were calculated via MCNP6 model of VVER-1000 mock-up, using mesh tally for the chosen part. TORT calculations were used for verification of MCNP results.

3.2.1. MCNP model

MCNP computation model is fixed source: where the neutron emission from fuel is defined directly with external sources and fission is dealt as absorption, this is mainly done for the computation time saving. The model geometry is maximally realistic and detailed including stainless steel components of the reactor core, baffle and barrel models up to the concrete biological shielding. This model is detailed described in [7]. The fixed source model uses directly defined neutron or gamma emission from the fuel. Neutron emission density was defined by means of fission density obtained in critical calculation in MCNP. The total gammas in selected part of baffle Fig. 2 are coming from the primary gamma from the fission and secondary. The primary gamma is absent in the case of fixed neutron source calculation. They are computed with additional calculation with gamma source in fuel. To calculate the available energy E_a from the Eq. 2, flux mesh tally (flux averaged over the grid cell) has to be modified by the damage energy production cross section library IRDF-2002 [12]. The Lindhard cut-off energy and displacement efficiency are chosen according to ref. [11]: E_d = 40 eV and k = 0.8 for major steel components. ENDF VII.1 [13] nuclear data library were used for the calculation model compilation.

3.2.2 TORT model

Used TORT model is composed of 1/6th energetic VVER-1000 reactor core, i.e. 28 fuel assembles of VVER-1000 reactor core and all components from baffle to concrete shielding are included. The *above-*mentioned TORT 3D model is currently used for neutron and gamma flux calculations for nuclear power plant (NPP). The cross section library used for presented calculations is BUGLE-96 (RSICC, 1999) 47-neutron-group/20-gamma-ray-group collapsed library based on ENDF/B-VI. This model is described in detail in [7].

4 Results

Presented results show the dpa and gamma heating maps in the chosen region of baffle. Calculated results with MCNP for mock-up were scaled to the VVER-1000 nominal power of 3000 MW thermal power of energetic nuclear power plant. As it is obvious from the Fig. 3 and 4 the distribution of dpa and gamma heating is very similar in both calculations. In both cases the distribution is taken from the middle of core height. TORT calculation is done for

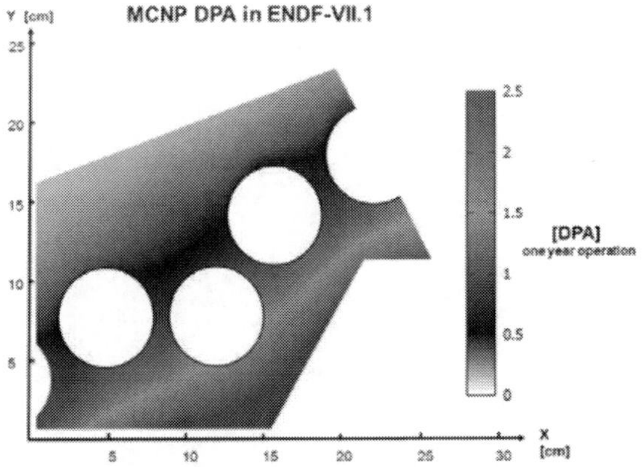

Fig. 3 Relative distribution of dpa in the selected part of baffle

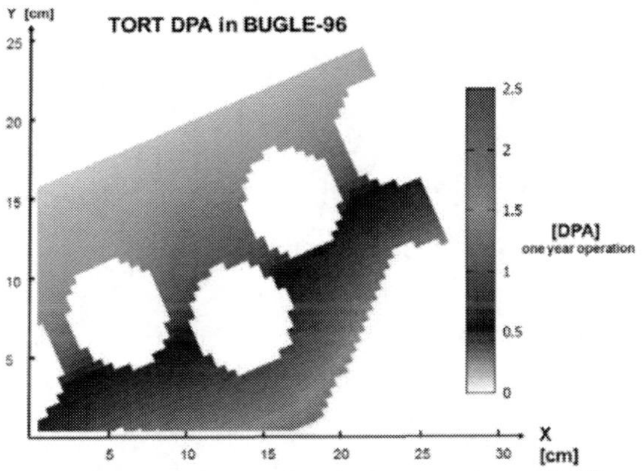

Fig. 4 Relative distribution of dpa in the selected part of baffle

the energetic reactor and there are several differences compared with MCNP in initial data, mainly composition of fuel (burn up included in case of TORT) fuel length: 125 cm in case of MCNP model and 350 cm in case of TORT model, length of fuel cycle and others described in [7].

Gamma heating map presented below Fig. 5 and 6 were obtained using same technique as in the case of dpa. In both cases the distribution is similar and the differences in absolute values are mainly because of above mentioned reasons.

5 Uncertainties and validation

Calculation uncertainty covers statistical uncertainty of calculation less than 1% MCNP or convergence criterion of 0.001 in case of TORT. The statistical uncertainties can be neglected, as they are negligible with regard to parameters uncertainties. For parameters uncertainties estimation the sensitive analysis were performed.

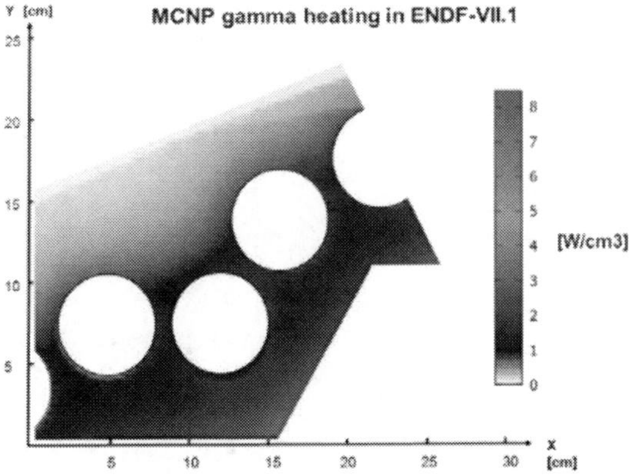

Fig. 5 Relative gamma heating map in the selected part of baffle

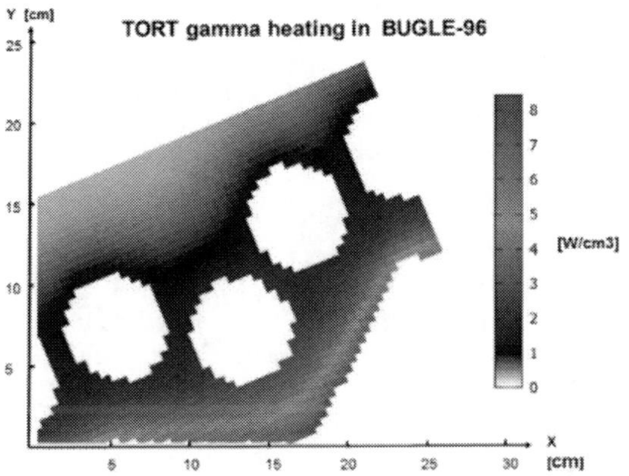

Fig. 6 Relative gamma heating map in the selected part of baffle

The main sources of calculation uncertainties are baffle steel density, boric acid concentration, fuel enrichment and gap between fuel pins and the baffle.
For dpa the main contributions to uncertainty are baffle steel density up to 1.4 %, boric acid concentration up to 0.12 %, fuel enrichment up to 0.25 % and gap distance up to 0.8 % giving total uncertainty in dpa 1.6 %.

For gamma heating the corresponding contributions to uncertainty are baffle steel density up to 2.7 %, boric acid concentration up to 0.15 %, fuel enrichment up to 1.4 % and gap distance up to 3.0 % giving total uncertainty in gamma heating values 4.3 %. The total standard uncertainty was determined by summing the values of square standard uncertainty for all of the parameters and taking the square root to convert the variance to a standard uncertainty.

DOI: 10.1051/epjconf/201715307019

Due to considerable differences in MCNP and TORT geometries the relative neutron fluxes are calculated in 3 specified points P1-P3, which are shown on the Fig. 1 and compared with measured spectra in the point P2 and P3. Fig. 7 shows relative neutron spectra for MCNP, TORT and experiment. Considering good agreement between the models and measurement the results of dpa and gamma heating calculations seem reasonable. The measurement uncertainties can be found in [10].

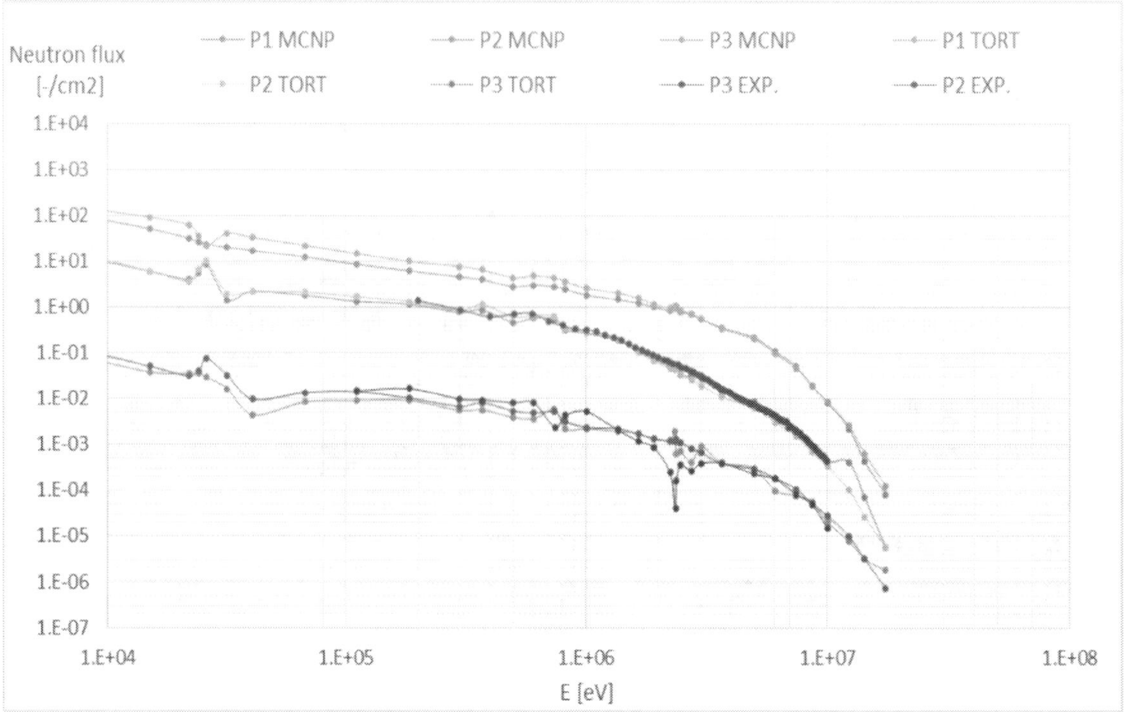

Fig. 7 Relative neutron spectra in position 1, 2 and 3

6 Conclusions

MCNP results for dpa and gamma heating in selected part of baffle are in an agreement with TORT results. Both distributions are comparable. Maximum dpa integrated over a year was estimated using MCNP as 2.5 and by TORT as 1.5 dpa. Maximum gamma heating in MCNP calculation is 4 W/cm3 compared with 8 W/cm3 in case of TORT. Used methods for absolute dpa and gamma heating estimation via MCNP calculations seems to be applicable for the energetic VVER-1000 reactor. The reason for a variance in the absolute values are due to the slight differences in nuclear parameters, geometry definition of computational models and by scaling process in MCNP. Dpa and gamma heating are the input data for void swelling estimation. In accordance with our experience we regard stochastic codes (MCNP) to be more reliable and with smaller margin of error, than codes using deterministic method.

Acknowledgement

The presented work was financially supported by the grant for Project SGS16/174/OHK4/2T/ 4, FNSPE of the CTU, department of Nuclear Reactors and the Ministry of Education, Youth and Sport Czech Republic Project LQ1603 (Research for SUSEN).

References

1. V.S. Neustroev, F.A. Garner / Journal of Nuclear Materials 386–388 (2009) 157–160.
2. B.Z. Margolin, I.P. Kursevich, A.A. Sorokin, V.S. Neustrojev: "The Relationship of Radiation Embrittlement and Swelling for Austenitic Steels for WWER Internals", Proceedings of the ASME 2009, Pressure Vessels and Piping conference, Prague, 2009.
3. J. L. Nelson et al., Proceedings of the 16th International Symposium on Environmental Degradation of Materials in Nuclear Power Systems - Water Reactors, Asheville, USA, 2013.
4. A.S. Kalchenko, V.V. Bryk, N.P. Lazarev, V.N. Voyevodin, F.A. Garner, Prediction of void swelling in the baffle ring of WWER-1000 reactors for service life of 30–60 years, Journal of Nuclear

Materials, Volume 437, Issues 1–3, June 2013, Pages 415-423, ISSN 0022-3115.

5 Norgett, M. J., Robinson, M. T. and Torrens, I. M. A proposed method of calculating v displacement dose rates. Nucl. Eng. Des. 33, 50–54 (1975).

6 R. S. Averback and T. Diaz de la Rubia, Displacement damage in irradiated metals and semiconductors, in Solid State Physics, edited by H. Ehrenfest and F. Spaepen, volume 51, pages 281--402, Academic Press, New York, 1998.

7 D. Harutyunyan, S. Vandlík, M. Schulc, M. Ruscak, E. Novak, Applicability of LR-0 mock-up results to VVER-1000 reactor pressure vessel issues, Annals of Nuclear Energy, Volume 98, December 2016, Pages 157-165, ISSN 0306-4549.

8 M. Košt'ál, V. Rypar, M. Švadlenková, The pin power distribution in the VVER-1000 mock-up on the LR-0 research reactor, Nucl. Eng. Des., 242 (2012), pp. 201–214 ISSN 0029-5493

9 M. Košt'ál, M. Švadlenková, J. Milčák, Absolute determination of power density in the VVER-1000 mock-up on the LR-0 research reactor, Appl. Radiat. Isot., 78 (2013), pp. 38–45 ISSN 0969-8043

10 M. Košt'ál, F. Cvachovec, V. Rypar, V. Juříček, Calculation and measurement of neutron flux in the VVER-1000 mock-up on the LR-0 research reactor, Annals of Nuclear Energy, Volume 40, Issue 1, February 2012, Pages 25-34, ISSN 0306-4549

11 Greenwood, L. R. and Smither, R. K. SPECTER; neutron damage calculations for materials irradiations. ANL/FPP/TM-197. Argonne National Laboratory (1985).

12 IAEA/NDS IRDF-2002 Damage Cross Section File 2002, https://www-nds.iaea.org/irdf2002/data/irdf2002_damage.dat

13 M.B. Chadwick, et al. s.l.: ENDF/B-VII.1 Nuclear data for science and technology: cross sections, covariances, fission product yields and decay data Nucl. Data Sheets, 112 (2011), pp. 2887–2996

Photon Shielding Features of Quarry Tuff

Hector Rene Vega-Carrillo[1,a], Luis Hernandez-Adame[2], Karen Arlete Guzman-Garcia[3], Arturo Agustin Ortiz-Hernandez[4], Jose Antonio Rodriguez-Rodriguez[5] and Cesar Antonio Juarez-Alvarado[5]

[1]*Universidad Autonoma de Zacatecas, UA de Estudios Nucleares, Cipres 10, Fracc. La Peñuela, 98068 Zacatecas, Zac., Mexico*
[2]*Laboratoire de Chimie Physique, UMR 8000-CNRS, Universite´ Paris-Sud, 91405 Orsay, France*
[3]*Universidad Politecnica de Madrid, Dept. Ingenieria Energetica, C. Jose Gutierrez Abascal 2, 28600 Madrid, Spain*
[4]*Instituto Politecnico Nacional, Campus Zacatecas, Blvd. El Bote, Zacatecas, Zac. Mexico*
[5]*Universidad Autonoma de Nuevo Leon, Fac. de Ingenieria Civil, C. Pedro de Alba s/n, San Nicolas de los Garza, NL, Mexico*

Abstract. Cantera is a quarry tuff widely used in the building industry; in this work the shielding features of cantera were determined. The shielding characteristics were calculated using XCOM and MCNP5 codes for 0.03, 0.07, 0.1, 0.3, 0.662, 1, 2, and 3 MeV photons. With XCOM the mass interaction coefficients, and the total mass attenuation coefficients, were calculated. With the MCNP5 code a transmission experiment was modelled using a point-like source located 42 cm apart from a point-like detector. Between the source and the detector, cantera pieces with different thickness, ranging from 0 to 40 cm were included. The collided and uncollided photon fluence, the Kerma in air and the Ambient dose equivalent were estimated. With the uncollided fluence the linear attenuation coefficients were determined and compared with those calculated with XCOM. The linear attenuation coefficient for 0.662 MeV photons was compared with the coefficient measured with a NaI(Tl)-based γ-ray spectrometer and a ^{137}Cs source.

1 Introduction

Due to the importance for radiation protection the shielding characteristics of man-made or natural occurring materials have been investigated. Thus, different shielding properties of building materials [1-3], ores [4], glasses [5, 6], plastics and polymers [7], Lunar soil samples [8], gemstones, like the Amethyst [9], Boron containing materials [10], alloys [11], nanoparticles of W and polyethylene dispersed in polymer [12], and concrete with different aggregates [13-16] have been reported.

Some of the shielding features include the effective atomic number, the effective electron density, the half value layer, the energy absorption and exposure buildup factors, the linear attenuation coefficient (μ), and the mass attenuation coefficient (μ/ρ) for photons.

In the cases of lunar soils the dose reduction and the protection against Galactic Cosmic Rays have been studied [8].

Another important shielding feature includes the materials behaviour against neutrons, thus Singh, Badiger and El-Khayatt [1] reported the shielding features against neutrons of building materials.

Also designed materials, like the Hormirad™, which is a high-density concrete has been evaluated as neutron shielding using experimental procedures with isotopic neutron sources and through Monte Carlo calculations [17]. Another designed material, known as NGS-concrete, polymers and standard cements mortars have been studied in their shielding features [18, 19]. In these studies, the half value layer and the ten value layer were evaluated.

The shielding characterization has been carried out through measurements, calculations or combining both procedures.

Tuffs are volcanic rocks made of an ash matrix with grain sizes ranging from fine clay minerals up to silt-sized material, containing lithic, vitric and crystal clasts. Tuffs are formed from pyroclastic flow, surge and fall deposits that combined with the primitive magma features, the deposition and compaction of the emitted material and the post depositional processes concur to find tuffs with different textures, and petrographic and chemical compositions [20].

Volcanic tuffs minerals have been proposed to improve the indoor environments reducing the microbial contaminants and the airborne fungal carcinogens [21].

Quarry tuffs are mostly soft and porous rocks used as building stones and for artwork because can be easily cut and reworked. Volcanic tuff stones are in different colours, they are used as covering materials for insulating and ornamental purposes on the exterior and interior of buildings [22-25].

[a] Corresponding author: fermineutron@yahoo.com

In the construction industry the quarry tuff (cantera) is used to coat the interior and exterior wall surfaces of buildings and houses avoiding the use of paint. Its use is part of the Mexico's stone heritage being present in pre-Hispanic, Colonial and modern architecture [26]. A radiometric analysis of this material from Turkey has been reported [24, 25], being in mostly of cases safe.

In Mexico, there are several cities where the cantera is widely used, it is found in prehispanic constructions, cathedrals, and modern buildings and houses. Cantera is used and the main construction material or as decorative item attached to the walls.

In several facilities having x-ray machines or sealed radioactive sources cantera is attached to the walls however, its features to shield x-or-γ photons are unknown; therefore in facilities with cantera hosting x-ray units or γ-ray sources the cantera is not accounted for shielding design or to evaluate the shielding effectiveness.

The aim of this work was to estimate the shielding characteristics of cantera using Monte Carlo methods and the XCOM code [27] for 0.03, 0.07, 0.1, 0.3, 0.662, 1, 2, and 3 MeV mono-energetic photons. The relative transmission of Kerma in air, Ka, and the Ambient dose equivalent, H*(10), was also determined.

For 0.662 MeV photons the calculated μ was compared with the μ measured using a NaI(Tl)-based γ-ray spectrometer and a ^{137}Cs source.

2 Materials and methods

From the local market a large piece of cantera was purchased and it was cut in 10x10 cm pieces with different thickness, ranging from 1 up to 40 cm. Each piece was weighted to determine the density, being 1.8 ± 2 % g cm^{-3}.

From the large piece of cantera several samples were taken in order to measure the chemical composition using Energy Dispersive X-ray Fluorescence. Cantera is mainly composed by SiO_2, Al_2O_3, K_2O, Na_2O, Fe_2O_3, and CaO, being in agreement with the chemical composition reported by Celik and Ergul [28]. Due to the total alkali (Na_2O+K_2O) versus silica (SiO_2) composition, this cantera tuff is Rhyolite [28, 29].

2.1 Calculations

The chemical composition of cantera was used to calculate the elemental concentration in weight fraction, being: Oxygen (0.4604 w/o), Silicon (0.2476 w/o), Aluminium (0.1022 w/o), Carbon (0.0651 w/o), Iron (0.0509 w/o), Potassium (0.0459 w/o), Sodium (0.0156 w/o), Magnesium (0.0074 w/o), and Calcium (0.0049 w/o). The elemental concentration was used to calculate the shielding features against photons using the XCOM and the Monte Carlo code MCNP5 [30].

2.1.1 XCOM

The XCOM code uses a large data base with Coherent scattering, Compton scattering (Incoherent scattering), Photoelectric absorption, and Pair production cross sections for several elements.

The chemical composition of cantera was input in the XCOM code and the mass interaction coefficients and the total mass attenuation coefficients, for 1 keV to 100 GeV photons, were calculated. The mass attenuation coefficient is calculated using equation 1.

$$\left(\frac{\mu}{\rho}\right) = \sum_i w_i \left(\frac{\mu}{\rho}\right)_i \qquad (1)$$

In equation 1, ρ is the cantera density, and w_i and $(\mu/\rho)_i$ are the weight fractions and mass attenuation coefficient of the constituent element i respectively.

The density of cantera and the total mass attenuation coefficients were used to calculate the linear attenuation coefficient, μ, for photons of 0.03, 0.07, 0.10, 0.3, 0.662, 1, 2 and 3 MeV, using equation 2.

$$\mu(E) = \rho \left(\frac{\mu}{\rho}\right)(E) \qquad (2)$$

Here, μ(E) is the linear attenuation coefficient of cantera for photons of E MeV, ρ is the density of cantera, and (μ/ρ)(E) is the mass attenuation coefficient of cantera for photons of E MeV.

2.1.2 MCNP5

In the MCNP5 calculations a model of a transmission experiment was built. Here, a point-like and isotropic photon source was used where the source term was mono-energetic, with photons of 30, 70, 100, and 300 keV in order to cover the x-rays for diagnosis, and 0.662, 1, 2 and 3 MeV to include gamma-ray sources.

A point-like detector was placed 42 cm from the source, between the source and the detector models of the cantera pieces where included. Calculations were carried out for 0, 0.5, 1, 2, …, 39, 40 cm-thick cantera pieces. The f5 tally was used to estimate the collided (total) and the uncollided photon fluence.

The uncollided photon fluence and the cantera thickness were fitted to the exponential function in order to calculate the linear attenuation coefficients that were compared with the μ calculated with the XCOM code.

The MCNP5 calculations also included the Kerma in air, Ka, and the Ambient dose equivalent, H*(10) that were calculated using the ICRP 74 [31] fluence-to-dose conversion coefficients. For each case the amount of histories used in the Monte Carlo calculations was large enough to have an uncertainty less than 5%.

2.2 Measurements

Using a 555 ± 5% MBq ^{137}Cs source in a lead shield with a 1 cm-diameter collimator, a transmission experiment was carried out using a narrow geometry array.

The transmitted photon spectrum was measured with gamma-ray spectrometer with a 7.62 Ø ×7.62 cm NaI(Tl) detector and a multichannel analyser. The detector was sited in a lead shield with 1 cm-diameter collimator. The centres of both collimators were aligned.

The area under the 0.662 MeV photopeak was measured as the cantera pieces with different thickness were placed between the source and the detector. The live-time of measurements was set to get 1% uncertainty in the photopeak net area.

For each cantera thickness the net count rates under the 0.662 photopeak were corrected due to background. Corrected count rates were normalized to the count rate measured without cantera and adjusted to an exponential function using the weighted regression [32]. The weight factors, ω, were calculated using equation 3.

$$\omega_i = \frac{1}{sc_i + sx_i} \quad (3)$$

Here, sc_i is the variance of count rates measured with the i-th cantera piece, sx_i is the variance of the i-th cantera thickness.

3 Results

3.1 Calculations

3.1.1 XCOM results

The cantera mass interaction coefficients, for 1 keV-100 GeV photons, calculated with the XCOM code, are shown in figure 1.

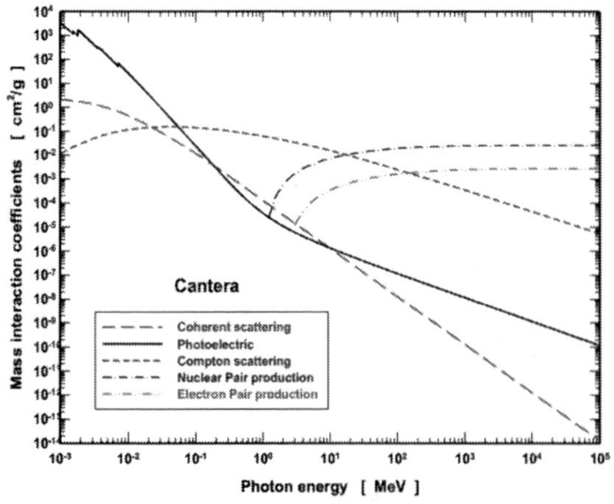

Figure 1. Photon mass interaction coefficients in cantera.

The interaction coefficients include coherent and incoherent (Compton) scattering, photoelectric effect, as well as, nuclear and electron pair production.

In cantera the photoelectric absorption is the main interaction mechanism for photons less than 60 keV, the best shielding performance is for low energy photons. For photons from 60 keV to approximately 20 MeV Compton scattering is the main interaction process with cantera where scattered photon, with lower energy, are produced.

The total mass attenuation coefficients of cantera for 10^{-3} to 10^{5} MeV photons, are shown in figure 2.

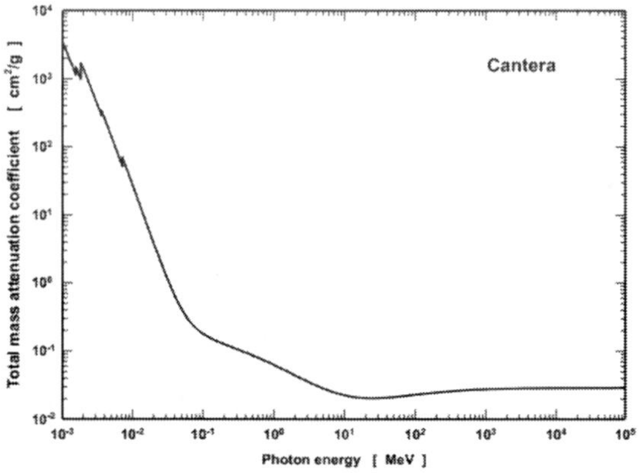

Figure 2. Total mass attenuation coefficients of cantera.

The total mass attenuation coefficients decreases sharply in the low energy region were discrete, small size, peaks appears. This behaviour is similar to the total mass attenuation coefficients reported for building materials [3]. In the low energy region there are seven resonances due to K absorption edge in Na, Mg, Al, Si, K, Ca, and Fe.

Using the mass attenuation coefficients and the density of cantera the linear attenuation coefficients for 0.03, 0.07, 0.1, 0.3, 0.662, 1, 2 and 3 MeV photons were calculated being included in Table 1.

Table 1. Linear attenuation coefficients of cantera.

Photon energy [MeV]	μ_{XCOM} [cm^{-1}]	μ_{MCNP5} [cm^{-1}]
0.030	2.2895	2.2920
0.070	0.4457	0.4455
0.100	0.3218	0.3219
0.300	0.1928	0.1926
0.662	0.1380	0.1379
1	0.1137	0.1135
2	0.0798	0.0797
3	0.0650	0.0649

3.1.2 Monte Carlo results

In Figure 3 are shown the relative uncollided photon transmission, estimated with the MCNP5 code for monoenergetic photon sources in function of the mass thickness of cantera.

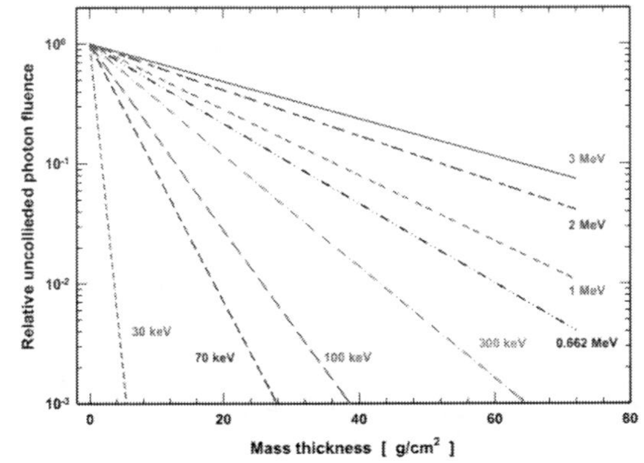

Figure 3. Uncollided photon transmission in cantera.

Data from Figure 3 were fitted into exponential functions and the linear attenuation coefficients were calculated.

In Table 1 are shown the μ values obtained with the XCOM and MCNP5 codes. There is a good agreement between the μ calculated using both codes.

The largest difference, noticed for 3 MeV photons, was 0.27 % being statistically insignificant. The Half value layer of cantera is 0.3, 1.6, 2.1, 3.6, 5.0, 6.1, 8.7, and 10.7 cm for 0.03, 0.07, 0.1, 0.3, 0.662, 1, 2, and 3 MeV photons respectively.

In Figure 4 the relative total photon (collided) transmission in cantera is shown.

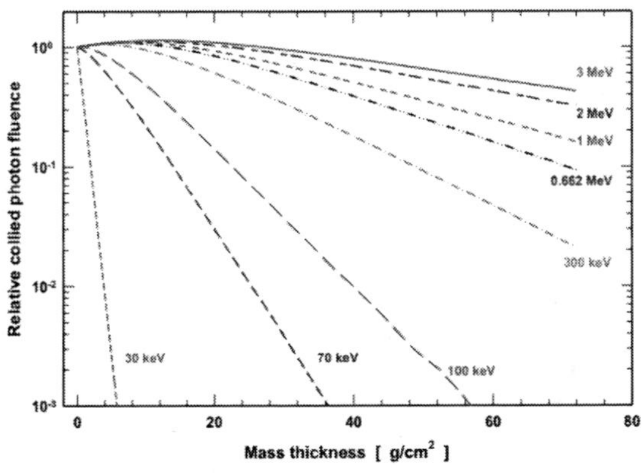

Figure 4. Collided photon transmission in cantera.

Total photon fluence includes the uncollided and the scattered photons. The differences between the photon transmission in figures 3 and 4 are due photon buildup in the detector. The photon buildup is the increase of photon fluence in the detector due to scattered photons by the cantera that reach the detector. Scattered photons have lower energies.

In Figure 5 the Ka transmission due to total photon fluence, in function of cantera mass thickness (x/ρ) for mono-energetic photons are presented. This feature is shown in terms of (x/ρ) because the density of cantera tuffs varies from 1.80 to 2.37 g cm^{-3} [22].

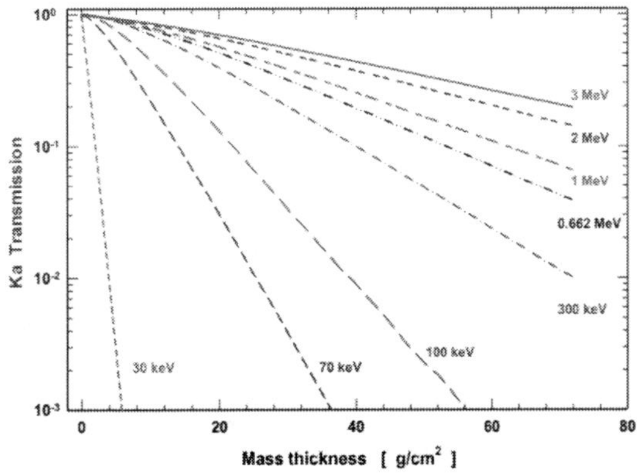

Figure 5. Kerma in air transmission in cantera.

The Ambient dose equivalent transmission of mono-energetic photons in cantera is shown in Figure 6.

Despite the fluence-to-H*(10) conversion coefficients are larger than the fluence-to-Ka conversion coefficients for photons from 0.03 to 3 MeV [31], for the same thickness of cantera the H*(10) transmission is slightly larger than Ka transmission. The probable explanation is because H*(10) and the Ka values were calculated using collided photons (uncollided and scattered) that reach the detector, therefore some scattered photons has energies where fluence-to-Ka conversion coefficients are larger than fluence-to-H*(10) conversion coefficients.

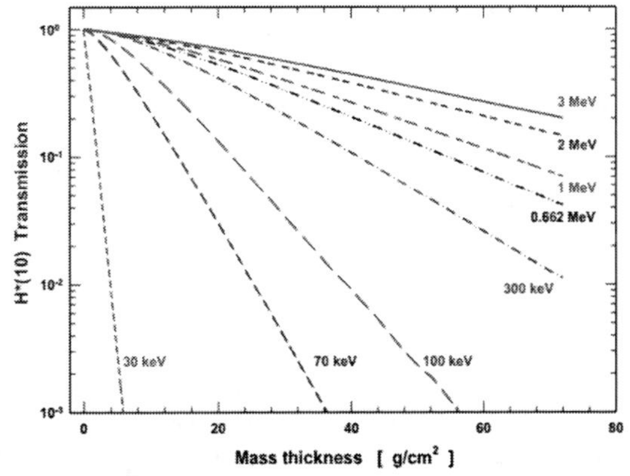

Figure 6. Ambient dose equivalent transmission in cantera.

3.2 Measurements

The linear attenuation coefficient of cantera for 0.662 MeV photons from ^{137}Cs source is 0.14 ± 0.01 cm^{-1}. This value was obtained through measurements carried out with the NaI(Tl) γ-ray spectrometer. The measured μ is in agreement with the μ-values for 0.662 MeV photons calculated with XCOM and MCNP5 codes, shown in Table 1.

The μ/ρ for 0.662 MeV is 0.076 $cm^2\ g^{-1}$ being similar to the value shown in building materials [2] and is larger than μ/ρ of aluminium (7.466×10^{-2} $cm^2\ g^{-1}$) and smaller than μ/ρ of lead (0.1101 $cm^2\ g^{-1}$) [33].

4 Conclusions

The cantera is widely used in the construction industry however, it's shielding features against x and γ-rays were unknown. Through calculations with XCOM and MCNP5 codes the shielding features of cantera have been estimated for 0.03, 0.07, 0.100, 0.300, 0.662, 1, 2, and 3 MeV monoenergetic photons. Using a ^{137}Cs and a NaI(Tl) γ-ray spectrometer the linear attenuation coefficient for 0.662 MeV photons was also measured.

Due to total alkali and silica content in this quarry tuff is defined as rhyolite having a density of 1.8 g cm^{-3}.

Linear attenuation coefficient of cantera varies from approximately 2.290 to 0.065 cm^{-1} for 30 keV to 3 MeV photons respectively.

For 30 keV photons the half value layer of cantera is 0.3 cm and 10.7 cm for 3 MeV γ-rays.

Photons with energy less than 60 keV are effectively shielded because the main interaction with cantera is through photoelectric interaction. Thus, the cantera is a good shielding material for rooms with x-ray units working below 60 kV. This x-ray equipment includes radio diagnostic x-rays for mammography or some few used for dental radiographies.

For photons with energy between 60 keV and 10 MeV the main interaction in cantera is Compton scattering. Here, the energy of the incoming photon is reduced, but scattered photons contribute with the photon and the dose build up that should be evaluated. Nevertheless in rooms with x-ray units working above 60 kV, like dental, diagnosis, interventional x-rays, or those with γ-ray sources, the cantera in the walls will help to shield photons and to reduce the dose.

For the same thickness of cantera the Ka and H*(10) transmission is almost the same.

A limitation of this work was the use of monoenergetic photons to represent x-rays, because x-ray tubes produce photons with a wide and continuous energy distribution together with few discrete peaks where just quite few photons have the largest energy. However, if the information here presented is used to calculate a shielding made with cantera will be conservative because the photon mean energy of actual x-rays is smaller than the energy of photons here used.

Another limitation of this work was that photons and dose buildup factors for cantera were not included.

References

1. V. P. Singh, N. M. Badiger and A. M. El-Khayatt, Radiat. Eff. Def. Solids **169**, 547 (2014)
2. K. S. Mann, B. Kaur, G. S. Sidhu, A. Kumar, Radiat. Phys. Chem. **87**, 16 (2013)
3. Ch. Singh, T. Singh, A. Kumar, G. S. Mudahar, Ann. Nucl. Energy **31**, 1199 (2004)
4. B. Oto, N. Yildiz, F. Akdemir, E. Kavaz, Prog. Nucl. Energy **85**, 391 (2015)
5. S. Kaewjang, U. Maghanemi, S. Kothan, H. J. Kim, P. Limkitjaroenporn and J. Kaewkhao, Nucl. Eng. Des. **280**, 21 (2014)
6. Y. Gedikoglu, G. Gedikoglu, G. Berkin, T. Ceyhan, M. A. Altinoz, Toxicol. Ind. Health **28**, 708 (2012)
7. K. S. Mann, A. Rani and M. S. Heer, Rad. Phys. Chem. **106**, 247 (2015)
8. J. Miller, L. Taylor, C. Zeitlin, L. Heilbronn, S. Guetersloh, M. DiGiuseppe, Y. Iwata, T. Murakami, Radiat. Meas. **44**, 163 (2009)
9. T. Korkut, H. Korkut, A. Karabulut, G. Budak, Ann. Nucl. Energy **38**, 56 (2011)
10. V. P. Singh, N. M. Badiger, Vacuum, **113**, 24 (2015)
11. D. N. Hopkins, M. Maqbool, M. S. Islam, Radiol. Phys. Technol. **5**, 229 (2012)
12. J. Kim, D. Seo, B. Ch. Lee, Y. S.Seo, W. H. Miller, Adv. Engr. Mat. **16**, 1083 (2014)
13. E.-S. A. Waly, M. A. Bourham, Ann. Nucl. Energy **85**, 306 (2015)
14. I. I. Bashter, Ann. Nucl. Energy **24**, 1389 (1997)
15. I. Akkurt, H. Akyildirim, B. Mavi, S. Kilincarslan, C. Basygit, Ann. Nucl. Energy **37**, 910 (2010)
16. A. M. El-Khayatt, Ann. Nucl. Energy **37**, 991 (2010)
17. E. Gallego, A. Lorente and H. R. Vega-Carrillo, Nucl. Tech. **162**, 399 (2009)
18. T. Piotrowski, D. B. Tefelski, J. J. Sokolowska, B. Jaworska, Acta Phys. Pol. A **128**, B-9 (2015)
19. T. Piotrowski, D. B. Tefelski, J. Skubalski, A. Zak, Acta Phys. Pol. A **128**, B-14 (2015)
20. P. Ciccioli, P. Plescia, D. Capitani, J. Phys. Chem. C **114**, 9328 (2010)
21. P. Yasaka, N. Pattanaboonmee, H. J. Kim, P. Limkitjaroenpron, J. Kaewkaho, Ann. Nucl. Energy **68**, 4 (2014)
22. R. López-Doncel, W. Wedekind, T. Leiser, S. Molina-Maldondo, A. Velasco-Sánchez, R. Dohrmann, A. Kral, A. Wittenborn, A. Aguillón-Robles, S. Siegesmund, Environ. Earth Sci. **75**, 212 (2016)
23. W. Wedekind, R. López-Doncel, R. Dohrmann, M. Kocher, S. Siegesmund, Environ. Earth Sci. **69**, 1203 (2013)
24. M. Degerlier, Radioprotection **48**, 215 (2013)
25. S. Turhan, E. Atici, A. Varinlioglu, Radioprotection **50**, 273 (2015)

26. N. A. Pérez, E. Lima, P. Bosch, J. Méndez-Vivar, J. Cult. Herit. **15**, 352 (2014)
27. M. J. Berger, J. H. Hubbell, S. M. Seltzer, J. Chang, J. S. Coursey, R. Sukumar, D. S. Zucker, K. Olsen, XCOM: Photon Cross Sections Database, version 1.5, [Online]. <http://physics.nist.gov/pml/data/xcom/data/xcom/index.cfm>. Gaithersburg, MD, USA.
28. M. Y. Celik, A. Ergul, Environ. Earth Sci. **74**, 3223 (2015)
29. M. J. Le Bas, R. W. Le Maitre, A. R. Woolley, Miner. Pet. **46**, 1 (1992)
30. X-5 Monte Carlo Team, Report LA-UR-03-1987 Los Alamos National Laboratory (2003)
31. ICRP74, Ann. ICRP **26**, 179 (1996)
32. H. R. Vega-Carrillo, Rev. Mex. Fis. **35,** 597 (1989)
33. T. E. Johnson. B. K. Birky, *Health Physics and Radiological Health* (Lippincott Williams & Wilkins, 2012)

Comparison of thick-target (alpha,n) yield calculation codes

Ana C. Fernandes[1,*], *Andreas* Kling[1] and *Gennadiy* N. Vlaskin[2]

[1]Centro de Ciências e Tecnologias Nucleares, Instituto Superior Técnico, Universidade de Lisboa, 2695-066 Bobadela, Portugal
[2]Innovation Technology Centre for the PRORYV Project, ROSATOM, 107140 Moscow, Russia

Abstract. Neutron production yields and energy distributions from (α,n) reactions in light elements were calculated using three different codes (SOURCES, NEDIS and USD) and compared with the existing experimental data in the 3.5-10 MeV alpha energy range. SOURCES and NEDIS display an agreement between calculated and measured yields in the decay series of ^{235}U, ^{238}U and ^{232}Th within ±10% for most materials. The discrepancy increases with alpha energy but still an agreement of ±20% applies to all materials with reliable elemental production yields (the few exceptions are identified). The calculated neutron energy distributions describe the experimental data, with NEDIS retrieving very well the detailed features. USD generally underestimates the measured yields, in particular for compounds with heavy elements and/or at high alpha energies. The energy distributions exhibit sharp peaks that do not match the observations. These findings may be caused by a poor accounting of the alpha particle energy loss by the code. A big variability was found among the calculated neutron production yields for alphas from Sm decay; the lack of yield measurements for low (~2 MeV) alphas does not allow to conclude on the codes' accuracy in this energy region.

1 Introduction

Alpha-induced neutron production yield data are required in numerous scenarios including the nuclear fuel cycle [1], nuclear safeguards and forensics [2], geological [3] and astrophysical studies [4], underground facilities [5], nuclear fusion [6] and radiation shielding [7]. In addition to thick-target production yields and neutron energy distributions most applications require small or at least reliable estimates for the uncertainty in the yield data.

Modern general-purpose radiation transport simulation codes [8-10] can model particle transport, production of secondaries and particle detection with an overall accuracy in the order of 10%. However their modelling of neutron production is either inaccurate or inexistent at the energy range relevant for most applications, and neutron yields and spectra are extracted more conveniently from evaluated cross-section data.

Various codes have been developed to derive the alpha-induced neutron spectra and production yields due to decay of various radionuclides. In this work we compare the data provided by three codes, and conclude on their accuracy upon comparison with experimental results.

2 Materials and methods

2.1 Overview of the tested codes

The evaluated codes are: (i) SOURCES-4C [11] developed at Los Alamos National Laboratory, USA and released by the Nuclear Energy Agency data bank; (ii) an application developed at the University of South Dakota, USA – hereafter the USD code – and available online [12]; (iii) the NEDIS-2.0 code [13], developed at the A.A. Bochvar Research Institute of Inorganic Materials, Moscow, Russia. A version of SOURCES-4A with an extended energy range that allows to include most decay alphas was developed at the University of Sheffield, UK and included in the evaluation [14].

All codes calculate thick-target yields by integrating the (α,n) reaction cross sections (σ) over the alpha particle range in the target material. The stopping power (s) of alpha particles in the material accounts for their energy loss during transport and the calculation consists of integrating σ(E)/s(E) between the threshold energy (E_{th}) for neutron production and the initial alpha energy (E_{α}).

When multiple alpha energies are involved e.g. from radioactive decay, the corresponding yields are summed to derive the total yield.

The yield in multi-element materials is derived from the elemental yields using the Bragg-Kleeman rule that expresses the mass stopping power of the compound or mixture as the sum of the elemental mass stopping

* Corresponding author: anafer@ctn.tecnico.ulisboa.pt

powers (weighted by the element mass fraction in the material).

The codes differ on their nuclear data libraries, namely for cross sections and stopping powers and on the integration algorithms. When deriving (α,n) energy distributions, different assumptions regarding the neutron angular distribution in the centre of mass system (CMS) are also found.

All codes except USD use the stopping power coefficients from Ziegler's tables/TRIM code [15] which enable to calculate energy-continuous atomic or mass stopping power. USD uses self-developed simulation procedures and extracts stopping powers from the ASTAR program [16]. The latter can be used to calculate alpha stopping power tables for many elements and some compounds based on values of the ICRU-49 report by the International Commission on Radiation Units and Measurements. The results for the stopping power values from ASTAR agree within a few percent with those of TRIM and with measurements [17]. Although ASTAR tables can be made with a very fine mesh, their size is in principle limited for integration in the USD code and may yield some steps or discontinuities.

With respect to cross sections, SOURCES-4C employs experimental data from various sources and calculated data from the nuclear physics code GNASH [18] for Ne and heavier elements. GNASH and therefore SOURCES-4C are limited to a maximum alpha energy of 6.5 MeV. However, as both the (α,n) cross sections and the average neutron energy rise with the alpha energy the 6.5 MeV cut reduces the total neutron yield from (α,n) reactions and artificially shifts the neutron spectrum to lower energies. This limitation was overcome in a modified version of SOURCES-4A (previously referred) in which the EMPIRE2.19 nuclear reaction code [19] is used to calculate the cross sections in an extended energy range and include more elements. The reliability of the EMPIRE data was evaluated upon comparison with experimental cross-section data, a few results being reported. A good (but unquantified) agreement is referred, especially at alpha energies below 10 MeV.

NEDIS-2.0 employs cross section data, complete and partial (on separate levels of the residual nuclei), from evaluated libraries. The cross sections of the code's standard database have been benchmarked for a vast number of light elements as detailed in Ref. 20, yielding an uncertainty smaller than 10%.

USD employs the nuclear reaction TALYS code to calculate the cross sections. Some benchmarking results and examples are reported for neutron- and proton-induced reactions [21], but we could not find any information regarding the code performance for (α,n) reactions.

The benchmarking status, as reported, for neutron yield calculations varies among the codes. Benchmarking of SOURCES-4C is reported in Ref. 14 and references therein; typical agreement between measured and calculated neutron production yields of 10-20% were found. NEDIS estimates an uncertainty below 10% for most materials due to the application of evaluated cross sections that nevertheless remain as the principal contributor to the yield uncertainty [20, 21]. USD quotes a good overall, but unquantified, agreement with experimental values.

As final note, both SOURCES-4C and NEDIS can provide (α,n) data for user-defined alpha energies or combination of alpha emitters, whereas USD is restricted to the ^{238}U and ^{232}Th series in secular equilibrium and to the Sm decay.

2.2 Evaluated quantities

In this work homogeneous mixtures are considered, in which the alpha-emitting atoms are embedded and directly adjacent to the target material. The quality of calculated yields was assessed upon comparison with experimental data. Both the total production yields and their energy distributions are investigated.

We considered mostly alpha energies in the 4-9 MeV range, which is representative of common particle accelerators and of actinide decay. The latter consisted mainly of the ^{235}U, ^{238}U and ^{232}Th series in equilibrium. Additionally, the low energy of ^{147}Sm decay (2.248 MeV, for which there are no yield measurements to our best knowledge) was also included due to the presence of Samarium in some detector components [12].

Only original yield measurements were considered: any cross section integration, yield conversion between elements and compound materials, or extrapolation beyond the experimental alpha energy range were excluded. The treatment of experimental data was restricted to cubic-splines interpolation when deriving yields for intermediate alpha energies, e.g., from isotopes' decay.

2.3 Experimental data

Extensive compilations of experimental (α,n) yields can be found and have been discussed in Ref. 22. In this work we have considered:

(i) thick-target yields measured by West and Sherwood [23] as a function of alpha energy (3.5-10 MeV) for various elements (Be, C, Mg, Al, Si, Fe) and compounds (BeO, BN, UC, UO_2);

(ii) thick-target yields measured by Bair and Gomez del Campo [24] as a function of alpha energy (3.5-7.5 MeV) for Li and B which complements the above elemental data. Measurements in PbF_2 and ZnF_2 were performed in order to derive the F yield (because the yield contribution from Pb and Zn is negligible). However, as the actual results were not reported measurements in PbF_2 (3.5-10 MeV) were extracted from Ref. 25;

(iii) thick-target yields measured by Feige et al. [26] as a function of alpha energy for various geological minerals (Na_2CO_3, K_2CO_3, CaF_2).

(iv) thick-target yields measured by Gorshov et al. [27] for a vast list of elements and compounds, in the emission spectra of ^{238}U and ^{232}Th decay series.

(v) neutron energy distributions measured by Jacobs and Liskien [28] as a function of alpha energy

(4-5.5 MeV) for C, Mg, Al and Si and furthermore BN, CaF_2, Al_2O_3, SiO_2 and UO_2. We note that data for BN, CaF_2 and UO_2 are representative of B, F and O, respectively;

(vi) neutron energy distributions from Am-Be and Am-Li sources [29, 30], representative of Be and Li yields at quasi-monoenergetic 5.480 MeV alphas.

3 Results

3.1 Yield dependence on alpha energy

Figure 1 shows calculated and measured yields as a function of alpha energy for various elements. The threshold energy (E_{th}) is indicated when the (α,n) reaction is endoenergetic.

Measurements from different works are plot allowing to perceive the consistency among the values. The data identified as SOURCES consists of SOURCES-4C results up to 6.5 MeV, and of the energy-extended SOURCES-4A in the 6.5-10 MeV energy range. In case the distributions do not match below 6.5 MeV, they are discriminated. Although Fe is not included in the standard version of NEDIS-2.0, yields were calculated using cross sections for the ^{54}Fe(α,n) reaction determined by Houck and by Levkovsii and available in the EXFOR database [31]. The cross section for ^{56}Fe was determined on the basis of the measured neutron yields from Ref. 23.

The results show that SOURCES-4A/C and NEDIS-2.0 are able to reproduce the measured yields for most elements within the experimental uncertainties or with small deviations from the experimental values. Important exceptions were found for C and Si, with a significant overestimation of the yield by SOURCES-4A. In the case of Silicon, SOURCES results start to diverge from measurements at 6.5 MeV, which evidences problems in the data employed by SOURCES-4A. For Carbon the SOURCES-4A and -4C yields split early at 5 MeV, meaning that the data employed by SOURCES-4C was not preserved by SOURCES-4A. For both, C and Si, the discrepancy with experimental data could be mitigated by normalising the EMPIRE-calculated data to yield measurements, e.g., from Ref. 23. NEDIS and SOURCES yields for Boron divide at ~8 MeV, with SOURCES data being coincident with that from Ref. 22 that extrapolates experimental values beyond 7.5 MeV.

Yield values for selected compound media are shown in Fig. 2. Besides UO_2 and PbF_2 which allow to evaluate the data quality for F and U, SiO_2 is included as an important material in nuclear waste vitrification and a main constituent of geological materials employed in shielding construction. In the case of PbF_2, the results of a recent effort to improve the accuracy of cross section data for the ^{19}F(α,n) reaction [32] were also employed in the NEDIS calculations. The results evidence a good agreement between calculated and experimental yields for O and F compounds; the SiO_2 yield overestimate by SOURCES reflects that in Si. The new ^{19}F(α,n) cross section data provides yield values 30-40% higher than standard NEDIS calculations (in the energy region >5.5 MeV) and exhibit a superior agreement with measurements.

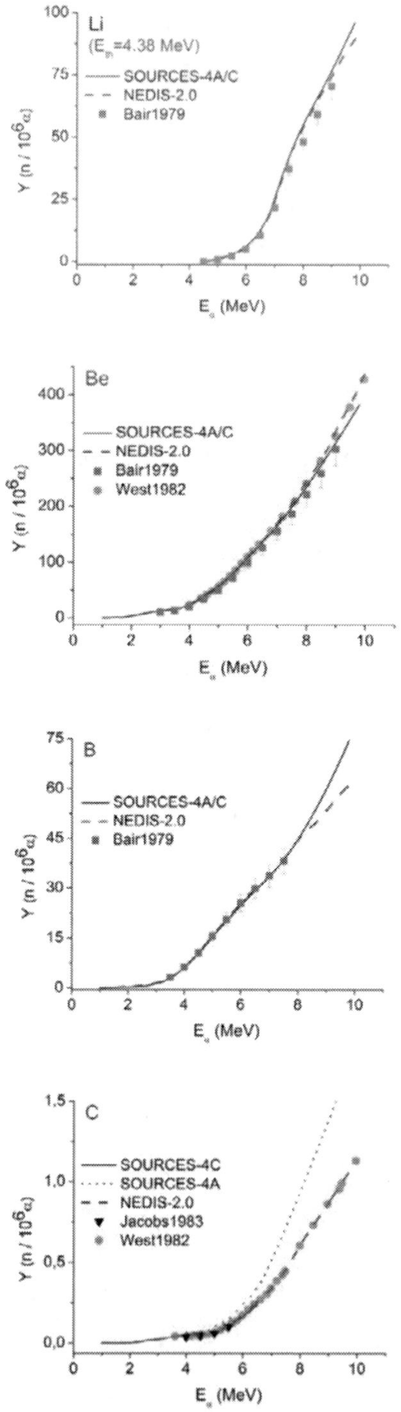

Fig. 1. Calculated and measured (α,n) yields in elemental materials, as a function of alpha energy. References of the experimental data are Bair1979 [24], Jacobs1983 [28], West1982 [23] and Feige1968 [26].

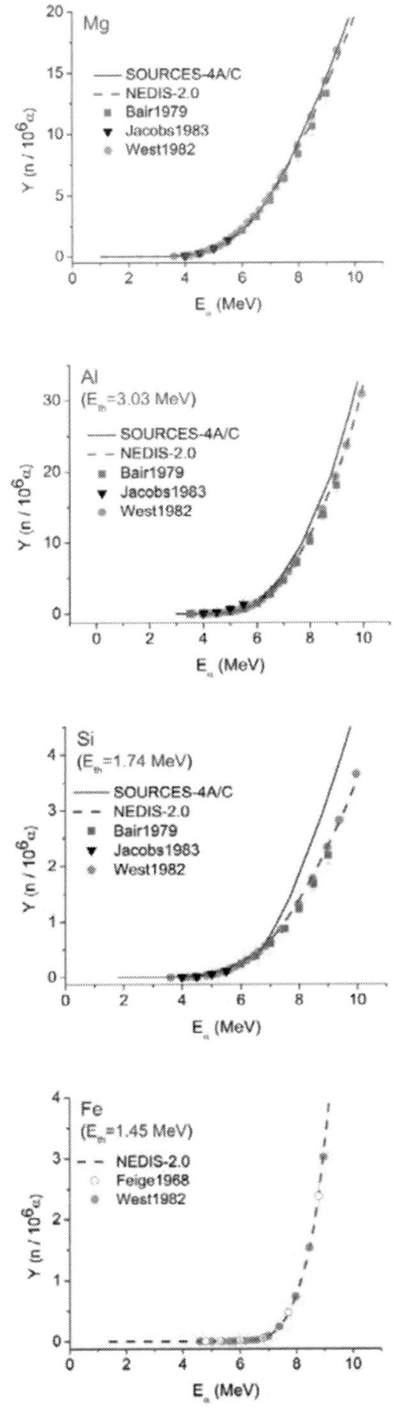

Fig. 1 (cont.). Calculated and measured (α,n) yields in elemental materials, as a function of alpha energy. References of the experimental data are Bair1979 [24], Jacobs1983 [28], West1982 [23] and Feige1968 [26].

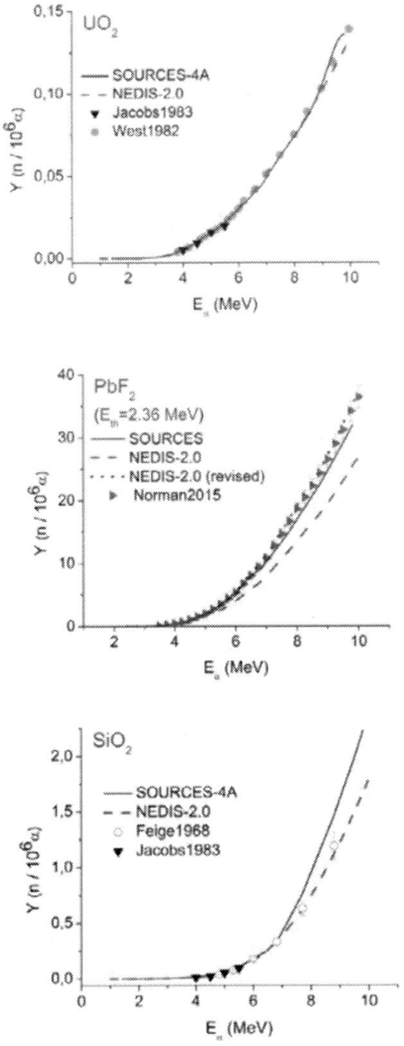

Fig. 2. Calculated and measured (α,n) yields in compound materials, as a function of alpha energy. References of the experimental data are Jacobs1983 [28], West1982 [23] and Feige1968 [26]. The revised NEDIS yield calculations employ cross section data from Ref. 32.

3.2 Yields for natural emitters

We now provide (α,n) yields for the emission spectra of ^{235}U, ^{238}U and ^{232}Th decay series in secular equilibrium, and for the decay of Sm. Evaluated decay data (alpha energies and intensities) [33] was used to weight the contribution of each emission line. The energy range and average energies of the decay series (considering emission intensities >1%) are: 4.22-7.39 MeV and 5.94 MeV for ^{235}U; 4.15-7.69 MeV and 5.36 MeV for ^{238}U; 3.95-8.78 MeV and 6.00 MeV for ^{232}Th. The ^{232}Th series is richer in high energies to which correspond therefore larger (α,n) yields. From a similar reasoning, the ^{238}U series displays the smaller yields.

Tables 1-3 contain measured and calculated yields for the referred decay series. The "experimental values" consist of either cubic-spline interpolation of the measured yields as a function of energy (Section 3.1) or the direct measurements from Ref. 27. Uncertainties are assumed to be at the 1-sigma level, and the following notation is used: 1.234 *56*=1.234±0.056 (i.e. the digits in italic indicate the uncertainty in the last digits of the quantity).

With respect to neutron production from Sm, only calculated yields in selected materials are reported in Table 4. Yield values using libraries other than the standard NEDIS (namely the Japanese Evaluated Neutron Data Library, JENDL and the Evaluated Neutron Data File, ENDF [34, 35]) are given for some targets.

Table 1. Measured and calculated neutron yields (n/$10^6\alpha$) for the spectrum of the ^{235}U decay series in secular equilibrium.

Target	Measured	SOURCES -4A	NEDIS -2.0
Li	9.82 *70* [24]	11.88	10.95 *77*
Be	100.0 *17* *	120.0	109 *3*
B	24.3 *16* [24]	26.61	24.5 *11*
C	0.1965 *28* [23]	0.3070	0.20 *1*
Mg	2.724 *35* *	2.796	2.7 *1*
Al	2.409 *28* *	2.832	2.45 *9*
Si	0.3258 *39* [23]	0.2286	0.32 *3*
BeO	40.42 *61* [23]	44.10	40.7 *15*
UC	0.03168 *48* [23]	0.04498	0.0295 *18*
CaF_2	9.8 *22* [26]	9.719	10.0 *7*
PbF_2	6.17 *31* [25]	6.207	6.38 *55*
Na_2CO_3	2.7 *4* [26]	2.805	2.34 *23*
SiO_2	0.22 *5* [26]	0.2365	0.20 *2*
K_2CO_3	0.22 *8* [26]	0.07523	0.14 *3*
UO_2	0.03230 *22* [23]	0.03169	0.032 *2*

* references:
Be: 100 *10* [24]; 110.2 *17* [23]
Mg: 2.732 *36* [23]; 2.50 *18* [24]; 2.67 *34* [26]
Al: 2.419 *29* [23]; 2.24 *11* [24]; 2.42 *35* [26]
Fe: 0.04651 [23]; 0.04992 [26]

Table 2. Measured and calculated neutron yields (n/$10^6\alpha$) for the spectrum of the ^{238}U decay series in secular equilibrium.

Target	Measured	SOURCES -4A	NEDIS -2.0	USD
Li	6.35 *44* [24]	7.375	7.19 *50*	6.40
Be	80.1 *13* *	79.93	79.6 *24*	39.5
B	21 *3* [27]	18.57	19.1 *8*	14.8
C	0.1397 *20* *	0.1977	0.137 *6*	0.120
Mg	1.728 *22* *	1.637	1.71 *6*	1.33
Al	1.514 *17* *	1.657	1.52 *5*	1.58
Si	0.2047 *25* *	0.2696	0.21 *2*	0.176
Fe	0.05913 *	-	0.056 *3*	0.0525
BeO	31.56 *30* *	29.37	29.3 *9*	14.3
H_3BO_3	3.1 *5* [27]	3.157	3.25 *12*	2.66
UC	0.02231 *33* [23]	0.02913	0.0205 *12*	5.76 $\times10^{-3}$
CaF_2	6.01 *92* *	5.802	6.48 *45*	4.68
PbF_2	4.08 *20* [25]	3.710	4.14 *40*	1.49
NaF	6.8 *21* [27]	7.369	7.7 *5*	6.05
Na_2CO_3	1.7 *3* [26]	1.660	1.44 *14*	0.783
Al_2O_3	0.8 *1* [27]	0.8437	0.78 *2*	0.860
SiO_2	0.13 *2* [26]	0.1550	0.134 *11*	0.116
K_2CO_3	0.12 *3* [26]	0.04783	0.097 *10*	0..265
UO_2	0.02312 *28* [23]	0.01997	0.0228 *11*	5.48 $\times10^{-3}$

* references:
Be: 80.1 *13* [23]; 72.1 *72* [24]; 92 *13* [27]
C: 0.1396 *20* [23]; 0.17 *4* [27]
Mg: 1.732 *23* [23]; 1.60 *11* [24]; 1.6 *2* [26]; 2.1 *4* [27]
Al: 1.521 *18* [23]; 1.409 *70* [24]; 1.47 *15* [26]; 1.5 *2* [27]
Si: 0.2045 *25* [23]; 0.26 *6* [27]
Fe: 0.07490 [23]; 0.06199 [26]
BeO: 29.35 *44* [23]; 33 *4* [27]
CaF_2: 5.7 *10* [26]; 7.0 *22* [27]

Table 3. Measured and calculated neutron yields (n/$10^6\alpha$) for the spectrum of the ^{232}Th decay series in secular equilibrium.

Target	Measured	SOURCES -4A	NEDIS -2.0	USD
Li	11.99 *84* [24]	12.25	15.5 *10*	5.60
Be	117.7 *20* *	103.7	118 *3*	35.9
B	27 *4* [27]	14.81	25.1 *11*	14.4
C	0.2305 *32* *	0.2991	0.23 *1*	0.132
Mg	3.280 *42* *	2.826	3.21 *12*	1.72
Al	3.086 *14* *	3.282	3.3 *1*	2.39
Si	0.4336 *52* *	0.4830	0.43 *3*	0.248
Fe	0.2657 *	-	0.263 *13*	0.203
BeO	43.40 *64* *	38.03	43.2 *14*	12.9
UC	0.03665 *55* [23]	0.04339	0.0379 *20*	6.38 x10^{-3}
CaF_2	10.58 *18* *	9.136	12.5 *8*	5.82
PbF_2	6.97 *35* [25]	5.813	7.97 *74*	1.85
NaF	12.7 *45* [27]	12.21	15.9 *9*	7.47
Na_2CO_3	3.3 *5* [26]	3.043	2.93 *27*	0.928
Al_2O_3	1.8 *2* [27]	1.651	1.67 *5*	1.29
SiO_2	0.27 *5* [26]	0.2641	0.25 *2*	0.157
K_2CO_3	0.25 *6* [26]	0.07032	0.21 *2*	0.0300
UO_2	0.03446 *41* [23]	0.02865	0.034 *2*	6.30 x10^{-3}

* references:
Be: 117.8 *20* [23]; 107 *11* [24]; 126 *15* [27]
C: 0.2303 *32* [23]; 0.31 *8* [27]
Mg: 3.289 *43* [23]; 3.00 *21* [24]; 3.1 *4* [26]; 3.6 *5* [27]
Al: 3.309 *40* [23]; 3.048 *15* [24]; 3.2 *3* [26]; 3.5 *4* [27]
Si: 0.4334 *52* [23]; 0.51 *11* [27]
Fe: 0.2636 [23]; 0.2677 [26]
BeO: 43.33 *65* [23]; 47 *5* [27]
CaF_2: 10.2 *22* [26]; 11.2 *32* [27]

Table 4. Calculated neutron yields (n/$10^6\alpha$) for the α spectrum from Sm decay.

Target	SOURCES -4C	NEDIS -2.0	USD
Be	5.510	5.50	7.89
B	0.2543	0.53	1.97
C	6.292x10^{-3}	7.3x10^{-3}	1.26x10^{-2}
Mg	2.437x10^{-4}	9.0x10^{-5}	1.87x10^{-4}
Si	1.987x10^{-5}	1.01x10^{-5}	1.18x10^{-6}
BeO	2.005	2.00	2.85
H_3BO_3	0.04621	0.095	0.351
UC	1.014x10^{-3}	1.08x10^{-3} *	6.06x10^{-4}
Na_2CO_3	9.019x10^{-4}	8.48x10^{-4}	1.65x10^{-3}
Al_2O_3	4.406x10^{-5}	5.89x10^{-5}	2.09x10^{-4}
SiO_2	8.096x10^{-5}	6.94x10^{-5}	2.36x10^{-4}
UO_2	4.412x10^{-5}	3.17x10^{-5} **	5.25x10^{-5}
K_2CO_3	7.425x10^{-4}	7.73x10^{-5} ***	1.25x10^{-3}

* standard: 1.18x10^{-3}; JENDL: 1.09x10^{-3}; ENDF: 9.69x10^{-4}
** standard: 3.55x10^{-5}; JENDL: 2.39x10^{-5}; ENDF: 3.58x10^{-5}
*** standard: 8.39x10^{-5}; JENDL: 7.8x10^{-5}; ENDF: 7.0x10^{-5}

In order to evaluate the accuracy of each code, calculated-to-experimental (C/E) yield ratios were plot for the ^{238}U and ^{232}Th series (Fig. 3). Elemental target materials were ordered along the x-axis by increasing order of atomic number (Z) in order to identify potential trends with stopping power. In the case of UO_2, Ca/PbF_2 and Na_2CO_3 the contributing element is indicated within brackets. The experimental uncertainty and, for NEDIS, the calculation uncertainty are included.

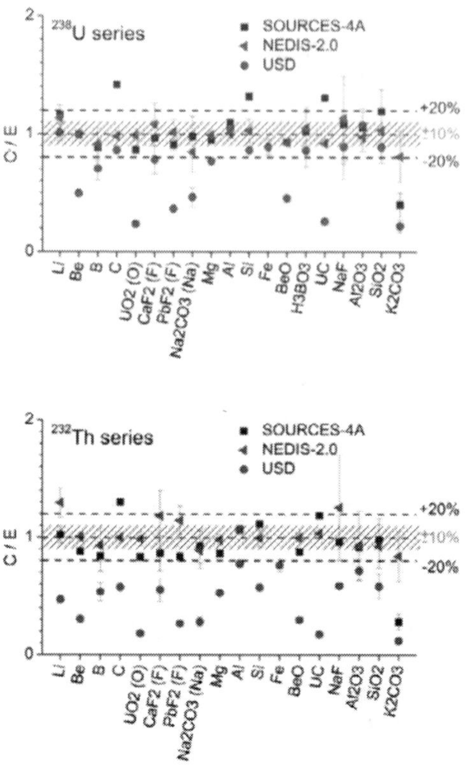

Fig. 3. Calculated-to-experimental (α,n) yield ratios in the alpha spectra of the ^{238}U and ^{232}Th decay series. The shaded area and dashed lines define the ±10% and ±20% limits, respectively.

For the ^{238}U decay spectrum, both SOURCES-4A and NEDIS agree with measurements within ±10% for most target materials. SOURCES-4A outliers correspond to yields in Si and C (for which SOURCES-4A fails to reproduce the measured yields as a function of energy, Fig. 1) and corresponding compounds, and also to the K compound which is not available as a target (i.e. only the contribution of C and O are accounted for). USD values for elemental materials lie within the ±20% limits except for Be. However for compounds with heavy elements (U, Pb) the agreement is degraded, which may indicate some difficulty in including the high stopping power in the yield calculation.

For the ^{232}Th series involving higher α energies the findings are not significantly altered, except for a noticeable overestimate of the Li yield by NEDIS. We attribute this to a systematic error in the experimental data, since all yields from the same work lie below those of other works (Fig.1). The accuracy of the USD code is clearly diminished, with an important underestimation of all yields.

The results show that the accuracy of SOURCES and NEDIS is generally in the ±10% range for decay spectra in the range of that of ^{238}U with a slight degradation to ±20% as alpha energies increase up to ~10 MeV.

3.3 Neutron energy distribution

The measured and calculated energy distributions of neutrons produced in (α,n) reactions are represented in Fig. 4 for target elements in the range Z=3-8. NEDIS is found to have a high performance in reproducing the structures in the spectra due to the application of evaluated cross sections with respect to the various excited states of the residual nucleus, as well as the anisotropic angular emission in the CMS in particular for ^{7}Li, ^{9}Be and ^{13}C. This is demonstrated in the neutron distribution for C and UO_2 where the low energy peak corresponds to the first and second excited states of ^{16}O and the first excited state of ^{20}Ne, respectively.

Some examples of the neutron energy distribution from compounds in the alpha decay spectra from U, Th and Sm are presented in Fig.5, allowing a comparison with results from the USD code. USD retrieves distributions with sharp features both at low and high alpha energies that are not realistic. The example of Sm-induced neutrons in UO_2 shows that USD does not convolute the contribution by the first excited states of ^{20}Ne and ^{21}Ne (both extending up to 1.0 MeV neutron energy) and the ground state of ^{21}Ne (maximum neutron energy of 1.4 MeV) leading us to suppose that the accounting of the slowing down process of alpha particles is insufficient.

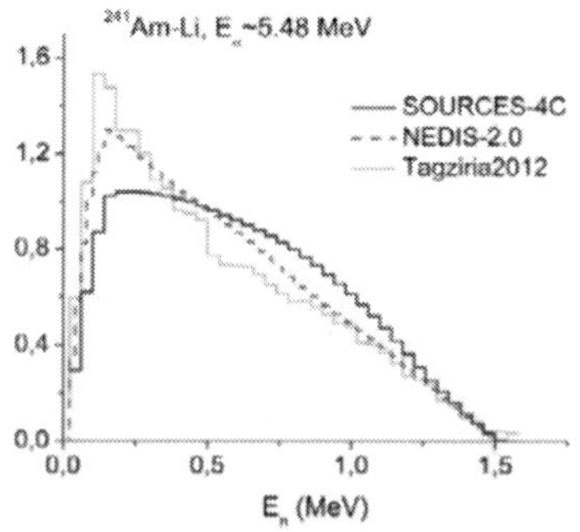

Fig. 4. Calculated and measured energy distributions of (α,n) neutrons induced by quasi-monoenergetic alphas in materials having a single element responsible for the neutron production.

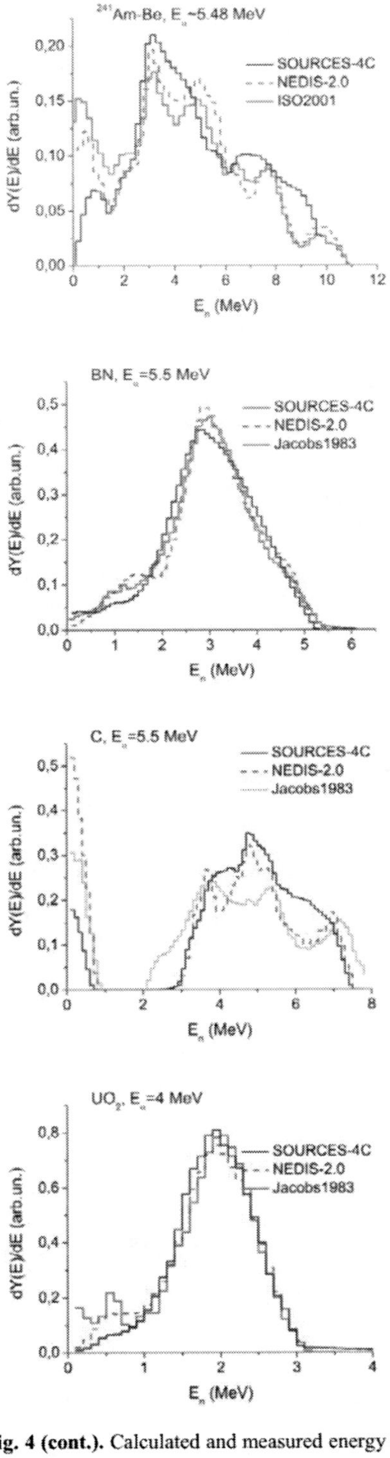

Fig. 4 (cont.). Calculated and measured energy distributions of (α,n) neutrons induced by quasi-monoenergetic alphas in materials having a single element responsible for the neutron production.

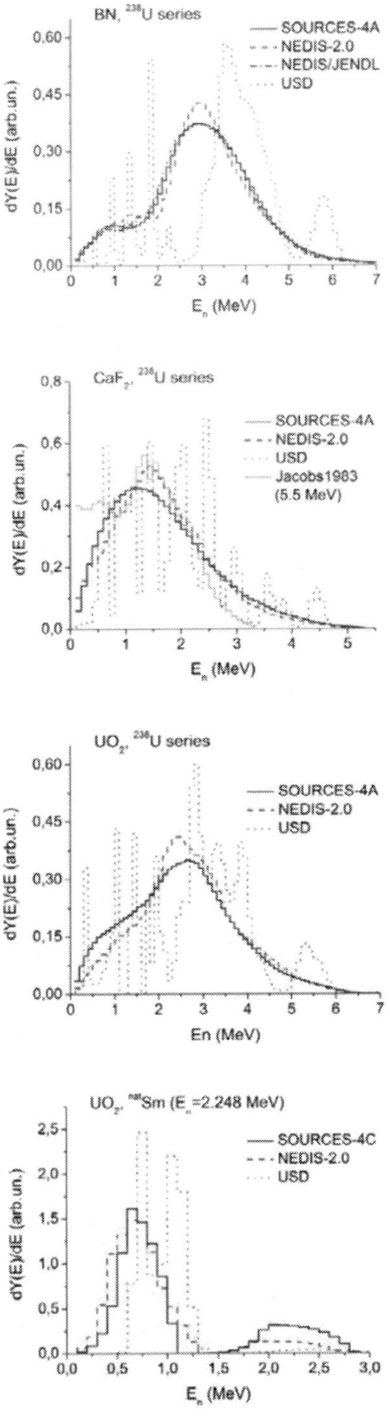

Fig. 5. Calculated energy distributions of (α,n) neutrons induced by the ^{238}U decay series and from Sm.

4 Conclusions

From the three (α,n) yield calculation codes evaluated in this work, SOURCES and NEDIS were found to reproduce total yield measurements within ±10-20% for most materials - larger limits corresponding to higher alpha energies. Beyond 6.5 MeV alphas, the extended-energy version of SOURCES-4A produces yields significantly larger than measured for Si and C. The neutron energy distributions calculated by the two codes agree well with measurements, with an improved performance of NEDIS regarding the description of the contribution by the various excited states of the residual nucleus.

For moderate (4-6 MeV) alpha energies, neutron yields extracted by the USD code for elemental materials are generally in good agreement (±20%) with measurements. The situation is deteriorated in compound materials with heavy elements, and for higher energies (up to 9 MeV), with an important underestimation of the neutron yields by the code. The neutron energy distributions display artefacts that are probably created by an insufficient accounting of the alpha particle slowing-down process. However, it may turn out that these features are not relevant to the process under investigation, namely when neutron moderation is important leading to their spreading [36].

A large discrepancy was found for the neutron production yield induced by Sm decay alphas.

The materials considered in this work are only those for which yield measurements were known. Although we believe that the conclusions from this work are generally applicable, the few exceptions that were found lead to the recommendation that distinct code+material+emitter combinations should be subject to an assessment by the user previous to application.

Acknowledgements

We thank Vitaly Kudryavtsev (University of Sheffield) for offering the version of SOURCES-4A evaluated in the present work.
This activity was supported by Fundação para a Ciência e Tecnologia (Portuguese Ministry of Science) via grant IF/00628/2012/CP0171/CT0008 and by projects UID/Multi/04349/2013 and PTDC/ EEI-ELC/2468/2014.

References

1. I. Crossland, *Nuclear Fuel Cycle Science and Engineering* (Woodhead, Cambridge, 2012)
2. D. Reilly, N. Ensslin, H. Smith, S. Kreiner, *Passive Nondestructive Assay of Nuclear Materials*, US Nuclear Regulatory Commission report NUREG/CR-5550 (US Government Printing Office, Washington DC, 1991)
3. *Isotope Methods for Dating Old Groundwater* (IAEA, Vienna, 2013)
4. A. Coc, F. Hammache, J. Kiener, Eur. Phys. J. A **51**, 34 (2015)
5. J.A. Formaggio, C.J. Martoff, Annu. Rev. Nucl. Part. S. **54**, 361 (2004)
6. A. Kaplan, J. Fusion En. 32, 382 (2013)
7. T. Nakamura, L. Heilbronn, Handbook on Secondary Particle Production and Transport by High Energy Heavy Ions (World Scientific, Singapore, 2006)
8. D.B. Pelowitz, *MCNPX User's Manual Version 2.7.0* LA-CP-11-00438 (Los Alamos National Laboratory, Los Alamos, 2011)
9. J. Allison, K. Amako, J. Apostolakis, et al., IEEE Trans. Nucl. Sci. **53**, 270 (2006)
10. T.T. Böhlen, F. Cerutti, M.P.W. Chin, et al., Nucl. Data Sheets **120**, 211 (2014)
11. W.B. Wilson, R.T. Perry, W.S. Charlton et al., Radiat. Prot. Dosim. **115**, 117 (2005)
12. D.-M. Mei, C. Zhang, A. Hime, Nucl. Instrum. Methods **A 606**, 651 (2009). Data available at http://neutronyield.usd.edu
13. G.N. Vlaskin, Yu.S. Khomyakov, V.I. Bulanenko, At. Energ. **117**, 357 (2015)
14. V. Tomasello, V.A. Kudryavtsev, M. Robinson, Nucl. Instrum. Methods **A 595**, 431 (2008)
15. J.F. Ziegler, J.P. Biersack, M.D. Ziegler, *SRIM: The Stopping and Range of Ions in Matter* (Srim, Chester, 2015). Data available at http://www.srim.org
16. M.J. Berger, J.S. Coursey, M.A. Zucker, J. Chang, *ESTAR, PSTAR, and ASTAR: Computer Programs for Calculating Stopping-Power and Range Tables for Electrons, Protons, and Helium Ions, v.1.2.3* (National Institute of Standards and Technology, Gaithersburg, 2005). Data available at http://physics.nist.gov/Star
17. H. Paul, A. Schinner, Nucl. Instrum. Methods **B 227**, 461 (2005)
18. P.G. Young, E.D. Arthur, M.B. Chadwick, Report LA-12343-MS (Los Alamos National Laboratory, Los Alamos, 1992)
19. M. Herman, R. Capote, B. Carlson et al., Nucl. Data Sheets **108**, 2655 (2007)
20. V.A. Vukolov, F.E. Chukreev, At. Énerg. **62**, 232 (1987)
21. www.talys.eu/documentation/
22. R. Heaton, H. Lee, P. Skensved, B.C. Robertson, Nucl. Instrum. Methods **A 276**, 529 (1989)
23. D. West and A.C. Sherwood, Ann. Nucl. Energy **9**, 551 (1982)
24. J.K. Bair, J. Gomez del Campo, Nucl. Sci. Eng. **71**, 18 (1979)
25. E.B. Norman, T.E. Chupp, K.T. Lesko, et al., Appl. Radiat. Isotopes **103**, 177 (2015)
26. Y. Feige, B.G. Oltman, J. Kastner, J. Geophys. Res. **73**, 3135 (1968)
27. G.V. Gorshov, O.S. Tsetkov, Sov. Atom. Energy **14**, 550 (1964)
28. G.J.H. Jacobs, H. Liskien, Ann. Nucl. Energy **10**, 541 (1983)
29. *Reference Neutron Radiations (Part 1)*, ISO 8529-1 (International Standards Organisation, Geneva, 2001)

30. H. Tagziria, M. Looman, Appl. Radiat. Isotopes **70**, 2395 (2012)
31. *Experimental Nuclear Reaction Data*. Available from, e.g., the IAEA Nuclear Data Section at https://www-nds.iaea.org/exfor/exfor.htm
32. W.A. Peters, M.S. Smith, S. Pittman et al., *A kinematically complete, interdisciplinary, and co-institutional measurement of the 19F(α,n) cross section for suclear safeguards science*, INL/EXT-16-38791 (Office of Scientific and Technical Information, Oak Ridge, TN, 2016). Available at https://www.osti.gov/scitech/servlets/purl/1263500
33. J.K. Tuli, *Nuclear Wallet Cards, 2011.* Available at http://www.nndc.bnl.gov/nudat2/
34. T. Murata, H. Matsunobu, K. Shibata, *Evaluation of the (α,xn) Reaction data for JENDL/AN-2005*, JAEA-Research 2006-052, Japanese Atomic Energy Agency, 2006.
35. M.B. Chadwick et al., Nuclear Data Sheets **112**, 2887 (2011)
36. A.C. Fernandes et al., this volume.

Worker dose under high-power operation of the J-PARC 3 GeV Rapid Cycling Synchrotron

Kazami Yamamoto[1,a]

[1]*J-PARC Center, Japan Atomic Energy Agency, 2-4 Shirakata, Tokai-mura, Naka-gun, Ibaraki-ken, 319-1195, Japan*

Abstract. The J-PARC 3 GeV Rapid Cycling Synchrotron (RCS) delivers a 1-MW, high-intensity beam to facilities downstream. In such high-intensity accelerators, the operational beam intensity is limited to keep worker exposure to the residual dose within acceptable tolerances. Therefore, we continue to pursue accelerator commissioning that reduces beam loss. In order to achieve further high-intensity operation, the J-PARC accelerator system has been drastically upgraded over the past two years. As a result, it was found that beam loss decreased, whereas output power increased; the residual doses were kept at the same level or decreased in RCS. A malfunction of a collimator occurred in April 2016, and we replaced it to a spare duct in a hurry. The broken collimator was higher activated, but exposure to workers was kept within the acceptable level.

1 Introduction

The J-PARC 3 GeV Rapid Cycling Synchrotron (RCS) was constructed to deliver a 1-MW, high-intensity beam to the Material and Life Science Experimental Facility (MLF) and the Main Ring (MR) [1]. In a high-intensity proton accelerator such as RCS, beam loss, which constitutes just a small fraction of the beam, causes large doses of prompt radiation and serious radio-activation, which may cause malfunctions in the accelerator components and higher exposure doses to maintenance workers. Therefore, the operational beam intensity is limited to keep workers' exposure to the residual dose within acceptable tolerances.

The J-PARC linac energy was upgraded to 400 MeV during the 2013 summer shutdown period [2], and the peak current of the ion source was increased to 50 mA during the summer shutdown period of 2014 [3]. Due to these upgrades, necessary equipment for 1-MW operation has been prepared. However, it is necessary to gradually increase the beam intensity while taking measures to minimize beam loss and residual dose during maintenance work. In this paper, we report the conditions of radio-activation and worker dose in the RCS after the accelerator upgrade.

2 History of beam power and residual doses

Figure 1 shows the history of the residual doses at typical loss points and the output power of the RCS from 2014 to summer of 2015. Here, we plot two graphs of the residual doses in the RCS. One is the dose on the charge-exchange foil chamber (which is the highest dose) and the other is the dose on the horizontal steering magnet chamber in the injection H^0 dump line. (the H^0 dump is used to abandon unexchanged H^- and H^0 beams during the injection period.)

Figures 2 and 3 show the higher-dose points. Due to a radioactive-material-leak accident from the Hadron experimental hall [4], the J-PARC facility was shut down for more than a half year after May 2013. During this long shutdown period, recovery work from the accident and linac acceleration-energy upgrade were performed. The user operation of the MLF resumed on February 17, 2014. Because the increment of the linac acceleration energy mitigated the space-charge effect, the beam loss of the RCS was reduced and we were able to deliver a 600-kW beam to the MLF [5]. However, we continued operation without any increase from the previous value of 300 kW. Because it had not yet been confirmed that a neutron target could accept a high-intensity beam, we had to limit the beam power from the viewpoint of target protection. Subsequently, the increment of the ion source and improvement of the target for 1-MW operation were completed during the summer shutdown period in 2014.

During this period, we investigated the radio-activation of the foil chamber and found that the radio-activation source was the interaction of the injection and circulating beams with the charge-exchange foil [6]. A method to decrease this activation is to reduce the interaction rate between the beams and the foil. Thus, since it was necessary to expand the painting-injection area to reduce the number of hits of the beams on the foil, we had to correct the modulation of the beta function caused by the edge effects of the injection bump magnet [7]. For this purpose, we installed an additional

[a] Corresponding author: kazami@post.j-parc.jp

quadrupole magnet system during the summer shutdown period in 2014. We also installed a new collimator in the H^0 dump line to reduce the neutron backscatter from the dump.

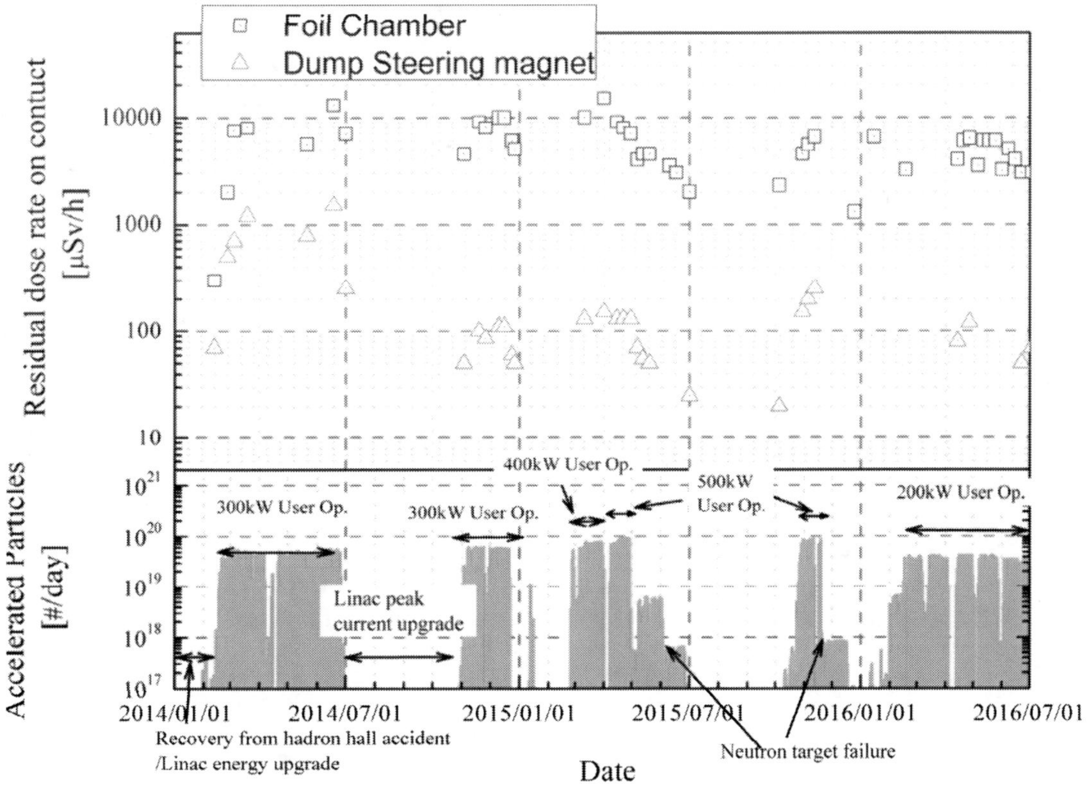

Figure 1. Beam power and residual dose histories of the RCS.

After these improvements, we resumed user operation at 300 kW from the end of February 2015. This was increased to 400 kW in the middle of March and further to 500 kW by the middle of April. Figure 4 shows the residual dose distributions of the RCS before and after the 500-kW operation. Before the operation, we corrected the modulation of the beta function with the new quadrupole magnet system and expanded the painting-injection area [8]. As a result, the dose rate of the foil chamber, which was above 10 mSv/h after a 400-kW operation, was decreased to less than 10 mSv/h. The residual dose of the H^0 dump line was also decreased by the new collimator. Based on these results, we intended to increase the output power to 600 kW in the middle of May; however, a cooling-water leak occurred in the neutron target at the MLF at the end of April, after which the supply to the MLF was stopped. Following this water-leak incident, operation to the MR only was conducted until July 1. The requirement for MR operation is less than 10% of that for the MLF, and the number of acceleration particles is less than 10%. Thus, all residual doses in the RCS were reduced.

The MLF user program resumed on October 27, but another water leak from the neutron target occurred on November 20 [9]. A spare target was installed and the user program was resumed on February 20 [10]. Since we did not have any spare neutron target, the output power was subsequently suppressed to less than 200 kW from the point of view of the target protection.

During the short maintenance period in April 2016, one of the ring collimators was broken [11]. Abnormal operation of the control system brought about a vacuum leak of the fifth secondary collimator chamber. Since the numerical simulation result indicated that the beam loss without this collimator would be within the acceptable level, we installed a straight duct instead of repairing the broken collimator. The repair period was reduced to 8 days by this countermeasure.

When we obtained a beam of more than 400 kW for the neutrino target after this malfunction, the residual dose on the replaced duct's surface exceeded 10 mSv/h. After the MR switched operation to the Hadron Experiment, the value reduced to 5 mSv at 6 hours after the beam had ceased operation. Furthermore, after a 2-month cooling-off period, it fell to less than 1 mSv at the time of maintenance work.

DOI: 10.1051/epjconf/201715307022

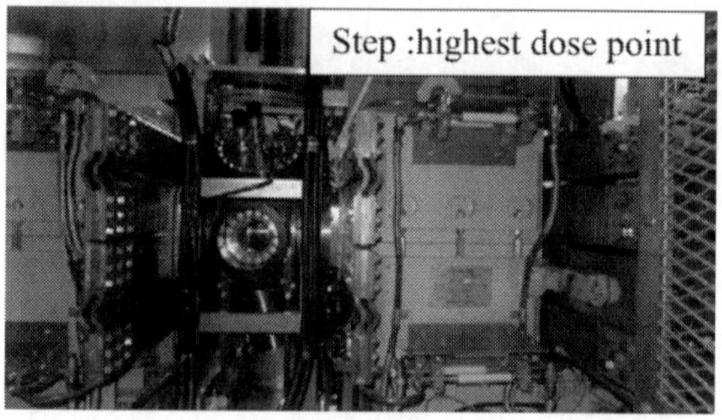

Figure 2: Photograph of the charge-exchange foil chamber.

It has a step between the foil chamber and the bump magnet chamber, and there is the highest dose point.

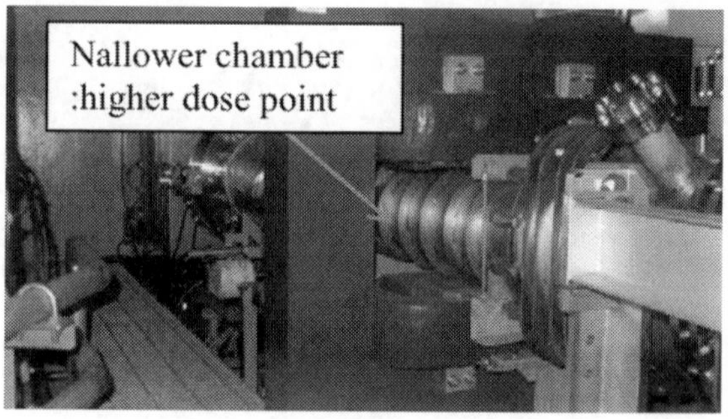

Figure 3: Photograph of the injection H^0 dump line.

The horizontal steering magnet chamber is higher dose point. This chamber is narrower than the upstream and downstream chambers.

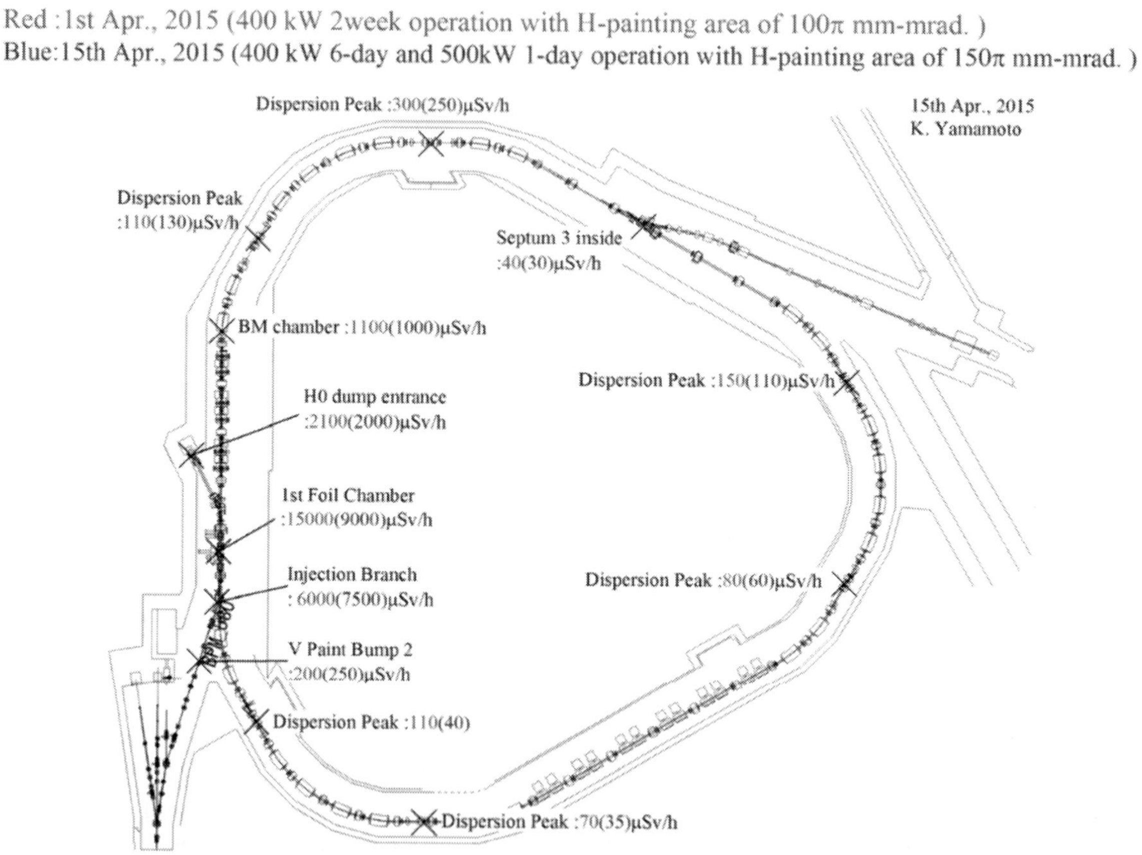

Figure 4. Residual dose distributions of the RCS before and after the 500-kW operation.

3 Worker doses during the maintenance period of 2014

A summary of the worker doses during maintenance work in the summer shutdown period of 2014 is given in Table 1. Since the RCS delivered a 300-kW beam to the MLF just before the summer shutdown of 2014, the residual dose on the foil chamber was slightly higher (7 mSv/h) just after beam shutdown. Maintenance work near the foil chamber commenced after 1.5 months of cooling, and the residual dose was about half of its previous value (just after beam shutdown). During the summer shutdown period in 2014, 53 workers were exposed to a dose of more than 0.01 mSv (this is the minimum dose that can be detected by an Alarm Pocket Dosimeter). A corrective dose to an RCS worker is 3.7 mSv. The number of workers who were exposed to more than 0.1 mSv was 12, and all of these worked near the foil chamber. The maximum dose for one worker was 0.31 mSv. This worker was involved in the replacing of a vacuum chamber near the foil chamber. Since the vacuum flange diameter of this chamber was very large (400 mm [12]) and there was insufficient work space, it was not easy to achieve an acceptable leak rate of less than 10^{-11} Pa m/s using this flange connection. Therefore, a great deal of time was spent reconnecting this flange near the foil chamber and the exposure dose was increased.

Table 1. Summary of the personal doses during maintenance work in the summer shutdown period of 2014.

Exposure dose [mSv]	Number of workers [#]
0.01–0.05	31
0.06–0.1	10
0.11–0.2	8
0.21-	4

4 Worker doses during the maintenance period of 2015

Since only beam operation to the MR was performed two months before the summer shutdown period of 2015, radioactive nuclides with short half-lives disappeared and the residual dose on the foil chamber was 3 mSv/h. Owing to nuclides with long lifetimes, this value did not decrease significantly during the maintenance period. During this summer shutdown period, 33 workers were exposed to a dose of more than 0.01 mSv, and the corrective dose to the RCS workers was 4.45 mSv. The number of workers who were exposed to more than 0.1 mSv was 22. The maximum dose for one worker was 0.42 mSv. The corrective dose and the number of workers who were exposed to higher doses increased in comparison with the previous year because of recovery work for a malfunction of the injection magnet. Since we removed the upper half of the magnet, which functioned as the radiation shielding, for maintenance, a duct with a high dose within the magnet was revealed and the exposure dose was increased.

The previous beam-commissioning results showed that the injection-beam parameter should be flexibly adjusted to a significant degree to reduce RCS beam loss. In order to secure sufficient flexibility, we needed to extend the physical aperture of the vacuum ducts in the magnets. The vertical apertures of the old ceramic ducts in the vertical bump magnets were adequate for the original beam condition, but the study results indicated that wider beam optics would be better for achieving the finer, low-loss matching condition. Thus, those ceramic ducts were replaced with wider ones (see Figure 5).

There was not such a high residual dose, even after operation at 500 kW. The residual dose distribution was similar to the result under a 300-kW operation in 2014 and higher radiation doses were concentrated in the injection area. Thus, significant doses were also caused by the work in the injection area.

Figure 5: Extension of the duct aperture.

Table 2. Summary of the worker doses during the maintenance work in the summer shutdown period of 2015.

Exposure dose [mSv]	Number of workers [#]
0.01–0.05	11
0.06–0.1	7
0.11–0.2	7
0.21–	8

5 Malfunction of the ring collimator

The beam-collimation system removes the beam halo and localizes the beam loss on itself in order to preserve other accelerator components [13]. The arrangement of this system is shown in Figure 6. It consists of one primary collimator, which scatters the halo particles, and five secondary collimators, which absorb those scattered particles [14]. Radiation shielding was designed to absorb a halo of at most 4 kW by the collimator system. Figure 7 shows the vacuum chamber and radiation shielding of the collimator. It was designed such that the radiation dose from the collimator block could be reduced to about 1/10,000th of its initial value by this shielding [15]. The capacity of the collimator was designed to be 4 kW, but in operation, the MLF output was 200 kW in June 2016, and we estimate that hundreds of Watts of beam loss occurred in the collimator system. In April 2016, a malfunction was caused in the VME system, i.e., the control system of the collimator. To fix this, we attempted a performance test, during which a vacuum leak occurred. Vacuum pumps stopped suddenly due to vacuum deterioration, and the part where a vacuum deterioration point was separated by a gate valve.

The investigation of the leak point revealed that the fifth secondary collimator was the source of the leak and it was quickly removed. Under these circumstances, the residual dose value on the collimator shielding was 5–10 μSv/h and that on the downstream chamber of the collimator was about 200 μSv/h. On the other hand, the residual dose of the collimator block, which absorbed the halo particles directly, was 40 mSv/h. We confirmed that the residual dose of the collimator was significantly reduced by the present shielding design.

We did not have a spare unit with which to replace the fifth secondary collimator. However, since the numerical simulation result indicated that the beam loss without this collimator would be at an acceptable level, we installed a straight duct instead of repairing the broken collimator. We spent 8 days on this recovery work. After resumption of user operation, the residual dose of the arc area was not as high as expected. However, there was a step in the new duct due to the difference in the upstream and downstream flange diameters, and a dose of maximum 10 mSv occurred at this step part.

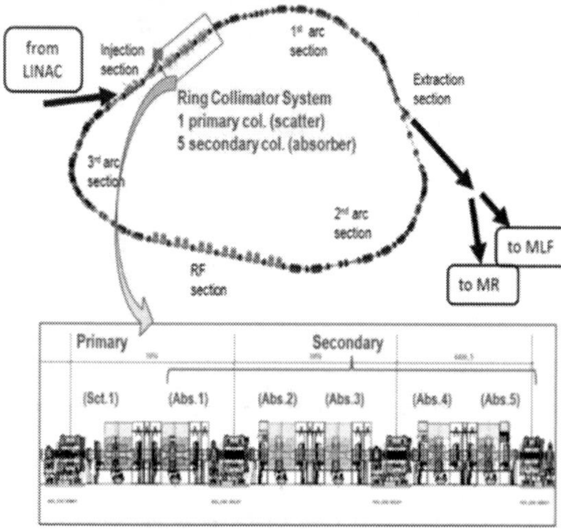

Figure 6: Layout of the collimator.

A total of 31 persons worked to replace the broken collimator and 24 workers were exposed to doses of more than 0.01 mSv. The corrective dose for the workers was 0.62 mSv. The maximum dose for one worker was 0.06 mSv and two workers were exposed to this dose. Their task was to replace and connect the collimator chamber.

Because the maintenance work varied over a range of tasks, coordinators were appointed for each task and the workers were separated. Therefore, we were able to average the worker doses and avoid concentrating on one worker. The exposure dose was less than 20% of that of year-round maintenance work.

Figure 7: Photograph of the collimator.

6 Conclusion

Beam commissioning of the J-PARC RCS was performed to achieve high-intensity operation. As a result, the residual dose around the accelerator component was maintained low enough for maintenance work, even after 500 kW of continuous operation. From a study of the beam under 1-MW output power, we confirmed that the beam loss was proportional to the output power [16]. However, to achieve an output power above 1 MW, we must lower the worker dose near the injection-foil chamber. The first way to reduce the activation of the foil chamber is mitigation of the interaction between the foil and the beam. We will try further optimization of the operational parameters of the correction quadrupole and injection-painting magnets. The second way is to replace the injection system. We will start developing a new injection system that simplifies maintenance work and includes radiation shielding. Regarding the broken collimator, we will investigate the cause of breakdown and install an improved model.

References

1. Y. Yamazaki ed., *"Accelerator Technical Design Report for High-Intensity Proton Accelerator Project, J-PARC"*, KEK Report 2002-13 and JAERI-Tech 2003-44 (2003)
2. K. Hasegawa, *"Commissioning of energy upgraded linac of J-PARC"*, Proc. LINAC14, TUIOB03 (2014)
3. T. Maruta, Y. Liu, K. Futatsukawa, T. Miyao, M. Ikegami, A. Miura, *"Commissioning of energy upgraded linac of J-PARC"*, Proc. IPAC15, THPF040 (2015)
4. T. Koseki, H. Hasegawa, *"Present Status of J-PARC -After the Shutdown Due to the Radioactive Material Leak Accident-"*, Proc. IPAC14, THPME061 (2014)

5. H. Hotchi, *"Commissioning the 400 MeV Linac at J-PARC and High Intensity Operation of the J-PARC RCS"*, Proc. IPAC14, TUXA01 (2014)
6. E. Yamakawa, M. Yoshimoto, M. Kinsho, *"Measurements and PHITS Monte Carlo Estimations of Residual Activities Induced by the 181 MeV Proton Beam in the Injection Area at J-PARC RCS Ring"*, JPS Conf. Proc. **8**, 012017 (2015)
7. H. Hotchi, N. Tani, Y. Watanabe, *"Numerical study for beam loss occurring for wide-ranging transverse injection painting and its mitigation scenario in the J-PARC 3-GeV RCS"*, NIM. Sect. A **778,** 102-114 (2015)
8. H. Hotchi, N. Tani, Y. Watanabe, H. Harada, S. Kato, K. Okabe, P. K. Saha, F. Tamura, M. Yoshimoto, *"Beam loss caused by edge focusing of injection bump magnets and its mitigation in the 3-GeV rapid cycling synchrotron of the Japan Proton Accelerator Research Complex"*, Phys. Rev. Accel. Beams **19,** 010401 (2016)
9. "*Information of the target problem at the MLF*", http://j-parc.jp/researcher/MatLife/en/news/20151127.html
10. "*On the restarting of Materials/Life Science Facility operation*", http://j-parc.jp/en/topics/20151225MLFtopics.html
11. K. Yamamoto et al., "*A malfunction of the Beam Collimator System in J-PARC 3 GeV Rapid Cycling Synchrotoron*", Proc. PASJ2013, MOP007 (in Japanese).
12. M. Kinsho, Y. Saito, Z. Kabeya, K. Tajiri, T. Nakamura, K. Abe, T. Nagayama, D. Nishizawa, N. Ogiwara, *"Development of alumina ceramics vacuum duct for the 3 GeV-RCS of the J-PARC project"*, Vacuum, **73,** 187-193 (2004)
13. J. B. Jeanneret, *Phys. Rev.* ST-AB 1, 081001 (1998).
14. K. Yamamoto, *Phys. Rev.* ST-AB 11, 123501 (2008).
15. K. Yamamoto, "The beam collimator system of J-PARC rapid cycling synchrotron", Proc. 42nd ICFA advanced beam dynamics workshop, HB2008 (2008).
16. H. Hotchi, *"Recent progress of J-PARC RCS beam commissioning –Toward realizing the 1-MW output beam power"*, Proc. IPAC15, TUBB3 (2015)

Shielding analyses of an AB-BNCT facility using Monte Carlo simulations and simplified methods

Bo-Lun Lai[1] and *Rong-Jiun* Sheu[2,*]

[1]Institute of Nuclear Engineering and Science, National Tsing Hua University, 101, Sec. 2, Kuang-Fu Road, Hsinchu 30013, Taiwan
[2]Department of Engineering and System Science, National Tsing Hua University, 101, Sec. 2, Kuang-Fu Road, Hsinchu 30013, Taiwan

Abstract. Accurate Monte Carlo simulations and simplified methods were used to investigate the shielding requirements of a hypothetical accelerator-based boron neutron capture therapy (AB-BNCT) facility that included an accelerator room and a patient treatment room. The epithermal neutron beam for BNCT purpose was generated by coupling a neutron production target with a specially designed beam shaping assembly (BSA), which was embedded in the partition wall between the two rooms. Neutrons were produced from a beryllium target bombarded by 1-mA 30-MeV protons. The MCNP6-generated surface sources around all the exterior surfaces of the BSA were established to facilitate repeated Monte Carlo shielding calculations. In addition, three simplified models based on a point-source line-of-sight approximation were developed and their predictions were compared with the reference Monte Carlo results. The comparison determined which model resulted in better dose estimation, forming the basis of future design activities for the first AB-BNCT facility in Taiwan.

1 Introduction

Boron neutron capture therapy (BNCT) is a form of targeted radiotherapy that relies on the thermal neutron capture reaction of boron-10, leading to two heavy charged particles to kill only boron-compound-bearing cells while sparing other surrounding tissues. It is an ideal selective form of radiotherapy if a high ratio of boron concentration between tumor and normal tissues could be achieved. The method could also be used to treat deeper-seated tumors in patients by using epithermal neutron beam. The Tsing Hua Open-Pool Reactor (THOR), a 2 MW research reactor located at the campus of National Tsing Hua University in Taiwan, has been successfully upgraded and renovated for BNCT application in 2004 [1]. Clinical trials for patients with locally recurrent head and neck cancer were started since August 1, 2010 [2]. Until now, more than 20 patients have been treated with BNCT and the overall response is encouraging.

THOR has been used for over 50 years and now is operating under its second licence extention. The BNCT group in Taiwan is seeking to establish an accelerator-based BNCT (AB-BNCT) facility to continue and expand this area of research. One major advantage of AB-BNCT over traditional reactor-based facilities is the public acceptance of sitting in a densely populated hospital. The proposed AB-BNCT facility in Taiwan is similar to the cyclotron-based neutron source at KURRI [3], where neutrons are produced from the bombardment of a beryllium target by 1-mA 30-MeV protons. To deliver high-quality epithermal neutron beam, the emitted fast neutrons from the target will be moderated by a specially designed beam shaping assembly (BSA) composed of iron, lead, polyethylene, bismuth and FLUENTAL™ [4]. This study aimed at investigating the characteristics of radiation field and bulk shielding requirements of such a facility.

2 Materials and methods

2.1 Facility and shielding layout

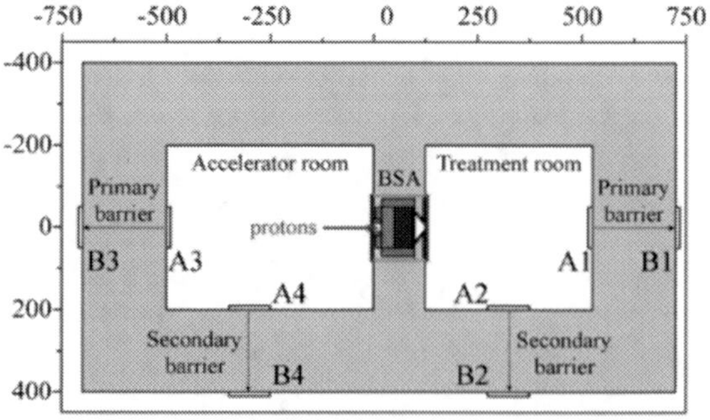

Fig. 1. Layout of a simple AB-BNCT facility, showing the bulk shielding structure and several locations of interest (A1-A4, B1-B4) inside and outside the concrete walls of the two rooms.

Fig. 1 shows a 2D slice of the bulk shielding structure of the proposed AB-BNCT facility, including an accelerator room and a patient treatment room. The

* Corresponding author: rjsheu@mx.nthu.edu.tw

abscissa and ordinate are the distance with unit of centimetre. During normal operation, a monoenergetic 30-MeV proton beam of 1-mA was launched to impinge on a beryllium target located inside the BSA. The cylindrical-shaped BSA with overall dimensions of 160 cm diameter and 125 cm length was entirely embedded in the partition wall (1.25 m thick) between the two rooms to reduce leakage radiation and simplify shielding design. The inner dimensions of the treatment room and accelerator room were 4×4×3 m^3 and 5×4×3 m^3, respectively. The thickness of the concrete wall was initially assumed to be 2 m, with a goal of reducing the transmitted dose rates to less than 10 μSv/h during normal operation. The primary barriers in Fig. 1 are referred to the walls directly facing the beam exit or entrance of the BSA where the radiation is expected to be the most intense, and the lateral concrete walls are called secondary barriers. Eight thin rectangles in Fig. 1 labelled A1 to A4 and B1 to B4 show the locations of scoring detectors, indicating typical radiation levels inside and outside the shielding structure during beam irradiation.

2.2 Calculation methods

The multi-particle Monte Carlo transport code MCNP6 [5] was used to simulate the proton-induced nuclear interactions and the subsequent transport of secondary particles. The major advantage of using MCNP6 in this study is the accuracy of the underlying cross sections for proton and neutron transport. Neutron and photon flux distributions were scored at locations of interest and converted to dose rates by folding with the ICRP-74 ambient dose equivalent conversion factors [6].

Two types of Monte Carlo shielding calculations were performed. The first type was a direct simulation of the radiation environment around the whole facility. The results were taken as the reference for comparison. The second type was a series of simulations of radiation transport in simplified configurations, aiming to generate shielding parameters for use with the point-source line-of-sight approximation [7, 8]. Eq. (1) is a classical two-parameter attenuation formula of the model, where H is the transmitted dose at the position of interest, E_p is the proton energy, θ is the angle with respect to the beam direction, d is the effective thickness of the shield, r is the distance between the target and the detector, H_0 and λ are the fitted parameters representing a pseudo-source term and the corresponding attenuation length under the specified conditions.

$$H(E_p,\theta,d/\lambda)=\frac{H_0(E_p,\theta)}{r^2}\exp(-\frac{d}{\lambda(E_p,\theta)}) \quad (1)$$

Applying a point-source line-of-sight model for estimating dose rates outside thick shields of an AB-BNCT facility was anticipated to be problematic because of the existence of the large-sized and complicated BSA enclosing the neutron production target. In addition, as shown in Fig. 1, the whole BSA including the target was embedded in the partition wall of two rooms. Considering these problems, this study took an approach of regarding the whole BSA as a radiation source in shielding analyses rather than a neutron-modifying device. In addition, as shown in Fig. 2, we considered two different geometries to generate shielding parameters, without and with the inclusion of the partition wall in the model. A coupling technique called the surface source write and read (SSW/SSR) in MCNP6 was implemented on the exterior surfaces of the BSA to reduce the computing burden of source generation in repeated Monte Carlo shielding simulations [5]. Note that the shielding parameters generated in this way were obviously the BSA dependent and intended only for use in shielding analyses of this facility.

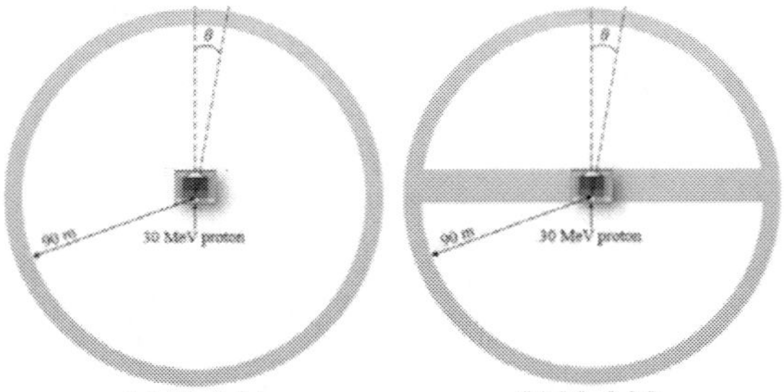

(a) Model 1 (b) Model 2

Fig. 2. Two simplified spherical geometries used to generate shielding parameters for the AB-BNCT facility in this study.

3 Results and discussion

3.1. Radiation field around the BSA

The design of BSA aims at production of an optimal epithermal neutron beam for BNCT purpose. A common definition for an epithermal energy range is 0.5 eV to 10 keV. The BSA not only is a critical component for the performance of neutron beam, but also plays an essential role in subsequent shielding analyses. In addition to the nearly collimated neutrons from the exit of the BSA, there are significant amounts of neutrons and photons emerging from all surfaces of the BSA. Appropriate shielding against these unwanted radiation sources is important that must be considered as an integral part of the overall facility design.

Fig. 3 shows neutron spectra emerged from the front, lateral and back surfaces of the BSA under the irradiation of 1-mA 30-MeV proton beam on the beryllium target. The neutron spectrum scored on the front surface indeed demonstrates the design purpose of the BSA, providing high-quality epithermal neutrons for patient treatment. Notably in Fig. 3, secondary neutrons emerged from the back and lateral surfaces, mainly fast and thermal neutrons respectively, are substantially more intense than those epithermal neutrons from the front surface. This characteristic poses a major challenge in radiation shielding for the accelerator room, resulting from a significant portion of fast neutrons produced in the target leaking immediately from the back surface of the BSA. In this study, both secondary neutrons and photons emerged from the BSA were considered and included in shielding calculations via the use of the SSW/SSR technique in MCNP6.

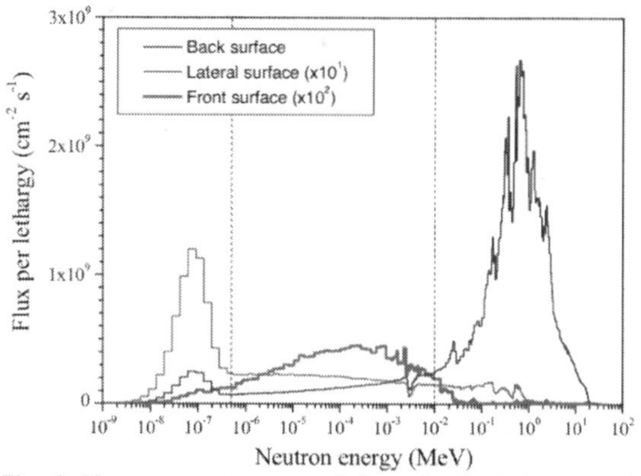

Fig. 3. Neutron spectra emerged from the front, lateral and back surfaces of the BSA under the irradiation of 1-mA 30-MeV proton beam on the beryllium target.

3.2 Monte Carlo simulations

Based on the shielding model in Fig. 1, Table 1 lists the MCNP6-predicted neutron and photon dose rates at several locations around the facility. Overall, the dose rates around the accelerator room are higher than those around the treatment room by about three orders of magnitude. This is because the radiation leaking from the back surface of the BSA is substantially more intense than that emerging from the front surface and more importantly with much higher neutron energies (Fig. 3). Inside the treatment room, neutron dose rates are higher than photon dose rates by approximately a factor of 4. This factor increases to ~40 in the accelerator room. Outside the shielding rooms, neutron dose rates are roughly comparable to photon dose rates. One of the advantages using SSW/SSR in this case was the saving of computing time while preserving the accuracies of full-scale Monte Carlo simulations. According to our experience, if only the shielding analysis of the treatment room is concerned, performing transport calculations based on the MCNP6-generated source term from the front surface of the BSA is good enough. Similarly, using only the source term from the back surface should be appropriate for shielding analyses of the accelerator room.

Fig. 4 shows the MCNP6-calculated depth dose profiles in primary and secondary barriers of the two rooms. From these trends, one can readily determine the required thicknesses of surrounding concrete walls to meet a prescribed dose limit. If a dose rate limit of 10 μSv/h is desired, the thicknesses of the primary and secondary barriers of the treatment room should be at least 128 cm and 114 cm, respectively. More shielding is required for the accelerator room because of the radiation distribution around the BSA. According to Fig. 4, at least 207 cm thick primary barrier and 192 cm thick secondary barrier are necessary to meet the same dose criterion. Although different thicknesses of concrete were required for these shielding walls, the attenuation lengths fitted for concrete thicker than 100 cm in Fig. 4 were found varying in a small range of 28–32 g/cm^2.

Table 1. MCNP6-predicted neutron and photon dose rates (μSv/h) at several locations around the facility (See Fig. 1).

Detector	Neutron	Photon
A1	$2.7\times10^5 \pm 3.3\%$	$6.5\times10^4 \pm 1.1\%$
A2	$2.5\times10^5 \pm 3.7\%$	$6.3\times10^4 \pm 1.1\%$
A3	$4.3\times10^8 \pm 1.4\%$	$1.1\times10^7 \pm 0.4\%$
A4	$4.5\times10^8 \pm 0.2\%$	$9.8\times10^6 \pm 0.1\%$
B1	$0.01 \pm 32.8\%$	$0.04 \pm 13.7\%$
B2	$0.007 \pm 100\%$	$0.02 \pm 3.5\%$
B3	$8.2 \pm 11.6\%$	$11.9 \pm 1.4\%$
B4	$2.5 \pm 17.8\%$	$5.5 \pm 1.8\%$

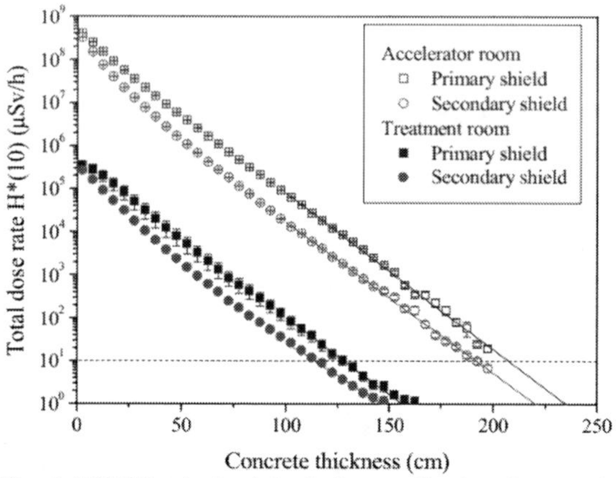

Fig. 4. MCNP6-calculated depth dose profiles in primary and secondary shields of the two rooms (See Fig. 1).

3.3 Point-source line-of-sight estimation

3.3.1 Source terms and attenuation lengths

The key for a successful use of a point-source line-of-sight model in shielding analysis relies on whether the source term and attenuation length are suitable to the problem under consideration. For the specific BSA designed for our facility, we have to generate associated shielding parameters by our own. As shown in Fig. 2, the authors adopted a similar calculation model used by Agosteo et al. [7]. The facility geometry was not modelled, but a simple spherical shell of concrete was placed around the BSA. MCNP6 was employed to simulate the proton-induced secondary radiation and their subsequent transport in surrounding concrete. Total dose rates including neutron and photon contributions were scored at given angular intervals and at various depths in concrete.

From the fitting of the MCNP6-calculated dose attenuation profiles in concrete, Fig. 5 summarizes the derived source terms and attenuation lengths as a function of angle for the two geometric models in Fig. 2. Notice that these parameters include dose contributions from both neutrons and photons for convenience in practical use. Comparing the shielding parameters

 DOI: 10.1051/epjconf/201715307023

generated by the two models revealed certain characteristics of the radiation field around the BSA. The source terms of Model 1 (without the partition wall) exhibited a general slightly increasing trend as the angle increases because of significant leakage radiation in backward directions. Due to the shielding effect of the partition wall against strong radiation in the accelerator room, the source terms of Model 2 were apparently smaller than those of Model 1 for the entire forward hemisphere, facing toward the treatment room.

The previous curve fitting was performed only for depths greater than 100 cm in shield that exhibited a better exponential attenuation in dose profiles. The results in Fig. 5 indicate a relatively constant attenuation length, ranging within 30–36 g/cm^2 for various angles and for the two geometrical models, with and without the partition wall. This suggests the thick-shield attenuation length in this case is mostly a material property rather than depending on the energy distribution of incident neutrons. This is because most neutrons in deep shielding regions have reached an approximate equilibrium spectrum.

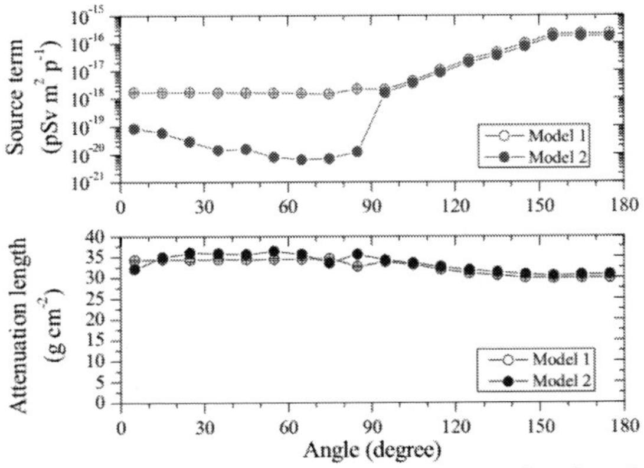

Fig. 5. Source terms and attenuation lengths as a function of angle for concrete shielding against 30-MeV protons hitting the beryllium target in the BSA.

3.3.2 Shielding requirements

Using Eq. (1) and the MCNP6-generated source terms and attenuation lengths, one can quickly estimate the transmitted dose rate at any position outside the shielding walls of the facility in Fig. 1. The results can be compared and verified with previous Monte Carlo predictions.

By using various approaches, Table 2 summarizes the estimated thicknesses of both primary and secondary barriers that are adequate to restrict dose-rates outside the facility to less than 10 μSv/h. Note that Models 1 and 2 estimated thicknesses of concrete walls based on different sets of shielding parameters, without and with considering the effect of the partition wall, respectively. Model 1 obviously overestimated the shielding requirements of the treatment room because of the inclusion of significant radiation backscattered from the accelerator room. As expected, Model 2 gave a more satisfactory prediction than Model 1 on the required thickness of the primary shield of the treatment room because of a more realistic geometry used in parameter generation. However, both Models 1 and 2 failed to predict the correct thicknesses of two secondary barriers. This was somewhat anticipated because the point-source line-of-sight approximation tends to underestimate the transmitted dose rates at large slant angles. It is a well-known caveat when using this approximation in dose rate assessment, resulting from the long slant path of direct penetration in shield and an inappropriate treatment of radiation scattering.

In order to partly address this issue, we developed Model 3, which is essentially a different application of the shielding parameters in Model 2. To estimate the dose rates outside the shielding walls, say at B1–B4, the dose rates on the inner surface of the shield, i.e. at A1–A4, were estimated first by directly attenuating the source term of the given angle with an inverse-square law. In addition to the geometric attenuation, the material attenuation was then taken into account only the perpendicular thickness of the concrete wall. For simplicity, the average attenuation length in Fig. 5 was always used in the dose estimation of Model 3 regardless of the angle at which the dose point views the target. The result turned out to be satisfactory if compared with the Monte Carlo prediction, as indicated in Table 2.

Table 2. Comparison of the estimated thicknesses (cm) of primary and secondary shields of the AB-BNCT facility by using Monte Carlo simulations and three simplified methods.

Method	Accelerator room		Treatment room	
	Primary	Secondary	Primary	Secondary
MCNP6	207	192	128	114
Model 1	209	157	166	145
Model 2	211	158	114	84
Model 3	198	203	129	120

4 Conclusions

Monte Carlo simulations are generally considered the most accurate method for solving radiation transport problems. However, it is time-consuming and its appropriate use requires knowledge and experience when dealing with a challenging shielding problem including complicated geometry and deep penetration. Approximate but simplified methods are often preferable in design phase, especially when the facility layout is expected to undergo several changes. Based on the point-source line-of-sight approximation and the MCNP6-generated shielding parameters, this study developed a suitable shielding model for our AB-BNCT facility, allowing us to readily estimate dose rates outside thick shields or, conversely, determine the barrier shielding. The estimated results were verified by comparing with the reference Monte Carlo results. Note that the shielding parameters used in this study were BSA-dependent and have to be regenerated if the BSA design changes. Nevertheless, the proposed shielding model and experience gained in this study will be useful for our future design work toward a practical AB-BNCT facility in Taiwan.

Acknowledgments

The work is financially supported by Industrial Technology Research Institute in Taiwan under contract no. 100A0303K8.

References

1. Y.W.H. Liu, T.T. Huang, S.H. Jiang, H.M. Liu, Appl. Radiat. Isotopes **61**, 1039 (2004).
2. L.W. Wang, S.J. Wang, P.Y. Chu, C.Y. Ho, S.H. Jiang, Y.H. Liu, Y.W.H. Liu, H.M. Liu, J.J. Peir, F.I. Chou, S.H. Yen, T.L. Lee, C.W. Chang, C.S. Liu, Y.W. Chen, K. Ono, Appl. Radiat. Isotopes **69**, 1803 (2011).
3. H. Tanaka, Y. Sakurai, M. Suzuki, S. Masunaga, Y. Kinashi, G. Kashino, Y. Liu, T. Mitsumoto, S. Yajima, H. Tsutsui, A. Maruhashi, K. Ono, Nucl. Instrum. Meth. B **267**, 1970 (2009).
4. S. Yang, *The beam shaping assembly and collimator design of epithermal neutron beam and in-phantom dose analysis for AB-BNCT* (National Tsing Hua University, Master Thesis, 2014)
5. T. Goorley, et al., Nucl. Technol. **180**, 298 (2012).
6. ICRP Publication 74, *Conversion coefficients for use in radiological protection against external radiation* (Pergamon Press, 1996).
7. S. Agosteo, M. Magistris, A. Mereghetti, M. Silari, Z. Zajacova, Nucl. Instrum. Meth. B **265**, 581 (2007).
8. B.L. Lai, R.J Sheu, U.T. Lin, Health Phys. **108**, S84 (2015).

Shutdown Dose Rate Analysis for the long-pulse D-D Operation Phase in KSTAR

Jin Hun Park[1], Jung-Hoon Han[1,a], D. H. Kim[2], K. S. Joo[3] and Y. S. Hwang[1,2]

[1] Center for Advance Research in Fusion Reactor Engineering (CARFRE), Seoul National University, Seoul, 08826, Korea
[2] Department of Nuclear Engineering, Seoul National University, Seoul, 08826, Korea
[3] Department of Physics, Myongji University, Yongin, 17058, Korea

Abstract. KSTAR is a medium size fully superconducting tokamak. The deuterium-deuterium (D-D) reaction in the KSTAR tokamak generates neutrons with a peak yield of $3.5x10^{16}$ per second through a pulse operation of 100 seconds. The effect of neutron generation from full D-D high power KSTAR operation mode to the machine, such as activation, shutdown dose rate, and nuclear heating, are estimated for an assurance of safety during operation, maintenance, and machine upgrade. The nuclear heating of the in-vessel components, and neutron activation of the surrounding materials have been investigated. The dose rates during operation and after shutdown of KSTAR have been calculated by a 3D CAD model of KSTAR with the Monte Carlo code MCNP5 (neutron flux and decay photon), the inventory code FISPACT (activation and decay photon) and the FENDL 2.1 nuclear data library.

1 Introduction

KSTAR started its operation from 2008 by creating first plasma and successfully completed its 1st phase of operation for 5 years by achieving H-mode plasma. Now it is in its 2nd phase operation aiming heating power of 28MW and developing Advanced Tokamak (AT) operation technology. At its 3rd operation stage from year 2018 with heating power of 28MW KSTAR is aiming for stable AT operation mode with full D-D plasma. At the last operation stage KSTAR would be utilized as a DEMO relevant facility.

The equally probable D-D fusion reactions are occurring in a deuterium plasma, one producing 2.45MeV neutrons and the other producing tritium, which are confined in the plasma and undergo D-T reactions, and produce 14.1MeV neutrons. Since the tritium population is quite small compared to deuterium ions, the 14.1MeV neutron production are about 2% of the 2.45MeV neutrons.

When these fusion reaction generated neutrons are irradiated onto the internal structure of tokamak all of the components are activated and give a serious radiation safety cause to human beings and delicate internal components. And as accurate as possible prediction of the amount of activation is very much required for operation licenses of KSTAR at its 3rd stage high power operation.

A 90° sector model developed at KAIST and POSTECH had a rather simplified Tokamak structure [1]. In particular, structurally simple PFC, vacuum vessel and port were adopted. Also material information used for in vessel components and most of structure were not reflecting most updated material component data. This research specifically reduce the calculation error by implementing a more refined and accurate structural data with updated tokamak structure material information.

2 Calculation

A Monte-Carlo model for KSTAR was developed for the analysis of neutron activation. A 3D CAD of the KSTAR [2] was converted into a geometrical input file for Monte-Carlo calculation through CAD interface program, MCAM [3].

This model is based on a 90° sector of KSTAR tokamak which contains the symmetry of both the vacuum vessel and in-vessel components. Smaller MCNP cells are added to have a neutron spectra at the various locations. Neutron analysis are performed for the central solenoid(CS) coil, toroidal field(TF) coil and poloidal field(PF) coil and nuclear heating.

Calculation of the shutdown dose rate are the so alled Rigorous two-step method (R2S) developed by FZK [4,5] and the second is the Direct one-step method (D1S) [6,7] developed by the ITER team and ENEA.

Both method, even if with different approches use the MCNP Monte Carlo Code[8], for transport calculation, and the FISPACT [9] inventory code.

In R2S, a regular neutron transport calculation provides the sparial distribution of the neutron flux spectra (MCNP, 1st step) the decay gamma source distribution is obtained through a nuclide inventory

* Corresponding author: junghoonhan@snu.ac.kr

DOI: 10.1051/epjconf/201715307024

calculation (FISPACT) using of the MCNP neutron flux spectra and the irradiation history, finally, a decay gamma transport calculation (MCNP, 2nd step) based on FISPACT decay gamma source distribution provieds the dose rate at the specified locations.

D1S method is based on the assumption that the decays gammas of the radioactive nuclides are promptly emitted ; hence, the neutrons and decay gammas are transported in a single MCNP run. The time correction factors required to take into account the build-up and the decay time of each radionuclide according to the real irradiaion history are calculated using FISPACT.

For these neutronics calculation FENDL 2.1 nuclear data library and special libraries for D1S are used.

The simulation assumed 100s pulse length for eachs hot with 30 minute shot to shot intervals and 10 shots per day during 7 days per week operation.

For a neutron source profile for MCNP calculation FTU neutron profile is adopted [10]. FTU neutron emission profile, obtained from ohmic and pellet fueled discharges, shows the neutron brightness profile, obtained by averaging data from 5 consecutive ohmic discharges with similar operation parameters. The solid line is a fit on the experimental data assuming a neutron emission profile of $S(r) = S_0(1 - \left(\frac{r}{a}\right)2)^\alpha$, where $\alpha = 8$.

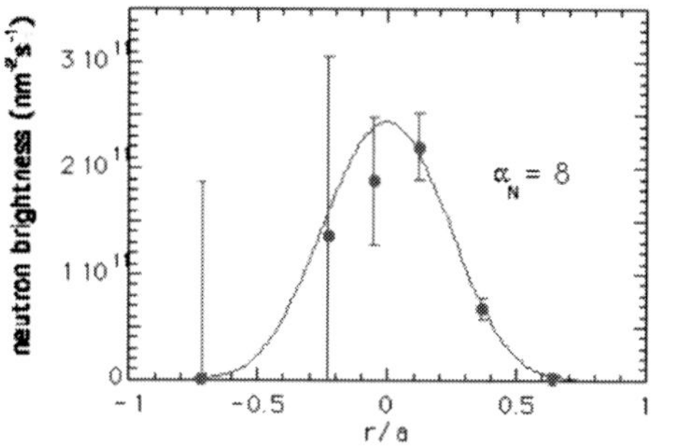

Fig. 1. Neutron emission profile resulted from ohmic discharge of FTU

3 Results

3.1. Central Solenoid coil (CS coil) shutdown dose rate calculation

Calculated the shutdown dose rate using the D1S and R2S methods.

Figure 2 is to show the shutdown dose rate of the CS coils. The measurement locations are 60cm (coil central) and 160cm (coil uppermost) from the center of plasma core. Although D1S and R2S results in a similar trend, they show a difference of approximately double. If the dose rate for the maintenance is 10^{-3} Sv/h, at the top of the CS coils, then it is estimated that it will be accessible for approximately fewer than three days after shutdown. However, according to the D1S calculations, it takes about a year to access the center of the coil. But, estimated that the R2S method will be accessible in about three days.

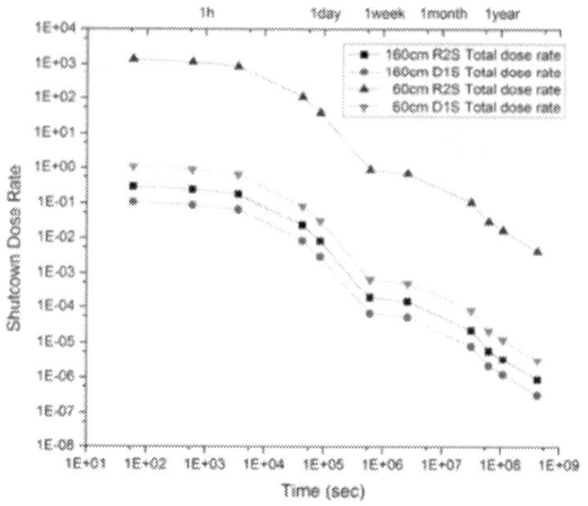

Fig. 2. CS coil shutdown dose rate of 60 and 160cm point from plasma center

The half-life of major radioactive materials is as follws : ^{60}Co 5.27years, ^{58}Co 70.86days, ^{51}Cr 27.7days, ^{94}Nb 2.03x10^4 years. Figure 3 and 4 represent of each radioactive material, depending on the distance from plasma center.

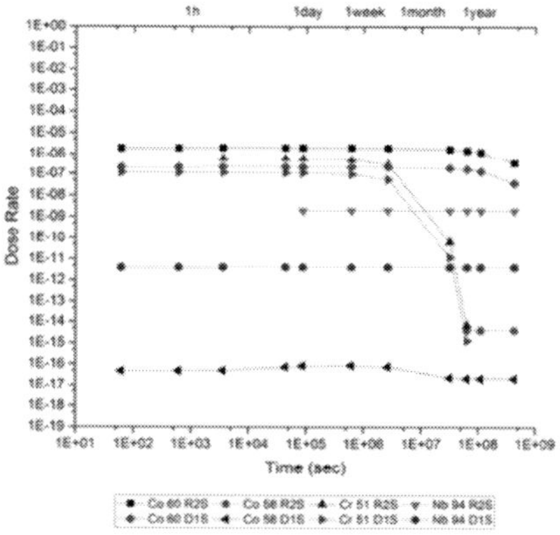

Fig. 3. The main nuclide dose rate of CS coil of 160cm point from the plasma center.

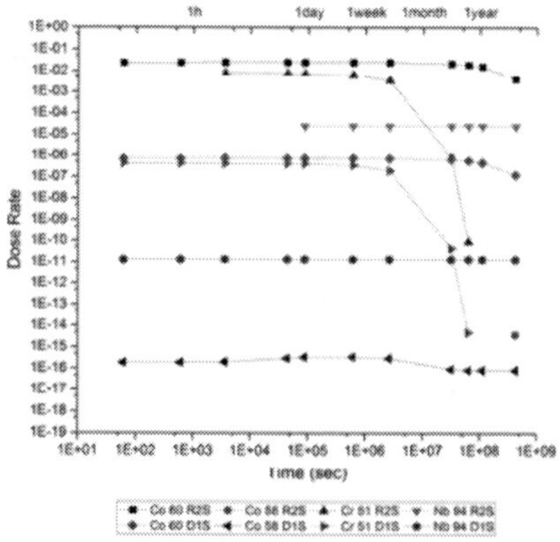

Fig. 4. The main nuclide dose rate of CS coil of 60cm point from the plasma center.

ITER uses Nb_3Sn and NbTi superconducting magnets such as KSTAR. The ^{60}Co, ^{58}Co, ^{51}Cr, and ^{94}Nb isotopes were produced by the reaction of the $NbSn_3$ superconducting magnet.

From Figure 3 and 4, can see that the ^{60}Co and ^{94}Nb have a long lasting dose after shutdown. ^{60}Co emits 0.324MeV beta radiation and 1.17MeV, 1.33MeV gamma rays and main contribute to the total shutdown dose rate after 1month, ^{51}Cr emits 320keV gamma ray and 5keV X-ray and disappear after 1-month. ^{60}Co and ^{51}Cr D1S and R2S method results error is relatively small. ^{58}Co and ^{94}Nb are small contributors to total shutdown dose rates. However, the dose rate difference varies considerably according to analysis methods. ^{58}Co emits 810.8keV gamma-ray and 474keV maximum beta+ decay. ^{94}Nb emits 2.045MeV beta- decay.

3.2. Toroidal Field coil (TF coil) shutdown dose rate calculation

The inner part of the CS and TF coils are located in a similar distance from the plasma. But TF coil is located close to plasma a little more than CS coil. The analytic locations are an inboard side of 45cm and outboard side of 150cm from the center of the plasma core. Closer the distance from plasma, it is possible to verify that the shutdown dose rate difference is increasing.

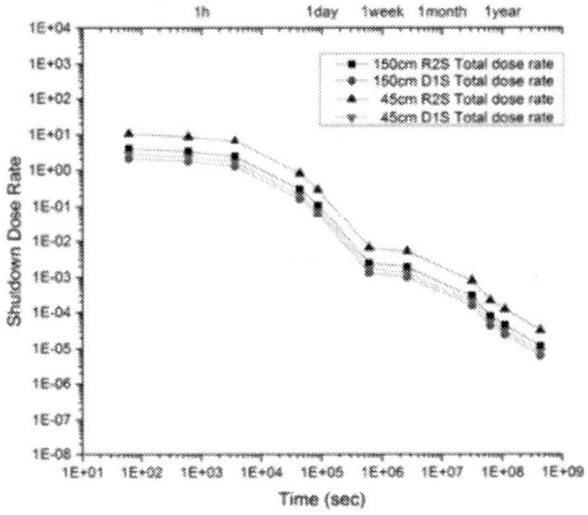

Fig. 5 TF coil shutdown dose rate of 45 and 150cm point from plasma center

Figure 5 is to show the shutdown dose rate of the TF coils. Depending on the calculation method the shutdown dose rate of 45cm points of out a difference of about 3 times, in the case of 150cm points difference of about 1.5 times. However, there is not much difference in shutdown dose rates compared to the overall CS coils. If the dose rate for the maintenance is 10^{-3} Sv/h, at the all of the TF coils, then it is estimated that it will be accessible for approximately fewer than three days after shutdown.

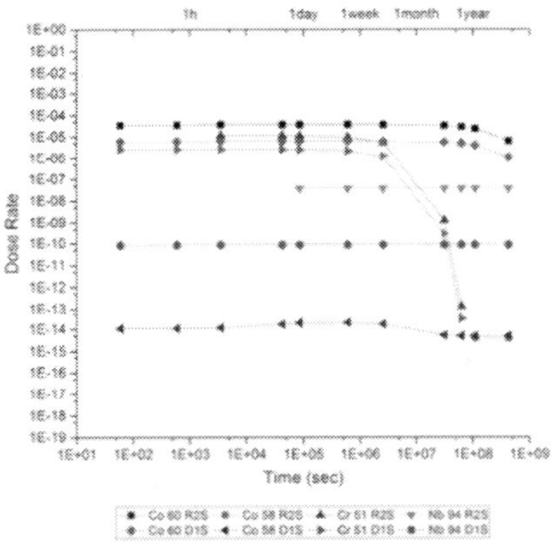

Fig. 6. The main nuclide dose rate of TF coil of 150cm point from the plasma center

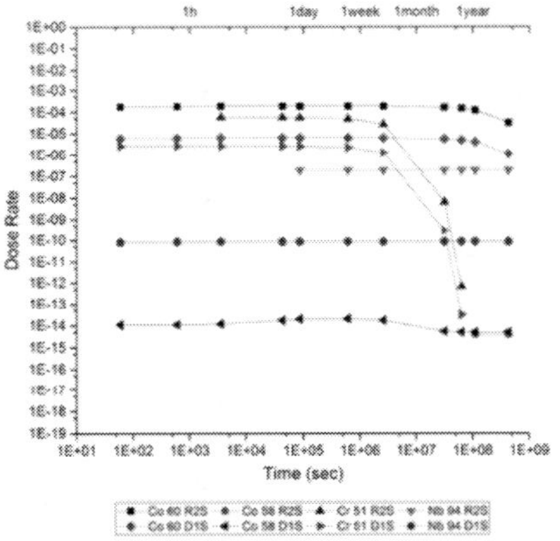

Fig. 7. The main nuclide dose rate of TF coil of 45cm point from the plasma center

Looking at the ^{94}Nb graph in figure 6 and 7, D1S and R2S method results difference are bigger than 160cm distance from plasma core the CS coil result. The reason for difference can be assumed to be dependent on the distance from the plasma. One more specific detail, ^{58}Co has almost same results using both methods. When comparing the CS coils and the TF coils, it is possible to ascertain that only 60cm distance from plasma core dose rate trend results of the CS coil is different.

3.3. Poloidal Field coil (PF coil) shutdown dose rate calculation

PF coils conductor consists of magnets stand of NbTi and jacket material of stainless steel (SS) 316LN. The results of shutdown dose rate D1S and R2S methods were shown in figure 8. The measurement location is 240cm distance upper side from plasma core.

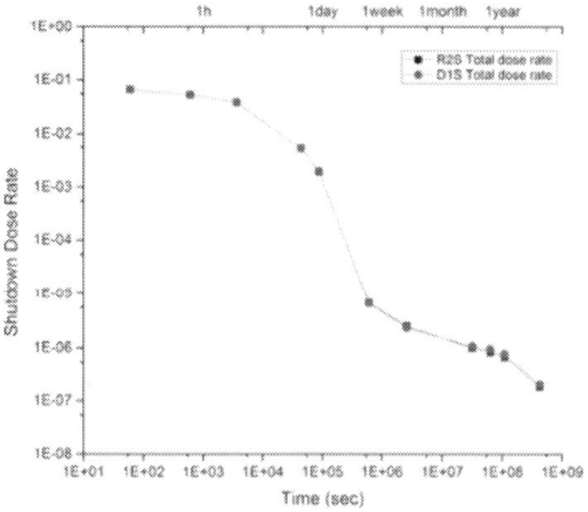

Fig. 8. TF coil shutdown dose rate of 240cm point from plasma center

The results of shutdown after one month are very similar to the two methods. After a month, shutdown dose rates vary slightly. However, both methods require approximately two days of access for maintenance. Figure 9 below shows the dose rate for the four major radioactive materials.

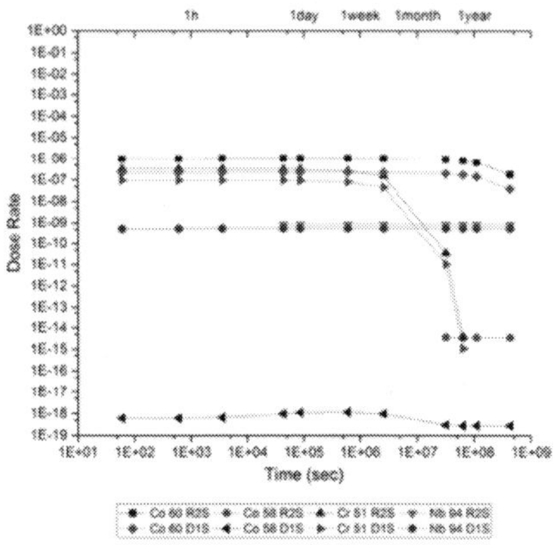

Fig. 9. The main nuclide dose rate of TF coil of 240cm point from the plasma center

The ^{60}Co, ^{94}Nb, ^{51}Cr three radioactive material dose rate results can be found to have small difference in the analysis methods. However, the ^{58}Co although the total dose rate contributes to small, the difference in dose rates can be found to be much better than the other coils. The total shutdown dose rate of the PF coils depends on ^{60}Co, like CS coils. But PF coils are more quickly stabilized than CS coils. Attention to the results of ^{94}Nb, we can show that low error rates at 1day after shutdown. This results seems a big difference in CS coils results. Especially CS coil is exposed to the high dose rates, we suspect the error in the high-dose rate Niobium libraries. Relatively low dose rate PF coils can be said to be relatively accurate results.

3.4. Nuclear heating

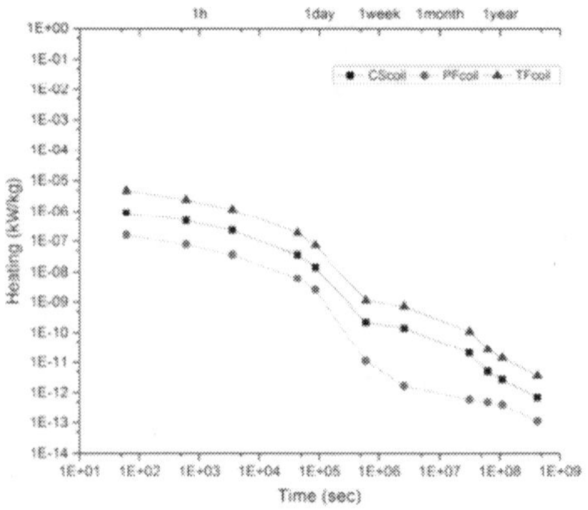

Fig. 10. Nuclear heating of main parts of Tokamak structure

Figure 10 shows time variation of nuclear heating for the major components KSTAR tokamak after shutdown. TF coils which are closer to the core plasma than CS coils shows less nuclear heating than that of CS coils due to its material composition difference. About 3 days after the nuclear heating appears to have been investigated by 10^{-9} kW/kg level. This means the temperature change of the nano-pico level, the heating of the superconducting magnet according to the nuclear heating can be seen that very little.

4 Summery and future work

By assuming KSTAR full D-D operation phase plan activation analysis are performed.

For a maintenance of KSTAR, from this analyses it take about a week after shutdown of tokamak operation to access tokamak core while for an access to the outside of KSTAR cryostat it needs 3 days after plasma operation.

For Niobium superconducting magnet there is increasing SDR results differences from two different analysis method as analysis points approaches to the plasma center. Especially for ^{94}Nb SDR results from two methods differences increasing as analysis points approaches to plasma center while for ^{58}Co trend is reversed. For these outcome further analyses are required by adopting different nuclear data libraries.

In addition since the above analyses excluded the effect of high dosage gamma rays from ^{16}N of cooling water it need further analyses which include the additional influences from the outside of magnet system.

Also it is very necessary to perform analyses for the in-vessel components of the KSTAR tokamak which might use the same material of ITER, such as Tungsten.

To assure for both human and machine safety further analyses using different data library and adopting various operation scenario, which gives different neutron source profile, are very much required.

 DOI: 10.1051/epjconf/201715307024

Acknowledgments

This work was supported by the Ministry of Science, ICT and Future Planning of the Republic of Korea under the Korean ITER project contract.

References

1. Hyunduk Kim et al., Radioactivity evaluation for the KSTAR TOKAMAK, Radiation Protection Dosimetry, Vol. **116**, No. 1-4 pp. 24-27 (2005)
2. M.S. Cheon et al., Diagnostic neutron activation system for KSTAR, 2nd International Workshop on Fast Neutron Detectors and Applications (2011)
3. FDS TEAM collaboration, Y. Wu, CAD-based interface program for fusion neutron transport simulation, Fusion engineering and Design 84 (2009)
4. Y. Chen, U. Fischer, ITER-FEAT, Shutdown dose rate analysis by Rigorous method final report on Contract EFDA/00-564 FZK Internet Bericht IRS Nr. 9/01, FUSION Nr. 177 (2001)
5. Y. Chen, U. Fischer, Rigorous MCNP based shutdown dose rate calculations: computational scheme, verification calculations and applications, ITER Fusion Eng. Des. 63/64 107-114 (2002)
6. H. Iida, D. Valenza, R. Plenteda, R. T. Santoro AND J. Dietz, Radiation Shielding for ITER to allow for Hands-on Maintenance inside the Cryostat, J. of Nuc. Sci. and Tech., Supplement 1. 235-242 (2000)
7. L. Petrizzi, H. Iida, D. Valenza and P. Batistoni, Improvement and benchmarking of the new shutdown dose estimation method by Monte Carlo code, Advanced Monte Carlo for Radiation Physics, Particle Transport Simulation and Applications, Proceedings of the MC2000 conference 23-26 Oct. 2000 Lisbon, Portugal. Springer 865-870 (2001)
8. J. F. Briestmaister, Ed, MCNPTM,--a general Monte Carlo N-particle transport code, Version 4C3, Los Alamos Nat. Lab. Report, LA13709-M (2000)
9. R. Forrest, FISPACT-2007 User Manual EASY 2007 Documentation Series UKAEA FUS 534 (2007)
10. ENEA Italian Agency for New Technologies Energy and the Environment Nuclear Fusion division, 1997 Progress report

Low energy and high energy dumps for ELI-NP accelerator facility: rational and Monte-Carlo calculations' results

A. Esposito[1], O. Frasciello[1,*], and M. Pelliccioni[1]

[1] *Istituto Nazionale di Fisica Nucleare - Laboratori Nazionali di Frascati, Via E. Fermi 40, 00044 Frascati, Italy.*

Abstract. ELI-NP will be a new international research infrastructure facility for laser-based Nuclear Physics to be built in Magurele, south west of Bucharest, Romania. For the machine to operate as an intense γ rays' source based on Compton back-scattering, electron beams are employed, undergoing a two stage acceleration to 320 MeV and 740 MeV (and, with an eventual energy upgrade, also to 840 MeV) beam energies. In order to assess the radiation safety issues, concerning the effectiveness of the dumps in absorbing the primary electron beams, the generated prompt radiation field and the residual dose rates coming from the activation of constituent materials, as well as the shielding of the adjacent environments against both prompt and residual radiation fields, an extensive design study by means of Monte Carlo simulations with FLUKA code was performed, for both low energy 320 MeV and high energy 720 MeV (840 MeV) beam dumps. For the low energy dump we discuss also the rational of the choice to place it in the building basement, instead of installing it in one of the shielding wall at the machine level, as it was originally conceived. Ambient dose equivalent rate constraints, according to the Rumenian law in force in radiation protection matter were 0.1 μSv/h everywhere outside the shielding walls and 1.4 μSv/h outside the high energy dump area. The dumps' placements and layouts are shown to be fully compliant with the dose constraints and environmental impact.

1 Introduction

The Extreme Light Infrastructure - Nuclear Physics (ELI-NP) constitutes one of the three arms of the European Extreme Light Infrastructure project, presently being in the building process. Using an high intensity laser up to 10^{24}W/cm^2 and a high energy low bandwidth gamma beam system (GBS), it will be a new international research infrastructure facility for laser based nuclear physics. The facility will provide a broad range of science covering new nuclear physics, astrophysics, fundamental high field physics as well as applications in nuclear materials, radioactive waste management, material science and life sciences.

The ELI-NP project, as defined by its promoter and beneficiary, the Romanian National Institute of Nuclear Physics and Engineering "Horia Hulubei" (IFIN-HH), will consist of ground-breaking research systems equipped with a 2x10 PW laser system and a 740 MeV (maximum operation energy) e^- linac, providing a tunable 0.2-19.5 MeV GBS. The γ rays are produced letting the relativistic e^- beam interact with a 2.4 eV laser pulse at a point. The ELI-NP GBS will allow the production of γ rays up to $\approx$ 19.5 MeV with a maximum e^- beam energy of 740 MeV.

An eventual energy upgrade of the e^- linac should raise the maximum energy up to 840 MeV. This operation mode has been used actually, in the prompt radiation and activation simulations, in order to assess results for the most conservative scenario.

According to the Technical Design Reports (TDRs), finalized and approved by ELI-NP International Scientific Advisory Board in June 2015, the ELI-NP Center will have eight experimental rooms, some of them being depicted in figure 1, namely: E1 for laser induced nuclear reactions; E2 for nuclear resonance fluorescence applications; E3 for positron source; E4/E5 for accelerated particle beams induced by High Power Laser System (HPLS) (0.1/1 PW) at high repetition rates; E6 for intense e^- and γ beams induced by high power laser beams; E7 for experiments with combined laser and γ beams; E8 for nuclear reactions induced by high energy γ beams [1].

The design of both the low energy (320 MeV) and high energy (840 MeV) line dumps has been worked out taking into consideration the main features of the systems, aiming to 1) avoid any interaction of the primary beam with the accelerator shielding walls, roof and floor included, 2) attenuate, at a reasonable cost and without compromising the utility of the apparatus for its design purpose, the prompt radiation at levels that are acceptable for humans outside the shielding walls, according with the Romanian law in force in radiation protection matter [2], 3) reduce the induced activity on the shielding walls, 4) reduce unwanted background in experimental detectors, 5) protect equipments against radiation damage and 6) reduce the cost of the shielding bulk.

*e-mail: oscar.frasciello@lnf.infn.it

 DOI: 10.1051/epjconf/201715307025

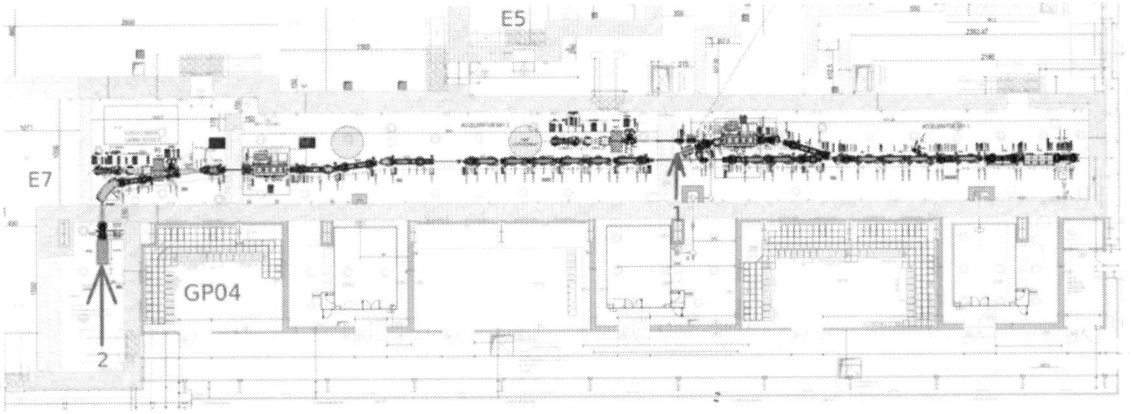

Figure 1. The ELI-NP machine layout and some experimental areas. Also indicated with red numbers are 1) the point where the low energy 320 MeV e^- beam is 90 degrees bended, in order to be delivered to the beam dump in the basement (as described in the text) and 2) the high energy 840 MeV beam dump and its ERA host room (Courtesy of A. Falone and V. Pettinacci).

The ELI-NP shielding wall calculations have been performed based on the criteria that the dose uptake for professional exposed personnel must be only 10% of the legal limits, according to the Romanian law. In more detail, the individual effective dose must be less than 2 mSv/y, the equivalent dose to the eye lens must be less then 15 mSv/y, the equivalent dose to the skin must be less then 50 mSv/y averaged over any cm^2 area of the skin and the equivalent dose to the extremities such as hands, forearms, feet and ankles must be less then 50 mSv/y [3]. Moreover, the H*(10) rate constraints dumps' design and shieldings must be compliant with, are 0.1 μSv/h everywhere outside main shielding walls, as the doses to the members of the public, and 1.4 μSv/h outside the high energy dump area, toward the GP04 room (refer to figure 1).

2 Monte Carlo simulations

All the simulations have been carried out with the FLUKA Monte Carlo code [4, 5], version 2011.2c. One of the major advantages of FLUKA, from the point of view of radiation safety applications, is the possibility to compute in the same simulation the ambient dose equivalent H*(10) due to all the components of the prompt radiation field and to the time evolution of the activation products of irradiated materials, being the code able to transport their emitted radiation. The code, indeed, is able to solve analytically the Bateman equation, describing the time evolution of a system irradiated with a given pattern:

$$\frac{dN_i}{dt} = -\sum_{j\neq i}\left[\lambda^d_{ji} + \bar{\sigma}_{ji}\bar{\phi}\right]N_i + \sum_{j\neq i}\left[\lambda^d_{ij} + \bar{\sigma}_{ij}\bar{\phi}\right]N_j \quad (1)$$

in which

$$\bar{\phi} = \int \phi(E)\mathrm{d}E \quad (2)$$

is the average spectrum of a particle fluence rate $\phi(E)$, which in equation 1 is kept constant during each time interval,

$$\bar{\sigma}_{ji} = \frac{1}{\bar{\phi}}\int \phi(E)\sigma_{ji}(E)\mathrm{d}E \quad (3)$$

is the particle induced cross section σ_{ij}, for the transmutation from isotope i to isotope j, averaged over the spectrum and λ^d_{ji} is the probability for radionuclide i to decay into radionuclide j. FLUKA can then generate and transport the residual radiation, thus allowing the estimation of the residual dose rates due to the decay of the produced residuals.

As will be shown in the next section, the dumps' constituent material blocks, on which the primary particles do impinge, have dimensions exceeding the 320 MeV and 840 MeV e^- range in the corresponding materials. For this reason those blocks must be treated as thick targets and, consequently, in our simulations the photo-nuclear physics was activated over the whole energy ranges for the relevant high Z elements in the dumps' models, namely Pb, W, Cu and concrete material.

In order to improve the statistics in the dose scoring regions, which are rather far apart from the interaction point of the primary particles of the beam with the concrete blocks inside the dumps (which are the first targets e^- do impinge upon), with several material regions in the middle, an importance biasing technique was used, increasing photons and neutrons importances up to a factor 512 toward the scoring regions. This way we were able to achieve calculated dose values affected by a statistical error not exceeding 10%, whereas the calculated values for residual volume specific activities had an acceptable statistical error for all the scored radionuclides, kept at the level of few percent.

3 Low energy 320 MeV beam dump

The low energy dump will be used to absorb the 320 MeV e^- beam during the machine operation to produce the low energy γ beam. The following assumptions subtend all the calculations for this dump:

- E(e^{-})= 320 MeV
- I=1.6 $\mu A \approx 9.9 \cdot 10^{12}\ e^{-}/s$
- Machine runtime= 3000 h/y.

As it was originally conceived, the low energy dump had to be placed at the accelerator level, horizontally arranged and mounted with additional concrete shields upon the building wall at the interface between accelerator bay 1 and 2, as shown in the 3D model of figure 2. This choice, however, posed several problems to good practice criteria for the concepts of machine installation and operation. It resulted firstly from preliminary simulations that in order to achieve acceptable dose rates levels outside the lateral 1.5 m thick shielding walls, several tons of concrete shield had to be installed around the dump, for a suitable protection of personnel present in the adjacent control room against the prompt radiation [6]. This represented a structural issue in terms of additional load for the building wall. Secondly, the unshielded backscattered radiation, to which neutrons contributed mostly, could represent a severe constraint to sustainable machine runs, operations and components' radiation hardness. So we proposed (and the results partially presented in this paper constituted the robust ground basis for the final acceptance of the proposal) to bend the low energy e^{-} beam at 90 degrees and to install its dump in the building basement area, under the accelerator floor, as shown in the 3D FLUKA model of figure 3, used for the simulations. Here the model is complete with the first eight nearest neighbours pillars, underground soil and water. The space sorrounding the dump is filled with air, with the standard composition of dry air at sea level. The dump itself 3D geometry is reported in figure 4 and does consist of a cylinder with diameter 150 cm, height 70 cm and internal core of cylinder smaller blocks of ordinary concrete, tungsten and lead, sorrounded by lead and ordinary concrete. Finally it lies over an additional 50 cm of ordinary concrete stand, thus reaching an overall final height of 120 cm.

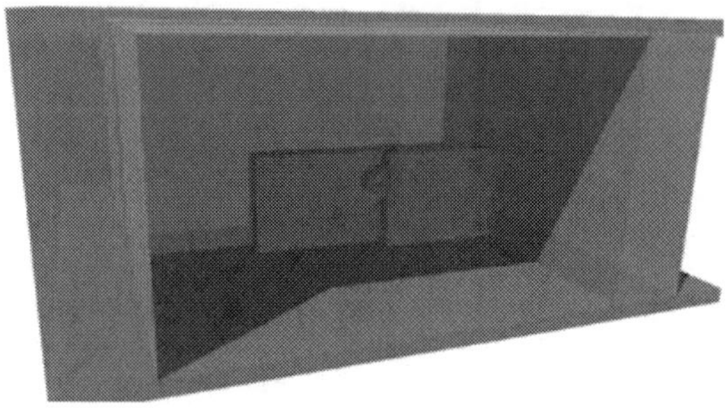

Figure 2. Low energy 320 MeV beam dump placed in the accelerator bay, complete model with the needed lateral shields.

3.1 Prompt radiation field

The interaction of the primary e^{-} beam with the dump materials produces secondary and tertiary radiations composed mainly of neutrons and photons. The computed H*(10) 3D distribution, in the building basement hosting the dump, due to the prompt radiation is reported in figure 5, the scale reporting values in units of pSv/e^{-}.

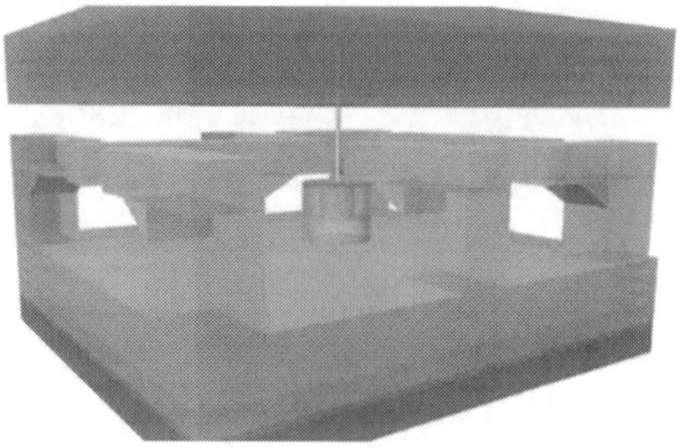

Figure 3. Low energy 320 MeV beam dump placed in the building basement, complete model with ancillary nearest neighbours pillars, underground soil and water.

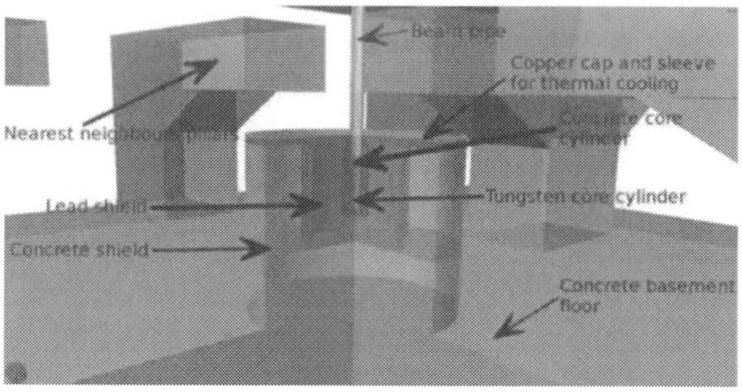

Figure 4. Low energy 320 MeV beam dump internal structure detail (clipped view).

Figure 5. Prompt radiation ambient dose equivalent rate 3D map (clipped view) for the 320 MeV beam dump. Values are reported in units of pSv/e^{-}.

The soil and ground-water beyond the concrete basement, as modeled in our simulations, receive an averaged H*(10) rate of $\approx$ 1.6μSv/h, as reported in figure 6. The average is taken over the single bins the indicated dose rate scoring volume was divided into. As a first result it fol-

Table 1. Radionuclides found in air and their specific activities after 3000 hs of beam dump irradiation, at 1 h cooling time. Acronyms PSN and PR stay for Parent Stable Nuclide and Production reaction, respectively.

Isotope	A	PSN	PR	$T_{1/2}$	$A_S\ \left[\frac{Bq}{cm^3}\right]$
Ar	41	^{40}Ar	(n, γ)	1.8 h	$8.46 \cdot 10^{-3}$
Ar	37	^{36}Ar	(n, γ)	35.1 d	$2.70 \cdot 10^{-4}$
C	14	^{14}N	(n, p)	5730 y	$2.35 \cdot 10^{-4}$
H	3	^{14}N	$(\gamma, ^3H)$	12.3 y	$4.96 \cdot 10^{-7}$
Ar	39	^{38}Ar	(n, γ)	269 y	$1.21 \cdot 10^{-8}$

lows that the prompt radiation field does not represent an issue from the point of view of the environmental impact.

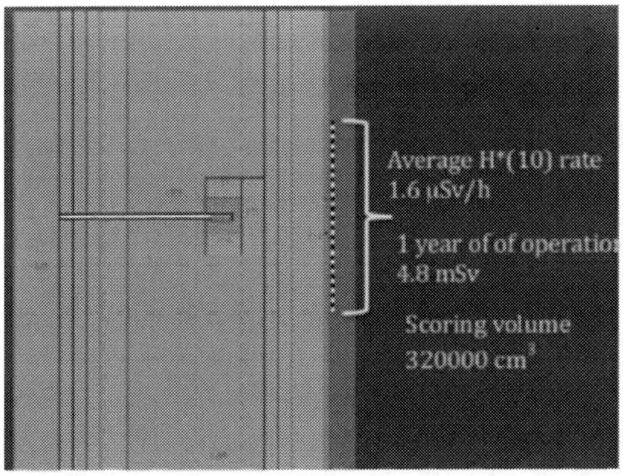

Figure 6. Average H*(10) rate delivered to soil and ground-water by the prompt radiation generated at low energy dump level.

3.2 Residual dose rates

Residual dose rates in soil and ground-water were found to be trivial and are not further addressed here. Special attention was paid on the estimation of the air activation in the dump area, both because there is and there could be no ventilation system and of the risk assessment for personnel in case of emergency access in the dump proximities. In our simulations we asked FLUKA to score the produced radionuclides and their volume specific activities in air, after an irradiation of 3000 h (equivalent to one year of full machine runtime) and at several cooling times, namely 1/2 h, 1h, 4h, 8h, 24 h. Just as an example we report in the table 1 the radionuclides scored and their specific activities at 1 h cooling time. It is clear that several radionuclides commonly produced in air by e^- beams, like ^{7}Be, ^{11}C, ^{13}N, ^{15}O, ^{16}N, ^{38}Cl, ^{39}Cl were not scored in our simulations. These radionuclides are produced via photo-nuclear reactions with specific thresholds energies, exceeding $\approx$ 10 MeV (actually the lowest threshold energy is 10.6 MeV for the (γ, n) reaction on ^{14}N, to produce ^{13}N). To establish the physical mechanisms justifying the absence of the above mentioned radionuclides in air in the dump region, the γ fluence in air was computed, for one beam shot at nominal intensity as reported in section 1, at 1 s and 1/2 h cooling times after 3000 h irradiation. The results are plotted in terms of isolethargic fluence, in figure 7. As it can be seen from the prompt radiation field

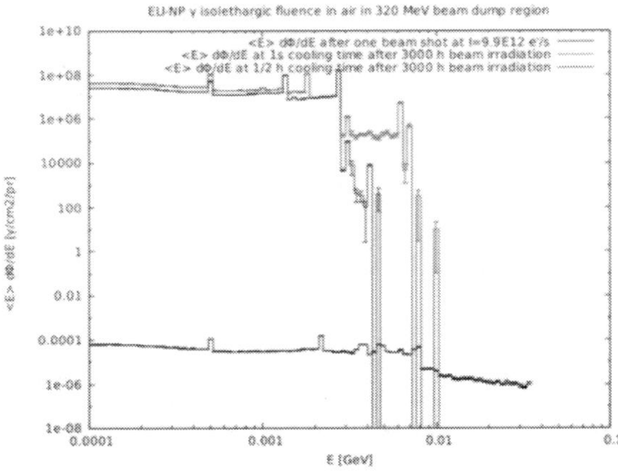

Figure 7. γ isolethergic fluence in air in 320 MeV beam dump area.

(that one which should lead to airborne activation) curve in black, very few photons ($\approx 10^{-6}$ γ/cm^2/pr) are scored above the threshold energy of 10.6 MeV, thus not sufficient to activate the radioisotopes of interest. For comparison, at 1 s cooling time and at 1/2 h cooling time all photons have energies well below that lowest threshold. That is why Be, C, N, O and Cl radioisotopes were not scored in our simulations. It is noteworthy to stress that the exemption level for ^{41}Ar in the Romanian law is 100 Bq/g.

In figure 8 the evolution of the residual H*(10) rate at 1 m from the lateral dump surface and at 1 m height from the floor is plotted, as a function of the cooling times, while the same quantity spatial distributions at 1/2 h and 24 h are shown in figure 9, with values in units of pSv/s. We can observe that already at 1 s after the end of irradiation an acceptable dose rate is found, $\approx$ 3 μSv/h, which decreases to $\approx$ 0.9 μSv/h at 24 h of cooling time.

The ^{41}Ar contribution to the residual H*(10) rate is found to be negligible, being ~ 10^{-5} μSv/h. The H*(10) rate due to the dump's materials and air activation, after the specified irradiation profile, at 1 m from the dump's surface, 5 minutes after machine shutdown, is $\approx$ 2.8 μSv/h. Thus 2.8 μSv is the upper bound effective dose a worker could receive if entering the area right after 5 minutes from the machine shutdown and resting at 1 m from the dump's surface for an hour. As a conclusion, the air activation radiation hazard in the low energy dump area was found to be negligible.

4 High energy 840 MeV beam dump

The high energy 740 MeV (840 MeV) dump will be used to absorb the e^- beam during the machine operation in the high energy γ beam production mode. The complete 3D

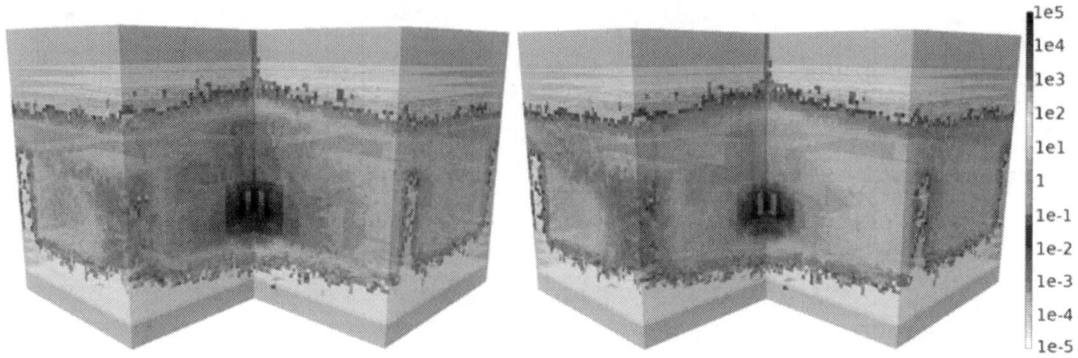

Figure 9. Low energy dump residual H*(10) rate 3D maps (clipped views), at 1/2 h (left) and 24 h (right) cooling times after 3000 h beam dump irradiation. Scale values are in units of pSv/s.

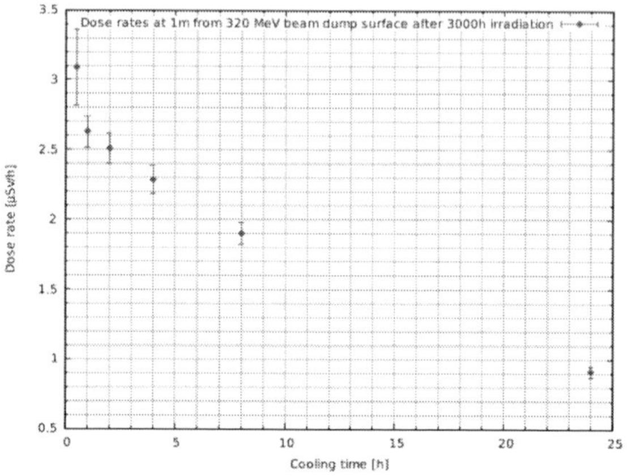

Figure 8. Low energy dump residual H*(10) rate at 1 m from the dump lateral surface, at 1 m height from the floor, after 3000 h of beam dump irradiation, as a function of cooling times.

models' geometry, as simulated in FLUKA, for the detailed position and structure of the dump are illustrated in figure 10 and 11, respectively. For this dump the simulations' assumptions are:

- E(e^-)= 840 MeV
- I=1.6 $\mu A \approx 9.9 \cdot 10^{12}\ e^-/s$
- Machine runtime= 3000 h/y.

Despite the low energy one, the high energy dump does benefit of a dedicated hosting room, as indicated in figure 1, called Electron Recovery Area (ERA). The dump features a geometry concept similar to the low energy one, with slight differences due to the horizontal and suspended positioning in the ERA. This allows for the thermal cooling system to be arranged quite differently, with the internal copper cladding terminating on the radiator cap which is placed at the end of the dump, along the beam direction. From the internal structure point of view, the dump is modeled as a cylinder of 160 cm in diameter and 200 cm in height, with inner core cylinders of ordinary concrete, tungsten and lead, sorrounded by lead and ordinary concrete.

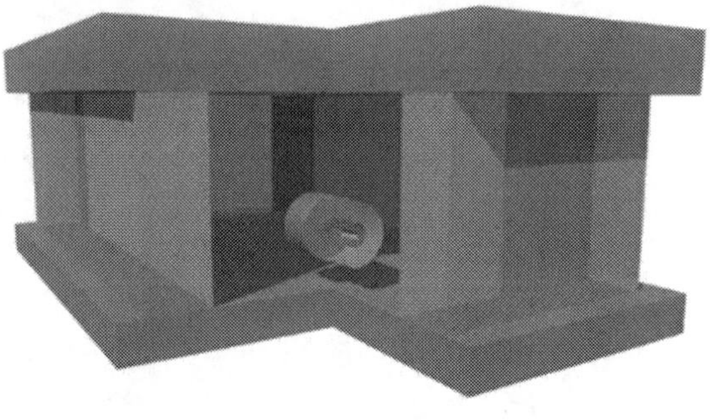

Figure 10. FLUKA 3D model (clipped view) for the position and orientation of the 840 MeV beam dump in the dedicated host room, complete with the building shielding walls.

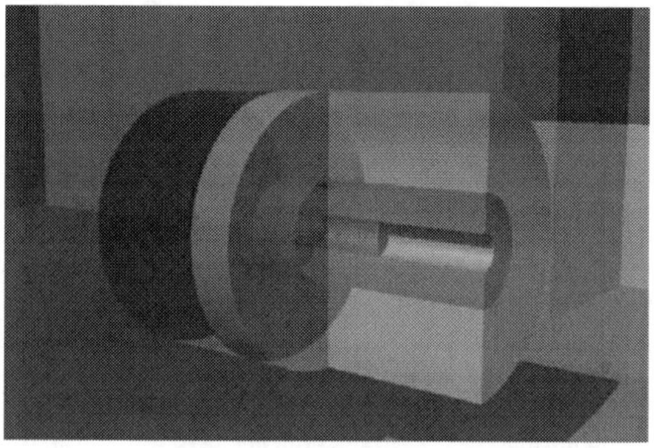

Figure 11. FLUKA 3D model (clipped view) for the structural details of the 840 MeV beam dump.

4.1 Prompt radiation field

As in the case of the low energy beam dump, the interaction of the e^- beam with dumps' materials does produce

DOI: 10.1051/epjconf/201715307025

secondary and tertiary radiation fields composed mainly by neutrons and photons. Prompt radiation ambient dose equivalent rate 3D map is reported in figure 12 for the 840 MeV dump. The H*(10) rate values, due to the prompt radiation field, found at 1 m from the dump's surface and at beam pipe level (dump longitudinal axis) are reported in the histogram of figure 13. Here we can see how the backscattered radiation determines a very high dose rate value of more then 1 Sv/h. While the dose level in the forward direction is the lower one, as expected, of ≈ 7 mSv/h, the levels on the left and right do access to ≈ 30 mSv/h. Moreover, in figure 14 the binned scoring regions of H*(10) outside the shielding walls (long pencils in black and white) and at 1 m from the dump's surface (white spots) are depicted, in overlay to the simulated geometry.

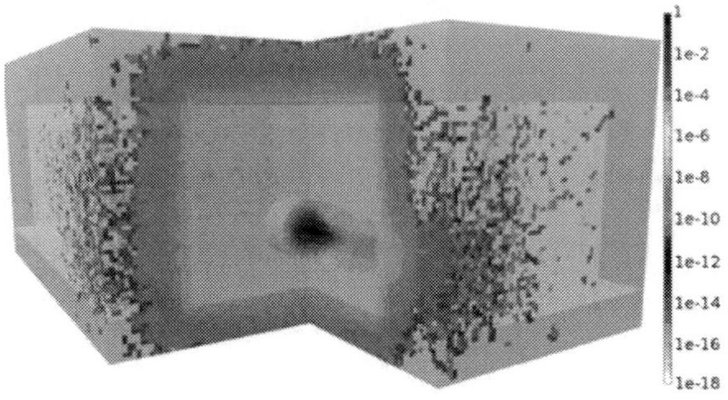

Figure 12. Prompt radiation ambient dose equivalent rate 3D map for the 840 MeV beam dump.

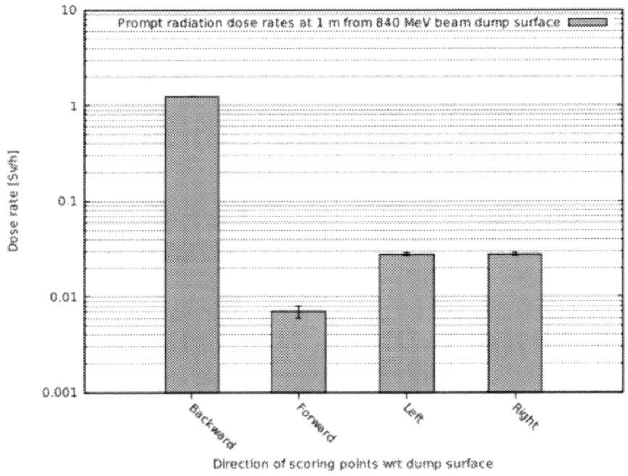

Figure 13. Ambient dose equivalent rates at 1 m from the dump's surface.

The H*(10) rate mean values, due to the prompt radiation, in the scoring volumes outside the ERA shielding walls, in the forward and lateral directions with respect to the beam direction (referring to figure 14 right and downside pencils, respectively), were found to be 0.67 μSv/h and 0.13 μSv/h respectively, both well below the 1.4 μSv/h constraint mentioned in section 1. For the other lateral

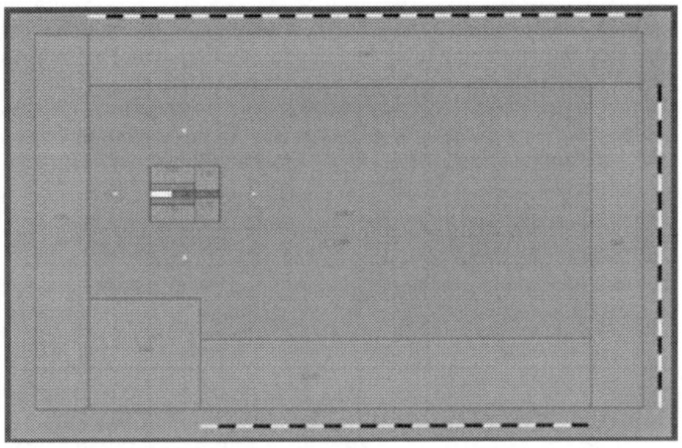

Figure 14. ERA FLUKA model, with the binned regions for the scoring of H*(10) (the pencils along the lateral shielding walls and the four spots at 1 m from the beam surface).

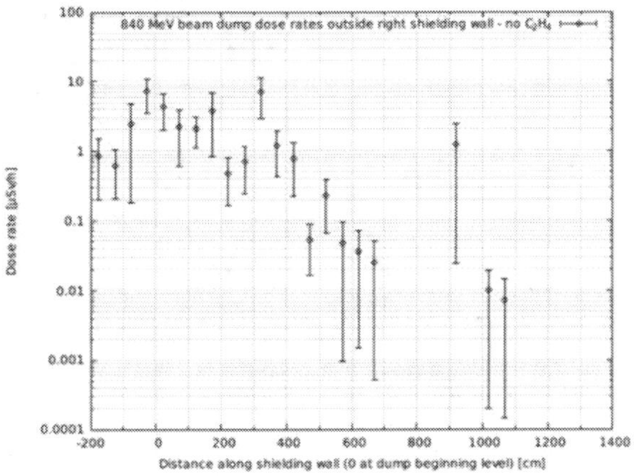

Figure 15. Ambient dose equivalent rates out of the lateral shielding wall (toward GP04 room), as a function of the distance along the wall.

shielding wall (upside in figure 14), the H*(10) rate as a function of the distance along the wall is plotted in figure 15. Here it is clearly visible that some values reach ≈ 7 μSv/h, which obviously was not acceptable in view of the constraints discussed above. Taking into account that the dominant contribution to the prompt radiation field in this region is due to neutrons, an asymmetric 25 cm thick$(C_2H_4)_n$ shell was added, as in figure 16. This countermeasure allowed a reduction of the scored dose rate by more than a factor of 10 at the dump beginning level, from 7 μSv/h to 0.14 μSv/h, thus allowing us to be compliant with the dose rate constraints.

4.2 Residual dose rates

The specific activities of the scored radionuclides in air in the ERA room are reported in table 2, for 3000 h of beam dump irradiation (equivalent to one year of full machine

 DOI: 10.1051/epjconf/201715307025

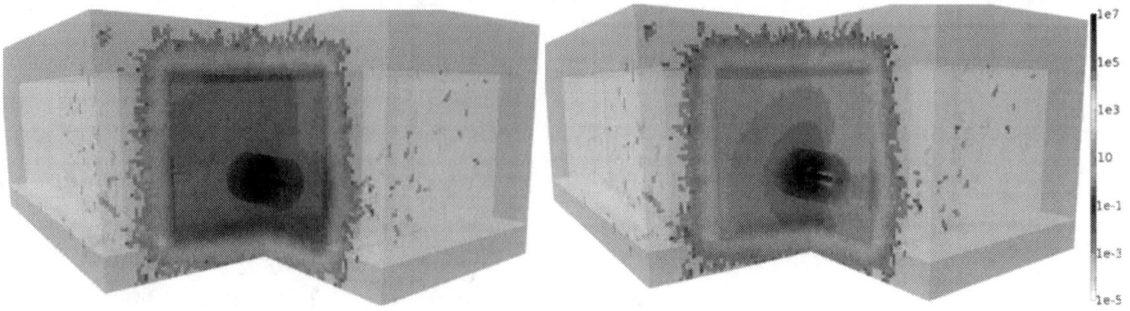

Figure 17. High energy dump residual H*(10) rate 3D maps (clipped views), at 1 s (left) and 24 h (right) cooling times after 3000 h beam dump irradiation. Scale values are in units of pSv/s.

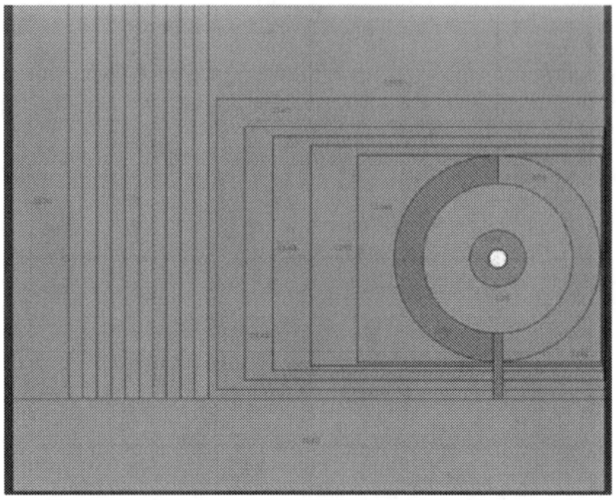

Figure 16. Asymmetric C_2H_4 shield geometry for the 840 MeV beam dump. The concentric rectangles around the dump are the used biasing regions for this set of simulations.

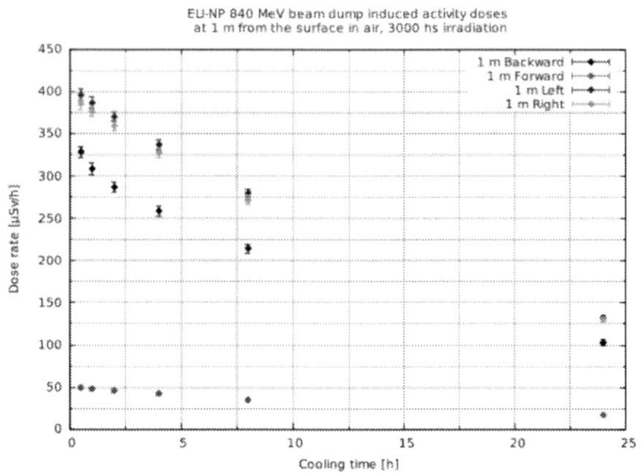

Figure 18. High energy dump residual H*(10) rate at 1 m from the dump's surface, after 3000 h of beam dump irradiation, as a function of cooling times.

runtime) and at 1 h of cooling time. Here we can find some of the above-mentioned radioisotopes commonly found at e^- accelerators, like ^{13}N and ^{15}O not scored in the low energy dump case, with specific activities more than 3 and 11 orders of magnitude, respectively, lower then the maximum scored one, that of ^{41}Ar.

In the case of high energy dump hosted in ERA room, the potential access by the personnel after the end of the irradiation is taken into account not for the emergency only, as it was for the low energy dump, but also for normal mantainance works. A ventilation system is thus foreseen for this area. In figure 17 the spatial distributions of H*(10) rate are shown, for 1 s and 24 h cooling times, with reported values in units of pSv/s. Consequently, the ambient dose equivalent rates, due to residual radioactivity, as a function of the cooling times, at 1 m from the dump's surface in the four directions are reported in the plot of figure 18. With respect to the low energy dump case, we can observe rather high residual dose rates in all directions (the lateral ones being the higher ones), still at 24 h after the end of irradiation. At 1 m from the dump's surface, on the left and on the right, indeed, residual H*(10) rate goes from $\approx$ 400 μSv/h at 1 s of cooling time to $\approx$ 130 μSv/h at 24 h of cooling time. The increase in the radiation hazard for personnel for this dump area, however, is strongly mitigated considering that working personnel could access in ERA only for a limited amount of time and, moreover, the presence of the ventilation system will operate in order to lower the intake risk from airborne activation. Under these circumstances no criticalities are advised from the radiological safety point of view.

Table 2. Radionuclides found in air in the ERA room and their specific activities after 3000 h of beam dump irradiation, at 1 h cooling time. Acronyms do stay for the same as in table 1.

Isotope	A	PSN	PR	$T_{1/2}$	A_S $\left[\frac{\text{Bq}}{\text{cm}^3}\right]$
Ar	41	^{40}Ar	(n,γ)	1.83 h	$7.32 \cdot 10^{-3}$
Ar	37	^{36}Ar	(n,γ)	35.1 d	$2.27 \cdot 10^{-4}$
C	14	^{14}N	(n,p)	5730 y	$2.17 \cdot 10^{-4}$
H	3	^{14}N	$(\gamma,^3\text{H})$	12.3 y	$5.66 \cdot 10^{-6}$
N	13	^{14}N	(γ,n)	6.9 d	$1.84 \cdot 10^{-6}$
Ar	39	^{38}Ar	(n,γ)	269 y	$3.00 \cdot 10^{-8}$
O	15	^{16}O	(γ,n)	122.24 s	$7.43 \cdot 10^{-14}$

5 Conclusions

All Monte Carlo calculations, performed by means of the FLUKA code under conservative conditions at maximum energy of 320 MeV and 840 MeV, with beam current of $9.9 \cdot 10^{12} e^-/s$, and 3000 h/y of machine operation are shown to be compliant with the dose constraints imposed by the Romanian law in force in radiation protection subject matter. The computed reference parameters ($H^*(10)$ rates, induced activity and radionuclides specific activities) for the dump in the basement area show that its environmental impact is trivial (i.e. negligible $H^*(10)$ rate delivered to soil and water and negligible activations of both materials). The air activation also, in the basement area, was found to be not a critical issue. Our simulations' results show, also, that the high energy dump is conceived in a proper way to satisfy the dose limit constraint to the members of the public, outside the building shield. The dose levels scored at 50 cm far away from the shielding wall toward the GP04 room, however, were not acceptable in terms of the constraints specified in section 1. As a countermeasure we modeled the dump with an additional asymmetric 25 cm thick $(C_2H_4)_n$ shell on its lateral surface, what resulted to be an effective remedy which allowed the dump to be compliant with the dose constraints. The studied models have already passed the engineering phase and are presently in the construction process.

References

[1] N. V. Zamfir, "Extreme Light Infrastructure Nuclear Physics (ELI-NP): Present status and perspectives", EPJ Web of Conferences **117**, 10001 (2016)

[2] Guvernul Romaniei, Cancelaria Primului Ministru, CNCAN, "Norme de Securitate Radiologica - Proceduri de Acceptare a Intraprinderilor Externe", **NSR-09**, (2003)

[3] I. O. Mitu, C. Ivan, F. Negoita, D. Aranghel, S. Bercea, E. Iliescu, C. Petcu, M. Gugiu, M. A. Popovici, C. A. Ur, S. Gales, N. V. Zamfir, "Radiation Protection and Safety at ELI-NP", Romanian Reports in Physics, **68** Supplement, S885-S945 (2016)

[4] A. Ferrari, P.R. Sala, A. Fasso', and J. Ranft, "FLUKA: a multi-particle transport code", CERN **2005-10** (2005), INFN/TC-05/11, SLAC-R-773

[5] T.T. Bohlen, F. Cerutti, M.P.W. Chin, A. Fasso', A. Ferrari, P.G. Ortega, A. Mairani, P.R. Sala, G. Smirnov, and V. Vlachoudis, "The FLUKA Code: Developments and Challenges for High Energy and Medical Applications", Nuclear Data Sheets **120**, 211-214 (2014)

[6] O. Frasciello, unpublished results

 DOI: 10.1051/epjconf/201715307026

The future SwissFEL facility – challenges from a radiation protection point of view

Claudia Strabel[1,*], *Albert* Fuchs[1], *Roman* Galev[1], *Eike* Hohmann[1], *Roland* Lüscher[1], *Elisa* Musto[1] and *Sabine* Mayer[1]

[1]Paul Scherrer Institute, CH-5232 Villigen, Switzerland

Abstract. The Swiss Free Electron Laser is a new large-scale facility currently under construction at the Paul Scherrer Institute. Accessible areas surrounding the 720 m long accelerator tunnel, together with the pulsed time structure of the primary beam, lead to new challenges for ensuring that the radiation level in these areas remains in compliance with the legal constraints. For this purpose an online survey system based on the monitoring of the ambient dose rate arising from neutrons inside of the accelerator tunnel and opportunely calibrated to indicate the total dose rate outside of the tunnel, will be installed. The presented study provides a conceptual overview of this system, its underlying assumptions and measurements so far performed to validate its concept.

1 Introduction

The Swiss Free Electron Laser (SwissFEL) [1] is the new large research facility at Paul Scherrer Institute (PSI), which will allow expanding the frontiers of knowledge in a wide range of scientific researches. It is foreseen to accelerate electrons up to an energy of 7 GeV with a pulsed-time structure of the primary beam in the range of a few ps, and to produce FEL pulses covering the wavelength range of 1 Å to 70 Å, i.e. 0.1–7 nm. The accelerator will be operated with a maximum charge of 800 pC per pulse and a maximum repetition rate of 100 Hz. The main beam parameters are summarised in Table 1.

The final layout of SwissFEL is illustrated in Figure 1. It comprises an injector area (gun and booster), three linear accelerating sections, two parallel undulator lines (Aramis and Athos) and several experimental areas, for a total length of 720 m. The facility is located underground, on the ground floor of a two-floor building, partly integrated in a regional recreation area (see Figure 2).

Table 1. Key beam parameters of SwissFEL.

Beam parameter	Range
Electron energy	max. 7 GeV
Bunch length	~ps
FEL wavelength	1 Å – 70 Å
Charge per pulse	800 pC
Repetition rate	max. 100 Hz

The normal SwissFEL operation will give rise to secondary radiation fields deriving from the interactions of the primary beam with the beam line components; these radiation fields have been investigated [2] by means of Monte Carlo simulations. During normal

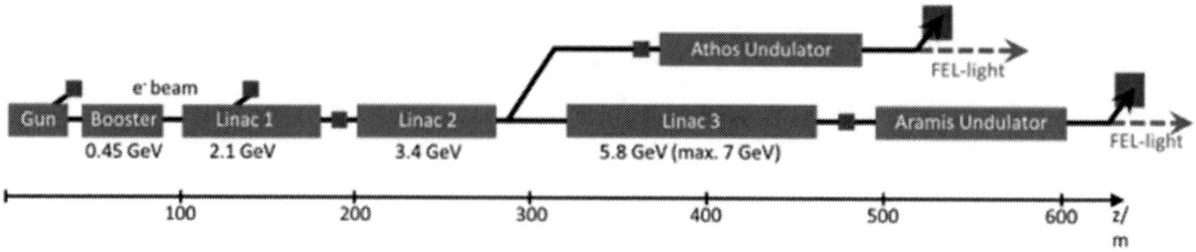

Fig. 1. The SwissFEL facility layout with the associated primary beam energies.

* Corresponding author: claudia.strabel@psi.ch

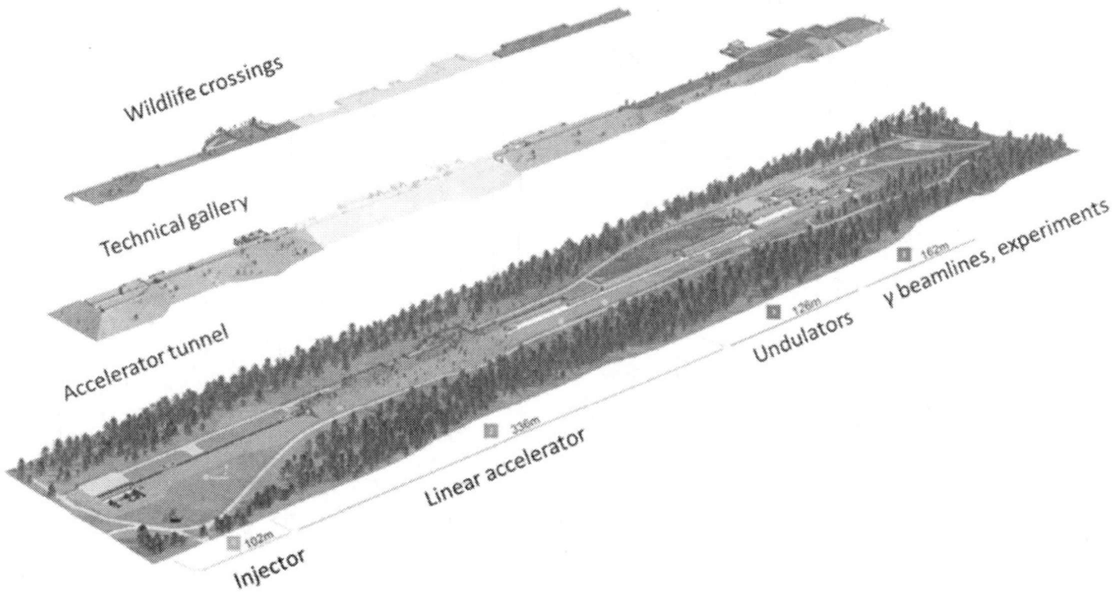

Fig. 2. The SwissFEL building site: With a total length of 720 m, the accelerator, composed of the injector and linear accelerator, and the undulators are located in the lower floor of a two-storey building. Most of the building will be underground, but the upper floor (technical gallery) will only be semi-covered with soil, to ensure good maintenance access. Wildlife crossings above of the technical gallery allow wildlife to pass from one side of the facility to the other. The experimental area is a single-storey building, which is naturally covered by the surrounding land.

operation, the shielding of the accelerator vault, designed in the early stages of the project, can reduce the radiation in the accessible areas to an acceptable level in compliance with legal constraints. It can also cope with a major accident scenario of a full beam loss for less than a few seconds. The radiation level will routinely be monitored by passive devices distributed along the areas neighbouring the accelerator vault, accumulating for time periods of three months. However, since the guideline values according to Swiss radiological protection regulation [3] can be exceeded in shorter time periods, it is foreseen to install an online dose rate monitoring system called Dose Rate Protection System (DRPS). Its aim, concept, calibration and preliminary measurements will be explained in the following sections.

2 The SwissFEL DRPS

2.1 Aim of the system

The aim of the DPRS is to provide the SwissFEL facility with a system capable of monitoring the dose rate in the different accessible areas inside and outside of the facility. The system shall raise warnings in case that the indicated dose rate will approach the predefined limits, and shall provide a signal, which will lead to an immediate shut down of the beam in case that a violation of the guideline values occurs. Furthermore, the DRPS shall allow for a continuous monitoring of the radiation level in absence of the beam such that it can be ensured that all of the beam components possibly leading to radiation are switched off during maintenance work inside of the tunnel.

2.2 Basic concept

The dose rate expected in the various accessible areas inside and outside of the facility is low. Therefore the associated measurement uncertainties in the respective areas are large. A systematic and reliable monitoring of the dose rate outside of the SwissFEL accelerator tunnel is hence not possible. This became the major constraint to be taken into account for the development of the DRPS concept. Instead of measuring the dose rate in the accessible areas, the design of the DRPS has been based on measuring the dose rate inside the accelerator tunnel and on the assumption that it can be employed to extrapolate with a satisfactory accuracy the dose rate outside of the tunnel. This assumption is mainly based on the fact that the impact of a partial or a total beam loss is independent on the loss position of the beam but results in a comparable stray field.

The different magnets of the beam-line used for alignment and manipulation of the beam allow only small angular changes of the beam with respect to the nominal beam direction. In case of misalignment of the beam or an unsatisfactory beam alignment, electrons impinging on the beam pipe will predominantly produce Bremsstrahlung. The latter is mainly peaked in the forward direction such that it will interact with the magnets and other beam-line elements, which are distributed along the full length of the accelerator tunnel. It produces then secondary neutrons, photons and electrons. The major contribution to the total dose rate outside of the accelerator tunnel is expected from secondary neutrons. The DRPS will therefore employ extended range neutron REM counters (see Section 2.4) placed along the full accelerator length, which will measure the neutron dose rate inside of the tunnel. With an adequate calibration of the detectors the measured dose rate can be extrapolated towards the outside of the accelerator tunnel taking also the electromagnetic component into account. Figure 3 schematically illustrates the concept described above.

This assumption has been validated by means of simulations performed with the Monte Carlo particle transport code FLUKA [4, 5]. The simulations included an electron beam of variable energy impinging on a cylindrical copper target with a length of 30 cm and a radius of 10 cm. These dimensions were chosen based on the assumption, that the beam can only be misaligned at small angles with respect to the beam direction and that the Bremsstrahlung arising from interactions with the beam pipe will be stopped when hitting a beam-line element. Finally, a simplified geometry with a concrete tunnel ceiling (thickness of 150 cm) was used. The simulations showed that for beam energies lower than 0.5 GeV, a full beam loss would lead to radiation levels outside of the accelerator tunnel that are in agreement with the guideline values, assuming a non-permanent stay (see Figure 4). For energies above 1 GeV the expected dose rate arising from photons outside of the accelerator vault contributes by less than 7% to the total dose rate, which is mainly created by secondary neutrons. In fact, high energy Bremsstrahlung is mainly peaked in forward direction whereas the distance and shielding is minimal under an angle of 90° with respect to the beam axis. In case of the latter, the spectral neutron distribution is almost independent of the energy of the primary electron beam for energies greater than 1 GeV (Figure 5, top) and the relative amount of neutrons with energies greater than 10 MeV is less than $(1.3 \pm 7.2)\%$ close to the ceiling and $(60.4 \pm 0.3)\%$ outside of the accelerator vault (Figure 5, bottom). As a result, the dose rate outside of the accelerator vault arises mainly from neutrons and can be measured inside with an adequate calibration.

Dedicated benchmark measurements of the DRPS concept will take place during the different commissioning phases of SwissFEL. Here, also gamma detectors will be employed in order to verify these numbers.

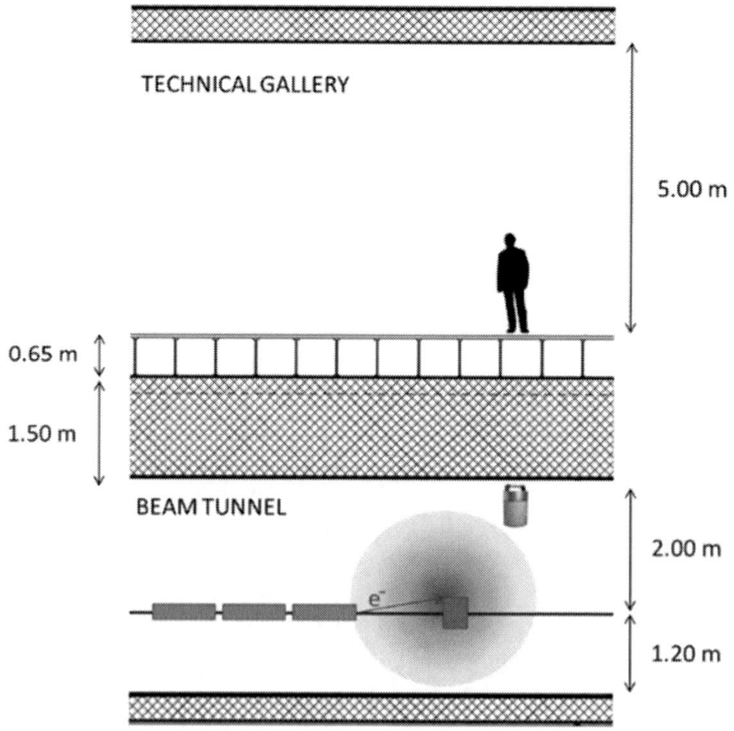

Fig. 3. Sketch of the DRPS concept. Misalignment of the beam results in Bremsstrahlung impinging on a magnet (presented by the red box). The total dose rate outside of the accelerator tunnel is monitored by a neutron detector, which is positioned inside of the accelerator tunnel and is calibrated such that it allows for the extrapolation of the total dose rate outside of the tunnel. See Section 2.4 for details about the LUPIN BF_3-NP detector manufactured by ELSE Nuclear [6, 7, 8] that is employed in the DRPS.

DOI: 10.1051/epjconf/201715307026

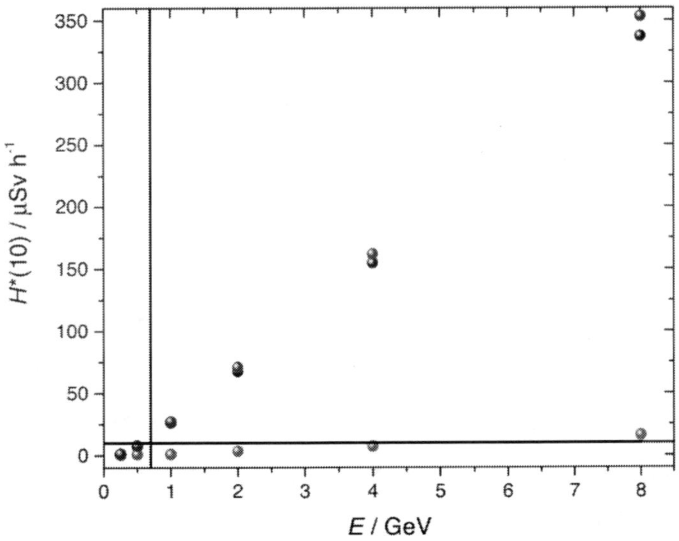

Fig. 4. Simulated dose rates outside of the accelerator vault, based on simplified assumptions and a full beam loss scenario arising from neutrons (black), photons (red) and total (blue). For electron energies less than 0.5 GeV (vertical black line), the calculated dose rates are not exceeding 10 µSv/h.

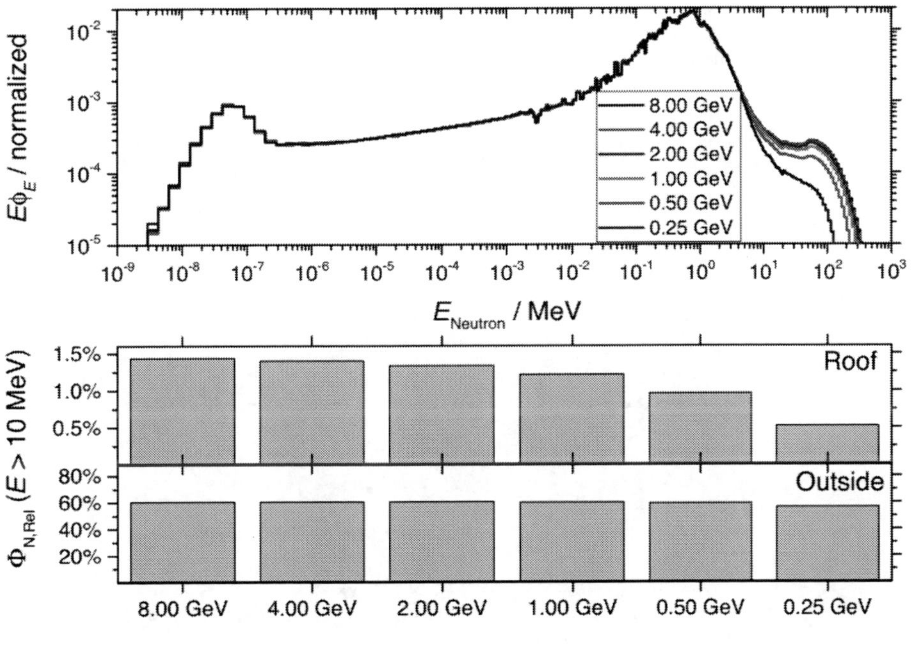

Fig. 5. Top: Simulated spectral neutron distribution near the roof of the SwissFEL vault for electrons impinging on a cylindrical copper target, for different electron energies. Bottom: Relative amount of neutron with energies greater than 10 MeV, for different electron energies close to the roof and outside of the accelerator vault.

DOI: 10.1051/epjconf/201715307026

2.3 Calibration procedure

The DRPS processes the dose rate measured inside the tunnel and extrapolates it to evaluate the dose rate in the various accessible areas surrounding the tunnel. To do this with a satisfactory accuracy, an adequate calibration is needed. In order to do so, a two-step procedure is foreseen.

In a first step, the accelerator tunnel is divided into different classes according to its shielding configuration. These classes comprise for example ventilation and cooling shafts as well as beam line crossings, which all exhibit different reductions of the concrete shielding of the surrounding accelerator tunnel. For each one of these classes a basic calibration factor is calculated using Monte Carlo methods, using a detailed geometric model. Figure 6 shows an example of such a shielding class by depicting a typical cooling shaft, through which the usually 1.5 m thick concrete ceiling of the accelerator tunnel is significantly reduced towards the above-lying technical gallery. In case of overlapping classes, the most conservative class will be used.

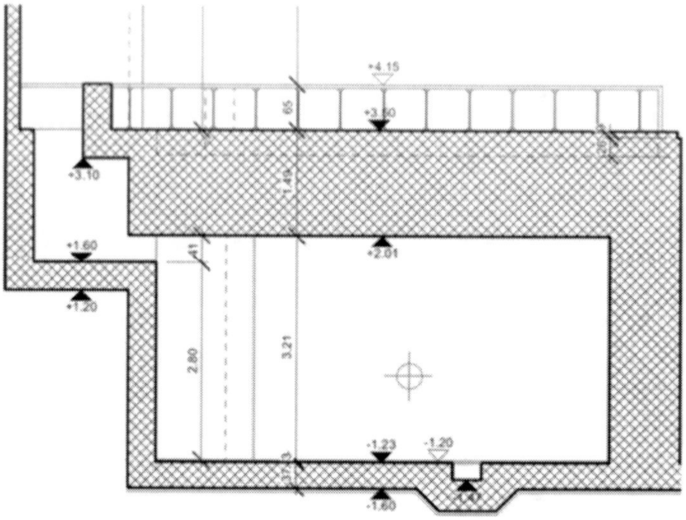

Fig. 6. Example of a calibration class with reduced shielding due to a cooling shaft with an opening of 41 cm from the accelerator tunnel towards the technical gallery above. The standard concrete shielding thickness towards the top gallery amounts to 1.5 m.

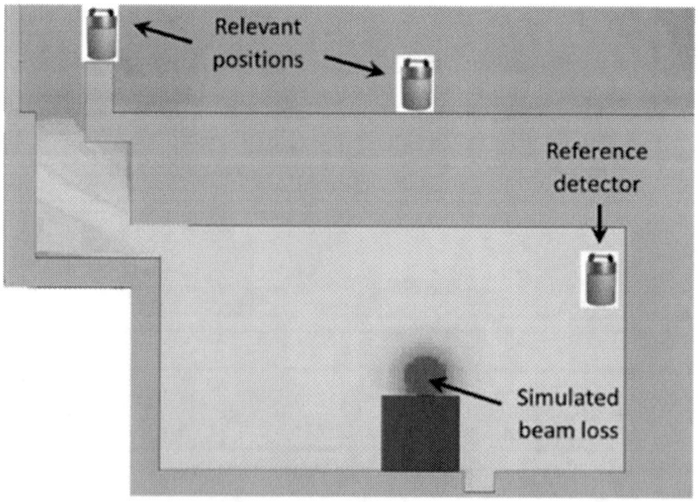

Fig. 7. Simulated dose rate generated from neutrons in case of a beam loss using FLUKA. The position of the reference detector inside of the accelerator tunnel is indicated on the right side. This detector will be used to extrapolate the neutron dose rate towards relevant positions in the accessible areas, such as the above-lying technical gallery. The extrapolation of the dose rate will be performed on the basis of the MC simulations and verified with dedicated benchmark measurements.

In a second step, FLUKA will be used to simulate a beam loss for each type of shielding class. The results will then be utilised to check the difference between the neutron dose rate at the position of the DRPS monitoring device inside of the accelerator tunnel and the total dose rate including the electromagnetic component in the accessible areas of the facility and its surrounding. This is illustrated in Figure 7. The difference between the neutron dose rate inside and the total dose rate outside of the tunnel will then determine the calibration factor.

To guarantee the respect of the dose rate limits even if the beam loss doesn't occur directly above a particular detector, a threshold according to a worst-case scenario will be applied. Since this option significantly reduces the threshold and therewith may interfere with the SwissFEL operation, the signal coming from the different detectors will be simultaneously acquired and opportunely combined.

Given the anisotropy of the secondary radiation field along the beamline due to the non-homogeneous shielding, the configuration in terms of interspace between the detector elements needs to be investigated, optimized and validated.

The correct calibration of the various detectors will finally be validated with dedicated measurement campaigns in which independent neutron detectors using a different method of data evaluation as well as gamma detectors will be deployed.

2.4 Preliminary measurements

Performing measurements inside of the SwissFEL tunnel is a challenge for the employed neutron survey instrumentation particularly due to the pulsed structure of the radiation and the large photon contribution.

The LUPIN (Long Interval, Ultra-wide dynamic, Pile-up free, Neutron rem-counter) [6, 7, 8] 5401 BF_3-NP detector manufactured by ELSE Nuclear was selected as main measurement device for the DRPS. It consists of a BF_3 neutron proportional counter, which is inserted in a cylindrical polyethylene moderator with a lead layer. It uses a front-end electronics based on a logarithmic amplifier. Its method allows the detection of every single radiation burst and is therefore ideally suited for the radiation fields at SwissFEL.

Next to the pulsed radiation fields, the detectors also have to cope with large radio-frequency (RF) fields, since they will be positioned above the beamline in proximity of the RF cavities responsible for providing the beam acceleration. To verify that the dose reading of the DRPS monitors is not affected by the RF fields, measurements with the LUPIN detector as well as other commercial available devices (LB6411 and WENDI-II) were performed at the test facility, where the SwissFEL RF cavities were commissioned. These measurements showed that there are no significant effects on the stability of the dose rate indication of the detectors due to the RF cavities.

3 Summary

The new large-scale facility SwissFEL currently under construction at PSI leads to challenges from a radiation protection point of view. An online survey system, the DRPS, will be used to monitor the ambient dose and ensure that the legal limits are respected. In this context, a brief overview of the system's concept, components and calibration has been provided.

References

1. R. Ganter (editor), SwissFEL CDR, http://www.psi.ch/swissfel/ (2012)
2. E. Hohmann, S. Reiche, A. Fuchs, R. Ganter, F. Le Pimpec, R. Lüscher, S. Mayer, T. Schietinger, PNST 4, 262-266 (2014)
3. Strahlenschutzverordnung, Der Schweizerische Bundesrat (2014)
4. T.T. Böhlen, F. Cerutti, M.P.W. Chin, A. Fassò, A. Ferrari, P.G. Ortega, A. Mairani, P.R. Sala, G. Smirnov and V. Vlachoudis, Nuclear Data Sheets 120, 211-214 (2014)
5. A. Ferrari, P.R. Sala, A. Fasso`, and J. Ranft, CERN-2005-10 (2005), INFN/TC_05/11, SLAC-R-773
6. M. Ferrarini, V. Varoli, A. Favalli, M. Caresana, B. Pedersen, Nucl. Instr. Meth. Phys. Res. Sec. **A 613**, 272 (2010)
7. M. Caresana, M. Ferrarini, G.P. Manessi, M.Silari, V. Varoli, Nucl. Instr. Meth. Phys. Res. Sec. **A 712**, 15 (2013)
8. M. Caresana, A. Denker, A. Esposito, M. Ferrarini, N. Golnik, E. Hohmann, A. Leuschner, M. Luszik-Bhadra, G. Manessi, S. Mayer, K. Ott, J. Röhrich, M. Silari, F. Trompier, M. Volnhals, M. Wielunski, Nucl. Instr. Meth. Phys. Res. Sec. **A** **737**, 203 (2014)

Computational study of radiation doses at UNLV accelerator facility

Matthew Hodges[1,*], *Alexander* Barzilov[1], *Yi-Tung* Chen[1], and *Daniel* Lowe[1]

[1]Department of Mechanical Engineering, University of Nevada, Las Vegas, 4505 S. Maryland Parkway, Las Vegas, NV, 89154, USA

Abstract. A Varian K15 electron linear accelerator (linac) has been considered for installation at University of Nevada, Las Vegas (UNLV). Before experiments can be performed, it is necessary to evaluate the photon and neutron spectra as generated by the linac, as well as the resulting dose rates within the accelerator facility. A computational study using MCNPX was performed to characterize the source terms for the bremsstrahlung converter. The 15 MeV electron beam available in the linac is above the photoneutron threshold energy for several materials in the linac assembly, and as a result, neutrons must be accounted for. The angular and energy distributions for bremsstrahlung flux generated by the interaction of the 15 MeV electron beam with the linac target were determined. This source term was used in conjunction with the K15 collimators to determine the dose rates within the facility.

1 Introduction

High energy (greater than 10 MeV) linear electron accelerators (linacs) often produce photoneutrons due to the energies of the generated bremsstrahlung photons being greater than that of the neutron binding energies of the materials they are likely to interact with [1]. These materials include the components of the linac itself in addition to the nearby structures within a shielded vault where the linac is housed. As an undesired secondary radiation, these photoneutrons contaminate the photon beam and increase the overall dose rates within the immediate surroundings. In order to accurately represent the radiation environment and understand the total dose rate due to the linac operation, it is necessary to not only determine the photon contribution but also that produced by neutrons.

The accelerator facility at UNLV was commissioned in 2015 and houses a Varian M6 linac. Previous research projects at UNLV include the radiation characterization of the facility due to the M6 operation [2]. A Varian K15 was considered for installation at the accelerator facility and as such, a radiation characterization of the vault must be performed. The K15 runs can be carried out in two modes: high energy and low energy. The high energy mode is capable of producing photons up to 15 MeV and the low energy mode is capable of producing photons up to 9 MeV. The K15 system uses a cone collimator in order to shape the emitted photons into a 30 degree cone. Other collimator shapes are possible, but only the conical collimator was studied in this research. This study was concerned with the development of computational model to determine the dose rate footprint throughout the UNLV facility for the K15 operation.

Fig. 1. Varian K15 linac [3].

2 Background

In order to determine the dose rates within the accelerator facility, first it was necessary to investigate the photon and neutron radiation flux generated in the linac's target and the irradiated samples. A cascade of bremsstrahlung photons is produced by the interaction of an electron beam striking upon a heavy metal target. The endpoint energy of this cascade is characteristic of the energy of the impacting electron beam. Photons emitted from the target interact with the atoms of matter as they are transported throughout the vault and the environment. If the energy of these photons is above the photoneutron threshold for the materials that are present within the accelerator or in the room, neutrons may also be produced.

* Corresponding author: Hodgesm@unlv.nevada.edu

 DOI: 10.1051/epjconf/201715307027

2.1 Photoneutron production

When the energy of the bremsstrahlung photons is greater than that of the neutron binding energy of the nuclei they interact with, a neutron is produced in the (γ,n) reaction. The neutron binding energy of a specific isotope is determined from the Q-value (difference in mass energies of the products and reactants). When the Q-value is negative, the target atom requires this input of energy from the projectile in order for the reaction to proceed. For a photonuclear reaction, this corresponds to the minimum threshold energy of the photon needed to induce the (γ,n) reaction. Values for (γ,n) threshold energies for materials commonly found in the linac and surrounding environment are shown in Table 1.

Table 1. Threshold energies of (γ,n) reaction.

Z	Isotope	Threshold energy (MeV)
1	^{2}H	2.2
5	^{11}B	11.5
6	^{12}C	18.7
7	^{14}N	10.6
8	^{16}O	15.7
11	^{23}Na	12.4
12	^{24}Mg	16.5
13	^{27}Al	13.1
25	^{55}Mn	10.2
26	^{56}Fe	11.2
27	^{59}Co	10.5
28	^{58}Ni	12.2
29	^{63}Cu	10.9
30	^{64}Zn	11.9

3 Computational model

The MCNPX code [4] was used in the computational study to determine the radiation dose rates within the UNLV accelerator facility. An MCNPX input deck requires several components in order to represent the model of the facility and the linac. The 3D geometry domain of the model (created through cell and surface cards) must be carefully defined in order to accurately represent the radiation transport processes.

The radiation source is the starting point of the model. If the source is incorrectly defined, this misrepresentation will propagate throughout the model resulting in unreliable results.

Tallies were used in order to calculate values of interest (i.e. radiation flux, dose, energy deposition). A tally must be specified within the computational model at specific geometric locations where these values are desired. Material cards define the matter within the model and must be properly represented. Physics models (i.e., electron, photon, and neutron interactions) must be properly defined for accurate solution to the problem.

3.1 Accelerator facility

The UNLV accelerator facility is built into a large mound of earth that stretches for about 20 meters north and 10 meters east of the building. The enveloping earthen berm acts as natural shielding to the environment from radiation produced by the linac.

An exclusion area on the top of the building is protected by fencing in order to keep out any animal life or people. The linacs in the accelerator facility are operated from a separate control room that is placed to the south of the facility building. The control room was not modeled in this study.

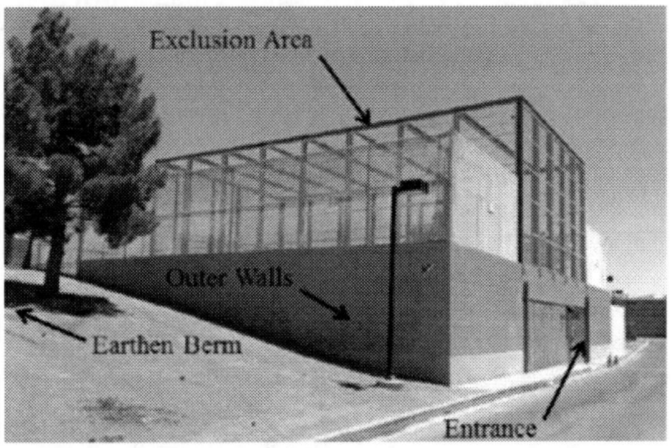

Fig. 2. UNLV accelerator facility.

The facility building consists of an entry room, a shielding maze, and an accelerator bay. There is a sample table at the northern end of the accelerator bay where samples may be placed for experiment. The geometry model of the facility was created within the MCNPX input deck according to on-site measurements [2]. Figure 3 details the layout and dimensions of the accelerator facility and room structures.

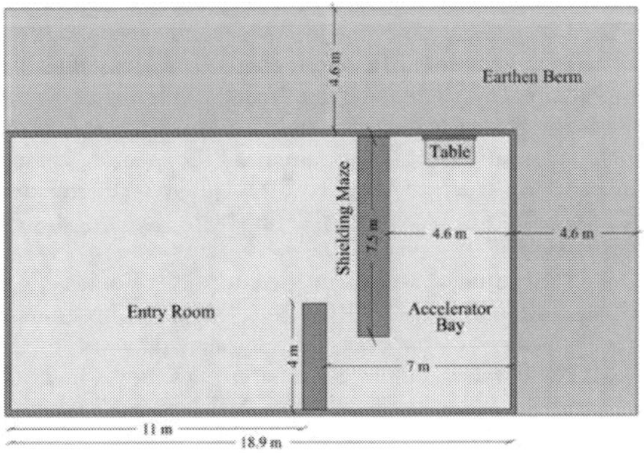

Fig. 3. Accelerator facility layout.

3.2 MCNPX model

The linac's target head and shielding components were defined in the computational model according to proprietary design provided by Varian. Material data used to define the room structure components (i.e. concrete, earth, air, etc.) were taken from the *Compendium of material composition data for radiation transport modeling* [5]. To account for photonuclear interactions, the *.70u* (LA150) cross section library was used.

Firstly, the simulation of an electron pencil beam (Gaussian in both energy and radial impact target location) impinging on the linac's target was carried out. The angular and energy distributions of the bremsstrahlung photon flux produced by this interaction were determined by placing F4 tallies on a series of conic segments 1 cm behind the linac's target head (see Fig .4). 100 equally spaced energy bins were used to determine the energy distribution of the photon flux. These tallies allowed for the determination of photon flux values at intervals of 10 degrees off the beam centerline.

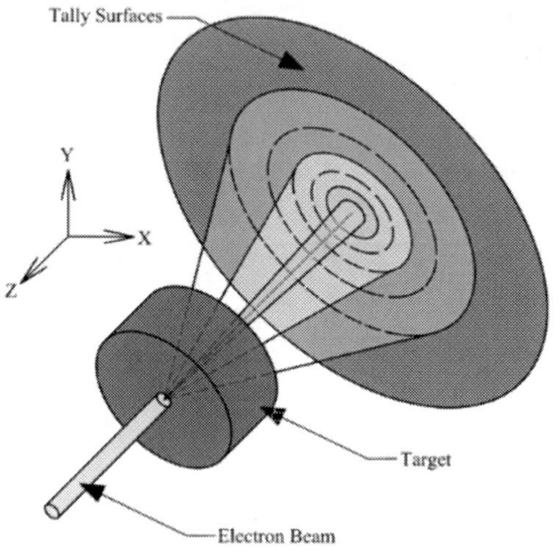

Fig. 4. Tallying the angular distribution of photons.

Using the bremsstrahlung source data, MCNPX was used to simulate the radiation transport throughout the facility. The RMESH card was used to determine the spatially-dependent radiation fluxes within the facility. A 2D mesh was placed at a height of 111 cm above the floor (positioned at the center of the photon conical beam) in order to determine the photon and neutron fluxes in 7.6-cm intervals (in both the x- and y-directions).

These fluxes were converted to the biological dose rates through the ANSI/ANS 6.1.1-1977 flux-to-dose rate factors [6] for both neutrons and photons. In the MCNPX model, this conversion was performed by specifying the dose conversion coefficient (*ic*=20) within the RMESH card.

A vertical profile of the photon dose rates at 1 meter north of the linac target head was determined through use of an F5 tally spaced 5 cm apart. The error associated with each tally was below 5% value. The neutron spectrum at 1 meter north of the linac's target head at a height of 1.11 m was also determined using an F5 tally. The MCNPX models were run for 500 million particle histories. This ensures that the statistical error associated with the respective F4 tallies was below the MCNPX recommendation of 10%. Once the calculations were completed, the MCNPX mesh output files were converted to Tecplot ready ASCII files through the use of the MCNPX command GRIDCONV. These files were then directly input into Tecplot for post processing of computational results.

4 Results

4.1 Bremsstrahlung photon spectra

The F4 tally results on the conic segments were used to create the spectral distributions of bremsstrahlung photons. The photon spectra at the respective energy modes are shown in Fig. 5 and Fig. 6. In each mode, the greatest fluxes of photons occur at the lowest energies. Additionally, the greatest photon fluxes occur within a conical angle of 10 degrees from the linac's target. As the angle increases and the cone widens, the photon flux decreases.

The spectra show similar trends between the two K15 energy modes. \For both energy modes, at 1 MeV, the photon fluxes occurring below 10 degrees are approximately 3 orders of magnitude greater than the photon fluxes at angles greater than 80 degrees. At 8 MeV, the differences between the two angles are 5 orders of magnitude.

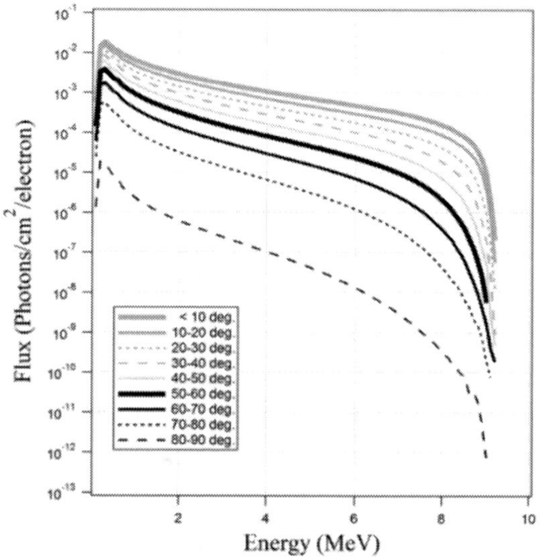

Fig. 5. Bremsstrahlung photon spectra, low energy mode.

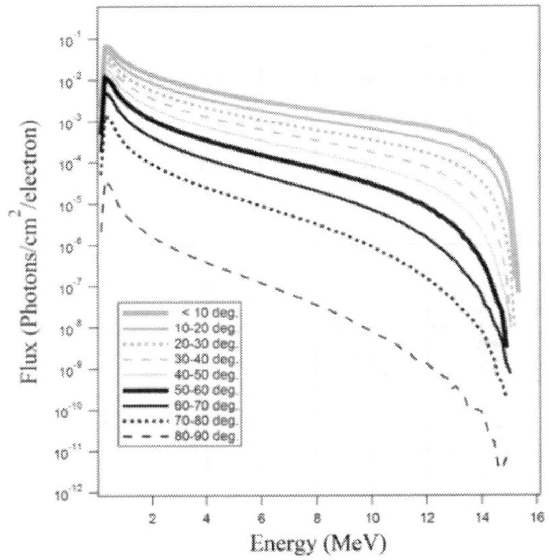

Fig. 6. Bremsstrahlung photon spectra, high energy mode.

4.2 Dose rates

Contour maps for photon and neutron dose rates were created in order to determine the radiation footprint within the building for the K15 operation in high and low energy modes. These maps were created for two different linac's operating conditions. The first was for operation of the K15 under normal conditions while the second was to determine the maximum dose rates within the facility if the K15 was operating in high energy mode without shielding and collimators. Comparison between the two scenarios results allowed for the determination of the effectiveness of the radiation shielding design in the reduction of dose rates outside the facility.

4.2.1 Normal operation dose rates

The photon dose rate contour maps for normal operation of the K15 system in high and low energy modes are shown in Fig. 7 and Fig. 8, respectively. The results show that photon fluxes and dose rates for both high and low energy modes are contained within the accelerator bay equipped with the concrete shielding maze.

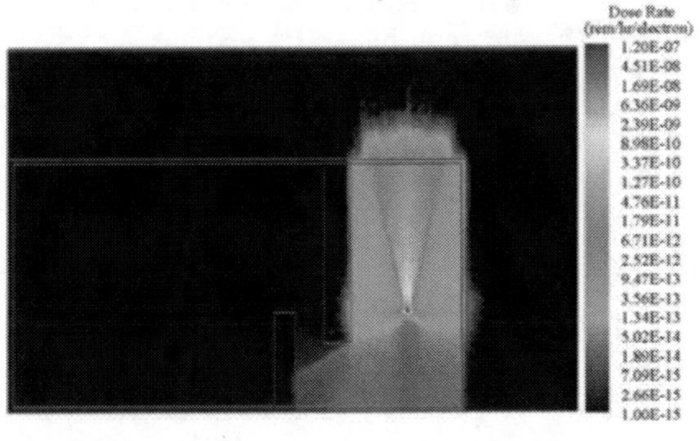

Fig. 7. Photon dose rate, normal operation in the high energy mode.

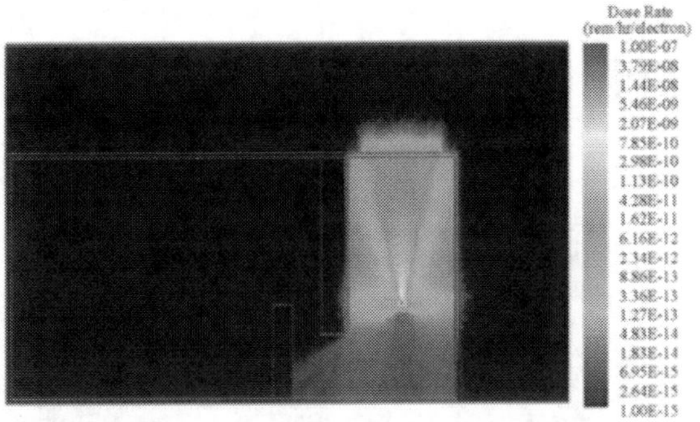

Fig. 8. Photon dose rate, normal operation in the low energy mode.

Dose rates in the entry way and shielding maze were determined to be 3 - 4 orders of magnitude lower than those in front of the linac assembly. The earthen berm to the north of the facility attenuates the photon fluxes behind the sample table within a few meters. The earthen berm to the east of the linac also minimizes the photon fluxes in less than a meter. In both instances, the dose rate was reduced to the limit set for the general public. The vertical profile of the photon dose rates at 1 meter north of the linac's target head is shown in Fig. 9.

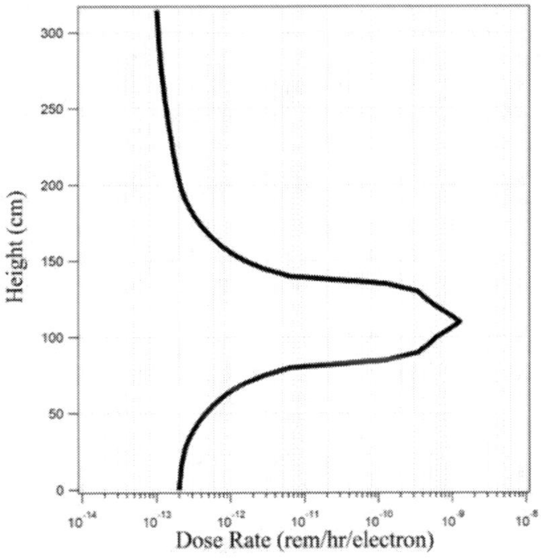

Fig. 9. Vertical profile of the photon dose rate, normal operation.

The results show that the largest photon dose rates occur at the beam height of 1.11 m from the floor. As the distance from the beam center increases, the photon dose rate decreases. The dose rates nearest the floor and ceiling are approximately four orders of magnitude lower than the maximum dose rate at the height of the beam. The largest photon dose rates occur at the height between 75 and 125 cm. This is due to being directly within the 30 degree conical collimator photon beam. Dose rates quickly decrease once out of the conical photon beam. The normal neutron dose rate contour map

DOI: 10.1051/epjconf/201715307027

for high energy operation is seen in Fig. 10. When operating in the low energy mode, the endpoint energy of the bremsstrahlung spectra is below the (γ,n) reaction thresholds of the materials used in the simulation and no photoneutrons were generated.

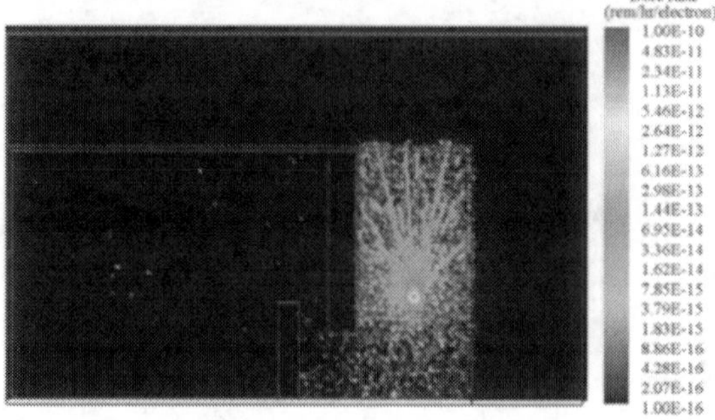

Fig. 10. Neutron dose rate, normal operation in the high energy mode.

The computed dose rates show that neutrons were primarily contained within the accelerator bay. The neutron dose rates to the north of the linac are largest due to the lack of polyethylene shielding directly behind the linac's target head, while the dose rates to the rear of the linac were minimized by the presence of neutron shielding.

These results also show that the neutron dose rates within the facility are several orders of magnitude lower than the photon dose rates. While the photon dose profile north of the linac's target head exhibits the shape created by the cone collimators, the neutron rate does not. This is best understood as the high-Z material in the collimators do not attenuate the neutron flux as they have low neutron scattering cross sections. The neutron spectrum at 1 m from the linac's target head is shown in Fig. 11. The results show that the largest flux of neutrons is in the 0.1 to 1 MeV energy range.

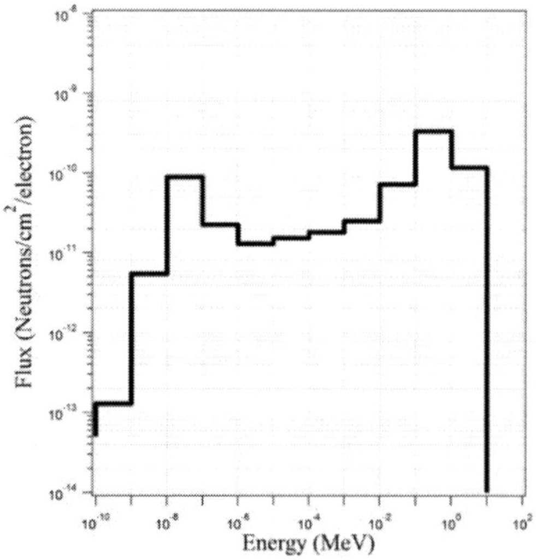

Fig. 11. Neutron spectrum.

4.2.2 Maximum dose rates

When the shielding and collimators were removed from the MCNPX model of the K15 system during the high energy mode operation, the dose rates within the accelerator facility increased. The results from this calculation help to understand the maximum dose rates present within the facility due to operation of the K15. Fig. 13 and Fig. 14 show the maximum dose rates within the building due to photons and neutrons, respectively.

Fig. 12. Maximum photon dose rates, high energy mode.

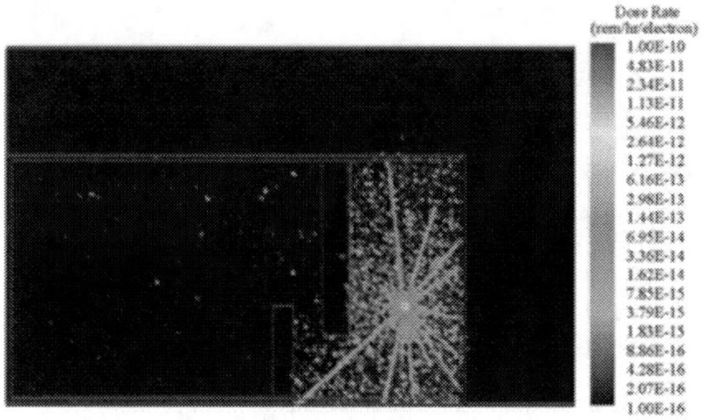

Fig. 13. Maximum neutron dose rate, high energy mode.

Without the linac shielding and collimators, the maximum photon dose rates increase throughout the building. The dose rate values were compared at the same pixel locations for the maximum and normal high energy modes. At 1 meter behind the linac's target head, the maximum dose is 238 times larger than the normal dose. In the center of the shielding maze, the maximum dose is 33 times larger than the normal dose. In the center of the entry room, the maximum dose is 7.5 times larger than the normal dose. The maximum neutron dose profile resembles that of the normal neutron dose rate in the northern half of the accelerator bay. The dose rates in the southern half of the accelerator bay have increased however, due to the removal of the neutron shielding.

5 Conclusions

The purpose of this study was to computationally characterize the dose rate footprint of the UNLV accelerator facility due to the operation of a Varian K15 linac. It was determined that when the K15 system is operated in the high energy mode, there is greater flux of bremsstrahlung photons than when it is operated in the low energy mode. Furthermore, when the linac is operated in the high energy mode, photoneutrons are produced whose contribution to dose rate cannot be ignored.

The linac's and facility's shielding components greatly reduce photon and neutron fluxes and dose rates outside of the facility, making it possible to comply with the regulatory limits (2 mrem/hr).

References

1. K. Kosako, K. Oishi, T. Nakamura, *Angular Journal of Nuclear Science and Technology*, **48**, 227-236 (2011)
2. M. Hodges, A. Barzilov, Y. Chen, D. Lowe, *Radiation Physics and Chemistry*, **127**, 72-77 (2016)
3. Varian Medical Systems. *Linatron K15*. Web. 10 Oct 2016. https://www.varian.com/x-ray-imaging-components/products/security-industrial-imaging/linear-accelerators/linatron-k15
4. D. Pelowitz (ed), *LA-CP-11-00438* (2011)
5. R. McConn, C. Gesh, R. Pagh, R. Rucker, R. Williams *PNNL-15780* (2011)
6. American National Standard, *ANSI/ANS 6.1.1-1977* (1977)

Neutron streaming studies along JET shielding penetrations

Ion E. Stamatelatos[1, a], *Theodora* Vasilopoulou[1], *Paola* Batistoni[2], *Barbara* Obryk[3], *Sergey* Popovichev[4], *Jonathan* Naish[4], and JET contributors*

EUROfusion Consortium, JET, Culham Science Centre, OX14 3DB, Abingdon, United Kingdom

[1] Institute of Nuclear and Radiological Sciences, Technology, Energy & Safety, NCSR "Demokritos", Athens, Greece
[2] ENEA – Fusion Technical Unit, Frascati, Rome, Italy
[3] Institute of Nuclear Physics, Polish Academy of Sciences, Kraków, Poland
[4]CCFE, Culham Science Centre, OX14 3DB, Abingdon, United Kingdom
*See the appendix of Romanelli F. et al, 2014, Proc. 25th IAEA Fusion Energy Conference 2014, St. Petersburg, Russia

Abstract. Neutronic benchmark experiments are carried out at JET aiming to assess the neutronic codes and data used in ITER analysis. Among other activities, experiments are performed in order to validate neutron streaming simulations along long penetrations in the JET shielding configuration. In this work, neutron streaming calculations along the JET personnel entrance maze are presented. Simulations were performed using the MCNP code for Deuterium-Deuterium and Deuterium-Tritium plasma sources. The results of the simulations were compared against experimental data obtained using thermoluminescence detectors and activation foils.

1 Introduction

The Joint European Torus (JET) is currently the largest tokamak in the world. The experiments and design studies performed by JET are consolidated to a large extent into the design of its successor ITER. Among others, benchmark experiments are being carried out at JET aiming to validate in a real fusion environment the neutronic codes and nuclear data applied in ITER nuclear analyses. In particular, measurements and calculations of the neutron fluence through the penetrations of the JET shielding aim to assess the capability of numerical tools to accurately predict neutron transport along the long paths and the complex geometries characterizing the ITER biological shield [1].

In the present work, neutron streaming through the JET Hall South West personnel entrance maze was evaluated. Monte Carlo calculations using the MCNP code were performed to predict neutron fluence and ambient dose equivalent along the maze. Deuterium-Deuterium (D-D) and Deuterium-Tritium (D-T) toroidal plasma sources were simulated. The results of the calculations were validated by comparison against measurements carried out using thermoluminescence detectors and activation foils.

2 Simulations

The personnel entrance maze is located at the South West (SW) corner of the JET Hall (Fig.1). The geometry and dimensions of the maze are shown in Figure 2.

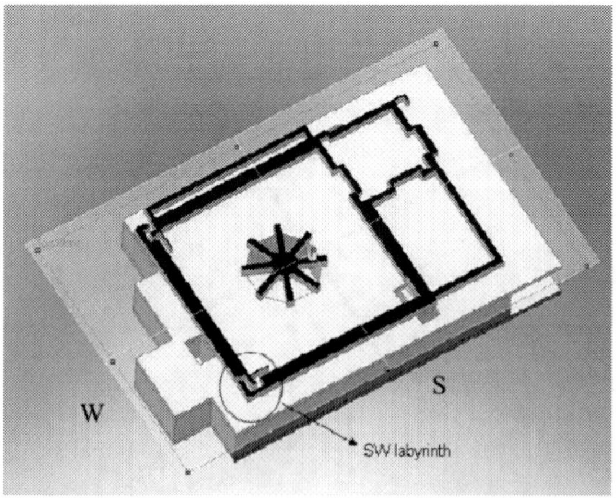

Figure 1. Ground view of JET Hall showing the location of the SW personnel entrance maze

[a] Corresponding author: ion@ipta.demokritos.gr

DOI: 10.1051/epjconf/201715307028

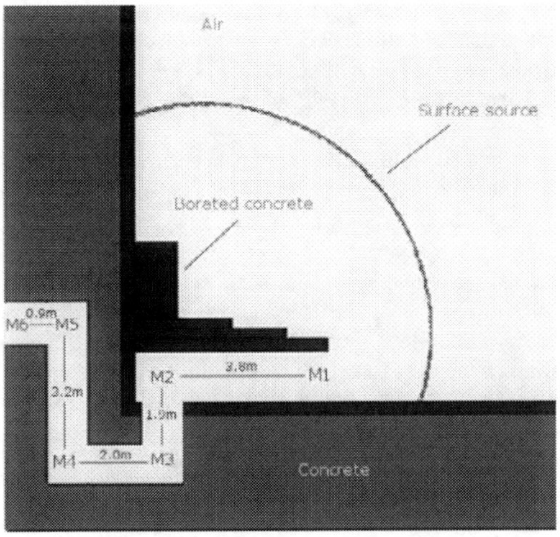

Figure 2. Cross sectional view of the SW entrance maze

The maze configuration provides four right-angle turns. The total length of the maze is 11.80 m and its height is 2.60 m. The maze width varies between 0.90 and 1.10 m and therefore the cross-sectional area ranges between 2.34 m^2 and 2.86 m^2. The thickness of the concrete wall is 2.50 m. The internal surface of the wall is covered by a layer of borated concrete (0.30 m in thickness). The densities of concrete and borated concrete were 2.43×10^3 $kg \cdot m^{-3}$ and 2.20×10^3 $kg \cdot m^{-3}$, respectively.

Simulations were performed for D-D and D-T toroidal plasma sources using Monte Carlo code MCNPX (version 2.5.0) [2]. Cross-section data were obtained from JEFF 3.1.2 and FENDL 2.1 libraries [3, 4]. All calculations were performed for the "as built" concrete compositions. The detailed concrete composition has been given elsewhere [5].

A two-stage simulation approach was employed. A detailed model of the JET Torus was used to produce a Surface Source Write file registering the neutrons directed towards the SW corner of the JET Hall. The model took into consideration the actual toroidal distribution of the neutrons produced in the JET plasma. The modeled geometry included the plasma facing components, the vacuum vessel, the magnetic coils, the shell, transformer limbs and concrete shielding walls. Outside the shell and before the walls, however, there are numerous substantial structures which have not been modeled in detail (diagnostic and heating systems, various equipments). Moreover, the vacuum vessel has ports through which neutrons may escape. These complications were approximated with a zone composed of iron, plastic and copper which has been adjusted in thickness until the JET external fission chamber measurements were consistent with the fluxes in the MCNP model [1]. This approximation was necessary since the complexity in the torus hall is too great for more detailed modeling on a realistic timescale. The Surface Source Write file registered neutrons on a spherical surface with center at the SW Hall corner (1.0 m above the floor level) and radius of 5.0 m (Fig. 2). The Surface Source Write file was used as Surface Source Read input file for the subsequent maze calculations performed in the context of this work.

Neutron fluence was calculated using track length estimate tallies of neutron flux in spherical cells of 0.3 m in radius positioned along the maze at 1.0 m height from the floor level. Moreover, ambient dose equivalent, H*(10), was calculated folding neutron flux by ambient dose equivalent to neutron fluence conversion factors as a function of neutron energy [6]. It is noted that the ambient dose equivalent H*(10) at a point of interest in a radiation field is the dose equivalent which would be generated in the associated oriented and expanded radiation field at a depth of 10 mm on the radius of the ICRU sphere which is oriented opposite to the direction of incident radiation. Statistical uncertainties were kept below 10% for all track length estimate tallies.

3 Results

Figures 3 & 4 show the MCNP predicted neutron energy spectrum at positions M1-M6 along the labyrinth for D-D and D-T JET plasma sources, respectively. Position M1 corresponds to the inner entrance of the maze (mouth) and position M6 to the exit door.

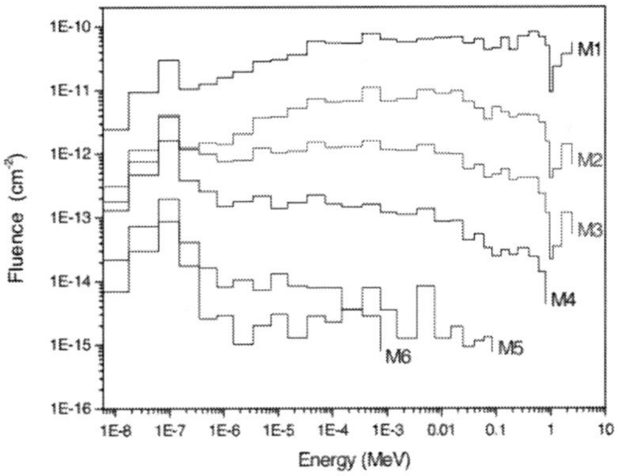

Figure 3. Neutron energy spectrum for the D-D source

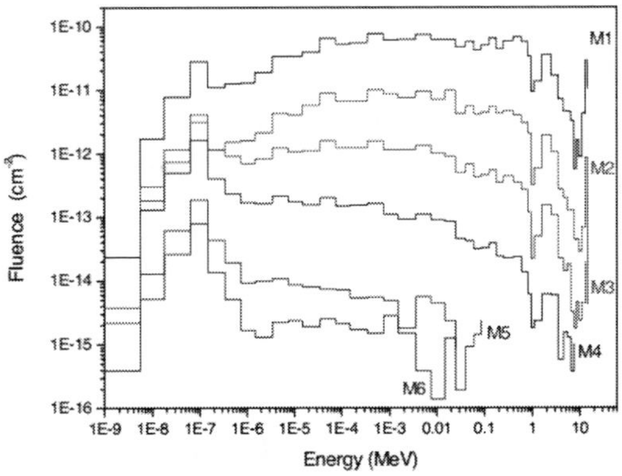

Figure 4. Neutron energy spectrum for the D-T source

The neutrons that are directed towards the labyrinth mouth have already been scattered by the machine, surrounding structures and wall materials one or more times and have lost a significant fraction of their kinetic energy. The mean energy of neutrons at the labyrinth mouth (M1) for the D-D and D-T source was approximately 270 keV and 650 keV, respectively. However, during their propagation along the maze, neutrons are elastically scattered by the wall materials and are further slowed down. The mean neutron energy at labyrinth position M3 for the D-D and the D-T source was 40 keV and 70 keV, respectively. At the exit (M6) neutron energy was in the eV region for both sources.

Figures 5 & 6 show the MCNP calculated neutron fluence along the maze, as a function of the summed centerline distance (L) from the inner maze entrance (mouth) to the exit door, divided by the square root of the maze cross-sectional area (A) for the D-D and D-T plasma sources, respectively. The results were normalized per JET source neutron. In these figures neutron fluence is presented in three energy groups: "thermal" (E<0.5 eV), "epithermal" (0.5 eV<E<0.1 MeV) and "fast" (E>0.1 MeV). The total neutron fluence (sum of the three groups) is also shown.

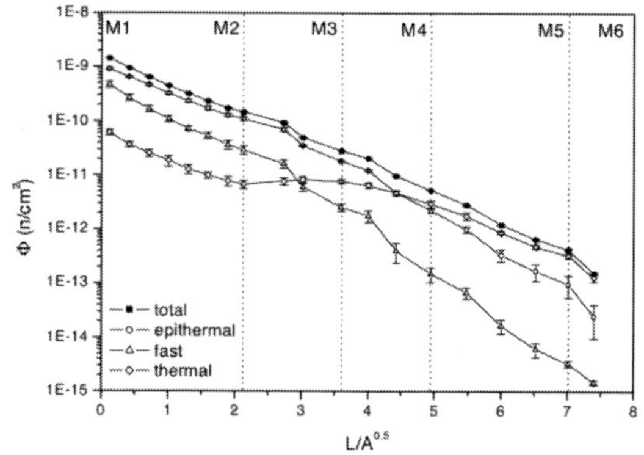

Figure 5. Neutron fluence for the D-D plasma source (normalization per JET neutron)

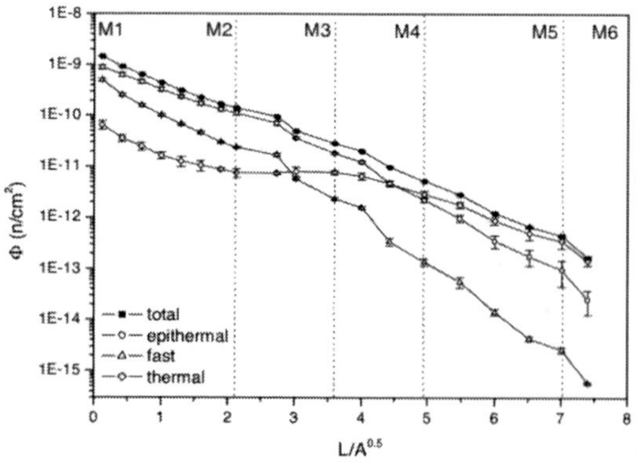

Figure 6. Neutron fluence for the D-T plasma source (normalization per JET neutron)

From Fig. 5 it can be seen that the total neutron fluence is attenuated along the maze (total length of 11.8 m) by about four orders of magnitude. The neutron fluence at the labyrinth exit was $(1.7 \pm 0.1)\times 10^{-13}$ cm^{-2} per JET neutron. In the first two legs of the labyrinth (sections M1-M2 and M2-M3) neutron fluence is dominated by epithermal neutrons. However, for $L/A^{0.5} \geq 4.5$ the thermal neutron group becomes the dominant one. At the labyrinth mouth (M1) the relative contribution of the fast, epithermal and thermal groups to the total neutron fluence was found to be 34%, 62% and 4%, respectively. At the second right-angle bend of the labyrinth (M3) the relative contribution to neutron fluence of the fast, epithermal and thermal groups was found to be 9%, 64% and 27%, respectively. At the exit door (M6), the relative contribution of fast, epithermal and thermal groups was 1%, 15% and 84%, respectively. Since the neutron fluence at the labyrinth exit is dominated by low energy neutrons, a further attenuation of the neutron fluence can easily be achieved by using a thermal neutron absorbing material (i.e. ^{6}Li-doped polyethylene) at the door shield. A similar behavior was also observed for the D-T plasma source (Fig. 6).

Figure 7 shows the MCNP calculated ambient dose equivalent, H*(10), along the maze as a function of the parameter $L/A^{0.5}$ for the D-D and D-T sources. The results are normalized per JET neutron.

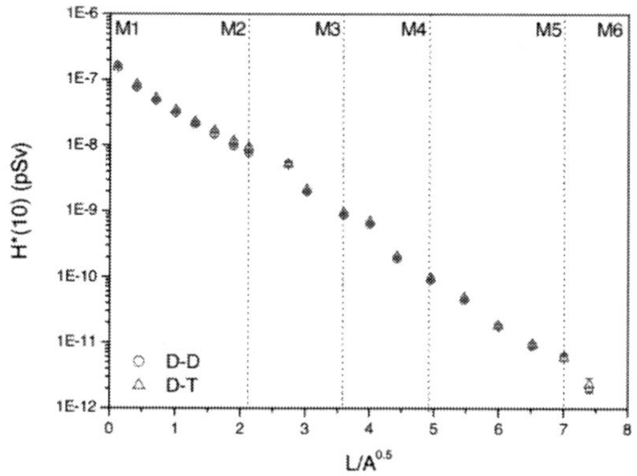

Figure 7. Neutron ambient dose equivalent, H*(10), as a function of $L/A^{0.5}$ for D-D and D-T plasma sources (per JET neutron)

As it can be seen from Fig. 7, in both source cases the ambient dose equivalent is decreasing along the total length of the maze. The H*(10) at the maze exit for the D-D and the D-T source was of $(2.0 \pm 0.2) \times 10^{-12}$ pSv and $(2.3 \pm 0.6) \times 10^{-12}$ pSv per JET neutron, respectively. The maze transmission factor was found to be 1.2×10^{-5} and 1.5×10^{-5} for the D-D and the D-T source, respectively. The maze transmission factor was defined as the ratio of the calculated ambient dose equivalent at the maze exit (M6) over the calculated dose at the maze mouth (M1) and is considered to be an index of the effectiveness of the maze in attenuating neutrons. The comparable calculated H*(10) values at the maze exit in D-D and D-T operation modes (per JET neutron) are attributed to the fact that neutrons entering the maze have already been slowed down due to multiple interactions in the Torus and wall materials and their propagation along the maze has become independent of their initial energy at the point of production (source). It is stressed that the results shown in Fig. 7 depend on the maze geometric configuration and shielding materials employed. Therefore, they cannot be readily generalized to other neutron sources and maze configurations as well.

4 Validation

A benchmark experiment was performed in the 2015-2016 JET D-D campaign in order to validate the results of the calculations. Measurements were performed using MCP-type thermoluminescence detectors (TLDs) and activation foils. The TLDs were produced and measured by the Institute of Nuclear Physics, Polish Academy of Sciences, Krakow, Poland. TLDs were positioned in rectangular and circular holders within large polyethylene cylindrical moderators (d=25 cm, H=25 cm) [7]. The activation foils used were Tantalum foil discs (bare and Cadmium covered). The foils were measured by the NCSR "Demokritos" group. The detectors were positioned at selected locations in the JET Hall and shielding penetrations (Fig. 8).

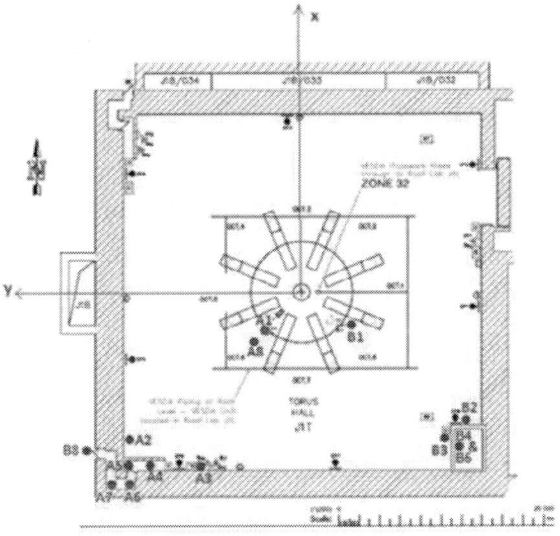

Figure 8. Detector locations in the Torus hall

To produce results comparable to those derived from the TLDs, the activation foils were positioned within large polyethylene moderator cylinders at selected locations within the JET hall, close to the moderators containing the TLDs (Fig. 9). More details on the experimental procedure have been discussed elsewhere [1].

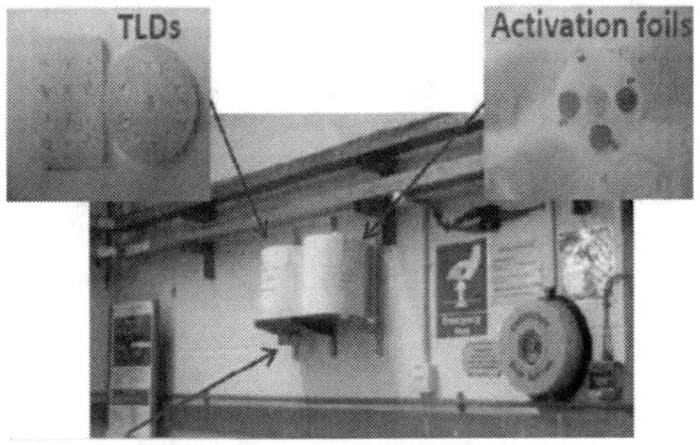

Figure 9. Detector positioning for irradiation in the Hall

The comparison of the simulations against the experimental measurements is presented in Table 1. In this table, the neutron fluence results derived using Tantalum foils and TLDs as well as the MCNP calculated values are given along with their % relative errors for five detector positions. One of these detectors (A1) was located near the tokamak, two detectors (A2 & A4) in the SW labyrinth region and two more detectors (B3 & B5) in the South East corner near the chimney (Fig. 8).

The neutron fluence values shown in Table 1 correspond to a total budget of 3.52E+18 neutrons, which is the total amount of neutrons produced at the source during the experimental campaign.

It is stressed that attention was paid in the development of the MCNP model and the subsequent calculations in order to ensure that the results will be comparable to those derived using the TLDs. For this reason, all associated parameters (such as the neutron self-shielding effect as well as the "shadowing" effect caused by the interference between individual TLDs and foils within the moderator) have been studied and taken into account in the calculations. A detailed description of the method has been presented [1, 7].

Table 1. Neutron fluence results for selected detector positions at the JET Hall

Detector Position	Neutron fluence (cm^{-2}) (% rel.error)				
	Ta foil	TLDs		MCNPX calculations	
		Circle	Square	Circle	Square
A1	2.7E+10 (16.4)	2.5E+10 (7.3)	2.7E+10 (6.0)	-	-
A2	1.0E+9 (20.0)	1.2E+9 (7.0)	1.3E+9 (6.6)	2.2E+9 (5.2)	2.3E+9 (5.2)
A4	2.2E+8 (33.2)	3.9E+8 (6.6)	5.1E+8 (7.8)	1.4E+9 (4.1)	1.4E+9 (4.1)
B3	9.6E+8 (18.9)	9.5E+8 (7.3)	1.0E+9 (6.3)	1.6E+9 (5.4)	1.7E+9 (5.4)
B5	2.1E+8 (43.3)	3.4E+8 (6.9)	4.0E+8 (7.2)	1.6E+8 (7.1)	1.7E+8 (7.1)

As it can be seen from Table 1, a good agreement was observed between TLDs and activation foil measurements for most of the positions studied. At detector positions A1, A2 and B3, the ratios of neutron fluence measured with TLDs over the one measured with Ta foil are within 0.9 to 1.3. At detector positions A4 (in the maze) and B5 (in the chimney), where lower fluence rates are measured, higher ratios -ranging from 1.6 to 2.3- are observed.

Furthermore, the calculations overestimate the measured fluences in all positions except B5 (in the chimney). The discrepancies are mainly attributed to approximations in the geometry model and in particular on the approximations used to describe the instruments and facilities surrounding the tokamak. These components were not possible to be modeled in detail on a realistic timescale, however they attenuate neutrons and therefore calculations over-estimated the neutron fluence in the Hall area. Nevertheless, taking into consideration the overall complexity of the geometry studied, the agreement between calculations and measurements is considered as satisfactory.

5 Discussion & Conclusions

Neutron streaming through tokamak ducts and channels has been studied by several workers [8-15]. Nevertheless, less attention has been given to neutron streaming through larger ducts or labyrinths in the tokamak biological shield. The methodologies generally used for the calculation of neutron streaming along multi-bent large ducts in accelerator facilities include application of the Monte Carlo technique and the use of predetermined "universal" analytical expressions [16, 17]. The advantage offered by the Monte Carlo simulation approach is the capability to explicitly model the source, geometry and materials configuration encountered.

The similar neutron transmission values obtained in this study for the D-D and the D-T JET plasma discharge sources were attributed to the fact that neutrons have been slowed down due to multiple interactions with the Torus and wall material and their propagation along the labyrinth has become independent of the initial energy of the neutron source and depend only on the labyrinth geometry and material configuration. Elastic neutron scattering in hydrogen and slow neutron absorption in boron in concrete have a significant effect on neutron streaming. Therefore, the results of this study confirmed that the accuracy of the neutron streaming simulations through the JET labyrinth depends on the exact knowledge of the labyrinth geometry and wall composition and is practically independent of the source neutron spectrum.

The calculations were validated by comparison against experimental measurements using both thermoluminescence detectors and activation foils. The results of the comparison showed that in most cases calculations overestimated the measured fluence. Nevertheless, the agreement between experimental results and calculations can be considered as satisfactory taking into consideration the overall complexity of the JET tokamak and shielding configuration studied. Moreover, Vasilopoulou et al [5] studied the effect of concrete composition on neutron streaming calculations along the maze and showed that the accurate knowledge of the hydrogen content in concrete is of outmost importance.

The results of this work contribute to the continuous operational radiation protection effort to minimize collective radiation dose to JET personnel during the current D-D and the forthcoming D-T experimental campaigns. Moreover, they provide important information from JET experience that may assist in the optimization and validation of the radiation shielding design methodology used in ITER.

Acknowledgments

This work has been started within EFDA and carried out within the framework of the EUROfusion Consortium and has received funding from the European Union's Horizon 2020 research and innovation programme under grant agreement No 633053. The views and opinions expressed herein do not necessarily reflect those of the European Commission.

References

1. P. Batistoni, S. Conroy, S. Lilley, J. Naish, B. Obryk, S. Popovichev, I.E. Stamatelatos, B.D. Syme, T. Vasilopoulou, Nucl. Fusion **55,** 053028 (2015)
2. W.G. McKinney et al., MCNPX overview, *Proceedings of the 2006 HSSW, FNAL, IL, LA-UR-06-6206, Sept. 2006* (2006)
3. A. Santamarina, D. Bernard, P. Blaise, M. Coste, A. Courcelle, T.D. Huynh, C. Jouanne, P. Leconte, O. Litaize, S. Mengelle, G. Noguère, J-M. Ruggiéri, O. Sérot, J. Tommasi, C. Vaglio, J-F. Vidal, The JEFF-3.1.1 Nuclear Data Library, JEFF Report 22, NEA No. 6807, OECD, ISBN 978-92-64-99074-6 (2009)
4. D.L. Aldama, A. Trkov, FENDL-2.1: Evaluated nuclear data library for fusion applications, INDC(NDS)-467 (2004)
5. T. Vasilopoulou, I.E. Stamatelatos, P. Batistoni, S. Conroy, B. Obryk, S. Popovichev, B.D. Syme, Radiat. Phys. Chem. **116,** 359-364 (2015)
6. International Commission on Radiological Protection (ICRP), The 2007 Recommendations of the International Commission on Radiological Protection, Publication 103, Ann. ICRP **37,** 2-4 (2007)
7. B. Obryk, P. Batistoni, S. Conroy, B.D. Syme, S. Popovichev, I.E. Stamatelatos, T. Vasilopoulou, P. Bilski, Fus. Eng. Des. **89,** 2235–2240 (2014)
8. M. Angelone, M., Batistoni, P., Petrizzi, L., Pillon, M., Fus. Eng. Des. **51–52,** 653-661 (2000)
9. Y. Chen, U. Fischer, S.P. Simakov, F.J. Wasastjern, Nucl. Mater. **367–370,** 1580-1585 (2007)
10. S. Sato, K. Maki, Fus. Eng. Des. **65** 501-524 (2003)
11. M. Yamauchi, K. Ochiai, Y. Morimoto, M. Wada, S. Sato, T. Nishitani, Radiat. Prot. Dosim. **116,** 542-546 (2005)

12. C. Konno, F. Maekawa, Y. Uno, Y. Kasugai, M. Wada, Y. Ikeda, H. Takeuchi, Fus. Eng. Des. **51–52,** 797-802 (2000)
13. K. Ochiai, S. Sato, M. Wada, N. Kubota, K. Kondo, M. Yamauchi, Y. Abe, T. Nishitani, C. Konno, Fus. Eng. Des. **82,** 2794-2798 (2007)
14. F. Moro, L. Petrizzi, G. Brolatti, B. Esposito, D. Marocco, R. Villari, Fus. Eng. Des. **84,** 1351-1356 (2009)
15. A. Serikov, U. Fischer, D. Leichtle, C.S. Pitcher, Fusion Eng. Des. **87,** 690-694 (2012)
16. H. Dinter, D. Dworak, K. Tesch, Nucl. Instr. Meth. Phys. Res. A **333,** 507-512 (1993)
17. E. Mauro, M. Silari, Nucl. Instr. Meth. Phys. Res. A, **608,** 28-36 (2009)

Light Ion Yields from Bombardment of Thick Targets by Protons, Helium-4 and Iron-56

Luis A. Castellanos[1], Natalie A. McGirl[1], Ashwin P. Srikrishna[1], Lawrence H. Heilbronn[1], Chiara La Tessa[2], Adam Rusek[2], Michael Sivertz[2], Steve Blattnig[3], Martha Clowdsley[3], Tony Slaba[3] and Cary Zeitlin[4]*

[1]Department of Nuclear Engineering, University of Tennessee, 1004 Estabrook Rd, Knoxville, TN, USA, 37996
[2]NASA Space Radiation Laboratory, Brookhaven National Laboratory, Upton, NY, USA, 11973
[3]National Aeronautics and Space Administration, Langley Research Center, Hampton, VA, USA, 23681
[4]Lockheed Martin, Information Systems and Global Solutions, Houston, TX, USA, 77285

Abstract. In March 2016 accelerator-based experiments colliding protons (0.4 and 0.8 GeV), helium (0.4 AGeV) and iron (0.4 and 0.8 AGeV) on thick aluminum targets with surface densities of 20, 40, and 60 g/cm2 were performed at the National Aeronautics and Space Administration Space Radiation Laboratory (NSRL) at Brookhaven National Laboratory. Two targets were utilized in each experimental configuration. Hydrogen and helium ions were detected using organic liquid scintillators in conjunction with thin plastic scintillators at 10°, 30°, 45°, 60°, 80°, and 135° from beam axis. Time-of-flight techniques and pulse shape discrimination were used to identify light ion species in order to generate double differential energy spectra of the light ion fragments. Comparisons of these measured yields were compared with Monte Carlo calculations generated by MCNP6. These yields will be used to quantify uncertainty in radiation transport codes utilized in risk assessment for spaceflight missions with prolonged exposures to galactic cosmic rays.

1 Introduction

The radiation environment beyond the confines of Earth's magnetosphere poses a significant risk for future manned missions. While exposure from Solar Energetic Particles (SEP) can be mitigated with increased shielding, the Galactic Cosmic Ray (GCR) component of dose is pervasive. Current calculations of Radiation Exposure Induced Death (REID) suggest that increased shielding has a negligible ability to reduce cancer risks [1]. These estimates of risk are based on particle fluences simulated by deterministic radiation transport codes, which are verified and validated by fits to measurements of GCR ion fluences from probes in Earth's atmosphere. The available database for comparison is limited, as the majority of data were collected at energies below 500 AMeV [2]. However, GCR at these energies are limited contributors to effective dose [3]. The limitations of the available data increase the uncertainties associated with transport calculations and subsequent risk assessments from GCR ions at higher energies. The purpose of the experiment is to provide measurements of double differential light ions yields through thick targets composed of materials commonly selected to shield against GCR. This information will be used to further benchmark transport codes at energies that are inadequately represented by prior measurement efforts.

2 Experiment Overview

Three separate ion species were delivered to the National Aeronautics and Space Administration (NASA) Space Radiation Laboratory (NSRL) target room over a 100-hour timespan from the Booster accelerator at Brookhaven National Laboratory (BNL). H-1, He-4, and Fe-56 beams were extracted at 0.4 AGeV, while H-1 and Fe-56 were extracted at 0.8 AGeV,. These projectiles were incident on the the center of the primary upstream target, composed of either aluminum with thicknesses of 20, 40, and 60 g/cm^2 and area 30 x 100 cm^2. A second downstream target was placed 3.5 m from the center of the upstream target. This configuration mirrors an enclosed space which was studied in work that identifies a local minimum in dose equivalent as a function of shielding thickness [4]. The downstream target matched the material of the upstream target with area 100 x 100 cm^2 and a fixed 60 g/cm^2 thickness, regardless of the upstream target's thickness.

Secondary light ions produced in the upstream target were detected with six cylindrical EJ-301 organic scintillator detectors, with active regions of dimensions 12.7- cm diameter and depth. In front of each detector was a thin NE-304 plastic scintillator to discriminate, or veto, charged and uncharged particles. The detectors were placed at 10°, 30°, 45°, 60°, 80°, and 135° from beam axis,

* Corresponding author: lcastell@vols.utk.edu

as illustrated in Figure 1. Total measured particle yields decreased with increasing angle and distance from target center. This influenced the choice of detector distances from target center, with care not to saturate the detection system and to accommodate shadow bars for neutron measurement. Timing information was initiated with a coincidence event between two thin EJ-228 plastic scintillators upstream from the primary target.

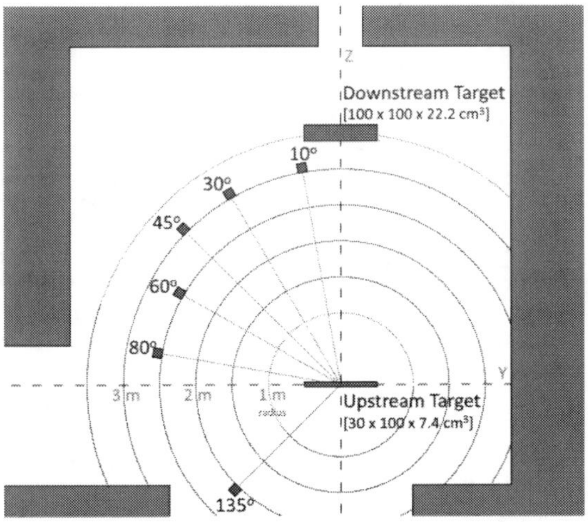

Fig. 1. NSRL target room schematic with experiment equipment illustrated, 20g/cm2 target shown.

This investigation will continue the efforts of the 18-hour commissioning experiment conducted in May 2015. Additional measurements will be conducted at NSRL to examine the yields resulting from other projectile species and energies, as well as other target materials and compositions in November 2016 and 2017.

3 Data Analysis Methodology

Light ions are identified by comparison of velocity and energy deposited in the detector. Some ions deposit their full kinetic energies in the detector. By interpolating from the continual slowing down approximation range tables provided by NIST [5], the full detector length of 12.7 cm is enough material to stop a proton of approximately 121.5 MeV. Particles with higher energies can escape or "punch through" the detector. The range (R) of a proton is related to the range of a different projectile (R_i) of mass (A) and charge number (Z) with the same velocity in an absorber of density (ρ) by equation 1. This relation is the key for determining the velocities of other ions

$$(R\rho)_i = (A/Z^2)_i \cdot (R\rho)_{proton} \quad (1)$$

An example of a histogram of light ion time of flight (ToF) and energy deposited (E) is presented in Figure 2. Detection events pertaining to the same ion are illustrated as pairs of lines which join at a peak. Peaks are indicative of particles losing full energies in the detector, with the line to the right of these peaks indicative of ions punching through the detection volume. Ions with higher mass deposit more energy and subsequently populate the upper parts of the histogram. The three lower pairs of lines represent hydrogen ions, the rightmost depicts proton detection events. This technique allows for discrimination between different secondary ion species' ToF spectra, which are next converted to energy spectra using the known distance between detector and the target.

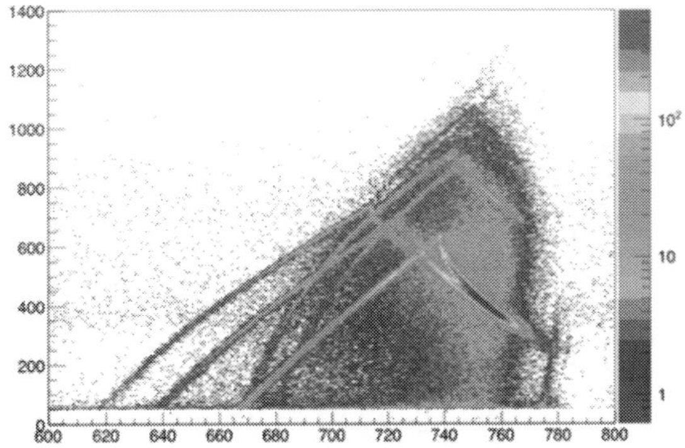

Fig. 2. Total energy deposited (Y) vs time-of-flight (X) histogram for Fe 0.4-AGeV projectiles at 10°.

Further discernment between particles was achieved utilizing pulse shape discrimination (PSD) techniques. For histograms, such as Figure 2, illustrating the detection of multiple ions and isotopes, there are points where the respective ions' charge deposition signatures overlap. PSD utilizes a feature of the data acquisition apparatus which integrates the charge deposition in two ways. One integrates the signal over the full 200-nanosecond "live" time of the acquisition system for a given detection event. The other is to integrate only the first 20 nanoseconds of the signal. Different ions produce signals in the detectors with different rise and fall times, and this feature of the scintillation allows for these two integration modes to serve as a tool for discerning differing ions from each other. A histogram charting these integrated values for the ions detected at the 10° position resulting from a 0.4-AGeV projectile is presented in Figure 3. The diagonal yellow line at the top right of the figure is indicative of helium ions. The two diagonal red-yellow bands at the lower left of the figure are indicative of protons, the lower of the two representative of protons depositing their full kinetic energies and while those populating the other line have enough energy to punch through. The ability to select events on this histogram allows one to most accurately attribute detection events to the correct ion at points where lines cross in the charge deposition vs ToF histograms.

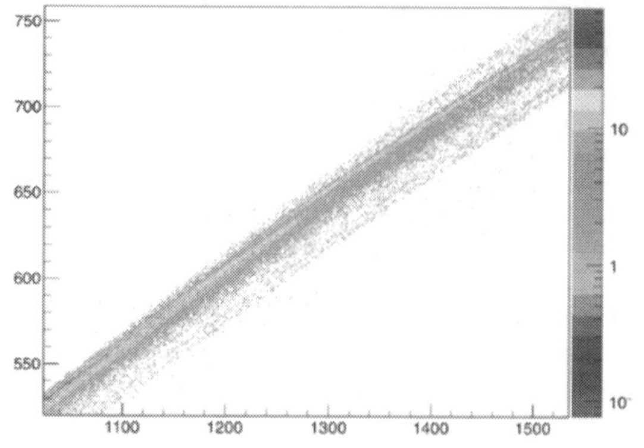

Fig. 3. Charge deposited in the first 20 ns of signal (Y) vs. total charge deposited (X) over the full 200ns integration gate.

4 Experiment Results

A summary of secondary ions detected organized by incident projectile is presented in Table 1. The angles listed are the widest angle of detection, thus detectors positioned at shallower or smaller angles than what is listed were also able to detect that respective ion. It was expected to not detect the heavier mass number greater than two ions from bombardments with proton projectiles.

Table 1. Summary of secondary ions detected resulting from bombardments incident projectiles and widest position of detection.

Projectile	H-1	H-2	H-3	He-3	He-4
0.4-AGeV Fe	135°	80°	45°	30°	30°
0.8-AGeV Fe	135°	80°	60°	30°	30°
0.4-GeV H	135°	45°	-	-	-
0.8-GeV H	135°	60°	-	-	-
0.4-AGeV He	135°	60°	45°	30°	30°

A small sample of the results collected will be presented in this paper to maintain brevity, while still showcasing the comprehensive detection capabilities of the experiment apparatus. Double differential yields of protons stemming from 0.4-AGeV iron on the three thicknesses of aluminium used are presented in Figure 4. The results exhibited predictable trends of yield as a function of angle. The relative rates of production per ion were also approximately what were expected, as supported in comparisons of the measured results to those simulated using MCNP6 [6]. An example of this comparison is presented in Figure 5.

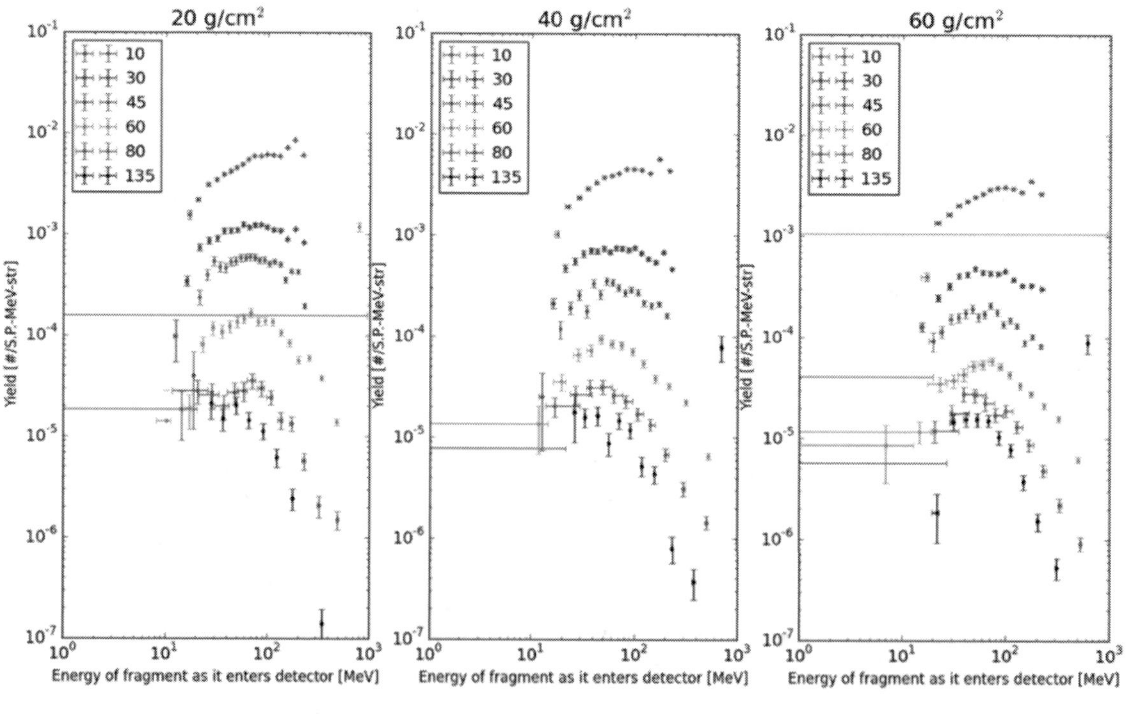

Fig. 4. Measured yields of protons resulting from the bombardment of aluminum upstream targets of three separate thicknesses by 0.4-AGeV Fe.

DOI: 10.1051/epjconf/201715307029

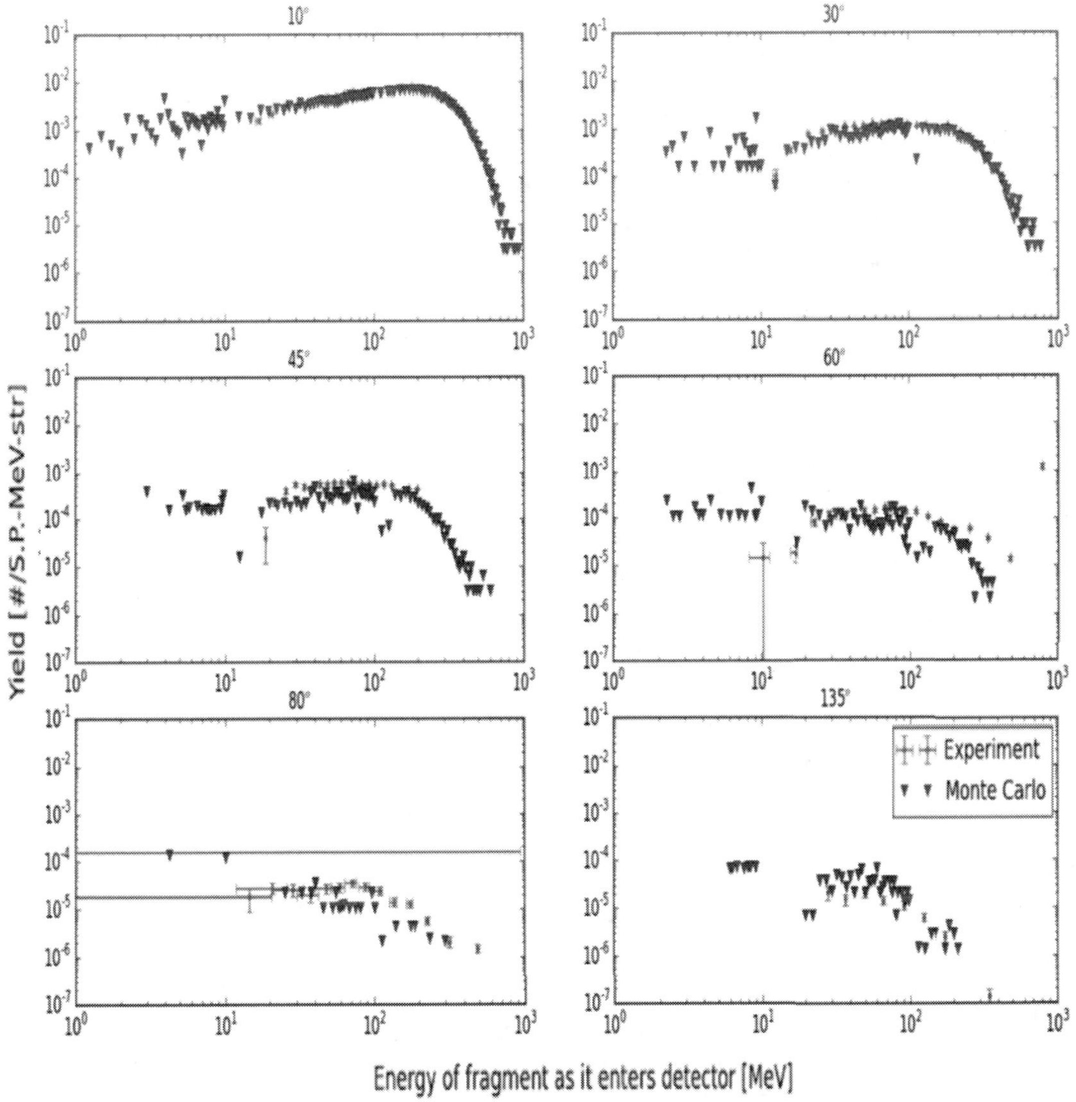

Fig. 5. Measured and simulated yields of protons resulting form the bombardment of $20 g/cm^2$ aluminum upstream target by 0.4-AGeV Fe.

The 0.4-AGeV iron experiment runs saw approximately 610 million projectile ions delivered with a 64% live acquisition time. This data rate produced proton yields with statistical uncertainties less than 10% for angles up to 60° save for a few low energy (less than 10 MeV/nucleon) bins. The wider angle detectors had uncertainties generally less than 20%.

MCNP6 simulation results are presented in this work. Other Monte Carlo tools will be utilized to benchmark these results in the future. Physics models selected to adhere to the default and suggested options as published in the software literature. Iron projectile simulations used 1.00 x 10^7 source particle histories and generated results with statistical errors similar to those of the measured yields. Currently the measured yields agree most closely with the forward-most angle simulation results, differing at most by a factor of two. The spectra shape of these tend to agree, as to where the drop-off in yield as a function of energy should occur, giving confidence in the agreement.

5 Summary

Double differential yields of secondary light ions produced by protons, helium and iron in thick targets of aluminium of 20, 40, and 60 g/cm^2 thickness were obtained from accelerator-based measurements performed at NSRL in March 2016. This data will supplement the database of measured ion fluences produced by space-flight probes. Ultimately, this information will help validate transport calculations of GCR through shielding materials used in the assessment of cancer risks associated with prolonged exposures that will occur in missions beyond Low Earth Orbit. These calculations are imperative in safely accomplishing the missions of furthering manned space flight.

Acknowledgements

This work was supported by Grants NNX12AL51A and NNX15AD89A from the United States National Aeronautics and Space Administration.

References

1. F. A. Cucinotta, M.-H. Y. Kim, L. Ren, Evaluating shielding effectiveness for reducing space radiation cancer risks, Radiat. Meas., **41**, 1173-1185 (2006)
2. T. C. Slaba, X. Xu, S. R. Blattnig, R. B. Norman, GCR environmental models III: GCR model validation and propagated uncertainties in effective dose, Space Weather, **12,** 233-245 (2014)
3. T. C. Slaba, S. R. Blattnig, GCR environmental models I: Sensitivity analysis for GCR environments, Space Weather, **12,** 217-224 (2014)
4. T. C. Slaba, C. J. Mertens, S. R. Blattnig, Radiation Shielding Optimization on Mars, NASA Technical Paper 2013- 217983 (2013)
5. PSTAR Stopping Power and Range Tables for Protons," National Institute of Standards and Technology (NIST): Physical Measurements Laboratory. [Online] Available: http://physics.nist.gov/PhysRefData/Star/Text/PSTAR.html
6. T. Goorley, et al., Initial MCNP6 Release Overview, Nuclear Technology, **180**, 298-315 (2012).

Neutron-gamma flux and dose calculations for feasibility study of DISCOMS instrumentation in case of severe accident in a GEN 3 reactor

Mariya Brovchenko[1] [a], *Isabelle* Duhamel[1], *Benjamin* Dechenaux[1]

[1]Institut de Radioprotection et de Sûreté Nucléaire (IRSN), Fontenay-aux-Roses, 92260, France

Abstract. The present paper presents the study carried out in the frame of the DISCOMS project, which stands for "DIstributed Sensing for COrium Monitoring and Safety". This study concerns the calculation of the neutron and gamma radiations received by the considered instrumentation during the normal reactor operation as well as in case of a severe accident for the EPR reactor, outside the reactor pressure vessel and in the containment basemat. This paper summarizes the methods and hypotheses used for the particle transport simulation outside the vessel during normal reactor operation. The results of the simulations are then presented including the responses for distributed Optical Fiber Sensors (OFS), such as the gamma dose and the fast neutron fluence, and for Self Powered Neutron Detectors (SPNDs), namely the neutron and gamma spectra. Same responses are also evaluated for severe accident situations in order to design the SPNDs being sensitive to the both types of received neutron-gamma radiation. By contrast, fibers, involved as transducers in distributed OFS have to resist to the total radiation gamma dose and neutron fluence received during normal operation and the severe accident.

1 Introduction

At Fukushima Daiichi, the loss of electricity power supplies has quickly led most of instrumentation to be inoperative and the operator (TEPCO) with no way to monitor the status and the progress of the accident. To overcome these important drawbacks, the DISCOMS project [1], which stands for "DIstributed Sensing for COrium Monitoring and Safety", considers taking advantage of the potential of distributed sensing technologies based on both "Optical Fiber Sensors" (Raman, Brillouin, and Rayleigh Reflectometry) and long-length "Self Powered Neutron Detectors" (SPNDs).

The goal consists in inquiring about the status of the third barrier of confinement and to define possible mitigation strategies in case of severe accident, namely: i) reactor pressure vessel breakthrough and corium relocation outside the vessel, ii) concrete floor erosion and iii) corium cooling. Such monitoring should consist in "sensing cables" inserted in basemat below the reactor vessel and interrogated from a rear base, where operators can work safely. Thus, optical fibers selected for their resistance to ionizing radiations and long-length SPNDs, both judiciously deployed within the reactor basemat, and the structures around it, will provide a useful real-time or on-demand monitoring, during normal operation as well as in accidental and post-accidental situations.

For GEN 3 reactors (e.g. EPR reactor), distributed sensors (OFS and SPND) could be implemented below the reactor pressure vessel close to the surface of the reactor building basemat concrete and, if technically feasible, embedded into the concrete. The SPNDs need a sufficient level of radiation to generate a signal, and could be placed on the surface of the reactor pit.

This paper discusses the contribution to the development of remote monitoring technologies to improve Nuclear Power Plants (NPPs) safety. The work presented in this paper aims to provide the radiation level in terms of fast neutron fluence and the gamma dose for the optical fibers in order to ensure their resistance during the reactor life. Additionally, the SPNDs should be optimized in order to be sensitive to neutron and/or gamma flux during the normal operation of the reactor and during the severe accident (i.e. accident involving serious core damage) by calculating neutron and gamma flux spectra.

2 Normal reactor operation

To evaluate the neutron and gamma fluxes outside the vessel and in the basemat during normal reactor operation, the Monte Carlo code MCNP [2], which allows a detailed 3D geometrical description of the reactor core, the internals in the vessel and the reactor pit, was used with the associated ENDF/B-VII.1 nuclear data library.

The radiation source in this problem being the fissions occurring in the core of the reactor, the flux of

[a] Corresponding author: mariya.brovchenko@irsn.fr

neutral particles is strongly attenuated, especially in the basemat. Therefore, such neutron-gamma shielding calculations require the use of variance reduction (VR) methods.

2.1 Variance reduction methods

The authors used the variance reduction techniques implemented in the MCNP code, based on the Weight Window and source biasing.

In such problems, the positions of the responses (side of the vessel, vessel bottom, within the basemat) and the physical quantities (gamma dose, fast neutrons and thermal neutrons) are multiple. Optimizing a unique Weight Window set of maps (neutrons and gammas for different energy groups) to all of the responses is a challenge.

For this reason, Kenneth W. Burn (ENEA, Bologna) assisted the authors in the calculation of the VR parameters with a detailed mathematical approach called the DSA (Direct Statistical Approach) that optimizes the VR parameters governing the sampling [3]. The approach has been extended to include, among other things, weight control and a multiresponse capability [4].

2.2 Source term

The only source term considered in these calculations relies on the neutrons coming from fissions. As can be observed from the results, the gamma radiation from the irradiated fuel can be easily neglected. Nevertheless, the activation of structures outside the core is not taken into account in this study and could be not negligible.

The fission neutrons source distribution used in this study is a flat distribution within each assembly. Thus, the decrease of the neutron flux on the border of the assemblies on the edge of the core, as shown in figure 1, is not modeled.

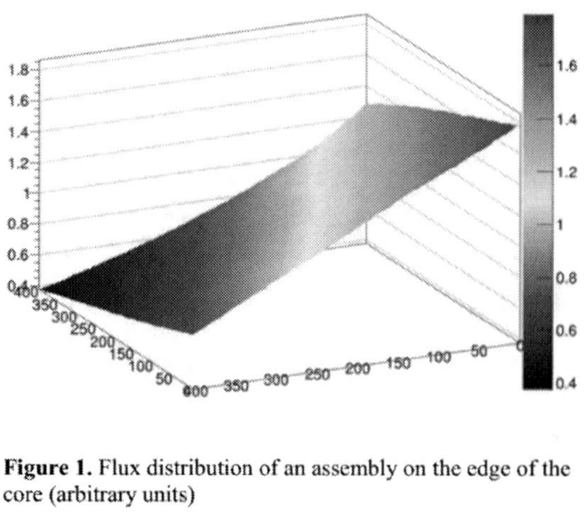

Figure 1. Flux distribution of an assembly on the edge of the core (arbitrary units)

Since only the first rows of assemblies on the edge of the core contribute to the flux estimation on or outside the vessel, the impact of this assumption was quantified. It leads to an overestimation of neutron and gamma fluxes outside the vessel of about 20% and 30% respectively.

2.3 Reactor model

Thanks to the MCNP code, detailed model of EPR reactor can be used for these calculations. The fuel compositions correspond to those of the end of an equilibrium cycle and are unique inside each assembly. The poisoned rods are treated separately.

No axial burn-up effect is taken into account. The geometry of the assemblies is detailed pin-by-pin (see figure 2).

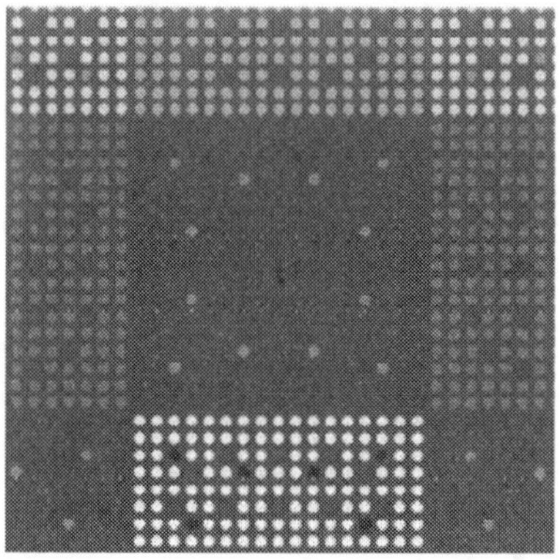

Figure 2. MCNP model of the pin-by-pin geometry of the assemblies

The geometry outside the core is modeled as precisely as possible. The positions of the responses of interest are presented in figures 3 and 4. The detectors are placed outside the vessel at the level of the active core. The responses are calculated with an axial segmentation, as detailed in figure 3.

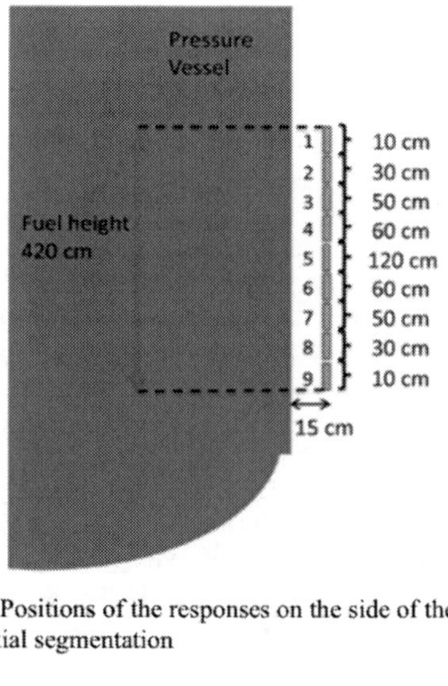

Figure 3. Positions of the responses on the side of the vessel with an axial segmentation

 DOI: 10.1051/epjconf/201715307030

The radiation is also studied under the vessel bottom on the basemat surface and at different depths inside the basemat. In the EPR reactor [5], it consists of a layer of sacrificial concrete (in green in figure 4) and a layer of refractory material (in blue in figure 4). As shown in figure 4, the reactor pit is divided into 3 rings.

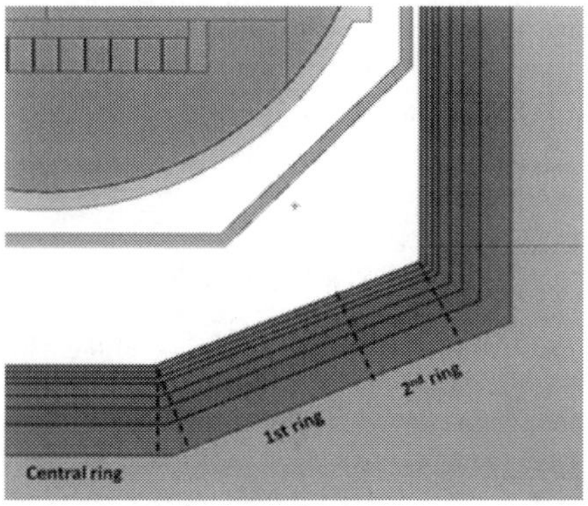

Figure 4. Positions of the responses in the reactor pit with a radial segmentation in rings

3 Severe accident situation

Several situations corresponding to different stages during a severe accident were considered in this study. The corium composition was calculated according to the hypotheses taken for each situation. Only the state with the spread corium in the reactor pit is presented here, in order to compare the radiation in the concrete to the one received during normal reactor operation.

3.1 Corium

The isotopic fuel composition is taken at the end of an equilibrium cycle, 3 hours after the reactor shut down, i.e. fissions stop. For sensitivity study, another corium composition was also considered supposing 49 hours after the reactor shut down. The fission products are supposed being not released from the corium. However, a sensitivity study was performed considering an important fission products release.

The corium model contains all the fuel mass, melted structural materials in the core and the vessel bottom. The composition is homogeneous and spread in the reactor pit, no corium-concrete interactions are considered.

3.2 Source term and transport

The source term considered here takes into account only the gamma rays issued from decay of instable isotopes found in the fuel composition. The spontaneous fissions activity was evaluated to be negligible (10^{-9} times lower than the total corium activity). The gamma rays source term is calculated using MCNP6 code and its new capabilities based on 25 energy groups emission data to generate the decay gamma rays. The decay heat evaluated by MCNP6 (tallying the energy deposition) was compared to the one given by the VESTA code [6] (using directly the decay energy from the JEFF 3.1 nuclear data library) resulting in a deviation of 13% on the decay heat. This deviation is acceptable for the purpose of its application here, since the calculation scheme is simplified: the source term and the transport can be performed in a single calculation.

The MCNP model, as presented in figure 5, is based on the same structures outside the vessel as used for the normal operation calculations. The positions of the responses in the reactor pit are the same as shown in figure 4. The corium is modeled as a cell with homogeneous composition and thus homogeneous source term.

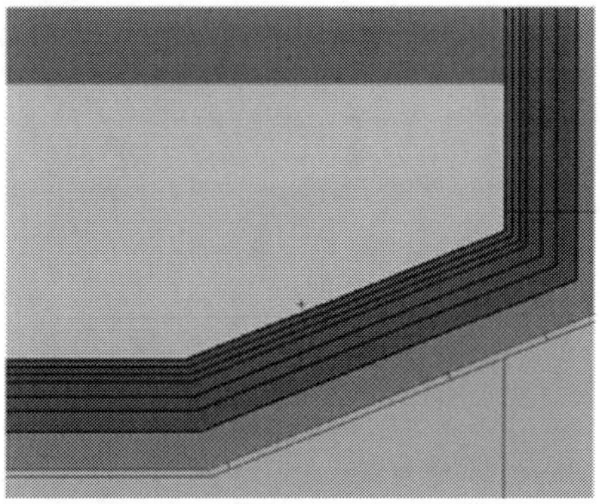

Figure 5. Corium (in beige) spread in the reactor pit on the sacrificial concrete (in red)

4 Results

4.1 Neutrons

The neutron responses were only calculated for the reactor full power operation state, since the spontaneous fission activity was negligible in the corium.

The responses were calculated using the VR parameters created with the DSA and converted to a set of classical Weight Windows. The transport calculation was performed using $3 \cdot 10^8$ source particles leading to a standard deviation of each response mostly lower than 1%. Results are reported in the figures below.

The neutron flux spectrum calculated at the different positions on the side of the vessel and on the basemat surface is presented in figure 6. The neutron flux varies by a factor 3 between the central axial segment and the bottom segment. The neutron energy spectrum is thereby unchanged.

The neutron flux on the surface of the basemat is more than one order of magnitude lower than on the side of the vessel, reaching 3 orders of magnitude for fast neutrons (E > 1 MeV). The neutron flux on the outer rings of the basemat surface is slightly higher than in the central ring, which is mainly due to the shielding by the bottom of the vessel. Indeed, as verified by flagging the surfaces in the MCNP tallies, the contribution of particles that crosses the bottom of the vessel is very low.

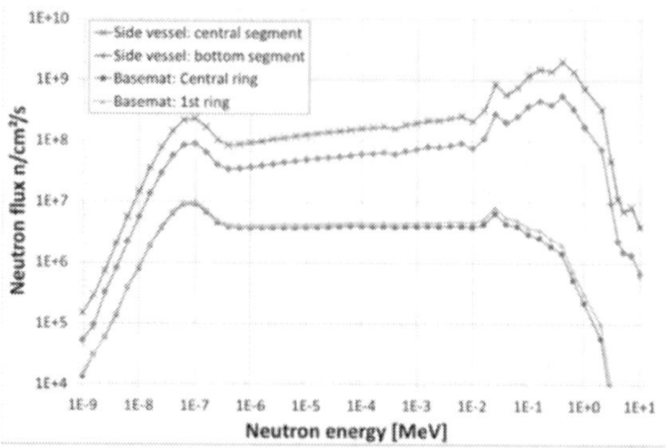

Figure 6. Neutron flux spectrum at different positions

Optical fiber sensors become opaque when subjected to neutron and gamma radiations. To protect them from the strong radiation in the nuclear reactor, the sensors can be embedded in the concrete or placed behind the refractory material layer.

To optimize the position, the neutron flux was evaluated at different depths in the reactor pit, considering 3 energy groups, as presented in the figure 7. Thereby, strong hypothesis was made: the flux, calculated here for the core corresponding to an end of the equilibrium cycle, is considered constant during the whole reactor lifetime. This hypothesis is nevertheless in accordance to the purpose of these evaluations to provide an order of magnitude of the radiation level at different positions.

As shown in figure 7, 50 cm of sacrificial concrete can provide an attenuation factor 100 for high (E > 1 MeV) and low energy (E < 0.1 MeV) neutron flux and a factor 10 000 for intermediate energy neutron flux (0.1 MeV < E < 1 MeV). 25 cm of refractory material provides an additional attenuation factor around 10 for all energy groups. It is to notice that the uncertainty becomes more important for values behind the refractory material, especially for the high energy neutrons.

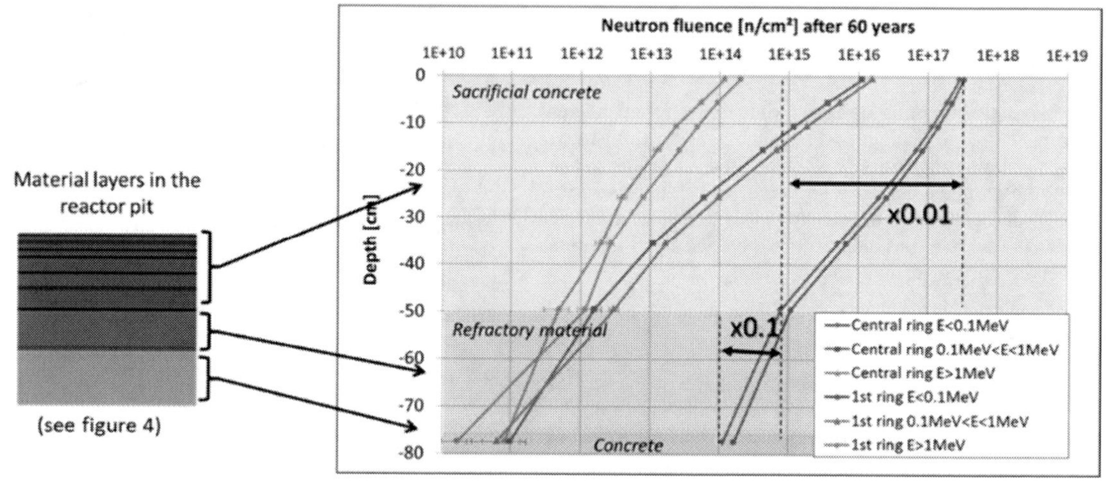

Figure 7. Neutron fluence attenuation considered for a constant flux during 60 years of reactor operation in the reactor pit divided in 3 energy groups in the different layers in the reactor pit

4.2 Gamma rays

For gamma radiation, the fluxes and doses were evaluated for the reactor operation at full power and the severe accident situations. The considered exposure time for the severe accident situation was 24 hours.

Figure 8 shows the gamma flux spectrum at different positions on the side of the vessel and on the surface of the basemat during normal reactor operation. Similarly to thermal neutrons, the gamma flux varies by a factor 3 between the central axial segment and the bottom segment and by more than one order of magnitude on the surface of the basemat. The gamma energy spectrum is almost unchanged even for the responses on the surface of the basemat.

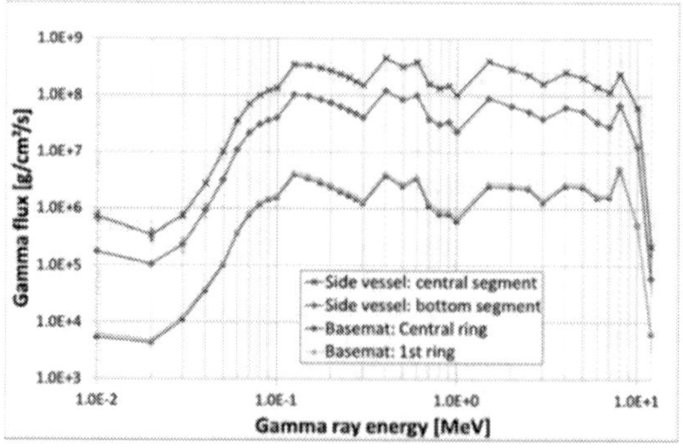

Figure 8. Gamma flux spectrum at different positions

To quantify the potential protection of the optical fibers from gamma radiation in the concrete the attenuation of the kerma in silica was calculated. The results are presented in figure 9. The attenuation is not exponential since the gamma rays are also produced by the neutrons directly in the basemat. Due to this, the sacrificial concrete provides only an attenuation of a factor 20. However, 25 cm of refractory material attenuates the kerma by an additional factor 100.

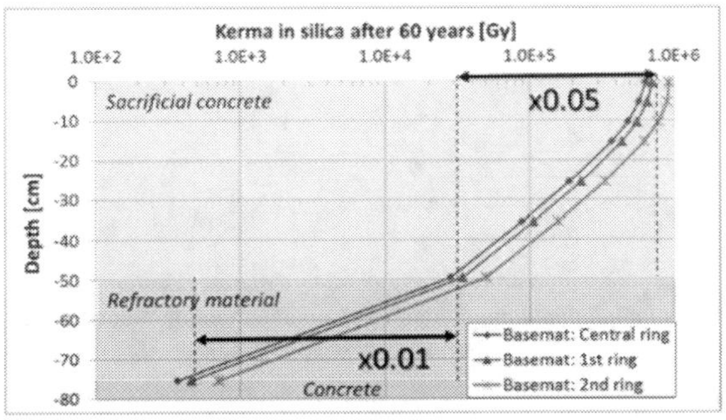

Figure 9. Gamma kerma in silica attenuation in the basemat considered for a constant flux during 60 years of reactor operation

For the severe accident situation, as described above, the same gamma responses in the basemat were calculated. The gamma ray spectrum was calculated on the concrete surface, directly in contact with the corium. The spectrum is compared to the one obtained during reactor operation at the same position and is presented in figure 10. The flux is 10 000 times higher in contact with the corium compared to the flux received during operation. The relative proportion of the high energy gamma rays (1-10 MeV) coming from the corium is low especially for gamma rays with E > 3 MeV. This part of the spectrum is very different from the one observed for gamma ray flux received during reactor operation due to the fact that the gamma rays issued from the neutron reactions have higher energies than the ones issued from the decays.

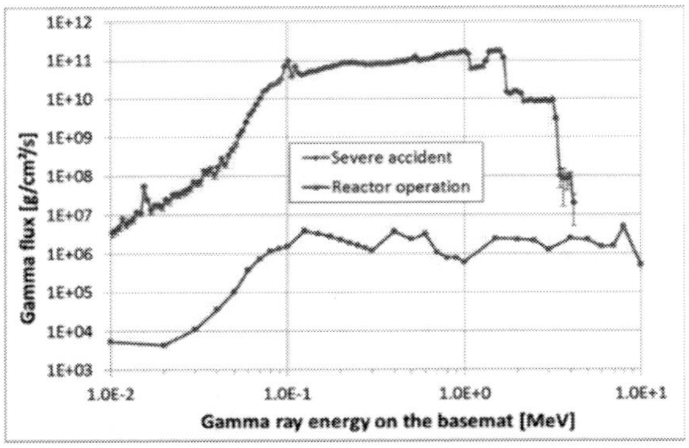

Figure 10. Gamma flux spectrum received in severe accident and during reactor operation on the surface of the basemat

The attenuation of the kerma in the basemat is shown in figure 11. For a gamma source, the concrete of 50 cm thickness provides a good protection with an attenuation of a factor 1000. An additional factor 1000 can also be provided by the refractory material (25 cm thickness).

With the hypothesis that the Optical Fiber Sensors have to resist at least for 24 hours after the accident onset, the choice of their position in the sacrificial concrete is driven by the gamma radiation received during normal reactor operation, as shown in figure 11. Indeed, in the first 5 cm of the concrete the gamma kerma from corium is more important than the one due to normal reactor operation, but at this distance from corium the sensor resistance to high temperatures is the dominant issue.

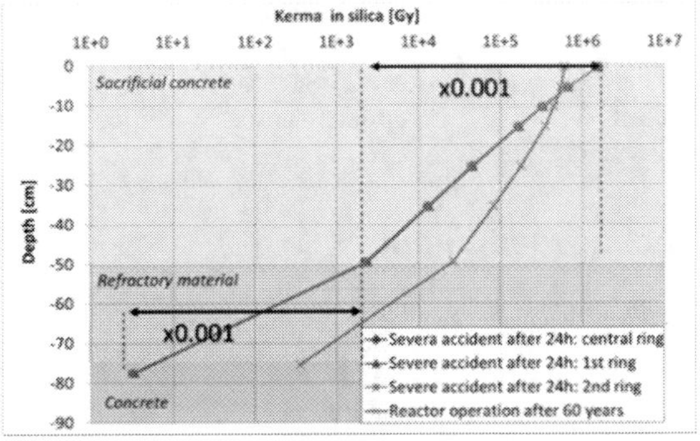

Figure 11. Gamma kerma in silica after 24 hours with corium on the basemat and after 60 years of normal operation

Due to the nature of a severe accident situation, assumptions must be made. In order to test the impact of these assumptions, some sensitivity calculations were performed.

The assumption for the corium isotopic composition of 3 hours "cooling time" after the reactor shut down was compared with 49 hours. The impact is a decrease of the kerma by around 30%, and a small change in the gamma ray spectrum.

The impact of the presence of volatile and half-volatile fission products in the corium was calculated. Two corium compositions were compared: one keeping all the fission products in the corium and another one considering an important release of volatile and semi-volatile fission products. The comparison showed a decrease of kerma by 30-35%, and a small change in the gamma ray spectrum.

Another important assumption is to consider that the corium composition is homogenous. Nevertheless, the impact of this hypothesis was not studied here.

5 Conclusion

In the framework of the DISCOMS project, the feasibility of the two types of instrumentation, Optical Fiber Sensors and SPNDs, is studied to contribute to the NPP safety improvement. The neutron and gamma fluxes and spectra were calculated at different positions outside the reactor vessel and on the basemat. The radiation was characterized for a state during normal reactor operation and severe accident situation. These results contribute to the design definition and the possible location choice for SPNDs with the aim that the SPNDs can provide a valuable signal for the both type of states.

To protect the optical fibers from radiation damages, fast neutron fluence and gamma kerma were calculated. For this, both types of situation were considered, the radiation received during reactor normal operation and the one received during the severe accident. As a result, the authors can conclude that the radiation damage to Optical Fiber Sensors will be mainly driven by the radiation received during reactor normal operation, regarding neutrons and gamma rays. In parallel to the radiation level calculations, the sensors designer should consider the temperature distribution and its evolution during the severe accident. The radiation and temperature "maps" should allow the DISCOMS project partners to select the most appropriate sensors' position and the instrumentation design.

Acknowledgment

This project is co-funded by the French "Programme d'Investissements d'Avenir (PIA)" referenced by n° ANR-11-RSNR-0007 and managed by the ANR (National Research Agency).

The authors would like to thank all the partners of the DISCOMS project, especially AREVA for the data supplied concerning the EPR modeling and instrumentation specification but also colleagues from the CEA involved in both Optical Fiber Sensors and SPNDs developments, for all the technical exchanges and fruitful discussions on these instrumentations.

Authors' particular thanks go to Kenneth W. Burn (ENEA), for his huge help in performing calculations of radiation during reactor operation, for his advice and for sharing his experience.

References

1. P. Ferdinand et al., "DISCOMS: DIstributed Sensing for COrium Monitoring and Safety", *CANSMART 2015: International Conference on Smart Materials and Structures, CANSMART, 15-17 July, Vancouver, BC, Canada* (2015)
2. T. Goorley et al., "Initial MCNP6 Release Overview – MCNP6 version 1.0", LA-UR-13-22934, (2013)
3. A. Dubi et al., "General Statistical Model for Geometrical Splitting in Monte Carlo – parts I and II", *Transp. Theory Stat. Phys.* **14**(2), 167-193 and 195-221 (1985)
4. K.W. Burn, "Optimizing Monte Carlo to multiple responses: the Direct Statistical Approach, 10 years on", *Nuclear Technology* **175**, 138–145 (2011)
5. M. Fischer, "The severe accident mitigation concept and the design measures for core melt retention of the European Pressurized Reactor (EPR)", *Nuclear Engineering and Design* **230** 169–180 (2004)
6. W. Haeck, "VESTA User's Manual - Version 2.1.0", IRSN DSU/SEC/T/2011-81, Institut de Radioprotection et de Sûreté Nucléaire, France (2012)

Evaluation of isotopic composition of fast reactor core in closed nuclear fuel cycle

Georgy Tikhomirov [1], Mikhail Ternovykh [1,a], Ivan Saldikov [1], Peter Fomichenko [2], and Alexander Gerasimov [3]

[1] National Research Nuclear University MEPhI (Moscow Engineering Physics Institute), Department of Theoretical and Experimental Physics of Nuclear Reactors, 31, Kashirskoye shosse, 115409, Moscow, Russia
[2] National Research Centre "Kurchatov Institute", Kurchatov Nuclear Technology Complex, Division of Fast and High Temperature Reactors, 1, Kurchatov sq, Moscow, 123183, Russia
[3] Institute for Theoretical and Experimental Physics, Department of safe nuclear reactors, 25, Bolshaya Cheremushkinskaya ul., 117218, Moscow, Russia

Abstract. The strategy of the development of nuclear power in Russia provides for use of fast power reactors in closed nuclear fuel cycle. The PRORYV (i.e. «Breakthrough» in Russian) project is currently under development. Within the framework of this project, fast reactors BN-1200 and BREST-OD-300 should be built to, inter alia, demonstrate possibility of the closed nuclear fuel cycle technologies with plutonium as a main source of energy. Russia has a large inventory of plutonium which was accumulated in the result of reprocessing of spent fuel of thermal power reactors and conversion of nuclear weapons. This kind of plutonium will be used for development of initial fuel assemblies for fast reactors. The closed nuclear fuel cycle concept of the PRORYV assumes self-supplied mode of operation with fuel regeneration by neutron capture reaction in non-enriched uranium, which is used as a raw material. Operating modes of reactors and its characteristics should be chosen so as to provide the self-sufficient mode by using of fissile isotopes while refueling by depleted uranium and to support this state during the entire period of reactor operation. Thus, the actual issue is modeling fuel handling processes. To solve these problems, the code REPRORYV (Recycle for PRORYV) has been developed. It simulates nuclide streams in non-reactor stages of the closed fuel cycle. At the same time various verified codes can be used to evaluate in-core characteristics of a reactor. By using this approach various options for nuclide streams and assess the impact of different plutonium content in the fuel, fuel processing conditions, losses during fuel processing, as well as the impact of initial uncertainties on neutron-physical characteristics of reactor are considered in this study.

Introduction

The strategy of the development of nuclear power in Russia provides for use of fast power reactors in closed nuclear fuel cycle[1,2,3]. The PRORYV (i.e. Breakthrough in Russian) project is currently under development. Within the framework of this project, fast reactors BN-1200 and BREST-OD-300 should be built to, inter alia, demonstrate possibility of the closed nuclear fuel cycle technologies with plutonium as a main source of power. Russia has a large inventory of plutonium which was accumulated in the result of reprocessing of spent fuel of thermal power reactors and conversion of nuclear weapons. This kind of plutonium will be used for development of initial fuel assemblies for fast reactors. The closed nuclear fuel cycle concept of the PRORYV assumes self-supplied mode of the reactors and regeneration by capture reaction in non-enriched uranium, which is used as a raw material.

1 Program code for simulating nuclide streams in fast reactors fuel cycle

Existing software codes for simulation of mass transfer in open and closed nuclear fuel cycles such as MEEMS, DANESS, ICECAT, NFCSim [4], REFCO83 [5] etc. usually consider a simplified scheme of stages of change in the composition of nuclear materials. Mathematical models of changes of the isotopic composition of fuel in an infinite cell can be considered as nuclear reactors in these programs. A reactor is often represented as a simple object with user-defined value of the energy release. Multi-group approximation calculation is used in some cases. However these programs do not always take into account the features of refueling in the reactor and the irradiation history of the fuel assemblies inside the core. This approach is justified frequently due to the fact that it does not require long calculation time to obtain the needed result. Such software in some cases consider the fuel cycle of a whole country as an object for analysis. Therefore, very detailed calculations are not necessarily,

[a] Corresponding author: myternovykh@mephi.ru

and such evaluations are sufficient to make recommendations for the further development of the energy complex of the country. Hence, simplifications underlying in these programs correspond to this kind of tasks. Codes COSI, DESAE [6], REBUS3/VARIANT8.0 [7] can also be considered as the codes of similar type.

The code REPRORYV (REcycle for PRORYV), which simulates the processes of loading, processing, and handling of fuel assemblies in the cores of BN-600, BN-800, BN-1200 [8, 9, 10] and other advanced fast reactors was developed to support the PRORYV project. This code can be used to analyze closed nuclear fuel cycles. It allows to study the possibility of the self-sufficient mode of operation in fast reactor. Similar work on the modeling of neutron-physical characteristics of the stationary states of the reactor core with detailed scheme of refueling of fuel assemblies was performed for the BN-600 reactor and was presented in [11]. A third-party software tool is used to calculate fuel irradiation and depletion on each step of fuel burnup. We can use these data to analyze external fuel cycle stages. Any code, which meets the requirements to reach the needed functionality, can be used as the third-party code.

Thus, REPRORYV is composed on modularity principle. In order to find the neutron-physical characteristics of a reactor, user can apply either precise methods (Monte-Carlo) or design methods (in this case diffusion approximation). Codes based on diffusion approximation run faster than those based on precise methods. This allows to evaluate quickly various input parameters of closed fuel cycle. After that more precise calculations of the best options can be made.

Russian code JARFR [12] is applied at present in the REPRORYV. The code is developed for evaluation of neutron-physical characteristics of fast reactors in the multi-group diffusion approximation. It is released with cross sections support based on the of, BNAB 93.01 nuclear data library [13]. The current version of the code JARFR is verified and used to study the characteristics of the project of Russian fast neutron reactors with sodium coolant, and with a variety of fuels - the current BN-600, introduced into the operation of the BN-800 and project of BN-1200.

JARFR can calculate full-scale reactor models with separate fuel assemblies. This program enables to set various properties of different fuel assemblies in the core of the considered reactor. Thus, the program JARFR allows to take into account the history of irradiation of fuel assemblies, their unique isotopic composition at any given moment of fuel lifetime. This is the main feature that distinguishes code REPRORYV from the programs mentioned above used for simulating mass transfer in closed nuclear fuel cycle.

Various schemes of formation of composition for nitride fuel (Pu-U-N) to be loaded into a fast reactor with are discussed in this paper. This type of fuel is planned to be used in BN-1200 reactor. The influence of plutonium relative to uranium fraction on the final result is evaluated. The breeding ratio and the amount of plutonium in the reactor during 60 years of operation are also calculated.

Special input file of JARFR code is required for correct operation of REPRORYV code. Results are calculated by JARFR code on each step based on this input file. To analyze the closed nuclear fuel cycle it is needed to specify some input parameters in REPRORYV input file: cooling time, reprocessing time, fuel reload layout, distribution of isotopes into different groups, and conditions of processing for each group.

Fuel reload layout indicates an order in which fuel assemblies need to be unloaded and replaced (Figure 1). The user can specify the part of the core which is unloaded each year. Detailed example of fuel reload layout is considered below.

```
                         0,
                        1,
                      2,  3,
                    4,  0,  5,
                  6,  1,  2,  3,
                4,  5,  6,  1,  0,
              2,  3,   0, 4,  5,  6,
            0,  1,  2,  3,  4,  5,  6,
          1,  2,  3,  4,  0,  5,  6,  1,
        2,  3,  4,  5,  6,  1,  2,  3,  4,
      5,  6,  1, 2,  3,  4,  5,   6,  1,  2,
    3,  4,  5,  6,  1,  2,  3,  4,  5,  6,  1,
  0,  0,  2,  3,  4,  5,  6,  1,  2,  3,  4,  0,
0,  0,  0,  0,  0,  0,  5,  6,  0,  0,  0,  0,  0,
0,  0,  0,  0,  0,  0,  0,  0,  0,  0,  0,  0,  0,  0,
0,  0,  0,  0,  0,  0,  0,  0,  0,  0,  0,  0,  0,
            0,  0,  0,  0,  0,  0,  0,
```

Figure 1. Fuel reload layout (60 degrees symmetry sector).

It is important to separate isotopes correctly to the groups and then operate with groups of isotopes on the stages of reprocessing. User may choose one of three options for each group as follows:

1. Do not change group concentrations, i.e. retain the mass of group as it was unloaded;
2. Remove the group after recycling (assign concentrations to zero);
3. Assign the initial value of the mass of the group, i.e. add or remove mass of the group so that the mass would be the same as at the first loading. When this option is selected, concentrations of each isotope can be changed while the total mass should be the same as at the initial step.

Thus, the user can allocate separate groups of plutonium isotopes (to change mass to its original value, "removing the excess of plutonium"), uranium (to add mass to the initial level by adding ^{238}U), structure materials (to leave them as they were) and actinides (remove them from reprocessed fuel, or leave them in case of the post-combustion scenario).

Various reload layouts can be introduced according to designer specifications. The effect of plutonium losses on fuel reprocessing stages is studied. Small loss of plutonium is allowable under equipment pollution in processes of separation of fuel fractions and fission products. To specify the input data of REPRORYV, the data from publications on the isotopic composition of fuel, energy release in the core composition and structural materials are used in this work [14, 15].

Uncertainties and errors of calculations

DOI: 10.1051/epjconf/201715307031

Advantage of the code REPRORYV is the possibility of direct calculation of the errors in the concentrations on the basis of errors in the input data. For this purpose, third-party code of nuclide kinetics BPSD (Burning and Poison calculation System with Deviation) [16] is used in REPRORYV code. The BPSD code was developed in the Institute of the Problems of the Development of Nuclear Power of the Russian Academy of Sciences (IBRAE RAS). This code allows to calculate any chains of nuclide transitions both for actinides and fission products. Library RUSFOND (IPPE) simulating up to about 1500 fission products and up to 40 actinides is used as database for BPSD code. Feature of the program is the direct calculation of the above-described errors: errors of the used numerical methods and errors of the input data, which are set inside the program. The result containing numerical values of the errors allows the user to take into account in their work the possible inaccuracies associated with the simulation, to get the confidence interval, the maximum deviation value, etc.

The user of the BPSD code has to assign the input data for the beginning of every step of burnup that is shown in figure 2. These input data are:

• Concentrations of actinides.

• Multi-group (26 groups) cross sections of fission, radioactive capture, (n,2n) and (n,3n) reactions.

• Neutrons spectrum (26 groups).

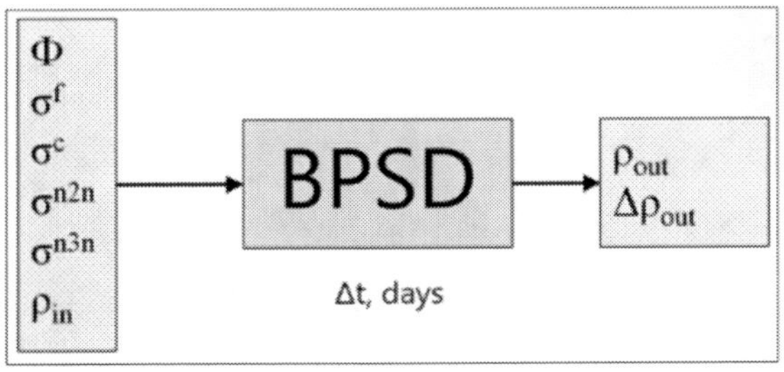

Figure 2. Simplified scheme of the BPSD program with the output of errors of the concentrations

At the moment the application of the BPSD code to estimate the uncertainty of the concentrations for the end of reactor operation, is under testing. This module will allow to include function of estimation of uncertainties of the mass for any time moment of reactor operation into the REPRORYV code. These uncertainties will allow to estimate the error of the resulting functionals, such as reactivity and density effects of reactivity.

Processing

Simulation of the recycling process in the REPRORYV code is based on the assumption that the discharged spent fuel should consist of two fractions: the product and the waste Scheme of processing in REPRORYV code taking into account uncertainties in isotopic composition is shown in figure 3.

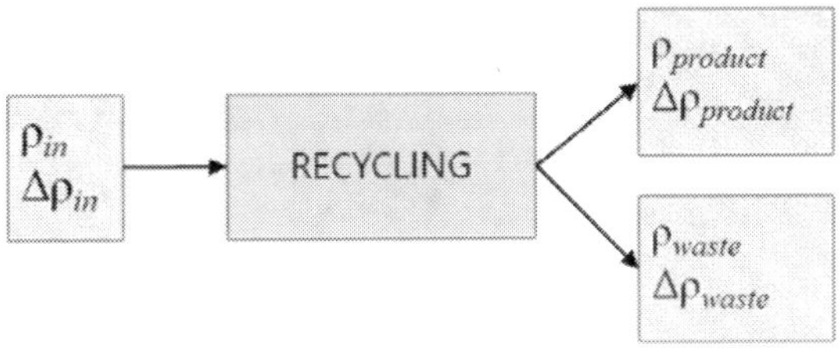

Figure 3. Scheme of processing in REPRORYV code taking into account uncertainties in isotopic composition.

In general case, the schema of processing is nonlinear and depends on many factors. Some of the software codes specifically designed to simulate the process of fuel reprocessing, separation of individual sections of processing lines, calculation of the processing coefficients relying on the chemical reactions, refining, etc. In the REPRORYV code, all processing steps are replaced by one common stage, after which it is already clear what part of the isotopes is the product and what part of the isotopes is waste. The general scheme of closed nuclear fuel cycle simulated in the task by the code REPRORYV is shown in figure 4.

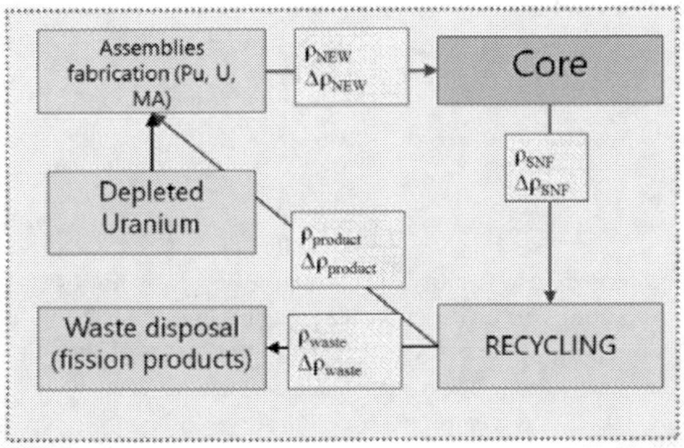

Figure 4. The general scheme of closed nuclear fuel cycle simulated in the task by the code REPRORYV

This simplified scheme allows us to specify vector of coefficients of processing for each isotope, removing the complexity of the scheme of spent fuel reprocessing. After processing the waste is directed to the burial, and the product is used to create new fuel loading for the reactor. Vector of coefficients of the separation of the product from the waste for the major isotopes (~20-30 isotopes) is assigned as the parameters of the separation process. Values of these coefficients of the separation were obtained with account of losses and errors.

DOI: 10.1051/epjconf/201715307031

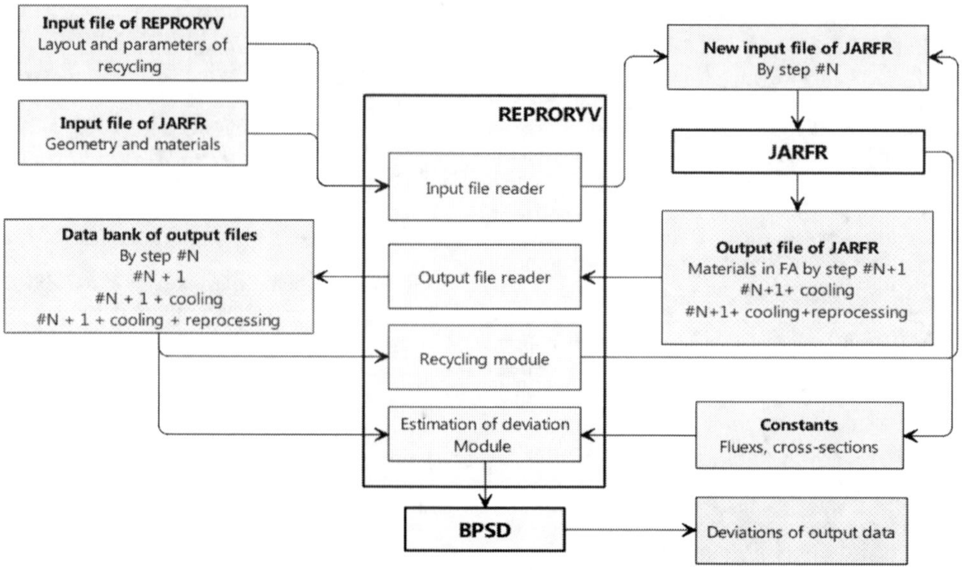

Figure 5. The general scheme of operation of the program REPRORYV

The following equations are used at each step of processing to obtain the new concentration values:

$$\vec{\rho}_{product} = \vec{k}_{recl} \times \vec{\rho}_{in}$$
$$\vec{\rho}_{waste} = \vec{k}_{waste} \times \vec{\rho}_{in}$$
$$\vec{k}_{recl} + \vec{k}_{waste} + \vec{d} = \vec{1}$$
$$|\vec{a}| > 0$$
$$|\vec{b}| > 0$$
$$|\vec{d}| > 0$$

where $\vec{\rho}_{in}$ is vector of concentrations of isotopes at the beginning of the procedure of processing;

$\vec{\rho}_{product}$ is vector of concentrations of isotopes in the product after processing;

$\vec{\rho}_{waste}$ is vector of concentrations of isotopes in the waste after processing;

$\vec{k}_{recl}$ is vector of coefficients of processing. Each component of this vector contains coefficients of processing for each of the major isotopes. Coefficient of processing means the share of the transition of each isotope in the product;

$\vec{k}_{waste}$ is vector of the coefficients of the transition to waste. Each component of this vector contains coefficients of transition to the waste for each of the major isotopes. Coefficient of transition to waste means the share of the transition of each isotope into the waste;

$\vec{d}$ is vector of the share of losses in the processing specified by the user.

It is obvious that each element of these vectors is greater than zero, and the sum of these vectors is equal to a unit vector (the law of conservation of mass).

There are certain difficulties in obtaining accurate values of losses at each operation of the processing. However, assignment of approximate values dives a possibility for the application of the algorithm with accurate data, which can be obtained experimentally directly in the processing plant.

After starting the program with all assigned required characteristics of the fresh fuel, irradiation parameters, cooling time, etc., the output file will contain the isotopic composition of the fuel for loading into a new reactor, its mass, neutron-physical characteristics of all active core and its parts, the errors of the obtained values.

The general scheme of operation of the program REPRORYV is presented in figure 5.

2 Core reloading layout

Reloading layout of the fuel assemblies in a BN-type reactor core was accepted as follows. Four different variants T1-T4 of plutonium mass content relative to uranium in the initial nitride fuel were considered. It was assumed that plutonium content was 12.6% (T1), 12.8% (T2), 13% (T3), and 14% (T4). Plutonium isotopic composition corresponded to that of discharged from the spent nuclear fuel of VVER-440 reactors [14,17] in all four cases. Initially fresh fuel assemblies were loaded in the core. In the first step of reloading after 330 days of operation, 12 fuel assemblies of 72 (1/6 part of the core) marked by index of "1" (first lot of fuel) in Figure 1 were unloaded and placed into intrareactor storage for cooling untill the activity of these fuel assemblies would decrease to the level when they could be reprocessed. This time of cooling was evaluated as two years in our case. New fuel assemblies with the same initial composition were placed to the positions of unloaded fuel assemblies during following 35 days. The time of operation during 330 days and time for reloading of 35 days were assumed for all steps of reloading. The total time was equal to one year. In the second step of reloadong after next 330 days of operation, fuel assemblies marked by index of "2"(second lot of fuel) were unloaded for cooling and replaced by new fuel assemblies during 35 days. Then reactor operated again during 330 days and so on. After two years of cooling, fuel assemblies of type "1" that have been loaded first, were sent for recycling. Several variants of reprocessing were considered. Actinides could be removed from the fuel (S1) or could be retained in the fuel (S2). In the process of separation of fission products, a small amount of plutonium could be extracted along with fission products. The amounts of plutonium as impurity to separated fission products of 0.1% (R1), 0.5% (R2), and 1% (R3) were considered. The time of reprocessing was assumed as one year. In the third step of reloading the fresh fuel assemblies were place to position "3" and the reactor operated for following 330 days. Finally in the fourth step of reloading, recycled fuel which was reprocessed over the last year was placed to position "4" (on the contrary to the fresh fuel as in the previous steps), composed of the first lot of fuel. Then, in following steps of reloading, fuel assemblies that have been unloaded and reprocessed after the second step (second lot) would be placed in position "5", after the third step - in position "6", etc. Thus, the reactor comes to steady state mode of operation, if the vector of plutonium (the fraction of each plutonium isotope in the mass of plutonium) at each following step of the refueling of the active core remains constant (self-sufficient plutonium mode)

3 Results

Geometry parameters and isotopic composition of the model of a BN-type reactor core were used as input for JARFR to obtain the results on multiplication factor and mass of plutonium at the end of the time interval. Fuel reloading layout and modes of recycling were specified in REPRORYV code. Results for T1, T2, T3, T4 variants were obtained using options S1 + R1. Results of different S and R variants were obtained using T3 variant. Each fuel assembly was homogenized in the active core. All results correspond to fuel density 11.5 g/cm^3. The reactor was set to the critical state (Keff =~ 1) by changing plutonium content in the active core only in the initial period of time. The tolerance of calculation of the multiplication factor in each case was equal to 10^{-5}. The rule "mass of the unloaded fuel assemblies should be equal to the mass of the loaded fuel assemblies in each of the cycles of refueling" was accepted for the formation of the new fuel composition by processing.

Figure 6 shows the dependence of the reactivity of the reactor core (in percent) on time (the lifetime of the reactor is 60 years) for different values of the plutonium share in the fuel in the initial moment of time. It is seen that all four types of fuel tend towards similar levels of reactivity. It should be noted that at the initial time moment the reactivity value for fuel with plutonium fraction of 14% is 3.8 times higher with respect to the value of reactivity for fuel with 12.6% of plutonium. Nevertheless we can conclude that the optimal value of a share of plutonium will be a value from 12% to 14%, because reactivity change during the reactor operation will be the smallest one just for this range of plutonium share. For this reason, there will be no need in high margin of reactivity in the control rods and their frequent moving at large amplitudes in the active core. This increases the overall safety level of the reactor plant.

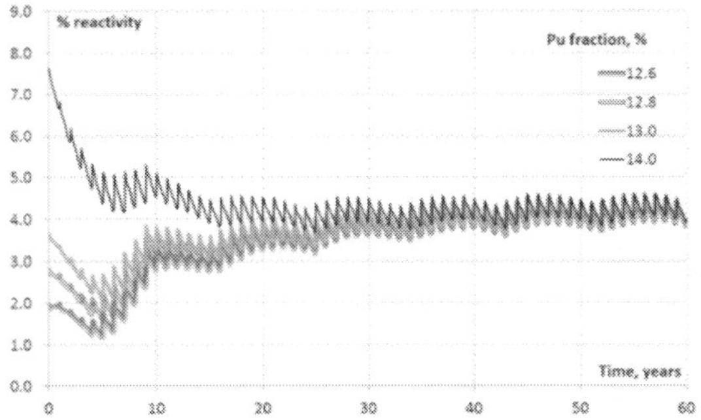

Figure 6. The dependence of reactivity of the core on time (the life cycle of the reactor is 60 years) for different values of plutonium share in fuel in the initial time

If we consider this dependence in the initial 3 years as shown in figure 7, we can see more smooth dependence of reactivity during one cycle (one year of reactor operation between refueling) for active core with initial share of 12.6% plutonium in fuel assemblies. Thus, it is possible to assume that if we remove accumulated excess of plutonium mass from the active core of the reactor, the excess reactivity in the core will be close to zero. Thus, after 60 years the difference between three types of plutonium loads actually disappears. Changing of reactivity between reloadings is the same. This can be seen in figure 8 for the end of reactor operation.

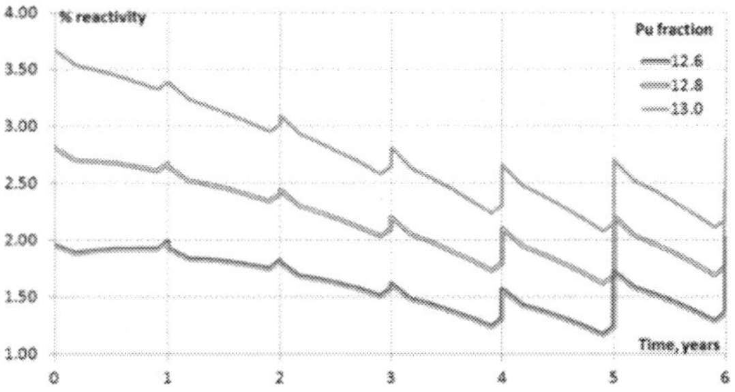

Figure 7. The dependence of reactivity of the reactor core on time in the beginning of the campaign (3 years) for different values of initial plutonium share in fuel

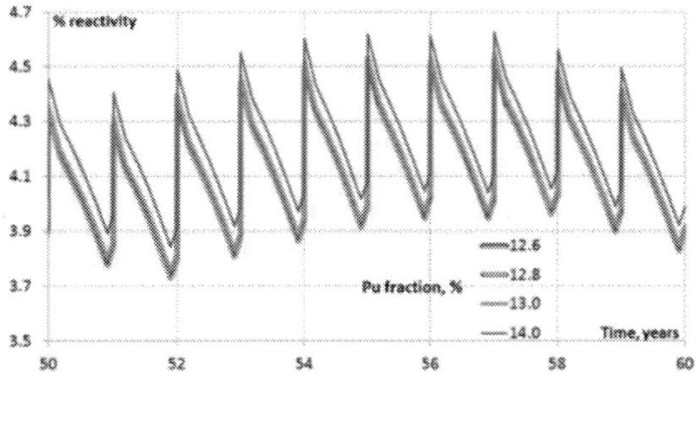

 DOI: 10.1051/epjconf/201715307031

Figure 8. The dependence of reactivity of the reactor core on time at the end of reactor operation (from 50 to 60 years) for different values of initial plutonium share in fuel

Figure 9 shows the dependence of the mass of plutonium in the reactor core at the end of each microcampaign depending on time for different loads of active core containing different shares of the plutonium in the total mass of fuel. It is seen that all the four active cores tend to some equilibrium value of the rate of plutonium accumulation. At the initial moment of time the ratio of the mass of plutonium when passing from 14% to 12.6% makes 1.1106. At the end of reactor operation this ratio is only 1.0019.

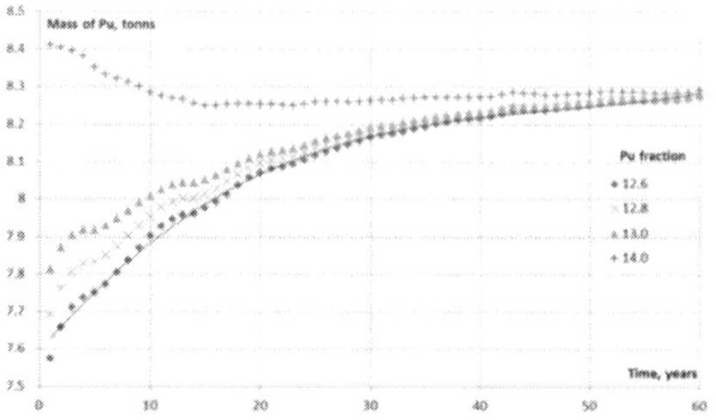

Figure 9. The dependence of the mass of plutonium in the reactor core at the end of each microcampaign on time for active cores containing different shares of plutonium in the total mass of the fuel.

Small fluctuations of plutonium mass are associated with the fact that in the first three microcampaign, fuel lots are loaded with initial content of plutonium (i.e. as it was the content at the beginning of reactor operation) instead of unloaded plutonium. Figure 10 shows the time-dependence of the plutonium mass loaded in each step from the same fuel lot. Since the time of irradiation of one fuel lot in the reactor is 6 years, followed by 2-year cooling in interim storage and 1-year processing, it is obvious that this fuel lot (without fission products and minor actinides) can be loaded again every 9 years. Therefore, the neutron-physical characteristics and the masses of isotopes will decrease shortly in every 9 years due to the fact that the plutonium is not added to reactor during the first three years. The black line in figure 12 shows the trend line. It is seen that, as for reactivity, the value of the mass of plutonium loaded in each fuel lot in the reactor, goes to the equilibrium value (mass does not change as much as at the initial time).

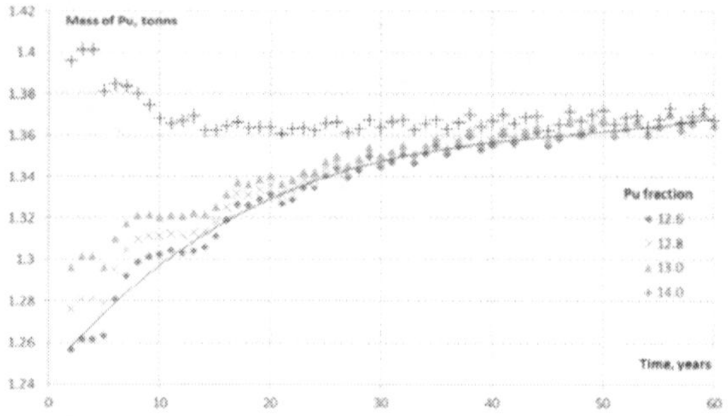

Figure 10. Dependence of plutonium mass downloaded at each step in one fuel lot on time for active cores containing different shares of plutonium in the total mass of fuel.

This rule is met also for fuel lots. In each fuel lot after 9 years it is necessary to add a smaller amount of plutonium to support the operation of the reactor (for the same mass of each fuel assembly, see table 1). From 4-th cycle the "excess" of plutonium accumulated in an amount of not more than 1%.

Table 1. Change (%) of the mass of plutonium during its production in each fuel lot during the reactor life cycle

	No of campaign						
Lot	1	2	3	4	5	6	7
1	-	-3.8	-2.1	-1.3	-0.7	-0.5	-0.3
2	-	-3.3	-1.8	-1.3	-0.6	-0.5	-0.2
3	-	-3.3	-1.9	-1.4	-0.6	-0.6	-0.2
4	-	-3.4	-2.2	-1.2	-0.7	-0.4	-0.3
5	-	-2.4	-1.8	-0.9	-0.7	-0.3	-0.3
6	-	-2.1	-1.6	-0.8	-0.7	-0.2	-
7	-	-2.1	-1.3	-0.8	-0.5	-0.3	-
8	-	-1.9	-1.0	-0.8	-0.3	-0.4	-
9	-	-2.1	-1.1	-0.8	-0.3	-0.4	-

To comply with the law of conservation of mass of unloaded and loaded fuel assemblies, the mass of depleted uranium added during the processing was varied. Since the mass of plutonium is increased even with the account of its burning due to fission reactions, the total mass of uranium in the active core decreases with time, see figure 11. Fluctuations in the uranium mass arise in this case for the same reason as fluctuations of plutonium.

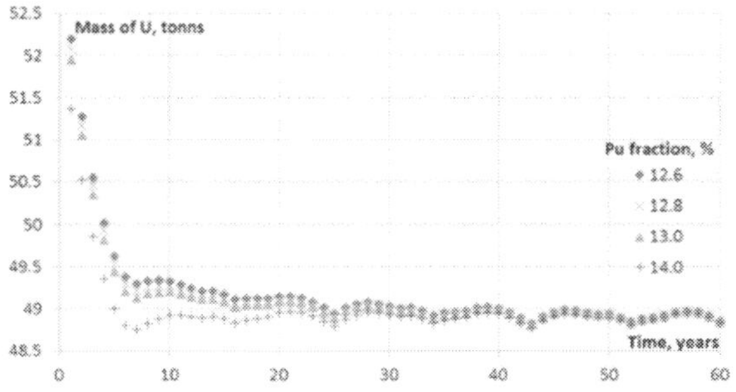

Figure 11. The dependence of uranium mass in the reactor core on time at the end of each microcampaign for active cores containing different shares of Pu in the total mass of the fuel.

Figure 12 presents dependence of reactivity of the reactor core on time for different values of plutonium loss during processing. In the initial time, dependence of the reactivity is the same for different values of plutonium loss. The influence of plutonium losses on neutron-physical characteristics will be maximum only at the end of reactor campaign. The ratio of the values of reactivity corresponding to loss of plutonium of 0.1% and 1.0% at the end of the reactor operation is 1.13.

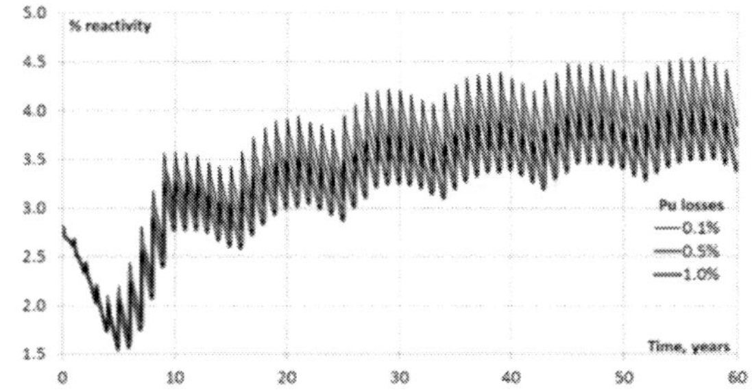

Figure 12. The dependence of reactivity of the reactor core on time (the life cycle of the reactor is 60 years) for different values of plutonium loss during processing.

Figure 13 shows the dependence of reactivity of the reactor core on time (campaign of the reactor is 60 years) when minor actinides are removed from the fuel or keep in fuel.

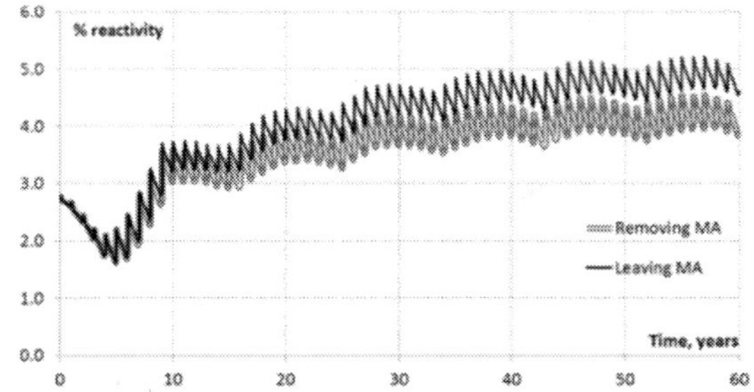

Figure 13. The dependence of reactivity of the core on time (the life cycle of the reactor is 60 years) for removal of minor actinides and leaving them in the fuel.

In the initial time, values of reactivity to these two variants coincide. For the end of reactor operation, actinides increase the reactivity compared to case of their complete removal during the processing by 1.18 times.

4 Conclusions

The paper deals with evaluation of fuel handling in fast reactors in the closed nuclear fuel cycle. To solve these problems, the REPRORYV program code is developed. It simulates the nuclide streams in out-of-reactor stages of the closed fuel cycle. Existed verified code JARFR is used for the calculation of neutron-physical characteristics of the core. Various options for nuclide streams are considered with a representative full-scale model of a fast reactor with sodium-cooled high-power. The changes of multiplication factor and mass of plutonium in the closed nuclear fuel cycle for a Russian BN-type fast reactor with use of the REPRORYV code are evaluated. Different scenarios of fuel reprocessing assuming the removing or retaining of actinides, taking into account various plutonium losses on reprocessing steps were considered. The questions of influence of the initial content of plutonium in nitride fuel (Pu-U-N) and the impact of the initial parameters to the possibility of fuel self-sufficient mode of the reactor during the whole period of its operation are studied.

References

1. V.P. Alferov, A.I. Radaev, M. V. Shchurovskaya, G. V. Tikhomirov, N.A. Hanan, F.A. Van Heerden, Ann. Nucl. Energy 77 (2015) 273–280.
2. A.S. Gerasimov, V.N. Kornoukhov, I.S. Sald'ikov, G.V. Tikhomirov, At. Energy 116 (2014).
3. B.R. Bergelson, V.V. Belonog, A.S. Gerasimov, G.V. Tikhomirov, At. Energy 107 (2009).
4. Physics and Safety of Transmutation Systems. A Status Report © OECD 2006 NEA No. 6090.
5. REFCO83, Nuclear Fuel Cycle Cost Economics Using Discounted Cash Flow Analysis. http://www.oecd-nea.org/tools/abstract/detail/nesc1065/
6. E. Andrianova., V. Davidenko, V. Tsibul'Skii.. Atomic Energy, 105 (6), 303-306 (2008).
7. REBUS3/VARIANT8.0, Code System for Analysis of Fast Reactor Fuel Cycles. http://www.oecd-nea.org/tools/abstract/detail/ccc-0653/.
8. V. Poplavsky, A. Tsiboulia, Yu. Khomyakov, V. Matveev, V. Eliseev, A. Tsikunov, B. Vasiliev, S. Belov, M. Farakshin. Core design and fuel cycle of advanced fast reactor with sodium coolant. *International Conference on Fast Reactors and Related Fuel Cycles - Challenges and Opportunities* - December 7-11, 2009, Kyoto, Japan.
9. B. Vasilyev, S. Shepelev, M. Ashirmetov, V. Poplavsky. BN-1200 Reactor Power Unit Design Development. *International Conference on Fast Reactors and Related Fuel Cycles: Safe Technologies and Sustainable Scenarios*. Paris, France, 4-7 March 2013.

10. A. Vasiliev, P. Fomichenko, P. Alekseev, V. Nevinnitsa, S. Belov, B. Vasiliev, V. Evseev, M. Farakshin. Preprint Kurchatov Institute IAE-6275/4 (2003).
11. A. I. Nevinitsa, V. I. Matveev, V. Yu. Stogov, V. A. Cherny, Problems of Atomic Science and Engineering, Series: Physics of Nuclear Reactors, 2, 53-57 (2011).
12. L.N. Yaroslavtseva. The JAR Code System for Nuclear Reactor Neutronics Calculations. VANT, issue 8 (37), 1983, pp. 41-43.
13. G. N. Manturov, M. N. Nikolayev, A. M. Tsybulya, Problems of Atomic Science and Engineering, Series: Nuclear Constants, 1, 4-12 (1996).
14. R. Prodanova. *Calculation of isotope composition of WWER- 440 spent fuel assembly by the NESSEL-NUKO code system on the basis of the ISTC burn up credit project data.* Institute for Nuclear Research and Nuclear Energy Bulgarian Academy of Sciences
15. S. B. Belov, A. V. Kiselyov, E. V. Marova, M. R. Farakshin, M. N. Zizin, P. A. Fomichenko, K. F. Raskach, M. Yu. Semyonov, Problems of Atomic Science and Engineering, Series: Physics of Nuclear Reactors, 4, 77-82 (2014).
16. A. Belov, E. Seleznev. Atomic Energy, 108, 321-324 (2010).
17. B.R. Bergelson, V.V. Belonog, A.S. Gerasimov, G.V. Tikhomirov, VVER nuclear fuel burnup with different absorbers, At. Energy. 109 (2011). doi:10.1007/s10512-011-9351-2.

Analysis of gamma ray dose for dried up pond storing low enriched UO_2 fuel

Yasushi Nauchi[1,*], and *Motomu* Suzuki[1]

[1]Central Research Institute of Electric Power Industry, Nuclear Risk Research Center, 2-6-1, Nagasaka, Yokosuka, Japan.

Abstract. Gamma ray dose is calculated for loss of coolant accident in spent fuel pond (SFP) storing irradiated fuels used in light water reactors. Influence of modelling of fuel assemblies, source distributions, and loading fraction of fuel assemblies in the fuel rack on the dose are investigated.

1 Introduction

In systems of light water reactors, discharged spent fuel assemblies (FAs) are stored in racks of a spent fuel ponds (SFP). A FA is stored until its radioactivity decays out below the limit for shipment. A SFP is filled with water which acts as coolant to remove the decay heat and shielding material against intense radiation from the FAs.

By the nuclear disaster occurred at Fukushima Dai-Ichi (1F) nuclear power station (NPS) on March 11, 2011, risk of loss of coolant accident in SFP (SFP LOCA) has been recognized. In 1F, the residual heat removal systems (RHR) in unit 1, 2, 3, and 4 were stopped due to the station black out. Even worse, coolant feeding systems were broken by hydrogen explosion in some of the units. Due to larger decay heat, the temperature of the coolant in unit 4 was raised even after the destruction of ceiling of reactor building (R/B). Therefore, decrement of the coolant level by evaporation was anticipated. In unit 4 of 1F, its decay heat removal function was recovered by feeding water from a concrete pumping vehicle prepared outside R/B and the SFP was prevented from LOCA, at last [1].

After the accidents, SFP-LOCA becomes one of the design base accidents (DBA) in Japanese NPS [2]. In addition to evaporation following to malfunction of RHR, leakage of coolant by siphon phenomenon initiated by destruction of outlet pipes of SFP cooling and cleanup systems is also lined up on the anticipated sequence to SFP LOCA. In the both cases, utilities have to assess what is happened when active region of FAs is exposed to air. The containment capability of fuel rod would be either intact or not depending on decay heat of FAs, flow of the dry air or vapour, and loading pattern of FAs in racks of SFP. The hazard of internal exposure due to intake of fission products must be estimated only for the case of fuel failure. Contrarily, the hazard of external exposure has to be estimated no matter whether the fuel is intact or not. However, systematic analyses of radiation dose have not been well performed for SFP LOCA.

In this study, we modelled FAs and a SFP and analysed radiation dose at the top height of SFP since the radiation dose at the level would be the source of skyshine radiation. Through the analyses, important parameters to be assessed for the radiation dose were outlined.

2 Base model and code

Under the condition where the containment capability of the claddings is intact, all radioactive materials are still remained in the fuel rods. Therefore, radiation transport from source inside the fuel pellets to the top height of the SFP is considered in this work. In this work, a design of PWR 17 x 17 type fuel assembly and a fuel rack was assumed based on published data. Source radiations were calculated for typical burn-up and cooling time. Gamma ray transport was calculated with a continuous energy Monte Carlo code.

2.1 Calculation geometry and component

Radiation dose at the top height of the SFP is determined by geometry of FA, the rack, and the SFP. The calculation models of them are described.

2.1.1 Fuel Assembly

Actual design information of PWR fuel assemblies are not published except for horizontal array of uranium oxide fuel, cladding, control rod guide tube. Currently, 3 dimensional neutronics calculations have been focused on in the reactor physics field and benchmark models are proposed for the calculations. Among them, BEAVRS benchmark model gives information of intermediate grid spacers, plenums, plenum springs, plugs of rods, and upper and lower nozzles [3]. Since shielding effects of those components are worth to be quantified in this study, the assembly model in the BEAVRS benchmark was

* Corresponding author: nauchi@criepi.denken.or.jp

adopted. The components are listed up in table 1 and upper part of the FA is shown in Fig. 1.

Table 1. Modelled components and material in fuel assembly

component	material
fuel	UO_2
cladding	Zry-4
end plug	Zry-4
Spacer	Inconel 718
	Zry-4
plenum	Helium
plenum spring	Inconel 718
Space	air
Lower/upper nozzle	SUS-304

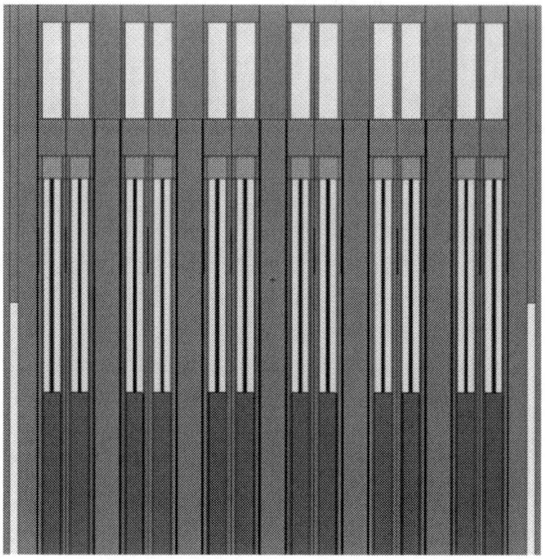

gray: dry air purple: Inconel-718
yellow: SUS-304 magenta: UO_2
Green: Zry-4 Pink: He gas

Fig. 1. Vertical model of top part of fuel assembly.

A fuel rod was modelled by combination of cylinders. A stack of uranium oxide (UO_2) fuel pellets were modelled by a cylinder rod of the active fuel height (366 cm), in which dishes or chamfers were smeared. As described later, only the gamma ray dose was considered in this work so that only atomic number densities of uranium and oxygen were assigned in this work. A fuel cylinder rod was enclosed in a rod made of Zircaloy – 4 (Zry-4). The plenum spring of spiral shape, actually, was modelled in a co-axial cylinder rod. The space in the rod was filled with helium gas. The both ends of rod were plugged.

In the PWR 17 x 17 type FA, 264 fuel rods, 24 control rod guide tubes, and a instrumentation guide tubes are assembled in square shaped array of which fuel rod pitch is 1.26cm. The pitch is fixed by the grid spacers. There are axially 8 grid spacers in a FA. They were modelled by square pipes. The top and bottom end of the spacers were modelled by Zry-4 and others were by Inconel-718.

The axial position of the spacers are fixed and suspended by control rod guide tubes. The tubes are fixed on upper and lower nozzles. The nozzles were also modelled in array of cylinder columns. Each column was set coaxially with a fuel rod.

2.1.2 SFP and rack

The FAs are put into cells of racks fixed on a floor of a SFP. The major role of the rack is to prevent from criticality incidents by keeping a distance between the adjacent assemblies. There are several types of rack employed in Japanese utilities. Among them, the rack made of stainless steel (SS) containing boron (B-SUS) has been popular after the re-racking strategy was taken to enhance capacities of SFPs. In the representative B-SUS rack, square pipes were arrayed in square shaped lattice of cell. Although detailed geometries of a rack and a SFP have not been published, we made model geometry as a base case referring to the literature [4].

In the model, the thickness of the wall of the square pipe was 4 mm and the wide of the pipe was 24 cm. The height of the rack was 4.4 m and the bottom of the fuel assemblies was assigned 30 cm above the bottom of the rack. The B-SUS plate was approximated as SUS-304 since contained boron influences little on the gamma ray transport. The horizontal cell pitch was 28 cm. The square pipes located at the center of the cells are separated by spacers equipped at bottom, intermediate, and top of height of the rack, actually. However, such kinds of spacer or joint fittings were assumed as air in the base model. The size of the SFP was set 9 m x 14 m, horizontally, referring to the literature [4]. Considering the cell pitch, the capacity of the rack was determined as 46 x 27 FAs. In the base case, all cells are filled with FAs.

red: concrete, gray: dry air
white: vacant cell, black: cell loaded with fuel

Fig. 2. Vertical model of fuel assembly and spent fuel pond.

The vertical geometry was shown in Fig. 2. The height of the SFP was set 12.4 m. The wall and bottom floor of SFP was concrete of 2 m in thickness. Metal liner on the inner surfaces of the SFP was neglected. The top height of SFP was located on the same level of the surface of the operating floor. The floor was made of concrete and its thickness was 30cm. In the Base model, the ceiling or the walls of R/B were not modelled and the operating floor was assumed to be extended to the horizontal calculation boundary. The boundary size was 360 m x 360 m around the center of SFP. Vertically, the lower boundary of the calculation was the bottom of the floor of the SFP and the upper boundary was 2 km above the operating floor. The SFP and the space above the operating floor were filled with dry air. The temperature of the all materials was assigned 293 K.

2.2 Source condition

2.2.1 Fuel Assembly

The source strength and its spectrum of gamma rays from FAs were calculated with the ORIGEN-ARP module of the SCALE-6.1 system [5]. In the calculation, the library for PWR 17 x 17 type FA (Westinghouse) was used. No gadolinium rod was considered. Initial ^{235}U enrichment was assigned 4.1% and the burn-up of the all assemblies were monochromatically determined 48 MWd/kgHM. The cycle length was set 420 days and the power of a FA was assumed constant during the 3 cycles. The cooling time was set 5 years. These kinds of FAs are widely used in Japanese utilities.

Calculated neutron and gamma ray yields were 8.24 x 10^8 /s/tHM and 1.56 x 10^{16} /s/tHM, respectively. The former is 7 ~ 8 orders of magnitude less than the latter. Accordingly, the neutron radiation dose from the spent FA is considered negligibly small. Hereafter, only the gamma ray dose was estimated for radiations by the SFP-LOCA. The gamma ray spectra were estimated in 47 energy groups structure as shown in Fig.3.

2.2.2 Gamma ray source distribution

In the present work, the number of gamma ray emission per FA and that per fuel pin were assumed constant over all FAs although burn-up of FAs and fuel pins varies in some extent. Inside the fuel pellet, gamma ray emission points were sampled uniformly.

Contrarily, the axial distribution of gamma ray emission was modelled in detail since the attenuation of gamma rays by UO_2 is significant in the estimation of radiation dose. In literature [6], radio activities of ^{134}Cs, ^{137}Cs and ^{106}Ru had been measured axially with fine spatial meshes of 2.5 cm for a spent PWR fuel rod. By averaging the count rates of gamma rays from the three kinds of nuclide, a model of axial distribution of gamma rays was evaluated in this work as shown in Fig. 4. The active length of fuels (366 cm) in the BEAVRS benchmark is longer than that of gamma ray measured length (345 cm). In this work, the measured gamma ray distribution was shifted upward so that the top part of the active region of the fuel in the base model radiate gamma rays. Oppositely, the bottom part of 20cm would not radiate gamma rays in the base model. The deviation of height of the grid spacers between the geometry model and the gamma ray source distribution one was neglected.

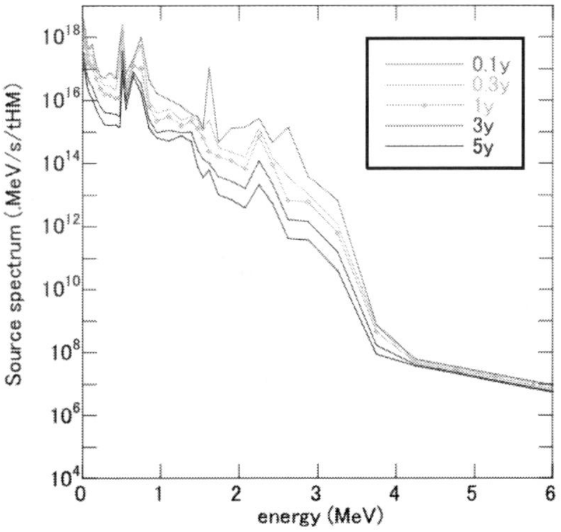

Fig. 3. Gamma ray spectrum from spent fuel in relation to cooling year. Base case is for cooled time of 5years.

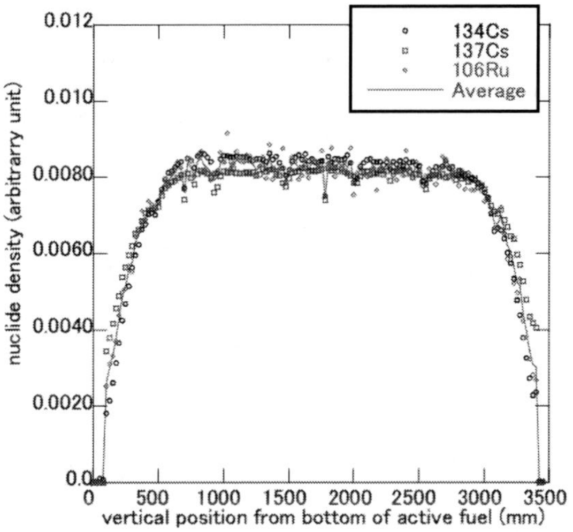

Fig. 4. Axial distribution of gamma ray emission in original literature [6].

2.3 Gamma ray transport calculation

Gamma ray transport calculations were performed with the MCNP-5.1.30 code [7]. The coupled transport calculations with either neutron or electron were neglected. The lower energy threshold of the gamma ray transport was set 10 keV. Above the energy, the transport was modelled in the "simple mode" where only the photo - electric effect, the Compton scattering, and the pair creation processes are considered. The mcplib04 library was used as the cross section [7].

Gamma ray source are assigned as described in section 2.2. The number of source employed was from 100 to 400 milion.

The surface type tally was adopted to calculate average gamma ray flux at the top height of SFP. No bias sampling techniques such as the weight window were applied. The flux to dose conversion factor was referred to ICRP pub 74 [8]. The calculated doses are given per source gamma ray emission. The relative errors of calculated doses were less than 1% for most of calculations in this work.

3 Gamma ray dose calculation and results

In this section, the radiation dose is calculated for the base case. Then, relative to the base case, sensitivities of parameters were studied.

3.1 Source model

3.1.1 Source spectrum

By MCNP calculation, doses are calculated per source gamma ray. The absolute doses are obtained by multiplying the MCNP results by the source intensity. Varying the coolant time, the source intensity and the spectra were calculated with the ORIGEN-ARP module. The spectra are shown in Fig. 3 together with for the base case which is cooled 5 years. With the spectra, the gamma ray doses were also calculated. The results are written in Table 2.

Table 2. Source intensity and spectrum effect on dose

Cooling year	*dose/source	*source intensity	*radiation dose
0.1	1.25	23.65	29.64
0.3	0.95	12.65	12.01
1	0.74	5.18	3.81
3	0.90	1.81	1.63
5	1.00	1.00	1.00

*data are normalized to base case (cooled 5 years)

As it is well known, the source intensity declines so rapidly with time during the 1st year after discharge. Even after a year, the intensity decreases from 520 % to 100 %. Therefore, the variation of intensity due to the decay time is essential for the dose estimation. Contrarily, the does per source varies not monotonically with decay time. As shown in Fig.3, the principle emitter of gamma rays varies within a short decay time and the spectrum varies so much. The cooling time effect on the dose per source during a period from 0.1 to 5 years was found about ±25%. The effect is not principle but significant for the precise estimation of the radiation dose.

3.1.2 Axial distribution of source

Sensitivity of the number of divisions for the axial distribution of the source was studied here. The sizes considered here were

1) Base case with the fine meshes of 2.5cm height
2) 24 coarse meshes of 15.24cm height
3) flat distribution along the active height

The distribution of model 2) was estimated by collapsing the fine distribution in model 1). The coarse mesh of the model 2) is widely used for estimating axial burn-up profiles in the reactor physics field. Estimation methods of the gamma ray yield and the spectra are easily to be implemented with a little effort for model 2). The model 3) is the easiest way to assume.

The calculated doses for model 2) and 3) are compared to that of model 1) in Table 3. In model 2), the gamma ray dose is increased by 33%. In model 3), radiation dose is done by 145%.

Table 3. Effect of axial distribution model of gamma ray emission on radiation dose

model	Relative Dose
Base Case	1.000
24 division	1.328
Flat	2.446

*data are normalized to base case

In order to study the source distribution effect, further analyses were performed. For the fine meshes applied in model 1), we put gamma ray sources at a certain height and estimated the dose. The calculated value can be regarded as the source importance at the height on the radiation dose. As shown in Fig. 5, the source importance decreases by an order of magnitude only within 15cm region from the top of active fuel (TAF). This is due to the fact that UO_2 is good shielding material for gamma ray. The slope in Fig. 5 becomes gentle as distance from TAF. In energy region where the pair creation is not dominant, the gamma ray attenuation length is shorter for the lower energy gamma rays. Within a short distance, lower energy gamma rays would be attenuated out. Accordingly, the gamma ray dose by a source at lower height would be dominated by higher energy gamma rays. The gentle slope in the lower region might be due to the attenuation of the higher energy gamma rays.

Due to the attenuation of gammas rays along vertical axis, the meshes of source of 15.24 cm length is considered still coarse. To take model 3) for the analysis of gamma ray dose during SFP-LOCA introduces extra conservative dose. If they can only take model 2) or 3) for the gamma ray source, some kinds of correction factor should be prepared.

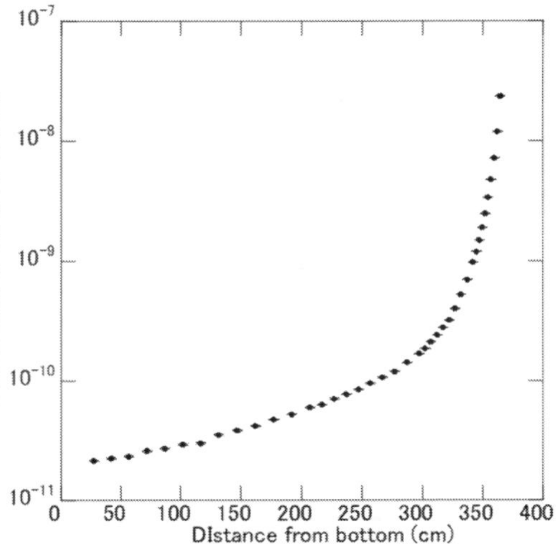

Fig. 5. Axial source importance on radiation dose.

3.2 Shielding effect of components

The shielding effect of grid spacers, upper materials above TAF, and walls in the rack cells were calculated for the SFP-LOCA condition. By changing each component in the base case to dry air, variation of the radiation dose was calculated. As listed in Table 4, the upper materials above TAF reduce the dose by 126% (2.263 →1) since the upper nozzle made of stainless steel, springs made of Inconel-716, upper end plugs and cladding tube made of Zry-4, have significant mass. Also, the effect of the rack cell wall is significant. In order to reduce the extra conservatism, those should be modelled. Contrarily, the effect of the grid spacers is not significant.

Table 4. Shielding effect of component

shielding material	Relative Dose
Base Case	1.000
Grid spacer	1.024
materials above TAF	2.263
rack cell wall	2.008

*data are normalized to base case

3.3 Partial loading

For SFP-LOCA, reduction of the loading fraction of FAs in the rack would be measure to reduce the maximum temperature of claddings to keep integrity of fuel rods. Generally, reduction of the fraction results in decrement of the total gamma ray source in SFP and increment of the source importance. In this section, the influence of the fraction was studied.

3.3.1 Random loading

For the rack of 46 x 27 cells in the base model, a loading fraction ξ_1 was uniformly sampled between 0 and 1. Then for the fraction ξ_1, (1242 x ξ_1) cells were filled with FAs and the other (1242 x (1 - ξ_1)) cells were kept vacant. The filled and vacant cells were also determined randomly as shown in Fig. 6. For tens of sets of random variables, loading patterns of FAs were determined and the radiation doses were calculated. The results are shown in Fig. 7. The gamma ray dose takes a maximal value when the loading fraction is 0.45~0.5. Although the number of gamma ray emission is proportional to the fraction, the gamma ray penetration probability increases with the decrement of the fraction. Compared to the base case, the gamma ray dose is increased up to 170%. For the conservative modelling, a kind of optimum search of the loading pattern would be required.

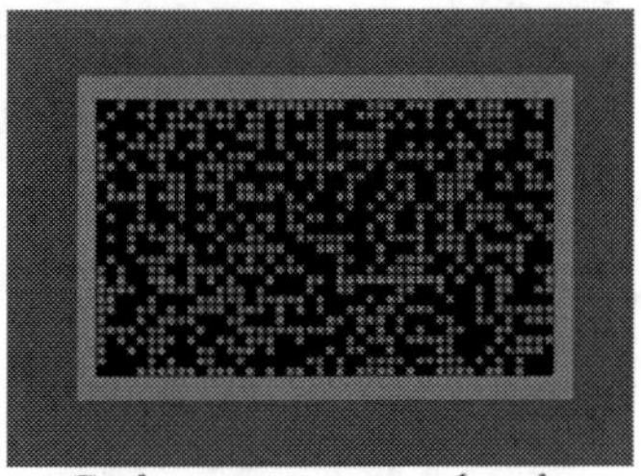

Red: concrete, gray: dry air
Gray in lattice: vacant cell, black: cell loaded with fuel

Fig. 6. Example of random loading of fuel assembly in spent fuel pond

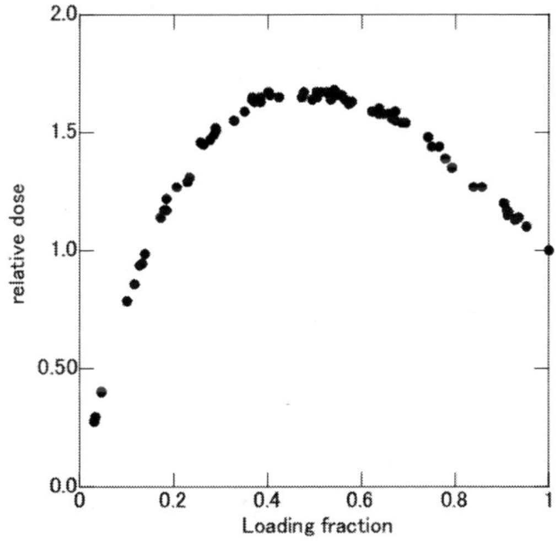

Fig. 7. Gamma ray dose in relation to the loading fraction of fuel assembly in rack. Loading pattern is randomly sampled.

3.3.2 Systematic loading

Beginning with the base case for the rack of 46 x 27 cells, other fuel loading patterns were made by discharging or charging FA systematically.

 DOI: 10.1051/epjconf/201715307032

1) Downsizing the fully charged area as 44 x 25 → 42 x 23→ 40 x 21→ …→20 x 1.
2) Discharging inner most layer of FAs as shown in Fig. 8.
Also, additionally two series of loading pattern were considered.
3) Checker board loading of FA by changing checker size.
4) Multi layer of vacant and FA rows.

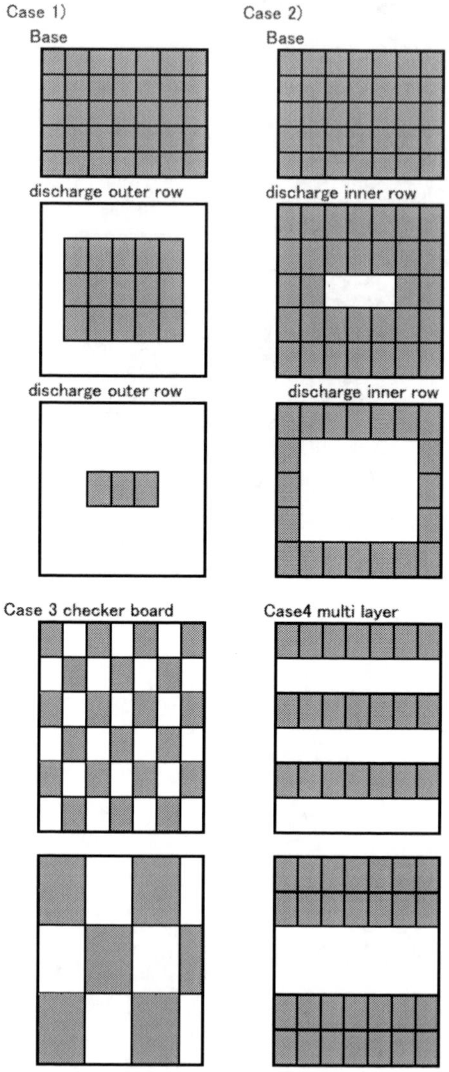

Fig. 8.Fractional loading pattern of fuel assemblies

In Fig. 9, calculate dose and dose per source are shown relative to base case for case 1) and case 2). In both cases, the dose per source increases as FAs are discharged since space is given for gamma ray streaming upward by discharging FAs. However, the increment does not well compensate the decrement of source by discharging. Thus, increment of gamma ray dose like the case of random discharging in Fig. 9 is not observed.

In table 5, calculated gamma ray doses for case 3) and 4) are listed. Contrarily to the case 1) and 2), radiation dose increases. Both in case 3) and 4), the gamma ray dose decreases as the size of condensed FAs or the number of rows of FAs are increased. As the size is increased, the number of FA facing 3 or 4 FAs is increased. Such inner FA would not contribute so much to radiation dose at the top height of SFP.

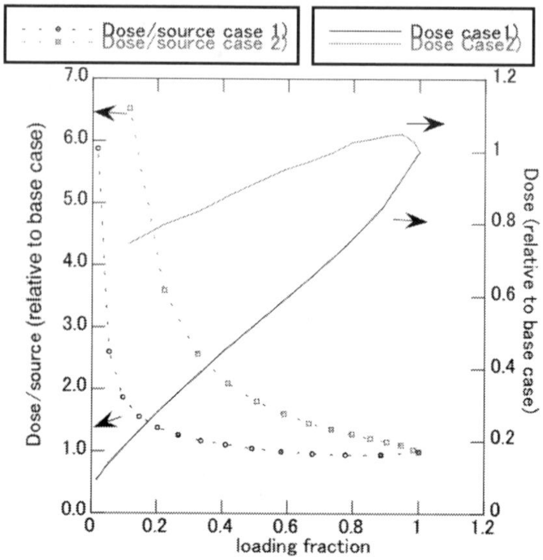

Fig. 9. Gamma ray dose for systematic loading of fuel assembly. Case 1) Downsizing of fuel array. Case 2) Inner most assemblies are discharged From Base case.

Table 5. Relative dose for partially loaded fuel assemblies

pattern	size	loading fraction	relative dose
checker board	1x1	0.519	1.745
	2x2	0.519	1.437
	3x3	0.519	1.257
	4x4	0.556	1.185
Multi Layer of rows	1row	0.500	1.872
	2row	0.501	1.741
	3row	0.500	1.548
	4row	0.502	1.430

*data are normalized to base case

The radiation doses by checker board or multi layer of rows of vacant and FA are greater than the case of random loading pattern. That is greater than base case by 87% at most. Accordingly, some kind of periodic patterns should be studied further to find conservative loading pattern of FAs.

3.3 Rack pitch

From a view point of the criticality safety, the rack pitch should be wider when the boron concentration in the square can is decreased. The pitch length and the concentration are chosen taking the cost and the benefit into account. Thus, the cell pitch of a rack varies with NPS. Here, the pitch length effect was studied.

Within the size of SFP in the current model, the cell pitch can be increased up to 30 cm for 46 x 27 cells. Here the case was called wide case. Oppositely, a narrow case was also made by assigning the pitch 26 cm. We calculate the relative dose for the case the rack cells are fully filled with the FAs. As shown in Table 6, the radiation dose increases from 0.77 to 1.31 by the increment of the pitch by 15%. For the variation of cell pitch, the effect of axial source modelling studied in section 3.1.2 was studied. As listed in Table 7, the effect is mitigated for wider cell pitch. By using wider pitch, gamma rays from source at lower height in a FA can stream upward between adjacent FAs. Accordingly, the radiation dose at the top height of SFP would be insensitive to the model of the axial distribution of gamma ray source.

Table 6. Pitch length of rack cell and radiation dose

cell pitch (cm)	26	28	30
radiation dose	0.766	1.000	1.308

*data are normalized to cel pitch = 28cm

Table 7. Effect of axial modelling of source on dose in relation to pitch length of rack cell

cell pitch (cm)	26	28	30
fine case	1.000	1.000	1.000
24 division	1.354	1.328	1.275
Flat	2.478	2.446	2.300

*data are normalized to fine case for each cell pitch

4 Summary and conclusion

Focusing on SFP-LOCA, calculation models for radiation dose at the top height of SFP were surveyed in the case where all radioactive materials are remained in fuel rods. A FA, rack, and SFP are modelled referring published data such as specifications of BEAVRS benchmark model. The dose calculation consists of evaluations of source and radiation transport.

It has been well known that cooling time of fuel assembly is essential for the radiation dose since it is major parameter to determine the source intensity. In this study, spectrum effect of source was additionally identified as ±25%.

One of the major findings here is the strong shielding effect of FA itself. Even to use axial distribution of gamma ray source in 24 divisions was found still coarse and overestimate the radiation dose by 33%. Conventional flat source model would give so conservative dose. The conservatism is comparative to neglect shielding effect of the wall of rack cell and components of FA equipped above the top of active fuel. The effect of the latter was found thanks to the BEAVRS benchmark model.

Significance of loading pattern in FAs was also identified. By reducing loading fraction by ~50%, the gamma ray dose would be increased by 87% although source intensity is reduced by the ~50%. That is due to enhancing gamma ray streaming upward by the discharging half of the FAs.

Generally, measures to mitigate SFP-LOCA take costs. Therefore, the authors consider the reasonable assessment of risk of radiation hazard is desirable. In the assessment, assumptions to cause extra conservatisms should be taken care of referring this work.

References

1. Tokyo Electric Power Company, "Fukushima Nuclear Accident Analysis Report (Interim Report)"HP http://www.tepco.co.jp/en/press/corp-com/release/betu11_e/images/111202e14.pdf, browse on Sep. 02,2016
2. https://www.nsr.go.jp/data/000034411.pdf,in Japanese, on internet, browse on Sep. 02,2016, for example.
3. N. Horelik, B. Herman, B. Forget, and K. Smith, "Benchmark for Evaluation and Validation of Reactor Simulations (BEAVRS)", Proceedings of International Conference on Mathematics and Computational Methods Applied to Nuclear Science and Engineering (M&C 2013), Sun Valley, Idaho, p.2986-2999, May 5-9 (2013).
4. Kansai Electric Power Company, press release, http://www.kepco.co.jp/corporate/pr/2004/0409-1_2j.html, in Japanese, on internet, browse on Aug. 31,2016.
5. I. C. Gauld, G. Radulescu, G. Ilas, et al., "Isotopic depletion and decay methods and analysis capabilities in SCALE," Nuclear Technology volume 174 p. 169~, May 2011.
6. A. Sasahara, et al., "Neutron and gamma ray source evaluation of LWR high burn-up UO2 and MOX spent fuels," Journal of Nuclear Science and Technology, vol. 41 (4), p.448-456, (2004).
7. X-5 Monte Carlo Team, MCNP – A General Monte Carlo N-Particle Transport Code, Version 5, LA-UR-03-1987 (2003).
8. "Conversion Coefficients for Radiological Protection Quantities for External Radiation Exposures," ICRP PUBLICATION 74.

Calculation of Neutron Production Rates and Spectra from Compounds of Actinides and Light Elements

Gennady Vlaskin[a] and Yuriy Khomiakov[1]

[1]ITCP, Innovation and Technology Centre for Proryv Project, Moscow, Russia

Abstract. The code NEDIS allows the calculation of neutron production rate and continuous energy spectra due to (α,n) reaction on Li, Be, B, C, O, F, Ne, Na Mg, Al, Si, P, S, Cl, Ar, K, and Ca. It accounts for anisotropic angular distribution of neutrons of (α,n) reaction in centre-of- mass system and dimensions of alpha emitting source material particles. Spontaneous fission spectra are calculated with evaluated half-life, spontaneous fission branching, ν-averaged per fission, and Watt spectrum parameters. The results of calculations by NEDIS can be used as input for Monte Carlo simulation for materials that will be used in radiation shielding and for underground neutron experiments

1 Introduction

The code NEDIS has been developed for simulation of neutron production rate and spectra due to (α,n)-reactions and spontaneous fission for using in non-destructive assay of plutonium in technology process of nuclear fuel fabrication and reprocessing [1,2]. The development has required the accumulation and evaluation of measured and calculated (α,n)-reaction cross-sections and angular neutron distributions for the individual levels of the residual nucleus, α-particle stopping cross-section data and the actinides decay constants. Today the code library includes (α,n)-reaction data for the light target nuclides in the Table 1 for alpha-particle energy less 9 MeV [3]. The code library has α-decay information for 58 heavy nuclides from ^{210}Pb to ^{254}Cf and 36 actinides have the spontaneous fission mode.

The neutron production code NEDIS2.0 was benchmarked using the following data: experimental measurements of thick-target yields of (α,n)-neutrons from Be, BeO, BN, C, UC, UO_2, Mg, Al, Si [4]; experimental data of thick-target neutron spectra of (α,n)-reaction from B, C, O, Mg, F, Al, Si, Al_2O_3, SiO_2 measured in energy-tunable alpha-particle accelerator [5]; experimental measurements of thick-target (α,n)-reaction yields and neutron spectra from compounds of actinides with light elements [6-8]. In this report we present some data of neutrons yields and spectra from (α,n)-reactions and spontaneous fission.

2 Calculation of neutron yields and energy spectra

The heavy nuclides in materials produce MeV α-particles. These α-particles interact with the nucleus in a thick target of light elements and yield neutrons. The neutron yield is calculated by:

$$Q(E_\alpha) = \int_{E_{th}}^{E_\alpha} \frac{\sigma_{tot}(E)}{\varepsilon(E)} dE \qquad (1)$$

where E_{th} –is the energy of threshold, E_α –is the initial energy of the α-particle, ε –is atomic stopping power [9], σ_{tot} total cross-section of the (α,n)-reaction. For an arbitrary primary α-particle spectrum the neutron yield is calculated by:

$$Q = \int_0^{E_{\alpha,max}} \frac{\sigma_{tot}(E_\alpha)}{\varepsilon(E_\alpha)} P(E_\alpha) dE_\alpha \qquad (2)$$

where $P(E_\alpha)$ number of α-particles with energy more than E_α.

2.1. Method of (α,n) neutron spectrum calculation

The differential neutron yield in the reaction (α,n) within the solid angle Ω^* in the centre-of-mass system (CMS) for an individual state of the residual nucleus is given by the expression

$$d^2Q = \frac{1}{\varepsilon}\left(\frac{\partial\sigma(E_\alpha,\Omega^*)}{\partial\Omega^*}\right) d\Omega^* dE_\alpha \qquad (3)$$

where $\frac{\partial\sigma(E_\alpha,\Omega^*)}{\partial\Omega^*}$ is the differential reaction cross section, which is expediently expressed as an expansion in Legendre polynomials

$$\frac{\partial\sigma(E_\alpha,\Omega^*)}{\partial\Omega^*} = \frac{\sigma^{int}(E_\alpha)}{4\pi}\left[\,1+\sum_\ell a_\ell(E_\alpha)P_\ell(\cos\theta^*)\,\right] \qquad (4)$$

[a] Corresponding author: vlasking@mail.ru

Putting (4) to the (3) and after transformation from solid angle in the CMS to energy E_n in laboratory system the differential neutron yield in the reaction (α,n) is obtained

$$d^2Q=\frac{1}{\varepsilon}\frac{\sigma^{int}(E_\alpha)}{4\pi}\left[1+\sum_\ell a_\ell(E_\alpha)P_\ell(\cos\theta^*)\right]I(E_\alpha)dE_n dE_\alpha \quad (5)$$

where $I(E_\alpha)$ Jacobian of the transformation is equal to

$$I(E_\alpha)=\frac{\partial\Omega^*}{\partial E_n}=2\pi\frac{d(\cos\theta^*)}{dE_n}=\frac{4\pi}{R(E_\alpha)} \quad (6)$$

as the energy of neutron and $\cos(\theta^*)$ is coupled by the expression

$$E_n=\frac{R(E_\alpha)}{4}\left(\gamma(E_\alpha)+\frac{1}{\gamma(E_\alpha)}+2\cos\theta^*\right) \quad (7)$$

where $R(E_\alpha)=E_n(E_\alpha,0°)-E_n(E_\alpha,180°)$, function $\gamma(E_\alpha)$ is the ratio of velocity CMS in laboratory system to velocity of neutron in CMS. After all appropriate transformation and integration with respect to E_α, the neutron spectrum of the (α,n)-reaction is obtained

$$\frac{dQ(E_n)}{dE_n}=\int_{X(E_n)}^{Y(E_n)}\frac{\sigma^{int}(E_\alpha)}{\varepsilon(E_\alpha)R(E_\alpha)}\left[1+\sum_{\ell\geq1}a_\ell(E_\alpha)P_\ell(\cos\theta^*)\right]dE_\alpha$$

(8)

The limits of integration $X(E_n)$ and $Y(E_n)$ are determined by the kinematics of the reaction [10].

The neutron spectrum of a mixture of elements may be determined from the formula for an arbitrary primary α-particle spectrum by

$$\frac{dQ(E_n)}{dE_n}=\sum_{i,j,k}\int_{X_{i,j,k}(E_n)}^{Y_{i,j,k}(E_n)}\frac{\sigma^{int}_{i,j,k}(E_\alpha)f_{i,j,k}(E_n,E_\alpha)RAT_{j,k}}{\left[\varepsilon_k(E_\alpha)+\sum_{\beta=1,\beta\neq k}^{L}\frac{N_\beta\varepsilon_\beta}{N_k}\right]R_{i,j,k}(E_\alpha)}P(E_\alpha)dE_\alpha$$

(9)

where i is the level of the residual nucleus as a result of reaction with the j-th isotope of the k-th element; j is the isotope of the k-th element participating in the (α,n)-reaction; k is the element of the j-th isotope which is involved in the (α,n)-reaction; β is a particular element in the mixture, $1\leq\beta\leq L$; L is the number of elements in the mixture; N_β, N_k are the numbers of atoms of the corresponding element in unit volume of mixture; $RAT_{j,k}$ is the ratio of the number of nuclei of isotope j which involved in the (α,n)-reaction to the total number of nuclei of element k or atomic abundance [11]; $P(E_\alpha)$ is the number of α-particles with energy more than E_α in an arbitrary primary α-particle spectrum; $f_{i,j,k}(E_n,E_\alpha)$ is the angle component expansion of cross-section in Legendre polynomials. The limits of integration are determined by the kinematics of the reaction to the ground state and each of the excited states of the residual nucleus.

Gauss method [12] with 20 knots is used with prescribed accuracy for the calculation equation (9) in the code NEDIS2.0. The code such as SOURCES4C [13] calculates the neutron yield of many sources assuming isotropy of neutron emission and sometimes homogeneity in constitution. SOURCES performs trapezoidal intergration in each alpha particle user discretised energy group to obtain the (α,n) spectrum contributions and total spectrum. It is restricted to alpha energies up to 6.5 MeV. The code as USD [14] did not compute the continuous retardation of the α-particles correctly.

Table 1. Light nuclides data available in NEDIS-2.0 library

Nuclide	Atomic abundance, [11]	Reaction	Reaction threshold, MeV	Number of levels, (last level energy)
^{6}Li	7.59	$^{6}Li(\alpha,n)^{9}B$	6.619	1(0.0)
^{7}Li	92.41	$^{7}Li(\alpha,n)^{10}B$	4.38	4(2.154)
^{9}Be	100	$^{9}Be(\alpha,n)^{12}C$	0	4(9.611)
^{9}Be	100	$^{9}Be(\alpha,n)3^{4}He$	2.272	2(0.42)
^{10}B	19.9	$^{10}B(\alpha,n)^{13}N$	0	9(8.00)
^{11}B	80.1	$^{11}B(\alpha,n)^{14}N$	0	11(7.03)
^{13}C	1.108	$^{13}C(\alpha,n)^{16}O$	0	5(7.117)
^{14}N	99.632	$^{14}N(\alpha,n)^{17}F$	6.088	2(0.495)
^{17}O	0.038	$^{17}O(\alpha,n)^{20}Ne$	0	5(5.618)
^{18}O	0.205	$^{18}O(\alpha,n)^{21}Ne$	0.851	7(5.337)
^{19}F	100	$^{19}F(\alpha,n)^{22}Na$	2.36	29(5.174)
^{21}Ne	0,27	$^{21}Ne(\alpha,n)^{24}Mg$	0	5(5.236)
^{22}Ne	9,25	$^{22}Ne(\alpha,n)^{25}Mg$	0,57	9(3.405)
^{23}Na	100	$^{23}Na(\alpha,n)^{26}Al$	3.485	29(4.22)
^{25}Mg	10.00	$^{25}Mg(\alpha,n)^{28}Si$	0	9(7.80)
^{26}Mg	11.01	$^{26}Mg(\alpha,n)^{29}Si$	0	5(3.069)
^{27}Al	100	$^{27}Al(\alpha,n)^{30}P$	3.027	20(4.80)
^{29}Si	4.6832	$^{29}Si(\alpha,n)^{32}S$	1.736	10(5.80)
^{30}Si	3.0872	$^{30}Si(\alpha,n)^{33}S$	3.959	11(4.21)
^{31}P	100	$^{31}P(\alpha,n)^{34}Cl$	6.380	1(0.0)
^{33}S	0.76	$^{33}S(\alpha,n)^{36}Ar$	2.244	2(1.970)
^{34}S	4.29	$^{34}S(\alpha,n)^{37}Ar$	5.176	1(0.0)
^{35}Cl	75.78	$^{35}Cl(\alpha,n)^{38}K$	6.530	1(0.0)
^{37}Cl	24.22	$^{37}Cl(\alpha,n)^{40}K$	4.292	21(3.52)
^{38}Ar	0.0629	$^{38}Ar(\alpha,n)^{41}Ca$	5.774	1(0.0)
^{40}Ar	99.604	$^{40}Ar(\alpha,n)^{43}Ca$	2.506	15(5.70)
^{41}K	6.73	$^{41}K(\alpha,n)^{44}Sc$	3.727	21(3.10)
^{42}Ca	0.647	$^{42}Ca(\alpha,n)^{45}Ti$	5.683	1(0.0)
^{43}Ca	0.135	$^{43}Ca(\alpha,n)^{46}Ti$	0.0	1(0.0)
^{44}Ca	2.086	$^{44}Ca(\alpha,n)^{47}Ti$	2.381	1(0.0)
^{46}Ca	0.004	$^{46}Ca(\alpha,n)^{49}Ti$	0.241	1(0.0)
^{48}Ca	0.178	$^{48}Ca(\alpha,n)^{51}Ti$	0.144	1(0.0)

2.2. Method of neutron spectrum calculation of spontaneous fission (SF)

Spontaneous fission spectra are calculated with evaluated half-life, spontaneous fission branching, ν-averaged per fission, and Watt spectrum parameters. The currently available nuclear data relevant to the spontaneous fission neutron yields are presented in [15]. Neutron multiplicities were similarly taken from [15].

The energy spectrum of neutrons emitted by the spontaneous fission of 36 actinides is described by a Watt formula: $e^{-E/a}\cdot \sinh(b\cdot E)^{1/2}$, where E is the neutron energy, and **a** and **b** are the radionuclide-specific fission parameters obtained from the NEDIS2.0 data library. The parameter **a** (temperature) deals with the neutron multiplicity by the Terrell formula [16] with our new coefficients for spontaneous fission

$$\mathbf{a}=0,48+0,2(1+\bar{\nu}_{SF})^{1/2} \quad (10)$$

The parameter **b** is determined by this temperature and average kinetic energy per nucleon of spontaneus fission products which is equal to E_f=0,76 MeV in NEDIS2.0 data library

$$\mathbf{b}=4E_f/\mathbf{a}^2=3,04/\mathbf{a}^2 \quad (11)$$

In this case the average energy of neuron spectrum of spontaneous fission is equal to

$$<E>=E_f+(3/2)*\mathbf{a}=1,48+0,3*(1+\bar{\nu}_{SF})^{1/2} \quad (12)$$

or

$$<E>=1,5a+0,25a^2b \quad (13)$$

The parameters of some nuclides are presented in the Tables 2 and 3 in comparison with [17-19].

Table 2. Radionuclide-specific fission parameters

Nuclide	The parameter **a**		
	NEDIS	SOURCES	[19]
U-238	0.827	0.6483	0.8366
Pu-240	0.8355	0.795	0.7989
Cm-248	0.8855	0.8084	-
Cf-252	0.916	1.025	1.025
Th-232	0.796	0.5934	-
	The parameter **b**		
U-238	4.445	6.811	4.339
Pu-240	4.335	4.689	4.903
Cm-248	3.877	4.536	-
Cf-252	3.623	2.926	2.9257
Th-232	4.755	8.030	-

Table 3. Average energy of SF spectrum

Nuclide	Average energy of spectrum			
	NEDIS	SOURCES	[17]	Exp.[18,19]
U-238	2.001	1.688	2.014	-
Pu-240	2.013	1.933	1.981	2.04±0.04
Cm-248	2.088	1.954	-	2.07±0.02
Cf-252	2.134	2.306	2.306	2.13±0.01
Th-232	1.954	1.597	-	-

3 Results of calculation of neutron yields and energy spectra

3.1. ^{227}Ac-^{13}C , ^{244}Cm-^{13}C and ^{210}Po-^{13}C sources of neutrons

Neutron spectral measurements have been reported [6] for two mixtures, one contains $^{227}AcO_2$ with daughter products and the second contains dioxide $^{244}CmO_2$ mixed with ^{13}C powder target material and [7] for ^{210}Po-^{13}C source of neutrons . The neutron yields calculated by NEDIS-2.0 are in satisfactory agreement with measurements for radius of $^{227}AcO_2$ ~1 μm and radius of $^{244}CmO_2$ ~12.8 μm. The spectra calculated by NEDIS-2.0 and measured in [6,7] are presented in Fig.1, Fig.2, and in Fig 3. Discrepancies in spectra are expected due to slowing-down of the neutrons in encapsulated sources and in the detector.

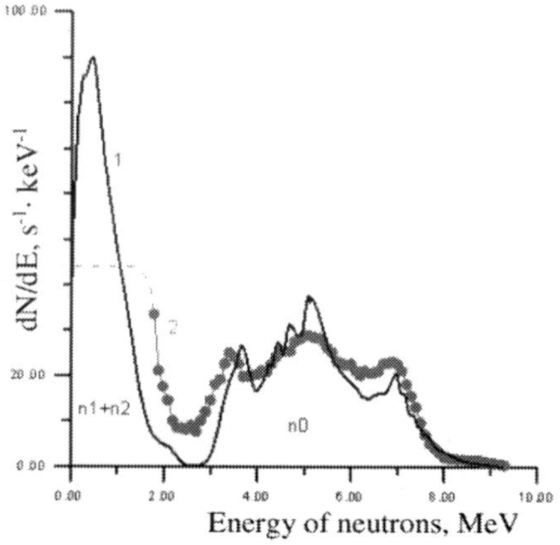

Fig. 1 - ^{227}Ac-^{13}C neutron spectra (1-calculated by NEDIS-2.0, 2 – measured)

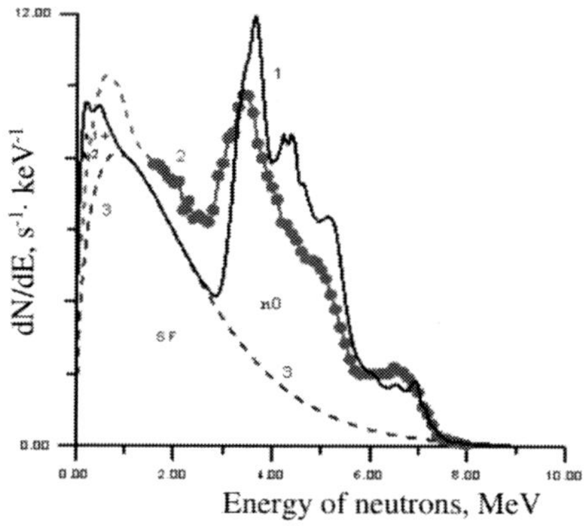

Fig. 2 - ^{244}Cm-^{13}C neutron spectra (1-calculated by NEDIS-2.0, 2 – measured, 3 – spontaneous fission spectrum)

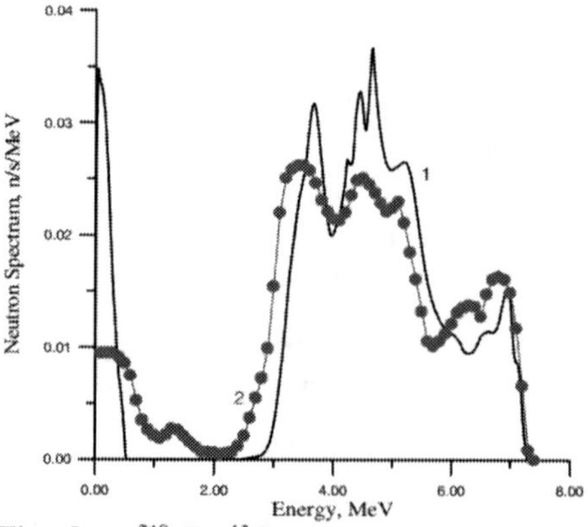

Fig. 3 – ^{210}Po-^{13}C neutron spectra (1-calculated by NEDIS-2.0, 2 – measured [7])

3.2 ^{241}Am(Au)-^{13}C plane source of neutrons for calibration in Daya Bay experiment

The NEDIS2.0 has been used to calculate the neutron yield and spectra from the source [8]. $P(E_\alpha)$ - the number of α-particles with energy more than E_α was calculated in continuous-retardation approximation, neglecting fluctuations in the α-particle energy loss due to scattering and inelastic collision in Am, in 1.1 μm gold and ^{13}C. Results are presented in Fig.4 and in Table 4, which are in agreement with an assay of [8].

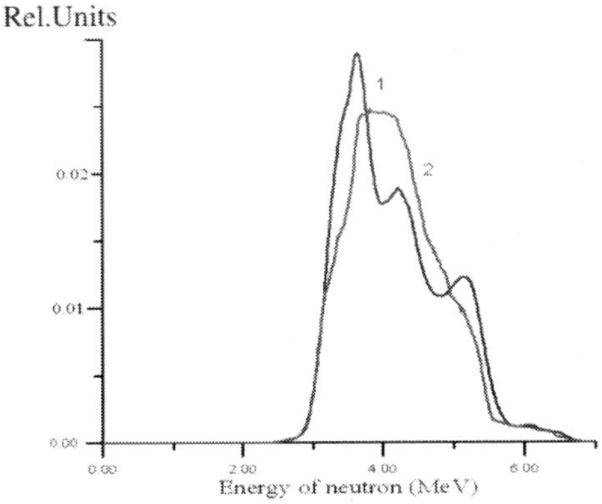

Fig. 4 – ^{241}Am(Au)-^{13}C neutron spectra (1-calculated by NEDIS-2.0, 2 –isotropic in CMS (α,n cross section from JENDL [20]).

Table 4. Neutron yields due to ^{13}C(α,n)^{16}O reaction.

Thickness of ^{241}Am disc(μm)	Deposit thickness of Au (μm)	Neutron yield(s^{-1})		
		Ground of ^{16}O	1-st (6,049 MeV)	2-nd (6,13 MeV)
1	0	3,62	0,400	0,048
4	0	2,52	0,109	0,0124
1	0,7	1,983	5,47e-03	3,43e-06
1	1	1,7	5,51e-07	0,0
1	1,1	1,622	0,0	0,0
2	1	1,47	2,75e-07	0,0
3	1	1,28	1,84e-07	0,0
4	1	1,11	1,38e-07	0,0
4	1,1	1,056	0,0	0,0
5	1	0,955	1,10e-07	0,0

3.3 Decay-induced neutrons

Table 5 presents earlier values not updated in a long-time of the neutron yield per 10^6 α-particle in the materials given by the experiments [21,22] and present-day ones by the NEDIS code which assumes secular equilibrium in the decay chains of ^{238}U and ^{232}Th. The results of calculation presented in the Table 5 are in satisfactory agreement with the measurements may be used to compare and benchmark new codes to estimate uncertainty of the (α,n) neutron yield calculations for a number of materials that are used and will be used for construction of dark matter detectors [14].

Table 5. Neutron yield (per 10^6 α-particles) due to presence of ^{238}U and ^{232}Th in secular equilibrium [21,22].

material	U-238		Th-232	
	Measurement	NEDIS	Measurement	NEDIS
Be	92±13	79,6±2,4	126±15	118±3
BeO	33±4	29,3±0,9	47±5	43,2±1,4
B	21±3	19,1±0,8	27±4	25,1±1,1
H_3BO_3	3,1±0,5	3,25±0,12	-	4,26±0,17
C	0,17±0,04	0,14±0,01	0,31±0,08	0,23±0,01
CaF_2	7,0±2,2	4,84±0,33	11,2±3,2	8,1±0,5
CaF_2	*5,7±1,0*	4,84±0,33	*10,2±2,2*	8,1±0,5
NaF	6,8±2,1	4,75±0,32	12,7±4,5	11,1±0,5
Na_2CO_3	*1,7±0,3*	1,44±0,14	*3,3±0,5*	2,93±0,3
Al	1,5±0,2	1,52±0,05	3,2±0,3	3,3±0,1
Al	*1,47±0,15*	1,52±0,05	*3,2±0,3*	3,3±0,1
Al_2O_3	0,8±0,1	0,78±0,04	1,8±0,2	1,67±0,08
Mg	2,1±0,4	1,71±0,06	3,6±0,5	3,21±0,12
Mg	*1,6±0,2*	1,71±0,06	*3,1±0,4*	3,21±0,12
SiO_2	*0,13±0,02*	0,134±0,011	*0,27±0,05*	0,25±0,02
Si	0,26±0,06	0,21±0,02	0,51±0,11	0,43±0,03
K_2CO_3	*0,12±0,03*	0,10±0,01	*0,25±0,06*	0,21±0,02

Neutron spectra due to presence of ^{238}U in secular equilibrium for Li,F,O,B,C and Be are presented in Fig.5.

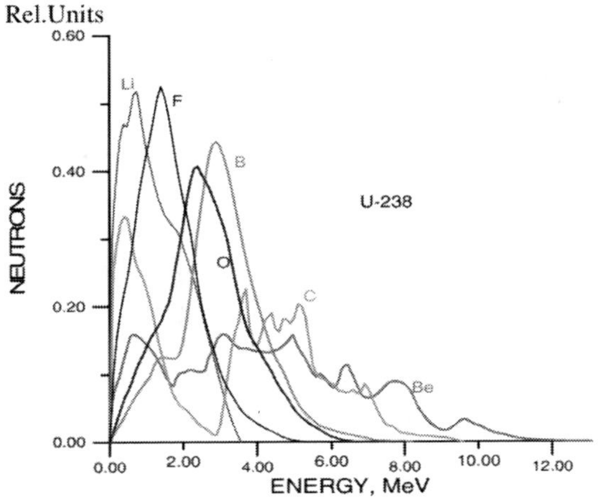

Fig. 5–Neutron spectra calculated by NEDIS-2.0.

3.4 Plutonium oxide

Measured total neutron yields from five plutonium oxide samples were collected in paper [23]. Two samples were measured at the Japan Atomic Energy Research Institute (JAERI) with using neutron coincidence counter and others at Oak Ridge National Laboratory (ORNL), Savannah River Plant (SRP), and at Pacific Northwest Laboratories (PNL). Results of calculations and measurements are given in Table 6. Measured and calculated neutron yields are in satisfactory agreement.

Table 6. Calculated and measured source magnitudes for several plutonium oxide samples

SAMPLE	NEUTRONS/S		
	MEASURED	SOURCES	NEDIS-2.0
JAERI-1	sf (184.7±4.8)	215	209.5
	α,n (157.8±9.8)	138	135.5
	tot (342.6±8.6)	353	345
JAERI-2	sf (42.1±1.1)	48	46.4
	α,n (51.3±2.5)	54	54.3
	tot (93.4±2.3)	102	100.7
ORNL	(868±9) 10^2	853 10^2	834 10^2
SRP	(855±13) 10^3	867 10^3	864 10^3
PNL	*2.3 10^5*	*2.39 10^5*	*2.34 10^5*

3.5 Borosilicate glass

The total neutron emission has been measured [24] at the Savanna River Laboratory from three glass samples doped with PuO_2 where mass fraction of ^{238}Pu ~ 90%. The results of measurements are compared with calculated neutron yields using NEDIS-2.0 and ORIGEN-S [25] codes in Table 7.

Table 7. Borosilicate glass results of emission rate

GLASS SAMPLE	MASS OF PU-238 (mg)	NEUTRONS/S		
		MEASURED	ORIGEN-S	NEDIS-2.0
1	10.8	4810	5571	**5682**
2	1.03	473	532	**542**
3	0.364	170	188	**192**

It was noted in [24] the uncertainty in the content of ^{238}Pu in the glass was about 5% and yields may be underestimated by as much as 10% due to the calibration with ^{252}Cf source (spectral differences). However, the dimensions of particles of PuO_2 in glass may be ~ 0.1-5 μm [26]. For 1.5 μm radius of particles NEDIS-2.0 outputs 4970, 474, 168 n/s and agreement with the measurements are ~ 3%. The predicted neutron source spectra from the glass sample 1 are illustrated in Fig.6.

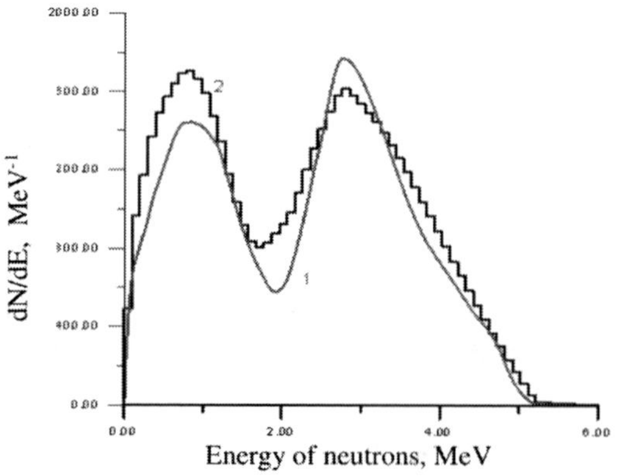

Fig. 6 - Energy dependent neutron source strength in glass sample 1 as calculated by 1-NEDIS-2.0 with account for radius of PuO_2 particles and 2- ORIGEN-S .

4 Conclusions

There is a growing need for high-accuracy (α,n)-yields and spectra from compounds of actinides and light elements in different nuclear technology mediums (as an example LiCl+KCl) due to interest in using neutron measurements for nuclear fuel fabrication and reprocessing plant safeguards and nuclear material accountancy. Improving these calculations will ultimately help safeguards professionals make better measurements with fewer resources. More broadly, computer codes such as MCNP are now being used to simulate shielding, require accurate continuous neutron spectrum and magnitude as input. The availability of accurate data is critical in these scenarios. The code NEDIS-2.0 is a useful and flexible computational tool for the accurate analysis of neutron emission rates and energy spectra of different materials of nuclear fuel cycle.

References

1. G.N. Vlaskin, E.V. Chvankin, Neutron spectrum and yield of accompanying photon radiation of lithium neutron sources. *Sov. At. Energy* 74, 129–134, 1993.
2. G. N. Vlaskin, "NEDIS2.0 code to calculate yields and spectra of neutrons produced in light element nuclei (α,n) reactions and spontaneous fission", Preprint VNIINM 06-1, 2006.
3. G.N.Vlaskin, Yu.S.Khomyakov, V.I.Bulanenko, Neutron Yield of the Reaction (α,n) on thick Targets comprised of Light Elements. *Atomic Energy*, Vol.117, No. 5, March, 2015 (Russian Original Vol.117,No. 5, November, 2014, pp. 287-293)
4. D. West, A.C. Sherwood. Measurements of Thick-Target (α,n) Yields from Light Elements, *Ann. Nucl. Enegy*, Vol. 9, pp. 551-577, 1982.
5. G.J.H. Jacobs, H. Liskien, Energy Spectra of Neutrons Produced by α-Particles in Thick Targets of Light Elements, *Ann. Nucl. Enegy*, Vol.10, pp. 541-552, 1983.
6. K.W. Geiger, L. Van-der-Zwan. Neutrons and γ-Rays from Radioactive ^{13}C(α,n) sources, *Nucl. Instr. Meth.*, 1978, Vol. 157, p. 199-201.
7. A.A.Chudotvortsev et.al., Energy spectrum of neutron and gamma radiation (α,n)-sources on ^{13}C, ^{18}O, ^{25}Mg. In book: Radiation Technique, V.18, M., Atomiztad,1978, p. 35.
8. J. Liu et al. , Neutron calibration sources in Daya-Bay experiment. Nuclear Instruments and Methods in Physics Research A 797 (2015) 260-264
9. J. Ziegler, Helium: Stopping Powers and Ranges in All Elements, Pergamon Press, New York (1977).
10. B. G. Whitmore and W. B. Baker, "The Energy Spectrum of Neutrons from a Po-Be Source", *Phys. Rev.*, 114, pp. 571-577(1959).
11. M. Berglund and M. Wieser, Isotopic Composition of the Elements 2009, (IUPAC Technical Report) *Pure Appl. Chem.* Vol.83, No.2, pp. 397-410, 2011
12. M.Abramovitz and I.Stegun(1974), *Handbook of*

Mathematical Functions (Dover, New York).
13. W.B. Wilson, R.T. Perry, W.S. Charlton, T.A. Parish and E.F. Shores, SOURCES: A code for calculating (alpha,n), spontaneous fission, and delayed neutron sources and spectra, *Radiation Protection Dosimetry* **115** (2005) 117-121.
14. D.-M. Mei, C. Zhang and A. Hime, Evaluation of (α,n) induced neutrons as a background for dark matter experiments, *Nuclear Instruments and Methods* **A 606** (2009)651-660. doi:10.1016/j.nima. 2009.04.032. Data available at http://neutronyield.usd.edu
15. S. Simakov, M. Verpelli, N. Otsuka. Update of the nuclear data for the neutron emissions for actinides of interest in safeguards. NDS, IAEA, 2015
16. J. Terrell, "Fission Neutron Spectra and Nuclear Temperatures", *Phys. Rev.* 113, pp. 527-541(1959).
17. D.G. Madland, "Calculation of Watt Distribution Parameters for Spontaneous and Neutron-Induced Fission," in Los Alamos National Laboratory report LA-8757-PR, p. 17 (March 1981).
18. V. Kalinin, V. Dushin, F.J.O. Hambsch, et al. Measurements of Prompt Neutron Multiplicity Distributions in Correlation with Mass-Energy Distribution of Fission Fragments in Spontaneous Fission of ^{252}Cf, ^{244}Cm and ^{248}Cm., J.of Nucl.Sci. &Techn.,Supl,2, Aug.,2002, Proc. Of Int. Conf. Nucl. Data for Sci. &Tech. Tsukuba, Japan, 2001, p. 250-253.
19. B. Gerasimenko, L. Drapchinsky, O. Kostochkin et al. Precision Measurements of Prompt Neutron Spectra in ^{240}Pu and ^{242}Pu Spontaneous Fission., J. of Nucl. Sci. &Techn.,Supl,2, Aug.,2002, Proc. Of Int. Conf. Nucl. Data for Sci. &Tech. Tsukuba, Japan, 2001, p. 362-365.
20. Murata T., Matsunobu H., Shibata K. Evaluation of the (α,xn) Reaction data for JENDL/AN-2005, // JAEA-Research 2006-052, 2006.
21. G. V. Gorshkov,V. A. Zyabkin, N. M. Lyatkovskaya and O. S. Tsvetkov, *Natural Neutron Background of the Atmosphere and Earth's Crust*, Atomizdat, Moscow (1966).
22. Y. Feige, B. Oltman, and J. Kastner, "Production rates of neutrons in soils due to natural radioactivity," *J. Geophys.Res.*, **73**, 3135–3142 (1968).
23. E.F. Shores. Plutonium oxide benchmark problems for the SOURCES code. Appl. Radiat. and Isot. 62(2005), 669-704
24. D.J. Pellarin,W.L. Matney and Bibler. (α,n) Neutron Emission from DWPF Glass, DPST-86-212, Memorandum to M.J. Plodinec, Savannah River Laboratory (1986)
25. I.G. Gauld, E.F. Shores, and R.T. Perry. New Neutron Source Algorithms in the ORIGEN-S code. In Proceeding of the 12th Biennial Topical Meeting of the Radiation Protection and Shielding Division of the American Nuclear Society, Spokane, WA, 17-21 September 2000 (2000).
26. J.D. Vienna, D.L. Alexander, M.J. Schweiger, D.K. Peeler, T.F. Meaker. Plutonium Dioxide Dissolution in Glass, PNNL-11346, 1996.

Determination of equilibrium fuel composition for fast reactor in closed fuel cycle

Mikhail Ternovykh[1a], Georgy Tikhomirov[1], Yury Khomyakov[2] and Igor Suslov[2]

[1] *National Research Nuclear University MEPhI (Moscow Engineering Physics Institute), Department of Theoretical and Experimental Physics of Nuclear Reactors, 31, Kashirskoe shosse, 115409, Moscow, Russia*
[2] *Institution «Innovation and Technology Center by «PRORYV» Project», State Atomic Energy Corporation « Rosatom», 2/8 Malaya Krasnoselskaya Street, 107140, Moscow, Russia*

Abstract. Technique of evaluation of multiplying and reactivity characteristics of fast reactor operating in the mode of multiple refueling is presented. We describe the calculation model of the vertical section of the reactor. Calculation validations of the possibility of correct application of methods and models are given. Results on the isotopic composition, mass feed, and changes in the reactivity of the reactor in closed fuel cycle are obtained. Recommendations for choosing perspective fuel compositions for further research are proposed.

1 Concept of the reactor project and criterion functionals considered

The developed algorithms and the results of this work are focused on the concept and design of high power reactors with sodium coolant, uranium-plutonium nitride fuel, high fuel burnup [1-3]. The increased diameter of the core and its significant flattening is typical for such reactors. Fuel assemblies have large fuel elements and increased volume share of fuel in the core. A feature of this concept is to refuse using separated blanket of fuel breeding in the radial and axial direction. Fuel assemblies with increased diameters of fuel rods are used in periferal rows of the core for flattering energy release. Initial fuel load of the same isotopic composition over the reactor is considered.

The reuse of spent fuel in the reactor after cooling and reprocessing is supposed. Fuel is made of the uranium and plutonium from the irradiated fuel assemblies after reprocessing with account of fixed losses during processing. Burnt fuel mass is compensated by depleted uranium and as needed by plutonium with isotopic composition of the starting load of the reactor.

We analyzed the possibility to achieve significant fuel burnup, required uranium and plutonium feeding mass, the change in reactivity between refueling when operating in the described fuel cycle.

2 Methods of estimation of reactor characteristics in the refueling mode

We considered the following scenario of refueling. The lifetime of the reactor was supposed as 60 years with maximum fuel lifetime of 6 years. Average fuel burnup at maximum fuel lifetime was 100 MWd/kg. Time of cooling and reprocessing of irradiated fuel was 3 years. The time between refuelings was 1 year (1 microcampaign). 1/6 part of the irradiated fuel assemblies is replaced in each refueling of the core. All fuel assemblies used in initial load and in 1-8 refuelings have the same isotopic composition. After that starting with the 9th refueling (first fuel reprocessing), fuel assemblies have another compositions. Thus, one cycle of fuel assembly is equal to 9 microcompaigns. We analyze the operation of the reactor during at least 60 microcompaigns in this paper.

Such refueling algorithm is extremely complicated, since it should take into account location of fuel assemblies with fresh anf reprocessed fuel in the core, possibility of fuel mix from different irradiated fuel assemblies during processing. Its application requires to keep real track of isotopic composition of many fuel assemblies in the core.

The proposed method of evaluation characteristics of the reactor in the refueling mode is based on calculated properties of fuel assemblies correspondind to average over the core parameters of fuel burnup. Reactor characteristics are predicted by the properties of the core cluster including six fuel assemblies with different properties. Different composition of the fuel assemblies in the cluster is taken into account for all lifetime of the reactor.

We use the term "cluster" in this paper. We mean that cluster is repeating system of elements in horizontal projection of active core larger than fuel assembly. The central element of calculation model of the cluster is assembly channel without fuel consisting of composition of structural materials and coolant. Cluster contains 6 types of fuel assemblies. Each one is a group of fuel

[a] Corresponding author: MYTernovykh@mephi.ru

assemblies that were loaded simultaneously in the active core after the next processing. Figure 1 shows the chart of refueling of reactor core (in symmetry of 60^0). Elements 0 are assembly channels in the reactor core and assemblies of the reactor protection. Elements 1-6 are groups of fuel assemblies which are reloaded simultaneously in active core. The selected clusters demonstrate repeated elements in different radial zones of the reactor.

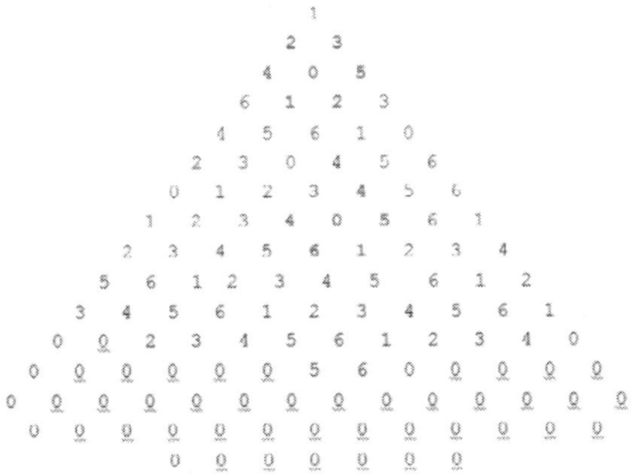

Fig. 1. Examples of marking of the calculating clusters of fuel assemblies in the active core

The application of this method and its testing was performed with SCALE, MCU, and JARFR program codes. The possibility of application of program code SCALE for solution of the tasks of fuel burnup in reactors with metal coolant is discussed in [4]. The task of changing the isotopic composition was solved using the model of the vertical section of the reactor core. Multi-zone 2-D model tooks into account the axial change of properties of fuel assemblies, structure elements (SE) above and below the core, end leakage from the reactor. Figure 2 presents the calculation model of fuel assembly that take into account the unevenness of change of the isotopic composition in the fuel zones 1-5. Calculations showed that the chosen model allowed us to predict correctly multiplication properties of the entire reactor.

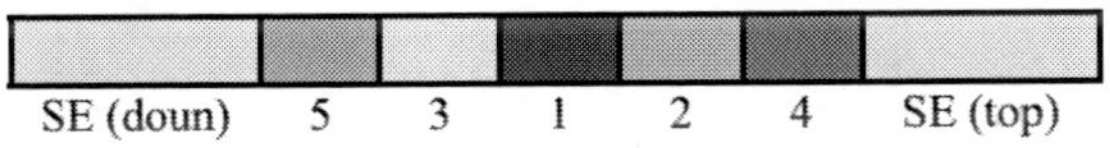

Fig. 2. Vertical section of the calculation model of fuel assembly.

To obtain isotopic composition and multiplying characteristics of the fuel assemblies during all time of reactor operation, one requires to perform calculations of 7 cycles of irradiation of fuel assemblies consisting of 9 microcampaigns. The numbering of states of fuel assemblies in microcampaigns of the reactor is presented in Table 1. Marked items in the table correspond to the mode of irradiation of the fuel assembly. The rest elements correspond to the mode of cooling and are necessary for modeling of processing of irradiated fuel and starting composition of fuel assembly of the next cycle. The reprocessing of irradiated fuel is modeled in the following way. All fission products and all isotopes except of uranium and plutonium are removed from the fuel. With account of parameters of the loss of uranium ε_U and plutonium ε_{Pu} during the processing, masses of isotopes of these elements used in the next cycle of fuel assembly are determined. Mass of plutonium in the fuel after processing is equal to the initial load of the reactor. The isotopic composition of plutonium feed is the same as in the starting load. The mass of uranium in feed is determined by the condition of conservation of equivalent fuel density.

Table 1. Accordance of fuel assembly state (year) to cycle number of fuel assembly and the number of microcampaign *n*

n	Cycle of fuel assembly						
	1	2	3	4	5	6	7
1	0	9	18	27	36	45	54
2	1	10	19	28	37	46	55
3	2	11	20	29	38	47	56
4	3	12	21	30	39	48	57
5	4	13	22	31	40	49	58
6	5	14	23	32	41	50	59
7	6	15	24	33	42	51	60
8	7	16	25	34	43	52	61
9	8	17	26	35	44	53	62

Modeling[5-8] of multiplying properties of cluster consisting of 6 fuel assemblies differing by the irradiation time in the reactor is carried out by averaging the multiplication factor of fuel assemblies of the cluster. This approach was justified by a series of direct calculations of realistic models of cluster consisting of 6 fuel assemblies. Analysis of the results was demonstrated that when modeling microcampaign of the reactor, the error in the multiplication factor and the masses of uranium and plutonium in the fuel does not exceed the uncertainty of the results related to the modeling of the geometry of the fuel assembly and irradiation of fuel. The principle of formation of properties of cluster of fuel assemblies during the period of reactor operation is shown in Table 2.

The proposed method includes evaluating of the possibility to provide criticality of the reactor during the whole time of its operation. Calculations were made for multi-zone 2-D model of the vertical section of the reactor core and full 3-D model of the reactor for several fuel types and burnups. Some results of calculations for three fuel compositions are shown in Figure 3. Parameter of adjustment was determined for prediction of multiplying properties of a reactor according to the results of 2-D calculations. It is shown that when using the parameter of adjustment to multiplication factor, error does not exceed 0.05 %, which is significantly less than the uncertainty of the results obtained according to various sources of neutron constants.

 DOI: 10.1051/epjconf/201715307034

Table 2. Status of fuel assembly in cluster (year) for microcampaign reactor *n*

n	Fuel assembly in cluster					
1	0	0	0	0	0	0
2	1	1	1	1	0	1
3	2	2	2	0	1	2
4	3	3	0	1	2	3
5	4	0	1	2	3	4
6	0	1	2	3	4	5
7	1	2	3	4	5	0
8	2	3	4	5	0	1
9	3	4	5	0	1	2
10	4	5	9	1	2	3
11	5	9	10	2	3	4
12	9	10	11	3	4	5
13	10	11	12	4	5	9
14	11	12	13	5	9	10
15	12	13	14	9	10	11
16	13	14	9	10	11	12
17	14	9	10	11	12	13
18	9	10	11	12	13	14
19	10	11	12	13	14	18
20	11	12	13	14	18	19
21	12	13	14	18	19	20
etc.						

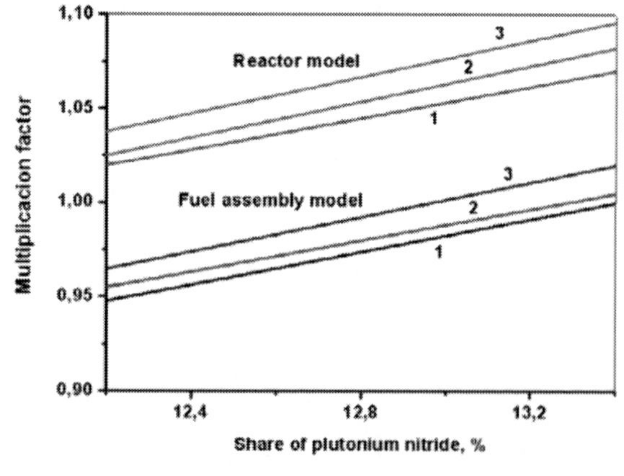

Fig. 3. Multiplication factor of reactor versus the equivalent fuel density for different start plutonium compositions.

It should be noted that this fact can be explained by the correct account of end leakage in multi-zone 2-D model for the reactor with significant flattening of the core.

3 Results and discussion

Method of evaluation of fast reactor characteristics in the refueling mode was used for the analysis of different fuel compositions. The share of plutonium nitride varied within 12-14%. Equivalent fuel density varied within 11,5-14 g/cm^3. Isotopic composition of plutonium corresponded to spent fuel of various thermal and fast reactors [9-13]. We additionaly analyzed refuelings with initial loads of the reactor on the basis of highly enriched nitride uranium or uranium-plutonium fuel with ^{239}Pu enrichment more than 90%.

Figure 4 presents the results of calculation of the multiplication factor of fuel assemblies of starting load during 6 years of irradiation and 3 years cooling before repeated use of fuel in the reactor. Isotopic composition of plutonium corresponded to real spent fuel of various reactors (curves 1-3). Figure 5 demonstrates the same information for fuel assemblies after the first processing of the fuel. Fuel assemblies of this type have to be loaded into the rector in 9 years of its operation. Note that for the considered fuel compositions, there are intervals of increase and decrease of reactivity during irradiation. This feature allows to compensate for the reactivity change in the reactor when operating in the mode of partial reloading.

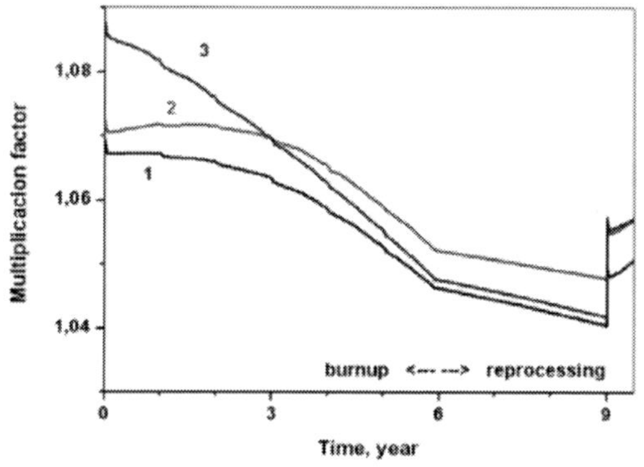

Fig. 4. Multiplication factor of fuel assembly versus the time of reactor operation for different start plutonium compositions.

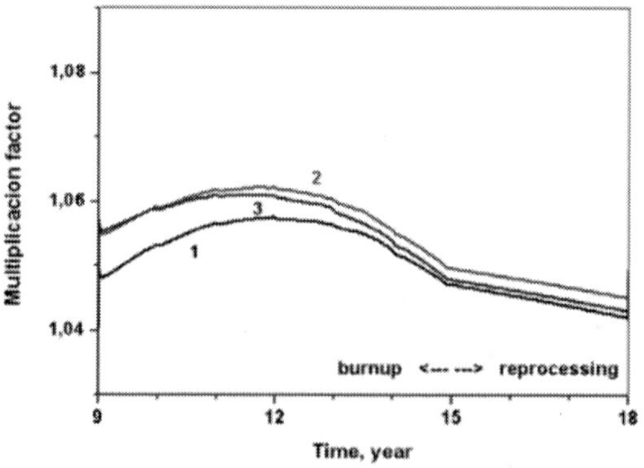

Figure 5 Multiplication factor of fuel assembly versus the time of reactor operation after the first reprocessing for different start plutonium compositions.

Examples of calculation results for further analysis are presented in Figures 6-7 for plutonium extracted from spent fuel of different reactors (curves 1-3) and for model plutonium (curve 4) and highly enriched uranium (curve 5).

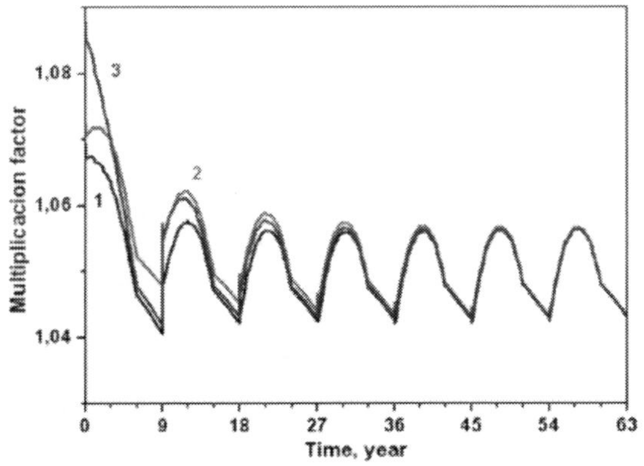

Fig. 6. Multiplication factor of fuel assembly versus the time of reactor operation in the refuelling mode for different plutonium compositions.

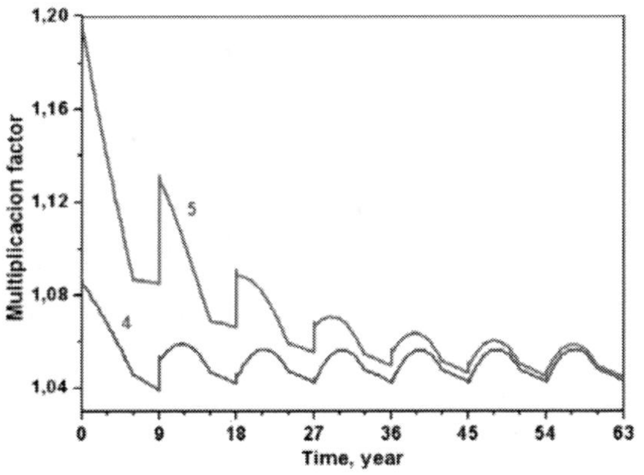

Fig. 7. Multiplication factor of fuel assembly versus the time of reactor operation in the refuelling mode for model plutonium and highly enriched uranium compositions.

Characteristics of fuel assemblies after 1-7 reprocessing are shown in Table 3. Negative values of feed by plutonium mean that excessive production of plutonium occurs in the considered fuel cycle.

The rate of transition of the fuel composition to the equilibrium in considered mode of processing was determined for different starting isotopic compositions of plutonium including compositions of irradiated fuel of different reactors and model composition with a high content of ^{239}Pu (curve 5). Share of ^{239}Pu in the composition for the considered compounds with the same proportion of the plutonium nitride in the fuel is presented in Figure 8. The mass of the plutonium feed in reprocessing is presented in Figure 9. Note that if the parameter of loss ε_{Pu} does not exceed 3 %, then for all starting compositions of plutonium from power reactors, fast neutron high power reactor with nitride fuel will not require plutonium feed but only depleted uranium feed during whole reactor operation.

Table 3. Feed by Uranium and Plutonium (M, g/kg), isotopic composition (X, %) after reprocessing with number N

N	M_U	M_{Pu}	X_{238}	X_{239}	X_{240}	X_{241}	X_{242}
1	110.4	-3.5	1.3	65.0	25.0	3.9	4.7
2	111.7	-5.7	0.7	65.1	27.1	3.2	3.9
3	112.1	-6.3	0.5	64.8	28.3	3.2	3.3
4	112.3	-6.5	0.4	64.5	28.9	3.2	2.9
5	112.3	-6.6	0.4	64.4	29.3	3.3	2.6
6	112.3	-6.7	0.4	64.4	29.5	3.3	2.4
7	112.3	-6.7	0.4	64.3	29.7	3.3	2.3

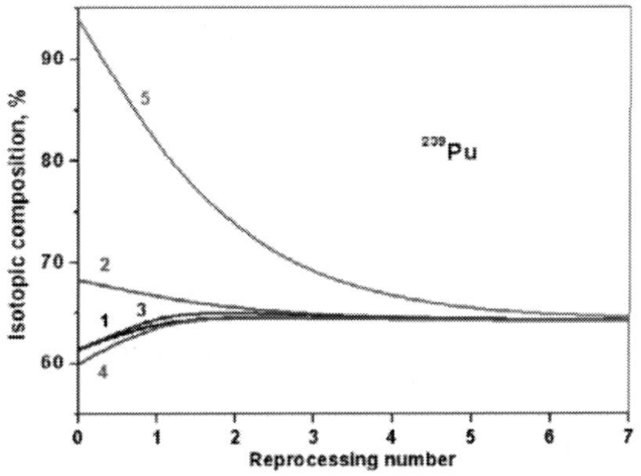

Fig. 8. Share of ^{239}Pu after reprocessing of fuel for different start plutonium compositions.

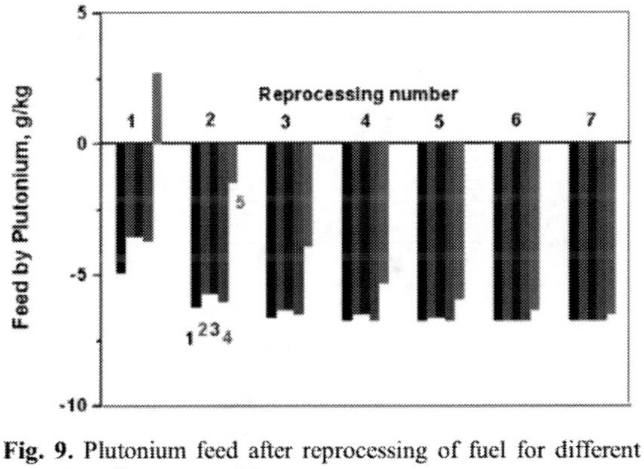

Fig. 9. Plutonium feed after reprocessing of fuel for different start plutonium compositions.

When choosing the starting fuel composition, there is a problem to provide criticality of the reactor during its operation. If necessary, the problem can be solved by increasing the share of plutonium nitride in the fuel. Figure 10 demonstrates this capability. It is seen that after the first processing, the change of the multiplication factor of fuel assembly has a similar form for different values of the share of plutonium nitride. There is an important feature: fuel assemblies of different groups compensate the increase and decrease of reactivity during the reactor operation.

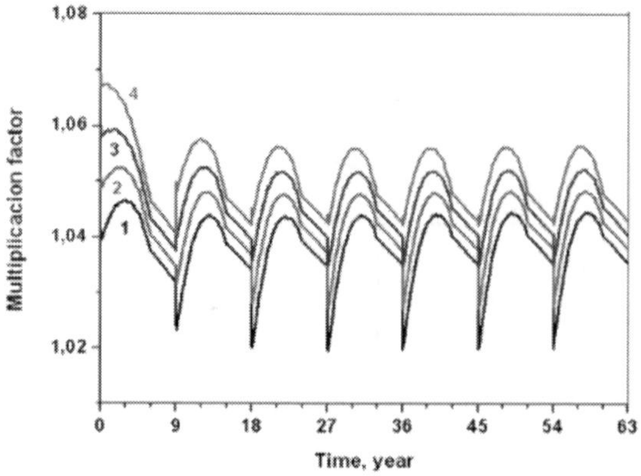

Fig. 10. Multiplication factor of fuel assembly versus the time of reactor operation in the refuelling mode for different share of plutonium nitride.

Additional research was made to evaluate the influence of non-uniformity of the energy release in the reactor and of the difference of compositions of simultaneously reprocessed fuel assemblies on the results of calculations. We considered variant of processing, in which plutonium was placed in the same fuel assembly without mixing with other fuel assemblies. At the same time fuel assembly after processing could be placed from a position with high power density in a position with lower power density and vice versa. Table 4 presents four variants of movement of fuel assemblies over the core under reloading. In these variants, the total fuel burnup has to be the same after 2, 5 and 7 cycle of fuel assembly.

Table 4. Ratio of power of fuel assembly to nominal power (%) for different variants of reloading *j*

j	Number of cycle of fuel assembly						
	1	2	3	4	5	6	7
1	100	100	100	100	100	100	100
2	125	75	125	75	100	125	75
3	75	125	75	100	125	75	125
4	75	125	100	125	75	125	75

Figure 11 and Table 5 demonstrate that in the mode of reloading characteristics of fuel assemblies depend on the sum fuel burnup and do not depend on their location in the reactor.

According to the results of variant calculations, we selected for the demonstration (Figure 12) starting fuel compositions, which provided the reactor to be critical during the whole time of its operation and provided fast achievement of equilibrium of the fuel composition.

We determined the amount of maximum change of reactivity of the reactor due to fuel burnup between reloading. This value is presented in Figures 13 and 14 in the form of continuous lines for intervals of non-equilibrium and equilibrium fuel compositions. Note that for the equilibrium fuel composition change of the reactor reactivity due to fuel burnup does not exceed the delayed neutron fraction and simplifies the regulation of the reactor.

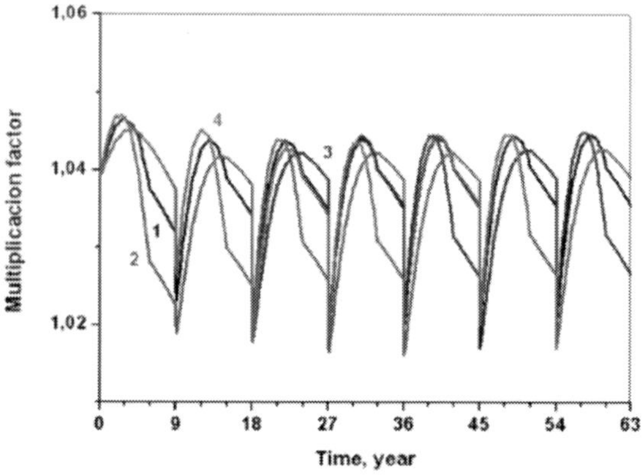

Figure 11. Multiplication factor of fuel assembly versus the time of reactor operation in the refuelling mode for different variants of reloading

Table 5. Multiplication factor of fuel assembly for different variants of reloading in the end of cycle with number *N*

N	1	2	3	4
2	1.0341	1.0380	1.0251	1.0251
5	1.0352	1.0354	1.0262	1.0385
7	1.0354	1.0390	1.0266	1.0390

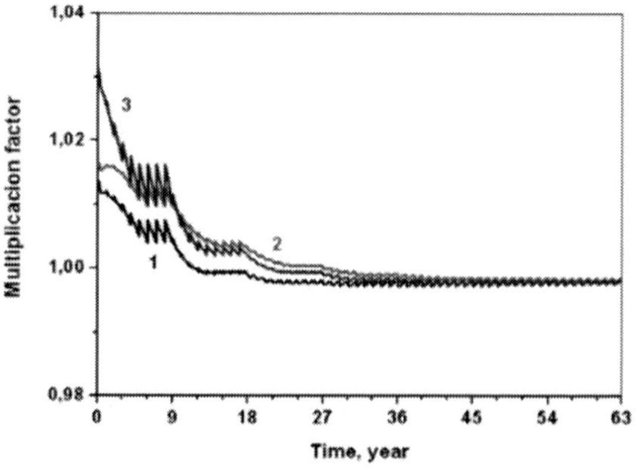

Figure 12. Multiplication factor of reactor versus the time of reactor operation in the refuelling mode for different plutonium compositions.

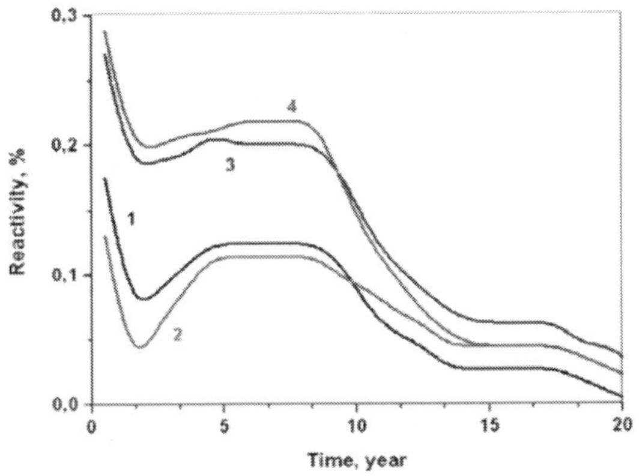

Figure 13. The maximum reactivity change of the reactor between the reloading during the first two cycles of the fuel assembly

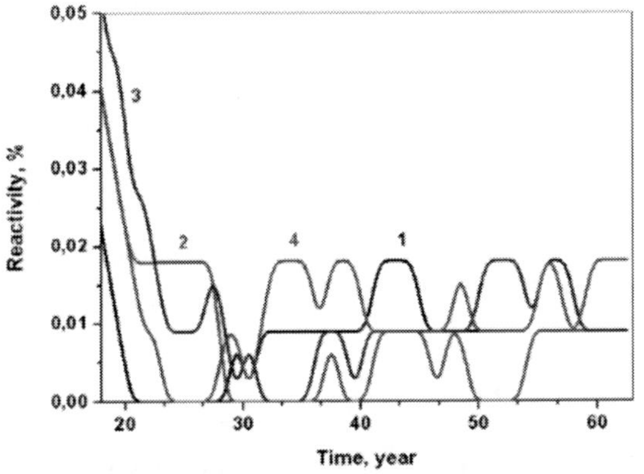

Figure 14. The maximum reactivity change between reloading after achievement of the equilibrium fuel composition

Note that calculation of the safety parameters of the reactor with equilibrium fuel composition was beyond the bounds of the purposes of this work.

4 Conclusions

These results demonstrate that fuel composition comes into equilibrium concentration in the multiple refueling reactor operation mode. If initial loads were based on plutonium from spent fuel of thermal and fast power reactors, equilibrium was achieved with twice repeated refueling. Initial fuel composition can be fit, which does not require further use of plutonium to compensate burned fuel mass. It is possible to find the mode of formation of secondary fuel and scheme of refuelinf, in which the change in reactivity between refuelings in the equilibrium mode does not exceed the fraction of delayed neutrons.

These conclusions do not apply to initial fuel loads based on highly enriched uranium nitride or uranium-plutonium nitride fuel with high enrichment of ^{239}Pu. For these fuels, equilibrium is reached after 4-5 refuelings.

References

1. V. I Rachkov, V. M. Poplavskii, A. M. Tsibulya, etc. Atomic Energy, **108**, 254-259 (2010).
2. V. M. Poplavskii, A. M. Tsibulya, Y. S. Khomyakov, etc. Atomic Energy, **108**, 260-266 (2010).
3. G. N. Vlaskin, V. I. Rachkov, Yu. S. Khomyakov, Atomic Energy, **116**, 311-315 (2014).
4. A. M. Sirotkin, M. Yu .Ternovykh, G. V. Tikhomirov, M. F. Khromova, Atomic Energy, **116**, 330-337 (2014).
5. V.P. Alferov, A.I. Radaev, M. V. Shchurovskaya, G. V. Tikhomirov, N.A. Hanan, F.A. Van Heerden, Ann. Nucl. Energy 77 (2015) 273–280.
6. A.M. Sirotkin, M.Y. Ternovykh, G.V. Tikhomirov, M.F. Khromova, At. Energy 116 (2014).
7. A.S. Gerasimov, V.N. Kornoukhov, I.S. Sald'ikov, G.V. Tikhomirov, At. Energy 116 (2014).
8. D.N. Skorokhodov, G.V. Tikhomirov, Phys. At. Nucl. 75 (2012).
9. A. N. Andrianov, V. G. Baranov, G. V. Tikhomirov, A. V. Khlunov, Atomic Energy, **104**, 463-469 (2008).
10. V. G. Baranov, M. Yu. Ternovykh, G. V. Tikhomirov, A. V. Khlunov, Atomic Energy, **105**, 391-396 (2008).
11. A. V. Moiseev, Yu. S. Khomyakov, M. Yu. Semenov, etc., *PHYSOR 2010*. **2**, 1577-1588 (2010).
12. I. R. Suslov, I. V. Tormyshev, O. G. Komlev, *Int. Conf. on Mathematics and Computational Methods Applied to Nuclear Science and Engineering*, **3**, 1655-1663 (2013).
13. A. D. Klimov, I. R. Suslov, I. V. Tormyshev, etc., Atomic Energy, **117**, 149-155 (2015).

Decay heat power of spent nuclear fuel of power reactors with high burnup at long-term storage

Mikhail Ternovykh [1], Georgy Tikhomirov [1,*], Ivan Saldikov [1] and Alexander Gerasimov [2]

[1]*National Research Nuclear University MEPhI (Moscow Engineering Physics Institute), Department of Theoretical and Experimental Physics of Nuclear Reactors, 31, Kashirskoye shosse, 115409, Moscow, Russia*
[2]*Institute for Theoretical and Experimental Physics, Department of safe nuclear reactors, 25, Bolshaya Cheremushkinskaya ul., 117218, Moscow, Russia*

Abstract. Decay heat power of actinides and fission products from spent nuclear fuel of power VVER-1000 type reactors at long-term storage is calculated. Two modes of storage are considered: mode in which single portion of actinides or fission products is loaded in storage facility, and mode in which actinides or fission products from spent fuel of one VVER reactor are added every year in storage facility during 30 years and then accumulated nuclides are stored without addition new nuclides. Two values of fuel burnup 40 and 70 MW·d/kg are considered for the mode of storage of single fuel unloading. For the mode of accumulation of spent fuel with subsequent storage, one value of burnup of 70 MW·d/kg is considered. Very long time of storage 10^5 years accepted in calculations allows to simulate final geological disposal of radioactive wastes. Heat power of fission products decreases quickly after 50-100 years of storage. The power of actinides decreases very slow. In passing from 40 to 70 MW·d/kg, power of actinides increases due to accumulation of higher fraction of ^{244}Cm. These data are important in the back end of fuel cycle when improved cooling system of the storage facility will be required along with stronger radiation protection during storage, transportation and processing.

1 Introduction

At present, safe nuclear thermal-neutron power reactors operating on uranium or uranium-plutonium fuel are used in nuclear power engineering of leading countries. Spent fuel is stored and accumulated in special storage facilities. During last tens years, advanced fuel with high burnup was developed and introduced in practice of nuclear power plants with light-water reactors. High burnup allows to increase economic efficiency of nuclear power plants and to decrease consumption of uranium. However application of nuclear fuel with high burnup influences strongly the back end of nuclear fuel cycle including interim and long-term storage of spent fuel, chemical processing, and ultimate disposal of radioactive wastes. Spent fuel of this type contains increased amount of high active actinides and fission products in comparison with common-type spent fuel. Therefore extended time of storage, improved cooling system of the storage facility will be required along with stronger radiation protection during storage, transportation and processing.

The main type of thermal-neutron power reactors in the world are light water reactors. Russian light water power reactors are named as VVER. They are the most commonly used in Russian nuclear power.

Typical burnup of fuel of VVER-1000 reactors was 40 MW·d/kg U during recent years. Now prospective fuel is studied which ensure increased burnup up to 70 MW·d/kg.

The promising area is going to the closed fuel cycle, in which the natural nuclear fuel and the artificial fissionable materials produced as a result of reactor operation (plutonium and the like) are used to a much more complete extent. Fast nuclear reactors should be the base of closed fuel cycle. The burnup of the nuclear fuel in fast reactors is several times higher than in thermal neutron reactors. This leads to a respective increase in the fission products and minor actinides buildup both in the reactor core and in the spent nuclear fuel. High-active actinides and fission products are accumulated in spent fuel in the process of burnup. They determine the decay heat power and gamma-radiation during subsequent storage, transportation, reprocessing. Short-lives nuclides decay during the short interim storage in several years after the discharge. Therefore nuclides with half-life longer than 10 years present the most radiation danger. These nuclides are ^{90}Sr, ^{137}Cs, ^{244}Cm, ^{238}Pu, ^{241}Am.

Decay heat power of actinides and fission products from spent nuclear fuel of power VVER-1000 type reactors at long-term storage is studied in this paper. Calculated data for two fuel burnups 40 and 70 MW·d/kg are discussed. Two modes of storage are considered: mode in which single portion of actinides or fission products is loaded in storage facility, and mode in which

* Corresponding author: GVTikhomirov@mephi.ru

actinides or fission products from spent fuel of one VVER reactor are added every year in storage facility during 30 years and then accumulated nuclides are stored without addition new nuclides.

2 Annual unloading of americium and curium isotopes from the VVER-1000 reactor

The masses of important fission products ^{90}Sr и ^{137}Cs along with long-lived ^{99}Tc and ^{126}Sn in annual discharge of the VVER are mainly the function of the power produced rather than of the mode of burnup. On the contrary, the quantity of the annually discharged different minor actinides increases in different ways with the burnup increase.

We calculated isotopic composition of spent fuel of VVER-1000 type reactor for burnup from 40 to 70 MW·d/kg. Initial enrichment 4.4% was accepted for burnup of 40 MW·d/kg with corresponding enrichment increase for higher burnup. Calculations were made using MCU, ORIGEN-SCALE codes [1-6]. Decay heat power in storage facility was calculated separately for actinides and fission products. Isotopes of neptunium, plutonium, americium, and curium were meant as actinides. We assumed that these isotopes were placed into storage facility as actinides, while uranium isotopes extracted from spent fuel in the chemical processing were returned into fuel cycle. Additionally, plutonium isotopes were produced as result of transformations of initial americium and curium isotopes. Power of alpha, beta, and gamma-radiation was taken for calculation of the decay heat power. Data on energy release in alpha and beta decays were taken from [7, 8].

Results of calculations of the main actinide masses from the annual discharge of the VVER-1000[9,10] reactor for the burnup of 40-70 MW·d/kg are presented in Table 1 and are illustrated in the Figure 1 for more wide limits of burnup.

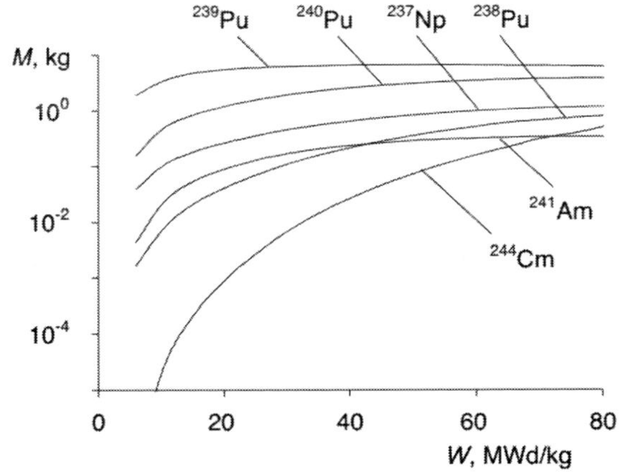

Fig. 1. Masses of main actinides in spent fuel of VVER-1000 reactor at different burnup of fuel

Table 1. Calculated data on annual unloading of actinides from VVER-1000 reactor at different burnup *B*, kg/year

Nuclide	*B*, MW·d/kg			
	40	50	60	70
^{237}Np	14.4	14.8	14.6	14.0
^{238}Pu	4.93	6.57	7.90	8.74
^{239}Pu	146	117	96.7	81.3
^{240}Pu	57.2	55.1	51.8	47.8
^{241}Pu	28.2	26.9	24.8	22.4
^{242}Pu	11.8	15.5	18.6	21.3
^{241}Am	5.25	5.10	4.73	4.27
^{243}Am	2.19	3.57	4.96	6.14
^{242}Cm	0.0029	0.0038	0.0044	0.0045
^{243}Cm	0.0063	0.0106	0.0144	0.0175
^{244}Cm	0.529	1.17	2.09	3.27

These data demonstrate that the increase of burnup from 40 to 70 MW·d/kg results in the corresponding increase of annual unloading of ^{238}Pu by 1.8 times while the mass of ^{244}Cm increases by 6.2 times.

3 Decay heat power of fission products and actinides from single unloading of spent fuel of VVER-1000 reactor in long-term storage

Calculated decay heat power of main fission products and actinides from 1 ton of spent fuel of VVER-1000 reactor with burnup of 40 and 70 MW·d/kg at long-term storage during the time *T* is presented in Tables 2 and 3. We assumed that initial moment of storage corresponded to the end of 3-year cooling after unloadung from the reactor. We considered ^{237}Np, ^{238}Pu, ^{239}Pu, ^{240}Pu, ^{241}Pu, ^{242}Pu, ^{241}Am, ^{243}Am, ^{242}Cm, ^{243}Cm, and ^{244}Cm as actinides initially loaded in storage facility. Analogously, initialy loaded fission products were ^{90}Sr, ^{99}Tc, ^{106}Ru, ^{125}Sb, ^{126}Sn, ^{137}Cs, ^{144}Ce, ^{154}Eu, ^{155}Eu.

We considered such very long time of storage 10^5 years which allows to simulate final geological disposal of radioactive wastes.

These data demonstrate that sum power of actinides in the initial period of storage increases by 4.6 times in going from burnup of 40 to burnup of 70 MW·d/kg. This is explained by the fact that power of actinides is first of all due to high-active nuclide ^{244}Cm, and secondly due to ^{238}Pu and ^{241}Am. Analogously sum power of fission products in the beginning of storage increases by 1.5 times when passing from 40 to 70 MW·d/kg. This is explained by the fact that power of fission products is due to respectively long-lived nuclides ^{90}Sr and ^{137}Cs with half-life about 30 years, and due to more short-lived nuclides ^{144}Ce and ^{106}Ru.

Table 2. Calculated decay heat power *P* of actinides and fission products FP at long-term storage for burnup of 40 MW·d/kg U

T, year	*P*, W		
	Actinides	FP	Total
1	258	1950	2210
10	264	960	1220
100	220	110	330
1000	64	0.03	64
10000	15	0.02	15
100000	0.85	0.01	0.86

Table 3. Calculated decay heat power *P* of actinides and fission products FP at long-term storage for burnup of 70 MW·d/kg U

T, year	*P*, W		
	Actinides	FP	Total
1	1180	2990	4170
10	890	1540	2430
100	440	180	620
1000	90	0.06	90
10000	21	0.04	21
100000	1.6	0.02	1.6

The power of fission products in the beginning of storage exceeds the power of actinides by 7 times for burnup of 40 MW·d/kg and by 2.5 times for burnup of 70 MW·d/kg. The power of fission products decreases quickly after 50-100 years of storage. So, for burnup of 40 MW·d/kg the power of fission products decreases by 3.3 times after 30 years, by 18 times after 100 years, by 1900 times after 300 years of storage. These data are about the same for burnup of 70 MW·d/kg.

The power of actinides decreases very slowly in the process of storage because of the long half-life of the main actinides. So, for burnup of 40 MW·d/kg the power of actinides is practically flat during the first 100 years of storage, decreases by 4 times after 1000 years, by 17 times after 10000 years of storage. For burnup of 70 MW·d/kg the power of actinides decreases by 2.7 times during the first 100 years of storage, by 13 times after 1000 years, by 56 times after 10000 years of storage.

The power of actinides makes equal to the power of fission products after 40-50 years of storage. At higher times of storage, the power of actinides far exceeds that of fission products.

Figures 2 and 3 demonstrate decay heat power of separate fission products for fuel burnup 40 and 70 MW·d/kg during long-term storage. Figures 4 and 5 present decay heat power of separate actinides for burnup 40 and 70 MW·d/kg. Figure 6 shows the power of gamma-radiation of actinides for burnup of 70 MW·d/kg.

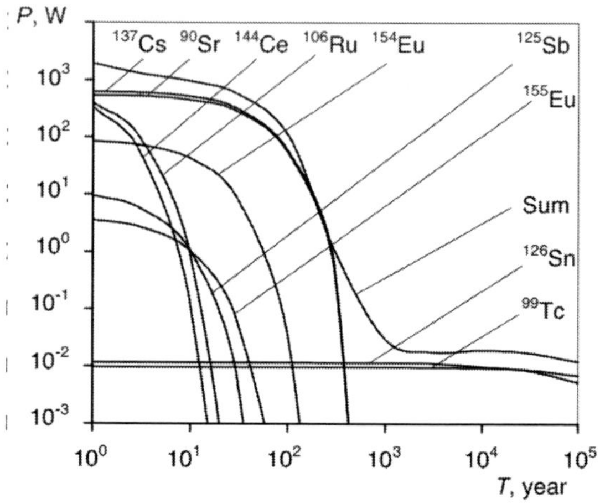

Fig. 2. Decay heat power of fission products from 1 ton of spent fuel of VVER-1000 reactor for fuel burnup 40 MW·d/kg at long-term storage

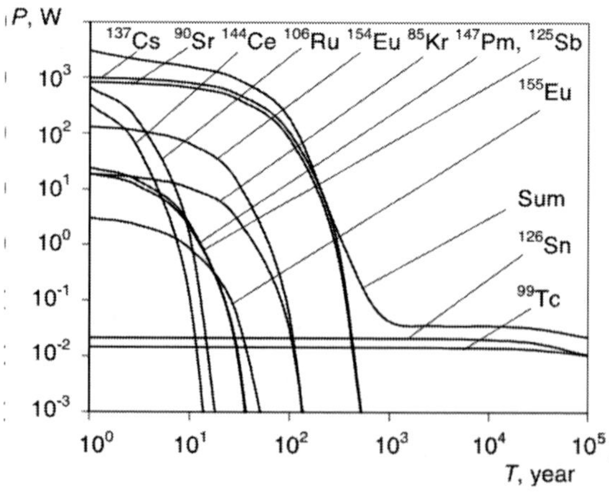

Fig. 3. Decay heat power of fission products from 1 ton of spent fuel of VVER-1000 reactor for fuel burnup 70 MW·d/kg at long-term storage

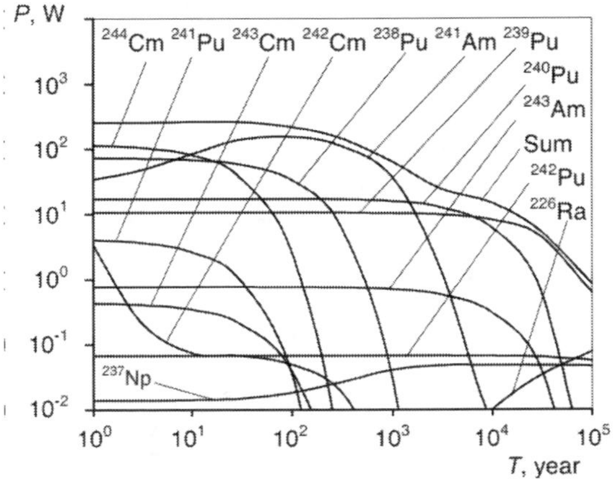

Fig. 4. Decay heat power of actinides from 1 ton of spent fuel of VVER-1000 reactor for fuel burnup 40 MW·d/kg at long-term storage

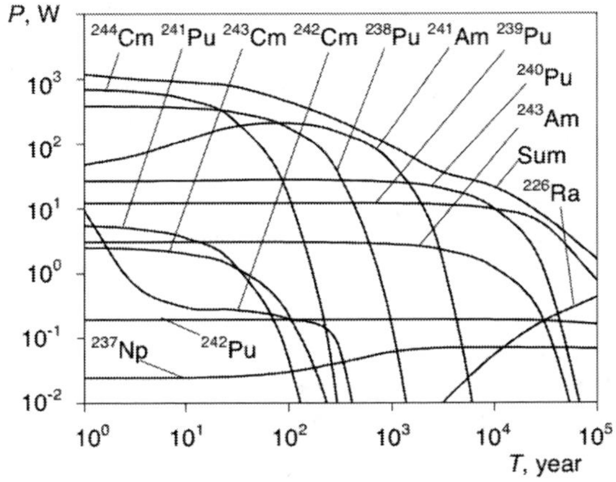

Fig. 5. Decay heat power of actinides from 1 ton of spent fuel of VVER-1000 reactor for fuel burnup 70 MW·d/kg at long-term storage

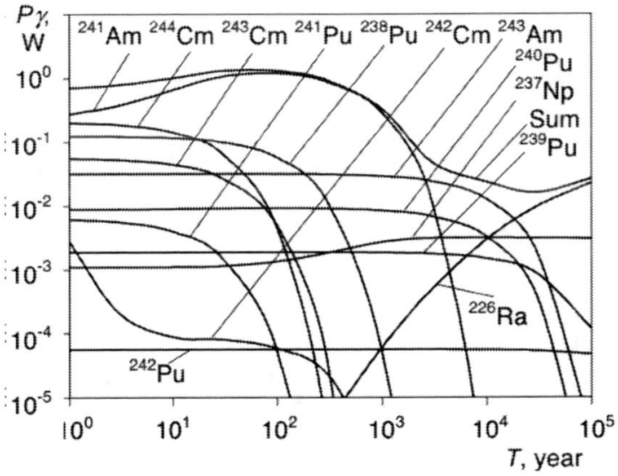

Fig. 6. Power of gamma-radiation of actinides from 1 ton of spent fuel of VVER-1000 reactor for fuel burnup 70 MW·d/kg at long-term storage

4 Decay heat power of fission products and actinides from spent fuel of VVER-1000 reactor for accumulation and subsequent long-term storage

Calculated data on decay heat power of separate fission products for the mode of accumulation and subsequent storage are presented on the Figure 7. Fuel burnup was equal to 70 MW·d/kg. Analogous data for separate actinides are shown in the figure 8. In the mode of accumulation, fission products or actinides from spent fuel of one VVER-1000 reactor are loaded into storage facility every year during 30 years. Then accumulation is finished. Accimilated nuclides are stored in the storage facility without addition other nuclides during the period of subsequent storage up to 100000 years. So very long time of storage 10^5 years allows to simulate final geological disposal of radioactive wastes.

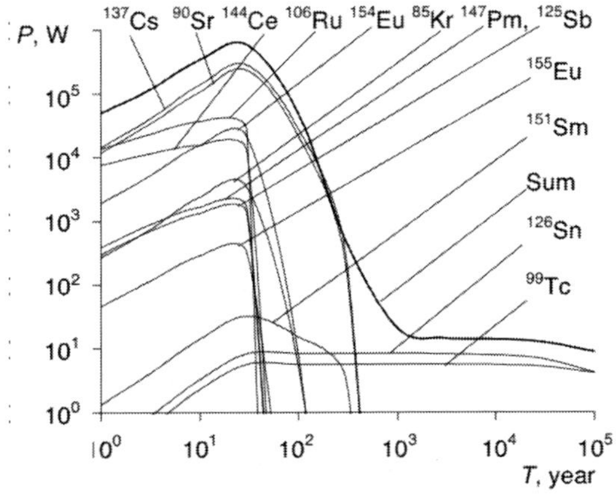

Fig. 7. Decay heat power of fission products in the mode of accumulation and subsequent long-term storage

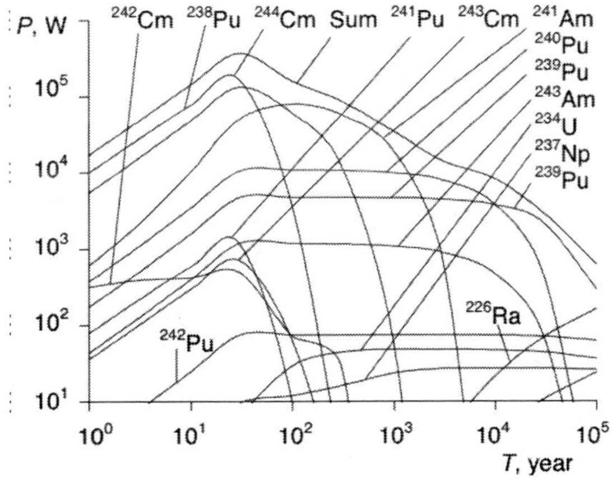

Fig. 8. Decay heat power of actinides in the mode of accumulation and subsequent long-term storage

The decay heat power of actinides for the case when only isotopes of neptunium, americium, and curium were placed into storage facility in the period of accumulation is presented in Figure 9. In this case we assumed that isotopes of uranium and plutonium were returned in fuel cycle while isotopes of neptunium, americium, and curium were directed to storage. Calculations were performed for fuel burnup of 70 MW·d/kg and for the time of accumulation of 30 years.

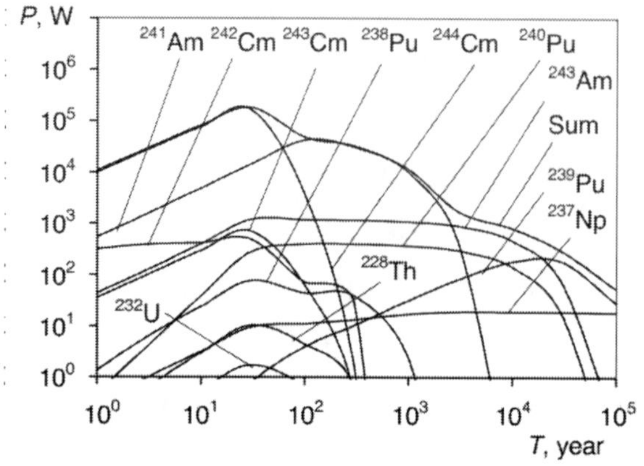

Fig. 9. Decay heat power of actinides in the mode of accumulation of neptunium, americium, and curium isotopes with subsequent long-term storage

5 The discussion of the results

The results presented above concerning to the long-term storage of fission products and actinides from single unloadung of spent fuel demonstrate the following.

Decay heat power of fission products (Figures 2 and 3) during the period of storage of about 30-300 years is mainly due to ^{90}Sr and ^{137}Cs with half-life periods of about 30 years. The other curves on Figures 2 and 3 correspond to more short-lived nuclides (except ^{99}Tc and ^{126}Sn). Their contribution to the decay heat power decreases quickly. Nuclides ^{90}Sr and ^{137}Cs decay after about 300 years. The power decreases by 4-5 orders of magnitude to the end of this period of time. After the decay of ^{90}Sr and ^{137}Cs, the long-lived ^{99}Tc and ^{126}Sn give the main share to the power.

When passing from fuel burnup 40 to 70 MW·d/kg, the decay heat power of fission products increases by 1.5-1.6 times for the period of storage less than 100 years (due to ^{90}Sr and ^{137}Cs), and by 2 times for the period of storage more than 1000 years (due to ^{99}Tc and ^{126}Sn).

Decay heat power of actinides (Figures 4 and 5) is mainly due to ^{244}Cm during the first 30 years. After that, the power is mainly due to ^{241}Am and ^{238}Pu in the period of storage of 30-100 years. The main nuclide is ^{241}Am in the period of 100-1000 years. Then the main nuclides are ^{240}Pu in the period of 10^3-10^4 years, and ^{239}Pu in more long period of time. Contribution of ^{237}Np is negligible during the whole period of storage.

The power of gamma-radiation of actinides (Figure 6) is due to ^{241}Am during 5000 years of storage. After 30000 years of storage, the power of gamma-radiation is due to ^{226}Ra, which was accumulated in storage as result of decays of other actinides.

Let's pass to the mode of accumulation and subsequent storage of fission products and actinides.

For the mode of accumulation of fission products in storage facility during 30 years, the power of fission products (Figure 7) icreases during the whole period of accumulation mainly due to ^{90}Sr and ^{137}Cs. Contribution of other short-lived fission products is much less. After the end of the period of accumulation, decay heat power recreases smoothly by 6-7 orders of magnitude during following 1000 years due to decay of ^{90}Sr and ^{137}Cs. After this period of time, the power decreases slowly due to very long lived nuclides ^{99}Tc and ^{126}Sn.

Decay heat power of actinides in the mode of accumulation with subsequent storage (Figure 8) demonstrates the same time dependensy as for long-term storage of a single unloading. In the period of accumulation the power increases mainly due to accumulation of ^{244}Cm and ^{238}Pu. After the end of the period of accumulation, the power decreases slowly due to decay of ^{238}Pu, and then due to decay of ^{241}Am, ^{240}Pu. The main nuclides are ^{238}Pu (50-100 years), ^{241}Am (100-2000 years), ^{240}Pu (2000-10000 years), ^{239}Pu (more than 10000 years).

Maximum decay heat power of fission products exceeds by 1.6 times the maximum decay heat power of actinides. It approximately corresponds to the end of the period of accumulation.

When we accumulate only isotopes of neptunium, americium, and curium (Figure 9) then maximum power is about 2 times lower with respect to the power with account of accumulation of ^{238}Pu (Figure 8). ^{241}Am makes basic contribution to total decay heat power of actinides in this case. After 3000 years of storage, the nain isotop is ^{243}Am, and then ^{239}Pu, produced in transformations of other actinides.

6 Conclusions

These data allow us to estimate the decay heat power and the gamma radiation level in long-term storage facility. Simple calculations for storage with a period of accumulation of nuclides in long-term storage facility can be used to describe the increase of energy release in the period of accumulation of radioactive waste and the decrease of energy release during period of storage due to decay of the nuclides. So long-term storage period of 100 000 years accepted for calculations allows to simulate final disposal of radioactive waste in geological formations.

In the period of accumulation in long-term storage, energy is determined mainly by fission products ^{90}Sr and ^{137}Cs. These nuclides decay during about 300 years after the end of the accumulation period, while the power of the actinides reduces slowly. During storage without the addition of spent fuel, the decay heat power determines mainly by the actinides. After 1000 years of storage, the total energy release is reduced only by 15 times in comparison with the maximum energy corresponding to the end of the period of accumulation in storage. Gamma radiation in the period of accumulation is determined mainly by fission products. The power of the gamma radiation of fission products is 3-4 times lower than the total power of energy release of the fission products, and gamma radiation of actinides by several orders of magnitude lower than their total decay heat power.

The maximum decay heat power of actinides from spent uranium fuel with a burnup of 70 MW·day/kg is about four times higher than that of for uranium fuel with burnup of 40 MW·d/kg due to higher accumulation of ^{244}Cm.

Thus, the use of fuel with high burnup may lead to a 50% increase in the total decay heat power of the fission products and actinides in long-term storage compared with common-type spent fuel. The fission products give a significant contribution to the total decay heat power only during the period of accumulation of spent fuel in long-term storage.

References

1. Kalugin, M.A., Oleynik, D.S., Shkarovsky, D.A. Annals of Nuclear Energy, Volume 82, 28 July 2015, Pages 54-62
2. SCALE: *A Modular Code System for Performing Standardized Computer Analyses for Licensing Evaluation*, Version 5, ORNL/TM-2005/39, **I-III**, ORNL (2005)
3. V.P. Alferov, A.I. Radaev, M. V. Shchurovskaya, G. V. Tikhomirov, N.A. Hanan, F.A. Van Heerden, Ann. Nucl. Energy 77 (2015) 273–280.
4. A.M. Sirotkin, M.Y. Ternovykh, G.V. Tikhomirov, M.F. Khromova, At. Energy 116 (2014).
5. A.S. Gerasimov, V.N. Kornoukhov, I.S. Sald'ikov, G.V. Tikhomirov, At. Energy 116 (2014).
6. D.N. Skorokhodov, G.V. Tikhomirov, Phys. At. Nucl. 75 (2012).
7. ICRP, 2008. Nuclear Decay Data for Dosimetric Calculations. ICRP Publication 107. Ann. ICRP 38 (3)
8. Brown, D.A. Nuclear Data Sheets. Volume 118, Issue 1, April 2014, Pages 98-103.
9. B.R. Bergelson, V.V. Belonog, A.S. Gerasimov, G.V. Tikhomirov, At. Energy 109 (2011).
10. B.R. Bergelson, V.V. Belonog, A.S. Gerasimov, G.V. Tikhomirov, At. Energy 107 (2009).

Consistent neutron-physical and thermal-physical calculations of fuel rods of VVER type reactors

Georgy Tikhomirov[1,*], *Ivan* Saldikov[1], *Mikhail* Ternovykh[1] , and *Alexander* Gerasimov[2]

[1]National Research Nuclear University MEPhI (Moscow Engineering Physics Institute), Department of Theoretical and Experimental Physics of Nuclear Reactors, 31, Kashirskoye shosse, 115409, Moscow, Russia.
[2]Institute for Theoretical and Experimental Physics, Department of safe nuclear reactors, 25, Bolshaya Cheremushkinskaya ul., 117218, Moscow, Russia.

Abstract. For modeling the isotopic composition of fuel, and maximum temperatures at different moments of time, one can use different algorithms and codes. In connection with the development of new types of fuel assemblies and progress in computer technology, the task makes important to increase accuracy in modeling of the above characteristics of fuel assemblies during the operation. Calculations of neutron-physical characteristics of fuel rods are mainly based on models using averaged temperature, thermal conductivity factors, and heat power density. In this paper, complex approach is presented, based on modern algorithms, methods and codes to solve separate tasks of thermal conductivity, neutron transport, and nuclide transformation kinetics. It allows to perform neutron-physical and thermal-physical calculation of the reactor with detailed temperature distribution, with account of temperature-depending thermal conductivity and other characteristics. It was applied to studies of fuel cell of the VVER-1000 reactor. When developing new algorithms and programs, which should improve the accuracy of modeling the isotopic composition and maximum temperature in the fuel rod, it is necessary to have a set of test tasks for verification. The proposed approach can be used for development of such verification base for testing calculation of fuel rods of VVER type reactors

1 Introduction

Simplified models of calculation are often used in traditional approach to modeling neutron-physical characteristics of fuel rods and isotopic composition in process of nuclear fuel burning[1-5]. For instance, in many cases neutron-physical characteristics of fuel depending on temperature are determined by use of average temperature of fuel rod, while the existing dependencies of the neutron cross sections on temperature are neglected. In thermal-physical calculations of the radial distribution of fuel temperature, constant average value of thermal conductivity factor is often used. However this factor depends really both on the temperature and on fuel burnup, which change with radius of fuel rod.

In this paper, mathematical models of consistent neutron-physical and thermal-physical calculation are presented. They are realized in codes designed in MEPhI and "Kurchatov Institute". This approach was applied to calculation investigation of fuel element of the power reactor of VVER-1000 type. Short description of benchmarks for verification of obtained results is given. It is demonstrated how the account of dependence of the thermal conductivity factor on temperature affects on the value of temperature in the center of fuel rod for the VVER type reactor.

2 Algorithms and methods

In order to obtain the dependence of temperature on radius of cylindrical fuel rod it is necessary to solve the equation of thermal conductivity with specified boundary conditions. If density of energy release and thermal conductivity factor are constant independent on radius of fuel rod, this problem has an analytical solution. However, thermal conductivity factor of oxide nuclear fuel depends on temperature and on fuel burnup. Energy release density and fuel burnup vary non-uniformly with radius of fuel rod in process of fuel burnup. Swelling of fuel pellet under deep burnup results in reduction of the gap between fuel pellet and cladding.

To value the contribution of each of these effects in change of maximum temperature of fuel rod, it is necessary to be able to calculate fuel burnup and change of isotopic composition, as well as distribution of energy release density on radius of fuel rod. For that it is necessary to solve multigroup equation of transport or diffusion of neutrons and system of equations of isotope kinetics. Remember that usually these equations are solved one after another. Firstly transport equation is solved with macroscopic constants calculated on the base of the known isotopic composition. Then one-group cross sections of all isotopes belonging to equations of

* Corresponding author: GVTikhomirov@mephi.ru

isotope kinetics are defined. Hereon the system of the equations of isotope kinetics is solved and new isotopic composition for the given moment of time is calculated. After that, we should solve once again transport equation since change of isotopic composition can result in change of macroscopic constants and of neutron spectrum.

Let's consider mathematical settings of the above described tasks.

2.1 Neutron diffusion equation

To obtain the one-grup cross sections and neutron fluxes, it is necessary to simulate neutron spectrum and hence to solve the neutron transport equation or diffusion equation. Usually multigroup approach is used for this purpose, in which one must solve a system of multigroup equations like (1). All symbols are traditional, g is the number of the group while G is a full number of groups.

$$\nabla D(\vec{r})\nabla\Phi^{g}(\vec{r})+\Sigma_{t}^{g}(\vec{r})\cdot\Phi^{g}(\vec{r})=\sum_{g'=1}^{G}\Sigma_{s}^{g'\to g}(\vec{r})\cdot\Phi^{g'}(\vec{r})+$$
$$+\frac{\chi^{g}}{4\pi}\frac{1}{K_{эфф}}\sum_{g'=1}^{G}\cdot\nu_{f}^{g'}\cdot\Sigma_{f}^{g'}(\vec{r})\cdot\Phi^{g'}(\vec{r}) \qquad (1)$$

The solution of these equations in particular geometry with specified boundary conditions allows to determine the neutron spectrum that can be used to obtain the one-group constants. To determine one-group neutron fluxes it is necassaru to calculate additionaly the spatial distribution of energy release density and know the full energy release in a simulated system.

We should keep in mind that diffusion coefficients $D(\vec{r})$ and macroscopic cross-sections $\Sigma^{g}(\vec{r})$ depend not only on the spatial variable, but also on the temperature in the spatial domain. In order to make correct system of multigroup equations with a purpose to prepare prepare one-group cross sections and fluxes, one need to set or be able to calculate the temperature in spatial areas.

2.2 Equation of isotope kinetics

System of equations of isotope kinetics is formulated for vector of isotope concentrations of nuclides $\vec{\rho}^{\,i}(t)$. Isotope concentrations in each spatial zone with number i are assumed to be the same in all points of spatial sone.

$$\frac{d\rho_{l}^{i}(t)}{dt}=\sum_{k}\lambda_{k\to l}\cdot\rho_{l}^{i}(t)-\lambda_{l}\cdot\rho_{l}^{i}(t)+$$
$$+\sum_{m}\rho_{m}^{i}(t)\cdot\sigma_{m\to l}^{i}\cdot\phi^{i}-\rho_{l}^{i}(t)\cdot\sigma_{l}^{i}\cdot\phi^{i} \qquad (2)$$

Each equation of the system of equations of isotope kinetics include the following values in addition to concentrations.

σ_{l}^{i} - one-group cross section of incineration of isotope with number l in spatial zone with number i due to nuclear reactions on l-th isotipe.

$\sigma_{m\to l}^{i}$ - one-group cross section of appearance of isotope with number l in spatial zone with number i due to nuclear reactions on m-th isotipe.

ϕ^{i} - one-group neutron flux in i-th spatial zone.

λ_{l} - decay constant of l-th isotope

$\lambda_{k\to l}$ - decay constant of k-th isotope resulting in appearance of l-th isotope.

Vector of concentrations changes in process of burning of nuclear fuel. Concentrations of isotopes may vary in different spatial locations of the fuel core. For example, area with modified microstructure, so-called rim-zone appeares in the outer layer of the fuel rod of thermal-neutron reactors. Its formation is associated with high concentration of fission products [6-8]. Therefore, application of constant concentrations over the whole fuel rod can lead to distortion of the simulation results. Along with isotope concentrations, other properties of fuel composition can vary with radius within fuel rod in the process of burnup and can influence on space distribution of the temperature. These variation of temperature don't take into account in the "classic" approach, and the process of formation of rim-zone can't be calculated.

In order to solve a system of equations (2), one must firstly calculate one-group fluxes ϕ^{i} and the cross section of the reactions σ_{l}^{i} and $\sigma_{m\to l}^{i}$ from the equation (1). It is also necessary to know the decay constants λ_{l} and $\lambda_{k\to l}$ for each nuclide and initial concentrations of isotopes ρ_{l}^{i}. After the solution of the equation system (2), the new value of the concentration vector $\vec{\rho}^{\,i}(t)$ is obtained, which corresponds to a certain moment of time.

2.3 Equation of thermal conductivity

In "classic" approach, the factor of thermal conductivity is considered to be independant on the radius within fuel rod, that allows to obtain an analytical decision, if radial distribution of the energy release density $q_{v}(r)$ is known.

However, really thermal conductivity factor of oxide nuclear fuel depends on the temperature T of this fuel and fuel burnup B [9] (Fig. 1).

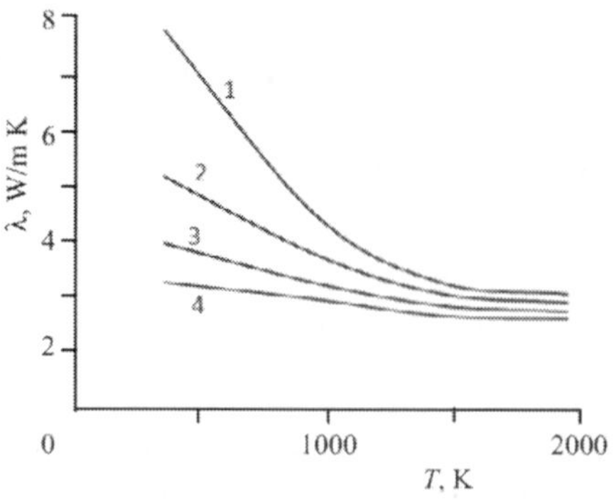

Fig. 1. Thermal conductivity factor of oxide fuel depending on temperature at burnup B = 0 (*1*), 30 (*2*), 60 (*3*), 90 (*4*) MW·d/kg U

Thus, equation of thermal conductivity is nonlinear one. In cylindrical geometry:

$$\frac{1}{r}\cdot\frac{\partial}{\partial r}\lambda(T(r),B)\cdot r\cdot\frac{\partial}{\partial r}T(r)+q_V(r)=0 \qquad (3)$$

where $\lambda(T(r),B)$ is thermal conductivity factor of fuel matherial depending on temperature and burnup, q_v is energy release density.

Several methods can be used to solve equation (3). One of them consists in dividing the fuel rod into annular zones. Factors of thermal conductivity remain constant within each zone : $\lambda(T(\mathrm{r})) = \{\lambda\,(T_i)\} = \{\lambda_i\}$.

Then the thermal conductivity equation becomes linear within each zone. However, in this case we need to use iterative procedure. Each iteration specifies the value of thermal conductivity factor in each layer. This allows to consider the dependence of thermal conductivity factor on temperature. Similarly, we can proceed with the dependence of thermal conductivity factor on fuel burnup : in each step of burnup to take from the table the new value of the factor of thermal conductivity and insert it into the equation to find the temperature.

3 Complex model of neutron-physics calculation of fuel cell

All of the above described relations and methods of their descriptions can be represented in a model of consistent thermal physical and neutron-physical calculations. In Fig. 2 the following designations are used: Eq.1 – transport equation, Eq.2 – equation of isotope kinetics, Eq.3 – equation of thermal conductivity. DB1 (data base) – dependence of microscopic cross sections from neutron energy, account of shielding, account of the temperature dependence of microscopic cross sections, DB2 – library of decay constant. Σ, σ are macroscopic and microscopic cross sections of interaction of neutrons with nuclei of the medium. **Φ** – absolute neutron flux (matched with the power and one-group cross sections), φ – multi-group neutron flux (the result of solving the transport equation). ρ is the vector of the nuclide concentration (the result of solving the system of equations of isotopic kinetics). q_v – energy release density. B – fuel burnup. T – temperature (the solution to the thermal conductivity equation). A1, A2, A3, A4, A5 – the algorithms that implement the ratio between the required values based on the model and experimental data :

A1. $\Sigma_{tot}(\vec{r},E,\overline{\rho}(t),\overline{\sigma}_{tot}(T,t)) = \overline{\sigma}_{tot}(T,t)\cdot\overline{\rho}(t)$

A2. Φ = C·φ, Φ – neutron flux, C – constant, φ – eigenvalues.

$\Sigma_{tot}(\vec{r},E,\overline{\rho}(t),\overline{\sigma}_{tot}(T,t))$ total macroscopic cross section of interaction of neutrons with energy E with nuclei in spatial point $\vec{r}$, which for each time moment t depends also on nuclei density $\overline{\rho}(t)$ and on microscopic cross sections of isotopes $\overline{\sigma}_{tot}(t)$.

A3. $q_v = \Sigma_f E_f\cdot\Phi + \Sigma_c\cdot E_c\cdot\Phi$ - energy release density.

A4. $B=\int q(t)dt$ - equation of burnup.

A5. Account of dependence of thermal conductivity on temperature and fuel burnup: $\lambda = \lambda(T, B)$.

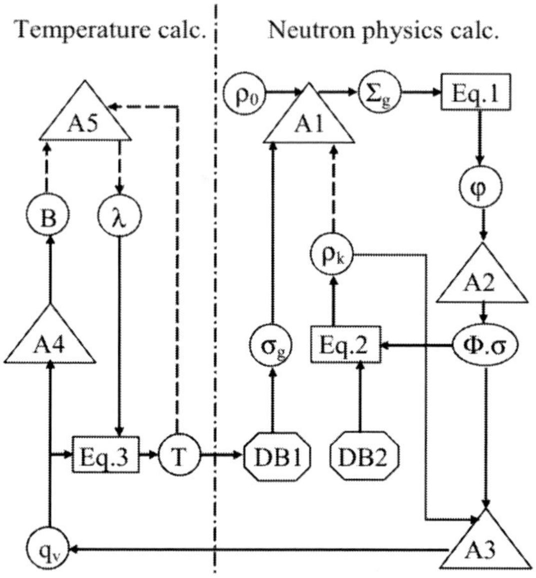

Fig. 2. Diagram of consistent thermal physical and neutron-physical calculations

Below we propose 4-steps sequence of calculations of neutron-physical characteristics of fuel rods based on diagram Fig. 2. It allows to simulate physical and neutron-physical characteristic of fuel more correctly and increases accuracy of the results.

1. We assign initial value of energy release density (q_v) and thermal conductivity factor λ. Then iterative procedure is made. Firstly, thermal conductivity equation (Eq.3) is solved. The result is temperature value in the cell. In case the cell is divided into zones, we obtain values of the temperature in each zone of the cell. Secondly we calculate new value of the thermal conductivity factor λ, corresponding to new value of the temperature. These two steps can be repeated many times to obtain required accuracy of the temperature. Simultaneously we get more accurate λ value.

2. This temperature value is used as input parameter for database (DB1) of the temperature dependencies of the microscopic cross sections σ. Using these data of σ together with initial values of nuclide concentrations (ρ_0), we obtain value of macroscopic cross section (Σ_g) by means of algorithm (A1).

3. With this macroscopic cross section we solve the transport equation (Eq.1). As a result, due to algorithm A2, we obtain group neutron fluxes in zones (Φ_g), that is to say spatial dependency of group neutron fluxes. At the same time, we save new values of the microscopic cross sections (Φ σ). We use the group neutron fluxes to calculate new value of energy release. We repeat the steps 1-3 as iterations over the energy release density (algorithm A3). Herewith we make more accurate the temperature distribution and take into account the temperature dependency of the thermal conductivity factor.

4. Using group neutron fluxes along with decay constants (DB2) and new values of the microscopic cross sections, we solve the equation of isotope kinetics (Eq.2). As a result we obtain isotope concentrations on the next step of burnup (ρ_k). With these new values of isotope concentrations and values of microscopic cross sections we obtain new value of macroscopic cross sections (algorithm A1). The steps 3-4 can be repeated several times as iterations.

We can perform iterations over the steps 1-4 to achieve required accuracy. Herewith we shall take into account the dependency of λ on fuel burnup.

Thereby, new complex approach allows to take into account:

– spatial distribution of energy release to determine neutron flux;

– dependence of the diffusion factor and macroscopic cross section not only on spatial value, but also on the temperature in given spatial area;

– presence of spatial-nonuniform structure in fuel rod (rim-zone);

– dependence of thermal conductivity factor of oxide fuel on the temperature of fuel;

– reduction of thermal conductivity factor of fuel in process of burnup.

In order to appreciate efficiency of the presented approach to calculation of temperature distribution over the radius of fuel rod, it is necessary to answer the following questions:

1. How do estimations of the maximum temperature of fuel change when turning from previous approaches to new approaches?

2. Is it necessary to change traditional schemes of calculation of isotopic composition in process of fuel burnup by including in them block of calculation of the temperature distribution?

3. How should we calculate fuel temperature if thermal conductivity factor depends on the temperature and fuel burnup?

4 Algorithms and codes used for calculations

Thermal-physical calculations in this paper were made using codes HEATING7, UNK_teplo, TempR_5.

HEATING7 is multipurpose code for solution of thermal conductivity tasks [10, 11]. It is included into code complex SCALE [12]. However it is not integrated code of the SCALE because it doesn't have standard format of input data and doesn't use in functional sequences of the SCALE.

Heatinh7 can solve stationary and nonstationary thermal conductivity tasks in one-, two- or three-dimensional square, cylindrical or spherical coordinates. Calculating model can include different materials. Thermal conductivity, density and energy release of each material can depend on time and temperature. Energy release can depend on time, temperature, space. Thermal conductivity can be anisotropic. Phase change can be taken into account. Thermal properties of materials can be included by user or can be taken from the own library of material thermal properties.

The energy release may depend on time, temperature, position, and boundary temperatures, which in turn may depend on time and coordinates. The boundary conditions may be assigned on the outer surface of the model or on the surface between parts of the model. They can be specified by temperatures or by any combination of heat flux, forced convection, natural convection and radiation. The parameters of the boundary conditions can depend on time and temperature. The spatial lattice for calculations can be arbitrary for any variable. HEATING uses the scheme of memory distribution during runtime, which allows to avoid the need of re-calculations and to meet the memory requirements for any tasks.The code UNK_teplo can solve the stationary thermal conductivity tasks in one-dimensional cylindrical coordinates with different materials. Density and thermal conductivity materials should be assigned for each layer. Numerical methods are used for calculation of temperature in each layer.

The code TempR_5 allows to calculate temperatures in cylindrical fuel element with piecewise constant thermal conductivity factor. Temperature distribution in each layer is obtained from analytical decision of thermal conductivity equation with assigned constant values of radius, thermal conductivity factor, and temperature. The results of calculation are average values of temperature over the layer, temperatures in points between layers and in the center of fuel element.

UNK_teplo is code for solution of thermal conductivity tasks. Code is included into program complex UNK [10, 13]

UNK_teplo can solve the stationary thermal conductivity tasks in one-dimensional cylindrical coordinates. Calculation model may include various materials. Thermal conductivity and density of each material is specified separately for each layer. Thermal properties of materials can be entered by the user.

Numerical methods are used to obtain the temperature values in each layer. The system of equations compiled according to the finite-difference scheme, then the equations are solved by sweep method.

The calculation of temperatures in the UNK complex is made in coordination with calculation of fuel burnup. Firstly, macro constants are prepared and the temperatures are calculated in different zones of the given geometry. After that, standard module of UNK starts to operate. It calculates the value of K_{eff} , the new macro constants, energy release in each zone. Then temperatures in zones are recalculated.

The code TempR_5 allows to calculate temperatures in cylindrical fuel element with piecewise constant thermal conductivity factor. Temperature distribution in each layer is obtained from analytical decision of thermal conductivity equation with assigned constant values of radius, thermal conductivity factor, and temperature. The results of calculation are average values of temperature over the layer, temperatures in points between layers and in the center of fuel element.

5 Results of test tasks calculations

Application of the proposed approach is illustrated by calculations of the temperature distribution along the radius of the fuel rod of the VVER-1000 reactor. These calculations can be considered as test tasks (benchmarks) to verify the developed algorithms. Test tasks will help to answer the above questions and to indicate the expediency of the complexity of the model for calculation of the fuel rod. We present below several results illustrating the capabilities of the algorithms. The tasks were solved in one-dimensional cylindrical geometry.

Model of the elementary cell of the VVER-1000 is taken as base model. The diameter of the central hole of the fuel pellet was taken as 1.5 mm, the external diameter of the pellet 7.55 mm; internal and external diameters of the cladding 7.72 and 9.17 mm; the pitch of the triangular lattice of rods 12.75 mm, temperature of coolant 575.7 K. For calculations, fuel rod was divided on 11 radial layers.

The temperature of the outer surface of the cladding was the same for all calculations. It was equal to the temperature of the coolant 575,7 K. The thermal conductivity factor was taken as 0.35 W/mK for the center hole and the gas gap between the fuel rod and cladding, and as 20.42 W/mK for cladding. These values were the same for all calculations.

First of all it is necessary to illustrate the advantage of the developed technique, which allows to take into account the dependence of thermal conductivity factor on temperature and on fuel burnup. Fig. 3 shows the results of calculations of the temperature inside the fuel element for four cases. The case 1 (curve *1*) corresponds to the traditional approach, in which λ was taken to be constant throughout the fuel rod. The λ value was calculated for the average temperature over the rod as 1000 K. The case 2 (curve *2*) corresponds to the application of the developed technique, however only dependence λ (T) for burnup B = 0 was considered. In cases 3 and 4, the dependence λ(T, B) was takes into account. Calculations were performed for B = 60 (curve *3*) and 120 MW·day/kgU (curve *4*). For all four cases, it was assumed that no gas gap was there between fuel rod and cladding, and energy release density q_v = 388 W/cm^3 was constant over the fuel rod.

The results presented in Fig.3 can be easily interpreted using data on λ(T, B) from Fig.1. In case 1 for λ = const, we obtain curve *1* in Fig.3. The temperature decreases from the center to the periphery of the fuel rod. Peripheral temperature is determined by the coolant temperature of 575 K.

In case 2 when we consider λ(T) at B = 0 (the upper curve *1* in Fig.1), the λ increases sharply due to temperature decrease in going from the center of the fuel rod to its periphery. Since the peripheral temperature is specified, the temperature at the center is lower than in case 1. The curve *2* lies below the curve *1*.

In case 3 (the dependence λ(T, B) corresponds to the curve *3* in Fig.1) the λ values are much less than in cases 1 and 2. The increase of the λ in going from the center to the periphery of fuel rod is slower than in cases 1 and 2. Since the energy release density is the same in all cases and peripheral temperature is specified, the temperature in the center of the rod is higher than in case 1. The curve *3* in Fig.3 lies above the curve *1*.

In case 4 (curve λ (T, B) is more flatter than curve *4* in Fig.1) similar arguments show that the temperature in the center of the rod is higher than in case 3. The curve *4* in Fig.3 lies above the curve *3*.

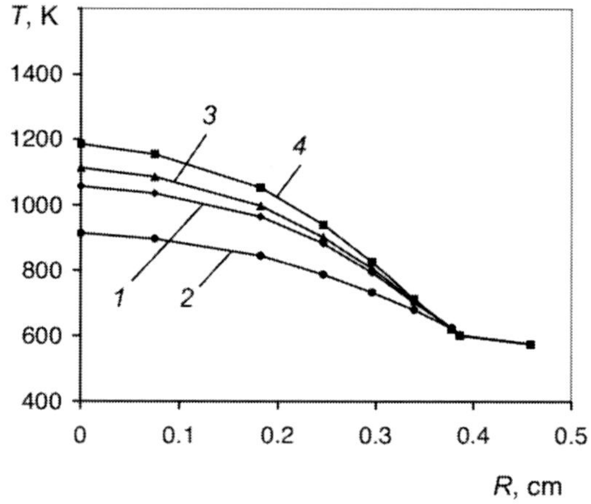

Fig. 3. λ = const (*1*), λ(T) is considered (*2*), λ(T, B) for B = 60 MWd/kg is considered (*3*), λ(T, B) for B = 120 MWd/kg is considered (*4*)

These data demonstrate the efficiensy of the developed methods of calculation of the temperature in the fuel element of the VVER reactor. Corrections to the value of the maximum temperature can reach 10-15%.

Let's now to turn to the results showing the capabilities of the algorithms.

In tasks 1 and 2 (Fig. 4 and 5), the uniform energy release density of 388 W/cm^3 was taken over the whole fuel rod. Thermal conductivity factor of the fuel rod λ was also taken constant. We considered three values of λ = 1.7 (curves *1* in figures 4 and 5), 2.1 (curves *2*) and 3.0 W/mK (curves *3*). The difference between the results shown in Figs.4 and 5 was that Fig.4 corresponded to fresh fuel with the gap between fuel rod and cladding, while Fig.5 corresponded to the "swollen" fuel. There was no gap in this case, and fuel was in thermal contact with cladding.

Results presented in Figs.4 and 5 demonstrate apparent dependence of the temperature curves for different values of λ. The lower λ, the higher is the curve *T*(*R*). All three curves converge on the outer surface of the fuel rod. Fig.4 clearly demonstrates sudden change in temperature in the gap between the outer surface of the fuel rod and cladding.

Fig.6 allows to compare the curves *1* of Figs.4 and 5.

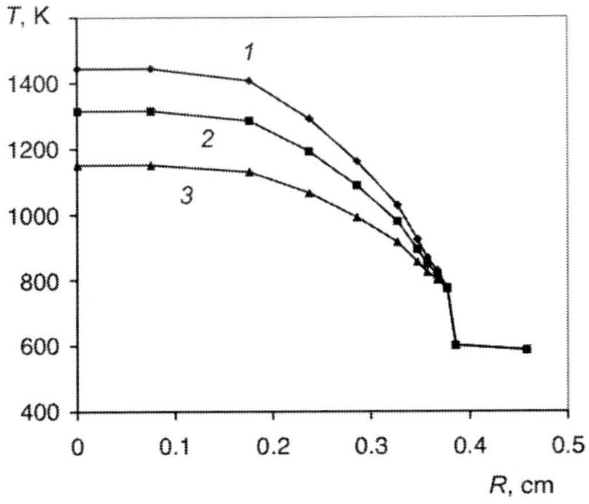

Fig. 4. Gap between fuel rod and cladding. Flat λ and power release

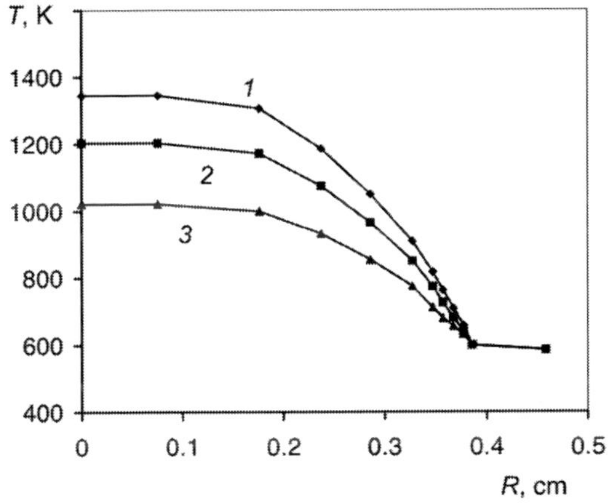

Fig. 5. No gap between fuel rod and cladding. Flat λ and power release

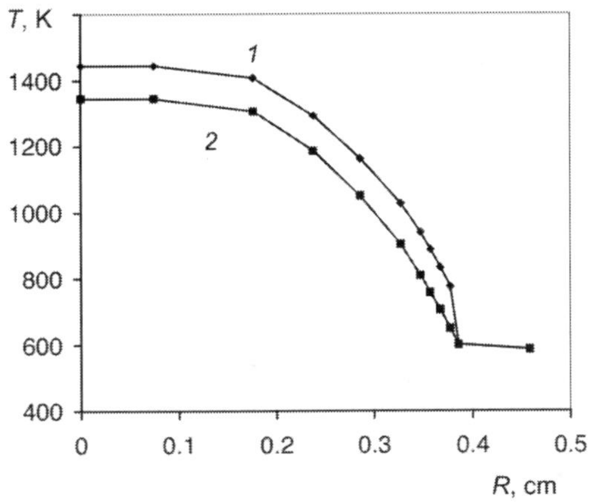

Fig. 6. Comparison of curves *1* from Figs. 3 and 4

In task 3 (Fig.7), thermal conductivity factor of fuel λ depended on the radius within the fuel rod. This dependence was the same for curves *1* and *2*. The λ value increased smoothly from 2.1 W/mK in the central layer to 6.5 W/mK in the outer layer of fuel rod. The energy release density was taken to be constant over the volume of the rod and was equal to 388 W/m^3 for curve *1* and 370 W/m^3 for curve *2*. Curve *1* corresponded to the presence of the gap between fuel rod and cladding, while curve 2 corresponded to the absence of the gap, and fuel was in thermal contact with cladding. The result is similar to the curves of Fig. 6, however in Fig. 7, the curves are more flat due to the increase of λ from the center to the periphery.

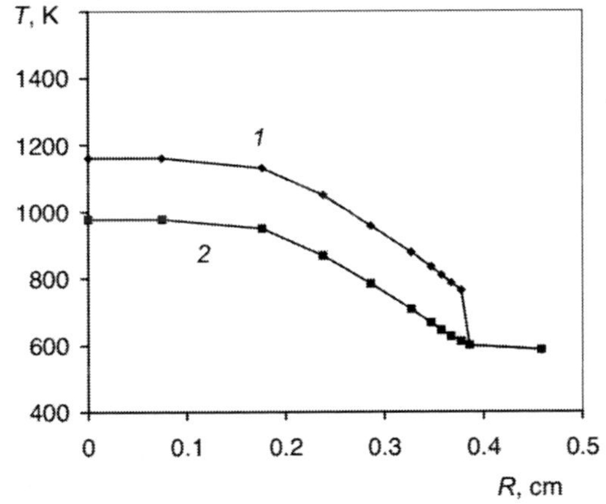

Fig. 7. λ increase with radius. Gap between fuel rod and cladding (*1*), no gap (*2*)

6 Conclusions

Thus, we present a consistent approach to complex calculation of nuclear fuel for VVER reactors. It is based on modern software and allows to obtain detailed distribution of power density and temperatures[14] in the fuel rod. The proposed approach can serve further for

DOI: 10.1051/epjconf/201715307036

development of verification base for testing of calculations of the fuel rods of VVER reactors.

References

1. V.P. Alferov, A.I. Radaev, M. V. Shchurovskaya, G. V. Tikhomirov, N.A. Hanan, F.A. Van Heerden, Ann. Nucl. Energy 77 (2015) 273–280.
2. A.M. Sirotkin, M.Y. Ternovykh, G.V. Tikhomirov, M.F. Khromova, At. Energy 116 (2014).
3. A.S. Gerasimov, V.N. Kornoukhov, I.S. Sald'ikov, G.V. Tikhomirov, At. Energy 116 (2014).
4. D.N. Skorokhodov, G.V. Tikhomirov, Phys. At. Nucl. 75 (2012).
5. B.R. Bergelson, V.V. Belonog, A.S. Gerasimov, G.V. Tikhomirov, At. Energy 109 (2011).
6. J. Piron, B. Bordin, G. Geoffroy, et al., *Proc. of Int. Topical Meeting on Light Water Reactor Fuel Performance.* USA, April 17-21, 321-327 (1994)
7. A. Shcheglov, V. Sidorenko, V. Proselkov, S. Boi'shagin. Atomic Energy, **80**, 222-224 (1996)
8. V. Baranov, M. Ternovykh, G. Tikhomirov, A. Khlunov, A. Tenishev, I. Kurina. *Proc. of the 9-th Int.Conf.on WWER Fuel Performance, Modelling and Experimental Support.* 17-24 September, Helena Resort, Bulgaria, 390-395 (2011)
9. I. Kurina, V. Popov, V. Rumyantsev, Atomic Energy, **101,** 802-808 (2006)
10. F. De Lillo, F. Cecconi, G. Lacorata, A. Vulpiani, EPL, **84** (2008) V. Davidenko, V. Tsibulsky. *Proc. of the Int. Conf. on the Physics of Nuclear Science and Technology,* Oct. 5–8, Long Island, New York, 1755-1760 (1998)
11. K. Childs, *HEATING 7.0 User's Manual,* K/CSD/INF/90-32 (Martin Marietta Energy Systems, Inc., July 1990)
12. SCALE: *A Modular Code System for Performing Standardized Computer Analyses for Licensing Evaluation*, Version 5, ORNL/TM-2005/39, **I-III**, ORNL (2005)
13. I. Belousov, V.Davidenko, V. Tsibulsky, Preprint IAE-6083/4, Moscow (1998)
14. V.I. Deev, K.V. Kutsenko, A.A. Lavrukhin, M.Y. Ternovykh, G.V. Tikhomirov, At. Energy 115 (2014).

Fission Product Appearance Rate Coefficients in Design Basis Source Term Determinations – Past and Present

Pedro B. Perez[1,*] and *John N.* Hamawi[2,**]

[1]AREVA SAS, Lyon, France
[2]AREVA INC, Marlborough, MA, U.S.A.

Abstract. Nuclear power plant radiation protection design features are based on radionuclide source terms derived from conservative assumptions that envelope expected operating experience. Two parameters that significantly affect the radionuclide concentrations in the source term are failed fuel fraction and effective fission product appearance rate coefficients. Failed fuel fraction may be a regulatory based assumption such as in the U.S. Appearance rate coefficients are not specified in regulatory requirements, but have been referenced to experimental data that is over 50 years old. No doubt the source terms are conservative as demonstrated by operating experience that has included failed fuel, but it may be too conservative leading to over-designed shielding for normal operations as an example. Design basis source term methodologies for normal operations had not advanced until EPRI published in 2015 an updated ANSI/ANS 18.1 source term basis document. Our paper revisits the fission product appearance rate coefficients as applied in the derivation source terms following the original U.S. NRC NUREG-0017 methodology. New coefficients have been calculated based on recent EPRI results which demonstrate the conservatism in nuclear power plant shielding design.

1 Introduction

Nuclear power plant (NPP) occupational doses are maintained ALARA by both design and operating practices. Operating plant occupational collective doses have decreased over the years demonstrating ALARA is being successfully practiced. NPPs are designed with radiation protection requirements that reflect the ICRP three fundamental principles of radiation protection [1]:

- Justification,
- Optimization of Protection, and
- Application of Dose Limits

Optimization is an important part in establishing design requirements for NPP radiation protection features. The design process includes consideration of source term, system layout, shielding material selection and ambient dose equivalent rate targets.

The design basis shielding source term has a direct impact on the plant design and cost associated with construction and ultimately decommissioning. Bulk-wall shield thicknesses are derived from the source term and dose rate targets. The end result could be an over-sized shield for normal operations impacting building dimensions. This would also raise a question if ALARA resources were properly allocated in the design.

Shielding source terms are derived to envelope expected operating experience. Assumptions are selected to be conservative and provide operational margins. Two variables that are particularly important are failed fuel fraction and fission product appearance rate coefficients. These two input variables significantly affect the radionuclide concentrations in the source term.

Failed fuel fraction may be a regulatory based assumption, such as the 0.25% in the U.S. Nuclear regulatory Commission (NRC) licensing basis [2] and [6]. Appearance rate coefficients are not specified as a regulatory requirement, but have been referenced to NUREG-0017 [3] in many U.S. Final Safety Analysis Reports (FSAR) and approved Design Certifications [4], as well as applications under review by the NRC.

Shielding methods have been improved over the years with exceptional advances in 3D Monte Carlo modelling capabilities. Yet the shielding design basis source terms did not advance at all over the past 50 years until EPRI published in December 2015 an updated ANSI/ANS 18.1 basis document [5]. New appearance rate coefficients can be derived from the recent EPRI data for current and future nuclear power plant shielding designs.

It should also be pointed out that failed fuel fraction assumption and fission product appearance rate coefficients are also applied in calculating initial source term for radiological consequences from design basis accidents. Another application is in the design of radioactive effluent processing systems. These examples are based on the U.S. NRC licensing basis; however,

* Corresponding author: pedro.perez@areva.com
** Retired, currently ENTECH Engineering, Westborough, MA, U.S.A.

other design applications may also benefit from an updated appearance rate coefficients.

2 Appearance Rate Coefficients

Fission product appearance rate coefficients express the transport of fission products from the fuel pellet to the reactor coolant. The physical processes have been studied experimentally and theoretically. Early work in the U.S. and Canada included perforated fuel rods operated in test loops, and fission products were measured in the coolant and purification media. Reference [8] is an example of the early work. These measurements and calculations provided a simplified "effective appearance rate" coefficient (k); that is fission product transfer from the pellet to the coolant including gap region interactions. These coefficients are publically documented in NUREG-0017 [3], for PWRs, and are provided in Table 1.

Table 1. NUREG-0017 PWR Appearance Rate Coefficients

Element	k (s^{-1})
Kr Xe gases	$6.5(10^{-8})$
I Br Rb Cs	$1.3(10^{-8})$
Mo Ag	$2.0(10^{-9})$
Te	$1.0(10^{-9})$
Sr Ba	$1.0(10^{-11})$
All others	$1.6(10^{-12})$

The coefficients (k) in Table 1 are called "effective" in the present work because they represent a single transfer mechanism. In fact, the symbol "k" is also used to avoid confusion with modern models.

Additional experiments and recent theoretical models have generally accepted that the fission product release rates involve two physical processes [9]:

- ν, pellet to gap escape rate (atoms s^{-1}), and
- ε, gap-to-coolant appearance rate (atoms s^{-1})

Experimentally, these two release rate coefficients can be individually measured. However, measurements in a commercial nuclear power plant can only provide radionuclide concentrations. This radiochemistry data along with fuel failure detection methodology can characterize the fuel rod damage.

3 RCS Source Terms

Multiple radiological source terms are used in a nuclear power plant design basis. The normal operations source term is often based on operating experience. ANSI/ANS 18.1 [10] has been the reference source term for U.S. plants. This source term, however, is outdated and reflects operating experience from early nuclear power plants. Recently EPRI has documented updated plant data reflecting improved fuel performance. Another source term is for shielding and following NRC guidance assumes a non-mechanistic 0.25% failed fuel fraction [6]. Other source terms used for accident analyses may also apply the appearance rates to establish an initial condition for reactor coolant concentrations [7]. Accident source terms are not treated in the present work.

Reactor coolant system (RCS) source terms used for shielding calculations are derived by applying a failed fuel fraction and effective appearance rate coefficients. A simplified expression is provided in ANSI/ANS 18.1 that is based on mass balance with radioactive decay:

$$C_i = R_i / M(\lambda_i + \beta) \quad (1)$$

Where: C_i = concentration of ith nuclide (Bq/g),
R_i = release rate of ith nuclide (Bq/s),
M = coolant mass (g)
λ_i = decay constant for i^{th} nuclide (s^{-1})
β = removal constant for i^{th} nuclide (s^{-1})

Removal constant (β) includes purification and any other removal mechanism.

Release rate (R_i) in Equation (1) is the product of the core inventory, failed fuel fraction and appearance rate coefficient. It is also referred to as the radionuclide appearance rate and is solved for all the fission products. Figure 1 provides a simplified diagram of a typical calculation with ELISA-2 [11].

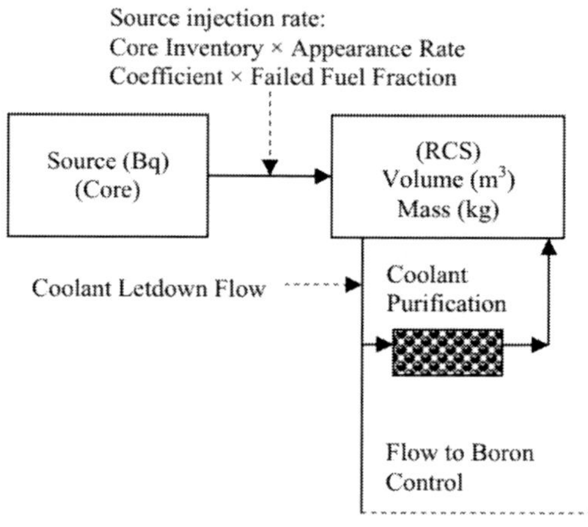

Figure 1. Schematic of ELISA-2 Model

Source terms for shielding have been based on older appearance rate coefficients that may be too conservative. These are applied in conjunction with a non-mechanistic fuel failure fraction and can result in oversizing shielding.

EPRI has recently published updated radionuclide concentrations from operating power plants. The data provides an opportunity to update the appearance rate coefficients and remain fully within the U.S. regulatory framework.

4 Updated Appearance Rate Coefficients

Appearance rate coefficients can now be re-evaluated. The EPRI report includes recent U.S. plant measured RCS concentrations and plant specific operating information. The raw data in the EPRI report was obtained from the EPRI Chemistry Monitoring and Assessment Database that included over 60 PWR plants. Ten (10) plants were selected for use in updating ANSI/ANS 18.1.

The recommended RCS concentrations for the reference plant, along with the reference-plant specific variables (i.e., volumes, flow rates, filtration efficiencies, . etc.) and the simplified model in Equation (1), were used in this work with ELISA-2 to back calculate "effective" appearance rate coefficients consistent with NUREG-0017 [3]. The first step was to group the elements for input to ELISA-2 according to the breakdown of the Periodic Table of the Elements shown in Figure 2 (from Los Alamos). This grouping can be modified based on the application; for example, radiological consequences of design basis accidents.

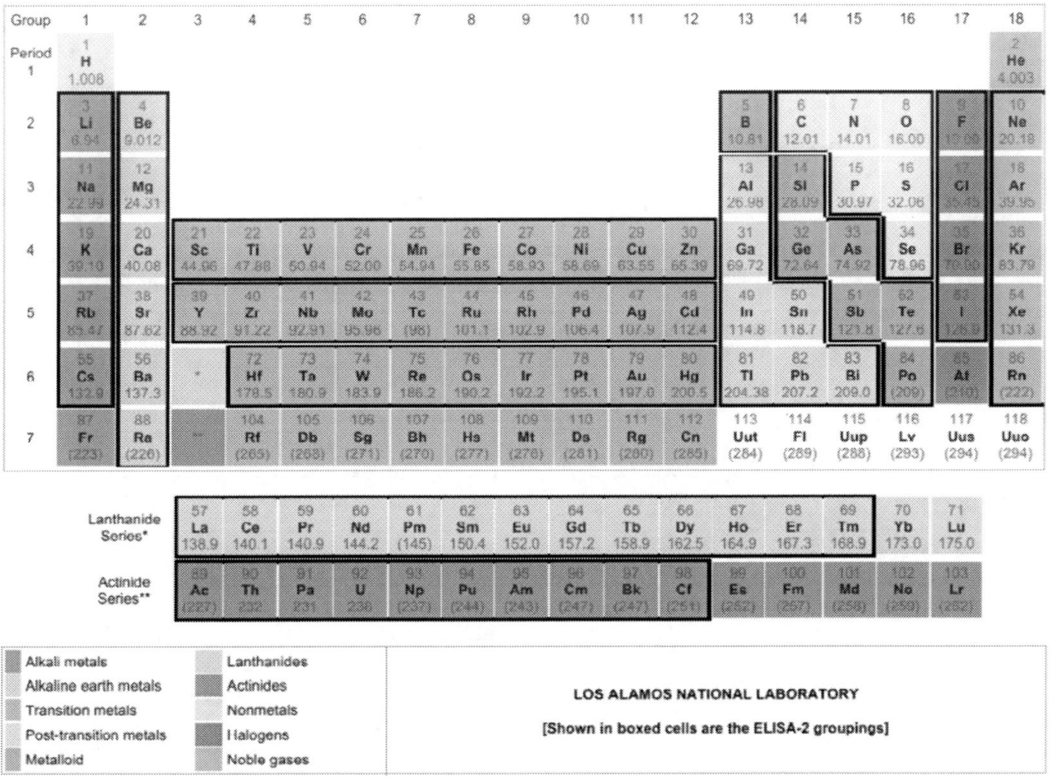

Figure 2. ELISA-2 Element Grouping

The second step was to prepare an ELISA2 computer run with the model in Figure 1. The assumed failed fuel fraction is consistent with the NUREG-0017 assumption of 0.12% [3]. This failed fuel assumption is necessary for the following two reasons: First, it allows a comparison of effective appearance rate coefficients based on updating the source term. Second, there is no failed fuel fraction information in the EPRI report [5].

The RCS concentrations calculated were compared to the measured concentrations in the EPRI report [5]. The third and final step was to adjust the effective appearance rate coefficients to arrive at calculated-to-measured (C/M) ratios close to unity by iterative ELISA-2 analyses. Our approach is simple and fully within the NRC regulatory guidance in SRP 11.1 [2].

The final EPRI based effective appearance rate coefficients obtained from ELISA-2 analyses with an assumed failed fuel fraction of 0.12% are compared to the NUREG-0017 based values in Table 2. Elements in groupings represented by **bold font** in Table 2 can be directly compared to NUREG-0017 results (i.e., Table 1).

The updated appearance rate coefficients in Table 2 can now be applied to demonstrate the impact on design basis primary coolant source terms.

5 Example Applications

The updated appearance rate coefficients in Table 2 can now be applied to demonstrate the impact on design basis primary coolant source terms. The ELISA-2 model depicted in Figure 1 is next applied to the reference plant described in Reference [5]. A full-core inventory based on 3340 MW-thermal, 5% U-235, 3 cycles is generated through use of ORIGEN-ARP. Table 3 lists the core inventory.

The two different sets of appearance rates (NUREG-0017 and ELISA-2, the latter based on reproduction of the current PWR measurements) were used as inputs to ELISA-2 runs on identical models assuming 0.12% failed fuel fraction consistent with original assumption. Table 4 lists the RCS inventory with updated coefficients.

Table 2. Appearance Rate Coefficients.

Element Grouping	**NUREG-0017**	**ELISA-2**
	k (s^{-1})	
Noble Gases Ne Ar **Kr** **Xe** Rn	$6.5(10^{-8})$	$6.7(10^{-10})$
Halogens F Cl **Br** **I**	$1.3(10^{-8})$	$3.1(10^{-10})$
Alkali Metals Li Na K **Rb** **Cs**	$1.3(10^{-8})$	$2.1(10^{-10})$
Alkali Earth Metals Be Mg Ca **Sr** **Ba** Ra	$1.0(10^{-11})$	$1.6(10^{-10})$ (a)
Transition Metals – Grp 1 Sc Ti V Cr Mn Fe Co Ni Cu Zn	$1.6(10^{-12})$	$1.5(10^{-8})$ (b)
Transition Metals – Grp 2 Y Zr Nb **Mo** Tc Ru Rh Pd Ag Cd	$2.0(10^{-9})$	$3.3(10^{-12})$
Transition Metals – Grp 3 Hf Ta W Re Os Ir Pt Au Hg	$1.6(10^{-12})$	$3.3(10^{-12})$ (c)
Post Transition Metals Al Ga In Sn Tl Pb Bi	$1.6(10^{-12})$	$9.2(10^{-10})$
Metaloids B Si Ge As Sb **Te** Po	$1.0(10^{-9})$	$1.6(10^{-10})$
Non-Metals C N O P S Se	$1.6(10^{-12})$	$1.6(10^{-10})$ (d)
Lanthanides La Ce Pr Nd Pm Sm Eu Gd Tb Dy Ho Er Tm	$1.6(10^{-12})$	$3.3(10^{-12})$(e)
Actinides Ac Th Pa U Np Pu Am Cm Bk Cf	$1.6(10^{-12})$	$6.7(10^{-15})$

(a) No data. Assumed to be the same as the Metaloid elements. Strontium and Barium elements are in the same group as Tellurium, per RG 1.183 [7].

(b) These are mostly activation products. The appearance rate coefficient is based on the calculated core inventory which includes structural-material activation. There is reasonably good comparison for all measured elements in this group.

(c) No data. Set the same as for Transition Metals Group 2.

(d) No data. Assumed to be the same as the Metaloid elements. Selenium is in the same group as Tellurium, per RG 1.183 [7].

(e) No data. Set the same as the Transition Metals - Group 2. Lanthanum is in the same group as Zirconium and Niobium in RG 1.183 [7].

Table 3. Core Inventory (MBq)

Ag110	3.67E+11	I129	9.77E+04	Pr144	4.37E+12	Tc99m	5.70E+12
Ag110m	7.88E+09	I130	5.22E+10	Pr144m	6.11E+10	Te123m	2.16E+07
Ag111	1.72E+11	I131	3.33E+12	Pu238	1.12E+10	Te125m	6.70E+09
Am241	5.62E+08	I132	4.88E+12	Pu239	1.24E+09	Te127	2.95E+11
Ar41	2.54E-01	I133	6.85E+12	Pu240	1.61E+09	Te127m	4.74E+10
Ba137m	3.89E+11	I134	7.77E+12	Pu242	6.03E+06	Te129	8.55E+11
Ba139	6.14E+12	I135	6.55E+12	Rb86	6.07E+09	Te129m	1.62E+11
Ba140	5.92E+12	In113m	5.03E+09	Rb88	2.45E+12	Te131	2.83E+12
Ba141	5.55E+12	Kr85	4.00E+10	Rb89	3.22E+12	Te131m	6.25E+11
Ba142	5.22E+12	Kr85m	9.07E+11	Rh103m	5.03E+12	Te132	4.74E+12
Br82	9.88E+09	Kr87	1.80E+12	Rh105	3.03E+12	Te133	3.69E+12
Br83	4.18E+11	Kr88	2.41E+12	Rh106	1.77E+12	Te133m	3.26E+12
Br84	7.70E+11	Kr89	3.02E+12	Ru103	5.03E+12	Te134	6.22E+12
Ce141	5.59E+12	La140	6.07E+12	Ru105	3.39E+12	Xe131m	4.33E+10
Ce143	5.25E+12	La141	5.59E+12	Ru106	1.61E+12	Xe133	6.88E+12
Ce144	4.33E+12	La142	5.40E+12	Sb122	4.00E+09	Xe133m	2.12E+11
Cm242	1.21E+11	Mn54	3.50E+09	Sb124	2.32E+09	Xe135	2.06E+12
Co58	2.10E+10	Mn56	9.95E+10	Sb125	3.12E+10	Xe135m	1.42E+12
Co60	2.70E+10	Mo99	6.29E+12	Sb127	2.90E+11	Xe137	6.22E+12
Cr51	7.18E+10	Na24	1.45E+07	Sb129	9.14E+11	Xe138	5.92E+12
Cs134	5.29E+11	Nb95	5.92E+12	Sb131	2.65E+12	Y90	3.24E+11
Cs136	1.72E+11	Nb95m	6.51E+10	Sn113	5.03E+09	Y91	4.37E+12
Cs137	4.11E+11	Nb96	1.06E+10	Sn119m	3.50E+10	Y91m	2.45E+12
Cs138	6.44E+12	Nb97	6.03E+12	Sn123	5.37E+09	Y92	4.55E+12
Cs139	6.03E+12	Nd147	2.22E+12	Sr89	3.34E+12	Y93	5.11E+12
Cu64	7.81E+06	Ni65	2.58E+09	Sr90	3.12E+11	Y95	5.55E+12
Eu154	2.37E+10	Np239	6.14E+13	Sr91	4.22E+12	Zn65	1.27E+03
Eu155	9.47E+09	Pm147	6.33E+11	Sr92	4.48E+12	Zn69	6.03E+06
Fe55	1.41E+10	Pm149	1.86E+12	Tc101	5.74E+12	Zn69m	2.85E+04
Fe59	1.07E+09	Pr143	5.25E+12	Tc104	4.11E+12	Zr95	5.88E+12
						Zr97	5.99E+12

DOI: 10.1051/epjconf/201715307037

Table 4. RCS Concentration with Updated Appearance Rate Coefficients (Bq/m^3)

Ag-110	2.15E+02	I-129	5.63E+00	Pr-144	2.66E+06	Tc-99m	3.13E+06
Ag-110m	4.84E+03	I-130	1.64E+06	Pr-144m	2.62E+04	Te-123m	6.41E+02
Ag-111	1.00E+05	I-131	1.83E+08	Pu-238	1.40E+01	Te-125m	2.00E+05
Am-241	7.03E-01	I-132	1.53E+08	Pu-239	1.55E+00	Te-127	8.26E+06
Ar-41	5.58E-06	I-133	2.68E+08	Pu-240	2.00E+00	Te-127m	1.42E+06
Ba-137m	1.51E+07	I-134	4.58E+07	Rb-86	2.32E+05	Te-129	1.10E+07
Ba-139	2.47E+07	I-135	1.47E+08	Rb-87	6.60E-07	Te-129m	4.80E+06
Ba-140	1.71E+08	In-113m	8.58E+05	Rb-88	8.28E+07	Te-131	8.75E+06
Ba-141	4.73E+06	Kr-85	2.61E+08	Rb-89	4.88E+06	Te-131m	1.39E+07
Ba-142	2.62E+06	Kr-85m	4.88E+07	Rh-103m	3.02E+06	Te-132	1.24E+08
Br-82	4.40E+05	Kr-87	2.76E+07	Rh-105	1.58E+06	Te-133	3.55E+06
Br-83	4.56E+06	Kr-88	8.24E+07	Rh-106	9.89E+05	Te-133m	7.97E+06
Br-84	2.17E+06	Kr-89	1.91E+06	Ru-103	3.05E+06	Te-134	1.17E+07
Ce-141	3.44E+06	La-140	3.78E+07	Ru-105	6.26E+05	Xe-131m	1.26E+08
Ce-143	2.46E+06	La-141	4.37E+06	Ru-106	9.89E+05	Xe-133	9.29E+09
Ce-144	2.66E+06	La-142	2.71E+06	Sb-122	1.03E+05	Xe-133m	1.44E+08
Cm-242	1.51E+02	Mn-54	9.75E+06	Sb-124	6.87E+04	Xe-135	4.63E+08
Co-58	5.82E+07	Mn-56	5.55E+07	Sb-125	9.28E+05	Xe-135m	4.39E+07
Co-60	7.54E+07	Mo-99	3.33E+06	Sb-127	7.77E+06	Xe-137	4.77E+06
Cr-51	1.97E+08	Na-24	3.36E+02	Sb-129	8.14E+06	Xe-138	1.68E+07
Cs-134	2.07E+07	Nb-95	3.63E+06	Sb-131	2.83E+06	Y-90	1.46E+06
Cs-136	6.49E+06	Nb-95m	3.99E+04	Sn-113	8.57E+05	Y-91	3.08E+06
Cs-137	1.60E+07	Nb-96	4.50E+03	Sn-119m	5.96E+06	Y-91m	3.28E+07
Cs-138	2.88E+07	Nb-97	2.43E+06	Sn-123	9.14E+05	Y-92	2.11E+07
Cs-139	3.47E+06	Nd-147	1.31E+06	Sr-89	9.88E+07	Y-93	1.55E+06
Cu-64	1.20E+04	Ni-65	1.41E+06	Sr-90	9.30E+06	Y-95	5.56E+04
Eu-154	1.46E+04	Np-239	6.46E+04	Sr-91	6.06E+07	Zn-65	3.54E+00
Eu-155	5.81E+03	Pm-147	3.89E+05	Sr-92	2.74E+07	Zn-69	1.44E+03
Fe-55	3.94E+07	Pm-149	9.58E+05	Tc-101	7.92E+04	Zn-69m	4.55E+01
Fe-59	2.96E+06	Pr-143	3.20E+06	Tc-104	7.27E+04	Zr-95	3.58E+06
						Zr-97	2.28E+06

Table 5 provides the RCS photon spectra (MeV/sec-m^3) for an RCS volume of 345.5 m^3 corresponding to all radionuclides including dissolved noble gases in Table 3. The RCS spectra in Table 4 can be used to demonstrate the different shielding requirements.

Table 5. RCS photon spectra (MeV/sec-m^3)

MeV	NUREG	ELISA-2	NUREG/ELISA2
0.01	4.18E+08	5.18E+06	80.6
0.025	3.74E+08	7.54E+06	49.7
0.0375	1.39E+10	1.50E+08	92.5
0.0575	1.12E+07	9.77E+05	11.4
0.0850	2.69E+10	2.89E+08	93.0
0.125	2.95E+08	3.04E+06	97.1
0.225	9.47E+09	1.62E+08	58.4
0.375	4.06E+09	9.91E+07	41.0
0.575	1.56E+10	4.47E+08	34.9
0.850	1.60E+10	4.75E+08	33.6
1.25	1.38E+10	5.50E+08	25.0
1.75	9.17E+09	2.35E+08	38.9
2.25	1.38E+10	1.77E+08	77.9
2.75	2.28E+09	3.33E+07	68.5
3.50	3.97E+08	5.86E+06	67.8
5.00	7.15E+07	7.65E+05	93.5

Figure 3 depicts a simple MicroShield™ Point Kernel model of a vessel. This hypothetical tank is 500 cm long, 38 cm in radius with a wall thickness of 2 cm. The content will be RCS liquid with concentrations corresponding to 0.12% failed fuel fraction.

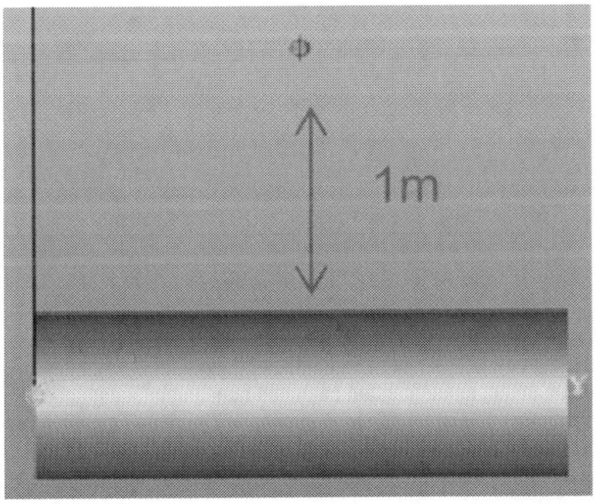

Figure 3. Shielding Model

The absorbed dose rate in air at 1 meter from the tank was calculated with the spectra in Table 5. Concrete shielding is placed between the source and receptor to reduce the absorbed dose rate to less than 10 μGy/h. The difference in shield thickness is a direct impact on a nuclear power plant civil works. Table 6 presents the dose rate results.

Table 6. Absorbed Dose in Air

Concrete Shielding	Absorbed Dose (μGy/h)	
	NUREG	ELISA2
None	1528	40
15 cm	340	8
50 cm	8	0.2

The difference in calculated dose rate due to updating appearance rate coefficients is immediately noticed when comparing unshielded absorbed dose rate. The RCS source term based on the NUREG-0017 photon spectra is 3.8 times greater.

A concrete shield of 50 cm would be required to lower the absorbed dose to 8 μGy/h with the NUREG-0017 spectra. On the other hand, only 15 cm of concrete would be required with the updated design basis radionuclide appearance rates calculated with ELISA2 in this present work.

The difference in concrete volume is significant when considering the civil works in a nuclear power plant. This example demonstrates the importance of source term assumptions in the design of nuclear power plants and supports updating ANSI/ANS 18.1.

6 Conclusions

Radionuclide source terms are used in many areas of designing nuclear power plants. The appearance rate of radionuclides and failed fuel fraction are two important design inputs used to generate source terms such as the shielding 0.25% failed fuel fraction in the U.S.N.R.C. Appearance rate coefficients currently applied in design work have not been updated since NUREG-0017 [3] was published in 1976.

Appearance rate coefficients can now be updated from a recent EPRI report [5] following the NUREG-0017 methodology. The present work demonstrates appearance rate coefficients predicted assuming 0.12% failed fuel fraction are generally much lower in comparison to historical values. The new coefficients demonstrate the conservatism in nuclear power plant source terms for shielding designs.

Radiation protection design features will continue to be ALARA when based on updated appearance rate coefficients reflecting modern fuel performance. The updated methodology would reduce the risk of over-sized shielding for normal operations impacting building dimensions and construction cost as well as the cost associated with ultimately decommissioning the nuclear power plant.

DOI: 10.1051/epjconf/201715307037

References

1. ICRP, 2007. *The 2007 Recommendations of the International Commission on Radiological Protection*. ICRP Publication 103. Ann. ICRP 37 (2-4) (2007).
2. U.S. Nuclear Regulatory Commission, NUREG-0800, Section 11.1 (2016).
3. U.S. Nuclear Regulatory Commission, NUREG-0017, Revision 0 (1976).
4. Westinghouse AP1000 Design Control Document, Revision 19, Tier 2, Chapter 11, Radioactive Waste Management (2011).
5. EPRI-3002005404, Advanced Nuclear Technology: Reactor Coolant Radiological Source Terms for Normal Operations – Updated and Revised Methodology. EPRI, Palo Alto, CA (2015).
6. U.S. Nuclear Regulatory Commission, Regulatory Guide 8.8, Revision 3 (1978).
7. U.S. Nuclear Regulatory Commission, Regulatory Guide 1.183 (2000).
8. J.D. Eichenberg, et al, WAPD-183, Bettis Plant, United States Atomic Energy Commission (1957).
9. EPRI-1019107, Fuel Reliability Monitoring and Failure Evaluation Handbook: Revision 2. EPRI, Palo Alto, CA (2010).
10. ANSI/ANS-18.1-1999, American Nuclear Society, Radioactive Source Term for Normal Operations of Light Water Reactors, an American National Standard (1999).
11. J.N. Hamawi, "ELISA-2 – Software Package for Radiological Evaluations at Light-Water Nuclear Power Plants (Source-Term Generation and Accident Analyses)", ENTECH Engineering, Westborough, MA, Technical Report P100-R25, Version 2.7 (2015).

Growth of micrometric oxide layers for the study of metallic surfaces decontamination by laser

Luisa Carvalho [a], Wilfried Pacquentin, Michel Tabarant, Hicham Maskrot and Alexandre Semerok

Den – Service d'Etudes Analytiques et de Réactivité des Surfaces (SEARS), CEA, Université Paris-Saclay, F-91191, Gif sur Yvette,

Abstract. The nuclear industry produces a wide range of radioactive waste in term of level of hazard, contaminants and material. For metallic equipment like steam generators, the radioactivity is mainly located in the oxide surface. In order to study and develop techniques for dismantling and for decontamination in a safe way, it is important to have access to oxide layers with a representative distribution of non-radioactive contaminants. We propose a method of formation of oxide layer on stainless steel 304L with europium (Eu) as contaminant marker. In this method, an Eu-solution is sprayed on the stainless steel samples. The specimen are firstly treated with a pulsed nanosecond laser and secondly the steel samples are exposed to a 600°C furnace for various durations in order to grow an oxide layer. The oxide structure and in-depth distribution of Eu in the oxide layer are analysed by scanning electron microscopy coupled with energy dispersive X-ray microanalyzer, and by glow discharge optical emission or mass spectrometry. The oxide layers were grown to thicknesses in the range of 200 nm to 4.5 µm regarding to the laser treatment parameters and the heating duration. These contaminated oxides have a 'duplex structure' with a mean weight percentage of 0.5% of europium in the volume of the oxide layer. It appears that europium implementation prevents the oxide growth by furnace but has no impact on laser heating. These oxide layers are used to study the decontamination of metallic surfaces such as stainless steel 304L using a nanosecond pulsed laser.

1 Introduction

In nuclear industry, the dismantling is a major issue. The cleaning of contaminated surfaces is currently based on chemical and mechanical processes. However few problems remain unsolved. These techniques usually generate secondary wastes and contaminated effluents which need long term storage and dismantling workers are exposed to occasional radiation.

In this context, the CEA (French Alternative Energies and Atomic Energy Commission) has developed a prototype for surfaces cleaning called Aspilaser and studied the cleaning of contaminated surfaces such as painted concrete [1-2] and stainless steel [3-4] by excimer or fiber lasers. In regard to metallic components such as pipe systems, the radionuclides are usually in the oxide layer which grows during the operations of the nuclear facility.

The principle of this technique is the laser ablation of the surface in order to eject the oxide layer and collect the radioactive particles with a HEPA filter for storage. In nuclear facilities, the use of a laser for the cleaning of contaminated surfaces has many advantages such as the possibility of remote treatment, a dry process and a minimum of secondary wastes.

In dismantling operations, the contaminated surfaces range widely in characteristics such as shape (pipes, plane surfaces ...), conditions of oxide growth (temperature, atmosphere ...) and nature of contaminants. A previous study [3] highlighted the need to create oxide layers in a controlled manner for the study the efficiency of laser decontamination and to optimize the cleaning process. In this paper, we present the recent results of the growth of micrometric oxide layer contaminated with non-radioactive element and the preliminary test of laser cleaning.

The aim of the growth of oxide layers on metallic surfaces with a non-radioactive contaminant is to enable the study and the development of decommissioning techniques without the problems of availability of radioactive samples and the constraints of work in hot laboratories. The goal of our work was also to find a way of formation of oxide layers with desired characteristics (thickness, Eu-concentration).

2 Oxidation of stainless steel 304L with non-radioactive marker

We investigate a method to form a reproducible and micrometric oxide layer on stainless steel AISI 304L (304L SS) with a volumic distribution of europium (Eu) in the layer thickness. This distribution is chosen to consider the Eu as a simulation of a radioactive element (actinide) Thus, the efficiency of decontamination can be determined

[a] Corresponding author: luisa.carvalho@cea.fr

by the measurement of Eu concentration before and after laser decontamination. The initial composition of the 304L SS samples was analyzed via glow discharge mass spectrometry (GDMS) and no trace of Eu were found with a detection limit of 100 ppb.

2.1 Procedure

The samples (30 mm x 30 mm x 3 mm) were cut by water jetfrom a cold-rolled 304L stainless steel sheet. The coupons were cleaned in an ultrasound bath with ethanol and the average surface roughness was Ra = 0.46 µm. The contamination is made before oxidation by deposition of a solution of $Eu(NO_3)_3$ in water at a concentration of 2000 or 4000 ppm in Eu depending of the studies. The solution is sprayed with a nebulizer and the flow is controlled by a pump device. The samples are heated using a hot plate at 200°C until the deposition is dried. We can estimate that the deposited amount of Eu is 0.5 up to 1.0 mg/cm² (2 – 4 × 10^{18} atoms/cm²). Some samples without contamination served as control samples.

Pre-oxidation is carried out by laser treatment in air at atmospheric pressure. The set up comprises an ytterbium fiber laser (Gaussian beam, λ = 1064 nm, mean power up to 20 W, frequency ν = 20 kHz, pulse duration τ = 110 ns at FWHM), a focusing lens (f = 420 mm) resulting in a laser beam waist radius ω = 84 ± 2 µm (at $1/e^2$ intensity level), a scanning system and an air aspiration system. For the oxidation regime, the samples were mounted on a micro control assembly and placed at the lens focal. The power of the beam varied between 5 and 18 W in order to study a range of fluence maximum per pulse (F) between 2.2 and 7.8 J/cm² (maximal fluence on the beam axis). The laser treatment is done by horizontal and vertical scanning (figure 1). The horizontal scanning is characterized by the scanning speed V in order to define the step Δx between two consequent laser spots. The vertical scanning is defined by the vertical step Δy between two parallel lines.

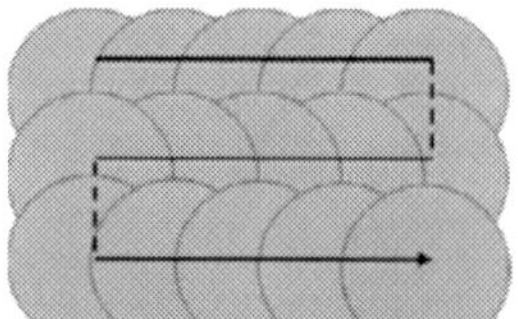

Figure 1. Scanning regime schema. The continuous arrow represents the laser beam path.

The scan speed and the distance between lines were chosen to provide a high spatial overlapping between two successive spots from 80 to 95 %. The overlapping percentage is calculated using the laser spot beam diameter at $1/e^2$. The laser treatment is defined by the maximum beam fluence F per pulse and the energy amount accumulated E_{acc} per surface unit of the sample after the treatment. F is calculated as follow:

$$F = E/S = 2P/\nu\pi\omega^2 \quad (1)$$

where E is the energy per pulse in J and S the surface of the laser beam in cm². The energy per pulse is the average power of the laser (P) divided by the repetition rate of 20 kHz. E_{acc} is determined as follow:

$$E_{acc}=P\Delta t/S_s \quad (2)$$

where Δt is the treatment duration and S_s is the treated surface in cm². In this work, we study oxide layers formed by an accumulated energy in a range from 50 to 600 J/cm².

After laser treatment, the oxide samples are exposed in the furnace at 600°C for a duration of 50 or 100 hours at in air at atmospheric pressure and the cooling down is done in the furnace switched off regime. Furnace treatment allows us to grow the oxide layer to a thickness depending of the furnace duration.

2.2 Oxide coatings characterization

The obtained oxide layers were characterized and analysed by glow discharge optical emission spectrometry (GD-OES), glow discharge mass spectrometry (GDMS), scanning electron microscopy (SEM) coupled with an energy dispersive X-ray analyser (EDX), X-ray diffraction (XRD) and micro-Raman spectroscopy. The oxidation procedure results in the formation of Eu-contaminated oxide layers with a thickness from 200 nm to 4.5 µm depending of the set of laser parameters and heating duration.

The Eu-contaminated layer formed after the pre oxidation by laser was analyzed by GD-OES and is presented on figure 2. SEM image (back scattering electrons) and GD-OES profile of a representative oxide layer (Eu initial contamination of 900 µg/cm², E_{acc}=600 J/cm², oxidation time in furnace of 50h) are shown on figure 3 and 4. One can see that the laser treatment results in a formation of an oxide layer (mostly Fe-oxide) of a thickness of 500 nm with Eu distribution in the whole oxide thickness and with an average concentration up to 5 % in weight.

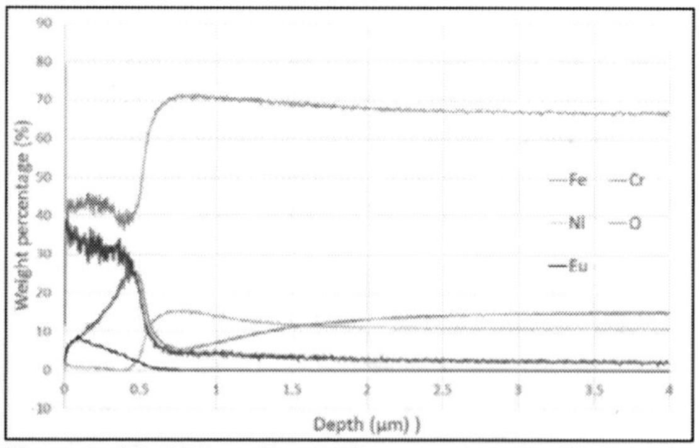

Figure 2. GD-OES profile of a Eu-contaminated oxide layer on 304L SS after laser preliminary treatment (9W, overlapping of 95%, E_{acc}=600 J/cm²).

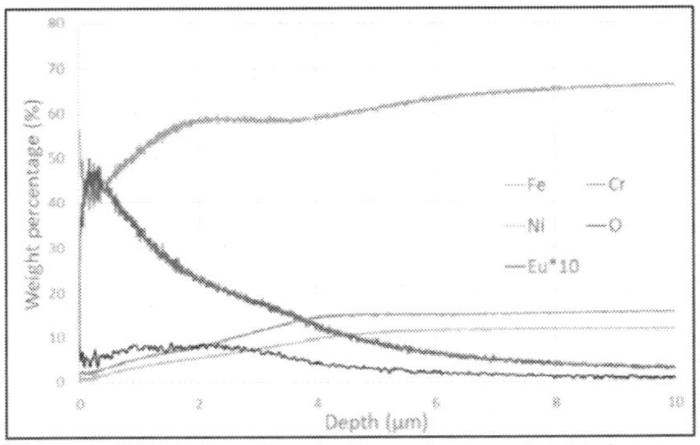

Figure 3. GD-OES profile of a Eu-contaminated oxide layer on 304L SS after laser preliminary treatment (9W, overlapping of 95%, E_{acc}=600 J/cm²) and an oxidation by furnace (50h, 600°C).

The oxidation by furnace allow us to have a layer of 4.5 µm with a contamination of 0.5% in weight of Eu in the whole thickness. In figure 3, the scale of Eu-distribution is multiplied by 10 in order to be seen in the profile. In both figure 2 and 3, the distribution of Eu follows the O-distribution. This means that the contamination is strictly in the oxide layer and has not been detected in the metal bulk, as was expected in order to use europium as contamination marker.

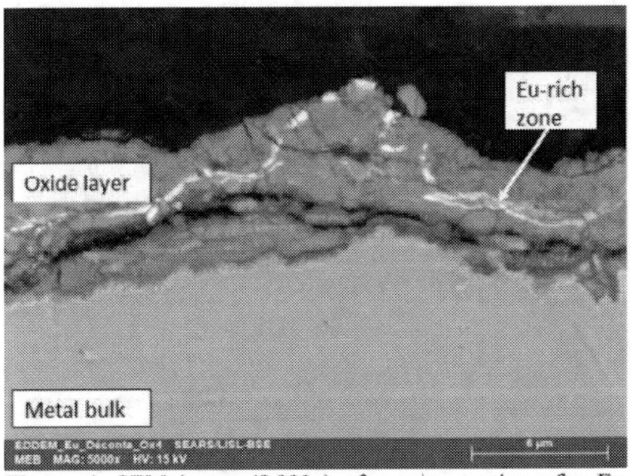

Figure 4. SEM image (5 000x) of a cross section of a Eu-contaminated oxide layer on 304L SS after laser preliminary treatment (9W, overlapping of 95%, E_{acc}=600 J/cm²) and an oxidation by furnace (50h, 600°C).

In the SEM image of the cross section (figure 4) the lightest parts match with Eu-rich zones (10% of weight by EDX analysis) and the rest is mostly ferrous oxide whereas the GD-OES analysis (figure 2 and 3) show that the contaminant is implemented in the whole oxide. The resulting contamination level in Eu appeared to be lower than the detection limit of EDX, XRD and micro-Raman methods. The monitoring of the Eu was made by GD-OES and GDMS which have detection limits at respectively 100 ppm and 100 ppb.

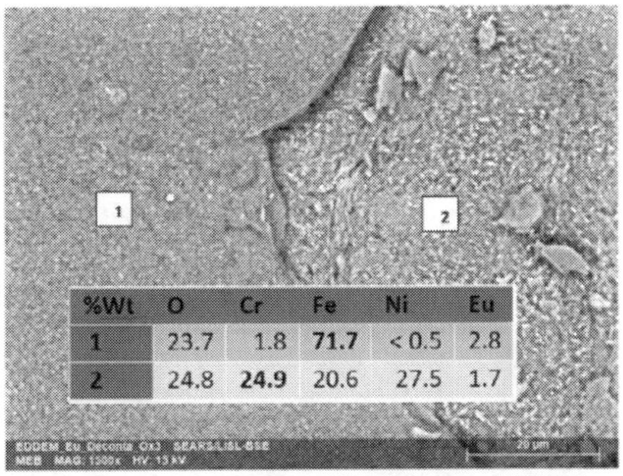

Figure 5. SEM image (1 500x) and EDS analysis (%Wt) of a cross section of a Eu-contaminated oxide layer on 304L SS after laser preliminary treatment (9W, overlapping of 95%, E_{acc}=600 J/cm²) and an oxidation by furnace (50h, 600°C). The peeling of the oxide during sectioning allowed us to access to an intern and an extern layer.

For the thick oxide layer, the SEM image (figure 5) and GD-OES profiles (figure 2) lead to the assumption of the presence of an outer thicker layer mainly composed by a Fe-rich oxide (zone 1 in figure 5) and an inner one enriched in Cr and Ni (zone 2 in figure 5). Nevertheless, the GD-OES profile (figure 3) did not confirm this hypothesis. The micro-Raman spectroscopy (laser wavelength $\lambda = 532$ nm) and XRD analysis (grazing incidence, $\theta = 1°$) of this sample showed mainly the formation of hematite Fe_2O_3 and we can assume the presence of a Fe-Cr spinel. This structure appeared to be similar to the oxide layers in industrial conditions [5]. The DRX analysis showed the same results for samples with and without Eu and were not able to reveal any specific phase with Eu. Our result is different than the one of a previous study of oxidation by furnace of 304L SS coupons contaminated by ytterbium [6]. In the study of Riffard *et al.*, the contaminated layer are thinner and the oxide is composed mainly a Cr-oxide, Cr_2O_3. Actually experiments and previous studies [6] [7] showed the influence of rare earths as inhibitors of oxidation. The advantage of our procedure is to implement the Eu by a first oxide layer in order to continue the growth of this layer by furnace despite the presence of Eu.

Our studies in order to find the optimal oxidation condition to control the thickness of the oxide layer allowed us to test different laser regime and furnace oxidation duration.

2.3 Effect of oxidation parameters

The first step of our study was to determine the influence of the laser parameters on the preliminary oxide layer. To do so, we varied the accumulated energy amount on the sample surface from 50 to 300 J/cm² by using several laser power (5, 9, 14 and 18 W) and 80 or 90 % of spatial overlap. We could notice that the continuous preliminary oxide thicknesses were from 20 to 140 nm thick (figure 6).

DOI: 10.1051/epjconf/201715307038

The oxide layers seem to grow linear with the accumulated energy amount.

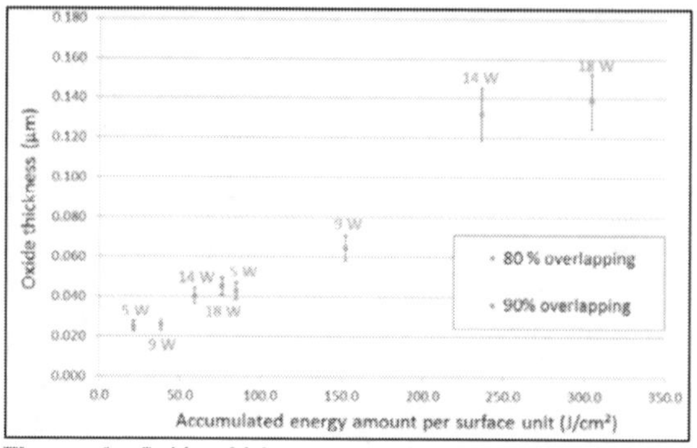

Figure 6. Oxide thicknesses resulting for different laser irradiations at 5, 9, 14 and 18 W and 80% overlap (in blue) and 90% overlap (in orange).

Moreover, in the case of the treatments at 18W/80% overlap and 5W/90% overlap, the oxides are quite similar in thickness although the fluences per pulse are different. This result highlights the fact that the principal oxidation parameter is the accumulated energy amount and not the fluence per pulse or the spatial overlap. The same experiments were performed without Eu-contamination and it appears that the presence of Eu did not interfere in the oxidation by laser.

In order to span a large variety of laser treatment, the laser fluence varied from 2.2 to 7.8 J/cm² per pulse. This laser treatment can lead to surface heating, melting or even ablation regarding to the irradiation regime. In the case of a nanosecond pulse, the laser wave is absorbed by the surface by the electrons excitation and transmitted to the metal lattice by relaxation. This implies the localized heating of the metal bulk, its oxidation and, possibly, its melting. The formation of the oxide during the laser scanning modifies the physical properties and the interaction regime between laser and the oxide surface. Interferences between in coming and out coming beam can occur because of the presence of this semi-transparent oxide and imply temperature oscillations [8]. According to [9], the theoretical laser melt threshold fluence can be estimated by:

$$\Phi_m = (\sqrt{\pi}/2)\kappa\theta_m(\tau/D)^{1/2} \qquad (3)$$

where κ=13 W/m.K is the thermal conductivity, θ=1859 K - the melting temperature, τ=110 ns is the laser pulse duration and D=3,8.10^{-6} m²/s, the thermal diffusivity [10]. This fluence represent the minimum fluence absorbed by the surface that leads to the surface melting. In our study, Φ_m = 0.36 J/cm². The absorptivity of the surface is A = 0.35 and so the melt threshold fluence can be estimated to 1 J/cm² which is lower than the laser fluences used in the preliminary treatment. SEM observations (figure 7) of single impacts of laser pulse in the pre oxidation conditions showed proof of fusion during the treatment.

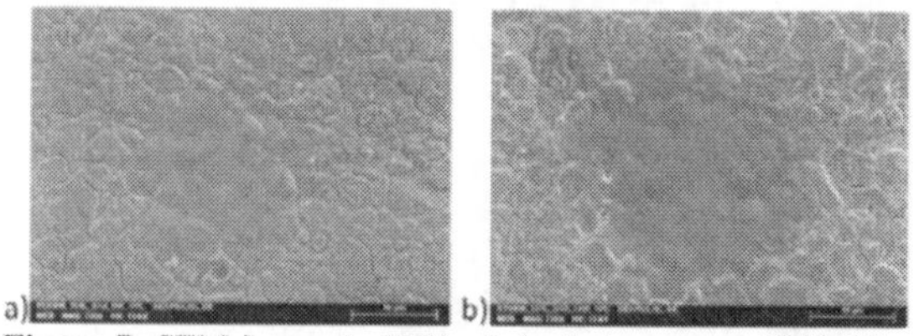

Figure 7. SEM image (x 700) of single impact on 304L SS a different laser treatment: (a) 9W (4.1 J/cm²), (b) 14W (6.3 J/cm²).

An analytical model [11] were used to simulate laser heating of a metallic surface by nanosecond pulse radiation. The temperature distribution spatial or evolution in time can be determined. The simulation of a single impact of a 1064 nm-pulse with a duration of 110 ns at a frequency of 20 kHz on a 304L stainless steel substrate were performed for different laser fluence. At a low fluence (1 J/cm²), we can see that the temperature at the surface rise to its maximum of 1650 K after 150 ns and after 500 ns, the metal temperature is 750 K (figure 8). We can assume that the increase of the laser pulse energy can lead to the melting temperature (θ=1859 K). The simulation condition (F=1J/cm²) is coherent with the melt threshold calculated in equation (3). When the maximum temperature is reached, the temperature profile in depth of the shows that the thermally affected zone is sub micrometric, like the resulting oxide layers.

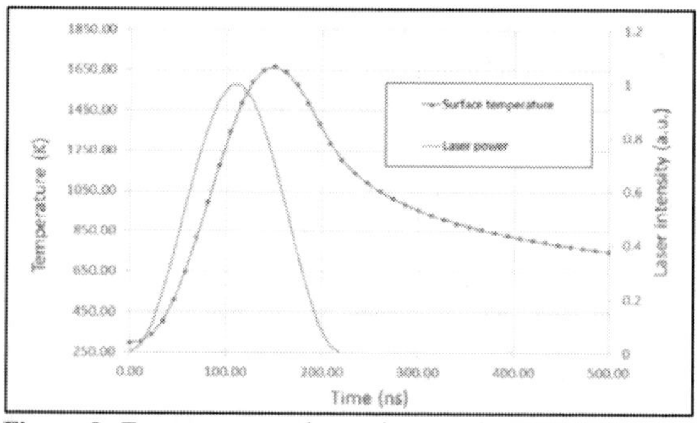

Figure 8. Temperature at the surface on the laser beam axis during a single laser pulse of 110 ns for a fluence of 1 J/cm².

Surface temperature calculated at 50 µs after laser shot with F=1 J/cm² is still at 330 K. Thus, the pulse energy accumulation at the surface metal has a weak influence on the final surface temperature after one single pulse. Nevertheless, the simulation of multi impacts is needed to determine this influence during the laser treatment. Moreover, during laser scanning, the step between two pulses is less than 10% of the beam dimension, we can consider that the treatment is performed on a surface already disturbed by the previous pulse. For the next simulations, we will have to take into account the oxidation by multi pulse on a substrate of 304L stainless steel with an oxide layer. More simulations need to be carried out but we can suppose that the oxidation is made from a liquid layer. This state allowed Eu to penetrate into the metal bulk and led to a fixed contamination of the sample surfaces.

In a second part, the influence of the treatment duration on the oxide layer characteristics was studied. Contaminated samples (900 µg/cm² of Eu) were first preliminary treated by laser (14W, 90% overlap) and oxidized in a furnace at 600°C for duration from 25 to 200h. The same experiments were performed with 304L samples without contamination and the preliminary oxide layer obtained by laser was 100 to 150 nm-thick. The figure 9 shows two main phenomena.

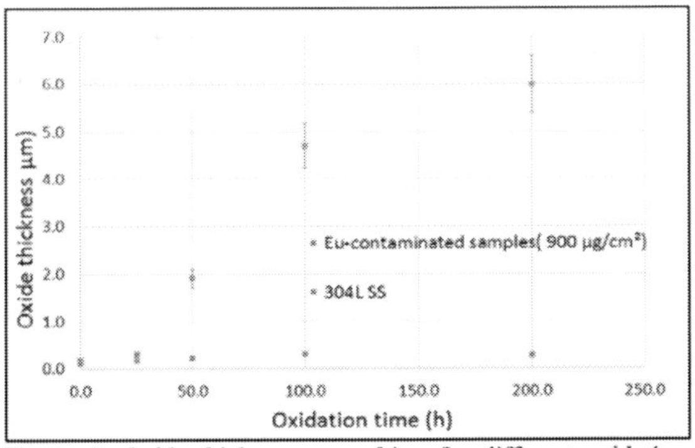

Figure 9. Oxide thicknesses resulting for different oxidation duration of Eu-contaminated samples (blue) and blank samples (orange).

On one hand the contamination of Eu stopped the oxidation as expected in the literature [6] [7]. On the other hand for blank samples, the oxide growths following a parabolic kinetics [6]. For the parabolic part of the curve, the parabolic rate is $k = 6.14 \pm 0.75 \times 10^5\ \mu m^2/s$. Unlike the laser treatment, the oxidation by furnace is influenced the Eu-contamination. Indeed the experiments presented table 1 shows that the increase of the concentration of the surface contamination implies the limitation of the thickness of the oxide.

Table 1. Oxide thickness resulting of a laser treatment (237 J/cm² accumulated) and a furnace oxidation (100h, 600°C) with different contamination concentration in Eu

Surface contamination of Eu before oxidation treatment (µg/cm²)	Oxide thickness (µm)
0	4.7 ± 0.5
500	1.6 ± 0.2
900	0.30 ± 0.01

Even with the behavior of oxidation inhibitor of the Eu, our procedure of oxidation allows us to carry on with the oxidation process. Knowing the influence of the main parameters (laser energy accumulated, furnace oxidation duration or Eu concentration), the thickness of the resulting oxide layer can be controlled and adjusted regarding to the study.

3 Laser decontamination of oxidized metal surfaces

The literature shows that decontamination of metallic surfaces by excimers [4] and ytterbium fiber laser [3] is successful with decontamination percentage up to 99 % for thick stainless steel. The oxide layer is removed efficiently for the first shots but in a second phase, the decontamination stagnates. We supposed that this limitation of the laser cleaning is the result of two phenomena. On one hand, the trapping of the oxide and contaminants into micro cracks of the materials prevent the laser beam to access full decontamination. And on the other hand, the heating of the surface during the laser treatment leads to the penetration of the contaminants in the metal bulk.

We proceeded to the decontamination of the contaminated oxide layer by laser. The decontamination set up is the same that the pre-oxidation one. The fluences applied to ablate the oxide layer is in the range of 4 to 8 J/cm² with a spot overlapping of 80 % (figure 10-11). The samples were analyzed by GD-OES in order to determine the oxide layer thickness but since the remaining Eu-concentration is near the detection limit of GD-OES, the GD-MS method was also used (figure 11). The figure 11 present the GD-MS profiles for an oxide layer obtained after laser preliminary treatment (14W, overlapping of 90%, E_{acc}=237 J/cm²) and an oxidation by furnace (50h, 600°C) With and without decontamination treatment. After oxidation, the oxide layer is 230 nm-thick and with an Eu-contamination of 0.55 % (5 500 ppm) and we can see that after 1 pass at 14W and 80% overlapping, the resulting Eu-concentration is less than 200 ppm. The decontamination treatment led also to the formation of an oxide layer of 300 nm.

Tests are carried out in order to compare the results to the literature and to find optimal laser parameters to decontaminate contaminated layers. Special care was taken to reduce thermal effects like oxidation or melting which can caused the penetration of Eu into the bulk.

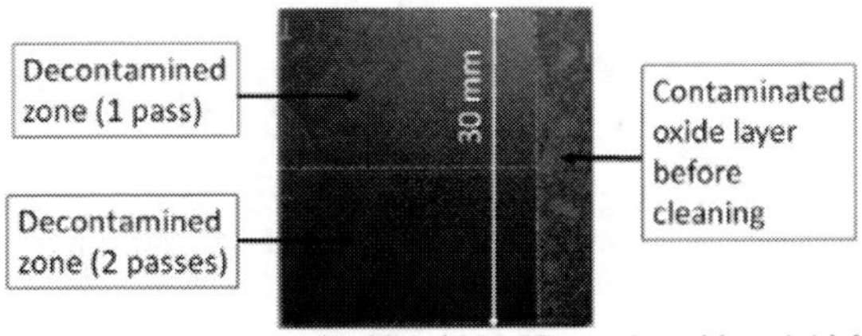

Figure 10. Overview of oxidized 304 SS samples with an initial layer thickness of 4.5 µm before and after laser cleaning (5.9 J/cm² per pulse and 80% of overlapping).

 DOI: 10.1051/epjconf/201715307038

a)

b)

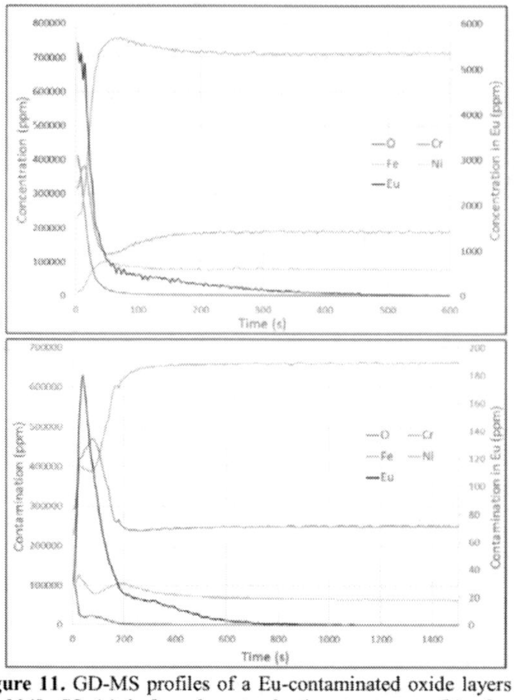

Figure 11. GD-MS profiles of a Eu-contaminated oxide layers on 304L SS (a) before decontamination treatment; (b) after decontamination treatment (14W, overlapping of 80%, F=5.9 J/cm² per pulse, 1 pass).

4 Conclusion

Our goal was to manage different way of Eu-implementation in oxide layer of 304L SS samples in order to carry out studies with non-radioactive contaminants and conventional equipment. The first results shown the Eu-contamination in volume of a micrometric oxide layer with similar characteristics as for metal samples oxidized in real conditions. These oxides layers were prepared with a contamination by Eu solution spraying, a pre-oxidation by laser and a final treatment by furnace to continue the oxidation. The thicknesses of the oxides are from 0.2 μm up to 4.5 μm with a mean Eu-concentration of 0.5 % of weight exclusively in the oxide layer. This method by furnace treatment simulates the oxidation of stainless steel exposed to high temperature.

The influence of oxide thermophysical and optical characteristics (composition, absorption coefficient …) on decontamination mechanism are studied by numerical simulation code that was developed in our laboratory. The experimental results will also be used to improve and validate the model of laser heating of micrometric layer on metallic substrate.

Finally, the formation of contaminated oxide layer with controlled characteristics is a first step for the study of laser cleaning of metallic surface with micrometric cracks.

References

1. F. Brygo, Ch. Dutouquet, D. Le Guern, R. Oltra, A. Semerok, J.M. Weurlesse, Appl. Surf. Sci. **252,** 2131–2138 (2006)
2. F. Champonnois, F. Beaumont, C. Lascoutouna, US Patent 20110315666 A1 (2009)
3. A. Leontyev, PhD thesis, Université Paris Sud - Paris XI, (2011)
4. Ph. Delaporte, M Gastaud, W. Marine, M. Sentis, O. Uteza, P. Thouvenot, J.L. Alcatraz, J.M. Le Samedy, D. Blin, Appl. Surf. Sci., **197-198**, 826-830 (2002)
5. Z. Homonnay, E. Kuzmann, K. Varga, Z. Nemeth, A. Szabo, K. Rado, K.E Mako, L. Köver, I. Cserny, D. Varga, J. Toth, J. Schunk, P. Tilky, G. Patek, J. Nucl. Mater., **348**, 191-204, (2006)
6. F. Riffard, H. Buscail, E. Caudron, R. Cueff, C. Issartel, S. Perrier, Appl. Surf. Sci., **252**, 3697-3706 (2006)
7. N. Karimi, PhD thesis, Université Blaise Pascal-Clermont-Ferrand II, (2007)
8. Wautelet, M. Appl. Phys. A, **50**, 131-139 (1990)
9. D. Bäuerle, *Chemical Processing with Lasers* (Springer, Berlin, 2000)
10. P. Lacombe, G. Béranger, B. Baroux, *Les aciers inoxydables,* (Les éditions de physique, 1990) (in French)
11. A. Semerok, S. V. Fomichev, J.-M. Weulersse, F. Brygo, P.-Y. Thro and C. Grisolia, J. Nucl. Mater., **420**, 198–206 (2012)

Particle track plotting in Visual MCNP6

Randy Schwarz[1,*]

[1]Visual Editor Consultants, PO Box 1308, Richland, WA 99352, USA

Abstract. A visual interface for MCNP6 has been created to allow the plotting of source points, collision points and particle tracks. The interface will allow users to visualize source and collision points based on parameters set by the user. This can be very valuable in understanding the physics of a problem and the important paths that contribute to a tally.

1 Introduction

The Visual Editor is a graphical user interface created for displaying geometry plots and results obtained from the MCNP computer code developed at Los Alamos National Laboratory. It consists of a graphical user interface written in C++ that is linked directly to the MCNP Fortran code.

The Initial creation of the graphical interface started in 1992 and it was first published in 1993 [1]. The original version of the graphical interface was created for Unix/Linux operating systems and was linked to the MCNP computer code [2]. The graphical interface was adapted to the Windows operating system in 1999 [3]. In 2008 the graphical interface was adapted to work with the MCNPX computer code [4].

In 2014 work began on creating a visual Interface for MCNP6 [5]. An initial release of this interface was created to plot geometries and tallies, including mesh tallies and weight window meshes [6]. This paper discusses the latest upgrade of the Visual Editor interface to add source generation point plotting and collision plotting with the option to add particle tracks.

Complete details on these capabilities along and their current status can be found at the Visual Editor website (www.mcnpvised.com).

2 Source point plotting

Sources in MCNP6 can be very complex and it is important to verify that the source geometry is correctly defined. MCNP6 will run even if the source geometry is not correct. Visual tools are required to verify that the source is correct.

The Visual Editor program is linked to the source generation routine in MCNP6 to obtain the source information so it can be plotted in the Visual Editor. Figure 1 shows a plot in the Visual Editor of a source defined by "SDEF rad=90" inside a sphere of 100 cm. This is a plot of the first 100,000 source points.

Figure 1. Source point plot for sdef rad=90

The user can set a number of parameters for plotting the source points including color and size and how much of the source to plot. Notice in Figure 1, the outside of the sphere is darker than the inside, this is because the source is a shell source and not a spherical volume source. This can be verified, by only plotting particles that are with 10 cm of the plot plane.

Figure 2 shows the same plot with the distance from the geometry plot plane set to +/- 10 cm. The size of the particle has also been increased so that it can be clearly seen that the source is a shell source and not a volume source.

* Corresponding author: randyschwarz@mcnpvised.com

DOI: 10.1051/epjconf/201715307039

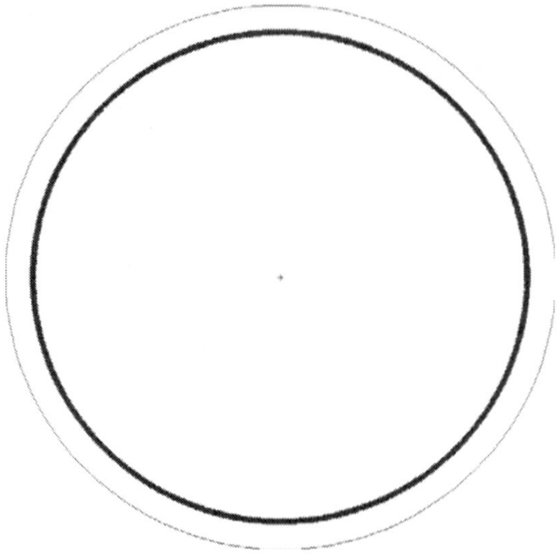

Figure 2. Distance from the plot plane set to +/- 10 cm.

It is a common mistake to incorrectly define a spherical volume source as a shell source. The Visual Editor clearly shows this source is a shell source.

Figure 3 shows a how a volume source should look. The source particles should be distributed normally when the distance from the plot plane is set to +/- 10 cm.

Figure 3. Spherical volume source defined correctly.

For complex source configurations it is essential that you have a visual confirmation that the source sampling is correct. Figure 4 shows a plot of the source for 3 spheres, 2 have a radius of 50 cm and one has a radius of 30 cm. The source is defined incorrectly because it does not use functional dependence so the two radii are used in all 3 positions. By plotting 100,000 source points the error becomes obvious.

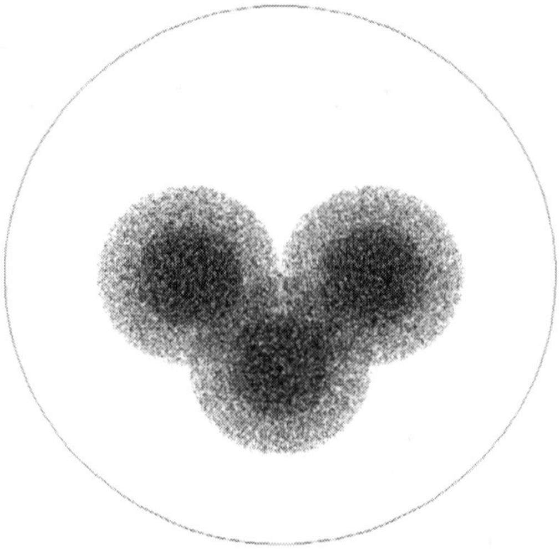

Figure 4. Spherical volume sources without functional dependence.

Without the Visual Editor, users typically look at the first 50 source particles to determine the validity of the source. Figure 5 shows the same source with only 50 source points plotted. From these points it is very hard to determine that the source is wrong.

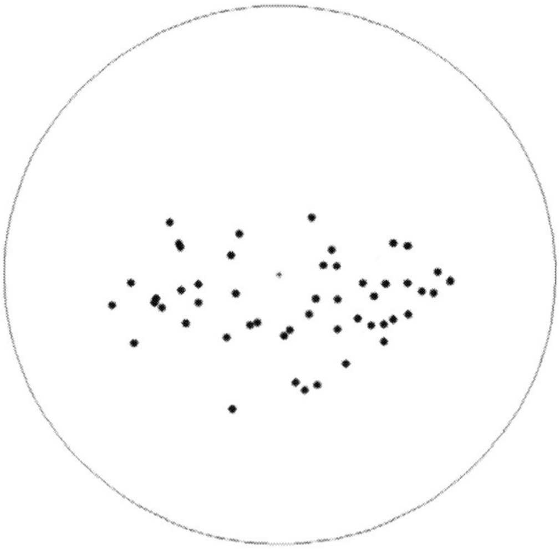

Figure 5. First 50 particles plotted for wrong source definition.

Figure 6 shows a plot of the correct source definition with functional dependence used to place the radii in the correct positions. From this plot it can easily be seen that the source is correct.

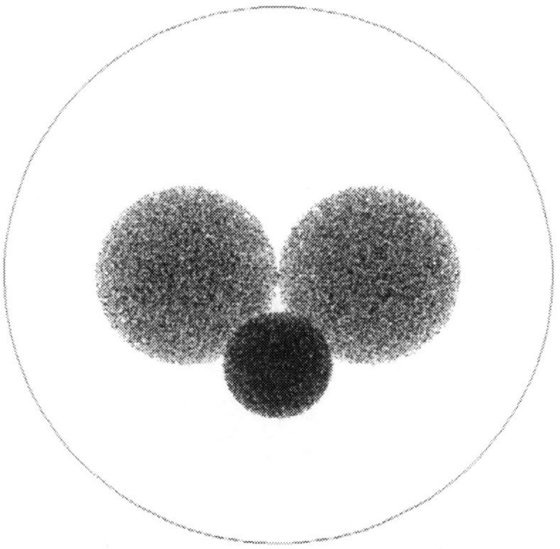

Figure 6. Three Spherical volume sources with functional dependence.

3 Collision point plotting

Because the Visual Editor has access to all of the particle information in MCNP6 it is also possible to plot particle collision points and optionally particle tracks.

Figure 7 shows a plot of collisions displayed from test input file 109 provided with the MCNP6 distribution. This input file has the "tropt" keyword and can only be run in MCNP6. Figure 7 shows the collisions from the first 5 source particles.

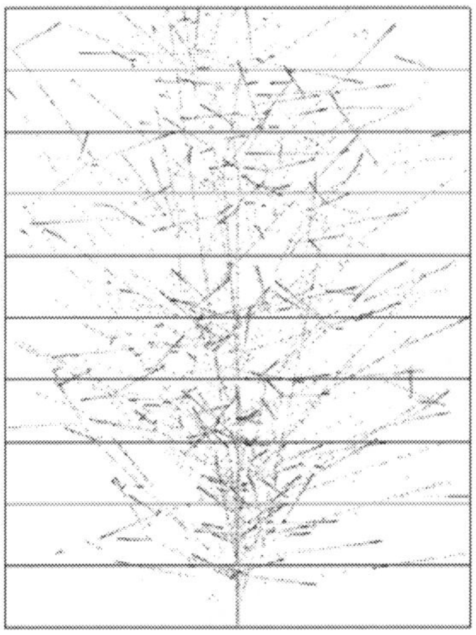

Figure 7. Collisions plotted for test input 109

Test input 109 has a proton source hitting a water cylinder producing secondary particles. The mode card has 29 different particle types as shown below

*mode n q h g / * z | ! k ? % ^ l b + _ - ~ x c y w o @ u < v >*

The color and size of each particle can be set inside the Visual Editor. Figure 8 shows the particle track plotting panel in the Visual Editor for this problem.

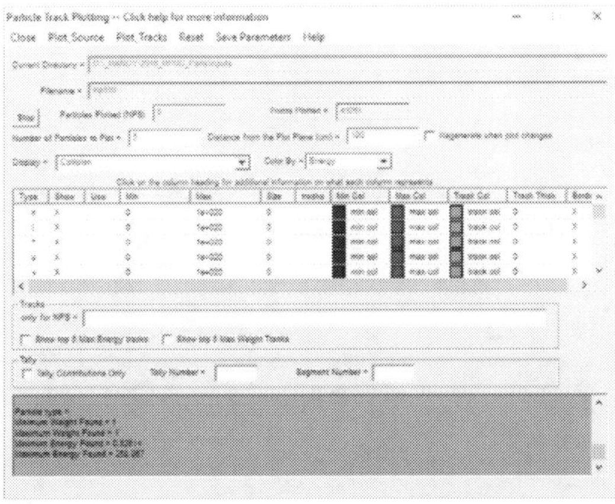

Figure 8. particle types listed in the Visual Editor for input 109.

In Figure 8 all particles have the same plot parameters making it difficult to differentiate the different particles plotted in Figure 7.

In Figure 9, we have set the size of the source protons to be larger (using the size column in Figure 8), to differentiate the source protons from the secondary particles created from the source protons.

The color of the particle represents the energy, with the source energy red and the color dropping down to blue for the lowest energy. The Visual Editor tabulates the energy of all particles that it displays and sets this energy range dynamically. The user can set a specific energy range by specifying a minimum and maximum value in the table shown in Figure 8.

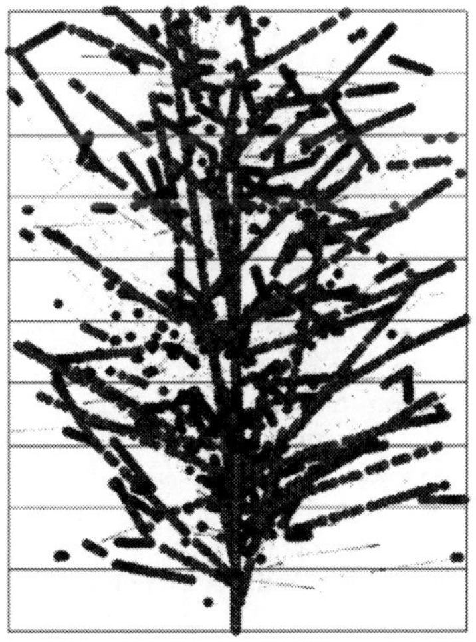

Figure 9. Proton source particles set larger.

 DOI: 10.1051/epjconf/201715307039

To determine the secondary particles being generated, we now turn off the plotting of the source protons by unselecting the "Show" column in Figure 8. We then set the size and color of the secondary particles to differentiate them as shown in the particle track plotting panel in Figure 10.

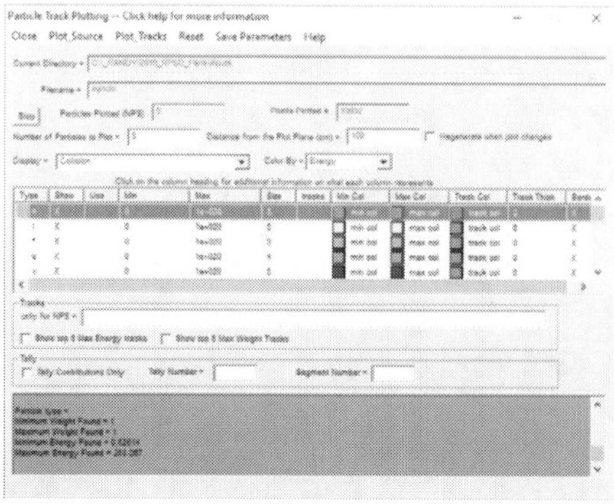

Figure 10. Secondary particles differentiated by color.

Figure 11 now shows the collisions colored by particle type for the first 5 particles.

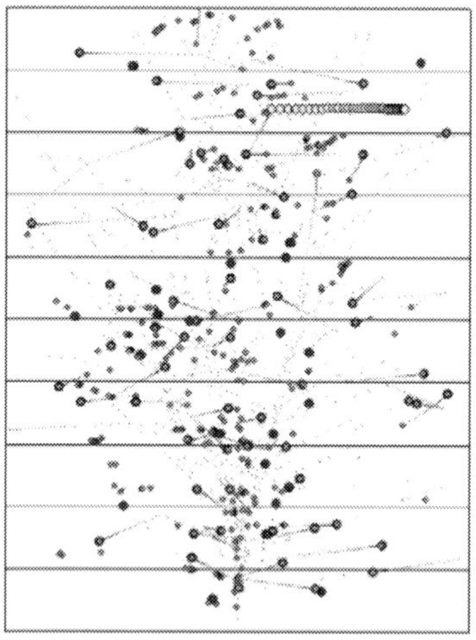

Figure 11. Secondary particles differentiated by size and color.

By comparing the table in Figure 10 with the plot in Figure 11, it is possible to now differentiate some of the secondary particles. The yellow particle moving to the right in the top center is a muon (particle type |). The small green dots are neutrons, The large purple dots are muon neutrinos (particle type v). The orange dots are electron neutrinos (particle type u).

The user may spend a significant amount of time setting particle size and color, so the option is provided to save this information for this input file by selecting the "Save Parameters" option in the menu shown in Figure 10.

To determine the path of the particles that cause the collisions, it is possible to turn on tracks as shown in Figure 12. Here tracks have been turned on for neutrons by selecting the tracks column in Figure 10. The color of the tracks has been set to magenta and the size has been slightly increased over the default size so that they show up better.

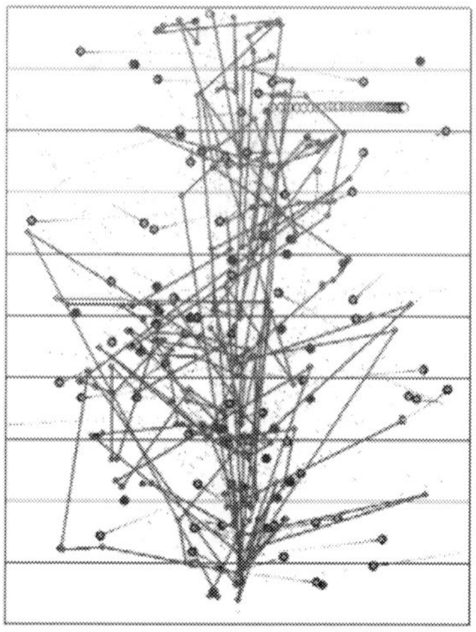

Figure 12. Particle tracks plotted for neutrons

4 Plotting collision points showing variance reduction

The ability to visually differentiate different particle types within the plot is critical for understanding the physics of the problem. However, just as important is the ability to plot particle tracks from variance reduction.

Figure 13 shows the source for a transportation cask geometry with a source inside a universe that occurs in 16 different locations. The source is biased within each pin and also biased axially to optimize the source points that will contribute to a detector located outside the cask. The source biasing in each pin can is shown in Figure 13.

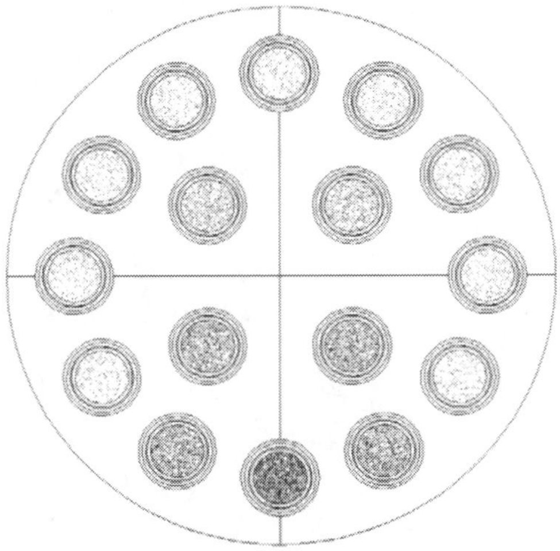

Figure 13. Source points plotted for a transportation cask showing source biasing.

Figure 14 shows the axial source biasing of the 16 pins. The source points are colored by weight. The under-biased source points at the bottom are red and the over-biased source particles at the top are blue.

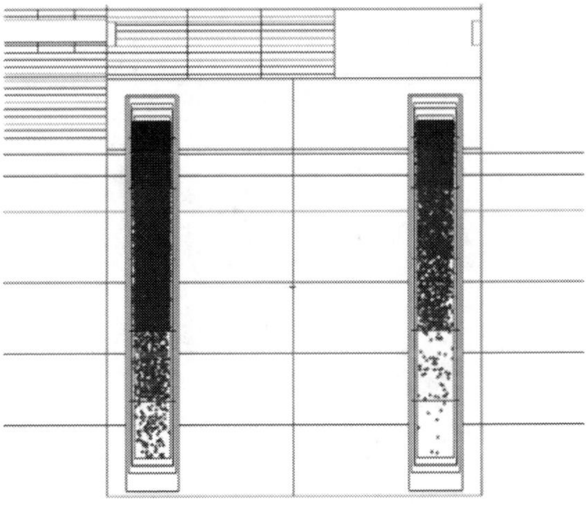

Figure 14. Source points plotted for a transportation cask showing axial source biasing.

Figure 14 shows both the presence of the radial biasing and the axial biasing. The distance from the plot plane has been set to 4 cm so the source points in the other pins do not show up in the plot.

Figure 15 shows an axial plot of collision points in the transportation cask shown previously in figures 13 and 14. The particles are being directed toward the tally volume outside the cask using variance reduction. The collision points are colored by weight, with the source particles being red and the low weight particles colored in blue. The effects of the variance reduction can be clearly seen.

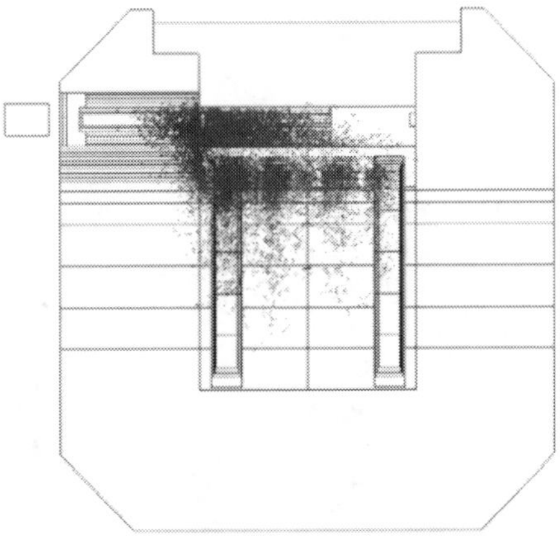

Figure 15. Collision points showing variance reduction.

The collisions can be plotted on top of any normal MCNP6 geometry plot, including color plots. Figure 16 shows the collision points plotted on top of a color plot of the importances. Since the collisions occur due to the presence of the importances, it can be very useful to verify the importance mapping along with the collision points.

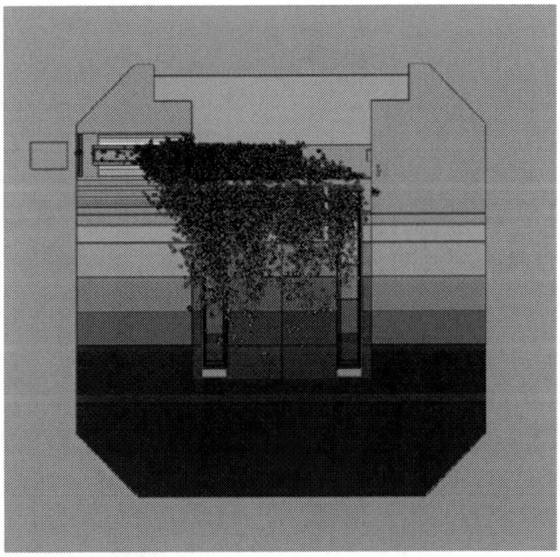

Figure 16. Collision points plotted with importance colors.

The most important collisions are those that contribute to the tally region outside the cask. The Visual Editor allows the user to show only those collisions from a particle that reaches a tally volume.

Figure 17 shows a plot similar to that in Figure 16, expect now only those particles that contribute to the tally are plotted. Since only a few tracks will contribute, it is necessary to plot a lot more tracks to see the important tracks for the tally.

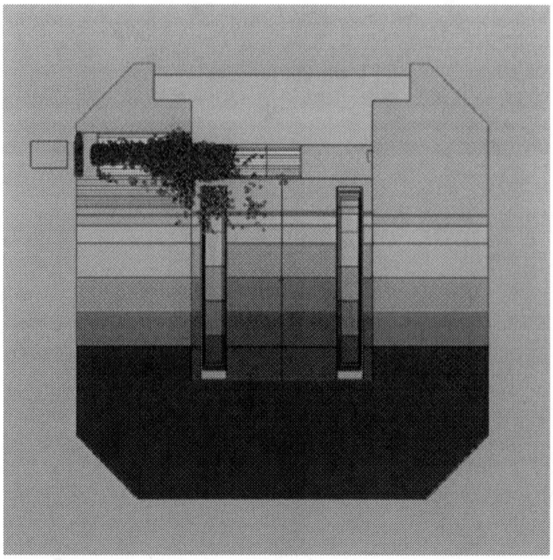

Figure 17. Collision that contribute to the tally.

Figure 18 shows a close up view of only those collisions that reach the tally volume. The track option has been turned on to show the path of the particles. The size and color of the track lines can be set by the user.

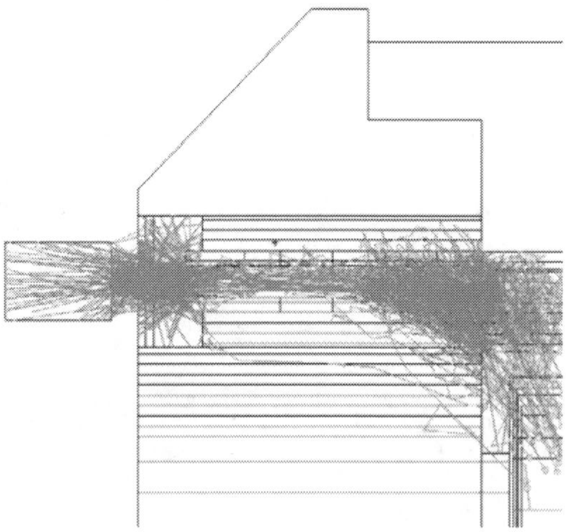

Figure 18. Particle tracks that contribute to the tally.

Particle tracks can also be shown for DXTRAN spheres. Figure 19 shows a shielding problem that consists of a hole in a concrete wall with a 1 MeV neutron source inside a sphere.

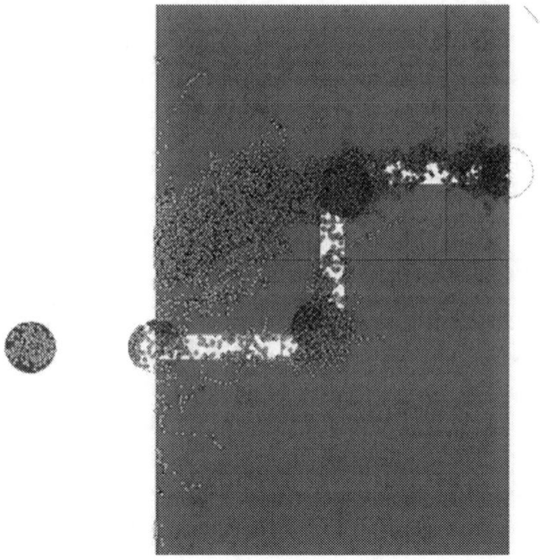

Figure 19. Tracks showing contributions to DXTRAN spheres.

This problem also includes weight windows since part of the contribution to the final tally is from particles moving through the concrete shield. In Figure 20, we plot the collisions on top of the weight window color mesh.

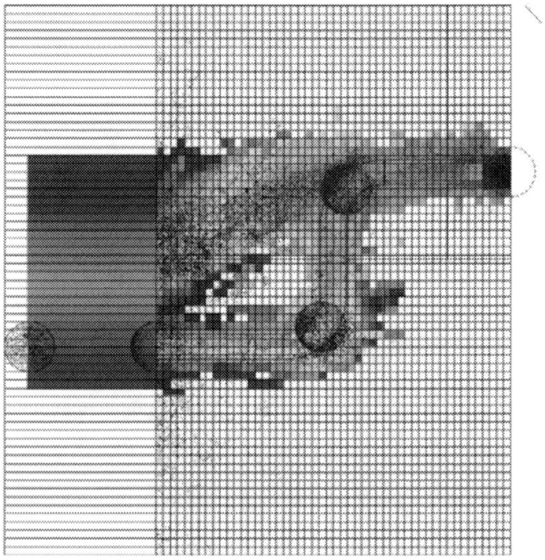

Figure 20. Collisions plotted on top of the weight windows mesh.

In Figure 21 we plot tracks instead of collisions. Two paths can be clearly seen, one going down the hole and one going through the shield. The plot legend has also been turned on to show the values for the weight window mesh.

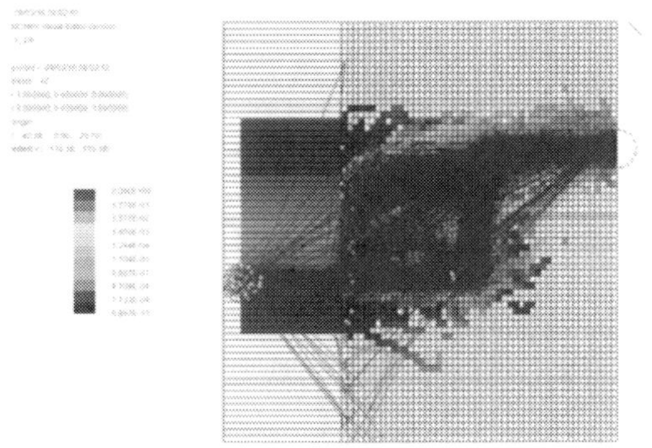

Figure 21. Tracks plotted on top of weight windows.

Sometimes it may be desirable to plot the collisions on top of a mesh tally. In Figure 22, we plot the collisions on top of a fmesh tally.

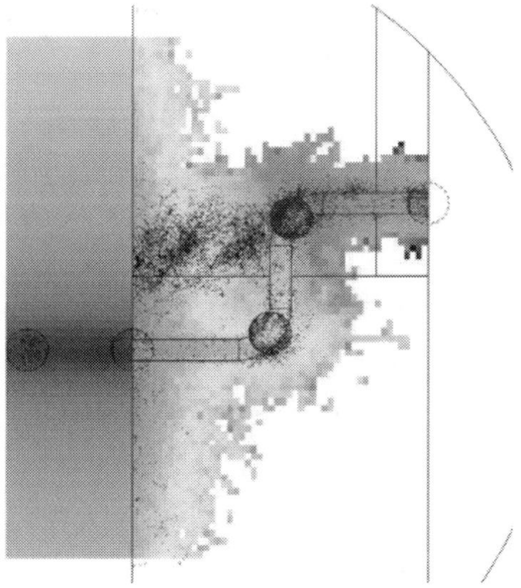

Figure 22. Collisions plotted on top of an fmesh tally.

5 Criticality source point plotting

For criticality calculations (KCODE source) it is essential to verify that all fissionable regions have been adequately sampled. There are calculational methods that can be used to help verify if the source is properly sampled, but they do not always work for complex critical systems.

The Visual Editor has the ability to plot the cycle-by-cycle source points for a criticality calculation so that the user can verify that the source has converged properly and that all points of interest have been sampled.

Figure 23 shows all the fission points sampled for a criticality calculation for a sub-critical system.

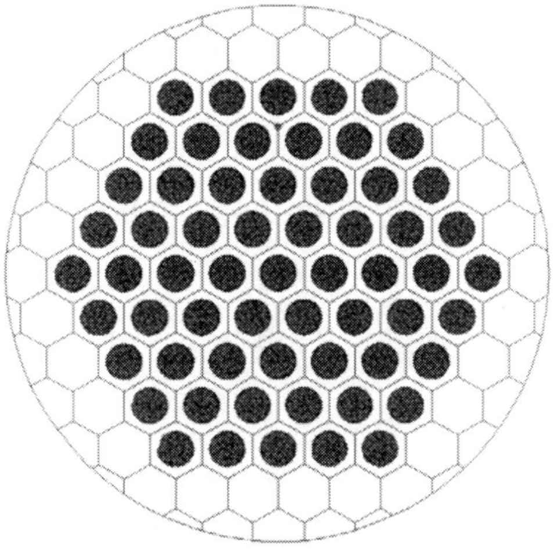

Figure 23. Fission points sampled for a KCODE calculation.

The fission points can be sampled for each cycle and an animation can be made of the points from cycle to cycle. Figure 24 shows the fission source distribution for the first cycle. The calculation was initiated with a single point at the center of the lattice which can be clearly seen in this plot.

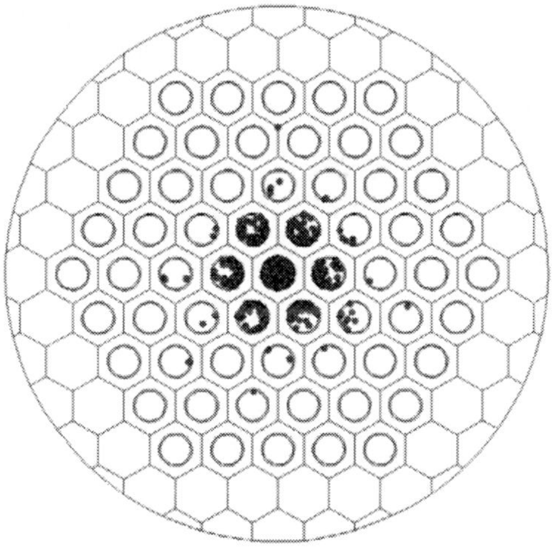

Figure 24. Fission points for the first cycle.

The ability to plot fission points is especially important for complex systems such as the one shown in Figure 25. The geometry is defined so that a single source point at the bottom of the geometry will not calculate the correct value of Keff because the fission points cannot travel up the small central region to reach the top portion that is super-critical. Figure 25 shows all of the fission points for the entire calculation, using 10,000 points per cycle.

Figure 25. Fission source points plotted for a non-converged criticality calculation.

This plot clearly shows that the most important fission volume at the top has not been sampled. Without the ability to visualize the location of the fission source points it is very hard to verify that the value of Keff is correct. The value calculated for this source location is 0.58, while the true value of Keff for the geometry which is dominated by the upper volume is 1.01.

The ability to plot the source point locations for a criticality calculation can provide valuable insight into the propagation of the fission points as a function of cycle.

Figure 26 shows the panel used for plotting the fission points. The user has a number of options including which cycles to plot and the size and color of the points to plot. For animations, the user can select the delay time (in milliseconds) between the different cycles.

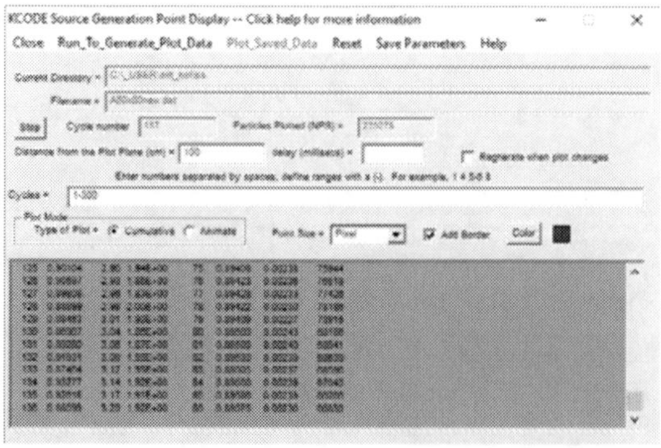

Figure 26. Panel for displaying criticality fission points.

To correctly sample the entire volume we use a cylindrical source to sample points up and down the central cylinder for the first cycle. Figure 27 shows the plot of the fission points for this calculation which will now calculate the correct value of Keff.

Figure 27. Fission source points, showing a converged source sampling.

4 Conclusion

The Visual Editor is a graphical user interface for MCNP6. The ability to plot the source and collisions allow the user to user to gain valuable insight into the source configuration and particle behavior. This is done in a user friendly easy to use interface.

References

1. R.A. Schwarz, L.L. Carter, T.P. Dole, S.M. Fredrickson, and B.M. Templeton, *Trans. Am. Nucl. Soc.*, **69**, 401 (1993)
2. J. F. Briesmeister, Ed., ``MCNP--A General Monte Carlo Code for Neutron and Photon Transport, Version 3A," Los Alamos National Laboratory report LA-7396-M, Rev. 2 (1986)
3. L.L. Carter, R.A. Schwarz, and J. Pfohl, *Trans. Am. Nucl. Soc.*, **81**, 256–257 (1999).
4. L.L. Carter, R.A. Schwarz, and A. L. Schwarz, NASA contract number NNL07AA10C (2008)
5. D. B. Pelowitz, Ed., ``MCNP6 Users Manual Version 1.0," Los Alamos National Laboratory report LA-CP-13-00634 (2013).
6. R.A. Schwarz, 18th Topical Meeting of the Radiation Protection and Shielding Division, Knoxville, TN, September 14-18 (2014)

Elaborate SMART MCNP Modelling Using ANSYS and Its Applications

Jaehoon Song*, *Han-bum* SURH, *Seung-jin* Kim and *Bonsueng* Koo

Korea Atomic Energy Research Institute, 989-111 Daedeok-daero, Yuseong-gu, Daejeon, 305-353, Korea

Abstract. An MCNP 3-dimensional model can be widely used to evaluate various design parameters such as a core design or shielding design. Conventionally, a simplified 3-dimensional MCNP model is applied to calculate these parameters because of the cumbersomeness of modelling by hand. ANSYS has a function for converting the CAD 'stp' format into an MCNP input in the geometry part. Using ANSYS and a 3-dimensional CAD file, a very detailed and sophisticated MCNP 3-dimensional model can be generated. The MCNP model is applied to evaluate the assembly weighting factor at the ex-core detector of SMART, and the result is compared with a simplified MCNP SMART model and assembly weighting factor calculated by DORT, which is a deterministic Sn code.

1 Introduction

The shielding design parameters for a safety analysis have been conventionally generated by a deterministic method such as DORT or TORT (Two-and Three-Dimensional Oak Ridge Discrete Ordinates Neutron/Photon Transport Code) [1]. These codes find the solution of the Boltzmann transport equation describing the given system. It uses the method of discrete ordinates to treat the directional variable and finite-difference methods to treat spatial variables. In addition, energy dependence is treated using a multi-group formulation. In order to treat these deterministic codes, a macroscopic cross section is required, such as BUGLE-96 [2], and to apply the cross section into DORT or TORT, the GIP [3] code is needed. BUGLE multi-group libraries are made on the basis of neutron or gamma energy spectrum coming from a commercial reactor. Therefore, there is a limitation to apply the macroscopic cross section into a new type of reactor with different vessel diameters like SMART. DORT or TORT requires a long time to make the geometry input part. GEOSHIELD [4] was developed to easily make a DORT geometry input. The code reads a CAD 'dxf' file and converts it into a DORT geometry input, and it can summarize the DORT output file to be easily read by users and provide visualized output summaries.

SMART is a small module type reactor of ~330MWth that is being developed at the Korea Atomic Energy Research Institute. It has 57 fuel assemblies, 8 steam generators, 4 coolant pumps, and a pressurizer vessel on top of the reactor. In particular, it has a flow mix header assembly for uniformly mixing S/G outlet coolant as well as neutron/gamma shielding and a flow skirt under the core, as shown in Fig. 1. The enrichment of the fuel has a range of 2wt% to 4.95wt% depending on the required thermal power, loading patterns, and cycle batch.

SMART has four ex-core detectors (power range) in the core shroud region. Its height is about 2 meters by covering the active core height. As a concept design, the 3 separate detectors are inserted into a hole axially.

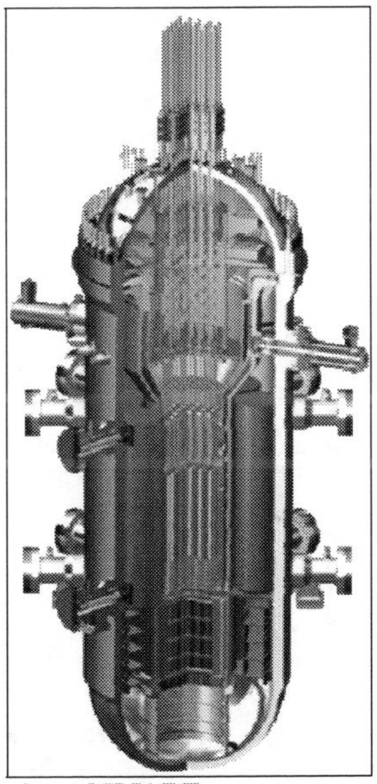

Fig. 1. Schmetic view of SMART

On the other hand, programs that convert a CAD 3-dimensional drawing file into an MCNP [5] input have been introduced since the 2000s, such as MCAM (CHINA) [6], ATTILA (US) [7] and VISED [8]. In

* Corresponding author: jh_song@kaeri.re.kr

order to use these programs, engineers should have a license for them. For example, MCAM has a good technology in that it can analyse overlapped surfaces, divide a complicated cell into several smaller cells, and sometimes find a failed cell and fix it. However, the MCAM sometimes does not give a result (MCNP geometry input with the coolant region) because the void region (coolant region) is generally not described in the CAD file used in the machinery division.

In order to figure out the mentioned problem, the coolant region is modelled before converting CAD into an MCNP input. In addition, the conversion is applied using ANSYS [9].

The ANSYS code is widely used in various engineering fields for an analysis of the structure mechanics, fluid dynamics, and so on. ANSYS also has a function to convert a CAD file ('*.stp') into an MCNP input in geometry. Therefore, it is easy to make the MCNP geometry input through cooperation with other engineering fields having an ANSYS code license.

In this paper, the procedures are introduced to generate an MCNP input using ANSYS, and the assembly weighting factor generated from the introduced procedure to be used as the input data in the core protection and monitoring system is evaluated and compared with previous generated design parameters as an application.

2 Procedures

First of all, an STP formatted CAD file is required. The CAD file can be obtained from the machinery division. The structural geometry in the STP file should be simplified by removing insignificant structures such as bolts, nuts, and slightly slanted edges in consideration of the shielding design. In addition, the remaining excluding reactor structures should be filled with coolant. Additionally, it is recommended that a complicated structure like a core support plate and fuel alignment plate should be divided into several parts. In addition, if some model has a reactor core like the SMART reactor model, the core should be modelled with a single cell. The reason for this is that if a complicated core with a fuel rod and cladding is modelled, it is not sure whether the converting is performed well or not. Moreover, it is more powerful and easier to use a repeated structure with a "fill" and "lattice" command in the MCNP for the reactor core modelling.

Secondly, read the STP format file on the design moduler in ANSYS. Then, apply 'EXPORT' to the MCNP format. The converted geometry input should be checked using VISED or MCNP plotting in order to confirm whether the geometry input is correctly made or not.

Thirdly, read the converted MCNP input in a text editor. Then, assign the material composition in each cell. In addition, put the core part, which is already manually made by the engineer in a single cell.

Fourthly, a particle transport calculation is done using a simple source definition, in which it is recommended that the starting particle be located in the centre position in the geometry with a lower material density in order to check the LOST particle from the incorrect modelling.

It is useful if this procedure be applied to other fields, a fusion reactor, a particle accelerator, and a medial system in view of reducing the geometry modelling time.

3 Model description

Fig 2 shows a 2-dimensional r-z view (axial direction) along with the red line in Fig. 3 (left), which is r-θ view in the model. The ex-core detectors are located between the core shroud and flow mixing header assembly. Because the detailed specifications for an ex-core detector is not yet determined, a single cylindrical cell is modelled. It covers the active core region in the axial direction. As shown in Fig. 2, the core shroud is composed of several layers. With this sophistic geometry, the effects of neutrons coming between the shroud layers can be considered. Fig. 3 (right) shows a complicated fuel alignment plate structure. This cell consists of 25 smaller cells and dozens of surfaces. Owing to the ANSYS converting function, this complicated geometry modelling can be performed.

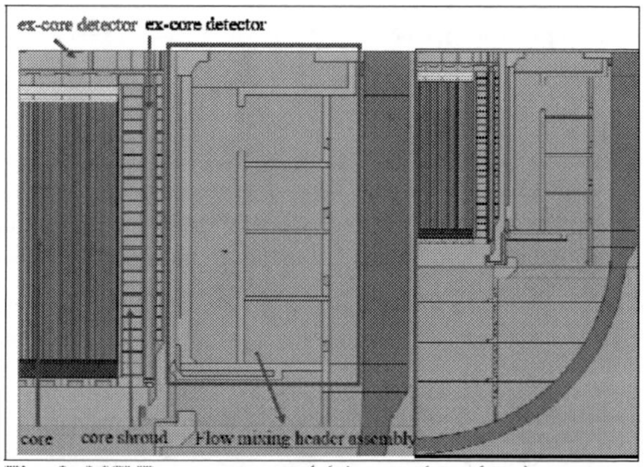

Fig. 2. MCNP geometry model in r-z view showing ex-core detector region (left) / MCNP whole core model(right)

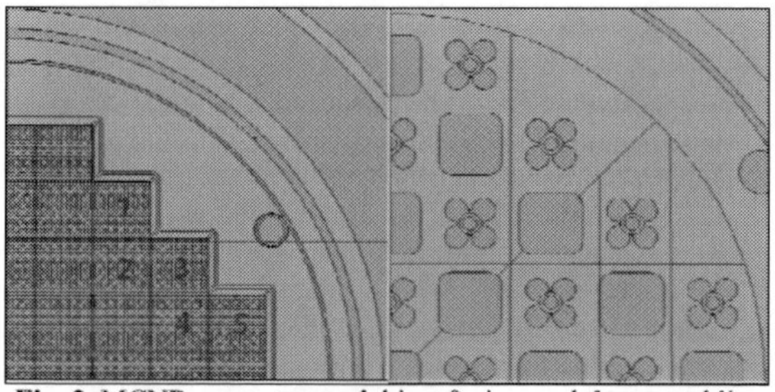

Fig. 3. MCNP geometry model in r-θ view and the assemblies with neutron source (left)/complicated fuel alignment plate model (right)

Table 1 shows the numbers of cells and surfaces used in the MCNP. Over 3000 surfaces are used in the model, and using these surfaces, 931 cells are made by ANSYS. The total time spent to make an MCNP geometry model containing 57 fuel assemblies using the introduced method is within about 40 hours.

Table 1. The number of cells and surfaces used in MCNP 3 dimensional model

	number
Cells	931
Surfaces	2697
Surface transformation	556

The neutron sources are assigned in 5 fuel assemblies in which they are most closed to ex-core detectors, due to a lower response of other assemblies with fission watt spectrum, as shown in Fig. 3(left). The number of histories used in this calculation is 4×10^{10}. Using the ft tally with the 'SCX' option, the neutrons coming from 5 assemblies with an assigned neutron source are counted in the cylindrical ex-core detector region cell. The reflective boundary conditions are applied in the y=0 and x=0 plane. The void boundary condition is applied in the radial direction owing to 1/4 core model.

4 Applications and Reference calculations

In this chapter, two types of reference calculations for AWF (assembly weighting factor) are introduced in order to compare the MCNP results generated from the introduced method using ANSYS with a conventional deterministic Sn method and with simplified MCNP 3 dimensional stochastic method. AWF is a relative response fraction for the ex-core detector of each fuel assembly. The AWF is generally used in the core protection and monitoring system (SCOMS/SCOPS) to make a core power distribution.

4.1. Dort

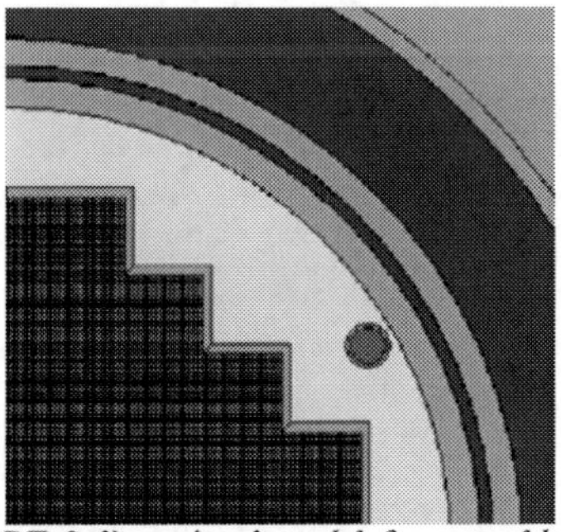

Fig. 4. DORT 2-dimensional model for assembly weighting factor

Fig. 4. Shows a DORT 2-dimensional r-θ model treated in reference [10]. The GIP [3] code and BUGLE-96 [2] library are used. The number of fine meshes in the radial and azimuth direction is 249 and 180, respectively. The number of scattering order is P5, and an S8 angular quadrature is used. The boundary conditions are the same with the above introduced MCNP model. The AWF result from the DORT model using adjoint equation is shown in Fig. 5.

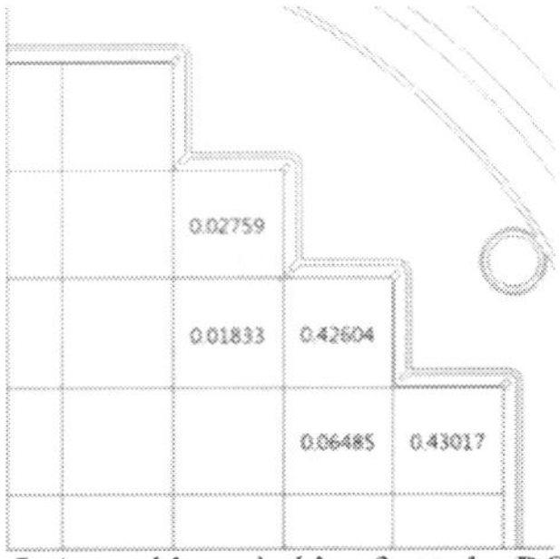

Fig. 5. Assembly weigthing factor by DORT

4.2 Simplified MCNP

This model has 677 surfaces, 140 cells, and 74 transformation cells. The model is not a 1/4 core but a whole core with 57 fuel assemblies. Therefore, the boundary conditions are all void conditions. The number of histories used is 1×10^{10}. It has an ex-core detector region, which is divided into 3 parts in the axial direction. The height of each detector is 15cm, and one is 35cm apart. For a comparison with the introduced model, the tally result on the central part is chosen. Fig. 6 shows the MCNP model and Fig. 7 shows the AWF result.

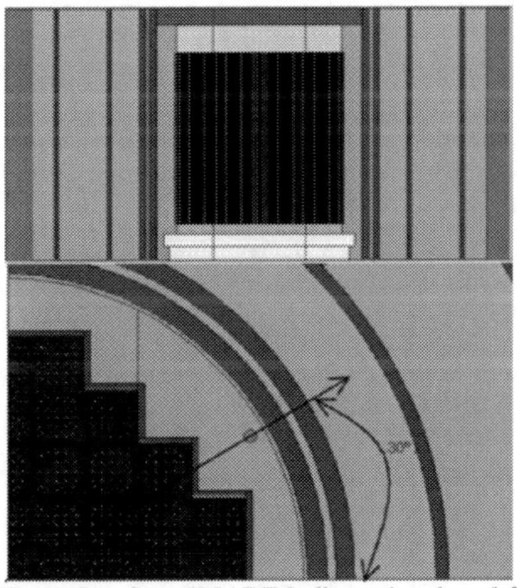

Fig. 6. Previous SMART 3-dimensional model

DOI: 10.1051/epjconf/201715307040

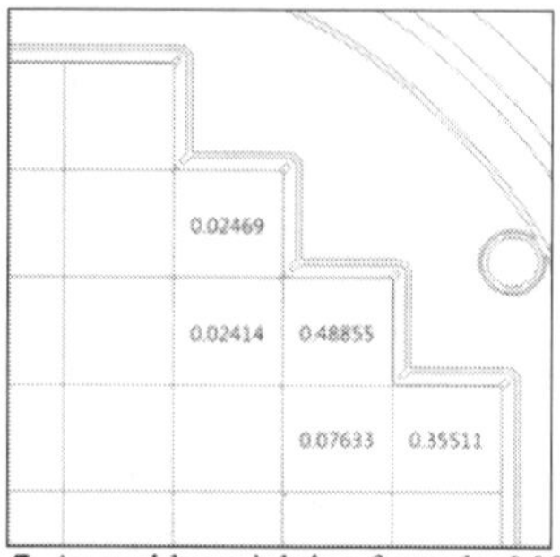

Fig. 7. Assembly weighting factor by MCNP

5 Comparison and conclusion

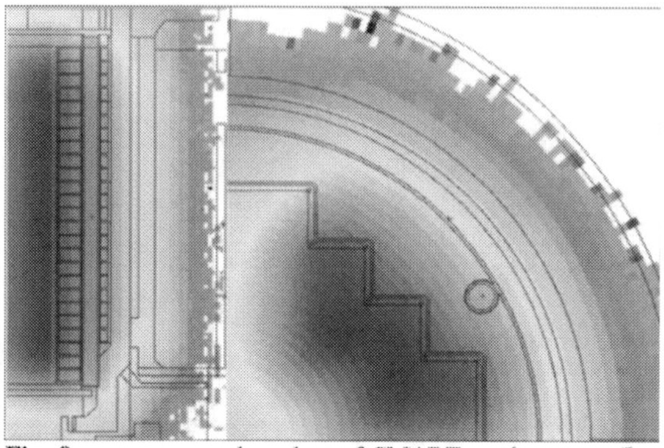

Fig. 8. x-z cross section view of SMART total neutron flux distribution by Kcode/Fmesh option using MCNP

Fig. 8 shows the fmesh tally result in the total energy range. The total relative error on the FT tally with the 'SCX' option is less than 0.05% in the ex-core detector region. The assembly weighting factors from the introduced MCNP model are compared with the DORT result and MCNP, which is a 3-dimensional simplified model, as shown in Fig. 9. In Fig. 9, the AWFs (red) by the introduced method are more similar than those by the simplified MCNP. The discrepancy between the introduced method and DORT is about 15%. In addition, the difference in both MCNP methods is over 25%. Actually, this percent is not important due to a small contribution with large difference assemblies.

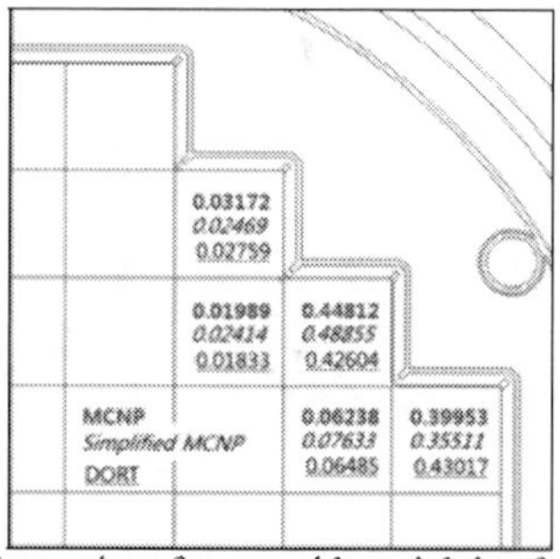

Fig. 9. Comparison for assembly weighting factors

Using ANSYS, the mcnp geometry input is converted and generated from an "STP" CAD 3-dimensional file. Before converting, the coolant region should be reflected in the "STP" file. Complicated structures are automatically divided into smaller cells in ANSYS. The AWFs from the introduced method are compared with AWFs from the previous evaluated simplified mcnp and with AWFs from DORT using an Sn deterministic method.

6 Further works

In order to calculate a more reliable assembly weighting factor by MCNP and compare it with DORT, the ex-core region will be expanded from one to three. Considering the back scattering effects from a flow mixing header assembly, a variance reduction technique will be attempted. The procedure introduced in reference [11] is a good approach for the case of a whole reactor model. And using reference [6], the assembly weighting factor of the MCAM-MCNP model will be compared with those of the ANSYS-MCNP model

This article was prepared with the help and support of HAN-BUM SURH in the machinery division and SEONG-JIN KIM and BON SEUNG KOO in the SMART PPE development division at the Korea Atomic Energy Research Institute. HAN-BUM SURH provided all sorted 'stp' files to convert in MCNP input. SEONG-JIN KIM provided the converted MCNP input using the 'stp' file and ANSYS.

References

1. RSICC CODE PACKAGE CCC-543, Two-and Three-Dimensional Discrete Ordinate Transport, Ver. 2.7.3.
2. RSICC DATA LIBRARY COLLECTION BUGLE-96, DLC-185, ORNL, March 28, 1996.
3. W. A. Rhodades, "The Gip Program for Preparation of Ground-Oranized Cross-sections", ORNL, APR. 13, (1978)
4. K. S. Kim, K. Y. KIM, H. Y. KIM, C. H. Lee and S. Q. Zee, "Development of the GEOSHIELD Program for the Automatic Particle Transport Calculation Using DORT," *Proc. 2006 Autumn Meeting of Korean Association for Radiation Protection*, Gyoengjoo, Korea, Nov. 23-24, 94 (2006)
5. D. B. Pelowitz, LA-CP-13-00634, Rev. 0, MCNP6 (TM) USER'S MANUAL Ver. 1.0 (2013)
6. Y. Wu, FDS Team, CAD-based interface programs for fusion neutron transport simulation, *Fusion Eng.* Des. 84, 1987 (2009)
7. M. Z. Youssef, P. Batistoni, L. Patrizzi, T. Wareing, I. M. Davis, *Fusion. Sci. Tech.*, 52, 801 (2007)
8. R.A. Schwarz, L.L. Carter, T P. Dole, S.M. Fredrickson, and B.M. Templeton, "Graphical User Input Interface for MCNP," *Trans. Am. Nucl. Soc.*, **69**, 401 (1993)
9. ANSYS v14.5 release, 14, Nov. 2012.
10. K. S. Kim, 002-NR301-031, Rev. 00, SMART Assembly Weighting Factor Evaluation, June 5, (2010)

11. S. Yun and S. J. Kim, A Cylinderical Shielding Design Concept for the Protype Gen-IV Sodium-cooled Fast Reactor, *Trans. of the KNS Spring Meeting*, (2014)

The Activation Detector Activity Calculations Using the Effective Source Method and Measurement

Vladimir Smutný[1], *Alena* Konečná[1], *Daniel* Sprinzl[1], *Vít* Klupák[2], and *Miroslav* Vinš[2]

[1]SKODA JS a.s., Calculation Department, Orlík 266/15, 316 00 Plzeň, Czech Republic
[2]Research Centre Rez, Hlavní 130, 250 68 Husinec-Rez, Czech Republic

Abstract. In the paper the application of effective source to the solution of activation detector activities in the reactor pressure vessel cavity of the VVER-1000 reactor is presented. The effective source method applies the Boltzmann transport operator to time integrated source data to obtain detector activities. Weighting the source data by time dependent depletion of the detector activity, the result of the calculation is the detector activity. The approach works because of the inherent linearity of radiation transport in non-multiplying time-invariant media. Integrated in this way, the source data are referred to as the 'effective source'. The effective source method thereby enables the analyst to replace numerous intensive transport calculations with a single transport calculation in which the time dependence and magnitude of the source are correctly represented. Detailed description of the effective source method is presented in previous works. First, there were performed neutron-physical calculations of few real VVER-1000 cycles using MOBY-DICK macrocode. Second, there follows 3-D transport calculation using the deterministic code TORT and the cross section library BUGLE-B7 and obtained results are presented. These calculation results of activation detector activities in the reactor cavity are compared with relevant activation detectors results of the ex-vessel measurement. The comparison between calculation and measurement of activation detectors activity in the reactor cavity is necessary to the calculation quality verifying for further fast neutron fluence onto the reactor pressure vessel credible calculation. The activation detectors positions are evident from Figs 1, 2, 3.

1 Introduction

The fast neutron fluence onto the reactor pressure vessel (RPV) is necessary quantity for radiation impact onto the RPV determination in general and particularly for the RPV material embrittlement analyses. There is usually requested the fast neutron fluence increment within 1 cycle for thresholds 0,5 MeV or 1,0 MeV. Values of the fast neutron fluence are obtained either by calculation or by deriving from activity measurement. In principle there are 2 main activity measurement positions towards the RPV: surveillance samples position before the RPV and activation detectors position behind the RPV. The activity is still uniquely assessed for cladding material taken directly from the RPV inner surface. The fast neutron fluence calculationaly derived from mentioned types of measurement is valuable especially in the points of measurement, while the calculation provides the neutron fluence in the whole RPV and particularly in unaccessible points for measurement – RPV inner surface and inner surface of RPV basic material. The approach to the neutron fluence determining is then following: the neutron fluence is prepared by calculation and measurement in certain points, and if the agreement is achieved then for other points of interest on the RPV the neutron fluence obtained by calculation is used.

In the paper the results of the activation detectors activity measurement are presented and compared with transport calculation using the effective source method.

2 The effective source method and application

The activities of the activation detectors in the reactor cavity were calculated with 3-D discrete ordinates deterministic transport code TORT [1] applying the effective source method. The effective source method used in forward transport calculations is based on following assumptions: the space and material arrangement of the RPV and its surrounding is unchangeable in time, the space and energy neutron distribution in the active core is known – fixed source problem. The assumption about time independent material composition holds because there are investigated quantities related only to the fast neutrons escaping from the active core.
There is possible to obtain the fast neutron fluence and the detector activity applying the Boltzmann transport equation operator to the time integral of the neutron source in the active core and to the time integral of time-weighted neutron source respectively, whereas the time-weighted function depends on decay constant of the detector.

Let's call the time integral of the neutron source as the effective source for neutron fluence and the time integral of the time-weighted neutron source as the effective source for detector activity [2-5].
The advantage of the effective source method is the demand of only one forward transport calculation for the time dependent neutron source (time-weighted neutron source) through the whole cycle.
The effective source method was applied to detectors activity calculation in the reactor cavity (detectors positions: Fig. 1, Fig. 2, Fig. 3).
The neutron source distribution on the pin by pin level in 360° was calculated using MOBY-DICK code [6] (3-D diffusion standard Czech code for the VVER-440 and VVER-1000 reactor core calculation) for the VVER-1000 reactor, NPP Temelin, Unit 1 cycle 12 and Unit 2 cycle 11. The effective sources were prepared for the activation detectors ^{46}Ti, ^{54}Fe, ^{55}Mn, ^{58}Ni, ^{63}Cu. The transport calculations with the effective sources for determination of the detectors activity in the reactor cavity were performed by the code TORT [1] in 120° symmetry, the angle approximation S_{16} and the anisotropic scattering P_3, using the broad group cross section library BUGLE-B7 [7] with 47 neutron groups. The activation cross sections taken from mentioned library were produced by groups collapsing using 1/4 RPV thickness weighting. As concerns the ^{46}Ti(n,p) reaction, our experience leads to taking the activation cross section from previous library version BUGLE-96 [8]. The calculation results are presented in Tab. 1, Tab. 2 and Fig. 4 - Fig. 11.

3 The measurement arrangement

The neutron fluence assessment in the cavity behind RPV is based on activation detector method. The measured activity values are used like input data for neutron fluence adjustment. The measurement in the reactor cavity is arranged according to Fig. 1, Fig. 2 and Fig. 3. The holder of activation detectors (red lines Fig. 3) is made from aluminum and it covers space between welds no. 3 and no. 4 and approximately 26 degree sector on RPV surface (sector C in Fig. 1 and green sector in Fig. 2). The holder supports activation detectors for spatial neutron fluence distribution and for neutron spectrum measurements. The spectrometric sets of activation detectors (metal disk foils with content of ^{46}Ti, ^{54}Fe, ^{55}Mn, ^{58}Ni, and ^{63}Cu isotopes) are placed into the small boxes which are located in positions P5, P6, P8 and P2 (Fig. 3). The copper and iron detectors are used for spatial activity distribution. These detectors are located with constant distance 100 mm on the vertical holder rod and 80 mm on the horizontal one. The green lines show positions of spatial distribution measurements in Fig. 3.

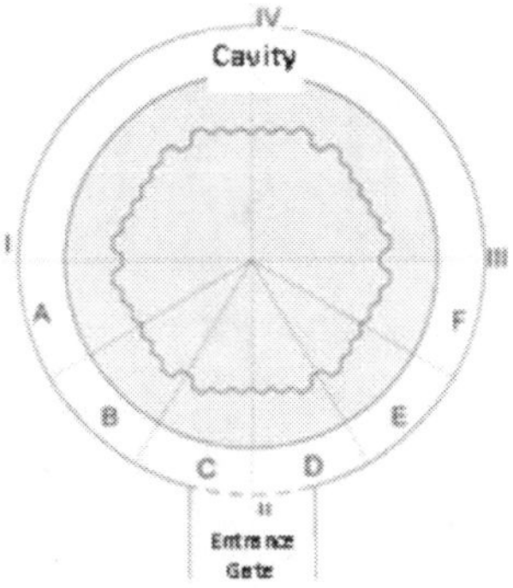

Fig. 1 Sectors C, D for holder placing in the RPV cavity

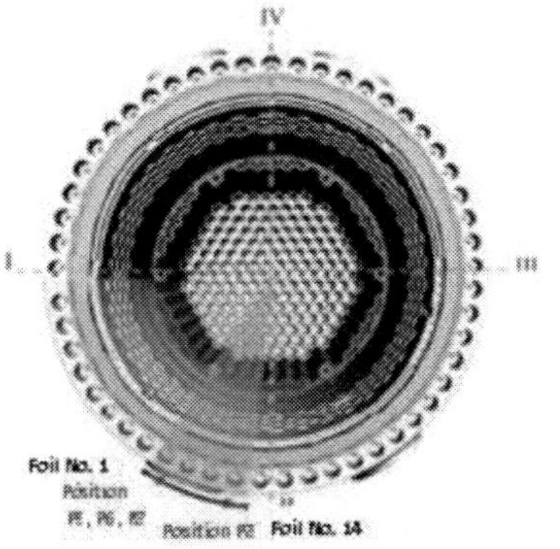

Fig. 2 Holder placing in the face of RPV cross section

Fig. 3 Holder placing, front sight to the RPV

4 The calculation and measurement comparison

The calculation and measurement results are compared in Tab. 1 for Unit 1 cycle 12 and in Tab. 2 for Unit 2 cycle 11 for activation detectors (^{46}Ti, ^{54}Fe, ^{55}Mn, ^{58}Ni, ^{63}Cu) located in positions P5, P6, P8, and P2 (Fig. 3).

Table 1. NPP Temelin Unit 1 cycle 12, activation detectors activities in the reactor cavity [1])

activation reaction	measurement (E) A [Bq/1 nucleus]	calculation (C) A [Bq/1 nucleus]	difference (C-E)/E [%]
box P5: RPV outer surface; azimuthally: 68,5°, axially: weld No. 3			
$^{54}Fe(n,p)^{54}Mn$	2,647E-17	3,365E-17	27,1
$^{58}Ni(n,p)^{58}Co$	8,015E-17	9,777E-17	22,0
$^{46}Ti(n,p)^{46}Sc$	1,099E-17	1,301E-17	18,4
$^{63}Cu(n,\alpha)^{60}Co$	7,470E-20	9,079E-20	21,5
$^{55}Mn(n,2n)^{54}Mn$	5,522E-19	5,616E-19	1,7
box P6: RPV outer surface; azimuthally: 68,5°, axially: active core midplane			
$^{54}Fe(n,p)^{54}Mn$	2,879E-17	3,445E-17	19,7
$^{58}Ni(n,p)^{58}Co$	8,378E-17	9,812E-17	17,1
$^{46}Ti(n,p)^{46}Sc$	1,089E-17	1,303E-17	19,7
$^{63}Cu(n,\alpha)^{60}Co$	7,941E-20	9,315E-20	17,3
$^{55}Mn(n,2n)^{54}Mn$	5,732E-19	5,733E-19	0,0
box P8: RPV outer surface; azimuthally: 68,5°, axially: weld No. 4			
$^{54}Fe(n,p)^{54}Mn$	2,818E-17	3,193E-17	13,3
$^{58}Ni(n,p)^{58}Co$	8,488E-17	9,359E-17	10,3
$^{46}Ti(n,p)^{46}Sc$	1,098E-17	1,230E-17	12,0
$^{63}Cu(n,\alpha)^{60}Co$	7,726E-20	8,439E-20	9,2
$^{55}Mn(n,2n)^{54}Mn$	5,676E-19	5,151E-19	-9,2
box P2: RPV outer surface; azimuthally: 82,5°, axially: active core midplane			
$^{54}Fe(n,p)^{54}Mn$	1,814E-17	2,159E-17	19,0
$^{58}Ni(n,p)^{58}Co$	5,301E-17	6,132E-17	15,7
$^{46}Ti(n,p)^{46}Sc$	7,497E-18	8,271E-18	10,3
$^{63}Cu(n,\alpha)^{60}Co$	5,516E-20	6,089E-20	10,4
$^{55}Mn(n,2n)^{54}Mn$	4,223E-19	4,043E-19	-4,3

1) Azimuthal variable, 0°: direction I in Fig. 1 and Fig. 2

Table 2. NPP Temelin Unit 2 cycle 11, activation detectors activities in the reactor cavity [1])

activation reaction	measurement (E) A [Bq/1 nucleus]	calculation (C) A [Bq/1 nucleus]	difference (C-E)/E [%]
box P5: RPV outer surface; azimuthally: 68,5°, axially: weld No. 3			
$^{54}Fe(n,p)^{54}Mn$	3,195E-17	4,047E-17	26,7
$^{58}Ni(n,p)^{58}Co$	9,070E-17	1,109E-16	22,3
$^{46}Ti(n,p)^{46}Sc$	1,238E-17	1,478E-17	19,3
$^{63}Cu(n,\alpha)^{60}Co$	9,949E-20	1,107E-19	11,2
$^{55}Mn(n,2n)^{54}Mn$	6,072E-19	6,577E-19	8,3
box P6: RPV outer surface; azimuthally: 68,5°, axially: active core midplane			
$^{54}Fe(n,p)^{54}Mn$	3,175E-17	4,146E-17	30,6
$^{58}Ni(n,p)^{58}Co$	8,866E-17	1,110E-16	25,2
$^{46}Ti(n,p)^{46}Sc$	1,193E-17	1,477E-17	23,8
$^{63}Cu(n,\alpha)^{60}Co$	1,006E-19	1,137E-19	13,1
$^{55}Mn(n,2n)^{54}Mn$	6,404E-19	6,728E-19	5,1
box P8: RPV outer surface; azimuthally: 68,5°, axially: weld No. 4			
$^{54}Fe(n,p)^{54}Mn$	3,023E-17	3,744E-17	23,8
$^{58}Ni(n,p)^{58}Co$	8,496E-17	1,043E-16	22,7
$^{46}Ti(n,p)^{46}Sc$	1,161E-17	1,370E-17	18,0
$^{63}Cu(n,\alpha)^{60}Co$	9,625E-20	1,001E-19	4,0
$^{55}Mn(n,2n)^{54}Mn$	5,879E-19	5,886E-19	0,1
box P2: RPV outer surface; azimuthally: 82,5°, axially: active core midplane			
$^{54}Fe(n,p)^{54}Mn$	1,961E-17	2,551E-17	30,1
$^{58}Ni(n,p)^{58}Co$	5,302E-17	6,795E-17	28,2
$^{46}Ti(n,p)^{46}Sc$	7,459E-18	9,154E-18	22,7
$^{63}Cu(n,\alpha)^{60}Co$	6,818E-20	7,277E-20	6,7
$^{55}Mn(n,2n)^{54}Mn$	4,536E-19	4,622E-19	1,9

The calculation and measurement results are compared graphically in the pictures Fig. 4 - Fig. 11 for activation detectors (^{54}Fe and ^{63}Cu) located with constant distance on vertical and horizontal holder rods.

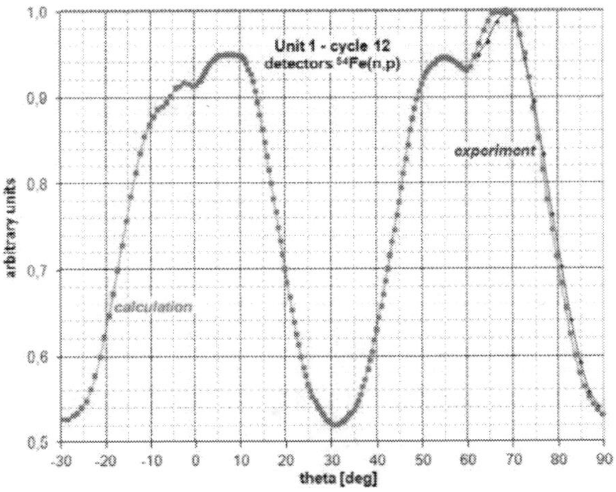

Fig. 4 ^{54}Fe detectors activity azimuthal distribution for Unit 1 cycle 12

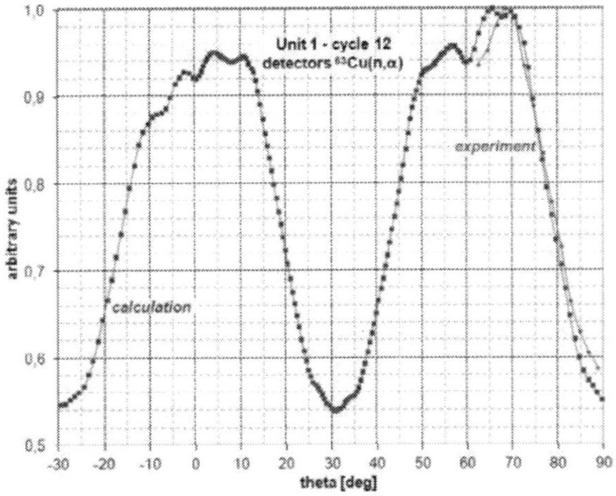

Fig. 5 ^{63}Cu detectors activity azimuthal distribution for Unit 1 cycle 12

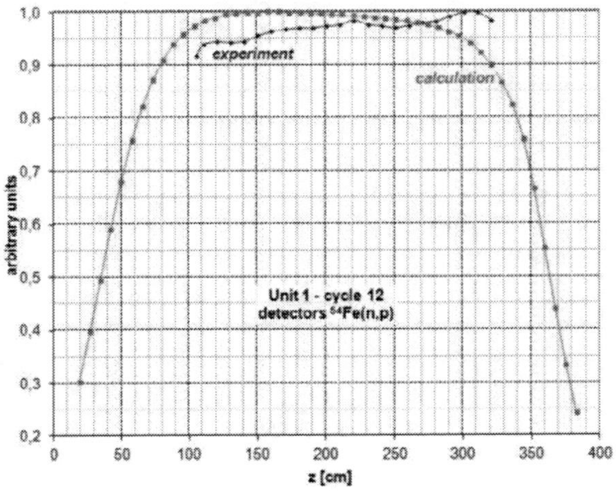

Fig. 6 ^{54}Fe detectors activity axial distribution for Unit 1 cycle 12

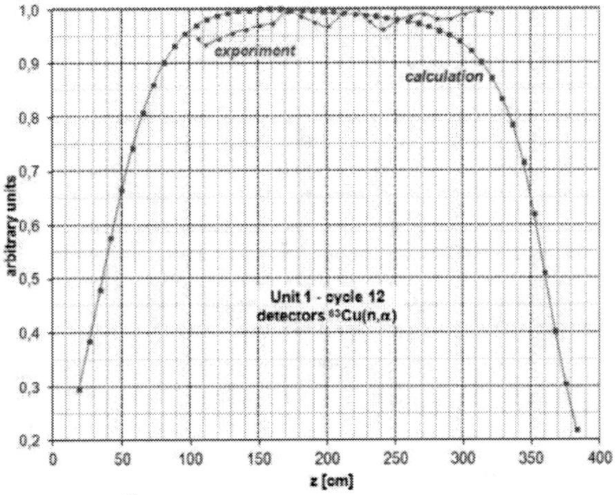

Fig. 7 ^{63}Cu detectors activity axial distribution for Unit 1 cycle 12

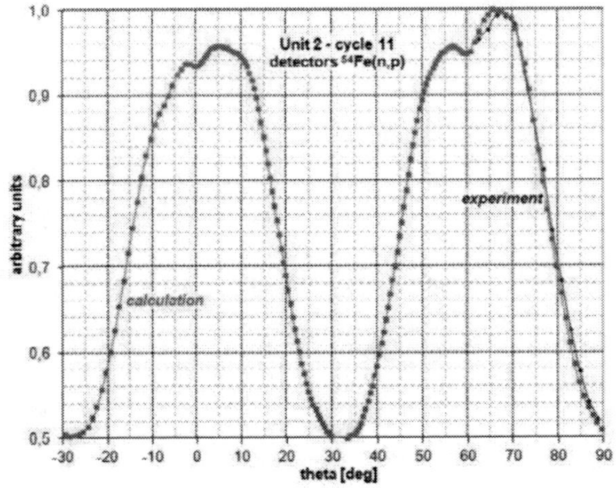

Fig. 8 ^{54}Fe detectors activity azimuthal distribution for Unit 2 cycle 11

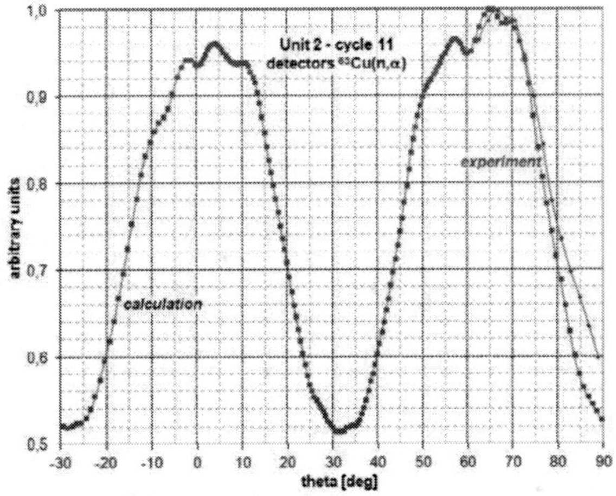

Fig. 9 ^{63}Cu detectors activity azimuthal distribution for Unit 2 cycle 11

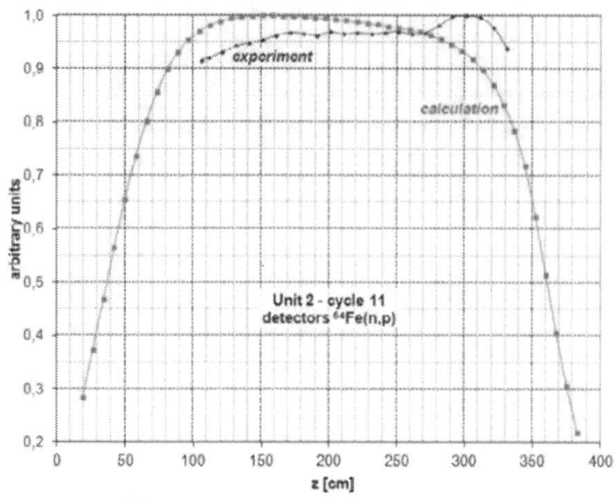

Fig. 10 ^{54}Fe detectors activity axial distribution for Unit 2 cycle 11

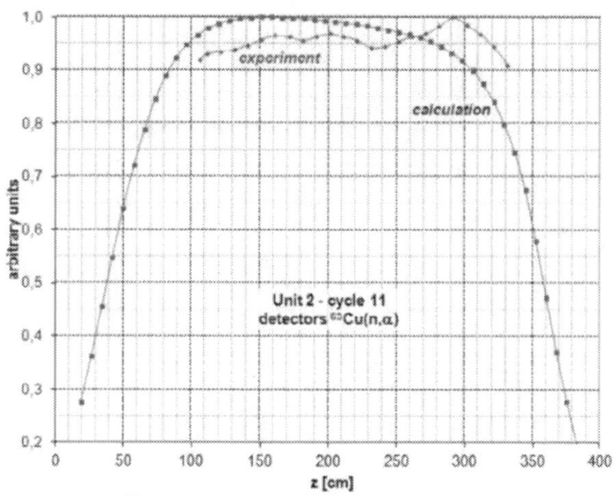

Fig. 11 ^{63}Cu detectors activity axial distribution for Unit 2 cycle 11

5 Results discussion

For both cases Unit 1 cycle 12 and Unit 2 cycle 11 (see Tab. 1 and Tab. 2) the calculation overestimates the measurement except one occurrence: ^{55}Mn in several RPV points for Unit 1 cycle 12. Generally, for ^{55}Mn detector is the difference between calculation and measurement results relatively low, but it should be remembered that the reaction energy threshold is relatively high. And though ^{55}Mn(n,2n) activation cross section shows significantly spectral dependence, the BUGLE-B7 library group system includes only 3 broad groups above the ^{55}Mn(n,2n) reaction threshold. The largest differences between calculation and measurement holds for ^{54}Fe detector: app. 20 – 30%. For other detectors (^{58}Ni, ^{46}Ti, ^{63}Cu) better agreement between calculation and measurement is achieved, whereas for Unit 1 cycle 12 the differences approximately 10 – 20% are displayed, while for Unit 2 cycle 11, 10 – 25%.

The comparisons of activity spatial distribution in arbitrary units are given in Fig. 4 - Fig. 11. There is good agreement between calculation and measurements for azimuthal distribution. Despite of azimuthal distribution, the shape of axial distributions obtained by calculation differs little from results of measurements, whereas the calculated curves show opposed trend to measured ones. If the assumption those measurements are correct, the possible explanation is strong dependence on neutron source definition used for transport calculation. The similar issue or situation for VVER-1000 reactor was described by Borodkin et al. [9].

It should be emphasized that the maximal differences between calculation and measurements in Fig. 10 and Fig. 11 are one tenth (10%) in arbitrary units and the uncertainties of activity are approximately 7% for copper foils and 5% for iron detector. The highest value of uncertainty 9% is for titanium. The main reason is low activity and statistical error of peak area during gamma spectroscopy.

References

1. DOORS 3.2a, *Program Package for One, Two and Three Dimensional Discrete Ordinates Transport* (CCC-650, RSICC, ORNL, 2003)
2. K. Ilieva, S. Antonov, S. Belousov, Nucl. Sci. & Eng. **122** No.1, 131-135 (1996)
3. K. Ilieva, T. Apostolov, I. Penev, S. Belousov, E. Taskaev, S. Antonov, T. Petrova, P. Tzokov, Z. Bojadjiev, Nucl. Engn. & Design **160**, 257-260 (1996)
4. A. Thornton, T.A. Lewis, J.R. Mossop, C.J. Page, S.A. Haddock, *10thInternational Symposium on Reactor Dosimetry* (Osaka, 1999)
5. J. Hep, A. Konečná, V. Krýsl, V. Smutný, *Reactor Dosimetry 14thInternational Symposium* (Bretton Woods, 2011)
6. V. Krýsl, P. Mikoláš, J. Šůstek, K. Vlachovský, *Description of program MOBY-DICK with changes for VVER-1000* (ŠKODA JS, 2015)
7. BUGLE-B7/VITAMIN-B7, *Broad-Group and Fine-Group and Coupled Neutron/Gama Cross-Section Libraries Derived from ENDF/B-VII.0 Nuclear Data* (DLC-245, RSICC, ORNL, 2011)
8. BUGLE-96, *Coupled 47 Neutron, 20 Gamma-Ray Group Cross Section Library Derived from ENDF/B-VI for LWR Shielding and Pressure Vessel Dosimetry Applications* (DLC-185, RSICC, ORNL, 1996)
9. P. Borodkin, G. Borodkin, N. Khrennikov, *Methodology of Fuel Burn Up Fitting in VVER-1000 Reactor Core by Using New Ex-Vessel Neutron Dosimetry and In-Core Measurements and its Application for Routine Reactor Pressure Vessel Fluence Calculations,* EPJ Web of Conferences, Volume **106** (2016), ISRD 15 – International Symposium on Reactor Dosimetry

Implementing Shared Memory Parallelism in MCBEND

Adam Bird[1,*], David Long[1,**], and Geoff Dobson[1,***]
[1]Amec Foster Wheeler, Kings Point House, Queen Mother Square, Poundbury, Dorchester, Dorset, DT1 3BW, United Kingdom

Abstract. MCBEND is a general purpose radiation transport Monte Carlo code from AMEC Foster Wheelers's ANSWERS® Software Service. MCBEND is well established in the UK shielding community for radiation shielding and dosimetry assessments. The existing MCBEND parallel capability effectively involves running the same calculation on many processors. This works very well except when the memory requirements of a model restrict the number of instances of a calculation that will fit on a machine. To more effectively utilise parallel hardware OpenMP has been used to implement shared memory parallelism in MCBEND. This paper describes the reasoning behind the choice of OpenMP, notes some of the challenges of multi-threading an established code such as MCBEND and assesses the performance of the parallel method implemented in MCBEND.

1 Introduction

MCBEND[1] is a well-established powerful Monte Carlo software tool for general radiation transport analysis for shielding and dosimetry applications. MCBEND is developed and licensed for use by AMEC Foster Wheeler's ANSWERS Software Service. The MCBEND package comprises not only the Monte Carlo code itself but also nuclear data libraries, user documentation, productivity tools of various kinds and user support services. Supporting geometry model visualisation and verification tools are also available.

The existing parallel capability in MCBEND known as the 'grid' option, effectively involves running the same calculation on many processors and combining the results. Because there is minimal communication required between processes the method scales almost linearly. MCBEND performs the combining of results and the user is presented with the output as if from a single run. When the accompanying user interface VisualWorkshop[2] is used the user is largely unaware of the activity. This system works well except when the memory requirements of the model are such that it reduces the number of instances that will fit on a workstation or node of a cluster.

To more effectively make use of parallel hardware the decision was taken to implement multi-threading in MCBEND in order to maximise the potential for shared-memory, and eliminating the memory constraints of the existing `grid' option. Multi-threading has been achieved within MCBEND using OpenMP[3]. The rationale for implementing a shared memory model using OpenMP is given in Section 1.1.

A description of the major design considerations and choices when implementing OpenMP are given in Section 2. Here, specific attention is given to the shared memory data management within a MCBEND calculation. Further implementation details are also given for the accumulation of scored quantities (Section 2.1) and the necessary requirement of re-producible results (Section 2.2) through the design of a new random number generator. Finally, the functional specification for the OpenMP multi-threaded parallel version of MCBEND is given in Section 2.3

1.1 Why OpenMP

When implementing a shared memory multi-threaded model there are several existing application technologies from which to choose. The following list gives the rationale behind the choice of OpenMP for MCBEND

- OpenMP is a mature technology with detailed documentation and advice on its implementation.
- OpenMP is implemented by the compiler and additional software is not required by clients to run on a given system.
- The multi-threaded behaviour is controlled by directives in comment lines within the Fortran source code. Since the implementation of OpenMP is applied on at coarse gain level over the main sample loop of a MCBEND calculation, see Section 2, load balancing considerations do not apply. Therefore, relatively few

* e-mail: Adam.Bird@amecfw.com
** e-mail: David.Long@amecfw.com
*** e-mail: Geoff.Dobson@amecfw.com

directives are required, making the implementation of OpenMP less intrusive on the existing code base and making future maintenance an easier task.

- OpenMP directives can be turned on or off, meaning that the same source code can be compiled as a sequential or parallel application. This is particularly useful when having a code base shared with other ANSWERS products and for debugging and maintenance purposes.

2 OpenMP Implementation.

In a simplified form a MCBEND calculation can be viewed in three stages:

Stage 1: Reading and set up, input data is read, data libraries are read and the data are pre-processed ready for the calculation to begin.

Stage 2: The calculation, basically a loop processing where individual particle histories are tracked from source through interaction events with the materials in the model until it escapes via absorption or moves out of the geometric or energy domain of the problem. During this processing, scores (tallies) are accumulated. The process is repeated until a specified number of histories are reached or a time limit is exceeded.

Stage 3: Post processing, where scored data are analysed and formatted for writing to output files.

Typically the reading, set up, post-processing and output, performed in stages 1 and 3, constitute <1% of the total run time for a MCBEND calculation. As such, there is very little to gain from the significant amount of effort needed to apply multi-threading to stages 1 and 3. Therefore, the multi-threaded parallelisation of MCBEND has only been applied to stage 2, the main calculation. The basis of this implementation can be seen by the simple 'fork join' model detailed in figure 1.

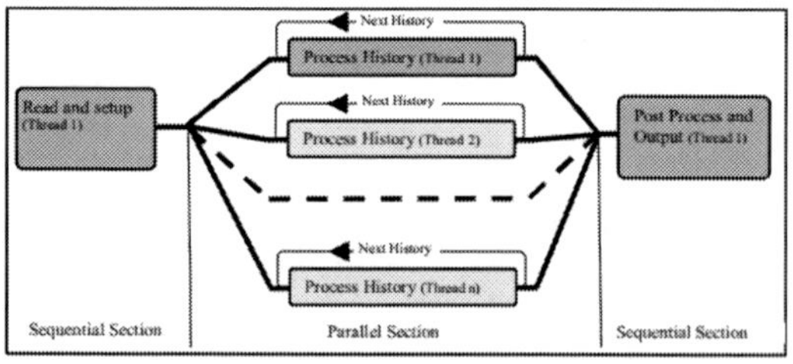

Figure 1. Multi-threading model for MCBEND.

Since each 'sample history' within stage 2 of a MCBEND shielding calculation is statistically independent from all the others, implementing multi-threading to the simple model presented in figure 1 is conceptually a simple problem involving a single coarse grain parallel region which encompasses the main sampling loop. A finer grain approach, that is using OpenMP to parallelise local sections of the code, for example, loops, was considered but it was decided that it would likely not deliver the performance or reduction of memory requirements desired.

When designing and implementing the coarse gain multi-threaded region within MCBEND, the most important consideration was for data to be correctly managed. The data within stage 2 of a MCBEND calculation fall into the following three categories.

Thread-private: Data which is private to a specific thread and should not be accessed, read from or write to, by any other thread running in the parallel region. Examples are the current attributes of individual sample histories such as the location, direction and energy.

Global read: Data which all threads within the parallel region can read, such as material cross-section values and data structures describing the geometry of the problem.

Global read-write: Data which all threads within the parallel region can read from and write to, such as scoring tally accumulators which are updated within the main sample loop when necessary.

With regards to data management it is a requirement that all thread-private data associated with a given sample should not be corrupted by other threads within the parallel region. Also, as global data cannot be specified within OpenMP as read-only, attention must be given to avoid the potential for so called 'data race' conditions when multiple threads within the parallel region have read and write access to the same variable. Violation of these conditions can lead to incorrect or irreproducible calculation results (reproducibility is necessary for verification, see Section 3).

Within the parallel region all thread-private data is duplicated for each individual thread. It was therefore critical to intelligently analyse the data present within the parallel region to identify a minimum set that needs to be thread-private. Having unnecessary amounts of thread-private data increases both the processing overhead of a multi-threaded application and the amount of memory used, which, in the limit that everything is declared thread-private, would mean that a multi-threaded MCBEND would have no advantage over the current 'grid' option.

The main effort in implementing multi-threading within MCBEND was to determine which data should be thread-private, shared globally but read only, or shared globally and writeable by all threads. In order to feasibly implement the correct assignment of all data within the parallel region, parts of the code and data structures were re-factored. The most challenging aspect of this process was to ensure that Fortran derived data types requiring to be thread-private were correctly initialised within the parallel region, as the OpenMP standard does not 'deep copy' such data structures, and therefore these had to be achieved explicitly.

Aside from the complexities of data management, the implementation of OpenMP within MCBEND has the following basic structure.

- Before entering the main parallel region, a global shared copy is made of all derived data type structures which are required to be thread-private.
- The main parallel region is opened.

- Thread-private copies of derived data types are created for each thread using globally shared initial values.
- The number of samples within the calculation is divided across the number of active threads using an OpenMP parallel 'DO' loop.
- Individual samples are tracked in parallel until all histories have been completed, with contributions from each sample to scoring tallies accumulated where necessary (see Section 2.1 for further details).
- Once tracking has been completed and the parallel 'DO' loop has been exited, any diagnostic data which has been accumulated within each individual thread is combined.
- The main parallel region is closed.
- Output files are generated sequentially.

In addition to the data management and basic structure described above, the final design of the OpenMP implementation required consideration of two remaining issues; the appropriate approach for accumulating scored quantities within the parallel region and the design of a new random number generator which generates reproducible calculation results.

2.1 Scoring Tallies

During the main calculation it is necessary to regularly update values from each sample to scoring tallies within the parallel region. For the OpenMP implementation, scoring is restricted to the Unified Tally (UT) feature of MCBEND, see Section 2.3. The use of UT scoring meshes can impose large memory usage on a MCBEND calculation, specifically when multiple fine resolution meshes are used. It is therefore impractical to define the data structures which define the UT scoring meshes as thread-private since large amounts of memory would be copied for each thread used.

Instead it was decided that the data structures responsible for storing accumulated scoring results would be shared globally, with each thread in the parallel region having write access. In order to avoid 'data race' conditions and ensure the validity of scored results, only one thread at a time is allowed to update values within a scoring mesh, achieved using the OpenMP 'ATOMIC' directive. When multiple threads attempt to synchronously update the same scoring mesh then any number of threads above unity are forced to wait idle (blocked) until the current thread has finished updating.

While using shared global data for UT scoring meshes does not require duplicating large amounts of memory, it has the potential to limit the performance of the code when increasing the number of processors used. When large numbers of threads are active in the parallel region, the likelihood that threads are blocked while updating scoring meshes is increased. This can result in a reduction of the performance gained from using additional processors, i.e. non-linear scaling.

While the issue of having shared global UT scoring mesh data potentially reducing the scaling performance is a genuine concern, it is overridden by the practical limitations duplicating thread-private UT meshes has on the memory requirements of typical system architectures used to perform MCBEND calculations. Since the maximum number of active threads for typical systems running MCBEND calculations is currently of the order of 10's, scaling issues relating to scoring within UT meshes are not expected to be fundamentally limiting to the performance of calculations. Looking to the future, as numbers of available cores increases further thought may need to be given to the treatment or scoring meshes.

2.2 Random Number Generator

The Monte-Carlo method of a MCBEND calculation uses random numbers to decide the outcome of every event that occurs within a sample history. In the sequential version of MCBEND each random number is provided by a combined lagged Fibonacci generator and linear congruential random number generator (RNG). The RNG provides a sequence of pseudo-random numbers which can be based on an initial user provided seed. If identical seeds are used for the same calculation then the same random number sequence will be generated enabling calculations to be exactly re-produced.

Unlike the sequential version of MCBEND, within the main sampling loop of the multi-threaded parallel region the exact sequence of samples is undetermined and will be different each time the same calculation is performed. Here, the nth random number using the sequential RNG's will be used for a different purpose, creating different results, making it impossible to re-produce a calculation exactly.

In order to provide re-producible calculation results a new RNG has been designed for use within the multithreaded parallel region. The new RNG is seeded by a user supplied value (if present) and the unique sample number of the history being tracked. This new 'thread-safe' RNG therefore produces a unique random sequence for each sample history, which is independent of the order in which samples are tracked. This allows a direct comparison between the results from a sequential calculation with those from a multi-threaded parallel calculation, or indeed multi-threaded parallel calculations generated with differing numbers of threads. The ability to generate such reproducible calculation results is essential for verification of the multi-threaded parallel version of MCBEND as detailed in Section 3. The thread-safe RNG is the default mode when performing multi-threaded parallel MCBEND calculations.

2.3 Functional Specification

The implementation of OpenMP does not cover all of the features contained within MCBEND; implementation has been limited to a subset of the more modern functional units associated with both neutron and gamma particle tracking and scoring. The main functionality omissions that will be included later are the IGES based CAD import and electron tracking. Other omissions include the DICE and multi group collision processing, legacy source options and some of the more advanced variance reduction methods. OpenMP multithreading has been applied to the following functionality within MCBEND;

- FG Simple body geometry, including polygon surface CAD import.
- Unified Tally.
- BINGO (Neutron and Gamma-ray collision processing).
- Splitting and Roulette (variance reduction).
- Dump and Restart.
- Hole Geometry (Woodcock tracking).
- Looping.

3 Verification

When creating the multi-threaded parallel version of MCBEND, verification that the developed code is correct is of the highest importance. The approach for verification was as follows;

- An 'intermediate' MCBEND was created that used the new random number generator, detailed in Section 2.2, with no other modifications.
- Each calculation in the test set, see Section 4, was performed using a sequential version of the 'intermediate' MCBEND.
- The results from the 'intermediate' MCBEND were compared against those from a standard MCBEND run. While the exact comparison of results for each test case is not possible they were confirmed to be statistically equivalent, which provides evidence that the new thread-safe RNG is producing the same distribution (to within acceptable limits) of random numbers as the RNG in the standard MCBEND.
- The results from the parallel OpenMP MCBEND were compared with identically seeded results from the 'intermediate' MCBEND and an exact match was expected.

Using this intermediate MCBEND allows us to prove (for the cases in the test set) that the introduction of multi-threading has not changed the basic calculation. Here, identical comparisons are only possible due to the new thread-safe RNG.

4 Test Set

For verification purposes the test set comprised 43 cases from the standard MCBEND verification set. The reduced functionality of the multi-threaded MCBEND necessitated this reduced set. Six of these models were used for performance testing of the multi-threaded version of MCBEND. This subset was chosen to cover the range of functionality that has been multi-threaded, as detailed in Section 2.3, including both FG and Hole geometry features, variance reduction through splitting and importance mesh generation, scoring of flux and response within UT meshes, different source options and collision type processing for neutron, gamma and coupled cases. A brief description of each test is given below.

- **311_bingo**: An infinite slab model with a monoenergetic uniform source. Scoring of flux occurs in a user supplied importance map mesh when boundaries are crossed.
- **fuel_flask_gamma**: A real world fuel flask example for gamma-ray dose rate calculations. Includes general, nest and array part FG features. Importance mesh is generated using the Calculate option and scoring occurs for both flux and response functions.
- **s17_t002**: Test of the Unified Tally unit for a gamma collision case with a line source.
- **s17_tab001**: Uses the pipe hole geometry with neutron collision processing. Uses looping functionality to produce tabular output for both flux and response function scored in UT mesh.
- **v2_aspis_onestep**: A coupled neutron-gamma case within a complex general part geometry. Scoring is for both flux and response function in a UT mesh.
- **v2_b2_hole**: Similar to the fuel_flask_gamma test case with the inclusion of Square and Plate hole geometry features.

5 Results

Performance testing of the multi-threaded version of MCBEND was performed on both Windows and Linux platforms. For Windows the testing was run on a standalone machine which has 12 processors. For Linux the testing was run on a single node of a HPC which has 16 processors. For each platform the multi-threaded version was compiled using the Intel Fortran 2015 compiler with the appropriate OpenMP option selected.

All test cases were run with the number of active threads in the parallel region ranging from unity to the maximum allowed for the specific platform, 12 for Windows and 16 for Linux. The results from all calculations were compared to those from the 'intermediate' version of MCBEND and were found to match identically.

Performance results were obtained by measuring the total computer processing unit time spent within the parallel region and comparing this with timings over the same code region (the main calculation sampling loop) for the 'intermediate' version. All performance timings were calculated from wall clock times obtained from calls to the Fortran system_clock function

5.1 Scaling

Scaling results for the v2_aspis_onestep test case from Windows and Linux are shown in figures 2 and 3 respectively. Here the scaling factor is plotted as a function of the number of threads used in the calculation. For each plot there are two scaling factors defined as,

$$\frac{REF_{seq}}{RUN} \quad \text{and} \quad \frac{REF_{1T}}{RUN}$$

DOI: 10.1051/epjconf/201715307042

where REF_{seq} is the reference time using the sequential 'intermediate' version of the code, REF_{1T} is the calculation time from the OpenMP version using a single thread and RUN is the calculation time from the OpenMP version for the number of threads being used

From figure 2 we find that on Windows the scaling factors when using the sequential reference times follow a linear trend based on the number of threads used, however with a constant offset from the exact ('perfect') linear scaling trend shown in black. Defining an offset factor as,

$$\left(\frac{Number\ of\ Threads}{Scaling\ Factor} - 1\right) * 100$$

the offset between the OpenMP results and the exact linear trend is approximately 50%, i.e. for 6 threads the speed up factor is 4 and for 12 threads the speed up factor is 8.

The linear trend seen in the scaling factors of figure 2 are a positive sign that OpenMP has been implemented correctly, and that performance issues related to the scoring (noted in Section 2.1) are not apparent when using the order of 10 threads. However, the relatively large offset factor noted above is less positive. While an offset factor is always expected when implementing OpenMP a value of 50% is greater than expected. A discussion of reasons behind this offset factor is given in Section 6.

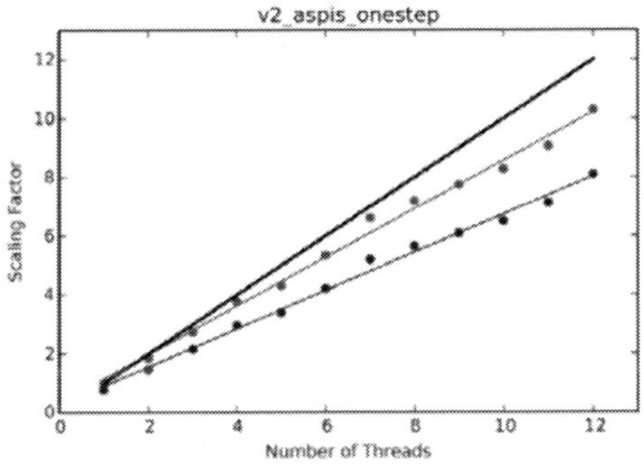

Figure 2. Scaling results for the v2_aspis_onestep test case on Windows. (blue) Scaling factor when the sequential model is used as reference run time. (red) Scaling factor when the OpenMP model with a single thread is used as the reference run time. (black) Exact linear scaling.

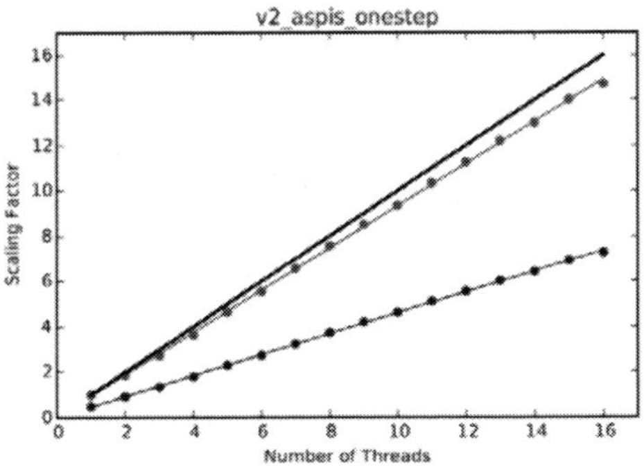

Figure 3. Scaling results for the v2_aspis_onestep test case on Linux. (blue) Scaling factor when the sequential model is used as reference run time. (red) Scaling factor when the OpenMP model with a single thread is used as the reference run time. (black) Exact linear scaling.

From figure 3 we find that the scaling results for Linux have the same form as those found for Windows. The scaling factors again follow a linear trend based on the number of threads used. However the offset factor seen when using the sequential reference times are larger than those found under Windows. Here the offset factor is of the order 120%, i.e using 8 threads results in a scaling factor of slightly less than 4 and using 16 threads results is a scaling factor of slightly less than 8.

It should be noted that all models within the test set exhibit scaling behaviour of the same form as that shown for the v2_aspis_onestep case, and as such these results are not displayed.

5.2 Test Set Performance

In addition to the scaling results presented above, the performance of all test set models under Windows and Linux are presented in tables 1 and 2 respectively. Here, SF_{SEQ} is the scaling factor when using sequential times as the reference, SF_{1T} is the scaling factor when using time from OpenMP with a single thread as the reference and the RUN times are those obtained with the maximum number of threads available on each platform,

Table 1. Windows test set performance results.

Test Name	REF_{seq}(s)	REF_{1T}(s)	RUN(s)	SF_{seq}	SF_{1T}
311_bingo	42.29	68.89	6.77	6.24	10.17
fuel_flask_gamma	94.64	119.12	12.54	7.54	9.49
S17_t002	32.51	46.69	4.6	7.06	10.14
S17_tab001	288.47	288.23	30.60	9.42	9.41
v2_aspis_onestep	100.76	128.09	12.43	8.10	10.30
v2_b2_holw	114.75	149.82	14.60	7.85	10.26

Table 2. Linux test set performance results.

Test Name	REF_{seq}(s)	REF_{1T}(s)	RUN(s)	SF_{seq}	SF_{1T}
311_bingo	14.71	35.15	2.37	6.2	14.81
fuel_flask_gamma	33.6	59.75	6.38	5.26	9.35
S17_t002	10.86	21.21	1.96	5.53	10.81
S17_tab001	100.12	180.56	19.24	5.2	9.38
v2_aspis_onestep	29.41	59.54	4.05	7.26	14.70
v2_b2_holw	39.95	70.11	7.99	4.99	8.76

From table 1 we find that the scaling factor when using the sequential reference times ranges from ~6-9.5 when using the maximum 12 threads. The offset factor for these results ranges from ~25-100% with an average offset factor of ~60%.

From table 2 we find that the scaling factor when using the sequential reference times ranges from ~5-7 when using the maximum 16 threads. The offset factor for these results ranges from ~120-200% with an average offset factor of ~180%. The performance results for the Linux platform are therefore consistently poorer than those seen under Windows.

DOI: 10.1051/epjconf/201715307042

6 Discussion and Conclusions

The magnitude of the difference in scaling factors observed when the OpenMP model with a single thread and the sequential model were used as the reference was not expected. The same source code was used to build the sequential and OpenMP versions that were used for performance testing. The only difference was the build option to enable OpenMP. What we have observed is that OpenMP running a single thread is approximately twice as slow as sequential code on Windows and between two and three times as slow on Linux.

An analysis of this performance issue using a simple test program revealed a problem associated with accessing data that is private to a thread but declared outside the scope of the parallel region, for example, Fortran module data declared using the OpenMP 'THREADPRIVATE' clause. For production code we use the Intel Fortran compiler and we have seen the problem using versions 13, 15 & 16 on the Windows and Linux operating systems. Intel has informed us that the performance issue is caused by additional run time checking that the data exists. Intel suggested not using THREADPRIVATE. This solution would require considerable re-factoring and is not a desirable option for us at this time.

Trials with version 6.1 of the Gfortran compiler on Linux do not show this problem and the difference between the OpenMP model with a single thread and the sequential case is of the order of a few percent. We are considering using an alternative compiler for the production version of the OpenMP MCBEND.

Another aspect of the performance is the memory use, especially as reducing the memory requirements of a parallel run is the main driver for this work. A multi-threaded case uses more memory than a sequential case but less than the equivalent 'grid' case. For example, a calculation using 12 threads uses four times the memory of a sequential calculation. This is a significant saving compared to the 'grid' option.

In conclusion we are very pleased with the linear nature of the scaling graphs which show, at least for the current test cases and current number of threads, that we have not yet reached a point where there is excessive interaction between threads. The main objective of reducing the memory requirements of a parallel calculation by using OpenMP has been met. The highest priority for us is to effectively remove the difference in offset factor between OpenMP with a single thread and the sequential case before we would consider a production version of MCBEND with multi-threading.

References

1. P. Cowan, G. Dobson, J. Martin, *Release of MCBEND11*, ICRS12-RPSD2012, Nara, Japan.
2. A. Bird and T. Fry, *Visual Workshop 2: A Model Viewer, Editor and Results Display Package for the Answers Shielding and Criticality Codes*, ICRS12-RPSD2012, Nara, Japan.
3. OpenMP Architecture Review Board OpenMP Application Program Interface Version 3.0, (2008).

Modeling of γ field around irradiated TRIGA fuel elements by R2S method

Ambrožič Klemen[1,*] and Snoj Luka[1,**]

[1] *Jožef Stefan Institute, Jamova cesta 39, SI-1000 Ljubljana, Slovenia*

Abstract. The JSI TRIGA reactor has several irradiation facilities with well characterized neutron fields. The characterization was performed by measurements and by utilizing Monte Carlo particle transport computational methods. Because of this, JSI TRIGA has become a reference center for neutron irradiation of detectors for ATLAS experiment (CERN). Thorough γ characterization of the reactor is however yet to be performed. Current Monte Carlo particle transport code only account for the prompt generation of neutron induced γ rays, which have been characterized, but are neglecting the time dependent delayed part, which may in some cases amount to more then 30 % of total γ flux in an operation reactor, and is the only source of γ-rays after reactor shutdown. Several common approaches of modeling delayed γ-rays , namely D1S and R2S exist.
In this paper an in-house developed R2S method code is described, coupling a Monte Carlo particle transport code MCNP6 and neutron activation code FISPACT-II, with intermediate steps performed by custom Python scripts.
An example of its capabilities is presented in terms of evaluation of utilization of JSI TRIGA nuclear fuel as a viable γ-ray source. In the model, fresh nuclear fuel is considered and a silicon pipe sample is modeled in. Fuel activities, dose and kerma rates on the sample, as well as emitted γ-ray spectra and isotopic contribution to the contact dose are calculated and presented.

1 Introduction

The Jožef Stefan Institute (IJS) TRIGA reactor is a 250 kW pool type reactor featuring numerous irradiation facilities with different characteristics in terms of size and neutron field properties [1]. Neutron fluxes and spectra in these facilities have been well characterized both computationally by Monte Carlo computational models [1] and experimentally validated [2]. Due to this, JSI TRIGA has become a reference center for neutron irradiation of detectors for ATLAS experiment (CERN) [3] within the AIDA project [4].
However this is not the case for γ-ray fields. Event though particle transport computer codes, such as Monte Carlo code MCNP support generation of neutron induced γ-rays, commonly only prompt γ-ray generation is supported by default [5]. Measurement with ionization chamber indicate that γ-ray fluxes due to decay of neutron activated isotopes (delayed γ-rays) contribute up to 30 % of total photon flux [6, 7]. Additionally these γ-ray fluxes are time dependent due to decay of fission and

*e-mail: klemen.ambrozic@ijs.si
**e-mail: luka.snoj@ijs.si

activation products and are also present after reactor shutdown. Up to date, only prompt γ fields have been characterized [5], with delayed part still to be evaluated.
Two approaches in Monte Carlo particle transport for modeling delayed γ-rays, namely D1S [8] and R2S [9, 10] are used. Initial steps on delayed γ ray characterization have been performed utilizing an in-house developed R2S method code on TRIGA nuclear fuel, to assess feasibility of its use as a viable γ-source. The code is written in Python 2.7 (Python) [11] and couples MCNP [12] particle transport capabilities with FISPACT-II [13] neutron activation and transmutation analysis computer code.
In the calculations fresh nuclear fuel is considered, and two activation time regimes are evaluated: short irradiation in terms of several hours at full reactor power which simulates intra-day reactor operation with emphasis on observing short-lived isotopes, and long term irradiation with fuel element burn-up of 1 MW d and 10 MW d, which is representative of several years of reactor operation and account for long-lived isotopes accomulation.
Calculations on JSI TRIGA fuel elements pose several advantages, due to the fact the operational history of every fuel element is known and easy acessibility of reactor core and fuel elements inside the reactor core, allowing for easy acess for experimental validation of the computational data, which will be performed in the next two years.
The current code is problem specific, but several of its modules could be used in a more generalized code, with possible cell under voxel approach [10], and as such offers significant insight in the method performance capabilities. A generalized version of the code is currently being developed and will be verified with an extensive experimental campaign.

2 Rigorous two-step method (R2S)

R2S [9] in reactor physics is a method of coupling particle transport capabilities with neutron activation calculations for characterization of delayed γ-ray emissions. The basic idea is, to divide the geometry of the computational model into small voxels, and calculate neutron flux and spectrum in each of them. This, along with initial material composition of the voxel is input into the activation analysis computer code, generating isotope inventory as well as γ-ray spectra respectively. The latter is then used as input into particle transport computer code again, so each voxel is represented in terms of a γ source respectively, calculating delayed photon field flux, spectrum and dose. Such approach allows user to calculate for example dose fields around activated or irradiated samples or in our case around irradiated nuclear fuel. The approach is commonly used in assessing radiation fields in ITER [9].
Our Python script couples MCNP6 [12] particle transport code with neutron activation and transmutation analysis code FIPACT-II [13] in the above mentioned fashion. MCNP model geometry is divided into voxels and in each voxel neutron flux and spectra are calculated using cell averaged track length neutron flux estimator (F4 in MCNP) and normalization to desirable reactor power [1].
Material composition and time dependent neutron flux in terms of normalized neutron spectra and total neutron fluxes are input into the activation calculations. Results of activation calculations are time dependent isotopic compositions of activated samples and gamma rays arising from activated material. The latter are then used as input source in MCNP model, where γ-ray transport is performed allowing user to calculate gamma flux and dose fields. Dose fields were calculated in term of ICRP-21 $H^{*}10$ dose [14] using flux-to-dose conversion factors and kerma using energy deposition over cell estimates (F6 in MCNP) are calculated. Several intermediate data conversion tasks are performed by the Python script for data input/output (I/O) compatibility between computer codes. A flow chart of script operation is presented in figure 1.

DOI: 10.1051/epjconf/201715307043

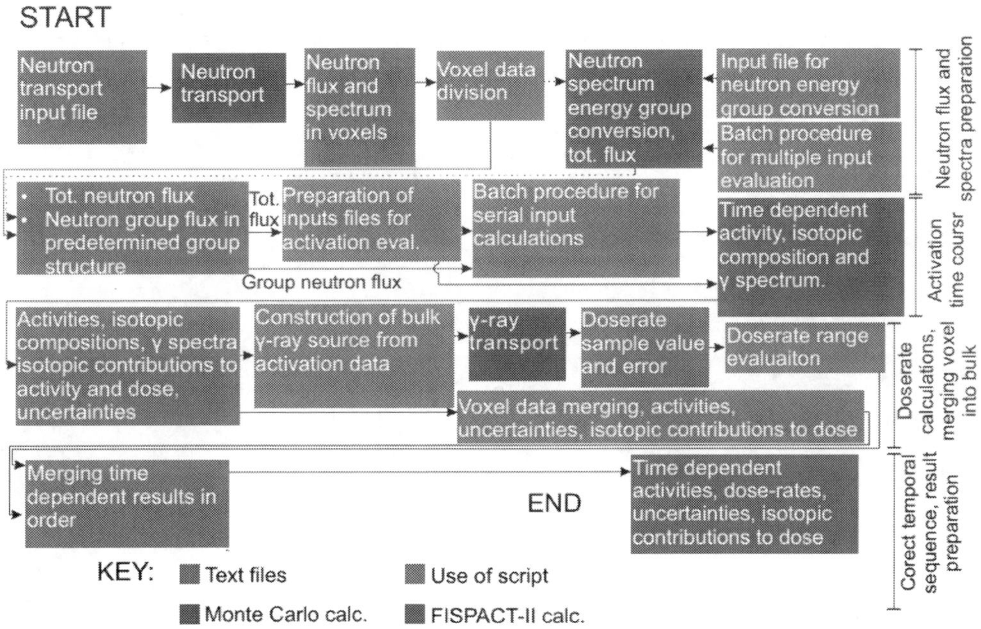

Figure 1: Flow chart of the R2S script operation.

3 JSI TRIGA fuel element R2S evaluation of delayed γ-ray dose.

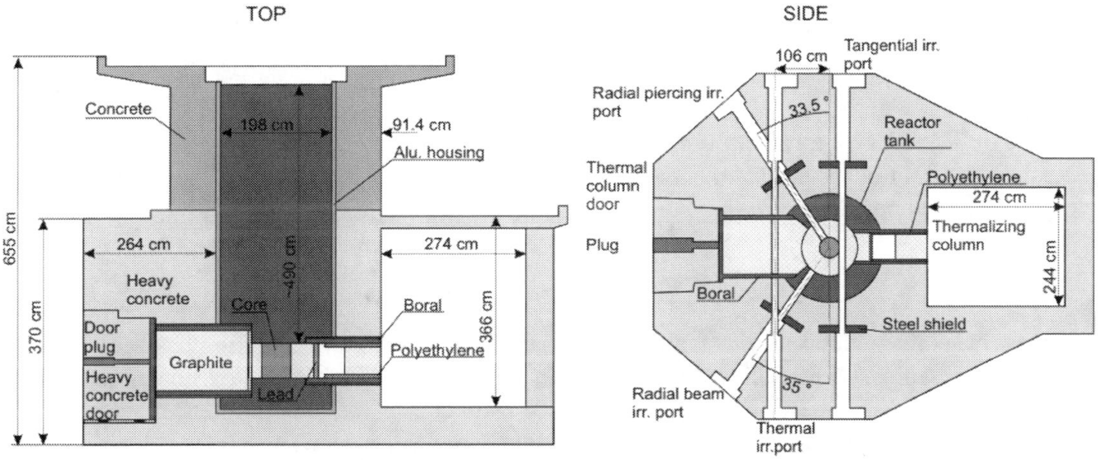

Figure 2: JSI TRIGA reactor, top and side schematic view.

Since activated nuclear fuel is highly radioactive, its utilization as a viable γ-ray source was calculated using R2S method. JSI TRIGA is a pool type, 250 kW reactor 2, with fuel elements arranged in an annular configuration in the reactor core 3. Each fuel element comprises of Zr rod, with homogeneous mixture of uranium-zirconium hydride alloy (U – ZrH) fuel/moderator and stainless-steel cladding 4. Several representative, standard fuel elements containing 12 wt % of U in U – ZrH (Uranium enrichment is 20 %) were selected for our calculations. U – ZrH part of each of the fuel elements was divided

DOI: 10.1051/epjconf/201715307043

into 250 small voxels (10 axially × 5 radially × 5 azimuthally) depicted on fig. 4, and the above mentioned R2S calculation method performed on them. Time dependency of γ-ray spectra and intensities was evaluated in each voxel, and a bulk photon source constructed respectively. Since an increased

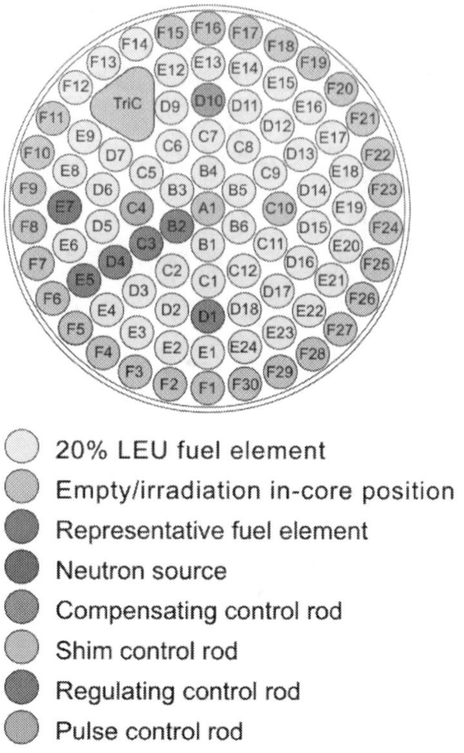

Figure 3: JSI TRIGA core with typical fuel element and control rod arrangement and in-core position annotation.

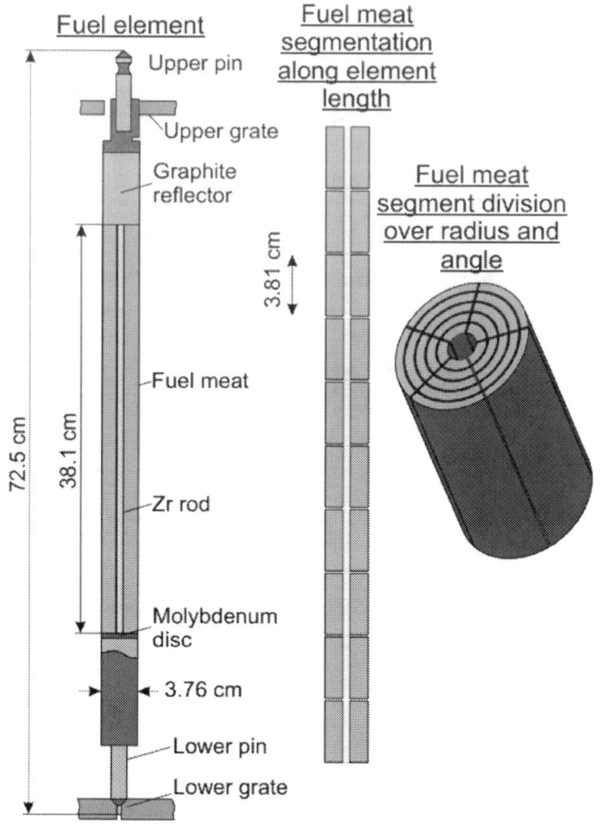

Figure 4: JSI TRIGA fuel element and fuel meat division in 250 voxels for R2S calculation.

interest in semiconductor testing against high dose and dose-rate of γ radiation has been observed, a pipe silicon sample surrounding the fuel element has been modeled in this particular computational model (fig. 6). A 3D mesh for track length-based estimator of the photon fluence, kerma and H^*10 dose using ICRP-21 [14] flux to dose conversion coefficients has been superimposed over the sample geometry. For the initial step, 100-group neutron spectra has been calculated in each of the fuel element meat voxels, to obtain the neutron activation software data (fig. 5).

Two distinct fresh TRIGA fuel irradiation time regimes were evaluated, for irradiation times of several hours and for 1 MWd and 10 MWd burn-ups. For irradiation times of several hours, operation at full reactor power for an appropriate amount of time is simulated. For longer irradiations, until a fuel element reaches burn-up of 1 MWd or 10 MWd, 40 h/week, 8 h d^{-1} operation at full power is simulated. The results are presented in terms of fuel element activities, H^*10 and kerma on the sample, as well as fuel element delayed γ spectra and isotopic contribution to contact dose vs. cooling time after activation completion respectively.

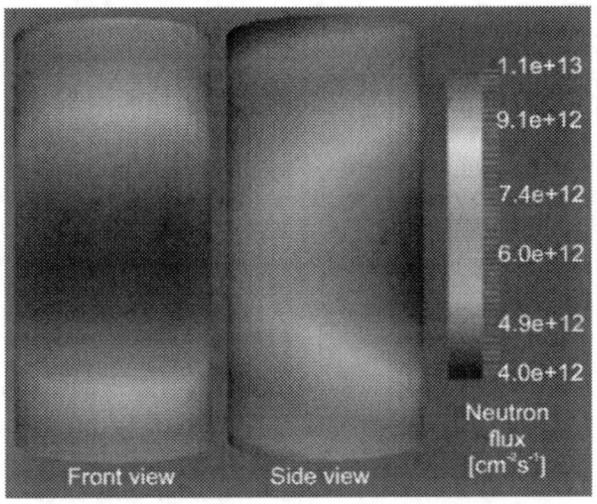

Figure 5: Total neutron flux distribution on the surface of the active part of the B2 position fuel element.

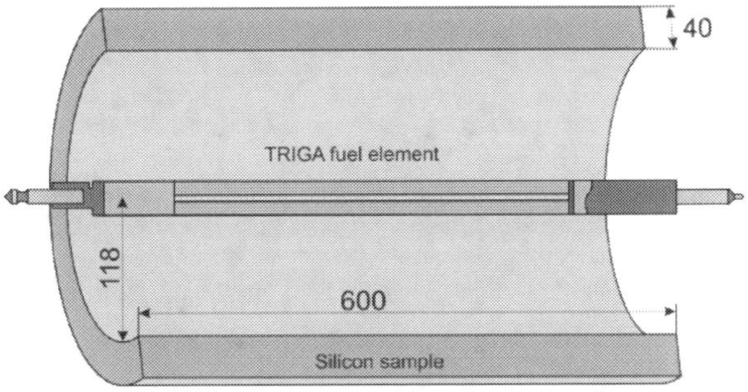

Figure 6: JSI TRIGA fuel element and the silicon pipe sample model schematic.

3.1 Irradiation times of several hours

The case of fuel element neutron irradiation for several hours corresponds to day to day reactor operations, which is the most convenient way of sufficiently activating the nuclear fuel, and use it as a viable γ source. Fresh nuclear fuel is considered. Kerma rate and dose rate to the silicon sample are given in range terms, due to large gradients of photon fields inside the silicon sample itself. The computational uncertainty (not accounting for the nuclear data uncertainty) is in all cases below 1 %.

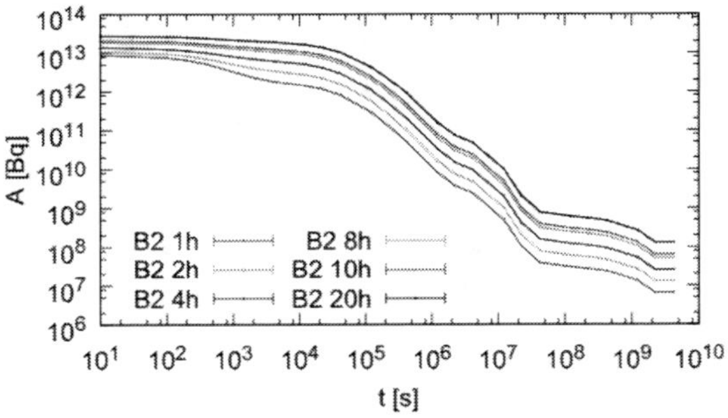

(a) B2 fuel element activity.

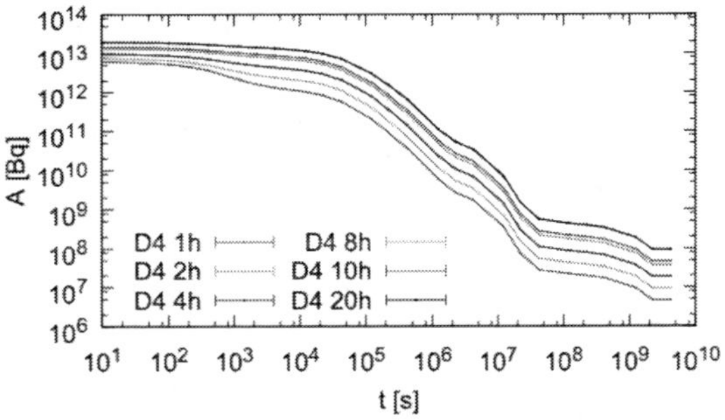

(b) D4 fuel element activity.

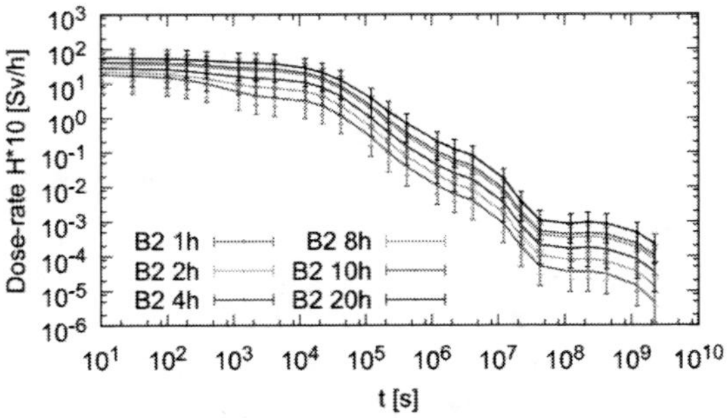

(c) B2 fuel element dose-rate to sample.

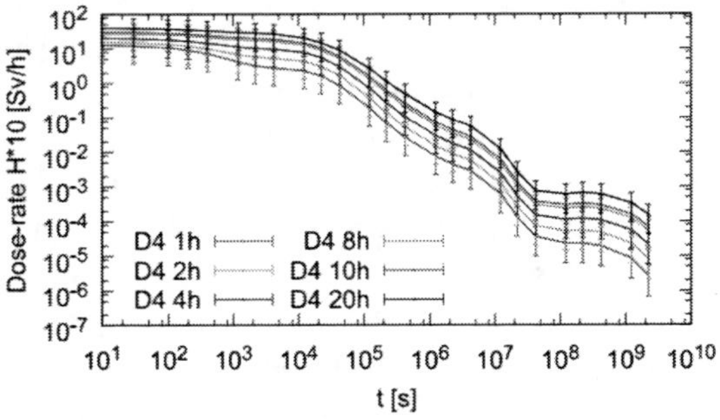

(d) D4 fuel element dose-rate to sample.

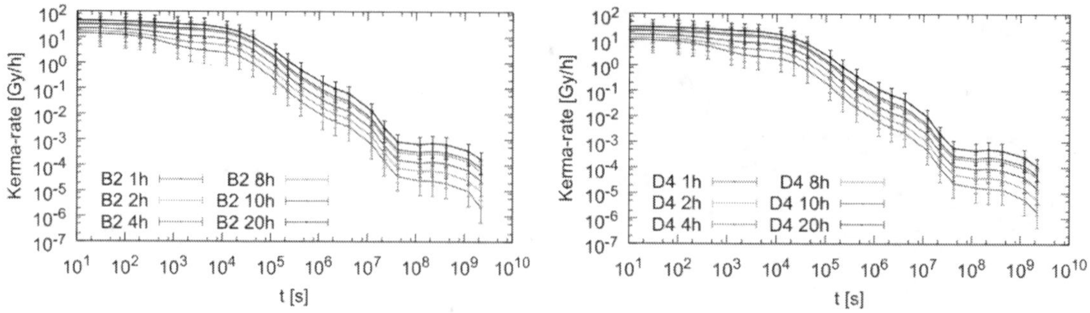

(e) B2 fuel element kerma-rate to sample.

(f) D4 fuel element kerma-rate to sample.

Figure 7: Two representative fuel elements, with different activation times. X-axis denotes cooling time.

In addition to the result presented in fig. 7, individual voxel spectra were summed up and renormalized to obtain an integral γ-spectrum of the fuel element just after the end of the irradiation (fig. 8). Results in terms of isotope contribution to the contact dose are also given, as this might be important, when dealing with decomposition of neutron activated materials (fig. 9).

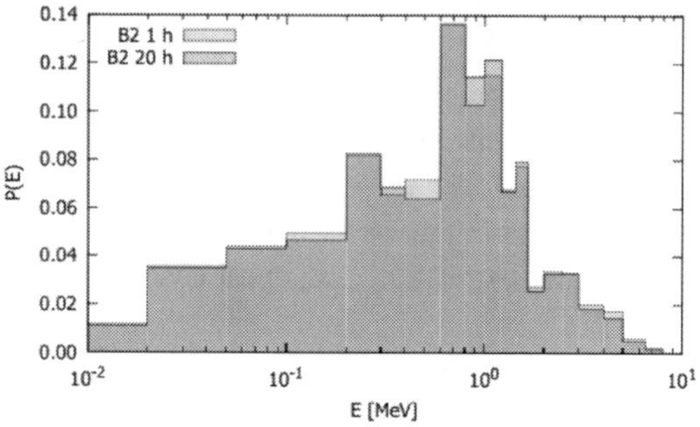

Figure 8: B2 fuel element γ-ray spectra after 1 h and 20 h of operation at full reactor power.

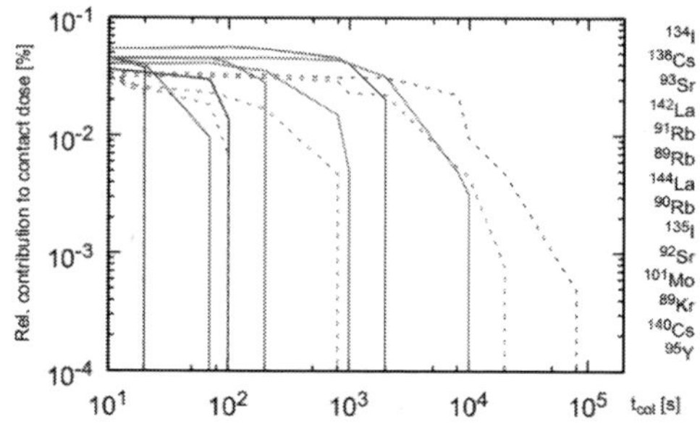

Figure 9: Isotope contributions to the contact dose in B2 element, after 20 h f operation at full reactor power.

3.2 Element burn-up of 1 MW d and 10 MW d

Longer irradiation of fresh nuclear fuel in terms of burn-up of 1 MW d and 10 MW d were also simulated, to account for accomulation of long lived isotopes, and to observe the bulk activity of the fuel elements, regardles of intraday operation. 40 h/w, 8 h d^{-1} at full reactor power operation was simulated on each individual representative fuel element, until a sufficient burn-up was achieved. Activity and γ-ray spectrum per fuel element, as well as kerma-rate and dose-rate on the silicon irradiation sample for each fuel element were calculated (fig. 10). Per element isotopic contribution to contact dose (fig. 12) and γ spectrum (fig. 11) were also calculated.
In fig. 10 we can observe the majority of long lived isotopic contribution, due to the fact, that the short-lived isotopes decay quickly, with all the elements having similar activities, kerma-rate and dose-rate contributions to the silicon sample after some 1×10^7 s. However, an $\times 10$ increase it that region with all elements being approximately equal is observed, when comparing two different burn-ups, showing a linear-rate of production of long-lived isotopes, which is consistent.

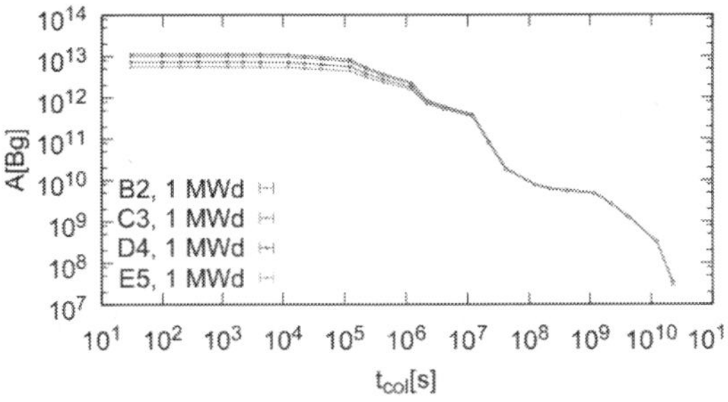

(a) Representative fuel element activity after 1 MW d burn-up.

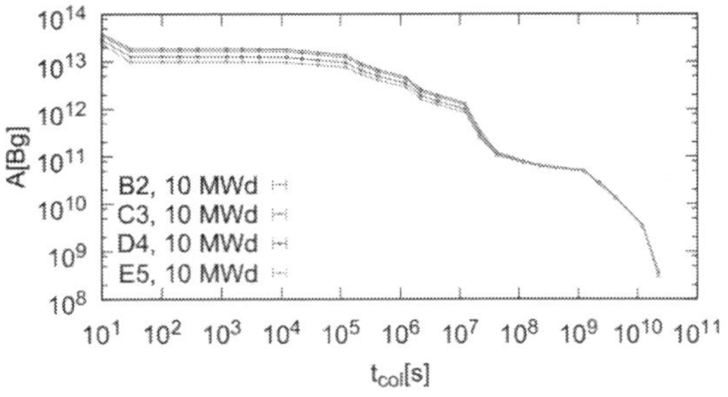

(b) Representative fuel element activity after 10 MW d burn-up.

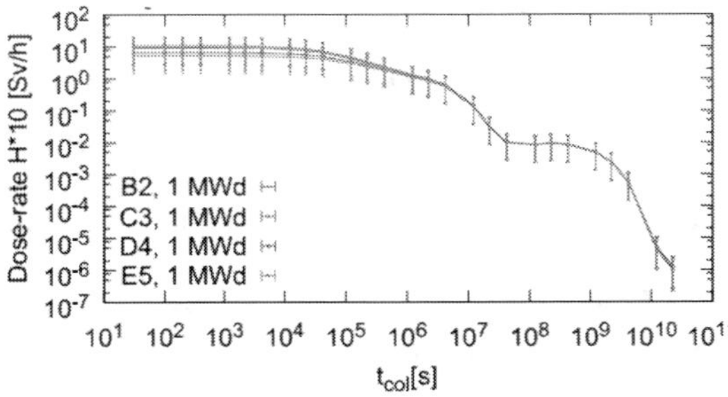

(c) H^*10 dose-rate on silicon sample after 1 MW d fuel element burn-up.

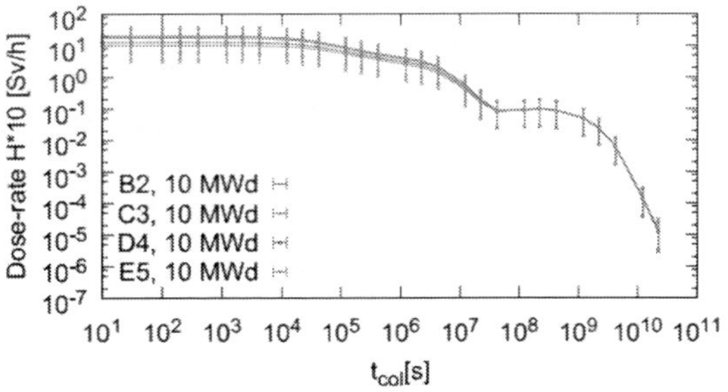

(d) H^*10 dose-rate on silicon sample after 10 MW d fuel element burn-up.

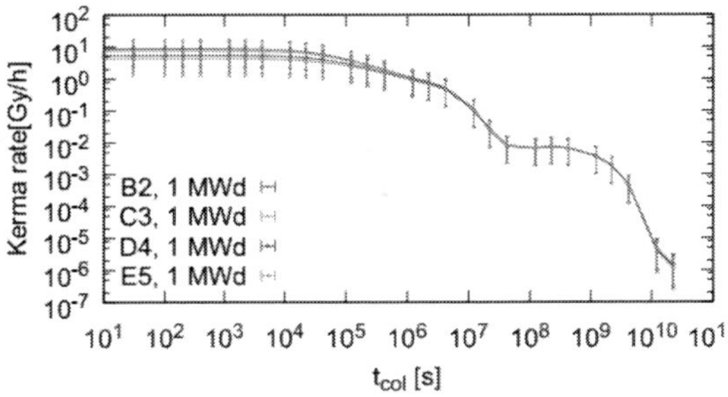

(e) Kerma-rate on silicon sample after 1 MW d fuel element burn-up.

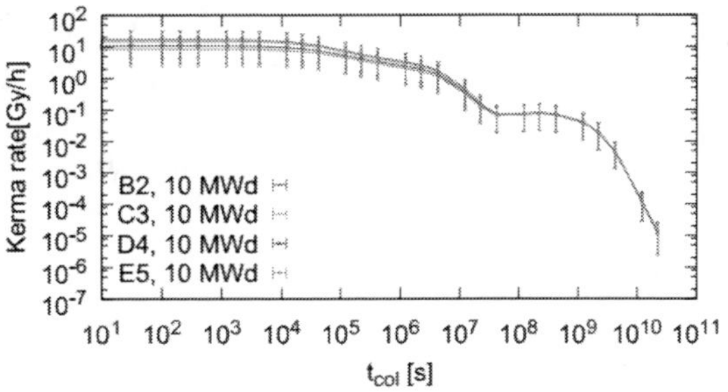

(f) Kerma-rate on silicon sample after 10 MW d fuel element burn-up.

Figure 10: Fuel elements activity, kerma-rate and dose-rate on the silicon sample after 1 MW d and 10 MW d burn-up.

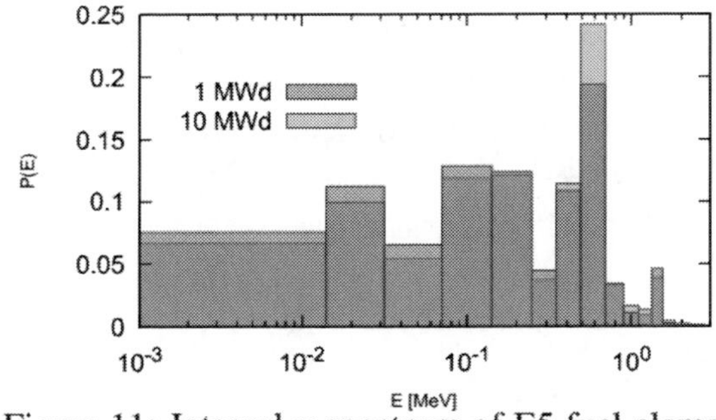

Figure 11: Integral γ spectrum of E5 fuel element after 1 MW d and 1 MW d irradiation.

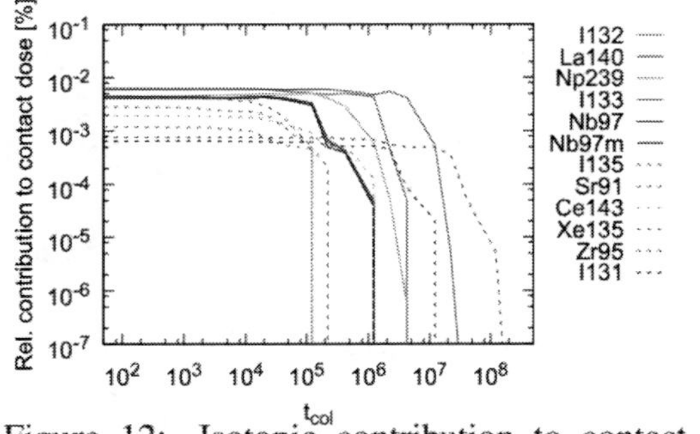

Figure 12: Isotopic contribution to contact dose after 1 MW d of B2 fuel element.

4 Summary

An R2S method script was developed for calculation of delayed γ-rays to improve photon characterization of irradiation facilities. JSI TRIGA nuclear fuel analysis on delayed gamma ray emission and dose deposition in was performed. Since then initial measurements on dose field around fuel element, irradiated for several hours were performed, showing agreement with dose calculations within the dose-rate distribution denoted on the graphs. To further evaluate on the code results, smaller samples around the irradiated elements should be used with precisely known positions and times after irradiation. Such measurements, utilizing different γ dozimeters, such as TLD, OSLD, RadFET, etc.. will be performed in the next two years, to evaluate high γ dose-rate measurement techniques inside mixed neutron and γ fields and evaluate computational results with measurements.
The script for the R2S is still in its initial stages of the development, as several steps have not yet been automated. Also a more general script version should be made, where neutron spectra and flux voxels will be subdivided into their respective cell parts. Also model segmentation should be performed in a way, that neutron spectrum and flux do not change significantly throughout the voxel. Estimates on mesh division have thus far been made by hand, instead a computer code would be developed for full automation and accurate representation of actual state. The code will be used for γ-field characterization of the TRIGA reactor allowing its utilization for γ radiation hardness studies. Moreover , the code could be later used to study activation of reactor components during it lifetime cycle for decomissioning purposes, to increase nuclear waste storage facilities utilization.

References

[1] L. Snoj, G. Žerovnik, A. Trkov, Applied Radiation and Isotopes **70**, 483 (2012)

[2] G. Žerovnik, T. Kaiba, V. Radulovic, A. Jazbec, S. Rupnik, L. Barbot, D. Fourmentel, L. Snoj, Applied Radiation and Isotopes **96**, 27 (2015)

[3] L. Snoj, B. Smodiš, *45 years of TRIGA Mark II in Slovenia*, in *Proceedings of the International Conference Nuclear Energy for New Europe* (2011)

[4] *Recent results of the ATLAS Upgrade Planar Pixel Sensors R&D Project* (Proceedings of science, 2011)

[5] K. Ambrožič, L. Snoj, G. Žerovnik (2016), to be published this year

[6] V. Radulović, D. Fourmentel, L. Barbot, J.F. Villard, T. Kaiba, Ž. Gašper, L. Snoj, Nuclear Instruments and Methods in Physics Research Section A: Accelerators, Spectrometers, Detectors and Associated Equipment **804**, 149 (2015)

[7] D. Fourmentel, V. Radulović, L. Barbot, J.F. Villard, G. Žerovnik, L. Snoj, M. Tarchalski, K. Pytel, F. Malouch, *Delayed Gamma Measurements in Different Nuclear Research Reactors Bringing Out the Importance of the Delayed Contribution in Gamma Flux Calculations*, in *Advancements in Nuclear Instrumentation Measurement Methods and their Applications, ANIMMA* (Instituto de Plasmas e Fusão Nuclear, Lisbon, 2015)

[8] R. Villari, U. Fischer, F. Moro, P. Pereslavtsev, L. Petrizzi, S. Podda, A. Serikov, Fusion Engineering and Design **89**, 2083 (2014), proceedings of the 11th International Symposium on Fusion Nuclear Technology-11 (ISFNT-11) Barcelona, Spain, 15-20 September, 2013

[9] A. Davis, Ph.D. thesis, University of Birmingham (2010)

[10] P. Sauvan, J.P. Catalán, F. Ogando, R. Juárez, J. Sanz, IEEE Transactions on Nuclear Science **63**, 375 (2016)

[11] G. Rossum, Tech. rep., Amsterdam, The Netherlands, The Netherlands (1995)

[12] T. Goorley, M. James, T. Booth, F. Brown, J. Bull, L. Cox, J. Durkee, J. Elson, M. Fensin, R. Forster et al., Nuclear Technology **180**, 298 (2012)
[13] J.C. Sublet C., J.W. Eastwood, J.G. Morgan, *The FISPACT-II User Manual*, UK Atomic Energy Authority, Culham center for fusion energy, issue 6 edn. (2016)
[14] J. Hale, Radiology **111**, 716 (1974)

On the development of radiation tolerant surveillance camera from consumer-grade components

Ambrožič Klemen[1,*], Snoj Luka[1,**], Öhlin Lars[2,***], Gunnarsson Jan[2,****], and Barringer Niklas[2,†]

[1] *Jožef Stefan Institute, Jamova cesta 39, SI-1000 Ljubljana, Slovenia*
[2] *ISEC-Industrial Security AB, Diabasgatan 12, SE-254 68 Helsingborg, Sweden*

Abstract. In this paper an overview on the process of designing a radiation tolerant surveillance camera from consumer grade components and commercially available particle shielding materials is given. This involves utilization of Monte-Carlo particle transport code MCNP6 and ENDF/B-VII.0 nuclear data libraries, as well as testing the physical electrical systems against γ radiation, utilizing JSI TRIGA mk. II fuel elements as a γ-ray sources. A new, aluminum, 20 cm × 20 cm × 30 cm irradiation facility with electrical power and signal wire guide-tube to the reactor platform, was designed and constructed and used for irradiation of large electronic and optical components assemblies with activated fuel elements. Electronic components to be used in the camera were tested against γ-radiation in an independent manner, to determine their radiation tolerance. Several camera designs were proposed and simulated using MCNP, to determine incident particle and dose attenuation factors. Data obtained from the measurements and MCNP simulations will be used to finalize the design of 3 surveillance camera models, with different radiation tolerances.

1 Introduction

In order to maintain video surveillance in radiation environment, the surveillance cameras have to be radiation tolerant. This can be achieved by selecting radiation tolerant (rad hard) camera components, or by adequate shielding of non rad hard components. This paper describes development of a camera, constructed from consumer grade electronic and optics components with the aid of publicly available rad-hard component databases [1] and a camera shielding enclosure design. Such a camera would be used inside NPP containment, spent nuclear fuel and waste facilities, in medical radiology, etc., and would therefore be exposed to both neutrons and γ-rays. Since consumer-grade parts usually do not exhibit heightened radiation tolerance, they have to be shielded from radiation.
The end result is to be 3 distinct surveillance camera models, with different levels of incident radiation tolerance requirements, supplied by ISEC. These levels were determined by close cooperation of ISEC and their customers and market demands and are given in terms of ^{60}Co dose in table 1.

*e-mail: klemen.ambrozic@ijs.si
**e-mail: luka.snoj@ijs.si
***e-mail: laoh@isec.se
****e-mail: jagu@isec.se
†e-mail: niba@isec.se

DOI: 10.1051/epjconf/201715307044

Table 1: Radiation tolerance goals for three different camera model versions

Camera model version	Dose-rate [Gy h^{-1}]	Cumulative dose [Gy]
High radiation tolerance	10^3	10^5
Medium radiation tolerance	10^3	10^4
Low radiation tolerance	10^1	10^3

Since radiation tolerance is not a required feature in consumer grade electronics, no data on radiation resistance is provided by the manufacturer and very little work on radiation tolerance assurance has been provided by third parties, due to low volume requirements of such components and consequently high production costs. Therefore the components to be used in the camera have to be tested against radiation tolerance, and their operational parameters monitored during the tests. This way a list of components radiation tolerances can be made. The most sensitive components or modules are then replaced by their rad hard counterparts or shielding requirements readjusted to most sensitive components. Strategical placement of components and modules inside shielding enclosure has a significant impact on overall radiation tolerance of the camera system.
The most appropriate tool for shielding design and analysis of doses are Monte Carlo neutron and photon transport codes, such as MCNP [2]. In the paper the above mentioned activities are presented.

2 Irradiation tests

Initially, camera modules, electronic and optics assemblies and components intended to be used in the camera models, were checked for their rad-hard counterparts in the available databases [1], and tested against radiation. The purpose of these tests is to determine the doses (both neutron and γ) that the components and modules can withstand. These tests were performed at the JSI TRIGA mark II reactor [3] (TRIGA) in the existing and newly constructed irradiation facilities. A gas-flow ionization chamber in close proximity of the test assembly was used for cumulative dose and dose-rate measurements during the irradiation procedure. A custom assembly-parameter-reader was constructed for each test assembly, which monitored several aspects of the assembly operation. A failure or malfunction would be noted on the parameter reader log, thus determining the radiation tolerance of the component.
Activated TRIGA fuel elements were utilized as a γ-ray source, either in an in-core Triangular channel (TriC) irradiation facility, or ex-core, by moving up to 6 TRIGA fuel elements to the fuel element rack on the edge of the reactor tank [4]. A new irradiation facility, an aluminum box with internal dimensions of 20 cm × 20 cm × 30 cm and a cable guide tube, leading to the reactor platform, was designed and constructed to utilize ex-core fuel elements as γ-ray sources. After the irradiations the logged data for both ionization chamber and for electronic components were evaluated and points of failure for each component determined. This process was repeated several times over, using different components and module designs, until a satisfactory radiation tolerance of the module was obtained. After the irradiations, the measurement data were post-processed, and points of failure determined. An example of an irradiation test data visualization is presented in figure 1.

3 Computational analysis

Monte Carlo particle transport calculations were carried out in order to assist in the development of a suitable shielding enclosure of the camera, as well as proposing additional shielding of the individual components. Several different stages of modeling the shielding enclosure took place, from scoping calculations on block shielding materials to progressively more complex geometrical models towards

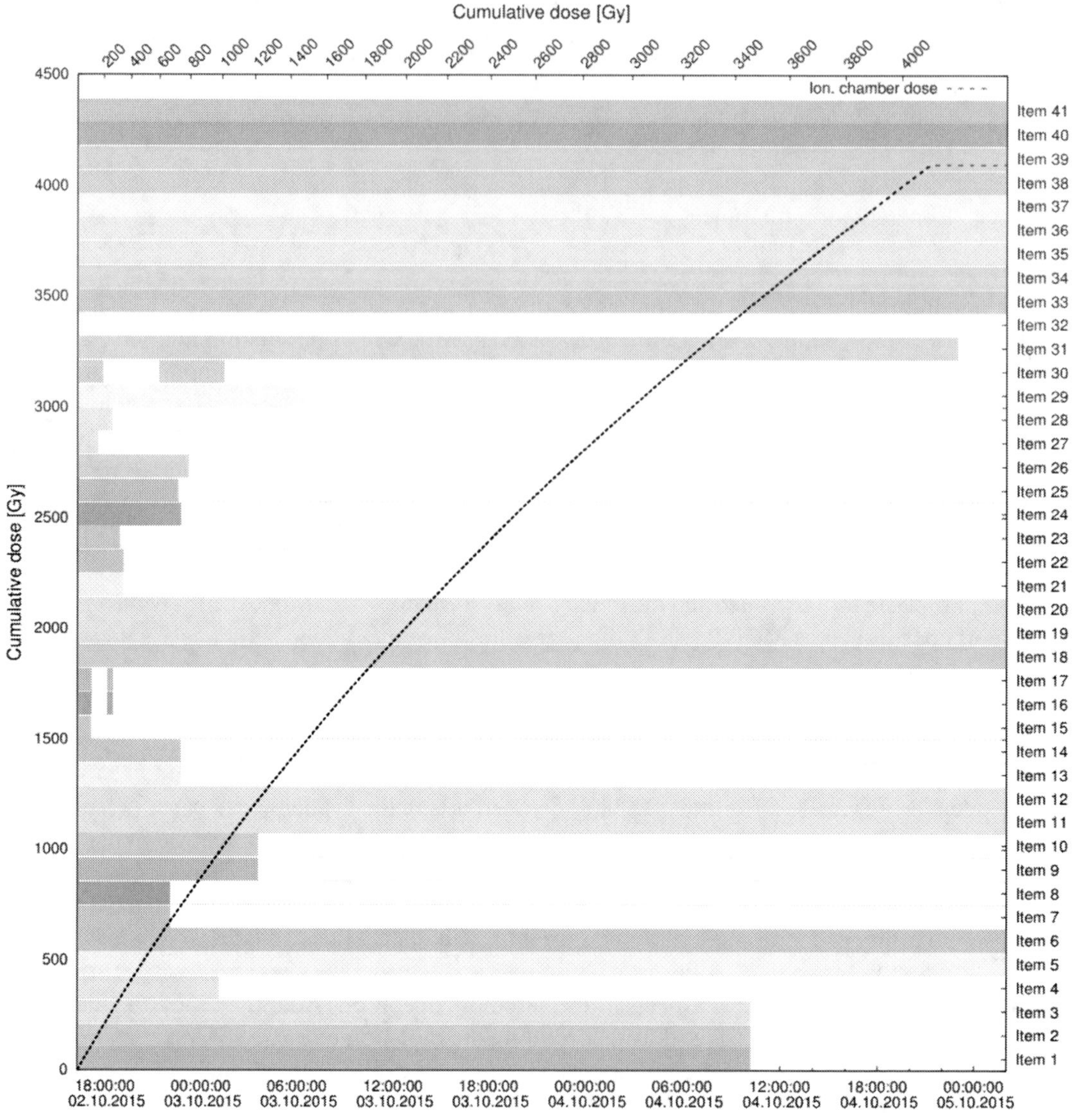

Figure 1: An example of electronic componets test data, with test items in the horizontal rows, and ionization chamber measurement data. Colored row means, the item is operational.

a final design. Several different neutron and photon spectra were used for radiation sources in the simulations (fig. 3). Photon spectra were previously calculated [5], and 6 MeV and 7 MeV γ lines added, to represent a percentage of whole spectrum, to simulate water activation γ lines.
Real world scenarios are taken into account in terms of radiation source geometries, source particle types and spectra. Two plane sources, one with particle emission directly towards the camera (fig. 2a), one with isotropic emission over half the solid angle (fig. 2b) which are representative of point sources at distance and close proximity, a half spherical (fig. 2c) and a a spherical source (fig. 2d), with isotropic particle emission from the half sphere / sphere surface over half the solid angle inwards,

which are representative of uniformly distributed radiation sources throughout the space, and account for backscattering from walls, should the camera be positioned close to the wall.

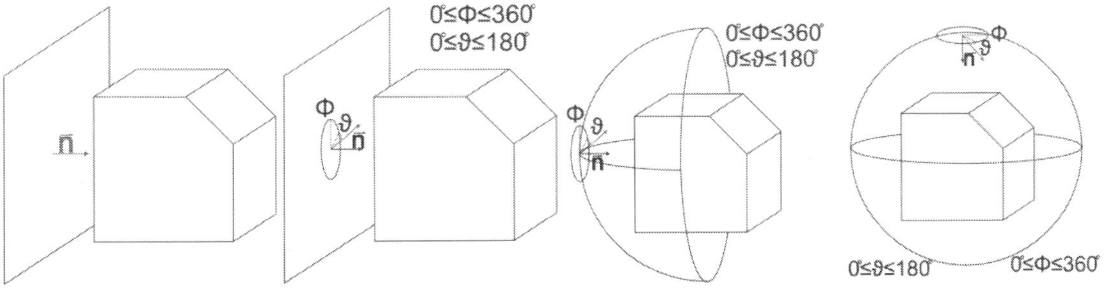

(a) Plane direct source geometry. (b) Plane isotropic source geometry. (c) Isotropic source geometry over half sphere shell inwards. (d) Isotropic source over sphere shell inwards.

Figure 2: Source geometries, used for simulations.

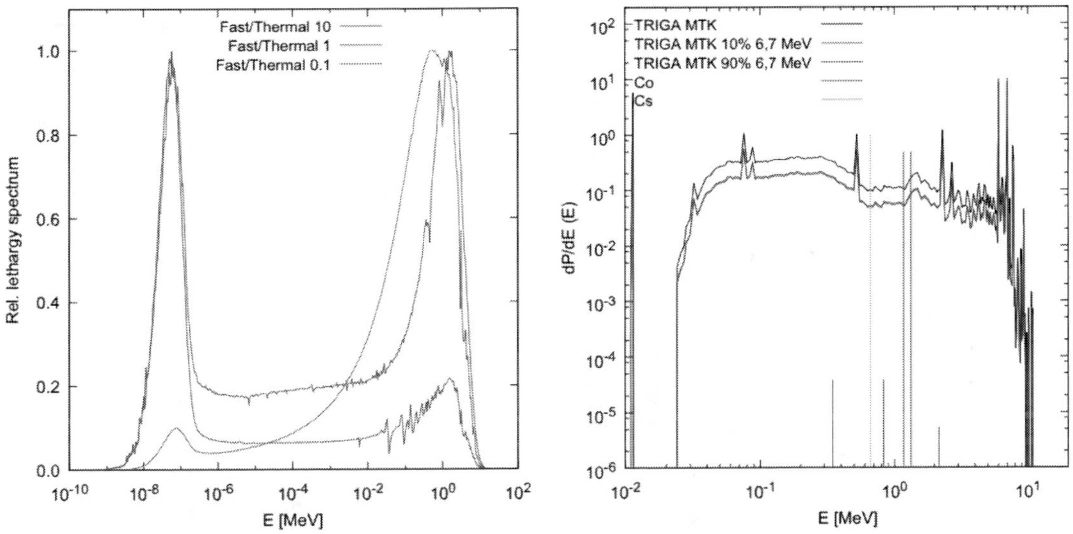

(a) Source neutron lethargy spectra, used in simulations. (b) Source γ-ray spectra used in simulations.

Figure 3: Source particle spectra, used for simulations.

Several different spectra, for both neutron and γ rays were used in simulations. Neutron spectra are of ^{235}U fission type, given in terms of ratios between fast (E_N >1 keV) and thermal (E_N <0.625 eV) neutron flux $\frac{F_N}{Th_N}$ and aim to represent current reactor spectra for thermal ($\frac{F_N}{Th_N} = 0.1$) and fast ($\frac{F_N}{Th_N} = 1$) fission reactors, as well as fusion devices ($\frac{F_N}{Th_N} = 10$).γ-ray spectra are spectra calculated inside TRIGA irradiation facilities, modified TRIGA irradiation facilities spectra (MTK: minor thermalization column), with added 6 MeV and 7 MeV γ lines arising from primary water activation (percentage denotes spectral fraction of 6 MeV and 7 MeV lines) due to (n,p) reactions on ^{16}O, and ^{60}Co and ^{137}Cs spectra.

Doses were calculated using either flux-to-dose conversion factors (ICRP-21 H*10 [6] for γ-rays,

ICRP-74 H*10 [7] for neutrons and ASTM E-722 [8] for neutron dose on silicon)(fig. 4), or direct kerma computations using MCNP track-length heating estimate (F6 tally in MCNP) [2].
Shielding material compositions data-sheets were supplied by the shielding manufacturers and isotopic compositions evaluated using MATSSF [9] computer code.
ENDF/B-VII.0 nuclear data libraries [10] were used in the calculations.

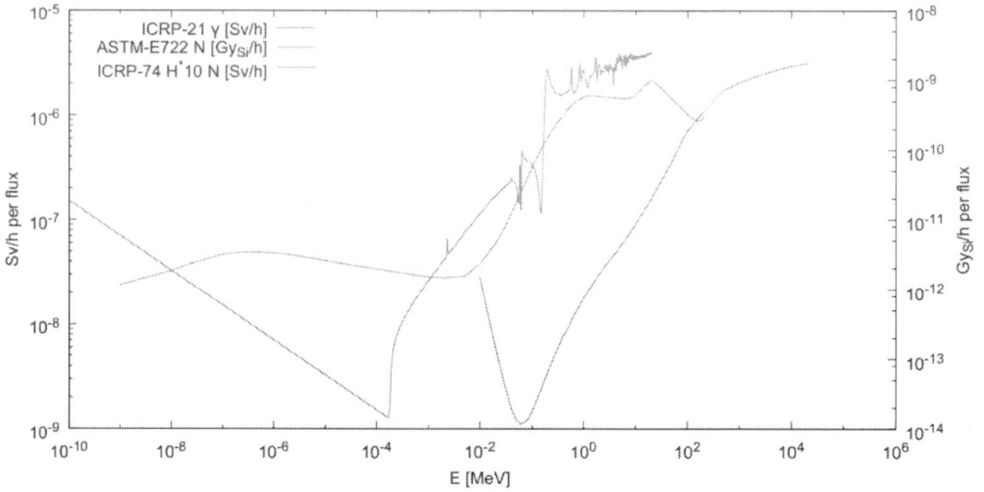

Figure 4: Particle flux to dose conversion factors: ICRP-21 for γ H^*10 [6], ICRP-74 for neutron H^*10 [7] and ASTM E-722 for silicon dose equivalent [8].

3.1 Bulk shielding analysis

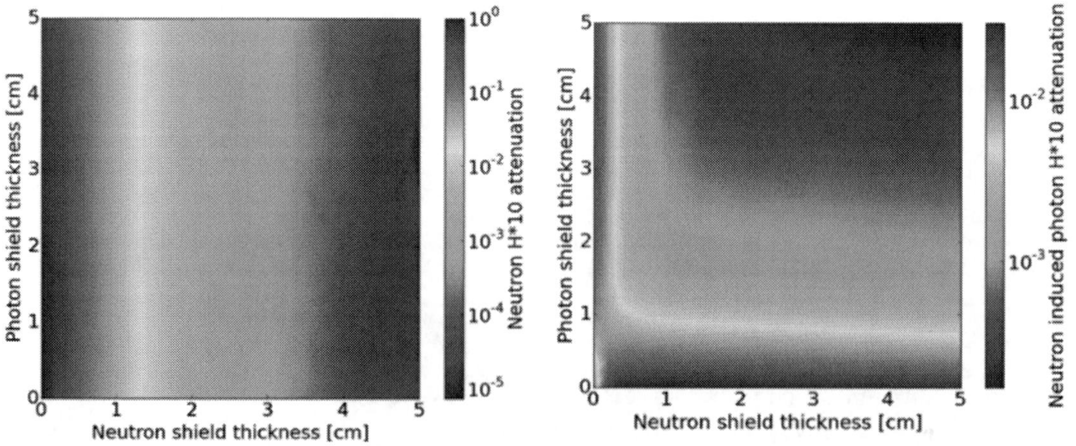

(a) Neutron H^*10 dose attenuation vs. photon and neutron shielding thickness.

(b) Neutron induced photon H^*10 dose attenuation vs. photon and neutron shielding thickness, normalized to incident neutron H^*10 dose.

Figure 5: Color map of neutron and neutron induced photon H^*10 dose attenuation vs. neutron and photon shielding thicknesses. Incident neutron spectrum (fig. 3a), $\frac{F}{Th}$ = 0.1, plane direct source.

Firstly scoping calculations to identify the best shielding materials were performed on slab shields. Lead and tungsten were chosen as γ-ray shielding material and several boron or lithium loaded polyethylene-based materials as neutron shielding.
In fig. 5, that in case of incident neutrons, neutron shielding induces γ-radiation due to neutron-induced production of photons. This should be taken into account, as the values of neutron induced γ-ray dose might exceed that of neutrons at some shielding thicknesses combinations.
Since the irradiation tests were going on in parallel, some desired attenuation estimates based on the initial measurement data and identification of least radiation resistant components were made.

3.2 Camera model design

An initial stage camera model CAD design was proposed, which had the camera module, electronic and optics components modeled in. An exact transcript into MCNP [2] geometry was made utilizing a Rhino® Grasshopper plug-in script [11]. Each electronic component was tallied for neutron and photon flux and dose. In addition a mesh was superimposed across the model geometry (MCNP FMESH card), and both neutron and photon fluxes and doses calculated in each of the mesh voxels. Material compositions of the electronics and optics components were supplied by manufacturers and isotopic composition calculated using MATSSF [9].

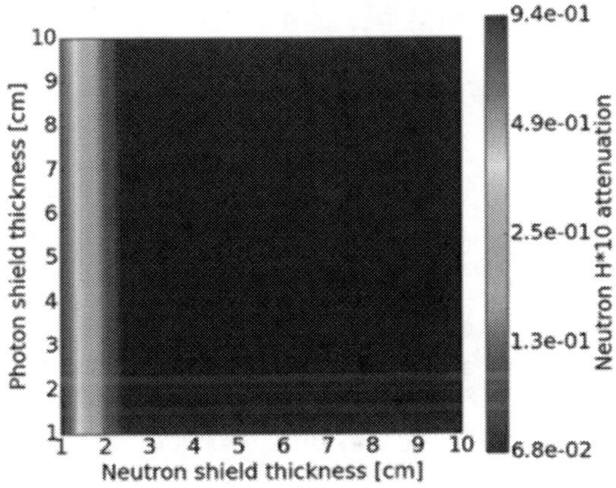

(a) Neutron H^*10 dose attenuation on an internal component vs. photon and neutron shielding thickness.

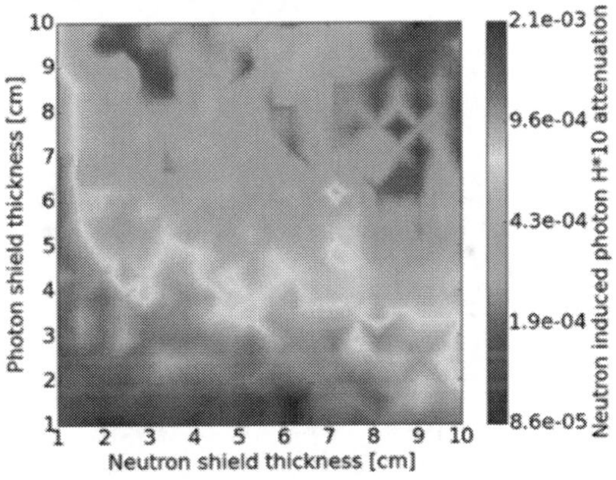

(b) Neutron induced photon H^*10 dose attenuation on an internal component vs. photon and neutron shielding thickness, normalized to incident neutron H^*10 dose.

Figure 6: Color plot of neutron and neutron induced photon H^*10 dose attenuation on an internal component vs. neutron and photon shielding thicknesses. Incident neutron spectrum (fig. 3a), $\frac{F}{Th}$ = 0.1, plane direct source.

A parametric analysis was performed, varying thicknesses of both neutron and γ shields, and neutron and photon source spectra and geometries. A Python v.2.7 script [12] (Python) was written in order to speed up the process and perform automatic input generation. Another, unshielded model was made, which served for normalization. It should be noted, that internal structure of the camera i.e. the components positions were changing during the parametric analysis, to accommodate several technical requirements, which in turn effect the shielding attenuation (figure 6b).

Results would be presented in terms of attenuation vs. shielding thicknesses on color map-type graphs for each component, incident particle type and spectrum, source geometry and selected shielding materials (figure 6).

3D colored dose and flux maps were also produced from superimposed mesh results (fig: 7), to

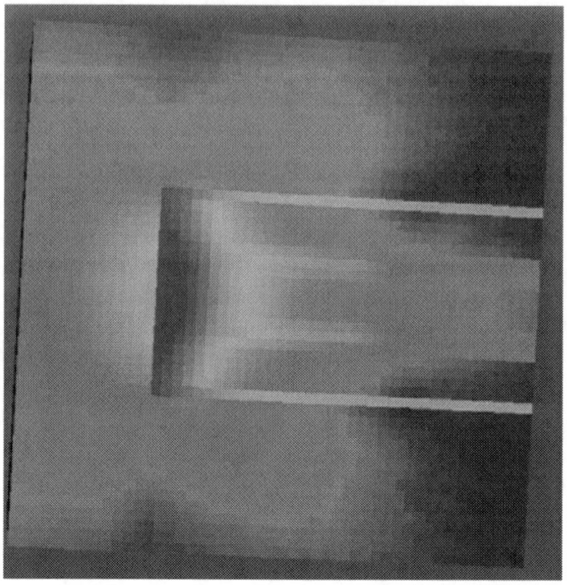

Figure 7: 3D ICRP-21 $H^{*}10$ dose map, with incident ^{60}Co spectra, emitted from plane isotropic source.

asses the significant gradients, locate main contributors to the dose on components inside shielding and propose improvements. This enabled us, to place additional shielding to some components, that could not be relocated.

Using the obtained results, one can assess the shielding thicknesses and shielding material combinations, and a narrower shielding thickness variation with all the different source spectra and geometries was performed and dose attenuation values tallied electronic components and modules (items on figure 8). An optimal shielding thickness was established this way, and minor position changes and shielding thickness changes on key positions were suggested, to conform with the requirements.

It is interesting to note, that in case of reactor γ-ray spectra, the doses inside the shielding enclosure are up to more then 30 % higher comparing to ^{60}Co spectrum. This is important, as the electronic components are commonly tested in ^{60}Co irradiation facilities, where test γ-spectra might significantly differ from γ-spectra during its intended operation, leading to an underestimation of shielding requirements (fig. 8).

4 Summary

When designing shielded radiation resistant electronic, a three step approach is proposed. Firstly the radiation tolerance of individual components should be determined, preferably by experiments in representative environment with respect to the intensity and spectrum of the incident particles, in our case gamma rays and neutrons Then the environmental source should be determined, i.e. the type of particles, their spectrum, angular flux, flux intensity, dose rate etc. When these boundary conditions

DOI: 10.1051/epjconf/201715307044

	Item 1	Item 2	Item 3	Item 4	Item 5	Item 6	Item 7
Plane, direct TRIGA MTK + 0.9 6,7 MeV	0.527	0.547	0.546	0.547	0.545	0.665	0.483
Plane, direct TRIGA MTK + 0.1 6,7 MeV	0.527	0.546	0.546	0.547	0.545	0.665	0.483
Plane, direct TRIGA MTK	0.523	0.551	0.552	0.551	0.550	0.670	0.484
Plane, direct Cs-137	0.390	0.446	0.444	0.445	0.441	0.600	0.443
Plane, direct Co-60	0.474	0.530	0.529	0.529	0.526	0.651	0.461
Plane, uniform over angle TRIGA MTK + 0.9 6,7 MeV	0.406	0.449	0.446	0.443	0.433	0.489	0.524
Plane, uniform over angle TRIGA MTK + 0.1 6,7 MeV	0.406	0.449	0.446	0.443	0.433	0.489	0.524
Plane, uniform over angle TRIGA MTK	0.406	0.448	0.446	0.442	0.432	0.488	0.522
Plane, uniform over angle Cs-137	0.277	0.321	0.319	0.316	0.306	0.386	0.440
Plane, uniform over angle Co-60	0.355	0.403	0.400	0.397	0.386	0.444	0.480
Spherical TRIGA MTK + 0.9 6,7 MeV	0.436	0.429	0.428	0.425	0.415	0.460	0.498
Spherical TRIGA MTK + 0.1 6,7 MeV	0.436	0.429	0.428	0.425	0.415	0.460	0.498
Spherical TRIGA MTK	0.435	0.428	0.428	0.425	0.415	0.460	0.498
Spherical Cs-137	0.308	0.302	0.302	0.299	0.290	0.352	0.383
Spherical Co-60	0.388	0.382	0.382	0.378	0.368	0.418	0.452

Figure 8: Kerma attenuation factors on items within the shielding enclosure, for different incident γ spectra and source geometries. Differences in attenuation between reactor, ^{60}Co and ^{137}Cs (latter two commonly used for testing) γ spectra may underestimate required shielding thicknesses.

are determined, the shielding design should be performed. In these activities computational support by Monte Carlo particle transport methods is essential. As the radiation field can change significantly from utility to utility all possible scenarios with respect to particles spectra and angular distribution are taken into account. For each of the scenarios the attenuation factor for the shield is calculated. It is important to note that γ dose due to neutron induced γ-rays in the shield can be significantly higher than the environmental gamma ray dose, hence special attention should be paid to neutron field determination, which is often difficult [13]. As γ radiation tolerance tests are commonly performed in dedicated facilities with ^{137}Cs or ^{60}Co sources, it is important to be aware that reactor γ spectrum leads to approx 30 % higher Si kerma at the same H^*10 environmental γ dose. One should also pay attention to dose definitions and measurement. H^*10 environment dose is commonly measured with probes, the damage on electronics however is commonly measured with RADFETs for γ-rays and Si PIN diodes for neutrons [14]. Their response is proportional to Si kerma for γ-rays and damage function [8] for neutrons. The differences and relations between different doses, measurements and

 DOI: 10.1051/epjconf/201715307044

environment should be considered and the most practical approach to asses this is by Monte Carlo neutron and photon transport modeling as presented in the paper.

References

[1] *ESCIES:European Space Components Information Exchange System*, https://escies.org/webdocument/showArticle?id=228&groupid=6, accessed: 2016-06-14

[2] T. Goorley, M. James, T. Booth, F. Brown, J. Bull, L. Cox, J. Durkee, J. Elson, M. Fensin, R. Forster et al., Nuclear Technology **180**, 298 (2012)

[3] L. Snoj, G. Žerovnik, A. Trkov, Applied Radiation and Isotopes **70**, 483 (2012)

[4] K. Ambrožič, L. Snoj, *Modeling of γ field around irradiated TRIGA fuel elements by R2S method*, in *Proceedings of the ICRS-13 & RPSD-2016* (2016)

[5] K. Ambrožič, Master's thesis, Fakulteta za Matematiko in Fiziko, UL (2015)

[6] J. Hale, Radiology **111**, 716 (1974)

[7] I.C. on Radiological Protection, I.C. on Radiation Units, Measurements, *ICRP Publication 74: Conversion Coefficients for Use in Radiological Protection Against External Radiation*, Annals of the ICRP (SAGE Publications, 1997), ISBN 9780080427393

[8] A. International, ASTM E722, ASTM International, West Conshohocken, Philadelphia, USA (2014)

[9] A. Trkov, G. Žerovnik, L. Snoj, M. Ravnik, Nuclear Instruments and Methods in Physics Research A **610**, 553 (2009)

[10] M. Chadwick, P. Obložinsky, M. Herman, N. Greene, R. McKnight, D. Smith, P. Young, R. MacFarlane, G. Hale, S. Frankle et al., Nuclear Data Sheets **107**, 2931 (2006), evaluated Nuclear Data File ENDF/B-VII.0

[11] B. Kos, L. Snoj, Master's thesis, Fakulteta za matematiko in fiziko, Univerza v Ljubljani (2015)

[12] G. Rossum, Tech. rep., Amsterdam, The Netherlands, The Netherlands (1995)

[13] I. Remec, M. Najžer, G. Pregl, Ph.D. thesis (1991), `https://books.google.si/books?id=QnVhOQAACAAJ`

[14] F. Ravotti, Ph.D. thesis, Université Montpellier II (2006)

DOI: 10.1051/epjconf/201715308001

Outline of the Fukushima Daiichi Accident. Lessons Learned and Safety Enhancements

Masashi Hirano*

Nuclear Regulation Authority, Japan

Abstract. On March 11, 2011, an earthquake and subsequent tsunamis off the Pacific coastline of Japan's Tohoku region caused widespread devastation in Japan. As of June 10, 2016, it is reported that a total of 15,894 people lost their lives and 2,558 people are still unaccounted for. In Fukushima Prefecture, approximately 100,000 people are still obliged to live away from their homes due to the earthquake and tsunami as well as the Fukushima Daiichi accident.

On the day, the earthquake and tsunami caused severe damages to the Tokyo Electric Power Company (TEPCO)'s Fukushima Daiichi Nuclear Power Station (NPS). All the units in operation, namely Units 1 to 3, were automatically shut down on seismic reactor protection system trips but the earthquake led to the loss of all off-site electrical power supplies to that site. The subsequent tsunami inundated the site up to 4 to 5 m above its ground level and caused, in the end, the loss of core cooling function in Units 1 to 3, resulting in severe core damages and containment vessel failures in these three units. Hydrogen was released from the containment vessels, leading to explosions in the reactor buildings of Units 1, 3 and 4. Radioactive materials were released to the atmosphere and were deposited on the land and in the ocean.

One of the most important lessons learned is an importance to prevent such large scale common cause failures due to extreme natural events. This leads to a conclusion that application of the defense-in-depth philosophy be enhanced because the defense-in-depth philosophy has been and continues to be an effective way to account for uncertainties associated with risks. From the human and organizational viewpoints, the final report from the Investigation Committee of the Government pointed out so-called "safety myth" that existed among nuclear operators including TEPCO as well as the government, that serious severe accidents could never occur in nuclear power plants in Japan.

After the accident, the Nuclear Regulation Authority (NRA) was established on September 19, 2012. The NRA very urgently developed and issued the new regulatory requirements on July 8, 2014, taking into the account the lessons learned from the accident. It is noted that the NRA issued the Statement of Nuclear Safety Culture on May 27, 2015 which clearly expressed the NRA's commitment to break with the safety myth.

This paper briefly presents the outline of the Fukushima Daiichi accident and summarizes the major lessons learned having been drawn and safety enhancements having been done in Japan for the purpose of giving inputs to the discussions to be taken place in the Special Invited Session "Fukushima, 5 years after".

* Corresponding author: masashi.hirano@nsr.go.jp

NRA, Japan
Nuclear Regulation Authority

Fukushima Special Session
"Fukushima, 5 years after"

Outline of the Fukushima Daiichi Accident, Lessons Learned and Safety Enhancements

Masashi HIRANO and **Harutaka HOSHI**
Nuclear Regulation Authority (NRA)

The joint conference of
the 13th Int. Conf. on Radiation Shielding (**ICRS-13**), and
19th Topical Meeting of the Radiation Protection and Shielding Division of ANS - 2016 (**RPSD-2016**)
October 3 - 6, 2016, Paris, France

NRA, Japan
Nuclear Regulation Authority

Contents

Current Status of Nuclear Regulation in Japan

NRA, Japan
Nuclear Regulation Authority

- NRA was established in Sep. 2012 and developed the new regulatory requirements for NPPs which came into force in July 2013.
  - All the 48 units had been shut down since Sep. 2013.
  - So far, a total of 26 units (17 PWRs and 9 BWRs) have applied for conformance review.
  - In Sep. 2014, NRA first approved the applications from Sendai Units 1 and 2.
  - As of Aug. 2016, NRA has granted permission to 7 PWRs of which four units have restarted (Sendai Units 1,2 and Takahama Units 3, 4)
- The IAEA IRRS mission's review was taken place in Jan. 2016.

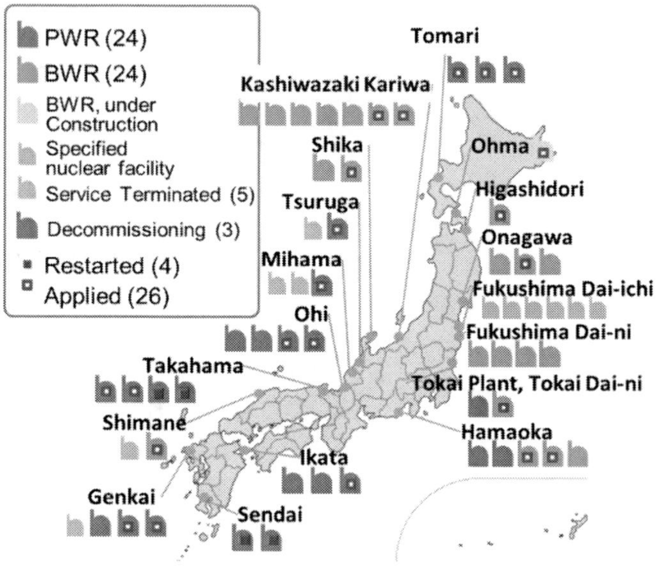

Y. Shimizu, IAEA Effective Nuclear Regulatory Systems (April 2016)

NRA, Japan
Nuclear Regulation Authority

Outline of the Accident

Tohoku-District off the Pacific Ocean Earthquake

- Occurred at **14:46** on **March 11, 2011**
- **9.0 Mw**: largest ever recorded in Japan
- Rupture duration: **170 sec**
- Epicenter: 38° 10''N and 142° 86''E, 23.7km in depth
- **Reverse fault type near the Japan Trench**

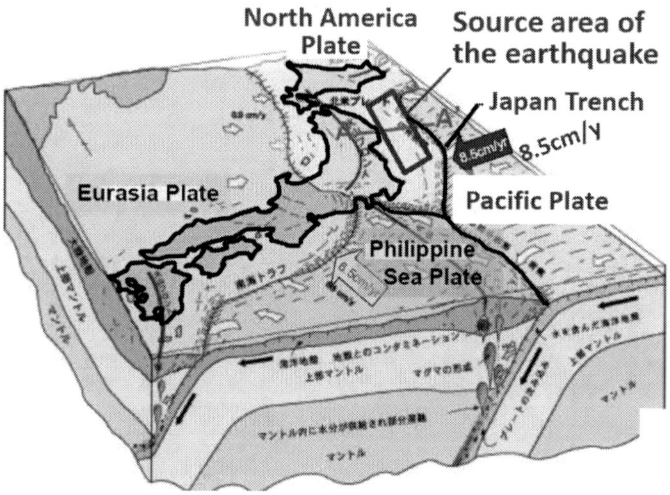

Edited to figure in "1st Government Report (June) 2011)"

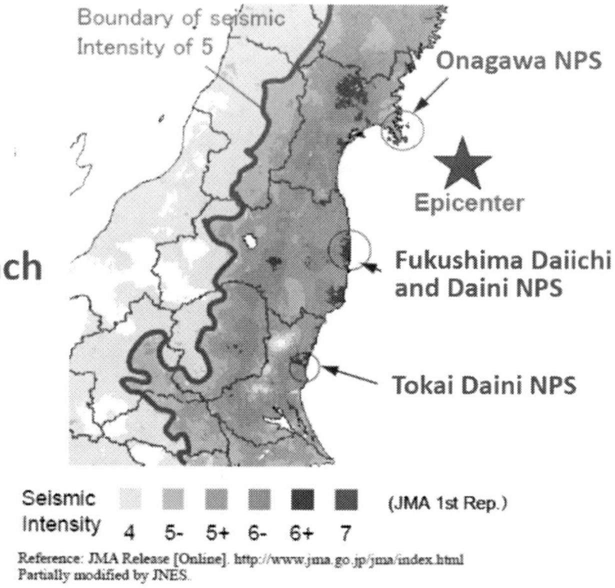

- Observed **max. acceleration** of seismic ground motion at **Fukushima-Daiichi** exceeded that of the **design basis by 26%**.

Accompanied Tsunami

- A **huge tsunami** was generated due to **a large slip**, more than 56m, **near the Japan Trench**.

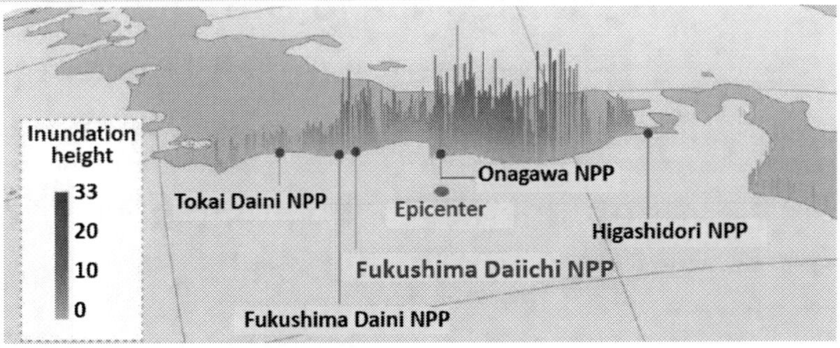

IAEA Fukushima report, Technical Volume 1/5 (2015)

Estimated slip distribution

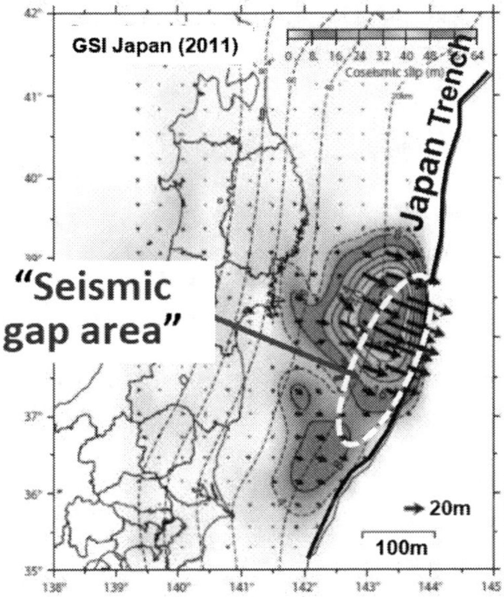

Sugino, et al., NRA, 24 Feb, 2016

- **Multi-segment rupture** including **"seismic gap area (no seismic record)"**

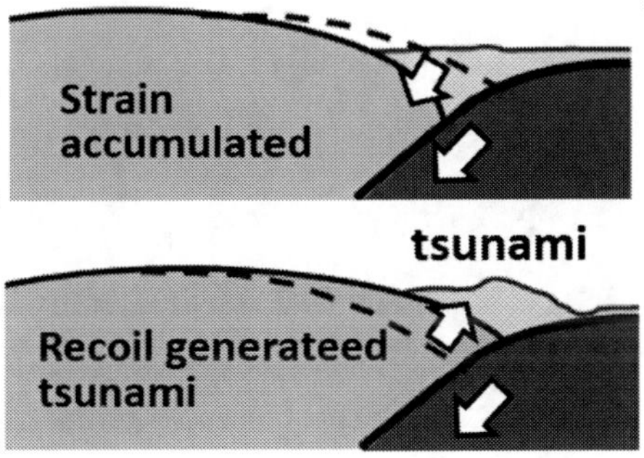

- Due to rupture of the **bonded interface**, the **North America Plate recoiled** and the strain energy was released, which generated the huge tsunami.

NRA, Japan
Nuclear Regulation Authority

Consequences of Natural Hazards and the Accident

- **Casualties due to earthquake/tsunami** (as of Dec.1, 2011)
  - **Deceased**: **15,840**
  - **Unaccounted for**: **3,547**
  - **Injured**: **5,951**
  - **Damaged buildings**: **1,009,074**

Interim Report from Government's Investigation Committee (Dec. 2011)

- As of June 2016, the number of evacuees has been reduced from **470,000** to **155,000**. Still **51,000 people** live in **temporary housing.**
- In **Fukushima Prefecture**, approx. **100,000** people are still **obliged to live away from their homes** due to the earthquake/tsunami as well as the Fukushima Daiichi accident. (website of the Fukushima Prefecture)
- **Evacuation order**: (1) Return is difficult (**red**), (2) Habitation is restricted (**yellow**), (3) Preparing for lifting orders (**Green**)
  - **Yellow and Green areas** will be lifted no later than March 2017. Current Status of Reconstruction and Challenges, June 2016, Reconstruction Agency

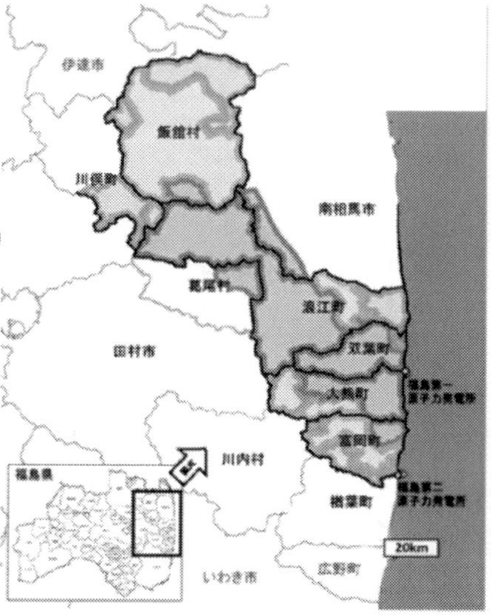

NRA, Japan
Nuclear Regulation Authority

Tsunami Heights at four Different Sites

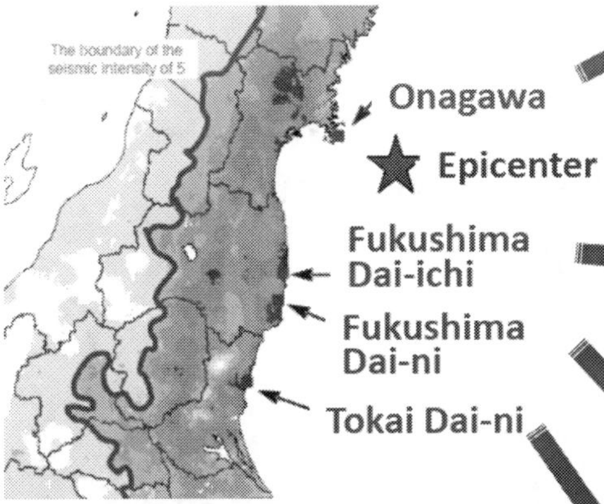

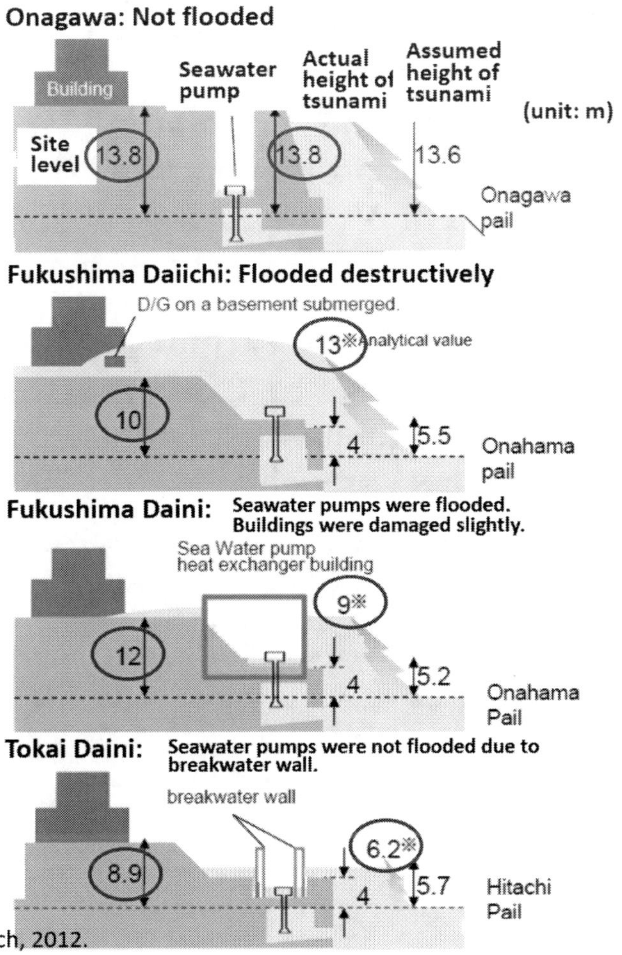

Tsunami at Fukushima Daiichi NPS

Yasui, METI, "Causes and Countermeasures", March, 2012.

NRA, Japan
Nuclear Regulation Authority

Loss of Offsite Power by Earthquake and Common-Cause Failures by Tsunami

- Units 1 to 3 automatically scrammed on seismic reactor protection trips. The **earthquake** led to **loss of offsite power**. All the **emergency diesel generators** (**EDGs**) automatically started.
- Then, **about 40 min. later**, the **tsunami** arrived at the site.
  - In Units **1 to 4**, **all the EDGs became inoperable** (SBO) together with loss of **DC** power.* It also caused **LUHS** as well.
  - In Unites **5 and 6**, an **air-cooled EDG of Unit 6 survived** and it supplied **AC power** to both units.

*DC power was available in Unit 3.

Seawater pumps for residual heat removal became inoperable
↓
Loss of Ultimate Heat Sink (LUHS)

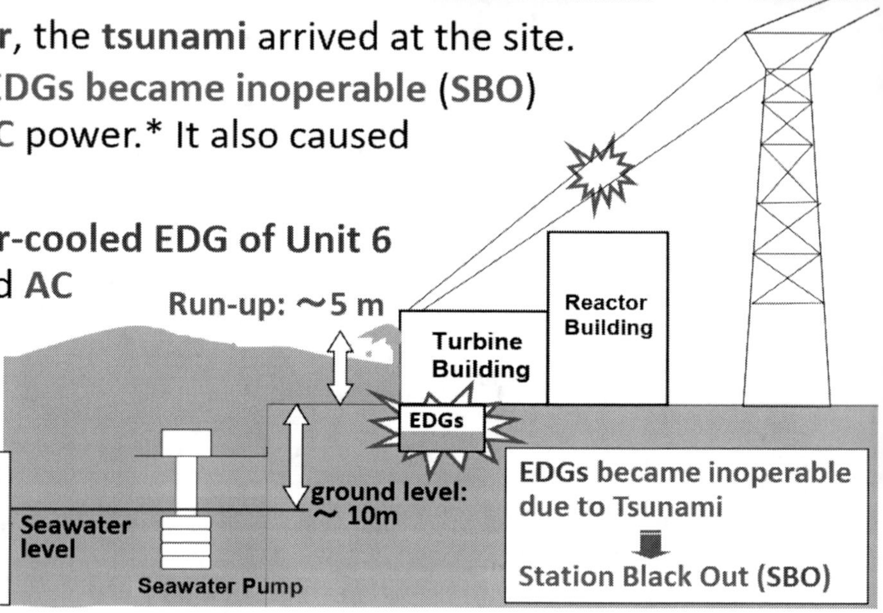

NRA, Japan
Nuclear Regulation Authority

Major Chronology of Events

Source: IAEA DG Report and Technical Volumes 1 (2015):

Unit 1	Unit 2	Unit 3
14:46 on Mar. 11: Shutdown due to **earthquake**		
LOOP: Loss of off-site power ➡ **EDGs started automatically**		
ICs (Isolation Condensers) operated manually	**RCIC** (Reactor Core Isolation Cooling System) started manually	**RCIC** started manually
15:27 on Mar. 11: Tsunami arrived		
Loss of AC and DC power		**Loss of AC** power (DC power available)
ICs and HPCI were not operable **March 12** • **PCV venting** at about **14:30** • **Hydrogen explosion at 15:36** • **Seawater injection at 19:04**	RCIC operated for about **70 hours** without DC power **March 14** • **RCIC stopped** until **~12: 30** • Operators **opened SRV** at **18:02** • **Seawater injection at 19:54** **March 15** • Possible PCV failure	**March 12** • RCIC stopped (**11:36**) and HPCI started (**12:35**) **March 13** • HPCI was **shut down** remote-manually from MCR at **2:42** • **SRVs opened at 9:08** (ADS conditions met). • **Fresh water injection at 9:25** **March 14** • **Hydrogen explosion at 11:01**

Hydrogen Explosion at Units 1, 3

- Hydrogen leaked from RPV to PCV and then to the RB, and exploded there.
- Analyses were done by the former JNES with **FLUENT** and **AUTODYN**.
- It was assumed that hydrogen leaked at:
  - **Top flange gasket at Unit 1**
  - **Sealing of hatch or penetrations at 1st floor at Unit 3**

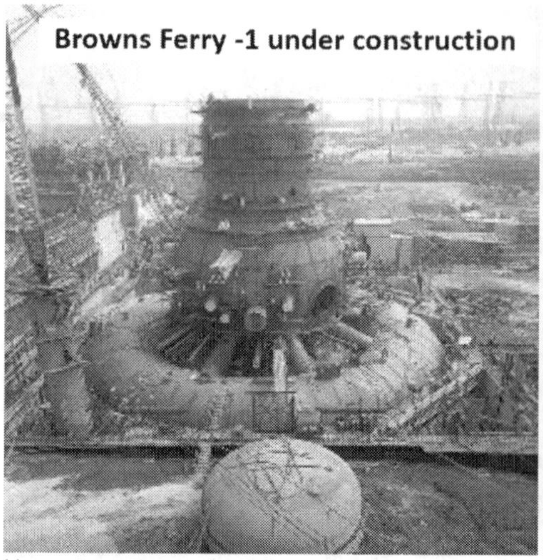
Browns Ferry -1 under construction

https://en.wikipedia.org/wiki/File:Browns_Ferry_Unit_1_under_construction.jpg

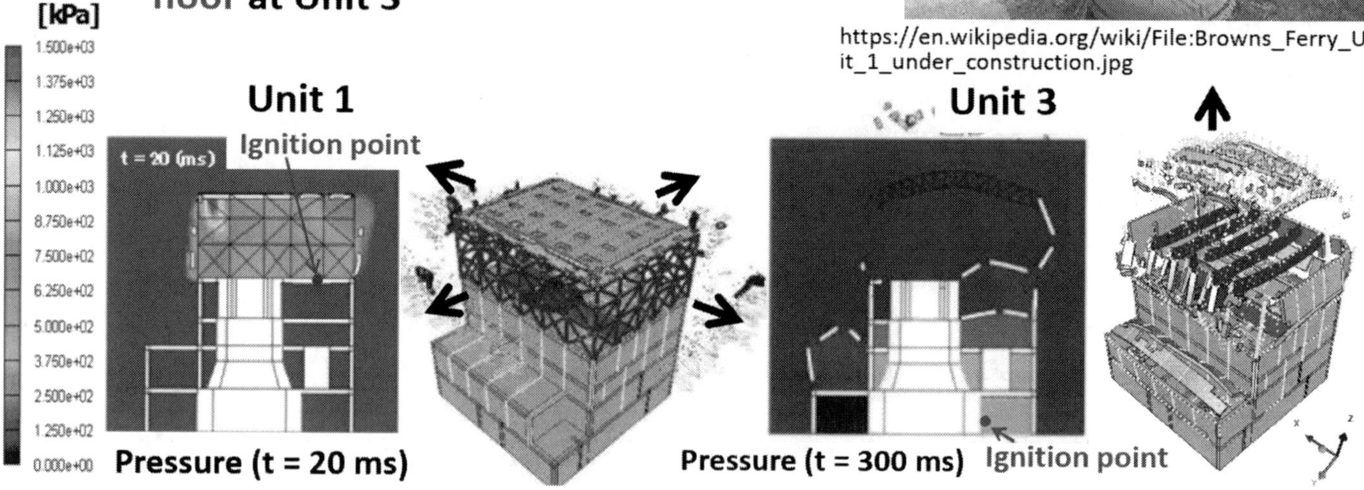

Hydrogen Explosion at Unit 4

NRA, Japan
Nuclear Regulation Authority

- **Hydrogen generated at Unit 3** might have been transported to Unit 4 **through SGTS piping and exhaust ducts.**
- The **exhaust stack was shared**. The valves of unit4 SGTS were open.

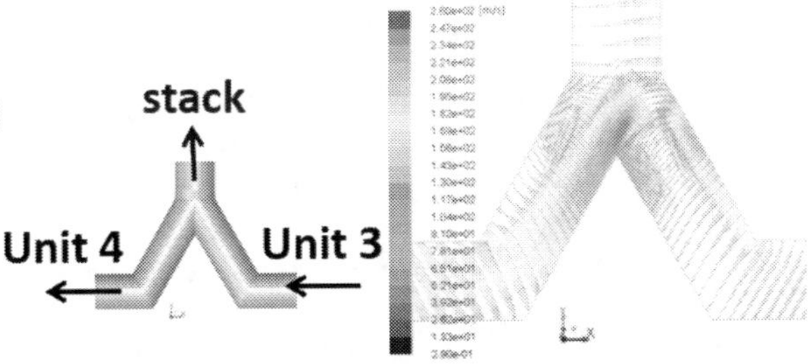

Analysis done by former JNES
WWW.nsr.go.jp/archive/jnes/content/000125907.pdf

Reactor Building of Unit 4

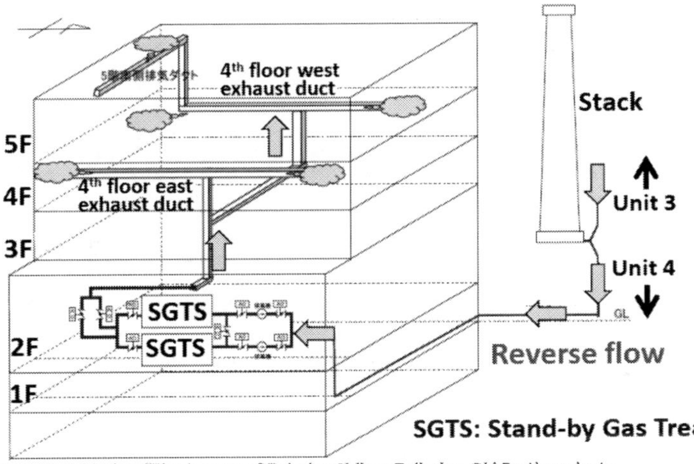

SGTS: Stand-by Gas Treatment System

Source: Added to "The impact of Tohoku-Chihou Taiheiyo-Oki Earthquake to Nuclear Reactor Facilities at Fukushima Daiichi Nuclear Power Station" (Sep. 9 2011, revised Sept. 28, 2011, Tokyo Electric Power Co., Inc.)

2nd Government Report

NRA, Japan
Nuclear Regulation Authority

End States of Units 1 to 3: OECD/NEA BSAF Project Phase I

OECD/NEA CSNI launched the BSAF (Benchmark Study of the Accident at the Fukushima Daiichi NPP) Project in Nov. 2012.

- 16 organizations from 8 countries (France, Germany, Korea, Russia, Spain, Switzerland, the United States, and Japan) participated.

Phase I Summary Report (March 2015)

- **Qualitative description of the plausible end states** after comparison of the best estimate case analyses

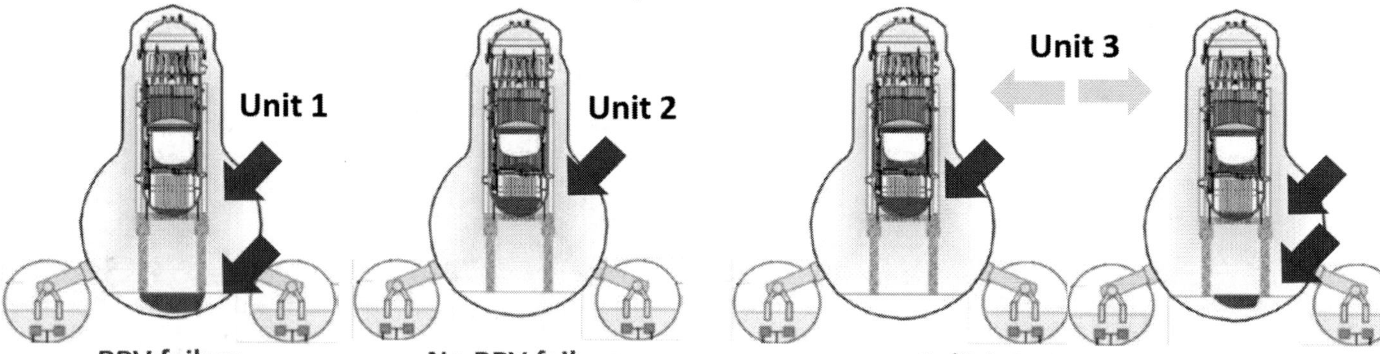

Unit 1: RPV failure. **Every calculation predicted that extensive damage occurred.**

Unit 2: No RPV failure. **The majority of calculations do not predict RPV failure**

Unit 3: Split into two groups:
a) No RPV failure
b) RPV failure and temporary MCCI scenario

NRA, Japan
Nuclear Regulation Authority

Source Term Evaluation

- The IAEA collected / compared the source terms evaluated by various organizations with different methods and assumptions.
- **Technical Volume 1 of the IAEA Fukushima report**: "**The total amount of aerosol based fission product releases** from all of the Fukushima Daiichi units was **about one order of magnitude smaller than that from the Chernobyl accident.**"

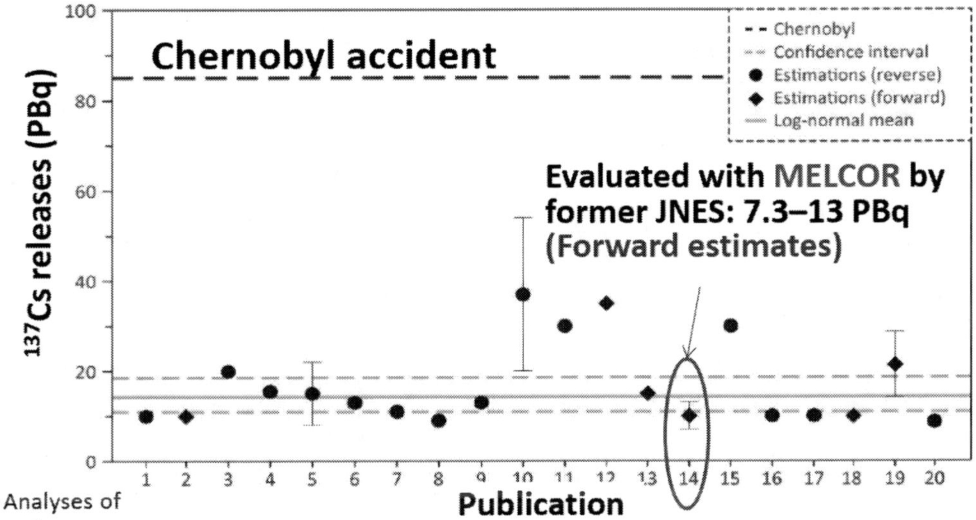

FIG. 1.4–4. Estimated atmospheric releases of 137Cs

IAEA Fukushima report, Technical Volume 1/5 (2015)

Publication [14]:
HOSHI, H., HIRANO, M., Severe Accident Analyses of Fukushima - Daiichi Units 1 to 3, Side event by Government of Japan at 56th IAEA General Conference, Vienna, 2012, JNES (2012),

NRA, Japan
Nuclear Regulation Authority

SA Progression and Source Terms

Hirano, Hoshi and Homma, Presented at U.S.NRC RIC, March 2013.

- The former JNES has conducted a SA progression analysis with **MELCOR** developed by U.S.NRC. The NRA continues to do it.
  - **Release timings** are in good agreement with the monitoring data.
  - **The largest peak in the morning on March 15** was probably caused by **PCV failure at Unit 2**

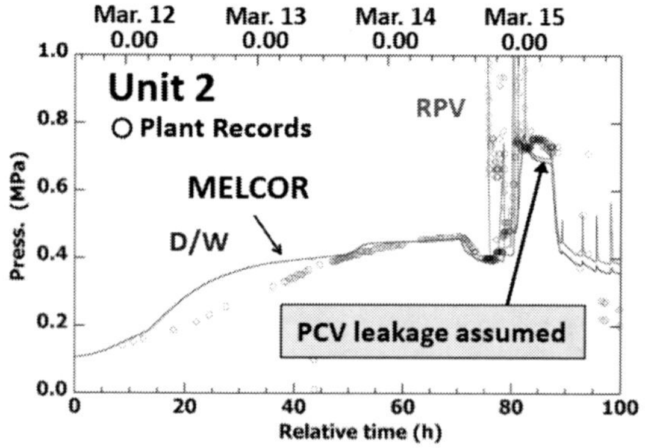

Comparison between the source terms by MELCOR and monitoring data

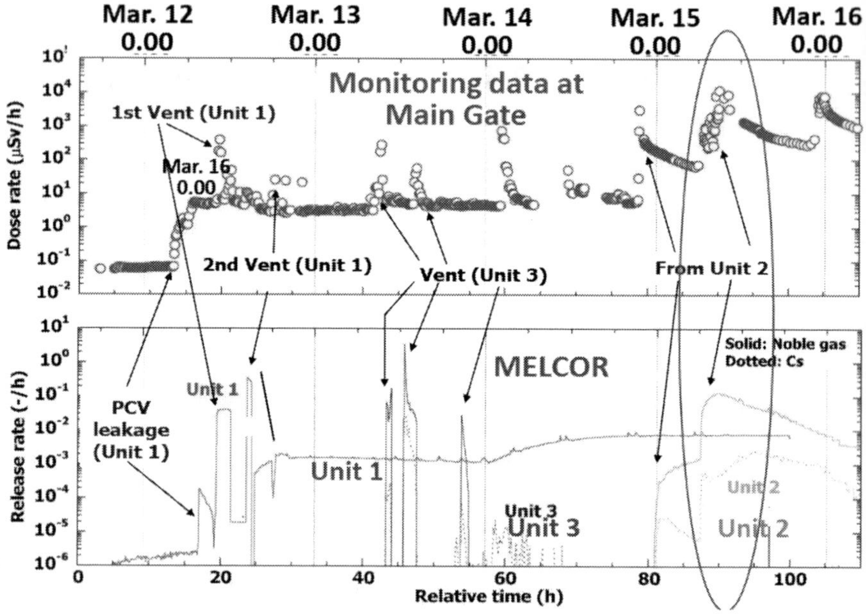

NRA, Japan
Nuclear Regulation Authority

Environmental Consequence Analysis

Hirano, Hoshi and Homma, Presented at U.S.NRC RIC, March 2013.

- By using the source terms with MELCOR, an environmental consequence analysis was done in JAEA with OSCAAR developed there.

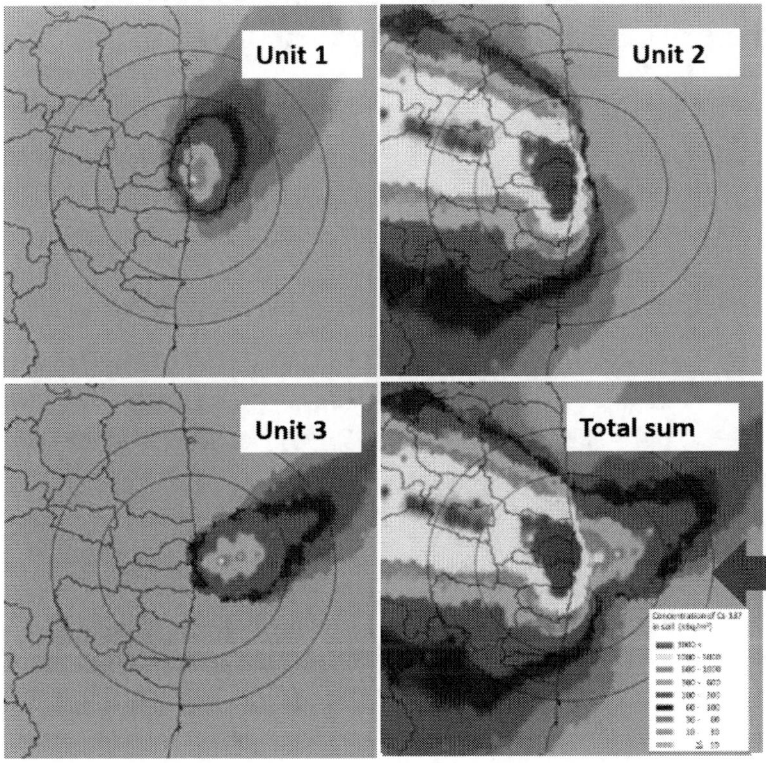

Major findings:

- The release from **Unit 2** on **March 15** had dominant contribution to the inland contamination.
- The release from **Units 1 and 3** mostly spread to the ocean.
- Sum of the release well reproduced the characteristics of the measured distribution of contamination.

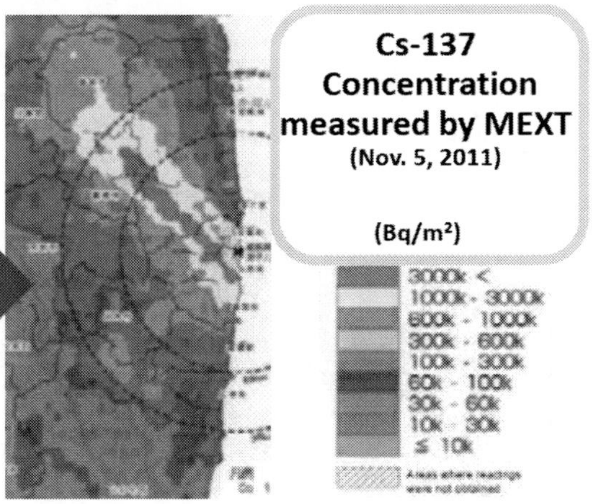

NRA, Japan
Nuclear Regulation Authority

Wind was whirling clockwise on March 15, 2011.

Ref.: Hirano, Presented at IAEA IEM9, 20–24 April 2015

- During the accident, the **SPEEDI** simulation had been done every hour assuming a **unit release (1 Bq/hour)** since the source terms were not available. ➡ The results were equivalent to the **meteorological data**.

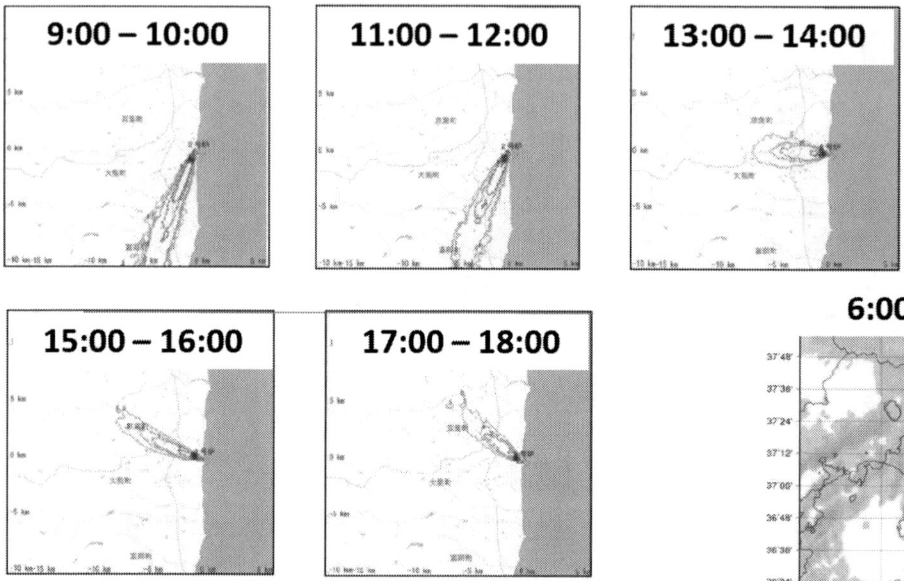

Final Report from Government's Investigation Committee (July 2012)

SPEEDI calculations Mar. 15, 2011

Source: M. Chino, JAEA, May 2013

Precipitation (Rainfall)

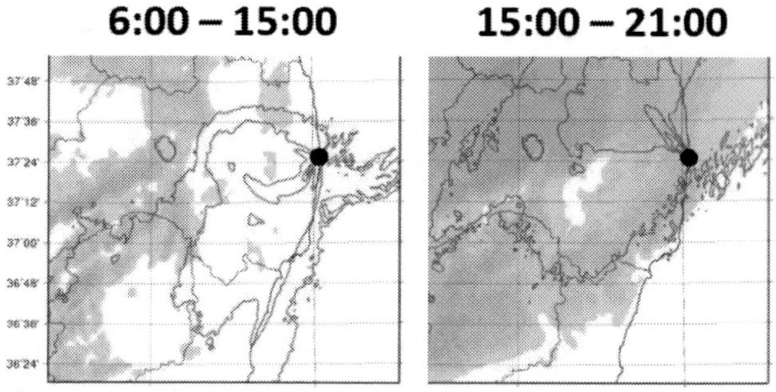

http://www.aec.go.jp/jicst/NC/iinkai/teirei/siryo2013/siryo18/siryo1.pdf

NRA, Japan
Nuclear Regulation Authority

Some Lessons Learned

NRA, Japan
Nuclear Regulation Authority

Lack of "Defense-in-Depth" against external hazards

- Large scale "common cause failures (CCFs)" by external hazards should have been prevented/mitigated.
  - We should have been prepared for "Low frequency, high consequence events" beyond design basis.
- **Actually, the Defense-in-Depth (DiD) had not been taken into consideration for external events.**
  - Watertight doors were not implemented, Seawater pumps were not protected. AM measures were not effective, ...
  - Training / drills assuming external events had not been conducted.
- **DiD has been and will continue to be an effective way to account for uncertainties.**
  - Effective independence between different layers of defense
  - More "diversity" and "independence" rather than "redundancy"

NRA, Japan
Nuclear Regulation Authority

"Tsunami risk had been recognized": Lack of Agility

Diet`s Report: NAIIC : The National Diet's Fukushima Nuclear Accident Independent Investigation Commission, Reported to the Diet in July 2012

- The Seismic Design Review Guide was revised in 2006 and a requirement was newly introduced against tsunami. So-called "back-check" was in progress, but its progress was very slow.
    - **"Uncertainty allows for wishful thinking" + "Safety myth"**
    - **Lack of agility, lack of safety culture**

Diet Report

- As the regulatory agency was aware of TEPCO's delaying of countermeasures, but did not follow up with any specific instructions. Nor did they properly supervise the back-check progress.
- When new findings indicate the possibility of a tsunami exceeding previous assumptions, the operator is required to **quickly implement countermeasures**, rather than taking time to clarify the scientific basis for that possibility through studies of sediment ....

TEPCP's "Trial Calculations" of Tsunami

IAEA Fukushima report, Technical Volume 2/5 (2015)

Seismicity Gap Area (no record of large earthquake):

- **Major opinion**: Majority of scientists believed the shallow plate boundary **along the Japan trench** was **not able to store a large amount of strain** (**coupling is weak**).

New findings / minor opinion

- In **2002, HERP*** pointed out the **possibility of a tsunami earthquake occurring anywhere along the Japan Trench including the gap area could not be ruled out.**
- **In 2008, TEPCO** did hypothetical calculation assuming a **M 8.2 earthquake** occurring **off the coast of Fukushima Pref.** ➡ **Max. ~15.7 m**
- TEPCO also conducted a **Jogan** (869)-**type tsunami** simulation ➡ **Max. ~9.2 m**
- However, **TEPCO** decided only on a plan to ask JSCE* for investigation such as **tsunami deposit survey, rather than take any immediate measures.**

*Headquarters for Earthquake Research Promotion

*JSCE : Japan Society of Civil Engineers

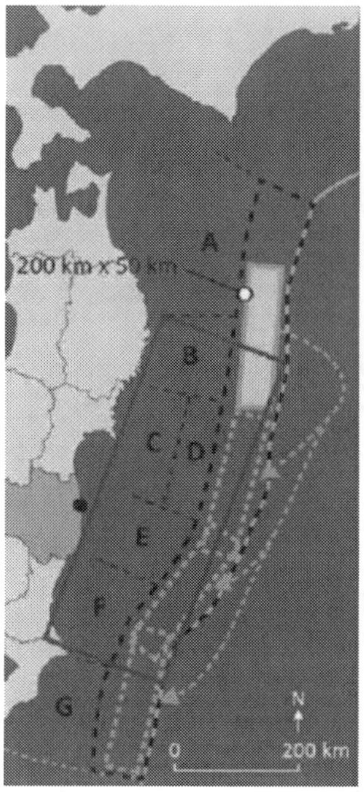

NRA, Japan
Nuclear Regulation Authority

Regulatory Aspects: "Regulatory Capture" Pointed out in the Diet's Report

Diet's Report (July 2012)

Message from Chairman

- ... this was a disaster "**Made in Japan.**" Its fundamental causes are to be found in the ... **Japanese culture**: our **reflexive obedience**; our **reluctance to question authority**; ... and our **insularity**.

Organizational issues ...

- ... actual relationship lacked independence and transparency, ... In fact, it was a typical example of "**regulatory capture**," ...

Lack of expertise

- ... the two incorporated technical agencies advising NISA, namely, **JNES and JAEA**, have been **too rigidly tied to NISA** ....

Conclusions

- ... **The** lack of expertise resulted in "**regulatory capture,**"... They **avoided their direct responsibilities** by **letting operators apply regulations on a voluntary basis.**

NRA, Japan
Nuclear Regulation Authority

Human / Organizational Aspects: "Myth of Safety" Pointed Out in the Government's Report

Final Report from Government's Investigation Committee (July 2012)

"Myth of Safety"

- … a fundamental problem of the inability to capture such crises as a reality that could happen in our lives; this, in turn, is the result of a **myth of safety** that existed among nuclear operators including TEPCO as well as the government, that serious severe accidents could never occur in NPPs in Japan.

"Basic Assumption" in the **IAEA DG Report**

- Because of the basic assumption that **NPPs in Japan were safe**, there was a tendency for organizations and their staff not to challenge the level of safety.
- The reinforced **basic assumption** among the stakeholders about the robustness of the technical design of NPPs resulted in a situation where safety improvements were not introduced promptly.

NRA, Japan
Nuclear Regulation Authority

"**Complacency**" together with "**Bureaucracy**" allowed "**Safety myth**" to prevail, having let "**Continuous improvement**" cease.

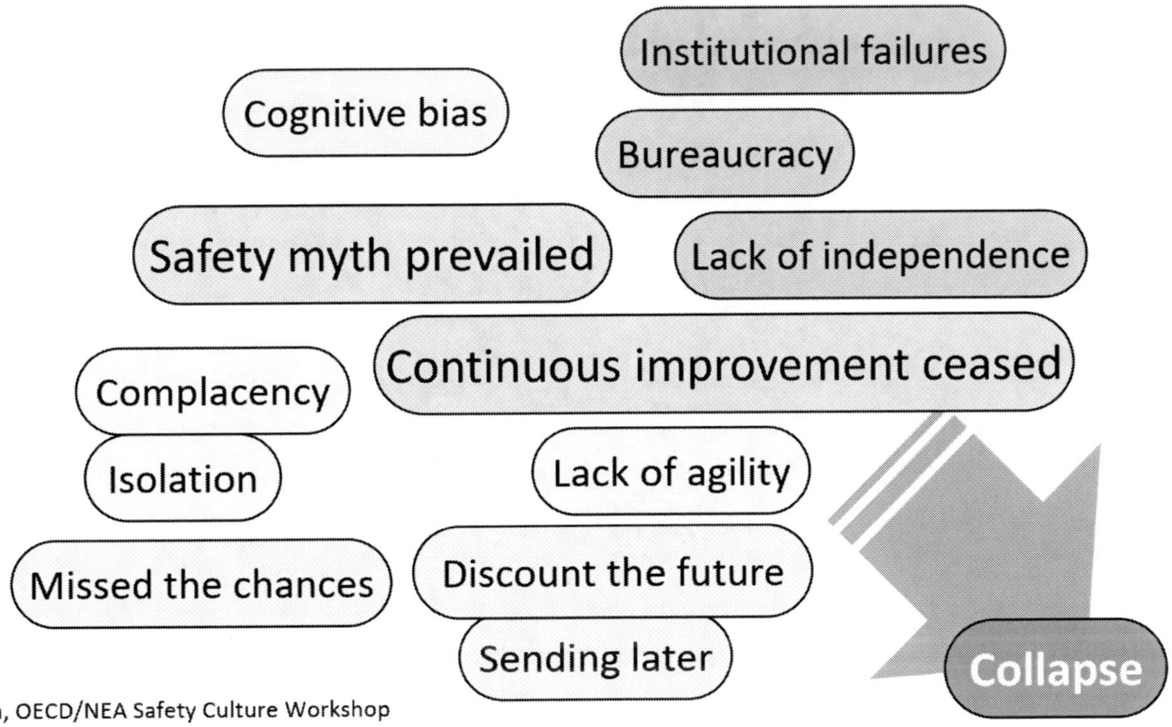

T. Fuketa, OECD/NEA Safety Culture Workshop (June 2015)

NRA , Japan
Nuclear Regulation Authority

Safety Enhancements

NRA, Japan
Nuclear Regulation Authority

NRA: Nuclear Regulation Authority

Established in Sept. 2012

■ Independence

Nuclear regulation and nuclear promotion were clearly separated, and the NRA was established as **an independent commission body** defined by law* affiliated with MOE (Minister of Environment).

* a council-system organization based on Article 3 of the National Government Organization Act, ensuring its independence without any control or supervision by other organizations.

■ Integrated

Nuclear regulation functions regarding safety, security, safeguards, radiation monitoring and radioisotopes were integrated into the NRA.

T. Fuketa, NRA, presented at U.S.NRC RIC2013, March 13, 2013: http://www.nsr.go.jp/data/000067048.pdf

Integrated and Independent

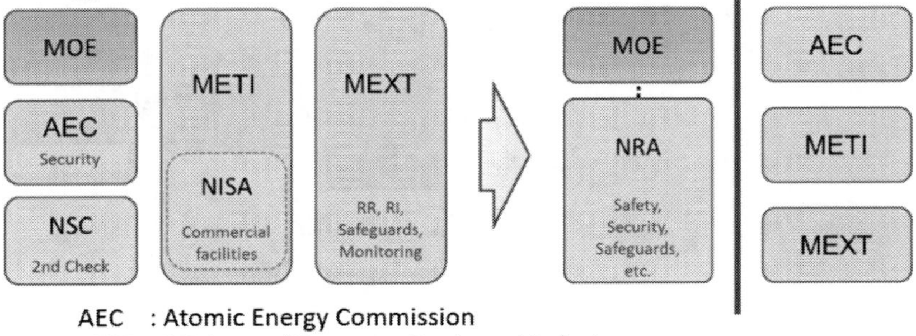

AEC : Atomic Energy Commission
METI : Ministry of Economy, Trade and Industry
MEXT : Ministry of Education, Culture, Sports, Science and Technology
MOE : Ministry of the Environment
NISA : Nuclear and Industrial Safety Agency (abolished)
NSC : Nuclear Safety Commission (abolished)

NRA, Japan
Nuclear Regulation Authority

New Regulatory Requirements: Structure

http://www.nsr.go.jp/data/000067048.pdf

Requirements for **B-DBA**

- **DEC**: Design extension conditions defined in IAEA SSR-2/1

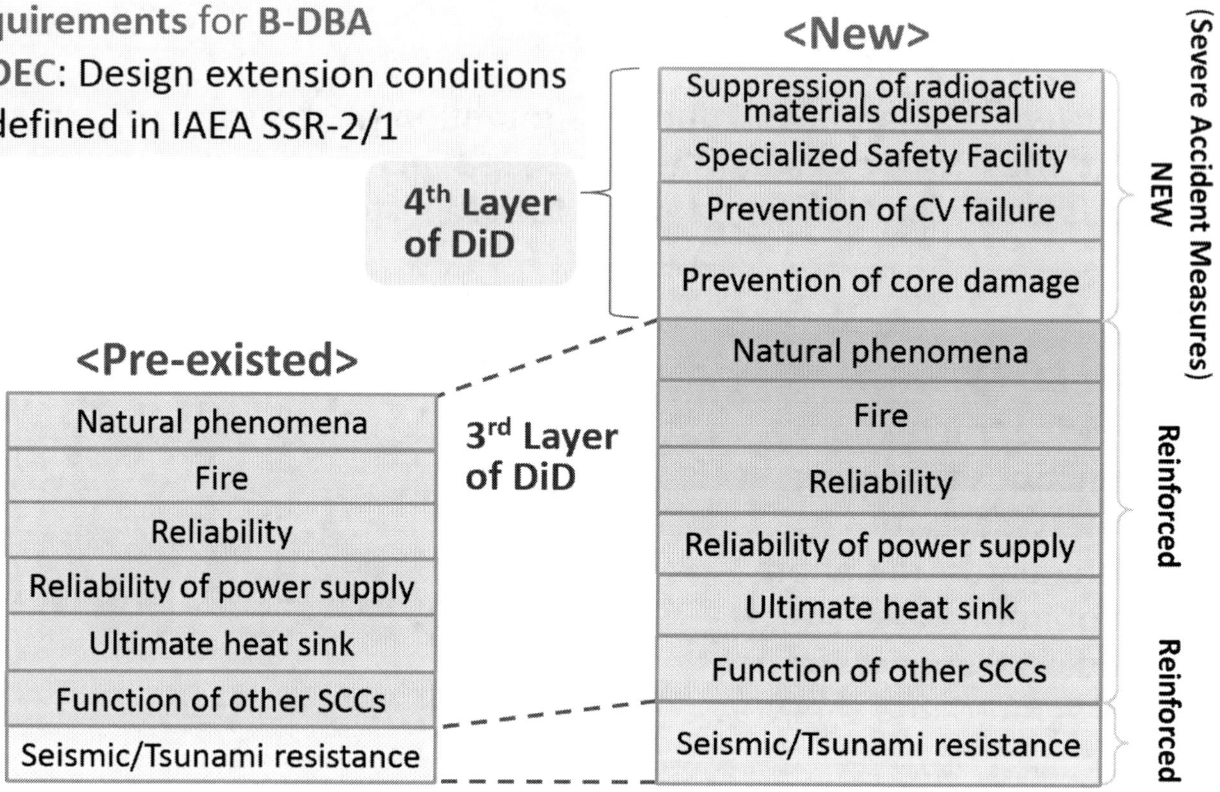

NRA, Japan Nuclear Regulation Authority

New Regulatory Requirements: **Enhanced Measures against Tsunami**

http://www.nsr.go.jp/data/000067048.pdf

More Stringent Standards on Tsunami ➡	It is required to define "design basis tsunami" that **exceeds the largest in the historical records** and to take **protective measures** such as **breakwater wall** based on it.
Enlarged Application of Higher Seismic Resistance ➡	SSCs for tsunami protective measures are **classified as Class S** equivalent to RPV etc. of **seismic design importance classification.**

Example of protective measures against tsunami (multiple measures)

- **Breakwater wall** for prevention of inundation to the site

- **Tsunami gate** for prevention of water penetration into the building

NRA, Japan
Nuclear Regulation Authority

New Regulatory Requirements:
Measures against Intentional Aircraft Crash, etc.

http://www.nsr.go.jp/data/000067048.pdf

"Specialized Safety Facility" is required to mitigate release of radioactive materials after core damage due to intentional aircraft crash.

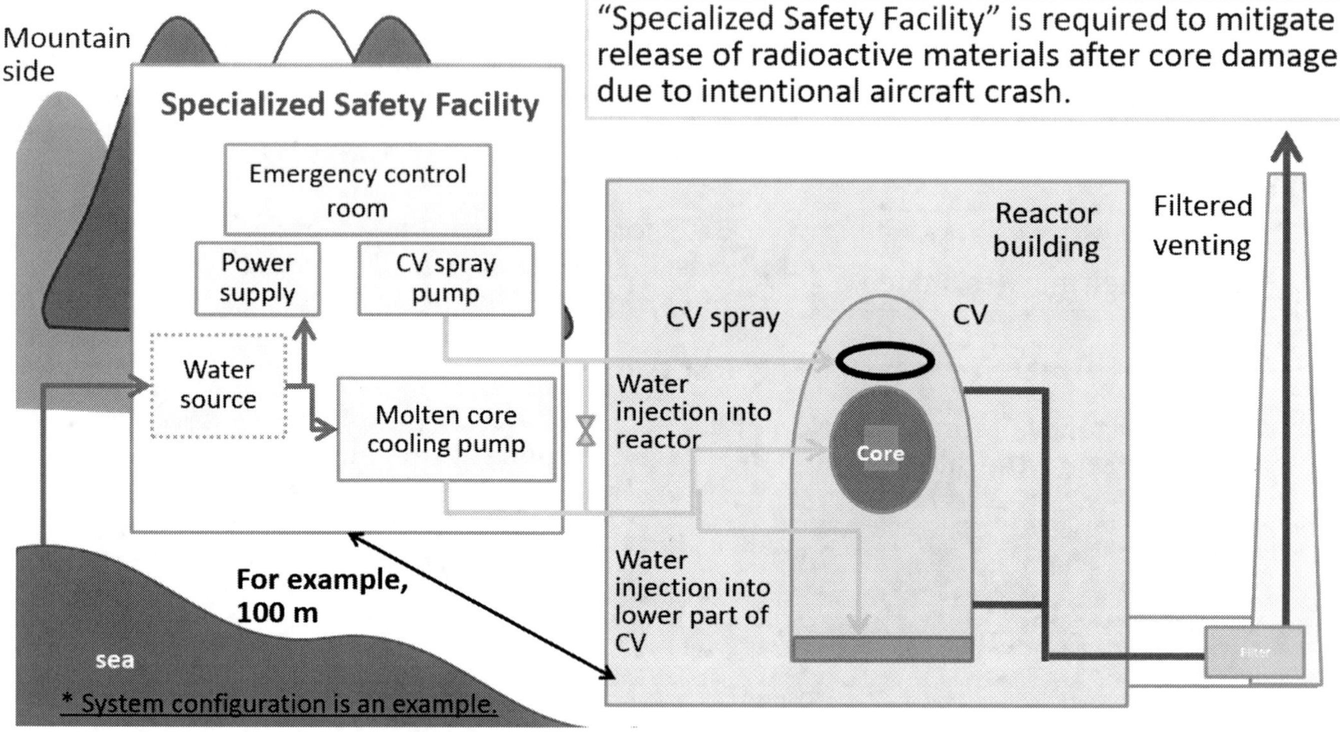

For BWR, one filtered venting for prevention of containment failure and another filtered venting of Specialized Safety Facility are acceptable solution.

NRA, Japan
Nuclear Regulation Authority

NRA's Statement of Nuclear Safety Culture

Issued on May 27, 2015

T. Fuketa, OECD/NEA Safety Culture Workshop (June 2015)

- NRA's Mission Statements

The NRA's Core Values and Principles

The Statement of Nuclear **Safety Culture**	The Code of Conduct on **Nuclear Security**

- **Eight Traits** of Nuclear Safety Culture in NRA's Statement
  1. Priority to Safety ➡ Break with "**safety myth**"
  2. Decision-making prioritized by safety
  3. Fostering, sustaining and strengthening safety culture
  4. Learning organization ➡ Seek out "**opportunities for improvement**"
  5. Communication ➡ Get rid of "**isolation**"/"**self-righteousness**"
  6. Questioning attitude ➡ Avoid "**complacency**"
  7. Rigorous and prudent judgment and action with agility
  8. Harmonization with nuclear security

NRA, Japan
Nuclear Regulation Authority

Summary

- One of the most important lessons learned is "**Lack of Defense-in-Depth against external hazards**". We need to continuously enhance DiD since **DiD has been and will continue to be an effective way to account for uncertainties.**
- We need to keep reminding that "**Complacency**" together with "**Bureaucracy**" allowed "**Safety myth**" to prevail, having let "**Continuous improvement" cease**.
- Commissioner Fuketa of NRA stated in the OECD/NEA SC Workshop*:
  - We, however, acknowledge we are oblivious. Numerous **sprouts of safety myth** reappear already.
  - We must incorporate lessons-learned into the "institutional memory" of the NRA. We must create an environment where **a gene letting us think "safety first" can survive**.

* T. Fuketa, OECD/NEA Safety Culture Workshop (June 2015)

References

- **1st Government Report (June 2011)**: Report of the Japanese Government to the IAEA Ministerial Conference on Nuclear Safety, Nuclear Emergency Response Headquarters, Government of Japan, June 2011
- **2nd Government Report (September 2011)**: Additional Report of the Japanese Government to the IAEA (Second Report), Nuclear Emergency Response Headquarters, Government of Japan, September 2011
- **Diet's Report (July 2012)**: The Official Report of the Fukushima Nuclear Accident Independent Investigation Commission (NAIIC), the National Diet of Japan, 2012
- **Interim Report from Government's Investigation Committee (Dec. 2011)**: Interim Report, Investigation Committee on the Accident at Fukushima NPS of TEPCO, Dec. 26, 2011
- **Final Report from Government's Investigation Committee (July 2012)**: Final report, Investigation Committee on the Accident at Fukushima NPS of TEPCO, July 23, 2012
- **IAEA DG Report and Technical Volumes 1 to 5 (2015)**: The Fukushima Daiichi Accident, Report by the Director General and Technical Volumes 1 to 5 (2015)
- **T. Fuketa, OECD/NEA Safety Culture Workshop (June 2015)**: OECD NEA/CNRA/ CSNI/CRPPH Joint Workshop on Challenges and Enhancements to Safety Culture of the Regulatory Body, Paris, June 3, 2015
- **Y. Shimizu, IAEA Effective Nuclear Regulatory Systems (April 2016)**: Y. Shimizu, "Lessons Learned from the Fukushima Daiichi Accident, Actions Taken and Challenges Ahead," presented at IAEA International Conference on Effective Nuclear Regulatory Systems, Vienna, Austria, April 11-15, 2016

Appendix

Key: It was difficult to open Air-Operated Valves (AOVs)

- For seawater injection by fire engines, it was necessary to depressurize RPV by opening SRVs. It was needed to open AOVs in the venting system to prevent PCV failure.
- AOVs need:
  - DC power to open the pilot valve (solenoid valve) ⬅ **Car batteries**
  - Compressed air ⬅ **Mobile compressor, nitrogen gas cylinders**

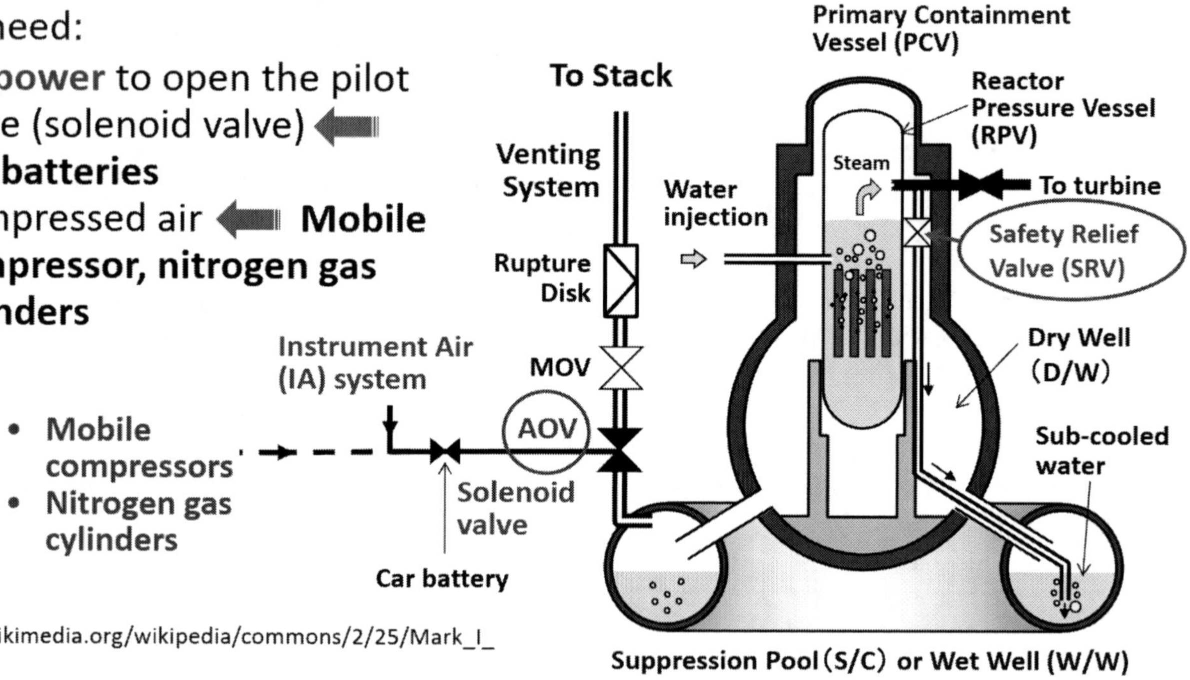

https://upload.wikimedia.org/wikipedia/commons/2/25/Mark_I_Containment.jpg

Appendix

Example of Use of Mobile Equipment:

Measures against Station Blackout (SBO)

✓ Install both mobile alternate power sources (power vehicle, batteries, etc.) and permanent alternate AC power sources.

- Batteries for 8 hours without load shedding + 16 hours with load shedding
- Alternate onsite AC power supply for 7 days
- External support from offsite by the 6th day
- Alternative MC, PC, MCC

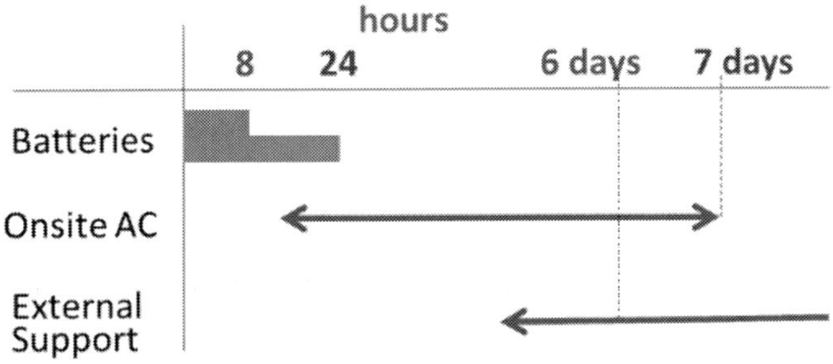

✓ It is also required to install the third station battery system.

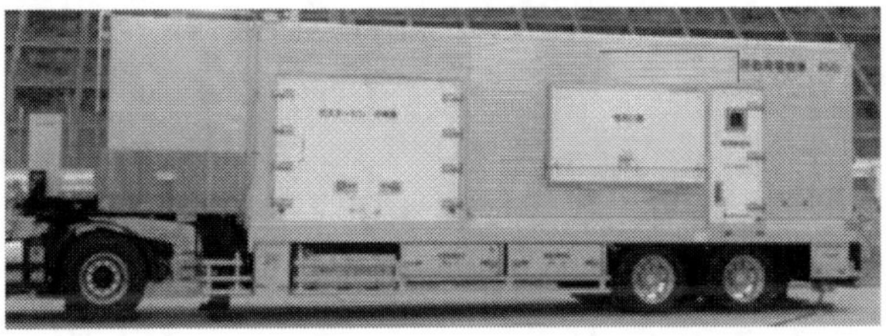

Alternative onsite AC power (Power Vehicle)

OPEN RADIATION: a collaborative project for radioactivity measurement in the environment by the public.

Jean-François Bottollier-Depois[1,*], *E.* Allain[2], *G.* Baumont[1], *N.* Berthelot[3], *I.* Clairand[1], *C.* Couvez[1], *G.* Darley[1], *B.* Henry[3], *T.* Jolivet[3], *P.* Laroche[4], *A.* Lebau-Livé[1], *V.* Lejeune[1], *J.* Miss[1], *W.* Monange[1], *F.* Quéinnec[1], *Y.* Richet[1], *C.* Simon[5], *F.* Trompier[1], *F.* Vayron[1]

1: *IRSN, Institute for radiological Protection and Nuclear Safety, 92262 Fontenay-aux-Roses, France*
2: *IFFoRME, Institut Français des Formateurs Risques Majeurs et protection de l'Environnement, 9 rue Jacques Louvel-Tessier, 75 010 Paris, France*
3: *Planète Sciences, 16, place Jacques Brel, 91130 Ris-Orangis, France*
4: *Agoralogie, 36 boulevard de la Bastille, 75012 Paris, France*
5: *UPMC, Pierre and Marie Curie University, 4, place Jussieu, 75005 Paris, France*

IRSN/PRP-HOM, France

Abstract. After the Fukushima accident, initiatives emerged from the public to carry out themselves measurements of the radioactivity in the environment with various devices, among which smartphones, and to share data and experiences through collaborative tools and social networks. Such measurements have two major interests, on the one hand, to enable each individual of the public to assess his own risk regarding the radioactivity and, on the other hand, to provide "real time" data from the field at various locations, especially in the early phase of an emergency situation, which could be very useful for the emergency management.

The objective of the OPENRADIATION project is to offer to the public the opportunity to be an actor for measurements of the radioactivity in the environment using connected dosimetric applications on smartphones. The challenge is to operate such a system on a sustainable basis in peaceful time and be useful in case of emergency. In "peaceful situation", this project is based on a collaborative approach with the aim to get complementary data to the existing ones, to consolidate the radiation background, to generate alerts in case of problem and to provide education & training and enhanced pedagogical approaches for a clear understanding of measures for the public. In case of emergency situation, data will be available "spontaneously" from the field in "real time" providing an opportunity for the emergency management and the communication with the public. ...

The practical objective is i) to develop a website centralising data from various systems/dosimeters, providing dose maps with raw and filtered data and creating dedicated areas for specific initiatives and exchanges of data and ii) to develop a data acquisition protocol and a dosimetric application using a connected dosimeter with a bluetooth connection.

This project is conducted within a partnership between organisms' representative of the scientific community and associations to create links with the public.

*Corresponding author: jeanfrancois.bottollier@irsn.fr

OPENRADIATION: a collaborative project for radioactivity measurements in the environment by the public

Jean-François BOTTOLLIER-DEPOIS (IRSN)
On behalf of the group

ICRS13 - RPSD2016
4 October 2016

E Allain[2], G Baumont[1], N Berthelot[3], I Clairand[1], C Couvez[1], G Darley[1], B Henry[3], T Jolivet[3], P Laroche[4], A Lebau-Livé[1], V Lejeune[1], J Miss[1], W Monange[1], F Quéinnec[1], Y Richet[1], C Simon[5], F Trompier[1], F Vayron[1]

[1] ***IRSN***, *Institute for Radiological Protection and Nuclear Safety, 92262 Fontenay-aux-Roses, France*

[2] ***IFFoRME***, *Institut Français des Formateurs Risques Majeurs et protection de l'Environnement, 75010 Paris, France*

[3] ***Planète Sciences***, *91130 Ris-Orangis, France*

[4] ***Agoralogie***, *75012 Paris, France*

[5] ***UPMC***, *Pierre and Marie Curie University, 75005 Paris, France*

open radiation

"Radiation dose to **researchers:**

- *a quantitative value of the **physical agent***

Radiation dose to **residents:**

- *a quantitative value of their **everyday life***

- Working together to support their self-help actions is the only way to help them to gain self confidence. Only then science has some meaning to people
- Provide the most accessible information
- Offer the possibility to each person to assess is own risk by doing himself measurements"

Prof. O. Niwa

open radiation

The context…

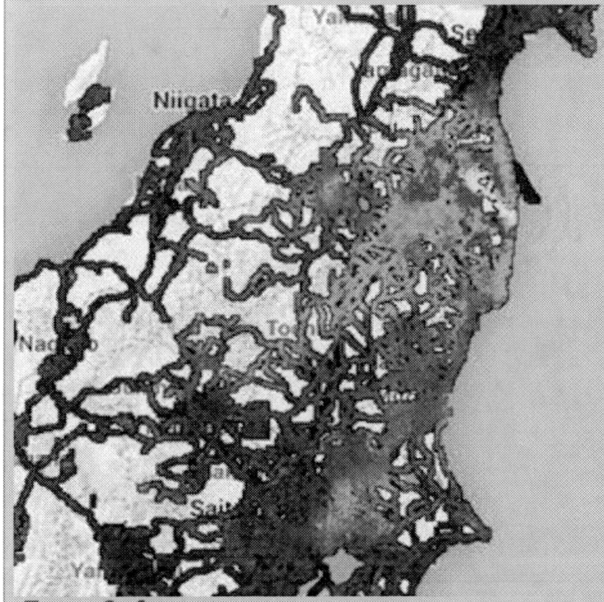

From Safecast

- After the Fukushima accident, **dosimetric applications** dedicated to the **public** appeared to realise **geo-localised measurements of the radioactivity** in the environment
- Today, such systems are used at a relatively large scale, in particular by the Japanese population, providing an interesting feedback

open radiation

… the context

In case of emergency situation: data from the population using various dosimetric applications

- ☺ Data coming **from the field in "real time"**
- ☺ Data provided **in mass…**
- ☹ … but with a **confident level not always very high**!
- ☹ **Important solicitation** of authorities/experts by the population foreseen

In « peaceful time »: data provided by the public in the frame of a collaborative approach & citizen science

- ☺ Data coming mainly from a **"motivated" public:** few% of the general population interesting by the topic, specific projects/studies (students, associations…)
- ☹ Data with a **confident level to be checked**
- ☹ **False alarms** may be generated

open radiation

What is the interest for the scientific community?

"Peaceful time"

- Offer to the public the possibility to contribute to a **citizen science** project on the **radioactivity measurement**
- Get **complementary data** to the existing ones and consolidate the background reference values
- Provide alerts in case of problem

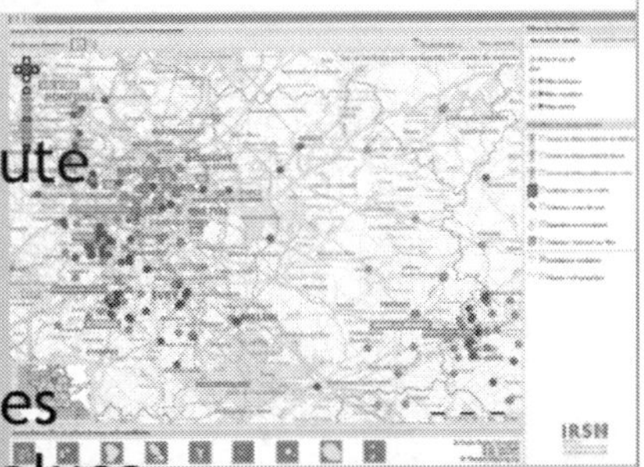

Emergency situation

- Get data contributing to the **management of emergency situations** (decision making, complementary to the existing measurements and the dispersion models…)
- Provide an **opportunity to communicate** with the public

open radiation

What is the interest for the public?

Collective use: "I perform measurements and send data for a collective use and to exchange information"

- *In normal situation*, by participating to a **collaborative project / citizen science**
- *In case of emergency*, by providing **spontaneously** data useful for the **stakeholders** involved in the **crisis management** and the **population**

Personal use: *"I perform measurements to assess my own risk, especially in case of emergency situation"*

➡ **These different modes have to be taken into account to develop an application for the public**

open radiation

What are the challenges?

Operate the system on a sustainable basis in peaceful time and useful in case of emergency situations

"Peaceful time"

- Set up a **pedagogic and sustainable system** used by the public
- Get data **robust** enough in **complement to the background** reference values
- Manage **alerts** in case of positive measurements
- Operate a **collaborative website**

Emergency situation

- Get data **robust** enough to contribute to the **management of emergency situations** (decision making…)
- Communicate with the **public**
- Operate the **website in "crisis" mode**

open radiation

The OPENRADIATION project

Collaborative project (open data & source) to measure the **radioactivity in the environment** involving the **public** and using **connected dosimetric applications** on smartphones

Develop an internet site

- **Centralise data** from **various** systems/dosimeters
- Provide dose maps with raw and "filtered" data
- Create **dedicated areas** for projects and information exchange

Develop a dosimetric application

- Develop a **connected dosimeter** using bluetooth (GM...)
- Develop a **smartphone application** to collect and transmit data

An open partnership

IRSN
INSTITUT
DE RADIOPROTECTION
ET DE SÛRETÉ NUCLÉAIRE

Institute for Radiological Protection and Nuclear Safety

Association involved in E&T for crisis management

Planete Sciences
une aventure pour les jeunes

Association involved citizen science for young people

UPMC
SORBONNE UNIVERSITÉS

University Pierre and Marie Curie

To be enlarged in the future:

Associations, academic partnerships, public representatives…

open radiation

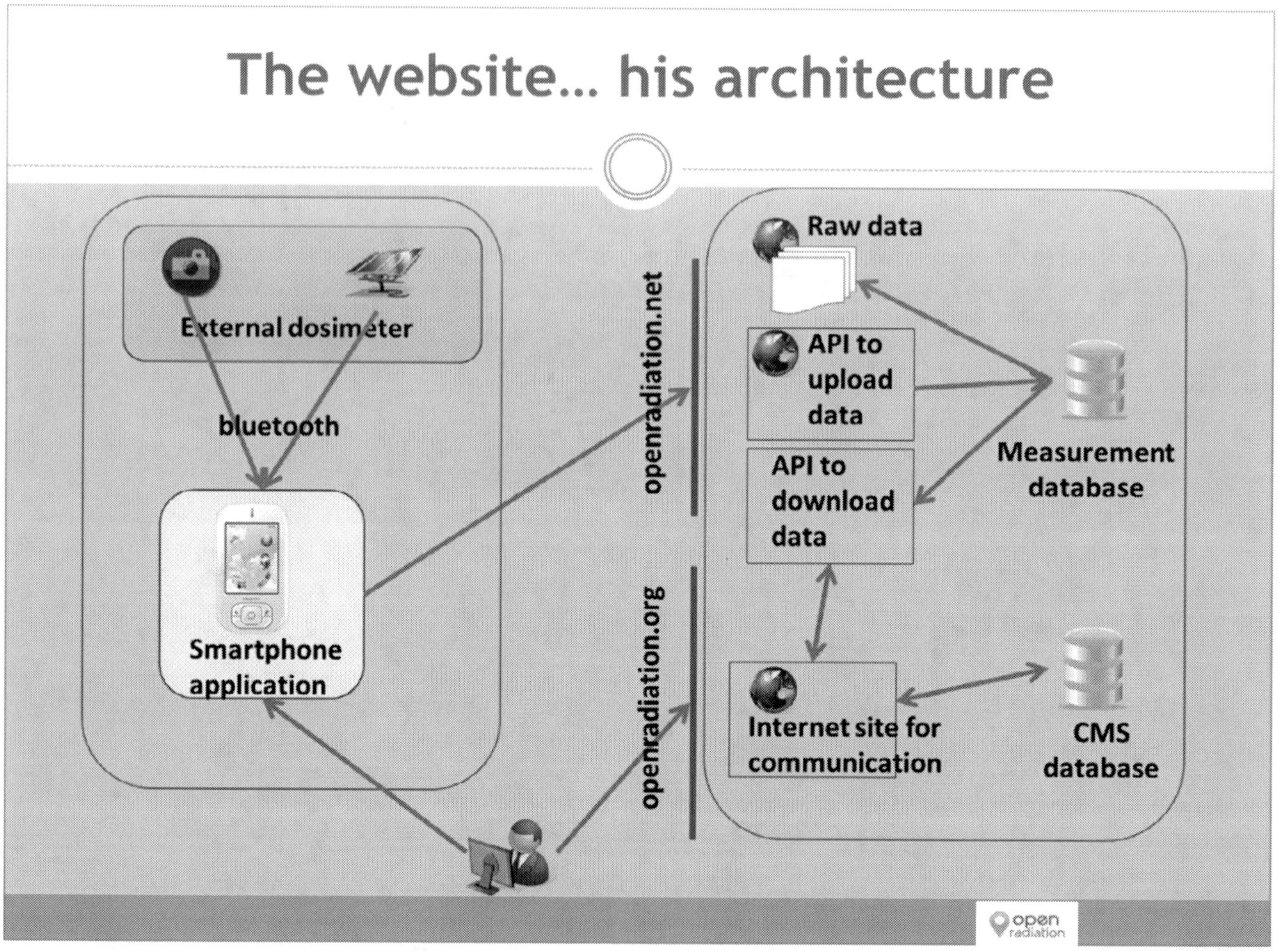
The website... his architecture
External dosimeter
bluetooth
Smartphone application
openradiation.net
openradiation.org
Raw data
API to upload data
API to download data
Measurement database
Internet site for communication
CMS database
open radiation

The website… under commissioning

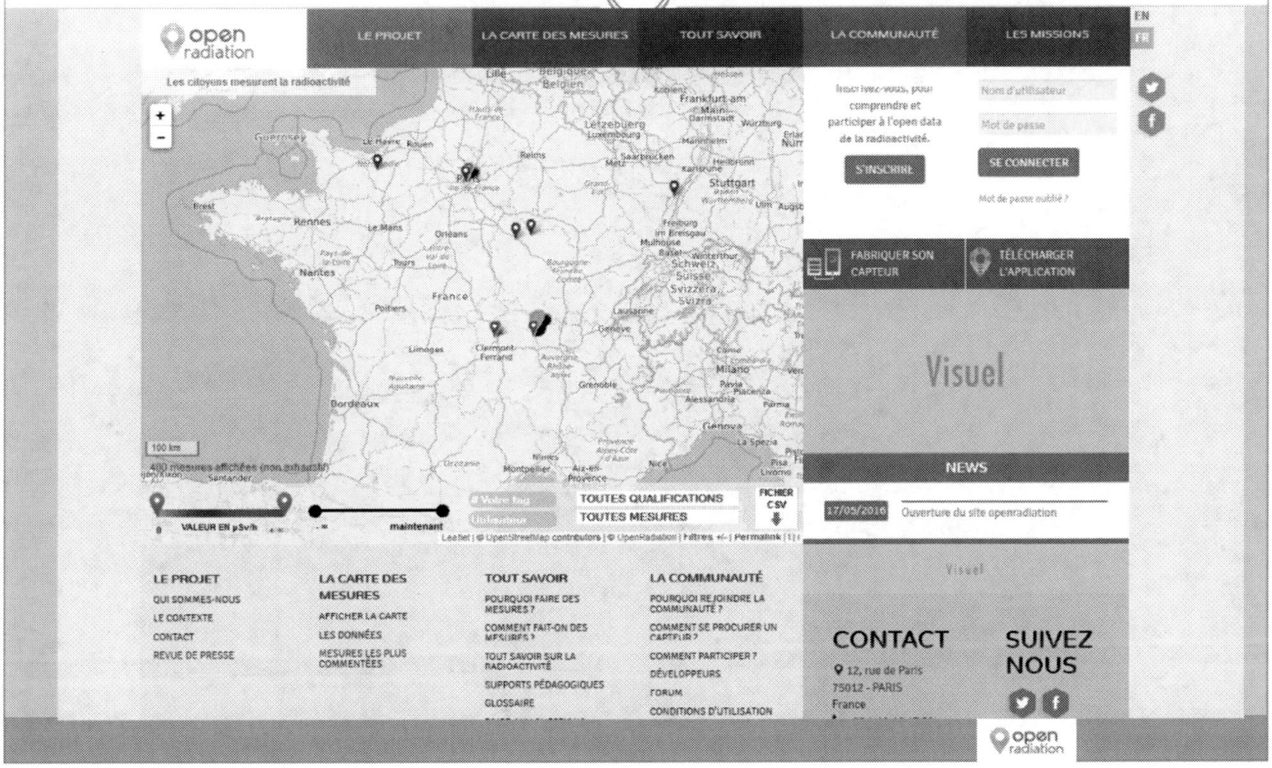

... the website

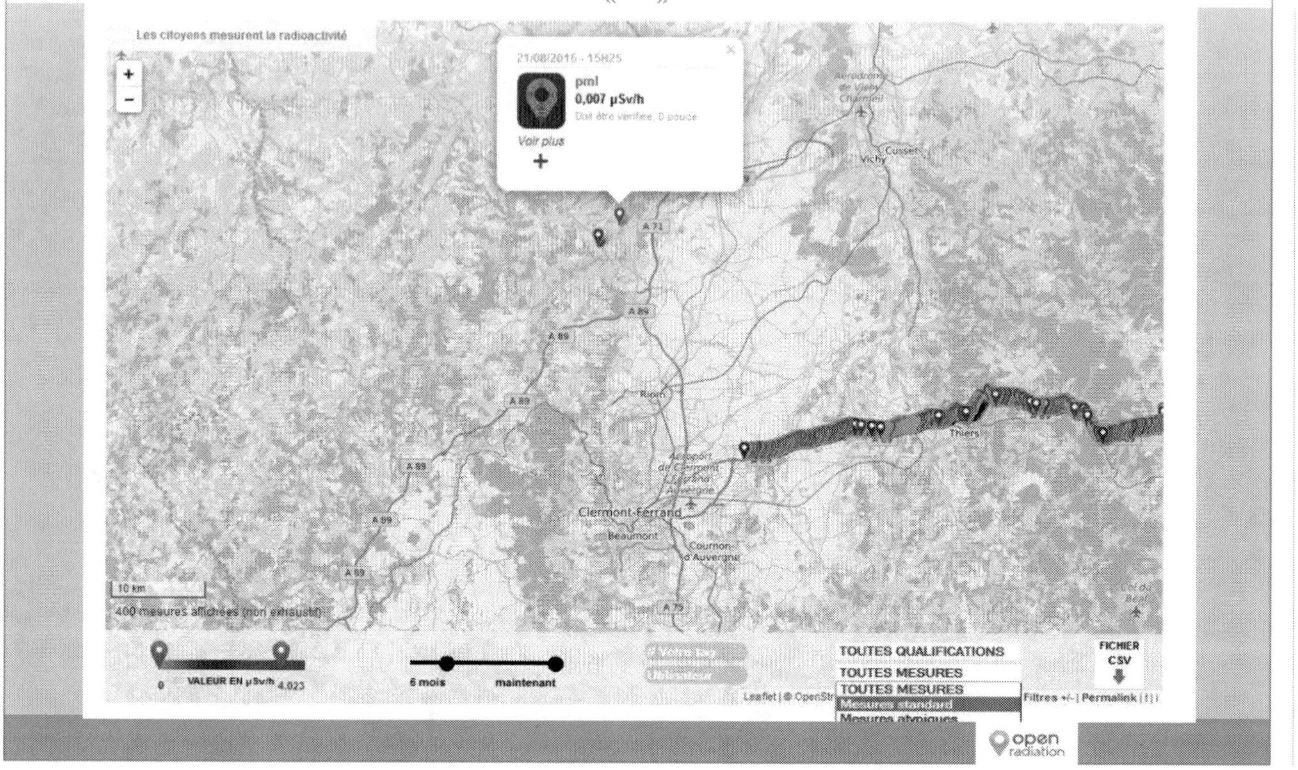
Les citoyens mesurent la radioactivité
21/08/2016 - 15h25
pml
0,007 µSv/h
Voir plus
Vichy
Cusset
Riom
Thiers
Clermont-Ferrand
Beaumont
Cournon-d'Auvergne
10 km
400 mesures affichées (non exhaustif)
VALEUR EN µSv/h
6 mois
maintenant
TOUTES QUALIFICATIONS
TOUTES MESURES
Mesures standard
FICHIER CSV
open radiation

... the website

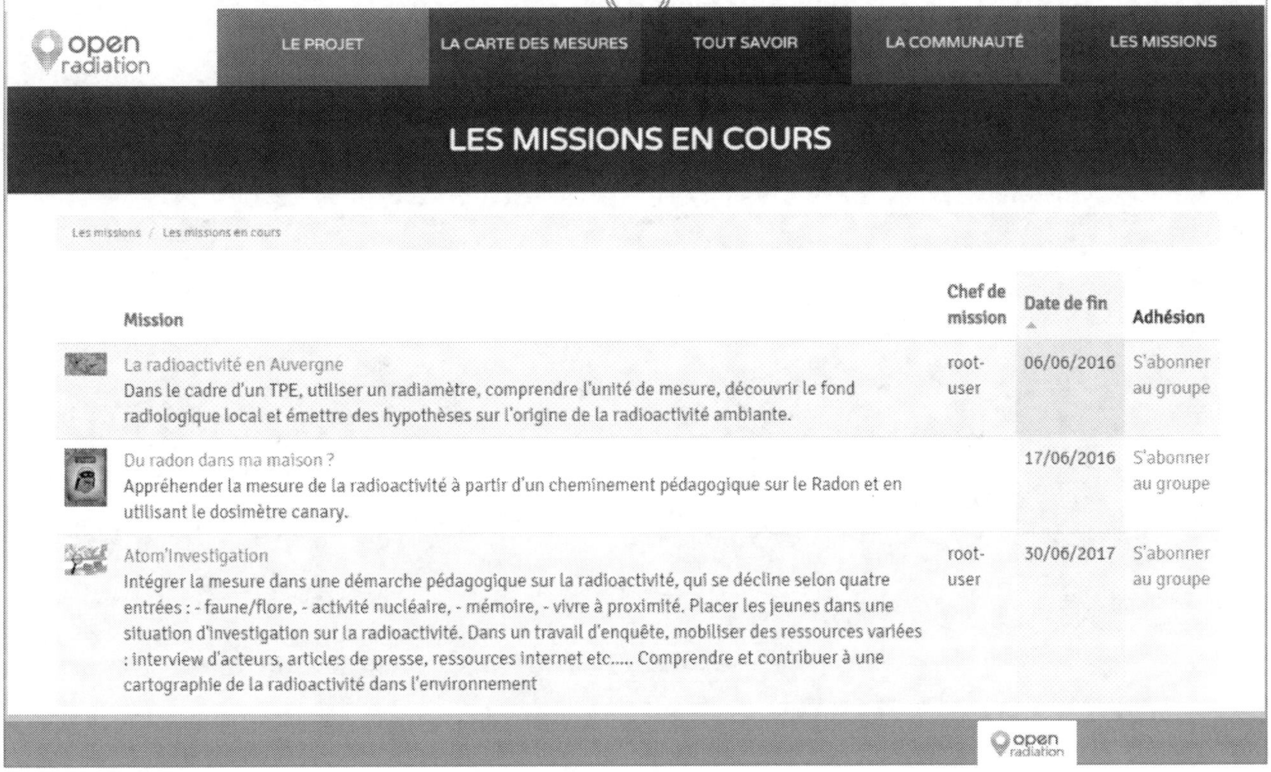

open radiation

LE PROJET | LA CARTE DES MESURES | TOUT SAVOIR | LA COMMUNAUTÉ | LES MISSIONS

LES MISSIONS EN COURS

Les missions / Les missions en cours

Mission	Chef de mission	Date de fin	Adhésion
La radioactivité en Auvergne Dans le cadre d'un TPE, utiliser un radiamètre, comprendre l'unité de mesure, découvrir le fond radiologique local et émettre des hypothèses sur l'origine de la radioactivité ambiante.	root-user	06/06/2016	S'abonner au groupe
Du radon dans ma maison ? Appréhender la mesure de la radioactivité à partir d'un cheminement pédagogique sur le Radon et en utilisant le dosimètre canary.		17/06/2016	S'abonner au groupe
Atom'Investigation Intégrer la mesure dans une démarche pédagogique sur la radioactivité, qui se décline selon quatre entrées : - faune/flore, - activité nucléaire, - mémoire, - vivre à proximité. Placer les jeunes dans une situation d'investigation sur la radioactivité. Dans un travail d'enquête, mobiliser des ressources variées : interview d'acteurs, articles de presse, ressources internet etc..... Comprendre et contribuer à une cartographie de la radioactivité dans l'environnement	root-user	30/06/2017	S'abonner au groupe

open radiation

The Open Geiger dosimeter

- GM counter (silicon diode in the future)
- Bluetooth connection
- Application on iPhone, Android and tablet
- Collaboration UPMC/IRSN/Planète Science
- 2 versions: "kit" and "packaged"

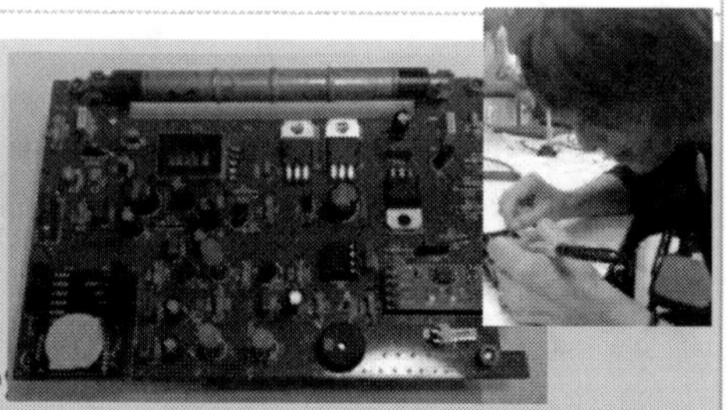

« kit » prototype

FabLab de Sorbonne Universités

« packaged » prototype

open radiation

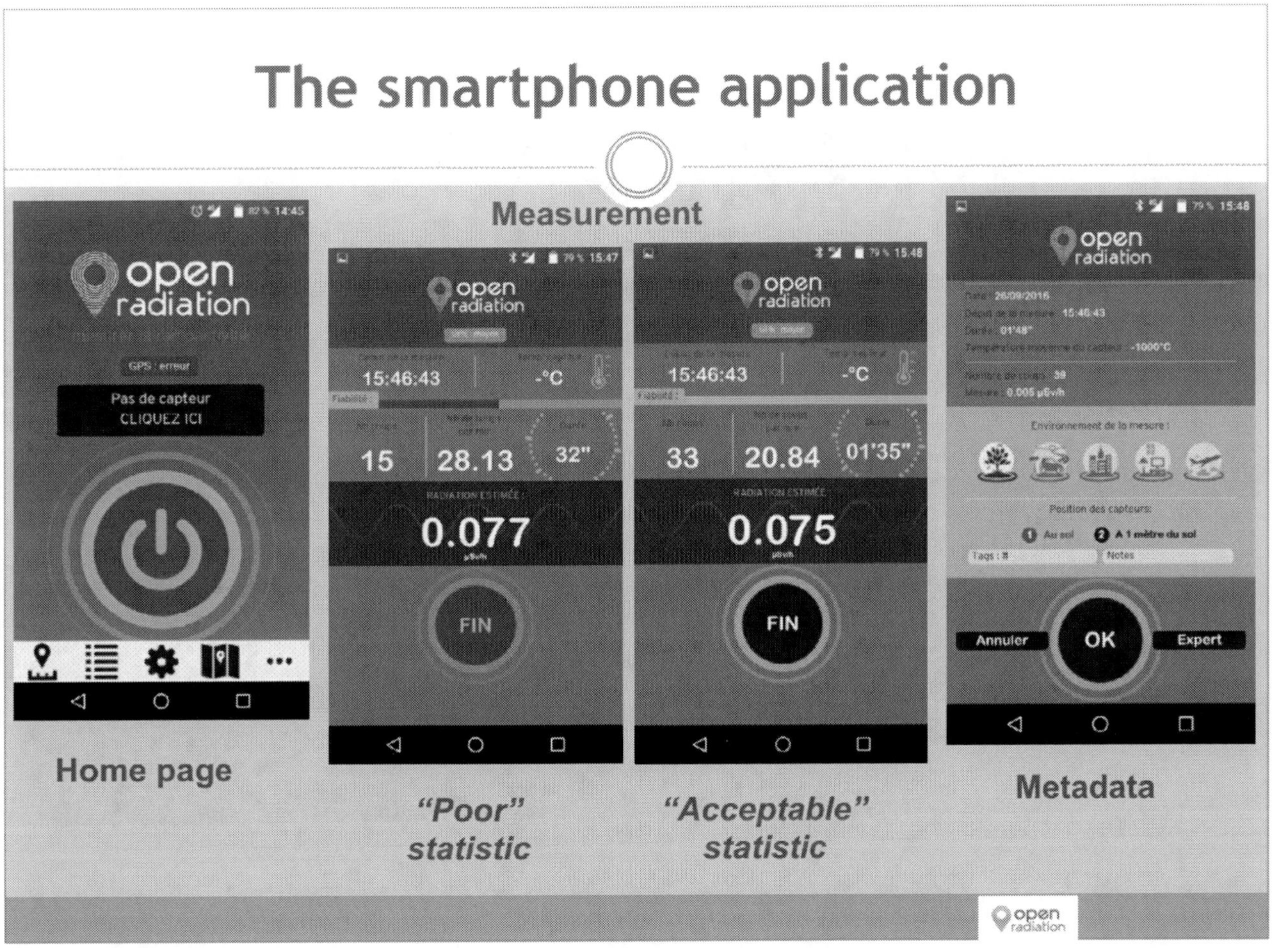
The smartphone application
Measurement
open radiation
Pas de capteur
CLIQUEZ ICI
Home page
15:46:43
-°C
15
28.13
32"
0.077
FIN
"Poor" statistic
15:46:43
-°C
33
20.84
01'35"
0.075
FIN
"Acceptable" statistic
open radiation
Annuler
OK
Expert
Metadata
open radiation

Other dosimeters

Other dosimeters can be used using a **specific interface** or enter **manually** measurements

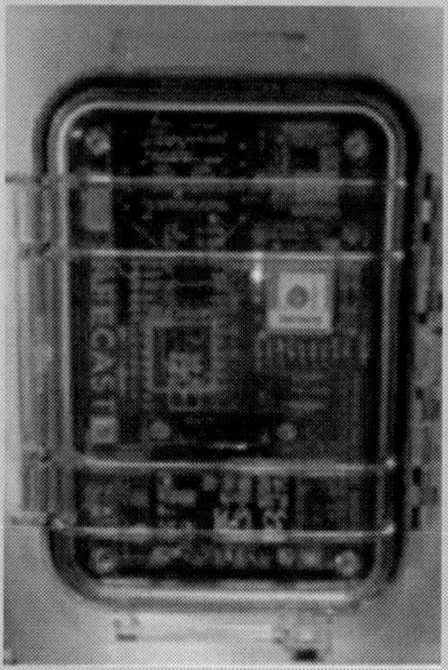

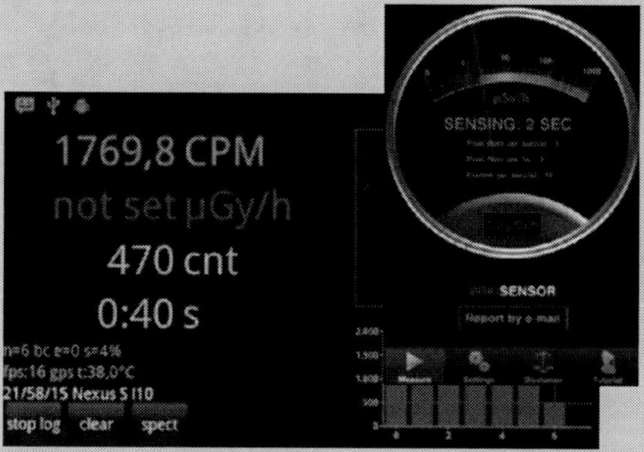

Specific interface: Safecast

open radiation

Around the project today

- **Projects with students:** Measurements with high schools from *Vichy et Villeneuve-sur-Lot, project « Atom' investigation » with high schools from Perpignan et de Dieppe, contact with Fukushima high school…*
- **Contact with other collaborative projects:** *Safecast, D-Shutle…*
- **Presentation of the project to potential users in France:** *ANCCLI (local committees), Paris district…*

Milestones and futures actions

- **February 2017:** *beta version available*
- **May-June 2017:** *site and dosimetric application available for the public*
- **2017-2018:** *definition and implementation of protocols for data analysis*
- **2017-2018:** *development of an application using the CMOS camera*

Thank you for your attention

ご清聴ありがとうございました

open radiation

Temporal Change of Environmental Contamination Conditions in Five Years after the Fukushima Accident

Kimiaki Saito[1,a]

[1]Japan Atomic Energy Agency, Fukushima Environmental Safety Center, 2-2-2 Uchisaiwai-cho, Chiyoda-ku, Tokyo 100-8577, Japan

Abstract. The temporal change of environmental contamination conditions after the Fukushima accident have been clarified based on large-scale environmental monitoring data repeatedly obtained in the 80 km zone. The decreasing tendency of air dose rates was confirmed to obviously depend on land uses. In human-related diverse environments the air dose rates have decreased much faster than the physical decay of radiocesium. The horizontal movement of radiocesium in undisturbed fields were found to be generally quite small, though it has gradually penetrated into the deeper parts of the ground.

1 Introduction

In the Fukushima accident, a large amount of radionuclides were released into the atmosphere and deposited over wide areas in East Japan. In order to obtain reliable and detailed environmental radiation data, large-scale environmental monitoring activities have been repeatedly performed in collaboration with many organizations in so-called mapping projects commissioned by the national government [1,2]. The data on air dose rates and radionuclide ground deposition accumulated in the projects have clarified the characteristics of the contamination conditions around the Fukushima site and their temporal changes. This paper summarizes the obtained results.

2 Large-scale environmental monitoring

2.1 Dose rate in air

Dose rates in air in terms of ambient dose equivalent rates have been measured in large scale mainly using the following five methods. The first four methods have been utilized in the mapping projects [1.2] and the last helicopter surveys have been conducted in other projects [3].

2.1.1 Fixed-location measurements above undisturbed flat fields

Air dose rates are measured with calibrated survey meters above undisturbed fields where the conditions are expected not to change for a long-time. The measurements aims to investigate natural weathering effects in reference locations. At about 6,500 locations in the 80 km zone, the measurements are being performed.

2.1.2 Car-borne survey on roads

Car-borne surveys are carried out over wide areas in East Japan [4] with a mobile survey system KURAM-II [5]. Simultaneous usage of more than one hundred systems in collaboration with municipalities enabled us to construct an air dose rate map over wide areas within a short time. A total distance of 30,000 – 80,000 km in East Japan have been covered in each survey.

2.1.3 Walk survey in human-related environments

In a walk survey, a person carrying a KURAMA-II system walks around in diverse environments related to human activities. Walk surveys provide important data from a viewpoint of exposure doses to inhabitants. About 600 areas of 1 km square in the 80 km zone are selected for walk surveys.

2.1.4 Unmanned helicopter survey

Unmanned helicopter surveys are employed to obtain data in the 5 km zone where crews are difficult to enter on land [6]. This method has better positional resolution than usual helicopter survey since the flying height is about 80 m.

2.1.5 Helicopter survey

Helicopter surveys have targeted for wide areas ranging from the 80 km zone to the whole Japan depending on the campaign. This method can obtain data over wide areas even though the data contain some uncertainties.

[a] Corresponding author:saito.kimiaki@jaea.go.jp

2.2 Radionuclides ground deposition

Radionuclide deposition densities in Bq/m^2 and depth profiles in Ba/kg for different soil layers have been investigated.

2.2.1 Radionuclide deposition density

In the first campaign, more than 11,000 soil samples were collected at more than 2,200 locations within the 100 km zone and the rest of Fukushima prefecture [7,8]. In principle, five samples were collected at one location. The collected soil samples were analysed at 22 institutes using Ge detectors for gamma-emitting radionuclides. Further, selected one hundred samples were analysed for alpha- and beta- emitting radionuclides.

Since the second campaign, *in situ* measurements with a portable Ge detector have been utilized which can determine the average deposition density of a location by detecting gamma rays from wide area [9].

2.2.2 Depth profile

Soil samples for different depth layers are collected using a scraper plate which can minimize cross contaminations among the samples [10]. The thickness of collected soil layer is changed from 5 mm to a few cm according to the condition. Collected samples are analysed using a Ge detector.

3 Initial contamination conditions

Among many radionuclides released from the Fukushima accident, the following radionuclides were detected in soil at a number of locations sufficient to show on a map: ^{134}Cs, ^{137}Cs, ^{131}I, ^{132m}Te, ^{129m}Ag, ^{89}Sr, ^{90}Sr, ^{238}Pu, $^{239+240}$Pu. A part of detected strontium and plutonium were judged to originate from the accident on the basis of their isotope ratios; however, the deposition densities were comparable to the levels of background radionuclides due to the nuclear weapon tests.

Effective doses for 50 years due to deposited radionuclides were conservatively evaluated from the observed maximum deposition densities using dose conversion coefficients shown in TECDOC-1162 [11]. Radiocesium were confirmed to be far more important than other radionuclides from a dosimetric point of view. The doses from radiostrontium and plutonium were negligible in comparison.

The total deposition amount of ^{137}Cs on the Japanese main island was estimated based on the results of the first and second campaigns as of June 2011 and March 2012, respectively. The first campaign targeted the region near Fukushima prefecture [7], while the second campaign covered wide areas in East Japan with in situ measurements [9]. These data were integrated together with helicopter survey data. It was estimated that 1.6x10^{15} Bq of ^{137}Cs was deposited within the 80 km zone and 2.0x10^{15} Bq on the main island. The total release of ^{137}Cs have been evaluated by several researchers; in most cases, the evaluated release was a little larger than 1x10^{16} Bq [12]. Therefore, it is considered that about 20% or a little less of the released ^{137}Cs was deposited on land in Japan.

The proportion of ^{137}Cs deposition for each land use to the total deposition was calculated within the 80 km zone referring to the National Land Numerical Information [13]. Forested areas occupy a large part of Fukushima prefecture. The proportion of ^{137}Cs deposition was found to be proportional to the area if we consider the whole 80 km zone: About 70% of ^{137}Cs was estimated to be deposited in forest; 20% in agricultural land; 5% in building site. The feature is different when we consider a smaller region. For example, in case of Fukushima city, the proportion of deposition in building site was as twice as that expected from the area.

4 Temporal changes

Temporal changes of contamination conditions were analysed in the 80 km zone having severe contamination on the basis of large-scale environmental monitoring data accumulated from 2011 to 2015 [1].

4.1 Dose rate in air

The decreasing tendency of dose rates in air was quite different due to the condition. Figure 1 indicates the time-dependent average air dose rates measured by different methods. Roughly speaking, the air dose rate above roads measured by car-borne surveys became 1/5 in four years since June 2011; that above undisturbed flat fields became 1/4; while, the air dose rate was estimated not to decrease up to 2/5 simply with the physical deay.

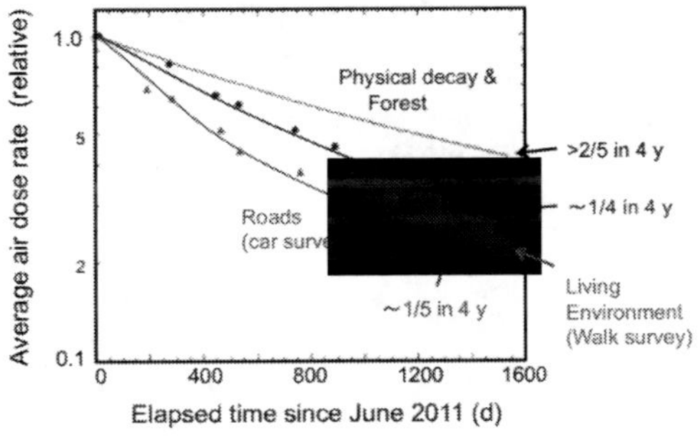

Figure 1. Decreasing tendency of the average air dose rate

Since walk surveys started in 2013, this kind of estimation is impossible; nevertheless, it has been made clear that air dose rates by walk surveys are higher than those above roads and lower than those above undisturbed flat fields on average. Thus, air dose rates in environments related to human activities were found to have decreased much faster than the physical decay of radiocesium.

In contrast, air dose rates in pure forest were found to have decreased nearly according to the physical decay as a whole, though the species of trees may slightly affect the decreasing tendency. The average air dose rate measured by helicopter surveys was reported to have

 DOI: 10.1051/epjconf/201715308003

become 1/3 in four years. This is because the helicopter surveys cover the whole area in the 80 km zone including wide forested areas.

Statistical analyses of car-borne data suggested two important factors affecting the reduction tendency of air dose rates: land uses and human activities. It is already well-known in the past studies that land uses affect the air dose rate reduction tendency. The knowledge obtained after the Chernobyl accident has been summarized as location factors which define time-dependent air dose rate reduction for different land uses [14].

Generally, the location factors have been determined by a limited number of data; therefor, the predicted air dose rates with current location factors have some uncertainties. In the studies performed after the Fukushima accident, massive car-borne survey data have been statistically analyzed, and this enabled to predict the air dose rate trend for each land use with the confidential intervals [15].

The air dose rate reduction tendency is obviously different between inside and outside the evacuation zone even if areas with the same land use are compared. This suggests that human activities are an important factor which decides air dose reduction; however, comprehensive quantitative analysis has not been done in connection with statistical quantities, which must be a future challenge.

4.2 Radionuclide deposition

4.2.1 Deposition density

Radiocesium deposition densities in undisturbed flat fields in terms of Bq/m^2 were analysed excluding the locations where decontamination activities were considered to be performed or ground conditions were changed somehow. Figure 2 gives the time-dependent average deposition densities for ^{134}Cs and ^{137}Cs. For the both nuclides, the deposition density has decreased according to the physical decay. It is inferred that the movement of radiocesium in horizontal directions in undisturbed fields is quite small.

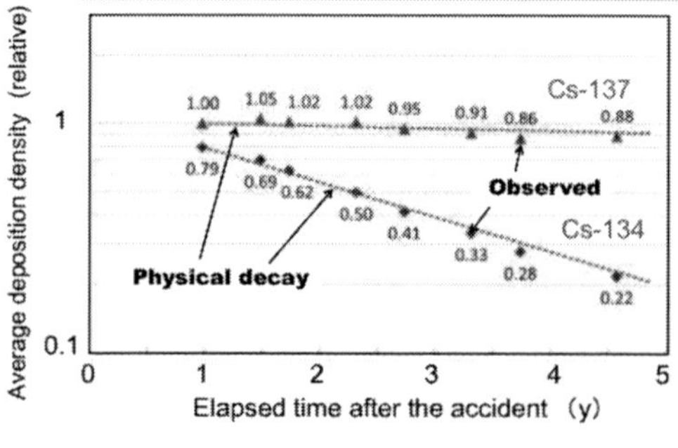

Figure 2. Time-dependent radiocesium deposition density

The knowledge obtained from radiocesium migration studies supports this tendency. It was found in the field measurements that soil erosion plays an important role in radiocesium migration in the natural environment but an annual radiocesium discharge rate is very small, generally an order of 0.1% or less [1].

4.2.2 Depth profile

Two kinds of typical depth profiles in undisturbed fields have been observed in the 80 km zone. The first one is an exponential distribution where the radiocesium concentration decreases monotonously with depth. The other is a distribution having a concentration peak at a certain depth. The latter distribution was found to be well approximated by a hyperbolic secant (sech) function [10]. Further, in some locations, disturbed distributions were found. Most of these disturbed distributions were considered due to decontamination activities or some human activities to change the ground conditions.

The 90% depth (L_{90}) defined as the depth up to which 90% of deposited radiocesium exist was calculated for the all depth profiles except disturbed distributions. Figure 3 indicates the observed 90% depths as a function of elapsed time. The whole data are shown by red or blue marks, and the average at each sampling time is by a green mark. Though the average L_{90} has increased with time, the average L_{90} in 2015 was still smaller than 5 cm. Decontamination by removing soil up to a depth of 5 cm is considered to be effective even now.

The penetration of deposited radiocesium into the ground has increased the shielding effect of soil for gamma rays. This is the main reason that air dose rates above undisturbed fields have reduced faster than the physical decay.

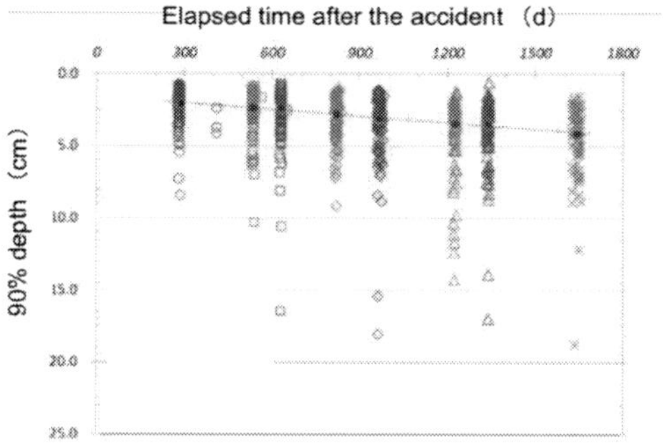

Figure 3. Increase of L_{90} with elapsed time

5 Conclusion

A huge amount of data periodically obtained in the mapping projects after the Fukushima accident have clarified the specific features of the time-dependent contamination conditions. The dose rate reduction was found to depend on land uses and human activities. The air dose rates in diverse environments connected with human living have decreased much faster than the physical decay of radiocesium. On the other hand, in pure forest environments, air dose rates have decreased according to the physical decay. The horizontal movement of radiocesium deposited on undisturbed fields

 DOI: 10.1051/epjconf/201715308003

is small while gradually penetrating into the ground. This results in air dose rate reduction faster than the physical decay because of increase in gamma-ray shielding effect by soil.

Acknowledgement

The data indicated in this paper are mainly from the mapping projects funded by the Ministry of Education, Sports, Science and Technology (MEXT) and by the Nuclear Regulation Authority (NRA).

References

1. NRA, http://radioactivity.nsr.go.jp/ja/list/338/list-1.html (in Japanese)
2. K. Saito, Y. Onda, J. Environ. Radioact. **139**, 240-249 (2015).
3. MEXT, http://fukushima.jaea.go.jp/initiatives/cat03/
4. M. Ando, Y. Nakahara, S. Tsuda, T. Yoshida, N. Matsuda, F. Takahashi, S. Mikami, N. Kinouchi, T. Sato, M. Tanigaki, K. Takamiya, N. Sato, R. Okumura, Y. Uchihori, K. Saito, J. Environ. Radioact. **139**, 266-280 (2015)
5. S. Tsuda, T. Yoshida, K.Saito, J. Environ. Radioact. **139**, 260-265 (2015)
6. Y. Sanada, T. Torii, J. Environ. Radioact. **139**, 294-299 (2015)
7. K. Saito, I. Tanihata, M. Fujiwara, T. Saito, S. Shimoura, T. Otsuka, Y. Onda, M. Hoshi, Y. Ikeuchi, F. Takahashi, N. Kinouchi, J. Saegusa, A. Seki, H. Takemiya, T. Shibata, J. Environ. Radioact. **139**, 308-319 (2015)
8. Y. Onda, H. Kato, M. Hoshi, Y. Takahashi, M. Nguyen, J. Environ. Radioact. **139**, 300-307 (2015)
9. S. Mikami, T. Maeyama, Y. Hoshide, R. Sakamoto, S. Sato, N. Okuda, S. Demongeot, R. Gurriaran, Y. Uwamino, H. Kato, M. Fujiwara, T. Sato, H. Takemiya, K. Saito, J. Environ. Radioact. **139**, 320-343 (2015).
10. N. Matsuda, S. Mikami, S. Shimoura, J. Takahashi, M. Nakano, K. Shimada, K. Uno, S. Hagiwara, K Saito, J. Environ. Radioact. **139**, 427-434 (2015).
11. IAEA, TECDOC-1162 (2000)
12. TEPCO, http://www.tepco.co.jp/cc/press/betu12_j/images/120524j0105.pdf (2012)
13. MLIT, http://nlftp.mlit.go.jp/ksj-e/index.html
14. EC, EUR 16541 EN (1996)
15. S. Kinase, T. Takahashi, S. Sato, R. Sakamoto, K. Saito, Radiat. Prot. Dosim. **160,** 318-324 (2014)

Modelling the dynamics of ambient dose rates induced by radiocaesium in the Fukushima terrestrial environment

Marc-André Gonze[1,*], *Christophe* Mourlon[1], *Philippe* Calmon[2], *Erwan* Manach[3], *Christophe* Debayle[3] and *Jean* Baccou[4]

[1] *IRSN, Institute for radiological Protection and Nuclear Safety, Laboratoire de Modélisation pour l'Expertise Environnementale (LM2E), 13115 St Paul-lez-Durance, France.*
[2] *IRSN, Institute for radiological Protection and Nuclear Safety, Laboratoire d'Études Radioécologiques en milieux Continental et Marin (LERCM), 13115 St Paul-lez-Durance, France.*
[3] *IRSN, Institute for radiological Protection and Nuclear Safety, Laboratoire de Surveillance Atmosphérique et d'Alerte (LS2A), 78116 Le Vésinet, France.*
[4] *IRSN, Institute for radiological Protection and Nuclear Safety, Laboratoire Incertitude et Modélisation des Accidents de Refroidissement (LIMAR), 13115 St Paul-lez-Durance, France.*

Abstract. Since the Fukushima accident, Japanese scientists have been intensively monitoring ambient radiations in the highly contaminated territories situated within 80 km of the nuclear site. The surveys that were conducted through mainly carborne, airborne and in situ gamma-ray measurement devices, enabled to efficiently characterize the spatial distribution and temporal evolution of air dose rates induced by Caesium-134 and Caesium-137 in the terrestrial systems. These measurements revealed that radiation levels decreased at rates greater than expected from physical decay in 2011-2012 (up to a factor of 2), and dependent on the type of environment (i.e. urban, agricultural or forest). Unlike carborne measurements that may have been strongly influenced by the depuration of road surfaces, no obvious reason can be invoked for airborne measurements, especially above forests that are known to efficiently retain and recycle radiocaesium.

The purpose of our research project is to develop a comprehensive understanding of the data acquired by Japanese, and identify the environmental mechanisms or factors that may explain such decays. The methodology relies on the use of a process-based and spatially-distributed dynamic model that predicts radiocaesium transfer and associated air dose rates inside/above a terrestrial environment (e.g., forests, croplands, meadows, bare soils and urban areas).

Despite the lack of site-specific data, our numerical study predicts decrease rates that are globally consistent with both aerial and in situ observations. The simulation at a flying altitude of 200 m indicated that ambient radiation levels decreased over the first 12 months by about 45% over dense urban areas, 15% above evergreen coniferous forests and between 2 and 12% above agricultural lands, owing to environmental processes that are identified and discussed. In particular, we demonstrate that the decrease over evergreen coniferous regions might be due the combined effects of canopy depuration (through biological and physical mechanisms) and the shielding of gamma rays emitted from the forest floor by vegetation. Our study finally suggests that airborne surveys might have not reflected dose rates at ground level in forest systems, which were predicted to slightly increase by 5 to 10% during the same period of time.

*Corresponding author: marc-andre.gonze@irsn.fr

IRSN
INSTITUT
DE RADIOPROTECTION
ET DE SÛRETÉ NUCLÉAIRE

Faire avancer la sûreté nucléaire

Modelling the dynamics of Ambient Dose Rates in Fukushima terrestrial environments

Gonze M.A., Mourlon C., Calmon P., Baccou J., Manach E. & Debayle C.

IRSN
CE Cadarache (France)

Field surveys of Ambient Dose Rates (ADRs)

Coordinated by the Japanese Nuclear Regulatory Authority / Atomic Energy Agency

Airborne (dec. 2012) along flying routes

Carborne (dec. 2012) along road network

In situ (dec. 2012) of undisturbed bare soils

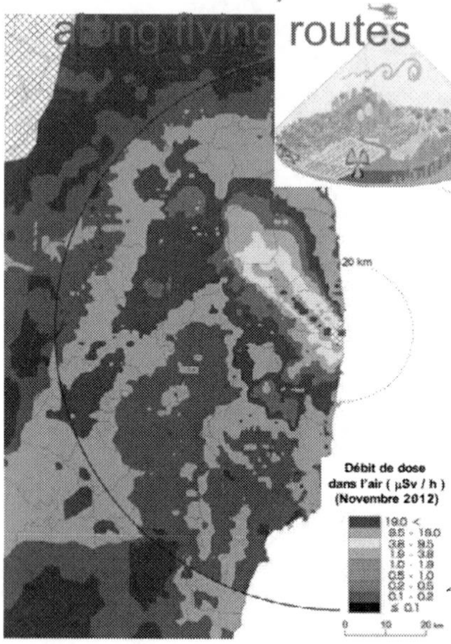

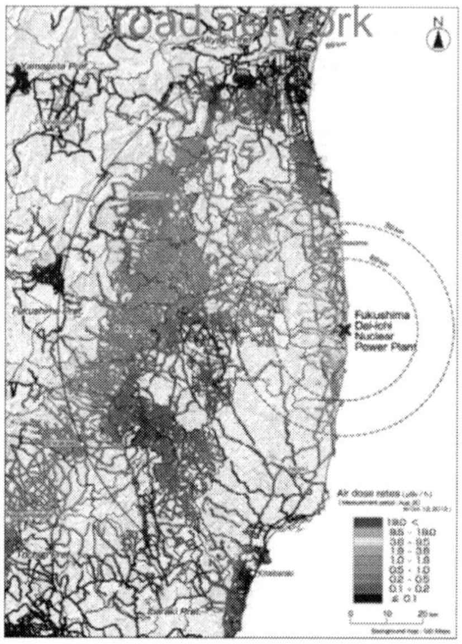

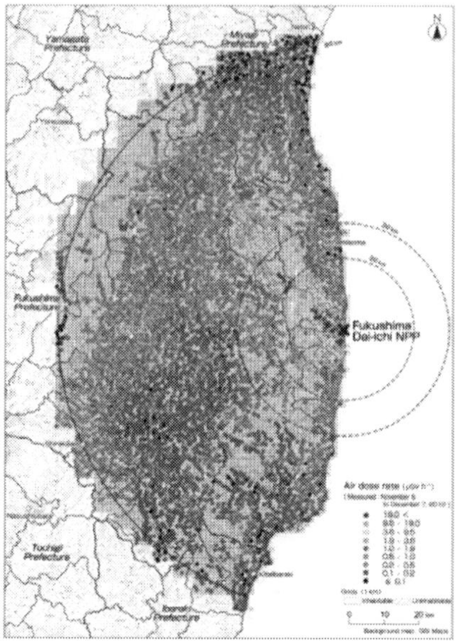

Sanada et al., 2012; Sanada et al. 2014a,b; Tsuda and Tsusumi, 2012; Tsuda et al. 2014; Tanigaki et al., 2013; Kinase et al., 2014; Andoh et al., 2014; Mikami et al., 2014a; Mikami et al., 2014b; Onda et al., 2014; Saito et al., 2014a; Saito et al., 2014b; Matsuda et al., 2014

Aerial *versus* in situ (Oct. 2011)

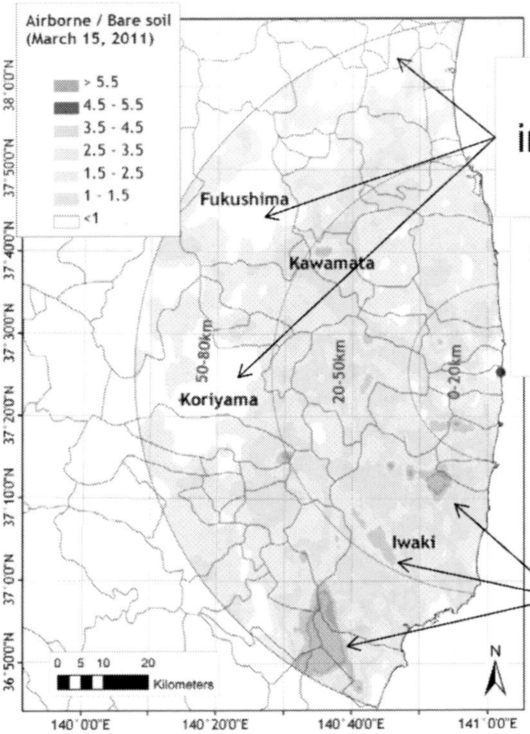

Landuse map

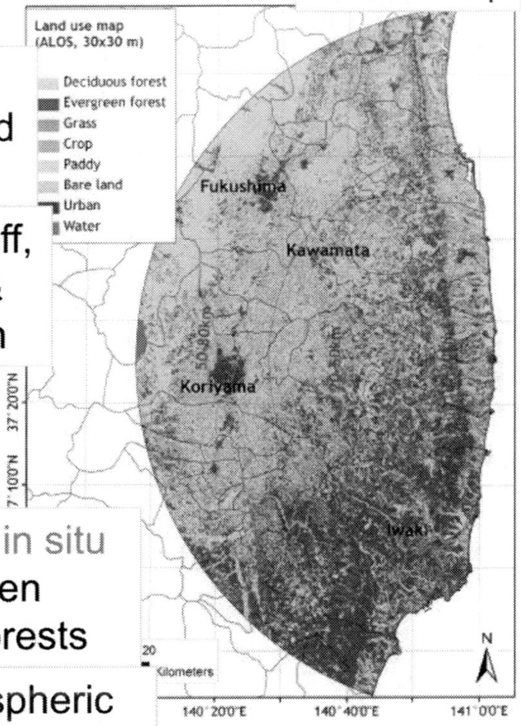

airborne < in situ
in densily inhabited areas

⇒ natural wash-off, traffic erosion & decontamination

airborne > 3× in situ
in evergreen coniferous forests

⇒ dry atmospheric deposition

Gonze et al., ES&T 2014

CONTEXT & OBJECTIVES MODELLING METHODOLOGY RESULTS & DISCUSSION

Time decrease of ADRs (1m height)

… averaged within 80 km & normalized by its value in June 2011

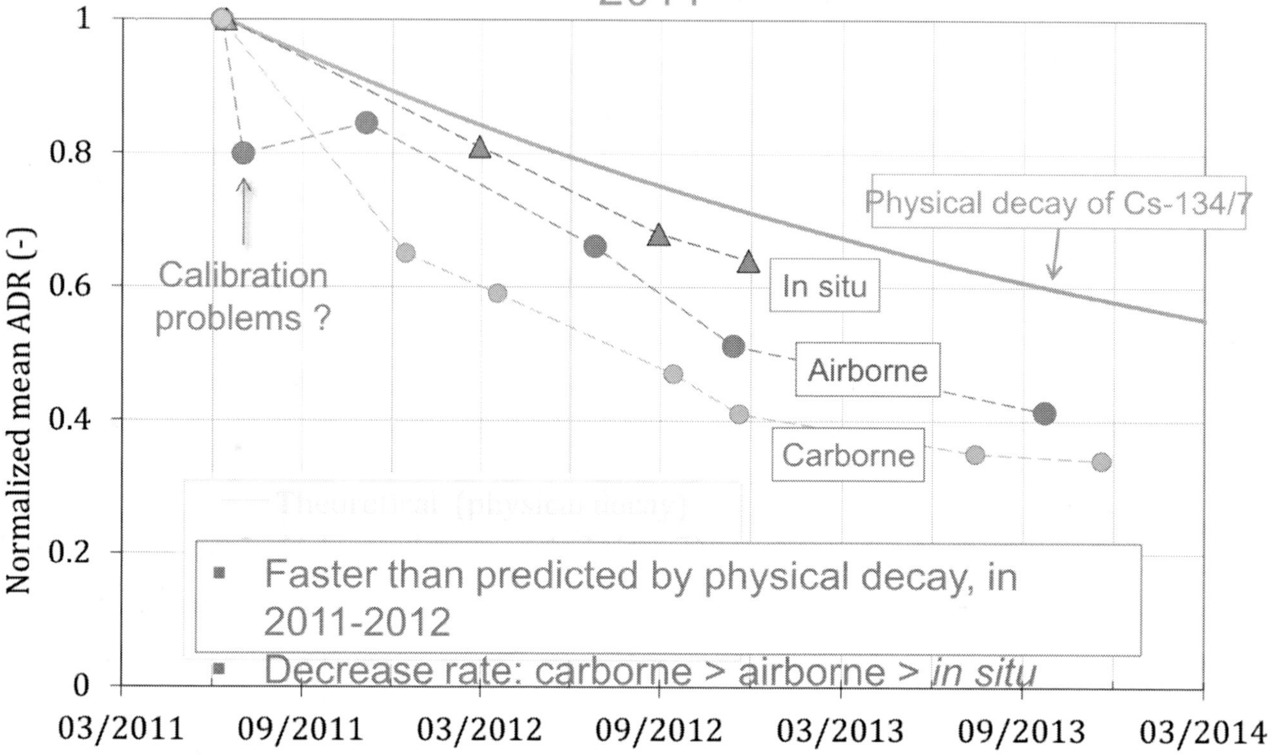

- Faster than predicted by physical decay, in 2011-2012
- Decrease rate: carborne > airborne > *in situ*

Time decrease of ADRs (1m height)

… averaged within 80 km & normalized by its value in June 2011

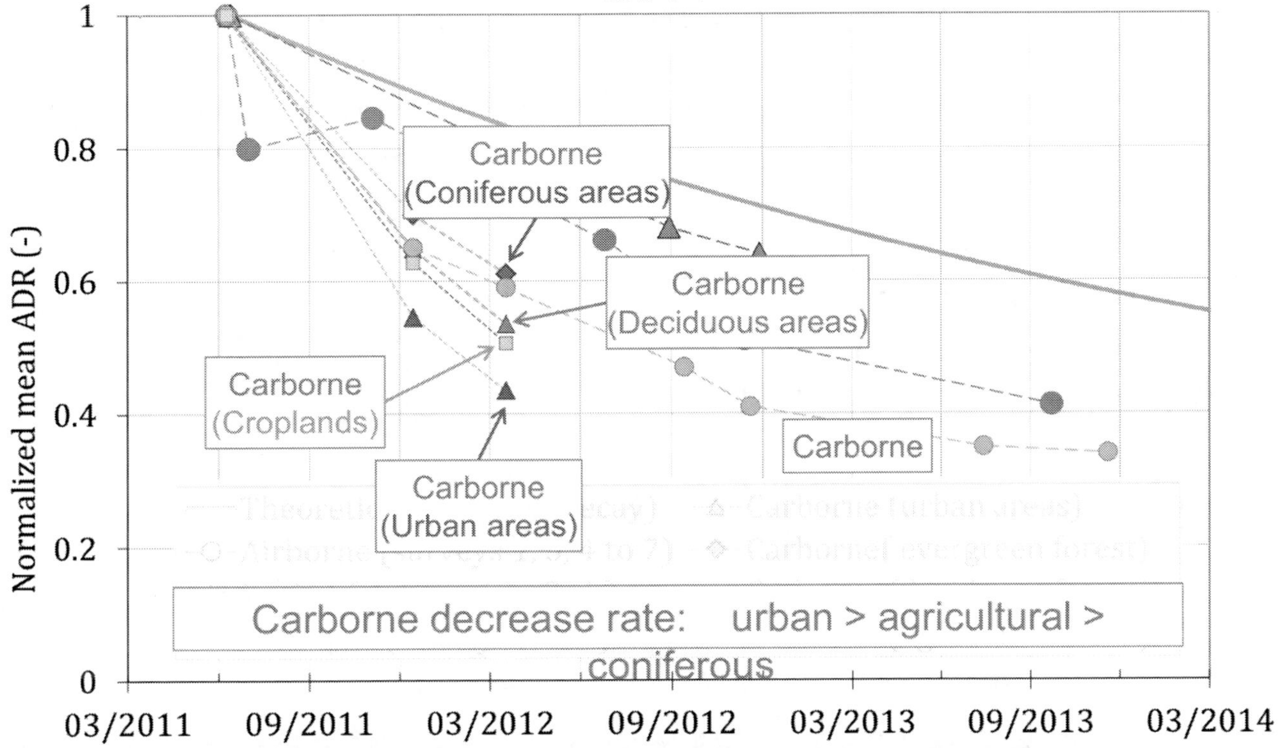

Ongoing research projects

Open questions

- Which environmental processes could explain such a variability in ADRs ?
- To what extent can forest vegetation modify ADRs inside/above forests ?
- Can we gain further understanding from process-based & dynamic models ?

AMORAD project (2013-2019)

Improvement of models for predicting radionuclide transfer in biosphere, with one focus put on forest systems

INVESTISSEMENTS D'AVENIR

French National Research Agency

EDOFU project (2014-2016)

Improvement of models for predicting γ ADRs in a multimedia and patchy environment

eDF

Electricité de France

http://www.irsn.fr/EN/Research/Research-organisation/Research-programmes/Pages/Research-programmes-2966.aspx

Modeling Cs transfer …

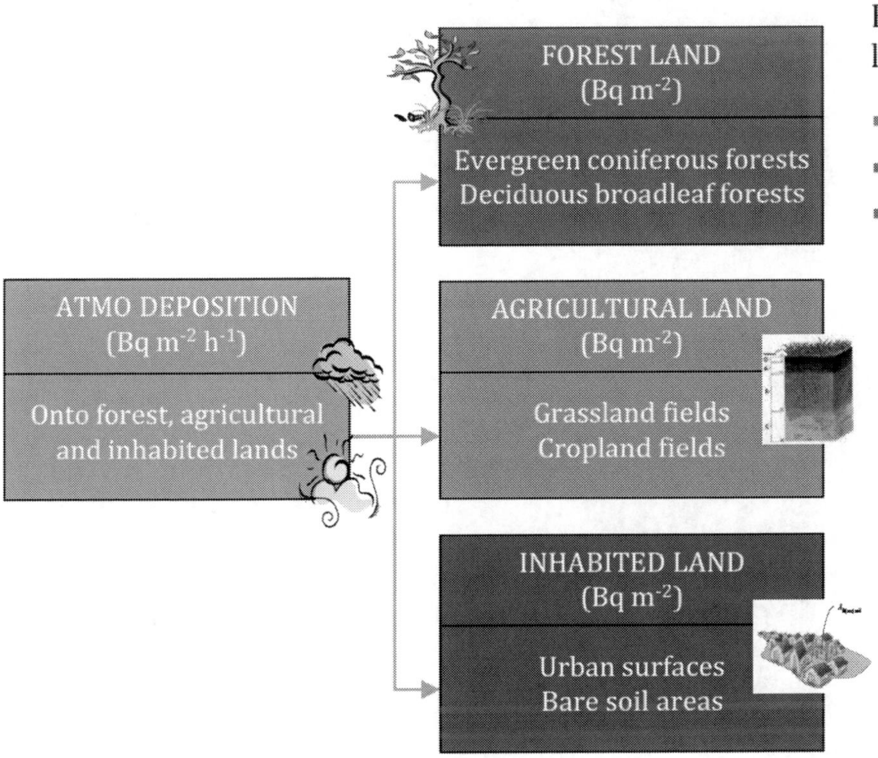

Key processes or factors that are likely to influence ADRs:

- Dry/wet deposition
- Interception by vegetation
- Depuration of vegetation

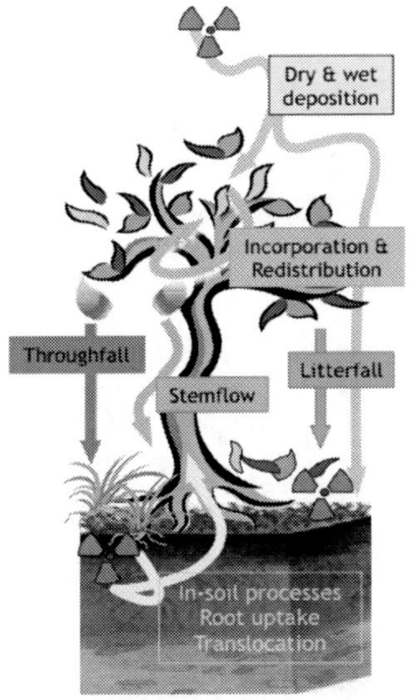

IRSN's modelling platform (SYMBIOSE)

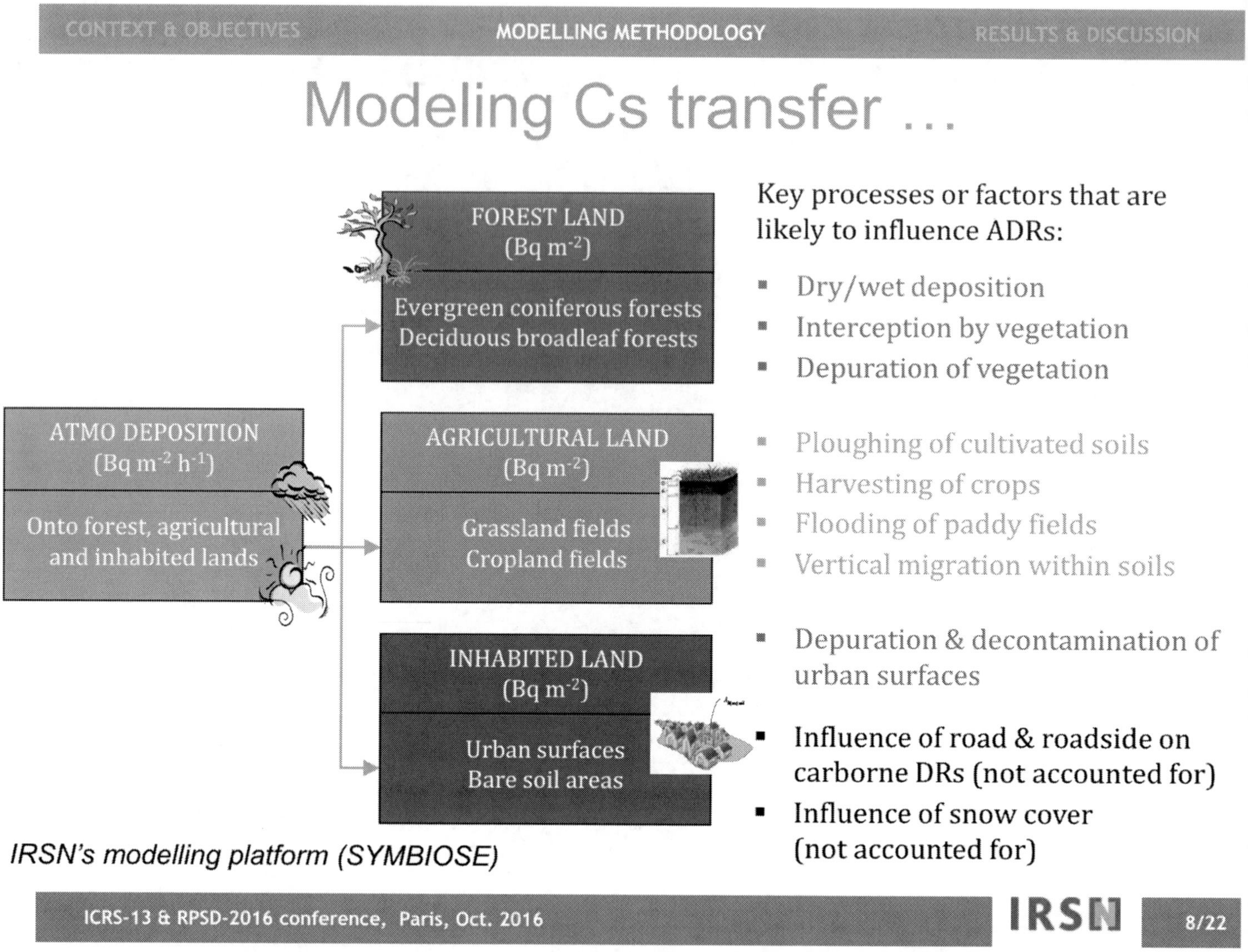
CONTEXT & OBJECTIVES
MODELLING METHODOLOGY
RESULTS & DISCUSSION
Modeling Cs transfer …
FOREST LAND
(Bq m-2)
Evergreen coniferous forests
Deciduous broadleaf forests
ATMO DEPOSITION
(Bq m-2 h-1)
Onto forest, agricultural and inhabited lands
AGRICULTURAL LAND
(Bq m-2)
Grassland fields
Cropland fields
INHABITED LAND
(Bq m-2)
Urban surfaces
Bare soil areas
IRSN's modelling platform (SYMBIOSE)
Key processes or factors that are likely to influence ADRs:
Dry/wet deposition
Interception by vegetation
Depuration of vegetation
Ploughing of cultivated soils
Harvesting of crops
Flooding of paddy fields
Vertical migration within soils
Depuration & decontamination of urban surfaces
Influence of road & roadside on carborne DRs (not accounted for)
Influence of snow cover (not accounted for)
ICRS-13 & RPSD-2016 conference, Paris, Oct. 2016
IRSN
8/22

… and ADRs

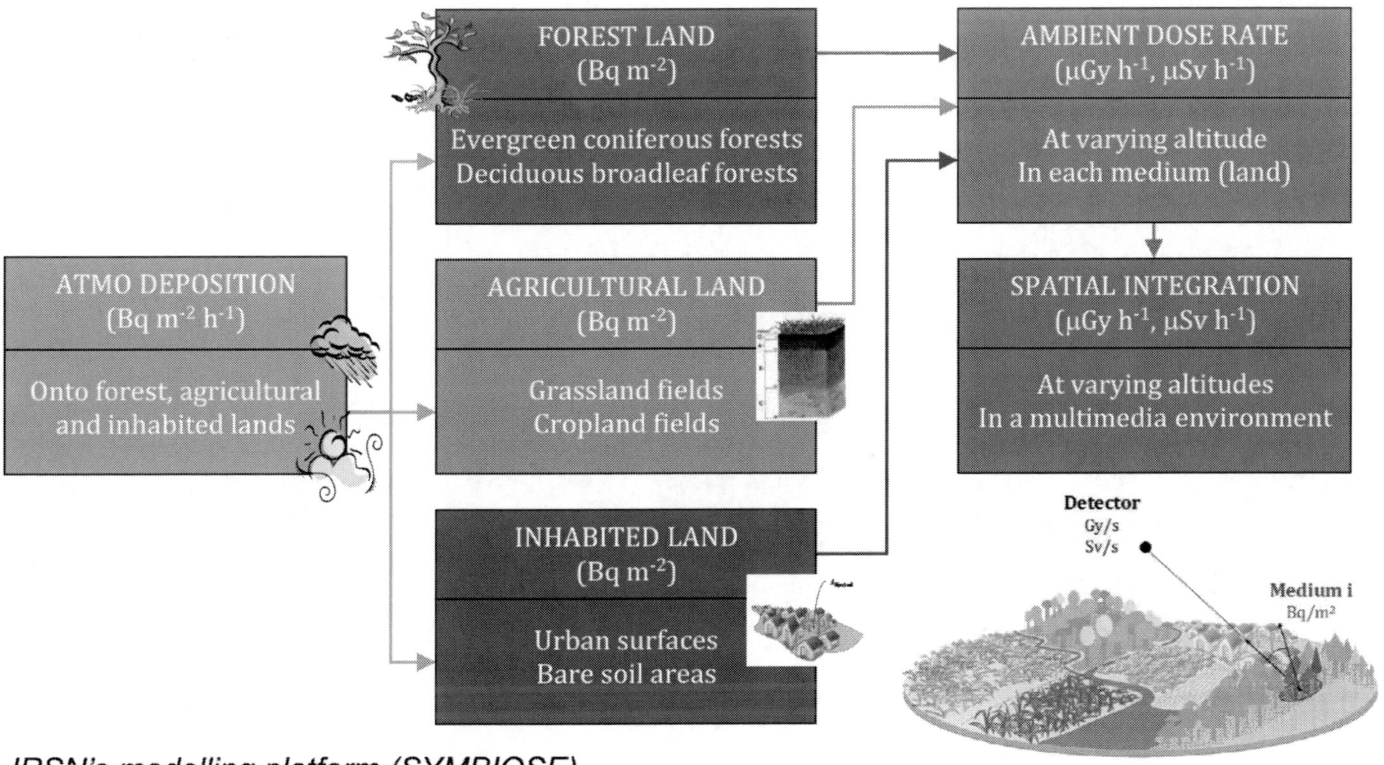

IRSN's modelling platform (SYMBIOSE)

ADR coefficients for a soil-vegetation medium

Use of :

- ADR coefficients expressed in (Gy/h)/(Bq/m^2)
- Pre-calculated with MCNP code
- Depending on detector altitude, depth of source and medium characteristics (e.g., geometry, density & elemental composition)

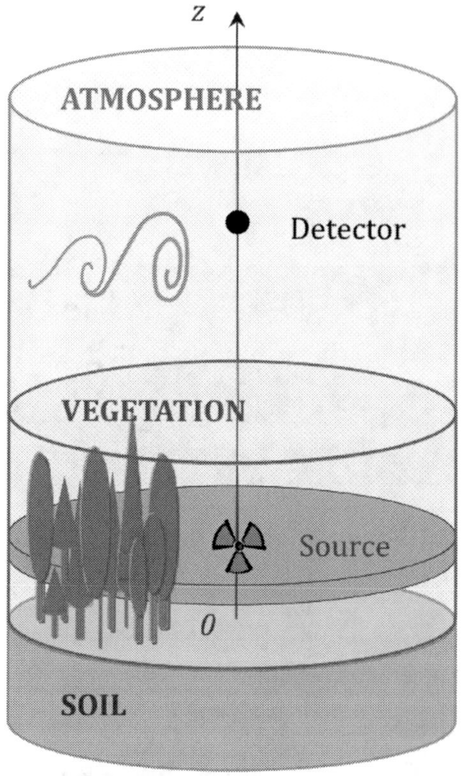

Equivalent Medium assumption

- Soil, vegetation & atmosphere layers with homogeneous properties
- Plane or volumetric radioactive source of homogeneous intensity

Regional simulations

- Spatio-temporal domain
  - 80 km from FDNPP
  - Computational regular grid
  - March 2011 to March 2014 (3 years)
- Environmental parameters
  - Mostly generic
  - A few site-specific
    - Land use map
    - Cs contamination maps (airborne & in situ)
    - Seasonal climatic data
    - Agricultural practices
    - Depuration/decontamination rates for urban surfaces
    - Depuration of forest vegetation
    - …

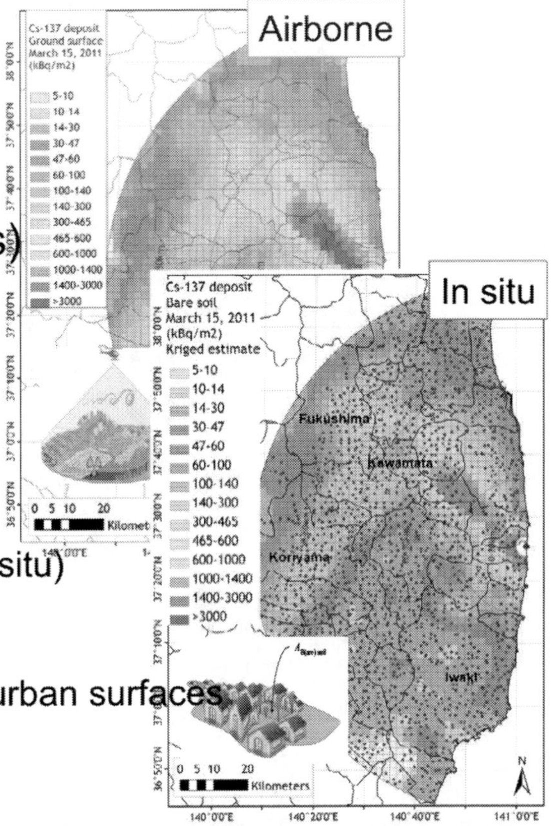

CONTEXT & OBJECTIVES MODELLING METHODOLOGY **RESULTS & DISCUSSION**

Tree depuration flux

at 21 evergreen coniferous sites

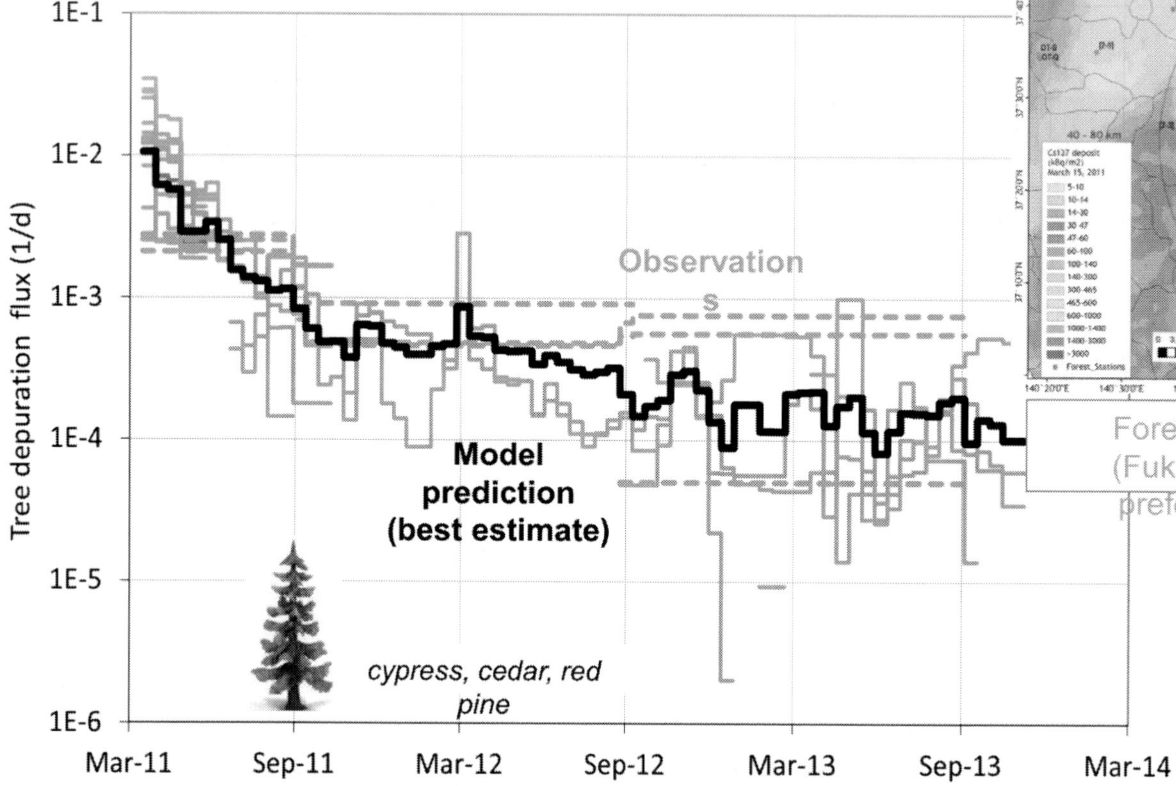

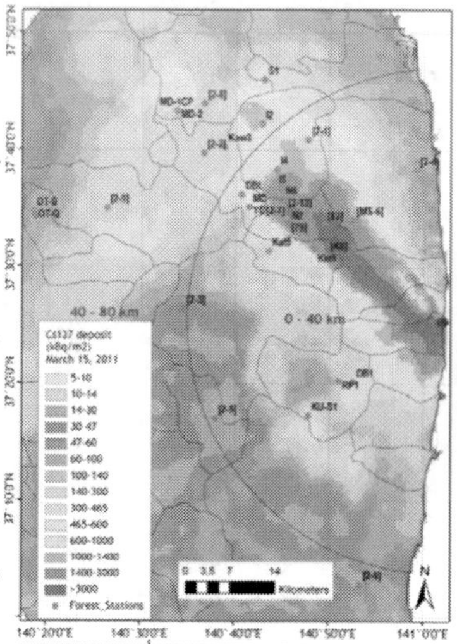

Forest sites (Fukushima prefecture)

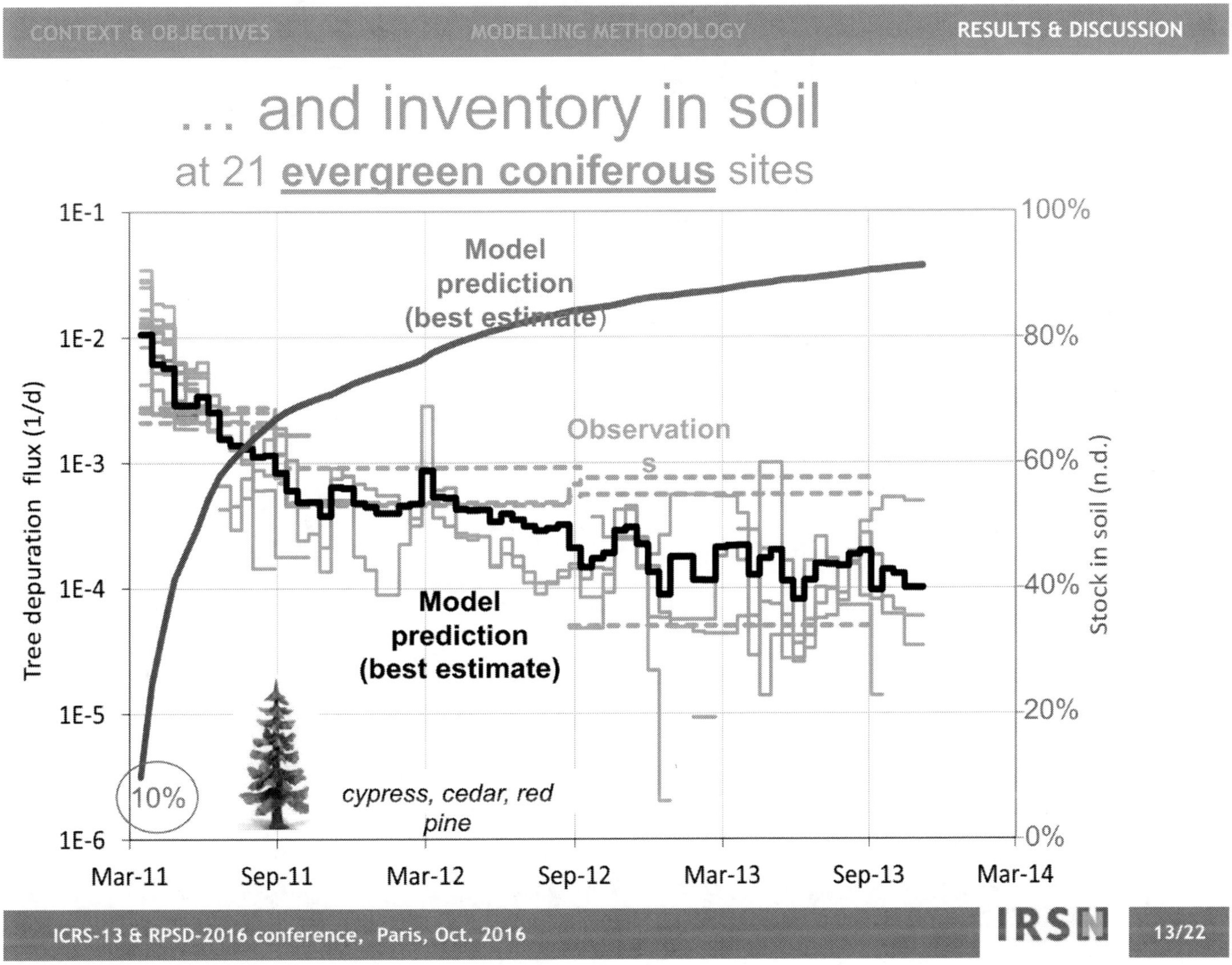
CONTEXT & OBJECTIVES
MODELLING METHODOLOGY
RESULTS & DISCUSSION
… and inventory in soil
at 21 evergreen coniferous sites
Model prediction (best estimate)
Observation s
Model prediction (best estimate)
Tree depuration flux (1/d)
Stock in soil (n.d.)
1E-1
1E-2
1E-3
1E-4
1E-5
1E-6
100%
80%
60%
40%
20%
0%
10%
cypress, cedar, red pine
Mar-11
Sep-11
Mar-12
Sep-12
Mar-13
Sep-13
Mar-14
ICRS-13 & RPSD-2016 conference, Paris, Oct. 2016
IRSN
13/22

Deciduous broadleaf forests (4 sites)

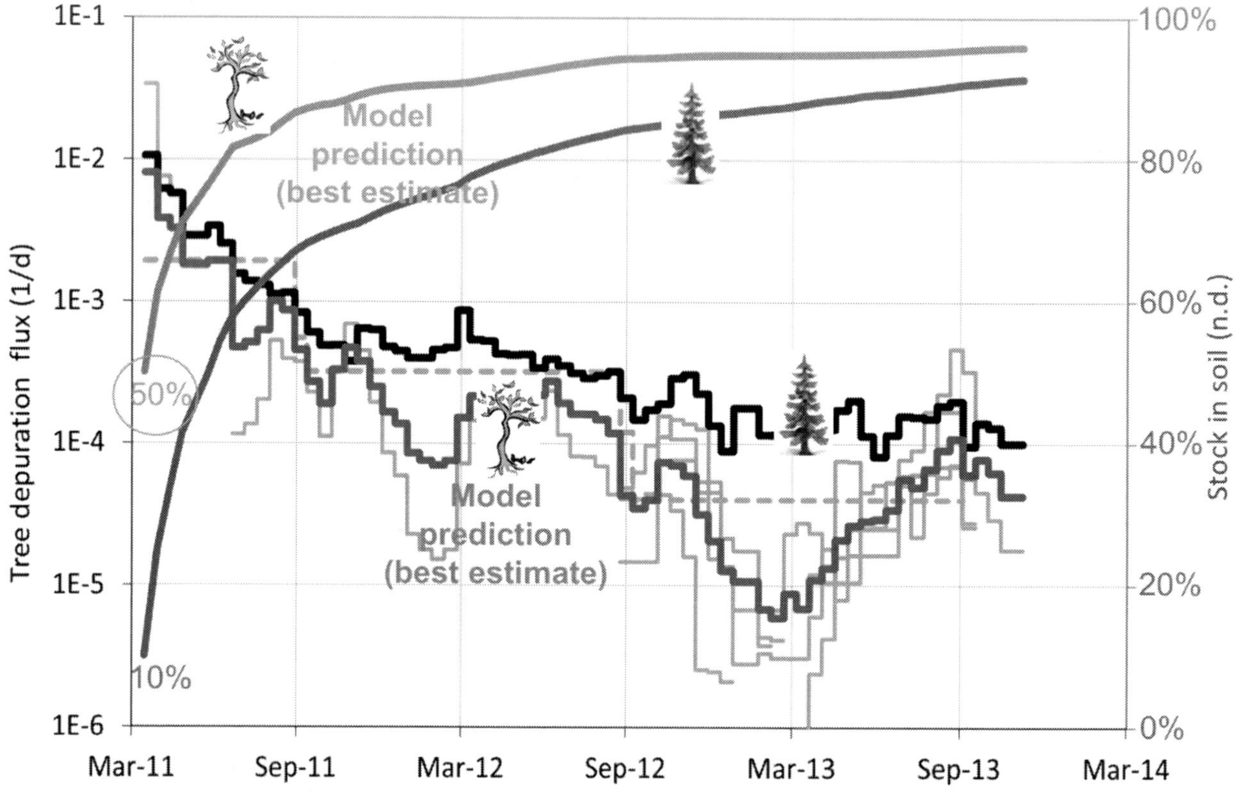

ADR coefficient in a coniferous forest

… for a **plane** source of ^{134}Cs of increasing height

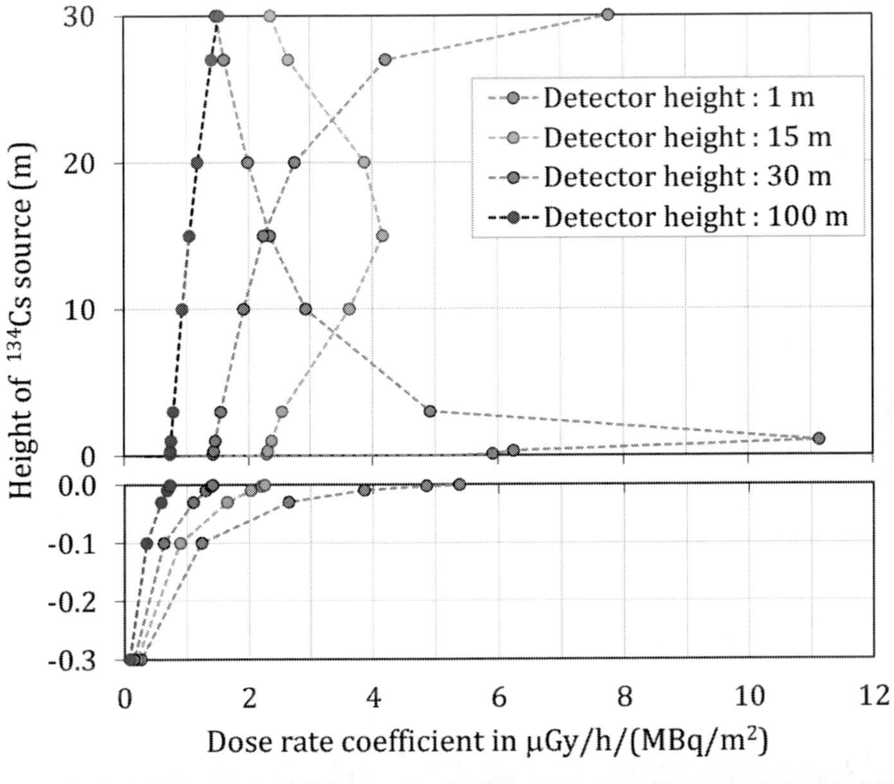

30 m tree layer

30 cm soil layer

ADR coefficients in a coniferous forest

for a **volumic** source of ^{134}Cs in the top 1cm soil layer or 30m tree layer

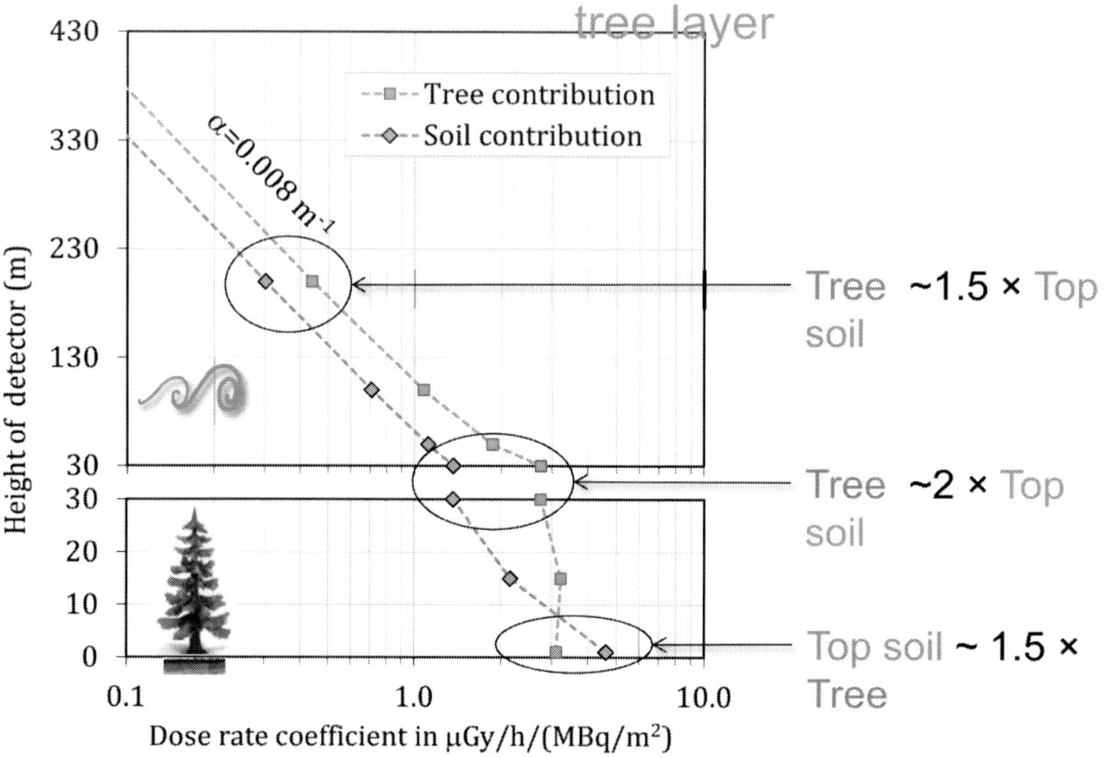

Predicted ADRs at ground level ($^{134+7}Cs$)…

Mean value within the 80 km region and normalized by its value in June 2011

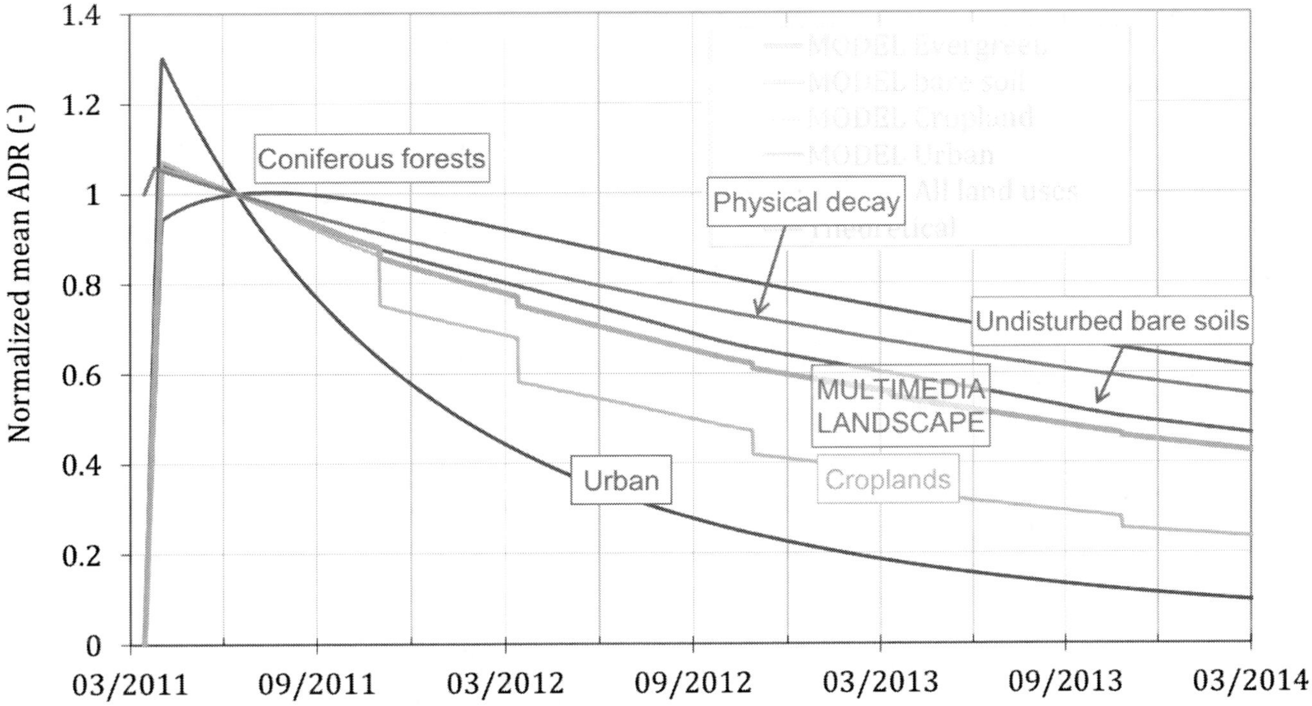

…*versus* in situ & carborne data

Mean value within the 80 km region and normalized by its value in June 2011

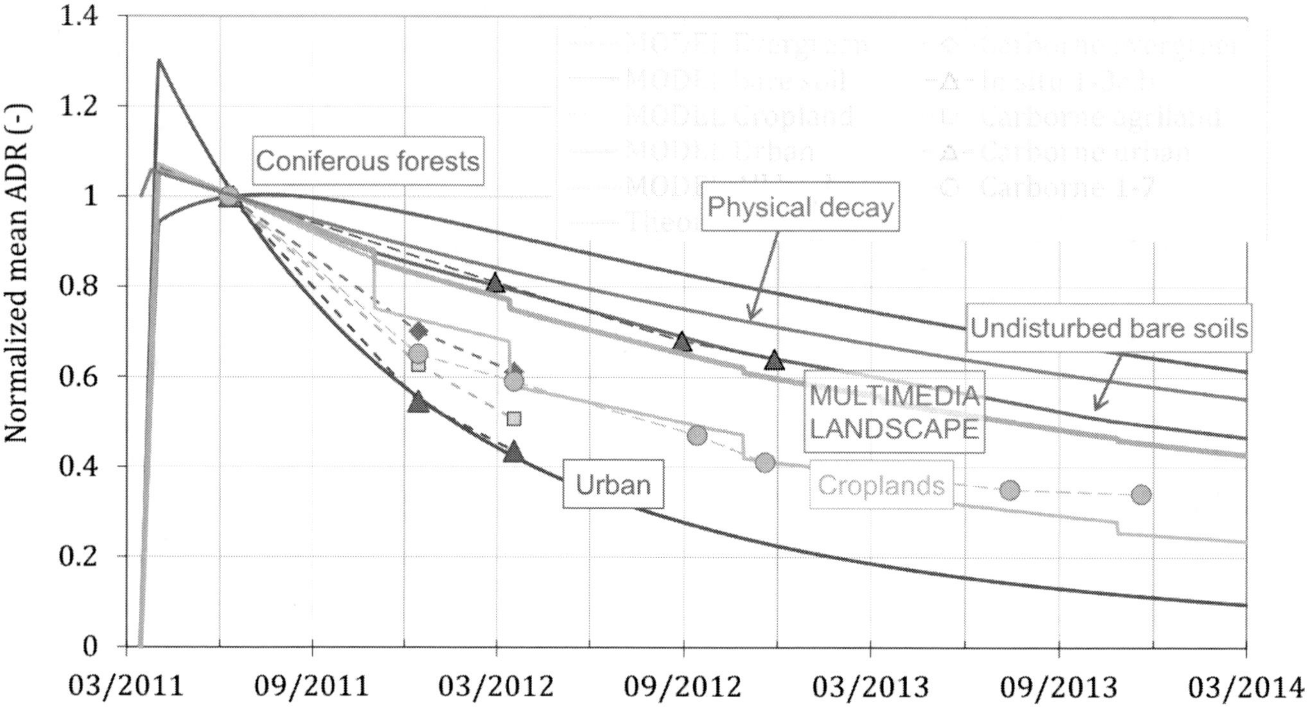

Predicted ADRs at 200m *versus* airborne data

Mean value within the 80 km region and normalized by its value in June 2011

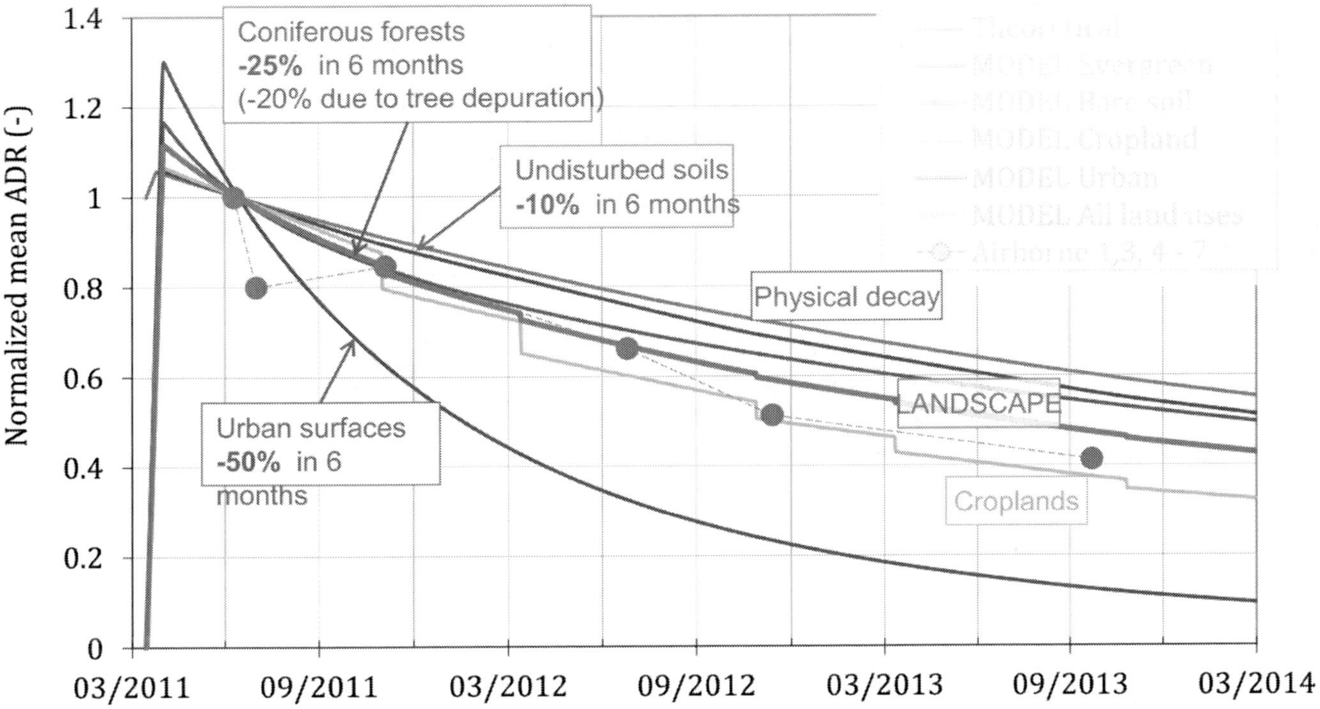

Further details in…

Science of the Total Environment 529 (2015) 30–39

Contents lists available at ScienceDirect

Science of the Total Environment

journal homepage: www.elsevier.com/locate/scitotenv

Modeling the early-phase redistribution of radiocesium fallouts in a evergreen coniferous forest after Chernobyl and Fukushima acciden…

P. Calmon, M.-A. Gonze [*], Ch. Mourlon

Institute of Radiation Protection and Nuclear Safety, CE Cadarache-Bat 153, BP3-13115 St-Paul-lez-Durance Cedex, France

HIGHLIGHTS

- Transfer of radiocesium atmospheric fallout in evergreen forests was modeled.
- The model was tested using observations from Chernobyl and Fukushima accidents.
- Model predictions of canopy interception and depuration agree with measurements.
- Unexpectedly high contribution of litterfall for the Japanese forest is discussed.
- Meteorological conditions and tree species characteristics are sensitive factors.

Environmental Science & Technology

Article

pubs.acs.org/est

Terms of Use

Assessment of Dry and Wet Atmospheric Deposits of Radioactive Aerosols: Application to Fukushima Radiocaesium Fallout

Marc-André Gonze,[*,†] Philippe Renaud,[†] Irène Korsakissok,[‡] Hiroaki Kato,[§] Thomas G. Hinton,[†] Christophe Mourlon,[†] and Marie Simon-Cornu[†]

[†]Institute for Radiological Protection and Nuclear Safety (IRSN), Environmental Research Division, BP 3–13115 St-Paul-lez-Durance Cedex, France

[‡]Institute for Radiological Protection and Nuclear Safety (IRSN), Nuclear & Radiological Emergency Division, BP 17–92262 Fontenay-aux-Roses Cedex, France

…ynamics, University of Tsukuba,1-1-1 Tennodai, Tsukuba, Ibaraki Japan

…ident led to massive … the land surfaces. The … Japanese authorities for … ne monitoring surveys … of bare soil areas (since … ences exist between the … se differences can be … lly based relationships … ology, which has been … n 80-km of the nuclear … d wet deposits that are … simulations from the … ute of Radioprotection

Journal of Environmental Radioactivity xxx (2015) 1–13

Contents lists available at ScienceDirect

Journal of Environmental Radioactivity

journal homepage: www.elsevier.com/locate/jenvrad

Modelling the dynamics of ambient dose rates induced by radiocaesium in the Fukushima terrestrial environment

Marc-André Gonze [a, *], Christophe Mourlon [a], Philippe Calmon [b], Erwan Manach [c], Christophe Debayle [c], Jean Baccou [d]

[a] Institut de Radioprotection et de Sûreté Nucléaire (IRSN), Laboratoire de Modélisation pour l'Expertise Environnementale (LM2E), Cadarache, … 159, 13115 St Paul-lez-Durance, France

[b] Institut de Radioprotection et de Sûreté Nucléaire (IRSN), Laboratoire d'Etudes Radioécologiques en milieux Continental et Marin (LERCM), C… Bâtiment 151, 13115 St Paul-lez-Durance, France

[c] Institut de Radioprotection et de Sûreté Nucléaire (IRSN), Laboratoire de Surveillance Atmosphérique et d'Alerte (LS2A), Le Vésinet, Bâtiment C8, 78116 Le Vésinet, France

[d] Institut de Radioprotection et de Sûreté Nucléaire (IRSN), Laboratoire Incertitude et Modélisation des Accidents de Refroidissement (LIMAR), Cadarache, Bâtiment 700, St Paul-lez-Durance 13115, France

Compilation & analysis of monitoring observations in Japanese forests to be published shortly

Summary

- Japanese data highly valuable for improving our understanding of post-accidental consequences in terrestrial systems
- Despite the lack of site-specific data & uncertainties on deposition characteristics, predicted ADRs globally consistent with field observations (thanks to some models improvement & calibration)
- This study suggests that:
  - Disagreement with carborne observations mostly attributed to road & roadside effects (not accounted for in our modelling study)
  - Airborne observations above vegetation may not be representative of ground ADRs, due to tree depuration & shielding of gamma rays
  - Influence of forest vegetation should be accounted for when converting counting rates (for especially low flying altitudes) => this requires assumptions on radionuclide distribution !

Predicted wet deposition ratio

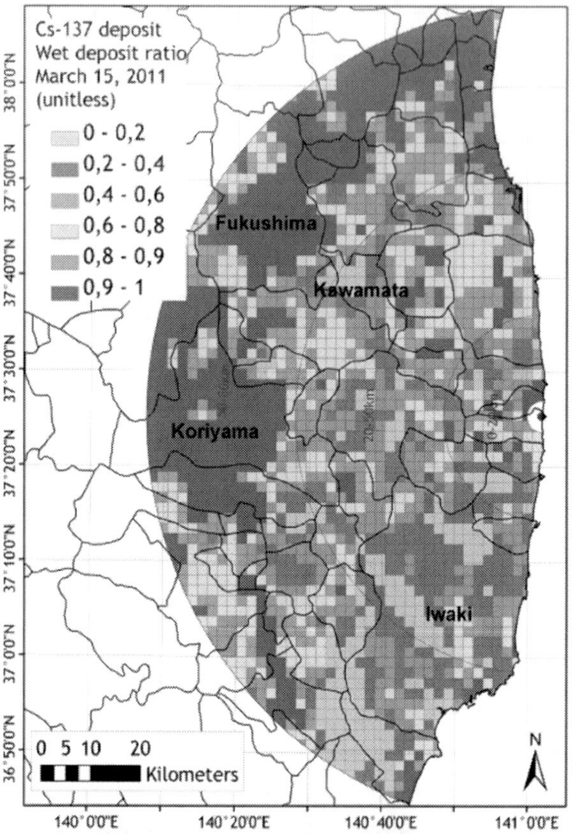

Wet fraction > 0.8
Intense wet deposition might have occurred in the Abukuma valley & to the NW of FDNPP

Wet fraction < 0.4
Noticeable dry deposition might have occurred in S/SW regions = high forested areas of the Abukuma mountains

Optimization of exposure and countryside waste management for different accidental radioactive release

Philippe Guétat*

Member of High Committee for Transparency and Information about Safety in Nuclear

Abstract. Since the Fukushima accident, Japanese scientists have been intensively monitoring ambient radiations in the highly contaminated territories situated within 80 km of the nuclear site. The surveys that were conducted through mainly carborne, airborne and in situ gamma-ray measurement devices, enabled to efficiently characterize the spatial distribution and temporal evolution of air dose rates induced by Caesium-134 and Caesium-137 in the terrestrial systems. These measurements revealed that radiation levels decreased at rates greater than expected from physical decay in 2011-2012 (up to a factor of 2), and dependent on the type of environment (i.e. urban, agricultural or forest). Unlike carborne measurements that may have been strongly influenced by the depuration of road surfaces, no obvious reason can be invoked for airborne measurements, especially above forests that are known to efficiently retain and recycle radiocaesium.

The purpose of our research project is to develop a comprehensive understanding of the data acquired by Japanese, and identify the environmental mechanisms or factors that may explain such decays. The methodology relies on the use of a process-based and spatially-distributed dynamic model that predicts radiocaesium transfer and associated air dose rates inside/above a terrestrial environment (e.g., forests, croplands, meadows, bare soils and urban areas).

Despite the lack of site-specific data, our numerical study predicts decrease rates that are globally consistent with both aerial and in situ observations. The simulation at a flying altitude of 200 m indicated that ambient radiation levels decreased over the first 12 months by about 45% over dense urban areas, 15% above evergreen coniferous forests and between 2 and 12% above agricultural lands, owing to environmental processes that are identified and discussed. In particular, we demonstrate that the decrease over evergreen coniferous regions might be due the combined effects of canopy depuration (through biological and physical mechanisms) and the shielding of gamma rays emitted from the forest floor by vegetation. Our study finally suggests that airborne surveys might have not reflected dose rates at ground level in forest systems, which were predicted to slightly increase by 5 to 10% during the same period of time.

* Corresponding author: philippe.guetat@cea.fr

Optimization of exposure and waste management for different accidental releases

Philippe GUETAT

Stackholder - France

- Paris 2016

1

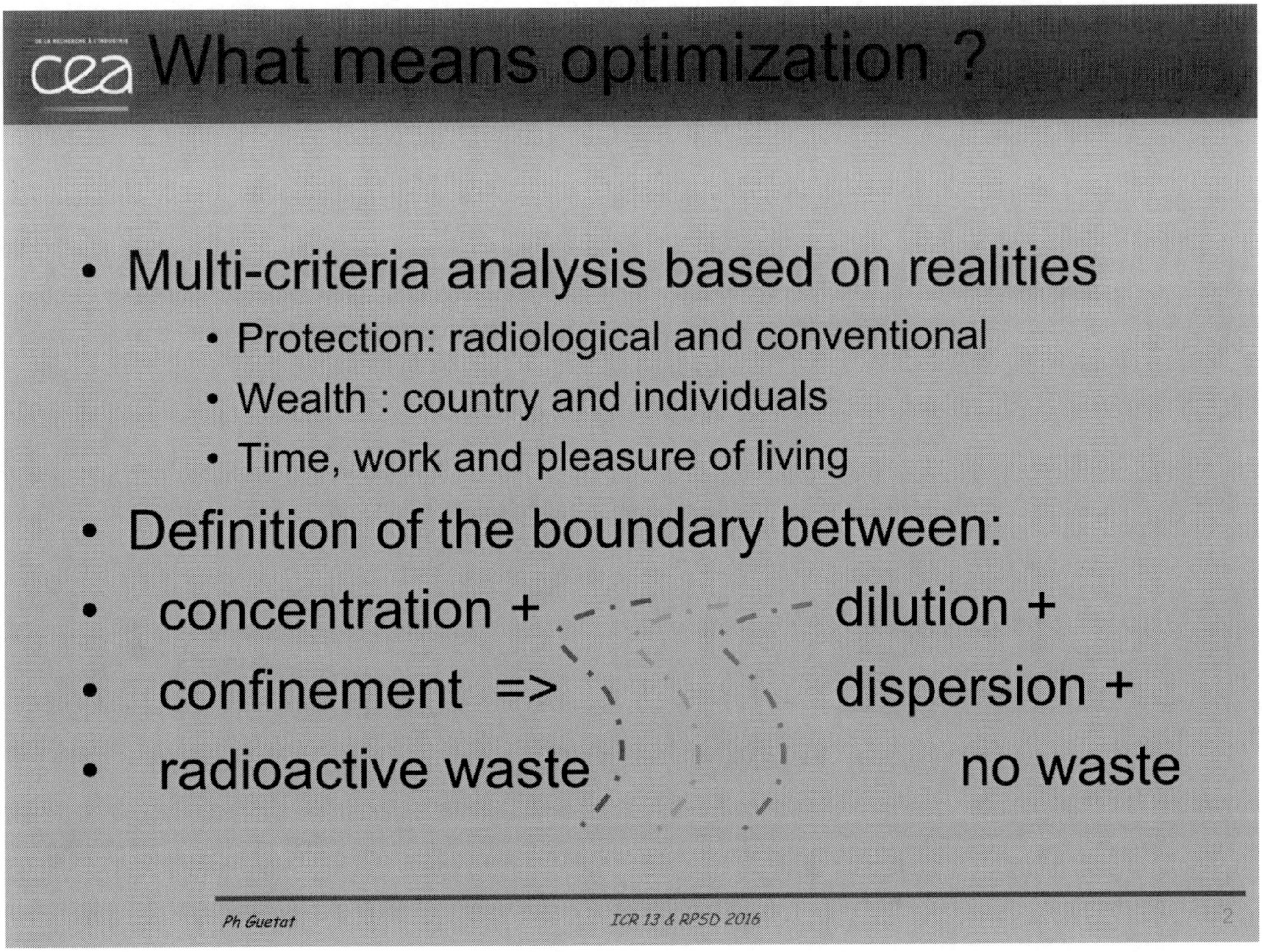
cea
What means optimization ?
• Multi-criteria analysis based on realities
• Protection: radiological and conventional
• Wealth : country and individuals
• Time, work and pleasure of living
• Definition of the boundary between:
• concentration +
• confinement =>
• radioactive waste
dilution +
dispersion +
no waste
Ph Guetat
ICR 13 & RPSD 2016
2

What to avoid
No, on the other side but here there is some light
You are sure to have lost some thing here ?
Ph Guetat
ICR 13 & RPSD 2016
3

I - REFERENCE VALUES

4

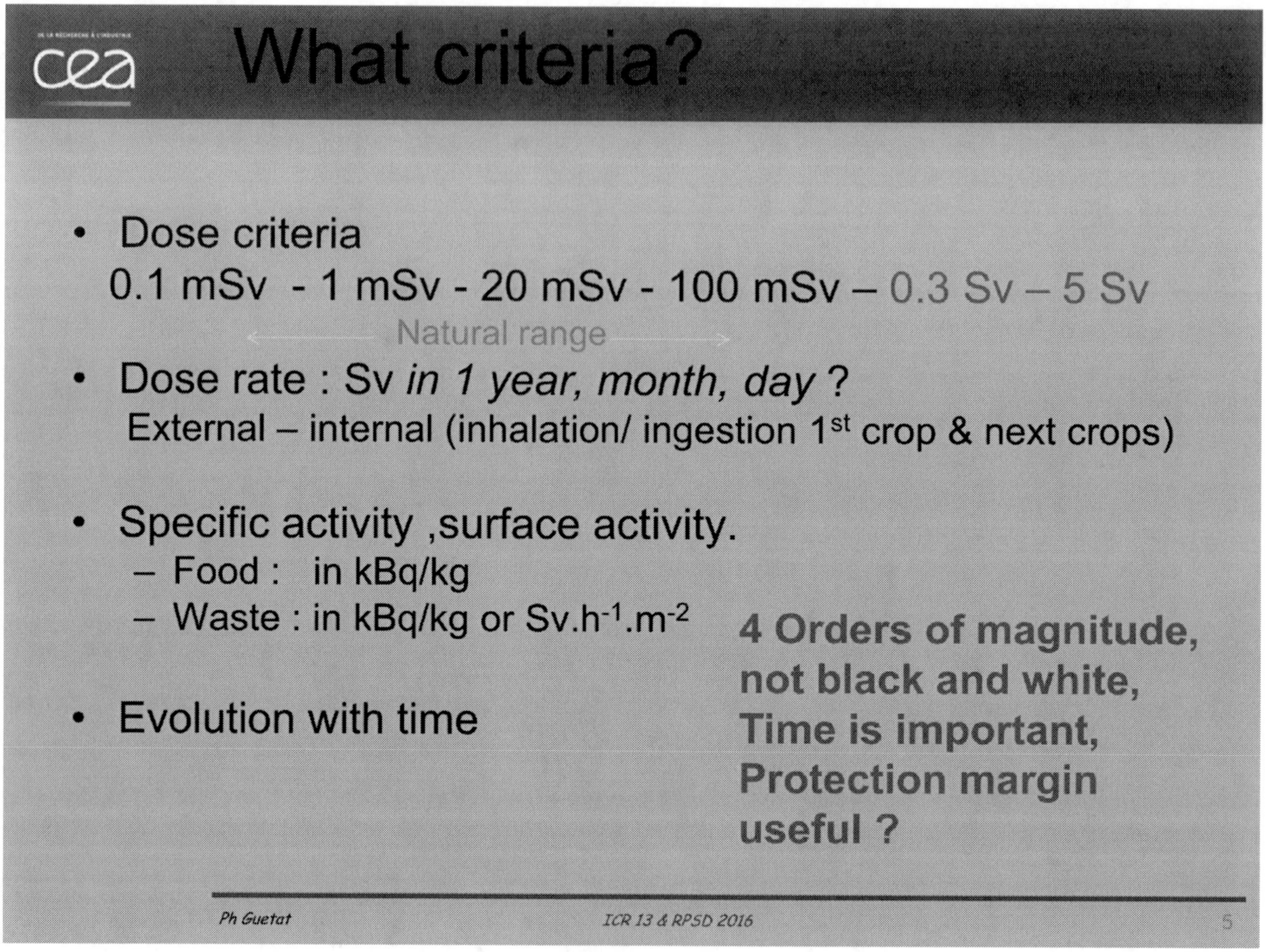
What criteria?
• Dose criteria
0.1 mSv - 1 mSv - 20 mSv - 100 mSv – 0.3 Sv – 5 Sv
Natural range
• Dose rate : Sv in 1 year, month, day ?
External – internal (inhalation/ ingestion 1st crop & next crops)
• Specific activity ,surface activity.
– Food : in kBq/kg
– Waste : in kBq/kg or Sv.h-1.m-2
• Evolution with time
4 Orders of magnitude, not black and white, Time is important, Protection margin useful ?
Ph Guetat
ICR 13 & RPSD 2016
5

What danger ?

1. Meltdown is necessarily associated to a significant release of energy. Releases are at a big height which brings an important protection to the neighborhood.
2. Delay before release = very short-lived nuclides disappear. Suppress effect on forest by β.
3. Melting is not explosion : Non-volatile Nuclides not emitted.

- Evacuation induces other more conventional but real dangers.
- The cure can be worse than the disease, especially if the evil is imaginary.

Ph Guetat ICR 13 & RPSD 2016 6

Reference values as elements of optimization

Relation between dose & relative risk for solid cancers and leukemia : Hiroshima Nagasaki

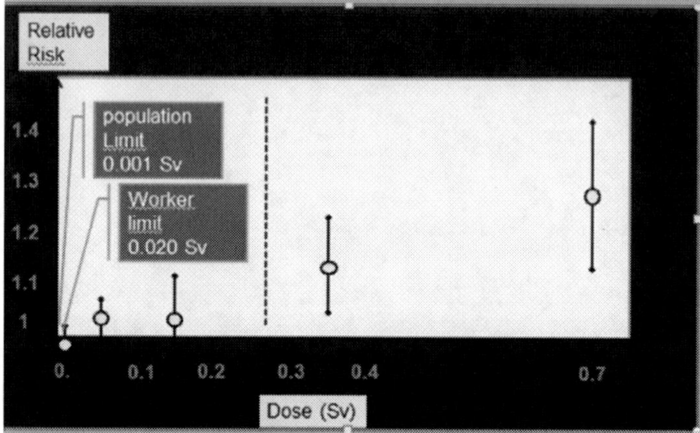

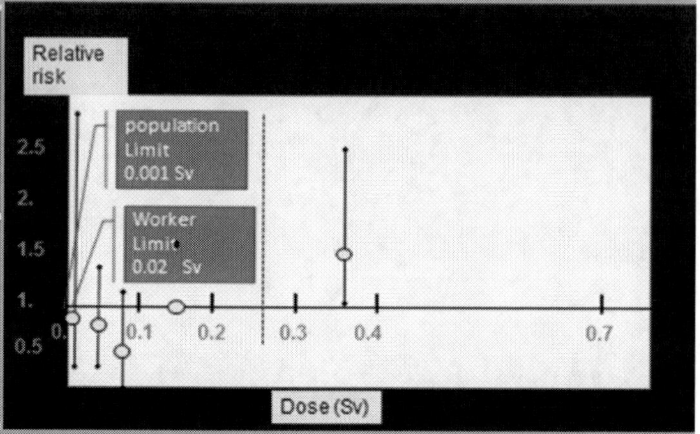

ICRP reference : 0.001, 0.02 and 0.1 Sv

In Japan: 0.3 Sv for workers = a good choice
No clinical effect – no detectable stochastic effect

DNA damages short term ? (weeks)

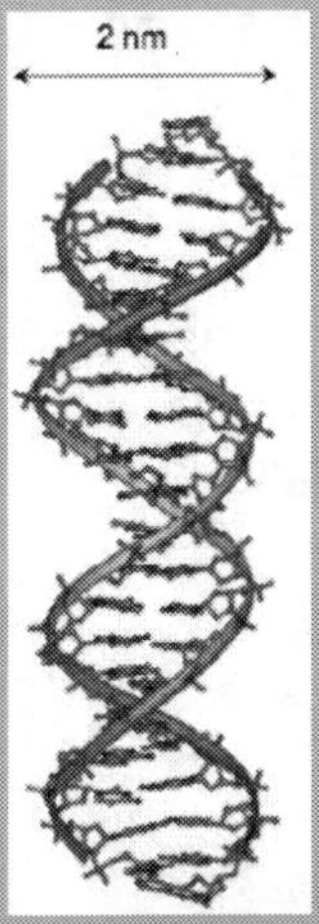

DNA Damage to **thyroid** cells	spontaneous over 1 month	Chernobyl young children	Fukushima young children
exposure	-	**3** Gy (2 and more) Inhal. & **ingest.**	Max **0.08** Gy
simple strand breaks	300 000 – 1 700 000	3000	80
Nucleobase loss	400 000	6000	160
Bases modification	100 000		
Double-strand breaks	**240**	120 (50%)	3 (1%)

no effect will be visible (exist) in Japan

Ph Guetat ICR 13 & RPSD 2016 8

DNA damages in long term ? (Years)

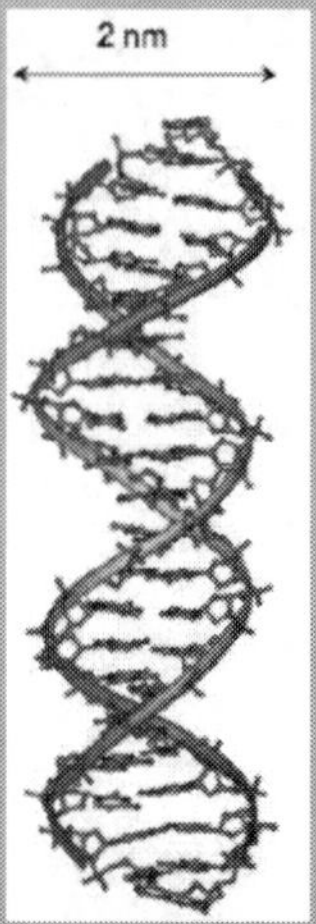

DNA damages per cell and per day	« spontaneous »	Radio-induced damages rate 100 milliGray/y
simple strand breaks	10 000 - 55 000	0.3
Nucleobase loss	12 600	0.6
Bases modification	3 200	
Double-strand breaks	8	0.012

(Burkart W et al. CR Acad Sci III 1999; 322:89-101;
Ward JF Prog Nucl Acids Res Mol Biol. 1988; 35: 95-125)

Reference values for food and feed

Bq/kg	Infant food	Dairy produce	Other major foods	Liquid food (CA), & *water (WHO)*	Minor food	Feed
I131	150 /*100*	500 / *100*	2000/ 100	125 / *10*	20000	
Cesium 134 - 137	400/ *1000*	1000 / *1000*	1250 / *1000*	1000 / *10*	12500	Pigs.............. 1350 Poultry, lambs, calves.......... 2500 Other.......... 5000
Tritium	- / *1 000*	- / *10 000*	/ *10 000*	- / *10 000*		*(As Organic molecules)*
Pu239	1 / *1*	20 / *10*	80 / *10*	20 / *1*	800	

1rst crop Food ban Fundamental

0.1 mSv : very low but today +/- unavoidable.

No reason to choose lower values

(Maximum, average, 95% ? M= Max/10)

Ph Guetat ICR 13 & RPSD 2016 10

Deposition – transfers – dose rate per hour and per year, specific activity

Case dry deposition for 1 $MBq.m^{-2}$ Total deposition					
External irradiation	infinite surface	Cs134	5.5 $\mu Sv.h^{-1}$	33 $mSv.an^{-1}$	Bq ^{134}Cs & ^{137}Cs not equivalent
		Cs137	2. $\mu Sv.h^{-1}$	12 $mSv.an^{-1}$	
	Homogeneous after ploughing	Cs134	0.9 $\mu Sv.h^{-1}$	5,3 $mSv.an^{-1}$	Fed. Guid n° 12 Depth 30 cm
	3 Bq/g	Cs137	0.33 $\mu Sv.h^{-1}$	2 $mSv.an^{-1}$	
Vegetable	Foliar deposition	aerosols	≈10^6 $Bq.kg^{-1}$	existing crop	1 to 5 $kg_{Fresh\ W}.m^{-2}$
	Soil-plant transfer	Cs 134 137	30 – 300 $Bq.kg^{-1}$	next crops	(Fresh weight)
Cow Milk	feed contamination	Cs 134 137	1.10^5 $Bq.L^{-1}$ max	short term	(Dynamic model)
Cow meat	soil> grass>meat	Cs137	50 – 500 $Bq.kg^{-1}$	Long term	
Exemption levels	BSS - Transport	Cs 134 137	10 $Bq.g^{-1}$	Japan 7 $Bq.g^{-1}$	local/national

2 - 3 $MBq.m^{-2}$ of Cs 137, a physical boundary

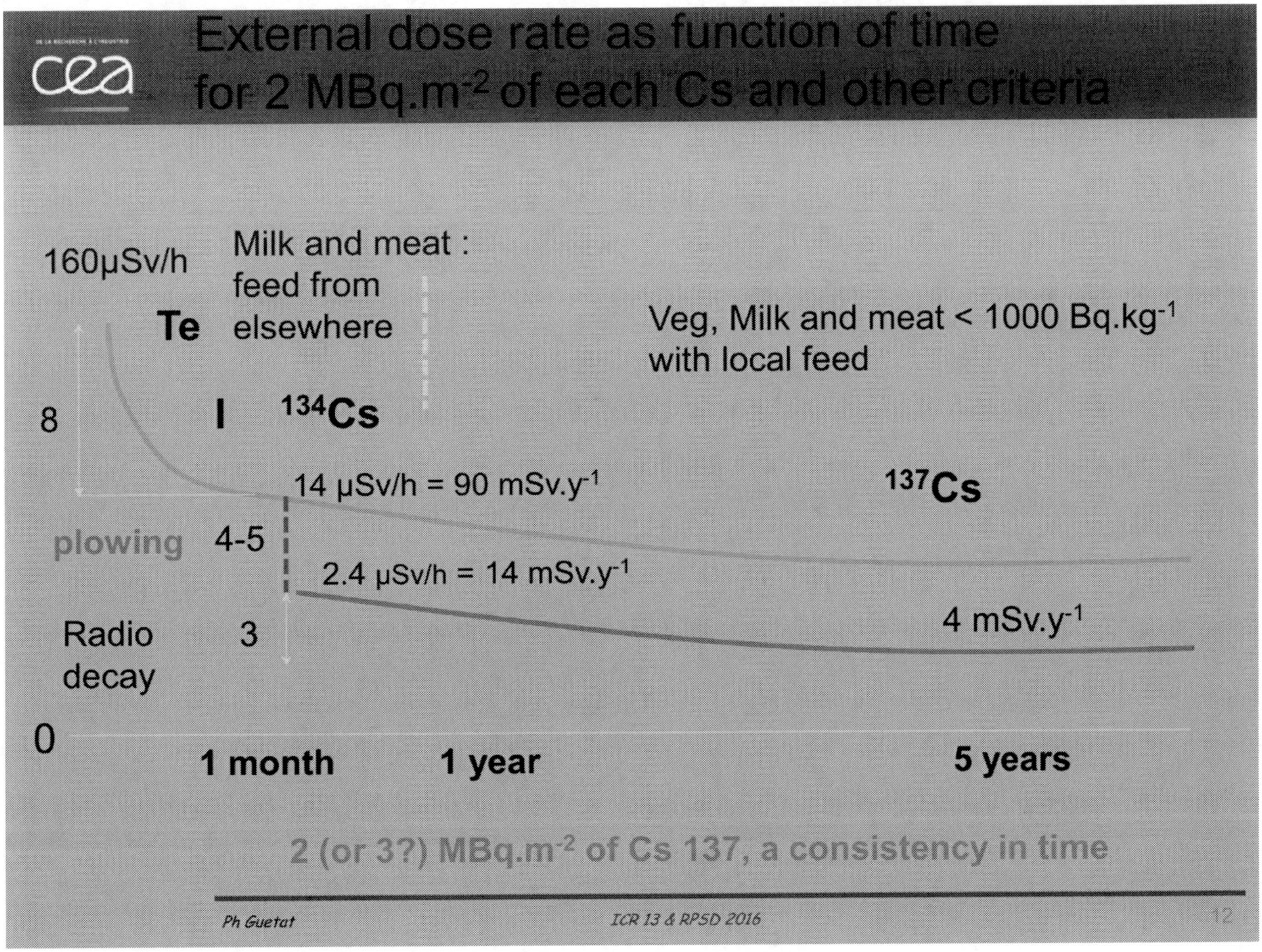
External dose rate as function of time
for 2 MBq.m-2 of each Cs and other criteria
160µSv/h
Milk and meat :
feed from
elsewhere
Te
Veg, Milk and meat < 1000 Bq.kg-1
with local feed
8
I
134Cs
14 µSv/h = 90 mSv.y-1
137Cs
plowing
4-5
2.4 µSv/h = 14 mSv.y-1
Radio
decay
3
4 mSv.y-1
0
1 month
1 year
5 years
2 (or 3?) MBq.m-2 of Cs 137, a consistency in time
Ph Guetat
ICR 13 & RPSD 2016
12

An illustration

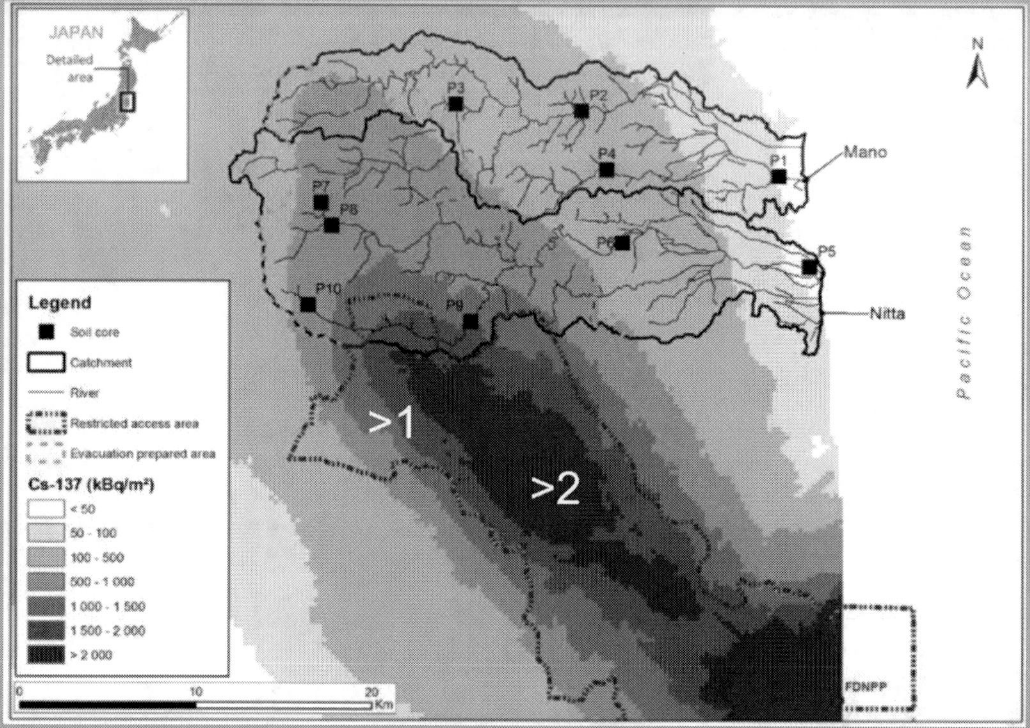

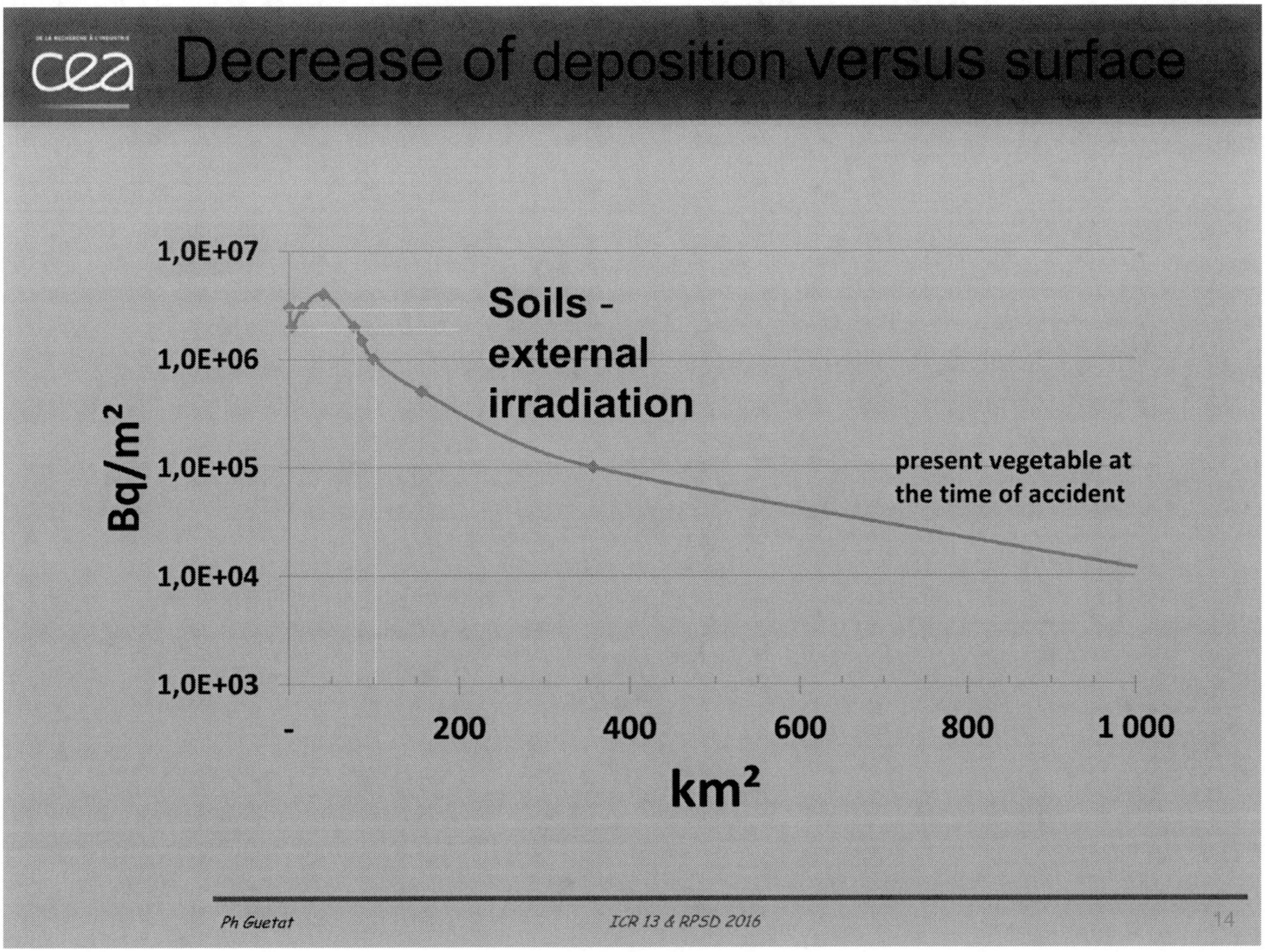
cea
Decrease of deposition versus surface
1,0E+07
1,0E+06
1,0E+05
1,0E+04
1,0E+03
Bq/m²
Soils -
external
irradiation
present vegetable at
the time of accident
-
200
400
600
800
1 000
km²
Ph Guetat
ICR 13 & RPSD 2016
14

II - WASTE

Remediation and Waste management

- The best waste is the waste that does not exist.
- Soil is a wealth to preserve
- Food is an important but short-term trouble.

To avoid waste

- Plowing, digging over
- Vegetable grubbing and burial by plowing, digging over.
- Distribution of uncontaminated Feed
- If contaminated milk, disposed on farm by spreading with urines/slurry

III – OTHER NUCLIDES

17

Tritium, tritiated water

- Is present in CANDU & ITER (risk = fire ?)
- water (HTO), organic (in environment, plants), (Gas HT: no risk), metallic hydrides.
- **HTO** : Very low radiotoxicity – rapid turnover 12 y period
- Very quick decrease in plants in 2 days (100 to 1000).
- No bioaccumulation, no bioconcentration. (major element)
- No health risk, but psychological risk.
- ***Food ban for some days, for few km for 10 g T.***
- Reference value 10^4 Bq/kg far too low at short term. Should be 10^6 10^7 Bq/kg for intervention.
- Dilution is absolutely the best, (especially in ocean).

Plutonium

- Fuel reprocessing
- Dense metal, non-volatile, no external irradiation, non-metabolized element.
- Transfer to plants and animals very small.
- Resuspension is of minor importance.
- For a standard 1g release, ***food restriction for leaf vegetable on few km***.
- Incineration is possible but useless.
- Standard impact assessment much over estimated. (density, veg. washing, dose factors)

Conclusions

- A major nuclear accident (with food ban) has presently no health effect, and environmental/waste effects are mainly psychological. (LNT law)
- Consistency between the criteria is necessary.
- Dilution has to be a large part of the post-accidental management.
- Food and Feed ban can be done at large scale. Soils should be preserved.
- Experts should not stress population and decision makers.

Estimation of Effective Dose from External Exposure in The Six Prefectures adjacent to Fukushima Prefecture

Hirokazu Miyatake[1,a], *Nobuaki* Yoshizawa[1], *Sachiko* Hirakawa[1], *Kana* Murakami[1], *Mari* Takizawa[1], *Masaki* Kawai[1], *Osamu* Sato[1], *Shunji* Takagi[1] and *Gen* Suzuki[2]

[1] *Mitsubishi Research Institute, Inc.: 2-10-3, Nagata-cho, Chiyoda-ku, Tokyo, Japan*
[2] *International University of Health and Welfare Clinic: 2600-6, Kitakanemaru, Otawara-city, Tochigi, Japan*

Abstract. The Fukushima Daiichi Nuclear Power Plant accident caused a release of radionuclides. Radionuclides were deposited on the ground not only in Fukushima prefecture but also in nearby prefectures. Since the accident, measurement of radiation in environment such as air dose rate and deposition density of radionuclides has been performed by many organizations and universities. In particular, Japan Atomic Energy Agency (JAEA) has been performing observations of air dose rate using a car-borne survey system continuously and over wide areas. In our study, using the data measured by JAEA, we estimated effective dose from external exposure in the six prefectures adjacent to Fukushima prefecture. Since car-borne survey was started a few months later after the accident, measured air dose rate in this method is mainly contributed by ^{137}Cs and ^{134}Cs whose half-lives are relatively long. Therefore, based on air dose rate of ^{137}Cs and ^{134}Cs and the ratio of deposition density of short-half-life nuclides to that of ^{137}Cs and ^{134}Cs, we also estimated effective dose contributed from not only ^{137}Cs and ^{134}Cs but also other short-half-life nuclides. We compared the effective dose estimated by the method above with that of UNSCEAR and measured data using personal dosimeters in some areas.

1 Introduction

The Fukushima Daiichi Nuclear Power Plant accident caused a release of radionuclides. Radionuclides were deposited on the ground not only in Fukushima prefecture but also in nearby prefectures. Since the accident, measurement of radiation in environment such as air dose rate and deposition density of radionuclides has been performed by many organizations and universities. In particular, Japan Atomic Energy Agency (JAEA) has been performing observations of air dose rate using a car-borne survey system continuously and over wide areas. There has been 11 times surveys performed as of March 2016 [1].

In our study, considering some factors such as time spent indoor and outdoor, background and radiation reduction by house, and using the data of air dose rate measured by JAEA [2], we estimated effective dose from external exposure in the six prefectures adjacent to Fukushima prefecture (Iwate pref., Miyagi pref., Ibaraki pref., Tochigi pref., Gunma Pref. and Chiba Pref.).

United Nations Scientific Committee on the Effects of Atomic Radiation has also estimated effective dose using measured date of radiation in environment [3]. In their method, density of radionuclides is used. There is a similarity between UNSCEAR's method and ours in terms of using measured data in environment. However, although the data that we use has been measured continuously, the data that UNSCEAR used was the one that was measured only one time.

There is also another approach such as measurement using personal dosimeters. This kind of survey was performed some places, such as Tochigi pref., Kashiwa city in Chba pref. and Marumori town in Miyagi pref [4] [5] [6] .

In the case of Marumori town in Miyagi pref., this town was divided into three areas (Area-A,B,C) and measurement of external dose was performed using personal dosimeter. There were a few hundred children as examinee, and the measurement had been continued for two years since September 2011. In this survey, other data such as time spent indoor and outdoor, body height and bodily weight. In the relevant survey, radiation reduction factor of typical types of house in the nearby areas was also measured [7]. For purpose of reference, typical measured data are shown in the Table. 1 for three areas.

Table. 1. Typical measured data of dose from external exposure of children in Marumori town in Miyagi pref. [5]

	Measurement period	Result
Area-A	2011.12~2012.9	2.25 mSv
Area-B	2012.6~2013.11	1.63 mSv
Area-C	2012.7~2013.11	1.26 mSv

Measurement using personal dosimeter has the advantage that the date measured contains various sorts of uncertainties such as time spend indoor and outdoor and type of houses, therefore it is not necessary to

[a] Corresponding author: himiyata@mri.co.jp

introduce any conversion coefficients or assumptions. On the other hand, examinees have to attach dosimeters to their bodies, and therefore it is difficult to keep measuring for years against many examinees.

In our study, we estimated effective dose from external exposure based on measured air dose rate by car-borne survey by JAEA, and compered its vale to the one estimated by UNSCEAR and the one measured using personal dosimeters in Marumori town in Miyagi Pref.

2 Methods

In our study, we annualized measured air dose rate with two models. By fitting with the models, time-dependent function of air dose rate. Maltiplying some conversion factors and subtracting background, air dose rate can be converted to effective dose rate contributed from nuclides that were released by The Fukushima Daiichi Nuclear Power Plant accident. Integrating this, accumulative effective dose from external exposure can be obtained (See Fig. 1).

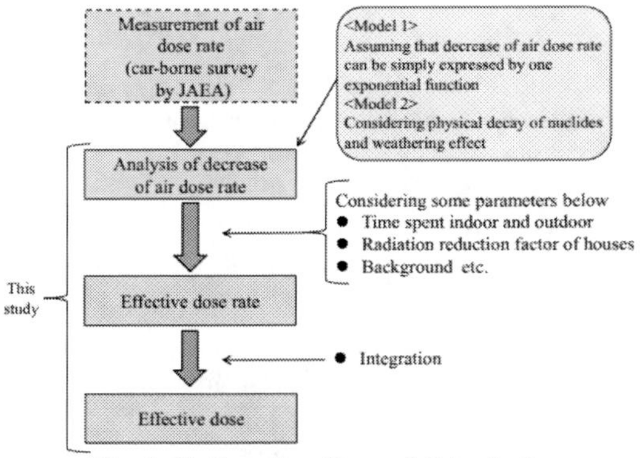

Fig. 1. Estimation flow of this study

2.1 Models of decrease of air does rate

Air dose rate decreases because of physical decay and weathering effect. In the case of physical decay, air dose rate decay at the pace determined by half-life of each nuclide. Weathering effect is a natural phenomenon where rainfall promotes nuclide to penetrate into the soil, therefore this depends on weather condition or soil characteristics of each area.

In this study, we use two models to analyze decrease of air dose rate. The first model is given by the equation (1). In this model, it is assumed that decrease of air dose rate can be simply expressed by one exponential function.

$$D_{\mathrm{air}}(t) = a\exp(-bt) + c \tag{1}$$

The second model is given by the equation (2) using the parameters listed in the Table. 2 [8]. In this model, decay of nuclide and weathering are considered.

$$D_{\mathrm{air}}(t) = (D_0 - D_{\mathrm{BG}}) \times \left\{ f_{\mathrm{fast}} \exp\left(-\frac{0.693}{T_{\mathrm{fast}}}t\right) + (1 - f_{\mathrm{fast}})\exp\left(-\frac{0.693}{T_{\mathrm{slow}}}t\right)\right\} \times \frac{k\exp(-\lambda_{134}t) + \exp(-\lambda_{137}t)}{k+1} + D_{\mathrm{BG}} \tag{2}$$

According to the previous studies, weathering effect can be expressed as the second term in equation (2) under the assumption that this effect can be divided into two elements whose contributions are normalized, where f_{fast} is fraction of fast element, and T_{fast} and T_{slow} are half-life of fast and slow elements [9]. The third term in equation (2) coresponds to physical decay of ^{137}Cs and ^{134}Cs, where k is ratio of ^{137}Cs to ^{134}Cs contributing to air dose under the assumption that the initial value of deposition density of each nuclide is same, and λ_{134} and λ_{137} are decay constants of each nuclide.

Table. 2. Parameters in the second model

Parameters	
D_0	Initial value of air dose rate (μSv/h)
D_{BG}	Background (μSv/h)
f_{fast}	Fraction of fast element of weathering effect (dimensionless)
T_{fast}	Half-life of fast element of weathering effect (y)
T_{slow}	Half-life of slow element of weathering effect (y)
k	Ratio of ^{137}Cs to ^{134}Cs contributing to air dose (dimensionless)
λ_{134}, λ_{137}	Decay constant of ^{134}Cs and ^{137}Cs (1/y)

2.2 Conversion to effective dose

Considering some factors such as time spent indoor and outdoor, background and radiation reduction of houses, air dose rate can be converted to effective dose rate as shown in equation (3). In this study, by substracting the background which was measured in the area of interest before the accident, we estimated the dose added by the nulides released by The Fukushima Daiichi Nuclear Power Plant accident the accident.

$$\frac{dE(t)}{dt} = (pD_{\mathrm{air}}(t) - D_{\mathrm{BG}}) \times (sT_{\mathrm{in}} + T_{\mathrm{out}}) \times \alpha \times f \tag{3}$$

In car-borne survey, measurement is conducted on streets with detectors set in the car. The measured air dose rate is the one on streets. Because of traffics of cars, nuclides deposited on streets are easier to be removed or penetrate into soil than in living environment, therefore air dose rate measured by car-borne survey tends to be smaller air dose rate measured in living environment. It is shown that the former is approximately 1.2 times lareger than the latter by man-borne survey conducted nearby the street where car-borne surevy was also conducted [10]. Considering this difference, the coefficient p is multiplied by $D_{\mathrm{air}}(t)$.

Air dose rate $D_{\mathrm{air}}(t)$ includes background. By subtracting background D_{BG}, the net contribution from nuclides released by The Fukushima Daiichi Nuclear

Power Plant accident is only considered. As background D_{BG}, we used the measured data in prefectural capitals of each prefecture before the accident (Table. 3) [11].

Table. 3. Background in prefectural capitals of each prefecture

Prefecture	D_{BG} (μSv/h)
Iwate pref.	0.062
Miyagi pref.	0.066
Ibaraki pref.	0.063
Tochigi pref.	0.055
Gunma pref.	0.065
Chiba pref.	0.059

After The Fukushima Daiichi Nuclear Power Plant accident, there was a statistical survey on time spent indoor and outdoor of Fukushima City office workers [12]. According to this, indoor workers spend 0.57 hours outdoor on average. In this study, although the targeted prefectures are not Fukushima prefecture, we used this value.

Radiations are reduced indoor because of shielding effect of houses. In this study, we adapted the value 0.4 for reduction factor s [13].

For conversion factor between air dose rate and effective dose, we adapted the value 0.6, where the target that we estimate is assumed to be adult [14].

Table. 4. Paremeters to conver frome air dose rate to effective dose rate.

Parameters	
p	Ratio of air dose rate in living atmosphere to that on streets (1.2)
D_{BG}	Background (see **Table. 3**)
s	Radiation reduction factor (0.4)
T_{in}	Time spent indoor (23.43 h/d)
T_{out}	Time spent outdoor (0.57 h/d)
α	Number of days per year (365 h/d)
f	Conversion factor between air dose rate and effective dose (0.6)

Finaly, by integrating according to equation (3), accumulated effective dose in the period (form year T_1 to T_2) can be estimated.

$$E(T_1, T_2) = \int_{T_1}^{T_2} \frac{dE(t)}{dt} dt \quad (4)$$

2.3 Effective dose considering short-half-life nuclides

Since car-borne survey was started a few months later after the accident, measured air dose rate in this method is mainly contributed by ^{137}Cs and ^{134}Cs whose half-lives are relatively long. Therefore, based on air dose rate of ^{137}Cs and ^{134}Cs and the ratio of deposition density of short-half-life nuclides to that of ^{137}Cs and ^{134}Cs, we also estimated effective dose contributed from not only ^{137}Cs and ^{134}Cs but also other short-half-life nuclides.

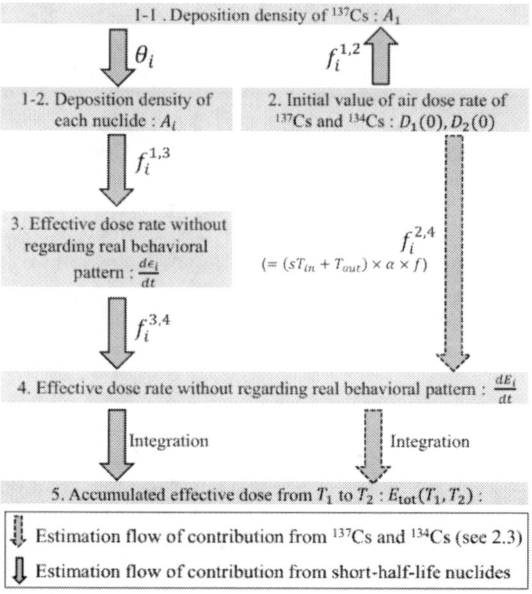

Fig. 2. Estimation flow of contribution of short-half-life nuclides

The initial value $D_{air}(0)$ can be obtained by plugging in 0 into t in eqution (1) or (2). This value can be considerd as sum of the contribution of ^{137}Cs, ^{134}Cs and background as shown in equation (5).

$$D_{air}(0) = D_1(0) + D_2(0) + D_{BG} \quad (5)$$

Where $D_1(0)$ is air dose rate contributed from ^{137}Cs in the soil and $D_2(0)$ is that of ^{134}Cs in the soil. As shown in the Table. 5, index i is defined to indicate each nuclide. Index $i = 1,2$ correspond ^{137}Cs and ^{134}Cs. By substructing back ground from $D_{air}(0)$, value $D_1(0) + D_2(0)$ can be obtained.

By defining deposition density for each nuclide i as A_i and conversion factor from deposition density A_i to air dose rate as $f_i^{1,2}$, for ^{137}Cs ($i = 1$) and ^{134}Cs ($i = 2$), the equation below is obtained.

$$D_1(0) + D_2(0) = f_1^{1,2} A_1 + f_2^{1,2} A_2 \quad (6)$$

Assumuing that the initial value of deposition densityof ^{137}Cs and ^{134}Cs is same ($A_1 = A_2$) [15], parameter A_2 can be eliminated, then the equation below is obtained.

$$A_1 = \frac{D_1(0) + D_2(0)}{f_1^{1,2} + f_2^{1,2}} \quad (7)$$

Defining the ratio of initial deposition density of each nuclide i to that of ^{137}Cs ($i = 1$) as ϑ_i, the equation below can be obtained.

$$A_i = \vartheta_i A_1 \quad (8)$$

Although this ratio has been studied in many previous studies, in this study, we adapted the same values as UNSCEAR (**Table. 5**) [15].

Table. 5. Deposition ratio and half-life for each nuclide i

i	Nuclide	ϑ_i	T_i^{half}
1	$^{137}Cs + ^{137m}Ba$	-	30 y
2	^{134}Cs	1.0	2.1 y
3	^{136}Cs	0.17	13 d
4	^{131}I	12	8.0 d
5	$^{129m+129}Te$	1.1	34 d
6	$^{132}Te + ^{132}I$	8.0	3.2 d
7	$^{110m+110}Ag$	2.8×10^{-3}	25 d

Defining the conversion factor from deoisition density A_i to effective dose rate $d\varepsilon_i/dt$ as $f_i^{1,3}$, effective dose ration contributed from each nuclide i can be expressed as the equation (9), where T_i^{half} is half-life for nuclide i.

$$\frac{d\varepsilon_i(t)}{dt} = A_i \exp\left(-\frac{\ln 2}{T_i^{half}}t\right) \quad (9)$$

As shown in 2.2, by multiplying some factors (defining its product as $f_i^{3,4}$ in equation (10)),

$$f_i^{3,4} = (sT_{in} + T_{out})\times\alpha \quad (10)$$

this can be converted to effective dose that considers time spent indoor and outdoor, and reduction factor is obtained as shown in equation (11).

$$\frac{dE_i(t)}{dt} = f_i^{3,4}\frac{d\varepsilon_i(t)}{dt} \quad (11)$$

Finally, by integrating equation (11), accumulated effective dose including short-half-life nucliide can be obtained.

$$E_{tot}(T_1,T_2) = E(T_1,T_2) + \sum_{i=3}^{7}\int_{T_1}^{T_2}\frac{dE_i(t)}{dt}dt \quad (12)$$

Since effective dose contributed from ^{137}Cs and ^{134}Cs ($E(T_1,T_2)$) can be obtaind in the way expressed in 2.2, the total effective dose is expresserd, the second term in equation (12) is expressed as the sam from $i=3$ to $i=7$.

Table. 6. Conversion factor $f_i^{1,2}$ and $f_i^{1,3}$ [15][16]

i	Nuclide	$f_i^{1,2}$	$f_i^{1,3}$
		($\times10^{-3}$ μSv/h per kBq/m^2)	
1	$^{137}Cs+^{137m}Ba$	2.3	-
2	^{134}Cs	6.3	-
3	^{136}Cs	-	4.7
4	^{131}I	-	0.85
5	$^{129m+129}Te$	-	0.15
6	$^{132}Te+^{132}I$	-	5.5
7	$^{110m+110}Ag$	-	6.1

3 Results

3.1 Time-dependent degrease of air dose rate

By fitting the average of measured data of air dose rate from the first the 7th car-borne survey with the two models shown in 2.1, we analysed time-dependent degrease of air dose rate in the 6 prefectures (Iwate pref., Miyagi pref., Ibaraki pref., Tochigi pref., Gunma Pref. and Chiba Pref.).

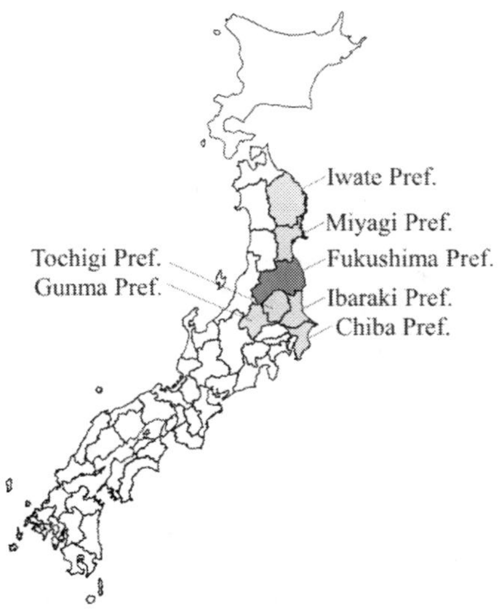

Fig. 3. Prefectures targeted in this study

In the first model, parameters a,b and c were free. In the second model, D_0,f_{fast} and T_{fast} were free, although T_{slow} was constan ($T_{slow}=92$ y) [7] and as D_{BG} the values in the **Table. 3** were used.

The parameters for Miyagi Pref., Tochigi Pref. and Ibaraki Pref. were derermined by fitting as shown in the **Table. 7** and **Table. 8**. For the three prefectures (Miyagi, Ibraki and Tochigi Pref.), averaged measured data and time-dependent degrease of air dose rate drawn by the determined parameters are shown in Fig. 4 ~Fig. 6.

Table. 7. Fitting paremeters for the first model (Miyagi, Ibraki and Tochigi Pref.)

	a (μSv/h)	b (1/y)	c (μSv/h)
Miyagi Pref.	0.48	3.0	0.075
Ibaraki Pref.	0.21	2.3	0.093
Tochigi Pref.	0.47	2.9	0.096

Table. 8. Fitting paremeters for the second model (Miyagi Pref., Ibraki Pref. and Tochigi Pref.)

	D_0 (μSv/h)	f_{fast}	T_{fast} (1/y)
Miyagi Pref.	0.58	0.97	3.0
Ibaraki Pref.	0.32	0.82	2.5
Tochigi Pref.	0.59	0.88	3.1

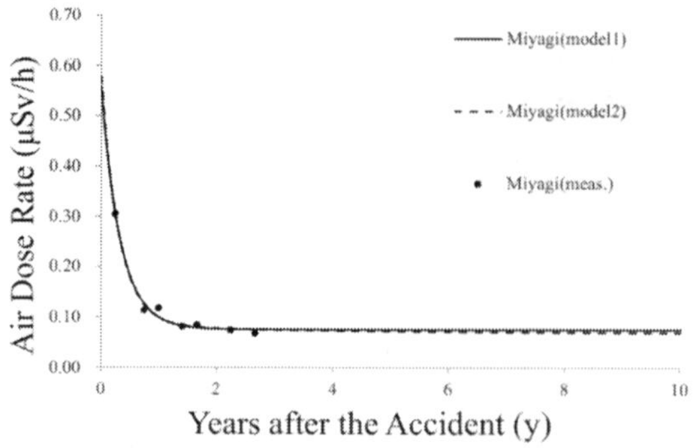

Fig. 4. Air does rate (Miyagi Pref.)

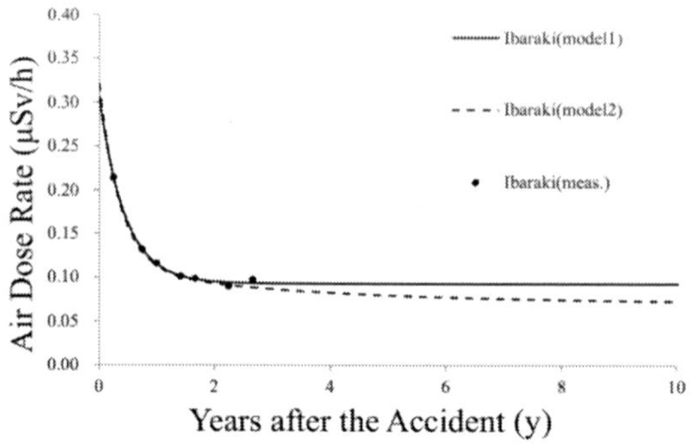

Fig. 5. Air does rate (Ibaralki Pref.)

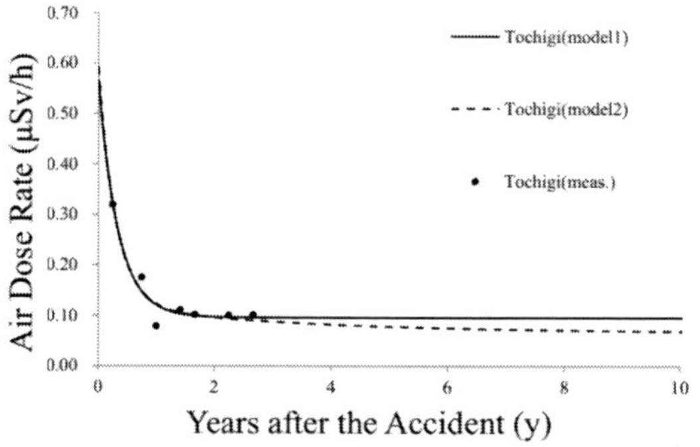

Fig. 6. Air does rate (Tochigi Pref.)

In the three prefectures (Miyagi, Ibraki and Tochigi Pref.) where measured data of air dose rate were relatively small, there were some cases where fitting parameters were not determined appropriately. In order to avoid this, we used the fitting parameters in the other prefecture (b in the first model, and T_{fast} in the second model) and fixed the parameter (f_{fast} in the second model). The parameters determined for these three prefectures are shown in Table. 9 and Table. 10.

Table. 9. Fitting paremeters for the first model (Iwate, Gunma and Chiba Pref.)

	a (μSv/h)	b (1/y)	c (μSv/h)
Iwate Pref.	0.17	0.81*1	0.030
Gunma Pref.	0.089	0.81	0.051
Chiba Pref.	0.13	0.81*1	0.053

*1 Using the parameter determined in Gunma Pref.

Table. 10. Fitting paremeters for the second model (Iwate, Gunma and Chiba Pref.)

	D_0 (μSv/h)	f_{fast}	T_{fast} (1/y)
Iwate Pref.	0.20	0.90*2	1.3*3
Gunma Pref.	0.15	0.90*2	1.3
Chiba Pref.	0.24	0.90*2	1.3*3

*2 Fixed with the typical value
*3 Using the parameter determined in Gunma Pref.

Averaged measured data and time-dependent degrease of air dose rate drawn by the determined parameters are shown in Fig. 7~Fig. 9.

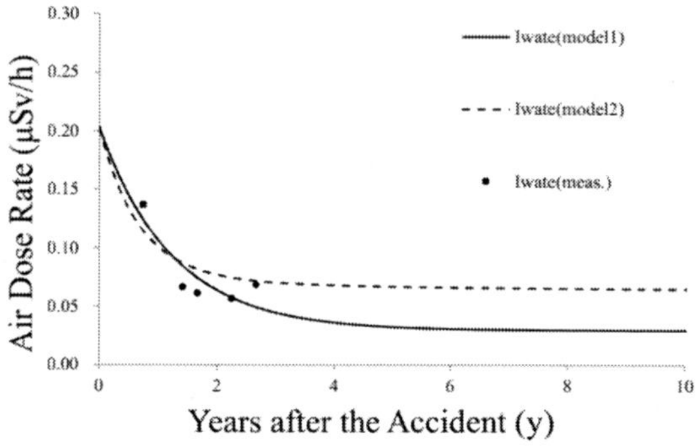

Fig. 7. Air does rate (Iwate Pref.)

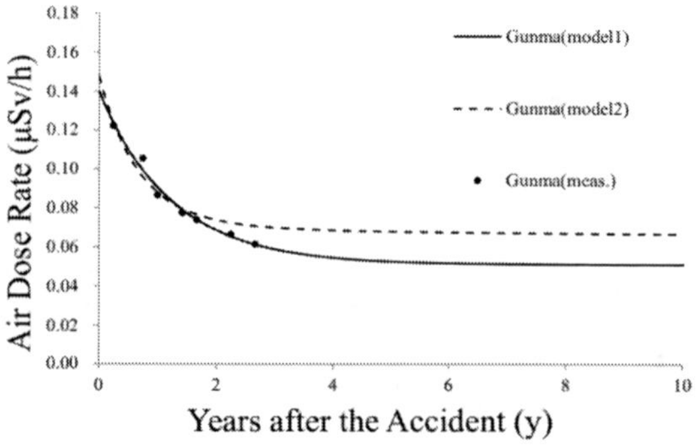

Fig. 8. Air does rate (Gunma Pref.)

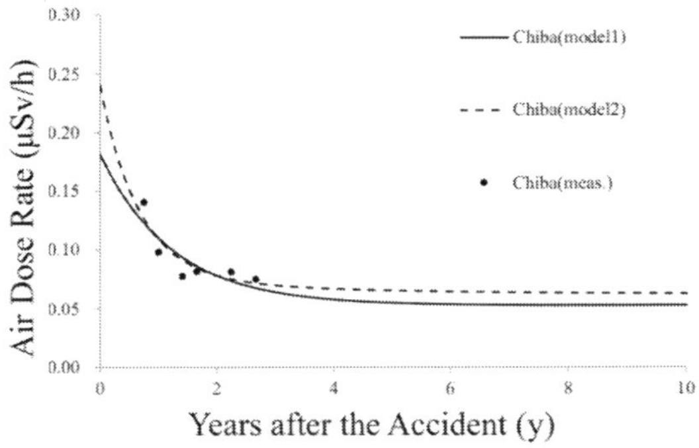

Fig. 9. Air does rate (Chiba Pref.)

3.2 Effective dose (considering ^{137}Cs and ^{134}Cs)

Based on the time-dependent degrease of air dose rate shown in 3.1, we estimated accumulated effective dose (considering only ^{137}Cs and ^{134}Cs) by equation (3) and (4). Fig. 10~Fig. 12 show the estimated results for Miyagi Pref., Ibaraki Pref. and Tochigi Pref.

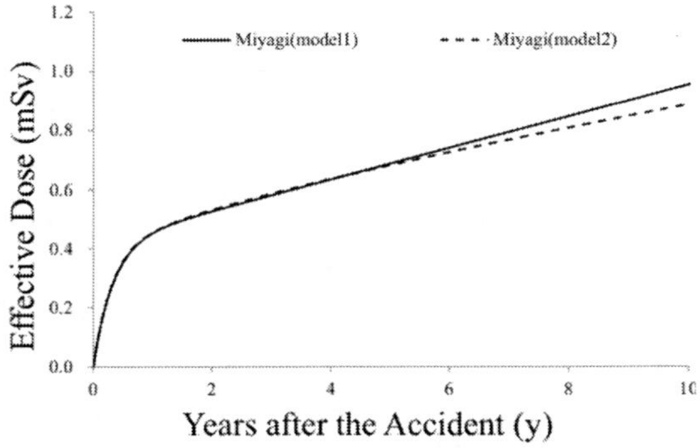

Fig. 10. Effective dose (considering ^{137}Cs and ^{134}Cs) (Miyagi Pref.)

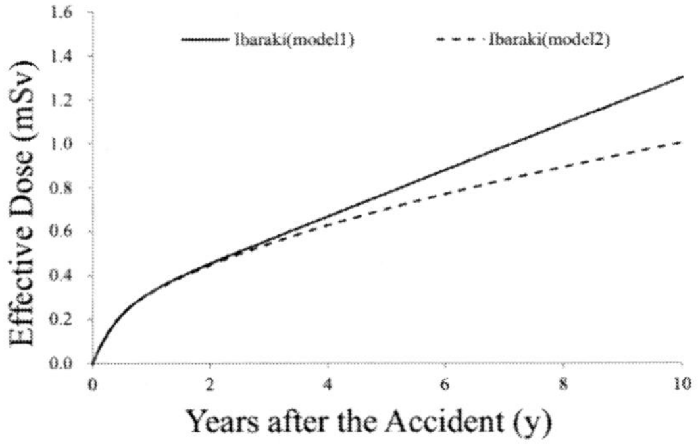

Fig. 11. Effective dose (considering ^{137}Cs and ^{134}Cs) (Ibaraki Pref.)

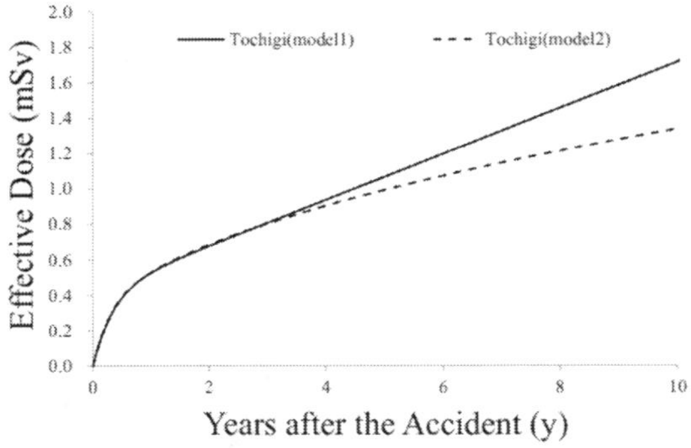

Fig. 12. Effective dose (considering ^{137}Cs and ^{134}Cs) (Tochigi Pref.)

For the three prefectures where some parameters were fixed with the values determined in the other prefecture or fixed with the typical value, the estimated results are shown in Fig. 13~Fig. 15. For the first model in Iwate Pref. and Gunma Pref., air dose rate decreases to the background level. In this case, when integrateing by equation (4), interval of integration was confined till air dose rate gets the background level.

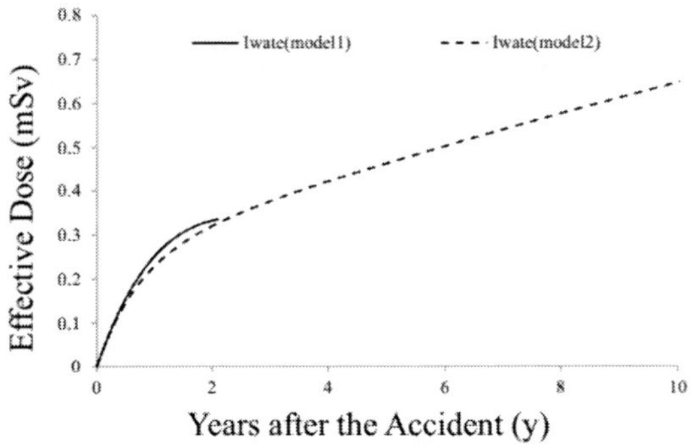

Fig. 13. Effective dose (considering ^{137}Cs and ^{134}Cs) (Iwate Pref.)

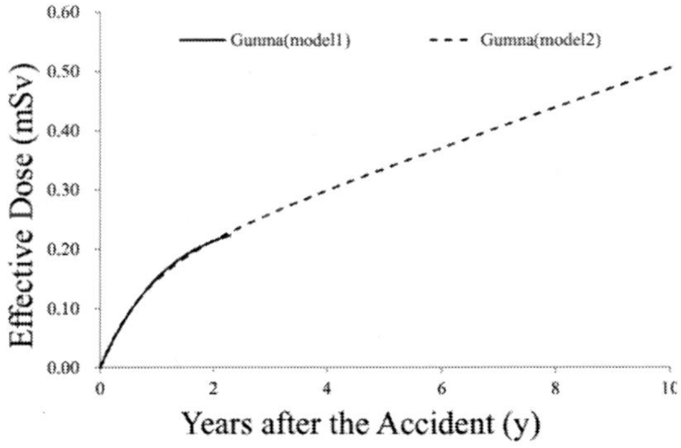

Fig. 14. Effective dose (considering ^{137}Cs and ^{134}Cs) (Gunma Pref.)

DOI: 10.1051/epjconf/201715308006

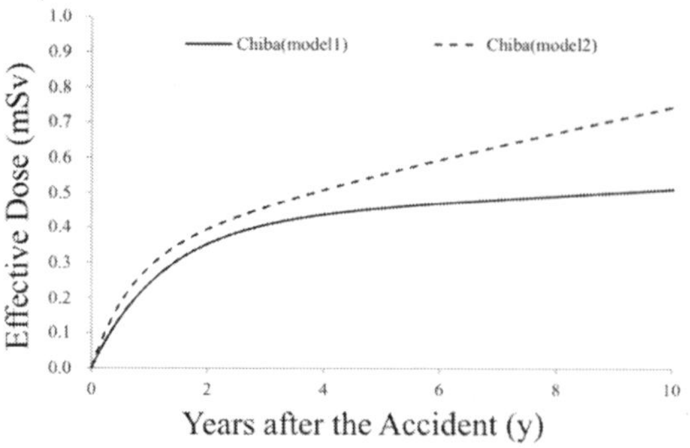

Fig. 15. Effective dose (considering ^{137}Cs and ^{134}Cs) (Chiba Pref.)

3.3 Effective dose (considering ^{137}Cs, ^{134}Cs and short-half-life nuclides)

The results shown in 3.2 do not include the contribution from short-half-life nuclides. By the method shown in 2.3, we also estimated accumulated effective dose considering short-half-life nuclides.

Estimation was conducted under the condition below.

- Only some prefectures (Miyagi Pref., Ibraki Pref. and Tochigi Pref.) were targeted.
- The interval of integration ranges from $T_1 = 0\text{y}$ to $T_2 = 3\ \text{y}$, because half-lives of other nuclides than ^{137}Cs and ^{134}Cs are relatively short (a few ten days at longest)
- The fast model was used because in this range, there is no significant deference between two models.

The estimated results are shown in Fig. 16~Fig. 18.

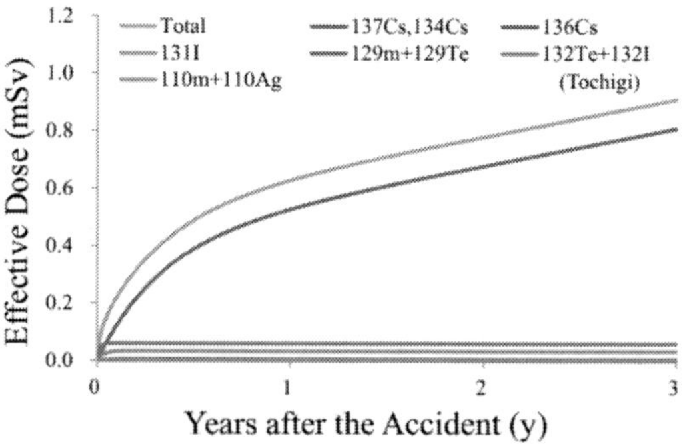

Fig. 16. Effective dose (considering ^{137}Cs, ^{134}Cs and short-half-life nuclides) (Tochigi Pref.)

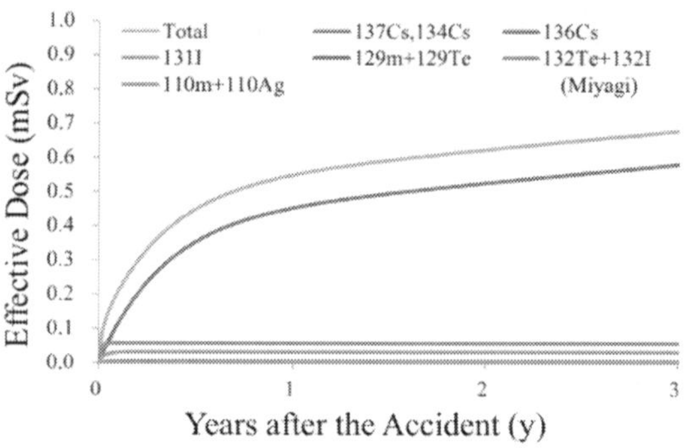

Fig. 17. Effective dose (considering ^{137}Cs, ^{134}Cs and short-half-life nuclides) (Miyagi Pref.)

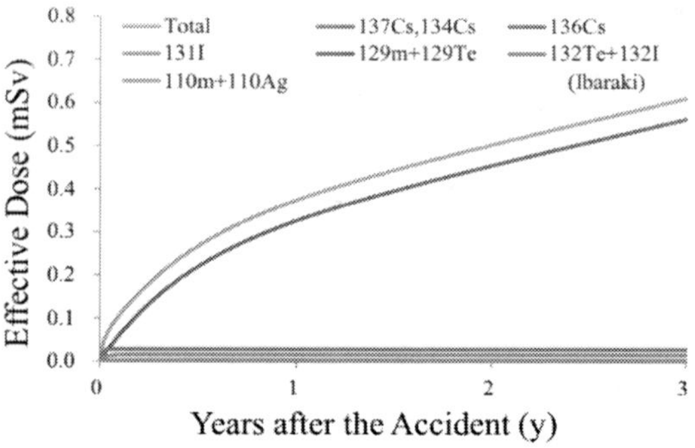

Fig. 18. Effective dose (considering ^{137}Cs, ^{134}Cs and short-half-life nuclides) (Ibaraki Pref.)

3.4 Estimation in Marumori Town in Miyagi Pref.

As shown in 1., there were some cases where measurement using personal dosimeters were conducted. In particular, in Marumori Town in Miyagi Pref., measurement had been conducted for relatively long [6]. In order to compare results, we estimated effective dose in this town.

First, we analysed averaged measured date of air dose rate in this town as the same way shown above, however here we used the first model here.

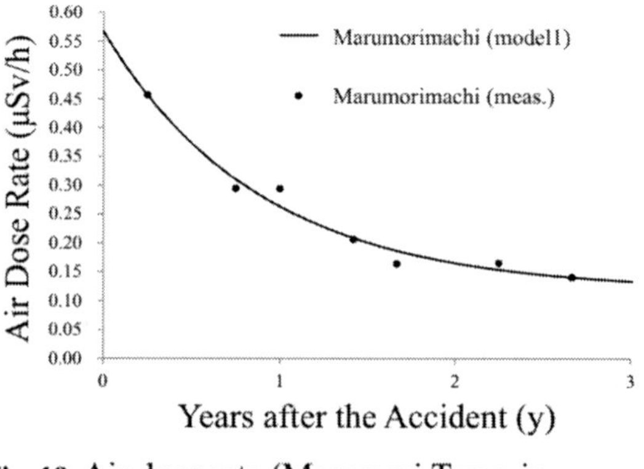

Fig. 19. Air does rate (Marumori Town in Miyagi Pref.)

Estimation of effective dose based on air dose rate was done almost the same way above exept for some small changes below.

- Distributions of parameters were considered as possible, and ten thouthands sets of randam numbers of the distributions were generated.
- Conversion factor was changed, because examinees in Marumori Town were children.
- Background was not sabstructed in equation (3) because accumulated value by personal dosimeters includes background.
- Interval of integration (T_1, T_2) in equation (4) was changed to adjust to the condition of measurement using personal dosimeters in Marumori Town.

The parameters used here are listed in **Table. 11**.

Table. 11. Parameters used in estimation for Marumori Town in Miyagi Pref.

Parameters		Value	Ref.
p	Lognormal distribution	Mode : 1.2	[10]
D_{BG}	-	-	-
s	Uniform distribution	Lower limit : 0.4 Upper limit : 1.0	[7]
T_{out}	Nomal distribution	Average : 3.0 h/d, SD : 1.5 h/d	[6]
α	Constan	365 h/d	-
f	Constant	0.7	[17]

Estimated results are shown in **Fig. 20~Fig. 22**. The values written in these figures with dashed lines are the typical value measured with personal dosimeter in each area in Marumori Town **(Table. 1)** [6].

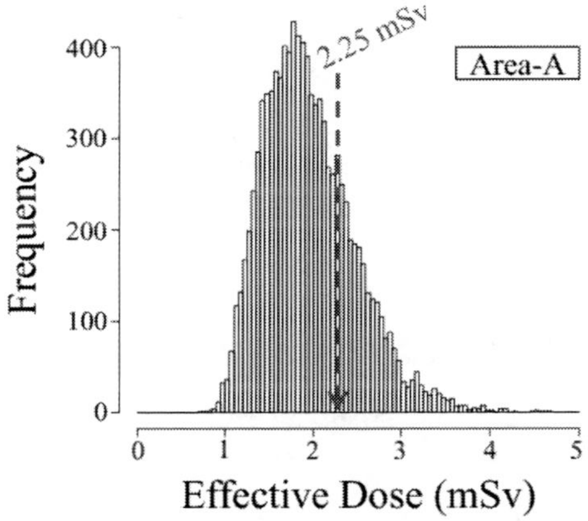

Fig. 20. Distribution of estimated effective dose including background in area A in Marumori Town (the value 2.25 mSv is the typical one measured in this area)

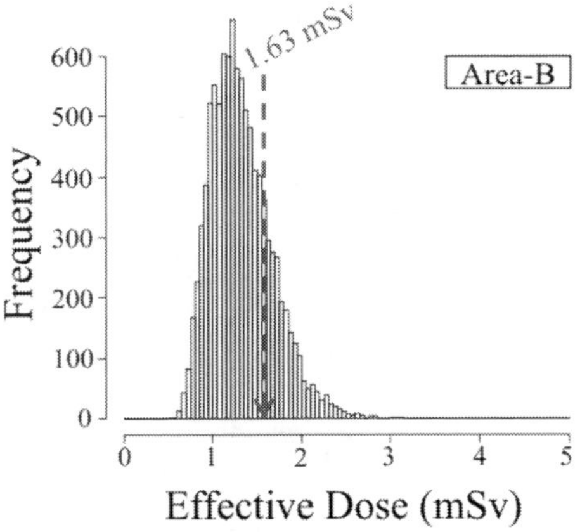

Fig. 21. Distribution of estimated effective dose including background in area B in Marumori Town (the value 1.63 mSv is the typical one measured in this area)

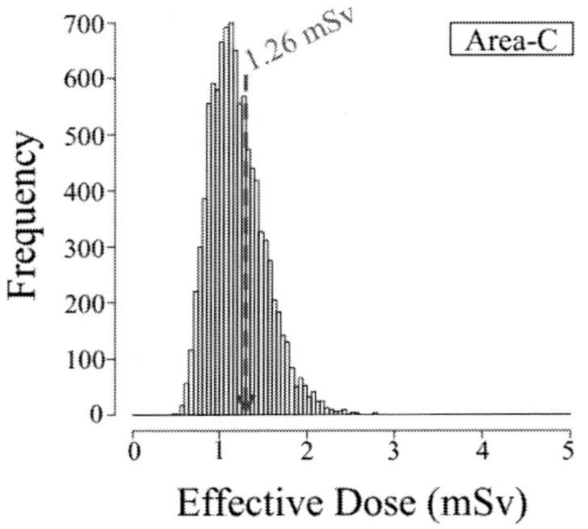

Fig. 22. Distribution of estimated effective dose including background in area C in Marumori Town (the value 1.26 mSv is the typical one measured in this area)

4 Discussions

4.1 Comparison with estimation of UNSCEAR

Estimated effective dose in this study and that of UNSCEAR are listed in Table. 12. In this list, these values does not include background. Although UNSCEAR estimated at the level of cities, towns and villages, in this study we estimated at the level of prefectures. Therefore, estimated values of UNSCEAR were shown minimum and maximum values given by cities, towns or villages belonging to each prefecture. [18][19].

Table. 12. Comparison of estimated results (mSv)

	This study			UNSCEAR	
	model	1 y.	10 y.	1 y.	10 y.
Iwate Pref.	model1	0.25	-	0.14	0.1
	model2	0.23	0.65	~0.31	~0.7
Miyagi Pref.	model1	0.45	0.95	0.05	0.1
	model2	0.45	0.89	~0.55	~1.3
Ibaraki Pref.	model1	0.33	1.30	0.06	0.1
	model2	0.32	1.00	~0.54	~1.3
Tochigi Pref.	model1	0.53	1.72	0.06	0.2
	model2	0.53	1.34	~1.07	~2.5
Gunma Pref.	model1	0.15	-	0.06	0.1
	model2	0.15	0.51	~0.50	~1.2
Chiba Pref.	model1	0.24	0.51	0.05	0.1
	model2	0.29	0.74	~0.76	~1.8

Table. 12 shows that the values we estimated are witin the range of those of UNSCEAR. However, it is necessary to see the maximum value given by estimation of UNSCEAR, because of the differnece of geographical unit for estimeation. The maximum values in Table. 12 are 1.07 mSv (for 1y. in Tochigi Pref.) and 2.5 mSv (for 10 y. in Tochigi Pref.). Tese values are given by Nasu Town in Tochigi Prefecture [18][19].

We estimated efective dose in Nasu Town by the same way as shown above. The resulut is shown in Fig. 23. In this figure, in order to compare with the estimation of UNSCEAR, the curve of UNSCEAR estimation that we calucurated based on the medhod of UNSCEAR in this study is also shown [15].

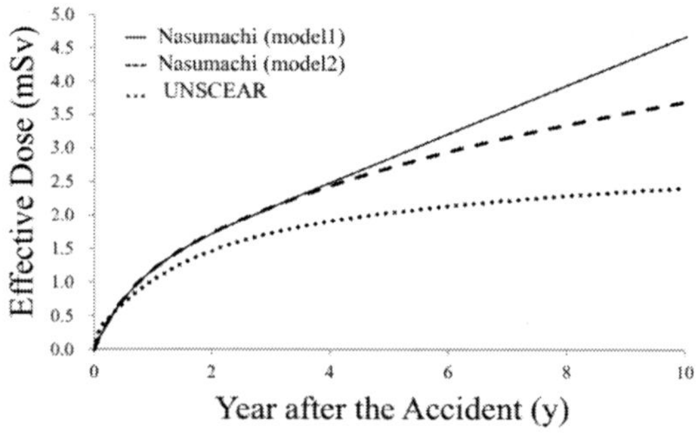

Fig. 23. Effective dose (Nasu Town in Tochigi Pref.)

Fig. 23shows that within a year after the accident the estimated values of both are almost same, however the longer time passes, the larger the estimated value in this study get.

In the UNSCEAR method, measured deposition density was used for input data. This measurement was conducted only one time within a few months after the accident. The number of measurement points was quite fewer than that of car-borne survey. In addition, some parameters used in UNSCEAR method are based on Chernobyl accident, therefore it is possible that environmental characteristics in Japan are not reflected enough.

4.2 Contribution from short-half-life nuclides

Fig. 24 shows the ratio of contribution from effective dose each nuclides to the total.

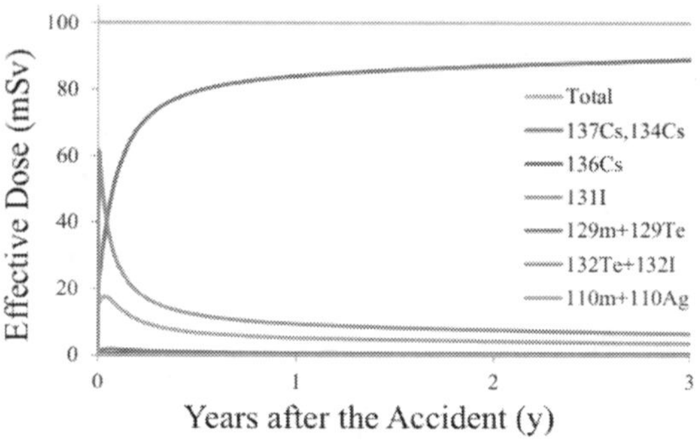

Fig. 24. Contribution from each nuclide to effective dose

This graph shows that within a few months after the accident the contribution from ^{132}Te, ^{132}I and ^{131}I is relatively large, and the contribution from ^{137}Cs and ^{134}Cs gets more dominant as time passes. At three years after the accident, the ratio of ^{137}Cs and ^{134}Cs to the total exceeds 80 percent. This suggests that estimated effective dose considering not only ^{137}Cs and ^{134}Cs but also other short-half-life nuclides is about 1.2 times larger than the one considering only ^{137}Cs and ^{134}Cs.

4.3 Comparison with measurement using personal dosimeters

Estimating effective dose in Marumori Town in Miyagi Pref., we considered distribution for some parameters. Therefore estimated value also has its distribution. Comparing the distributions of estimated value with typical measured value using personal dosimeters for three cases, it is shown that the latter value is within the distribution and corresponds to about 1.2 to 1.3 times lager value than the mode of each distribution. Although the distribution and typical measured values are in good agreement, it is necessary to be aware that they are basically different. There are some assumptions and some conversion factors that have uncertainty behind estimated values, although measured values using personal dosimeters are reflected by real behavioural pattern of each examinee.

REFERENCES

1. JAEA, Report in Japanese (2016), pp 102-119 http://radioactivity.nsr.go.jp/ja/list/564/list-1.html
2. JAEA, Database for Radioactive Substance Monitoring Data, http://emdb.jaea.go.jp/emdb/
3. UNSCEAR, UNSCEAR 2013 Report Volume 1 (2013)
4. Report written in Japanes, http://www.pref.tochigi.lg.jp/e04/documents/documents/documents/report.pdf
5. T. Iimoto et al., Measures of a Local Government in Tokyo Metropolitan Area on Environmental Radioactive Conteamination due to Fukushima Dai-ichi Nuclear Power Plant Accident and Related

 DOI: 10.1051/epjconf/201715308006

Discussion, Radiation Biology Research Communications **48** (1) , pp. 15-38 (2013)
6. H. Yoshida et al., Survey Report on Personal Dose Equivalent and Indoor and Outdoor Staying Time for Children in the Southern Miyagi Prefecture after the Fukushima Daiichi Nuclear Power Plant Accident, RADIOISOTOPES Vol. **64** No. **5**, pp. 319-333 (2015)
7. H. Yoshida et al., Effect of Shieldings on Ambient Equivalent Dose Rate Reduction Inside Resident's House after the Fukushima Daiichi Nuclear Power Plant Accident, RADIOISOTOPES Vol. **62** No. **4**, pp. 203-210 (2013)
8. JAEA, Report in Japanese (2016), http://radioactivity.nsr.go.jp/ja/contents/12000/11995/33/part2.pdf
9. H.J.Gale, D.L.O.Humphreys and E.M.R.Fisher, Weathering of Caesium-137 of soil, Nature, 4916, 257-261 (1964)
10. JAEA, Report in Japanese (2014), http://radioactivity.nsr.go.jp/ja/list/504/list-1.html
11. The Japan Chemical Analysis Center, Environmental Radioactivity and Radiation in Japan, http://www.kankyo-hoshano.go.jp/study_menu.html
12. Takahara, S., Iijima M., Shimada K., Kimura M. and Homma T., Probabilistic Assessment of Doses to the Public Living in Areas Contaminated by the Fukushima Daiichi Nuclear Power Plant Accident, Radiation Monitoring and Dose Estimation of the Fukushima Nuclear Accident, pp. 197–214 (2014)
13. IAEA, Generic procedures for assessment and response during a radiological emergency, IAEA-TECDOC-**1162**, pp.1-181 (2000)
14. ICRP, Conversion Coefficients for use in Radiological Protection against External Radiation, ICRP Publication **74**, Ann. ICRP 26 pp.3-4 (1996)
15. UNSCEAR, METHODOLOGY FOR THE ASSESSMENT OF DOSE FROM EXTERNAL EXPOSURE AND INHALATION OF RADIOACTIVE MATERIAL, UNSCEAR 2013 Report, ATTACHMENT C-12 (2013)
16. Saito K, Ambient dose equivalent conversion coefficients for radionuclides exponentially distributed in the ground, Journal of Nuclear Science and Technology, **51**, pp. 1274-1287(2014)
17. K. Akahane et al., NIRS external dose estimation system for Fukushima residents after the Fukushima Dai-ichi NPP accident., Sci Rep. 2013;**3** pp.1670 (2013)
18. UNSCEAR , EFFECTIVE DOSES IN JAPAN FOR THE FIRST YEAR, UNSCEAR 2013 Report, ATTACHMENT C-14 (2013)
19. UNSCEAR , TOTAL EFFECTIVE DOSES IN JAPAN FOR 10-YEAR AND LIFETIME EXPOSURE, UNSCEAR 2013 Report, ATTACHMENT C-19 (2013)

Determination of kinetic parameters for 123-I thyroid uptake in healthy Japanese

Hiroyuki Kusuhara[1,a], and Kazuya Maeda[1]

[1]*Laboratory of Molecular Pharmacokinetics, Graduate School of Pharmaceutical Sciences, the University of Tokyo, Japan*

Abstract. The purpose of this study was to compare the kinetic parameters for iodide thyroid accumulation in Japanese today with previously reported values. We determined the thyroid uptake of 123-I at 24 hours after the oral administration in healthy male Japanese without any diet restriction. The mean value was 16.1±5.4%, which was similar or rather lower than those previously reported in Japan (1958-1972). Kinetic model analysis was conducted to obtain the clearance for thyroid uptake from the blood circulation. The thyroid uptake clearance of 123-I was 0.540±0.073 ml/min, which was almost similar to those reported previously. There is no obvious difference in the thyroid uptake for 24 hours, and kinetic parameters in healthy Japanese for these 50 years. The fraction of distributed to the thyroid gland is lower than the ICRP reference man, and such difference must be taken into consideration to estimate the radiation exposure upon Fukushima accident in Japan.

1 Introduction

It is well known that thyroid accumulation of radio-iodide depends on the ingestion of stable iodides because of non-linearity in the thyroid accumulation. Because Japanese food contains content of stable iodides, Japanese accumulates radio iodide upon exposure with lower magnitude than the ICRP reference man. Indeed, in previously studies (conducted in 1958-1967, the mean value of the thyroid uptake of radio-iodide for 24 hours after administration was 21±8 (mean±SD) %.

After Fukushima accident, we have a concern if the thyroid uptake and kinetic parameters of radio-iodide in Japanese today remains to be similar to those previously reported values. To address this issue, we designed clinical study in healthy subjects with no diet restriction to determine the thyroid uptake of 123-I in Japanese subjects with normal thyroid function.

2 Clinical studies in healthy subjects

2.1 Study Design

Clinical study was conducted in Nagasaki University Hospital PET/Molecular Imaging Center (Nagasaki, Japan). Fifteen Japanese male subjects were enrolled in this study (Age 41.8±0.4 years old, body weight 68.9±8.0 kg, BMI 23.5±0.5, and normal thyroid function). The subjects received no diet restriction. Following oral dose of 123-I (Sodium Iodide Capsules, Nihon Medi-Physics, Co., Ltd, Japan; 4.53-7.98 MBq), the blood was collected at 20 min, 40 min, 1, 1.5, 2, 4, 6, 9 and 24 hours after administration, and urine was collected for 10 hours after administration. The total radioactivity associated with blood and urine specimens were measured by gamma-counter. The thyroid uptake of 123-I was determined at 20 min, 40 min, 1, 1.5, 2, 4, 6, 9 and 24 hours after administration. This study was approved by the Ethics Review Boards of Faculty of Pharmaceutical Sciences, The University of Tokyo, and Faculty of Pharmaceutical Sciences, Nagasaki University.

The area under the blood concentration time profile (AUC) was calculated as trapezoidal approximation method. The renal clearance (CL_r) of 123-I was determined by dividing the amount of total radioactivity excreted into the urine by the area under the blood concentration time profile.

2.2 Result

Thyroid uptake of 123-I were increased along with time, and reached to 16.1±5.4% (mean±SE) of the oral dose after 24hours in average (ranging from 9.1-26%). The total radioactivity recovered for 10 hours after administration was 49.7±6.2%. CL_r of the total radioactivity was 36.1±8.4 ml/min.

[a] Corresponding author: kusuhara@mol.f.u-tokyo.ac.jp

 DOI: 10.1051/epjconf/201715308007

2.3 Comparison of the thyroid uptake for 24 hours

The thyroid uptake of 123-I for 24 hours determined in this study was compared with previously reported values which were collected from literatures (include published in Japanese). As shown in Figure 1, the mean value of the thyroid uptake was similar or rather slightly lower in the Japanese subjects today. Kunii Y et al also conducted similar study in another area in Japan (Tokyo) where the mean value (12.8±5.7%, [1]) was similar to our study. These values were lower than the corresponding value in ICRP reference man (30%).

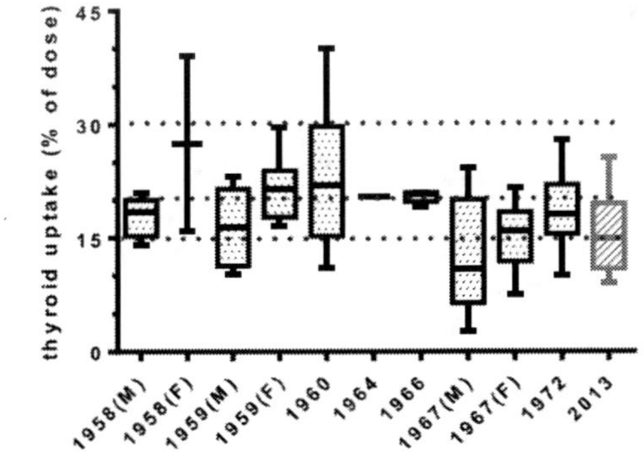

Figure 1 Comparison of the 24 hour-thyroid uptake of radioiodide in Japanese

3 Model analyses

3.1 Kinetic models

One- or two compartment model was employed in the analysis of blood concentration time profiles, and thyroid accumulation. The kinetic parameters such as distribution volume, and thyroid uptake clearance were obtained by iterative non-linear least square method using Napp software [2] on Macintosh computer.

3.2 Result of model analysis

Blood concentration time profiles followed one or two compartment model depending on subjects; in 9 of 15 subjects, the blood concentration time profiles were accounted for by the 1 compartment model where 1st order absorption was considered.

The distribution volume in the central compartment was 21.2±4.1 l, and thyroid uptake clearance was 0.540±0.073 (mean±SE) l/hr.

3.3 Comparison of thyroid uptake clearance in Japanese population

Nagataki S et al reported the thyroid uptake clearance in healthy subjects in 1967 [2]. The mean and standard error were compared with those in our study (Figure 2). Our data was almost comparable with previously reported value.

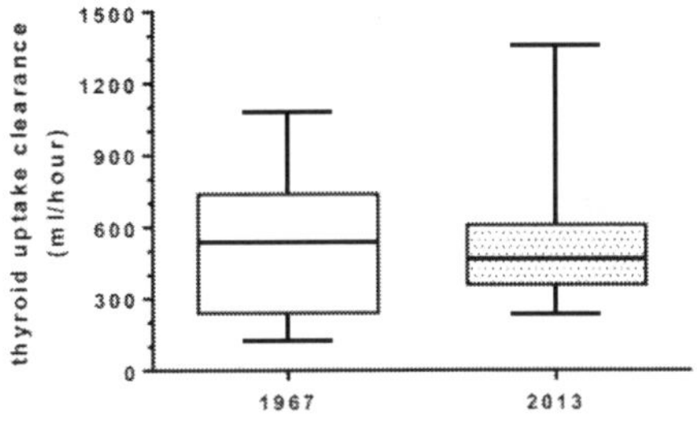

Figure 2 Comparison of the thyroid uptake clearance of radio iodide in Japanese

Conclusion

For these 50 years, no significant difference was observed in the thyroid accumulation of radio iodide for 24 hours, and thyroid uptake clearance. The present study confirmed that thyroid accumulation of radio iodide is lower in Japanese than the ICRP reference man. This was also confirmed in another study independently conducted in Tokyo. Such difference must be taken into consideration to estimate the radiation exposure upon Fukushima accident in Japan.

This study also determined the distribution volume and renal clearance in Japanese population. Using the kinetic model, and parameters, the thyroid accumulation of radio iodide will be estimated according to the exposure scenarios during evacuation. The kinetic parameters in virtual subjects can be generated with means and coefficient of variations. Monte Carlo method may allow estimation of the likely distribution of thyroid accumulation based on each exposure scenario.

References

[1]Y. Kunii, H. et al. Abstract (P2-03-03) in Nippon Naibunpi Gakkai zasshi, 88, 526-526 (2012)
[2]A. Hisaka, Y. Sugiyama, J. Pharmacokinet. Biopharm. **26**, 495-519 (1998)
[3]S. Nagataki, K. Shizume, K. Nako, J. Clin. Endocrinol. Metab. 27, 638-647 (1967)

Acknowledgement

Clinical study was conducted as collaboration with Dr Shuichi Yamashita, and Dr. Takashi Kudo (Nagasaki University, Japan). The authors would like to thank Akihiro Inano (Fukushima Medical University), and Sawako Nakajima and Masako Tsurumaru (Nagasaki University Hospital) for their support.

This study was conducted by funds from Ministry of the Environment.

Development of a tool for calculating early internal doses in the Fukushima Daiichi nuclear power plant accident based on atmospheric dispersion simulation

Osamu Kurihara[1,*], Eunjoo Kim[1], Naoaki Kunishima[1,2], Kotaro Tani[1], Tetsuo Ishikawa[3], Kazuo Furuyama[1], Shozo Hashimoto[1] and Makoto Akashi[1]

[1] *National Institutes for Quantum and Radiological Sciences and Technology, National Institute of Radiological Sciences, 4-9-1 Anagawa Inage-ku, Chiba city, Chiba, Japan*
[2] *Self-Defense Forces Central Hospital, 1-2-24 Ikejiri, Setagaya-ku, Tokyo, Japan*
[3] *Fukushima Medical University, 1-Hikarigaoka, Fukushima city, Fukushima, Japan*

Abstract. A tool was developed to facilitate the calculation of the early internal doses to residents involved in the Fukushima Nuclear Disaster based on atmospheric transport and dispersion model (ATDM) simulations performed using Worldwide version of System for Prediction of Environmental Emergency Information 2nd version (WSPEEDI-II) together with personal behavior data containing the history of the whereabouts of individul's after the accident. The tool generates hourly-averaged air concentration data for the simulation grids nearest to an individual's whereabouts using WSPEEDI-II datasets for the subsequent calculation of internal doses due to inhalation. This paper presents an overview of the developed tool and provides tentative comparisons between direct measurement-based and ATDM-based results regarding the internal doses received by 421 persons from whom personal behavior data available.

1 Introduction

A huge amounts of radionuclides was released into the environment as a consequence of the accident at the Tokyo Electric Power Company's (TEPCO's) Fukushima Daiichi Nuclear Power Plant (FDNPP) in 2011 [1]. The Japanese Government ordered residents living near the FDNPP to evacuate shortly after the accident in order to minimize their radiation exposure. It was reported that most of the residents evacuated in accordance with the evacuation orders [2].

Needless to say, it is of great importance to evaluate the levels of radiation exposure in the populations living in the radiologically affected areas. Much effort has been made on this task up to the present; however, there still remain many difficulties in estimating the early internal dose due to the intake of short-lived radionuclides represented by iodine-131 (^{131}I), a significant contributor of internal thyroid doses in particular for small children. Neither human nor environmental measurements could be adequately performed shortly after the accident. According to the published research [3-6], only around 1,300 human measurements were performed within the first two month after the accident. Extensive direct (*in vivo*) measurements using whole-body counters (WBCs) were started in late June, 2011[7, 8]; however, ^{131}I could not have been detected at this point of time.

One of the possible solutions to estimating the early internal dose is to make use of atmospheric transport and dispersion model (ATDM) simulations of the radionuclides. The authors have reported the feasibility of this method based on the results of the Worldwide version of System for Prediction of Environmental Emergency Dose Information 2nd version (WSPEEDI-II) developed by the Japan Atomic Energy Agency (JAEA) [9]. In the study, the early internal doses calculated from the time-series ground level air concentration maps (^{131}I, ^{137}Cs) produced by WSPEEDI-II were compared to those obtained from direct measurement results on the assumption that the individuals remained outdoors at all time at representative places or behaved in accordance with a representative evacuation pattern. As a result, it was found that the former doses were several times higher than the 90th percentile values of the latter doses. This discrepancy was supposed to be minimized by taking into account actual personal behaviors after the accident.

Afterwards, personal behavior data became partially available to the authors through the permission of research ethics committees of the National Institute of Radiological Sciences (NIRS) and Fukushima Medical University (FMU). These data were produced based on a self-administrative questionnaire about each individual's whereabouts for the first four months after the accident and have been used for calculating the external doses to residents of Fukushima Prefecture in the first four months after the accident [10]. The authors were provided with the personal behavior data of those for whom direct measurement results were available. The present paper describes the development of a tool for facilitating internal dose calculations using the results of WSPEEDI-II and shows tentative comparisons between the direct-measurement-based and ATDM-based internal

* Corresponding author: kurihara.osamu@qst.go.jp

doses to 421 persons for whom personal behavior data were available.

2 Materials and methods

2.1 ATDM simulations

The developed tool used datasets from the time-series ground-level air concentration maps produced using WSPEEDI-II (Table 1). The original simulations that produced these datasets have been described elsewhere [11, 12]. The results obtained by Terada et al were cited by the United Nations Scientific Committee on the Effect of Atomic Radiation (UNSCEAR) [13], and the findings of Katata et al were obtained by utilizing an updated source term based on newly obtained environmental data and an improved deposition scheme. Each dataset contains hourly-averaged air concentration maps of the radionuclides from March 12 to May 1 (Terada-Regional) or to April 1 (Katata-Regional, Katata-Local), 2011. The physicochemical form of ^{131}I was not considered in the simulations by Terada et al, whereas it was separately calculated for elemental iodine (I_2), methyl iodine (CH_3I) and particulate iodine in the simulations by Katata et al. The calculation domains of each simulation are shown in Figure 1. The nominal spatial resolutions of the regional and local models are 3 km x 3 km and 1 km x 1 km, respectively.

Table 1. Simulation datasets used in the tool.

Dataset	Terada-Regional	Katata-Regional	Katata-Local
Calculation domains (Covered area)	681 km x 951 km (East Japan)	381 km x 561 km (East Japan)	187 km x 187 km (Fukushima Prefecture)
Spatial resolution (nominal)	3 km x 3 km	3 km x 3 km	1 km x 1 km
Time resolution	1 h		
Period of calculation in 2011	from 6:00 on Mar. 12 to 0:00 on May. 1 (JST*)	from 6:00 on Mar. 12 to 5:00 on April. 1 (JST*)	from 6:00 on Mar. 12 to 5:00 on April. 1 (JST*)
Data size	about 64 GB	about 14.4 GB	about 21.6 GB
Radionuclide	^{131}I, ^{137}Cs	^{131}I (particulate, elemental, methyl iodide), ^{132}Te, ^{134}Cs, ^{137}Cs	

* Japan Standard Time

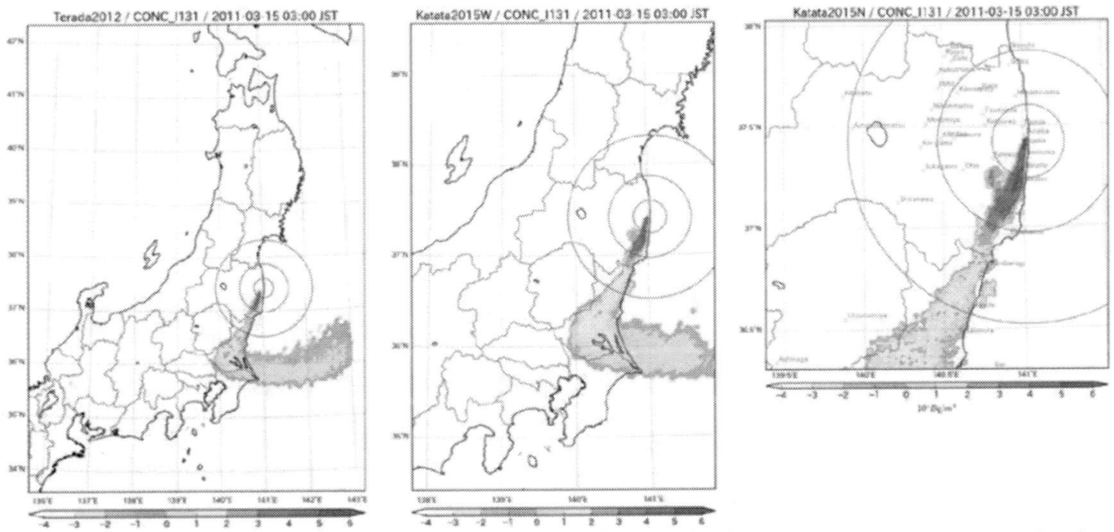

Figure 1. Calculation domains of the datasets used in tool (left: Terada-Regional, middle: Katata-Regional, right: Katata-Local).

2.2 Personal behaviour data

The personal behavior data provided include the whereabouts (the place name and its latitude and longitude) and time spent indoors, outdoors or in migration during the first four months after the accident. All these data were stored in digital files. The data were provided in different formats depending on the period of time (Table 2). The information on the whereabouts was obtained hourly for the first and second periods (March 12 to March 25, 2011), whereas was obtained in terms of daily representative places (mostly one place) where the subject stayed for the last period (March 26 to July 11, 2011). However, complete information is not necessarily available because some data were missing. Additionally, the subject's whereabouts during migration from one place to another is basically unknown, which could result in a large uncertainty in the internal dose estimation, in particular for those who migrated over a long distance and for a long period of time.

For this study, the authors received 422 personal behavior data from FMU. These persons comprised 310 of 1,080 subjects (children) of the thyroid exposure screening campaign conducted by the Nuclear Emergency Response Local Headquarters at the end of March, 2011 and 112 of 174 subjects of the pilot survey by means of a WBC at the NIRS during the period between June 27 and July 28, 2011. The internal doses of the 421 persons (one subject of the screening campaign was not identified) were already evaluated using the direct measurement results in the previous studies [8, 14] and were compared with those obtained in this study. However, this comparison is rather tentative because the former doses were evaluated assuming a simple intake scenario, namely acute inhalation on March 15 for the subjects of the screening campaign or March 12 for the subjects of the pilot survey. This scenario would somewhat overestimate the internal doses.

Table 2. Formats of personal behavior data.

Period in 2011	Data format
Mar. 12 – Mar. 16	Hourly place data (staying, departure and destination) are written on each column.
Mar. 17 – Mar. 25	Place data (staying, departure and destination) are written on each row. The time of departure from the place where the person was present in the morning is calculated from the sum of the time duration of the migration.
Later Mar. 26	Place where the person was present in the morning and visited temporarily.

2.3 Development of the tool

The tool was designed assuming that it would need to manipulate a huge amount of simulation data (with a number of items per one original simulation dataset being on the giga order) generated by WSPEEDI-II and also possibly additional personal behavior data in the future. From this point of view, a commercial package of database software was not suitable because of the limitated data manipulation that is possible.

The tool has been developed using the C# language with the .NET Framework 4.6 (Microsoft Corporation, US) and is operable on personal computers (64 bits) with Windows 7 or later. In this tool, all data used in the analyses were stored in databases that were originally designed to enable traverse access. As a result, users can instantly access the desired information after the data are loaded into the PC memory. The tool operations can be performed by both the Command Prompt in Windows and the Graphic User Interface (GUI). The main tasks of the tool are as follows.

● Interpolation of personal behavior data

As described above, the personal behavior data were provided in the different formats depending on the period of time. To facilitate the subsequent dose calculations, the tool interpolates these data to be arrayed hourly as same as in WSPEEDI-II. The first individual's whereabouts on each day was assigned to every hour of the day during the period on or after March 26, 2011 (Table 2), because of the lack of information on the exact times at which the subjects were engaged in specific behaviors. When data were missing on a certain day, it was assumed that the person continued to stay at the place given last in the personal behavior data.

● Extraction of the air concentration data

The tool extracts the air concentration data from the simulation grid (and its surrounding eight grids) nearest the place where the person stayed at a certain time from the datasets of WSPEEDI-II. In the case of a period of migration, the air concentration data were extracted for the place of departure and the destination. The extraction was performed hourly until the end of the simulations using the results of the previous task.

● Visualization of the outputs

The extracted air concentration data can be displayed in multiple graphs as a function of time, which is useful as it allows one to see when the person was most significantly exposed to radioactive plumes. The tool can also create animations of the radioactive plumes of the radionuclide selected and the movement of the persons selected using Google Earth.

2.4 Internal dose calculations

The function of the internal dose calculations has not been incorporated in the current version of the tool, but it is easy to perform this function using the extracted air concentration data and a commercial spreadsheet software package such as Microsoft EXCEL. In this study, the internal dose was calculated using the following equation.

$$D_{i,j} = e_{i,j} \times I_{i,j} = e_{i,j} \times \sum_k C_{i,nearest}(t_k) \cdot b_j \cdot F(t_k) \quad (1)$$

where, suffixes *i*, *j* and *k* denote the radionuclide, the age group and the chronological number of the air concentration map, respectively, *e* is the dose per unit intake (DPUI) (Sv Bq^{-1}), $C_{nearest}(t)$ is the hourly averaged air concentration at the time *t* at the nearest simulation grid (Bq m^{-3}) (or is the average of the air concentrations at the place of departure and the destination in the case of migration), b is the ventilation volume per hour (m^3) and F is the ratio of indoor to outdoor air concentration.

Regarding ^{131}I, the internal dose calculations were separately performed for each physicochemical form in the case of the datasets by Katata et al, whereas they were performed using the weighted DPUI of ^{131}I (60% of elemental iodine and 40% of particulate iodine) in the case of the dataset of Terada et al. The composition of the weighted DPUI was determined based on observations at the FDNPP and three JAEA sites in Ibaraki Prefecture (located south of Fukushima Prefecture) [15]. The parameters used in Equation (1) are provided in Table 3 [16, 17]

Table 3. Parameters used in internal dose calculations.

Age-group	infant	1-y	5-y	10-y	15-y	adult
Ventilation volume (m^3 d^{-1})	2.86	5.16	8.72	15.3	20.1	22.2
Dose coefficient (Sv Bq^{-1})						
Effective dose of $^{134}Cs^*$	1.1E-08	7.3E-09	5.2E-09	5.3E-09	6.3E-09	6.6E-09
Effective dose of $^{137}Cs^*$	8.8E-09	5.4E-09	3.6E-09	3.7E-09	4.4E-09	4.6E-09
Thyroid equivalent dose of ^{131}I						
(CH_3I)	2.6E-06	2.5E-06	1.5E-06	7.4E-07	4.8E-07	3.1E-07
(I_2)	3.3E-06	3.2E-06	1.9E-06	9.5E-07	6.2E-07	3.9E-07
(Particulate)	1.4E-06	1.4E-06	7.3E-07	3.7E-07	2.2E-07	1.5E-07

* Type F compunds

3 Results and discussion

3.1 Overview of the tool

Figure 2 shows the start-up screen of the tool. The functions of each panel on the screen can be explained as follows. Figure 2a depicts the time-series of the air concentrations from all the WSPEEDI-II datasets at the gird nearest the latitude and longitude that a user designates. The air concentrations at the surrounding eight grids can be also displayed on sub panels. Figure 2b and Figure 2c depict graphs similar to those in the previous figure, but for the person(s) selected. The former can compare all the WSPEEDI-II datasets in terms of one selected person and the latter can compare multiple persons in terms of one selected dataset. The air concentrations displayed on these panels are those extracted based on the personal behavior data. Whether the average of the air concentrations at the place of departure and the destination during migration is used or not can also be chosen on these panels. Figure 2d is a screenshot of Google Earth with the simulation data and the marks of the places where the selected persons stayed. Figure 2e is the screen for controlling the animation on Google Earth. The numbers of subjects who correspond to the search criteria (i.e., subject group, age group, gender, municipality, date) are also displayed in the same panel.

3.2 Examples of output from the tool

The time-series of the air concentrations at the places where the persons stay (Figures 2b and 2c) correspond to the relative intake amount rate via inhalation, which is one of the modifiers of the internal dose evaluated from direct measurements. Figure 3 demonstrates the comparison of the ^{137}Cs air concentration between Terada-Regional (Figure 3a) and Katata-Regional (Figure 3b) for the 112 subjects of the pilot survey. As shown, the largest intake for most of the subjects occured on different days in the two simulations: March 15 for Terada-Regional and March 12 for Katata-Regional. About a half of the 112 subjects are from Namie town. Their personal behavior data suggested that most of them evacuated within a few days after the accident. On the other hand, this period is of great importance in determining the early internal dose because of the abundance of short-lived radionuclides other than ^{131}I. The dose contributions of ^{132}Te ($T_{1/2}$: 3.2 d)-^{132}I ($T_{1/2}$: 2.3 h) and ^{133}I ($T_{1/2}$: 20.8 h) were estimated to be as great as 30-40% of the internal thyroid dose from ^{131}I in the case of inhalation intake occurring as early as March 12 [18].

Figure 4 demonstrates the time-trend of the whereabouts of the 310 subjects of the screening campaign conducted in Kawamata town, Iwaki city and Iitate village at the end of March, 2011. It is noted that no evacuation orders were issued in these municipalities (excluding areas within the 30 km-radius of the FDNPP) before the campaign. Thus, the result suggested that a significant percentage of the persons voluntarily evacuated to other places (partly outside Fukushima Prefecture) and returned at the time of the measurements. It was also found that especailly for those examined in Iite village and Iwaki city, the total number of evacuees reaches to the maximum on March 21. As described later, the thyroid doses received by the 309 subjects were small, so it is difficult to determine if the voluntary evacuation was effective for reducing the doses. This should be further analyzed in addition to the issues addressed in the subjects of the pilot survey.

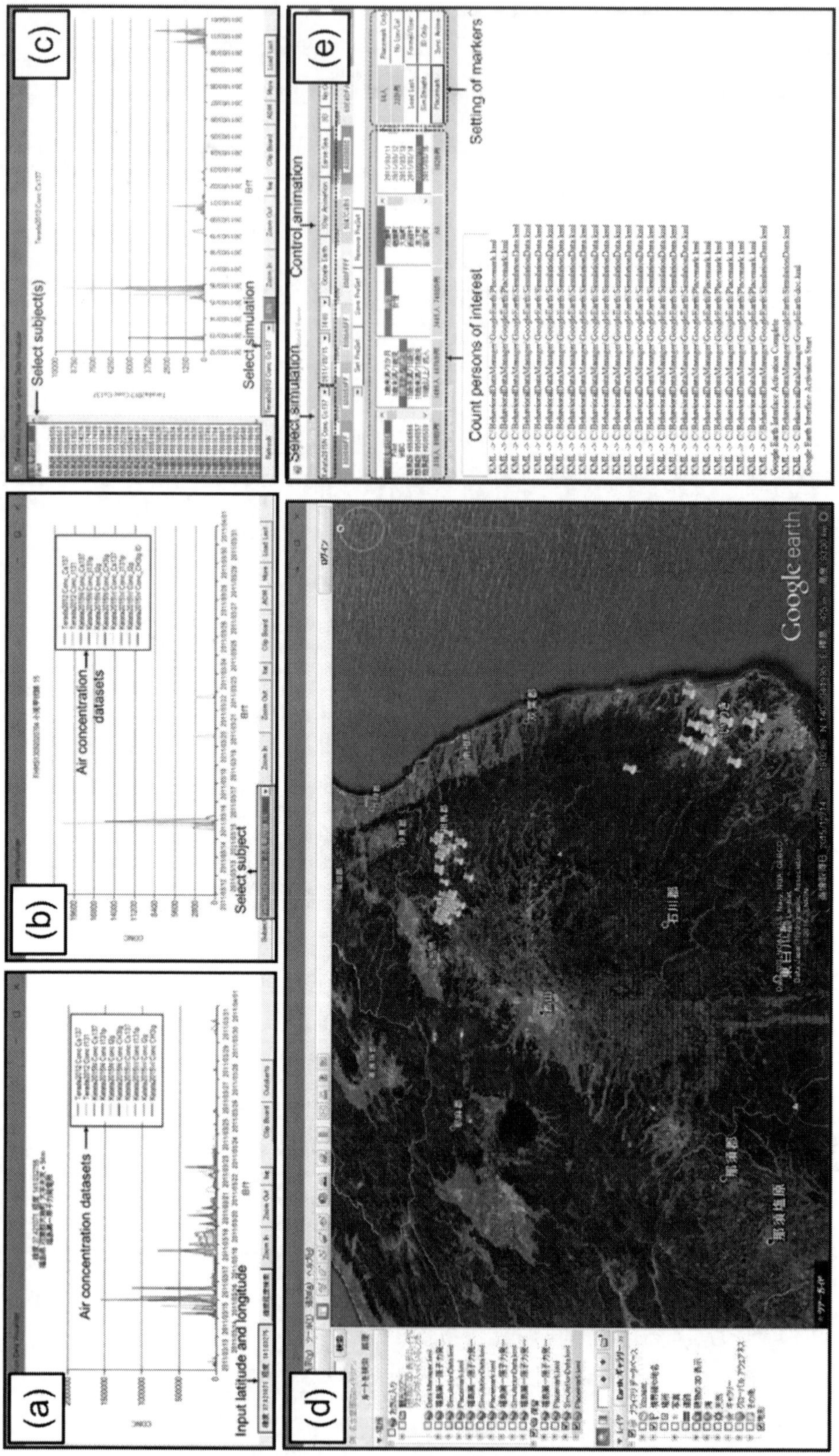

Figure 2. Start-up screen of the tool with explanations.

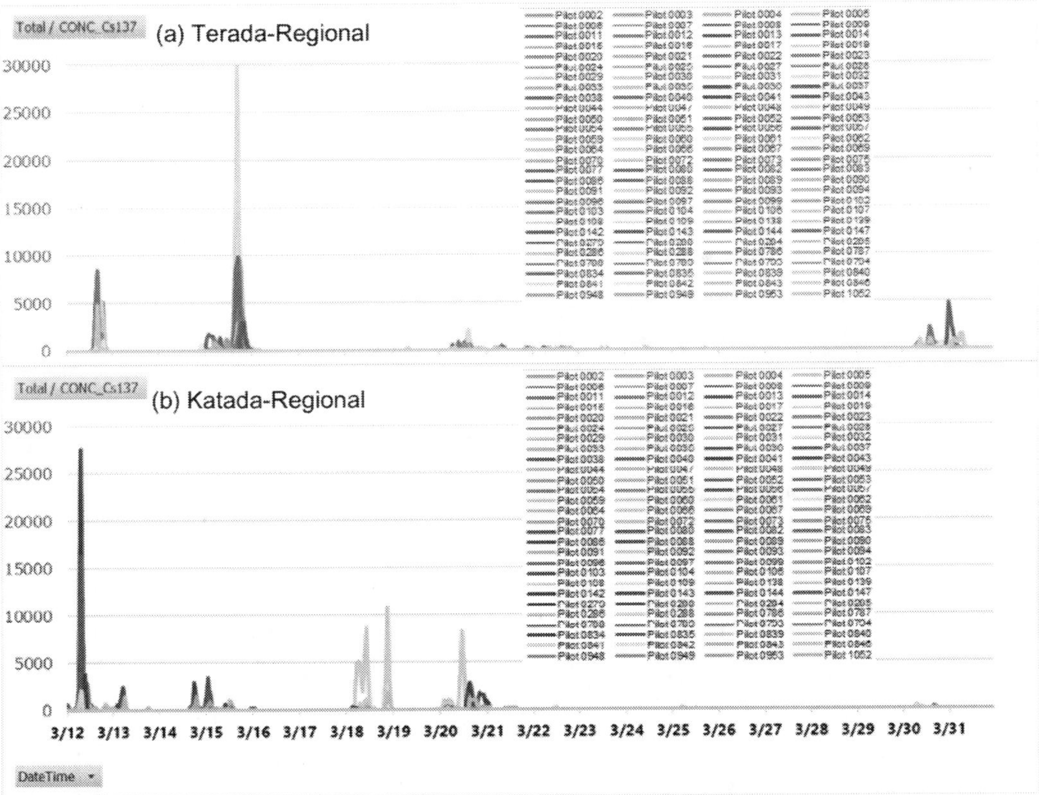

Figure 3. Comparison of ^{137}Cs air concentration between (a) Terada-Regional and (b) Katata-Regional for 112 subjects of the pilot survey.

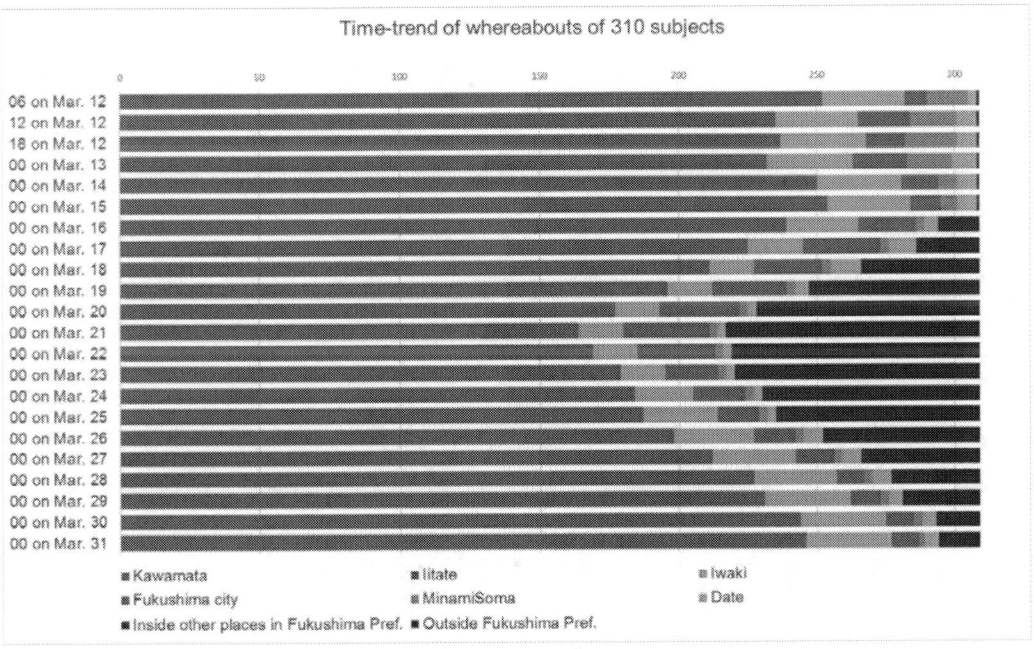

Figure 4. Whereabouts of 310 subjects of the screening campaign.

3.3 Comparison between measured and simulated doses

Figure 5a compares the thyroid equivalent doses (^{131}I) received by the 309 subjects from the direct measurements (screening campaign) with those determined in the simulation (Terada-Regional). For the latter doses, the rate of indoor to outdoor air concentration, F, was set at 0.5. Figure 5b is the same figure, but it compares the direct measurements with another simulation (Katata-Regional). As shown in the figure, the simulations generally tended to give higher doses than the direct measurements, even taking into account the indoor dose reduction. Figure 6 compares the histograms of the thyroid equivalent doses shown above for each municipality where the screening campaign was conducted. The figure demonstrates that the overestimation is more marked in Katata-Regional than in Terada-Regional in the case of Iwaki city and Iitate village, whereas the overestimation in Karata-Regional is reduced compared to Terada-Regional in the case of Kawamata town.

Figure 7a compares the effective doses (^{134}Cs and ^{137}Cs) received by the 112 subjects from the direct measurements (pilot survey) with those determined in the simulation (Terada-Regional). Figure 7b is the same figure, but it compares the direct measurements with another simulation (Katata-Regional). The rate of indoor to outdoor air concentration was set at the same level as above. In the case of Terada-Regional, the air concentration of ^{134}Cs was assumed to be the same as that of ^{137}Cs because of the lack of data. Figure 8 compares the histograms of the effective doses determined from the direct measurements with those determined in the simulations. From these figures, it still seems difficult to reproduce the individual doses. Analyses of the persons whose dose were considerably overestimated or underestimated indicated that either the simulations or their personal behavior data were not satisfactorily accurate. Protracted intake until the direct measurements were made can also be considered as one possible reason for the underestimation, if any occurred. On the other hand, the dose distribution from the direct measurements agrees reasonably well with that in the simulations. Further studies are necessary to improve the internal dose estimation based on ATDM simulations and the developed tool would be useful for this future task.

4 Summary

Estimation of the early internal dose received by residents in the FDNPP accident based on ATDM simulations is neceaasry because of the shortage of direct measurements of individuals. In the present study, a tool was developed to facilitate this estimation based on the simulations by WSPEEDI-II and the available personal behavior data. The tool generates hourly-averaged air concentration data at the simulation grids nearest to places where the person stayed from the huge datasets of WSPEEDI-II. Using the outputs from the tool, the internal doses received by 421 persons for whom both direct measurement data and personal behavior data were available were calculated and compared with those from the measurements, suggesting a need for further studied to improve the internal dose estimation based on the ATDM simulation. The developed tool would be useful for this future task.

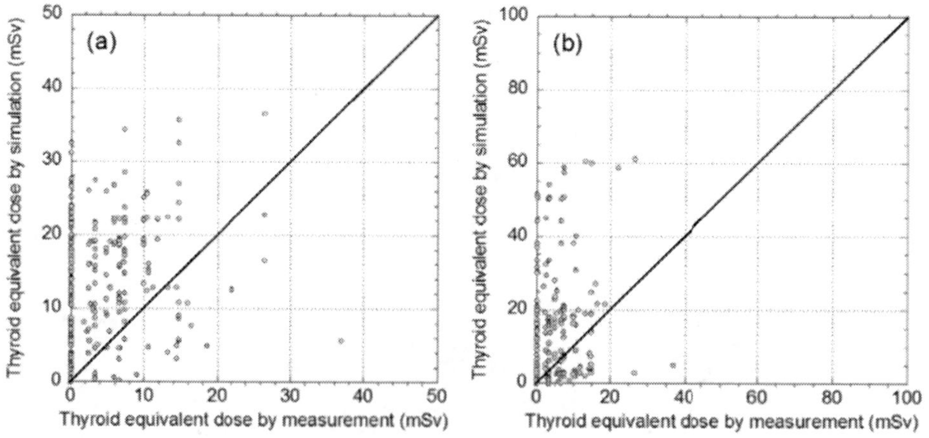

Figure 5. Comparison between direct measurement-based and simulation-based thyroid equaivalent doses (^{131}I) received by 309 subjects (a: Terada-Regional, b: Katata-Regional).

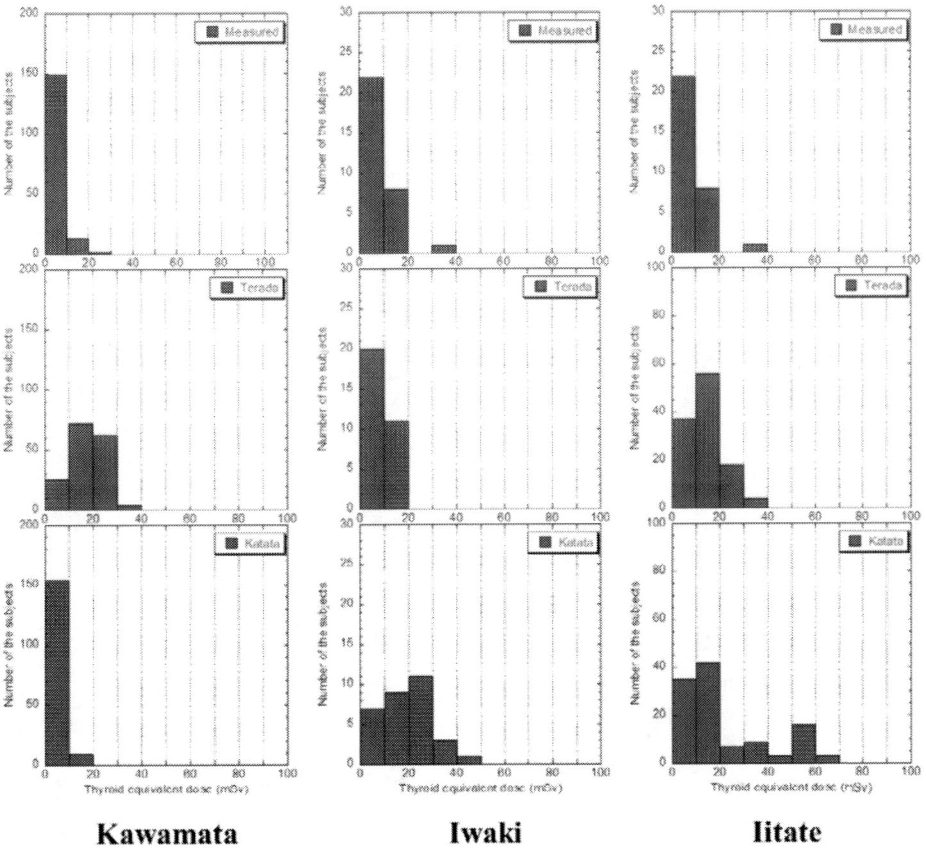

Figure 6. Histograms of thyroid equivalent doses (upper row: measured, middle row:Terada-Regional, lower row: Katata-Regional; left column: Kawamata town, middle column: Iwaki city, right column: Iitate village).

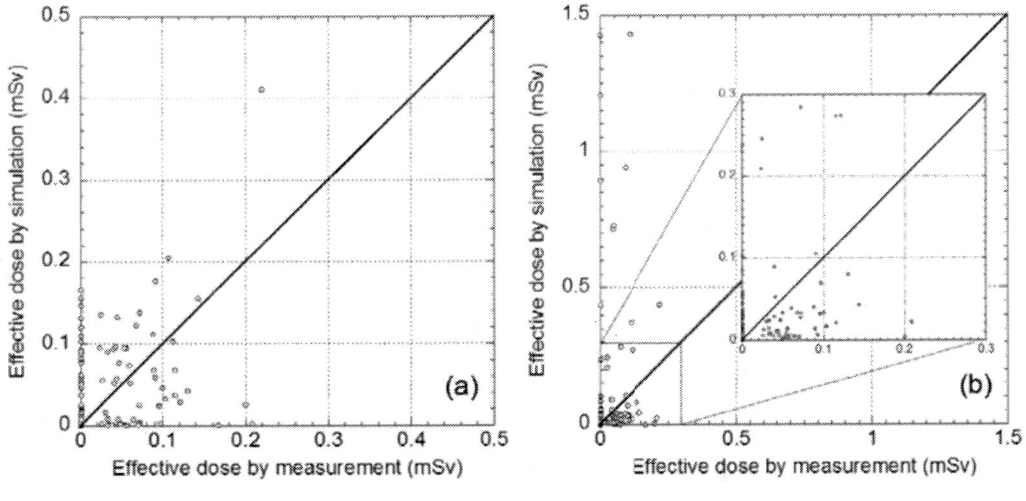

Figure 7. Comparison between direct measurement-based and simulation-based effective doses (^{134}Cs and ^{137}Cs) received by 112 subjects (a: Terada-Regional, b: Katata-Regional).

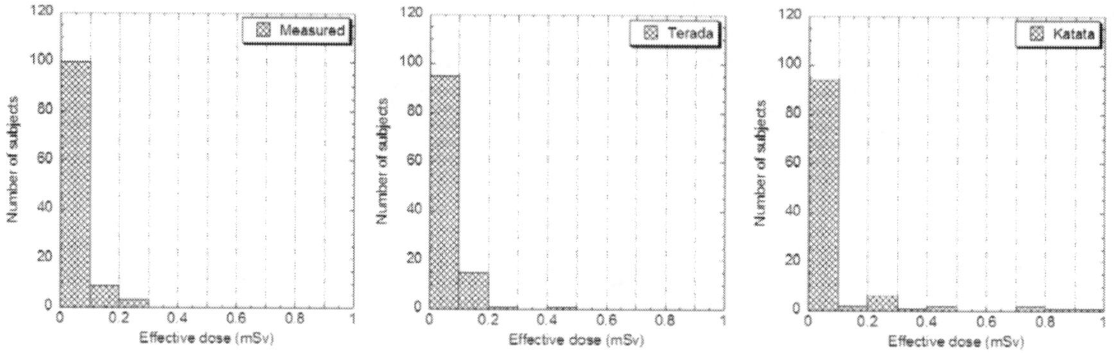

Figure 8. Histograms of effective doses (left: measured, middle: Terada-Regional, right: Katata-Regional).

Acknowledgement

The authors wish to express their thanks to Dr. Haruyasu Nagai of the Japan Atomic Energy Agency and his colleagues for kindly providing datasets from WSPEEDI-II. This study was conducted as part of the comprehensive study on the dose estimation of resiednts in the Fukushima Daiichi Nuclear Power Plant accident funded by the Ministry of the Environment, Japan (FY. 2014-2016) and also as a collaborative study between the NIRS and Fukushima Medical University.

References

1. Nuclear Emergency Response Headquarters Government of Japan, *Report of Japan Government to the IAEA Ministerial Conference on Nuclear Safety – The accident at TEPCO's Fukushima Nuclear Power Stations June 2011*, Available at http://japan.kantei.go.jp/kan/topics/201106/iaea_houkokusho_e.html. Accessed on 24 August 2016.
2. The National Diet of Japan. *The official report of the Fukushima Nuclear Accident Independent Investigation Commission. 2012*, Available at http://warp.da.ndl.go.jp/info:ndljp/pid/3856371/naiic.go.jp. Accessed on 24 August 2016 (in Japanese).
3. E Kim, O Kurihara, T Suzuki, M Matsumoto, K Fukutsu, Y Yamada, N Sugiura, M Akashi, *Proc. of the first NIRS symposium on reconstruction of early internal dose in the TEPCO Fukushima Daiichi nuclear power station accident*, **NIRS-M-252**(2012), Available at http://www.nirs.qst.go.jp/publication/irregular/pdf/nirs_m_252.pdf. Accessed on 24 August 2016.
4. S Tokonami, M Hosoda, S Akiba, A Sorimachi, I Kashiwakura and M Balonov, *Sci. Rep.* **2**, 507(2012)
5. N Matsuda, A Kumagai, A Ohtsuru, N Morita, M Miura, M Yoshida, T Kudo, N Takamura and S Yamashita, *Radiat. Res.* **179**, 663 (2013)
6. N Kamada, O Saito, S Endo, A Kimura, K Shizuma, *J. Environ. Radioact.* **110**, 84 (2012)
7. T Momose, C Takada, T Nakagawa, K Kanai, O Kurihara, N Tsujimura, Y Ohi, T Murayama, T Suzuki, Y Uezu and S Furuta, *Proceedomgs of the first NIRS symposium on reconstruction of early internal dose in the TEPCO Fukushima Daiichi nuclear power station accident*, **NIRS-M-252** (2012)
8. E Kim, O Kurihara, N Kunishima, T Nakano, K Tani, M Hachiya, T Momose, T Ishikawa, Sh Tokonami, M Hosoda, and M Akashi, *Health Phys* (to be published)
9. H Terada, H Nagai, A Furuno, T Kakefuda. T Harayama, M Chino. *WSPEEDI 2nd version. J. At. Energy Soc.* Jpn. **7**, 257(2008). (in Japanese)
10. T Ishikawa, S Yasumura, K Ozasa, K Kobashi, H Yasuda, M Miyazaki, K Akahane, Sh Yonai, A Ohtsuru, A Sakai, R Sakata, K Kamiya and M Abe, *Sci. Rep.* **5**, 12712(2015)
11. H Terada, G Katata, M Chino and H Nagai, *J. Environ. Radioact.* **112,** 141(2012)
12. G Katata, M Chino, T Kobayashi, H Terada, M Ota, H Nagai, M Kajino, R Draxler, M C Hort, A Malo, T Torii and Y Sanada, *Atoms. Chem. Phys.* **15**, 1029 (2015)
13. United Nation Scientific Committee on the Effect of Atomic Radiation (UNSCEAR). *UNSCEAR 2013 report volume 1*, **Scientific Annex A** (2014)
14. E Kim, O Kurihara, N Kunishima, T Momose, T Ishikawa, M Akashi, *J. Radiat. Res.* **57(S1)**, i118 (2016)
15. O Kurihara, E Kim, K Fukutsu, M Matsumoto, S Soheigh., K Akahane, K Sakai. *Radiation Monitoring and Dose Estimation of the Fukushima Nuclear Accident,* Takahashi S ed., Speinger Open,177 (2014)
16. International Commission on Radiological Protection, *Age-dependent doses to the members of the public from intake of radionuclides: Part 4 inhalation dose coefficients.* **ICRP Publication 71. Ann. ICRP 25 (3-4)**, (1995)
17. International Commission on Radiological Protection. Compendium of dose coefficients based on ICRP Publication 60. **ICRP Publication 119. Ann. ICRP 41 (Suppl.)**, (2012)
18. S M Shinkarev, K V Kotenko, E O Granovskaya et al., *Radiat. Prot. Dosim.* **164,** 51(2015)

Estimation of thyroid equivalent doses during evacuation based on body surface contamination levels in the nuclear accident of FDNPS in 2011

Takashi Ohba[1,2], *Arifumi* Hasegawa[3], *Yoshitaka* Kohayakawa[4], *Hisayoshi* Kondo[4], *Gen* Suzuki[5,*]

[1] Department of Radiation Health Management, Fukushima Medical University, 1 Hikarigaoka Fukushima, Japan
[2] Barcelona Institute for Global Health (ISGlobal), 88 Doctor Aiguader Barcelona, Spain
[3] Department of Radiation Disaster Medicine, Fukushima Medical University, 1 Hikarigaoka Fukushima, Japan
[4] National Hospital Organization Disaster Medical Center, 3256 Midorimachi Tachikawa, Japan
[5] International University of Health and Welfare Clinic, 2600-1 Kitakanemaru Ohtawara, Japan

Abstract. To reduce uncertainty in thyroid dose estimation, residents' radiation protection behavior should be reflected in the estimation. Screening data of body surface contamination provide information on exposure levels during evacuation. Our purpose is to estimate thyroid equivalent doses based on body surface contamination levels using a new methodology. We obtained a record of 7,539 residents/evacuees. Geiger-Mueller survey meter measurement value in cpm was translated into Bq/cm^2 according to the nuclides densities obtained by measuring clothing from two persons by germanium γ-spectrometer. The measurement value of body surface contamination on head was adjusted by a natural removal rate of 15 hours and radionuclides' physical half-life. Thyroid equivalent dose of 1-year-old children by inhalation was estimated by two-dimensional Monte Carlo simulation. The proportions of evacuees/residents with measurement value in cpm of Namie and Minamisoma groups were higher than those of other groups during both periods ($p<0.01$, Kruskal-Wallis). During 12-14 March period, 50 and 95 percentiles of thyroid equivalent doses by inhalation were estimated as 2.7 and 86.0 mSv, respectively, for Namie group, and 4.2 and 17.2 mSv, respectively, for Minamisoma group, 0.1 and 1.0 mSv, respectively, for Tomioka/Okuma/Futaba/Naraha group, and 0.2 and 2.1 mSv, respectively, for the other group. During 15-17 March period, 50 and 95 percentiles of thyroid equivalent doses by inhalation were 0.8 and 15.7 mSv, respectively, for Namie group, and 1.6 and 8.4 mSv, respectively, for Minamisoma group, 0.2 and 13.2 mSv, respectively, for Tomioka/Okuma/Futaba/Naraha group, and 1.2 and 12.7 mSv, respectively, for the other group. It was indicated that inhalation dose was generally higher in Namie and Minamisoma groups during 12-14 March than those during 15-17 March might reflect different self-protective behavior to radioactive plumes from other groups.

1 Introduction

The risk of childhood thyroid cancer is a major concern after the nuclear accident at the Fukushima Daiichi Nuclear Power Station (FDNPS) in March 2011 [1]. Because of paucity in direct thyroid measurements after the accident [2,3], thyroid equivalent doses were estimated by an atmospheric transfer, dispersion and deposition model (ATDM) using a source term by Terada in UNSCEAR 2013 Report [4]. However, the UNSCEAR's estimates of settlement-averaged thyroid doses were two to five times higher than the corresponding values based on direct measurements. This discrepancy might be, at least in part, due to the difficulty in estimating temporal and spatial nuclides concentrations in the air by ATDM and individual self-protective behavior during evacuation.

Most residents living in the Fukushima Prefecture were screened for their body surface contaminations by Geiger-Mueller (GM) survey meter [5,6]. Screening survey was started from 12 March and conducted about 43,000 residents until 17 March [5]. These residents began evacuating within a radius of 20 km from the FDNPS at 18:25 (JST) on 12 March [4]. Since only a small fraction of radioactive plume blew inland, the time and direction of plumes that blew inland and residents' behavior at the occasion were important factors for exposure. A radioactive plume released at evening on 12 March directed to north-northwest direction [7], many evacuees who evacuated to this direction might have been exposed. In addition, a radioactive plume released from the FDNPS on 15 and 16 March blew inland from south-southwest direction in the morning and changing the direction clock wisely into north-west [7], many

* Corresponding author: gensuzki@iuhw.ac.jp

residents might have been exposed to plume but at rather remote places from the FDNPS.

To reduce uncertainty in thyroid dose estimation, residents' radiation protection behavior should be considered in the estimation. The levels of body surface contamination might reflect the exposure levels during evacuation. In the present study, we introduced a new methodology for estimating thyroid doses based on the levels of body surface contamination.

2 Methods

2.1 Principals of analysis

2.1.1 Calculation of radionuclide density (Bq/cm²) from screening survey using GM survey meter

Body surface contamination density $S_i(T)$ (Bq/cm²) of radionuclide-i was calculated by the following formula in (1), where N and N_b represent GM measurements (count per minute; cpm) of body surface and background, respectively, $\varepsilon_{e,i}$ and $\varepsilon_{s,i}$ are instrument efficiency and emission efficiency of β-ray, W (cm²) is detector area, and ER_i is emission rate of β- ray [8].

$$S_i(T) = \frac{(N-N_b)}{\varepsilon_{e,i} \cdot W \cdot \varepsilon_{s,i} \cdot ER_i} \quad (1)$$

2.1.2 Conversion of β-ray into radionuclide densities

GM survey meter measures almost all β emissions from radionuclides. In order to estimate thyroid equivalent dose, one has to know the composition of β-ray from radionuclides on cloths and body surface. Radionuclides' composition on clothing of two persons was measured by a whole body counter (WBC) equipped with germanium (Ge) γ-spectrometer (AccuScan by CANBERRA Industries Inc.).

We utilized this composition for body surface contamination by a plume on 12 March. As to radionuclides composition on body surface during 15-17 March, we adopted air dust sampling data by Japan Atomic Energy Agency (JAEA) in Tokai Village [9]. Major radionuclides contributing screening survey, *i.e.*, ^{131}I, ^{132}I, ^{133}I, ^{135}I, ^{132}Te, ^{134}Cs, and ^{137}Cs, were determined in Becquerel (Bq). As relative radionuclides' composition on body surface was known, the β-ray proportion from each radionuclide was calculated using equipment efficiency in each β-ray energy and major β-ray emission rate in each radionuclide using equation (1).

2.1.3 Natural removal rate

The natural removal rate of sub-micron particles from human skin was reported to be 14.7 h [10]. In the analyses, we used 15 h for particulate radionuclides as well as for elemental radioiodine. The average of multiple measurements on head and neck, at most 6, was considered as the representative value of individual, as there were no statistically significant differences in contamination levels among head, hand and clothing in the same individual [6].

Body surface contamination density $S_i(at\ plume)$ (Bq/cm²) of radionuclide-i at the time of exposure of the plume was calculated by a following formula in (2), where T_{scr} (h) is the time from exposure of the plume to measurement of body surface contamination, HL_i (h) is a half-life of each radionuclide, and 15 (h) is a half-life of the natural remove.

$$S_i(at\ plume) = S_i(T) \cdot \left(0.5^{\frac{T_{scr}}{HL_i}} \times 0.5^{\frac{T_{scr}}{15.0}}\right) \quad (2)$$

We assumed that the first exposure ending time to plume was set up at 19:00 on 12 March. Natural removal rate was calculated from the first exposure ending time to screening survey time at 12:00 on 13 March, or at 12:00 on 14 March. The second exposure ending time to plume was set up at 9:00 on 16 March because of the plume to south-southwest direction began blowing in the morning on 15 March, and changing the direction clock wisely into north-west, and finally to east direction in the morning on 16 March [7]. The natural removal rate was calculated from the second exposure ending to screening survey time at 12:00 on 16 March, or at 12:00 on 17 March. As to evacuees/residents who were measured on 15 March, we did not apply natural removal rate.

2.1.4 Thyroid dose estimation model by inhalation based on body surface contamination level

On the assumption that radionuclide-i deposits on body surface with a dry deposition velocity of v_i (cm/s), then cumulative body surface contamination density $S_i(at\ plume)$ (Bq/cm²) during evacuation from time 0 to time T (hour), as shown in Figure 1., would be

$$S_i(at\ plume) = 3.6 \times 10^{-3} \int_0^T v_i\, A_i(t) dt \quad (3)$$

, where $A_i(t)$ represents radionuclide-i concentration in the air at time t (Bq/m³).

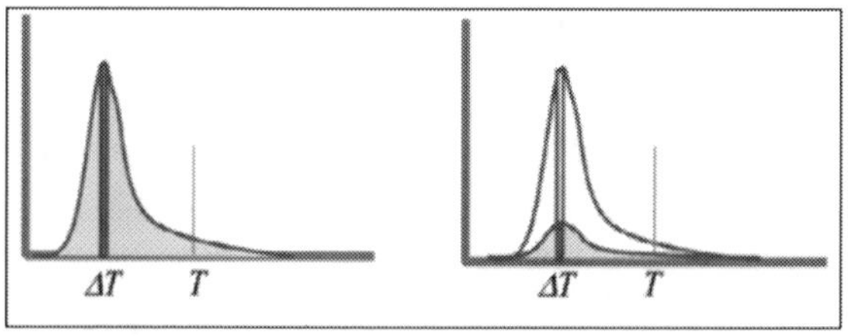

Fig. 1. The contents of a cumulative body surface contamination density *Si(at plume)* (Bq/cm²) during evacuation from time 0 to time T (hour).

On the other hand, inhalation thyroid equivalent dose on (4), $ED_{i\text{-}thy}(at\ plume)$, (Sv) during times 0 to T hour is expressed as

$$ED_{i-thy}(at\ plume) = B \cdot T \cdot f_i \frac{\int_0^T A_i(t)\, dt}{T}$$

$$= \frac{B \cdot f_i \cdot S_i(at\ plume)}{3.6\times10^{-3} \cdot v_i} \quad (4)$$

, where B (m^3/h) is age-specific ventilation parameter at light exercise (1-year-old), and fi is an age-specific thyroid dose coefficient (1-year-old) of ICRP Publication 71 [11]. As $S_i(at\ plume)$ can be calculated using measurement value $S_i(T)$ in equation (1) and equation (2), $ED_{i\text{-}thy}(at\ plume)$ can be estimated if v_i is given.

2.1.5 Physicochemical forms of radionuclide

Radioiodine (^{131}I) was assumed to be composed of 20% elementary iodine, 30% methyl-iodine, and 50% particulate iodine, which was referred from Katata et al. [7]. On the other hand, radioiodine (^{132}I, ^{133}I) were assumed to be composed of 50% elementary iodine, and 50% particulate iodine [9]. ^{132}Te, and ^{137}Cs were assumed to be composed of 100% particulate form.

By comparing ^{132}Te/^{131}I ratio on cloths in our data and that of air dust samplings on 12 March at 20km-zone from the FDNPS [12], relative deposition velocity of ^{132}Te was estimated to be 1.2-times faster than that of ^{131}I.

2.1.6 Two-dimensional Monte Carlo simulation

Two-dimensional Monte Carlo (2D-MC) simulation was conducted using a Monte Carlo sampling method by Crystal Ball (release 11.1.2.3.500, Kozo Keikaku Engineering Inc. Tokyo, Japan) on the assumption that v_i followed a uniform distribution ranging from 0.1 to 0.5 cm/s, and body surface contamination levels followed district-specific lognormal-, exponential- or Weibll-distribution, of which parameters were automatically selected by Crystal Ball software based on Anderson-Darling (A-D) statistics. The upper and lower bounds of the uniform distribution were adopted from reports on the FDNPS accident, the Windscale accident, and the SL-1 accident [13-17]. 2D-MC simulation was performed 200-times for uncertain v_i part combined with 2000-times for variable part, *i.e.*, cpm distribution. Deposition velocity of ^{132}Te was modified by 1.2-times based on the difference of ^{132}Te/^{131}I ratios between dust sampling data on 12 March and WBC data of two persons' clothing.

2.2 Analysis of data

2.2.1 GM survey meter screening

Body surface contamination was measured in cpm at 5 cm apart from body surface with a GM survey meter equipped with ϕ50mm-probe (TGS-133, TGS-136, and TGS-146; Hitachi-Aloka Medical, Ltd. and so on.) wrapped with a plastic cling film. Multiple measurements were done in the same individual from head to feet and results were written in an individual formatted sheet. The upper limit of measurement was 100,000 cpm. A detailed method of screening survey was reported elsewhere [5,6].

From the local government of Fukushima Prefecture, we obtained 7,291 records, as of May 2016. There were 6 completely blank records and 7,285 records were included in the present analysis. In addition, 254 residents/evacuees were screened by GM survey meter in Fukushima Medical University (FMU) hospital, and we pooled those 7,285 and 254 records. Thus, 7,539 evacuees/residents were the subject in the present analysis. Each record contained personal information including residence town, age, gender, date of screening, screening place, and measurements data in cpm on a body map.

This study was approved by the Ethics Committee of International University of Health and Welfare (Approval number: 13-B-79) as well as by the Ethics Committee of FMU (Approval number: No. 2136).

2.2.2 Data cleaning and Statistical analysis

There were 4 categories of records, such as "BG (background)", "Measurement value in cpm (including over 13,000 cpm)", "No measurement value recorded (no description about the measurement value of body surface contamination)", and "Specific symbols (-(hyphen), o (circle), / (slash), ✔, OK, None, "<10,000 cpm", Normal, and so on)". We interviewed several screening teams on these symbols, but as of the end of August 2016, we could not determine whether these symbols represented "Value over BG but less than screening level". Thus, in the present study, we categorized data into "BG", and "Measurement value in cpm", "No measurement value recorded (over background but less than screening level, *i.e.*, 13,000 cpm)", and "Specific Symbols". In the analyses, measured values were subtracted by 60 cpm in order to estimate thyroid equivalent dose at 1-year-old by inhalation. Finally, the value was log-transformed in the analyses.

According to evacuation directions, the evacuees/residents were classified into 4 district-groups; Tomioka/Okuma/Futaba/Naraha group, Namie group, Minamisoma group, and the other group. The other group consisted of residents living in Hirono Town, Tamura City, Kawauchi Village, Katsurao Village, Iitate Village, other 21 municipalities in Fukushima Prefecture, and outside Fukushima area. Since radioactive plumes flew over inland on 12 March and on 15 and 16 March [7], the distribution of surface contamination of residents was classified by 2 different periods: during 12-14 March period, and during 15-17 March period.

IBM SPSS Statistics for Windows ver. 23.0.0.2 software package (IBM Corp., Armonk, NY, USA) was used for statistical analyses. x^2-Test was used to compare proportions of over 13,000 cpm in 4 district-groups (Tomioka/Okuma/Futaba/Naraha group, Namie group, Minamisoma group, and the other group) using *post hoc* z-Test contributed Bonferroni collection, Kruskal-Wallis to compare the measurement value (cpm) in 4 district-groups, and *post hoc* analyses were performed automatically multiple comparison analysis in SPSS.

The p-values of <0.05 were considered statistically significant.

3 Results

3.1 The summary of contamination level in residents

The demographic feature of screened subjects were 41.9 ± 23.5 (mean ± SD) years old (ranged 0-100, excluding 353 missing data), and 49.1 % males (excluding 721 missing data). Table 1 showed subject numbers in each contamination category. The proportion of "BG", "Measurement value in cpm", "No measurement value recorded", and "Specific symbols" were 2.4 %, 27.7 %, 33.7 %, and 36.2 %, respectively. Among evacuees/residents who had data about body surface contamination, only 0.4% of evacuees/residents showed contamination levels over 13,000 cpm. The proportion of over 13,000 cpm in Namie group was higher than any other group during both periods ($p<0.001$, x^2-Test).

Table 1. The proportion of categorical contamination levels.

Subjects' living area on 11th March 2011	BG	Measurement value in cpm [Over 13,000 cpm]	No measurement value recorded	Specific symbols	Total
Tomioka/Okuma/Futaba/Naraha	81	893 [10]	937	705	2,626
(%)	(3.1)	(34.0) (0.4)	(35.7)	(26.8)	
Namie	31	328 [15]	272	321	967
(%)	(3.2)	(33.9) (1.6)	(28.1)	(33.2)	
Minamisoma	20	589 [0]	781	688	2,078
(%)	(1.0)	(28.3) (0.0)	(37.6)	(33.1)	
The other*	46	250 [2]	552	1,018	1,868
(%)	(2.5)	(13.4) (0.1)	(29.5)	(54.5)	
Total	178	2,060 [27]	2,542	2,732	7,539
(%)	(2.4)	(27.3) (0.4)	(33.7)	(36.2)	

* The address of the other group included Hirono Town, Tamura City, Kawauchi Village, Katsurao Village, Iitate Village, the other 21 municipalities in Fukushima Prefecture, and outside Fukushima area.

3.2 Body surface contamination level in residents after modifying natural removal rate

The box-and-whisker plots of body surface contamination levels are shown in Figure 2.

During 12-14 March periods, the levels of measurement values in cpm were almost same between Namie and Minamisoma groups, but statistically higher than those of Tomioka/Okuma/Futaba/Naraha group and the other group ($p<0.01$, Kruskal-Wallis). (Figure 2). During 15-17 March period, the contamination levels of Tomioka/Okuma/Futaba/Naraha group was statistically the lowest in all groups ($p<0.001$, Kruskal-Wallis). It was noted that the contamination levels in the other group, that included Kawamata, Iitate, Katurao villages located north-west direction from FDNPS, increased and became comparable to the levels in Namie and Minamisoma groups (Figure 2).

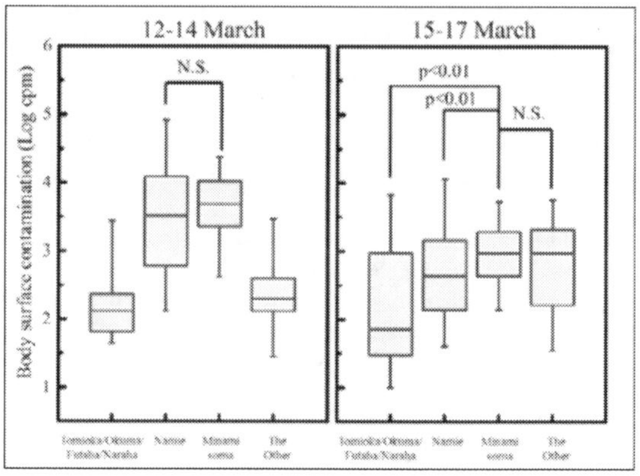

Fig. 2. The box-and-whisker plots of body surface contamination after natural remove collection. N.S. is no significant.

3.3 Radionuclides' ratio in the air and clothing

Table 2 depicted the temporal pattern of radionuclides' ratios in the air based on WBC and air dust samplings. It was demonstrated that $^{132}Te/^{131}I$ ratio by air dust sampling within 20 km radius from the FDNPS on 12 March was almost the same as those by WBC measurements of cloths of two persons exposed on the same day after modifying physical decay. On 12 March, $^{132}Te/^{131}I$ ratio was 1.50 by source term, 1.93 by air dust samplings (N=3) within 20 km radius from the FDNPS, and 1.79 by the estimated air radioactive concentration using conversion factor of GM survey meter. On 15 March, $^{132}Te/^{131}I$ ratio was 0.20 for source term, 0.50 for air dust samplings of average data (N=5) in JAEA, and 0.36 by the estimation using conversion factor of GM survey meter.

Table 2. The temporal pattern of radionuclides' ratios in the air on 12 March and 15 March.

On 12 March	$^{132}Te/^{131}I$	$^{133}I/^{131}I$	$^{137}Cs/^{131}I$
Source term by Katata et al. at 12:00	1.50	1.10	0.10
Air dust sampling average data (N=3) at 20 km-zone from Fukushima Prefecture at 9:00-13:30	1.93		0.06
Estimated air concentration from measured by clothing in FMU at 12:00*	1.79	1.14	0.08

On 15 March	$^{132}Te/^{131}I$	$^{133}I/^{131}I$	$^{137}Cs/^{131}I$
Source term by Katata et al. at 20:00	0.20	0.10	0.01
Air dust sampling average data (N=5) at JAEA (Tokai Village) at 3:00-21:00	0.50	0.12	0.08
Air dust samplings in JAEA using our conversion factor of GM survey meter at 12:00*	0.36	0.11	0.06

Radionuclide ratios were based on source term by Katata, et al. [7], air dust sampling data within 20 km radius the FDNPS by the local government of Fukushima Prefecture on 12 March [12], and air dust sampling by JAEA located in Tokai Village on 15-17 March [9]. * Estimated nuclides ratios by simulation were based on the nuclides compositions of clothing on 12 March or those of air dust samplings by JAEA on 15 March.

3.4 2D-MC simulation of estimated thyroid equivalent dose at 1-year-old children by inhalation using body surface contamination

We assumed that the distribution of contamination among evacuees/residents having measurement values in cpm would be representative of other evacuees/residents in the same district who had only categorical values, *i.e.*, "No measurement value recorded", or "Specific Symbols". Using distributions in Figure 2, thyroid equivalent doses of 1-year-old children by inhalation were estimated by 2D-MC simulation in 12-14 March period (Table 3) and 15-17 March period (Table 4). Judging from A-D scores, the fitness of body surface contamination to group-specific log normal, exponential or Weibll distribution was practically satisfactory in Namie, Minamisoma, and the others groups. On the other hand, the distribution of Tomioka/Okuma/Futaba/Naraha group was heterogeneous and fitting to a log normal distribution was rather poor.

In 12-14 March period (Table 3), 50 percentile thyroid equivalent dose at 1-year-old children was highest in Minamisoma group (4.2 mSv), while 95 percentile dose was highest in Namie group (86.0 mSv). In 15-17 March period (Table 4), 50 percentile thyroid equivalent dose was less than 1.6 mSv, and 95 percentile dose was less 16 mSv in all groups.

If the same evacuee/resident exposed 3-times on 12, 15 and 16 March, his/her 50 percentile inhalation dose would be the sum of 50 percentile doses in Tables 3 and 4, for example, 4.3 mSv for Namie group.

Table 3. Estimated thyroid equivalent dose of 1-year-old children by inhalation using 2D-MC simulation during 12-14 March period.

4 district-group	50 %-tile (90% U.I.)	90%-tile (90% U.I.)	95%-tile (90% U.I.)
Tomioka/Okuma/ Futaba/Naraha	0.1 mSv (0.1 - 0.3)	0.6 mSv (0.3 - 1.3)	1.0 mSv (0.5 - 2.0)
Namie	2.7 mSv (1.4 - 5.5)	40.8 mSv (20.9 - 85.2)	86.0 mSv (80.1- 91.9)
Minamisoma	4.2 mSv (2.2 - 8.8)	13.4 mSv (6.9 - 28.0)	17.2 mSv (8.8 - 35.8)
The Others	0.2 mSv (0.1 - 0.4)	1.3 mSv (0.7 - 2.6)	2.1 mSv (1.1 - 4.5)

The distributions (A-D scores) of body surface contamination in 4 district-groups were log normal (23.2), log normal (0.84), exponential (0.49), and log normal (2.21) for Tomioka/Okuma/Futaba/Naraha, Namie, Minamisoma, and the others, respectively. The U.I. is uncertainty interval.

Table 4. Estimated thyroid equivalent dose of 1-year-old children by inhalation using 2D-MC simulation during 15-17 March period.

4 district-group	50 %-tile (90% U.I.)	90%-tile (90% U.I.)	95%-tile (90% U.I.)
Tomioka/Okuma/ Futaba/Naraha	0.2 mSv (0.1 - 0.3)	5.4 mSv (2.8 - 11.3)	13.2 mSv (6.8 - 27.6)
Namie	0.8 mSv (0.4 - 1.7)	8.2 mSv (4.2 - 17.3)	15.7 mSv (8.1 - 32.9)
Minamisoma	1.6 mSv (0.8 - 3.3)	5.9 mSv (3.0 - 12.3)	8.4 mSv (4.3 - 17.5)
The Others	1.2 mSv (0.6 - 2.5)	8.3 mSv (4.3 - 17.4)	12.7 mSv (6.5 - 26.5)

The distributions (A-D scores) of body surface contamination in 4 district-groups were log normal (4.91), log normal (1.44), log normal (1.03), and Weibll (1.29) for Tomioka/Okuma/Futaba/Naraha, Namie, Minamisoma, and the others, respectively. The U.I. is uncertainty interval.

4 Discussions

In the present study, we proposed a new methodology for estimating thyroid equivalent dose by inhalation based on the levels of body surface contamination. Since the exposure levels would differ by the timing and route of evacuation, as well as by sheltering status at the time of radioactive plume, body surface contamination was a reasonable indicator of exposure, which reflected individual self-protective behavior.

Evacuation order was delivered to residents living in the 20 km zone from the FDNPS at evening on 12 March. About 93% of residents living in Tomioka, Okuma, Futaba and Naraha Towns evacuated to evacuation centers in Kawauchi Village, Tamura City, Kawamata Town, and Iwaki City as of midnight on 13 March [18]. Since these evacuation centers were not attacked by a plume released on 12 March, residents might expose to plume before or during evacuation. In contrast, many residents in Namie Town evacuated to an evacuation center on Tsushima area in Namie Town, which placed on borderline of 20 km zone from FDNPS. Although about 81% of residents in Namie Town evacuated as of midnight on 13 March [18], a plume on 12 March blew inland towards Namie Town and Minamisoma City. As to residents in Minamisoma City, evacuation order was delivered to only Odaka area, and about 30 % of residents evacuated as of midnight on 13 March [18]. These differences in evacuation timing and route may be responsible for difference in body surface contamination levels and estimated thyroid doses among 4 district-groups.

In the present study, we estimate thyroid equivalent dose of 1-year old children by inhalation based on body surface contamination levels. In order to convert GM survey meter measurement in cpm into radionuclides densities in Bq/cm^2, one has to know radionuclides' composition on body surface. Since we had simultaneously analyzed radionuclides' composition on clothing who evacuated on 12 March, and because the radionuclides' composition of air dust samplings at 20 km-zone from FDNPS on the same day was published [12], we could accomplish the task. As to radionuclides' composition on body surface on 15-17 March, we assumed relative deposition velocity of ^{132}Te to ^{131}I would be same through 12 March to 17 March, and estimated nuclides composition on the body surface would be proportional to those composition in air at JAEA on 15 March. A total number of subjects having measurement values in cpm were 343 in Namie Town, 589 in Minamisoma City and these numbers were sufficient for illustrating the district-specific distribution of body surface contamination. On the other hand, these numbers were too small to illustrate a district-specific distribution in other area, and we had to pool subjects into Tomioka/Okuma/Futaba/Naraha group and the other group. By this reason, the distributions of body surface contamination in the latter district-groups were rather heterogeneous and uncertain in the present 2D-MC simulation. However, our estimates were very close to estimates based on thyroid measurements or whole body measurements.

In the UNSCEAR 2013 Fukushima Report, settlement-average thyroid doses during evacuation in 1-year-old infants was estimated as follows: 0-46 mGy for Tomioka/Okuma/Futaba/Naraha Town, 37 mGy and/or 59 mGy for Namie Town, 6.4 mGy or 45 mGy for Minamisoma City, and 0-63 mGy for the other [4]. It was noted that UNSCEAR's estimates were a sum of inhalation and ingestion doses and based on an atmospheric transfer and dispersion model with Terada's source term [19]. It was said that the variability of thyroid inhalation dose could be from 30-50% of settlement-average thyroid doses up to about two to three times higher than settlement-average thyroid doses.

Tokonami et al. estimated thyroid doses of children based on direct thyroid measurements of 46 evacuees/residents, most of them from Namie Town. The median thyroid equivalent dose and the maximum dose for children were 4.2 and 23 mSv, respectively [3]. In the estimation, the authors hypothesized that subjects were exposed on 15 March. Recently, Kim et al. reported internal thyroid dose estimation at 1-year-old children mostly from inhalation during 15-23 March, 2011. The authors estimated thyroid dose by 1,080 direct thyroid measurements, about 3,000 radio-cesium measurements by WBC and the simulation of behavior pattern by atmospheric dispersion [2]. In order to estimate thyroid doses from WBC data, the authors assumed ^{131}I/^{137}Cs intake ratio being 3, which was 2- to 3-times lower than those in the air. The 90 percentile of internal thyroid dose at 1-year-old children was 10-30 mSv for Tomioka/Okuma/Futaba/Naraha Town, 20 mSv for Namie Town, and 20 mSv for Minamisoma City [2].

In the present report, inhalation thyroid doses were higher in Namie and Minamisoma groups than Tomioka/Okuma/Futaba/Naraha group during 12-14 March period as well as 15-17 March period. In addition, inhalation doses were higher during 12-14 March period than those during 15-17 March period. This was a new finding that a plume on 12 March had suffered people living Namie Town and Minamisoma City than expected before. Our 50 and 90 percentiles of thyroid equivalent dose at 1-year-old children in Namie group were 2.7 mSv and 40.8 mSv during 12-14 March periods, respectively, and were within a 2-times difference range of Tokonami's estimates. Our 90 percentile estimate was also higher than those by Kim, *i.e.* 20 mSv. While, our estimates for Tomioka/Okuma/Futaba/Naraha group were lower than Kim's estimates: 0.6 mSv during 12-14 March periods and 5.4 mSv during 15-17 March periods vs. 10-30 mSv. Since radio-cesium detected by WBC could be derived from acute inhalation during early phase of the accident but also from chronic ingestion, radio-iodine uptake scenario could influence the estimation.

We proposed in the present study a new methodology for estimating thyroid equivalent dose using district-specific body surface contamination in cpm. However, it is still in a developmental stage and has several sources of uncertainty. First, the natural removal rate (half-life time) was chosen as 15 hours based on one report [10]. However, it is not certain whether radioactive particles or elemental iodine behave like submicron silica particles. In addition, natural removal rate may be influenced meteorological condition, human activity and physicochemical property of skin, hair and clothing. Second, we assumed deposition velocity following a uniform distribution ranging from 0.1 to 0.5 cm/s. The deposition velocity in experimental conditions was reported to be quite heterogeneous and changed along with the physicochemical property of surface and meteorological condition [15,16]. In the present study, we adopted values reported in the FDNPS accident, the Windscale accident, and the SL-1 accident [13,14,16]. However, it is not certain whether these values can be applied to human skin, hair or clothing. Third, the chemical form of radio-iodine was uncertain. In the present study, we adopted 20 % elementary iodide, 30% methyl-iodine and 50% particulate form as Katata reported [7]. If the fraction of methyl iodine increases with time, the deposition velocity of radio-iodine will decrease, and then, our estimation will tend to underestimate. Fourth, we chose the constant respiration rate (0.35 m^3/h) at 1-year-old children of lightly exercise during evacuation, and a thyroid uptake rate of 30%. These parameters are reported in ICRP Publication 68 [20]. Since the variation of thyroid uptake rate is known in Japanese, a thyroid uptake rate may become one of uncertainty in our report.

As to our obtained data from the Fukushima government office, there were only rounded address and age, date of screening, screening place, and measurements data in cpm or "Specific symbols". It was not possible to correlate our data with the resident personal behavior record of the Fukushima Health Management Survey [21] or measurement data such as

WBC measurement [22]. Furthermore, we could not classify an affiliation of a person who was screened body surface contamination survey. If first responders or TEPCO workers of the early phase in the FDNPS accident were contented in our data list, their contamination level might be quite different from those of residents [23]. We are planning to correlate GM survey meter data to WBC measurements and resident personal behavior record in order to clarify these issues. There were categorical data in our data set, such as "No measurement value recorded", and "Specific symbols". In order to clarify their meaning, we began conducting questionnaire survey to surveyors who conducted screening survey for body surface contamination during the first 6 days of the FDNPS accident. It is still premature to conclude, but several surveyors answered "Specific symbols" represented a condition where body surface contamination was over BG but less than predetermined screening level, *i.e.*, 13,000 cpm.

5 Conclusion

Thyroid equivalent dose of 1-year-old children by inhalation was estimated by 2D-MC simulation using four district-specific distributions of body surface contamination in cpm. Inhalation dose was higher in Namie and Minamisoma groups than other districts due to exposure to a plume on 12 March, 2011. Our estimates were lower than UNSCEAR's estimates, partly because ingestion dose was not included.

Acknowledgment

This study was supported by the funding from the Ministry of Environment, Japan.

References

1. S. Midorikawa, S. Suzuki, and A. Ohtsuru, Science **352** (6286), 666 (2016)
2. E. Kim, O. Kurihara, N. Kunishima, T. Momose, T. Ishikawa, and M. Akashi, J. Radiat. Res. **57 Suppl 1**, i118 (2016)
3. S. Tokonami, M. Hosoda, S. Akiba, A. Sorimachi, I. Kashiwakura, and M. Balonov, Sci. Rep. **2**, 507 (2012)
4. United Nations Scientific Committee on the Effects of Atomic Radiation, *UNSCEAR 2013 Report Annex A: Levels and effects of radiation exposure due to the nuclear accident after 2011 great east-Japan earthquake and tsunami* (United Nations, New York, 2014)
5. H. Kondo, J. Shimada, C. Tase, T. Tominaga, H. Tatsuzaki, M. Akashi, K. Tanigawa, Y. Iwasaki, T. Ono, M. Ichihara, Y. Kohayakawa, and Y. Koido, Health Phys. **105** (1), 11 (2013)
6. T. Ohba, M. Miyazaki, H. Sato, A. Hasegawa, M. Sakuma, T. Yusa, F. Shishido, and A. Ohtsuru, Health Phys. **107** (1), 10 (2014)
7. G. Katata, M. Chino, T. Kobayashi, H. Terada, M. Ota, H. Nagai, M. Kajino, R. Draxler, M. C. Hort, A. Malo, T. Torii, and Y. Sanada, Atmos. Chem. Phys. Discuss. **15** (2), 1029 (2015)
8. H. Ogino, T. Ichiji, and T. Hattori, Radiat. Prot. Dosimetry **151** (1), 36 (2012)
9. S. Furuta, S. Sumiya, H. Watanabe, M. Nakano, K. Imaizumi, M. Takeyasu, A. Nakada, H. Fujita, T. Mizutani, M. Morisawa, Y. Kokubun, T. Kono, M. Nagaoka, H. Yokoyama, T. Hokama, T. Isozaki, M. Nemoto, Y. Hiyama, T. Onuma, C. Kato, and, T. Kurachi, *Results of the Environmental Radiation Monitoring Following the Accident at the Fukushima Daiichi Nuclear Power Plant -Interim Report (Ambient Radiation Dose Rate, Radioactivity Concentration in the Air and Radioactivity Concentration in the Fallout)* (JAEA-Review, Tokai, 2011)
10. H. Hession, M. Byrne, S. Cleary, K. G. Andersson, and J. Roed, J. Environ. Radioact. **85** (2-3), 196 (2006)
11. International Commission on Radiological Protection, *Age-dependent Doses to Members of the Public from Intake of Radionuclides - Part 4 Inhalation Dose Coefficients* (ICRP Publication 71, Oxford, 1995)
12. The local government of Fukushima Prefecture, *The result of gamma-ray radionuclide analysis about air dust sampling in emergency monitoring* (Japanese) [online]. Available at: http://www.pref.fukushima.lg.jp/uploaded/attachment/166539.pdf. Accessed 10 August 2016.
13. H. Amano, M. Akiyama, B. Chunlei, T. Kawamura, T. Kishimoto, T. Kuroda, T. Muroi, T. Odaira, Y. Ohta, K. Takeda, Y. Watanabe, and T. Morimoto, J. Environ. Radioact. **111**, 42 (2012)
14. A. C. Chamberlain, Q. J. R. Meteorol. Soc. **85** (366), 362 (1959)
15. G. A. Sehmel, Atmos. Environ. **14** (9), 983 (1980)
16. F. A. Gifford Jr. and D. H. Pack, Surface deposition of airborne material, *Technical Progress Review, Nuclear Safety* **3** (4), 205 (1963)
17. N. G. Stewert and R. N. Crooks, Nature **182** (4636), 627 (1958)
18. The National Diet of Japan, *The official report of The Fukushima Nuclear Accident Independent Investigation Commission* (The National Diet of Japan-The Fukushima Nuclear Accident Independent Investigation Commission, Tokyo, 2012)

19. H. Terada, G. Katata, M. Chino, and H. Nagai, J. Environ. Radioact. **112C**, 141 (2012)

20. International Commission on Radiological Protection, *Dose Coefficients for Intakes of Radionuclides by Workers* (ICRP Publication 68, Oxford, 1994)

21. S. Yasumura, M. Hosoya, S. Yamashita, K. Kamiya, M. Abe, M. Akashi, K. Kodama, and K. Ozasa, Journal of Epidemiology **22** (5), 375 (2012)

22. K. Kamiya, T. Ishikawa, S. Yasumura, A. Sakai, T. Ohira, H. Takahashi, A. Ohtsuru, S. Suzuki, M. Hosoya, M. Maeda, H. Yabe, K. Fujimori, S. Yamashita, H. Ohto, and M. Abe, Radiat. Prot. Dosimetry, in press (2016)

23. T. Nakano, E. Kim, K. Akahane, T. Tominaga, H. Tatsuzaki, O. Kurihara, and N. Sugiura, Direct measurement for highly-exposed TEPCO workers and NIRS first responders involved in the Fukushima NPS accident, *The 1st NIRS symposium on reconstruction of early internal dose in the TEPCO Fukushima Daiichi nuclear power station accident* (National Institute of Radiological Sciences, Chiba, 2012)

Investigation of Main Radiation Source above Shield Plug of Unit 3 at Fukushima Daiichi Nuclear Power Station

Hideo Hiratama[1,2*], Kenjiro Kondo[1,2], Seishiro Suzuki[1], Yoshihiko Tanimura[1], Kohei Iwanaga[1], Hiroshi Nagata[1]

[1]*Nuclear Regulation Authority, 1-9-9, Roppongi, Minato-ku, Tokyo, Japan 106-8450*
[2] *High Energy Accelerator Research Organization, 1-1, Oho, Tsukuba-shi, Ibaraki, Japan 305-0801*
*Corresponding Author: hideo.hirayama@kek.jp

Abstract

Pulse height distributions were measured using a CdZnTe detector inside a lead collimator to investigate main source producing high dose rates above the shield plugs of Unit 3 at Fukushima Daiichi Nuclear Power Station. It was confirmed that low energy photons are dominant. Concentrations of Cs-137 under 60 cm concrete of the shield plug were estimated to be between 8.1E+9 and 5.7E+10 Bq/cm^2 from the measured peak count rate of 0.662 MeV photons. If Cs-137 was distributed on the surfaces of the gaps with radius 6m and with the averaged concentration of 5 points, 2.6E+10 Bq/cm^2, total amount of Cs-137 is estimated to be 30 PBq.

KEYWORDS: Unit 3, Fukushima Daiichi, CdZnTe, pulse height distribution, EGS5

I. Introduction

Ambient dose equivalent rates at the operation floor of Fukushima Daiichi Nuclear Power Station Unit 3 were very high. It was order of Sv/h at maximum at 5 m from the floor surface. Maximum dose rate was reduced to about 200 mSv/h by removing debris, cutting and sucking of concrete surface using remote devices. It was estimated that about 5 mm of concrete surface was removed by cutting and sucking. The reduction rate of about 20 % is smaller from the fact that contamination by Cs-134 and Cs-137 attached at concrete superficially. We supposed considering the construction of shield plugs shown in Figure 1 that Cs-134 and Cs-137 attached between gaps of 3 layers of shield plugs at which gases including high density Cs-134 and Cs-137 leaked through at the time of accident. If this estimation is correct, the main component of the photon field above the shield plug is scattered photons within 60 cm concrete. To confirm this estimation, measurements of pulse height distribution were performed using a CdZnTe detector inside a lead collimator with enough shielding.

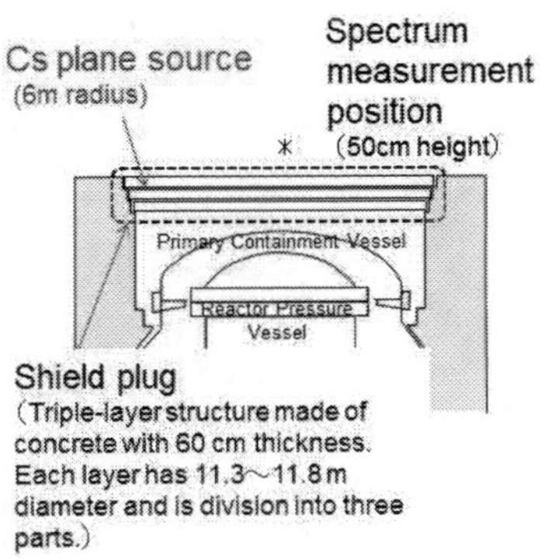

Figure 1: Construction of shield plug.

II. Detector and collimator

The collimator shown in Figure 2 was used to measure pulse height distribution of photons emitted from small region outside the shield plug etc. The collimator consists with a 30 cm x 30 cm x 30 cm lead block. This lead block was set inside a 2 cm thick iron container. A 0.8 cm diameter hole is opening at a center of the lead block and a 1 cm x 1 cm x 1 cm CdZnTe detector (GR-1, Kromek[1)]) was set at 12.5 cm from bottom of the lead block.

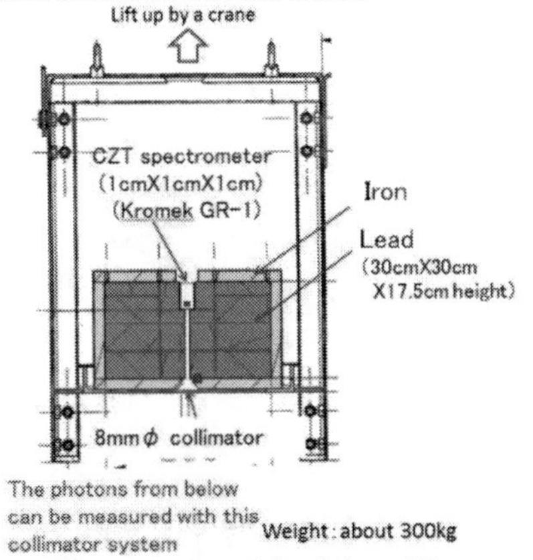

Figure 2: Schematic drawing of the collimator system

If 1 photon/cm^2 toward the detector uniformly enters the bottom of the collimator, numbers of unscattered photons become as follows ;

0.662 MeV	: through hole	0.503
	through lead	0.000515
0.3 MeV	: through hole	0.503
	through lead	1.26E-20.

DOI: 10.1051/epjconf/201715308010

From this result, it is clear that contributions from photons transmitting lead can be neglected.

Weight of the collimator system is about 300 kg. This system was set at the position that the detector surface was 50 cm from the surface of the floor by using a huge remote operating crane.

III. Measurement points and results

24 points were selected for pulse height distribution measurements from various environments; the shield plug of the reactor, iron shield on the floor, the spent fuel pool etc. The measurement points are densely-arranged above the shield plugs and their joints because high level contamination is expected.

Measurement points and measured pulse height distributions are shown in Figure 3 and 4, respectively. As the full energy absorption peaks, corresponding to gamma-rays from Cs-134 and Cs-137 only are detected. All pulse height distributions show high count rates at low energy region mainly corresponding to scattered photons.

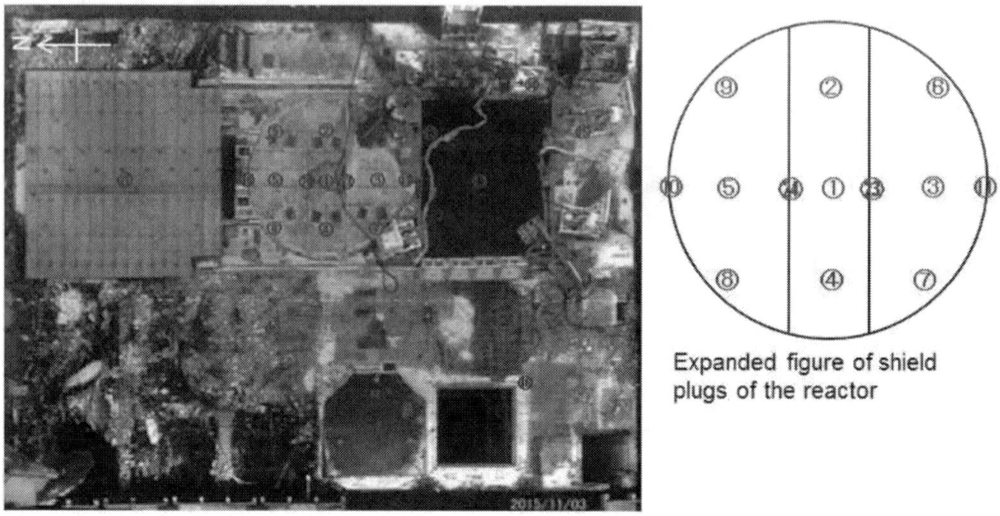

Figure 3: **Measurement points**

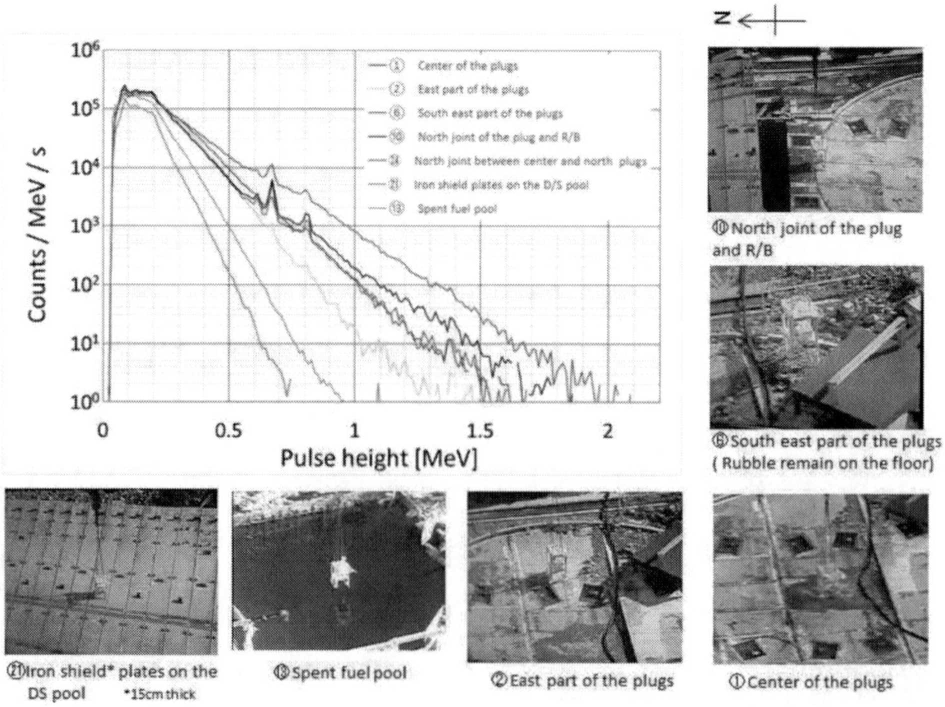

Figure 4 :Measured pulse height distributions

IV. Estimation of concentration of Cs-137 at the gap of the shield plug

1. Conversion factor from peak count rates to a concentration density

Numbers of unscattered photons from radial segmented plane sources of Cs-137 entering to the CdZnTe detector inside the collimator system were calculated with EGS5 Monte Carlo code[2)]. The calculated 0.662 MeV photons entering on the detector surface became 7.5E-09 photons/sec per Bq/cm^2. The contribution of Cs-137 outside 7.2 cm from a center of the collimator was about 1 %. From this result, an effective source radius for this measurement can be treated as 7 cm.

A peak efficiency of the detector for 0.662 MeV photon was 0.078 by calculating with EGS5 supposing that photons entered normally on the detector surface and uniformly within the collimator radius .

The conversion coefficient from peak count rates of 0.662 keV photons to a concentration density became 1.7E+09 Bq/cm^2 per cps.

2. Estimation of concentration

Concentrations were estimated from the measured peak counts rates and the obtained conversion coefficient at the point from No.1 to No.5 which were above the shield plugs except their joints. Results are shown in Table 1. Concentrations varied almost factor 5 between these points from 8.0E+09 to 5.7E+10 Bq/cm^2. If Cs-137 was distributed on the surfaces of the gaps of three shield plugs with radius 6m and with the averaged concentration of these 5 points, 2.6E+10 Bq/cm^2, total amount of Cs-137 is estimated to be 30 PBq and is about 10 % of the estimated total inventory of Cs-137 at Unit 3[3)].

IV. Conclusion

From our measurement, it can be expected that larger amount of Cs-137 than that emitted to environment at the time of accident exists between the gaps of the shield plugs. This must be considered at future processes of decommissioning of the reactor.

Acknowledgment

We have to express out appreciation to High Energy Accelerator Research Organization for providing the collimator system. We are also immensely grateful to Fukushiam Daiichi Nuclear Power Station related Tokyo Electric Power Company, Toshiba Corporation and Kajima Corporation JV for their helping at Unit 3 measurement.

Table 1 Estimated Cs-137 concentration

Point	Peak count rates (cps)	Concentration (Bq/cm^2)
No. 1	32.8	5.7E+10
No. 2	7.19	1.2E+10
No. 3	23.4	4.0E+10
No.4	4.72	8.1E+09
No. 5	8.67	1.5E+10

References

1) GR1-Gamma Ray Spectrometer, Kromek Group plc <http://www.kromek.com/index.php/products/all-nuclear-products/gr1-gamma-ray-spectrometer> (confirmed 27, January, 2016)

2) H. Hirayama, Y. Namito, A. F. Bielajew, S. J. Wilderman and W. R. Nelson, "The EGS5 Code System", SLAC-R-730 (2005) and KEK Report 2005-8 (2005).

3) Kenji Nishihara, Hiroki Iwamoto, and Kenya Suyama, " Estimation of Fuel Compositions in Fukushima-Daiichi Nuclear Power Plant", JAEA-Data/Code 2012-018(2012).

 DOI: 10.1051/epjconf/201715308011

Investigation on Main Radiation Source at Operation Floor of Fukushima Daiichi Nuclear Power Station Unit 4

Hideo Hirayama[1,2*], Kenjiro Kondo[1,2], Seishiro Suzuki[1], Shimpei Hamamoto[1,3], Kohei Iwanaga[1]

[1]*Nuclear Regulation Authority, 1-9-9, Roppongi, Minato-ku, Tokyo, Japan 106-8450*
[2] *High Energy Accelerator Research Organization, 1-1, Oho, Tsukuba-shi, Ibaraki, Japan 305-0801*
[3]*Japan Atomic Energy Agency, (765-1 Funaishikawa, Tokai-mura, Naka-gun, Ibaraki 319-1184, Japan (Current affiliation)*

*Corresponding Author: hideo.hirayama@kek.jp

Abstract

Pulse height distributions were measured using a $LaBr_3$ detector set in a 1 cm lead collimator to investigate main radiation source at the operation floor of Fukushima Daiichi Nuclear Power Station Unit 4. It was confirmed that main radiation source above the reactor well was Co-60 from the activated steam dryer in the DS pool (Dryer-Separator pool) and that at the standby area was Cs-134 and Cs-137 from contaminated buildings and debris at the lower floor. Full energy peak count rate of Co-60 was reduced about 1/3 by 12mm lead sheet placed on the floor of the fuel handling machine.

KEYWORDS: Unit 4, Fukushima Daiichi, $LaBr_3$, pulse height distribution

I. Introduction

Ambient dose equivalent rates at the operation floor of Fukushima Daiichi Nuclear Power Station Unit 4 were far lower than those at Unit 1, 2 and 3 due to the fact that Unit 4 was under a periodical inspection at the time of the accident, March 2011. However, there were several places at which ambient dose equivalent rates exceeded several hundreds μSv/h even in this situation. This level is still higher at the position of the manual fuel removal operation. Before our measurement, Tokyo Electric Power Company had a plan to set a shielding wall at the north end of the cover of the operation floor by regarding that photons at the operation floor of Unit 4 mainly came from the contamination at the operation floor of Unit 3. If this assumption was correct, dose rates were supposed to decrease gradually depending on the distance from the north end on the operation floor of Unit 4, by considering the distance from the Unit 3. However, dose rate decreased rapidly depending on the distance. From this tendency, we supposed that there were other radiation sources around Unit 4. Pulse height distribution measurements were performed to investigate radiation sources for the dose at various places on the operation floor.

II. Measurements

A $LaBr_3$ detector with 2 inches diameter and 2 inches length (Inspector 1000[1)]) was used for pulse height distribution measurements. Detector set inside a lead box with 1 cm thick and 1 cm diameter hole at the center was used to measure at the places at which dose rates were over 100 μSv/h and to get information of photon direction. Bird-eye view of the Unit 4 reactor building operation floor and the points of measurement are shown in Figure 1.

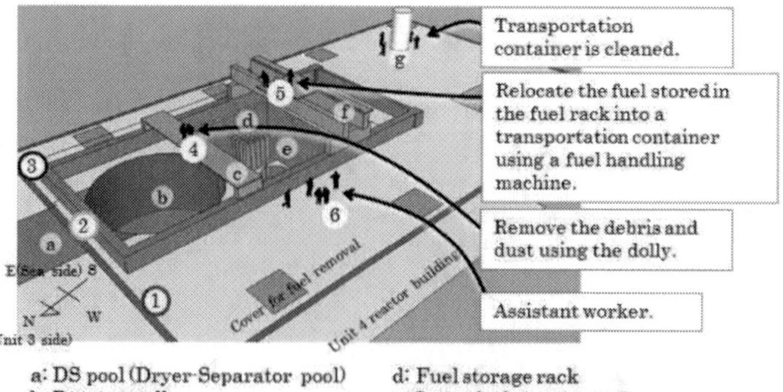

a: DS pool (Dryer-Separator pool)
b: Reactor well
c: Dolly
d: Fuel storage rack
e: Spent fuel storage pool
f: Fuel handling machine (FHM)
g: On-premise transportation container

Figure 1: Bird-eye view of the Unit 4 reactor building operation floor

III. Results and discussions

1. Pulse height distribution

Measured pulse height distributions at point 1, 2 and 3 are shown in Figure 2. At the point 2, full energy absorption peaks of Co-60 were clearly seen and those of Cs-134 and Cs-137 were very small. On the other hand, full energy absorption peaks of Cs-134 and Cs-137 were dominant and those of Co-60 were relatively small at other points.

At Unit 4, the steam dryer was set on the separator in the DS pool (Dryer-Separator pool) for replacing the shroud and the top of the steam dryer cropped out from water surface. From this configuration, the steam dryer which was heavily contaminated by Co-60 could be seen via water in the reactor well. The steam dryer could not be seen directly from other 2 points. From this result, it can be confirmed that source of Co-60 is the steam dryer in the DS pool.

Ambient dose equivalent rates at 5 m height at the operation floor of Unit 3 was maximum at the center of the shield plug[2)]. If photons emitted from Cs-134 and Cs-137 came from Unit 3 with similar distribution at the operation floor of Unit 3, peak count rates at the point 2 must be higher than those at other 2 points. The small peak count rates of Cs-134 and Cs-137 at point 2 can be explained if photons from Cs-134 and Cs-137 came from depositions at the lower floors outside the North end cover of Unit 4 and were shielded by water in the DS pool.

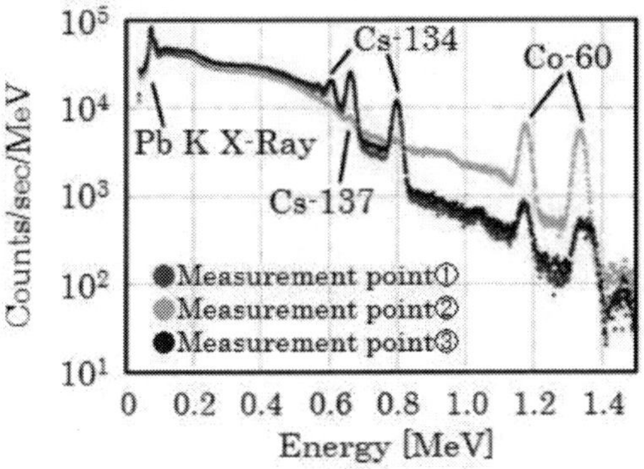

Figure 2: Comparison of pulse-height distribution

The decrease of photons from Co-60 at points 1 and 3 was supposed due to shielding effect by the operation floor.

2. Floor shielding effect

The shielding wall at the north end was supposed to be effective for photons from Cs-134 and Cs-137 contaminated debris at the lower floors. On the other hand, this wall was considered not to be effective for Co-60 photons from the steam dryer because they penetrate water under the wall and reach the working place.

We investigated the shielding effect of lead sheet set at the FHM floor and its dolly floor. Figure 3 shows the study point of the shielding effect. Full energy peak count rates of photons from Co-60, Cs-134 and Cs-137 were compared with or without a 12 mm lead sheet at the points 4, 5 and 6. The obtained results are shown in Figure 4.

At the FHM floor and the dolly floor, peak count rates of Co-60 were reduced to about 1/3 by the 12 mm lead sheet. This result shows that the floor shield is effective to reduce exposur dose for workers above the FHM or the dolly.

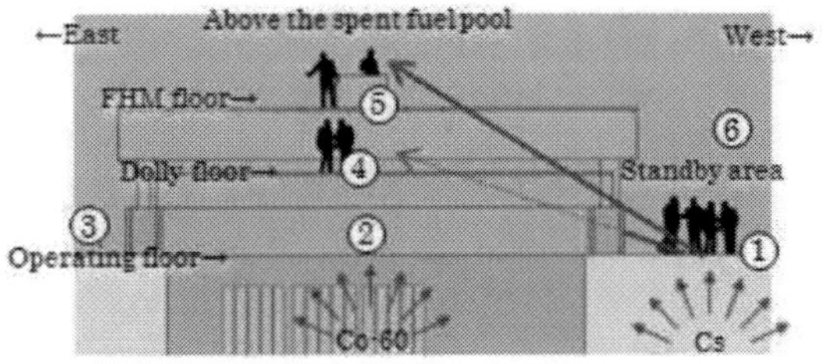

Figure 3: Study point of floor shielding effect

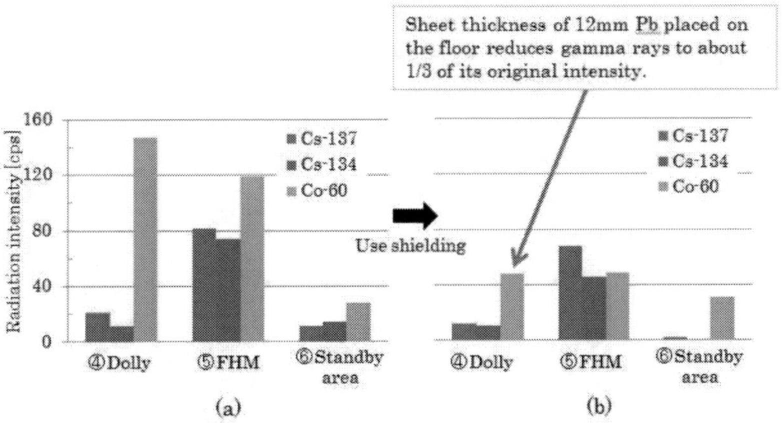

Figure 4: Reduction of peak count rates due to 12 mm lead

3. Ambient dose equivalent contribution from Unit 3

From pulse height distributions, it became clear that contributions from the operation floor of Unit 3 were small at least for unscattered photons. High contaminations at the operation floor of Unit 3 were supposed to increase ambient dose equivalent rate at the operation floor of Unit 4 in the form of skyshine. These skyshine photons were supposed to be well shielded by the wall and the ceiling board of the operation floor due to their lower energy.

IV. Conclusion

From our measurements, it became clear that photons emitted from Cs-134 and Cs-137 came mainly from contaminated buildings and debris at the lower floors of Unit 4 and those from Co-60 came from the steam dryer in the DS pool. At the FHM floor and the dolly floor, peak count rates of Co-60 were reduced to about 1/3 by the 12 mm lead sheet and these floor shields are effective to reduce exposure doses for workers above the FHM or the dolly.

References

1) CANBERRA Cooperate, InSpector™ 1000 Digital Hand-Held Multichannel Analyzer, <http://www.canberra.com/products/hp_radioprotection/inspector-1000.asp> (Confirmed January 27, 2016).

2) METI, First Secretariat meeting of the Team for Countermeasures for Decommissioning and Contaminated Water Treatment, Material 4-5, December 26,2013, "Results of dose rate measurement after removing small debris above the reactor well", <http://www.meti.go.jp/earthquake/nuclear/pdf/131226/131226_01hh.pdf> (Confirmed January 27, 2016) (in Japanese).

DOI: 10.1051/epjconf/201715308012

Updating source term and atmospheric dispersion simulations for the dose reconstruction in Fukushima Daiichi Nuclear Power Station Accident

Haruyasu Nagai[1,*], *Hiroaki* Terada[1], *Katsunori* Tsuduki[1], *Genki* Katata[1], *Masakazu* Ota[1], *Akiko* Furuno[1], and *Shusaku* Akari[2]

[1]Nuclear Science and Engineering Center, Japan Atomic Energy Agency, 2-4 Shirakata, Tokai, Ibaraki, 319-1195 Japan
[2]Makuhari System Laboratory, Fujitsu Limited, 1-9-3 Nakase, Mihama-ku, Chiba, 261-8588 Japan

Abstract. In order to assess the radiological dose to the public resulting from the Fukushima Daiichi Nuclear Power Station (FDNPS) accident in Japan, especially for the early phase of the accident when no measured data are available for that purpose, the spatial and temporal distribution of radioactive materials in the environment are reconstructed by computer simulations. In this study, by refining the source term of radioactive materials discharged into the atmosphere and modifying the atmospheric transport, dispersion and deposition model (ATDM), the atmospheric dispersion simulation of radioactive materials is improved. Then, a database of spatiotemporal distribution of radioactive materials in the air and on the ground surface is developed from the output of the simulation. This database is used in other studies for the dose assessment by coupling with the behavioral pattern of evacuees from the FDNPS accident. By the improvement of the ATDM simulation to use a new meteorological model and sophisticated deposition scheme, the ATDM simulations reproduced well the ^{137}Cs and ^{131}I deposition patterns. For the better reproducibility of dispersion processes, further refinement of the source term was carried out by optimizing it to the improved ATDM simulation by using new monitoring data.

1 Introduction

The Fukushima Daiichi Nuclear Power Station (FDNPS) accident in Japan, which was triggered by the magnitude 9.0 earthquake and resulting tsunami on March 11, 2011, caused a month-long discharge of significant amount of radioactive materials into the atmosphere. In order to assess the radiological dose to the public resulting from this release, there are some issues to be solved. One is for the internal dose due to short lived radionuclides such as radioiodine during the early phase of the accident, which cannot be assessed by using measured data. Another is for the external dose due to the direct radiation from radioactive plume and the ground shine at the point without measurement, which were considered to have large temporal variation caused by several plume passages. To solve these problems, the atmospheric transport, dispersion and deposition model (ATDM) simulation that can reproduce the spatial and temporal distribution of radioactive materials is useful. By simulating atmospheric dispersion conditions to be consistent with measurements, calculation results can complement spatiotemporally discrete measurements for dose assessment.

The source term of radioactive materials released into the atmosphere during the FDNPS accident is essential to evaluate the environmental distribution of radioactive materials. We have estimated the source term by comparing measurements of air concentration of radioactive materials or dose rate in the environment with calculated results by ATDM [1-4]. The ATDM used in these studies is WSPEEDI (Worldwide version of System for Prediction of Environmental Emergency Dose Information) [5] developed by the Japan Atomic Energy Agency (JAEA). Our source term has been validated by comparing with other estimations or by using in simulations of ATDMs [4, 6, 7]. The United Nations Scientific Committee on the Effects of Atomic Radiation (UNSCEAR) has used our source term [4] for estimating levels of radioactive material in the terrestrial environment and doses to the public, because they were derived from, and the models were already optimized to fit measurements of radioactive material in the environment [8]. However, UNSCEAR also pointed out in its 2013 report that the source term has large uncertainties and needs to be improved in future studies, in particular as more information becomes available on the progression of the accident, greater use is made of measurements in the environment, and improved assessment methods are implemented [8].

To improve our source term, we recently made detailed source term estimation by using additional monitoring data near the FDNPS and WSPEEDI including new deposition scheme [9]. The simulation of modified WSPEEDI and the new source term

* Corresponding author: nagai.haruyasu@jaea.go.jp

DOI: 10.1051/epjconf/201715308012

successfully reproduced the local and regional deposition pattern of ^{131}I and ^{137}Cs derived from airborne monitoring by the United States Department of Energy (US-DOE) and Ministry of Education, Culture, Sports, Science and Technology (MEXT) [10, 11].

In this study, new source term and updated WSPEEDI simulations are used to reconstruct the spatial and temporal distribution of radioactive materials in the environment during the FDNPS accident. Then, a database of spatiotemporal distribution of radioactive materials in the air and on the ground surface is developed from the output of the WSPEEDI simulations. For the better reproducibility of dispersion processes provided by the database, further refinement of the source term is also carried out by optimizing it to the improved ATDM simulation. This database is used for the comprehensive dose assessment by coupling with the behavioral pattern of evacuees from the FDNPS accident. It can also provide basic information and simulation techniques to understand environmental impacts by the accident and to predict future conditions.

2 Materials and method

2.1 New source term

The source term, which is necessary for ATDM simulations to set up release condition of radioactive materials, was refined by considering the particle size and chemical form of radionuclides, and giving temporal change with short time step. For this purpose, the source term estimated by Katata et al. [9] was the most appropriate one in the published source terms, and was used in this study. The temporal change in release rate of ^{131}I from 12 to 24 March 2011 estimated by Katata et al. [9] is shown in Fig. 1. This source term (new source term) is the latest update after the series of source term estimations by the combined analysis of environmental monitoring data and ATDM simulations at JAEA [1-4]. The major improvements from the source term used by UNSCEAR [4] are as follows:

- Air dose rates from automatic monitoring posts of Fukushima Prefecture [12], which are located within 20 km from FDNPS, during the period from March 12 -16, 2011 were used.
- Deposition distributions of ^{131}I and ^{137}Cs on April 3, 2011 derived from the US-DOE and MEXT airborne survey [11] were used.
- The estimation method with coupled simulations of atmospheric and oceanic dispersion models using concentrations of ^{134}Cs and ^{137}Cs in marine surface over the Pacific Ocean observed during the period from April 2 to May 17, 2011 [13] was applied to correct the release rate during the period when the plume flowed toward the Pacific Ocean.
- The atmospheric dispersion model was modified to include sophisticated deposition scheme [9].

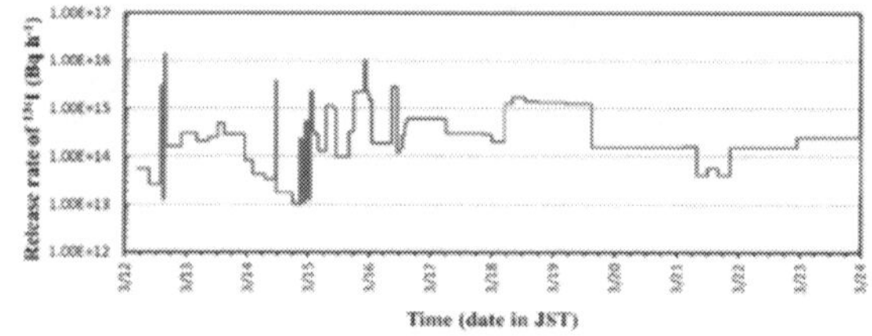

Fig. 1. Temporal change in estimated release rate of ^{131}I.

2.2 Improvement of ATDM simulation

The ATDM used in this study is WSPEEDI [5] developed by JAEA. The original WSPEEDI consists of non-hydrostatic mesoscale atmospheric model MM5 [14] and Lagrangian particle dispersion model GEARN [5]. MM5 is a community model having many users all over the world. The Lagrangian particle dispersion model GEARN calculates the atmospheric dispersion of radionuclides by tracing the trajectories of a large number (typically a million) of marker particles discharged from a release point. By using meteorological field predicted by MM5, it calculates the movement of each particle affected by both the advection due to mean wind and subgrid scale turbulent eddy diffusion. A part of the radioactivity in the air is deposited on the ground surface by turbulence (dry deposition) and precipitation (wet deposition).

Both meteorological and dispersion models have been changed in this study. For the meteorological model, a new meteorological model (WRF) [15] and data assimilation method (WRF-DA) [16] are introduced. WRF is the updated model of MM5, and has many new physical options and functions including advanced data assimilation method WRF-DA. GEARN have been modified to use a sophisticated deposition scheme. The new scheme deals with dry and fog-water deposition, cloud condensation nuclei (CCN) activation, and subsequent wet scavenging due to mixed-phase cloud microphysics (in-cloud and below-cloud scavenging) for radioactive iodine gas (I_2 and CH_3I) and other particles (CsI, Cs, and Te) as shown in Fig. 2 [9].

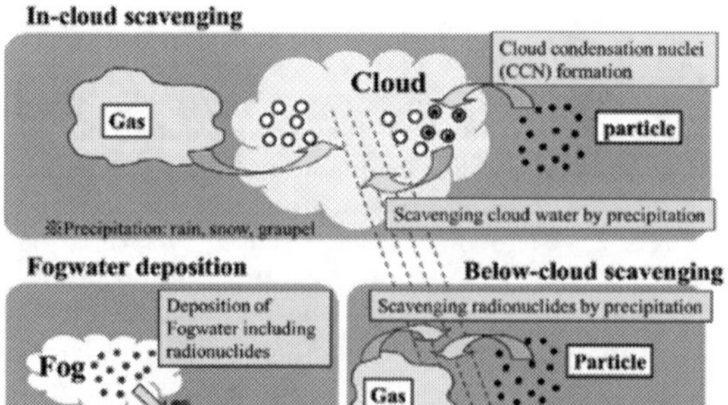

Fig. 2. Processes considered in the new wet deposition scheme.

ATDM simulations by the new model using the new source term were carried out. Simulation results by various calculation conditions including results by the old model and source term used in UNSCEAR 13 report (UNSCEAR source term) were compared with measured surface deposition, and the reproducibility of new model is evaluated.

2.3 Development of database

A database of spatiotemporal distribution of radioactive materials in the air and on the ground surface has been constructed by developing calculation and analysis methods for database as shown in Fig. 3. At first, meteorological calculations are carried out by using MM5, WRF, and WRF-DA to make dataset of meteorological fields. Then, the dispersion calculation by using the modified dispersion model (GEARN-new) is conducted with unit release condition for a time segment (one hour) of release period. This calculation is carried out for each combination of meteorological field, nuclide type considering deposition property, release height, and release time segment, and matrix outputs for every calculation conditions are made.

The spatiotemporal distribution of air concentration and surface deposition of radioactive materials for any condition of source term can be reproduced immediately by linear combination of matrix outputs. Therefore, it is easy to compare results applying many kinds of source term with monitoring data, and find out the optimum source term with which the difference between calculations and measurements can be minimized. The optimum source term and resulting spatiotemporal distribution of radioactive materials in the environment are used for the assessment of radiological dose to the public.

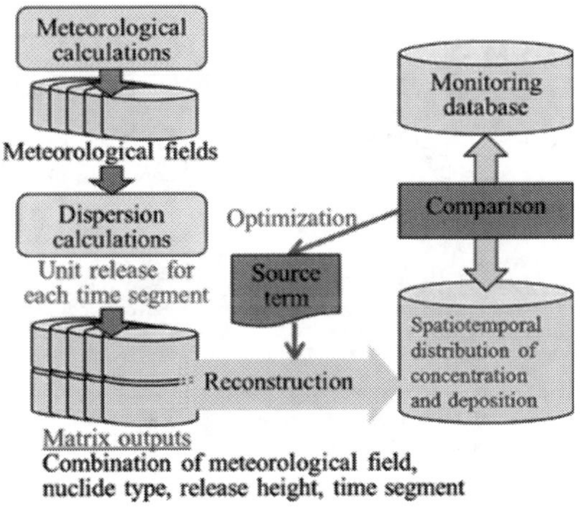

Fig. 3. Calculation and analysis methods for database.

2.4 Optimization of the source term using the database and new monitoring data

In the database, dataset by ATDM simulations with combinations of the meteorological model (MM5, WRF, and WRF-DA) and modified dispersion model (new GEARN) are stored. The spatiotemporal distribution of air concentration and surface deposition of radioactive materials can be reproduced by applying the source term to the dataset of each ATDM simulation. In this study, the new source term [9] was used as the most appropriate one in the published source terms. Since this source term was estimated by using the meteorological model MM5 and dispersion model (new GEARN) [9], it is considered to provide the optimum result by applying to the dataset of ATDM simulation with MM5 and new GEARN. Although the new meteorological model WRF and data assimilation method WRF-DA can possibly provide better meteorological fields, the difference of meteorological field from one by MM5 can cause the increase of discrepancy between the measurements and dispersion calculation results using the new source term. In addition to this, new monitoring data that have hourly atmospheric ^{134}Cs and ^{137}Cs radioactivity concentrations derived by analysing suspended particulate matter (SPM) collected at air pollution monitoring station [17] have been available after the estimation of the new source term. Therefore, optimization of the source term for ATDM simulations with new meteorological field is necessary.

The optimization of the source term was carried out by comparing hourly air concentration of ^{137}Cs between simulations and new measurements derived from SPM samples [17]. SPM samples at 99 sampling points have been analysed and published as the dataset. In this dataset 71 points using glass fiber sampling tape, which is less likely to have cross-contamination due to the apposition than polytetrafluoroethylene (PTFE) sampling tape, and one point near the FDNPS using PTFE sampling tape were used for the comparison. These 72 sampling points used for the comparison are shown in Fig. 4.

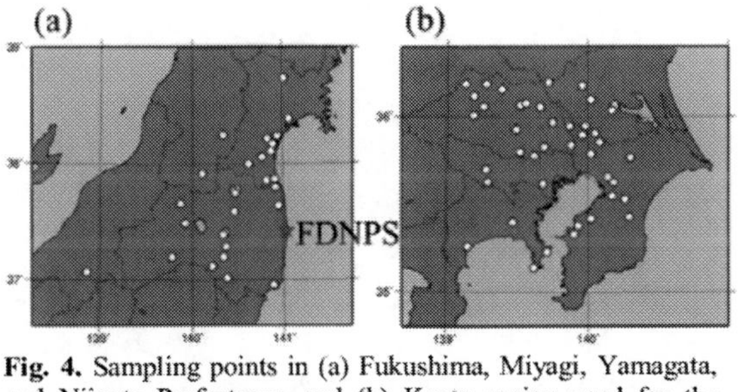

Fig. 4. Sampling points in (a) Fukushima, Miyagi, Yamagata, and Niigata Prefectures, and (b) Kanto region used for the comparison between simulated and measured air concentrations of ^{137}Cs.

3 Results and discussion

3.1. Performance test of ATDM

Simulated deposition patterns of ^{137}Cs by combination of models (MM5 or WRF, old GEARN or new GEARN) and source terms (UNSCEAR source term [4] or new source term [9]) are shown in Fig. 5. Also, deposition patterns of ^{137}Cs by the airborne monitoring by MEXT [10] and simulation with the combination of WRF-DA, new GEARN, and new source term are shown in Fig. 6.

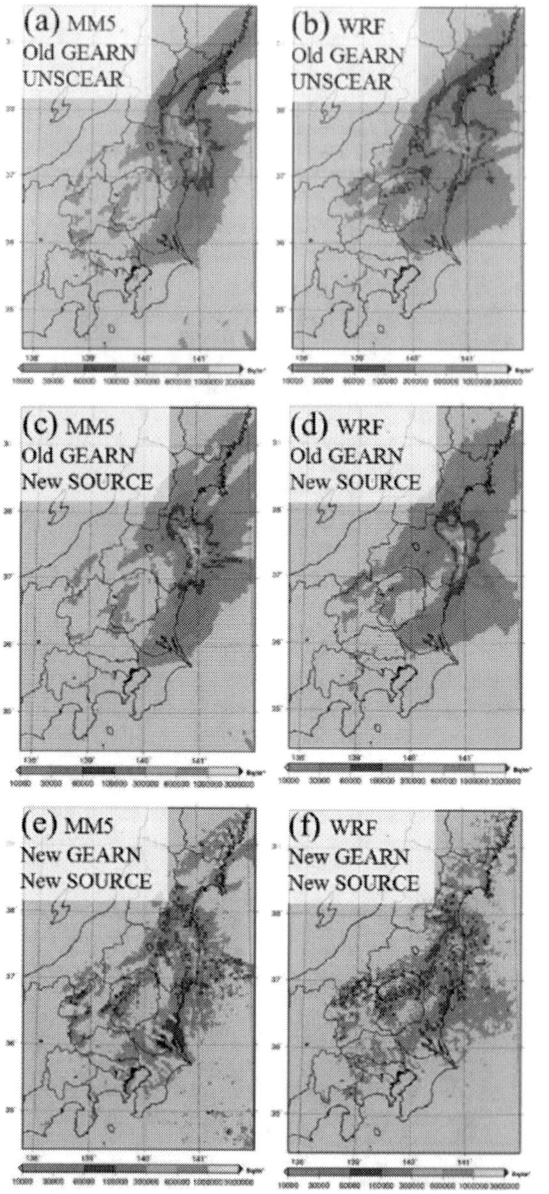

Fig. 5. Deposition patterns of ^{137}Cs by simulations with combination of (a) MM5, old GEARN, and UNSCEAR source term, (b) WRF, old GEARN, and UNSCEAR source term, (c) MM5, old GEARN, and new source term, (d) WRF, old GEARN, and new source term, (e) MM5, new GEARN, and new source term, and (f) WRF, new GEARN, and new source term.

In the cases using the former meteorological model MM5 (Fig. 5(a), (c), (e)), the reproducibility of ^{137}Cs deposition pattern was increased by using the new source term and new GEARN, especially around southwestern and northern areas from FDNPS, where the original WSPEEDI (Fig. 5(a)) under- and over-estimated, respectively, compared with the ^{137}Cs deposition of airborne monitoring (Fig. 6(a)).

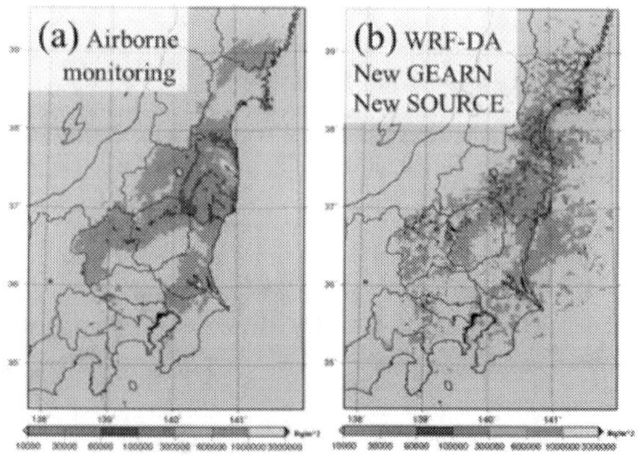

Fig. 6. Deposition patterns of ^{137}Cs of (a) the airborne monitoring by MEXT, and (b) simulation with combination of WRF-DA, new GEARN, and new source term.

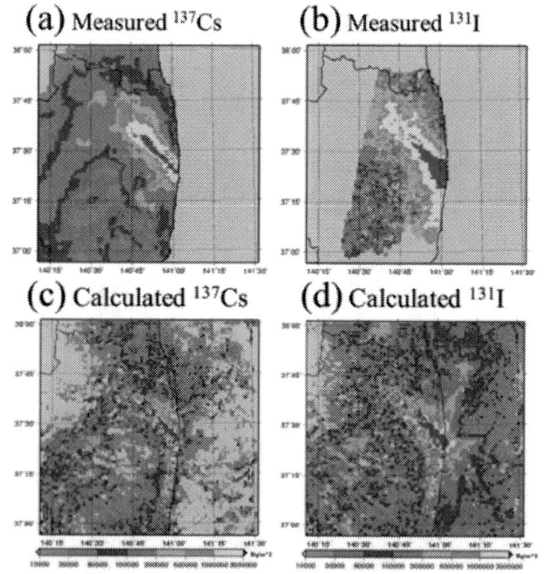

Fig. 7. Local deposition patterns of (a) ^{137}Cs and (b) ^{131}I by the airborne monitoring by MEXT, and (c) ^{137}Cs and (d) ^{131}I by simulation with combination of WRF-DA, new GEARN, and new source term.

The utilization of new meteorological model WRF had both positive and negative effects on the reproducibility of ^{137}Cs deposition pattern (Fig. 5(b), (d), (f)). For the deposition pattern of the eastern Japan scale, simulations using WRF agreed better with measurement than those using MM5. This indicates that WRF has higher performance in meteorological field prediction than MM5. However, the reproducibility of ^{137}Cs deposition pattern in the vicinity of FDNPS was decreased by using WRF. It is considered that the source term was optimized to the ATDM simulation with MM5 and a small difference in meteorological field between MM5 and WRF may cause a large difference in timing of plume passage and, consequently, deposition pattern.

By using the meteorological field by WRF-DA, the reproducibility of ^{137}Cs deposition pattern became higher than that by WRF (Fig. 6(b)). It is considered that WRF-DA successfully simulated actual meteorological field more appropriately than MM5 and WRF by assimilating observed meteorological data. Although MM5 simulation also used the data assimilation, the data assimilation method of WRF-DA is more advanced and can provide better meteorological field. The good performance of ATDM simulation with WRF-DA is also shown in the comparison of local deposition patterns of ^{137}Cs and ^{131}I (Fig. 7). Therefore, it is considered that ATDM simulation with WRF-DA and new GEARN can provide better results of spatiotemporal distribution of air concentration and surface deposition of radioactive materials if the new source term is optimised to this dataset.

3.2 Database

A prototype of database of spatiotemporal distribution of radioactive materials in the air and on the ground surface has been developed (Fig. 8). The analysis method to reproduce spatiotemporal distribution of air concentration and surface deposition of radioactive materials for any condition of source term by linear combination of matrix outputs was validated by using sample outputs. While it took several hours to execute usual ATDM simulation, the same outputs were reproduced from matrix outputs by using the analysis function of database within a few minutes. With this result, we confirmed the effectiveness of the analysis method to optimize the new source term and provide results of atmospheric dispersion analysis to other studies for the dose assessment.

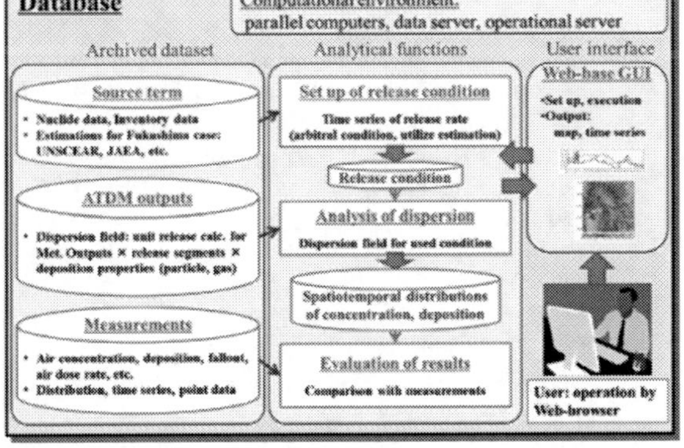

Fig. 8. Schematic design of database of spatiotemporal distribution of radioactive materials.

3.3 Optimization of the source term

The optimization of the source term for ATDM simulation with WRF-DA and new GEARN was carried out by comparing hourly air concentration of ^{137}Cs between simulations and measurements [17]. Before the optimization of the source term, statics of the comparison: the percent within a factor of two, five, and ten (FA2, FA5, FA10), correlation coefficient (CC), are summarized in Table 1. In this comparison, statistics were calculated for fractions of samples within specific area and period regarding the plume passage as well as for the total samples. The source term was modified to reduce the discrepancy between simulations and measurements especially for the area and period with low scores (North of FDNPS, Mar. 12-14 and Mar. 18-19), where the impact by this discrepancy on the dose estimation is significant. After some modification of the source term, statistics of the comparison became to have higher scores for most cases. Although lower scores for some cases were existed, the reproducibility of atmospheric dispersion processes by the ATDM simulation was improved as a whole by this optimization of the source term. Simulations with the new source term (Katata 2015 source) [9] and optimum source term are compared for the regional ^{137}Cs deposition (Fig. 9), local ^{137}Cs deposition (Fig. 10), and local ^{131}I deposition (Fig. 11). Although differences between the simulations were small, the deposition patterns by using optimum source term agreed better with measured ones.

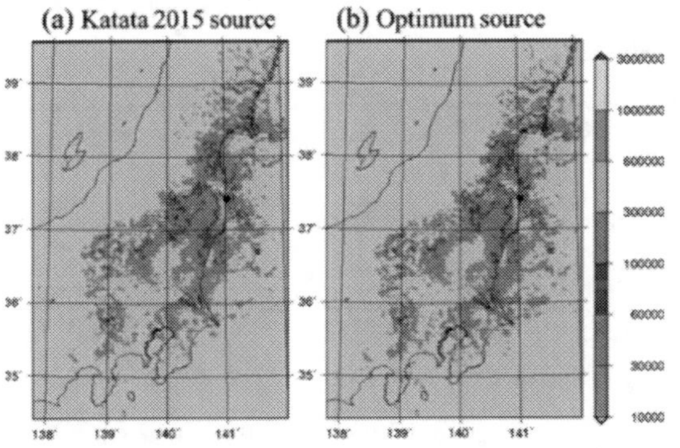

Fig. 9. Deposition patterns of ^{137}Cs of simulations with (a) the Katata 2015 source term, and (b) optimum source term.

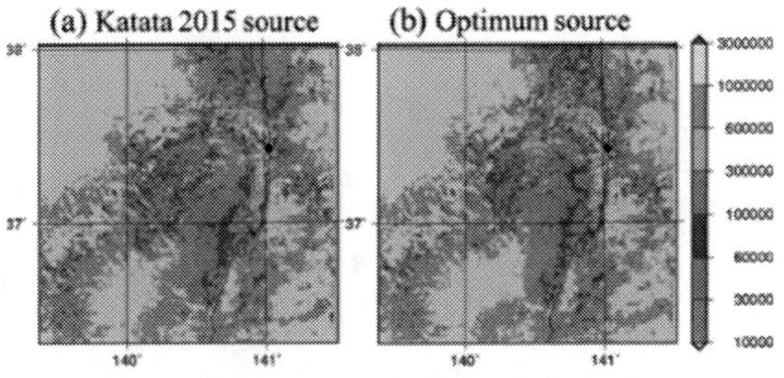

Fig. 10. Local deposition patterns of ^{137}Cs of simulations with (a) the Katata 2015 source term, and (b) optimum source term.

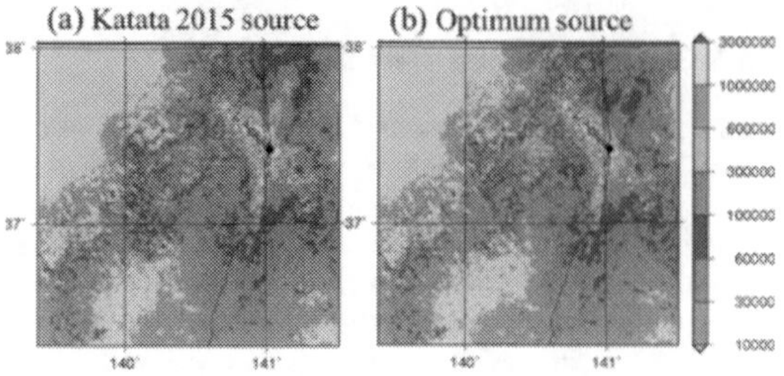

Fig. 11. Local deposition patterns of ^{131}I of simulations with (a) the Katata 2015 source term, and (b) optimum source term.

Table 1. Statics of the comparison: the percent within a factor of two, five, and ten (FA2, FA5, FA10), correlation coefficient (CC), of hourly air concentration of ^{137}Cs between simulations before the optimization and measurements. Statistics were calculated for fractions of samples within specific area and period regarding the plume passage as well as for the total samples.

Area and period	FA2 (%)	FA5 (%)	FA10 (%)	CC
North of FDNPS Mar. 12-14	12.0	24.0	42.0	-0.20
Kanto Mar. 15-16	16.2	37.3	49.0	0.19
Fukushima Mar. 15-16	13.5	35.4	44.8	0.27
North of FDNPS Mar. 18-19	11.4	25.7	34.3	-0.09
Southern Tohoku* Mar. 20-21	13.0	34.1	42.2	0.22
Kanto Mar. 20-21	22.7	51.4	66.3	0.69
Total	17.8	41.7	54.3	0.08

* Southern Tohoku: Fukushima, Miyagi, and Yamagata Prefectures

Table 2. Statics of the comparison: the percent within a factor of two, five, and ten (FA2, FA5, FA10), correlation coefficient (CC), of hourly air concentration of ^{137}Cs between simulations after the optimization and measurements. Statistics were calculated for fractions of samples within specific area and period regarding the plume passage as well as for the total samples.

Area and period	FA2 (%)	FA5 (%)	FA10 (%)	CC
North of FDNPS Mar. 12-14	12.0	26.0	48.0	-0.16
Kanto Mar. 15-16	16.2	37.3	49.0	0.19
Fukushima Mar. 15-16	15.6	34.4	46.9	0.27
North of FDNPS Mar. 18-19	14.3	34.3	45.7	0.33
Southern Tohoku* Mar. 20-21	12.4	28.1	39.5	0.23
Kanto Mar. 20-21	24.1	52.0	66.9	0.71
Total	18.3	41.7	54.7	0.26

* Southern Tohoku: Fukushima, Miyagi, and Yamagata Prefectures

4 Conclusion

In this study, the source term of radioactive materials discharged into the atmosphere was refined and the ATDM simulation was improved to use a new meteorological model and sophisticated deposition scheme for the accurate assessment of the radiological dose to the public resulting from the FDNPS accident. By using the ATDM simulations, a prototype of database of spatiotemporal distribution of radioactive materials in the air and on the ground surface is also developed. The new ATDM simulation reproduced the ^{137}Cs deposition pattern for the eastern Japan scale better than the former ATDM simulation. Furthermore, the source term was optimized to fit the new ATDM simulation by using new monitoring data that have hourly atmospheric ^{137}Cs concentrations derived by analysing SPM collected at air pollution monitoring station [17]. As the result, the reproducibility of atmospheric dispersion processes by the ATDM simulation was improved, and it can provide better information for the comprehensive dose assessment by the collaborative studies.

The authors express their gratitude to Prof. Hiromi Yamazawa of Nagoya University, Dr. Matthew Hort of UK Met Office, Dr. Werner Rühm of Helmholtz Center Munich, and Dr. André Bouville of National Cancer Institute, NIH, DHHS for their helpful comments and suggestions. This study was conducted in the framework funded and commissioned by Ministry of Environment of Japan.

References

1. M. Chino, H. Nakayama, H. Nagai, H. Terada, G. Katata, H. Yamazawa, J. Nucl. Sci. Technol., **48**, 1129 (2011)
2. G. Katata, H. Nagai, H. Terada, M. Chino, J. Environ. Radioactiv., **111**, 2 (2012)
3. G. Katata, M. Ota, H. Terada, M. Chino, H. Nagai, J. Environ. Radioactiv., **109**, 103 (2012)
4. H. Terada, G. Katata, M. Chino, and H. Nagai, J. Environ. Radioactiv., **112**, 141 (2012)
5. H. Terada and M. Chino, J. Nucl. Sci. Technol., **45**, 920 (2008)
6. Y. Morino, T. Ohara, M. Watanabe, S. Hayashi, M. Nishizawa, Environ. Sci. Technol., **47**, 2314 (2013)
7. R. Draxler, D. Arnold, M. Chino, S. Galmarini, M. Hort, A. Jones, S. Leadbetter, A. Malo, C. Maurer, G. Rolph, K. Saito, R. Servranckx, T. Shimbori, E. Solazzo, G. Wotawa, J. Environ. Radioactiv., **139**, 172 (2015)
8. UNSCEAR (United Nations Scientific Committee on the Effects of Atomic Radiation), *UNSCEAR 2013 Report: Sources, Effects and Risks of Ionizing Radiation, Vol. I*, United Nations, New York, USA (2014)
9. G. Katata, M. Chino, T. Kobayashi, H. Terada, M. Ota, H. Nagai, M. Kajino, R. Draxler, M. Hort, A. Malo, T. Torii, Y. Sanada, Atmos. Chem. Phys., **15**, 1029 (2015)
10. NRA (Nuclear Regulation Authority), *Results of the (i) Fifth Airborne Monitoring Survey and (ii)*

Airborne Monitoring Survey Outside 80km from the Fukushima Dai-ichi NPP, http://radioactivity.nsr.go.jp/en/contents/6000/5790/24/203_0928_14e.pdf (2012, last access: 20 January 2016)

11. T. Torii, T. Sugita, C. E. Okada, M. S. Reed, D. J. Blumenthal, Health Phys., **105**, 192 (2013)
12. Fukushima Prefecture, *Results of air dose rate measurement on March 2011 (data retrieved from monitoring posts in Fukushima Prefecture)*, http://www.atom-moc.pref.fukushima.jp/monitoring/monitoring201103/201103_mpdata.html, (2012, in Japanese, last access: 12 November 2014)
13. T. Kobayashi, H. Nagai, M. Chino, H. Kawamura, J. Nucl. Sci. Technol., **50**, 255 (2013)
14. G. Grell, J. Dudhia, D. Stauffer, *A Description of the Fifth-Generation Penn State/NCAR Mesoscale Model (MM5)*, NCAR Tech, Note NCAR/TN-3921STR, National Center for Atmospheric Research, Boulder, Colorado, USA (1994)
15. W. C. Skamarock, J. B. Klemp, J. Dudhia, et al., *A description of the Advanced Research WRF Version 3*, NCAR Tech. Note NCAR/TN-475STR, National Center for Atmospheric Research, Boulder, Colorado, USA (2008)
16. D. Barker, X-Y Huang, Z. Liu, et al., Bulletin of American Meteorological Society, **93**, 831 (2012)
17. Y. Oura, M. Ebihara, H. Tsuruta, T. Nakajima, T. Ohara, M. Ishimoto, H. Sawahata, Y. Katsumura, W. Nittae, J. Nucl. Radiochem. Sci., **15**, 15 (2015)

Internal Dose from Food and Drink Ingestion in the Early Phase after the Accident

Masaki Kawai[1,*], *Nobuaki* Yoshizawa[1], *Sachiko* Hirakawa[1], *Kana* Murakami[1], *Mari* Takizawa[1], *Osamu* Sato[1], *Shunji* Takagi[1], *Hirokazu* Miyatake[1], *Tomoyuki* Takahashi[2] and *Gen* Suzuki[3]

[1]Mitsubishi Research Institute, Inc., 2-10-3, Nagata-cho, Chiyoda-ku, Tokyo, Japan
[2]Kyoto University Research Reactor Institute, Asashironishi2-chome, Kumatori-machi, Sennan-gun, Osaka, Japan
[3]International University of Health and Welfare Clinic, 2600-6, Kitakanemary, Otawara-city, Tochigi, Japan

Abstract. Activity concentrations in food and drink, represented by water and vegetables, have been monitored continuously since the Fukushima Daiichi Nuclear Power Plant accident, with a focus on radioactive cesium. On the other hand, iodine-131 was not measured systematically in the early phase after the accident. The activity concentrations of iodine-131 in food and drink are important to estimate internal exposure due to ingestion pathway. When the internal dose from ingestion in the evacuation areas is estimated, water is considered as the main ingestion pathway. In this study, we estimated the values of activity concentrations in water in the early phase after the accident, using a compartment model as an estimation method. The model uses measurement values of activity concentration and deposition rate of iodine-131 onto the ground, which is calculated from an atmospheric dispersion simulation. The model considers how drinking water would be affected by radionuclides deposited into water. We estimated the activity concentrations of water on Kawamata town and Minamisouma city during March of 2011 and the committed effective doses were 0.08 mSv and 0.06 mSv. We calculated the transfer parameters in the model for estimating the activity concentrations in the areas with a small amount of measurement data. In addition, we estimated the committed effective doses from vegetables using atmospheric dispersion simulation and FARMLAND model in case of eating certain vegetables as option information.

1 Introduction

Radioactive materials were discharged into the atmosphere after the accident at Fukushima Daiichi Nuclear Power Plant of Tokyo Electric Power Company, resulted from the Great East Japan Earthquake which occurred on March 11, 2011. For health risk assessment to local residents, it is important to estimate how much radiation they were exposed to by radioactive materials.

There are two typical reports in the systematic dose estimation due to the accident by the following international institutions. World Health Organization (WHO) conducted a temporary evaluation using information obtained at that time and released a report with the aim of considering measures to reduce health effects of exposure to radiation through emergency response measures after the accident in 2012[1]. United Nations Scientific Committee on the Effects of Atomic Radiation (UNSCEAR), taking into consideration information within about two years after the accident, released a report in 2014. The report estimated more closely to the actual situation than that provided by WHO[2]. However, also in the evaluation of UNSCEAR, there are some factors of overestimation or underestimation due to uncertainty in source terms, lack of measured data on iodine-131 following the accident, etc. In order to obtain the accident impact regarding the internal dose from food ingestion, the survey of food supply in the early phase after the accident in Fukushima prefecture was investigated to provide information of the actual situation[3]. The investigation proves that there was less possibility of fresh vegetables or milk being widely supplied. The investigation also shows that tap water was used both for drinking and cooking. The report by WHO shows the estimation of internal dose in the early phase using the measured concentrations of foods. These data included the results of food samples collected just for monitoring purpose; therefore it might have led to an overestimation of representative dose. Internal dose due to tap water was not considered with the entire dose because the dose estimated with maximum detected levels was low in comparison with dose from other pathways. The report by UNSCEAR shows the estimation of the internal dose in the early phase after the accident using the activity concentrations of foods in IAEA/FAO database. The concentration data also included the results for identification of foods with the high concentrations. It appears to lead to overestimation of representative dose. Internal dose from tap water was estimated with the monitored data and consideration of restriction in the report.

Also, various studies of dose estimation from food after the accident have been published. Harada et al. estimated internal doses for residents in three areas: village of Kawauchi, Tamano area of Soma and Haramachi area of Minamisoma, during August-

* Corresponding author: makawai@mri.co.jp

September 2012 by food-duplicate survey[4]. The internal dose estimated from dietary intake of radioactive cesium averaged 0.0058, 0.019 and 0.0088 mSv/y. The authors pointed out that the survey cannot consider day to day variation in individuals as an uncertainty and the result does not include the dose from iodine-131 in the early phase. Tsutsumi et al. estimated the committed effective dose from ingestion using market basket method in three prefectures: Tokyo, Miyagi and Fukushima[5]. The food for dose estimation was collected in each prefecture in autumn 2011 and the estimated doses of radioactive cesium were 0.021, 0.017 and 0.019 mSv/year. They noted that the doses were significantly lower than the maximum permissible dose (1mSv/year) determined by the Ministry of Health, Labour and Welfare, Japan. Yamaguchi et al. reported the results of dose estimation based on the monitoring data of activity concentration in food from the early phase after the accident through December 2012[6]. They estimated the committed effective dose from iodine-131 and radioactive cesium in food by age group. The highest median value was 0.19 mSv in children 13-18 years of age. They mentioned some factors of overestimation or underestimation such as effectiveness restrictions to food.

Taking into account the international reports and these studies, we considered an estimation method of Iodine-131 concentrations in tap water for dose estimation. As for vegetables, we considered an estimation method of concentration to calculate the internal dose from the intake of vegetables which might have been consumed.

In this study, the estimation mainly covered near the evacuation areas, where estimating the dose was considered to be particularly important in the early phase. For reference, we used the results of WSPEEDI established by the Japan Atomic Energy Agency (JAEA), National Research and Development Agency, in order to obtain deposition rate[7].

2 Methods

2.1 Estimation Method for Iodine-131 Concentration of Tap Water

To estimate iodine-131 concentration of tap water, we used a compartment model where radionuclides contained in tap water are deemed to attenuate at a fixed rate to their concentration, and increase at every new depositions. In estimating iodine-131 concentration of tap water through the compartment model, changes of activity concentration in tap water are represented by the rate of radionuclides newly deposited into water and effective attenuation of radionuclides contained in water,

$$\frac{\mathrm{d}C}{\mathrm{d}t} = p - \lambda C \qquad (1)$$

Where C is the iodine-131 concentration in tap water (Bq/kg), p is the rate of deposition into water (Bq/kg·hour), and λ is the effective attenuation coefficient (l/hour).

Formula (1) can be represented as follows, provided the rate of deposition into water is replaced by the rate of radionuclides deposition onto the ground, which is calculated using an atmospheric dispersion simulation and conversion coefficient.

$$\frac{\mathrm{d}C}{\mathrm{d}t} = ap' - \lambda C \qquad (2)$$
$$(p = ap')$$

Where p'=rate of iodine-131 deposition onto the ground, which is calculated with an atmospheric dispersion simulation (Bq/m^2·hour), and a=conversion coefficient (m^2/kg).

Conversion coefficient and effective attenuation coefficient are determined using the least squares method and finding the best fit, substituting measurement values of iodine-131 concentration in tap water. The optimal value of the concentration can be estimated with the parameters.

Based on this estimation method, data necessary for estimating iodine-131 concentration in water are iodine-131 concentration values actually measured in water and the deposition rate onto the ground calculated from an atmospheric dispersion simulation. The deposition rate onto the ground was calculated by total ground depositions at a time step and a next time step by WSPEEDI. As for measurement values of iodine-131 concentration, we collected data contained in publicly-available information, which was obtained at water purifying facilities or drinking spaces of running water at local municipal offices[8]. In addition, tap water travels until it reaches each drinking space, starting from the source of tap water, then on to a purifying facility and a distribution facility. Radionuclides in the atmosphere are more likely to deposit into water at the source and at some parts of a water purifying facility, via these relay points. Water distributed through a water purifying facility to each household, on the other hand, seems less affected by radionuclides deposition because most of the tap water piping goes under the ground or indoor. Taking into account these conditions, the water source is deemed to be the place where radionuclides are deposited, when the deposition rate onto the ground is calculated from an atmospheric dispersion simulation. When the location of water source is not identified, a purifying facility is substituted. It is assumed that it takes twelve hours for tap water to be taken in after the deposition of radionuclides. In other words, the activity concentration in tap water at a time affected by the depotion rate onto the ground in the water source at 12 hours before.

2.2 Dose Estimation from Intake of Tap water

The effective dose of iodine-131 was estimated as;

$$Dose(Sv) = A \times B \times \sum_{t} \overline{C}_t \qquad (3)$$

 DOI: 10.1051/epjconf/201715308013

where A is the coefficient for iodine-131 (Sv/Bq), B is the daily consumption of water (kg/day), t is the number of days and $\overline{C}$ is the daily mean concentration of iodine-131 in water at t (Bq/kg).

The coefficient is reference value from the publications of the International Commission on Radiological Protection (ICRP)[9]. Assuming the water intake volume in the early phase after the accident, the impact of the earthquake disaster should be considered. In some cases, the evacuees obtained bottled water for drink in this phase[10]. The bottled water can be regarded as unconcentrated water; therefore the bottled water volume was not added for dose estimation in this study. On the other hand, tap water which may be utilized for cooking and some amount of drink should be considered in water intake volume for dose estimation. The media volume of water intake for evacuees in Fukushima prefecture was assumed be 1.2 L/day based on hearing survey results. The median volume for non-evacuees was assumed to be 1.8 L/day based on existing inquiry results[11].

Activity concentrations in water have been generally measured since the accident. The restriction was requested according to the index values for the restriction on the intake of food and beverages set out by the Nuclear Safety Commission when the activity concentration was high. For example, on March 21, 2011, the activity concentration in tap water of the small-scale water supply in Iitate village exceeded the index value for restriction and Fukushima prefecture informed the residents using the tap water to refrain from drinking water[12]. In consideration of these actual situations about water, the activity concentration in water was determined by following the chart below.

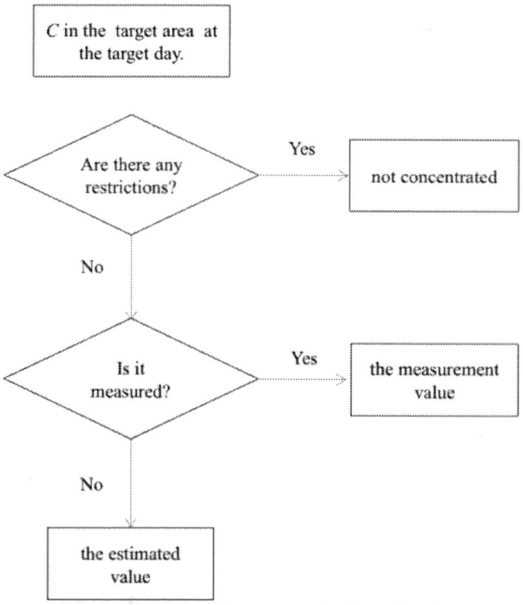

Fig. 1. Flow chart for concentration of water

2.3 Estimation Method for Concentration of Vegetables

In order to estimate the internal dose from intake of vegetables immediately after the accident, activity concentrations of vegetables were estimated using FARMLAND model. FARMLAND model is utilized to predict the transfer of radionuclides into foods such as green vegetables, root vegetables, grain and cow milk, by deposition onto the ground[13]. In this model, the movement of radionuclides within each module is represented by transfers between interconnected compartments and within each module it is assumed that first-order kinetics apply. The model set the specific parameter in each model of vegetables such as green vegetables and root vegetables. In this study, the activity concentrations of vegetables were estimated by these parameters[14]. The depositions onto the ground and the vegetables were calculated using the deposition rates by WSPEEDI and interception factor. Fig. 2 shows the concept for estimation of activity concentration in vegetables.

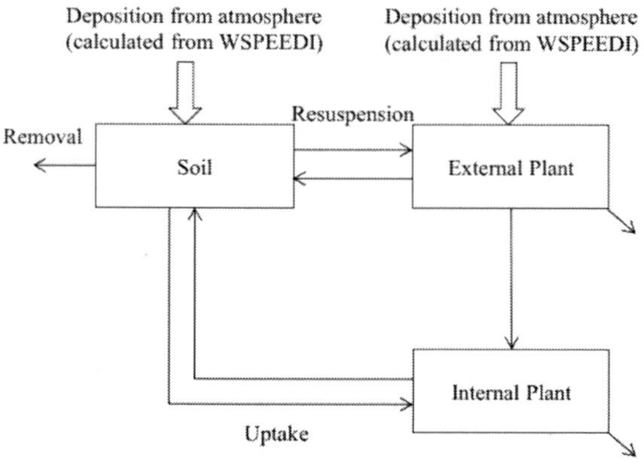

Fig. 2. Concept for estimation of concentration in vegetables

Ministry of Health, Labour and Welfare publishes activity concentrations in food including vegetables measured by various facilities immediately after the accident[15]. We estimated the concentrations on the areas where there were measurements, to discuss the result of this model. However, it is difficult to obtain the details of the locations. Therefore, the location of each municipal office was substituted for that of planting and the deposition onto the ground of the location by WSPEEDI was used for estimation.

2.4 Dose Estimation from Intake of Vegetables

Considering the dose estimation from intake of vegetables in the early phase after the accident, the impact of the earthquake disaster should be considered. In particular, evacuees did not generally take the concentrated vegetables in refuge life. However, any residents may take the outdoor-grown vegetables. In this case, if the intake amount of the vegetables is calculated by general intake amounts of Japanese, the amount does not reflect the situation in the term. Therefore, we provide the dose estimated by an amount of radionuclides per 100 g of a vegetable taking as option information. The effective dose from certain vegetables per 100 g was estimated as;

$$Dose(Sv/kg) = A_k \times C_{k,i,t} \quad (4)$$

where k is the radionuclide, A_k is the coefficient for radionuclide k (Sv/Bq), t is the number of days, i is the kind of vegetable and $C_{k,i,t}$ is the concentration of radionuclide k at t.

3 Results

3.1 Tap Water

We estimated the activity concentrations of water in the areas where there had been measurements in the early phase after the accident. Fig. 3 and Fig. 4 show the estimation results using the measurement values on Kawamata town and Minamisouma city.

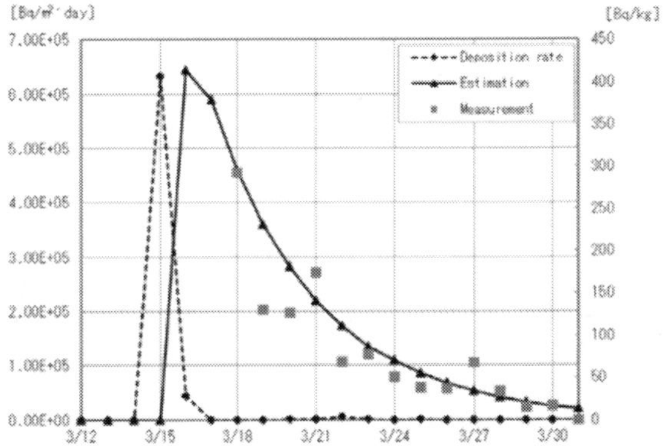

Fig. 3. Estimation Result of Tap Water on Kawamata town. The points indicate the measument values, the line the esitamed values and the broken line the deposition rate calculated by WSPEEDI.

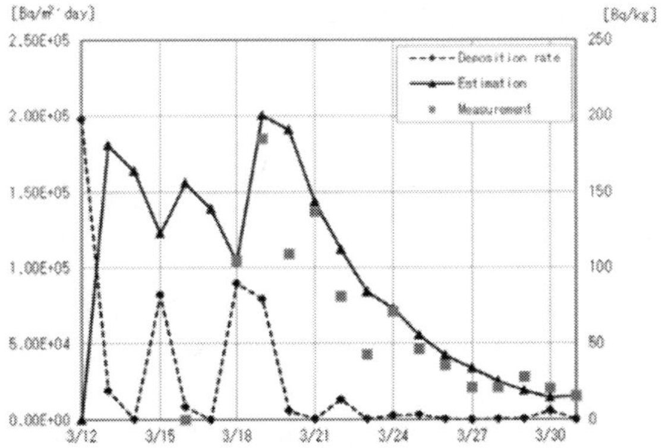

Fig. 4. Estimation Result of Tap Water on Minamisoma city. The points indicate the measument values, the line the esitamed values and the broken line the deposition rate calculated by WSPEEDI.

Table 1 shows the committed effective dose of adult evacuees and adult non-evacuees from water.

Table 1. Committed effective dose from water based on the measured concentration, the estimated concentration and the assumed intake volume on Kawamata town and Minamisoma city during March of 2011 (mSv).

	Dose for evacuee	Dose for non-evacuee
Kawamata town	0.05	0.08
Minamisoma city	0.04	0.06

Fig. 5 and Fig. 6 show the parameters of conversion coefficients and of effective attenuation coefficients in the areas where the activity concentrations was measured.

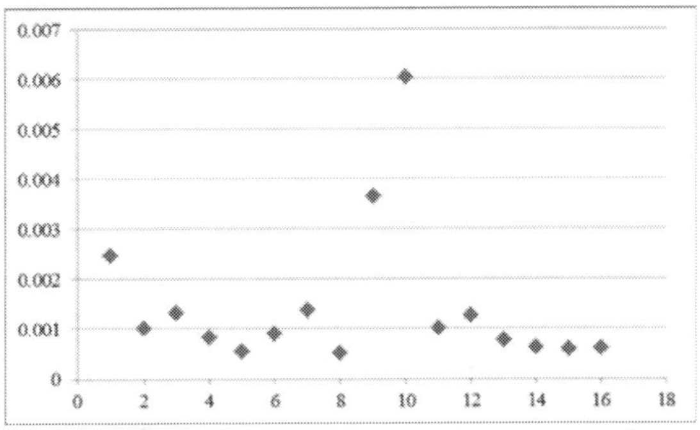

Fig. 5. Parameters of conversion coefficients on various areas.

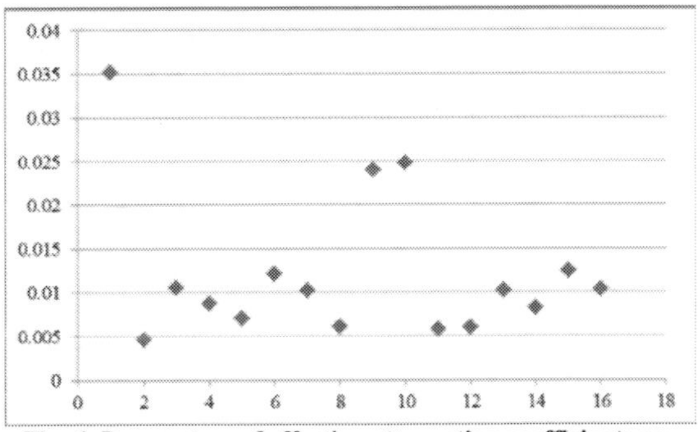

Fig. 6. Parameters of effective attenuation coefficients on various areas.

The mean value of conversion coefficients and of effective attenuation coefficients are 0.0015 and 0.012.

3.2 Vegetables

We estimated activity concentrations of vegetables using deposition rates obtained by WSPEEDI. Table 2 and Table 3 show the concentrations of spinach and cabbage measured on March 28, 2011 and the estimated concentrations.

Table 2. Activity concentration for spinach (Bq/kg).

Location	Activity concentration (Measured value)		Activity concentration (Estimated value)	
	^{131}I	^{137}Cs	^{131}I	^{137}Cs
Izumizaki village	3500	4700	26000	18000
Otama village	5900	17000	2500	1200
Ono town	5100	2300	6000	3200
Tanagura town	2700	3100	7100	3800
Tamura city	8400	8400	21000	13000
Nakajima village	2300	1100	17000	11000
Hanawa town	2100	830	4000	2500
Hirata village	5300	460	9000	31
Yabuki Town	1600	320	18000	12000
Yabuki Town	2000	400	18000	12000

Table 3. Activity concentration for cabbage (Bq/kg).

Location	Activity concentration (Measured value)		Activity concentration (Estimated value)	
	^{131}I	^{137}Cs	^{131}I	^{137}Cs
Asakawa town	1100	1400	2600	1800
Ishikawa town	12	0	2100	1600
Minamisouma city	140	51	8000	6800

Table 4 shows the committed effective dose from cabbage and spinach by 100 g ingestion in Table 2 and Table 3. The dose were estimated by contributions of ^{131}I, ^{134}Cs and ^{137}Cs.

Table 4. Estimation of committed effective dose from spinach (mSv/100g).

Location	Committed effective dose (By measured concentration)	Committed effective dose (By estimated concentration)
Izumizaki village	0.02	0.1
Otama village	0.07	0.009
Ono town	0.02	0.02
Tanagura town	0.02	0.03
Tamura city	0.05	0.09
Nakajima village	0.008	0.07
Hanawa town	0.007	0.02
Hirata village	0.01	0.03
Yabuki Town	0.004	0.08
Yabuki Town	0.006	0.08

Table. 5 Estimation of committed dose from cabbage (mSv/100g)

Location	Committed effective dose (By measured concentration)	Committed effective dose (By esitmated concentration)
Asakawa town	0.007	0.01
Ishikawa town	0.00003	0.01
Minamisouma city	0.0005	0.04

4 Discussions

The estimation results of activity concentration in tap water was in good agreement with the concentration measured in Kawamata town and Minamisouma city. These results proved that the activity concentration in tap water can be estimated by the rate of iodine-131 deposition onto the ground from an atmopspheric dispersion simlation. The committed doses were lower than the maximum permissible dose 1 mSv/year, however, considering the results of the other researches for dose estimation, the dose cannot be ignored. For the areas without actual measurement data of activity concentration in water, though, either conversion coefficients or effective attenuation coefficients cannot be determined by the measured concentration. In this case, the average values of conversion coefficients and effective attenuation coefficients, which were obtained from those of areas with measured concentration, were used for supposition of coefficients to estimate activity concentration for the areas without measured data. As a factor of overestimation or underestimation, the water intake volumes for evacuee and non-evacuee vary, though we set the values of both. In addition, the facility effect of a water purifying facility is factored into parameters such as conversion coefficients in this model using measurement values, as some reports mentioned that it can reduce the influence of iodine-131.

For vegetables, the result shows that the activity concentration and the dose can be estimated by this model at a certain day in the early phase after the accident. These vegetables measured for the activity concentration were non-distributed food because of the ristriction of distribution. Due to that, the vegetables evaluated in this study were seemed to be rarely eaten. It is neccessary to adjust the parameters in the model because the concentrations of cabbage seemed to be overestimated in comparison with the measured concentration. Though we estimated two kinds of vegetables in this estimation, the estimations will be improved more appropriate to the reality by hearing survey about what the evacuees ate.

The values of activity concentrations in water and vegetables will change according to the improvement of the atmospheric dispersion simulation, WSPEEDI. Therefore, if WSPEEDI is improved, we will provide more accurate doses in the early phase. On the other hand, the estimated values may change according to different distance, because the deposition rate after the accident varies considerably, despite being merely several kilometers apart. The errors of these results vary due to the errors of the deposition rates by WSPEEDI, therefore we will consider the uncertainty for dose estimation including that of WSPEEDI. Finally, the results may help the dose estimation of the actual situation in the early phase after the accident by reducing these uncertainties.

References

1. WHO, Preliminary dose estimation from the nuclear accident after the 2011 Great East Japan Earthquake and Tsunami (2012).
2. UNCEAR, UNSCEAR 2013 Report to the General Assembly with scientific annexes (2014).
3. Hirakawa S., et al., "Surveys of Food Distribution Just after the Nuclear Accident and Challenges", Journal of Japan Society for Safety Engineering, 53(3), p167-p172, 2014(in japanese).
4. Harada, K.H., T. Niisoe, M. Imanaka et al., "Radiation dose rates now and in the future for residents neighboring restricted areas of the Fukushima Daiichi Nuclear Power Plant." Proc Natl Acad Sci U S A 111(10): E914-923 (2014).
5. Tsutumi T., et al., "Estimation of the Committed Effective Dose of Radioactive Cesium and Potassium by the Market Basket Method", Journal of the Food Hygienic Society of Japan, 54(1), p7- 13, 2013(in japanese).
6. Yamaguchi I., et al., "Dose estimation from food intake due to the Fukushima Daiichi nuclear power plant accident", 62(2), p138-143, 2013 (in japanese).
7. Terada H., et al., "Development of worldwide version of system for prediction of environmental emergency dose information: WSPEEDI 2nd version", Journal of the Atomic Energy Society of Japan, 7, p257–p267, 2008 (in japanese).
8. Ministry of Health, Labour and Welfare, "The Survey Results of Radioactive Materials in Tap Water" (in japanese) Available at : <http://www.mhlw.go.jp/file/06-Seisakujouhou-10900000-Kenkoukyoku/1_houshasei_110719_m1.pdf> [Accessed 23 August 2016]
9. ICRP, "Compendium of Dose Coefficients based on ICRP Publication 60", ICRP Publication 119. 2012.
10. Iitate Village, "Record of Devastation from the Great East Japan Earthquake and TEPCO Fukushima Daiichi Nuclear Power Station Accident (First Edition) – Progress Made in the First Two Years after the Disaster", 2013(in japanese). Available at : <http://archive.vill.iitate.fukushima.jp/dsearch/ics/view_data.php?dataId=0005639> [Accessed 23 August 2016]
11. Health, labour and science research grant "Comprehensive study on risk evaluation and control of water quality in tap water" Contribution section meeting report(in japanese).
12. Ministry of Health, Labour and Welfare press release, 2011. Available at : <http://www.mhlw.go.jp/english/policy/health/water_supply/dl/7-1_march_21_02.pdf> [Accessed 23 August 2016]
13. Brown J. and Simmonds J. R., "FARMLAND: a dynamic model for the transfer of radionuclides through terrestrial foodchains. National Radiological Protection Board.", NRPB-R273, 1995.
14. Sherwood J. (private communication).
15. Ministry of Health, Labour and Welfare, "Database of radioactive substances in food". Available at : <http://www.radioactivity-db.info/search.aspx> [Accessed 23 August 2016]

Retrospective reconstruction of Iodine-131 distribution through the analysis of Iodine-129

Hiroyuki Matsuzaki[1,a], Yasuyuki Muramatsu[2], Takeshi Ohno[2], Wei Mao[1]

[1]*MALT, The University of Tokyo, 2-11-16 Yayoi Bunkyo-ku, Tokyo, Japan*
[2]*EGakushuin University, 1-5-1 Mejiro, Toshima-ku, Tokyo, Japan*

Abstract. Iodine-131 distribution released from the Fukushima Dai-ichi Nuclear Power Plant accident was reconstructed through the iodine-129 measurements. From nearly 1,000 surface soil samples iodine was extracted by the pyro hydrolysis method. Extracted iodine was then mixed with carrier, purified and finally collected as silver iodide. Silver iodide sample was pressed into the cathode holder and set at the ion source of the MALT facility, The University of Tokyo. The isotopic ratio $^{129}I/^{127}I$ was measured by means of Accelerator Mass Spectrometry. From ^{129}I data obtained, ^{131}I deposition map was constructed. There observed various fine structures in the map which could not estimated neither by the simulation nor ^{137}Cs distribution.

1 Introduction

Among various radioactive nuclides released from the Fukushima Daiichi Nuclear Power Plant (FDNPP) accident, Iodine-131 showed high radioactivity just after the accident. If taken into human body, Iodine-131 concentrates in the thyroid and may cause the thyroid cancer. The recognition about the risk of Iodine-131 dose originated from the experience of the Chernobyl accident based on the epidemiological study [1]. It is thus important to investigate the detailed deposition distribution of I-131 to evaluate the radiation dose due to I-131 and watch the influence on the human health. However I-131 decays so rapidly (half life = 8.02 d) that it cannot be detected several months after the accident. Long lived radioactive isotope I-129 (half life = 1.57E+7 yr,), which is also a fission product as well as I-131, is ideal proxy for I-131 because they are chemically identical. Near 1,000 soil samples collected around FDNPP site have been analyzed and I-131 map was constructed.

2 Experimental procedure

Soil samples were selected from the collection made by Ministry of Education, Culture, Sports, Science & Technology (MEXT), Japan. Experimental procedure is summarized in Figure 1. After homogenization, iodine was extracted by pyro hydrolysis method and trapped into an alkaline solution. Adding iodine carrier, purified by solvent extraction, and finally silver iodide (AgI) precipitation was made. Isotopic ratio of $^{129}/^{127}I$ was measured by Accelerator Mass Spectrometry at MALT, The University of Tokyo. The iodine concentration was determined by ICP-MS performed at Gakushuin University and MALT.

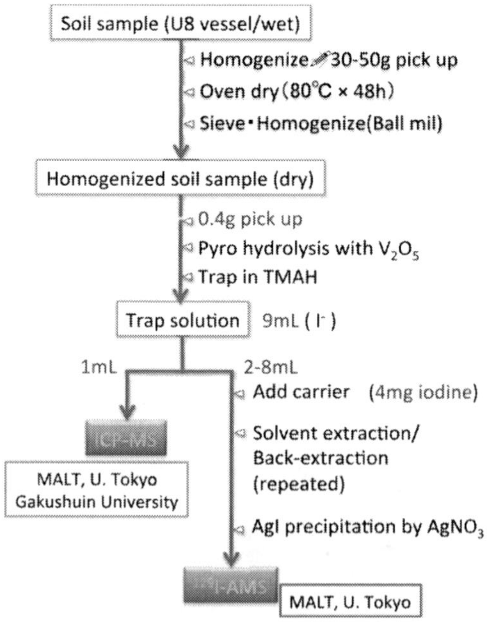

Figure 1. Experimental procedure.

The ^{129}I concentration in soil was calculated according to Equation (1):

[a] Corresponding author: hmatsu@um.u-tokyo.ac.jp

$$n_{129} = \frac{1}{w} \; R \; w n_{127} \; \frac{V_1 \;\; V_q}{V_1} \quad b \; R_C b \; \frac{V_1}{V_1 \;\; V_q}, \qquad (1)$$

here, the each parameter denotes:

- w: Dry weight of soil treated.
- n_{127}: Stable iodine (^{127}I) concentration in soil.
- n_{129}: ^{129}I concentration in soil.
- v_1: Weight of trap solution (TMAH-base).
- v_q: Aliquot weight for trap solution for ICP-MS.
- b: Carrier amount added (as Iodine weight).
- R: AMS result ($^{129}I/^{127}I$ ratio = 10^{-12} - 10^{-10} range).
- R_C: $^{129}I/^{127}I$ of the carrier ($=1.7\times10^{-13}$).

The overall analytical error was determined by the AMS measurement which showed typically 1-3%.

3 Results and discussion

3.1 Reconstructed ^{131}I distribution map

Near 1,000 samples were analysed and made distribution map for iodine-131. For the conversion from ^{131}I to ^{129}I the estimated isotopic ration of $^{129}I/^{131}I$ in the FDNPP reactor was used [2, 3]. The reconstructed map is shown in Figure 2, 3.

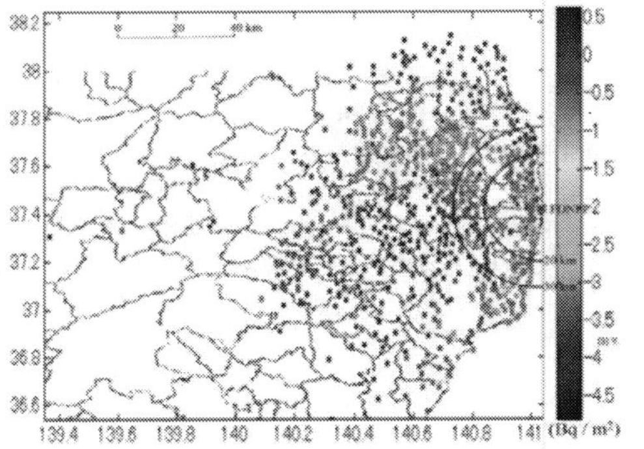

Figure 2. Reconstructed ^{131}I map.

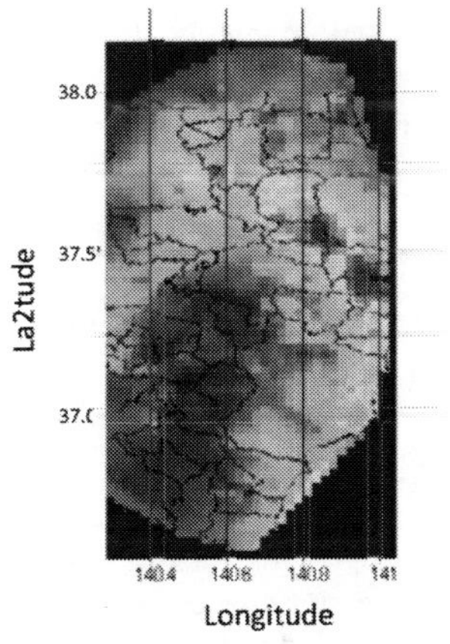

Figure 3. Reconstructed ^{131}I map converted to the mesh distribution for the convenience for the comparison with the simulation. The concentration scale is common with Figure 2.

3.2 Estimation of overall uncertainty

There are several factors that introduce uncertainty to the reconstructed value other than analytical error. Figure 4 shows the factors and introduced uncertainty. It was found that overall uncertainty should be less than 30%.

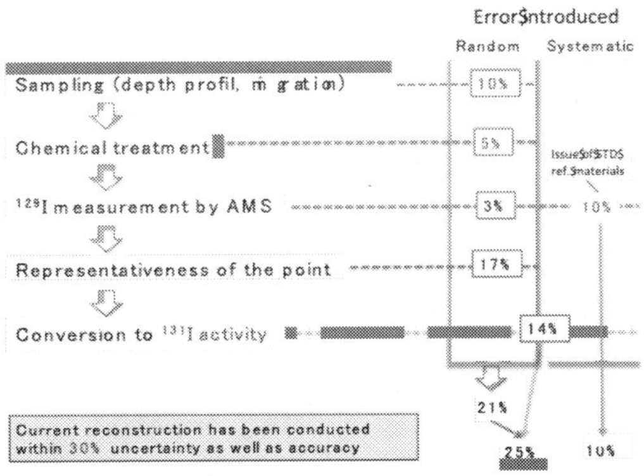

This is not bad compared to the ^{137}Cs map constructed by the direct measurements.

Figure 4. Estimation of overall uncertainty for the reconstruction of ^{131}I.

In Figure 4, the evaluation of the uncertainty accompanied with "sampling" referred to Honda et al. (2015) [4]. Schnetger and Muramats (1996) was referred for the chemical treatment especially for the extraction yield of iodine at pyro hydrolysis [5]. Issue of the standard reference material for ^{129}I-AMS is described in Matsuzaki et al. (2015) [6]. The "representativeness" is related to the local inhomogeneity of deposition density. In actual measurement, there is always occurred case that even two points close to each other show quite different activity. This causes difficult to evaluate the radiation dose of the residents.

References

1. Y. Nikiforov and D. R. Gnepp, Cancer, Vol. **47**, 748 (1994)
2. Y. Miyake, et al., Geochem. J., Vol. **46**, 327 (2012)
3. Y. Miyake, et al., NIM **B361**, 627 (2015)
4. M. Honda, et al., JER **146**, 35 (2015)
5. B. Schnetger and Y. Muramatsu, Analyst **11**, 1627 (1996)
6. H. Matsuzaki, et al. NIM **B361**, 63 (2015)

SEE cross section calibration and application to quasi-monoenergetic and spallation facilities

Rubén García Alía[1,a], Markus Brugger[1], Salvatore Danzeca[1], Julien Mekki[2], and Adam Thornton[1]

[1]CERN, CH-1211, Genève, Switzerland
[2]Centre National d'Etudes Spatiales (CNES), 18 Avenue Edouard Belin, 31400 Toulouse, France

Abstract. We describe an approach to calibrate SEE-based detectors in monoenergetic fields and apply the resulting semi-empiric responses to more general mixed-field cases in which a broad variety of particle species and energy spectra are involved. The calibration of the response functions is based both on experimental proton and neutron data and considerations derived from Monte Carlo simulations using the FLUKA code. The application environments include the quasi-monoenergetic neutrons at RCNP, the atmospheric-like VESUVIO spallation spectrum and the CHARM high-energy accelerator test facility.

1 Introduction

In a mixed radiation field (i.e. one composed of different particles species and energies) the number of SEEs N in a given time period will correspond to the convolution of the individual differential particle fluences $\frac{d\phi_i}{dE}(E)$ and SEE cross sections $\sigma_i(E)$ summed over all the different particle species i, as expressed in Eq. 1.

$$N = \sum_i \int \frac{d\phi_i}{dE}(E)\sigma_i(E)dE \qquad (1)$$

The cross sections as a function of energy can be expressed as the product of a certain constant cross section value and a weighting function $\omega_i(E)$ as shown in Eq. 2, therefore Eq. 1 can be rewritten as Eq. 3 where ϕ_i^{eq} is defined as the equivalent fluence for the particle species i.

$$\sigma_i(E) = \sigma_i \cdot \omega_i(E) \qquad (2)$$

$$N = \sum_i \sigma_i \int \frac{d\phi_i}{dE}(E)\omega_i(E)dE = \sum_i \sigma_i \phi_i^{eq} \qquad (3)$$

For the radiation fields that will be treated in this paper, the particle species contributing to SEEs can be divided into two categories according to their interaction mechanisms: thermal neutrons and high energy hadrons (HEH) [1]. Therefore, Eq. 3 can be developed into two terms and resolved for the so-called mixed-field HEH cross section σ^*_{HEH} as shown in Eq. 4. For a given mixed-field measurement, this value will depend on the number of measured SEEs, the associated equivalent fluences and the sensitivity to thermal neutrons.

$$\sigma^*_{HEH} = \frac{N - \sigma_{th}\phi_{th}^{eq}}{\phi_{HEH}^{eq}} \qquad (4)$$

In the case of the thermal neutrons, SEEs are known to be induced by the neutron capture in ^{10}B and the production of ^7Li and ^4He as ionizing particles. Therefore, the thermal neutron SEE cross section can be expressed as shown in Eq. 5 where σ_{th} is the SEE cross section at 0.025 eV and $\omega_{th}(E)$ is defined as shown in Eq. 6.

$$\sigma_{th} = \int \omega_{th}\frac{d\phi_n}{dE}(E)dE \qquad (5)$$

$$\omega_{th}(E) = \left[\frac{0.025\ eV}{E(eV)}\right]^{\frac{1}{2}} \qquad (6)$$

As to what concerns the HEH contribution and according to experimentally supported nuclear interaction physical considerations that will be developed in the final version of the paper, we assume that for both protons and neutrons, the respective cross sections can be expressed as shown in Eq. 2 where i is either protons or neutrons and $\omega_i(E)$ is a three-parameter Weibull function as shown in Eq. 7 where E_o in the onset energy and W and s are the shape parameters. In this case the constant cross section term σ_i corresponds to the saturation value.

$$w(E) = 1 - e^{-((E-E_o)/W)^s} \qquad (7)$$

Therefore, the σ^*_{HEH} value extracted in a mixed-field through Eq. 4 can be compared to the value retrieved in the monoenergetic measurement set in Eq. 2 which is considered as the calibrated value.

In the work here presented, a set of mixed-field HEH SEE cross sections are measured and benchmarked against the calibrated value. We will use the ratio R between

[a]e-mail: ruben.garcia.alia@cern.ch

DOI: 10.1051/epjconf/201715308015

the mixed-field and calibrated HEH cross sections as a means of quantifying how well the given environment is described in terms of SEE induction.

2 SEE Cross Section Calibration

2.1 PSI: protons between 30 and 230 MeV

The ESA Monitor SEU cross section was measured during a set of test campaigns at PSI between September 2011 and April 2014. The Proton Irradiation Facility (PIF) beam line at the Paul Scherrer Institute (PSI) is used extensively by the space community as well as by research groups in other disciplines [2]. Since 2007, the COMET cyclotron is in operation designed to produce a 1000 nA, 250 MeV proton beam which is currently used in three Gantries and the PIF experimental station. The initial proton energy delivered by the cyclotron during the ESA Monitor calibration tests was 230 MeV, which could be degraded through the use of copper plates of different thicknesses.

At PSI, the beam intensity monitoring system consists of an ionization chamber located downstream the degrader. A plastic scintillator placed in the location of the test sample is used to measure the flux before the test runs and establish the conversion factors between the counts from the ionization chamber and the actual proton flux. The same scintillators can be used to extract the beam profile by moving it horizontally and vertically in millimeter steps. The profile is typically homogeneous within ~10% in a diameter of 5 cm. As to what regards the beam current and respective fluxes, the maximum value is roughly 5 nA, corresponding to a flux of $\sim 2 \cdot 10^8\ p/cm^2/s$ at 230 MeV and $\sim 4 \cdot 10^7\ p/cm^2/s$ at 30 MeV. These values can be reduced linearly with intensity down to a lower limit of roughly 0.1 nA.

In order to evaluate the sensitivity spread among individual detectors, eight different monitor were tested at 230 MeV, yielding an average cross section value as shown in Eq. 8 and a relative 2σ deviation of 7%. This spread was assumed to originate from the sensitivity differences among the detectors, as the count statistics error was significantly smaller in every case (at least $5 \cdot 10^3$ SEU counts were accumulated during each run, corresponding to a 2σ spread of $\sim$ 3%). Moreover, this spread was assumed to be constant for the different test conditions considered. In addition, a 10% error in the fluence value provided by the facility is assumed as typically reported.

$$\sigma_{HEH} = (2.63 \pm 0.32) \cdot 10^{-14}\ cm^2/bit \qquad (8)$$

The ESA Monitor SEU cross section results were normalized to the 230 MeV value and fitted to a three-parameter Weibull function $\omega_p(E)$ such as that shown in Eq. 7. The normalized experimental data and the respective response function $\omega_p(E)$ are plotted in Fig. 1. In addition to the uncertainly related to the σ_{HEH} value (deriving from the sensitivity spread and the fluence measurement) an extra 15% error is considered for the $\omega_p(E)$ function in order to account for the actual energy dependence of the response as opposed to the considered fit.

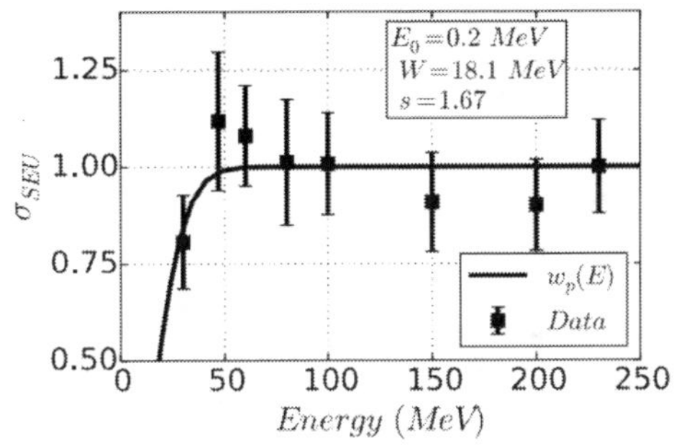

Figure 1. Normalized PSI proton cross section data together with fitted Weibull response function.

2.2 PTB: Neutrons between 5 and 15 MeV

The PTB Neutron Reference Fields (PIAF) are quasi-monoenergetic neutron reference beams in the energy range from thermal to 200 MeV [3]. The main purpose of such beams is the energy response calibration of instruments used for neutron monitoring and dosimetry. The principle behind the production of intermediate and high energy quasi-monoenergetic fields is that of light ions (protons, deuterium) accelerated in a proton or Van-de-Graaf (VgG) accelerator impinging on gas or solid low-Z targets (deuterium, tritium, ^7Li). The reactions and field properties of the energies used in the ESA Monitor calibration campaign are shown in Table 1.

Table 1. PTB calibration for different energies in the 1-20 MeV range. Flux rates correspond to a distance of 1 m from the source. Ti(T) stands for tritiated titanium. The energy spread is represented by the FWHM. The relative contribution of neutrons scattered in the target is indicated by Φ_{sc}/Φ

Reaction	$< E_n >$ (MeV)	$FWHM_{E_n}$ (MeV)	Target	Flux ($/cm^2/s$)	Φ_{sc}/Φ (%)
$^2H(d,n)^3He$	5.0	0.2	D_2-gas	$5.2 \cdot 10^3$	< 1.0
$^2H(d,n)^3He$	8.0	0.2	D_2-gas	$1.9 \cdot 10^4$	< 1.0
$^3H(d,n)^4He$	14.8	0.4	Ti(T)	$1.3 \cdot 10^4$	3.0

The measurements of the neutron peak fluence are performed relative to the differential n-p scattering cross section by means of a proton recoil proportional counter (PRPC) or a recoil proton telescope (RCT) depending on the energy. Likewise, the measurement of the spectral fluence is carried out using pulsed beams and the Time-Of-Flight (TOF) technique with scintillators and fission ionization chambers. This detection technique cannot be used for low energy neutrons at PTB (24 keV - 19 MeV) due to the beam pulse frequency, therefore their spectral characterization is performed through the unfolding of Bonner sphere readings.

At PTB the ESA SEU Monitor was characterized with the energies reported in Table 1. These correspond to a range in which the SEE cross section is known to have a strong dependence with energy and that can have an important impact on the overall high-energy accelerator SEE rate [1].

The respective normalized SEU cross section values and Weibull response fit $w_n(E)$ are shown in Fig. 2 in logarithmic scale and along with the proton response $\omega_p(E)$. Similarly to the proton case, a 15% error is attributed to the use of an analytical fit $\omega_n(E)$ as opposed to the actual response. These results are regarded as monoenergetic for SEU calibration purposes as (i) the proportion of scattered neutrons is at the percent level (ii) the cross section decreases rapidly with energy in this range.

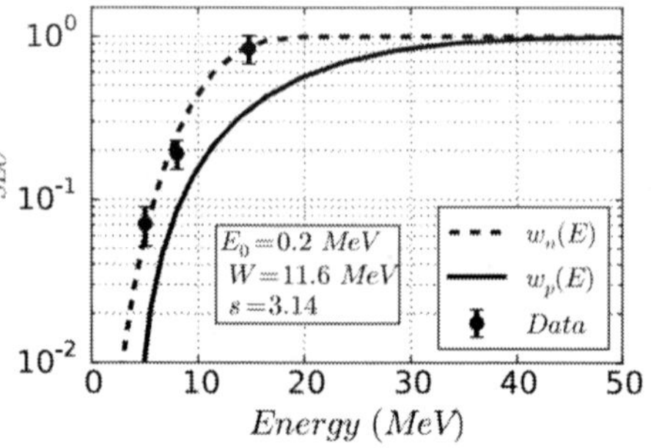

Figure 2. Normalized PTB neutron cross section data together with the fitted Weibull response function.

3 Application to quasi-monoenergetic and spallation facilities

3.1 RCNP: quasi-monoenergetic neutrons at 100 and 300 MeV

At RCNP, quasi-monoenergetic neutron beams in the 80-400 MeV range are provided through the $^7Li(p, xn)$ reaction [4]. A 1 cm thick enriched Li target is used to produce the neutrons. An NE213 liquid scintillator was used to detect the generated neutron spectra using a TOF technique. The proton energy during the ESA Monitor calibration experiments performed in November 2014 was determined to be 100 and 296 MeV, yielding neutron peak energies of 96 and 293 MeV respectively. As a figure-of-merit of the contribution of scattered neutrons to the total spectrum, the ratio between the peak and total above 3 MeV fluences is used, $\Phi_{peak}/\Phi_{>3\ MeV}$, which was reported to be equal to 0.41 for the 100 MeV case and 0.44 for the 296 MeV case. The beam intensity is monitored using a Faraday cup collecting the protons after interacting with the production target.

The measured spectra as reported by the facility are shown in Fig. 3. The response function $w_n(E)$ is convoluted with the neutron fluxes $\phi_n(E)$ in order to yield the ϕ_{HEH}^{eq} from Eq. 3 in which the thermal neutron contribution is assumed to be negligible. The resulting cross sections are shown in Table 2 for the quasi-monoenergetic RCNP neutron beams with their associated uncertainty. The latter considers the count statistics, the sensitivity spread, a 10% margin in the neutron flux measurement and a 15% uncertainty related to the use of $w_n(E)$ as a response function. The normalized cross section results are plotted in Fig. 4. As can be seen, the 96 and 293 MeV neutron values are 27% and 5% larger than the saturation value, respectively; and therefore within the considered statistical uncertainty. This is compatible with the fact that as will be shown in the final version of the paper through Monte Carlo simulations, in virtue of their similar nuclear reaction cross section protons and neutrons are expected to yield equivalent SEE cross sections above roughly 50 MeV.

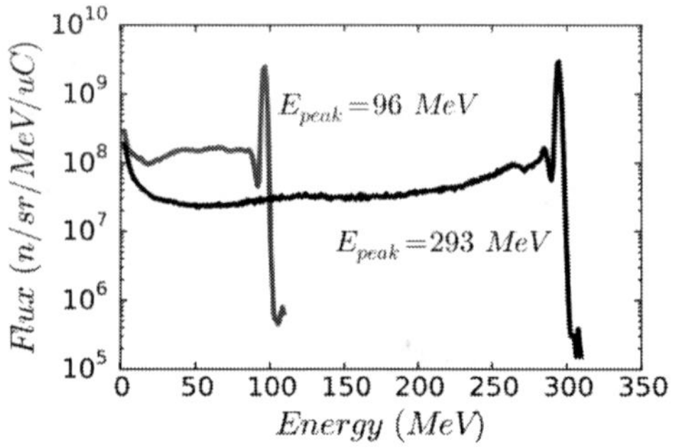

Figure 3. RCNP spectra for 100 and 296 MeV as measured through the NE213 TOF technique.

Table 2. SEU cross section summary for the ESA SEU Monitor at RCNP.

Energy (MeV)	σ_{SEU} $(\cdot 10^{-14}\ cm^2/bit)$	$\pm 2\sigma$ (%)
96	3.36	±26
293	2.77	±26

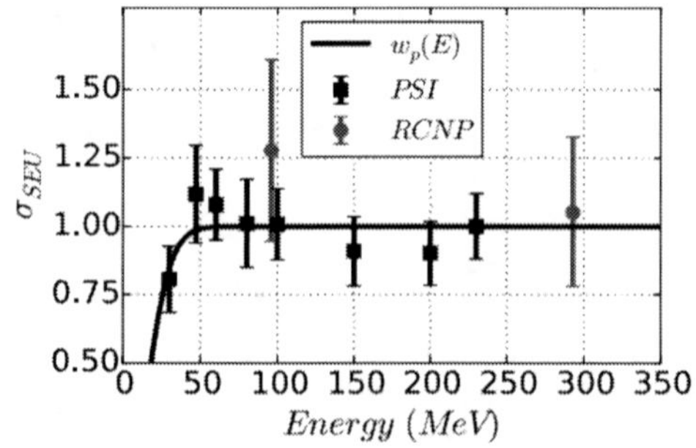

Figure 4. Normalized RNCP neutron quasi-monoenergetic cross section data together with the fitted Weibull proton response function.

3.2 VESUVIO: Atmospheric-like neutron spectrum

The VESUVIO neutron beam is part of the ISIS-STFC laboratory in Oxford, UK. Despite its main use as a condensed matter research instrument [5]. Benchmark measurements have been performed proving the VESUVIO provides a neutron spectrum similar to the ambient at sea level. Neutrons are generated through the interaction of a 800 MeV, 2 μA proton beam with a spallation target. The proton beam is accelerated in a synchrotron as two 100 ns long pulses with a frequency of 50 Hz. The VESUVIO beamline is at 60° with respect to the initial proton beamline. The neutron flux obtained above 10 MeV is $\sim 5.8 \cdot 10^4/cm^2/s$ therefore roughly an order of magnitude lower than those available at TRIUMF or LANSCE. The neutron spectrum is calculated using the MCNPX Monte Carlo simulation tool and benchmarked against TOF measurements performed with different detectors including Bonner spheres, activation foils, CCD devices and Thin Film Breakdown Counters (TFBC). The flux measurement during the VESUVIO experiments relies on the benchmarked MCNPX calculations scaled with the beam current.

Concerning the ESA SEU Monitor measurements performed in March 2014, two configurations were used: one with the primary neutron VESUVIO beam and one with a 1.5 mm Cd foil surrounding the detector in order to absorb the thermal neutrons by means of the roughly 7000 b associated capture cross section. The two concerned spectra can be seen in Fig. 5 as reported by the facility through benchmarked MCNPX simulations.

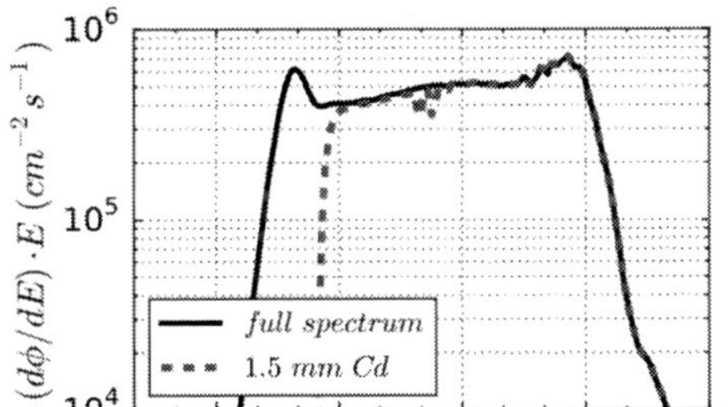

Figure 5. VESUVIO lethargy flux spectrum with and without the Cadmium absorber.

As the ESA Monitor is known to be sensitive to thermal neutrons and provided the high energy neutron spectrum remains unaltered when introducing the cadmium absorber, the thermal neutron SEU cross section σ_{th} can be extracted through the ratio of the differences of the SEU rates (SER) and the respective equivalent thermal neutron fluxes ϕ_{th}^{eq}, without and with absorber, as shown in Eq. 9.

$$\sigma_{th} = \frac{SER_{full} - SER_{Cd}}{\phi_{th}^{eq(full)} - \phi_{th}^{eq(Cd)}} = (2.30 \pm 0.62) \cdot 10^{-15}\ cm^2/bit \quad (9)$$

The HEH cross section can therefore be derived from Eq. 3 yielding the result shown in Eq. 10.

$$\sigma^*_{HEH} = (2.79 \pm 0.60) \cdot 10^{-14}\ cm^2/bit \quad (10)$$

When divided by σ_{HEH} the resulting ratio is 1.06 ± 0.27 therefore the VESUVIO cross section value is fully compatible with the calibration.

3.3 CHARM: high-energy accelerator like mixed field

In the final version of the paper, results for the CHARM accelerator-like mixed field facility at CERN will be shown for different particle spectra, concentrating on the impact of the hadron species (mainly protons, neutrons and pions) as well as their energy distribution.

4 Conclusions and Outlook

Results for the monoenergetic calibration of the ESA SEU Monitor are shown and applied to various mixed-field cases. For all three conditions (100 and 300 MeV quasi-monoenergetic neutron fields at RCNP and spallation neutron spectrum at VESUVIO) the figure-of-merit *R* defined as the ratio between the mixed-field and reference HEH cross sections is compatible with 1. Therefore, this method provides both a general approach and specific calibration results that can be used to determine the degree of accuracy in which a mixed radiation fields is described in terms of capability of inducing SEEs.

References

[1] K. Roed *et al.*, *IEEE Trans. Nucl. Sci.* **59**, 1040 (2012)
[2] W. Hajdas *et al.*, *Radiation Effects Data Workshop* 160 (2002)
[3] S. Röttger *et al.*, *4th International Workshop on Nuclear Fission and Fission-Product Spectroscopy* **1175** 375 (2009)
[4] Y. Iwamoto *et al.*, *Nuclear Instruments and Methods in Physics Research Section A: Accelerators, Spectrometers, Detectors and Associated Equipment* **804**, 50 (2015)
[5] C. Andreani *et al.*, *Applied Physics Letters* **92**, 114101 (2008)

Shielding experiments of concrete and iron for the 244 MeV and 387 MeV quasi-mono energetic neutrons using a Bonner sphere spectrometer (at RCNP, Osaka Univ.)

Tetsuro Matsumoto[1,a], Akihiko Masuda[1], Jun Nishiyama[1,2], Hiroshi Iwase[3,4], Yosuke Iwamoto[5], Daiki Satoh[5], Masayuki Hagiwara[3,4], Hiroshi Yashima[6], Tatsushi Shima[7], Takashi Nakamura[8,9], Hideki Harano[1], Atsushi Tamii[7], Kichiji Hatanaka[7]

[1]*National Institute of Advanced Industrial Science and Technology (AIST), 1-1-1 Umezono, Tsukuba, Ibaraki 305-8568, Japan*
[2]*Tokyo Institute of Technology, 2-12-1 O-okayama, Meguro-ku, Tokyo, 152-8350 Japan*
[3]*High Energy Accelerator Research Organization (KEK), 1-1Oho, Tsukuba, Ibaraki 305-0801, Japan*
[4]*Department of Accelerator Science, Graduate University for Advanced Studies (SOKENDAI), 1-1Oho, Tsukuba, Ibaraki 305-0801, Japan*
[5]*Japan Atomic Energy Agency (JAEA), 2-4 Shirakata,Tokai, Naka, Ibaraki 319-1195, Japan*
[6]*Research Reactor Institute, Kyoto University, 2-1010 Asashiro-nishi, Kumatori, Sennan, Osaka 590-0494, Japan*
[7]*Research Center for Nuclear Physics (RCNP), Osaka University, 10-1Mihogaoka, Ibaraki, Osaka 567-0047, Japan*
[8]*Shimizu Corporation, Etchujima 3-4-17, Koto-ku, Tokyo 135-8530, Japan*
[9]*Cyclotron and Radioisotope Center, Tohoku University, 6-3 Aramaki, Aoba, Sendai 980-8578, Japan*

Abstract. Neutron energy spectra behind concrete and iron shields were measured for quasi-monoenergetic neutrons above 200 MeV using a Bonner sphere spectrometer (BSS). Quasi-monoenergetic neutrons were produced by the ^{7}Li(p,xn) reaction with 246-MeV and 389-MeV protons. Shielding materials are concrete blocks with thicknesses from 25 cm to 300 cm and iron blocks with thicknesses from 10 cm to 100 cm. The response function of BSS was also measured at neutron energies from 100 MeV to 387 MeV. In data analysis, the measured response function was used and the pingpong scattering effect between the BSS and the shielding material was considered. The neutron energy spectra behind the concrete and iron shields were obtained by the unfolding method using the MAXED code. Ambient dose equivalents were obtained as a function of a shield thickness successfully.

1 Introduction

Recently, high-energy and intense beam accelerators, such as the Japan Proton Accelerator Research Complex (J-PARC) are used for various studies and industries. In these facilities, secondary neutrons with energies above 100 MeV are produced around accelerators and beam lines by nucleon-nucleus and nucleus-nucleus reactions. For accurate shielding design of high energy accelerators, experiment data are indispensable to construction of these facilities from the point of view of radiation protection. High-energy neutron penetration data for main shielding materials (concrete and iron) are very important. However, the shielding experimental data for high energy neutrons above 100 MeV are still insufficient both in quality and in quantity as compared with those below 100 MeV [1-3].

In this study, we measured the neutron energy spectra behind concrete and iron shields using a Bonner sphere spectrometer (BSS) for quasi-monoenergetic neutrons produced by the ^{7}Li(p,xn) reaction with 246-MeV and 389-MeV protons at the Research Center for Nuclear Physics (RCNP) of the Osaka University. Moreover, the response function of the BSS was also measured for energy range from 100 MeV to 400 MeV at RCNP. The measured response function of BSS was used in data analysis of the shielding experiments.

2 Experiments

2.1 Shielding experiments

Figure 1 shows a schematic view of a typical experimental arrangement. Shielding experiments were performed at the time-of-flight (TOF) beam course of RCNP. Quasi-monoenergetic neutrons with peak energies of 243.5 MeV and 386.6 MeV were produced by the ^{7}Li(p,xn) reaction by bombarding a 1-cm thick Li

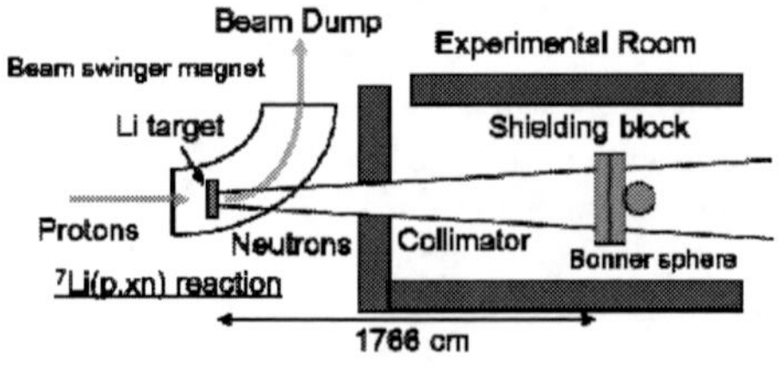

Figure 1. Schematic view of typical experimental arrangement at RCNP.

[a] Corresponding author: t-matsumoto@aist.go.jp

target with 246-MeV and 389-MeV protons from the AVF cyclotron and the ring cyclotron of RCNP. A the neutron beam emitted in the forward direction was extracted into the TOF room through an iron collimator of 12-cm wide and 10-cm high aperture embedded inside a concrete wall with a thickness of 15 cm. The proton beam penetrated through the target was guided to a beam dump using a beam swinger magnet. Neutron source spectra in the TOF room were measured with liquid scintillators by means of the TOF method [4, 5].

Shielding materials are concrete blocks (2.33 g/cm^3) with thicknesses from 25 cm to 300 cm and iron blocks (7.87 g/cm^3) with thicknesses from 10 cm to 100 cm. The components of concrete are given in **Table 1**. The distance between the Li target and upstream surface of the shielding materials was 17.7 m. The BSS was contact with the backward surface of shielding material.

Neutron energy spectra behind the shields were measured with the BSS [6]. The BSS consists of a ^3He spherical proportional counter (CENTRONIC LTD, SP9, gas pressure: 21.3 kPa) and polyethylene (PE, 0.95 g/cm^3) moderators with diameters from 7.62 cm to 24.1 cm. In addition, measurements were performed using inserting metal shells made of lead (457p: 10.2-cm-diameter PE sphere + 1.27-cm thick Pb shell + 2.54-cm thick PE shell) and copper (457c: 10.2-cm-diameter PE sphere + 1.27-cm-thick Cu shell + 2.54-cm thick PE shell) [7].

2.2 Response measurements of BSS

The responses of BSS were measured at neutron energies from 100 MeV to 387 MeV in neutron feilds of RCNP. The responses for mono-energetic neutrons were evaluated by the two-angle differential measurement method, which was described in detail elsewhere [6]. The responses at neutron energies of 144 keV, 565 keV, 5.0 MeV and 14.8 MeV were also calibrated in the mono-energetic neutron standard fields [10] of the National Institute of Advanced Industrial Science and Technology.

The response function of BSS was also simulated with the MCNPX code [8]. The JENDL-HE[9] file was used in the simulation.

3 Data Analysis and Results

The measured and the calculated response results of BSS are in good agreement within measurement uncertainties above 100 MeV for the BSS with the polyethylene moderator. However, there is the difference between the measured and calculated results for the BSS with the Pb or Cu shell. The measured responses were reflected in data analysis of shielding experiments.

The BSS was contacted with the surface of shielding block, actual response functions were affected by the neutron multi-scattering effect between the moderator of BSS and the shielding block, called ping-pong effect. The response function with the ping pong effect was also evaluated with the MCNPX code. The response function including the ping-pong effect was used in the data analysis to obtain more precise results.

Neutron energy spectra behind the shields were obtained by unfolding method using the MAXED code [11]. From the neutron energy spectra with the BSS, the ambient dose equivalent was obtained using conversion coefficients based on the International Commission on Radiological Protection recommendations in 1990 [12, 13]. **Figure 2** shows results of neutron ambient dose equivalent per proton beam charge for the 244 MeV experiment of the concrete shielding. Figure 2 shows the results obtained using the response function of BSS with and without the ping pong effect.

In the shielding experiments, other measurements with a 25.4-cm-diameter and 25.4-cm-thick NE213 scintillator were also performed. The final results of neutron ambient dose equivalents and the attenuation length will be obtained after comparing with the experimental data obtained by the NE213 scintillator in fast energy region and calculated results in all energy region.

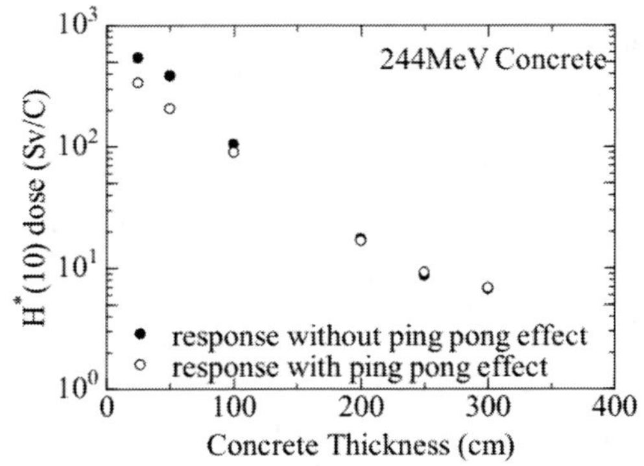

Figure 2. Ambient dose equivalents as a function of the concrete shield. The results were obtained using the response function with and without the ping pong effect.

References

1. T. Ishikawa, Y. Miyama and T. Nakamura, Neutron penetration through iron and concrete shields with the use of 22.0- and 32.5-MeV quasi-monoenergetic sources, *Nucl. Sci. Eng.* **116**, 278 (1994).
2. N. Nakao, H. Nakashima, T. Nakamura, S. Tanaka, S. Tanak, K. Shin, M. Baba, Y. Sakamoto and Y. Nakane, Transmission through shields of quasi-monoenergetic neutrons generated by 43- and 68-MeV protons – I: Concrete shielding experiment and calculation for practical application, *Nucl. Sci. Eng.* **124**, 228 (1996).
3. H. Yashima, H. Iwase, M. Hagiwara, Y. Kirihara, S. Taniguchi, H. Yamakawa, K. Oishi, Y. Iwamoto, D. Satoh, Y. Nakane, H. Nakashima, T. Itoga, N. Nakao, T. Nakamura, A. Tamii and K. Hatanaka, Benchmark experiment of neutron penetration through iron and concrete shields for hundreds-of-MeV quasi-monoenergetic neutrons – I: Measurements of neutron spectrum by a multimoderator spectrometer, *Nucl. Technol.* 168, **298** (2009).

4. Y. Iwamoto, M. Hagiwara, D. Satoh, H. Iwase, H. Yashima, T. Itoga, T. Sato, Y. Nakane, H. Nakashima, Y. Sakamoto, T. Matsumoto, A. Masuda, J. Nishiyama, A. Tamii, K. Hatanaka, C. Theis, E. Feldbaumer, L. Jaegerhofer, C. Pioch, V. Mares and T. Nakamura, Quasi-monoenergetic neutron energy spectra for 246 and 389 MeV ^{7}Li(p,n) reactions at angles from 0 to 30, *Nucl. Instr. Methods* **A629**, 43 (2011).
5. Y. Iwamoto, M. Hagiwara, D. Satoh, S. Araki, H. Yashima, T. Sato, A. Masuda, T. Matsumoto, N. Nakao, T. Shima, T. Kin, Y. Watanabe, H. Iwase, T. Nakamura, Nuclear Instrument and Methods A **804**, 50 (2015).
6. A. Masuda, T. Matsumoto, H. Harano, J. Nishiyama, Y. Iwamoto, M. Hagiwara, D. Satoh, H. Iwase, H. Yashima, T. Nakamura, T. Sato, T. Itoga, Y. Nakane, H. Nakashima, Y. Sakamoto, C. Theis, E. Feldbaumer, L. Jaeferhofer, C. Pioch, V. Mares, A. Tamii and K. Hatanaka, Response measurement of a Bonner sphere spectrometer for high-energy neutrons, *IEEE trans. Nucl. Sci.* 59 (1), 161 (2012).
7. B. Weigel and A. V. Alevra, NEMUS – the PTB Neutron Multisphere Spectrometer: Bonner sphere and more, *Nucl. Inst, Meth.* **A476**, 36 (2002).
8. D. B. Pelowitz, *MCNPX use's manual Ver 2.5.0*, Los Alamos National Laboratory, LA-CP-05-0369 (2005).
9. Y. Watanabe, K. Kosako, S. Kunieda, S. Chiba, R. Fujimoto, H. Harada, M. Kawai, F. Maekawa, T. Murata, H. Nakashima, K. Niita, N. Shigyo, S. Shimakawa and N. Yamano, Status of JENDL High Energy File, *J. Korean Physical Society* **59 (2)**, 1046 (2011).
10. H. Harano, T. Matsumoto, Y. Tanimura, Y. Shikaze, M. Baba and T. Nakamura, Monoenergetic and quasi-monoenergeti neutron reference fields in Japan, *Radiat. Meas.* **45**, 1076 (2010).
11. M. Reginatto, P. Goldhagen and S. Neumann, Spectrum unfolding, sensitivity analysis and propagation of uncertainties with the maximum entropy deconvolution code MAXED, *Nucl. Inst. Methods* **A476**, 242 (2002).
12. *1990 Recommendations of the International Commission on Radiological Protection*, ICRP Publication 60, International Commission on Radiological Protection (1991).
13. T. Sato, A. Endo, M. Zankl, N. Petoussi-Henss, H. Yasuda and K. Niita, Fluence-to-dose conversion coefficients for aircrew dosimetry based on the new ICRP Recommendations, *Prog. Nucl. Sci. Technol.* **1**, 134 (2011).

EPJ Web of Conferences 153, 08017 (2017) DOI: 10.1051/epjconf/201715308017
ICRS-13 & RPSD-2016

Calibration of the radiation monitors from DESY and SPring-8 at the quasi-mono-energetic neutron beams using 100 and 300 MeV ^{7}Li(p,n) reaction at RCNP in Osaka Japan in November 2014

Albrecht Leuschner[1,*], *Yoshihiro* Asano[2] and *Alfred* Klett[3]

[1]Deutsches Elektronen-Synchrotron, Notkestr. 85, 22607 Hamburg, Germany
[2]SPring-8/RIKEN, Sayo, Hyogo 679-5148, Japan
[3]formerly Berthold Technologies, Calmbacher Str. 22, 75323 Bad Wildbad, Germany

Abstract. At the ring cyclotron facility of the Research Center for Nuclear Physics (RCNP) Osaka University, Osaka, Japan a series of measurement campaigns had been continued with quasi mono-energetic neutron beams in November 2014. A ^{7}Li target was bombarded with 100 and 300 MeV protons and the generated neutron beams were directed into a long time-of-flight tunnel at 0 and 25 degrees deflection angle with respect to the proton beam. At a distance of 41 m the cross section of the neutron beam was large enough for the illumination of square meter sized objects like extended range rem-counters. The research institutes SPring-8/RIKEN, Japan, and DESY, Germany, participated in this campaign for the calibration of 4 different types of active ambient dose rate monitors: LB 6411, LB 6411-Pb, LB 6419 and LB 6420. The measurements of their responses are reported and compared with the calculated values.

1 Introduction

At the ring cyclotron facility of the Research Center for Nuclear Physics (RCNP) Osaka University, Osaka, Japan a series of measurement campaigns [1] had been continued with quasi mono-energetic neutron beams in November 2014. A ^{7}Li target was bombarded with 100 and 300 MeV protons and the generated neutron beams were directed into a long time-of-flight tunnel at 0 and 25 degrees deflection angle with respect to the proton beam. At a distance of 41 m the cross section of the neutron beam was large enough for the illumination of square meter sized objects like extended range rem-counters. The research institutes SPring-8/RIKEN, Japan, and DESY, Germany, participated in this campaign for the calibration of 4 different types of active ambient dose rate monitors: LB 6411, LB 6411-Pb, LB 6419 and LB 6420.

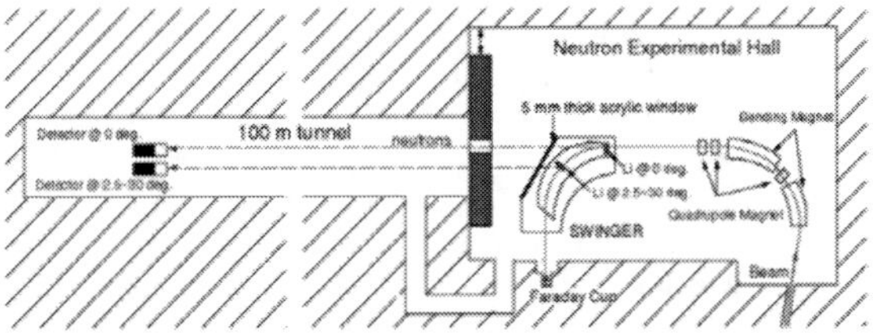

Figure 1: Geometry, proton beam: red, neutron beam: blue.

Figure 1 shows a sketch of the neutron experimental facility at RCNP. For the 25° beam conditions the Lithium target had to be moved inside the vacuum chamber of the swinger magnet and the shielding wall (black) had to be moved to the right position as it is indicated by the arrows in figure 1. The neutron beam is characterized by time-of-flight measurements [2] and Bonner Sphere measurements [3]. The detailed characterization of the neutron beams are reported elsewhere [4]. The beam parameters are summarized in table 1. In the first column the 4 beam settings are listed.

Table 1: Neutron beam qualities for position 41 m derived from Bonner Sphere measurements [3]

Designation		Mean neutron energy [MeV]	Dose per proton charge [Sv/C]
Experimental setup	Spectral correction		
100MeV-25°		44	0.45
100MeV- 0°		62	0.57
	100MeV (0.839)	**98**	**0.20**
300MeV-25°		138	0.26
300MeV- 0°		205	0.39
	300MeV (0.855)	**292**	**0.17**

The second column contains the 100 MeV and 300 MeV mono-energetic neutron spectra after the spectral correction by the proposed subtraction method [1]. The mean neutron energy (3rd column) and the dose per proton charge (4th column) are derived from the dose spectrum of the Bonner Sphere measurement [3] by the inverse-square law scaling (35 m → 41 m).

* Corresponding author: Albrecht.Leuschner@desy.de

2 Method

The dose rate monitors are calibrated in terms of ambient dose equivalent H*(10). Here, it is just called dose. For the calculation of the dose response as well for the dose spectrum of the neutron beam a common set of dose conversion coefficients H*(10)/Φ(E) [5,6] is used where Φ is the neutron fluence . It is shown in figure 2 above 0.1 MeV up to 1 GeV. H*(10)/Φ(E) decreases from 0.1 MeV towards lower energies and is almost constant around 10 pSv cm^2 below 0.001 MeV down to thermal energies. This energy region is excluded as it doesn't contribute to the dose responses of the dose rate monitors from the high-energetic neutron beam.

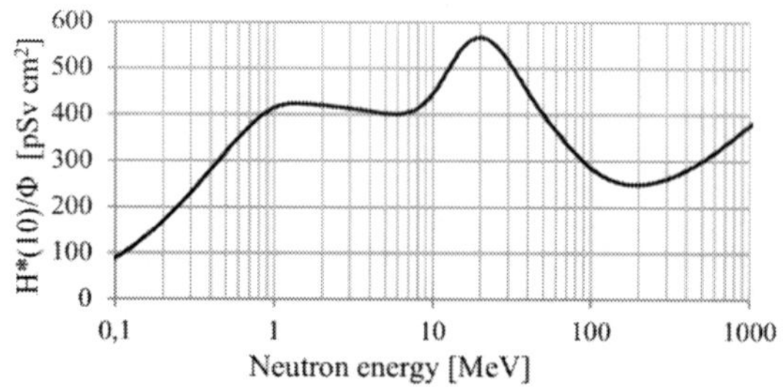

Figure 2 : Dose conversion coefficient H*(10)/Φ(E)

The dose spectra are taken from the Bonner Sphere measurements [3] using the H*(10)/Φ(E) from above. As an example the 100 MeV spectra are shown in figure 3. The spectral correction is done with the equation

$$H(E) = H_0(E) - c \times H_{25}(E) \quad (1)$$

where H(E) is the dose in the energy bin with the energy E and $H_0(E)$, $H_{25}(E)$ are the measured spectra at 0 and 25 degrees emission angle. The parameter c is obtained such that the dose integral between 0.1 MeV and 70 MeV is zero. The value of the parameter c is given in the 2nd column of table 1. The measured and corrected spectra are shown in figure 3 for the 100 MeV case. The corrected spectrum shows a few negative parts having no physical meaning. Nevertheless, any integrated quantity like mean energy or total dose (see table 1) are correct. The folding with a flat detector response gives a correct result. It still works well if the detector response doesn't change much within the energy range of a positive and negative wiggle. The equation (1) is applied to the dose responses of the detectors as well.

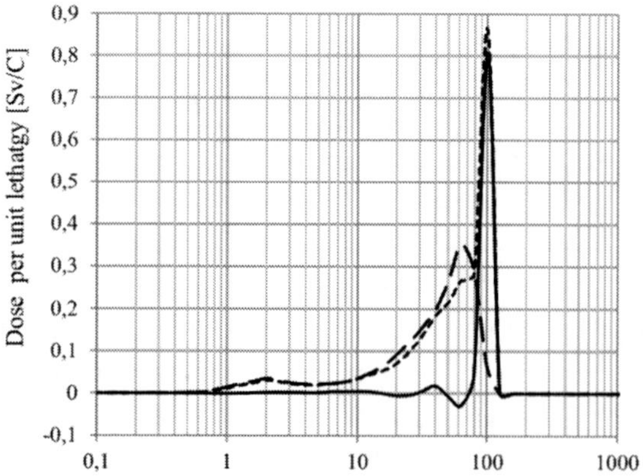

Figure 3 : Dose spectra per proton charge on target at the proton energy of 100 MeV. Angle 0° (dotted line), angle 25° (dashed line) and corrected spectrum (solid line) for position 41 m.

We are calling neutrons with energies below 20 MeV low-energetic neutrons and neutrons above 20 MeV high-energetic neutrons. The corresponding doses are designated H_{LEN} (low-energetic neutron dose), H_{HEN} (high-energetic neutron dose) and H_N for the entire spectrum:

$$H_N = H_{LEN} + H_{HEN} \quad (2)$$

The calculations of the detector responses were performed with the Monte-Carlo-Code FLUKA [7, 8] (FLUKA2011 versions 2b.6 and 2c.2). The devices are fully illuminated by a mono-energetic and parallel neutron beam with a rectangular cross section A. The scored quantities are given per one primary neutron. The reference dose H_{ref} at the neutron energy E is then

$$H_{ref}(E) = H^*(10)/\Phi(E) \times 1/A\,. \quad (3)$$

The respective response is the scored number of tritons N(E) in the sensitive volume of the ^{3}He – proportional counter for example. The response dose H_{res} is calculated by multiplying N(E) with the calibration factor (H/C). For all devices N tritons are assumed to produce C=N counts, i.e. the tube is 100 % efficient and the calibration factor for calculation and measurement is the same.

$$H_{res}(E) = H/C \times N(E) \quad (4)$$

3 Results on detectors

3.1 LB 6411 and LB 6411- Pb

The LB 6411 is a spherical moderated rem-counter with a ^{3}He – proportional counter. The response R_{He} is obtained by counting the reaction products of the nuclear reaction ^{3}He(n,p)T. The LB 6411 is designed [9, 10] to measure the low-energetic neutron dose H_{LEN}. It has been used all over the world for decades.

In order to extend the energy range the dose rate monitor LB 6411 can be equipped by an outer lead shell (10 mm Pb). The combined system is designated LB 6411-Pb [11] and measures the total neutron dose H_N. The calibrations factor is kept unchanged. In figure 4 calculated and measured responses are shown. The overresponse of the LB 6411 can be explained by contributions from the 1 MeV region of the neutron spectrum. It is the same absolute amount for the LB 6411-Pb, but due to the higher response its proportion is smaller.

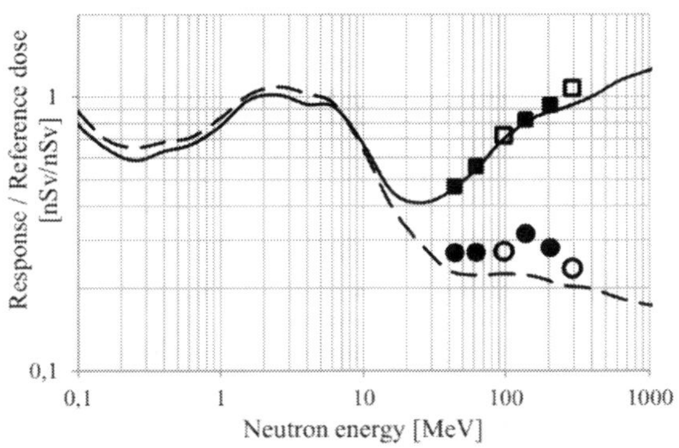

Figure 4 : Response dose divided by the reference dose of the LB 6411 (dashed, circles) and LB 6411-Pb (solid, squares); lines: calculated values; filled markers: measured values; open markers: corrected values by eq.(1).

3.2 LB 6419

The LB 6419 was designed to measure pulsed and continuous neutron and gamma radiation at high energy accelerators [12, 13]. As the proton bunch spacing of this facility is too small the capabilities for the measurement of pulsed radiation could not be tested.

The LB 6419 comprises a cylindrical moderated rem-counter with a ^{3}He – proportional counter and a plastic scintillator. The response of the former, R_{He}, is obtained by counting the reaction products of the nuclear reaction ^{3}He(n,p)T. Its moderator is made of polyethylene and does not contain neither any response-shaping absorbers like Cd or B nor converters like Pb. So it measures H_{LEN}.

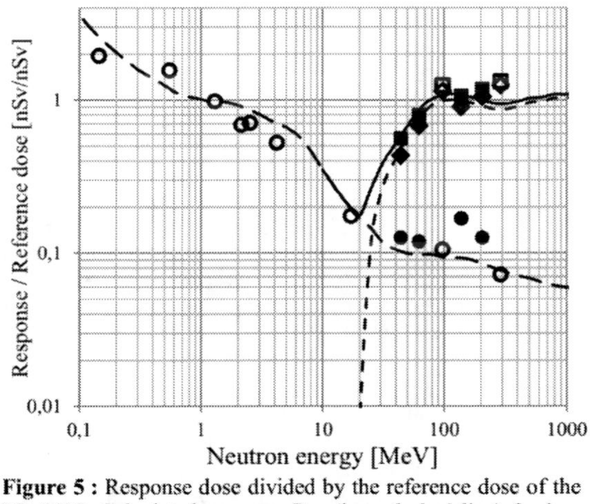

Figure 5 : Response dose divided by the reference dose of the LB 6419. Calculated response R_{He} (long dashed line) for low energy neutrons and R_{Sc} (short dashed line) for high energy neutrons. Sum of them (solid line) for the total neutron dose. Measured raw data (filled markers); measured corrected values by eq.(1) (open markers).

The response of the latter R_{Sc} is obtained by collecting scintillator light above a 20 MeV threshold. Any responses from electro-magnetic radiation such as γ, $e^{\pm}$, $\mu^{\pm}$ are discriminated. The response comes from the energy deposition of charged products from neutron scattering on hydrogen nuclei of the scintillator H(n,n)p and on carbon nuclei as C(n,p) and C(n,α). It [14] measures H_{HEN}. The total neutron dose H_N is yielded by summing up both of the responses R_{He} and R_{Sc}.

The LB 6419 has been used at DESY since 2009. The figure 5 shows the responses of the LB 6419 detector versus the neutron energy. The response R_{He} for low energy neutrons (long dashed line) is met by the measurement (open circles) very well. Earlier measurements at the NPL [12, 15] at energies up to 20 MeV are marked with open circles. Here, low energetic stray radiation was corrected by means of shadow cones.

The measured response R_{Sc} for high energy neutrons (diamonds) comes very close to the reference value at and above 100 MeV. Its calibration was performed by measurements at CERF in 2010 and 2012 [16]. The short dashed line is a Monte Carlo calculation for the response of the scintillator. The measured sums for the total neutron dose H_N are marked with squares. The solid line illustrates the total neutron dose response above 10 MeV.

3.3 LB 6420

The LB 6420 is a derivative of the LB 6419 and a prototype up to now. The improvements pertain to the angular and the energy dependence of the response R_{He} by changing the shape and the structure of the moderator only. The main consequence is that the response R_{He} is assigned to the total neutron dose H_N here. The response R_{Sc} is for the high energy neutrons and loses its importance by becoming additional information, only.

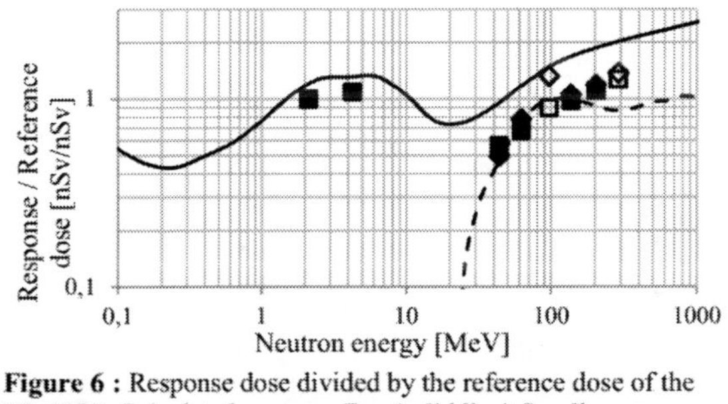

Figure 6 : Response dose divided by the reference dose of the LB 6420. Calculated response R_{He} (solid line) for all neutrons and R_{Sc} (dashed line) for high energy neutrons. Measured raw data (filled markers); measured corrected values by eq.(1) (open markers).

The figure 6 shows the responses of the LB 6420. The calibration measurements with neutron sources (^{252}Cf, ^{241}AmBe) are marked with filled squares. The measured responses R_{He} (squares) and R_{Sc} (diamonds) fit together very well and agree with the reference. But the calculated response R_{He} for the total neutron dose is a factor of 2 too high in the 100 MeV region. This discrepancy needs further investigation.

4 Results on beam

4.1 Distance scaling on beam axis

The parameters of the neutron beams as given in table 1 were measured at 35m distance from the target. But during the measurements the detectors were located at 41 m distance. Therefore the parameters have to be scaled according to the inverse-square-law $(35\ m / 41\ m)^2 = 0.72873$. This scaling factor was verified by the measurement with the LB 6419 responses for the 300 MeV, 0 degree setting. The results are given in table 2. The deviations Δ of the measured values from the inverse-square-law are on the level of a few percent.

Table 2 : Measured doses by the LB 6419 for the positions 35 m and 41 m (3rd and 4th column) compared with the inverse-square law (2nd column).

Quantity	Positions		Factor	Δ
Distance[m]	35	41	0.7287	0 %
H_{LEN} [Sv/C]	0.0719	0.0510	0.7093	-2.7 %
H_{HEN} [Sv/C]	0.539	0.397	0.7365	1.1 %

The other way round, the distance from the detector to the neutron source can be calculated by using the measured scaling factor together with the 6 m base between the positions 35 m and 41 m. The scaling factor for H_{HEN} points to the position -1 m while the scaling factor for H_{LEN} points to the position +3 m. Taking into account that 55 % of the response H_{LEN} stem from low-energetic neutrons their source is estimated to be at +6 m. The former may correspond to the target and the latter to the collimator in the shielding wall. The real geometry (see fig. 1) is very well illustrated by these data.

The different scaling for low-energetic neutrons and high-energetic neutrons leads to a small (approx. 7 %) overestimation of the low-energetic part of the Bonner Sphere dose spectrum at the position 41 m. Mostly, this effect can be neglected. But it should be considered for monitors with high response for low-energetic neutrons and low response for high-energetic neutrons like the H_{LEN} of LB 6411 and LB 6419.

4.2 Off axis neutron stray radiations

The stray radiation was measured with the LB 6419 on the tunnel floor close to the tunnel wall at the longitudinal position of approx. 44 m while the LB 6420 was in the beam at the 300 MeV, 25 degrees setting. The ratio of the stray radiation by the radiation on axis amounts to 5 % and 7 % for the high-energetic and low-energetic neutron dose, respectively.

4.3 Carbon-11 method

Since the early days of nuclear physics the Carbon-11 method has been used to measure high-energetic neutrons. It is based on the inelastic nuclear reaction $^{12}C(n,2n)^{11}C$ in a plastic scintillator. The clean β^+ decay pattern of ^{11}C with a half-life of 20 min leads to the response from high-energetic neutrons.

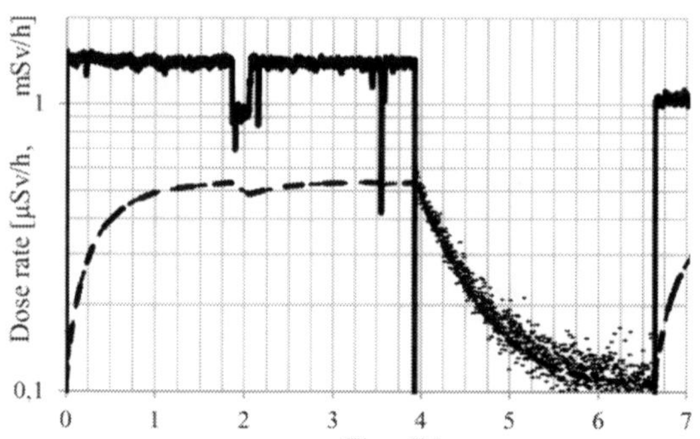

Figure 7 : History of the high-energetic neutron dose H_{HEN} of the LB 6419 (solid line) in mSv/h. The gamma dose after irradiation in µSv/h is marked by dots. The dashed line shows the calculated gamma dose from the ^{11}C decay in µSv/h.

In figure 7 the history of the parasitic measurement of the high-energetic neutron dose by the C-11 method is shown. Fitting the gamma dose of the LB 6419 to the decay pattern after the irradiation (here in the period between 4 to 6.5 h) leads to a dose response from high-energetic neutrons.

The ^{11}C half-life and the dose response due to the decay of one ^{11}C nucleus are the fixed parameters while the natural background dose rate (^{40}K, cosmic muons) and the number of produced ^{11}C nuclei per H_{HEN} are fitted.

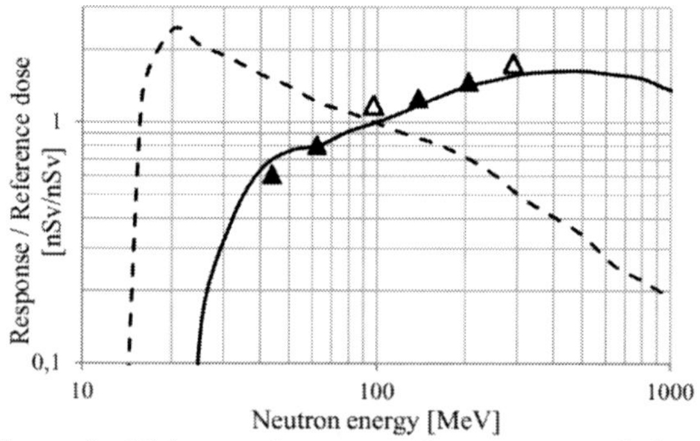

Figure 8: High-energetic neutron dose response of the LB 6419 by means of the ^{11}C method. Solid line: FLUKA calculation for $^{12}C(n,2n)^{11}C$. Dashed line: FLUKA calculation for $^{12}C(n,p)^{12}B$. Filled triangles: measured raw data. Open triangles: measured corrected values by eq.(1).

The measurements were performed at the position 41 m in parasitic modes with the HMGU [3] and CERN [17] group, simultaneously. In figure 8 experimental results (triangles) and calculated values (solid line) are compared. They agree very well with each other. The calibration factor was chosen such that the calculated response function is correct at 100 MeV which is the typical neutron peak energy behind shieldings of high energy accelerators. Its value is set to be 5.43 nSv per ^{11}C nucleus produced in the scintillator.

 DOI: 10.1051/epjconf/201715308017

Some accelerators like the electron linear accelerators FLASH, XFEL and laser plasma accelerators like LUX are operated at repetition rates of 10 Hz and lower. Here, the high-energetic neutron dose can be measured by making use of the inelastic nuclear reaction $^{12}C(n,p)^{12}B$. The ^{12}B nuclei decay with a half-life of 20 ms. Thus, there are more than 5 half-lives between successive pulses. Its response is shown in figure 8 as the dashed line. At low repetition rates the Boron-12 method can be used during exposure but the Carbon-11 cannot.

4.4 Muon and gamma contributions

The response of the LB 6419 plastic scintillator due to any electro-magnetic radiation like γ, $e^{\pm}$ and $\mu^{\pm}$ with energies above 8 MeV contributes to the so-called muon peak at 8 MeV. This corresponds to its characteristic dimension of 4 cm. The muon peak only appears in the absorbed dose – energy spectrum in lethargy representation as shown in figure 9. The background measurement at DESY (dotted line) clearly shows the muon peak at 8 MeV. Its energy is a bit higher than that of the CERF measurement (dashed line) because of the different angular distributions of the incoming muons. The spectrum at 300 MeV, 0 degree (solid line) as well as those at the other settings do not show any hints from electro-magnetic radiation. The muon peak is missing.

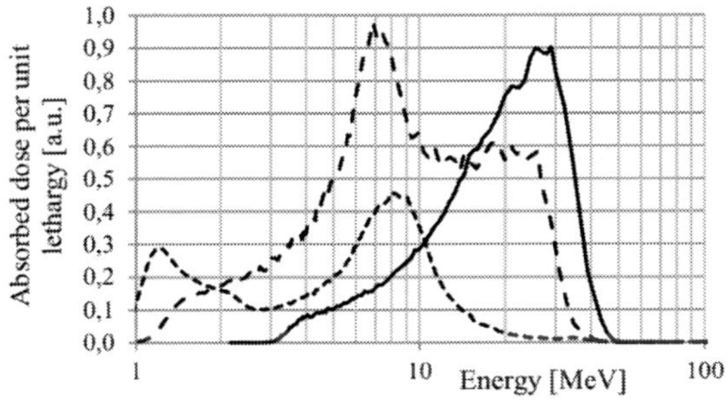

Figure 9 : Absorbed dose spectrum in lethargy representation by the plastic scintillator of the LB 6419. Solid line: RCNP, 300 MeV, 0 degree data. Dashed line: CERF, 120 GeV, CT7+2m. Dotted line: DESY, natural background.

For the experimental setup as shown in figure 1 no charged particles escape from the target chamber on the neutron beam axis as it sits in a magnetic dipole. The only candidates travelling on beam axis are photons from the π^0 decay. Another source of pions could be the neutron beam itself hitting the collimator. But again, their decay products do not appear in the spectra measured at RCNP. Their effect can be considered to be negligible.

5 Conclusions

The active neutron dose monitors LB 6411, LB 6411-Pb, LB 6419 and LB 6420 were tested in the high energy neutron beam facility at the RCNP. All the measurements give a very consistent picture with respect to the reference doses obtained from time-of-flight and Bonner Sphere spectroscopy. The LB 6411-Pb, the LB 6419 and the LB 6420 are suited to measure high energetic neutron dose. The final results are summarized in table 3.

Table 3 : Dose responses H_N of the monitors in Sv/C proton charge on target compared to the reference BSS measurement.

Seup	BSS	LB 6411-Pb	LB 6420	LB 6419	LB 6419 C-11
100-25	0.45	0.21	0.25	0.25	0.27
100-00	0.57	0.32	0.38	0.45	0.46
100-cor	**0.20**	**0.14**	**0.17**	**0.24**	**0.23**
300-25	0.26	0.21	0.25	0.28	0.33
300-00	0.39	0.37	0.43	0.46	0.48
300-cor	**0.17**	**0.18**	**0.22**	**0.23**	**0.30**

Thanks to Prof. Tatsushi Shima, et al. from RCNP for hospitality and stable beam conditions. Thanks to Dr. Yosuke Iwamoto from JAEA and Drs. Vladimir Mares and Sebastian Trinkl from HMGU for unfolding neutron spectra and delivering reference doses. We are grateful for all the support we got from DESY, thanks to Dr. Norbert Tesch.

References

1. Yosuke Iwamoto, *Characterization of quasi-mono-energetic neutron field and radiation instruments with spectrometry using 100 and 300 MeV ^{7}Li(p,n) reaction.* Proposal for Experiment at RCNP (2013).
2. Y. Iwamoto, M. Hagiwara, D. Satoh, S. Araki, H. Yashima et al. *Characterization of high-energy quasi-mono-energetic neutron energy spectra and ambient dose equivalents of 80-389 MeV ^{7}Li(p,n) reactions using time-of-flight method.* Nucl. Instr. and Meth. A, **804**, 50–58 (2015)
3. V. Mares, S. Trinkl et al. *Neutron spectrometry and dosimetry in 100 and 300 MeV quasi-mono-energetic neutron field at RCNP, Osaka University, Japan.* This session. ICRS13, (2016)
4. Y. Iwamoto et al., *Experimental analysis of neutron and background gamma-ray energy spectra of 80-400 MeV ^{7}Li(p,n) reactions under quasi-mono-energetic neutron field at RCNP, Osaka University.* This session. ICRS13 (2016).
5. M. Pelliccioni, Overview of fluence-to-effective dose and fluence-to-ambient dose equivalent conversion coefficients for high energy radiation calculated using the FLUKA code, Radiat. Prot. Dosim. (2000); **88**:279–297.
6. International Commission on Radiological Protection (ICRP), *Conversion coefficients for use in radiological protection against external radiation.* ICRP Publication 74. Pergmon (1997).
7. T.T. Bohlen, F. Cerutti, M.P.W. Chin. A. Fassò, A. Ferrari, A. Mairani, P.R. Sala, G. Smirnov,

DOI: 10.1051/epjconf/201715308017

V. Vlachoudis, *The FLUKA Code: Developments and Challenges for High Energy and Medical Applications*, Nuclear Data Sheets, **120**, 211-214 (2014).

8. A. Ferrari, P.R. Sala, A. Fassò, and J. Ranft, *FLUKA: a multi-particle transport code*, CERN-2005-10 (2005), INFN/TC_05/11, SLAC-R-773.
9. A. Klett and B. Burgkhardt, *The New Remcounter LB6411: Measurement of Neutron Ambient Dose Equivalent H*(10) according to ICRP60 with High Sensitivity*, IEEE Transactions on Nuclear Science, Vol. **44**, No.3, pp. 757-759, June 1997
10. A. Burgkhardt, G. Fieg, A. Klett, A. Plewnia, B.R.L. Siebert, *The Neutron Fluence and H*(10) Response of the New LB6411 Rem Counter,* Neutron Dosimetry, Proceedings of the Eighth Symposium, Paris, November 13-17 1995, Radiation Protection Dosimetry, Vol. **70**, Nos. 1-4, (1997), pp. 361-364.
11. A. Klett, S. Mayer, C. Theis, H. Vincke, *A Neutron Dose Rate Monitor For High Energies,* Radiation Measurements, Elsevier, Special Issue, Volume **41**, Supplement 2, (2007) pp. 279-282
12. A. Klett, A. Leuschner, N. Tesch, *A dose meter for pulsed neutron fields,* Radiation Measurements **45** (2010), 1242 – 1244.
13. A. Klett, A. Leuschner, *A pulsed neutron dose monitor,* Nuclear Science Symposium Conference Record, 2007. NSS '07. IEEE, Year: 2007, Volume: **3**, 1982 – 1983.
14. H. Dinter, B. Racky and K. Tesch, *Neutron dosimetry at high energy accelerators*, NIM A **376** (1996) 104-114.
15. National Physical Laboratory (NPL), London, UK, http://www.npl.co.uk/science-technology/neutron-metrology/products-and-services/monoenergetic-neutron-fluence-and-dose-standards
16. CERF-EU, Genf, http://cerf-dev.web.cern.ch/
17. R. Garcia Alia et al. *SEE cross section calibration and application to quasi-mono-energetic and spallation facilities.* This session. ICRS13, (2016)

Characterization of the PTW 34031 ionization chamber (PMI) at RCNP with high energy neutrons ranging from 100 – 392 MeV

C. Theis[1a], P. Carbonez, E. Feldbaumer, D. Forkel-Wirth, L. Jaegerhofer, M. Pangallo, D. Perrin, C. Urscheler, S. Roesler and H. Vincke, M. Widorski[1]

Y. Iwamoto[2], M. Hagiwara[3], D. Satoh[2], H. Iwase[3], H. Yashima[4], T. Matsumoto[5], A. Masuda[5], J. Nishiyama[5], H. Harano[5], T. Itoga[6], T. Nakamura[7], T. Sato[2], Y. Nakane[2], H. Nakashima[2], Y. Sakamoto[2], S. Taniguchi[8], N. Nakao[9], A. Tamii[10], T. Shima[10], K. Hatanaka[10]

[1]*CERN, HSE-RP, Geneva 23, CH-1211 Geneva, Switzerland*
[2]*JAEA, Tokai-mura, Japan*
[3]*KEK, Tsukuba, Japan*
[4]*Kyoto University, Kyoto, Japan*
[5]*AIST, Tsukuba, Japan*
[6]*RIKEN, Tsukuba, Japan*
[7]*Tohoku University, Sendai, Japan*
[8]*Spring-8 synchrotron radiation facility, Hyogo, Japan*
[9]*Institute of Technology, Shimizu Corporation, Tokyo, Japan*
[10]*RCNP, Osaka University, Osaka, Japan*

Radiation monitoring at high energy proton accelerators poses a considerable challenge due to the complexity of the encountered stray radiation fields. These environments comprise a wide variety of different particle types and span from fractions of electron-volts up to several terra electron-volts. As a consequence the use of Monte Carlo simulation programs like FLUKA is indispensable to obtain appropriate field-specific calibration factors. At many locations of the LHC a large contribution to the particle fluence is expected to originate from high-energy neutrons and thus, benchmark experiments with mono-energetic neutron beams are of high importance to verify the aforementioned detector response calculations. This paper summarizes the results of a series of benchmark experiments with quasi mono-energetic neutrons of 100, 140, 200, 250 and 392 MeV that have been carried out at RCNP - Osaka University, during several campaigns between 2006 and 2014.

[a] Corresponding author: christian.theis@cern.ch

1 Introduction

Radiation monitoring at high-energy proton accelerators poses a considerable challenge due to the complexity of the encountered stray radiation fields. On one hand they span over an energy range comprising energies from fractions of electron-volts (eV) up to several TeV. On the other hand also the composition of the fields is very complex as it comprises many different particle types (hadrons, leptons, photons). At CERN gas-filled ionization chambers are commonly used in such mixed fields as they are sensitive to many different particle types which directly or indirectly cause ionization of the active detection medium. However, standard calibration sources like ^{137}Cs, ^{60}Co, Am-Be and Pu-Be neither cover the full energy range nor the mixed particle composition of the radiation environments in which these detectors are eventually utilized. As a consequence the use of Monte Carlo simulation programs like FLUKA (1)(2) is indispensable to study and obtain an appropriate monitor calibration depending on the radiation field (3). As shown in Ref. (4) calibration-source based calibration can be applicable behind thick lateral shielding but this depends on the detector as well as the actual radiation field. In any case a thorough characterization and understanding of the detector response to mixed radiation fields is of high importance.

At CERN an air-filled ionization chamber under atmospheric pressure PTW-34031, which internally is named PMI and manufactured by PTW, is utilized for the measurement of dose in the LHC experiment caverns during the operation of the machine as well as the determination of the residual dose after the beam has been turned off. The response of this monitor to mixed fields has been studied with the help of Monte Carlo simulations as well as experimental benchmarks (5). The results generally confirmed good agreement of the FLUKA calculations and the experiments but due to the fact that the detector yields integral results, independent of particle species and energy, no detailed benchmark-based conclusion could be drawn for individual particle types.

In the radiation fields present during accelerator operation at CERN, a large contribution to the particle fluence is expected to originate from high-energy neutrons. Hence, benchmark experiments with mono-energetic neutron beams are indispensable to verify the highly important neutron part of the aforementioned detector response calculations. Around the LHC a significant contribution of neutrons with energies above 100 MeV is to be expected and thus, a series of benchmark experiments with quasi mono- energetic neutrons of 100, 140, 200, 250 and 392 MeV have been carried out during several campaigns between 2006 and 2014. This paper includes the results of the latest experiment conducted at the end of 2014 as well as an overall summary of the whole campaign.

2 Benchmark experiments at RCNP

2.1 Setup

The Research Center for Nuclear Physics (RCNP) at the University of Osaka houses a unique chain of cyclotrons which is used mainly for Quantum Chromodynamics (QCD) studies. An AVF cyclotron is acting as an injector to a ring cyclotron which can deliver proton beams of up to 400 MeV to various experimental halls (6) as depicted in Figure 1. Among these facilities there is a time of flight (TOF) tunnel of 100 m length which is available for experiments with quasi-mono-energetic neutrons. In our experiment protons of 100, 140, 200, 250 and 392 MeV were directed at a 1 cm thick ^{7}Li target to produce quasi mono-energetic neutrons at the respective energies. All charged particles were deflected by a beam swinger magnet. The remaining beam of neutral particles is collimated by an iron collimator of 1.5 m with an opening of 10 cm (horizontally) x 12 cm (vertically) before entering the TOF tunnel.

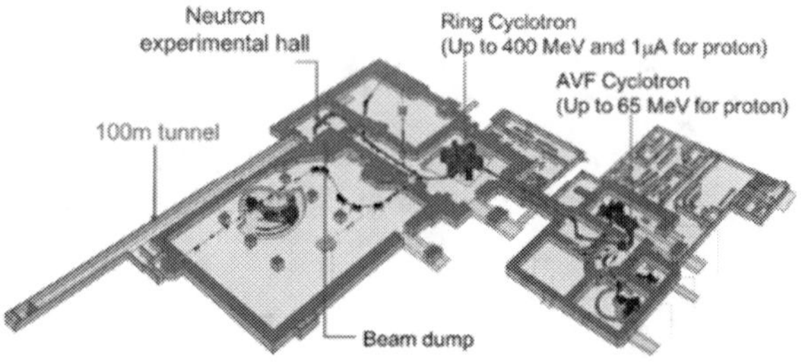

Figure 1: Schematic of the layout of the RCNP facility.

For the experiment two identical detectors were used which allowed for checking the equipment for any kind of malfunction. As illustrated in Figure 2 the monitors were placed on in the TOF tunnel at a certain distance from the target during the irradiation. This distance was determined by the prerequisite that the whole detector should be covered homogeneously by the neutron beam which, coming from the target, is divergent. Taking the beam extensions at the end of the collimator and the detector geometry as well as the dimensions into account it was found that for the PMI a distance of at least 18 m with respect to the target safely fulfills this requirement. With exception of 2014 the detectors were always placed in a distance of 25 m due to the fact that the space in front was occupied by other material and equipment that was tested during the same campaigns. In 2014 the chambers were moved closer and located at a distance of 18 m as the space behind our detectors was made available for parasitic irradiation tests of electronics. In order to compare the results all values have been scaled and normalized to the same distance of 18 m using the $1/r^2$ law.

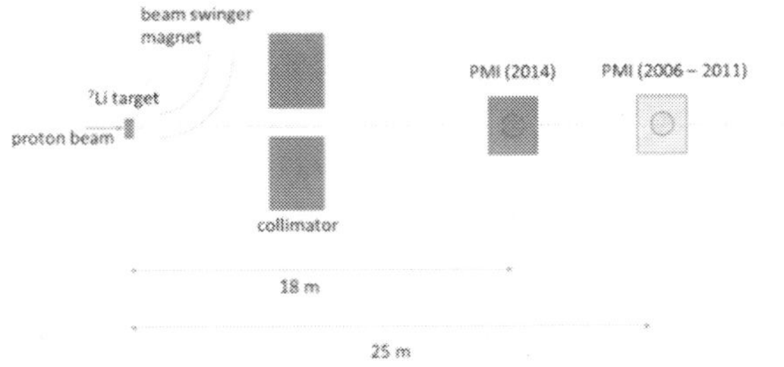

Figure 2: Sketch of the experimental setup at RCNP used in the years 2006 – 2011, when the PMI was located in a distance of 25 m from the target, as well as 2014 when the detector was placed closer to allow for parasitic measurements behind it.

2.2 Simulation of the experiment using quasi-monoenergetic neutrons from 100 MeV up to 392 MeV peak energy

In order to be as accurate as possible the actual neutron fluence spectrum should be used as a source term in the calculations instead of mono-energetic neutrons. Over the years a complete set of measurements were conducted covering the energy range from 100 up to 392 MeV. For each of those scenarios the neutron fluence spectra have been determined by Iwamoto et al. (7)(8) (see Figure 3) and were used in the subsequent simulation of the experiment to characterize the PMI's response.

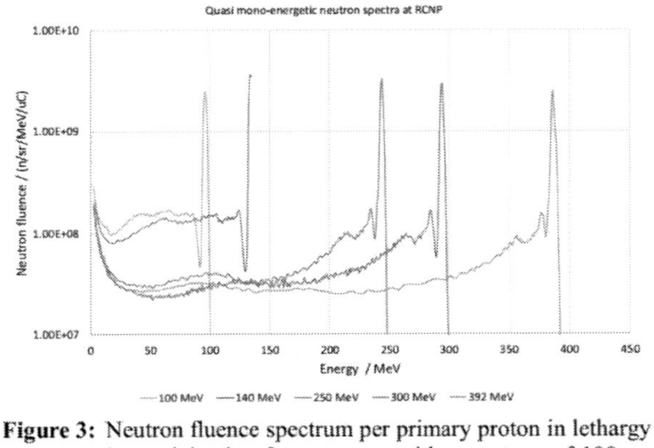

Figure 3: Neutron fluence spectrum per primary proton in lethargy representation, originating from protons with an energy of 100 up to 392 MeV impinging on a 1 cm thick ^{7}Li target. The spectra were measured at RCNP Osaka between 2009 and 2014, using an organic NE213 liquid scintillator (7)(8).

For this purpose the detector's response function to neutrons was convoluted with the measured neutron spectrum and the resulting detector current was normalized to 1 uA proton current on the Lithium target. These results were subsequently compared to the measured values.

3 Comparison of simulation and measurements

First measurements were conducted in 2006 with neutrons of 250 and 392 MeV peak energy. First comparisons to simulations, which used neutron spectra that had been determined prior to these measurements, yielded a discrepancy that was not completely understood. Therefore, these measurements were repeated in 2009 and the respective neutron spectra were measured during the same campaign by Iwamoto et al. (7). During the second experiment the measurements of the first tests could be reproduced and verified. Yet, the discrepancy with respect to the simulations remained. At 250 MeV an underestimation of the calculations of about 15% was found whereas at 392 MeV an underestimation of even 38% could be observed.

One has to keep in mind that during the (p,n) reaction of protons impinging on a ^{7}Li target not only neutrons but also other particles are emitted. All charged particles are deflected by the beam swinger magnet but a photon component remains. The spectra measurements that were performed with the NE213 scintillator discriminated the photon component, but for ionization chambers this is not possible. First assumptions that the contribution might generally be negligibly small were proven false, as the measured charge for the neutron energy of 392 MeV was significantly higher than the values obtained from simulations which assumed a pure neutron beam. Unfortunately the neutron TOF facility at RCNP has not been characterized with respect to photons and thus, no detailed information about the photon spectra was available. Therefore, it was attempted to calculate the respective spectra using FLUKA simulations of the proton interaction with the ^{7}Li target and the results are shown in Figure 4.

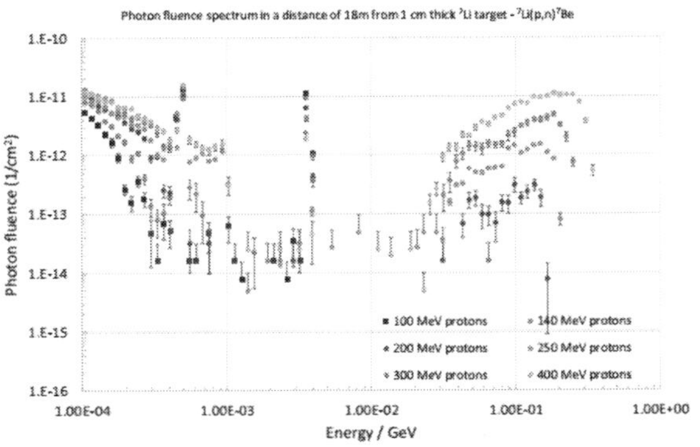

Figure 4: Photon fluence spectra per primary particle obtained from FLUKA calculations of protons with energies ranging between 100 and 392 MeV impinging on a 1 cm thick ^{7}Li target.

Figure 4 nicely illustrates the distinction between low-energetic photons originating from the de-excitation of ^{7}Be as well as a notable high-energy component above 10 MeV.

On one hand these findings corroborated the hypothesis that there is a non-negligible photon component which needs to be taken into account when using ionization chambers in the field at RCNP. On the other hand it also triggered a more detailed investigation of the simulation of the ^{7}Li target reaction using FLUKA. As discussed in detail in (3) the process is quite intricate to model as there is a low-energy photon component (< 1 MeV) which originates from excited states of ^{7}Be as well as a high-energy contribution that can be attributed to the decay of π^0 produced according to inelastic nucleon-nucleon models (9). The energy threshold for pion production is at 290 MeV for a fixed target configuration and for incident proton energies of 392 MeV this threshold is easily overcome and reliable result are obtained. At proton energies of 250 MeV which is below the production threshold of pions one would not expect their creation and thus, no occurrence of high-energy photons due to π^0 decay. However, one has to take into account that nucleons are not at rest but subject to Fermi motion which in turn allows for pion production even at energies below 290 MeV. It should be kept in mind that in FLUKA the radial dependence of the Fermi motion is implemented using the approximated shell model and density distributions. Consequently, these simplifications may have a notable impact on the obtained pion production for ^{7}Li and incident proton energies of 250 MeV.

There is also some uncertainty regarding the accuracy of the prediction of the low-energetic photons (< 1 MeV) due to the use of a so-called optical potential to model scattering processes in Lithium. As the nuclear levels are widely spaced the statistical assumptions of the models can be too coarse for detailed simulations (10). However, as shown in (3) for the 250 MeV scenario 67% of the photon response of the PMI can be attributed to high energetic gammas >> 1 MeV and this value even rises to 93% considering 392 MeV protons impinging on the target. In the latter case the influence of the approximations due to the Fermi motion are expected to be of lower importance as the primary energy of the incident protons is clearly beyond the pion production threshold, whereas this is not the case for the 250 MeV scenario.

Looking once more at Figure 4 it can be seen that the high-energy photon fluence starts to significantly diminish with decreasing energy of the primary protons and at 140 MeV and below it has completely vanished due to the absence of the π^0 decay. It was found that for primary proton energies of 200 MeV and below the photon contribution of the ^{7}Li target reaction to the simulated detector response is below 1% (11). Thus, it can be considered as negligible whereas at energies of 250 MeV or higher their contribution cannot be neglected (3) and must be taken into account. A detailed summary of all measured as well as simulated results can be found in Table 1.

Table 1 Summary of the measured as well as simulated detector response normalized per unit proton current on the Lithium target (3)(11). The simulations also include explicitly the photon contribution based on the spectra shown in Figure 4. It should be noted that all values have been scaled to reflect a setup at the same distance of 18 m from the target, whereas in 2009 and 2011 measurements had to be conducted at 25 m. The uncertainties include statistical fluctuations of the measurements & simulations as well as a systematic error of 10% for the beam intensity monitoring as well as an overall uncertainty of 5% for the neutron fluence measurements.

Energy (MeV)	Measured current per beam intensity (pA/μA)	Calculated current per beam intensity (pA/μA) (neutrons only)	Calculated current per beam intensity (pA/μA) (neutrons + γ)
100	40.8 +/- 15%	33.4 +/- 5%	33.4 +/- 5%
140	36.3 +/- 10%	35.1 +/- 9%	33.4 +/- 5%
200	30.1 +/- 11%	29.7 +/- 11%	29.7 +/- 11%
250	28.9 +/- 3%	24.5 +/- 7%	30.9 +/- 7%
300	28.6 +/- 12%	24.6 +/- 5%	36.5 +/- 6%
392	36.7 +/- 3%	22.7 +/- 11%	36.3 +/- 3%

Looking at the values given in Table 1 it can be clearly seen that at energies of 250 MeV and higher the detected photon contribution from the target is not anymore negligible. This conclusion is also corroborated by looking at the ratios of simulation (with and without photon component) in comparison to the measurement results depicted in Figure 5.

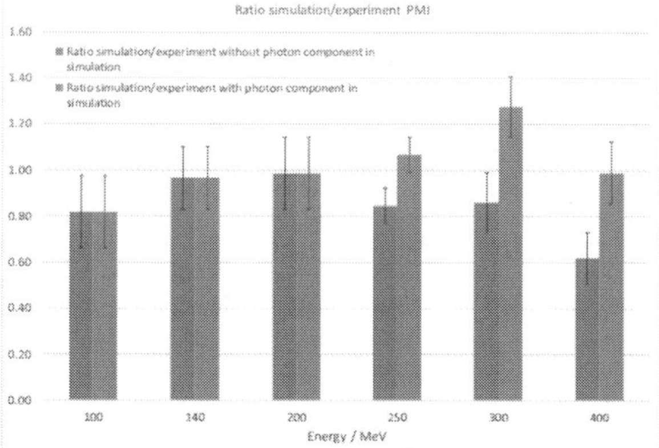

Figure 5 Comparison of the simulated PMI detector response with respect to the measurement values for quasi mono-energetic neutron energies ranging from 100 MeV to 392 MeV. For all cases the simulation results with and without the additionally simulated photon component is shown.

As can be seen for energies of 200 MeV and below the gamma contribution from the target reaction to the detector signal can be considered to be negligible. However, at 250 MeV and above a notable discrepancy can be observed if it is omitted from the simulations. At an energy range of 250 – 300 MeV a slight overestimation of the total signal could be suspected from the graph. This might be explained by the fact that in this energy range the approximation of the Fermi motion in the modeling of the nucleus has a greater impact than at 392 MeV where clearly defined reaction channels for the production and decay of π^0 particles are present. However, considering all uncertainties the overall agreement can still be considered as quite good.

At the lowest energy of 100 MeV a slight underestimation of the simulation can be observed in Figure 5. This might be explained by the fact that at such low energies the measurement of the proton current on the target with a Faraday cup becomes less accurate. However, this would need to be studied in detail for confirmation. In general, very good agreement of the FLUKA calculations of the PMI response to the benchmark measurements was observed. The experiment also highlighted the need to include in the simulations the complete neutron fluence spectra with the continuum as well as a thorough treatment of the photon component of the target reaction. While the latter can be ignored for detectors that are sensitive to neutrons only it becomes important for ionization chambers, especially at energies above 200 MeV.

4 Conclusion

Between 2006 and 2014 a number of measurement campaigns with quasi mono-energetic neutrons has been conducted at RCNP, Osaka University. The goal of these experiments was to verify the characterization and response simulations of PMI ionization chambers that had previously been carried out with the FLUKA Monte Carlo code. As neutrons represent a strong component in the radiation fields typically encountered at CERN's accelerators such tests in well-defined conditions represent an important complement to previous experiments that had already been carried out in mixed radiation fields at CERN.

In general very good agreement was found comparing the FLUKA simulations and the measurements. Yet, it should be emphasized that even with quasi mono-energetic neutrons the full energy spectrum has to be taken into account in the simulations to obtain reliable simulation results. In addition it was found that at proton energies of 250 MeV and above a non-negligible gamma component originating from the target $^7Li(p,n,)^7Be$ reaction was observed that had to be taken into account in the simulations in order to achieve satisfying agreement. The additional photon fluence mainly originates from the creation and subsequent decay of high-energetic π^0 particles.

Acknowledgements
The authors would like to express their gratitude towards Alfredo Ferrari for helpful discussions and his support regarding the understanding of FLUKA's physics models.

References

1) A. Fassò, A. Ferrari, J. Ranft, and P.R. Sala, *FLUKA: a multi-particle transport code*, CERN Yellow Report, INFN/TC_05/11, SLAC-R-773, (2005).
2) T.T. Böhlen, F. Cerutti, M.P.W. Chin, A. Fassò, A. Ferrari, P.G. Ortega, A. Mairani, P.R. Sala, G. Smirnov and V. Vlachoudis, *The FLUKA Code: Developments and Challenges for High Energy and Medical Applications*, Nuclear Data Sheets **120**, 211-214 (2014).
3) C. Theis, *The radiation environment in underground workplaces of the LHC*, PhD thesis, TU-Graz, (2007).
4) C. Theis, D. Forkel-Wirth, M. Fuerstner, S. Mayer, Th. Otto; S. Roesler, H. Vincke, *Field calibration studies for ionisation chambers in mixed high-energy radiation fields*, Radiation Protection Dosimetry Advance Access published on August 1, 2007, DOI 10.1093/rpd/ncm062. Radiation Protection Dosimetry **126**, pp- 299-305, (2007).
5) H. Vincke, D. Forkel-Wirth, D. Perrin, C. Theis, *Simulation and measurements of the response of an air ionisation chamber exposed to a mixed high-energy radiation field*, Radiation Protection Dosimetry **116** (1-4): 380-386. doi: 10.1093/rpd/nci088, (2005).
6) H. Sakai et al., *Facility for the (p,n) polarization transfer measurement*, Nucl. Instrum. Meth. Phys. Res. A 369, 120, (1996).
7) Y. Iwamoto, M. Hagiwara, D. Satoh, H. Iwase, H. Yashima, T. Matsumoto, A. Masuda, J. Nishiyama, T. Nakamura, T. Sato, Y. Nakane, H. Nakashima, Y. Sakamoto, C. Theis, E. Feldbaumer, L. Jaegerhofer, C. Pioch, V. Mares, A. Tamii, K. Hatanaka, T. Itoga, *Benchmark experiment of neutron penetration through iron and concrete shields using 138, 243 and 387 MeV quasi-monoenergetic neutrons part-II:Measurements of neutron energy spectra*, Proceedings of the SATIF 10 conference held at CERN, Switzerland, 2 – 4 June (2010).
8) Y. Iwamoto, M. Hagiwara, D. Satoh, S. Araki, H. Yashima, T. Sato, A. Masuda, T. Matsumoto, N. Nakao, T. Shima, T. Kin, Y. Watanabe, H. Iwase, T. Nakamura, *Characterization of high-energy quasi-monoenergetic neutron energy spectra and ambient dose equivalents of 80–389 MeV 7Li(p,n) reactions using a time-of-flight method*, Nuclear Instrument and Methods A **804**, 50 (2015).
9) A. Ferrari and P.R. Sala, *The physics of high energy reactions*, Proceedings of Workshop on nuclear reaction data and nuclear reactors physics, design and safety, Trieste, (1996).
10) S. G. Mashnik, M. B. Chadwick, H. G. Hughes, R. C. Little, R. E. MacFarlane, L. S. Waters, P.G. Young, *$^7Li(p,n)$ nuclear data library for incident proton energies to 150 MeV*, Proceedings 2000 ANS/ENS International Meeting, Nuclear Applications of Accelerator Technology (AccApp00), (2000).
11) C. Urscheler, Radiological studies for the luminosity upgrade of the CMS detector, PhD thesis at the ETH Zurich, Diss. ETH No. 20578, (2012).

Experimental analysis of neutron and background gamma-ray energy spectra of 80-400 MeV ^{7}Li(p,n) reactions under the quasi-monoenergetic neutron field at RCNP, Osaka University

Yosuke Iwamoto[1,a], Tatsuhiko Sato[1], Daiki Satoh[1], Masayuki Hagiwara[2,3], Hiroshi Yashima[4], Akihiko Masuda[5], Tetsuro Matsumoto[5], Hiroshi Iwase[2,3], Tatsushi Shima[6], Takashi Nakamura[7,8]

[1]*Japan Atomic Energy Agency (JAEA), 2-4 Shirakata,Tokai, Naka, Ibaraki 319-1195, Japan*
[2]*High Energy Accelerator Research Organization (KEK), 1-1Oho, Tsukuba, Ibaraki 305-0801, Japan*
[3]*Department of Accelerator Science, Graduate University for Advanced Studies (SOKENDAI), 1-1Oho, Tsukuba, Ibaraki 305-0801, Japan*
[4]*Research Reactor Institute, Kyoto University, 2-1010 Asashiro-nishi, Kumatori, Sennan, Osaka 590-0494, Japan*
[5]*National Institute of Advanced Industrial Science and Technology, 1-1-1 Umezono, Tsukuba, Ibaraki 305-8568, Japan*
[6]*Research Center for Nuclear Physics (RCNP), Osaka University, 10-1Mihogaoka, Ibaraki, Osaka 567-0047, Japan*
[7]*Shimizu Corporation, Etchujima 3-4-17, Koto-ku, Tokyo 135-8530, Japan*
[8]*Cyclotron and Radioisotope Center, Tohoku University, 6-3 Aramaki, Aoba, Sendai 980-8578, Japan*

Abstract. To develop the 100-400 MeV quasi-monoenergetic neutron field, we measured neutron and unexpected gamma-ray energy spectra of the ^{7}Li(p,n) reaction with 80-389 MeV protons in the 100-m time-of-flight (TOF) tunnel at the Research Center for Nuclear Physics (RCNP) cyclotron facility. Neutron energy spectra with energies above 3 MeV were measured by the TOF method, which had been reported in our previous papers, and photon energy spectra with energies above 0.1 MeV were measured by the automatic unfolding function of the radiation dose monitor DARWIN. For neutron spectra, the contribution of peak intensity to the total intensity integrated with energies above 3 MeV varied between 0.38 and 0.48 in the proton energy range of 80–389 MeV. For gamma-ray spectra, high-energetic gamma-rays at around 70 MeV originated from the decay of π^0 were observed with proton energies higher than 200 MeV. For the 246-MeV proton incident reaction, the contribution of gamma-ray dose to neutron dose is negligible because the ratio of gamma-ray dose to neutron dose is 0.014.

1 Introduction

The expected neutron fluence spectra in such as high-energy particle accelerators and flight altitudes environments typically have a peak at approximately 100 MeV, with maximum energies reaching the GeV range. For development of high-energy neutron monitors employed in these environments, it is important to calibrate instruments in well-defined high-energy mono-energetic neutron fields during response validation.

In high-energy regions above 100 MeV, quasi-monoenergetic neutron fields [1] generated using ^{7}Li(p,n)^{7}Be (g.s. + 0.429 MeV) have been developed at 0° and 16° by iThemba [2] in energy regions of up to 200 MeV. Furthermore, the Research Center for Nuclear Physics (RCNP) cyclotron facility at Osaka University has been used to generate neutron fields in energy regions of up to 400 MeV, and neutron beams are available in the angular range of 0°–25° in the 100-m time-of-flight (TOF) tunnel. Our previous study has investigated the characterization of neutron energy spectra for 80-, 100-, 137-, 200-, 246-, 300- and 389-MeV protons in RCNP's neutron field [3,4]. The RCNP's neutron field has been widely used for various studies such as development of radiation monitors [5,6], neutron shielding experiments [7] and measurement of neutron induced activation cross sections [8]. It has been reported in the International Atomic Energy Agency (IAEA) [9] and the European Radiation Dosimetry group (EURADOS) publications [10] as a well-characterized high-energy monoenergetic neutron field.

In the high-energy neutron field, it is also important to assess the contribution of high-energetic gamma-rays generated from π^0 decay. To create a π^0 with rest energy 135 MeV, it is necessary to give the incoming proton at least 290 MeV of kinetic energy for a fixed nucleon in a target. One should note that Fermi motion of nucleons allows for π^0 production even at energies below 290 MeV [11]. Thus, it is important for calibration of radiation monitors to characterize not only neutron spectra but also high-energy gamma-ray spectra in the high-energy neutron field.

In this paper, we reported measurements of gamma-ray energy spectra and doses in the RCNP quasi-monoenergetic neutron field. Measurements of neutron energy spectra was also reported to compare neutron data with gamma-ray spectra and doses.

[a] Corresponding author: iwamoto.yosuke@jaea.go.jp

2 Experiments

The experiments were carried out in the 100-m TOF tunnel of the RCNP ring cyclotron. A schematic view of the experimental arrangement is shown in **Fig. 1**. This section briefly describes the experiments. Details for neutron measurements were in our previous paper [4].

Proton beams extracted from the ring cyclotron at 80-, 100-, 137-, 200-, 246-, 300- and 389-MeV were made to strike a 1.0-cm-thick lithium target (^{7}Li 99.88%) placed in a swinger located in a vacuum chamber. Neutron energies were measured by a TOF method.

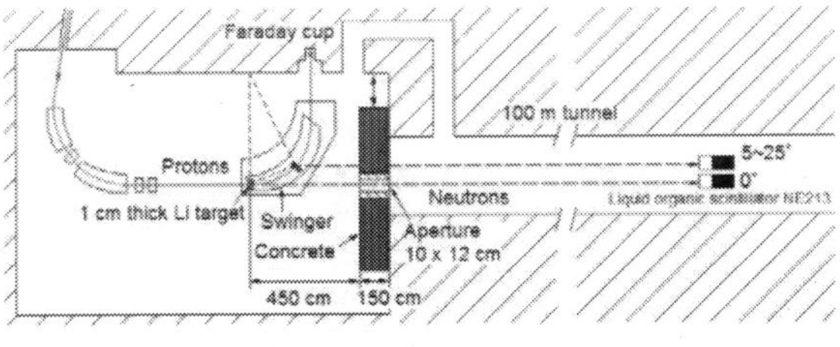

Figure 1. Illustration of experimental setup in the 100-m tunnel at RCNP cyclotron facility [4].

For neutron measurements at 0°, the target was set at the entrance of the swinger. Measurements at angles between 5° and 25° (0°, 5°, 10°, 15°, 20°, 25°) were made by moving the target downward along the curve trajectory of the proton beam in the swinger. To measure proton beam intensity, we used a swinger magnet to drive protons into a Faraday cup after they passed through the target. Neutrons produced at the target entered the 100-m tunnel through a 5-mm-thick acrylic window at the exit side of the swinger and a 10 cm × 12 cm aperture in a 150-cm-thick movable iron collimator embedded in a concrete wall located 4.5 m from the target. A clearing magnet equipped in the movable collimator served to remove charged particles from the neutron beam. The movable collimator and the swinger magnet allowed neutron emissions to be measured at angles between 0° and 25°.

The neutron TOF measurements with energies above 3 MeV were made at angles between 0° and 25° using NE213 organic liquid scintillators of two sizes (12.7 cm × 12.7 cm and 5.08 cm× 5.08 cm (in diameter × length)) with different flight path lengths between the target and detector surfaces. The neutron-detection efficiency of NE213 was calculated using the SCINFUL-QMD code [12, 13]. Finally, the energy spectra were determined on the basis of the detection efficiency, detector solid angle, and proton beam current.

Gamma-ray energy spectra with energies above 0.1 MeV were measured at 0° for 100-, 246-, 300- and 389-MeV by the automatic unfolding function of the radiation dose monitor DARWIN composed of a phoswitch-type scintillation detector, which consists of liquid organic scintillator BC501A coupled with ZnS(Ag) scintillation sheets [14].

3 Results

Figure 2 compares the measured neutron energy spectra at the RCNP quasi monoenergetic neutron filed [3]. Peak neutrons covered the range from transitions to the ground state and the first excited state of ^{7}Be to transitions to the ground state of ^{6}Be. The peak neutron energy is lower than the proton energy because of energy loss in the lithium target, and it is lower than the 1.88 MeV threshold energy of the ^{7}Li(p,n) reaction. The neutron peak intensity is 0.9–1.1 × 10^{10} neutrons/sr/μC. The contribution of peak intensity to the total intensity integrated with energies above 3 MeV varied between 0.38 and 0.48 in the incident proton energy range of 80–389 MeV.

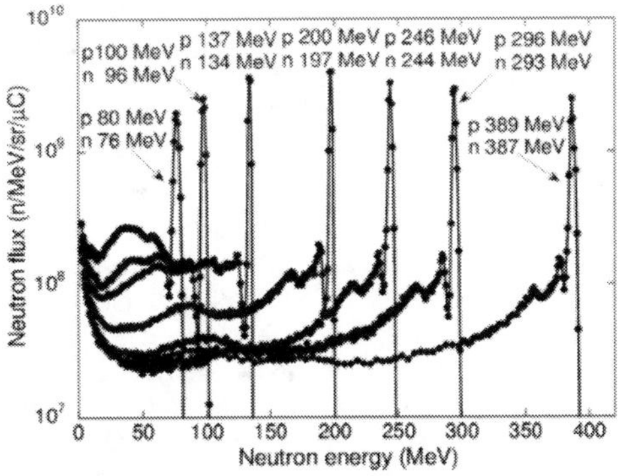

Figure 2. Neutron energy spectra at 0° for 80-, 100-, 137-, 200-, 246-, 296-, and 389-MeV Li(p,xn) reactions and 1-cm-thick lithium target [4].

Figure 3 shows gamma-ray and neutron energy spectra at 0° for 246-MeV ^{7}Li(p,xn) reactions. The unfolding function of DARWIN is useful for investigating gamma-ray and neutron spectra during the dose measurement, though it cannot clearly reproduce quasi-mono-energetic neutron peak as shown in **Fig. 3**. For gamma-ray spectra, low energy photons with energies below 20 MeV can be attributed to de-excitation of ^{7}Be, whereas high-energetic gamma-rays at around 70 MeV originate from the decay of π^0, which are created according to non-elastic nucleon-nucleon interactions [4]. The ratio of gamma-ray flux above 20 MeV to total gamma-ray flux is 0.31. At neutron and gamma-ray energies around 70 MeV, the gamma-ray flux is almost same with neutron flux. On the other hands, the contribution of gamma ray ambient dose to neutron dose is negligible because the ratio of gamma-ray dose to neutron dose is 0.014. High-energetic gamma-rays were also observed for 300- and 389-MeV protons, though it was not observed for 100-MeV proton incidence.

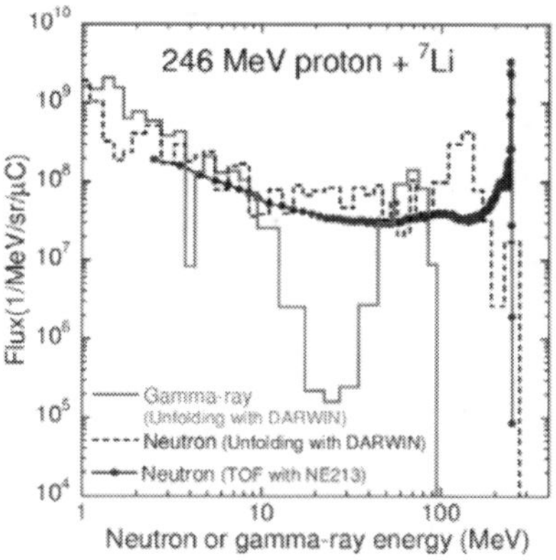

Figure 3. Neutron and gamma-ray energy spectra at 0° for 246-MeV ^{7}Li(p,xn) reactions. Solid line is gamma-ray spectrum and dashed line is neutron spectrum measured by the unfolding method with DARWIN, and full circle shows neutron TOF spectrum measured with an NE 213 liquid organic scintillator.

4 Summary

We measured neutron and gamma-ray energy spectra of the ^{7}Li(p,n) reaction with 80-389 MeV protons in the 100-m TOF tunnel at RCNP cyclotron facility. Neutron energy spectra with energies above 3 MeV were measured by the TOF method and gamma-ray energy spectra with energies above 0.1 MeV were measured by the automatic unfolding function of DARWIN. For neutron spectra, the contribution of peak intensity to the total intensity integrated with energies above 3 MeV varied between 0.38 and 0.48 in the proton energy range of 80–389 MeV. For gamma-ray spectra, high-energetic gamma-rays at around 70 MeV originate from the decay of π^0 were observed with proton energies higher than 200 MeV. For the 246-MeV proton incident reaction, the ratio of gamma-ray flux above 20 MeV to total gamma-ray flux is 0.31 and the gamma-ray flux is almost same with neutron flux at neutron and gamma-ray energies around 70 MeV. On the other hands, the contribution of gamma ray ambient dose to neutron dose is negligible because the ratio of gamma-ray dose to neutron dose is 0.014. The experimental data will be useful to consider the contribution of high-energetic gamma-rays on the response measurement of the neutron monitor in the RCNP quasi-monoenergetic neutron field.

In future work, we will analyze the experimental gamma-ray data of NE213 liquid organic scintillators with the unfolding method to deduce the spectra for all proton energies and various angles.

Acknowledgements

The authors wish to thank the staff at RCNP for providing the beams and for their assistance during the experiments.

References

1. H. Harano, R. Nolte, Metrologia **48**, S202 (2011).
2. M. Mosconi, E. Musonza, A. Buffler, R. Nolte, S. Röttger, F.D. Smit, Radiation Measurements **45**, 1342 (2010).
3. Y. Iwamoto, M. Hagiwara, T. Matsumoto, et al., Nucl. Inst. Method. in Phys. Res. A **690**, 10 (2012).
4. Y. Iwamoto, M. Hagiwara, D. Satoh, S. Araki, H. Yashima, T. Sato, A. Masuda, T. Matsumoto, N. Nakao, T. Shima, T. Kin, Y. Watanabe, H. Iwase, T. Nakamura, Nucl. Inst. Method. in Phys. Res. A **804**, 50 (2015).
5. V. Mares, C. Pioch, W. Rühm, H. Iwase, Y. Iwamoto et al., IEEE Trans. Nucl. Sci., **60**, 299 (2013).
6. A. Masuda, T. Matsumoto, H. Harano, J. Nishiyama, Y. Iwamoto et al., IEEE Trans. Nucl. Sci., **59**, 161 (2012).
7. M. Hagiwara, H. Iwase, Y. Iwamoto, D. Satoh, T. Matsumoto, Prog. Nucl. Sci. Technol., **4**, 327 (2014).
8. H. Yashima, S. Sekimoto, K. Ninomiya, Y. Kasamatsu, Radiat. Prot. Dosim., **161**, 139 (2014).
9. Y. Iwamoto, H. Yashima, D. Datoh, M. Hagiwara, H. Iwase et al., IAEA-TECDOC-1743 (CD-ROM), 177 (2014).
10. S. Pomp, D.T. Bartlett, S. Mayer et al., EURADOS Repost 2013-02 (2013).
11. A. Ferrari, P.R. Sala, "The physics of high energy reactions," Proceedings of Workshop on nuclear reaction data and nuclear reactors physics, design and safety, Trieste, (1996).
12. D. Satoh, T. Sato, A. Endo, Y. Yamaguchi, M. Takada, K. Ishibashi, J. Nucl. Sci. Technol. **43**, 714 (2006).
13. D. Satoh, T. Sato, N. Shigyo, K. Ishibashi, "SCINFUL-QMD: Monte Carlo based computer code to calculate response function and detection efficiency of a liquid organic scintillator for neutron energies up to 3 GeV," JAEA-Data/Code, 2006-023.
14. T. Sato, D. Satoh, A. Endo, Y. Yamaguchi, J. Nucl. Sci. Technol. **42**, 768 (2005).

Neutron spectrometry and dosimetry in 100 and 300 MeV quasi-mono-energetic neutron field at RCNP, Osaka University, Japan

Vladimir Mares[1,a], Sebastian Trinkl[1,2], Yosuke Iwamoto[3], Akihiko Masuda[4], Tetsuro Matsumoto[4], Masayuki Hagiwara[5], Daiki Satoh[3], Hiroshi Yashima[6], Tatsushi Shima[7] and Takashi Nakamura[8]

[1]Helmholtz Zentrum München (HMGU), 85764 Neuherberg, Germany
[2]Technische Universität München, Physik-Department, 85748 Garching, Germany
[3]Japan Atomic Energy Agency (JAEA), Tokai, Japan
[4]National Metrology Institute of Japan, National Institute of Advanced Industrial Science and Technology (NMIJ/AIST), Tsukuba, Japan
[5]High Energy Accelerator Research Organization (KEK), Tsukuba, Japan
[6]Research Reactor Institute, Kyoto University, 2-1010 Asashiro-nishi, Kumatori, Sennan, Osaka 590-0494, Japan
[7]Research Center for Nuclear Physics (RCNP), Osaka University, Mihogaoka, Ibaraki, Japan
[8]Tohoku University, Sendai, 980-0872, Japan

Abstract. This paper describes the results of neutron spectrometry and dosimetry measurements using an extended range Bonner Sphere Spectrometer (ERBSS) with ^{3}He proportional counter performed in quasi-mono-energetic neutron fields at the ring cyclotron facility of the Research Center for Nuclear Physics (RCNP), Osaka University, Japan. Using 100 MeV and 296 MeV proton beams, neutron fields with nominal peak energies of 96 MeV and 293 MeV were generated via ^{7}Li(p,n)^{7}Be reactions. Neutrons produced at 0° and 25° emission angles were extracted into the 100 m long time-of-flight (TOF) tunnel, and the energy spectra were measured at a distance of 35 m from the target. To deduce the corresponding neutron spectra from thermal to the nominal maximum energy, the ERBSS data were unfolded using the MSANDB unfolding code. At high energies, the neutron spectra were also measured by means of the TOF method using NE213 organic liquid scintillators. The results are discussed in terms of ambient dose equivalent, $H^*(10)$, and compared with the readings of other instruments operated during the experiment.

1 Introduction

The high-energy neutrons, which are e.g. present at flight altitudes or behind the shielding of particle accelerators or proton and ion therapy treatment facilities, need particular attention because, for example, at typical flight altitudes or outside the shielding of particle accelerators about half of the neutron ambient dose equivalent, H*(10), originates from neutrons with energies above 20 MeV [1]. Therefore, it is necessary to determine neutron dose rates associated to high-energy neutron fields as precisely as possible.

For the high-energy neutron monitors the proper calibration is very important. The only way of the instrument calibration is measurement performed in preferably mono-energetic neutron fields with known spectral fluence rate distribution. One of the most unique facilities worldwide which provides quasi-mono-energetic neutron fields with energies up to 400 MeV using the ^{7}Li(p,n)^{7}Be reaction is the cyclotron facility at the Research Center for Nuclear Physics (RCNP), Osaka University, Japan [2].

2 Materials and methods

2.1 RCNP facility and experimental setup

At the RCNP in Osaka the protons are pre-accelerated in an AVF cyclotron and then injected into a ring cyclotron where they are accelerated up to energies of 400 MeV. For the measurements presented in this work, proton beams with nominal energies of 100 and 296 MeV were generated and focused on a 10 mm thick natLi (7.6% ^{6}Li and ^{7}Li 92.4%) target situated in a beam swinger magnet inside a vacuum chamber (see Fig. 1). The neutrons were extracted at two different emission angles, 0° and 25°, to the 100 m long TOF tunnel through an iron collimator with 10x12 cm^2 aperture size embedded in a concrete wall of 1.5 m thickness at distance of 4.5 m to target. The resulting neutron beam homogenously covered a minimum solid angle of 2.18x10^{-4} sr assuming a rotationally symmetric beam with radius 5.0 cm at the end of the collimator. At the position of the ERBSS measurement, i.e. 35 m from the target, the radius of the neutron beam was about 29 cm, which was broad enough

[a] Corresponding author: mares@helmholtz-muenchen.de

 DOI: 10.1051/epjconf/201715308020

to fully illuminate the largest sphere with its radius of 19.05 cm.

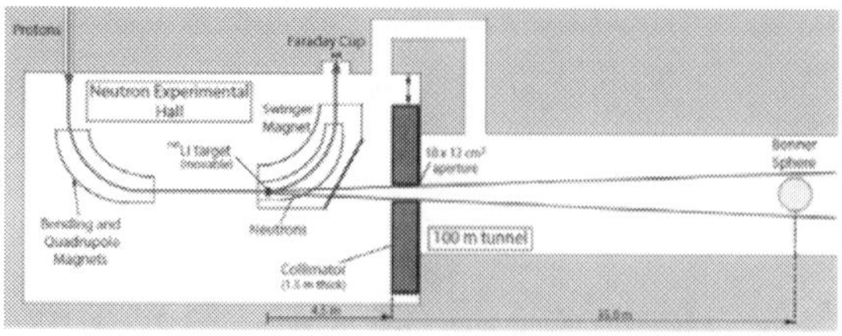

Figure 1. Sketch of the neutron experiment facility and experimental setup at the RCNP [3].

2.2 Extended range Bonner sphere spectrometer

The HMGU ERBSS system is composed of a 3.3 cm diameter spherical ^{3}He (partial pressure of 172 kPa) proportional counter type SP9 (Centronic Ltd.) and 15 polyethylene (PE) spheres of different diameters ranging from 2.5" (6.35 cm) to 15" (38.10 cm). Additionally, two 9" (22.86 cm) PE spheres are designed with lead shells of 0.5" (1.27 cm) and 1" (2.54 cm) to extend the detection range to neutrons above 10 MeV [4, 5]. A bare ^{3}He proportional counter without any surrounding material is also used as the 18th measuring channel. This system enables accurate measurements of neutron fluence spectra and ambient dose equivalent, $H^*(10)$, values since the fluence response functions (HEMA99) of all spheres with ^{3}He proportional counter in their center were calculated by means of Monte Carlo (MC) simulations [4, 5] and experimentally validated at 13 neutron energies between thermal and 14.8 MeV [6, 7] as well as with quasi-mono-energetic neutron fields with peak energies at 244 MeV and 387 MeV [8].

3 Results and discussion

Using the 18 measured count rates and the HEMA99 response function, the spectral neutron fluence distribution was unfolded with the MSANDB unfolding code [9]. The resulting neutron spectra in the lethargy representation generated by protons with energy of 100 MeV and 296 MeV and emission angle of 0° are shown in Figures 2 and 3. The neutron spectra are normalized to the proton beam current measured by a Faraday Cup (FC) inside a beam dump (see Fig. 1). The neutron spectral distributions at 0° exhibit five regions with different fluence intensity: a) a Maxwell-Boltzmann peak, b) flat epithermal region, c) an evaporation peak, d) a breakup and spallation continuum, and e) a peak region. There two the most important high-energy areas of these quasi-mono-energetic neutron spectra - continuum and peak region. The neutron spectra at 25° do not include the peak region because these neutrons produced in target – moved along the proton beam trajectory and turned 25 degree - cannot enter the TOF tunnel through aperture in 150 cm thick iron collimator embedded in a 150 cm thick concrete wall (see Fig. 1).

The high-energy neutrons produced in Li target were also measured by means of the time-of-flight (TOF) method using 3 scintillator detectors [2]. The energy resolution obtained with TOF method was 1 MeV in energy region above detection threshold of 3 MeV. To allow comparison with the ERBSS spectra in Figures 2 and 3, the TOF spectra were rebinned to 10 log-equidistant bins per decade, which is the typical binning structure used in unfolding.

The both ERBSS and TOF neutron spectra were folded with the same fluence-to-dose $h^*(10)$ conversion coefficients to estimate the neutron ambient dose equivalent, $H^*(10)$. The conversion coefficients from ICRP 74 [10] extended to high energies with data from Pelliccioni [11] were used. These data were interpolated in the energy range between 1 meV and 10 GeV with 130 logarithmically equidistant energy bins, i.e. 10 bins per decade, for ERBSS spectra, and with 1 MeV bins between 3 MeV and neutron peak energy for TOF spectra.

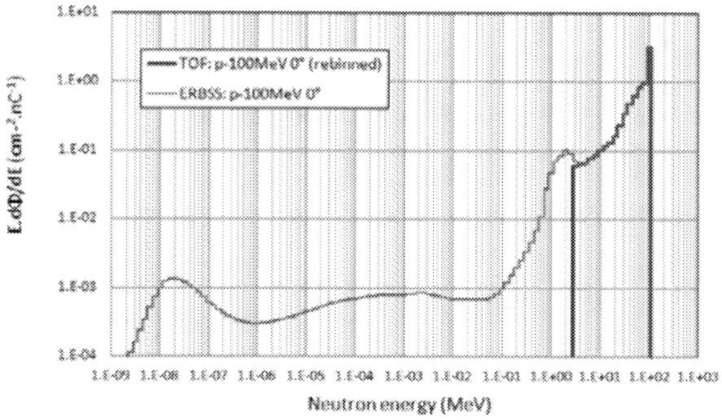

Figure 2. Neutron energy spectra at a distance of 35 m from the Li target bombarded by 100 MeV protons measured with ERBSS (thin blue line) and the TOF (thick red line) method (rebinned), at 0°.

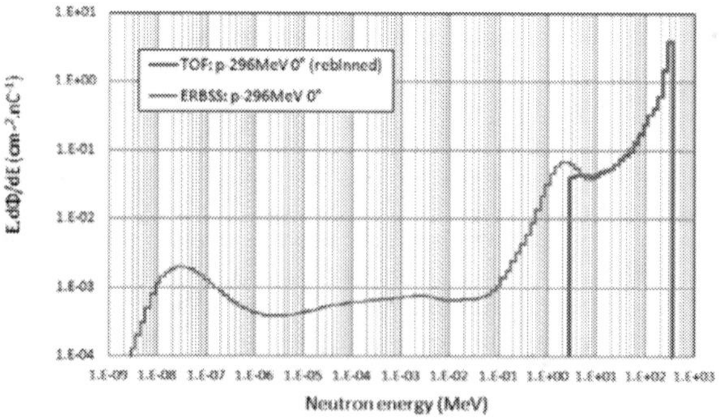

Figure 3. The same as in Fig. 2 but for 296 MeV protons.

In Table 1, the ERBSS $H^*(10)$ values are compared to to those deduced from the TOF spectra, at 0° and 25°. Note that TOF doses were calculated in energy region above detection threshold (3 MeV). This is the reason why TOF doses are about 10 – 30 % lower than ERBSS doses. The data in Table 1 should be treated as preliminary results.

Table 1. Neutron ambient dose equivalent, $H^*(10)$, in Sv/C, at the distance of 35 m from the target. (*contribution integrated for neutrons above 3 MeV.)

Proton energy [MeV]	Emission angle	ERBSS	TOF*
100	0°	0.82	0.62
100	25°	0.64	0.50
296	0°	0.55	0.51
296	25°	0.36	0.33

4 Conslusions

The agreement between ERBSS and TOF neutron spectra above 5 MeV is very good. The evaporation peak around 2 MeV was only identified in ERBSS measurements because of 3 MeV detection threshold of TOF method. Comparison in terms of ambient dose equivalent, $H^*(10)$ between ERBSS and TOF values for both proton energies and emission angles shows very good agreement.

Acknowledgements

The authors would like to thank the staff of the RCNP, Osaka University, Japan for providing technical support, and all measurement campaign participants for their helpful cooperation and kind support during the experiment.

References

1. B. Wiegel et al., Radiat. Meas. **44**, 660-672 (2009)
2. Y. Iwamoto, M. Hagiwara, D. Satoh, S. Araki, H. Yashima et al., Nucl. Instr. and Meth. A **804**, 50–58 (2015)
3. C.D. Pioch, PhD Thesis, TU Munich (2012)
4. V. Mares, G. Schraube, H. Schraube, Nucl. Instr. Meth. A **307**, 398–412 (1991)
5. V. Mares, A. Sannikov, H. Schraube, SATIF-3 Proceedings, Sendai, Japan, 237-248 (1998)
6. A.V. Alevra, M. Cosack, J.B. Hunt, D.J. Thomas, H. Schraube, Radiat. Prot. Dosim. **40**, 91-10 (1992)
7. D.J. Thomas, A.V. Alevra, J.B. Hunt, H. Schraube Radiat. Prot. Dosim. **54**, 25-31 (1994)
8. V. Mares et al., IEEE Trans. Nucl. Sc. **60**, 299-304 (2013)
9. M. Matzke, Private communication later integrated into the neutron metrology file IAEA1279 ZZ-NMF-90 available from NEA Databank (1987)
10. ICRP, International Commission on Radiological Protection, ICRP Publication 74. Pergamon (1997).
11. M. Pelliccioni, Radiat. Prot. Dosim. **88**, 279–297 (2000).

Shielding experiments of concrete and iron for the 244 MeV and 387 MeV quasi-mono energetic neutrons using an organic scintillator (at RCNP, Osaka Univ.)

Masayuki Hagiwara[1,2,a], Hiroshi Iwase[1,2], Yosuke Iwamoto[3], Daiki Satoh[3], Tetsuro Matsumoto[4], Akihiko Masuda[4], Hiroshi Yashima[5], Yoshihiro Nakane[3], Hiroshi Nakashima[3], Yukio Sakamoto[3], Tatsushi Shima[6], Atsushi Tamii[6], Kichiji Hatanaka[6], Takashi Nakamura[7,8]

[1]*High Energy Accelerator Research Organization (KEK), 1-1Oho, Tsukuba, Ibaraki 305-0801, Japan*
[2]*Department of Accelerator Science, Graduate University for Advanced Studies (SOKENDAI), 1-1Oho, Tsukuba, Ibaraki 305-0801, Japan*
[3]*Japan Atomic Energy Agency (JAEA), 2-4 Shirakata,Tokai, Naka, Ibaraki 319-1195, Japan*
[4]*National Institute of Advanced Industrial Science and Technology (AIST), 1-1-1 Umezono, Tsukuba, Ibaraki 305-8568, Japan*
[5]*Research Reactor Institute, Kyoto University, 2-1010 Asashiro-nishi, Kumatori, Sennan, Osaka 590-0494, Japan*
[6]*Research Center for Nuclear Physics (RCNP), Osaka University, 10-1Mihogaoka, Ibaraki, Osaka 567-0047, Japan*
[7]*Shimizu Corporation, Etchujima 3-4-17, Koto-ku, Tokyo 135-8530, Japan*
[8]*Cyclotron and Radioisotope Center, Tohoku University, 6-3 Aramaki, Aoba, Sendai 980-8578, Japan*

Abstract. A shielding benchmark experiment has been performed using a quasi-monoenergetic ^{7}Li(p,n) neutron source with the peak energies of 244 and 387 MeV at the Research Center for Nuclear Physics (RCNP) of Osaka University, in order to assess the accuracy of nuclear data and calculation codes used in high-energy accelerator shielging design. Energy spectra behind bulk shields of 10- to 100-cm-thick iron, 25- to 300-cm-thick concrete and their composite are measured using a NE213 organic liquid scintillator with a diameter and thickness of 25.4 cm each with a time-of-flight and an unfolding method. The neutron attenuation lengths are illustared for iron and concrete as a function of the incident energy.

1 Introduction

In recent years, accelerators have been used for various purposes of not only fundamental research fields such as nuclear physics, material and life science but also industrial and medical applications. The developments of recent accelerator technologies are rapidly increasing the maximum energies and intensities that can be handled. In high-energy accelerator facilities, the shielding of high-energy neutrons is of prime importance for keeping the radiation level of the working area and surrounding environment below the dose limits determined by law and the facilities. Adequate shielding designs are necessary for saving of the building costs as well as radiation safety issue.

The shielding benchmark experiment using the p-^{7}Li quasi-monoenergetic neutron sources have been performed for 40 and 65 MeV in Takasaki Ion Accelerators for Advanced Radiation Application (TIARA), Japan Atomic Research Agency [1, 2] and for 137 MeV in the Research Center for Nuclear Physics (RCNP), Osaka University, Japan [3, 4], because the data are very useful to investigate the attenuation length of monoenergetic neutrons and the accuracy of nuclear data libraries and nuclear reaction models implemented in the simulation codes used for accelerator shielding designs.

In order to extend the shielding benchmark experiment using p-^{7}Li quasi-monoenergetic neutron sources to higher energies, we have developed the irradiation field of p-^{7}Li quasi-monoenergetic neutron sources with the peak energies of hundreds of MeV in RCNP. [3, 5] In 2009, we have performed a seris of benchmark experiments using a NE213 scintillator of 25.4-cm in diameter and 25.4-cm in thickness [6] and a Bonner sphere spectrometer [7] for the hundreds of MeV p-^{7}Li quasi-monoenergetic neutron sources, and obtained the pleliminary neuron spectra in time and energy behind iron and concrete shields with several thicknesses from 10- to 300-cm for the p-^{7}Li quasi-monoenergetic neutron sources with the peak energies of 244 and 387 MeV. The attenuation lengths for neutrons with energies of 244 and 387 MeV were estimated by fitting the attenuation profile of peak neutrons as a function of shield depth with single exponential form. Some difference between measured and calculated spectra are observed especially for thick shields and for the 387 MeV neutron source, due primarily to counting of the charged hadrons produced in the shields.

We have reanalized the neutron spectra transmitted through the iron, concrete and their composite shield with thickness from 10- to 300-cm, which were measured with a TOF and an unfolding method by a large NE213

[a] Corresponding author: masayuki.hagiwara@kek.jp

scintillator of 25.4 cm in diameter and 25.4 cm in thickness. The energy spectra measured by the NE213 scintillator are compared with those measured by a Booner sphere spectrometer. The neutron attenuation length were obtained for iron and concrete as a function of neutron incident energies.

2 EXPERIMENT AND ANALYSIS

2.1 Experimental facility and setup

The experiment was performed at the neutron time-of-flight beam course in RCNP. Figure 1 illustrates the experimental setup used in this study. Proton beams of 246 and 389 MeV extracted from the ring cyclotron were transported to the experimental room and hit a 1.0-cm-thick ^{nat}Li target installed in a vacuum chamber. Protons passing through the target were bended towards the beam dump by the swinger magnet to measure the proton beam intensity with a Faraday cup. Neutrons produced at 0-degree from the target were extracted into the TOF tunnel through a 150-cm-thick iron collimator with 10 × 12 cm aperture embedded in a 150-cm-thick concrete wall located 4.5 m away from the target, while charged particles were rejected by a vertical bending magnet located in the collimator. The neutron beam size was 30 × 36 cm at 18 m from the target, which was determined by the 10 × 12 cm aperture. The details of the p-^{7}Li quasi-monoenergetic source neutrons were described in Ref. 5.
We used 118 × 118 × 10 cm-thick iron blocks and 120 × 120 × 25 cm-thick concrete blocks as shielding materials. In the shielding experiments, these blocks were assembled with thickness from 10 to 100 cm for the iron shield, from 25 to 300 cm for the concrete shield and for the composite shield, 70 cm-thick iron before 200 cm-thick concrete were used. The shielding assemblies were placed ~18 m away from the target. The density of the iron and concrete used in this experiment was 7.87 and 2.33 g·cm^{-3}, respectively, and the atomic compositions of the concrete are listed in Ref. 4.

2.2. Measurement of neutron spectra behind Shields

We used a large NE213 scintillator of 25.4 cm in diameter and 25.4 cm in thickness for the neutron detector [8], because the NE213 can stop recoil proton up to 180 MeV and therefore has good energy resolution for high energy neutrons. The NE213 was placed on the neutron beam axis in contact with the back surface of the shields. The distance between the target and the NE213 were changed from 18 m to 21 m with increasing the thickness of the shields. The proton beam was used of intensity between 1 nA to 1 μA. The data taking system was the same as that used in the previous measurements. [3, 5]

In order to deduce the energy spectra of neutrons transmitted through the shields, we applied a time-of-flight (TOF) method and an unfolding method based on the FORIST code [3] for the peak energy region and the continuous energy region, respectively, because it was difficult to obtain the energy spectra by the unfolding method alone due to oscillation caused by the sharp monoenergetic peak of the p-^{7}Li neutron source and the poor accuracy of responses for high-energy neutrons above 180 MeV which is the upper energy that recoil protons can be stopped in the detector and deposit their full energy. Thus, we separately analyzed the energy spectra for the peak energy region and the continuous energy region using the TOF and unfolding method, respectively. The TOF method was applied above the boundary of the times corresponding to 231 and 361 MeV in the TOF spectra for the incident proton energies of 246 MeV and 389 MeV, respectively. In order to deduce the energy spectra including the scattered neutrons for the TOF peak region, we used the relation between the TOF and the energy spectra calculated by the PHITS code. [9] The energy spectra for the TOF peak region were normalized with the measured fluence of the TOF peak region considering the detector efficiency on the calculated energy spectra. The energy spectra for the continuous energy region were deduced by the FORIST unfolding code coupled with the response matrix which was prepared by the calculation with the SCINFUL-QMD code [10] based on the measured response data [8]. The energy spectra separately analyzed for the two TOF regions (peak and continuous region) were combined by summing them in each energy bin. The details of this technique are described in the Ref. 3. The energy calibration of the light output was performed using the Compton edge of 4.43 MeV γ rays from ^{241}Am-Be source and the recoiled proton edges. The proton energy was converted to electron energy of the equivalent light output (MeVee) by using the calibration value from Ref. 10.

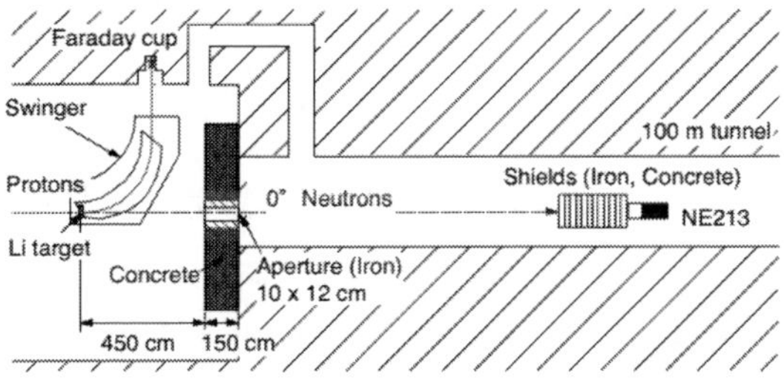

Figure 1. Illustration of the experimental setup used in this study

3 Results and discussion

3.1. Energy spectra

Figure 3 shows the typical resluts of energy spectra of neutrons transmitted through the iron measured by an NE213 with energies above 30 MeV, in comparison with tthose measured by a Bonner sphere spectrometr [7]. The both measured spectra are generally in good agreement in the overlap energy region, although the NE213 data are generally several tens % higher than the Bonner data. This difference might come from counting protons

DOI: 10.1051/epjconf/201715308021

produced in the shields. This effect will be corrected in the final papaer.

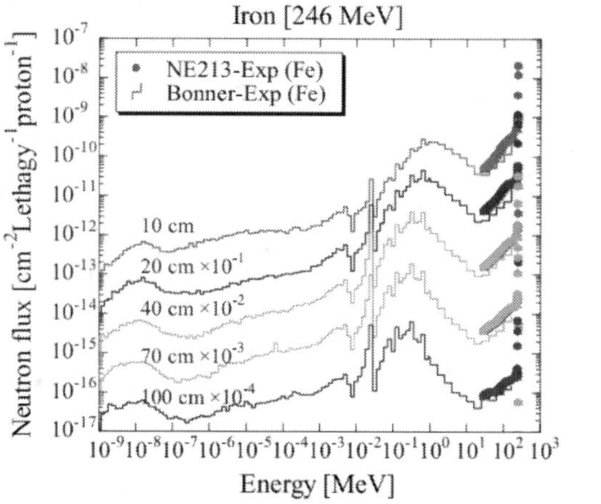

Figure 2. Comparison of the neutron energy spectra transmitted through the iron with various thicknesses for the 246 MeV p-^{nat}Li quasi-monoenergetic neutrons between experimental data measured by a NE213 scintillator (indicated by symbols) and a Bonner sphere spectrometer (indicated by lines).

3.2. Attenuation length

The attenuation profile of the peak neutron flux for iron and concrete shields are fitted by a single exponential curve as a function of depth of the shields. The measured attenuation lengths give good agreement with the calculation results by PHITS within 6 %.[6] Figure 3 shows attenuation length of neutrons transmitted through iron and concrete shields as a function of incident neutron energies. The plotted data for neutrons energies less than 140 MeV were obtained from the Refs. 1-3. The attenuation lengths show signs of the plateau above 250 MeV.

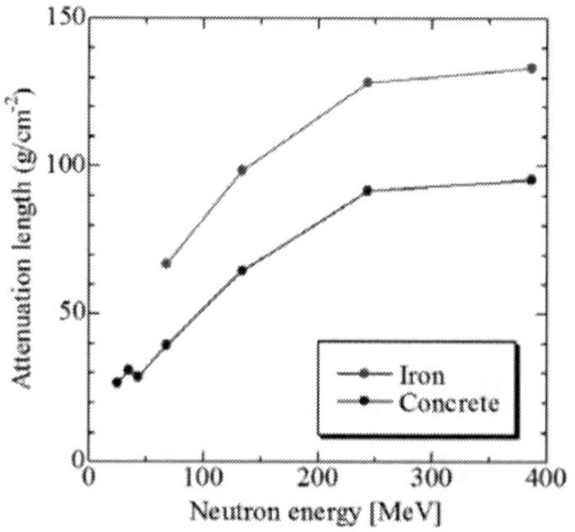

Figure 3. Attenuation length of neutrons transmitted through the iron and concrete shields as a function of incident neutron energies. The plotted data for neutrons energies less than 100 MeV were obtained from the Refs 1-3..

4 CONCLUSION

Neutrons transmitted through 10- to 100-cm-thick iron, 25- to 300-cm-thick concrete and their composite shield (70-cm-iron before 200-cm-concrete) were measured in the energy spectra with a 25.4 diameter × 25.4 thick NE213 scintillator using the 244 and 387 MeV p-^{7}Li quasi-mono-energetic neutron sources at RCNP. The comparison between the experimental data measured by a NE213 scintilator and a Bonner sphere spectrometer shows generally good agreement in the overlap energy region. The neutron attenuation length were obtained for iron and concrete as a function of neutron incident energies.The both attenuation lengths show signs of the plateau above 250 MeV.

References

1. N. Nakao, H. Nakashima, T. Nakamura, Sh. Tanaka, Su. Tanaka, K. Shin, M. Baba, Y. Sakamoto and Y. Nakane, Transmission through shields of quasi-monoenergetic neutrons generated by 43- and 68-MeV protons: part I - concrete shielding experiment and calculation for practical application, Nucl. Sci. Eng. 124 (1996), pp. 228-242.
2. H. Nakashima, N. Nakao, Sh. Tanaka, T. Nakamura, K. Shin, Su. Tanaka, S. Meigo, Y. Nakane, H. Takada, Y. Sakamoto and M. Baba, Transmission through shields of quasi-monoenergetic neutrons generated by 43- and 68-MeV protons: part II - iron shielding experiment and analysis for investigating calculation methods and cross-section data, Nucl. Sci. Eng. 124 (1996), pp. 243-257.
3. M. Hagiwara, H. Iwase, Y. Kirihara, H. Yashima, Y. Iwamoto, D. Satoh, Y. Nakane, H. Nakashima, T. Nakamura, A. Tamii, and K. Hatanaka, Benchmark experiment of neutron penetration through iron and concrete shields for hundreds-of-MeV quasi-monoenergetic neutrons - II: Measurements of neutron spectrum by an organic liquid scintillator, Nucl. Tech. 168 (2009), pp. 304-309.
4. H. Yashima, H. Iwase, M. Hagiwara, Y. Kirihara, S. Taniguchi, H. Yamakawa, K. Oishi, Y. Iwamoto, D. Satoh, Y. Nakane, H. Nakashima, T. Itoga, N. Nakao, T. Nakamura, A. Tamii, and K. Hatanaka, Benchmark experiment of neutron penetration through iron and concrete shields for hundreds-of-MeV quasi-monoenergetic neutrons - 1: Measurements of neutron spectrum by a multimoderator spectrometer, Nucl. Tech., 168 (2009), pp. 298-303.
5. Y. Iwamoto, M. Hagiwara, D. Satoh, H. Iwase, H. Yashima, T. Itoga, T. Sato, Y. Nakane, H. Nakashima, Y. Sakamoto, T. Matsumoto, A. Masuda, J. Nishiyama, A. Tamii, K. Hatanaka, C. Theis, E. Feldbaumer, L. Jaegerhofer, C. Pioch, V. Mares, and T. Nakamura, Quasi-monoenergetic neutron energy spectra for 246 and 389 MeV ^{7}Li(p,n) reactions at angles from 0° to 30°, Nucl. Instr. and Meth. A 629 (2011), pp. 43-49.
6. M. Hagiwara, H. Iwase, Y. Iwamoto, D. Satoh, T. Matsumoto, A. Masuda, H. Yashima, Y. Nakane, H. Nakashima, Y. Sakamoto, A. Tamii, K. Hatanaka

and T. Nakamura, Shielding benchmark experiment using hundreds of MeV quasi-monoenergetic neutron source by a large organic scintillator, Progress in Nuclear Science and Technology 4, 327-331 (2014).
7. T. Matsumoto, A. Masuda, J. Nishiyama, H. Harano, H. Iwase, Y. Iwamoto, M. Hagiwara, D. Satoh, H. Yashima, Y. Nakane, H. Nakashima, Y. Sakamoto, C. Pioch, V. Mares, A. Tamii, K. Hatanaka and T. Nakamura, Measurement of neutron energy spectra behind shields for quasi-monoenergetic neutrons generated by 246-MeV and 389-MeV protons using a Bonner sphere spectrometer, Progress in Nuclear Science and Technology 4, 332-336 (2014).
8. S. Taniguchi, N. Nakao, T. Nakamura, H. Yashima, Y. Iwamoto, D. Satoh, Y. Nakane, H. Nakashima, T. Itoga, A. Tamii, and K. Hatanaka, Development of a quasi-monoenergetic neutron field using the 7Li(p,n)7Be reaction in the energy range from 250 to 390 MeV at RCNP, Rad. Prot. Dosim. 126, 1-4 (2007), pp. 23-27.
9. T. Sato, K. Niita, N. Matsuda, S. Hashimoto, Y. Iwamoto, S. Noda, T. Ogawa, H. Iwase, H. Nakashima, T. Fukahori, K. Okumura, T. Kai, S. Chiba, T. Furuta and L. Sihver, Particle and Heavy Ion Transport Code System PHITS, Version 2.52, J. Nucl. Sci. Technol. 50:9, 913-923 (2013).
10. D. Satoh, T. Sato, N. Shigyo, and K. Ishinbashi, SCINFUL-QMD: Monte Carlo based computer code to calculate response function and detection efficiency of a liquid organic scintillator for neutron energies up to 3 GeV, JAEA-Data/Code, 2006-023, Japan Atomic Energy Agency, (2006) .

Dose Measurements through the Concrete and Iron Shields under the 100 to 400 MeV Quasi-Monoenergetic Neutron Field (at RCNP, Osaka Univ.)

Yoshihiro Nakane[1*], Yosuke Iwamoto[1], Masayuki Hagiwara[2], Hiroshi Iwase[2], Tatsuhiko Sato[1], Akihiko Masuda[3], Tetsuro Matsumoto[3], Tomoya Nunomiya[4], Hiroshi Yashima[5], Daiki Satoh[1], Hiroshi Nakashima[1], Tatsushi Shima[6], Atsushi Tamii[6], Kichiji Hatanaka[6] and Takashi Nakamura[7]

[1]*Japan Atomic Energy Agency*, [2]*High Energy Accelerator Research Organization (KEK)*, [3]*National Institute of Advanced Industrial Science and Technology*, [4]*Fuji Electric Co., Ltd.*, [5]*Kyoto University*, [6]*Osaka University*, [7]*Tohoku University*
*Corresponding Author: nakane.yoshihiro@jaea.go.jp

Shielding benchmark experiments are useful to verify the accuracy of calculation methods for the radiation shielding designs of high-energy accelerator facilities. In the present work, the benchmark experiments were carried out for 244- and 387-MeV quasi-monoenergetic neutron field at RCNP of Osaka University. Neutron dose rates through the test shields, 100-300 cm thick concrete and 40-100 cm thick iron, were measured by four kinds of neutron dose equivalent monitors, three kinds of wide-energy range monitors applied to high-energy neutron fields above 20 MeV and a conventional type rem monitor for neutrons up to 20 MeV, placed behind the test shields. Measured dose rates were compared one another. Measured results with the wide-energy range monitors were in agreement one another for both the concrete and the iron shields. For the conventional type rem monitor, measured results are smaller than those with the wide-energy range monitors for the concrete shields, while that are in agreements for the iron shields. The attenuation lengths were obtained from the measurements. The lengths from all the monitors are in agreement on the whole, though some differences are shown. These results are almost same as those from others measured at several hundred MeV neutron fields.

KEYWORDS: 100-400MeV, Shielding Experiments, Concrete, Iron, Neutron Dose

I. Introduction

In the radiation shielding designs and safety analyses, calculation methods were used(1) for high-energy and high-intensity proton accelerator facilities. In order to verify the accuracy of calculation methods for the dose estimations of shielding designs, shielding benchmark experiments for several hundred MeV neutrons are useful, especially for those performed at the fields that the intensity and the energy spectrum of source are well known.

In the present work, neutron dose rates through the concrete and the iron test shields were measured by four kinds of neutron dose equivalent monitors for 244- and 387-MeV quasi-monoenergetic neutrons generated from the ^{7}Li(p,n) reactions at the Research Center of Nuclear Physics (RCNP) ring cyclotron of Osaka University(2). Dose rates measured by the monitors were compared with one another. The attenuation lengths were obtained from the measurements for the concrete and the iron shields, and compared.

II. Experiments

1. Experimental Setup

Experiments were carried out at the neutron time-of-flight (TOF) beam course in the RCNP of Osaka University. Figure 1 shows a schematic view of the experimental arrangement. The experimental room is a tunnel of 100-m in length, 4-m in width and 3.5-m in height. Quasi-monoenergetic source neutrons, the peak energies were 244- and 387-MeV, were produced in a 1.0-cm thick ^{7}Li target(99.9%) bombarded with 246- and 389-MeV protons, respectively. Source neutrons emitted in the forward direction reached an experimental room through a 10.0-cm-wide and 12.0-cm-height aperture, and 150-cm-long iron collimator embedded in a concrete wall. The protons penetrating through the target were bent towards a beam dump by the swinger magnet to measure the proton beam intensity with a Faraday cup inside the dump. The test shields, 100-, 200- and 300-cm thick concrete and 40-, 70- and 100-cm thick iron, were placed on the neutron beam at the position of 17.66 m from the Li target. Atomic densities of the concrete and the iron test shields used in the

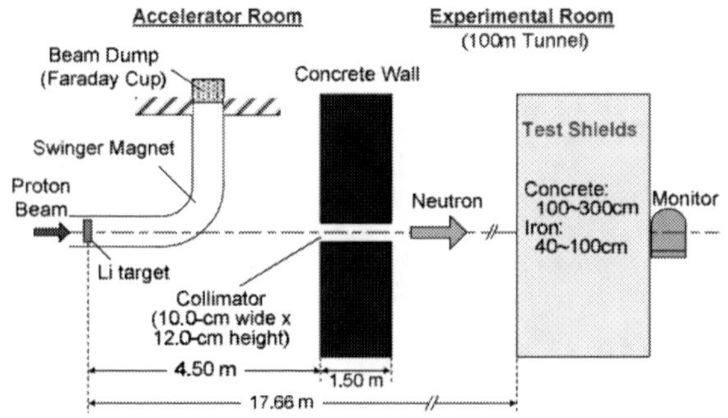

Figure 1: Schematic View of the Experimental Arrangement at RCNP

DOI: 10.1051/epjconf/201715308022

experiments were listed in Table 1. Neutron dose rates were measured with four kinds of neutron monitors placed behind the test shields.

Material	Density ($g\,cm^{-3}$)	Element	Atomic density ($10^{22}\,cm^{-3}$)
Concrete	2.31	Hydrogen	1.498
		Oxygen	4.188
		Sodium	0.123
		Magnesium	0.062
		Aluminum	0.312
		Silicon	1.110
		Potassium	0.038
		Calcium	0.430
		Iron	0.141
Iron	7.87	Iron	8.487

Table 1: Atomic Densities of the Test Shields

The energy spectra of source neutrons were shown in Figure 2. The spectra have been measured[(3)] by the TOF methods with three sizes of NE213 liquid organic scintillation detectors (25.4 x 25.4 cm2 for neutrons above 100 MeV, 12.7 x 12.7 cm2 for neutrons from 10 to 100 MeV, and 5.08 x 5.08 cm2 for neutrons below 10 MeV in diameter and length).

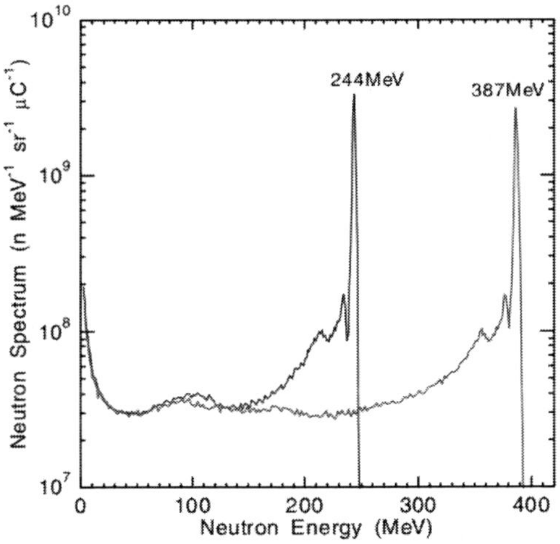

Figure 2: Energy Spectra of Source Neutrons[(3)]

2. Neutron Dose Equivalent Monitors

Table 2 shows the neutron dose equivalent monitors used in the present work. Four monitors, two kinds of modified Andersson-Braun type[(4)] monitors with lead or tungsten breeder(JAEA wide range[(5)] and WENDI[(6)]), a liquid organic scintillator type monitor with the response functions applied for high-energy neutron dose(DARWIN[(7)]), and a conventional Andersson-Braun type rem monitor(ALNOR), were used for the measurements. In our previous work, energy responses of these monitors have been measured[(8)] for 134-, 197-, 244- and 387-MeV quasi-monoenergetic neutron fields at the RCNP. Three monitors except the conventional rem monitor were wide-energy range type, and designed for applying to measure dose rate for high-energy neutrons above 20MeV. The JAEA wide range monitor was developed for applying to neutron monitoring for high-energy proton accelerator facility, J-PARC[(9)], and products of the monitor have been stationed in the facility for monitoring neutron dose rates. Because the monitor responses do not underestimate dose rates at high-energy neutron fields, the monitor was designed that the energy responses reproduced the neutron flux-to-dose conversion coefficients for ambient dose equivalent, $H^*(10)/\varphi$, for the energy below 20 MeV, and effective dose[(10)], E/φ, for the energy above 20 MeV.

III. Results and discussions

1. Comparisons of measured results

Figure 3 shows the measured results of dose rates through the concrete shields for the 244- and the 387-MeV sources. Measured results of neutron dose were normalized to proton beam charge (Coulomb) of beam dump shown in Figure 1. In the figure 3, lines show the interpolation of measured results. It was found from the figure that the measured results of the conventional type rem monitor (ALNOR) are smaller than those of all the wide-energy range monitors, about an order smaller at the 100-cm-thck concrete shields for both the 244- and the 387-MeV sources. For the comparison with the wide-energy range monitors, the results are in agreements one another in less than two. The tendency of attenuations for the measurements of the JAEA wide range monitor was quite larger than those of other wide-energy range monitors (DARWIN and WENDI). This is ascribed to the differences of reproducing the flux-to-dose conversion coefficients on the design for neutrons above 20

Monitor	Detector type	Data of Andersson-Braun type: Counter	Breeder for high-energy neutron	Shape	Wide-energy range	Reproduced dose conversion coefficients: ≤ 20MeV	> 20MeV
JAEA wide range	Modified Andersson-Braun	^3He	Lead	Spherical	yes	H*(10)	*E*
WENDI	Modified Andersson-Braun	^3He	Tungsten	Cylindrical	yes	H*(10)	H*(10)
DARWIN	Liquid organic scintillator	---	---	---	yes	H*(10)	H*(10)
Alnor(Studsvik) 2202D	Andersson-Braun	BF_3	none	Cylindrical	no	H*(10)	---

Table 2: Neutron Dose Equivalent Monitors Used in the Present Work

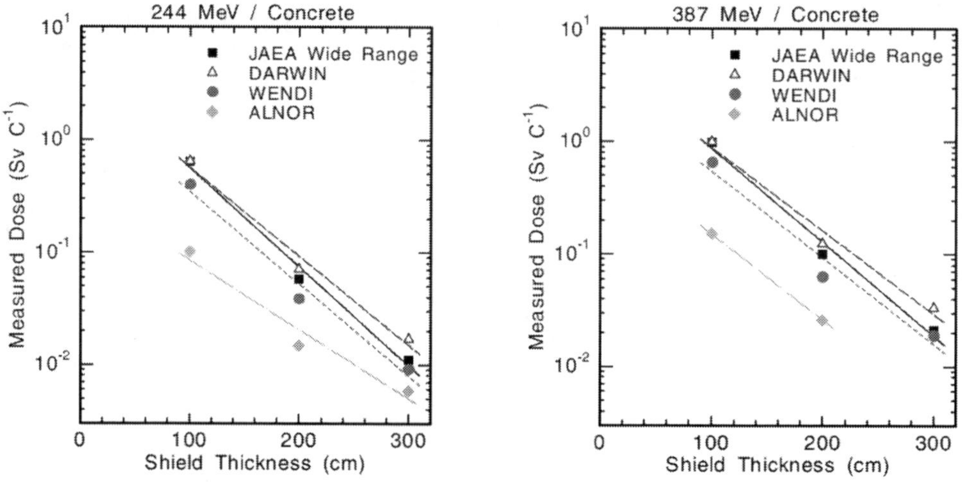

Figure 3: **Measured Results of Dose Rates through the Concrete Test Shields**

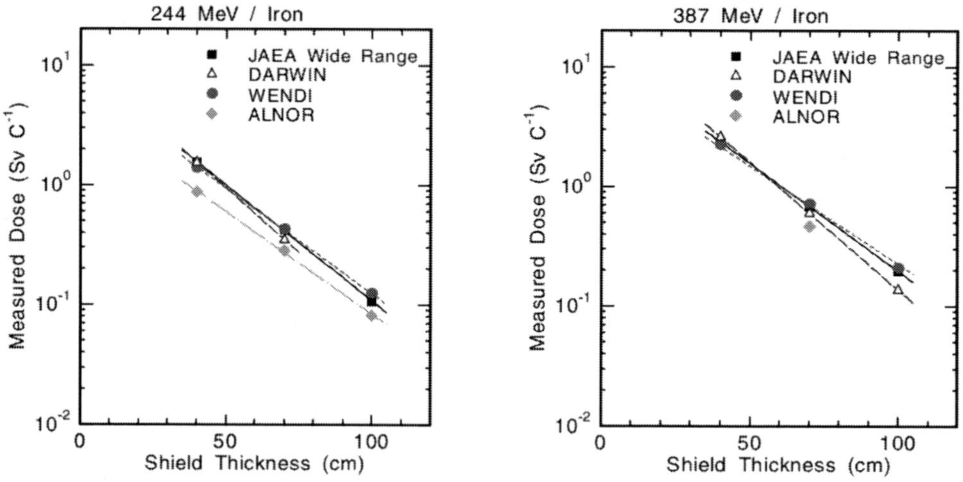

Figure 4: **Measured Results of Dose Rates through the Iron Test Shields**

MeV shown in Table 2.

Figure 4 shows the results of dose rates through the iron test shields for the 244- and the 387-MeV sources. The results of the wide-energy range monitors are in good agreement one another within 30% on the whole. The tendency of attenuations for the measurements of DARWIN was quite larger than those of other wide-energy range monitors (JAEA wide range and WENDI). The results of the conventional type monitor are about half of those of the wide-energy range monitors, not depend on the thickness of the shields and the energy of source neutrons. For the measurements through the iron shields, the differences between the conventional type monitor and the wide-energy range monitors are smaller than those for the concrete shields. It is because that low energy neutrons are dominant for the energy spectra through the iron shields.

2. Attenuation Lengths

The attenuation lengths obtained from the measured results with the four monitors are listed in Table 3. In the table, the attenuation lengths obtained from only two data are shown in italic letter without a standard deviation. For the concrete shields, the attenuation lengths obtained from the measurements of the three wide-energy range monitors are 114~127 and 120~136 (g cm^{-2}) for the 244- and the 387-MeV source, respectively. For the conventional type rem monitor, the length is larger than those of the wide-energy range monitors for the 244-MeV source. On the other hand, that is in agreement for the 387-MeV source. It is guessed that the length is obtained from the measurements only for 100-cm and 200-cm thick shields for the 387-MeV source, of which the tendency of the attenuations of the conventional type monitor is in agreement with that of the wide-energy range monitors. For the iron shields, the attenuation lengths for all the monitors are 159~200 and 160~209 (g cm^{-2}) for the 244- and the 387- MeV source, respectively. The lengths of ALNOR, JAEA wide range and WENDI are in agreement, while that of DARWIN is smaller than those of other monitors. For the comparison with the concrete and the iron shields, the attenuation lengths for the iron shields are larger than those for the concrete shields. These values are almost same as those from others measured(11-13) at CERN, KEK and ISIS for several hundred MeV neutron fields.

Monitors	Attenuation Length (g cm^{-2})			
	Concrete Shield		Iron Shield	
	244 MeV	387 MeV	244 MeV	387 MeV
JAEA Wide Range	114 ± 12	120 ± 13	177 ± 3	190 ± 0
WENDI	122 ± 17	130 ± 24	194 ± 2	209 ± 11
DARWIN	127 ± 16	136 ± 18	*159*	160 ± 0
ALNOR	163 ± 34	*132*	200 ± 5	---

Table 3: Attenuation Lengths of Doses Measured with the monitors

V. Conclusion

Shielding benchmark experiments were carried out for the 244- and the 387-MeV quasi-monoenergetic neutron fields at RCNP of Osaka University. Neutron dose rates through the concrete and the iron test shields were measured by four kinds of neutron dose equivalent monitors placed behind the test shields. The results measured with the wide-energy range monitors are in agreements one another both for the concrete and the iron test shields. For the conventional type rem monitor, the results are smaller than those with the wide-energy range monitors for the concrete shields, while those are in agreement for the iron shields. The attenuation lengths were obtained from the measurements for the concrete and the iron shields, and the results are almost same as those from others for several hundred MeV neutron fields.

References

1) e.g., Y. Nakane, T. Abe and H. Nakashima, "Monte Carlo Calculations for the Shielding Design of Beam Injection and Extraction Areas at the 3-GeV Synchrotron in J-PARC," Proc. of ICRS-11, Nuclear Technology, **168**, 519-523 (2009).
2) H. Sakai, H. Okamura, H. Otsu, T. Wakasa, S. Ishida, N. Sakamoto, T. Uesaka, Y. Satou, S. Fujita and K. Hatanaka, "Facility for the (p,n) polarization transfer measurement," Nucl. Instrum. Methods Phys. Res., **A369,** 120-134 (1996).
3) Y. Iwamoto, M. Hagiwara, D. Satoh, H. Iwase, H. Yashima, T. Itoga, T. Sato, Y. Nakane, H. Nakashima, Y. Sakamoto, T. Matsumoto, A.Masuda, J. Nishiyama, A. Tamii, K. Hatanaka, C. Theis, E. Feldbaumer, L. Jaegerhofer, C. Pioch, V.Mares and T. Nakamura, "Quasi-monoenergetic neutron energy spectra for 246 and 389 MeV ^{7}Li(p,n) reactions at angles from 0 to 30 degree," Nucl. Instrum. Methods Phys. Res., **A629,** 43-49 (2011).
4) C. Birattari, A. Ferrari, C. Nuccetelli, M. Pelliccioni and M. Silari, "An extended range neutron rem counter," Nucl. Instrum. Methods Phys. Res., **A297,** 250-257 (1990).
5) Y. Nakane, Y. Harada, Y. Sakamoto, T. Oguri, M. Yoshizawa, F. Takahashi, T. Ishikura, T. Fujimoto, S. Tanaka and N. Sasamoto, *Evaluation of Energy Response of Neutron Rem Monitor Applied to High-energy Accelerator Facilities*, JAERI-Tech 2003-011, Japan Atomic Energy Research Institute (2003). [in Japanese]
6) R.H. Olsher, H.H. Hsu, A. Beverding, J.H. Kleck, W.H. Casson, D.G. Vasilik and R.T. Devine, "WENDI: An improved neutron rem meter," Health Physics, **79(2)**, 170-181 (2000).
7) T. Sato, D. Satoh, A. Endo and Y. Yamaguchi, "Darwin: dose monitoring system applicable to various radiations with wide energy ranges," Radiation Protection Dosimetry, **126**, 501-505 (2007).
8) Y. Nakane, M. Hagiwara, Y. Iwamoto, H. Iwase, D. Satoh, T. Sato, H. Yashima, T. Matsumoto, A. Masuda, T. Nunomiya, Y. Sakamoto, H. Nakashima, A. Tamii, K. Hatanaka and T. Nakamura, "Response Measurement of Various Neutron Dose Equivalent Monitors in 134-387 MeV Neutron Fields," Proc. of ICRS-13, Progress in Nuclear Science and Technology, **4**, 704-708 (2014).
9) URL: http://j-parc.jp/
10) *Radiation Dose Conversion Coefficients for Radiation Shielding Calculations: 2004*, AESJ-SC-R002:2004, Atomic Energy Society of Japan, (2004). [in Japanese]
11) G.R. Stevenson, K.L. Liu and R.H. Thomas, "Determination of Transverse Shielding for Proton Accelerators Using the Moyer Model," Health Phys., **43**, 13-29 (1982).
12) S. Ban, H. Hirayama and k.Katoh, "Measurement of Secondary Neutron Fluxes around Beam Stop for 500 MeV Protons", Nucl. Instrum. Methods, **184**, 409-412 (1981).
13) T. Nunomiya, N. Nakao, P. Wright, T. Nakamura, E. Kim, T. Kurosawa, S. Taniguchi, M. Sasaki, H. Iwase, Y. Uwamino, T. Shibata, S. Ito and D.R. Perry, "Measurements of Attenuation Lengths Through Concrete and Iron for Neutrons Produced by 800-MeV Proton on Tantalum Target at ISIS", Nucl. Instrum. Methods Phys. Res., **A476,** 85-89 (2002).

AUTHOR INDEX

AUTHOR INDEX

AUTHOR INDEX

AUTHOR INDEX

AUTHOR INDEX

AUTHOR INDEX

AUTHOR INDEX

AUTHOR INDEX

EDP Sciences – Web of Conferences
17, Avenue du Hoggar
Parc d'Activité de Courtabouf
BP 112
F-91944 Les Ulis Cedex A
France

ISBN 978-1-5108-4812-2

CURRAN ASSOCIATES INC.
proceedings.com

9 781510 848122